W9-BNQ-018

EVOLUTIONARY ANALYSIS

FOURTH EDITION

Scott Freeman
University of Washington

Jon C. Herron
University of Washington

Upper Saddle River, NJ 07458

Library of Congress Cataloging-in-Publication Data

Freeman, Scott
Evolutionary analysis / Scott Freeman, Jon C. Herron. — 4th ed.

p. cm.

ISBN 0-13-227584-8
1. Evolution (Biology)--Textbooks. I. Herron, Jon C., II. Title.
QH366.2.F73 2007
576.8--dc22

2006034384

Senior Editor: Andrew Gilfillan
Assistant Editor: Jessica Berta
Senior Media Editor: Patrick Shriner
Editor in Chief, Development: Carol Trueheart
Development Editor: Ann Scanlan-Rohrer
Production Editor: Debra A. Wechsler
Executive Managing Editor: Kathleen Schiaparelli
Assistant Managing Editor: Beth Sweeten
Media Production Editor: Gina Cheselka; Robert Merenoff
Director of Marketing: Patrick Lynch
Manufacturing Buyer: Alan Fischer
Manufacturing Manager: Alexis Heydt-Long
Director of Creative Services: Paul Belfanti
Creative Director: Juan R. López
Art Director: Kenny Beck
Interior Design: Suzanne Behnke
Cover Design: Kenny Beck
Senior Managing Editor, Art Production and Management: Patricia Burns
Manager, Production Technologies: Matthew Haas
Managing Editor, Art Management: Abigail Bass
Art Production Editor: Rhonda Aversa
Illustrations: RMBlue Studios; Jon C. Herron
Manager, Formatting: Allyson Graesser
Formatting Assistance: Jacqueline Ambrosius
Director, Image Resource Center: Melinda Reo
Manager, Rights and Permissions: Zina Arabia
Interior Image Specialist: Beth Brenzel
Cover Image Specialist: Karen Sanatar
Image Permission Coordinator: Debbie Latronica
Photo Researcher: Jerry Marshall
Editorial Assistant: Lisa Tarabokjia
Cover Image: The Earth—Digital Vision/Getty Images; Snowy owl—Winfried Wisniewski/Jupiter Images; Leafcutter ant—Tim Flach/Getty Images; *Homo ergaster* skull—©David Brill; HIV virus—Geostock/Getty Images; Zebra—Alan & Sandy Carey/ Getty Images; *Mimulus lewisii* flower—Mark Turner/Jupiter Images. Back cover illustrations: RMBlue Studios

Published by Pearson Education, Inc.
Pearson Prentice Hall
Pearson Education, Inc.
Upper Saddle River, NJ 07458

Printed in the United States of America
10 9 8 7 6 5 4 3 2 1

ISBN: 0-13-227584-8

Pearson Education LTD., *London*
Pearson Education Australia PTY, Limited, *Sydney*
Pearson Education Singapore, Pte. Ltd
Pearson Education North Asia Ltd., *Hong Kong*
Pearson Education Canada, Ltd., *Toronto*
Pearson Educación de Mexico, S.A. de C.V.
Pearson Education—Japan, *Tokyo*
Pearson Education Malaysia, Pte. Ltd

Brief Contents

PART I

INTRODUCTION 1

CHAPTER 1 A Case for Evolutionary Thinking: Understanding HIV 3
CHAPTER 2 The Pattern of Evolution 37
CHAPTER 3 Darwinian Natural Selection 73
CHAPTER 4 Estimating Evolutionary Trees 111

PART II

MECHANISMS OF EVOLUTIONARY CHANGE 141

CHAPTER 5 Mutation and Genetic Variation 143
CHAPTER 6 Mendelian Genetics in Populations I: Selection and Mutation as Mechanisms of Evolution 169
CHAPTER 7 Mendelian Genetics in Populations II: Migration, Genetic Drift, and Nonrandom Mating 223
CHAPTER 8 Evolution at Multiple Loci: Linkage and Sex 281
CHAPTER 9 Evolution at Multiple Loci: Quantitative Genetics 319

PART III

ADAPTATION 361

CHAPTER 10 Studying Adaptation: Evolutionary Analysis of Form and Function 363
CHAPTER 11 Sexual Selection 401
CHAPTER 12 Kin Selection and Social Behavior 447
CHAPTER 13 Aging and Other Life History Characters 483
CHAPTER 14 Evolution and Human Health 529
CHAPTER 15 Phylogenomics and the Molecular Basis of Adaptation 575

PART IV

THE HISTORY OF LIFE 603

CHAPTER 16 Mechanisms of Speciation 605
CHAPTER 17 The Origins of Life and Precambrian Evolution 639
CHAPTER 18 The Cambrian Explosion and Beyond 689
CHAPTER 19 Development and Evolution 725
CHAPTER 20 Human Evolution 753

Contents

Preface xi

PART I

INTRODUCTION 1

CHAPTER 1

A Case for Evolutionary Thinking: Understanding HIV 3

1.1 The Natural History of the HIV/AIDS Epidemic 4
1.2 Why Does AZT Work in the Short Run, But Fail in the Long Run? 11
Box 1.1 *Can understanding how resistance evolves help researchers design better treatments?* 16
1.3 Why is HIV Fatal? 16
1.4 Why Are Some People Resistant to HIV? 22
1.5 Where Did HIV Come From? 25
Box 1.2 *When did HIV move from chimpanzees to humans?* 28
Summary 30 • Questions 30
Exploring the Literature 31 • Citations 32

CHAPTER 2

The Pattern of Evolution 37

Box 2.1 *A brief history of ideas on evolution* 39
2.1 Evidence of Change through Time 40
2.2 Evidence of Common Ancestry 50
Box 2.2 *Homology and model organisms* 59
2.3 The Age of Earth 60
Box 2.3 *A closer look at radiometric dating* 64
2.4 Is There Necessarily a Conflict between Evolutionary Biology and Religion? 65
Summary 68 • Questions 68
Exploring the Literature 69 • Citations 70

CHAPTER 3

Darwinian Natural Selection 73

3.1 Artificial Selection: Domestic Animals and Plants 74
3.2 Evolution by Natural Selection 76
3.3 The Evolution of Flower Color in an Experimental Snapdragon Population 78
3.4 The Evolution of Beak Shape in Galápagos Finches 80
Box 3.1 *Issues that complicate how heritabilities are estimated* 84
3.5 The Nature of Natural Selection 90
3.6 The Evolution of Darwinism 94
3.7 The Debate over "Scientific Creationism" and Intelligent Design Creationism 97
Summary 105 • Questions 106
Exploring the Literature 107 • Citations 107

CHAPTER 4

Estimating Evolutionary Trees 111

4.1 The Logic of Phylogeny Inference 112
4.2 The Phylogeny of Whales 119
Box 4.1 *A note on distance methods* 126
4.3 Using Phylogenies to Answer Questions 130
Summary 136 • Questions 137
Exploring the Literature 139 • Citations 139

PART II

MECHANISMS OF EVOLUTIONARY CHANGE 141

CHAPTER 5

Mutation and Genetic Variation 143

5.1 Where New Alleles Come From 144
5.2 Where New Genes Come From 152
5.3 Chromosome Mutations 156
5.4 Measuring Genetic Variation in Natural Populations 160
Summary 166 • Questions 166
Exploring the Literature 167 • Citations 168

CHAPTER 6

Mendelian Genetics in Populations I: Selection and Mutation as Mechanisms of Evolution 169

6.1 Mendelian Genetics in Populations: The Hardy–Weinberg Equilibrium Principle 170
Box 6.1 *Combining probabilities* 175
Box 6.2 *The Hardy–Weinberg equilibrium principle with more than two alleles* 180
6.2 Selection 182
Box 6.3 *A general treatment of selection* 186
Box 6.4 *Spongy brain diseases* 189
Box 6.5 *Statistical analysis of allele and genotype frequencies using the χ^2 (chi-square) test* 192
Box 6.6 *Predicting the frequency of the CCR5-Δ32 allele in future generations* 194
6.3 Patterns of Selection: Testing Predictions of Population Genetic Theory 194 
Box 6.7 *An algebraic treatment of selection on recessive and dominant alleles* 198
Box 6.8 *Stable equilibria with heterozygote superiority and unstable equilibria with heterozygote inferiority* 202
6.4 Mutation 210
Box 6.9 *A mathematical treatment of mutation as an evolutionary mechanism* 212
Box 6.10 *Allele frequencies under mutation-selection balance* 215
Box 6.11 *Estimating mutation rates for recessive alleles* 216
Summary 218 • Questions 219
Exploring the Literature 220 • Citations 221

CHAPTER 7

Mendelian Genetics in Populations II: Migration, Genetic Drift, and Nonrandom Mating 223

7.1 Migration 225
Box 7.1 *An algebraic treatment of migration as an evolutionary process* 227
Box 7.2 *Selection and migration in Lake Erie water snakes* 229
7.2 Genetic Drift 232
Box 7.3 *The probability that a given allele will be the one that drifts to fixation* 241
Box 7.4 *Effective population size* 245
Box 7.5 *The rate of evolutionary substitution under genetic drift* 250
7.3 Genetic Drift and Molecular Evolution 251
7.4 Nonrandom Mating 264
Box 7.6 *Genotype frequencies in an inbred population* 269
7.5 Conservation Genetics of the Illinois Greater Prairie Chicken 273
Summary 276 • Questions 276
Exploring the Literature 278 • Citations 279

CHAPTER 8

Evolution at Multiple Loci: Linkage and Sex 281

8.1 Evolution at Two Loci: Linkage Equilibrium and Linkage Disequilibrium 282
Box 8.1 *The coefficient of linkage disequilibrium* 285
Box 8.2 *Hardy–Weinberg analysis for two loci* 286

Box 8.3 Sexual reproduction reduces linkage disequilibrium 290
8.2 Practical Reasons to Study Linkage Disequilibrium 295
Box 8.4 Estimating the age of the GBA–84GG mutation 297
8.3 The Adaptive Significance of Sex 302
Summary 313 • Questions 313
Exploring the Literature 315 • Citations 316

CHAPTER 9
Evolution at Multiple Loci: Quantitative Genetics 319

9.1 The Nature of Quantitative Traits 319
9.2 Identifying Loci That Contribute to Quantitative Traits 324
Box 9.1 QTL mapping 328
9.3 Measuring Heritable Variation 333
Box 9.2 Additive genetic variation versus dominance genetic variation 336
9.4 Measuring Differences in Survival and Reproductive Success 338
Box 9.3 The selection gradient and the selection differential 340
9.5 Predicting the Evolutionary Response to Selection 341
Box 9.4 Selection on multiple traits and correlated characters 344
9.6 Modes of Selection and the Maintenance of Genetic Variation 346
9.7 The Bell-Curve Fallacy and Other Misinterpretations of Heritability 350
Summary 355 • Questions 355
Exploring the Literature 357 • Citations 358

PART III
ADAPTATION 361

CHAPTER 10
Studying Adaptation: Evolutionary Analysis of Form and Function 363

10.1 All Hypotheses Must Be Tested: Oxpeckers Reconsidered 364
10.2 Experiments 367
Box 10.1 A primer on statistical testing 371
10.3 Observational Studies 372
10.4 The Comparative Method 376
Box 10.2 Calculating phylogenetically independent contrasts 380
10.5 Phenotypic Plasticity 380
10.6 Trade-Offs and Constraints 383
10.7 Selection Operates on Different Levels 392
10.8 Strategies for Asking Interesting Questions 395
Summary 396 • Questions 397
Exploring the Literature 398 • Citations 399

CHAPTER 11
Sexual Selection 401

11.1 Sexual Dimorphism and Sex 402
11.2 Sexual Selection on Males: Competition 408
Box 11.1 Alternative male mating strategies 412
11.3 Sexual Selection on Males: Female Choice 415
Box 11.2 Runaway sexual selection in stalk-eyed flies? 426
11.4 Sexual Selection on Females 429
Box 11.3 Extra-pair copulations and multiple mating 430
11.5 Sexual Selection in Plants 434
11.6 Sexual Dimorphism in Body Size in Humans 438
Summary 440 • Questions 441
Exploring the Literature 443 • Citations 444

CHAPTER 12
Kin Selection and Social Behavior 447

12.1 Kin Selection and the Evolution of Altruism 448
Box 12.1 Calculating coefficients of relatedness 449

Box 12.2 Kin recognition 454
12.2 Evolution of Eusociality 459
Box 12.3 The evolution of the sex ratio 462
12.3 Parent–Offspring Conflict 467
12.4 Reciprocal Altruism 471
Box 12.4 Prisoner's dilemma: Analyzing cooperation and conflict using game theory 473
Summary 477 • Questions 478
Exploring the Literature 479 • Citations 480

CHAPTER 13

Aging and Other Life History Characters **483**

13.1 Basic Issues in Life History Analysis 485
13.2 Why Do Organisms Age and Die? 487
Box 13.1 A trade-off between cancer risk and aging 490
13.3 How Many Offspring Should an Individual Produce in a Given Year? 502
Box 13.2 Is there an evolutionary explanation for menopause? 503
13.4 How Big Should Each Offspring Be? 509
13.5 Conflicts of Interest between Life Histories 514
13.6 Life Histories in a Broader Evolutionary Context 517
Summary 524 • Questions 524
Exploring the Literature 526 • Citations 527

CHAPTER 14

Evolution and Human Health **529**

14.1 Evolving Pathogens: Evasion of the Host's Immune Response 531
14.2 Evolving Pathogens: Antibiotic Resistance 538
14.3 Evolving Pathogens: Virulence 541
14.4 Tissues as Evolving Populations of Cells 546
Box 14.1 Genetic sleuthing solves a medical mystery 547
14.5 The Adaptationist Program Applied to Humans 550
14.6 Adaptation and Medical Physiology: Fever 556
14.7 Adaptation and Human Behavior: Parenting 561
Box 14.2 Is cultural evolution Darwinian? 562
Summary 569 • Questions 569
Exploring the Literature 571 • Citations 571

CHAPTER 15

Phylogenomics and the Molecular Basis of Adaptation **575**

15.1 Transposable Elements and Levels of Selection 576
Box 15.1 Categories of transposable elements 578
15.2 Lateral Gene Transfer 584
15.3 The Molecular Basis of Adaptation 591
15.4 Frontiers in Phylogenomics 596
Summary 600 • Questions 600
Exploring the Literature 601 • Citations 601

PART IV

THE HISTORY OF LIFE **603**

CHAPTER 16

Mechanisms of Speciation **605**

16.1 Species Concepts 605
Box 16.1 What about bacteria and archaea? 607
16.2 Mechanisms of Genetic Isolation 611
16.3 Mechanisms of Divergence 616
16.4 Secondary Contact 623
16.5 The Genetics of Speciation 629
Summary 633 • Questions 634
Exploring the Literature 635 • Citations 635

CHAPTER 17

The Origins of Life and Precambrian Evolution 639

17.1 What Was the First Living Thing? 640
17.2 Where Did the First Living Thing Come From? 651
Box 17.1 *The Panspermia Hypothesis* 652
17.3 What Was the Last Common Ancestor of All Extant Organisms and What Is the Shape of the Tree of Life? 660
17.4 How Did the Last Common Ancestor's Descendants Evolve into Today's Organisms? 675
Summary 680 • Questions 681
Exploring the Literature 683 • Citations 684

CHAPTER 18

The Cambrian Explosion and Beyond 689

18.1 The Nature of the Fossil Record 690
18.2 The Cambrian Explosion 694
18.3 Macroevolutionary Patterns 702
18.4 Mass Extinctions 709
Summary 720 • Questions 721
Exploring the Literature 722 • Citations 722

CHAPTER 19

Development and Evolution 725

19.1 The Foundations of Evo-Devo Research 726
19.2 Homeotic Genes and Diversification in Animal Body Plans 728
19.3 Deep Homology and Diversification in Animal Limbs 735
19.4 Homeotic Genes and the Evolution of the Flower 742
19.5 Frontiers in Evo-Devo Research 747
Summary 749 • Questions 749
Exploring the Literature 750 • Citations 750

CHAPTER 20

Human Evolution 753

20.1 Relationships among Humans and Extant Apes 754
Box 20.1 *Genetic differences between humans, chimpanzees, and gorillas* 762
20.2 The Recent Ancestry of Humans 764
20.3 The Origin of the Species *Homo sapiens* 773
Box 20.2 *Genetic diversity among living humans* 776
Box 20.3 *Using linkage disequilibrium to date the divergence between African and non-African populations* 784
20.4 The Evolution of Uniquely Human Traits 786
Summary 791 • Questions 792
Exploring the Literature 793 • Citations 795

Glossary 799

Illustration Credits 806

Index 816

Preface

Evolutionary Analysis is for undergraduates majoring in the life sciences. We assume that readers have completed much or all of their introductory coursework and are beginning to explore in more detail particular areas of biology that are relevant to their personal and professional lives.

We expect our readers to pursue careers in a diversity of fields, including medicine, education, environmental management and conservation, journalism, biotechnology, and research. We therefore attempt, throughout the book, to show the relevance of evolution to all of biology and to real-world problems.

Our primary goal is to encourage readers to think like scientists. We present evolutionary biology not as a collection of facts but as an ongoing research effort. When exploring an issue, we begin with questions. Where did HIV come from? Why did prairie chicken populations continue to decline despite successful efforts to restore their habitat? How closely related are humans and chimpanzees? We use such questions to motivate discussions of background information and theory. These discussions enable us to frame alternative hypotheses, consider how they can be tested, and make predictions. We then present and analyze data, consider its implications, and highlight new questions for future research. The analytical and technical skills readers learn from this approach are broadly applicable, and will stay with them long after the details of particular examples have faded.

Consistent with our presentation of evolutionary biology as a dynamic research enterprise, we have tried to keep our coverage as up-to-date as possible. This has made the fourth edition as exciting to work on—and as daunting—as the first three. Many of the fields we cover are advancing at a rate we would not have dreamed possible just a few years ago. More than once our editors have had to tear chapter manuscripts from our hands while there were still more changes we wanted to make.

There is something new in every chapter. Among the new examples we are most excited about are these:

- Evidence that disease progression in HIV patients results, in part, from evolution of the viral population toward greater competititve fitness (Chapter 1).
- Population-genetic data indicating that heterozygotes survived an epidemic of spongy brain disease at a higher rate than homozygotes (Chapter 6).
- A demonstration that the substitution of one allele for another at a single locus can dramatically alter a flower's appeal to different pollinators (Chapter 9).
- Data suggesting that the frequency of Apert syndrome in human populations reflects selection at the level of stem cells in tissues acting in oppostion to selection at the level of individuals in populations (Chapter 10).
- Results showing that female crickets scent-mark their mates to avoid copulating twice with the same male (Chapter 11).

- Common-garden experiments revealing that an invasive plant evolved greater weediness when released from a life-history trade-off (Chapter 13).
- Documentation that a change in the phenotype of a plant resulted from a change in gene regulation (Chapter 15).
- Phylogenetic evidence that viruses played a key role in the transition from RNA World to DNA world, and in the origin of the three domains of life (Chapter 17).
- New insights into the evolution of the tetrapod limb from both developmental genetics and a recently discovered transistional fossil (Chapter 19).

We encourage readers to check the literature for new developments that have been reported since the book went to press; there are sure to be many.

Two trends we have witnessed since the first edition are reflected in changes to the table of contents. First, phylogenies have grown so central to research in evolution that we felt we had to introduce tree thinking much earlier in the book. We have therefore moved Estimating Evolutionary Trees into Unit 1, where it has become Chapter 4. Second, the Human Genome Project and the technologies it fostered have ignited an explosion of genomic data. Comparative analyses of whole genomes have yielded startling insights into the evolutionary process. To convey some of the excitement of this new frontier, we have added a chapter on Phylogenomics and the Molecular Basis of Adaptation. As Chapter 15, it concludes Unit 3.

There are four units in all:

- **Part I, Introduction**, shows that evolution is relevant outside books and classrooms, establishes the fact of evolution with a mixture of classical and recent evidence, presents natural selection as an observable process, and develops modern methods for reconstructing evolutionary trees.
- **Part II, Mechanisms of Evolutionary Change**, develops the theoretical underpinnings of modern evolutionary biology by exploring how mutation, selection, migration, and drift produce evolutionary change.
- **Part III, Adaptation**, introduces a variety of methods for studying adaptation, and offers detailed accounts of research in sexual selection, kin selection, life history evolution, and Darwinian medicine.
- **Part IV, The History of Life**, begins with an analysis of speciation. It then considers the origin of life, the universal phylogeny, and major events in the history of multicellular organisms. Because of its importance for understanding macroevolution, our chapter on Development and Evolution is part of this unit. The unit and book end with human evolution.

Most chapters include boxes that cover special topics or methods, provide more detailed analyses, or offer derivations of equations. All chapters end with a set of questions that encourage readers to review the material, apply concepts to new issues, and explore the primary literature.

Additional Resources for Instructors and Students

The new edition of *Evolutionary Analysis* offers an Instructor Resource Center on CD containing all of the line art, tables, and all photos from the book in both JPEG and PowerPoint® files. All presentation art has been carefully modified for optimal visibility when projected. An unlabeled version of the line art is also included to facilitate customization.

The student Companion Website for *Evolutionary Analysis* has again been revised and updated. This website is accessible through the book's homepage at:

www.prenhall.com/freeman

It features carefully crafted chapter study quizzes that offer elaborate instructional feedback. These quizzes are designed to increase understanding of the underlying concepts of each chapter as well as prepare students for taking tests. Activities such as simulations and case studies challenge students to pose questions, formulate hypotheses, design experiments, analyze data, and draw conclusions. Many of these activities accompany downloadable software programs that allow students to conduct their own virtual investigations. The Companion Website also offers the answers to the end-of-chapter questions and weblinks to other evolution-related sites.

Acknowledgments

We owe the effectiveness and success of *Evolutionary Analysis* to the generosity, creativity, energy, and support of the many colleagues and students who have helped us write a better book. They have reviewed chapters, shared their data and photographs, answered our questions, emailed us with suggestions, sent us reprints, and talked with us at meetings. It is a privilege to spend time with this remarkable community, and we thank them for their collaboration.

In preparing the fourth edition we have been guided by thoughtful, detailed, and constructive critiques by:

Butch Brodie, *Indiana University*
George W. Gilchrist, *College of William & Mary*
David Gray, *California State University, Northridge*
Andy Jarosz, *Michigan State University*
Nicole Kime, *Edgewood College*
Martin Morgan, *Washington State University*
Leslee A. Parr, *University of Wisconsin*
Andy Peters, *University of Wisconsin*
Thomas Ray, *University of Oklahoma*
David Ribble, Trinity University
Peter Tiffin, *University of Minnesota*
Robert S. Wallace, *Iowa State University*
Yufeng Wang, *University of Texas, San Antonio*
Paul Wilson, *California State University*

If deficiencies remain, the fault is ours for not following more closely their excellent advice.

The following colleagues read chapters in proof. Their keen eyes and thoughtful feedback applied considerable polish to the manuscript:

Lynda Delph, *Indiana University*
Stephen Freeland, *University of Maryland, Baltimore County*
Tamra Mendelson, *Lehigh University*
Sara Via, *University of Maryland*
Helen Young, *Middlebury College*

RMBlue Studios helped develop the figure program, prepared the beautiful new illustrations, and revised the existing art. Kathleen Hunt revised the thought-provoking end-of-chapter questions and exploring the literature items. Brooks Miner assisted with library research, helped plan chapter revisions, and made invaluable suggestions on the manuscript.

The editorial and production team at Pearson Prentice Hall have been, as ever, superb. We thank them for their guidance, support, collaboration, and friendship. ESM President Paul Corey has been steadfastly committed to this project from the beginning. Editorial Assistant Lisa Tarabokjia arranged for the reviews. Senior Media Editor Patrick Shriner and Assistant Editor Jessica Berta developed the media components. Art Director Kenny Beck designed the book. Jacqueline Ambrosius oversaw the formatting and composition of the text. Prentice Hall's remarkable team of sales representatives are, even as we put the final touches on this edition, disseminating the book to professors everywhere.

Production Editor Debra Wechsler is outstanding at her job and a joy to work with. How she has put up with us, in the midst of overseeing a million last-minute details, we do not understand. We cannot thank her enough.

Finally, we thank the two extraordinary editors with whom it has been our great fortune to collaborate. Former Publisher Sheri Snavely willed this project into existence and devoted herself wholeheartedly to its success through its first three editions. It will always be her book. After getting us started on the fourth edition she passed the project into the highly capable hands of Acquistions Editor Andrew Gilfillan. He has been a pillar of wisdom and support. Now it is his book too.

Jon C. Herron
Scott Freeman
Seattle, Washington

PART I

INTRODUCTION

Where did Earth's organisms come from? Why are there so many different kinds? How did they come to be so apparently well-designed? These are the fundamental questions of evolutionary biology. The answers are found in both the pattern and mechanism of evolution. The pattern is descent with modification from common ancestors. The primary mechanism is natural selection.

Our first goal in Part I (Chapters 1–4) is to introduce the pattern and process of evolution. In Chapter 1 we explore an example, the evolution of HIV. In Chapter 2 we look at the pattern of evolution and at evidence of common ancestry. In Chapter 3 we focus on the mechanism of evolution. Natural selection is evolutionary biology's organizing principle; its simplicity is among the discipline's charms. Natural selection is widely misconstrued, however. Understanding it requires moving beyond slogans like "survival of the fittest." In Chapter 4 we cover methods for reconstructing evolutionary history.

Our second goal is to introduce the experimental and analytical methods used by the biologists who study evolution. These methods are a prominent theme throughout the text. We emphasize them to help readers learn how to ask questions, design experiments, analyze data, and critically review scientific papers. The detailed examples we present make the general concepts of evolutionary biology clear and also provide insight into how we know what we know. ■

Bobobos are, along with common chimpanzees, our closest living relatives. Here, a female stretches.

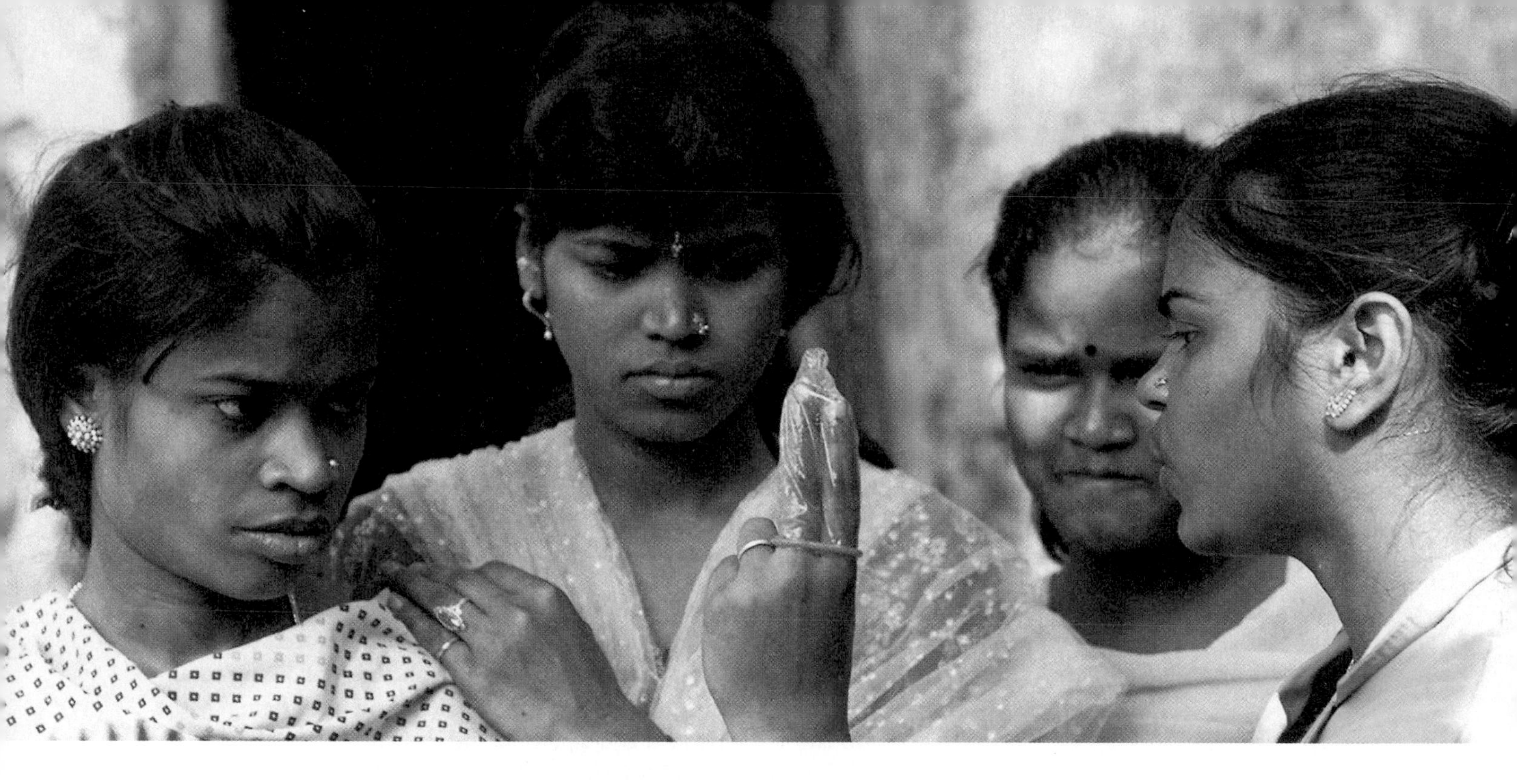

Prostitutes in the Songachi redlight district in Calcutta, India, learn from a health worker about the benefits of using condoms. In Songachi, an aggressive campaign to educate sex workers, madams, and pimps to distribute condoms, and to encourage condom use, has kept the prevalance of HIV among prostitutes below 12%. In other redlight districts, HIV prevalence has climbed above 50% (Cohen 2004).

1

A Case for Evolutionary Thinking: Understanding HIV

Why study evolution? Although he scarcely mentioned it in *On the Origin of Species* (1859), one of Charles Darwin's motives was that understanding evolution can help us understand ourselves. "Light will be thrown," Darwin wrote, "on the origin of man and his history." For Theodosius Dobzhansky (1973), an architect of the modern view of evolution we present in this text, the reward was that evolutionary biology is the conceptual foundation that supports all the life sciences. "Nothing in biology makes sense," he declared, "except in the light of evolution." For some readers the incentive may be that a course in evolution is required for the completion of their degree.

Here we suggest yet another reason to study evolution: The tools and techniques of evolutionary biology offer crucial insights into matters of life and death. To justify this claim, we explore the evolution of human immunodeficiency virus (HIV). HIV is the virus that causes acquired immune deficiency syndrome (AIDS).

A deep look at this prominent contemporary issue will introduce the scope of evolutionary analysis. It will illustrate the kinds of questions evolutionary biologists ask, show how an evolutionary perspective can inform research throughout the biological sciences, and introduce concepts that we will explore in detail elsewhere in the book.

HIV makes a compelling case study because it raises issues almost certain to influence the personal and professional life of every reader. HIV exemplifies pressing public health issues: It is an emerging virus. It rapidly evolves drug resistance. And it is deadly. AIDS already qualifies as one of the most devastating epidemics our species has experienced.

Here are the questions we address:

- Why did early AIDS treatments, like the drug azidothymidine (AZT), look promising when first used but prove ineffective in the long run?
- Why does HIV kill people?
- Why are some people resistant to becoming infected or to progressing to disease once they are infected?
- Where did HIV come from?

As a case study, HIV will demonstrate how evolutionary biologists study adaptation and diversity.

Some of these questions may not sound as if they have anything to do with evolutionary biology. But evolutionary biology is the science devoted to understanding two things: (1) how populations change through time following modifications in their environment, and (2) how new species come into being. More formally, evolutionary biologists study adaptation and diversity. These are exactly the issues targeted by our questions about HIV and AIDS. Before we tackle them, however, we need to delve into some background biology.

1.1 The Natural History of the HIV/AIDS Epidemic

The worst epidemic in human history, judging by the number of deaths, was probably the influenza of 1918. It swept the globe in a matter of months, killing 50 to 100 million people (Johnson and Mueller 2002). The second worst was likely the Black Death, caused by a highly virulent pathogen whose identity remains controversial (see Raoult et al. 2000; Gilbert et al. 2004; Christakos and Olea 2005; Duncan and Scott 2005). It ravaged Europe from 1347 to 1352, taking 30 to 50% of the population—roughly 25 million lives (Derr 2001). More localized outbreaks over the next 300 years killed millions more. Also worthy of mention is the New World smallpox epidemic unleashed around 1520 by European conquistadores. Its death toll is harder to reckon, but over the succeeding decades it decimated Native American populations across two continents (Roberts 1989; Snow 1995; Patterson and Runge 2002).

AIDS is among the worst epidemics in human history.

The AIDS epidemic, first recognized by medical professionals in 1981, has rapidly earned a place among this grim company (UNAIDS 2005). HIV has so far infected more than 65 million people. Twenty-five million have already died of the opportunistic infections that characterize AIDS. Among the rest, many are gravely ill. And many are still spreading the disease. The Joint United Nations Programme on HIV/AIDS has estimated that by 2020 the AIDS epidemic will have claimed a total of almost 90 million lives (UNAIDS 2002a).

Figure 1.1 summarizes the global pattern of the AIDS epidemic. In the map in Figure 1.1a, regions are shaded to show the prevalence of HIV infection among adults, and labeled to indicate the total number of adults and children infected with HIV and the sex ratio among infected adults. The bar graphs in Figure 1.1b document the growth of the epidemic over time in different parts of the world.

Every day roughly 13,400 people are newly infected with HIV and 8,500 people die of AIDS (UNAIDS 2005). According to the World Health Organiza-

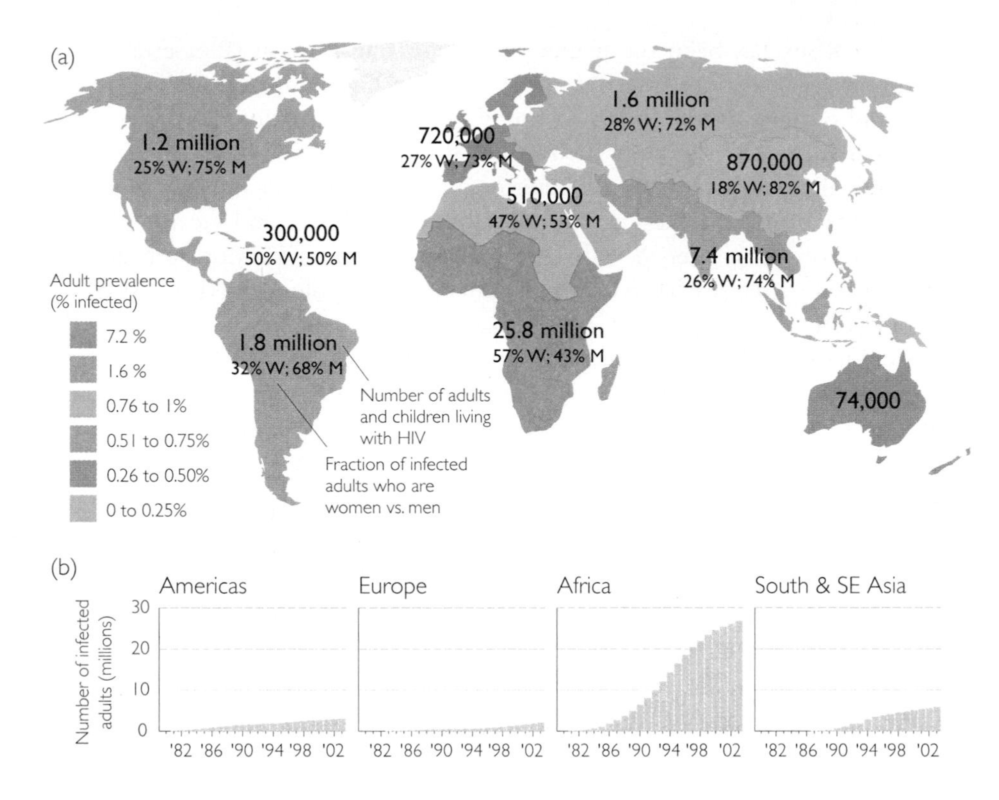

Figure 1.1 The HIV/AIDS pandemic (a) This map shows the geographic distribution of HIV infections. Each region is shaded to indicate the prevalence of infection among adults. In addition, regions are labeled with the total number of individuals living with HIV and the sex ratio among infected adults. More than three-fifths of the people infected with HIV live in sub-Saharan Africa; another one-fifth live in south and southeast Asia. Data from UNAIDS (2005). (b) These bar graphs illustrate the growth in the number of adults living with HIV since the pandemic began in the early 1980s. Redrawn from WHO (2004).

tion, AIDS is now responsible for about 4.9% of all deaths worldwide (WHO 2004). AIDS causes a smaller fraction of deaths than cancer (12.5%), heart attacks (12.6%), strokes (9.7%), and lower respiratory tract infections (6.8%)—common causes of death among the elderly. But AIDS causes more deaths than tuberculosis (2.7%), malaria (2.2%), car accidents (2.1%), homicides (1%), and wars (0.3%).

The epidemic has wrought its greatest devastation, by far, in sub-Saharan Africa (see Piot et al. 2001). The average prevalence of HIV among adults there is 7.2% (UNAIDS 2005). Worst hit is Swaziland, with an adult prevalence of 38.8%, followed by Botswana at 37.3%; Lesotho, 28.9%; and Zimbabwe, 24.6% (UNAIDS 2004). In Lesotho, an individual who turned 15 in the year 2000 has a 74% chance of contracting HIV by age 50 (UNAIDS 2002a). In Botswana, the AIDS epidemic has dragged the average life expectancy from 65 years to 40 and is expected to pull it lower still (Figure 1.2).

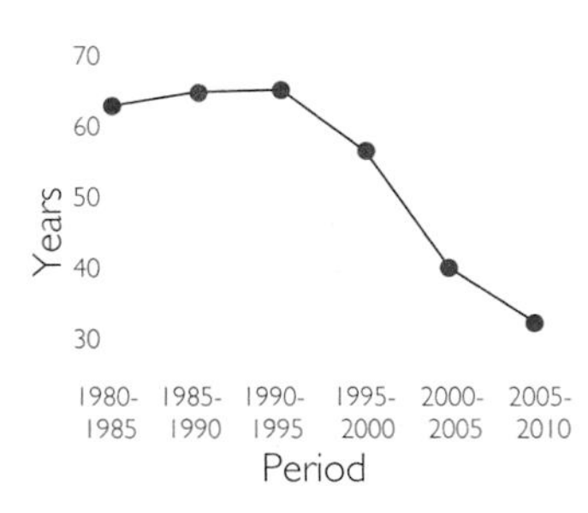

Figure 1.2 Life expectancy in Botswana This graph shows the estimated life expectancy at birth for individuals born between 1980 and 2000, and projected life expectancy for individuals born between 2000 and 2010. The decline after 1990 is due to the AIDS epidemic. Redrawn from Figure 12 in UNAIDS (2004).

In the industrialized countries of North America and western Europe, overall infection rates are much lower than in sub-Saharan Africa (UNAIDS 2004, 2005). In western Europe the adult prevalence of HIV infection is just 0.3%. In Canada the adult prevalence is also 0.3%, and in the United States it is 0.6%. For certain risk groups, however, infection rates rival those in Africa's most devastated regions. Among men who have sex with men, the infection rate is 18% in New York City, 19% in Los Angeles, 24% in San Francisco, and 40% in Baltimore (CDC 2005). Among injection drug users the infection rate is 18% in Chicago and about 25% in New York City (Piot et al. 2001).

HIV establishes a new infection when a bodily fluid holding the virus, usually blood or semen, carries it from an infected person directly onto a mucous membrane or into the bloodstream of an uninfected person. The virus can be passed during heterosexual sex, homosexual sex, oral sex, needle sharing, transfusion with contaminated blood products, childbirth, and breastfeeding. The virus has spread via different routes in different regions. In sub-Saharan Africa and in India

An HIV infection can be acquired only from someone else who already has it.

heterosexual sex has been the primary mode of transmission (Piot et al. 2001; Schmid et al. 2004; Lopman et al. 2005—but see Gisselquist et al. 2002, 2004; Brody and Potterat 2005). In China the virus spread first among injection drug users, then among blood donors whose plasma was collected in an unsafe manner, and finally among partners in heterosexual sex (Kaufman and Jing 2002). In the United States and western Europe homosexual sex and needle sharing among injecting drug users have been the most common transmission routes, although heterosexual sex has recently played a growing role (UNAIDS 2005).

Programs to curb the spread of HIV have seen success (Figure 1.3). After the AIDS epidemic arrived in Thailand in the late 1980s and began to accelerate in the early 1990s, the Ministry of Health launched a campaign to encourage young people to reduce risky sexual practices and use condoms (Nelson et al. 2002). In less than a decade, the incidence of HIV infection among military draftees dropped from over 11% to less than 3%, in concert with an increase in condom use during visits to commercial sex workers (and a decrease in the frequency of such visits). A sexual health education program for sex workers in Ivory Coast contributed to a similarly dramatic drop in HIV infection rates, again coincident with an increase in condom use (Ghys et al. 2002).

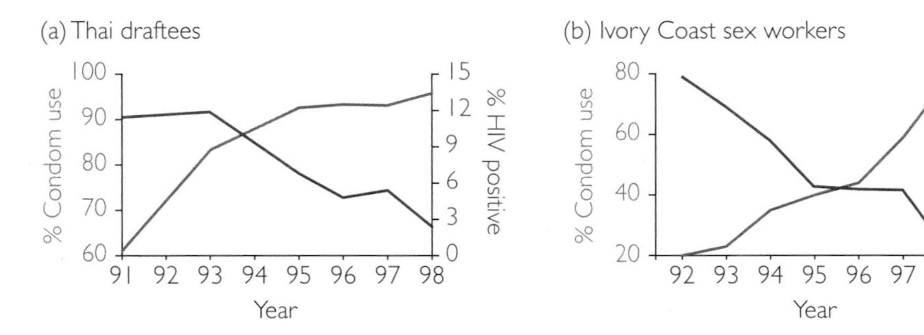

Figure 1.3 Successful HIV/AIDS prevention These graphs document the success of HIV prevention programs in (a) Thailand and (b) Ivory Coast. As condom use went up, the incidence of HIV infection went down. Drawn from data in Nelson et al. (2002) and Ghys et al. (2002).

There is no room for complacency, however. The graph in Figure 1.4 shows that around the year 2000 the rate of new HIV infections began to rise, in parallel with infection rates for other sexually transmitted diseases, among men who have sex with men in London. The same thing is happening in San Francisco and elsewhere (Kellogg, McFarland, and Katz 1999; Hamers and Downs 2004; Giuliani et al. 2005). It appears that the introduction of effective long-term drug therapies, which for some individuals have at least temporarily transformed HIV into a manageable chronic illness, may also have prompted an increase in risky sexual behavior (Katz et al. 2002; Chen et al. 2002; Crepaz, Hart, and Marks 2004). Additional cause for concern is the increasingly widespread abuse of methamphetamine, which is associated with risky behavior and a greater chance of contracting HIV (Buchacz et al. 2005).

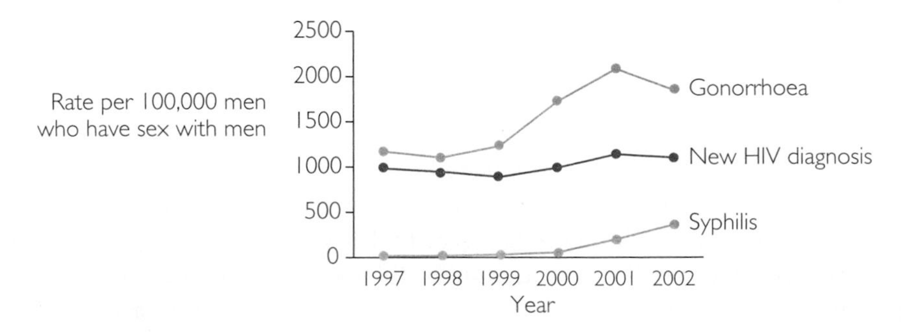

Figure 1.4 Rates of new HIV diagnosis and other sexually transmitted diseases among men who have sex with men in London This graph documents recent increases in the incidence of gonorrhoea and syphilis, and an increase in the rate of new HIV diagnosis, among men who have sex with men in London. From Macdonald et al. (2004).

What is HIV?

Like all viruses, HIV is an intracellular parasite that cannot reproduce on its own. HIV invades specific types of cells in the human immune system. The virus uses the enzymatic machinery and energy of these cells to make copies of itself, killing the host cells in the process.

HIV is a parasite that afflicts cells in the human immune system.

Figure 1.5 outlines HIV's life cycle in more detail. The life cycle includes an extracellular phase and an intracellular phase. During the extracellular, or infectious phase, the virus moves from one host cell to another, and can be transmitted from host to host. The extracellular form of a virus is called a virion or virus particle. During the intracellular, or parasitic phase, the virus replicates.

HIV initiates its replication phase by latching onto two proteins on the surface of a host cell. After adhering first to CD4, found on the surface of certain immune system cells, HIV attaches to a second protein, called a coreceptor. This fuses the virion's envelope with the host's cell membrane and spills the contents of the virion into the cell. These contents include the virus's diploid genome (two copies of a single-stranded RNA molecule) and three proteins: reverse

HIV virions enter host cells by binding to proteins on their surface, then use the host cell's own machinery to make new virions.

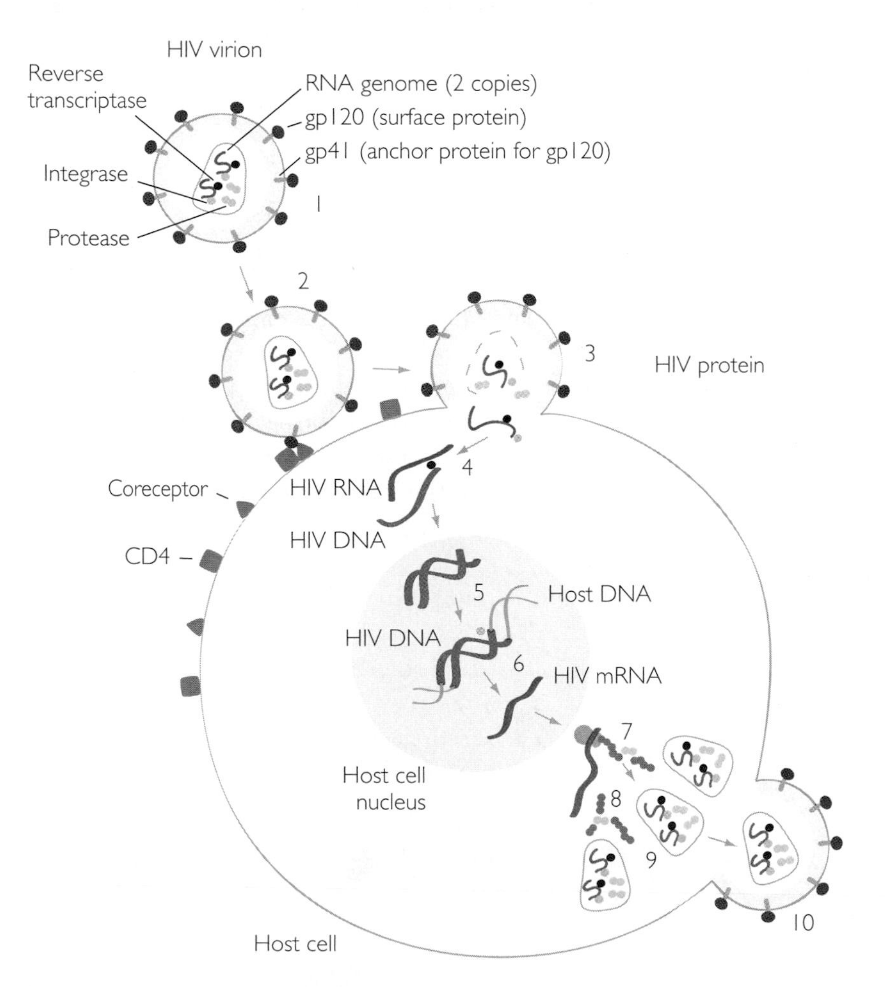

1) HIV's extracellular form, known as a virion, encounters a host cell

2) HIV's gp120 protein binds to CD4 and coreceptor on host cell

3) HIV's RNA genome, reverse transcriptase, integrase, and protease enter host cell

4) Reverse transcriptase synthesizes HIV DNA from HIV's RNA template

5) Integrase splices HIV DNA into host genome

6) HIV DNA is transcribed to HIV mRNA by the host cell's RNA polymerase

7) HIV mRNA is translated to HIV precursor proteins by host cell's ribosomes

8) Protease cleaves precursors into mature viral proteins

9) New generation of virions assembles inside host cell

10) New virions bud from host cell's membrane

Figure 1.5 The life cycle of HIV An HIV virion (1) invades a host cell by binding to two proteins on the cell's surface (2), allowing the virion to spill its contents into the cell (3). Once inside the host cell, HIV's reverse transcriptase makes a DNA copy of the viral genome (4). HIV's integrase inserts this DNA copy into the host cell's genome (5). The host cell's RNA polymerase transcribes the viral genome into messenger RNA (6), and the host cell's ribosomes transcribe the viral mRNA into precursor proteins (7). HIV's protease cleaves the precursors, yielding mature viral proteins (8). New virions assemble in the host cell's cytoplasm (9), then bud from the host cell's membrane (10).

transcriptase, which transcribes the virus's RNA genome into DNA; integrase, which splices the DNA genome into the host cell's genome; and protease, which plays a role in the preparation of new viral proteins.

Note that in HIV, as in other retroviruses, the flow of genetic information is different than in cells and in viruses with DNA genomes. In retroviruses, genetic information does not follow the familiar route from DNA to mRNA to proteins. Instead, information flows from RNA to DNA, then to mRNA to proteins. It is that first step, featuring a backward flow of information, that inspired the *retro* in retrovirus and the *reverse* in reverse transcriptase.

Once HIV's genome has been inserted into the host cell's chromosomes, the host cell's RNA polymerase transcribes the viral genome into mRNA, and the host cell's ribosomes synthesize viral proteins. New virions assemble in the host cell cytoplasm, then bud off the cell membrane and enter the bloodstream. There, the new virions may find another cell to infect in the same host, or they may be transmitted to a new host.

A notable feature of HIV's life cycle is that the virus uses the host cell's own enzymatic machinery—its polymerases, ribosomes, and tRNAs—in almost every step. This is why HIV, and viral diseases in general, are so difficult to treat. Drugs that interrupt the virus's life cycle are almost certain to also interfere with the host cell's enzymatic functions and therefore cause debilitating side effects.

How Does HIV Cause AIDS?

Despite a quarter century of intensive research, the mechanisms whereby HIV infection leads to immune deficiency remain incompletely understood (Brenchley et al. 2006; Grossman et al. 2006). The short version is this: HIV parasitizes immune system cells, particularly helper T cells. After a long battle against the virus, the immune system's supply of helper T cells is badly depleted. Because helper T cells play a crucial role in the response to invading pathogens (Figure 1.6), this leaves the host vulnerable to a variety of secondary infections.

Evidence of hidden complexity lurking behind this short version comes from a study by Guido Silvestri and colleagues (2005). These researchers used SIVsm as

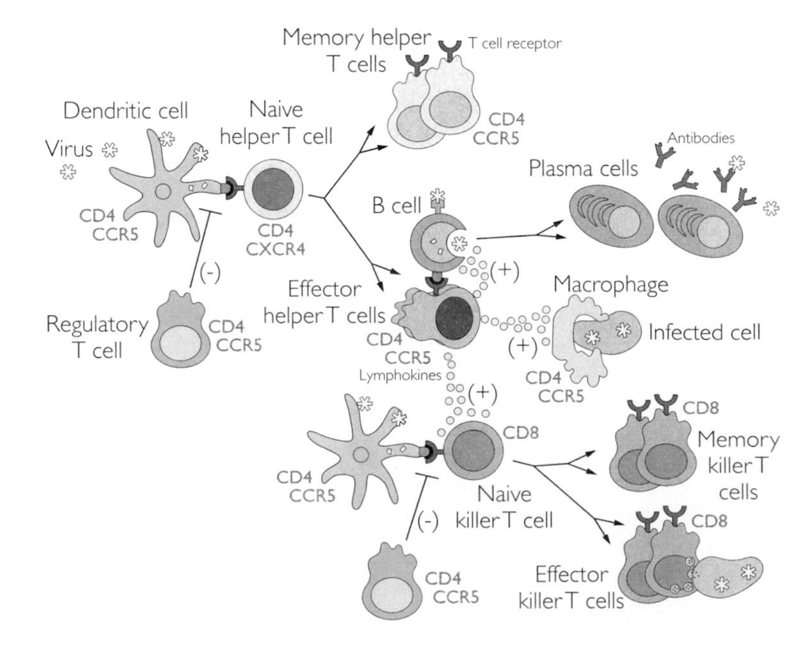

Figure 1.6 How the immune system fights a viral infection Dendritic cells (green) take up the virus and present bits of its proteins to naive helper T cells. Once activated by a bit of viral protein that fits its T cell receptor, a helper T cell divides to produce memory cells (yellow) and effector cells (red). Memory helper T cells sit out the current battle but remain ready to trigger a quick reaction should the same virus invade again. Effector helper T cells join the current fight. In part by releasing signalling molecules called chemokines, they stimulate B cells to mature into plasma cells that produce antibodies that bind the virus. Effector helper T cells also stimulate macrophages to ingest infected cells and help activate naive killer T cells. Activated killer T cells divide to produce memory cells and effector cells. Effector killer T cells identify and kill cells infected with the invading virus. The immune response is kept under control by regulatory T cells. The orange labels identify cell-surface proteins, some of which are exploited by HIV to gain entry into cells. Modified from NIAID (2003).

a model for HIV. SIVsm is a simian immunodeficiency virus that is related to HIV, but infects monkeys. SIVsm's natural host, the sooty mangabey, tolerates infection with SIVsm without getting sick. Rhesus macaques infected with SIVsm, however, typically develop AIDS. Silvestri and colleagues infected three sootey mangabeys and three rhesus macaques with SIVsm from the same source, then monitored the battle between the virus and the hosts' immune systems.

The virus sustained high levels of replication in all six hosts. Two of the three rhesus macaques, but none of the sooty mangabeys, showed chronically high immune activation evidenced by abundant proliferation of T cells. It was, paradoxically, these two rhesus macaques—the ones whose immune systems reponded most aggressively to the infection—that developed AIDS. It seems that the host's own immune response contributes to the development of immunodeficiency.

Consistent with this conclusion, Paolo Rizzardi and colleagues (2002) found, in a small clinical trail, that human HIV patients treated with antiretroviral drugs plus the immunosuppressant cyclosporin maintained higher helper T cell counts than control patients treated with antiretrovirals alone.

To decipher these results, we need to look at the T cell life cycle. T cells derive from stem cells in the bone marrow (Figure 1.7a). These stem cells generate precursors that mature into naive T cells in the thymus. Naive T cells are activated in lymph nodes. An activated T cell undergoes a burst of proliferation, yielding effector and memory cells. These circulate in the blood and move through tissues. A large fraction of the body's memory cells reside in lymphoid tissue associated with mucus membranes lining the nose, mouth, lungs, and especially the gut.

Naive T cells and memory T cells are long lived (Figure 1.7b). In contrast, effector cells, which actively engage in the fight against invaders, are short lived (Moulton and Farber 2006). Furthermore, any given T cell lineage has a finite capacity for replication—a capacity that is reduced with each cell division. This means that each burst of replication within a T cell lineage brings that lineage closer to exhaustion. As we will see shortly, these patterns may help explain how sustained immune activation during HIV infection can utimately deplete the body's supply of helper T cells and lead to the collapse of the host's defenses.

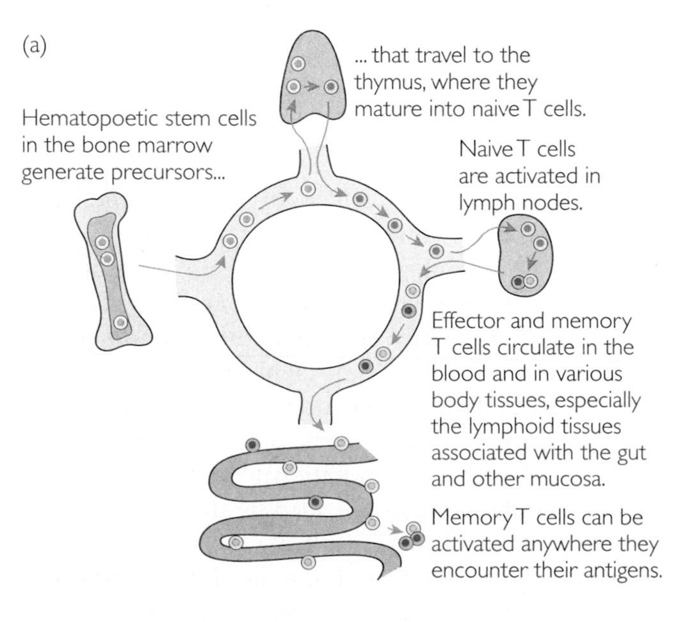

(a)
Hematopoetic stem cells in the bone marrow generate precursors...
... that travel to the thymus, where they mature into naive T cells.
Naive T cells are activated in lymph nodes.
Effector and memory T cells circulate in the blood and in various body tissues, especially the lymphoid tissues associated with the gut and other mucosa.
Memory T cells can be activated anywhere they encounter their antigens.

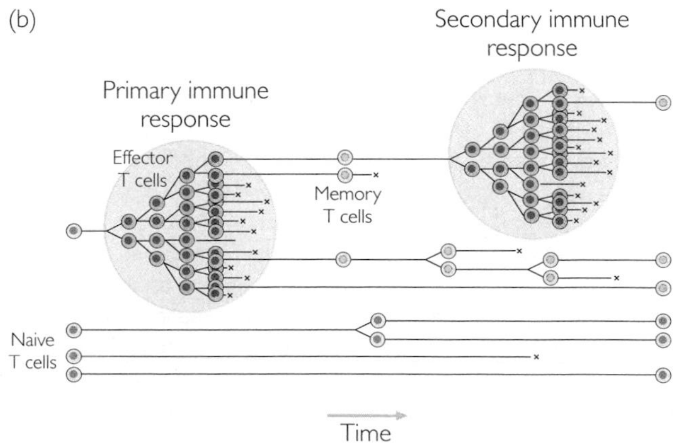

Figure 1.7 The life history of T cells (a) T cells derive from stem cells in bone marrow, mature in the thymus, and are activated in lymph nodes. (b) Naive and memory T cells are long lived; effector T cells are short lived. A given T cell lineage has a finite capacity to replicate. Modified from Grossman et al. (2002).

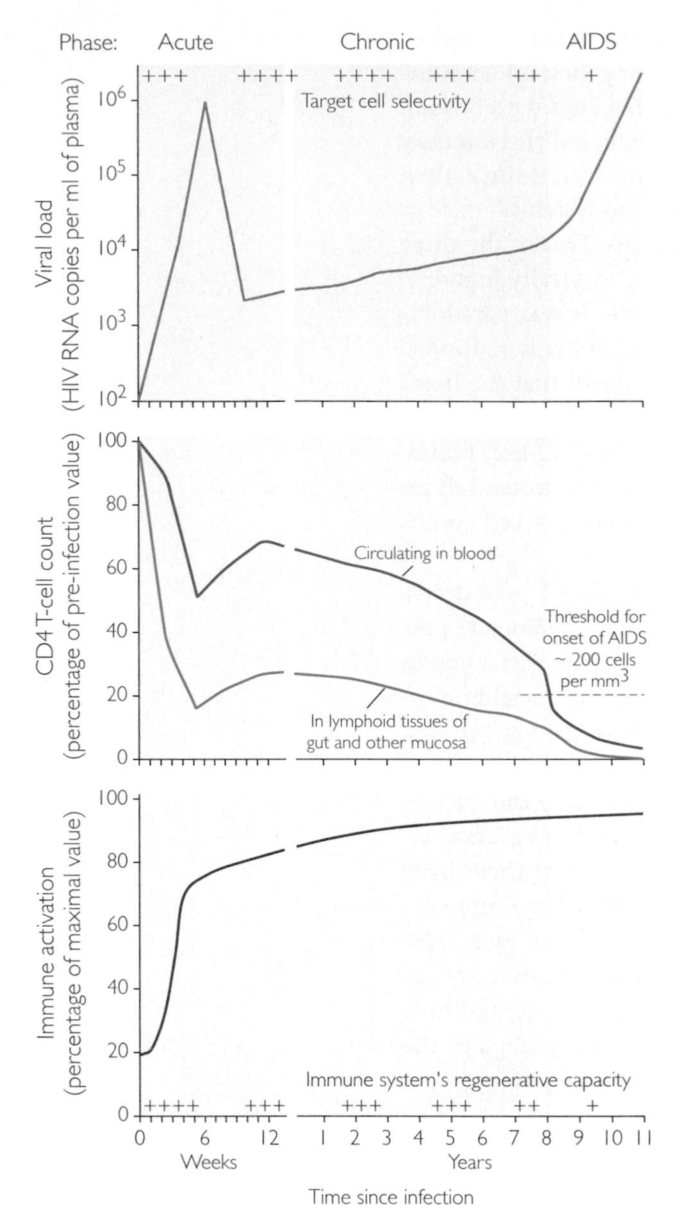

Figure 1.8 The general pattern of progression of an untreated HIV infection An untreated HIV infection typically has three phases: an acute phase, in which the host may show general symptoms of a viral infection; a chronic phase in which the host is largely asymptomatic; and an AIDS phase, in which the host's immune system collapses, leaving the host vulnerable to the opportunistic infections. The viral load (top) spikes during the acute phase, then falls as the host mobilizes an immune response. The immune response fails to halt viral replication, however, and during the acute and AIDS phases the viral load climbs again. Late in the infection, the viral population often evolves the capacity to infect a greater variety of host cells. The patient's CD4 T-cell counts (center) fall during the acute phase, then recover somewhat. During the chronic and AIDS phases, they fall again. The host's immune system remains highly activated (bottom) throughout. This may help fight the virus, but it also provides cells in which the virus can replicate and ultimately exhausts the immune system's capacity to regenerate. After Bartlett and Moore (1998), Brenchley et al. (2006), Grossman et al. (2006).

An untreated HIV infection exhibits distinct phases, in which the loss of helper T cells happens at different rates and appears to be driven by different mechanisms (Douek et al. 2003; Derdeyn and Silvestri 2005; Brenchley et al. 2006; Grossman et al. 2006). The graphs in Figure 1.8 track the viral load (top), helper T cell counts (center), and level of immune activation (bottom) in a typical host as his or her infection progresses through the acute phase, the chronic phase, and the terminal AIDS phase.

In the acute, or initial, phase, HIV virions enter the host's body and begin to replicate. As was shown in Figure 1.5 (page 7), HIV gains entry into a host cell by first latching onto the cell-surface protein CD4, then binding to a coreceptor. Look back at Figure 1.6. The presence of CD4 and other cell surface proteins on various immune system cells is indicated by the orange labels. The coreceptor used by most of the HIV strains responsible for new infections is CCR5. These viral strains can thus infect dendritic cells, macrophages, regulatory T cells, and especially memory and effector helper T cells.

HIV replicates explosively, and the concentration of virions in the blood climbs steeply. At the same time, the concentrations of CD4 T cells plummet, largely because HIV kills them while replicating. Hardest hit are the memory helper T cells in the lymphoid tissues of the gut (Guadalupe et al. 2003; Brenchley et al. 2004; Mehandru et al. 2004). Because the gut is both large and vulnerable to penetration by pathogens, the loss of these T cells is a severe blow to the body's defenses.

The acute phase ends when viral replication slows and the concentration of virions in the blood drops. One reason may be that the virus simply runs short of host cells it can easily invade. In addition, however, the immune system mobilizes against the infection and killer T cells begin to target host cells infected with HIV. The host's CD4 T cell counts recover somewhat.

HIV has been slowed, but it has not been stopped. As the chronic phase begins, the immune system struggles to recover from its initial losses while continuing to fight the virus. Throughout the chronic phase, the immune system remains highly activated. The reasons are not fully understood. To some extent the chronic activation is due to the ongoing effort to control the HIV infection. Additional causes may include stimulation by proteins encoded by the virus, the destruction of regulatory T cells by the virus (Oswald-Richter et al. 2004), and the need to fight other pathogens slipping past the weakened gut defenses.

The chronically activated state of the immune system may enhance some aspects of the host's response to HIV. However, it also generates a steady supply of activated CD4 T cells in which HIV can replicate. And it burns through the host's supply of naive and memory helper T cells by stimulating them to divide and differentiate into short-lived effector cells (Deeks and Walker 2004; Garber et al. 2004). Replacement of lost helper T cells ultimately depends on the production of new naive T cells by the thymus. Thymic output declines with age, however, and is impaired by HIV infection. HIV infection also damages the bone marrow and lymph nodes. As the battle wears on, the immune system's capacity to regenerate steadily erodes. The viral load climbs again and the CD4 T cell counts fall. The chronic phase ends when the concentration of helper T cells in the blood drops below about 200 cells per cubic millimeter.

With so few helper T cells left, the immune system can no longer function. The patient develops AIDS. The syndrome is characterized by opportunistic infections with bacterial and fungal pathogens that rarely cause problems for people with robust immune systems. In the absence of effective anti-HIV drug therapy, an HIV-infected individual who has begun showing symptoms of AIDS typically can expect to live two or three more years.

AIDS begins when HIV infection has progressed to a point where the immune system does not function properly.

Having covered the virus's basic biology, we are ready to explore questions about HIV's evolution. The first issue long frustrated everyone involved in fighting the epidemic: Why was it so hard to develop drugs capable of combating HIV? It was certainly not for lack of trying; governments and private companies committed hundreds of millions of dollars to AIDS research and drug development. The story of AZT, one of the first anti-AIDS drugs, turned out to be typical. Early on, AZT looked promising, but ultimately it proved a disappointment. To explain why, we need to introduce evolution by natural selection.

1.2 Why Does AZT Work in the Short Run, But Fail in the Long Run?

To combat viral infections, researchers look for drugs that are capable of inhibiting enzymes special to the virus. For example, a drug that blocks reverse transcription should kill retroviruses with minimal side effects. This is the rationale behind azidothymidine, or AZT.

Figure 1.9 shows how reverse transcription works. HIV's reverse transcriptase uses the viral RNA as a template to construct a complementary strand of

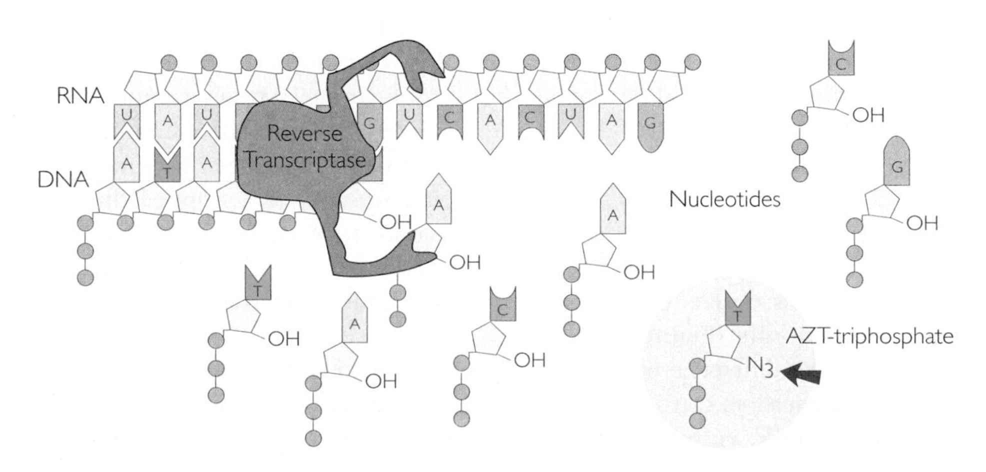

Figure 1.9 How AZT blocks reverse transcriptase HIV's reverse transcriptase enzyme uses nucleotides from the host cell to build a DNA strand complementary to the virus's RNA strand. AZT mimics a normal nucleotide well enough to fool reverse transcriptase, but it lacks the attachment site for the next nucleotide in the chain.

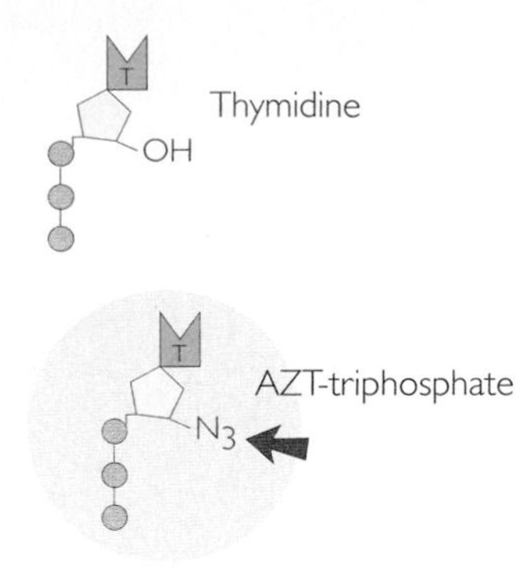

Figure 1.10 Thymidine versus AZT

DNA. Reverse transcriptase makes the DNA using building blocks—nucleotides—stolen from the host cell.

The figure also shows how AZT stops reverse transcription. Note the *thymidine* in AZT's name: AZT is similar in its chemical structure to the normal nucleotide thymidine—so similar that AZT fools reverse transcriptase into picking it up and incorporating it into the growing DNA strand. Note also, however, that there is a crucial difference between normal thymidine and AZT (Figure 1.10). Where thymidine has a hydroxyl group (—OH), AZT has an azide group ($—N_3$). The hydroxyl group that AZT lacks is precisely where reverse transcriptase would attach the next nucleotide to the growing DNA molecule. Reverse transcriptase is now stuck. Unable to add more nucleotides, it cannot finish its job. AZT thus interrupts the pathway to new viral proteins and new virions.

In early tests, AZT worked. It effectively halted the loss of macrophages and T cells in AIDS patients. AZT also caused serious side effects because it sometimes fools DNA polymerase and interrupts DNA synthesis in host cells. But it did promise to inhibit, or at least slow, the progression of the disease. By 1989, however, after only a few years of use, patients stopped responding to treatment. Their CD4 cell counts again began to decline. Why?

In principle, AZT could lose its effectiveness in either or both of two ways. One way is that the patient's own cellular physiology could change. After it enters a cell, AZT has to be phosphorylated by the cell's own thymidine kinase enzyme to become biologically active. Perhaps long-term exposure to AZT causes a cell to make less thymidine kinase. If so, AZT would become less effective over time. Patrick Hoggard and colleagues (2001) tested this hypothesis by periodically checking the intracellular concentrations of phosphorylated AZT in a group of patients taking the same dose of AZT for a year. The data refute the hypothesis. The concentrations of phosphorylated AZT did not change over time.

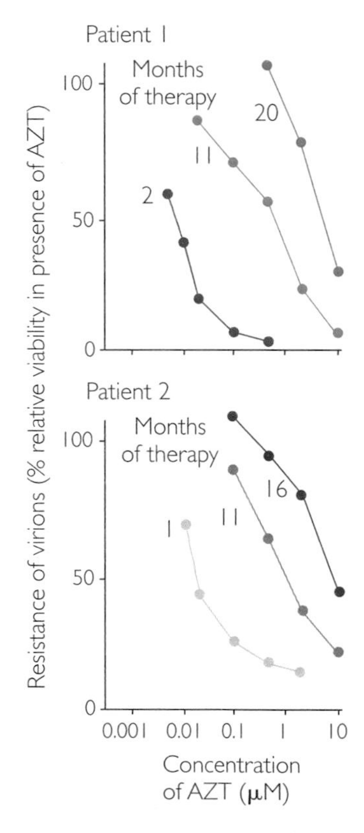

Figure 1.11 **HIV populations evolve resistance to AZT within individual patients** As therapy continued in these two patients, higher concentrations of AZT were required to curtail the replication of viruses sampled from the patients' blood. Redrawn from Larder et al. (1989).

The other way AZT could lose its effectiveness is that the population of virions living inside the patient could change so that the virions themselves would be resistant to disruption by AZT. To find out whether populations of virions become resistant to AZT over time, Brendan Larder and colleagues (1989) took samples of HIV from patients and grew the virus on cultured cells in Petri dishes. Figure 1.11 shows data for two patients that the researchers monitored for many months. Each curve in the graphs falls, showing how rapidly HIV's ability to replicate is curbed by increasing concentrations of AZT. Examine the three curves for Patient 1. Virions sampled from this patient after he had been taking AZT for 2 months were still susceptible to the drug. The virions lost their ability to replicate almost entirely at moderate concentrations of AZT. Virions sampled from the patient after 11 months on AZT were partially resistant. They could be stopped, but it took about 10 times as much AZT to do it. Virions taken after 20 months on AZT were highly resistant. They were completely unaffected by AZT concentrations that stopped the first sample and could still replicate fairly well at concentrations that stopped the second sample. The data for Patient 2 tell the same story. Populations of virions within individual patients change to become resistant to AZT. In other words, the populations evolve. In most patients, the evolution of AZT-resistant HIV takes just 6 months (Figure 1.12).

What is the difference between a resistant virion versus a susceptible one? To answer this question, consider a thought experiment. If we wanted to genetically engineer an HIV virion capable of replicating in the presence of AZT, what would we do? The simplest answer might be to change the active site in the reverse transcriptase enzyme, making it less likely to mistake AZT for the normal nucleotide. The cartoon in Figure 1.13a shows how this might work in principle.

In practice, we could use a mutagenic chemical or ionizing radiation to generate strains of HIV with altered nucleotide sequences in their genomes and thus altered amino acid sequences in their proteins. If many mutants were generated, at least a few would carry changes in the part of the reverse transcriptase molecule that recognizes and binds to normal thymidine. A model of the actual structure of reverse transcriptase's binding site appears in Figure 1.13b. If one of the reverse transcriptases with an altered binding site were less likely to mistake AZT for the normal nucleotide, then the mutant variant of HIV would be able to continue replicating in the presence of the drug. In populations of HIV virions treated with AZT, strains unable to replicate in the presence of AZT would decline in numbers, and the new form would come to dominate the HIV populations.

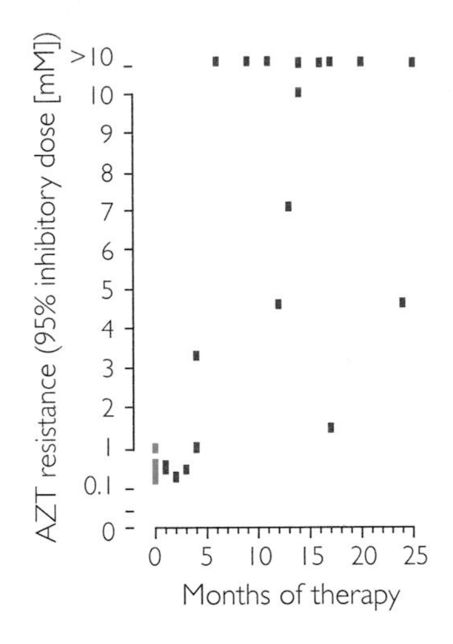

Figure 1.12 In most patients, AZT resistance evolves within six months This graph plots resistance in 39 patients checked at different times. Redrawn from Larder et al. (1989).

The steps involved in this thought experiment are just what happens inside the bodies of HIV patients like the ones followed by Larder and colleagues. How do we know? In studies similar to Larder's, researchers took repeated samples of HIV virions from patients receiving AZT. In each sample, the researchers sequenced the reverse transcriptase gene. They found that viral strains present late in treatment were genetically different from viral strains that had been present before treatment in the same host individuals. The mutations associated with AZT resistance were often the same from patient to patient (St. Clair et al. 1991; Mohri et al. 1993; Shirasaka et al. 1993) and were located in the active site of reverse transcriptase (Figure 1.13c). Researchers have directly observed the evolution of AZT resistance in dozens of AIDS patients. In each individual, mutations in the HIV genome led to specific amino acid substitutions in the active site of reverse transcriptase. These genetic changes allowed the mutant strains of the virus to replicate in the presence of AZT. Unlike the situation in our thought experiment, however, no conscious manipulation took place. How, then, did the change in the viral strains occur?

Some mutations in the active site of reverse transcriptase make the enzyme less likely to add AZT instead of thymidine.

The answer is that reverse transcriptase is error prone, and HIV's genome has no instructions for making error-correcting enzymes. As a result, over half of the DNA transcripts produced by reverse transcriptase contain at least one mistake in their

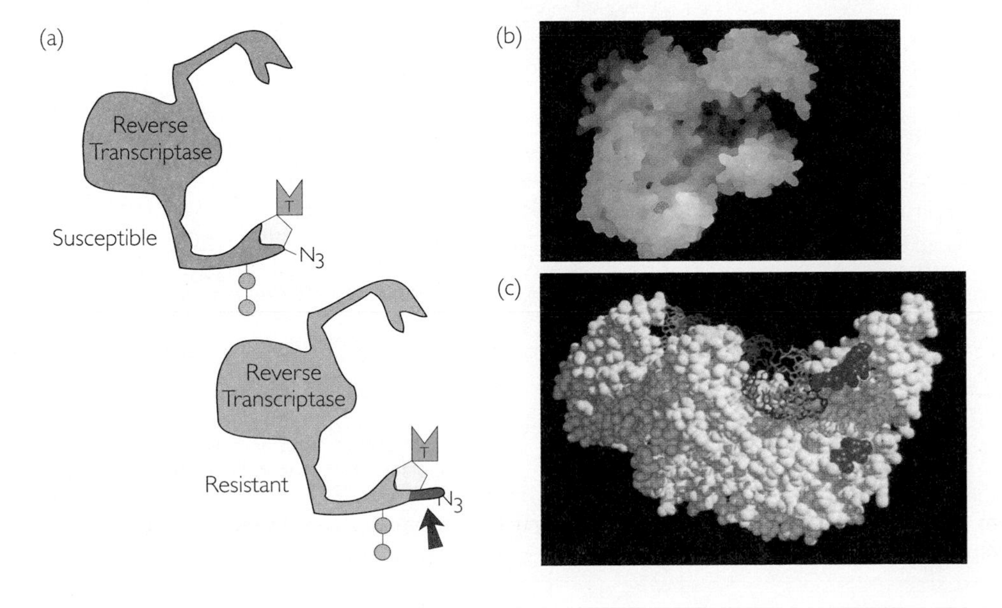

Figure 1.13 The difference between AZT-sensitive versus AZT-resistant reverse transcriptases (a) This cartoon shows how a change in reverse transcriptase's active site might enable it to recognize AZT as an imposter. (b) This image shows the large groove in the reverse transcriptase enzyme where the substrate (RNA) binds. (c) The red spheres on this image indicate the locations of amino acid substitutions correlated with resistance to AZT. Note that they are localized in the enzyme's groove, or active site. From Cohen (1993).

nucleotide sequence, also known as a mutation (Hübner et al. 1992; Wain-Hobson 1993). In fact, HIV has the highest mutation rate of any virus or organism observed to date. Because thousands of generations of HIV replication take place within each patient during the course of an infection, a single strain of HIV can produce hundreds of different reverse transcriptase variants over time.

Simply because of their numbers, it is a virtual certainty that one or more of these variants contains an amino acid substitution that lessens reverse transcriptase's affinity for AZT. If the patient takes AZT, the replication of unaltered HIV variants is suppressed, but the resistant mutants will still be able to synthesize some DNA and produce new virions. As the resistant virions reproduce and the nonresistant virions fail to propagate, the fraction of the virions in the patient's body that are resistant to AZT increases over time. Furthermore, each

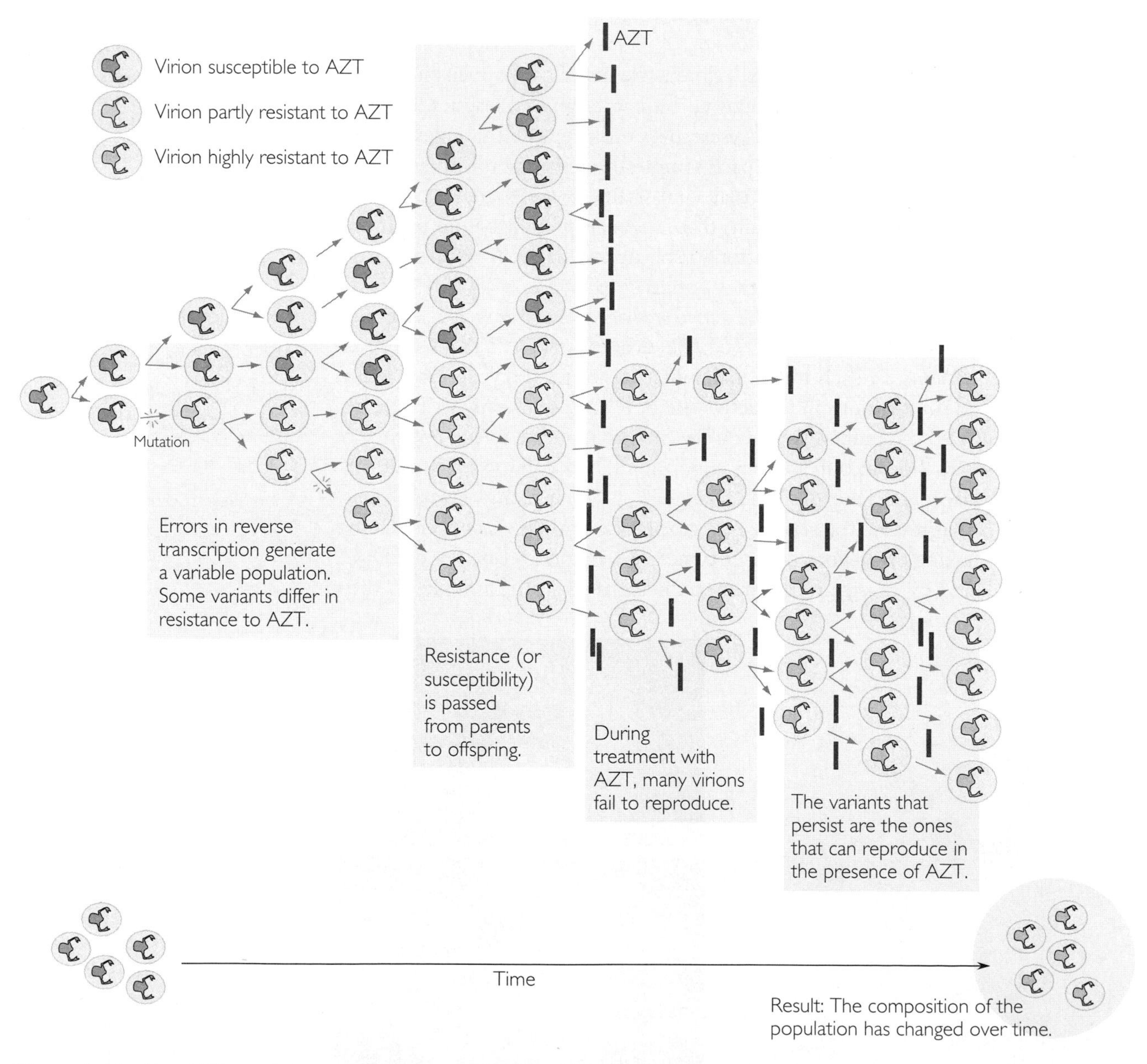

Figure 1.14 How HIV populations evolve resistance to AZT Variation due to mutation, inheritance, and differences in survival due to AZT results in a change in the composition of the population over time.

new generation in the viral population is likely to contain virions with additional new mutations. Some of these additional mutations may further enhance the ability of reverse transcriptase to function in the presence of AZT. Because they reproduce faster, the virions that carry these new mutations will also increase in frequency at the expense of their less-resistant contemporaries.

This process of change over time in the composition of the viral population is called evolution by natural selection. It has occurred so consistently in patients taking AZT that the use of AZT alone has been abandoned as an AIDS therapy. (We discuss more advanced therapies in Box 1.1.)

Changes in the genetic makeup of HIV populations over time have led to increased drug resistance. This is an example of evolution by natural selection.

Now consider a slightly different question. We have been following what happens to virions carrying different versions of the reverse transcriptase gene when AZT is present. What happens when AZT is absent? Are the mutant HIV strains also more efficient at reproducing when host cells do not contain AZT? No. When AZT therapy has been stopped, the proportion of AZT-resistant virions in the viral population has fallen back toward what it was before AZT treatment began. Back-mutations that restored reverse transcriptase's amino acid sequence to its original configuration became common because the virons that carried them out-reproduced AZT-resistant forms (St. Clair et al. 1991). The viral strain that increases in frequency is the one that replicates fastest *in the current environment*. Without AZT present, natural selection favors nonmutant virions; with AZT present, natural selection favors mutant virions. Is evolution by natural selection unidirectional and irreversible? It is not.

Note that the process we have described involves four steps (Figure 1.14):

1. Transcription errors produce mutations in the reverse transcriptase gene. Virions carrying mutant reverse transcriptase genes produce versions of the reverse transcriptase enzyme that vary in their resistance to AZT.
2. The mutant virions pass their reverse transcriptase genes, and thus their AZT resistance or susceptibility, to their offspring. In other words, AZT resistance is heritable.
3. During treatment with AZT, some virions are better able to survive and reproduce than others.
4. The virions that persist in the presence of AZT are the ones with mutations in their reverse transcriptase genes that confer resistance.

Heritable traits that lead to survival and reproductive abundance spread in populations; heritable traits that lead to reproductive deficit disappear. This is evolution by natural selection.

The result is that the composition of the viral population within the host changes over time. Virons resistant to AZT comprise an ever larger fraction of the population; virions susceptible to AZT become rare. There is nothing mysterious or purposeful about evolution by natural selection; it just happens. It is an automatic consequence of simple, cold arithmetic.

Because evolution by natural selection is an automatic consequence of cold arithmetic, it can happen in any population in which the four steps occur. That is, it can happen in any population in which there is heritable variation in reproductive success. We will see many examples in the chapters to come.

One measure of whether we understand a process is whether we can control it. If we truly understand the mechanism of evolution by natural selection as it operates inside the bodies of HIV patients, we should be to find a way to stop it—or at least slow it down. For a discussion of how researchers have used their understanding of the mechanism of resistance evolution to devise more effective therapies, see Box 1.1.

Box 1.1 Can understanding how resistance evolves help researchers design better treatments?

Researchers have developed a variety of antiretroviral drugs that, like AZT, target processes unique to viral enzymes and proteins (see Figure 1.5, page 7; Pomerantz and Horn 2003; Pommier et al. 2005). Drugs in use or under development include:

- **Reverse transcriptase inhibitors.** Some, like AZT, inhibit reverse transcriptase by mimicking the normal building blocks of DNA. Others inhibit reverse transcriptase by directly blocking the enzyme's active site.
- **Protease inhibitors.** These prevent HIV's protease enzyme from cleaving viral precursor proteins to produce mature components for new virions.
- **Fusion inhibitors.** These bar HIV from entering host cells in the first place by interfering with HIV's gp120 or gp41 proteins or by blocking proteins on the surface of the host cell that HIV latches onto.
- **Integrase inhibitors.** These block HIV's integrase from inserting HIV DNA into the host genome, thereby preventing the transcription of new viral RNAs.

Experience so far indicates that when any antiretroviral drug is used alone, the outcome will be the same as we have seen with AZT. The virus population in the host quickly evolves resistance (see, for example, St. Clair et al. 1991; Condra et al. 1996; Ala et al. 1997; Deeks et al. 1997; Doukhan and Delwart 2001).

With any single drug, as we saw with AZT, just one or a few mutations in the gene for the targeted protein can render the virus resistant. With its high mutation rate, short generation time, and large population size, HIV generates so many mutant genomes that one with the crucial combination of mutations is likely to appear within a fairly short time. When there is genetic variation for replication in the presence of the drug, and the drug is present, then the population inevitably evolves.

What we need is a way to increase the number of mutations that must be present in a virion's genome to render the virion resistant. The more mutations needed for resistance, the lower the probability that the mutations will occur together in a single virion. In other words, we need a strategy to reduce the genetic variation for resistance to zero. Without genetic variation, the viral population cannot evolve.

The simplest way to increase the number of mutations required for resistance is to use two or more drugs at once. Resistance to the drugs must be conferred by different mutations. Ideally, mutations that make HIV resistant to one of the drugs will also make the virus susceptible to another of the drugs (see St. Clair et al. 1991).

The good news is that treatment cocktails using combinations of drugs have proven effective. For example, Roy Gulick and colleagues (1997) found that in many patients a cocktail of two reverse transcriptase inhibitors (AZT and dideoxy-3′-thiacytidine, or 3TC), plus a protease inhibitor (indinavir), can reduce the number of HIV virions in blood plasma to undetectable levels for at least a year. Results such as these have earned multidrug treatments the collective nickname Highly Active Anti-Retroviral Therapy, or HAART (Cohen 2002a; for more on drug combinations used in HAART, see Kalkut 2005).

Frank Palella and colleagues (2002) followed nearly 1,800 patients on various HAART regimens for six years. With the advent of HAART in 1996, mortality rates among the patients dropped dramatically (Figure 1.15a), as did the incidence of opportunistic infections typical of AIDS (Figure 1.15b). Understanding how resistance evolves has helped researchers save lives.

1.3 Why Is HIV Fatal?

One of the keys to becoming an evolutionary biologist is to learn "selection thinking." The idea is that evolution by natural selection, as outlined in Section 1.2, is an automatic process that simply happens whenever a population shows the necessary heritable variation in survival and reproductive success. Traits con-

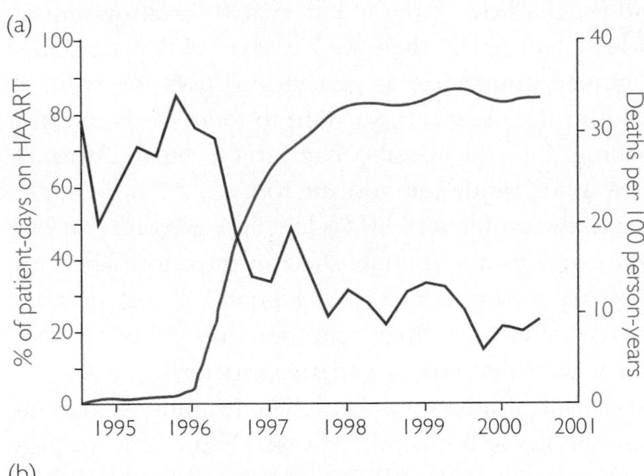

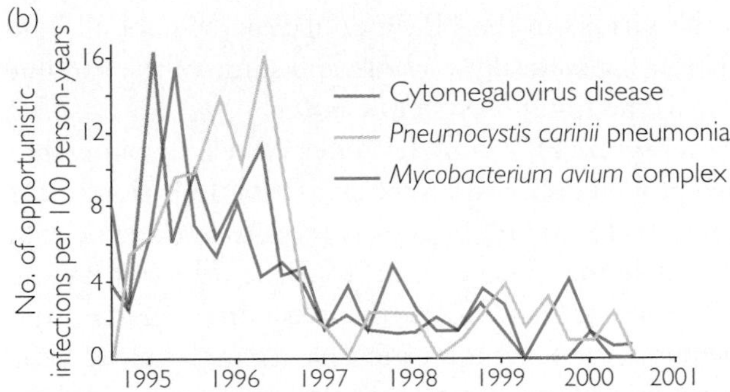

Figure 1.15 Successes of highly active antiretroviral therapy In a sample of 1,800 patients, the introduction of multiple-drug cocktails led to a dramatic reduction in both the AIDS death rate (a) and the incidence of secondary infections characteristic of AIDS (b). From Palella et al. (2002).

The bad news is that multidrug cocktails do not cure HIV infection. A reservoir of viable HIV genomes remains in the body, hidden in the chromosomes of resting white blood cells and possibly other tissues (Chun et al. 1997; Finzi et al. 1997; Wong et al. 1997b). As a result, when patients go off HAART their viral loads climb rapidly (Chun et al. 1999; Davey et al. 1999; Oxenius et al. 2002; Kaufmann et al. 2004). The hidden HIV reservoir may persist for decades (Finzi et al. 1999). Researchers are experimenting with therapies that might deplete it, but it is unclear whether it will ever be possible to drain the reservoir completely (Lehrman et al. 2005; Smith 2005).

A crucial question is whether the virions in the hidden reservoir are dormant or replicating. It appears that in some patients, HAART suppresses all replication and only dormant virions persist (see, for example, Finzi et al. 1997; Wong et al. 1997b; Zhang et al. 1999). So long as all virions are dormant, then the viral population will not evolve.

In other patients, however, some virions continue to replicate (see, for example, Günthard et al. 1999; Ramratnam et al. 2000; Sharkey et al. 2000; Frost et al. 2001). Ongoing replication suggests that the viral population harbored at least some variation in resistance before therapy started. As long as partially resistant virions are continuing to reproduce, there is an opportunity for mutations conferring additional resistance to appear and, under selection imposed by the drugs, accumulate within viral strains (Kristiansen et al. 2005). Several teams of researchers have documented the evolution of HIV strains that are simultaneously resistant to multiple drugs, including both reverse transcriptase inhibitors and protease inhibitors (Wong et al. 1997a; Gallago et al. 2001; Grant et al. 2002; Evans et al. 2005; Markowitz et al. 2005).

A further disappointment is that many patients taking multidrug cocktails suffer side effects that are difficult or impossible to tolerate (Cohen 2002a). Nausea, anemia, and a variety of metabolic disorders make it hard to adhere to prescribed therapy (Sabundayo et al. 2006). Patients maintain lower concentrations of the antiretroviral drugs, increasing the probability that partially resistant virions will be able to reproduce and hence that viral populations will evolve.

The bottom line is that the high activity of HAART carries an expiration date for most patients (Chen et al. 2003; Mocroft et al. 2004). In Palella's study—the study that produced the dramatic data presented in Figure 1.15—few HAART regimens remained effective for more than three years.

Anti-HIV therapies that are easily tolerated and that permanently suppress viral replication, and viral evolution, remain a goal of ongoing research.

ducive to surviving and reproducing spread throughout the population; traits conducive to dying without issue disappear. If we want to understand why a particular trait is common in a particular population, a good first step is to try to understand how the trait might affect the survival and reproductive success of individuals. In this section, we apply selection thinking to a puzzling feature of HIV infections: Untreated, they are almost always fatal.

(a)

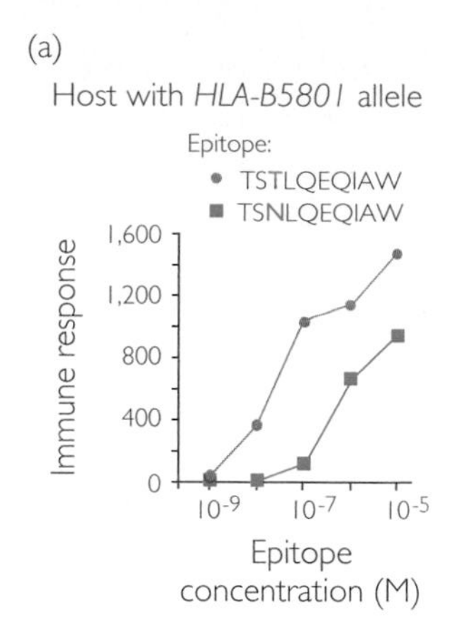

(b)

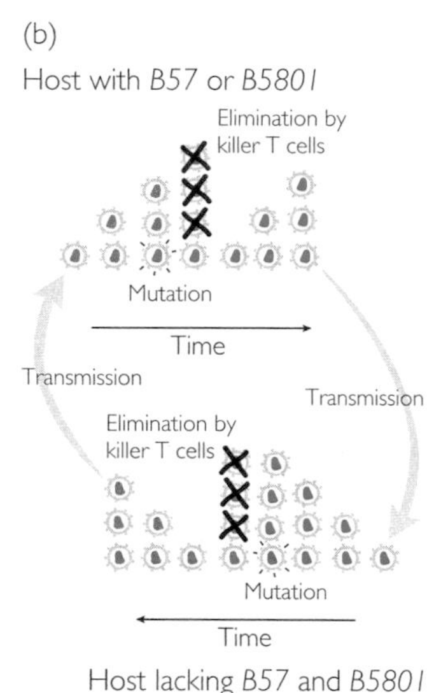

Figure 1.16 An HIV escape mutation (a) This graph shows the vigor of the immune response by an HIV patient's white blood cells as a function of the concentration of the protein fragment, or epitope, being tested. The two frangments are variants of a small portion of the p24 protein. Each letter represents an amino acid: T = threonine; S = serine; N = asparginine; and so on. The units of immune response are number of cells, per million, producing interleukin-gamma. From Leslie and colleagues (2004). (b) In hosts carrying either HLA allele *B57* or *B5801*, the HIV population evolves toward high frequencies of the N variant. In hosts with other genotypes, the HIV population evolves toward high frequencies of the T variant.

Short-Sighted Evolution

Dying of AIDS is clearly bad for the host. If there is heritable variation among humans in the resistance to HIV and AIDS, then we can expect that resistance will spread throughout the human population as generations pass. We will explore this issue in Section 1.4. But the organism we want to focus on here is not the host; it is the virus. Is killing the host not also bad for the virus? After all, when the host dies, the virions living inside the host die too.

To apply selection thinking to the problem of HIV's lethality, imagine that one or a few virions have invaded a new host and established an infection. The virions in the new population are replicating rapidly (see Figure 1.8, top, on page 10). As they use reverse transcriptase to copy their genomes, they generate many mutations. The growing population is developing genetic variation.

Now the host's body mobilizes its immune response. The immune system attacks the HIV virions with antibodies and killer T cells (see Figure 1.6 on page 8). These eliminate many of the virions in the HIV population, but not all. The reason is that the HIV population is genetically variable, and some of the variants are less susceptible than others to the immune system's assault.

Antibodies and killer T cells recognize HIV and HIV-infected cells by binding to **epitopes**—short pieces of viral protein displayed on the surface of the virion or the infected cell. These epitopes are encoded in HIV's genes. Mutations in the genes can change the epitopes and may enable the mutant virion to evade detection by the host's current arsenal of antibodies and killer T cells. As the infection progresses from the acute phase to the chronic phase, the HIV population has already evolved. Variants easily recognized by the first wave of the immune attack have disappeared; variants less easily recognized persist (Price et al. 1997; Allen et al. 2000).

Figure 1.16 provides an example of a mutation that helps HIV virions evade the immune response in some patients. The mutation affects an epitope in a protein, called p24, that is a component of the capsule that surrounds the core of the HIV virion. Infected host cells display this epitope on their surface, along with a host protein called a human leucocyte antigen, or HLA. When a killer T-cell recognizes the foreign epitope alongside the self HLA protein, it destroys the infected cell.

In a survey of virions from more than 300 patients, A. J. Leslie and colleagues (2004) found that in most strains of HIV the third amino acid in the eptitope is threonine. However, in most HIV strains from patients who carry either of two particular alleles of HLA-B—*B5801* or *B57*—the third amino acid is asparginine.

Experiments in test tubes showed why. Leslie and colleagues took white blood cells from a patient with the *B5801* allele and exposed them to different versions of the p24 epitope (Figure 1.16a). The patient's cells reacted much more strongly to the version with threonine than to the version with asparginine. White blood cells from patients with the *B57* allele showed a similar pattern.

Leslie and colleagues found several cases in which an individual with either *B5801* or *B57* had become infected with HIV from a host lacking both alleles. By periodically sampling the viral population in the new host, they were able to document the population's evolution. Early in the infection, the virions all had threonine in position three of the p24 epitope (Figure 1.16b). Soon, however, mutant virions appeared with asparginine in position three. Eventually, virions with threonine went extinct, and only virions with asparginine remained. The researchers also found cases in which individuals lacking *B5801* and *B57* became infected with HIV from a host with one of these alleles. Periodic sampling from these patients showed that their viral populations evolved in the opposite direction.

Because the immune system never completely curtails HIV's replication, the HIV population inside a host evolves throughout the chronic phase of infection. The HIV population produces as many as 10 million to 100 million new virions each day (Ho et al. 1995; Wei et al. 1995). As they replicate, the virions accidentally generate mutations that change their epitopes. Some of the mutant virions enjoy unfettered reproduction until the immune system develops antibodies and killer T cells that recognize their altered proteins. Then these mutants disappear, and a new generation of virions with novel epitopes automatically takes their place.

Raj Shankarappa and colleagues (1999), working in the laboratory of James Mullins, documented continuous evolution in the HIV population during the chronic phase of infection in several individual patients. The data for one of the patients appear in Figure 1.17. Look first at Figure 1.17a. The scientists periodically harvested HIV virions from the patient's blood and read the sequence of nucleotides in a portion of the gene for gp120. This protein sits on HIV's outer envelope, where it initiates fusion with host cells by binding to CD4 and a coreceptor (see Figure 1.5 on page 7). The portion of the gene the scientists studied determines which coreceptor the virion uses and contains an epitope targeted by the host's immune system. The team noted the nucleotide sequence in the first sample they took from the patient and compared all subsequent samples to it. Over the first seven years the researchers followed the patient, the sequences went from nearly identical to the reference sequence to differing at almost 8% of their nucleotides.

Now notice what happened between year six and year eight. The trend line stopped climbing and flattened out. In other words, the rate of evolution slowed dramatically. Why? Did the virus population stop generating the genetic variation that fuels evolution by natural selection? Probably not. Figure 1.17b shows that the concentration of virions was high at the time. With so many virions replicating, the population was certainly continuing to churn out mutant genomes at a furious rate. Did the way in which viral genotype influences survival and reproduction change? Probably so. Until year seven, virions with genotypes that gave them novel epitopes were more likely to survive and proliferate (see Ross and Rodrigo 2002); after year seven, this advantage appears to have vanished.

Figure 1.17c shows that around the time the rate of virus evolution slowed, the patient's CD4 T-cell count plunged. At year six, it was about 1,200 cells per cubic

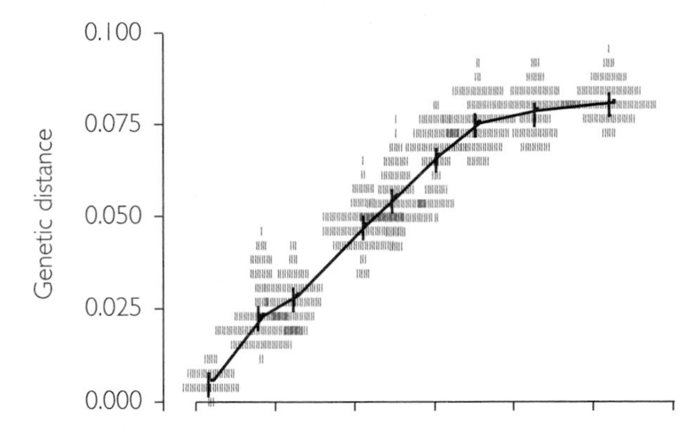

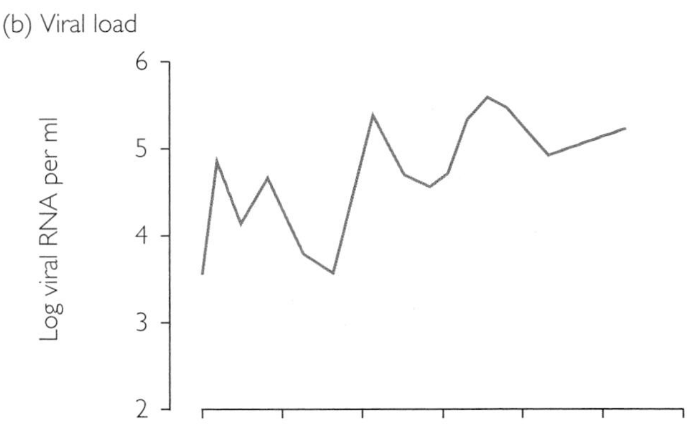

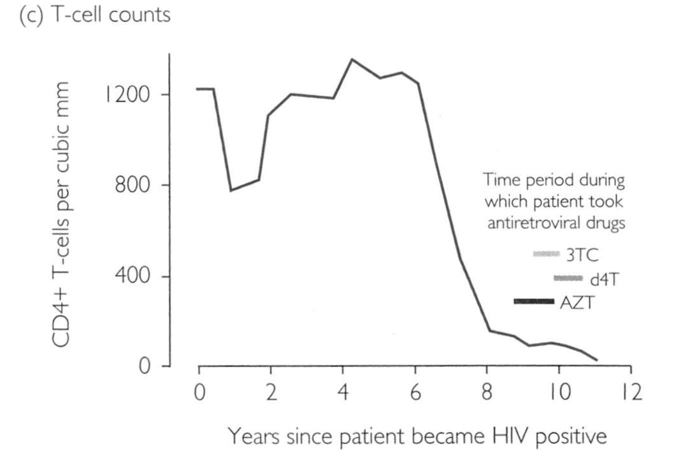

Figure 1.17 Evolution of the HIV population within an individual patient (a) Each blue tick represents a virion sampled from the patient during the course of the infection; its horizontal position indicates when it was sampled, and its vertical position indicates how genetically different it was from the first sample. The black line shows the trend: Virions sampled later had diverged more. (b) The patient's viral load increased over time. (c) The patient's CD4 T-cell count stayed fairly high for several years, then dropped precipitously. From Shankarappa et al. (1999).

Pathogen populations evolve inside individual hosts in response to selection imposed by the immune system.

millimeter; by year eight it was less than 200. The patient's immune system was collapsing. The collapse of the immune system meant that the patient's body was no longer producing new kinds of antibodies and new kinds of killer T cells. This freed the HIV population from the selective agent that was forcing it to evolve. There was no longer any benefit to having novel epitopes. Instead, the strains most capable of rapid replication simply spread, and those less capable became rare (see Williamson et al. 2005).

The evolution of the HIV population appears to contribute to the collapse of the immune system in at least three ways. First, it is the continuous evolution toward novel epitopes that enables the viral population to stay far enough ahead of the immune response to keep replicating in high numbers. Eventually, as described in Section 1.1, the continuously replicating viral population burns through the host's supply of naive and memory T cells and destroys the body's ability to replace them.

Second, the viral population within most hosts evolves toward ever more aggressive replication. Ryan Troyer and colleagues (2005) took sequential HIV samples from several untreated patients. The researchers grew the virions from each sample on white blood cells from an uninfected donor. To each culture dish the researchers added one of four control strains of HIV against which the virions from the patient would have to compete. In each dish, the viral strain that could replicate most efficiently became numerically predominant. Troyer and colleagues assessed the competitive fitness of the virions from the patients' samples based on their overall performance against the four control strains. The results appear in Figure 1.18. Each color represents the sequential samples from a particular patient. In seven of eight cases, the competitive fitness of the patient's virions steadily increased over time. With two of their patients, Troyer and colleagues also competed strains sampled early against strains sampled later. The strains later sampled always won. The longer a patient harbors an HIV population, the more damaging the virions in the population become.

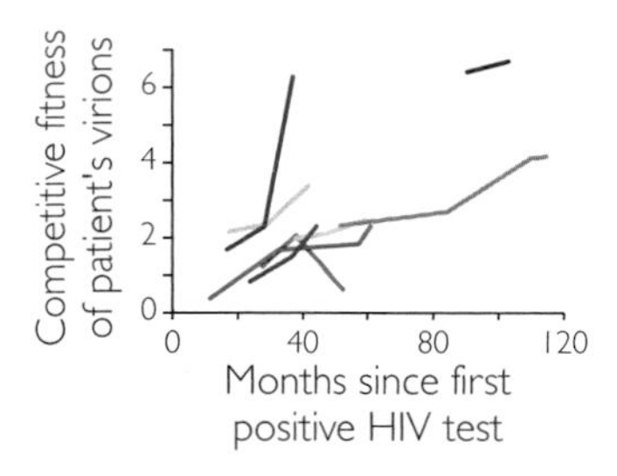

Figure 1.18 HIV populations in most hosts evolve toward more aggressive replication Each color represents sequentially sampled virions from a particular host. Competitive fitness reflects the virions' ability to replicate in white blood cells from an uninfected donor in the presence of control strains. In seven of eight patients, the HIV population's ability to persist in the fact of competition increased over time. Rerendered from Troyer et al. (2005).

Third, in at least half of all hosts—and possibly many more—strains of HIV evolve that can infect naive T cells (Shankarappa et al. 1999; Moore et al. 2004). An HIV virion's ability to infect a given cell type is determined by the coreceptor the virion uses. The coreceptor, as shown in Figure 1.5 on page 7, is the second of two proteins HIV latches onto to infiltrate a host cell. Early in most HIV infections, most virions in the HIV population use as their coreceptor a protein called CCR5. CCR5 is found on dendritic cells, macrophages, and regulatory, resting, and effector T cells (see Figure 1.6 on page 8). As the infection progresses and the HIV population evolves, virions often emerge that exploit a different coreceptor, a protein called CXCR4. Such virions, called X4 virions, may even become numerically predominant. This is what happened in the patient whose infection is detailed in Figure 1.17. X4 virions were absent early in the infection, became strongly predominant between years 5 and 8, and then became fairly rare again by year 11. CXCR4 is found on naive T cells.

The rapid evolution of the HIV population inside a host hastens the collapse of the host's immune system. This evolution is short-sighted because it also hastens the extinction of the HIV population.

Because naive T cells are the progenitors of memory and effector T cells, the emergence of virions that can infect and kill naive T cells is typically bad news for the host. Hetty Blaak and colleagues (2000) sampled the viral populations of 16 HIV patients to determine whether they contained X4 virions. Then, for a time span running from a year before to a year after the date of this sample, the researchers calculated the average helper T cell counts in the blood of patients with X4 viruses versus the patients without. The results appear in Figure 1.19.

The average T cell counts in the patients without X4 viral strains held fairly steady over time; the average counts in the patients with X4 strains fell. When virions arise that undermine the immune system's ability to replenish its stock of T cells, they apparently accelerate the immune system's demise.

The evolution of the HIV population within a host is short-sighted (Levin and Bull 1994; Levin 1996). The virions do not look to the future and anticipate that as their population evolves it will ultimately kill its host and thereby cause its own extinction. The virions can't look to the future; they are just tiny, thoughtless molecular machines. Evolution by natural selection does not look to the future either. It can't; it is just a mathematical process that happens automatically. As a result, the HIV population in any particular host ultimately evolves itself right out of existence.

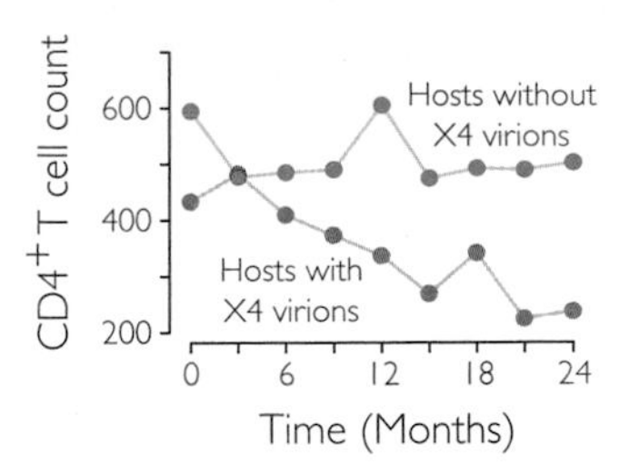

Figure 1.19 HIV strains that use CXCR4 as a coreceptor hasten the collapse of their hosts' immune systems Helper T cell counts fall faster in patients that harbor X4 strains, apparently because X4 strains infect and kill naive T cells and thereby undermine their hosts' ability to generate new memory and effector T cells. Rerendered from Blaak et al. (2000).

The short-sighted nature of HIV evolution is particularly clear in the case of X4 virions. Comparisons of the virions present in sequentially infected hosts show that HIV strains that use CXCR4 as a coreceptor do not get transmitted to new hosts (Zhu et al. 1993; Clevestig et al. 2005). Pathogens that cannot move to new hosts cannot survive for long. Even if they played no role in destroying their hosts' immune systems, X4 strains would thus be destined for certain extinction.

In summary, selection thinking leads us to the conclusion that HIV infection is fatal, at least in part, because of the short-sighted evolution of the HIV population inside the host. Lethal strains of HIV become predominant in the host because they enjoy a short-term advantage in survival and reproduction.

A Correlation between Lethality and Transmission?

Short-sighted evolution may not be the only reason HIV infections are fatal. The evidence for this claim is that rare strains of HIV exist that kill their hosts more slowly than common strains, if at all (Geffin et al. 2000; Rhodes et al. 2000; Tobiume et al. 2002). The best-known attenuated strain of HIV spread from a blood donor to eight transfusion recipients in Australia in the early 1980s (Deacon et al. 1995; Learmont et al. 1999; Birch et al. 2001; Churchill et al. 2006). This group of patients is known as the Sydney Bloodbank Cohort. The donor and four of the recipients have been followed for up to 25 years since they became infected. They are getting old enough that two have died of age-associated conditions unrelated to HIV. Two members of the cohort eventually developed AIDS, one 17 years and the other 18 years after infection, but both responded well to antiretroviral therapy. One member had detectable levels of HIV in his blood but remained asymptomatic. The remaining two have had normal T cell counts and viral loads so as to be undetectable. Overall, the cohort has fared much better than a comparison group infected with normal strains of HIV.

What accounts for the slower progression to AIDS in the Syndey Bloodbank cohort? N. J. Deacon and colleagues (1995) examined the genome of the HIV strain infecting the cohort. The cohort's strain, they discovered, is missing part of the gene for a viral protein called Nef. The mechanisms through which the Nef protein works are not entirely understood (Fackler and Baur 2002), but researchers have shown that Nef helps HIV virions gain entry into host cells (Schaeffer et al. 2001; Papkalla et al. 2002); boosts viral replication (Aiken and Trono 1995; Linnemann et al. 2002); and helps HIV-infected cells evade the host's immune system (Swann et al. 2001). Loss or reduction of these functions apparently limits the damage HIV does to the host's immune system, perhaps in part because they result in lower viral loads.

Relatively benign strains of HIV are known to exist.

The important point is that alterations to HIV's genome can render the virus less damaging to the host, but still able to survive in the host's body. If the host lives longer, then the HIV population persists longer too. That would seem to be good for the virus. Why, then, are HIV strains bearing such genetic alterations so rare?

Selection thinking will suggest an answer if we recognize that even for a pathogen that is completely benign, all hosts eventually die. To persist beyond the life span of the host, a viral population must at some point colonize new hosts. This means that there is a second level of natural selection acting on HIV. The first level of selection is the one we have already explored: There are differences among virions in their ability to survive and reproduce within a given host. The second level of selection occurs when there are differences among viral strains in their ability to move from one host to another. Strains that are good at getting transmitted to new hosts will become more common over time; strains that are bad at getting transmitted will disappear.

Relatively benign strains of HIV may be transmitted from host to host at low rates.

A reasonable hypothesis is that HIV strains with mutant *nef* genes are rare because they are seldom transmitted from one host to another. Poor transmission could be due to the reduced ability of Nef-deficient virions to invade new cells. It could also be partially explained by lower viral loads maintained in hosts by Nef-deficient viral populations. With fewer virions present in bodily fluids, the chance of a successful transfer during a given bout of sex or needle sharing is reduced.

A similar scenario may explain why HIV-2 is less common worldwide than HIV-1. The HIV we have been discussing to this point is HIV-1. HIV-2 is a related virus that humans acquired from a different primate species (see Section 1.5). HIV-2 is less damaging to its hosts than HIV-1. Individuals infected with HIV-2 do progress to AIDS, but they do so much more slowly than individuals infected with HIV-1 (Marlink et al. 1994). The slower course of HIV-2 infection appears to be related to the fact that viral loads are lower in hosts infected with HIV-2 than in hosts infected with HIV-1 (Popper et al. 1999). Among the reasons for these lower viral loads may be that HIV-2's Nef protein has a function that was lost in HIV-1's ancestors before they began infecting humans: It prevents chronic immune activation (Schindler et al. 2006). However, in addition to being less harmful to its hosts HIV-2 is also transmitted at lower rates than HIV-1 (Kanki et al. 1994). Differences in transmission rate may explain why HIV-2 has remained largely confined to West Africa (Bock and Markovitz 2001), while HIV-1 has spread across the globe.

A second reason HIV infections are fatal may be that traits that predispose HIV to kill also enhance its ability to infect new hosts.

To summarize, a second reason why HIV infection is fatal is that the traits that predispose HIV populations to ultimately kill their hosts—traits like the ability to replicate rapidly, evade the immune system, and maintain large population sizes—may also enhance HIV's ability to colonize new hosts. Selection at the level of host-to-host transmission may favor such traits, even at the expense of killing individual hosts more quickly.

In the next section we will continue to practice selection thinking, but we will shift our focus from the virus back to the host. Doing so will suggest a third reason why HIV infections are fatal: Perhaps the human population has not had time to evolve an adequate defense.

1.4 Why Are Some People Resistant to HIV?

We have already mentioned, in Section 1.1, that HIV has relatives, called the simian immunodeficiency viruses (SIVs), that infect various species of primates.

The natural hosts of these SIVs typically tolerate infection without getting sick. In wild populations of African green monkeys, for example, over half of all adults are infected with SIVagm but there is no evidence that they suffer disease as a result (Kuhmann et al. 2001). When rhesus macaques are infected with SIVagm, however, they often develop AIDS. Such results suggest that the natural hosts of SIVs have evolved effective defenses that novel hosts lack.

Could humans evolve similar defenses against HIV? If so, why have we not? Recall that for a population to evolve, the individuals must vary, and the variation must be passed genetically from parents to offspring. Is there heritable variation among humans in susceptibility to HIV?

For a population to evolve, it must harbor genetic differences among individuals.

In the early 1990s, work from several laboratories demonstrated that some people remain uninfected even after repeated exposure to HIV, and that some people who are infected with the virus survive many years longer than expected (see Cao et al. 1995). In the mid 1990s, a team led by Edward Berger identified the coreceptor molecules that allow HIV to enter host cells (see Feng et al. 1996; Alkhatib et al. 1996). Soon after, Rong Liu and coworkers (1996) and Michel Samson and associates (1998) suggested that resistant individuals might have unusual forms of the coreceptor molecules and that these mutant proteins might thwart HIV's entry into host cells.

To test this hypothesis, Samson and colleagues sequenced the gene that codes for the coreceptor CCR5 from three HIV-infected individuals who were long-term survivors. As predicted, one of the individuals had a mutant form of the gene. Because this allele is distinguished by a 32-base-pair deletion in the normal sequence of DNA, Samson and coworkers named it the *Δ32* allele (*Δ* is the Greek letter delta). Then they showed that HIV cannot enter cells that have the *Δ32* form of CCR5. This experiment confirmed that the *Δ32* allele protects individuals from infection.

Modifications in the expression or structure of CCR5 turn out to be a common defense among the hosts of SIVs as well (Chen et al. 1998; Palacios et al. 1998; Kuhmann et al. 2001; Veazey et al. 2003). Many African green monkeys, for example, carry a CCR5 gene encoding an amino acid substitution that makes it harder for SIV to enter their cells. Most red capped mangabeys carry a CCR5 gene with a 24-base-pair deletion that has the same effect. Some sooty mangabeys carry a CCR5 gene with a different deletion that renders the protein nonfunctional.

We can conclude that human populations do, in fact, harbor genetic variation for resistance to HIV infection and disease progression. And that this raw material for evolution involves physiological mechanisms similar to those that confer resistance in monkeys. Has the genetic variation in humans been molded by natural selection imposed by the AIDS pandemic?

In human populations, some individuals carry alleles that make them resistant to infection with HIV.

To find out how common the *Δ32* allele is in various human populations, Samson and colleagues (1998) took DNA samples from a large number of individuals of northern European, Japanese, and African heritage, examined the gene for CCR5 in each individual, and calculated the frequency of the normal and *Δ32* alleles in each population. A strong pattern emerged: The mutant allele was present at a relatively high frequency of 9% in Europeans but is completely absent in individuals of Asian or African descent. Further research has confirmed this result. The *CCR5-Δ32* allele is common in northern Europe and declines dramatically in frequency to both the south and the east (Figure 1.20).

Comparing the map of *Δ32* frequency in Figure 1.20 with the map of HIV prevalence in Figure 1.1 on page 5 reveals a striking disconnect. The *Δ32* allele

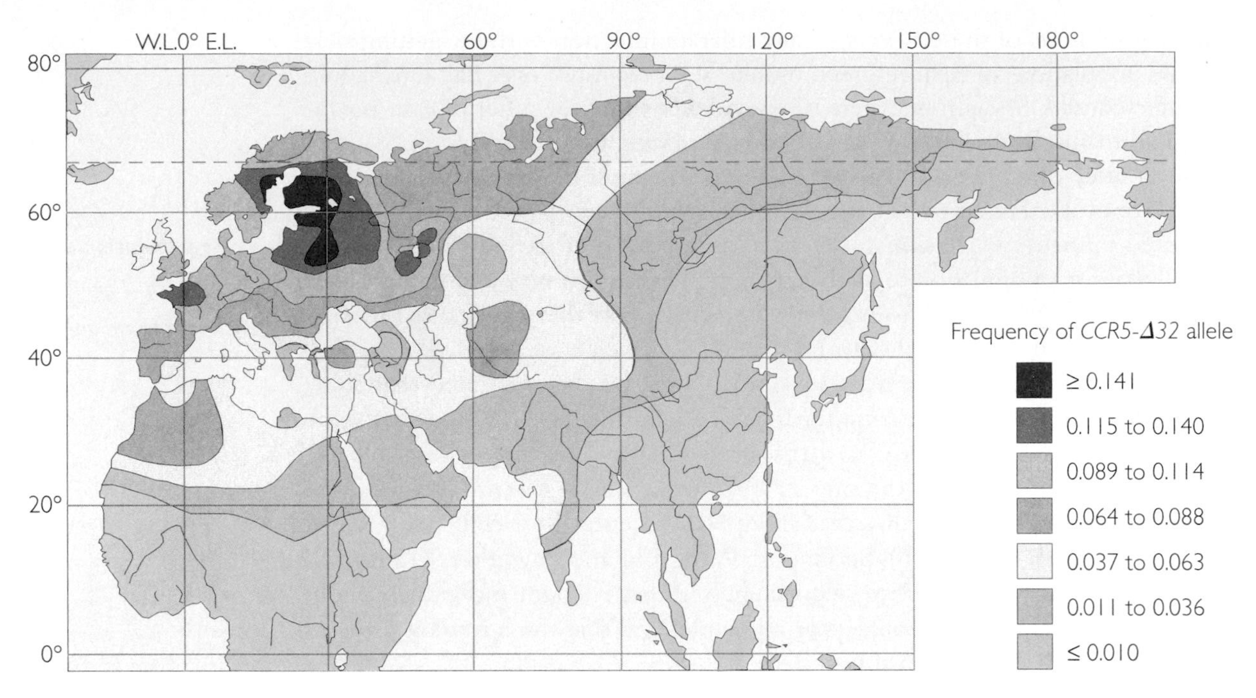

Figure 1.20 The frequency of the *CCR5-Δ32* allele in the Old World The *Δ32* allele is at highest frequency in northern Europe. From there its frequency declines to the south and to the east. From Limborskaa et al. (2002).

Curiously, the frequency of the best-known protective allele is highest in regions with low rates of HIV infection.

is common in a part of the world where HIV infection is relatively rare. And HIV infection is rampant in parts of the world where the *Δ32* allele is rare or absent. Perhaps we should not have expected the geographic distribution of the *Δ32* allele to reflect the distribution of HIV. As we will see in the next section, not only is the current HIV pandemic young, but HIV itself is a new human disease. It takes generations for a population to change as a result of natural selection, and there has not been enough time for HIV to change human populations—yet.

Researchers comparing uninfected exposed individuals and long-term survivors versus individuals who develop AIDS have by now discovered many other genes at which different alleles confer different susceptibility or resistance to HIV (see O'Brien and Nelson 2004; Gao et al. 2005; Gonzales et al. 2005; Modi et al. 2006). Evolutionary biologists are measuring how common protective alleles are in various populations and predicting how their frequencies may change as the epidemic continues (Schliekelman et al. 2001; Sullivan et al. 2001; Ramaley et al. 2002). We will consider some of these predictions in detail in Chapter 6.

It appears that HIV is too new a human disease to have triggerd subustantial evolutionary change in human populations. The pattern in the map above remains to be explained.

If the AIDS pandemic does not explain the geographic pattern in Figure 1.20, what does? Samson and coworkers offered two explanations: (1) The *CCR5-Δ32* allele may have been recently favored by natural selection in European populations; or (2) The allele could have risen to high frequency by chance, in a process called genetic drift. In line with the natural selection hypothesis, researchers have suggested that the *Δ32* allele confers protection against a pathogen other than HIV, such as bubonic plague (Stephens et al. 1998) or smallpox (Lalani et al. 1999). Under these scenarios, the *Δ32* allele would have risen to high frequency because of the survival advantage it offered during devastating epidemics that swept Europe during the past millennium. In a scheme related to the genetic drift hypothesis, another biologist has proposed that the *Δ32* allele first appeared and achieved a high frequency among the Vikings, and then was disseminated across Europe during the Viking raids of the 8th, 9th, and 10th centuries (Lucotte 2001). Researchers have also recently begun discovering costs associated with the *Δ32* allele. For example, homozygotes are more susceptible to West Nile virus (Glass et al. 2006). This suggests that natural selection against the allele will also have to considered. We will revisit the puzzle of *CCR5-Δ32's* history and geographic distribution in Chapters 4, 6, and 8.

1.5 Where Did HIV Come From?

We noted in Section 1.1 that AIDS was first recognized in 1981. It is a new disease for humans. Its cause, HIV, is a new pathogen. Viruses, like other organisms, only come from reproduction of their kind. Where did the first HIV virions come from?

The first clue is that, as we have mentioned, HIV's genome and life cycle are similar to those of the SIVs, a family of viruses that infect a variety of primates. Like HIV, the simian immunodeficiency viruses infect their hosts' immune systems. Unlike HIV, they do not appear to cause serious disease.

A logical hypothesis is that HIV is derived from one of the SIVs and that the global AIDS epidemic started when this SIV moved from its primate host into humans. Which SIV is the ancestor of HIV? To find out, evolutionary biologists have reconstructed the evolutionary history of the viruses in the SIV/HIV family.

How Do Researchers Reconstruct Evolutionary History?

Just as the historical relationships among individuals are described by their genealogy, the historical relationships among populations or species are described by their **phylogeny**. A picture of these evolutionary relationships shows the family tree of a group of species or populations and is called a **cladogram** or **phylogenetic tree**. The methodology for reconstructing phylogenies is complex in its details (all of Chapter 4 is devoted to this topic), but the basic logic is simple. In general, more closely related species should be more similar than distantly related forms. In the case of HIV, researchers infer the historical relationships among strains by comparing the nucleotide sequences of their genes. The premise is that strains with similar nucleotide sequences shared a common ancestor more recently than strains with different sequences.

A phylogenetic tree shows the historical relationships among a group of viruses or organisms.

The Origin of HIV

Beatrice Hahn and colleagues sequenced the gene that codes for reverse transcriptase in several SIVs and compared them to the sequences found in a variety of HIV strains (Gao et al. 1999; Hahn et al. 2000). Using their data to estimate the relationships among these viruses, the researchers produced the phylogeny shown in Figure 1.21a. In this tree, the lengths of the horizontal lines indicate the percentage of nucleotides that are different between the genes of viral strains. Short branches between species mean that their sequences are similar; longer branches mean that their sequences are more divergent. Because sequences diverge as a result of mutations that accumulate over years, the length of the horizontal branches on this tree correlate roughly with time. (In contrast, the lengths of the vertical lines are arbitrary. They are simply adjusted to make the tree more readable.)

To read the tree and understand what it implies about the history of HIV, start at the orange arrow on the lower left. The branching point, or **node**, at this arrow represents the common ancestor of all viruses included in the tree. Note that each

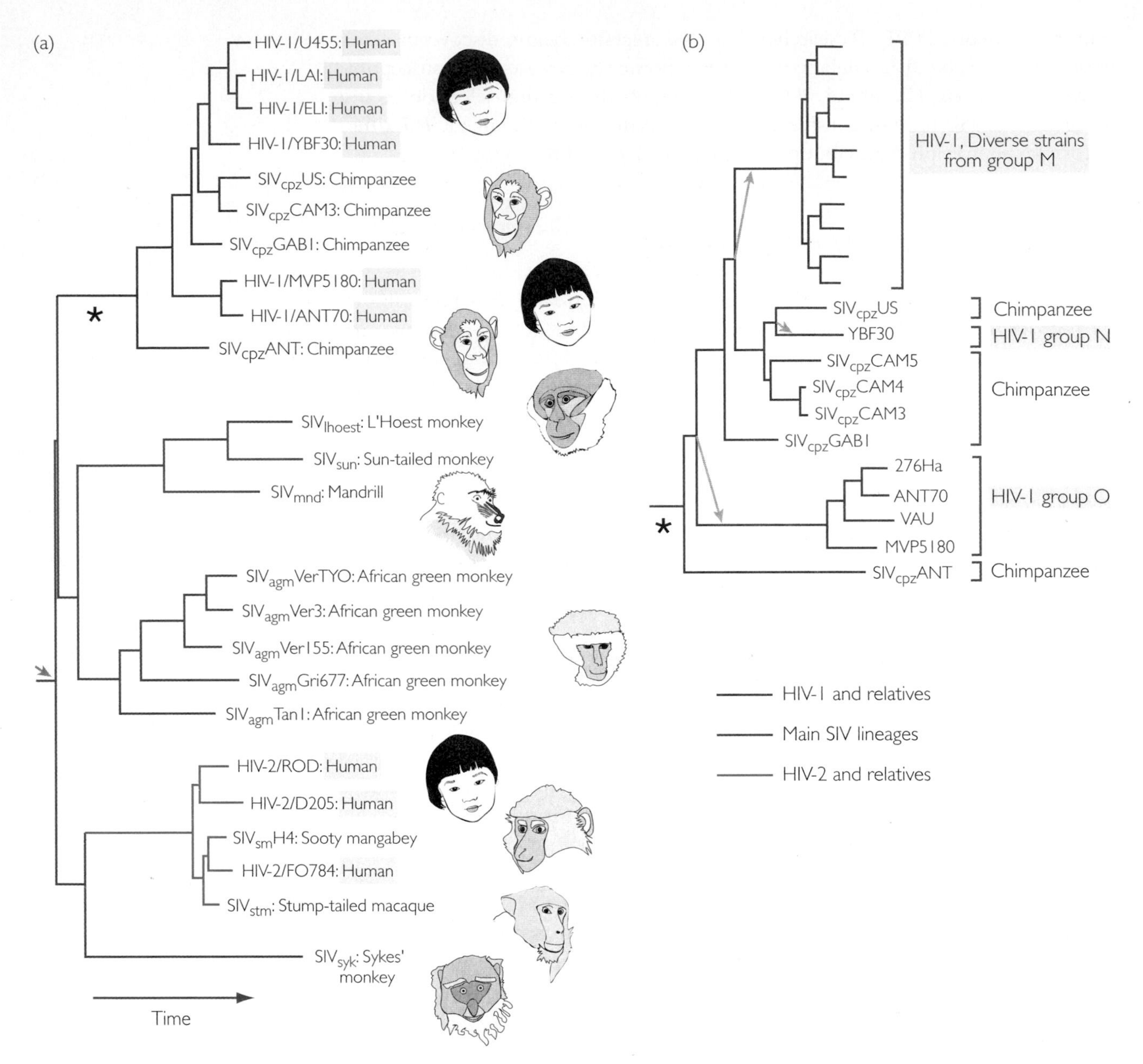

Figure 1.21 The family tree of HIV and related viruses (a) This tree shows the evolutionary relationships among the two major forms of HIV, called HIV-1 and HIV-2, and the immunodeficiency viruses that afflict nonhuman primates. Note that the viruses that branch off near the orange arrow at the base of the tree parasitize monkeys. Based on this observation, researchers conclude that virus strains jumped from monkeys to humans. (b) This tree shows a more detailed analysis performed by Hahn et al. (2000). (The asterisk marks the same branch point on both trees.) The arrows indicate the places on the tree where immunodeficiency viruses were transmitted from chimpanzees to humans. According to this tree, each major strain of HIV-1 originated in a different transmission event from a chimpanzee host, as represented by the gray arrows. Redrawm from Hahn et al. (2000).

of the distinct groups, or **lineages**, that branch off from the ancestral population lead to viruses that infect monkeys or chimpanzees. The blue branches diversified into viruses that infect a variety of nonhuman primates, while the red and green branches begat viruses that parasitize both human and nonhuman hosts.

Where did the human immunodeficiency viruses come from? Find the viruses designated HIV-2, near the bottom on the tree, and note that they share a limb on the tree with a virus that infects a species of monkey called the sooty mangabey. HIV-2 is prevalent in West Africa and is less virulent than HIV-1, the virus that is causing the global AIDS epidemic. Because sooty mangabeys are hunted for food and kept as pets in West Africa, and because their viral gene sequences are so closely related to HIV-2, researchers concur that the virus was probably transmitted from sooty mangabeys to humans in the recent past. Once it had moved into humans, evolution by natural selection led to the strain known as HIV-2. (The reader may notice that a virus called SIV_{stm} is also a close relative of HIV-2. This strain was obtained from a captive stump-tailed macaque that received its infection from a sooty mangabey.)

The two main types of HIV, HIV-2 and HIV-1, were transmitted to humans from different sources. HIV-2 originated in sooty mangabeys, while HIV-1 was originally transmitted to humans from chimpanzees.

Now look at the red lineage at the top of the tree. This lineage diversified into strains that infect humans and chimpanzees. These populations include HIV-1, the virus that is causing the AIDS epidemic. Because chimpanzees are hunted for food in Africa and because their gene sequences are so closely related to HIV-1, Hahn and colleagues inferred that the SIV that infects chimpanzees (SIV_{cpz}) was transmitted from chimps to humans, where it evolved into HIV-1.

To examine this transmission event more closely, Hahn and coworkers compared HIV-1 and SIV_{cpz} sequences from the gene that codes for the proteins found on the surface of the virions. The tree based on these data gives a more detailed view of the relationships among these viruses and is reproduced in Figure 1.21b. Note that HIV strains form three distinct clusters, which HIV researchers call subgroups M, N, and O. Each HIV subgroup is closely related to a different strain of SIV_{cpz}, This is evidence that the virus jumped from chimps to humans on at least three different occasions. HIV-1 was transmitted to humans from chimps not once, but multiple times. More recent reconstructions using viruses sampled from wild chimp populations has established that the natural reservoir from which all three HIV-1 subgroups are derived is a subspecies of chimpanzee known as *Pan trogodytes troglodytes* (Keele et al. 2006).

Each major subgroup of HIV-1 originated in an independent transmission event from chimpanzees to humans.

When did SIV move from chimpanzees to humans? Work on this question has focused on the group M branch of HIV, at the top of the tree in Figure 1.20b. It is group M that is responsible for the bulk of the global AIDS epidemic. Several groups of researchers have used sequence data from various group M strains to estimate the age of their last common ancestor (see Box 1.2). There is considerable uncertainty, but the best estimate is that the last common ancestor of the group M HIV-1 viruses lived in the 1930s. This common ancestor could, in principle, have lived in either a chimpanzee or a human. However, the available evidence is most consistent with a human host (Hillis 2000; Rambaut et al. 2001; Sharp et al. 2001). The implication is that the group M strains of HIV-1 originated in a transfer of SIV from chimps to humans that happened more than 60 years ago.

One medical lesson of the fact that HIV-1 is derived from SIV_{cpz} is that chimpanzees are an important animal to study. Crucial questions yet to be answered include: How common is SIV_{cpz} in the wild? How is it transmitted? And perhaps most important, why does this virus not make chimpanzees ill?

HIV Diversity and the Difficulty of Developing a Vaccine

Another medical lesson from HIV's evolutionary trees is a clue as to why it has been so hard for AIDS researchers to develop an effective vaccine. Vaccines have been responsible for the great success stories in controlling viral diseases from polio to smallpox. The difficulty of designing antiviral drugs, together with the rate at which HIV has evolved drug resistances, has made vaccine development an urgent priority for the AIDS research community. Is it possible to design a vaccine that would make people immune to HIV?

Box 1.2 When did HIV move from chimpanzees to humans?

Here we outline the method used by Bette Korber and colleagues (2000) to estimate the date of the common ancestor of the group M strains of HIV-1. The researchers analyzed nucleotide sequences from 159 different samples of HIV-1.

First, Korber and associates reconstructed an evolutionary tree from their sequence data. The tree, shown in Figure 1.22a, is unrooted, so it looks a bit different from the trees we saw before in Figure 1.21. Each twig on the tree represents a particular sequence. The distance along the tree from the tip of one twig to the tip of another indicates the genetic difference between two sequences. The tree is divided into several distinct branches. The HIV strains on these branches are referred to as subtypes and are designated by letters. The branch point in the center of the tree, highlighted in orange, represents the common ancestor of the 159 sequences on the tips of the twigs.

Next, Korber and colleagues prepared a scatterplot showing the genetic difference between each virion on their tree versus the common ancestor as a function of the year in which the virion was collected (Figure 1.22b). Individual virions are represented on the scatterplot by colored letters corresponding to their subtype. As we saw in Figure 1.17a (page 19) for virions evolving within a single patient, the virions analyzed in Figure 1.22b show increasing divergence with time. That is, the later a sample was collected, the greater its genetic difference versus the common ancestor. The orange line is the statistical best-fit line through the data points.

Finally, Korber and colleagues extrapolated the best-fit line back in time to estimate the year in which a sample would have to have been collected to have a genetic difference of zero versus the common ancestor (Figure 1.22c). In other words, they extrapolated back to the date of the common ancestor itself. The best-fit line hits zero at 1931. Extrapolating so far beyond the data is a bit dangerous, and there may be biases in the data due to sampling error. The true relationship between sequence divergence and time could be anywhere between the gray lines in the figure. Korber and associates estimate with 95% confidence that the common ancestor of their group M virions lived sometime between 1915 and 1941, indicated by the red bar on the horizontal axis. Additional analyses by the same and other teams have produced similar estimates (Salemi et al. 2001; Sharp et al. 2001; Yusim et al. 2001).

Vaccines work by priming the immune system to respond quickly to an infection. To respond to bacterial and viral infections, T cells have to identify a protein from the pathogen as foreign, or non-self. A fragment of foreign protein that is recognized as non-self and that triggers a response by T cells is called, as we noted in Section 1.3, an epitope. Vaccines consist of epitopes from killed or weakened virions. Although no actual infection occurs after a vaccination, the immune system responds by activating cells that recognize the epitopes presented. If an authentic infection starts later, the immune system is prepared to respond more quickly than it otherwise could have. Usually, the invader is eliminated before the infection causes disease.

In the case of HIV, most epitopes presented to the immune system are derived from the protein called gp120 that coats the virion's surface (see Figure 1.5, page 7). To be effective, then, a vaccine would have to contain epitopes from the gp120 proteins found in many different strains of HIV. The evolutionary trees we have examined reveal that many different subgroups of HIV-1 exist as a result of independent transmission events from chimpanzees to humans. The resulting diversity of HIV strains poses a challenge to vaccine development. Furthermore,

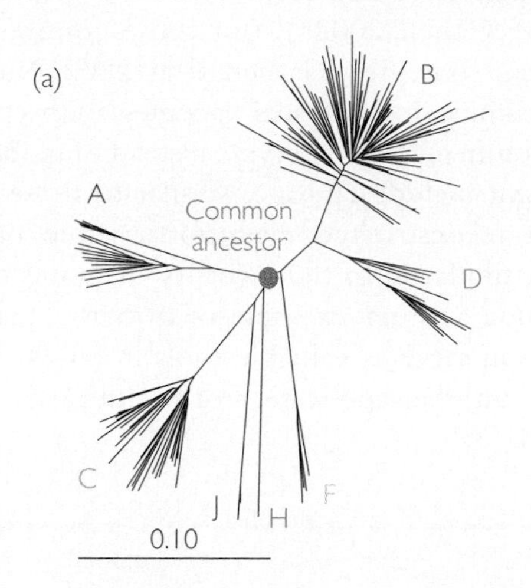

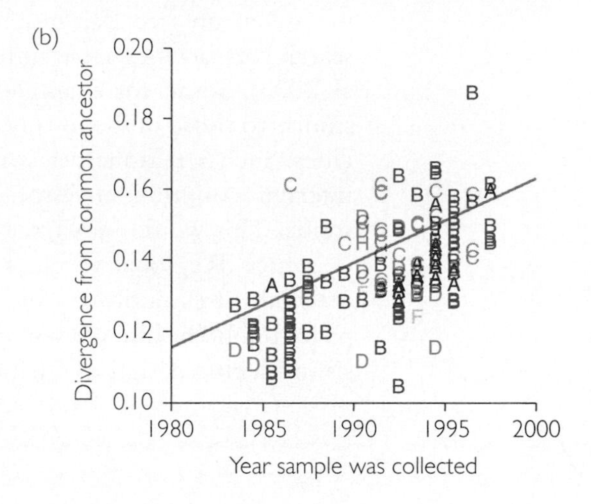

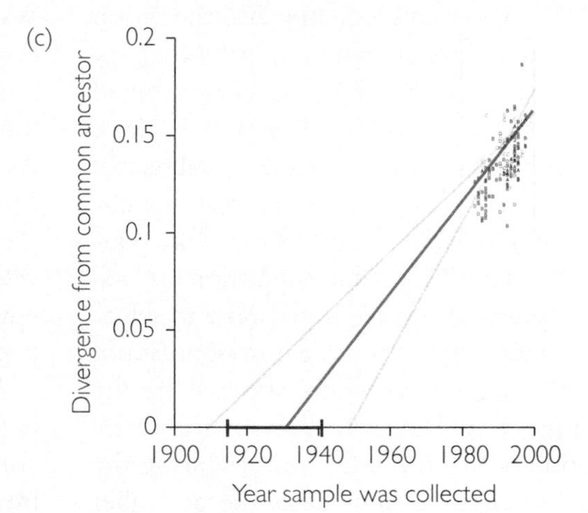

Figure 1.22 Dating the common ancestor of HIV-1 strains in group M (a) An unrooted evolutionary tree for 159 group M HIVs. The tip of each twig represents a virion; the distance traveled from one tip to another represents the genetic difference between the two virions. The orange dot marks the common ancestor of all group M strains. (b) This scatterplot shows the genetic difference between each HIV sample in (a) versus the common ancestor as a function of the date the sample was collected. The statistical best-fit line is in orange. (c) Extrapolating the best-fit line in (b) back to zero genetic difference gives an estimate of the date at which the common ancestor lived. From Korber et al. (2000).

because transmission of SIVs to humans has happened repeatedly in the past, it is likely to continue in the future.

The branch lengths in Figure 1.21b and Figure 1.22a suggest that sequence divergence is high even within subgroups of HIV-1. Indeed, the envelope proteins of HIV strains in the same subtype can differ by as much as 20%, and in strains from different subtypes they can differ by as much as 35% (Gaschen et al. 2002). Research by Tuofo Zhu and coworkers (1998) also drives this point home. Zhu and associates sequenced HIV-1 genes found in a blood sample taken from a Congolese man in 1959. This is the earliest sample of HIV so far discovered. The researchers' analysis shows that the 1959 sample is strikingly different from contemporary strains. In the researchers' own words (p. 596), "The diversification of HIV-1 in the past 40–50 years portends even greater viral heterogeneity in the coming decades." The rapid evolution of HIV, like the rapid genetic change commonly observed in flu and cold viruses, makes vaccine design difficult.

The results of the first large study of an AIDS vaccine were announced in early 2003 (Cohen 2003a, b). The vaccine, AIDSVAX, made by VaxGen, failed to protect subjects who took it any better than a placebo. Many researchers are pessimistic about the prospects for ever finding an effective AIDS vaccine (Korber et al. 1998; Letvin 1998; Baltimore and Heilman 1998). Others have not given up (see Baltimore 2000; Cohen 2002b). But those continuing to search for vaccines are refining their strategies (Gashen et al. 2002; Nickle et al. 2003). Some, for example, are aiming for regional vaccines, with epitopes similar to those of the locally predominant HIV subtype, instead of global vaccines. Such a regional vaccine might include epitopes similar to those of the inferred common ancestor on a reconstructed evolutionary tree of local strains. This would maximize their similarity to the diversity of strains now in existence. Regional vaccines provide another example of how the analytical methods of evolutionary biology can serve as valuable tools in efforts to improve public health. We will discuss another application of evolutionary analysis in vaccine design in Chapter 14.

Summary

Each time an HIV virion invades a host cell the virion reverse transcribes its RNA genome into a DNA copy that serves as the template for the next generation of virus particles. Because reverse transcription is error prone, an HIV population quickly develops substantial genetic diversity. Some genetic variants replicate rapidly while others die. As a result, the composition of the population changes over time. That is, the population evolves.

Just as HIV populations evolve in response to selection imposed by their hosts, so too the host populations may evolve in response to selection imposed by the virus. Some human populations harbor genetic variation for susceptibility to HIV infection. If, during the AIDS epidemic, susceptible individuals die at higher rates than resistant individuals, then genetic composition of these populations will change over time.

HIV's potential for rapid evolution has profound consequences for individuals and for public health. Within infected individuals HIV populations quickly evolve resistance to any single antiretroviral drug and can even evolve resistance to multiple-drug cocktails. Without effective antiretroviral therapy, HIV populations also continuously evolve to evade the host's immune response, a process that ultimately contributes to the collapse of the immune system and the onset of AIDS. Among infected individuals, HIV diversifies so rapidly and to such an extent that an effective broad-spectrum vaccine will be difficult or impossible to develop. Our best hope for slowing the global AIDS epidemic remains education aimed at encouraging individuals to practice safer sex and use clean needles.

HIV belongs to a family of viruses that infect a variety of primates. Evolutionary trees based on genetic comparisons reveal that HIV-2 jumped to humans from sooty mangabeys. HIV-1 jumped to humans from chimpanzees, and has done so more than once.

By focusing in this chapter on adaptation and diversification in HIV, we introduced topics that will resonate throughout the text: mutation and variation, competition, natural selection, phylogeny reconstruction, lineage diversification, and applications of evolutionary theory to scientific and human problems.

Questions

1. When did HIV enter the human population, and from what source? How do we know?
2. Review the process by which the HIV population inside a human host evolves resistance to the drug AZT. How might a similar scenario explain the evolution of antibiotic resistance in a population of bacteria?
3. In the early 1990s, researchers began to find AZT-resistant strains of HIV-1 in recently infected patients who had never received AZT. How can this be?
4. What traits of HIV contribute to its rapid evolution?
5. Given the risk of evolution of resistance, why do you think the two patients shown in Figure 1.11 were not given high doses of AZT immediately, rather than starting them with low doses?
6. The idea behind multiple-drug therapy for HIV is to increase the number of mutations required for resistance and therefore reduce the amount of genetic variation in the viral population for survival in the presence of drugs. Could we achieve the same effect by using antiretroviral drugs in sequence instead of simultaneously? Why or why not?

7. Some physicians have advocated "drug holidays" as a way of helping HIV patients cope with the side effects of multiple-drug therapy. Under this plan, every so often the patient would stop taking drugs for a while. From an evolutionary perspective, does this seem like a good idea or a bad idea? Justify your answer.
8. Recall that we discussed two different types of selection in this chapter: selection of different virus strains within one host and selection of those virus strains that are able to transmit themselves from host to host. Now consider the hypothesis, traditionally championed by biomedical researchers, that disease-causing agents naturally evolve into more benign forms as the immune systems of their hosts evolve more efficient responses to them. Is the evidence we have reviewed on the evolution of HIV within and among hosts consistent with this hypothesis? Why or why not?
9. Science fiction authors often make interesting statements about evolution:
   a. Respond to the following quote, from the Mr. Spock character in *Star Trek*: "A truly successful parasite is commensal, living in amity with its host, or even giving it positive advantages, as, for instance, the protozoans who live in the digestive system of your termites and digest for them the wood they eat. A parasite that regularly and inevitably kills its host cannot survive long, in the evolutionary sense, unless it multiplies with tremendous rapidity; it is not prosurvival."
   b. HIV is a tiny, robotic, molecular machine. Many science fiction books describe robots that evolve to become intelligent and conscious (and, usually, seek freedom, develop emotions, and start wars with humans). Under what conditions could robots actually evolve? Is it necessary that the robots reproduce, for example?
10. How does HIV-2 illustrate the trade-off between virulence (damage to the current host) and transmission (moving to new hosts)?
11. Some researchers expect that human populations will evolve in response to the AIDS epidemic because alleles that confer resistance to HIV infection should increase in frequency in the population over time. Do you agree with this prediction? If so, when and where do you think this will happen first? How would you design a study to test your predictions?
12. Suppose that HIV were the ancestor of the SIVs, instead of the other way around. If immunodeficiency viruses were originally transmitted from humans to monkeys and chimpanzees, make a sketch of what Figure 1.21a would look like.
13. Not all viruses are dangerous. (The common cold is an example.) HIV, however, is nearly 100% lethal. Describe three major hypotheses for why HIV is so highly lethal.

Exploring the Literature

14. Drug resistance has evolved in a wide variety of viruses, bacteria, and other parasites. The following papers describe evolution of drug resistance in the hepatitis B virus (HBV) and in the bacterium that causes tuberculosis:

    Blower, S. M., and T. Chou. 2004. Modeling the emergence of "hot zones": tuberculosis and the amplification dynamics of drug resistance. *Nature Medicine* 10: 1111–1116.

    Shaw, T. A., A. Bartholomeusz, and S. Locarnini. 2006. HBV drug resistance: mechanisms, detection and interpretation. *Journal of Hepatology* 44: 593–606.
15. For documentation of the contingent nature of natural selection in the context of drug resistance in HIV, see:

    Devereux, H. L., V. C. Emery, M. A. Johnson, and C. Loveday. 2001. Relative fitness in vivo of HIV-1 variants with multiple drug resistance-associated mutations. *Journal of Medical Virology* 65: 218–224.
16. This paper shows that the mutation that renders HIV resistant to the antiretroviral drug 3TC also make reverse transcriptase less error-prone:

    Wainberg, M. A., W. C. Drosopoulos, H. Salomon, et al. 1996. Enhanced fidelity of 3TC-selected mutant HIV-1 reverse transcriptase. *Science* 271: 1282–1285.
17. Stanley Trask and colleagues (2002) hypothesize that most HIV-1 transmissions in sub-Saharan Africa occur between married couples. That is, the husband acquires HIV and then passes it to his wife, or the wife acquires HIV and then passes it to her husband. The researchers then use a reconstructed evolutionary tree to test their hypothesis. Think about how this test might work. What would the tree look like if the hypothesis is true? If it is false? Then look up Trask et al.'s paper:

    Trask, S. A., C. A. Derdeyn, U. Fideli, et al. 2002. Molecular epidemiology of human immunodeficiency virus type 1 transmission in a heterosexual cohort of discordant couples in Zambia. *Journal of Virology* 76: 397–405.

    Also see an interesting update of this topic, which found that risk of acquiring HIV from an infected partner is higher if the two partners share certain immune-system alleles:

    Dorak, M., J. Tang, J. Penman-Aguilar, et al. 2004. Transmission of HIV-1 and HLA-B allele-sharing within serodiscordant heterosexual Zambian couples. *Lancet* 363: 2137—2139.
18. In Box 1.1, we discussed several new classes of HIV drugs, including fusion inhibitors. Some fusion inhibitors

work by sticking to CCR5, preventing HIV from binding to the protein and using it as a coreceptor. Imagine treating an HIV infection with one of these CCR5 antagonists alone. How might the HIV population evolve in response? That is, predict the kinds of mutations that might rise to high frequency because they confer on HIV the ability to replicate in the presence of the drug. Then see:

Mosier, D. E., G. R. Picchio, R. J. Gulizia, et al. 1999. Highly potent RANTES analogues either prevent CCR5-using human immunodeficiency virus type 1 infection in vivo or . . . [remainder of title truncated to avoid giving away an answer]. *Journal of Virology* 73: 3544–3550.

Trkola, A., S. E. Kuhmann, J. M. Strizki, et al. 2002. HIV-1 escape from a small molecule, CCR5-specific entry inhibitor does not involve . . . [remainder of title truncated to avoid giving away an answer]. *Proceedings of the National Academy of Sciences USA* 99: 395–400.

For more information on this class of drugs, see:

Krambovitis, E., F. Porichis, and D. A. Spandidos. 2005. HIV entry inhibitors: a new generation of antiretroviral drugs. *Acta Pharmacologica Sinica* 26: 1165–1173.

19. See the following paper for a review of recent efforts to develop an HIV vaccine:

Girard, Marc P., S. K. Osmanov, and M. P. Kieny. 2006. A review of vaccine research and development: The human immunodeficiency virus (HIV). *Vaccine* 24: 4062–4081.

20. AIDS has generated a number of controversial fringe theories. Some deny the link between HIV and AIDS, contending that HIV is a harmless opportunistic infection and that AIDS itself is caused by other factors such as drug abuse. These hypotheses have been progressively discredited in the past two decades, but are still widely believed by many people, including many minority gay men and some public health officials (notably in South Africa). Another fringe hypothesis contends that HIV originated not from wild chimpanzees but from an experimental oral polio vaccine that was derived from chimpanzee cell cultures and was administered to many Africans during the late 1950s.

Read more about the history and current status of these hypotheses in the articles "AIDS reappraisal" and "OPV AIDS hypothesis" at the publicly edited encyclopedia Wikipedia (*www.wikipedia.org*). In your opinion, have these hypotheses been adequately tested?

The OPV hypothesis has been the subject of recent research. Recently, researchers have managed to obtain samples of the polio vaccine that was used in Africa in the 1950s. By sequencing ribosomal lRNAs present in the vaccines they were able to test whether the species used to prepare the vaccine was really chimpanzee, as proposed. In addition, new information is available on whether and where there are wild chimpanzee populations that harbor the SIV type that is most closely related to HIV. See:

Berry, N., A. Jenkins, J. Martin, et al. 2005. Mitochondrial DNA and retroviral RNA analyses of archival oral polio vaccine (OPV CHAT) materials: evidence of [rest of title deleted to avoid giving away the answer]. *Vaccine* 23: 1639–1648.

Keele, B. F., F. van Heuverswyn, Y. Li, et al. 2006. Chimpanzee reservoirs of pandemic and nonpandemic HIV-1. *Science Express Reports* (*www.sciencemag.org*), 10.1126/science.1126531.

21. See the following articles and websites for recent information about the HIV pandemic:

Stover, J., S. Bertozzi, J-P Gutierrez, et al. 2006. The global impact of scaling up HIV/AIDS prevention programs in low- and middle-income countries. *Science* 311: 1474–1476.

HIV guide and review of recent literature from the magazine *New Scientist*: *http://www.newscientist.com/channel/health/hiv*.

A complete medical textbook on HIV, available for free PDF download: *http://www.hivmedicine.com/*.

National Institute of Health's AIDS website: *http://www.niaid.nih.gov/daids/*.

A detailed article on HIV from the publicly edited encyclopedia Wikipedia, with references, links, and recent news: *www.wikipedia.org/wiki/hiv*.

Citations

Aiken, C., and D. Trono. 1995. Nef stimulates human immunodeficiency virus type 1 proviral DNA synthesis. *Journal of Virology* 69: 5048–5056.

Ala, P. J. et al. 1997. Molecular basis of HIV-1 protease drug resistance: Structural analysis of mutant proteases complexed with cyclic urea inhibitors. *Biochemistry* 36: 1573–1580.

Alkhatib, G., C. Combadiere, C.C. Broder, Y. Feng, P. E. Kennedy, P. M. Murphy, and E. A. Berger. 1996. CC CKR5: A RANTES, MIP-1α, MIP-1β receptor as a fusion cofactor for macrophage-tropic HIV-1. *Science* 272: 1955–1958.

Allen, T. M., D. H. O'Connor, P. Jing, et al. 2000. Tat-specific cytotoxic T lymphocytes select for SIV escape variants during resolution of primary viraemia. *Nature* 407: 386–390.

Baltimore, D. 2002. Steering a course to an AIDS vaccine. *Science* 296: 2297.

Baltimore, D., and C. Heilman. 1998. HIV vaccines: prospects and challenges. *Scientific American* 279 (July): 98–103.

Bartlett, J. G., and R. D. Moore. 1998. Improving HIV therapy. *Scientific American* 279 [July]: 84–89.

Bestilny, L. J., M. J. Gill, C. H. Mody, and K. T. Riabowol. 2000. Accelerated replicative senescence of the peripheral immune system induced by HIV infection. *AIDS* 14: 771–780.

Birch, M.R., J. C. Learmont, W. B. Dyer, et al. 2001. An examination of signs of disease progression in survivors of the Sydney Blood Bank Cohort (SBBC). *Journal of Clinical Virology* 22: 263–270.

Blaak, H., A. B. van't Wout, M. Brouwer, et al. 2000. In vivo HIV-1 infection of $CD45RA^+CD4^+$ T cells is established primarily by syncytium-inducing variants and correlates with the rate of $CD4^+$ T cell decline. *Proceedings of the National Academy of Sciences USA* 97: 1269–1274.

Bock, P. J., and D. M. Markovitz. 2001. Infection with HIV-2. *AIDS* 15: S35–S45.

Brenchley, J. M., T. W. Schacker, L. E. Ruff, et al. 2004. $CD4^+$ T cell depletion during all stages of HIV disease occurs predominantly in the gastrointestinal tract. *Journal of Experimental Medicine* 200: 749–759.

Brenchley, J. M., D. A. Price, and D. C. Douek. 2006. HIV disease: fallout from a mucosal catastrophe? *Nature Immunology* 7: 235–239.

Brody, S., and J. J. Potterat. 2005. HIV Epidemiology in Africa: Weak Variables and Tendentiousness Generate Wobbly Conclusions. *PLoS Medicine* 2: e137.

Buchaz, K., W. McFarland, et al. 2005. Amphetamine use is associated with increased HIV incidence among men who have sex with men in San Francisco. *AIDS* 19: 1423–1424.

Cao, Y., L. Qin, L. Zhang, J. Safrit, and D. D. Ho. 1995. Virologic and immunologic characterization of long-term survivors of human immunodeficiency virus type 1 infection. *New England Journal of Medicine* 332: 201–208.

Carrington, M. et al. 1999. *HLA* and HIV-1: Heterozygote advantage and *B*35-CW*04* disadvantage. *Science* 283: 1748–1752.

Catania, J. A., D. Osmond, R. D. Stall, et al. 2001. The continuing HIV epidemic among men who have sex with men. *American Journal of Public Health* 91: 907–914.

CDC. 2005. HIV prevalence, unrecognized infection, and HIV testing among men who have sex with men—five U.S. cities, June 2004–April 2005. *Morbitity and Mortality Weekly Report* 54 (24): 597–601.

Chen, R. Y., A. O. Westfall, et al. 2003. Duration of highly active antiretroviral therapy regimens. *Clinical Infectious Diseases* 37: 714-722.

Chen, S. Y., S. Gibson, M. H. Katz, et al. 2002. Continuing increases in sexual risk behavior and sexually transmitted diseases among men who have sex with men: San Francisco, Calif., 1999–2001. *American Journal of Public Health* 92: 1387.

Chen, Z., D. Kwon, et al. 1998. Natural infection of a homozygous *Δ24* CCR5 red-capped mangabey with an R2b-tropic simian immunodeficiency virus. *Journal of Experimental Medicine* 188: 2057–2065.

Christakos, G., and R. A. Olea. 2005. New space-time perspectives on the propagation characteristics of the Black Death epidemic and its relation to bubonic plague. *Stochastic Environmental Research and Risk Assessment* 19: 307–314.

Chun, T.-W., R. T. Davey, Jr., D. Engel, et al. 1999. Re-emergence of HIV after stopping therapy. *Nature* 401: 874–875.

Chun, T.-W., L. Stuyver, S. B. Mizell, et al. 1997. Presence of an inducible HIV-1 latent reservoir during highly active antiretroviral therapy. *Proceedings of the National Academy of Sciences USA* 94: 13193–13197.

Churchill, M. J., D. I. Rhodes, et al. 2006. Longitudinal analysis of human immunodeficiency virus type 1 *nef*/long terminal repeat sequences in a cohort of long-term survivors infected from a single source. *Journal of Virology* 80: 1047–1052.

Cohen, J. 1993. AIDS Research: The Mood is Uncertain. *Science* 260: 1254–1261.

Cohen, J. 2002a. Confronting the limits of success. *Science* 296: 2320–2324.

Cohen, J. 2002b. Monkey puzzles. *Science* 296: 2325–2326.

Cohen, J. 2003a. AIDS vaccine trial produces disappointment and confusion. *Science* 299: 1290–1291.

Cohen, J. 2003b. Vaccine results lose significance under scrutiny. *Science* 299: 1495.

Cohen, J. 2004. Songachi sex workers stymie HIV. *Science* 304: 506.

Condra, J. H., D. J. Holder, and W. A. Schleif, et al. 1996. Genetic correlates of *in vivo* viral resistance to indinavir, a human immunodeficiency virus type 1 protease inhibitor. *Journal of Virology* 70: 8270–8276.

Crepaz, N., T. A. Hart, and G. Marks. 2004. Highly active antiretroviral therapy and sexual risk behavior:a meta-analytic review. *JAMA* 292: 294–236.

Darwin, C. 1859. *On the Origin of Species by Means of Natural Selection, Or the Preservation of the Favoured Races in the Struggle for Life.* London: John Murray.

Davenport, M. P., J. J. Zaunders, M. D. Hazenberg, et al. 2002. Cell turnover and cell tropism in HIV-1 infection. *Trends in Microbiology* 10: 275–278.

Davey, R. T. Jr., N. Bhat, C. Yoder, et al. 1999. HIV-1 and T cell dynamics after interruption of highly active antiretroviral therapy (HAART) in patients with a history of sustained viral suppression. *Proceedings of the National Academy of Sciences USA* 96: 15109–15114.

Deacon, N. J. et al. 1995. Genomic structure of an attenuated quasi species of HIV-1 from a blood transfusion donor and recipients. *Science* 270: 988–991.

Deeks, S. G., M. Smith, M. Holodniy, and J. O. Kahn. 1997. HIV-1 protease inhibitors. *Journal of the American Medical Association* 277: 145–153.

Deeks, S. G., and B. D. Walker. 2004. The immune response to AIDS virus infection: good, bad, or both? *Journal of Clinical Investigation* 113: 808–810.

Derdeyn, C. A., and G. Silvestri. 2005. Viral and host factors in the pathogenesis of HIV infection. *Current Opinion in Immunology* 17: 366–373.

Derr, M. 2001. New theories link Black Death to ebola-like virus. *The New York Times* 2 October Late Edition: F4.

Dobzhansky, T. 1973. Nothing in biology makes sense except in the light of evolution. *American Biology Teacher* 35: 125–129.

Douek, D. C., L. J. Picker, and R. A. Koup. 2003. T cell dynamics in HIV-1 infection. *Annual Review of Immunology* 21: 265–304.

Doukhan, L. and E. Delwart. 2001. Population genetic analysis of the protease locus of human immunodeficiency virus type 1 quasispecies undergoing drug selection, using a denaturing gradient-heteroduplex tracking assay. *Journal of Virology* 75: 6729–6736.

Dukers, N. H. T. M., J. Spaargaren, R. B. Geskus, et al. 2002. HIV incidence on the increase among homosexual men attending an Amsterdam sexually transmitted disease clinic: Using a novel approach for detecting recent infections. *AIDS* 16: F19–F24.

Duncan, C. J., and S. Scott. 2005. What caused the Black Death? *Postgraduate Medical Journal* 81: 315–320.

Evans, B., P. Cane, et al. 2005. Estimating HIV-1 drug resistance in antiretroviral-treated individuals in the United Kingdom. *Journal of Infectious Diseases* 192: 967–973.

Fackler, O. T., and A. S. Baur. 2002. Live and let die: Nef functions beyond HIV replication. *Immunity* 16: 493–497.

Feng, Y., C. C. Broder, P. E. Kennedy, and E. A. Berger. 1996. HIV-1 entry cofactor: Functional cDNA cloning of a seven-transmembrane, G protein-coupled reactor. *Science* 272: 872–877.

Finzi, D., J. Blankson, J. D. Siliciano, et al. 1999. Latent infection of $CD4^+$ T cells provides a mechanism for lifelong persistence of HIV-1, even in patients on effective combination therapy. *Nature Medicine* 5: 512–517.

Finzi, D., M. Hermankova, T. Pierson, et al. 1997. Identification of a reservoir for HIV-1 in patients on highly active antiretroviral therapy. *Science* 278: 1295–1300.

Frost, S. D. W., H. F. Günthard, J. K. Wong, et al. 2001. Evidence for positive selection driving the evolution of HIV-1 *env* under potent antiviral therapy. *Virology* 284: 250–258.

Gallago, O., C. de Mendoza, J. J. Pérez-Elías, et al. 2001. Drug resistance in patients experiencing early virological failure under a triple combination including indinavir. *AIDS 2001* 15: 1701–1706.

Gao, F., E. Bailes, D. L. Robertson, Y. Chen, C. M. Rodenburg, S. F. Michael, L. B. Cummins, L. O. Arthur, M. Peeters, G. M. Shaw, P. M. Sharp, and B. H. Hahn. 1999. Origin of HIV-1 in the chimpanzee *Pan troglodytes troglodytes. Nature* 397: 436–441.

Gao, X., A. Bashirova, et al. 2005. AIDS restriction HLA allotypes target distinct intervals of HIV-1 pathogenesis. *Nature Medicine* 11: 1290–1292.

Garber, D. A., G. Silvestri, A. P. Barry, et al. 2004. Blockade of T cell costimulation reveals interrelated actions of $CD4^+$ and $CD8^+$ T cells in control of SIV replication. *Journal of Clinical Investigation* 113: 836–845.

Gaschen, B., J. Taylor, K. Yusim, et al. 2002. Diversity considerations in HIV-1 vaccine selection. *Science* 296: 2354–2360.

Geffin, R., D. Wolf, R. Müller, et al. 2000. Functional and structural defects in HIV type 1 *nef* genes derived from pediatric long-term survivors. *AIDS Research and Human Retroviruses* 16: 1855–1868.

Ghys, P. D., M. O. Diallo, V. Ettiègne-Treoré, et al. 2002. Increase in condom use and decline in HIV and sexually transmitted diseases among female sex workers in Abidjan, Côte d'Ivoire, 1991–1998. *AIDS* 16: 251–258.

Gilbert, M. T. P., J. Cuccui, et al. 2004. Absence of Yersinia pestis-specific DNA in human teeth from five European excavations of putative plague victims. Mircobiology 150: 341–354.

Gisselquist, D., R. Rothenberg, J. Potterat, and E. Drucker. 2002. HIV infections in sub-Saharan Africa not explained by sexual or vertical transmission. *International Journal of STD and AIDS* 13: 657–666.

Gisselquist, D., Potterat, J. J., et al. 2004. Does selected ecological evidence give a true picture of HIV transmission in Africa? *International Journal of STD and AIDS* 15: 434–439.

Giuliani, M., A. D. Carlo, et al. 2005. Increased HIV incidence among men who have sex with men in Rome. *AIDS* 19: 1429–1431.

Glass, W. G., D. H. McDermott, et al. 2006. CCR5 deficiency increases risk of sympotmatic West Nile virus infection. *Journal of Experimental Medicine* 203: 35–40.

Gonzalez, E., H. Kulkarni, et al. 2005. The influence of *CCL3L1* gene-containing segmental duplications on HIV-1/AIDS susceptibility. *Science* 307: 1434–1440.

Goulder, P. J. R., C. Brander, Y. Tang, et al. 2001. Evolution and transmission of stable CTL escape mutations in HIV infection. *Nature* 412: 334–338.

Grant, R. M., F. M. Hecht, M. Warmerdam, et al. 2002. Time trends in primary HIV-1 drug resistance among recently infected persons. *Journal of the American Medical Association* 288: 181–188.

Grossman, Z., M. Meier-Schellersheim, et al. 2002. $CD4^+$ T-cell depletion in HIV infection: Are we closer to understanding the cause? *Nature Medicine* 8: 319–323.

Grossman, Z., M. Meier-Schellersheim, et al. 2006. Pathogenesis of HIV infection: what the virus spares is as important as what it destroys. *Nature Medicine* 12: 289–295.

Guadalupe, M., E. Reay, S. Sankaran, et al. 2003. Severe $CD4^+$ T-cell depletion in gut lymphoid tissue during primary human immunodeficiency virus type 1 infection and substantial delay in restoration following highly active antiretroviral therapy. *Journal of Virology* 77: 11708–11717.

Gulick, R. M., J. W. Mellors, D. Havlir, et al. 1997. Treatment with indinavir, zidovudine, and lamivudine in adults with human immunodeficiency virus infection and prior antiretroviral therapy. *The New England Journal of Medicine* 337: 734–739.

Günthard, H. F., S. W. Frost, A. J. Leigh-Brown, et al. 1999. Evolution of envelope sequences of human immunodeficiency virus type 1 in cellular reservoirs in the setting of potent antiviral therapy. *Journal of Virology* 73: 9404–9412.

Hahn, B. H., G. M. Shaw, K. M. De Cock, and P. M. Sharp. 2000. AIDS as a zoonosis: Scientific and public health implications. *Science* 287: 607–614.

Hamers, F. F., and A. M. Downs. 2004. The changing face of the HIV epidemic in western Europe: what are the implications for public health policies? *The Lancet* 364: 83–94.

Hillis, D. M. 2000. AIDS. Origins of HIV. *Science* 288: 1757–1759.

Ho, D. D., A. U. Neumann, A. S. Perelson, et al. 1995. Rapid turnover of plasma virions and CD4 lymphocytes in HIV-1 infection. *Nature* 373: 123–126.

Hoggard, P. G., J. Lloyd, S. H. Khoo, et al. 2001. Zidovudine phosphorylation determined sequentially over 12 months in human immunodeficiency virus-infected patients with or without previous exposure to antiretroviral agents. *Antimicrobial Agents and Chemotherapy* 45: 976–980.

Hübner, A., M. Kruhoffer, F. Grosse, and G. Krauss. 1992. Fidelity of human immunodeficiency virus type 1 reverse transcriptase in copying natural RNA. *Journal of Molecular Biology* 223: 595–600.

Jekle, A., B. Schramm, P. Jayakumar, et al. 2002. Coreceptor phenotype of natural human immunodeficiency virus with Nef deleted evolves in vivo, leading to increased virulence. *Journal of Virology* 76: 6966–6973.

Johnson, N. P. A. S., and J. Mueller. 2002. Updating the accounts: Global mortality of the 1918–1920 "Spanish" influenza pandemic. *Bulletin of the History of Medicine* 76: 105–115.

Kalkut, G. 2005. Antiretroviral therapy: an update for the non-AIDS specialist. *Current Opinion in Oncology* 17: 479-484.

Kanki P. J., K. U. Travers, S. MBoup, et al. 1994. Slower heterosexual spread of HIV-2 than HIV-1. *Lancet* 343: 943–946.

Katz M. H., S. K. Schwarcz, T. A. Kellogg, et al. 2002. Impact of highly active antiretroviral treatment on HIV seroincidence among men who have sex with men: San Francisco. *American Journal of Public Health* 92: 388–394.

Kaufman, J., and J. Jing. 2002. China and AIDS—the time to act is now. *Science* 296: 2339–2340.

Kaufmann, D. E., M. Lichterfeld, et al. 2004. Limited durability of viral control following treated acute HIV infection. *PLoS Medicine* 2: e36.

Keele, B. F., F. Van Heuverswyn, et al. 2006. Chimpanzee reservoirs of pandemic and nonpandemic HIV-1. *Science* 313: 523–526.

Kellogg, T., W. McFarland, and M. Katz. 1999. Recent increases in HIV seroconversion among repeat anonymous testers in San Francisco. *AIDS 1999* 13: 2303–2304.

Korber, B., M. Muldoon, J. Theiler, et al. 2000. Timing the ancestor of the HIV-1 pandemic strains. *Science* 288: 1789–1796.

Korber, B., J. Theiler, and S. Wolinsky. 1998. Limitations of a molecular clock applied to considerations of the origin of HIV-1. *Science* 280: 1868–1870.

Kristiansen, T. B., A. G. Pedersen, et al. 2005. Genetic evolution of HIV in patients remaining on a stable HAART regimen despite insufficient viral suppression. *Scandinavian Journal of Infectious Diseases* 37: 890–901.

Kuhmann, S. E., D. Madani, et al. 2001. Frequent substitution polymorphisms in African green monkey CCR5 cluster at critical sites for infections by simian immunodeficiency virus SIVagm, implying ancient virus-host coevolution. *Journal of Virology* 75: 8449–8460.

Lalani, A. S., J. Masters, W. Zeng, et al. 1999. Use of chemokine receptors by poxviruses. *Science* 286: 1968–1971.

Larder, B. A., G. Darby, and D. D. Richman. 1989. HIV with reduced sensitivity to Zidovudine (AZT) isolated during prolonged therapy. *Science* 243: 1731–1734.

Learmont, J. C., A. F. Geczy, J. Mills, et al. 1999. Immunologic and virologic status after 14 to 18 years of infection with an attenuated strain of HIV-1. A report from the Sydney Blood Bank Cohort. *New England Journal of Medicine* 340: 1715–1722.

Lehrman, G., I. B. Hogue, et al. 2005. Depletion of latent HIV-1 infection in vivo: a proof of concept study. *The Lancet* 366: 549–555.

Leslie, A. J., K. J. Pfafferott, et al. 2004. HIV evolution: CTL escape mutation and reversion after transmission. *Nature Medicine* 10: 282–289.

Letvin, N. L. 1998. Progress in the development of an HIV-1 vaccine. *Science* 280: 1875–1880.

Levin, B. R. 1996. The evolution and maintenance of virulence in microparasites. *Emerging Infectious Diseases* 2: 93–102.

Levin, B. R., and J. J. Bull. 1994. Short-sighted evolution and the virulence of pathogenic microorganisms. *Trends in Microbiology* 2: 76–81.

Limborska, S. A., O. P. Balanovsky, E. V. Balanovskaya, et al. 2002. Analysis of *CCR5-Δ32* geographic distribution and its correlation with some climatic and geographic factors. *Human Heredity* 53: 49–54.

Linnemann, T., Y.-H. Zheng, R. Mandic, and B. M. Peterlin. 2002. Interaction between Nef and phosphatidylinositol-3-kinase leads to activation of p21-activated kinase and increased production of HIV. *Virology* 294: 246–255.

Liu, R. et al. 1996. Homozygous defect in HIV-1 coreceptor accounts for resistance of some multiply-exposed individuals to HIV-1 infection. *Cell* 86: 367–377.

Lopman, B. A., G. P. Garnett, et al. 2005. Individual Level Injection History: A Lack of Association with HIV Incidence in Rural Zimbabwe. *PLoS Medicine* 2: e37.

Lucotte, G. 2001. Distribution of the CCR5 gene 32-basepair deletion in West Europe. A hypothesis about the possible dispersion of the mutation by Vikings in historical times. *Human Immunology* 62: 933–936.

Macdonald, N., S. Dougan, et al. 2004. Recent trends in diagnoses of HIV and other sexually transmitted infections in England and Wales among men who have sex with men. *Sexually Transmitted Infections* 80: 492–497.

Markowitz, M., H. Mohri, et al. 2005. Infection with multidrug resistant, dual-tropic HIV-1 and rapid progression to AIDS: a case report. *The Lancet* 365: 1031–1038.

Marlink, R., P. Kanki, I. Thior, K. Travers, G. Eisen, T. Siby, I. Traore, C-C. Hsieh, M. C. Dia, E-H. Gueye, J. Hellinger, A. Gueye-Ndiaye, J-L. Sankalé, I. Ndoye, S. Mboup, and M. Essex. 1994. Reduced rate of disease development after HIV-2 infection as compared to HIV-1. *Science* 265: 1587–1590.

McCune, J. M. 2001. The dynamics of $CD4^+$ T-cell depletion in HIV disease. *Nature* 410: 974–979.

Mehandru, S., M. A. Poles, et al. 2004. Primary HIV-1 infection is associated with preferential depletion of $CD4^+$ T lymphocytes from effector sites in the gastrointestinal tract. *Journal of Experimental Medicine* 200: 761–770.

Mocroft, A., B. Ledergerber, et al. 2004. Time to virological failure of 3 classes of antiretrovirals after initiation of highly active antiretroviral therapy: results from the EuroSIDA Study Group. *Journal of Infectious Diseases* 190: 1947–56.

Modi, W. S., J. Lautenberger, et al. 2006. Genetic variation in the *CCL18-CCL3-CCL4* chemokine gene cluster Influences HIV type 1 transmission and AIDS disease progression. *American Journal of Human Genetics* 79: 120–128.

Mohri, H., M. K. Singh, W. T. W. Ching, and D. D. Ho. 1993. Quantitation of zidovudine-resistant human immunodeficiency virus type 1 in the blood of treated and untreated patients. *Proceedings of the National Academy of Sciences USA* 90: 25–29.

Moore, C. B., M. John, I. R. James, et al. 2002. Evidence of HIV-1 adaptation to HLA-restricted immune responses at a population level. *Science* 296: 1439–1443.

Moore, J. P., S. G. Kitchen, et al. 2004. The CCR5 and CXCR4 coreceptors—central to understanding the transmission and pathogenesis of human immunodeficiency virus type 1 infection. *AIDS Research and Human Retroviruses* 20: 111–126.

Mosier, D. E., G. R. Picchio, R. J.Gulizia, et al. 1999. Highly potent RANTES analogues either prevent CCR5-using human immunodeficiency virus type 1 infection in vivo or rapidly select for CXCR4-using variants. *Journal of Virology* 73: 3544–3550.

Moulton, V. R., and D. L. Farber. 2006. Committed to memory: lineage choices for activated T cells. *Trends in Immunology*: In press.

National Institute of Allergy and Infectious Diseases (NIAID). 2003. *Understanding the Immune System: How It Works.* NIH Publication No. 03-5423. Available online at: *http://www.niaid.nih.gov/publications/immune/the_immune_system.pdf.*

Nelson, K. E., S. Eiumtrakul, D. D. Celentano, et al. 2002. HIV infection in young men in northern Thailand, 1991–1998: Increasing role of injection drug use. *Journal of Acquired Immune Deficiency Syndromes* 29: 62–68.

Nickle, D. C., M. A. Jensen, et al. 2003. Consensus and ancestral state HIV vaccines. *Science* 299: 1515–1518.

O'Brien, S. J., and G. W. Nelson. 2004. Human genes that limit AIDS. *Nature Genetics* 36: 565–574.

Oswald-Richter, K., S. M. Grill, et al. 2004. HIV infection of naturally-occurring and genetically reprogrammed human regulatory T-cells. *PLoS Biology* 2: e198.

Oxenius, A., D. A Price, et al. 2002. Stimulation of HIV-specific cellular immunity by structured treatment interruption fails to enhance viral control in chronic HIV infection. *Proceedings of the National Academy of Sciences USA* 99: 13747–13752.

Palacios, E., L. Digilio, et al. Parallel evolution of CCR5-null phenotypes in humans and in a natural host of simian immunodeficiency virus. *Current Biology* 8: 943–946.

Palella, F. J., Jr., J. S. Chmiel, A. C. Moorman, et al. 2002. Durability and predictors of success of highly active antiretroviral therapy for ambulatory HIV-infected patients. *AIDS* 16: 1617–1626.

Papkalla, A., J. Münch, C. Otto, and F. Kirchhoff. 2002. Nef enhances human immunodeficiency virus type 1 infectivity and replication independently of viral coreceptor tropism. *Journal of Virology* 76: 8455–8459.

Patterson, K. B., and T. Runge. 2002. Smallpox and the Native American. *American Journal of the Medical Sciences* 323: 216–222.

Penn, M. L., J.-C. Grivel, B. Schramm, et al. 1999. CXCR4 utilization is sufficient to trigger $CD4^+$ T cell depletion in HIV-1-infected human lymphoid tissue. *Proceedings of the National Academy of Sciences USA* 96: 663–668.

Piot, P., M. Bartos, et al. 2001. The global impact of HIV/AIDS. *Nature* 410: 968–973.

Pomerantz, R. J., and D. L. Horn. 2003. Twenty years of therapy for HIV-1 infection. *Nature Medicine* 9: 867–873.

Pommier, Y., A. A. Johnson, and C. Marchand. 2005. Integrase inhibitors to treat HIV/AIDS. *Nature Reviews Drug Discovery*: 4: 236–248.

Popper S. J., A. D. Sarr, K. U. Travers, et al. 1999. Lower human immunodeficiency virus (HIV) type 2 viral load reflects the difference in pathogenicity of HIV-1 and HIV-2. *Journal of Infectious Diseases* 180: 1116–1121.

Price, D. A., P. J. R. Goulder, P. Klenerman, et al. 1997. Positive selection of HIV-1 cytotoxic T lymphocyte escape variants during primary infection. *Proceedings of the National Academy of Sciences USA* 94: 1890–1895.

Quillent, C. et al. 1998. HIV-1-resistance phenotype conferred by combination of two separate inherited mutations of CCR5 gene. *Lancet* 351: 14–18.

Ramaley, P. A., N. French, P. Kaleebu, et al. 2002. Chemokine-receptor genes and AIDS risk. *Nature* 417: 140.

Rambaut, A., D. L. Robertson, O. G. Pybus, et al. 2001. Human immunodeficiency virus. Phylogeny and the origin of HIV-1. *Nature* 410: 1047–1048.

Ramratnam, B., J. E. Mittler, L. Q. Zhang, et al. 2000. The decay of the latent reservoir of replication-competent HIV-1 is inversely correlated with the extent of residual viral replication during prolonged antiretroviral therapy. *Nature Medicine* 6: 82–85.

Raoult, D., G. Aboudharam, E. Crubézy, et al. 2000. Molecular identification by "suicide PCR" of *Yersinia pestis* as the agent of medieval Black Death. *Proceedings of the National Academy of Sciences USA* 97: 12800–12803.

Rhodes, D. I., L. Ashton, A. Solomon, et al. 2000. Characterization of three *nef*-defective human immunodeficiency virus type 1 strains associated with long-term nonprogression. *Journal of Virology* 74: 10581–10588.

Rizzardi, G. P., A. Harari, et al. 2002. Treatment of primary HIV-1 infection with cyclosporin A coupled with highly active antiretroviral therapy. *Journal of Clinical Investigation* 109: 681–688.

Roberts, L. 1989. Disease and death in the new world. *Science* 246: 1245–1247.

Ross, H. A., and A. G. Rodrigo. 2002. Immune-mediated positive selection drives human immunodeficiency virus type 1 molecular variation and predicts disease duration. *Journal of Virology* 76: 11715–11720.

Sabundayao, B. P., J. H. McArthur, et al. 2006. High frequency of highly active antiretroviral therapy modifications in patients with acute or early human immunodeficiency virus infection. *Pharmacotherapy* 26: 674–681.

St. Clair, M. H., J. L. Martin, G. Tudor-Williams, M. C. Bach, C. L. Vavro, D. M. King, P. Kellam, S. D. Kemp, and B. A. Larder. 1991. Resistance to ddI and sensitivity to AZT induced by a mutation in HIV-1 reverse transcriptase. *Science* 253: 1557–1559.

Salemi, M., K. Strimmer, W. H. Hall, et al. 2001. Dating the common ancestor of SIV_{cpz} and HIV-1 group M and the origin of HIV-1 subtypes using a new method to uncover clock-like molecular evolution. *The FASEB Journal* 15: 276–278.

Samson, M. et al. 1998. Resistance to HIV-1 infection in caucasian individuals bearing mutant alleles of the CCR-5 chemokine receptor gene. *Nature* 382: 722–725.

Schaeffer, E., R. Geleziunas, and W. C. Greene. 2001. Human immunodeficiency virus type 1 Nef functions at the level of virus entry by enhancing cytoplasmic delivery of virions. *Journal of Virology* 75: 2993–3000.

Schindler, M., J. Münch, et al. 2006. Nef-mediated suppression of T cell activation was lost in a lentiviral lineage that gave rise to HIV-1. *Cell* 125: 1055-1067.

Schliekelman, P., C. Garner, and M. Slatkin. 2001. Natural selection and resistance to HIV. Nature 411: 545–546.

Schmid, G. P., A. Buvé, et al. 2004. Transmission of HIV-1 infection in sub-Saharan Africa and effect of elimination of unsafe injections. *The Lancet* 363: 482–488.

Shankarappa, R., J. B. Margolick, S. J. Gange, et al. 1999. Consistent viral evolutionary changes associated with the progression of human immunodeficiency virus type 1 infection. *Journal of Virology* 73: 10489–10502.

Sharkey M. E., I. Teo, T. Greenough, et al. 2000. Persistence of episomal HIV-1 infection intermediates in patients on highly active anti-retroviral therapy. *Nature Medicine* 6: 76–81.

Sharp, P. M., E. Bailes, R. R. Chaudhuri, et al. 2001. The origins of acquired immune deciency syndrome viruses: where and when? *Philosophical Transactions of the Royal Society of London* B 356: 867–876.

Shirasaka, T., R. Yarchoan, M. C. O'Brien, R. N. Husson, B. D. Anderson, E. Kojima, T. Shimada, S. Broder, and H. Mitsuya. 1993. Changes in drug sensitivity of human immunodeficiency virus type 1 during therapy with azidothymidine, dideoxycytidine, and dideoxyinosine: An in vitro comparative study. *Proceedings of the National Academy of Sciences USA* 90: 562–566.

Silvestri, G., A. Fedanov, et al. 2005. Divergent host responses during primary simian immunodeficiency virus SIVsm infection of natural sooty mangabey and nonnatural rhesus macaque hosts. *Journal of Virology* 79: 4043–4054.

Smith, M. W. et al. 1997. Contrasting genetic influence of CCR2 and CCR5 variants on HIV-1 infection and disease progression. *Science* 277: 959–965.

Smith, S. M. 2005. Valproic acid and HIV-1 latency: beyond the sound bite. *Retrovirology* 2: 56–58.

Snow, D. R. 1995. Microchronology and demographic evidence relating to the size of pre-Columbian North American Indian populations. *Science* 268: 1601–1604.

Stephens, J. C., D. E. Reich, D. B. Goldstein, et al. 1998. Dating the origin of the *CCR5-Δ32* AIDS-resistance allele by the coalescence of haplotypes. *American Journal of Human Genetics* 62: 1507–1515.

Stephenson, J. 2002. Researchers explore new anti-HIV agents. *Journal of the American Medical Association* 287: 1635–1637.

Strizki, J. M., S. Xu, N. E. Wagner, et al. 2001. SCH-C (SCH 351125), an orally bioavailable, small molecule antagonist of the chemokine receptor CCR5, is a potent inhibitor of HIV-1 infection *in vitro* and *in vivo*. *Proceedings of the National Academy of Sciences USA* 98: 12718–12723.

Sullivan, A. D., J. Wigginton, and D. Kirschner. 2001. The coreceptor mutation *CCR5-Δ32* influences the dynamics of HIV epidemics and is selected for by HIV. *Proceedings of the National Academy of Sciences USA* 98: 10214–10219.

Swann, S. A., M. Williams, C. M. Story, et al. 2001. HIV-1 Nef blocks transport of MHC class I molecules to the cell surface via a PI3-kinase-dependent pathway. *Virology* 282: 267–277.

Tobiume, M. M. Takahoko, T. Yamada, et al. 2002. Inefficient enhancement of viral infectivity and CD4 downregulation by human immunodeficiency virus Type 1 Nef from Japanese long-term nonprogressors. *Journal of Virology* 76: 5959–5965.

Trkola, A., S. E. Kuhmann, J. M. Strizki, et al. 2002. HIV-1 escape from a small molecule, CCR5-specific entry inhibitor does not involve CXCR4 use. *Proceedings of the National Academy of Sciences USA* 99: 395–400.

Troyer, R. M., K. R. Collins, et al. 2005. Changes in human immunodeficiency virus type 1 fitness and genetic diversity during disease progression. *Journal of Virology* 79: 9006–9018.

UNAIDS. 2002a. Report on the global HIV/AIDS epidemic 2002. Available at *http://www.unaids.org*.

UNAIDS. 2002b. A global view of HIV infection. Available at *http://www.unaids.org/barcelona/presskit/graphics.html#global*.

UNAIDS. 2004. Report on the global AIDS epidemic. Available at *http:/www.unaids.org/bangkok2004/report.html*.

UNAIDS. 2005. AIDS epidemic update: Special report on HIV prevention. December 2005. Available at: *http://www.unaids.org/epi/2005/doc/report_pdf.asp*.

Veazey, R., B. Ling, et al. 2003. Decreased CCR5 expression on CD4+ T cells of SIV-infected sooty mangabeys. *AIDS Research and Human Retroviruses* 19: 227–233.

Vila-Coro, A. J., M. Mellado, A. M. de Ana, et al. 2000. HIV-1 infection through the CCR5 receptor is blocked by receptor dimerization. *Proceedings of the National Academy of Sciences USA* 97: 3388–3393.

Wain-Hobson, S. 1993. The fastest genome evolution ever described: HIV variation in situ. *Current Opinion in Genetics and Development* 3: 878–883.

Weber, J. 2001. The pathogenesis of HIV-1 infection. *British Medical Bulletin* 58: 61–72.

Wei, X., S. K. Ghosh, M. E. Taylor, et al. 1995. Viral dynamics in human immunodeficiency virus type 1 infection. *Nature* 373: 117–122.

Williamson, S., S. M. Perry, et al. 2005. A statistical characterization of consistent patterns of human immunodeficiency virus evolution within infected patients. *Molecular Biology and Evolution* 22: 456-468.

Wong, J. K., H. F. Günthard, D.V. Havlir, et al. 1997a. Reduction of HIV-1 in blood and lymph nodes following potent antiretroviral therapy and the virologic correlates of treatment failure. *Proceedings of the National Academy of Sciences USA* 94: 12574–12579.

Wong, J. K., M. Hezareh, H. F. Günthard, et al. 1997b. Recovery of replication-competent HIV despite prolonged suppression of plasma viremia. *Science* 278: 1291–1295.

World Health Organization (WHO) 2001. The world health report 2001. Available at *http://www.who.int/whr/*.

World Health Organization (WHO) 2004. The world health report 2004: Changing history. Available at *http://www.who.int/whr/*.

Yusim, K., C. Kesmir, B. Gaschen, et al. 2002. Clustering patterns of cytotoxic T-lymphocyte epitopes in human immunodeficiency virus type 1 (HIV-1) proteins reveal imprints of immune evasion on HIV-1 global variation. *Journal of Virology* 76: 8757–8768.

Yusim, K., M. Peeters, O. G. Pybus, et al. 2001. Using human immunodeficiency virus type 1 sequences to infer historical features of the acquired immune deficiency syndrome epidemic and human immunodeficiency virus evolution. *Philosophical Transactions of the Royal Society of London* B 356: 855–866.

Zhang L., B. Ramratnam, K. Tenner-Racz, et al. 1999. Quantifying residual HIV-1 replication in patients receiving combination antiretroviral therapy. *New England Journal of Medicine* 340: 1605–1613.

Zhu, T., H. Mo, N. Wang, et al. 1993. Genotypic and phenotypic characterization of HIV-1 in patients with primary infection. *Science* 261: 1179–1181.

Zhu, T., B. T. Korber, A. J. Nahmias, E. Hooper, P. M. Sharp, and D. D. Ho. 1998. An African HIV-1 sequence from 1959 and implications for the origin of the epidemic. *Nature* 391: 594–597.

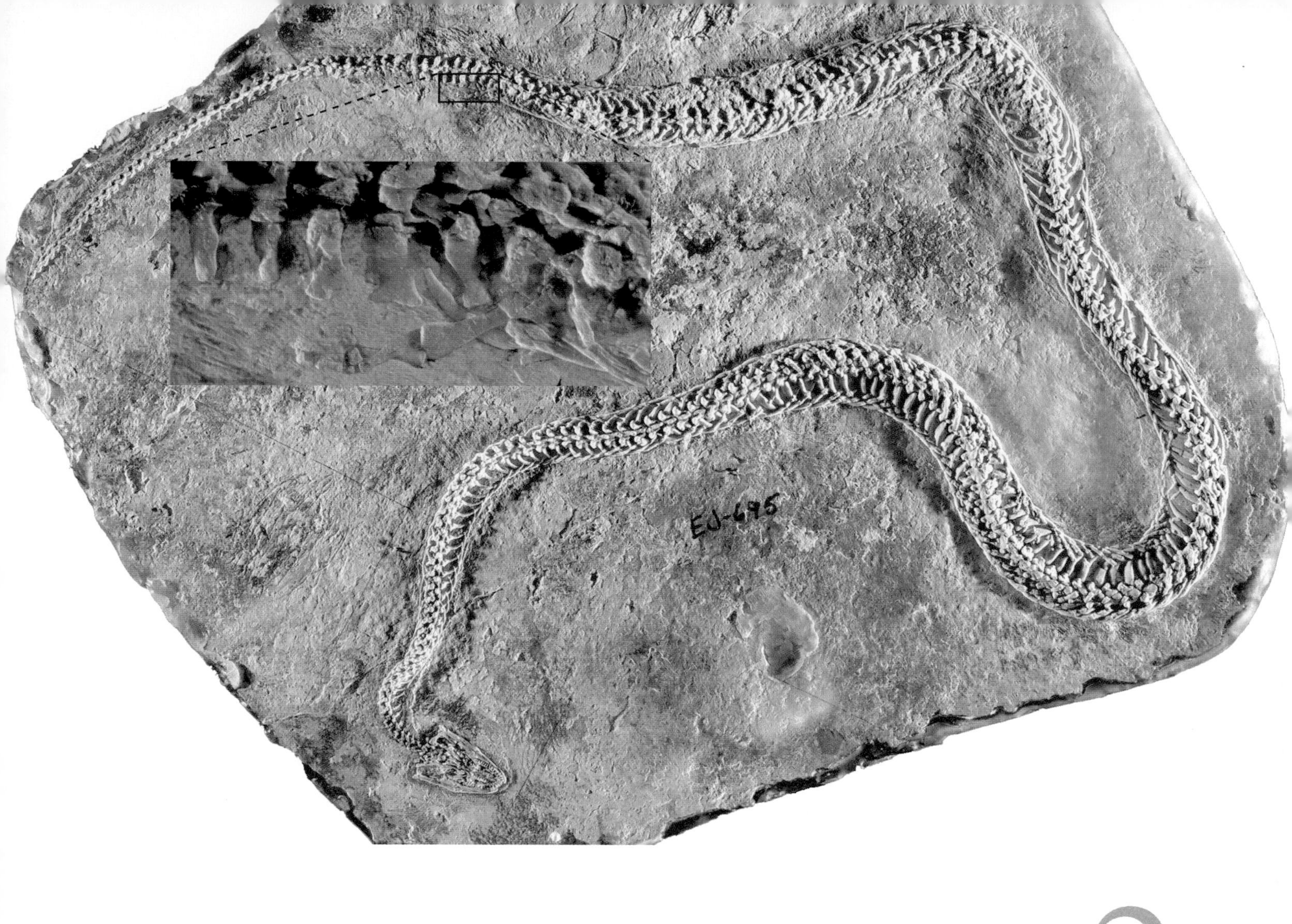

2
The Pattern of Evolution

A snake with legs. This 95-million-year-old fossil snake has small but fully formed hind limbs. Called Haas's Holy Land snake (*Haasiophis terrasanctus*), it documents the previous existence of snakes leggier than any alive now. See Tchernov et al. (2000); Rieppel et al. (2003).

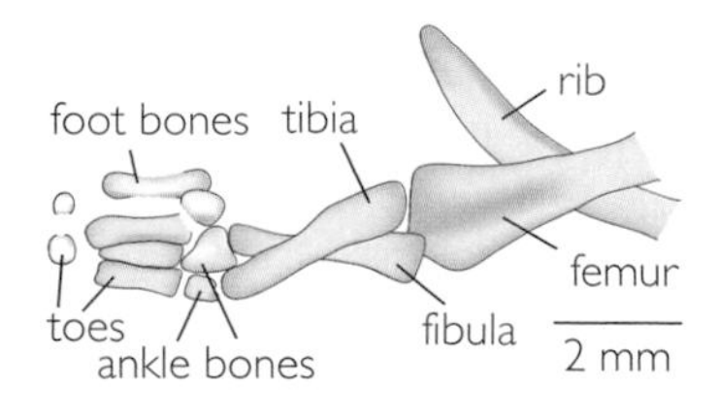

Where do we come from, we humans and the staggering variety of other organisms with which we share our planet? Members of the general public are, in the United States at least, divided over the answer. When, in a recent survey, 2,000 adult Americans were asked to choose a phrase to complete the statement "Life on Earth has . . . "

42% picked ". . . existed in its present form since the beginning of time,"

48% picked ". . . evolved over time," and

the remaining 10% said they did not know (PEW Research Center 2005).

The first option derives from a literal reading of the Bible's Book of Genesis (1:1–2:4). In this version of life's history, all organisms were created by God during the six days of creation. The ideal types formed by this special process, including Adam and Eve, were the progenitors of all organisms. Species are unchanged since their creation, or immutable, and variation within each type is limited. As expressed by John Ray (1686), the first scientist to give a biological definition of species, "... [O]ne species never springs from the seed of another."

Among the most literal interpreters of the Book of Genesis was James Ussher, the Archbishop of Armagh. In 1650 Ussher, drawing on anstronomical cycles, Old Testament geneologies, and other references, fixed the beginning of time "upon the entrance of the night preceeding the twenty-third day of October" in 4004 B.C. (reprinted in English in Ussher 1658, page 1). Starting in 1701, Ussher's figure for the year of creation appeared, without attribution, in a marginal note in English Bibles printed by Clarendon Press, Oxford. It was taken as authoritative by many readers for centuries thereafter (see Brice 1982).

Scientific theories often have two components. The first is a statement about a pattern that exists in the natural world; the second is a process that explains the pattern.

Note that the Theory of Special Creation has two components. The first component is a set of assertions—claims about the pattern of life's history. These claims are: (1) Species do not change through time; (2) they were created independently of one another; and (3) they were created recently. The second component identifies the process that is responsible for producing the pattern: separate and independent acts of creation by a designer.

When the English naturalist Charles Darwin began to study biology seriously, as a college student in the early 1820s, the Theory of Special Creation had been challenged on its details by some scholars, but was still the leading explanation in Europe for the origin of species. By the time Darwin began working on the problem in the 1830s, however, dissatisfaction with the theory had begun to grow in earnest. Research in the biological and geological sciences was advancing rapidly, and the data clashed with special creation's central claims. Hints of the new theory that would replace it were in the air (for a brief history, see Box 2.1).

The Theory of Special Creation and the Theory of Descent with Modification make different assertions about whether species can change, where species came from, and the age of the Earth and life.

The scientist who brought this new theory to fruition was, of course, Darwin himself. Drawing on his own work and that of others, Darwin marshalled evidence that the pattern of life's history is different than is claimed by Special Creation (Figure 2.1). First, species are not immutable, but change through time. Second, species are derived not independently, but from common—that is, shared—ancestors. "I should infer ...," Darwin said in *On the Origin of Species* (1859, p. 484), "... that probably all the organic beings which have ever lived on this earth have descended from some one primordial form, into which life was first breathed." Third, the Earth and life are considerably more than 6,000 years old.

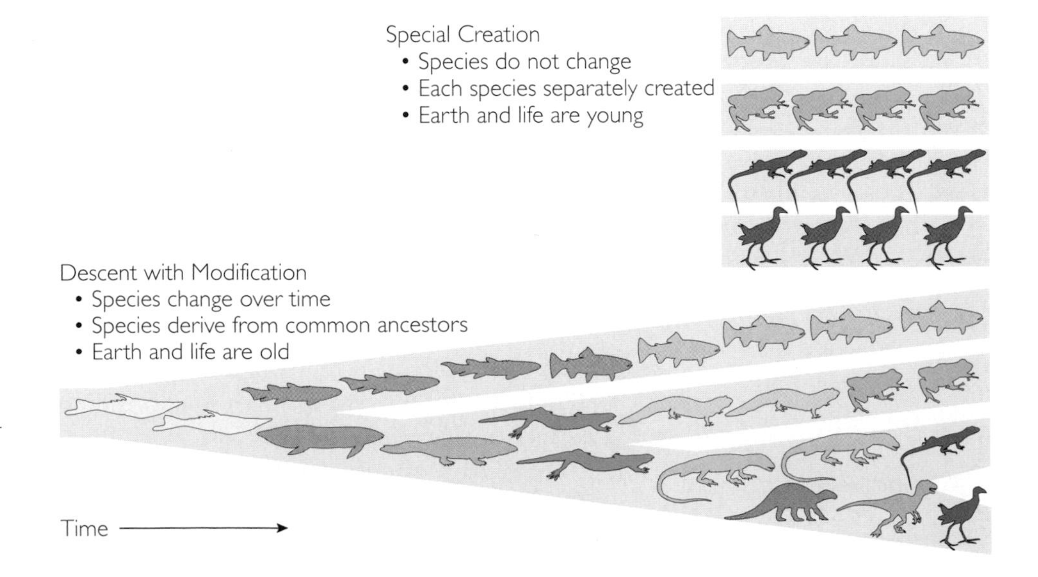

Figure 2.1 Two views of the history of life These cartoons illustrate the contrasting claims made by the Theory of Special Creation versus the Theory of Descent with Modification.

Box 2.1 A brief history of ideas on evolution

In his most famous book, *On the Origin of Species by Means of Natural Selection*, first published in 1859, Charles Darwin set himself two objectives: To marshal evidence on the fact of evolution and to identify natural selection as the mechanism responsible. Neither idea was unique to Darwin, but he understood them more clearly and treated them much more comprehensively than his forerunners. Furthermore, he anticipated where they would lead with an accuracy that continues to startle modern biologists. As a result, it is Darwin's name that is today most closely associated with the Theory of Evolution.

The fact of evolution

The fact of evolution had been proposed by several workers in the late 1700s and early 1800s, including Comte de Buffon, Erasmus Darwin (Charles' grandfather), and the great French biologist, Jean-Baptiste Lamarck (Eiseley 1958; Desmond and Moore 1991). Darwin himself (1872) cited Lamarck as the first writer "whose conclusions on the subject excited much attention." In works published in 1809 and 1815, Lamarck advanced the notion that all species, including humans, are derived by gradual evolution from other species. This process was driven, according to Lamarck, by the inheritance of acquired characteristics and by an inherent tendency for all organisms to progress from simple to complex form. To explain the continued existence of simple life forms, Lamarck suggested that they are continuously replenished by spontaneous generation from nonliving matter.

Between 1844 and 1853, Robert Chambers published 10 editions of a popular book called *The Vestiges of the Natural History of Creation*. Darwin found Chambers's scientific reasoning confused and his evidence inadequate but credited Chambers for promoting the idea of evolution and for "removing prejudice" against it.

Although the idea of evolution had been under discussion for decades, it was Darwin who convinced the scientific community that it was true—that Earth's species are the products of descent with modification from a common ancestor (Mayr 1964). Darwin had worked on the material in *The Origin* for more than 20 years before publishing it and had gathered an overwhelming collection of detailed evidence from a variety of fields of biology. His masterful presentation of this evidence was persuasive. Within a decade of *The Origin's* initial publication, the fact of evolution had achieved general acceptance. Darwin's mechanism of evolution, on the other hand, did not at first fair so well.

The mechanism of evolution

At least two authors discovered natural selection well before Darwin did (Darwin 1872). In 1813 W. C. Wells used it to explain how human populations on different continents came to differ in their physical appearance and resistance to disease. In 1831, Patrick Matthew discussed it in a treatise on farming trees for lumber with which to build ships. Neither work was widely read, and neither came to Darwin's attention until after he had published the first edition of *The Origin*.

Alfred Russel Wallace independently discovered natural selection while Darwin was incubating his ideas. Indeed, it was Darwin's receipt of a manuscript sent to him by Wallace that finally prompted Darwin to go public. Wallace's paper, and one by Darwin, were read before the Linnean Society of London in 1858, and Darwin published his book the following year.

In contrast to the rapid acceptance of the fact of descent with modification, natural selection was not widely accepted as the mechanism of adaptive evolution until the 1930s. Instead, Lamarckism, and a variety of other mechanisms, remained popular. There were many reasons for the prolonged debate (see Mayr and Provine 1980; Gould 1982; Bowler 2002). Among them was that natural selection depends on genetic variation, and no one, save Gregor Mendel, whose work on garden peas and the mechanism of inheritance was ignored by virtually everyone, understood genetics. Mendel's work was finally rediscovered in 1900, leading to the development of population genetics over the next three decades. Finally, population genetics was merged with natural selection, and the combination was used to explain gradual evolution, speciation, and macroevolution. Many of the researchers who participated in this "Modern Synthesis" viewed T. G. Dobzhansky's *Genetics and the Origin of Species*, published in 1937, as book that marked the establishment of modern evolutionary biology.

This chapter reviews evidence that supports Darwin's view of life's history. This evidence—some of it presented by Darwin, much of it accumulated since—has conviced virtually all scientists who study life that Darwin was right. Darwin called the pattern he saw "descent with modification." It has since come to be known as evolution.

Darwin also identified a process to explain the pattern. This process, natural selection, is the subject of Chapter 3.

The first three sections of this chapter explore data that challenge each of the assertions made by the Theory of Special Creation—that species are immutable, independent, and recent—and instead support the Theory of Descent with Modification. The final section briefly discusses whether Darwin's refutation of special creation necessarily places evolutionary biology in conflict with religion.

2.1 Evidence of Change through Time

The Theory of Special Creation asserts that species, once created, are immutable. This claim is challenged by several lines of evidence that support the alternative hypothesis that populations of organisms change through time. The data we review here come from living species as well as from extinct forms preserved in the fossil record.

Evidence from Living Species

The living evidence for descent with modification comes in two forms. First, by monitoring natural populations, we can directly observe small scale change, or **microevolution**. Second, if we examine the bodies of living organisms, we can find evidence of dramatic change, or **macroevolution**.

Direct Observation of Change through Time

During the past several decades, biologists have documented evolutionary change in hundreds of different species. As an example, consider work by Scott Carroll and colleagues on the soapberry bug, *Jadera haematoloma*, an insect native to the southern United States.

Soapberry bugs feed by using their long beaks to attack the inflated, balloon-like fruit capsules of their host plants (Carroll et al. 2005). The bugs probe the the seams between the capsule's panels, trying to reach the seeds, which are held in the center of the capsule far away from the walls (Figure 2.2). When a bug manages to reach a seed, it pierces the seed coat, liquefies its contents, then sucks them up. Prior to 1925, soapberry bugs in Florida lived exclusively on their native host, the round-capsuled balloon vine. The balloon vine occurs primarily in the southern tip of Florida and the Florida Keys, with a few individuals scattered across central Florida. Starting in 1926 and gaining momentum in the 1950s, gardeners in central Florida began planting as an ornamental an Asian relative of the balloon vine called the flat-podded golden rain tree. As its name suggests, the flat-podded golden rain tree has flat-capsuled fruits. Soapberry bugs in central Florida began exploiting the new import, and their populations grew.

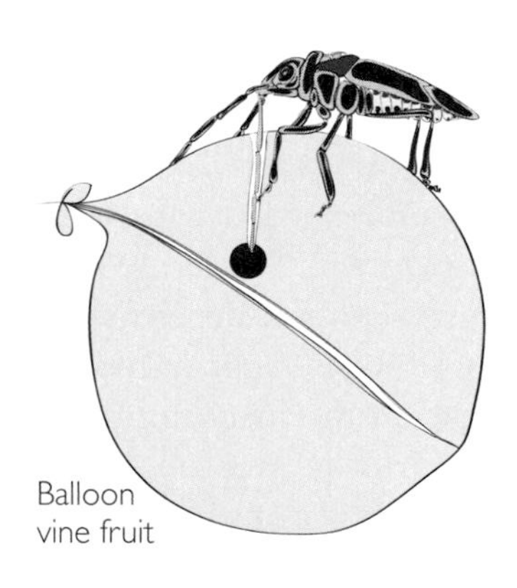

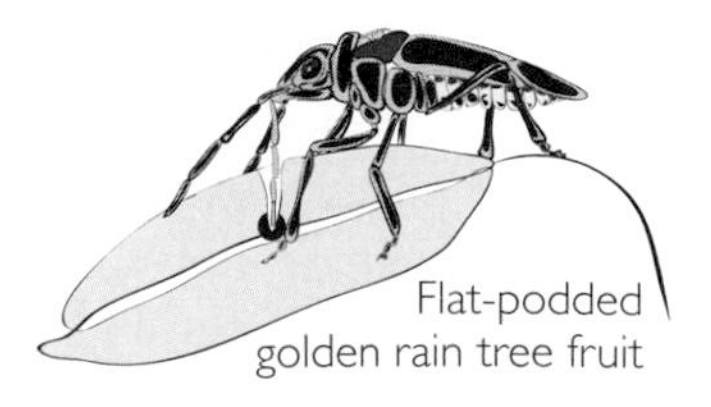

Figure 2.2 Soapberry bugs After Fig. 1 in Carroll and Boyd (1992).

Carroll and Christian Boyd (1992) collected soapberry bugs living on balloon vines on Key Largo and on flat-podded golden rain trees at Lake Wales. The researchers measured the lengths of the bugs' beaks. They found that, on average, the population of bugs living on the flat-capsuled host had much shorter beaks

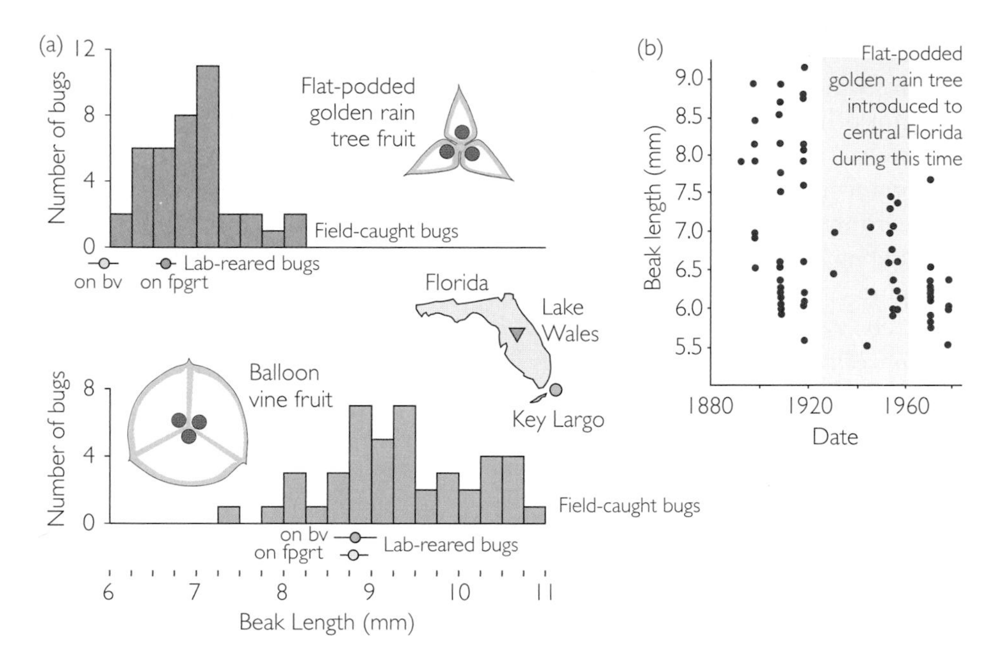

Figure 2.3 Evolutionary change in soapberry bugs
(a) The bar graphs show that the soapberry bugs found in nature on flat-podded golden rain trees have, on average, shorter beaks than bugs found on balloon vines. The balloon vine is native to southern Florida, while the flat-podded golden rain tree was introduced in the late 1920s from Asia. The data points below the horixontal axes show the mean beak length (±standard error) of bugs reared in the lab on fruits from balloon vine (bv) versus flat-podded golden rain tree (fpgrt). (b) The scatterplot shows the beak lengths of female soapberry bugs from Florida in museum collections (each data point represents one individual). See text for explanation. Rendered from Figs. 3, 4, and 6 in Carroll and Boyd (1992) and Fig. 3 in Carroll et al. (1997).

than the population living on the round-capsuled host (Figure 2.3a). The short-beaked bugs living on the flat-capsuled host are descendants of long-beaked bugs that lived on the round-capsuled host (Carroll and Boyd 1992; Carroll et al. 2001). The data thus imply that the Lake Wales soapberry bug population evolved following their switch to a new host.

There is, however, an alternative interpretation. Perhaps, as they grow, soapberry bugs develop a beak just long enough to reach the seeds of the fruits they find themselves feeding on. To rule out this possibility, Carroll and colleagues (1997) reared soapberry bugs from both Lake Wales and Key Largo on fruits from both host plants. The average beaks lengths of these lab-reared bugs are indicated below the horizontal axes in the histograms in Figure 2.3a. The soapberry bugs from the Key Largo population developed long beaks regardless of whether they grew up on flat fruits or round ones. Likewise, the bugs from the Lake Wales population developed short beaks regardless. These results show that the bugs from the two populations are genetically different. The short-beaked Lake Wales bugs have descended with modification from long-beaked ancestors.

Observations on living organisms provide direct evidence of microevolution by showing that populations and species change over time.

Carroll and Boyd (1992) measured the beaks of soapberry bugs that had been collected in Florida and preserved in museum collections. The data from the museum specimens document the reduction in beak length that followed the introduction of the flat-podded golden rain tree (Figure 2.3b).

Carroll and colleagues (2005) reported an additional episode to the soapberry story. The balloon vine, the soapberry bug's native host in southern Florida, has been spreading as an invasive weed in Australia for some 80 years. There it has been colonized by the Australian soapberry bug, *Leptocoris tagalicus*. The Australian bug's native host is a small-podded tree called the woolly rambutan, and the Australian bug consequently has a relatively short beak. Populations of Australian bugs that have been living for many generations on balloon vines, however, have evolved significantly longer beaks than their ancestors.

The characteristics of soapberry bugs are not immutable. Instead, they have changed substantially over time. Many more examples of change over time directly observed in populations of living organisms appear throughout the book.

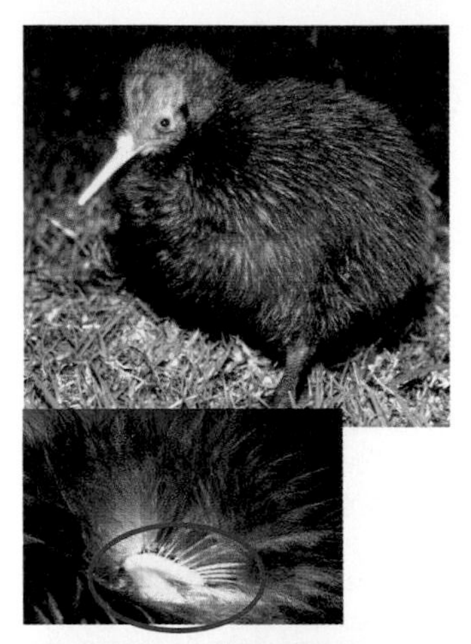

Figure 2.4 Vestigial structural traits Top, the brown kiwi, a flightless bird, has tiny, useless wings. Bottom, the rubber boa has a tiny remnant hind limb, called a spur, on either side of its vent.

Vestigial Organs

By the time Darwin began working on "the species question," comparative anatomists had described a number of curious traits called **vestigial structures**. A vestigial structure is a useless or rudimentary version of a body part that has an important function in other, closely allied, species. Darwin argued that vestigial traits are inexplicable under the Theory of Special Creation, but readily interpretable under the Theory of Evolution.

Figure 2.4 shows examples of vestigial structural traits. The North Island brown kiwi, *Apteryx mantelli*, a flightless bird, has tiny, stubby wings. The rubber boa, *Charina bottae*, has remnant hind limbs, represented internally by rudimentary hips and leg bones and externally by minute spurs. The evolutionary interpretation of these vestigial structures is that kiwis and boas are descended, with modification, from ancestors in which the wings or hind legs were fully formed and functional.

Humans, too, have vestigial structures. We have, for example, a tiny tailbone called the coccyx (Figure 2.5a). We also have muscles attached to our hair follicles that contract to make our body hair stand on end when we are cold or frightened (Figure 2.5b). If we were hairy, like chimpanzees, then the contraction of these arrector pili muscles would increase the loft of our fur to keep us warm or make us look bigger and more intimidating to enemies (Figure 2.5c). But we are not hairy, so we just get goosebumps. Our goosebumps imply that we are descended from ancestors that were hairier than we are. Likewise, our tiny tailbones imply that we are descended from ancestors with tails.

Figure 2.6 illustrates a vestigial developmental trait. Chickens have three digits in the "hands" of their wings and four digits in their feet. But when chicken embryos are treated with a stain to mark the tissues that initiate bone development, an additional digit—marked with an arrow in the figure—appears in each. These extra digits later disappear, leaving no trace in the adult. Why? The evolutionary explanation is that birds are descended from ancestors that, like most tetrapods, had five digits in all of their limbs. The modifications in development that transformed these ancestral limbs into wings and birds' feet take place after the fourth and fifth digits begin to form, but before the structure of the limb is fully defined. The transient extra digit is a vestige of the ancestral developmental program.

Vestigial traits also occur at the molecular level. Humans have one on chromosome 6. It is a DNA sequence that looks like a gene for the enzyme CMAH (CMP-*N*-acetylneuraminic acid hydroxylase), except that it is disabled by a 92-base-pair deletion (Chou et al. 1998). Most mammals, including chimpanzees, bonobos, gorillas, and orangutans, make CMAH in abundance, but humans cannot (Chou et al. 2002). CMAH converts an acidic sugar, destined for display on

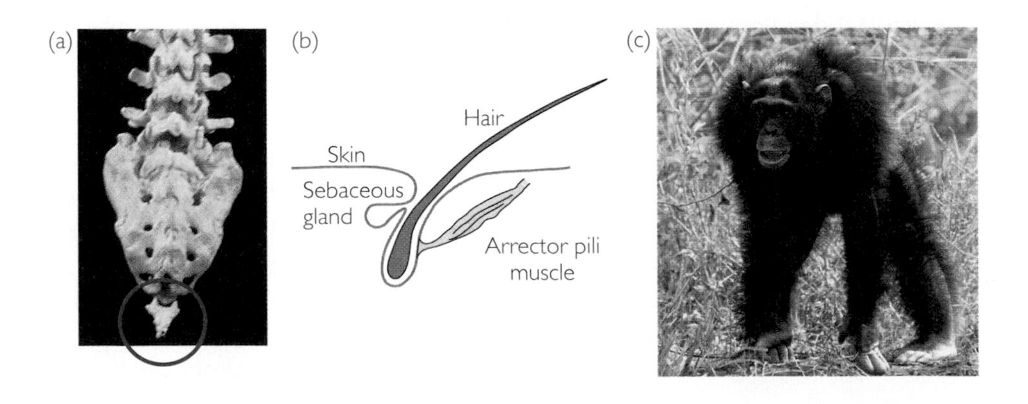

Figure 2.5 Human vestigial traits (a) Humans have a rudimentary tailbone, called the coccyx. (b) Humans have a muscle, the arrector pili, at the base of each hair follicle. When it contracts, producing a goosebump, the hair stands up. (c) If we were hairy, like this chimpanzee, then contraction of our arrector pili muscles would increase the loft of our fur to keep us warm, or to make us look more intimidating.

the surface of the cells that make it, from one form to another. As a result of our inability to make CMAH, we humans have a different biochemical signature on our cell membranes. This appears to explain why humans and chimpanzees are largely immune to each other's malaria parasites (Martin et al. 2005). Our possession of a nonfunctional gene is difficult to reconcile with the notion that humans were created in their present form. But it is readily explicable if humans are descended with modification from ancestors that made CMAH.

We have interpreted the vestigial traits we have mentioned as evidence of evolution. In some cases, this interpretation can be tested. The threespine stickleback, *Gasterosteus aculeatus,* is a small fish that lives in coastal waters throughout the Northern Hemisphere and readily invades freshwater (Bell and Foster 1994). Marine sticklebacks carry heavy body armor, featuring bony plates protecting their sides and pelvic fins modified into spines (Figure 2.7a). Freshwater sticklebacks, however, often carry light armor. They have fewer bony plates, and their pelvic strutures are reduced to vestiges or absent altogether [Figure 2.7b and (c)]. William Cresko and colleagues (2004) studied lightly armored freshwater sticklebacks in Alaska. The researchers suspected that these fish were descended from marine ancetors that had invaded lakes left by melting glaciers at the end of the last ice age.

In the lab, Cresko mated heavily armored marine fish with lightly armored freshwater fish (generation P, Figure 2.7d). The offspring were healthy, fertile, and heavily armored (generation F_1, Figure 2.7d). Mated with each other, these fish produced four kinds of progeny: heavily armored; fully plated with vestigial pelvic structure; lightly plated with full pelvic structure; and lightly armored (generation F_2, Figure 2.7d). The counts of these phenotypes approximated the 9:3:3:1 ratio expected for a dihybrid cross. These results suggest that the marine and freshwater sticklebacks are closely related, and that the differences in their body armor are largely (though not completely) controlled by the alleles of two Mendelian genes.

If the recessive alleles for light plating and a reduced pelvis exist in marine stickleback populations—hidden in heterozygotes—then marine populations invading freshwater might evolve toward the freshwater form rapidly enough to watch it happen. Michael Bell and colleagues (2004) documented just such a swift transition. Loberg Lake, Alaska, was poisoned in 1982 so that it could be restocked with trout and salmon for recreational fishing. By 1988 the lake had been invaded by marine sticklebacks from nearby Cook Inlet. Bell monitored the Loberg population from 1990 through 2001. In just a dozen years, the composition of the population changed from over 95% fully plated fish to over 75% lightly plated. In Loberg Lake, and probably elsewhere, freshwater sticklebacks with vestigial armor are, indeed, the modified descendants of heavily armored marine fish.

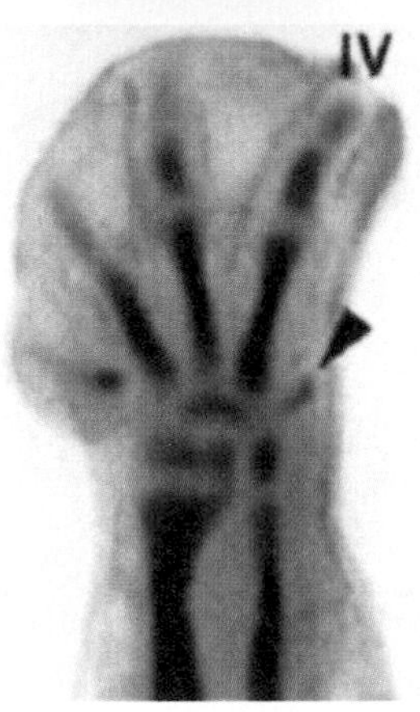

Figure 2.6 Developmental vestigial traits Adult chickens have three digits in their wings and four in their feet. But during development, an extra digit appears for a short time in the "hand" (top) and foot (bottom). From Burke and Feduccia (1997).

Reduced or useless body parts are evidence of both microevolution and macroevolution.

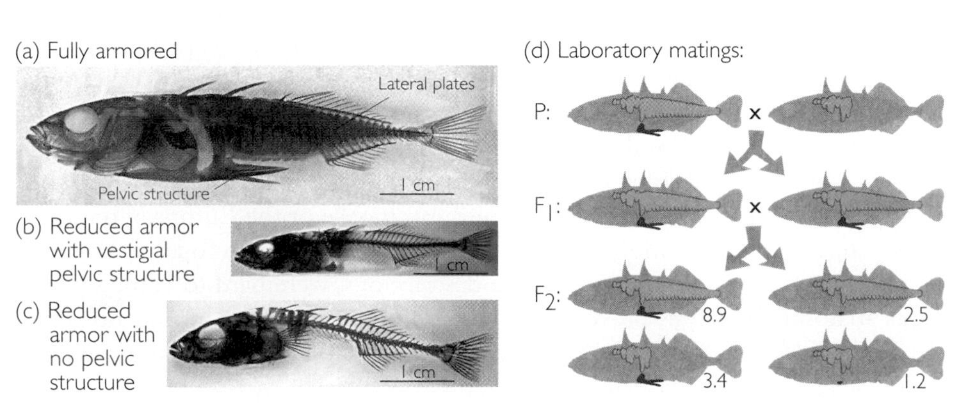

Figure 2.7 Reduced armor in freshwater sticklebacks (a) A marine fish heavily armored with bony plates and pelvic spines. (b) A freshwater fish with light plating and a vestigial pelvic structure. (c) a freshwater fish with no pelvic structure at all. (d) Matings show that these differences are largely genetically determined. Photos from Fig. 1, and drawings based on Fig. 4 in Cresko et al. (2004).

Figure 2.8 The Irish elk confirms the fact of extinction
Cuvier confirmed that fossils of the huge ice-age deer called the Irish elk represented an extinct species.

Evidence from the Fossil Record

A **fossil** is a trace of any organism that lived in the past. The total, worldwide collection of fossils, scattered among thousands of different institutions and individuals, is called the **fossil record**.

The simple fact that fossils exist, and that the vast majority of fossil forms are unlike species that are living today, argues that life has changed through time. Three specific observations about the fossil record helped Darwin and other 19th century scientists make the case that fossils are evidence of evolution.

The Fact of Extinction

The fact that many species have gone extinct suggests that Earth's flora and fauna have changed over time.

In 1801, the comparative anatomist Baron Georges Cuvier published a list of 23 species that were no longer in existence. This list was a direct challenge to the widely accepted hypothesis that unusual forms in the fossil record would eventually be found as living species, once European scientists had visited all parts of the globe. Cuvier pointed to mastodons and other enormous creatures that had been excavated from the rocks of the Paris basin. These species were so large, he argued, that it was unlikely they were still alive but had simply escaped detection.

Controversy over the fact of extinction ended after 1812, when Cuvier published a careful examination of fossils of the Irish elk—the huge ice-age deer shown in Figure 2.8. Fossils of this deer had been found throughout northern Europe and the British Isles. Other scientists had suggested that the Irish elk belonged to a living species, such as the American moose or the European reindeer. These suggestions were more reasonable than they might seem today. For species like the moose, specimens, even reliable descriptions, were hard to come by. Cuvier's anatomical analysis proved that the Irish elk was neither moose, nor reindeer, nor did it belong to any other living species (Gould 1977). It was a species unto itself, and it was extinct.

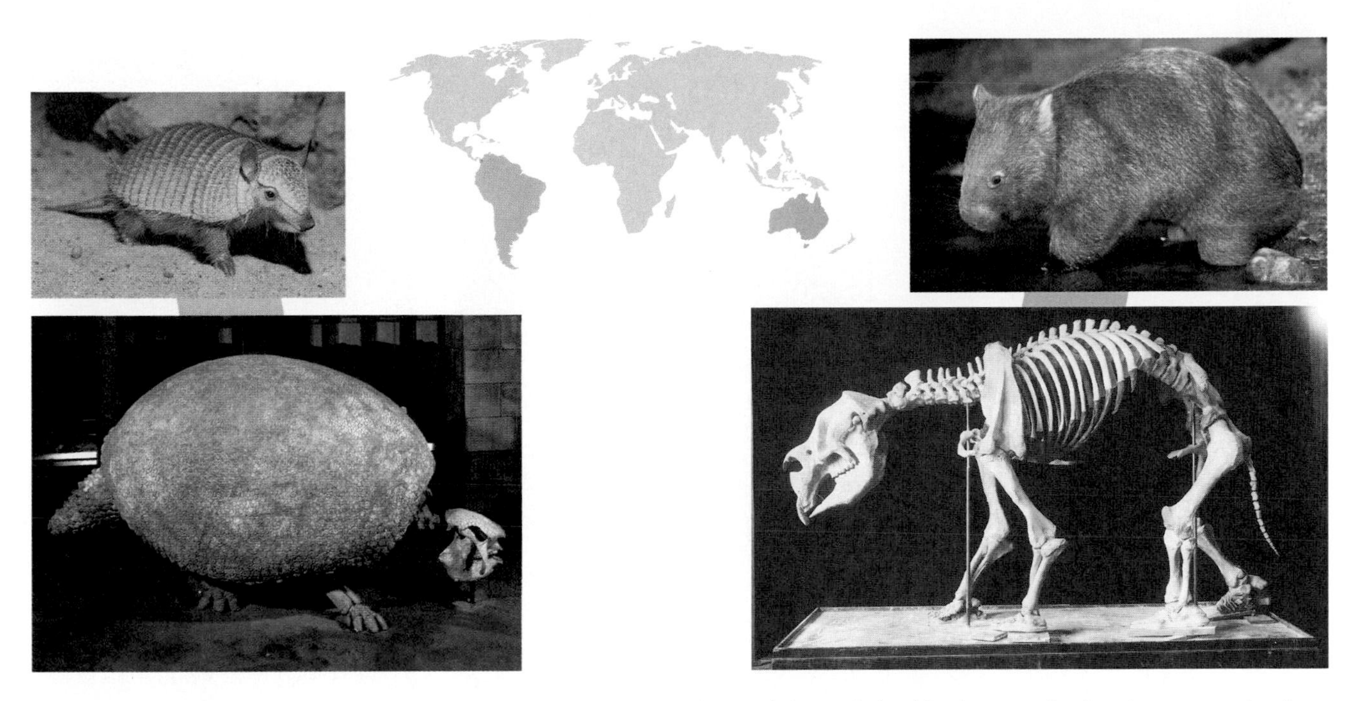

Figure 2.9 **The law of succession** Early researchers so routinely observed close relationships between fossil and extant species from the same geographical area and between fossil forms in adjacent rock strata that the pattern became known as the law of succession. Darwin noted the similarities between the contemporary pygmy armadillo (*Zaedyus pichiy*) (top left) and the fossil glyptodont (bottom left) of Argentina. Richard Owen confirmed a pattern first recognized by William Clift when Owen identified the extinct Australian mammal *Diprotodon* (lower right) as a marsupial similar to the wombats (upper right) that live in Australia today (Dugan 1980).

By the time Darwin wrote *On The Origin of Species*, extinct plants and animals were being found in rock layers that had been formed in many different times and places. Creationists contended that these species had perished in a series of floods akin to the biblical event at the time of Noah. In contrast, Darwin and other biologists interpreted extinct species as the relatives of living organisms. They pointed to the fact of extinction as evidence that Earth's flora and fauna have changed through time.

The Law of Succession

An early 18th century paleontologist named William Clift was the first to publish an observation later confirmed and expanded upon by Darwin (Darwin 1859; Dugan 1980; Eiseley 1958). Fossil and living organisms in the same geographic region are related to each other and are distinctly different from organisms found in other areas (Figure 2.9). Clift worked on the extinct mammals of Australia and noted that they were marsupials, closely related to forms alive in Australia today. Darwin analyzed the armadillos of Argentina and their relationship to the fossil glyptodonts he excavated there. The mammalian faunas of the two continents are obviously different, yet each continent's extant fauna is strikingly similar to the continent's recent fossil forms. The general pattern of correspondence between fossil and living forms from the same locale came to be known as the **law of succession**. The law is supported by analyses from a wide variety of locations and taxonomic groups. Darwin's theory of evolution gives the law a straightforward explanation. Today's species are descended with modification from ancestors that lived in the same region; it is to be expected that they would bear a stronger resemblance to their recent ancestors than to their more distantly related kin in other parts of the world.

The resemblance between living and fossil forms in the same region suggests that living organisms are descended with modficiation from earlier species.

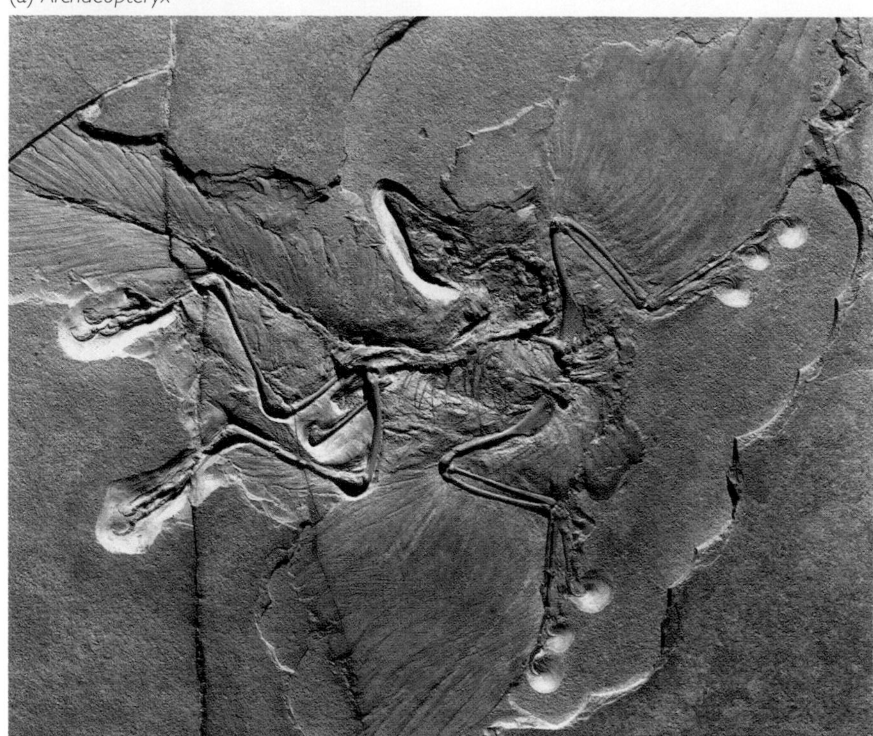

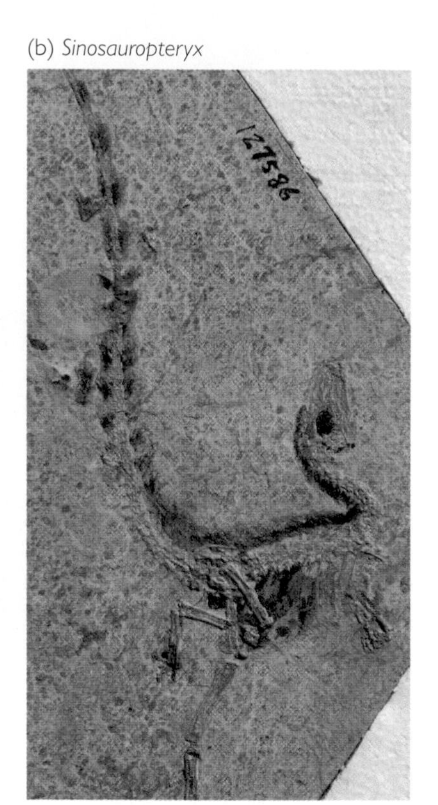

Figure 2.10 A bird with a dinosaur's skeleton and a dinosaur with feathers (a) *Archaeopteryx*, a bird with modern feathers and a dinosaur-like skeleton. (b) *Sinosauropteryx prima*, a dinosaur with bristly structures on its neck, back, flanks, and tail that many paleontologists believe are down-like feathers. From Chen et al. (1998).

Transitional Forms

Darwin asserted that species are descended with modification from earlier forms and that fossils represent past populations, some of which were ancestors of creatures alive today. If Darwin was right, then the fossil record should capture evidence of transformations in progress: transitional species showing a mix of features, including traits typical of ancestral populations and novel traits seen later in descendants. Few transitional forms had been discovered by Darwin's time, and he took pains to explain that they should be rare in the fossil record. Since Darwin's day, however, many transitional fossils have been found.

The most famous transitional form, *Archaeopteryx*, was discovered shortly after Darwin published *On the Origin of Species* (Figure 2.10a; see Christiansen and Bonde 2004). This crow-sized animal lived 145 to 150 million years ago in what is now Germany. The fact that it had feathers and, apparently, at least some ability to fly identifies it as a bird (Padian and Chiappe 1998; Alonso et al. 2004). Its skeleton, however, was so reptilian—with teeth, three-clawed hands, and a long, bony tail—that specimens of *Archaeopteryx* have been mistaken for remains of the dinosaur *Compsognathus* (see Wellnhofer 1988). Darwin's friend and champion Thomas Henry Huxley (1868) was among the first to recognize the skeletal similarities between dinosaurs and birds and to suggest that *Archaeopteryx* documents an evolutionary transition from one to the other.

When we call *Archaeopteryx* a transitional fossil, we are not claiming that it was on the direct line of descent from dinosaurs to modern birds. Instead, *Archaeopteryx* likely represents an extinct side branch on the evolutionary tree that connects dinosaurs to birds. We call *Archaeopteryx* a transitional fossil be-

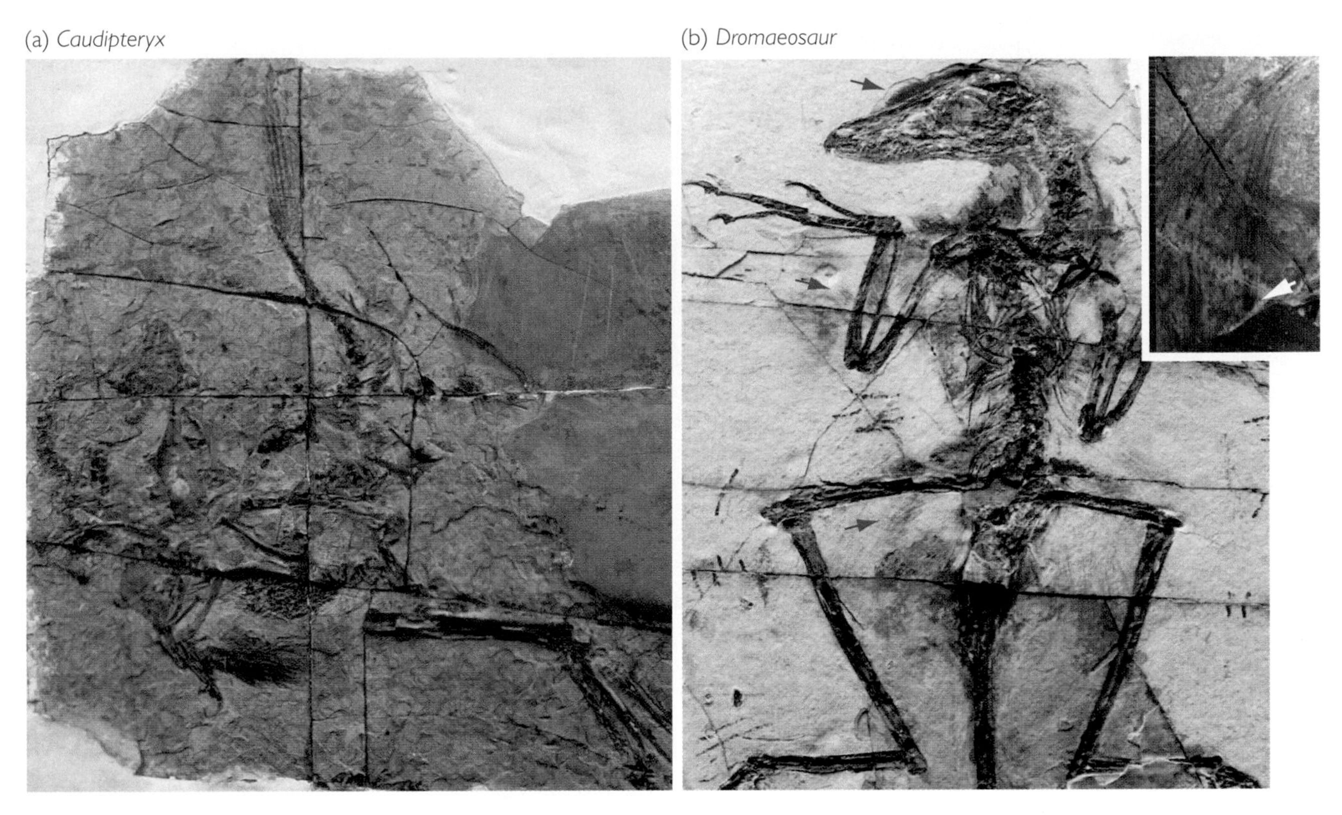

Figure 2.11 More feathered dinosaurs (a) *Caudipteryx zoui*, a dinosaur with elongated feathers on its arms and tail. From Ji et al. (1998). (b) A young dromaeosaur, probably *Sinornithosaurus millenii*, with simple feathers; arrows mark the most visible impressions. From Ji et al. (2001). The inset shows a feather from another *S. millenii;* the arrow marks the base of the tuft of filaments. From Xu et al. (2001).

cause it demonstrates the past existence of species intermediate in form between dinosaur and bird. *Archaeopteryx* indicates, with its fully modern feathers and dinosaurian skeleton, that birds evolved their birdness piecewise. Feathers came first, before the skeletal and muscular modifications associated with modern powered flight (Garner et al. 1999).

If feathers were among the first evolutionary steps on the path from dinosaurs to birds, then the fossil record should hold another kind of transitional form: dinosaurs with feathers in various stages of evolution (Unwin 1998). Feathered dinosaurs are most likely to lurk among the theropods—the bipedal carnivores, including *Compsognathus* and *Tryannosaurus rex,* with which birds share the largest number of evolutionary innovations (Gauthier 1986; Prum 2002).

In recent years, paleontologists excavating fossil beds in China's Liaoning Province have unearthed several feathered theropods (see Norell and Xu 2005). The *Sinosauropteryx prima* specimen shown in Figure 2.10b, roughly the size of a chicken, is exquisitely preserved (Chen et al. 1998). Many paleontologists believe that the bristly structures on its neck, back, flanks, and tail are simple feathers (Chen et al. 1998; Unwin 1998; Currie and Chen 2001; but see Geist and Feduccia 2000). More convincing are the fossils in Figure 2.11. *Caudipteryx zoui,* the turkey-sized theropod in Figure 2.11a, had sprays of long feathers on its hands and tail (Ji et al. 1998). Nearly the entire body of the 60-cm-long dromaeosaur in Figure 2.11b is covered with filamentous structures (Ji et al. 2001). The tuft in the inset is from a closely related specimen (Xu et al. 2001). Paleontologists interpret these structures as feathers because they bear key features that are today found only in feathers and because they match intermediate stages

(a) *Dromaeosaur* feathers

(b) *Microraptor gui*

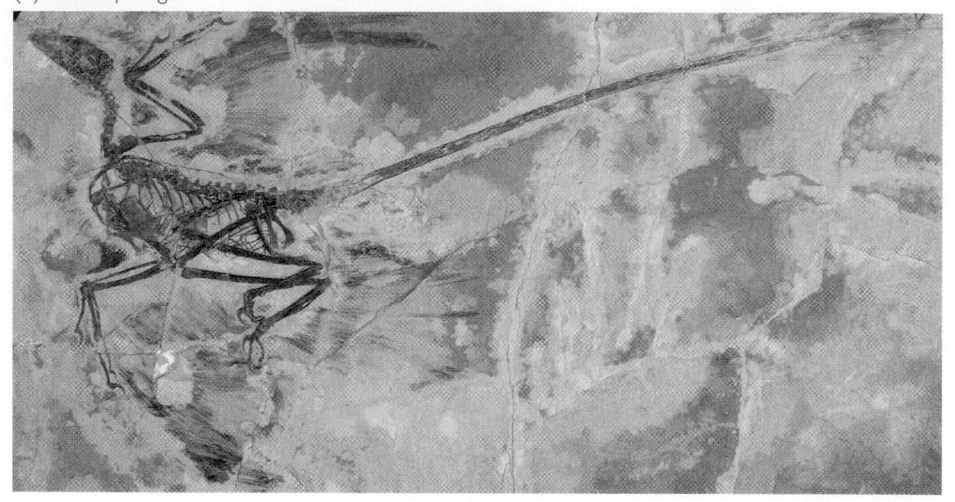

Figure 2.12 Dinosaur feathers (a) Modern feathers, with filaments branching from a central shaft, from a dromaeosaur. From Norell et al. (2002). (b) *Microraptor gui*, a dromaeosaur with flight feathers on all four limbs—that is, a four-winged dinosaur. From Xu et al. (2003).

predicted by a model of feather evolution based on how feathers develop (Ji et al. 2001; Sues 2001; Xu et al. 2001; Prum and Brush 2002). Finally, Figure 2.12a shows modern feathers, complete with filaments branching from a central shaft, that adorn another fossil dromaeosaur (Norell et al. 2002). Modern feathers also grace both the arms and legs of *Microraptor gui*, the 80-cm-long four-winged dinosaur in Figure 2.12b (Xu et al. 2003).

These and other feathered dinosaurs support Huxley's contention that birds evolved from dinosaurs. Indeed, they have made it difficult to say just what a bird is and what distinguishes it from an ordinary theropod. It used to be easy: If it had feathers, it was a bird. Under that denfinition, however, *Sinosauropteryx*, *Caudipteryx*, and even some *Tyrannosaurs* were birds (Xu et al. 2004). A more restrictive, but reasonable, definition is that if it has feathers and can fly, or if it is descended from an animal that had feathers and could fly, then it is a bird. Even by this criterion it may turn out that dromaeosaurs like *Velociraptor*, a predator much loved by movie makers, was a bird (see Makovicky 2005; Perkins 2005).

Transitional fossils document the past existence of species displaying mixtures of traits typical of what are today distinct groups of organisms. They are evidence for macroevolution.

For another suite of transitional forms, consider the whales pictured in Figure 2.13. Because the earliest mammal fossils represent terrestrial species, biologists infer that the ancestors of whales also lived on land. Consistent with this idea, some modern whales still have vestigial pelvis and leg bones (Figure 2.13a). Between the terrestrial ancestors and modern whales, then, there should be intermediate forms that have functioning limbs, as well as features that identify them as ocean-dwelling species. Two such transitional fossils are shown. *Basilosaurus isis,* analyzed by Philip Gingerich and colleagues (1990), lived approximately 38 million years ago (mya). *Basilosaurus* was clearly an exclusively aquatic animal but had tiny and fully formed hind limbs. Gingerich contends that the limbs of *Basilosaurus* were too reduced to function in swimming and may instead have served as grasping organs during copulation. *Ambulocetus natans,* discovered and described by

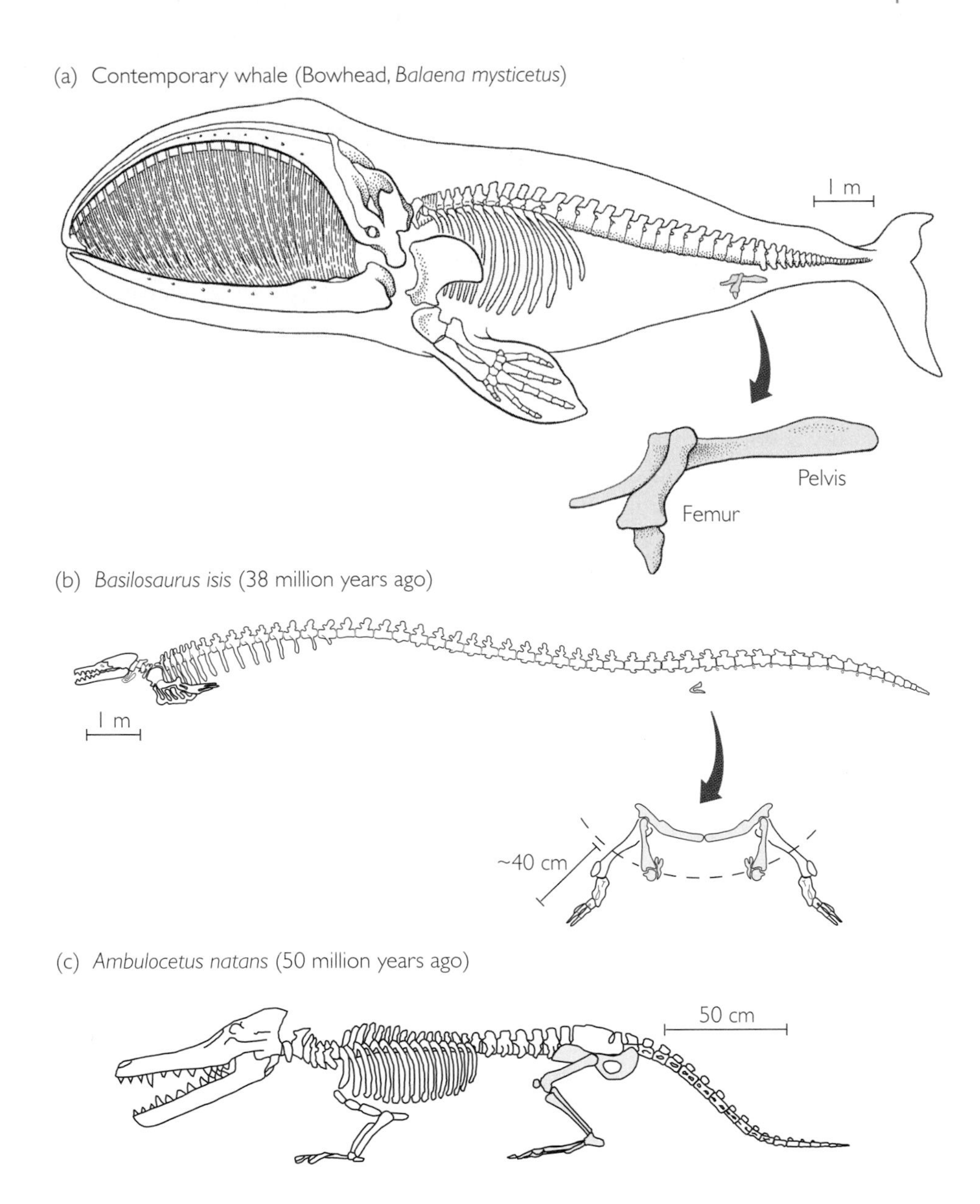

Figure 2.13 Transitional fossils documenting the evolution of whales from leggy ancestors (a) Some contemporary whales have a vestigial femur and pelvis. (b) *Basilosaurus isis* fossils are about 38 million years old. They have reduced hind limbs that probably did not function in swimming. Instead, the legs may have been used as a grasping structure during copulation. From Gingerich et al. (1990). (c) *Ambulocetus natans* (literally translated, means "walk-whale swimming") fossils are about 50 million years old. They have functional hind limbs that were probably used as paddles in swimming. From Thewissen et al. (1994).

J. G. M. Thewissen and co-workers (1994), lived about 50 mya. It had enormous hind limbs that made it awkward on land. From an analysis of how the limbs articulated with the body, however, Thewissen suggests that *Ambulocetus* was an excellent swimmer that used its limbs much as modern otters do. These fossils mark a major evolutionary transition. As predicted, their form is intermediate between limbed ancestors and limbless descendants.

An additional example of a transitional fossil appears on page 37.

Evidence of Descent with Modification

We have reviewed evidence from both living and fossil organisms. The evidence shows that species change over time on a small, or microevolutionary scale—as when soapberry bugs evolved shorter beaks. And it shows that species change over time on a large, or macroevolutionary scale—as when birds evolved from dinosaurs. That gets us half the way from the Theory of Special Creation to Darwin's view of life's history. Next, we consider evidence of common ancestry.

2.2 Evidence of Common Ancestry

In contrast to the Theory of Special Creation, Darwin's theory of life's history holds that species are not independent, but are connected by descent from a shared ancestor. This means that species have genealogical relationships analogous to the family trees of individual humans. Before we introduce evidence supporting Darwin's view, it will be worth spending a few paragraphs introducing the graphical tools evolutionists use to think about relationships among species.

An Introduction to Tree Thinking

We begin with a thought experiment concerning events of a sort that will, if Darwin's theory is correct, have happened often during the history of life. Imagine a population of snails living on an island (Figure 2.14a, left). From time to time novel traits appear in the population and, if they confer an ability to survive and reproduce at higher rates, become prevalent (Figure 2.14b, left). In addition, snails occasionally cross the water—perhaps by rafting on floating vegetation—to establish new populations on uninhabited islands (Figure 2.14c–g, left). After a series of colonizations, evolutionary changes, and one extinction, we have three distinct populations, all decended from the common ancestral population that originally inhabited Island 1 (Figure 2.14g, left).

The illustrations in the rightmost column of Figure 2.14 show how we can draw a diagram—called an **evolutionary tree**, or **phylogenetic tree**, or **phylogeny**—to record the history of our snails. The original population on Island 1 is represented by line segment that will be the root of our tree (Figure 2.14a, right). We place on the root a picture of a typical snail from the ancestral population. As time passes, our tree grows from left to right. We record the evolution of coiled shells in the Island 1 population with a bar across the tree labeled with the novel trait that has become newly prevalent (Figure 2.14b, right). We also add a new picture to the tree showing that a typical snail now has a coiled shell. When snails from Island 1 invade Island 2, the effect is that our ancestral population has split. We show this in our tree by dividing the root into two branches (Figure 2.14c, right). We add the rest of the evolutionary transitions and population subdivisions to our tree in the same way (Figure 2.14d–g, right). When the population on Island 2 goes extinct, its branch stops growing (Figure 2.14f, right).

Phylogenetic trees are a visual representation of descent with modification from a common ancestor.

The completed phylogeny at lower right neatly summarizes the entire sequence of events depicted in the seven cartoons in the first column. To read the history of our snails, we start at the root on the left-hand side and read across from left to right. (It may seem that we should start at the Island 1 population on the right side of the tree, but this population of banded snails is not the ancestor of the snails on Islands 3 and 4. Instead, all three populations are descended from the simple-shelled snails that lived on Isand 1 at the start.) Scanned from left to right, the tree shows that coiled shells evolved before pink shells, which in turn preceeded high spires and spikes. It also shows that at the end of our history the populations on Islands 3 and 4 are more closely related to each other than either is to the population on Island 1, because they share a more recent common ancestor: the population of pink high spired snails that once lived on Island 3.

Figure 2.14 (opposite) Evolutionary trees describe histories of descent with modification The left column depicts the history of a population of imaginary snails as they evolved and spread across a chain of four islands. The right column encodes this history in a growing evolutionary tree.

(a) A population of snails on Island 1.

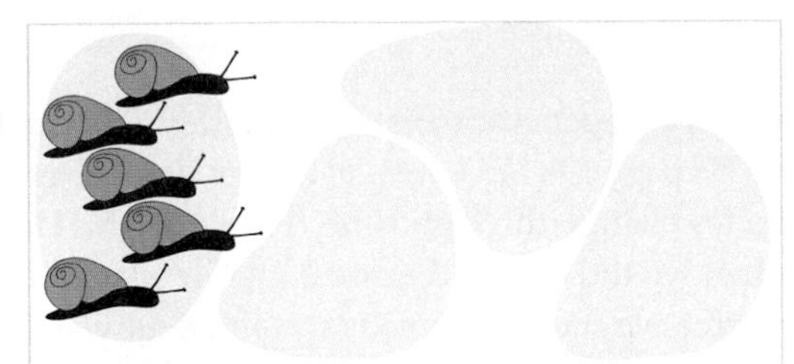

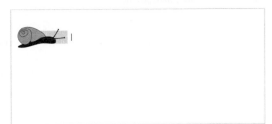

(b) Elongated shells appear and become common, replacing uncoiled shells.

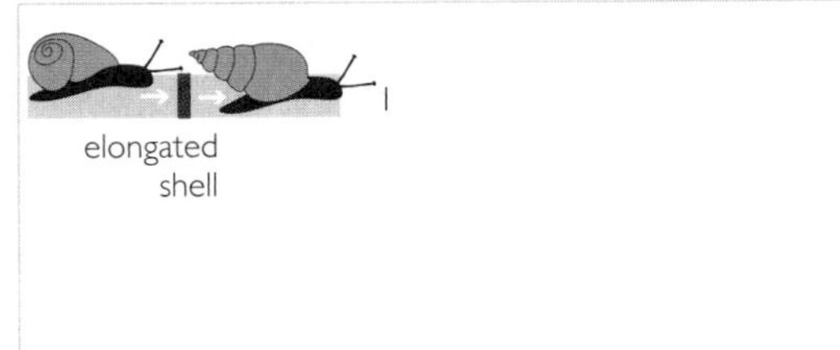

(c) Snails from Island 1 colonize Island 2.

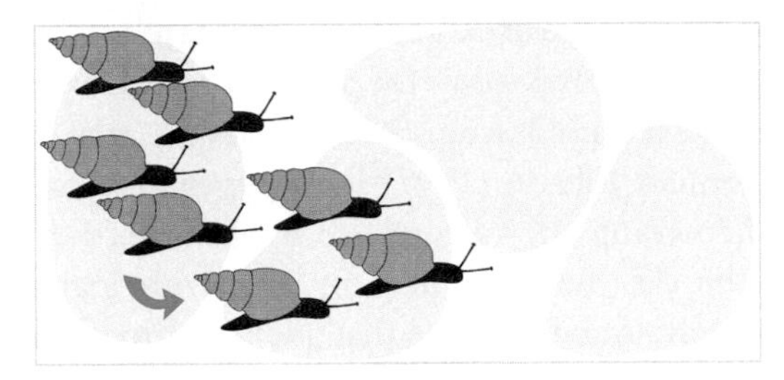

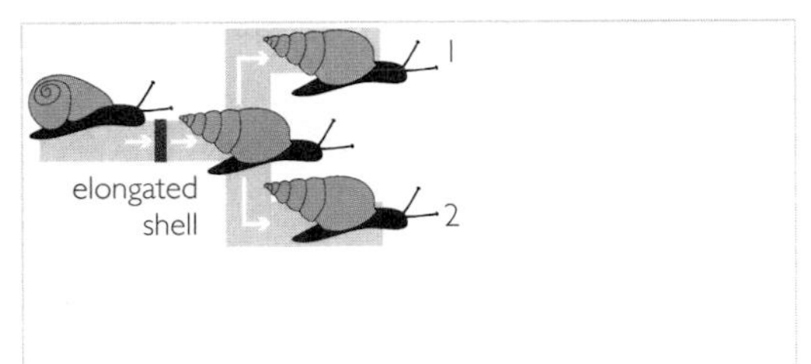

(d) Island 2 population evolves pink shells. Later, Island 2 snails colonize Island 3.

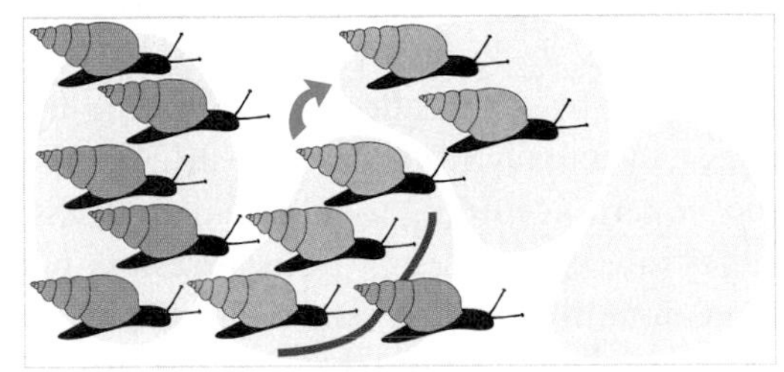

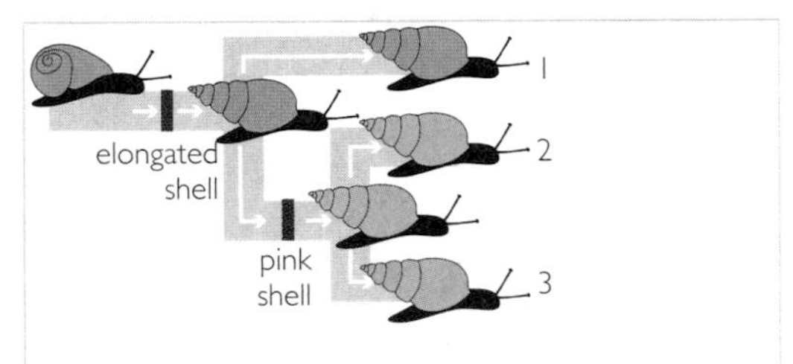

(e) Island 1 snails evolve banded shells; Island 3 snails evolve high spire.

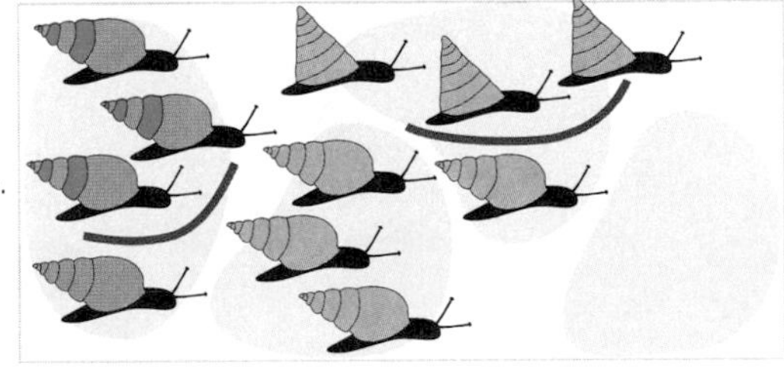

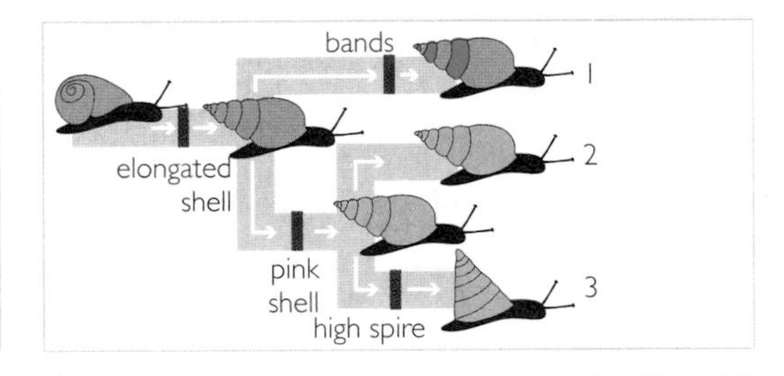

(f) Island 2 population goes extinct; Island 3 snails colonize Island 4.

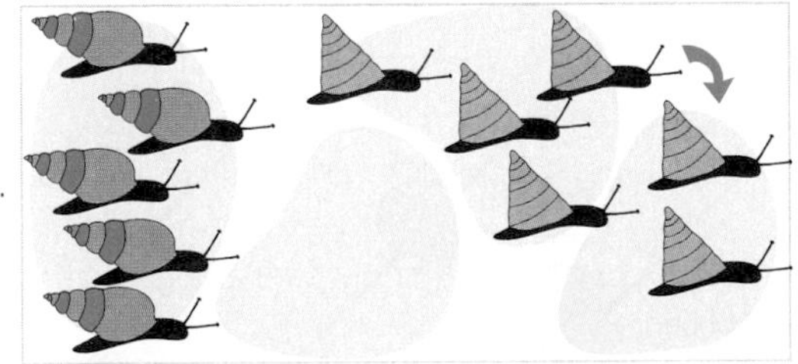

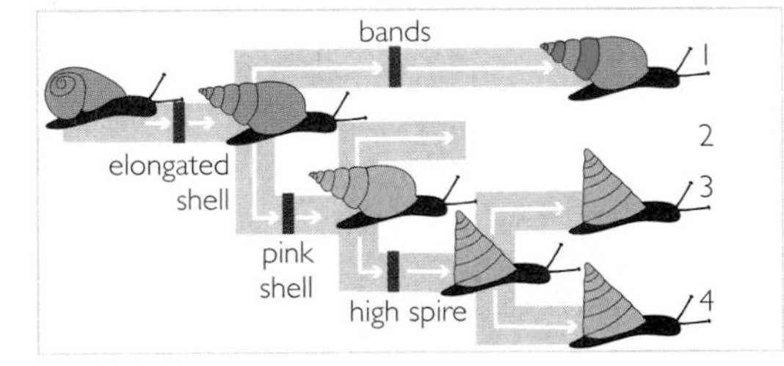

(g) Island 3 snails evolve spiked shells.

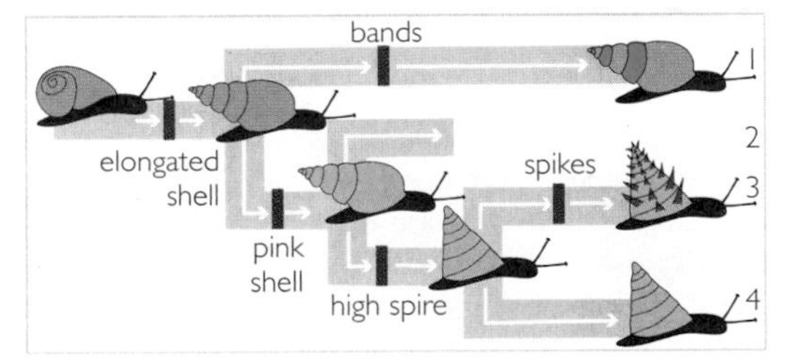

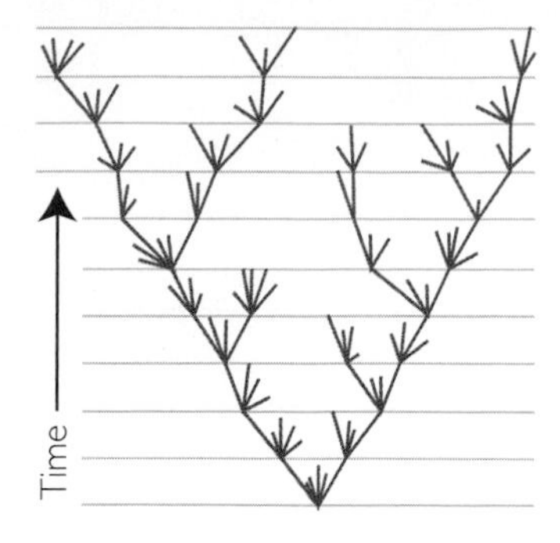

Figure 2.15 Darwin's evolutionary tree The root of this tree is at the bottom, so we read the history it records from bottom to top.

The inventor of phylogenetic trees was Charles Darwin himself. The only illustration in his 490-page *On the Origin of Species* was a diagram that presented his view of how species change through time (Figure 2.15). Darwin has oriented his tree differently than we did ours in Figure 2.14. He placed the root at the bottom. To read Darwin's tree we start at the root and move upward. Both orientations are common in scientific literature. All trees, however, show populations or species diversifying over time. That means all are read from the narrowest part, the root, toward the broadest part, the tips. Darwin also represented the splitting of populations with shallower angles than the 90° turns we used in ours. Again, both styles are common in the literature. It is the order of the branchings that matters, not the style in which they are drawn. If the branch lengths on a particular tree are proportional to time or to the amount of genetic change that has occurred since taxa diverged, then a scale or labeled axis is provided. Otherwise, branch lengths are arbitrary and are arranged so as to improve readability.

Figure 2.16 shows a phylogeny for a group of real organisms: several species of big cats. It is excerpted from a much larger tree reconstructed by Lars Werdelin and Lennart Olsson (1997) for a paper they called "How the leopard got its spots." According the researchers' hypothesis, the most recent common ancestor of all the cats, near the bottom of the phylogeny, had a flecked coat. Reading up the right branch, we see that after the divergence of the lineage leading to the jaguarundi, but before the diversification of snow leopards, tigers, jaguars, lions, and leopards, flecks were modified to become rosettes. A rosette is a central fleck surrounded by smaller flecks. Continuing up the rightmost branch of the phylogeny, we eventually reach the leopard, whose spots are also arranged in rosettes. So the leopard, according to Werdelin and Olsson, got its spots by descent with modification, most recently from ancestors with coats like the leopard's own and more distantly from ancestors with flecks. As an exercise, the reader may wish to use the tree to trace how the tiger got its stripes.

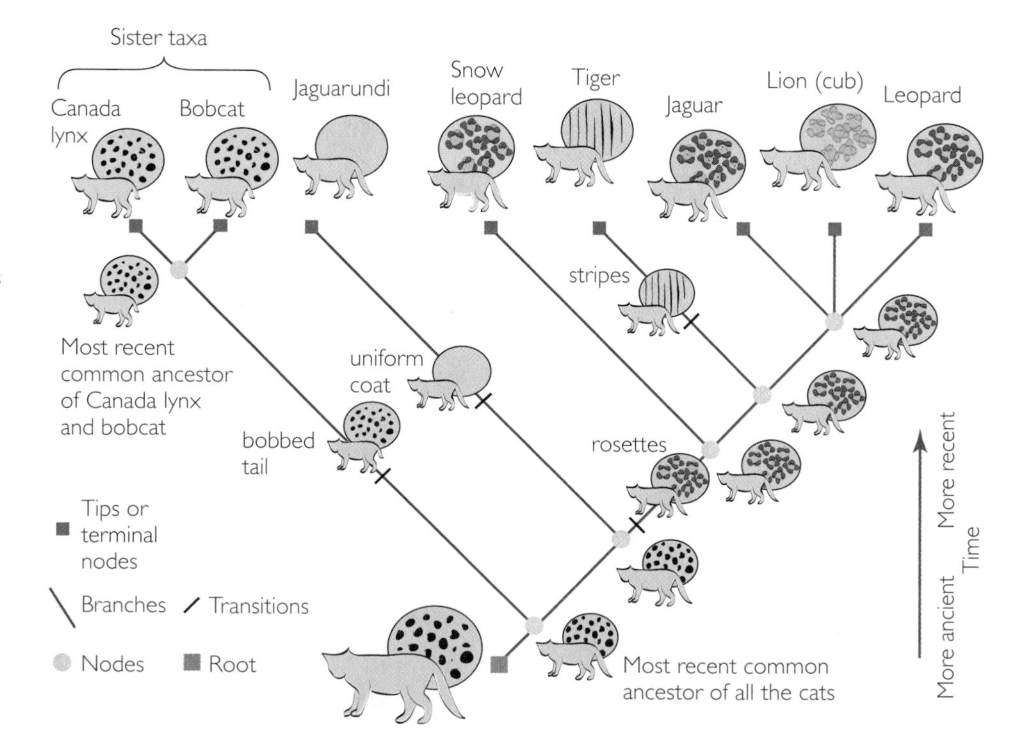

Figure 2.16 An evolutionary tree for eight species of cats According to this tree, the common ancestor of all extant cats had a flecked coat pattern. The tree shows the evolutionary transitions leading to the extant cats' divergent patterns. The key names the parts of an evolutionary tree. The branch tips, or terminal nodes, represent the most recent species—typically extant forms. The root represents the common ancestor of all other species on the tree. The branches trace the history of descent. The transitions mark modifications. The nodes denote points where one species split into two or more descendent species. Sister taxa are each other's closest relatives. After Figs. 3 and 4 in Werdelin and Olsson (1997).

The reader may have noticed that we referred to the phylogeny in Figure 2.16 as a hypothesis. This is because neither Werdelin and Olsson, nor we, nor anyone else, knows the true evolutionary history of the leopard and its kin. Not only do we not know the true history of coat patterns, we do not know the true pattern of branching on the tree from the common ancestor to today's cats. The best we can do is use the available data to identify the most plausible scenarios about evolutionary history. We will discuss techniques for doing so in Chapter 4. But it is always useful to keep in mind the distinction between our hypotheses and the (almost always) unknowable truth. This distinction helps explain why researchers using different data sets have inferred somewhat different phylogenies for the cats. Warren Johnson and Stephen O'Brien (1997), for example, reconstructed a phylogeny in which lions and tigers are each other's closest relatives. Michelle Mattern and Deborah McLennan (2000) reconstructed a tree in which tigers and jaguars are each other's closest relatives. And Warren Johnson and colleagues (2006) reconstructed a tree in which jaguars and lions are each other's closest kin.

Having introduced tree thinking, we are ready to explore the evidence that phylogenies portray the pattern of life's history—that is, the evidence that Earth's various life forms are all related.

Ring Species

The first kind of evidence one might seek for common ancestry is documentation that one species can split into two. Ring species, explified by the Siberian greenish warbler (*Phylloscopus trochiloides*), provide such evidence (Figure 2.17). The greenish warbler's geographic range forms a ring surrounding the Tibetan Plateau. Although the songs the warblers sing increase in complexity from south to north around both sides of the ring, individuals recognize each other as members of the same species by interbreeding everywhere they meet (Irwin et al. 2001a; Wake 2001). The exception is in central Siberia, where the northeastern form encounters the northwestern form. There the two varieties refuse to mate.

Some species appear to be in the process of splitting into two. Freely interbreeding populations connect the entire species, but members of certain populations do not interbreed.

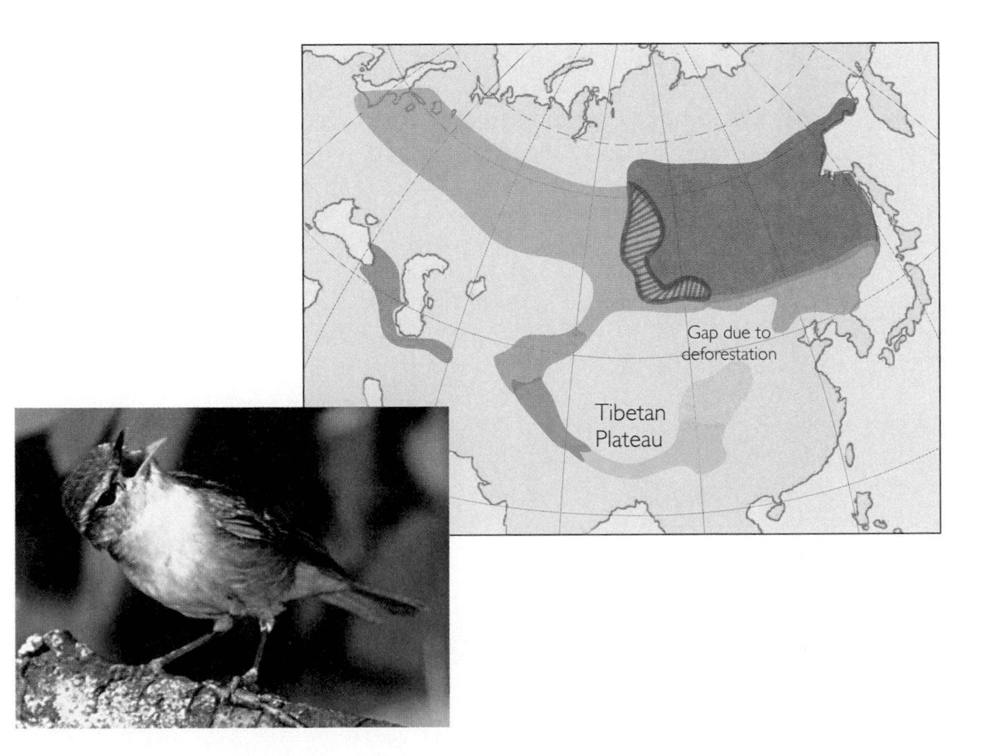

Figure 2.17 Evidence that one species can split into two Left, a Siberian greenish warbler (*Phylloscopus trochiloides*). Right, a map showing the ranges of the greenish warbler's geographic variants. The birds interbreed everywhere they meet around the Tibetan Plateau, except where the northwestern form meets the northeastern form in the region indicated by hash marks. There, the birds behave as different species. Photo by D. Irwin, from Wake (2001); map from Irwin et al. (2005).

Darren Irwin and colleagues (2005) presented genetic evidence that there are no other biological boundaries, aside from central Siberia, between one form of greenish warbler and another. In other words, all greenish warblers are members of a single large population that loops around on itself. Irwin argues that this population originated in the south, expanding northward from there in two directions. By the time the two fronts reconnected many generations later, the birds were sufficiently modified as to be mutually disinterested in romance.

Greenish warbers show that with space and time one species can gradually divide into two. For a review of additional examples of ring species, see Irwin (2001b). With this evidence of common ancestry on a small scale, we now turn to evidence of common ancestry on larger scales.

Homology

As the fields of comparative anatomy and comparative embryology developed in the early 1800s, one of the most striking results to emerge was that fundamental similarities underlie the obvious physical differences among species. Early researchers called the phenomenon **homology**—literally, the study of likeness. Richard Owen, Britain's leading anatomist, defined homology as "the same organ in different animals under every variety of form and function."

Structural and Developmental Homology

Organisms show curious similarities in structure and development unrelated to function. These similarities are difficult to explain under the Theory of Special Creation, but easy to explain under the Theory of Evolution.

A famous example of homology comes from work by Owen and by Baron Georges Cuvier of Paris, the founder of comparative anatomy. They described extensive similarities among vertebrate skeletons and organs. A few of these are illustrated in Figure 2.18.

Referring to Owen and Cuvier's work, Darwin (1859, p. 434) wrote,

> "What could be more curious than that the hand of a man, formed for grasping, that of a mole for digging, the leg of the horse, the paddle of the porpoise, and the wing of the bat, should all be constructed on the same pattern, and should include the same bones, in the same relative positions?"

His point was that the underlying design of these vertebrate forelimbs is similar, even though their function and appearance are different. This makes the similarity

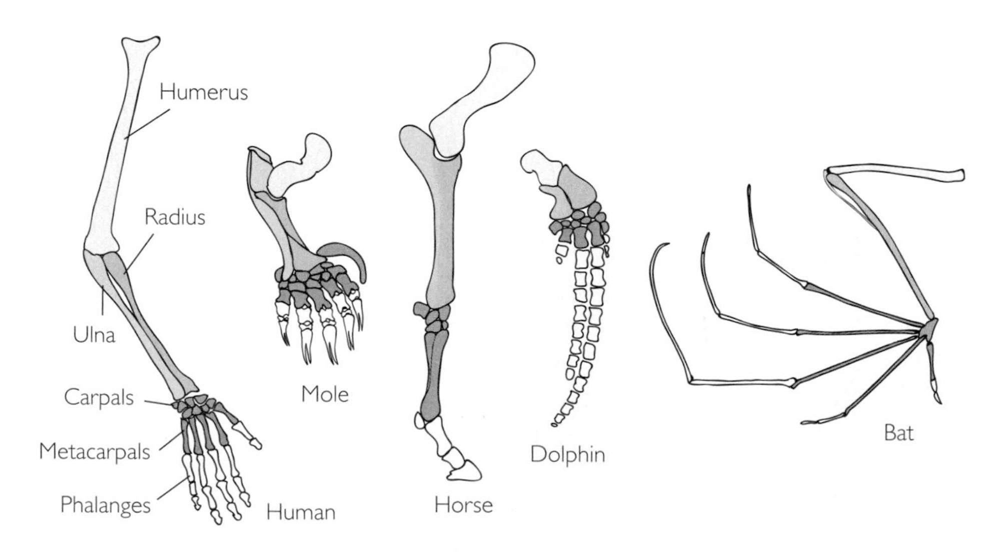

Figure 2.18 Structural homologies These vertebrate forelimbs are used for different functions, but have the same sequence and arrangement of bones. In this illustration, homologous bones are colored in the same way and are labeled on the human arm.

in design among vertebrate forelimbs different from, say, the similarity in design between a shark and a whale (Figure 2.19). Both shark and whale have a streamlined shape, short fins or flippers for steering, and a powerful tail for propulsion. These similarities in form make sense in light of their function: fast movement through water. Human engineers incorporate the same features into watercraft. In contrast, the internal similarity between forelimbs with radically different functions seems arbitrary. Would an engineer design implements for grasping, digging, running, swimming, and flying using the same set of structural elements in the same arrangement? Based on this observation, Darwin concluded that the similarity among vertebrate forelimbs is hard to explain under the Theory of Special Creation. But the similarity does make sense if all vertebrates descended from a common ancestor, from which they inherited the fundamental design of their limbs. According to Darwin, homology supports the Theory of Evolution.

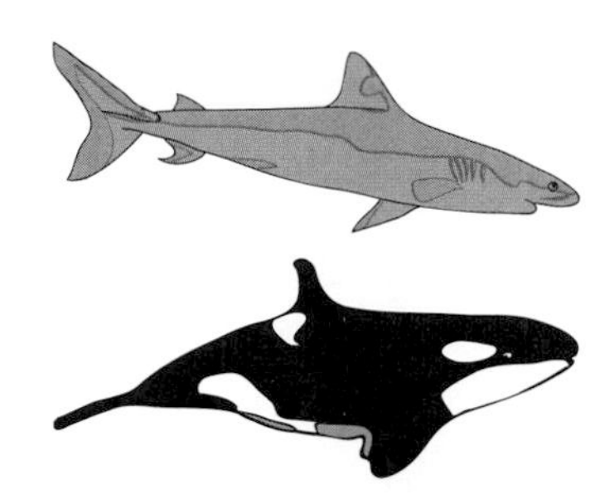

Figure 2.19 Nonhomologous similarities This shark and Orca both have streamlined shapes, powerful tails, and short fins or flippers, even though one is a fish and the other a mammal. These similarities all make sense in light of their function and are not homologous.

The examples of homology known in Darwin's time ranged beyond adult forms and vertebrates. The naturalist Louis Agassiz was among many who observed that the embryos of a great variety of vertebrates bear some striking similarities, especially early in development (Figure 2.20). Darwin himself (1862) analyzed the anatomy of orchid flowers and showed that, even though they are diverse in shape and in the pollinators they attract, they are actually constructed from the same set of component pieces. Like vertebrate forelimbs, the flowers in Figure 2.21 have the same parts in the same relative positions.

What causes these similarities? Darwin argued that descent from a common ancestor is the most logical explanation. He contended that the embryos in Figure 2.20 are similar because all vertebrates evolved from the same common ancestor and because some developmental stages have remained similar as reptiles, birds, and mammals diversified over time. Likewise, he argued that the orchids in Figure 2.21 are similar because they share a common ancestor.

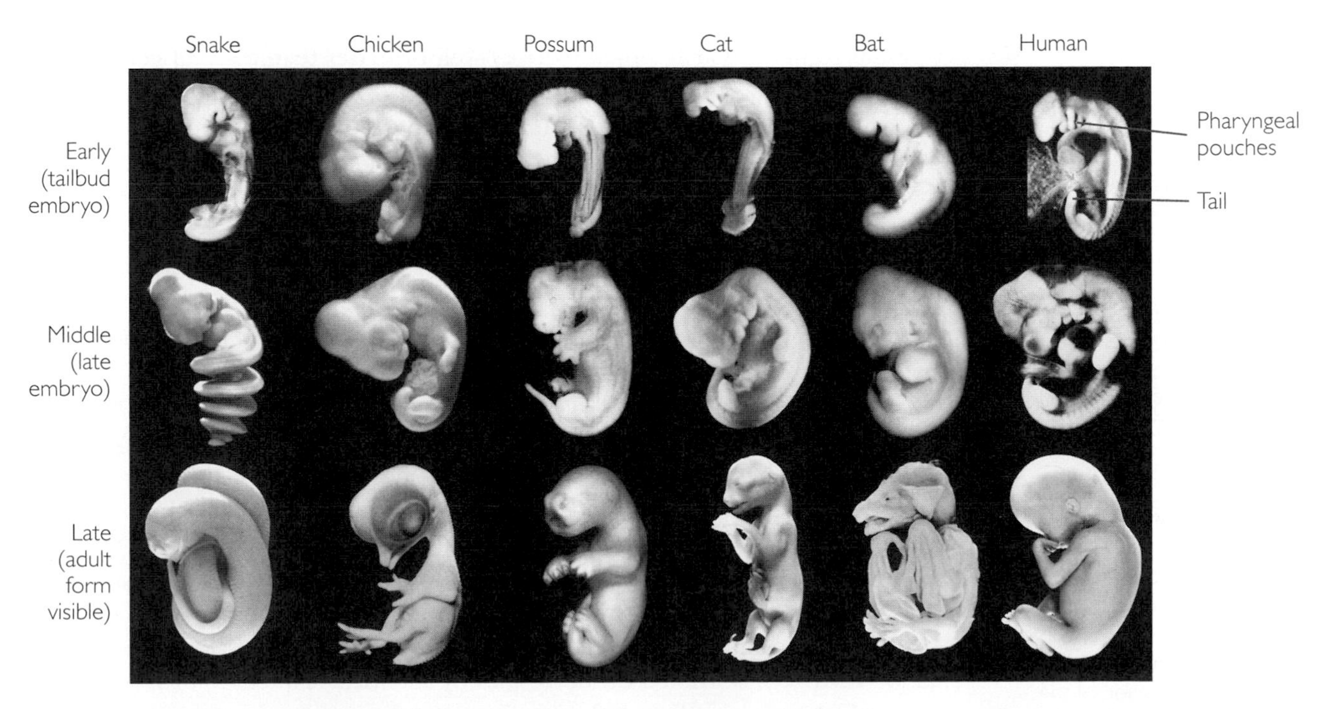

Figure 2.20 Developmental homologies Embryos from different vertebrates show striking similarities early in development. Note that the early embryos shown here all have pharygeal pouches and a tail. From Richardson, et al. (1998).

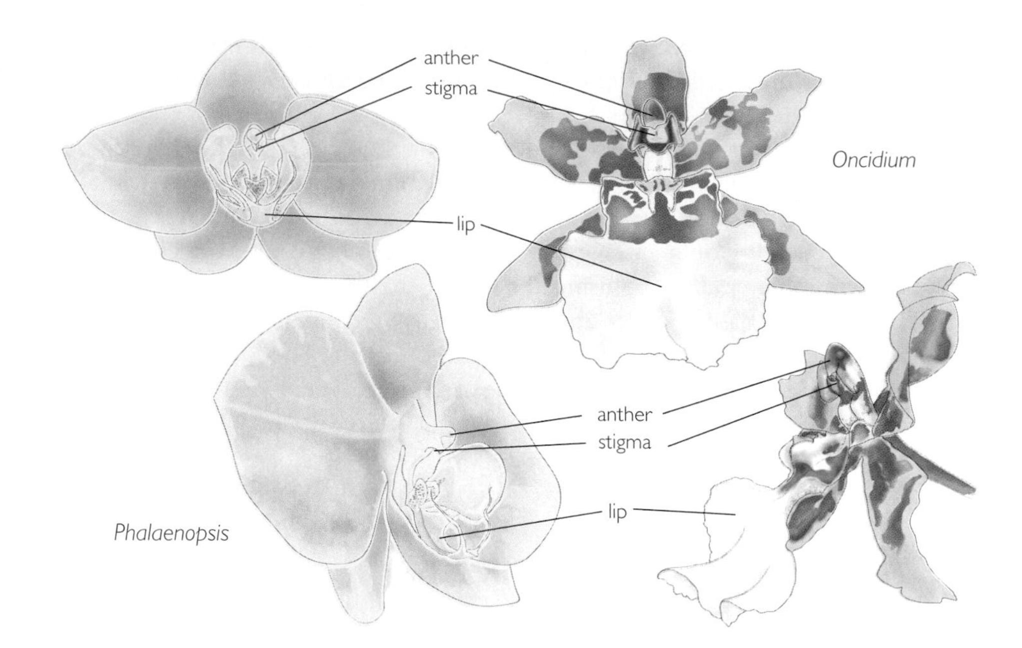

Figure 2.21 More structural homologies Orchid flowers are diverse in size and shape, but are comprised of elements that are similar in structure and orientation. From Darwin (1862).

Molecular Homology

Advances in molecular genetics have revealed other fundamental similarities among organisms. Prominent among these is the genetic code. With a few minor exceptions, all organisms studied to date use the same nucleotide triplets, or codons, to specify the same amino acids to be incorporated into proteins (Figure 2.22). The particular assignment of codons to amino acids in the genetic code reduces the deleterious effects of point mutations and translation errors (Freeland et al. 2000). However, an enormous number of alternative codes is theoretically possible, some of which would work as well or better than the real genetic code (Judson and Haydon 1999). Furthermore, having a unique genetic code might offer distinct advantages. For example, if humans used a different genetic code from chimpanzees, we would not have been susceptible to the chimpanzee virus that jumped to humans and became HIV (see Chapter 1). When the virus attempted to replicate inside human cells, its proteins would have been garbled during translation. If alternative genetic codes are possible, and if using them would be advantageous, then why do virtually all organisms use the same one? Darwinism provides a logical answer: All organisms inherited their genetic code from a common ancestor.

First base	Second base G (RNA Codon)	Amino acid	Abbreviation	Third base
U	UGU	Cysteine	C	U
	UGC	Cysteine	C	C
	UGA	Stop		A
	UGG	Tryptophan	W	G
C	CGU	Arginine	R	U
	CGC	Arginine	R	C
	CGA	Arginine	R	A
	CGG	Arginine	R	G
A	AGU	Serine	S	U
	AGC	Serine	S	C
	AGA	Arginine	R	A
	AGG	Arginine	R	G
G	GGU	Glycine	G	U
	GGC	Glycine	G	C
	GGA	Glycine	G	A
	GGG	Glycine	G	G

Figure 2.22 A genetic homology: the genetic code In almost every organism, the same nucleotide triplets, or codons, specify the same amino acids to be incorporated into proteins. This chart shows a portion of the code. The entire code appears in Chapter 5, Figure 5.3.

Our second example of molecular homology involves a genetic flaw found on chromosome 17 in the human genome. Shared flaws are especially useful in distinguishing between special creation versus descent from a common ancestor. The reason is familiar to any instructor who has caught a student cheating on an exam. If A sat next to B and wrote identical correct answers, it tells us little. But if A sat next to B and wrote identical wrong answers, our suspicions rise. Likewise, shared flaws in organisms suggest common ancestry.

On chromosome 17 the gene for a protein called peripheral myelin protein-22, or PMP-22, is flanked on both sides by identical sequences of DNA, called the CMT1A repeats (Figure 2.23a). This situation arose when the distal repeat, which contains part of the gene for a protein called COX10, was duplicated and

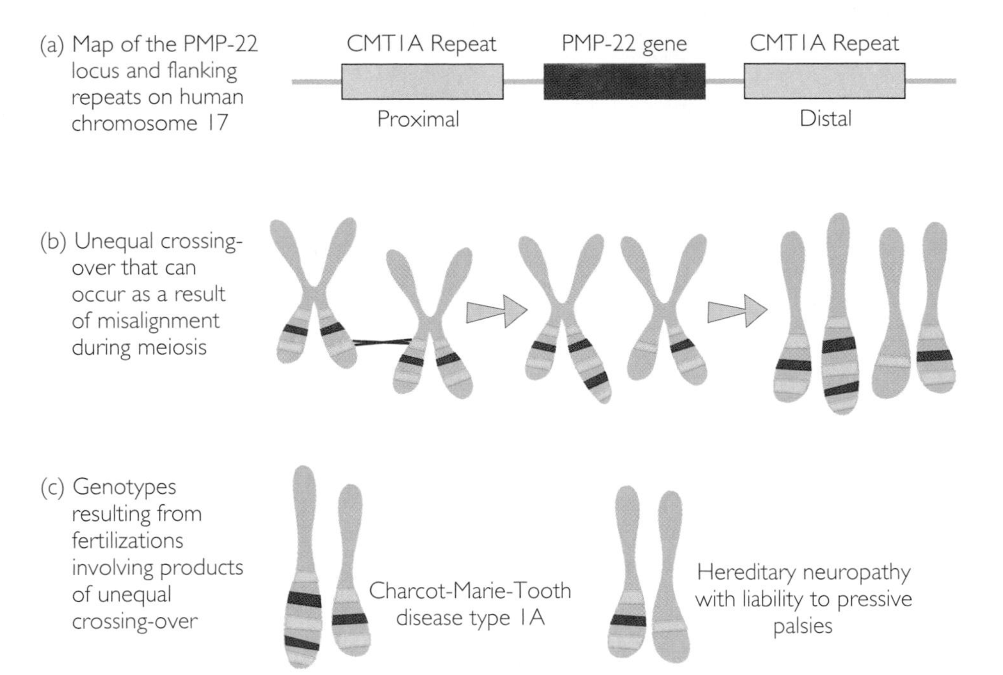
(a) Map of the PMP-22 locus and flanking repeats on human chromosome 17
CMT1A Repeat
PMP-22 gene
CMT1A Repeat
Proximal
Distal
(b) Unequal crossing-over that can occur as a result of misalignment during meiosis
(c) Genotypes resulting from fertilizations involving products of unequal crossing-over
Charcot-Marie-Tooth disease type 1A
Hereditary neuropathy with liability to pressive palsies

Figure 2.23 A genetic flaw that humans share with chimpanzees (a) The proximal CMT1A repeat, near the gene for PMP-22, is a duplication of the distal repeat on the other side of the gene. (b) The proximal repeat can align with the distal repeat during meiosis, resulting in unequal crossing-over. (c) The genotypes that result from the unequal crossing-over are associated with neurological disorders.

inserted on the other side of the PMP-22 gene (Reiter et al. 1997). The presence of the proximal CMT1A repeat has to be considered a genetic flaw because it occasionally lines up with the distal repeat during meiosis, resulting in unequal crossing-over (Figure 2.23b; Lopes et al. 1998). Among the products are a chromosome with two copies of PMP-22 and a chromosome that is missing the PMP-22 gene altogether. If either of these abnormal chromosomes participates in a fertilization, the resulting zygote is predisposed to neurological disease (Figure 2.23c). Individuals with three copies of PMP-22 suffer from a condition called Charcot-Marie-Tooth disease type 1A. Individuals with only one copy of PMP-22 suffer from hereditary neuropathy with liability to pressive palsies.

Motivated by the hypothesis that humans share a more recent common ancestor with the chimpanzees than either humans or chimps do with any other species, Marcel Keller and colleagues (1999) examined the chromosomes of common chimpanzees, bonobos (also known as pygmy chimpanzees), gorillas, orangutans, and several other primates. Both common chimps and bonobos share with us the paired CMT1A repeats that can induce unequal crossing-over. The proximal repeat is absent, however, in gorillas, orangutans, and all other species the researchers examined. This result is difficult to explain under the view that humans and chimpanzees were separately created. But it makes sense under the hypothesis that humans are a sister species to the two chimpanzees. All three species inherited the proximal repeat from a recent common ancestor, just as the Island 3 and Island 4 snails in Figure 2.14 inherited pink shells.

Curious similarities among organisms occur at the molecular level as well.

Our third example of molecular homology concerns another kind of genetic quirk that might be considered a flaw: **processed pseudogenes**. Before we explain what processed pseudogenes are, note that most genes in the human genome consist of small coding bits, or **exons**, separated by noncoding intervening sequences, or **introns**. After a gene is transcribed into messenger RNA, the introns have to be spliced out before the message can be translated into protein. Note also that the human genome is littered with **retrotransposons**,

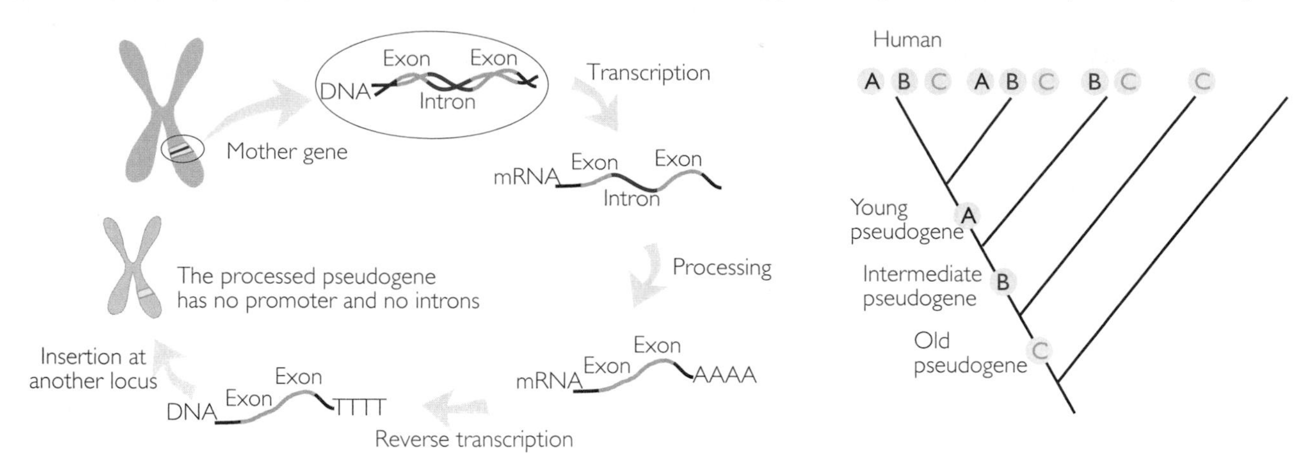

(c) Distribution of six human pseudogenes of various ages

Pseudogene	Estimated age	Human	Chimp	Gorilla	Orangutan	Rhesus monkey	Capuchin monkey	Hamster
α-Enolase Ψ_1	11 mya	●	●	●				
AS Ψ_7	16 mya	●	●		●			
CALM II Ψ_2	19 mya	●	●	●	●			
AS Ψ_1	21 mya	●	●	●	●	●		
AS Ψ_3	25 mya	●	●	●	●	●		
CALM II Ψ_3	36 mya	●	●	●	●	●	●	

Figure 2.24 **Processed pseudogenes used to test Darwin's hypothesis of common ancestry** (a) Processed pseudogenes arise when processed messenger RNAs are reverse transcribed and inserted into the genome; biologists can estimate their age by the number of mutations they have accumulated. (b) If Darwin's hypothesis of common ancestry is correct, then older, processed pseudogenes should occur in a broader range of species. (c) The taxonomic distributions of these six processed pseudogenes are consistent with this prediction.

retrovirus-like genetic elements that jump from place to place in the genome via transciption to RNA, reverse transcription to DNA, and insertion at a new site (see Luning Prak and Kazazian 2000). Some of the retrotransposons in our genome are active and encode functional reverse transcriptase.

Now we can explain that processed pseudogenes are nonfunctional copies of normal genes that originate when processed mRNAs are accidentally reverse transcribed to DNA by reverse transcriptase, then inserted back into the genome at a new location (Figure 2.24a). Processed pseudogenes are readily distinguished from their mother genes because they lack both introns and promoters.

For our purposes, the useful feature of processed pseudogenes is that it is possible to estimate their ages. Because processed pseudogenes have no function, they tend to accumulate mutations. The older a processed pseudogene, the more mutations it will have accumulated. By comparing the sequence of a processed pseudogene with that of its mother gene, we can estimate the number of mutations the pseudogene has accumulated. And from the number of mutations we can estimate the pseudogene's age.

By combining what we know about processed pseudogenes with the tree thinking introduced at the beginning of this section, we can devise a test of Darwin's view of life's history. If Darwin is correct—if species are related by descent from a common ancestor—then older processed pseudogenes should be shared by a greater variety of species. The logic behind this claim is illustrated in Figure 2.24b. The earlier the ancestor in which a processed pseudogene appeared, the more descendent species will have inherited it. Some of the descendents may have lost the pseudogene by deletion of the entire sequence, but if we examine enough species the overall pattern should be clear.

Felix Friedberg and Allen Rhoads (2000), estimated the ages of six processed pseudogenes in the human genome. The ages ranged from 11 million years to 36 million years. The researchers then looked for the same six processed pseudogenes in the genomes of the chimpanzee, gorilla, orangutan, rhesus monkey, black-capped capuchin monkey, and hamster. The results, shown in Figure 2.24c, are consistent with our prediction. Humans share the youngest of the six pseudogenes only with the African great apes (chimpanzee and gorilla). We share the four pseudogenes of intermediate age with an increasing diversity of primates (although the 16 million-year-old pseudogene appears to have been lost in gorillas). Finally, we share the oldest with the African great apes, the Asian great ape (orangutan), the Old World monkey (Rhesus), and the New World monkey (capuchin). These processed pseudogenes are molecular homologies, whose distribution among primates is evidence for common ancestry.

The Modern Concept of Homology

Darwin's interpretation of homology has become deeply embedded in biological thinking. So deeply, in fact, that the interpretation has become the definition. Under Owen's definition, homology referred to curious similarity in structure despite differences in function. Now many biologists define homology as similarity due to the inheritance of traits from a common ancestor (Abouheif 1997; Mindell and Meyer 2001). Box 2.2 points out that homology underlies the use of model organisms in biomedical research and drug testing. In other words, much of modern biomedical research is founded on the assumption that humans are related to the rest of Earth's organisms by descent from a common ancestor. The tremendous successes of this research effort can thus be taken as powerful evidence for evolution.

Darwin's interpretation of homology has become the definition: similarity due to common ancestry.

Box 2.2 Homology and model organisms

Homology may appear to be an abstract concept, but it is actually the guiding principle behind most biomedical research. Homology is the reason medical researchers can obtain valid results when testing the safety of new drugs in mice or studying the molecular basis of disease in rats. The results can be extrapolated to humans if the molecular or cellular basis of the phenomenon being studied is homologous.

Researchers choose a study organism—also called a "model" organism—based on the degree of homology required to study a particular process or disease. In psychiatry and the behavioral sciences, for example, monkeys and apes are often the preferred experimental subject because aspects of their behavior and brain structures are homologous with those of humans. Because some of the genes involved in more basic processes, such as the cell cycle, are homologous between even distant relatives, researchers can use baker's yeast (*Saccharomyces cerevisiae*) to study why certain malfunctioning genes cause cancer in humans. At an even more fundamental level, the genes involved in DNA repair are homologous between humans and the bacterium *Escherichia coli*. Primates, yeast, and bacteria share these characteristics with humans because they inherited them from a common ancestor.

Relationships among Species

Darwin's recognition of relationship through shared descent extended to phenomena other than homology. His trip to the Galápagos Islands had a strong influence on his thinking about the relationships among species. While he was aboard the HMS *Beagle* during a five-year mapping and exploratory mission, Darwin collected and cataloged the flora and fauna encountered during the voyage. He was especially impressed by the mockingbirds he found during his work in the Galápagos, because several islands had distinct populations. Although they were all similar in color, size, and shape—and thus clearly related to one another—each mockingbird population seemed distinct enough to be classified as a separate species. This was confirmed later by a taxonomist colleague of Darwin's back in England. Darwin and others followed up on this result with studies showing the same pattern in Galápagos tortoises and finches: The various islands hosted different, but closely related, species (see Desmond and Moore 1991).

To explain this pattern, Darwin hypothesized that a small population of mockingbirds colonized the Galápagos from South America long ago. His thesis was that the population expanded in the new habitat and that subpopulations subsequently colonized different islands in the group. Once mockingbird populations had become physically isolated from one another in this way, they diverged enough to become distinct species.

Species that are extremely similar to one another tend to be clustered geographically. This suggests that they were not created independently, but are descended from a common ancestor that lived in the same area.

Like structural homologies, the existence of closely related forms in island groups was a logical outcome of descent with modification. In contrast, both patterns were inconsistent with special creation, which predicted that organisms were created independently. Under special creation, no particular patterns are expected in the design or geographic relationships of organisms.

Evidence of Common Ancestry

In their genes, in their development, and in their structure, Earth's organisms resemble each other to a remarkable degree. Some of these resemblances make functional sense, like the resemblance in shape between a shark and a whale. Such resemblances can be explained under either special creation or evolution. Other resemblances, however, like the bones of the forelimb in various vertebrates or the processed pseudogenes in humans and primates, make little or no sense for function. They are much more easily explained under Darwin's view that organisms are descended from a common ancestor.

2.3 The Age of Earth

The young science of geology confirmed that Earth had existed for vast stretches of time. Evolution is a time-dependent process, but special creation is not.

By the time Darwin began working on the origin of species, data from geology had challenged a linchpin in the Theory of Special Creation: that Earth was only about 6,000 years old. Evidence was mounting that Earth was ancient.

Much of this evidence was grounded in a principle called **uniformitarianism**, first articulated by James Hutton in the late 1700s. Uniformitarianism is the claim that geological processes taking place now operated similarly in the past. It was proposed in direct contrast to a hypothesis called catastrophism. This was the proposal that today's geological formations resulted from catastrophic events, like the biblical flood, which occurred in the past on a scale never observed today.

The assumption of uniformitarianism and rejection of catastrophism led Hutton, and later Charles Lyell, to infer that Earth was unimaginably old in human terms. This conclusion was driven by data. These early geologists measured the rate of ongoing rock-forming processes such as the deposition of mud, sand, and gravel at beaches and river deltas and the accumulation of marine shells (the precursors of limestone). Based on these observations, it was clear that vast stretches of time were required to produce the immense rock formations being mapped in the British Isles and Europe by these researchers.

The Geologic Time Scale

When Darwin began his work, Hutton and followers were already in the midst of a 50-year effort to put the major rock formations and fossil-bearing strata of Europe in a younger-to-older sequence. Their technique was called relative dating because its objective was to determine how old each rock formation was relative to other strata. Relative dating was an exercise in logic based on the following assumptions:

- Younger rocks are deposited on top of older rocks (this is called the principle of superposition).
- Lava and sedimentary rocks, like sandstones, limestones, and mudstones, were originally laid down in a horizontal position. As a result, any tipping or bending events in these types of rocks must have occurred after deposition (principle of original horizontality).
- Rocks that intrude into seams in other rocks, or as dikes, are younger than their host rocks (principle of cross-cutting relationships).
- Boulders, cobbles, or other fragments found in a body of rock are older than their host rock (principle of inclusions).
- Earlier fossil life forms are simpler than more recent forms, and more recent forms are most similar to existing forms (principle of faunal succession).

Using these rules, geologists established the chronology of relative dates known as the geologic time scale (Figure 2.25). They also created the concept of the geologic column, which is a geologic history of Earth based on a composite, older-to-younger sequence of rock strata. (There is no one place on Earth where all rock strata that have formed through time are still present. Instead, there are always gaps where some strata have eroded completely away. But by combining data from different locations, geologists are able to assemble a complete record of geologic history.)

The principle of uniformitarianism, the geologic time scale, and the geologic column furnished impressive evidence for an ancient Earth. Geologists began working in time scales of tens of millions of years, instead of a few thousand years, long before Darwin published his ideas on change through time and descent with modification. These data were important to the Theory of Evolution. Special creation is an instantaneous process, but evolutionary change required long periods of time to produce the diversity of life seen today.

Included with the geologic time scale in Figure 2.25 are ages now known from radiometic dating (the time scale is non-linear) and an evolutionary tree showing the currently accepted relationships among a few familiar extant organisms and some important fossils. The divergence times noted on the phylogeny are estimates based on genetic data (Hedges and Kumar 2003). The older ones, in particular, are controversial (see Graur and Martin 2004; Hedges and Kumar 2004; Reisz and Müller 2004a, b; Glazko et al. 2005).

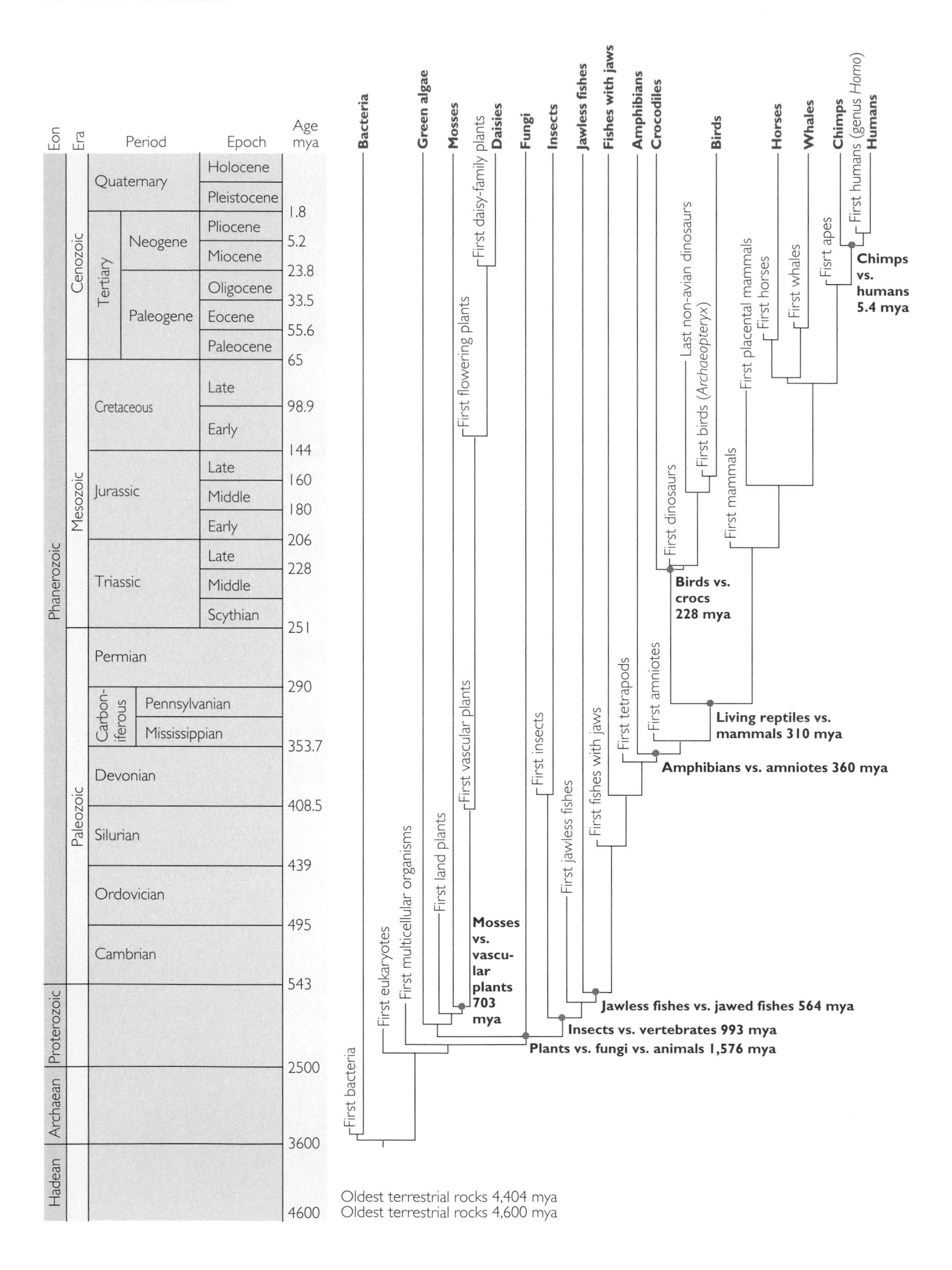

Eon
Era
Period
Epoch
Age mya
Phanerozoic
Proterozoic
Archaean
Hadean
Cenozoic
Mesozoic
Paleozoic
Quaternary
Tertiary
Neogene
Paleogene
Cretaceous
Jurassic
Triassic
Permian
Carbon-iferous
Pennsylvanian
Mississippian
Devonian
Silurian
Ordovician
Cambrian
Holocene
Pleistocene
Pliocene
Miocene
Oligocene
Eocene
Paleocene
Late
Early
Late
Middle
Early
Late
Middle
Scythian
1.8
5.2
23.8
33.5
55.6
65
98.9
144
160
180
206
228
251
290
353.7
408.5
439
495
543
2500
3600
4600
Bacteria
Green algae
Mosses
Daisies
Fungi
Insects
Jawless fishes
Fishes with jaws
Amphibians
Crocodiles
Birds
Horses
Whales
Chimps
Humans
First bacteria
First eukaryotes
First multicellular organisms
First land plants
First vascular plants
First flowering plants
First daisy-family plants
First insects
First jawless fishes
First fishes with jaws
First tetrapods
First amniotes
First dinosaurs
Last non-avian dinosaurs
First birds (Archaeopteryx)
First mammals
First placental mammals
First horses
First whales
Fisrt apes
First humans (genus Homo)
Chimps vs. humans 5.4 mya
Birds vs. crocs 228 mya
Living reptiles vs. mammals 310 mya
Amphibians vs. amniotes 360 mya
Mosses vs. vascular plants 703 mya
Jawless fishes vs. jawed fishes 564 mya
Insects vs. vertebrates 993 mya
Plants vs. fungi vs. animals 1,576 mya
Oldest terrestrial rocks 4,404 mya
Oldest terrestrial rocks 4,600 mya

Figure 2.25 (opposite) The geologic time scale The sequence of eons, eras, periods, and epochs shown on the left part of this diagram was established through the techniques of relative dating. Each named interval of time is associated with a distinctive fossil flora and fauna. The absolute ages included here were added much later, when radiometric dating systems became available. The abbreviation mya stands for millions of years ago. The evolutionary tree on the right includes some important fossils (in black) and a few familar extant organsisms (in green). See Hedges (2002) for a detailed overview of the tree of life. The divergence times marked by red dots are estimates based on genetic data (Hedges and Kumar 2003).

Radiometric Dating

By the mid-19th century, Hutton, Lyell, and their followers had established, beyond a reasonable doubt, that Earth was old. But how old? How much time has passed since life on Earth began?

Marie Curie's discovery of radioactivity in the early 1900s gave scientists a way to answer these questions. Using a technique called radiometric dating, physicists and geologists began to assign absolute ages to the relative dates established by the geologic time scale.

The technique for radiometric dating uses unstable isotopes of naturally occurring elements. These isotopes decay, meaning that they change into either different elements or different isotopes of the same element. Each isotope decays at a particular and constant rate, measured in a unit called a half-life. One half-life is the amount of time it takes for 50% of the parent isotope present to decay into its daughter isotope (Figure 2.26). The number of decay events observed in a rock sample over time depends only on how many radioactive atoms are present. Decay rates are not affected by temperature, moisture, or any other environmental factor. As a result, radioactive isotopes function as natural clocks. For more detail, see Box 2.3.

Because of their long half-lives, potassium–argon and uranium–lead systems are the isotopes of choice for determining the age of Earth. Using these systems, what rocks can be tested to determine when Earth first formed? Current models of Earth's formation predict that the planet was molten for much of its early history, which makes answering this question difficult. If we assume that all of the components of our solar system were formed at the same time, however, two classes of candidate rocks become available to date the origin of Earth: moon rocks and meteorites. Both uranium–lead and potassium–argon dating systems

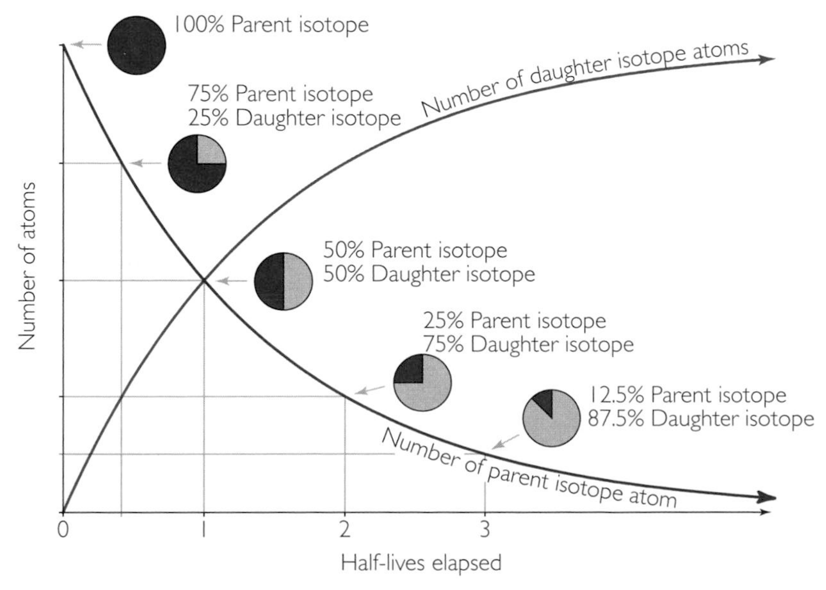

Figure 2.26 Radioactive decay Many radioactive isotopes decay through a series of intermediates until a stable daughter isotope is produced. Researchers measure the ratio of parent isotope to daughter isotope in a rock sample, then use a graph like this to convert the measured ratio to the number of half-lives elapsed. Multiplying the number of half-lives that have passed by the number of years it takes for a half-life to elapse yields an estimate for the absolute age of the rock.

Box 2.3 | A closer look at radiometric dating

Radiometric dating allows geologists to assign absolute ages to rocks. Here is how the technique works: First, the half-life of a radioactive isotope is determined by putting a sample in an instrument that records the number of decay events over time. For long-lived isotopes, of course, researchers must extrapolate from data collected over a short time interval. Then the ratio of parent-to-daughter isotopes in a sample of rock is measured, often with an instrument called a mass spectrometer. Once the half-life of the parent isotope and the current ratio of parent-to-daughter isotopes are known, the number of years that have passed since the rock was formed can be calculated.

A critical assumption here is that the ratio of parent-to-daughter isotope when the rock was formed is known. This assumption can be tested. Potassium–argon dating, for example, is an important system for dating rocks of volcanic origin. We can predict that, initially, there will be zero of the daughter isotope, argon-40, present. That is because argon-40 is a gas that bubbles out of liquid rock and only begins to accumulate after solidification. Observations of recent lava flows confirm that this is true: Expressed as a ratio of percentages, the ratio of potassium-40 to argon-40 in newly minted basalts, lavas, and ashes is, as predicted, 100:0 (see Damon 1968, Faure 1986).

Of the many radioactive atoms present in Earth's crust, the isotopes listed in Table 2.1 are the most useful. Not only are they common enough to be present in measurable quantities, but they are also stable in terms of not readily migrating into or out of rocks after their initial formation. If the molecules did move, it would throw off our estimate of the surrounding rock's age.

To choose an isotope suitable for dating rocks and fossils of a particular age, geochronologists and paleontologists look for a half-life short enough to allow a measurable amount of daughter isotope to accumulate, but long enough to ensure that a measurable quantity of parent isotope is still left. In many instances more than one isotope system can be used on the same rocks or fossils, providing an independent check on the date.

Table 2.1 Parent and daughter isotopes used in radiometric dating

Method	Parent isotope	Daughter isotope	Half-life of parent (years)	Effective dating range (years)	Materials commonly dated
Rubidium-strontium	Rb-87	Sr-87	47 billion	10 million–4.6 billion	Potassium-rich minerals such as biotite, potassium, muscovite, feldspar, and hornblende; volcanic and metamorphic rocks
Uranium-lead	U-238	Pb-206	4.5 billion	10 million–4.6 billion	Zircons, uraninite, and uranium ore such as pitchblende; igneous and metamorphic rock
Uranium-lead	U-235	Pb-207	71.3 million	10 million–4.6 billion	Same as above
Thorium-lead	Th-232	Pb-208	14.1 billion	10 million–4.6 billion	Zircons, uraninite
Potassium-argon	K-40	Ar-40	1.3 billion	100,000–4.6 billion	Potassium-rich minerals such as biotite, muscovite, and potassium feldspar; volcanic rocks
Carbon-14	C-14	N-14	5,730	100–100,000	Any carbon-bearing material, such as bones, wood, shells, charcoal, cloth, paper, and animal droppings

place the age of the moon rocks brought back by the Apollo astronauts at 4.53 billion years. Also, virtually every meteorite found on Earth that has been dated yields an age of 4.6 billion years. Therefore, scientists can infer that the planet is about 4.6 billion years old.

How long has life on Earth been evolving? Paleontologists have found fossils of organisms that appear to be cyanobacteria and eukaryotic algae that are about 2 billion years old (Golubic and Hofmann 1976; Han and Runnegar 1992). They have also reported apparent fossils as much as 3.4 to 3.5 billion years old (Knoll and Barghoorn 1997; Schopf 1993), but some of these reports are controversial (Brasier 2002; Dalton 2002; Schopf et al. 2002). Geochemists have found biological molecules preserved in rocks that are 2.7 billion years old (Brocks et al. 1999), and chemical evidence suggestive of life that is more than 3.7 billion years old (Mojzsis et al. 1996; Rosing 1999). Again, however, some of the reports of earlier life are controversial (Fedo and Whitehouse 2002; Kerr 2002; van Zullen et al. 2002). Together, the data indicate that life has been evolving on Earth for at least 2.7 billion years and possibly for more than 3.7 billion years.

Radiometric dating confirms that the Earth and life are billions of years old.

The Age of Earth

In the 19th century, relative dating suggested that Earth was much older than the 6,000 years predicted by Bishop Ussher. In the 20th century, absolute dating confirmed that life has existed at least 450,000 times longer than suggested by the Theory of Special Creation.

2.4 Is There Necessarily a Conflict between Evolutionary Biology and Religion?

We have discussed evidence that conflicts with the Theory of Special Creation, a view of life's history derived from a literalist interpretation of the Bible's Book of Genesis. And we have argued that this evidence is consistent with Darwin's alternative, the theory of descent with modification from a common ancestor. Does this mean that evolutionary biology is necessarily in conflict with Christianity or with religion in general? As with the other questions we address throughout the book, we will treat this issue as an empirical matter. First we will briefly consider some theory—in this case drawn from the philosophy of science. Then we will look at some evidence.

Methodological Naturalism and Ontological Naturalism

Scientists seek to understand the natural world: what it contains, how it arrived at its current state, and the laws that regulate its behavior. As a fundamental rule governing how they conduct their inquiries, modern scientists have adopted a position that philosophers of science call **Methodological Naturalism** (Pennock 1996). Under this position, the only hypotheses researchers propose to account for natural phenomena, and the only explanations they accept, are hypotheses and explanations that involve strictly natural causes.

The adoption of Methodological Naturalism by scientists can be justified on purely practical grounds: It is the only way to make any progress. If we allow ourselves the option of appealing to supernatural explanations, then we have no way of knowing when we should keep struggling with a hard problem and when we should simply give up and declare the phenomenon in question a miracle. Through the discipline of never giving up, scientists have made great progress in solving mysteries that previous generations deemed beyond the reach of rational understanding. Darwin's triumph in explaining the origin of species is just one example among many.

Scientists reject supernatural explanations for natural phenomena. This does not mean that all scientists reject the existence of the supernatural.

While Methodological Naturalism has proven to be an invaluable guiding principle for learning about the world, it must be distinguished from **Ontological Naturalism**. Ontological Naturalism is the position that the natural world is all there is. As we have said, a Methodological Naturalist *assumes*, for the sake of inquiry and argument, that nothing else exists. An Ontological Naturalist goes further and actually *believes* that nothing else exists. Whether the reader will think that making this distinction is splitting hairs or merely stating the obvious will likely depend on his or her own philosophical stance. The point for our purposes is that committing oneself to Methodological Naturalism does not necessarily, either logically or emotionally, also involve committing oneself to Ontological Naturalism. In less precise but plainer language, the fact that a person refrains from discussing God at work does not necessarily mean that he or she is an atheist at home.

Are There Religious Evolutionary Biologists?

That is the theory; what about the data? What do evolutionary biologists actually believe about the existence of the supernatural?

Certainly many evolutionary biologists are Ontological Naturalists. Here are samples from the responses of two, when asked to explain in writing why they are secular humanists (Bonner et al. 1997). The first is from Richard Dawkins:

> "We are on our own in the universe. Humanity can expect no help from outside, so our help, such as it is, must come from our own resources. As individuals we should make the most of the short time we have, for it is a privilege to be here. We should seize the opportunity presented by our good fortune and fill our brief minds, before we die, with understanding of why, and where, we exist."

The second is from Edward O. Wilson:

> "... scientific materialism explains vastly more of the tangible world, physical and biological, in precise and useful detail, than the Iron-Age theology and mysticism bequeathed us by the modern great religions ever dreamed. It offers an epic view of the origin and meaning of humanity far greater, and I believe more noble, than conceived by all the prophets of old combined. Its discoveries suggest that, like it or not, we are alone. We must measure and judge ourselves, and we will decide our own destiny."

Other evolutionary biologists, however, reject Ontological Naturalism. And some of them are deeply religious. Again, we offer two writing samples. The first is from Kenneth Miller (1999, p. 267):

> "A nonbeliever, of course, puts his or her trust in science and finds no value in faith. And I certainly agree that science allows believer and nonbeliever alike to investigate the natural world through a common lens of observation, exper-

iment, and theory. The ability of science to transcend cultural, political, and even religious differences is part of its genius, part of its value as a way of knowing. What science cannot do is assign either meaning or purpose to the world it explores. This leads some to conclude that the world as seen by science is devoid of meaning and absent of purpose. It is not. What it does mean, I would suggest, is that our human tendency to assign meaning and value must transcend science and, ultimately, must come from outside it."

The second is from Loren Eiseley (1946, p. 210):

> "I would say that if 'dead' matter has reared up this curious landscape of fiddling crickets, song sparrows, and wondering men, it must be plain even to the most devoted materialist that the matter of which he speaks contains amazing, if not dreadful powers, and may not impossibly be, as Hardy has suggested, 'but one mask of many worn by the Great Face behind.' "

To determine the frequency of religious belief among American scientists, Edward Larson and Larry Witham (1997, 1998, 1999) surveyed several hundred mathematicians, astronomers, physicists, and biologists, asking the subjects whether they believe in: (1) a personal God to whom one prays and expects to receive an answer, and (2) personal immortality that transcends death. Larson and Witham were attempting to replicate, as closely as possible, a survey performed in 1914 by James Leuba. Leuba phrased his questions to reflect what he saw as the core teachings of traditional Christian churches. In 1996, as in 1914, some of the subjects objected that the survey's definitions of God and of the afterlife were too restrictive. Nonetheless, about 40% of all scientists Leuba surveyed, including about 30% of the biologists, reported belief in God. Roughly the same number reported belief in personal immortality. Leuba had predicted that the frequency of religious belief among scientists would decline over time. In fact, however, Larson and Witham found that the frequency of belief had changed little over the 80 years between surveys.

For many scientists and theologians there is no necessary conflict between science and religion.

A textbook on evolution is not the place for a discussion of theology, but we will also mention before leaving this issue that most theologians see no conflict between evolution and religion. As the General Assembly of the United Presbyterian Church in the United States put it in a resolution they adopted in 1982 (see National Center for Science Education 2000; see also Alters and Alters 2001; Pennock 2001):

> "... the imposition of a fundamentalist viewpoint about the interpretation of Biblical literature—where every word is taken with uniform literalness and becomes an absolute authority on all matters, whether moral, religious, political, historical or scientific—is in conflict with the perspective on Biblical interpretation characteristically maintained by Biblical scholars and theological schools in the mainstream of Protestantism, Roman Catholicism and Judaism. Such scholars find that the scientific theory of evolution does not conflict with their interpretation of the origins of life found in Biblical literature."

Religious belief is less common among scientists than among nonscientists (Bishop 1998). But for many people, including many evolutionary biologists, there is no contradiction between accepting evolution and believing in God.

Summary

The pattern component of the Theory of Evolution contends that species have changed through time and are related to one another through descent from a common ancestor. Darwin argued forcefully for this theory in his book *On the Origin of Species*, published in 1859. At that time, one leading explanation for the history of life was the Theory of Special Creation, which maintained that species were created independently and recently and that they do not change through time.

Data sets from living and fossil species refute the hypothesis that species do not change through time. The presence of rudimentary structures, transitory developmental stages, and vestigial DNA sequences in contemporary organisms is readily understood as a result of change through time. Change in important characteristics, like the beak length of soapberry bugs, has also been observed directly in hundreds of different species. The hypothesis of change through time is further supported by the extensive extinctions, the law of succession, and transitional forms documented in the fossil record.

Several lines of evidence argue that species were not created independently. For example, extensive structural, developmental, and genetic homologies exist among organisms. These similarities are most logically explained as the product of descent from a common ancestor. Similarly, closely related groups of species in the same geographic area, such as the mockingbirds, finches, and tortoises that Darwin observed on the Galápagos islands, are readily interpreted as the descendants of populations that colonized the area in the past.

By the mid-1800s, the principle of uniformitarianism and the completion of the geologic time scale had persuaded most scientists that Earth is much older than the few thousand years posited by the Theory of Special Creation. This result was verified in the early 1900s by radiometric dating. The best data available suggest that Earth formed about 4.6 billion years ago. The first fossil evidence for life is 3.7 billion years old.

The Theory of Evolution is successful because it provides a logical explanation for a wide variety of observations and because it makes predictions that can be tested and verified.

Questions

1. Review the evidence for evolution analyzed in Sections 2.1–2.3. List the sources of evidence that were available to Darwin and those that appeared later. Indicate which evidence you consider strongest and which you consider weakest. Explain why.
2. Suppose you were a typical English citizen of 1859, reading Darwin's new book *On the Origin of Species*. Given the data that Darwin had available (see your answer for question 1), would you have been convinced by his arguments for common ancestry? For natural selection? If not, do you think the evidence finally swung in the favor of either proposal when: *Archeopteryx* was discovered (1861); Mendel's rules of inheritance became widely known (1900); Dobzhansky and others showed how genetics are related to natural selection (1937–onward); the molecular structure of DNA was determined (1959); fossils of very ape-like hominids were discovered (1970s–present); DNA sequence information became available (1990s–present); fossils of feathered dinosaurs and legged whales were discovered (1990s–present); or, not at all?
3. As evidence of change over time, Figure 2.3a presents two populations of soapberry bugs that differ in their beak lengths and are descended from a common ancestor. How might different breeds of dogs, or different breeds of cats, be used to construct a similar argument? How could you test that all breeds of dogs, or all breeds of cats, are descended from a common ancestor?
4. Figures 2.10 and 2.13 show examples of transitional fossils. If Darwin's theory of evolution is correct, and all organisms are descended with modification from a common ancestor, predict some other examples of transitional forms that should have existed and that might have produced fossils. If such fossils are someday found, will that strengthen the hypothesis that such transitional species once existed? Conversely, if such fossils have not been found, does this weaken the hypothesis that the transitional species once existed?
5. The transitional fossils in Figure 2.10 through 2.12 demonstrate that dinosaurs evolved feathers long before they evolved flight. Clearly, feathers did not evolve for their aerodynamic advantages. What else, besides aerodynamics, do feathers do for birds today? What advantages might feathers have offered for dinosaurs? Can you think of a way to test your hypothesis?
6. Section 2.2 presented two definitions of homology: the classical definition articulated by Richard Owen (p. 54) and the modern definition favored by many contemporary biologists (p. 59). Look at the vestigial organs shown in Figure 2.4. Is the tiny wing of a brown kiwi

homologous to the wing of an eagle? Are the spurs of a rubber boa homologous to the hind limbs of a kangaroo? By which definition of homology?

7. Analogy and homology are important concepts used in comparing species. Traits are homologous if they are derived, evolutionarily and developmentally, from the same source structure. Traits are analogous if they have similar functions but are derived, evolutionarily and developmentally, from different source structures. A classic example of analogous structures is insect wings and bat wings. Which of the following pairs of structures are analogous and which are homologous?
   a. The dorsal fins of a porpoise and a salmon
   b. The flippers of a porpoise and the pectoral fins (front fins) of a salmon
   c. The jointed leg of a ladybird beetle and a robin
   d. A rhesus monkey's tail and a human's coccyx
   e. The bright red bracts (modified leaves) of a poinsettia and the green leaves of a rose
   f. The bright red bracts of a poinsettia and the red petals of a rose
8. Draw a simple phylogenetic tree showing what the relationships among five living species might be. Then draw a genealogy of your family or a friend's family, starting with the oldest and continuing to the youngest generation. Label the parts of each diagram. How are phylogenetic trees and pedigrees similar? How are they different?
9. According to the evolutionary tree shown in Figure 2.16, are jaguarundis more closely related to tigers or to bobcats? Why?
10. In the early 20th century, radiometric dating allowed geologists to assign absolute ages to most fossil-bearing strata. The absolute dates turned out to be entirely consistent with the relative dating done in the early 19th century. What does this result say about the assumptions behind relative dating listed on page 61?
11. Based on the assumption that extinctions were caused by catastrophic, worldwide floods of the type described in the Bible, what predictions does the Theory of Special Creation make about the nature of the fossil record? What predictions does the Theory of Evolution make about the nature of the fossil record? What evidence exists to confirm or refute your predictions?

Exploring the Literature

12. Darwin's classic book still stands as one of the most influential books written in the past two centuries. It should be read by any serious student of biology. The full text of *On the Origin of Species* (full title: *On the Origin of Species by Means of Natural Selection, or, the Preservation of Favoured Races in the Struggle for Life*), is available free online at: *http://www.talkorigins.org/faqs/origin.html*
13. Figure 2.3 documents the rapid evolution of soapberry bugs following their switch to a new host plant. For a similar example of swift evolution—this time in a bird—see:

    Smith, T. B., L. A. Freed, et al. 1995. Evolutionary consequences of extinctions in populations of a Hawaiian honeycreeper. *Conservation Biology* 9: 107-113.

    Is there an alternative explanation, besides genetic evolution, that could explain the change in beak size that Smith documented? (Hint: Look back at the discussion of soapberry bugs on pages 40–41). What experiment would have to be done to rule out this alternative, and why was it impossible for Smith to do? How did the authors of the following papers, which also document rapid evolution, rule out the alternative explanation?:

    Karban, R. 1989. Fine-scale adaptation of herbivorous thrips to individual host plants. *Nature* 340: 60-61.

    Magurran, A. E., B. H. Seghers, et al. Behavioral consequences of an artificial introduction of guppies (*Poecilia reticulata*) in N. Trinidad: Evidence for the evolution of anti-predator behavior in the wild. *Proceedings of the Royal Society of London, Biological Sciences* 248: 117-122.
14. Figure 2.7 presents evidence that in sticklebacks the presence or absence of bony armor is determined largely by the alleles of a single gene, and that the same is true for pelvic spines. For a report on the identification of the gene controlling bony armor, see:

    Colosimo, P. F., K. E. Hosemann, et al. 2005. Widespread parallel evolution in sticklebacks by repeated fixation of ectodysplasin alleles. *Science* 307: 1928-1933.

    For the discovery of the gene controlling pelvic spines, see:

    Cole, N. J, M. Tanaka, et al. 2003. Expression of limb initiation genes and clues to the morphological diversification of threespine stickleback. *Current Biology* 13: R951-R952.

    Shapiro, M. D., M. E. Marks, et al. 2004. Genetic and developmental basis of evolutionary pelvic reduction in threespine sticklebacks. *Nature* 428: 717-723.
15. For two different hypotheses about how dinosaurs might have used their wings during the evolution of flight, see:

    Xu, X., and F. Zhang. 2005. A new maniraptoran dinosaur from China with long feathers on the metatarsus. *Naturwissenschaften* 92 (4): 173–177.

    Dial, K. P. 2003. Wing-assisted incline running and the evolution of flight. *Science* 299: 402–404.
16. For additional recent fossil evidence on the kinship between birds and dinosaurs, including a new specimen of *Archaeopteryx*, see:

    Mayr, G., B. Pohl, and D.S. Peters. 2005. A well-preserved *Archaeopteryx* specimen with theropod features. *Science* 310: 1483–1486.

Schweitzer, M. H., J. L. Wittmeyer, and J. R. Horner. 2005. Gender-specific reproductive tissue in ratites and *Tyrannosaurus rex*. *Science* 308: 1456–1460.

Xu, X., and Mark A. Norell. 2004. A new troodontid dinosaur from China with avian-like sleeping posture. *Nature* 431: 838–841.

Zhou, Z. 2004. The origin and early evolution of birds: discoveries, disputes, and perspectives from fossil evidence. *Naturwissenschaften*. 91(10): 455–471.

17. For another example of a genetic design flaw we humans share with chimpanzees, see:

Kawaguchi, H., C. O'hUigin, and J. Klein. 1992. Evolutionary origin of mutations in the primate cytochrome p450c21 gene. *American Journal of Human Genetics* 50: 766–780.

18. We mentioned in Section 2.2 that different Galápagos islands have distinct but closely related species of giant tortoises. For a phylogenetic analysis of the origin and relationships among Galápagos tortoises, see:

Caccone, A., J. P. Gibbs, V. Ketmaier, et al. 1999. Origin and evolutionary relationships of giant Galápagos tortoises. *Proceedings of the National Academy of Sciences USA* 96: 13223–13228.

19. The following represent a suite of independent data sets that combine to corroborate an evolutionary view of the history of life: phylogenies that are estimated from morphological data, phylogenies that are estimated from molecular data, radiometric dating, and the fossil record. Research on the evolution of tetrapods presents a good example of combining data sets. The following citations will help you get started on this literature:

Daeschler, E.B., N.H. Shubin, and F.A. Jenkins, Jr. 2006. A Devonian tetrapod-like fish and the evolution of the tetrapod body plan. *Nature* 440 (7085): 757–763.

Ahlberg, P. E., and Z. Johanson. 1998. Osteolepiforms and the ancestry of tetrapods. *Nature* 395: 792–794.

Hedges, S. B., and L. L. Poling. 1999. A molecular phylogeny of reptiles. *Science* 283: 998–1001.

Shubin, N. 1998. Evolutionary cut and paste. *Nature* 394: 12–13.

Shubin, N., E. B. Daeschler, and F. A. Jenkins, Jr. 2006. The pectoral fin of *Tiktaalik roseae* and the origin of the tetrapod limb. *Nature* 440: 764–771.

Zardoya, R., and A. Meyer. 1996. Evolutionary relationships of the coelacanth, lungfishes, and tetrapods based on the 28S ribosomal RNA gene. *Proceedings of the National Academy of Sciences USA* 93: 5449–5454.

20. We reported in Section 2.4 that most theologians, and many evolutionary biologists, see no necessary conflict between religion and evolutionary biology. For extended discussions by a theologian and a biologist, see:

Clouser, Roy. 2001. Is theism compatible with evolution? Chapter 21 (pages 513–536) in Pennock, R. T., ed. 2001. *Intelligent Design Creationism and its Critics*. Cambridge, MA: The MIT Press.

Miller, Kenneth R. 1999. *Finding Darwin's God: A Scientist's Search for Common Ground between God and Evolution*. New York: Cliff Street Books.

For an extended discussion by an evolutionary biologist who feels there *is* a conflict see:

Dawkins, R. 2006. *The God Delusion*. Boston: Houghton Mifflin.

21. Based largely on the strength of the evidence that Darwin compiled, there has been little, if any, scientific debate about the fact of evolution since the 1870s. Biologists continue to debate the evidence for evolution with nonscientists, however. To participate in the discussion, explore *http://www.talkorigins.org*.

Citations

Abouheif, E. 1997. Developmental genetics and homology: A hierarchical approach. *Trends in Ecology and Evolution* 12: 405–408.

Alonso, P. D., A. C. Milner., R. A. Ketcham, et al. 2004. The avian nature of the brain and inner ear of *Archaeopteryx*. *Nature* 430: 666-669.

Alters, B. J., and S. M. Alters. 2001. *Defending Evolution: A Guide to the Creation/Evolution Controversy*. Sudburry, MA: Jones and Bartlett.

Beddall, B. G. 1957. Historical notes on avian classification. *Systematic Zoology* 6: 129-136.

Bell, M. A., W. E. Aguirre, and N. J. Buck. 2004. Twelve years of contemporary armor evolution in a threespine stickleback population. *Evolution* 58: 814-824.

Bell, M. A., and S. A. Foster, eds. 1994. *The Evolutionary Biology of the Threespine Stickleback*. Oxford: Oxford University Press.

Bishop, G. 1998. The religious worldview and American beliefs about human origins. *The Public Perspective* 9: 39–44.

Brasier, M. D., O. R. Green, A. P. Jephcoat, et al. 2002. Questioning the evidence for Earth's oldest fossils. *Nature* 416: 76–81.

Brice, W. R. 1982. Bishop Ussher, John Lightfoot and the age of creation. *Journal of Geological Education* 30: 18-26.

Bonner, Y., H. Bondi, T. Nasrin, et al. 1997. Why I am a secular humanist. *Free Inquiry* 18: 18–23.

Bowler, P. J. 2002. Evolution: history. In *Encyclopedia of Life Sciences*. Macmillan Publishers Ltd. Nature Publishing Group: *http://www.els.net*.

Brocks, J. J., G. A. Logan, R. Buick, R. E. Summons. 1999. Archean molecular fossils and the early rise of eukaryotes. *Science* 285: 1033–1036.

Burke, A. C., and A. Feduccia. 1997. Developmental patterns and the identification of homologies in the avian hand. *Science* 278: 666–668.

Carroll, S. P., and C. Boyd. 1992. Host race radiation in the soapberry bug: Natural history with the history. *Evolution* 46: 1052–1069.

Carroll, S. P., H. Dingle, T. R. Famula, and C. W. Fox. 2001. Genetic architecture of adaptive differentiation in evolving host races of the soapberry bug, *Jadera haematoloma*. *Genetica* 112–113: 257–272.

Carroll, S. P., H. Dingle, and S. P. Klassen. 1997. Genetic differentiation of fitness-associated traits among rapidly evolving populations of the soapberry bug. *Evolution* 51: 1182–1188.

Carroll, S. P., J. E. Loye, et al. 2005. And the beak shall inherit—evolution in response to invasion. *Ecology Letters* 8: 944-951.

Chen, P.-J., Z. M. Dong, and S. N. Zhen. 1998. An exceptionally well-preserved theropod dinosaur from the Yixian Formation of China. *Nature* 391: 147–152.

Chou, H.-H., T. Hayakawa, S. Diaz, et al. 2002. Inactivation of CMP-N-acetylneuraminic acid hydroxylase occurred prior to brain expasion during human evolution. *Proceedings of the National Academy of Science USA* 99: 11736-11741.

Chou, H.-H., H. Takematsu, S. Diaz, et al. 1998. A mutation in human CMP-sialic acid hydroxylase occurred after the *Homo-Pan* divergence. *Proceedings of the National Academy of Sciences USA* 95: 11751-11756.

Cresko, W. A., A. Amores, C. Wilson, et al. 2004. Parallele genetic basis for repeated evolution of armor loss in Alaskan threespine stickleback populations. *Proceedings of the National Academy of Sciences USA* 101: 6050-6055.

Christiansen, P., and N. Bonde. 2004. Body plumage in *Archaeopteryx*: a review and new evidence from the Berlin specimen. *Comptes Rendus Palevol* 3: 99-118.

Currie, P.J., and P.J. Chen. 2001. Anatomy of *Sinosauropteryx prima* from Liaoning, northeastern China. *Canadian Journal of Earth Sciences* 38: 1705–1727.

Dalton, R. 2002. Squaring up over ancient life. *Nature* 417: 782–784.

Damon, P. E. 1968. Potassium-argon dating of igneous and metamorphic rocks with applications to the basin ranges of Arizona and Sonora. In Hamilton, E. I., and R. M. Farquhar, eds. *Radiometric Dating for Geologists*. London: Interscience Publishers.

Darwin, C. 1859. *On the Origin of Species by Means of Natural Selection, First Edition*. London: John Murray.

Darwin, C. 1862. *The Various Contrivances by Which Orchids are Fertilized by Insects*. London: John Murray.

Darwin, C. 1872. *On the Origin of Species by Means of Natural Selection*, 6th ed.

Desmond, A. and J. Moore. 1991. *Darwin*. New York: W.W. Norton.

Dugan, K. G. 1980. Darwin and *Diprotodon*: The Wellington Caves fossils and the law of succession. *Proceedings of the Linnean Society of New South Wales* 104: 265–272.

Eiseley, L. 1946. *The Immense Journey*. New York: Vintage Books.

Eiseley, L. 1958. *Darwin's Century*. Garden City, NY: Anchor Books.

Faure, G. 1986. *Principles of Isotope Geology*. New York: John Wiley & Sons.

Fedo, C. M., and M. J. Whitehouse. 2002. Metasomatic origin of quartz-pyroxene rock, Akilia, Greenland, and implications for Earth's earliest life. *Science* 296: 1448–1452.

Freeland, S. J., R. D. Knight, L. F. Landweber, and L. D. Hurst. 2000. Early fixation of an optimal genetic code. *Molecular Biology and Evolution* 17: 511–518.

Friedberg, F., and A. R. Rhoads. 2000. Calculation and verification of the ages of retroprocessed pseudogenes. *Molecular Phylogenetics and Evolution* 16: 127–130.

Garner, J. P., G. K. Taylor, and A. L. R. Thomas. 1999. On the origins of birds: The sequence of character acquisition in the evolution of avian flight. *Proceedings of the Royal Society of London* B 266: 1259–1266.

Gauthier, J. A. 1986. Saurischian monophyly and the origin of birds. *Memoires of the California Academy of Sciences* 8: 1–55.

Geist, N. R., and A. Feduccia. 2000. Gravity-defying behaviors: Identifying models for protaves. *American Zoologist* 40: 664–675.

Gingerich, P. D., B. H. Smith, and E. L. Simons. 1990. Hind limbs of Eocene Basilosaurus: Evidence of feet in whales. *Science* 249: 154–156.

Glazko, G.V., E.V. Koonin, and I. B. Rogozin. 2005. Molecular dating: ape bones agree with chicken entrails. *Trends in Genetics* 21: 89-92.

Golubic, S., and H. J. Hofmann. 1976. Comparison of Holocene and mid-Precambrian Entophysalidaceae (Cyanophyta) in stromatolitic algal mats—cell-division and degradation. *Journal of Paleontology* 50: 1074–1082.

Gould, S. J. 1977. *Ever Since Darwin: Reflections in Natural History*. New York: Norton.

Gould, S. J. 1982. Introduction to reprinted edition of Dobzhansky, T. G. 1937. *Genetics and the Origin of Species*. New York: Columbia University Press.

Graur, D., and W. Martin. 2004. Reading the entrails of chickens: molecular timescales of evolution and the illusion of precision. *Trends in Genetics* 20: 80-86.

Han, T.-M., and B. Runnegar. 1992. Megascopic eukaryotic algae from the 2.1-billion-year-old Negaunee Iron-Formation, Michigan. *Science* 257:232–235.

Hedges, S. B. 2002. The origin and evolution of model organisms. *Nature Reviews Genetics* 3: 838-849.

Hedges, S. B., and S. Kumar. 2003. Genomic clocks and evolutionary timescales. *Trends in Genetics* 19: 200-206.

Hedges, S. B., and S. Kumar. 2004. Precision of molecular time estimates. *Trends in Genetics* 20: 242-247.

Huxley, T. H. 1868. On the animals which are most nearly intermediate between birds and reptiles. *Geological Magazine* 5: 357-365.

Irwin, D. E., S. Bensch, and T. D. Price. 2001a. Speciation in a ring. *Nature* 409: 333-337.

Irwin, D. E., J. H. Irwin, and T. D. Price. 2001b. Ring species as bridges between microevolution and speciation. *Genetica* 112-113: 223-243.

Irwin, D. E., S. Bensch, et al. 2005. Speciation by distance in a ring species. *Science* 307: 414-416.

Ji., Q., P. J. Currie, et al. 1998. Two feathered dinosaurs from northeastern China. *Nature* 393: 753-761.

Ji, Q., M. A. Norell, K. Q. Gao, et al. 2001. The distribution of integumentary structures in a feathered dinosaur. *Nature* 410: 1084–1088.

Johnson, W. E., and S. J. O'Brien. 1977. Phylogenetic reconstruction of the Felidae using 16S rRNA and NADH-5 mitochondrial genes. *Journal of Molecular Evolution* 44: S98–S116.

Johnson, W. E., E. Eizirik, et al. 2006. The late Miocene radiation of modern Felidae: A genetic assessment. *Science* 311: 73-77.

Judson, O. P., and D. Haydon. 1999. The genetic code: What is it good for? An analysis of the effects of selection pressures on genetic codes. *Journal of Molecular Evolution* 49: 539–550.

Keller, M. P., B. A. Seifried, and P. F. Chance. 1999. Molecular evolution of the CMT1A-REP region: A human- and chimpanzee-specific repeat. *Molecular Biology and Evolution* 16: 1019–1026.

Kerr, R. A. 2002. Reversals reveal pitfalls in spotting ancient and E.T. life. *Science* 296: 1384–1385.

Knoll, A. H., and E. S. Barghoorn. 1977. Archean microfossils showing cell division from the Swaziland System of South Africa. *Science* 198: 396–398.

Larson, E. J., and L. Witham. 1997. Scientists are still keeping the faith. *Nature* 386: 435–436.

Larson, E. J., and L. Witham. 1998. Leading scientists still reject God. *Nature* 394: 313.

Larson, E. J., and L. Witham. 1999. Scientists and religion in America. *Scientific American* 281(September): 88–93.

Lopes, J., N. Ravisé, A. Vandenberghe, et al. 1998. Fine mapping of de novo CMT1A and HNPP rearrangements within CMT1A-REPs evidences two distinct sex-dependent mechanisms and candidate sequences involved in recombination. *Human Molecular Genetics* 7: 141–148

Luning Prak, E. T., and H. H. Kazazian, Jr. 2000. Mobile elements and the human genome. *Nature Reviews Genetics* 1: 135–144.

Makovicky, P. J., S. Apestuguía, and F. L. Angolín. 2005. The earliest dromaeosaurid theropod from South America. *Nature* 437: 1007-1011.

Martin, M. J., J. C. Rayner, P. Gagneux, et al. 2005. Evolution of human-chimpanzee differences in malaria susceptibility: Relationship to human genetic loss of N-glycolylneuraminic acid. *Proceedings of the National Academy of Sciences USA* 102: 12819-12824.

Mattern, M.Y., and D. A. McLennan. 2000. Phylogeny and speciation of felids. *Cladistics* 16: 232–253.

Mayr, E. 1964. *Introduction to: Darwin, C., 1859. On the Origin of Species: A Facsimile of the First Edition*. Cambridge, MA: Harvard University Press.

Mayr, E., and W. B. Provine, eds. 1980. *The Evolutionary Synthesis: Perspectives on the Unification of Biology*. Cambridge, MA: Harvard University Press.

Miller, Kenneth R. 1999. *Finding Darwin's God*. New York: HarperCollins. See Also: Miller, K.R., 1999. *Finding Darwin's God. Brown Alumni Magazine* 100 (Nov.–Dec.). (Available online at *http://www.brown.edu/Administration/Brown_Alumni_Magazine/00/11-99/index.html)*

Mindell, D. P., and A. Meyer. 2001. Homology evolving. *Trends in Ecology and Evolution* 16: 434–440.

Mojzsis, S. J., G. Arrhenius, K. D. McKeegan, T. M. Harrison, A. P. Nutman, and C. R. L. Friend. 1996. Evidence for life on Earth before 3,800 million years ago. *Nature* 384: 55–59.

National Center for Science Education. 2000. Voices for evolution. Available online at *http://www.ncseweb.org/article.asp?category=2*

Norell, M., Q. Ji, K. Gao, et al. 2002. 'Modern' feathers on a non-avian dinosaur. *Nature* 416: 36–37.

Norell, M. A., and Xing Xu. 2005. Feathered dinosaurs. *Annual Review of Earth and Planetary Sciences* 33: 277-299.

Padian, K., and L. M. Chiappe. 1998. The origin and early evolution of birds. *Biological Reviews* 73: 1-42.

Pennock, R. T. 1996. Naturalism, evidence, and creationism: The case of Phillip Johnson. *Biology and Philosophy* 11: 543–549. (Reprinted in Pennock 2001.)

Pennock, R. T., ed. 2001. *Intelligent Design Creationism and its Critics*. Cambridge, MA: The MIT Press.

Perkins, S. 2005. Raptor line: Fossil finds push back dinosaur ancestry. *Science News* 168: 243,

PEW Research Center for the People & The Press. 2005. Religion a strength and weakness for both parties: Public divided on origins of life. Survey report released 30 August. http://people-press.org/reports/pdf/254.pdf

Prum, R. O. 2002. Why ornithologists should care about the theropod origins of birds. *The Auk* 119: 1-17.

Prum, R. O., and A. H. Brush. 2002. The evolutionary origin and diversification of feathers. *Quarterly Review of Biology* 77: 261–295.

Ray, J. 1686. *Historia Plantarum*. London. (Vol. 1, p. 40; translated by E. T. Silk, as reproduced in Beddall 1957.)

Reisz, R. R., and J. Müller. 2004. Molecular timescales and the fossil record: a paleontological perspective. *Trends in Genetics* 20: 237-241.

Reisz, R. R., and J. Müller. 2004. The comparative method for evaluating fossil calibration dates: a reply to Hedges and Kumar. *Trends in Genetics* 20: 596-597.

Reiter, L. T., T. Murakami, T. Koeuth, et al. 1997. The human COX10 gene is disrupted during homologous recombination between the 24 kb proximal and distal CMT1A-REPs. *Human Molecular Genetics* 6: 1595–1603.

Richardson, M. K., J. Hanken, L. Selwood, et al. 1998. Haeckel, embryos, and evolution. *Science* 280: 983.

Rieppel, O., H. Zaher, E. Tchernov, and M. J. Polcyn. 2003. The anatomy and relationships of *Haasiophis terrasanctus*, a fossil snake with well-developed hind limbs from the mid-cretaceous of the Middle East. *Journal of Paleontology* 77: 536-558.

Rosing, M. T. 1999. Depleted carbon microparticles in >3700-Ma seafloor sedimentary rocks from West Greenland. *Science* 283: 674–676.

Schopf, J. W. 1993. Microfossils of the early Archean Apex Chert: New evidence of the antiquity of life. *Science* 260: 640–646.

Schopf, J. W., A. B. Kudryavtsev, D. G. Agresti, et al. 2002. Laser-Raman imagery of Earth's earliest fossils. *Nature* 416: 73–76.

Sues, H.-D. 2001. Ruffling feathers. *Nature* 410: 1036–1037.

Tchernov, E., O. Rieppel, H. Zaher, et al. A fossil snake with limbs. *Science* 287: 2010-2012.

Thewissen, J. G. M., S. T. Hussain, and M. Arif. 1994. Fossil evidence for the origin of aquatic locomotion in archaeocete whales. *Science* 263: 210–212.

Unwin, D. M. 1998. Feathers, filaments, and theropod dinosaurs. *Nature* 391: 119–120.

Ussher, James. 1650. *Annales Veteris Testamenti*. London: J. Flesher and L. Sadler. (Pubished in English as Ussher 1658).

Ussher, James. 1658. *The Annals of the World*. London: Printed by E. Tyler for J. Crook and G. Bedell.

van Zullen, M. A., A. Lepland, and G. Arrhenius. 2002. Reassessing the evidence for the earliest traces of life. *Nature* 418: 627–630.

Wake, D. B. 2001. Speciation in the round. *Nature* 409: 299-300.

Wellnhofer, P. 1988. A new specimen of *Archaeopteryx*. *Science* 240: 1790-1792.

Werdelin, L., and L. Olsson. 1997. How the leopard got its spots: a phylogenetics view of the evolution of felid coat patterns. *Biological Journal of the Linnean Society* 62: 383–400.

Xu, X., Z. Zhou, and R. O. Prum. 2001. Branched integumental structures in *Sinornithosaurus* and the origin of feathers. *Nature* 410: 200-204

Xu, X., Z. Zhou, X. Wang, et al. 2003. Four-winged dinosaurs from China. *Nature* 421: 335-340.

3

Darwinian Natural Selection

Survival of the fattest. The photo shows a pack of African wild dogs bringing down an impala. Data gathered by Alistair Pole and colleagues (2003) show that wild dogs prey on the skinniest, weakest impala. The graph shows, for different times of year, the mean amount of stored fat in the bone marrow of impala taken by wild dogs versus a random sample of impala.

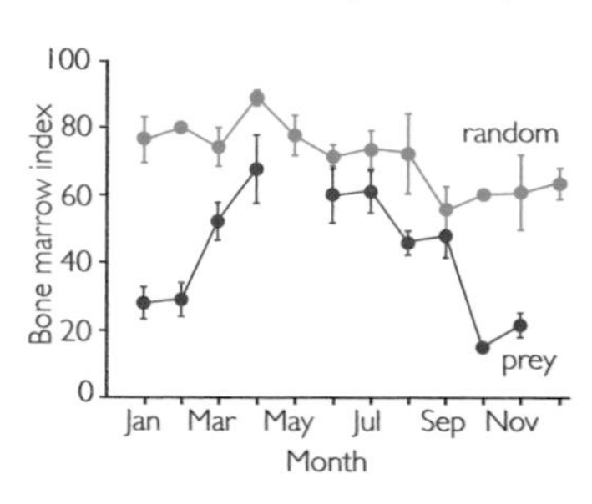

"It is quite conceiveable," Darwin wrote in his introduction to *On the Origin of Species* (1859, p. 3) "that a naturalist, reflecting on the mutual affinities of organic beings, on their embryological relations, their geographical distribution, geological succession, and other such facts, might come to the conclusion that each species had not been independently created, but had descended . . . from other species."

This assertion concerns the pattern of life's history. A growing body of evidence, amassed by Darwin and the early evolutionists who were his intellectual forebears, indicated that both fossilized and living organisms were derived with modification from either a single common ancestor or a few. The evidence was indirect and the interpretation startling, but Darwin's argument was so compelling that scientific debate over descent with modification virtually ended by the mid-1870s. Evolution was, and is, an established fact.

"Nevertheless," he continued, "such a conclusion, even if well founded, would be unsatisfactory, until it could be shown how the innumerable species inhabiting this world have been modified. . . . "

Darwin knew as well as anyone that the mere recognition of a pattern does not amount to a complete scientific theory. If we are to claim any understanding of life's history, we must explain not only what happened, but how. What is the mechanism that produces the pattern we call evolution? Chapter 2 focused on the evidence for descent with modification; this chapter introduces the process, natural selection, that Darwin asserted produces the pattern.

3.1 Artificial Selection: Domestic Animals and Plants

To increase the frequency of desirable traits in their stocks, plant and animal breeders employ artificial selection.

To understand the mechanism of evolution in nature, Darwin studied the mechanism of evolution under domestication. That is, he studied the method plant and animal breeders use to modify their crops and livestock. Darwin's favorite domestic organism was the pigeon. Darwin became a pigeon breeder himself to learn the experts' techniques. To refine a particular breed of pigeon so that, for example, the birds' tail feathers fan more spectacularly, or their body feathers curl more elegantly, breeders employ artificial selection. They scrutinize their flocks and select the individuals with the most desirable traits. These birds the breeders mate with each other to produce the next generation. If the desirable traits are passed from parents to offspring, then the next generation, consisting of the progeny of only the selected birds, will show the desirable traits in a higher proportion than existed in last year's flock.

Our favorite domestic organism is the tomato. The domestic tomato, *Solanum lycopersicum,* occurs around the world, both in cultivation and as a weedy escapee. It is closely related to, and can interbreed with, several species of wild tomatoes, all found in western South America (Spooner et al. 2005). The domestic tomato was first cultivated by Native Americans before Europeans arrived in the New World (Tanksley 2004). It traveled back to Europe with the early explorers and spread around the globe from there (Albala 2002).

The power of artificial selection is evident in Figure 3.1. All species of wild tomato have small fruit like the currant tomato on the left, typically less than a centimeter across and weighing just a few grams (Frary et al. 2000). The ancestor of the domestic tomato probably had similarly tiny fruit. Modern varieties of domestic tomato, like the Red Giant on the right, have fruit 15 cm or more across that can weigh more than a kilogram. Descent with modification, indeed.

Figure 3.1 Wild and domestic tomatoes Wild tomatoes have tiny fruit, like that of the currant tomato on the left. Domestic tomatoes are descended from tiny-fruited ancestors, but as a result of artificial selection have large fruit, like that of the Red Giant on the right. From Frary et al. (2000).

Research by molecular biologists allows us to understand at least part of what happened during the domestication of tomatoes at the level of individual genes. Tomatoes carry, on chromosome 2, a gene called *fw2.2* (Tanksley 2004). The gene encodes a protein made during early fruit development (Frary et al. 2000). The protein's job is to repress cell division; the more of the protein a plant makes, the smaller its fruit (Liu et al. 2003). Changes in the nucleotide sequence in the *fw2.2* promoter—the gene's on-off switch—alter the timing of production and the total amount of protein made (Cong et al. 2002; Nesbitt and Tanksley 2002).

Every wild tomato ever tested has carried alleles of *fw2.2* associated with high production of the repressor protein and small fruit (Tanksley 2004). Every cultivated tomato has carried alleles associated with low production of the protein and large fruit. Anne Frary and collegues (2000), working in the laboratory of Steven Tanksley, used genetic engineering to place copies of a small-fruit allele into domestic tomatoes. The fruit on the left in Figure 3.2 is from an unmanipulated plant; the fruit on the right is from a sibling of the unmanipulated plant that has been genetically modified to carry the wild, small-fruit allele of *fw2.2*. The fruits differ in size by about 30%.

Figure 3.2 A genetically determined difference in fruit size These tomatoes are from sibling plants. The one on the left carries only domestic alleles of the *fw2.2* gene. The one on the right carries, in addition, copies of the wild allele. The *fw2.2* gene encodes a protein that represses fruit growth. From Frary et al. (2000).

Tanksley enivisions a scenario in which early tomato farmers noticed variation in fruit size among their plants (Nesbitt and Tanksley 2002; Tanksley 2004). Some of this variation was due to the plants' possession of different alleles of the *fw2.2* gene. Large fruit alleles might have been present as rare variants prior to domestication, or they might have arisen as new mutations in cultivated populations. Because the farmers preferred larger tomatoes, year after year they planted their fields with seeds from the largest fruit of the previous crop. By this discipline the farmers eventually eliminated small-fruited alleles from their stocks.

Farmers practicing artificial selection can change more than size. The domesticated vegetables shown in Figure 3.3—broccoli, brussels sprouts, cauliflower, kale, and kohlrabi—are strikingly different in achitecture. Yet all can readily interbreed, and are classified by botanists as varieties of wild cabbage, *Brassica oleracea,* from which they are derived.

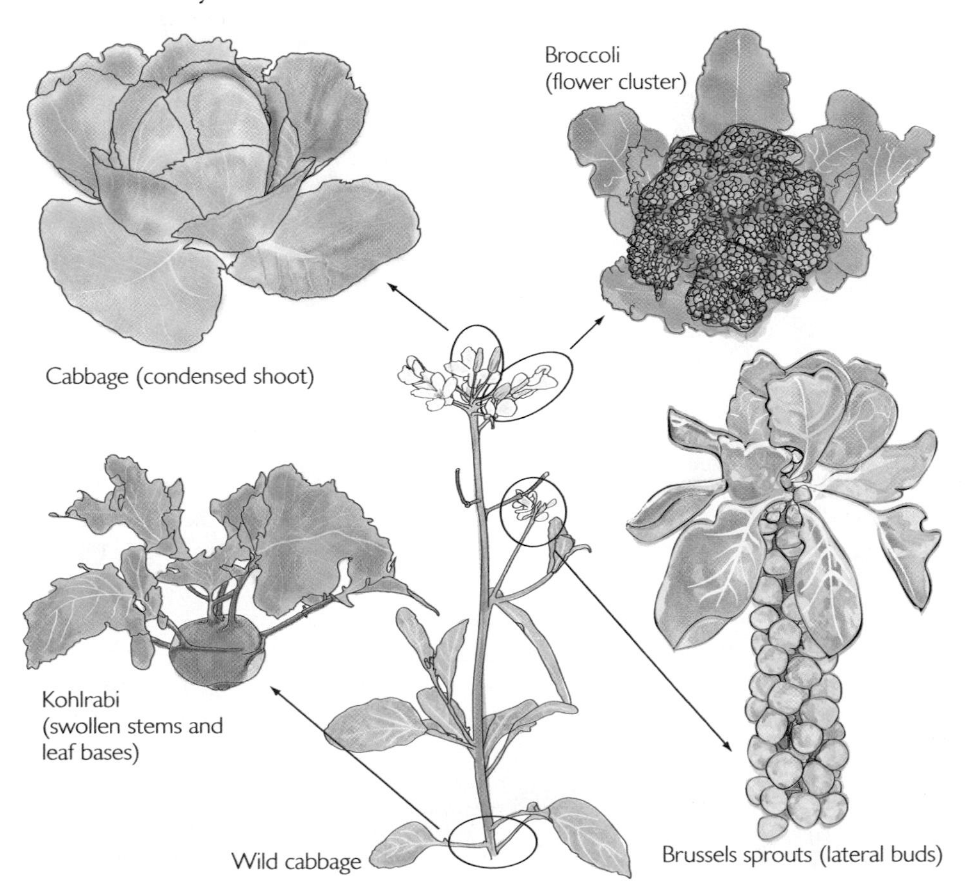

Figure 3.3 Wild and domestic varieties of *Brassica oleracea* Cauliflower (*Brassica oleracea botrytis*), broccoli (*Brassica oleracea italica*), brussels sprouts (*Brassica oleracea gemmifera*), kale (*Brassica oleracea acephala*), and kohlrabi (*Brassica oleracea gongylodes*) are all derived from wild cabbage (*Brassica oleracea oleracea*). After Niklaus (1997).

3.2 Evolution by Natural Selection

Darwin realized that a process much like artificial selection happens in nature. His Theory of Evolution by Natural Selection holds that descent with modification is the logical outcome of four postulates, which he laid out in his introduction to *On the Origin of Species by Means of Natural Selection*. Darwin (1859, p. 459) considered the rest of the book one long argument in their support. Darwin's postulates, claims about the nature of populations, are as follows:

Darwin and Wallace realized that a process similar to artificial selection happens automatically in nature.

1. Individuals within populations are variable.
2. The variations among individuals are, at least in part, passed from parents to offspring.
3. In every generation, some individuals are more successful at surviving and reproducing than others.
4. The survival and reproduction of individuals are not random; instead they are tied to the variation among individuals. The individuals with the most favorable variations, those who are better at surviving and reproducing, are naturally selected.

If these four postulates are true, then the composition of the population changes from one generation to the next. Figure 3.4 shows how Darwin's theory might play out in a population of chilies eaten by packrats.

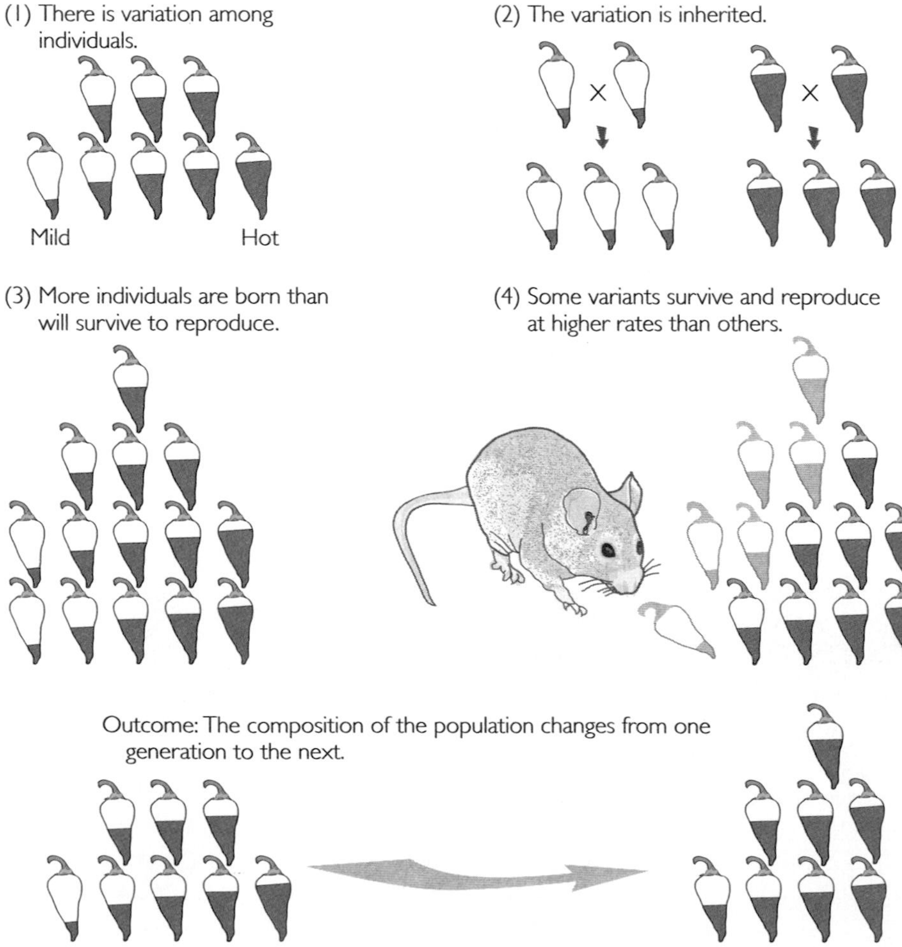

(1) There is variation among individuals.

Mild

Hot

(2) The variation is inherited.

(3) More individuals are born than will survive to reproduce.

(4) Some variants survive and reproduce at higher rates than others.

Outcome: The composition of the population changes from one generation to the next.

Figure 3.4 Darwin's Theory of Evolution by Natural Selection Darwin's theory consists of four claims about populations of organisms and a logical outcome that follows, as a matter of simple mathematics, if the four postulates are true. These cartoons show how the theory might work in a population of chili plants whose fruits are attacked by packrats. If the chilies vary in the spiciness of their fruit, and if packrats prefer milder chilies, and if the hot survivors pass their spiceness to their offspring, then the population will show a higher proportion of hot-fruited chilies each generation. Inspired by Tewksbury and Nabhan (2001).

The logic is straightforward: If there are differences among the individuals in a population that can be passed on to offspring, and if there is differential success among those individuals in surviving and/or reproducing, then some traits will be passed on more frequently than others. As a result, the characteristics of the population will change slightly with each succeeding generation. This is Darwinian evolution: gradual change in populations over time.

Natural selection is a process that produces descent with modification, or evolution.

Note that while the logic is straightforward it contains a subtlety that can cause confusion. To understand how natural selection works, we have to think statistically. The selection itself—the surviving and reproducing—happens to individuals, but what changes is populations. Recall the HIV virions discussed in Chapter 1. Because of differences in the amino acid sequences of the reverse transcriptase active site, individual virions within the same host varied in their ability to synthesize DNA in the presence of AZT. Virions with mutant forms of reverse transcriptase that were less likely to bind AZT reproduced more successfully. When they reproduced, they passed their reverse transcriptase mutations to their offspring. In the next generation, then, a higher percentage of virions carried the modified form of reverse transcriptase than in the generation before. This change in the population is evolution by natural selection.

Darwin referred to the individuals who are better at surviving and reproducing, and whose offspring make up a greater percentage of the population in the next generation, as more fit. In so doing he gave the everyday English words *fit* and *fitness* a new meaning. **Darwinian fitness** is the ability of an individual to survive and reproduce in its environment.

An important aspect of fitness is its relative nature. Fitness refers to how well an individual survives and how many offspring it produces compared to other individuals of its species. Biologists use the word **adaptation** to refer to a trait or characteristic of an organism, like a modified form of reverse transcriptase, that increases its fitness relative to individuals without the trait.

An adaptation is a characteristic that increases the fitness of an individual compared to individuals without the trait.

Darwin's mechanism of evolution was, incidentally, discovered independently by a colleague of Darwin's named Alfred Russel Wallace. Though trained in England, Wallace had been making his living in Malaysia by selling natural history specimens to private collectors. While recuperating from a bout with malaria in 1858, he wrote a manuscript explaining natural selection and sent it to Darwin. Darwin, who had written his first draft on the subject in 1842 but never published it, immediately realized that he and Wallace had formulated the same theory. Brief papers by Darwin and by Wallace were read together before the Linnean Society of London, and Darwin then rushed *On the Origin of Species* into publication (17 years after he had written the first draft). Today, Darwin's name is more prominently associated with the Theory of Evolution by Natural Selection for two reasons: He had clearly thought of it before Wallace, and his book provided a full exposition of the idea, along with massive documentation.

One of the most attractive aspects of the Darwin–Wallace theory is that each of the four postulates and their logical consequence can be verified independently. That is, the theory is testable. There are neither hidden assumptions nor anything that has to be accepted uncritically. In the next two sections, we examine each of the four assertions, and Darwin's predicted result, by reviewing two studies: a recent experiment on snapdragons and an ongoing study of finches in the Galápagos Islands off the coast of Ecuador. These studies show that the Theory of Evolution by Natural Selection be tested rigorously, by direct observation.

The Theory of Evolution by Natural Selection is testable.

3.3 The Evolution of Flower Color in an Experimental Snapdragon Population

Kristina Niovi Jones and Jennifer Reithel (2001) wanted to know whether natural selection by bumblebees could influence the evolution of a floral trait controlled by alleles of a single gene. To find out, they established an experimental population of 48 snapdragons in which they made sure that Darwin's postulates 1 and 2 were true. Then they monitored the plants and their offspring to see whether postulates 3 and 4, and the predicted outcome, were true as well.

Postulate 1: There Is Variation among Individuals

The snapdragons in Jones and Reithel's population varied in flower color. Three-quarters of the plants had flowers that were almost pure white, with just two spots of yellow on the lower lip. The rest had flowers that were yellow all over.

Postulate 2: Some of the Variation Is Heritable

The variation in color among Jones and Reithel's plants was due to differences in the plants' genotypes for a single gene. The gene has two alleles, which we will call *S* and *s*. Individuals with either genotype *SS* or *Ss* have white flowers with just two spots of yellow. Individuals with genotype *ss* are yellow all over. Among the 48 plants in the experimental population, 12 were *SS*, 24 were *Ss*, and 12 were *ss*. Figure 3.5a shows the variation in phenotype among Jones and Reithel's snapdragons, and the variation in genotype responsible for it.

Testing Postulate 3: Do Individuals Vary in Their Success at Surviving or Reproducing?

Although Jones and Reithel ran their experiment in a meadow in Colorado, they kept their snapdragons in pots and made sure all of the plants survived.

When researchers set up a plant population in which postulates 1 and 2 were true, they found that postulate 3 was true as well . . .

The researchers did not intervene, however, to help the snapdragons reproduce. Instead, they let free-living bumblebees pollinate the plants. To gauge the plants' success at reproducing by exporting pollen, Jones and Reithel tracked the number of times bees visited each flower. To gauge the plants' success at reproducing by making seeds, the researchers counted the seeds produced from each fruit. Consistent with Darwin's third postulate, the plants showed considerable variation in reproductive success, both as pollen donors and as seed mothers.

Testing Postulate 4: Is Reproduction Nonrandom?

Jones and Reithel expected that one color would attract more bees than the other, but they did not know which color it would be. The yellow spots on otherwise white snapdragons are thought to serve as nectar guides, helping bumblebees find the reward the flower offers. All-yellow flowers lack nectar guides and so might be less attractive to bees, or they might be more visible against the background vegetation and thus *more* attractive. Jones and Reithel found that white flowers attracted twice as many bee visits as yellow flowers (Figure 3.5b, left).

. . . as were postulate 4 . . .

Reproductive success through seed production was less strongly associated with color than was success through pollen donation. Nonetheless, the white plants were somewhat more robust than the yellow plants and so produced, on average, slightly more seeds per fruit (Figure 3.5b, right).

(a) Composition of parental population

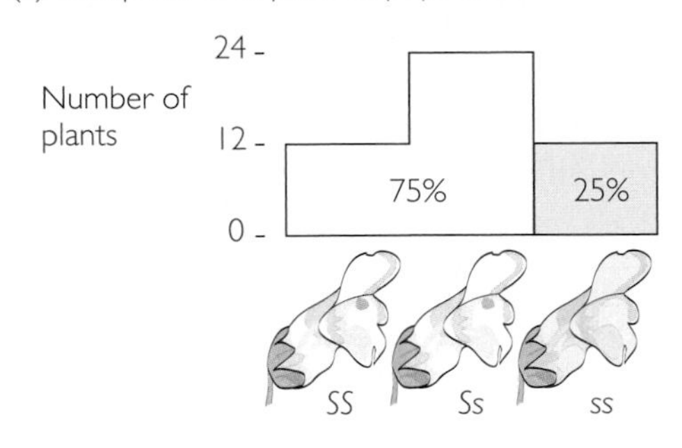

(b) Differences in reproductive success through male function (left) and female function (right)

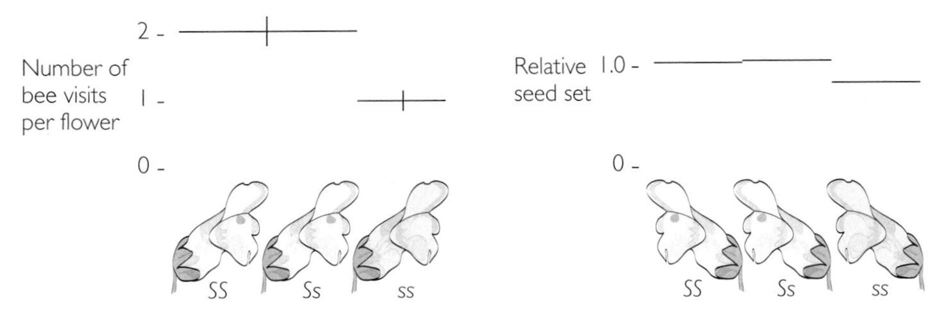

(c) Composition of offspring population

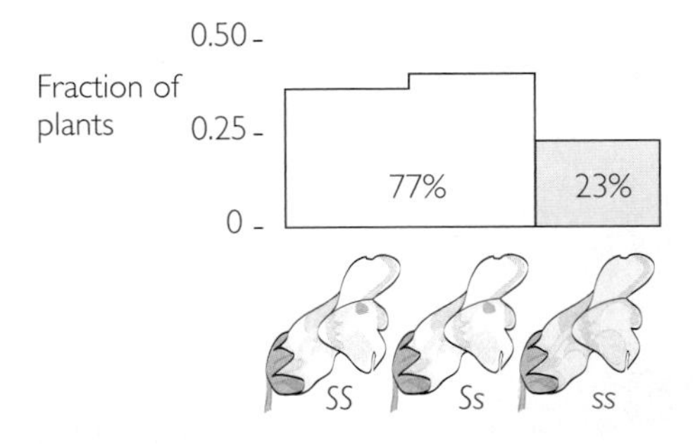

Figure 3.5 Darwin's Theory of Evolution by Natural Selection demonstrated in an experimental population of snapdragons (a) The plants in the parental population vary in flower color. This variation in phenotype is due to variation in genotype. The graph shows the number of plants in the population with each of the three possible genotypes. (b) The white plants are more successful at reproducing. They are visited by bumblebees twice as often (left), and make more seeds (right). (c) Because plants with white flowers are more successful at passing on their genes, they occupy a larger fraction of the population in the next generation. Prepared from data in Jones and Reithel 2001. [In (b) left, the vertical bars show the size of the standard error; they indicate the accuracy of the researchers' estimate of the mean number of bee visits. In (b) right, the values for relative seed set were calculated as the fraction of seeds actually produced by plants with a particular genotype divided by the fraction of seeds expected based on the frequencies of the genotypes.]

Consistent with Darwin's fourth postulate, reproductive success was not random. Through both pollen donation and seed production, white plants had higher reproductive success than yellow plants.

Testing Darwin's Prediction: Did the Population Evolve?

The bumblebees that volunteered to participate in Jones and Reithel's experiment played the same role that Darwin did in breeding pigeons: They selected particular individuals and granted them high reproductive success. Since white plants had higher reproductive success than yellow, and since flower color is determined by genes, the next generation of snapdragons should have had a higher proportion of white flowers.

. . . and Darwin's prediction that the population would evolve as a result.

Indeed, the next generation did have a higher proportion of white flowers (Figure 3.5c). Among the plants in the starting population, 75% had white flowers; among their offspring, 77% had white flowers. The snapdragon population evolved as predicted. An increase of two percentage points in the proportion of white flowers might not seem like much. But modest changes can accumulate over many generations. With Jones and Reithel's population evolving at this rate, it would not take many years for white flowers to all but take over.

Jones and Reithel's experiment shows that Darwin's theory works, at least in experimental populations when researchers have made certain that Darwin's first two postulates hold. But does the theory work in completely natural populations, in which researchers have manipulated nothing? To find out, we turn to research on Darwin's finches in the Galápagos Islands.

3.4 The Evolution of Beak Shape in Galápagos Finches

Peter Grant and Rosemary Grant and their colleagues have been studying finches in the Galápagos Archipelago since 1973 (see P. R. Grant 1999; B. R. Grant and P. R. Grant 1989, 2003; P. R. Grant and B. R. Grant 2002a, 2002b, 2005. 2006; B. R. Grant 2003). Collectively called Darwin's finches, the birds are derived from a small flock of grassquits that invaded the archipelago from Central or South America some 2.3 million years ago (Sato et al. 2001). The descendents of this flock today comprise 13 species that live in the Galápagos, plus a 14th that lives on Cocos Island. Close examination of the evolutionary tree in Figure 3.6 reveals that all of these species are closely related. The deepest split on the tree separates two lineages of warbler finches that still recognize each other as potential mates and are thus classified (despite each having its own name) as belonging

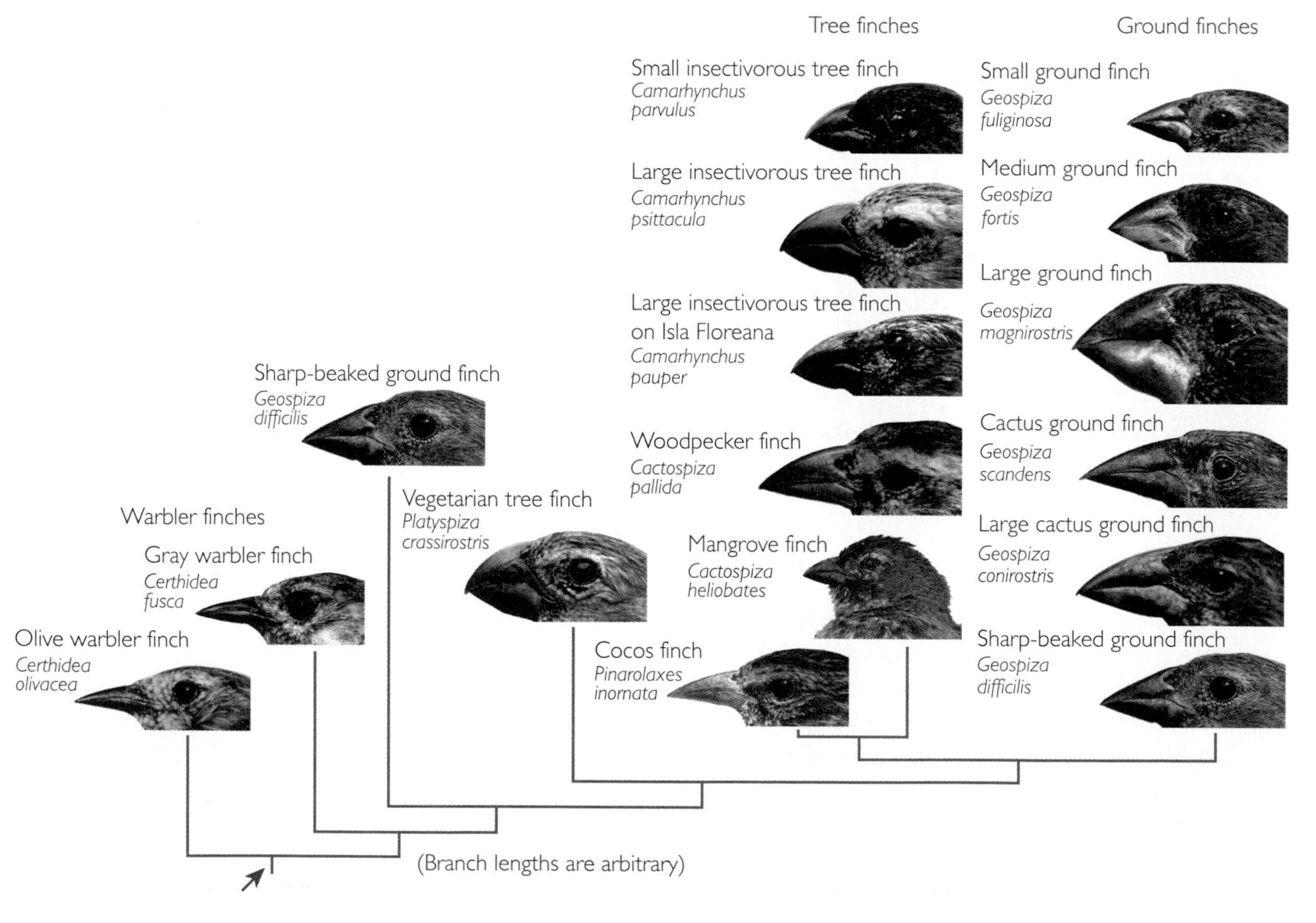

Figure 3.6 **Diversity in Darwin's finches** These finches are all descended from a common ancestral population (red arrow) that traveled from Central or South America to the Galápagos Archipelago. The evolutionary tree, estimated from similarities and differences in DNA sequences by Kenneth Petren and colleagues (2005), shows the sometimes complex relationships among the major groups. The photos, from Petren et al. (1999) and Grant and Grant (1997), show the extensive variation among species in beak size and shape.

to a single species. The next deepest split separates two lineages of sharp-beaked ground finches that are likewise considered a single species. Consistent with their close kinship, all species of Darwin's finches are similar in size and coloration. They range from 4 to 6 inches in length and from brown to black in color. They do, however, show remarkable variation the size and shape of their beaks.

The beak is the primary tool used by birds in feeding, and the enormous range of beak morphologies among the Galápagos finches reflects the diversity of foods they eat. The warbler finches (*Certhidea olivacea* and *Certhidea fusca*) feed on insects, spiders, and nectar; woodpecker and mangrove finches (*C. pallida* and *C. heliobates*) use twigs or cactus spines as tools to pry insect larvae or termites from dead wood; several ground finches in the genus *Geospiza* pluck ticks from iguanas and tortoises in addition to eating seeds; the vegetarian finch (*Platyspiza crassirostris*) eats leaves and fruit.

For a test of the Theory of Evolution by Natural Selection, we focus on data Grant and Grant and colleagues have gathered on the medium ground finch, *Geospiza fortis,* on Isla Daphne Major (Figure 3.7).

Daphne Major's size and location make it a superb natural laboratory. Like all of the islands in the Galápagos, it is the top of a volcano (Figure 3.8). The island is tiny. It rises from the sea to a maximum elevation of just 120 meters. It has one

Figure 3.7 **The medium ground finch, *Geospiza fortis*** (top) An adult male; (bottom) an adult female.

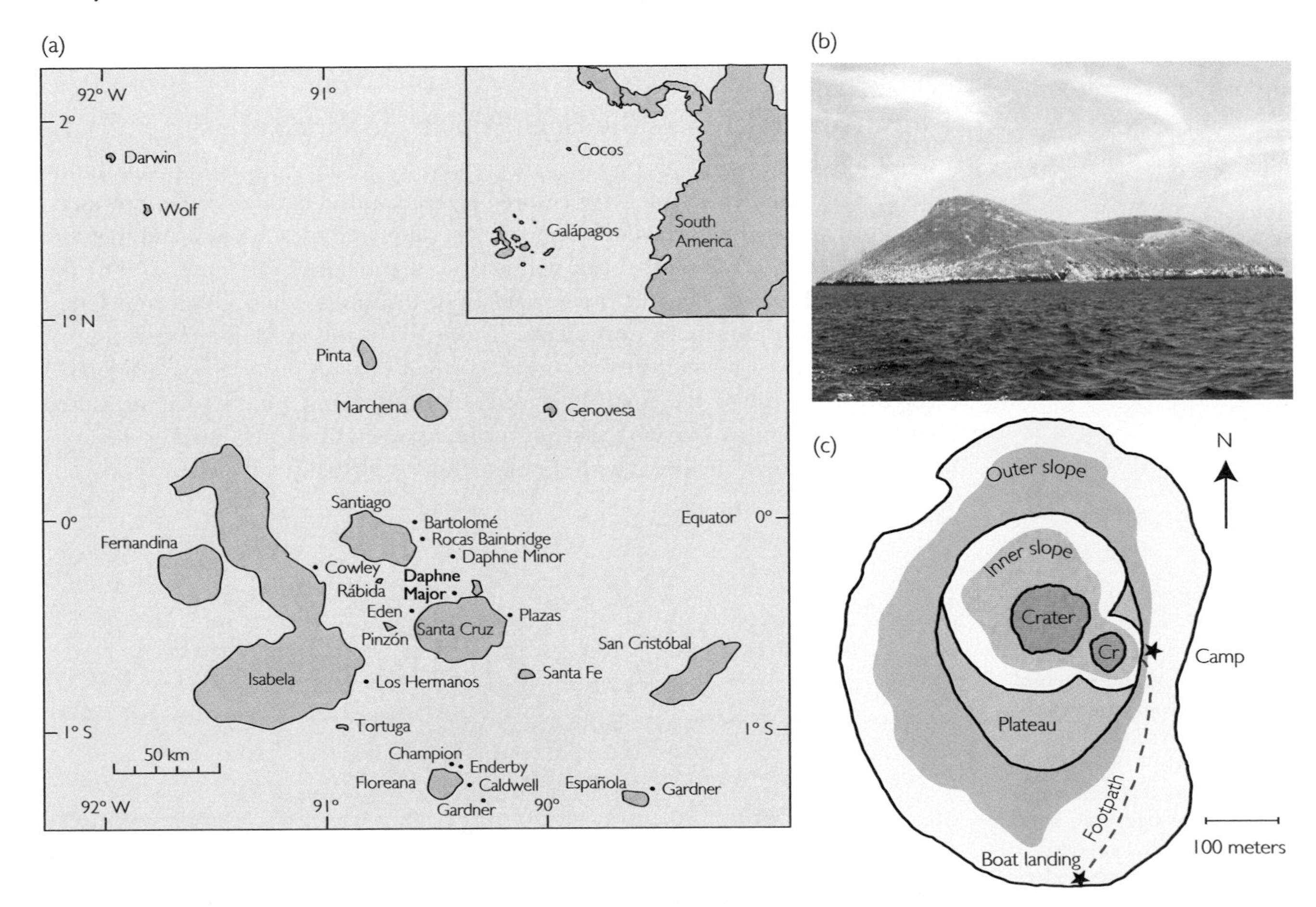

Figure 3.8 **The Galápagos Archipelago and Isla Daphne Major** (a) Cocos Island and the Galapágos Archipelago, home of Darwin's finches. Isla Daphne Major is a tiny speck between Santa Cruz and Santiago. (b) Isla Daphne Major, seen from a boat approaching the island. Visible as a faint white line running upward from left to right is the footpath that runs from the boat landing (at the waterline) to the campsite (on the rim of the crater). (c) A map of Daphne Major. Note the island's tiny size. Rerendered from Boag and Grant (1984a).

main crater, with a small secondary crater adjacent to it. There is only one spot on the island that is both flat enough and large enough to pitch a camp. It takes just 20 minutes to walk from the campsite all the way around the main crater's rim and back to camp. The climate is seasonal even though the location is equatorial. A warmer, wetter season from January through May alternates with a cooler, drier season from June through December. The vegetation consists of dry forest and scrub, with several species of cactus.

The medium ground finches on Daphne Major make an ideal study population. Few finches migrate onto or off of the island, and the population is small enough to be studied exhaustively. In an average year, there are about 1,200 individual finches on the island. By 1977, Grant and Grant's team had captured and marked more than half of them; since 1980, virtually 100% of the population has been marked. Medium ground finches live up to 16 years (Grant and Grant 2000). Their generation time is 4.5 years (Grant and Grant 2002).

Medium ground finches are primarily seed eaters. The birds crack seeds by grasping them at the base of the bill and then applying force. Grant and Grant and their colleagues have shown that both within and across finch species, beak size is correlated with the size of seeds harvested. In general, birds with bigger beaks eat larger seeds, and birds with smaller beaks eat smaller seeds. This is because birds with different beak sizes are able to handle different sizes of seeds more efficiently (Bowman 1961; Grant et al. 1976; Abbott et al. 1977; Grant 1981b).

Testing Postulate 1: Is the Finch Population Variable?

Some Geospiza fortis have beaks that are only half as deep as other individuals.

The researchers mark every finch they catch by placing on its legs one numbered aluminum band and three colored plastic bands. This allows them to identify individual birds in the field. The scientists also weigh each finch and measure its wing length, tail length, beak width, beak depth, and beak length. All of the traits they have investigated are variable. For example, when Grant and Grant plotted measurements of beak depth in the Isla Daphne Major population of medium ground finches, the data indicated that beak depth varies considerably (Figure 3.9). All of the finch characteristics Grant and Grant have measured clearly conform to Darwin's first postulate. As we will see in Chapter 4, variation among the individuals within populations is virtually universal.

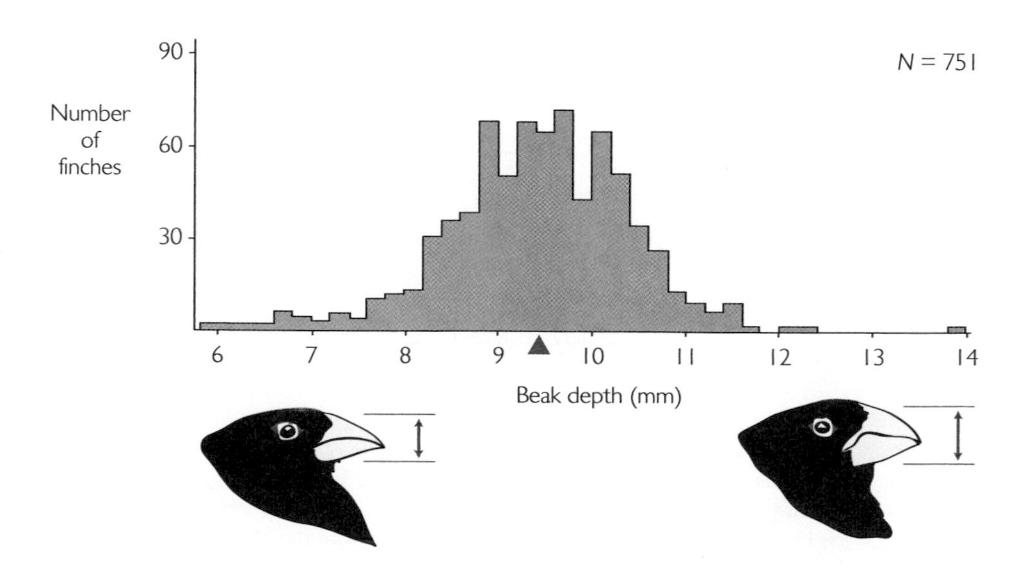

Figure 3.9 Variation in beak depth in medium ground finches This histogram shows the distribution of beak depth in medium ground finches on Daphne Major in 1976. A few birds have shallow beaks; a few birds have deep beaks; most birds have medium beaks. (*N* stands for sample size; the blue arrow along the *x*-axis indicates the mean, or average.) Rerendered from Boag and Grant (1984b).

Testing Postulate 2: Is Some of the Variation among Individuals Heritable?

Within the Daphne Major population, individual finches could vary in beak depth because the environments they have experienced are different or because their genotypes are different, or both. There are several ways that environmental variation could cause the variation in beak depth documented in Figure 3.9. Variation in the amount of food that individual birds happened to have received as chicks can lead to variation in beak depth among adults. Injuries or abrasion against hard seeds or rocks can also affect beak size and shape.

To determine whether at least part of the variation among finch beaks is genetically based, and thus passed from parents to offspring, Peter Boag, a colleague of Peter Grant and Rosemary Grant, estimated the **heritability** of beak depth.

The heritability of a trait is defined as the proportion of the variation observed in a population that is due to variation in genes. Because it is a proportion, heritability varies between 0 and 1. We will develop the theory behind how heritability is estimated more fully in Chapter 9. For now, we point out that if the differences among individuals are due to differences in the alleles they have inherited, then offspring will resemble their parents.

Boag compared the average beak depth of families of *G. fortis* young after they had attained adult size to the average beak depth of their mother and father. Boag's data reveal a strong correspondence between relatives. As the plot in Figure 3.10 shows, parents with shallow beaks tend to have chicks with shallow beaks, and parents with deep beaks tend to have chicks with deep beaks. This is evidence that a large proportion of the observed variation in beak depth is genetically based and can be passed to offspring (Boag and Grant 1978; Boag 1983).

In finches, the beak depths of parents and offspring are similar. This observation suggests that some alleles tend to produce shallow beaks, while other alleles tend to produce deeper beaks.

Boag himself would be the first to say that caution is warranted in interpreting his data. Environments shared by family members, maternal effects, conspecific nest parasitism, and misidentified paternity can cause graphs like the one in Figure 3.10 to exaggerate, or to underplay, the heritability of traits (see Box 3.1). However, Lukas Keller and colleagues (2001) have used modern genetic analyses to eliminate most of these confounding factors (Box 3.1). It is clear that Darwin's second postulate is true for the medium ground finches on Daphne Major: A substantial fraction of the variation in beak size is due to variation in genotype.

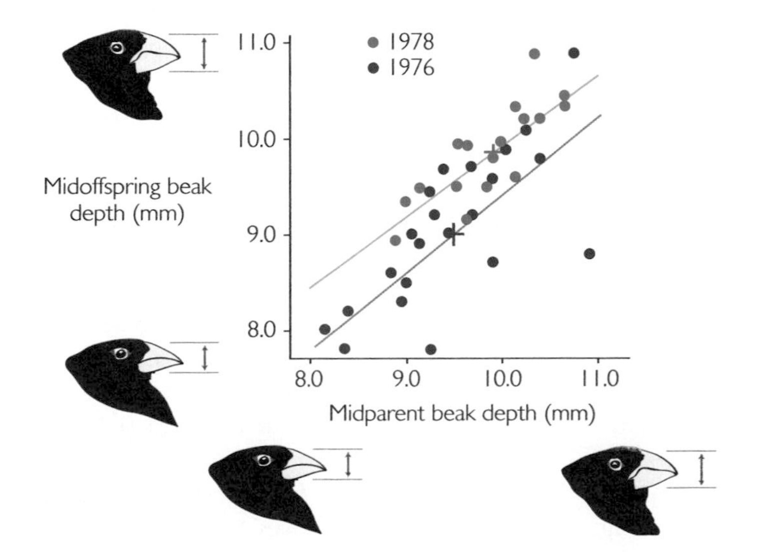

Figure 3.10 Heritability of beak depth in *Geospiza fortis* This graph shows the relationship between the beak depth of parents and their offspring. Midparent value is the average of the maternal and paternal measurements; midoffspring value is the average of the offspring measurements. The lines in the graph are statistical best-fit lines. The green line and circles are from 1978 data, and the blue line and circles are from 1976 data. Both years show a strong relationship between the beak depth of parents and their offspring. Rerendered from Boag (1983).

Box 3.1 Issues that complicate how heritabilities are estimated

Heritabilities are estimated by measuring the similarity of traits among closely related individuals. The idea is that genes run in families. If the variation in phenotype among individuals is due in part to variation in genotype, then relatives will tend to resemble one another. But a number of confounding issues can complicate this approach. We will consider four such issues here: misidentified paternity, conspecific nest parasitism, shared environments, and maternal effects.

Misidentified paternity In many species of birds, even socially monogamous birds like medium ground finches, females sometimes have extrapair sex. This means that a chick's social father is not always its biological father. If researchers simply assume that the social father at a nest is the biological father of all the chicks, they may underestimate the heritability. Although it is expensive and time consuming, misidentified paternity can be avoided by using genetic paternity tests.

Conspecific nest parasitism In some species of birds, females sneak into each other's nests and lay extra eggs. This means that even the social mother at a nest might not be the biological parent of all the chicks. Again, researchers may underestimate the heritability. As with misidentified paternity, this problem can be avoided by using genetic tests.

Shared environments Relatives share their environment as well as their genes, and any correlation that is due to their shared environment inflates the estimate of heritability. For example, it is well known that birds tend to grow larger when they have abundant food as chicks. But the most food-rich breeding territories are often claimed and defended by the largest adults in the population. Young from these territories will tend to become the largest adults in the next generation. As a result, a researcher might measure a strong relationship between parent and offspring beak and body size, and claim a high heritability for these traits, when in reality there is none. In this case, the real relationship is between the environments that parents and their young each experienced as chicks.

In many species, this problem can be circumvented by performing what are called cross-fostering, common garden, or reciprocal-transplant experiments. In birds, these experiments involve taking eggs out of their original nest and placing them in the nests of randomly assigned foster parents. Measurements in the young, taken when they are fully grown, are then compared with the data from their biological parents. This experimental treatment removes any bias in the analysis created by the fact that parents and offspring share environments.

Maternal effects Even cross-fostering experiments cannot remove environmental effects that are due to differences in the nutrient stores or hormonal contents of eggs. These are called maternal effects. They can be largely avoided by estimating heritabilities from the resemblance between offspring and their fathers only.

Lukas Keller and colleagues (2001) have made the most painstaking estimates to date of the heritability of morphological traits in Daphne Major's medium ground finches. The researchers performed genetic analyses to confirm the parentage of all the chicks in their sample. They found no evidence of conspecific nest parasitism, but they did find that 20% of the chicks had been fathered by extrapair males. Excluding these chicks from their data set, Keller and colleagues estimated that the heritability of beak depth is 0.65 (with a standard error of 0.15). In other words, about 65% of the variation among finches in beak depth appears to be due to differences in genes. This estimate is uncontaminated by extrapair paternity, conspecific nest parasitism, and maternal effects. It might, however, contain some error due to shared environments.

It has not been possible for the Galápagos researchers to perform a cross-fostering experiment on Darwin's finches. Because the Galápagos are a national park, experiments that manipulate individuals beyond catching and marking are forbidden. But the finches themselves have conducted a sort of cross-fostering experiment: As we mentioned above, about 20% of the chicks have been raised by males who are not their biological fathers. If some of the resemblance between parents and offspring is due to shared environments, then these chicks should resemble their social fathers. Using data on the social fathers and their foster offspring, Keller and colleagues calculated the "heritability" of beak depth. It was less than 0.2 and was not statistically distinguishable from zero. This suggests that shared environments have little influence on the resemblance among relatives' beaks.

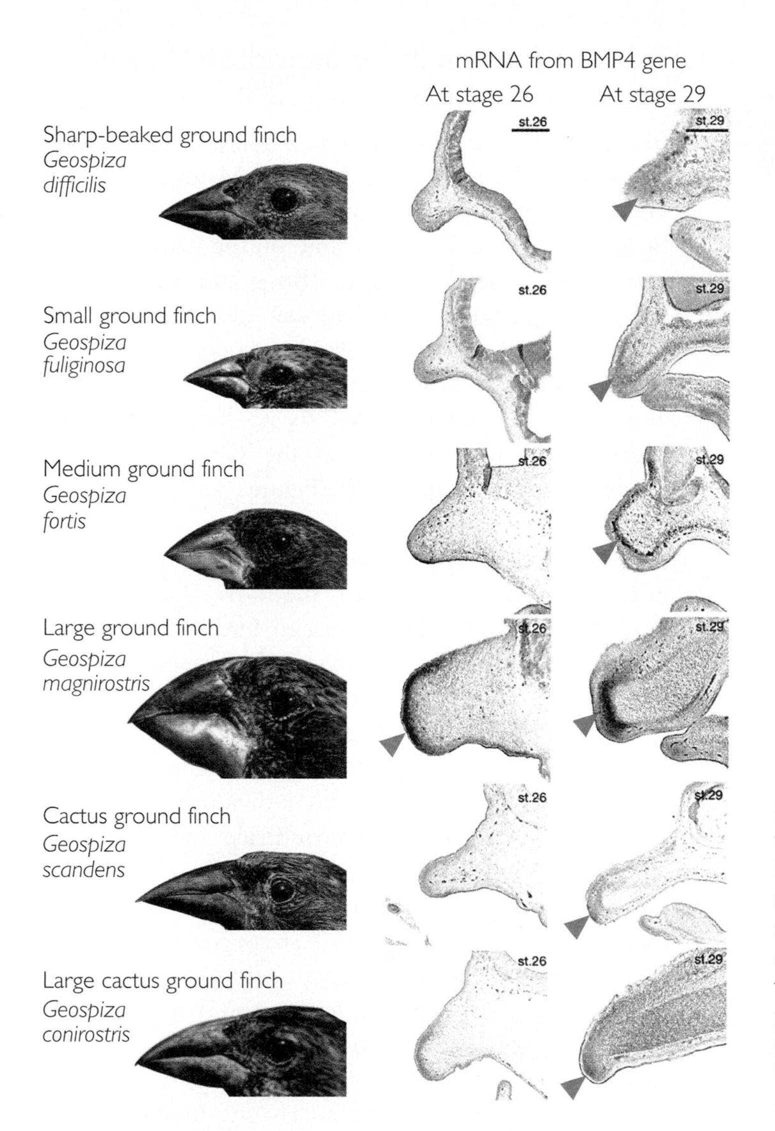

Figure 3.11 **Bone morphogenic protein 4 and beak development in Darwin's ground finches** The first column illustrates the differences in beak size and shape among the six species of ground finches. The second and third columns show cross sections of the upper beak bud in embryos of each species at two stages of development. The cross sections have been treated with a probe that stains mRNA made from the gene for bone morphogenic protein 4, or BMP4. The stained mRNA appears as dark areas indicated by arrowheads. Adult finch photos from Petren et al. (1999); embryos from Abzhanov et al. (2004).

We do not know the identity of the specific genes reponsible for variation in beak size in medium ground finches. However, Arhat Abzhanov and colleagues (2004), working in the laboratory of Clifford Tabin, discovered a tantalizing clue. These researchers focussed on growth factors known to be active during embryonic development. Among them was bone morphogenic protein 4, or BMP4, a signalling molecule that helps sculpt the shape of bird beaks (Wu et al. 2004). For all six species of ground finches, Abzhanov and colleagues treated embryos of different ages with a probe that stains messenger RNA made by the gene that encodes BMP4. As the photos in Figure 3.11 show, ground finch species with larger beaks make BMP4 mRNA (and presumably BMP4) earlier and in larger quantities than species with smaller beaks. The large ground finch, *Geospiza magnirostris*, for example, has by far the biggest beak; it is also the only species that begins making BMP4 mRNA at stage 26 of development. Abzhanov and colleagues suggest that the different species of ground finches harbor alternate versions of one or more of the genes that determine when, where, and how strongly the BMP4 gene is activated. A reasonable hypothesis would be that a similar genetic mechanism is responsible for some of the variation among individuals in the medium ground finch population on Daphne Major.

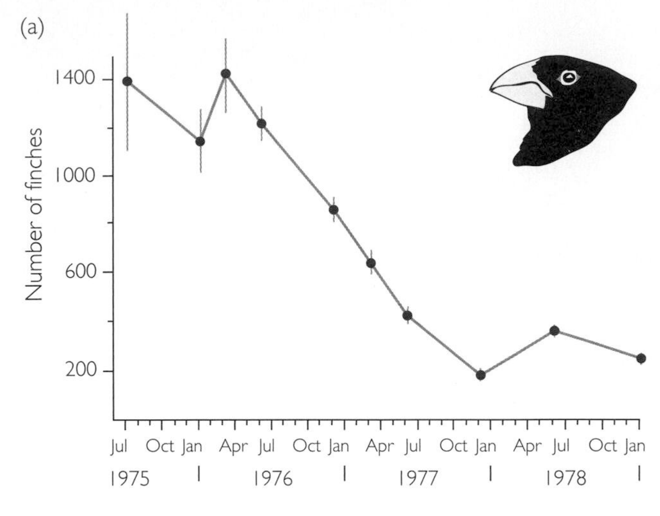

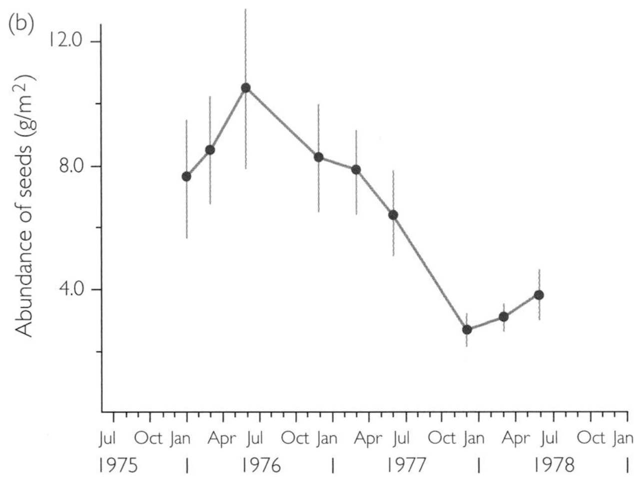

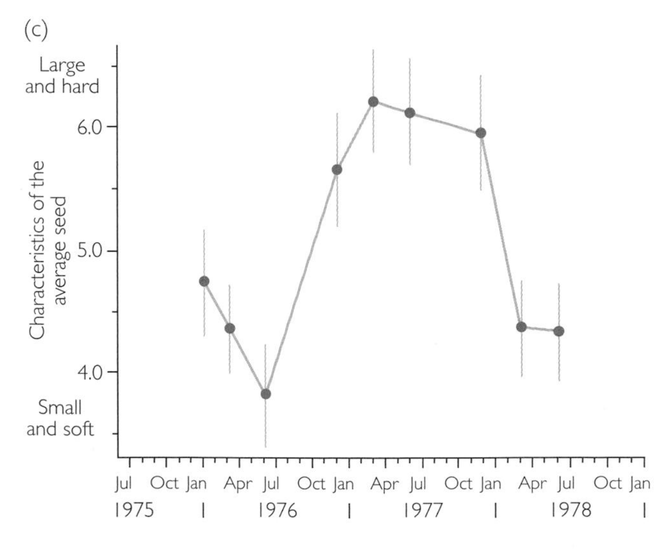

Testing Postulate 3: Do Individuals Vary in Their Success at Surviving or Reproducing?

Because Grant and Grant and their colleagues have been monitoring the finches on Daphne Major every year since 1973, two members of the research team, Peter Boag and Laurene Ratcliffe, were on the island in 1977 to witness a terrible drought (Boag and Grant 1981; Grant 1999). Instead of the normal 130 mm of rainfall during the wet season, the island got only 24 mm. The plants made few flowers and few seeds. The medium ground finches did not even try to breed. Over the course of 20 months, 84% of the *Geospiza fortis* on Daphne Major disappeared (Figure 3.12a). The researchers inferred that most died of starvation. The decline in population size was simultaneous with a decline in the availability of the seeds the birds depend on for food (Figure 3.12b); 38 emaciated birds were actually found dead, and none of the missing birds reappeared the following year. It is clear that only a fraction of the population survived to reproduce. This sort of mortality is not unusual. For example, Rosemary Grant has shown that 89% of *Geospiza conirostris* individuals die before they breed (Grant 1985). Trevor Price and coworkers (1984) determined that an additional 19% and 25% of the *G. fortis* on Daphne Major died during subsequent drought events in 1980 and 1982, respectively.

In fact, in every natural population studied, more offspring are produced each generation than survive to breed. If a population is not increasing in size, then each parent will, in the course of its lifetime, leave an average of one offspring that survives to breed. But the reproductive capacity (or biotic potential) of organisms is astonishing. Darwin (1859) picked the elephant to illustrate this point, because it was the slowest breeder then known among animals. He calculated that if all the descendants of a single pair survived and reproduced, then after just 750 years there would be 19 million of them. The numbers are even more startling for rapid breeders. Dodson (1960) calculated that if all the descendents of a pair of starfish survived and reproduced, then after just 16 years they would exceed 10^{79}, the estimated number of electrons in the visible universe.

Figure 3.12 Decline of ground finch population and available seeds during the 1977 drought (a) This graph shows the number of ground finches found on Daphne Major before, during, and after the drought. The vertical lines through each data point represent a quantity called the standard error, which indicates the amount of variation in census estimates. The lines in this graph are simply drawn from point to point to make the trend easier to see. (b) This graph shows the abundance of seeds on Daphne Major before, during, and after the drought. (c) This graph shows the characteristics of the average seed available as food to medium ground finches before, during, and after the drought. The hardness index plotted on the *y*-axis is a special measure created by Boag and Grant (1981).

Similarly, data show that in most populations some of the individuals that survive to breed are more successful at mating and producing offspring than others. Just as variation in survival does, variation in reproductive success represents selection. Darwin's third postulate is universally true.

Testing Postulate 4: Are Survival and Reproduction Nonrandom?

Darwin's fourth claim was that the individuals who survive and go on to reproduce, or who reproduce the most, are those with certain, favorable variations. Did a nonrandom, or selected, subset of the medium ground finch population survive the 1977 drought? The answer is yes.

As the drought wore on, not only the number, but also the types of seeds available changed dramatically (Figure 3.12c). The finches on Daphne Major eat seeds from a variety of plants. The seeds range from small and soft to large and hard. The small, soft seeds, easy to crack, are the birds' favorites. During the drought, as at other times, the finches ate the small, soft seeds first. Once most of the small, soft seeds were gone, the large, hard fruits of an annual plant called *Tribulus cistoides* became a key food item. Only large birds with deep, narrow beaks can crack and eat *Tribulus* fruits successfully. The rest of the finches were left to turn over rocks and scratch the soil in search of the few remaining smaller seeds.

During the drought, finches with larger, deeper beaks had an advantage in feeding, and thus in surviving.

The top graph in Figure 3.13 is from Figure 3.9 on page 82. It shows the beak sizes of a large and random sample of the birds living on Daphne Major the year before the drought. The bottom graph in Figure 3.13 shows the beak sizes of a random sample of 90 birds who survived the drought. The average survivor had a deeper beak than the average nonsurvivor. Because deep beaks and large body sizes are positively correlated, and because large birds tend to win fights over food, the average survivor had a larger body size too.

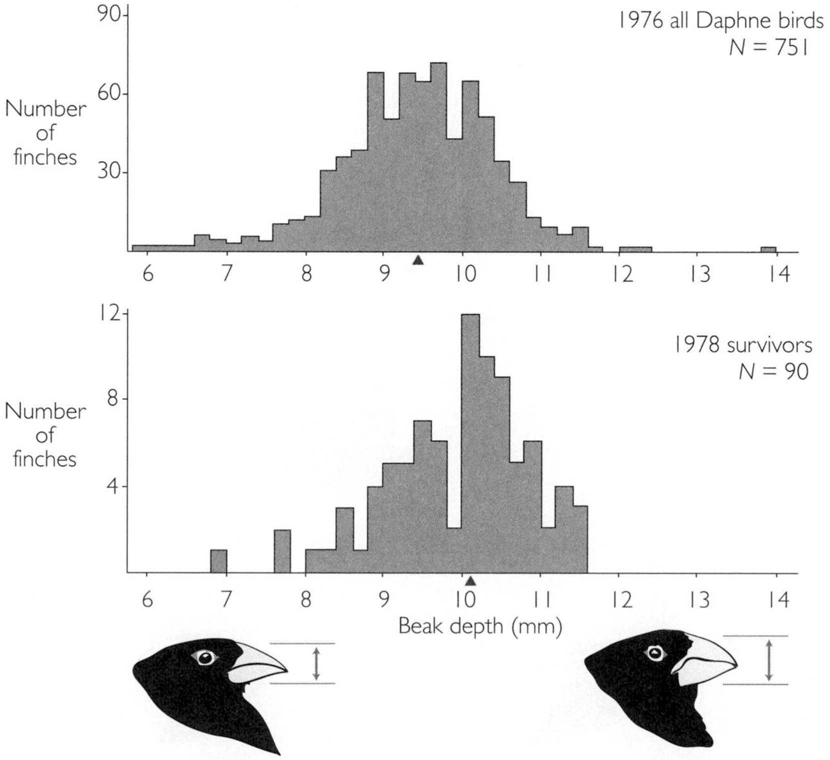

Figure 3.13 Beak depth before and after natural selection These histograms show the distribution of beak depth in medium ground finches on Daphne Major, before and after the drought of 1977. The blue triangles indicate the population means. Rerendered from Boag and Grant (1984b).

The 1977 selection event, as dramatic as it was, was not an isolated occurrence. In 1980 and 1982 there were similar droughts, and selection again favored individuals with large body size and deep beaks (Price et al. 1984). Then, in 1983, an influx of warm surface water off the South American coast, called an El Niño event, created a wet season with 1,359 mm of rain on Daphne Major. This dramatic environmental change (almost 57 times as much rain as in 1977) led to a superabundance of small, soft seeds and, subsequently, to strong selection for smaller body size (Gibbs and Grant 1987). After wet years, small birds with shallow beaks survive better and reproduce more because they harvest small seeds much more efficiently than large birds with deep beaks. Larger birds were favored in drought conditions, but smaller birds were favored in wet years. Natural selection—as we pointed out in our analysis of HIV evolution in Chapter 1—is dynamic.

Testing Darwin's Prediction: Did the Population Evolve?

As a result of the drought, the finch population evolved. Selection occurs within generations; evolution occurs between generations.

All four of Darwin's postulates are true for the medium ground finch population on Daphne Major. Darwin's theory therefore predicts a change in the composition of the population from one generation to the next. When the deep-beaked birds who survived the drought of 1977 bred to produce a new generation, they should have passed their genes for deep beaks to their offspring. Figure 3.14 confirms that they did. The chicks hatched in 1978, the year after the drought, had deeper beaks, on average, than the birds hatched in 1976, the year before the drought. The population evolved.

Peter Grant and Rosemary Grant and their colleagues have continued to monitor the Daphne Major finch population since the 1970s. As a result of unpredictable changes in the climate and bird community, and consequent changes in the Daphne Major plant community, the researchers have seen selection events

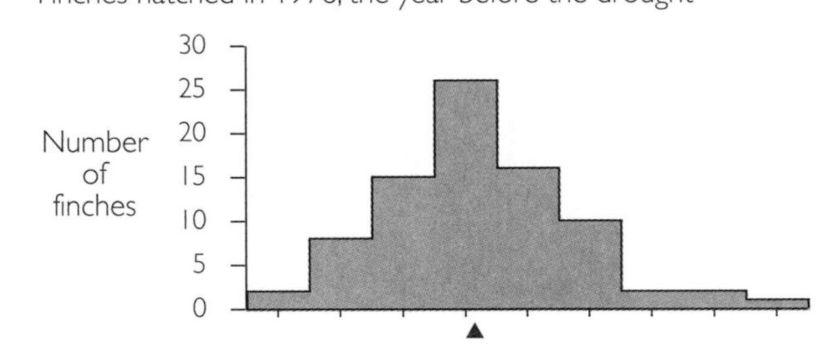

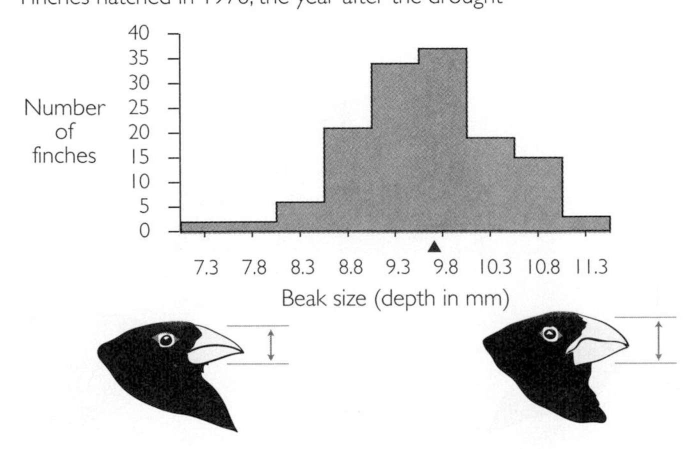

Figure 3.14 Beak depth in the finches hatched the year before the drought versus the year after the drought The red triangles represent population means. Redrawn from Grant and Grant (2003).

in which deep-beaked birds were more likely to survive and selection events in which shallow-beaked birds were more likely to survive.

Figure 3.15 shows the pattern of change in the population averages for three traits across three decades. Each of the three traits is a statistical composite of measurable traits, like beak depth. For example, "PC1 beak size" (Figure 3.15a) combines beak depth, beak length, and beak width. The evolutionary changes that occurred as a result of 1977 drought are highlighted in red.

Figure 3.15a shows, first, what we have already seen: During the drought of 1977 the finch population evolved a significantly larger average beak size. In addition, the figure shows that the population remained at this large mean beak size until the mid-1980s, then evolved back to the mean beak size it started with. There the population stayed for many years, until another drought struck.

The drought of 2003 and 2004 was as bad as the drought of 1977 (Grant and Grant 2006). Once again the medium ground finches ran short of food and many perished. This time, however, the medium ground finches faced an additional challenge: competition from a substantial population of large ground finches (*Geospiza magnirostris*) that had become established on the island. The large ground finches dominated access to, and consumed, the *Tribulus* fruits on which the large-beaked medium ground finches had survived in 1977. As a result, the medium ground finches with large beaks died at higher rates than the ones with small beaks did, and the population evolved toward smaller beak size.

In its mean beak shape and mean body size the medium ground finch population also showed substantial evolution [Figure 3.15b and (c)]. The average bird in 2001 had a significantly sharper beak, and was significantly smaller than the average bird in the mid-1970s (Grant and Grant 2002).

Grant and Grant's long-term study demonstrates that Darwin's mechanism of evolution can be documented in natural populations. When all four of Darwin's

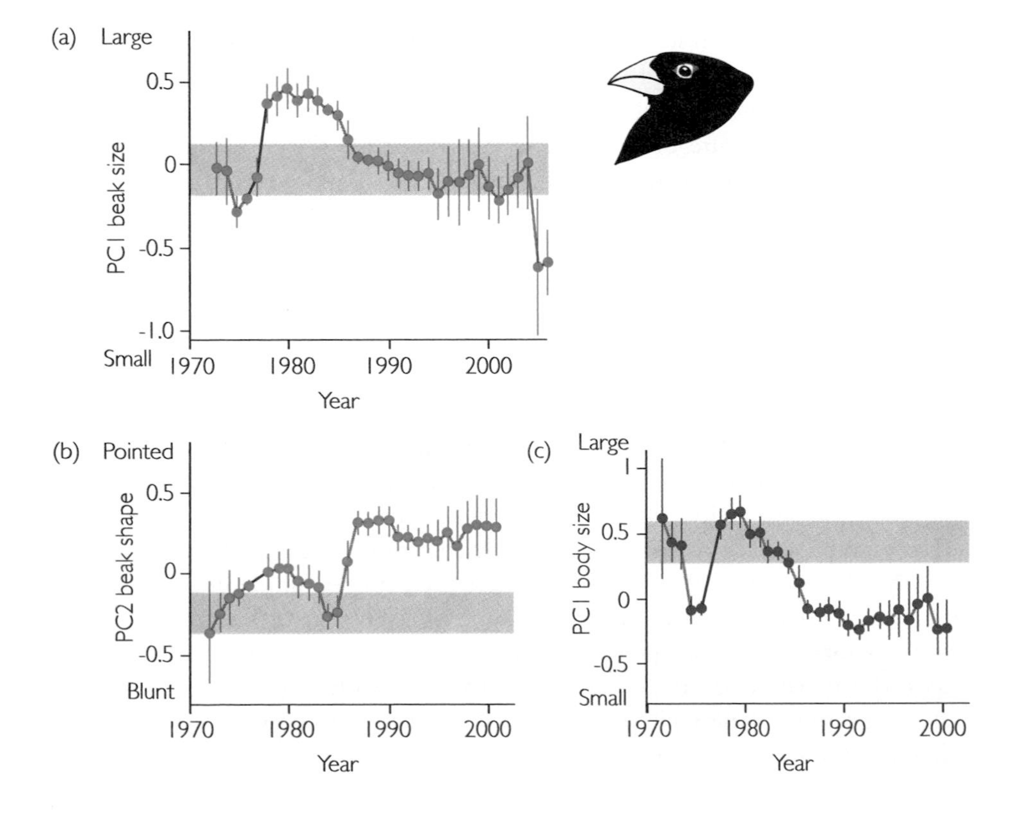

Figure 3.15 Thirty years of evolution in the medium ground finch population on Isla Daphne Major These graphs track the average adult values for beak size, beak shape, and body size among the *Geospiza fortis* on Daphne Major from the early 1970s into the 2000s. The vertical whiskers represent the 95% confidence interval for the estimated mean. If there had been no evolution, the confidence intervals for all dots would have overlapped the tan band—the 95% confidence interval for 1973, the first year with complete data. The changes that occurred during the drought of 1977 are indicated in red. In (a) the change that occurred during the drought of 2004 is indicated in orange. The population showed significant evolution in all three traits. (a) Rerendered from Grant and Grant (2006). (b and c) Rerendered from Grant and Grant (2002).

postulates are true in a population, the population evolves. The study also shows that small evolutionary changes over short time spans can accumulate into larger changes over longer time spans.

3.5 The Nature of Natural Selection

Although the Theory of Evolution by Natural Selection can be stated concisely, tested rigorously in natural populations, and validated, it can be difficult to understand thoroughly. One reason is that under Darwin's theory descent with modification is essentially a statistical process: a change in the trait distributions of populations. Statistical thinking does not come naturally to most people, and there are a number of widely shared ideas about natural selection that are incorrect. Our goal in this section is to cover some key points about how selection does and does not operate.

Natural Selection Acts on Individuals, but Its Consequences Occur in Populations

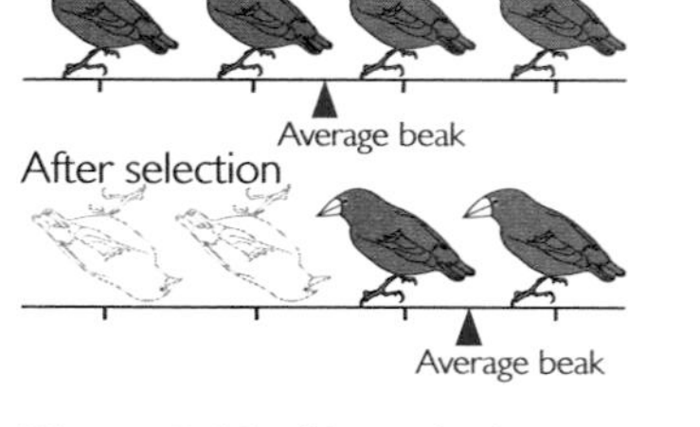

Figure 3.16 Natural selection happens to individuals, but what changes is populations During the drought on Daphne Major individuals did not change their beak depths; they simply lived or died. What changed was the average beak depth, a characteristic of the population.

When HIV strains were selected by exposure to AZT, or finch populations were selected by changes in seed availability, none of the selected individuals (virions or finches) changed in any way. They simply lived through the selection event while others died or reproduced more than competing virions or birds. What changed after the selection process was the characteristics of the populations of virions and finches, not the affected individuals themselves. Specifically, a higher frequency of HIV virions in the population were able to replicate in the presence of AZT, and a higher proportion of finches had deep beaks.

To state this point another way, the effort of cracking *Tribulus* seeds did not make finch beaks become deeper and their bodies larger, and the effort of transcribing RNA in the presence of AZT did not change the amino acid composition of the reverse transcriptase active site. Instead, the average beak depth and body size in the finch population increased because more smaller finches died than larger ones (Figure 3.16), and the average active site sequence in reverse transcriptase changed because certain mutants did a better job of making new virions.

Natural Selection Acts on Phenotypes, but Evolution Consists of Changes in Allele Frequencies

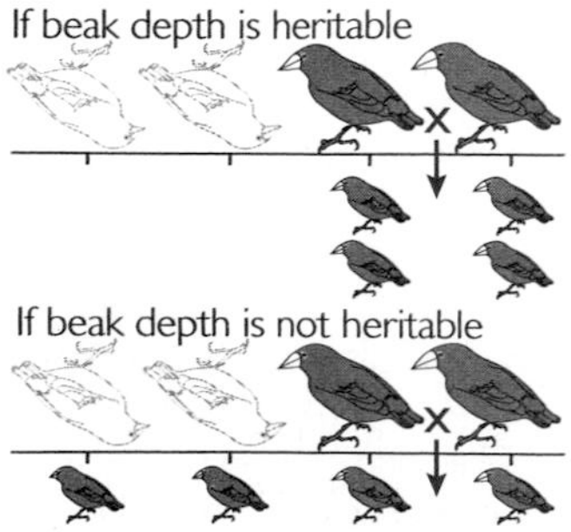

If beak depth is heritable

If beak depth is not heritable

Figure 3.17 Populations evolve only if traits are heritable If variation is due to differences in genotype, then the survivors of selection pass their successful phenotypes to their offspring.

Finches with large bodies and deep beaks would have survived at higher rates during the drought even if all of the variation in the population had been environmental in origin (that is, if heritabilities had been zero). But no evolution would have occurred. Selection would have altered the frequencies of the phenotypes in the population, but in the next generation the phenotype distribution might have gone back to what it was before selection occurred (Figure 3.17).

Only when the survivors of selection pass their successful phenotypes to their offspring, via genotypes that help determine phenotypes, does natural selection cause populations to change from one generation to the next. On Daphne Major, the variation in finch phenotypes that selection acted on had a genetic basis. As a result, the new phenotypic distribution seen among the survivors persisted into the next generation.

Natural Selection Is Not Forward Looking

Each generation is descended from the survivors of selection by the environmental conditions that prevailed in the generation before. The offspring of the HIV virions and finches that experienced selection are better adapted to environments dominated by AZT and drought conditions, respectively, than their parents' generation was. If the environment changed again during the lifetime of these offspring, however, they might not be adapted to the new conditions.

Natural selection adapts populations to conditions that prevailed in the past, not conditions that might occur in the future.

There is a common misconception that organisms can be adapted to future conditions, or that selection can look ahead in the sense of anticipating environmental changes during future generations. This is impossible. Evolution is always a generation behind any changes in the environment.

New Traits Can Evolve, Even Though Natural Selection Acts on Existing Traits

Differences in survival or reproduction—that is, natural selection—can only occur among variants that already exist. Death by starvation of small-beaked finches, for example, does not instantaneously create birds with big beaks optimal for cracking *Tribulus* fruits. It merely winnows the breeding population down to the largest-beaked of the birds already living.

This may seem to imply that new traits cannot evolve by natural selection. But the evolution of new traits is, in fact, possible for two reasons. First, during reproduction in all species, mutations produce new alleles. Second, during reproduction in sexual species, meiosis and fertilization recombine existing alleles into new genotypes. Mutation and recombination yield new suites of traits for selection to sort.

Consider, for example, an artificial selection study run at the University of Illinois (Moose et al. 2004). Since they started in 1896 with 163 ears of corn, researchers have been sowing for next year's crop only seeds from the plants with the highest oil content in their kernels. In the starting population, oil content ranged from 4–6% by weight. After 100 generations of selection, the average oil content in the population was about 20% (Figure 3.18). That is, a typical plant in the present population has over three times the oil content of the most-oil rich plant in the founding population. Mutation, recombination, and selection together produced a new phenotype.

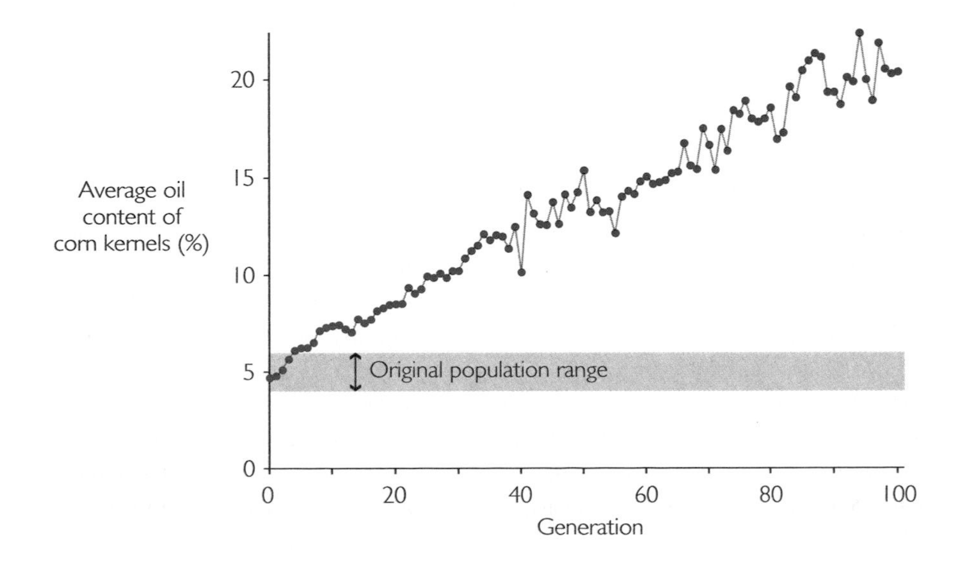

Figure 3.18 **Persistent long-term selection can result in dramatic changes in traits** These data, from the Illinois Long-Term Selection Experiment, document the increase in oil content in corn kernels during 100 generations of artificial selection. The average for the 100th generation lies far outside the range of the founding generation. Modified from Moose et al. (2004).

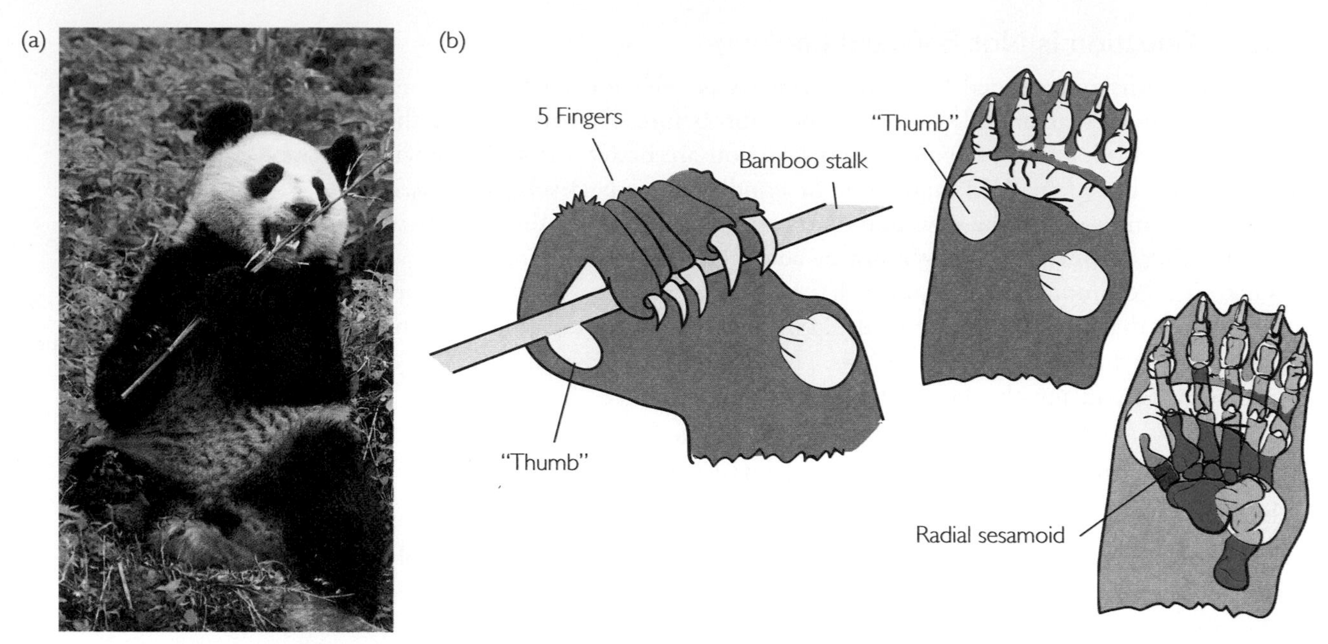

Figure 3.19 The panda's thumb (a) Giant pandas can grasp and manipulate bamboo stalks in their paws. (b) These drawings show how the panda's thumb, actually a modified wrist bone, helps clamp a stalk in the animal's curled fingers. After Endo et al. (1999a, b; 2001).

Persistent natural selection can lead to the evolution of entirely new functions for existing behaviors, structures, or genes. The giant panda's thumb provides an example (Gould 1980; Endo et al. 1999a, 1999b, 2001). Pandas use this structure to get a firm grip on the stalks of their favorite food, bamboo (Figure 3.19a). Anatomically, the "thumb" is actually an enlarged and modified radial sesamoid bone, which in closely related species is part of the wrist (Figure 3.19b). Knowing how natural selection works in contemporary populations, we surmise that when pandas first began eating bamboo, there was variation among individuals in the length of the radial sesamoid. Pandas with bigger radial sesamoids had firmer grips, fed more efficiently, and consequently had more offspring. As a result of continued selection over many generations, the average size of the bone increased in the population until it reached its present dimensions.

A trait that is used in a novel way and is eventually elaborated by selection into a completely new structure, like the radial sesamoid of the ancestral panda, is known as a **preadaptation**. An important point about preadaptations is that they represent happenstance. A preadaptation improves an individual's fitness fortuitously—not because natural selection is conscious or forward looking.

Natural Selection Does Not Lead to Perfection

The previous paragraphs argue that populations evolving by natural selection become better adapted over time. It is equally important, however, to realize that evolution does not result in organisms that are perfect.

Consider the male mosquito fish (*Gambusia affinis*), whose anal fin is modified to serve as a copulatory organ, or gonopodium. Brian Langerhans and colleages (2005) found that females prefer males with larger gonopodia. But when predators attack, a big gonopodium is literally a drag, slowing a male's escape. A perfect male would be irresistable to females and fleet enough to evade any predator. Alas, no male can be both. Instead, each population evolves a phenotype that strikes a compromise between opposing agents of selection (Figure 3.20).

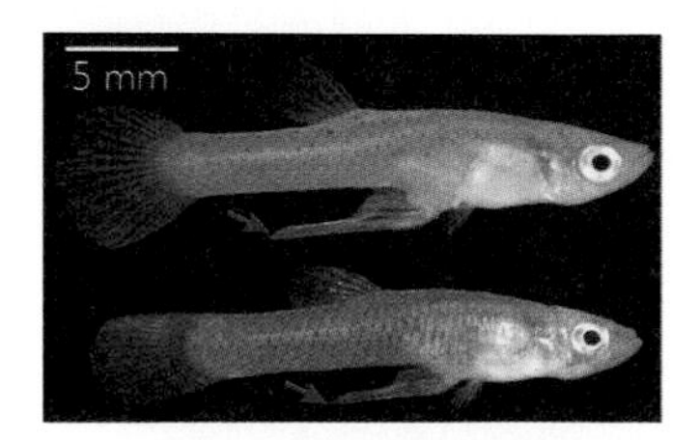

Figure 3.20 No guy is perfect These males sport gonopodia that attract mates but hinder escape. The male at bottom is from a high-predation population.

Natural selection cannot simultaneously optimize all traits. It leads to adaptation, not perfection.

Natural Selection Is Nonrandom, but It Is Not Progressive

Evolution by natural selection is sometimes characterized as a random or chance process, but nothing could be further from the truth. Mutation and recombination, the processes that generate genetic variation, are random with respect to the changes they produce in phenotypes. But natural selection, the mechanism that sorts among variant phenotypes and genotypes, is the opposite of random. It is, by definition, the nonrandom superiority at survival and reproduction of some variants over others. This is why evolution by means of natural selection is nonrandom, why, instead, it increases adaptation to the environment.

As the HIV, finch, and panda examples show, however, nonrandom selection as it occurs in nature is completely free of any entity's conscious intent. Darwin came to regret using the phrase "naturally selected," because people thought it implied a conscious act or choice by some entity. Nothing of the sort happens.

Also, although evolution has tended toward increases in the complexity, degree of organization, and specialization of organisms over time, it is not progressive in the sense of leading toward some predetermined goal. Evolution makes populations "better" only in the sense of increasing their average adaptation to their environment. There is no inexorable trend toward more advanced forms of life. For example, contemporary tapeworms have no digestive system and have actually evolved to be simpler than their ancestors. Snakes evolved from ancestors that had limbs. The earliest birds in the fossil record had teeth.

Unfortunately, a progressivist view of evolution dies hard. Even Darwin had to remind himself to "never use the words higher or lower" when discussing evolutionary relationships. It is true that some organisms are the descendants of ancient lineages and some are the descendants of more recent lineages, but all organisms in the fossil record and those living today were adapted to their environments. They are all able to survive and reproduce. None is "higher" or "lower" than any other.

There is no such thing as a higher or lower plant or animal.

Fitness is Not Circular

The Theory of Evolution by Natural Selection is often criticized by nonbiologists as tautological, or circular in its reasoning. That is, after reviewing Darwin's four postulates, one could claim, "Of course individuals with favorable variations are the ones that survive and reproduce because the theory defines favorable as the ability to survive and reproduce."

The key to resolving the issue is to realize that the word "favorable," although a convenient shorthand, is misleading. The only requirement for natural selection is for certain heritable variants to do better than others, as opposed to random ones. As long as a nonrandom subset of the population survives better and leaves more offspring, evolution will result. In the examples we have been analyzing, research not only determined that nonrandom groups survived a selection event, but also uncovered why those groups did better than others.

It should also make sense by now that Darwinian fitness is not an abstract quantity. Fitness can be measured in nature. This is done by counting the offspring that individuals produce, or by observing their ability to survive a selection event, and comparing each individual's performance to that of others in the population. These are independent and objective criteria for assessing fitness.

Selection Acts on Individuals, Not for the Good of the Species

One of the most pervasive misconceptions about natural selection, especially selection on animal behavior, is that individual organisms will perform actions for the good of the species. Self-sacrificing, or altruistic, acts do occur in nature. Prairie dogs give alarm calls when predators approach, which draws attention to themselves. Lion mothers sometimes nurse cubs that are not their own. But traits cannot evolve by natural selection unless they increase the bearer's fitness relative to competing individuals. If an allele existed that produced a truly altruistic behavior—that is, a behavior that reduced the bearer's fitness and increased the fitness of others—it would quickly disappear from the population. As we will see in Chapter 12, every altruistic behavior that has been studied in detail has been found to increase the altruist's fitness, either because the beneficiaries of the behavior are close genetic relatives (as in prairie dogs) or because the beneficiaries reciprocate (as in nursing lions) or both.

Individuals do not do things for the good of the species. They behave in a way that maximizes their individual fitness.

The idea that animals will do things for the good of the species is so ingrained, however, that we will make the same point a second way. Consider lions again. Lions live in social groups called prides. Coalitions of males fight to take over prides. If a new group of males defeats the existing pride males in combat, the newcomers quickly kill all of the pride's nursing cubs. These cubs are unrelated to them. Killing the cubs increases the new males' fitness because pride females become fertile again sooner and will conceive offspring by the new males (Packer and Pusey 1983, 1984). Infanticide is widespread in animals. Clearly, behavior like this does not exist for the good of the species. Rather, infanticide exists because, under certain conditions, it enhances the fitness of the individuals who perform the behavior relative to individuals who do not.

3.6 The Evolution of Darwinism

Because evolution by natural selection is a general organizing feature of living systems, Darwin's theory ranks as one of the great ideas in intellectual history. Its impact on biology is analogous to that of Newton's laws on physics, Copernicus's Sun-Centered Theory of the Universe on astronomy, and the Theory of Plate Tectonics on geology. In the words of evolutionary geneticist Theodosius Dobzhansky (1973), "Nothing in biology makes sense except in the light of evolution."

For all its power, though, the Theory of Evolution by Natural Selection was not universally accepted by biologists until some 70 years after it was initially proposed. There were three serious problems with the theory, as originally formulated by Darwin, that had to be resolved.

1. Because Darwin knew nothing about mutation, he had no idea how variability was generated in populations. As a result, he could not answer critics who maintained that the amount of variability in populations was strictly limited and that natural selection would grind to a halt when variability ran out. It was not until the early 1900s, when geneticists such as Thomas Hunt Morgan began experimenting with fruit flies, that biologists began to appreciate the continuous and universal nature of mutation. Morgan and colleagues showed that mutations occur in every generation and in every trait.

2. Because Darwin knew nothing about genetics, he had no idea how variations are passed on to offspring. It was not until Mendel's experiments with peas were rediscovered and verified, 35 years after their original publication, that biologists understood how parental traits are passed on to offspring. Mendel's laws of segregation and independent assortment confirmed the mechanism behind postulate 2, which states that some of the variation observed in populations is heritable.

Until then, many biologists proposed that genes acted like pigments in paint. Advocates of this hypothesis, called **blending inheritance**, argued that favorable mutations would simply merge into existing traits and be lost. In 1867, a Scottish engineer named Fleeming Jenkin published a mathematical treatment of blending inheritance, along with a famous thought experiment concerning the offspring of light-skinned and dark-skinned people. For example, if a dark-skinned sailor became stranded on an equatorial island inhabited by light-skinned people, Jenkins' model predicted that no matter how advantageous dark skin might be (in reducing skin cancer, for example), the population would never become dark-skinned because traits like skin color blended. If the dark-skinned sailor had children by a light-skinned woman, their children would be brown-skinned. If they, in turn, had children with light-skinned people, their children would be light-brown-skinned, and so on. Conversely, if a light-skinned sailor became stranded on a northern island inhabited by dark-skinned people, blending inheritance argued that, no matter how advantageous light skin might be (in facilitating the synthesis of vitamin D with energy from UV light, for example), the population would never become light. Under blending inheritance new variants are swamped, and new mutations diluted, until they cease to have a measurable effect. For natural selection to work, favorable new variations have to be passed on to offspring intact, and remain discrete.

We understand now, of course, that phenotypes blend in some traits, like skin color, but genotypes never do. Jenkins's hypothetical population would, in fact, become increasingly darker or lighter skinned if selection were strong and mutation continually added darker- or lighter-skinned variants to the population via changes in the genes involved in regulating the production of melanin (Figure 3.21).

Darwin himself struggled with the problem of inheritance, and eventually adopted an entirely incorrect view based on the work of Jean-Baptiste Lamarck. Lamarck was a great French biologist of the early 19th century who proposed that species evolve through the inheritance of changes wrought in individuals. Lamarck's idea was a breakthrough: It recognized that species have changed through time and proposed a mechanism to explain how. His theory was wrong, however, because offspring do not inherit phenotypic changes acquired by their parents. If people build up muscles lifting weights, their offspring are not more powerful; if giraffes stretch their necks reaching for leaves in treetops, it has no consequence for the neck length of their offspring.

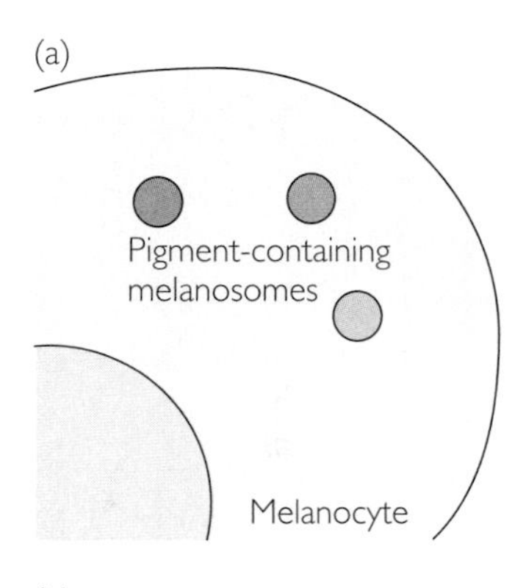

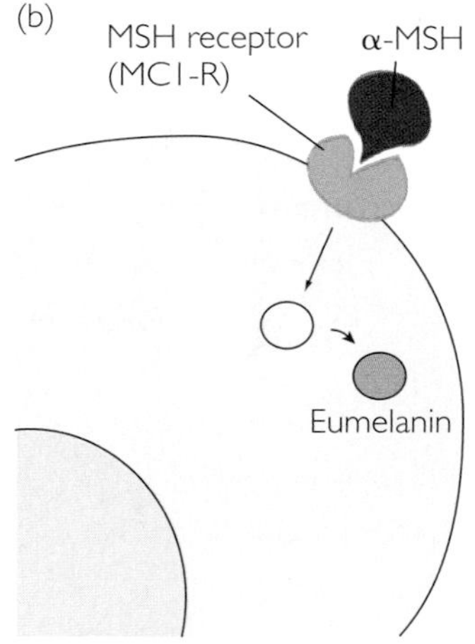

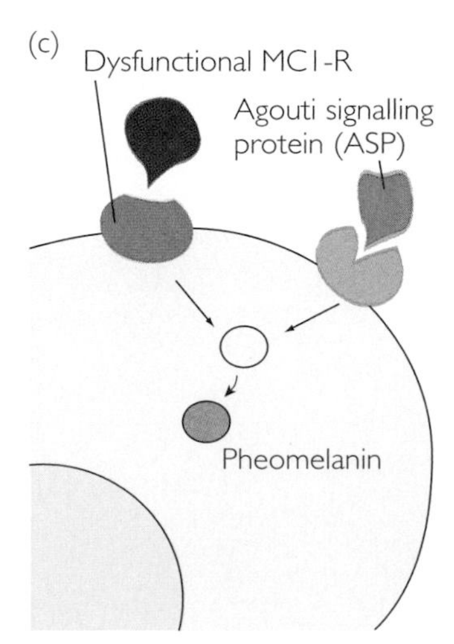

Figure 3.21 Why blending inheritance does not occur (a) Skin (and hair) color in mammals is largely determined by the production of pigments in cells called melanocytes. (b) When alpha melanocyte-stimulating hormone (α–MSH) binds to the melanocortin 1 receptor (MC1-R), it stimulates melanocytes to make eumelanin, which is brownish black. (c) When MC1-R is dysfunctional, or when it is blocked by agouti signalling protein (ASP), melanocytes make pheomelanin, which is reddish yellow.

Variation in human coloration has been tied to allelic variation in both the gene for MC1-R and the gene for ASP (Harding et al. 2000; Schaffer and Bolognia 2001; Kanetsky et al. 2002). For example, homozygotes for the *Arg151Cys* allele of the MC1-R gene almost always have red hair and fair skin (Smith et al. 1998). The effects of alleles may blend in determining the phenotype. An individual with just one copy of the *Arg151Cys* allele, for instance, may have intermediate coloration. But the alleles themselves are passed on intact to offspring, and two *Arg151Cys* heterozygotes can have a homozygous red-haired offspring. Thus inheritance is particulate, not blending. After Schaffer and Bolognia 2001.

3. Lord Kelvin, the foremost physicist of the 19th century, published an important series of papers in the early 1860s estimating the age of Earth at 15–20 million years. Kelvin's analyses were based on measurements of the Sun's heat and the temperature of Earth. Because fire was the only known source of heat at the time, Kelvin assumed that the Sun was combusting like an enormous lump of coal. This had to mean that the Sun was gradually burning down, releasing progressively less heat with each passing millennium. Likewise, both geologists and physicists believed that the surface of Earth was gradually cooling. This was based on the assumption that Earth was changing from a molten state to a solid one by radiating heat to the atmosphere, a view apparently supported by measurements of progressively higher temperatures deeper down in mineshafts. These data allowed Kelvin to calculate the rate of radiant cooling.

The bottom line from Kelvin's calculations was that the transition from a hot to cold Sun and hot to cold Earth created a narrow window of time when life on Earth was possible. The window was clearly too narrow to allow the gradual changes of Darwinism to accumulate, and thus strongly supported a role for instantaneous and special creation in explaining adaptation and diversity.

The discovery of radioactive isotopes early in the 20th century changed all that. Kelvin's calculations were correct, but his assumptions were wrong. Scientists confirmed that Earth's heat is a by-product of radioactive decay, not radiant cooling, and that the Sun's heat is from nuclear fusion, not combustion.

The Modern Synthesis

The Modern Synthesis resolved decades of controversy over the validity of evolution by natural selection.

Understanding variability, inheritance, and time was so difficult that the first 70 years of evolutionary biology were characterized by turmoil (see Provine 1971; Mayr 1980, 1991). But between 1932 and 1953 a series of landmark books were published that successfully integrated genetics with Darwin's four postulates and led to a reformulation of the Theory of Evolution. This restatement, known as the Modern Synthesis or the Evolutionary Synthesis, was a consensus grounded in two propositions:

- Gradual evolution results from small genetic changes that are acted upon by natural selection.
- The origin of species and higher taxa, or macroevolution, can be explained in terms of natural selection acting on individuals, or microevolution.

With the synthesis, Darwin's original four postulates and their outcome could be restated along the following lines:

1. As a result of mutation creating new alleles, and segregation and independent assortment shuffling alleles into new combinations, individuals within populations are variable for many traits.
2. Individuals pass their alleles on to their offspring intact.
3. In every generation, some individuals are more successful at surviving and reproducing than others.
4. The individuals that survive and reproduce, or who reproduce the most, are those with the alleles and allelic combinations that best adapt them to their environment.

The outcome is that alleles associated with higher fitness increase in frequency from one generation to the next.

This View of Life

Darwin ended the introduction to the first edition of *On the Origin of Species* with a statement that still represents the consensus view of evolutionary biologists (Darwin 1859, p. 6): "Natural Selection has been the main but not exclusive means of modification." We now think of modification in terms of changes in the frequencies of the alleles responsible for traits like beak depth and AZT resistance. We are more keenly aware of other processes that cause evolutionary change in addition to natural selection. (Chapters 6 and 7 explore these processes in detail.) But the Darwinian view of life, as a competition between individuals with varying abilities to survive and reproduce, has proven correct in almost every detail.

As Darwin wrote in his concluding sentence (1859, p. 490): "There is grandeur in this view of life, with its several powers, having been originally breathed into a few forms or into one; and that, whilst this planet has gone cycling on according to the fixed law of gravity, from so simple a beginning endless forms most beautiful and most wonderful have been, and are being, evolved."

3.7 The Debate over "Scientific Creationism" and Intelligent Design Creationism

Scientific controversy over the fact of evolution ended in the late 1800s, when the evidence reviewed in Chapter 2 simply overwhelmed the critics. Whether natural selection was the primary process responsible for both adaptation and diversity was still being challenged until the 1930s, when the works of the Modern Synthesis provided a mechanistic basis for Darwin's four postulates and unified micro- and macroevolution. Evolution by natural selection is now considered the great unifying idea in biology. Although scientific discourse about the validity of evolution by natural selection ended well over a half-century ago, a political and philosophical controversy in the United States and Europe still continues (Holden 1995; Kaiser 1995). What is this debate, and why is it occurring?

Creationists want the Theory of Special Creation to be taught in public schools, even though it was dismissed as a viable alternative to the Theory of Evolution by Natural Selection over a century ago.

History of the Controversy

The Scopes Trial of 1925, popularly known as the Monkey Trial, is perhaps the most celebrated event in a religious debate that has raged since Darwin first published *On the Origin of Species* (see Gould 1983, essay 20; Larson 1997). John Scopes (Figure 3.22) was a biology teacher who gave his students a reading assignment about Darwinian evolution. This was a clear violation of the State of Tennessee's Butler Act, which prohibited the teaching of evolution in public schools. William Jennings Bryan, a famous politician and a fundamentalist orator, was the lawyer for the prosecution; Clarence Darrow, the most renowned defense attorney of his generation, led Scopes's defense. Although Scopes was convicted and fined $100, the trial was widely perceived as a triumph for evolution because Bryan had suggested, while on the stand as a witness, that the six days of creation described in Genesis 1:1–2:4 may each have lasted far longer than 24 hours. This was considered a grave inconsistency, and therefore a blow to the integrity of the creationist viewpoint. But far from ending the debate over teaching evolution in U.S. schools, the Scopes trial was merely a way station.

Figure 3.22 **Scopes on Trial** John Scopes, right, confers with a member of his defense team.

The Butler Act, in fact, stayed on the books until 1967; it was not until 1968, in *Epperson v. Arkansas,* that the U.S. Supreme Court struck down laws that prohibit the teaching of evolution. The court's ruling was made on the basis of the U.S. Constitution's separation of church and state. In response, fundamentalist religious groups in the United States reformulated their arguments as "creation science" and demanded equal time for what they insisted was an alternative theory for the origin of species. By the late 1970s, 26 state legislatures were debating equal-time legislation (Scott 1994). Arkansas and Louisiana passed such laws only to have them struck down in state courts. The Louisiana law was then appealed all the way to the U.S. Supreme Court, which decided in 1987 (*Edwards v. Aquillard*) that because creationism is essentially a religious idea, teaching it in the public schools was a violation of the first amendment. Two justices, however, formally wrote that it would still be acceptable for teachers to present alternative theories to evolution (Scott 1994).

One response from opponents of evolution has been to drop the words creation and creator from their literature and call either for equal time for teaching that no evolution has occurred, or for teaching a proposal called Intelligent Design Theory, which infers the presence of a designer from the perfection of adaptation in contemporary organisms (Scott 1994; Schmidt 1996). In the fall of 2005 the case of *Kitzmiller et al. v. Dover Area School District* was tried in Dover, Pennsylvania. The school district had enacted a policy requiring that students in biology classes "be made aware of gaps/problems in Darwin's Theory and of other theories of evolution including, but not limited to, intelligent design." A group of parents sued the school district on the grounds, again, that the policy violates the first ammendment. The court agreed (Goodstein 2005; Jones 2005).

The complexity and perfection of organisms is a time-worn objection to evolution by natural selection. Darwin was aware of it; in his *Origin* he devoted a section of the chapter titled "Difficulties on Theory" to "Organs of extreme perfection." How can natural selection, by sorting random changes in the genome, produce elaborate and integrated traits like the vertebrate eye?

Perfection and Complexity in Nature

The English cleric William Paley, writing in 1802, promoted the Theory of Special Creation with a now-classic argument. If a person found a watch and discovered that it was an especially complex and accurate instrument, they would naturally infer that it had been made by a highly skilled watchmaker. Paley then drew a parallel between the watch and the perfection of the vertebrate eye and asked his readers to infer the existence of a purposeful and perfect Creator. He contended that organisms are so well-engineered that they have to be the work of a conscious designer. This logic, still used by creationists today, is called the Argument from Design (Dawkins 1986).

The Argument from Design contends that adaptations must result from the actions of a conscious entity.

Because we perceive perfection and complexity in the natural world, evolution by natural selection seems to defy credulity. There are actually two concerns here. The first is how random changes can lead to order. Mutations are chance events, so the generation of variation in a population is random. But the selection of those variants, or mutants, is nonrandom: It is directed in the sense of increasing fitness. And adaptations—structures or behaviors that increase fitness—are what we perceive as highly ordered, complex, or even perfect in the natural world. But there is nothing conscious or intelligent about the process. The biologist Richard Dawkins captured this point by referring to natural selection as a blind watchmaker.

A second, and closely related, concern is: How can complex, highly integrated structures, like the vertebrate eye, evolve through the Darwinian process of gradual accumulation of small changes? Each evolutionary step would have to increase the fitness of individuals in the population. Darwinism predicts that complex structures have evolved through a series of intermediate stages, or graded forms. Is this true? For example, when we consider a structure like the eye, do we find a diversity of forms, some of which are more complex than others?

The answer to these questions is yes. In some unicellular species there are actually subcellular organelles with functions analogous to the eye. The eyespots of a group of protozoans called euglenoids, for example, contain light-absorbing molecules that are shaded on one side by a patch of pigment. When these molecules absorb light, they undergo structural changes. Because light can reach them from one side only, a change in the light-absorbing molecule contains useful information about where light is coming from. Some dinoflagellates even have a subcellular, lenslike organelle that can concentrate light on a pigment cup. It is unlikely that these single-celled protists can form an image, however, because they are not capable of neural processing. Rather, their eye probably functions in transmitting information about the cell's depth in the water column, helping the cell orient itself and swim toward light.

More complex eyes have a basic unit called the photoreceptor. This is a cell that contains a pigment capable of absorbing light. The simplest type of multicellular eye, consisting of a few photoreceptor cells in a cup or cuplike arrangement, is shown in Figures 3.23a and 3.23b. This type of eye is found in a wide diversity of taxa, including flatworms, polychaetes (segmented worms in the phylum Annelida), some crustaceans (the shrimps, crabs, and allies), and some vertebrates. These organs are used in orientation and day-length monitoring (Willson 1984; Brusca and Brusca 1990). Slightly more complex eyes, like those illustrated in Figure 3.23c, have optic cups with a narrow aperture acting as a lens and may be capable of forming images in at least some species. These are found in a few nemerteans (ribbon worms) and annelids (segmented worms), copepod crustaceans, and abalone and nautiloids (members of the phylum Mollusca). The most complex eyes (Figure 3.23d) fall into two functional categories based on whether the photoreceptor cells are arrayed on a retina that is concave, like the eyes of vertebrates and octopuses, or convex, like the compound eyes of insects and other arthropods (Goldsmith 1990). These eyes have lenses, and in most cases are capable of forming images.

It is important to recognize that the simpler eyes we have just reviewed do not themselves represent intermediate forms on the way to more advanced structures. The eyespots, pigment cups, and optic cups found in living organisms are contemporary adaptations to the problem of sensing light. They are not ancestral forms. It is, however, sensible to argue that the types of eyes discussed here form an evolutionary pathway (Gould 1983, essay 1). That is, it is conceivable that eyes like these formed intermediate stages in the evolution of the complex eyes found in vertebrates, octopuses, and insects. This is exactly what Darwin argued in his section on organs of extreme perfection. (To learn more about the evolution of the eye, see Salvini-Plawen and Mayr 1977; Nilsson and Pelger 1994; Quiring et al. 1994; Dawkins 1994; Donner and Maas 2004; Gehring 2004; Fernald 2004.)

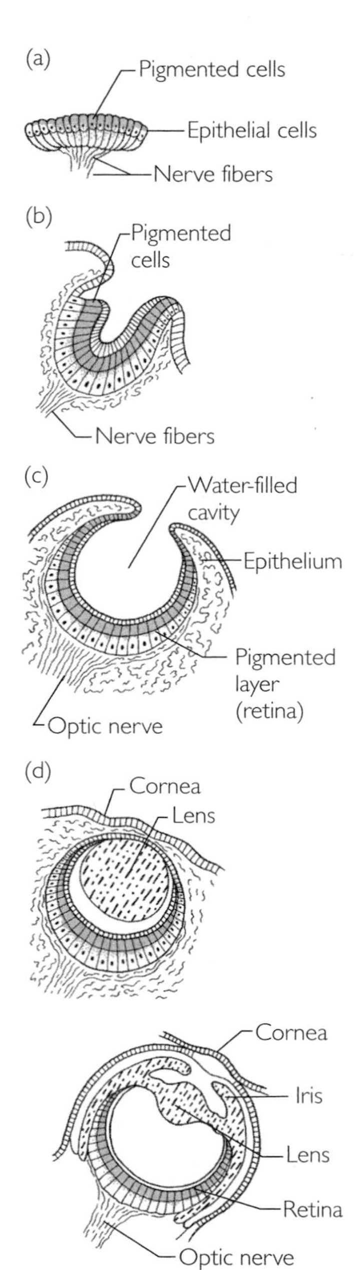

Figure 3.23 Variation in mollusc eyes (a) A pigment spot; (b) a simple pigment cup; (c) the simple optic cup found in abalone; (d) the complex lensed eyes of a marine snail called *Littorina* and the octopus. Pigmented cells are shown in color. From Brusca and Brusca 1990.

The Argument from Biochemical "Design"

Summarizing his views on perfection and complexity in nature, Darwin wrote (1859, p. 189):

> If it could be demonstrated that any complex organ existed, which could not possibly have been formed by numerous, successive, slight modifications, my theory would absolutely break down. But I can find out no such case.

Creationist Michael Behe (1996), believes he has found a profusion of such cases. Behe claims that many of the molecular machines found inside cells are irreducibly complex, and could not have been built by natural selection. Behe writes (p. 39):

> By *irreducibly complex* I mean a single system composed of several well-matched, interacting parts that contribute to the basic function, wherein the removal of any one of the parts causes the system to effectively cease functioning.

Among the examples Behe offers is the eukaryotic cilium (also known, when it is long, as a flagellum).

Figures 3.24a and 3.24b show a cross section of the stalk, or axoneme, of one of these cellular appendages. Its main structural components are microtubules, made of proteins called α-tubulin and β-tubulin. At the core of the axoneme are two singlet microtubules, held together by a protein bridge. Surrounding the central pair are nine doublet microtubules. The doublet microtubules are connected to the central pair by protein spokes. Neighboring doublets are also connected to each other by an elastic protein called nexin. The cilium is powered by the dynein motors on the doublet microtubules. As the motors on each doublet crawl up their neighboring doublet, they cause the entire axoneme to bend.

Here is Behe again (1998):

> Cilia are composed of at least a half dozen proteins: alpha-tubulin, beta-tubulin, dynein, nexin, spoke protein, and a central bridge protein. These combine to perform one task, ciliary motion, and all of these proteins must be present for the cilium to function. If the tubulins are absent, then there are no filaments to

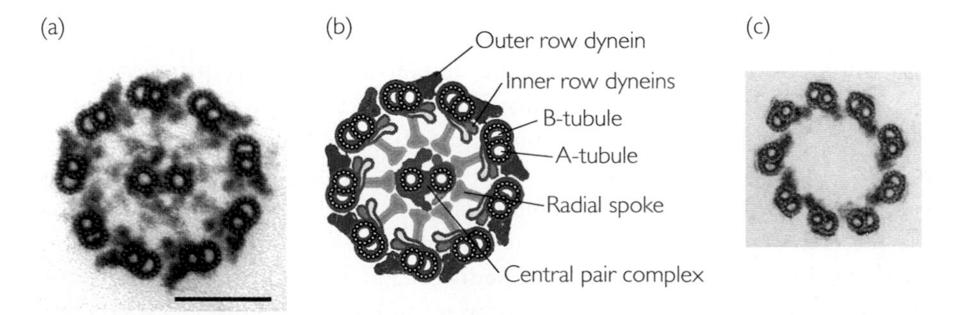

Figure 3.24 Eukaryotic flagella (a) An electron micrograph showing a cross section through a flagellum of the single-celled alga *Chlamydomonas*. Scale bar = 100 nm (b) An interpretive drawing showing the individual components of the flagellum in (a). (c) A cross section through the flagellum of an eel sperm. See text for explanation. Parts (a) and (b) are from Mitchell (2000); (c) is from Woolley (1997).

> slide; if the dynein is missing, then the cilium remains rigid and motionless; if nexin or the other connecting proteins are missing, then the axoneme falls apart when the filaments slide.

Behe finds it implausible that the cilium could have arisen by natural selection, a stepwise process in which each step involves an incremental improvement over what came before:

> [S]ince the complexity of the cilium is irreducible, then it can not have functional precursors. Since the irreducibly complex cilium can not have functional precursors it can not be produced by natural selection, which requires a continuum of function to work. Natural selection is powerless when there is no function to select.

Once he has concluded that the cilium cannot have arisen by natural selection, Behe infers that it must have been designed.

Intelligent Design Theory is a modern version of the Argument from Design.

The first thing we can say about Behe's argument is that the eukaryotic cilium is not, in fact, irreducibly complex. It is certainly not irreducibly complex in an evolutionary sense. This is demonstrated by organisms with cilia that are simpler in structure than the one pictured in Figure 3.24a and (b) (see Miller 1999). Figure 3.24c, for example, shows a cross section of an eel sperm's flagellum. It is fully functional, even though it lacks the central pair of singlet microtubules, the spokes, and the outer row of dynein motors. The cilium is not even irreducibly complex in a mechanical sense. This is shown by a mutation in the single-celled alga *Chlamydomonas* known as *pf14*. The flagella of cells carrying this mutation lack spokes. Although the *pf* in *pf14* stands for paralyzed flagella, the flagella of mutant cells do still function under the right chemical conditions or on the right genetic backgrounds (Frey et al. 1997).

The second thing we can say about Behe's argument is that even if the cilium were irreducibly complex, he would still be wrong to conclude that it cannot have functional precursors and cannot have been built by natural selection. Behe's argument assumes that evolution by natural selection builds molecular machines and their components from scratch, and that the individual component proteins are useless until the entire structure has been assembled in its final form. In fact, evolution by natural selection cobbles molecular machines together from preexisting and functional component proteins that it co-opts for new roles (True and Carrol 2002). If the components of complex molecular machines are recruited from other jobs, then we no longer have to explain how the components were maintained by selection while the machine evolved from scratch.

Richard Lenski and colleagues (2003) showed that evolution by natural selection can, in fact, build complex machines in just this way by studying populations of digital organisms. A digital organism is a self-replicating computer program. Each of the organisms in Lenski et al.'s virtual world has a genome composed of a series of simple instructions—low-level scraps of computer code. There are some two dozen possible instructions in all, which can be strung together in any order and repeated any number of times. Most possible sequences of instructions do nothing. Some allow an organism to copy itself. Still others allow an organism to take numbers as inputs, perform logical functions on them, and produce meaningful outputs. The researchers started with a large population of identical organisms whose modest-sized genomes allowed them to replicate themselves but not to perform logical functions. Replication was imperfect, meaning that occasionally one or more of the instructions in the genome was replaced with another chosen at random, or an instruction was inserted or deleted at random. The organisms had to compete for the chance to run their instructions and reproduce. If an organism appeared that could correctly perform one or more logical functions, it was rewarded with additional running time.

The capacity to perform simple logical functions evolved first. Complex functions evolved later, building on the simple ones and co-opting them for new purposes. In genomes capable of performing the most complex function, many of the individual instructions were crucial; deleting them destroyed the organism's ability to perform the function. Intriguingly, some of the mutations on the path to the most complex function were initially harmful. That is, they disrupted the machinery for one or more simple functions. But they set the stage for later mutations that helped assemble new and more complex functions from old.

A striking demonstration of gene co-option in real organisms comes from the crystallins of animal eye lenses (True and Carrol 2002). Crystallins are water-soluble proteins that form densely packed, transparent, light-refracting arrays constituting about a third of the mass of the lens. Animal eyes contain an astonishing diversity of crystallins (Figure 3.25). Some, such as the α and $\beta\gamma$ crystallins, are widely distributed across the vertebrates and must have evolved early. These ancient crystallins evolved from duplicate copies of genes for proteins with other functions. Other crystallins are unique to particular taxa and must have evolved recently. Most of these recently evolved crystallins are similar or identical to enzymes that function outside the eye. Some, in fact, *are* enzymes that function outside the eye. That is, in some cases a single gene encodes a single protein that functions as an enzyme in some tissues and as a crystallin in the lens. The $\in$ crystallin in chickens, for instance, is a metabolic enzyme called lactate dehydrogenase B. The antifreeze proteins in the blood of Arctic and Antarctic marine fishes provide additional examples of proteins co-opted for new functions (Baardsnes and Davies 2001; Fletcher et al. 2001).

Crystallins and antifreeze proteins have simple jobs as proteins go. They have switched roles during their evolutionary history, but have not been incorporated into complex molecular machines. However, most of the components of the molecular machines Behe cites are homologous to proteins with other cellular functions. The microtubules and dyneins of the eukaryotic cilium, for example, are similar to components of the spindle apparatus employed in cell division. And work on simple examples such as crystallins and antifreeze proteins has paved the way for progress on more challenging problems. Researchers have begun reconstructing the evolutionary origins of complex molecular machines and metabolic pathways. Examples include the Krebs citric acid cycle (Meléndez-Hevia, et al. 1996; Huynen, et al. 1999), the cytochrome *c* oxidase proton pump (Musser and Chan 1998), the blood-clotting cascade (Krem and Di Cera 2002), and various bacterial flagella (Pallen and Matzke 2006).

Behe is right that we have not yet worked out in detail the evolutionary histories of the molecular machines he takes as examples of irreducible complexity. He would have us give up and attribute them all to miracles. But that is no way to make progress. Ironically, Behe began claiming that the origins of cellular biochemistry would never be deciphered just as the techniques and data required to do so were becoming available. Among these are automated DNA sequencers and the whole-genome sequences they are providing. We predict that in the coming decades all of Behe's examples of irreducible complexity will yield to evolutionary analysis.

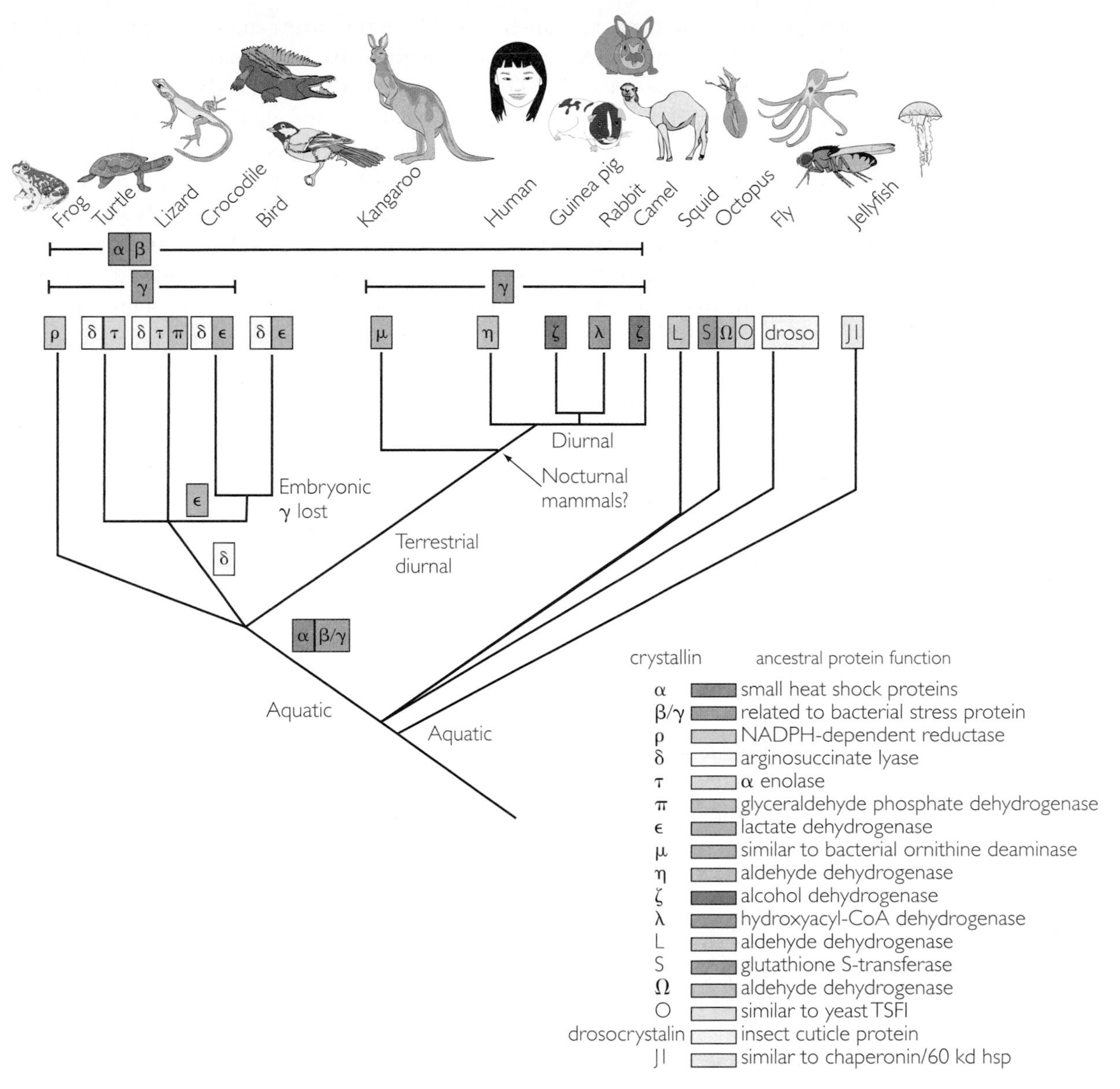

Figure 3.25 **Gene co-option in the crystallins of animal eye lenses** Crystallin proteins are major components of the lenses in animal eyes. All are derived from proteins with other functions. In some cases crystallins are encoded by duplicates of the genes for the proteins they are derived from; in other cases crystallins are encoded by the same genes. This phylogeny shows the evolutionary relationships among a variety of animals. The color-coded Greek letters indicate the crystallins found in the lenses of each animal. The table lists the proteins the various crystallins are derived from. Redrawn from True and Carroll 2002.

Other Objections

Here are four additional arguments that creationists use regularly, with responses from an evolutionary perspective (see Gish 1978; Kitcher 1982; Futuyma 1983; Gould 1983 essays 19, 20, 21; Dawkins 1986; Swinney 1994):

1. Evolution by natural selection is unscientific because it is not falsifiable and because it makes no testable predictions. Each of Darwin's four postulates is independently testable, so the theory meets the classical criterion that ideas must be falsifiable to be considered scientific. Also, the claim that evolutionary biologists do not make predictions is not true. Paleontologists routinely (and correctly) predict which strata will bear fossils of certain types (a spectacular example was that fossil marsupial mammals would be found in Antarctica); Peter Grant and Rosemary Grant have used statistical techniques based on evolutionary theory to correctly predict the amount and direction of change in finch characteristics during selection events in the late 1980s and early 1990s (Grant and Grant 1993, 1995). Scientific creationism, on the other hand, amounts to an oxymoron; in the words of one of its leading advocates, Dr. Duane Gish (1978, p. 42): "We cannot discover by scientific investigations anything about the creative processes used by God."

2. Because Earth was created as little as 6,000–8,000 years ago, there has not been enough time for Darwinian evolution to produce the adaptation and diversity observed in living organisms. Creation scientists present short-Earth theories and argue that most geological landforms and strata resulted from the flood during the time of Noah. (For example, see Gish 1978 and Swinney 1994.) Most simply disbelieve the assumptions behind radiometric dating and deny the validity of the data. The assumption of uniformitarianism in the evolution of life and landforms is also rejected by creation scientists. Again, we quote Gish (1978, p. 42): "We do not know how God created, what processes He used, *for God used processes which are not now operating anywhere in the natural universe*" (emphasis original).

The assumptions of radiometric dating have been tested, however, and demonstrated to be correct. Radiometric dating has demonstrated that rock strata differ in age, and that Earth is about 4.6 billion years old.

3. Because organisms progress from simpler to more complex forms, evolution violates the Second Law of Thermodynamics. Although the Second Law has been stated in a variety of ways since its formulation in the late 19th century, the most general version is: "Natural processes tend to move toward a state of greater disorder" (Giancoli 1995). The Second Law is focused on the concept of entropy. This is a quantity that measures the state of disorder in a system. The Second Law, restated in terms of entropy, is "The entropy of an isolated system never decreases. It can only stay the same or increase" (Giancoli 1995).

The key to understanding the Second Law's relevance to evolution is the word "isolated." The Second Law is true only for closed systems. Organisms, however, live in an open system: Earth, where photosynthetic life-forms capture the radiant energy of the Sun and convert it to chemical energy that they and other organisms can use. Because energy is constantly being added to living systems, the Second Law does not apply to their evolution.

A similar objection is William Dembski's (2002) assertion that natural selection cannot lead the evolution of complex genetic information because it is no better than a random search. He stakes this claim on a set of results in theoretical computer science called the No Free Lunch Theorems. These show that averaged over all possible problems, no set of rules for finding a solution is better than any other, including random trial and error. But as Allen Orr (2002) points out, the No Free Lunch Theorems do not apply to Darwinian evolution, because Darwinian evolution is not a search for a predefined target. It is, instead, "sheer cold demographics." Genomes that make more copies of themselves become more common; genomes that make fewer copies disappear.

4. No one has ever seen a new species formed, so evolution is unproven. And because evolutionists say that speciation is too slow to be directly observed, evolution is unprovable and thus based on faith. Although speciation is a slow process, it is ongoing and can be studied. In Chapter 2 we discussed an example: Eastern versus western greenish warblers have diverged as they expanded their range around the Tibetan Plateau, to the point that the two forms act like different species where they meet in the north. Chapter 16 will cover other experimental and observational studies of speciation in action.

Also, it is simply incorrect to claim that the only way to prove something happened is to observe it directly. Here is a rather contrived example: Imagine that you and two friends are stranded on an otherwise deserted island. You find one friend face down with a knife in his back, and you know that you did not do it. Although you did not directly observe the murder, you can infer the identity of the guilty party. We make inferences of this sort all the time in everyday life. They are common in science as well. We cannot observe atoms directly, for example, but there is considerable evidence on which to infer that they exist.

What Motivates the Controversy?

For decades, evolution by natural selection has been considered one of the best-documented and most successful theories in the biological sciences. Many scientists see no conflict between evolution and religious faith (Easterbrook 1997; Scott 1998), and many Christians agree. In 1996, for example, Pope John Paul II acknowledged that Darwinian evolution was a firmly established scientific result and stated that accepting Darwinism was compatible with traditional Christian understandings of God.

If the fact of evolution and the validity of natural selection are utterly uncontroversial, and if belief in evolution is compatible with belief in God, then why does the creationist debate continue?

During a discussion about whether material on evolution should be included in high school textbooks, a member of the Alabama State School Board named David Byers said, "It's foolish and naive to believe that what children are taught about who they are, how they got here, doesn't have anything to do with what they conclude about why they are here and what their obligations are, if, in fact, they have any obligations, and how they should live" (National Public Radio 1995). This statement suggests that, for some creationists, the controversy is not about the validity of the scientific evidence or its compatibility with religion. Instead, the concern is about what evolution means for human morality and behavior.

Creationists and evolutionists, it is safe to say, share the desire that children should grow up to become morally responsible adults. Creationists fight evolution because they believe it is morally dangerous. Evolutionary biologists, on the other hand, tend to believe that morality and moral guidance derive from sources outside of biology, and that children should learn what science shows us about how we and Earth's other living things came to be.

Summary

Before Darwin began to work on the origin of species, many scientists had become convinced that species change through time. The unique contribution made by Darwin and Wallace was to realize that the process of natural selection provided a mechanism for this pattern, which Darwin termed descent with modification.

Evolution by natural selection is the logical outcome of four facts: (1) Individuals vary in most or all traits; (2) some of this variation is genetically based and can be passed on to offspring; (3) more offspring are born than can survive to breed, and of those that do breed, some are more successful than others; and (4) the individuals that reproduce the most are a nonrandom, or more fit, subset of the general population. This selection process causes changes in the genetic makeup of populations over time, or evolution.

Questions

1. In everyday English, the word "adaptation" means an adjustment to environmental conditions. How is the evolutionary definition of adaptation different from the everyday English sense?

2. a. Describe Darwin's four postulates in your own words. What would have happened in the snapdragon experiment if any of the four had *not* been true?
   b. If Darwin's four postulates are true for a given population, is there any way that evolution can not happen? What does this imply about whether evolution is or is not occurring in most populations today?

3. Think about how the finch bill data demonstrate Darwin's postulates.
   a. What would Figure 3.9 have looked like if bill depth was not variable?
   b. What would Figure 3.14 look like if bill depth was variable but the variation was not heritable?
   c. In Figure 3.10, why is the line drawn from 1978 data, after the drought, higher on the *y*-axis than the line drawn from 1976 data, before the drought?

4. According to the text, it is legitimate to claim that most finches died from starvation during the 1977 drought because "there was a strong correspondence between population size and seed availability." Do you accept this hypothesis? If so, why don't the data in Figure 3.12 show a perfect correspondence between when seed availability started declining and when population size started declining?

5. A common creationist criticism of the finch study is, "But it's just a little change in beak shape. Nothing really new has evolved." Or put a different way, "It's just microevolution and not macroevolution." The finch team continues to spend a great deal of effort on their project—traveling thousands of miles to the remote Galápagos every year, just to try to band an entire population of birds and all their nestlings and measure their bills. How would you respond to the creationists' criticisms? Do you think the ongoing 30-year-effort of the finch bill project has been worthwhile? Is it useful to try to document microevolution, and does it tell us anything about how macroevolution might work?

6. Suppose that you are starting a long-term study of a population of annual, flowering plants isolated on a small island. Reading some recent papers has convinced you that global warming will probably cause long-term changes in the amount of rain the island receives. Outline the observations and experiments you would need to do in order to document whether natural selection occurs in your study population over the course of your research. What traits would you measure, and why?

7. At the end of an article on how mutations in variable number tandem repeat (VNTR) sequences of DNA are associated with disease, Krontiris (1995, p. 1683) writes: "the VNTR mutational process may actually be positively selected; by culling those of us in middle age and beyond, evolution brings our species into fighting trim." This researcher proposes that natural selection on humans favors individuals who die relatively early in life. His logic is that the trait of dying from VNTR mutations is beneficial and should spread because the population as a whole becomes younger and healthier as a result. Can this hypothesis be true, given that selection acts on individuals? Explain.

8. Describe three major objections to Darwin's theory in the 19th century that were eventually resolved by discoveries by other scientists in the 20th century. What does this tell us about the utility of a theory that cannot yet answer all questions but that appears to be better than all alternative theories?

9. Many working scientists are relatively uninterested in the history of their fields. Did the historical development of Darwinism, reviewed in Section 3.6, help you understand the theory better? Why or why not? Do you think it is important for practicing scientists to spend time studying history?

10. a. Describe Behe's argument of "irreducible complexity." Is it a logical argument? How does it apply to the bacterial flagellum or the vertebrate eye?
    b. Opponents of intelligent design refer to irreducible complexity as an "argument from personal incredulity" (i.e., "I personally can't imagine how this could have evolved, so it must not have evolved."). What is the logical flaw of an argument from personal incredulity? Do you think it is fair to characterize irreducible complexity in this way?

11. In 1995, the Alabama School Board, after reviewing high school biology texts, voted to require that this disclaimer be posted on the inside front cover of the approved book (National Public Radio 1995):

    > This textbook discussed evolution, a controversial theory some scientists present as a scientific explanation for the origin of living things, such as plants, animals, and humans. No one was present when life first appeared on Earth; therefore, any statement about life's origins should be considered as theory, not fact.

    Do you accept the last sentence in this statement? Does the insert's point of view pertain to other scientific theories, such as the Cell Theory, the Atomic Theory, the Theory of Plate Tectonics, and the Germ Theory of Disease?

12. In the final opinion on the Dover School Board intelligent design trial of 2005 (*Kitzmiller v. Dover*), District Court Judge John E. Jones wrote (in part): "To be sure, Darwin's theory of evolution is imperfect. However, the fact that a scientific theory cannot yet render an explanation on every point should not be used as a pretext to thrust an untestable alternative hypothesis grounded in religion into the science classroom or to misrepresent well-established scientific propositions."

Do you agree with Judge Jones? Why or why not? [See #18 for more information on this trial.]

13. As discussed in Chapter 2, a 2005 poll of U.S. adults found that 42% of the respondents believe that life on earth "has existed in its present form since the beginning of time". Given the evidence for evolution by natural selection, comment on why so few people in the United States accept it.

Exploring the Literature

14. During the past 50 years, hundreds of viruses, bacteria, fungi, and insects have evolved resistance to drugs, herbicides, fungicides, or pesticides. These are outstanding examples of evolution in action. In several of these cases, we know the molecular mechanisms of the evolutionary changes involved. To explore this topic further, look up the following papers. Think about how the evidence from these studies compares with the evidence for evolution in Darwin's finches and HIV.

Anthony, R. G., T. R. Waldin, J. A. Ray, S. W. J. Bright, and P. J. Hussey. 1998. Herbicide resistance caused by spontaneous mutation of the cytoskeletal protein tubulin. *Nature* 393: 260–263.

Cohen, M. L. 1992. Epidemiology of drug resistance: Implications for a post-antimicrobial era. *Science* 257: 1050–1055.

Davies, J. 1994. Inactivation of antibiotics and the dissemination of resistance genes. *Science* 264: 375–382.

Van Rie, J., W. H. McGaughey, D. E. Johnson, B. D. Barnett, and H. Van Melleart. 1990. Mechanism of insect resistance to the microbial insecticide *Bacillus thuringiensis*. *Science* 247: 72–74.

15. It seems unlikely that selection of traits "for the good of the species" can occur. However, it now appears that under certain conditions, such as small group size and very low migration, group selection of altruistic behaviors may in fact be possible. Look up the following papers to learn more about this topic:

Avilés, L., and P. Tufino. 1998. Colony size and individual fitness in the social spider *Anelosimus eximius*. *American Naturalist* 152: 403–418.

Borrello, M.E. 2005. The rise, fall, and resurrection of group selection. *Endeavour* 29 (1):43–47.

Ono, S., and K. Misawa, and K. Tsuji. 2003. Effect of group selection on the evolution of altruistic behavior. *J. Theor. Biol.* 220 (1):55–66.

16. For detailed critical discussions of Intelligent Design Creationism, see:

Miller, K. R. 1999. *Finding Darwin's God: A Scientist's Search for Common Ground Between God and Evolution.* New York: Cliff Street Books.

Pennock, R.T., ed. 2001. *Intelligent Design Creationism and Its Critics.* Cambridge, MA: The MIT Press. (See especially Chapter 10 by M. J. Behe, Chapter 11 by Philip Kitcher, and Chapter 12 by M. J. Brauer and D. R. Brumbaugh.)

Young, M., and T. Edis (editors). 2004. *Why Intelligent Design Fails: A Scientific Critique of the New Creationism.* New Jersey: Rutgers University Press.

17. For new hypotheses about the evolution of Behe's two most famous examples of "irreducibly complex" systems, the flagellum and the blood clotting cascade, see:

Aird, W.C. 2003. Hemostasis and irreducible complexity. *J. Thromb. Haemost.* 1 (2): 227–230.

Hanumanthaiah, R., K. Day, and P. Jagadeeswaran. 2002. Comprehensive analysis of blood coagulation pathways in teleostei: evolution of coagulation factor genes and identification of zebrafish factor VIIi. *Blood Cells Mol. Dis.* 29 (1): 57–68.

Davidson, C.J., R.P. Hirt, K. Lal, P. Snell, G. Elgar, E.G. Tuddenham, and J.H. McVey. 2003. Molecular evolution of the vertebrate blood coagulation network. *Thromb. Haemost.* 89 (3): 420–428.

Li, J.Y., and C.F. Wu. 2005. New symbiotic hypothesis on the origin of eukaryotic flagella. *Naturwissenschaften* 92 (7): 305–309.

Mitchell, D.R. 2004. Speculations on the evolution of 9+2 organelles and the role of central pair microtubules. *Biol. Cell* 96 (9): 691–696.

18. For the full court decision of the 2005 Dover School Board intelligent design case (Kitzmiller vs. Dover School District), see:

www.pamd.uscourts.gov/kitzmiller/kitzmiller_342.pdf

Citations

Abbott, I., L. K. Abbott, and P. R. Grant. 1977. Comparative ecology of Galápagos ground finches (*Geospiza* Gould): Evaluation of the importance of floristic diversity and interspecific competition. *Ecological Monographs* 47: 151–184.

Abzhanov, A., M. Protas, B. R. Grant, et al. 2004. *Bmp4* and morphological variation of beaks in Darwin's finches. *Science* 305: 1462–1465.

Albala, K. 2002. *Eating Right in the Renaissance.* Berkeley: University of California Press. Page 237.

Baardsnes, J., and P. L. Davies. 2001. Sialic acid synthetase: The origin of fish type III antifreeze protein? *Trends in Bichemical Sciences* 26: 468–469.

Behe, M. 1996. *Darwin's Black Box: The Biochemical Challenge to Evolution.* New York: Free Press/Simon and Schuster.

Behe, M. 1998. Molecular machines: Experimental support for the design inference. *Cosmic Pursuit* (Spring): 27–35. Available online at: *http://www.discovery.org/viewDB/index.php3?program=CRSC&command=view&id=54.*

Boag, P. T. 1983. The heritability of external morphology in Darwin's ground finches (*Geospiza*) on Isla Daphne Major, Galápagos. *Evolution* 37: 877–894.

Boag, P. T., and P. R. Grant. 1978. Heritability of external morphology in Darwin's finches. *Nature* 274: 793–794.

Boag, P. T., and P. R. Grant. 1981. Intense natural selection in a population of Darwin's finches (Geospizinae) in the Galápagos. *Science* 214: 82–85.

Boag, P. T., and P. R. Grant. 1984a. Darwin's finches (*Geospiza*) on Isla Daphne Major, Galápagos: Breeding and feeding ecology in a climatically variable environment. *Ecological Monographs* 54: 463–489.

Boag, P. T., and P. R. Grant. 1984b. The classical case of character release: Darwin's finches (*Geospiza*) on Isla Daphne Major, Galápagos. *Biological Journal of the Linnean Society* 22: 243–287.

Bowman, R. I. 1961. Morphological differentiation and adaptation in the Galápagos finches. *University of California Publications in Zoology* 58: 1–302.

Brusca, R. C., and G. J. Brusca. 1990. *Invertebrates.* Sunderland, MA: Sinauer.

Cong, B., J. Liu, and S. D. Tanksley. 2002. Natural alleles at a tomato fruit size quantitative trait locus differ by heterchronic regulatory mutations. *Proceedings of the National Academy of Sciences USA* 99: 13606–13611.

Darwin, C. 1859. *On the Origin of Species by Means of Natural Selection.* London: John Murray.

Dawkins, R. 1986. *The Blind Watchmaker.* Essex: Longman Scientific.

Dawkins, R. 1994. The eye in a twinkling. *Nature* 368: 690–691.

Dembski, W. A. 2002. *No Free Lunch: Why Specified Complexity Cannot Be Purchased Without Intelligence.* Lanham, MA: Rowman & Littlefield.

Dobzhansky, T. 1973. Nothing in biology makes sense except in the light of evolution. *American Biology Teacher* 35: 125–129.

Dodson, E. O. 1960. *Evolution: Process and Product.* New York: Reinhold Publishing.

Donner, A. L., and R. L. Mass. 2004. Conservation and non-conservation of genetic pathways in eye specification. *International Journal of Developmental Biology* 48: 743–753.

Easterbrook, G. 1997. Science and God: A warming trend? *Science* 277: 890–893.

Endo, E., Y. Hayashi, D. Yamagiwa, et al. 1999a. CT examination of the manipulation system in the giant panda (*Ailuropoda melanoleuca*). *Journal of Anatomy* 195: 295–300.

Endo, H., D. Yamagiwa, Y. Hayashi, et al.. 1999b. Role of the giant panda's pseudo-thumb. *Nature* 397: 309–310.

Endo, H., M. Sasaki, Y. Hayashi, et al. 2001. Carpal bone movements in gripping action of the giant panda (*Ailuropoda melanoleuca*). *Journal of Anatomy* 198: 243–246.

Fernald, R. D. 2004. Evolving Eyes. *International Journal of Developmental Biology* 48: 701–705.

Fletcher, G. L., C. L. Hew, and P. L. Davies. 2001. Antifreeze proteins of teleost fishes. *Annual Review of Physiology* 63: 359–90.

Frary, A., T. C. Nesbitt, et al. 2000. *fw2.2*: A quantitative trait locus key to the evolution of tomato fruit size. *Science* 289: 85–88.

Frey, Erica, C. J. Brokaw, and C. K. Omoto. 1997. Reactivation at low ATP distinguishes among classes of *paralyzed flagella* mutants. *Cell Motility and the Cytoskeleton* 38: 91–99.

Futuyma, D. J. 1983. *Science on Trial: The Case for Evolution.* New York: Pantheon.

Gehring, W. J. 2004. Historical perspective on the development and evolution of eyes and photoreceptors. *International Journal of Developmental Biology* 48: 707–717.

Giancoli, D. C. 1995. *Physics: Principles with Applications.* Englewood Cliffs, NJ: Prentice Hall.

Gibbs, H. L., and P. R. Grant. 1987. Oscillating selection on Darwin's finches. *Nature* 327: 511–513.

Gish, D. T. 1978. *Evolution: The Fossils Say No!* San Diego: Creation-Life Publishers.

Goldsmith, T. H. 1990. Optimization, constraint, and history in the evolution of eyes. *Quarterly Review of Biology* 65: 281–322.

Goodstein, L. 2005. Judge bars 'Intelligent Design' from Pa. classes. *The New York Times* 20 December.

Gould, S. J. 1980. *The Panda's Thumb.* New York: W. W. Norton.

Gould, S. J. 1983. *Hen's Teeth and Horse's Toes.* New York: W. W. Norton.

Grant, B. R. 1985. Selection on bill characters in a population of Darwin's finches: *Geospiza conirostris* on Isla Genovesa, Galápagos. *Evolution* 39: 523–532.

Grant, B. R. 2003. Evolution in Darwin's finches: A review of a study on Isla Daphne Major in the Galápagos Archipelago. *Zoology* 106: 255–259.

Grant, B. R., and P. R. Grant. 1989. *Evolutionary Dynamics of a Natural Population.* Chicago: University of Chicago Press.

Grant, B. R., and P. R. Grant. 1993. Evolution of Darwin's finches caused by a rare climatic event. *Proceedings of the Royal Society of London* B 251: 111–117.

Grant, B. R., and P. R. Grant. 2003. What Darwin's finches can teach us about the evolutionary origin and regulation of biodiversity. *BioScience* 53 (10): 965–975.

Grant, P. R. 1981a. Speciation and adaptive radiation on Darwin's finches. *American Scientist* 69: 653–663.

Grant, P. R. 1981b. The feeding of Darwin's finches on *Tribulus cistoides* (L.) seeds. *Animal Behavior* 29: 785–793.

Grant, P. R. 1991. Natural selection and Darwin's finches. *Scientific American* October: 82–87.

Grant, P. R. 1999. *Ecology and Evolution of Darwin's Finches,* 2nd ed. Princeton: Princeton University Press.

Grant, P. R., and B. R. Grant. 1995. Predicting microevolutionary responses to directional selection on heritable variation. *Evolution* 49: 241–251.

Grant, P. R., and B. R. Grant. 2000. Non-random fitness variation in two populations of Darwin's finches. *Proceedings of the Royal Society of London* B 267: 131–138.

Grant, P. R., and B. R. Grant. 2002a. Unpredictable evolution in a 30-year study of Darwin's finches. *Science* 296: 707–711.

Grant, P. R., and B. R. Grant. 2002b. Adaptive radiation of Darwin's finches. *American Scientist* 90(2): 130–139.

Grant, P. R., and B. R. Grant. 2005. Darwin's finches. *Current Biology* 15: R614–R615.

Grant, P. R., and B. R. Grant. 2006. Evolution of character displacement in Darwin's finches. *Science* 313: 224–226.

Grant, P. R., B. R. Grant, J. N. M. Smith, I. J. Abbott, and L. K. Abbott. 1976. Darwin's finches: Population variation and natural selection. *Proceedings of the National Academy of Sciences USA* 73: 257–261.

Harding, R. M., E. Healy, et al. 2000. Evidence for variable selective pressures at MC1R. *American Journal of Human Genetics* 66: 1351–1361.

Holden, C. 1995. Alabama schools disclaim evolution. *Science* 270: 1305.

Huynen, M. A., T. Dandekar, and P. Bork. 1999. Variation and evolution of the citric-acid cycle: A genomic perspective. *Trends in Microbiology* 7: 281–291.

Jones, K. N., and J. S. Reithel. 2001. Pollinator-mediated selection on a flower color polymorphism in experimental populations of *Antirrhinum* (Scrophulariaceae). *American Journal of Botany* 88: 447–454.

Jones, J. E., III. 2005. *Tammy Kitzmiller v. Dover Area School District, Memorandum Opinion.* US District Court for the Middle District of Pennsylvania, Case No. 04cv2688.

Kaiser, J. 1995. Dutch debate tests on evolution. *Science* 269: 911.

Kanetsky, P. A., J. Swoyer, et al. 2002. A polymorphism in the agouti signaling protein gene is associated with human pigmentation. *American Journal of Human Genetics* 70: 770–775.

Keller, L. F., P. R. Grant, B. R. Grant, and K. Petren. 2001. Heritability of morphological traits in Darwin's finches: Misidentified paternity and maternal effects. *Heredity* 87: 325–336.

Kitcher, P. 1982. *Abusing Science: The Case Against Creationism.* Cambridge, MA: MIT Press.

Krem, M. M., and E. Di Cera. 2002. Evolution of enzyme cascades from embryonic development to blood coagulation. *Trends in Biochemical Sciences* 27: 67–74.

Krontiris, T. G. 1995. Minisatellites and human disease. *Science* 269: 1682–1683.

Langerhans, R. B., C. A. Layman, and T. J. DeWitt. 2005. Male genital size reflects a tradeoff between attracting mates and avoiding predators in two live-bearing fish species. *Proceedings of the National Academy of Sciences USA* 102: 7618–7623.

Larson, E. J. 1997. *Summer for the Gods: The Scopes Trial and America's Continuing Debate Over Science and Religion*. Cambridge, MA: Harvard University Press.

Lenski, R. E., C. Ofria, R. T. Pennock, and C. Adami. 2003. The evolutionary origin of complex features. *Nature* 423: 139–144.

Liu, J., B. Cong, and S. D. Tanksley. 2003. Generation and analysis of an artificial gene dosage series in tomato to study the mechanisms by which the cloned quantitative trait locus *fw2.2* controls fruit size. *Plant Physiology* 132: 292–299.

Mayr, E. 1980. Prologue. In Mayr, E., and W. B. Provine, eds. *The Evolutionary Synthesis.* Cambridge, MA: Harvard University Press.

Mayr, E. 1991. *One Long Argument: Charles Darwin and the Genesis of Modern Evolutionary Thought.* Cambridge, MA: Harvard University Press.

Meléndez-Hevia, E., T. G. Waddell, and M. Cascante. 1996. The puzzle of the Krebs citric acid cycle: Assembling the pieces of chemically feasible reactions, and opportunism in the design of metabolic pathways during evolution. *Journal of Molecular Evolution* 43: 293–303.

Miller, K. R. 1999. *Finding Darwin's God: A Scientist's Search for Common Ground Between God and Evolution.* New York: Cliff Street Books.

Mitchell, D. R. 2000. *Chlamydomonas* flagella. *Journal of Phycology* 36: 261–273.

Moose, S. P., J. W. Dudley, and T. R. Rocheford. 2004. Maize selection passes the century mark: a unique resource for 21st century genomics. *Trends in Plant Science* 9: 358–364.

Musser, S. M., and S. I. Chan. 1998. Evolution of the cytochrome *c* oxidase proton pump. *Journal of Molecular Evolution* 46: 508–520.

National Public Radio. 1995. Evolution disclaimer to be placed in Alabama textbooks. Morning Edition, Transcript #1747, Segment #13.

Nesbitt, T. C., and S. D. Tanksley. 2002. Comparative sequencing in the genus Lycopersicon: Implications for the evolution of fruit size in the domestication of cultivated tomatoes. *Genetics* 162: 365–379.

Niklaus, K. J. 1997. *The Evolutionary Biology of Plants.* Chicago: University of Chicago Press.

Nilsson, D.-E., and S. Pelger. 1994. A pessimistic estimate of the time required for an eye to evolve. *Proceedings of the Royal Academy of London* B 256: 53–58.

Orr, H. A. 2002. The Return of Intelligent Design (review of No Free Lunch: Why Specified Complexity Cannot Be Purchased without Intelligence by William A. Dembski). *Boston Review* Summer: 53–56.

Packer, C., and A. E. Pusey. 1983. Adaptations of female lions to infanticide by incoming males. *American Naturalist* 121: 716–728.

Packer, C., and A. E. Pusey. 1984. Infanticide in carnivores. In G. Hausfater and S. B. Hrdy, eds. *Infanticide.* New York: Aldine Publishing Company, pp. 31–42.

Pallen, M. J., and N. J. Matzke. 2006. From *The Origin of Species* to the origin of bacterial flagella. *Nature Reviews Microbiology* 4:784–790.

Petren, K., B. R. Grant, and P. R. Grant. 1999. A phylogeny of Darwin's finches based on microsatellite DNA length variation. *Proceedings of the Royal Society of London* B 266: 321–329.

Petren, K., P. R. Grant, B. R. Grant, and L. F. Keller. 2005. Comparative landscape genetics and the adaptive radiation of Darwin's finches: the role of peripheral isolation. *Molecular Ecology* 14: 2943–2957.

Pole, A., I J. Gordon, and M. L. Gorman. 2003. African wild dogs test the 'survival of the fittest' paradigm. *Proceedings of the Royal Society of London* B (Supplement) 270: S57.

Price, T. D., P. R. Grant, H. L. Gibbs, and P. T. Boag. 1984. Recurrent patterns of natural selection in a population of Darwin's finches. *Nature* 309: 787–789.

Provine, W. B. 1971. *The Origins of Theoretical Population Genetics.* Chicago: University of Chicago Press.

Quiring, R., U. Walldorf, U. Kloter, and W. J. Gehring. 1994. Homology of the *eyeless* gene of *Drosophila* to the *small eye* gene in mice and *aniridia* in humans. *Science* 265: 785–789.

Root-Bernstein, R. S. 1995. Darwin's rib. *Discover* (September) 38–41.

Salvini-Plawen, L. V., and E. Mayr. 1977. On the evolution of photoreceptors and eyes. *Evolutionary Biology* 10: 207–263.

Sato, A., H. Tichy, C. O'hUigin, et al. 2001. On the Origin of Darwin's finches. *Molecular Biology and Evolution* 18: 299–311.

Schaffer, J. V., and J. L. Bolognia. 2001. The melanocortin-1 receptor: Red hair and beyond. *Archives of Dermatology* 137: 1477–1485.

Schmidt, K. 1996. Creationists evolve new strategy. *Science* 273: 420–422.

Scott, E. C. 1994. The struggle for the schools. *Natural History* 7: 10–13.

Scott, E. C. 1998. Two kinds of materialism. *Free Inquiry* 18: 20.

Smith, R., E. Healy, et al. 1998. Melanocortin 1 receptor variants in an Irish population. *Journal of Investigative Dermatology* 111: 119–122.

Spooner, D. M., I. E. Peralta, and S. Knapp. 2005. Comparison of AFLPs with other markers for phylogenetic inference in wild tomatoes [*Solanum* L. section *Lycopersicon* (Mill.) Wettst.] *Taxon* 54: 43–61.

Swinney, S. 1994. *Evolution: Fact or Fiction.* Kansas City, MO: 1994 Staley Lecture Series, KLJC Audio Services.

Tanksley, S. D. 2004. The genetic, developmental, and molecular bases of fruit size and shape variation in tomato. *The Plant Cell* 16: S181–S189.

Tewksbury, J. J., and G. P. Nabhan. 2001. Directed deterrence by capsaicin in chillies. *Nature* 412: 403–404.

True, J. R., and S. B. Carroll. 2002. Gene co-option in physiological and morphological evolution. *Annual Review of Cell and Developmental Biology* 18: 53–80.

Willson, M. F. 1984. *Vertebrate Natural History.* Philadelphia: Saunders.

Woolley, D. M., 1997. Studies on the eel sperm flagellum. I. The structure of the inner dynein arm complex. *Journal of Cell Science* 110: 85–94.

Wu, P., T.-X. Jiang, S. Suksaweang, et al. 2005. Molecular shaping of the beak. *Science* 305: 1465–1466.

4

Estimating Evolutionary Trees

This solitary tree is an appropriate symbol for the tree of life—the evolutionary tree that describes the relationships among all species. This chapter introduces the tools that biologists use to analyze the tree of life.

You might recall from Chapter 2 that the evolutionary history of a group of species is called its **phylogeny**, and that a **phylogenetic tree** is a graphical summary of this history. An evolutionary tree describes the pattern, and in some cases the timing, of events that occurred as species diversified. It records the sequence in which lineages appeared and documents which organisms are more closely or distantly related.

Because we do not have direct knowledge of evolutionary history in the vast majority of cases, phylogenetic trees have to be inferred from data. What types of data do biologists use to estimate trees like the ones we used in Chapters 1 and 2 to infer where HIV came from and how leopards got their spots? How do researchers know that they have inferred the most accurate tree implied by the data? This chapter focuses on these questions. It introduces how evolutionary trees are put together and evaluated. It also explores how they are used in more detail.

The chapter opens with an introduction to the basic principles that biologists use to infer phylogenetic trees, and continues with an analysis of the phylogenetic relationships among some major vertebrate groups and then between whales and other mammals. These case studies illustrate how researchers choose data for a phylogenetic problem, use the principle of parsimony in phylogeny estimation, and evaluate the reliability of a particular phylogeny. This presentation parallels the sequence of decisions that faces researchers when estimating phylogenies

(Swofford et al. 1996). The chapter closes with a series of examples that illustrate how phylogenetic thinking is applied in contemporary research. Topics in this last section range widely, from how we should classify the diversity of life to the evolution of human influenza strains that have caused deadly pandemics. If you have a thorough understanding of evolution by natural selection and become comfortable with tree thinking, you will be well on the way to thinking like an evolutionary biologist.

4.1 The Logic of Phylogeny Inference

At its most basic level, the logic of estimating evolutionary relationships is simple: The most closely related taxa should have the most traits in common. Naively, we would say that any traits that have a genetic basis and that vary among the taxa involved can be assessed for similarity and help us reconstruct who evolved from whom. Many types of characters could qualify: the sequence of nucleotides in a particular gene, the presence or absence of specific skeletal elements or flower parts, or the mode of embryonic or larval development.

Grouping species by their similarities and distinguishing groups by their differences would seem to be a fairly straightforward way of inferring their evolutionary relationships. Unfortunately, phylogeny inference is anything but simple in practice. Let's first consider which types of similar traits are informative, and then delve into processes that complicate phylogeny inference.

Synapomorphies Identify Monophyletic Groups

The most fundamental principle of phylogeny inference is that only certain types of homologous characters are useful in estimating phylogenetic trees. Recall from Chapter 2 that homology is defined as similarity in traits that is due to descent from a common ancestor. The types of homologies that are useful in estimating phylogenies are called synapomorphies. A **synapomorphy** is a homologous trait that is shared among certain species and is similar because it was modified in a common ancestor. Synapomorphies are shared, derived traits.

To get a better feel for this concept, consider the relationship between synapomorphies and other types of homologies, using the genetic code as an example. The genetic code is a homologous trait that is shared by all organisms alive today. It is a synapomorphy that groups all living organisms into a single lineage that descended from the same common ancestor. Any group that includes an ancestor and all of its descendants is called a monophyletic group (or clade or lineage; Figure 4.1). The key point to recognize is that the genetic code helps identify bacte-

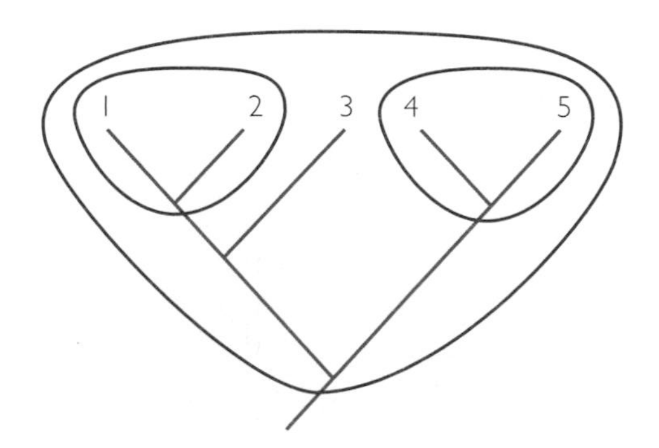

Figure 4.1 Monophyletic groups are comprised of an ancestor and all of Its descendants Monophyletic groups are also called clades or lineages. The groups circled here are all monophyletic. The group described by species 1, 2, 3, and their closest common ancestor is also monophyletic.

ria and mammals as members of the same monophyletic group, but it does not help us distinguish bacteria from eukaryotes. Instead, bacteria and mammals each have synapomorphies that identify them as distinct monophyletic groups, within the monophyletic group consisting of all species alive today. Bacteria are identified by synapomorphies such as cell walls that contain a compound called peptidoglycan; eukaryotes have synapomorphies such as the nuclear envelope.

To state this concept another way, all synapomorphies are homologous traits, but not all homologous traits are synapomorphies. Synapomorphies can be identified at whatever taxonomic level a researcher might be interested in: populations, species, genera, phyla, and so on.

A synapomorphy is a trait that is similar among species because the common ancestor of those species also had the trait. Synapomorphies are important because they identify monophyletic groups.

Two ideas are key to understanding why evolutionary relationships can be inferred by analyzing synapomorphies. The first is that synapomorphies identify evolutionary branch points. As Chapter 16 will show, the process called speciation starts when two populations become genetically isolated, meaning that gene flow is reduced or absent. When genetic separation occurs and species begin evolving independently, some of the homologous traits in each population undergo changes due to mutation, selection, and drift. These changed traits are synapomorphies that identify the populations belonging to the two independent, descendant lineages (Figure 4.2a). The second key idea is that synapomorphies are nested. That is, as you move through time and trace a tree from its root to its tips, each branching event adds one or more shared, derived traits (Figure 4.2b). As a result, the hierarchy described by synapomorphies also describes the hierarchy of branching events.

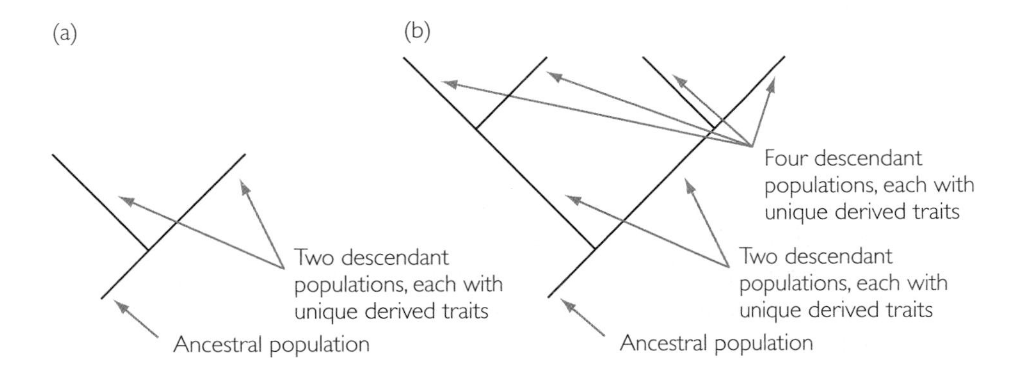

Figure 4.2 Synapomorphies arise in ancestral populations, and are passed on to descendants (a) Speciation leads to the creation of two independent populations. Each acquires unique traits by mutation, selection, and genetic drift but they share traits inherited from their common ancestor. (b) As you go up a tree, synapomorphies create a nested hierarchy. Each successive monophyletic group can be identified by synapomorphies that arose in its ancestors.

These theoretical insights are due to the German entomologist Willi Hennig (1979), who began writing on phylogeny inference methods in the 1950s. Phylogeny inference methods that use these principles are called cladistic methods.

To implement a cladistic approach to inferring a phylogeny, researchers have to determine which traits are more ancient and which are more derived. There are several ways to do this. One of the most basic and reliable methods is called outgroup analysis (Maddison et al. 1984). In outgroup analysis, the character state in the group of interest (the ingroup) is compared to the state in a very close relative that clearly branched off earlier (the outgroup). Finding an appropriate outgroup, in turn, involves borrowing conclusions from other phylogenetic analyses or confirming an earlier appearance in the fossil record.

Each monophyletic group on an evolutionary tree can be identified by one or more synapomorphies.

Then each branch on the tree corresponds to one or more synapomorphies that distinguish the derived groups. A phylogenetic tree inferred by clustering synapomorphies in this way is called a **cladogram**.

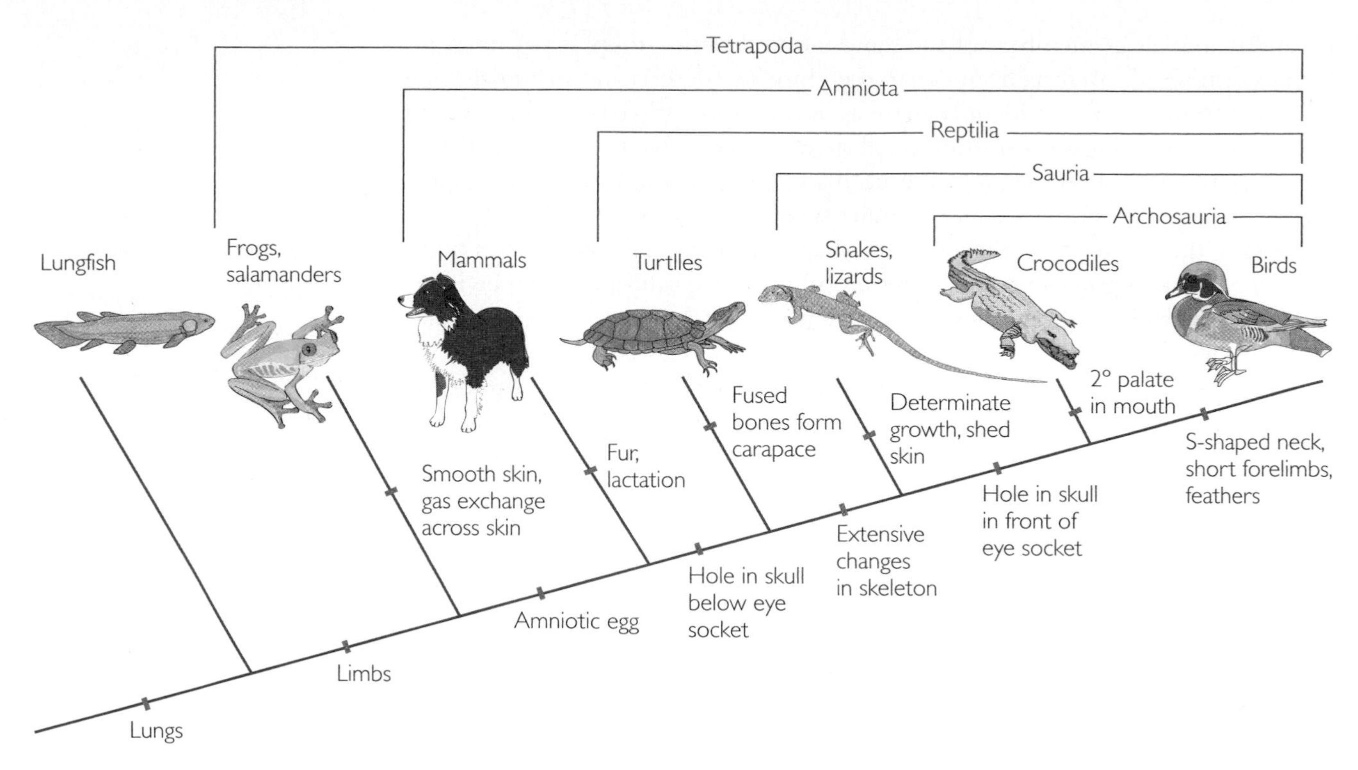

Figure 4.3 Synapomorphies reveal the relationships among tetrapods The traits that are labeled at each hash mark on this tree are synapomorphies shared by the descendant species above that point. For example, birds have feathers and other shared, derived traits that identify them as birds. But they also have four limbs that identify them as a member of the monophyletic group called Tetrapoda, amniotic eggs that identify them as members of the clade called Amniota, and so on.

By convention, synapomorphies are indicated on cladograms with bars across the branches and then described in an accompanying key or labels. The synapomorphies marked on the tree in Figure 4.3 are traits that allowed researchers to estimate the evolutionary relationships among the tetrapods, or four-footed vertebrates. Note that the evolution of limbs in vertebrates is a synapomorphy that identifies the clade called Tetrapoda; moist, scaleless skin and the ability to exchange gases across the skin are some of the synapomorphies that link the lineage called Amphibia; the amniotic egg is a synapomorphy that distinguishes the monophyletic group called Amniota; and so on. The groups are ordered this way based on data indicating that limbs are derived from the limbs in the common ancestor of tetrapods and lungfish, that amphibian skin is derived from the scaled skin of fish-like ancestors, and that the amniotic egg is derived from eggs similar to those observed in living species of lungfish and Amphibia.

Problems in Reconstructing Phylogenies

To reconstruct a phylogenetic tree accurately, researchers have to analyze homologous traits and identify characteristics that qualify as synapomorphies. Unfortunately, it is entirely possible for similar traits to evolve independently in different groups of species. In cases like these, species share similar traits that were not derived from a common ancestor. Instead of serving as synapomorphies that identify monophyletic groups, some similar traits may actively mislead efforts to reconstruct evolutionary history. Let's consider why these problems arise, and then look at how biologists address them.

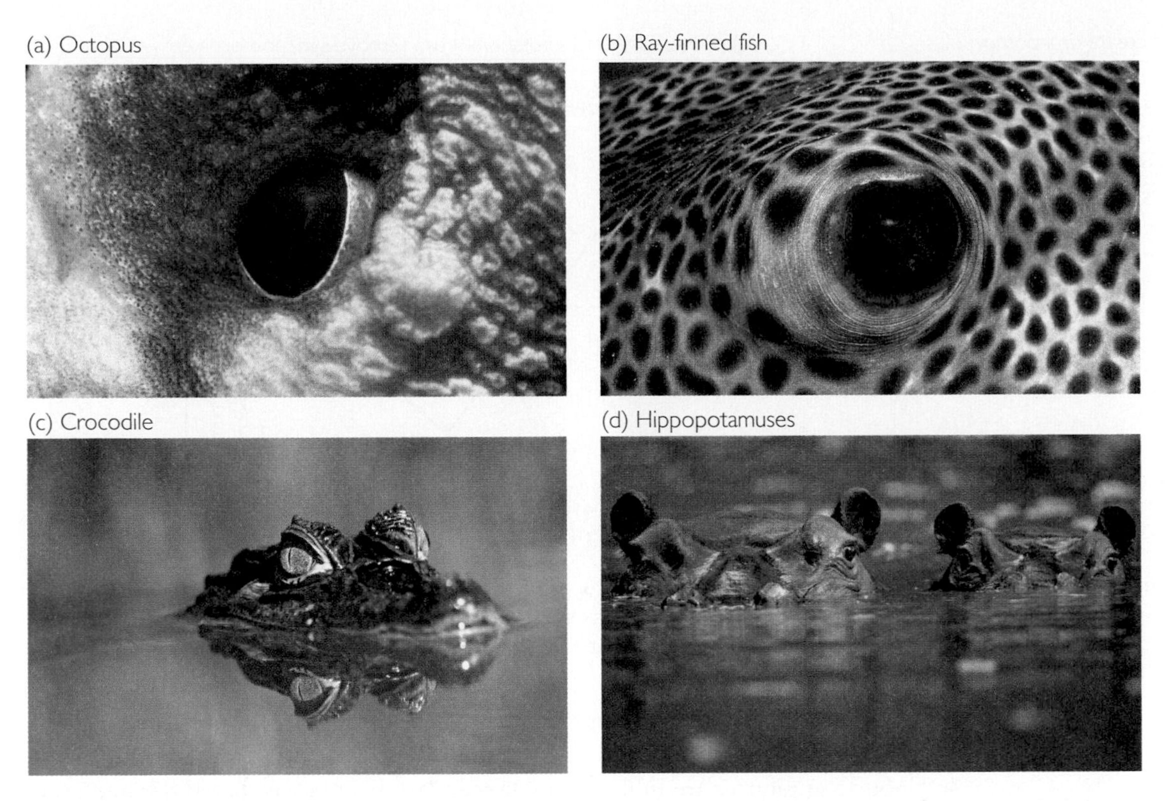

Figure 4.4 Similar traits may not be homologous The pairs of species shown have similar traits even though they are not closely related. The octopus (a) and ray-finned fish (b) have camera eyes. The crocodile (c) and hippos (d) have skulls in which the eyes sit on top. Theses similarities are due to convergent evolution, not common ancestry.

Not All Similar Traits Are Homologous

Octopuses share a large number of derived characteristics with squid, clams, mussels, and other mollusks. Yet they also have an image-forming "camera eye" similar in structure and function to the eyes found in ray-finned fish and other vertebrate animals [Figure 4.4a, (b)]. Within vertebrates, both crocodiles and hippopotamuses have eyes that are located at the tops of their skulls instead of at the side [Figure 4.4c, (d)]. Yet crocodiles have an array of synapomorphies that identify them as reptiles, while hippos share fur, lactation, and other derived traits with mammals.

Morphological similarities like the eyes and skulls pictured in Figure 4.4 evolve independently in different lineages due to **convergent evolution**, which occurs when natural selection favors similar structures as solutions to problems posed by similar environments. Octopuses and vertebrates are not hypothesized to have inherited their sophisticated camera eyes from a common ancestor. Instead, the two lineages independently evolved camera eyes because both depend on eyesight to find food and avoid danger. Likewise, hippopotamuses and crocodiless do not have eyes on the tops of their heads because they inherited the trait from a common ancestor. Hippos and crocs both spend large parts of the day submerged in water, so having eyes located at the tops of their skulls is hypothesized to help them watch for food or predators while keeping their heads cool and hidden. Other examples of convergent evolution include the wings of bats and birds, the streamlined shapes of sharks and whales, and the elongated, limbless bodies of snakes and legless lizards.

If convergent evolution has occurred, then similar traits are not homologous and do not qualify as synapomorphies.

(a) Mutations can create synapomorphies

DNA sequences in descendants:

TGCTATT TGCTTTT TGCTTTT

DNA sequence changes to: TGCTTTT

Ancestral DNA sequence: TGCTATT

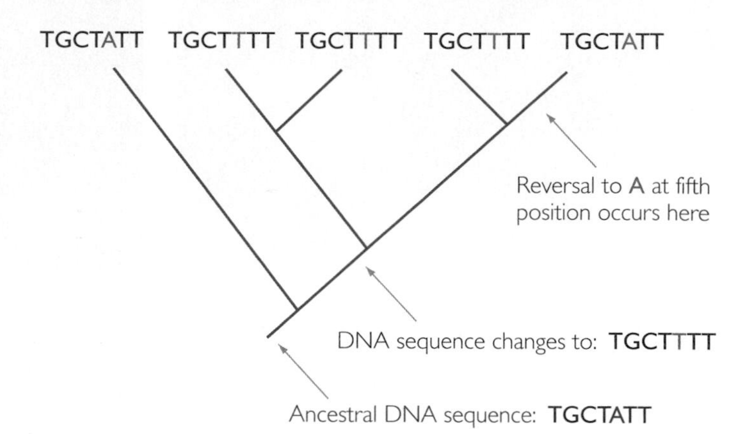

Figure 4.5 Reversals complicate phylogeny inference (a) Read this tree up from the root, and notice that a change in the fifth position of this DNA sequence creates a shared, derived character in the descendant populations. (b) If a reversal changed the fifth position back to the ancestral state later in the evolution of this group, it would make it much more difficult to infer the correct phylogeny.

The same types of similarities can arise at the molecular level. Evolution at the molecular level entails changes in nucleotide sequence due to mutation. Novel nucleotide sequences are inherited by descendant lineages as shared derived traits (Figure 4.5a). But species can share nucleotide sequences for reasons other than common ancestry. To see how this can happen, put your finger at the base of the tree in Figure 4.5b. As you begin to move your finger up the tree, notice the 7-base-pair DNA sequence that was present in the ancestral population. This same sequence is still found in species A today. After species A split off, a change occurred in the DNA found in the ancestor to species B–E. Specifically, a mutation occurred that changed the fifth base in the sequence from an A to a T. Three of the descendant species, B–D, have this changed sequence. But in the ancestor of species E, a mutation occurred that changed the base in the fifth position back to A. As a result, species A and species E have an identical DNA sequence. They did not inherit this sequence from the same ancestor, however. Instead, species E inherited the sequence from the ancestor at the top arrow, and species A inherited the sequence from the ancestor at the base of the tree. In species A and E, the similar bases in the fifth position are not homologous.

If a reversal has occurred, then similar traits are not homologous and do not qualify as synapomorphies.

In DNA sequence data, a change like the one illustrated in Figure 4.5 is termed a **reversal**. Reversals are common in DNA data because there are only four possible states for each base in a sequence. Other things being equal, there is a 25% chance that a reversal to the previous state will occur each time a change occurs at a particular site in DNA.

Convergence and reversal are lumped under the term **homoplasy**. If similarities in traits are not due to homology, then they are due to homoplasy. How do biologists tell the difference?

Distinguishing Homology from Homoplasy

It is one thing to recognize that homoplasy occurs, and another thing to distinguish it from homology and keep it from leading to incorrect conclusions about which species are most closely related.

The most efficient way to distinguish homology from homoplasy is to analyze many traits in reconstructing evolutionary relationships instead of just one or a few. For example, ray-finned fish and other vertebrates have a bony skeleton and a wide array of other traits that distinguish them from octopuses and other mollusks. Grouping species only on the basis of eye structure would suggest that octopuses and vertebrates are closely related, but this hypothesis quickly runs into trouble when a larger set of characteristics are examined. If you insisted that octopus eyes and vertebrate eyes were homologous, you would need to explain how their hearts, skeletons, and other characteristics got to be so different.

The essence of phylogeny inference is to maximize the use of reliable information while minimizing the impact of misleading information. Synapomorphies are reliable; homoplasy is misleading.

To drive this point home, consider the evolutionary trees shown in Figure 4.6, which were estimated on the basis of a large number of traits. The tree in part (a) assumes that the octopus eye and vertebrate eye are homologous. Under this hypothesis, a camera eye would have evolved at the point indicated at the base of the tree and would then have been lost at all the points marked in lineages derived from that point. In contrast, the tree in part (b) assumes that the octopus eye and vertebrate eye resulted from convergent evolution. Under this hypothesis, a camera eye would have had to evolve at the two points marked. The tree in part (a) implies six changes while the tree in part (b) implies just two changes. Under **parsimony**, biologists consider the hypothesis of convergence as the most likely.

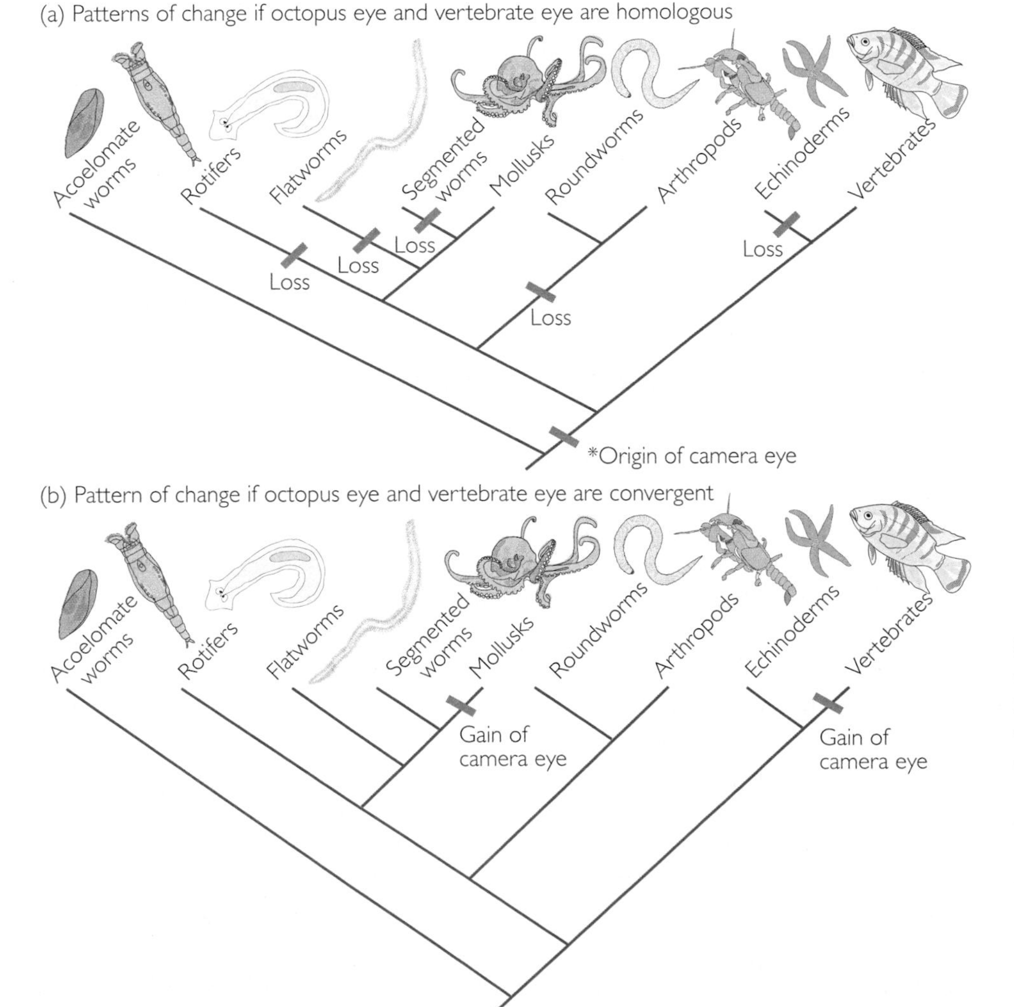

Figure 4.6 Using parsimony to distinguish homology from homoplasy The trees shown were estimated using a large number of synapomorphies in DNA sequences. (a) If the camera eyes of octopuses and vertebrates are homologous, then six evolutionary changes occurred, as shown. (b) If the camera eyes of octopuses and vertebrates are convergent, then two evolutionary changes occurred, as shown.

In addition, careful study usually shows that homoplasius morphological traits are not as similar as they initially appear. Octopuses and vertebrates each have a layer of light-sensitive cells in their eyes; but in octopuses these cells are oriented toward the opening where light enters the eye, while in vertebrates the light-sensitive cells are oriented in the opposite direction. Muscles in the octopus eye move the lens back and forth and change the shape of the eyeball in order to focus the image. But in vertebrates, focusing is accomplished by muscles that change the shape of the lens.

If you think back to the structural, developmental, and genetic homologies described in Chapter 2 for a moment, it should make sense that homoplasius morphological traits are not identical. Structural similarities that are due to homoplasy have evolved independently. As a result, it is almost certain that different alleles and developmental pathways are responsible for the traits. But morphological traits that are similar due to shared ancestry should develop in similar ways in embryos and depend on the products of similar alleles. The logic here is the developmental pathways and alleles observed in different species will be similar if they were present in a common ancestor. Structural homologies result from developmental homologies, which result from genetic homologies.

Unfortunately, it is rarely possible to eliminate homoplasius characters from the datasets used to infer phylogenetic trees. For example, it is not possible to analyze the underlying developmental or genetic traits responsible for morphological homologies in fossil taxa or in rare or poorly studied species living today, and DNA sequence data and other types of molecular characters are almost guaranteed to contain reversals.

Homoplasy is a fact of life in phylogeny inference. It represents "noise" in the data sets used to reconstruct evolutionary history. Homoplasius traits are analogous to the bad or misleading measurements that are present in almost every dataset used in science. Homoplasius characters are troublesome in phylogeny inference because they suggest that species are closely related, when in fact they are not. If homoplasy is inevitable, how can trees still be estimated accurately?

Resolving Conflicts in Data Sets: The Role of Parsimony

Parsimony provides one way to identify which branching pattern, among the many that are possible, minimizes the confusing effects of homoplasy and most accurately reflects actual evolutionary history. Parsimony is a general logical criterion. Under parsimony, simpler explanations are preferred over more complex explanations. When parsimony is applied to phylogeny inference, the preferred tree is the one that minimizes the total amount of evolutionary change that has occurred.

The rationale for invoking parsimony in phylogeny inference is simple and compelling. In many instances, it is valid to assume that convergence and reversal will be rare relative to similarity that is due to modification from a common ancestor (but see Felsenstein 1978, 1983). Reversals and convergence both require multiple evolutionary changes. It makes sense, then, that the tree that minimizes the total amount of change implied by the data will also be the one that minimizes the amount of homoplasy. The most parsimonious tree should therefore be the best estimate of the actual phylogenetic relationships among the species being studied.

Parsimony is a logical way to distinguish homology from homoplasy and identify synapomorphies. It is not infallible, however.

In some cases, though, parsimony may not work well. Researchers who set out to reconstruct a phylogenetic tree also have to make decisions about which traits to use. Their goal is to measure the characteristics that are least subject to homoplasy and thus the most reliable source of synapomorphies. To see how evolutionary biologists address these questions and cope with other issues in phylogeny inference, let's delve into a case history: the effort to reconstruct the evolutionary history of whales.

4.2 The Phylogeny of Whales

The whales, dolphins, and porpoises share an array of features that are unusual for mammals, the most spectacular of which is a lack of posterior limbs. Whales are so highly adapted to aquatic life, in fact, that it has been extremely difficult to figure out which mammal group is their closest living relative. The oldest fossils that can be recognized as whales come from rocks in the Himalayas that are about 53.5 million years old. These whales had hind limbs and resembled an extinct group of amphibious mammals called the mesonychians (Thewissen and Hussain 1993; Thewissen et al. 1994; Bajpai and Gingerich 1998). Thus, the fossil record supports the hypothesis that whales evolved from ancestors that had hind limbs. Until recently, though, biologists lacked fossils with synapomorphies that clearly linked the early whales with mammals from a particular group still living today. As a result, researchers have had to analyze traits found in living groups of mammals and try to find synapomorphies that are shared with whales.

Choosing Characters: Morphology and Molecules

In launching a phylogenetic analysis, the first task is to choose characters to use as data. The phylogeny of whales, like many other phylogenetic problems, has been studied using two very different types of characters: (1) skeletal features and other morphological characteristics, and (2) DNA sequences and other molecular traits.

Which type of character is best? There is no hard and fast answer to this question, because morphological and molecular datasets each have pros and cons. Morphological traits are essential when studying species that exist only as fossils, and using morphological traits becomes especially compelling when homoplasy can be distinguished from homology—usually by examining living representatives of the group in question and documenting that the structures being studied develop from the same populations of cells present in embryos or from homologous genes. On the negative side, evaluating the form of a single morphological trait in a group of species often requires slow, painstaking work by a highly trained expert.

Molecular characters have other advantages and disadvantages. Thanks to technological advances, the cost of generating large amounts of sequence data has declined drastically. A large number of nucleotides in DNA from a variety of genes may now be analyzed fairly rapidly. In addition, evolutionary biologists have developed sophisticated models to analyze how different types of DNA sequences should change through time. If used properly, these models make it possible to minimize the impact of homoplasy and accurately estimate the phylogeny implied by the data. And because reversals represent two changes at a particular site in a sequence, researchers can search for genes that change slowly relative to the groups in question and thus minimize the probability that reversals have taken place. It is almost impossible to avoid homoplasy entirely, though, because just

Morphological and molecular traits each have advantages and disadvantages as data for inferring evolutionary relationships.

four character states exist at each site in DNA (A, C, G, and T). Homoplasy can also be difficult to recognize in molecular data.

Because pros and cons exist for each type of character, researchers often try to analyze both morphological and molecular traits. This is exactly what has been done in the case of whales.

Finding the Best Tree Implied by the Data

Parsimony with a Single Morphological Character

Based on analyses of skeletal characters, several studies placed cetaceans as a close relative of the ungulates, or hoofed mammals (Flower 1883; Simpson 1945; Novacek 1993). The ungulates, in turn, consist of two major groups. The horses and rhinos form the perissodactyls; the cows, deer, hippos, pigs, peccaries, and camels form the artiodactyls. The perissodactyls and artiodactyls are both monophyletic groups—each consists of a common ancestor and all of its descendants.

In morphological data sets, several synapomorphies identify the Artiodactyla as monophyletic. Although these synapomorphies include some skull and dental characteristics, the most notable shared, derived feature is found in an ankle bone called the astragalus (Prothero et al. 1988; Milinkovich and Thewissen 1997). In artiodactyls, the astragalus has an unusual shape: both ends of the bone are smooth and pulley shaped (Figure 4.7). This shape allows the foot to rotate in a wide arc around the end of the ankle and contributes to the long stride and strong running ability observed in many artiodactyls.

The question is, how closely related are the whales to the artiodactyls? Are they a closely related outgroup or are they actually part of the clade itself? Specifically, how likely is it that whales share a recent common ancestor with hippos—the only ungulate that spends a great deal of time in water? If whales are closely related to hippos, it suggests a logical evolutionary scenario. After the extinction of the dinosaurs and the large marine reptiles, some mammals began making their living in shallow water habitats. Whales became fully aquatic over time, while hippos remained semi-aquatic. But a whale-hippo relationship also means that species like dolphins and porpoises are closely related to completely terrestrial forms like deer, cows, and pigs.

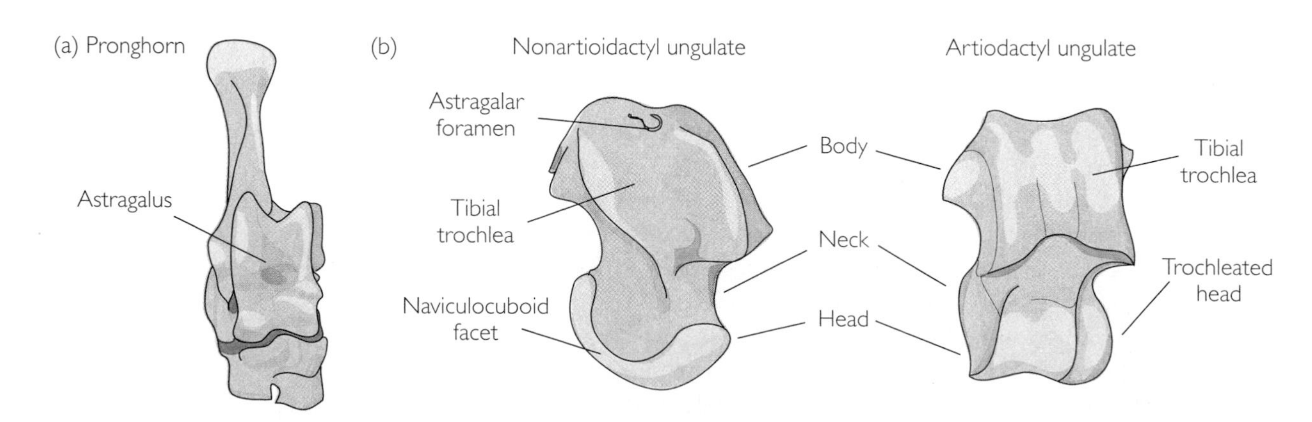

Figure 4.7 The astragalus is a synapomorphy that defines artiodactyls (a) The astragalus is the highest bone in the ankle, around which the foot rotates to extend forward or backward. (b) The astragalus of a nonartiodactyl ungulate (left) and an artiodactyl ungulate (right). In the artiodactyl, both ends of the astragalus are pulley-shaped. From Schaeffer (1948). Copyright © 1948 *Evolution*. Reprinted by permission of *Evolution*.

As a key shared, derived character, the presence of a pulley-shaped astralagus argues that hippos and whales are not closely related (Luckett and Hong 1998). The logic here is straightforward. If hippos and other Artiodactyla form a monophyletic group, then the pulley-shaped astragalus evolved just once without subsequent changes. To visualize this prediction, examine Figure 4.8a and draw a black bar across the branch that leads to the common ancestor of all artiodactyls. But if whales are a closely related "sister" group to the hippos, then the origin of the pulley-shaped astragalus was followed by the loss of this synapomorphy in the lineage leading to whales. To visualize this, put a black bar across the branch that leads to all the artiodactyls in Figure 4.8b and an open bar along the branch leading to whales. The black bars you've put on the figures represent the gain of a trait and the open bar represents a loss. The figure should now show that the whale + hippo hypothesis in Figure 4.8b is less parsimonious than the Artiodactyla hypothesis in Figure 4.8a because it implies one extra step in evolution. This kind of inference is the heart of a phylogenetic analysis based on parsimony. You have compared two alternative trees and concluded that the tree implying the fewest evolutionary changes is most likely to be correct, given the available data.

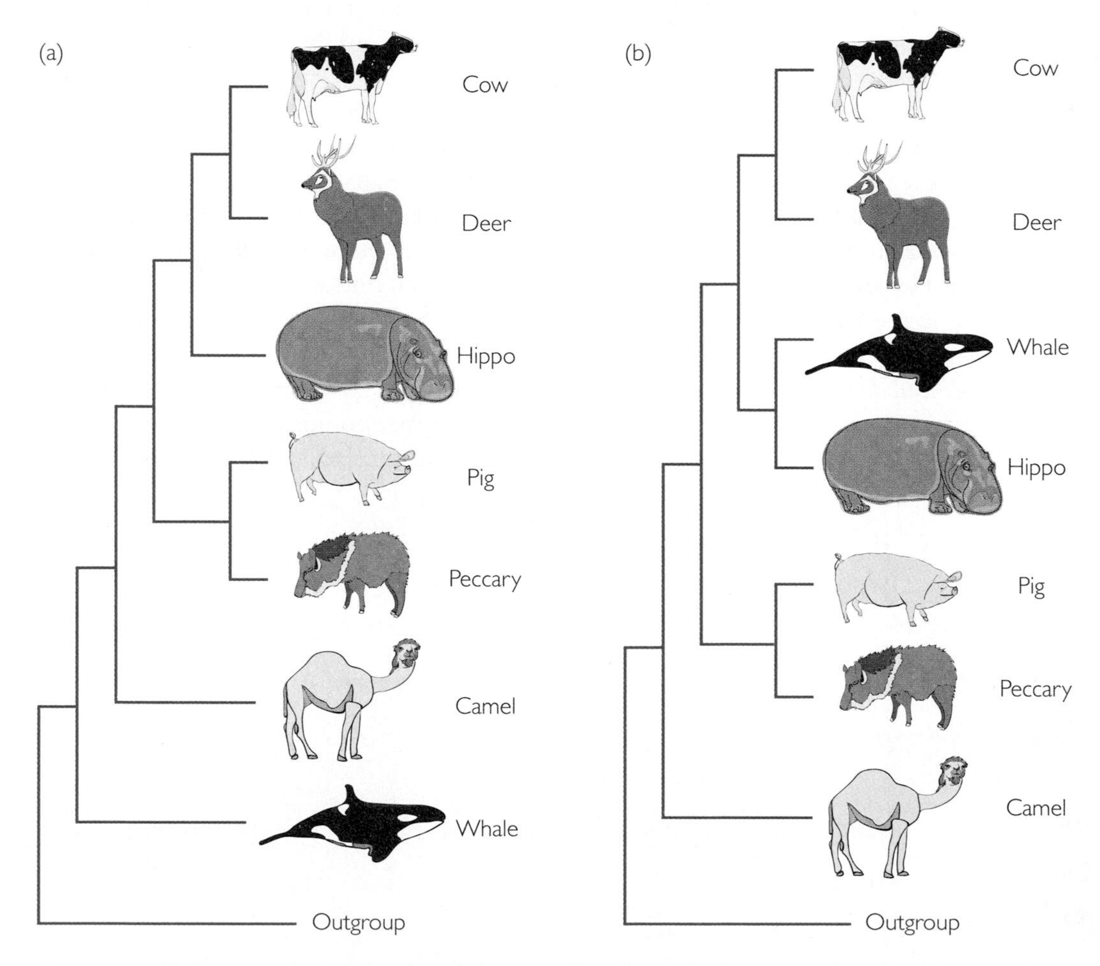

Figure 4.8 **Phylogenetic hypotheses for whales and other mammals** The tree in (a) shows the Artiodactyla hypothesis: Whales and dolphins are related to the ungulates, possibly as the sister group to the artiodactyls (represented by cows, deer, hippos, pigs, peccaries, and camels). The outgroup to these species is from the ungulate group called Perissodactyla (horses and rhinos). The tree in (b) shows the whale + hippo hypothesis. It is identical to (a) with one exception: The branch leading to whales is moved so that whales are the sister group to the hippos.

Living whales have no ankles, though, so the shape of the whale astragalus as a possible artiodactyl trait cannot be assessed. Fortunately, some fossil whales have hind limbs (Gingerich et al. 1990). Johannes Thewissen and Sandra Madar (1999) found fossil ankle bones in the same deposits containing the oldest whale fossils and compared these to the ankles of living and extinct artiodactyls. They concluded that some features of the pulley-shaped astragalus are, in fact, found in the earliest whales. Their data support the hypothesis that whales are descended from an artiodactyl ancestor. Their finding was controversial, however. Critics suggested that the bones in question might actually have belonged to some other artiodactyl—not to a whale—and emphasized that a more general analysis of many morphological traits excluded the whales from the Artiodactyla (Luckett and Hong 1998; O'Leary and Geisler 1999).

When results are inconclusive or controversial, biologists try to resolve the conflict by re-analyzing the existing data, analyzing additional data, or looking for new types of data.

Faced with conflicting data like these, evolutionary biologists have only one course of action: Evaluate other types of data.

Parsimony with Multiple Molecular Characters

John Gatesy and colleagues (1999) set out to analyze the evolutionary history of whales by assembling DNA sequence data from four whale species, eight species of artiodactyl including a hippo, and an outgroup from the Perissodactyla. Figure 4.9 shows 60 of the characters in their dataset, for eight of the taxa involved. The data are from nucleotide sites 141–200 in the DNA sequences for a milk protein gene called beta-casein. The discussion that follows uses these eight taxa and 60 characters to illustrate how multiple characters are analyzed under parsimony.

To find the best tree implied by sequence data or other types of traits, a computer generates many or all of the tree shapes that are possible for the number of species in question, and then maps each character, one at a time, onto each tree. The most parsimonious pattern of character change is noted for each character on each tree. Although these steps sound complex, you've just completed them. Recall that you mapped a single trait—astralagus shape—onto the two alternative trees in Figure 4.8 and recorded the most parsimonius pattern of change for this character. Because you were considering just a single character and a small

Figure 4.9 Sequence data for parsimony analysis These data are 60 nucleotides of aligned sequence from a milk-protein gene in six artiodactyls, a whale (the dolphin *Lagenorhynchus obscurus*), and a perissodactyl as an outgroup. An X at a site indicates an ambiguously identified nucleotide. Some of the invariant or uninformative sites are shaded blue; sites that provide synapomorphies are shaded orange. The phylogeny is based on a parsimony analysis of these nucleotide synapomorphies.

number of species, you were able to do this by hand. When a computer is evaluating a large data set in the same way, it sums the number of changes in all of the characters in the data set for each of the possible trees. Under parsimony, the best tree is the one that implies the fewest character state changes across all characters.

The first thing to note about the 60 characters in Figure 4.9 is that only 15 contain usable phylogenetic information. Many of the characters do not vary at all among the species in the study. For example, all of the species involved have G at site 142. The most likely explanation for this observation is that all of the species in the study inherited a G at this position from their common ancestor. The site is analogous to the presence of the same genetic code in bacteria and mammals—an observation that does not help us identify bacteria and mammals as distinct groups.

Other sites in the data set are variable but uninformative. This situation is illustrated by site 192, where all taxa have C except the camel. To serve as a synapomorphy and thus be useful in inferring a phylogeny, a character has to group two or more taxa. It may be that a C at this site serves as a synapomorphy that helps distinguish members of the camel family as a monophyletic group. If so, then additional study should show that llamas, guanacos, and other species in the camel family share this trait.

On the basis of their entire data set, Gatesey and coworkers claimed that whales and hippos are each other's closest relatives. To understand why their parsimony analysis supported this claim, consider site 162. Note that the cow, deer, whale, and hippo all have a T at this site. Because the other artiodactyls and the outgroup have C at site 162, the T probably represents a shared, derived trait. Similarly, site 166 provides the only synapomorphy in these 60 bases for a monophyletic group consisting of whales and hippos. It's critical to note, though, that not all of the informative characters support the same groupings. Site 177 provides a synapomorphy for a clade consisting of whales, hippos, pigs, and peccaries, which conflicts with the information at site 162. Clearly, reversal or convergence has resulted in homoplasy. Either site 177 or 162 does not reflect the actual evolutionary history of artiodactyls.

What happens when a computer maps each of the informative characters onto the two trees presented in Figure 4.8? The Artiodactyla hypothesis in Figure 4.8a implies a total of 47 nucleotide changes while the whale + hippo hypothesis in Figure 4.8b implies just 41. The difference is due to nucleotide sites 151, 162, 166, 176, 177, and 194. For each of these characters, the whale + hippo hypothesis implies fewer changes than the Artiodactyla hypothesis does. As an exercise, try to find the most parsimonious reconstruction for each of these six characters on both trees, just as you did for the astragalus in Figure 4.8. This is the best way to satisfy yourself that the Artiodactyla hypothesis implies a tree that is, in fact, six steps longer than the tree implied by the whale + hippo hypothesis. For these 60 characters and eight taxa, the whale + hippo hypothesis is more parsimonious and therefore the preferred tree. Based on this conclusion, the research group dubbed their collection of molecular characters the WHIPPO-1 data set (for WHale-hIPPO).

Under parsimony, it is logical to conclude that similarities in DNA sequence data from whales and hippos are synapomorphies that link them in the same monophyletic group.

Searching among Possible Trees

In reality, researchers have to evaluate many possible trees to determine which is most parsimonius given the data—not just two trees as we have done. Even in a moderately sized study, the number of tree topologies that are possible becomes astonishingly large. When four species are included, only three different branching patterns are possible. Adding a fifth species to the data set makes the number of possible topologies jump from 3 to 15; a sixth leads to 105 and a seventh to 945. For the eight species in Figure 4.8 there are 10,395 possible trees. If this sounds immense, consider that it is routine now for studies to include 50 or more taxa. With large analyses like these, an incomprehensibly large number of different trees is possible. Fortunately, several different approaches can be used to have a computer search among all of these possible trees to find the most parsimonious one.

When the number of taxa in a study is relatively low—typically fewer than 11—a computer program can evaluate all of the possible trees. This strategy is called an exhaustive search. Because it guarantees that the optimal tree implied by a particular data set will be found, the approach is called an exact method. An exhaustive search of all 10,395 trees for the data in Figure 4.8 produced a single shortest phylogeny of 41 steps, identical to the tree in Figure 4.8b.

However, the original WHIPPO-1 data set consists of 13 taxa and about 8,000 molecular characters. Because an exhaustive search based on this data set is prohibitively slow, Gatesy et al. (1999) used two other methods for searching among the possible trees. These methods take advantage of some logical or computational shortcuts. They end up searching only some parts of the landscape of all possible trees, while maximizing the probability of finding the most parsimonious tree.

It is critical to evaluate a large number of possible trees to choose the best one implied by the data at hand.

In this case, both search methods produced two most-parsimonious trees, and both of the resulting trees were consistent with the phylogeny in Figure 4.8b. This agreement among search methods indicates that the close relationship between whales and hippos is not an artifact of incomplete searching or of the inability of search methods to find the most parsimonious solutions. Results like these help reassure researchers that they have found the most accurate tree implied by the data.

Evaluating the Best Tree

Having compared several or all of the possible trees, the question becomes: Just how good is the "best" tree? Is the most parsimonius tree significantly better than trees that are only slightly less parsimonius? To answer this question, researchers usually evaluate a series of highly parsimonius trees implied by the data in the study, and create a consensus that represents the branching pattern supported by all of the nearly optimal trees.

Using Other Methods Besides Parsimony: Maximum Likelihood and Bayesian Inference

In addition to evaluating the best trees implied by a parsimony analysis, most researchers analyze their data sets with a phylogenetic method that computes a probability or likelihood that alternative trees are supported by the data. These approaches are called **maximum likelihood** (**ML**) and **Bayesian Markov Chain Monte Carlo** (**BMCMC**) methods (Felsenstein 1981; Huelsenbeck et al. 2001).

With DNA sequence data, the essence of the likelihood approach is to ask the question: Given a mathematical formula that describes the probability that different types of nucleotide substitution will occur, and given a particular phylogenetic tree with known branch lengths, how likely am I to obtain this particular set of DNA sequences? A computer program can answer this question by evaluating each of the possible tree topologies and computing the probability of producing the observed data, given the specified model of character change. This probability is reported as the tree's likelihood. The criterion for accepting or rejecting competing trees is to choose the one with the highest likelihood. In addition, likelihood methods allow investigators to evaluate the hypothesis that similar trees are just as likely as the one with highest likelihood. As a result, the statistical tests that are possible with ML analyses give researchers an objective criterion for deciding just how good a particular tree is.

Bayesian approaches are similar to likelihood methods, except that the principle is to ask what the probability is of a particular tree being correct, given the data and a model of how the traits in question change over time. A computer analysis can then evaluate trees and find the one with the highest probability, given the data and the model of character change.

Likelihood and Bayesian approaches have become increasingly popular recently, in part because they provide an objective criterion for evaluating how much better certain trees are than others. In addition, when investigators have used a computer to simulate evolution in a group of hypothetical organisms and then used parsimony, ML, and BMCMC methods to evaluate the data, they often find that ML or BMCMC methods do best in terms of recovering the actual pattern of evolution (e.g., Hall 2005). Box 4.1 introduces a third general approach to inferring phylogenies, called distance methods.

Evolutionary biologists continue to investigate when parsimony, ML/BMCMC, or distance methods function best (Felsenstein 2004; Kolaczkowski and Thornton 2004; Steel 2005) and new inference methods continue to be developed (e.g., Tamura et al. 2004). The current consensus is that ML/BMCMC approaches are both reliable and powerful in analyzing DNA sequence data or other characters where the pattern of trait change over time is understood reasonably well. But most researchers agree that it is wise to use a combination of approaches, with parsimony, ML, and BMCMC being the current methods of choice. This strategy provides several ways of evaluating how good a particular tree is: (1) producing a consensus tree under parsimony, (2) using statistical tests to evaluate the best trees under ML and BMCMC, and (3) comparing the best trees under parsimony, ML, and BMCMC to see how consistent they are (Huelsenbeck and Hillis 1993; Hillis et al. 1994). If the results are consistent, researchers are confident that they have indeed found the best tree supported by the data.

Most researchers use a variety of criteria to identify the best tree implied by their data. The trees identified under different criteria can then be compared.

Evaluating Particular Branches: Bootstrapping

Suppose a research team analyzes a data set using parsimony, ML, and BMCMC approaches, and that the results make them confident that they have found the best tree implied by their data. How confident can they be that particular nodes and branches within that tree are supported by the data? For example, how confident are we that the data strongly support the node joining whales with hippos? This question is analogous to asking about the reliability of measuring height in a group of people. If you measure some people and compute the average of the sample, you need to ask whether your sample average is an accurate representation of the true average for the entire population. To answer this question, you use the measurements you made to compute a statistic that quantifies the amount of variation around the mean. If the variation is high, you become less confident that your sample average is reliable.

Box 4.1 A note on distance methods

Distance methods offer a radically different approach to phylogeny inference than parsimony, likelihood, or Bayesian methods. Here the idea is to convert discrete character data, such as the presence or absence of a morphological trait or the identity of a nucleotide at a homologous location in a gene, into a distance value (Swofford et al. 1996). For example, two species are separated by a genetic distance of 10% if an average of 10 nucleotides have changed per 100 bases in a particular gene. Converting discrete characters to a single distance measure in this way results in a loss of specific information about which traits have changed, but attempts to capture the overall degree of similarity between taxa.

It's important to recognize that to convert discrete data into an estimate of distance, an investigator has to assume a specific model of character change over time. When computing genetic distances from data on DNA sequence divergence, researchers prefer formulas that correct for multiple substitutions at the same site and for differences in the frequency of transition and transversion substitutions (Kimura 1980; Wakely 1996).

To estimate a phylogeny from distance data, computer programs cluster taxa so that the most similar forms are found close to one another on the resulting tree. Small genetic distances should indicate recent divergence from a common ancestor and a close phylogenetic relationship. Currently the most widely used clustering algorithm is the neighbor-joining method developed by Naruya Saitou and Masatoshi Nei (1997; see also Levy et al. 2006).

Biologists refer to the general strategy of grouping taxa according to their similarities as a phenetic approach (Sneath and Sokal 1973). The preferred tree is the one that minimized the total distance among taxa. In their overall logic, then, phenetic methods contrast with the cladistic approach introduced earlier in the chapter, and with likelihood and Bayesian probability approaches.

An example of a phenetic analysis is shown in Figure 4.10. Note that the left part of the figure shows the matrix of pairwise genetic distances among the sequences in Figure 4.9. These distances were calculated with a formula that accounts for multiple substitutions at the same site. Notice that the smallest genetic distances are between the whale and hippo sequence and between the deer and cow sequences, and that the genetic distances among these four sequences are also small. The clustering analysis on the right side of the figure groups whales and hippos as sister taxa based on these genetic distances, with cows and deer as their next closest relatives.

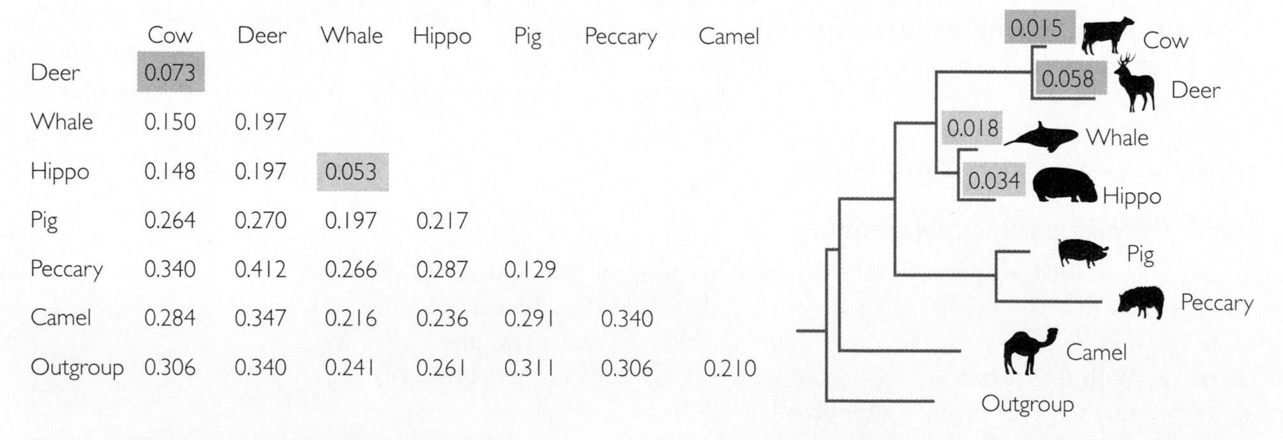

	Cow	Deer	Whale	Hippo	Pig	Peccary	Camel
Deer	0.073						
Whale	0.150	0.197					
Hippo	0.148	0.197	0.053				
Pig	0.264	0.270	0.197	0.217			
Peccary	0.340	0.412	0.266	0.287	0.129		
Camel	0.284	0.347	0.216	0.236	0.291	0.340	
Outgroup	0.306	0.340	0.241	0.261	0.311	0.306	0.210

Figure 4.10 Genetic distances for cluster analysis Each entry in this table is a genetic distance between a pair of taxa, calculated from the sequence data in Figure 4.9. The phylogeny here was produced by a clustering analysis of these genetic distances. Notice that pairs of taxa with low genetic distances are grouped as sister taxa, such as the cow and deer (blue) or the whale and hippo (orange). The lengths of the branches are proportional to the expected proportion of nucleotide differences between groups, and are shown numerically for several branches.

To answer questions about the reliability of particular branches in a phylogenetic tree, investigators have to evaluate those branches statistically (see Bremer 1994; Swofford et al. 1996; Huelsenbeck and Rannala 1997). With trees based on ML or BMCMC analyses, you can compare trees with and without a particular branch and compute which one is more likely (see Felsenstein 2004). With trees based on a parsimony analysis, the most popular way to evaluate particular branches is called **bootstrapping** (Felsenstein 1985).

In bootstrapping, a computer creates a new data set from the existing one by repeated sampling. For example, if there are 300 base pairs of sequence in a study, the computer begins the bootstrapping process by randomly selecting one of the sites and using it as the first entry in a new data set. Then, it randomly selects another site, which becomes the second data point in the new data set. (There is a 1/300 chance that this second point will be the same site as the first.) The computer keeps resampling like this until the new data set has 300 base pairs of data, representing a random selection of the original data. This new data set is then used to estimate the phylogeny. By repeating this process many times, the investigator can say that particular branches occur in 50%, 80%, or 100% of the trees estimated from the resampled data sets. The more times a branch occurs in the bootstrapped estimates, the more confidence we have that the branch actually exists. If bootstrap support for a particular branch is low, say under 50%, an investigator will usually conclude that the branching pattern is uncertain in that part of the tree and collapse that particular branch into a **polytomy**, or a point of uncertainty, in the published tree.

Bootstrapping is a technique for evaluating which branches on a particular tree are more well-supported than others.

Resampling the data in Figure 4.9 through bootstrapping indicates strong support for the whale + hippo clade. Of 1,000 resampled data sets, 71% included a whale + hippo clade. When all taxa and all molecular characters in the WHIPPO-1 data set were analyzed, bootstrap support for this node approached 100% (Hillis 1999). In some analyses of known phylogenies—where investigators bred laboratory organisms and split their populations to create lineages of known relatedness—bootstrap support of around 70% or greater was usually associated with the true phylogeny (Hillis and Bull 1993). Based on bootstrapping, the close relationship between whales and hippos seems sound.

Resolving Conflict

What happens when trees produced by different phylogeny inference methods or by analyses of different characters conflict? For example, the results of the morphological analyses in Figure 4.8 and the molecular studies in Figure 4.9 do not agree. Trees that are based on different traits or on different phylogeny inference methods are treated as competing estimates of the group's actual evolutionary history. In cases of conflict, researchers have more confidence in trees that are estimated with larger data sets, from characters that are less subject to homoplasy, and with inference methods that are most appropriate to the data in question. But in many cases the most prudent position is to wait for additional data that are independent of the traits analyzed to date. In the case of whale evolution, one possibility is that newly discovered fossil artiodactyls and whales might have characteristics that clearly support one of the two hypotheses. Alternatively, new molecular data might be able to resolve the conflict. In the case of whales, both events occurred.

New molecular data that clarified the phylogeny of whales emerged before the recent fossil finds. The data in question are the presence or absence of DNA sequences that occasionally insert themselves into new locations in a genome. The genes involved are called SINEs and LINEs, for Short or Long INterspersed Elements. The presence or absence of a particular SINE or LINE at a homologous location in the genomes of two different species can be used as a trait in phylogeny inference.

Traits that are unlikely to converge or to reverse over time are ideal for inferring evolutionary relationships.

David Hillis (1999) outlined the potential advantages of using SINEs and LINEs in phylogeny inference. It is well-established that transposition events, in which the parastitic sequence inserts itself in a new location in the host genome, are relatively rare. As a result, it is extremely unlikely that two homologous SINEs would insert themselves into two independent host lineages at exactly the same location. This kind of convergence is possible but clearly highly improbable. Reversal to the ancestral condition is also unlikely, because the loss of a SINE or LINE can usually be detected. When SINEs and LINEs are lost, it is common to also observe the associated loss of part of the host genome. As a result, researchers can usually tell if a particular parasitic gene is absent or has been lost. If convergence and reversal are rare or can be identified, then homoplasy is unlikely. SINEs and LINEs should be extraordinarily reliable characters to use in phylogeny inference.

What do SINEs and LINEs have to say about whale evolution? Masato Nikaido and colleagues (1999) answered this question by analyzing 20 different SINEs and LINEs found in the genomes of artiodactyls. The data for the taxa we have been considering are given in Figure 4.11, along with a tree that shows how these data map onto the whale + hippo tree. Look at each of the 20 genes in turn, and note that the presence or absence of each SINE or LINE acts as a

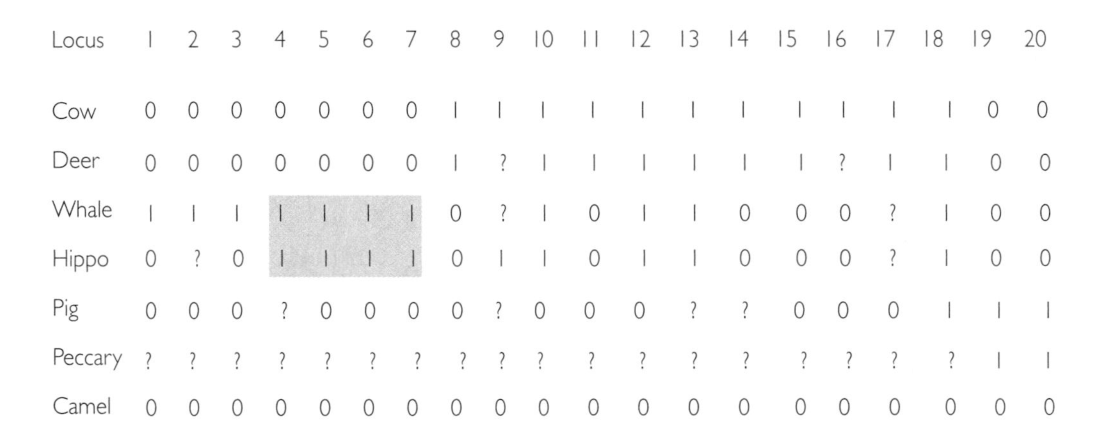

Locus	1	2	3	4	5	6	7	8	9	10	11	12	13	14	15	16	17	18	19	20
Cow	0	0	0	0	0	0	0	1	1	1	1	1	1	1	1	1	1	1	0	0
Deer	0	0	0	0	0	0	0	1	?	1	1	1	1	1	1	?	1	1	0	0
Whale	1	1	1	1	1	1	1	0	?	1	0	1	1	0	0	0	?	1	0	0
Hippo	0	?	0	1	1	1	1	0	1	1	0	1	1	0	0	0	?	1	0	0
Pig	0	0	0	?	0	0	0	0	?	0	0	0	?	?	0	0	0	1	1	1
Peccary	?	?	?	?	?	?	?	?	?	?	?	?	?	?	?	?	?	?	1	1
Camel	0	0	0	0	0	0	0	0	0	0	0	0	0	0	0	0	0	0	0	0

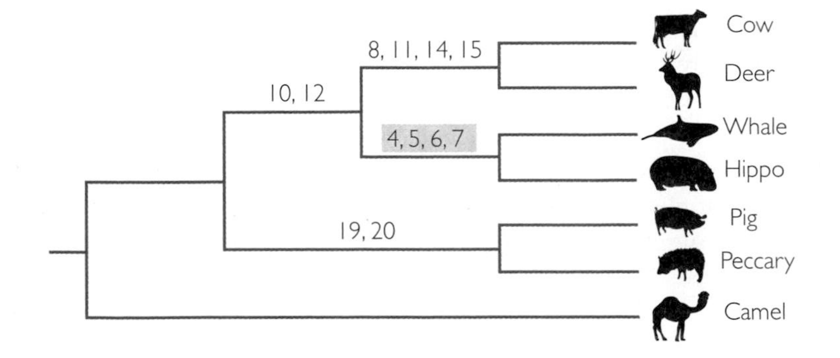

Figure 4.11 Nearly perfect phylogenetic characters? This table shows the presence (1) or absence (0) of a SINE or LINE at 20 loci in the genomes of six artiodactyls and a whale (Baird's beaked whale, *Berardius bairdii*). Question marks (?) indicate loci that are questionable in some taxa. Data are from Nikaido et al. (1999). The phylogenetic tree was produced by a parsimony analysis of these 20 characters. The presence of a SINE or LINE at loci 4–7 defines a clade of whales and hippos.

synapomorphy that identifies exactly one clade in the phylogeny. Stated another way, there is no homoplasy at all in the data set and thus no conflicts among the characters when they are mapped onto the tree. The analysis is remarkably clean and strongly corroborates the conclusion from the DNA sequence studies.

Not long after Nikaido and coworkers published their conclusions, two research teams simultaneously announced fossil finds that were characterized as "one of the most important events in the past century of vertebrate palaeontology" (de Muizon 2001, p. 260). The oldest of the fossils came from 48 million-year-old rocks and represented two species: the wolf-sized *Pakicetus attocki* and the fox-sized *Ichthyolestes pinfoldi*. Both were long-legged, long-tailed creatures that were clearly terrestrial. Both species have synapomorphies in the size and shape of ear bones that clearly identify them as whales, as well as a pulley-like astralagus that mark them as artiodactyls (Thewissen et al. 2001). The same two characteristics are clearly present in two slightly more recent species, *Rodhocetus kasrani* and *Artiocetus clavis*, dated to 47 million years ago (Gingerich et al. 2001). Taken together, the suite of new fossils describes the transition from terrestrial to aquatic life in a satisfying way (Figure 4.12; see also Figure 2.13). The fossil record is now able to confirm that this transition took place in a lineage of artiodactyls that became today's whales, dolphins, and porpoises. Recent analyses of the fossil record have also identified an extinct group of

Biologists become more confident that a phylogenetic tree is correct when results from different types of analyses and traits agree.

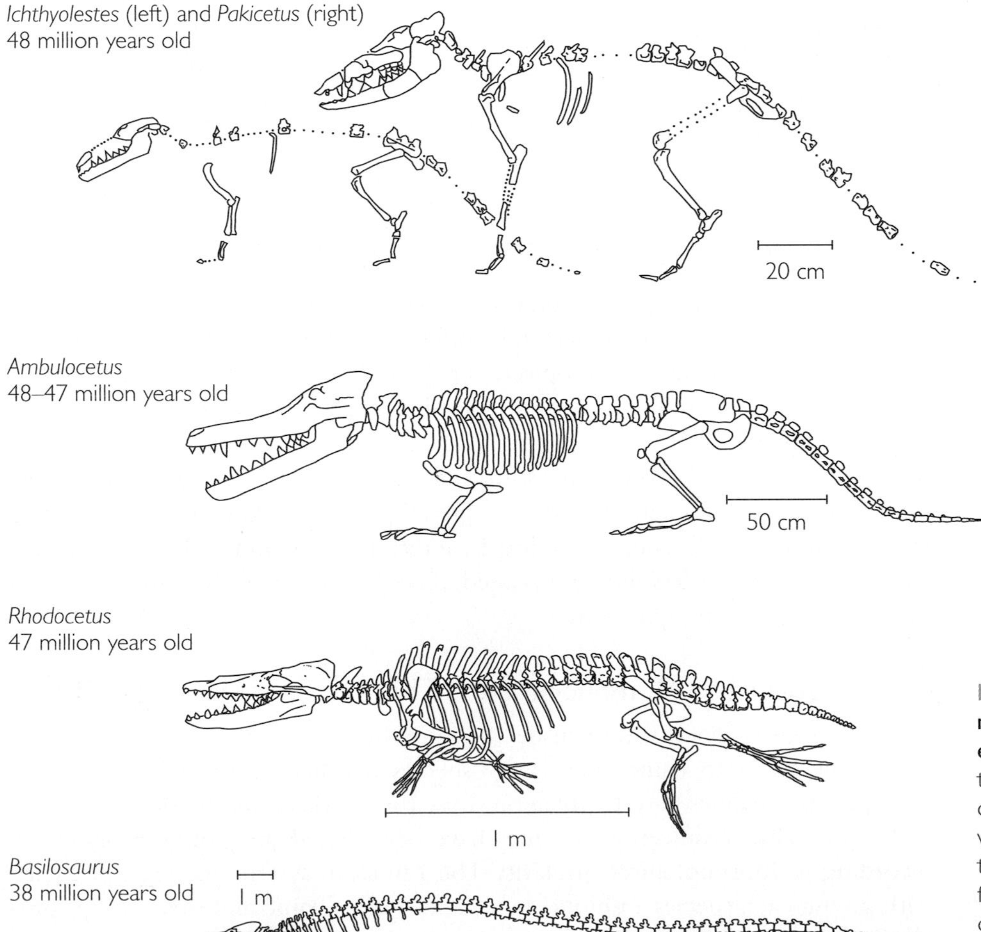

Fgiure 4.12 Early whales moved from terrestrial to aquatic environments The fossils pictured here illustrate some of the changes that occurred early in whale evolution, as members of this lineage made the transition from land to water. A comparison of some of these fossils to a modern whale is shown in Figure 2.13.

semi-aquatic artiodactyls as the sister group to hippos (Boisserrie et al. 2005). This report creates a link between the ancestors of today's hippopotamuses and the ancestors of today's whales, and suggests that both may have evolved from the same semi-aquatic ancestor.

Take-Home Messages

If the whale evolution story has a moral, it is that inferring phylogenies presents a series of challenges: choosing homologous characteristics to study that provide synapomorphies and minimize homoplasy, searching among the many trees that are possible to find the one that is either the most parsimonius or the most likely under a particular model of character evolution, and using bootstrapping or other techniques to evaluate how well the data support particular branches in the tree. In addition, it is common for different research teams to come up with conflicting results by analyzing different traits, and common for conflicts to be resolved by additional analyses—often using new types of traits or phylogeny inference methods.

The past 25 years have seen a virtual explosion in the size and quality of the data sets used in estimating phylogenetic trees and in the sophistication of phylogeny inference techniques. Work continues, however. Evolutionary biologists are beginning the use the presence and absence of particular genes, diagnosed from genome sequencing data, as traits in phylogeny inference (Korbel et al. 2002; Shedlock et al. 2004). There is more interest in using a combination of morphological and molecular characters in the same analyses (e.g., Wahlberg et al. 2005) and DNA sequences from a wide array of genes instead of just one or a few (Rokas et al. 2003). Year by year and study by study, biologists are getting a clearer and clearer picture of the Tree of Life's size and shape.

4.3 Using Phylogenies to Answer Questions

The first two sections of this chapter have focused on methods. Their goal was to show that estimating evolutionary relationships requires a series of careful decisions about which data are appropriate for the task and how they should be analyzed. Now we turn to the issue of how evolutionary trees can be used to answer interesting questions.

Research programs based on estimating and interpreting evolutionary trees are thriving. In the remainder of this chapter, the goal is to sample the diversity of applications for phylogenetic thinking. In addition to revealing which species and lineages are more or less distantly related, phylogenetic trees can directly answer an array of interesting questions.

Classification and Nomenclature: Is There Such a Thing as a Fish?

Classifying the diversity of organisms is one of the most basic tasks faced by biologists. The effort to name and classify species is called **systematics**. What do phylogenetic analyses have to say about how this work should be done?

Traditionally, classification schemes have been based on grouping organisms according to morphological similarity. The Linnaean system, for example, starts with giving each species a unique genus and species name and then groups progressively more similar species into kingdoms, classes, orders, families, and genera. Classification schemes that depend on analyses of similarity like this are called

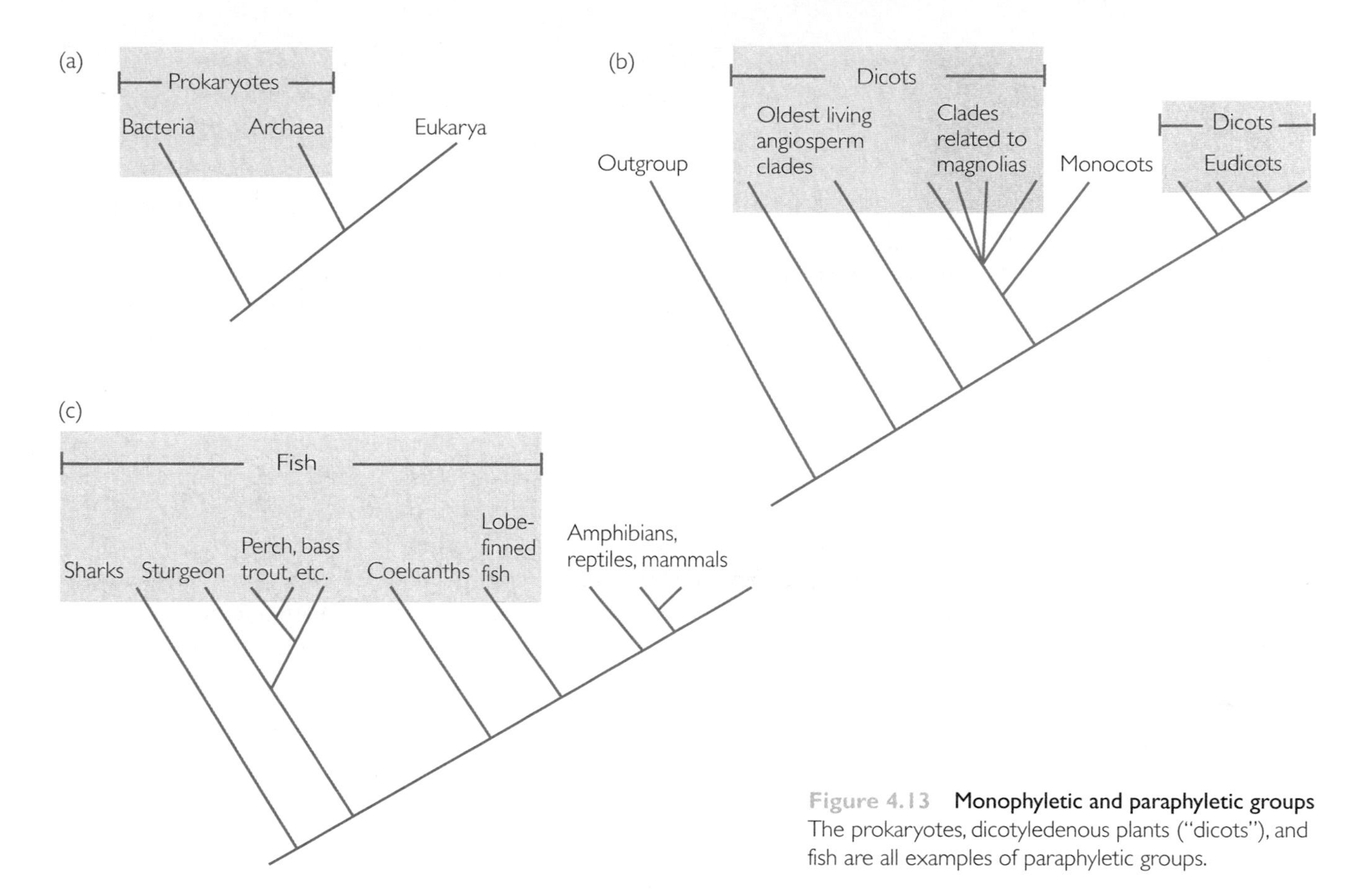

Figure 4.13 Monophyletic and paraphyletic groups The prokaryotes, dicotyledenous plants ("dicots"), and fish are all examples of paraphyletic groups.

phenetic approaches. In contrast, naming schemes that are based on evolutionary relationships are referred to as phylogenetic or **cladistic** approaches. According to phylogenetic systematics, classification systems should be tree based, with names and categories that reflect the actual sequence of branching events. More specifically, only **monophyletic groups**, which include all descendants of a common ancestor, are named. **Paraphyletic groups**, which include some but not all descendants of a common ancestor, should not be named.

To appreciate how phylogenetic naming schemes affect classification, consider the phylogenies in Figure 4.13. As these trees show, the groups called prokaryotes, dicots, and fish are all paraphyletic. Under a cladistic naming scheme, none of these familiar names would be used. Under a tree-based scheme for naming lineages, there is no such thing as a fish. A group called "the fish" would have to include the tetrapods (terrestrial vertebrates including amphibians, mammals, and reptiles) as well as the ray-finned fishes, lobe-finned fishes, and lungfishes.

Similarly, a cladistic classification would nest the whales as a subgroup within the Artiodactyla, which in turn are a subgroup of the Mammalia. In this case, the contrast with traditional classification would be to reduce the Cetacea from the high-level taxon called an order to a lower-level group related to hippos.

Phylogenetic systematics represents an important philosophical break from classical approaches to taxonomy. An increasing number of taxonomists and systematists are calling for a complete overhaul of the traditional phenetic naming scheme, with the goal of creating an explicitly phylogenetic classification (de Queiroz and Cantino 2001). This viewpoint is increasingly being put into practice. For example, recent papers on whale evolution refer to the lineage described in Figure 4.8b as the Cetartiodactyls, for "cetaceans + artiodactyls."

If species and groups are classified in a way that reflects their evolutionary history, only monophyletic groups would be given names.

Using Molecular Clocks: When Did Humans Start Wearing Clothes?

Some of our most basic questions about the history of life concern when major events occurred and how rapidly they took place. Chapter 18 introduces a few of the many evolutionary events that have been dated from the fossil record. But what options are available when fossil data are missing, as they are for many groups? In at least some instances, it should be possible to address questions about the timing and rate of evolution by analyzing molecular traits that change at a steady rate. This hypothesis, called the **molecular clock**, originated with Emile Zuckerkandl and Linus Pauling (1962).

There are good theoretical reasons to expect that at least some types of DNA sequences change in a clocklike fashion. Specifically, many mutations change an individual's DNA but not its phenotype. In most cases, mutations like these are not exposed to natural selection. Instead of being favored or eliminated by selection, these "neutral" changes respond to a random process called genetic drift. As Chapter 7 will show, the neutral theory of molecular evolution predicts that neutral changes in DNA should accumulate in populations at a rate equal to the mutation rate. If the mutation rate does not change much over time, and if generation times remain similar, then the number of neutral molecular differences between two taxa should be proportional to the age of their most recent common ancestor. By documenting the number of different neutral mutations observed in two species and multiplying by a calibration rate, representing how frequently neutral changes occur per million years, researchers can estimate when the two species diverged.

Molecular clocks offer the possibility of dating events that are not documented in the fossil record.

Although the possibility of dating events from estimates of genetic divergence is compelling, there are several important caveats. For example, it is critical to realize that the mutation rate to neutral alleles will vary from gene to gene and lineage to lineage, and even from base to base. For reasons that are explained in Chapter 5, silent site changes in the third positions of codons are much more likely to be neutral with respect to fitness and thus to accumulate at a clocklike rate than replacement changes that occur at the first and second positions in codons. And if allele frequencies change rapidly due to strong selection at a particular gene, it is highly unlikely that the mutations involved are accumulating in a clocklike fashion. Finally, rates of change that are calibrated for a particular gene and lineage are unlikely to work for other groups, which may have different generation times and selection histories (Martin et al. 1992; Martin and Palumbi 1993; Hillis et al. 1996).

Even if clocklike change occurs in a particular gene and lineage, how can the rate be determined? To answer this question, investigators have to turn to the fossil or geological records. The idea is to measure the genetic distance between two taxa whose divergence date is known from fossil or geological data and then to use this calibration to date the divergence times of groups that have no fossil record.

As an example of how researchers can use molecular clocks to date events, consider work by Ralf Kittler and colleagues (2003) on the origin of human body lice. Body lice (*Pedicularis corporus*) are distantly related to pubic lice (*Pthirus pubis*) but are extremely similar to head lice (*Pedicularis capitus*). Human body lice feed on the body but live in clothing, while head lice live in hair and feed on the scalp. Both species are restricted to humans—chimpanzees and our other close relatives have their own specialized species of lice.

Kittler et al. reasoned that if human body lice are adapted to live in clothing, then they must have diverged from human head lice about the time that humans began wearing clothes. Based on sequence data from a large number of head and body lice collected from humans at 12 different locations around the world, they estimated the average percentage of bases that differed between head lice and body lice.

To convert this estimate of genetic divergence into a time of divergence, the biologists analyzed homologous sequences in lice that parasitize chimps. The fossil record documents that the common ancestor of humans and chimps lived about 5.5 million years ago. By assuming that the head lice of humans and chimps diverged at the same time their host species did, the group was able to estimate the percentage of bases that change per million years. When they multiplied this rate by the amount of divergence observed between human head lice and body lice, they got an estimate of 72,000 +/− 42,000 years ago for the origin of body lice, and thus clothing.

Applications of other molecular clocks, summarized in Chapter 20, suggest that our species originated between 100,000–200,000 years ago. Both fossil and genetic data agree that modern humans originated in Africa and then moved throughout the world. If body lice and clothing originated between 114,000–30,000 years ago, it is possible that people left Africa wearing clothes.

Analyzing Phylogeography: How Did Chameleons Get from Africa to India?

The effort to understand where organisms live and how they came to be there is called biogeography. Biogeographers ask questions about why certain species are found in certain parts of the world and how geographic distributions have changed through time. When researchers turn to phylogenies for help in answering these types of questions, the research program is called **phylogeography**. We will use phylogeographic approaches to study the origin and radiation of human populations in Chapter 20. Here we introduce the strategy by considering a classical problem: How the breakup of the supercontinent Gondwana, starting about 200 million years ago, affected the ranges of species that are alive today.

Our specific focus is on the 134 species in the lizard family Chamaeleonidae (Figure 4.14). Chameleons are found throughout Africa, Madagascar, and parts of India, as well as on Indian Ocean islands such as the Seychelles, the Comoros, and Reunion. Many of these areas were created by the breakup of Gondwana. When seafloor spreading caused the supercontinent to break apart, various continental plates moved apart to form today's continents of South America, Africa,

Figure 4.14 **Chameleons are a clade of lizards** More than 130 species of chameleons live in Africa, Madagascar, islands in the Indian Ocean, and India.

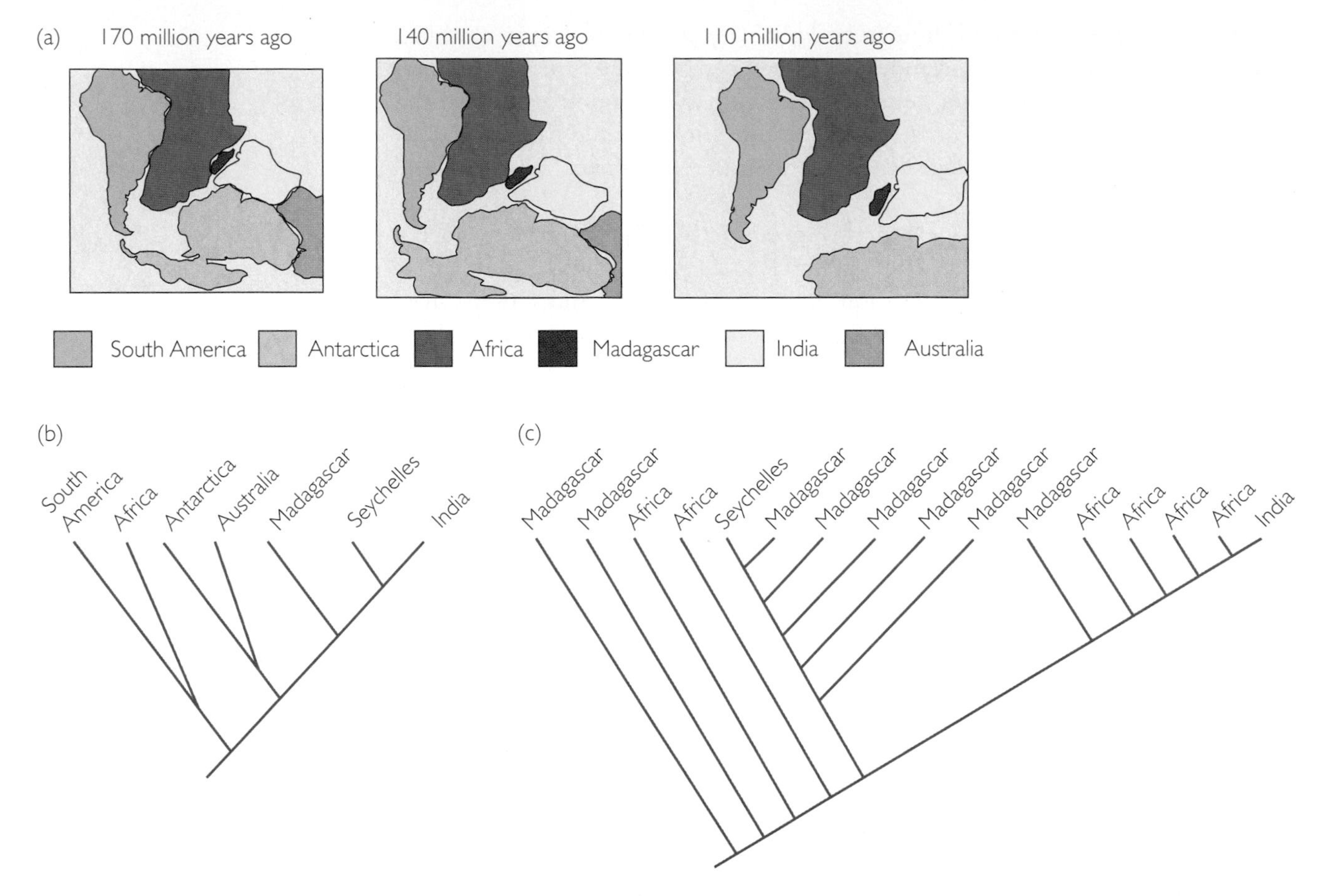

Figure 4.15 Did chameleons speciate when Gondwanaland broke up? (a) These drawings show the sequence of events that occurred as the supercontinent Gondwanaland broke up. (b) This tree shows the relationships among chameleon groups predicted by the Gondwanaland breakup hypothesis. The Seychelles Islands are not formed of continental plate material (they are volcanic or coral. (c) This tree shows the actual relationships among chameleons.

Analyzing phylogenies can be central to understanding why certain species are found in particular parts of the world.

Antarctica, and Australia and islands such as Madagascar and the Seychelles. In addition, the Indian subcontinent split off and began moving to the north and east, where it eventually collided with the Asian landmass (Figure 4.15a).

The question is: Did chameleon populations float into their current geographic positions, riding pieces of Gondwana? If so, then the populations living today should have the phylogenetic relationships shown in Figure 4.15b. Alternatively, chameleons could have diversified well after the supercontinent had broken up, through transoceanic dispersal of small populations from one landmass to another. To test these contrasting hypotheses, C. J. Raxworthy and colleagues (2002) estimated the phylogeny of chameleons by analyzing homologous traits in species from throughout the group's present range. The traits they studied included morphological and behavioral characteristics as well as DNA sequences.

Their results are shown in Figure 4.15c. Note that instead of following the sequence of events that occurred as the supercontinent broke up, the phylogeny suggests that chameleons have diversified via dispersal to new habitats. Specifically, the tree implies that populations have dispersed from Madagascar to Africa on several occasions, as well as from Madagascar to the Seychelle Islands and from Africa to India. The dispersal hypothesis also has to be invoked to explain the origin of chameleons on the Comoros Islands and Reunion, which are volcanic in origin and have never been in contact with continental plates.

The phylogeographic analysis indicates that chameleons did not simply ride chunks of Gondwana into their present positions. Instead, small groups of chameleons happened on to small rafts made of floating vegetation that dispersed from Madagascar and Africa to new habitats in the north and east.

Co-Speciation: When New Species of Aphids Form, What Happens to the Bacteria That Live inside Their Cells?

Predation, parasitism, mutualism, and other types of interactions between species are common in nature. When natural selection occurs during these interactions and produces adaptations in both species involved, **coevolution** is said to take place. In some human populations, for example, certain alleles have increased in frequency because they confer protection against specific strains of the malaria parasite. But in response, new strains of the parasite have evolved that are able to thrive in hosts who have these alleles (Gilbert et al. 1998). In this way, host and parasite species are continuously coevolving.

How can phylogenetic thinking be used in coevolutionary studies? One of the most dramatic cases of coevolution is called cospeciation. As Chapter 16 will show, speciation occurs when a population splits into two groups that become genetically isolated and then begin to diverge genetically. When cospeciation occurs, the speciation process occurs in two interacting species simultaneously.

As an example of how cospeciation might happen, consider the association between aphids and the bacteria that live inside specialized aphid cells. Aphids are small insects that make their living by sucking phloem sap from plants (Figure 4.16a). Many have cells called bacteriocytes that house symbiotic bacteria (Figure 4.16b). The relationship is considered to be mutualistic because it is thought to be beneficial for both parties and is termed an endosymbiosis. The aphids supply the bacterial cells with nutrients and a safe habitat while the bacteria produce amino acids that are not present in phloem sap and that aphids are not able to synthesize on their own. The bacteria are transmitted from one generation of aphids to the next via egg cells.

When co-speciation occurs, species that interact extensively—like parasites and hosts—speciate together.

The question is: When aphid populations diverge to form distinct species, do their symbiotic bacteria speciate along with them? The answer would seem to be yes, except that a plausible alternative to this cospeciation hypothesis exists. Aphids are parasitized by wasps, which insert a structure called an ovipositor into the bodies of their victims and lay eggs. Thus, it is reasonable to suppose that as female wasps move from one host to another, new types of bacterial cells are at least occasionally introduced to each aphid population. If so, then new species of

(a)

(b)

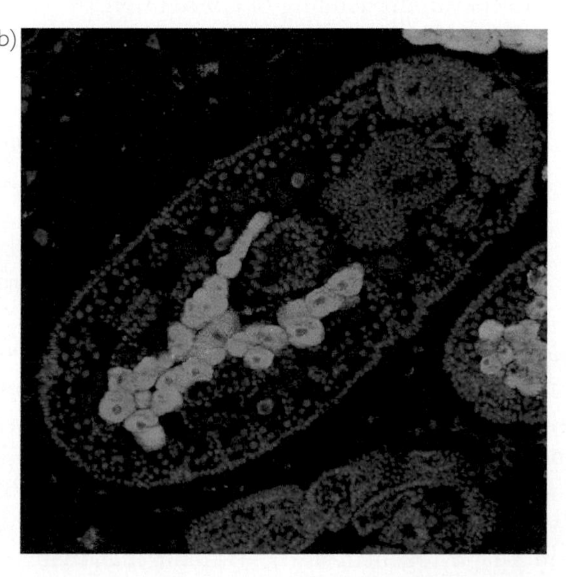

Figure 4.16 Aphids and their bacterial endosymbionts
(a) Aphids insert an organ called a stylet into plant vascular tissues and suck sap. (b) Specialized aphid cells are stained blue in this micrograph; the bacterial cells inside are stained green.

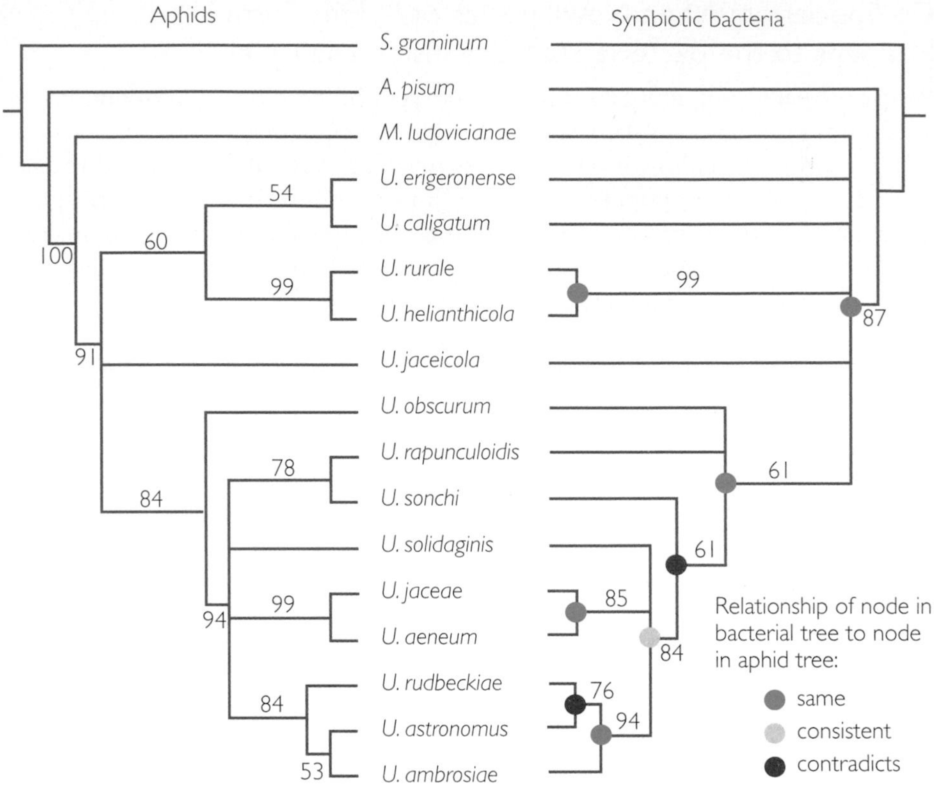

Figure 4.17 Did aphids and their bacterial endosymbionts speciate together? The tree on the left is a phylogeny of aphid species that have endosymbiotic bacteria. The tree on the right is a phylogeny of bacteria in the genus *Buchnera* that live inside aphid cells. Numbers above branches indicate the bootstrap support for that branch; branches that were poorly supported by bootstrapping were collapsed into polytomies. Aphid genus names are: *S* = *Schizaphis*; *A.* = *Acrythosiphon*; *M.* = *Macrosiphoniella*; *U.* = *Uroleucon*.

aphids might host bacterial cells that are not closely related to those of their ancestral species, but were acquired from a wasp.

To test these hypotheses, Marta Clark and coworkers (2000) used DNA sequence data to estimate the phylogeny of 17 aphid species and their bacterial symbionts. The trees that resulted from this analysis are shown in Figure 4.17. Note that in every case but two, the branching pattern in the aphid tree is the same as the branching pattern in the bacterial tree, or at least consistent with it. This is strong evidence that aphids and the bacteria they house routinely cospeciate. The aphid-bacteria data are a good example of how phylogenies can lead to a deeper understanding of coevolution.

In combination with the HIV and leopard analyses in Chapters 1 and 2, the four examples provided here should give you a good feel for the variety of questions that can be answered by estimating and evaluating phylogenies. If you are more comfortable with tree thinking and convinced that analyzing evolutionary trees can lead to some exciting science, then this chapter has been a success.

Summary

Recent conceptual and technological advances have revolutionized our ability to estimate phylogenies accurately. Research on methods to infer phylogenies is advancing rapidly, and phylogenetic thinking is beginning to pervade biology.

The first step in estimating a phylogeny is to select and measure characters that can be phylogenetically informative. The molecular or morphological characters employed in phylogeny inference have to be homologous, variable among the taxa in the study, and resistant to homoplasy. In addition, only synapomorphies—homologous traits that are shared among species because they were derived from a common ancestor—are useful in identifying monophyletic groups.

The second step in phylogeny inference is to decide whether a parsimony, maximum likelihood, Bayesian, or distance method is most appropriate for analyzing the data in hand. Because each of these approaches can work well with different types of traits or situations, researchers regularly use several with each data set. Parsimony approaches are implemented by finding the tree that implies the smallest number of character state changes in the traits being studied. Several different computer algorithms can be employed to search among the very large number of trees that are possible and evaluate them according to a parsimony, likelihood, Bayesian probability, or minimum-distance criterion.

A third step in phylogeny inference is to evaluate the quality of the best tree implied by each of the inference methods employed. A variety of statistical techniques can be used to quantify the degree of support for the whole tree when compared with other trees, or for specific lineages within a best tree. When conflicts arise between trees that were based on different data sets or estimated through different inference approaches, researchers seek to resolve the conflict by analyzing data that are less subject to homoplasy.

Phylogenetic thinking is being applied to a wide variety of problems in evolutionary biology, from systems for classifying the diversity of life to the origins of disease-causing agents. Informative uses of phylogenies include dating events that are poorly documented in the fossil record, analyzing the geographic distribution of species, and studying coevolution.

Questions

1. What is a synapomorphy?
2. What is homoplasy? Explain the difference between convergence and reversal.
3. Why does homoplasy make it difficult to estimate the evolutionary history of species accurately?
4. Why is it logical to use parsimony as a way to minimize the impact of homoplasy in phylogeny inference?
5. What are some of the advantages and disadvantages of using morphological and molecular traits to infer phylogenies?
6. How can researchers defend the claim that certain traits are homologous?
7. How can researchers defend the claim that certain traits are derived or ancestral?
8. Explain why traits might be similar due to convergent evolution. Give an example of morphological traits that are convergent.
9. Explain the difference between monophyletic and paraphyletic groups. Give an example of each.
10. Why are evolutionary relationships only revealed by analyzing synapomorphies?
11. Explain the difference between cladistic and phenetic approaches to inferring phylogenies.
12. In the case of the whale astralagus, why was it misleading to use parsimony when inferring the phylogeny of artiodactyls?
13. What is a molecular clock? When and why should molecular clocks work?
14. In Figure 4.9, site 192 was considered "variable but uninformative." Why is it uninformative?
15. High-crowned teeth that are well suited for grazing are found in some rodents, rabbits and hares, most cloven-hoofed animals, horses, and elephants. Examine Figure 4.18, which shows the relationships of these and other mammalian orders. Are high-crowned teeth a synapomorphy or a product of convergent evolution?

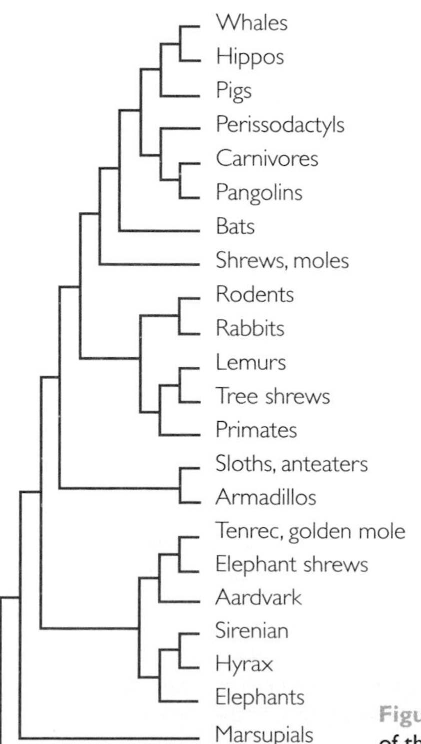

Figure 4.18 Phylogeny of the mammals From Murphy, W.J. et al. (2001).

16. Using the information in Figure 4.3, explain why the bones found in bird wings and bat wings are homologous. Then explain why the use of the forelimb for powered flight is a convergent trait in birds and bats.

17. According to Figure 4.18, are the ungulates—which consist of the horses + tapirs + rhino clade and the whales + hippos + pigs clade—a monophyletic group? Draw arrows on the tree indicating where the hoofed-foot trait was gained and/or lost. In your opinion, should the name ungulate still be used?

18. Examine the three primate phylogenies shown in Figure 4.19. Do the three phylogenies show the same relationships and the same order of branching? Do they give different impressions of whether there was a "goal" of primate evolution or what the "highest" primate is? Explain.

19. Figure 4.20 shows the phylogeny of some ant species that farm fungi for food, along with the phylogeny of the fungal species that they have domesticated. Does cospeciation occur in these interacting species? Explain.

20. Darwin maintained that there is no such thing as a higher or lower animal or plant. Explain what he meant.

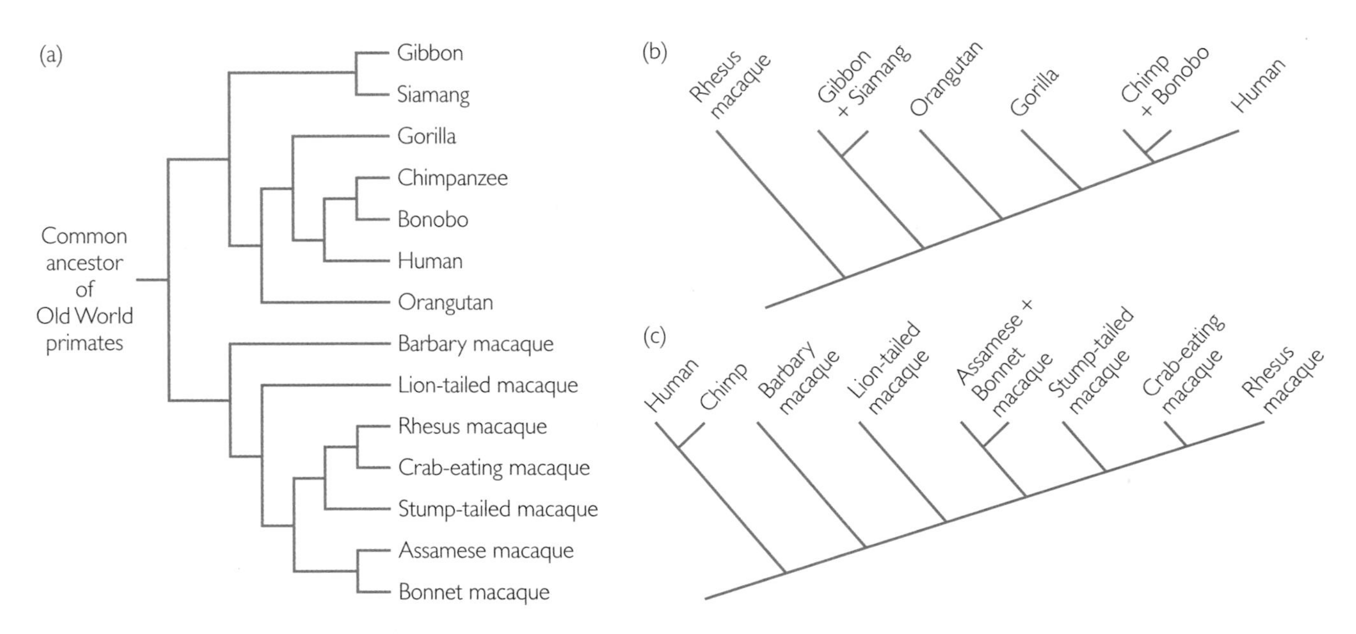

Figure 4.19 Phylogenies showing the relationships of some Old World primates (Branch lengths are not scaled.)

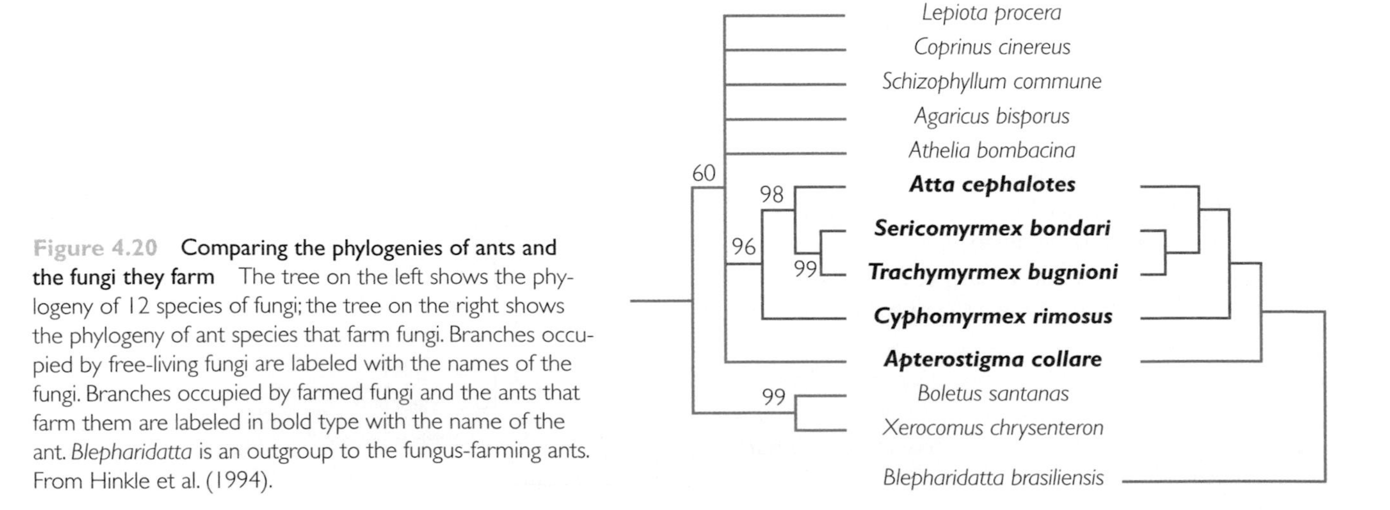

Figure 4.20 **Comparing the phylogenies of ants and the fungi they farm** The tree on the left shows the phylogeny of 12 species of fungi; the tree on the right shows the phylogeny of ant species that farm fungi. Branches occupied by free-living fungi are labeled with the names of the fungi. Branches occupied by farmed fungi and the ants that farm them are labeled in bold type with the name of the ant. *Blepharidatta* is an outgroup to the fungus-farming ants. From Hinkle et al. (1994).

Exploring the Literature

21. Leading researchers are now maintaining World Wide Web sites that provide trees for all forms of life. One major effort is called The Tree of Life (*http://tolweb.org/tree/phylogeny.html*). For background information and the addresses of some other key sites, see the Web site associated with this text.

22. Phylogenetic studies of coevolution are a booming research area. For an introduction to this literature, see:

Brower, A.V. Z. 1996. Parallel race formation and the evolution of mimicry in Heliconius butterflies: A phylogenetic hypothesis from mitochondrial DNA sequences. *Evolution* 50: 195–221.

Currie, C.R. et al. 2003. Ancient tripartite coevolution in the attine ant-microbe symbiosis. *Science* 299: 386-388.

Farrell, B. D., A. S. Sequeira, B. C. O'Meara, B. B. Normark, J. H. Chung, and B. H. Jordal. 2001. The evolution of agriculture in beetles (Curculionidae: Scolytinae and Platypodinae). *Evolution* 55: 2011–2027.

Gargas, A., P. T. DePriest, M. Grube, and A. Tehler. 1995. Multiple origins of lichen symbioses in fungi suggested by SSU rDNA phylogeny. *Science* 268: 1492–1495.

Mant, J. G, F. P. Schiestl, R. Peakall, and P. H. Weston. 2002. A phylogenetic study of pollinator conservatism among sexually deceptive orchids. *Evolution* 56: 888–898.

23. Methods for extracting and analyzing DNA from fossils are now sophisticated enough that researchers have been able to infer the phylogeny of some extinct forms. For examples, see:

Baker, A.J., L.J. Huynen, O. Haddrath, C.D. Millar, and D.M. Lambert. 2005. Reconstructing the tempo and mode of evolution in an extinct clade of birds with ancient DNA: The giant moas of New Zealand. *Proceedings of the National Academy of Sciences USA* 102: 8257–8262.

Karanth, K.P., T. Delefosse, B. Rakotosamimanana, T.J. Parsons, and A.D. Yoder. 2005. Ancient DNA from giant extinct lemurs confirms single origin of Malagasy primates. *Proceedings of the National Academy of Sciences USA* 102: 5090–5095.

Citations

Bajpai, S., and P. D. Gingerich. 1998. A new Eocene archaeocete (Mammalia, Cetacea) from India and the time of origin of whales. *Proceedings of the National Academy of Science, USA* 95: 15464–15468.

Boisserie, J.-R., F. Lihoreau, and M. Brunet. 2005. The position of Hippopotamidae within Cetartiodactyla. *Proceedings of the National Academy of Sciences USA* 102: 1537-1541.

Bremer, K. 1994. Branch support and tree stability. *Cladistics* 10: 295–304.

Chapela, I.H., S.A. Rehner, T.R. Schultz, and U.G. Mueller. 1994. Evolutionary history of the symbiosis between fungus-growing ants and their fungi. *Science* 266: 1691-1694.

Clark, M. A., N. A. Moran, P. Baumann, and J. J. Wernegreen. 2000. Cospeciation between bacterial endosymbionts (*Buchnera*) and a recent radiation of aphids (*Uroleucon*) and pitfalls of testing for phylogenetic congruence. *Evolution* 54: 517–525.

de Muizon, D. 2001. Walking with whales. Nature 413: 259–260.

de Queiroz, K., and P.D. Cantino. 2001. Phylogenetic nomenclature and the Phylocode. *Bulletin of Zoological Nomenclature* 58: 254–271.

Felsenstein, J. 1978. Cases in which parsimony or compatibility methods will be positively misleading. *Systematic Zoology* 27: 401–410.

Felsenstein, J. 1981. Evolutionary trees from DNA sequences: A maximum likelihood approach. *Journal of Molecular Evolution* 17: 368–376.

Felsenstein, J. 1983. Parsimony in systematics: Biological and statistical issues. *Annual Review of Ecology and Systematics* 14: 313–333.

Felsenstein, J. 1985. Confidence limits on phylogenies: An approach using the bootstrap. *Evolution* 39: 783–791.

Felsenstein, J. 2004. *Inferring Phylogenies*. Sinauer: Sunderland, MA.

Flower, W. H. 1883. On whales, present and past and their probable origin. *Proceedings of the Zoological Society of London* 1883: 466–513.

Gatesy, J., M. Milinkovitch, V. Waddell, and M. Stanhope. 1999. Stability of cladistic relationships between Cetacea and higher-level artiodactyl taxa. *Systematic Biology* 48: 6–20.

Gilbert, S. C. et al. 1998. Association of malaria parasite population structure, HLA, and immunological antagonism. *Science* 279: 1173–1177.

Gingerich, P. D., B. H. Smith, and E. L. Simons. 1990. Hind limbs of Eocene *Basilosaurus*: Evidence of feet in whales. *Science* 249: 154–157.

Gingerich, P. D., M. ul Haq, I. S. Zalmout, I. H. Khan, and M. S. Malkani. 2001. Origin of whales from early artiodactyls: hands and feet of Eocene Protocetidae from Pakistan. *Science* 293: 2239–2242.

Hall, B.G. 2005. Comparison of the accuracies of several phylogenetic methods using protein and DNA sequences. *Molecular Biology and Evolution* 22: 792:802.

Hayasaka, K., K. Fujii, K. and S. Horai. 1996. Molecular phylogeny of macaques: Implications of nucleotide sequences from an 896-base pair region of mitochondrial DNA. *Molecular Biology and Evolution* 13: 1044-1053.

Hennig, W. 1979. *Phylogenetic Systematics*. Urbana: University of Illinois Press.

Hillis, D. M. 1999. SINEs of the perfect character. *Proceedings of the National Academy of Sciences USA* 96: 9979–9981.

Hillis, D. M., and J. J. Bull. 1993. An empirical test of bootstrapping as a method for assessing confidence in phylogenetic analysis. *Systematic Biology* 42: 182–192.

Hillis, D. M., J. P. Huelsenbeck, and C. W. Cunningham. 1994. Application and accuracy of molecular phylogenies. *Science* 264: 671–677.

Hillis, D. M., B. K. Mable, and C. Moritz. 1996. Applications of molecular systematics: The state of the field and a look to the future. In D. M. Hillis, C. Moritz, and B.K. Mable, eds. *Molecular Systematics*. Sunderland, MA: Sinauer, 515–543.

Hinkle, G., J. K. Wetterer, et al. 1994. Phylogeny of the Attine ant fungi based on analysis of small subunit ribosomal RNA gene sequences. *Science* 266: 1695–1697.

Huelsenbeck, J. P., and D. M. Hillis. 1993. Success of phylogenetic methods in the four-taxon case. *Systematic Biology* 42: 247–264.

Huelsenbeck, J. P., and B. Rannala. 1997. Phylogenetic methods come of age: Testing hypotheses in an evolutionary context. *Science* 276: 227–232.

Huelsenbeck, J.P., F. Ronquist, R. Nielsen, and J.P. Bollback. 2001. Bayesian inference of phylogeny and its impact on evolutionary biology. *Science* 294: 2310–2314.

Kimura, M. 1980. A simple method for estimating rates of base substitution through comparative studies of nucleotide sequences. *Journal of Molecular Evolution* 16: 111–120.

Kittler, R., M. Kayser, and M. Stoneking. 2003. Molecular evolution of *Pedicularis humanus* and the origin of clothing. *Current Biology* 13: 1414–1417.

Kolaczkowski, B. and J.W. Thornton. 2004. Performance of maximum parsimony and likelihood phylogenetics when evolution is heterogeneous. *Nature* 431: 980–984.

Korbel, J.O., B. Snel, M.A. Huynen, and P. Bork. 2002. SHOT: a web server for the construction of genome phylogenies. *Trends in Genetics* 18: 158–162.

Levy, D. R. Yoshida, and L. Pachter. 2006. Beyond pairwise distances: neighbor-joining with phylogenetic diversity estimates. *Molecular Biology and Evolution* 23: 491–498.

Luckett, W. P., and N. Hong. 1998. Phylogenetic relationships between the orders Artiodactyla and Cetacea: A combined assessment of morphological and molecular evidence. *Journal of Mammalian Evolution* 5: 127–182.

Maddison, W. P., M. J. Donoghue, and D. R. Maddison. 1984. Outgroup comparison and parsimony. *Systematic Zoology* 33: 83–103.

Martin, A. P., G. J. P. Naylor, and S. R. Palumbi. 1992. Rates of mitochondrial DNA evolution in sharks are slow compared with mammals. *Nature* 357: 153–155.

Martin, A. P., and S. R. Palumbi. 1993. Body size, metabolic rate, generation time, and the molecular clock. *Proceedings of the National Academy of Sciences USA* 90: 4087–4091.

Milinkovitch, M.C. and J.G.M. Thewissen. 1997. Even-toed fingerprints on whale ancestry. *Nature* 388: 622-624.

Murphy, W. J., E. Eizirik, et al. 2001. Resolution of the early placental mammal radiation using Bayesian phylogenetics. *Science* 294: 2348–2351.

Nikaido, M., A. P. Rooney, and N. Okada. 1999. Phylogenetic relationships among cetartiodactyls based on insertions of short and long interspersed elements: Hippopotamuses are the closest extant relatives of whales. *Proceedings of the National Academy of Sciences USA* 96: 10261–10266.

Novacek, M. J. 1993. Reflections on higher mammalian phylogenetics. *Journal of Mammalian Evolution* 1: 3–30.

O'Leary, M. A., and J. H. Geisler. 1999. The position of Cetacea within Mammalia: Phylogenetic analysis of morphological data from extinct and extant taxa. *Systematic Biology* 48: 455–490.

Prothero, D. R., E. M. Manning, and M. Fischer. 1988. The phylogeny of the ungulates. In M. J. Benton, ed. *The Phylogeny and Classification of the Tetrapods, Volume 2-Mammals*. Oxford: Clarendon Press, 201–234.

Raxworthy, C. J., M. R. J. Forstner, and R.A. Nussbaum. 2002. Chameleon radiation by oceanic dispersal. *Nature* 415: 784–787.

Rokas, A., B.L. Williams, N. King, and S.B. Carroll. 2003. Genome-scale approaches to resolving incongruence in molecular phylogenies. *Nature* 425: 798–804.

Saitou, N. and M. Nei. 1987. The neighbor-joining method: A new method for reconstructing evolutionary trees. *Molecular Biology and Evolution* 4: 406–425.

Shedlock, A.M., K. Takahashi, and N. Okada. 2004. SINEs of speciation: tracking lineages with retrotransposons. *Trends in Ecology and Evolution* 19: 545–553.

Simpson, G. G. 1945. The principles of classification and a classification of mammals. *Bulletin of the American Museum of Natural History* 85: 1–350.

Sneath, P. H. A., and R. R. Sokal. 1973. *Numerical Taxonomy: The Principles and Practice of Numerical Classification*. San Francisco: Freeman.

Steel, M. 2005. Should phylogenetic models be trying to 'fit an elephant'? *Trends in Genetics* 21: 307–3009.

Swofford, D. L., G. J. Olsen, P. J. Waddell, and D. M. Hillis. 1996. Phylogenetic inference. In D. M. Hillis, C. Moritz, and B. K. Mable, eds. *Molecular Systematics*. Sunderland, MA: Sinauer, 407–514.

Tamura, K., M. Nei, and S. Kumar. 2004. Prospects for inferring very large phylogenies by using the neighbor-joining method. *Proceedings of the National Academy of Sciences USA* 101: 11030–11035.

Thewissen, J. G. M., and S. T. Hussain. 1993. Origin of underwater hearing in whales. *Nature* 361: 444–445.

Thewissen, J. G. M., S. T. Hussain, and M. Arif. 1994. Fossil evidence for the origin of aquatic locomotion in archaeocete whales. *Science* 263: 210–212.

Thewissen, J. G. M., and S. I. Madar. 1999. Ankle morphology of the earliest cetaceans and its implications for the phylogenetic relations among ungulates. *Systematic Biology* 48: 21–30.

Thewissen, J. G. M., E. M. Williams, L. J. Roe, and S. T. Hussain. 2001. Skeletons of terrestrial cetaceans and the relationship of whales to artiodactyls. *Nature* 413: 277–281.

Wahlberg, N., M.F. Braby, et al. 2005. Synergistic effects of combining morphological and molecular data in resolving the phylogeny of butterflies and skippers. *Proceedings of the Royal Society B* 272: 1577-1586.

Wakely, J. 1996. The excess of transitions among nucleotide substitutions: New methods of estimating transition bias underscore its significance. *Trends in Ecology and Evolution* 11: 158–163.

Zuckerkandl, E., and L. Pauling. 1962. Molecular disease, evolution and genic heterogeneity. In M. Kash and B. Pullman, eds. *Horizons in Biochemistry*. New York: Academic Press.

PART II

MECHANISMS OF EVOLUTIONARY CHANGE

Genetic variation is the raw material for evolution by natural selection. The photo on the right shows a typical fruit fly (*Drosophila*). The remaining photos show mutants: a bar-eyed fly, a four-winged fly, and a white-eyed fly.

The first four chapters of this text focused on natural selection as a mechanism of evolutionary change. By the end of Part I we had defined evolution as a change in allele frequencies within populations and probed how natural selection functions as an agent of evolutionary change.

Natural selection is not the only process that alters allele frequencies and causes evolution, however. Mutation, migration, and genetic drift also change allele frequencies in populations.

To gain a more thorough understanding of the mechanisms of evolutionary change, it's logical to begin with a chapter devoted to mutation—a process that continually introduces new alleles, and occasionally new genes, into populations. Using a combination of algebraic models and experimental tests, Chapters 6 and 7 go on to probe how selection, mutation, migration, and drift act on this variation to produce evolutionary change. Chapter 7 also investigates how inbreeding and other forms of nonrandom mating affect the fate of alleles in populations. Chapters 8 and 9 conclude the unit by focusing on interactions that occur among genes as the four forces act and by exploring how biologists study evolutionary change in traits that are molded by large numbers of genes. ■

5

Mutation and Genetic Variation

The zebra standing second from left is darkly colored or "melanistic," due to a mutation in a gene involved in the synthesis or transport of dark-colored melanin pigments.

The genetic variation that natural selection and other evolutionary forces act on originates in mutation. Meiosis also creates genetic variation in populations that reproduce sexually, because crossing over leads to new groupings of alleles on individual chromosomes and independent assortment leads to new mixtures of chromosomes in the resulting daughter cells. But meiosis reshuffles *existing* alleles into new combinations. Mutation is the only process that creates completely *new* alleles and new genes. As a result, biologists refer to mutation as the ultimate source of genetic variation. Once this variation is produced, then selection, drift, and migration can act. Mutations are the raw material of evolution.

This chapter has two goals: to investigate the mechanisms responsible for generating new alleles and new genes, and to explore how biologists quantify the amount of genetic variation that exists in natural populations. We begin by reviewing how single-base mutations and other types of small-scale changes occur in DNA sequences. These processes produce new alleles. Later in the chapter we consider larger-scale changes that can produce new genes, change the organization of individual chromosomes, or alter the number of chromosomes in a species. The chapter closes by considering how researchers analyze genetic variation within species.

5.1 Where New Alleles Come From

The instructions for making and maintaining an organism are encoded in its hereditary material—the molecule called deoxyribonucleic acid, or DNA. As Figure 5.1a shows, DNA is made up of smaller molecules called deoxyribonucleotides. The four deoxyribonucleotides found in DNA are similar in structure: Each contains the 5-carbon sugar called deoxyribose, a phosphate group, and a distinctive nitrogen-containing base. As the illustration shows, the five carbons in deoxyribose are each numbered in a specific way. These atoms are referred to as the 1′ carbon, 2′ carbon, and so forth; the prime symbol indicates that they are part of the deoxyribose subunit and not the attached nitrogenous base. The four bases, in turn, belong to two discrete chemical groups: Cytosine and thymine are pyrimidines, whereas adenine and guanine are purines. The four deoxyribonucleotides are routinely abbreviated to C, T, A, and G.

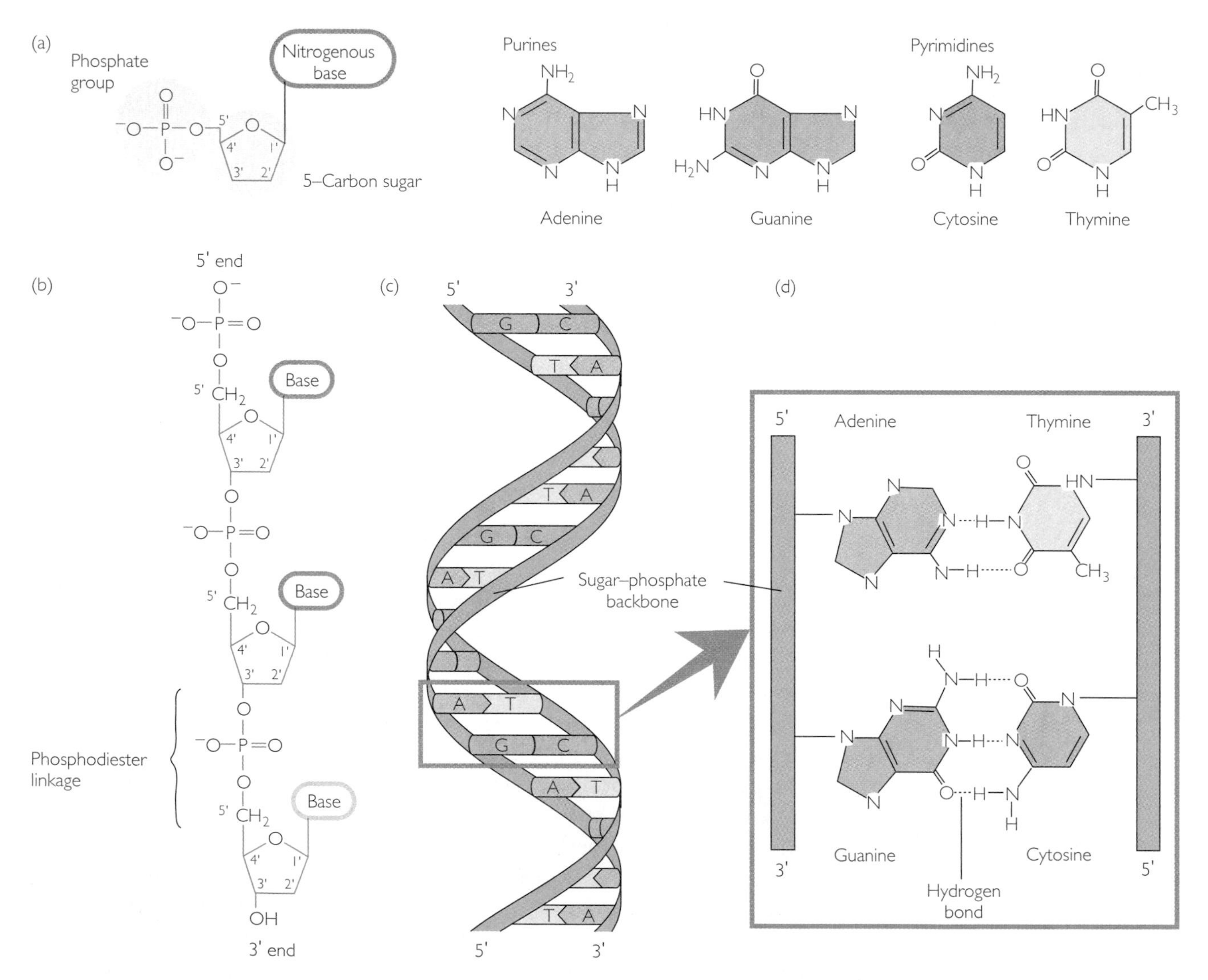

Figure 5.1 **The structure of the genetic material** (a) The diagram on the left shows the generalized form of a nucleotide. Note that hydrogen and oxygen atoms bonded to the numbered carbons are not shown. The diagrams on the right show the structure of the four nitrogenous bases. (b) Nucleic acids form via phosphodiester linkages between the 5′ carbon on one nucleotide and the 3′ carbon on another. (c) When complementary bases on opposite DNA strands form hydrogen bonds, the molecule twists into a double helix like the one shown here. (d) Adenine and thymine form two hydrogen bonds; cytosine and guanine form three.

Figure 5.1b illustrates how these molecules are linked into long strands by phosphodiester bonds that form between the 5′ carbon of one deoxyribonucleotide and the 3′ carbon of another. A single strand of DNA, then, consists of a sequence of bases attached to sugar-phosphate groups that form a "backbone." The DNA found in cells normally consists of two such strands. The two strands are wound around one another to produce the double helix diagrammed in Figure 5.1c. This structure is stabilized in part by hydrophobic interactions among atoms in the interior of the helix and in part by hydrogen bonds that form between the nitrogenous bases on either strand. Due to the geometry of the bases and the amount of space available inside the helix, hydrogen bonds only form when adenine and thymine (A—T) or guanine and cytosine (G—C) bases line up on opposite strands. These purine–pyrimidine combinations are called **complementary base pairs**. As Figure 5.1d shows, three hydrogen bonds form between G and C. Only two hydrogen bonds are made between A and T.

The Nature of Mutation

A mutation is any change in DNA. Genes are made of DNA, so changes in DNA create changes in genes.

Once James Watson and Francis Crick (1953) had deduced the double-helical structure of DNA shown in Figure 5.1c, they immediately realized that complementary base pairing provided a mechanism for copying the hereditary material. As Figure 5.2 illustrates, one strand serves as the template for making a copy of the other strand. In 1960, Arthur Kornberg succeeded in isolating the first of several proteins, called DNA polymerases, that are responsible for copying the DNA in cells.

In the late 1950s and early 1960s, a series of experiments succeeded in clarifying how the sequence of bases in DNA encodes information and how the genetic information is used to synthesize the ribonucleic acids (RNAs) and proteins that form the cell's structures and control the chemical reactions

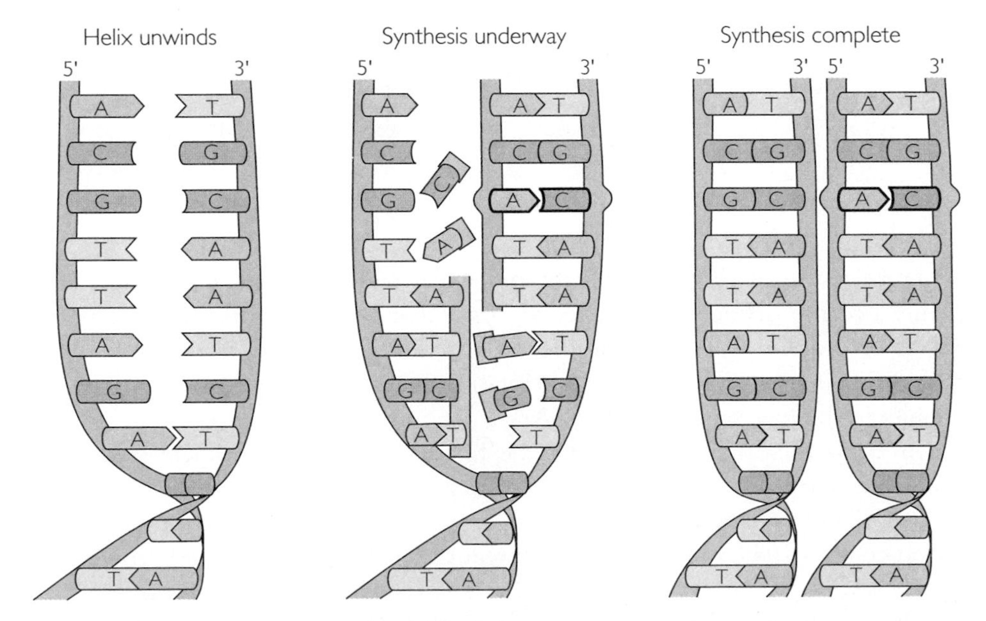

Figure 5.2 **DNA forms a template for its synthesis** Because of complementary base pairing, each strand in a DNA molecule forms a template for the synthesis of the complementary strand. If DNA polymerase inserts the wrong base, as in the strand at the far right, it results in a mismatched pair that must be repaired. If the repair is not made, a mutation results.

occurring inside. The central result was that DNA is transcribed into messenger RNA (mRNA), which is then translated into protein (Figure 5.3a). Researchers also established that the genetic code is read in triplets called codons. Each of the 64 different codons specifies a particular amino acid. Because only 20 amino acids need to be specified by the 64 codons, the genetic code is highly redundant—meaning that the same amino acid can be specified by more than one codon. Figure 5.3b shows the genetic code that is common to virtually all organisms living today.

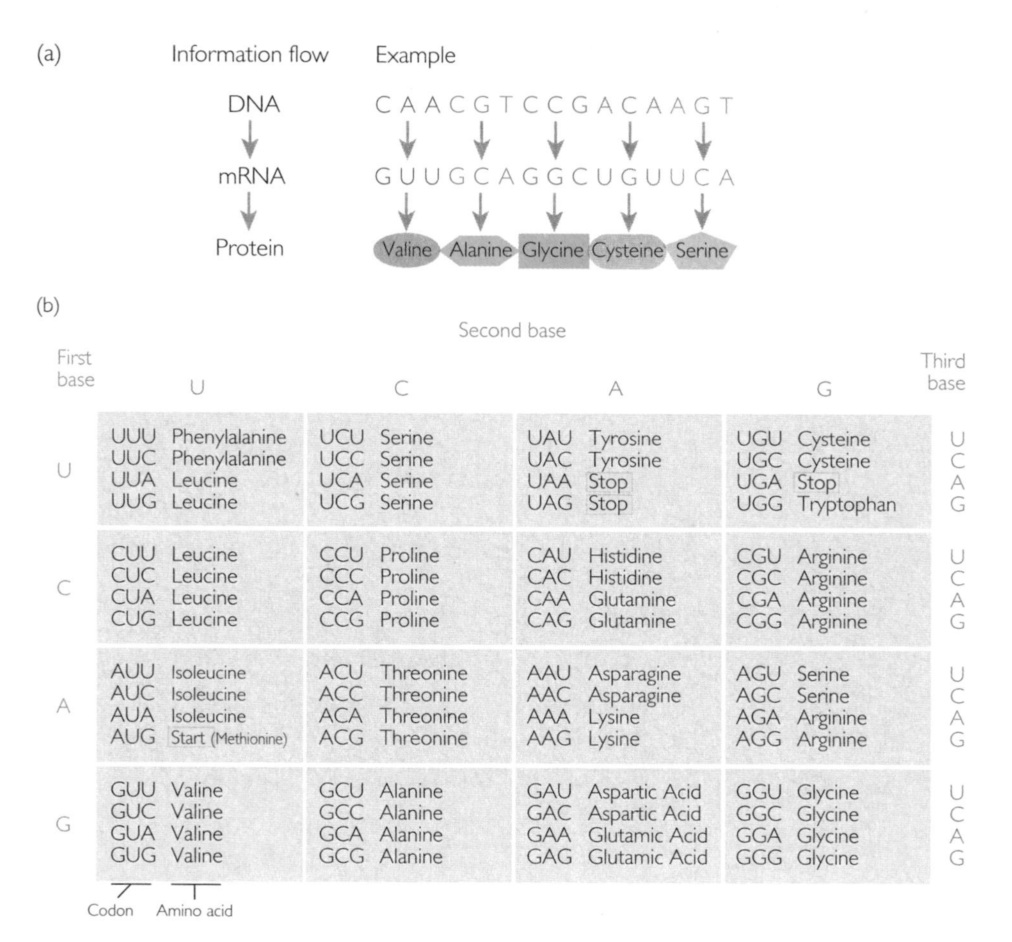

First base	U	C	A	G	Third base
U	UUU Phenylalanine	UCU Serine	UAU Tyrosine	UGU Cysteine	U
	UUC Phenylalanine	UCC Serine	UAC Tyrosine	UGC Cysteine	C
	UUA Leucine	UCA Serine	UAA Stop	UGA Stop	A
	UUG Leucine	UCG Serine	UAG Stop	UGG Tryptophan	G
C	CUU Leucine	CCU Proline	CAU Histidine	CGU Arginine	U
	CUC Leucine	CCC Proline	CAC Histidine	CGC Arginine	C
	CUA Leucine	CCA Proline	CAA Glutamine	CGA Arginine	A
	CUG Leucine	CCG Proline	CAG Glutamine	CGG Arginine	G
A	AUU Isoleucine	ACU Threonine	AAU Asparagine	AGU Serine	U
	AUC Isoleucine	ACC Threonine	AAC Asparagine	AGC Serine	C
	AUA Isoleucine	ACA Threonine	AAA Lysine	AGA Arginine	A
	AUG Start (Methionine)	ACG Threonine	AAG Lysine	AGG Arginine	G
G	GUU Valine	GCU Alanine	GAU Aspartic Acid	GGU Glycine	U
	GUC Valine	GCC Alanine	GAC Aspartic Acid	GGC Glycine	C
	GUA Valine	GCA Alanine	GAA Glutamic Acid	GGA Glycine	A
	GUG Valine	GCG Alanine	GAG Glutamic Acid	GGG Glycine	G

Figure 5.3 **In organisms, information flows from DNA to RNA to proteins** (a) In cells, the sequence of bases in DNA is transcribed to a sequence of bases in a strand of messenger RNA, which is then translated into a sequence of amino acids in a protein. Note that RNA contains a nitrogenous base called uracil instead of thymine. An adenine in DNA specifies a uracil in RNA. (b) This is the genetic code. Each of the 64 mRNA codons shown here specifies an amino acid or the start or end of a transcription unit. Note that in many instances, changing the third base in a codon does not change the message.

These results inspired an explicitly molecular view of the gene and mutation. Genes became defined as stretches of DNA that code for a distinctive type of RNA or protein product. Alleles became defined as versions of the same gene that differ in their base sequence. Mutations were understood as any type of change in the base sequence of DNA.

To apply these insights, consider the first mutation ever characterized on a molecular level: the change in the human gene for hemoglobin that results in sickle-cell disease. Hemoglobin is the oxygen-carrying protein found in red blood cells. In 1949, Linus Pauling's lab reported that people suffering from sickle-cell disease had a form of hemoglobin different from that of people without the disease. In 1958, Vernon Ingram showed that the difference between normal and sickle-cell hemoglobin was due to a single amino acid change at position number 6 in the protein chain, which is 146 amino acids long. Instead of having glutamic acid at this position, the sickling allele has valine. Further work established that the amino acid replacement is caused by a single base substitution in the hemoglobin gene. The mutant allele has an adenine instead of a thymine at nucleotide 2 in the codon for amino acid 6. A change like this is called a **point mutation** because it alters a single point in the base sequence of a gene.

Point mutations create a change in a single base pair in DNA. They occur when errors in DNA replication or repair are not fixed properly.

Point mutations are caused by one of two processes: random errors in DNA synthesis or random errors in the repair of sites damaged by chemical mutagens or high-energy radiation. Both types of changes result from reactions catalyzed by DNA polymerase. Look back at Figure 5.2, and note that DNA polymerase inserted the wrong deoxyribonucleotide opposite a cytosine in the far-right strand. If this mismatch is not repaired, then a DNA molecule with a point mutation will be created the next time that the strands are copied. An error like this resulted in the sickling allele in human hemoglobin.

If DNA polymerase mistakenly substitutes a purine (A or G) for another purine, or a pyrimidine (T or C) for another pyrimidine, during normal synthesis or the synthesis that occurs during a repair, the resulting point mutation is called a **transition** (Figure 5.4). If a purine is substituted for a pyrimidine or a pyrimidine for a purine, the mutation that results is called a **transversion**. Of the two kinds of point mutation, transitions are much more common. In a recent study of mutation types in the roundworm *Caenorhabditis elegans*, transitions outnumbered transversions by 1.6:1. The leading hypothesis to explain this observation is that transitions cause much less disruption in the DNA helix during synthesis, so are less likely to be recognized as an error and therefore less likely to be immediately corrected by DNA polymerase or later by mismatch repair enzymes.

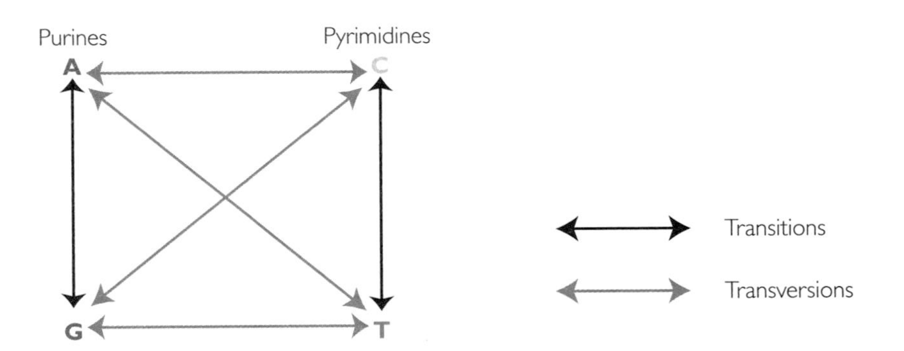

Figure 5.4 Transitions and transversions

If either type of base substitution occurs in the coding region of a gene, the mutation changes the codon read by the protein, called RNA polymerase, that synthesizes RNA from the DNA template. For example, the substitution of an A for a T in the gene for hemoglobin is a transversion that changes the message in codon number 6. More specifically, the change from A to T occurs at the second position in codon 6. Instead of specifying GAA or GAG as the sixth codon in the mRNA for hemoglobin, the mutant gene results in an mRNA containing GUA or GUG. A quick look at the genetic code in Figure 5.3b should convince you that this change results in a valine instead of glutamic acid in the resulting protein.

Spend a moment more analyzing the genetic code and you'll note an important general pattern: Changes in the first or second position of a codon almost always change the amino acid specified by the resulting mRNA. But because of the redundancy in the genetic code, changes in the third position frequently produce no change at all. Point mutations that result in an amino acid change are called **replacement** (or **nonsynonymous**) **substitutions**; those that result in

Point mutations may or may not produce a change in the protein encoded by a gene. Mutations produce new phenotypes when they change the gene product.

no change are called **silent site** (or **synonymous**) **substitutions**. Both types of point mutations create new alleles. Now the question is: How often do mutations occur?

Mutation Rates

How often are new alleles formed? Traditionally, biologists had to estimate mutation rates by studying large populations of organisms and counting the number of offspring that had observable mutant phenotypes in each generation (e.g., Drake et al. 1998; Grogan et al. 2001). In most cases, these observable mutant phenotypes were due to knock-out or **loss-of-function mutations**—meaning, changes in DNA that inactivate a gene, leading to a complete lack of gene product. In humans, for example, the traits called achondroplasia (a type of dwarfism) and hemophilia A (impaired blood-clotting) are due to loss-of-function mutations.

Researchers recognized that collecting data on observable phenotypes results in an underestimate of the actual mutation rate. This is because silent site mutations and the vast majority of replacement substitutions have effects that are much more subtle than a complete loss of function. As a result, these types of mutations are not detected when researchers assess offspring phenotypes. The new alleles created by these types of mutations go completely uncounted in traditional surveys.

Still, traditional methods of estimating mutation rates uncovered an intriguing pattern: When mutation rates were scaled for generation time and reported on a per-genome, per-cell-division basis, organisms as diverse as bacteria, archaea, fungi, plants, and animals appeared to have roughly equal mutation rates. This result suggested that the mutation rate per cell division is approximately equal in most or all organisms, and led to the hypothesis that natural selection had led to a single, common mutation rate.

Recent Work: Direct Estimates of Mutation Rates

The view that organisms share a common mutation rate has been challenged recently, in part by data sets where the mutation rate was estimated directly. As an example, consider work done by Dee Denver and colleagues (2004) on the roundworm *Caenorhabditis elegans*. These researchers started their work with a single individual. Because *C. elegans* have both male and female gonads and can self-fertilize, the researchers were able to establish many independent lines, or families, from this individual's offspring. Each of these lines, in turn, was propagated for a large number of generations by rearing the next generation of offspring and then randomly selecting a single individual as the progenitor of that line's subsequent generation. Each family line was maintained in the most benign environment possible, with optimal temperature and humidity, minimal crowding of individuals, abundant food, and no predators or parasites. The purpose of this protocol was to minimize the impact of natural selection and maximize the probability that every mutation that occurred would be passed on to the next generation—even mutations that would lead to low fitness in a normal environment and thus be eliminated by natural selection.

Researchers have begun to estimate mutation rates directly, by sequencing DNA from the same study population over time.

To estimate the mutation rate, the group sequenced tens of thousands of base pairs of DNA from the individual at the start of the experiment and from each of over 50 family lines derived from that individual at generation number 280, 353, and 396. They found a total of 30 mutations, yielding an estimated mutation rate of 2.1×10^{-8} mutations per site per generation. More than half of the 30 mutations consisted of **indels**, or small numbers of bases that had been inserted into or deleted from the genome (see Garcia-Diaz and Kunkel 2006). Overall, this mutation rate is equivalent to 2.1 mutations per genome per generation. This means that each new worm carries an average of about 2.1 new mutations in its nuclear genome. The result carries an important punchline: Mutation introduces a great deal of genetic variation into populations in every generation.

Denver et al. (2000) used the same protocol to study the mutation rate in the mitochondrial DNA of *C. elegans*. Mitochondria are organelles that have a circular chromosome independent of the nuclear chromosomes. In this case the analysis identified a total of 26 new mutations. Of these, 16 were base substitutions, with 13 transitions and 3 transversions. The remaining 10 mutations were indels. Based on the number of base pairs sequenced, the overall mutation rate was estimated to be 1.6×10^{-7} mutations per site per generation. This is about an order of magnitude higher than the rate observed for nuclear DNA in *C. elegans*. This result is logical, because mitochondria lack some of the DNA repair enzymes found in the nucleus. Because fewer errors are repaired in mitochondrial DNA, more mutations occur.

Having more precise estimates of mutation rates is helpful, but what biologists would really like to know is how rates vary among populations and species. Baer et al. (2005) attempted to address this question using the same rearing protocol as the previous two studies. In this study, however, the researchers created the experimental lines from two different populations of each of three roundworm species. Although sequence data are not yet available from this experiment, an indirect method of calculating the overall mutation rate suggested that each population and each species has a different mutation rate. Although this conclusion is still tentative, it is in stark contrast to the conclusion from early work indicating that organisms may share a common mutation rate. Do populations of different species actually have distinct mutation rates, shaped by natural selection to maximize fitness given that organism's way of life and environment?

Early results suggest that mutation rates (1) may exceed two new mutations per individual per generation, and (2) vary among populations and species.

Investigating Natural Selection on Mutation Rate

Although a substantial number of new alleles appear in most populations every generation, it is important to recognize that on a per-site basis DNA synthesis is astonishingly accurate. From the data on *C. elegans* nuclear DNA, for example, it appears that an incorrect base is inserted about once in every 100 million nucleotides. This is a remarkable observation. It implies that the enzymes responsible for copying and repairing DNA have been under intense selection.

Frances Gillin and Nancy Nossal (1976a and b) were the first to document that DNA polymerases vary in their accuracy. They did this by investigating single-base substitutions in the DNA polymerase of a virus called bacteriophage T4. Some of the mutations that Gillin and Nossal isolated decreased the rate at which polymerase made errors during DNA replication and reduced the overall mutation rate. Other mutations in polymerase increased the error rate and heightened the overall mutation rate. Since their work was published, investigators have been able to isolate and characterize DNA polymerases from the bacterium *Escherichia coli*, HIV, and other organisms or viruses that also vary in accuracy (for examples see Shinkai and Loeb 2001; Gutiérrez-Rivas and Menéndez-Arias 2001; Minnick et al. 2002). Working with a virus that has an RNA genome, Furio et al. (2005) showed that the polymerase responsible for copying its genes exhibits a fundamental trade-off between accuracy and speed: Mutants that increased accuracy were slower than more error-prone mutants.

Point mutation rates also depend on how efficiently mistakes in DNA synthesis are corrected. In *E. coli*, the bacterium *Salmonella enteritidis*, and *C. elegans*, mutations in the genes responsible for repairing base-pair mismatches and damaged DNA result in mutation rates that are 100 to 1,000 times higher than normal (LeClerc et al. 1996; Matic et al. 1997; Denver et al. 2005). The efficiency of DNA mismatch repair, like the error rate of DNA polymerase, is clearly a trait with heritable variation.

If mutation rate is a trait that varies among individuals within populations, then when are lower or higher mutation rates advantageous? Thanks to recent studies on *E. coli*, an answer to this question is starting to emerge. Arjan de Visser and colleagues (1999) compared how fitness changed over time in *E. coli* populations with normal and high mutation rates in laboratory environments that were either novel or identical to the culture conditions normally experienced by the cells. They found that cells with elevated mutation rates had higher fitness than normal cells only when they were grown in novel environments. Similarly, Antoine Giraud and coworkers (2001) found that when they injected bacteria-free mice with equal numbers of *E. coli* from populations with high and low mutation rates, the "mutator" cells initially had a large fitness advantage. This advantage disappeared, however, as time passed and the cells became adapted to the novel environment.

Mutation rates vary due to variation in the structure of enzymes involved in DNA replication and repair. High mutation rates may be selectively advantageous in novel or rapidly changing environments.

The message of these studies is that higher mutation rates may be adaptive when organisms colonize new environments to which they are poorly adapted, or when rapid copying of genes is advantageous. Otherwise, individuals with lower mutation rates should have higher fitness. A key idea here is that if individuals are well adapted to their environment, then most mutations are likely to be deleterious—meaning that they lower fitness. In contrast, individuals with high mutation rates appear to be favored when a population is in a novel or rapidly changing environment, where mutations are more likely to be beneficial.

What do the data have to say about these ideas? Is it true that in a normal environment, most mutations are indeed deleterious?

The Fitness Effects of Mutations

Some of the best data on the fitness effects of mutations come from the mutation-accumulation lines that Denver et al. studied in *Caenorhabditis elegans*. Recall that these individuals were allowed to accumulate mutations in a benign environment where natural selection was minimal or nonexistent. When Larissa Vassilieva and associates (2000) measured the fitness of individuals from these lines over time, they found that longevity, the production of offspring, and other key aspects of fitness declined steadily relative to control populations. Control populations were maintained in normal conditions of laboratory culture, where large numbers of individuals competed for resources and a large sample of individuals were randomly chosen as the progenitors of the next generation. In the control populations, then, individuals with deleterious mutations would produce relatively few offspring and be less likely to be represented in the next generation. The results for survival to sexual maturity, graphed in Figure 5.5a, are typical. These results support the hypothesis that the vast majority of mutations are deleterious. If they are allowed to accumulate, fitness declines.

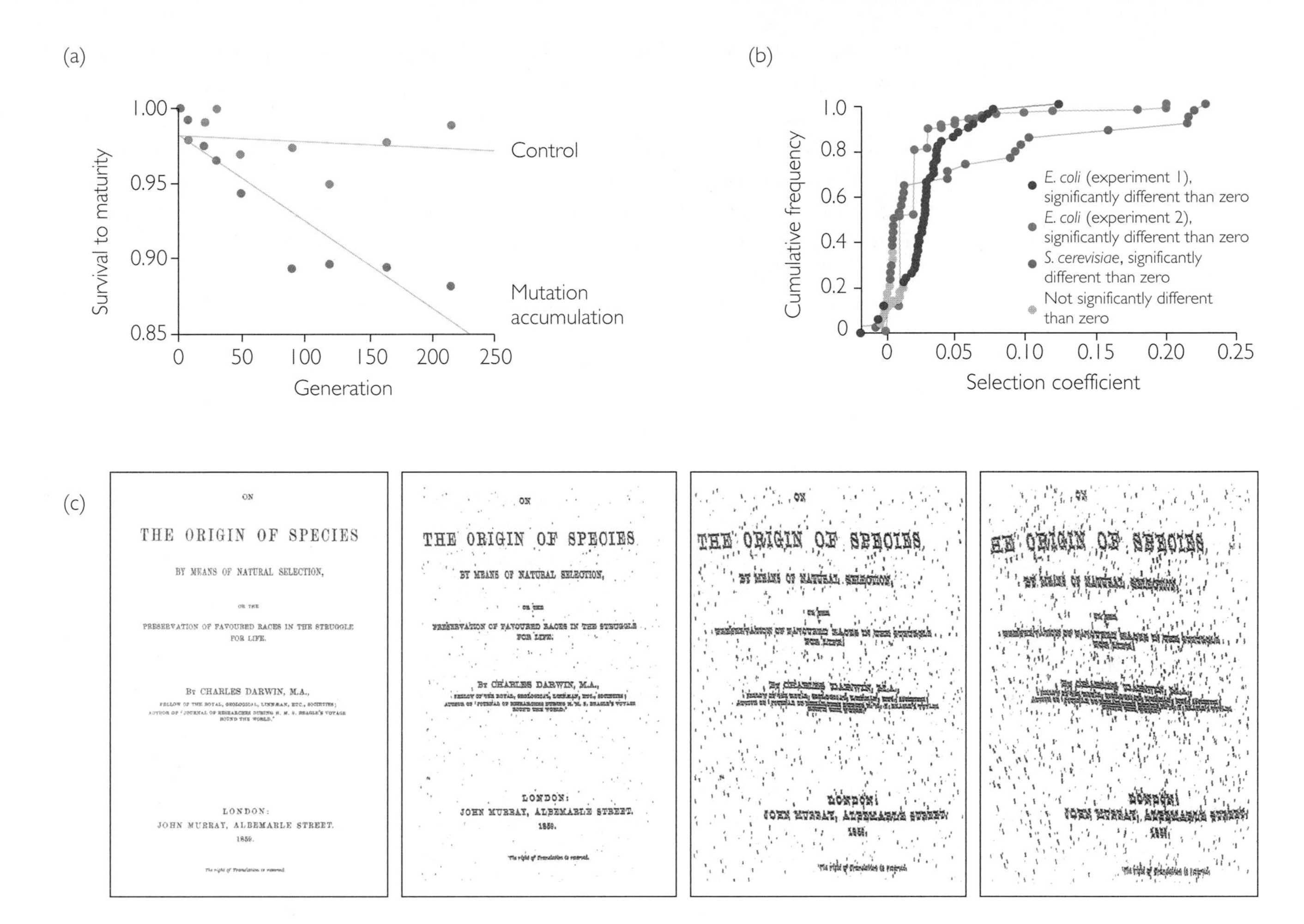

Figure 5.5 How do most mutations effect fitness? (a) This graph shows the percentage of individuals that survived to adulthood through time in populations that were allowed to accumulate all mutations versus control lines where natural selection eliminated most deleterious mutations. (b) To create the data plotted here, researchers inserted DNA sequences into random locations in the genomes of *Escherichia coli* or the yeast *Saccharomyces cerevisiae*. (c) These images were created from the original on the far left by making a copy of a copy 20, 50, and 100 times.

The roundworm data are also consistent with a large body of more indirect evidence, from mutation accumulation experiments on *Drosophila melanogaster* and other organisms, that most mutations are only mildly deleterious—meaning that each mutation reduces fitness by only about 2% when heterozygous (see review by Lynch et al. 1999). Consider, for example, the results graphed in Figure 5.5b for *E. coli* and the yeast *Saccharomyces cerevisiae*. To collect these data, researchers manipulated bacterial or yeast cells so that each would receive a large insertion mutation in a random location. Then they cultured many offspring from these experimental cells and measured the growth rate of each population relative to controls, which were identical except for the insertion mutation. The **selection coefficient** plotted along the x-axis represents the difference in fitness between each experimental population and the control. Note that upwards of 70% of the mutations have selection coefficients less than 2%. Because these mutations consisted of large insertions that were likely to knock out genes, their average effect on fitness should be much greater than that of an average base substitution.

For a population through time, then, the situation is analogous to the one illustrated in Figure 5.5c. Note that the first panel in the figure shows the title page of Darwin's book *The Origin of Species*. This page is equivalent to the *C. elegans* population at the start of the mutation accumulation experiment. To create the page illustrated in the second panel, a biologist reproduced the original page on an exceptionally good quality copy machine—one that makes very few errors per letter or block of white space. He then made a copy of this copy, and then a copy of the copied copy, and then a copy of that copy, and so on, for a total of 100 generations. Note that random errors in the print or white space have begun to noticeably change the appearance of the page by generation 20. Eventually, the page is entirely unreadable. Like the *C. elegans* populations, its "fitness" has declined dramatically due to the accumulation of many mutations of small effect. The message of these analyses is that most of the new alleles created by mutation are quickly eliminated from the population by natural selection.

The vast majority of mutations reduce fitness slightly or are neutral with respect to fitness.

It is important to recognize, though, that a few mutations are beneficial and that many mutations have little to no detectable effect on fitness. Alleles that have no effect on fitness are said to be **neutral**. (Silent site substitutions in DNA, for example, do not alter gene products and thus are not subject to natural selection based on protein or RNA function. As Chapter 7 will show, though, even silent site substitutions can be subject to natural selection. This phenomenon is called codon bias, and occurs when it is more efficient for DNA polymerase to transcribe certain codons for the same amino acid rather than others.)

To summarize the growing body of literature on mutation rates and the fitness effects of mutations, it appears that mutations are surprisingly frequent when considered on a per genome per generation basis, that the mutation rate may vary considerably among populations and species, and that the vast majority of mutations are either neutral or slightly deleterious. It is not surprising that most of the random changes in the amino acid sequences of proteins do not improve their ability to function, because most proteins have been under selection for millions of years. You would not expect a random change to improve a protein's function any more than you would expect a random change in a computer's circuitry to improve processing performance. Mutation continually introduces new alleles to populations, but most of these new alleles have slightly deleterious effects. Understanding the fate of deleterious, neutral, and beneficial alleles is a major focus of Chapters 6 and 7.

5.2 Where New Genes Come From

Several kinds of mutations can create entirely new genes, just as several types of mutations can create new alleles. **Gene duplications** are probably the most important source of new genes, however.

Most duplications result from one of two processes. The first is called retrotransposition. The term combines the word root retro, meaning backward, with the word transpose, meaning to change position. The choice is appropriate because the process begins when a processed messenger RNA, which lacks introns but has a poly (A) tail, is reverse-transcribed by the enzyme reverse transcriptase (see Chapter 1) to form a double-stranded segment of DNA. If this molecule becomes integrated into one of the main chromosomes, then the genome has acquired a duplicated copy of the original gene.

A more common source of duplicated genes is a phenomenon known as unequal cross-over. **Unequal cross-over** is a chance mistake caused by the proteins involved in managing the genetic recombination that occurs during meiosis. As Figure 5.6 shows, the event begins when homologous chromosomes do not synapse correctly during prophase of meiosis I. It results in one chromosome that contains a deletion and one chromosome that contains a redundant stretch of

Gene duplication can create new genes because it creates new DNA.

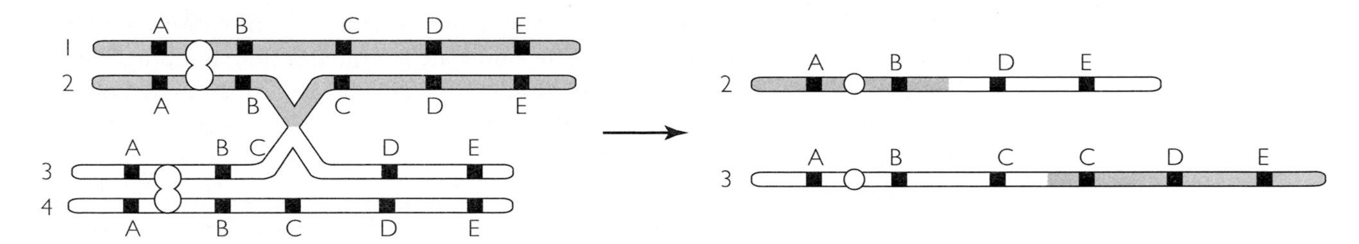

Figure 5.6 **Unequal cross-over and the origin of gene duplications** The letters and bars on each chromosome in the diagram indicate the location of genes; the open circles indicate the location of the centromere. The chromosomes on the left have synapsed, but cross-over has occurred at nonhomologous points. As a result, one of the cross-over products (chromosome #2) lacks gene C and one (chromosome #3) has a duplication of gene C.

DNA. In the latter case, the resulting chromosome has an extra copy of the sequences involved in the unequal cross-over. The extra copy is a duplicated segment.

Retrotransposition and unequal crossover leave distinctive "footprints" in the genome. Retrotransposed genes lack introns and lack the nearby regulatory sequences found in the original gene. They also code for a poly (A) tail and are usually found far from the original gene. In contrast, genes that were duplicated during unequal cross-over are found back-to-back, or in tandem. The duplicated segments tend to have at least some of the regulatory sequences that flank the original gene, and contain the same introns. Now the question is: How often do these events occur?

Rates of Gene Duplication

The best data on rates of gene duplication come from work by Michael Lynch and John Conery (2000), who analyzed the genomes of nine species of eukaryotes ranging from humans to baker's yeast. By documenting the number of genes that had been duplicated in each species and comparing it to the age of each species, they estimated an average rate of genes that duplicate and increase to high frequency in populations as 0.01 per gene per million years. This rate is similar to the rate that point mutations occur per base pair in individual genes. It means that in species with a genome containing 10,000 genes, a gene is duplicated and increases to high frequency every 10,000 years on average.

This conclusion was bolstered by recent work from Ze Cheng and colleagues (2005), who compared the complete genome sequences of common chimpanzees and humans and identified DNA segments that had been duplicated in each species. Humans and chimps shared a common ancestor about 7.5 million years ago, and their data indicate that since then about 2.7% of the genomes have been duplicated in one species but not the other. This represents a rate of about 4.4 megabases of DNA being added to each genome per million years.

Gene duplications occur often enough to be an important source of genetic variation in populations over time.

The message of these analyses is that in addition to a constant introduction of new alleles in every individual in every generation, it is common to find duplicated genes rising to high frequency in populations over time, creating an additional source of genetic variation. What happens to these duplicated segments?

The Fate of Duplicated Genes

When a DNA sequence is duplicated via retrotransposition or unequal cross-over, the original gene should continue to produce a normal product. The duplicated sequences, in contrast, are redundant and may accumulate mutations without consequences to the phenotype. The duplicated sequence might even change function over time and become an entirely new gene instead of an extra copy of an existing gene. This is an important point. Because it creates additional DNA, gene duplication results in entirely new possibilities for gene function.

To illustrate this point, consider the globin gene family in humans. These genes are found in two major clusters of sequences that code for the protein subunits of hemoglobin. The groups are the α-like cluster on chromosome 16 and the β-like cluster on chromosome 11 (α and β are the Greek letters alpha and beta). A completed hemoglobin molecule is made up of an iron-binding heme group surrounded by four protein subunits—two coded by genes in the α-like cluster and two coded by genes in the β-like cluster.

The data plotted in Figure 5.7 show that each gene in the α- and β-like families is expressed at a different time in development. In first-trimester human embryos, for example, hemoglobin is made up of two ζ (zeta) chains and two ε (epsilon) chains, while in adults it is made up of two α chains and two β chains. (The sickle-cell mutation occurs in one of the β chains.) Different combinations of globin polypeptides result in hemoglobin molecules with important functional differences. For example, fetal hemoglobin has a higher affinity for oxygen molecules than adult hemoglobin. As a result, oxygen molecules are efficiently transferred from the mother to the fetus during pregnancy.

The genes that make up the globin family are thought to be a product of gene duplication events. This hypothesis is supported by the remarkable correspondence in the length and position of the exons and introns observed among globin genes (see Figure 5.8). The logic here is that it is extremely unlikely that such high structural resemblance could occur in genes that do not share a recent common ancestor. The duplication hypothesis is also supported by the observation of high sequence similarity among globin genes and by their similar function.

The general model, then, is that an ancestral globin gene was duplicated several times during the course of vertebrate evolution. In several of these new genes,

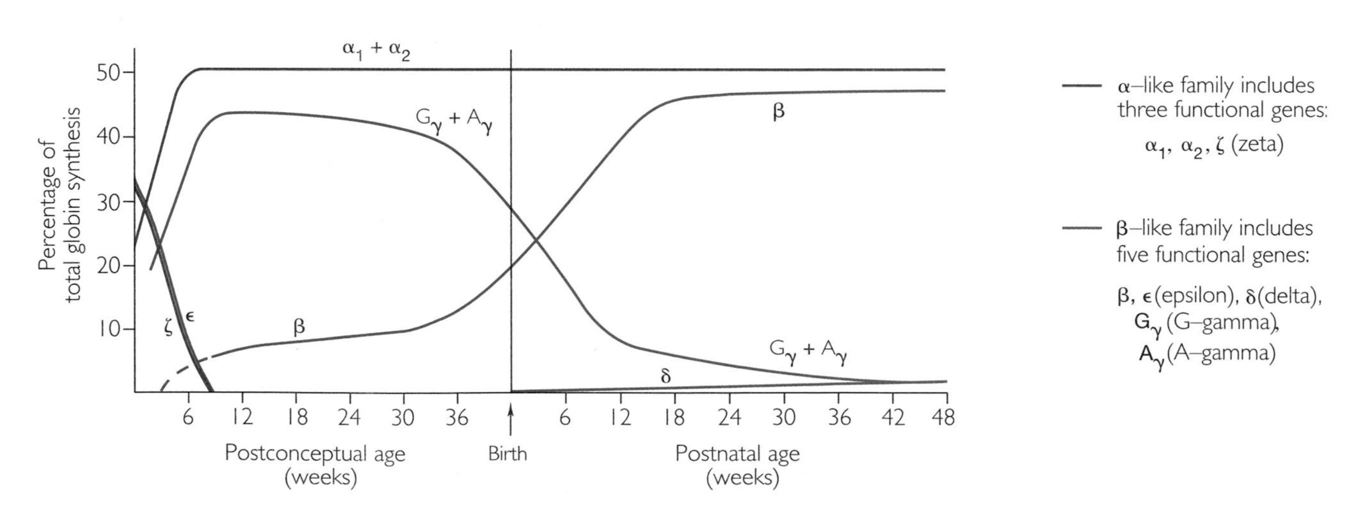

Figure 5.7 **Timing of expression differs among members of the globin gene families** This graph shows changes in the expression of genes from the α- and β-globin families in humans during pregnancy and after birth. In embryos, hemoglobin is made up of ζ-globin from the α-like gene cluster and ε-globin from the β-like gene cluster. In the fetus, hemoglobin is made up of α-globin from the α-like gene cluster and γ-globin from the β-like gene cluster. In adults, most hemoglobin is made up of α-globin from the α-like gene cluster and β-globin from the β-like gene cluster; a small number contain δ-globin. Each of these hemoglobins has important functional differences.

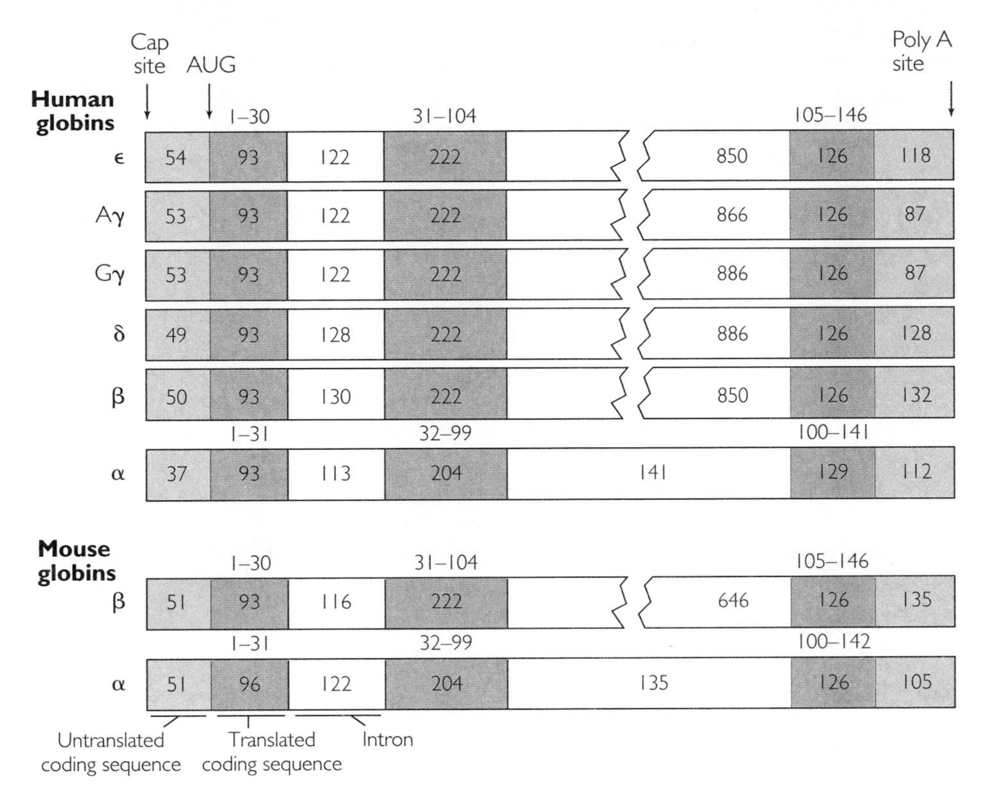

Figure 5.8 Transcription units in the globin gene family
In these diagrams, the numbers inside the boxes denote the number of nucleotides present in the primary transcript, while the numbers above the boxes give the amino acid positions in the resulting polypeptide. AUG is the start codon. The lengths and positions of introns and exons in loci throughout the α- and β-like clusters are virtually identical.

mutations changed the function of the duplicated region's protein product in a way that was favored by natural selection, leading to the formation of the gene family. Because the α- and β-like clusters also contain nonfunctional loci called **pseudogenes**, which are not transcribed, biologists infer that some duplicated loci were rendered functionless by loss-of-function mutations. These mutations may be indels that disrupt key sequences or point mutations that create a stop codon and truncate the gene product.

Duplicated genes can (1) retain their original function and provide an additional copy of the parent gene, (2) gain a new function through mutation and selection, or (3) become functionless pseudogenes.

It is important to note, however, that not all gene duplications result in sequences that evolve different functions or no function at all. In some important cases, like rRNA genes, multiple copies of the same gene have an identical or nearly identical base sequence and produce a product with the same function as the ancestral sequence.

Whether duplicated sequences have no function, the same function, or a different function, duplicated genes are homologous with respect to each other because they are derived from the same common ancestral sequence. But biologists also recognize that homologous genes come in one of two types (Fitch 1970). Genes that are duplicated and then diverge in sequence, like the members of the α- and β-globin clusters in humans, are said to be **paralogous**. Paralogs can be contrasted with genes that are homologous but have diverged after a speciation event, like the β-globin genes found in mice and in humans (see Figure 5.8). Homologs found in different species are said to be **orthologous**.

Recent analyses of whole-genome sequencing data have shown that most organisms contain an array of gene families similar to the globin genes found in humans and other mammals. In the species studied to date, duplicated genes are common and the size of gene families varies a great deal. Table 5.1 summarizes current estimates of the number and sizes of gene families found in yeast, fruit flies, and roundworms. Note that each of these organisms contains hundreds of

Table 5.1 Sizes of gene families

These data are from analyses of protein-coding regions in whole-genome sequences. In most cases, genes were designated as members of the same family if at least 30% of the amino acids in the resulting protein are identical. Modified from Gu et al. 2002.

	Number of families		
Size of gene family	**Yeast**	**Fruit flies**	**Roundworms**
2	415	404	665
3	56	113	188
4	23	46	93
5	9	21	71
6–10	19	52	104
>10	8	38	98

gene families comprised of 2 genes, and dozens of gene families that are made up of more than 10 genes. Analyses of the human genome have produced similar results (Li et al. 2001). Thus far, the largest gene family described is found in the mouse and consists of 1,296 paralogous genes involved in olfactory reception (Zhang and Firestein 2002).

To summarize, segments of DNA can be duplicated by several different mechanisms. These processes occur often enough to represent an important source of genetic variation in populations. Duplicated segments may be knocked out by mutations and have no function, they may retain the same function as the original sequence, or they may gain a new function. Paralogs form gene families and are an important aspect of genome structure and function, as Chapter 15 will show.

5.3 Chromosome Mutations

The mutations discussed thus far occur on the scale of a single base pair in DNA to segments containing tens of thousands of base pairs. But the amount of DNA affected by point mutations and gene duplications pales in comparison to mutations that alter the gross morphology of chromosomes. Some of these mutations affect only gene order and organization; others produce duplications or deletions that affect the total amount of genetic material. They can also involve the entire DNA molecule or segments of varying sizes. Here we focus on two types of chromosome alterations that are particularly important in evolution.

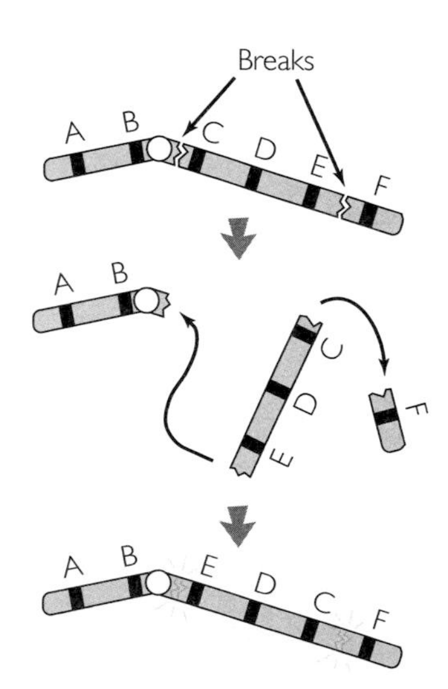

Figure 5.9 Chromosome inversion Inversions result when a chromosome segment breaks in two places, flips, and reanneals. Note that after the event, the order of the genes labeled C, D, and E is inverted.

Inversions

Chromosome inversions often result from a multistep process that starts when radiation causes two double-strand breaks in a chromosome. After breakage, a chromosome segment can detach, flip, and reanneal in its original location. As Figure 5.9 shows, gene order along the chromosome is now inverted.

In addition to involving much larger stretches of DNA than point mutations and gene duplications, inversions produce very different consequences. Inversions affect a phenomenon known as genetic **linkage**. Linkage is the tendency for alleles of different genes to assort together at meiosis. For example, genes on the same chromosome tend to be more tightly linked (that is, more likely to be inherited together) than genes on nonhomologous chromosomes. Similarly, the closer together genes are on the same chromosome, the tighter the linkage. Crossing over at meiosis, on the other hand, breaks up allele combinations and reduces linkage (see Chapter 8).

When inversions are heterozygous, meaning that one chromosome contains an inversion and the other does not, the inverted sequences cannot align properly when homologs synapse during prophase of meiosis I. When inversions are heterozygous, then, successful crossing-over events are extremely rare. The result is that alleles inside the inversion are locked so tightly together that they are inherited as a single "supergene."

Inversions change gene order and lessen the frequency of crossing over between homologous segments of chromosomes. As a result, the alleles inside inversions tend to be inherited as a unit.

Inversions are common in *Drosophila*—the most carefully studied of all insects. Are they important in evolution? To answer this question, consider a series of inversions found in populations of *Drosophila subobscura*. This fruit fly is native to western Europe, North Africa, and the Middle East, and has six chromosomes. Five of these chromosomes are **polymorphic** for at least one inversion (Prevosti et al. 1988), meaning that chromosomes with and without the inversions exist. Biologists have known since the 1960s that the frequencies of these inversions vary regularly with latitude and climate. This type of regular change in the frequency of an allele or an inversion over a geographic area is called a **cline**. Several authors have argued that different inversions must contain specific combinations of alleles that function well together in cold, wet weather or hot, dry conditions. But is the cline really the result of natural selection on the supergenes? Or could it be an historical accident, caused by differences in the founding populations long ago?

A natural experiment has settled the issue. In 1978 *D. subobscura* showed up in the New World for the first time, initially in Puerto Montt, Chile, and then four years later in Port Townsend, Washington, USA. Several lines of evidence argue that the North American population is derived from the South American one. For example, of the 80 inversions present in Old World populations, precisely the same subset of 19 is found in both Chile and Washington State. Within a few years of their arrival on each continent, the *D. subobscura* populations had expanded extensively along each coast and developed the same clines in inversion frequencies found in the Old World (Figure 5.10). The clines are even correlated with the same general changes in climate type: from wet marine environments, to mediterranean climates, to desert and dry steppe habitats (Prevosti et al. 1988; Ayala et al. 1989). This is strong evidence that the clines result from natural selection and are not due to historical accident.

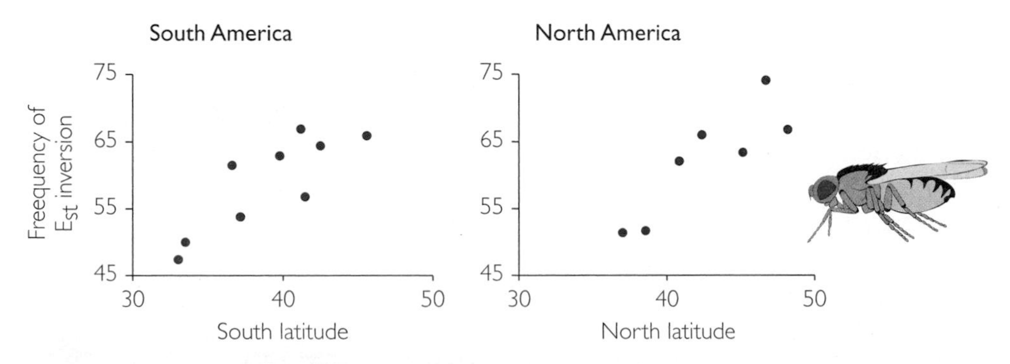

Figure 5.10 Inversion frequencies form clines in *Drosophila subobscura* These graphs plot the frequency of an inversion called E_{st} in South American and North American populations of *Drosophila subobscura*. From data in Prevosti et al. 1988; see also Balanya et al. 2003.

Which genes are locked in the inversions, and how do they affect adaptation to changes in climate? In the lab, *D. subobscura* lines that are bred for small body size tend to become homozygous for the inversions found in the dryer, hotter part of the range (Prevosti 1967). Recent research has confirmed that pronounced and parallel clines in body size exist in fly populations from North America, South America, and Europe (Gilchrist et al. 2004). These results hint that alleles inside the inversions affect body size, with natural selection favoring large flies in cold, wet climates and small flies in hot, dry areas. Research into this natural experiment is continuing. In the meantime, the fly study illustrates a key point about inversions: They are an important class of mutations because they affect selection on groups of alleles—a topic that will be explored in detail in Chapter 8.

Genome Duplication

The final type of mutation considered here occurs at the largest scale possible: entire sets of chromosomes. For example, if homologous chromosomes fail to segregate during meiosis I or if sister chromatids do not separate properly during meiosis II, the resulting cells may have double the number of chromosomes of the parent cell. In plants, similar mutations can occur during the mitotic cell divisions that lead up to gamete formation. Mutations like these can lead to the formation of a diploid gamete in species where gametes are normally haploid.

Figure 5.11 shows one possible outcome of a chromosome-doubling mutation. In the diagram, the individual that produces diploid gametes contains both male and female reproductive structures and can self-fertilize. When it does so, a tetraploid (4n) offspring results. If this offspring self-fertilizes when it matures, or if it mates with its parent or a tetraploid sibling that also produces diploid gametes, then a population of tetraploids can become established.

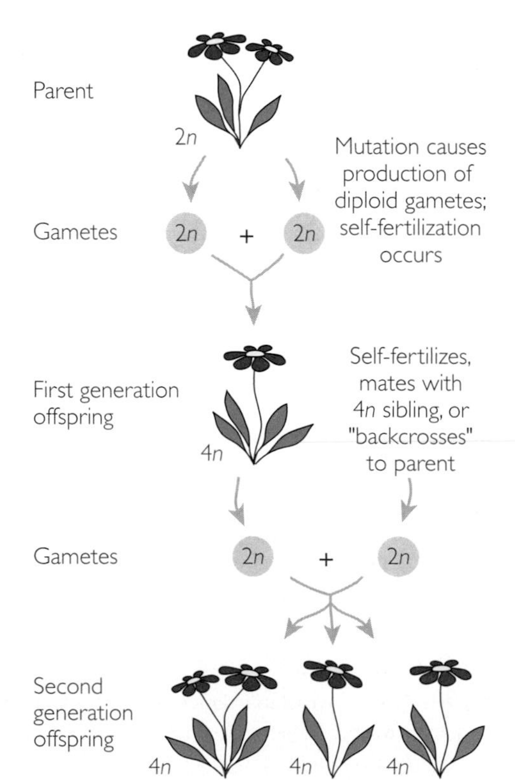

Figure 5.11 How are tetraploid individuals produced in plants? See text for explanation.

Organisms that have more than two chromosome sets are said to be **polyploid**. Polyploid organisms can be tetraploid (4n), hexaploid (6n), octoploid (8n), or higher. The phenomenon is common in plants and rare in animals—probably because self-fertilization is much more common in plants than animals. Nearly half of all angiosperm (flowering plant) species and the vast majority of the ferns are descended from ancestors where polyploidization occurred. In animals, polyploidy occurs in taxa like earthworms and some flatworms where individuals contain both male and female gonads and can self-fertilize. It is also present in animal groups that are capable of producing offspring without fertilization, through a process called parthenogenesis. In some species of beetles, sow bugs, moths, shrimp, goldfish, and salamanders, a type of parthenogenesis occurs that can lead to chromosomal doubling.

Polyploid individuals have one or more extra copies of every gene in the genome.

Now let's consider two reasons why polyploidy is an important type of mutation in evolution: It can lead to new species being formed, and it represents an enormous and rapid increase in the total amount of genetic material present.

Polyploidy and Speciation

To understand why a genome duplication event can lead to a new species being formed, imagine the outcome of matings between individuals in a tetraploid population and the most closely related diploid population. If individuals from the two populations mate, they produce triploid offspring. When these individuals mature and meiosis occurs, the homologous chromosomes cannot synapse correctly because they are present in an odd number. As a result, the vast majority of the gametes produced by triploids end up with the wrong number of chromosomes and fail to survive. Triploid individuals have extremely low fertility.

In contrast, recall from Figure 5.11 that when tetraploid individuals continue to self-fertilize or mate among themselves, then fully fertile tetraploid offspring will result. In this way, natural selection should favor polyploids that are reproductively isolated from their parent population. Diploid and tetraploid populations that are genetically isolated are on their way to becoming separate species.

Polyploid individuals may represent a new species, because it is difficult for them to breed with individuals from the original (non-polyploid) population.

Genome Duplication and Genetic Innovation

Doubled chromosome sets, like the duplicated individual genes analyzed earlier in the chapter, are free to gain new function as a result of mutation and natural selection. Polyploidy is a key source of genetic variation because it produces hundreds or thousands of duplicated genes.

How often does genome duplication occur? Justin Ramsey and Douglas Schemske (1998) addressed this question by calculating how frequently tetraploids are formed in angiosperms. According to published studies, flowering plant species typically produce diploid gametes at a frequency of 0.00465 per generation. Consequently, tetraploid offspring should be produced at a rate of $0.00465 \times 0.00465 = 2.16 \times 10^{-5}$. In each generation, then, about 2 out of every 100,000 offspring are tetraploid. The take-home message of this analysis is striking: In flowering plants, polyploid formation occurs about as frequently as point mutations in individual genes.

Because genome duplication doubles the amount of hereditary material present, biologists have wondered whether there might be a correlation between genome duplication events and bursts of evolution during the history of life. To address this hypothesis, researchers have scanned the genomes of an array of organisms to find evidence of past duplication events. This search is framed by the realization that instead of being polyploid, descendant species might retain only a portion of the originally duplicated genes—presumably, the portions that have acquired important new functions. If so, then past genome duplication events can be recognized by the presence of many large duplicated regions, in the same 5′ to 3′ orientation, on many different chromosomes.

For example, researchers have compared a large series of genes found in ray-finned fish to homologous genes in land-dwelling vertebrates, and concluded that in most cases, ray-finned fish have more copies of each gene (Chen et al. 2004; Jaillon et al. 2004). Based on these data, it is logical to infer that a genome duplication event occurred early the evolution of the ray-finned fish. This result is intriguing, because ray-finned fish are far and away the most species-rich lineage of vertebrates, and perhaps the most morphologically and ecologically diverse as well.

Genome duplication events have been an important source of genetic variation over the course of evolution. Early results suggest that genome duplications have been correlated with the origin of particularly species-rich and rapidly changing groups.

The idea that there might be a correlation between important bursts of evolution and an increase in the total amount of genetic material present has been bolstered by recent work indicating that genome duplication events occurred early in the evolution of angiosperms, which represent over 80% of land plants living today (DeBodt et al. 2005), and early in the evolution of vertebrates (Panopoulou and Poustka 2005). It is important to be cautious in interpreting these studies, however. Conclusions may change as more complete genome sequences become available, and the evidence available to date is correlational in nature. It has yet to be shown that a genome duplication event actually caused the evolution of a particularly species-rich and diverse lineage.

Although work on patterns of genome duplication continues, it is clear that chromosome-level mutations are important in evolution. Inversions may lock groups of alleles into tightly linked "supergenes," polyploidization can lead to the formation of new species, and genome duplication provides massive amounts of redundant gene sequences that may then diversify.

5.4 Measuring Genetic Variation in Natural Populations

Table 5.2 summarizes all of the various types of mutation reviewed thus far. Because these processes generate new alleles, genes, and chromosomes, they create the genetic variation that is the raw material for evolution. The purpose of this chapter's concluding section is to introduce the methods that biologists use to measure the overall amount of genetic variation present in natural populations, with a focus on measurements of allelic variation at individual genes.

The data analyzed earlier in rates of point mutation, gene duplication, and genome duplication should have convinced you that populations are constantly acquiring significant amounts of genetic variation. This is a stark contrast, however, with classical views of genetic variation in populations, which held that populations should contain little genetic variation. Biologists had come to this conclusion by reasoning that because in any population inhabiting a particular environment, one allele of each gene should confer higher fitness than all other alleles of that gene. Natural selection should preserve the allele most conducive to survival and reproduction and eliminate the rest. The one best allele was called the wild type; any other alleles present were considered mutants and were expected to be extremely rare.

Table 5.2 Types of mutation with significant evolutionary impact: A summary

Name	Description	Mechanism	Significance
Point mutation	Base pair substitutions in DNA sequences	Chance errors during DNA synthesis or during repair of damaged DNA	Creates new alleles
Chromosome inversion	Flipping of a chromosome segment, so order of genes along the chromosome changes	Breaks in DNA caused by radiation or other insults	Alleles inside the inversion are likely to be transmitted together, as a unit
Gene duplication	Duplication of a short stretch of DNA, creating an extra copy of the sequence	Unequal crossing-over during meiosis or retrotransposition	Redundant new genes may acquire new functions, by mutation
Genome duplication	Addition of a complete set of chromosomes	Errors in meiosis or (in plants) mitosis	May create new species; massive gene duplication

Pioneering work by Harris (1966) and Lewontin and Hubby (1966), who analyzed the proteins encoded by alleles, initiated a dramatic break with the classical view of limited genetic variation. These early studies found that instead of containing one wild-type allele at extremely high frequency, most populations routinely harbor an array of alleles. The deeper biologists have looked for allelic variation in the decades since, the more they have found. Today, evolutionary biologists recognize that the vast majority of natural populations harbor substantial genetic variation.

Determining Genotypes

The first step in measuring the diversity of alleles present at a particular gene is to determine the genotypes of a large sample of individuals in a population. To do so, biologists usually look directly at the DNA of the alleles themselves.

As an example of how researchers do this work, consider the gene in humans called *CCR5*. You might recall from Chapter 1 that this gene codes for a cell surface protein found on white blood cells, and that the CCR5 protein is used as a coreceptor by most sexually transmitted strains of HIV-1. One *CCR5* allele has a 32-base-pair deletion in the gene sequence. As a result of the deletion, the gene's protein product is severely shortened and thus nonfunctional. Individuals who are homozygous for the allele with the deletion lack CCR5 on the surfaces of their T cells. As a result, HIV-1 virions cannot bind to the cells to initiate an infection. Homozygous individuals do not get infected with HIV, even if they are repeatedly exposed to the virus.

Biologists can now measure the extent of genetic variation in natural populations directly by determining the genotypes of a large sample of individuals.

We will call the functional allele *CCR5+*, or just +, and the allele with the 32-base-pair deletion *CCR5-Δ32*, or just *Δ32*. Individuals with genotype

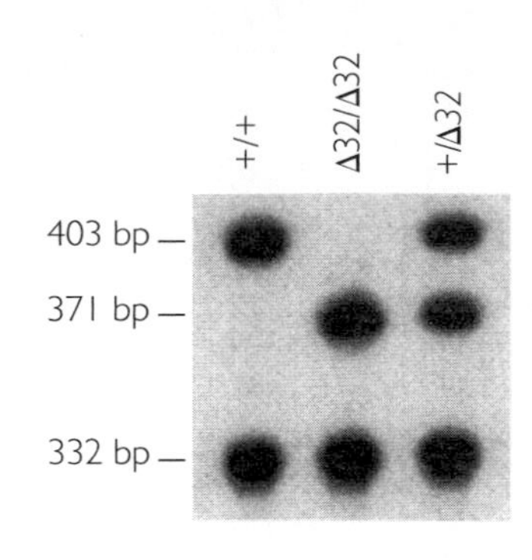

Figure 5.12 Determining *CCR5* genotypes by electrophoresis of DNA Each lane of this gel contains DNA fragments prepared from the *CCR5* alleles of a single individual. The locations of the dark spots, or bands, on the gel indicate the sizes of the fragments. Each genotype yields a unique pattern of bands. From Samson et al. (1996). Reprinted with permission from *Nature*. © 1996, Macmillan Magazines Ltd.

+/+ are susceptible to infection with HIV-1, individuals with genotype +/*Δ32* are susceptible, but may progress to AIDS more slowly, and individuals with genotype *Δ32/Δ32* are resistant to most sexually transmitted strains of the virus.

Upon learning of the *CCR5-Δ32* allele, AIDS researchers immediately wanted to know how common it is. Michel Samson and colleagues (1996) developed a genotype test that works as follows. Researchers first extract DNA from a sample of the subject's cells. Then the researchers use a polymerase chain reaction (PCR) to make many copies of a region of the gene, several hundred base pairs long, that contains the site of the 32-base-pair deletion. (PCR duplicates a targeted sequence many times over by employing a test-tube DNA replication system in combination with specifically tailored primer sequences that direct the DNA polymerase to copy just the locus of interest.) Finally, the researchers cut the duplicated DNA sequences with a restriction enzyme and run the resulting fragments on an electrophoresis gel.

The results appear in Figure 5.12. Both alleles yield two DNA fragments. The fragments from a *CCR*5-+ allele are 332 and 403 base pairs long. The fragments from a *CCR5-Δ32* allele are 332 and 371 base pairs long. Homozygotes have just two bands in their lane on the gel, whereas heterozygotes have three bands.

Several laboratories have completed surveys of *CCR5* genotypes in various indigenous populations from around the world. Data excerpted from a survey by Jeremy Martinson and colleagues (1997) appear in Table 5.3.

Table 5.3 Diversity of *CCR5* genotypes in various populations

Population	Number of people tested	Number with each genotype			Allele frequency (%)	
		+/+	+/*Δ32*	*Δ32/Δ32*	*CCR5*-+	*CCR5-Δ32*
Europe						
Ashkenazi	43	26	16	1	79.1	20.9
Iceland	102	75	24	3	85.3	14.7
Britain	283	223	57	3		
Italy	91	81	10	0		
Middle East and Asia						
Saudi Arabia	241	231	10	0		
Yemen	34	34	0	0		
Russia (Udmurtia)	46	38	7	1		
Pakistan	34	32	2	0		
Hong Kong	50	50	0	0		
Mongolia	59	59	0	0		
Philippines	26	26	0	0		
Africa						
Nigeria	111	110	1	0		
Central African Republic	52	52	0	0		
Kenya	80	80	0	0		

Excerpted from Martinson et al. 1997. Copyright © 1997, *Nature Genetics*. Reprinted by permission of the Nature Publishing Group, New York, NY.

Calculating Allele Frequencies

We have noted that a pressing question concerning the *CCR5-Δ32* allele is: How common is it? To answer this question precisely, we need to use the data on genotypes in Table 5.3 to calculate the frequency of the *Δ32* allele in the various populations tested. The frequency of an allele is its fractional representation among all the alleles present in a population.

As an example, we will calculate the frequency of the *Δ32* allele in the Ashkenazi population in Europe from the data in the first row of Table 5.3. The simplest way to calculate allele frequencies is to count allele copies. Martinson and colleagues tested 43 individuals. Each individual carries 2 alleles, so the researchers tested a total of 86 alleles. Of these 86 alleles, 18 were copies of the *Δ32* allele: 1 from each of the 16 heterozygotes, and 2 from the single homozygote. Thus the frequency of the *Δ32* allele in the Ashkenazi sample is

$$\frac{18}{86} = 0.209$$

or 20.9%. We can check our work by calculating the frequency of the + allele. It is

$$\frac{(52 + 16)}{86} = 0.791$$

or 79.1%. If our calculations are correct, the frequencies of the 2 alleles should sum to one, which they do.

An alternative method of figuring the allele frequencies in the Ashkenazi population is to calculate them from the genotype frequencies. Martinson and colleagues tested 43 individuals, so the genotype frequencies are as follows:

+/+	**+/Δ32**	**Δ32/Δ32**
$\frac{26}{43} = 0.605$	$\frac{16}{43} = 0.372$	$\frac{1}{43} = 0.023$

The frequency of the *Δ32* allele is the frequency of *Δ32/Δ32* plus half the frequency of +/*Δ32*:

$$0.023 + \frac{1}{2}(0.372) = 0.209$$

This is the same value we got by the first method.

Allele frequencies are filled in for the first two rows in Table 5.3. The rest of the rows are left to you to practice calculating allele frequencies. Readers who do so will discover an intriguing pattern. The *CCR5-Δ32* allele is common in populations of northern European extraction, with frequencies as high as 21%. As we move away from northern Europe, both to the east and to the south, the frequency of the *Δ32* allele declines. Outside of Europe, the Middle East, and western Asia, the *Δ32* allele is virtually absent. (To visualize this result, look at the map of *Δ32* frequency in Figure 1.20.) We will return to this pattern in Chapters 6 and 8.

How Much Genetic Diversity Exists in a Typical Population?

Studies on allelic diversity, similar to the work on the frequency of the *Δ32* allele in humans, have been done in a wide variety of populations and genes. Biologists use two summary statistics to summarize these types of data: the mean heterozygosity, and the percentage of polymorphic genes. The mean heterozygosity can be interpreted in two equivalent ways: as the average frequency of heterozygotes across loci, or as the fraction of genes that are heterozygous in the genotype of the average individual. The percentage of polymorphic loci is the fraction of genes in a population that have at least two alleles.

Early efforts to study allelic diversity in populations were based on a technique called allozyme electrophoresis. It involved isolating proteins from a large sample of individuals, separating the proteins in an electrophoresis gel, and then staining the gel to visualize the proteins produced by a particular gene. If the alleles present in a population were different enough that their protein products had different sizes or charges, then the proteins would migrate differently in the gel and would show up as different bands.

Allozyme electrophoresis studies demonstrated that most natural populations harbor substantial genetic variation. Figure 5.13 summarizes data on mean heterozygosities from invertebrates, vertebrates, and plants. As a broad generalization, in a typical natural population, between 33 and 50% of the genes that code for enzymes are polymorphic, and the average individual is heterozygous at 4–15% of its genes (Mitton 1997).

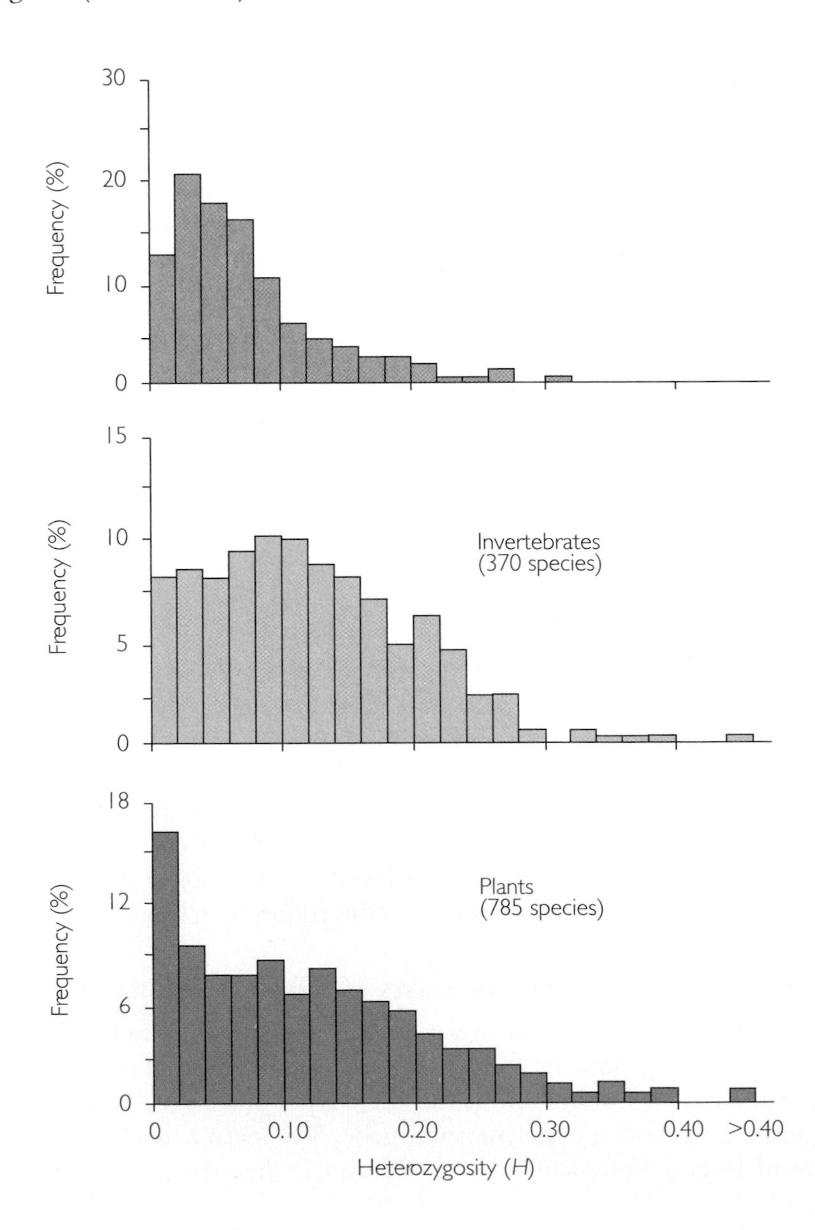

Figure 5.13 Analysis of proteins reveals that most populations harbor considerable genetic diversity These histograms show the distribution of enzyme heterozygosities among species of animals and plants. For example, about 7% of all plant species have a heterozygosity between 0.10 and 0.12. From Fig. 2.2, p. 19, of Avise (1994). © 1994, Chapman and Hall. Reprinted by permission of Springer Science and Business Media.

Methods that directly examine the DNA of alleles are much more powerful at revealing genetic diversity, however. This is because not every change in the DNA sequence at a gene produces an electrophoretically distinguishable protein. Among the most intensively studied gene to date is the gene associated with cystic fibrosis in humans. This gene encodes a protein called the cystic fibrosis transmembrane conductance regulator (CFTR). CFTR is a cell surface protein expressed in the mucus membrane lining of the intestines and lungs. Gerald Pier and colleagues (1997) demonstrated that one of CFTR's key functions is to enable the cells of the lung lining to ingest and destroy *Pseudomonas aeruginosa* bacteria. Individuals homozygous for loss-of-function mutations in the CFTR gene have cystic fibrosis. One of the symptoms of this disease is chronic infection with *P. aeruginosa*, eventually leading to severe lung damage. Molecular geneticists have examined the DNA sequence in the CFTR alleles of more than 15,000 cystic fibrosis patients, for a total of more than 30,000 copies of disease alleles. They have discovered more than 500 different loss-of-function mutations at this one locus (Figure 5.14). Although we will return to the CFTR gene in Chapter 6, the message of this work and similar studies is clear: The amount of genetic variation in most populations is extremely high. At most genes in most populations, dozens or even hundreds of different alleles are present.

Recent studies have shown that in most populations, many alleles are present at every gene in the genome. Genetic variation is extensive.

The classical view of genetic diversity, under which little diversity was expected in most populations, is clearly wrong. But why is so much allelic diversity present in most populations? Two modern views have replaced the traditional idea. According to the balance, or selectionist theory, genetic diversity is maintained by natural selection—in favor of rare individuals, in favor of heterozygotes, or in favor of different alleles at different times and places. In contrast, the neutral theory claims that most of the alleles at most polymorphic loci are functionally and selectively

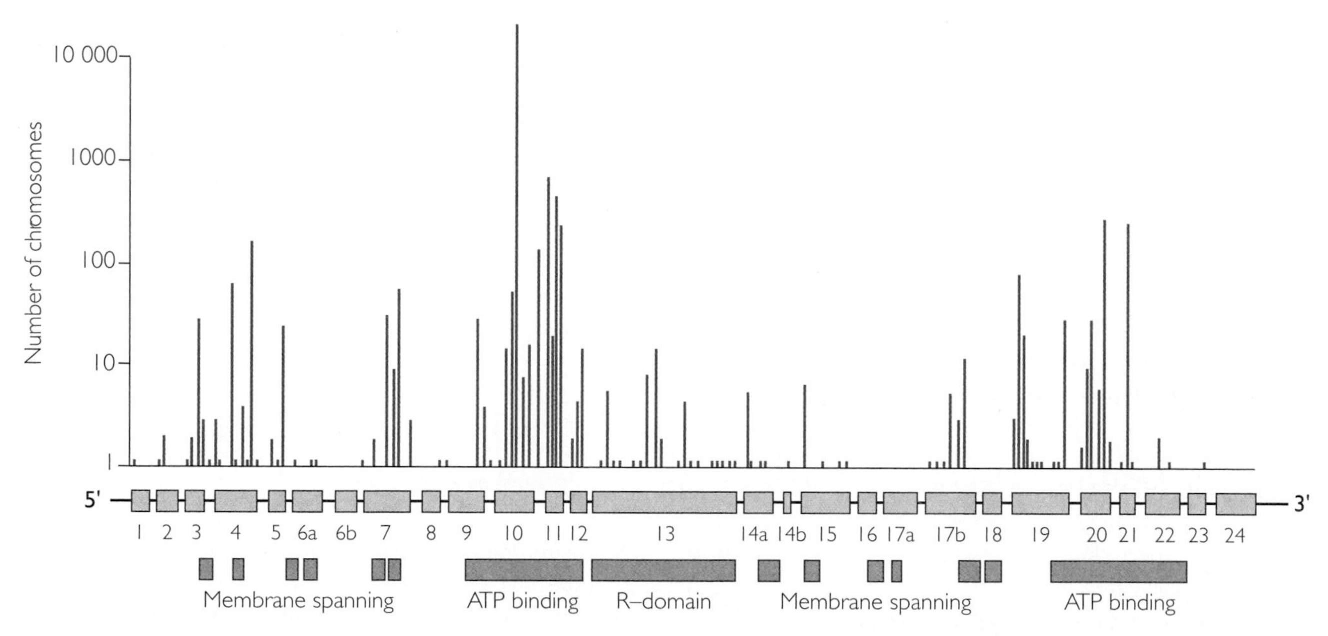

Figure 5.14 Sequencing studies have revealed enormous genetic diversity at the cystic fibrosis locus in humans This graph shows the abundance and location of the loss-of-function mutations discovered in an analysis of over 30,000 disease-causing alleles at the cystic fibrosis locus. The histogram shows the number of copies of each mutation found. The genetic map below it, in which the boxes represent exons, shows the location of each mutation within the CFTR gene. The boxes at the bottom of the graph give the functions of the coding regions of the gene. From Fig. 2, p. 395, in Tsui (1992). Copyright © 1992, Elsevier Science. Reprinted with permission of Elsevier Science.

equivalent and are maintained by genetic drift. In effect, the neutral theory holds that genetic diversity is maintained because it is not eliminated by selection. We will consider the selectionist and neutral theories in more detail in Chapters 6 and 7. Now let's delve into how drift, selection, and other processes act on the impressive genetic variation present in populations.

Summary

Mutations range from single base-pair substitutions to the duplication of entire chromosome sets and vary in impact from no amino acid sequence change to single amino acid changes to gene creation and genome duplication.

Point mutations result from errors made by DNA polymerase during DNA synthesis or errors made by DNA repair enzymes. Point mutations in first and second positions of codons frequently result in replacement substitutions that lead to changes in the amino acid sequence of proteins. Point mutations in third positions of codons usually result in silent substitutions that do not lead to changes in the amino acid sequence of proteins. Both types of point mutations create new alleles.

Point mutations and small insertions and deletions, or indels, are surprisingly frequent when considered on a per genome per generation basis. Recent studies confirm that the vast majority of mutations are either neutral or slightly deleterious in terms of their affect on fitness. The mutation rate is a trait that is subject to evolution by natural selection, because the speed and accuracy of DNA polymerase and the many proteins involved in repairing mismatched bases vary among individuals. A growing body of data suggests that the mutation rate varies considerably among populations and species, and that individuals with high mutation rates might be favored in novel or rapidly changing environments.

The most common sources of new genes are duplication events that result from errors during crossing-over. Over time, a duplicated gene may remain unchanged and produce additional copies of the same gene product, or diverge from its parent sequence to acquire a distinctive function, or become a nonfunctional pseudogene.

Chromosome alterations form a large class of mutations. Chromosome inversions have interesting evolutionary implications because they reduce the frequency of recombination between genes inside the inversion. As a result, alleles within inversions tend to be inherited as a group instead of independently. When genomes duplicate and most chromosomes are retained over time, the resulting population is polyploid. Polyploidy is particularly common in plants and is important because polyploid individuals are genetically isolated from the population that gave rise to them. Recent data also indicate that genome duplication followed by extensive chromosome loss has occurred in an array of lineages, and suggest that genome duplications may be correlated with particularly rapid bursts of evolution, resulting in species-rich and ecologically diverse groups.

Evolutionary biologists typically measure allelic diversity in populations by sampling large numbers of individuals and analyzing the proteins encoded by various alleles or the DNA of the alleles themselves. Such studies have revealed that most natural populations contain substantial genetic diversity.

Questions

1. What is the difference between a silent site mutation and a replacement mutation?
2. How do chromosome inversions occur?
3. Diagram the sequence of events that leads to the formation of second-generation polyploid individuals in plants that can self-fertilize.
4. How does unequal cross-over lead to the duplication of DNA sequences?
5. According to the data available to date, are most mutations deleterious, beneficial, or neutral? On average, are their fitness effects large, small, or nonexistent?
6. What is a transition? What is a transversion? Which is more common?
7. Why are the terms "point mutation" and "gene duplication" appropriate?
8. Why is genetic variation important in evolution?

9. Discuss why mutation rates vary among individuals within populations, and among species.
10. Compare and contrast the evolutionary effects of point mutations, chromosome inversions, gene duplications, and polyploidization.
11. What are the possible fates of a gene sequence that becomes duplicated due to unequal crossing-over? What events lead to each of these consequences?
12. What evidence do researchers use to defend the hypothesis that two or more stretches of DNA sequence are related by gene duplication events?
13. Suppose a silent site mutation occurs in an exon that is part of the β-globin gene in a human. Has a new allele been created? Defend your answer.
14. The amino acid sequences encoded by the red and green visual pigment genes found in humans are 96% identical (Nathans et al. 1986). These two genes are found close together on the X chromosome, while the gene for the blue pigment is located on chromosome 7. Among primates, only Old World monkeys, the great apes, and humans have a third pigment gene—New World monkeys have only one X-linked pigment gene. Comment on the following three hypotheses:
    - One of the two visual pigment loci on the X chromosome originated in a gene duplication event.
    - The gene duplication event occurred after New World and Old World monkeys had diverged from a common ancestor, which had two visual pigment genes.
    - Human males with a mutated form of the red or green pigment gene experience the same color vision of our male primate ancestors.
15. Chromosome number can evolve by smaller-scale changes than duplication of entire chromosome sets. For example, domestic horses have 64 chromosomes per diploid set while Przewalski's horse, an Asian subspecies, has 66. Przewalski's horse is thought to have evolved from an ancestor with $2n = 64$ chromosomes. The question is: Where did its extra chromosome pair originate? It seems unlikely that an entirely new chromosome pair was created de novo in Przewalski's horse. To generate a hypothesis explaining the origin of the new chromosome in Przewalski's horse, examine the adjacent figure. The drawing shows how certain chromosomes synapse in the hybrid offspring of a domestic horse–Przewalski's horse mating (Short et al. 1974). The remaining chromosomes show a normal 1:1 pairing. Do you think this sort of gradual change in chromosome number involves a change in the actual number of genes present, or just rearrangement of the same number of genes?

Exploring the Literature

16. In mammals, sperm cells are produced by parent cells (spermatogonia) that undergo constant cell division throughout life, whereas egg cells are produced only during fetal development. Do you think the average number of mutations per gamete might differ in males vs. females? Why or why not? Compare your ideas to the hypotheses and data in:

Makova, K. D. and W.-H. Li. 1993. Strong male-driven evolution of DNA sequences in humans and apes. *Nature* 416: 624–626.

17. Some evolutionary geneticists have suggested that the genetic code has been shaped by natural selection to minimize the deleterious consequences of mutations. For an entry into the literature on this issue, see:

Caporaso, J.G., M. Yarus, and R. Knight. 2005. Error minimization and coding triplet/binding site associations are independent features of the canonical genetic code. *Journal of Molecular Evolution* 61: 597–607.

Freeland S.J., and L. D. Hurst. 1998. Load minimization of the genetic code: History does not explain the pattern. *Proceedings of the Royal Society London B* 265: 2111–2119.

Freeland S.J., and L. D. Hurst. 1998. The genetic code is one in a million. *Journal of Molecular Evolution* 47: 238–248.

Knight, R. D., S. J. Freeland, and L. F. Landweber. 1999. Selection, history and chemistry: The three faces of the genetic code. *Trends in Biochemical Sciences* 24: 241–247.

18. Several new members of the globin family of genes have recently been discovered and characterized. To review this literature, see:

Burmester, T., B. Welch, S. Reinhardt, and T. Hankeln. 2000. A vertebrate globin expressed in the brain. *Nature* 407: 520–523.

Burmester, T., B. Ebner, B. Weich, and T. Hankeln. 2002. Cytoglobin: A novel globin type ubiquitously expressed in vertebrate tissues. *Molecular Biology and Evolution* 19: 416–421.

Sun, Y., K. Jin, X.O. Mao, Y. Zhu, and D.A. Greenberg. 2001. Neuroglobin is upregulated by and protects neurons from hypoxic-ischemic injury. *Proceedings of the National Academy of Sciences USA* 98: 15306–15311.

Vinogradov, S.N., D. Hoogewijs, et al. 2005. Three globin lineages belonging to two structural classes in the genomes of the three kingdoms of life. *Proceedings of the National Academy of Sciences USA* 102: 11385–11389.

Watts, R.A., P.W. Hunt, A.N. Hvitvad, M.S. Hargrove, W.J. Peacock, and E.S. Dennis. 2001. A hemoglobin from plants homologous to truncated hemoglobins from microorganisms. *Proceedings of the National Academy of Sciences USA* 98: 10119–10124.

Citations

Avise, John C. 1994. *Molecular Markers, Natural History and Evolution.* New York: Chapman & Hall.

Ayala, F. J., L. Serra, and A. Prevosti. 1989. A grand experiment in evolution: The *Drosophila subobscura* colonization of the Americas. *Genome* 31: 246–255.

Balaynà, J., L. Serra, G.W. Gilchrist, R.B. Huey, M. Pascual, F. Mestres, and E. Solé. 2003. Evolutionary pace of chromosomal polymorphism in colonizing populations of *Drosophila subobscura*: An evolutionary time series. *Evolution* 57: 1837–1845.

Chen, W.-J., G. Orti, and A. Meyer. 2004. Novel evolutionary relationship among four fish model systems. *Trends in Genetics* 20: 424–1431.

Cheng, Z. M.Ventura, et al. 2005. A genome-wide comparison of recent chimpanzee and human segmental duplications. *Nature* 437: 88–193.

De Bodt, S., S. Maere, and Y. Van de Peer. 2005. Genome duplication and the origin of angiosperms. *Trends in Ecology and Evolution* 20: 591–1597.

de Visser, J. A. G. M., C. W. Zeyl, P. J. Gerrish, J. L. Blanchard, and R. E. Lenski. 1999. Diminishing returns from mutation supply rate in asexual populations. *Science* 283: 404–406.

Denver, D. R., K. Morris, M. Lynch, L. L. Vassilieva, and W. K. Thomas. 2000. High direct estimate of mutation rate in the mitochondrial genome of *Caenorhabditis elegans*. *Science* 289: 2342–2344.

Denver, D.R., K. Morris, M. Lynch, and W.K. Thomas. 2004. High mutation rate and predominance of insertions in the *Caenorhabditis elegans* nuclear genome. *Nature* 430: 679-682.

Denver, D.R., S. Feinberg, S. Estes, W.K. Thomas, and M. Lynch. 2005. Mutation rates, spectra, and hotspots in mismatch repair-deficient *Caenorhabditis elegans*. *Genetics* 170: 107-113.

Drake, J. W., B. Charlesworth, D. Charlesworth, and J. F. Crow. 1998. Rates of spontaneous mutation. *Genetics* 148: 1667–1686.

Fitch, W. 1970. Distinguishing homologous from analogous proteins. *Systematic Zoology* 19: 99–113.

Furio, V., A. Moya, and R. Sanjuan. 2005. The cost of replication fidelity in an RNA virus. *Proceedings of the National Academy of Sciences USA* 102: 10233–10237.

Garcia-Diaz, M. and T.A. Kunkel. 2006. Mechanism of a genetic glissando: structural biology of indel mutations. *Trends in Biochemical Sciences* 31: 206-214.

Gilchrist, G.W., R.B. Huey, J. Balanyà, M. Pascual, and L. Serra. 2004. A time series of evolution in action: A latitudinal cline in wing size in South American *Drosophila subobscura*. *Evolution* 58: 768-780.

Gillin, F. D., and N. G. Nossal. 1976a. Control of mutation frequency by bacteriophage T4 DNA polymerase I. The ts CB120 antimutator DNA polymerase is defective in strand displacement. *Journal of Biological Chemistry* 251: 5219–5224.

Gillin, F. D., and N. G. Nossal. 1976b. Control of mutation frequency by bacteriophage T4 DNA polymerase II. Accuracy of nucleotide selection by L8 mutator, CB120 antimutator, and wild type phage T4 DNA polymerases. *Journal of Biological Chemistry* 251: 5225–5232.

Giraud, A., I. Matic, O. Tenaillon, A. Clara, M. Radman, M. Fons, and F. Taddei. 2001. Costs and benefits of high mutation rates: Adaptive evolution of bacteria in the mouse gut. *Science* 291: 2606–2608.

Grogan, D. W., G. T. Carver, and J. W. Drake. 2001. Genetic fidelity under harsh conditions: Analysis of spontaneous mutations in the thermoacidophilic archaeon *Sulfolobus acidocaldarius*. *Proceedings of the National Academy of Sciences USA* 98: 7928–7933.

Gu, Z., A. Cavalcanti, F.-C. Chen, P. Bouman, and W.-H. Li. 2002. Extent of gene duplication in the genomes of *Drosophila*, nematode, and yeast. *Molecular Biology and Evolution* 19: 256–262.

Gutiérrez-Rivas, M. and L. Menéndez-Arias. 2001. A mutation in the primer grip region of HIV-1 reverse transcriptase that confers reduced fidelity of DNA synthesis. *Nucleic Acids Research* 29: 4963–4972.

Harris, H. 1966. Enzyme polymorphisms in man. *Proceedings of the Royal Society London B* 164: 298–310.

Ingram, V. M. 1958. How do genes act? *Scientific American* 198: 68–76.

Jaillon, O., J.-M. Aury, et al. 2004. Genome duplication in the teleost fish *Tetraodon nigroviridis* reveals the early vertebrate proto-karyotype. *Nature* 431: 946–957.

LeClerc, J. E., B. Li, W. L. Payne, and T. A. Cebula. 1996. High mutation frequencies among *Escherichia coli* and *Salmonella* pathogens. *Science* 274: 1209–1211.

Lewontin, R. C., and J. L. Hubby. 1966. A molecular approach to the study of genetic heterozygosity in natural populations. II. Amount of variation and degree of heterozygosity in natural populations of *Drosophila pseudoobscura*. *Genetics* 54: 595–609.

Li, W.-H., Z. Gu, H. Wang, and A. Nekrutenko. 2001. Evolutionary analysis of the human genome. *Nature* 409: 847-849.

Lynch, M., J. Blanchard, D. Houle, T. Kibota, S. Schultz, L. Vassilieva, and J. Willis. 1999. Perspective: spontaneous deleterious mutation. *Evolution* 53: 645–1663.

Lynch, M. and J. S. Conery. 2000. The evolutionary fate and consequences of duplicate genes. *Science* 290: 1151–1155.

Martinson, J. J., N. H. Chapman, et al. 1997. Global distribution of the CCR5 gene 32-base-pair deletion. *Nature Genetics* 16: 100–1103.

Matic, I., M. Radman, F. Taddei, B. Picard, C. Doit, E. Bingen, E. Denamur, and J. Elion. 1997. Highly variable mutation rates in commensal and pathogenic *Escherichia coli*. *Science* 277: 1833–1834.

Minnick, D. T., L. Liu, N. D. F. Grindley, T. A. Kunkel, and C. M. Joyce. 2002. Discrimination against purine-pyrimidine mispairs in the polymerase active site of DNA polymerase I: A structural explanation. *Proceedings of the National Academy of Sciences USA* 99: 1194–1199.

Mitton, J. B. 1997. *Selection in Natural Populations.* Oxford: Oxford University Press.

Nathans, J., D. Thomas, and D. S. Hogness. 1986. Molecular genetics of human color vision: The genes encoding blue, green, and red pigments. *Science* 232: 193–202.

Panopoulou, G. and A.J. Poustka. 2005. Timing and mechanism of ancient vertebrate genome duplications-the adventure of a hypothesis. *Trends in Genetics* 21: 559–567.

Pauling, L., H. A. Itano, S. J. Singer, and I. C. Wells. 1949. Sickle-cell anemia, a molecular disease. *Science* 110: 543–548.

Pier, G. B., M. Grout , and T. S. Zaidi. 1997. Cystic fibrosis transmembrane conductance regulator is an epithelial cell receptor for clearance of *Pseudomonas aeruginosa* from the lung. *Proceedings of the National Academy of Science USA*. 94: 12088–12093.

Prevosti, A. 1967. Inversion heterozygosity and selection for wing length in *Drosophila subobscura*. *Genetical Research Cambridge* 10: 81–93.

Prevosti, A., G. Ribo, L. Serra, M. Aguade, J. Balaña, M. Monclus, and F. Mestres. 1988. Colonization of America by *Drosophila subobscura*: Experiment in natural populations that supports the adaptive role of chromosomal-inversion polymorphism. *Proceedings of the National Academy of Science USA* 85: 5597–5600.

Ramsey, J., and D.W. Schemske. 1998. Pathways, mechanisms, and rates of polyploid formation in flowering plants. *Annual Review of Ecology and Systematics* 29: 467–501.

Samson, M., F. Libert, et al. 1996. Resistance to HIV-1 infection in caucasian individuals bearing mutant alleles of the CCR5 chemokine receptor gene. *Nature* 382: 722–725.

Shinkai, A. and L.A. Loeb. 2001. In vivo mutagenesis by *Escherichia coli* DNA polymerase I. *Journal of Biological Chemistry* 276: 46759–46764.

Short, R. V., A. C. Chandley, R. C. Jones, and W. R. Allen. 1974. Meiosis in interspecific equine hybrids. II. The Przewalski horse/domestic horse hybrid. *Cytogenetics and Cell Genetics* 13: 465–478.

Tsui, L.-C. 1992. The spectrum of cystic fibrosis mutations. *Trends in Genetics* 8: 392–398.

Vassilieva, L. L., A. M. Hook, and M. Lynch. 2000. The fitness effects of spontaneous mutations in *Caenorhabditis elegans*. *Evolution* 54: 1234–1246.

Watson, J. D., and F. H. C. Crick 1953. A structure for deoxyribose nucleic acid. *Nature* 171: 737–738.

Zhang, X. and S. Firestein. 2002. The olfactory receptor gene superfamily of the mouse. *Nature Neuroscience* 5: 124-133.

The flour beetles in this population vary in color: Some are red, others black.

6

Mendelian Genetics in Populations I: Selection and Mutation as Mechanisms of Evolution

In Chapter 3 we considered the logical structure of Darwin's Theory of Evolution by Natural Selection and reviewed evidence that the theory provides an accurate mechanistic explanation of descent with modification. As Darwin himself recognized, however, the theory is incomplete without an accurate understanding of the mechanism of inheritance. Mendelian and molecular genetics, which we reviewed in Chapter 5, have provided that understanding. We now have the tools we need to develop a more complete model of the mechanism of evolution.

Population genetics, the subject of Chapters 6, 7, and 8, integrates Darwin's Theory of Evolution by Natural Selection with Mendelian genetics (for a history, see Provine 1971). The crucial insight of population genetics is that changes in the relative abundance of traits in a population can be tied to changes in the relative abundance of the alleles that influence them. From a population–genetic perspective, evolution is a change across generations in the frequencies of alleles. Population genetics provides the theoretical foundation for much of our modern understanding of evolution.

In this chapter we introduce the fundamental aspects of population genetic theory, then explore the role of natural selection and mutation in evolution. Throughout the chapter we use the theory we develop to address practical issues concerning human diseases and human evolution.

6.1 Mendelian Genetics in Populations: The Hardy–Weinberg Equilibrium Principle

Most people are susceptible to HIV. Their best hope of avoiding infection is to avoid contact with the virus. There are, however, a few individuals who remain uninfected despite repeated exposure. In 1996, AIDS researchers discovered that at least some of this variation in susceptibility has a genetic basis (see Chapters 1 and 5). The gene responsible encodes a cell surface protein called CCR5. CCR5 is the handle exploited by most sexually transmitted strains of HIV-1 as a means of binding to white blood cells. A mutant allele of the CCR5 gene, called *CCR5-Δ32*, has a 32-base-pair deletion that destroys the encoded protein's ability to function. Individuals who inherit two copies of this allele have no CCR5 on the surface of their cells and are therefore resistant to HIV-1. Given that individuals homozygous for *CCR5-Δ32* are much less likely to contract HIV, we might ask whether the global AIDS epidemic will cause an increase in the frequency of the *Δ32* allele in human populations. If so, how fast will it happen?

Population genetics begins with a model of what happens to allele and genotype frequencies in an idealized population. Once we know how Mendelian genes behave in the idealized population, we will be able to explore how they behave in real populations.

Before we can hope to answer such questions, we need to understand how the *CCR5-Δ32* allele would behave without the AIDS epidemic. In other words, we need to develop a null model for the behavior of genes in populations. This null model should specify, under the simplest possible assumptions, what will happen across generations to the frequencies of alleles and genotypes. The model should apply not just to humans but to any population of organisms that are both diploid and sexual. In this first section of the chapter, we develop such a model and explore its implications. In the next section we will add natural selection to the model, which will enable us to address our questions about the AIDS epidemic and the *CCR5-Δ32* allele.

We will develop our model by scaling Mendelian genetics up from the level of families, where the reader has used it until now, to the level of entire populations. A **population** is a group of interbreeding individuals and their offspring (Figure 6.1). The crucial events in the life cycle of a population are these: The adults produce gametes, the gametes combine to make zygotes, the zygotes develop into juveniles, and the juveniles grow up to become the next generation of adults. We want to track the fate, across generations, of Mendelian genes in such a population. That is, we want to know whether a particular allele or genotype will become more common or less common over time, and why.

Imagine that the mice in Figure 6.1 have in their genome a particular Mendelian locus, the A locus, with two alleles: *A* and *a*. We can begin tracking these alleles at any point in the life cycle. We will then follow them through one complete turn of the cycle, from one generation to the next, to see if their frequencies change.

A Simulation

Our task of following alleles around the life cycle will be simplest if we start with the gametes produced by the adults when they mate. We will assume that the

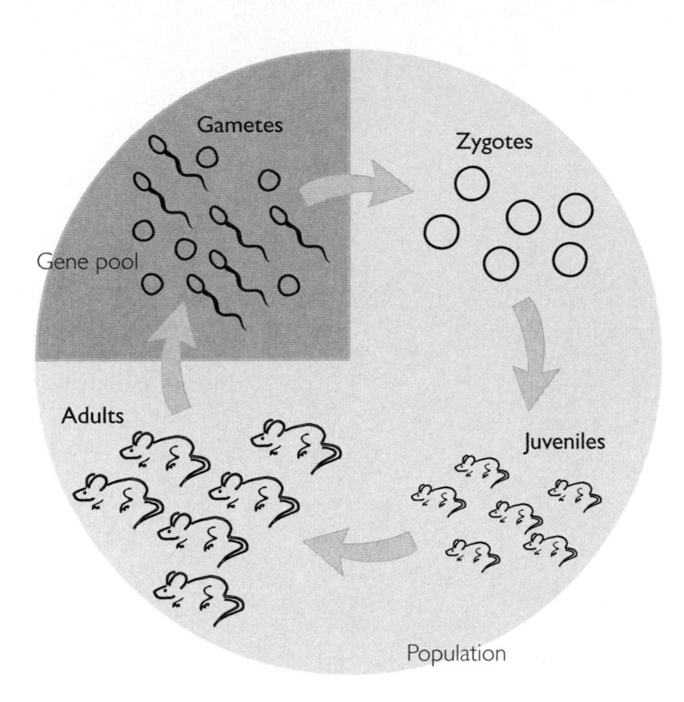

Figure 6.1 The life cycle of an imaginary population of mice, highlighting the stages that will be important in our development of population genetics

adults choose their mates at random. A useful mental trick is to picture random mating happening like this: We take all the eggs and sperm produced by all the adults in the population, dump them together in a barrel, and stir. This barrel is known as the **gene pool**. Each sperm in the gene pool swims about at random until it collides with an egg, whereupon the egg and sperm fuse to make a zygote. Something rather like this actually happens in sea urchins and other marine creatures that simply release their gametes onto the tide. For other organisms, like mice and humans, it is obviously a simplification.

The adults in our mouse population are diploid, so each carries two copies of the A locus. But the adults made their eggs and sperm by meiosis. Following Mendel's law of segregation, each gamete received just one copy of the A locus. Imagine that 60% of the eggs and sperm received a copy of allele *A*, and 40% received a copy of allele *a* (Figure 6.2). In other words, the frequency of allele *A* in the gene pool is 0.6, and the frequency of allele *a* is 0.4.

What happens when eggs meet sperm? For example, what fraction of the zygotes they produce has genotype *AA*? And once these zygotes develop into juveniles, grow up, and spawn, what are the frequencies of alleles *A* and *a* in the next generation's gene pool?

Starting with the eggs and sperm that constitute the gene pool, our model tracks alleles through zygotes and adults and into the next generation's gene pool.

One way to find out is by simulation. We can close our eyes and put a finger down on Figure 6.2 to choose an egg. Perhaps it carries a copy of allele *A*. Now we close our eyes and put down a finger to choose a sperm. Perhaps it carries a copy of allele *a*. If we combine these gametes we get a zygote with genotype *Aa*. We encourage the reader to carry out this process to make a large sample of zygotes—50, say, or even 100. We have paused to do so as we write. Among the 100 zygotes we made, 34 had genotype *AA*, 57 had *Aa*, and 9 had *aa*.

Now let us imagine that all these zygotes develop into juveniles, and that all the juveniles survive to adulthood. Imagine, furthermore, that when the adults reproduce, they all donate the same number of gametes to the gene pool. We can choose any number of gametes we like for the standard donation, so we will choose 10 to make the arithmetic easy. We will not worry

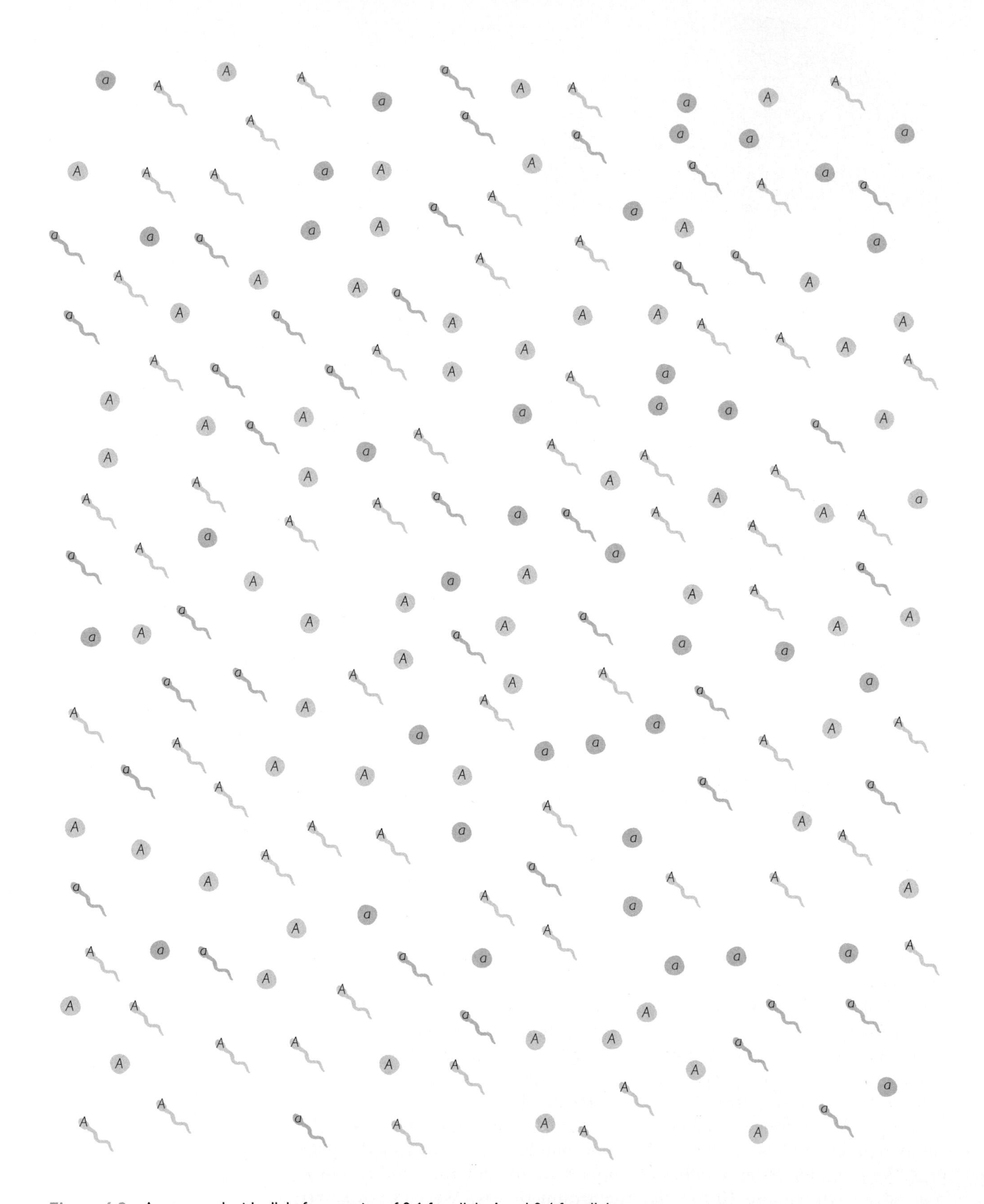

Figure 6.2 A gene pool with allele frequencies of 0.6 for allele *A* and 0.4 for allele *a*

about whether a particular adult makes eggs or sperm; instead we will simply count gametes:

The 34 *AA* adults together make a total of 340 gametes: 340 carry allele *A*; none carry allele *a*.

The 57 *Aa* adults together make a total of 570 gametes: 285 carry allele *A*; 285 carry allele *a*.

The 9 *aa* adults together make a total of 90 gametes: none carry allele *A*; 90 carry allele *a*.

Summing the gametes carrying copies of each allele we get 625 carrying *A* and 375 carrying *a*, for a total of 1,000. The frequency of allele *A* in the new gene pool is 0.625; the frequency of allele *a* is 0.375.

We have followed the alleles around one complete turn of the population's life cycle and found that their ending frequencies are somewhat different from their starting frequencies (Figure 6.3). In other words, our population has evolved.

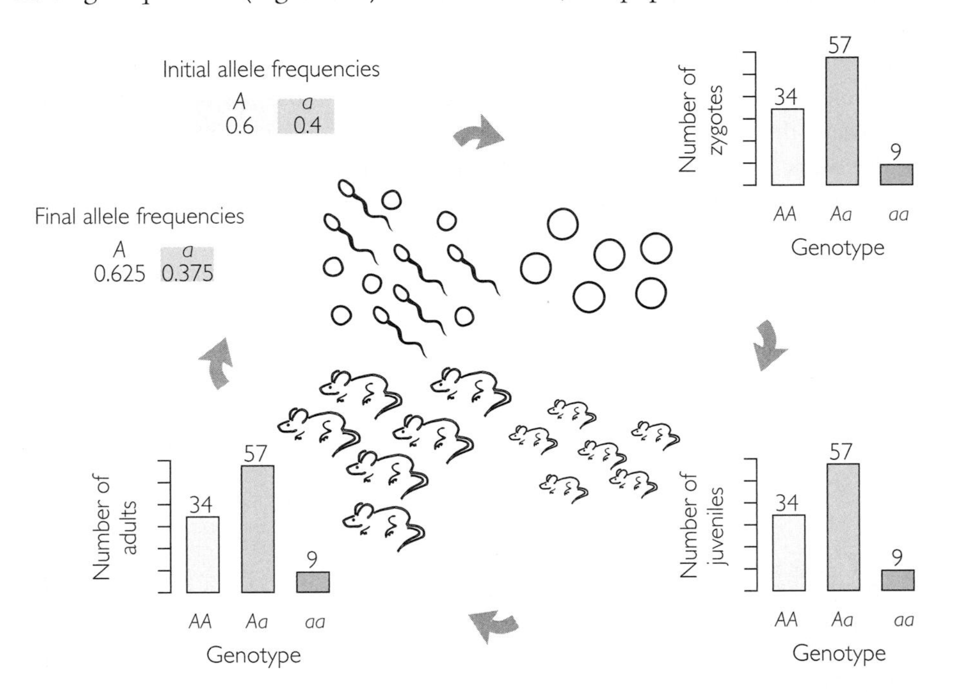

Figure 6.3 Allele and genotype frequencies throughout the life cycle in a numerical simulation We made the zygotes by picking gametes at random from the gene pool in Figure 6.2, and assumed that all of the zygotes survived. The reader's results, on repeating this exercise, will be somewhat different.

The genotype frequencies among the zygotes in the reader's sample, and the frequencies of the alleles in the reader's next generation, will almost certainly be somewhat different from ours. Indeed, we carried out the simulation two more times ourselves and got different results each time. In our second simulation we got, among the zygotes, 41% *AA*, 44% *Aa*, and 15% *aa*. The allele frequencies in the next generation's gene pool were 0.63 for *A* and 0.37 for *a*. In our third simulation we got, among the zygotes, 34% *AA*, 49% *Aa*, and 17% *aa*. The allele frequencies in the next generation were 0.585 for *A* and 0.415 for *a*.

In simulated populations, allele frequencies change somewhat across generations. This is evolution resulting from blind luck.

Our three results are not wildly divergent, but they are not identical either. In two cases the frequency of allele *A* rose, whereas in one case it fell. The reason we got a different result each time is that in each simulation blind luck in pulling gametes from the gene pool gave us a different number of zygotes with each genotype. The fact that blind luck can cause a population to evolve unpredictably is an important result of population genetics. This mechanism of evolution is called **genetic drift**. We will return to genetic drift as a mechanism of evolution in Chapter 7. For now, however, we are interested not in whether evolution is sometimes unpredictable but whether it is ever predictable. We want to consider what would have happened in our simulations if blind luck had played no role at all.

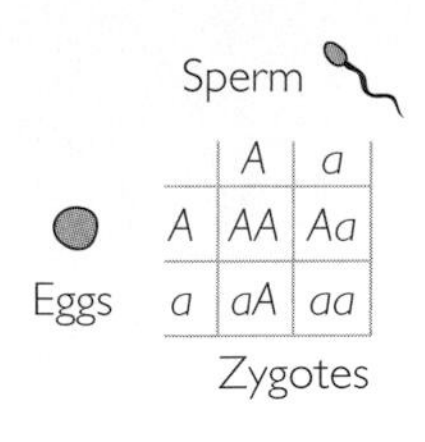

Figure 6.4 Punnett square for a cross between two heterozygotes This device makes accurate predictions about the genotype frequencies among the zygotes because the genotypes of the eggs and sperm are represented in the proportions in which the parents produce them.

A Numerical Calculation

We can discover the luck-free result of combining eggs and sperm to make zygotes by using a Punnett square. Punnett squares, invented by Reginald Crundall Punnett, are more typically used in Mendelian genetics to predict the genotypes among the offspring of a particular male and female. Figure 6.4, for example, shows the Punnett square for a mating between an *Aa* female and an *Aa* male. We write the genotypes of the eggs made by the female, in the proportions we expect her to make them, along the side of the square. We write the genotypes of the sperm made by the male, in the proportions we expect him to make them, along the top. Then we fill in the boxes in the square to get the genotypes of the zygotes. This Punnett square predicts that among the offspring of an *Aa* female and an *Aa* male, one-quarter will be *AA*, one-half *Aa*, and one-quarter *aa*.

We can use the same device to predict the genotypes among the offspring of an entire population (Figure 6.5a). The trick is to write the egg and sperm genotypes along the side and top of the Punnett square in proportions that reflect their frequencies in the gene pool. Sixty percent of the eggs carry copies of allele *A* and 40% carry copies of allele *a*, so we have written six *A*'s and four *a*'s along the side of the square. Likewise, for the sperm, we have written six *A*'s and four *a*'s along the top. Filling in the boxes in the square, we find that among 100 zygotes in our population, we can expect 36 *AA*'s, 48 *Aa*'s, and 16 *aa*'s.

The Punnett square in Figure 6.5a suggests that we could also predict the genotype frequencies among the zygotes by multiplying probabilities. Figure 6.5b shows the four possible combinations of egg and sperm, the zygotes they produce, and a calculation specifying the probability of each (see also Box 6.1).

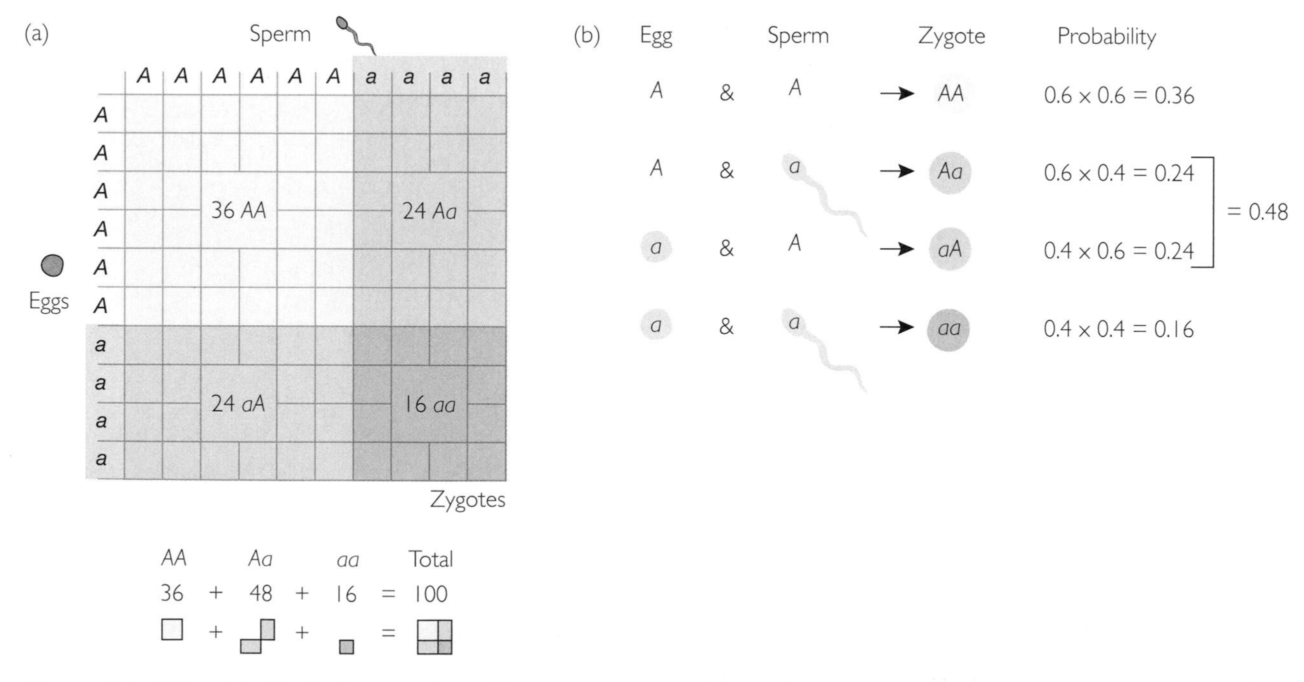

Figure 6.5 When blind luck plays no role, random mating in the gene pool of our model mouse population produces zygotes with predictable genotype frequencies (a) A Punnett square. The genotypes of the gametes are listed along the left and top edges of the box in proportions that reflect the frequencies of *A* and *a* eggs and sperm in the gene pool. The shaded areas inside the box represent the genotypes among 100 zygotes formed by random encounters between gametes in the gene pool. (b) We can also calculate genotype frequencies among the zygotes by multiplying allele frequencies. (See Box 6.1.)

Box 6.1 Combining probabilities

The combined probability that two independent events will occur together is equal to the product of their individual probabilities. For example, the probability that a tossed penny will come up heads is $\frac{1}{2}$. The probability that a tossed nickel will come up heads is also $\frac{1}{2}$. If we toss both coins together, the outcome for the penny is independent of the outcome for the nickel. Thus the probability of getting heads on the penny and heads on the nickel is

$$\frac{1}{2} \times \frac{1}{2} = \frac{1}{4}$$

The combined probability that either of two mutually exclusive events will occur is the sum of their individual probabilities. When rolling a die we can get a one or a two (among other possibilities), but we cannot get both at once. Thus, the probability of getting either a one or a two is

$$\frac{1}{6} + \frac{1}{6} = \frac{1}{3}$$

For example, if we look into the gene pool and pick an egg to watch at random, there is a 60% chance that it will have genotype A. When a sperm comes along to fertilize the egg, there is a 60% chance that the sperm will have genotype A. The probability that we will witness the production of an AA zygote is therefore

$$0.6 \times 0.6 = 0.36$$

If we watched the formation of all the zygotes, 36% of them would have genotype AA. The calculations in Figure 6.5b show that random mating in the gene pool produces zygotes in the following proportions:

AA	Aa	aa
0.36	0.48	0.16

(The Aa category includes heterozygotes produced by combining either an A egg with an a sperm or an a egg with an A sperm.) Notice that

$$0.36 + 0.48 + 0.16 = 1$$

This confirms that we have accounted for all of the zygotes.

We now let the zygotes grow to adulthood, and we let the adults produce gametes to make the next generation's gene pool. When chance plays no role, will the frequencies of alleles A and a in the new gene pool change from one generation to the next?

If we assume, as we did above, that 100 adults each make 10 gametes, then:

The 36 AA adults together make a total of 360 gametes: 360 carry allele A; none carry allele a.

The 48 Aa adults together make a total of 480 gametes: 240 carry allele A; 240 carry allele a.

The 16 aa adults together make a total of 160 gametes: none carry allele A; 160 carry allele a.

Summing the gametes carrying each allele we get 600 carrying copies of A and 400 carrying copies of a, for a total of 1,000. The frequency of allele A in the new gene pool is 0.6; the frequency of allele a is 0.4.

We can also calculate the composition of the new gene pool using frequencies. Because adults of genotype AA constitute 36% of the population, they will make 36% of the gametes. All of these gametes carry copies of allele A. Likewise, adults of genotype Aa constitute 48% of the population, and will make

48% of the gametes. Half of these gametes carry copies of allele *A*. So the total fraction of the gametes in the gene pool that carry copies of *A* is

$$0.36 + \left(\frac{1}{2}\right)0.48 = 0.6$$

Figure 6.6a shows this calculation graphically. The figure also shows a calculation establishing that the fraction of the gametes in the gene pool that carry copies of allele *a* is 0.4. Notice that

$$0.6 + 0.4 = 1$$

This confirms that we have accounted for all of the gametes. Figure 6.6b shows a geometrical representation of the same calculations.

Numerical examples show that when blind luck plays no role, allele frequencies remain constant from one generation to the next.

We have come full circle, and this time, unlike in our simulations, we have arrived precisely where we began (Figure 6.7). We started with allele frequencies of 60% for *A* and 40% for *a* in our population's gene pool. We followed the alleles through zygotes, juveniles, and adults and into the next generation's gene pool. The allele frequencies in the new gene pool are still 60% and 40%. When blind luck plays no role, the allele frequencies for *A* and *a* in our population are in equilibrium: They do not change from one generation to the next. The population does not evolve.

The first biologist to work a numerical example, tracing the frequencies of Mendelian alleles from one generation to the next in an ideal population, was G. Udny Yule in 1902. He started with a gene pool in which the frequencies of two alleles were 0.5 and 0.5 and showed that in the next generation's gene pool the allele frequencies were still 0.5 and 0.5. The reader may want to reproduce Yule's calculations as an exercise.

Like us, Yule concluded that the allele frequencies in his imaginary population were in equilibrium. Yule's conclusion was both groundbreaking and correct, but he took it a bit too literally. He had worked only one example, and he believed

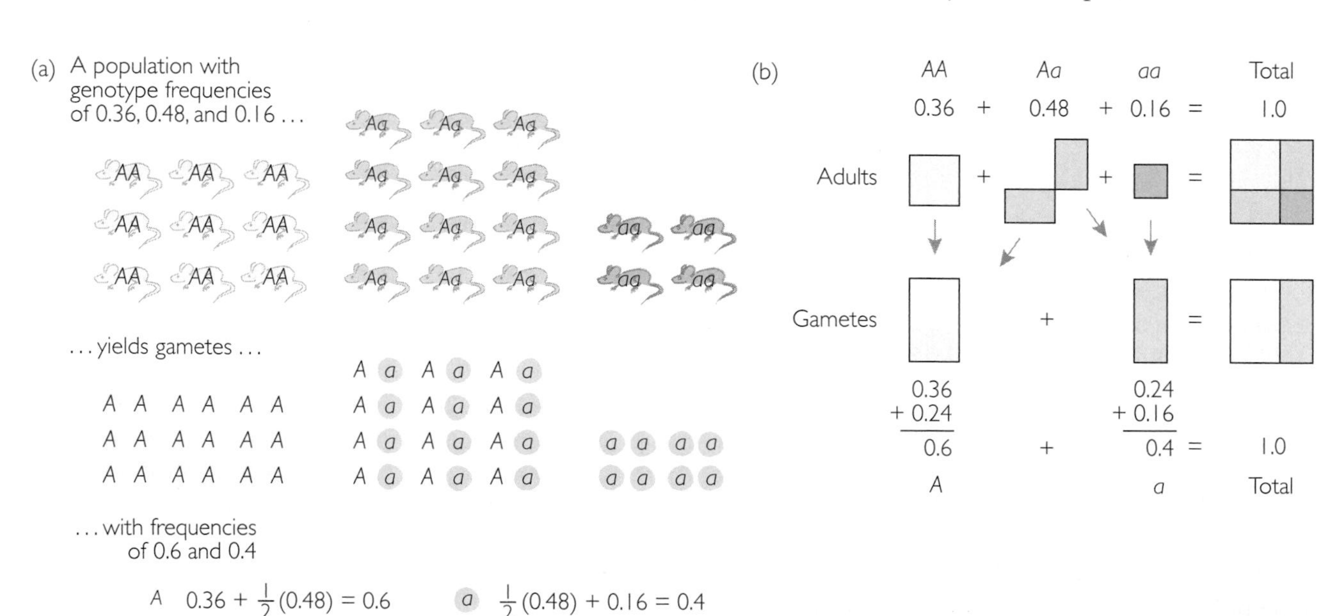

Figure 6.6 When the adults in our model mouse population make gametes, they produce a gene pool in which the allele frequencies are identical to the ones we started with a generation ago (a) Calculations using frequencies. (b) A geometrical representation. The area of each box represents the frequency of an adult or gamete genotype. Note that half the gametes produced by *Aa* adults carry allele *A*, and half carry allele *a*.

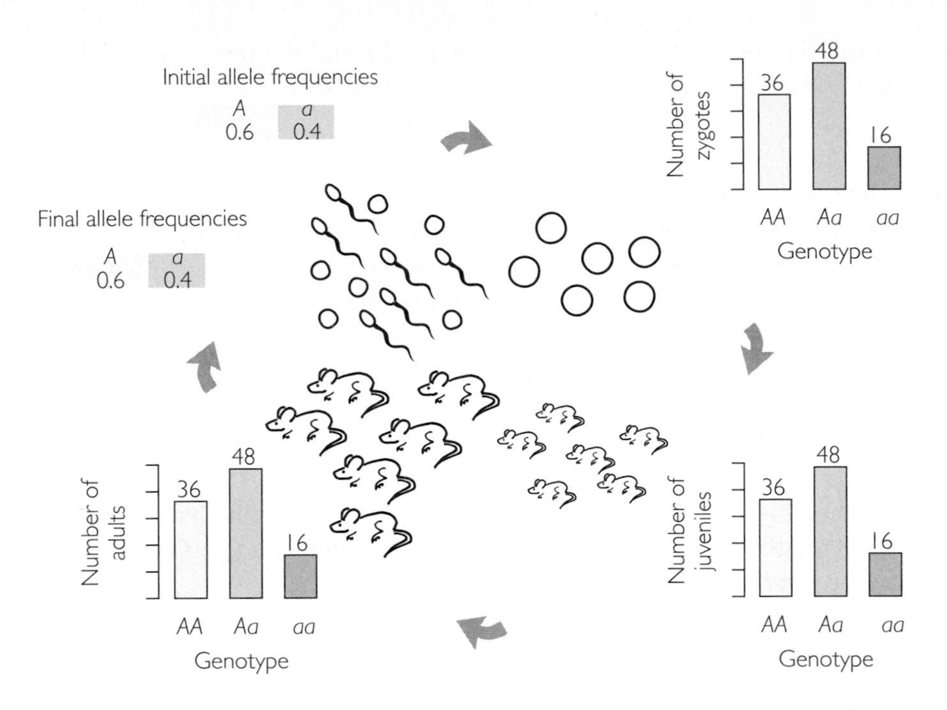

Figure 6.7 When blind luck plays no role in our model population, the allele frequencies do not change from one generation to the next We made the zygotes with the Punnett square in Figure 6.5, and assumed that all the zygotes survived.

that allele frequencies of 0.5 and 0.5 represented the only possible equilibrium state for a two-allele system. For example, Yule believed that if a single copy of allele A appeared as a mutation in a population whose gene pool otherwise contained only copies of a, then the A allele would automatically increase in frequency until copies of it constituted one-half of the gene pool. Yule argued this claim during the discussion that followed a talk given in 1908 by none other than Reginald Punnett. Punnett thought that Yule was wrong, but he did not know how to prove it.

We have already demonstrated, of course, that Punnett was correct in rejecting Yule's claim. Our calculations showed that a population with allele frequencies of 0.6 and 0.4 is in equilibrium too. What Punnett wanted, however, is a general proof. This proof should show that any allele frequencies, so long as they sum to 1, will remain unchanged from one generation to the next.

Punnett took the problem to his mathematician friend Godfrey H. Hardy, who produced the proof in short order (Hardy 1908). Hardy simply repeated the calculations that Yule had performed, using variables in place of the specific allele frequencies that Yule had assumed. Hardy's calculation of the general case indeed showed that any allele frequencies can be in equilibrium.

The General Case

The challenge now is to prove algebraically that there was nothing special about our numerical examples. Any allele frequencies will remain constant from generation to generation.

For our version of Hardy's general case, we will again work with an imaginary population. We are concerned with a single locus with two alleles: A_1 and A_2. We use capital letters with subscripts because we want our calculation to cover cases in which the alleles are codominant as well as cases in which they are dominant and recessive. The three possible diploid genotypes are A_1A_1, A_1A_2, and A_2A_2.

As in our simulations and numerical example, we will start with the gene pool and follow the alleles through one complete turn of the life cycle. The gene pool will contain some frequency of A_1 gametes and some frequency of A_2 gametes.

We will call the frequency of A_1 in the gene pool p and the frequency of A_2 in the gene pool q. There are only two alleles in the population, so

$$p + q = 1$$

The first step is to let the gametes in the gene pool combine to make zygotes. Figure 6.8a shows the four possible combinations of egg and sperm, the zygotes they produce, and a calculation specifying the probability of each. For example, if we pick an egg to watch at random, the chance is p that it will have genotype A_1. When a sperm comes along to fertilize the egg, the chance is p that the sperm will have genotype A_1. The probability that we will witness the production of an A_1A_1 zygote is therefore

$$p \times p = p^2$$

If we watched the formation of all the zygotes, p^2 of them would have genotype A_1A_1. The calculations in Figure 6.8a show that random mating in our gene pool produces zygotes in the following proportions:

A_1A_1	A_1A_2	A_2A_2
p^2	$2pq$	q^2

Figure 6.8b shows a Punnett square that yields the same genotype frequencies. The Punnett square also shows geometrically that

$$p^2 + 2pq + q^2 = 1$$

This confirms that we have accounted for all the zygotes. The same result can be demonstrated algebraically by substituting $(1 - p)$ for q in the expression $p^2 + 2pq + q^2$, then simplifying.

We have gone from the allele frequencies in the gene pool to the genotype frequencies among the zygotes. We now let the zygotes develop into juveniles, let the juveniles grow up to become adults, and let the adults produce gametes to make the next generation's gene pool.

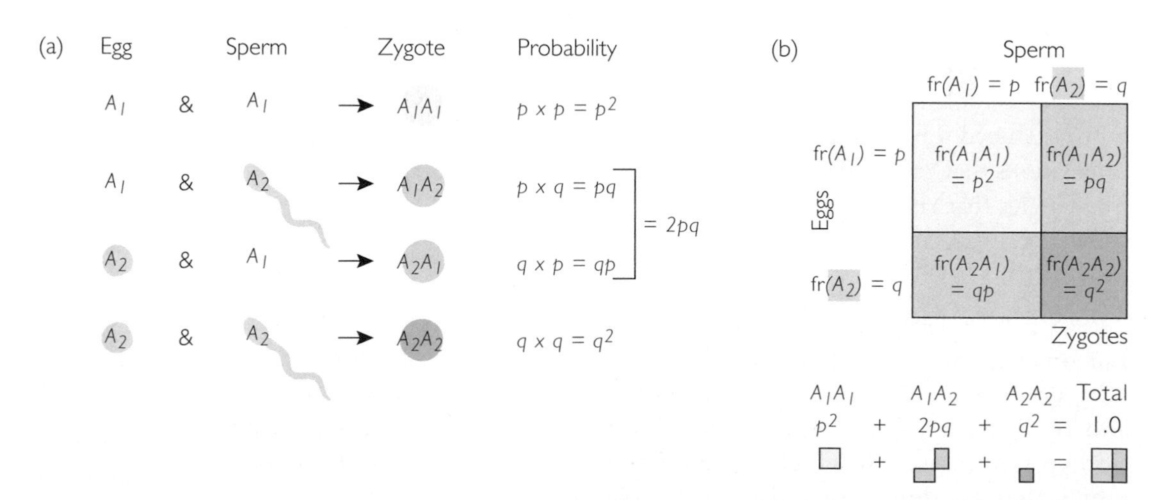

Figure 6.8 The general case for random mating in the gene pool of our model mouse population (a) We can predict the genotype frequencies among the zygotes by multiplying the allele frequencies. (b) A Punnett square. The variables along the left and top edges of the box represent the frequencies of *A* and *a* eggs and sperm in the gene pool. The expressions inside the box represent the genotype frequencies among zygotes formed by random encounters between gametes in the gene pool.

We can calculate the frequency of allele A_1 in the new gene pool as follows. Because adults of genotype A_1A_1 constitute a proportion p^2 of the population, they will make p^2 of the gametes. All of these gametes carry copies of allele A_1. Likewise, adults of genotype A_1A_2 constitute a proportion $2pq$ of the population, and will make $2pq$ of the gametes. Half of these gametes carry copies of allele A_1. So the total fraction of the gametes in the gene pool that carry copies of A_1 is

$$p^2 + \left(\frac{1}{2}\right)2pq = p^2 + pq$$

We can simplify the expression on the right by substituting $(1 - p)$ for q. This gives

$$\begin{aligned} p^2 + pq &= p^2 + p(1 - p) \\ &= p^2 + p - p^2 \\ &= p \end{aligned}$$

Figure 6.9 shows this calculation graphically. The figure also shows a calculation establishing that the fraction of the gametes in the gene pool that carry copies of allele A_2 is q. We assumed at the outset that p and q sum to 1, so we know that we have accounted for all of the gametes.

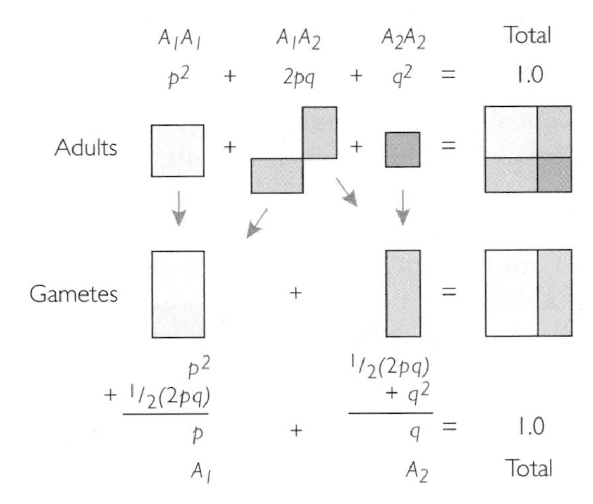

Figure 6.9 A geometrical representation of the general case for the allele frequencies produced when the adults in our model population make gametes The area of each box represents the frequency of an adult or gamete genotype.

Our model has shown that our idealized population does not evolve. This conclusion is known as the Hardy–Weinberg equilibrium principle.

Once again we have come full circle, and arrived back where we started. We started with allele frequencies of p and q in our population's gene pool. We followed the alleles through zygotes and adults and into the next generation's gene pool. The allele frequencies in the new gene pool are still p and q. The allele frequencies p and q can be stable at any values at all between 0 and 1, as long as they sum to 1. In other words, *any* allele frequencies can be in equilibrium, not just $p = q = 0.5$ as Yule thought.

This is a profound result. At the beginning of the chapter we defined evolution as change in allele frequencies in populations. The calculations we just performed show, given simple assumptions, that in populations following the rules of Mendelian genetics, allele frequencies do not change.

We have presented this result as the work of Hardy (1908). It was derived independently by Wilhelm Weinberg (1908) and has become known as the Hardy–Weinberg equilibrium principle. (Some evolutionary biologists refer to it as the Hardy–Weinberg–Castle equilibrium principle, because William Castle

[1903] worked a numerical example and stated the general equilibrium principle nonmathematically five years before Hardy and Weinberg explicitly proved the general case [see Provine 1971].) The Hardy–Weinberg equilibrium principle yields two fundamental conclusions:

- **Conclusion 1:** The allele frequencies in a population will not change, generation after generation.
- **Conclusion 2:** If the allele frequencies in a population are given by p and q, the genotype frequencies will be given by p^2, $2pq$, and q^2.

We get an analogous result if we generalize the analysis from the two-allele case to the usual case of a population containing many alleles at a locus (see Box 6.2).

What Use Is the Hardy–Weinberg Equilibrium Principle?

It may seem puzzling that in a book about evolution we have devoted so much space to a proof apparently showing that evolution does not happen. Evolution does, of course, happen—we saw it happen in this chapter in our own simulations. What makes the Hardy–Weinberg equilibrium principle useful is that it rests on a specific set of simple assumptions. When one or more of these assumptions is violated, the Hardy–Weinberg conclusions no longer hold.

We left some of the assumptions unstated when we developed our null model of Mendelian alleles in populations. We can now make them explicit. The crucial assumptions are as follows:

1. **There is no selection.** All members of our model population survived at equal rates and contributed equal numbers of gametes to the gene pool. When this assumption is violated—when individuals with some genotypes survive and reproduce at higher rates than others—the frequencies of alleles may change from one generation to the next.
2. **There is no mutation.** In the model population, no copies of existing alleles were converted by mutation into copies of other existing alleles, and no new alleles were created. When this assumption is violated, and, for example, some alleles have higher mutation rates than others, allele frequencies may change from one generation to the next.

Box 6.2 The Hardy–Weinberg equilibrium principle with more than two alleles

Imagine a single locus with several alleles. We can call the alleles A_i, A_j, A_k, and so on, and we can represent the frequencies of the alleles in the gene pool with the variables p_i, p_j, p_k, and so on. The formation of a zygote with genotype A_iA_i requires the union of an A_i egg with an A_i sperm. Thus the frequency of any homozygous genotype A_iA_i is p_i^2. The formation of a zygote with genotype A_iA_j requires either the union of an A_i egg with an A_j sperm, or an A_j egg with an A_i sperm. Thus, the frequency of any heterozygous genotype A_iA_j is $2p_ip_j$.

For example, if there are three alleles with frequencies p_1, p_2, and p_3. such that

$$p_1 + p_2 + p_3 = 1$$

then the genotype frequencies are given by

$$(p_1 + p_2 + p_3)^2 = p_1^2 + p_2^2 + p_3^2 + 2p_1p_2 + 2p_1p_3 + 2p_2p_3$$

and the allele frequencies do not change from generation to generation.

3. There is no migration. No individuals moved into or out of the model population. When this assumption is violated, and individuals carrying some alleles move into or out of the population at higher rates than individuals carrying other alleles, allele frequencies may change from one generation to the next.

4. There are no chance events that cause individuals with some genotypes to pass more of their alleles to the next generation than others. That is, blind luck plays no role. We saw the influence of blind luck in our simulations. We avoided it in our analysis of the general case by assuming that the eggs and sperm in the gene pool collided with each other at their actual frequencies of p and q, with no deviations caused by chance. Another way to state this assumption is that the model population was infinitely large. When this assumption is violated, and by chance some individuals contribute more alleles to the next generation than others, allele frequencies may change from one generation to the next. This mechanism of allele frequency change is called, as we said earlier, **genetic drift**.

The Hardy–Weinberg equilibrium principle becomes useful when we list the assumptions we made about our idealized population. By providing a set of explicit conditions under which evolution does not happen, the Hardy–Weinberg analysis identifies the mechanisms that can cause evolution in real populations.

5. Individuals choose their mates at random. We explicitly set up the gene pool to let gametes find each other at random. In contrast to assumptions 1 through 4, when this assumption is violated—when, for example, individuals prefer to mate with other individuals of the same genotype—allele frequencies do not change from one generation to the next. Genotype frequencies may change, however. Such shifts in genotype frequency, in combination with a violation of one of the other four assumptions, can influence the evolution of populations.

By furnishing a list of specific ideal conditions under which populations will not evolve, the Hardy–Weinberg equilibrium principle identifies the set of events that can cause evolution in the real world (Figure 6.10). This is the sense in which the Hardy–Weinberg equilibrium principle serves as a null model. Biologists can measure allele and genotype frequencies in nature, and determine whether the Hardy–Weinberg conclusions hold. A population in which conclusions 1 and 2 hold is said to be in **Hardy–Weinberg equilibrium**. If a population is not in Hardy–Weinberg equilibrium—if the allele frequencies change from generation to generation or if the genotype frequencies cannot, in fact, be predicted by multiplying the allele frequencies—then one or more of the Hardy–Weinberg model's assumptions is being violated. Such a discovery does

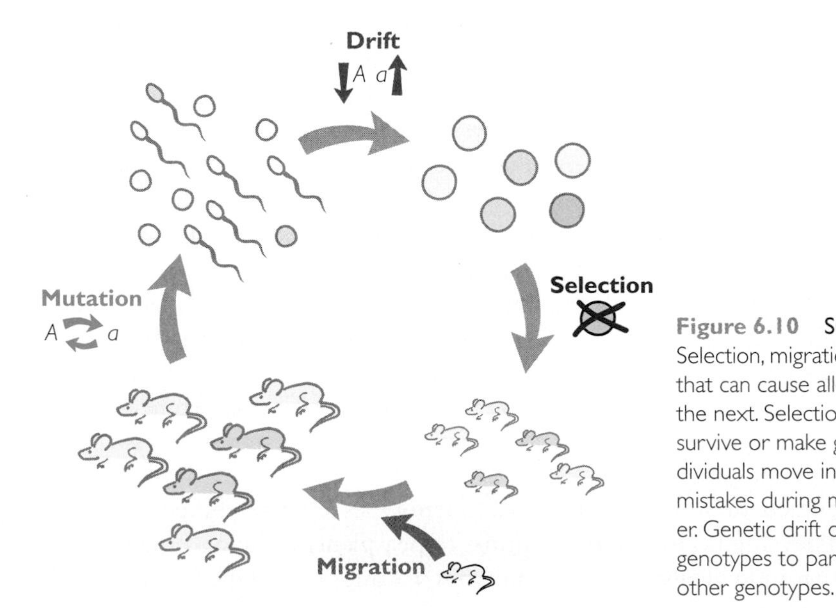

Figure 6.10 Summary of the mechanisms of evolution Selection, migration, mutation, and genetic drift are the four processes that can cause allele frequencies to change from one generation to the next. Selection occurs when individuals with different genotypes survive or make gametes at different rates. Migration occurs when individuals move into or out of the population. Mutation occurs when mistakes during meiosis turn copies of one allele into copies of another. Genetic drift occurs when blind chance allows gametes with some genotypes to participate in more fertilizations than gametes with other genotypes.

not, by itself, tell us which assumptions are being violated, but it does tell us that further research may be rewarded with interesting discoveries.

In the following sections of Chapter 6, we consider how violations of assumptions 1 and 2 affect the two Hardy–Weinberg conclusions, and we explore empirical research on selection and mutation as mechanisms of evolution. In Chapter 7, we consider violations of assumptions 3, 4, and 5.

Changes in the Frequency of the *CCR5-Δ32* Allele

We began this section by asking whether we can expect the frequency of the *CCR5-Δ32* allele to change in human populations. Now that we have developed a null model for how Mendelian alleles behave in populations, we can give a partial answer. As long as individuals of all CCR5 genotypes survive and reproduce at equal rates, as long as no mutations convert some CCR5 alleles into others, as long as no one moves from one population to another, as long as populations are infinitely large, and as long as people choose their mates at random, then no, the frequency of the *CCR5-Δ32* allele will not change.

This answer is, of course, thoroughly unsatisfying. It is unsatisfying because none of the assumptions will be true in any real population. We asked the question in the first place precisely because we expect *Δ32/Δ32* individuals to survive the AIDS epidemic at higher rates than individuals with either of the other two genotypes. In the next two sections, we will see that our null model, the Hardy–Weinberg equilibrium principle, provides a framework that will allow us to assess with precision the importance of differences in survival.

6.2 Selection

Our analysis in Section 6.1 was motivated by a desire to predict whether the frequency of the *CCR5-Δ32* allele will change as a result of the AIDS epidemic. We started by scaling Mendelian genetics up from single crosses to whole populations. This is was the first step in integrating Mendelism with Darwin's Theory of Evolution by Natural Selection. The next step is to add differences in survival and reproductive success. Doing so will make the algebra a bit more complicated. But it will also let us glimpse the predictive strength of population genetics.

First on the list of assumptions about our idealized population was that individuals survive at equal rates and have equal reproductive success. We now explore what happens to allele frequencies when this assumption is violated.

In the population model we used to derive the Hardy–Weinberg equilibrium principle, first on our list of assumptions was that all individuals survive at equal rates and contribute equal numbers of gametes to the gene pool. Systematic violations of this assumption are examples of **selection**. Selection happens when individuals with particular phenotypes survive to reproductive age at higher rates than individuals with other phenotypes, or when individuals with particular phenotypes produce more offspring during reproduction than individuals with other phenotypes. The bottom line in either kind of selection is differential reproductive success: Some individuals have more offspring than others. Selection can lead to evolution when the phenotypes that exhibit differences in reproductive success are heritable—that is, when certain phenotypes are associated with certain genotypes.

Population geneticists often assume that phenotypes are determined strictly by genotypes. They might, for example, think of pea plants as being either tall or short, such that individuals with the genotypes *TT* and *Tt* are tall and individuals with the genotype *tt* are short. Such a view is at least roughly accurate for some traits, including the examples we use in this chapter.

When phenotypes fall into discrete classes that appear to be determined strictly by genotypes, we can think of selection as if it acts directly on the genotypes. We can then assign a particular level of lifetime reproductive success to each genotype. In reality, most phenotypic traits are not, in fact, strictly determined by genotype. Pea plants with the genotype *TT*, for example, vary in height. This variation is due to genetic differences at other loci and to differences in the environments in which the pea plants grew. We will consider such complications in Chapter 9. For the present, however, we adopt the simple view.

When we think of selection as if it acts directly on genotypes, its defining feature is that some genotypes contribute more alleles to future generations than others. In other words, there are differences among genotypes in fitness.

Our task in this section is to incorporate selection into the Hardy–Weinberg analysis. We begin by asking whether selection can change the frequencies of alleles in the gene pool from one generation to the next. In other words, can violation of the no-selection assumption lead to a violation of conclusion 1 of the Hardy–Weinberg equilibrium principle?

Adding Selection to the Hardy–Weinberg Analysis: Changes in Allele Frequencies

We start with a numerical example that shows that selection can indeed change the frequencies of alleles. Imagine that in our population of mice there is a locus, the B locus, that affects the probability of survival. Assume, as we did for the A locus in Figure 6.2, that the frequency of allele B_1 in the gene pool is 0.6 and the frequency of allele B_2 is 0.4 (Figure 6.11). After random mating, we get genotype frequencies for B_1B_1, B_1B_2, and B_2B_2, of 0.36, 0.48, and 0.16. The rest of our

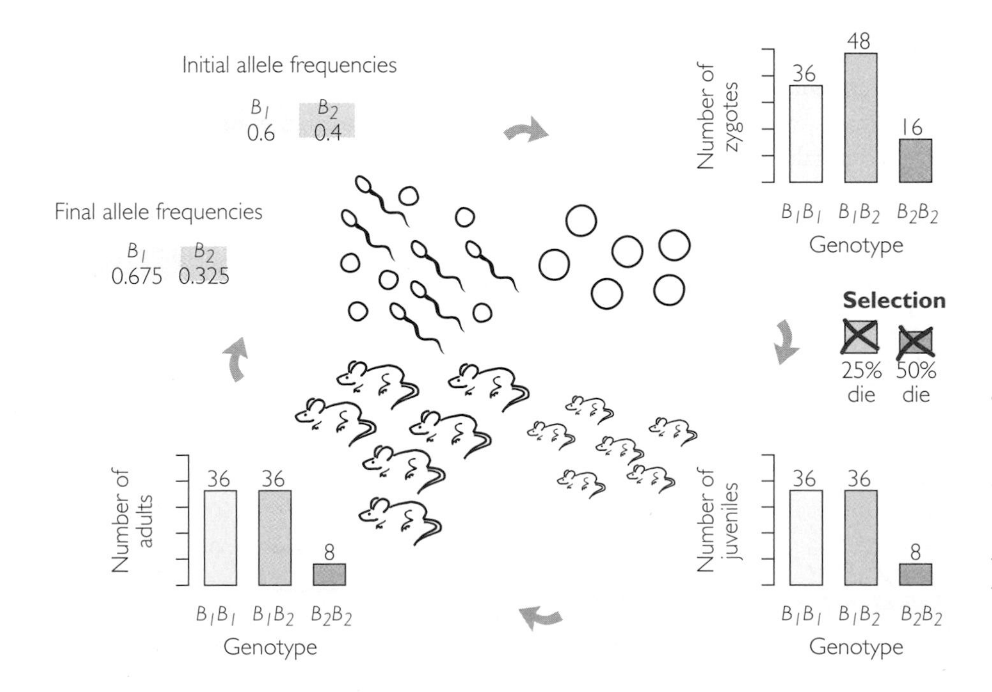

Figure 6.11 Selection can cause allele frequencies to change across generations This figure follows our model mouse population from one generation's gene pool to the next generation's gene pool. The bar graphs show the number of individuals of each genotype in the population at any given time. Selection, in the form of differences in survival among juveniles, causes the frequency of allele B_1 to increase.

calculations will be simpler if we give the population of zygotes a finite size, so imagine that there are 100 zygotes:

B_1B_1	B_1B_2	B_2B_2
36	48	16

These zygotes are represented by a bar graph on the upper right in the figure. We will follow the individuals that develop from these zygotes. Those that survive to adulthood will breed to produce the next generation's gene pool.

We incorporate selection by stipulating that the genotypes differ in their rates of survival. All of the B_1B_1 individuals survive, 75% of the B_1B_2 individuals survive, and 50% of the B_2B_2 individuals survive. As shown in Figure 6.11, there are now 80 adults in the population:

B_1B_1	B_1B_2	B_2B_2
36	36	8

If we assume that each surviving adult donates 10 gametes to the next generation's gene pool, then

The 36 B_1B_1 adults together make a total of 360 gametes: 360 carry allele B_1; none carry allele B_2.

The 36 B_1B_2 adults together make a total of 360 gametes: 180 carry allele B_1; 180 carry allele B_2.

The 8 B_2B_2 adults together make a total of 80 gametes: none carry allele B_1; 80 carry allele B_2.

A numerical example shows that when individuals with some genotypes survive at higher rates than individuals with other genotypes, allele frequencies can change from one generation to the next. In other words, our model shows that natural selection causes evolution.

Summing the gametes carrying copies of each allele we get 540 carrying copies of B_1 and 260 carrying copies of B_2, for a total of 800. The frequency of allele B_1 in the new gene pool is $540/800 = 0.675$; the frequency of allele B_2 is $260/800 = 0.325$. The frequency of allele B_1 has risen by an increment of 7.5 percentage points. The frequency of allele B_2 has dropped by the same amount.

Violation of the no-selection assumption has resulted in violation of conclusion 1 of the Hardy–Weinberg analysis. The population has evolved in response to selection.

We used strong selection to make a point in our numerical example. Rarely in nature are differences in survival rates large enough to cause such dramatic change in allele frequencies in a single generation. If selection continues for many generations, however, even small changes in allele frequency in each generation can add up to substantial changes over the long run. Figure 6.12 illustrates the cumulative change in allele frequencies that can be wrought by selection. The figure is based

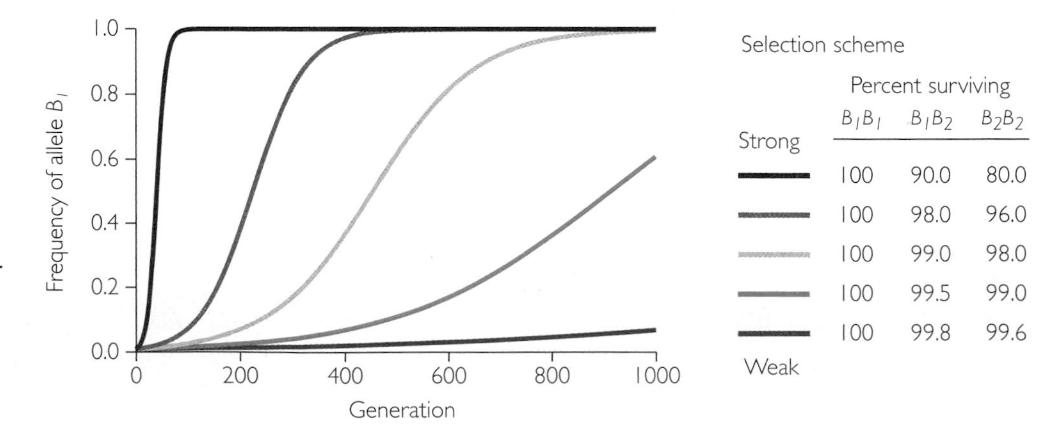

Figure 6.12 Persistent selection can produce substantial changes in allele frequencies over time Each curve shows the change in allele frequency over time under a particular selection intensity.

on a model population similar to the one we used in the preceding numerical example, except that the initial allele frequencies are 0.01 for B_1 and 0.99 for B_2. The red line shows the change in allele frequencies when the survival rates are 100% for B_1B_1, 90% for B_1B_2, and 80% for B_2B_2. The frequency of allele B_1 rises from 0.01 to 0.99 in less than 100 generations. Under weaker selection schemes, the frequency of B_1 rises more slowly, but still inexorably. (See Box 6.3 for a general algebraic treatment incorporating selection into the Hardy–Weinberg analysis.)

Empirical Research on Allele Frequency Change by Selection

Douglas Cavener and Michael Clegg (1981) documented a cumulative change in allele frequencies over many generations in a laboratory-based natural selection experiment on the fruit fly (*Drosophila melanogaster*). Fruit flies, like most other animals, make an enzyme that breaks down ethanol, the poisonous active ingredient in beer, wine, and rotting fruit. This enzyme is called alcohol dehydrogenase, or ADH. Cavener and Clegg worked with populations of flies that had two alleles at the ADH locus: Adh^F and Adh^S. (The F and S refer to whether the protein encoded by the allele moves quickly or slowly through an electrophoresis gel.)

The scientists maintained two experimental populations of flies on food spiked with ethanol and two control populations of flies on normal, nonspiked food. The researchers picked the breeders for each generation at random. This is why we are calling the project a natural selection experiment: Cavener and Clegg set up different environments for their different populations, but the researchers did not themselves directly manipulate the survival or reproductive success of individual flies.

Every several generations, Cavener and Clegg took a random sample of flies from each population, determined their ADH genotypes, and calculated the allele frequencies. The results appear in Figure 6.13. The control populations showed no large or consistent long-term change in the frequency of the Adh^F allele. The experimental populations, in contrast, showed a rapid and largely consistent increase in the frequency of Adh^F (and, of course, a corresponding decrease in the frequency of Adh^S). Hardy–Weinberg conclusion 1 appears to hold true in the control populations, but is clearly not valid in the experimental populations.

Empirical research on fruit flies is consistent with our conclusion that natural selection can cause allele frequencies to change.

Can we identify for certain which of the assumptions of the Hardy–Weinberg analysis is being violated? The only difference between the two kinds of populations is that the experimentals have ethanol in their food. This suggests that it is

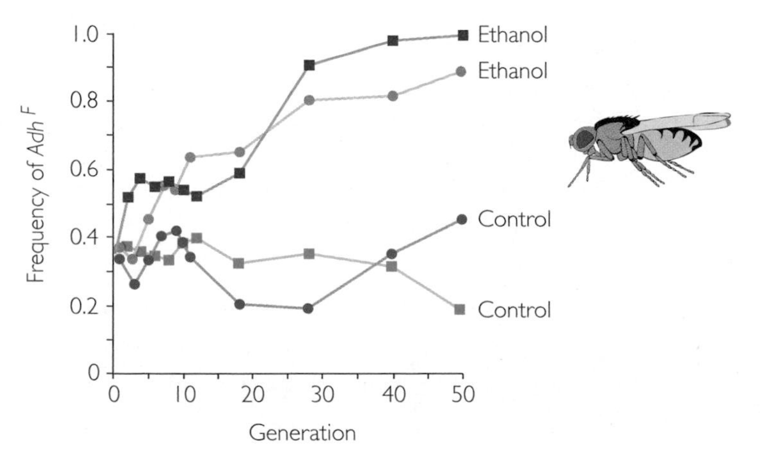

Figure 6.13 Frequencies of the Adh^F allele in four populations of fruit flies over 50 generations The blue and green dots and lines represent control populations living on normal food; the red and orange dots and lines represent experimental populations living on food spiked with ethanol. From Cavener and Clegg (1981).

Box 6.3 A general treatment of selection

Here we develop equations that predict allele frequencies in the next generation, given allele frequencies in this generation and the fitnesses of the different genotypes. We start with a population that has a gene pool in which allele A_1 is at frequency p and allele A_2 is at frequency q. We allow gametes to pair at random to make zygotes of genotypes A_1A_1, A_1A_2, and A_2A_2 at frequencies p^2, $2pq$, and q^2, respectively. We incorporate selection by imagining that A_1A_1 zygotes survive to adulthood at rate w_{11}, A_1A_2 zygotes survive at rate w_{12}, and A_2A_2 zygotes survive at rate w_{22}. All individuals that survive produce the same number of offspring. Therefore, a genotype's survival rate is proportional to the genotype's lifetime reproductive success, or fitness. We thus refer to the survival rates as fitnesses. The average fitness for the whole population, $\overline{w}$, is given by the expression:

$$\overline{w} = p^2w_{11} + 2pqw_{12} + q^2w_{22}$$

[To see this, note that we can calculate the average of the numbers 1, 2, 2, and 3 as $\frac{(1 + 2 + 2 + 3)}{4}$ or as $(\frac{1}{4} \times 1) + (\frac{1}{2} \times 2) + (\frac{1}{4} \times 3)$. Our expression for the average fitness is of the second form: We multiply the fitness of each genotype by its frequency in the population and then sum the results.]

We now calculate the genotype frequencies among the surviving adults (right before their gametes go into the gene pool). The new frequencies of the genotypes are

A_1A_1	A_1A_2	A_2A_2
$\frac{p^2w_{11}}{\overline{w}}$	$\frac{2pqw_{12}}{\overline{w}}$	$\frac{q^2w_{22}}{\overline{w}}$

(We have to divide by the average fitness in each case to ensure that the new frequencies still sum to 1.)

Finally, we let the adults breed, and calculate the allele frequencies in the gene pool:

- For the A_1 allele: A_1A_1 individuals contribute $\frac{p^2w_{11}}{\overline{w}}$ of the gametes, all of them A_1. A_1A_2 individuals contribute $\frac{2pqw_{12}}{\overline{w}}$ of the gametes, half of them A_1. So the new frequency of A_1 is

$$\frac{p^2w_{11} + pqw_{12}}{\overline{w}}$$

- For the A_2 allele: A_1A_2 individuals contribute $\frac{2pqw_{12}}{\overline{w}}$ of the gametes, half of them A_2; A_2A_2 individuals contribute $\frac{q^2w_{22}}{\overline{w}}$ of the gametes, all of them A_2. So the new frequency of A_2 is

$$\frac{pqw_{12} + q^2w_{22}}{\overline{w}}$$

The reader should confirm that the new frequencies of A_1 and A_2 sum to 1.

It is instructive to calculate the change in the frequency of allele A_1 from one generation to the next. This value, Δp, is the new frequency of allele A_1 minus the old frequency of A_1;

$$\begin{aligned}\Delta p &= \frac{p^2w_{11} + pqw_{12}}{\overline{w}} - p \\ &= \frac{p^2w_{11} + pqw_{12}}{\overline{w}} - \frac{p\overline{w}}{\overline{w}} \\ &= \frac{p^2w_{11} + pqw_{12} - p\overline{w}}{\overline{w}} \\ &= \frac{p}{\overline{w}}(pw_{11} + qw_{12} - \overline{w})\end{aligned}$$

The final expression is a useful one, because it shows that the change in frequency of allele A_1 is proportional to $(pw_{11} + qw_{12} - \overline{w})$. The quantity $(pw_{11} + qw_{12} - \overline{w})$ is sometimes called the **average excess** of allele A_1. It is equal to the average fitness of allele A_1 when paired at random with other alleles $(pw_{11} + qw_{12})$ minus the average fitness of the population $(\overline{w})$. When the average excess of allele A_1 is positive, A_1 will increase in frequency. In other words, if the average A_1-carrying individual has higher-than-average fitness, then the frequency of allele A_1 will rise.

The change in the frequency of allele A_2 from one generation to the next is

$$\begin{aligned}\Delta q &= \frac{pqw_{12} + q^2w_{22}}{\overline{w}} - q \\ &= \frac{q}{\overline{w}}(pw_{12} + qw_{22} - \overline{w})\end{aligned}$$

the no-selection assumption that is being violated in the experimental populations. Flies carrying the Adh^F allele appear to have higher lifetime reproductive success (higher fitness) than flies carrying the Adh^S allele when ethanol is present in the food. Cavener and Clegg note that this outcome is consistent with the fact that alcohol dehydrogenase extracted from Adh^F homozygotes breaks down ethanol at twice the rate of alcohol dehydrogenase extracted from Adh^S homozygotes. Whether flies with the Adh^F allele have higher fitness because they have higher rates of survival or because they produce more offspring is unclear.

Adding Selection to the Hardy–Weinberg Analysis: The Calculation of Genotype Frequencies

The calculations and example we have just discussed show that selection can cause allele frequencies to change across generations. Selection invalidates conclusion 1 of the Hardy–Weinberg analysis. We now consider how selection affects conclusion 2 of the Hardy–Weinberg analysis. In a population under selection, can we still calculate the genotype frequencies by multiplying the allele frequencies?

Often, we cannot. As before, we use a population with two alleles at a locus affecting survival: B_1 and B_2. We assume that the initial frequency of each allele in the gene pool is 0.5 (Figure 6.14). After random mating, we get genotype frequencies for B_1B_1, B_1B_2, and B_2B_2, of 0.25, 0.5, and 0.25. The rest of our calculations will be simpler if we give the population of zygotes a finite size, so imagine that there are 100 zygotes:

B_1B_1	B_1B_2	B_2B_2
25	50	25

These zygotes are represented by a bar graph on the upper right in the figure. We will follow the individuals that develop from these zygotes. Those that survive to adulthood will breed to produce the next generation's gene pool.

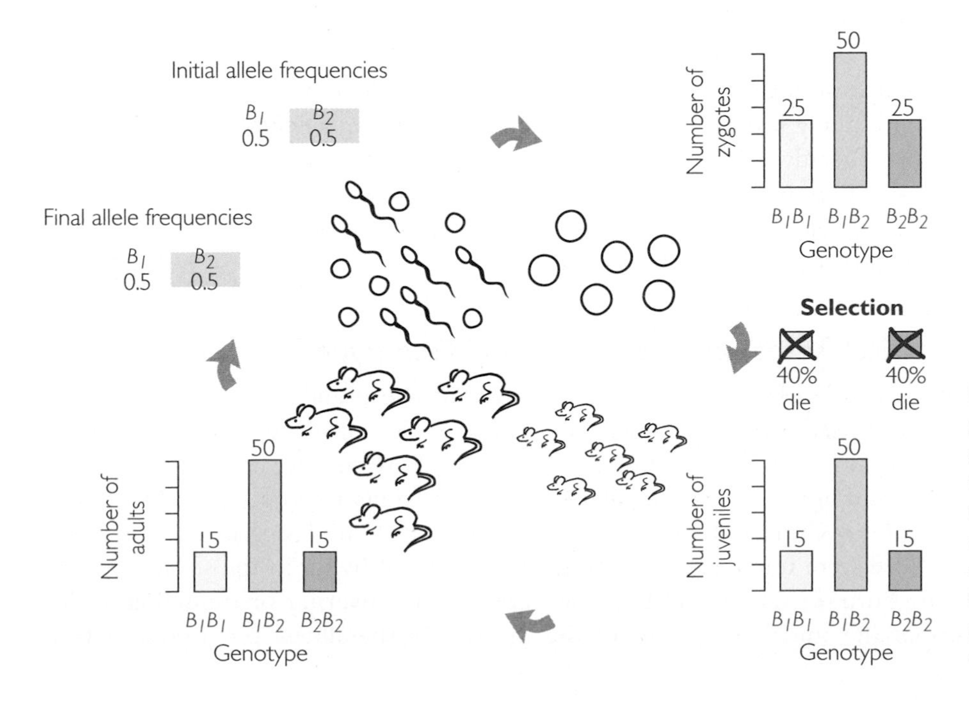

Figure 6.14 Selection can change genotype frequencies so that they cannot be calculated by multiplying the allele frequencies When 40% of the homozygotes in this population die, the allele frequencies do not change. But among the survivors, there are more heterozygotes then predicted under Hardy–Weinberg equilibrium.

As in our first selection example, we incorporate selection by stipulating that the genotypes differ in their rates of survival. This time, 60% percent of the B_1B_1 individuals survive, all of the B_1B_2 individuals survive, and 60% of the B_2B_2 individuals survive. As shown in Figure 6.14, there are now 80 adults in the population:

B_1B_1	B_1B_2	B_2B_2
15	50	15

If we assume that each surviving adult donates 10 gametes to the next generation's gene pool, then

The 15 B_1B_1 adults together make a total of 150 gametes: 150 carry allele B_1; none carry allele B_2.

The 50 B_1B_2 adults together make a total of 500 gametes: 250 carry allele B_1; 250 carry allele B_2.

The 15 B_2B_2 adults together make a total of 150 gametes: none carry allele B_1; 150 carry allele B_2.

Summing the gametes carrying each allele we get 400 carrying B_1 and 400 carrying B_2, for a total of 800. Both alleles are still at a frequency of 0.5. Despite strong selection against homozygotes, the frequencies of the alleles have not changed; the population has not evolved.

But let us calculate frequencies of the three genotypes among the surviving adults. These frequencies are as follows:

B_1B_1	B_1B_2	B_2B_2
15/80	50/80	15/80
= 0.1875	= 0.625	= 0.1875

Natural selection can also drive genotype frequencies away from the values predicted under the Hardy–Weinberg equilibrium principle.

These genotype frequencies reveal that violation of the no-selection assumption *has* resulted in violation of conclusion 2 of the Hardy–Weinberg analysis. We can no longer calculate the genotype frequencies among the adult survivors by multiplying the frequencies of the alleles. For example:

$$\begin{array}{ccc} \textit{Frequency of } B_1B_1 & & (\textit{Frequency of } B_1)^2 \\ 0.1875 & \neq & (0.5)^2 = 0.25 \end{array}$$

We used strong selection in our numerical example to make a point. In fact, selection is rarely strong enough to produce, in a single generation, such a large violation of Hardy–Weinberg conclusion 2. Even if it does, a single bout of random mating will immediately put the genotypes back into Hardy–Weinberg equilibrium. Nonetheless, researchers sometimes find violations of Hardy–Weinberg conclusion 2 that seem to be the result of selection.

Empirical Research on Selection and Genotype Frequencies

Our example comes from research by Simon Mead and colleagues (2003) on genetic variation for resistance to kuru. Kuru is a fatal neurological disorder known only from an epidemic that struck the Foré people during the last century. The Foré are a traditional society that inhabits remote jungle hamlets in Papua New Guinea's Eastern Highlands Province. Until less than a century ago, the Foré were uncontacted by Westerners and still living in the stone age. The name kuru, in the language of the Foré, means shivering or trembling; it describes the initial symptoms of the disease. As the illness progresses, victims begin to stagger while walking, then develop trouble talking, chewing, and swallowing. The final stages are coma and death.

Although restricted to the Foré, kuru belongs to a group of maladies called spongiform encephalopathies, or spongy brain diseases (see Box 6.4 for more information). Other spongy brain diseases of humans include Creutzfeldt-Jakob disease, Gerstmann-Sträussler-Scheinker disease, and fatal familial insomnia. Among other mammals known to suffer spongy brain diseases are sheep and goats (scrapie), deer and elk (chronic wasting disease), and, most famously, cattle (bovine spongiform encephalopathy, or mad cow disease).

Humans can contract spongy brain disease through surgery with contaminated instruments or transplantation of infected tissue (see Prusiner 1997; Will 2003; Aguzzi and Polymenidou 2004). More frighteningly, many species that suffer spongy brain diseases can acquire them by eating affected tissue from an earlier victim. A large outbreak of mad cow disease among cattle in the 1980s was apparently sparked when contaminated meat and bone meal was fed to calves. Ingestion can carry spongy brain diseases across species boundaries. More than a hundred humans have gotten a spongiform encephalopathy called variant Creutzfeldt-Jakob disease by eating meat from cattle that died of bovine spongiform encephalopathy.

During the worst known human epidemic of spongy brain disease, thousands of Foré people unwittingly gave themselves kuru by eating the bodies of relatives who had died of the disease. The Foré had begun practicing ritual funereal

Box 6.4 Spongy brain diseases

The spongy brain diseases get their name from the pattern of neurological degeneration they share, which turns brain tissues into something that, under a microscope, looks like a sponge (Prusiner 1997; Aguzzi and Polymenidou 2004). These degenerating tissues are choked with globs of a protein called PrP, short for *prion protein*. The primary structure of PrP is encoded in the host's genome, and the protein is made by the host's own cells—particularly bone marrow cells that divide to produce blood cells, and central nervous system cells destined to become neurons. In its normal conformation, PrP sits on the surface of cells and seems to help direct cell maturation (Couzin 2006; Steele et al. 2006; Zhang et al. 2006). But it can adopt abnormal conformations as well. It is these misfolded versions of the protein that stick together to form globs, also called amyloid plaques.

Whether and how the accreting plaques of PrP kill brain cells is uncertain. But two lines of evidence implicate PrP in the development of disease. First, mutations in the PrP gene are known to cause heritable spongy brain diseases, including familial Creutzfeldt-Jakob disease, Gerstmann-Sträussler-Scheinker disease, and fatal familial insomnia (Goldfarb et al. 1991, 1992; Dloughy et al. 1992). Second, mice that cannot make PrP do not get spongy brain diseases (Büeler et al. 1993; Prusiner et al. 1993).

As mentioned in the main text, a curious thing about spongy brain diseases is that while some are caused by mutations passed genetically from parents to offspring and some are sporadic, still others are transmissible (see Will 2003). A few researchers still expect that a virus or bacterium will eventually prove to be the transmissible agent (Manuelidis and Lu 2003; Arjona et al 2004; Bastain 2005). Most, however, believe the infectious agent in transmissible spongiform encephalopathies is the misfolded protein itself (Prusiner 1996; Legname et al. 2004; Castilla et al. 2005; Shorter and Lindquist 2005). Under one version of this hypothesis a misfolded PrP can, by contact, induce a normal PrP to adopt the misfolded shape. Under another version, a prexisting clump of misfolded proteins accretes and stabilizes newly misfolded proteins that arise spontaneously.

cannibalism just a few decades earlier. At the height of the tragedy, in the late 1950s, kuru was killing roughly 1% of the Foré population each year. Since then the Foré have stopped eating their dead and the epidemic has gradually waned.

Mead and colleagues, working in the laboratory of John Collinge, knew that an individual's genotype for the PrP gene on chromosome 20 influences his or her susceptibility to variant Creutzfeldt-Jakob disease (Will 2003). Two common alleles of the gene differ in the amino acid specified for position 129 of the encoded protein. In one allele the amino acid is valine; in the other allele it is methionine. Among the general Caucasian population, about 39% have genotype *Met/Met*, 50% have *Met/Val*, and 11% have *Val/Val*. To date, however, every victim of variant CJD has had genotype *Met/Met*. Mead and colleagues wanted to know whether there was similar genetic variation for resistance to kuru among the Foré.

The researchers answered their question by using the Hardy–Weinberg equilibrium principle. They focussed on Foré women and girls because women and girls were the most common victims of kuru. During mortuary feasts women and girls ate the internal organs, including the brains, of their deceased kin. The brain and spinal cord are the most infectious tissues. When men joined in the ritual they tended to eat just the muscles, and fewer of them got sick (Cooke 1999).

Mead and colleagues first determined the allele frequencies for the PrP gene among 140 Foré females who were too young to have practiced ritual cannibalism and who were thus never exposed to kuru (see Hedrick 2003 for the data). The frequencies were:

Met	***Val***
0.48	0.52

If the population is in Hardy–Weinberg equilibrium, then multiplying these allele frequencies will allow us to predict the genotype frequencies:

Met/Met	***Met/Val***	***Val/Val***
$(0.48)^2 = 0.23$	$2(0.48)(0.52) = 0.5$	$(0.52)^2 = 0.27$

These are, in fact, close to the actual genotype frequencies among the unexposed females:

Met/Met	***Met/Val***	***Val/Val***
$\frac{31}{140} = 0.22$	$\frac{72}{140} = 0.51$	$\frac{37}{140} = 0.26$

The allele and genotype frequencies among the unexposed females thus conform to conclusion 2 of the Hardy-Weinberg analysis.

Mead and colleagues then determined the allele frequencies among 30 older Foré women who had participated in mortuary feasts but had never contracted kuru. The frequencies were nearly the same as among the younger females:

Met	***Val***
0.52	0.48

The discovery that genotype frequencies in a population are not in Hardy–Weinberg equilibrium may be a clue that natural selection is at work.

If this segment of the population is, like their younger counterparts, in Hardy–Weinberg equilibrium, then multiplying the genotype frequencies will again allow us to predict the genotype frequencies:

Met/Met	***Met/Val***	***Val/Val***
$(0.52)^2 = 0.27$	$2(0.52)(0.48) = 0.5$	$(0.48)^2 = 0.23$

This time, the predicted values are a poor fit to the actual genotype frequencies:

Met/Met	*Met/Val*	*Val/Val*
$\frac{4}{30} = 0.13$	$\frac{23}{30} = 0.77$	$\frac{3}{30} = 0.10$

The survivors of the kuru epidemic are in violation of conclusion 2 of the Hardy–Weinberg analysis. There is a striking excess of heterozygotes, and a corresponding deficit of homozygotes. This discrepancy between our prediction the data is statistically significant (see Box 6.5).

Mead and colleagues believe the most plausible explanation for the missing homozygotes is that they were present in the population when the now-elderly women were young, but that they contracted kuru and died. In other words, it appears that homozygotes are susceptible to kuru whereas heterozygotes are resistant. Hedrick (2003) estimated that *Met* homozygotes and *Val* homozygotes survived the epidemic at rates of about 40% and 25%, respectively, compared to heterozygtoes.

Changes in the Frequency of the *CCR5-Δ32* Allele Revisited

We are now in a position to give a more satisfying answer to the question we raised at the beginning of Section 6.1: Will the AIDS epidemic cause the frequency of the *CCR5-Δ32* allele to increase in human populations? The AIDS epidemic could, in principle, cause the frequency of the allele to increase rapidly, but at present it appears that it will probably not do so in any real population. This conclusion is based on the three model populations depicted in Figure 6.15 (See Box 6.6 for the algebra.) Each model is based on different assumptions about the initial frequency of the *CCR5-Δ32* allele and the prevalence of HIV infection. Each graph shows the predicted change in the frequency of the *Δ32* allele over 40 generations, or approximately 1,000 years of evolution.

The model population depicted in Figure 6.15a offers a scenario in which the frequency of the *Δ32* allele could increase rapidly. In this scenario, the initial frequency of the *CCR5-Δ32* allele is 20%. One-quarter of the individuals with genotype +/+ or +/*Δ*32 contract AIDS and die without reproducing, whereas all of the *Δ32/Δ32* individuals survive. The 20% initial frequency of *Δ32* is approximately equal to the highest frequency reported for any population, a sample of Ashkenazi Jews studied by Martinson et al. (1997). The mortality rates approximate the situation in Botswana, Namibia, Swaziland, and Zimbabwe, where up to 25% of individuals between the ages of 15 and 49 are infected with HIV (UNAIDS 1998). In this model population, the frequency of the *Δ32* allele increases by as much as a few percentage points each generation. By the end of 40 generations, the allele is at a frequency of virtually 100%. Thus, in a human population that combined the highest reported frequency of the *Δ32* allele with the highest reported rates of infection, the AIDS epidemic could cause the frequency of the allele to increase rapidly.

At present, however, no known population combines a high frequency of the *Δ32* allele with a high rate of HIV infection. In northern Europe, many populations have *Δ32* frequencies between 0.1 and 0.2 (Martinson et al. 1997; Stephens et al. 1998), but HIV infection rates are under 1% (UNAIDS 1998). A model population reflecting these conditions is depicted in Figure 6.13b. The initial frequency of the *Δ32* allele is 0.2, and 0.5% of the +/+ and +/*Δ32* individuals contract

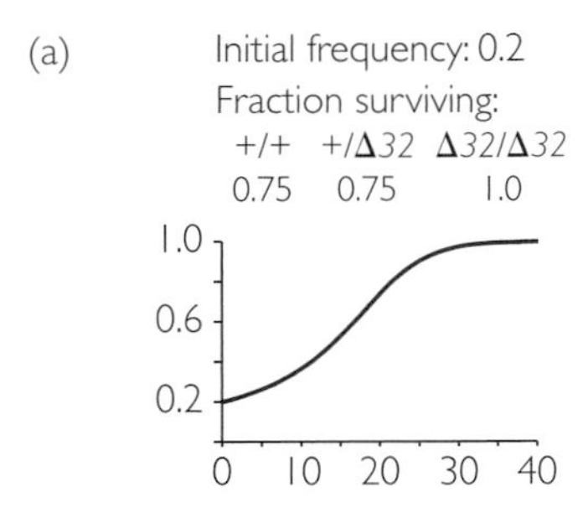

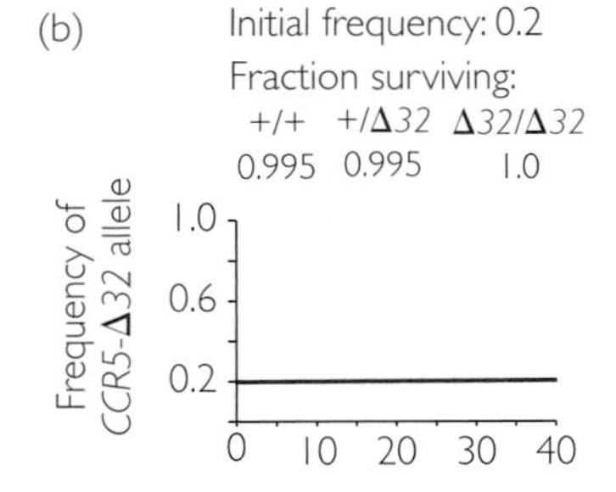

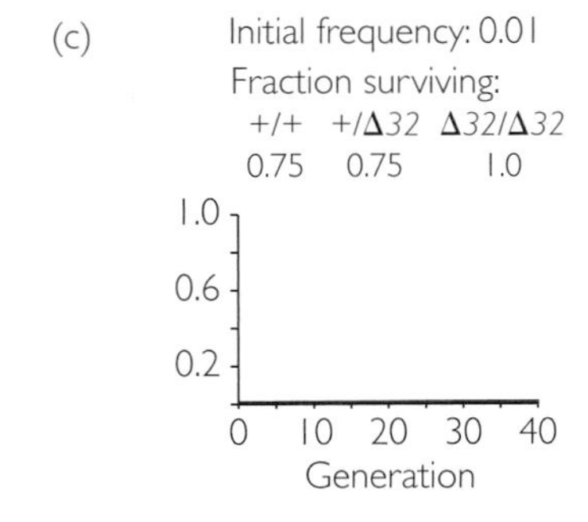

Figure 6.15 Predicted change in allele frequencies at the CCR5 locus due to the AIDS epidemic under three different scenarios (a) When the initial frequency of the *CCR5-Δ32* allele is high and a large fraction of the population becomes infected with HIV, the allele frequencies can change rapidly. However, no real population combines these characteristics. (b) In European populations allele frequencies are high, but only a small fraction of individuals become infected. (c) In parts of Africa there are high infection rates, but allele frequencies are low.

Box 6.5 Statistical analysis of allele and genotype frequencies using the χ^2 (chi-square) test

Here we use the data from Mead and colleagues (2003; see Hedrick 2003 for the numbers) to illustrate one method for determining whether genotype frequencies deviate significantly from what they would be under Hardy–Weinberg equilibrium. The researchers surveyed a population of 30 females who had eaten dead kin, yet survived the kuru epidemic without getting sick. The numbers of individuals with each genotype were as follows:

Met/Met	***Met/Val***	***Val/Val***
4	23	3

From these numbers, we will calculate the allele frequencies of *Met* and *Val* and determine whether the genotype frequencies observed are those we would expect according to the Hardy–Weinberg equilibrium principle. There are five steps:

1. Calculate the allele frequencies. The sample of 30 individuals is also a sample of 60 gene copies. All 8 copies carried by the 4 *Met/Met* women are *Met*, as are 23 of the copies carried by the 23 *Met/Val* women. Thus, the frequency of the *Met* allele is

$$\frac{(8 + 23)}{60} = 0.52$$

and the frequency of the *Val* allele is

$$\frac{(23 + 6)}{60} = 0.48$$

2. Calculate the genotype frequencies expected under the Hardy–Weinberg principle, given the allele frequencies calculated in Step 1. According to the Hardy–Weinberg equilibrium principle, if the frequencies of two alleles are p and q, then the frequencies of the genotypes are p^2, $2pq$, and q^2. Thus, the expected frequencies of genotypes among the survivors of the kuru epidemic are

Met/Met	***Met/Val***	***Val/Val***
$(0.52)^2$	$2(0.52)(0.48)$	$(0.48)^2$
$= 0.27$	$= 0.5$	$= 0.23$

3. Calculate the expected number of individuals of each genotype under Hardy–Weinberg equilibrium. This is simply the expected frequency of each genotype multiplied by the number of individuals in the sample, 54. The expected values are

Met/Met	***Met/Val***	***Val/Val***
$(0.27)(30)$	$(0.5)(30)$	$(0.23)(30)$
$= 8$	$= 15$	$= 7$

The numbers of individuals expected are different from the number of individuals actually observed (4, 23, and 3). The actual sample contains more heterozygotes and fewer homozygotes than expected. Is it plausible that this large a difference between expectation and reality could arise by chance? Or is the difference statistically significant? Our null hypothesis is that the difference is simply due to chance.

4. Calculate a test statistic. We will use a test statistic that was devised in 1900 by Karl Pearson. It is

AIDS and die without reproducing. The frequency of the $\Delta 32$ allele hardly changes at all. Selection is too weak to cause appreciable evolution in such a short time.

In parts of sub-Saharan Africa, as many as a quarter of all individuals of reproductive age are infected with HIV. However, the $\Delta 32$ allele is virtually absent (Martinson et al. 1997). A model population reflecting this situation is depicted in Figure 6.13c. The initial frequency of the $\Delta 32$ allele is 0.01, and 25% of the $+/+$ and $+/\Delta 32$ individuals contract AIDS and die without reproducing. Again, the frequency of the $\Delta 32$ allele hardly changes at all. When the $\Delta 32$ allele is at low frequency, most copies are in heterozygotes. Because heterozygotes are susceptible to infection, these copies are hidden from selection.

Our exploration of natural selection has given us tools we can use to predict the future of human populations.

called the chi-square (χ^2). The chi-square is defined as

$$\chi^2 = \sum \frac{(\text{observed} - \text{expected})^2}{\text{expected}}$$

where the symbol $\sum$ indicates a sum taken across all the classes considered. In our data there are three classes: the three genotypes. For our data set:

$$\chi^2 = \frac{(4-8)^2}{8} + \frac{(23-15)^2}{15} + \frac{(3-7)^2}{7} = 8.55$$

5. Determine whether the value of the test statistic is signficant. The chi-square is defined in such a way that the chi-square gets larger as the difference between the observed and expected values gets larger. How likely is it that we could get a chi-square as large as 8.55 by chance? Most statistical textbooks have a table that provides the answer. In Zar's (1996) book this table is called "Critical values of the chi-square distribution."

To use this table, we need to calculate a number called the degrees of freedom for the test statistic. The number of degrees of freedom for the chi-square is equal to the number of classes minus the number of independent values we calculated from the data for use in determining the expected values. For our chi-square there are three classes: the three genotypes. We calculated two values from the data for use in determining the expected values: the total number of individuals, and the frequency of the *Val* allele. (We also calculated the frequency of the *Met* allele, but it is not independent of the frequency of the *Val* allele, because the frequency of *Met* is one minus the frequency of *Val*.) Thus the number of degrees of freedom is 1. (Another formula for calculating the degrees of freedom in chi-square tests for Hardy–Weinberg equilibrium is

$$df = k - 1 - m$$

where k is the number of classes and m is the number of independent allele frequencies estimated from the data.)

According to the table, the critical value of chi-square for one degree of freedom and $P = 0.05$ is 3.841. This means that there is a 5% chance under the null hypothesis of getting $\chi^2 \geq 3.841$. The probability under the null hypothesis of getting $\chi^2 \geq 8.55$ is therefore (considerably) less than 5%. We reject the null hypothesis and assert that our value of chi-square is statistically significant at $P < 0.05$. (In fact, in this case, $P < 0.0034$.)

The chi-square test tells us that the alleles of the gene for PrP in the population of individuals who survived the kuru epidemic are not in Hardy–Weinberg equilibrium. This indicates that one or more of the assumptions of the Hardy–Weinberg analysis has been violated. By itself, however, it does not tell us which assumptions are being violated, or how.

The analysis we have just described is based on a number of simplifying assumptions. We have assumed, for example, that all HIV-infected individuals die without reproducing. In fact, however, many HIV-infected individuals have children. We have also assumed that the death rate is the same in heterozygotes as in +/+ homozygotes. In reality, although heterozygotes are susceptible to HIV infection, they appear to progress more slowly to AIDS (Dean et al. 1996). As a result, the fitness of heterozygotes may actually be higher than that of +/+ homozygotes. We challenge the reader to explore the evolution of human populations under a variety of selection schemes, to see how strongly our simplifying assumptions affect the predicted course of evolution. For analyses of more complex models of human evolution in response to selection imposed by AIDS,

Box 6.6 Predicting the frequency of the *CCR5-Δ32* allele in future generations

Let q_g be the frequency of the *CCR5-Δ32* allele in the present generation. Based on Box 6.3, we can write an equation predicting the frequency of the allele in the next generation, given estimates of the survival rates (fitnesses) of individuals with each genotype. The equation is

$$q_{g+1} = \frac{(1 - q_g)q_g w_{+\Delta} + q_g^2 w_{\Delta\Delta}}{(1 - q_g)^2 w_{++} + 2(1 - q_g)q_g w_{+\Delta} + q_g^2 w_{\Delta\Delta}}$$

where q_{g+1} is the frequency of the *Δ32* allele in the next generation, w_{++} is the fitness of individuals homozygous for the normal allele, $w_{+\Delta}$ is the fitness of heterozygotes, and $w_{\Delta\Delta}$ is the fitness of individuals homozygous for the *CCR5-Δ32* allele.

After choosing a starting value for the frequency of the *Δ32* allele, we plug it and the estimated fitnesses into the equation to generate the frequency of the *Δ32* allele after one generation. We then plug this resulting value into the equation to get the frequency of the allele after two generations, and so on.

see models by Schliekelman et al. (2001) (but also Ramaley et al. 2002) and by Sullivan et al. (2001).

6.3 Patterns of Selection: Testing Predictions of Population Genetic Theory

In the 1927 case of *Buck v. Bell*, the United States Supreme Court upheld the state of Virginia's sterilization statute by a vote of eight to one. Drafted on the advice of eugenicists, the sterilization law was intended to improve the genetic quality of future generations by allowing the forced sterilization of individuals afflicted with hereditary forms of insanity, feeblemindedness, and other mental defects. The court's decision in *Buck v. Bell* reinvigorated a compulsory sterilization movement dating from 1907 (Kevles 1995). By 1940, 30 states had enacted sterilization laws, and by 1960 over 60,000 people had been sterilized without their consent (Reilly 1991; Lane 1992). In hindsight, the evidence that these individuals suffered from hereditary diseases was weak. But what about the evolutionary logic behind compulsory sterilization? If the genetic assumptions had been correct, would sterilization have been an effective means of reducing the incidence of undesirable traits?

Before we try to answer this question, it will be helpful to address a more general one. How well does the theory of population genetics actually work? We developed this theory in Sections 6.1 and 6.2. The final product is a model of how allele frequencies change in response to natural selection (Figures 6.11 and 6.12, Boxes 6.3 and 6.6). If our model is a good one, it should accurately predict the direction and rate of allele frequency change under a variety of selection schemes. It should work, for example, whether the allele favored by selection is dominant or recessive, common or rare. It should work whether selection favors heterozygotes or homozygotes. It should even predict what will happen when a particular allele is favored by selection under some circumstances and disfavored in others.

In this section we will find out how well our model works. We will use the theory we have developed to predict the course of evolution under different patterns of selection, and we will compare our predictions to empirical data from experimental populations. We will then return to our question about eugenic sterilization.

Selection on Recessive and Dominant Alleles

For our first test, we will see whether our theory accurately predicts changes in the frequencies of recessive and dominant alleles. Our example comes from the work of Peter Dawson (1970). Dawson had been studying a laboratory colony of flour beetles (*Tribolium castaneum*) and had identified a gene we will call the l locus. This locus has two alleles: + and *l*. Individuals with genotype +/+ or +/*l* are phenotypically normal, whereas individuals with genotype *l*/*l* do not survive. In other words, *l* is a recessive lethal allele.

Dawson collected heterozygotes from his beetle colony and used them to establish two new experimental populations. Because all the founders were heterozygotes, the initial frequencies of the two alleles were 0.5 in both populations. Because *l*/*l* individuals have zero fitness, Dawson expected his populations to evolve toward ever lower frequencies of the *l* allele and ever higher frequencies of the + allele. He let his two populations evolve for a dozen generations, each generation measuring the frequencies of the two alleles.

Dawson used the equations derived in Box 6.3 and the method described in Box 6.6 to make a quantitative prediction of the course of evolution in his populations. We can reproduce this prediction with a straightforward numerical calculation like the ones we performed in Figures 6.11 and 6.12. Imagine a gene pool in which alleles + and *l* are both at a frequency of 0.5. If we combine gametes at random to make 100 zygotes, we get the three genotypes in the following numbers:

+/+	+/*l*	*l*/*l*
25	50	25

Now we imagine that all the *l*/*l* individuals die and that everyone else survives to breed. Finally, imagine that each of the survivors donates 10 gametes to the new gene pool:

The 25 +/+ survivors make a total of 250 gametes: 250 carry allele +; none carry allele *l*.

The 50 +/*l* survivors make a total of 500 gametes: 250 carry allele +; 250 carry allele *l*.

This gives us 500 copies of the + allele and 250 copies of the *l* allele for a total of 750. In this new gene pool, the frequency of the + allele is 0.67, and the frequency of the *l* allele is 0.33. We have gone from the gene pool in generation zero to the gene pool in generation one. The frequency of the + allele has risen, and the frequency of the *l* allele has fallen.

To get from generation one's gene pool to generation two's gene pool we just repeat the exercise. We combine the gametes in generation one's gene pool at random to make 100 zygotes—45 +/+, 44 +/*l*, and 11 *l*/*l*—and so on. The only problem with using pencil-and-paper numerical calculations to predict evolution is that chasing the alleles around and around the life cycle all the way to generation 12 is a tedious job.

With a computer, however, predicting how Dawson's population will evolve is quick and easy. We can use a spreadsheet application to set up the required calculations ourselves (see Boxes 6.3 and 6.6), or we can use any of a variety of population genetics programs that are already set up to do the calculations for us. Such programs take starting allele frequencies and genotype fitnesses as input and use the model we have developed in this chapter to produce predicted allele frequencies in future generations as output. We encourage the reader to get one of these programs and experiment with it.

The prediction for Dawson's experiment appears as two gray curves in Figure 6.16. The curve in the top graph predicts the falling frequency of the *l* allele; the curve in the bottom graph predicts the rising frequency of the + allele. The curves predict that evolution will be rapid at first but will slow as the experiment proceeds.

Empirical research on flour beetles shows that predictions made with population genetic models are accurate, at least under laboratory conditions.

Dawson's data appear in the graphs as colored circles and triangles. They match our theoretical predictions closely. This tight fit between prediction and data may seem unsurprising, even mundane. It should not. It should be astonishing. We used a simple model of the mechanism of evolution, combining the fundamental insights of Gregor Mendel with those of Charles Darwin, to predict how a population would change over 12 generations. If the creatures in question had been humans instead of flour beetles, it would have meant forecasting events that will happen in 300 years. And Dawson's data show that our prediction was spot on. If we had a theory that worked like that for picking stocks or race horses—well, we could have retired years ago. Our model has passed its first test.

Dawson's experiment shows that dominance and allele frequency interact to determine the rate of evolution. When a recessive allele is common (and a dominant allele is rare), evolution by natural selection is rapid. In contrast, when a recessive allele is rare, and a dominant allele is common, evolution by natural selection is slow. The Hardy–Weinberg equilibrium principle explains why. First

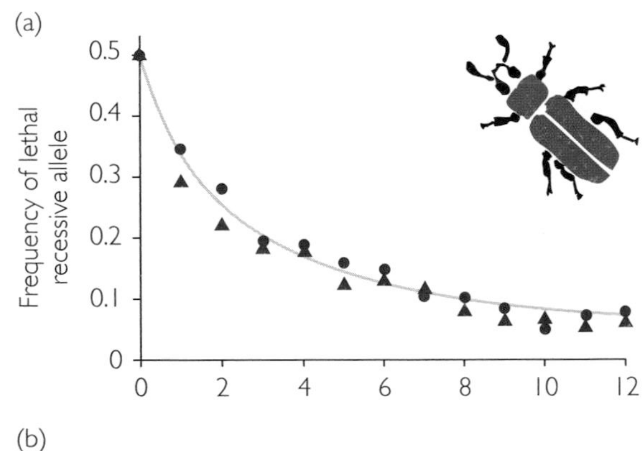

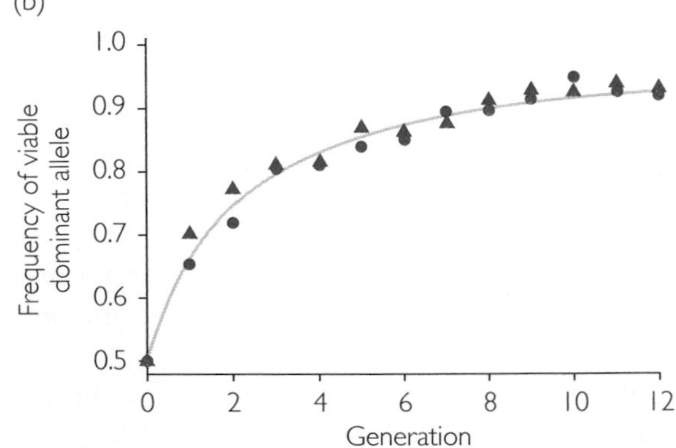

Figure 6.16 Evolution in laboratory populations of flour beetles (a) The decline in frequency of a lethal recessive allele (red symbols) matches the theoretical prediction (gray curve) almost exactly. As the allele becomes rare, the rate of evolution slows dramatically. (b) This graph plots the increase in frequency of the corresponding dominant allele. Redrawn from Dawson (1970).

imagine a recessive allele that is common: Its frequency is, say, 0.95. The dominant allele thus has a frequency of 0.05. By multiplying the allele frequencies, we can calculate the genotype frequencies:

AA	***Aa***	***aa***
$(0.05)^2$	$2(0.05)(0.95)$	$(0.95)^2$
$= 0.0025$	$= 0.095$	$= 0.9025$

Roughly 10% of the individuals in the population have the dominant phenotype, while 90% have the recessive phenotype. Both phenotypes are reasonably well represented, and if they differ in fitness, then the allele frequencies in the next generation may be substantially different. Now imagine a recessive allele that is rare: Its frequency is 0.05. The dominant allele thus has a frequency of 0.95. The genotype frequencies are

Natural selection is most potent as a mechanism of evolution when it is acting on common recessive alleles (and rare dominant alleles). When a recessive allele is rare, most copies are hidden in heterozygotes and protected from selection.

AA	***Aa***	***aa***
$(0.95)^2$	$2(0.95)(0.05)$	$(0.05)^2$
$= 0.9025$	$= 0.095$	$= 0.0025$

Approximately 100% of the population has the dominant phenotype, while approximately 0% has the recessive phenotype. Even if the phenotypes differ greatly in fitness, there are so few of the minority phenotype that there will be little change in allele frequencies in the next generation. In a random mating population, most copies of a rare recessive allele are phenotypically hidden inside heterozygous individuals and thereby immune from selection. For an algebraic treatment of selection on recessive and dominant alleles see Box 6.7.

Selection on Heterozygotes and Homozygotes

In our next two tests we will see whether our model can accurately predict what happens when selection favors heterozygotes or homozygotes. Both tests will use data on laboratory populations of fruit flies (*Drosophila melanogaster*).

Selection Favoring Heterozygotes

Our first example comes from research by Terumi Mukai and Allan Burdick (1959). Like Dawson, Mukai and Burdick studied evolution at a single locus with two alleles. We will call the alleles *V*, for viable, and *L* for lethal. This is because flies with genotype *VV* or *VL* are alive, whereas flies with genotype *LL* are dead. The researchers used heterozygotes as founders to establish two experimental populations with initial allele frequencies of 0.5. They allowed the populations to evolve for 15 generations, each generation measuring the frequency of the viable allele.

So far Mukai and Burdick's experiment sounds just like Dawson's. If it is, then our theory predicts that the *V* allele will rise in frequency—rapidly at first, then more slowly. By generation 15 it should reach a frequency of over 94%. But that is not what happened.

Mukai and Burdick's data appear in Figure 6.18 (p.200), represented by the red symbols. As expected, the frequency of the viable allele increased rapidly over the first few generations. However, in both populations the rate of evolution slowed long before the *V* allele approached a frequency of 0.94. Instead, the viable allele seemed to reach an equilibrium, or unchanging state, at a frequency of about 0.79.

Box 6.7 An algebraic treatment of selection on recessive and dominant alleles

Here we develop equations that illuminate the differences between selection on recessive versus dominant alleles. Imagine a single locus with two alleles. Let p be the frequency of the dominant allele A, and let q be the frequency of the recessive allele a.

Selection on the recessive allele

Let the fitnesses of the genotypes be given by

w_{AA}	w_{Aa}	w_{aa}
1	1	$1 - s$

where s, called the **selection coefficient**, represents the strength of selection against homozygous recessives relative to the other genotypes. (Selection in favor of homozygous recessives can be accomodated by choosing a negative value for s.)

Based on Box 6.3, the following equation gives the frequency of allele a in the next generation, q', given the frequency of a in this generation and the fitnesses of the three genotypes:

$$q' = \frac{pqw_{Aa} + q^2w_{aa}}{\overline{w}} = \frac{pqw_{Aa} + q^2w_{aa}}{p^2w_{AA} + 2pqw_{Aa} + q^2w_{aa}}$$

Substituting the fitness values from the table above, and $(1 - q)$ for p, then simplifying, gives

$$q' = \frac{q(1 - sq)}{1 - sq^2}$$

If a is a lethal recessive, then s is equal to 1. Substituting this value into the preceding equation gives

$$q' = \frac{q(1 - q)}{1 - q^2} = \frac{q(1 - q)}{(1 - q)(1 + q)} = \frac{q}{(1 + q)}$$

A little experimentation shows that once a recessive lethal allele becomes rare, further declines in frequency are slow. For example, if the frequency of allele a in this generation is 0.01, then in the next generation its frequency will be approximately 0.0099.

Selection on the dominant allele

Let the fitnesses of the genotypes be given by:

w_{AA}	w_{Aa}	w_{aa}
$1 - s$	$1 - s$	1

where s, the selection coefficient, represents the strength of selection against genotypes containing the dominant allele relative to homozygous recessives. (Selection in favor of genotypes containing the dominant allele can be accomodated by choosing a negative value of s.)

Based on Box 6.3, we can write an equation that predicts the frequency of allele A in the next generation, p', given the frequency of A in this generation and the fitness of the three genotypes:

$$p' = \frac{p^2w_{AA} + pqw_{Aa}}{\overline{w}} = \frac{p^2w_{AA} + pqw_{Aa}}{p^2w_{AA} + 2pqw_{Aa} + q^2w_{aa}}$$

Substituting the fitnesses from the table, and $(1 - p)$ for q, then simplifying, gives

$$p' = \frac{p(1 - s)}{1 - 2sp + sp^2}$$

If A is a lethal dominant, s is equal to 1. Substituting this value into the foregoing equation shows that a lethal dominant is eliminated from a population in a single generation.

Selection on recessive alleles versus selection on dominant alleles

Selection on recessive alleles and selection on dominant alleles are opposite sides of the same coin. Selection against a recessive allele is selection in favor of the dominant allele and vice versa.

Figure 6.17a (left) shows 100 generations of evolution in a model population under selection against a recessive allele and in favor of the dominant allele. At first, the frequencies of the alleles change rapidly.

How could this happen? An equilibrium frequency of 0.79 for the viable allele means that the lethal allele has an equilibrium frequency of 0.21. How could natural selection maintain a lethal allele at such a high frequency in this population? Mukai and Burdick argue that the most plausible explanation is **heterozygote superiority**, also known as **overdominance**. Under this hypothesis,

(a) Selection against a recessive allele (s = 0.5) and for a dominant allele

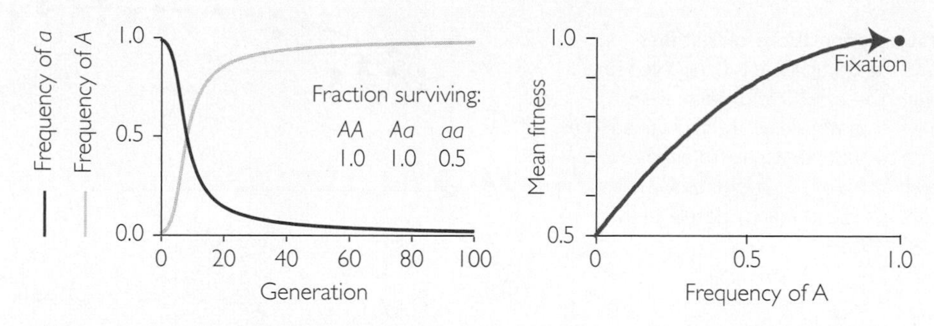

(b) Selection for a recessive allele and against a dominant allele ($s = 0.6$)

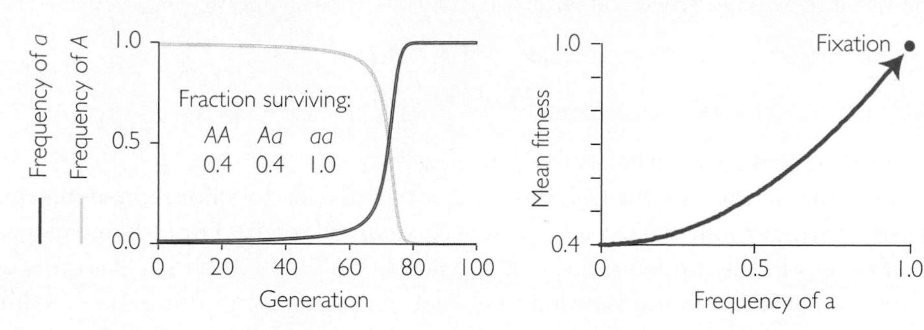

Figure 6.17 Evolution in model populations under selection on recessive and dominant alleles Graphs on the left show changes in allele frequencies over time. Graphs on the right show adaptive landscapes: Changes in population mean fitness as a function of allele frequencies.

As the recessive allele becomes rare, however, the rate of evolution slows dramatically. When the recessive allele is rare, most copies in the population are in heterozygous individuals, where they are effectively hidden from selection.

The figure also shows (right) the mean fitness of the population (see Box 6.3) as a function of the frequency of the dominant allele. As the dominant allele goes from rare to common, the mean fitness of the population rises. Mean fitness is maximized when the favored allele reaches a frequency of 100%. Graphs of mean fitness as a function of allele frequency are often referred to as adaptive landscapes.

Figure 6.17b (left) shows 100 generations of evolution in a model population under selection in favor of a recessive allele and against the dominant allele. At first, the frequencies of the alleles change slowly. The recessive allele is rare, most copies present are in heterozygotes, and selection cannot see it. However, as the recessive allele becomes common enough that a substantial fraction of homozygotes appear, the rate of evolution increases dramatically. Once the pace of evolution accelerates, the favorable recessive allele quickly achieves a frequency of 100%. That is, the recessive allele becomes fixed in the population.

The figure also shows (right) the mean fitness of the population (see Box 6.3) as a function of the frequency of the recessive allele. As the recessive allele goes from rare to common, the mean fitness of the population rises. Mean fitness is maximized when the favored allele reaches a frequency of 100%.

heterozygotes have higher fitness than either homozygote. At equilibrium, the selective advantage enjoyed by the lethal allele when it is in heterozygotes exactly balances the obvious disadvantage it suffers when it is in homozygotes.

A little experimentation with a computer should allow the reader to confirm that Mukai and Burdick's hypothesis explains their data nicely. The red curve in

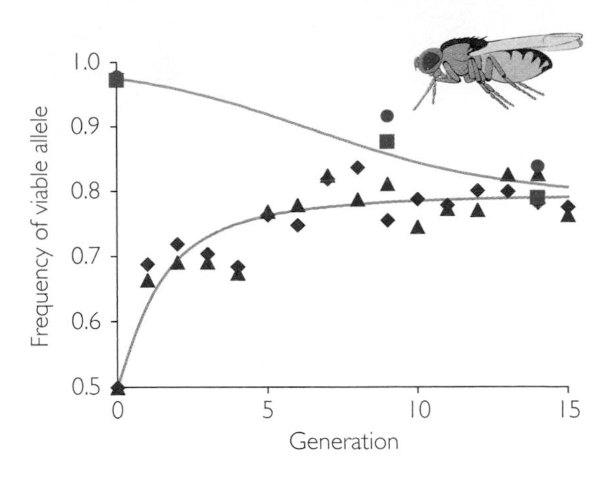

Figure 6.18 Evolution in four laboratory populations of fruit flies In homozygous state, one allele is viable and the other allele is lethal. Nonetheless, the populations that started with a frequency of 0.5 for both alleles (red) evolved toward an equilibrium in which both alleles are maintained. The likely explanation is that heterozygotes enjoy superior fitness to either homozygote. The blue populations represent a test of this hypothesis. The data (circles and squares) match the theoretical prediction (line) closely. Drawn from data presented in Mukai and Burdick (1959).

Figure 6.18 represents evolution in a model population in which the fitnesses of the three genotypes are as follows:

VV	*VL*	*LL*
0.735	1.0	0

This theoretical curve matches the data closely.

Note that in this case the fit between theory and data does not represent a rigorous test of our model. That is because we examined the data first, then tweaked the fitnesses in the model to make its prediction fit. That is a bit like shooting at a barn and painting a target around the bullet hole. Mukai and Burdick's flies did, however, provide an opportunity for a test of our model that is rigorous. And Mukai and Burdick performed it.

Research on fruit flies shows that natural selection can act to maintain two alleles at a stable equilibrium. One way this can happen is when heterozygotes have superior fitness.

The researchers established two more experimental populations, this time with the initial frequency of the viable allele at 0.975. If the genotype fitnesses are, indeed, those required to make our model fit the red data points in Figure 6.18, then this time our model predicts that the frequency of the *V* allele should fall. As before, it should ultimately reach an equilibrium near 0.79. The predicted fall toward equilibrium is shown by the blue curve in Figure 6.18. Mukai and Burdick's data appear in the figure as blue symbols. The data match the prediction closely. Our model has passed its second test.

Mukai and Burdick's flies have shown us something new. In all our previous examples, selection has favored one allele or the other. Under such circumstances our model predicts that sooner or later the favored allele will reach a frequency of 100%, and the disfavored allele will disappear. By keeping a population at an equilibrium in which both alleles are present, however, heterozygote superiority can maintain genetic diversity indefinitely. For an algebraic treatment of heterozygote superiority, see Box 6.8.

Selection Favoring Homozygotes

Our second example comes from work by G. G. Foster and colleagues (1972). These researchers set up experiments to demonstrate how populations evolve when heterozygotes have lower fitness than either homozygote. Foster and colleagues used fruit flies with compound chromosomes. Compound chromosomes are homologous chromosomes that have swapped entire arms, so that one homolog has two copies of one arm, and the other homolog has two copies of

(a) A normal pair of homologous chromosomes (each has one blue arm and one green arm).

(b) A pair of compound chromosomes (one has two blue arms, the other has two green arms).

(c) Gametes made by an individual with compound chromosomes may contain both chromosomes, one, or neither.

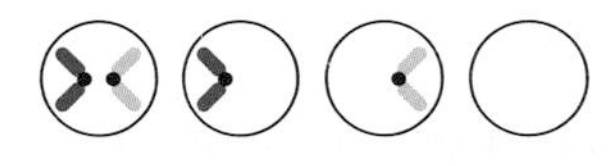

(d) When individuals with compound chromosomes mate, one quarter of their zygotes are viable.

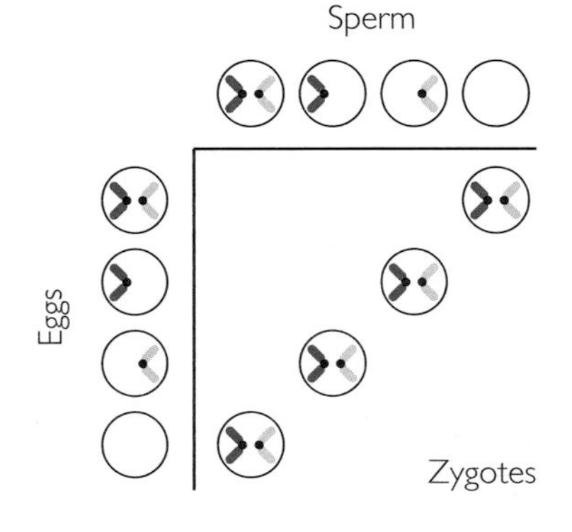

(e) When an individual with compound chromosomes mates with an individual with normal chromosomes, none of their zygotes are viable.

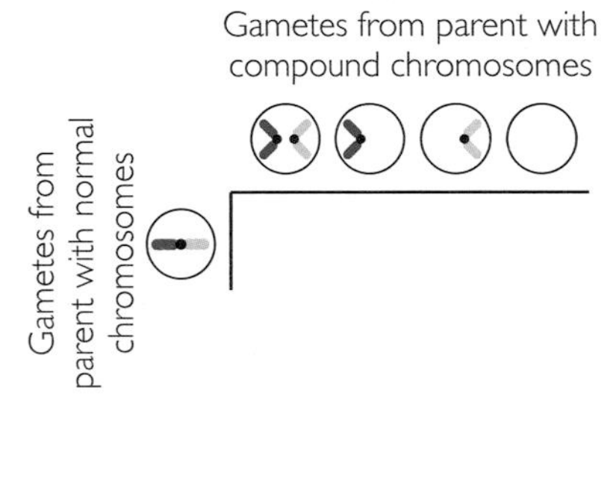

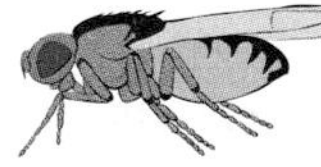

(f) Left: Evolution in 11 populations of *Drosophila melanogaster* containing compound second chromosomes [*C(2)*] and compound third chromosomes [*C(3)*]. The initial frequency of *C(2)* ranged from 0.4 to 0.65. Right: Evolution in 13 populations of *Drosophila melanogaster* containing a mixture of compound second chromosomes [*C(2)*] and normal second chromosomes [*N(2)*]. The initial frequency of *C(2)* ranged from 0.71 to 0.96.

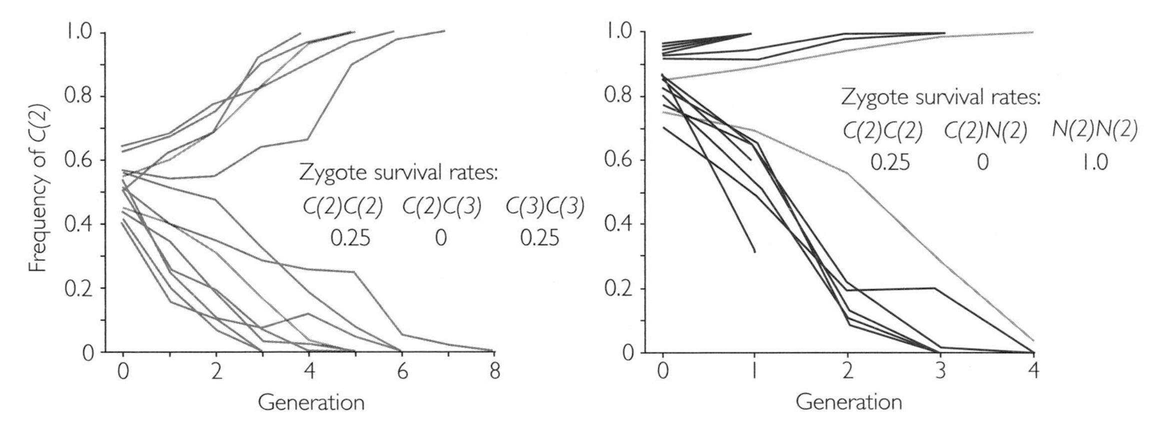

Figure 6.19 An experiment designed to show how populations evolve when heterozygotes have lower fitness than either homozygote (a–e) The experimental design makes clever use of compound chromosomes. (f) The data (orange and red) match the theoretical predictions (gray). Redrawn with permission from Foster et al. (1972).

the other arm [Figure 6.19a and (b)]. During meiosis, compound chromosomes may or may not segregate. As a result, four kinds of gametes are produced in equal numbers: gametes with both homologous chromosomes, gametes with just one member of the pair, gametes with the other member of the pair, and gametes with neither member of the pair (Figure 6.19c). When two flies with compound chromosomes mate with each other, one-quarter of their zygotes have every chromosome arm in the correct dose and are thus viable. The other

It is also possible for heterozygotes to have inferior fitness.

Box 6.8 Stable equilibria with heterozygote superiority and unstable equilibria with heterozygote inferiority

Here we develop algebraic and graphical methods for analyzing evolution at loci with overdominance and underdominance. Imagine a population in which allele A_1 is at frequency p and allele A_2 is at frequency q. In Box 6.3, we developed an equation describing the change in p from one generation to the next under selection:

$$\Delta p = \frac{p}{\overline{w}}(pw_{11} + qw_{12} - \overline{w})$$
$$= \frac{p}{\overline{w}}(pw_{11} + qw_{12} - p^2w_{11} - 2pqw_{12} - q^2w_{22})$$

Substituting $(1 - q)$ for p in the first and third terms in the expression in parentheses gives

$$\Delta p = \frac{p}{\overline{w}}[(1 - q)w_{11} + qw_{12} - (1 - q)^2w_{11} - 2pqw_{12} - q^2w_{22}]$$

which, after simplifying and factoring out q, becomes

$$\Delta p = \frac{pq}{\overline{w}}(w_{12} + w_{11} - qw_{11} - 2pw_{12} - qw_{22})$$

Now, by definition, the frequency of allele A_1 is at equilibrium when $\Delta p = 0$. The equation above shows that $\Delta p = 0$ when $p = 0$ or $q = 0$. These two equilibria are unsurprising. They occur when one allele or the other is absent from the population. The equation also gives a third condition for equilibrium, which is

$$w_{12} + w_{11} - qw_{11} - 2pw_{12} - qw_{22} = 0$$

Substituting $(1 - p)$ for q and solving for p gives

$$\hat{p} = \frac{w_{22} - w_{12}}{w_{11} - 2w_{12} + w_{22}}$$

where $\hat{p}$ is the frequency of allele A_1 at equilibrium. Finally, let the genotype fitnesses be as follows:

A_1A_1	A_1A_2	A_2A_2
$1 - s$	1	$1 - t$

Positive values of the selection coefficients s and t represent overdominance; negative values represent underdominance. Substituting the fitnesses into the previous equation and simplifying gives

$$\hat{p} = \frac{t}{s + t}$$

For example, when $s = 0.4$ and $t = 0.6$, heterozygotes have superior fitness, and the equilibrium frequency for allele A_1 is 0.6. When $s = -0.4$ and $t = -0.6$, heterozygotes have inferior fitness, and the equilibrium frequency for allele A_1 is also 0.6.

Another useful method for analyzing equilibria is to plot Δp as a function of p. Figure 6.20a shows such a plot for the two numerical examples we just calculated. Both curves show that $\Delta p = 0$ when $p = 0$, $p = 1$, or $p = 0.6$.

The curves in Figure 6.20a also allow us to determine whether an equilibrium is stable or unstable. Look at the red curve; it describes a locus with heterozygote superiority. Notice that when p is greater than 0.6, Δp is negative. This means that when the frequency of allele A_1 exceeds its equilibrium value, the population will move back toward equilibrium in the next generation. Likewise, when p is less than 0.6, Δp is positive. When the frequency of allele A_1 is below its equilibrium value, the population will move back toward equilibrium in the next genera-

three-quarters have too many or too few of copies of one or both chromosome arms and are thus inviable (Figure 6.19d). When a fly with compound chromosomes mates with a fly with normal chromosomes, none of the zygotes they make are viable (Figure 6.19e).

Foster and colleagues established two sets of laboratory populations. In the first set of populations, some of the founders had compound second chromosomes [*C(2)*] and others had compound third chromosomes [*C(3)*]. Note that if

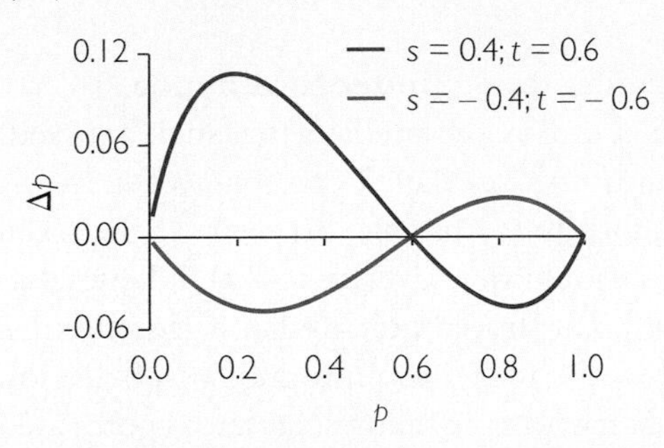

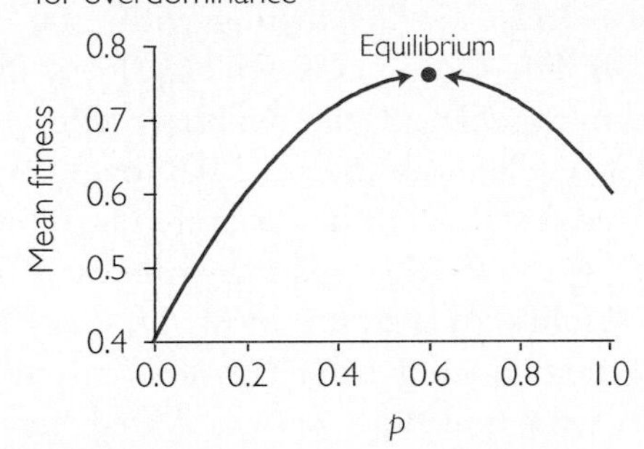

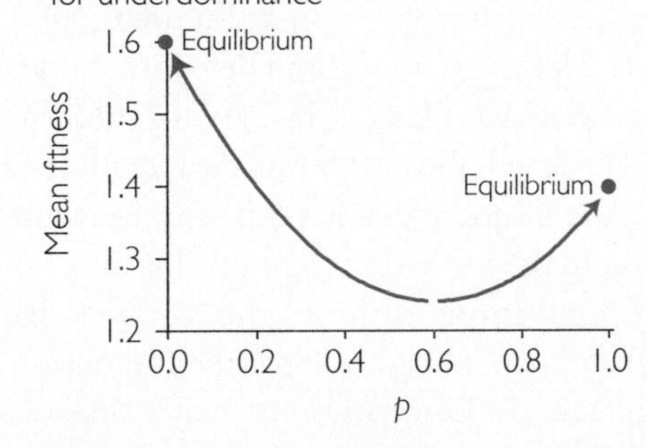

Figure 6.20 A graphical analysis of stable and unstable equilibria at loci with overdominance and underdominance (a) A plot of Δp as a function of p. (b) and (c) Adaptive landscapes.

tion. The "internal" equilibrium for a locus with heterozygote superiority is stable.

Figure 6.20b shows an adaptive landscape for a locus with heterozygote superiority. The graph plots population mean fitness as a function of the frequency of allele A_1. Mean fitness is low when A_1 is absent, and relatively low when A_1 is fixed. As the allele frequency moves from either direction toward its stable equilibrium, the population mean fitness rises to a maximum.

Now, look at the blue curve in Figure 6.20a. It describes a locus with heterozygote inferiority. If p rises even slightly above 0.6, p will continue to rise toward 1.0 in subsequent generations; if p falls even slightly below 0.6, p will continue to fall toward 0 in subsequent generations. The internal equilibrium for a locus with heterozygote inferiority is unstable.

Figure 6.20c shows an adaptive landscape for a locus with heterozygote inferiority. Population mean fitness is lowest when the frequency of allele A_1 is at its unstable internal equilibrium. As the allele frequency moves away from this equilibrium in either direction, mean fitness rises.

A comparison of the adaptive landscape in Figure 6.20c with those in Figure 6.20b and Figure 6.17 offers a valuable insight. As a population evolves in response to selection, the mean fitness of the individuals in the population tends to rise. Selection does not, however, always maximize mean fitness in a global sense. Depending on the initial allele frequencies, the population depicted in Figure 6.20c may evolve toward either fixation or loss of A_1. If the allele becomes fixed, the population will be at a stable equilibrium, but the population's mean fitness will be substantially lower than it would be if the allele were lost.

two flies with compound second chromosomes mate, one-quarter of their offspring survive. Likewise, if two flies with compound third chromosomes mate, one-quarter of their offspring survive. But if a fly with compound second chromosomes (and normal third chromosomes) mates with a fly with compound third chromosomes (and normal second chromosomes), none of their offspring survive. For purposes of analysis, then, we can treat the second and third chromosome as though they are alleles of a single locus. Thus, the founders consisted of *C(2)C(2)* homozygotes and *C(3)C(3)* homozygotes. Based on the zygote viabilities we just described, the fitnesses of the possible offspring genotypes in the mixed population are

$C(2)C(2)$	$C(2)C(3)$	$C(3)C(3)$
0.25	0	0.25

In other words, the genotypes exhibit strong **underdominance**.

The algebraic analysis described in Box 6.8 predicts that such a mixed population will be in genetic equilibrium, with both alleles present, when the frequency of *C(2)* is exactly 0.5. This equilibrium is unstable, however. If the frequency of *C(2)* ever gets above 0.5, then it should quickly rise to 1.0. Likewise, if the frequency of *C(2)* ever dips below 0.5, it should quickly fall to zero. Experimentation with a computer should allow the reader to reproduce this behavior.

Intuitively, the reason for the behavior is as follows. Heterozygotes are inviable, so the adults in the population are all homozygotes. Imagine first that *C(2)C(2)* individuals are common and *C(3)C(3)* individuals are rare. If the flies mate at random, then most matings will involve *C(2)C(2)* flies mating with each other, or *C(2)C(2)* flies mating with *C(3)C(3)* flies. Only rarely will *C(3)C(3)* flies mate with their own kind. Consequently, most *C(3)C(3)* flies will have zero reproductive success, and the frequency of *C(2)* will climb toward 1.0. Now imagine that *C(3)C(3)* individuals are common and *C(2)C(2)* individuals are rare. Under random mating, most matings involve *C(3)C(3)* flies mating with each other, or *C(3)C(3)* flies mating with *C(2)C(2)* flies. As a result, most *C(2)C(2)* flies will have zero reproductive success, and the frequency of *C(2)* will fall toward zero.

Foster and colleagues set up 11 mixed populations, with *C(2)* frequencies ranging from about 0.4 to about 0.65, then monitored their evolution for up to eight generations. Predictions for the evolution of populations with initial *C(2)* frequencies of 0.45 and 0.55 appear as gray lines in the left graph in Figure 6.19f. The data from Foster et al.'s flies appear as orange lines. There is some deviation between prediction and result, probably due to genetic drift. That is, in a few of the experimental populations the frequency of *C(2)* started above 0.5 but ultimately fell to zero. In all 11 populations, however, once the frequency of *C(2)* had moved substantially away from 0.5 it continued moving in the same direction until it hit zero or one.

In the researchers' second set of populations, some of the founders had compound second chromosomes [*C(2)*] and others had normal second chromosomes [*N(2)*]. If two flies with compound second chromosomes mate, one-quarter of their offspring are viable. If a fly with compound second chromosomes mates with a fly with normal second chromosomes, none of their offspring is viable. If two flies with normal second chromosomes mate, all of their offspring are viable. Again, for purposes of analysis, we can treat each chromosome as though it were a single allele. Thus, the founders consisted of *C(2)C(2)* homozygotes and *N(2)N(2)* homozygotes. The fitnesses of the genotypes in the mixed population are

$C(2)C(2)$	$C(2)N(2)$	$N(2)N(2)$
0.25	0	1.0

As in the first set of populations, the genotypes exhibit strong **underdominance**. This time, however, one kind of homozygote has much higher fitness than the other.

The algebraic analysis described in Box 6.8 predicts an unstable equilibrium when the frequency of *C(2)* is exactly 0.8. If the frequency of *C(2)* ever gets above 0.8, then it should quickly rise to 1.0. Likewise, if the frequency of *C(2)* ever dips below 0.8, it should quickly fall to zero. Experimentation with a computer should allow the reader to reproduce this prediction.

The intuitive explanation is as follows. Heterozygotes are inviable, so the adults in the population are all homozygotes. Imagine first that *C(2)C(2)* individuals are common and *N(2)N(2)* individuals are rare. If the flies mate at random, then almost all matings will involve *C(2)C(2)* flies mating with each other, or *C(2)C(2)* flies mating with *N(2)N(2)* flies. Only very rarely will *N(2)N(2)* flies mate with their own kind. Consequently, most *N(2)N(2)* flies will have zero reproductive success, and the frequency of *C(2)* will climb to 1.0. Now imagine that there are enough *N(2)N(2)* flies present that appreciable numbers of them *do* mate with each other. These matings will produce four times as many offspring as matings between *C(2)C(2)* flies. Consequently, the frequency of *N(2)* will climb to 1.0 and the frequency of *C(2)* will fall to zero.

Foster and colleagues set up 13 mixed populations, with *C(2)* frequencies ranging from 0.71 to 0.96, then monitored their evolution for up to four generations. Predictions for the evolution of populations with initial *C(2)* frequencies of 0.75 and 0.85 appear as gray lines in the right graph in Figure 6.19f. The data appear as red lines. Qualitatively, the outcome matches the theoretical prediction nicely. In populations with higher initial *C(2)* frequencies, *C(2)* quickly rose to fixation, while in populations with lower initial *C(2)* frequencies, *C(2)* was quickly lost. The exact location of the unstable equilibrium turned out to be approximately 0.9 instead of 0.8. Foster and colleagues note that their *C(2)C(2)* flies carried recessive genetic markers that the biologists had bred into them to allow for easy identification. They suggest that these markers reduced the relative fitness of the *C(2)C(2)* flies below the value of 0.25 inferred solely on the basis of their compound chromosomes.

Our model's predictions were not as accurate for Foster et al.'s experiments as they were for Dawson's and Mukai and Burdick's. Nonetheless the model performed well. It predicted something we have not seen before: an unstable equilibrium above which the frequency of an allele would rise and below which it would fall. It predicted that the unstable equilibrium would be higher in Foster et al.'s second set of populations than in their first. And its predictions about the rate of evolution were roughly correct. Our model has passed its third test.

When heterozygotes have inferior fitness, one allele tends to go to fixation while the other allele is lost. However, different populations may lose different alleles.

Foster et al.'s experiments demonstrate that heterozygote inferiority leads to a loss of genetic diversity within populations. By driving different alleles to fixation in different populations, however, heterozygote inferiority may help maintain genetic diversity among populations.

Frequency-Dependent Selection

For our fourth and final test of population genetic theory we will see whether our model can predict the evolutionary outcome when the fitness of individuals with a particular phenotype depends on their frequency in the population. Our example, from the work of Luc Gigord, Mark Macnair, and Ann Smithson (2001), concerns a puzzling color polymorphism in the Elderflower orchid (*Dactylorhiza sambucina*).

Elderflower orchids come in two colors: yellow and purple (Figure 6.21a). Populations typically include both colors, with yellow usually somewhat more common. The flowers attract bumblebees, which serve as the orchid's primary pollinator. But the bumblebees that visit Elderflower orchids are always disappointed. To the bees the orchid's colorful flowers appear to advertise a reward, but in fact they offer nothing. The puzzle Gigord and colleagues wanted to solve was this: How can two distinct deceptive advertisements persist together in Elderflower orchid populations?

The researchers' hypothesis grew out of earlier observations by Smithson and Macnair (1997). When naive bumblebees visit a stand of orchids to sample the flowers, they tend to alternate between colors. If a bee visits a purple flower first and finds no reward, it looks next in a yellow flower. Finding nothing there either, it tries another purple one. Disappointment sends it back to a yellow, and so on, until the bumblebee gives up and leaves. Because bumblebees tend to visit equal numbers of yellow and purple flowers, orchids with the less common of the two colors receive more visits per plant. If more pollinator visits translates into higher

Figure 6.21 Frequency-dependent selection in Elderflower orchids (a) A mixed population. Some plants have yellow flowers, others have purple flowers. (b) Through male function, yellow flowers have higher fitness than purple flowers when yellow is rare, but lower fitness than purple flowers when yellow is common. (c) Through female function, yellow flowers have higher fitness than purple flowers when yellow is rare, but lower fitness than purple flowers when yellow is common. The dashed vertical lines show the predicted frequency of yellow flowers, which matches the frequency in natural populations. From Gigord et al. (2001).

reproductive success, then the rare-color advantage could explain why both colors persist. Selection by bumblebees favors yellow until it becomes too common, then it favors purple. This is an example of **frequency-dependent selection**.

To test their hypothesis Gigord and colleagues collected and potted wild orchids, then placed them in the orchids' natural habitat in 10 experimental arrays of 50 plants each. The frequency of yellow flowers varied among arrays, with two arrays at each of five frequencies: 0.1, 0.3, 0.5, 0.7, and 0.9. The researchers monitored the orchids for removal of their own pollinia (pollen-bearing structures), for deposition of pollinia from other individuals, and for fruit set. From their data, Gigord and colleagues estimated the reproductive advantage of yellow flowers, relative to purple, via both male and female function.

Selection can also maintain two alleles in a population if each allele is advantageous when it is rare.

The resulting estimates of relative reproductive success, plotted as a function of the frequency of yellow flowers, appear in Figure 6.21b and (c). Consistent with the researchers' hypothesis, yellow-flowered orchids enjoyed higher reproductive success than purple-flowered plants when yellow was rare and suffered lower reproductive success when yellow was common.

Gigord and colleagues calculated the relative reproductive success of yellow orchids as

$$RRS_y = \frac{2(RS_y)}{RS_y + RS_p}$$

where RS_y and RS_p are the absolute reproductive success of yellow and purple orchids. The relationship between relative reproductive success via male function and the frequency of yellow flowers is given by the best-fit line in Figure 6.21b. It is

$$RRS_y = -0.66F_y + 1.452$$

where F_y is the frequency of yellow flowers.

We can incorporate this relationship into a population–genetic model. We might imagine, for example, that flower color is determined by two alleles at a single locus, with yellow recessive to purple. We set the starting frequency of the yellow allele to an arbitrary value. We assign fitnesses to the three genotypes as we have before, except that the fitnesses change each generation with the frequency of yellow flowers. When we use a computer to track the evolution of our model population, we discover that the frequency of the yellow allele moves rapidly to equilibrium at an intermediate value. This value is precisely the allele frequency at which yellow flowers have a relative fitness of 1. We get the same result if we imagine that yellow flowers are dominant. Again the equilibrium value for the yellow allele is the frequency at which yellow and purple flowers have equal fitness.

The dashed vertical lines in Figure 6.21b and (c) indicate the predicted equilibrium frequencies Gigord and colleagues calculated for each of their fitness measures. The predictions are 61%, 69%, and 72% yellow flowers. The researchers surveyed 20 natural populations in the region where they had placed their experimental arrays. The actual frequency of yellow flowers, 69 ± 3%, is in good agreement with the predicted frequency. Our model has passed its fourth test.

Gigord et al.'s study of Elderflower orchids demonstrates that frequency-dependent selection can have an effect similar to heterozygote superiority. Both patterns of selection can maintain genetic diversity in populations.

Compulsory Sterilization

We can use population genetic models to evaluate whether eugenic sterilization could have accomplished the aims of its proponents, had their assumptions about the heritability of traits been correct. The answer depends on the frequency of the alleles in question, and on the criteria for success.

The theory of population genetics, despite its simplifying assumptions, allows us to predict the course of evolution. Our four tests show that the model we have developed works remarkably well. So long as we know the starting allele frequencies and genotype fitnesses, the model can predict how allele frequencies will change, under a variety of selection schemes, many generations into the future. The requisite knowledge is easiest to get, of course, for experimental populations living under controlled conditions in the lab. But Gigord et al.'s study of Elderflower orchids shows that the model can even make fairly accurate predictions in natural populations. Given its success in the four tests, it is reasonable to use our model to consider the evolutionary consequences of a eugenic sterilization program. The proponents of eugenic sterilization sought to reduce the fitness of particular genotypes to zero and thereby to reduce the frequency of alleles responsible for undesirable phenotypes. Would their plan have worked?

The phenotype that caught the eugenicists' attention perhaps more than any other was feeblemindedness. The Royal College of Physicians in England defined a feebleminded individual as "One who is capable of earning his living under favorable circumstances, but is incapable from mental defect existing from birth or from an early age (a) of competing on equal terms with his normal fellows or (b) of managing himself and his affairs with ordinary prudence" (see Goddard 1914). Evidence presented in 1914 by Henry H. Goddard, who was the director of research at the Training School for Feebleminded Girls and Boys in Vineland, New Jersey, convinced many eugenicists that strength of mind behaved like a simple Mendelian trait (see Paul and Spencer 1995). Normalmindedness was believed to be dominant and feeblemindedness recessive.

A recessive genetic disease is not a promising target for a program that would eliminate it by sterilizing affected individuals. As Figures 6.16a and 6.17a show, rare recessive alleles decline in frequency slowly, even under strong selection. On the other hand, eugenicists did not believe that feeblemindedness was especially rare (Paul and Spencer 1995). Indeed, they believed that feeblemindedness was alarmingly common and increasing in frequency. Edward M. East (1917) estimated the frequency of feeblemindedness at three per thousand. Henry H. Goddard reported a frequency of 2% among New York schoolchildren. Tests of American soldiers during World War I suggested a frequency of nearly 50% among white draftees.

We will assume a frequency for feeblemindedness of 1% and reproduce a calculation reported by R. C. Punnett (1917) and revisited by R. A. Fisher (1924). Let f be the purported allele for feeblemindedness, with frequency q. If 1% of the population has genotype ff, then, by the Hardy–Weinberg equilibrium principle, the initial frequency of f is

$$q = \sqrt{0.01} = 0.1$$

If all affected individuals are sterilized, then the fitness of genotype ff is zero (or, equivalently, the selection coefficient for genotype ff is 1). Using the equation developed in Box 6.7, we can calculate the value of q in successive generations, and from q we can calculate the frequency of genotype ff.

The result appears in Figure 6.22. Over 10 generations, about 250 years, the frequency of affected individuals declines from 0.01 to 0.0025.

Whether a geneticist saw this calculation as encouraging or discouraging depended on whether he or she saw the glass as partially empty or partially full. Some looked at the numbers, saw that it would take a very long time to completely eliminate feeblemindedness and argued that compulsory sterilization was such a hopelessly slow solution that it was not worth the effort. Others, such as Fisher, dismissed this argument as "anti-eugenic propaganda." Fisher noted that after just one generation, the frequency of affected individuals would drop from 100 per 10,000 to 82.6 per 10,000. "In a single generation," he wrote, "the load of public expenditure and personal misery caused by feeblemindedness . . . would be reduced by over 17 percent." Fisher also noted that most copies of the allele for feeblemindedness are present in heterozygous carriers rather than affected individuals. Along with East, Punnett, and others, Fisher called for research into methods for identifying carriers.

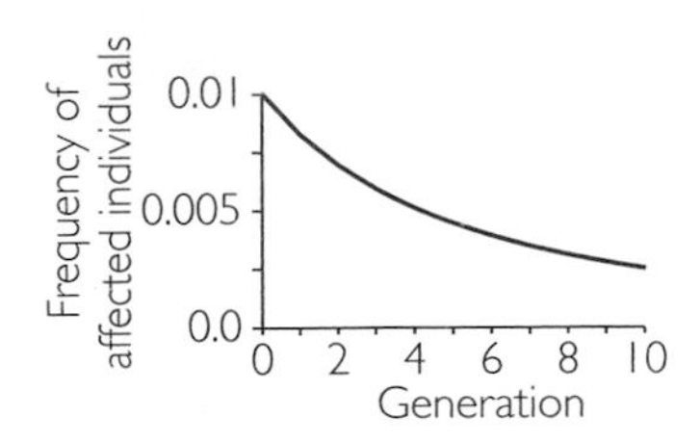

Figure 6.22 Predicted change in the frequency of homozygotes for a putative allele for feeblemindedness under a eugenic sterilization program that prevents homozygous recessive individuals from reproducing

While their evolutionary logic was sound, the eugenicists' models were built on dubious genetic hypotheses. It is not entirely fair to use modern standards to criticize Goddard's research on the genetics of feeblemindedness. Mendelian genetics was in its infancy. Nonetheless, looking back after nearly a century, we can see that Goddard's evidence was deeply flawed. We will consider three problems.

First, the individuals whose case studies he reports are a highly diverse group. Some have Down syndrome; some have other forms of mental retardation. At least one is deaf and appears to be the victim of a woefully inadequate education. Some appear to have been deposited at Goddard's training school by widowed fathers who felt that children from a prior marriage were a liability in finding a new wife. Some may just have behaved differently than the directors of the school thought they should. Concluding the first case report in his book, Goddard writes of a 16-year-old who has been at the school for seven years:

> "Gertrude is a good example of that type of girl who, loose in the world, makes so much trouble. Her beauty and attractiveness and relatively high [intelligence] would enable her to pass almost anywhere as a normal child and yet she is entirely incapable of controlling herself and would be led astray most easily. It is fortunate for society that she is cared for as she is."

Second, Goddard's methods for collecting data were prone to distortion. He sent caseworkers to collect pedigrees from the families of the students at the training school. The caseworkers relied on hearsay and subjective judgments to assess the strength of mind of family members—many of whom were long since deceased.

Third, Goddard's method of analysis stacked the cards in favor of his conclusion. He first separated his 327 cases into a variety of categories: definitely hereditary cases; probably hereditary cases; cases caused by accidents; and cases with no assignable cause. He apparently placed cases in his "definitely hereditary" group only when they had siblings, recent ancestors, or other close kin also classified as feebleminded. When he later analyzed the data to determine whether feeblemindedness was a Mendelian trait, Goddard analyzed only the data from his "definitely hereditary" group. Given how he had filtered the data ahead of time, it is not too surprising that he concluded that feeblemindedness is Mendelian.

Although feeblemindedness is not among them, many genetic diseases are now known to be inherited as simple Mendelian traits. Yet eugenic sterilization has few advocates. One reason is that most serious genetic diseases are recessive and very rare; sterilization of affected individuals would have little impact on the frequency at which new affected individuals are born. A second reason is that mainstream attitudes about reproductive rights have changed to favor individual autonomy over societal mandates (Paul and Spencer 1995). A third reason is that, as we will discuss in the next section, there is a growing list of disease alleles that are suspected or known to be maintained in populations by heterozygote superiority. It would be futile and possibly ill-advised to try to reduce the frequency of such alleles by preventing affected individuals from reproducing.

6.4 Mutation

Cystic fibrosis is among the most common serious genetic diseases among people of European ancestry, affecting approximately 1 newborn in 2,500. Cystic fibrosis is inherited as an autosomal recessive trait (see Chapter 5). Affected individuals suffer chronic infections with the bacterium *Pseudomonas aeruginosa* and ultimately sustain severe lung damage (Pier et al. 1997). At present, most individuals with cystic fibrosis live into their thirties or forties (Elias et al. 1992), but until recently few survived to reproductive age. In spite of the fact that cystic fibrosis was lethal for most of human history, in some populations as many as 4% of individuals are carriers. How can alleles that cause a lethal genetic disease remain this common?

Our consideration of heterozygote superiority in the previous section hinted at one possible answer. Another potential answer is that new disease alleles are constantly introduced into populations by mutation. Before we can evaluate the relative merits of these two hypotheses for explaining the persistence of any particular disease allele, we need to discuss mutation in more detail.

In Chapter 5, we presented mutation as the source of all new alleles and genes. In its capacity as the ultimate source of all genetic variation, mutation provides the raw material for evolution. Here, we consider the importance of mutation as a mechanism of evolution. How rapidly does mutation cause allele frequencies to change over time? How strongly does mutation affect the conclusions of the Hardy–Weinberg analysis?

Adding Mutation to the Hardy–Weinberg Analysis: Mutation as an Evolutionary Mechanism

Second on the list of assumptions for the Hardy–Weinberg equilibrium principle was that there are no mutations. We now explore what happens to allele frequencies when this assumption is violated.

Mutation by itself is generally not a rapid mechanism of evolution. To see why, return to our model population of mice. Imagine a locus with two alleles, A and a, with initial frequencies of 0.9 and 0.1. A is the wild-type allele, and a is a recessive loss-of-function mutation. Furthermore, imagine that copies of A are converted by mutation into new copies of a at the rate of 1 copy per 10,000 per generation. This is a very high mutation rate, but it is within the range of mutation rates known. Back mutations that restore function are much less common than loss-of-function mutations, so we will ignore mutations that convert copies of a into new copies of A. Finally, imagine that all mutations happen while the adults are making gametes to contribute to the gene pool.

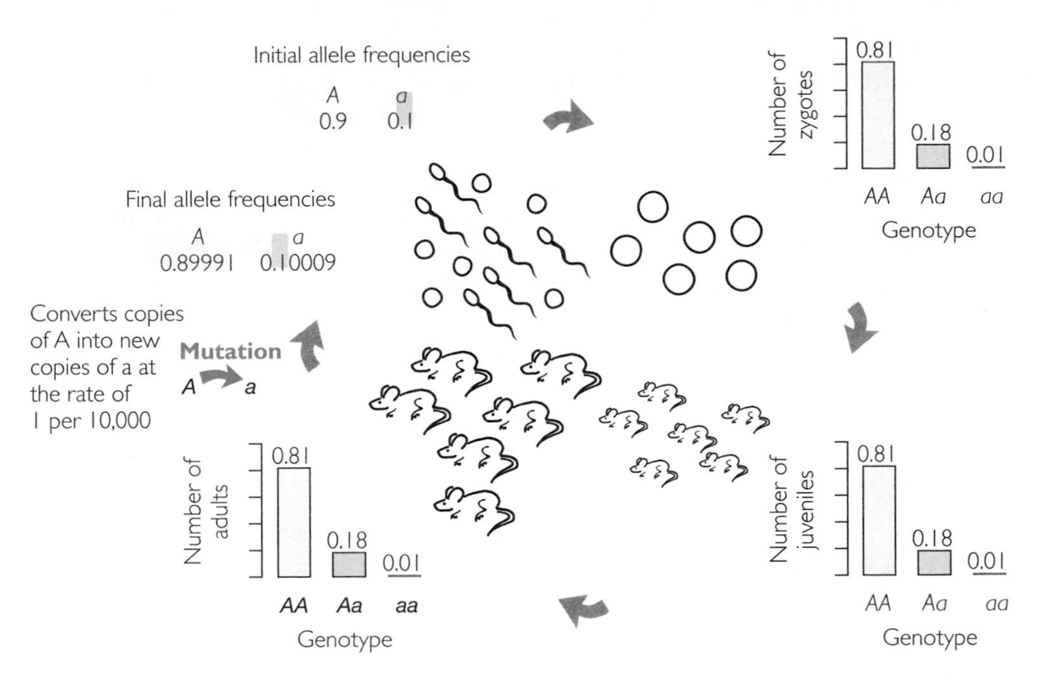

Figure 6.23 **Mutation is a weak mechanism of evolution** In a single generation in our model population, mutation produces virtually no change in allele and genotype frequencies.

Figure 6.23 follows the allele and genotype frequencies through one turn of the life cycle. Among the zygotes, juveniles, and adults the genotypes are in the Hardy–Weinberg proportions:

AA	Aa	aa
0.81	0.18	0.01

Now the adults make gametes. Were it not for mutation, the allele frequencies in the new gene pool would be

A	a
0.9	0.1

But mutation converts 1 of every 10,000 copies of allele A into a new copy of allele a. The frequency of A after mutation is given by the frequency before mutation minus the fraction lost to mutation; the frequency of a after mutation is given by the frequency before mutation plus the fraction gained by mutation. That is,

A	a
$0.9 - (0.0001)(0.9)$	$0.1 + (0.0001)(0.9)$
$= 0.89991$	$= 0.10009$

The new allele frequencies are almost identical to the old allele frequencies. As a mechanism of evolution, mutation has had virtually no effect.

But virtually no effect is not the same as exactly no effect. Could mutation of A into a, occurring at the rate of 1 copy per 10,000 every generation for many generations, eventually result in an appreciable change in allele frequencies? The graph in Figure 6.24 provides the answer (see Box 6.9 for a mathematical

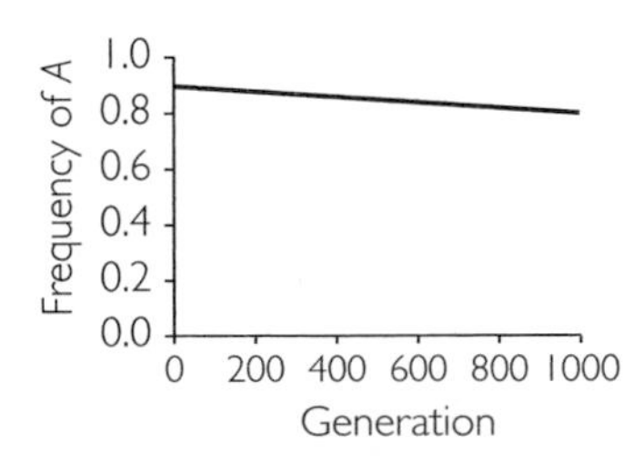

Figure 6.24 **Over very long periods of time, mutation can eventually produce appreciable changes in allele frequency**

Box 6.9 A mathematical treatment of mutation as an evolutionary mechanism

Imagine a single locus with two alleles: a wild-type allele, A, and a recessive loss-of-function mutation, a. Let μ be the rate of mutation from A to a. Assume that the rate of back mutation from a to A is negligible. If the frequency of A in this generation is p, then its frequency in the next generation is given by

$$p' = p - \mu p$$

If the frequency of a in this generation is q, then its frequency in the next generation is given by

$$q' = q + \mu p$$

The change in p from one generation to the next is

$$\Delta p = p' - p$$

which simplifies to

$$\Delta p = -\mu p$$

After n generations, the frequency of A is approximately

$$p_n = p_0 e^{-\mu n}$$

where p_n is the frequency of A in generation n, p_0 is the frequency of A in generation 0, and e is the base of the natural logarithms.

Readers familiar with calculus can derive the last equation as follows. First, assume that a single generation is an infinitesimal amount of time, so that we can rewrite the equation $\Delta p = -\mu p$ as

$$\frac{dp}{dg} = -\mu p$$

Now divide both sides by p, and multiply both sides by dg to get

$$\left(\frac{1}{p}\right)dp = -\mu dg$$

Finally, integrate the left side from frequency p_0 to p_n and the right side from generation 0 to n; then solve for p_n.

Hardy–Weinberg analysis shows that mutation is a weak mechanism of evolution.

treatment). After 1,000 generations, the frequency of allele A in our model population will be about 0.81. Mutation can cause substantial change in allele frequencies, but it does so slowly.

As mutation rates go, the value we used in our model, 1 per 10,000 per generation, is very high. For most genes, mutation is an even less efficient mechanism of allele frequency change.

Mutation and Selection

Although mutation alone usually does not cause appreciable changes in allele frequencies, this does not mean that mutation is unimportant in evolution. In combination with selection, mutation becomes a crucial piece of the evolutionary process. This point is demonstrated by an experiment conducted in Richard Lenski's lab (Lenski and Travisano 1994; Elena et al. 1996). Lenski and coworkers studied the evolution of a strain of *Escherichia coli* that is incapable of recombination (here, recombination means conjugation and exchange of DNA among cells). For *E. coli* populations of this strain, mutation is the only source of genetic variation. The researchers started several replicate populations with single cells placed in a glucose-limited, minimal salts medium—a demanding environment for these bacteria. After allowing each culture to grow to about 5×10^8 cells, Lenski and colleagues removed an aliquot (containing approximately 5 million cells) and transferred it to fresh medium. The researchers performed these transfers daily for 1,500 days, or approximately 10,000 generations.

At intervals throughout the experiment, the researchers froze samples of the transferred cells for later analysis. Because *E. coli* are preserved but not killed by

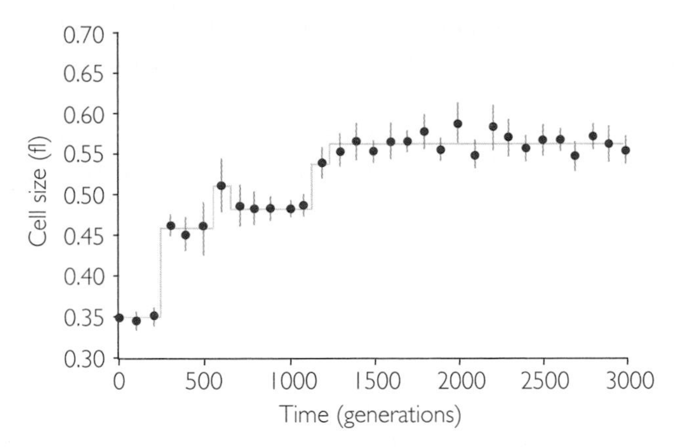

Figure 6.25 Change over time in cell size of an experimental *E. coli* population Each point on the plot represents the average cell size in 10 replicate assays of the population. The vertical lines are error bars; 95% of the observations fall within the range indicated by the bars. Reprinted with permission from Elena et al. (1996).

freezing, Lenski and colleagues could take ancestors out of the freezer and grow them in a culture flask with an equivalent number of cells from descendant populations. These experiments allowed the team to directly measure the relative fitness of ancestral and descendant populations, as the growth rate of each under competition. In addition to monitoring changes in fitness over time in this way, the Lenski team measured cell size.

During the course of the study, both fitness and cell size increased dramatically in response to natural selection. The key point for our purposes is that within any given population these increases occurred in jumps (Figure 6.25). The steplike pattern resulted from a simple process: the occurrence of beneficial mutations that swept rapidly through the population. Each new mutation enabled the bacteria that carried it to divide at a faster rate. The frequency of the mutants quickly increased as they out-reproduced the other members of the population. Eventually, each new mutation became fixed in the population. The time from the appearance of each mutation to its fixation in the population was so short that we cannot see it in the figure. Most of the beneficial mutations caused larger cell size. Thus the plot of cell size over time also shows abrupt jumps. Between the appearance of one beneficial mutation and the next, the population stood still. Why larger cells were beneficial in the nutrient-poor laboratory environment is a focus of ongoing research (Mongold and Lenski, 1996; Lenski et al. 1998).

Research with bacteria illustrates that while mutation itself is only a weak mechanism of evolution, it nonetheless supplies the raw material on which natural selection acts.

The experiment by Lenski and colleagues reinforces one of the messages of Chapter 5. Without mutation, evolution would eventually grind to a halt. Mutation is the ultimate source of genetic variation.

Mutation-Selection Balance

Unlike the minority of mutations that led to increased cell size and higher fitness in Lenski et al.'s *E. coli* populations, most mutations are deleterious. Selection acts to eliminate such mutations from populations. Deleterious alleles persist, however, because they are continually created anew. When the rate at which copies of a deleterious allele are being eliminated by selection is exactly equal to the rate at which new copies are being created by mutation, the frequency of the allele is at equilibrium. This situation is called **mutation-selection balance**.

What is the frequency of the deleterious allele at equilibrium? If the allele is recessive, its equilibrium frequency, $\hat{q}$, is given by

$$\hat{q} = \sqrt{\frac{\mu}{s}}$$

where μ is the mutation rate, and s, the selection coefficient, is a number between 0 and 1 expressing the strength of selection against the allele (see Box 6.10 for a derivation). This equation captures with economy what intuition tells us about mutation-selection balance. If the selection coefficient is small (the allele is only mildly deleterious) and the mutation rate is high, then the equilibrium frequency of the allele will be relatively high. If the selection coefficient is large (the allele is highly deleterious) and the mutation rate is low, then the equilibrium frequency of the allele will be low.

Research by Brunhilde Wirth and colleagues (1997) on patients with spinal muscular atrophy provides an example. Spinal muscular atrophy is a neurodegenerative disease characterized by weakness and wasting of the muscles that control voluntary movement. It is caused by deletions in a locus on chromosome 5 called the telomeric survival motor neuron gene (*telSMN*). In some cases, the disease may be exacerbated by additional mutations in a nearby gene. Spinal muscular atrophy is, after cystic fibrosis, the second most common lethal autosomal recessive disease in Caucasians (McKusick et al. 1999).

Collectively, the loss-of-function alleles of *telSMN* have a frequency of about 0.01 in the Caucasian population. Wirth and colleagues estimate that the selection coefficient is about 0.9. With such strong selection against them, we would expect that disease-causing alleles would slowly but inexorably disappear from the population. How, then, do they persist at a frequency of 1 in 100?

At the same time selection removes deleterious alleles from a population, mutation constantly supplies new copies. In some cases, this balance between mutation and selection may explain the persistence of deleterious alleles in populations.

One possibility is that the disease alleles are being kept in the population by a balance between mutation and selection. If we substitute the allele frequency and selection coefficient for $\hat{q}$ and s in the equation above, and then solve for μ, we find that this scenario requires a mutation rate of about 9.0×10^{-5} ($=0.9 \times 10^{-4}$) mutations per *telSMN* allele per generation. Wirth et al. analyzed the chromosomes of 340 individuals with spinal muscular atrophy, and the chromosomes of their parents and other family members. They found that 7 of the 340 affected individuals carried a new mutation not present in either parent. These numbers allowed the scientists to estimate directly the mutation rate at the *telSMN* locus (see Box 6.11). Their estimate is 1.1×10^{-4}. This directly measured mutation rate is in good agreement with the rate predicted under the hypothesis of a mutation-selection balance. Wirth et al. conclude that mutation-selection balance provides a sufficient explanation for the persistence of spinal muscular atrophy alleles.

Are the Alleles That Cause Cystic Fibrosis Maintained by a Balance between Mutation and Selection?

Cystic fibrosis is caused by recessive loss-of-function mutations in a locus on chromosome 7 that encodes a protein called the cystic fibrosis transmembrane

Box 6.10 Allele frequencies under mutation-selection balance

Here we derive equations for predicting the equilibrium frequencies of deleterious alleles under mutation-selection balance. Imagine a single locus with two alleles, A_1 and A_2, with frequencies p and q. A_1 is the wild type; A_2 is deleterious. Let μ be the rate at which copies of A_1 are converted into copies of A_2 by mutation. Assume that the rate of back mutation is negligible.

Selection will continuously remove copies of A_2 from the population, while mutation will continuously create new copies. We want to calculate the frequency of A_2 at which these processes cancel each other. Following Felsenstein (1997), we will perform our calculation in a roundabout way. We will develop an equation in terms of p that describes mutation-selection balance for allele A_1. Then we will solve the equation for q to get the equilibrium frequency of A_2. This approach may seem perverse, but it greatly simplifies the algebra.

Mutation-selection balance for a deleterious recessive allele

Imagine that A_2 is a deleterious recessive allele, such that the genotype fitnesses are given by

w_{11}	w_{12}	w_{22}
1	1	$1 - s$

where the selection coefficient s gives the strength of selection against A_2.

First, we will write an equation for p^*, the frequency of allele A_1 after selection has operated, but before mutations occur. From Box 6.3, this equation is

$$p^* = \frac{p^2 w_{11} + pqw_{12}}{p^2 w_{11} + 2pqw_{12} + q^2 w_{22}}$$

Substituting the fitnesses from the table above, and $(1 - p)$ for q, then simplifying gives

$$p^* = \frac{p}{1 - s(1 - p)^2}$$

Next we will write an expression for p', the frequency of allele A_1 after mutations occur. These mutations convert a fraction μ of the copies of A_1 into copies of A_2, leaving behind a fraction $(1 - \mu)$. Thus

$$p' = (1 - \mu)p^* = \frac{(1 - \mu)p}{1 - s(1 - p)^2}$$

Finally, when mutation and selection are in balance, p' is equal to p, the frequency of allele A_1 that we started with:

$$\frac{(1 - \mu)p}{1 - s(1 - p)^2} = p$$

This simplifies to

$$(1 - p)^2 = \frac{\mu}{s}$$

Substituting q for $(1 - p)$ and solving for q yields an equation for $\hat{q}$, the equilibrium frequency of allele A_2 under mutation-selection balance:

$$\hat{q} = \sqrt{\frac{\mu}{s}}$$

If A_2 is a lethal recessive, then $s = 1$, and the equilibrium frequency of A_2 is equal to the square root of the mutation rate.

Mutation-selection balance for a lethal dominant allele

Imagine that A_2 is a lethal dominant allele, such that the genotype fitnesses are given by

$\mathbf{w}_{11}$	$\mathbf{w}_{12}$	$\mathbf{w}_{22}$
1	0	0

Now the expression for p^* simplifies to

$$p^* = 1$$

which makes sense because, by definition, selection removes all copies of the lethal dominant A_2 from the population. Now the expression for p' is

$$p' = 1 - \mu$$

and the equilibrium condition is

$$1 - \mu = p$$

Substituting $(1 - q)$ for p and simplifying gives

$$\hat{q} = \mu$$

In other words, the equilibrium frequency of A_2 is equal to the mutation rate.

Box 6.11 Estimating mutation rates for recessive alleles

Here, we present the method used by Brunhilde Wirth and colleagues (1997) to estimate mutation rates for recessive alleles. The key information required is the fraction of affected individuals that carry a brand-new mutant allele. With modern molecular techniques, this fraction can be obtained by direct examination of the chromosomes of affected individuals and their relatives.

Let q be the frequency of recessive loss-of-function allele a. Ignoring the extremely rare individuals with two new mutant copies, there are two ways to be born with genotype aa:

1. An individual can be the offspring of two carriers. The probability of this outcome for a given birth is the product of: (a) the probability that an offspring of two carriers will be affected; (b) the probability that the mother is a carrier; and (c) the probability that the father is a carrier. This probability is given by

$$\left[\frac{1}{4}\right] \times [2q(1-q)] \times [2q(1-q)]$$

2. An individual can be the offspring of one carrier and one homozygous dominant parent and can receive allele a from the affected parent and a new mutant copy of a from the unaffected parent. The probability of this outcome for a given birth is the product of (a) the probability that an offspring of one carrier will receive that carrier's mutant allele; (b) the probability that the mother is a carrier; (c) the probability that the father is a homozygous dominant; and (d) the mutation rate *plus* the same probability for the scenario in which the father is the carrier and the mother is the homozygous dominant:

$$\left\{\left[\frac{1}{2}\right] \times [2q(1-q)] \times [(1-q)^2] \times [\mu]\right\}$$
$$+ \left\{\left[\frac{1}{2}\right] \times [2q(1-q)] \times [(1-q)^2] \times [\mu]\right\}$$
$$= [2q(1-q)] \times [(1-q)^2] \times [\mu]$$

With these probabilities, we can write an expression for r, the fraction of affected individuals that carry one new mutant allele. This is the second probability divided by the sum of the second probability and the first. Simplified just a bit, we have

$$r = \frac{2q(1-q)(1-q)^2\mu}{2q(1-q)(1-q)^2\mu + q(1-q)q(1-q)}$$

Simplifying further yields

$$r = \frac{2(1-q)\mu}{2(1-q)\mu + q}$$

Finally, assume that q is small, so that $(1-q)$ is approximately equal to one. This assumption gives

$$r = \frac{2\mu}{2\mu + q}$$

which can be solved for μ:

$$\mu = \frac{rq}{2 - 2r}$$

The mutation rate for spinal muscular atrophy

In Caucasian populations, spinal muscular atrophy affects about 1 infant in 10,000, implying that the frequency of the mutant allele is

$$q = \sqrt{0.0001} = 0.01$$

Wirth and colleagues examined the chromosomes of 340 affected patients and their family members. The researchers discovered that 7 of their patients had a new mutant allele not present in either parent. Thus,

$$r = \frac{7}{340} = 0.021$$

Substituting these values for q and r into the equation for μ gives the estimate

$$\mu = \frac{(0.021)(0.01)}{2 - 2(0.021)} = 0.00011$$

The mutation rate for cystic fibrosis

In Caucasian populations, cystic fibrosis affects about 1 infant in 2,500. Wirth and colleagues cite data from other authors to establish that only 2 of about 30,000 cystic fibrosis patients studied proved to have a new mutant allele not present in either parent. These figures give an estimated mutation rate of

$$\mu = 6.7 \times 10^{-7}$$

conductance regulator (CFTR). CFTR is a cell surface protein expressed in the mucus membrane lining the intestines and lungs. Gerald Pier and colleagues (1997) demonstrated that one of CFTR's key functions is to enable cells of the lung lining to ingest and destroy *Pseudomonas aeruginosa* bacteria. These bacteria cause chronic lung infections in individuals with cystic fibrosis, eventually leading to severe lung damage (Figure 6.26). Selection against the alleles that cause cystic fibrosis appears to be strong. Until recently, few affected individuals survived to reproductive age; those that do survive are often infertile. And yet the alleles that cause cystic fibrosis have a collective frequency of approximately 0.02 among people of European ancestry.

Could cystic fibrosis alleles be maintained at a frequency of 0.02 by mutation-selection balance? If we assume a selection coefficient of 1 and use the equation derived in Box 6.10, the mutation rate creating new disease alleles would have to be 4×10^{-4}. The actual mutation rate for cystic fibrosis alleles appears to be considerably lower than that: approximately 6.7×10^{-7} (see Box 6.11). We can conclude that a steady supply of new mutations cannot, by itself, explain the maintenance of cystic fibrosis alleles at a frequency of 0.02.

In other cases, the frequency of a deleterious allele may be too high to explain by mutation-selection balance. This may be a clue that heterozygotes have superior fitness.

Our discussion of heterozygote superiority suggests an alternative explanation (Figure 6.18 and Box 6.8). Perhaps the fitness cost suffered by cystic fibrosis alleles when they are in homozygotes is balanced by a fitness advantage they enjoy when they are in heterozygotes.

Gerald Pier and colleagues (1998) hypothesized that cystic fibrosis heterozygotes might be resistant to typhoid fever and therefore have superior fitness. Typhoid fever is caused by *Salmonella typhi* bacteria (also known as *Salmonella enterica* serovar *typhi*). The bacteria initiate an infection by crossing the layer of epithelial cells that line the gut. Pier and colleagues suggested that *S. typhi* bacteria infiltrate the gut by exploiting the CFTR protein as a point of entry. If so, then heterozygotes, which have fewer copies of CFTR on the surface of their cells, should be less vulnerable to infiltration.

Pier and colleagues tested their hypothesis by constructing mouse cells with three different CFTR genotypes: homozygous wild-type cells; heterozygotes with one functional CFTR allele and one allele containing the most common human cystic fibrosis mutation, a single-codon deletion called *ΔF508*; and homozygous *ΔF508* cells. The researchers exposed these cells to *S. typhi*, then measured the number of bacteria that got inside cells of each genotype. The results

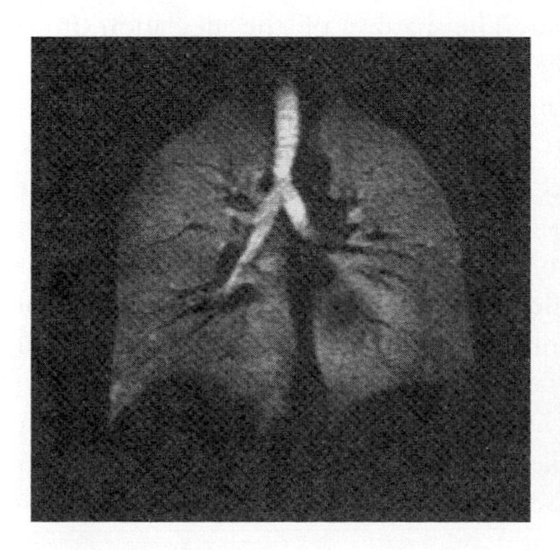

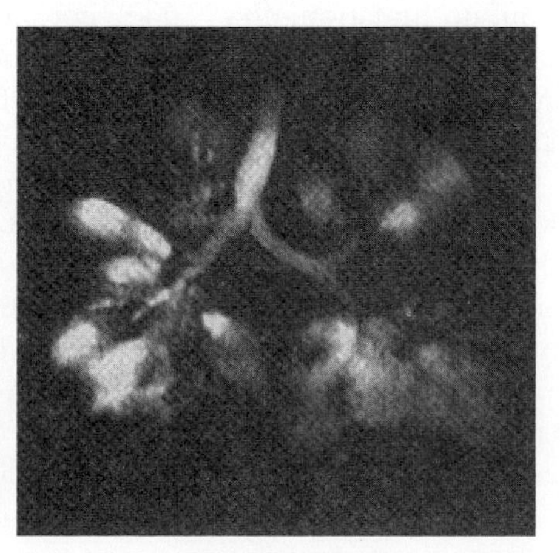

Figure 6.26 A normal lung (left) versus a lung ravaged by the bacterial infections that accompany cystic fibrosis (right)

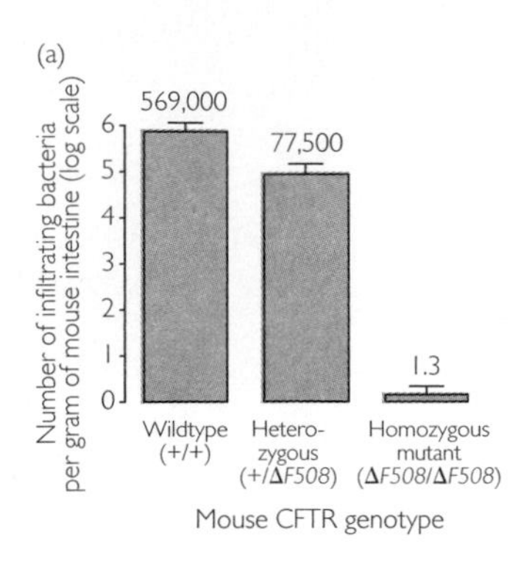

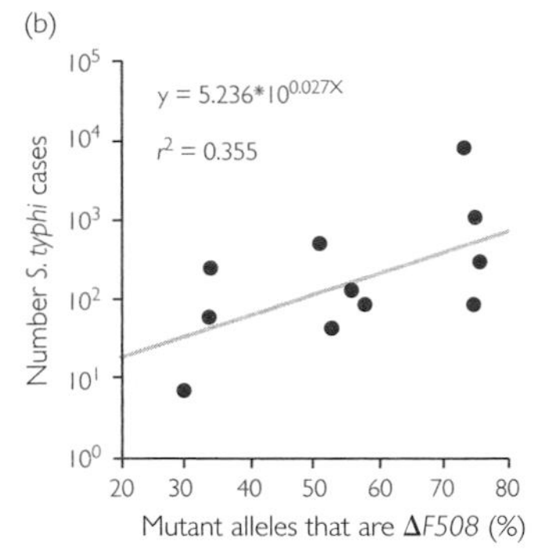

Figure 6.27 Heterozygotes for the *ΔF508* allele are resistant to typhoid fever (a) Cultured mouse cells heterozygous for cystic fibrosis show substantial resistance to infiltration by the bacteria that cause typhoid fever. Cells homozygous for *ΔF508*, the most common human disease allele, are almost totally resistant. From Pier et al. (1998). (b) Data from 11 European countries suggest that *S. typhi* selects for carriers of the *ΔF508* allele. The frequency of the allele, among cystic fibrosis mutations, in the generation following a typhoid fever outbreak increases with the severity of the outbrreak. From Lyczak et al. (2002).

were dramatic (Figure 6.27a). As the researchers predicted, homozygous *ΔF508* cells were almost totally resistant to infiltration by *S. typhi,* while homozygous wild-type cells were highly vulnerable. Heterozygous cells were partially resistant; they accumulated 86% fewer bacteria than did the wild-type cells. These results are consistent with the hypothesis that cystic fibrosis disease alleles are maintained in human populations because heterozygotes have superior fitness during typhoid fever epidemics.

Also consistent with the hypothesis are two more recent discoveries by Pier and coworkers. First, Jeffrey Lyczak and Pier (2002) found that *S. typhi* bacteria manipulate the gut cells of their hosts, causing the cells to display more CFTR protein on their membranes and easing the bacteria's entry. This helps explain why cells that cannot make CFTR are resistant to invasion. Second, Lyczak, Carolyn Cannon, and Pier (2002), using data compiled from the literature, found an apparent association across 11 European countries between the severity of typhoid fever outbreaks and the frequency a generation later, among CFTR mutations, of the *ΔF508* allele (Figure 6.27b).

Pier et al.'s research serves as another example in which an evolutionary analysis has proved valuable in biomedical research.

Summary

Population genetics represents a synthesis of Mendelian genetics and Darwinian evolution and is concerned with the mechanisms that cause allele frequencies to change from one generation to the next. The Hardy–Weinberg equilibrium principle is a null model that provides the conceptual framework for population genetics. It shows that under simple assumptions—no selection, no mutation, no migration, no genetic drift, and random mating—allele frequencies do not change. Furthermore, genotype frequencies can be calculated from allele frequencies.

When any one of the first four assumptions is violated, allele frequencies may change across generations. Selection, mutation, migration, and genetic drift are thus the four mechanisms of evolution. Nonrandom mating does not cause allele frequencies to change and is thus not a mechanism of evolution. It can, however, alter genotype frequencies and thereby affect the course of evolution.

Population geneticists can measure allele and genotype frequencies in real populations. Thus, biologists can test whether allele frequencies are stable across generations and whether the genotype frequencies conform to Hardy–Weinberg expectations. If either of the conclusions of the Hardy–Weinberg analysis is violated, it means that one or more of the assumptions does not hold. The nature of the deviation from Hardy–Weinberg expectations does not, by itself, identify the faulty assumption. We can, however, often infer which mechanisms of evolution are at work based on other characteristics of the populations under study.

Selection occurs when individuals with different genotypes differ in their success at getting copies of their genes into future generations. It is a powerful mechanism of evolution. Selection can cause allele frequencies to change from one generation to the next and can take genotype frequencies away from Hardy–Weinberg equilibrium. Some patterns of selection tend to drive alleles to fixation or to loss; other patterns of selection serve to maintain allelic diversity in populations. Population genetic theory allows us to make accurate predictions about both the direction and the rate of evolution under a variety of patterns of selection.

Alone, mutation is a weak evolutionary mechanism. Mutation does, however, provide the genetic variation that is the raw material for evolution. In some cases a steady supply of new mutant alleles can counterbalance selection against those same alleles and thereby serve to hold allele frequencies at equilibrium.

Questions

1. List the five conditions that must be true for a population to be in Hardy–Weinberg equilibrium. Why is it useful to know the conditions that will prevent evolution from occurring? For each of the five, specify whether violation of that assumption will result in changes in genotype frequencies, allele frequencies, or both.

2. Why was it important that Hardy used variables in his mathematical treatment of allele frequencies, instead of working several more examples with specific allele frequencies?

3. Name the phenomenon being described in each (hypothetical) example, and describe how it is likely to affect allele frequencies in succeeding generations.

a. A beetle species is introduced to an island covered with dark basaltic rock. On this dark background, dark beetles, *TT* or *Tt*, are much more resistant to predation than are light-colored beetles, *tt*. The dark beetles have a large selective advantage. Both alleles are relatively common in the group of beetles released on the new island.

b. Another beetle population, this time consisting of mostly light beetles and just a few dark beetles, is introduced onto a different island with a mixed substrate of light sand, vegetation, and black basalt. On this island, dark beetles have only a small selective advantage.

c. A coral reef fish has two genetically determined types of male, one of which is much smaller and "sneaks" into larger males' nests to fertilize their females' eggs. When small males are rare, they have a selective advantage over large males. However, if there are too many small males, large males switch to a more aggressive strategy of nest defense, and small males lose their advantage.

d. In a tropical plant, *CC* and *Cc* plants have red flowers and *cc* plants have yellow flowers. However, *Cc* plants have defective flower development and produce very few flowers.

e. In a species of bird, individuals with genotype *MM* are susceptible to avian malaria, *Mm* birds are resistant to avian malaria, and *mm* birds are resistant to avian malaria, but the *mm* birds are also vulnerable to avian pox.

4. In Mead et al.'s study of the Foré, why was it important that they studied the group of younger women? Can you think of any other possible explanations for their data?

5. Black color in horses is governed primarily by a recessive allele at the A locus. *AA* and *Aa* horses are nonblack colors such as bay, while *aa* horses are black all over. (Other loci can override the effect of the A locus, but we will ignore that complication.) In the internet group "rec.equestrian", one person asked why there are relatively few black horses of the Arabian breed. One response was, "Black is a rare color because it is recessive. More Arabians are bay or gray because those colors are dominant." What is wrong with this explanation? (Assume that the *A* and *a* alleles are in Hardy–Weinberg equilibrium, which was probably true at the time of this discussion.) Generally, what does the Hardy–Weinberg model show us about the impact that an allele's dominance or recessiveness has on its frequency?

6. In humans, the *COL1A1* locus codes for a certain collagen protein found in bone. The normal allele at this locus is denoted with *S*. A recessive allele *s* is associated with reduced bone mineral density and increased risk of fractures in both *Ss* and *ss* women. A recent study of 1,778 women showed that 1,194 were *SS*, 526 were *Ss*, and 58 were *ss* (Uitterlinden et al. 1998). Are these two alleles in Hardy–Weinberg equilibrium in this population? How do you know? What information would you need to determine whether the alleles will be in Hardy–Weinberg equilibrium in the next generation?

7. We used Figure 6.13 as an example of how the frequency of an allele (in fruit flies) does not change in unselected (control) populations but does change in response to selection. However, look again at the unselected control lines in Figure 6.13. The frequency of the allele in the two control populations did change a little, moving up and down over time. Which assumption of the Hardy–Weinberg model is most probably being violated? If this experiment were repeated, what change in experimental design would reduce this deviation from Hardy–Weinberg equilibrium?

8. Most animal populations have a 50:50 ratio of males to females. This does not have to be so; it is theoretically possible for parents to produce predominantly male offspring or predominantly female offspring. Imagine a population with a male-biased sex ratio, say, 70% males and 30% females. Which sex will have an easier time finding a mate? As a result, which sex will probably have higher average fitness? Which parents will have higher fitness—those that produce mostly males or those that produce mostly females? Now imagine the same population with a female-biased sex ratio, and answer the same questions. What sort of selection is probably maintaining the 50:50 sex ratio seen in most populations?

9. Discuss how each of the following recent developments may affect the frequency of alleles that cause cystic fibrosis (CF).
   a. Many women with CF now survive long enough to have children. (CF causes problems with reproductive ducts, but many CF women can bear children nonetheless. CF men are usually sterile.)
   b. Typhoid fever in developed nations has declined to very low levels since 1900.
   c. In some populations, couples planning to have children are now routinely screened for the most common CF alleles.
   d. Drug-resistant typhoid fever has recently appeared in several developing nations.

10. Kerstin Johannesson and colleagues (1995) studied two populations of a marine snail living in the intertidal zone on the shore of Ursholmen Island. Each year, the researchers determined the allele frequencies for the enzyme aspartate aminotransferase (don't worry about what this enzyme does). Their data are shown in the graphs in Figure 6.28. The first year of the study was 1987. In 1988, a bloom of toxic algae (pink bands) killed all of the snails in the intertidal zone across the entire island. That is why there are no data for 1988 and 1989. Although the snails living in the intertidal zone were exterminated by the bloom, snails of the same species living in the splash zone just above the intertidal survived unscathed. By 1990, the intertidal zone had been recolonized by splash-zone snails. Your challenge in this question is to develop a coherent explanation for the data in the graphs. In each part, be sure to name the evolutionary mechanism involved (selection, mutation, migration, or drift).
   a. Why was the frequency of the Aat^{120} allele higher in both populations in 1990 than it was in 1987? Name the evolutionary mechanism, and explain.
   b. Why did the allele frequency decline in both populations from 1990 through 1993? Name the evolutionary mechanism, and explain.
   c. Why are the curves traced by the 1990–1993 data for the two populations generally similar but not exactly identical? Name the evolutionary mechanism, and explain.
   d. Predict what would happen to the allele frequencies if we followed these two populations for another 100 years (assuming there are no more toxic algal blooms). Explain your reasoning.

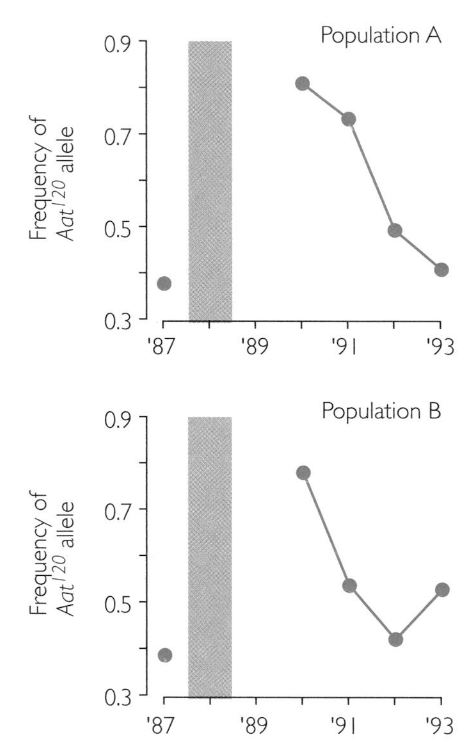

Figure 6.28 **Changes over time in the frequency of an allele in two intertidal populations of a marine snail** From Johannesson et al. (1995).

Exploring the Literature

11. Very often, the first step in any study of genetic variation is evaluation of any deviations from Hardy-Weinberg equilibrium. Read the following paper to explore how careful examination of Hardy-Weinberg equilibrium is necessary for assessing gene-disease associations in humans.

Trikalinos, T.A., G. Salanti, M.J. Khoury, and J.P. Ioannidis. 2006. Impact of violations and deviations in Hardy–Weinberg equilibrium on postulated gene-disease associations. *Am. J. Epidemiol.* 163 (4): 300–309.

12. In the example of the Elderflower orchid, we saw that frequency-dependent selection tends to maintain the presence of both yellow and purple flowers in mixed populations. See the following references for some interesting cases of possible frequency-dependent selection in other species. How plausible do you find each scenario?

Dugatkin, L.A., M. Perlin, J.S. Lucas, and R. Atlas. 2005. Group-beneficial traits, frequency-dependent selection and genotypic diversity: an antibiotic resistance paradigm. *Proceedings of the Royal Society of London B* 272 (1558): 79–83.

Faurie, C. and M. Raymond. 2005. Handedness, homicide and negative frequency-dependent selection. *Proceedings of the Royal Society of London B* 272: 25–28.

Hori, M. 1993. Frequency-dependent natural selection in the handedness of scale-eating cichlid fish. *Science* 260: 216–219.

Sinervo, B., and C. M. Lively. 1996. The rock-paper-scissors game and the evolution of alternative male strategies. *Nature* 380: 240–243.

13. The version of the adaptive landscape presented in Box 6.7 and Box 6.8, in which the landscape is a plot of mean fitness as a function of allele frequency, is actually somewhat different from the original version of the concept that Sewall Wright presented in 1932. Furthermore, there is even a third common interpretation of the adaptive landscape idea. For a discussion of the differences among the three versions, see Chapter 9 in:

Provine, W. B. 1986. *Sewall Wright and Evolutionary Biology.* Chicago: University of Chicago Press.

For Sewall Wright's response to Provine's history, see:

Wright, S. 1988. Surfaces of selective value revisited. *The American Naturalist* 131: 115–123.

Wright's original 1932 paper is reprinted in Chapter 11 of:

Wright, S. 1986. *Evolution: Selected Papers.* William B. Provine, ed. Chicago: University of Chicago Press.

14. If your library has the earliest volumes of the *Journal of Heredity*, read:

Bell, Alexander Graham. 1914. How to improve the race. *Journal of Heredity* 5: 1–7.

Keep in mind that population genetics was in its infancy; Mendelism had yet to be integrated with natural selection. What was accurate and inaccurate in Bell's understanding of the mechanisms of evolution? Would the policy Bell advocated actually have accomplished his aims? Why or why not? If so, would it have done so for the reasons Bell thought it would?

15. For an example in which strong natural selection caused rapid change in allele frequencies in wild populations, see:

Rank, N. E., and E. P. Dahlhoff. 2002. Allele frequency shifts in response to climate change and physiological consequences of allozyme variation in a montane insect. *Evolution* 56: 2278–2289.

16. Patients with cystic fibrosis (CF) are chronically infected with *Pseudomonas aeruginosa* bacteria. Their immune systems are engaged in a constant battle with the bacteria. In addition, they take powerful antibiotics to help keep the bacterial populations under control. Consider the consequences for the bacteria. How would you expect a *P. aeruginosa* population to evolve in the environment found inside a CF patient's lungs? What novel traits would you expect to appear? Make some predictions, then see the following paper (we are withholding the full title to avoid giving too much away):

Oliver, A., R. Cantón, et al. 2000. High frequency of . . . in cystic fibrosis lung infection. *Science* 288: 1251–1253.

17. As discussed in this chapter, the chemokine receptor CCR5 is the major means by which HIV gains entry to human white blood cells. CCR5 is also important in susceptibility to other important diseases. One example is described in the following article. Consider how CCR5's multiple role in different emerging diseases may affect its evolution, and the implications for medical treatments.

Glass, W. G., D. H. McDermott, J. K. Lim, S. Lekhong, S. F. Yu, W. A. Frank, J. Pape, R. C. Cheshier, and P.M. Murphy. 2006. CCR5 deficiency increases risk of symptomatic West Nile virus infection. *Journal of Experimental Medicine* 203(1):35–40.

Citations

Much of the population genetics material in this chapter is modeled after presentations in the following:

Crow, J. F. 1983. *Genetics Notes.* Minneapolis, MN: Burgess Publishing.

Felsenstein, J. 1997. *Theoretical Evolutionary Genetics.* Seattle, WA: ASUW Publishing, University of Washington.

Griffiths, A. J. F., J. H. Miller, D. T. Suzuki, R. C. Lewontin, and W. M. Gelbert. 1993. *An Introduction to Genetic Analysis.* New York: W. H. Freeman.

Templeton, A. R. 1982. Adaptation and the integration of evolutionary forces. In R. Milkman, ed., *Perspectives on Evolution.* Sunderland, MA: Sinauer, 15–31.

Here is the list of all other citations in this chapter:

Aguzzi, A., and M. Polymenidou. 2004. Mammalian prion biology: Once century of evolving concepts. *Cell* 116: 313-327.

Arjona, A., L. Simarro, et al. 2004. Two Creutzfeldt–Jakob disease agents reproduce prion protein-independent identities in cell cultures. *Proceedings of the National Academy of Sciences USA* 101: 8768-8773.

Bastain, F. O. 2005. Spiroplasma as a candidate agent for the transmissible spongiform encephalopathies. *Journal of Neuropathology and Experimental Neurology* 64: 833-838.

Büeler, H.R., A. Aguzzi, et al. 1993. Mice devoid of PrP are resistant to scrapie. Cell 73: 1339–1347.

Castilla, J., P. Saá, et al. 2005. In vitro generation of infectious scrapie prions. Cell 121: 195-206.

Castle, W. E. 1903. The laws of heredity of Galton and Mendel, and some laws governing race improvement by selection. *Proceedings of the American Academy of Arts and Sciences* 39: 223–242.

Cavener, D. R., and M. T. Clegg. 1981. Multigenic response to ethanol in *Drosophila melanogaster. Evolution* 35: 1–10.

Cooke, J. 1999. Once were cannibals. *Sydney Morning Herald* 28 August: 40.

Couzin, J. 2006. The prion protein has a good side? You bet. *Science* 311: 1091.

Dawson, P. S. 1970. Linkage and the elimination of deleterious mutant genes from experimental populations. *Genetica* 41: 147–169.

Dean, M., M. Carrington, et al. 1996. Genetic restriction of HIV-1 infection and progression to AIDS by a deletion allele of the CKR5 structural gene. *Science* 273: 1856–1862.

Dlouhy, S. R., K. Hsiao, et al. 1992. Linkage of the Indiana kindred of Gerstmann-Straussler-Scheinker disease to the prion protein gene. *Nature Genetics* 1: 64-67.

East, E. M. 1917. Hidden feeblemindedness. *Journal of Heredity* 8: 215–217.

Elena, S. F., V. S. Cooper, and R. E. Lenski. 1996. Punctuated evolution caused by selection of rare beneficial mutations. *Science* 272: 1802–1804.

Elias, S., M. M. Kaback, et al. 1992. Statement of The American Society of Human Genetics on cystic fibrosis carrier screening. *American Journal of Human Genetics* 51: 1443–1444.

Fisher, R. A. 1924. The elimination of mental defect. *The Eugenics Review* 16: 114–116.

Foster, G. G., M. J. Whitten, T. Prout, and R. Gill. 1972. Chromosome rearrangements for the control of insect pests. *Science* 176: 875–880.

Gigord, L. D. B., M. R. Macnair, and A. Smithson. 2001. Negative frequency-dependent selection maintains a dramatic flower color polymorphism in the rewardless orchid *Dactylorhiza sambucina* (L.) Soò. *Proceedings of the National Academy of Sciences USA* 98: 6253–6255.

Goddard, H. H. 1914. *Feeblemindedness: Its Causes and Consequences.* New York: The Macmillan Company.

Goldfarb, L. G., P. Brown, et al. 1991. Transmissible familial Creutzfeldt-Jakob disease associated with five, seven, and eight extra octapeptide coding repeats in the PRNP gene. *Proceedings of the National Academy of Sciences USA* 88: 10926-10930.

Goldfarb, L. G., R. B. Petersen, et al. 1992. Fatal familial insomnia and familial Creutzfeldt-Jakob disease: disease phenotype determined by a DNA polymorphism. *Science* 258: 806-808.

Hardy, G. H. 1908. Mendelian proportions in a mixed population. *Science* 28: 49–50.

Hedrick, P. W. 2003. A heterozygote advantage. *Science* 302: 57.

Johannesson, K., B. Johannesson, and U. Lundgren. 1995. Strong natural selection causes microscale allozyme variation in a marine snail. *Proceedings of the National Academy of Sciences USA* 92: 2602–2606.

Kevles, D. J. 1995. *In the Name of Eugenics: Genetics and the Uses of Human Heredity.* Cambridge, MA: Harvard University Press.

Lane, H. 1992. *The Mask of Benevolence: Disabling the Deaf Community.* New York: Vintage Books.

Legname, G., I. V. Baskakov, et al. 2004. Synthetic mammalian prions. *Science* 305: 673-676.

Lenski, R. E., J. A. Mongold, et al. 1998. Evolution of competitive fitness in experimental populations of *E. coli*: What makes one genotype a better competitor than another? *Antonie Van Leeuwenhoek International Journal of General and Molecular Microbiology* 73: 35–47.

Lenski, R. E., and M. Travisano. 1994. Dynamics of adaptation and diversification: A 10,000-generation experiment with bacterial populations. *Proceedings of the National Academy of Sciences USA* 91: 6808–6814.

Lyczak, J. B., C. L. Cannon, and G. B. Pier. 2002. Lung infections associated with cystic fibrosis. *Clinical Microbiology Reviews* 15: 194–222.

Lyczak, J. B., and G. B. Pier. 2002. *Salmonella enterica* serovar *typhi* modulates cell surface expression of its receptor, the cystic fibrosis transmembrane conductance regulator, on the intestinal epithelium. *Infection and Immunity* 70: 6416–6423.

Manuelidis, L., and Z. Y. Lu. 2003. Virus-like interference in the latency and prevention of Creutzfeldt–Jakob disease. *Proceedings of the National Academy of Sciences USA* 100: 5360-5365.

Martinson, J. J., N. H. Chapman, et al. 1997. Global distribution of the CCR5 gene 32-base-pair deletion. *Nature Genetics* 16: 100–103.

McKusick, Victor A., et al. 1999. Spinal muscular atrophy I. Record 253300 in Online Mendelian Inheritance in Man. Center for Medical Genetics, Johns Hopkins University (Baltimore, MD) and National Center for Biotechnology Information, National Library of Medicine (Bethesda, MD). World Wide Web URL: *http://www.ncbi.nlm.nih.gov/omim/*.

Mead, S., M. P. H. Stumpf, et al. 2003. Balancing selection at the prion protein gene consistent with prehistoric kurulike epidemics. *Science* 300: 640-643.

Mongold, J. A., and R. E. Lenski. 1996. Experimental rejection of a nonadaptive explanation for increased cell size in *Escherichia coli*. *Journal of Bacteriology* 178: 5333–5334.

Mukai, T., and A. B. Burdick. 1959. Single gene heterosis associated with a second chromosome recessive lethal in *Drosophila melanogaster. Genetics* 44: 211–232.

Paul, D. B., and H. G. Spencer. 1995. The hidden science of eugenics. *Nature* 374: 302–304.

Pier, G. B., M. Grout, and T. S. Zaidi. 1997. Cystic fibrosis transmembrane conductance regulator is an epithelial cell receptor for clearance of *Pseudomonas aeruginosa* from the lung. *Proceedings of the National Academy of Sciences USA* 94: 12088–12093.

Pier, G. B., M. Grout, et al. 1998. *Salmonella typhi* uses CFTR to enter intestinal epithelial cells. *Nature* 393: 79–82.

Provine, W. B. 1971. *The Origins of Theoretical Population Genetics.* Chicago: The University of Chicago Press.

Prusiner, S. B., D. Groth, et al. 1993. Ablation of the prion protein (PrP) gene in mice prevents scrapie and facilitates production of anti-PrP antibodies. *Proceedings of the National Academy of Sciences USA* 90: 10608–10612.

Prusiner, S. B. 1996. Molecular biology and pathogenesis of prion diseases. *Trends in Biochemical Sciences* 21: 482–487.

Prusiner, S. B. 1997. Prion diseases and the BSE crisis. *Science* 278: 245–251.

Punnett, R. C. 1917. Eliminating feeblemindedness. *Journal of Heredity* 8: 464–465.

Ramaley, P. A., N. French, P. Kaleebu, et al. 2002. Chemokine-receptor genes and AIDS risk. *Nature* 471: 140.

Reilly, P. 1991. *The Surgical Solution: A History of Involuntary Sterilization in the United States.* Baltimore, MD: Johns Hopkins University Press.

Schliekelman, P., C. Garner, and M. Slatkin. 2001. Natural selection and resistance to HIV. *Nature* 411: 545–546.

Shorter, J., and S. Lindquist. 2005. Prions as adaptive conduits of memory and inheritance. *Nature Reviews Genetics* 6: 435-450.

Smithson, A., and M. R. Macnair. 1997. Negative frequency-dependent selection by pollinators on artificial flowers without rewards. *Evolution* 51: 715–723.

Steele, A. D., J. G. Emsley, et al. 2006. Prion protein (PrPc) positively regulates neural precursor proliferation during developmental and adult mammalian neurogenesis. *Proceedings of the National Academy of Sciences USA* 103: 3416–3421.

Stephens, J. C., D. E. Reich, et al. 1998. Dating the origin of the *CCR5-Δ32* AIDS-resistance allele by the coalescence of haplotypes. *American Journal of Human Genetics 62*: 1507–1515.

Sullivan, A. D., J. Wigginton, and D. Kirschner. 2001. The coreceptor mutation *CCR5-Δ32* influences the dynamics of HIV epidemics and is selected for by HIV. *Proceedings of the National Academy of Sciences USA* 98: 10214–10219.

Uitterlinden A. G., H. Burger, et al. 1998. Relation of alleles of the collagen type IA1 gene to bone density and the risk of osteoporotic fractures in postmenopausal women. *New England Journal of Medicine* 338: 1016–1021.

UNAIDS. 1998. AIDS epidemic update: December 1998. (Geneva, Switzerland). World Wide Web URL: *http://www.unaids.org*.

United States Supreme Court. 1927. *Buck v. Bell*, 274 U.S. 200.

Weinberg, W. 1908. Ueber den nachweis der vererbung beim menschen. *Jahreshefte des Vereins für Vaterländische Naturkunde in Württemburg* 64: 368–382. English translation in Boyer, S. H. 1963. *Papers on Human Genetics*. Englewood Cliffs, NJ: Prentice Hall.

Will, R. G. 2003. Acquired prion disease: iatrogenic CJD, variant CJD, kuru. British Medical Bulletin 66: 255–265.

Williams, A. G., M. P. Rayson, et al. 2000. The ACE gene and muscle performance. *Nature* 403: 614.

Wirth, B., T. Schmidt, et al. 1997. De novo rearrangements found in 2% of index patients with spinal muscular atrophy: Mutational mechanisms, parental origin, mutation rate, and implications for genetic counseling. *American Journal of Human Genetics 61*: 1102–1111.

Yule, G. U. 1902. Mendel's laws and their probable relations to intra-racial heredity. *New Phytologist 1*: 193–207; 222–238.

Zar, J. H. 1996. *Biostatistical Analysis*, 3d ed. Upper Saddle River, NJ: Prentice Hall.

Zhang, C. C., et al. 2006. Prion protein is expressed on long-term repopulating hematopoietic stem cells and is important for their self-renewal. *Proceedings of the National Academy of Sciences USA* 103: 2184-2189.

7

Mendelian Genetics in Populations II: Migration, Genetic Drift, and Nonrandom Mating

Waterleaf plants show greater inbreeding depression during their second year of life than during their first. This suggests that the seed mother's influence on offspring phenotype through the provisioning of seeds can, for a time, mask the ill effects of deleterious recessive alleles.

The Greater Prairie Chicken, *Tympanuchus cupido pinnatus*, is a 2-pound bird with a 10-pound mating display (Figure 7.1). Each spring, the males congregate in communal breeding areas, called leks, where they stake out small territories and advertise for mates. They spread their tail feathers, stomp their feet, and inflate the bright orange air sacs on their throats. As the birds draw air into the sacs it makes a booming noise that is audible for miles—like the sound created when a person blows air across the mouth of a large empty bottle, but much louder (Thomas 1998). Females visit the lek, inspect the displaying males, and choose a mate.

Two hundred years ago, the state of Illinois was almost entirely covered with prairie (Figure 7.2a) and was home to millions of greater prairie chickens. In 1837, however, the steel plow arrived (Thomas 1998). With the first blade that could break through the dense roots of prairie plants, the steel plow allowed farmers to convert prairie into cropland. As the Illinois prairie shrank, the range of the Illinois greater prairie chicken dwindled with it (Figure 7.2b–d). And as the bird's range contracted, its numbers fell: to just 25,000 in 1933; 2,000 in 1962; 500 in 1972; 76 in 1990

Figure 7.1 A greater prairie chicken This male has inflated his air sacs and fanned his feathers as part of his courtship display.

(Westemeier et al. 1991; Bouzat et al. 1998). By 1994, there were fewer than 50 greater prairie chickens left in Illinois (Westemeier et al. 1998). These remaining birds belonged to two remnant populations—one in Marion County and the other in Jasper County.

Efforts to save remnant populations of greater prairie chickens by restoring prairie habitat appeared at first to be succeeding. Soon, however, population sizes resumed their decline. Why?

Efforts to save the Illinois greater prairie chicken began with a ban on hunting in 1933 (Thomas 1998). In 1962 and 1967, respectively, the habitats occupied by the Jasper County and Marion County populations were established as sanctuaries and as sites for the restoration and management of grasslands (Westemeier et al. 1998). Figure 7.3 tracks the number of males seeking mates on leks in Jasper County from 1963 to 1997. From the mid-1960s through the early 1970s, the number of birds increased steadily. The conservation measures appeared to be working. In the mid-1970s, however, the population began to crash again. The population hit its all-time low of five or six males in 1994, despite the fact that there was now more managed grassland available to the birds than there had been in 1963.

Why did the Jasper County prairie chicken population continue to decline from the mid-1970s to the mid-1990s, even though the amount of habitat available was increasing? And what did wildlife managers do to finally reverse the decline?

The answers to these questions involve three phenomena introduced in Chapter 6 but not discussed there: migration, genetic drift, and nonrandom mating. We identified these processes as potentially important factors in the evolution of populations when we developed the Hardy–Weinberg equilibrium principle. When an ideal population has no selection, no mutation, no migration, and an infinitely large size, and when individuals choose their mates at random, then (1) the allele frequencies do not change from one generation to the next, and (2) the genotype frequencies can be calculated by multiplying the allele frequencies. In Chapter 6 we looked at what happens when we relax the

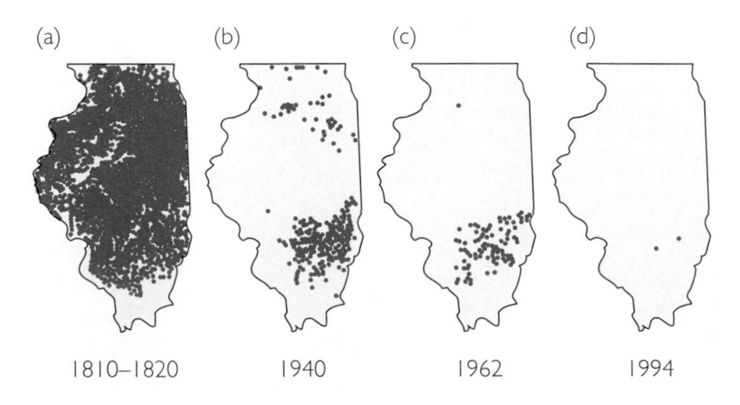

Figure 7.2 Habitat destruction and the shrinking range of Illinois greater prairie chickens Map (a) shows the extent of prairie in Illinois before the introduction of the steel plow. Maps (b), (c), and (d) show the distribution of the greater prairie chickens in 1940, 1962, and 1994. From Westemeier et al. (1998).

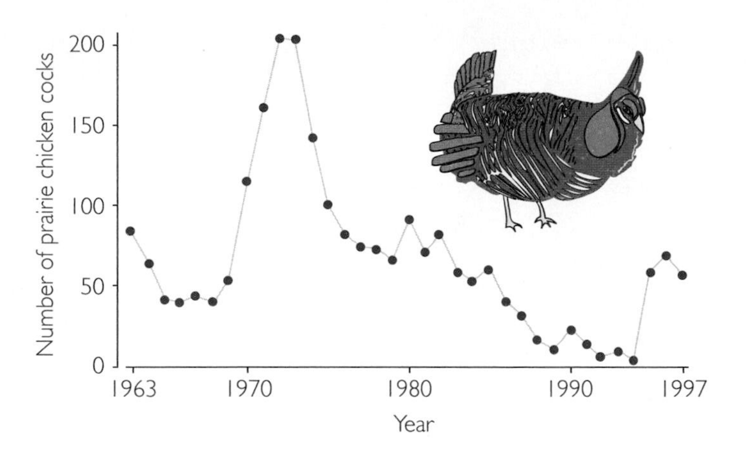

Figure 7.3 A greater prairie chicken population in danger of extinction This graph plots the number of male prairie chickens displaying each year on booming grounds in Jasper County, Illinois, from 1963 to 1997. Redrawn with permission from Westemeier et al. (1998).

assumptions of no selection and no mutation. In this chapter we will explore what happens when we relax the assumptions of no migration, infinite population size, and random mating. We will then return to the case of the Illinois greater prairie chicken and address the questions it poses.

7.1 Migration

Migration, in an evolutionary sense, is the movement of alleles between populations. This use of the term migration is distinct from its more familiar meaning, which refers to the seasonal movement of individuals. To evolutionary biologists, migration means gene flow: the transfer of alleles from the gene pool of one population to the gene pool of another population. Migration can be caused by anything that moves alleles far enough to go from one population to another. Mechanisms of gene flow range from the occasional long-distance dispersal of juvenile animals to the transport of pollen, seeds, or spores by wind, water, or animals. The actual amount of migration among populations in different species varies enormously, depending on how mobile individuals or propagules are at various stages of the life cycle.

Adding Migration to the Hardy–Weinberg Analysis: Migration as a Mechanism of Evolution

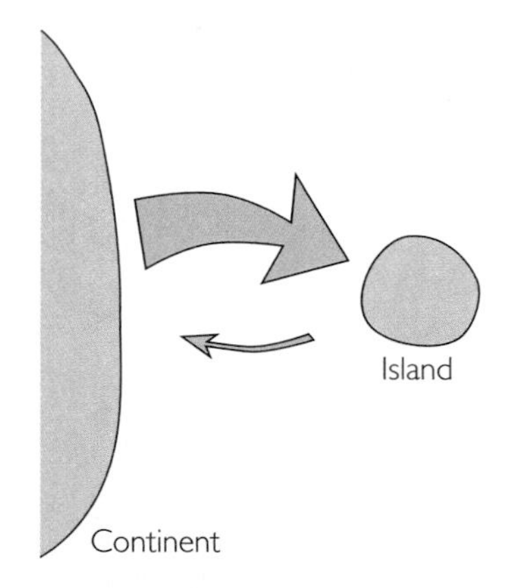

Figure 7.4 The one-island model of migration The arrows in the diagram show the relative amount of gene flow between the island and continental populations. Alleles arriving on the island from the continent represent a relatively large fraction of the island gene pool, whereas alleles arriving on the continent from the island represent a relatively small fraction of the continental gene pool.

To investigate the effects of migration on the two conclusions of the Hardy–Weinberg analysis, we consider a simple model of migration, called the one-island model. Imagine two populations: one on a continent, and the other on a small island offshore (Figure 7.4). Because the island population is so small relative to the continental population, any migration from the island to the continent will be inconsequential for allele and genotype frequencies on the continent. So migration, and the accompanying gene flow, is effectively one way, from the continent to the island. Consider a single locus with two alleles, A_1 and A_2. Can migration from the continent to the island take the allele and genotype frequencies on the island away from Hardy–Weinberg equilibrium?

To see that the answer is yes, imagine that before migration, the frequency of A_1 on the island is 1.0 (that is, A_1 is fixed in the island population—see Figure 7.5). When gametes in a gene pool in which A_1 is fixed combine at random to make zygotes, the genotype frequencies among the zygotes are 1.0 for A_1A_1, 0 for A_1A_2, and 0 for A_2A_2. Our calculations will be simpler if we give our

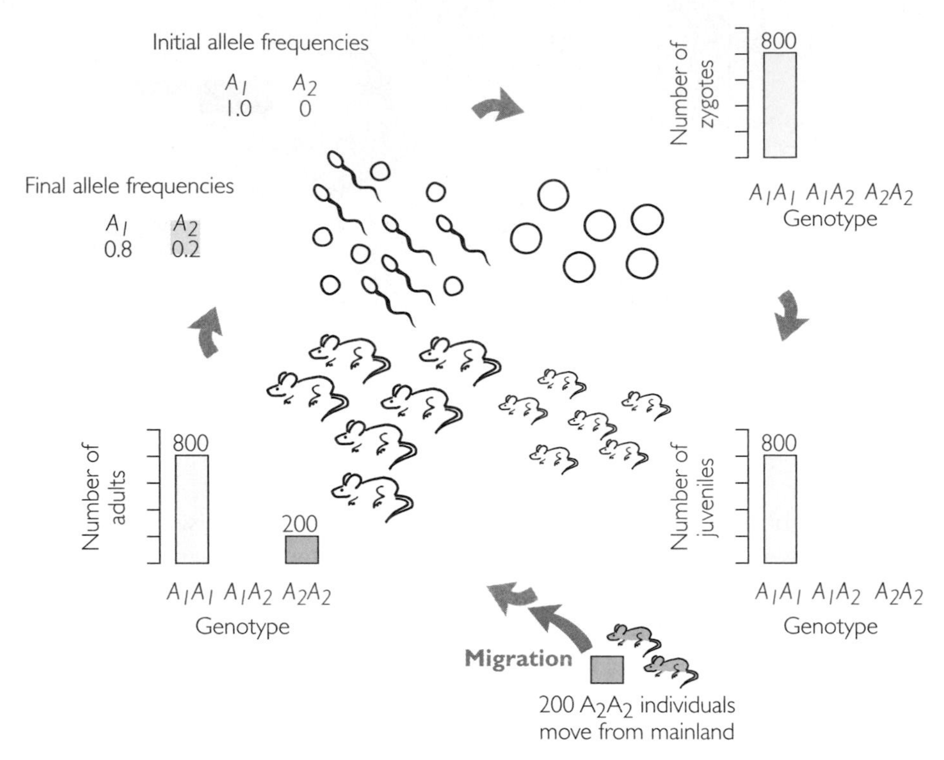

Figure 7.5 Migration can alter allele and genotype frequencies This diagram follows an imaginary island population of mice from one generation's gene pool (initial allele frequencies) to the next generation's gene pool (final allele frequencies). The bar graphs show the number of individuals of each genotype in the population at any given time. Migration, in the form of individuals arriving from a continental population fixed for allele A_2, increases the frequency of allele A_2 in the island population.

population a fixed size, so imagine that there are 800 zygotes, which we will let develop into juveniles and grow up to become adults.

Now suppose that the continental population is fixed for allele A_2 and that before the individuals on the island reach maturity, 200 individuals migrate from the continent to the island. After migration, 80% of the island population is from the island, and 20% is from the continent. The new genotype frequencies are 0.8 for A_1A_1, 0 for A_1A_2, and 0.2 for A_2A_2. When individuals on the island reproduce, their gene pool will have allele frequencies of 0.8 for A_1 and 0.2 for A_2.

Migration is a potent mechanism of evolution. In practice, migration is most important in preventing populations from diverging.

Migration has changed the allele frequencies in the island population, violating Hardy–Weinberg conclusion 1. Before migration, the island frequency of A_1 was 1.0; after migration, the frequency of A_1 is 0.8. The island population has evolved as a result of migration. (For an algebraic treatment of migration as a mechanism of allele frequency change, see Box 7.1).

Migration has also produced genotype frequencies among the adults on the island that are not consistent with Hardy–Weinberg conclusion 2. Under the Hardy–Weinberg equilibrium principle, a population with allele frequencies of 0.8 and 0.2 should have genotype frequencies of 0.64, 0.32, and 0.04. Compared to these expected values, the postmigration island population has an excess of homozygotes and a deficit of heterozygotes. A single bout of random mating will, of course, put the population back into Hardy–Weinberg equilibrium for genotype frequencies.

Box 7.1 An algebraic treatment of migration as an evolutionary process

Let p_I be the frequency of allele A_1 in an island population, and p_C be the frequency of A_1 in the mainland population. Imagine that every generation a group of individuals moves from the mainland to the island, where they constitute a fraction m of the island population. We want to know how the frequency of allele A_1 on the island changes as a result of migration and whether there is an equilibrium frequency for A_1 at which there will be no further change even if migration continues.

We first write an expression for p_I', the frequency of A_1 on the island in the next generation. A fraction $(1 - m)$ of the individuals in the next generation were already on the island. Among these individuals, the frequency of A_1 is p_I. A fraction m of the individuals in the next generation came from the mainland. Among them, the frequency of A_1 is p_C. Thus the new frequency of A_1 in the island population is a weighted average of the frequency among the residents and the frequency among the immigrants:

$$p_I' = (1 - m)(p_I) + (m)(p_C)$$

We can now write an expression for Δp_I, the change in the allele frequency on the island from one generation to the next:

$$\Delta p_I = p_I' - p_I$$

Substituting our earlier expression for p_I' and simplifying gives

$$\Delta p_I = (1 - m)(p_I) + (m)(p_C) - p_I = m(p_C - p_I)$$

Finally, we can determine the equilibrium frequency of allele A_1 on the island. The equilibrium condition is no change in p_I. That is,

$$\Delta p_I = 0$$

If we set our expression for Δp_I equal to zero, we have

$$m(p_C - p_I) = 0$$

This expression shows that the frequency of A_1 will remain constant on the island if there is no migration $(m = 0)$, or if the frequency of A_1 on the island is already identical to its frequency on the mainland $(p_I = p_C)$. In other words, without any opposing mechanism, migration will eventually equalize the frequencies of the island and mainland populations.

Empirical Research on Migration as a Mechanism of Evolution

The water snakes of Lake Erie (Figure 7.6) provide an empirical example of migration from a mainland population to an island population. These snakes (*Nerodia sipedon*) live on the mainland surrounding Lake Erie and on the islands in the lake. Individuals vary in appearance, ranging from strongly banded to unbanded. To a rough approximation, color pattern is determined by a single locus with two alleles, with the banded allele dominant over the unbanded allele (King 1993a).

On the mainland virtually all the water snakes are banded, whereas on the islands many snakes are unbanded (Figure 7.7). The difference in the composition of mainland versus island populations appears to be the result of natural selection caused by predators. On the islands, the snakes bask on limestone rocks at the water's edge. Following up on earlier work by Camin and Ehrlich (1958), Richard B. King (1993b) showed that among very young snakes unbanded individuals are more cryptic on island rocks than are banded individuals. The youngest and smallest snakes are presumably most vulnerable to predators. King (1993b) used mark-recapture studies, among other methods, to show that on the islands unbanded snakes indeed have higher rates of survival than banded snakes.

Migration of individuals from the mainland to islands appears to be preventing the divergence of island versus mainland populations of Lake Erie water snakes.

If selection favors unbanded snakes on the islands, then we would expect that the island populations would consist entirely of unbanded snakes. Why is this not the case? The answer, at least in part, is that every generation several banded

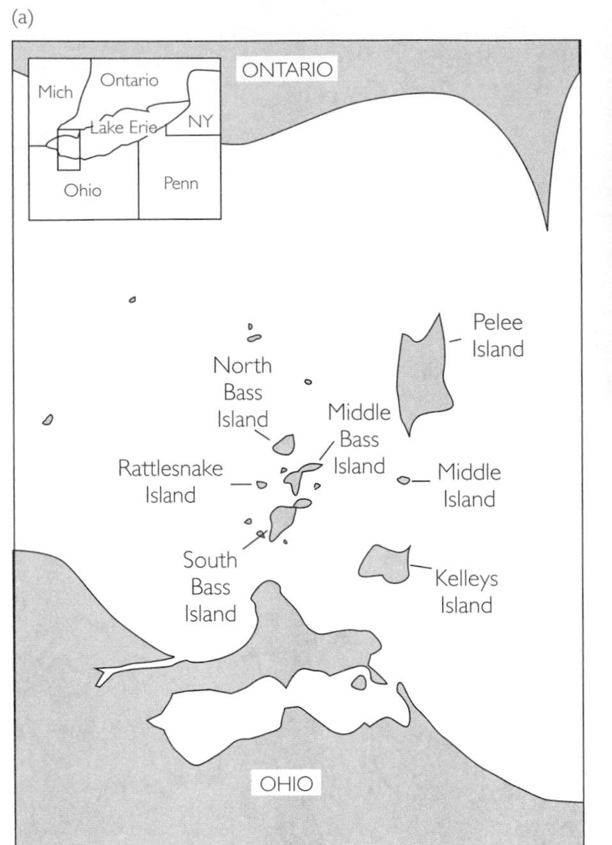

Figure 7.6 **Water snakes and where they live** The map in (a) shows the island and mainland areas in and around Lake Erie where Richard King and colleagues studied migration as a mechanism of evolution in water snakes. The photo in (b) shows unbanded, banded, and intermediate forms of the Lake Erie water snake (*Nerodia sipedon*). From King and Lawson (1995).

snakes move from the mainland to the islands. The migrants bring with them copies of the allele for banded coloration. When the migrant snakes interbreed with the island snakes, they contribute these copies to the island gene pool. In this example, migration is acting as an evolutionary mechanism in opposition to natural selection, preventing the island population from becoming fixed for the unbanded allele. (For an algebraic treatment of the opposing influences of selection and migration on the Lake Erie water snakes, see Box 7.2.)

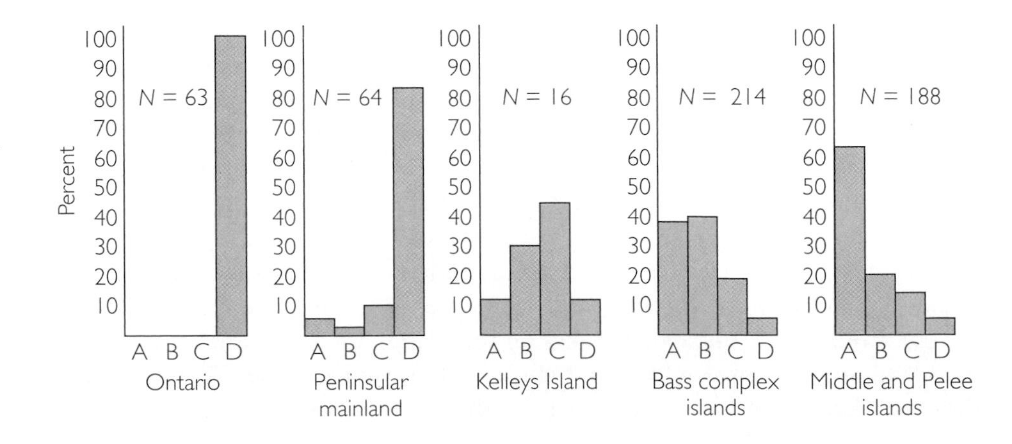

Figure 7.7 **Variation in color pattern within and between populations** These histograms show frequency of different color patterns in various populations. Category A snakes are unbanded; category B and C snakes are intermediate; category D snakes are strongly banded. Snakes on the mainland tend to be banded; snakes on the islands tend to be unbanded or intermediate. From Camin and Ehrlich (1958).

Box 7.2 Selection and migration in Lake Erie water snakes

As described in the main text, the genetics of color pattern in Lake Erie water snakes can be roughly approximated by a single locus, with a dominant allele for the banded pattern and a recessive allele for the unbanded pattern (King 1993a). Selection by predators on the islands favors unbanded snakes. If the fitness of unbanded individuals is defined as 1, then the relative fitness of banded snakes is between 0.78 and 0.90 (King and Lawson 1995). So why has selection not eliminated banded snakes from the islands? Here we calculate the effect that migration has when it introduces new copies of the banded allele into the island population every generation.

King and Lawson (1995) lumped all the island snakes into a single population, because snakes appear to move among islands much more often than they move from the mainland to the islands. King and Lawson used genetic techniques to estimate that 12.8 snakes move from the mainland to the islands every generation. The scientists estimated that the total island snake population is between 523 and 4,064 individuals, with a best estimate of 1,262. This means that migrants represent a fraction of 0.003 to 0.024 of the population each generation, with a best estimate of 0.01.

With King and Lawson's estimates of selection and migration, we can calculate the equilibrium allele frequencies in the island population at which the effects of selection and migration exactly balance each other. Let A_1 represent the dominant allele for the banded pattern, and A_2 the recessive allele for the unbanded pattern. Let p represent the frequency of A_1, and q the frequency of A_2. Following Box 6.3, we create individuals by random mating, then let selection act. After selection (but before migration), the new frequency of allele A_2 is

$$q^* = \frac{pqw_{12} + q^2w_{22}}{\overline{w}}$$

where w_{12} is the fitness of A_1A_2 heterozygotes, w_{22} is the fitness of A_2A_2 homozygotes, and $\overline{w}$ is the mean fitness of all the individuals in the population, given by $(p^2w_{11} + 2pqw_{12} + q^2w_{22})$.

For our first calculation, we will use $w_{11} = w_{12} = 0.84$, and $w_{22} = 1$. A relative fitness of 0.84 for banded snakes is the midpoint of the range within which King and Lawson (1995) estimated the true value to fall. This gives

$$q^* = \frac{pq(0.84) + q^2}{[p^2(0.84) + 2pq(0.84) + q^2]}$$

Substituting $(1 - q)$ for p gives

$$q^* = \frac{(1 - q)q(0.84) + q^2}{[(1 - q)^2(0.84) + 2(1 - q)q(0.84) + q^2]}$$
$$= \frac{0.84q + 0.16q^2}{0.84 + 0.16q^2}$$

Now we allow migration, with the new migrants representing, in this first calculation, a fraction 0.01 of the island's population (King and Lawson's best estimate). None of the new migrants carry allele A_2, so the new frequency of A_2 is

$$q' = (0.99)\frac{0.84q + 0.16q^2}{0.84 + 0.16q^2}$$

The change in q from one generation to the next is

$$\Delta q = q' - q = (0.99)\frac{0.84q + 0.16q^2}{0.84 + 0.16q^2} - q$$

Plots of Δq as a function of q appear in Figure 7.8. Look first at the green curve (b). This curve is for the function we just calculated. It shows that if q is greater than 0.05 and less than 0.93 in this generation, then q will be larger in the next generation (Δq is positive). If q is less than 0.05 or greater than 0.93 in this generation, then q will be smaller in the next generation (Δq is negative). The points where the curve crosses the horizontal axis, where $\Delta q = 0$, are the equilibrium points. The upper equilibrium point is stable: if q is less than 0.93, then q will rise in the next generation; if q is greater than 0.93, then it will fall in the next generation. Thus a middle-of-the-road prediction, given King and Lawson's estimates of selection and gene flow, is that the equilibrium frequency of the unbanded allele in the island population will be 0.93.

Box 7.2 (Continued)

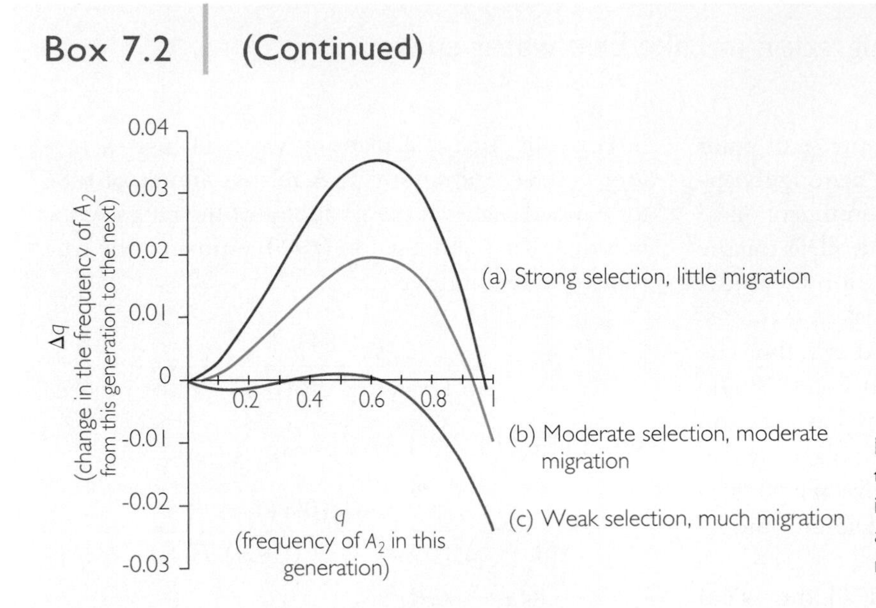

Figure 7.8 The combined effects of selection and migration on allele frequencies in island water snakes The curves show Δq as a function of q for different combinations of migration and selection. See text for details.

Curve (a) is a high-end estimate; it uses fitnesses of 0.78 for A_1A_1, 0.78 for A_1A_2, and 1 for A_2A_2, and a migration rate of 0.003 (0.3% of every generation's population are migrants). It predicts an equilibrium at $q = 0.99$. Curve (c) is a low-end estimate; it uses fitnesses of 0.90 for A_1A_1, 0.90 for A_1A_2, and 1 for A_2A_2, and a migration rate of 0.024 (2.4% of every generation's population are migrants). It predicts an equilibrium at $q = 0.64$.

King and Lawson's best estimate of the actual frequency of A_2 is 0.73. This value is toward the low end of our range of predictions. Our calculation is a relatively simple one, and leaves out many factors, including recent changes in the population sizes of both the water snakes and their predators, as well as recent changes in the frequencies of banded versus unbanded snakes. For a detailed treatment of this example, see King and Lawson (1995) and Hendry et al. (2001).

Migration as a Homogenizing Evolutionary Process across Populations

Migration of water snakes from the mainland to the islands makes the island population more similar to the mainland population than it otherwise would be. This is the general effect of migration: It tends to homogenize allele frequencies across populations. In the water snakes, this homogenization is opposed by selection.

How far would the homogenization go if selection did not oppose it? The algebraic model developed in Box 7.1 shows that gene flow from a continent to an island will eventually drive the allele frequency on the island to a value exactly equal to what it is on the continent. In other words, if allowed to proceed unopposed by any other mechanism of evolution, migration will eventually homogenize allele frequencies across populations completely.

Barbara Giles and Jérôme Goudet (1997) documented the homogenizing effect of gene flow on populations of red bladder campion, *Silene dioica*. Red bladder campion is an insect-pollinated perennial wildflower (Figure 7.9a). The populations that Giles and Goudet studied occupy islands in the Skeppsvik Archipelago, Sweden. These islands are mounds of material deposited by glaciers during the last ice age and left underwater when the ice melted. The area on which the islands sit is rising at a rate of 0.9 centimeters per year. As a result of this geological uplift, new islands are constantly rising out of the water. The Skeppsvik Archipelago thus contains dozens of islands of different ages.

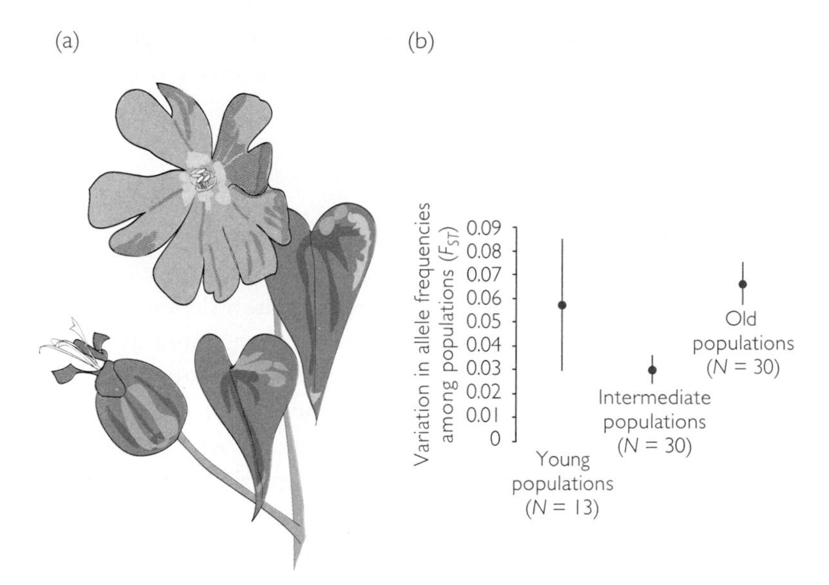

Figure 7.9 Variation in allele frequencies among populations of red bladder campion *Silene dioica* (a) Red bladder campion, a perennial wildflower. (b) Giles and Goudet's (1997) measurements of variation in allele frequencies among populations. The red dots represent values of F_{ST} (see text); the vertical red lines represent standard errors (larger standard errors represent less certain estimates of F_{ST}). There is less variation in allele frequencies among intermediate age populations than among young populations ($P = 0.05$). There is more variation in allele frequencies among old populations than among intermediate populations ($P = 0.04$). After Giles and Goudet (1997).

Red bladder campion seeds are transported by wind and water, and the plant is among the first to colonize new islands. Campion populations grow to several thousand individuals. There is gene flow among islands as a result of seed dispersal and the transport of pollen by insects. After a few hundred years, campion patches are invaded by other species of plants and by a pollinator-borne disease. Establishment of new seedlings stops, and populations dwindle as individuals die.

Giles and Goudet predicted that young populations, having been founded by the chance transport of just a few seeds, would vary in their allele frequencies at a variety of loci. (We will consider why in more detail in Section 7.2.) Populations of intermediate age should be more homogeneous in their allele frequencies as a result of migration—that is, as a result of gene flow among populations via seed dispersal and pollen transport. Finally, the oldest populations, no longer exchanging genes and thus structured mainly by the fortuitous survival of a few remaining individuals, should again become more variable in their allele frequencies.

The researchers tested their predictions by collecting leaves from many individual red bladder campions on 52 islands of different ages. By analyzing proteins in the leaves, Giles and Goudet determined each individual's genotype at six enzyme loci. They divided their populations by age into three groups: young, intermediate, and old. For each of these groups, they calculated a test statistic called F_{ST}. A value for F_{ST} refers to a group of populations and reflects the variation in allele frequencies among the populations in the group. The value of F_{ST} can be anywhere from 0 to 1. Larger values represent more variation in allele frequency among populations. Populations homogenized by gene flow should have similar allele frequencies and thus low F_{ST} values.

The results confirm Giles and Goudet's predictions (Figure 7.9b). There is less variation in allele frequencies among populations of intermediate age than among young and old populations. The low diversity among intermediate populations probably reflects the homogenizing influence of gene flow. The higher diversity of young and old populations probably represents genetic drift, the subject of the next section.

In summary, migration is the movement of alleles from population to population. Within a participating population, migration can cause allele frequencies to change from one generation to the next. For small populations receiving immigrants from large source populations, migration can be a potent mechanism of evolution. Across groups of populations, gene flow tends to homogenize allele frequencies. Thus migration tends to prevent the evolutionary divergence of populations.

7.2 Genetic Drift

In Chapter 3, we refuted the misconception that evolution by natural selection is a random process. To be sure, Darwin's mechanism of evolution depends on the generation of random variation by mutation. The variation generated by mutation is random in the sense that when mutation substitutes one amino acid for another in a protein, it does so without regard to whether the change will improve or damage the protein's ability to function. But natural selection itself is anything but random. It is precisely the nonrandomness of selection in sorting among mutations that leads to adaptation.

We are now in a position to revisit the role of chance in evolution. Arguably, the most important insight from population genetics is that natural selection is not the only mechanism of evolution. Among the nonselective mechanisms of evolution, there is one that is absolutely random. That mechanism is genetic drift. We first encountered genetic drift in Chapter 6. When we simulated drawing gametes from a gene pool to make zygotes, we found that blind luck produced different outcomes in different trials. Genetic drift does not lead to adaptation, but it does lead to changes in allele frequencies. In the Hardy–Weinberg model, genetic drift results from violation of the assumption of infinite population size.

A Model of Genetic Drift

To begin our exploration of how genetic drift works, we will return to a simulation like the one we used in Chapter 6. Imagine an ideal population that is finite—in fact, small—in size. As usual, we will focus on a single locus with two alleles, A_1 and A_2, Imagine that in the present generation's gene pool, allele A_1 is at frequency 0.6, and allele A_2 is at frequency 0.4 (Figure 7.10a, upper left). We will let the gametes in this gene pool combine at random to make exactly 10 zygotes. These 10 zygotes will constitute the entire population for the next generation.

We can simulate the production of 10 zygotes from our gene pool with a physical model. The gene pool appears in Figure 7.10b. It includes 100 gametes. Sixty of these eggs and sperm carry allele A_1; 40 carry A_2. We make each zygote by closing our eyes and putting a finger down to chose a random egg, then closing our eyes and putting a finger down to choose a random sperm. (The chosen gametes remain in the gene pool and can be chosen again. We are imagining that our gene pool is much bigger than what we can see in the illustration, and that removing a few gametes has no effect on the allele frequencies.) We are pausing to choose gametes as we write. The genotypes of the 10 zygotes are

$$\begin{array}{ccccc} A_2A_1 & A_1A_1 & A_1A_1 & A_1A_1 & A_2A_2 \\ A_1A_1 & A_2A_2 & A_1A_2 & A_1A_1 & A_1A_1 \end{array}$$

Counting the genotypes, we have A_1A_1 at a frequency of 0.6, A_1A_2 at a frequency of 0.2, and A_2A_2 at a frequency of 0.2 (Figure 7.10a). Counting the allele

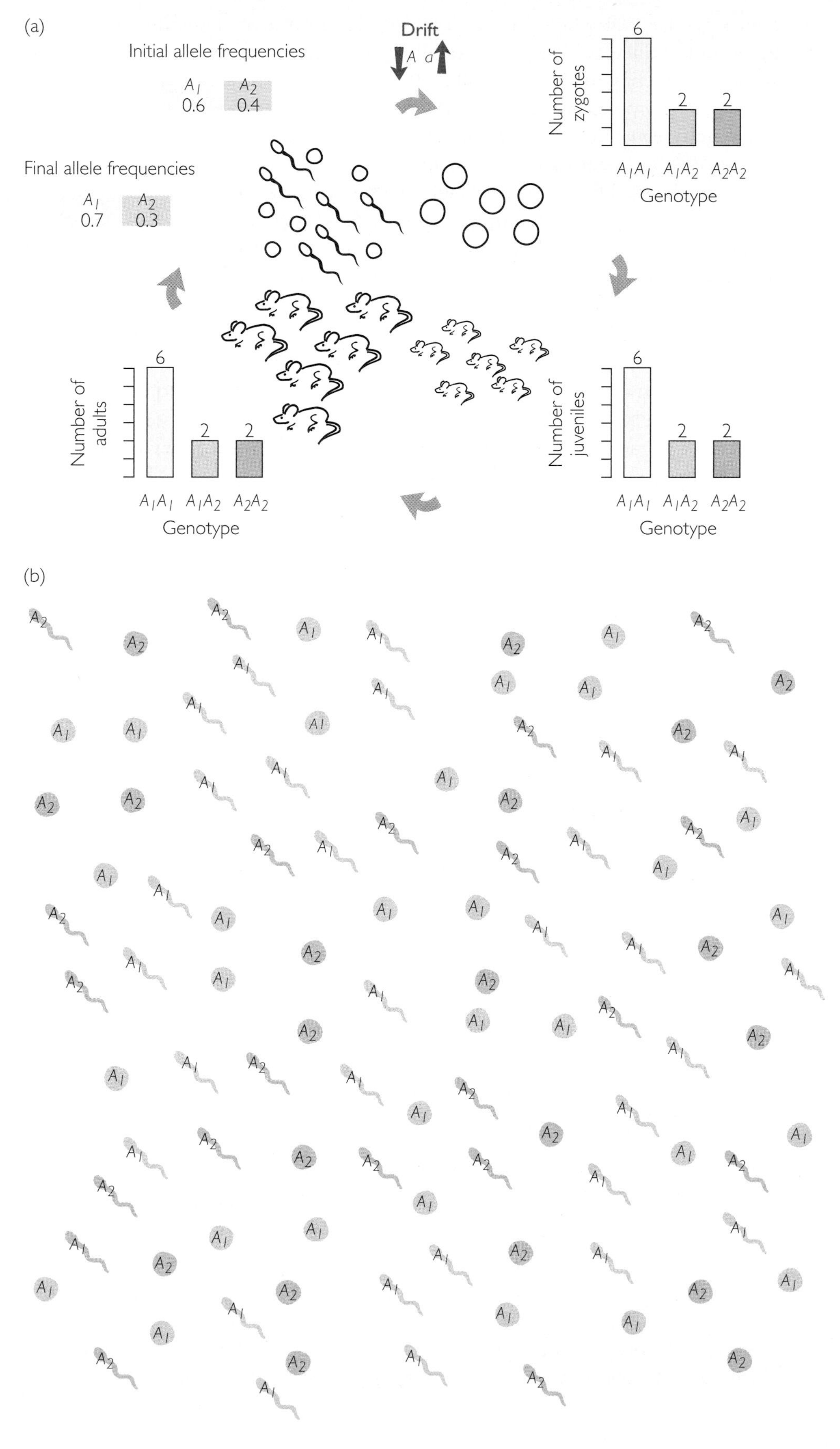

Figure 7.10 Chance events can alter allele and genotype frequencies (a) This diagram follows an imaginary population of 10 mice from one generation's gene pool (initial allele frequencies) to the next generation's gene pool (final allele frequencies). The bar graphs show the number of individuals of each genotype in the population at any given time. Genetic drift, in the form of sampling error in drawing gametes from the initial gene pool (b) to make zygotes, increases the frequency of allele A_1. Note that many other outcomes are also possible.

copies, we see that when these zygotes develop into juveniles, which then grow up and reproduce, the frequency of allele A_1 in the new gene pool will be 0.7, and the frequency of allele A_2 will be 0.3 (Figure 7.10a).

In populations of finite size, chance events—in the form of sampling error in drawing gametes from the gene pool—can cause evolution.

We have completed one turn of the life cycle of our model population. Nothing much seems to have happened, but note that both conclusions of the Hardy–Weinberg equilibrium principle have been violated. The allele frequencies have changed from one generation to the next, and we cannot calculate the genotype frequencies by multiplying the allele frequencies.

The reason our population has failed to conform to the Hardy–Weinberg principle is simply that the population is small. In a small population, chance events produce outcomes that differ from theoretical expectations. The chance events in our simulated population were the blind choices of gametes to make zygotes. We picked gametes carrying copies of A_1 and A_2 not in their exact predicted ratio of 0.6 and 0.4 but in a ratio that just happened to be a bit richer in A_1 and a bit poorer in A_2. This kind of random discrepancy between theoretical expectations and actual results is called **sampling error**. Sampling error in the production of zygotes from a gene pool is **genetic drift**. Because it is nothing more than cumulative effect of random events, genetic drift cannot produce adaptation. But it can, as we have seen, cause allele frequencies to change. Blind luck is, by itself, a mechanism of evolution.

Sometimes it is difficult to see the difference between genetic drift and natural selection. In our simulated small population, copies of allele A_1 were more successful at getting into the next generation than copies of allele A_2. Differential reproductive success is selection, is it not? In this case, it is not. If it had been selection, the differential success of alleles in our model population would have been explicable in terms of the phenotypes the alleles confer on the individuals that carry them. Individuals with one or two copies of A_1 might have been better at surviving, finding food, or attracting mates. In fact, however, individuals carrying copies of allele A_1 were none of these things. They were just lucky; their alleles happened to get drawn from the gene pool more often. Selection is differential reproductive success that happens for a reason. Genetic drift is differential reproductive success that just happens.

Selection is differential reproductive success that happens for a reason; genetic drift is differential reproductive success that just happens.

Another way to see that genetic drift is different from selection is to recognize that the genotype and allele frequencies among our 10 zygotes could easily have been different from what they turned out to be. To prove it, we can repeat the exercise of drawing gametes from our gene pool to make 10 zygotes. This time, the genotypes of the zygotes are

$$
\begin{array}{ccccc}
A_1A_1 & A_1A_1 & A_1A_1 & A_2A_1 & A_1A_2 \\
A_2A_2 & A_1A_2 & A_1A_1 & A_2A_1 & A_2A_2
\end{array}
$$

Among this set of zygotes the genotype frequencies are 0.4 for A_1A_1, 0.4 for A_1A_2, and 0.2 for A_2A_2. The allele frequencies are 0.6 for A_1 and 0.4 for A_2.

Repeating the exercise a third time produces these zygotes:

$$
\begin{array}{ccccc}
A_1A_1 & A_1A_1 & A_1A_1 & A_1A_2 & A_1A_1 \\
A_1A_2 & A_2A_1 & A_2A_2 & A_2A_2 & A_2A_2
\end{array}
$$

Now the genotype frequencies are 0.4 for A_1A_1, 0.3 for A_1A_2, and 0.3 for A_2A_2, and the allele frequencies are 0.55 for A_1 and 0.45 for A_2.

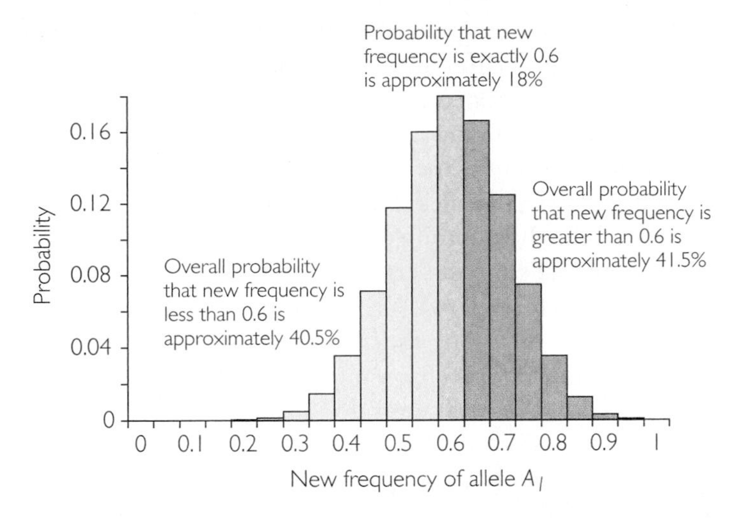

Probability that new frequency is exactly 0.6 is approximately 18%
Overall probability that new frequency is greater than 0.6 is approximately 41.5%
Overall probability that new frequency is less than 0.6 is approximately 40.5%
Probability
0.16
0.12
0.08
0.04
0
0 0.1 0.2 0.3 0.4 0.5 0.6 0.7 0.8 0.9 1
New frequency of allele A_1

Figure 7.11 The range of possible outcomes in our model population of ten mice This graph shows the possible outcomes, and the probability of each, when we make 10 zygotes by drawing alleles from a gene pool in which alleles A_1 and A_2 have frequencies of 0.6 and 0.4. The single most probable outcome is that the allele frequencies will remain unchanged. However, the chance of this happening is only about 18%.

Here is a summary of the results from our model population:

	Frequency A_1
In the gene pool	0.6
In the first set of 10 zygotes	0.7
In the second set of 10 zygotes	0.6
In the third set of 10 zygotes	0.55

The three sets of zygotes have shown us that if we start with a gene pool in which allele A_1 is at a frequency of 0.6 and make a population of just 10 zygotes, the frequency of A_1 may rise, stay the same, or fall. In fact, the new frequency of A_1 among a set of 10 zygotes drawn from our gene pool could turn out to be anywhere from 0 to 1.0, although outcomes at the extremes of this range are not likely. The graph in Figure 7.11 shows the theoretical probability of each possible outcome. Overall, there is about an 18% chance that the frequency of allele A_1 will stay at 0.6, about a 40.5% chance that it will drop to a lower value, and about a 41.5% chance that it will rise to a higher value. The reader should not just take our word for it, but should use the gene pool in Figure 7.10b to make a few batches of zygotes. Again, the point is that genetic drift is evolution that simply happens by chance.

Genetic Drift and Population Size

Genetic drift is fundamentally the result of finite population size. If we draw gametes from our gene pool to make a population of more than 10 zygotes, the allele frequencies among the zygotes will get closer to the values predicted by the Hardy–Weinberg equilibrium principle. Closing our eyes and pointing at a book quickly becomes tedious, so we used a computer to simulate drawing gametes to make not just 10, but 250 zygotes (Figure 7.12a). As the computer drew each gamete, it gave a running report of the frequency of A_1 among the zygotes it had made so far. At first this running allele frequency fluctuated wildly. For example, the first zygote turned out to have genotype A_2A_2, so the running frequency of allele A_1 started at zero. The next several zygotes were mostly A_1A_1 and A_1A_2, which sent the running frequency of allele A_1 skyrocketing to 0.75. As the cumulative number of zygotes made increased, the frequency of allele A_1 in the new

Genetic drift is most important in small populations.

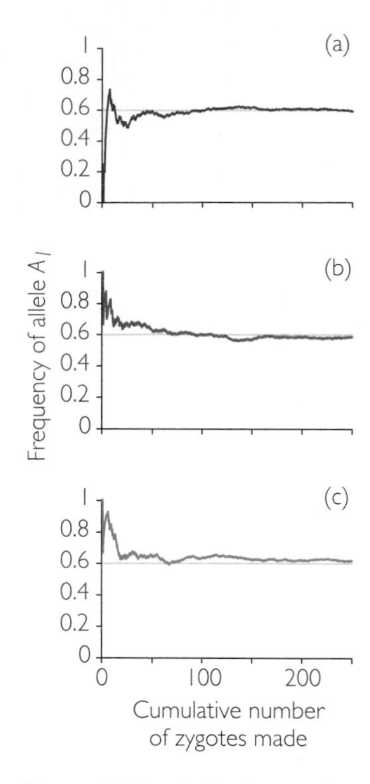

Figure 7.12 A simulation of drawing alleles from a gene pool, run three times At first the new frequency of allele A_1 fluctuates considerably, in a unique trajectory for each run. As the number of zygotes made increases, however, the new frequency of A_1 settles toward the expected value of 0.6.

generation bounced around less and less, gradually settling toward the expected value of 0.6. The deviations from expectation that we see along the way to a large number of zygotes are random, as illustrated by the graphs in Figure 7.12b and (c). These graphs show two more sets of draws to make 250 zygotes. In each, the allele frequency in the new generation fluctuates wildly at first, but in a unique pattern. As in the first graph, however, the allele frequency in the new generation always eventually settles toward the theoretically predicted value of 0.6.

Our simulations demonstrate that sampling error diminishes as sample size increases. If we kept drawing gametes forever, to make an infinitely large population of zygotes, the frequency of allele A_1 among the zygotes would be exactly 0.6. Genetic drift is a powerful evolutionary mechanism in small populations, but its power declines in larger populations. We will return to this point in later sections.

Empirical Research on Sampling Error as a Mechanism of Evolution: The Founder Effect

If we want to observe genetic drift in nature, the best place to look is in small populations. Populations are often small when they have just been founded by a group of individuals that have moved, or been moved, to a new location. The allele frequencies in the new population are likely, simply by chance, to be different from what they were in the source population. This is called the **founder effect**.

The founder effect is a direct result of sampling error. For example, if 35 different alleles are present at a single locus in a continental population of lizards, but just 15 individuals are on a mat of vegetation that rafts to a remote island (see Censky et al. 1998), the probability is zero that the new island population will contain all of the alleles present on the continent. If, by chance, any of the founding individuals are homozygotes, allele frequencies in the new population will have shifted even more dramatically. In any founder event, some degree of random genetic differentiation is almost certain between old and new populations. In other words, the founding of a new population by a small group of individuals typically represents not only the establishment of a new population but also the instantaneous evolution of differences between the new population and the old one.

Sonya Clegg and colleagues (2002) used a molecular genetic analysis to look for evidence of founder effects in island populations of silvereyes (*Zosterops lateralis*). Silvereyes are small songbirds native to Australia and Tasmania (Figure 7.13a). Five times in recent history naturalists documented the migration of silvereyes to new islands to establish new populations (Figure 7.13b). First, in 1830, birds from Tasmania colonized New Zealand's South Island. In 1856 some of their descendants moved on to both Chatham Island and Palmerston North, at the south end of North Island. In 1865 descendants of the Palmerston North birds relocated to Auckland, at the north end of North Island. Finally, in 1904, silvereyes from Auckland settled on Norfolk Island.

Clegg and colleagues caught silvereyes in mist nets on the Australian mainland, on Tasmania, and in all five recently colonized island locations. The researchers took blood samples from the birds, then released them. After extracting DNA from the blood samples, Clegg and colleagues determined each individual's genotype at each of six microsatellite loci. Microsatellites are regions of noncoding DNA with numerous easily identifiable alleles, distinguished by the number of times a short sequence of nucleotides is repeated. For each of the seven populations, Clegg and colleagues calculated the average number of alleles per locus.

(a)

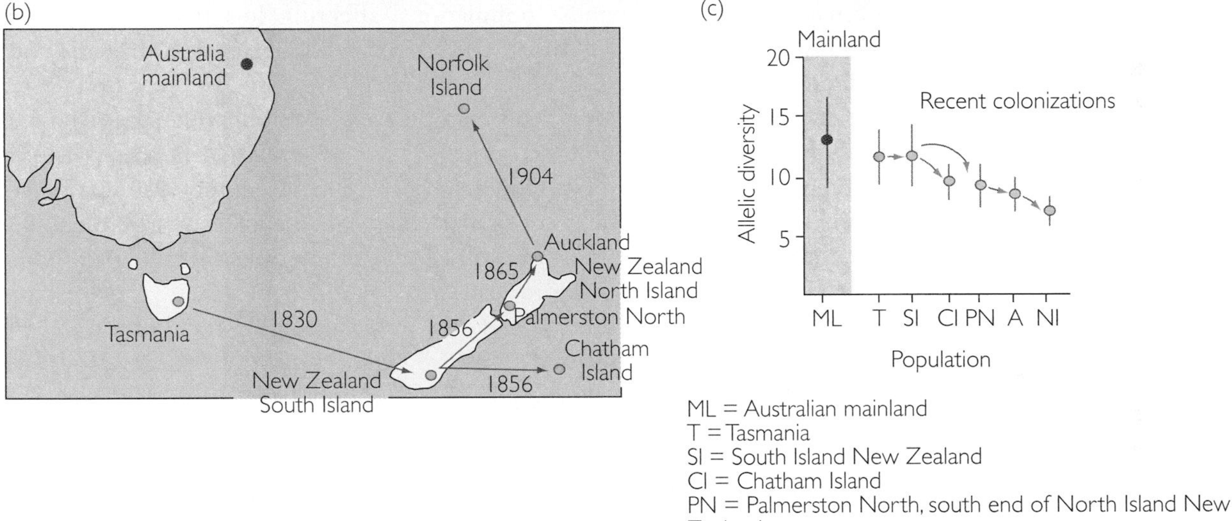

Figure 7.13 **The founder effect in an island-hopping bird** (a) A silvereye, *Zosterops lateralis*. (b) Silvereyes have been documented to colonize new islands in recent history. (c) Allelic diversity has declined along the silvereye's route of travel. From Clegg et al. (2002).

If founder effects have accompanied the establishment of new silvereye populations, then each new population will contain a randomly chosen subset of the alleles that were present in the source population that the founders came from. As the birds hopped from island to island, the allelic diversity in their populations should have fallen. Today, the allelic diversity of populations should decline along the birds' routes of travel: from mainland Australia to Tasmania, South Island, then Chatham Island and also from South Island to Palmerston North, Auckland, then Norfolk Island.

When a new population is founded by a small number of individuals, it is likely that chance alone will cause the allele frequencies in the new population to be different from those in the source population. This is the founder effect.

The data, graphed in Figure 7.13c, confirm this prediction. The decline in allelic diversity from any one population to the next is so small as to be statistically undetectable, probably because each new population was founded by somewhere between 20 and 200 migrants (see Estoup and Clegg 2003). But because the founding events were sequential, these small declines in diversity add up. The Norfolk Island population harbors just 60% of the allelic diversity present in Tasmania and just over half that present in the mainland population. Evolution occurred, not through selection, but by random sampling error. This is genetic drift in the form of the founder effect.

Founder effects are often seen in genetically isolated human populations. For example, the Pingelapese people of the Eastern Caroline Islands, located about 2,700 miles southwest of Hawaii, are descended from 20 survivors of a typhoon and subsequent famine that devastated Pingelap Atoll in about 1775 (Sheffield 2000). Among the survivors was a heterozygous carrier of a recessive loss-of-function allele of the *CNGB3* gene (Sundin et al. 2000). This gene, located on chromosome eight, encodes one component of a protein crucial to the function of cone cells, the photoreceptors in the retina that give us color vision. We know this survivor was a carrier because four generations after the typhoon homozygotes for the mutant allele began to appear among his descendants. These individuals have achromatopsia, a condition characterized by complete color blindness, extreme sensitivity to light, and poor visual acuity. Achromatopsia is rare in most populations, affecting less than 1 person in 20,000 (Winick et al. 1999). Among today's 3,000 Pingelapese, however, about 1 in 20 are achromats.

The high frequency of the achromatopsia allele among the Pingelapese is probably not due to any selective advantage it confers on either heterozygotes or homozygotes. Instead, the high frequency of the allele is simply due to chance. Sampling error by the typhoon, a founder effect, left the allele at a frequency of at least 2.5%. Further genetic drift in subsequent generations carried it still higher, to its current frequency of more than 20%.

Our examples from silvereyes and the Pingelapese illustrate not only the founder effect, but the cumulative nature of genetic drift. In the next section we will consider the cumulative consequences of genetic drift in more detail.

Random Fixation of Alleles and Loss of Heterozygosity

We have seen that genetic drift can change allele frequencies in a single generation, and that drift is even more powerful as a mechanism of evolution when its effects are compounded over many generations. We can further investigate the cumulative effects of genetic drift with the same physical model we have used before: closing our eyes and picking gametes from a paper gene pool. Our starting point will be the gene pool in Figure 7.14a, with alleles A_1 and A_2 at frequencies of 0.6 and 0.4. We will call the parents who produced this gene pool generation zero. As we did before, we now blindly select gametes to simulate the production of 10 zygotes by random mating. This time, the allele frequencies among the newly formed zygotes turn out to be 0.5 for A_1 and 0.5 for A_2. We will call these zygotes generation one.

To continue the simulation for another generation, we need to set up a new gene pool, with alleles A_1 and A_2 at frequencies of 0.5 and 0.5 (Figure 7.14b). Drawing gametes from this gene pool, we get the zygotes for generation two. Generation two's allele frequencies happen to be 0.4 for A_1 and 0.6 for A_2.

We now set up a gene pool with alleles A_1 and A_2 at frequencies of 0.4 and 0.6 (Figure 7.14c) and draw zygotes to make generation three. Generation three's allele frequencies are 0.45 for A_1 and 0.55 for A_2.

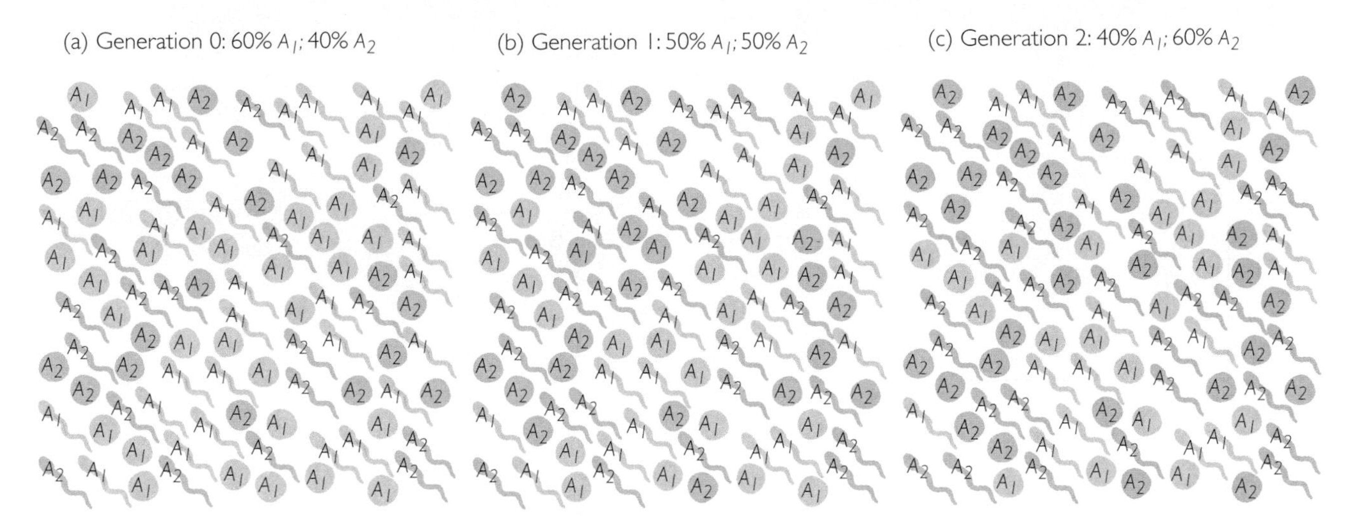

Figure 7.14 A simulation of the cumulative effects of genetic drift The gametes that make each generation's zygotes are drawn, with sampling error, from the previous generation's gene pool.

Now we need a gene pool with alleles A_1 and A_2 at frequencies of 0.45 and 0.55, and so on. The advantage of using a computer to simulate drawing gametes from gene pools is rapidly becoming apparent. We can have the computer run the simulation for us generation after generation for as long as we like, then plot graphs tracing the frequency of allele A_1 over time.

Graphs (a), (b), and (c) in Figure 7.15 show the results of 100 successive generations of genetic drift in simulated populations of different sizes. Each graph tracks allele frequencies in eight populations. Every population starts with allele frequencies of 0.5 for A_1 and 0.5 for A_2. The populations tracked in graph (a) have just 4 individuals each, the populations tracked in graph (b) have 40 individuals each, and the populations tracked in graph (c) have 400 individuals each. Three patterns are evident:

1. Because the fluctuations in allele frequency from one generation to the next are caused by random sampling error, every population follows a unique evolutionary path.
2. Genetic drift has a more rapid and dramatic effect on allele frequencies in small populations than in large populations.
3. Given sufficient time, genetic drift can produce substantial changes in allele frequencies even in populations that are fairly large.

Under genetic drift, every population follows a unique evolutionary path. Genetic drift is rapid in small populations and slow in large populations.

Note that if genetic drift is the only evolutionary mechanism at work in a population—if there is no selection, no mutation, and no migration—then sampling error causes allele frequencies to wander between 0 and 1. This wandering is particularly apparent in the population whose evolution is highlighted in the graph in Figure 7.15b. During the first 25 generations, allele A_1's frequency rose from 0.5 to over 0.9. Between generations 25 and 40 it dropped back to 0.5. Between generations 40 and 80 the frequency bounced between 0.5 and 0.8. Then the frequency of A_1 dropped precipitously, so that by generation 85 it hit 0 and A_1 disappeared from the population altogether. The wandering of allele frequencies produces two important and related effects: (1) Eventually alleles drift to fixation or loss, and (2) the frequency of heterozygotes declines.

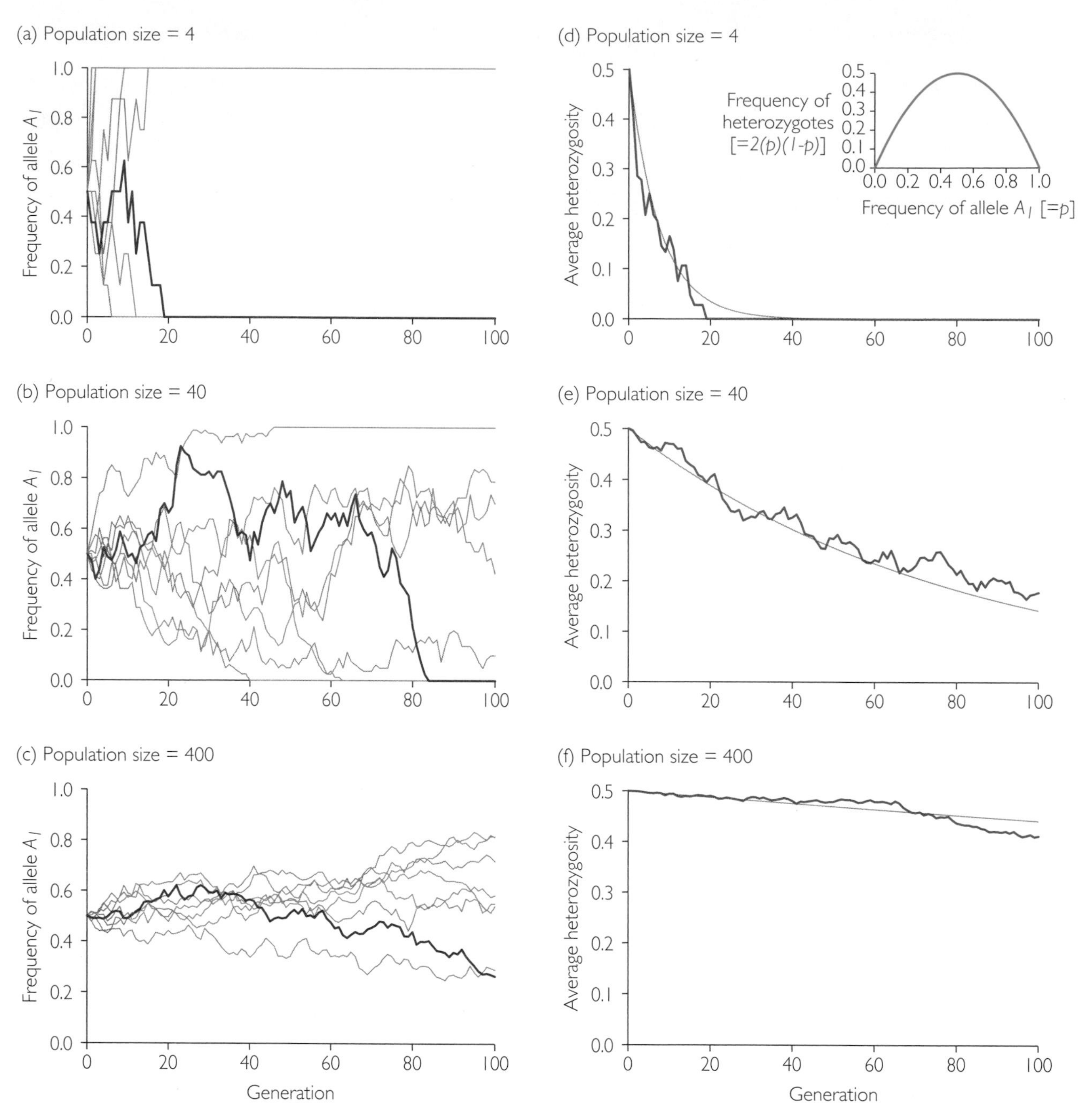

Figure 7.15 Simulations of genetic drift in populations of different sizes Plots (a), (b), and (c) show the frequency of allele A_1 over 100 generations. Eight populations are tracked in each plot, with one population highlighted in red. Plots (d), (e), and (f) show the average frequency of heterozygotes over 100 generations in the same sets of simulated populations. The gray curves represent the rate of decline predicted by theory. The inset in plot (d) shows the frequency of heterozygotes in a population, calculated as $2(p)(1 - p)$, as a function of p, the frequency of allele A_1. Collectively, the graphs in this figure show that genetic drift leads to random fixation of alleles and loss of heterozygosity and that drift is a more potent mechanism of evolution in small populations.

Random Fixation of Alleles

As any allele drifts between frequencies of 0 and 1.0, sooner or later the allele will meet an inevitable fate: Its frequency will hit one boundary or the other. If the allele's frequency hits 0, then the allele is lost forever (assuming that mutation or migration do not reintroduce it). If the allele's frequency hits 1, then the allele is said to be fixed, also forever. Among the eight populations tracked in Figure 7.15a, allele A_1 drifted to fixation in five and to loss in three. Among the eight populations tracked in Figure 7.15b, A_1 drifted to fixation in one and to loss in three. It is just a matter of time before A_1 will become fixed or lost in the other populations as well. As some alleles drift to fixation and others drift to loss, the allelic diversity present in a population declines.

Now imagine a finite population in which there are several alleles present at a particular locus: A_1, A_2, A_3, A_4, and so on. If genetic drift is the only evolutionary mechanism at work, then eventually one of the alleles will drift to fixation. At the same moment one allele becomes fixed, the last of the other alleles will be lost.

If genetic drift is the only evolutionary process at work, eventually one allele will drift to a frequency of 1 (that is, to fixation) and all other alleles will be lost.

We would like to be able to predict which alleles will meet which fate. We cannot do so with certainly, but we can give odds. Sewall Wright (1931) proved that the probability that any given allele in a population will be the one that drifts to fixation is equal to that allele's initial frequency (see Box 7.3). If, for example, we start with a finite population in which A_1 is at a frequency of 0.73, and A_2 is at a frequency of 0.27, there is a 73% chance that the allele that drifts to fixation will be A_1.

Loss of Heterozygosity

As allele frequencies in a finite population drift toward fixation or loss, the frequency of heterozygotes in the population decreases. Graphs (d), (e), and (f) in Figure 7.15 show the decline in the frequency of heterozygotes in our simulated populations.

To see why the frequency of heterozygotes declines, look first at the inset in graph (d). The inset plots the frequency of heterozygotes in a random mating

Box 7.3 The probability that a given allele will be the one that drifts to fixation

Sewall Wright (1931) developed a detailed theory of genetic drift. Among many other results, he showed that the probability that a given allele will be the one that drifts to fixation is equal to that allele's initial frequency. Wright's model of genetic drift is beyond the scope of this book, but we can provide an intuitive explanation of fixation probabilities.

Imagine a population of N individuals. This population contains a total of $2N$ gene copies. Imagine that every one of these gene copies is a unique allele. Assume that genetic drift is the only mechanism of evolution at work.

At some point in the future, one of the $2N$ alleles will drift to fixation, and all the others will be lost. Each allele must have an equal chance of being the one that drifts to fixation; that is what we meant when we assumed that drift is the only mechanism of evolution at work. So we have $2N$ alleles, each with an equal probability of becoming fixed. Each allele's chance must therefore be $\frac{1}{2N}$.

Now imagine that instead of each allele being unique, there are x copies of allele A_1, y copies of allele A_2, and z copies of allele A_3. Each copy of allele A_1 has a $\frac{1}{2N}$ chance of being the one that drifts to fixation. Therefore, the overall probability that a copy of allele A_1 will be the allele that drifts to fixation is

$$x \times \frac{1}{2N} = \frac{x}{2N}$$

Likewise, the probability that a copy of allele A_2 will be the allele that drifts to fixation is $\frac{y}{2N}$, and the probability that a copy of allele A_3 will be the allele that drifts to fixation is $\frac{z}{2N}$.

Notice that, $\frac{x}{2N}$, $\frac{y}{2N}$, and $\frac{z}{2N}$ are also the initial frequencies of A_1, A_2, and A_3 in the population. We have shown that the probability that a given allele will be the one that drifts to fixation is equal to that allele's initial frequency.

population as a function of p, the frequency of allele A_1. Because the individuals mate at random, we can calculate the frequency of heterozygotes as $2(p)(1 - p)$. The frequency of heterozygotes has its highest value, 0.5, when A_1 is at frequency 0.5. As the frequency of A_1 drops toward 0 or rises toward 1, the frequency of heterozygotes falls. And, of course, if the frequency of A_1 reaches 0 or 1, the frequency of heterozygotes falls to 0.

Now look at graphs (a), (b), and (c). In any given generation, the frequency of A_1 may move toward or away from 0.5 in any particular population (so long as A_1 has not already been fixed or lost). Thus the frequency of heterozygotes in any particular population may rise or fall. However, the overall trend across all populations is for allele frequencies to drift away from intermediate values and toward 0 or 1. So the average frequency of heterozygotes, across populations, should tend to fall.

Now look at graphs (d), (e), and (f). The heavy blue line in each graph tracks the frequency of heterozygotes averaged across the eight populations in question. The frequency of heterozygotes does indeed tend to fall, rapidly in small populations and slowly in large populations. Eventually one allele or the other will become fixed in every population, and the average frequency of heterozygotes will fall to 0.

The frequency of heterozygotes in a population is sometimes called the population's **heterozygosity**. We would like to be able to predict just how fast the heterozygosity of finite populations can be expected to decline. Sewell Wright (1931) showed that, averaged across many populations, the frequency of heterozygotes obeys the relationship

As alleles drift to fixation or loss, the frequency of heterozygotes in the population declines.

$$H_{g+1} = H_g\left[1 - \frac{1}{2N}\right]$$

where H_{g+1} is the heterozygosity in the next generation, H_g is the heterozygosity in this generation, and N is the number of individuals in the population. The value of $\left[1 - \frac{1}{2N}\right]$ is always between $\frac{1}{2}$ and 1, so the expected frequency of heterozygotes in the next generation is always less than the frequency of heterozygotes in this generation. The gray curves in Figure 7.15d, (e), and (f) show the declines in heterozygosity predicted by Wright's equation.

To appreciate just one of the implications of the inevitable loss of heterozygosity in finite populations, imagine that you are responsible for managing a captive population of an endangered species. Suppose there are just 50 breeding adults in zoos around the world. Even if you could arrange the shipment of adults or semen to accomplish random mating, you would still see a loss in heterozygosity of 1% every generation due to genetic drift.

An Experimental Study on Random Fixation and Loss of Heterozygosity

Our discussion of random fixation and loss of heterozygosity has so far been based on simulated populations and mathematical equations. Peter Buri (1956) studied these phenomena empirically, in small laboratory populations of the fruit fly, *Drosophila melanogaster*. Adopting an approach that had been used by Kerr and Wright (1954), Buri established 107 populations of flies, each with eight females and eight males. All the founding flies were heterozygotes for alleles of an eye-color gene called *brown*. All the flies had the same genotype: bw^{75}/bw. Thus, in all 107 populations, the initial frequency of the bw^{75} allele was 0.5. Buri maintained these populations for 19 generations. For every population in every generation, Buri kept the population size at 16 by picking eight females and eight males at random to be the breeders for the next generation.

What results would we predict? If neither allele bw^{75} nor allele bw confers a selective advantage, then we expect the frequency of allele bw^{75} to wander at random by genetic drift in every population. Nineteen generations should be enough, in populations of 16 individuals, for many populations to become fixed for one allele or the other. Because allele bw^{75} has an initial frequency of 0.5, we expect this allele to be lost about as often as it becomes fixed. As the bw^{75} allele is drifting toward fixation or loss in each population, we expect the average heterozygosity across all populations to decline. The rate of decline in heterozygosity should follow Wright's equation, given in the previous section.

Buri's results confirm these predictions. Each small graph in Figure 7.16 is a histogram summarizing the allele frequencies in all 107 populations in a particular generation. The horizontal axis represents the frequency of the bw^{75} allele, and the vertical axis represents the number of populations showing each frequency. The frequency of bw^{75} was 0.5 in all populations in generation zero, which is not shown in the figure. After one generation of genetic drift, most populations still had an allele frequency near 0.5, although one population had an allele frequency as low as 0.22 and another had an allele frequency as high as 0.69. As the frequency of the bw^{75} allele rose in some populations and fell in others, the distribution of allele frequencies rapidly spread out. In generation four, the frequency of bw^{75} hit 1 in a population for the first time. In generation six, the frequency of bw^{75} hit 0 in a population for the first time. As the allele frequency reached 0 or 1 in more and more populations, the distribution of frequencies became U-shaped. By the end of the experiment, bw^{75} had been lost in 30 populations and had become fixed in 28. The 30:28 ratio of losses to fixations is very close to the 1:1 ratio we would predict under genetic drift. During Buri's experiment there was dramatic evolution in nearly all 107 of the fruit fly populations, but natural selection had nothing to do with it.

The genetic properties of the *brown* locus were such that Buri could identify all three genotypes from their phenotypes. Thus Buri was able to directly assess the frequency of heterozygotes in each population. The frequency of heterozygotes in generation zero was 1, so the heterozygosity in generation one was 0.5. Every generation thereafter, Buri noted the frequency of heterozygotes in each population, then took the average heterozygosity across all 107 populations. Figure 7.17 tracks these values for average heterozygosity over the 19 generations of the experiment. Look first at the red dots, which represent the actual data. Consistent with our theoretical prediction, the average frequency of heterozygotes steadily declined.

Empirical studies confirm that under genetic drift alleles become fixed or lost and the frequency of heterozygotes declines. Indeed, these processes often happen faster than predicted.

The fit between theory and results is not perfect, however. The dashed gray curve in the figure shows the predicted decline in heterozygosity, using Wright's equation and a population size of 16. The actual decline in heterozygosity was more rapid than expected. The solid gray curve shows the predicted decline for a population size of 9; it fits the data well. Buri's populations lost heterozygosity as

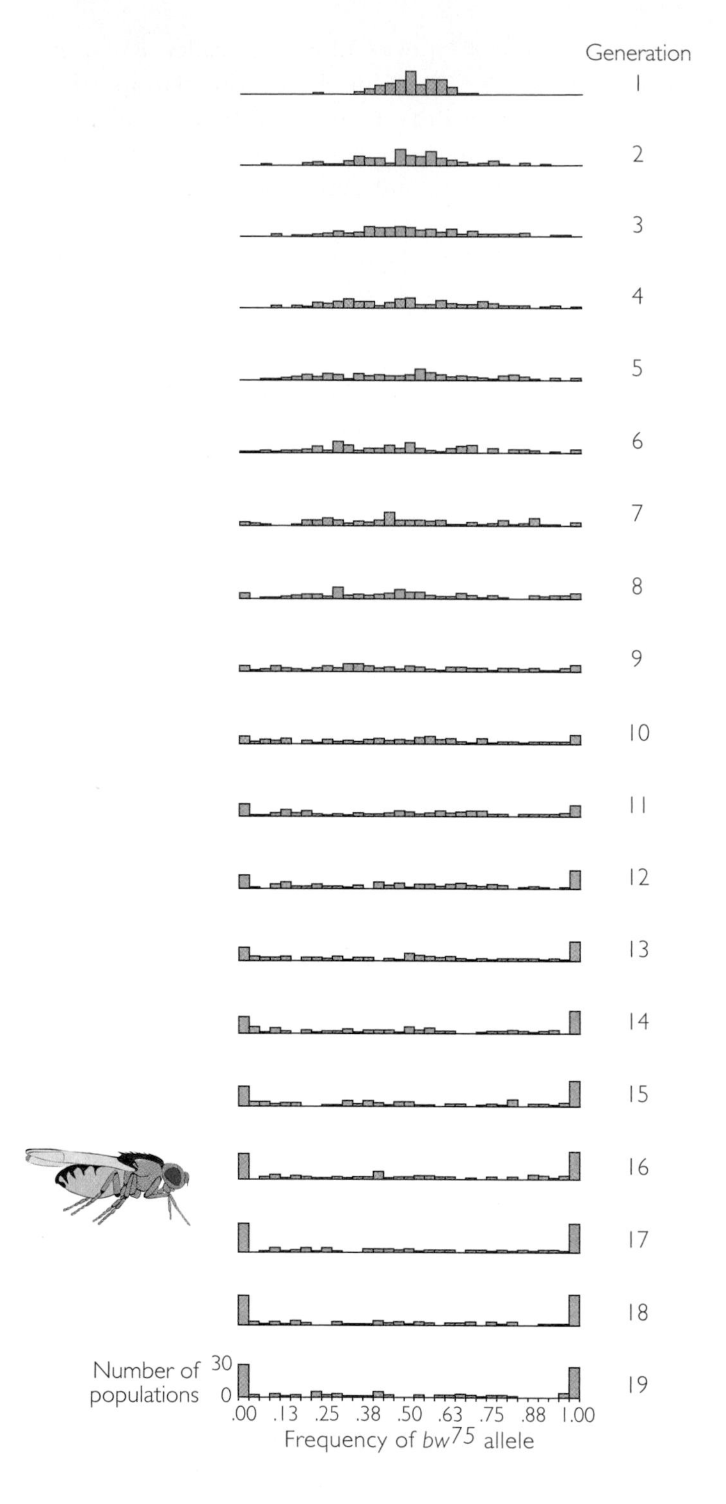

Figure 7.16 Nineteen generations of genetic drift in 107 populations of 16 fruit flies Each line is a histogram summarizing the allele frequencies in all 107 populations in a particular generation. The horizontal axis represents the frequency of the bw^{75} allele, and the vertical axis represents the number of populations showing each frequency. The frequency of bw^{75} was 0.5 in all populations in generation zero (not shown). By the end of the experiment, 30 populations had become fixed at a frequency of 0, and 28 had become fixed at a frequency of 1 (bottom line). Throughout the experiment, however, the distribution of allele frequencies remained symmetrical around 0.5. From data in Buri (1956), after Ayala and Kiger (1984).

though they contained only 9 individuals instead of 16. In other words, the **effective population size** in Buri's experiment was 9 (see Box 7.4). Among the explanations are that some of the flies in each population may have died due to accidents before reproducing, or some males may have been rejected as mates by the females.

Buri's experiment with fruit flies shows that the theory of genetic drift allows us to make accurate qualitative predictions, and reasonably accurate quantitative

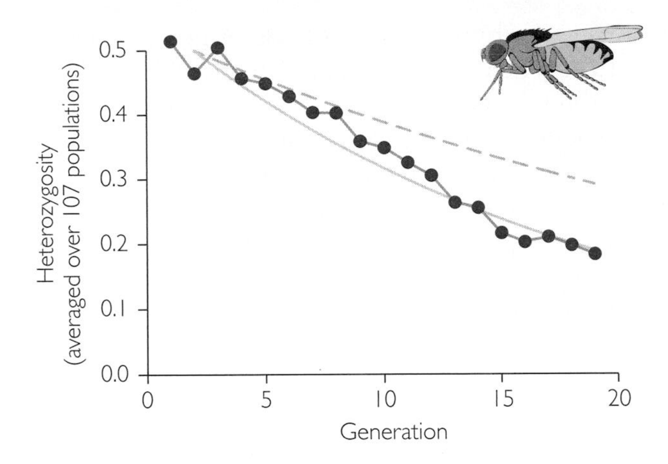

Figure 7.17 The frequency of heterozygotes declined with time in Buri's experimental populations The red dots show the frequency of heterozygotes in each generation, averaged across all 107 populations. The dashed gray curve shows the theoretical prediction for a population of 16 flies. The solid gray curve shows the theoretical prediction for a population of 9 flies. The graph demonstrates that (1) heterozygosity decreases across generations in small populations, and (2) although all the populations had an actual size of 16 flies, their effective size was roughly 9. Replotted from data in Buri (1956), after Hartl (1981).

predictions, about the behavior of alleles in finite populations—at least in the lab. In the next section we will consider evidence on random fixation of alleles and loss of heterozygosity in natural populations.

Random Fixation and Loss of Heterozygosity in Natural Populations

Alan Templeton and colleagues (1990) tested predictions about the random fixation of alleles by documenting the results of a natural experiment in Missouri's Ozark Mountains. Although now largely covered in oak–hickory forest, the Ozarks were part of a desert during an extended period of hot, dry climate that lasted from 8,000 to 4,000 years ago. The desert that engulfed the Ozarks was contiguous with the desert of the American Southwest. Many southwestern

Box 7.4 Effective population size

The effective population size is the size of an ideal theoretical population that would lose heterozygosity at the same rate as an actual population of interest. The effective population size is virtually always smaller than the actual population size. In Buri's experiment, two possible reasons for the difference in effective versus actual population size are that (1) some of the flies in each bottle died (by accident) before reproducing and (2) fruit flies exhibit sexual selection by both male–male combat and female choice (see Chapter 11)—either of which could have prevented some males from reproducing.

The effective population size is particularly sensitive to differences in the number of reproductively active females versus males. When there are different numbers of each sex in a population, the effective population size N_e can be estimated as

$$N_e = \frac{4N_m N_f}{(N_m + N_f)}$$

where N_m is the number of males and N_f is the number of females.

To see how strongly an imbalanced sex ratio can reduce the effective population size, use the formula to show that when there are 5 males and 5 females, $N_e = 10$; when there is 1 male and 9 females, $N_e = 3.6$; and when there is 1 male and 1,000 females, $N_e = 4$. Consider the logistical problems involved in maintaining a captive breeding program for a species in which the males are extremely aggressive and will not tolerate each other's presence.

desert species expanded their ranges eastward into the Ozarks. Among them was the collared lizard (Figure 7.18a). When the warm period ended, the collared lizard's range retracted westward and the Ozarks were largely overgrown with savannas. Within these mixed woodlands and grasslands, however, on exposed rocky outcrops, were small remnants of desert habitat called glades. Living in these glades were relict populations of collared lizards.

Every five years or so, wildfires swept the Ozark savannahs (Templeton et al. 2001). This periodic burning was essential to the maintenance of the savanna plant community. We know this because of what happened after European settlers arrived. First the Europeans clearcut the Ozark woodlands. Then, starting in about 1950, they suppressed all fires. These interventions allowed the oak-hickory forest that covers the area today to invade the savannas. And they allowed eastern red cedars to invade the glades.

The invasion of the glades by eastern red cedars, and the other woody plants that followed, was bad news for collared lizards (Templeton et al. 2001). The cedars partially overgrew many of the glades, drastically reducing their size. The oak–hickory forest between the glades was even worse. Its dense understory prevented the lizards from migrating from one glade to another. Most of the lizard populations, even some separated by just 50 meters of oak–hickory forest, were sufficiently isolated from each other that there was little or no gene flow among them. And the relict populations in the few remaining glades were tiny; most harbored no more than a few dozen lizards.

Because of the small size and genetic isolation of the glade populations, Templeton and colleagues (1990) predicted that the collared lizards of the Ozarks

(a)

Collared lizard (*Crotaphytus collaris*)

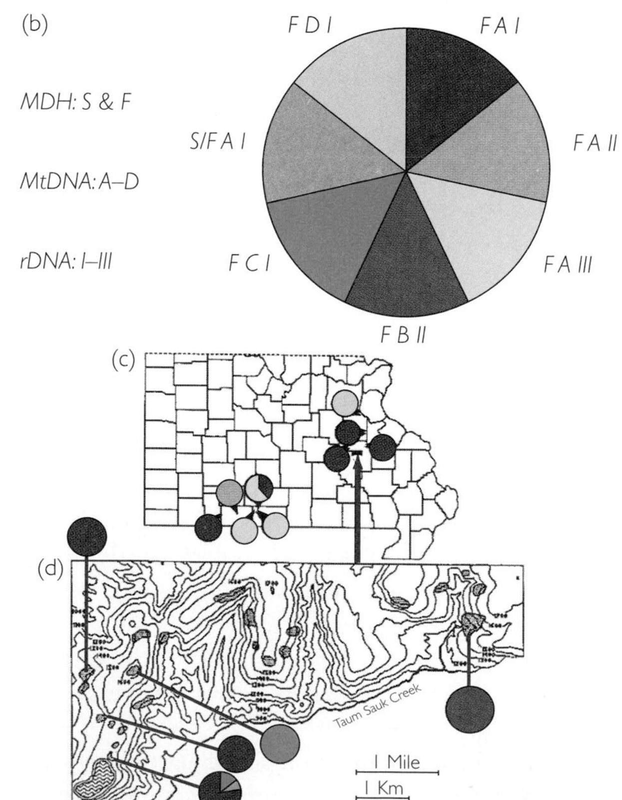

Figure 7.18 Genetic variation in Ozark glade populations of the collared lizard (a) A collared lizard. (b) This pie diagram gives a key to the seven distinct multilocus genotypes that Templeton et al. (1990) found in Ozark collared lizards. Each multilocus genotype is characterized by a malate dehydrogenase (MDH) genotype [the two alleles are "slow" (*S*) and "fast" (*F*)], a mitochondrial DNA haplotype (designated *A–D*), and a ribosomal DNA genotype (designated *I–III*). (c) This is a map of southern Missouri, showing the locations and genetic compositions of nine glade populations. The shading of each pie diagram represents the frequency in a single population of each multilocus genotype present. (d) This is an expanded map of a small piece of the map in (c). It gives the locations and genetic compositions of five more glade populations. From Templeton et al. (1990).

would bear a strong imprint of genetic drift. Within each population, most loci should be fixed for a single allele, and genetic variation should be very low. Which allele became fixed in any particular population should be a matter of chance, however, so there should be considerable genetic diversity among populations.

Templeton and colleagues (1990) assayed several glade populations for genetic variation. The researchers screened lizards for their genotypes at a variety of enzyme loci, for their ribosomal DNA genotypes, and for their mitochondrial DNA genotypes. The researchers identified among the lizards seven distinct multilocus genotypes (Figure 7.18b). Confirming the predicted consequences of isolation and small population size, most glade populations were fixed for a single multilocus genotype, with different genotypes fixed in different populations [Figure 7.18c and (d)].

Empirical data from a natural experiment confirm that small isolated populations lose their genetic diversity as a result of drift.

Templeton and colleagues (2001) believed that the nearly complete loss of genetic diversity in the glade populations had doomed the Ozark collared lizards to extinction. This extinction would happen one glade at a time, with any of a number of proximate causes. If a pathogen appeared that could infect and kill one of the lizards in a glade, it could infect and kill all lizards in the glade—because they were virtually identical. As the biological and physical environment changed, the lizard populations would be unable to evolve in response—because genetic variation is the raw material for adaptive evolution. And if an adaptation did evolve in one of the populations, it would be unable to spread to other glades—because the lizards were unable to cross the oak–hickory forests that divided them. Templeton and colleagues surveyed 130 Ozark glades. Consistent with their expectations, two-thirds of them were already devoid of collared lizards.

If Templeton and colleagues were right, simple measures could save the Ozark collared lizards. One is the relocation of lizards to repopulate the empty glades. In the 1980s, with the cooperation of the Missouri Department of Conservation, Templeton and colleagues established three new populations in the Stegall Mountain Natural Area, a former ranch with many glades but no lizards. The lizards in the new populations thrived but did not migrate, neither from population to population nor to any of the empty glades. As long as the oak–hickory forest was still in the way, the populations would remain isolated and suffer the long-term consequences of genetic drift. Starting in 1994, the Missouri Department of Conservation and the United States Forest Service began using controlled burns to clear the oak–hickory forest at Stegall Mountain. The lizards responded almost immediately, moving from population to population and colonizing many of the empty glades. This should restore the genetic diversity of the glade populations and dramatically improve the collared lizard's prospects for long-term survival.

Jennifer Brisson, Jared Strasburg, and Templeton (2003) monitored the results of a controlled experiment at Taum Sauk Mountain State Park, 80 km from Stegall Mountain. They compared the collared lizard populations occupying glades in an area that had been treated with a series of controlled burns versus populations occupying glades in an unburned area. Consistent with observations at Stegall Mountain, the burned area supported a much larger population of lizards, and the lizards there moved from glade to glade and colonized empty glades at much higher rates.

In their research on collared lizards, Templeton and colleagues documented the random fixation of alleles and loss of heterozygosity in small populations. Andrew Young and colleagues (1996), in contrast, looked for evidence of these processes among populations of various sizes. The researchers compiled data from the literature on three flowering herbs and a tree. From these data they plotted two measures of overall genetic diversity against population size. The first measure was genetic polymorphism, the fraction of loci within the genome that have at least two alleles with frequencies higher than 0.01. The second was allelic richness, the average number of alleles per locus. Both of these measures are related to heterozygosity. Imagine a single locus in a randomly mating population. As the number of alleles at the locus increases and as the fraction of those alleles that have substantial frequencies increases, the frequency of heterozygotes at the locus increases as well. If, on the other hand, the locus is fixed for a single allele, then no individual in the population will be a heterozygote. Genetic polymorphism, allelic richness, and heterozygosity rise and fall together. Because genetic drift is more pronounced in small populations than in large ones, and because genetic drift results in the loss of heterozygosity, Young and colleagues predicted that small populations would have lower levels of polymorphism and allelic richness. Young et al.'s plots appear in Figure 7.19. Consistent with their prediction, in almost every case smaller populations did indeed harbor less genetic diversity.

The studies by Templeton et al. and Young et al. show that in at least some natural populations genetic drift leads, as predicted, to random fixation and reduced heterozygosity. The loss of genetic diversity in small populations is of particular concern to conservation biologists, for two reasons. First, genetic diversity is the raw material for adaptive evolution. Imagine a species reduced to a few remnant

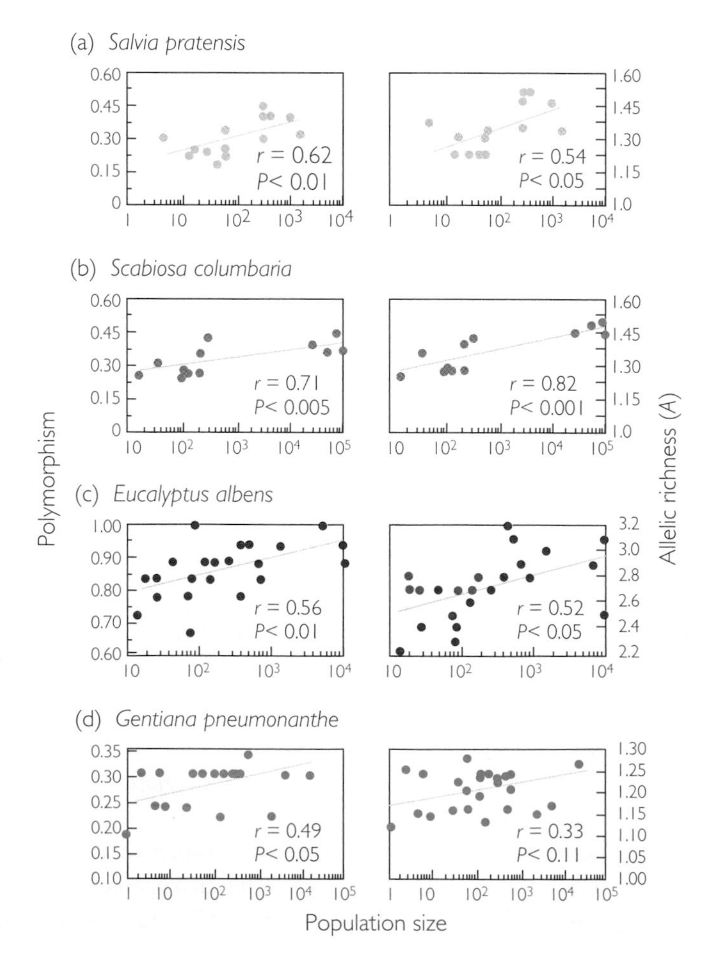

Figure 7.19 Population size and genetic diversity Each data point on these scatterplots represents a population of flowering plants. Polymorphism, plotted on the vertical axis of the graphs at left, is the proportion of allozyme loci at which the frequency of the most common allele in the population is less than 0.99. In other words, polymorphism is the fraction of alleles that are substantially polymorphic. Allelic richness, plotted on the vertical axis of the graphs at right, is the average number of alleles per locus. The statistic r is a measure of association, called the Pearson correlation coefficient, which varies from 0 (no association between variables) to 1 (perfect correlation). P specifies the probability that the correlation coefficient is significantly different from zero. *Salvia pratensis*, *Scabiosa columbaria*, and *Gentiana pneumonanthe* are all flowering herbs; *Eucalyptus albens* is a tree. [The blue dots in (c) right represent small populations isolated by less than 250 m.] Reprinted from Young et al. (1996).

populations by habitat destruction or some other environmental change. Genetic drift may rob the remnant populations of their potential to evolve in response to a changing environment at precisely the moment the environment is changing most drastically. Second, a loss of heterozygosity also entails an increase in homozygosity. Increased homozygosity often leads to reduced fitness in experimental populations (see, for example, Polans and Allard 1989; Barrett and Charlesworth 1991). Presumably this involves the same mechanism as inbreeding depression: It exposes deleterious alleles to selection. We will consider inbreeding depression in Section 7.4.

The Rate of Evolution by Genetic Drift

The theory and experiments we have discussed in this section establish that sampling error can be an important mechanism of evolution. The next aspect of drift that we will consider is the rate of evolution when genetic drift is the only process at work.

First, we need to define what we mean by the rate of evolution at a single locus. We will take the rate of evolution to be the rate at which new alleles created by mutation are substituted for other alleles already present. Figure 7.20 illustrates the process of **substitution** and distinguishes substitution from mutation. The figure follows a gene pool of 10 alleles for 20 generations. Initially, all of the alleles are identical (white dots). In the fourth generation, a new allele appears (light orange dot), created by a mutation in one of the original alleles. Over several generations, this allele drifts to high frequency. In generation 15, a second new allele appears (blue dot), created by a mutation in a descendant of the first light orange allele. In generation 19, the last of the white alleles is lost. At this point, we can say that the light orange allele has been substituted for the white allele. Thus, by evolutionary substitution, we mean the fixation of a new mutation, with or without additional mutational change.

When genetic drift is the only mechanism of evolution at work, the rate of substitution is simply equal to the mutation rate (see Box 7.5). This is true regardless of the population size, because two effects associated with population size cancel each other out: More mutations occur in a larger population, but in a large population each new mutation has a smaller chance of drifting to fixation. Under genetic drift, large populations generate and maintain more genetic variation than small populations, but populations of all sizes accumulate substitutions at the same rate.

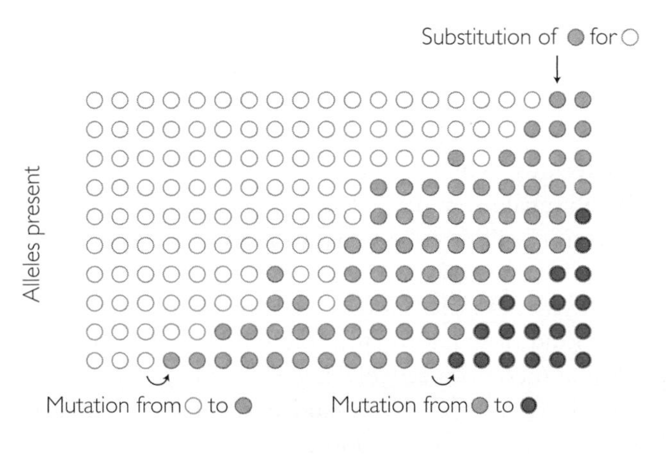

Figure 7.20 Mutation is the creation of a new allele; substitution is the fixation of the new allele, with or without additional mutational change This graph shows the 10 alleles present in each of 20 successive generations, in a hypothetical population of five individuals. During the time covered in the graph, the light orange allele was substituted for the white allele. The blue allele may ultimately be substituted for the light orange allele, or it may be lost.

Box 7.5 The rate of evolutionary substitution under genetic drift

Here we show a calculation establishing that when genetic drift is the only mechanism of evolution at work, the rate of evolutionary substitution is equal to the mutation rate (Kimura 1968).

Imagine a diploid population of size N. Within this population are $2N$ alleles of the locus of interest, where by "alleles" we mean copies of the gene, regardless of whether they are identical or not. Let v be the rate of selectively neutral mutations per allele per generation, and assume that each mutation creates an allele that has not previously existed in the population. Then every generation, there will be

$$2Nv$$

new alleles created by mutation. Because by assumption all new alleles are selectively neutral, genetic drift is the only process at work. Each new allele has the same chance of drifting to fixation as any other allele in the population. That chance, equal to the frequency of the new allele, is

$$\frac{1}{2N}$$

Therefore, each generation the number of new alleles that are created by mutation and are destined to drift to fixation is

$$2Nv \times \frac{1}{2N} = v$$

The same argument applies to every generation. Therefore, the rate of evolution at the locus of interest is v substitutions per generation.

Mutation, selection, and drift in molecular evolution

It will be useful for our discussion of molecular evolution to explore in more detail what we mean by v, the rate of neutral mutations. Imagine that the locus of interest is a gene encoding a protein that is L amino acids long. Let u be the rate of mutations per codon per generation. The overall rate of mutation at our locus is given by

$$\mu = uL(d + a + f) = uLd + uLa + uLf$$

where d is the fraction of codon changes that are deleterious, a is the fraction that are selectively advantageous, f is the fraction that are selectively neutral, and $d + a + f = 1$. Note that the rightmost term, uLf, is equal to our earlier v.

In showing that the rate of substitution is equal to v, we assumed that d and a are both equal to zero. In any real population, of course, many mutations are deleterious and d is not zero. This does not change our calculation of the substitution rate. Deleterious alleles are eliminated by natural selection and do not contribute to the rate of evolutionary substitution.

Proponents of the neutral theory hold that a is approximately equal to zero and that f is much larger than a. Therefore, they predict that evolutionary substitution will be dominated by neutral mutations and drift and will occur at the rate $v = uLf$, as we have calculated.

Proponents of the selectionist theory hold that a is too large to ignore and that the rate of evolutionary substitution will be significantly influenced by the action of natural selection in favor of advantageous alleles.

Of course, mechanisms of evolution other than drift are often at work. We can allow some natural selection into our model and still get a similar result. Imagine that some mutations are deleterious, while others are selectively neutral. The deleterious mutations will be eliminated by natural selection and will never become fixed. The rate of substitution will then be equal to the rate at which neutral mutations occur.

Evolutionary biologists are divided on the relevance of this calculation to real populations. All agree that one kind of mutation and one kind of selection have been left out (see Box 7.5). Some mutations are selectively advantageous and are swept to fixation by natural selection more surely and much faster than drift would ever carry them. Evolutionists are of two minds, however, over how often this happens.

Proponents of the **neutral theory**, long championed by Motoo Kimura (1983), hold that advantageous mutations are exceedingly rare and that most alleles of most genes are selectively neutral. Neutralists predict that for most genes in most populations, the rate of evolution will, indeed, be equal to the neutral mutation rate.

Proponents of the **selectionist theory**, most strongly championed by John Gillespie (1991), hold that advantageous mutations are common enough that they cannot be ignored. Selectionists predict that for many genes in most populations, the rate of substitution will reflect the action of natural selection on advantageous mutations.

The neutralist–selectionist debate has largely been fought in the arena of molecular evolution, so that is where we will go, in the next section, to explore it.

In summary, genetic drift is a nonadaptive mechanism of evolution. Simply as a result of chance sampling error, allele frequencies can change from one generation to the next. Genetic drift is most powerful in small populations and when compounded over many generations. Ultimately, genetic drift leads to the fixation of some alleles and the loss of others and to an overall decline in genetic diversity.

When mutation, genetic drift, and selection interact, three processes occur: (1) Deleterious alleles appear and are eliminated by selection; (2) neutral mutations appear and are fixed or lost by chance; and (3) advantageous alleles appear and are swept to fixation by selection. The relative importance of (2) and (3) in determining the overall substitution rate is a matter of debate.

7.3 Genetic Drift and Molecular Evolution

The field of molecular evolution was launched in the mid-1960s, when biochemists succeeded in determining the amino acid sequences of hemoglobin, cytochrome *c*, and other particularly abundant and well-studied proteins found in humans and other vertebrates. These data sets provided the first opportunity for evolutionary biologists to compare the amount and rate of molecular change among species.

Early workers in the field made several striking observations about these data sets. Foremost among them were calculations made by Motoo Kimura (1968). Kimura took the number of sequence differences observed in the well-studied proteins of humans and horses and scaled them for time using divergence dates from the fossil record. He then extrapolated these rates of molecular evolution to all of the protein-coding loci in the genome. His calculations implied that as the two lineages diverged from their common ancestor, novel mutations leading to amino acid replacements had, on average, risen to fixation once every two years. Given that most mutations should be deleterious, this rate seemed far too high to be due to natural selection. Beneficial mutations fixed by natural selection should be extremely rare.

A second observation, contributed by Emil Zuckerkandl and Linus Pauling (1965), was that the rate of amino acid sequence change in certain proteins appeared to be constant through time, or clocklike, during the diversification of vertebrates. This result also seemed to be inconsistent with the action of natural selection, which should be episodic in nature and correlated with changes in the environment rather than with time.

Early analyses of molecular evolution suggested that rates of change were high and constant through time. These conclusions appeared to be in conflict with what might be expected under natural selection.

In short, early data on molecular evolution did not match the expectation that natural selection was responsible for most evolutionary change. The results raised an important question: If natural selection does not explain evolution at the molecular level, then what process is responsible for the observation of rapid, clocklike change? Many researchers believe that the answer is genetic drift.

The Neutral Theory of Molecular Evolution

The neutral theory models the fate of new alleles that were created by mutation and whose frequencies change by genetic drift. It claims to explain most evolutionary change at the level of nucleotide sequences.

Kimura (1968, 1983) formulated the Neutral Theory of Molecular Evolution to explain the observed patterns of amino acid sequence divergence. This theory claims that the vast majority of nucleotide changes that become fixed in populations are neutral with respect to fitness and that genetic drift dominates evolution at the level of DNA sequences. Kimura held that natural selection on beneficial mutations is largely inconsequential as an explanation for the differences among species observed at the molecular level. Based on the calculation in Box 7.5, Kimura postulated that the rate of molecular evolution is equal to the mutation rate. Kimura's theory was astonishing to many evolutionary biologists, for two reasons:

1. The size of the population has no role. The models and experiments we reviewed in Section 7.2 showed that genetic drift is far more effective at changing the frequency of alleles in small populations than in large ones. But Kimura showed that, for strictly neutral mutations, the rate of fixation of novel alleles due to drift does not depend on population size.
2. Positive natural selection is excluded. The theory's central claim is that the vast majority of base substitutions are neutral. The overall rate of evolution will be equal to the neutral mutation rate only when this proposition is true. Proponents of the neutral theory go further, however, by pointing out that even if a small proportion of nonneutral mutations occur in a population, they are likely to be deleterious and rapidly eliminated by natural selection. Thus, the mutation rate v will represent the maximum rate of evolutionary change measured.

Although Kimura's theory appeared to explain why the amino acid sequences of hemoglobin, cytochrome *c*, and other proteins change steadily through time, the theory was inspired by fairly limited amounts of protein-sequence data. How did the neutral theory hold up, once large volumes of DNA sequence data became available?

Patterns in DNA Sequence Divergence

During the late 1970s and 1980s researchers mined growing databases of DNA sequences to analyze the amounts and rates of change in different loci. To discern meaningful patterns in the data, it became routine to create categories defined by the type of sequence being considered. The most basic distinction was between coding and noncoding sequences. Coding sequences contain the instructions for tRNAs, rRNAs, or proteins; noncoding sequences include introns, regions that flank coding regions, regulatory sites, and the unusual loci called pseudogenes that were introduced in Chapters 2 and 5. What predictions does the neutral theory make about the rate and pattern of change in these different types of sequences? Have these predictions been verified or rejected?

Pseudogenes Establish a Canonical Rate of Neutral Evolution

The evolution of pseudogenes conforms to the assumptions and predictions of the neutral theory.

Pseudogenes are functionless stretches of DNA that result from gene duplication events (see Chapter 5). Because they do not code for proteins, mutations in pseudogenes should be completely neutral with respect to fitness and should increase to fixation solely as a result of genetic drift. For this reason, pseudogenes are considered a paradigm of neutral evolution (Li et al. 1981). As predicted by the neutral theory, the divergence rates recorded in pseudogenes—which should be equal to the neutral mutation rate v—are among the highest observed among loci and sites in nuclear genomes (Li et al. 1981, Li and Graur 2000). This finding provided strong support for the neutral theory as an explanation for evolutionary change at the molecular level. It also quantified the rate of evolution due to drift. How do rates of change in other types of sequences compare to this standard, or canonical, rate of evolution caused by drift?

Silent Sites Change Faster than Replacement Sites in Most Coding Loci

Chapter 5 pointed out that two basic types of point mutations occur in the coding regions of genes. To review, recall that bases in DNA are read in groups of three, called codons, and that only 20 different amino acids need to be specified by the 64 codons in the genetic code. This means that the genetic code contains considerable redundancy. In the portion of the code shown in Figure 7.21a, for example, two codons specify phenylalanine, two specify leucine, and four code for serine. As a result of the code's redundancy, base-pair changes may or may not lead to amino acid sequence changes (Figure 7.21b). DNA sequence changes that do not result in amino acid changes are called **silent-site** (or **synonymous**) **mutations**; sequence changes that do result in an amino-acid change are called **replacement** (or **nonsynonymous**) **mutations**.

Figure 7.21c presents data on the rate of replacement versus silent substitution in a gene of the influenza virus, based on comparisons of a 1968 flu virus with samples collected over the subsequent 20 years (Gojobori et al. 1990). Both kinds of substitution accumulated in a linear, clocklike fashion, but the rate of silent changes is much higher than the rate of replacement changes.

This pattern accords with the neutral theory in important ways. Silent changes are not exposed to natural selection on protein function because they do not

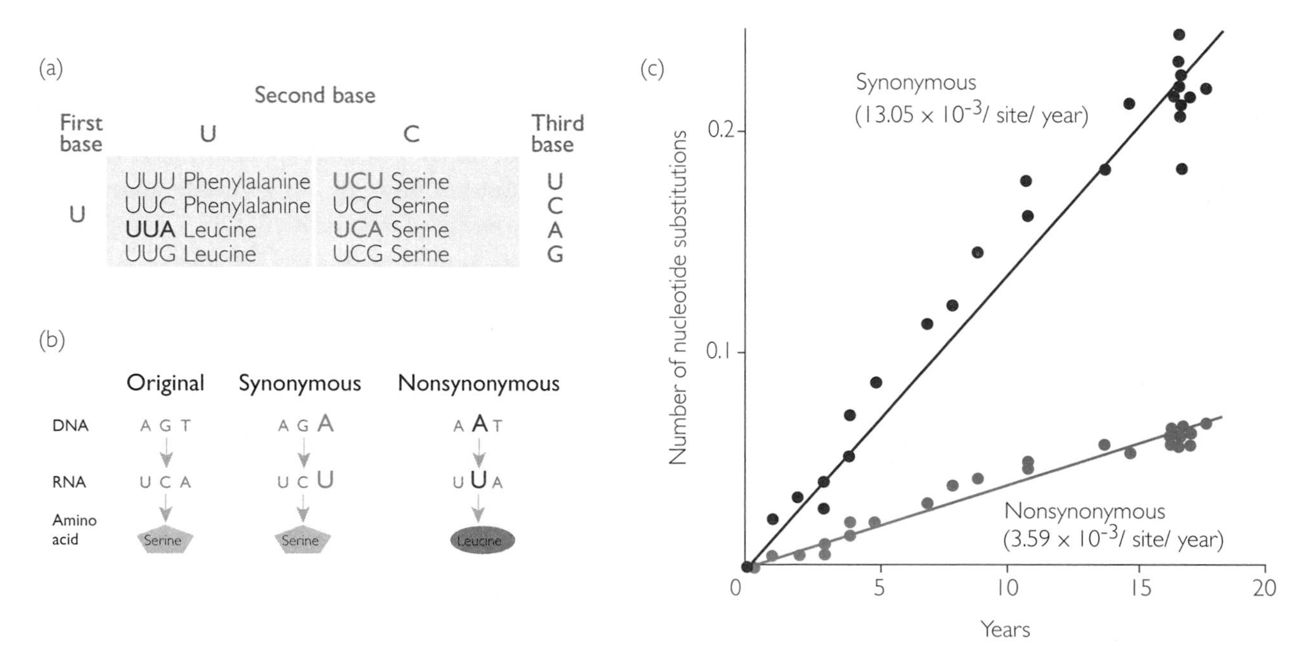

Figure 7.21 **Molecular evolution in influenza viruses is consistent with the neutral theory.** Because the genetic code is redundant (a), there are two kinds of point mutations (b). The neutral theory predicts that both kinds of substitution will accumulate in populations by genetic drift, but that synonymous, or silent, substitutions will accumulate faster, as happens in flu viruses (c). From Gojoburi et al. (1990).

Natural selection against deleterious mutations is called negative selection. Natural selection favoring beneficial mutations is called positive selection.

In most coding sequences, substitution rates are higher at silent sites than at replacement sites. This is consistent with the notion that molecular evolution is dominated by drift and negative selection.

alter the amino acid sequence. As a result, new alleles created by silent mutations should increase or decrease in frequency through time, largely as a result of drift. Replacement mutations, in contrast, change the amino acid sequences of proteins. If most of these alterations are deleterious, then they should be eliminated by natural selection. (This type of natural selection is called **negative** or **purifying selection**, as opposed to **positive selection** on beneficial mutations.) Less frequently, replacement mutations occur that have no effect on protein function and may be fixed by drift.

Molecular biologists have compared the rate of silent versus replacement substitutions in a great variety of coding loci (Table 7.1). In the vast majority, the rate of silent changes is far higher than the rate of replacement changes.

In a similar vein, Austin Hughes and colleagues (2003) examined the DNA of 102 ethnically diverse humans to quantify the amount of standing genetic diversity at 1,442 single nucleotide polymorphism sites—that is, points in the genome at which some individuals have one nucleotide and other individuals have another. The researchers found lower standing diversity, measured as the fraction of individuals who are heterozygotes, for polymorphisms that involve amino acid changes versus polymorphisms that do not. The implication is that most single-nucleotide mutations that swap one amino acid for another in a protein are deleterious and held at low frequency by negative selection.

These observations are consistent with the patterns predicted if most mutations are either deleterious or neutral and drift dominates molecular evolution. As a result, they support the central tenet of the neutral theory.

Variation among Loci: Evidence for Functional Constraints

The data in Table 7.1 contain another important pattern. When homologous coding sequences from humans and rodents are compared, some loci are found to be nearly identical, while others have undergone rapid divergence. This result turns out to be typical. Rates of molecular evolution vary widely among loci.

The key to explaining this pattern lies in the following observation: Genes that are responsible for the most vital cellular functions appear to have the lowest rates of replacement substitutions. Histone proteins, for example, interact with DNA to form structures called nucleosomes. These protein–DNA complexes are a major feature of the chromatin fibers in eukaryotic cells. Changes in the amino acid sequences of histones disrupt the structural integrity of the nucleosome, with negative consequences for DNA transcription and synthesis. In contrast, genes that are less vital to the cell, and thus under less stringent functional constraints, show more rapid rates of replacement substitutions. When functional constraints are lower, a larger percentage of replacement mutations are neutral with respect to fitness and may be fixed by drift.

The Nearly Neutral Model

Although the neutral theory appeared to account for several important patterns in DNA sequence data, data sets that indicated clocklike change in proteins compared across species presented a problem. The issue was that the neutral mutation rate v should vary among species as a function of generation time. Over any given time interval, more neutral mutations should occur in species with short generation times than in species with long generation times. Contrary to expectation, at least some protein sequence comparisons appeared to undergo clock-

Table 7.1 Rates of nucleotide substitution vary among genes and among sites within genes

These data report rates of replacement and silent substitutions in protein-coding genes compared between humans and either mice or rats. The data are expressed as the average number of substitutions per site per billion years, plus or minus a statistical measure of uncertainty called the standard error. L is the number of codons compared.

Gene	L	Replacement rate ($\times 10^9$)	Silent rate ($\times 10^9$)
Histones			
Histone 3	135	0.00 ± 0.00	6.38 ± 1.19
Histone 4	101	0.00 ± 0.00	6.12 ± 1.32
Contractile system proteins			
Actin a	376	0.01 ± 0.01	3.68 ± 0.43
Actin b	349	0.03 ± 0.02	3.13 ± 0.39
Hormones, neuropeptides, and other active peptides			
Somatostatin-28	28	0.00 ± 0.00	3.97 ± 2.66
Insulin	51	0.13 ± 0.13	4.02 ± 2.29
Thyrotropin	118	0.33 ± 0.08	4.66 ± 1.12
Insulin-like growth factor II	179	0.52 ± 0.09	2.32 ± 0.40
Erythropoietin	191	0.72 ± 0.11	4.34 ± 0.65
Insulin C-peptide	35	0.91 ± 0.30	6.77 ± 3.49
Parathyroid hormone	90	0.94 ± 0.18	4.18 ± 0.98
Luteinizing hormone	141	1.02 ± 0.16	3.29 ± 0.60
Growth hormone	189	1.23 ± 0.15	4.95 ± 0.77
Urokinase plasminogen activator	435	1.28 ± 0.10	3.92 ± 0.44
Interleukin I	265	1.42 ± 0.14	4.60 ± 0.65
Relaxin	54	2.51 ± 0.37	7.49 ± 6.10
Hemoglobins and myoglobin			
α-globin	141	0.55 ± 0.11	5.14 ± 0.90
Myoglobin	153	0.56 ± 0.10	4.44 ± 0.82
β-globin	144	0.80 ± 0.13	3.05 ± 0.56
Apolipoproteins			
E	283	0.98 ± 0.10	4.04 ± 0.53
A-I	243	1.57 ± 0.16	4.47 ± 0.66
A-IV	371	1.58 ± 0.12	4.15 ± 0.47
Immunoglobulins			
Ig V_H	100	1.07 ± 0.19	5.66 ± 1.36
Igγ1	321	1.46 ± 0.13	5.11 ± 0.64
Igκ	106	1.87 ± 0.26	5.90 ± 1.27
Interferons			
α1	166	1.41 ± 0.13	3.53 ± 0.61
β1	159	2.21 ± 0.24	5.88 ± 1.08
γ	136	2.79 ± 0.31	8.59 ± 2.56

Source: Li and Graur (1991)

like change in absolute time—independent of differences in generation time among the species being compared.

To account for this observation, Tomoko Ohta and Motoo Kimura (1971; Ohta 1972, 1977) developed mathematical models exploring how drift and selection would affect mutations that are slightly deleterious, instead of being strictly neutral. Ohta's work showed that mutations are effectively neutral—meaning that they are fixed or eliminated by drift instead of selection—when

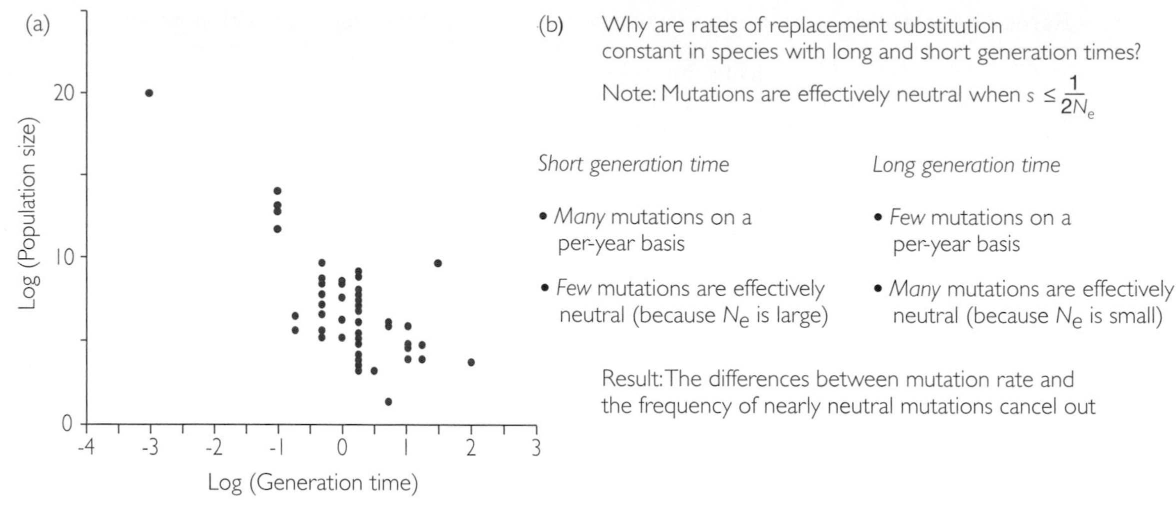

Figure 7.22 Generation time, population size, and nearly neutral mutations (a) This graph plots the logarithm of population size versus the logarithm of generation time. Statistical tests confirm that there is a strong negative correlation between the two variables. From L. Chao and D. E. Carr (1993). (b) Differences in generation time and population size can lead to clocklike change in replacement substitutions that are nearly neutral with respect to fitness.

$s \leq \frac{1}{2N_e}$, where s is the selection coefficient and N_e is the effective population size (the number of breeding adults).

The nearly neutral model explains why, in some cases, rates of sequence change correlate with absolute time instead of generation time.

How does this nearly neutral model explain the observation of molecular clocks in absolute time? As Lin Chao and David Carr (1993) have shown, there is a strong negative correlation between average population size in a species and its generation time. Species with short generation times tend to have large populations; species with long generation times tend to have small populations (Figure 7.22a). This is important because, according to Ohta's model, drift fixes a larger percentage of mutations in organisms with small population sizes. The upshot is that an increase in evolutionary rate due to the fixation of nearly neutral mutations in these small-population, long-generation species offsets the higher mutation rate in short-generation species and results in the molecular clock (Figure 7.22b). Consistent with this view, most studies have shown that replacement substitutions show relatively small differences in rates among mammalian lineages with different generation times (see Li and Tanimura 1987; Li et al. 1987). As predicted by the neutral theory, silent substitutions in mammals show much more pronounced generation-time effects.

The Neutral Theory as a Null Hypothesis: Detecting Natural Selection on DNA Sequences

The neutralist-selectionist controversy is a debate about the relative importance of drift and positive selection in explaining molecular evolution.

Since their inception, the neutral and nearly neutral theories have been controversial (see Berry 1996; Ohta and Kreitman 1996). Discussion has focused on the claims by Kimura (1983) and King and Jukes (1969) that the number of beneficial mutations fixed by positive natural selection is inconsequential compared to the number of mutations that change in frequency under the influence of drift. Is this claim accurate? How can we determine that natural selection has been responsible for changes observed at the molecular level?

When researchers compare homologous DNA sequences among individuals and want to explain the differences they observe, they routinely use the neutral theory as a null hypothesis. The neutral theory specifies the rates and patterns of sequence change that occur in the absence of natural selection. If the changes that are actually observed are significantly different from the predictions made by the neutral theory, and if a researcher can defend the proposition that the sequences in question have functional significance for the organism, then there is convincing evidence that natural selection has caused molecular evolution.

Here we examine a few of the strategies being used to detect molecular evolution due to natural selection. We begin with studies of replacement changes, then explore evidence that many silent-site mutations are also under selection.

Selection on Replacement Mutations

We noted above that according to the neutral theory silent mutations are expected to evolve largely by genetic drift. Replacement mutations are expected to be deleterious, in which case they are eliminated by negative selection and we will not see them, or to be neutral, in which case they, too, evolve by drift. If the neutral theory is wrong for a particular gene, however, and replacement mutations are advantageous, then they will be rapidly swept to fixation by positive selection. Thus, to find out whether replacements within a particular gene are deleterious, neutral, or advantageous, we can compare two sequences and calculate the rate of nonsynonymous subsitutions per site (d_N) and the rate of synonymous substitutions per site (d_S). If we take their ratio, we will get

$$\frac{d_N}{d_S} < 1 \text{ when replacements are deleterious,}$$

$$\frac{d_N}{d_S} = 1 \text{ when replacements are neutral, and}$$

$$\frac{d_N}{d_S} > 1 \text{ when replacements are advantageous.}$$

When sequences evolve by drift and negative selection, synonymous substitutions outnumber replacement substitutions. When sequences evolve by drift and positive selection, replacement substitutions outnumber synonymous substitutions.

Austin Hughes and Masatoshi Nei (1988) tested the neutral theory by estimating the ratio of replacement to silent substitutions in genes vital to immune function. When mammalian cells are infected by a bacterium or a virus, they respond by displaying pieces of bacterial or viral protein on their surfaces. Immune system cells then kill the infected cell. (This prevents the bacterium or virus inside the cell from replicating.) The membrane proteins that display pathogen proteins are encoded by a cluster of genes called the major histocompatibility complex, or MHC. The part of an MHC protein that binds to the foreign peptide is called the antigen recognition site (ARS). Hughes and Nei (1988) studied sequence changes in the ARS of MHC loci in humans and mice.

When Hughes and Nei compared alleles from the MHC complexes of 12 different humans and counted the number of differences observed in silent versus replacement sites, they found significantly more replacement site than silent site changes. The same pattern occurred in the ARS of mouse MHC genes, although the differences were not as great. This pattern could only result if the replacement changes were selectively advantageous. The logic here is that positive selection causes replacement changes to spread through the population much more quickly than neutral alleles can spread by chance.

It is important to note, however, that Hughes and Nei found this pattern only in the ARS. Other exons within the MHC showed more silent than replacement changes or no difference. At sites other than the ARS, then, they could not rule out the null hypothesis that sequence change is dominated by drift.

Many examples have been found in which replacement substitutions outnumber synonymous substitutions, a signature of positive selection.

Research by Gavin Huttley and colleagues (2000) on *BRCA1*, a gene associated with breast cancer, provides another example. These researchers sequenced exon 11 from the BRCA1 genes of a variety of mammals, then inferred the rates of nonsynonymous and synonymous substitution along the branches of the evolutionary tree that connects the extant species to their common ancestors (Figure 7.23). Along most branches of the phylogeny the value of $\frac{d_N}{d_S}$ was less than one, consistent with the neutral theory. On the branches connecting humans and chimpanzees to their common ancestor, however, $\frac{d_N}{d_S}$ was significantly greater than one. This suggests that the sequence of exon 11 has been under positive selection in the ancestors of today's humans and chimps. The selective agent responsible remains unknown.

Comparing Silent and Replacement Changes within and between Species. The research by Hughes and Nei and by Huttley and colleagues provides clear examples of gene segments where neutral substitutions do not predominate. Thanks to the efforts of numerous researchers, many other loci have been found where replacement substitutions outnumber silent substitutions.

Even though the $\frac{d_N}{d_S}$ criterion for detecting positive selection has been useful Paul Sharp (1997) points out that it is extremely conservative. Replacement substitutions will only outnumber silent substitutions when positive selection has been very strong. In a comparison of 363 homologous loci in mice and rats, for example, only one showed an excess of replacement over silent changes. But as Sharp notes (1997: 111), "it would be most surprising if this were the only one of these genes that had undergone adaptive changes during the divergence of the two species." Are more sensitive methods for detecting natural selection available?

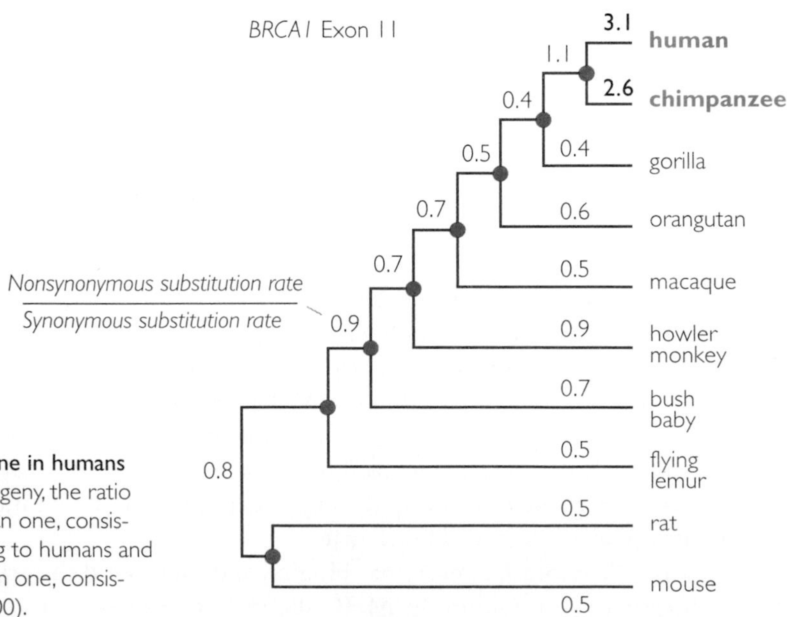

Figure 7.23 Positive selection on the *BRCA1* gene in humans and chimpanzees On most branches of this phylogeny, the ratio of replacement to silent substitution rates is less than one, consistent with neutral evolution. On the branches leading to humans and chimps, however, the ratio is significantly greater than one, consistent with positive selection. From Huttley et al. (2000).

John McDonald and Martin Kreitman (1991) invented a test for natural selection that is widely used. The McDonald–Kreitman, or MK, test is based on an important corollary to the neutral theory's prediction that silent substitutions occur more rapidly than replacement substitutions. According to the neutral theory, the ratio of replacement to silent-site substitutions in any particular locus should be constant through time. Based on this proposition, McDonald and Kreitman predicted that the ratio of replacement to silent-site substitutions in between-species comparisons should be the same as the ratio of nonsynonymous to synonymous polymorphisms in within-species comparisons.

Their initial test of this prediction compared sequence data from the alcohol dehydrogenase (*Adh*) gene of 12 *Drosophila melanogaster*, 6 *D. simulans*, and 12 *D. yakuba* individuals. *Adh* was an interesting locus to study for two reasons: Fruit flies feed on rotting fruit that may contain toxic concentrations of ethanol, and the alcohol dehydrogenase enzyme catalyzes the conversion of ethanol to a nontoxic product. Because of the enzyme's importance to these species, and because ethanol concentrations vary among food sources, it is reasonable to suspect that the locus is under strong selection when populations begin exploiting different fruits.

In an attempt to sample as much within-species variation as possible, the individuals chosen for the study were from geographically widespread locations. McDonald and Kreitman aligned the *Adh* sequences from each individual in the study and identified sites where a base differed from the most commonly observed nucleotide, or what is called the consensus sequence. The researchers counted differences as fixed if they were present in all individuals from a particular species, and as **polymorphisms**—or allele differences within species—if they were present in only some individuals from a particular species. Differences that were fixed in one species and polymorphic in another were counted as polymorphic.

McDonald and Kreitman found that 29% of the differences that were fixed between species were replacement substitutions. Within species, however, only 5% of the polymorphisms in the study represented replacements. Rather than being the same, these ratios show an almost sixfold, and statistically significant, difference ($P = 0.006$). This is strong evidence against the neutral model's prediction. McDonald and Kreitman's interpretation is that the differences in replacement mutations fixed in different species are selectively advantageous. They suggest that these mutations occurred after *D. melanogaster*, *D. simulans*, and *D. yakuba* had diverged and spread rapidly to fixation due to positive selection in the differing environments occupied by these species.

Using the MK test, natural selection has now been detected in loci from plants, protists, and a variety of animals (Escalante et al. 1998; Purugganen and Suddith 1998). With an extension of the MK test applied to 35 genes in *D. simulans* and *D. yakuba*, Nick Smith and Adam Eyre-Walker (2002) estimated that 45% of all amino acid substitutions between the genomes of the two species were fixed by positive selection. With an extension applied to the genomes of humans and chimpanzees, Carlos Bustamante and colleagues (2005) identified 304 human genes that have evolved under positive selection.

Which Loci Are under Strong Positive Selection? Thanks to studies employing the Hughes and Nei analysis, the MK test, and other strategies, generalizations are beginning to emerge concerning the types of loci where positive natural selection has been particularly strong (Yang and Bielawki 2000; Vallender and Lahn

2004; Nielsen 2005; Nielsen et al. 2005). Replacement substitutions appear to be particularly abundant in loci involved in arms races between pathogens and their hosts, in loci with a role in reproductive conflicts such as sperm competition and egg-sperm interactions, and in recently duplicated genes that have attained new functions. Positive selection has also been detected in genes involved in sex determination, gametogenesis, sensory perception, interactions between symbionts, tumor suppression, and programmed cell death, as well as in genes that code for certain enzymes or regulatory proteins. Table 7.2 lists a few specific examples.

As data accumulate from genome-sequencing projects in closely related species, such as humans and chimpanzees, the number and quality of comparative studies are exploding. Even before the era of genome sequencing began, however, it became clear that silent substitutions, as well as replacement changes, are subject to natural selection.

Selection on "Silent" Mutations

The term silent mutation was coined to reflect two aspects of base changes at certain positions of codons: They do not result in a change in the amino acid sequence of the protein product, and they are not exposed to natural selection. The second proposition had to be discarded, however, in the face of data on phenomena known as codon bias, hitchhiking, and background selection. How can mutations that do not alter an amino acid sequence be affected by natural selection?

Direct Selection on Synonymous Mutations: Codon Bias and Other Factors. Most of the 20 amino acids are encoded by more than one codon. We have emphasized

Table 7.2 Studies that confirm positive selection on replacement mutations

Although this table lists just a few examples, it underscores a general point: Evidence for positive selection is particularly strong in genes that code for proteins involved in disease resistance, reproductive conflict, interactions between symbionts, and the development of novel traits.

Gene	Species	Rationale	Reference
MHC Class II	Humans	Strong selection for divergence among antigen-recognition proteins	Hughes, A.L., and M. Nei. 1989. *Proceedings of the National Academy of Sciences USA* 86: 958–962.
Semenogelin II gene	Primates	Strong selection for semen coagulation in species with intense sperm competition	Dorus, S. D., et al. 2004. *Nature Genetics* 12: 1326-1329.
Lysin protein and receptor	Abalone	Strong selection on species-specific egg-recognition proteins on sperm	Swanson, W.J, and V.D. Vacquier. 1998. *Science* 281: 710–712.
Eosinophil cationic protein	Primates	Strong selection on a recently duplicated gene involved in disease resistance	Zhang, J., H.F. Rosenberg, and M. Nei, 1998. *Proceedings of the National Academy of Sciences USA* 95: 3708–3713.
Self-incompatibility loci	Tomato family plants	Strong selection for divergence among proteins involved in self-fertilization	Clark, A.G., and T.-H. Kao. 1991. *Proceedings of the National Academy of Sciences USA* 88: 9823–9827.
Multicolored fluorescent proteins	Reef-building corals	Evolving algal sybionts impose selection on ability of host corals to regulate them.	Field, S. F., et al. 2006. *Journal of Molecular Evolution* 62: 332-339.
ASPM	Humans and great apes	Selection for increased brain size.	Evans, P. D., et al. 2004. *Human Molecular Genetics* 13: 489-494.

that changes among redundant codons do not cause changes in the amino-acid sequences of proteins, and have implied that these silent-site changes are neutral with respect to fitness. If this were strictly true, we would expect that codon usage would be random, and that each codon in a suite of synonymous codons would be present in equal numbers throughout the genome of a particular organism. But early sequencing studies confirmed that codon usage is highly nonrandom (Table 7.3). This phenomenon is known as **codon bias**.

Several important patterns have emerged from studies of codon bias. In every organism studied to date, codon bias is strongest in highly expressed genes—such as those for the proteins found in ribosomes—and weak to nonexistent in rarely expressed genes. In addition, the suite of codons that are used most frequently correlates strongly with the most abundant species of tRNA in the cell (Figure 7.24).

The leading hypothesis to explain these observations is natural selection for translational efficiency (Sharp and Li 1986; Sharp et al. 1988; Akashi 1994). The logic here is that if a "silent" mutation in a highly expressed gene creates a codon that is rare in the pool of tRNAs, the mutation will be selected against. The selective agent is the speed and accuracy of translation. Speed and accuracy are especially important when the proteins encoded by particular genes are turning over rapidly and the corresponding genes must be transcribed continuously. It is reasonable, then, to observe the strongest codon bias in highly expressed genes.

Codon bias suggests that some synonymous mutations are not selectively neutral.

Selection against certain synonymous mutations represents a form of negative selection; it slows the rate of molecular evolution. As a result, codon bias may explain

Table 7.3 Codon bias

This table reports the relative frequencies of codons found in genes from three different species the bacterium *Escherichia coli*, baker's yeast (*Saccharomyces cerevisiae*), and the fruit fly *Drosophila melanogaster*. If each codon were used equally in each genome, the relative frequencies would all be 1. Deviations from 1.00 indicate codon bias. The amino acids listed are leucine, valine, isoleucine, phenylalanine, and methionine. "High" and "Low" differentiate data from highly transcribed versus rarely transcribed genes. In every case reported here, codon bias is more extreme in highly expressed genes.

		Escherichia coli		*Saccharomyces cerevisiae*		*Drosophila melanogaster*	
Amino acid	**Codon**	**High**	**Low**	**High**	**Low**	**High**	**Low**
Leu	UUA	0.06	1.24	0.49	1.49	0.03	0.62
	UUG	0.07	0.87	5.34	1.48	0.69	1.05
	CUU	0.13	0.72	0.02	0.73	0.25	0.80
	CUC	0.17	0.65	0.00	0.51	0.72	0.90
	CUA	0.04	0.31	0.15	0.95	0.06	0.60
	CUG	5.54	2.20	0.02	0.84	4.25	2.04
Val	GUU	2.41	1.09	2.07	1.13	0.56	0.74
	GUC	0.08	0.99	1.91	0.76	1.59	0.93
	GUA	1.12	0.63	0.00	1.18	0.06	0.53
	GUG	0.40	1.29	0.02	0.93	1.79	1.80
Ile	AUU	0.48	1.38	1.26	1.29	0.74	1.27
	AUC	2.51	1.12	1.74	0.66	2.26	0.95
	AUA	0.01	0.50	0.00	1.05	0.00	0.78
Phe	UUU	0.34	1.33	0.19	1.38	0.12	0.86
	UUC	1.66	0.67	1.81	0.62	1.88	1.14
Met	AUG	1.00	1.00	1.00	1.00	1.00	1.00

Source: Sharp et al. (1988)

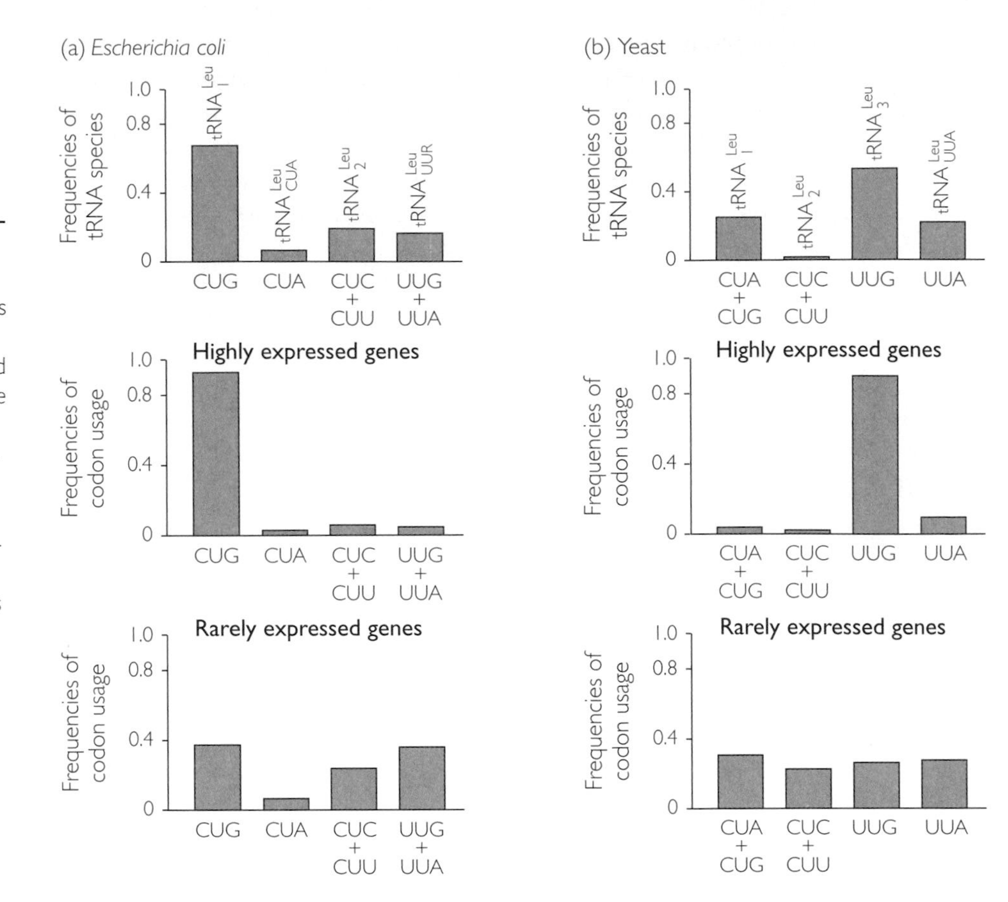

Figure 7.24 Codon bias correlates with the relative frequencies of tRNA species The bar chart in the top row of both (a) and (b) shows the frequencies of four different tRNA species that carry leucine in *E. coli* (a) and the yeast *Saccharomyces cerevisiae* (b). The bar charts in the middle and bottom rows report the frequency of the mRNA codons corresponding to each of these tRNA species in the same organisms. The mRNA codons were measured in two different classes of genes: those that are highly transcribed (middle) and those that are rarely transcribed (bottom). The data show that codon usage correlates strongly with tRNA availability in highly expressed genes, but not at all in rarely expressed genes. From Li and Graur (1991).

the observation that silent changes do not accumulate as quickly as base changes in pseudogenes. Other synonymous mutations may experience selection as a result of their effects on mRNA stability or exon splicing (see Chamary et al. 2006). The general message here is that not all redundant sequence changes are "silent" with respect to natural selection.

Indirect Effects on Synonymous Mutations: Hitchhiking and Background Selection. Another phenomenon that affects the rate and pattern of change at silent sites is referred to as **hitchhiking**, or a **selective sweep**. Hitchhiking can occur when strong positive selection acts on a particular amino acid change. As a favorable mutation increases in frequency, neutral or even slightly deleterious mutations closely linked to the favored site will increase in frequency along with the beneficial locus. These linked mutations are swept along by selection and can actually "hitchhike" to fixation. Note that this process only occurs when recombination fails to break up the linkage between the hitchhiking sites and the site that is under selection.

Perhaps the best example of hitchhiking discovered to date was found on the fourth chromosome of fruit flies. The fourth chromosome in *Drosophila* is remarkable because no recombination occurs along its entire length. As a consequence, the entire chromosome represents a linkage group and is inherited like a single gene.

Andrew Berry and colleagues (1991) sequenced a 1.1-kb region of the fourth chromosome in 10 *Drosophila melanogaster* and 9 *D. simulans*. This chromosome region includes the introns and exons of a gene expressed in fly embryos, called *cubitus interruptus Dominant* (*ciD*). Berry et al. found a remarkable pattern in the sequence data: No differences were observed among the *D. melanogaster* individuals surveyed. The entire 1.1 kb of sequence was identical in the 10 individuals. Furthermore, only one base difference was observed among the *D. simulans* in the study. This means that there was almost no polymorphism (variation within species) in this region. In contrast, a total of 54 substitutions were found when the same sequences were compared between the two species. A key observation here is that genes on other chromosomes surveyed in the same individuals showed normal amounts of polymorphism. These latter data serve as a control and confirm that the lack of variation observed in and around the *ciD* locus is not caused by an unusual sample of individuals. Rather, there is something unusual about the fourth chromosome in these flies.

Berry et al. suggest that selective sweeps recently eliminated all or most of the variation on the fourth chromosome in each of these two species. They argue that an advantageous mutation anywhere on the fourth chromosome would eliminate all within-species polymorphism as it increased to fixation. New variants, like the one polymorphism observed in the *D. simulans* sampled, will arise only through mutation. In this way, selective sweeps create a "footprint" in the genome: a startling lack of polymorphism within linkage groups. Similar footprints have been found in other chromosomal regions where the frequency of recombination is low, including the ZFY locus of the human Y chromosome (Dorit et al. 1995) and a variety of loci in *D. melanogaster* and other fruit flies (for example, see Nurminsky et al. 1998).

Has hitchhiking produced all of these regions of reduced polymorphism? The answer is probably not. Another process, called background selection, can produce a similar pattern (Charlesworth et al. 1993). Background selection results from negative selection against deleterious mutations, rather than positive selection for advantageous mutations. Like hitchhiking, it occurs in regions of reduced recombination. The idea here is that selection against deleterious mutations removes closely linked neutral mutations and produces a reduced level of polymorphism.

Although the processes called hitchhiking and background selection are not mutually exclusive, their effects can be distinguished in at least some cases. Hitchhiking results in dramatic reductions in polymorphism as an occasional advantageous mutation quickly sweeps through a population. Background selection causes a slow, steady decrease in polymorphism as frequent deleterious mutations remove individuals from the population. The current consensus is that hitchhiking is probably responsible for the most dramatic instances of reduced polymorphism in linked regions—for example, where sequence variation is entirely eliminated—while background selection causes the less extreme cases.

The Current Status of the Neutral Theory

The Neutral Theory of Molecular Evolution explains the clocklike evolution of nucleotide sequences that we saw Figure 7.21. It also explains the fact that silent substitutions outnumber replacement substitutions in most genes, as we saw in Figure 7.21 and Table 7.1. And the Neutral Theory serves as a null hypothesis that allows researchers to identify examples of positive selection on nucleotide sequences, as illustrated in Figure 7.23 and reviewed in Table 7.2. By all these criteria, the Neutral Theory of Molecular Evolution is extraordinarily useful.

As a null hypothesis for detecting positive selection in molecular evolution, the neutral theory has been highly successful.

Data are now accumulating that will allow researchers to evaluate the Neutral Theory's claim that most molecular evolution is dominated by negative selection and drift.

What of the Neutral Theory's fundamental claim that the vast majority of nucleotide changes that become fixed in populations are selectively neutral and that molecular evolution is largely the result of genetic drift? What we need to assess this claim is: (1) data for as many substitutions as possible in as many species as possible, and (2) a breakdown of the proportion of substitutions that are neutral versus deleterious versus beneficial. The data that we need are now accumulating. We have already mentioned Smith and Eyre-Walker's (2002) estimate that 45% of the amino acid substitutions distinguishing two species of fruit flies were fixed by positive selection. In similar studies on primates and fruit flies, Justin Fay and colleagues (2001, 2002) have likewise concluded that positive selection is a more important mechanism of molecular evolution than the Neutral Theory claims. Peter Andolfatto (2005) extended the McDonald-Kreitman test to analyze the evolutionary divergence between *Drosophila melanogaster* and *D. simulans* in various kinds of non-coding DNA. He estimated that 20% of the nucleotide differences in DNA located between genes and in introns, and 60% of the nucleotide differences in the untranslated portions of mature mRNAs, evolved as a result of positive selection. These observations suggest that the Neutral Theory's most provacative claim was overstated. It is still too early, however, for definitive general pronouncements on the relative importance of drift versus selection in molecular evolution.

7.4 Nonrandom Mating

We have so far considered what happens in populations when we relax the assumptions of no migration and no genetic drift. The final assumption of the Hardy–Weinberg analysis is that individuals in the population mate at random. In this section we relax that assumption and allow individuals to mate nonrandomly. Nonrandom mating does not, by itself, cause evolution. Nonrandom mating can nonetheless have profound indirect effects on evolution.

The most common type of nonrandom mating, and the kind we will focus on here, is inbreeding. Inbreeding is mating among genetic relatives. The effect of inbreeding on the genetics of a population is to increase the frequency of homozygotes compared to what is expected under Hardy–Weinberg assumptions.

Inbreeding decreases the frequency of heterozygotes and increases the frequency of homozygotes compared to expectations under Hardy–Weinberg assumptions.

To show how this happens, we will consider the most extreme example of inbreeding: self-fertilization, or selfing. Imagine a population in Hardy–Weinberg equilibrium with alleles A_1 and A_2 at initial frequencies of 0.5 each. The frequency of A_1A_1 individuals is 0.25, the frequency of A_1A_2 individuals is 0.5, and the frequency of A_2A_2 individuals is 0.25 (Figure 7.25a). Imagine that there are 1,000 individuals in the population: 250 A_1A_1, 500 A_1A_2, and 250 A_2A_2. If all the individuals in the population reproduce by selfing, homozygous parents will produce all homozygous offspring, while heterozygous parents will produce half homozygous and half heterozygous offspring. Among 1,000 offspring in our population, there will be 375 A_1A_1, 250 A_1A_2, and 375 A_2A_2. If the population continues to self for two more generations, then, among every 1,000 individuals in the final generation, there will be 468.75 homozygotes of each type and 62.5 heterozygotes (Figure 7.25b). The frequency of heterozygotes has been halved every generation, and the frequency of homozgyotes has increased.

Conclusion 2 of the Hardy–Weinberg analysis is violated when individuals self: We cannot predict the genotype frequencies by multiplying the allele frequencies. Note that in generation three, in Figure 7.25b, the allele frequencies are

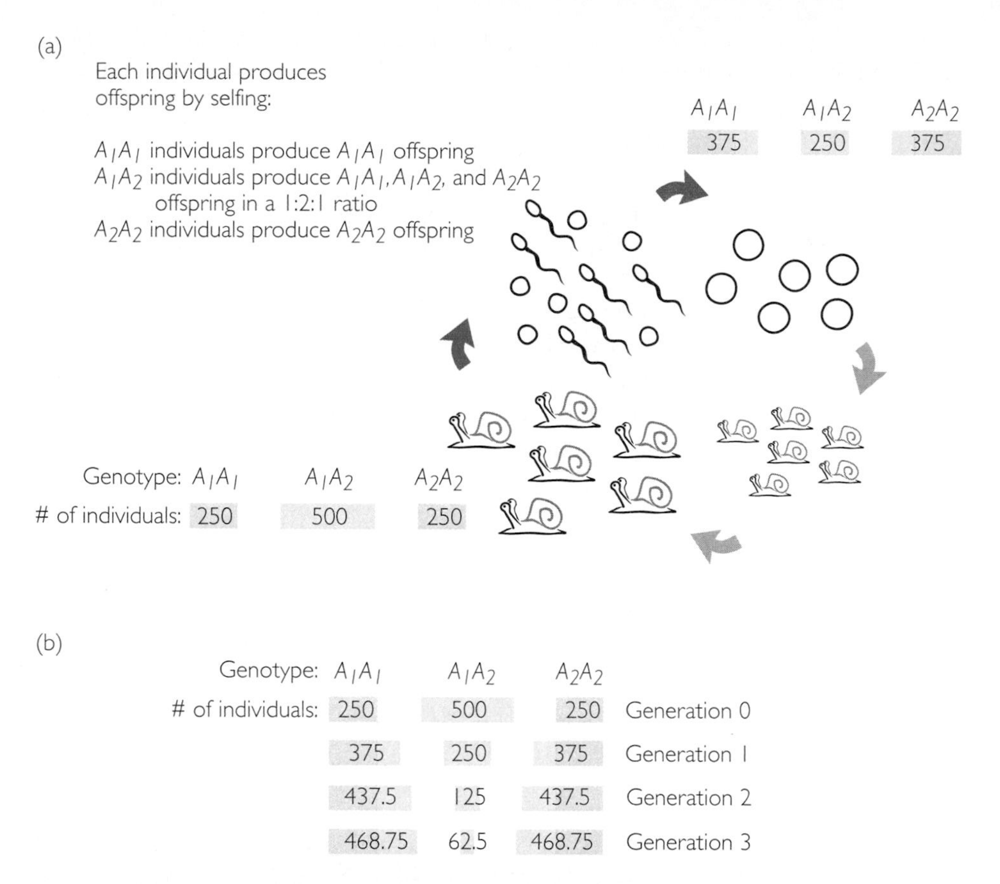

Figure 7.25 Inbreeding alters genotype frequencies (a) This figure follows the genotype frequencies in an imaginary population of 1,000 snails from one generation's adults (lower left) to the next generation's zygotes (upper right). The frequencies of both allele A_1 and A_2 are 0.5. The colored bars show the number of individuals with each genotype. Every individual reproduces by selfing. Homozygotes produce homozygous offspring and heterozygotes produce both heterozygous and homozygous offspring, so the frequency of homozygotes goes up, and the frequency of heterozygotes goes down. (b) These bar charts show what will happen to the genotype frequencies if this population continues to self for two more generations. The orange portions of the bars show the decrease in heterozygosity and the increase in homozygosity due to inbreeding.

still 0.5 for A_1 and 0.5 for A_2. Yet the frequency of heterozygotes is far less than $2(0.5)(0.5)$. Compared to Hardy–Weinberg expectations, there is a deficit of heterozygotes and an excess of homozygotes. The general case under selfing is shown algebraically in Table 7.4.

What about Hardy–Weinberg conclusion 1? Do the allele frequencies change from generation to generation under inbreeding? They did not in our numerical example. We can check the general case by calculating the frequency of allele A_1 in the gene pool produced by the population shown in the last row of Table 7.4. The frequency of allele A_1 in the gene pool is equal to the frequency of

Table 7.4 Changes in genotype frequencies with successive generations of selfing

The frequency of allele A_1 is p and the frequency of allele A_2 is q. Note that allele frequencies do not change from generation to generation—only the genotype frequencies. After Crow (1983).

Generation	Frequency of A_1A_1	Frequency of A_1A_2	Frequency of A_2A_2
0	p^2	$2pq$	q^2
1	$p^2 + (pq/2)$	pq	$q^2 + (pq/2)$
2	$p^2 + (3pq/4)$	$pq/2$	$q^2 + (3pq/4)$
3	$p^2 + (7pq/8)$	$pq/4$	$q^2 + (7pq/8)$
4	$p^2 + (15pq/16)$	$pq/8$	$q^2 + (15pq/16)$

A_1A_1 adults in the population $\left(= p^2 + \frac{15pq}{16}\right)$ plus half the frequency of $A_1A_2\left(= \frac{1}{2}\left[\frac{pq}{8}\right]\right)$. That gives

$$p^2 + \frac{15pq}{16} + \frac{1}{2}\left[\frac{pq}{8}\right] = p^2 + \frac{15pq}{16} + \frac{pq}{16} = p^2 + pq$$

Now substitute $(1 - p)$ for q to give $p^2 + p(1 - p) = p$. This is the same frequency for allele A_1 that we started out with at the top of Table 7.4. Although inbreeding does cause genotype frequencies to change from generation to generation, it does not cause allele frequencies to change. Inbreeding by itself, therefore, is not a mechanism of evolution. As we will see, however, inbreeding can have important evolutionary consequences.

Empirical Research on Inbreeding

Because inbreeding can produce a large excess of homozygotes, Hardy–Weinberg analysis can be used to detect inbreeding in nature. As an example, we consider research on California sea otters (*Enhydra lutris*). Sea otters (Figure 7.26) were once abundant along the West Coast of North America from Alaska to Baja California. They were nearly wiped out, however, by the 18th and 19th century fur

Figure 7.26 A sea otter feeding in a kelp bed off Monterey, California.

trade. At its lowest point, the sea otter population in California numbered fewer than 50 individuals (Lidicker and McCollum 1997). The good news is that since they were placed under protection in 1911 the California otters have been making a comeback. By the end of the 20th century, there were some 1,500 of them.

As a result of the bottleneck, the California otter population harbors less genetic diversity than before the fur hunters arrived (Larson et al. 2002). William Lidicker and F. C. McCollum (1997) investigated whether the reduced size and density of the otter population also led to inbreeding.

Lidicker and McCollum determined the genotypes of a number of California otters for each of 31 allozyme loci. One of them was the PAP locus (1-phenylalanyl-1-proline peptidase), for which the otter population harbored two alleles: *S* (for slow) and *F* (for fast). Among a sample of 33 otters, the number of individuals with each genotype were:

SS	***SF***	***FF***
16	7	10

The sample of 33 otters includes 66 alleles. The frequencies of *S* and *F* are:

$$S: \quad \frac{2(16) + 7}{66} \approx 0.6 \qquad F: \quad \frac{7 + 2(10)}{66} \approx 0.4$$

Data revealing a deficit of heterozygotes and an excess of homozygotes may be evidence of inbreeding.

If the otter population were in Hardy-Weinberg equilibrium, the genotype frequencies would be:

$$SS: (0.6)^2 = 0.36 \qquad SF: 2(0.6)(0.4) = 0.48 \qquad FF: (0.4)^2 = 0.16$$

The actual frequencies, however, were:

$$SS: \frac{16}{33} = 0.485 \qquad SF: \frac{7}{33} = 0.212 \qquad FF: \frac{10}{33} = 0.303$$

There are more homozygotes and fewer heterozygotes than expected in a population in which individuals are mating at random. Lidicker and McCollum also determined the PAP genotypes of six sea otters from Alaska, where the population experienced a less severe bottleneck. Their sample size was small, but the Alaskan otters showed no evidence of missing heterozygotes (1 had genotype *SS*; 3 had *SF*; 2 had *SS*).

Table 7.5 gives the mean frequencies of heterozygotes across all 31 loci for all the otters Lidicker and McCollum examined. The overall results are consistent the results for the PAP locus. The California otter population shows a substantial deficit of heterozygotes. This is consistent with inbreeding.

Strictly speaking, the excess of homozygotes shows only that one or more of the Hardy–Weinberg assumptions is being violated in the otter population. In principle, a defecit of heterozygotes could result from selection against them and in favor of homozgotes. The appearance of a heterozygote deficit could also arise if the California otters, which Lidicker and McCollum treated as single population, actually comprise two separate populations with different allele frequencies. Lidicker and McCollum consider these alternative explanations, however, and conclude that inbreeding is more plausible. They recommend that recovering

Table 7.5 The observed and expected number of heterozygotes for California and Alaska sea otters

The numbers given here are means across 31 loci for 74 otters from California and 9 otters from Alaska. For each population, the observed number of individuals with a particular kind of genotype is compared to the number expected under Hardy–Weinberg conditions of random mating and no mutation, selection, migration, or genetic drift.

	California	Alaska
Heterozygotes observed	**4.6%**	**6.8%**
Heterozygotes expected	**7.2%**	**7.7%**

Source: Lidicker and McCollumn (1997).

otter populations be monitored for evidence of inbreeding depression, a phenomenon we will discuss later in this section.

General Analysis of Inbreeding

So far our treatment of inbreeding has been limited to self-fertilization and sibling mating. But inbreeding can also occur as matings among more distant relatives, such as cousins. Inbreeding that is less extreme than selfing produces the same effect as selfing—it increases the proportion of homozygotes—but at a slower rate. For a general mathematical treatment of inbreeding, population geneticists use a conceptual tool called the **coefficient of inbreeding**. This quantity is symbolized by F, and is defined as the probability that the two alleles in an individual are identical by descent (meaning that both alleles came from the same ancestor allele in some previous generation). Box 7.6 shows that in an inbred population that otherwise obeys Hardy–Weinberg assumptions, the genotype frequencies are

A_1A_1	A_1A_2	A_2A_2
$p^2(1 - F) + pF$	$2pq(1 - F)$	$q^2(1 - F) + qF$

The reader can verify these expressions by substituting the values $F = 0$, which gives the original Hardy–Weinberg genotype ratios, and $F = 0.5$, which represents selfing and gives the ratios shown for generation 1 in Table 7.4.

The same logic applies when many alleles are present in the gene pool. Then, the frequency of any homozygote A_iA_i is given by

$$p_i^2(1 - F) + p_iF$$

and the frequency of any heterozygote A_iA_j is given by

$$2p_ip_j(1 - F)$$

where p_i is the frequency of allele A_i and p_j is the frequency of allele A_j.

The last expression states that the fraction of individuals in a population that are heterozygotes (that is, the population's heterozygosity) is proportional to

Box 7.6 Genotype frequencies in an inbred population

Here we add inbreeding to the Hardy–Weinberg analysis. Imagine a population with two alleles at a single locus: A_1 and A_2, with frequencies p and q. We can calculate the genotype frequencies in the next generation by letting gametes find each other in the gene pool, as we would for a random mating population. The twist added by inbreeding is that the gene pool is not thoroughly mixed. Once we have picked an egg to watch, for example, we can think of the sperm in the gene pool as consisting of two fractions: a fraction $(1 - F)$ carrying alleles that are not identical by descent to the one in the egg, and the fraction F carrying alleles that are identical by descent to the one in the egg (because they were produced by relatives of the female that produced the egg). The calculations of genotype frequencies are as follows:

- **A_1A_1 homozygotes:** There are two ways we might witness the creation of an A_1A_1 homozygote. The first way is that we pick an egg that is A_1 (an event with probability p) and watch it get fertilized by a sperm that is A_1 by chance, rather than by common ancestry. The frequency of unrelated A_1 sperm in the gene pool is $p(1 - F)$, so the probability of getting a homozygote by chance is

$$p \times p(1 - F) = p^2(1 - F)$$

The second way to get a homozygote is to pick an egg that is A_1 (an event with probability p) and watch it get fertilized by a sperm that is A_1 because of common ancestry (an event with probability F). The probability of getting a homozygote this way is pF. The probability of getting an A_1A_1 homozygote by either the first way or the second way is the sum of their individual probabilities:

$$p^2(1 - F) + pF$$

- **A_1A_2 heterozygotes:** There are two ways to get an A_1A_2, heterozygote. The first way is to pick an egg that is A_1 (an event with probability p) and watch it get fertilized by an unrelated sperm that is A_2. The frequency of A_2 unrelated sperm is $q(1 - F)$, so the probability of getting a heterozygote this first way is $pq(1 - F)$. The second way is to pick an egg that is A_2 (probability: q) and watch it get fertilized by an unrelated sperm that is A_1 [probability: $p(1 - F)$]. The probability of getting a heterozygote the second way is $qp(1 - F)$. The probability of getting a heterozygote by either the first way or the second way is the sum of their individual probabilities:

$$pq(1 - F) + qp(1 - F) = 2pq(1 - F)$$

- **A_2A_2 homozygotes:** We can get an A_2A_2 homozygote either by picking an A_2 egg (probability: q) and watching it get fertilized by an unrelated A_2 sperm [probability: $q(1 - F)$], or by picking an A_2 egg (probability: q) and watching it get fertilized by a sperm that is A_2 because of common ancestry (probability: F). The overall probability of getting an A_2A_2 homozygote is

$$q^2(1 - F) + qF$$

The reader may wish to verify that the genotype frequencies sum to 1.

$(1 - F)$. If we compare the heterozygosity of an inbred population, H_F, with that of a random mating population, H_0, then the relationship will be

$$H_F = H_0(1 - F)$$

Anytime F is greater than 0, the frequency of heterozygotes is lower in an inbred population than it is in a random mating population.

Computing *F*

To measure the degree of inbreeding in actual populations, we need a way to calculate F. Doing this directly requires a pedigree—a diagram showing the genealogical relationships of individuals. Figure 7.27 shows a pedigree leading to a female who is the daughter of half-siblings. There are two ways this female could receive alleles that are identical by descent. One is that she could receive two copies of her grandmother's "red triangle" allele (Figure 7.27a). This will happen if the grandmother passes the triangle allele to her daughter and to her son, the daughter passes it to the granddaughter, and the son passes it to the granddaughter. The total probability of this scenario is $\frac{1}{16}$. The second way is that she could receive two copies of her grandmother's "blue diamond" allele (Figure 7.27b). The total probability of this scenario is $\frac{1}{16}$. The probability that the daughter of half-siblings will have two alleles identical by descent by either the first scenario or the second scenario is $\frac{1}{16} + \frac{1}{16} = \frac{1}{8}$. Thus, F for an offspring of half-siblings is $\frac{1}{8}$.

Inbreeding Depression

Inbreeding may lead to reduced mean fitness if it generates offspring homozygous for deleterious alleles.

Although inbreeding does not directly change allele frequencies, it can still affect the evolution of a population. Among the most important consequences of inbreeding for evolution is inbreeding depression.

Inbreeding depression usually results from the exposure of deleterious recessive alleles to selection. To see how this works, consider the extreme case illustrated by loss-of-function mutations. These alleles are often recessive, because a single wild-type allele can still generate enough functional protein, in most instances, to produce a normal phenotype. Even though they may have no fitness consequences at all in heterozygotes, loss-of-function mutations can be lethal in homozygotes. By increasing the proportion of individuals in a population that are homozygotes, inbreeding increases the frequency with which deleterious recessives affect phenotypes. Inbreeding depression refers to the effect these alleles have on the average fitness of offspring in the population.

Studies on humans have shown that inbreeding does, in fact, expose deleterious recessive alleles, and data from numerous studies consistently show that children of first cousins have higher mortality rates than children of unrelated parents

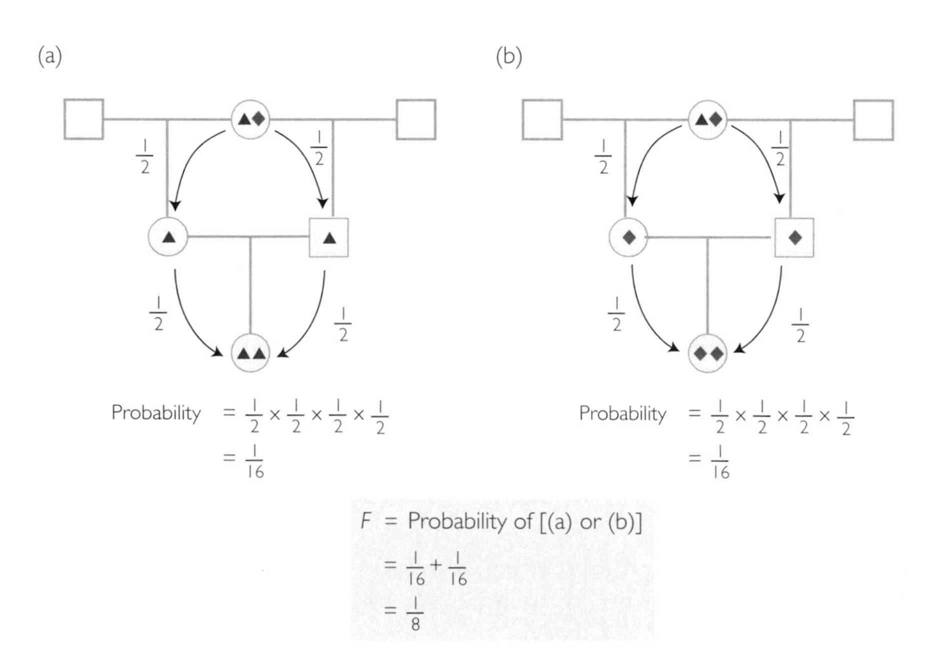

Figure 7.27 Calculating *F* from a pedigree In parts (a) and (b), squares represent males; circles represent females; arrows represent the movement of alleles from parents to offspring via gametes. The red triangles and blue diamonds represent alleles at a particular locus.

(Figure 7.28). Strong inbreeding depression has also been frequently observed in captive populations of animals (for example, Hill 1974; Ralls et al. 1979).

Perhaps the most powerful studies of inbreeding depression in natural populations concern flowering plants, in which the inbreeding can be studied experimentally. In many angiosperms, selfed and outcrossed offspring can be produced from the same parent through hand pollination. In experiments like these, inbreeding depression can be defined as

$$\delta = 1 - \frac{w_s}{w_o}$$

where w_s and w_o are the fitnesses of selfed and outcrossed progeny, respectively. This definition makes levels of inbreeding depression comparable across species. Three patterns are starting to emerge from experimental studies.

First, inbreeding effects are often easiest to detect when plants undergo some sort of environmental stress. For example, when Michele Dudash (1990) compared the growth and reproduction of selfed and outcrossed rose pinks (*Sabatia angularis*), the plants showed some inbreeding depression when grown in the greenhouse or garden, but their performance diverged more strongly when they were planted in the field. Lorne Wolfe (1993) got a similar result with a waterleaf (*Hydrophyllum appendiculatum*): Selfed and outcrossed individuals had equal fitness when grown alone, but differed significantly when grown under competition. And in the common annual called jewelweed (*Impatiens capensis*), McCall et al. (1994) observed the strongest inbreeding effects on survival when an unplanned insect outbreak occurred during the course of their experiment.

Second, inbreeding effects are much more likely to show up later in the life cycle (not, for example, during the germination or seedling stage). This pattern is striking (Figure 7.29). Why does it exist? Wolfe (1993) suggests that maternal effects—specifically, the seed mother's influence on offspring phenotype through provisioning of seeds—can mask the influence of deleterious recessives until later in the life cycle.

Third, inbreeding depression varies among family lineages. Michele Dudash and colleagues (1997) compared the growth and reproductive performance of

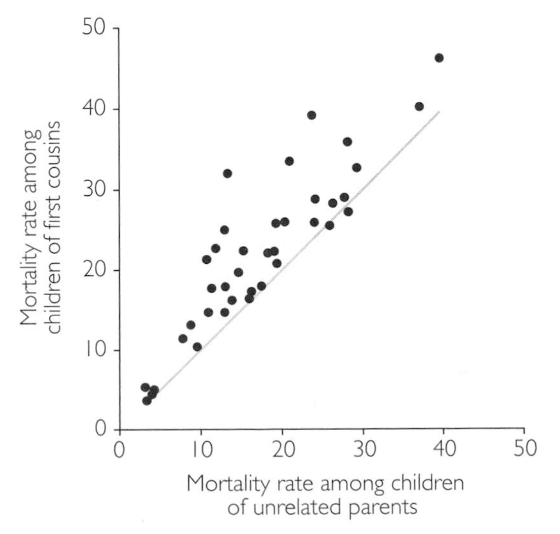

Figure 7.28 Inbreeding depression in humans Each dot on this graph represents childhood mortality rates for a human population. The horizontal axis represents mortality rates for children of unrelated parents; the vertical axis represents mortality rates for children of first cousins. The gray line shows where the points would fall if mortality rates for the two kinds of children were equal. Although childhood mortality rates vary widely among populations, the mortality rate for children of cousins is almost always higher than the rate for children of unrelated parents—usually by about four percentage points. Plotted from data in Bittles and Neel (1994).

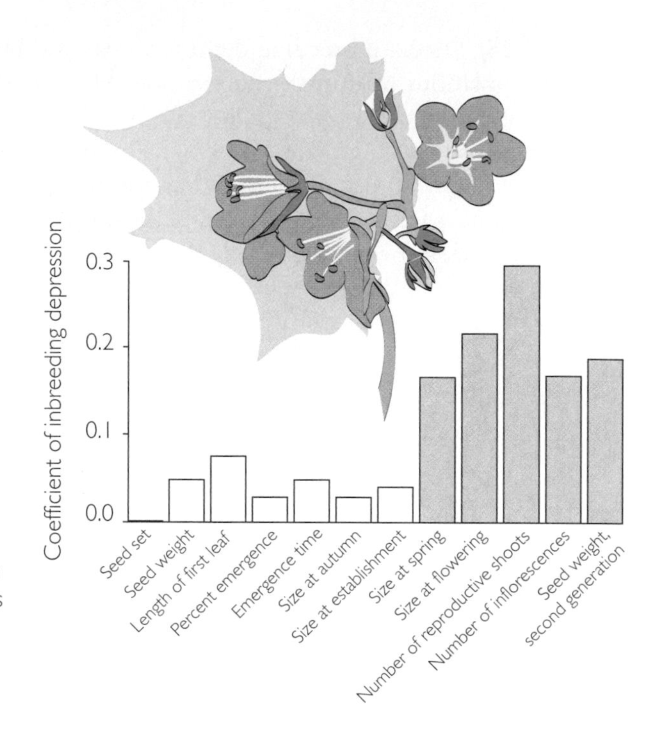

Figure 7.29 **Inbreeding depression in flowering plants increases as individuals age** These data are for waterleaf, a biennial. The open bars show data from the first year of growth; the filled bars indicate traits expressed in the second year (when the plants mature, flower, and die). Inbreeding depression is much more pronounced in the second year than the first. Redrawn from Wolfe (1993).

inbred versus outcrossed individuals from each of several families in two annual populations of the herb *Mimulus guttatus*. Some families showed inbreeding depression; others showed no discernible effect of type of mating; still others showed improved performance under inbreeding.

Inbreeding depression has been documented in natural populations of animals as well. Long-term studies in two separate populations of a bird called the great tit (*Parus major*) have shown that inbreeding depression can have strong effects on reproductive success. When Paul Greenwood and coworkers (1978) defined inbred matings as those between first cousins or more closely related individuals, they found that the survival of inbred nestlings was much lower than that of outbred individuals. Similarly, A. J. van Noordwijk and W. Scharloo (1981) showed that in an island population of tits, there is a strong relationship between the level of inbreeding in a pair and the number of eggs in a clutch that fail to hatch (Figure 7.30). More recently, Keller et al. (1994) found that outbred individuals in a population of song sparrows in British Columbia, Canada, were much more likely than inbred individuals to survive a severe winter.

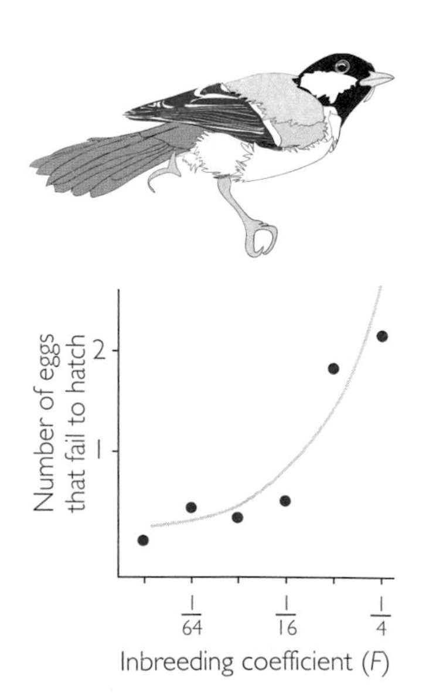

Figure 7.30 **Inbreeding increases egg failure in great tits** From van Noordwijk and Scharloo (1981).

Given the theory and data we have reviewed on inbreeding depression, it is not surprising that animals and plants have evolved mechanisms to avoid it. Mechanisms of inbreeding avoidance include mate choice, genetically controlled self-incompatibility, and dispersal. But under some circumstances, inbreeding may be unavoidable. In small populations, for example, the number of potential mates for any particular individual is limited. If a population is small and remains small for many generations, and if the population receives no migrants from other populations, then eventually all the individuals in the population will be related to each other even if mating is random. Thus, small populations eventually become inbred, and the individuals in them may suffer inbreeding depression. This can be a problem for rare and endangered species, and it creates a challenge for the managers of captive breeding programs, as we will see in Section 7.5.

In summary, nonrandom mating does not, by itself, alter allele frequencies. It is not, therefore, a mechanism of evolution. Nonrandom mating does, however, alter the frequencies of genotypes. It can thereby change the distribution of phenotypes in a population and alter the pattern of natural selection and the evolution of the population. For example, inbreeding increases the frequency of homozygotes and decreases the frequency of heterozygotes. This can expose deleterious recessive alleles to selection, leading to inbreeding depression.

7.5 Conservation Genetics of the Illinois Greater Prairie Chicken

We opened this chapter with the case of the Illinois greater prairie chicken (Figure 7.1), a once abundant bird that in the mid-1990s appeared to be destined for extinction. Like a great many other vulnerable and endangered species, the prairie chicken's worst enemy is habitat destruction (Figure 7.2). Before the introduction of the steel plow, prairie covered more than 60% of Illinois; today less than one hundredth of a percent of that prairie remains (Westemeier et al. 1998). Yet habitat destruction is not the prairie chicken's only problem. Beginning in the early 1960s, conservationists established prairie chicken reserves and worked to restore and maintain prairie habitats. From the late 1960s to the early 1970s, their efforts appeared to be working, as the prairie chicken's numbers began to rebound. But the apparent success was short-lived: By the mid-1970s, the prairie chicken population fell once again into a steady decline (Figure 7.3). Something else was now threatening the survival of the Illinois greater prairie chicken, but what? Our discussion of migration, genetic drift, and nonrandom mating has given us the tools to understand the likely answer.

Ronald Westemeier and colleagues (1998) developed a hypothesis that runs as follows: Destruction of the prairie did two things to the prairie chicken population. First, it directly reduced the size of the birds' population. Second, it fragmented the population that remained. By 1980, the few prairie chickens that survived in Illinois were trapped on small islands of prairie in a sea of farmland. Each island had its own small population of birds. These small populations were geographically isolated from each other and from populations in other states.

Small populations with little or no gene flow are precisely the setting in which genetic drift is most powerful. And genetic drift results in random fixation and declining heterozygosity. If some of the alleles that become fixed are deleterious recessives, then the average fitness of individuals will be reduced. A reduction in fitness due to genetic drift is reminiscent of inbreeding depression. In fact, it *is* inbreeding depression. Reduced heterozygosity due to drift and increased homozygosity due to inbreeding are two sides of the same coin. In a small population all individuals are related, and there is no choice but to mate with kin.

Michael Lynch and Wilfried Gabriel (1990) have proposed that an accumulation of deleterious recessives (a phenomenon known as genetic load) can lead to the extinction of small populations. They noted that when exposure of deleterious mutations produces a reduction in population size, the effectiveness of drift is increased. The speed and proportion of deleterious mutations going to fixation subsequently increases, which further decreases population size. Lynch and Gabriel termed this synergistic interaction between mutation, population size, and drift a "mutational meltdown."

Westemeier and colleagues suggested that the remnant populations of Illinois greater prairie chickens were trapped in just such a scenario. As the populations lost their genetic diversity, the birds began to suffer inbreeding depression. This inbreeding depression reduced individual reproductive success and caused the remnant populations to continue their decline even as the amount of suitable habitat increased. The continued decline in population size led to even more drift, which led to worse inbreeding depression, and so on. The birds had fallen into an "extinction vortex" (see Soulé and Mills 1998).

Early efforts to conserve remnant populations of greater prairie chickens apparently failed because the birds were suffering from inbreeding depression.

To test their hypothesis, the researchers first used data from a long-term study of the Jasper County population to look for evidence of inbreeding depression. The researchers plotted the hatching success of prairie chicken eggs, a measure of individual fitness, as a function of time (Figure 7.31). Throughout the 1960s, over 90% of greater prairie chicken eggs in Jasper County hatched. This rate is comparable to what it was in the 1930s and to what it is today in larger prairie chicken populations in other states. By 1970, however, a steady decline in hatching success had begun. By the late 1980s, the hatching rate dipped below 80%. The all-time low came in 1990, with fewer than 40% of the eggs hatching. The decline in hatching success is statistically significant, and represents a substantial reduction in individual fitness. In other words, it looks like inbreeding depression.

If the decline in hatching success in Jasper County greater prairie chickens was, in fact, due to inbreeding depression caused by genetic drift, then it should be accompanied by a genetic signature. The Jasper County population should show less genetic diversity than larger populations in other states, and less genetic diversity now than it had in the past. Juan Bouzat and colleagues (1998) analyzed the DNA of a number of greater prairie chickens from Illinois, Kansas, Minnesota, and Nebraska, and determined each bird's genotype at six selectively neutral loci (these were noncoding regions with variable numbers of short tandem repeats). As predicted, the Illinois birds had an average of just 3.67 alleles per

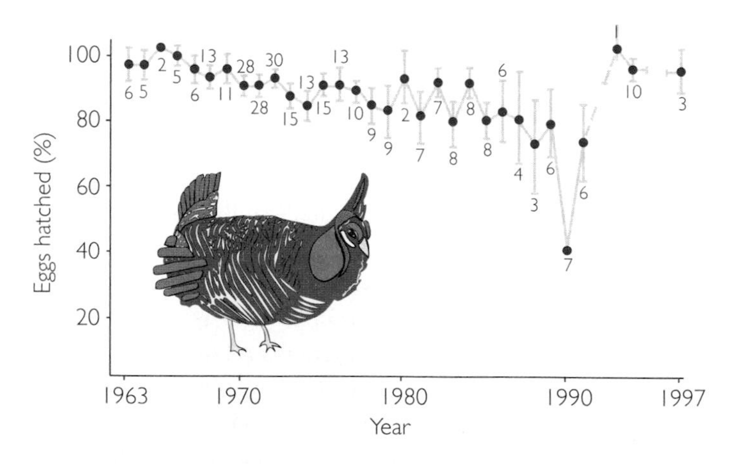

Figure 7.31 Declining hatching success in a greater prairie chicken population This graph plots, for greater prairie chickens in Jasper County, Illinois, the fraction of eggs hatching each year from 1963 to 1997. The small number below each data point indicates the number of nests followed; the whiskers indicate ±1 standard error (a statistical measure of uncertainty in the estimated hatching rate). The decline in hatching success from the mid-1960s to the early 1990s appears to reflect inbreeding depression. Redrawn with permission from Westemeier et al (1998).

locus, significantly fewer than the 5.33 to 5.83 alleles per locus shown by the other populations (Table 7.6). The researchers were even able to extract DNA from 10 museum specimens that had been collected in Jasper County in the 1930s, plus 5 from the 1960s. As shown in the last column in Table 7.6, Bouzat and colleagues used the data from the museum specimens to estimate that the Jasper County population once had an average of at least 5.12 alleles per locus. Consistent with the extinction vortex hypothesis, the greater prairie chickens of Jasper County are genetically depauperate, compared to both their own ancestral population and other present-day populations.

Inbreeding depression in remnant populations of greater prairie chickens was caused by a loss of allelic diversity under genetic drift.

The final test of the extinction vortex hypothesis was to use it to develop a practical conservation strategy. If the problem for the Jasper County prairie chicken population is reduced genetic diversity, then the solution is gene flow. Migrants from other populations should carry with them the alleles that have been lost in Jasper County. Reintroduction of these lost alleles should reverse the effects of drift and eliminate inbreeding depression. Natural migration of greater prairie chickens into Jasper County ceased long ago. But in 1992, conservation biologists began trapping greater prairie chickens in Minnesota, Kansas, and Nebraska, and moving them to Jasper County. The plan seems to be working. Westemeier and colleagues (1998) report that in 1993, 1994, and 1997, hatching rates in Jasper County were over 90%—higher than they had been in 25 years (Figure 7.31). And the Jasper County population is growing (Figure 7.3).

All of the data we have presented on Illinois greater prairie chickens come from observational studies, so it is always possible that some uncontrolled and unknown environmental variable is responsible for the variation in hatching success. On present evidence, however, Westemeier et al.'s extinction vortex hypothesis—involving migration, genetic drift, and nonrandom mating—appears to be the best explanation.

Migration, in the form of birds transported by biologists, appears to be restoring genetic diversity to remnant populations and alleviating inbreeding depression.

Table 7.6 Number of alleles per locus found in each of the current populations of Illinois, Kansas, Minnesota, and Nebraska and estimated for the Illinois prebottleneck population

Locus	Illinois	Kansas	Minnesota	Nebraska	Illinois prebottleneck*
ADL42	3	4	4	4	3
ADL23	4	5	4	5	5
ADL44	4	7	8	8	4
ADL146	3	5	4	4	4
ADL162	2	5	4	4	6
ADL230	6	9	8	10	9
Mean	3.67	5.83	5.33	5.83	5.12
SE	0.56	0.75	0.84	1.05	0.87
Sample size	32	37	38	20	15

Note:

- SE indicates standard error of mean number of alleles per locus. The Illinois population in column 1 shows signficantly less allelic diversity than the rest of the populations ($P < 0.05$).
- Number of alleles in the Illinois prebottleneck population include both extant alleles that are shared with the other populations and alleles detected in the museum collection.

Source: From Bouzat et al. (1998).

Summary

Among the important implications of the Hardy—Weinberg equilibrium principle is that natural selection is not the only mechanism of evolution. In this chapter, we examined violations of three assumptions of the Hardy–Weinberg analysis first introduced in Chapter 6 and considered their effects on allele and genotype frequencies.

Migration, in its evolutionary meaning, is the movement of alleles from one population to another. When allele frequencies are different in the source population than in the recipient population, migration causes the recipient population to evolve. As a mechanism of evolution, migration tends to homogenize allele frequencies across populations. In doing so, it may tend to eliminate adaptive differences between populations that have been produced by natural selection.

Genetic drift is evolution that occurs as a result of sampling error in the production of a finite number of zygotes from a gene pool. Just by chance, allele frequencies change from one generation to the next. Genetic drift is more dramatic in smaller populations than in large ones. Over many generations, drift results in an inexorable loss of genetic diversity. If some of the alleles that become fixed are deleterious recessives, genetic drift can result in a reduction in the fitness of individuals in the population.

The Neutral Theory of Molecular Evolution suggests that genetic drift is the most important mechanism of evolution at the level of DNA sequences. The Neutral Theory explains the clocklike evolution observed in some genes, and serves as a null hypothesis for detecting the action of positive natural selection.

Nonrandom mating does not directly change allele frequencies and is thus not, strictly speaking, a mechanism of evolution. However, nonrandom mating does influence genotype frequencies. For example, inbred populations have more homozygotes and fewer heterozygotes than otherwise comparable populations in which mating is random. An increase in homozygosity often exposes deleterious recessive alleles and results in a reduction in fitness known as inbreeding depression.

As illustrated by the case of the Illinois greater prairie chicken, the phenomena discussed in this chapter find practical application in conservation efforts. Drift can rob small remnant populations of genetic diversity, resulting in inbreeding depression and greater risk of extinction. Migration can sometimes restore lost genetic diversity, improving a population's chances for long-term survival.

Questions

1. Conservation managers often try to purchase corridors of undeveloped habitat so that larger preserves are linked into networks. Why? What genetic goals do you think the conservation managers are aiming for?
2. The graph in Figure 7.32 shows F_{ST}, a measure of genetic differentiation between populations as a function of geographic distance. The data are from human populations in Europe. Genetic differentiation has been calculated based on loci on the autosomes (inherited from both parents), the mitochondrial chromosome (inherited only from the mother), and the Y chromosome (inherited only from the father). Note that the patterns are different for the three different kinds of loci. Keep in mind that migration tends to homogenize allele frequencies across populations. Develop a hypothesis to explain why allele frequencies are more homogenized across popula-

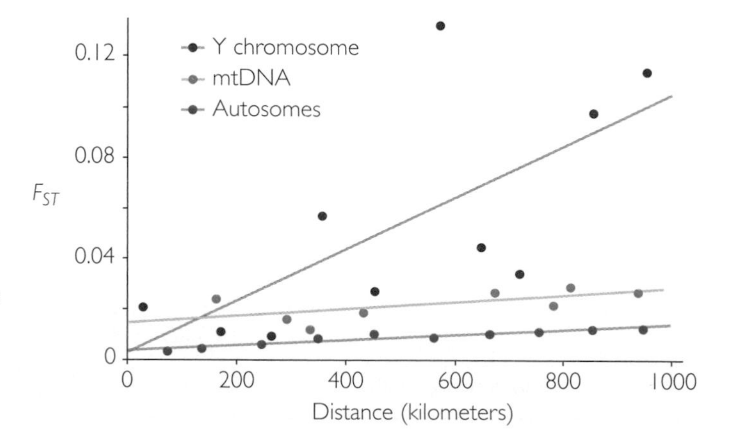

Figure 7.32 Genetic distance between human populations as a function of geographic distance Genetic distance (F_{ST}) is a measure of genetic differentiation among populations. Here it has been calculated based on autosomal loci (blue), mitochondrial loci (green), and Y-chromosome loci (red). From Seielstad et al. (1998).

tions for autosomal and mitochondrial loci than for Y-chromosome loci. Then go to the library and look up the following paper, to see if your hypothesis is similar to the one favored by the biologists who prepared the graph:

Seielstad, M. T., E. Minch, and L. L. Cavalli-Sforza. 1998. Genetic evidence for. . . . in humans. *Nature Genetics* 20:278–280. [Part of title deleted to encourage readers to develop their own hypotheses.]

3. Consider three facts: (i) Loss of heterozygosity may be especially detrimental at MHC loci, because allelic variability at these loci enhances disease resistance; (ii) Microsatellite loci show that the gray wolves on Isle Royale, Michigan, are highly inbred (Wayne et al. 1991); (iii) This wolf population crashed during an outbreak of canine parvovirus during the 1980s. How might these facts be linked? What other hypotheses could explain the data? How could you test your ideas?

4. If you were a manager charged with conserving the collared lizards of the Ozarks, one of your tasks might be to reintroduce the lizards into glades in which they have gone extinct. When reintroducing lizards to a glade, you will have a choice between using only individuals from a single extant glade population or from several extant glade populations. What would be the evolutionary consequences of each choice, for both the donor and recipient populations? Which strategy will you follow, and why?

5. Bodmer and McKie (1995) review several cases, similar to achromatopsia in the Pingelapese, in which genetic diseases occur at unusually high frequency in populations that are, or once were, relatively isolated. An enzyme deficiency called hereditary tyrosinemia, for example, occurs at an unusually high rate in the Chicoutimi region north of Quebec City in Canada. A condition called porphyria is unusually common in South Africans of Dutch descent. Why are genetic diseases so common in isolated populations? What else do these populations all have in common?

6. Remote oceanic islands are famous for their endemic species—unique forms that occur nowhere else (see Quammen 1996 for a gripping and highly readable account). Consider the roles of migration and genetic drift in the establishment of new species on remote islands.
   a. How do plant and animal species become established on remote islands? Do you think island endemics are more likely to evolve in some groups of plants and animals than others?
   b. Consider a new population that has just arrived at a remote island. Is the population likely to be large or small? Will founder effects, genetic drift, and additional waves of migration from the mainland play a relatively large or a small role in the evolution of the new island population (compared to a similar population on an island closer to the mainland)? Do your answers help explain why unusual endemic species are more common on remote islands than on islands close to the mainland?

7. By using the start codon AUG as a guidepost, researchers can determine whether substitutions in pseudogenes correspond to silent changes or replacement changes. In contrast to most other loci, the rate of silent and replacement changes is identical in pseudogenes. Explain this observation in light of the neutral theory of evolution.

8. When researchers compare a gene in closely related species, why is it logical to infer that positive natural selection has taken place if replacement substitutions outnumber silent substitutions?

9. What is codon bias? Why is the observation of nonrandom codon use evidence that certain codons might be favored by natural selection? If you were given a series of gene sequences from the human genome, how would you determine whether codon usage is random or nonrandom?

10. Draft sequences are now available for both the human and the chimpanzee genomes. Outline how you would analyze homologous genes in the two species to determine which of the observed sequence differences result from drift and which result from selection.

11. Recall that the fourth chromosome of *Drosophila melanogaster* does not recombine during meiosis. The lack of genetic polymorphism on this chromosome has been interpreted as the product of a selective sweep. If the fourth chromosome had normal rates of recombination, would you expect the level of polymorphism to be different? Why?

12. As we have seen, inbreeding can reduce offspring fitness by exposing deleterious recessive alleles. However, some animal breeders practice generations of careful inbreeding within a family, or "line breeding," and surprisingly many of the line-bred animals, from champion dogs to prize cows, have normal health and fertility. How can it be possible to continue inbreeding for many generations without experiencing inbreeding depression due to recessive alleles? (*Hint:* Consider some of the differences between animal breeders and natural selection in the wild.) Generally, if a small population continues to inbreed for many generations, what will happen to the frequency of the deleterious recessive alleles over time?

13. In the mid-1980s, conservation biologists reluctantly recommended that zoos should not try to preserve captive populations of all the endangered species of large cats. For example, some biologists recommended ceasing efforts to breed the extremely rare Asian lion, the beautiful species seen in Chinese artwork. In place of

the Asian lion, the biologists recommended increasing the captive populations of other endangered cats, such as the Siberian tiger and Amur leopard. By reducing the number of species kept in captivity, the biologists hoped to increase the captive population size of each species to several hundred, preferably at least 500. Why did the conservation biologists think that this was so important as to be worth the risk of losing the Asian lion forever?

14. In this chapter we saw that in many cases, gene frequencies in small populations change at different rates than in large populations. As a review, state how the following processes tend to vary in speed and effects in small versus large populations. (Assume the typical relationship of population size and generation time.)

    Selection
    Migration
    Genetic drift
    Inbreeding
    New mutations per individual
    New mutations per generation in the whole population
    Substitution of a new mutation for an old allele
    Fixation of a new mutation

Exploring the Literature

15. For a paper that explores migration as a homogenizer of allele frequencies among human populations, see:

    Parra, E. J., A. Marcini, et al. 1998. Estimating African-American admixture proportions by use of population-specific alleles. *American Journal of Human Genetics* 63: 1839–1851.

16. Newly available human genome sequences are being examined with a variety of new techniques to assess the role of positive natural selection in recent human evolution. For a start on this literature, see:

    Voight, B.F., S. Kudaravalli, X. Wen, and J.K. Pritchard. 2006. A map of recent positive selection in the human genome. *PLoS Biology* 4 (3): e72.

17. For another example like the research on collared lizards by Templeton and colleagues (1990), in which biologists took advantage of a natural experiment to test predictions about the effect of genetic drift on genetic diversity, see:

    Eldridge, M. D. B., J. M. King, et al. 1999. Unprecedented low levels of genetic variation and inbreeding depression in an island population of the black-footed rock-wallaby. *Conservation Biology* 13: 531–541.

18. We mentioned in Section 7.4 that inbreeding depression is a concern for biologists trying to conserve endangered organisms with small population sizes. Inbreeding depression turns out to vary among environments and among families. In addition, new genetic techniques are enabling more precise measures of inbreeding in wild populations that have unknown genealogies. For more information, see:

    Spielman, D., B.W. Brook, and R. Frankham. 2004. Most species are not driven to extinction before genetic factors impact them. *Proceedings of the National Academy of Sciences USA* 101 (42): 15261–15264.

    Hedrick, P. W., and S. T. Kalinowski. 2000. Inbreeding depression in conservation biology. *Ann. Rev. Ecol. Syst.* 31: 139–162.

    Liberg, O., H. Andren, H-C Pedersen, H. Sand, D. Sejberg, P. Wabakkan, M. Akesson, and S. Bensch. 2005. Severe inbreeding depression in a wild wolf (*Canis lupus*) population. *Biology Letters* 1: 17–20

19. An essential step of any conservation program is to determine the minimum population size necessary to make the extinction of a species unlikely over the long term. The following papers explore this question:

    Lande, R. 1995. Mutation and conservation. *Conservation Biology* 9: 782–791.

    Lynch, M. 1996. A quantitative genetic perspective on conservation issues. In J. C. Avise and J. Hamrick, eds. *Conservation Genetics: Case Histories from Nature.* New York: Chapman and Hall, 471–501.

20. Cheetahs have long been cited as a classic example of a species whose low genetic diversity put it at increased risk of extinction. Other researchers have debated the validity of this view. For a start on the literature, see:

    Menotti-Raymond, M., and S. J. O'Brien. 1993. Dating the genetic bottleneck of the African cheetah. *Proceedings of the National Academy of Sciences USA* 90: 3172–3176

    Merola, M. 1994. A reassessment of homozygosity and the case for inbreeding depression in the cheetah, *Acinonyx jubatus*: Implications for conservation. Conservation Biology 8: 961–971.

21. In animals, the rate of sequence change appears to vary as a function of metabolic rate as well as generation time. Gillooly and colleagues have recently attempted to unify these data with the original classic neutral model of evolution. According to their model, the molecular clock ticks at one substitution "per unit of mass-specific metabolic energy" rather than per unit time. Here is Gillooly's paper, along with two of the original papers that raised the metabolic-rate issue:

    Gillooly, J.F., A.P. Allen, G.B. West, and J.H. Brown. 2005. The rate of DNA evolution: effects of body size and temperature on the molecular clock. *Proceedings of the National Academy of Sciences USA* 102 (1): 140–145.

    Martin, A. P., G. J. P. Naylor, and S. R. Palumbi. 1992. Rates of mitochondrial DNA evolution in sharks are slow compared with mammals. *Nature* 357: 153–155.

    Martin, A. P. and S. R. Palumbi. 1993. Body size, metabolic rate, generation time, and the molecular clock. *Proceedings of the National Academy of Sciences USA* 90: 4087–4091.

Citations

Much of the population genetics material in this chapter is modeled after presentations in the following:

Crow, J. F. 1983. *Genetics Notes.* Minneapolis, MN: Burgess Publishing.

Felsenstein, J. 1997. *Theoretical Evolutionary Genetics.* Seattle, WA: ASUW Publishing, University of Washington.

Griffiths, A. J. F., J. H. Miller, D. T. Suzuki, R. C. Lewontin, and W. M. Gelbert. 1993. *An Introduction to Genetic Analysis.* New York: W. H. Freeman.

Maynard Smith, J. 1998. *Evolutionary Genetics.* Oxford: Oxford University Press.

Roughgarden, J. 1979. *Theory of Population Genetics and Evolutionary Ecology: An Introduction.* New York: Macmillan Publishing.

Templeton, A. R. 1982. Adaptation and the integration of evolutionary forces. In R. Milkman, ed., *Perspectives on Evolution.* Sunderland, MA: Sinauer, 15–31.

Here is the list of all other citations in this chapter:

Akashi, H. 1994. Synonymous codon usage in *Drosophila melanogaster:* natural selection and translational accuracy. *Genetics* 144: 927–935.

Andolfatto, P. 2005. Adaptive evolution of non-coding DNA in *Drosophila. Nature* 437: 1149-1152.

Ayala, F. J., and J. A. Kiger, Jr. 1984. *Modern Genetics.* Menlo Park, CA: Benjamin/Cummings.

Barrett, S. C. H., and D. Charlesworth. 1991. Effects of a change in the level of inbreeding on the genetic load. *Nature* 352: 522–524.

Berry, A., J. W. Ajioka, and M. Kreitman. 1991. Lack of polymorphism on the *Drosophila* fourth chromosome resulting from selection. *Genetics* 129: 1111–1117.

Berry, A. 1996. Non-non-Darwinian evolution. *Evolution* 50: 462–466.

Bittles, A. H., and J.V. Neel. 1994. The costs of human inbreeding and their implications for variations at the DNA level. *Nature Genetics* 8: 117–121.

Brisson, J. A., J. L. Strassburg, and A. R. Templeton. 2003. Impact of fire management on the ecology of collared lizard (*Crotaphytus collaris*) populations living on the Ozark Plateau. *Animal Conservation* 6: 247–254.

Bodmer, W., and R. McKie. 1995. *The Book of Man.* New York: Scribner.

Bouzat, J. L., H. A. Lewin, and K. N. Paige. 1998. The ghost of genetic diversity past: Historical DNA analysis of the greater prairie chicken. *American Naturalist* 152: 1–6.

Buri, P. 1956. Gene frequency in small populations of mutant *Drosophila. Evolution* 10: 367–402.

Bustamante, C. D., A. Fledel-Alon, et al. 2005. Natural selection on protein-coding genes in the human genome. *Nature* 437: 1153-1157.

Camin, J. H., and P. R. Ehrlich. 1958. Natural selection in water snakes (*Natrix sipedon* L.) on islands in Lake Erie. *Evolution* 12: 504–511.

Censky, E. J., K. Hodge, and J. Dudley. 1998. Over-water dispersal of lizards due to hurricanes. *Nature* 395: 556.

Chao, L. and D. E. Carr. 1993. The molecular clock and the relationship between population size and generation time. *Evolution* 47: 688–690.

Chamary, J.V., J. L. Parmley, and L. D. Hurst. 2006. Hearing silence: Non-neutral evolution at synonymous sites in mammals. *Nature Reviews Genetics* 7: 98-108.

Charlesworth, B., M. T. Morgan, and D. Charlesworth. 1993. The effects of deleterious mutations on neutral molecular variation. *Genetics* 134: 1289–1303.

Clegg, S. M., S. M. Degnan, J. Kikkawa, et al. 2002. Genetic consequences of sequential founder events by an island-colonizing bird. *Proceedings of the National Academy of Sciences, USA* 99: 8127–8132.

Dorit, R. L., H. Akashi, and W. Gilbert. 1995. Absence of polymorphism at the *ZFY* locus on the human Y chromosome. *Science* 268: 1183–1185.

Dudash, M. R. 1990. Relative fitness of selfed and outcrossed progeny in a self-compatible, protandrous species, *Sabatia angularis* L. (Gentianaceae): A comparison in three environments. *Evolution* 44: 1129–1139.

Dudash, M. R., D. E. Carr, and C. B. Fenster. 1997. Five generations of enforced selfing and outcrossing in *Mimulus guttatus:* Inbreeding depression variation at the population and family level. *Evolution* 51: 54–65.

Escalante, A. A., A. A. Lal, and F. J. Ayala. 1998. Genetic polymorphism and natural selection in the malaria parasite *Plasmodium falciparum. Genetics* 149: 189–202.

Estoup, A. and S. M. Clegg. 2003. Bayesian inferences on the recent island colonization history by the bird *Zosterops lateralis lateralis. Molecular Ecology* 12: 657–674.

Fay, J. C., G. J. Wyckoff, and C.-I Wu. 2001. Positive and negative selection on the human genome. *Genetics* 158: 1227–1234.

Fay, J. C., G. J. Wyckoff, and C.-I Wu. 2002. Testing the neutral theory of molecular evolution with genomic data from *Drosophila. Nature* 415: 1024–1026.

Giles, B. E., and J. Goudet. 1997. Genetic differentiation in *Silene dioica* metapopulations: Estimation of spatiotemporal effects in a successional plant species. *American Naturalist* 149: 507–526.

Gillespie, J. H. 1991. *The Causes of Molecular Evolution.* New York: Oxford University Press.

Gojoburi, T., E. N. Etsuko, and M. Kimura. 1990. Molecular clock of viral evolution, and the neutral theory. *Proceedings of the National Academy of Sciences USA* 87: 10015–10018.

Greenwood, P. J., P. H. Harvey, and C. M. Perrins. 1978. Inbreeding and dispersal in the great tit. *Nature* 271: 52–54.

Hamilton, W. D. 1967. Extraordinary sex ratios. *Science* 156: 477–488.

Hamilton, W. D. 1979. Wingless and fighting males in fig wasps and other insects. In M. S. Blum and N. A. Blum, eds., *Sexual Selection and Reproductive Competition in Insects.* New York: Academic Press, 167–220.

Hartl, D. L. 1981. *A Primer of Population Genetics.* Sunderland, MA: Sinauer.

Hendry, A. P., T. Day, and E. B. Taylor. 2001. Population mixing and the adaptive divergence of quantitative traits in discrete populations: A theoretical framework for empirical tests. *Evolution* 55: 459–466.

Hill, J. L. 1974. *Peromyscus:* Effect of early pairing on reproduction. *Science* 186: 1042–1044.

Hughes, A. L., and M. Nei. 1988. Pattern of nucleotide substitution at major histocompatibility complex class I loci reveals overdominant selection. *Nature* 335: 167–170.

Hughes, A. L., B. Packer, et al. 2003. Widespread purifying selection at polymorphic sites in human protein-coding loci. *Proceedings of the National Academy of Sciences USA* 100: 15754-15757.

Huttley, G. A., E. Easteal, M. C. Southey, et al. 2000. Adaptive evolution of the tumor suppressor *BRCA1* in humans and chimpanzees. *Nature Genetics* 25: 410–413.

Keller, L., P. Arcese, J. N. M. Smith, W. M. Hochachka, and S. C. Stearns. 1994. Selection against inbred song sparrows during a natural population bottleneck. *Nature* 372: 356–357.

Kerr, W. E., and S. Wright. 1954. Experimental studies of the distribution of gene frequencies in very small populations of *Drosophila melanogaster.* I. Forked. *Evolution* 8: 172–177.

Kimura, M. 1968. Evolutionary rate at the molecular level. *Nature* 217: 624–626.

Kimura, M. 1983. *The Neutral Theory of Molecular Evolution.* New York: Cambridge University Press.

King, J. L., and T. H. Jukes. 1969. Non-Darwinian evolution. *Science* 164: 788–798.

King, R. B. 1987. Color pattern polymorphism in the Lake Erie water snake, *Nerodia sipedon insularum. Evolution* 41: 241–255.

King, R. B. 1993a. Color pattern variation in Lake Erie water snakes: Inheritance. *Canadian Journal of Zoology* 71: 1985–1990.

King, R. B. 1993b. Color-pattern variation in Lake Erie water snakes: Prediction and measurement of natural selection. *Evolution* 47: 1819–1833.

King, R. B., and R. Lawson. 1995. Color-pattern variation in Lake Erie water snakes: The role of gene flow. *Evolution* 49: 885–896.

King, R. B., and R. Lawson. 1997. Microevolution in island water snakes. *BioScience* 47: 279–286.

Larson, S., R. Jameson, et al. 2002. Loss of genetic diversity in sea otters (*Enhydra lutris*) associated with the fur trade of the 18th and 19th centuries. *Molecular Ecology* 11: 1899–1903.

Li, W.-H., T. Gojobori, and M. Nei. 1981. Pseudogenes as a paradigm of neutral evolution. *Nature* 292: 237–239.

Li, W.-H., and D. Graur. 1991. *Fundamentals of Molecular Evolution*. Sunderland, MA: Sinauer.

Li, W.-H. and M. Tanimura. 1987. The molecular clock runs more slowly in man than in apes and monkeys. *Nature* 326: 93–96.

Li, W.-H., M. Tanimura, and P. M. Sharp. 1987. An evaluation of the molecular clock hypothesis using mammalian DNA sequences. *Journal of Molecular Evolution* 25: 330–342.

Lidicker, W. Z., and F. C. McCollum. 1997. Allozymic variation in California sea otters. *Journal of Mammology* 78: 417-425.

Lynch, M., and W. Gabriel. 1990. Mutation load and the survival of small populations. *Evolution* 44: 1725–1737.

McCall, C., D. M. Waller, and T. Mitchell-Olds. 1994. Effects of serial inbreeding on fitness components in *Impatiens capensis*. *Evolution* 48: 818–827.

McDonald, J. H., and M. Kreitman. 1991. Adaptive protein evolution at the *Adh* locus in *Drosophila*. *Nature* 351: 652–654.

Nielsen, R. 2005. Molecular signatures of natural selection. *Annual Review of Genetics* 39: 197-218.

Nielsen, R., C. Bustamante, et al. 2005. A scan for positively selected genes in the genomes of humans and chimpanzees. *PLoS Biology* 3: e170.

Nurminsky, D. I., M. V. Nurminskaya, D. De Aguiar, and D. L. Hartl. 1998. Selective sweep of a newly evolved sperm-specific gene in *Drosophila*. *Nature* 396: 572–575.

Ohta, T. 1972. Evolutionary rate of cistrons and DNA divergence. *Journal of Molecular Evolution* 1: 150–157.

Ohta, T. 1977. Extension to the neutral mutation random drift hypothesis. In M. Kimura, ed. *Molecular Evolution and Polymorphism*. Mishima, Japan: National Institute of Genetics, 148–176.

Ohta, T., and M. Kimura. 1971. On the constancy of the evolutionary rate of cistrons. *Journal of Molecular Evolution* 1:18–25.

Ohta, T., and M. Kreitman. 1996. The neutralist-selectionist debate. *BioEssays* 18: 673–683.

Paul, R. E. L., M. J. Packer, M. Walmsley, M. Lagog, L. C. Ranford-Cartwright, R. Paru, and K. P. Day. 1995. Mating patterns in malaria parasite populations of Papua, New Guinea. *Science* 269: 1709–1711.

Polans, N. O., and R. W. Allard. 1989. An experimental evaluation of the recovery potential of ryegrass populations from genetic stress resulting from restriction of population size. *Evolution* 43: 1320–1324.

Purugganen, M. D. and J. I. Suddith. 1998. Molecular population genetics of the *Arabidopsis* CAULIFLOWER regulatory gene: Nonneutral evolution and naturally occurring variation in floral homeotic evolution. *Proceedings of the National Academy of Sciences USA* 95: 8130–8134.

Quammen, D. 1996. *The Song of the Dodo.* New York: Touchstone.

Ralls, K., K. Brugger, and J. Ballou. 1979. Inbreeding and juvenile mortality in small populations of ungulates. *Science* 206: 1101–1103.

Read, A. F., A. Narara, S. Nee, A. E. Keymer, and K. P. Day. 1992. Gametocyte sex ratios as indirect measures of outcrossing rates in malaria. *Parasitology* 104: 387–395.

Seielstad, M. T., E. Minch, and L. L. Cavalli-Sforza. 1998. Genetic evidence for a higher female migration rate in humans. *Nature Genetics* 20: 278–280.

Sharp, P. M. 1997. In search of molecular Darwinism. *Nature* 385: 111–112.

Sharp, P. M., E. Cowe, D. G. Higgins, D. Shields, K. H. Wolfe, and F. Wright. 1988. Codon usage patterns in *Escherichia coli, Bacillus subtilis, Saccharomyces cerevisiae, Schizosaccharomyces pombe, Drosophila melanogaster,* and *Homo sapiens:* A review of the considerable within-species diversity. *Nucleic Acids Research* 16: 8207–8211.

Sharp, P. M., and W.-H. Li. 1986. An evolutionary perspective on synonymous codon usage in unicellular organisms. *Journal of Molecular Evolution* 24: 28–38.

Sheffield, V. C. 2000. The vision of Typhoon Lengkieki. *Nature Medicine* 6: 746–747.

Smith, N. G. C., and A. Eyre-Walker. 2002. Adaptive protein evolution in *Drosophila*. *Nature* 415: 1022–1024.

Soulé, M. E., and L. S. Mills. 1998. No need to isolate genetics. *Science* 282: 1658–1659.

Sundin, O. H., J.-M. Yang, Y. Li, et al. 2000. Genetic basis of total colourblindness among the Pingelapese islanders. *Nature Genetics* 25: 289–293.

Templeton, A. R., K. Shaw, E. Routman, and S. K. Davis. 1990. The genetic consequences of habitat fragmentation. *Annals of the Missouri Botanical Garden* 77: 13–27.

Templeton, A. R., R. J. Robertson, et al. 2001. Disrupting evolutionary processes: The effect of habitat fragmentation on collared lizards in the Missouri Ozarks. *Proceedings of the National Academy of Sciences USA* 98: 5426–5432.

Thomas, J. 1998. A bird's race toward extinction is halted. *The New York Times* 29 December: D3.

Vallender, E. J., and B. T. Lahn. 2004. Positive selection on the human genome. *Human Molecular Genetics* 13: R245-R254.

van Noordwijk, A. J., and W. Scharloo. 1981. Inbreeding in an island population of the great tit. *Evolution* 35: 674–688.

Wayne, R. K., N. Lehman, D. Girman, P. J. P. Gogan, D. A. Gilbert, K. Hansen, R. O. Peterson, U. S. Seal, A. Eisenhawer, L. D. Mech, and R. J. Krumenaker. 1991. Conservation genetics of the endangered Isle Royale gray wolf. *Conservation Biology* 5: 41–51.

Westemeier, R. L., J. D. Brawn, et al. 1998. Tracking the long-term decline and recovery of an isolated population. *Science* 282: 1695–1698.

Westemeier, R. L., S. A. Simpson, and D. A. Cooper. 1991. Successful exchange of prairie-chicken eggs between nests in two remnant populations. *Wilson Bulletin* 103:717–720.

Winick, J. D., M. L. Blundell, B. L. Galke, et al. 1999. Homozygosity mapping of the achromatopsia locus in the Pingelapese. *American Journal of Human Genetics* 64: 1679–1685.

Wolfe, L. M. 1993. Inbreeding depression in *Hydrophyllum appendiculatum*: Role of maternal effects, crowding, and parental mating history. *Evolution* 47: 374–386.

Wright, S. 1931. Evolution in Mendelian populations. *Genetics* 16: 97–159.

Yang, Z., and J. P. Bielawski. 2000. Statistical methods for detecting molecular adaptation. *Trends in Ecology and Evolution* 15: 496–503.

Young, A., T. Boyle, and T. Brown. 1996. The population genetic consequences of habitat fragmentation for plants. *Trends in Ecology and Evolution* 11: 413–418.

Zuckerkandl, E. and L. Pauling. 1965. Evolutionary divergence and convergence in proteins. In V. Bryson and H. J. Vogel, eds. *Evolving Genes and Proteins*. New York: Academic Press, 97–165.

8

Evolution at Multiple Loci: Linkage and Sex

An asexual lizard. Female desert grasslands whiptails (*Cnemidophorus uniparens*) lay unfertilized eggs that hatch into clones of their mother. There are no males.

In Chapters 6 and 7 we introduced basic population genetics, built on the Hardy–Weinberg equilibrium principle. The models we discussed are elegant and powerful. Figure 6.16 (page 196), for example, documents a case in which a researcher used population genetics to accurately predict the course of evolution 12 generations into the future. In human terms, that is equivalent to accurately predicting what will happen 300 years from now. Population genetics is an extraordinarily successful theory.

As with many theories, however, basic population genetics buys its elegance at the price of simplification. The models we have used until now track allele frequencies at just one locus at a time. We have only been able to consider the evolution of traits that are (or appear to be) controlled by a single gene. The genomes of real organisms, of course, contain hundreds or thousands of loci. And many traits are determined by the combined influence of numerous genes.

In Chapter 8 we will take our models of the mechanics of evolution closer to real organisms by considering two or more loci simultaneously. Our first step in that direction, the subject of Section 8.1, will be an extension of the Hardy–Weinberg analysis that follows two loci at a time. The two-locus model will tell us when we can use the single-locus models developed in Chapters 6 and 7 to make predictions and when we must take into account the confounding influence of selection at other loci.

Our discussion of the two-locus version of Hardy–Weinberg analysis, which uses terms like *linkage disequilibrium*, may at first seem hopelessly abstract. But effort invested in understanding it will produce two surprising payoffs. These are the subjects of Sections 8.2 and 8.3. First, the two-locus model provides tools we can use to reconstruct the history of genes and populations. We will use these tools to address, among other issues, an unresolved question from the discussion in Chapters 1, 5, and 6 of *CCR5-Δ32*, the allele that protects against HIV: Where did the *Δ32* allele come from and why does it occur only in Europe? Second, the two-locus model provides insight into the adaptive significance of one of the most striking and puzzling characteristics of organisms: sexual reproduction.

8.1 Evolution at Two Loci: Linkage Equilibrium and Linkage Disequilibrium

In this section we will expand the one-locus version of Hardy–Weinberg analysis to consider two loci simultaneously. In principle, we could focus on any pair of loci in an organism's genome. Our discussion will be easier to understand, however, if we focus on a pair of loci located sufficiently close together on the same chromosome that crossing-over between them is rare. That is, we will consider two loci that are physically linked (Figure 8.1). We will imagine that locus A has two alleles, *A* and *a*, and that locus B has two alleles, *B* and *b*.

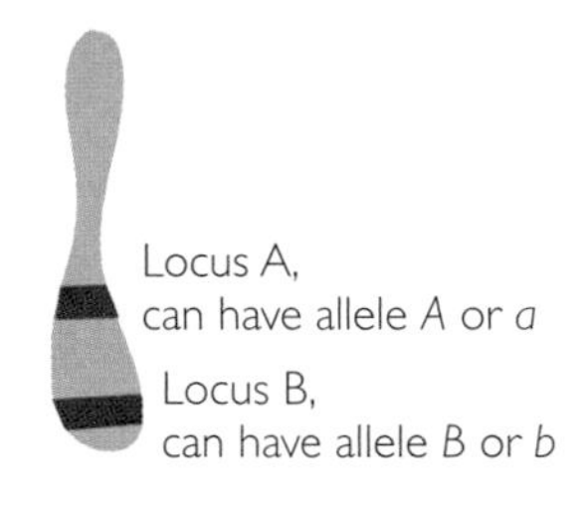

Figure 8.1 A pair of linked loci

In the single-locus version of Hardy–Weinberg analysis, we were concerned primarily with tracking allele frequencies. In the two-locus version, we will be concerned with tracking both allele frequencies and chromosome frequencies. Note that the assumptions we made in the previous paragraph allow four different chromosome genotypes: *AB, Ab, aB,* and *ab*. The multilocus genotype of a chromosome or gamete is sometimes referred to as its **haplotype** (a term that comes from the contraction of *haploid genotype*).

When we use population-genetic models to analyze evolution at a particular locus, do we need to worry about the effects of selection at other loci? Only if the locus of interest and the other loci are in linkage disequilibrium.

Our main goal will be to determine whether selection at the A locus will interfere with our ability to use the models of Chapters 6 and 7 to make predictions about evolution at the B locus. The answer will be: sometimes—depending on whether the loci are in linkage equilibrium or linkage disequilibrium. We will define linkage equilibrium and linkage disequilibrium shortly.

A Numerical Example

A numerical example will illustrate key concepts and help us define terms. Figure 8.2 shows two hypothetical populations, each with a gene pool containing 25 chromosomes. In studying the genetic structure of these populations, the first thing we might do is calculate allele frequencies. In the top population, for example, 15 of the 25 chromosomes carry allele *A* at locus A. Thus the frequency of allele *A* is $15/25 = 0.6$. The same is true for the bottom population. In fact, the allele frequencies at both loci are identical in the two populations. If we were studying locus A only, or locus B only, we would conclude that the two populations are identical.

But the populations are not identical. This we discover when we calculate the chromosome frequencies. In the top population, for example, 12 of the 25 chro-

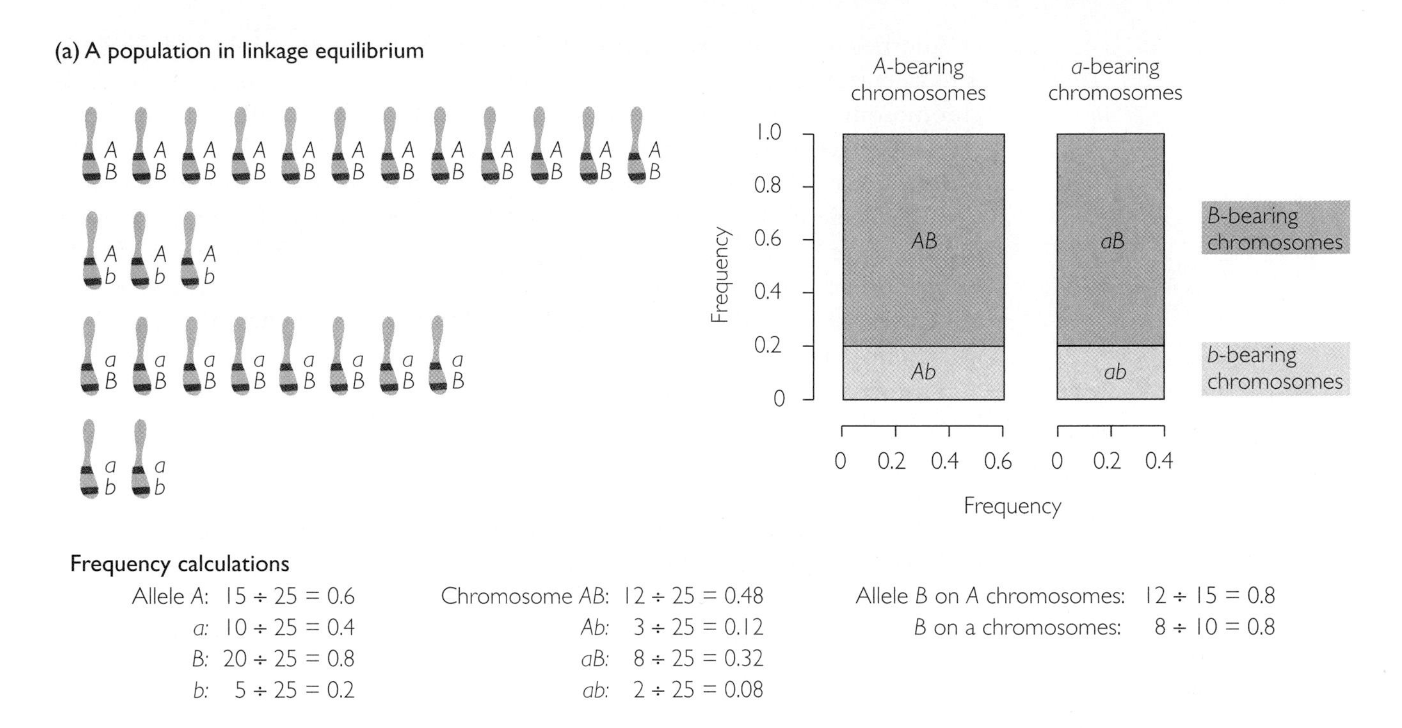

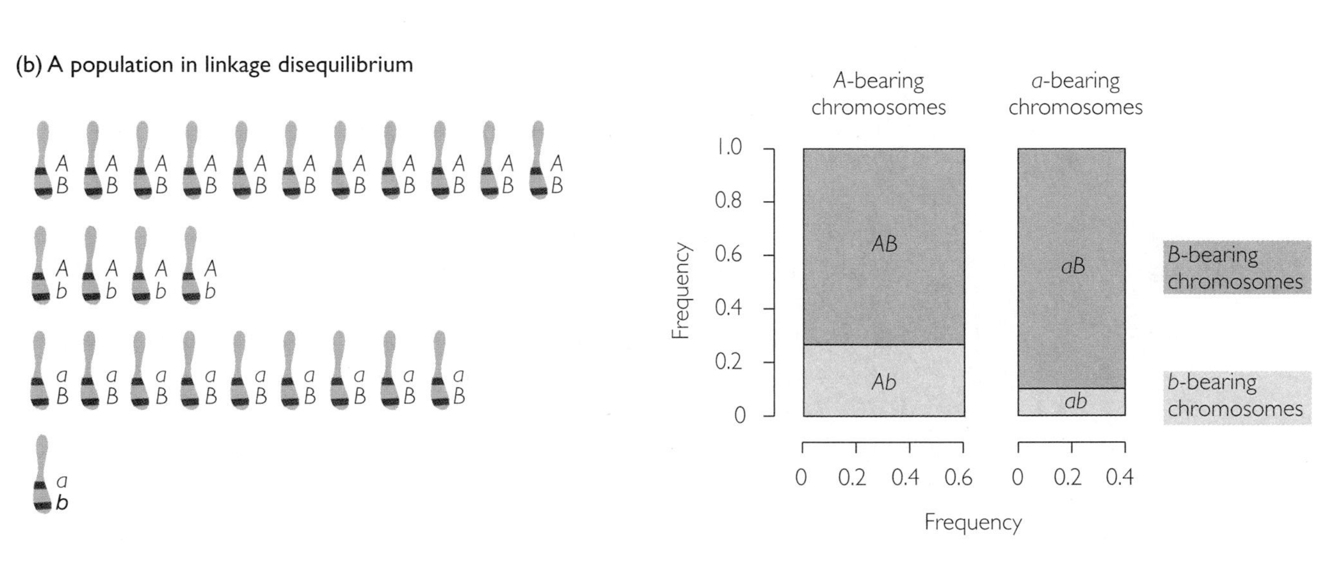

Frequency calculations

Allele *A*: 15 ÷ 25 = 0.6
a: 10 ÷ 25 = 0.4
B: 20 ÷ 25 = 0.8
b: 5 ÷ 25 = 0.2

Chromosome *AB*: 11 ÷ 25 = 0.44
Ab: 4 ÷ 25 = 0.16
aB: 9 ÷ 25 = 0.36
ab: 1 ÷ 25 = 0.04

Allele *B* on *A* chromosomes: 11 ÷ 15 = 0.73
B on a chromosomes: 9 ÷ 10 = 0.9

Figure 8.2 Two populations: one in linkage equilibrium; one in linkage disequilibrium

mosomes carry haplotype *AB*, giving this haplotype a frequency of 0.48. In the bottom population, on the other hand, the frequency of *AB* chromosomes is 11 of 25, or 0.44. This is the first lesson of two-locus Hardy–Weinberg analysis: A pair of populations can have identical allele frequencies but different chromosome frequencies.

Another way to see the difference between the two populations in Figure 8.2 is to calculate the frequency of allele *B* on chromosomes carrying allele *A* versus chromosomes carrying allele *a*. In the top population there are 15 chromosomes carrying allele *A*, 12 of which carry allele *B*. The frequency of *B* on *A* chromosomes is thus 12/15 = 0.8. In the same population there are 10 chromosomes carrying allele *a*, 8 of which carry allele *B*. The frequency of *B* on *a* chromosomes is thus 8/10 = 0.8. In the top population, then, the frequency of allele *B* is the same on chromosomes carrying *A* as it is on chromosomes carrying *a*. The same is not true for the bottom population. There, the frequency of *B* is 0.73 on *A* chromosomes, but 0.9 on *a* chromosomes.

To understand linkage disequilibrium, it is helpful to recognize that when we consider two linked loci at once, populations can have identical allele frequencies, but different chromosome (that is, haplotype) frequencies.

The bar graphs in Figure 8.2 provide a visual representation of the difference between the two populations. The widths of the two bars in each graph represent the frequencies of *A*-bearing chromosomes versus *a*-bearing chromosomes. Note that the combined widths of the two bars must equal 1, so if one bar gets wider, the other must get narrower. The darkly shaded versus lightly shaded portion of each bar represents the frequency of allele *B* versus allele *b* on the chromosomes in question. The bar graphs let us see at a glance what we discovered by calculation in the previous paragraph. In the top population the frequency of *B* is the same on *A* chromosomes as on *a* chromosomes—the same fraction is shaded in both bars. In the bottom population the frequency of *B* is lower on *A* chromosomes than on *a* chromosomes.

Linkage Disequilibrium Defined

In the top population in Figure 8.2, locus A and locus B are in linkage equilibrium. In the bottom population, the loci are in linkage *dis*equilibrium. Two loci in a population are in **linkage equilibrium** when the genotype of a chromosome at one locus is independent of its genotype at the other locus. This means that knowing the genotype of the chromosome at one locus is of no use at all in predicting the genotype at the other. Two loci are in **linkage disequilibrium** when there is a nonrandom association between a chromosome's genotype at one locus and its genotype at the other locus. If we know the genotype of a chromosome at one locus, it provides a clue about the genotype at the other.

These definitions are rather abstract. More concretely, the following conditions are true for a pair of loci if, and only if, they are in linkage equilibrium:

When genotypes at one locus are independent of genotypes at another locus, the two loci are in linkage equilibrium. Otherwise, the loci are in linkage disequilibrium.

1. The frequency of *B* on chromosomes carrying allele *A* is equal to the frequency of *B* on chromosomes carrying allele *a*.
2. The frequency of any chromosome haplotype can be calculated by multiplying the frequencies of the constituent alleles. For example, the frequency of *AB* chromosomes can be calculated by multiplying the frequency of allele *A* and the frequency of allele *B*.
3. The quantity *D*, known as the coefficient of linkage disequilibrium, is equal to zero. *D* is calculated as

$$g_{AB}g_{ab} - g_{Ab}g_{aB}$$

where g_{AB}, g_{ab}, g_{Ab}, and g_{aB} are the frequencies of *AB*, *ab*, *Ab*, and *aB* chromosomes (see Box 8.1).

We have already established, by calculation and with bar graphs, that the first condition is true for the top population in Figure 8.2 but false for the bottom

Box 8.1 The coefficient of linkage disequilibrium

The coefficient of linkage disequilibrium, symbolized by D, is defined as

$$g_{AB}g_{ab} - g_{Ab}g_{aB}$$

where g_{AB}, g_{ab}, g_{Ab}, and g_{aB} are the frequencies of AB, ab, Ab, and aB chromosomes.

To see why D is called the coefficient of linkage disequilibrium, recall that when two loci are in linkage equilibrium, the allele frequencies at one locus are independent of allele frequencies at the other locus. Let p and q be the frequencies of A and a, and let s and t be the frequencies of B and b. If a population is in linkage equilibrium, then $g_{AB} = ps$, $g_{Ab} = pt$, $g_{aB} = qs$, and $g_{ab} = qt$. And furthermore,

$$D = psqt - ptqs = 0$$

If, on the other hand, the population is in linkage disequilibrium, then $g_{AB} \neq ps$, $g_{Ab} \neq pt$, $g_{aB} \neq qs$, and $g_{ab} \neq qt$. And $D \neq 0$.

The maximum value that D can assume is 0.25, when AB and ab are the only chromosomes present and each has a frequency of 0.5. The minimum value that D can assume is -0.25, when Ab and aB are the only chromosomes present and each is at a frequency of 0.5. Thus calculating D is a useful way to quantify the degree of linkage disequilibrium in a population.

population. The reader should verify that the second and third conditions are likewise true for the top population but false for the bottom one.

The Two-Locus Version of Hardy–Weinberg Analysis

We can perform a two-locus version of Hardy–Weinberg analysis analogous to the single-locus version we performed in Chapter 6. We assume Hardy–Weinberg conditions of no selection, no mutation, no migration, infinite population size, and random mating, and we follow chromosome frequencies through one complete turn of our population's life cycle, from gametes in the gene pool to zygotes to juveniles to adults and back to gametes in the gene pool. This calculation is given in Box 8.2. It provides our first piece of evidence that linkage equilibrium is important in evolution. If the two loci in our ideal population are in linkage equilibrium, then under Hardy–Weinberg conditions chromosome frequencies will not change from one generation to the next. If, instead, the loci are in linkage disequilibrium, then the chromosome frequencies will change.

Under Hardy–Weinberg assumptions, chromosome frequencies remain unchanged from one generation to the next, but only if the loci in question are in linkage equilibrium. If the loci are in linkage disequilibrium, the chromosome frequencies move closer to linkage equilibrium each generation.

What Creates Linkage Disequilibrium in a Population?

Three mechanisms can create linkage disequilibrium in a random-mating population: selection on multilocus genotypes, genetic drift, and population admixture. We will consider each of these mechanisms in turn. As we mentioned earlier, the mechanisms that create linkage disequilibrium may be easier to visualize if the reader imagines how they would apply to a pair of loci that are not assorting independently because they are physically linked.

In random mating populations, three mechanisms create linkage disequilibrium: selection on multilocus genotypes, genetic drift, and population admixture.

Selection on Multilocus Genotypes Can Create Linkage Disequilibrium

To see how selection on multilocus genotypes can create linkage disequilibrium, start with the population whose gene pool is shown in Figure 8.2a. Locus A and locus B are in linkage equilibrium. Imagine that the gametes in this gene pool combine at random to make zygotes. The 10 kinds of zygotes produced, and

Box 8.2 Hardy–Weinberg analysis for two loci

Here we develop the two-locus version of the Hardy–Weinberg equilibrium principle. We will show that when an ideal population is in linkage equilibrium, the chromosome frequencies do not change from one generation to the next.

In the single-locus version of Hardy–Weinberg analysis, introduced in Chapter 6, we followed allele frequencies around a complete turn of the life cycle of a population, from one generation's gene pool into zygotes, from zygotes into juveniles, from juveniles into adults, and from adults into the next generation's gene pool. We will use a similar strategy here, except that we will track not allele frequencies but chromosome frequencies. The chromosomes in our organisms contain two loci: the A locus, with alleles *A* and *a;* and the B locus, with alleles *B* and *b.* (We do not intend these symbols to necessarily imply a dominant/recessive relationship between alleles. We use them only because they make the equations easier to read than alternative notations.) There are four kinds of chromosomes: *AB, Ab, aB,* and *ab.*

Imagine an ideal population in whose gene pool chromosomes *AB, Ab, aB,* and *ab* are present at frequencies g_{AB}, g_{Ab}, g_{aB}, and g_{ab}, respectively. If the gametes in the gene pool combine at random to make zygotes, among the possible zygote genotypes is *AB/AB.* Its frequency will equal the probability that a randomly chosen egg contains an *AB* chromosome multiplied by the probability that a randomly chosen sperm contains an *AB* chromosome, or $g_{AB} \times g_{AB}$. Another possible zygote genotype is *AB/Ab.* Its frequency will be $2 \times g_{AB} \times g_{Ab}$. This expression contains a 2 because there are two ways to make an *AB/Ab* zygote: An *AB* egg can be fertilized by an *Ab* sperm, or an *Ab* egg can be fertilized by an *AB* sperm. Overall, there are 10 possible zygote genotypes. Their frequencies are

AB/AB	**Ab/Ab**	**aB/aB**	**ab/ab**	**AB/Ab**
$g_{AB}g_{AB}$	$g_{Ab}g_{Ab}$	$g_{aB}g_{aB}$	$g_{ab}g_{ab}$	$2g_{AB}g_{Ab}$
AB/aB	**AB/ab**	**Ab/aB**	**Ab/ab**	**aB/ab**
$2g_{AB}g_{aB}$	$2g_{AB}g_{ab}$	$2g_{Ab}g_{aB}$	$2g_{Ab}g_{ab}$	$2g_{aB}g_{ab}$

If we allow these zygotes to grow to adulthood without selection, then the genotype frequencies among the adults will be the same as they are among the zygotes.

We have followed the chromosome frequencies from gene pool to zygotes to juveniles to adults. We can now calculate the chromosome frequencies in the next generation's gene pool. Consider chromosome *AB.* Gametes containing *AB* chromosomes can be produced by 5 of the 10 adult genotypes. The adults that can make *AB* gametes, together with the allotment of *AB* gametes they contribute to the new gene pool, are

Adult	*Allotment of AB gametes contributed*	*Notes*
AB/AB	$g_{AB}g_{AB}$	
AB/Ab	$\left(\frac{1}{2}\right)(2g_{AB}g_{Ab})$	
AB/aB	$\left(\frac{1}{2}\right)(2g_{AB}g_{aB})$	
AB/ab	$(1 - r)\left(\frac{1}{2}\right)(2g_{AB}g_{ab})$	where r = recombination rate
Ab/aB	$(r)\left(\frac{1}{2}\right)(2g_{Ab}g_{aB})$	where r = recombination rate

The first row in this table is straightforward: *AB/AB* adults constitute a fraction $g_{AB}g_{AB}$ of the population and thus contribute $g_{AB}g_{AB}$ of the gametes in the gene pool, all of them *AB.* The second row is also straightforward: *AB/Ab* adults constitute a fraction $2g_{AB}g_{Ab}$ of the population and thus contribute $2g_{AB}g_{Ab}$ of the gametes in the gene pool, half of them *AB.* The third row is straightforward as well. It is the last two rows of the table that require explanation.

Adults of genotype *AB/ab* will produce gametes containing *AB* chromosomes only when meiosis occurs *without* crossing-over between the A locus and the B locus. When no crossing-over occurs, half of the gametes produced by *AB/ab* adults carry *AB* chromosomes. If r is the rate of crossing-over, or recombination, between the A locus and the B locus, then the allotment of *AB* gametes contributed to the gene pool by *AB/ab* individuals is $(1 - r)\left(\frac{1}{2}\right)(2g_{AB}g_{ab})$.

Adults of genotype *Ab/aB* produce gametes containing *AB* chromosomes only when meiosis occurs *with* crossing-over between the A locus and the B locus. When crossing-over occurs, half of the gametes produced by *Ab/aB* adults carry *AB* chromosomes. If r is the rate of crossing-over, then the allotment of *AB* gametes contributed to the gene pool by *Ab/aB* individuals is $(r)\left(\frac{1}{2}\right)(2g_{Ab}g_{aB})$.

Box 8.2 (Continued)

We can now write an expression for g_{AB}', the frequency of AB chromosomes in the new gene pool:

$$\begin{aligned} g_{AB}' &= g_{AB}g_{AB} + \left(\tfrac{1}{2}\right)(2g_{AB}g_{Ab}) + \left(\tfrac{1}{2}\right)(2g_{AB}g_{aB}) \\ &\quad + (1-r)\left(\tfrac{1}{2}\right)(2g_{AB}g_{ab}) + (r)\left(\tfrac{1}{2}\right)(2g_{Ab}g_{aB}) \\ &= g_{AB}g_{AB} + g_{AB}g_{Ab} + g_{AB}g_{aB} \\ &\quad + g_{AB}g_{ab} - rg_{AB}g_{ab} + rg_{Ab}g_{aB} \\ &= g_{AB}(g_{AB} + g_{Ab} + g_{aB} + g_{ab}) \\ &\quad - r(g_{AB}g_{ab} - g_{Ab}g_{aB}) \end{aligned}$$

We can simplify this expression further by noting that $(g_{AB} + g_{Ab} + g_{aB} + g_{ab}) = 1$, and that $g_{AB}g_{ab} - g_{Ab}g_{aB}$ is D, defined in the text and Box 8.1. This gives us

$$g_{AB}' = g_{AB} - rD$$

We leave it to the reader to derive the expressions for the other three chromosome frequencies, which are:

$$g_{Ab}' = g_{Ab} + rD;\ g_{aB}' = g_{aB} + rD;\ g_{ab}' = g_{ab} - rD$$

The expressions for g_{AB}', g_{Ab}', g_{aB}', and g_{ab}' show that when a population is in linkage equilibrium—when $D = 0$—the chromosome frequencies do not change from one generation to the next. When, on the other hand, the population is in linkage disequilibrium—when $D \neq 0$—the chromosome frequencies do change from one generation to the next. The first population geneticist to report this result was H. S. Jennings (1917).

We should note that allele frequencies at a pair of loci can be in linkage disequilibrium even when the loci are on different chromosomes. For loci on different chromosomes, it is appropriate to speak of gamete frequencies rather than chromosome frequencies. The Hardy–Weinberg analysis for such a situation is identical to the one we have developed here, except that r is always equal to exactly $\left(\tfrac{1}{2}\right)$.

their expected frequencies, appear in the grid in Figure 8.3a. Because 32% of the eggs are *aB*, for example, and 32% of the sperm are *aB*, we predict that the frequency of *aB*/*aB* zygotes will be $0.32 \times 0.32 = 0.1024$. Now let the zygotes develop into adults and assign phenotypes as follows: Individuals with genotype *ab*/*ab* have a size of 10. For other genotypes, every copy of *A* or *B* adds 1 unit to

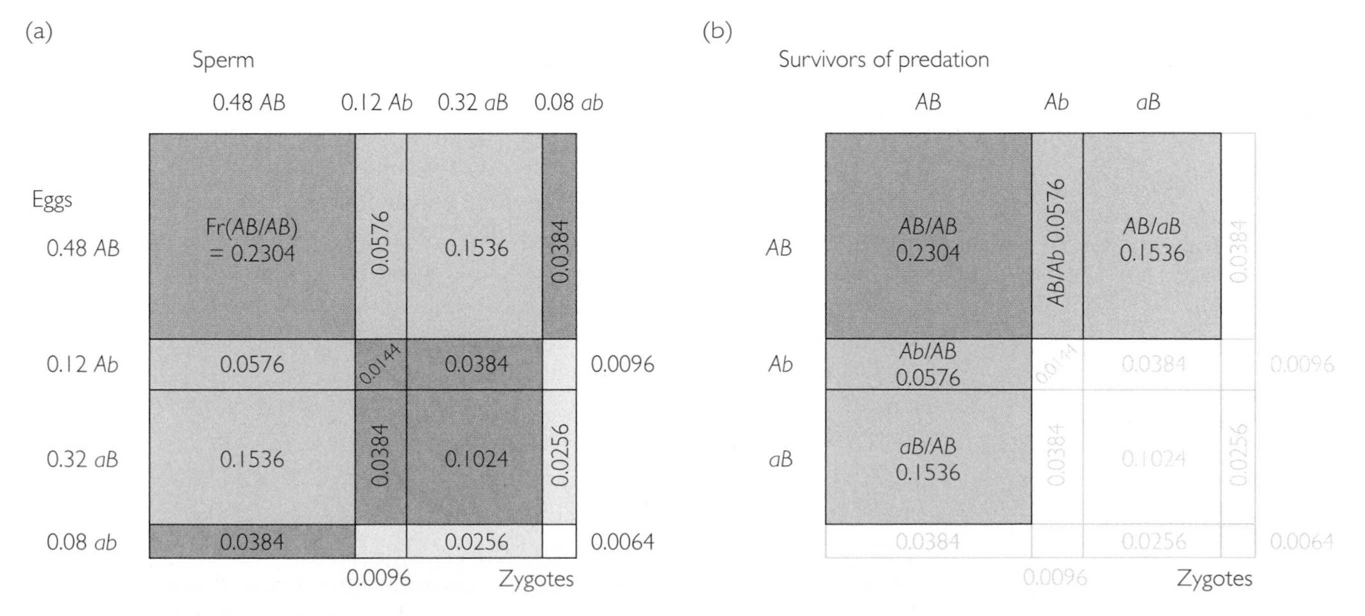

Figure 8.3 Selection on multilocus genotypes can create linkage disequilibrium Diagram (a) shows the expected frequencies of zygotes produced by random mating in the linkage-equilibrium population from Figure 8.2a. Diagram (b) shows the genotypes that survive after predators kill all individuals with fewer than three capital-letter alleles in their genotype. The population of survivors is in linkage disequilibrium because some possible genotypes are missing.

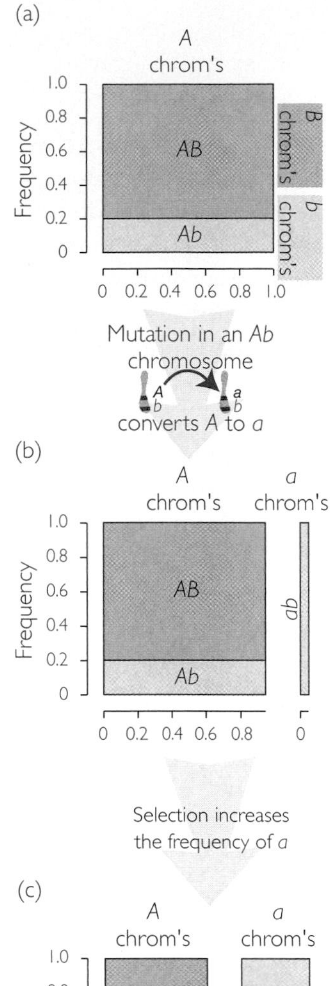

Figure 8.4 Genetic drift can create linkage disequilibrium (a) Chromosome frequencies in a finite population in which only one allele, *A*, is present at locus A; (b) chromosome frequencies after a mutation creates a single copy of allele *a*; (c) chromosome frequencies after selection in favor of *a* increases the frequency of *ab* chromosomes. The population in (c) is in linkage disequilibrium. Drift is the crucial mechanism that created the disequilibrium, because this scenario could only happen in a finite population.

the individual's size. For example, *aB*/*aB* individuals have a size of 12, and *AB*/*Ab* individuals have a size of 13. Finally, imagine that predators catch and eat every individual whose size is less than 13. The survivors, which represent 65.28% of the original population, appear in the grid in Figure 8.3b.

In the population of survivors, locus A and locus B are in linkage disequilibrium. Perhaps the easiest way to see this is to calculate the frequency of allele *a* and allele *b*. One way to calculate the frequency of *a* is this: Of the survivors, the fraction carrying copies of allele *a* is $(0.1536 + 0.1536)/0.6528 \approx 0.47$. All of these carriers of allele *a* are heterozygotes. Therefore, the frequency of allele *a* in the population of survivors is $0.5 \times 0.47 \approx 0.24$. The frequency of *b* is approximately 0.09. If our two loci were in linkage equilibrium, then, by criterion 2 on our list, the frequency of *ab* chromosomes among the survivors would be $0.24 \times 0.09 \approx 0.02$. In fact, the frequency of *ab* chromosomes is 0. Because a nonrandom subset of multilocus genotypes survived, our two loci are in linkage disequilibrium. As an exercise, the reader should demonstrate that the loci are in linkage disequilibrium by criteria 1 and 3 as well.

Genetic Drift Can Create Linkage Disequilibrium

To see how genetic drift can create linkage disequilibrium, look at the scenario diagrammed in Figure 8.4. This scenario starts with a gene pool in which the only chromosomes present are *AB* and *Ab* (Figure 8.4a). In other words, copies of allele *a* do not exist in this population. Locus A and locus B are in linkage equilibrium. Now imagine that in a single *Ab* chromosome, a mutation converts allele *A* into allele *a*. This creates a single *ab* chromosome (Figure 8.4b). It also puts the population in linkage disequilibrium because there is now a possible chromosome haplotype—*aB*—that is missing. The missing haplotype could be created by another mutation or by recombination during meiosis in an *AB*/*ab* diploid, but it may be many generations before either happens. Finally, imagine that selection favors allele *a* over allele *A*, so that *a* increases in frequency and *A* decreases (Figure 8.4c). This increases the degree of linkage disequilibrium between locus A and locus B.

The reader may wonder why we are ascribing the linkage disequilibrium created in this scenario to genetic drift, when the key events seem to be mutation and selection. The reason is that the scenario, as we described it, could only happen in a finite population. In an infinite population, the mutation converting allele *A* into allele *a* would happen not once, but many times each generation, on both *AB* and *Ab* chromosomes. At no point would *aB* chromosomes be missing. Selection favoring *a* over *A* would simultaneously increase the frequency of both *ab* and *aB* chromosomes. Locus A and locus B would never be in linkage disequilibrium. Because our scenario can only create linkage disequilibrium in a finite population, the crucial evolutionary mechanism at work is genetic drift. It was sampling error that caused the mutation creating allele *a* to happen only once, and in an *Ab* chromosome.

Population Admixture Can Create Linkage Disequilibrium

Finally, to see how population admixture can create linkage disequilibrium, imagine two gene pools (Figure 8.5). In one, there are 60 *AB* chromosomes, 20 *Ab* chromosomes, 15 *aB* chromosomes, and 5 *ab* chromosomes. In the other, there are 10 *AB*, 40 *Ab*, 10 *aB*, and 40 *ab* chromosomes. Locus A and locus B are in linkage equilibrium in both gene pools, as the top two bar graphs in Figure 8.5 show. Now combine the two gene pools. This produces a new gene pool in which there are 70 AB, 60 Ab, 25 aB, and 45 ab chromosomes. In this new gene pool, locus A and locus B are in linkage disequilibrium.

Selection on multilocus genotypes, genetic drift, and population admixture can all create linkage disequilibrium because they can all produce populations in which some chromosome haplotypes are underrepresented and others over-represented, compared to what their frequencies would be under linkage equilibrium. In our multilocus selection scheme, for example, selection acted more strongly against *ab* than any other haplotype because no individual containing an *ab* chromosome survived. In our drift scenario, a chance event led to the creation of an *ab* chromosome but no *aB* chromosome. In our population admixture example, a simple combination of populations with different allele and chromosome frequencies created a new population with an excess of *AB* and *ab* chromosomes.

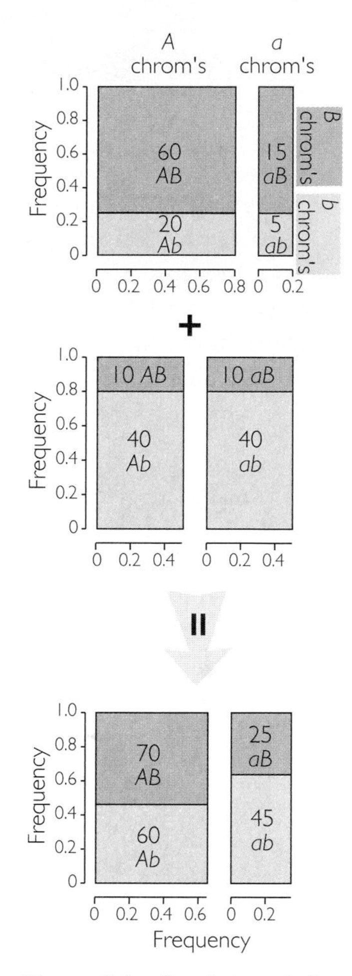

Figure 8.5 Population admixture can create linkage disequilibrium The top two bar charts represent chromosome frequencies in two distinct populations, each in linkage equilibrium. Mixed together, these two populations yield the population shown in the bottom bar chart; it is in linkage disequilibrium.

What Eliminates Linkage Disequilibrium from a Population?

At the same time that selection, drift, and admixture may be creating linkage disequilibrium in a population, sexual reproduction inexorably reduces it. By sexual reproduction, we mean meiosis with crossing-over and outbreeding. The union of gametes from unrelated parents brings together chromosomes with different haplotypes. When the zygotes grow to adulthood and themselves reproduce, crossing-over during meiosis breaks up old combinations of alleles and creates new ones. The creation of new combinations of alleles during sexual reproduction is called **genetic recombination**. Because genetic recombination tends to randomize genotypes at one locus with respect to genotypes at another, it tends to reduce the frequency of overrepresented chromosome haplotypes and to increase the frequency of underrepresented haplotypes. In other words, genetic recombination reduces linkage disequilibrium.

The action of sexual reproduction in reducing linkage disequilibrium is demonstrated algebraically in Box 8.3. The analysis in the box shows that under Hardy–Weinberg assumptions the rate of decline in linkage disequilibrium between a pair of loci is proportional to the rate of recombination between them. Predictions of the rate of decline for several different rates of recombination appear in Figure 8.6.

Michael Clegg and colleagues (1980) documented the decay of linkage disequilibrium in experimental populations of fruit flies. Every population they studied harbored two alleles at each of two loci on chromosome 3. One locus encodes the enzyme esterase-c; we will call it locus A, and its alleles A and a. The other locus encodes the enzyme esterase-6; we will call it locus B, and its alleles B and b.

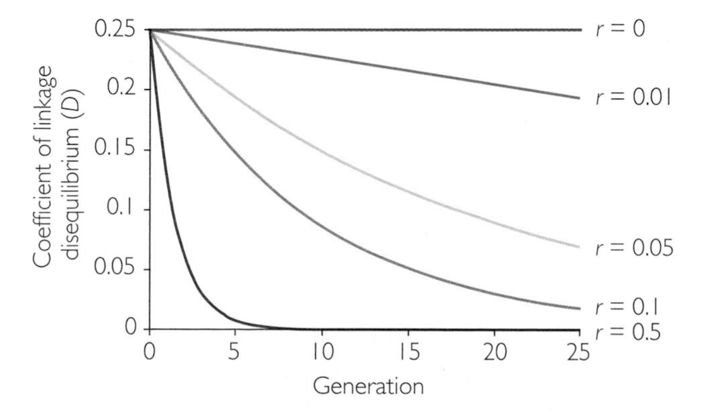

Figure 8.6 With sexual reproduction and random mating, linkage disequilibrium falls over time This graph shows the level of linkage disequilibrium between two loci over 25 generations in a random-mating population. The population starts with linkage disequilibrium at its maximum possible value, 0.25. Each curve shows the decline in linkage disequilibrium, according to the equation $D' = D(1 - r)$, for a different value of r. With $r = 0.5$, which corresponds to the free recombination of loci on different chromosomes, the population reaches linkage equilibrium in less than 10 generations. With $r = 0.01$, which corresponds to closely linked loci, linkage disequilibrium persists for many generations. After Hedrick (1983).

Box 8.3 Sexual reproduction reduces linkage disequilibrium

Here we show that the level of linkage disequilibrium inexorably declines in a random-mating sexual population. We do so by starting with the definition of D, given in the text and Box 8.1, and deriving an expression for D', the coefficient of linkage disequilibrium in the next generation.

By the definition of D,

$$D' = g_{AB}'g_{ab}' - g_{Ab}'g_{aB}'$$

Substituting the expressions for g_{AB}', g_{ab}', g_{Ab}', and g_{aB}' that were derived in Box 8.2 gives

$$\begin{aligned} D' &= [(g_{AB} - rD)(g_{ab} - rD)] \\ &\quad - [(g_{Ab} + rD)(g_{aB} + rD)] \\ &= [g_{AB}g_{ab} - g_{AB}rD - g_{ab}rD + (rD)^2] \\ &\quad - [g_{Ab}g_{aB} + g_{Ab}rD + g_{aB}rD + (rD)^2] \\ &= g_{AB}g_{ab} - g_{AB}rD - g_{ab}rD + (rD)^2 \\ &\quad - g_{Ab}g_{aB} - g_{Ab}rD - g_{aB}rD - (rD)^2 \end{aligned}$$

Canceling and rearranging terms gives

$$\begin{aligned} D' &= g_{AB}g_{ab} - g_{Ab}g_{aB} - g_{AB}rD \\ &\quad - g_{ab}rD - g_{Ab}rD - g_{aB}rD \\ &= (g_{AB}g_{ab} - g_{Ab}g_{aB}) - rD(g_{AB} \\ &\quad + g_{ab} + g_{Ab} + g_{aB}) \end{aligned}$$

Finally, the expression $(g_{AB}g_{ab} - g_{Ab}g_{aB})$ is equal to D, and the expression $(g_{AB} + g_{ab} + g_{Ab} + g_{aB})$ is equal to 1, so we have

$$D' = D - rD = D(1 - r)$$

Recall that r is the rate of recombination during meiosis, which is always between 0 and $\frac{1}{2}$. This means that $(1 - r)$ is always between $\frac{1}{2}$ and 1. Thus, unless there is no recombination at all between a pair of loci, the linkage disequilibrium between them will move closer to 0 every generation. The higher the rate of recombination between the loci, the faster the population reaches linkage equilibrium.

Clegg and colleagues set up fruit fly populations with only AB and ab chromosomes, each at a frequency of 0.5. The researchers also set up populations with only Ab and aB chromosomes, again each at a frequency of 0.5. Thus every population was initially in complete linkage disequilibrium, either with $D = 0.25$, or with $D = -0.25$.

The researchers maintained their fly populations for 48 to 50 generations, at sizes of approximately 1,000 individuals, and let the flies mate as they pleased. Every generation or two, the researchers sampled each population to determine the frequencies of the four chromosome haplotypes and calculated the level of linkage disequilibrium between the two loci. For reasons beyond the scope of our discussion, the researchers measured linkage disequilibrium not with D, but with a related statistic called the correlation of allelic state. The correlation of allelic state, r, is defined as follows:

$$r = \frac{D}{\sqrt{pqst}}$$

where p and q are the frequencies of A and a and s and t are the frequencies of B and b. There is no one-to-one relationship between values of D and the correlation of allelic state, but as a general rule we can say that as linkage disequilibrium in a population declines, and as D moves from 0.25 or -0.25 toward 0, the correlation of allelic state declines as well, moving toward 0 from 1.0 or -1.0. Clegg and colleagues predicted that this is just what they would see in their freely mating fruit fly populations.

The results appear in Figure 8.7. The smooth gray curves show the predicted pattern of decline; the jagged colored lines show the data. As predicted, crossing-over during meiosis created the missing chromosome haplotypes, and the linkage

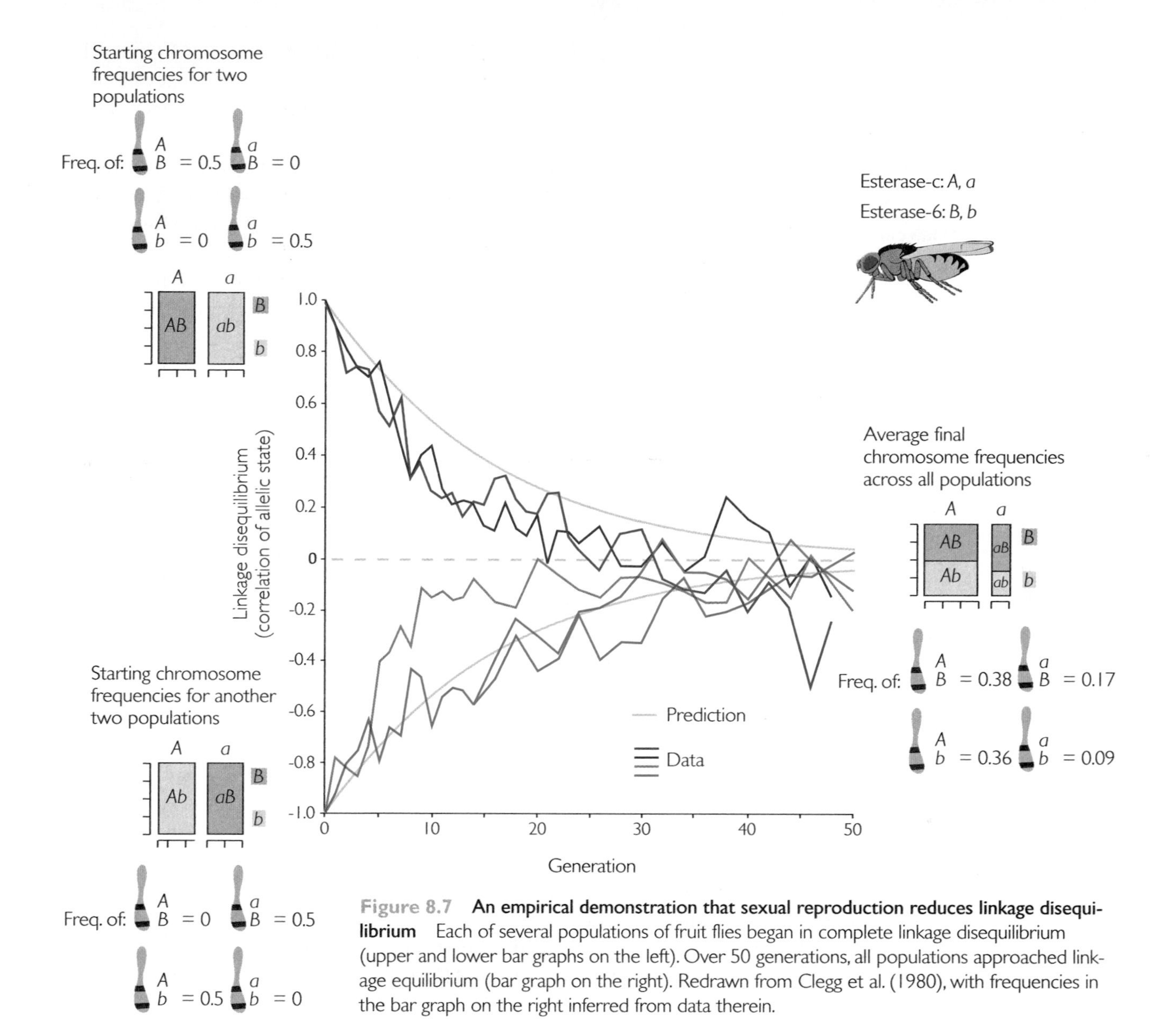

Figure 8.7 An empirical demonstration that sexual reproduction reduces linkage disequilibrium Each of several populations of fruit flies began in complete linkage disequilibrium (upper and lower bar graphs on the left). Over 50 generations, all populations approached linkage equilibrium (bar graph on the right). Redrawn from Clegg et al. (1980), with frequencies in the bar graph on the right inferred from data therein.

disequilibrium between the loci declined. Indeed, linkage disequilibrium declined somewhat faster than predicted. Clegg and colleagues believe that the more-rapid-than-expected decline was the result of heterozygote superiority at the enzyme loci they were studying. Heterozygote superiority would increase the frequency of individuals heterozygous for both loci, and thus provide more opportunities for crossing-over to break down nonrandom associations between alleles at one locus and alleles at the other.

One mechanism reduces linkage disequilbrium: genetic recombination resulting from meiosis and outbreeding (that is, sex).

Why Does Linkage Disequilibrium Matter?

We have defined linkage disequilibrium as a nonrandom association between genotypes at different loci. We have identified multilocus selection, genetic drift, and population admixture as evolutionary mechanisms that can create it. We have seen that sexual reproduction reduces it, restoring populations to a state of linkage equilibrium. And we have demonstrated that in an ideal Hardy–Weinberg population that is in linkage equilibrium, chromosome frequencies do not change from one generation to the next. We have not, however, addressed what we said at the outset was to be our primary question: Can selection at one locus interfere with our ability to use single-locus models to predict the course of evolution at other loci? We are ready to address this question now.

When a pair of loci are in linkage disequilibrium, selection at one locus can change allele frequencies at the other locus. This means that single-locus models may make inaccurate predictions.

The bad news is that if locus A and locus B are in linkage disequilibrium, then selection at locus A changes the frequencies of the alleles at locus B. This means that a single-locus population genetic model looking only at locus B will make inaccurate predictions about evolution.

Figure 8.8a illustrates how selection on locus A can change allele frequencies at locus B. Before selection, allele *B* is at high frequency. Most copies of *B* are on *aB* chromosomes. Selection in favor of allele *A* decreases the frequency of *aB* chromosomes. As *aB* chromosomes disappear, they take copies of *B* with them. Because the frequency of *B* is much lower among *A*-bearing chromosomes than among *a*-bearing chromosomes, many of the lost copies of *B* are replaced by copies of *b*. The end result is that the frequency of *B* declines.

Note that in the scenario we just described, selection was acting only at locus A, not at locus B. Genotypes at locus B had no effect on fitness. Instead, the frequency of allele *B* dropped simply because it got dragged along for the ride. But if we were monitoring locus B only and watching the frequency of allele *B* decline over time, we might erroneously conclude that the target of selection was locus B itself. This is the most depressing lesson of the two-locus version of Hardy–Weinberg analysis: Single-locus studies can be derailed by linkage disequilibrium.

Is this bad news merely hypothetical, or does it alter the course of evolution in nature? Todd Schlenke and David Begun (2004) discovered a case in which selection favoring a new allele at one locus appears, as a result of linkage disequilibrium, to have altered allele frequencies at nearby loci. In a sample of *Drosophila simulans* fruit flies collected in California, Schlenke and Begun assessed the variation in nucleotide sequence at various loci along a stretch of chromosome 2. The researchers found a 100-kb-long region in which there is essentially no genetic variation at all (Figure 8.9). For each of the region's 22 known and putative genes, every fly in the researchers' sample was a homozygote for exactly the same allele.

Schlenke and Begun assayed the variation at three loci within the 100-kb region in a sample of *D. simulans* from Africa. The African fly population showed

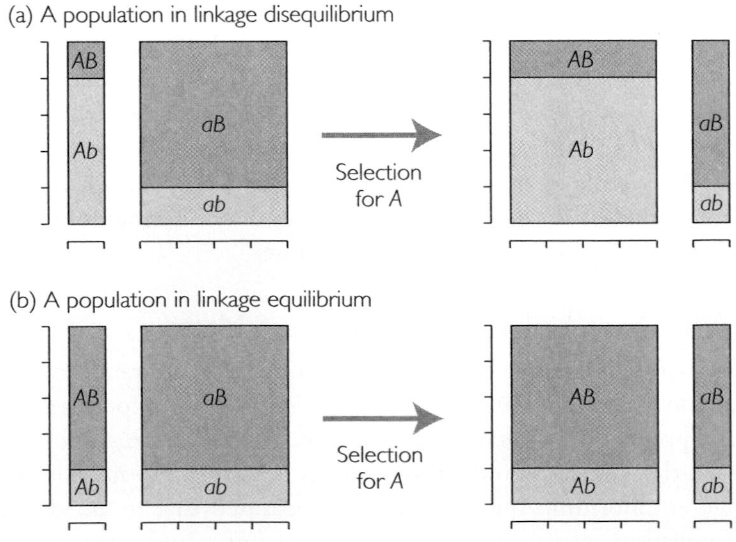

(a) A population in linkage disequilibrium

AB
Ab
aB
ab
Selection for A
AB
Ab
aB
ab

(b) A population in linkage equilibrium

AB
Ab
aB
ab
Selection for A
AB
Ab
aB
ab

Figure 8.8 Linkage equilibrium, selection at a single locus, and allele frequencies at a linked locus In a population in linkage disequilibrium, as shown in (a), selection in favor of allele *A* at locus A causes a decline in the frequency of allele *B* at locus B. In a population in linkage equilibrium, as shown in (b), selection in favor of allele *A* has no effect on the frequency of allele *B*.

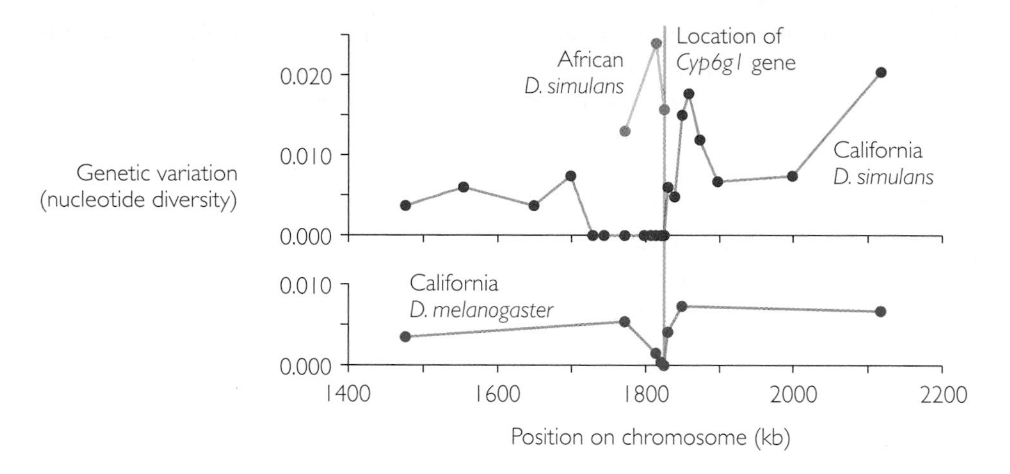

Figure 8.9 A selective sweep in a fruit fly population allowed physically linked alleles at nearby loci to hitchike to high frequency The red dots report the amount of variation in nucleotide sequence at loci along a swath of chromosome 2 in a California population of *Drosophila simulans*. In the center of the swath is a 100-kb-long stretch with virtually no variation. The orange dots report sequence variation in a African population of *D. simulans*. What caused the reduction in diversity in California but not Africa? The blue dots provide a clue; they report sequence variation in a California population of *D. melanogaster*. The point of low diversity corresponds to a locus at which a high-expression allele of the *Cyp6g1* gene has become fixed in the population, apparently because it confers insecticide resistance. A similar high-expression allele has become fixed in the California *D. simulans* population. As this allele rose to fixation, physically linked alleles at nearby loci hitchhiked to high frequency along with it. Plotted from data in Table 1 in Schlenke and Begun (2004).

normal amounts of genetic diversity. The reduction in diversity within the 100-kb region, whatever its cause, apparently happened only in North America.

Schlenke and Begun next checked a sample of a closely related species, *Drosophila melanogaster,* collected in California. Like California *D. simulans, D. melanogaster* showed drastically reduced genetic diversity, but in a much smaller region; only, in fact, in the immediate vicinity of the gene *Cyp6g1*. This gene encodes a cytochrome P450 enzyme that breaks down toxic molecules.

Drosophila melanogaster from California—and from many other parts of the world—carry a transposable element, called *Accord*, that has inserted itself upstream of *Cyp6g1* (Daborn et al. 2002; Catania et al. 2004). The presence of the *Accord* sequence in front of *Cyp6g1* results in a dramatic increase in gene expression. This increase in *Cyp6g1* expression confers resistance to insecticides.

Schlenke and Begun sequenced the *Cyp6g1* gene in *D. simulans* from California and from Africa. The Californian flies, but not the African ones, carry a copy of a different transposable element, called *Doc,* that has inserted itself upstream of *Cyp6g1* in almost exactly the same location as *Accord* in *D. melanogaster*. As with *Accord* in *D. melanogaster,* the presence of *Doc* in *D. simulans* results in a doubling of *Cyp6g1* expression and appears to confer resistance to insecticides. This is a remarkable case of parallel evolution.

Schlenke and Begun believe the best explanation for their data is this. In the recent past, inside a single *D. simulans* fly somewhere in California, a single copy of *Doc* inserted itself in front of the *Cyp6g1* gene in a single chromosome 2. This resulted in a new allele of *Cyp6g1*. Because the new allele conferred resistance to insecticides, it quickly rose to fixation in the population. At the moment of its birth the new allele of *Cyp6g1* was in linkage disequilibrium with alleles at nearby loci, through the mechanism diagrammed in Figure 8.4 on page 288. The alleles that just happened to be on the same individual chromosome as the first copy of the new *Cyp6g1* allele—be they neutral, advantageous, or slightly deleterious—hitchhiked to fixation before recombination could swap any copies of them out for other alleles. A worrisome implication is that if we had looked at one of the nearby loci in isolation, and seen a particular allele at high frequency in California and low frequency elsewhere, we might have mistakenly concluded that natural selection in California had favored that particular allele.

The good news is that if linkage disequilibrium is absent—if locus A and locus B are in linkage equilibrium—then selection on locus A has no effect whatsoever on allele frequencies at locus B. Look at Figure 8.8b. Selection in favor of allele A again eliminates many aB chromosomes. But because the frequency of B is the same among A-bearing chromosomes as among a-bearing chromosomes, every copy of allele B that is lost is replaced by another copy of B. If selection at locus A has no effect on allele frequencies at locus B, then it will not interfere with our use of single-locus models to analyze locus B's evolution.

When a pair of loci are in linkage equilibrium, selection at one locus has no effect on allele frequencies at the other, and we can use single-locus models with confidence. Fortunately, sex is so good at reducing linkage disequilibrium that most pairs of loci are in linkage equilibrium most of the time.

Still better news is that in random-mating populations sex is so good at eliminating linkage disequilibrium that most pairs of loci are in linkage equilibrium most of the time. Research by Elisabeth Dawson and colleagues (2002) illustrates this claim. Dawson and colleagues surveyed linkage disequilibrium among 1,504 marker loci on human chromosome 22 in a population of European families. The marker loci were sites showing allelic variation in single nucleotides or small insertions or deletions. Figure 8.10a shows that nearby loci do tend to be in linkage disequilibrium with each other in some regions of chromosome 22. However, Figure 8.10b shows that the amount of linkage disequilibrium among 18,736 pairs of loci falls rapidly with the physical distance between the loci in the pairs.

Gavin Huttley and colleagues (1999) surveyed the entire human genome for linkage disequilibrium among short tandem-repeat loci. A short tandem-repeat locus is a spot where a short nucleotide sequence is repeated several times; such loci typically have several alleles. Huttley and colleagues conducted 200,000 pairwise tests of linkage disequilibrium involving 5,000 loci distributed across all 22 autosomes. Like Dawson, Huttley and colleagues found several places where neighboring loci exhibit substantial linkage disequilibrium. One such region, known from earlier studies, is an area on chromosome 6 containing the human leukocyte antigen (HLA) loci. The HLA loci encode proteins the immune system uses to recognize foreign invaders. The HLA loci are under strong selection, and the disequilibrium among them is probably due to selection on multilocus genotypes. Overall, however, pairs of loci exhibiting linkage disequilibrium were in the minority. The pairs most likely to show disequilibrium were closely linked physically—that is, situated near enough to each other on the same chromosome that crossing-over between them is rare. Huttley and colleagues focused on pairs of loci close enough that crossing-over occurs between them in 4% or fewer of meiotic cell divisions. Of these pairs, just 4% exhibited linkage disequilibrium.

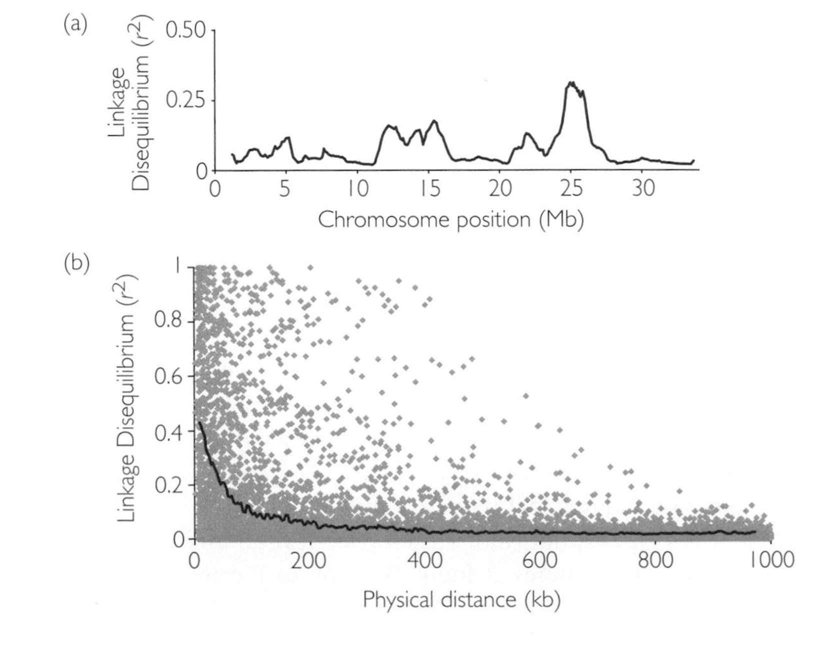

Figure 8.10 On human chromosome 22, most pairs of loci are in linkage equilibrium (a) This graph, which plots the average linkage disequilibrium among nearby loci (squared correlation of allelic state) along human chromosome 22, shows that linkage disequilibrium does exist in the human genome. (b) This graph, however, shows that the further apart a pair of loci are, the less likely it is that they will be in linkage disequilibrium. From Figures 1 and 2 in Dawson et al. (2002).

The International HapMap Consortium (2005) has assembled a database of over one million loci in the human genome with allelic variation at single nucleotides, and has determined the complete genotypes at these sites for 269 individuals in four populations. Researchers have used these data to analyze patterns of linkage disquilibrium across the entire genome in detail (see McVean et al. 2005 for an overview). They have confirmed the general findings of earlier studies. There are blocks of sequence throughout the human genome in which nearby loci are in linkage disequilibrium, but on a larger scale linkage disquilibrium falls rapidly with distance between loci (see De La Vega et al. 2005 Supplemental Figure 1; International HapMap Consortium 2005 Supplementary Figure 6).

In a similar study Magnus Nordborg and colleagues (2005) surveyed the genome of the plant *Arabidopsis thaliana*, a member of the mustard family. One might expect the *Arabidopsis* genome to harbor considerable linkage disequilibrium, even among loci far apart or on different chromosomes. This is because *Arabidopsis* typically self-fertilizes, which leads to increased homozygosity and reduced opportunity for recombination. For each of 96 plants from around the world, Nordborg and colleagues determined the sequences of short fragments of 876 loci scattered throughout the genome. They found that linkage disequilibrium between pairs of loci declines rapidly with the distance between them (Figure 8.11). The researchers concluded that *Arabidopsis* outcrosses enough that its genome resembles that of other sexually reproducing species. Even a relatively small amount of recombination goes a long way toward reducing linkage disequilibrium.

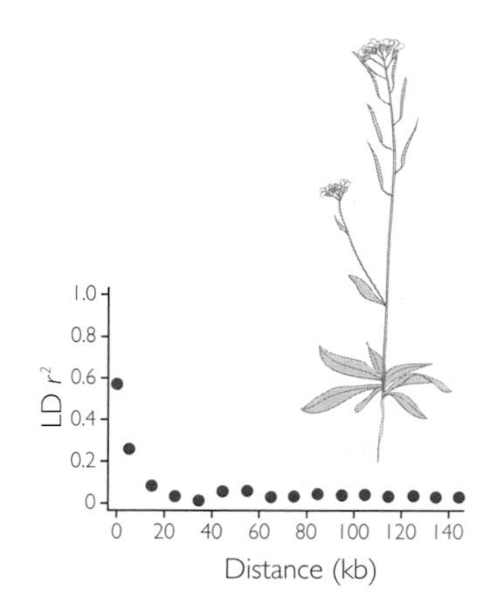

Figure 8.11 Linkage disequilibrium in a highly selfing plant Even in *Arabidopsis thaliana*, a plant thought to reproduce by self-fertilization almost but not quite all of the time, linkage disequilibrium falls rapidly with the distance between loci. From Nordborg et al. (2005).

We can summarize the take-home message in our exploration of two-locus Hardy–Weinberg analysis as follows. Population geneticists need to be aware that any particular locus of interest may be in linkage disequilibrium with other loci, especially other loci located nearby. If the locus of interest is, in fact, in linkage disequilibrium with another locus, then single-locus population genetic models may yield inaccurate predictions. Nonetheless, in freely mating populations, most pairs of loci can be expected to be in linkage equilibrium. In general, we can expect that single-locus models will work well most of the time.

8.2 Practical Reasons to Study Linkage Disequilibrium

In the introduction to this chapter, we promised rewards awaiting readers who mastered the abstractions of Section 8.1. Two such rewards are these: Measurements of linkage disequilibrium provide clues that are useful in reconstructing the history of genes and populations; and linkage disequilibrium can be used to identify alleles that have recently been favored by positive selection.

Reconstructing the History of Genes and Populations

Type 1 Gaucher disease is caused by mutations in the gene encoding the enzyme glucocerebrosidase, also known as acid β-glucosidase, or GBA (Beutler 1993). GBA is found inside lysosomes within cells, where it breaks down the lipid glucocerebroside for recycling. When GBA activity is low or absent, glucocerebroside accumulates within cells. For this reason, Gaucher disease is categorized as a lysosomal storage disorder. Its symptoms include enlargment of the liver and spleen, anemia, and fragile bones.

While Gaucher disease occurs worldwide, it is most common among Ashkenazi Jews, of whom 1 in 19 is a carrier (Strom et al. 2004). The majority of Ashkenazi carriers harbor a nonsynonymous substitution altering amino acid

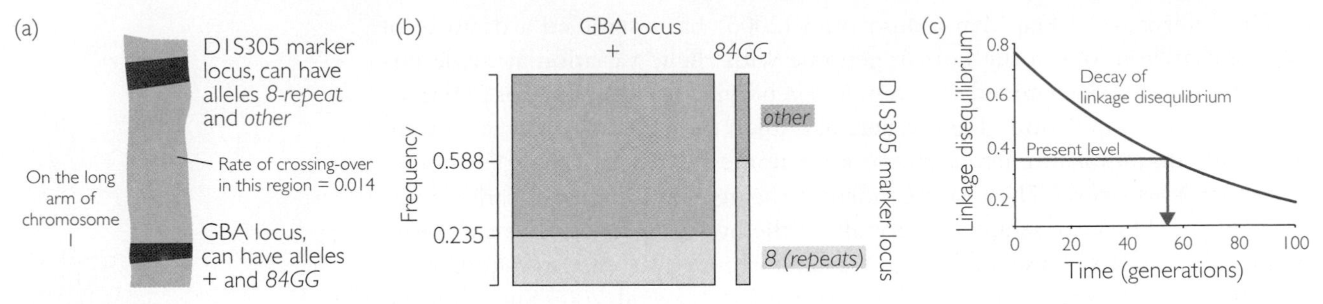

Figure 8.12 Dating the most recent common ancestor of extant copies of the *GBA-84GG* allele (a) The GBA locus is on the long arm of chromosome 1. Nearby is a short tandem repeat locus called D1S305. (b) The GBA locus and D1S305 are in linkage disequilibrium. The frequency of the *8-repeat* allele is higher on chromosomes carrying *84GG* than on normal chromosomes. This suggests that the most recent common ancestor of extant copies of *84GG* was on a chromosome carrying the *8-repeat* allele. (c) We can quantify the linkage disequilibrium between the GBA locus and D1S305 as the difference between the frequency of the *8-repeat* allele on *84GG* chromosomes versus normal chromsomes. The blue curve shows the decay in linkage disequilibrium since the time of the most recent common ancestor. The red lines show the present level of linkage disequilibrium and the inferred number of generations elapsed since the most recent common ancesotor. If the generation time is 25 years, then the most recent common ancestor of extant *84GG* alleles existed about 1,375 years ago. Based on data and calculations in Diaz et al. (2000).

370 in the GBA protein (Beutler et al. 1991). This allele, called *N370S*, is found in other populations as well. A substantial minority of Ashkenazi carriers, however, harbor an allele that is exclusive to the Ashkenazim. This allele, called *84GG*, has an extra guanine inserted at nucleotide position 84. The insertion results in a frame shift and complete loss of GBA function.

George Diaz and colleagues (2000) wanted to know how long the *GBA-84GG* mutation has been circulating among the Ashkenazim. They identified a marker locus on chromosome 1 that is in linkage disquilibrium with the GBA locus (Figure 8.12a). This locus, called D1S305, is a short tandem repeat polymorphism. About 59% of chromosomes carrying the *84GG* allele at the GBA locus carry the *8-repeat* allele at D1S305, whereas only about 24% of chromosomes carrying the normal allele at the GBA locus carry the *8-repeat* allele at D1S305 (Figure 8.12b).

How did this linkage disequilibrium arise? As discussed in Section 8.1, there are three possibilities: selection on multilocus genotypes, genetic drift, and population admixture. Selection on multilocus genotypes is an unlikely candidate, because D1S305 is a noncoding locus and its alleles appear to be selectively neutral. Population admixture is also unlikely, because it would require a source population in which the frequency of the *84GG* allele is much higher than it is in the Ashkenazim, and no such population exists. That leaves genetic drift.

Diaz and colleagues believe that, similar to the scenario in Figure 8.4, all copies of *84GG* now circulating among the Ashkenazim are derived from a single common ancestor that carried the *8-repeat* allele at D1S305. The ancestral copy of *84GG* might have been a new mutation that arose in an Ashkenazi individual. It might have been a mutation carried into the Ashkenazi population by a single heterozygous migrant from a population in which *84GG* was subsequently lost. Or it might have been the only copy to survive the genetic drift that accompanied a bottleneck, or reduction in the size of the Ashkenazi population.

When there was only one copy of *84GG* in the population, the linkage disequilibrium between the GBA locus and D1S305 was complete. The frequency of the *8-repeat* allele on chromosomes carrying *84GG* was 100%. Soon, however, *84GG* had produced multiple descendants and the disequilbrium began to break down. Crossing-over swapped other alleles at the short tandem repeat locus onto chromosomes carrying *84GG*. Eventually the frequency of the *8-repeat* allele will be the same on *84GG* chromosomes as it is on normal chromosomes. Diaz and colleagues used the rate of crossing-over between the GBA locus and D1S305 to estimate the rate at which their linkage disequilibrium is decaying (Figure 8.12c, blue curve). Then they used the rate of decay and the present level of linkage disequilbrium to calculate that the most recent common ancestor ancestor of all extant *84GG* alleles existed between 750 and 2,325 years ago, with a best estimate of 1,375 years ago (Figure 8.12c, red lines; see Box 8.4 for details).

Box 8.4 Estimating the age of the *GBA-84GG* mutation

Here we outline the calculation George Diaz and colleagues (2000) used to estimate the age of the most recent common ancestor of all copies of the *GBA-84GG* mutation in the Ashkenazim (see also Slatkin and Rannala 2000). The GBA locus is on chromosome 1 (Figure 8.12). For our purposes it has two alleles: the normal allele, +, and *84GG*. Nearby is the short tandem repeat locus D1S305, with alleles *8* (for the number of repeats) and *other*.

The GBA locus and D1S305 are in linkage disequilibrium: The frequency of allele *8* is higher on chromosomes carrying *84GG* than on chromosomes carrying +. Assuming that this linkage disequilibrium is not being maintained by selection on multilocus genotypes, it will be in the process of decaying as a result of recombination. Diaz and colleagues developed an equation that describes this decay.

They started with an equation that predicts the frequency of *8* on *84GG* chromosomes in any given generation from its frequency in the generation before. The researchers assumed that *84GG* chromosomes are rare enough that they virtually always pair with + chromosomes, that the frequency of *8* on + chromosomes is constant over time, and that there is no mutation. The equation is

$$X_t = (1 - c)X_{t-1} + cY,$$

where X_t is the frequency of *8* on *84GG* chromosomes in generation t, Y is the frequency of *8* on + chromosomes, and c is the rate of crossing-over between the GBA locus and the short tandem repeat locus. The first term on the right accounts for the *84GG–8* chromosomes that do not experience recombination. The second term on the right accounts for the *84GG* chromosomes that receive a copy of the *8* as a result of recombination.

Subtracting Y from both sides gives

$$X_t - Y = (1 - c)X_{t-1} - Y + cY.$$

Factoring $-Y$ out of the two right-most terms gives

$$X_t - Y = (1 - c)X_{t-1} - Y(1 - c).$$

And factoring $(1 - c)$ out of the two terms on the right gives

$$X_t - Y = (1 - c)(X_{t-1} - Y).$$

We can think of $(X_t - Y)$ as a measure of linkage disequilibrium, because if the GBA locus and D1S305 were in linkage equilibrium $(X_t - Y)$ would be zero. Our equation thus describes the decay of linkage disquilibrium. Compare it to the equation we derived in Box 8.3.

According to our equation, each generation the difference between the frequency of *8* on *84GG* versus + chromosomes declines by a factor of $(1 - c)$. This implies that

$$X_t - Y = (1 - c)^t(X_0 - Y),$$

where X_0 is the frequency of *8* on *84GG* chromosomes in the generation in which the *84GG* mutation last appeared in the population as just a single gene copy. Note that when the population contained only one copy of *84GG* (which was on an *84GG–8* chromosome) the frequency of *8* on *84GG* chromosomes was 100%. In other words, $X_0 = 1$, thus

$$X_t - Y = (1 - c)^t(1 - Y).$$

Dividing both sides by $(1 - Y)$ and taking the natural logarithm of both sides allows us to solve for t:

$$\ln\left[\frac{(X_t - Y)}{(1 - Y)}\right] = \ln[(1 - c)^t]$$

$$\ln\left[\frac{(X_t - Y)}{(1 - Y)}\right] = t\ln(1 - c)$$

$$t = \frac{1}{\ln(1 - c)} \times \ln\left[\frac{(X_t - Y)}{(1 - Y)}\right]$$

All we have to do to estimate t is plug estimates of X_t, Y, and c into this equation.

Based on a sample of 85 + chromosomes and 58 *84GG* chromosomes, Diaz and colleagues estimated that $X_t = 0.588$ and $Y = 0.235$. Their estimate of the recombination rate, c, was 0.014. These values put the time of the last common ancestor of all copies of *84GG* at 55 generations ago. If the generation time is 25 years, the last common ancestor existed about 1,375 years ago. This estimate is sensitive to the recombination rate, which, being a small number, is hard to measure accurately. Allowing for error in the recombination rate, Diaz and colleagues concluded that the *84GG* carrier of the ancestral allele probably lived between 750 and 2,325 years ago.

The Ashkenazim carry at unusually high frequency alleles causing a number of other genetic disorders, including Tay-Sachs disease, Fanconi anemia type C, factor XI deficiency, and elevated risk of breast cancer. Neil Risch and colleagues (2003) reviewed efforts, including that of Diaz and colleagues, to estimate the ages of the most recent common ancestors of eleven Ashkenazi disease alleles. The ages fall roughly into three categories: about 12 generations old, about 50 generations old, and more than 100 generations old.

Because the list of mechanisms that create linkage disequilibrium is short, the presence of linkage disequilibrium in a population provides clues to the population's past.

These dates are broadly consistent with the history of the Ashkenazi population. The Ashkenazim trace their recent ancestry to eastern and central Europe and their more ancient ancestry to the Middle East. The group of disease alleles whose most recent common ancestors are over 100 generations old can be explained by a genetic drift in the form of a founder effect accompanying the departure of Jewish populations from the Middle East 2,000 to 3,000 years ago (Risch et al. 2003). The group whose most recent common ancestors are about 50 generations old can be explained by a founder effect associated with the arrival of the Ashkenazim in central Europe 1,000 to 1,500 years ago. And the group whose most recent common ancestors are about 12 generations old can be explained by a founder effect associated with the arrival of the Ashkenazim in Lithuania within the last 400 years.

While they generally confirm accounts of Ashkenazi history, the genetic analyses suggest that the migrations of this population have entailed more severe bottlenecks than might otherwise have been appreciated. There is ongoing debate over whether genetic drift is sufficient to explain the high frequencies of all Ashkenazi disease alleles. Some researchers have concluded that it is (Behar et al. 2004; Slatkin 2004); others invoke selection favoring heterozygotes for at least some categories of alleles (Cochran et al. 2006).

Although not associated with a genetic disease, another allele that is surprisingly common among the Ashkenazim, as well as other European populations, is *CCR5-Δ32*. This allele, which we discussed in Chapters 1, 5, and 6, is a loss-of-function mutation at the CCR5 locus. It protects homozygotes against sexually transmitted strains of HIV-1.Unresolved issues from our earlier discussions are where the *Δ32* allele came from and why is it common only in Europe.

J. Claiborne Stephens and colleagues (1998) addressed these questions with an analysis similar to that by Diaz and colleagues on *GBA-84GG*. Stephens and colleagues found that the CCR5 locus is in strong linkage disequilibrium with two nearby marker loci, both noncoding and apparently neutral. Most chromosomes carrying the *Δ32* allele at the CCR5 locus also carry a specific haplotype at the marker loci. This suggests that the *Δ32* allele arose just once, as a unique mutation. As the *Δ32* allele subsequently rose to high frequency, the marker alleles that happened to be linked with it came along for the ride.

The linkage disequilibrium between the CCR5 locus and the marker loci is no longer perfect. Since the *Δ32* allele first appeared, recombination and/or additional mutations have put *Δ32* into new haplotypes. Stephens and colleagues used estimates of the rates of crossing-over and mutation to calculate how fast the linkage disequilibrium would be expected to break down, then used this calculation to estimate the age of the last common ancestor of all extant copies of the *Δ32* allele. The researchers concluded that the common ancestor lived between 275 and 1,875 years ago, with a best estimate of about 700 years ago.

This result was tantalizing. It implied that the frequency of the *Δ32* allele in Europe had risen from virtually zero to 15% or more in roughly 30 generations.

Such a rapid climb can be explained most readily by strong natural selection. What might have been the selective agent that gave the *Δ32* allele such an advantage? The most obvious suspects were epidemic diseases: the Black Death (Stephens et al. 1998), which swept Europe during the 14th century and killed a third of the population; and smallpox (Lalani et al. 1999; Galvani and Slatkin 2003).

Pardis Sabeti and colleagues (2005) pointed out, however, that Stephens and colleagues had based their calculations on a genetic map that has turned out to be flawed. The marker loci Stephens used are closer to the CCR5 locus, and the recombination rates between the markers and CCR5 are therefore lower, than previously thought. Using a larger set of genetic markers, Sabeti and colleagues calculated that the common ancestor of all extant copies of the *Δ32* allele lived between 3,150 and 7,800 years ago, with a best estimate of 5,075 years ago. On this and other evidence, Sabeti and colleagues argued that the current frequency and distribution of the *Δ32* allele could be explained by genetic drift.

But the story does not end there. Using PCR, Susanne Hummel and colleagues (2005; see also Hedrick and Verrelli 2006) were able to recover DNA sequences from the CCR5 locus in the skeletons of 17 Bronze Age Europeans. These individuals lived 2,900 years ago in what is now northwestern Germany, and were buried in the Lichtenstein Cave. Four of them were heterozygous carriers of the *Δ32* allele. This confirms that the allele is at least a few thousand years old. It also puts the Bronze Age frequency of the allele at about 12%—well within the modern range. Together, Sabeti's age estimate and Hummel's discovery suggest that the *Δ32* allele rose to its present frequency within 200 generations of its appearance by mutation.

John Novembre, Alison Galvani, and Montgomery Slatkin (2005) developed a population-genetic simulation of how a unique allele might increase in frequency as it spread across Europe. Their model includes parameters describing the distance individuals move in a lifetime and the strength of selection. The model indicates that the *Δ32* allele could not have achieved its present distribution as quickly as it apparently did without the aid of natural selection.

The answer to our questions about the origin of the *Δ32* allele appear to be as follows: The allele was created by a unique mutation that occurred in Europe within the past several thousand years. And the allele does not occur outside Europe either because the mutation creating it has never occurred in a non-European population, or because when the mutation *has* occurred outside Europe, it has not been favored by selection. The agent of selection remains unknown.

Detecting Positive Selection

We have seen that a unique mutation, by the mere fact of its birth, puts its locus in linkage disequilibrium with nearby markers. This linkage disequilibrium immediately begins to break down. This means that when we find a locus in linkage disequilibrium with nearby markers, we suspect that the locus harbors a young allele. If a young allele is at high frequency, we suspect that during its short life the allele has been favored by positive natural selection. Using this logic, Pardis Sabeti, David Reich, and colleagues (2002) developed a general method for identifying alleles recently favored by selection. Sabeti and colleagues demonstrated their method by applying it to the G6PD locus in humans.

The G6PD locus, located on the X chromosome, encodes an essential housekeeping enzyme called glucose-6-phosphate dehydrogenase (Ruwende and Hill

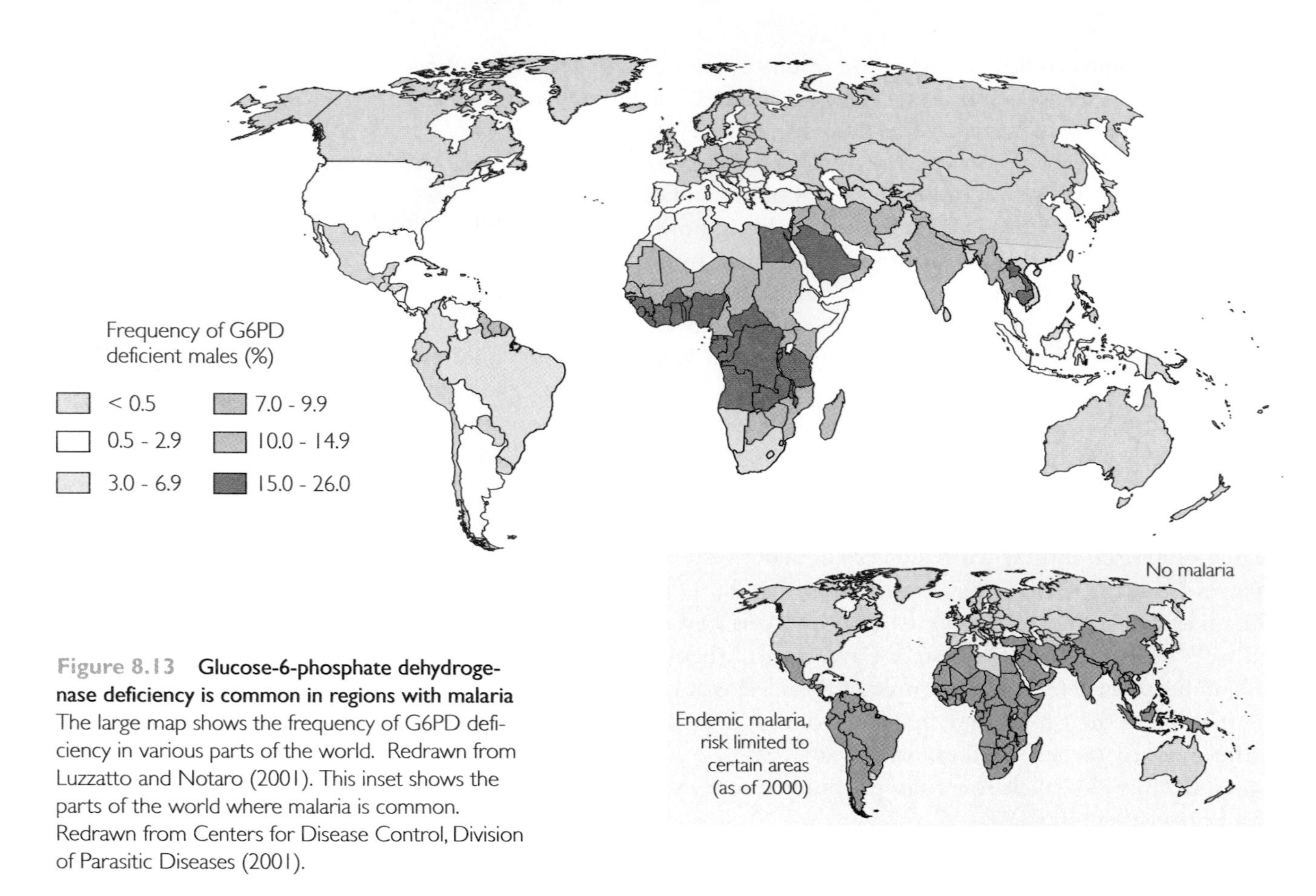

Figure 8.13 Glucose-6-phosphate dehydrogenase deficiency is common in regions with malaria The large map shows the frequency of G6PD deficiency in various parts of the world. Redrawn from Luzzatto and Notaro (2001). This inset shows the parts of the world where malaria is common. Redrawn from Centers for Disease Control, Division of Parasitic Diseases (2001).

1998). The locus is highly variable. Hundreds of alleles are known, distinguishable by the encoded protein's biochemical properties. Dozens of these variants reach frequencies of 1% or more. And many common alleles have reduced enzymatic activity (Figure 8.13a). Indeed, with 400 million people affected worldwide, G6PD deficiency is the most common enzyme deficiency known. Individuals with mild G6PD deficiency often have no symptoms, but individuals with more severe deficiencies can suffer episodes in which their red blood cells rupture, a condition known as acute hemolytic anemia. Why is this potentially serious genetic condition so common? The geographic distribution of G6PD deficiency suggests that it confers some resistance to malaria [Figure 8.13a and (b)]. This inference is supported by epidemiological evidence. For example, individuals carrying the allele *G6PD-202A*, a reduced-activity variant common in Africa, have a substantially lower risk of suffering severe malaria.

Sabeti, Reich, and colleagues reasoned that if the *G6PD-202A* allele confers resistance to malaria, then it should bear the signature of recent positive selection. What would the signature of recent positive selection look like? Imagine a new mutant allele appearing in a finite population. The allele is unique and is consequently in complete linkage disequilibrium with other loci. If it is neutral, meaning that its frequency evolves by genetic drift, our new allele can anticipate one of three fates. It may disappear, in which case we will not have it to study. It may persist but remain rare. Or it may persist and gradually drift to high frequency. If our allele persists, its linkage disequilibrium with other loci will break down under the influence of recombination. The further away the other loci are, the more rapidly the linkage disequilibrium will break down. Thus, in a population evolving by mutation and genetic drift we can expect to find three kinds of alleles: alleles that are rare and (because they are young) have high linkage disequilibrium with other loci; alleles that are rare and (because they are old) have low linkage disequilibrium; and alleles that are common and (because they are old) have low linkage disequilibrium. What we do not expect to find is alleles that are common and have high linkage disequilibrium with other loci. The signature of recent positive selection is thus a high frequency combined with high linkage disequilibrium with other loci. The higher the frequency, and the further the linkage disequilibrium extends from the locus of interest, the stronger the selection must have been.

When an allele at a coding locus is in linkage disequilibrium with alleles at nearby neutral marker loci, we can infer that the coding allele is relatively young. When a young allele is at high frequency, we can infer that it has recently been favored by positive selection.

We already know that the *G6PD-202A* allele has a high frequency—about 18% in the three African populations Sabeti and colleagues studied. To assess *G6PD-202A*'s linkage disequilibrium, the researchers examined the X chromosomes of a total of 230 men. First they looked at the G6PD gene on each chromosome. They found a total of nine alleles of the gene, distinguishable by the combination of nucleotides present at 11 SNP loci within the gene itself. SNP is an acronym for single-nucleotide polymorphism. An SNP locus is a particular base pair in the genome where different individuals have different nucleotides. One of the nine alleles the researchers were able to distinguish by SNP haplotype was *G6PD-202A*.

The researchers next determined each X chromosome's genotype for 14 SNPs outside the G6PD gene, located at distances ranging up to 413,000 base pairs. To characterize each G6PD allele's linkage disequilibrium with these 14 outside loci, Sabeti and colleagues calculated a quantity they call the **extended haplotype homozygosity**, or **EHH**. An allele a's extended haplotype homozygosity to a particular outside distance x is defined as the probability that two randomly chosen chromosomes carrying allele a will also have the same genotype at all marker loci out to x. The higher an allele's EHH, the greater its linkage disequilibrium with nearby loci. Figure 8.14a plots the EHH values for *G6PD-202A* (red) and the other eight alleles (black) out to the locations of each of the outside SNP loci. The *G6PD-202A* allele clearly has the highest linkage disequilibrium, extending further away from the G6PD gene.

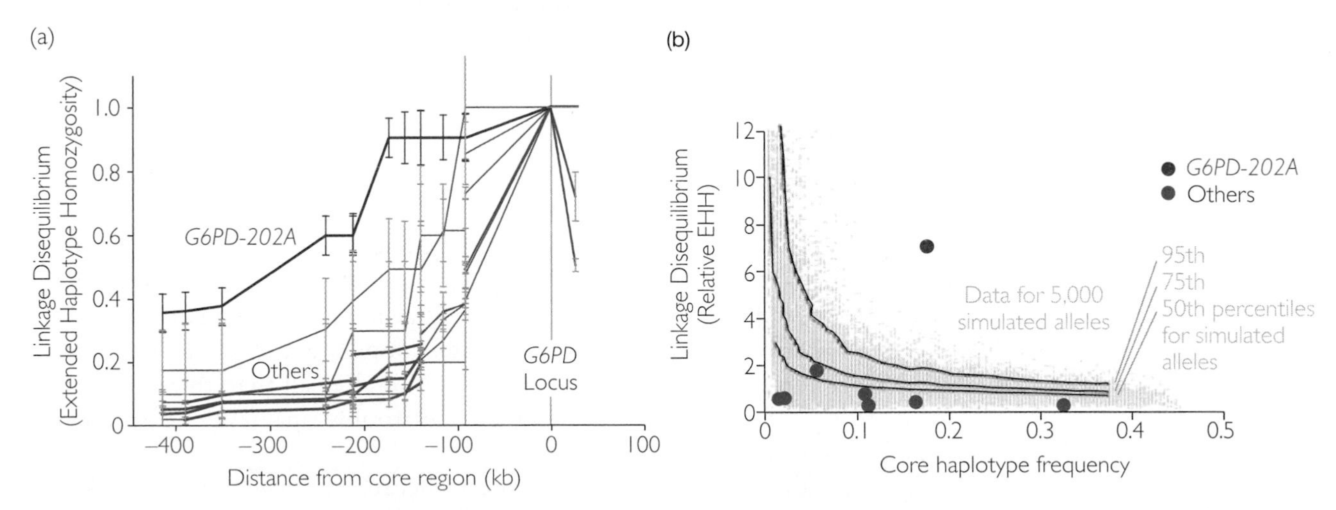

Figure 8.14 The signature of positive selection (a) This graph shows the linkage disequilibrium (extended haplotype homozygosity) of alleles at the G6PD locus at various distances. Allele *G6PD-202A* (red) has higher linkage disequilibrium, extending farther along the chromosome, than the other eight alleles (black). (b) This graph shows that *G6PD-202A*'s high linkage disequilibrium, in combination with its high frequency, clearly distinguish it from other alleles at the G6PD locus and from neutral alleles in simulated populations evolving by genetic drift. An allele's relative EHH is its EHH at the most distant marker divided by the average EHH at that same marker of all other alleles. See text for more explanation. From Sabeti et al. (2002).

So now we know that *G6PD-202A* has both a high frequency and high linkage disequilibrium. Is this a strong enough signature of positive selection to rule out genetic drift? To find out, Sabeti and colleagues ran computer simulations of genetic drift, producing several thousand replicates of their actual data set. The frequencies of the alleles in these simulated data sets, and their EHH values at 413 kb, are plotted in the graph in Figure 8.14b. They form a gray cloud that clings to the horizontal and vertical axes. Consistent with the verbal argument we made above, neutral alleles evolving by drift can have a high frequency, or high linkage disequilibrium, but not both. *G6PD-202A* and the eight other alleles from the actual data set are also plotted in the graph. The eight other alleles, shown in blue, fall well within the gray cloud. Their numbers are easily explainable by drift. But *G6PD-202A*, shown in red, is clearly an outlier. Sabeti and colleagues concluded that *G6PD-202A* has recently been favored by natural selection, a finding consistent with the hypothesis that the allele confers partial resistance to malaria.

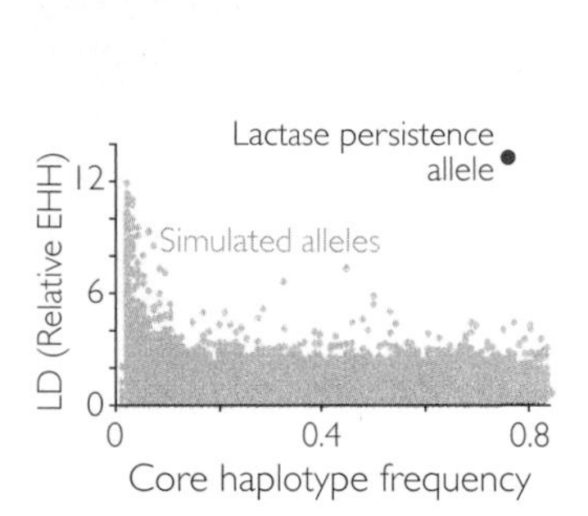

Figure 8.15 Evidence of positive selection on the ability to digest milk sugar in Europeans This graph, similar to Figure 8.14b on the previous page, shows that an allele associated with persistent production of lactase in adulthood has a combination of high linkage disequilibrium with nearby markers and high frequency that would be extraordinarily unusual among simulated alleles evolving by genetic drift. From Bersaglieri et al. (2004).

The method developed by Sabeti and colleagues is readily applicable to other loci. Todd Bersaglieri, Sabeti, and colleagues (2004) used it to show that an allele associated with the persistence of lactase production beyond the age of weaning has recently been favored by positive selection in Europeans (Figure 8.15). It is this allele that gives individuals in populations with a history of dairying the ability, rare among humans, to digest milk sugar in adulthood. Benjamin Voight, Jonathan Pritchard, and colleagues (2006) used an extension of the method to scan the entire human genome for loci showing evidence of recent positive selection in East Asians, Europeans, and West Africans (Yoruba). Among the loci bearing, in one or more of these groups, the signature of positive selection—in the form of high frequency and high linkage disequilibrium—are genes involved in sperm motility and fertilization, olfaction, skin color, skeletal development, and carbohydrate metabolism.

We have shown that an understanding of linkage disequilibrium yields powerful tools for reconstructing the history of alleles and for detecting positive selection. An additional reward we promised readers was that understanding linkage disequilibrium would help us understand the adaptive significance of sexual reproduction. The mystery of sex is the subject of the next section.

8.3 The Adaptive Significance of Sex

Sexual reproduction is complicated, costly, and dangerous. Searching for a mate takes time and energy, and may increase the searcher's risk of being killed by a predator. Once found, a potential mate may demand additional exertion or investment before agreeing to cooperate. Sex itself may expose the parties to sexually transmitted diseases. And after all that, the mating may prove to be infertile. Why not avoid all the trouble and risk and simply reproduce asexually instead?

Many species are capable of both sexual and asexual reproduction.

This question sounds odd to human ears, because we have no choice: We inherited from our ancestors the inability to reproduce by any other means than sex. But many organisms do have a choice, at least in a physiological sense: They are capable of both sexual and asexual reproduction and regularly switch between the two. Most aphid species, for example, have spring and summer populations composed entirely of asexual females. These females feed on plant juices

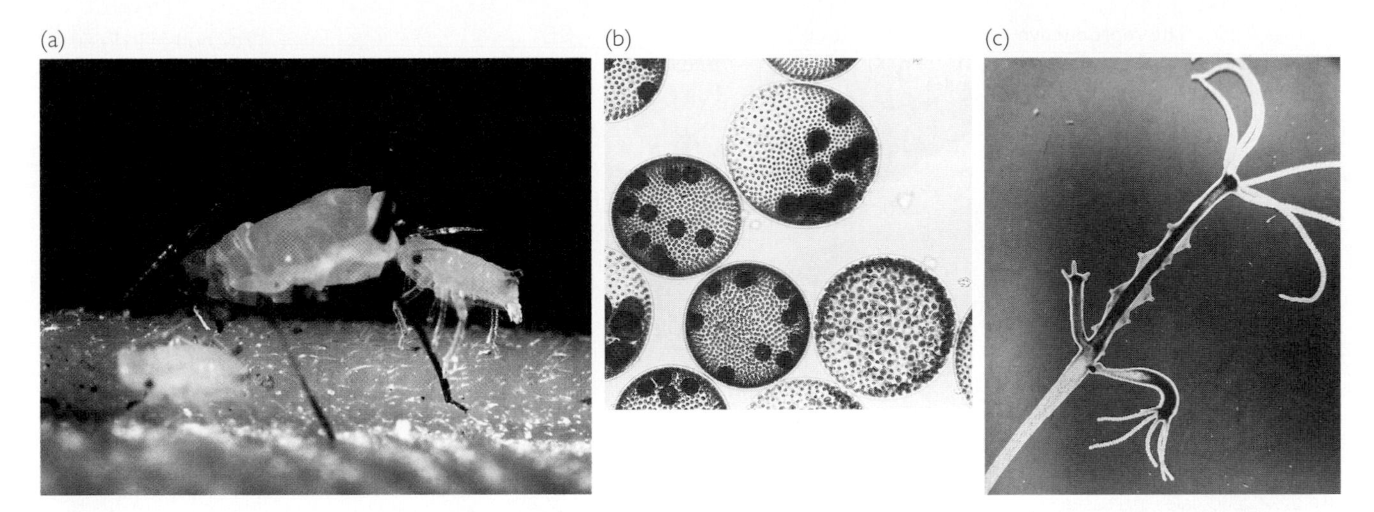

Figure 8.16 Organisms with two modes of reproduction (a) Asexual reproduction in an aphid. The large aphid is giving birth to a daughter, produced by parthenogenesis, that is genetically identical to its mother. In the fall, the aphids will switch to sexual reproduction. (b) *Volvox aureus*, a freshwater alga. Each large sphere is a single adult individual. Before maturity, any individual has the potential to develop as a sexual male, a sexual female, or an asexual. The (entirely visible) individual at lower right is a male. The randomly oriented disks are packets of sperm. The large individual above and to his left is a female. Each of the dark fuzzy spheres inside her is an encysted zygote. The individual directly to the left of the male is an asexual. Each of the dark spheres inside it is an offspring, developing by mitosis into a clone of the parent. (c) A hydra. This individual is reproducing both sexually and asexually. The crown of tentacles at the upper left surround the hydra's mouth. Along the body below the mouth are rows of testes. Below the testes are two asexual buds.

and, without the participation of males, produce live-born young genetically identical to their mothers (Figure 8.16a). This mode of reproduction, in which offspring develop from unfertilized eggs, is called **parthenogenesis**. In the fall aphids change modes, producing males and sexual females. These mate, and the females lay overwintering eggs from which a new generation of parthenogenetic females hatch the following spring.

Many other organisms are capable of both sexual and asexual reproduction. *Volvox*, for example, like aphids, alternate between sexual and asexual phases (Figure 8.16b). Hydra can reproduce sexually and asexually at the same time (Figure 8.16c). So can the many species of plants that reproduce both by sending out runners and by developing flowers that exchange pollen with other individuals.

Which Reproductive Mode Is Better: Sexual or Asexual?

The existence of two different modes of reproduction in the same population raises the question of whether one mode will replace the other over time. John Maynard Smith (1978) approached this question by developing a null model. The null model explores, under the simplest possible assumptions, the evolutionary fate of a population in which some females reproduce sexually and others reproduce asexually. Maynard Smith made just two assumptions:

1. A female's reproductive mode does not affect the number of offspring she can make.
2. A female's reproductive mode does not affect the probability that her offspring will survive.

Maynard Smith also noted that all the offspring of a parthenogenetic female are themselves female, whereas the offspring of a sexual female are a mixture, typically with equal numbers of daughters and sons.

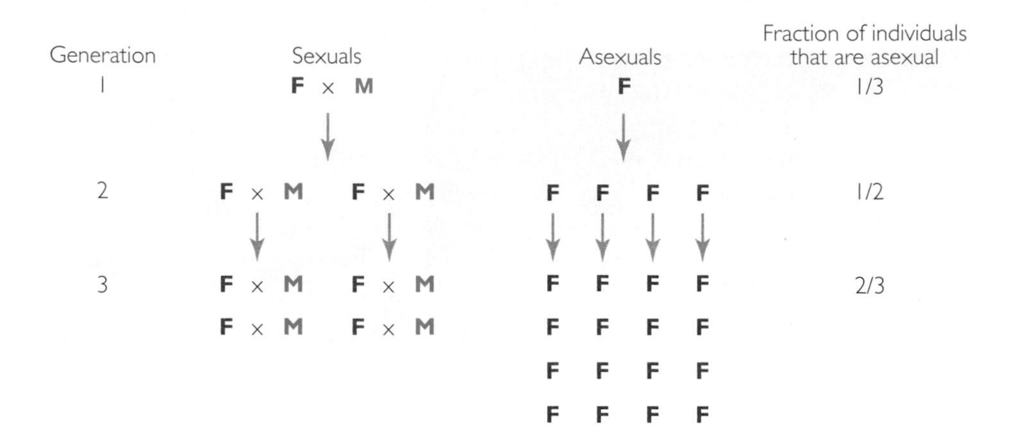

Figure 8.17 The reproductive advantage of asexual females Imagine a population founded by three individuals: a sexual female, a sexual male, and an asexual female. Every generation, each female produces four offspring, after which the parents die. All offspring survive to reproduce. Half of the offspring of sexual females are female; the other half are male. All of the offspring of asexual females are, of course, female. Under these simple assumptions, the fraction of individuals in the population that are asexual females increases every generation. After John Maynard Smith (1978).

In a population conforming to Maynard Smith's assumptions, asexual females produce twice as many grandchildren as sexual females (Figure 8.17). This means that asexual females will constitute a larger and larger fraction of the population every generation. Ultimately, asexual females should completely take over. In principle, all that would be required is for a mutation to produce a single asexual female in an otherwise exclusively sexual population. From the moment the mutation occurred, the population would be destined to be overwhelmed by asexuals.

The persistence of sex is a paradox, because a simple model shows that asexual females should rapidly take over any population.

And yet such asexual takeovers do not seem to have happened very often. The vast majority of multicellular species are sexual, and there are many species, like aphids, *Volvox*, and hydra, in which sexual and asexual reproduction stably coexist. Maynard Smith's model demonstrates, as he intended it to, that these facts represent a paradox for evolutionary theory.

Obviously, sex must confer benefits that allow it to persist in spite of the strong reproductive advantage offered by parthenogenesis. But what are these benefits? The mathematical logic of Maynard Smith's model is correct, so the benefits of sex must lie in the violation of one or both of the assumptions. This is the model's greatest value. By making a short list of explicit assumptions, Maynard Smith focused the inquiry on just a few essential facts of biology.

The first assumption, that the number of offspring a female can make does not depend on whether she is sexual or asexual, is violated in species in which fathers provide resources or other forms of parental care essential for producing young. With no male to provide help, asexual females are likely to produce fewer offspring. Species in which female reproductive success is limited by male parental care certainly exist. Examples include humans, many birds, and pipefish (see Chapter 11). However, species with male parental care are in the minority. In most species—most mammals and most insects, for example—males contribute only genes. A general advantage to sex is thus more likely to be found in the violation of the second assumption, that the probability that a female's offspring will survive does not depend on whether she produces them asexually or sexually.

R. L. Dunbrack and colleagues (1995) tested this second assumption experimentally. They showed that it is wrong, at least under the conditions of their experiments. Dunbrack and colleagues studied laboratory populations in the flour beetle *Tribolium castaneum*. In each of a series of trials, the researchers established a mixed population founded with equal numbers of red beetles and black beetles. The beetles of one color were designated the "sexual" strain, and the beetles of the other color were designated the "asexual" strain. For example, in half of the trials the red beetles were the sexual strain, and the black beetles the asexual strain. We will stay with these designations for the rest of our description of the protocol.

Flour beetles are not actually capable of asexual reproduction, so the researchers had to manage the black population so that it was numerically and evolutionarily equivalent to a population in which individuals really do reproduce asexually. Every generation, the researchers counted the adults of the black strain and threw them out. They then replaced each one of the discarded black adults with three new black adults from a reservoir population of pure-bred black beetles unexposed to competition with reds. This procedure effectively gave the black strain a threefold reproductive advantage over the red strain, but prevented them from adapting to the new environment. Thus, the black strain was analogous to an asexual subpopulation in which every generation is genetically identical to the generation before, but in which individuals enjoy an even greater reproductive edge than the twofold advantage that an actual asexual strain would have in nature. Because every generation of the black population was (except for drift) genetically identical to the generation before, the black strain could not evolve in response to selection imposed by competition with red.

The red adults were allowed to breed among themselves and remain in the experimental culture. Thus they constituted a sexual population that could evolve in response to competitive interactions with the black (asexual) strain.

Dunbrack and colleagues added an environmental challenge for their beetle populations by spiking the flour in which the beetles lived with the insecticide malathion. This imposed selection on the flour beetle populations that favored the evolution of insecticide resistance. Finally, the researchers used a clever procedure, the details of which need not concern us here, to prevent the red and black beetles from mating with each other. The researchers maintained the experiment for 30 generations, which took two years.

Maynard Smith's null model predicts that, in each trial, the asexual strain should occupy an ever-increasing fraction of the population, until eventually the sexual strain is eliminated altogether. The model's first assumption is built into the experiment; in fact, asexual individuals in the experiment produced *more* offspring each generation than sexual individuals. The only way asexuals could fail to take over is if the model's second assumption is incorrect.

Dunbrack and colleagues performed eight replicates of their experiment. Four were as we described, with red as the sexual (evolving) strain and black as the asexual (nonevolving) strain. Each of these replicates used a different concentration of malathion. The other four replicates used black as the sexual (evolving) strain and red as the asexual (nonevolving) strain. Again, each replicate used a different concentration of malathion. The researchers also performed a control for each of the eight replicates. In the controls, neither the red nor the black beetles were allowed to evolve, but one color or the other had a threefold reproductive advantage.

The results appear in Figure 8.18. In the control cultures [Figure 8.18b and (d)], the outcome was always consistent with Maynard Smith's null model: The strain with the threefold reproductive advantage quickly eliminated the other strain. In the experimental cultures, however [Figure 8.18a and (c)], the outcome was always contrary to the prediction of the null model. Initially, the asexual strain appeared to be on its way to taking over, but within about 20 generations, depending on the concentration of malathion, the evolving sexual strain recovered. Eventually, the evolving sexual strain completely eliminated the nonevolving asexual strain, despite the asexual strain's threefold reproductive advantage.

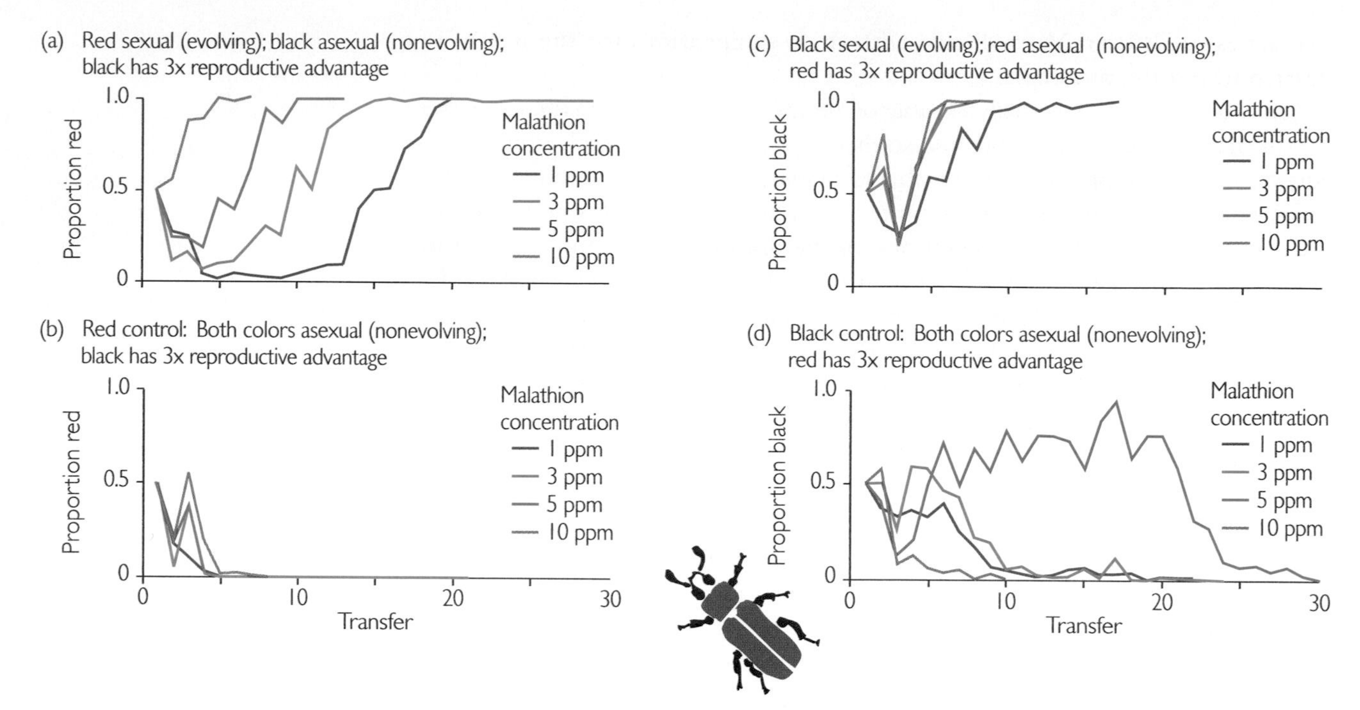

Figure 8.18 **An experimental test of assumption 2 of Maynard Smith's null model** Each panel shows the relative frequency, in a mixed population, of the flour beetle strain that has been placed at a reproductive disadvantage but [in (a) and (c)] reproduces sexually and evolves. The four time-series in each panel represent cultures treated with different concentrations of malathion (ppm = parts per million). A transfer is analogous to a generation. (a) Red is at a reproductive disadvantage, but has sexual reproduction and evolves. Red eventually eliminates black. (b) Red is at a reproductive disadvantage and does not evolve. Red is quickly eliminated by black. (c) Black is at a reproductive disadvantage, but has sexual reproduction and evolves. Black quickly eliminates red. (d) Black is at a reproductive disadvantage and does not evolve. Black is eventually eliminated by red. Redrawn from Dunbrack et al. (1995).

Notice, too, that the speed with which the evolving sexual strain eliminated the nonevolving asexual strain depended on the concentration of malathion. Higher concentrations of malathion meant stronger selection for insecticide resistance. And the stronger selection meant a greater advantage for sex.

At least under some conditions, descendants produced by sexual reproduction achieve higher fitness than descendants produced by asexual reproduction.

We can conclude that assumption 2 of the null model is incorrect. Over time spans of just a few generations, descendants produced by sexual reproduction achieve higher fitness than descendants produced by asexual reproduction.

The next question is, Why? The only inherent difference between offspring that a female makes sexually versus asexually is that asexual offspring are genetically identical to their mother and to each other, whereas sexual offspring are genetically different from their mother and from each other. Most theories about the benefits of sex are concerned with reasons why females that produce genetically diverse offspring will see more of them survive and reproduce than will females that produce genetic copies of themselves.

We should note at this point that there is a tremendous diversity of theories about the advantages of sex. We have space here to focus only on population genetic models and tests, and a small number of them at that. For a more comprehensive overview of the field, see Michod and Levin (1988).

Sex in Populations Means Genetic Recombination

When population geneticists talk about sex, what they usually mean, and what we mean here, is reproduction involving (1) meiosis with crossing-over and (2) matings between unrelated individuals, such as occur during random mating. The consequence of these processes acting in concert is genetic recombination. If we follow a particular allele through several generations of a pedigree, every generation the allele will be part of a different multilocus genotype. For example, a particular allele for blue eyes may be part of a genotype that includes genes for blond hair in one generation and part of a genotype that includes genes for brown hair in the next generation.

Sex, to a population geneticist, means genetic recombination. Selection—on any sort of trait—seems to favor recombination.

There is experimental evidence that selection, in general, favors recombination. For example, Sarah Otto and Thomas Lenormand (2002) reviewed experiments in which researchers subjected populations to artificial selection for traits that have nothing to do with sex or recombination, then assessed whether the rate of genetic recombination during meiosis had changed (see also Otto and Barton 2001). In many of the experiments artificial selection on other traits had, indeed, produced a statistically significant change in the rate of recombination (Figure 8.19). And in the majority of such cases, the change was an increase.

How can we explain these results? In a population-genetic analysis, genetic recombination reduces linkage disequilibrium. It does so by shuffling multilocus genotypes. This was a central conclusion of Section 8.1.

In a population-genetic analysis, sex has exactly one effect: It reduces linkage disequilibrium.

In fact, the reduction of linkage disequilibrium is the *only* consequence of sex at the level of population genetics (Felsenstein 1988). In a population that is already in linkage equilibrium, sex has no effect. If sex has no effect, it can confer no benefits. Therefore, any population-genetic model for the evolutionary benefits of sex must, at minimum, include two things. First, the model must include a mechanism that eliminates particular multilocus genotypes or produces an excess of others, thereby creating linkage disequilibrium. Second, the model must include a reason why genes that tend to reduce linkage disequilibrium—by promoting sex—are favored.

Based on this analysis, Joe Felsenstein (1988) neatly divides nearly all population-genetic models for the benefit of sex into two general theories. These general theories are distinguished by the evolutionary mechanism they posit for the creation of linkage disequilibrium. Some models posit genetic drift as the factor that creates linkage disequilibrium; other models posit selection on multilocus genotypes.

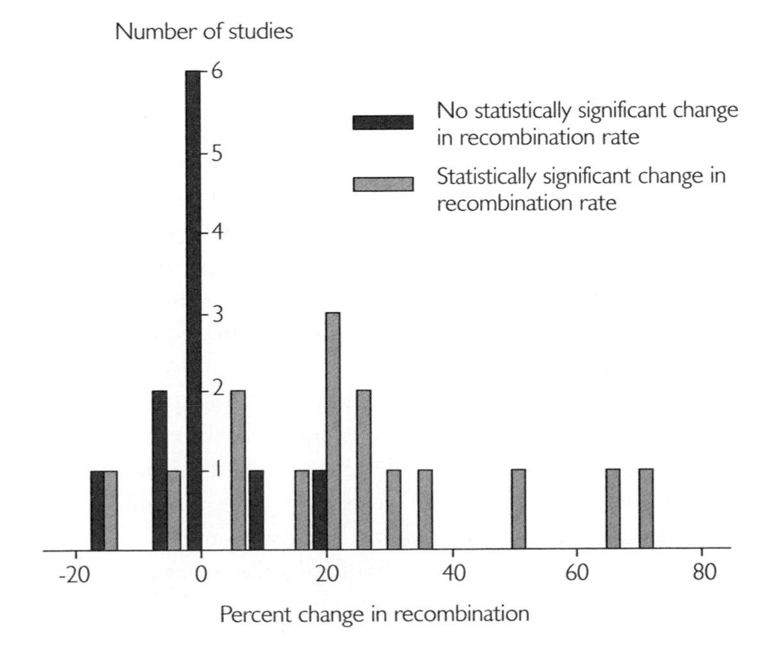

Figure 8.19 **Artificial selection favors increased genetic recombination** This graph summarizes the results of experiments in which researchers artificially selected for traits unrelated to sex and recombination rate, then assessed changes in rate of recombination during meiosis. Among the experiments in which selection produced a significant change in recombination rate (brown bars), the majority of the changes were increases. Rerendered from Otto and Lenormond (2002).

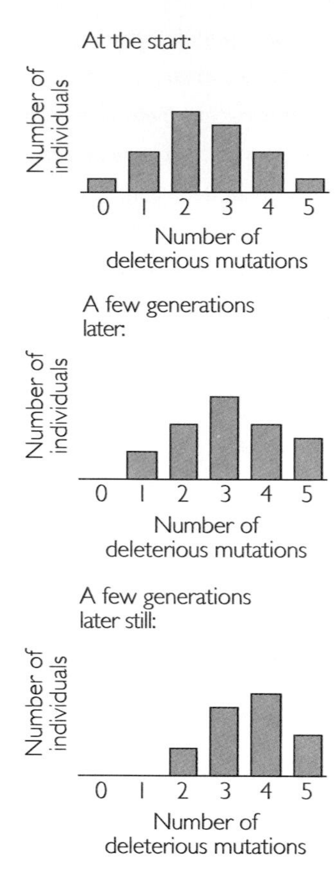

Figure 8.20 Muller's ratchet: Asexual populations accumulate deleterious mutations Each histogram shows a snapshot of a finite asexual population. In any given generation, the class with the fewest deleterious mutations may be lost by drift. Because forward-mutation to deleterious alleles is more likely than back-mutation to wild-type alleles, the distribution slides inexorably to the right. After Maynard Smith (1988).

Genetic Drift, in Combination with Mutation, Can Make Sex Beneficial

According to the drift theory of sex, mutation and drift create problems that sex can solve. Imagine, for example, that an asexual female sustains a deleterious genetic mutation in her germ cells. She will pass the mutation to all of her offspring, which in turn will pass it to all of their offspring. The female's lineage is forever hobbled by the deleterious mutation. The only hope of escape is if one of her descendants is lucky enough to experience either a back-mutation or an additional mutation that compensates for the first. If the female were sexual, however, she could produce mutation-free offspring immediately, simply by mating with a mutation-free male.

The role of drift in this scenario becomes clear when we scale the model up to the level of populations. The most famous drift model is called Muller's ratchet; it argues that asexual populations are doomed to accumulate deleterious mutations. H. J. Muller (1964) imagined a finite asexual population in which individuals occasionally sustain deleterious mutations. Because the mutations envisioned by Muller are deleterious, they will be selected against. The frequency of each mutant allele in the population will reflect the mutation rate, the strength of selection, and genetic drift (see Chapters 6 and 7).

At any given time, Muller's population may include individuals that carry no mutations, individuals that carry one mutation, individuals that carry two mutations, and so on. Because the population is asexual, we can think of these groups as distinct subpopulations and plot the relative number of individuals in each subpopulation in a histogram (Figure 8.20). The number of individuals in each group may be quite small, depending on the size of the entire population and on the balance between mutation and selection (see Chapter 6). The group with zero mutations is the one whose members, on average, enjoy the highest fitness; but if this group is small, then in any given generation chance events may conspire to prevent the reproduction of all individuals in the group. If this happens just once, then the zero-mutation subpopulation is lost, and the members of the one-mutation group are now the highest-fitness individuals. The only way the zero-mutation group will reappear is if a member of the one-mutation group sustains the back-mutation that converts into a zero-mutation individual.

With the demise of the zero-mutation group, the members of the one-mutation subpopulation enjoy the highest mean fitness. But this group may also be quite small and may be lost by chance in any given generation. Again, the loss of the group by drift is much easier than its re-creation by back-mutation. As the ratchet clicks away, and highest-fitness group after highest-fitness group is lost from the population, the average fitness of the population declines over time. The burden imposed by the accumulating mutations is known as the **genetic load**. Eventually, the genetic load carried by the asexual population becomes so high that the population goes extinct.

Sex breaks the ratchet. If the no-mutation group is lost by chance in any given generation, it can be quickly reconstituted by outcrossing and recombination. If two individuals mate, each carrying a single copy of a deleterious mutation, one-quarter of their offspring will be mutation free. In Muller's view, the genes responsible for sex are maintained in populations because they help to create zero-mutation genotypes. As these zero-mutation genotypes increase in frequency, the genes for sex increase in frequency with them—in effect going along for the ride.

In Muller's scenario, linkage disequilibrium is created by drift. Particular multilocus genotypes are at lower-than-linkage-equilibrium frequencies because chance events have eliminated them. These missing multilocus genotypes are the zero-mutation genotype, then the one-mutation genotype, and so on. Sex reduces linkage disequilibrium by re-creating the missing genotypes.

Haigh (1978; reviewed in Maynard Smith 1988) developed and explored an explicit mathematical model of Muller's ratchet. Not surprisingly, the most critical parameter in the model is population size. In populations of 10 or fewer individuals, drift is a potent mechanism of evolution and the ratchet turns rapidly. In populations of more than 1,000, drift is a weak mechanism of evolution and the ratchet does not turn at all. Also important are the mutation rate and the impact of deleterious mutations. The ratchet operates fastest with mildly deleterious mutations. This is because severely deleterious mutations are eliminated by selection before drift can carry them to fixation.

Dan Andersson and Diarmid Hughes (1996) tested Muller's ratchet experimentally with populations of the bacterium *Salmonella typhimurium*. From a standard wild-type strain, Andersson and Hughes set up 444 cultures, each established from a single individual (Figure 8.21). After letting them grow overnight, the researchers propagated the cultures, again from a single individual. After another night of growth, they propagated the cultures again, and so on. The bacteria reproduced asexually by binary fission. The daily bottlenecks, during which the population size of each culture was reduced to one, exposed the cultures to genetic drift. The researchers maintained the experiment for two months, giving drift about 1,700 generations to operate. Based on Muller's ratchet, Andersson and Hughes predicted that the bacterial cultures would accumulate deleterious mutations.

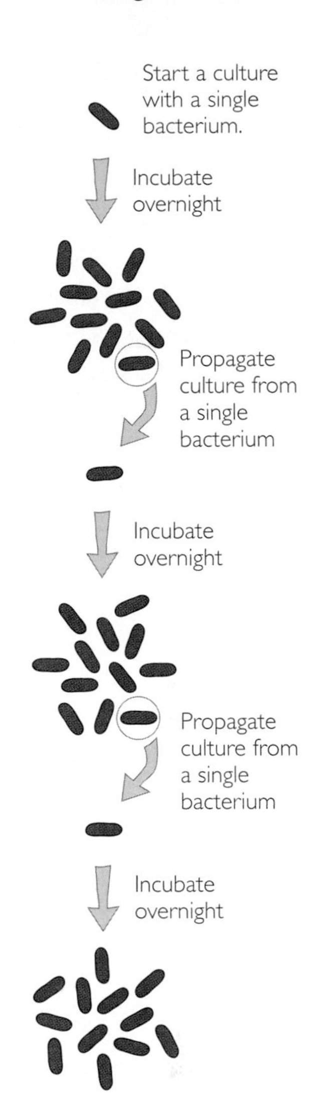

Figure 8.21 A bacterial population subjected to periodic bottlenecks The bacterial population is propagated from a single individual (the bottleneck), allowed to grow and divide to produce a large colony, then propagated again from a single individual. The bottlenecks provide an opportunity for genetic drift to operate.

Andersson and Hughes checked this prediction by comparing the fitness of each experimental culture to the wild-type strain that was their common ancestor. The researchers assessed fitness by measuring population growth rate. Among their 444 cultures, Andersson and Hughes found 5, or 1%, with significantly reduced fitness. The generation time of these 5 cultures ranged from 25.0 to 47.5 minutes, compared to 23.2 minutes for the wild-type ancestor. None of the 444 cultures had higher fitness than the wild-type. These results are consistent with Muller's ratchet.

J. David Lambert and Nancy Moran (1998) took advantage of a natural experiment to assess whether Muller's ratchet operates in nature. Lambert and Moran studied nine species of bacteria that live inside the cells of insects. These bacteria are obligate endosymbionts—they live *only* inside the cells of insects. The bacteria are transmitted from mother to offspring by traveling in the cytoplasm of the egg, just as mitochondria are. Note that the endosymbiotic bacteria in Lambert and Moran's study are propagated in a manner analogous to the protocol Andersson and Hughes used in their lab study. The difference is that, in Lambert and Moran's study, the bacteria had been propagated that way for millions of years.

The nine species of bacteria that Lambert and Moran studied represent at least four separate inventions of the endosymbiotic lifestyle, and each group of independently derived endosymbionts has close relatives that are free-living. Lambert and Moran checked whether, compared to their free-living relatives, the endosymbionts have accumulated deleterious mutations. The researchers focused on small-subunit ribosomal RNA (rRNA) genes. From the sequence of each species' rRNA gene, Lambert and Moran calculated the thermal stability of the encoded rRNA. Stability is a good thing in rRNAs, and deleterious mutations in an rRNA gene would reduce stability. In every case, Lambert and Moran found that the endosymbiotic bacteria have rRNAs that are 15 to 25% less stable than their free-living relatives. Again, this result is consistent with Muller's ratchet.

Sex may be advantageous because it re-creates favorable multilocus genotypes that have been lost to drift. The genes for sex then ride to high frequency in the high-fitness genotypes they help to create.

Thus Muller's ratchet works, both in principle and in practice. By counteracting the ratchet, sex can confer benefits. Working with a more general model of disequilibrium arising from deleterous mutations, Peter Keightley and Sarah Otto (2006) found sex beneficial even in—indeed, more so in—large populations.

There is a problem, however, with models that explain sex by positing drift as the source of linkage disequilibrium. The benefits conferred by sex in these models accumulate only over the long term. If an asexual female appeared in a sexual population, it would take many generations for Muller's ratchet to catch up with her descendants and lower their fitness enough to drive them to extinction. In the meantime, the asexual female's descendants would enjoy the twofold reproductive benefit identified by Maynard Smith. Yet the rarity of asexual species suggests, and the experiment by Dunbrack and colleagues demonstrates, that the advantage of sex accrues over just a few generations. This reasoning has prompted a search for short-term benefits of sex.

Selection Imposed by a Changing Environment Can Make Sex Beneficial

To see the logic of the changing-environment theories for sex, first imagine an asexual female and a sexual female living in a constant environment. If these females themselves survived to reproduce, and their offspring live in the same environment, then the offspring of the asexual female will probably survive to reproduce also. After all, they will receive exactly their mother's already-proven genotype. The genetically diverse offspring of the sexual female, however, may or may not survive to reproduce, depending on the nature of the genetic differences between their mother and themselves. By this reasoning, in a constant environment asexual reproduction is a safer bet.

In a variable environment, however, all bets are off. If the environment is changing in such a way that the asexual female herself might not survive in the new conditions, then her offspring will have poor prospects too. If the environment changes for a sexual female, however, there is always a chance that some of her diverse offspring will have genotypes that allow them to thrive in the new conditions. Some changing-environment theories focus on changes in the physical environment, whereas others focus on changes in the biological environment. Note that all changing-environment theories assume trade-offs, such that genotypes that do relatively well in some environments necessarily do relatively poorly in others.

Sex may be advantageous because it re-creates favorable multilocus genotypes that were recently eliminated by selection. Again, the genes for sex then ride to high frequency in the high-fitness genotypes they help to create.

A. H. Sturtevant and K. Mather (1938, reviewed in Felsenstein 1988) were the first to consider an explicit population-genetic model of varying selection. They imagined that selection favors some multilocus genotypes in some generations (say, genotypes *AABB* and *aabb* in a two-locus model) and other genotypes (*AAbb* and *aaBB*) a few generations later. The environment would alternate between a selection regime that generates linkage disequilibrium with positive values of D and a regime that generates linkage disequilibrium with negative values of D (see Box 8.1). Under these conditions, sex could be favored for its ability to re-create genotypes that were recently eliminated by selection but are now favored. As in Muller's ratchet, the genes for sex get a ride to high frequency in the high-fitness multilocus genotypes they help to create.

The variable pattern of selection required by the changing-environment theories can be imposed either by physical factors in the environment or by biological interactions. Currently, the most popular changing-environment theory of sex, sometimes called the *Red Queen hypothesis*, involves evolutionary arms races between parasites and their hosts (for reviews see Seger and Hamilton 1988; Lively 1996). Parasites and their hosts are locked in a perpetual struggle, with the host evolving to defend itself and the parasite evolving to evade the host's defenses. It is easy to imagine that a population of parasites would select in favor of some multilocus host genotypes in some generations and in favor of other multilocus host genotypes in other generations. Figure 8.22 presents a scenario for such an ongoing evolutionary interaction.

Curtis Lively (1992) investigated whether parasites, in fact, select in favor of sex in their hosts. Lively studied the freshwater snail *Potamopyrgus antipodarum*. This snail, which lives in lakes and streams throughout New Zealand, is the host of over a dozen species of parasitic trematode worms. The trematodes typically castrate their host by eating its gonads. In an evolutionary sense, castration is equivalent to death: It prevents reproduction. Trematodes thus exert strong selection on snail populations for resistance to infection. Most populations of the snail contain two kinds of females: obligately sexual females that produce a mixture of male and female offspring, and obligately parthenogenetic females whose daughters are clones of their mother. (Note that both kinds of female must have an ovary to reproduce; the difference is that the eggs of the parthenogenetic females do not have to be fertilized.) The proportion of sexuals versus asexuals varies from population to population. So does the frequency of trematode infection. If an evolutionary arms race between the snails and the trematodes selects in favor of sex in the snails (see Figure 8.22), then sexual snails should be more common in the populations with higher trematode infection rates.

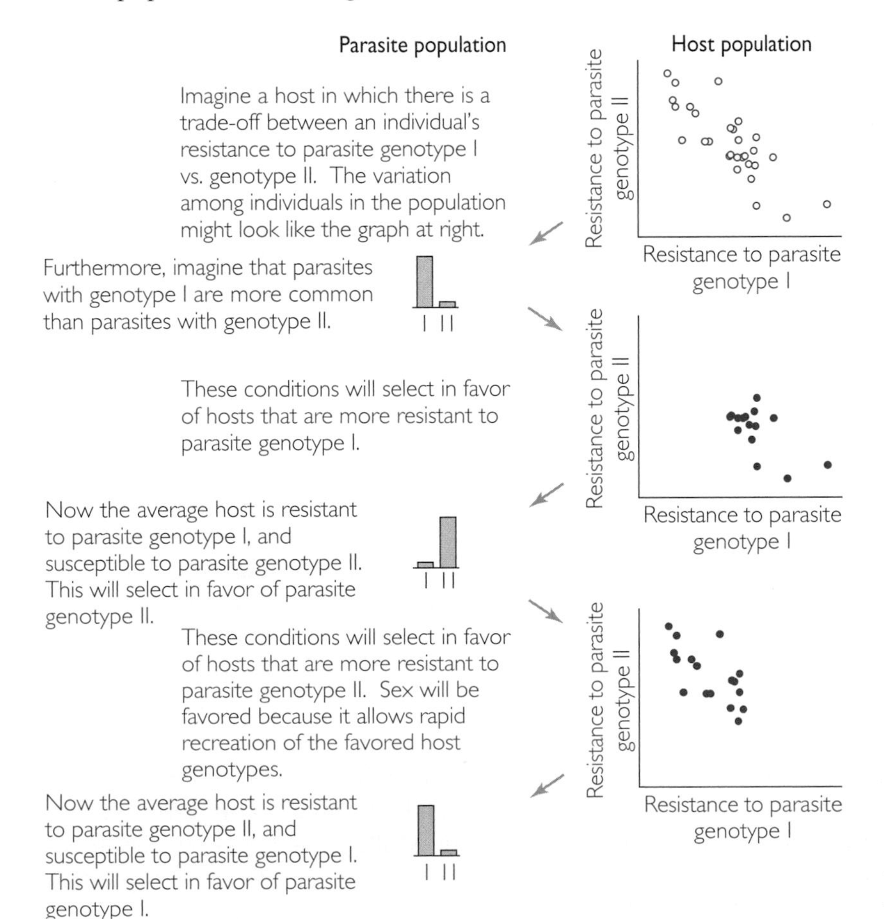

Figure 8.22 A host parasite arms race can make sex beneficial Hosts resistant to parasite genotype I are necessarily susceptible to parasite genotype II, and vice versa. As the parasite population evolves in response to the hosts, it first selects for hosts resistant to parasite genotype I, then for hosts resistant to parasite genotype II. Genes for sex ride to high frequency in the currently more-fit genotypes they help to create.

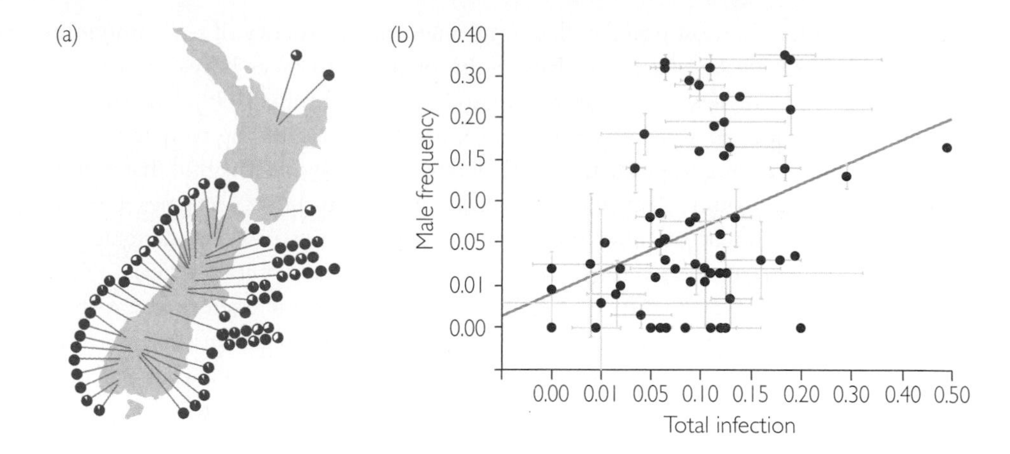

Figure 8.23 The frequency of sexual individuals in populations of a host snail is positively correlated with the frequency of its trematode parasites (a) The map shows the locations of the 66 lakes Lively sampled. In each population's pie diagram, the white slice represents the frequency of males. (b) The scatterplot shows the frequency of males in each population versus the proportion of snails infected with trematodes. The graph includes a best-fit line. Males are more frequent in populations in which more snails are infected. From Lively (1992).

Lively took samples of snails from 66 lakes and determined the sex of each snail and whether it was infected with parasites. Lively used the frequency of males in each population as an index of the frequency of sexual females, on the logic that males are produced only by sexual females. Lively found that a higher proportion of the females are indeed sexual in more heavily parasitized populations (Figure 8.23). This result is consistent with the varying-selection theory of sex.

Lively noted that because his study was observational, alternative explanations for the association he found should be considered. For example, if

1. trematode infection rates are higher in more dense populations of snails, because high host density facilitates parasite transmission, and
2. the frequency of parthenogenetic females is higher in less dense populations of snails, because the real benefit of parthenogenesis is that it allows females to reproduce even when mates are hard to find,

then these two effects in combination would produce a positive association between the frequency of sexuals and the frequency of trematode infection. Lively rejected this alternative explanation by showing that although there is a positive correlation between infection rate and snail density (effect 1 is true), there is also a positive correlation between the frequency of parthenogenetic females and snail density (effect 2 is false). After considering this and other alternative explanations, Lively concluded that the simplest explanation for the pattern he found is that the trematodes indeed select in favor of sexual reproduction by the snails.

In summary, in the context of population genetics the effect of sex is to reduce linkage disequilibrium. A population-genetic model for the adaptive value of sex must therefore have two components: a mechanism for the creation of linkage disequilibrium, and a reason why selection favors traits that tend to reduce linkage disequilibrium. There are two classes of models for sex. In the first class, genetic drift creates linkage disequilibrium. Sex is then favored because it helps to re-create high-fitness genotypes lost by drift. In the second class, natural selection creates linkage disequilibrium. Then the pattern of selection changes, and sex is favored because it helps to re-create the now-favorable genotypes recently selected against. The various scenarios favoring sex are mutually compatible with each other. It is likely that varying selection imposed by changes in both the biological and physical environments combines with Muller's ratchet to create an advantage for sex that is greater than the advantage any one factor can produce alone (Howard and Lively 1998).

Summary

The single-locus models of Chapters 6 and 7 are powerful, but potentially oversimplified. Extension of the Hardy–Weinberg analysis to two loci reveals complications. When genotypes at one locus are nonrandomly associated with genotypes at the other, the loci are in linkage disequilibrium. Even under Hardy–Weinberg assumptions, chromosome frequencies change across generations. Furthermore, selection on one locus can alter allele frequencies at the other, and single-locus models may make inaccurate predictions. When genotypes at one locus are independent of genotypes at the other locus, however, the loci are in linkage equilibrium. Chromosome frequencies do not change across generations. Selection on one locus has no effect on allele or genotype frequencies at the other, and we can use single-locus models to make predictions about evolution.

In a random-mating population, linkage disequilibrium can be created by selection on multilocus genotypes, genetic drift, and population admixture. All three mechanisms create an excess of some chromosome haplotypes and a deficit of others. Linkage disequilibrium is reduced by sexual reproduction. Sex brings together chromosomes with different haplotypes, and crossing-over during meiosis allows the chromosomes to exchange genes. This genetic recombination tends to break up overrepresented haplotypes and create underrepresented haplotypes.

Measurements of linkage disequilibrium are useful in inferring the history of alleles. If an allele is in linkage disequilibrium with nearby neutral marker loci, we can infer that the allele is relatively young. If we have an estimate of the rate at which disequilibrium between the allele and the nearby neutral marker loci breaks down, then we can use the strength of the persisting disequilibrium to estimate the allele's age. If an allele is both young, as indicated by linkage disequilibrium, and present at high frequency, then we can infer that the allele has recently been favored by positive natural selection.

The fact that sexual reproduction reduces linkage disequilibrium provides a key to understanding why sexual reproduction persists in populations. Simple theoretical arguments suggest that asexual reproduction should sweep to fixation in any population in which it appears. Empirical observations and experiments indicate, however, that sex confers substantial benefits. These benefits can be found in the population-genetic consequences of sex. When drift or selection has reduced the frequency of particular multilocus genotypes below their expected levels under linkage equilibrium, sexual reproduction can be favored because it re-creates the missing genotypes.

Questions

1. Describe the three mathematical consequences of linkage equilibrium. That is, what three equations about genotype and chromosome frequencies will be true if a population is in linkage equilibrium? What is D and how is it calculated?
2. Figure 8.3 presented an example of selection favoring certain multilocus genotypes. The chapter text demonstrated that, after selection, the population failed criterion #2 for linkage equilibrium. Now test the same population in some different ways:
   - **a.** What is the frequency of B on chromosomes that are carrying allele A? What is the frequency of B on chromosomes carrying allele a? Does the population meet criterion #1?
   - **b.** What is D, the coefficient of linkage disequilibrium? Does the population meet criterion #3?
   - **c.** From the postselection population in Figure 8.3b, develop a bar graph like the ones in Figure 8.2. Does this bar graph confirm that the postselection population is in linkage disequilibrium?
3. In horses, the basic color of the coat is governed by the E locus. *EE* and *Ee* horses can make black pigment, while *ee* horses are a reddish chestnut. A different locus, the R locus, can cause *roan*, a scattering of white hairs throughout the basic coat color. However, the roan allele has a serious drawback: *RR* embryos always die during fetal development. *Rr* embryos survive and are roan, while *rr* horses survive and are not roan. The E locus and the R locus are tightly linked.

   Suppose that several centuries ago a Spanish galleon with a load of conquistadors' horses was shipwrecked by a large grassy island. Just by chance, the horses that survived the shipwreck and swam to shore were 20 chestnut roans (*eeRr*) and 20 nonroan homozygous blacks (*EErr*). On the island, they interbred with each other and established a wild population.

The island environment exerts no direct selection on either locus.

a. What was D, the coefficient of linkage disequilibrium, in the initial population of 20 horses? Was the initial population in linkage equilibrium or not? If not, what chromosomal genotypes were underrepresented?

b. Do you expect the frequency of the chestnut allele, e, to increase or decrease in the first crop of foals? Would your answer be different if the founding population had been just 10 horses (5 of each color)? Explain your reasoning.

c. If you could travel to this island today, can you predict what D would be now? Do you have predictions about whether more horses will be roan versus nonroan, or chestnut versus black? If not, explain what further information you would need.

4. Imagine a population of pea plants that is in linkage equilibrium for two linked loci, flower color (P = purple, p = pink) and pollen shape L = long, l = round).

a. What sort of selection event would create linkage disequilibrium? For example, will selection at just one locus (e.g., all red-flowered plants die) create linkage disequilibrium? How about selection at two loci (e.g., red-flowered plants die, and long-pollen plants die)? How about selection on a certain combination of genotypes at two loci (e.g., only plants that are both red-flowered and have long pollen grains die)?

b. Now imagine a population that is already in linkage disequilibrium for these two loci. Will selection for purple flowers affect evolution of pollen shape? How is your answer different from your answer to part a, and why?

5. Populations of rats exposed to the poison warfarin rapidly evolve resistance. The gene for warfarin resistance is located on rat chromosome 1. Michael Kohn and colleagues (2000) surveyed rats in five German rat populations known to vary in their recent exposure to warfarin and in their resistance. The researchers determined the genotype of each rat at a number of marker loci near the warfarin resistance gene. For each population, the researchers calculated the average heterozygosity (H) among the marker loci, the fraction of loci that were out of Hardy–Weinberg equilibrium (HWE), and the fraction of marker-locus pairs that were in linkage disequilibrium (LD). Their results appear in Figure 8.24. Based on these graphs, rank the five populations in order, from lowest to highest, for exposure to warfarin and resistance. Explain your reasoning.

6. a. In the beetle evolution experiment (Figure 8.18), Dunbrack et al. did not actually use sexual beetles and asexual beetles. Instead, they used two colors of sexual beetles, but forced one color of beetles to grow in population size as if it were asexual. They also did the experiment again with the other color as "asexual." Why was it important that the researchers run the experiment both ways? Compare the graphs of the two different sets of experiments (red asexual, and black asexual). Did the two strains of beetles perform differently?

b. Actual asexual beetles would reproduce twice as fast as sexual beetles, but in the experiment the authors made the asexual beetles reproduce three times as fast. Why do you think they did this?

c. The researchers' simulated asexual population was not allowed to evolve at all in response to selection imposed by competition. Is this different from what would have happened in an actual population of asexual beetles? Do you think Dunbrack et al.'s experiment is a valid test of asexual reproduction versus sexual reproduction? Briefly describe the next experiment that you think Dunbrack et al. should do to follow up on this topic.

7. In 1992, Spolsky, Phillips, and Uzzell reported genetic evidence that asexually reproducing lineages of a sala-

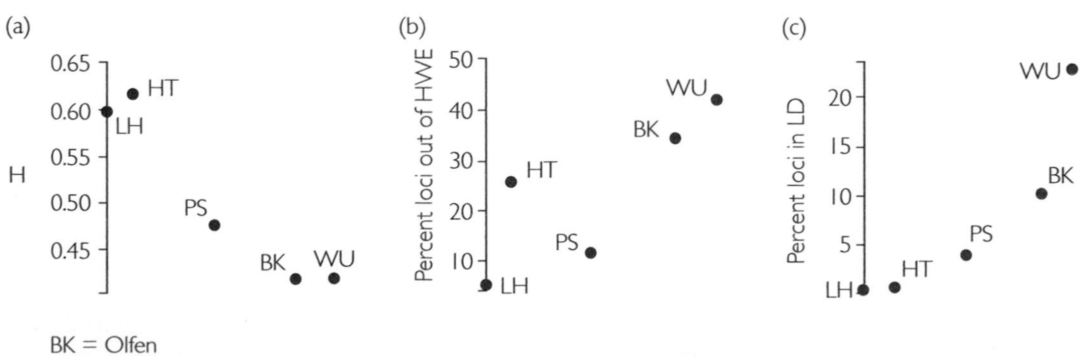

Figure 8.24 Population genetic data on five rat populations Rerendered from Kohn et al. (2000).

mander species have persisted for about 5 million years, an unusually long time. Is this surprising? Why or why not? Speculate about what sort of environment these asexual salamanders live in, and whether their population sizes are typically small (say, less than 100) or large (say, more than 1,000).

8. How can you identify an allele that has experienced recent strong positive selection?

9. Describe the major hypotheses for the cause of high frequency of the *CCR5-Δ32* allele among European populations. Why is the age of the allele relevant for distinguishing among the hypotheses? Do we know how old this allele is, and if so, what the evidence is?

10. *Volvox* (Figure 8.16b) are abundant and active in lakes during the spring and summer. During winter they are inactive, existing in a resting state. During most of the spring and summer, *Volvox* reproduce asexually; but at times they switch and reproduce sexually instead. When would you predict that *Volvox* would be sexual: spring, early summer, or late summer? Explain your reasoning.

11. In mammals, sex is determined by the X and Y chromosomes. Females are XX; males are XY. The Y chromosome contains a gene that causes development of testes, which then causes the mammal embryo to become male. The Y chromosome does not undergo crossing-over with the X during spermatogenesis in males, but the two X's cross-over with each other during oogenesis in females.

a. The Y chromosome is thought to have once been the same size as the large, fully functional X chromosome. But during the evolution of the mammals, the Y chromosome seems to have accumulated an enormous number of deleterious mutations. It has also lost almost all of its genes and has shrunk to a rudimentary chromosome containing just the testis-determining gene, a few other genes, and some nonfunctional remnants of other genes. Why has this occurred?

b. Birds use a reverse system, in which females have two different chromosomes (called WZ in birds) and males have two of the same kind of chromosome (ZZ). In birds, sex is determined by a gene on the W chromosome that causes ovary formation, which then causes the bird embryo to become female. Would you predict one of these chromosomes might have accumulated mutations in the same way that the Y chromosome has? If so, which one?

c. Some plants also have genetically determined sex but are polyploid. Should their sex chromosomes show accumulation of mutations?

Exploring the Literature

12. To read more about the accumulation of mutations in sex chromosomes, see:

Berlin, S., and H. Ellegren. 2006. Fast accumulation of nonsynonymous mutations on the female-specific W chromosome in birds. *Journal of Molecular Evolution* 62: 66–72.

Gerrard, D.T., and D.A. Filatov. 2005. Positive and negative selection on mammalian Y chromosomes. *Molecular Biology and Evolution* 22: 1423–1432.

13. If the endosymbiotic bacteria studied by Lambert and Moran (1998) suffer reduced fitness as a result of Muller's ratchet, then mitochondria should too. Do they? See:

Lynch, M. 1996. Mutation accumulation in transfer RNAs: Molecular evidence for Muller's ratchet in mitochondrial genomes. *Molecular Biology and Evolution* 13: 209–220.

Lynch, M. 1997. Mutation accumulation in nuclear, organelle, and prokaryotic transfer RNA genes. *Molecular Biology and Evolution* 14: 914–925.

14. Many human pathogens, including bacteria and eukaryotes, are capable of both asexual reproduction and genetic recombination (that is, sex in the population-genetic sense). The frequency of recombination in a pathogen population can have medical implications. (Think, for example, about how fast resistance to multiple antibiotics will evolve in a population of bacteria that does have recombination versus a poplation that does not.) How can we tell whether a given pathogen population is engaging in genetic recombination or is predominantly clonal? Genetic recombination is such a powerful mechanism for reducing linage disequilibrium that the amount of disequilibrium in a pathogen population provides a clue. See:

Xu, J. 2004. The prevalence and evolution of sex in microorganisms. 2004. *Genome* 47: 775–780.

Maynard Smith, J., N. H. Smith, M. O'Rourke, and B. G. Spratt. 1993. How clonal are bacteria? *Proceedings of the National Academy of Sciences USA* 90: 4384–4388.

Burt, A., D.A. Carter, G. L. Koenig, T. J. White, and J. W. Taylor. 1996. Molecular markers reveal cryptic sex in the human pathogen *Coccidioides immitis*. *Proceedings of the National Academy of Sciences USA* 93: 770–773.

Go, M. F., V. Kapur, D.Y. Graham, and J. M. Musser. 1996. Population genetic analysis of *Helicobacter pylori* by multilocus enzyme electrophoresis: Extensive allelic diversity and recombinational population structure. *Journal of Bacteriology* 178: 3934–3938.

Gräser, Y., et al. 1996. Molecular markers reveal that population structure of the human pathogen *Candida albicans* exhibits both clonality and recombination. *Proceedings of the National Academy of Sciences USA* 93: 12473–12477.

Jiménez, M., J. Alvar, and M. Tibayrenc. 1997. *Leishmania infantum* is clonal in AIDS patients too: Epidemiological implications. *AIDS* 11: 569–573.

15. Most evolutionary biologists assume that the genetic recombination that occurs in bacteria, through processes such as conjugation and transduction, are equivalent to eukaryotic sex—and that they are favored by selection for similar reasons. For a beautifully written argument that recombination in bacteria is nothing like eukaryotic sex and evolved for entirely different reasons, see:

    Redfield, R. J. 2001. Do bacteria have sex? *Nature Reviews Genetics* 2: 634–639.

16. Asexual populations must deal with the accumulation of deleterious mutations due to Muller's ratchet, but they do manage to evolve nonetheless. See:

    Bachtrog, D., and I. Gordo. 2004. Adaptive evolution of asexual populations under Muller's ratchet. *Evolution* 58 (7): 1403–1413.

17. One way to learn about why sex is maintained in the vast majority of eukaryotes is to study the exceptions—taxa of that have persisted for millions of years without sex. Are there any? If so, what might they tell us? See:

    Judson, O. P., and B. B. Normark. 1996. Ancient asexual scandals. *Trends in Ecology and Evolution* 11: 41–46.

    Judson, O. P., and B. B. Normark. 2002. Sinless originals. *Science* 288: 1185–1186.

    Mark Welch, D., and M. Meselson. 2002. Evidence for the evolution of Bdelloid rotifers without sexual reproduction or genetic exchange. *Science* 288: 1211–1215.

    Schon, I., and K. Martens. 2003. No slave to sex. *Proceedings of the Royal Society of London B* 270: 827–823.

18. We have seen that parasites can promote sexual reproduction in their hosts. Do the hosts also affect the parasites in the same way? See:

    Howard, R. S., and C. M. Lively. 2002. The ratchet and the Red Queen: The maintenance of sex in parasites. *Journal of Evolutionary Biology* 15 (4): 648—656.

Citations

Please note that much of the population genetics in this chapter is modeled after presentations in the following:

Cavalli-Sforza, L. L., and W. F. Bodmer. 1971. *The Genetics of Human Populations.* San Francisco: W. H. Freeman and Company.

Felsenstein, J. 1997. *Theoretical Evolutionary Genetics.* Seattle, WA: ASUW Publishing, University of Washington.

Felsenstein, J. 1988. Sex and the evolution of recombination. In R. E. Michod and B. R. Levin, eds. *The Evolution of Sex.* Sunderland, MA: Sinauer, 74–86.

Hartl, D. L. 1981. *A Primer of Population Genetics.* Sunderland, MA: Sinauer.

Maynard Smith, J. 1998. *Evolutionary Genetics,* 2nd ed. Oxford: Oxford University Press.

Here is the listing of all citations in this chapter:

Andersson, D. I., and D. Hughes. 1996. Muller's ratchet decreases fitness of a DNA-based microbe. *Proceedings of the National Academy of Sciences USA* 93: 906–907.

Behar, D., M. F. Hammer, et al. 2004. MtDNA evidence for a genetic bottleneck in the early history of the Ashkenazi Jewish population. *European Journal of Human Genetics* 12: 355-364.

Bersaglieri, T., P. C. Sabeti, et al. 2004. Genetic signatures of recent positive selection at the lactase gene. *American Journal of Human Genetics* 74: 1111-1120.

Beutler, E. 1993. Gaucher disease as a paradigm of current issues regarding single gene mutations of humans. *Proceedings of the National Academy of Sciences USA* 90: 5384-5390.

Beutler, E., T. Gelbart, et al. 1991. Identification of the second common Jewish Gaucher disease mutation makes possible population-based screening for the heterozygous state. *Proceedings of the National Academy of Sciences USA* 88: 10544-10547.

Catania, F., M. O. Kauer, et al. 2004. World-wide survey of an *Accord* insertion and its association with DDT resistance in *Drosophila melanogaster. Molecular Ecology* 13: 2491-2405.

Centers for Disease Control, Division of Parasitic Diseases. 2001. *Malaria.* Published online at *http://www.dpd.cdc.gov/dpdx/HTML/Malaria.asp?body=Frames/M-R/Malaria/body_Malaria_page2.htm.*

Clegg, M. T., J. F. Kidwell, and C. R. Horch. 1980. Dynamics of correlated genetic systems. V. Rates of decay of linkage disequilibria in experimental populations of *Drosophila melanogaster. Genetics* 94: 217–234.

Cochran, G., J. Hardy, and H. Harpending. 2006. Natural history of Ashkenazi intelligence. *Journal of Biosocial Science* 38: 659–693.

Daborn, P. J., J. L. Yen, et al. 2002. A single P450 allele associated with insecticide resistance in *Drosophila. Science* 297: 2253-2256.

Dawson, E., G. R. Abecasis, S. Bumpstead, et al. 2002. A first-generation linkage disequilibrium map of human chromosome 22. *Nature* 418: 544–548.

De La Vega, F. M., H. Isaac, et al. 2005. The linkage disequilibrium maps of three human chromosomes across four populations reflect their demographic history and a common underlying recombination pattern. *Genome Research* 15: 454-462.

Diaz, G. A., B. D. Gelb, et al. 2000. Gaucher disease: The origins of the Ashkenazi Jewish *N370S* and *84GG* acid β-glucosidase mutations. *American Journal of Human Genetics* 66: 1821-1832

Dunbrack, R. L., C. Coffin, and R. Howe. 1995. The cost of males and the paradox of sex: An experimental investigation of the short-term competitive advantages of evolution in sexual populations. *Proceedings of the Royal Society of London* B 262: 45–49.

Felsenstein, J. 1988. Sex and the evolution of recombination. In R. E. Michod and B. R. Levin, eds. *The Evolution of Sex.* Sunderland, MA: Sinauer, 74–86.

Galvani, A. P., and M. Slatkin. 2003. Evaluating plague and smallpox as historical selective pressures for the *CCR5-Δ32* HIV-resistance allele. *Proceedings of the National Academy of Sciences USA* 100: 15276-15279.

Haigh, J. 1978. The accumulation of deleterious mutations in a population: Muller's ratchet. *Theoretical Population Biology* 14: 251–267.

Hedrick, P. W. 1983. *Genetics of Populations.* Boston: Science Books International.

Hedrick, P. W., and B. C. Verrelli. 2006. "Ground truth" for selection on *CCR5-Δ32. Trends in Genetics* 22: 293-296.

Howard, R. S., and C. M. Lively. 1998. The maintenance of sex by parasitism and mutation accumulation under epistatic fitness functions. *Evolution* 52: 604–610.

Hummel, S., D. Schmidt, et al. 2005. Detection of the *CCR5-Δ32* HIV resistance gene in Bronze Age skeletons. *Genes and Immunity* 6: 371-374.

Huttley, G. A., M. W. Smith, et al. 1999. A scan for linkage disequilibrium across the human genome. *Genetics* 152: 1711–1722.

International HapMap Consortium, The. 2005. A haplotype map of the human genome. *Nature* 437: 1299-1320.

Jennings, H. S. 1917. The numerical results of diverse systems of breeding, with respect to two pairs of characters, linked or independent, with special relation to the effects of linkage. *Genetics* 2: 97–154.

Keightley, P. D., and S. P. Otto. 2006. Interference among deleterious mutations favors sex and recombination in finite populations. *Nature* 443: 89–92.

Kohn, M. H., H.-J. Pelz, and R. K. Wayne. 2000. Natural selection mapping of the warfarin-resistance gene. *Proceedings of the National Academy of Science USA* 97: 7911–7915.

Lalani, A. S., J. Masters, et al. 1999. Use of chemokine receptors by poxviruses. *Science* 286: 1968–1971.

Lambert, J. D., and N. A. Moran. 1998. Deleterious mutations destabilize ribosomal RNA in endosymbiotic bacteria. *Proceedings of the National Academy of Sciences USA* 95: 4458–4462.

Lively, C. M. 1992. Parthenogenesis in a freshwater snail: Reproductive assurance versus parasitic release. *Evolution* 46: 907–913.

Lively, C. M. 1996. Host-parasite coevolution and sex. *BioScience* 46: 107–114.

Luzzatto, L., and R. Notaro. 2001. Protecting against bad air. *Science* 293: 442–443.

Maynard Smith, J. 1978. *The Evolution of Sex*. Cambridge: Cambridge University Press.

Maynard Smith, J. 1988. The evolution of recombination. In R. E. Michod and B. R. Levin, eds. *The Evolution of Sex*. Sunderland, MA: Sinauer, 106–125.

McVean, G., C. C. A. Spencer, and R. Chaix. 2005. Perspectives on Human Genetic Variation from the HapMap Project. *PLoS Genetics* 1: e54.

Michod, R. E., and B. R. Levin, eds. 1988. *The Evolution of Sex*. Sunderland, MA: Sinauer.

Muller, H. J. 1964. The relation of recombination to mutational advance. *Mutation Research* 1: 2–9.

Nordborg, M., T. T. Hu, et al. 2005. The pattern of polymorphism in *Arabidopsis thaliana*. *PLoS Biology* 3: e196.

Novembre, J., A. P. Galvani, and M. Slatkin. 2005. The geographic spread of the *CCR5-Δ32* HIV-resistance allele. *PLoS Biology* 3: e339.

Otto, S. P., and N. H. Barton. 2001. Selection for recombination in small populations. *Evolution* 55: 1921–1931.

Otto, S. P., and T. Lenormand. 2002. Resolving the paradox of sex and recombination. *Nature Reviews Genetics* 3: 252–261.

Risch, N., H. Tang, et al. 2003. Geographic distribution of disease mutations in the Ashkenazi Jewish population supports genetic drift over selection. *American Journal of Human Genetics* 72: 812-822.

Ruwende, C. and A. Hill. 1998. Glucose-6-phosphate dehydrogenase deficiency and malaria. *Journal of Molecular Medicine* 76:581–588.

Sabeti, P. C., D. E. Reich, J. M. Higgins, et al. 2002. Detecting recent positive selection in the human genome from haplotype structure. *Nature* 419: 832–837.

Sabeti, P. C., E. Walsh, et al. 2005. The case for selection at *CCR5-Δ32*. *PLoS Biology* 3: e378.

Schlenke, T. A., and D. J. Begun. 2004. Strong selective sweep associated with a transposon insertion in *Drosophila simulans*. *Proceedings of the National Academy of Sciences USA* 101: 1626-1631.

Seger, J., and W. D. Hamilton. 1988. Parasites and sex. In R. E. Michod and B. R. Levin, eds. *The Evolution of Sex*. Sunderland, MA: Sinauer, 176–193.

Slatkin, M. 2004. A population-genetic test of founder effects and implications for Ashkenazi Jewish diseases. *American Journal of Human Genetics* 75: 282-293.

Slatkin, M., and B. Rannala. 2000. Estimating allele age. *Annual Review of Genomics and Human Genetics* 1: 225-249.

Spolsky, C. M., C. A. Phillips, and T. Uzzell. 1992. Antiquity of clonal salamander lineages revealed by mitochondrial DNA. *Nature* 356: 706–710.

Stephens, J. C., D. E. Reich, et al. 1998. Dating the origin of the *CCR5-Δ32* AIDS-resistance allele by the coalescence of haplotypes. *American Journal of Human Genetics* 62: 1507–1515.

Strom, C. M., B. Crossley, et al. 2004. Molecular screening for diseases frequent in Ashkenazi Jews: Lessons learned from more than 100,000 tests performed in a commercial laboratory. *Genetics in Medicine* 6: 145-152.

Sturtevant, A. H., and K. Mather. 1938. The interrelations of inversions, heterosis, and recombination. *American Naturalist* 72: 447–452.

Voight, B. F., Sridhar Kudaravalli, et al. 2006. A map of recent positive selection in the human genome. *PLoS Biology* 4: e72.

9 Evolution at Multiple Loci: Quantitative Genetics

The population genetics models of Chapters 6, 7, and 8 allow us to understand, sometimes with considerable precision, how and why populations evolve. However, we can use these models only when we are content to analyze the evolution of just one or two loci at a time. That is often not good enough, because many interesting traits are determined by the combined influence of alleles at many loci. When studying such traits, we often do not know the identities of the particular loci involved. This chapter introduces **quantitative genetics**, the branch of evolutionary biology that provides tools for analyzing the evolution of multilocus traits. As in Chapter 8, the material we present will contain some abstract ideas, but there will be a surprising practical payoff for readers who master them. Our discussion of quantitative genetics will allow us to debunk erroneous claims about the cause of differences in IQ scores among ethnic groups.

Japanese flounder do not fall into discrete color categories. Instead, they show continuous variation. The histogram below shows that color intensity among these fish is normally distributed. From Shikano (2005).

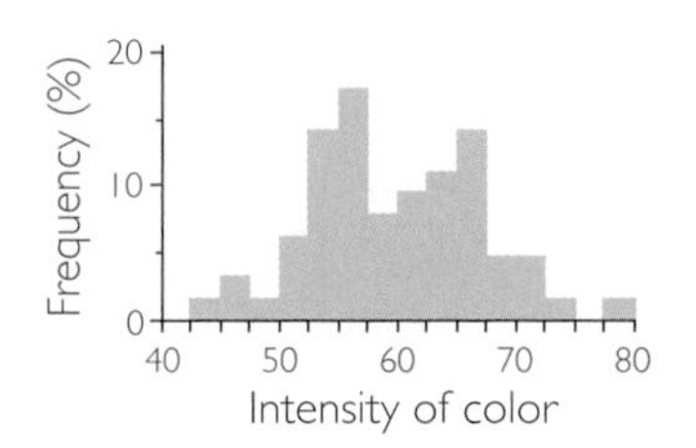

9.1 The Nature of Quantitative Traits

Throughout our coverage of evolutionary genetics, we have been discussing traits for which the phenotypes come in discrete categories. A flour beetle is either alive or dead; a person either has cystic fibrosis or does not. We might call

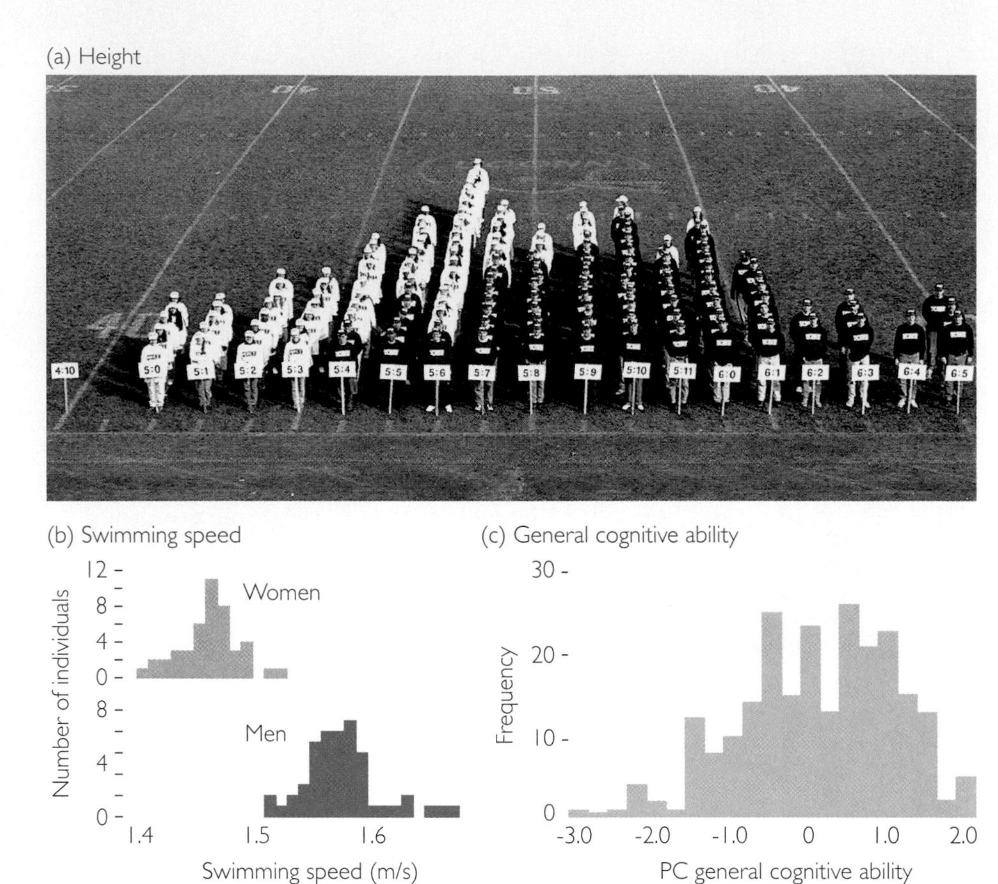

Figure 9.1 Some quantitative traits in humans (a) These students and faculty at the University of Connecticut have formed a living histogram by arranging themselves into columns by height. The women are wearing white shirts; the men are wearing blue. (b) These graphs show the variation in average speed over a distance of 1,500 meters for swimmers who competed in the finals at the 2002 Phillips 66 Summer Nationals. Plotted from data found at http://www.usaswimming.org/fast_times/template.pl?opt=results&eventid=471. (c) This graph shows the variation in general cognitive ability, assessed as a statistical composite of scores from a variety of tests, for Swedish participants in a twin study. For more details see Twins in Section 9.3. Rerendered from McClearn et al. (1997).

characters such as these **qualitative traits**, because we can assign individuals to categories just by looking at them, or perhaps by conducting a simple genetic test.

Traits with discrete phenotypes are special examples; most traits in most organisms show continuous variation. Examples of human traits with continuous variation are height, athletic ability, and intelligence (Figure 9.1). Examples in other organisms include beak length in soapberry bugs (Figure 2.3, page 41) and bill depth in medium ground finches (Figure 3.9, page 82). For traits with continuous variation we cannot assign individuals to discrete phenotypic categories by simple inspection. Instead, we have to take measurements. For this reason, characters with continuously distributed phenotypes are called **quantitative traits**. Quantitative traits are determined by the combined influence of (1) the genotype at many different loci, and (2) the environment.

Quantitative traits are traits for which the distribution of phenotypes is continuous rather than discrete.

Early in the 20th century there was considerable debate among biologists over whether Gregor Mendel's model of genetics can be applied to quantitative traits (see Provine 1971). Among the first researchers to provide convincing affirmative evidence was Edward East (1916). East worked with populations of longflower tobacco (*Nicotiana longiflora*). The trait he studied was the length of the corolla, the part of the flower formed by the petals. In longflower tobacco the corolla is shaped like a tube.

East started with two pure-breeding strains of *Nicotiana*, one with short corollas and the other with long corollas. He crossed individuals from these parental strains to produce F_1 hybrids, then let the F_1 hybrids self-fertilize to produce an F_2 generation. Before looking at East's data on corolla length in the F_1s and F_2s, let us first make predictions using Mendel's model of genetics.

The simplest Mendelian model we might devise is one in which corolla length is determined by a single locus with two alleles. We will imagine that the alleles are codominant, so there will be three phenotypes. The genotype *aa* will produce short flowers, *aA* will produce medium flowers, and *AA* will produce long flowers (Figure 9.2a). By this model, East's first cross is between *aa* and *AA* parents. All the F_1s will have genotype *aA*, and all will have medium flowers. When the F_1s self-fertilize, the F_2s they produce will have genotypes *aa*, *aA*, and *AA* in the proportions $\frac{1}{4}:\frac{1}{2}:\frac{1}{4}$. The reader can check this prediction with a 2×2 Punnett square.

East knew from his prior experience with longflower tobacco that this model is much too simple. In most populations corolla length is highly variable, and the variations form a continuum, not three distinct phenotypes. However, a straightforward modification of our simple Mendelian model will improve its prospects for predicting the results of East's crosses.

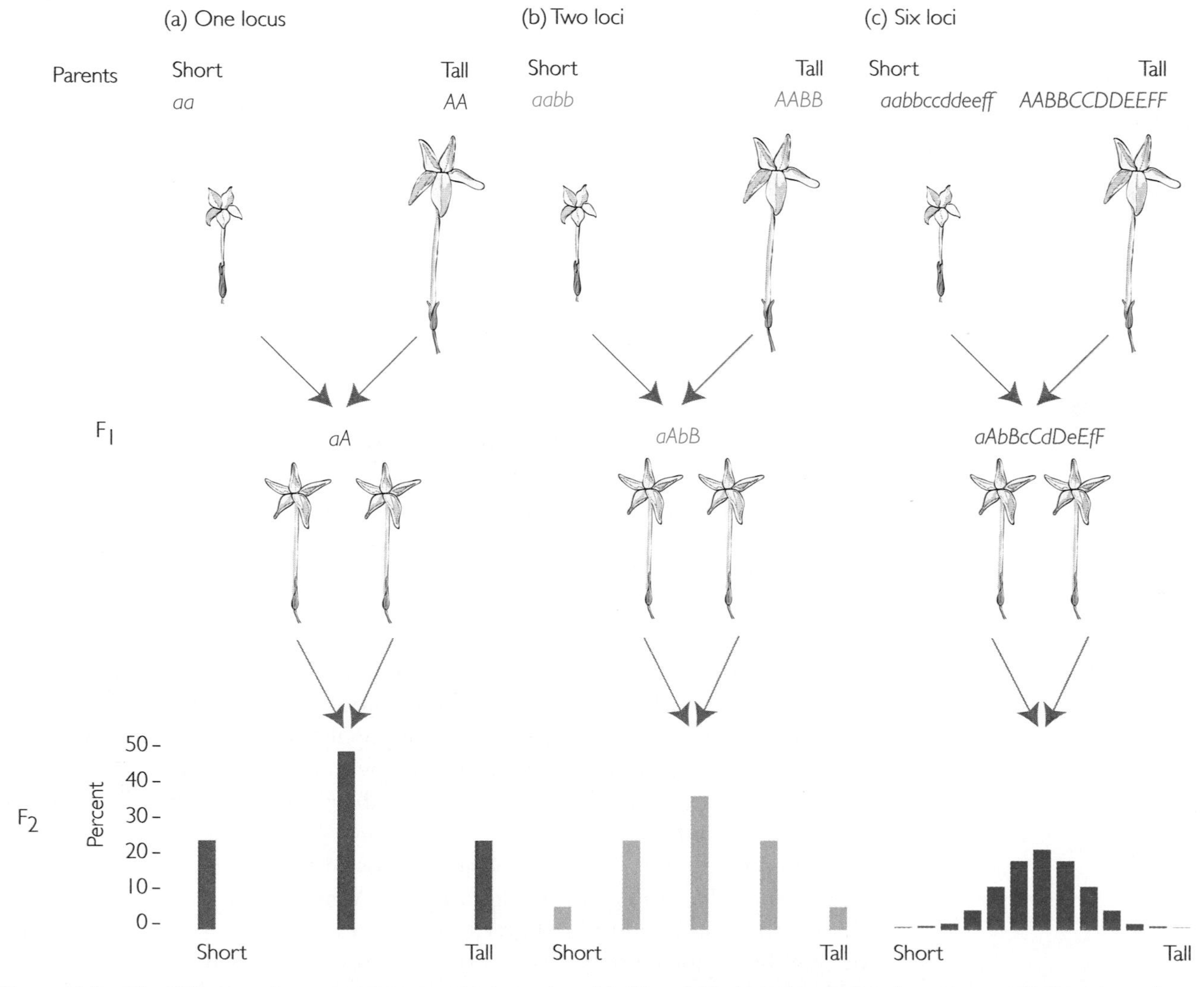

Figure 9.2 **Mendelian genetics can explain quantitative traits** (a), (b), and (c) show the predicted genotypes and phenotypes for parental, F_1, and F_2 plants under Mendelian models in which corolla length is determined by the alleles at one, two, and six loci. See text for more details.

Instead of imagining that corolla length is determined by the alleles at a single locus, we will imagine it is determined by the alleles at two loci (Figure 9.2b). We still have our first locus, with alleles *a* and *A*, but now we add a second locus, with alleles *b* and *B*. As with the first locus, each copy of an allele designated by a capital letter independently contributes to a longer corolla. In our new model there are five phenotypes, associated with genotypes that have 0, 1, 2, 3, or 4 capital letter alleles. East's parental cross, *aabb* × *AABB*, produces F_1s with genotype *aAbB* and medium corollas. When the F_1s self-fertilize, the F_2s they produce will have phenotypes ranging from short to long in the proportions $\frac{1}{16}:\frac{4}{16}:\frac{6}{16}:\frac{4}{16}:\frac{1}{16}$. The reader can check this prediction with a 4 × 4 Punnett square (the gametes made by an *aAbB* plant are *ab*, *aB*, *Ab*, and *AB*).

The two-locus model is a step in the right direction, but it still produces F_2s with discrete phenotypes. We can fix this by skipping ahead to a model in which corolla length is determined not by one or two loci, but by several. Figure 9.2c shows the predictions for a model with six loci. This model yields 13 phenotypes, associated with genotypes that have 0 to 12 capital letter alleles. East's parental cross in this model is *aabbccddeeff* × *AABBCCDDEEFF*. The F_1s will have genotype aAbBcCdDeEfF and medium corollas. The F_2s will have phenotypes ranging from short to long in the proportions $\frac{1}{4096}:\frac{12}{4096}:\frac{66}{4096}:\frac{220}{4096}:\frac{495}{4096}:\frac{792}{4096}:\frac{924}{4096}:\frac{792}{4096}:\frac{495}{4096}:\frac{220}{4096}:\frac{66}{4096}:\frac{12}{4096}:\frac{1}{4096}$. This prediction can be checked with a 64 × 64 Punnett square, but we suspect the reader will just take our word for it.

Quantitative traits are consistent with Mendelian genetics. They are influenced by the combined effects of the genotype at many loci.

By the time there are 13 different phenotypes they begin to grade into one another, and with real plants it will be difficult to assign individuals to particular categories without using a ruler. In other words, in our six-locus model corolla length is a quantitative trait.

Now we come to East's two key predictions. Note that in the one-locus and two-locus models, substantial numbers of F_2 plants have phenotypes identical to those of the parental strains. In the six-locus model there are also F_2 plants with phenotypes identical to the parental strains, but not very many of them. Just 1 in 4,096 F_2 plants, for example, has genotype *aabbccddeeff* and the shortest possible corollas. East's first prediction was that unless we breed and measure thousands of plants, the range of variation we will see in the F_2s will not extend all the way to the original parental phenotypes. Note also that just because parental phenotypes do not appear in a population of a few hundred F_2s does not mean they have been lost forever. The alleles necessary to produce genotypes *aabbccddeeff* and *AABBCCDDEEFF* are still present in the population. They are just all in heterozygotes. East's second prediction was that with a few generations of selective breeding for short or tall corollas we should be able to recover the original parental phenotypes.

East's data appear in Figure 9.3. When he crossed short- and long-flowered parents they produced F_1s with medium flowers. When he let the F_1s self-fertilize, the F_2s they produced showed greater variation in phenotype than the F_1s. But because he examined only 454 F_2 plants, not several thousand, he found no F_2s with phenotypes approaching the extremes of the parental generation. Finally, starting with the F_2 plants, East selectively bred for short corollas and long corol-

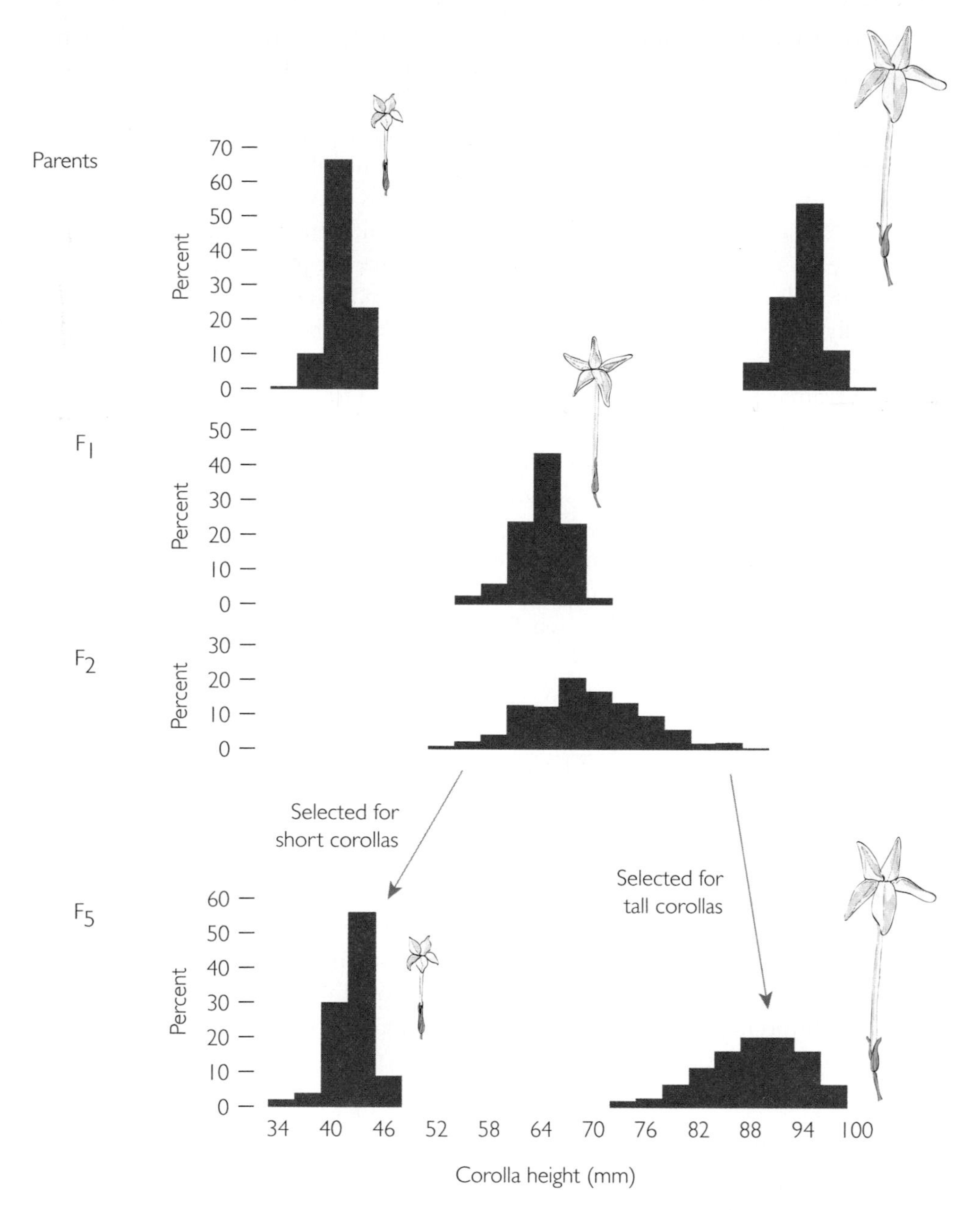

Figure 9.3 Edward East's data confirm the predictions of the Mendelian model in Figure 9.2c East crossed short and long-flowered parental plants to produce medium-flowered F_1s. He then selfed the F_1s, which produced an F_2 generation that was more variable than the F_1 generation, but did not approach the extremes of the parental strains. Finally, East recovered the parental phenotypes by selectively breeding from the F_2 plants. Drawn from data in Table 1 of East (1916); after Ayala (1982).

las. By the time he reached the F_5 generation, most of the plants in his selected lines had corolla lengths within the ranges of the original parents. East's data confirm his predictions. His experiment, and others like it, established that quantitative traits are determined by the combined influence of Mendelian alleles at many loci.

Quantitative traits are also influenced by the environment.

The fit between East's data and his Mendelian model is impressive, but there is one respect in which it is not perfect. Under a strict interpretation of even the

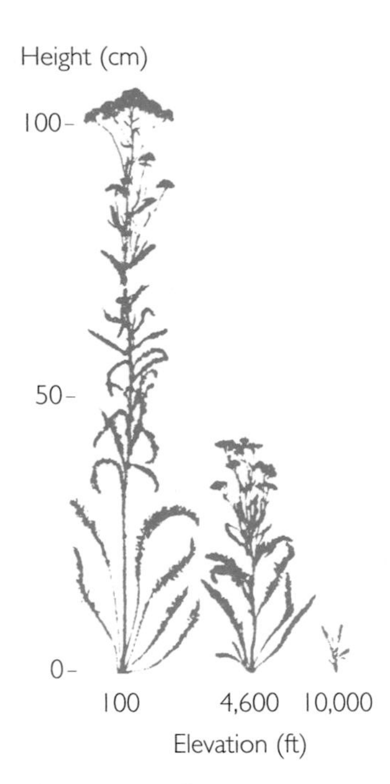

Figure 9.4 Quantitative traits are influenced by the environment as well as genotype These three yarrow plants were grown from cuttings of the same individual, and are thus genetically identical. Reared at different altitudes, they show dramatic differences in height. Reprinted from Clausen, Keck, and Hiesey (1948).

six-locus model, the short parental plants should all have been exactly the same. Likewise, the long parental plants and the F_1s should all have been exactly the same. In each of these groups all the plants have the same genotype, so they should all have the same phenotype. But, of course, they do not. There is some phenotypic variation even among genetically identical plants. The reason is that each plant, even in East's experimental garden, was exposed to a unique environment. Some got a little more water, others got a little more sun. These small differences in environment produced small differences in phenotype.

The influence of environmental differences on quantitative phenotypes is especially clear in Figure 9.4. The three yarrow plants shown in the figure are genetically identical. Jens Clausen, David Keck, and William Hiesey (1948) grew them from cuttings of a single individual, collected from a population living at an elevation of 50 feet. They reared the clones in gardens at elevations of 100, 4,600, and 10,000 feet. The environmental differences associated with elevation had a dramatic impact on the plants' heights.

The data presented in this section substantiate a claim we made at the outset. Quantitative traits are determined by the combined influence of the genotype at many different loci and the environment. In the next section we consider modern techniques for identifying some of the particular loci that underlie quantitative variation.

9.2 Identifying Loci That Contribute to Quantitative Traits

The loci that influence quantitative traits are called, appropriately enough, **quantitative trait loci**, or **QTLs**. Often we would like to identify the QTLs behind an interesting quantitative trait. Modern genetic and statistical methods make it possible to do so. We will review two such methods: QTL mapping, and investigation of candidate loci. Our examples come from a study of the genetics of adaptation in monkeyflowers, a study of human personality, and a study of genetic factors contributing to a human disease.

QTL Mapping

QTL mapping is the collective name for a suite of related techniques that employ marker loci to scan chromosomes and identify regions containing genes that contribute to a quantitative trait. We will illustrate QTL mapping with an example from research by H. D. Bradshaw, Jr. and colleagues (1998) on two species of monkeyflowers, *Mimulus cardinalis* and *Mimulus lewisii* (Figure 9.5).

Mimulus cardinalis and *M. lewisii* are sister species that hybridize readily in the lab and produce fertile offspring. They also have overlapping ranges in the Sierra Nevada Mountains of California. But no hybrids have ever been found in the field. The reason is that the two monkeyflowers attract different pollinators. *Mimulus cardinalis* is pollinated by hummingbirds, whereas *M. lewisii* is pollinated by bees.

The difference in pollinators between *M. lewisii* versus *M. cardinalis* is reflected in the dramatic differences in their floral morphology. Bees do not see well in the red part of the visible spectrum and need a platform to land on before walking into a flower and foraging. Hummingbirds, in contrast, see red well, have long, narrow beaks, and hover while harvesting nectar. *M. lewisii* has a prominent land-

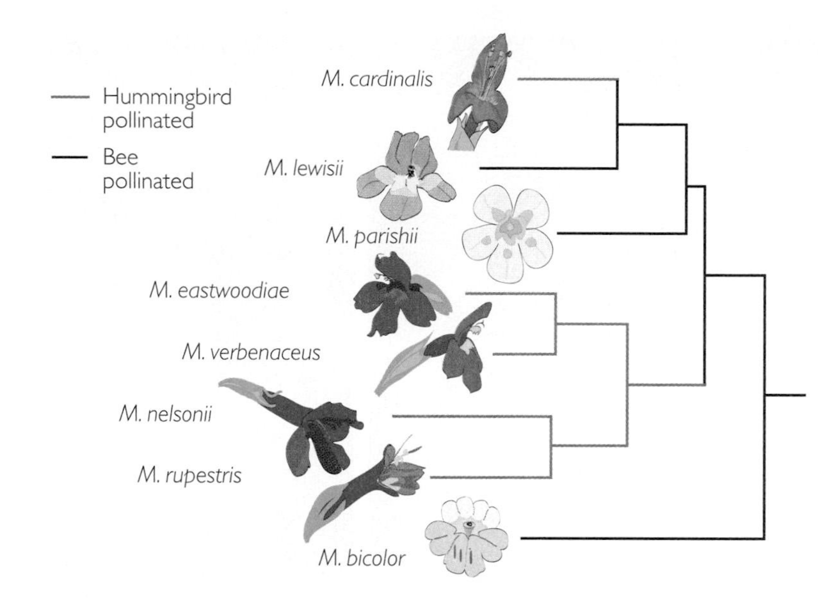

Figure 9.5 A phylogeny of *Mimulus cardinalis*, *Mimulus lewisii*, and kin The common ancestor of these species was pollinated by bees. Pollination by hummingbirds evolved twice: once in the common ancestor of *M. eastwoodiae* and kin, and once in *M. cardinalis*. What genes are involved and what are their effects? After Beardsley et al. (2003).

ing pad, while *M. cardinalis* has an elongated tube with a nectar reward at the end. Indeed, the flowers of *M. lewisii* and *M. cardinalis* conform to classical bee- and bird-pollinated colors and shapes.

As the phylogeny in Figure 9.5 shows, the most recent common ancestor of *M. lewisii* and *M. cardinalis* was pollinated by bees (Beardsley et al. 2003). This implies that many characteristics of *M. lewisii*'s bee-pollinated flower are ancestral and that *M. cardinalis*'s more tubular, reddish, hummingbird-pollinated flower is derived. The questions are: What genes are responsible for the radical makeover of *M. cardinalis*'s flower? How many of them are there? How strong are their effects? QTL mapping offers a way to find out.

Bradshaw and colleagues' interest in these questions was motivated in part by a fascination with the plants themselves and in part by a debate in theoretical evolutionary genetics. Starting with Ronald Fisher (1930) most evolutionary geneticists have held that the alleles driven to fixation by natural selection, and consequently responsible for the adaptive differences between species, were virtually all extremely subtle in their effects on the phenotype. A minority of theoreticians, however, most prominently H. Allen Orr, have contended that some of the alleles fixed during adaptive evolution are obvious in their phenotypic effects (Orr and Coyne 1992; Orr 1998, 1999). Bradshaw and coworkers saw that monkeyflowers were an ideal test case. If Orr's view is correct, the researchers would find QTLs with strong effects on floral phenotype; if Fisher's view is correct, they would find only QTLs with subtle effects.

Bradshaw and colleagues crossed *M. lewisii* and *M. cardinalis* to produce F_1 hybrids(Figure 9.6a–c). They then crossed the F_1s to produce 465 F_2 individuals. The F_2 individuals show an astonishing diversity of floral phenotypes (Figure 9.6d–l). This result is similar to that of Edward East (depicted in Figure 9.3), and it has a similar genetic explanation. The parental species, *M. lewisii* and *M. cardinalis*, were essentially homozygous at all loci that influence floral appearance. As a result, the F_1s were all heterozygous. The F_2s are the product of genetic recombination among the F_1 heterozygotes. At any given locus, any given F_2 may be a homozygote for the *M. lewisii* allele, a heterozygote, or a homozygote for the *M. cardinalis* allele. Bradshaw and colleagues scored all 465 F_2

Figure 9.6 ***Mimulus cardinalis, Mimulus lewisii,*** **and their F_1 and F_2 descendents** Photo (a) shows *M. lewisii*, photo (b) shows an F_1 hybrid, and photo (c) shows *M. cardinalis*. The remaining photos (d–l) show F_2 hybrids produced by crosses between F_1s. The F_2s show wide variation in their floral characters. Reprinted from Schemske and Bradshaw (1999).

plants for each of a dozen floral characters that differ between the two species (Table 9.1).

Bradshaw and colleagues also determined the genotype of each F_2 individual at each of 66 marker loci randomly distributed throughout the monkeyflower genome. A marker locus is a site in the genome where the nucleotide sequence varies among chromosomes and where a simple genetic test will identify different alleles. Bradshaw and colleagues chose marker loci at which *M. cardinalis* are all homozygous for one allele and *M. lewisii* are all homozygous for another. This meant that the F_1 plants were all heterozygous, and that the F_2s could be homozygous for the *M. lewisii* allele, heterozygous, or homozygous for the *M. cardinalis* allele.

Table 9.1 Flowers of *Mimulus cardinalis* versus *Mimulus lewisii*

Traits scored by Bradshaw et al. (1998) are listed in the first column, grouped by function in pollination. The direction of the difference between species is indicated in the second and third columns. After Bradshaw et al. (1998).

Characteristic	*M. cardinalis*	*M. lewisii*	Notes
Pollinator attraction			
Purple pigment (anthocyanins) in petals	high	low	
Yellow pigment (carotenoids) in petals	high	low	• The yellow pigment in *M. lewisii* petals is arranged in stripes called nectar guides, which are interpreted as a "runway" for bees as they land on the wide petals.
Lateral petal width	high	low	
Corolla width	low	high	
Corolla projected area	low	high	
Upper petal reflexing	high	low	
Lateral petal reflexing	high	low	
Pollinator reward			
Nectar volume	high	low	• Nectar volume is probably higher in bird flowers simply because birds drink more than bees.
Pollinator efficiency			
Stamen (male structure) length	high	low	• The difference in stamen and pistil length is important: In *M. cardinalis* these structures extend beyond the flower and touch the hummingbird's forehead as it feeds.
Pistil (female structure) length	high	low	
Corolla aperture width	low	high	
Corolla aperture height	high	low	

To see the logic of QTL mapping, imagine a quantitative trait locus that influences one of the monkeyflower floral traits. We will call the *cardinalis* allele Q_C and the *lewisii* allele Q_L. Imagine also a marker locus at which the *cardinalis* allele is M_C and the *lewisii* allele is M_L. Consider first a case in which the QTL and the marker locus are physically linked—that is, close together on the same chromosome. The *M. cardinalis* parental plant had genotype Q_CM_C/Q_CM_C and the *M. lewisii* parent had genotype Q_LM_L/Q_LM_L, where Q_CM_C indicates a two-locus genotype on a single chromosome. The F_1 plants all had genotype Q_CM_C/Q_LM_L. There will be rare cases in which there is crossing-over between the QTL and the marker locus, but except for these, the F_2 population will consist of plants with three genotypes: Q_CM_C/Q_CM_C, Q_CM_C/Q_LM_L, and Q_LM_L/Q_LM_L. Plants homozygous for the *cardinalis* marker allele will tend toward the *cardinalis* phenotype, heterozygotes will have intermediate phenotypes, and plants homozgyous for the *lewisii* marker allele will tend toward the *lewisii* phenotype (Figure 9.7, first column). In the language of Chapter 8, the marker locus and the QTL are in linkage disequilibrium. Of the four possible chromosome genotypes, only two are present. This linkage disequilibrium reveals itself in a nonrandom association between the genotype at the marker locus and the phenotype influenced by the QTL.

We can detect the presence and location of loci influencing a quantitative trait by crossing parents from populations with fixed differences. Among the grandoffspring, we look for associations between phenotype and genotype at marker loci.

Consider now a case in which the QTL and the marker locus are unlinked. The *M. cardinalis* parental plant had genotype $Q_C/Q_CM_C/M_C$ and the *M. lewisii* parent had genotype $Q_L/Q_LM_L/M_L$. The F_1s all had genotype $Q_C/Q_L\ M_C/M_L$. Because the QTL and the marker are unlinked, the F_2 population will include plants with nine genotypes: $Q_C/Q_C\ M_C/M_C$, $Q_C/Q_L\ M_C/M_C$, $Q_L/Q_L\ M_C/M_C$, and so on. Among the F_2s, there will be no association between genotype at the marker locus and phenotype for the trait influenced by the QTL (Figure 9.7, second column).

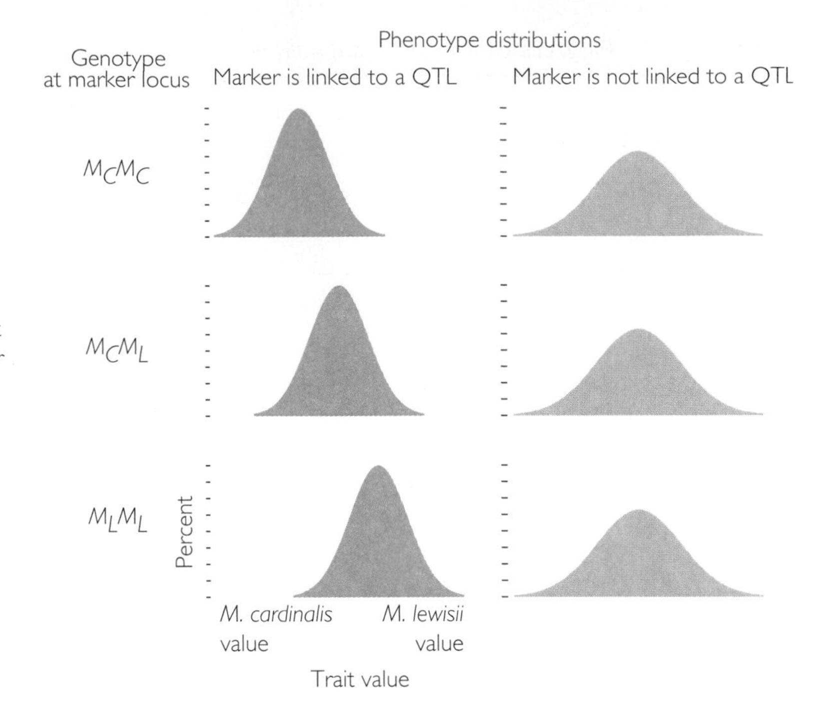

Figure 9.7 **The logic of QTL mapping** In QTL mapping, researchers start with parents from different species, cross them to produce F_1 hybrids, then self or intercross the F_1 hybrids to produce a large population of F_2s (see Figure 9.6). For each F_2 individual the researchers measure the phenotype for the quantitative trait of interest and the genotype at marker loci distributed across the genome. Finally, examining the entire F_2 population, researchers compare individuals with different genotypes at each marker locus. If phenotypes differ among individuals with different genotypes at a particular marker locus, as shown here on the left, then we can infer that the marker locus sits near a locus that contributes to the quantitative trait.

In QTL mapping, then, researchers examine an F_2 population for statistical associations between genotype at marker loci and phenotype. If phenotype is associated with genotype at a particular marker locus, we can interpret the association as evidence that a QTL influencing the trait of interest is located near the marker (see Box 9.1 for more details). By using multiple marker loci, researchers can estimate both the location of QTLs and the strength of their influence on phenotype.

Box 9.1 QTL mapping

Here we use a simplified example to illustrate the statistical reasoning employed in maximum-likelihood QTL mapping. We will consider a *qualitative* phenotypic trait controlled entirely by a single locus with two codominant alleles. We want to know whether the locus for this trait is linked to a particular marker locus.

Let the alleles at the locus affecting phenotype be P and p, and the alleles at the marker locus be M and m. Imagine that we have crossed a parent homozygous for allele P at the trait locus and allele M at the marker locus with a parent homozygous for allele p and allele m. The F_1 offspring of this cross are heterozygous for both loci. We then crossed two F_1 individuals to produce an F_2 individual, which turned out to be homozygous for both allele P and allele M. We want to know whether this outcome constitutes evidence that the trait locus and the marker locus are linked. To do so, we will calculate the **likelihood** of producing a double homozygote from our F_1 cross. The likelihood of a particular outcome is its probability given a model of the process that produced it. We will consider two models: one in which the loci are linked with a recombination frequency of 0.1, and one in which they are not linked.

Linkage model with $r = 0.1$: Under this model the loci are linked with the distance between them such that 10% of the gametes meiosis produces are recombinants. The genotype for our F_1 individuals under this model was MP/mp. These individuals produce gametes in the following proportions: 45% MP, 5% Mp, 5% mP, and 45% mp. Thus the probability that they would produce an offspring of genotype MP/MP is $0.45 \times 0.45 = 0.2025$.

Free-recombination model: Under this model the loci are unlinked. As a result, they recombine during meiosis 50% of the time. The genotype for our F_1 individuals under this model was $M/m\ P/p$. These individuals produce gametes in the following proportions: 25% MP, 25% Mp, 25% mP,

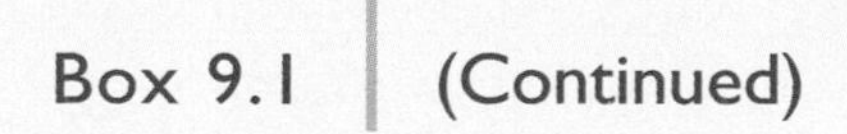

Box 9.1 (Continued)

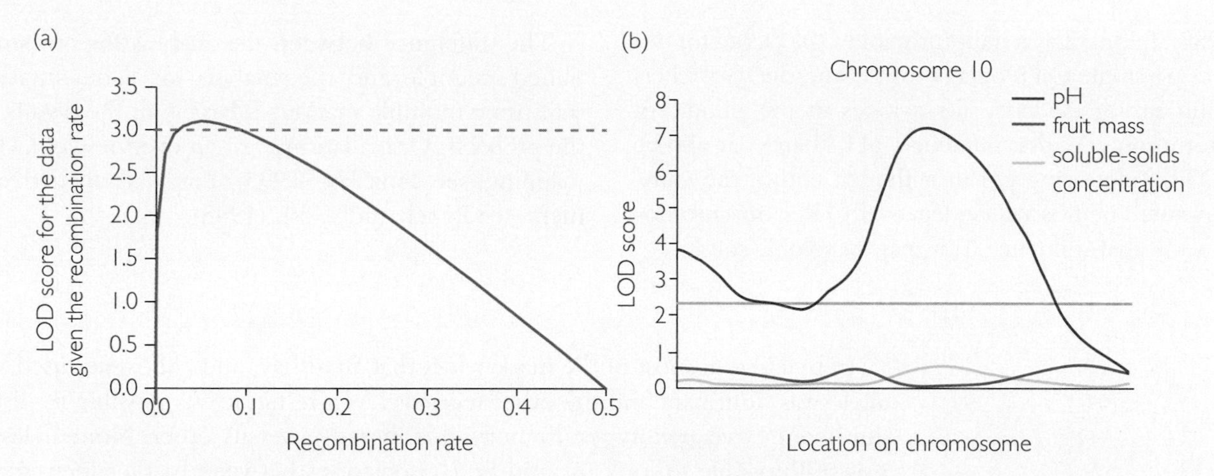

Figure 9.8 Results from QTL mapping studies are often summarized with plots of LOD score LOD score measures the degree to which we can better explain the data with a model in which a locus influencing phenotype is linked to the marker loci examined versus a model in which no such linkage exists. Where the LOD score crosses a threshold chosen by the researchers, the researchers conclude that there is evidence of linkage. (a) This graph shows LOD score as a function of recombination rate for our hypothetical example. (b) This graph shows LOD score as a function of chromosomal location for three quantitative traits in tomatoes. Rerendered from Paterson et al. (1988).

and 25% *mp*. Thus the probability that they would produce an offspring of genotype $M/M\ P/P$ is $0.25 \times 0.25 = 0.0625$.

If we divide the likelihood under the linkage model by the likelihood under the free recombination model we get the **likelihood ratio**, which is 3.24. This calculation shows that our double homozygote F_2 offspring is over three times more likely under the linkage model than under the free recombination model. In other words, our F_2 offspring provides evidence that the trait locus and the marker locus are linked.

But it does not provide much evidence; it is only a single individual. We need to look at many F_2 offspring and assess the strength of the evidence they provide collectively. To do this we take the logarithm of the likelihood ratio, which gives us a value called the **LOD** score, short for *logarithm of the odds*. The LOD score for our first F_2 individual is 0.511. Now we can calculate the LOD scores for other individuals and sum them to get an overall LOD score. Imagine that we examine nine more F_2s and find that two have genotype *MMPP*, four have genotype *MmPp*, one has genotype *Mmpp*, and two have genotype *mmpp*. The LOD scores are 0.511 for each *MMPP* individual, 0.215 for each *MmPp* individual, −0.444 for the *Mmpp* individual, and 0.511 for each *mmpp* individual. The overall LOD score for our 10 F_2s is 2.97.

Many geneticists consider a LOD score of 3.0 to be the threshold for concluding that a trait locus is linked to a marker locus. Our overall LOD score falls just shy of this threshold. Thus we cannot conclude that the loci are linked with a recombination frequency of 0.1. But what about other recombination frequencies? We could just as well have chosen a recombination frequency of 0.05 for our linkage model, or 0.2, or any value between 0 and 0.5. Figure 9.8a plots the overall LOD score for our data under all possible linkage models. The model under which the LOD score reaches its maximum value, 3.10, is one in which the recombination rate is 0.05. We can conclude that the trait locus and the marker locus are linked, and that the best estimate for the recombination rate between them is 0.05.

Analyzing genotypes for multiple marker loci, researchers can scan chromosomes for evidence of loci that influence the phenotype for quantitative traits. Often they summarize their analyses with plots of LOD score as a function of the chromosomal location of the QTL under the linkage model. Figure 9.8b, for example, shows LOD score as a function of location on chromosome 10 for three quantitative traits in tomatoes—fruit mass, concentration of soluble solids, and pH—in a study by Andrew Paterson and colleagues

Box 9.1 (Continued)

(1988). Taking 2.4 as their threshold LOD score for detecting a single QTL on a chromosome, the researchers found strong evidence for a locus in the middle of chromosome 10 that influences pH. There may also be a QTL influencing pH near the left end of the chromosome. There is no evidence of QTLs on chromosome 10 that influence fruit mass or soluble solids.

The difference between the analysis for our simplified example and the analysis for a quantitative trait using multiple markers is largely in the details of the probability calculations. For an overview of QTL mapping, see Tanksley (1993). For a detailed treatment, see Lynch and Walsh (1998).

In practice, at most of the marker loci that Bradshaw and colleagues used, one allele was dominant and the other recessive. As a result, it was possible to distinguish only two genotypes: homozygous recessive versus other. Nonetheless, it was still possible to look for, and find, associations between marker locus genotypes and floral phenotypes. For each of the dozen floral traits the researchers scored, they found between one and six QTLs that influence flower phenotype.

Some of the QTLs Bradshaw and colleagues found had pronounced effects on the appearance of the flowers (Figure 9.9). For 9 of the 12 floral characters, there was at least one QTL at which differences in genotype explained over 25% of the variation in flower phenotype.

To confirm that the QTLs that Bradshaw and colleagues identified were, indeed, the loci subject to selection during the diversification of the two species, Douglas Schemske and Bradshaw (1999) reared a large series of F_2 individuals in the greenhouse and recorded the amount of purple pigment, yellow pigment, and nectar in their flowers, along with overall flower size. Then they planted the individuals in a natural habitat where both species of monkeyflowers naturally coexist, and recorded which pollinators visited which flowers. Their data revealed a strong trend. Bees prefer large flowers and avoid flowers with a high concentration of yellow pigments. Hummingbirds, in contrast, tended to visit the most nectar-rich flowers and those with the highest amounts of purple pigment.

By collecting tissues from each F_2 individual planted in the field and determining which QTL markers they contained, the researchers were able to calculate that an allele associated with increased concentration of yellow pigments reduced bee visitation by 80%, while an allele responsible for increasing nectar production doubled hummingbird visitation. It is reasonable to surmise that changes in the frequencies of these alleles, driven by differential success in at-

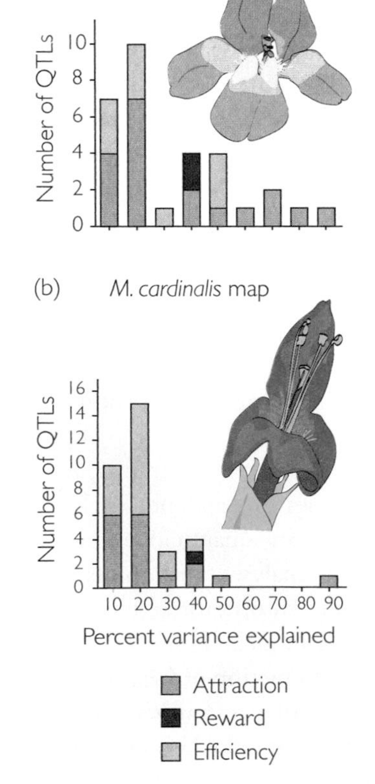

Figure 9.9 QTLs for floral traits in *Mimulus lewisii* and *Mimulus cardinalis*, sorted by the strength of their effects on the phenotype Bradshaw and colleagues (1998) found between one and six QTLs for each of the dozen floral traits they mapped. Differences in genotype for the majority of these QTLs explained only a modest amount of the variation in flower phenotype, as indicated by the higher bars on the left side of these graphs. However, for 9 of the 12 floral traits, there was at least one QTL at which differences in genotype explained more than 25% of the variation in phenotype. These QTLs with large phenotypic effects are indicated by the bars in the center and right of these graphs. There are separate graphs for *M. lewisii* and *M. cardinalis* because most of the marker loci used by Bradshaw and colleagues can only be detected in one species or the other. However, the QTLs themselves almost certainly occur in both genomes. Rerendered from Bradshaw and colleagues (1998).

tracting hummingbirds as pollinators, was the mechanism behind the diversification of the two monkeyflowers.

Bradshaw and Schemske (2003) reproduced an event that might have been the first step in the evolution of *M. cardinalis* from a bee-pollinated ancestor. They bred plants that had the *M. lewisii* genotype across virtually the entire genome, except that they carried the *M. carinalis* genotype at a QTL, mentioned in the previous paragraph, that strongly influences the amount of yellow pigment in the petals. This single genetic change turns *M. lewisii's* petals from purplish-pink to pale yellow-orange (Figure 9.10). When they monitored these plants in a natural habitat, the researchers found that the *M. lewisii* plants with the novel genotype were considerably less attractive to bees than the wild-type, but dramatically more attractive to hummingbirds. The yellow-orange *M. lewisii* were still much less attractive to hummingbirds than are wild-type *M. cardinalis,* but they had been given a good start down the path toward switching pollinators. The results of Bradshaw and Schemske's monkeyflower research are consistent with Orr's view that adaptive evolution often involves the selective fixation of alleles with large effects; it is inconsistent with Fisher's view.

(a) Wild-type *M. lewisii*

(b) *M. lewisii* with *M. cardinalis* genotype at the *YUP* locus

Figure 9.10 A novel allele at a single locus can dramatically alter a flower's attractiveness to different pollinators The monkeyflowers shown here are full siblings. Genetically they are virtually identical, except that they carry different alleles at the *YUP* locus. The flower in (a) is about 6 times as attractive to bees than the flower in (b). The flower in (b) is roughly 70 times more attractive to hummingbirds than the flower in (a).

QTL mapping can reveal the number of loci that influence a quantitative trait, the magnitude of their effects on phenotype, and their location in the genome. It cannot, however, tell us the identity of the loci and the proteins they encode. To determine these, researchers must evaluate candidate loci.

Candidate Loci

To detect a QTL and find its location, we look for an association between the phenotype and the genotype at a marker locus. To learn the identity of a QTL and the protein it encodes, we have to look for associations between the genotype at a coding locus, the structure and function of its gene product, and the phenotype. Sometimes we know to evaluate a particular coding locus because we already know something about the function of its gene product and suspect that it may play a role in the phenotype. Other times we know to evaluate a particular coding locus because its location matches that of a QTL we have mapped using markers.

Our first example comes from a study of human personality, most aspects of which are quantitative traits. Jonathan Benjamin and colleagues (1996) were interested in a personality trait called novelty seeking. Novelty seeking, assessed with questionnaires, is highly variable among individuals and has the familiar bell-shaped distribution. People with high novelty-seeking scores tend to be more impulsive, excitable, and exploratory, whereas people with low scores tend to be reflective, stoic, and rigid (Ebstein et al. 1996). Benjamin and colleagues had reason to suspect that some of the variation in novelty seeking might be associated with allelic variation in the gene for the D4 dopamine receptor, or D4DR.

D4DR is a neurotransmitter receptor. It sits on the surface of neurons in the brain, waiting to receive messages from other cells in the form of dopamine, a neurotransmitter. Benjamin and colleagues knew that neurons employing D4DR as a receptor participate in thought and emotion in humans and in exploratory behavior in animals. They also knew that one of the coding regions of the D4DR gene contains a 48-base-pair tandem repeat. The number of repeats varies from two to eight, and different repeats have distinguishable physiological properties.

We can confirm that a particular locus influences a quantitative trait by looking for associations between genotype and phenotype.

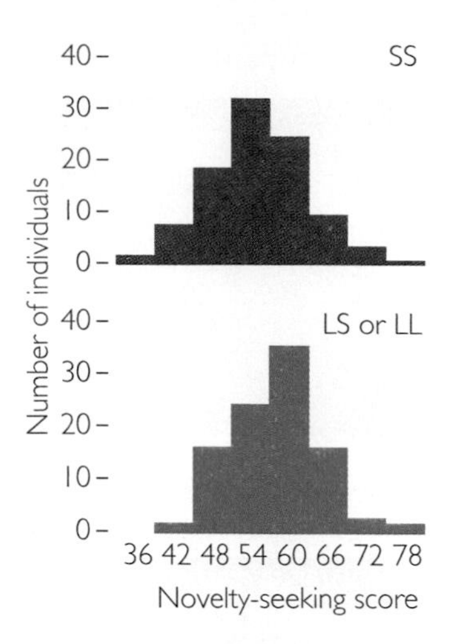

Figure 9.11 Identification of a quantitative trait locus influencing a personality trait Sequence variation at the D4 dopamine receptor locus can be reduced to two categories of alleles: short (S) and long (L). Individuals with genotype LS or LL tend to score slightly but significantly higher on psychological tests of novelty seeking. Redrawn from Benjamin et al. (1996).

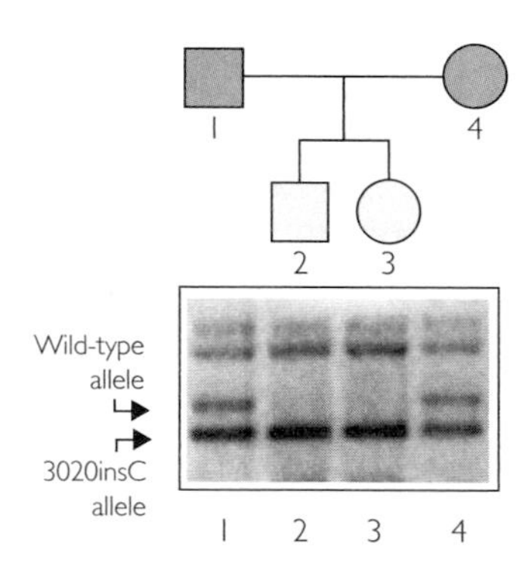

Figure 9.12 Transmission of a suspect allele in a family with Crohn's disease The pedigree shows that in this family the parents are unaffected and both children have Crohn's disease. The electrophoresis gel shows DNA fragments from PCR amplification of a portion of the NOD2 gene. Both parents are carriers of the *3020insC* allele; both children are homozygotes. These data are consistent with the hypothesis that *3020insC* homozygotes are at higher risk for Crohn's disease. From Ogura et al. (2001).

To see if the D4DR gene is a QTL influencing novelty seeking, Benjamin and colleagues determined the novelty-seeking scores and D4DR genotypes of 315 people. The researchers divided the D4DR alleles they found into two categories: short (S) for alleles with two to five repeats versus long (L) for alleles with six to eight repeats. They divided their subjects into two categories by their D4DR genotypes: a group with two short alleles (SS) versus a group with at least one long allele (SL or LL). When the researchers compared the novelty-seeking scores for the two groups, they found that people with at least one long allele scored three points higher, on average, than people with two short alleles (Figure 9.11). This difference is slight; it explains only 3–4% of the variation in novelty seeking. There are some SS individuals who score much higher than some SL or LL individuals. But genotype at the D4DR locus does appear to exert some influence on this personality trait (see also Schinka et al. 2002).

Our second example comes from a study of Crohn's disease, a chronic digestive disorder characterized by inflammation of the intestine. Crohn's disease tends to run in families, suggesting that it may be caused, in part, by genetic factors. Furthermore, its frequency has increased in recent decades, suggesting that it may be caused, in part, by environmental factors, such as bacterial pathogens (Hugot et al. 2001) or even reduced exposure to intestinal worms (Moreels and Pelckmans 2005; Summers et al. 2005a, 2005b). The severity of Crohn's disease ranges from mild to extreme. We can think of Crohn's disease as a quantitative trait.

Yasunori Ogura and colleagues (2001) knew, from mapping studies using marker loci, that a QTL influencing susceptibility to Crohn's disease is located in a particular region of chromosome 16. This region contains several coding loci, some of which had been evaluated previously and shown to play no role in the disease. Ogura and colleagues investigated a gene called NOD2. NOD2 encodes a protein that helps regulate the immune response to bacteria in the intestine (Kobayashi et al. 2005). It makes sense that genotypes at the NOD2 locus might be associated with the risk of developing Crohn's disease.

Ogura and colleagues sequenced the NOD2 genes from a dozen patients with Crohn's disease. In three patients they found a frameshift mutation, a single nucleotide insertion that creates, downstream, a premature stop codon. This allele, called *3020insC*, encodes a protein that is shorter than normal and has compromised function.

To find out whether the *3020insC* allele can increase the risk of Crohn's disease, the researchers performed what is called a transmission disequilibrium test (see Lynch and Walsh 1998). They screened a large sample of Crohn's disease patients and found 56 independent cases in which a patient had a parent who carried one copy of the *3020insC* allele and thus had had the opportunity to pass the allele to the patient (see Figure 9.12 for an example). If the allele plays no role in Crohn's disease, then we would expect the heterozygous parent to have transmitted the allele to the patient in half the cases. If, on the other hand, the allele does play a role, we would expect the parents to have transmitted the allele in more than half the cases. In fact, the parent transmitted the allele in 39 cases and failed to transmit it in only 17. A chi-square test shows that this result is statistically significant.

The *3020insC* allele is clearly not the sole cause of Crohn's disease. There are many Crohn's patients who do not carry the allele, and many individuals who carry the allele who do not have Crohn's. But the allele does appear to increase a person's risk of contracting the disease.

The techniques and examples we have discussed in this section show that it is possible to trace a quantitative trait to the Mendelian loci that influence it. This effort, however, is time consuming and expensive. We need tools that allow us to analyze and understand the genetics and evolution of quantitative traits, even when we do not know the identities of the many specific genes involved. These tools are the subject of the next three sections of the chapter.

9.3 Measuring Heritable Variation

Quantitative genetics allows us to analyze evolution by natural selection in traits controlled by many loci.

Recall the basic tenets of Darwin's theory of evolution by natural selection: If there is heritable variation among the individuals in a population, and if there are differences in survival and/or reproductive success among the variants, then the population will evolve. Quantitative genetics includes tools for measuring heritable variation, tools for measuring differences in survival and/or reproductive success, and tools for predicting the evolutionary response to selection. In this section we focus on the first challenge, measuring heritable variation.

Imagine a population of organisms in which there is continuous variation among individuals in some trait. For example, imagine a population of humans in which there is continuous variation among individuals in height. Continuously variable traits are typically normally distributed, so that a histogram of a trait has the familiar bell-curve shape. Assuming our human population follows this pattern, a few people are very short, many people are more or less average in height, and a few people are very tall (Figure 9.1a, page 320). We want to know: Is height heritable?

It is worth thinking carefully about exactly what this question means. Questions about heritability are often expressed in terms of nature versus nurture. But such questions are meaningful only if they concern comparisons among individuals. It makes no sense to focus only on a woman on the far left of Figure 9.1a and ask, without reference to the other individuals, whether this woman is 5 feet tall because of her genes (nature) or because of her environment (nurture). She had to have both genes and an environment to be alive and of any height at all. She did not get 3 feet of her height from her genes, and 2 feet from her environment, so that $3 + 2 = 5$. Instead, she got all of her 5 feet from the activity of her genes operating within her environment. Within this single student we cannot, even in principle, disentangle the influence of nature and nurture.

The first step in a quantitative genetic analysis is to determine the extent to which the trait in question is heritable. That is, we must partition the total phenotypic variation (V_P) into a component due to genetic variation (V_G) and a component due to environmental variation (V_E).

The only kind of question it makes sense to ask is a comparative one: Is the shortest woman shorter than the tallest woman because they have different genes, because they grew up in different environments, or both? This is a question we can answer. In principle, for example, we could take an identical twin of the short woman and raise her in the environment experienced by the tall woman. If this twin still grew up to be 5 feet tall, then we would know that the difference between the shortest and tallest women is due entirely to differences in their genes. If the twin grew up to be 6 feet 2 inches tall, then we would know that the difference between the shortest and tallest women is due entirely to differences in their environments. In fact, the twin would probably grow up to be somewhere between 5 feet and 6 feet 2 inches. This would indicate that the difference between the two women is partly due to differences in their genes and partly due to differences in their environments. Considering the whole population, rather than just two individuals, we can ask: What fraction of the variation in height among the students is due to variation in their genes, and what fraction is due to variation in their environments?

The fraction of the total variation in a trait that is due to variation in genes is called the **heritability** of the trait. The total variation in a trait is referred to as the **phenotypic variation**, and is symbolized by V_P. Variation among individuals that is due to variation in their genes is called **genetic variation** and is symbolized by V_G. Variation among individuals due to variation in their environments is called **environmental variation**, and is symbolized by V_E. Thus, we have

$$\text{heritability} = \frac{V_G}{V_P} = \frac{V_G}{V_G + V_E}$$

More precisely, this fraction is known as the **broad-sense heritability**, or degree of genetic determination. We will define the narrow-sense heritability shortly. Heritability is always a number between 0 and 1.

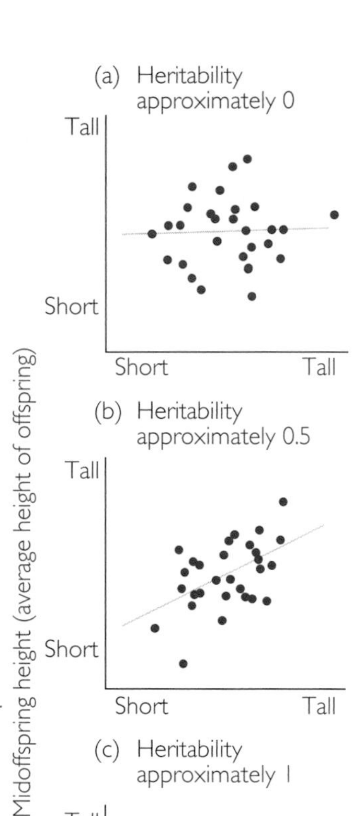

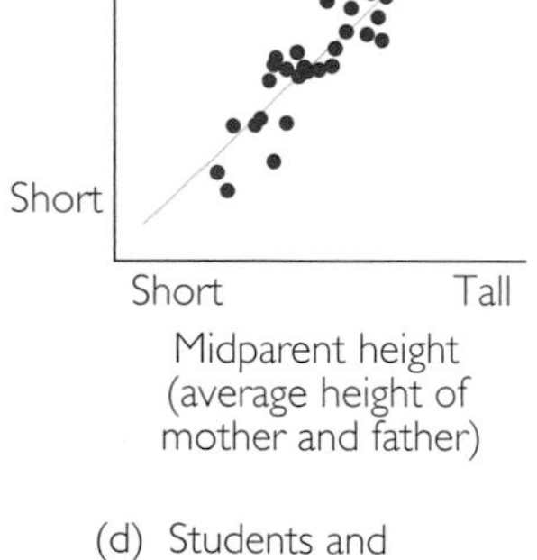

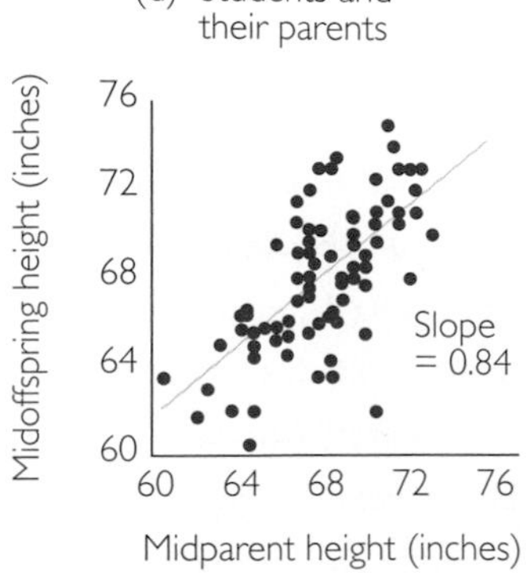

Estimating Heritability from Parents and Offspring

Before wading any deeper into symbolic abstractions, we note the simple truth that if the variation among individuals is due to variation in their genes, then offspring will resemble their parents. It is easy, in principle, to check whether they do. We first make a scatterplot with offspring's trait values represented on the y-axis, and their parents' trait values on the x-axis (Figure 9.13). We have two parents for every offspring, so we use the midparent value, which is the average of the two parents. If we have more than one offspring for each family, we use a midoffspring value as well. We then draw the best-fit line through the data points. If offspring do not resemble their parents, then the slope of the best-fit line through the data will be near 0 (Figure 9.13a); this is evidence that the variation among individuals in the population is due to variation in their environments, not variation in their genes. If offspring strongly resemble their parents, the slope of the best-fit line will be near 1 (Figure 9.13c); this is evidence that variation among individuals in the population is due to variation in their genes, not variation in their environments. Most traits in most populations fall somewhere in the middle, with offspring showing a moderate resemblance to their parents (Figure 9.13b); this is evidence that the variation among individuals is partly due to variation in their environments and partly due to variation in their genes. Figure 9.13d shows data for an actual population of students. (For another example, look back at the scatterplot in Figure 3.10, page 83, which analyzes the heritability of beak depth in Darwin's finches.)

The examples in Figure 9.13 illustrate that the slope of the best-fit line for a plot of midoffspring versus midparents is a number between 0 and 1 that reflects the degree to which variation in a population is due to variation in genes. In other words, we can take the slope of the best-fit line as an estimate of the heritability. If we determine the best-fit line using the method of least-squares linear regression, which minimizes the sum of the squared vertical distances between

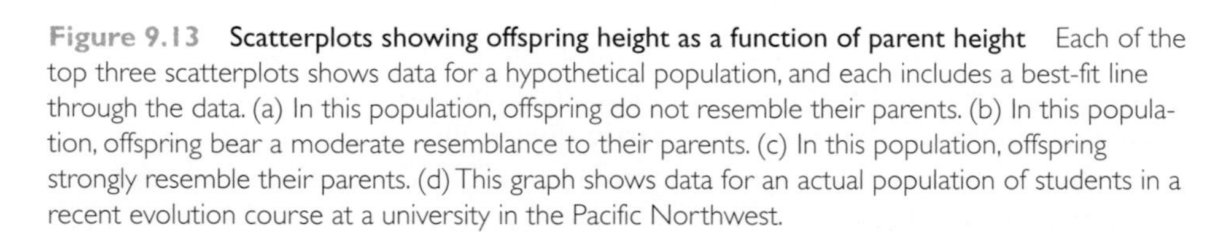

Figure 9.13 Scatterplots showing offspring height as a function of parent height Each of the top three scatterplots shows data for a hypothetical population, and each includes a best-fit line through the data. (a) In this population, offspring do not resemble their parents. (b) In this population, offspring bear a moderate resemblance to their parents. (c) In this population, offspring strongly resemble their parents. (d) This graph shows data for an actual population of students in a recent evolution course at a university in the Pacific Northwest.

the points and the line, then the slope represents a version of the heritability symbolized by h^2 and known as the **narrow-sense heritability**. Least-squares linear regression is the standard method taught in introductory statistics texts and used by statistical software packages to determine best-fit lines. (For readers familiar with statistics, it may prevent some confusion if we note here that h^2 is *not* the fraction of the variation among the offspring that is explained by variation in the parents. That quantity would be r^2. Instead, h^2 is an estimate of the fraction of the variation among the *parents* that is due to variation in their genes.)

To explain the difference between narrow-sense heritability and broad-sense heritability, we need to distinguish between two components of genetic variation: additive genetic variation versus dominance genetic variation. Additive genetic variation (V_A) is variation among individuals due to the additive effects of genes, whereas dominance genetic variation (V_D) is variation among individuals due to gene interactions such as dominance (see Box 9.2). The total genetic variation is the sum of the additive and dominance genetic variation: $V_G = V_A + V_D$. The broad-sense heritability, defined earlier, is V_G/V_P. The narrow-sense heritability, h^2, is defined as follows:

The heritability, h^2, is a measure of the (additive) genetic variation in a trait.

$$h^2 = \frac{V_A}{V_P} = \frac{V_A}{V_A + V_D + V_E}$$

When evolutionary biologists mention heritability without noting whether they are using the term in the broad or narrow sense, they almost always mean the narrow-sense heritability. We will use the narrow-sense heritability in the rest of this discussion. It is the narrow-sense heritability, h^2, that allows us to predict how a population will respond to selection.

When estimating the heritability of a trait in a population, it is important to keep in mind that offspring can resemble their parents for reasons other than the genes the offspring inherit (see Box 3.1, page 84). Environments run in families too. Among humans, for example, some families exercise more than others, and different families eat different diets. Our estimate of heritability will be accurate only if we can make sure that there is no correlation between the environments experienced by parents and those experienced by their offspring. We obviously cannot do so in a study of humans. In an animal study, however, we could collect all the offspring at birth, then foster them at random among the parents. In a plant study, we could place seeds at random locations in a field.

James Smith and André Dhondt (1980), for example, studied song sparrows (*Melospiza melodia*) to determine the heritability of beak size. They collected young from natural nests, sometimes as eggs and sometimes as hatchlings, and moved them to the nests of randomly chosen foster parents. When the chicks grew up, Smith and Dhondt calculated midoffspring values for the chicks and midparent values for both the biological and foster parents. Graphs of offspring beak depth versus parental beak depth appear in Figure 9.14. The

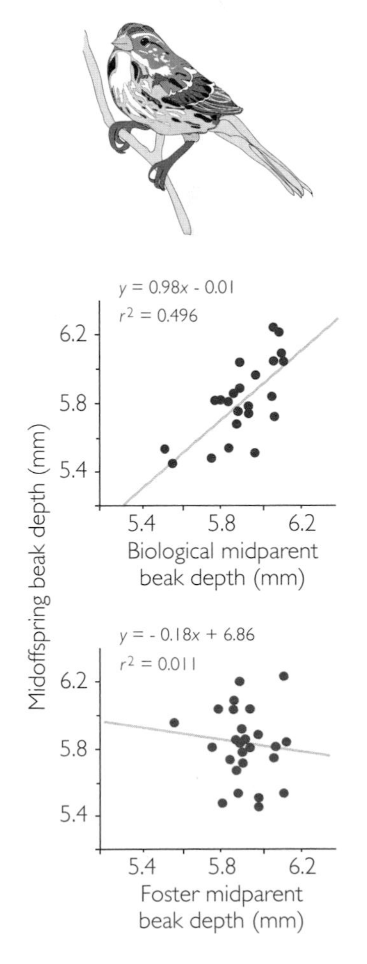

Figure 9.14 A field experiment on the heritability of beak size in song sparrows The top scatterplot shows the relationship between midoffspring beak size (mean family beak depth) and the biological parents' beak size (biological midparent beak depth). The bottom scatterplot shows the relationship between midoffspring beak size (mean family beak depth) and the foster parents' beak size (foster midparent beak depth). Chicks resemble their biological parents strongly, and their foster parents not at all. From Smith and Dhondt (1980).

Box 9.2 Additive genetic variation versus dominance genetic variation

Here we use a numerical example to distinguish between additive genetic variation and dominance genetic variation. To simplify the discussion, we will analyze genetic variation at a single locus with two alleles as though we were analyzing a quantitative trait. We will assume that there is no environmental variation in the trait in question: An individual's phenotype is determined solely and exactly by its genotype. The alleles at the locus are A_1 and A_2, each has a frequency of 0.5, and the population is in Hardy–Weinberg equilibrium. We will consider two situations: (1) the alleles are codominant; (2) allele A_2 is dominant over allele A_1.

Situation (1): Alleles A_1 and A_2 are codominant.

A_1A_1 individuals have a phenotype of 1. In A_1A_2 and A_2A_2 individuals, each copy of allele A_2 adds 0.5 to the phenotype. At the left in Figure 9.15a is a histogram showing the distribution of phenotypes in the population. At the center and right are scatterplots that allow us to analyze the genetic variation in the population. The *x*-axis represents the genotype, calculated as the number of copies of allele A_2. The *y*-axis represents the phenotype. The horizontal gray line shows the mean phenotype for the population (= 1.5). The plot at center shows that the total genetic variation V_G is a function of the deviations of the data points from the population mean (green arrows). We can quantify V_G by calculating the sum of the squared deviations. The plot at right shows the best-fit line through the data points (red). The additive genetic variation V_A is defined as that fraction of the total genetic variation that is explained by the best-fit line (blue arrows). In this case, the best-fit line explains all of the genetic variation, so $V_G = V_A$. There is no dominance genetic variation.

Situation (2): Allele A_2 is dominant over allele A_1.

This time, A_1A_1 individuals again have a phenotype of 1. The effects of substituting copies of A_2 for copies of A_1 are not strictly additive, however: The first copy of A_2 (which makes the genotype A_1A_2) changes the phenotype from 1 to 2. The second copy of A_2 (which makes the genotype A_2A_2) does not alter the phenotype any further. At left in Figure 9.15b is a histogram showing the distribution of phenotypes in the population. At center and right are scatterplots that allow us to analyze the genetic variation in the population. The plot at center shows that the total genetic variation V_G is a function of the deviations of the data points (green arrows) from the population mean (gray line; = 1.75). The plot at right shows the best-fit line through the data points (red). The additive genetic variation V_A is that fraction of the total genetic variation that is explained by the best-fit line (blue arrows). The dominance genetic variation V_D is that fraction of the total genetic variation left unexplained by the best-fit line (yellow arrows). In this case, the best-fit line explains only part of the genetic variation, so $V_G = V_A + V_D$.

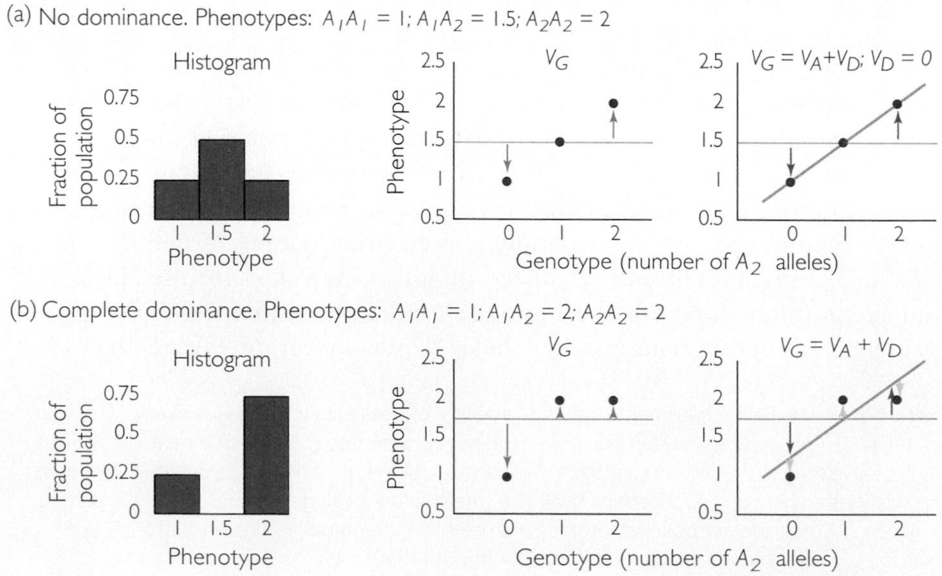

(a) No dominance. Phenotypes: $A_1A_1 = 1$; $A_1A_2 = 1.5$; $A_2A_2 = 2$

(b) Complete dominance. Phenotypes: $A_1A_1 = 1$; $A_1A_2 = 2$; $A_2A_2 = 2$

Figure 9.15 Additive genetic variation versus dominance genetic variation in a trait controlled by two alleles at a single locus

chicks resembled their biological parents strongly, and their foster parents not at all. These results show that virtually all the variation in beak depth in this population is due to variation in genes. Smith and Dhondt estimated that the heritability of beak depth is 0.98.

Estimating Heritability from Twins

There are a variety of other methods for estimating heritability besides calculating the slope of the best-fit line for offspring versus parents. For example, studies of twins can be used. The logic of twin studies works as follows (Figure 9.16). Monozygotic (identical) twins share their environment and all of their genes, whereas dizygotic (fraternal) twins share their environment and half of their genes. If heritability is high, and variation among individuals is due mostly to variation in genes, then monozygotic twins will be more similar to each other than are dizygotic twins. If heritability is low, and variation among individuals is due mostly to variation in environments, then monozygotic twins will be as different from each other as dizygotic twins.

Gerald McClearn and colleagues (1997) used a twin study to estimate the heritability of general cognitive ability, a measure of intelligence, in a Swedish population. Figure 9.1c (page 320) shows the distribution of this trait among 110 pairs of monozygotic twins and 130 pairs of same-sex dizygotic twins. Even though all the participants in the study were at least 80 years old, and environmental variation had had ample time to exert its influence, the monozygotic twins resembled

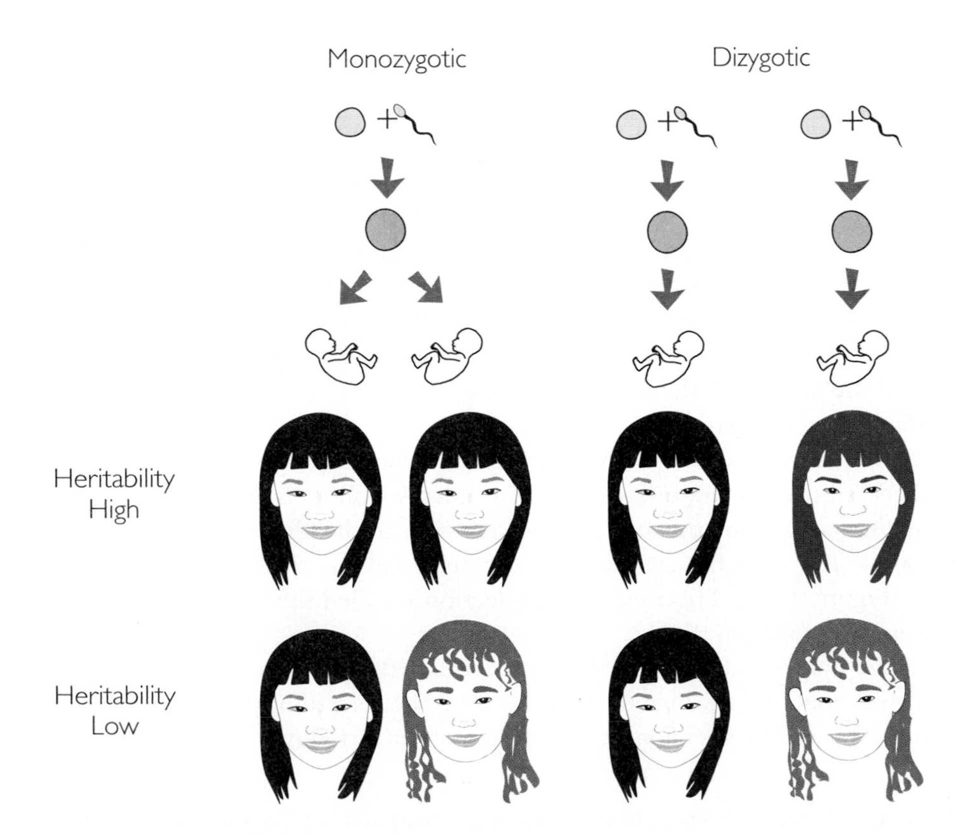

Figure 9.16 **Estimating heritability from twin studies** Monozygotic twins develop from a single zygote, and thus share all their genes. Dizygotic twins develop from separate zygotes, and share half their genes. If the heritability of a trait is high, monozygotic twins will resemble each other more strongly than dizygotic twins.

each other substantially more strongly than the dizygotic twins. McClearn and colleagues estimate that the heritability of general cognitive ability in their study population is 0.62.

For detailed treatments of methods for measuring heritability, see Falconer (1989) and Lynch and Walsh (1998). Data on the heritability of traits are frequently misinterpreted, particularly when the species under study is humans. We will return to this issue in Section 9.7.

9.4 Measuring Differences in Survival and Reproductive Success

In the preceding section we developed techniques for measuring the heritable variation in quantitative traits, the first tenet of Darwin's theory of evolution by natural selection. The next tenet of Darwin's theory is that there are differences in survival and/or reproductive success among individuals. We now discuss techniques for measuring differences in success—that is, for measuring the strength of selection. Once we can measure both heritable variation and the strength of selection, we will be able to predict evolutionary change in response to selection.

The kind of differences in success envisioned by Darwin's theory are systematic differences. On average, individuals with some values of a trait survive at higher rates, or produce more offspring, than individuals with other values of a trait. To measure the strength of selection, we first note who survives or reproduces and who fails to do so. Then we quantify the difference between the winners and the losers in the trait of interest.

The second step in a quantitative genetic analysis is to measure the strength of selection on the trait in question. One measure is the selection differential, S, equal to the difference between the mean of the selected individuals and the mean of the entire population.

In selective-breeding experiments, the strength of selection is easy to calculate. Consider, for example, an experiment conducted by R. J. Di Masso and colleagues (1991). These researchers set out to breed mice with longer tails. They wanted to know how the developmental program that constructs a mouse's tail would change under selection. Would a mouse embryo make a longer tail by elongating the individual vertebrae, or by adding extra ones? Every generation, the researchers measured the tails of all the mice in their population. Then they picked the mice with the longest tails, and let them breed among themselves to produce the next generation.

To see how to quantify the strength of selection, suppose the researchers picked as breeders the one-third of the mice whose tails are the longest. The simplest measure of the strength of selection is the difference between the mean tail length of the breeders and the mean tail length of the entire population (Figure 9.17a). This measure of selection is called the selection differential and is symbolized by S.

A second (and related) measure of the strength of selection is the selection gradient.

There is a second measure of the strength of selection that is useful because of its broad applicability. This measure is called the selection gradient (Lande and Arnold 1983). The selection gradient is calculated as follows:

1. Assign absolute fitnesses to the mice in the population. We will think of fitness as survival to reproductive age. In our population, one-third of the mice survived long enough to reproduce. (This does not necessarily mean that the short-tailed mice were actually killed, just that they were removed from the

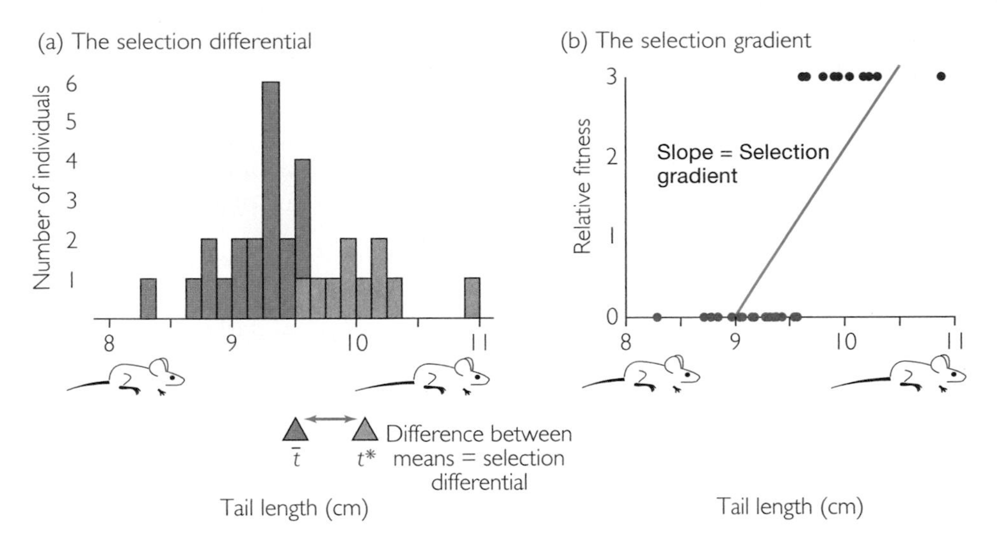

Figure 9.17 Measuring the strength of selection (a) The histogram shows the variation in tail length in a fictional population of lab mice. The red bars represent the mice chosen as breeders for the next generation. The gray triangle indicates the average tail length for the entire population; the red triangle indicates the average tail length for the breeders. The difference between these two averages is the selection differential. (b) A scatterplot for the same fictional population of mice showing relative fitness (see text) as a function of tail length. Red dots represent mice chosen as breeders for the next generation. The scatterplot includes the best-fit line (green). The slope of the best-fit line is the selection gradient.

breeding population; as far as our breeding population is concerned, the short-tails did not breed so they did not survive long enough to reproduce.) The short-tailed two-thirds of the mice have a fitness of 0, and the long-tailed one-third have a fitness of 1.

2. Convert the absolute fitnesses to relative fitnesses. The mean fitness of the population is 0.33 (if, for example, there are 30 mice in the population, the mean is $[\{20 \times 0\} + \{10 \times 1\}]/30 = 0.33$). We calculate each mouse's relative fitness by dividing its absolute fitness (0 or 1) by the mean fitness (0.33). The short-tailed mice have a relative fitness of 0; the long-tailed mice have a relative fitness of 3.
3. Make a scatterplot of relative fitness as a function of tail length, and calculate the slope of the best-fit line (Figure 9.17b). The slope of the best-fit line is the selection gradient.

It may not appear at first glance that the selection gradient and the selection differential have much to do with each other. In fact, they are closely related, and each can be converted into the other. If we are analyzing selection on a single trait like tail length, then the selection gradient is equal to the selection differential divided by the variance in tail length (see Box 9.3). An advantage of the selection gradient is that we can calculate it for any measure of fitness, not just survival. We might, for example, measure fitness in a natural population of mice as the number of offspring weaned. If we first calculate each mouse's relative fitness (by dividing its number of offspring by the mean number of offspring), then plot relative fitness as a function of tail length and calculate the slope of the best-fit line, then that slope is the selection gradient.

The selection gradient, β, for trait t is equal to the selection differential, S, divided by the variance:

$$\beta = \frac{S}{var(t)}$$

Box 9.3 The selection gradient and the selection differential

The selection differential is an intuitively straightforward measure of the strength of selection: It is the difference between the mean of a trait among the survivors and the mean of the trait among the entire population. The selection gradient, while more abstract, has several advantages. Among them is that the selection gradient can be calculated for a wider variety of fitness measures. The fact that the selection gradient is closely related to the selection differential lends the former some of the intuitive appeal of the latter.

Here we show that in our example on tail length in mice (Figure 9.15), the selection gradient for tail length t is equal to the selection differential for tail length divided by the variance of tail length. Imagine that we have 30 mice in our population. First, note that the selection differential is

$$S = t^{\star} - \bar{t}$$

where $t^{\star}$ is the mean tail length of the 10 mice we kept as breeders, and $\bar{t}$ is the mean tail length of the entire population of 30 mice.

The selection gradient is the slope of the best-fit line for relative fitness w as a function of tail length. The slope of the best-fit line in linear regression is given by the covariance of y and x divided by the variance of x:

$$\text{slope} = \frac{\text{cov}(y, x)}{\text{var}(x)}$$

The covariance of y and x is defined as

$$\text{cov}(y, x) = \frac{1}{n} \sum_{i=1}^{n} (y_i - \bar{y})(x_i - \bar{x})$$

and the variance of x is defined as

$$\text{var}(x) = \frac{1}{n} \sum_{i=1}^{n} (x_i - \bar{x})^2$$

where n is the number observations, $\bar{y}$ is the mean value of y, and $\bar{x}$ is the mean value of x. The selection gradient for t is therefore:

$$\text{selection gradient} = \frac{\text{cov}(w, t)}{\text{var}(t)}$$

Thus, what we need to show is that $\text{cov}(w, t) = t^{\star} - \bar{t}$

Because (by definition) the mean relative fitness is 1, we can write

$$\begin{aligned}
\text{cov}(w, t) &= \frac{1}{30} \sum_{i=1}^{30} (w_i - 1)(t_i - \bar{t}) \\
&= \frac{1}{30} \sum_{i=1}^{30} (w_i t_i) - \frac{1}{30} \sum_{i=1}^{30} (w_i \bar{t}) \\
&\quad - \frac{1}{30} \sum_{i=1}^{30} (t_i) + \frac{1}{30} \sum_{i=1}^{30} (\bar{t}) \\
&= \frac{1}{30} \sum_{i=1}^{30} (w_i t_i) - \bar{t} - \bar{t} + \bar{t} \\
&= \frac{1}{30} \sum_{i=1}^{30} (w_i t_i) - \bar{t} \\
&= t^{\star} - \bar{t}
\end{aligned}$$

The last step may not be transparent. To see that

$$\frac{1}{30} \sum_{i=1}^{30} (w_i t_i) = t^{\star}$$

note that for the first 20 mice $w_i = 0$, and for the last 10 mice $w_i = 3$. This means that

$$\begin{aligned}
\frac{1}{30} \sum_{i=1}^{30} (w_i t_i) &= \frac{1}{30} \sum_{i=21}^{30} (3 t_i) \\
&= \frac{3}{30} \sum_{i=21}^{30} (t_i) \\
&= \frac{1}{10} \sum_{i=21}^{30} (t_i) = t^{\star}
\end{aligned}$$

In their mice, Di Masso et al. selected for long tails in 18 successive generations. The mice in the 18th generation had tails more than 10% longer than mice in a control population. The long-tailed mice had 28 vertebrae in their tails, compared to 26 or 27 vertebrae for the controls. The developmental program had been altered to make more vertebrae, not to elongate the individual vertebrae.

9.5 Predicting the Evolutionary Response to Selection

Once we know the heritability and the selection differential, we can predict the evolutionary response to selection. Here is the equation for doing so:

$$R = h^2S$$

where R is the predicted response to selection, h^2 is the heritability, and S is the selection differential.

The logic of this equation is shown graphically in Figure 9.18. This figure shows a scatterplot of midoffspring values as a function of midparent values, just like the scatterplots in Figure 9.13. The scatterplot in Figure 9.18 represents tail lengths in a population of 30 families of mice. The plot includes a best-fit line, whose slope estimates the heritability h^2.

Look first at the x-axis. $\overline{P}$ is the average midparent value for the entire population. $P^\star$ is the average of the 10 largest midparent values. The difference between $P^\star$ and $\overline{P}$ is the selection differential (S) that we would have applied to this population had we picked as our breeders only the 10 pairs of parents with the largest midparent values.

Now look at the y-axis. $\overline{O}$ is the average midoffspring value for the entire population. $O^\star$ is the average midoffspring value for the 10 pairs of parents with the largest midparent values. The difference between $O^\star$ and $\overline{O}$ is the evolutionary response (R) we would have gotten as a result of selecting as breeders the 10 pairs of parents with the largest midparent values.

The slope of a line can be calculated as the rise over the run. If we compare the population averages with selection versus without selection, we have a rise of $(O^\star - \overline{O})$ over a run of $(P^\star - \overline{P})$ so

$$h^2 = \frac{(O^\star - \overline{O})}{(P^\star - \overline{P})} = \frac{R}{S}$$

In other words, $R = h^2S$.

Once we know h^2 and S, we can use them to predict the response to selection as $R = h^2S$.

We now have a set of tools for studying the evolution of multilocus traits under natural selection. We can estimate how much of the variation in a trait is due to variation in genes, quantify the strength of selection that results from differences in survival or reproduction, and put these two together to predict how much the population will change from one generation to the next.

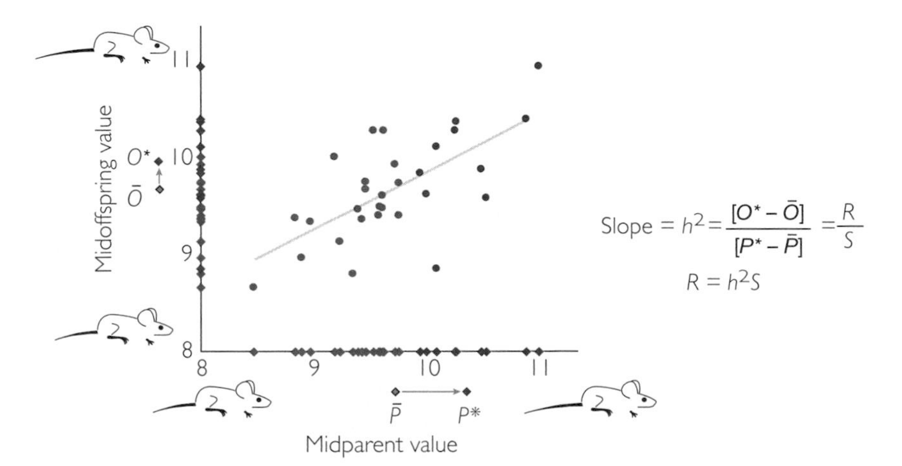

Figure 9.18 The response to selection is equal to the heritability multiplied by the selection differential The midoffspring and midparent values are indicated both as dots on the scatterplot and as diamonds on the y- and x-axes. The red symbols represent the 10 families with the largest midparent values. $\overline{P}$ is the average midparent value for the entire population; P^* is the average midparent value of the families with the largest midparent values. $\overline{O}$ is the average midoffspring value for the entire population; O^* is the average midoffspring value for the families with the largest midparent values. After Falconer (1989).

(a)

(b)

Figure 9.19 An alpine skypilot and a bumblebee (a) Alpine skypilot (*Polemonium viscosum*). (b) Bumblebee (*Bombus* sp.).

Alpine Skypilots and Bumblebees

As an example of the questions biologists answer with quantitative genetics we review Candace Galen's (1996) research on flower size in the alpine skypilot (*Polemonium viscosum*), a perennial Rocky Mountain wildflower (Figure 9.19a). Galen studied populations on Pennsylvania Mountain, Colorado, including populations growing at the timberline and populations in the higher-elevation tundra. At the timberline, skypilots are pollinated by a diversity of insects, including flies, small solitary bees, and some bumblebees. In the tundra, they are pollinated almost exclusively by bumblebees (Figure 9.19b). The flowers of tundra skypilots are, on average, 12% larger in diameter than those of timberline skypilots. Previously, Galen (1989) had documented that larger flowers attract more visits from bumblebees and that skypilots that attract more bumblebees produce more seeds.

Galen wanted to know whether the selection on flower size imposed by bumblebees is responsible for the larger flowers of tundra skypilots. If it is, she also wanted to know how long it would take for selection by bumblebees to increase the average flower size in a skypilot population by 12%—the difference between tundra and timberline flowers.

Galen worked with a small-flowered timberline population. First, she estimated the heritability of flower size. She measured the flower diameters of 144 skypilots and collected their seeds. She germinated the seeds and planted the 617 resulting seedlings at random in the same habitat their parents had lived in. Seven years later, 58 had matured and Galen could measure their flowers. This let her plot offspring flower diameter (corolla flare) as a function of maternal, or seed-parent, flower diameter (Figure 9.20). The slope of the best-fit line is approximately 0.5. For reasons beyond the scope of this discussion, the slope of the best-fit line for offspring versus a single parent (as opposed to the midparent) is an estimate of $\frac{1}{2}h^2$ (see Falconer 1989). Thus, the heritability of flower size in the timberline skypilots is roughly $2 \times 0.5 = 1$. Note, however, that the scatter in Figure 9.20 makes the true slope uncertain. Galen's statistical analysis indicated that she could safely conclude only that the heritability of flower size is between 0.2 and 1. That is, at least 20% of the phenotypic variation in skypilot flower size is due to additive genetic variation.

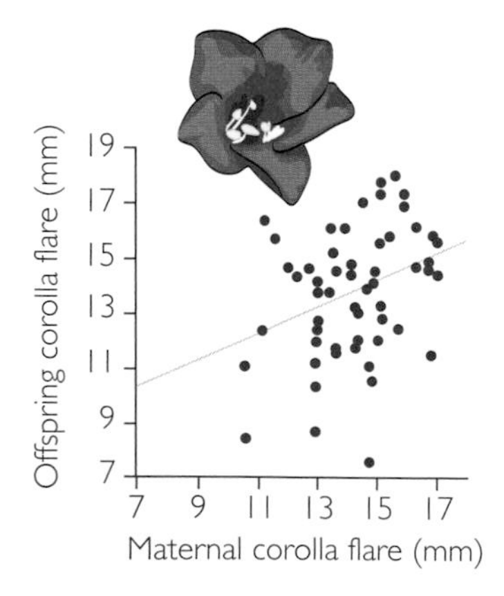

Figure 9.20 Estimating the heritability of flower size (corolla flare) in alpine skypilots This scatterplot shows offspring corolla flare as a function of maternal plant corolla flare for 58 skypilots. The slope of the best-fit line is 0.5. Redrawn from Galen (1996).

Next Galen estimated the strength of the selection imposed by bumblebee pollinators. Recall that bumblebees prefer to visit larger flowers and that more bumblebee pollinators means more seeds. Galen built a screen-enclosed cage at her study site, moved 98 soon-to-flower skypilots into it, and added bumblebees. The cage kept the bumblebees in and all other pollinators out. When the caged skypilots flowered, Galen measured their flowers. Later she collected their seeds, germinated them in the lab, and planted the seedlings at random back out in the parental habitat. Six years later, Galen counted the surviving offspring that had been produced by each of the original caged plants. Using the number of surviving 6-year-old offspring as her measure of fitness, Galen plotted relative fitness as a function of flower size and calculated the slope of the best-fit line (Figure 9.21). This slope, 0.13, is the selection gradient resulting from bumblebee pollination. Multiplying the selection gradient by the variance in flower size, 5.66, gives the selection differential: $S = 0.74$ mm. The average flower size was 14.2 mm. Thus the selection differential can also be expressed as $\frac{0.74}{14.2} = 0.05$, or 5%. Roughly, this means that when skypilots compete to reproduce by enticing bumblebees to visit them, the plants that win have flowers 5% larger than those of the average plant.

Galen performed two control experiments to confirm that bumblebees select in favor of larger flowers. In one control, she pollinated skypilots by hand (without regard to flower size); in the other, she allowed skypilots to be pollinated by all other natural pollinators except bumblebees. In neither control was there any relationship between flower size and fitness; only bumblebees select for larger flowers.

Galen's data allowed her to predict how the population of timberline skypilots should respond to selection by bumblebees. The scenario she imagined was that a population of timberline skypilots that had been pollinated by a variety of insects moved (by seed dispersal) to the tundra, where the plants are now pollinated exclusively by bumblebees. Using the low-end estimate that $h^2 = 0.2$, and the estimate that $S = 0.05$, Galen predicted that the response to selection would be $R = h^2S = 0.2 \times 0.05 = 0.01$. Using the high-end estimate that $h^2 = 1$, and the estimate that $S = 0.05$, Galen predicted that the response to selection would be $R = h^2S = 1 \times 0.05 = 0.05$. In other words, a single generation of selection by bumblebees should produce an increase of 1–5% in the average flower size of a population of timberline skypilots moved to the tundra.

Galen's prediction was, therefore, that flower size would evolve rapidly under selection by bumblebees. Is this prediction correct? Recall the experiment described earlier, in which Galen reared offspring of timberline skypilots that had been pollinated by hand and offspring of timberline skypilots that been pollinated exclusively by bumblebees. Galen calculated the mean flower size of each group and found that the offspring of bumblebee-pollinated skypilots had flowers that were, on average, 9% larger than those of hand-pollinated skypilots (Figure 9.22). Her prediction was correct: Skypilots show a strong and rapid response to selection. In fact, the response is even larger than Galen predicted.

Galen concluded that the 12% difference in flower size between timberline and tundra skypilots can be plausibly explained by the fact that timberline skypilots are pollinated by a diversity of insects, whereas tundra skypilots are pollinated almost exclusively by bumblebees. Timberline skypilots can set seed even if bumblebees avoid them, but tundra skypilots cannot. Furthermore, it would take only a few generations of bumblebee-only pollination for a population of timberline skypilots to evolve flowers that are as large as those of tundra skypilots.

In Sections 9.3, 9.4, and 9.5, we have used the tools of quantitative genetics to analyze the evolution of just one trait at a time. The tools we have developed can be generalized, however, to analyze the simultaneous evolution of multiple traits. Box 9.4 provides an introduction.

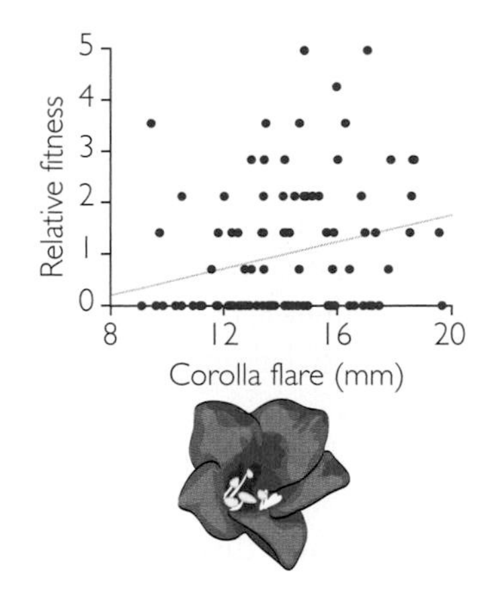

Figure 9.21 Estimating the selection gradient in alpine skypilots pollinated by bumblebees This scatterplot shows relative fitness (number of surviving 6-year-old offspring divided by average number of surviving 6-year-old offspring) as a function of maternal flower size (corolla flare). The slope of the best-fit line is 0.13. Prepared with data provided by Candace Galen.

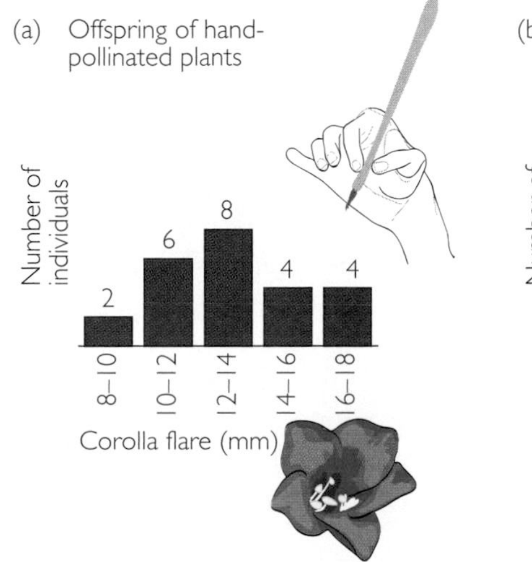

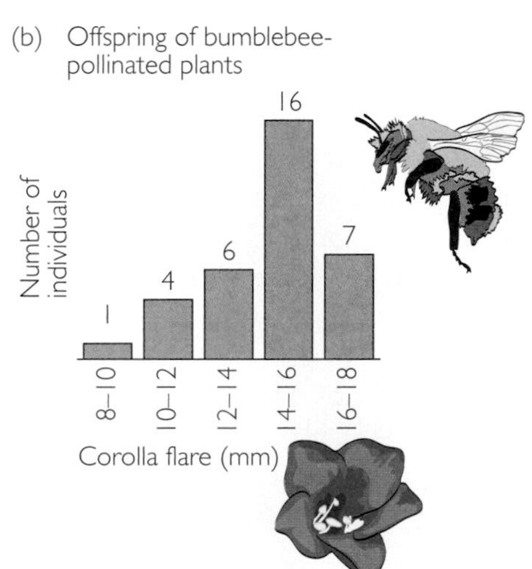

Figure 9.22 Measuring the evolutionary response to selection in alpine skypilots These histograms show the distribution of flower size (corolla flare) in the offspring of hand-pollinated skypilots (a; average = 13.1 mm) and bumblebee-pollinated skypilots (b; average = 14.4 mm). Redrawn from Galen (1996).

Box 9.4 Selection on multiple traits and correlated characters

In the main text we analyze selection on just one quantitative trait at a time. Selection in nature, however, often acts on several traits at once. Here we provide a brief introduction to how the techniques of quantitative genetics can be extended to analyze selection on multiple traits. For mathematical details, see Lande and Arnold (1983), Phillips and Arnold (1989), and Brodie, Moore, and Janzen (1995).

In Chapter 3 we discussed natural selection on beak size in Darwin's finches. During the drought of 1976–1977 on Daphne Major Island, medium ground finches with deeper beaks survived at higher rates. Beak depth is heritable, so the population evolved. Peter Grant and Rosemary Grant have reanalyzed the data from this selection episode, looking at selection on several traits simultaneously. We will discuss their analysis of two traits: beak depth and beak width. We will present a qualitative overview only; for the numbers see Grant and Grant (1995).

The medium ground finches of Daphne Major vary both in beak depth and beak width. These two traits are strongly correlated. Deep beaks tend to be wide, and shallow beaks tend to be narrow. The reasons why this might be true are beyond the scope of this discussion. We will just take it as given that it is difficult or impossible to build a finch with a beak that is deep and narrow or a beak that is shallow and wide.

During the drought of 1976–1977, when food was scarce and many finches starved, selection acted on both beak depth and beak width. If we were looking at just one of these characteristics, we could measure the strength of selection as the slope of the best-fit line relating fitness to beak size. This is the selection gradient introduced in the main text. To look at both characteristics at once, we can measure the strength of selection as the slope of the best-fit *plane* relating fitness to both beak depth *and* beak width. This slope is the two-dimensional selection gradient.

Look at the three-dimensional graph in Figure 9.23a. Beak depth is represented on one horizontal axis; beak width is represented on the other horizontal axis. Fitness is represented on the vertical axis. The surface given by the blue grid is the best-fit plane. The fitness at each corner of the best-fit plane, or selection surface, is marked with a blue triangle. Selection favored birds with beaks that were both deep *and* narrow. The bird with the highest chance of surviving the drought would be a bird with a very deep, very narrow beak—located at the right rear corner of the selection surface.

Recall that beak depth and width are correlated, however. It is impossible to build the perfect bird. The correlation between depth and width is represented by the double-headed black arrow on the floor of the three-dimensional graph. During the drought, selection pushed most strongly in the direction indicated by the wide dark blue arrow. This would have taken the population average from the center of the graph directly up the steepest path along the selection surface toward the best possible beak shape. But because of the correlation between beak depth and beak width, the population could not move in the direction selection pushed; it could only move along the double-headed black arrow. Selection favored deep beaks more strongly than it favored narrow beaks. As a result, the population average moved toward a beak that was deeper and *wider* than it was before the drought. This change is represented by the wide green arrow.

Three-dimensional graphs can be difficult to interpret, so we have included Figures 9.23b, (c), and (d), which illustrate the same analysis with two-dimensional graphs. Figure 9.23b shows the selection gradient on beak depth, with beak width held constant. Selection favored birds with deeper beaks. Figure 9.23c shows the selection gradient on beak width, with beak depth held constant. Selection favored narrower beaks. Figure 9.23d shows the correlation between beak width and beak depth (double-headed black arrow), with fitness represented by the intensity of blue color across the graph. Selection pushed the population average toward a bird with a deep, narrow beak (wide dark blue arrow), but because of the correlation between depth and width, the population could not go there. Selection favored increased depth more strongly than decreased width, so the population average moved toward a deeper, wider beak (wide green arrow).

Grant and Grant's analysis of selection on finch beaks illustrates the advantages of looking at several

Box 9.4 (Continued)

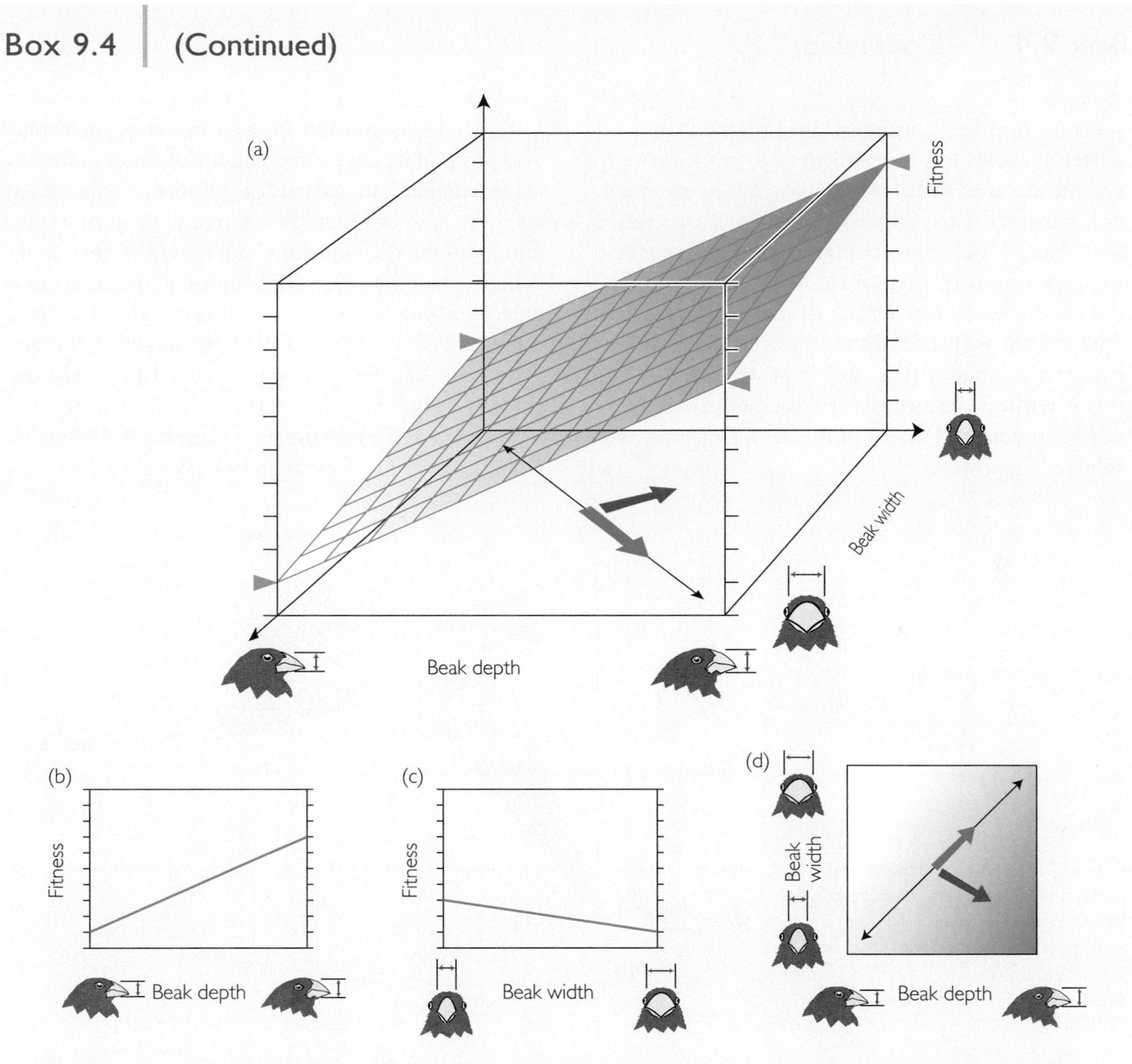

Figure 9.23 A multidimensional analysis of selection on beak size in medium ground finches (a) The grid plane shows the relationship between fitness and both beak depth and beak width. Birds with deep and narrow beaks had highest fitness. (b), (c), and (d) show the same scenario in two-dimensional graphs.

traits at once, of using selection gradients to measure the strength of selection, and of recognizing that traits may be correlated with each other. Imagine that we were to look only at beak width, and calculate the selection differential. The average survivor had a wider beak than did the average bird alive before the drought. The selection differential, the difference between the population mean before and after selection, would suggest that selection favored wider beaks. But the multidimensional analysis reveals that this was not the case. Beak width was selected against, but was dragged along for the ride anyway as a result of stronger selection on beak depth.

Grant and Grant assumed for their analysis that the relationship between beak depth, beak width, and fitness was linear, as shown by the planar selection surface in Figure 9.23a. However, the relationship between a pair of traits and fitness is not always linear. Work by Edmund Brodie (1992) provides an example. Brodie monitored the survival of

Box 9.4 (Continued)

several hundred individually marked juvenile garter snakes. He estimated the effect on fitness of two traits that help the snakes evade predators: color pattern (striped versus unstriped or spotted) and escape behavior (straight-line escape versus many reversals of direction). Brodie's analysis produced the selection surface shown in Figure 9.24. The snakes with the highest rates of survival were those with stripes that fled in a straight line, and those without stripes that performed many reversals of direction. Other combinations of traits were selected against.

Given a selection surface, we can follow the evolution of a population by tracking the position of the average individual. In general, a population is expected to evolve so as to move up the steepest slope from its present location. As Grant and Grant's study of finch beaks demonstrated, however, correlations among traits may prevent a population from following this predicted route. Selection surfaces like those shown in Figures 9.23a and 9.23 are sometimes referred to as adaptive landscapes, but this term has a complex history and several different meanings (see Chapter 9 in Provine 1986; Chapter 11 of Wright 1986; Wright 1988).

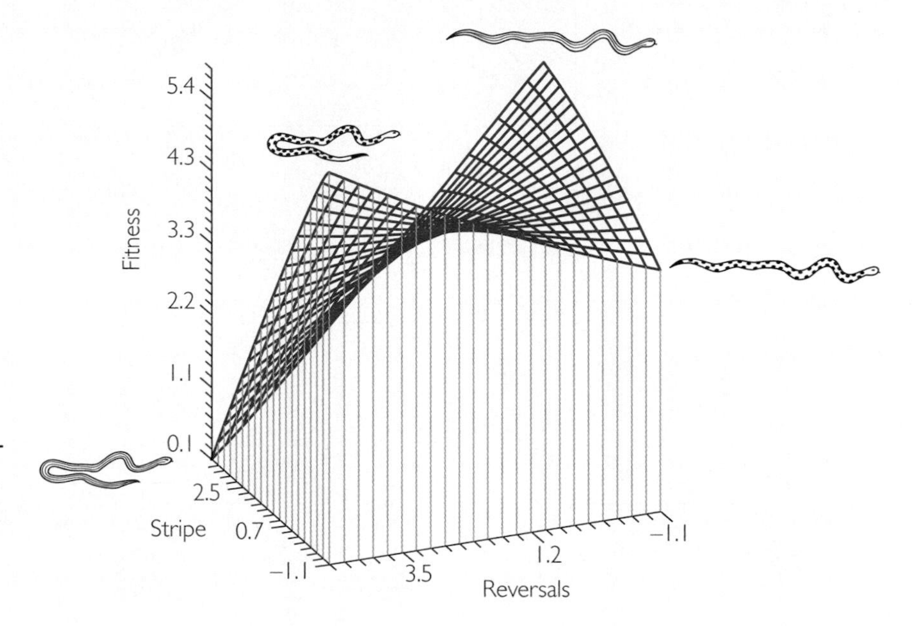

Figure 9.24 A multidimensional analysis of selection on antipredator defenses in garter snakes The grid surface shows the relationship between fitness and both color pattern and evasive behavior. From Fig. 1 in Brodie (1992).

9.6 Modes of Selection and the Maintenance of Genetic Variation

Selection on a population may take any of a variety of forms. Directional selection and stabilizing selection tend to reduce the amount of variation in a population; disruptive selection tends to increase the amount of variation.

In our discussions of selection on quantitative traits, we have assumed that the relationship between phenotype and fitness is simple. In our mice, long tails were better than short tails; in skypilots, bigger flowers were better than smaller flowers. Before leaving the topic of selection on quantitative traits, we note that the relationship between phenotype and fitness may be complex. A variety of patterns, or modes of selection, is possible.

Figure 9.25 shows three distinct modes of selection acting on a hypothetical population. Each column represents a different mode. The histogram in the top row shows the distribution of values for a phenotypic trait before selection. The graph in the middle row shows the relationship between phenotype and fitness, plotted as the probability of surviving as a function of phenotype. The histogram in the bottom row shows the distribution of phenotypes among the survivors. The triangle and bar below each histogram show the mean and variation in the population. (The bar representing variation encompasses ±2 standard deviations around the mean, or approximately 95% of the individuals in the population.)

In **directional selection**, fitness consistently increases (or decreases) with the value of a trait (Figure 9.25, first column). Directional selection on a continuous trait changes the average value of the trait in the population. In the hypothetical population shown in Figure 9.25, the mean phenotype before selection was 6.9, whereas the mean phenotype after selection was 7.4. Directional selection also reduces the variation in a population, although often not dramatically. In our hypothetical population, the standard deviation before selection was 1.92, whereas the standard deviation after selection was 1.89.

In **stabilizing selection**, individuals with intermediate values of a trait have highest fitness (Figure 9.25, middle column). Stabilizing selection on a continuous trait does not alter the average value of the trait in the population. Stabilizing selection does, however, trim off the tails of the trait's distribution, thereby reducing the variation. In our hypothetical population, the standard deviation before selection was 1.92, whereas the standard deviation after selection was 1.04.

In **disruptive selection**, individuals with extreme values of a trait have the highest fitness (Figure 9.25, last column). Disruptive selection on a continuous trait does not alter the average value of the trait in the population. Disruptive

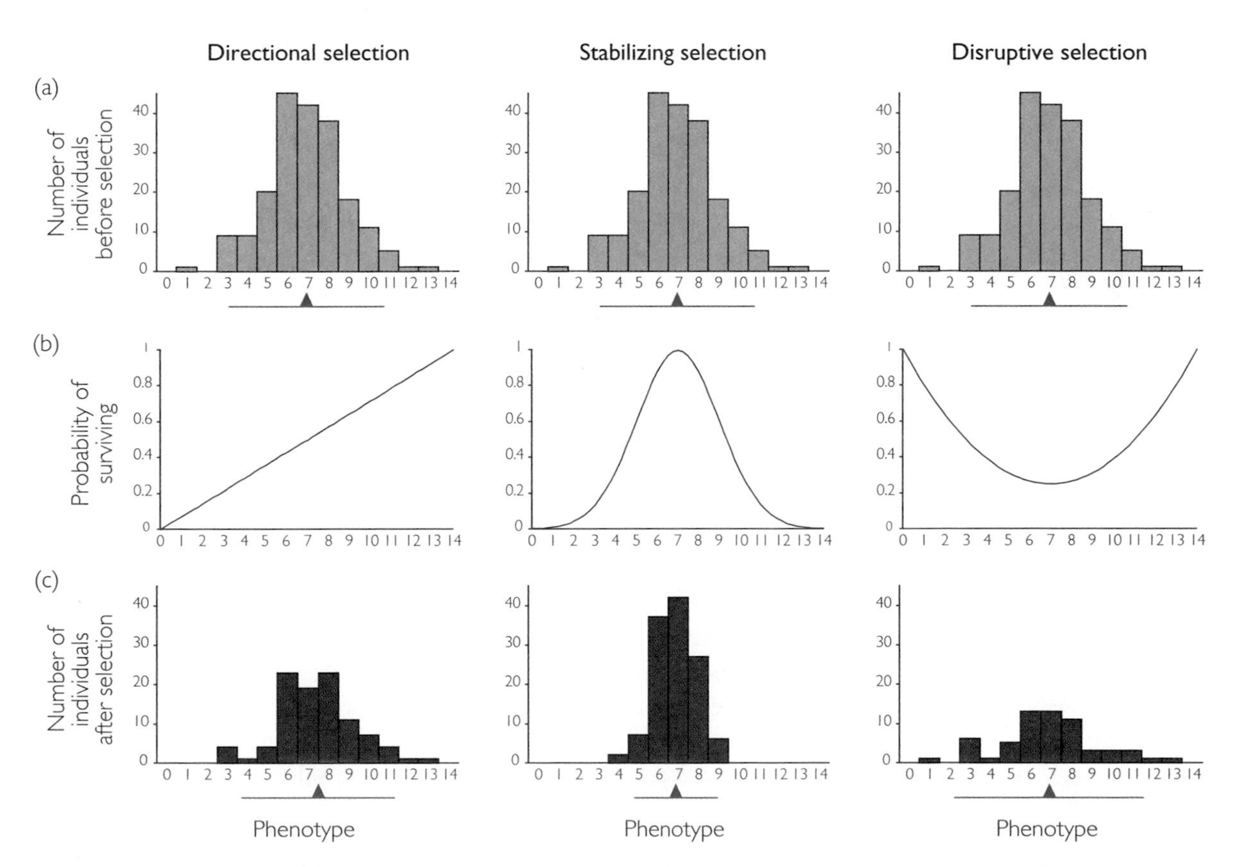

Figure 9.25 Three modes of selection Each column represents a mode of selection. The graphs in row (a) are histograms showing the distribution of a phenotypic trait in a hypothetical population before selection. The graphs in row (b) show different patterns of selection; they plot the probability of survival (a measure of fitness) as a function of phenotype. The graphs in row (c) are histograms showing the distribution of the phenotypic trait in the survivors. The blue triangle under each histogram shows the mean of the population. The blue bar under each histogram shows the variation ±2 standard deviations from the mean). After Cavalli-Sforza and Bodmer (1971).

selection does, however, trim off the top of the trait's distribution, thereby increasing the variance. In our hypothetical population, the standard deviation before selection was 1.92, whereas the standard deviation after selection was 2.33.

All three modes of selection eliminate individuals with low fitness and preserve individuals with high fitness. As a result, all three modes of selection increase the mean fitness of the population.

We have already seen examples of directional selection. In alpine skypilots pollinated by bumblebees, for instance, larger flowers have higher fitness. And in medium ground finches, the drought of 1976–1977 on Daphne Major selected for birds with larger beaks (see Chapter 3).

Research by Arthur Weis and Warren Abrahamson (1986) provides an elegant example of stabilizing selection. Weis and Abrahamson studied a fly called *Eurosta solidaginis*. The female in this species injects an egg into a bud of the tall goldenrod, *Solidago altissima*. After hatching, the fly larva digs into the stem and induces the plant to form a protective gall. As it develops inside its gall, the larva may fall victim to two kinds of predators. First, a female parasitoid wasp may inject *her* egg into the gall, where the wasp larva will eat the fly larva. Second, a bird may spot the gall and break it open, again to eat the larval fly. Weis and Abrahamson established that genetic variation among the flies is partly responsible for the variation in the size of the galls they induce. The researchers also collected several hundred galls and determined, by dissecting them, the fate of the larva inside each one.

Weis and Abrahamson discovered that parasitoid wasps impose on the gall-making flies strong directional selection favoring larger galls (Figure 9.26a). Nearly all larvae in galls under about 16 mm in diameter were killed by wasps, whereas larvae in larger galls had at least a fighting chance to survive. However, the researchers also found that birds impose on the gall makers strong directional selection favoring *smaller* galls (Figure 9.26b). Together, selection by wasps and selection by birds add up to stabilizing selection on gall size. Figure 9.26c shows the distribution of sizes among the galls before and after selection.

Research by Thomas Bates Smith (1993) provides an example of disruptive selection. Bates Smith studied an African finch called the black-bellied seedcracker. Birds in this species exhibit two distinct beak sizes: large and small. The birds in the two groups specialize on different kinds of seeds. Bates Smith followed the fate of over 200 juvenile birds. The graphs in Figure 9.27 show the distribution of beak sizes among all juveniles and among juveniles that survived to adulthood. The graphs reveal disruptive selection: The survivors were the birds with bills that were either relatively large or relatively small. Birds with beaks of intermediate size did not survive. (Note that an element of stabilizing selection appears to be at work here too: Except in the case of birds with extremely long bills, the birds with the most extreme phenotypes did not survive.)

Evolutionary biologists generally assume that directional selection and stabilizing selection are common, whereas disruptive selection is rare. If the prepon-

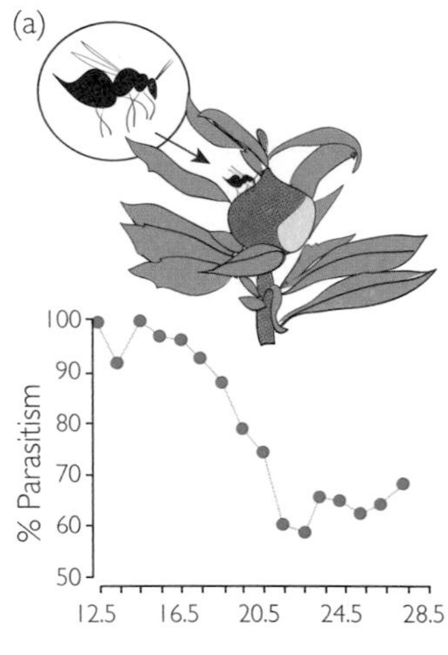

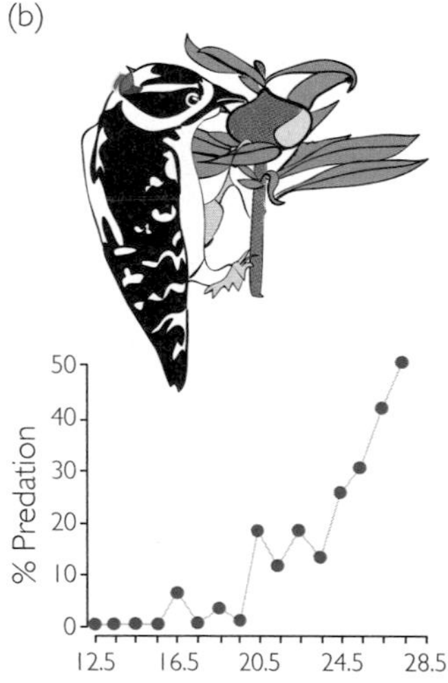

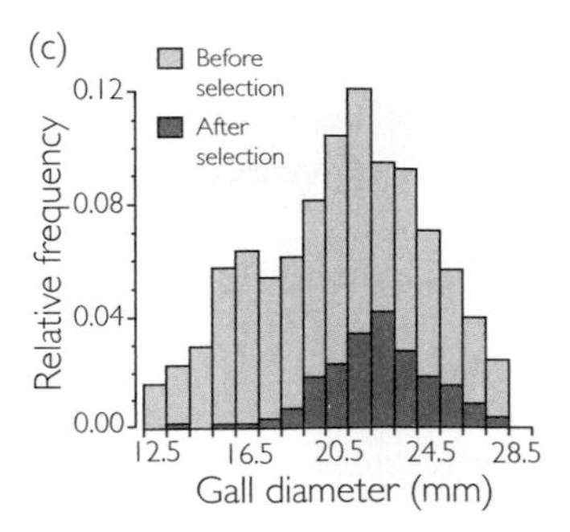

Figure 9.26 Stabilizing selection on a gall-making fly (a) Parasitoid wasps kill fly larvae inside small galls at higher rates than they kill larvae inside large galls. (b) Birds kill fly larvae inside large galls at higher rates than they kill larvae inside small galls. (c) The distribution of gall sizes before (tan + red portion of bars) and after (red portion of bars) selection by parasitoids and birds. Overall, fly larvae inside medium-sized galls survived at the highest rates. From Weis and Abrahamson (1986).

derance of directional and stabilizing selection is real, however, it creates a puzzle. Recall from Figure 9.25 that both directional and stabilizing selection reduce the phenotypic variation present in a population. If the trait in question is heritable, then these modes of selection will reduce the genetic variation in the population as well. Eventually, the genetic variation in any trait related to fitness should be eliminated altogether, and the population should reach an equilibrium at which the mean value of the trait, the variation in the trait, and the mean fitness of the population will all cease to change. The puzzle is that populations typically exhibit substantial genetic variation, even in traits closely related to fitness. How is this genetic variation maintained?

Here are three possible solutions to the puzzle of how genetic variation for fitness is maintained:

1. Most populations are not in evolutionary equilibrium with respect to directional and/or stabilizing selection. In any population there is a steady, if slow, supply of new favorable mutations creating genetic variation for fitness-related traits. While favorable mutations are rising in frequency, but have not yet become fixed, the population will exhibit genetic variation for fitness. We will call this the "Fisher's Fundamental Theorem Hypothesis." It was Ronald Fisher who first showed mathematically that the rate at which the mean fitness of a population increases is proportional to the additive genetic variation for fitness, a result he called the Fundamental Theorem of Natural Selection.
2. In most populations there is a balance between deleterious mutations and selection. In any population, there is a steady supply of new deleterious mutations. In Chapter 6 we showed that unless the mutation rate is high or selection is weak, selection will keep any given deleterious allele at low frequency. But quantitative traits are determined by the combined influence of many loci of small effect. Selection on the alleles at any single locus affecting a quantitative trait may be very weak, allowing substantial genetic variation to persist at the equilibrium between mutation and selection.
3. Disruptive selection, or patterns of selection with similar effects, may be more common than is generally recognized. Other patterns of selection that can maintain genetic variation in populations include frequency-dependent selection in which rare phenotypes (and genotypes) have higher fitness than common phenotypes and selection imposed by a fluctuating environment.

All of these hypotheses are controversial, and have been the subject of considerable theoretical and empirical research (see, for example, Barton and Turelli 1989). A detailed discussion is beyond the scope of this text. We can, however, provide a brief review of an intriguing experiment by Santiago Elena and Richard Lenski.

Elena and Lenski (1997) studied six populations of the bacterium *Escherichia coli*. These populations were established from a common ancestral culture, so they were closely related to each other. Each population was founded by a single bacterium, so within any given culture all genetic variation had arisen as a result of new mutations. The six populations had been evolving in a constant lab environment for 10,000 generations. During this time, the mean fitness of each population, assessed via competition experiments, had increased by over 50% relative to

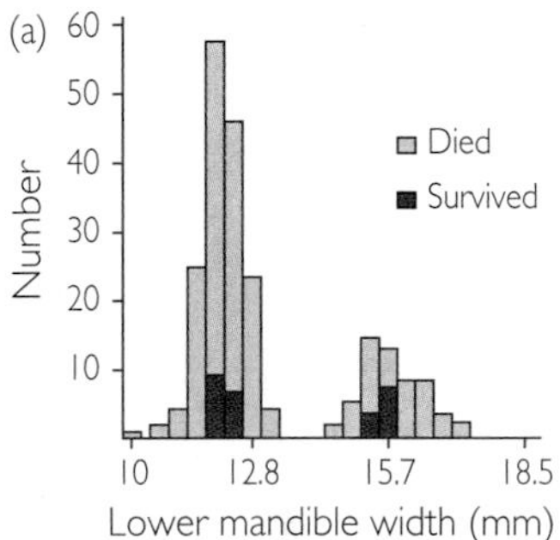

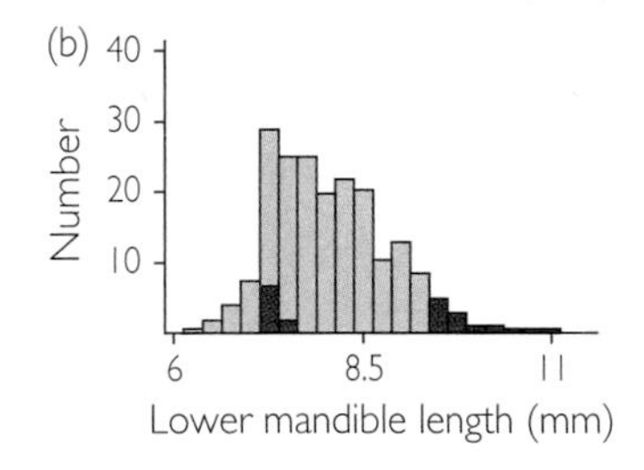

Figure 9.27 Disruptive selection on bill size in the black-bellied seedcracker (*Pyrenestes o. ostrinus*) Each graph shows the distribution of lower bill widths (a) or lengths (b) in a population of black-bellied seedcrackers, an African finch. The light-colored portion of each bar represents juveniles that did not survive to adulthood; the dark-colored portion represents juveniles that did survive. The survivors were those individuals with bills that were either relatively large or relatively small. Rerendered from Bates Smith (1993).

their common ancestor. However, much of the increase in fitness had occurred in the first few thousand generations. After 10,000 generations, the populations appeared to have arrived at an evolutionary equilibrium. Elena and Lenski assessed the genetic variation in fitness among the various strains present in each of their six populations and found it to be significant. On average, two strains selected from the same population differed in fitness by about 4%.

Elena and Lenski tested the Fisher's Fundamental Theorem hypothesis by using the standing genetic variation in fitness within each population to predict how much additional improvement in fitness should occur in a further 500 generations of evolution. Depending on the assumptions used, the researchers predicted an additional increase in fitness of between 4 and 50%. In fact, between generations 10,000 and 10,500, none of the six populations showed any significant increase in mean fitness. Elena and Lenski concluded that the genetic variation in fitness within their *E. coli* populations is not the result of a continuous supply of new favorable mutations in the process of rising to fixation.

Elena and Lenski tested the mutation-selection balance hypothesis by noting that two of their six populations had evolved extraordinarily high mutation rates—on the order of 100 times higher than the other four populations and the common ancestor. If the genetic variation in fitness within each population is maintained by mutation-selection balance, then the two populations with high mutation rates should show by far the highest standing genetic variation in fitness. One of the two high-mutation populations did exhibit much higher genetic variation for fitness than the four low-mutation populations. But the other high-mutation population did not. Elena and Lenski concluded that the genetic variation in fitness within their *E. coli* populations is probably not the result of a balance between deleterious mutations and selection.

Finally, Elena and Lenski tested the hypothesis that frequency-dependent selection is maintaining the genetic variation for fitness in their populations. The researchers used competition experiments to determine whether the various strains present in each population enjoyed a fitness advantage when rare. They found that a typical *E. coli* strain, when rare, did indeed have a fitness edge of about 2% relative to its source population. Furthermore, across the six bacterial populations, the intensity of frequency-dependent selection was significantly correlated with the amount of standing genetic variation for fitness. Elena and Lenksi noted that the three hypotheses they tested are mutually compatible. Nonetheless, the researchers concluded that the best explanation for the variation in fitness in their populations is frequency-dependent selection. Whether this conclusion applies to other populations and other organisms remains to be seen.

9.7 The Bell-Curve Fallacy and Other Misinterpretations of Heritability

We promised, in the introduction to this chapter, that our discussion of quantitative genetics would allow us to debunk erroneous claims about differences in IQ scores among ethnic groups. We are now ready to make good on that promise.

A key point is that the formula for heritability includes both genetic variation, V_G, and environmental variation, V_E. Any estimate of heritability is, therefore, specific to a particular population living in a particular environment. As a result, heritability tells us nothing about the causes of differences between populations that live in different environments.

We can illustrate this point with data from an experiment by Jens Clausen, David Keck, and William Hiesey (1948). Clausen, Keck, and Hiesey studied *Achillea*, a perennial wildflower. *Achillea* will grow from cuttings, making it possible to create duplicates—clones—of a single individual. The researchers collected seven plants from a wild population and took two cuttings from each.

Clausen, Keck, and Hiesey grew one cutting from each plant in an experimental garden at Mather, California, (Figure 9.28, top row). As the Mather cuttings grew up side by side, they lived in the same soil, got the same amount of water, the same amount of sunlight, and so on. Because the plants experienced the virtually same environment, differences among them in height at maturation are almost entirely due to genetic variation. The heritability of size in the Mather population is approximately 1.

Clausen, Keck, and Hiesey grew the second cutting from each plant in an experimental garden at Stanford, California (Figure 9.28, bottom row). As the Stanford cuttings grew up side by side, they lived in the same soil, got the same amount of water, the same amount of sunlight, and so on. Because the plants experienced the virtually same environment, differences among them in height at

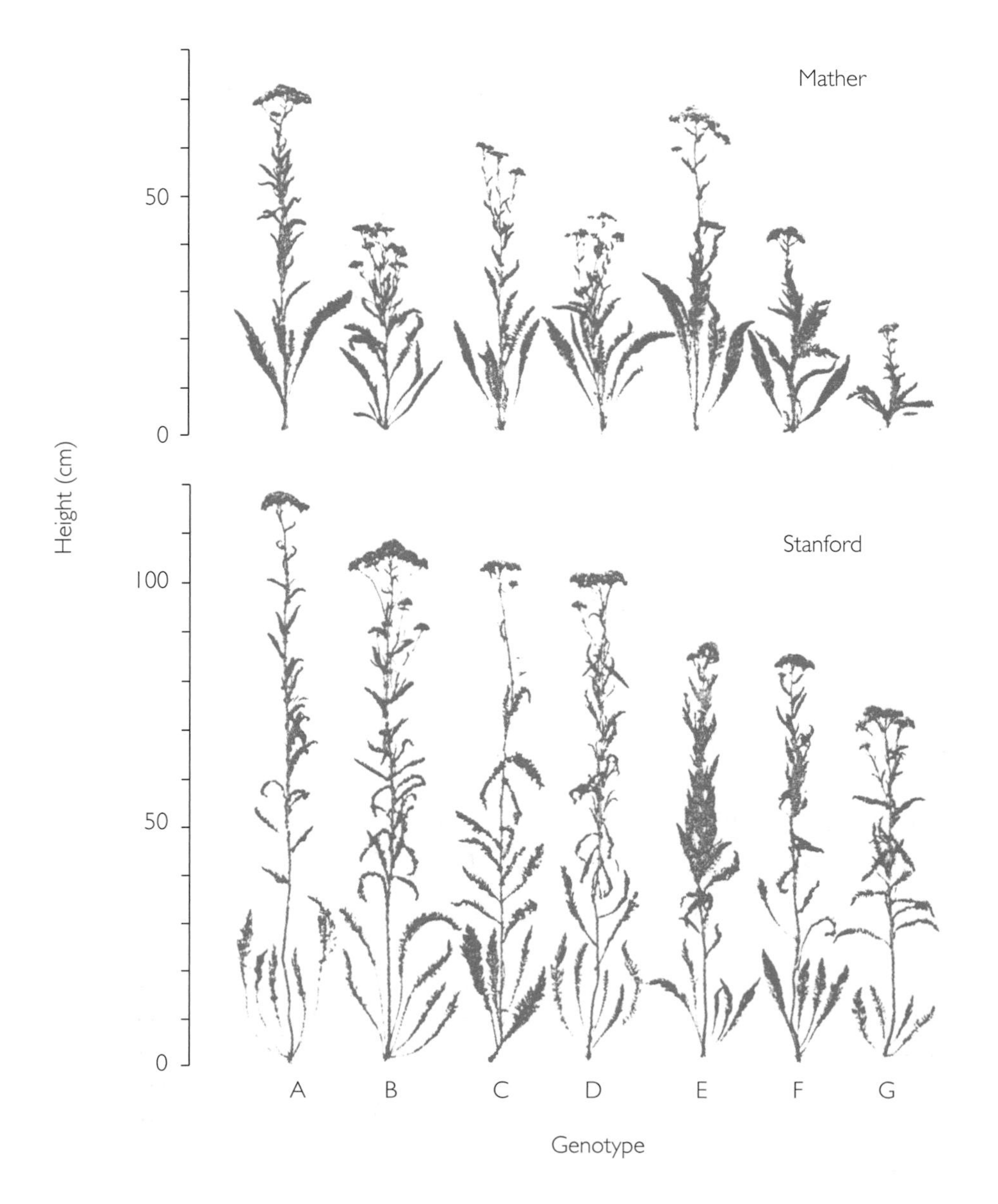

Figure 9.28 High heritability within populations tells us nothing about the cause of differences between populations We know the variation in height among the plants within each of these populations is entirely due to differences in their genes, because the plants grew in experimental common gardens where all experienced the same environment. The plants in the Stanford population are taller, on average, than the plants in the Mather population. Does this mean that the Stanford population is genetically superior to the Mather population? No: We know these two populations are genetically identical because they were grown from cuttings of the same seven plants. Reprinted from Clausen, Keck, and Hiesey (1948).

maturation are almost entirely due to genetic variation. The heritability of size in the Stanford population is approximately 1.

Studies of heritability are often misinterpreted as implying that differences between populations are due to differences in genes.

Notice the plants in the Stanford population are, on average, taller than the plants in the Mather population. We have high heritability in both populations, and a difference in mean height between populations. Does this mean that the Stanford population is genetically superior to the Mather population with respect to height? Of course not; Clausen, Keck, and Hiesey set up the populations to be identical in genetic composition. *The fact that heritability is high in each population tells us nothing about the cause of differences between the populations, because the populations were reared in different environments.*

The mistaken notion that heritability can tell us something about the causes of differences between populations has been particularly persistent in studies of human intelligence. In 1994, Charles Murray and Richard J. Herrnstein sold thousands of copies of their book, *The Bell Curve.* Murray and Herrnstein claimed that the difference in average IQ scores between African Americans and European Americans is due to genetic differences between these groups (our analysis of their argument is based on an extract of their book published in *The New Republic* [Murray and Herrnstein 1994]).

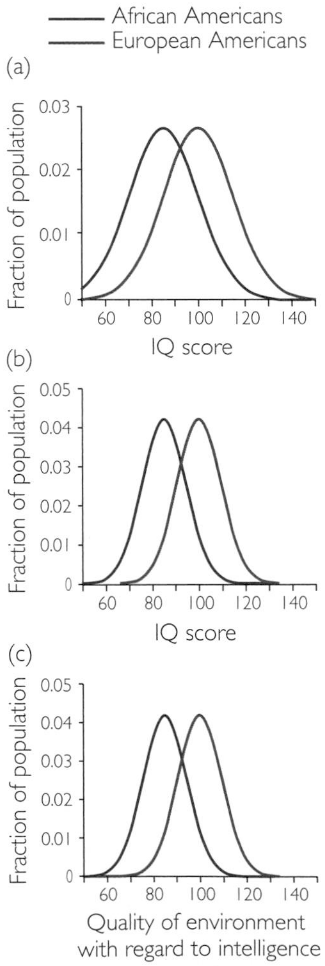

Murray and Herrnstein take note of the point we just made with *Achillea.* They state, "Most scholars accept that I.Q. in the human species as a whole is substantially heritable, somewhere between 40 percent and 80 percent, meaning that much of the observed variation in I.Q. is genetic. And yet this information tells us nothing for sure about the origin of the differences between groups." Having said that, however, Murray and Herrnstein proceed to develop the following erroneous argument.

Based on various sources, Murray and Herrnstein assume that: (1) the average IQ of African Americans is 85; (2) that the average IQ of European Americans is 100; and (3) that the variance (a statistical measure of the variation) in each group is 225. Bell curves representing these assumptions appear in Figure 9.29a. Murray and Herrnstein further assume that the heritability of IQ in each group is 0.6. There are reasons to dispute each of Murray and Herrnstein's assumptions, but we will grant them here—for the sake of argument only. (There are also reasons to dispute whether IQ tests measure anything meaningful at all, but we leave that argument to others.)

Next, Murray and Herrnstein imagine what the bell curves for IQ would look like if all the *genetic* variation among individuals within each population were removed. In other words, they imagine that all African Americans have been made genetically identical to the average African American, and all European Americans have been made genetically identical to the average European American. On the assumption that 60% of the variation within each group is due to genetic variation, this leaves 40% of the original variation within each group. Bell curves representing this thought experiment appear in Figure 9.29b.

Figure 9.29 Illustrations of Murray and Herrnstein's erroneous argument concerning IQ and ethnicity See text for explanation.

Now Murray and Herrnstein consider the proposition that the difference between the average IQ of the African Americans and the average IQ of the European Americans in Figure 9.29b is solely due to differences in environment. Under this proposition, Murray and Herrnstein say, we could replace the label "IQ score" with the label "Quality of environment with regard to intelligence," as shown in Figure 9.29c. Murray and Herrnstein find it implausible that the difference in the quality of the environment experienced by African

Americans versus European Americans is as great as that shown Figure 9.29c. They conclude that at least part of the difference between the mean IQ score of African Americans versus European Americans must be due to genetic differences between the groups.

There are at least two serious flaws in Murray and Herrnstein's argument. First, by replacing the label on the horizontal axis in Figure 9.29b with the label in Figure 9.29c, Murray and Herrnstein are implicitly assuming that there is a linear relationship between environment and IQ. This assumption is almost certainly wrong.

Second, Murray and Herrnstein's argument from their own incredulity amounts to rhetorical technique, not science. A scientific approach to Murray and Herrnstein's hypothesis would be to conduct a common garden experiment: Rear European Americans and African Americans together in an environment typically experienced by European Americans, and then compare their IQ scores. This design, and the reciprocal experiment, in which everyone is reared in an environment typically experienced by African Americans, is shown in Figure 9.30.

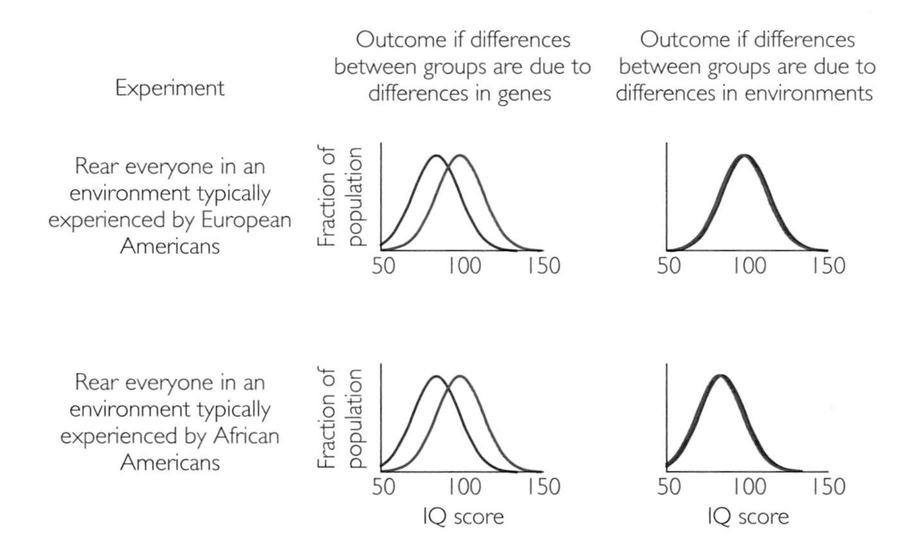

Figure 9.30 An experiment that would test Murray and Herrnstein's claim The left column describes two experimental treatments. The middle and right columns show the predicted outcomes under the hypothesis that differences between groups are due to differences in genes versus the predicted outcomes under the hypothesis that differences between groups are due to differences in environments.

We cannot do this experiment with humans. It might be suggested that we could approximate the experiment by studying European American and African American children that have been adopted into similar families. But the children would still differ in appearance, and might therefore be treated differently by their parents, their teachers, their peers, and so on. In other words, even though they lived in similar families, the children might experience very different environments. Because we cannot do the definitive experiment, we simply have no way to assess whether genetics has anything to do with the difference in IQ score between ethnic groups.

But experiments like the one in Figure 9.30 have been done with plants and animals, and it is instructive to look at the results. For example, Clausen, Keck, and Hiesey (1948) conducted a series of common garden experiments with the plant *Achillea* (Figure 9.31). Plants in this genus collected from low-altitude populations make more stems than plants collected from high-altitude populations (Figure 9.31a). Is the difference between the low- versus high-altitude plants due to differences in their genes or differences in their environments? When plants from low altitude and plants from high altitude are grown together at low altitude, the low-altitude plants make more stems (Figure 9.31b). This

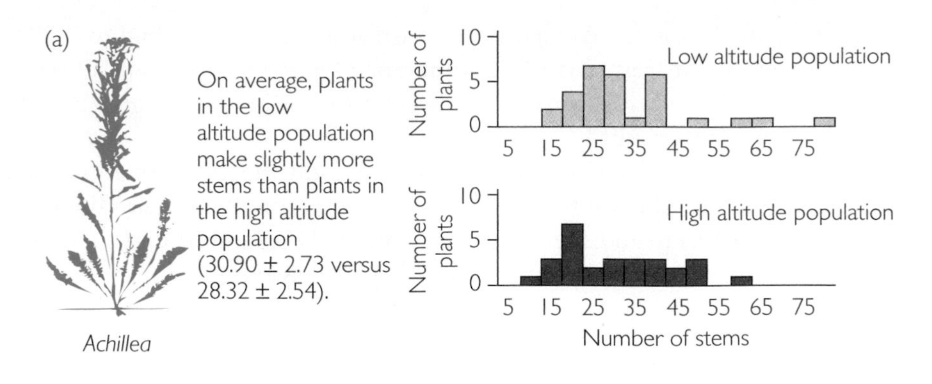

(b)

When Clausen, Keck, and Hiesey (1948) grew Achillea from both populations in a garden at low altitude, the plants from the low altitude population made more stems (30.90 ± 2.73 versus 7.21±1.08).

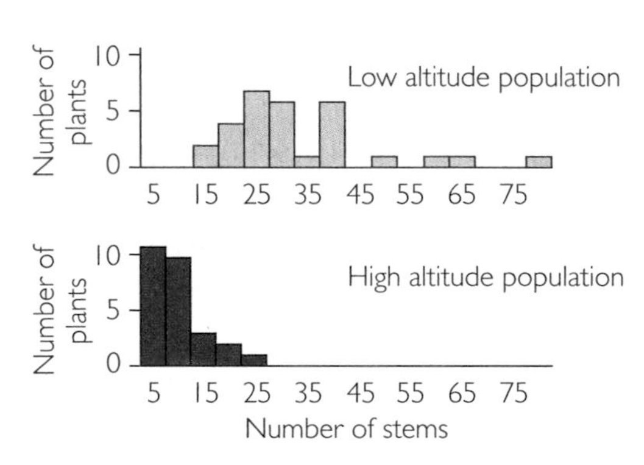

(c)

When the researchers grew cuttings from the same plants in a garden at high altitude, the high altitude plants made more stems (19.89 ± 2.26 versus 28.32 ± 2.54).

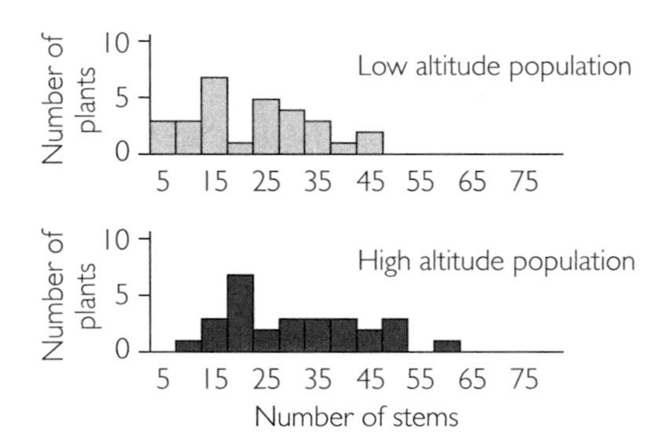

Figure 9.31 Data from experiments by Clausen, Keck, and Hiesey (1948) (a) A comparison between Achillea populations from low altitude (San Gregorio, California) versus high altitude (Mather, California). (b) Plants from low and high altitude grown in a common garden at low altitude (Stanford, California). (c) Plants from low and high altitude grown in a common garden at high altitude (Mather, California).

The only way to determine the cause of differences between populations is to rear individuals from each of the populations in identical environments.

result is consistent with the hypothesis that plants from low altitude are genetically programmed to make more stems. When plants from low altitude and plants from high altitude are grown together at high altitude, however, the *high-altitude* plants make more stems (Figure 9.31c). This result was wholly unanticipated in the experimental design. It reveals genetic differences between low- and high-altitude plants in the way each responds to the environment. It also reveals that each population of plants is superior in its own environment of origin. This unanticipated outcome demonstrates that hypothetical claims about the causes of differences between populations are no substitute for experimental results.

What would happen if we could do this kind of experiment with African Americans and European Americans? No one has the slightest idea. It is misleading to say that high heritabilities for IQ within groups tell us "nothing for sure about the origin of the differences between groups" (Murray and Herrnstein 1994). In fact, high heritabilities within groups tell us *nothing at all* about the origin of differences between groups.

Finally, it is worth noting that heritability also tells us nothing about the role of genes in determining traits that are shared by all members of a population. There is no variation among humans in number of noses. The heritability of nose number is undefined, because $V_A/V_P = 0/0$. This obviously does not mean that our genes are not important in determining how many noses we have.

So what good *does* it do us to measure the heritability of a trait? Precisely and only this: It allows us to predict whether selection on the trait will cause a population to evolve.

Summary

Quantitative traits show continuous variation among individuals. They are influenced by the genotype at many loci as well as the environment. Sometimes we can identify loci that contribute to a quantitative trait. We start with phenotypically distinct parental strains or species in which we have identified marker loci at which different alleles are fixed in each parental population. We then generate a large population of F_2 individuals and look for associations between the genotype at the marker loci versus phenotype. Such associations indicate that the marker is linked to a locus that influences the trait of interest. If known protein-encoding genes are nearby, and if their function is plausibly connected to the phenotype, they may warrant further investigation.

Often we do not know the identity of the loci that influence a quantitative trait. Quantitative genetics gives us tools for analyzing the evolution of such traits, anyway. Heritability can be estimated by examining similarities among relatives. The strength of selection can be measured by analyzing the relationship between phenotypes and fitness. When we know both the heritability of a trait and the strength of selection on the trait, we can predict how the population will evolve in response to selection.

Selection on quantitative traits can follow a variety of patterns, including directional selection, stabilizing selection, and disruptive selection. Directional selection and stabilizing selection reduce genetic variation in populations. Nonetheless, genetic variation persists in most populations, even for traits closely related to fitness. Variation may persist because most populations are not in equilibrium, because there is a balance between mutation and selection, or because disruptive selection (and related patterns, like frequency-dependent selection) are more common than generally recognized.

Estimates of heritability are often misinterpreted. Estimates of heritability depend on both the genetic composition of the study population and the environment it was studied in. As a result, heritabilities tell us nothing about the cause of differences between populations.

Questions

1. Degree of antisocial behavior is a quantitative trait in human males. Avshalon Caspi and colleagues (2002) used data on several hundred men to investigate the relationship between antisocial behavior and two factors. The first factor was genotype at the locus that encodes the enzyme monoamine oxidase A (MAOA). MAOA acts in the brain, where it breaks down a variety of the neurotransmitters nerve cells use to communicate with each other. The gene for MAOA is located on the X chromosome. As a result of genetic differences in the gene's promoter, some men have low MAOA activity and others have high MAOA activity. The second factor was the experience of maltreatment during childhood. Based on a variety of evidence, the researchers determined whether each man had experienced no maltreatment, probable maltreatment, or severe maltreatment. The data are summarized in Figure 9.32.

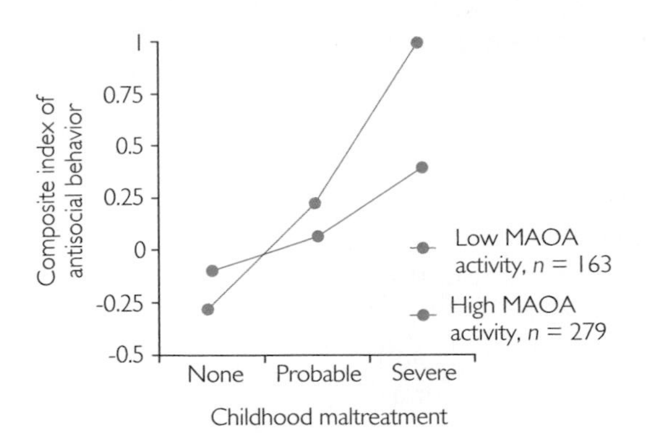

Figure 9.32 Degree of antisocial behavior among men with different levels of MAOA activity as a function of childhood maltreatment Reprinted from Caspi, et al. (2002).

a. Is the variation among men in antisocial behavior at least partly due to differences in genotype? Explain.
b. Is the variation among men in antisocial behavior at least partly due to differences in environment? Explain.
c. Do men with different genotypes respond the same way to changes in the environment? Explain.
d. Is antisocial behavior heritable? Explain.
e. Do these data influence your opinion about how men who exhibit antisocial behavior should be treated and/or punished?

2. The serotonin transporter is a cell-surface protein that recycles the neurotransmitter serotonin after it has been used to carry a message between nerve cells in the brain. There are two alleles of the serotonin transporter gene: *s* and *l*. Klaus-Peter Lesch and colleagues (1996) found that people with genotypes *ls* and *ss*, score slightly, but significantly, higher than people with genotype *ll* on psychological tests of neuroticism (see Figure 9.33).

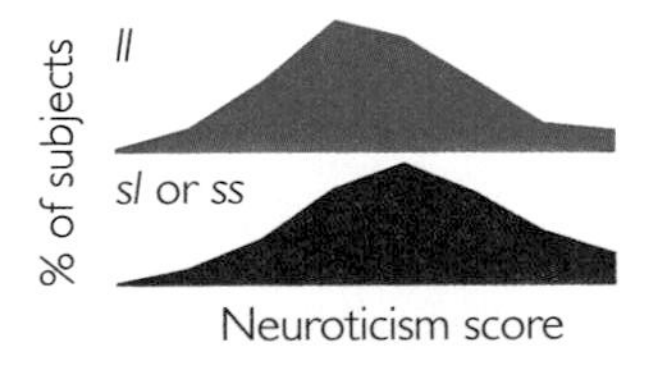

Figure 9.33 Distribution of neuroticism score among people with different genotypes of the serotonin transporter gene Redrawn from Lesch et al. (1996).

a. Are these data consistent with the hypothesis that the serotonin transporter gene is a QTL that influences neuroticism? Explain.
b. Is the serotonin transporter gene *the* gene for neuroticism? Explain.
c. Can you think of another plausible explanation, in which the serotonin transporter gene plays no role at all in neuroticism? Explain.

3. A group of owners of Thoroughbred racehorses asks you how they can identify some of the particular loci and alleles that distinguish winners from losers. Describe, in as much detail as possible, a research program that might reveal this information.

4. Suppose you're telling your roommate that you learned in biology class that within any given human population, height is highly heritable. Your roommate, who is studying nutrition, says, "That doesn't make any sense, because just a few centuries ago most people were shorter than they are now, and it's clearly because of diet. If most variation in human height is due to genes, how could diet make such a big difference?" Your roommate is obviously correct that poor diet can dramatically affect height. How do you explain this apparent paradox to your roommate?

5. Now consider heritability in more general terms. Suppose heritability is extremely high for a certain trait in a certain population.
a. First, can the trait be strongly affected by the environment despite the high heritability value? To answer this question, suppose that all the individuals within a certain population have been exposed all their lives to the same level of a critical environmental factor. Will the heritability value reflect the fact that the environment is very important?
b. Second, *can the heritability value change* if the environment changes? To answer this question, imagine that the critical environmental factor changes, such that different individuals are now exposed to different levels of this environmental factor. What happens to variation in the trait in the whole population? What happens to the heritability value?

6. A dog breeder has asked you for advice. The breeder keeps Alaskan huskies, which she races in sledding events. She would like to breed huskies that run faster. The table on the next page gives data on the running speeds (m/s) of 15 families of dogs in the breeder's kennel.
a. Use a piece of graph paper to prepare a scatterplot of midoffspring values versus midparent values. Approximately what is the heritability of running speed in the breeder's dog population?
b. If she selectively breeds her dogs, will the next generation run substantially faster than the dogs she has now?
c. What else would you suggesst this breeder should try if she wants to win more races?

Family	*Midparent*	*Midoffspring*
1	12.7	10.8
2	7.6	8.0
3	14.4	8.0
4	4.3	9.7
5	11.3	6.6
6	12.5	6.2
7	8.9	12.5
8	8.2	7.4
9	6.3	3.4
10	12.7	6.7
11	13.9	7.9
12	7.3	13.6
13	5.9	7.4
14	12.8	12.1
15	12.5	11.3

7. Imagine that the dog breeder in Question 6 were to pick just the five pairs of parents with the highest midparent values and use them as breeders for the next generation of dogs.

a. Calculate the selection differential and the selection gradient the breeder has imposed on her population of dogs.

b. Use your estimate of the heritability from Question 6 and the selection differential you just calculated to predict the response to selection. What is the predicted average running speed of the dogs in the next generation?

c. How does your predicted speed of the next generation compare to the actual average running speed of offspring of the fastest five families? Discuss.

8. In our discussion of Weis and Abrahamson's work on goldenrod galls (data plotted in Figure 9.26), we mentioned that the researchers established that there is heritable variation among flies in the size of the galls they induce. How do you think Weis and Abrahamson did this? Describe the necessary experiment in as much detail as possible.

9. Given the strength of selection that bumblebees exert on alpine skypilots, why haven't flower corollas in the tundra population evolved to be even larger than they are now? Develop at least two hypotheses, and describe how you could you test your ideas.

10. a. Describe, in your own words, the three major modes of selection and their general effects on population means and on population variation.

b. Which mode of selection is at work on gall size of the gall-making flies?

c. If parasitoid wasps became extinct, which mode of selection would affect the next generation of gall-making flies? Predict what would happen to average gall size in subsequent generations.

Exploring the Literature

11. For a study on aphids similar to the plant study pictured in Figure 9.26, see:

Via, S. 1991. The genetic structure of host plant adaptation in a spatial patchwork: Demographic variability among reciprocally transplanted pea aphid clones. *Evolution* 45: 827–852.

12. For a study in which researchers used the tools of quantitative genetics to predict how behavior in a bird population might evolve in response to global warming, see:

Pulido, F., P. Berthold, et al. 2001. Heritability of the timing of autumn migration in a natural bird population. *Proceedings of the Royal Society of London B* 268: 953–959.

13. How far and how fast can directional selection on a quantitative trait shift the distribution of the trait in a population? For one answer, see:

Weber, K. E. 1996. Large genetic change at small fitness cost in large populations of *Drosophila melanogaster* selected for wind tunnel flight: Rethinking fitness surfaces. *Genetics* 144: 205–213.

14. As genome-wide mapping techniques become increasingly sophisticated and cost-effective, QTLs are rapidly being identified for many human diseases. For some examples, see:

Kissebah, A. H., G. E. Sonnenberg, J. Myklebust, et al. 2000. Quantitative trait loci on chromosomes 3 and 17 influence phenotypes of the metabolic syndrome. *Proceedings of the National Academy of Sciences USA* 97: 14478–14483.

Arya, R., R. Duggirala, C. P. Jenkinson, L. Almasy, J. Blangero, P. O'Connell, and M. P. Stern. 2004. Evidence of a novel quantitative-trait locus for obesity on chromosome 4p in Mexican Americans. *American Journal of Human Genetics* 74: 272–282.

Nyholt, D. R., K. I. Morley, M. A. R. Ferreira, S. E. Medland, D .I. Boomsma, A. C. Heath, K. R. Merikangas, G. W. Montgomery, and N. G. Mart. 2005. Genomewide significant linkage to migrainous headache on chromosome 5q21. *American Journal of Human Genetics* 77: 500–512.

Majumder, P. P., and S. Ghogh. 2005. Mapping quantitative trait loci in humans: achievements and limitations. *Journal of Clinical Investigation* 115: 1419–1424.

15. It is difficult to estimate heritability accurately in humans because it is so hard to disentangle the shared environment and the shared genes within families. (Twin studies, while useful, have certain limitations.) Recall that siblings are assumed to share one-half of their genes. This is true on average, but any given pair of siblings may theoretically share between 0% and 100% of their genes, due to independent assortment of chromosomes and crossing-over during meiosis. (The actual percentage is almost always between 35% and 65%.) Recently it has become possible to measure the actual percentage of shared genes. See the following paper for an innovative approach to heritability estimates that relies on measurement of the actual percentage of genes shared by family members.

Visscher, P. M., S. E. Medland, M. A. R. Ferreira, K. I. Morley, G. Zhu, B. K. Cornes, G. W. Montgomery, and N. G. Martin. 2006. Assumption-free estimation of heritability from genome-wide identity-by-descent sharing between full siblings. *PLoS Genetics* 2: e41.

16. For further information on genetic and environmental effects on human intelligence, see:

Greenwood, P .M., and R. Parasuraman. 2003. Normal genetic variation, cognition, and aging. *Behavioral and Cognitive Neuroscience Reviews* 2: 278–306.

Posthuma, D., M. Luciano, E. J. C. de Geus, M. J. Wright, P. I. Slagboom, G. W. Montgomery, D. I. Boomsma, and N. G. Martin. 2005. A genomewide scan for intelligence identifies quantitative trait loci on 2q and 6p. *American Journal of Human Genetics* 77: 318–326.

17. See the following book for an interesting and insightful tour of twin studies, and other aspects of twin biology:

Segal, N. 2000. *Entwined Lives: Twins and What They Tell Us about Human Behavior.* New York: Plume.

Citations

Please note that much of the quantitative genetics in this chapter is modeled after presentations in the following:

Ayala, F. J. 1982. *Population and Evolutionary Genetics: A Primer.* Menlo Park, CA: Benjamin/Cummings.

Falconer, D. S. 1989. *Introduction to Quantitative Genetics.* New York: John Wiley & Sons.

Felsenstein, J. 1997. *Theoretical Evolutionary Genetics.* Seattle, WA: ASUW Publishing, University of Washington.

Jorde, L. B., J. C. Carey, et al. 1999. *Medical Genetics*, 2nd ed. St. Louis: Mosby.

Lynch, M., and B. Walsh. 1998. *Genetics and Analysis of Quantitative Traits.* Sunderland, MA: Sinauer.

Maynard Smith, J. 1998. *Evolutionary Genetics,* 2nd ed. Oxford: Oxford University Press.

Here is the listing of all citations in this chapter:

Barton, N. H., and M. Turelli. 1989. Evolutionary quantitative genetics: How little do we know? *Annual Review of Genetics* 23: 337–370.

Bates Smith, T. 1993. Disruptive selection and the genetic basis of bill size polymorphism in the African finch *Pyrenestes. Nature* 363: 618–620.

Beardsley, P. M., A. Yen, and R. G. Olmstead. 2003. AFLP phylogeny of *Mimulus*, section *Erythranthe* and the evolution of hummingbird pollination. *Evolution* 57: 1397-1410.

Benjamin, J., L. Li, C. Patterson, et al. 1996. Population and familial association between the D4 dopamine receptor gene and measures of novelty seeking. *Nature Genetics* 12: 81–84.

Bradshaw, H. D., Jr., K. G. Otto, et al. 1998. Quantitative trait loci affecting differences in floral morphology between two species of monkeyflower (*Mimulus*). *Genetics* 149: 367–382.

Bradshaw, H. D., Jr., and D. W. Schemske. 2003. Allele substitution at a flower colour locus produces a pollinator shift in monkeyflowers. *Nature* 426: 176-178.

Brodie, E. D., III. 1992. Correlational selection for color pattern and antipredator behavior in the garter snake *Thamnophis ordinoides. Evolution* 46: 1284–1298.

Brodie, E. D., III, A. J. Moore, and F. J. Janzen. 1995. Visualizing and quantifying natural selection. *Trends in Ecology and Evolution* 10: 313–318.

Caspi, A., J. McClay, et al. 2002. Role of genotype in the cycle of violence in maltreated children. *Science* 297: 851–854.

Cavalli-Sforza, L. L., and W. F. Bodmer. 1971. *The Genetics of Human Populations.* San Francisco: W. H. Freeman and Company.

Clausen, J., D. D. Keck, and W. M. Hiesey. 1948. *Experimental Studies on the Nature of Species. III. Environmental Responses of Climatic Races of Achillea.* Washington, D.C.: Carnegie Institution of Washington Publication No. 581, 45–86.

Di Masso, R. J., G. C. Celoria, and M. T. Font. 1991. Morphometric traits and femoral histomorphometry in mice selected for body conformation. *Bone and Mineral* 15: 209–218.

East, E. M. 1916. Studies on size and inheritance in *Nicotania. Genetics* 1: 164–176.

Ebstein, R. P., O. Novick, R. Umansky, et al. 1996. Dopamine D4 receptor (D4DR) exon III polymorphism associated with the human personality trait of novely seeking. *Nature Genetics* 12: 78–80.

Elena, S. F., and R. E. Lenski. 1997. Long-term experimental evolution in *Escherichia coli*. VII. Mechanisms maintaining genetic variability within populations. *Evolution* 51: 1058–1067.

Falconer, D. S. 1989. *Introduction to Quantitative Genetics.* New York: John Wiley & Sons.

Fisher, R. A. 1930. *The Genetical Theory of Natural Selection.* Oxford, UK: Clarendon Press.

Galen, C. 1989. Measuring pollinator-mediated selection on morphometric floral traits: Bumblebees and the alpine sky pilot, *Polemonium viscosum. Evolution* 43: 882–890.

Galen, C. 1996. Rates of floral evolution: Adaptation to bumblebee pollination in an alpine wildflower, *Polemonium viscosum. Evolution* 50: 120–125.

Grant, P. R., and B. R. Grant. 1995. Predicting microevolutionary responses to directional selection on heritable variation. *Evolution* 49: 241–251.

Hedrick, P. W. 1983. *Genetics of Populations.* Boston: Science Books International.

Hugot, J.-P., M. Chamaillard, H. Zouali, et al. 2001. Association of NOD2 leucine-rich repeat variants with susceptibility to Crohn's disease. *Nature* 411: 599–603.

Jennings, H. S. 1917. The numerical results of diverse systems of breeding, with respect to two pairs of characters, linked or independent, with special relation to the effects of linkage. *Genetics* 2: 97–154.

Kobayashi, K. S., M. Chamaillard, et al. 2005. Nod2-dependent regulation of innate and adaptive immunity in the intestinal tract. *Science* 307: 731–734.

Lande, R., and S. J. Arnold. 1983. The measurement of selection on correlated characters. *Evolution* 37: 1210–1226.

Lesch, K.-P., D. Bengel, A. Heils, et al. 1996. Association of anxiety-related traits with a polymorphism in the serotonin transporter gene regulatory region. *Science* 274: 1527–1531.

Lively, C. M. 1996. Host-parasite coevolution and sex. *BioScience* 46: 107–114.

Lynch, M., and B. Walsh. 1998. *Genetics and Analysis of Quantitative Traits*. Sunderland, MA: Sinauer.

McClearn, G. E., B. Johansson, S. Berg, et al. 1997. Substantial genetic influence on cognitive abilities in twins 80 or more years old. *Science* 276: 1560–1563.

Moreels, T. G., and P. A. Pelckmans. 2005. Gasterointestinal parasites: potential therapy for refractory inflammatory bowel diseases. *Inflammatory Bowel Disease* 11: 178-184.

Murray, C., and R. J. Herrnstein. 1994. Race, genes and I.Q.—An apologia. *The New Republic* 211 (October 31): 27–37.

Ogura, Y., D. K. Bonen, N. Inoara, et al. 2001. A frameshift mutation in *NOD2* associated with susceptibility to Crohn's disease. *Nature* 411: 603–606.

Orr, H. A. 1998. The population genetics of adaptation: The distribution of factors fixed during adaptive evolution. *Evolution* 52: 935–949.

Orr, H. A. 1999. The evolutionary genetics of adaptation: A simulation study. *Genetical Research, Cambridge* 74: 207–214.

Orr, H. A., and J. A. Coyne. 1992. The genetics of adaptation: A reassessment. *American Naturalist* 140: 725–742.

Paterson, A. H., E. S. Lander, et al. 1988. Resolution of quantitative traits into Mendelian factors by using a complete linkage map of restriction fragment length polymorphisms. *Nature* 335: 721–726.

Phillips, P. C., and S. J. Arnold. 1989. Visualizing multivariate selection. *Evolution* 43: 1209–1222.

Provine, W. B. 1971. *The Origins of Theoretical Population Genetics*. Chicago: University of Chicago Press.

Provine, W. B. 1986. *Sewall Wright and Evolutionary Biology*. Chicago: University of Chicago Press.

Schemske, D. W., and H. D. Bradshaw, Jr. 1999. Pollinator preference and the evolution of floral traits in monkeyflowers (*Mimulus*). *Proceedings of the National Academy of Science USA* 96: 11910–11915.

Schinka, J. A., E. A. Letsch, and F. C. Crawford. 2002. DRD4 and novelty seeking: Results of meta-analyses. *American Journal of Medical Genetics (Neuropsychiatric Genetics)* 114: 643–648.

Shikano, T. 2005. Marker-based estimation of heritability for body color variation in Japanese flounder *Paralichthys olivaceus*. *Aquaculture* 249: 95-105.

Smith, J. M. N., and A. A. Dhondt. 1980. Experimental confirmation of heritable morphological variation in a natural population of song sparrows. *Evolution* 34: 1155–1160.

Summers, R. W., D. E. Elliot, et al. 2005a. *Trichuris suis* therapy in Crohn's disease. *Gut* 54: 87–90.

Summers, R. W., D. E. Elliot, et al. 2005b. *Tichuris suis* therapy for active ulcerative colitis: a randomized controlled trial. *Gasteroenterology* 128: 825–832.

Tanksley, S. D. 1993. Mapping polygenes. *Annual Review of Genetics* 27: 205–233.

Weis, A. E., and W. G. Abrahamson. 1986. Evolution of host-plant manipulation by gall makers: Ecological and genetic factors in the *Solidago-Eurosta* system. *American Naturalist* 127: 681–695.

Wright, S. 1986. *Evolution: Selected papers*. William B. Provine, editor. Chicago: University of Chicago Press.

Wright, S. 1988. Surfaces of selective value revisited. *American Naturalist*. 131: 115–121.

PART III ADAPTATION

The proboscis of this tangle-veined fly and the floral tubes of the mountain drumstick it is sipping nectar from are well-matched.

The population genetic models and experiments that we introduced in Part II revealed four mechanisms of evolution: selection, mutation, migration, and genetic drift. Of the four, natural selection is the only evolutionary process that results in adaptation. An adaptation is a trait that allows an individual to leave more offspring than it would if it lacked the trait. It is natural selection that explains how organisms came to be apparently well designed.

In Part III we explore adaptation in depth. We begin, in Chapter 10, by surveying techniques evolutionary biologists use to study adaptation. How can a researcher test the hypothesis that a particular trait is adaptive? In Chapter 11 we consider how selection acts on morphological traits and behaviors that allow individuals to attract mates. Sexual selection can produce striking differences in the behavior and appearance of males and females. In Chapter 12 we ask how social interactions affect the fitness of individuals. We explore the evolutionary basis of altruism and introduce kin selection. In Chapter 13 we seek to understand how natural selection shapes aging, the timing of reproduction, and investment in offspring. In Chapter 14, we apply the lessons of evolutionary biology to practical issues in human health. Finally, in Chapter 15 we look at how selection acts at the level of genes and genomes. Together, the chapters in Part III explore the consequences of natural selection in all of its forms. ■

10

Studying Adaptation: Evolutionary Analysis of Form and Function

Pigeons shorn of the hooks on the tips of their beaks suffer worsening infestations of lice. Pigeons allowed to regrow their hooks (red) fare better than continuously trimmed pigeons (blue). Redrawn from Clayton et al. (2005).

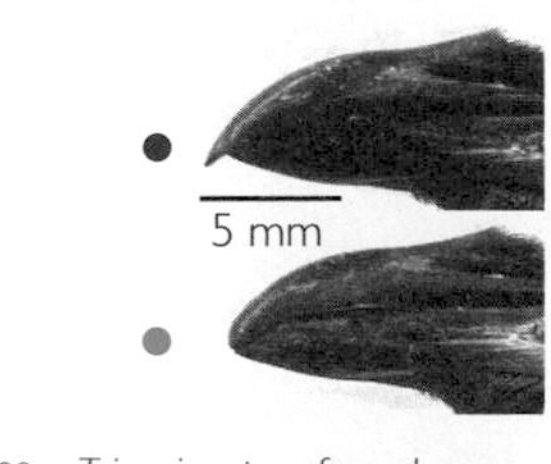

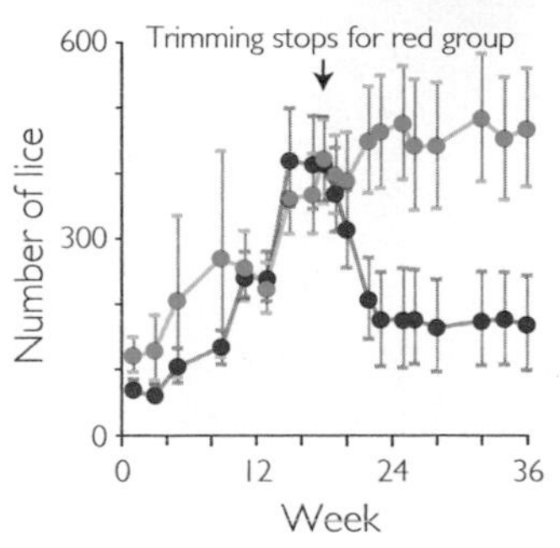

Why do rock pigeons carry tiny hooks on the tips of their beaks? Dale Clayton and colleagues (2005) thought they knew the answer. Rock pigeons often suffer infestations of feather-eating lice. In previous generations, the individuals with tiny hooks on their beaks were better able to kill the lice while preening, and consequently maintained superior feathers. These well-feathered birds were better than their hookless peers at keeping warm, evading predators, and attracting mates. Over a lifetime, they had more offspring.

Clayton and his coworkers are careful, however, about testing their ideas. They once thought that shedding of lice is among the reasons pigeons molt, or drop and regrow their feathers. But when the researchers induced pigeons to molt and compared them to unmolted controls, they found no difference in ectoparasite loads (Moyer et al. 2002).

Clayton and colleagues (2005) tested their hunch about beak hooks by catching 26 pigeons and trimming the tips of their beaks. Over 18 weeks, the hookless pigeons grew ever lousier. The researchers then allowed 13 of the birds, chosen at random, to regrow their hooks. Over subsequent weeks the newly re-hooked birds enjoyed falling parasite loads, while the hookless birds remained as heavily infested as ever. About beak hooks, it appears, Clayton and colleagues were right.

The explanation of organismal design is among the triumphs of the theory of evolution by natural selection. Individuals in previous generations varied in their design, and the ones with the best designs passed on their genes in greater numbers. A trait, or integrated suite of traits, that increases the fitness of its possessor is called an **adaptation** and is said to be **adaptive**.

Demonstrating that the traits of organisms are indeed adaptations has been one of the major activities of evolutionary biology since the time of Darwin (Mayr 1983). This research effort is sometimes called the adaptationist program. Roughly speaking, in order to demonstrate that a trait is an adaptation, we need first to determine what a trait is for and then show that individuals possessing the trait contribute more genes to future generations than individuals lacking it.

The adaptive significance of some traits is obvious: Eyes are manifestly devices for detecting objects at a distance by gathering and analyzing light; in many animal species, individuals with good eyesight will be better able to find food and avoid predators than individuals with poor eyesight. Other traits offer more subtle advantages, and understanding their adaptive significance requires effort.

A plausible hypothesis about the adaptive value of a trait is the beginning of a careful study, not the end.

Obvious explanations, in particular, can be dangerously seductive. As we shall see in the first section of the chapter, conventional wisdom is sometimes wrong. No explanation for the adaptive value of a trait should be accepted simply because it is plausible and charming (Gould and Lewontin 1979). All hypotheses must be tested. Hypotheses can be tested by using them to make predictions, then checking to see if the predictions are correct.

This chapter explores a variety of methods evolutionary biologists use to test hypotheses about adaptations, including experiments, observational studies, and the comparative method. In the last sections of the chapter we also explore complexities of biological form and function that continue to make the adaptationist program a challenging and active area of research.

10.1 All Hypotheses Must Be Tested: Oxpeckers Reconsidered

Oxpeckers deserve an explanation (Figure 10.1). Why do these birds flock to large mammals? Why do their hosts tolerate them? Most readers will either have

Figure 10.1 Oxpeckers on a forest buffalo The association between oxpeckers and large mammals has traditionally been consitered mutually beneficial. The oxpeckers get an easy meal of ticks and a safe place to eat it; their hosts receive a free cleaning. But does the traditional view stand up to careful scrutiny?

heard, or will quickly conceive, the traditional answer. The oxpeckers are looking for an easy meal of ticks and a safe place to eat it. Their hosts are happy to oblige in return for a free cleaning. This mutually advantageous association sometimes runs even deeper, as when the birds apparently minister to their mammalian benefactors by cleaning their wounds.

The trouble with this traditional answer is that on careful observation neither the oxpeckers nor their hosts seem to believe it. This we know from Paul Weeks (1999), who spent a year in Zimbabwe watching red-billed oxpeckers feed on domestic cattle. Weeks was able to establish that oxpeckers do sometimes eat ticks, because he found tick parts in the pellets of undigestible material that the birds occasionally regurgitate. But he seldom witnessed oxpeckers eating ticks. More often, he watched them ignore ticks that Weeks himself could plainly see. Instead, the birds devoted more than 85% of their feeding time to three activities: licking blood from open wounds; probing the hosts' ears, apparently for wax; and scissoring their beaks through the hosts' hair, apparently gleaning and eating dead skin. The hosts, in turn, seemed anything but pleased to have the oxpeckers around. When oxpeckers were worrying their wounds or poking in their ears, the cattle tried, once or twice per minute and with limited success, to shoo the birds away.

To better understand what oxpeckers do, Weeks (2000) set up an oxpecker-exclusion experiment. He divided a small herd of cattle, at random, into two groups. One group he allowed oxpeckers to visit as usual. The other group he protected from oxpeckers by paying an assistant to chase the birds away. Weeks ran the experiment for a month, then switched the treatments and ran the experiment for another month. Finally, he shuffled the cattle to form two new groups and ran the experiment for a third month. At the beginning and end of each month, Weeks counted the ticks on each ox.

The results appear in Figure 10.2. The graph shows, for each month-long trial, the change in tick load for cattle with oxpeckers versus cattle without. If oxpeckers serve their hosts by eating ticks, the change in tick load should be worse—more positive or less negative—for cattle without oxpeckers than for cattle with oxpeckers. This is what happened in trials one and three. But the opposite happened in trial two, and there was no significant difference between cattle with versus without oxpeckers in any of the trials. In other words, oxpeckers have no discernable effect on their hosts' tick loads.

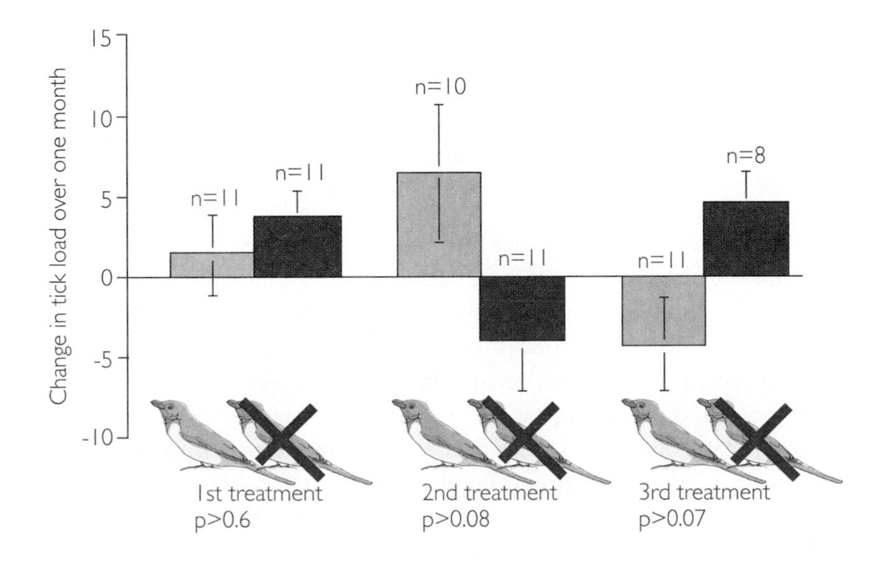

Figure 10.2 Red-billed oxpeckers have no effect on the tick loads on cattle This graph shows, for each of three month-long trials, the change in tick load from the beginning of the month to the end for domestic cattle exposed versus unexposed to oxpeckers. There is no clear pattern in the data, and none of the comparisons is statistically significant. The n's above the bars indicate numbers of cattle. Redrawn from Weeks (2000).

Before and after each month-long trial, Weeks also counted the number of open wounds on each ox. The results appear in Figure 10.3, and this time they are clear. Cattle exposed to oxpeckers have, on average, many more open wounds than cattle protected from the birds. We have already mentioned Weeks's observation that oxpeckers spend a considerable fraction of their feeding time drinking blood from open wounds. He also saw the oxpeckers enlarge existing wounds, and found that wounds took considerably longer to heal when oxpeckers were present.

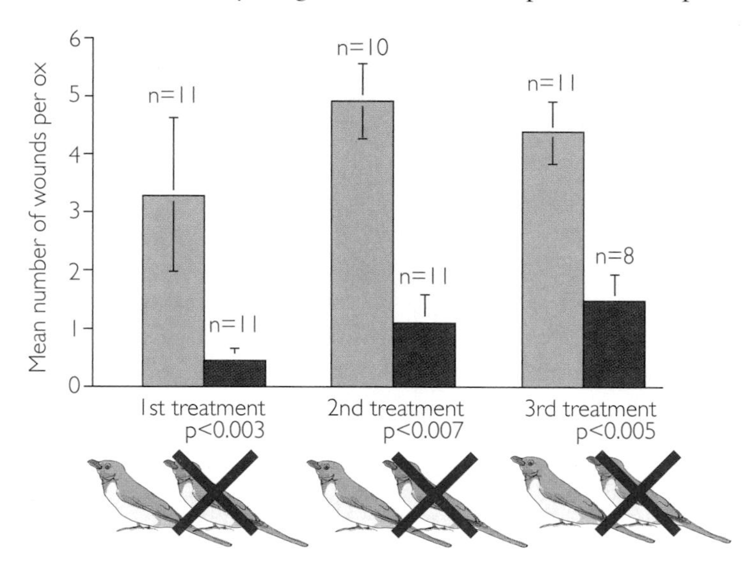

Figure 10.3 Red-billed oxpeckers maintain open wounds on their hosts This graph shows, for each of three month-long trials, the mean number of open wounds per individual on cattle exposed versus unexposed to oxpeckers. Redrawn from Weeks (2000).

Finally, Weeks scored the amount of wax in the ears of each ox. Again the results, which appear in Figure 10.4, are clear. Cattle exposed to oxpeckers have considerably less earwax. Whether this is good or bad for the cattle is unclear.

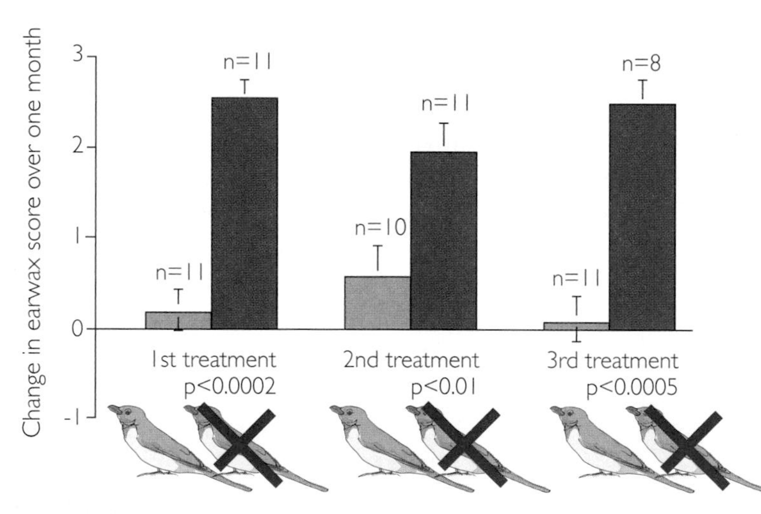

Figure 10.4 Red-billed oxpeckers remove their hosts' earwax This graph shows, for each of three month-long trials, the change in the amount of wax in the ears of cattle exposed versus unexposed to oxpeckers. Whether having their earwax removed is good or bad for the hosts is unclear. Redrawn from Weeks (2000).

Weeks concluded that oxpeckers are vampires and eaters of earwax. Even when they do eat ticks, oxpeckers prefer adult females that have already engorged themselves with blood—that is, ticks that have already done their damage to the host. Weeks acknowledges the possibility that oxpeckers may eat enough ticks to provide a benefit for other hosts or in other environments. For the cattle Weeks studied, however, oxpeckers appear to be less mutualists than parasites.

One flaw in Weeks's study is that domestic cattle are not among the oxpeckers' native hosts. Alan McElligott and colleagues (2004) watched oxpeckers feeding on a pair of black rhinoceros. Consistent with their behavior in Weeks's study, the birds spent most of their foraging time at open wounds. Moreover, the majority of the wounds on the two rhinos were injuries the researchers watched the oxpeckers create. On the available evidence, it appears that the conventional wisdom about oxpeckers and their hosts is wrong.

The oxpecker example demonstrates that we cannot uncritically accept a hypothesis about the adaptive significance of a behavior, or any other trait, simply because it is plausible. We must subject all hypotheses to rigorous tests.

All hypotheses must be tested.

Here are some other caveats to keep in mind when studying adaptations:

- Differences among populations or species are not always adaptive. There are two species of oxpecker; one has red bills, the other yellow. It is possible that each color is adaptive for the species that wears it. But it is also possible that the difference is not adaptive at all. Mutations causing different colors may have become fixed in the two oxpeckers by genetic drift (see Chapter 7). At the molecular level, much of the variation among individuals, populations, and species may be selectively neutral (see Chapter 7).
- Not every trait of an organism, or every use of a trait by an organism, is an adaptation. While feeding on large mammals, oxpeckers may sometimes meet a potential mate. This does not necessarily mean that feeding on large mammals evolved because it creates mating opportunities.
- Not every adaptation is perfect. Feeding on the blood and earwax of large mammals may provide oxpeckers with high-quality meals. But because many large mammals migrate long distances, it may also expose oxpeckers to the risk of an unpredictable food supply.

Alternative explanations must also be considered.

In the next three sections, we review three methods evolutionary biologists use to test hypotheses about the adaptive significance of traits. The first of these sections concerns experiments, the second looks at observational studies, and the third explores the comparative method.

10.2 Experiments

Experiments are among the most powerful tools in science. A well-designed experiment allows us to isolate and test the effect that a single, well-defined factor has on the phenomenon in question. We have already reviewed a variety of experiments in earlier chapters. Figures 6.13 (page 185) and 8.18 (page 306), for example, illustrate the results of experiments on laboratory insect populations. Here our focus is on the process of planning and interpreting experiments. We have chosen our example because it illustrates several aspects of good experimental design.

Experiments are the most powerful method for testing hypotheses. A good experiment restricts the difference between study groups to a single variable.

What Is the Function of the Wing Markings and Wing-waving Display of the Tephritid Fly *Zonosemata*?

The tephritid fly *Zonosemata vittigera* has distinctive dark bands on its wings. When disturbed, the fly holds its wings perpendicular to its body and waves them up and down. Entomologists had noticed that this display seems to mimic the leg-waving, territorial threat display of jumping spiders (species in the family Salticidae). These entomologists suggested that, because jumping spiders are fast and have a nasty bite, a fly mimicking a jumping spider might be avoided by a wide variety of other predators. Erick Greene and colleagues (1987) had a

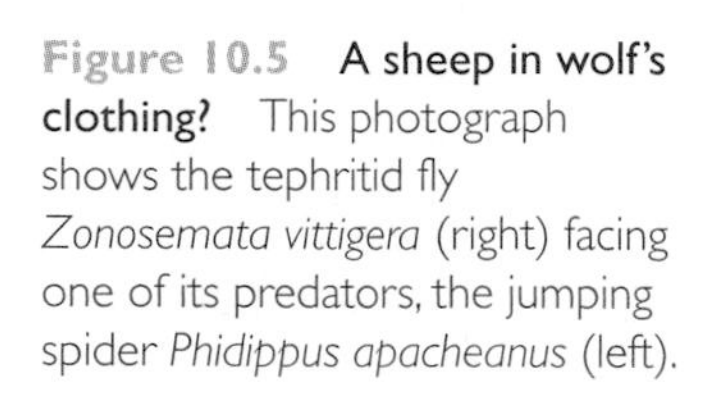

Figure 10.5 A sheep in wolf's clothing? This photograph shows the tephritid fly *Zonosemata vittigera* (right) facing one of its predators, the jumping spider *Phidippus apacheanus* (left).

different idea. Because jumping spiders are *Zonosemata*'s major predators (Figure 10.5), Greene and colleagues proposed that the fly uses its wing markings and wing-waving display to intimidate the jumping spiders themselves. The fly, in other words, is a sheep in wolf's clothing. Mimicry of a predator's behavior by its own prey had never before been recorded.

Both mimicry explanations are plausible hypotheses about the adaptive value of the fly's wing-waving display, but unless we test them they are just good stories. Can these hypotheses be tested rigorously? Greene and his co-workers (1987) set out to do so with an experiment.

The first step in any evolutionary analysis is to phrase the question as precisely as possible. In this case, Do the wing markings and the wing waving of *Zonosemata vittigera* mimic the threat displays that jumping spiders use on each other, and thereby allow the flies to escape predation? Stating a question precisely makes it easier to design an experiment that will provide a clear answer.

The researchers' next step was to list alternative explanations for the behavior. Good experiments test as many competing hypotheses as possible (Platt 1964). Note that each of the following is a biologically realistic explanation, not an implausible straw man proposed just to give the impression of rigor.

Good experimental designs test the predictions made by several alternative hypotheses.

Hypothesis 1: The flies do not mimic jumping spiders. This is a distinct possibility, because other fly species have dark wing bands and wing-flicking displays that do not deter predators. In many species, the flies use their markings and displays during courtship.

Hypothesis 2: The flies mimic jumping spiders, but the flies behave like spiders to deter other, nonspider predators. Other fly predators that might be intimidated by a jumping spider, or a jumping spider mimic, include other kinds of spiders, assassin bugs, preying mantises, and lizards.

Hypothesis 3: The flies mimic jumping spiders, and this mimicry functions specifically to deter predation by the jumping spiders themselves.

In an ideal experiment, the control and experimental groups are treated identically except for exactly one factor.

To test these alternatives, Greene and colleagues needed flies with some parts, but not all, of the *Zonosemata* display. The biologists found they could cut the wings off a *Zonosemata* fly and glue them back on with household glue. And they could cut the wings off a *Zonosemata* fly and replace them with the wings of a housefly (*Musca domestica*), which are clear and unmarked. Remarkably, the surgically altered *Zonosemata* still waved their wings normally and could even fly.

Greene and colleagues created a total of five experimental groups of flies (Figure 10.6). The five treatments distinguish among the three hypotheses, because each hypothesis makes a different suite of predictions about what will happen in encounters between predators and flies. The treatments also allow the

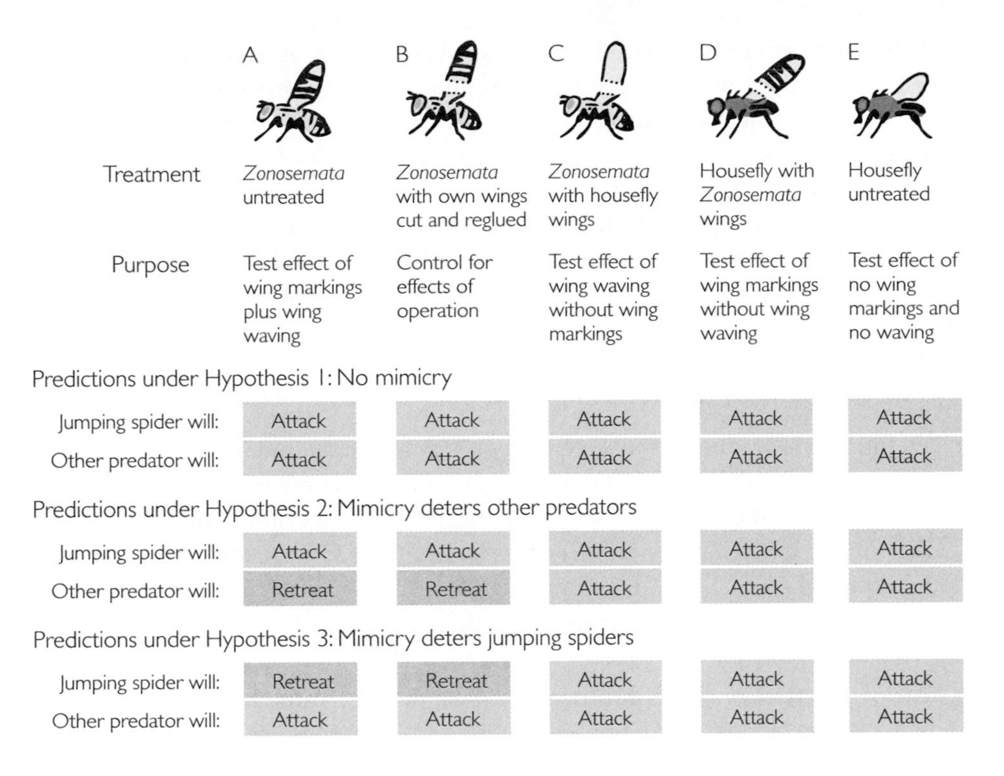

	A	B	C	D	E
Treatment	*Zonosemata* untreated	*Zonosemata* with own wings cut and reglued	*Zonosemata* with housefly wings	Housefly with *Zonosemata* wings	Housefly untreated
Purpose	Test effect of wing markings plus wing waving	Control for effects of operation	Test effect of wing waving without wing markings	Test effect of wing markings without wing waving	Test effect of no wing markings and no waving
Predictions under Hypothesis 1: No mimicry					
Jumping spider will:	Attack	Attack	Attack	Attack	Attack
Other predator will:	Attack	Attack	Attack	Attack	Attack
Predictions under Hypothesis 2: Mimicry deters other predators					
Jumping spider will:	Attack	Attack	Attack	Attack	Attack
Other predator will:	Retreat	Retreat	Attack	Attack	Attack
Predictions under Hypothesis 3: Mimicry deters jumping spiders					
Jumping spider will:	Retreat	Retreat	Attack	Attack	Attack
Other predator will:	Attack	Attack	Attack	Attack	Attack

Figure 10.6 Surgical treatments used in experiments testing the function of wing-waving display This table shows the predicted outcomes when different predators encounter flies with different treatments. Note that each hypothesis makes a unique suite of predictions. (The predictions listed for hypotheses 2 and 3 assume that both *Zonosemata's* wing markings and wing waving are necessary for effective mimicry.)

researchers to determine whether both the wing markings and the wing-waving display are important in mimicry. This is a powerful experimental design.

To run the experiment, Greene et al. had to measure the responses of jumping spiders and other predators to the five types of experimental flies. When confronted with a test fly, would the spiders retreat, stalk and attack, or kill? The researchers starved 20 jumping spiders from 11 different species for two days. Then they presented one of each of the five experimental fly types to each spider, in random order. The researchers made these presentations in a test arena and recorded each jumping spider's most aggressive response during a five-minute interval. There was a clear difference: Jumping spiders tended to retreat from flies that gave the wing-waving display with marked wings, but attacked flies that lacked either wing markings, wing waving, or both (Figure 10.7).

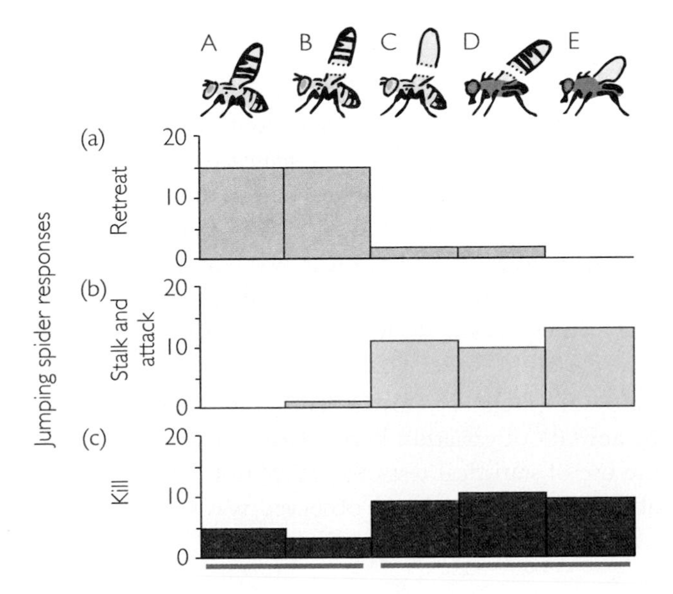

Figure 10.7 Tephritid flies mimic jumping spiders to avoid predation These graphs show how jumping spiders reacted to the fly types listed in Figure 10.6. The hieght of each bar represents the number of jumping spiders, out of 20, that showed each type of maximum response to each type of fly. Thus, for group A flies, 15 of the spiders retreated from their test flies [graph (a)], and five killed their test flies [graph (c)]. For group B flies, 15 of the spiders retreated from their flies, 2 stalked and attacked but did not kill their flies, and 3 killed their flies. From Greene et al. (1987).

Most of the flies that waved marked wings (groups A and B) survived their encounters with spiders unscathed, whereas most of the flies that lacked markings or waving or both were attacked, and many were killed. The red bars at bottom identify treatment groups where the spider responses were statistically indistinguishable from one another. Groups A and B were indistinguishable from each other, but different from groups C, D, and E. Because each spider was presented with one fly of each type, the sample size in each treatment group was 20.

When the researchers tested treatments A, C, and E against other predators (nonsalticid spiders, assassin bugs, mantises, and whiptail lizards), all of the test flies were captured and eaten. In fact, when Greene et al. placed flies before these nonsalticid predators, there was not even an appreciable difference in time-to-capture among the three treatment groups.

Comparison of Figures 10.6 and 10.7 shows that these results are consistent with hypothesis 3, but inconsistent with hypotheses 1 and 2. Thus, Greene et al.'s experiment provides strong support for the hypothesis that tephritid flies mimic their own jumping-spider predators to avoid being eaten by them (see also Mather and Roitberg 1987).

The Greene study illustrates important points about experimental design:

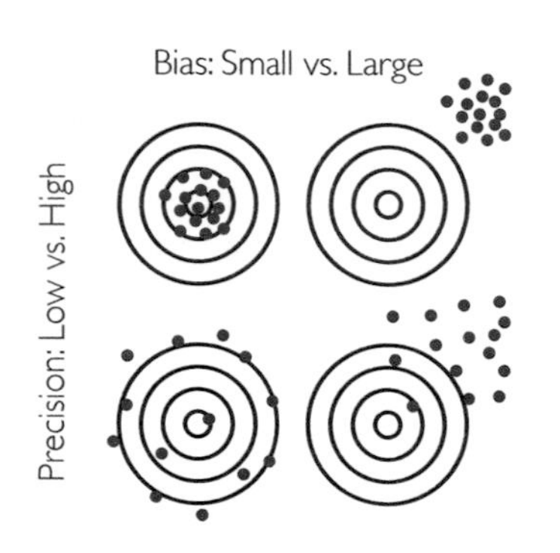

Figure 10.8 Bias and precision When designing an experiment or set of observations with the goal of estimating some quantity, it is important to minimize bias and maximize precision. In this illustration the quantity being estimated in a study is represented by the bull's-eye, and the data points collected are represented by red dots. Techniques like standardizing the experimental conditions and randomizing other factors help to minimize bias and increase precision. Note that our ability to measure precision depends on having a large number of data points.

- Defining and testing effective control groups is critical. In Greene et al.'s study, groups A and B (Figures 10.6 and 10.7) served as controls. These individuals demonstrated that the wing surgery itself had no effect on the behavior of the flies or the spiders. When the *Zonosemata* in group C were attacked and eaten by jumping spiders, Green and colleagues could be sure that this was because the flies no longer had markings on their wings, not simply because their wings had been cut and glued.
- All of the treatments (controls and experimentals) must be handled exactly alike. It was critical that Greene et al. used the same test arena, the same time interval, and the same definitions of predator response in each test. Using standard conditions allows a researcher to avoid bias and increase the precision of the data (Figure 10.8). Think about the problems that could arise if a different test arena were used for each of the five treatment groups.
- Randomization is a key technique for equalizing other, miscellaneous effects among control and experimental groups. In essence, it is another way to avoid bias. Greene and colleagues presented the different kinds of test flies to the spiders and other predators in random order. What problems could arise if they had presented the five types of flies in the same sequence to each spider?
- Repeating the test on many individuals is essential. It is almost universally true in experimental (and observational) work that larger sample sizes are better. This is because the goal of an experiment is to estimate a quantity. In this case, the quantity was the likelihood that a fly will be attacked by jumping spiders as a function of its ability to wave marked wings.

Replicated experiments or observations do two things:

Large sample sizes are better but researchers have to trade off the costs and benefits of collecting ever more data.

- They reduce the amount of distortion in the estimate caused by unusual individuals or circumstances. For example, 4 of the 10 *Zonosemata* with marked wings that were attacked were pounced on and killed before they even had a chance to display (groups A and B in Figures 10.6 and 10.7). Because Greene and colleagues were using standardized conditions, it was not acceptable to simply throw out these data points, even though they might represent bad luck. If events like this really do represent bad luck, they will be rare, and will not bias the result as long as the sample size is large.
- Replicated experiments allow researchers to understand how precise their estimate is by measuring the amount of variation in the data. Knowing how precise the data are allows the use of statistical tests. Statistical tests, in turn, allow us to quantify the probability that the result we observed was simply due to chance (see Box 10.1).

In sum, Greene et al.'s experimental design was successful because it allowed independent tests of the effect that predator type, wing type, and wing display have on the ability of *Zonosemata* flies to escape predation. Experiments are the most powerful means of testing hypotheses about adaptation. In the next section, we consider how careful observational studies can sometimes be nearly as good as experiments.

Box 10.1 A primer on statistical testing

The fundamental goal of many experimental and observational studies is to estimate the value of some quantity in two groups, such as a treatment group and a control group, and to determine if there is a difference in the quantity between the groups. In the studies we have reviewed thus far, researchers have estimated quantities like the depth of finch beaks (Chapter 3) and how frequently flies are attacked by spiders (this chapter). The groups we want to compare in these examples consist of finches before and after the drought and flies with marked or unmarked wings.

As our example, we will focus on a comparison between the flies in groups B and C of Greene et al.'s experiment (Figures 10.6 and 10.7). To simplify the discussion, we will lump together the outcomes *stalked and attacked* and *killed* to form a single category, *attacked*. When the researchers presented a group B fly to each of 20 jumping spiders, 15 of the spiders retreated and 5 attacked the fly. In contrast, when the researchers presented to each of the same 20 spiders a group C fly, 1 of the spiders retreated and 19 attacked the fly. It certainly looks as though jumping spiders are less aggressive toward flies waving marked wings (group B) than toward flies waving unmarked wings (group C).

Once we have measured the quantity in each group and have observed a difference between groups, the statistical question becomes, Is the difference real, or could it simply be due to random variation? It is conceivable that if we tested a much larger population of spiders and flies, we would discover that, in fact, spiders respond the same way to flies waving unmarked wings as they do to flies waving marked wings. Under this scenario, the apparent difference we observed in the experiment was just a chance result.

An analogy can be found in tossing coins. Imagine you have two fair coins. It is conceivable that you could toss the first coin 20 times and get 15 heads and 5 tails and then toss the second coin 20 times and get 1 head and 19 tails. It would appear that the coins are different, but the truth is that if you tossed both coins a very large number of times, you would discover that the coins are the same. (Note that this analogy is imperfect: The true rate at which a fair coin gives heads is 0.5, whereas the true rate at which spiders attack flies might be anywhere from 0 to 1.) What we want to know is, What is the probability that we could get a difference in spider behavior as large as the one we observed if the truth were that spiders actually respond the same way to both kinds of fly?

Answering this question requires a statistical test. The first step in a statistical test is to specify the null hypothesis. This is the hypothesis that there is actually no difference between the groups. In our example, the null hypothesis is that the presence or absence of wing markings does not affect the way jumping spiders respond to flies. According to this hypothesis, the true frequency of attack is the same for flies with markings on their wings as for flies without markings on their wings.

The second step is to calculate a value called a test statistic. A test statistic is a number that characterizes the magnitude of the difference between the groups. More than one test statistic might be appropriate for Greene et al.'s data. Greene and colleagues chose a test statistic that compares the actual rates of retreat and attack observed in the experiment to the rates of retreat and attack that would have been expected if the null hypothesis were true.

The third step is to determine the probability that chance alone could have made the test statistic as large as it is. In other words, if the null hypothesis were true, and we did the same experiment many times, how often would we get a value for the test

Box 10.1 (Continued)

statistic that is larger than the one we actually got? The answer comes from a reference distribution. This is a mathematical function that specifies the probability, under the null hypothesis, of each of all the possible values of the test statistic. Often, it is possible to look up the answer in a statistical table in a book or to have a computer calculate it. For Greene et al.'s data and test statistic, the probability that chance alone would have made the test statistic this large is considerably less than 0.01. In other words, if the null hypothesis were true, and if Greene et al. repeated their experiment many times, they would have gotten a value of the test statistic larger than the one they actually got in fewer than one in 100 experiments. This means that the null hypothesis is probably wrong and that there *is* a real difference in how jumping spiders respond to flies waving marked versus unmarked wings.

The fourth and final step is to decide whether to consider the outcome of the experiment statistically significant. By convention, scientists generally consider the value of a test statistic significant if its probability under the null hypothesis is less than 1 in 20, or 0.05. By this criterion, Greene et al.'s result is significant with room to spare. In other words, when Greene and colleagues claimed to have demonstrated that flies must have markings on their wings to deter jumping-spider attacks, the researchers were taking a chance of less than 1 in 100 of later being proved wrong by someone else who might repeat their experiment. That chance was low enough for them to claim that their result is statistically significant. If there is more than a 5% probability that the difference observed is due to chance, the convention is to accept the null hypothesis of no real difference between groups.

In scientific papers the probability of finding the observed differences by chance is reported as a P value, where the P stands for probability. In the original paper published by Greene et al., for example, the caption to the figure that is our Figure 10.6 includes the phrase "$P < 0.01$."

Statistical tests are based on explicit models of the processes that produced the data and of the design of the experiment. Many different types of random processes are modeled statistically; when analyzing data, it is essential to know enough about statistics to be able to choose a model appropriate to the data collected in a particular study.

10.3 Observational Studies

When an experiment is impractical, a careful observational study may be the next best method for evaluating a hypothesis.

Some hypotheses about adaptations are difficult or impossible to test with experiments. It is hard to imagine, for example, how we could do a controlled experiment to test alternative hypotheses about why giraffes have long necks. To do so, we would have to be able to make giraffes that are identical in all respects except the lengths of their necks. Experiments may also be inappropriate when a hypothesis makes predictions about how organisms will behave in nature. When experiments are impractical or inappropriate, careful observations can sometimes yield sufficient information to evaluate a hypothesis.

Behavioral Thermoregulation

The vast majority of organisms are ectothermic, which means that their body temperatures are determined by the temperatures of their environments. As Figure 10.9 demonstrates for desert iguanas (*Dipsosaurus dorsalis*), body temperature has a profound effect on an ectotherm's physiological performance. Desert iguanas can survive short exposures to body temperatures as low as 0°C and as high as about 47°C, but they can function only between about 15°C and 45°C. Within this narrower range, cold iguanas run and digest slowly, tire quickly, and hear poorly. As they get warmer, they run and digest more

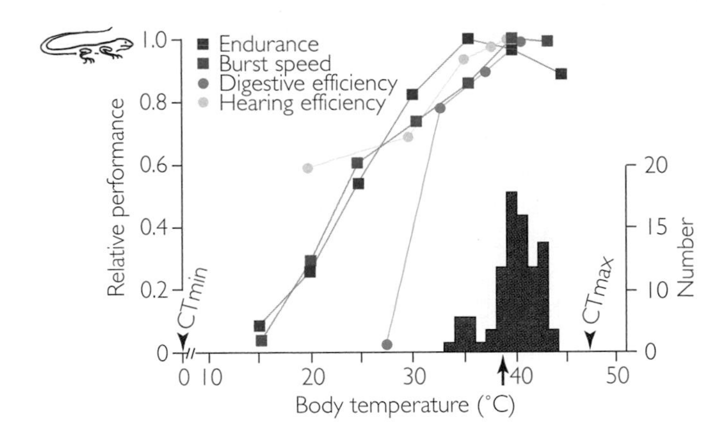

Figure 10.9 Physiological abilities of the desert iguana (*Dipsosaurus dorsalis*) as a function of body temperature The colored squares and circles show locomotor endurance, burst (sprint) speed, digestive efficiency, and hearing efficiency of iguanas as a function of body temperature. The shaded region is a histogram showing the distribution of body temperatures for active iguanas captured in nature. The black arrow indicates the body temperature chosen by iguanas in the lab. CTmax is the critical thermal maximum, that is, the temperature above which the lizard cannot right itself. CTmin is the critical thermal minimum. Reprinted from Huey and Kingsolver (1989).

quickly, tire more slowly, and hear more keenly. The iguanas' physiological capacities reach a plateau in the mid-to-high 30s. Above about 45°C, the iguanas are too hot and collapse.

The relationship between physiological performance and temperature is called a thermal performance curve. The shape of the desert iguana's thermal performance curves is typical of those of a variety of physiological processes in a diversity of organisms. Given the sensitivity of physiological function to temperature, we can predict that ectotherms will exhibit behavioral thermoregulation. That is, ectotherms should move around in the environment so as to maintain themselves at or near the temperature at which they perform the best.

As the temperature of their environment changes, for example, desert iguanas do not just passively accept the consequences. Instead, they regulate their body temperature by moving into the sun to warm up or into the shade to cool off. The iguanas prefer to maintain themselves at body temperatures in the high 30s (Figure 10.9, arrow). This is the center of the range of temperatures at which the iguanas perform best. The iguana's temperature preferences are not surprising. After all, an iguana never knows when it will need to run away from a predator. In nature, of course, iguanas may not always have a sufficient range of environmental temperatures to move among to maintain themselves at exactly their preferred temperature. As the shaded histogram of field body temperatures in Figure 10.9 shows, however, desert iguanas do reasonably well.

Note that although we have asserted that desert iguanas thermoregulate, the mere fact that iguanas captured in nature are usually at or near their optimal body temperature does not, by itself, prove that they are active in maintaining those temperatures. It could be that the environments in which they live are always in the mid-to-high 30s. To prove behavioral thermoregulation, we must show (1) that the animal in question is choosing particular temperatures more often than it would encounter those temperatures if it simply moved at random through its environment, and (2) that its choice of temperatures is adaptive.

Do Garter Snakes Make Adaptive Choices When Looking for a Nighttime Retreat?

Ray Huey and colleagues (1989b) made a detailed study of the thermoregulatory behavior of the garter snake (*Thamnophis elegans*) at Eagle Lake, California. Garter snakes are affected by temperature in the same way as desert iguanas, except that for garter snakes, the optimal temperature, preferred temperature, and maximum survivable temperature are all a few degrees lower than the corresponding temperatures for iguanas. Huey et al. surgically implanted several snakes with miniature radio transmitters. Each implanted transmitter emits a beeping signal that allows a biologist with a handheld receiver and directional antenna to find the implanted snake, even when the snake is hiding under a rock or in a burrow. In addition, the transmitter reports the snake's temperature by changing the rate at which it beeps.

Garter snakes in the lab prefer to stay at temperatures between 28°C and 32°C. Huey and colleagues found that snakes in nature do a remarkable job of thermoregulating in the same range. Figure 10.10 shows the body temperatures of two of the implanted snakes, each over the course of a 24-hour day. Both snakes kept their temperature within or near the preferred range. How do the garter snakes manage to thermoregulate so well? The two snakes shown in Figure 10.10 spent the day under or near rocks. Other options include moving up and down a burrow and staying on the surface of the ground while moving back and forth from sunshine to shade.

A good observational study seeks to find circumstances in nature that resemble an experiment.

Huey and colleagues compared the relative merits of each of these thermoregulatory strategies by monitoring the environmental temperature under rocks of various sizes, and at various depths in a burrow, and by monitoring the temperature of a model snake left on the surface in the sun or shade (Figure 10.11). For a snake under a rock, the thickness of the rock proves critical. A snake under a thin rock (Figure 10.11a) would not only get dangerously cold at night, but would overheat in the daytime. (As Huey says, "The snake would die by 11 a.m., and remain dead until at least 6 p.m.") A snake under a thick rock (Figure 10.11b) would remain safe all day, but would never reach its preferred temperature. Rocks of medium thickness are just right (Figure 10.11c). By moving around under the rock, a snake under a rock of medium thickness can stay close to or within its preferred temperature range for the entire day. A snake moving up and down a burrow could do reasonably well (Figure 10.11d), but would get colder at night than a snake under a medium-sized rock. Finally, a snake on the surface could thermoregulate effectively in the daytime by moving between sun and shade, but would get dangerously cold at night (Figure 10.11e). Putting these observations together, it appears that snakes have many options for thermoregulation during the daytime, as long as they avoid thin rocks or direct sun in the afternoon. At night, however, it appears that the best place for a snake to be is under a rock of medium thickness.

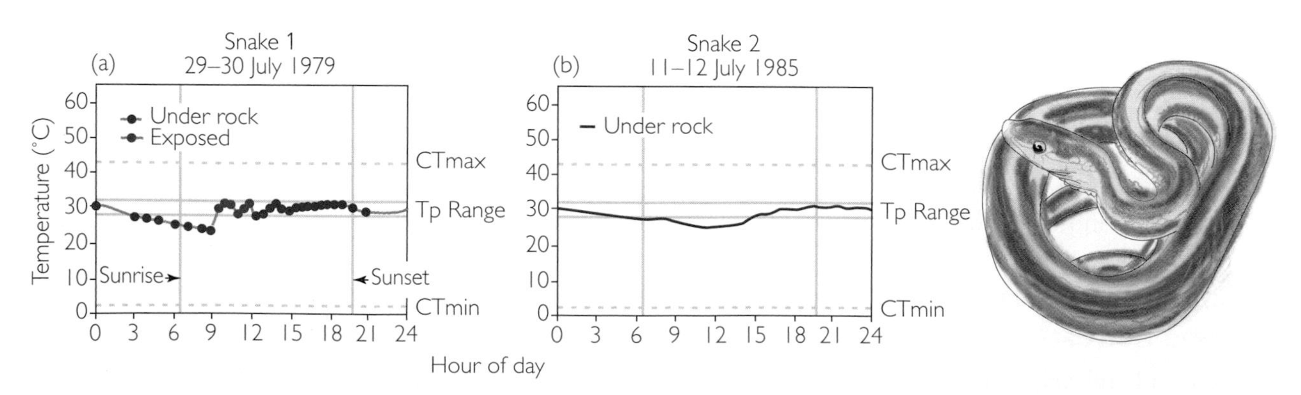

Figure 10.10 Body temperatures of garter snakes in nature (a) Snake 1 spent part of the day under a rock (red dots and line), and part of the day in the sun (blue dots and line). (b) Snake 2 spent the entire day under a rock. Tp is the preferred temperature range, measured in the lab. CTmax and CTmin are defined in Figure 10.9. Both snakes kept their temperature near 30°C for the entire day. From Huey et al. (1989b).

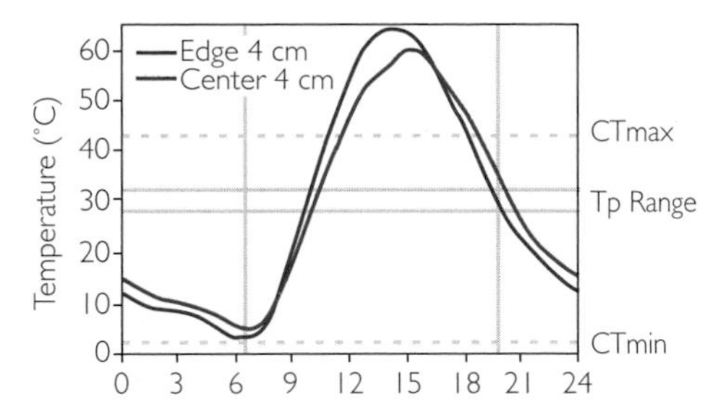

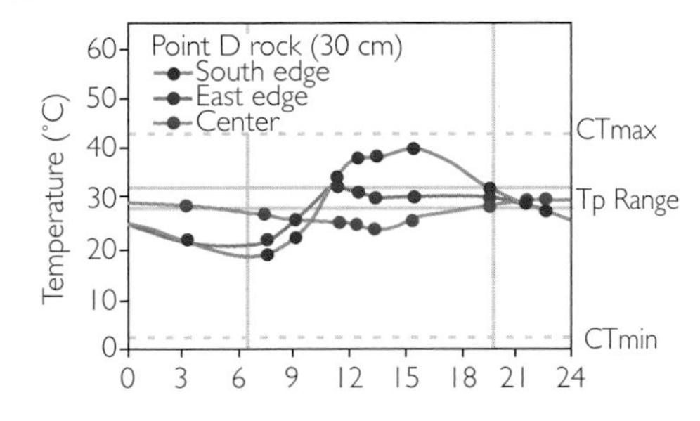

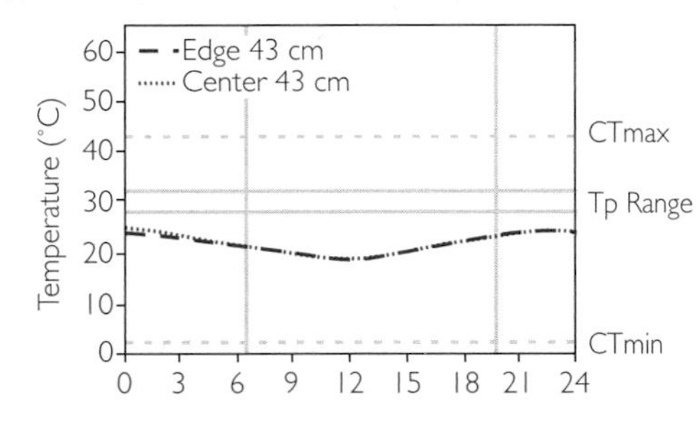

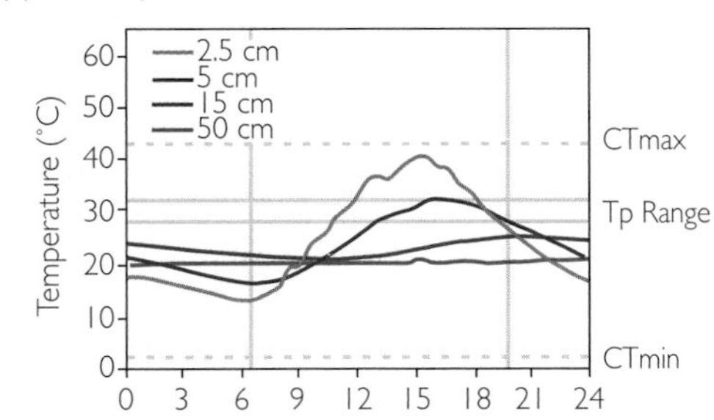

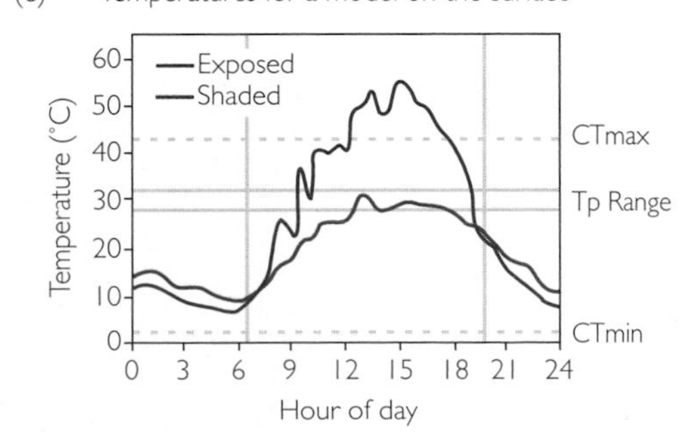

Figure 10.11 Environmental temperatures available to garter snakes at Eagle Lake The graphs show the daily cycle of temperatures in various places a snake might go. Under a thin (4 cm) rock (a), it is cold at night and hot during the day. Under a thick (43 cm) rock (c), it is cool all the time. Under a medium (30 cm) rock (b), there is virtually always a spot within the range of temperatures preferred by snakes. In a burrow (d), it is cool at night and cool to warm in the daytime, with the exact temperature depending on depth within the burrow. On the surface (e), it is cold at night and just right to hot in the daytime, depending on whether a snake is in shade or direct sunlight. From Huey et al. (1989b).

Most garter snakes do, in fact, retreat under rocks at night. Under the hypothesis of behavioral thermoregulation, Huey and colleagues predicted that snakes would choose their nighttime retreats adaptively. That is, they predicted that snakes would preferentially select rocks of medium thickness. Huey et al. tested their prediction by comparing the availability of rocks of different sizes at Eagle Lake to the sizes of the rocks actually chosen as nighttime retreats by radio-implanted snakes (Table 10.1). Thin, medium, and thick rocks are equally available, so if the snakes chose their nocturnal retreats at random, they should be found equally often under rocks of each size. In fact, however, the garter snakes are almost always found under medium rocks or thick rocks. The fact that snakes avoid thin rocks is good evidence that the snakes are active behavioral thermoregulators.

What made the observational study by Huey and colleagues effective in testing the hypothesis that garter snakes thermoregulate is the care with which the researchers monitored the snakes' environment. By determining the options available to snakes, and measuring the frequency of each option in the environment, the researchers were able to show that the snakes they observed were not simply picking their retreats at random, but were instead making an adaptive choice. In the next section, we consider a kind of observational study that looks at adaptations on a broader scale. Biologists using the comparative method evaluate hypotheses by looking at patterns of evolution among species.

Table 10.1 Distributions of rocks available versus rocks chosen by snakes

Thin, medium, and thick rocks are equally abundant at Eagle Lake, but garter snakes retreating under rocks at night show a strong preference for medium ($P < 0.05$; chi-square test with thin and thick rocks combined because of small expected values).

	Thin (<20 cm)	Medium (20–40 cm)	Thick (>40 cm)
Rocks available to snakes	**32.4%**	**34.6%**	**33%**
Rocks chosen by snakes	**7.7%**	**61.5%**	**30.8%**

Source: From Table 1 in Huey et al. (1989b).

10.4 The Comparative Method

In Sections 10.2 and 10.3 we considered how experiments and observations on individuals within populations can be used to test hypotheses about adaptation. Here we examine how comparisons among species can be used to study the evolution of form and function. Our example comes from a group of bats called the Megachiroptera, which includes the fruit bats and the flying foxes (Figure 10.12a).

Why Do Some Bats Have Bigger Testes than Others?

The comparative method seeks to evaluate hypotheses by testing for patterns across species, such as correlations among traits, or correlations between traits and features of the environment.

Males in some of these bat species have larger testes for their body size than others. Based on work on a variety of other animals, David Hosken (1998) hypothesized that large testes are an adaptation for sperm competition. Sperm competition occurs when a female mates with two or more males during a single estrus cycle, and the sperm from the different males are in a race to the egg. One way a male can increase his reproductive success in the face of sperm competition is to produce larger ejaculates. By entering more sperm into the race, he increases his odds of winning. And the way to produce larger ejaculates is to have larger testes.

To evaluate the sperm competition hypothesis, Hosken needed to use it to develop a testable prediction. Hosken knew that fruit bats and flying foxes roost in groups, and that the size of a typical group varies dramatically among species, from two or three individuals to tens of thousands. Hosken reasoned that females living in larger groups would have more opportunities for multi-

(a)

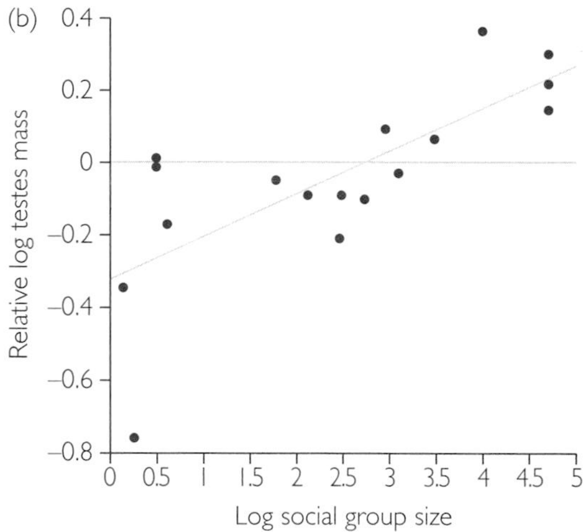

Figure 10.12 Variation in testis size among fruit bats and flying foxes (a) A grey-headed flying fox (*Pteropus poliocephalus*). (b) Relative testes size (that is, testes size adjusted for body size) as a function of roost group size for 17 species of fruit bats and flying foxes. From Hosken (1998).

ple matings, and that males living in larger groups would thus experience greater sperm competition. Hosken predicted that whenever a bat species evolves a preference for roosting in larger groups, its males will also evolve larger testes for their body size.

The simplest way to test this hypothesis is to gather data for a variety of species and prepare a scatterplot showing relative testes size as a function of roost group size. When Hosken did this, he found that the two variables are strongly correlated (Figure 10.12b). Bat species that live in larger groups have larger testes for their body size. As Hosken knew, however, there may be less evidence in this graph than meets the eye.

Figure 10.13 illustrates why. Imagine, for simplicity, that we have plotted a graph for only six species. We will call these species A, B, C, D, E, and F. Figure 10.13a shows a scatterplot for relative testes size versus group size. Like the real scatterplot in Figure 10.12b, this graph shows a positive correlation between the two traits. Now imagine that the evolutionary relationships among our six species are as shown in the phylogeny in Figure 10.13b. Species A, B, and C are all closely related to each other, as are species D, E, and F. It may be that species A, B, and C all inherited their small group sizes and their small testes from their common ancestor (green arrow). Likewise, it may be that species D, E, and F all inherited their large group sizes and large testes from their common ancestor (orange arrow). The possibility that our six species inherited their traits from just two common ancestors deflates the strength of our evidence considerably.

When we prepare a scatterplot and use it as the basis for claims about nature, we want all the data points to be independent of each other. If they are independent, then each data point makes a separate statement for or against our claim. Furthermore, independence of the data points is a requirement for traditional statistical tests. To make sure our scatterplot accurately reflects the nature of the evidence, we should therefore replace the points for species A, B, and C with a single point representing their common ancestor, and we should do the same with the points for species D, E, and F.

The graph in Figure 10.13c shows the result. It may, in fact, be true that group size and testes size evolve together, and it may be true that sperm competition is the reason. However, a scatterplot with only two data points is weak evidence on which to base such a claim.

Joe Felsenstein (1985) developed a better way to evaluate cross-species correlations among traits. What we look at in Felsenstein's method are patterns of

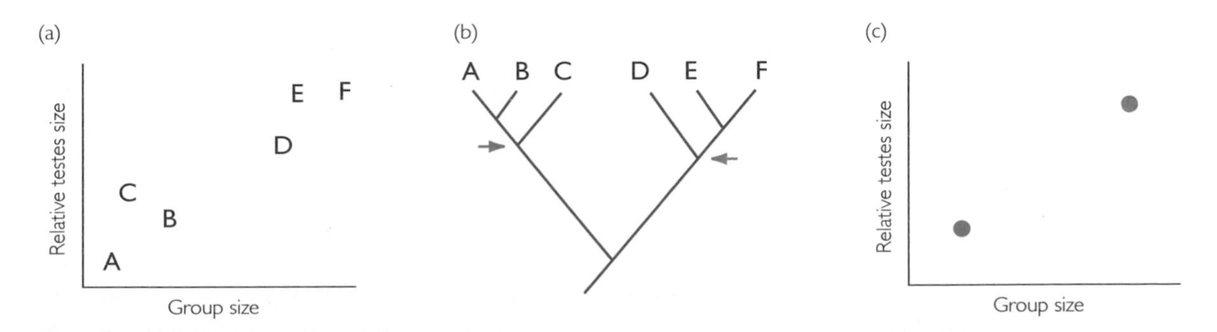

Figure 10.13 A simple scatterplot may provide only weak evidence that two traits evolve in tandem See text for explanation. After Lauder et al. (1995).

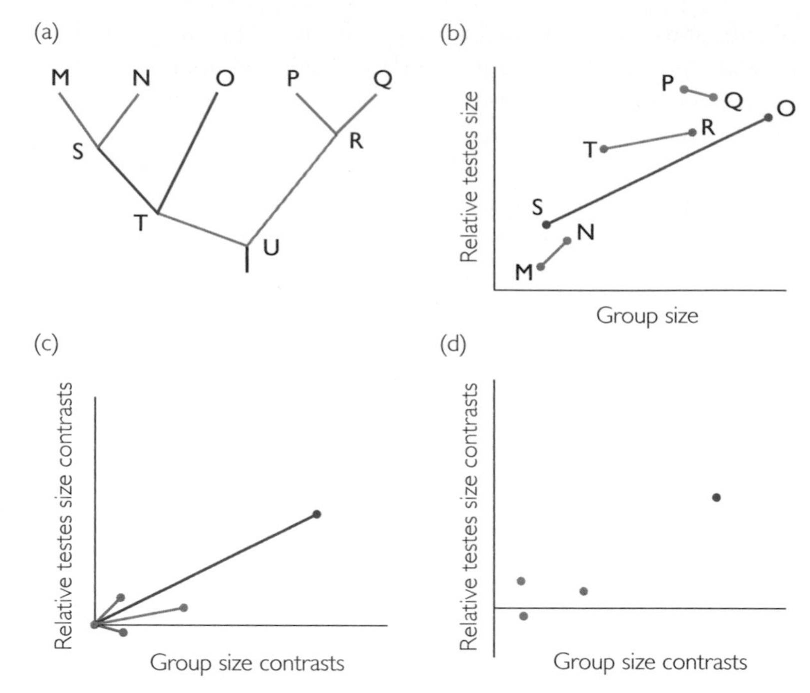

Figure 10.14 **A graphical interpretation of the basic procedure in Felsenstein's method for evaluating phylogenetically independent contrasts** See text for explanation.

divergence as sister species evolve independently away from their common ancestors. Figure 10.14 shows a graphical interpretation of the method's basic approach.

The first thing we need is a phylogeny for the species we are studying. Figure 10.14a shows a hypothetical phylogeny for five extant species. We will call these species M through Q. The phylogeny also includes the common ancestors that lived at all the nodes on the tree. These are species R, S, T, and U. Note that there are four places on this phylogeny where sister species diverged from a common ancestor; each is indicated by a different color. For example, M and N are sister species that diverged from common ancestor S. Likewise, S and O are sister species that diverged from common ancestor T. What we want to know is this: When species diverge from a common ancestor, does the species that evolves larger group sizes also evolve larger testes?

Proper application of the comparative method requires knowledge of the evolutionary relationships among the species under study.

We can answer this question by first plotting all of the pairs of sister species on a scatterplot, with lines connecting their data points (Figure 10.14b). We then slide each pair (without stretching or tilting their connector) until the left point rests on the origin (Figure 10.14c). Finally, we can erase the points at the origin and the connecting lines. We are left with a scatterplot with four data points (Figure 10.14d). Each data point represents the divergence, or contrast, that arose between a pair of sister species as they evolved away from their common ancestor. If the contrasts are correlated with each other, then we can conclude that when a species evolved a larger group size than its sister species, it also tended to evolve larger testes. In practice, we must make adjustments to the data before we can do statistical tests to evaluate the strength of any patterns. These adjustments are described in Box 10.2.

Hosken (1998) repeated his analysis of testes size and group size in bats using Felsenstein's method, which is called the method of phylogenetically independent contrasts. Figure 10.15a shows a phylogeny of the 17 bat species whose data Hosken analyzed. Figure 10.15b shows the association between relative testes size and roost group size, uncorrected for the influence of phylogenetic history. Figure 10.15c shows a plot of the contrasts in relative testes size versus the con-

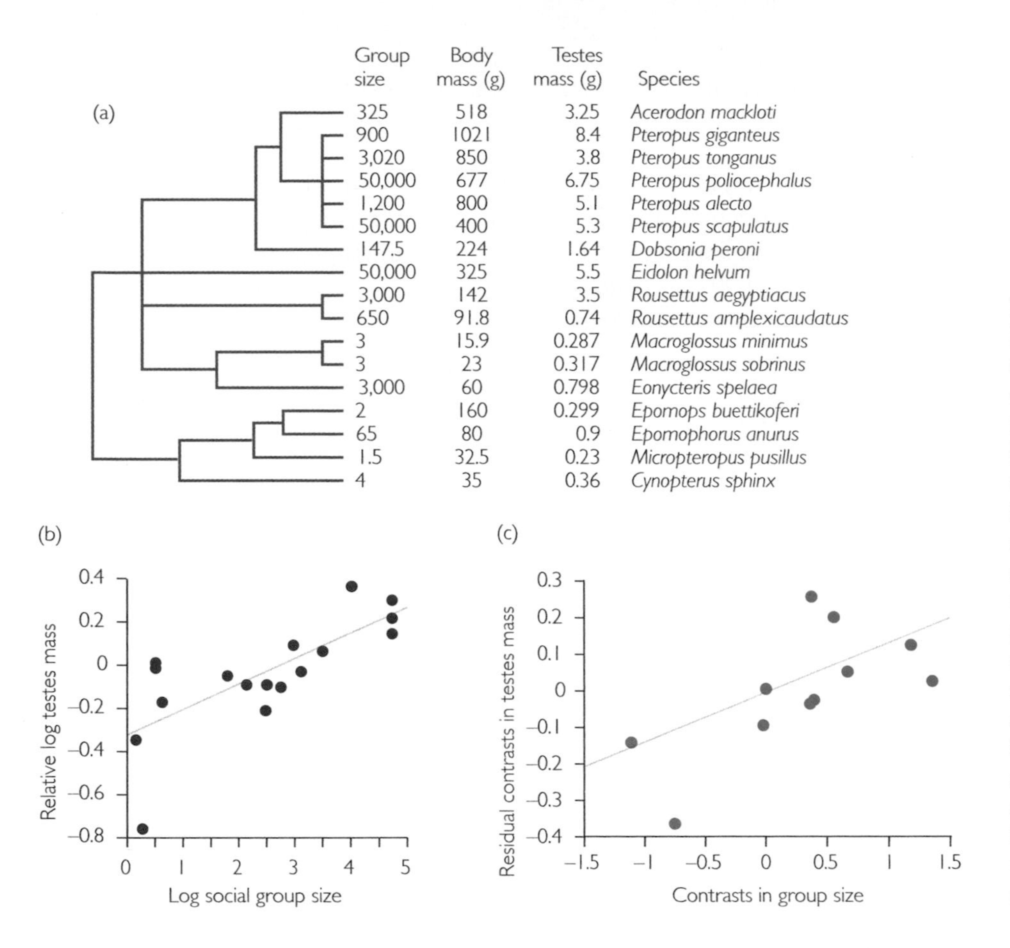

(a)
Group size
Body mass (g)
Testes mass (g)
Species
325 518 3.25 Acerodon mackloti
900 1021 8.4 Pteropus giganteus
3,020 850 3.8 Pteropus tonganus
50,000 677 6.75 Pteropus poliocephalus
1,200 800 5.1 Pteropus alecto
50,000 400 5.3 Pteropus scapulatus
147.5 224 1.64 Dobsonia peroni
50,000 325 5.5 Eidolon helvum
3,000 142 3.5 Rousettus aegyptiacus
650 91.8 0.74 Rousettus amplexicaudatus
3 15.9 0.287 Macroglossus minimus
3 23 0.317 Macroglossus sobrinus
3,000 60 0.798 Eonycteris spelaea
2 160 0.299 Epomops buettikoferi
65 80 0.9 Epomophorus anurus
1.5 32.5 0.23 Micropteropus pusillus
4 35 0.36 Cynopterus sphinx
(b)
Relative log testes mass
Log social group size
(c)
Residual contrasts in testes mass
Contrasts in group size

Figure 10.15 Correlated evolution of group size and testes size in fruit bats and flying foxes (a) A phylogeny for 17 species of bats, showing roost group size, body mass, and testes mass for each species. (b) The correlation between relative testes size and roost group size among the 17 species shown in the phylogeny, uncorrected for the influence of phylogenetic history. (c) Independent contrasts for relative testes size versus group size. The points on this graph show that when a bat species evolved larger (or smaller) group sizes than its sister species, it also tended to evolve larger (or smaller) testes ($P = 0.027$). From Hosken (1998).

trasts in group size. There is a significant positive correlation among the contrasts. In other words, when analyzed correctly the data show that when a bat species evolved larger roosting group sizes than its sister species, it also tended to evolve larger testes for its body size. Hosken concluded that the evidence from flying foxes and fruit bats is consistent with the hypothesis that large testes are an adaptation to sperm competition.

With data on 57 species representing all kinds of bats, Scott Pitnick and colleagues (2006) used phylogenetically independent contrasts to show that across species there is a negative association between testis size and brain size. That is, bat species that have evolved larger testes have also tended to evolve smaller brains. This association does not tell us whether the evolution of one of these traits drove the evolution of the other, or whether both traits were driven by some third factor. Pitnick and colleagues suggest that because brains and testes are metabolically expensive to grow and operate, bat species that invest heavily in one have only limited resources to invest in the other. The negative correlation does not appear to hold for fruit-eating bats, however, perhaps because—for a variety of reasons—their energy budgets are not as tight as those of other bats.

We have now considered three methods evolutionary biologists use to evaluate hypotheses about adaptation. In the next two sections of the chapter, we turn to complexities in organismal form and function that are active areas of current research. In the examples we discuss, researchers use experiments, observational studies, and the comparative method to investigate hypotheses about phenotypic plasticity (Section 10.5) and trade-offs and constraints on adaptation (Section 10.6).

When formulating and testing hypotheses about adaptation, biologists must keep in mind that organisms, and the lives they live, are complex.

Box 10.2 Calculating phylogenetically independent contrasts

Here we use an example from Garland and Adoph (1994) to illustrate the calculation of independent contrasts from a phylogeny (see also: Felsenstein 1985; Martins and Garland 1991; Garland et al. 1999; Garland et al. 2005). Figure 10.16 shows the phylogeny we will use. It shows the relationships among polar bears, grizzly bears, and black bears, and gives the body mass and home range of each. We will calculate independent contrasts for both traits among the bears. The steps are as follows:

1. Calculate the contrasts for pairs of sibling species at the tips of the phylogeny. In our three-species tree, there is just one pair of sibling species in which both species reside at the tips: polar bears and grizzly bears. The polar bear–grizzly bear contrast for body mass is:

$$265 - 251 = 14$$

The polar bear–grizzly bear contrast for home range is:

$$116 - 83 = 33$$

2. Prune each contrasted pair from the tree, and estimate the trait values for their common ancestor by taking the weighted average of the descendants' phenotypes. When calculating the weighted average, weight each species by the reciprocal of the branch length leading to it from the common ancestor. In our example, we are pruning polar bears and grizzlies from the tree and estimating the body mass and home range of their common ancestor A. The branch lengths from A to its descendants are both two units long. Thus, the weighted average for body mass is:

$$\text{Body mass of species A} = \frac{\left(\frac{1}{2}\right)265 + \left(\frac{1}{2}\right)251}{\left(\frac{1}{2}\right) + \left(\frac{1}{2}\right)} = 258$$

The weighted average for home range is:

$$\text{Home range of species A} = \frac{\left(\frac{1}{2}\right)116 + \left(\frac{1}{2}\right)83}{\left(\frac{1}{2}\right) + \left(\frac{1}{2}\right)} = 99.5$$

3. Lengthen the branch leading to the common ancestor of each pruned pair by adding to it the product of the branch lengths from the common ancestor to its descendants, divided by their sum. In our example, we are lengthening the branch leading to species A. The new branch length is:

$$3 + \frac{2 \times 2}{2 + 2} = 4$$

10.5 Phenotypic Plasticity

Throughout much of this book, we treat phenotypes as though they were determined solely and immutably by genotypes. We know, however, that phenotypes are often strongly influenced by the environment as well. Chapter 9 included a section on estimating how much of the phenotypic variation among individuals is due to variation in genotypes and how much is due to variation in environments. Here, we focus on the interplay between genotype, environment, and phenotype.

Another way to say that an individual's phenotype is influenced by its environment is to say that its phenotype is plastic. When phenotypes are plastic, individuals with identical genotypes may have different phenotypes if they live in different environments. Phenotypic plasticity is itself a trait that can evolve, and it may or may not be adaptive. As with the other traits we have discussed, to demonstrate that an example of phenotypic plasticity is adaptive, we must first determine what it is for, then show that individuals who have it achieve higher fitness than individuals who lack it.

4. Continue down the tree calculating contrasts, estimating the phenotypes of the common ancestors, and lengthening the branches leading to the common ancestors. In our example, the only remaining contrast is between species A and black bears. We do not need to estimate the phenotype of species B, or lengthen the branch leading to it, because species B is at the root of our tree. The species A–black bear contrast for body mass is:

$$258 - 93 = 165$$

The species A–black bear contrast for home range is:

$$99.5 - 57 = 42.5$$

5. Divide each contrast by its standard deviation to yield the standardized contrasts. The standard deviation for a contrast is the square root of the sum of its (adjusted) branch lengths. The standard deviation for the polar bear–grizzly bear contrast is:

$$\sqrt{2 + 2} = 2$$

The standard deviation for the species A–black bear contrast is:

$$\sqrt{4 + 5} = 3$$

The standardized contrasts for our example are given in Figure 10.16.

Once we have calculated the standardized contrasts, we can use them to prepare a scatterplot and to perform traditional statistical tests.

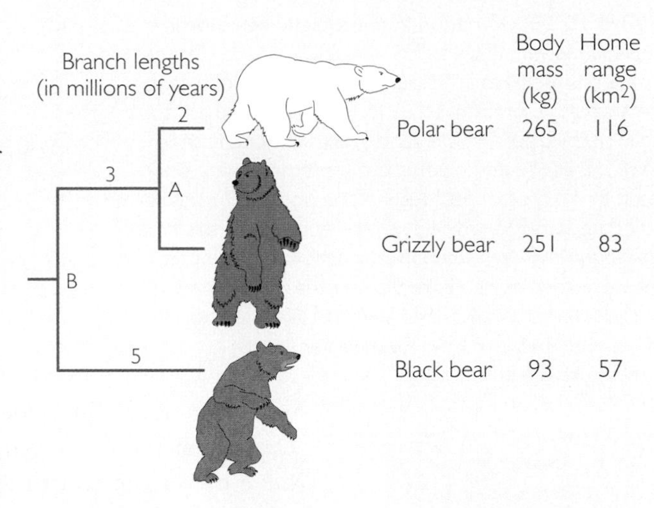

Contrast	Value for body mass	Standard deviation	Standardized contrast
Polar – Grizzly	265 – 251 = 14	2	7
A – Black bear	258 – 93 = 165	3	55

Contrast	Value for home range	Standard deviation	Standardized contrast
Polar – Grizzly	116 – 83 = 33	2	16.5
A – Black bear	99.5 – 57 = 42.5	3	14.17

Figure 10.16 An example showing how the data are adjusted when calculating phylogenetically independent contrasts From Garland and Adolph (1994).

Phenotypic Plasticity in the Behavior of Water Fleas

To illustrate phenotypic plasticity, we present the water flea, *Daphnia magna*. *Daphnia magna* is a tiny filter-feeding crustacean that lives in freshwater lakes (Figure 10.17). Conveniently for evolutionary biologists, *Daphnia* reproduce asexually most of the time. In other words, *Daphnia* clone themselves. This makes them ideal for studies of phenotypic plasticity, because researchers can grow genetically identical individuals in different environments and compare their phenotypes.

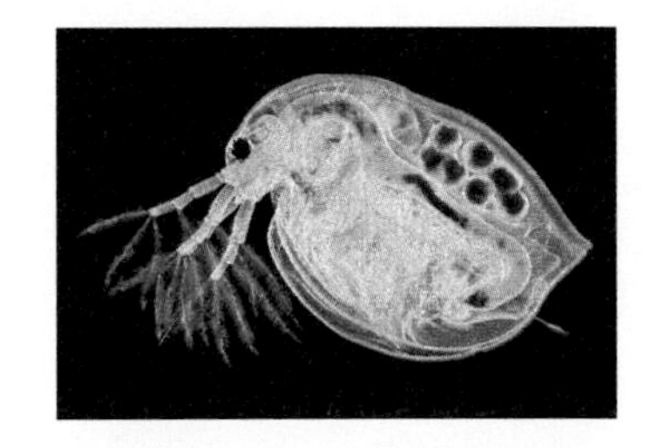

Figure 10.17 A water flea, *Daphnia magna* The branched appendages are antennae; the water flea uses them like oars for swimming. The dark object nearby is an eyespot. Also visible through the transparent carapace are the intestine and other internal organs. Enlarged about 10X.

Luc De Meester (1996) studied phenotypic plasticity in *D. magna*'s phototactic behavior. An individual is positively phototactic if it swims toward light and negatively phototactic if it swims away from light. De Meester measured the phototactic behavior typical of different genotypes of *D. magna*. In each single test, De Meester placed 10 genetically identical individuals in a graduated cylinder, illuminated them from above, gave them time to adjust to the change in environment, then watched to see where in the column they swam. De Meester summarized the results by calculating an index of phototactic behavior. The

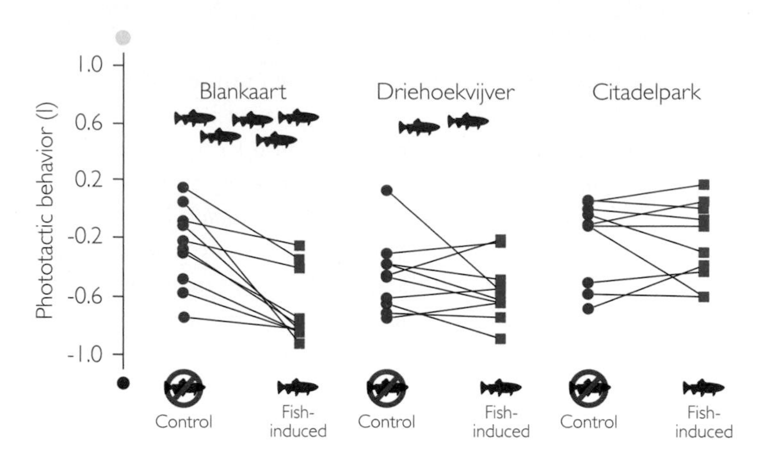

Figure 10.18 Variation in phototactic behavior in *Daphnia magna* Blankaart, Driehoekvijver, and Citadelpark are three lakes in Belgium. Each red dot represents the average result from three to five tests of the phototactic behavior of a single genotype (described in main text). The connected blue square represents the average result from three or four tests of the phototactic behavior of the same genotype. The difference is that this time the *Daphnia* were tested in water that had previously been occupied by fish. Lake Blankaart is home to many fish; Lake Driehoekvijver has few fish; Lake Citadelpark has no fish. Redrawn from De Meester (1996).

index can range in value from −1 to +1. A value of −1 means that all the *Daphnia* in the test swam to the bottom of the column, away from the light. A value of +1 means that all the *Daphnia* in the test swam to the top of the column, toward the light. An intermediate value indicates a mixed result.

De Meester measured the phototactic behavior of 10 *Daphnia* genotypes (also called clones) from each of three lakes. The results, indicated by the red dots in Figure 10.18, show that most *Daphnia* tend somewhat to avoid light. They also show that each lake harbors considerable genetic variation in phototactic behavior.

Genetically identical individuals reared in different environments may be different in form, physiology, or behavior. Such individuals demonstrate phenotypic plasticity.

De Meester also measured the phototactic behavior of the same 30 *Daphnia* genotypes in water that had been previously occupied by fish. The results are indicated by the blue squares in Figure 10.18. The red dot and blue square for each genotype are connected by a line. These lines are called **reaction norms**; they show a genotype's change in phenotype across a range of environments. *Daphnia magna*'s phototactic behavior is phenotypically plastic. In Lake Blankaart, in particular, most *Daphnia* genotypes score considerably lower on the phototactic index when tested in the presence of chemicals released by fish.

Finally, and most importantly, De Meester's results demonstrate that phenotypic plasticity is a trait that can evolve. Recall that a trait can evolve in a population only if the population contains genetic variation for the trait. Each of the *Daphnia* populations De Meester studied contains genetic variation for phenotypic plasticity. That is, some genotypes in each population alter their behavior more than others in the presence versus the absence of fish (Figure 10.18). Genetic variation for phenotypic plasticity is called **genotype-by-environment interaction**.

Has phenotypic plasticity evolved in the *Daphnia* populations De Meester studied? It apparently has. The average genotype in Lake Blankaart shows considerably more phenotypic plasticity than the average genotype in either of the other lakes. Blankaart is the only one of the lakes with a sizeable population of fish. Fish are visual predators, and they eat *Daphnia*. A reasonable interpretation is that predation by fish selects in favor of *Daphnia* that avoid well-lit areas when fish are present.

Christophe Cousyn, De Meester, and colleagues (2001) tested this hypothesis by taking advantage of the fact that *Daphnia* produce resting eggs that remain viable even after being buried in sediment for decades. The researchers took sediment cores from Oud Heverlee Pond, a small man-made lake constructed in 1970. From sediments of three different depths, representing distinct episodes in the history of the pond, the researchers hatched *Daphnia* clones. Each set of clones is a sample from the population's past. The researchers measured the phototactic behavior of the reawakened genotypes in the presence and absence of chemicals released by fish.

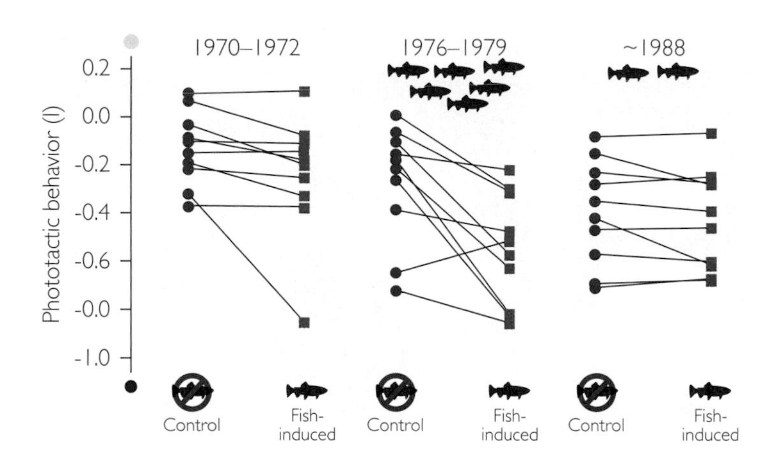

Figure 10.19 Evolution of phototactic behavior in *Daphnia magna* As in Figure 10.18, each pair of symbols connected by a line represents the phototactic behavior of a single genotype in the absence versus presence of chemicals released by fish. The three sets of genotypes come from samples of resting eggs produced during distinct episodes in the history of Oud Heverlee Pond. The earliest sample is from before the pond was stocked with plankivorous fish. The middle sample is from the period of heavy stocking. The last sample is from a period of reduced stocking. The degree of phenotypic plasticity shown by the population changed over time. Clones from the period of heavy stocking stay out of the light when they smell fish. Redrawn from Cousyn et al. (2001).

The people who built Oud Heverlee Pond began stocking it with planktivorous fish in 1973. They stocked it heavily until the mid-1980s, then less heavily through the late-1980s. Cousyn, De Meester, and colleagues predicted that the *Daphnia* population in the pond would have evolved in response to fish predation, and that genotypes preserved in resting resting eggs from the period of heavy stocking would show greater phenotypic plasticity in phototactic behavior than earlier or later genotypes.

When there is genetic variation for the degree or pattern of phenotypic plasticity, plasticity itself can evolve. Plasticity is adaptive when it allows individuals to adjust their phenotype so as to increase their fitness in the particular environment in which they find themselves.

The results appear in Figure 10.19. As predicted, the water flea population in Oud Heverlee changed over time. Clones from the period of heaviest fish stocking show the greatest shift in behavior in across environments. They stay out of the light when they smell predators.

Phenotypic plasticity is widespread, and perhaps underappreciated as an adaptation. As Theodosius Dobzhansky pointed out in 1937 (page 170), "Selection deals not with the genotype as such, but with its dynamic properties, its reaction norm, which is the sole criterion of fitness in the struggle for existence."

10.6 Trade-Offs and Constraints

It is impossible for any population of organisms to evolve optimal solutions to all selective challenges at once. We have mentioned examples of trade-offs in passing. In Section 10.4, for example, we noted that large testes help bats win at sperm competition but appear to impose metabolic costs that lead to the evolution of smaller and less energetically demanding brains. In Chapter 3, page 92, we lamented the mosquito fish whose large gonopodium entices mates but slows his escape from predators. In this section, we explore additional factors that limit adaptive evolution. These include trade-offs, functional constraints, and lack of genetic variation.

It is impossible to build a perfect organism. Organismal design reflects a compromise among competing demands.

Female Flower Size in a Begonia: A Trade-Off

The tropical plant *Begonia involucrata* is **monoecious**—that is, there are separate male and female flowers on the same plant. The flowers are pollinated by bees. As the bees travel among male flowers gathering pollen, they sometimes also transfer pollen from male flowers to female flowers. The male flowers offer the bees a reward, in the form of the pollen itself. The female flowers offer nothing; instead they get pollinated by deceit (Ågren and Schemske 1991). Not surprisingly, bees make more and longer visits to male flowers than to female flowers.

Figure 10.20 ***Begonia involucrata*** (a) Male (left) and female (right) flowers. The flowers lack true petals. Instead, each has a pair of petaloid sepals. The sepals are white or pinkish. In the center of each flower is a cluster of yellow anthers or stigmas. The stigmas of female flowers resemble the anthers of males. (b) An inflorescence, or stalk bearing many flowers. Each inflorescence makes both male and female flowers. Typically, the male flowers open first, and the female flowers open later. The inflorescence shown is unusual in having flowers of both sexes open at once.

The female flowers resemble the male flowers in color, shape, and size (Figure 10.20a). This resemblance is presumably adaptive. Given that bees avoid female flowers in favor of male flowers, the rate at which female flowers are visited should depend on how closely they mimic male flowers. The ability to attract pollinators should, in turn, influence fitness through female function, because seed set is limited by pollen availability. Presumption is not evidence, however. There are other possibilities. Doug Schemske and Jon Ågren (1995) sought to distinguish between two hypotheses about how bees might select on female flower size:

Hypothesis 1: The more closely female flowers mimic typical male flowers, the more often they will trick bees into visiting. Selection on female flowers is stabilizing, with best phenotype for females identical to the mean phenotype of males (Figure 10.21a, left).

Hypothesis 2: The more closely female flowers mimic the most rewarding male flowers, the more often they will succeed in duping bees. If larger male flowers offer bigger rewards, then selection on female flowers is directional, with bigger flowers always favored over smaller flowers (Figure 10.21a, right).

Schemske and Ågren made artificial flowers of three different sizes (Figure 10.21b), arrayed equal numbers of each in the forest, and watched to see how often bees approached and visited them. The results were clear: The larger the flower, the more bee approaches and visits it attracted (Figure 10.21c). Selection by bees on female flowers is strongly directional.

Taken at face value, Schemske and Ågren's results suggest that female flower size in *Begonia involucrata* is maladaptive. Selection by bees favors larger flowers, yet female flowers are no bigger than male flowers. Why are female flowers not huge? One solution to this paradox is that *B. involucrata* simply lacks genetic variation for female flowers that are substantially larger than male flowers. Schemske and Ågren have no direct evidence on this suggestion; *B. involucrata* is a perennial that takes a long time to reach sexual maturity, so quantitative genetic experiments are difficult to do.

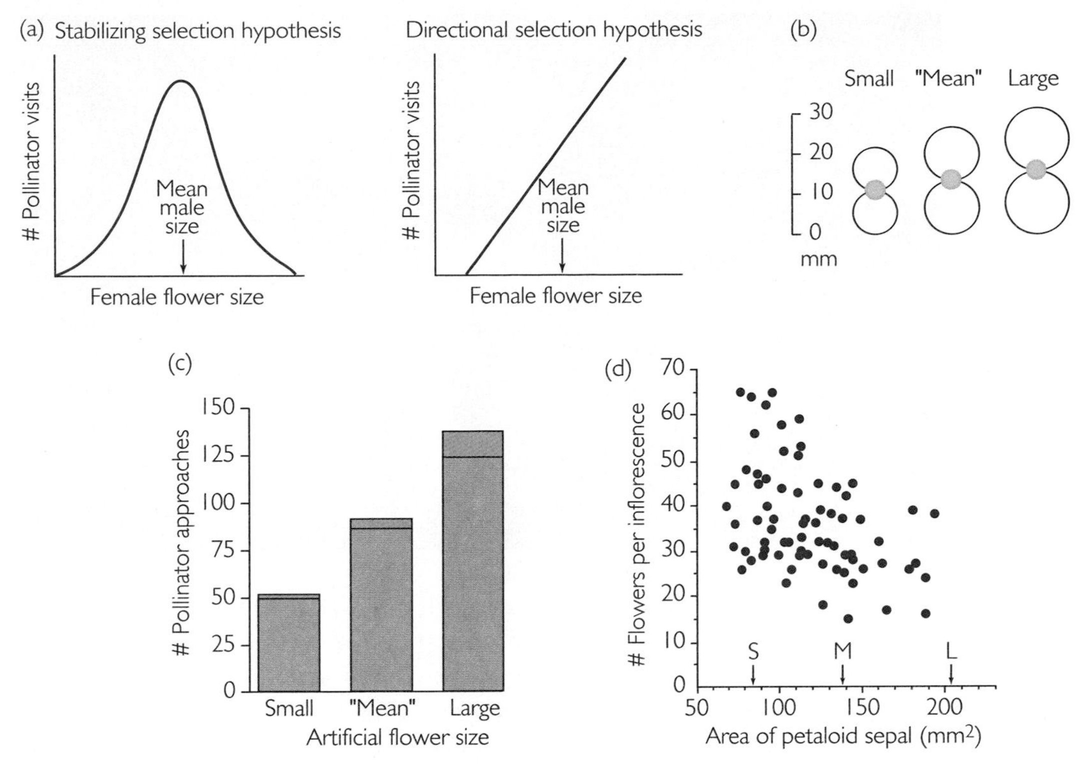

Figure 10.21 **An analysis of selection on female flower size in *Begonia involucrata*** (a) The two hypotheses investigated by Schemske and Ågren (1995). See text for more details. (b) Schemske and Ågren's three size classes of artificial flowers. The "mean" size class is the same size as the mean size of natural male flowers. (c) Pollinator preference as a function of flower size. The blue bars represent the number of bees that approached the artificial flowers; the brown bars represent the number of pollinators that actually visited the artificial flowers. Schemske and Ågren placed equal numbers of each size flower in the forest, but larger flowers attracted significantly more approaches and significantly more visits from bees. (d) Number of female flowers per inflorescence as a function of flower size. There is a statistically significant trade-off between flower size and flower number. From Schemske and Ågren (1995).

Another solution to the paradox is that focusing on individual female flowers gives us too narrow a view of selection. Schemske and Ågren expanded their focus from individual flowers to inflorescences (Figure 10.20b). The researchers measured the size and number of the female flowers on 74 inflorescences. They discovered a trade-off: The larger the female flowers on an inflorescence, the fewer flowers there are (Figure 10.21d). Such a trade-off makes intuitive sense. If an individual plant has a finite supply of energy and nutrients to invest in flowers, it can slice this pie into a few large pieces or many small pieces but not into many large pieces. Inflorescences with more flowers may be favored by selection for two reasons. First, bees may be more attracted to inflorescences with more flowers. Second, more female flowers mean greater potential seed production. Schemske and Ågren hypothesize that female flower size in *B. involucrata* has been determined, at least in part, by two opposing factors: directional selection for larger flower and the trade-off between flower size and number.

Resources devoted to one body part or function may be resources stolen from another part or function.

Flower Color Change in a Fuchsia: A Constraint

Fuchsia excorticata, also known as the Kotukutuku, is a bird-pollinated tree endemic to New Zealand (Delph and Lively 1989). Its flowers hang downward like

Figure 10.22 ***Fuchsia excorticata*** This bird-pollinated tree is native to New Zealand. Why do its flowers change color?

bells (Figure 10.22). The ovary is at the top (10.23a). The body of the bell consists of the hypanthium, or floral tube, and the sepals. The style resembles an elongated clapper. It is surrounded by shorter stamens and a set of reduced petals.

The hypanthium and sepals are the most conspicuously showy parts of the flower. They remain green for about 5.5 days after the flower opens, then begin to turn red (Figure 10.23b). The transition from green to red lasts about 1.5 days, at the end of which the hypanthium and sepals are fully red. The red flowers remain on the tree for about five days. The red flowers then separate from the ovary at the abscission zone and drop from the tree.

Pollination occurs during the green phase and into the intermediate phase, but it is complete by the time the flowers are fully red. The flowers produce nectar on days 1 through 7 (Figure 10.23b). Most flowers have exported more than 90% of their pollen by the end of that time. The stigmas are receptive to pollen at least until the second day of the fully red phase, but rarely does pollen arriving after the first day of the red phase actually fertilize eggs. Not surprisingly, bellbirds and other avian pollinators strongly prefer green flowers and virtually ignore nectarless red flowers (Delph and Lively 1985).

Why do the flowers of this tree change color? A general answer, supported by research in a variety of plants, is that color change serves as a cue to pollinators,

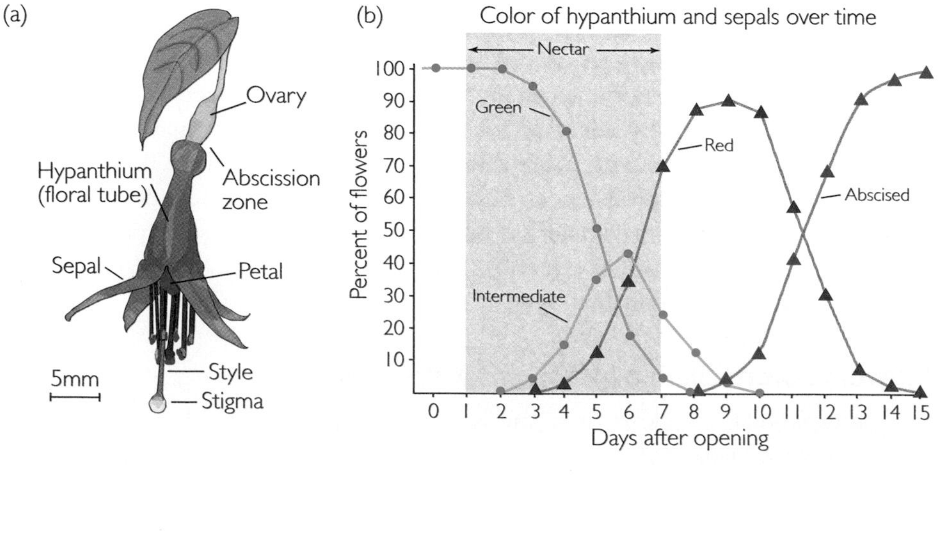

(a) Ovary, Hypanthium (floral tube), Abscission zone, Sepal, Petal, Style, Stigma, 5mm

(b) Color of hypanthium and sepals over time — Nectar; Green; Red; Abscised; Intermediate; Percent of flowers; Days after opening

Figure 10.23 **Flower color change in *Fuchsia excorticata*** (a) A *Fuchsia excorticata* flower. (b) The horizontal axis shows flower age, in days after opening. The vertical axis and graph lines show the percentage of flowers that are in each color phase at each age. From Delph and Lively (1989).

alerting them that the flowers are no longer offering a reward (for a review see Delph and Lively 1989). By paying attention to this cue, pollinators can increase their foraging efficiency; they do not waste time looking for nonexistent rewards. Individual plants benefit in return, because when pollinators forage efficiently they also transfer pollen efficiently. They do not deposit viable pollen on unreceptive stigmas, and they do not deposit nonviable pollen on receptive stigmas.

This answer is only partially satisfying, however. Why does *F. excorticata* not just drop its flowers immediately after pollination is complete? Dropping the flowers would give an unambiguous signal to pollinators that a reward is no longer being offered, and it would be metabolically much cheaper than maintaining the red flowers for several days. Retention of the flowers beyond the time of pollination seems maladaptive.

Lynda Delph and Curtis Lively (1989) consider two hypotheses for why *F. excorticata* keeps its flowers (and changes them to red) instead of just dropping them. The first is that red flowers may still attract pollinators to the tree displaying them, if not to the red flowers themselves. Once drawn to the tree, pollinators could then forage on the green flowers still present. Thus, retention of the red flowers could increase the overall pollination efficiency of the individual tree retaining them. If this hypothesis is correct, then green flowers surrounded by red flowers should receive more pollen than green flowers not surrounded. Delph and Lively tested this prediction by removing red flowers from some trees but not from others, and from some branches within trees but not from others. The researchers then compared the amount of pollen deposited on green flowers in red-free trees and branches versus red-retaining trees and branches. They found no significant differences. The pollinator-attraction hypothesis does not explain the retention of the red flowers in *F. excorticata*.

The second hypothesis Delph and Lively consider is that a physiological constraint prevents *F. excorticata* from dropping its flowers any sooner than it does. This physiological constraint is the growth of pollen tubes. After a pollen grain lands on a stigma, the pollen germinates. The germinated pollen grain grows a tube down through the style to the ovary. The pollen grain's two sperm travel through this tube to the ovary, where one of the sperm fertilizes an egg. The growth of pollen tubes takes time, especially in a plant like *F. excorticata,* which has long styles. If the plant were to drop its flowers before the pollen tubes had time to reach the ovaries, the result would be the same as if the flowers had never been pollinated at all. Delph and Lively pollinated 40 flowers by hand. After 24 hours, they plucked 10 of the flowers, dissected them, and examined them under a microscope to see whether the pollen tubes had reached the ovary. After 48 hours, they plucked and dissected 10 more flowers, and so on. The results appear in Table 10.2. It takes about 3 days for the pollen tubes to reach the ovary.

This result is consistent with the physiological constraint hypothesis. *F. excorticata* cannot start the process of dropping a flower until about 3 days after the flower is finished receiving pollen. Dropping a flower involves forming a structure called an abscission zone between the ovary and the flower (Figure 10.23a).

Traits or behaviors that would appear to be adaptive may, in fact, be physiologically or mechanically problematic.

Table 10.2 Pollen tube growth in *Fuchsia excorticata*

Days since pollination	1	2	3	4
Percentage of 10 flowers with pollen tubes in ovary	**0**	**20%**	**100%**	**100%**

Source: After Delph and Lively (1989).

The abscission zone consists of several layers of cells that form a division between the ovary and the flower. In *F. excorticata,* the growth of the abscission layer takes at least 1.5 days. The plant is therefore constrained to retain its flowers for at least 4.5 days after pollination ends. In fact, the plant retains its flowers for about 5 days. Delph and Lively suggest that flower color change in *F. excorticata* is an adaptation that evolved to compensate for the physiological constraints that necessitate flower retention. Given that the plant had to retain its flowers, selection favored individuals offering cues that allow their pollinators to distinguish the receptive versus unreceptive flowers on their branches. The pollinators deposit the incoming pollen onto receptive stigmas only, and they carry away only outgoing pollen that is viable.

Host Shifts in an Herbivorous Beetle: Constrained by Lack of Genetic Variation?

In several previous chapters, we have made the point that genetic variation is the raw material for evolution by natural selection. Because natural selection is the process that produces adaptations, genetic variation is also the raw material from which adaptations are molded. Conversely, populations of organisms may be prevented from evolving particular adaptations simply because they lack the necessary genetic variation to do so.

Here is an extreme example: Pigs have not evolved the ability to fly. We can imagine that flying might well be adaptive for pigs. It would enable them to escape from predators and to travel farther in search of their favorite foods. Pigs do not fly, however, because the vertebrate developmental program lacks genetic variation for the growth of both a trotter and a wing from the same shoulder. Other vertebrates have evolved the ability to fly, of course. But in bats and in birds, the developmental program has been modified to convert the entire forelimb from a leg to a wing; in neither group does an entirely new limb sprout from the body. Too bad for pigs.

Pig flight makes a vivid example, but in the end it is a trivial one. The wished-for adaptation is too unrealistic. Douglas Futuyma and colleagues have sought to determine whether lack of genetic variation has constrained adaptation in a more realistic and meaningful example (Funk et al. 1995; Futuyma et al. 1995; references therein). Futuyma and colleagues studied host plant use by herbivorous leaf beetles in the genus *Ophraella*. Among these small beetles, each species feeds, as larvae and adults, on the leaves of one or a few closely related species of composites (plants in the sunflower family, the Asteraceae). Each species of host plant makes a unique mixture of toxic chemicals that serve as defenses against herbivores. For the beetles, the ability to live on a particular species of host plant is a complex adaptation that includes the ability to recognize the plant as an appropriate place to feed and lay eggs, as well as the ability to detoxify the plant's chemical defenses.

An estimate of the phylogeny for 12 species of leaf beetle appears in Figure 10.24. The figure also lists the host plant for each beetle species. The evolutionary history of the beetle genus has included several shifts from one host plant to another. Four of the host shifts were among relatively distantly related plant species: They involved switches from a plant in one tribe of the Asteraceae to a plant in another tribe. These shifts are indicated in the figure by changes in the shading of the phylogeny. Other shifts involved movement to a new host in the same genus as the ancestral host, or in a genus closely related to that of the ancestral host.

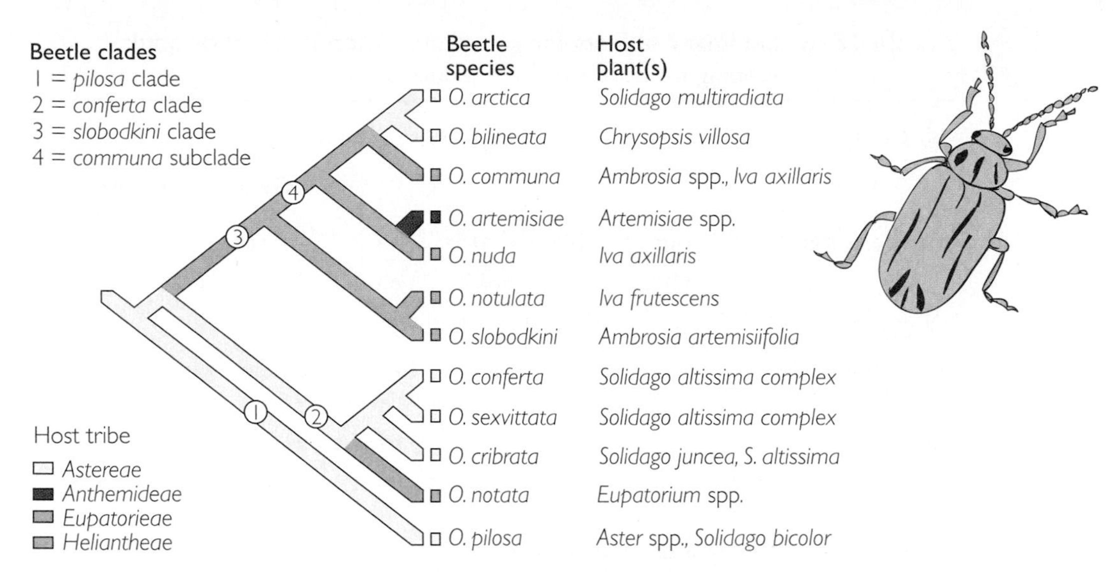

Figure 10.24 **Phylogeny of the leaf beetles, genus *Ophraella*** The numbers on the tree define the major branches (clades) of beetles. The shading of branches indicates the tribes of host species. The evolutionary history of the beetle genus has included four host shifts across tribes. From Futuyma et al. (1995).

Each combination of a beetle species and the host plant used by one of its relatives represents a plausible evolutionary scenario for a host shift that might have happened, but did not. For example, the beetle *Ophraella arctica* might have switched to the host *Iva axillaris*. Futuyma and colleagues have attempted to elucidate why some host shifts have actually happened while others have remained hypothetical. Here are two hypotheses:

Hypothesis 1: All host shifts are genetically possible. That is, every beetle species harbors sufficient genetic variation in its feeding and detoxifying mechanisms to allow at least some individuals to feed and survive on every potential host species. If a few individuals can feed and survive, they can be the founders for a new population of beetles that will evolve to become well-adapted to the new host. Because all host shifts are genetically possible, the pattern of actual host shifts has been determined by ecological factors and by chance. Ecological factors might include the abundance of the various host species within the geographic ranges of the beetle species, and the predators and competitors associated with each host species.

Hypothesis 2: Most host shifts are genetically impossible. That is, most beetle species lack sufficient genetic variation in their feeding and detoxifying mechanisms to allow any individuals to feed and survive on any but a few of the potential host species. The pattern of actual host shifts has been largely determined by what was genetically possible. Genetically possible host shifts have happened; genetically impossible host shifts have not.

We have presented these hypotheses as mutually exclusive. In fact, the truth is almost certainly that the actual pattern of host shifts has resulted from a mixture of genetic constraints, ecological factors, and chance. What Futuyma and colleagues were looking for was concrete evidence that genetic constraints have been at least part of the picture.

Futuyma and colleagues used a quantitative genetic approach (see Chapter 9) to determine how much genetic variation the beetles harbor for feeding and surviving on other potential hosts. The researchers examined various combinations

Table 10.3 Summary of tests for genetic variation in larval or adult feeding on potential host plants

(a) Tests for genetic variation in larval or adult feeding, by relationship among host plants

	Genetic variation?	
Beetle tested for feeding on a plant that is . . .	**Yes**	**No**
. . . in the same tribe as the beetle's actual host	7	1
. . . in a different tribe than the beetle's actual host	14	17

Conclusion: Genetic variation for feeding is more likely to be found when a beetle is tested on a potential host that is closely related to its actual host.

(b) Tests for genetic variation in larval or adult feeding, by relationship among beetles

	Genetic variation?	
Beetle tested for feeding on a plant that is . . .	**Yes**	**No**
. . . the host of a beetle in the same major clade	12	4
. . . the host of a beetle in a different major clade	9	14

Conclusion: Genetic variation for feeding is more likely to be found when a beetle is tested on a potential host that is the actual host of a closely related beetle.

Source: From Table 7 in Futuyma et al. (1995).

of four of the beetle species listed in Figure 10.24 with six of the host plants. Their tests revealed that there is little genetic variation in most beetle species for feeding and surviving on most potential host species. In 18 of 39 tests of whether larvae or adults of a beetle species would recognize and feed on a potential host plant, the researchers found no evidence of genetic variation for feeding. In 14 of 16 tests of whether larvae could survive on a potential host plant, the researchers found no evidence of genetic variation for survival.

Populations sometimes lack the genetic variation that would provide the raw material to evolve particular adaptations.

These results suggest that hypothesis 2 is at least partially correct. Many otherwise-plausible host shifts appear to be genetically impossible. Futuyma and colleagues performed an additional test of hypothesis 2 by looking for patterns in their data on genetic variation for larval and adult feeding. If hypothesis 2 is correct, then a beetle species is more likely to show genetic variation for feeding on a potential new host if the new host is a close relative of the beetle's present host. Futuyma et al.'s data confirm this prediction (Table 10.3a). Likewise, if hypothesis 2 is correct, then a beetle species is more likely to show genetic variation for feeding on a potential new host if the new host is the actual host of one of the beetles' close relatives. Futuyma et al.'s data also confirm this prediction (Table 10.3b). Futuyma and colleagues conclude that hypothesis 2 is at least partially correct. The history of host shifts in the beetle genus *Ophraella* has been constrained by the availability of genetic variation for evolutionary change.

Host Shifts in Feather Lice: Constrained by Dispersal Ability?

In the study we have just discussed, Futuyma and colleagues sought to show that host shifts are at least sometimes constrained by lack of genetic variation. The alternative explanations for why some host shifts have happened and others have not are ecological factors and chance. Dale Clayton and Kevin Johnson (2003) have identified a case in which host shifts appear to be constrained by an ecological factor.

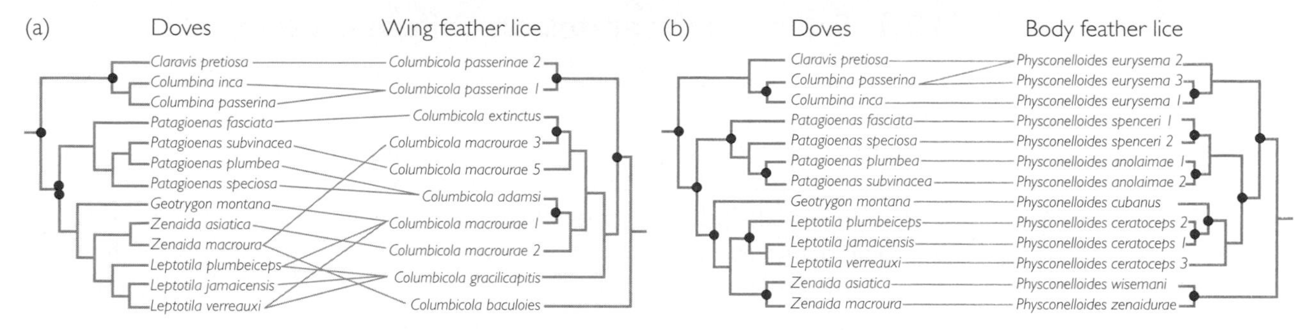

Figure 10.25 Phylogenetic congruence and discord for doves and their feather lice (a) The tree on the left is doves; the tree on the right is for their wing feather lice (genus *Columbicola*). Lines connect the parasite species to the bird species they infect. The many crossing lines indicate frequent host shifts in the evolutionary history of the lice. (b) The tree on the left is for doves; the tree on the right is for their body feather lice (genus *Physconelloides*). Lines connect the parasite species to the bird species they infect. The absence of crossing lines indicates that the lice have not changed hosts. Instead, they have gone along for the ride, splitting into new lineages when their hosts have. Redrawn from Clayton and Johnson (2003).

Clayton and Johnson analyzed the history of host shifts in the feather lice that infest doves. These ectoparasites include lice that live on wing feathers (genus *Columbicola*) and lice that live on body feathers (genus *Physconelloides*). Figure 10.25a compares the evolutionary trees for several dove species versus their wing feather lice. The phylogenies are not congruent, indicating that body feather lice have switched host species frequently. Figure 10.25b compares the evolutionary trees for the same dove species versus their body feather lice. This time the phylogenies are highly congruent, indicating that body feather lice have not switched host species. Instead, they have simply gone along for the ride, speciating only when their hosts have speciated.

Why have wing feather lice switched host species often while body feather lice have not? Experiments in which Clayton and colleagues (2003) transferred feather lice to novel hosts suggest that many host switches are genetically possible. Transplanted lice attach and feed on novel hosts. They can also evade the host's preening as long as their new host is similar in body size to their native host. Instead of being constrained by lack of variation for the ability to survive on novel hosts, Clayton and Johnson think that body feather lice simply have fewer chances to switch host species. This is because body feather lice disperse among individual hosts less readily than wing feather lice do. Field observations by Noah Whiteman and colleagues (2004) support this contention. These researchers looked for wing and body lice from Galápagos doves on Galápagos hawks. The two parasite species are equally common on doves, their native host, but on hawks dove wing lice are much more common than dove body lice.

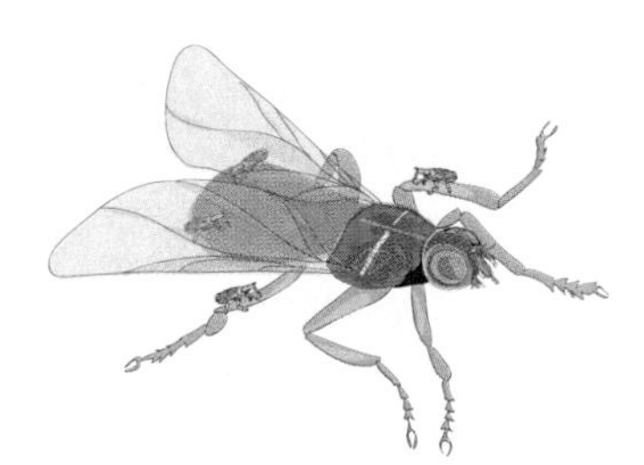

Figure 10.26 Dispersal via a lousy fly This drawing, based on a live example, shows three wing feather lice hitching a ride on the legs of a parasitic fly. After Clayton et al. (2004).

One way feather lice move from one host to another is via direct bodily contact between the two birds. Another way is by hitching a ride on the legs of a parasitic hippoboscid fly, as shown in Figure 10.26. The flies are less host-specific than lice, so a stowaway louse may find itself deposited on a novel host. Published records suggest that wing feather lice hitch rides on flies much more often than body feather lice. Apparently the reason body feather lice have so rarely switched host species is that they could not get a lift.

In this section and the previous one we have examined complications of organismal form and function that must be taken into account when studying adaptation. In the next section, we consider another kind of complication that must sometimes be taken into account—a complication in the action of natural selection itself.

10.7 Selection Operates on Different Levels

In the examples we have discussed so far, both in this chapter and in the book, we have been concerned with natural selection operating at the level of individuals within populations. At this level it is the birth, reproduction, and death of individual organisms that determines which alleles become common and which disappear. If an allele influences phenotype such that the average individual carrying the allele has greater than average reproductive success, then the allele's frequency will rise; otherwise the allele's frequency will fall (see Box 6.3, page 186).

Selection can act at other levels as well. In this section we first describe an experiment demonstrating that selection can act at the level of organelles within cells. We then discuss a genetic disease in humans that may be maintained at unexpectedly high frequency due to a similar phenomenon. Selection may favor the causative allele at the level of cells within individuals despite strong selection against the allele at the level of individuals within populations.

A Demonstration That Selection Acts at Different Levels

Douglas Taylor and colleagues (2002) used yeast (*Saccaromyces cerevisiae*) and their mitochondria to demonstrate that selection can act simultaneously at different levels. The researchers chose yeast because normal yeast cells can harvest energy in two ways: by fermentation and by respiration. This means that for yeast, unlike for most eukaryotes, the ability to respire is not essential for life.

Respiration in yeast cells, as in other eukaryotes, is carried out by mitochondria. Mitochondria occasionally sustain large deletions in their genomes that render them unable to respire. These non-respiratory mitochondria are intracellular parasites. They extract energy and material from their host cells and provide nothing in return. For most eukaryotic cells, having exclusively parasitic mitochondria would be fatal. For yeast cells, however, it is merely disadvantageous. It limits the yeast cells to harvesting energy by fermentation, but it does not kill them.

Organisms harbor populations of cells, organelles, and nucleotide sequences. Selection can operate within these populations.

Taylor and colleagues established yeast populations in which the founding individuals contained both normal and parasitic mitochondria. That is, each individual cell in the yeast population was itself home to a genetically variable population of mitochondrial genomes. These mitochondrial genomes replicate independently of the host cell's nuclear genome. Some mitochondrial genomes may replicate more rapidly than others. This means that the population of mitochondrial genomes within a cell can evolve by natural selection just as any other population can. Mitochondrial genomes that replicate rapidly will become common; mitochondrial genomes that replicate slowly will become rare.

At the level of mitochondrial genomes within a yeast cell, selection favors parasites over normal mitochondria. This is because parasitic mitochondrial genomes can replicate faster. If we stayed inside a yeast cell and tracked the evolution of its mitochondrial population, we would expect parasitic mitochondria to inexorably increase in frequency.

At the level of yeast cells within a petri dish, however, selection favors the ability to respire. Yeast cells that can respire can harvest energy more quickly and thus replicate faster. If we stayed inside a petri dish and tracked the evolution of its yeast population, we would expect respiration-competent yeast cells to inexorably increase in frequency.

Selection at the level of mitochondria within yeast cells is thus in opposition to selection at the level of yeast cells within petri dishes. What is the ultimate outcome? That depends on the relative intensities of selection at the two different levels.

Taylor and colleagues maintained their yeast cultures for 150 generations at three population sizes: small (about 10 yeast cells), medium (about 250 yeast cells), and large (about 18,000 cells). In small populations, selection among yeast cells is largely insignificant. Sampling error during propagation is the primary determinant of which yeast lineages persist and which disappear. In small populations, therefore, Taylor and colleagues predicted that selection at the level of mitochondria within cells would dominate. They expected that at the end of their experiment most yeast cells in their small cultures would contain exclusively parasitic mitochondria. In large populations, however, selection among yeast cells is important. Sampling error during propagation is negligible, and speed of growth and replication is the primary determinant of which yeast lineages persist and which disappear. In large populations, therefore, the researchers expected that yeast cells containing exclusively parasitic mitochondria would be rare.

As a control, the researchers established yeast cultures in which the founding individuals contained both chloramphenicol-susceptible and chloramphenicol-resistant mitochondria. Chloramphenicol susceptibility was selectively neutral under the conditions of the experiment. Averaged across populations, Taylor and colleagues expected the frequency of yeast cells containing exclusively susceptible mitochondria to remain at intermediate frequency.

The results, shown in Figure 10.27, confirm the researchers' predictions. In small populations most yeast cells contained exclusively parasitic mitochondria,

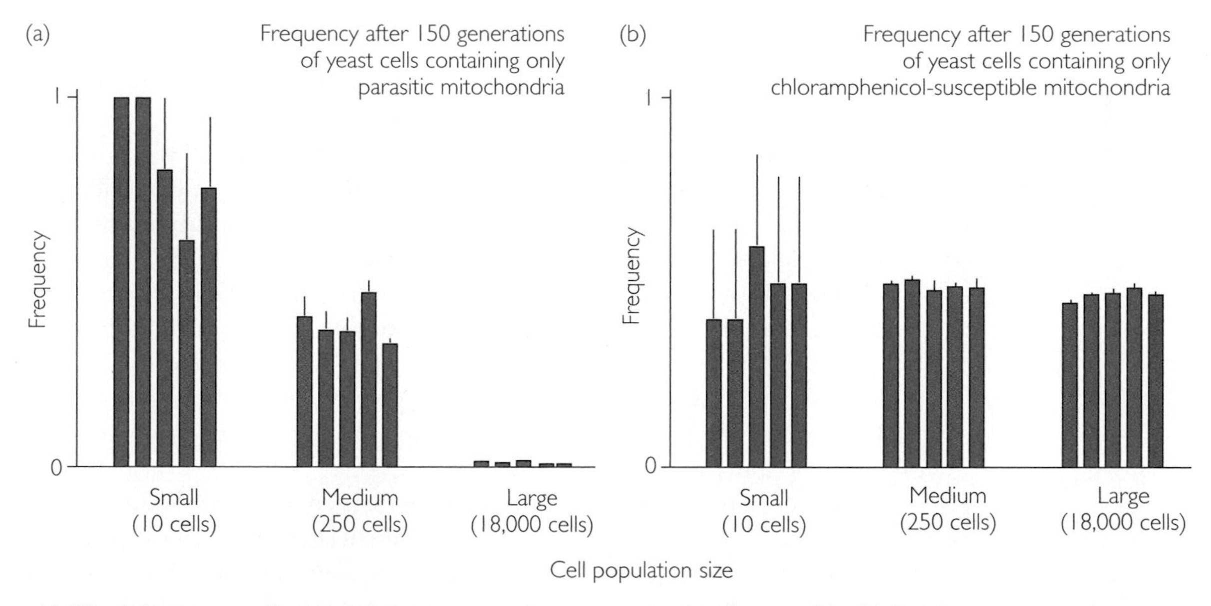

Figure 10.27 Selection on cells in populations versus selection on mitochondria in cells (a) Each bar represents the average of five experimental populations started with yeast cells containing a mixture of normal versus parasitic mitochondria. Among mitochondria within yeast cells, selection favors parasites, because they replicate faster. This selective advantage was constant across experiments. Among yeast cells within populations, selection favors yeast containing normal mitochondria, because they can harvest energy by respiration as well as fermentation. This selective advantage varies among yeast cultures maintained at different population sizes; it is weakest in small populations and strongest in large populations. Parasitic mitochondria thrive in small yeast populations but fall to low frequency in large yeast populations. (b) Each bar represents the average of four or five control populations started with yeast cells containing a mixture of chloramphenicol-resistant versus chloramphenicol-susceptible mitochondria. Chloramphenicol resistance is selectively neutral at both levels of selection. From Taylor et al. (2002).

whereas in large populations few yeast cells contained exclusively parasitic mitochondria. In control cultures the frequency of yeast cells containing exclusively chloramphenicol-susceptible mitochondria, averaged across populations, remained at intermediate levels.

This experiment demonstrates that when selection among yeast cells is relatively weak, selection among mitochondria within yeast cells can lead to the fixation of traits that decrease the mean fitness of the yeast population. Had we not recognized that selection acts on different levels, we might have mistakenly concluded that loss of the ability to respire is somehow adaptive for yeast cells living in small populations. Instead, we can see that a trait that is maladaptive for the organism that carries it may be driven to fixation by selection at lower levels. Taylor and colleagues believe their work will also provide insight into the progression of degenerative genetic diseases associated with the accumulation, within tissues, of mutant mitochondria.

Just as a yeast cell is home to an evolving population of self-replicating mitochondrial genomes, an animal is home to an evolving population of self-replicating cellular genomes. Recent research has suggested that evolution of cell populations can explain puzzling features of a human genetic disease.

Multilevel Selection in Apert Syndrome?

Selection at the level of cells, organelles, or sequences may be in opposition to selection at the level of whole organisms.

Apert sydrome is a genetic disease caused by a mutation in the gene for fibroblast growth factor receptor 2 (FGFR2). The manifestations are severe, including premature fusion of the suture joints in the skull, facial malformation, and fusion of the fingers and toes. Given that the condition is dominant and affected individuals have low fitness, it is not surprising that most cases are caused by new mutations. Less expected is that the mutant allele virtually always comes from the father. The risk of Apert syndrome increases with the father's age, and the incidence among newborns suggests a rather high mutation rate.

To better understand these facts, Anne Goriely and colleagues (2003) used modern molecular techniques to assess the frequency of mutant alleles among the sperm produced by men of different ages. The most common Apert mutation is a single nucleotide substitution at nucleotide 755 in the FGFR2 gene, the middle position in codon 252. The normal codon reads TCG, specifying the amino acid serine. There are three possible substitutions for C, yielding three different codons. TGG encodes tryptophan and causes Apert syndrome. TTG encodes leucine and results in either a normal phenotype or a condition known as Crouzon syndrome. TAG is a stop codon and has not been documented as a germline mutation.

Goriely and colleagues determined the frequency of each substitution among the sperm and blood cells of men of different ages. The data appear in Figure 10.28. Look first at graph (a) and focus on the diamonds, which represent the frequencies of the Apert mutation among the sperm of different men. There is considerable variation but a general pattern of increase with age. We might explain this pattern by positing that the mutation occassionally occurs among the stem-cell lineages that divide to produce sperm, and that as men age they accumulate more mutant lineages. The same argument might apply to the Crouzon mutation (10.28b). What this explanation does not explain, however, is that the frequency of the nonsense mutation does not increase with age (10.28c). It also fails to explain why none of the frequencies change with age in blood cells, as shown by the squares in all three graphs.

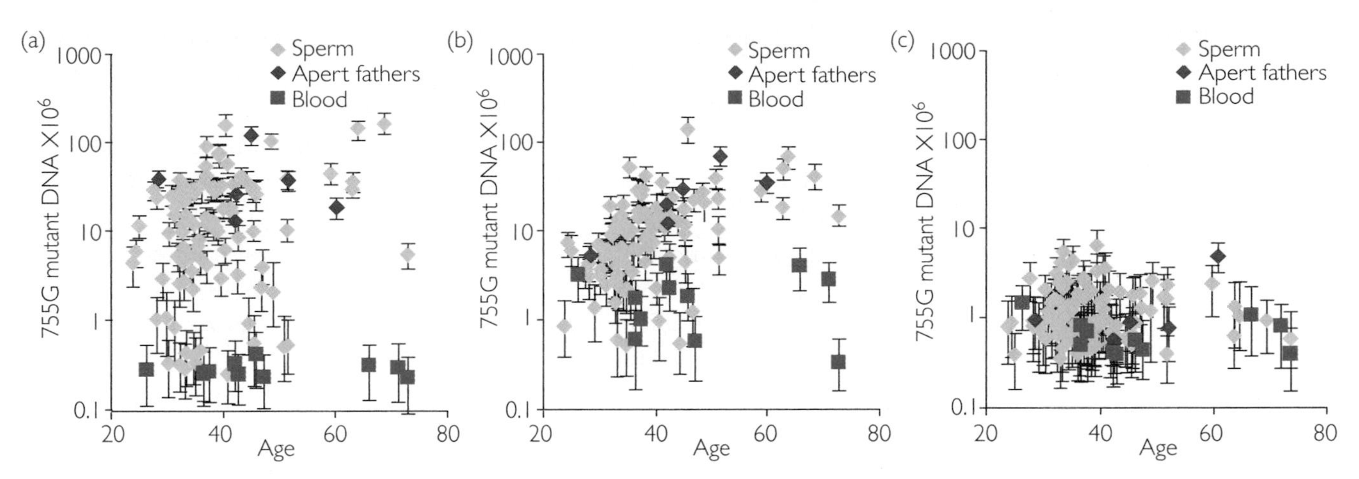

Figure 10.28 **Change in the frequencies of three mutations among cells in two tissues as a function of age in men** (a) The frequency of the mutation causing Apert syndrome increases with age among sperm but not among blood cells. (b) The frequency of the mutation causing Crouzon syndrome increases with age among sperm but not among blood cells. (c) The frequency of a nonsense mutation does not change with age in either tissue. All three mutations are single-nucleotide substitutions at position 755 in the gene for FGFR2. These patterns make sense if the Apert and Crouzon mutations induce increased proliferation in the stem cells that produce sperm. From Goriely et al. (2003).

Goriely and colleagues think there is a better explanation, one that accounts for all of these puzzling features (see also Crow 2003). They suggest that all three substitutions occur only rarely in the stem cell lineages that yield both sperm and blood. When either the Apert mutation or the Crouzon mutation occurs in spermatogonia, however, it enables the cells that carry it to divide more rapidly, essentially giving rise to small cancers that continue to produce sperm. Because the mutant cells enjoy higher reproductive success, the population of spermatogonia—which now harbors genetic variation—evolves. The frequency of mutant stem cells, and thus the frequency of mutant sperm, increases over time. Consistent with this explanation is that the Apert and Crouzon mutations are both gain-of-function mutations. They enable the FGR2 protein to bind its ligands more strongly and to bind a greater variety of molecules. Also consistent is that FGR2 is expressed spermatogonial stem cells (Goriely et al. 2005).

If Goriely and colleagues are right, then the frequency of Apert syndrome in the human population is determined by a balance between opposing patterns of selection acting at different levels. When the Apert mutation occurs within the testes of an individual man, selection acting at the level of cells within tissues causes the frequency of the mutation to increase. When babies are born with Apert syndrome, however, selection at the level of individuals within populations causes the frequency of the mutation to fall.

10.8 Strategies for Asking Interesting Questions

We began this chapter with a review of approaches evolutionary biologists use when testing hypotheses about organismal form and function. However, testing a hypothesis is the second half of a good research project. The first half is formulating a hypothesis in the first place. Formulating hypotheses worthy of testing means asking interesting questions, then making educated guesses about the answers. We close the chapter with a brief list of strategies for asking good questions about evolution:

- Study natural history. Descriptive studies can lead to the discovery of new patterns that need explanation. Some things in nature just leap out and demand

Learning how to ask good questions is as important as learning how to answer them.

explanation, such as the oxpecker's odd habits, or the wing-waving display of *Zonosemata*. Some of the most compelling science happens when a researcher simply picks an organism and sets out to learn about it.

- Question conventional wisdom. It is often untested. What makes Week's work on oxpeckers so captivating is that it undermines an adaptive scenario long accepted as fact.
- Question the assumptions underlying a popular hypothesis or research technique. Felsenstein's development of an improved method of comparative analysis grew out of the recognition that the traditional approach to comparative research was violating its own assumptions.
- Draw analogies that transfer questions from field to field or taxon to taxon. If fruit bats and flying foxes that evolve larger group sizes also evolve larger testes for their body size, might not the same be true of other kinds of animals?
- Ask why not. The studies reviewed on trade-offs and constraints were motivated by researchers who thought their study organisms were failing to do something that might be adaptive.

Summary

Among the major activities of evolutionary biology is analyzing the form and function of organisms to determine whether and why particular traits are adaptive. To establish that a trait is adaptive, researchers must formulate hypotheses about how the trait is used and why individuals possessing the trait have higher fitness than individuals lacking it. Then, because no hypothesis should be accepted simply because it is plausible, researchers must put their hypotheses to test. Researchers test hypotheses by using them to make predictions, then collecting data to determine whether the predictions are correct.

Researchers use a variety of approaches in collecting data to test hypotheses. The most powerful method is the controlled experiment. Controlled experiments involve groups of organisms that are identical but for a single variable of interest. The experimental variable can then be confidently identified as the cause of any differences in survival and reproductive success among the groups. When experiments are impractical, careful observational studies can yield data valuable for testing hypotheses. Finally, comparisons among species can be used to confirm or refute predictions, so long as researchers take into account the shared evolutionary history of the species under study.

When analyzing adaptations, we do well to keep in mind that organisms are complicated. Individuals may be phenotypically plastic, so that genetically identical individuals reared in different environments have different phenotypes. The function of a particular trait may change over evolutionary time, and it may reflect a compromise among competing environmental or physiological demands. Finally, populations may simply lack the genetic variation required to become perfectly adapted to their environments.

We also do well to keep in mind that selection can act at multiple levels. When selection at the level of organisms within populations is relatively weak, selection at the level of organelles within cells or cells within tissues can drive the evolution of traits that are maladaptive for the organisms that carry them. These and other complications are the subjects of current research by evolutionary biologists.

Questions

1. Describe in your own words the difference between an experimental study, an observational study, and a comparative study. What sorts of questions are they each suited for (i.e., why don't researchers always use the experimental method)? Give an example of each type of study from this chapter.

2. What were Futuyma et al.'s two hypotheses for why leaf beetles have not colonized all possible species of host plants? What did they do to test the hypotheses? How do their results illuminate the general question of whether all traits are adaptive?

3. For which of the following studies would you recommend the use of Felsenstein's method of phylogenetically independent contrasts? Why?

a. A comparison of feather parasite burden and beak shape in different species of birds.

b. An experiment that tests whether birds whose beak shape are experimentally altered will end up with greater parasite loads (similar to Clayton et al.'s study).

c. An observational study that measures the correlations among beak shape of individual birds, with their preening behavior, and with their parasite loads.

4. What is an evolutionary trade-off? Why do they occur? Give two examples. How does the occurrence of trade-offs illuminate the general question of whether all traits are adaptive?

5. What is an evolutionary constraint? Why do they occur? Give two examples. How does their occurrence illuminate the general question of whether all traits are adaptive?

6. How does Apert syndrome explain why some traits occur that may be maladaptive?

7. As a review, list all of the reasons you can think of that may cause a given trait to not be adaptive, despite the action of positive natural selection on the trait.

8. a. Why was it important that Greene and colleagues tested tephritid flies whose wings had been cut off and then glued back on?

b. Why did they do the wing-cutting experiments at all? For example, why didn't they just compare intact tephritid flies to houseflies?

c. Why was it important that the five types of flies were presented to each spider in random order?

9. In Huey et al.'s experiment, snakes often chose thick rocks despite the associated risk of being too cool. Outline two hypotheses for why snakes sometimes choose thick rocks. Are your hypotheses testable? Do both hypotheses assume that the behavioral trait of choosing thick rocks is adaptive?

10. Geckos are unusual lizards in that they are active at night instead of during the day. Describe the difficulties a gecko would face in trying to use its behavior to regulate its temperature at night. Would you predict that geckos have an optimal temperature for sprinting that is the same, higher, or lower than that of a typical diurnal lizard? Huey et al. (1989a) found that the geckos they studied had optimal temperatures that are the same as those of typical diurnal lizards (a finding in conflict with the researchers' own hypothesis). Can you think of an explanation?

11. Suppose that fish were introduced into Lake Citadelpark, one of the lakes De Meester studied. What do you predict will happen to phenotypic plasticity in the *Daphnia* of Lake Citadelpark? Outline the observations you would need to make to test your prediction.

12. Throughout this chapter, we have stressed this fundamental question: How can we test whether a given trait is adaptive or not? As a further exercise, think about the costs and benefits of being a certain body size. For example, a mouse can easily survive a 30- foot fall. A human falling 30 feet would probably be injured, and an elephant falling 30 feet would probably be killed. Finally, a recent study of the bone strength of *Tyrannosaurus rex* revealed that if a fast-running *T. rex* ever tripped, it would probably die (Farlow, Smith, and Robinson 1995). Given these costs, why has large body size ever evolved? Can you think of some costs of small body size? How would you test your ideas?

13. Imagine you are an explorer who has just discovered two previously unknown large islands. Each island has a population of a species of shrub unknown elsewhere. On Island A, the shrubs have high concentrations of certain poisonous chemicals in their leaves. On Island B, the shrubs have nonpoisonous, edible leaves. The islands differ in many ways—for instance, Island A has less rainfall and a colder winter than Island B, and it has some plant-eating insects that are not found on Island B. Island A also has a large population of muntjacs, a small tropical deer that loves to eat shrubs. You suspect the muntjacs have been the selective force that has caused evolution of leaf toxins. How could you test this hypothesis? What alternative hypotheses can you think of? What data would disprove your hypothesis, and what data would disprove the other hypotheses?

14. Consider skin color in humans. Does this trait show genetic variation? Phenotypic plasticity? Genotype-by-environment interaction? Give examples documenting each phenomenon. Could phenotypic plasticity for skin color evolve in human populations? How?

15. The example on *Begonias* (Section 10.6) illustrated that organisms are frequently caught between opposing forces of selection. Each of the following examples also illustrates a tug-of-war between several forces of natural selection. For each example, hypothesize about what selective forces may maintain the trait described and what selective forces may oppose it.

a. A male moose grows new antlers, made of bone, each year.

b. Douglas fir trees often grow to more than 60 feet tall.

c. A termite's gut is full of cellulose-digesting microorganisms.

d. Maple trees lose all of their leaves in the autumn.

e. A male moth has huge antennae, which can detect female pheromones.

f. A barnacle attaches itself permanently to a rock when it matures.

16. Schemske and Ågren (1995) used artificial flowers instead of real flowers in their experiment (Section 10.6). What were the advantages of using artificial flowers? (There are at least two important ones.) What were the disadvantages?

17. P1 is a virus that infects bacteria. It often resides for long periods of time as a plasmid inside a bacterial cell, replicating and being transmitted to descendants of the original cell. Among the genes on P1 are two loci that form what is called an addiction module. One of the genes (called "doc," which stands for "death on cure") encodes a small protein that is both highly poisonous to bacterial cells and chemically stable. The other gene (called "phd," which stands for "prevent host death") encodes a protein that serves as an antidote to the poison. This antidote is chemically unstable because it is degraded by a bacterial protease. If a bacterial cell containing the P1 plasmid divides and produces a daughter cell that does not contain the plasmid, that daughter cell's protoplasm still contains the poison and the antidote. The antidote breaks down quickly, while the poison persists. The daughter cell then dies. Explain carefully why it is selectively advantageous for the P1 virus to carry an addiction module. How does selection at the level of the virus and its genes affect selection at the level of the bacteria that harbor them? (For more information on this example, see Lehnherr et al. 1993; Lehnherr and Yarmolinsky 1995.)

18. An exercise used in some graduate programs is to have students list 20 questions that they would like to answer. Groups of students then discuss the questions and help each other sort out which would be the most interesting to pursue. There are many criteria for deciding that a question is interesting. Is it new? Does it address a large or otherwise important issue? Would pursuing it lead to other questions? Is it feasible, or would development of a new technique make it feasible? Try this exercise yourself.

Exploring the Literature

19. An important aspect of evaluating scientific papers is to consider other explanations for the data that the authors might have overlooked. See if you can think of alternate explanations for the data presented in the following papers:

Benkman, C. W., and A. K. Lindholm. 1991. The advantages and evolution of a morphological novelty. *Nature* 349: 519–520.

Soler, M., and A. P. Møller. 1990. Duration of sympatry and coevolution between the great spotted cuckoo and its magpie host. *Nature* 343: 748–750.

Finally, see the following review of Soler and Møller's work for an example of how scientific criticism can result in better science by all involved:

Lotem, A., and S. I. Rothstein. 1995. Cuckoo-host coevolution: From snapshots of an arms race to the documentation of microevolution. *Trends in Ecology and Evolution* 10: 436–437.

20. Male sticklebacks sometimes steal eggs from other males' nests to rear as their own. Sievert Rohwer suggested that egg stealing is a courtship strategy. Males often eat eggs out of their own nests, in effect robbing the reproductive investment made by their mates and using the proceeds to fund their own reproductive activities. Females, in consequence, should prefer to lay eggs in nests already containing the eggs of other females, thereby reducing the risk to their own. A female preference for nests already containing eggs would mean that males without eggs in their nests could increase their attractiveness by stealing eggs from other males. See:

Rohwer, S. 1978. Parent cannibalism of offspring and egg raiding as a courtship strategy. *American Naturalist* 112: 429–440.

For a related phenomenon in birds, see:

Gori, D. F., S. Rohwer, and J. Caselle. 1996. Accepting unrelated broods helps replacement male yellow-headed blackbirds attract females. *Behavioral Ecology* 7: 49–54.

21. For a dramatic example of phenotypic plasticity in which an herbivorous insect uses the chemical defenses of its host as a cue for the development of defenses against its predators, see:

Greene, Erick. 1989. A diet-induced developmental polymorphism in a caterpillar. *Science* 243: 643–646.

See this paper for a discussion of an organism's "choice" about *when* to begin changing phenotype:

Nishimura, Kinya. 2006. Inducible plasticity: Optimal waiting time for the development of an inducible phenotype. *Evolutionary Ecology Research* 8 (3): 553–559.

22. Among the challenges faced by parasites is moving from one host to another. This challenge is a particularly potent agent of selection for parasites in which every individual must spend different parts of its life cycle in different hosts. What adaptations might you expect to find in parasites to facilitate dispersal from host to host? For dramatic examples in which parasites manipulate their hosts' behavior or appearance, see:

Tierney, J. F., F. A. Huntingford, and D. W. T. Crompton. 1993. The relationship between infectivity of *Schistocephalus solidus* (Cestoda) and antipredator behavior of its intermediate host, the three-spined stickleback, *Gasterosteus aculeatus*. *Animal Behavior* 46: 603–605.

Lafferty, K. D., and A. K. Morris. 1996. Altered behavior of parasitized killifish increases susceptibility to predation by bird final hosts. *Ecology* 77: 1390–1397.

Bakker, T. C. M., D. Mazzi, and S. Zala. 1997. Parasite-induced changes in behavior and color make *Gammarus pulex* more prone to fish predation. *Ecology* 78: 1098–1104.

23. Brown-headed cowbirds (*Molothrus ater*) lay their eggs in other birds' nests, a behavior called nest parasitism. When this strategy succeeds, the host birds accept the cowbird egg as one of their own and rear the cowbird chick. When the strategy fails, the host birds recognize the cowbird egg as an imposter and eject it from the nest. Why do any host species accept cowbird eggs in their nests? Given the obvious costs involved in rearing a chick of another species, acceptance seems maladaptive. Evolutionary biologists have proposed two competing hypotheses to explain why some host species accept cowbird eggs. The evolutionary lag hypothesis posits that species that accept cowbird eggs do so simply because they have not yet evolved ejection behavior. Either the host species lack genetic variation that would allow them to evolve ejection behavior, or the host species have been exposed to cowbird nest parasitism only recently and therefore have not had sufficient time for such behavior to evolve. The evolutionary equilibrium hypothesis posits that host species that accept cowbird eggs do so because they face a fundamental mechanical constraint: Their bills are too small to allow them to grasp a cowbird egg, and if they tried to puncture the cowbird egg they would destroy too many of their own eggs in the process. Given this constraint, host species have evolved a strategy that makes the best of a bad situation. Think about how you would test each of the competing hypotheses. Then see:

Rohwer, S., and C. D. Spaw. 1988. Evolutionary lag versus bill-size constraints: A comparative study of the acceptance of cowbird eggs by old hosts. *Evolutionary Ecology* 1988: 27–36.

Rohwer, S., C. D. Spaw, and E. Røskaft. 1989. Costs to northern orioles of puncture-ejecting parasitic cowbird eggs from their nests. *Auk* 106: 734–738.

Røskaft, E., S. Rohwer, and C. D. Spaw. 1993. Cost of puncture ejection compared with costs of rearing cowbird chicks for northern orioles. *Ornis Scandinavica* 24: 28–32.

Sealy, S. G. 1996. Evolution of host defenses against brood parasitism: Implications of puncture-ejection by a small passerine. *Auk* 113: 346–355.

Given that some host species eject cowbird eggs by first puncturing the egg, then lifting it out of the nest, what adaptations would you expect to find in cowbird eggs? Would these adaptations carry any costs? See:

Spaw, C. D., and S. Rohwer. 1987. A comparative study of eggshell thickness in cowbirds and other passerines. *Condor* 89: 307–318.

Picman, J. 1997. Are cowbird eggs unusually strong from the inside? *Auk* 114: 66–73.

24. For additional recent examples in which evolutionary biologists used a comparative approach employing independent contrasts to address interesting questions, see:

Downes, S. J., and M. Adams. 2001. Geographic variation in antisnake tactics: The evolution of scent-mediated behavior in a lizard. *Evolution* 55: 605-615.

Iwaniuk, A. N., S. M. Pellis, and I. Q. Whishaw. 1999. Brain size is not correlated with forelimb dexterity in fissiped carnivores (Carnivora): A comparative test of the principle of proper mass. *Brain, Behavior, and Evolution* 54: 167–180.

Citations

Ågren, J., and D. W. Schemske. 1991. Pollination by deceit in a Neotropical monoecious herb, *Begonia involucrata*. *Biotropica* 23: 235–241.

Clayton, D. H., and K. P. Johnson. 2003. Linking coevolutionary history to ecological process: doves and lice. *Evolution* 57: 2335–2341.

Clayton, D. H., S. E. Bush, B. M. Goates, and K. P. Johnson. 2003. Host defense reinforces host-parasite cospeciation. *Proceedings of the National Academy of Sciences USA* 100: 15694–15699.

Clayton, D. H., S. E. Bush, and K. P. Johnson. 2004. Ecology of congruence: past meets present. *Systematic Biology* 53: 165-173.

Clayton, D. H., B. R. Moyer, et al. 2005. Adaptive significance of avian beak morphology for ectoparasite control. *Proceedings of the Royal Society of London B* 272: 811-817.

Cousyn, C., L. De Meester, et al. 2001. Rapid, local adaptation of zooplankton behavior to changes in predation pressure in the absence of neutral genetic changes. *Proceedings of the National Academy of Sciences USA* 98: 6256-6260.

Crow, J. F. 2003. There's something curious about paternal-age effects. *Science* 301: 606-607.

Delph, L. F., and C. M. Lively. 1985. Pollinator visits to floral colour phases of *Fuchsia excorticata*. *New Zealand Journal of Zoology* 12: 599–603.

Delph, L. F., and C. M. Lively. 1989. The evolution of floral color change: Pollinator attraction versus physiological constraints in *Fuchsia excorticata*. *Evolution* 43: 1252–1262.

De Meester, L. 1996. Evolutionary potential and local genetic differentiation in a phenotypically plastic trait of a cyclical parthenogen, *Daphnia magna*. *Evolution* 50: 1293–1298.

Dobzhansky, T. 1937. *Genetics and the Origin of Species.* New York: Columbia University Press.

Farlow, J. D., M. B. Smith, and J. M. Robinson. 1995. Body mass, bone "strength indicator," and cursorial potential of *Tyrannosaurus rex*. *Journal of Vertebrate Paleontology* 15: 713–725.

Felsenstein, J. 1985. Phylogenies and the comparative method. *The American Naturalist* 125: 1–15.

Funk, D. J., D. J. Futuyma, G. Ortí, and A. Meyer. 1995. A history of host associations and evolutionary diversification for *Ophraella* (Coleoptera: Chrysomelidae): New evidence from mitochondrial DNA. *Evolution* 49: 1008–1017.

Futuyma, D. J., M. C. Keese, and D. J. Funk. 1995. Genetic constraints on macroevolution: The evolution of host affiliation in the leaf beetle genus *Ophraella*. *Evolution* 49: 797–809.

Garland, T., Jr., and S. C. Adolph. 1994. Why not do two-species comparative studies: Limitations on inferring adaptation. *Physiological Zoology* 67: 797–828.

Garland, T., Jr., P. E. Midford, and A. R. Ives. 1999. An introduction to phylogenetically based statistical methods, with a new method for confidence intervals on ancestral values. *American Zoologist* 39: 374–388.

Garland, T., Jr., A. F. Bennett, and E. L. Rezende. 2005. Phylogenetic approaches in comparative physiology. *Journal of Experimental Biology* 208: 3015–3035.

Goriely, A., G. A. T. McVean, et al. 2003. Evidence for selective advantage of pathogenic FGFR2 mutations in the male germ line. *Science* 301: 643–646.

Goriely, A., G. A. T. McVean, et al. 2005. Gain-of-function amino acid substitutions drive positive selection of FGFR2 mutations in human spermatogonia. *Proceedings of the National Academy of Sciences USA* 102: 6051–6056.

Gould, S. J., and R. C. Lewontin. 1979. The spandrels of San Marco and the Panglossian paradigm: A critique of the adaptationist programme. *Proceedings of the Royal Society of London B* 205: 581–598.

Greene, E., L. J. Orsak, and D. W. Whitman. 1987. A tephritid fly mimics the territorial displays of its jumping spider predators. *Science* 236: 310–312.

Hosken, D. J. 1998. Testes mass in megachiropteran bats varies in accordance with sperm competition theory. *Behavioral Ecology and Sociobiology* 44: 169–177.

Huey, R. B., and J. G. Kingsolver. 1989. Evolution of thermal sensitivity of ectotherm performance. *Trends in Ecology and Evolution* 4: 131–135.

Huey, R. B., P. H. Niewiarowski, J. Kaufmann, and J. C. Herron. 1989a. Thermal biology of nocturnal ectotherms: Is sprint performance of geckos maximal at low body temperatures? *Physiological Zoology* 62: 488–504.

Huey, R. B., C. R. Peterson, S. J. Arnold, and W. P. Porter. 1989b. Hot rocks and not-so-hot rocks: Retreat-site selection by garter snakes and its thermal consequences. *Ecology* 70: 931–944.

Lauder, G. V., R. B. Huey, et al. 1995. Systematics and the study of organismal form and function. *BioScience* 45: 696–704.

Lehnherr, H., E. Maguin, S. Jafri, and Y. B. Yarmolinsky. 1993. Plasmid addiction genes of bacteriophage P1: doc, which causes death on curing of prophage, and phd, which prevents host death when prophage is retained. *Journal of Molecular Biology* 233: 414–428.

Lehnherr, H., and Y. B. Yarmolinsky. 1995. Addiction protein Phd of plasmid prophage P1 is a substrate of the ClpXP serine protease of *Escherichia coli*. *Proceedings of the National Academy of Sciences USA* 92: 3274–3277.

Martins, E. P., and T. Garland, Jr. 1991. Phylogenetic analyses of the correlated evolution of continuous characters: A simulation study. *Evolution* 45: 534–557.

Mather, M. H., and B. D. Roitberg. 1987. A sheep in wolf's clothing: Tephritid flies mimic spider predators. *Science* 236: 308–310.

Mayr, E. 1983. How to carry out the adaptationist program? *American Naturalist* 121: 324–334.

McElligott, A. G., I. Maggini, et al. 2004. Interactions between red-billed oxpeckers and black rhinos in captivity. *Zoo Biology* 23: 347–354.

Moyer, B. R., D. W. Gardiner, and D. H. Clayton. 2002. Impact of feather molt on ectoparasites: looks can be deceiving. *Oecologia* 131: 203–210.

Pitnick, S., K. E. Jones, and G. S. Wilkinson. 2006. Mating system and brain size in bats. *Proceedings of the Royal Society B* 273: 719–724.

Platt, J. R. 1964. Strong Inference. *Science* 146: 347–353.

Schemske, D. W., and J. Ågren. 1995. Deceit pollination and selection on female flower size in *Begonia involucrata:* An experimental approach. *Evolution* 49: 207–214.

Taylor, D. R., C. Zeyl, and E. Cooke. 2002. Conflicting levels of selection in the accumulation of mitochondrial defects in *Saccharomyces cerevisiae*. *Proceedings of the National Academy of Sciences USA* 99: 3690–3694.

Weeks, P. 1999. Interactions between red-billed oxpeckers, *Buphagus erythrorhynchus,* and domestic cattle, *Bos taurus,* in Zimbabwe. *Animal Behaviour* 58: 1253–1259.

Weeks, P. 2000. Red-billed oxpeckers: vampires or tickbirds? *Behavioral Ecology* 11: 154–160.

Whiteman, N. K., D. Santiago-Alarcon, et al. 2004. Differences in straggling rates between two genera of dove lice (Insecta: Phthiraptera) reinforce population genetic and cophylogenetic patterns. *International Journal for Parasitology* 34: 1113–1119.

11

Sexual Selection

Male collared lizards (*Crotaphytus collaris,* above) wear brighter colors than females (below). They also attract more attention from predators. Jerry Husak and colleagues (2006) left painted clay models of males and females in the Wichita Mountains of Oklahoma and checked them over several days for bite marks left by mammals, snakes, and birds. Of 20 models of each sex, predators attacked 14 males and 0 females.

Males and females often differ strikingly in size, appearance, and behavior. In marine iguanas, for example, males weigh twice as much as females. Males become intensely territorial during the breeding season, while females remain gregarious throughout the year. In long-tailed widowbirds, the adults of the two sexes have plumage so distinct that it would be easy to mistake them for different species. Males are jet black, carry tail feathers several times the length of their own bodies, and have red and yellow shoulder patches. Females are colored a cryptic brown, with short tail feathers and no shoulder patches. In gray tree frogs, males have dark throats and sing. Females have white throats and are silent. In stalk-eyed flies, both sexes wear their eyes on the ends of long thin stalks, but males have longer eyestalks than females. In some species of pipefish, females have blue stripes and skin folds on their bellies. Males lack these ornaments. The photos of males and females in Figure 11.1 provide additional examples.

In humans, too, females and males are conspicuously different. Our differences exceed the obvious and essential ones in genitalia and reproductive organs. They are found in the appearance of our faces, the sound of our voices, the distribution of our body fat and body hair, and our size. The size difference between women and men is documented in Figure 11.2.

(a) Red deer

(b) Guppies

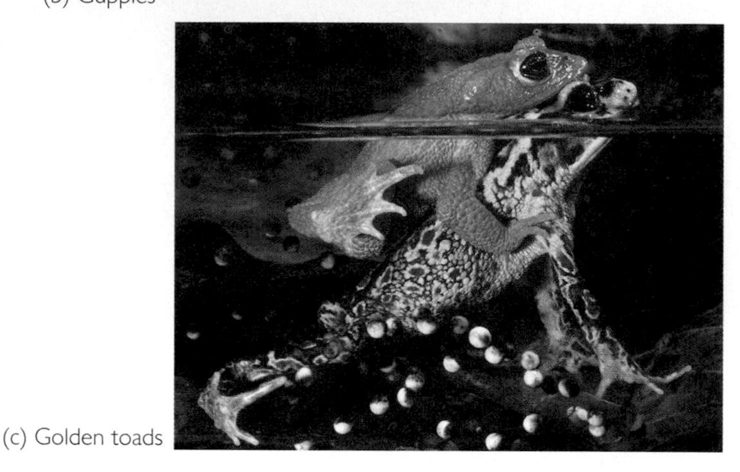
(c) Golden toads

Figure 11.1 **The differences between males and females (the sexual dimorphism) in red deer (*Cervus elaphus*), guppies (*Poecilia reticulata*), and golden toads (*Bufo periglenes*)** In (a), the male is on the left; in (b) and (c), the male is on the top.

A difference between the sexes is called a sexual dimorphism.

A difference between the males and females of a species is called a **sexual dimorphism**. In this chapter, we ask why sexual dimorphism occurs in such a great variety of organisms. It is a question Charles Darwin (1871) wrote half a book about, and it has captivated evolutionary biologists ever since.

11.1 Sexual Dimorphism and Sex

In previous chapters, we have explained the traits of organisms with the theory of evolution by natural selection. In Chapter 3, for example, we saw how natural selection shapes the beaks of medium ground finches on Daphne Major. When the finches face a drought and small soft seeds are rare, big beaks confer an advantage. Big beaks help finches survive by enabling them to open large, hard seeds and

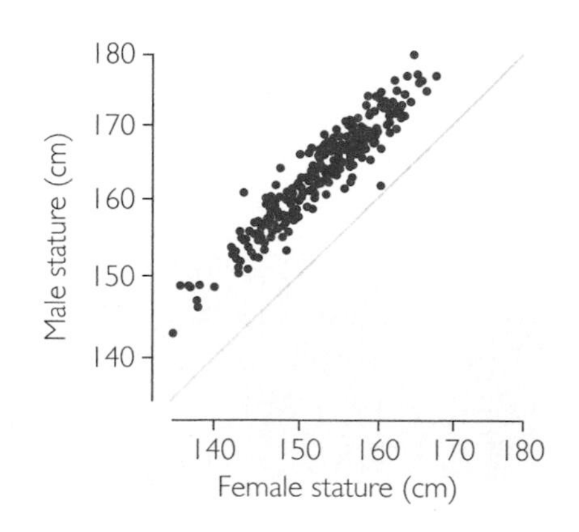

Figure 11.2 **Women and men differ in height** For each of more than 200 human societies, the average height of the men is plotted against the average height of the women. The diagonal line shows where the points would fall if men and women were of equal height. People vary widely in height from society to society: In the shortest society, the average man is about 143 cm tall (about 4 feet, 8 inches) and the average woman about 135 cm (~4′5″); in the tallest society, the average man is about 180 cm tall (~5′11″), and the average woman about 165 cm (~5′5″). But in every society the average man is taller than the average woman, usually by about 10%. From Rogers and Mukherjee (1992).

thus get more to eat. When the big-beaked survivors reproduce, they pass genes for big beaks to their offspring. In a similar way, natural selection can explain a great variety of other traits, from the bars on fly wings (see Figure 10.7, page 369) to the hiding places chosen by garter snakes (see Figure 10.11, page 375). But differences between the sexes of a species, like those shown in Figure 11.1, are often not among them.

Sexual dimorphism is often a puzzle, because natural selection cannot always explain it.

To see why, try to imagine how we might use evolution by natural selection to explain the tail streamers in male long-tailed widowbirds (Figure 11.3). Two problems arise. First, if long tail feathers can improve the survival or fecundity of a widowbird, then why do only males have them? Second, how could enormously long tail feathers improve the survival, or the fecundity, of widowbirds in the first place? As with bright colors in collared lizards, long tail feathers probably make male widow birds easier for predators to find and catch. Furthermore, long tail feather requires considerable energy to grow, maintain, and drag around (see Pryke and Anderson 2005; Walther and Clayton 2005). Energy spent on feathers is energy that cannot be spent on making offspring. It appears that the theory of evolution by natural selection can explain neither why male and female widow birds are different nor why the birds' most striking trait, long tail feathers, exists at all.

Figure 11.3 The sexual dimorphism in long-tailed widowbirds (*Euplectes progne*) The male is black with long tail feathers and red and yellow shoulder patches; the female is brown and cryptic.

As Darwin himself was the first to recognize, sex provides a solution to the puzzle of sexual dimorphism. To see why, consider life without sex. For organisms that reproduce without sex (see Chapter 8), getting genes into the next generation is straightforward, if not always easy. The two big challenges are surviving long enough to reproduce, then reproducing. Sex adds a third major challenge: finding a member of the opposite sex and persuading him or her to cooperate.

Charles Darwin realized that individuals vary not only in their success at surviving and reproducing, but also in their success at persuading members of the opposite sex to mate. About birds, for example, Darwin wrote,

> "Inasmuch as the act of courtship appears to be with many birds a prolonged and tedious affair, so it occasionally happens that certain males and females do not succeed during the proper season, in exciting each other's love, and consequently do not pair" (1871, page 107).

In its evolutionary consequences, failing to mate is the same as dying young. The victim makes no genetic contribution to future generations. Darwin had already applied the label natural selection to differential reproductive success due to variation among individuals in survival and reproduction. Differential reproductive success due to variation among individuals in success at getting mates he called **sexual selection**. We can develop a theory of evolution by sexual selection that is logically equivalent to the theory of evolution by natural selection: If there is heritable variation in a trait that affects the ability to obtain mates, then variants conducive to success will become more common over time.

Asymmetries in Sexual Reproduction

If sexual selection is to explain differences between the sexes, it will have to act on the sexes differently. Angus John Bateman (1948) argued that it can. The logic he developed to support his claim was later refined by Robert Trivers (1972). It hinges on a simple observation: In many animals, eggs (or pregnancies) are more expensive than ejaculates.

In more general terms, mothers typically make a larger parental investment in each offspring than fathers. By **parental investment** we mean energy and time

Figure 11.4 Orangutan mothers invest considerably more time and energy in each offspring than orangutan fathers

expended both in constructing an offspring and in caring for it. Ultimately, parental investment is measured in fitness. Parental investment increases the reproductive success of the offspring receiving it. At the same time, it decreases the remaining reproductive success that the investing parent may achieve in the future by way of additional offspring.

Consider the parental investments made by male and female orangutans (Figure 11.4). Adult orangutans of opposite sex tolerate each other's company for one purpose only (Nowak 1991). After a brief tryst, including a copulation that lasts about 15 minutes, the male and female go their separate ways. If a pregnancy results, then the mother, who weighs about 40 kg, will carry the fetus for 8 months, give birth to a 1-kg baby, nurse it for about 3 years, and continue to protect it until it reaches the age of 7 or 8. For the father, who weighs about 70 kg, the beginning and end of parental investment is a few grams of semen, which he can replace in a matter of hours or days. In their pattern of parental investment, orangutans are typical mammals. In more than 90% of mammal species, females provide substantial parental care and males provide little or none (Woodroffe and Vincent 1994).

Because mammalian mothers provide such intensive parental care, mammals present a somewhat extreme example of disparity in parental investment. In most animal species, neither parent cares for the young. Mated pairs of parents just make eggs, fertilize them, and leave them. But in these species, too, females usually make a larger investment in each offspring than males. Eggs are typically large and yolky, with a big supply of stored energy and nutrients. Think of a sea turtle's eggs, some of which are as large as a hen's eggs. Most sperm, on the other hand, are little more than DNA with a propeller. Even when a single ejaculate delivers hundreds of millions of sperm, the ejaculate seldom represents more than a fraction of the investment contained in a clutch of eggs.

The key to explaining sexual dimorphism is in recognizing that sexual reproduction imposes different selection pressures on females versus males.

When eggs are more expensive than ejaculates—when mothers make a larger parental investment than fathers— the factors limiting lifetime reproductive success will often be different for males versus females. A female's potential reproductive success will be relatively small, and her realized reproductive success is likely to be limited more by the number of eggs she can make (or pregnancies she can carry) than by the number of males she can convince to mate with her. In contrast, a male's potential reproductive success will be relatively large, and his realized reproductive success is likely to be limited more by the number of females he can convince to mate with him than by the number of ejaculates he can make. Access to mates will be a limiting resource for males, but not for females. Under such circumstances, Bateman and Trivers predicted that sexual selection—variation in mating success—will be a more potent force in the evolution of males than in the evolution of females.

When one parent invests more than the other in each offspring, the reproductive success of the heavily investing parent is often limited by resources and time. In contrast, the reproductive success of the lightly investing parent is limited by number of mates.

Bateman (1948) tested this prediction in laboratory populations of the fruit fly, *Drosophila melanogaster.* He found, as he predicted, that number of mates had a larger impact on the reproductive success of males than on the reproductive success of females. His results were not as clean as they might have been, however, and in hindsight his study can be criticized on methodological grounds (see Tang-Martinez and Ryder 2005). Nonetheless Bateman's key insight, that to understand sexual dimorphism we must quantify the relationship between number of mates and reproductive success for both males and females, is central to the theory of sexual selection (see Arnold 1994; Arnold and Duvall 1994). It will therefore be useful to consider in some detail two recent replications of Bateman's study using other species.

Asymmetric Limits on Reproductive Success in Newts and Pipefish

Adam Jones and colleagues (2002) quantified the relative strength of sexual selection on male and female rough-skinned newts (*Taricha granulosa*). Male rough-skinned newts gather in ponds in early winter to wait for females. Females saunter in during January and February, easily attract amorous males (Figure 11.5a), and finish mating shortly after arriving. As a result, at any given time more males than females are prowling the pond. After mating, females lay 300 or more eggs, one at a time, over several weeks or months. Neither parent cares for the young. The cost of eggs and sperm thus accounts for the entirety of parental investment. The investment per offspring is larger for females than for males.

Jones and colleagues captured all the newts from a single pond when the newts had finished mating and the females were laying eggs. The biologists housed the females in individual containers, induced them to lay the rest of their eggs by injecting them with hormones, and reared the eggs to hatching. The researchers used genetic tests to identify each hatchling's father. This gave Jones and colleagues sufficient information to determine the number of mates, and the number of offspring, for all the adults in their sample.

The majority of the males, it turned out, failed to mate (Figure 11.5b, upper left). Of those that succeeded, most mated just once or twice. In sharp contrast, all of the females mated at least once, and most mated two or three times (Figure 11.5b, lower left).

Not surprisingly, the males showed a pronounced variation in number of offspring (Figure 11.5b, upper right). Those who failed to mate, of course, fathered no offspring, while the lucky few fathered as many as 300 or more. All of the females had offspring, most between 100 and 300 of them (Figure 11.5b, lower right).

Of greatest interest is the extent to which access to mates determined reproductive success. Figure 11.5c plots number of offspring versus number of mates for both sexes, along with best-fit lines showing the average effect of a change in mating success on reproductive success (Arnold and Duvall 1994). Look first at the data for females (red dots and line). It appears that mating with more than one male may have carried some benefit. The slope of the best-fit line, however, is not statistically distinguishable from zero. For males, however, the association between number of mates and number of offspring was strong and highly significant (blue dots and line). For males, more mates meant more offspring.

(a)

Rough-skinned newt (*Taricha granulosa*)

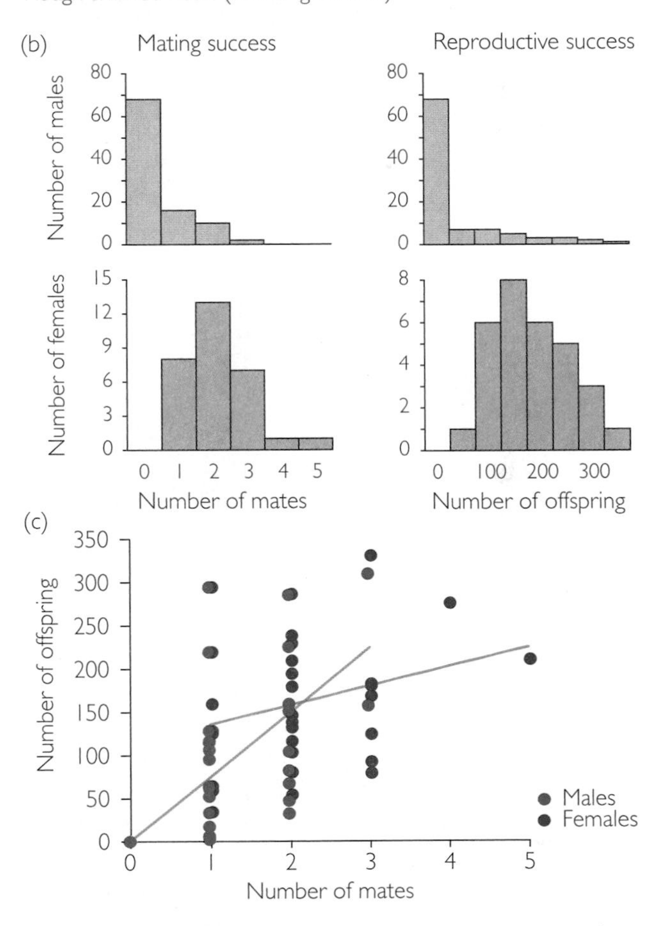

Figure 11.5 Asymmetries in sexual reproduction in rough-skinned newts (a) A mating pair. The male is above, the female below. (b) The distributions of number of mates and number of offspring for male (blue) and female (red) newts in a natural population. (c) Reproductive success versus number of mates for male (blue) and female (red) rough-skinned newts in a natural population. The slope of the best-fit line, or Bateman gradient, is steeper for males than for females ($P < 0.001$). Plotted from data provided by Adam G. Jones, Georgia Institute of Technology.

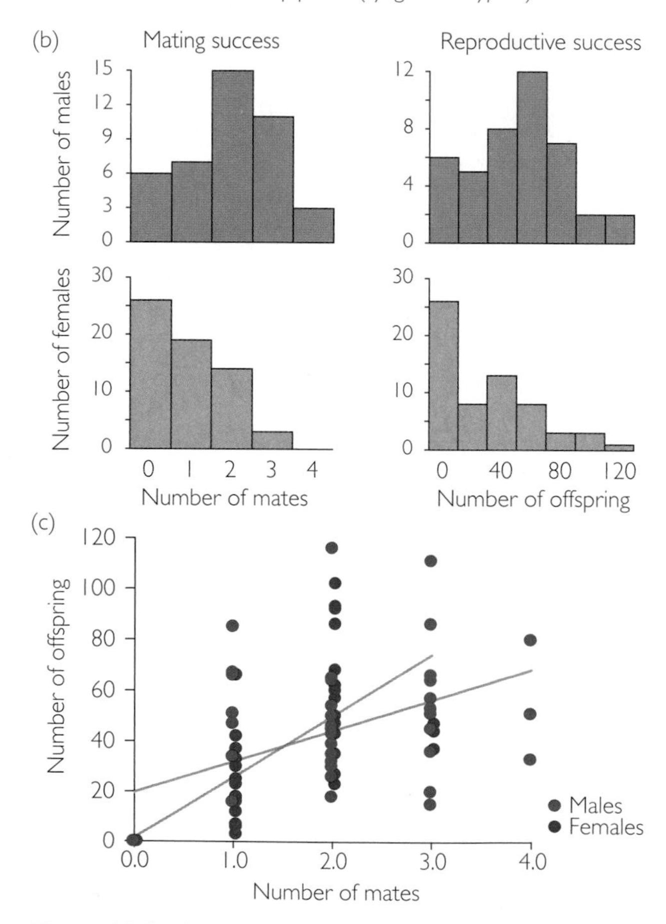

Figure 11.6 Asymmetries in sexual reproduction in broad-nosed pipefish (a) A mating pair. The male is on the left, the female on the right. (b) Reproductive success versus number of mates for male (blue) and female (red) broad-nosed pipefish in a laboratory experiment. (c) Reproductive success versus number of mates for male (blue) and female (red) broad-nosed pipefish in a laboratory experiment. The slope of the best-fit line is steeper for females than for males ($P = 0.004$). Graphs plotted from data provided by Adam G. Jones, Georgia Institute of Technology.

This result is consistent with Bateman's prediction. In rough-skinned newts, sexual selection is a more potent force in the evolution of males than in the evolution of females. This means that heritable traits that are associated, in males, with failure to mate will tend to disappear, while heritable traits associated with mating success will tend to become common. Male rough-skinned newts develop crests on their tails during the breeding season, and Jones and colleagues found that males who found mates displayed significantly taller crests than males that failed. We can infer that tail crests evolved as a result of sexual selection.

The pattern we have seen in rough-skinned newts is common. For males more than females in many species, fitness is determined by access to mates. But the pattern is by no means universal. This we can see from data on sexual selection in broad-nosed pipefish.

Broad-nosed pipefish (*Syngnathus typhle*) live in eelgrass beds (Figure 11.6a). In pipefish families, as in their kin, the seahorses, the father provides all the parental care. The male has a brood pouch into which the female lays her eggs. The male carries the eggs, protects them, and provides them with oxygen and nutrients until they hatch (Figure 11.7). Adam Jones and colleagues (2000; see also Jones et al. 2005) caught pipefish off the coast of Sweden before the breeding season began and let them mate in barrels in the lab. The data in Figure 11.6 come from two experiments. In the first, each barrel contained four males and four females. In the second, each barrel contained two males and six females. The second experiment probably mimics natural conditions more closely than the first. It takes a female less time to produce a clutch of eggs than it takes a male to rear them to hatching (Berglund et al. 1989). As a result, at any given time there are more females with eggs to lay than males with space to accept them.

Jones and colleagues used genetic tests to determine each offspring's mother. As in the newt study, this enabled the biologists to determine the number of mates and the number of offspring for each adult.

The results for pipefish are similar to those for newts, except that the roles of the sexes are reversed. In pipefish, more females than males failed to mate (Figure 11.6b, upper and lower left). Consequently, the reproductive success of females was more skewed than the reproductive success of males, with a sharper distinction between the winners and the losers (Figure 11.6b, upper and lower right). Most importantly, it was in females that reproductive success depended most strongly on mating success (Figure 11.6c). In broad-nosed pipefish, sexual selection is a more potent force in the evolution of females than it is in the evolution of males. Heritable traits that are associated, in females, with failure to mate will tend to disappear, while heritable traits associated with mating success will become more common. We need to keep this result in mind as we consider the behavioral consequences of asymmetrical limits on fitness.

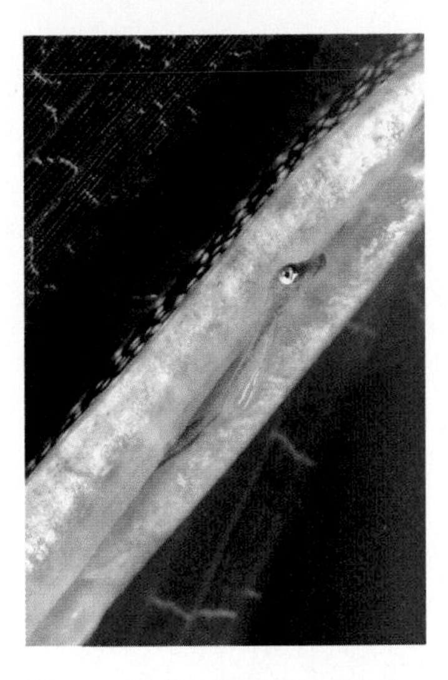

Figure 11.7 **A pregnant male pipefish, *Syngnathus typhle*, gives birth** This hatchling is emerging from his father's brood pouch, where it developed under its father's care after its mother deposited her eggs there.

Behavioral Consequences of Asymmetric Limits on Fitness

An asymmetry in the factors that limit reproductive success for females versus males allows us to predict differences in the mating behavior of the two sexes. Consider the pattern seen in Jones et al.'s rough-skinned newts. For males reproductive success is limited by access to mates, and at any given time there are more males than females in the pond looking for love. Under such circumstances, we can predict that males will compete amongst themselves for opportunities to fertilize eggs. For females, in contrast, reproductive success is limited by capacity to make eggs, mating involves the commitment of a large investment, and there is an excess of willing partners. We can expect that females will be selective about which partners they accept and which they reject.

More generally, when sexual selection is strong for one sex and weak for the other we can predict that:

- **Members of the sex subject to strong sexual selection will be competitive.**
- **Members of the sex subject to weak sexual selection will be choosy.**

These predictions have been confirmed in a great variety of animal species. We will look at some examples shortly.

In making these general predictions, we have used inclusive language for a reason. It is easy to get carried away with generalities, as Bateman and many who followed appear, in hindsight, to have done (see Knight 2002). Bateman thought that greater sexual selection on males than on females is inherent in maleness and femaleness as such. He and others therefore assumed that the optimal strategy for males, in virtually any species, would be to mate with as many females as possible, and that the optimal strategy for females would be to choose one male and mate with him only. These assumptions have often turned out to be wrong. We will see, later in the chapter, that males often have good reasons to stick with one mate and that females often have good reasons to be promiscuous.

Furthermore, as Jones et al.'s pipefish study shows, greater sexual selection on males than on females is not inherent in the identity of the sexes themselves. When access to mates is limiting for females instead of males, we predict that females will compete with each other over access to males and that males will be choosy.

Theory predicts that when one sex is subject to sexual selection and the other is not, the members of the sex experiencing selection will compete over mates and the members of the other sex will be choosy.

Competition for mates in one sex and choosiness in the other can play out in two ways. First, members of the competitive sex may fight amongst themselves, head-to-head, claw-to-claw, or antler-to-antler. Sometimes they fight over direct control of mates, sometimes they fight over control of a resource vital to mates, and sometimes they just fight. The members of the other sex then mate with the winners. This form of sexual selection is called **intrasexual selection**, because the key event that determines reproductive success (the fighting) involves interactions among the members of a single sex. Second, instead of fighting the members of the competitive sex may advertise for mates by singing, dancing, or showing off bright colors. The members of the other sex then choose the individual with the best display. This form of sexual selection is called **intersexual selection**, because the key event that determines reproductive success (the choosing) involves an interaction between members of the two sexes.

In the next two sections, we will look at examples of intrasexual and intersexual selection on males. We then will look at sexual selection on females.

11.2 Sexual Selection on Males: Competition

Sexual selection by male–male competition often occurs when individual males can monopolize access to females. Males may monopolize females through direct control of the females themselves or through control of some resource important to females, such as feeding territory or nest sites. Male–male competition can also occur for no apparent reason beyond simply impressing females. In this section, we consider examples of research into three forms of male–male competition: outright combat, sperm competition, and infanticide.

Combat

Male–male competition can take the form of combat over access to females.

Outright combat is the most obvious form of male–male competition for mates. Intrasexual selection involving male–male combat over access to mates can favor morphological traits including large body size, weaponry, and armor. Male–male combat also selects for tactical cleverness.

Our example of male–male combat comes from the marine iguanas (*Amblyrhynchus cristatus*) of the Galápagos Islands (Figure 11.8). Marine iguanas have a lifestyle unique among the lizards. They make their living grazing on algae in the intertidal zone. Between bouts of grazing, they bask on rocks at the water's edge. Basking warms the iguanas, which aids digestion and prepares them for their next foray into the cold water. Marine iguanas grow to different sizes on different islands, but, as we mentioned earlier, on any given island the males get larger than the females (Figure 11.9a).

Figure 11.8 **A Galápagos marine iguana** These unusual lizards make their living foraging on algae in the intertidal zone.

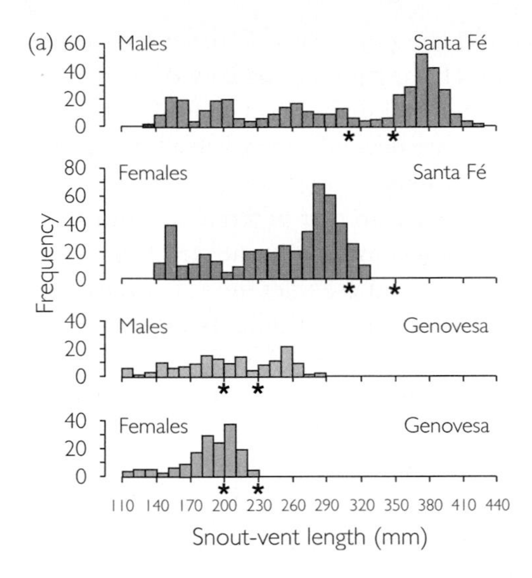

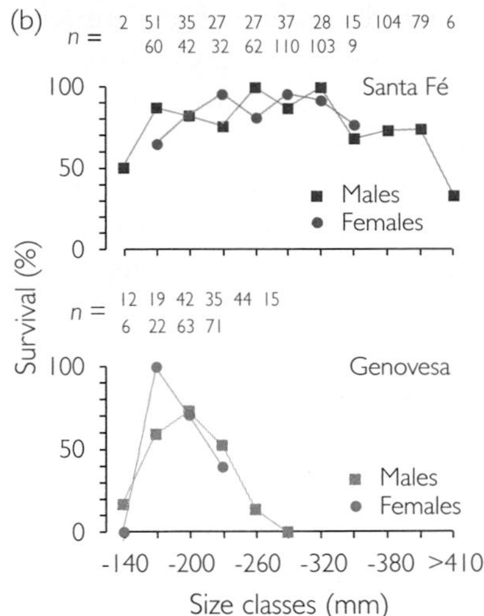

Figure 11.9 Natural selection on body size in marine iguanas (a) Size distributions of male and female marine iguanas on two different Galápagos Islands, Genovesa and Santa Fé. Asterisks mark the maximum sizes at which iguanas were able to maintain their weight in two different years (1991–1992 and 1992–1993). From Wikelski et al. (1997). (b) Survival rates of marked individuals of different sizes (snout–vent length, mm) from March 1991 to March 1992 on Genovesa and from February 1990 to February 1992 on Santa Fé. The sample sizes, or number of individuals in each group, are given by *n*. From Wikelski and Trillmich (1997).

The sexual size dimorphism in marine iguanas is an excellent example for the study of sexual selection, because we know a great deal about how marine iguana size is affected by natural selection (Wikelski et al. 1997; Wikelski and Trillmich 1997). Martin Wikelski and Fritz Trillmich documented natural selection on iguana body size by monitoring the survival of marked individuals on two islands over one to two years. Natural selection was much harsher on Genovesa than on Santa Fé, but it was clearly at work on both islands. Moreover, selection was stabilizing. Medium-sized iguanas survived at higher rates than either small iguanas or large iguanas (Figure 11.9b).

Potential agents of this natural selection on body size are few. Marine iguanas do not compete with other species for food and have virtually no predators. Other than reproduction, about all the iguanas have to contend with is competition for food among themselves.

Larger iguanas can harvest more algae, and thus gather more energy, but they also expend more energy on metabolism. Wikelski and colleagues (1997) found during two different years that small iguanas ran a net energy surplus, but large iguanas ran a net energy deficit. Consistent with the hypothesis that the availability of food limits body size, the largest iguanas on Santa Fé and Genovesa lost weight during both 1991–1992, a bad year for algae, and 1992–1993, a fairly good year (see also Wikelski and Thom 2000). The largest sizes at which iguanas were able to maintain their weight are indicated by the asterisks in Figure 11.9a.

Now compare Figure 11.9a with Figure 11.9b. The maximum sizes at which iguanas could sustain their weight are close to the optimal sizes for survival. Furthermore, the largest females on each island are near the optimal size for survival, but the largest males are much larger than the optimal size. The large body size of male marine iguanas is thus an evolutionary puzzle. We apparently cannot explain it by natural selection, because Wikelski and Trillmich have shown that natural selection acts against it. It is exactly the kind of puzzle for which Darwin invoked sexual selection.

As we discussed earlier, a crucial issue in sexual selection is the relative parental investment per offspring made by females versus males. In marine iguanas, the parental investment by females is much larger. Each female digs a nest on a beach away from the basking and feeding areas, buries her eggs, guards the nest for a few days, and then abandons it (Rauch 1988). Males provide no parental care at all. So parental investment by females consists mostly of producing eggs, and parental investment by males consists entirely of producing ejaculates. Females lay a single clutch of one to six eggs each year, into which they put about 20% of their body mass (Rauch 1985; Rauch 1988; Wikelski and Trillmich 1997). Compared to the female investment, the cost of the single ejaculate needed to fertilize all the eggs in a clutch is paltry. This difference in investment suggests that the maximum potential reproductive success of males is much higher than that of females. Number of mates will limit the lifetime reproductive success of males, but not females.

The iguanas' mating behavior is consistent with these inferences. Females copulate only once each reproductive season. Martin Wikelski, Silke Bäurle, and their field assistants followed several dozen marked females on Genovesa. The researchers watched the females from dawn to dusk every day during the entire month-long mating season in 1992–1993 and 1993–1994 (Wikelski and Bäurle 1996). They also watched the marked females from dawn to dusk every day during the subsequent nesting seasons. Every marked female that dug a nest and laid eggs had been seen copulating, but no marked female had been seen copulating more than once. Male iguanas, in contrast, attempt to copulate many times with many different females. But the opportunity to copulate with females is a privilege a male iguana has to fight for (Figure 11.10).

Figure 11.10 Male marine iguanas in combat Note the number painted on the individual on the right; he is participating in a study.

Prior to the mating season each year, male iguanas stake out territories on the rocks where females bask between feeding bouts. In these small, densely packed territories (Figure 11.11a), males attempt to claim and hold ground by ousting male interlopers. Confrontations begin with head-bobbing threats and escalate to chases and head pushing. If neither male backs down, fights can end with bites leaving serious injuries on the head, neck, flanks, and legs (Trillmich 1983). Males that hold territories are more attractive to females than males that do not (Trillmich 1983; Rauch 1985; Partecke et al. 2002). Because only some males manage to claim territories, because some males manage to maintain their claims for a longer period than others, and because females prefer some territories and territory holders over others, there is extreme variation among males in the number of copulations obtained (Figure 11.11b).

Because claiming and holding a territory involves combat with other males, bigger males tend to win. In the iguana colony that Krisztina Trillmich (1983) studied on Camaaño Islet, the male that got 45 copulations (Figure 11.11b), far more than any other male, was iguana 59 (his territory is shown in Figure 11.11a). His neighbor, iguana 65, was the second most successful with 10 copulations. Both of their territories were females' favorite early-morning and late-afternoon basking places. Trillmich reported that iguana 59 was the largest male in the colony; that to claim his territory, he had to eject four other males who tried to take it; and that during

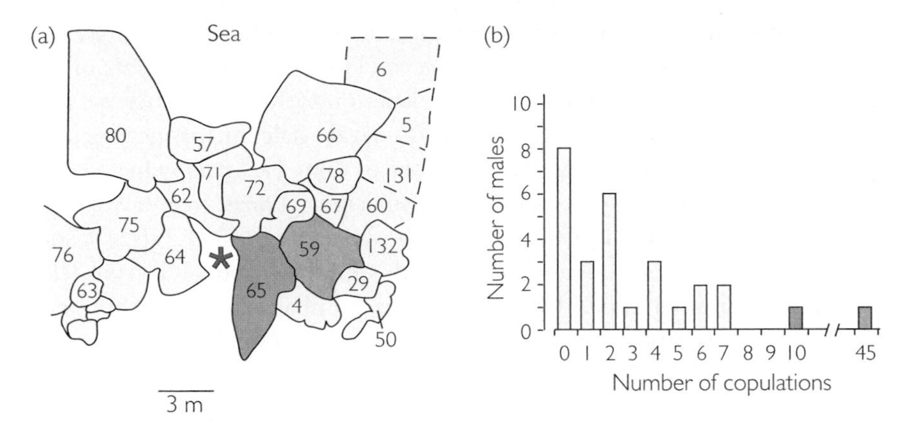

Figure 11.11 Mating success in male marine iguanas (a) A cluster of iguana mating territories on Camaaño Islet, Galápagos. Lines show boundaries of mating territories on January 16, 1978; numbers identify territory owners. As the scale bar for this map shows, mating territories are only a few square meters in size. The dark blue asterisk indicates where Krisztina Trillmich sat to watch the iguanas. (Camaaño Islet has only 880 m of shoreline and supports a population of nearly 2,000 iguanas.) From Trillmich (1983). (b) Histogram showing variation in number of copulations obtained by male iguanas on mating territories shown in (a). Note the break in the horizontal scale; the most successful male, iguana 59, got more than four times as many copulations as any of his rivals. The histogram includes only males that claimed a territory for at least a short time during the mating season. From Trillmich (1983).

his tenure, he lost parts of it to four neighboring males who were pushing their territories in from the sides. Wikelski and coworkers studied iguana colonies on Genovesa and Santa Fé (Wikelski et al. 1996; Wikelski and Trillmich 1997). Consistent with Krisztina Trillmich's observations, these researchers found that the mean size of males that actually got to copulate was significantly larger than the mean size of all males that tried to copulate (Table 11.1).

Male marine iguanas fight over territories where females congregate. Large iguanas win more fights, claim better territories, and thus get to copulate with more females. This pattern of sexual selection has led to the evolution of large body size in males.

Table 11.1 Sexual selection differentials for male body size in marine iguana colonies on Santa Fé and Genovesa

Body size is given as snout–vent length (SVL). The standardized selection differential (see Chapter 9) is the difference between the average body size of all males that copulated at least once and the average body size of all males that tried to copulate, expressed in standard deviations of the distribution of body sizes of all males that tried to copulate. (The standard deviation is the square root of the variance.) Both standardized selection differentials are positive ($P < 0.05$), indicating that males that got to copulate were larger on average than males that tried to copulate. From Wikelski and Trillmich (1997).

	N	Average size (mm SVL)	Standard deviation	Standardized selection differential
Santa Fé				
Males that copulated	253	401	13	0.42
All males that tried to	343	390	26	
Genovesa				
Males that copulated	25	243	26	0.77
All males that tried to	147	227	21	

If we assume that body size is heritable in marine iguanas, then we have variation, heritability, and differential mating success. These are the elements of evolution by sexual selection. We thus have an explanation for why male marine iguanas get so much bigger than the optimal size for survival. Male iguanas get big because bigger males get more mates and pass on more of their big-male genes.

Male–male combat, analogous to that in marine iguanas, happens in a great variety of species, including the red deer shown in Figure 11.1 (see Clutton-Brock 1985). In addition to large body size, this kind of sexual selection leads to the evolution of other traits that are assets in combat, such as weaponry and armor. Male–male combat can also lead to the evolution of alternative male mating strategies (see Box 11.1).

Box 11.1 Alternative male mating strategies

Victory in male–male combat typically goes to the large, strong, and well armed. But what about the smaller males? Is their only chance at fitness to survive until they are large enough to win fights? Often small males attempt to mate by employing alternative strategies. Sometimes they succeed.

In marine iguanas, small adult males are ousted from the mating territories on the basking grounds. But many do not give up; they continue trying to convince females to copulate with them. The small males are not terribly successful, but they do get about 5% of the matings in the colony (Wikelski et al. 1996). Small males attempting to mate with females are often harassed by other males. This happens to large territorial males too, but it happens more often to small males. Furthermore, copulations by small males are more likely to be disrupted before the male has time to ejaculate (Figure 11.12).

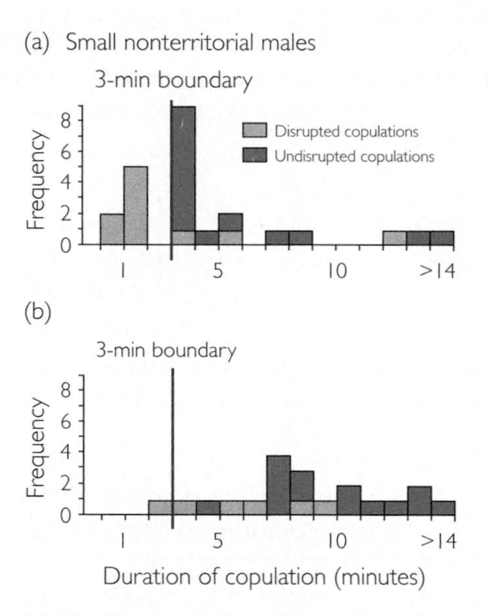

Figure 11.12 Duration of copulations by male marine iguanas Histograms showing the distribution of copulation durations for (a) 24 small nonterritorial males, and (b) 20 large territorial males. Pink areas indicate copulations that were disrupted by other males. The red vertical line at 3 minutes marks the approximate amount of time a male must copulate before he can ejaculate. Large territorial males had copulations that were significantly longer and less likely to be disrupted before the 3-minute boundary. From Wikelski and Bäurle (1996).

The small males solve the problem of disrupted copulations by ejaculating ahead of time (Wikelski and Bäurle 1996). They use the stimulation of an attempted copulation, or even of seeing a female pass by, to induce ejaculation. The males then store the ejaculate in their cloacal pouches. If he gets a chance to mate, a small male transfers his stored ejaculate to the female at the beginning of copulation. Wikelski and Bäurle examined the cloacae of a dozen females caught immediately after copulations that had lasted less than three minutes. None of these females had copulated earlier that mating season, but 10 of the 12 females had old ejaculates in their cloacae that must have been transferred during the short copulation.

The sperm in these old ejaculates were viable. From dawn to dusk every day for about a month, Wikelski and Bäurle watched five of the females until they laid their eggs. None of the five copulated again, but all laid fertilized eggs.

Prior ejaculation appears to be a strategy practiced more often by small nonterritorial males than by

Sperm Competition

Male–male competition does not necessarily stop when copulation is over. The real determinant of a male's mating success is not whether he copulates, but whether his sperm fertilize eggs. If an animal has internal fertilization, and if a female mates with two or more different males within a short period, then the sperm from the males will be in a race to the eggs. Indeed, females may produce litters or clutches in which different offspring are fathered by different males. Batches of offspring with multiple fathers have been documented in a variety of animals, including squirrels (Boellstorff et al. 1994), bears (Schenk and Kovacs 1995), birds (Gibbs et al. 1990), lizards (Olsson et al. 1994), and spiders (Watson

Male–male competition can take the form of sperm competition. If a female mates with two or more males, the male whose sperm win the race to the eggs has higher reproductive success.

large territorial males. Wikelski and Bäurle caught 13 nonterritorial and 13 territorial males at random; 85% of the nonterritorial males had stored ejaculates in their cloacal pouches, versus only 38% of the territorial males ($P < 0.05$). This difference is unlikely to result from more frequent copulation by territorial males, because even territorial males copulate only about once every 6 days (Wikelski et al. 1996).

Alternative, or sneaky, male mating strategies have also evolved in a variety of other species. In coho salmon, *Oncorhynchus kisutch,* for example, males return from the sea to spawn and die at one or the other of two distinct ages (Gross 1984; Gross 1985; Gross 1991). One group, called hooknoses, returns at 18 months. They are large, armed with enlarged hooked jaws, and armored with cartilaginous deposits along their backs. The other group, called jacks, returns at 6 months. They are small, poorly armed, and poorly armored.

When a female coho is ready to mate, she digs a nest and then lays her eggs in it. As she prepares the nest, males congregate. The males use one of two strategies in trying to fertilize the female's eggs (Figure 11.13). Some males fight for a position close to the female. These fighters quickly sort themselves out by size, making a line downstream of the nest. When the female lays her eggs, the males spawn over them in order. The first male to spawn fertilizes the most eggs. Males using the other strategy do not fight for position but instead look for a hiding place near the female. When the female lays her eggs, these sneakers attempt to dart out and spawn over the eggs.

Among hooknoses, those that adopt the fighting strategy are more successful. Among jacks, those that adopt the sneaky strategy are more successful. The relative fitness of hooknoses versus jacks depends, in part, on the frequency of each type of male in the breeding population.

There is an important distinction between the iguana example and the coho example. In marine iguanas, the small nonterritorial males appear to be making the best of a bad situation while they grow to a large enough size to successfully fight for a territory. In coho, a male irreversibly becomes either a hooknose or a jack. Which strategy a male coho pursues depends on a mixture of environmental and genetic factors.

Figure 11.13 Alternative mating strategies in coho salmon This figure shows a coho mating group. The large fish at the right (upstream) is a female that has built a nest and is ready to lay her eggs. Downstream from the female are four males that have opted for the fighting strategy. These males, three hooknoses and a jack, have sorted themselves out by size. Two other jacks have opted for the sneaky strategy. They have found hiding places near the female, one behind a rock, the other in a shallow. After Gross (1991).

Figure 11.14 A Mediterranean fruit fly, *Ceratitis capitata*

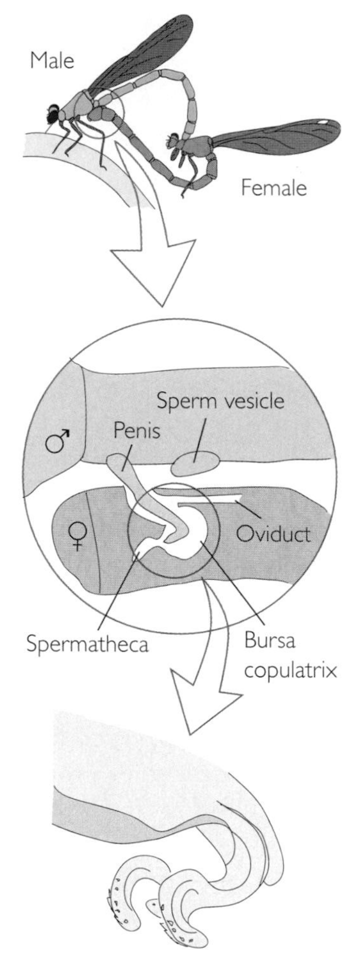

Figure 11.15 Sperm competition in damselflies During copulation (top), the male uses the barbed horns on his penis (bottom) to remove sperm left by the female's previous mates. Redrawn from Waage (1984).

1991). It happens in humans too; Smith (1984) reviews several reports of twins with different fathers.

Given sperm competition, what traits contribute to victory? One useful trait might simply be the production of large ejaculates containing many sperm. If sperm competition is something of a lottery, then the more tickets a male buys, the better his chances of winning. This hypothesis has been tested by Matthew Gage (1991) with the Mediterranean fruit fly, *Ceratitis capitata* (Figure 11.14). Gage's experiment was based on the observation that, although ejaculates are cheaper than eggs, they are not free (see, for example, Nakatsuru and Kramer 1982). Gage reasoned that if male Mediterranean fruit flies are subject to any constraints on sperm production, they might benefit from conserving their sperm, using during each copulation only the minimum number necessary to ensure complete fertilization of the female's eggs. But if larger ejaculates contribute to victory in sperm competition, males whose sperm are at risk of competition should release more sperm during copulation than males whose sperm are not at risk. If the number of sperm released is unimportant to the outcome of competition, then males should release the same number of sperm regardless of the risk of competition.

Gage raised and mated male medflies under two sets of conditions. One group of 20 he raised by themselves and allowed to mate in private; the other group of 20 he raised in the company of another male and allowed to mate in the presence of that second male. Immediately after each mating, Gage dissected the females and counted the number of sperm the males had released. Males raised and mated in the presence of a potential rival ejaculated more than 2 1/2 times as many sperm (average $\pm$ standard error $=$ 3,520 $\pm$ 417) as males raised and mated in isolation (1,379 $\pm$ 241), a highly significant difference ($P < 0.0001$). Gage's interpretation was that large ejaculates do contribute to victory in sperm competition and that male medflies dispense their sperm to balance the twin priorities of ensuring successful fertilization and conserving sperm.

In addition to large ejaculates, sperm competition has apparently led to various other adaptations. Males may guard their mates, prolong copulation, deposit a copulatory plug, or apply pheromones that reduce the female's attractiveness (Gilbert 1976; Beecher and Beecher 1979; Sillén-Tullberg 1981; Thornhill and Alcock 1983; Schöfl and Taborsky 2002). During copulation in many species of damselflies, the male uses special structures on his penis to scoop out sperm left by the female's previous mates (Figure 11.15; Waage 1984, 1986). R. E. Hooper and M. T. Siva-Jothy (1996) used genetic paternity tests to show that this strategy is highly effective. In the damselfly species they studied, the second male to mate with a female fertilized nearly all of the eggs produced during her first postcopulatory bout of oviposition.

Infanticide

In some species of mammals, competition between males continues even beyond conception. One example, discovered by B. C. R. Bertram (1975) and also studied by Craig Packer and Anne Pusey (reviewed in Packer et al. 1988), happens in lions. The basic social unit of lions is the pride. The core of a pride is a group of closely related females—mothers, daughters, sisters, nieces, aunts, and so on—and their cubs. Also in the pride is a small group of adult males; two or three is a typical number. The males are usually related to each other but not to the adult females. This system is maintained because females reaching sexual maturity stay in the pride they were born into, whereas newly mature males move to another pride.

The move for young adult males from one pride to another is no stroll in the park. The adult males already resident in the new pride resist the invaders. That is why males stay with their other male kin: Each group, the residents and the newcomers, forms a coalition. The residents fight the newcomers, sometimes violently, over the right to live in the pride. If the residents win, they stay in the pride and the newcomers search for a different pride to take over. If the residents lose, they are evicted, and the newcomers have exclusive access to the pride's females—exclusive, that is, until another coalition of younger, stronger, or more numerous males comes along and kicks them out. Pusey and Packer found that the average time a coalition of males holds a pride is a little over two years. Because residence in a pride is the key to reproductive success in lions, males in a victorious coalition quickly begin trying to father cubs. One impediment to quick fatherhood, however, is the presence of still-nursing cubs fathered by males of the previous coalition. That is because females do not return to breeding condition until after their cubs are weaned.

How can the males overcome this problem? They frequently employ the obvious, if grisly, solution: They kill any cubs in the pride that have not been weaned (Figure 11.16). Packer and Pusey have shown that this strategy causes the cubs' mothers to return to breeding condition an average of eight months earlier than they otherwise would. Infanticide by males is the cause of about 25% of all cub deaths in the first year of life and over 10% of all lion mortality.

Male–male competition can take the form of infanticide. By killing other males' cubs, male lions gain more opportunities to mate.

Figure 11.16 Lion infanticide In this photo by George B. Schaller, a male lion has just killed another male's cub, which it now carries in its mouth.

Infanticide improves the males' reproductive prospects, but is obviously detrimental to the reproductive success of the females. The females try to protect their own interests in this bad situation (Packer and Pusey 1983). They defend their cubs from infanticidal males, occasionally at the cost of their own lives. Nonetheless, Packer and Pusey report that young cubs rarely survive more than 2 months in the presence of a new coalition of males. With this shift in focus to female reproductive strategy, we leave the subject of male–male conflict and move to the other side of sexual selection on males: female choice.

11.3 Sexual Selection on Males: Female Choice

There is a great variety of species in which male reproductive success is limited by opportunities to mate but in which males are unable to monopolize either females themselves or any resource vital to females. In many such species, the males advertise for mates. Females typically inspect advertisements of several males before they choose a mate. Sexual selection by mate choice leads to the evolution of elaborate courtship displays.

When males cannot monopolize access to females, they often compete by advertising for mates. Although biologists were long skeptical that females discriminate among the advertising males, female choice is now well established.

Charles Darwin first asserted that female choice is an important mechanism of selection in 1871, in *The Descent of Man, and Selection in Relation to Sex*. Although widely accepted today, the notion that females actively discriminate among individual males was controversial for several decades. Most evolutionary biologists thought that female discrimination was limited to choosing a male of the right species (see Trivers 1985). Beyond allowing females to identify a mate of the right species, male courtship displays were thought to function primarily in overcoming a general female reluctance to mate. Once ready to mate, a female would accept any male at hand.

We begin this section by describing two sets of experiments that demonstrate that females are in fact highly selective, actively choosing particular males from among the many available. We then consider the functions of female choosiness. Potential benefits to a choosy female include the acquisition of good genes for her offspring and the acquisition of resources offered by males. Alternatively, females may prefer male displays that exploit preexisting sensory biases built into the females' nervous systems.

Female Choice in Red-Collared Widowbirds

Our first experiment demonstrating active female choice comes from the work Sarah Pryke and Steffan Andersson (2005) on the red-collared widowbird (*Euplectes ardens*), a close relative of the long-tailed widowbird pictured in Figure 11.3 on page 403. During the breading season, red-collared widowbirds are highly dimorphic. Adult males are jet black with long tail feathers and a crimson collar (Figure 11.17). Adult females are streaked with yellow and brown, and have normal-length tail feathers.

Figure 11.17 A male red-collared widowbird During the breeding season, adult males wear jet black plummage, with long tail feathers and a red collar.

Inspired by a classic study by Malte Andersson (1982) on long-tailed widowbirds, Pryke and Andersson captured 120 males before they had claimed nesting territories and assigned every other one to the control group or the experimental group. The researchers trimmed the tail feathers of the control birds to a maximum length of 20 cm, just 2 cm below the population average. They trimmed the tail feathers of the experimental birds to a maximum length of 12.5 cm. The researchers weighed the birds and measured their tarsi. The researchers then put a unique set of colored ankle bands on each of the birds and released them.

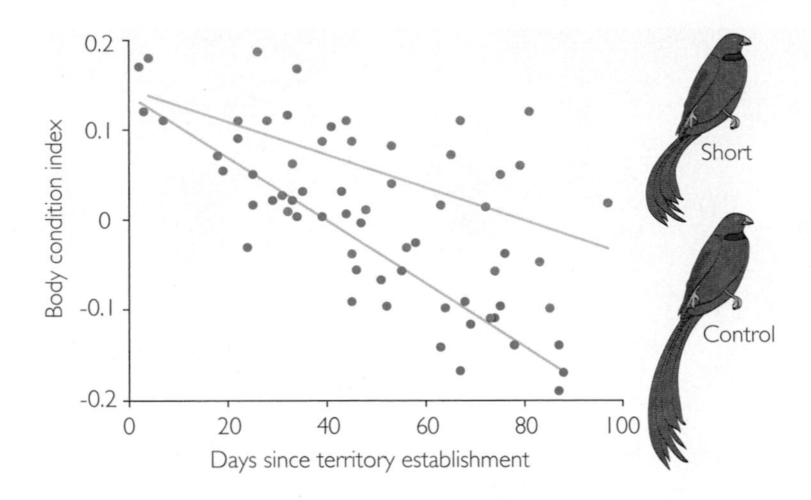

Figure 11.18 **Long tail feathers are a ball-and-chain for male red-collared widowbirds** This graph shows the decline in body condition (weight relative to linear size) throughout the breeding season for males with experimentally shortened tail feathers versus controls. The control males lost weight much more quickly. From Pryke and Andersson (2005).

Pryke and Andersson recaptured about half of the males at various times during the breeding season and reweighed them. For each one they calculated an index of body condition that reflects the bird's mass relative to its tarsus length. The data appear in Figure 11.18. The control males, who retained their long tails, lost weight much faster than the experimental males. This in spite of the fact that the control males spent less time flying around and showing off their tails to females. It appears that long tail feathers are simply expensive to drag around and maintain. In other words, natural selection probably acts against them. As with large body size in male marine iguanas, long tail feathers in widowbirds are just the sort of dimorphic trait Darwin invoked sexual selection to explain.

Could long tail feathers be a signal that males use to intimidate their rivals in the competition to claim territories? Apparently not. Forty-eight of the shortened males were able to acquire territories versus 43 of the control males, roughly the same success rate for each group. And the territories the two groups claimed were of similar size and quality. Pryke and Andersson concluded that long tail feathers did not evolve as a result of intrasexual selection.

Why, then, do male red-collared widowbirds have long tail feathers? Pryke and Andersson suspected from the beginning that the reason was that female red-collared widowbirds think long tails are sexy. The researchers monitored each male's territory throughout the breeding season, noting the number of females he enticed to nest there. As shown in Figure 11.19, on average the control males attracted nearly three times as many mates as the experimentally shortened males. We would need genetic tests to be certain, but presumably the control males enjoyed higher reproductive success as well.

Prkye and Andersson's experiment corroborates Darwin's contention that females are choosy. They actively discriminate among males of their own species and select particular kinds for their mates.

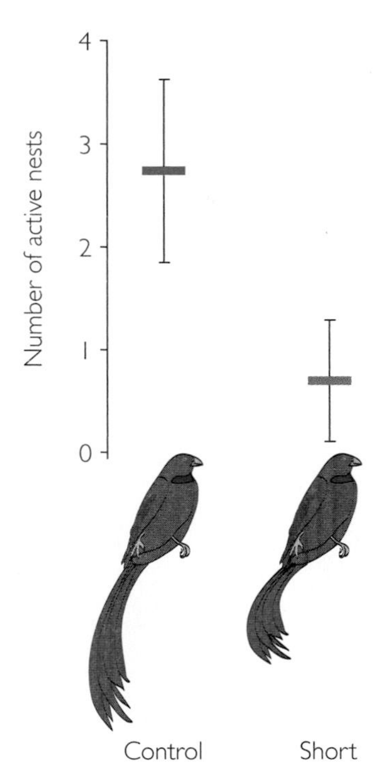

Figure 11.19 **Female red-collared widowbirds prefer long-tailed males** This graph shows the mean (± standard deviation) number of females that nested in the territories of control versus shorted males. From Pryke and Andersson (2005).

Female Choice in Gray Tree Frogs

Our second set of experiments demonstrating active female choice comes from the work of H. Carl Gerhardt and colleagues on gray tree frogs. Gray tree frogs, *Hyla versicolor,* live in woodlands in the eastern United States. During the breeding season, the males serenade the females (Figure 11.20). The love song of the gray tree frog is a series of calls, each call consisting of a number of pulses, or trills. Some males give long calls, with many trills, while other males give short calls,

Figure 11.20 A male gray tree frog (*Hyla versicolor*) singng to attract a mate

with few trills. Gerhardt and colleagues suspected, for at least two reasons, that female gray tree frogs discriminate among potential mates on the basis of their songs. First, they knew that when a male hears others joining him to make a chorus, he sometimes increases the length of his calls. Second, several times in the field the researchers had watched females go right past one singer to mate with a more distant one. The researchers hypothesized that females prefer to mate with longer-calling males.

Gerhardt, Miranda Dyson, and Steven Tanner (1996) captured female gray tree frogs in nature and tested their preferences in the laboratory. In one experiment, the researchers released females between a pair of loudspeakers (Figure 11.21a). Each speaker played a computer-synthesized mating call. To make their experiment conservative, the researchers made the call they expected to be less attractive louder, either by increasing the volume on that speaker, or by releasing the female closer to it. Then they waited to see which speaker the female would approach. They found that 30 of 40 females (75%) preferred long calls to short calls, even when the short calls were louder. In another experiment, Gerhardt, Dyson, and Tanner released female frogs facing two loudspeakers (Figure 11.21b). The closer speaker played short calls, while the more distant speaker played long calls. The researchers found that 38 of 53 females (72%) went past the short-calling speaker to approach the long-calling speaker.

Sarah Bush, Gerhardt, and Johannes Schul (2002) quantified female tastes more precisely. By placing females in front of a single loudspeaker, playing calls, and noting how enthusiastically each female responded to experimental calls versus a standard control call, the researchers learned that females are particularly disdainful of unusually short calls (Figure 11.21c). As calls get longer, females are less choosy—although they do still prefer songs with more trills. This result is consistent with observations made under more natural conditions by Joshua Schwartz, Bryant Buchanan, and Gerhardt (2001). In the midst of a chorus of singing males and other background noise, females are less able to make fine distinctions among males. But they do continue to discriminate, especially against short calls, and call length continues to explain a modest but statistically significant fraction of the variation in mating success among males.

The experiments of Gerhardt and colleagues show that female gray tree frogs are choosy, preferring, as predicted, males giving longer calls. And they show that longer-calling males are more likely to find mates.

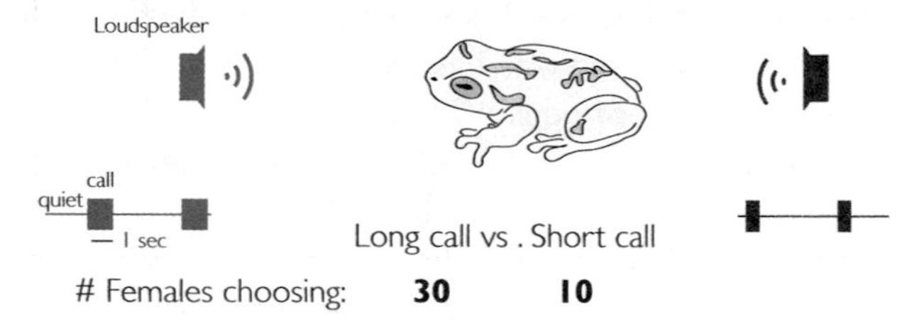

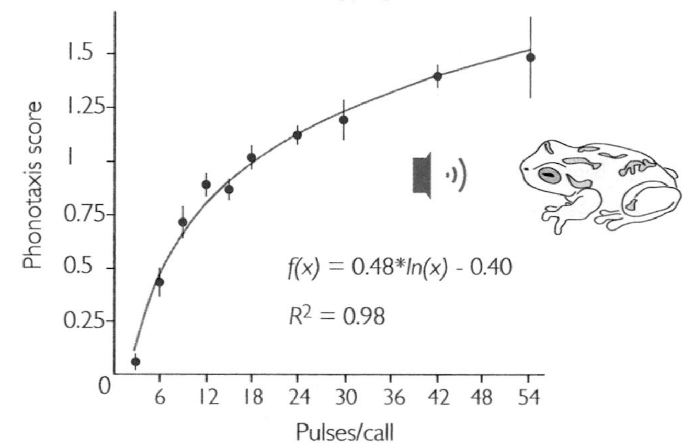

Figure 11.21 **Gerhardt et al.'s data on the preferences of female gray tree frogs** (a) Most females prefer long calls to short calls, even when the short calls are initially louder ($P < 0.001$). After Gerhardt el al. (1996). (b) Most females will pass a loudspeaker playing short calls to approach a loudspeaker playing long calls ($P < 0.001$). After Gerhardt el al. (1996). (c) Females discriminate most strongly against short calls. A female's "phonotaxsis score" for a particular test call is the time it took her to approach a control call with 18 pulses per second divided by the time it took her to approach the test call. Higher scores indicate a stronger preference for the test call relative to the control. Each data point is the average score of 10 females; the whiskers show ± 1 standard error. From Bush et al. (2002).

Female choice, as illustrated by red-collared widowbirds and gray tree frogs, is thought to be the selective agent responsible for the evolution of a great variety of male advertisement displays—from the gaudy tail feathers of the peacock, to the chirping of crickets, to the leg tufts of wolf spiders. Some male displays, like those of peacocks, are loud and clear; others are more subtle. In barn swallows, for example, a mere 2 cm added to—or subtracted from—a male's tail feathers can alter his attractiveness enough to dramatically influence his reproductive success (Møller 1988). Why should the females care about such a small difference? And for that matter, why should females care about any of the advertisements, even the loud ones, that males use to attract mates? We will consider three explanations.

Choosy Females May Get Better Genes for Their Offspring

A variety of factors has been suggested to explain female preferences.

One possibility is that the displays given by males are indicators of genetic quality. If males giving more attractive displays are genetically superior to males giving less attractive displays, then choosy females will secure better genes for their offspring (Fisher 1915, Williams 1966, Zahavi 1975).

Allison Welch and colleagues (1998) used an elegant experiment to investigate whether male gray tree frogs giving long calls are genetically superior to males giving short calls (Figure 11.22). During two breeding seasons, the researchers collected unfertilized eggs from wild females. They divided each female's clutch into separate batches of eggs, then fertilized one batch of eggs with sperm from a long-calling male and the other batch of eggs with sperm from a short-calling male. They reared some of the tadpoles from each batch of eggs on a generous diet and the others on a restricted diet.

This experimental design allowed Welch and colleagues to compare the fitness of tadpoles that were maternal half-siblings—that is, tadpoles with the same mother, but different fathers. When comparing tadpoles fathered by long-calling males versus short-calling males, the researchers did not have to worry about uncontrolled differences in the genetic contribution of the mothers, because the mothers were the same.

Welch and colleagues measured five aspects of offspring performance related to fitness: larval growth rate (faster is better); time to metamorphosis (shorter is better); mass at metamorphosis (bigger is better); larval survival; and post-metamorphic growth (faster is better). The results of their comparisons appear in Table 11.2. In 18 comparisons between the offspring of long-calling males versus short-calling males, there was either no significant difference, or better performance by the offspring of long callers. The offspring of short callers never did better. Overall, the data indicate that the offspring of long-calling males have significantly higher fitness. This result is consistent with the good genes hypothesis. At least one genetic difference between long-calling males versus short-calling males is that they feed more voraciously as tadpoles,

Choosy females sometimes get better genes for their offspring.

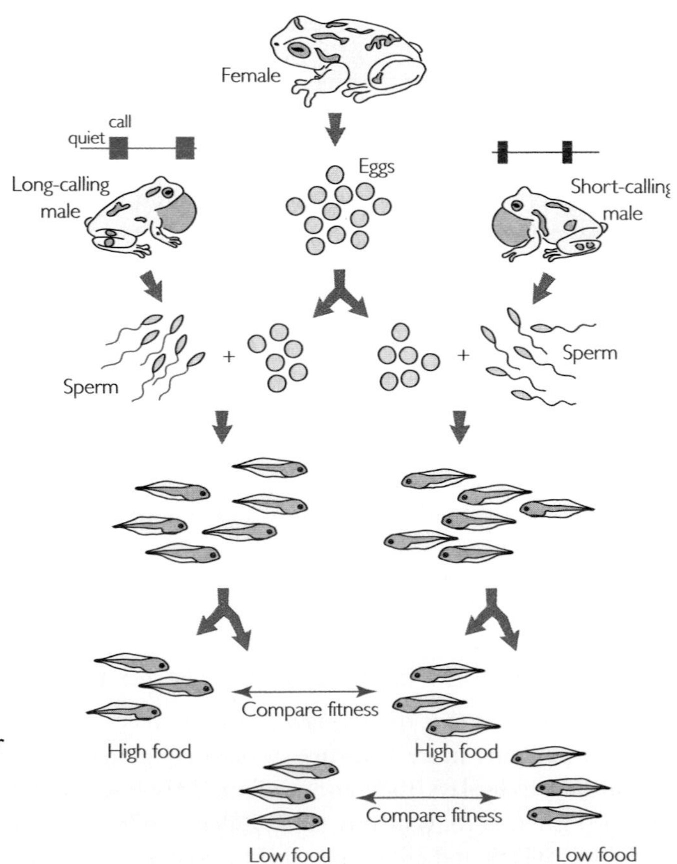

Figure 11.22 Welch et al.'s experiment to determine whether male gray tree frogs that give long calls are genetically superior to males that give short calls Overall, the experiment included batches of eggs from 20 different females fertilized with sperm from 25 different pairs of males.

Table 11.2 Fitness of the offspring of long-calling male frogs vs. short-calling male frogs

NSD = no significant difference; LC better = offspring of long-calling males performed better than offspring of short-calling males; — = no data taken. The overall result: Offspring fathered by long-calling males had significantly higher fitness than their maternal half-sibs fathered by short-calling males ($P < 0.0008$).

	1995		1996	
Fitness measure	**High food**	**Low food**	**High food**	**Low food**
Larval growth	NSD	**LC better**	**LC better**	**LC better**
Time to metamorphosis	**LC better**	NSD	**LC better**	NSD
Mass at metamorphosis	NSD	**LC better**	NSD	NSD
Larval survival	**LC better**	NSD	NSD	NSD
Postmetamorphic growth	—	—	NSD	**LC better**

even when a predator is present (Doty and Welch 2001). The nature of other genetic differences between long-calling frogs versus short-calling frogs is a subject for future research.

Choosy Females May Benefit Directly through the Acquisition of Resources

In many species the males provide food, parental care, or some other resource that is beneficial to the female and her young. If it is possible to distinguish good providers from poor ones, then choosy females reap a direct benefit in the form of the resource provided. Such is the case in the hangingfly (*Bittacus apicalis*), studied by Randy Thornhill (1976). Hangingflies live in the woods of eastern North America, where they hunt for other insects. After a male catches an insect, he hangs from a twig and releases a pheromone to attract females. When a female approaches, the male presents his prey. If she accepts it, the pair copulates while she eats (Figure 11.23).

Choosy females sometimes get food from their mates.

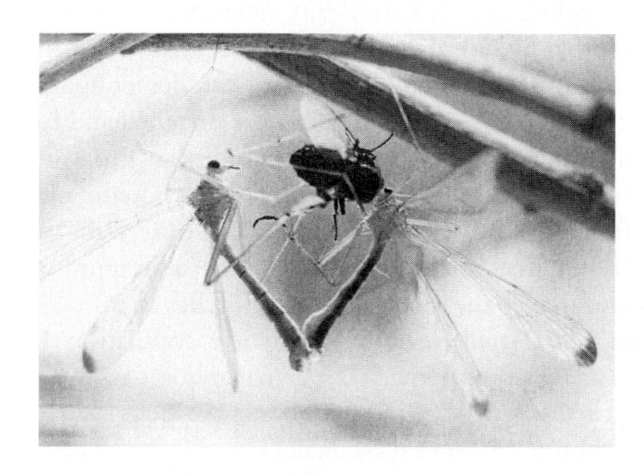

Figure 11.23 **Courtship and mating in hangingflies** A female (right) copulates with a male while eating a blowfly he has captured and presented to her.

The larger the prey, the longer it takes her to eat it, and the longer the pair copulates (Figure 11.24a). The longer the pair copulates, the more sperm the female accepts from the male (Figure 11.24b). If she finishes her meal in less

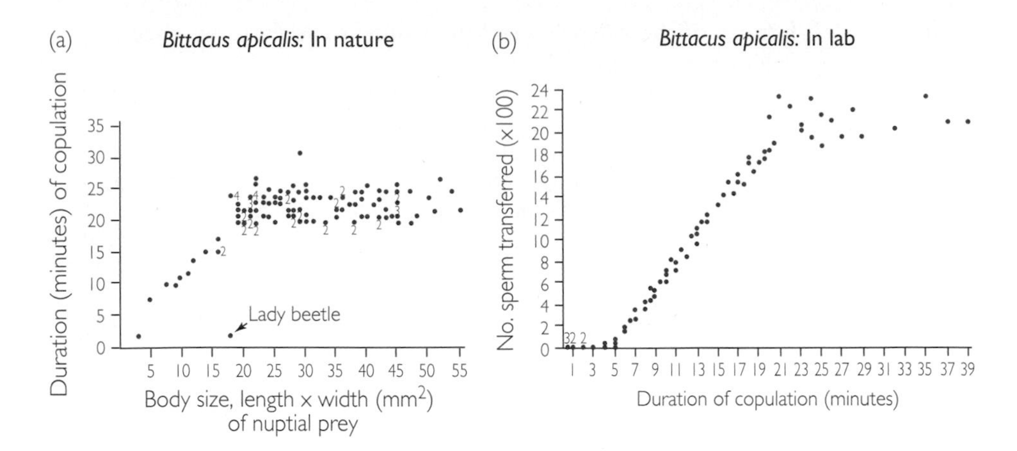

Figure 11.24 Courtship and mating in hangingflies (a) The larger the gift the male presents to the female, the longer the pair copulates. Copulation ends after about 20 minutes, even if the female is still eating. A lady beetle presented by one male is an exception to the general pattern. Even though the beetle was fairly large, the female hangingfly rejected it and broke off copulation almost immediately. (b) The longer a pair copulates, the more sperm the female allows the male to transfer. The male must present a gift that takes at least five minutes to eat or the female breaks off the copulation without accepting any sperm. From Thornhill (1976).

than 20 minutes, the female breaks off the copulation and flies away looking for another male and another meal. The female's preference for males bearing large gifts benefits her in two ways: (1) It provides her with more nutrients, allowing her to lay more eggs; and (2) it saves her from the need to hunt for herself. Hunting is dangerous, and males die in spider webs at more than twice the rate of females. The males behave in accord with the same kind of economic analysis: If the female is still eating after accepting all the sperm she can, the male grabs his gift back and flies off to look for a second female to share it with.

Choosy Females May Have Preexisting Sensory Biases

Females use their sensory organs and nervous systems for many other purposes than just discriminating among potential mates. It is possible that selection for such abilities as avoiding predators, finding food, and identifying members of the same species may result in sensory biases that make females particularly responsive to certain cues (see Enquist and Arak 1993). This may in turn select on males to display those cues, even if the cues would otherwise have no relation to mating or fitness. In other words, the preexisting bias, or sensory exploitation, hypothesis holds that female preferences evolve first and that male mating displays follow.

Research by Heather Proctor on the water mite *Neumania papillator* illustrates possible sensory exploitation (1991; 1992). Members of this species are small freshwater animals that live amid aquatic plants and make their living by ambushing copepods. Water mites have simple eyes that can detect light but cannot form images. Instead of vision, water mites rely heavily on smell and touch. Both males and females hunt copepods by adopting a posture that Proctor calls net-stance. The hunting mite stands on its four hind legs on an aquatic plant, rears up, and spreads its four front legs to form a sort of net. The mite waits until it detects vibrations in the water that might be produced by a swimming copepod, then turns toward the source of the vibrations and clutches at it.

Mating in *Neumania papillator* does not involve copulation. Instead, the male attaches sperm-bearing structures called spermatophores to an aquatic plant, then attempts to induce the female to accept them. He does this by fanning water across the spermatophores toward the female. The moving water carries to the female pheromones released by the spermatophores. When the female smells the pheromones, she may pick up the spermatophores.

Male water mites search for females by moving about on aquatic vegetation. When a male smells a female, he walks in a circle while lifting and trembling his

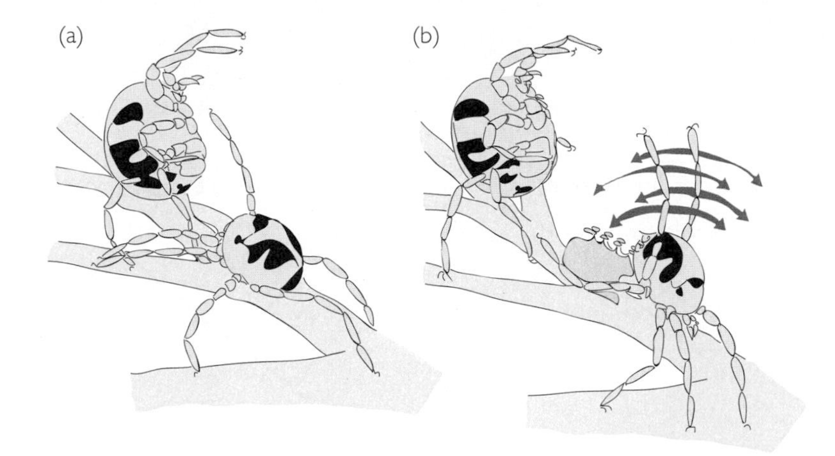

Figure 11.25 **Courtship in the water mite, *Neumania papillator*** (a) The female (on the left) is in net-stance, waiting to ambush a copepod; the male has found her and is now trembling his legs. (b) The female has turned toward the male in response to the trembling. The male has deposited spermatophores and is now fanning water across them. The sausage-shaped objects on top of the spermatophores are sperm packets. Redrawn from Proctor (1991).

front legs (Figure 11.25a). If the male has detected a female that is still there, not just the scent of one that has recently left, the female typically turns toward the trembling male. Often she also clutches at him. At this point, the male deposits his spermatophores and begins to fan (Figure 11.25b).

Proctor suspected that male leg-trembling during courtship evolved in *N. papillator* because it mimics the vibrations produced by copepods and thereby elicits predatory behavior from the female. She tested this hypothesis with a series of experiments in which she watched water mites under a microscope. First, Proctor measured the frequency of vibrations produced by trembling males, and compared it to the frequency of vibrations produced by copepods. Water mites tremble their legs at frequencies of 10 to 23 cycles per second, well within the copepod range of 8 to 45 cycles per second. Second, Proctor observed the behavior during net-stance of female water mites when they were alone, when they were with copepods, and when they were with males. Females in net-stance rarely turned and never clutched unless copepods or males were present, and the behavior of females toward males was similar to their behavior toward copepods. Third, Proctor observed the responses to male mites of hungry females versus well-fed females. Hungry females turned toward males, and clutched them, significantly more often than well-fed females. All of these results are consistent with the hypothesis that male courtship trembling evolved to exploit the predatory behavior of females.

Sometimes choosy females may simply be responding to courting males as though the males were prey.

Males employing leg trembling during courtship probably benefit in several ways. First, males appear to use the female response to trembling to determine whether a female is actually present. Proctor observed that a male that has initiated courtship by trembling is much more likely to deposit spermatophores if the female clutches him than if she does not. Second, trembling appears to allow males to distinguish between receptive females versus unreceptive ones. Proctor observed that a male has a strong tendency to deposit spermatophores for the first female he encounters that remains in place after he initiates courtship, but that virgin females are more likely to remain in place than are nonvirgins. Third, males appear to use the female response to trembling to determine which direction the female is facing. Proctor observed that males deposit their spermatophores in front of the female more often than would be expected by chance. These benefits mean that a male that trembles should get more of his spermatophores picked up by females than would a hypothetical male that does not tremble. In other words, a male that trembles would enjoy higher mating success. This is consistent with the hypothesis that trembling evolved by sexual selection.

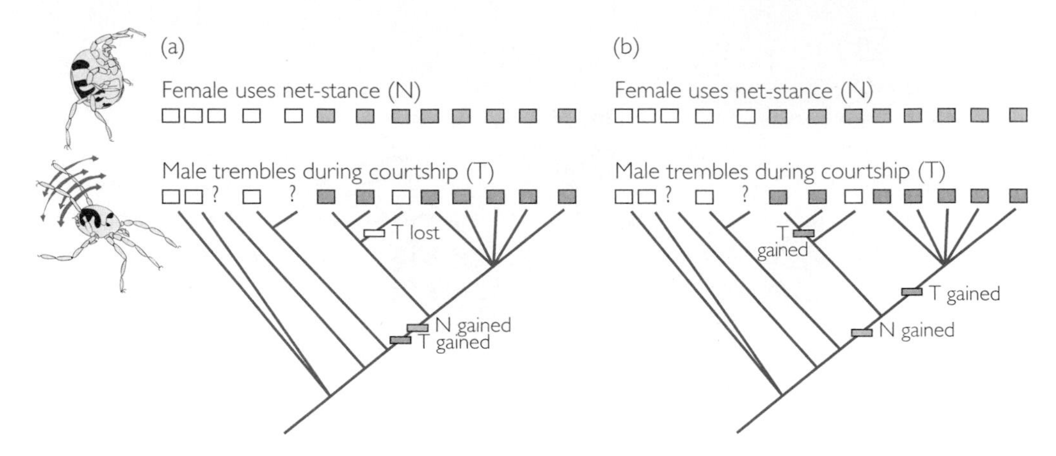

Figure 11.26 A phylogeny of the water mite *Neumania papillator* and several related species The boxes above the tips of the branches indicate which species have net-stance and which species have male courtship trembling. A colored box indicates the trait is present; an open box indicates the trait is absent. The two versions of the phylogeny show the two most likely scenarios for the evolution of these two traits. Redrawn from Proctor (1992).

A key prediction of the sensory exploitation hypothesis is that net-stance evolved before male trembling. Proctor tested this hypothesis by using a suite of morphological characters to estimate the phylogeny of *Neumania papillator* and several related water mites. (Methods for estimating phylogenies were discussed in Chapter 4.) She noted which species have net-stance and which species have male courtship trembling. She then inferred the places on the phylogeny at which net-stance and courtship trembling are most likely to have evolved, based on the assumption that simpler evolutionary scenarios are more probable. Proctor concluded that one of two evolutionary scenarios is most likely to be correct (Figure 11.26). In the first scenario, net-stance and courtship trembling both evolved at the base of the branch that includes all species with either or both of these traits, and trembling was subsequently lost once (Figure 11.26a). In the second scenario, net-stance evolved at the base of the branch, and courtship trembling subsequently evolved twice (Figure 11.26b). The first scenario supplies insufficient evidence to test the prediction that net-stance evolved before trembling. We simply cannot tell, under this scenario, whether one trait evolved before the other, or the two traits evolved simultaneously. The second scenario is consistent with the prediction that net-stance evolved first. Which scenario is closest to the truth remains unknown. However, given the phylogenetic evidence in combination with the data from her observations of water mite behavior, Proctor concludes that sensory exploitation is the best explanation for the evolution of courtship trembling.

Figure 11.27 A male fiddler crab ready to attract a mate This male has constructed a sand hood over his burrow, which will help entice females to visit.

For another example in which male courtship displays appear to have evolved to exploit preexisting female sensory biases, see the work of John Christy and colleagues (2003) on fiddler crabs. Males in some species build sand hoods over the burrows from which they court females (Figure 11.27). Christy has evidence that these hoods are attractive to females because they look like good places to seek shelter from predators.

Other Explanations for Female Choice

We have considered three explanations for female choice that provide good reasons why females might prefer some male displays over others. It is also possible, however, that female preferences are essentially arbitrary. One version of this idea is the sexy-son hypothesis. According to this hypothesis, once a particular male advertisement display is favored by a majority of females, selection on females will automatically reinforce a preference for the fashionable trait. The reason is that females choosing fashionable mates will have more fashionable sons, and therefore more grandchildren, than females choosing unfashionable mates. Selection on the females to maintain the preference and selection on males—by the females because of the preference—form a self-perpetuating loop.

The spotted cucumber beetle, *Diabrotica undecimpunctata howardi,* provides an example. The female in this insect discriminates among males to some extent by how they smell (Brodt et al. 2006). For the most part, however, she decides whether to allow any particular male to fertilize her eggs based on how he behaves during copulation (Figure 11.28). While they copulate, the male uses his antennae to stroke the female (Tallamy et al. 2002). The faster his movements, the more likely she is to relax the muscle that allows him to deposit his sperm in her reproductive tract.

Figure 11.28 Spotted cucumber beetles in love The male is on the left.

To find out what female spotted cucumber beetles gain by accepting sperm from fast stroking males only, Douglas Tallamy and colleagues (2003) assigned females at random to be paired with either a fast male or a slow male. Because they had no choice, most of the females eventually accepted sperm from their assigned mate. The researchers then looked for differences in fitness between the two groups of females and their offspring.

Females mated with fast-stroking males and females mated with slow-stroking males laid similar numbers of eggs, and saw their eggs hatch at similar rates. The offspring of the two groups of females developed at similar rates, reached similar sizes, and enjoyed simliar fecundities. The offspring sired by fast males survived to adulthood at a somewhat higher rate than the offpring sired by slow males, but the difference was not statistically significant. The only significant differences Tallamy and colleagues could find between the offspring of fast versus slow males were that the sons of fast-stroking males grew up to be fast strokers themselves, and that—apparently as a result—the sons of fast-stroking males had their sperm accepted by females at over twice the rate as sons of slow-stroking males. The researchers concluded that what female spotted cucumber beetles gain by being choosy is sexy sons and that the benefit of having sexy sons is substantial. How a preference for fast-stroking males became established in the first place is anyone's guess. But once fast males were fashionable, females who prefered them had more grandchildren.

Theory also suggests that female preferences can be completely arbitrary.

A more technically detailed version of the notion that sexual selection by mate choice can reinforce arbitrary preferences, called the runaway selection hypothesis, is discussed in Box 11.2.

We should note that all the explanations for female choice that we have discussed are mutually compatible. It is possible, at least in principle, for a single courtship display to indicate genetic quality, to predict direct benefits the female is likely to receive, to exploit biases in the female's sensory system, and to be reinforced by selection in favor of sexy sons. Much contemporary research on sexual selection is focused on determining the relative importance of these different factors in the evolution of female preferences. One approach is to test several alternative hypotheses in a single species (see Box 11.2 for an example). Another approach is to use the alternative hypotheses to develop testable predictions about how sexually selected traits will vary among closely related species on an evolutionary tree (see Prum 1997 for an example).

Box 11.2 Runaway sexual selection in stalk-eyed flies?

Runaway selection is an idea first elaborated by Ronald Fisher in 1915, and perhaps traceable to a remark made by T. H. Morgan in 1903 (reviewed in Anderson 1994). The idea is worth explaining in some detail, because it illustrates useful concepts in evolutionary genetics. We will discuss runaway selection in the context of research by Gerald Wilkinson and colleagues on the stalk-eyed flies of Southeast Asia.

Stalk-eyed flies carry their eyes on the ends of long thin appendages. In both sexes bigger flies have longer eyestalks, but males have longer stalks for their size than females. By day the flies are solitary and forage for rotting plants. In the evening, the flies congregate beneath overhanging stream banks, where they cling in small groups to exposed root hairs and spend the night (Figure 11.29). At dawn and dusk, the flies roosting together on a root hair often mate with each other. Neither sex cares for the young, so a female's investment in each offspring is larger than a male's. Not surprisingly, males attempt to evict each other from the root-hair roosts in order to be the only male in the group at mating time. In male–male confrontations, the male with longer eyestalks typically wins, so male–male competition may partially explain the evolution of eyestalks (Burkhardt and de la Motte 1983; Burkhardt and de la Motte 1987; Panhuis and Wilkinson 1999). As we will see, however, there is evidence that female choice has also played a role.

To see how the runaway selection hypothesis works, imagine a population of stalk-eyed flies in which both males and females are variable, males in the lengths of their eyestalks and females in their mating preferences for stalk length. These two patterns of variation should combine to produce assortative mating; that is, the females that prefer the longest stalks will mate with the longest-stalked males, and the females that prefer the shortest stalks will mate with the shortest-stalked males (Figure 11.30a). Assume, furthermore, that in both sexes the variation is heritable—that is, that at least part of the variation in stalk length, and part of the variation in preference, is due to variation in genes (see Chapter 9). Under these assumptions, offspring that receive genes for long eyestalks from their fathers tend to also receive genes for a preference for long-stalked mates from their mothers. In other words, if the assortative mating persists for some generations, then it will establish a genetic correlation (linkage disequilibrium) between the stalk-length genes and the preference genes (see Chapter 8). If we were to take a group of males, mate each with a number of randomly chosen females, and then examine their sons and daughters, we would find that sons with long eyestalks tend to have sisters with a preference for long-stalked males (Figure 11.30b). This association means that if we conduct artificial selection on the stalk lengths of the males (and only on the males), we should get a correlated evolutionary response in the preferences of the females (Figure 11.30c).

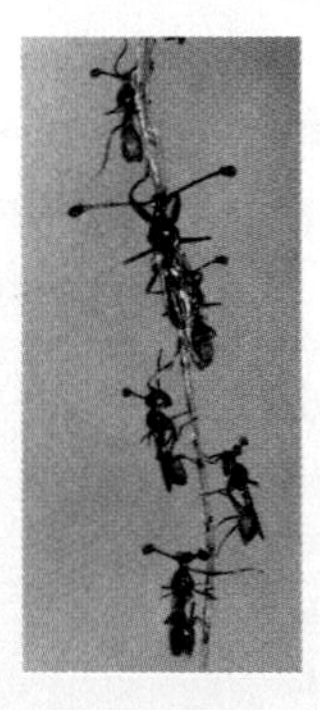

Figure 11.29 A group of Malaysian stalk-eyed flies (*Cyrtodiopsis whitei*) gathered on a root hair to spend the night The largest fly is a male; the others are females.

Wilkinson and Paul Reillo (1994) tested the prediction that selection on male stalk length will produce a correlated response in female preferences. Wilkinson and Reillo collected stalk-eyed flies (*Cyrtodiopsis dalmanni*) in Malaysia and used them to establish three laboratory populations. In each population, the researchers separated the males and females immediately after the adult flies emerged from their pupae and kept them apart for two to three months. Wilkinson and Reillo then chose breeders for the next generation of each population as follows. For the control line, they used 10 males and 25 females picked at random. For the long-selected line, they used the 10 males with longest eyestalks from a pool of 50 males picked at random and 25 females picked at random. For the short-selected line, they

Box 11.2 (Continued)

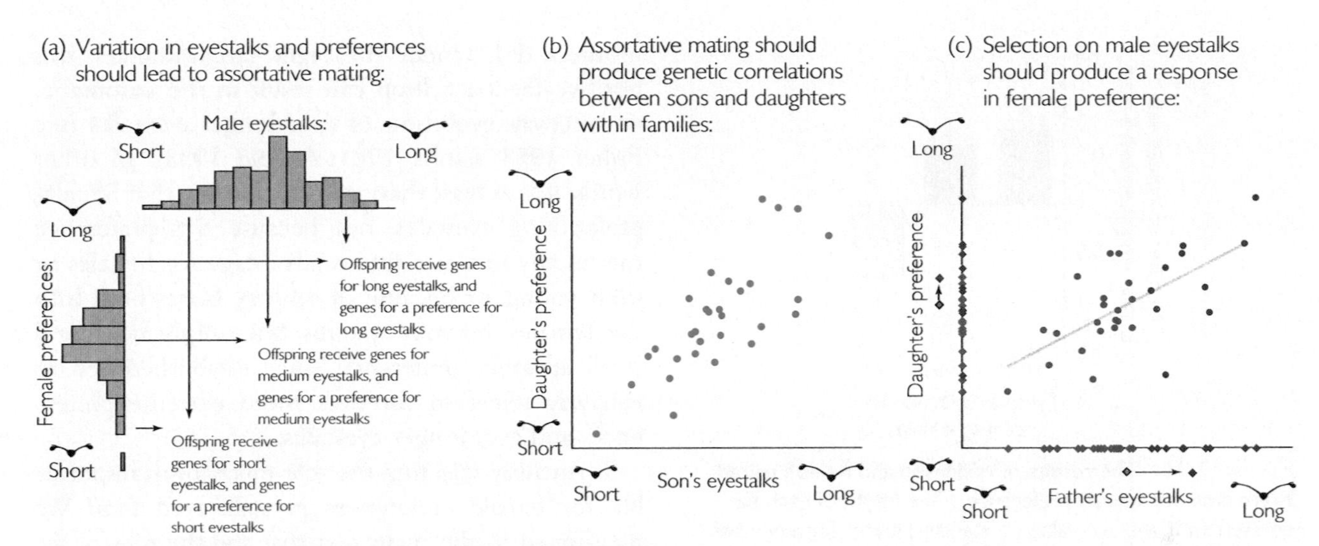

Figure 11.30 How runaway selection works in theory (a) Assortative mating. Females with different preferences choose among males with different eyestalk lengths. If both traits are heritable, then offspring receiving genes for long eyestalks also tend to receive genes for "long" preferences. (b) A genetic correlation between male stalk length and female preference. Each point represents the average value of the offspring of a male mated with each of a number of randomly chosen females. Males whose sons have long eyestalks also tend to have daughters that seek long eyestalks in their mates. After Arnold (1983). (c) Female preference evolves as a correlated response to selection on male stalk length. Each circle represents the stalk length and preference, respectively, of a father–daughter pair. The fathers are also represented by diamonds on the horizontal axis and the daughters by diamonds on the vertical axis. If we select the longest-stalked males as breeders (red diamonds on the horizontal axis and red circles), we should see a response in the daughters. The arrows indicate the selection differential and predicted response (see Chapter 9). The gray diamond below the horizontal axis marks the average of all fathers in the population, and the red diamond marks the average of selected fathers. The gray diamond to the left of the vertical axis marks the average of all daughters, and the red diamond marks the average of daughters of selected males. After Falconer (1989).

used the 10 males with shortest eyestalks from a pool of 50 males picked at random and 25 females picked at random. After 13 generations the populations had diverged substantially in eyestalk length. Wilkinson and Reillo then performed paired-choice tests to assay female preferences in each population.

In each test, Wilkinson and Reillo placed two males in a cage with five females. The males had the same body size, but one was from the long-selected line, and thus had eyestalks that were long, and the other was from the short-selected line, and thus had eyestalks that were short (but still longer than the eyestalks of females from any line). The two males were separated by a clear plastic barrier, and each had his own artificial root hair on which to roost. In the center of the plastic barrier was a hole, just large enough to allow the females to pass back and forth but too small for the males, with their longer eyestalks, to fit through. Wilkinson and Reillo then watched to see which male attracted more females. The scientists performed 15 to 25 tests for each of the three lines. In both the control and the long-selected lines, more females chose to roost for the night with the long-stalked male. In the short-selected line, however, more females chose to roost for the night with the short-stalked male (Figure 11.31). Artificial selection for short eyestalks in males had changed the mating preferences of females.

This result neatly accomplishes several things at once:

- It demonstrates that female stalk-eyed flies are choosy.
- It demonstrates that both male eyestalk length and female preference are heritable.
- It illustrates that selection on one trait can produce an evolutionary response in another trait (see Chapter 9).

Box 11.2 (Continued)

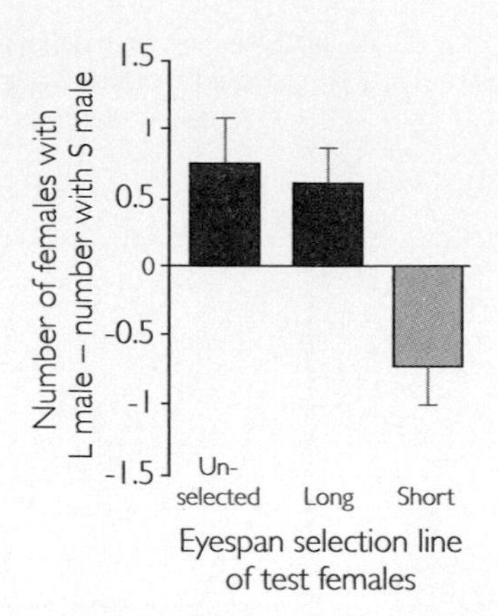

Figure 11.31 The results of Wilkinson and Reillo's paired choice tests for female preference The height of each bar represents the average value (± standard error), for a number of trials, of the difference between the number of females that preferred the long-stalked male and the number that preferred the short-stalked male. Positive values indicate that more females preferred long-stalked males. Unselected females preferred long-stalked males ($P = 0.0033$), as did females from the long-selected line ($P = 0.005$). Females from the short-selected line preferred short-stalked males ($P = 0.023$). From Wilkinson and Reillo (1994).

- It is consistent with Fisher's 80-year-old prediction that sexual selection by female choice produces genetic correlations between male traits and female preferences.

The scenario for the evolution of long eyestalks by runaway selection is as follows. At some time in the past, eyestalks were much shorter than they are now. At some point, a situation arose in which a majority of the females preferred longer-than-average eyestalks. Perhaps a preference for long stalks was favorable for females, because males with longer stalks were genetically superior, or perhaps the female preference was the result of genetic drift. Whatever the cause of the initial female preference, the consequence was that the males with long eyestalks left more offspring. As Fisher first noted, this can create a positive feedback loop, because as Wilkinson and Reillo's experiment showed, selection on males for longer eyestalks produces a correlated response in female preferences. Each generation's males have longer eyestalks than their fathers had, but each generation's females prefer longer stalks than their mothers did. Under the right circumstances, this positive-feedback loop can result in the automatic, or runaway, evolution of ever-longer eyestalks (see Fisher 1958; Lande 1981; Arnold 1983). In other words, it is at least theoretically possible that females prefer long eyestalks, not because this preference carries any intrinsic fitness advantage for females or their young, or because of sensory biases built into the females' nervous systems, but simply because a small arbitrary preference, once established, led to runaway selection for ever more extreme preferences and ever longer eyestalks.

Is runaway selection the sole mechanism responsible for female preferences in stalk-eyed flies? We mentioned in the main text that the theories of female choice we have discussed are mutually compatible. Wilkinson and various colleagues have continued their research on stalk-eyed flies, looking for evidence of other mechanisms selecting on female preferences. Because males provide no parental care and offer no gifts to females, it seems unlikely that choosy females receive direct benefits—unless long-stalked males lay claim to better nighttime roosts. And Wilkinson, Heidi Kahler, and Richard Baker (1998) found no evidence that females had preexisting sensory biases favoring long eyestalks before long eyestalks evolved in males. There is evidence, however, that choosy females get better genes for their offspring.

Choosy female stalk-eyed flies appear to get better genes for their offspring for at least two reasons. First, stalk length in males is correlated with a trait biologists refer to as condition (David et al. 1998). Roughly speaking, condition is general health and vigor as demonstrated by the ability to gather and store energy. Wilkinson and Mark Taper (1999) found that condition is genetically variable. By choosing a male with long eyestalks, a female fly can give her offspring better genes for condition. Second, some male stalk-eyed flies carry an allele on their X-chromosome that causes them to have more daughters than sons, and some males carry an allele on their Y-chromosome that counteracts the allele on the X-chromosome and causes the males to have more sons than daughters. Wilkinson, Daven Presgraves, and Lili Crymes (1998) found that the frequency of the X-chromosome allele

Box 11.2 (Continued)

does not vary across males with different stalk lengths, but the frequency of the Y-chromosome allele does. It is higher among males with long eyestalks. This matters because, in wild populations of stalk-eyed flies, females outnumber males and females can expect more grandoffspring through their sons than through their daughters. By choosing a mate with long eyestalks, a female fly can increase her chances of producing many sons.

In summary, long eyestalks appear to have evolved in stalk-eyed flies in response to a combination of male–male competition, a female preference for mates with good genes, and possibly a female preference reinforced by runaway selection. This combination of forces favoring long eyestalks raises a new question: Why are the males' eyestalks not even longer than they are now—for example, twice the length of the flies' bodies, or three times? One hypothesis is that if the eyestalks were any longer they would be a serious impediment to survival. As far as we know, this hypothesis has not been tested.

11.4 Sexual Selection on Females

We mentioned earlier that biologists were slow to accept the idea that females can actively discriminate among their suitors. Biologists were also slow to recognize that females themselves can experience sexual selection. Part of the problem was conceptual. Given that a single insemination typically delivers vastly more sperm than a female needs to fertilize her eggs, and that mating exposes a female to diseases and other dangers, biologists could see little that a female might gain from mating more than once, let alone with more than one male. But the problem was also practical. Few biologists could watch their research subjects closely enough and continuously enough to witness multiple matings. Modern genetic tests that would reveal multiple paternity within clutches or litters had not yet been developed. There were hints, however. There were species of birds and fish, such as the greater painted snipe (see Knight 2002) and the broad-nosed pipefish, in which the females are showier and the males provide all the parental care. And there were biologists, such as Sarah Hrdy (1979), who watched as females in the species they studied pursued multiple mates. When biologists began using genetic tests, they discovered that multiple mating is rampant (see Box 11.3). We begin this section with two examples.

Sexual selection acts on females too.

Polyandry: Multiple Mating by Females

John Hoogland (1998) studied polyandry in Gunnison's prairie dogs. His study is unusual, and valuable, because it relied on careful observation instead of genetic tests. Genetic tests can identify females that have conceived with more than one male. Direct observation can, in addition, identify matings that did not result in conception or birth. Prairie dogs mate in burrows, so Hoogland did not witness copulations. But he could infer from their above-ground behavior what the prairie dogs were doing. Among other things, before or after copulating, male prairie dogs often bark a unique call.

During 15,000 person-hours of observation over a 7-year study, Hoogland and his assistants gathered data on the mating behavior and reproductive success of more than 200 female prairie dogs. Hoogland found that 65% of these females

Box 11.3 Extra-pair copulations and multiple mating

Biologists once thought that polyandry—multiple mates per female—was rare. Researchers seldom witnessed it, and knew no reason to suspect it. That, however, was before they developed methods of genetic analysis that enable them to indirectly estimate the rate of extra-pair copulations—or at least extra-pair conceptions. Figure 11.32 shows two such tests performed on red-winged blackbirds.

Figure 11.32a shows paternity analysis using a restriction-fragment length polymorphism. The photo in the figure depicts an electrophoresis gel. Each lane in this electrophoresis gel contains DNA that has been extracted from an individual bird, cut with a restriction enzyme, and labeled with a probe that recognizes a sequence of DNA that occurs at a single locus. Bands on the gel are inherited as simple Mendelian alleles. Individual M1 (center lane) is an adult male red-wing who had two mates on his territory, F1 and F2. M2 and M3 are adult male neighbors of M1, F1, and F2. The numbers 1, 2, and 3 represent chicks in the nest shared by M1 and F1. Chick 1 has a band in its lane (arrow) that is present in neither its mother (F1) nor its social father (M1). This band is present, however, in M2. We can infer that F1 had an extra-pair copulation with M2 (or with an unknown male with the same genotype). The numbers 4, 5, and 6 represent chicks in the nest shared by M1 and F2. Chick 6 has a band in its lane (arrow) that is present in neither its mother (F2) nor its social father (M1). This band is present, however, in M3. We can infer that F2 had an extra-pair copulation with M3.

Figure 11.32b shows a paternity analysis of the same families using DNA fingerprints. Each lane contains DNA that has been extracted from an individual bird, cut with a restriction enzyme, and labeled with a probe that recognizes a sequence of DNA that occurs at many loci. Bands on the gel are inherited as simple Mendelian alleles, although we do not know which band corresponds to an allele at which locus. The DNA fingerprints confirm the same cases of extra-pair copulation we inferred from the gel in Figure 11.32a.

How common is extra-pair copulation in red-winged blackbirds? Elizabeth Gray (1997) used DNA fingerprints to assess the frequency of extra-pair copulation in a population in eastern Washington state. She estimated that in a given breeding season between 50% and 64% of all nests contained at least one chick sired by a male other than its social parent. Red-winged blackbirds are not unusual. By using genetic paternity tests, biologists have discovered that many socially monogamous birds engage in frequent extra-pair copulations.

Genetic analyses have, in fact, revealed that females in a great many animal species mate with more than one male (see Knight 2002). True monogamy, it turns out, is the mating system that is rare.

mated with more than one male, and a few mated with as many as four or five. The females have good reason to seek multiple mates. The probability of getting pregnant and giving birth was 92% for females that had only one or two mates, but 100% for females that had three or more. Hoogland thinks the reason is that some males are either permanently sterile or temporarily depleted of sperm. Female prairie dogs are only fertile once a year, and for a very short time. By mating with more than one male, a female increases her chances of receiving enough viable sperm to fertilize all her eggs.

Among the females that gave birth, litter size on the day the pups first came above ground increased with number of mates (Figure 11.33). This pattern may in part be due to associations with a third factor, the female's body size. Larger females can support larger pregancies. They may also attract more attention from amorous males, or be better able to resist one mate's attempts to prevent them from seeking other mates. But associations with the female's body size are not the

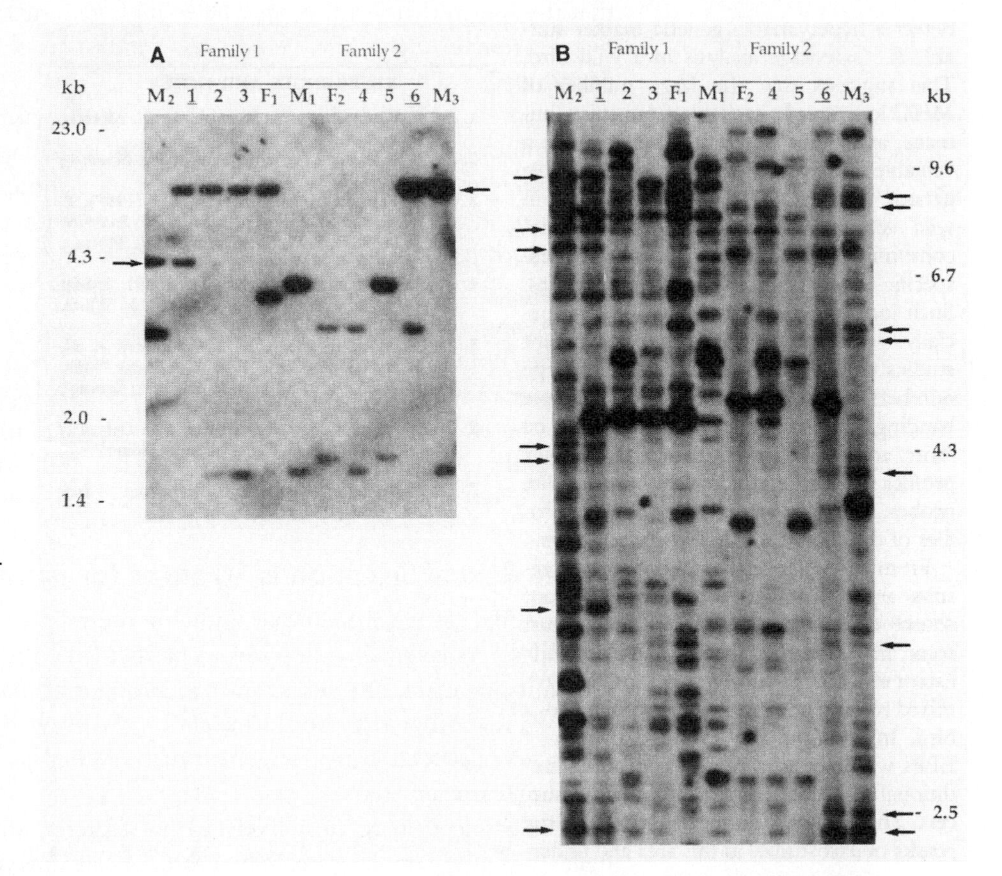

Figure 11.32 Genetic analyses demonstrating extra-pair copulations in red-winged blackbirds (a) A paternity analysis using a traditional restriction-fragment length polymorphism. (b) A paternity analysis of the same families using DNA fingerprints. Reprinted from H. L. Gibbs et al. (1990).

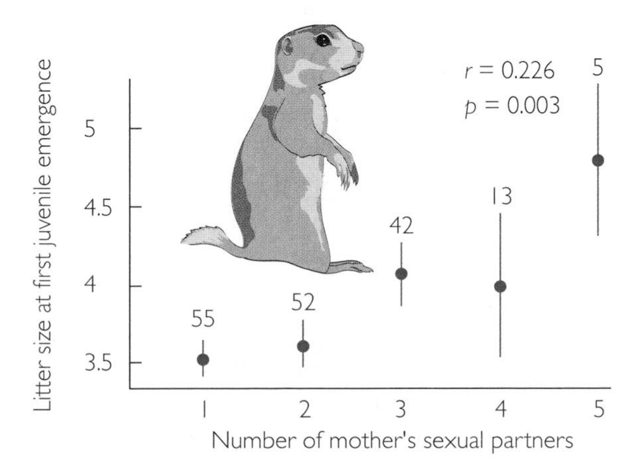

Figure 11.33 Female Gunnison's prairie dogs (*Cynomys gunnisoni*) increase their reproductive success by mating with more than one male Data points show average litter sizes, on the day the pups came above ground for the first time, for females who mated with different numbers of males. The lines above and below symbols show ±1 standard error; the numbers above the data points show the number of females in each category. From Hoogland (1998).

whole story. When Hoogland used statistical methods to remove the associations with body size, females with more mates still had larger litters. This might be due to increased genetic diversity among litters of pups with different fathers. The more genetically diverse a female's pups, the less likely they will all be killed by the same disease or parasite. Or perhaps sperm competition increases the chance that the sperm that fertilize eggs are of high genetic quality. Although the precise reasons remain to be discovered, Hoogland's data show that female Gunnison's prairie dogs experience sexual selection.

Females often benefit from mating with more than one male.

In some species, benefits garnered by mating with multiple males have led to the evolution of special mechanisms that help females avoid mating twice with the same partner. For example, when given a choice, female crickets (*Gryllodes sigillatus*), show a clear preference for mating with novel males versus previous mates. Do the females avoid mating twice with the same male by remembering what he smells like? Or do the females recognize their own lingering scent? Tracie Ivy and colleagues (2005) let female crickets choose between a novel male and the inbred brother of a male they had already mated with. The females did not discriminate between the males. But when the researchers gave females a choice between a novel male and a male who had mated with the female's own inbred sister, most chose the novel male. Ivy and colleagues concluded that female crickets scent-mark their mates so that they can avoid them in the future.

When Sexual Selection Is Stronger for Females than for Males

Species in which males invest more in each offspring, and are thus a limiting resource for females, provide a valuable opportunity to test the rules of sexual selection.

In Section 11.1 we argued that knowing the relative strength of sexual selection on males versus females allows us to predict how the sexes will approach an opportunity to mate. The sex subject to stronger sexual selection should be competitive; the other sex should be choosy. We saw these predictions confirmed when sexual selection is stronger for males. What happens when sexual selection is stronger for females?

Sexual selection is most likely to be stronger for females than males when males provide parental care. When fathers care for young, male parental investment per offspring may be comparable to, or even greater than, female parental investment. Species with male parental care include humans, many fish, about 5% of frogs, and over 90% of birds. When males actually do invest more per offspring than females, access to mates will be a limiting resource for females.

In Figure 11.6 (page 406) we examined data on a species in which sexual selection is a stronger for females than for males: the broad-nosed pipefish, *Syngnathus typhle*. We return to pipefish now. The work we will review was done on *S. typhle* and another pipefish, *Nerophis ophidion,* by Gunilla Rosenqvist, Anders Berglund, and their colleagues.

Recall that in pipefish males provide all the parental care. In *N. ophidion,* the male has a brood patch on his belly; in *S. typhle,* the male has a brood pouch. In both species, the female lays her eggs directly onto or into the male's brood structure. The male supplies the eggs with oxygen and nutrients until they hatch.

Although the extensive parental care provided by male pipefish requires energy, the pivotal currency for pipefish reproduction is not energy but time (Berglund et al. 1989). Females of both *N. ophidion* and *S. typhle* can make eggs faster than males can rear them to hatching. As a result, access to male brood space limits female reproductive success. If the theory of sexual selection we have developed is correct, then in these pipefish the females should compete over access to mates, and the males should be choosy.

In *N. ophidion,* the females are larger than the males and have two traits the males lack: dark blue stripes and skin folds on their bellies. These traits appear to function primarily as advertisements for attracting mates. For example, females develop skin folds during the breeding season and lose them after, and in captivity females develop skin folds only when males are present (Rosenqvist 1990). In paired-choice tests (Figure 11.34a), *N. ophidion* males are choosy, preferring larger females (Figure 11.34b) and females with larger skin folds (Figure 11.34c). Females, in contrast, appear to be less choosy. In paired-choice tests, females showed no tendency to discriminate between males of different sizes (Berglund and Rosenqvist 1993).

In *S. typhle,* the males and females are similar in size and appearance. Females, however, can change their color to intensify the zigzag pattern on their sides

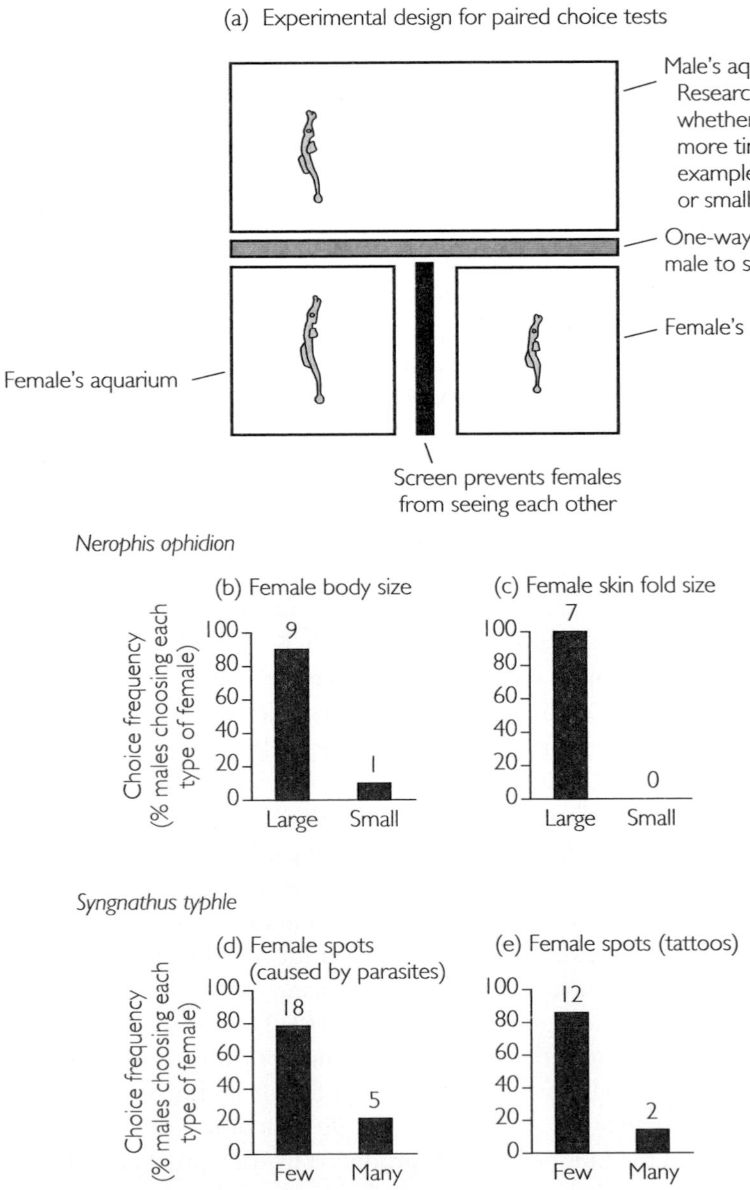

Figure 11.34 Male choice in pipefish (a) In paired-choice tests, researchers place a male pipefish in an aquarium from which he can see two females. The researchers infer which female the male would prefer as a mate from where he spends more of his time. In (b–d) the numbers above the bars indicate the number of males tested. After Rosenqvist and Johansson (1995). (b) Given a choice between large or small females, male pipefish prefer large females ($P = 0.022$). Replotted from Rosenqvist (1990). (c) Given a choice between females with large or small skin folds, male pipefish prefer large-folded females ($P = 0.016$). Replotted from Rosenqvist (1990). (d) Given a choice between females with many black spots (caused by a parasite) or females with few black spots, male pipefish prefer females with few spots ($P < 0.05$). Replotted from Rosenqvist and Johansson (1995). (e) Males still prefer females with few spots, even when the spots are tattooed onto parasite-free females ($P < 0.01$). Replotted from Rosenqvist and Johansson (1995).

(Berglund et al. 1997; Bernet et al. 1998). The females compete with each other over access to males (Berglund 1991) and while doing so display their dark colors. Females initiate courtship and mate more readily than males (Berglund and Rosenqvist 1993). Males are choosy (Rosenqvist and Johansson 1995). In paired-choice tests (Figure 11.34a), male *S. typhle* prefer females showing fewer of the black spots that indicate infection with a parasitic worm, whether the black spots were actually caused by parasites (Figure 11.34d) or were tattooed onto the females (Figure 11.34e). This choosiness benefits the males directly, because females with fewer parasites lay more eggs for the males to fertilize and rear.

The mating behavior of pipefish males and females is consistent with the theory of sexual selection. Other examples of "sex-role reversed" species whose behavior appears to support the theory include moorhens (Petrie 1983), spotted sandpipers (Oring et al. 1991a,b; 1994), giant waterbugs (see Anderson 1994), and some species of katydids (Gwynne 1981; Gwynne and Simmons 1990).

11.5 Sexual Selection in Plants

Plants are sometimes sexually dimorphic (Renner and Ricklefs 1995). Orchids in the genus *Catasetum* provide the most dramatic example. Individual plants produce male and female flowers at different times. The flowers of the opposite sex are so different that early orchid systematists placed individuals with male flowers (Figure 11.35a) in one genus and individuals with female flowers (Figure 11.35b) in another. The herb *Wurmbea dioica,* from Australia, provides a more typical example. The males make larger flowers than the females (Figure 11.35c). We have seen that sexual selection can explain sexual dimorphism in animals. Can it also explain sexual dimorphism in plants?

Sexual selection theory can be applied to plants as well as animals.

Many of the ideas we have developed about sexual selection in the context of animal mating can, in fact, be applied to plants (Bateman 1948, Willson 1979; but see Grant 1995). In plants, mating involves the movement of pollen from one individual to another. The recipient of the pollen, the seed parent, must produce a fruit. As a result, the seed parent may make a larger reproductive investment per seed than the pollen donor, which must only make pollen. When pollen is transported from individual to individual by animals, a plant's access to mates is a function of its access to pollinators. Based on the principles of sexual selection in animals, we can hypothesize that access to pollinators limits the reproductive success of pollen donors to a greater extent than it limits the reproductive success of seed parents.

Maureen Stanton and colleagues (1986) tested this hypothesis in wild radish (*Raphanus raphanistrum*). Wild radish is a self-incompatible annual herb that is pollinated by a variety of insects, including honeybees, bumblebees, and butterflies. Many natural populations of wild radish contain a mixture of white-flowered and yellow-flowered individuals. Flower color is determined by a single locus: White (*W*) is dominant to yellow (*w*). Stanton and colleagues set up a study population with eight homozygous white plants (*WW*) and eight yellow plants (*ww*). The scientists monitored the number of pollinator visits to plants of each color, then measured reproductive success through female and male function.

Measuring reproductive success through female function was easy: The researchers just counted the number of fruits produced by each plant of each color. Measuring reproductive success through male function was harder; in fact, it was not possible at the level of individual plants. Note, however, that a yellow seed

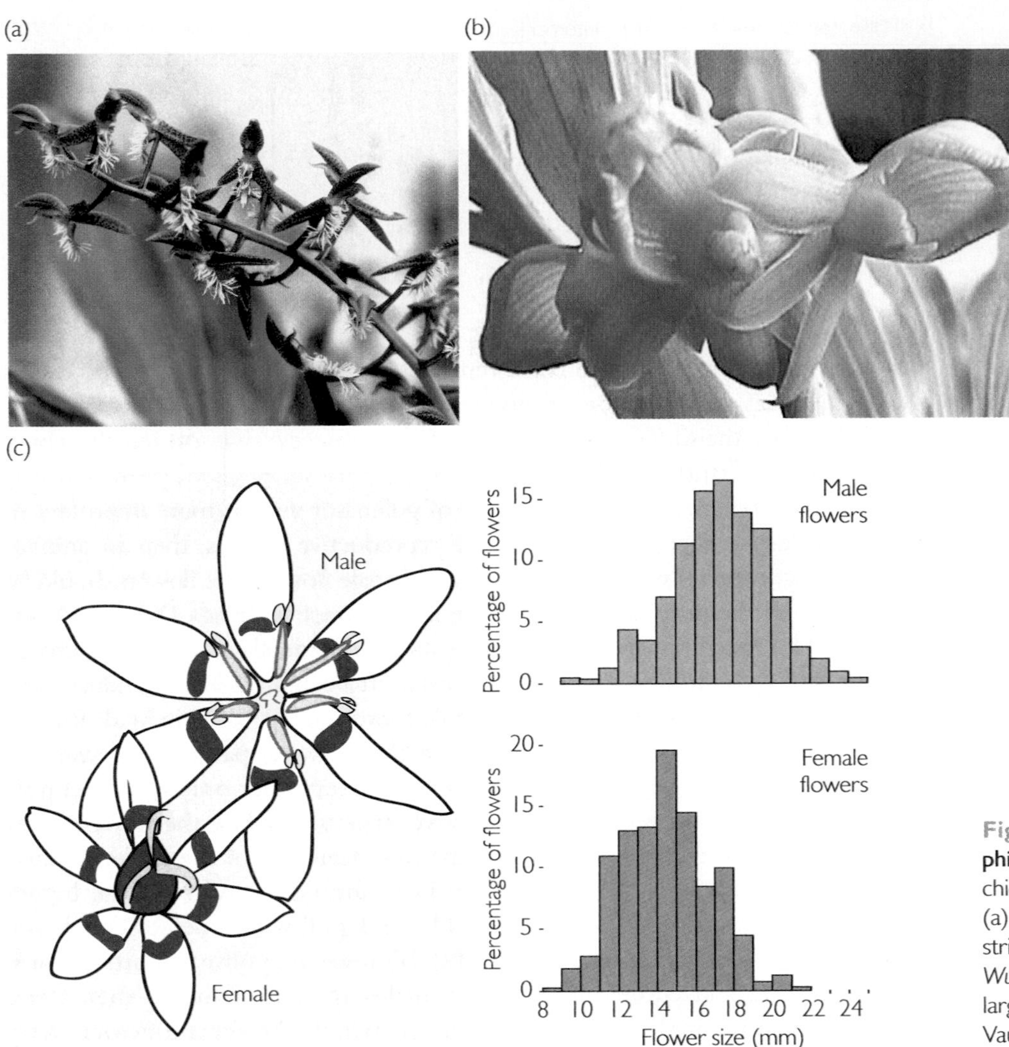

Figure 11.35 Sexual dimorphism in plants (a) In the orchid *Catasetum barbatum*, male (a) and female (b) flowers are strikingly different. In the herb *Wurmbea dioica* (c), males make larger flowers. Redrawn from Vaughton and Ramsey (1998).

parent (*ww*) will produce yellow offspring (*ww*) if it mated with a yellow pollen donor (*ww*), but white offspring (*Ww*) if it mated with a white pollen donor (*WW*). Thus, by rearing the seeds produced by the yellow seed parents and noting the color of their flowers, Stanton and colleagues could compare the population-level reproductive success of white versus yellow pollen donors. The relative reproductive success of pollen donors through yellow seed parents should be a reasonable estimate of the pollen donors' relative reproductive success through seed parents of both colors. The scientists repeated their experiment three times.

As Stanton and colleagues expected from previous research, the yellow-flowered plants got about three-quarters of the pollinator visits (Figure 11.36a). If reproductive success is limited by the number of pollinator visits, then the yellow-flowered plants should also have gotten about three-quarters of the reproductive success. This was true for reproductive success through pollen donation (Figure 11.36c), but not for reproductive success through seed production (Figure 11.36b). Reproductive success through seed production was simply proportional to the number of plants of each type. These results are consistent with the typical pattern in animals: The reproductive success of males is more

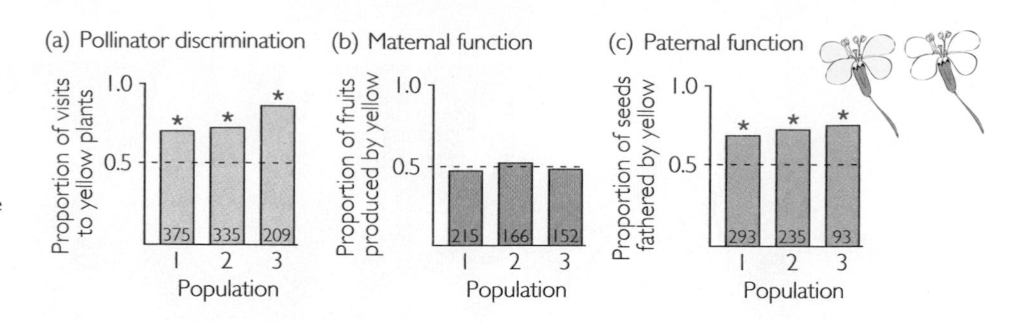

Figure 11.36 Reproductive success through pollen donation is more strongly affected by the number of pollinator visits than is reproductive success through the production of seeds The numbers inside each bar represent the number of pollinator visits (a), the number of fruits (b), and the number of seeds (c) examined. Bars marked with an asterisk have heights significantly different from 0.5 ($P < 0.0001$). (a) In populations with equal numbers of white and yellow flowers, the yellow-flowered plants got most of the pollinator visits. (b) Despite the inequality in pollinator visits, white- and yellow-flowered plants produced equal numbers of fruits. (c) The majority of the seeds, however, were fathered by yellow plants. From Stanton et al. (1986).

limited by access to mates than is the reproductive success of females. The results also suggest that the evolution of showy flowers that attract pollinators has been driven more by their effect on male reproductive success than on female reproductive success (Stanton et al. 1986).

If it is true, in general, that the number of pollinator visits is more important to male reproductive success than to female reproductive success, then in animal-pollinated plant species with separate male and female flowers, the flowers should be dimorphic and the male flowers should be more attractive. Lynda Delph and colleagues (1996) tested this hypothesis with a survey of animal- and wind-pollinated plants, including both dioecious species (separate male and female individuals) and monoecious species (separate male and female flowers on the same individual).

Delph and her coauthors first noted that the showiest parts of a flower, the petals and sepals that together form the perianth, serve not only to attract pollinators, but also to protect the reproductive structures when the flower is developing in the bud. If protection were the only function of the perianth, then the sex that has the bigger reproductive parts should always have the bigger perianth. This was indeed the case in all 11 wind-pollinated species Delph and colleagues measured (Figure 11.37a, right). If, however, pollinator attraction is also important, and more important to males than to females, then there should be a substantial number of species in which the female flowers have bigger reproductive parts, but the male flowers have bigger perianths. This was the case in 29% of the 42 animal-pollinated plants Delph and colleagues measured (Figure 11.37a, left). Furthermore, in species that are sexually dimorphic, male function tends to draw a greater investment in number of flowers per inflorescence and in strength of floral odor, although not in quantity of nectar (Figure 11.37b). These results are consistent with the hypothesis that sexual selection, via pollinator attraction, is often stronger for male flowers than for female flowers.

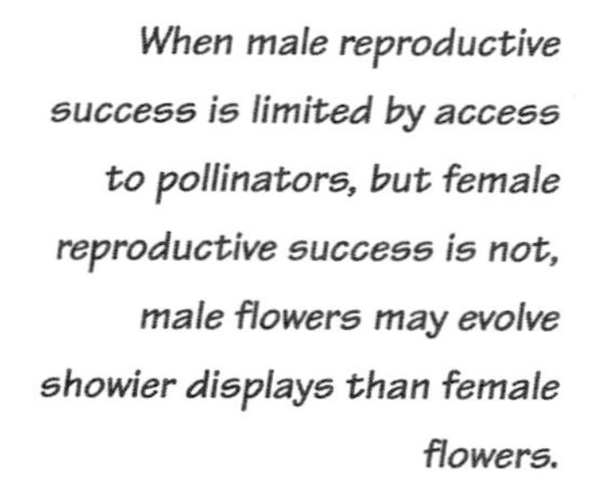
When male reproductive success is limited by access to pollinators, but female reproductive success is not, male flowers may evolve showier displays than female flowers.

Can sexual selection explain the particular examples of sexual dimorphism we introduced at the beginning of this section? Recall that in the herb *Wurmbea dioica,* males make larger flowers than females. This plant is pollinated by bees, butterflies, and flies. Glenda Vaughton and Mike Ramsey (1998) found that bees and butterflies visit larger flowers at higher rates than smaller flowers. As a result, pollen is removed from large flowers more quickly than from small flowers. Males with large flowers may benefit from exporting their pollen more quickly if a head start allows their pollen to beat the pollen of other males in the race to females' ovules. In addition, larger male flowers make more pollen, giving the pollen donor more chances to win. For females, larger flowers probably do not confer any benefit. Female flowers typically receive more than four times the pollen needed to fertilize all their ovules, and seed production is therefore not limited by pollen. These pat-

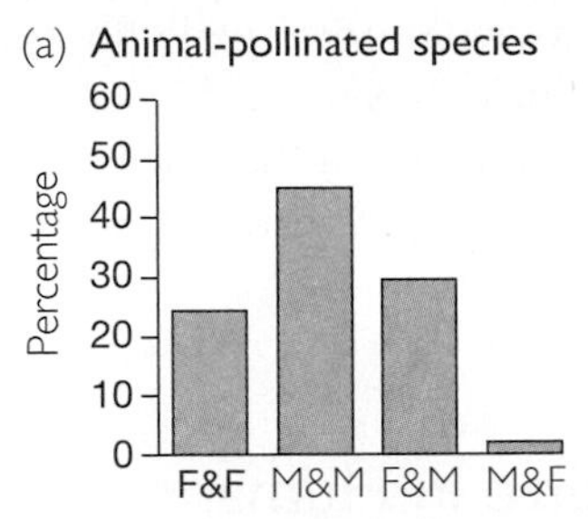

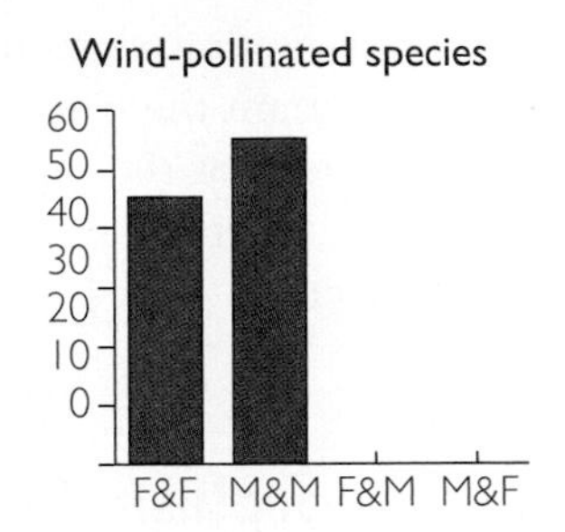

F&F: Female flower has larger reproductive parts and the larger perianth.
M&M: Male flower has larger reproductive parts and the larger perianth.
F&M: Female flower has larger reproductive parts, but male has the larger perianth.
M&F: Male flower has larger reproductive parts, but female has the larger perianth.

(b)

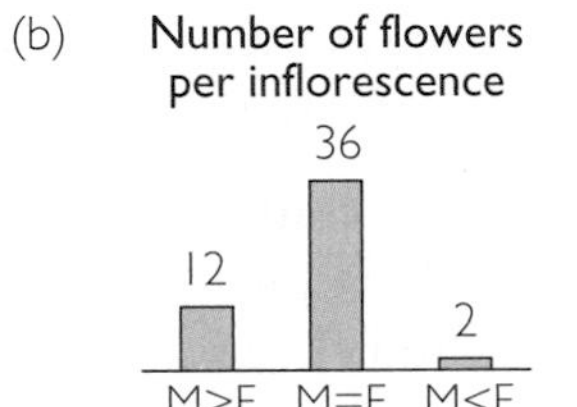

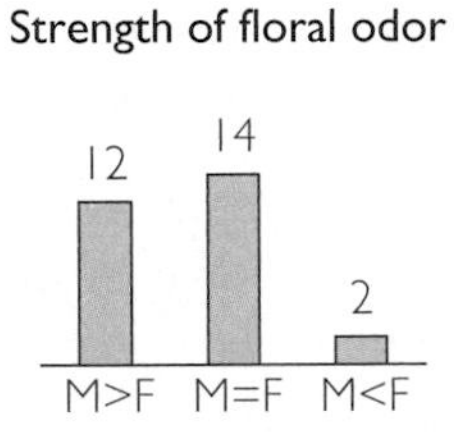

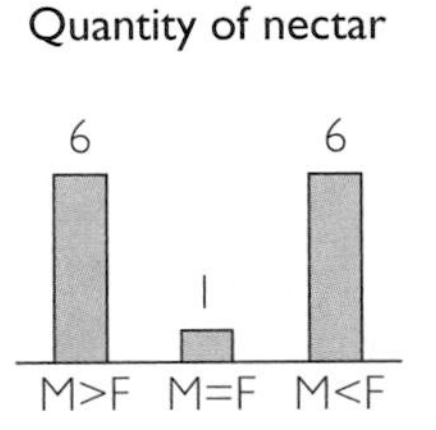

M>F: Male investment is greater than female investment.
M=F: Male investment and female investment are equal.
M<F: Male investment is less than female investment.

Figure 11.37 Patterns of sexual dimorphism in plants with separate male and female flowers (a) In all 11 wind-pollinated species measured (right), the sex with larger reproductive parts has the larger perianth. In 29% of the 42 animal-pollinated species measured (left), the female has larger reproductive parts but the male has the larger perianth, a pattern significantly different from that for wind-pollinated species ($P < 0.001$). From Delph et al. (1996). (b) When animal-pollinated plants have flowers that are sexually dimorphic for investment in pollinator attraction, investment by males tends to be larger for two of the three traits studied ($P < 0.01$; $P = 0.01$; N.S.). Drawn from data in Delph et al. (1996).

terns are consistent with the hypothesis that sexual selection on males is responsible for the sexual dimorphism in flower size.

Orchids in the genus *Catasetum,* the plants with the dramatically dimorphic flowers, have an unusual pollination system. They are pollinated exclusively by male euglossine bees. The orchids attract the bees with fragrant chemicals, such as cineole, which the bees collect and use in their own attempts to attract mates. The flowers of male *Catasetum* orchids are loaded with a single pollen-bearing structure, called a pollinarium. When a bee trips the trigger in a male flower, the flower shoots the pollinarium at the bee like a rubber band off a finger. The pollinarium sticks to the back of the bee with an adhesive that makes it impossible for the bee to remove. When the bee later visits a female flower, one of the pollen masses on the pollinarium lodges in the receptive structure on the female flower, called the stigmatic cleft, and is torn from the pollinarium. The stigmatic cleft quickly swells shut. This means that a female flower typically receives pollen from just one male.

Gustavo Romero and Craig Nelson (1986) observed bees pollinating the flowers of *Catasetum ochraceum*. The researchers found that after being shot with a pollinarium by one male flower, the bees avoided visiting other male flowers but continued to forage in female flowers. Romero and Nelson offer the following scenario to account for the sexual dimorphism in the flowers of *C. ochraceum*. At any given time, there are many more male flowers blooming than female flowers. This, in combination with the fact that female flowers accept pollen from only one male, means that there is competition among male flowers over opportunities to mate. The competition is further intensified by the fact that a second pollinarium attached to a bee would probably interfere with the first. A male flower that has attracted a bee and loaded it with a pollinarium would be at a selective advantage if it could prevent the bee from visiting another male flower. It is therefore adaptive for male flowers to train bees to avoid other male flowers, so long as they do not also train the bees to avoid female flowers. If this scenario is correct, forcible attachment of the pollinarium to the bee and sexually dimorphic flowers make sense together, and both are due to competition for mates—that is, to sexual selection.

11.6 Sexual Dimorphism in Body Size in Humans

One of the examples of sexual dimorphism that we cited at the beginning of this chapter was body size in humans (Figure 11.2, page 402). We now ask whether the sexual dimorphism in human size is the result of sexual selection. It is a difficult question to answer because sexual selection concerns mating behavior. The evolutionary significance of human behavior is hard to study for at least two reasons:

- Human behavior is driven by a complex combination of culture and biology. Studies based on the behavior of people in any one culture provide no means of disentangling these two influences. Cross-cultural studies can identify universal traits or broad patterns of behavior, either of which may warrant biological explanations. Cultural diversity is rapidly declining, however, and some biologists feel that it is no longer possible to do a genuine cross-cultural study.
- Ethical and practical considerations prohibit most of the kinds of experiments we might conduct on individuals of other species. This means that most studies of human behavior are observational. Observational studies can identify correlations between variables, but they offer little evidence of cause and effect.

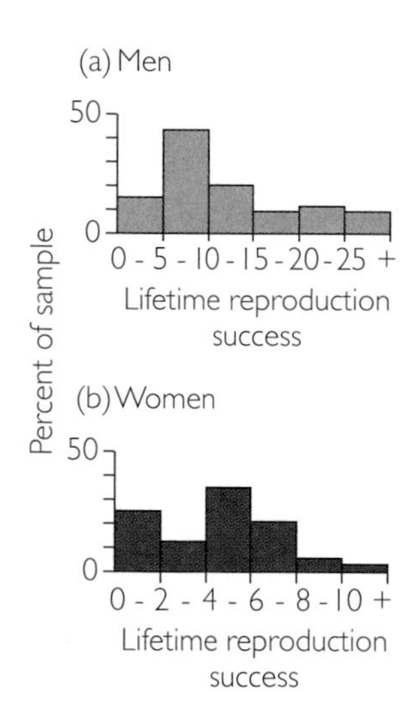

Figure 11.38 Variation in lifetime reproductive success among Kipsigis men and women (a) For the men, the height of each bar represents the percentage of men who had 0 to 5 children, 6 to 10 children, and so on. (b) For the women, the height of each bar represents the percentage of women who had 0 to 2 children, 3 or 4 children, 5 or 6 children, and so on. Some of the men had more than 25 children; few of the women had more than 10. From Borgerhoff Mulder (1988).

Human behavior is inherently fascinating, however, and we therefore proceed, with caution, to briefly consider the question of sexual selection and body size in humans.

The most basic knowledge of human reproductive biology indicates that the opportunity for sexual selection is greater in men than in women. Data from a single culture will suffice to illustrate this point. Research by Monique Borgerhoff Mulder (1988) on the Kipsigis people of southwestern Kenya revealed that the men with highest reproductive success had upwards of 25 children, while the most prolific women rarely had more than 10 (Figure 11.38). In Kipsigis culture, it appears that the reproductive success of men was limited by mating opportunities to a greater extent than was the reproductive success of women. But is there any evidence that reproductive competition, either via male–male interactions or female choice, selects for larger body size in men?

The most obvious kind of sexual selection to look at is male–male competition, because it drives the evolution of large male size in a great variety of other species. Men do, on occasion, compete among themselves over access to mates, but so do women. Do men compete more intensely? On the reasoning that homicide is an unambiguous indication of conflict, and that virtually all homicides are reported to the police, Martin Daly and Margo Wilson (1988) assembled data on rates of same-sex homicide from a variety of modern and traditional cultures. In all of these cultures, men kill men at much higher rates than women kill women. In the culture with the most balanced rates of male–male versus female-female killings, men committed 85% of the same-sex homicides. In several cultures, men committed all of the same-sex homicides.

Data from the United States and Canada show that the majority of perpetrators, and victims, of male–male homicides are in their late teens, twenties, and early thirties. On these and other grounds, Daly and Wilson interpret much male–male homicide as a manifestation of sexually selected competition among men.

If Daly and Wilson's interpretation is correct, then men who are more successful in male–male combat should have higher mating success and higher fitness, at least in pre-modern cultures without formal police and criminal justice systems. Napoleon Chagnon (1988) reported data on the Yąnomamö that confirm this prediction, at least for one culture. The Yąnomamö are a pre-modern people that live in the Amazon rain forest in Venezuela and Brazil. They take pride in their ferocity. Roughly 40% of the adult men in Chagnon's sample had participated in a homicide, and roughly 25% of the mortality among adult men was due to homicide. The Yąnomamö refer to men who have killed as *unokais*. Chagnon's data show that *unokais* have significantly more wives, and significantly more children, than non-*unokais* (Figure 11.39).

The Yąnomamö fight with clubs, arrows, spears, machetes, and axes. It would be reasonable to predict that *unokais* are larger than non-*unokais*. Chagnon (1988) reports, however, that "Personal, long-term familiarity with all the adult males in this study does not encourage me to conclude at this point that they could easily be sorted into two distinct groups on the basis of obvious biometric characters, nor have detailed anthropometric studies of large numbers of Yąnomamö males suggested this as a very likely possibility." Data on the relationship between male–male competition, body size, and mating success in other cultures are scarce.

B. Pawlowski and colleagues (2000) investigated the hypothesis that the sexual dimorphism in human body size is a result of female choice. The researchers gathered data from the medical records of 3,201 Polish men. They used statistical techniques to remove the effects of a variety of confounding variables, including residence in cities versus rural areas, age, and education. Pawlowski and

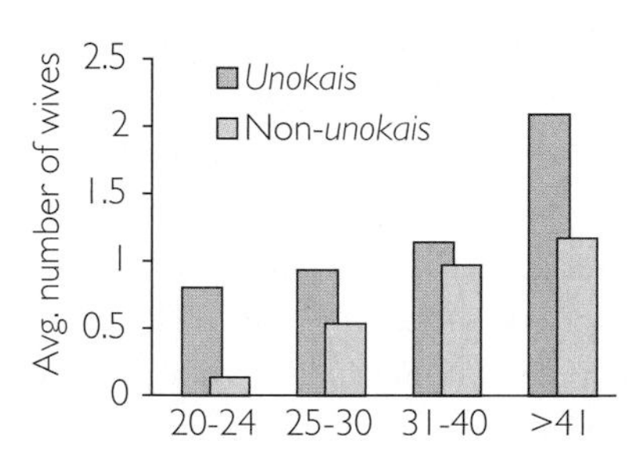

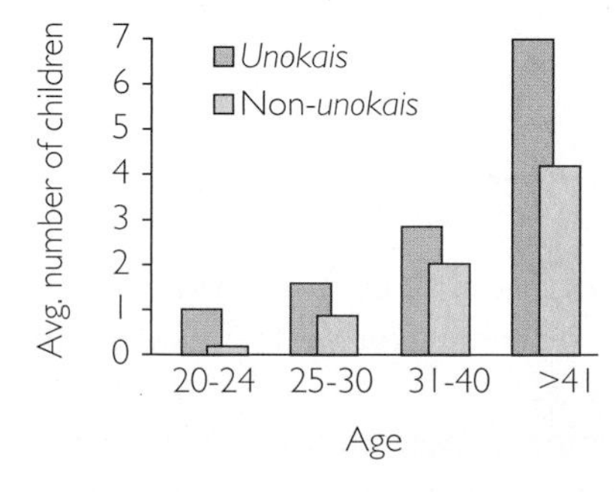

Figure 11.39 Mating and reproductive success of *unokais* (killers) versus non-*unokais* among Yąnomamö men These graphs show the average number of wives (top) and children (bottom) for adult men of various ages. Taken together, the data show that *unokais* are more successful than non-*unokais* ($P < 0.00001$). Plotted from data in Chagnon (1988).

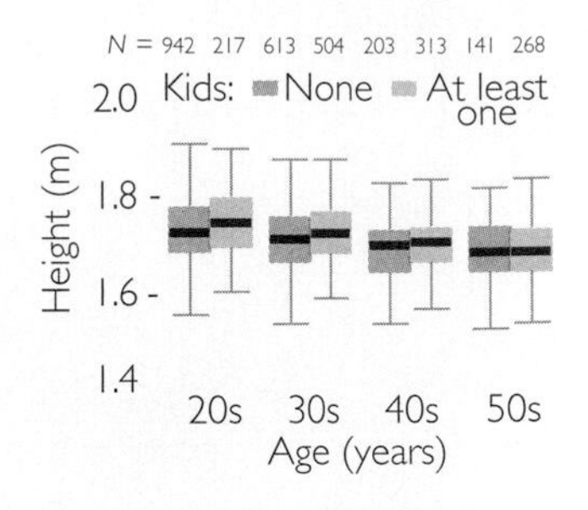

Figure 11.40 Men with children are taller, on average, than childless men. The heavy horizontal bars show, for a sample of Polish men, the average height of individuals in each age class, by whether or not they have children. The colored boxes show ±1 standard deviation around the mean; the whiskers show the range about the mean that includes 95% of the men. *N* is the number of men in each category. Though small, the differences between men with versus without children are statistically significant for men in their twenties ($P = 0.005$), thirties ($P = 0.001$), and forties ($P = 0.002$). The difference is not significant for men in their fifties ($P = 0.863$). From Pawlowski et al. (2000).

colleagues then compared bachelors to married men. The married men were taller by a slight but statistically significant margin. In addition, men with one or more children were significantly taller than childless men (Figure 11.40). The exception to this pattern was the group of men in their fifties, within which there was no difference in height between fathers versus childless men. Pawlowski and colleagues note that the men in their fifties reached marrying age shortly after World War II, when the ratio of women to men in Poland was unusually high. The researchers speculate that the men in their fifties had experienced less intense sexual selection, through female choice, than is the norm.

Additional evidence suggesting that female choice favors tall men comes from a study by Ulrich Mueller and Allan Mazur (2001). Mueller and Mazur surveyed members of the class of 1950 from the United States Military Academy at West Point. Among these career officers, unlike in many other more diverse populations, height was not associated with social status or socioeconomic success. Height was, however, associated with reproductive success. The tallest men had, over their lifetimes, more wives, and younger second wives, than other men. As a result, the tallest men had more children.

Daniel Nettle (2002) examined the relationship between height and reproductive success in women. Analyzing data from a large national health survey in Britain, Nettle found a weak but significant effect. Unlike in men, selection on women is stabilizing. Women of slightly less than average height had more children than either shorter or taller women. The cause appears to be that women of moderate height were healthier, on average, than extremely short or extremely tall women. That is, the higher fitness of slightly shorter-than-average women is due to natural selection.

On the data we have reviewed, then, there is some evidence that sexual selection, primarily through female choice, is responsible for the fact that men are taller than women. It is best, however, to consider the evolutionary significance of sexual size dimorphism in humans unresolved. The studies we have reviewed are observational, and the associations they documented are small. As a result, the evidence they provide about causation is only suggestive. It is possible that we humans simply inherited our sexual size dimorphism from our ancestors, who were more sexually dimorphic in size than we are (McHenry 1992). What is really needed to settle the issue is data from a larger number of cultures on the relationship between body size, number of mates, survival, and reproductive success for both women and men. Preferably, the data would come from hunter–gatherer cultures, whose members live the lifestyle ancestral for our species. The most technically challenging factor to measure accurately is the reproductive success of men. Modern techniques for genetic analysis have made it feasible, in principle, to collect such data (Figure 11.32, page 431). However, the research remains to be done.

It is unclear whether sexual selection helps maintain the sexual dimorphism in body size in humans. Males compete for mates, but larger males do not necessarily win. Females are choosy, and limited data suggest a slight preference for taller men.

Summary

Sexual dimorphism, a difference in form or behavior between females and males, is common. The difference often involves traits, like the enormous tail feathers of the peacock, that appear to be opposed by natural selection. To explain these puzzling traits, Darwin invoked sexual selection. Sexual selection is differential reproductive success resulting from variation in mating success.

Mating success is often a more important determinant of fitness for one sex than for the other. Often, but by no means always, it is males whose reproductive success is limited by mating opportunities, and females whose reproductive success is limited by resources rather than matings.

The members of the sex experiencing strong sexual selection typically compete among themselves over access to mates. This competition may involve direct combat, gamete competition, infanticide, or advertisement.

The members of the sex whose reproductive success is limited by resources rather than matings are typically choosy. This choosiness may provide the chooser with direct or indirect benefits, such as food or better genes for its offspring, or it may be the result of a preexisting sensory bias.

The theory of sexual selection was developed to explain sexual dimorphism in animals, but it applies to plants as well. In plants, access to pollinators is sometimes more limiting to reproductive success via pollen donation than to reproductive success via seed production. This can lead to the evolution of sexual dimorphism in which male flowers are showier than female flowers.

Questions

1. Under what conditions will sexual selection produce *different* traits in the two sexes (i.e., sexual dimorphism)? Why is one sex often "choosy" while the other is "showy"?
2. What is the difference between *inter*sexual selection and *intra*sexual selection? What kinds of traits do they each tend to produce? Give three examples of each.
3. In marine iguanas and long-tailed widowbirds, what evidence is there that sexual selection acts contrary to natural selection? That is, what is the evidence that the sexually selected trait may reduce survival? What does this imply about survival rates of "attractive" males in many species, as compared to less attractive or less competitive males?
4. What are four reasons that females may choose males with particular traits and reject other males? That is, how does the female benefit from her choice? Give one example for each.
5. Which parts of a flower are under more intensive selection for "showiness" (size, scent, color), the seed-producing parts or the pollinator-producing parts? Why? In plants that have separate genders, are the showier flowers found on male plants or female plants? In the sex that has less showy flowers, why does it have flowers at all?
6. In our discussion of rough-skinned newts, we inferred that tail crests in males evolved by sexual selection. Why is this a reasonable inference? Do you think the mechanism of sexual selection was male–male competition or female choice? Why? Design an experiment to find out.
7. Figure 11.6 shows the results of the experiment by Jones et al. (2000), in which broad-nosed pipefish mated in barrels in the lab. Each barrel contained either four males and four females, or two males and six females. Jones and colleagues also did experiments in which each barrel contained six males and two females. What do you think the analogous graphs from these experiments looked like? Why?
8. The graphs in Figure 11.41 show the variation in lifetime reproductive success of male versus female elephant seals (Le Boeuf and Reiter 1988). Note that the scales on the

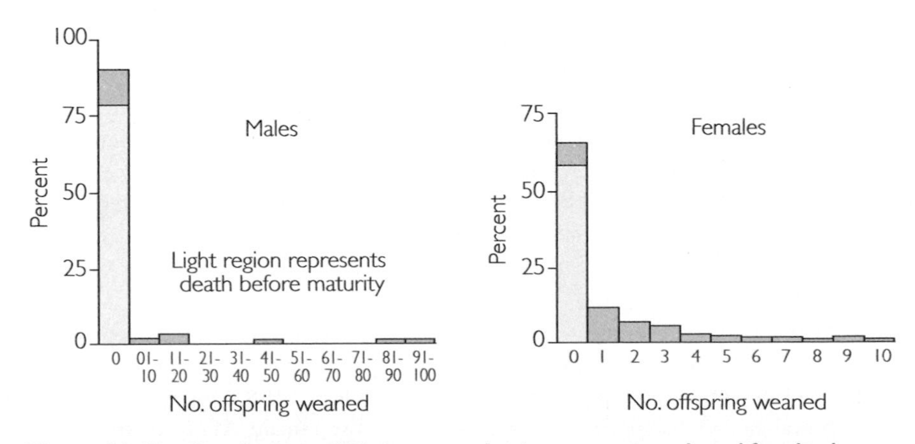

Figure 11.41 Distributions of lifetime reproductive success in male and female elephant seals From Le Boeuf and Reiter (1988).

horizontal axes are different. Why is the variation in reproductive success so much more extreme in males than females? Draw a graph showing your hypothesis for the relationship between number of mates and reproductive success for male and female elephant seals. Why do you think male elephant seals are four times larger than female elephant seals? Why aren't males even bigger?

9. Males in many species often attempt to mate with strikingly inappropriate partners. Ryan (1985), for example, describes male túngara frogs clasping other males. Some orchids mimic female wasps and are pollinated by amorous male wasps—who have to be fooled twice for the strategy to work. Would a female túngara or a female wasp make the same mistake? Why or why not? (Think of general explanations that are applicable to a wide range of species.)

10. What sex would you guess is the sage grouse pictured in Figure 11.42? What is it doing and why? Do you think this individual provides parental care? What else can you guess about the social system of this species?

Figure 11.42 **Sage grouse**

11. Male butterflies and moths commonly drink from puddles, a behavior known as puddling. Scott Smedley and Thomas Eisner (1996) report a detailed physiological analysis of puddling in the moth *Gluphisia septentrionis*. A male *G. septentrionis* may puddle for hours at a time. He rapidly processes huge amounts of water, extracting the sodium and expelling the excess liquid in anal jets (see Smedley and Eisner's paper for a dramatic photo). The male moth will later give his harvest of sodium to a female during mating. The female will then put much of the sodium into her eggs. Speculate on the role this gift plays in the moth's mating ritual and in the courtship roles taken by the male and the female. How would you test your ideas?

12. The scatterplot in Figure 11.43 shows the relationship between the importance of attractiveness in mate choice (as reported by subjects responding to a questionnaire) and the prevalence of six species of parasites (including leprosy, malaria, and filaria) in 29 cultures

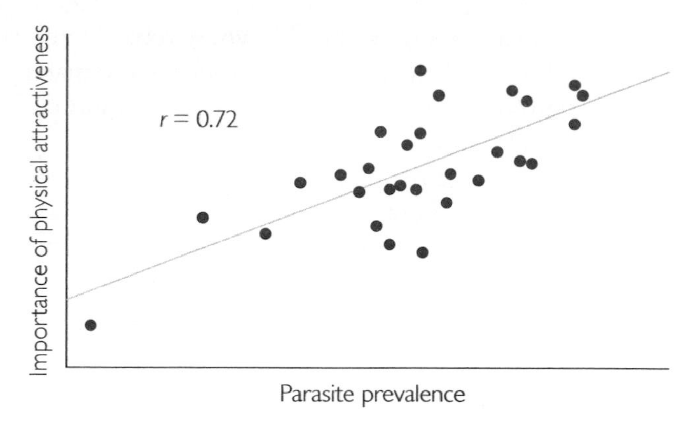

Figure 11.43 **Importance of physical attractiveness in mate choice versus parasite prevalence in 29 human cultures**

(Gangestad 1993; Gangestad and Buss 1993). (Statistical techniques have been used to remove the effects of latitude, geographic region, and mean income.) What is the pattern in the graph? Does this pattern make sense from an evolutionary perspective? One of the parasitic diseases is schistosomiasis. There is evidence that resistance to schistosomiasis is heritable (Abel et al. 1991). What do women gain (evolutionarily) by choosing an attractive mate? What do men gain (evolutionarily) by choosing an attractive mate? Can you offer a cultural explanation that could also account for this pattern?

13. In many katydids, the male delivers his sperm to the female in a large spermatophore which contains nutrients the female eats (for a photo, see Gwynne 1981). The female uses these nutrients in the production of eggs. Darryl Gwynne and L.W. Simmons (1990) studied the behavior of caged populations of an Australian katydid under low-food (control) and high-food (extra) conditions. Some of their results are graphed in Figure 11.44. (The graph shows the results from four sets of replicate cages; calling males = number of males calling at any given time; matings/female = number of times each female mated; % reject by M = fraction of the time a female approached a male for mating and was rejected; % reject by F = fraction of the time a female approached a male but then rejected him before copulating; % with F–F comp = fraction of matings in which one or more females were seen fighting over the male.) When were the females choosy and the males competitive? When were the males choosy and the females competitive? Why?

14. In some species of deep-sea anglerfish, the male lives as a symbiont permanently attached to the female (see Gould 1983, essay 1). The male is tiny compared to the female. Many of the male's organs, including the eyes, are reduced, though the testes remain large. Oth-

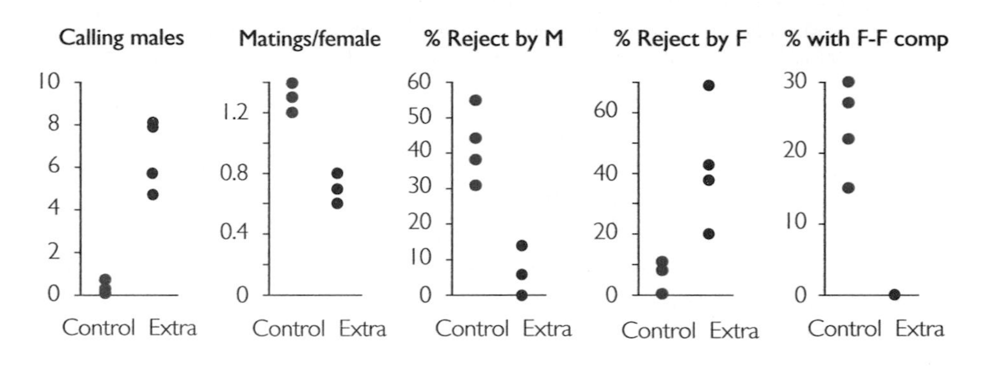

Figure 11.44 Behavior of male and female katydids under control versus extra-food conditions

ers, such as the jaws and teeth, are modified for attachment to the female. The circulatory systems of the two sexes are fused, and the male receives all of his nutrition from the female via the shared bloodstream. Often, two or more males are attached to a single female. What are the costs and benefits of the male's symbiotic habit for the male? For the female? What limits the lifetime reproductive success of each sex—the ability to gather resources, or the ability to find mates? Do you think that the male's symbiotic habit evolved as a result of sexual selection or natural selection? (It may be helpful to break the male symbiotic syndrome into separate features, such as staying with a single female for life, physical attachment to the female, reduction in body size, and nutritional dependence on the female.)

Exploring the Literature

15. If a single insemination provides all the sperm necessary to fertilize an entire clutch of eggs, then what do females gain by engaging in extra-pair copulations? Earlier we presented data showing that female Gunnison's prairie dogs that mate with multiple males are more likely to get pregnant and have larger litters. For more hypotheses and tests, see:

Kempenaers, B., G. R. Verheyn, M. van den Broeck, T. Burke, C. van Broeckhoven, and A. A. Dhondt. 1992. Extra-pair paternity results from female preference for high-quality males in the blue tit. *Nature* 357: 494–496.

Madsen, T., R. Shine, J. Loman, and T. Hakansson. 1992. Why do female adders copulate so frequently? *Nature* 355: 440–441.

Gray, E.M. 1996. Female control of offspring paternity in a western population of red-winged blackbirds (*Agelaius phoeniceus*). *Behavioral Ecology and Sociobiology* 38: 267–278.

Gray, E. M. 1997. Do female red-winged blackbirds benefit genetically from seeking extra-pair copulations? *Animal Behaviour* 53: 605–623.

Gray, E. M. 1997. Female red-winged blackbirds accrue material benefits from copulating with extra-pair males. *Animal Behaviour* 53: 625–629.

16. Why do females sometimes copulate more than once with the same male?

Petrie, M. 1992. Copulation frequency in birds: Why do females copulate more than once with the same male? *Animal Behaviour* 44: 790–792.

17. In many species, males "guard" females after copulation, following her closely, apparently in an effort to prevent sperm competition from other males that she may mate with later. In many insects, the male may even remain attached to the female in a copulatory position. See this paper for an interesting case of a grasshopper in which the male spends as long as 17 days mounted on the female.

Cueva del Castillo, R. 2003. Body size and multiple copulations in a neotropical grasshopper with an extraordinary mate-guarding duration. *Journal of Insect Behavior* 16: 503–522.

18. In plants, the analog of sperm competition is a race among pollen tubes to reach the ovules. For an exploration of whether the pollen of some donors is consistently superior in competition with the pollen from other donors, see:

Snow, A. A., and T. P. Spira. 1991. Pollen vigor and the potential for sexual selection in plants. *Nature* 352: 796–797.

Snow, A. A., and T. P. Spira. 1996. Pollen-tube competition and male fitness in *Hibiscus moscheutos*. *Evolution* 50: 1866–1870.

19. Peacocks are among the most famous animals with an elaborate male mating display. For research on sexual selection in peacocks, see:

Loyau, A., M. S. Jalme, C. Cagniant, and G. Sorci. 2005. Multiple sexual advertisements honestly reflect health status in peacocks (*Pavo cristatus*). *Behavioral Ecology and Sociobiology* 58: 552–557.

Loyau, A., M. S. Jalme, and G. Sorci. 2005. Intra-and intersexual selection for multiple traits in the peacock (*Pavo cristatus*). *Ethology* 111: 810–820.

Petrie, M. 1992. Peacocks with low mating success are more likely to suffer predation. *Animal Behaviour* 44: 585–586.

Petrie, M., T. Halliday, and C. Sanders. 1991. Peahens prefer peacocks with elaborate trains. *Animal Behaviour* 41: 323–331.

20. Figure 11.9 (page 409) shows that iguanas of intermediate size survive bad years at higher rates. For evidence of a surprising adaptation that may help big iguanas survive bad years, see:

Wikelski, M., and C. Thom. 2000. Marine iguanas shrink to survive El Niño. *Nature* 403: 37–38.

21. For evidence that sperm competition can lead to an escalating battle between the sexes, see:

Crudgington, H. S., and M. T. Siva-Jothy. 2000. Genital damage, kicking and early death: The battle of the sexes takes a sinister turn in the bean weevil. *Nature* 407: 855–856.

Holland, B., and W. R. Rice. 1999. Experimental removal of sexual selection reverses intersexual antagonistic coevolution and removes a reproductive load. *Proceedings of the National Academy of Sciences USA* 96: 5083–5088.

22. Our discussion of sexual selection in plants focused mostly on the plants' perspective. See this paper for the perspective of the pollinators, who have their own preferences about which flowers they want to visit:

Abraham, J. N. 2005. Insect choice and floral size dimorphism: Sexual selection or natural selection? *Journal of Insect Behavior* 18: 743–756.

23. Does the "sexy son" hypothesis have any evidence to support it? See these two papers for two very different stories:

Gustafsson, L., and A. Qvarnstrom. 2006. A test of the "sexy son" hypothesis: Sons of polygynous collared flycatchers do not inherit their fathers' mating status. *American Naturalist* 167: 297–302.

Gwinner, H., and H. Schwabl. 2005. Evidence for sexy sons in European starlings (*Sturnus vulgaris*). *Behavioral Ecology and Sociobiology* 58: 375–382.

24. We have seen that in most cases, the more "showy" sex will be the one that provides less parental care. See this paper for an interesting exception to the rule:

Heinsohn, R., S. Legge, and J. A. Endler. 2005. Extreme reversed sexual dichromatism in a bird without sex role reversal. *Science* 309: 617–619.

25. The "runaway selection" hypothesis posits that male ornaments may evolve in response to a preexisting sensory bias of females. But why would females have a preexisting "sensory bias" in the first place? See these three papers for evidence that male ornaments and courtship behavior can evolve in response to female sensory biases that evolved originally for foraging for orange and red fruits and for grapes, respectively:

Grether, G. F., G. R. Kolluru, F. H. Rodd, J. de la Cerda, J., and K. Shimazaki. 2005. Carotenoid availability affects the development of a colour-based mate preference and the sensory bias to which it is genetically linked. *Proceedings of the Royal Society of London B* 272: 2181–2188.

Smith, C., I. Barber, R. J. Wootton, and L. Chittka. 2004. A receiver bias in the origin of three-spined stickleback mate choice. *Proceedings of the Royal Society of London B* 271: 949–955.

Madden, J. R., and K. Tanner. 2003. Preferences for coloured bower decorations can be explained in a nonsexual context. *Animal Behaviour* 65: 1077–1083.

Citations

Abel, L., F. Demenais, et al. 1991. Evidence for the segregation of a major gene in human susceptibility/resistance to infection by *Schistosoma mansoni*. *American Journal of Human Genetics* 48: 959–970.

Andersson, M. 1982. Female choice selects for extreme tail length in a widowbird. *Nature* 299: 818–820.

Andersson, M. 1994. *Sexual Selection*. Princeton, NJ: Princeton University Press.

Arnold, S. J. 1983. Sexual selection: The interface of theory and empiricism. In P. Bateson, ed. *Mate Choice*. Cambridge: Cambridge University Press, 67–107.

Arnold, S. J. 1994. Bateman's principles and the measurement of sexual selection in plants and animals. *American Naturalist* 144: S126–149.

Arnold, S. J., and D. Duvall. 1994. Animal mating systems: A synthesis based on selection theory. *American Naturalist* 143: 317–348.

Bateman, A. J. 1948. Intra-sexual selection in *Drosophila*. *Heredity* 2: 349–368.

Beecher, M. D., and I. M. Beecher. 1979. Sociobiology of bank swallows: Reproductive strategy of the male. *Science* 205: 1282–1285.

Berglund, A. 1991. Egg competition in a sex-role reversed pipefish: Subdominant females trade reproduction for growth. *Evolution* 45: 770–774.

Berglund, A., and G. Rosenqvist. 1993. Selective males and ardent females in pipefish. *Behavioral Ecology and Sociobiology* 32: 331–336.

Berglund, A., G. Rosenqvist, and P. Bernet. 1997. Ornamentation predicts reproductive success in female pipefish. *Behavioral Ecology and Sociobiology* 40: 145–150.

Berglund, A., G. Rosenqvist, and I. Svensson. 1989. Reproductive success of females limited by males in two pipefish species. *American Naturalist* 133: 506–516.

Bernet, P., G. Rosenqvist, and A. Berglund. 1998. Female-female competition affects female ornamentation in the sex-role reversed pipefish *Syngnathus typhle*. *Behaviour* 135: 535–550.

Bertram, C. R. 1975. Social factors influencing reproduction in wild lions. *Journal of Zoology* 177: 463–482.

Boellstorff, D. E., D. H. Owings, M. C. T. Penedo, and M. J. Hersek. 1994. Reproductive behavior and multiple paternity of California ground squirrels. *Animal Behaviour* 47: 1057–1064.

Borgerhoff Mulder, M. 1988. Reproductive success in three Kipsigis cohorts. In T.H. Clutton-Brock, ed. *Reproductive Success.* Chicago: University of Chicago Press, 419–435.

Brodt, J. F., D. W. Tallamy, and J. Ali. 2006. Female choice by scent recognition in the spotted cucumber beetle. *Ethology* 112: 300–306.

Burkhardt, D., and I. de la Motte. 1983. How stalk-eyed flies eye stalk-eyed flies: Observations and measurements of the eyes of *Cyrtodiopsis whitei* (Dopsidae, Diptera). *Journal of Comparative Physiology* 151: 407–421.

Burkhardt, D., and I. de la Motte. 1987. Physiological, behavioural, and morphometric data elucidate the evolutive significance of stalked eyes in Diopsidae (Diptera). *Entomologia Generalis* 12: 221–233.

Bush, S. L., H. C. Gerhardt, and J. Schul. 2002. Pattern recognition and call preferences in treefrogs (Anura: Hylidae): a quantitative analysis using a no-choice paradigm. *Animal Behaviour* 63: 7–14.

Chagnon, N. A. 1988. Life histories, blood revenge, and warfare in a tribal population. *Science* 239: 985–992.

Christy, J. H., P. R. Y. Backwell, and U. Schober. 2003. Interspecific attractiveness of structures built by courting male fiddler crabs: Experimental evidence of a sensory trap. *Behavioral Ecology and Sociobiology* 53: 84–91.

Clutton-Brock, T. H. 1985. Reproductive Success in Red Deer. *Scientific American* 252 (February): 86–92.

Daly, M., and M. Wilson. 1988. *Homicide.* New York: Aldine de Gruyter.

Darwin, C. 1871. *The Descent of Man, and Selection in Relation to Sex.* London: John Murray.

David, P., A. Hingle, et al. 1998. Male sexual ornament size but not asymmetry reflects condition in stalk-eyed flies. *Proceedings of the Royal Society of London B* 265: 2211–2216.

Delph, L. F., L. F. Galloway, and M. L. Stanton. 1996. Sexual dimorphism in flower size. *American Naturalist* 148: 299–320.

Doty, G. V., and A. M. Welch. 2001. Advertisement call duration indicates good genes for offspring feeding rate in gray tree frogs (*Hyla versicolor*). *Behavioral Ecology and Sociobiology* 49: 150–156.

Enquist, M., and A. Arak. 1993. Selection of exaggerated male traits by female aesthetic senses. *Nature* 361: 446–448.

Falconer, D. S. 1989. *Introduction to Quantitative Genetics.* New York: John Wiley & Sons.

Fisher, R. A. 1915. The evolution of sexual preference. *Eugenics Review* 7: 184–192.

Fisher, R. A. 1958. *The Genetical Theory of Natural Selection,* 2nd ed. New York: Dover.

Gage, M. J. G. 1991. Risk of sperm competition directly affects ejaculate size in the Mediterranean fruit fly. *Animal Behaviour* 42: 1036–1037.

Gangestad, S. W. 1993. Sexual selection and physical attractiveness: Implications for mating dynamics. *Human Nature* 4: 205–235.

Gangestad, S. W., and D. M. Buss. 1993. Pathogen prevalence and human mate preferences. *Ethology and Sociobiology* 14: 89–96.

Gerhardt, H. C., M. L. Dyson, and S. D. Tanner. 1996. Dynamic properties of the advertisement calls of gray tree frogs: Patterns of variability and female choice. *Behavioral Ecology* 7: 7–18.

Gibbs, H. L., P. J. Weatherhead, P. T. Boag, B. N. White, L. M. Tabak, and D. J. Hoysak. 1990. Realized reproductive success of polygynous red-winged blackbirds revealed by DNA markers. *Science* 250: 1394–1397.

Gilbert, L. E. 1976. Postmating female odor in Heliconius butterflies: A male-contributed anti-aphrodisiac? *Science* 193: 419–420.

Gould, S. J. 1983. *Hen's Teeth and Horse's Toes.* New York: W. W. Norton & Company.

Grant, V. 1995. Sexual selection in plants: Pros and cons. *Proceedings of the National Academy of Sciences USA* 92: 1247–1250.

Gray, E. M. 1997. Do female red-winged blackbirds benefit genetically from seeking extra-pair copulations? *Animal Behaviour* 53: 605–623.

Gross, M. R. 1984. Sunfish, salmon, and the evolution of alternative reproductive strategies and tactics in fishes. In G. W. Potts and R. J. Wootton, eds. *Fish Reproduction: Strategies and Tactics.* London: Academic Press, 55–75.

Gross, M. R. 1985. Disruptive selection for alternative live histories in salmon. *Nature* 313: 47–48.

Gross, M. R. 1991. Salmon breeding behavior and life history evolution in changing environments. *Ecology* 72: 1180–1186.

Gwynne, D. T. 1981. Sexual difference theory: Mormon crickets show role reversal in mate choice. *Science* 213: 779.

Gwynne, D. T., and L. W. Simmons. 1990. Experimental reversal of courtship roles in an insect. *Nature* 346: 172–174.

Hoogland, J. L. 1998. Why do female Gunnison's prairie dogs copulate with more than one male? *Animal Behaviour* 55: 351–359.

Hooper, R. E., and M. T. Siva-Jothy. 1996. Last male sperm precedence in a damselfly demonstrated by RAPD profiling. *Molecular Ecology* 5: 449–452.

Hrdy, S. B. 1979. Infanticide among animals: A review, classification, and examination of the implications for the reproductive strategies of females. *Ethology and Sociobiology* 1: 13-40.

Husak, J. F., J. M. Macedonia, et al. 2006. Predation cost of conspicuous male coloration in collared lizards (*Crotaphytus collaris*): An experimental test using clay-covered model lizards. *Ethology* 112: 572–580.

Ivy, T. M., C. B. Weddle, and S. K. Sakaluk. 2005. Females use self-referent cues to avoid mating with previous mates. *Proceedings of the Royal Society of London B* 272: 2475-2478.

Jones, A. G., J. R. Arguello, and S. J. Arnold. 2002. Validation of Bateman's principles: A genetic study of sexual selection and mating patterns in the rough-skinned newt. *Proceedings of the Royal Society of London B* 269: 2533–2539.

Jones, A. G., G. Rosenqvist, A. Berglund, et al. 2000. The Bateman gradient and the cause of sexual selection in a sex-role-reversed pipefish. *Proceedings of the Royal Society of London B* 267: 677–680.

Jones, A. G., G. Rosenqvist, et al. 2005. The measurement of sexual selection using Bateman's principles: An experimental test in the sex-role-reversed pipefish *Syngnathus typhle. Integrative and Comparative Biology* 45: 874-884.

Knight, J. 2002. Sexual stereotypes. *Nature* 415: 254–256.

Lande, R. 1981. Models of speciation by sexual selection on polygenic traits. *Proceedings of the National Academy of Sciences USA* 78: 3721–3725.

Le Boeuf, B. J., and J. Reiter. 1988. Lifetime reproductive success in northern elephant seals. In T.H. Cutton-Brock, ed. *Reproductive Success.* Chicago: University of Chicago Press, 344–362.

McHenry, H. M. 1992. Body size and proportions in early hominids. *American Journal of Physical Anthropology* 87: 407–431.

Møller, A. P. 1988. Female choice selects for male sexual tail ornaments in the monogamous swallow. *Nature* 332: 640–642.

Møller, A. P. 1991. Sexual selection in the monogamous barn swallow (*Hirundo rustica*). I. Determinants of tail ornament size. *Evolution* 45: 1823–1836.

Mueller, U., and A. Mazur. 2001. Evidence of unconstrained directional selection for male tallness. *Behavioral Ecology and Sociobiology* 50: 302–311.

Nakatsuru, K., and D. L. Kramer. 1982. Is sperm cheap? Limited male fertility and female choice in the lemon tetra (Pisces, Characidae). *Science* 216: 753–755.

Nettle, D. 2002. Women's height, reproductive success and the evolution of sexual dimorphism in modern humans. *Proceedings of the Royal Society of London B* 269: 1919–1923.

Nowak, R. M. 1991. *Walker's Mammals of the World.* Baltimore: Johns Hopkins University Press.

Olsson, M., A. Gullberg, and H. Tegelstrom. 1994. Sperm competition in the sand lizard, *Lacerta agilis. Animal Behaviour* 48: 193–200.

Oring, L. W., M. A. Colwell, J. M. Reed. 1991a. Lifetime reproductive success in the spotted sandpiper (*Actitis macularia*)—sex-differences and variance-components. *Behavioral Ecology and Sociobiology* 28: 425–432.

Oring L. W., J. M. Reed, et al. 1991b. Factors regulating annual mating success and reproductive success in spotted sandpipers (*Actitis macularia*). *Behavioral Ecology and Sociobiology* 28: 433–442.

Oring, L. W., J. M. Reed, and S. J. Maxson. 1994. Copulation patterns and mate guarding in the sex-role reversed, polyandrous spotted sandpiper, *Actitis macularia. Animal Behaviour* 47: (5) 1065–1072.

Packer, C., L. Herbst, A. E. Pusey, J. D. Bygott, J. P. Hanby, S. J. Cairns, and M. Borgerhoff Mulder. 1988. Reproductive success of lions. In T. H. Clutton-Brock, ed. *Reproductive Success: Studies of Individual Variation in Contrasting Breeding Systems.* Chicago: University of Chicago Press, 263–283.

Packer, C., and A. E. Pusey. 1983. Adaptations of female lions to infanticide by incoming males. *American Naturalist* 121: 716–728.

Panhuis, T. M., and G. S. Wilkinson. 1999. Exaggerated male eye span influences contest outcome in stalk-eyed flies (Diopsidae). *Behavioral Ecology and Sociobiology* 46: 221–227.

Partecke, J., A. von Haeseler, and M. Wikelski. 2002. Territory establishment in lekking marine iguanas, *Amblyrhynchus cristatus:* Support for the hotshot mechanism. *Behavioral Ecology and Sociobiology* 51: 579–587.

Pawlowski, B., R. I. M. Dunbar, and A. Lipowicz. 2000. Tall men have more reproductive success. *Nature* 403: 156.

Petrie, M. 1983. Female moorhens compete for small fat males. *Science* 220: 413–415.

Proctor, H. C. 1991. Courtship in the water mite, *Neumania papillator:* Males capitalize on female adaptations for predation. *Animal Behaviour* 42: 589–598.

Proctor, H. C. 1992. Sensory exploitation and the evolution of male mating behaviour: A cladistic test using water mites (Acari: Parasitengona). *Animal Behaviour* 44: 745–752.

Prum, Richard O. 1997. Phylogenetic tests of alternative intersexual selection mechanisms: Trait macroevolution in a polygynous clade (Aves: Pipridae). *American Naturalist* 149: 668–692.

Pryke, S. R., and S. Andersson. 2005. Experimental evidence for female choice and energetic costs of male tail elongation in red-collared widowbirds. *Biological Journal of the Linnean Society* 86: 35-43.

Rauch, N. 1985. Female habitat choice as a determinant of the reproductive success of the territorial male marine iguana (*Amblyrhynchus cristatus*). *Behavioral Ecology and Sociobiology* 16: 125–134.

Rauch, N. 1988. Competition of marine iguana females *Amblyrhynchus cristatus* for egg-laying sites. *Behavior* 107: 91–106.

Renner, S. S., and R. E. Ricklefs. 1995. Dioecy and its correlates in the flowering plants. *American Journal of Botany* 82: 596–606.

Rogers, A. R., and A. Mukherjee. 1992. Quantitative genetics of sexual dimorphism in human body size. *Evolution* 46: 226–234.

Romero, G. A., and C. E. Nelson. 1986. Sexual dimorphism in Catasetum orchids: Forcible pollen emplacement and male flower competition. *Science* 232: 1538–1540.

Rosenqvist, G. 1990. Male mate choice and female-female competition for mates in the pipefish *Nerophis ophidion. Animal Behaviour* 39: 1110–1115.

Rosenqvist, G., and K. Johansson. 1995. Male avoidance of parasitized females explained by direct benefits in a pipefish. *Animal Behaviour* 49: 1039–1045.

Ryan, M. J. 1985. *The Túngara Frog: A Study in Sexual Selection and Communication.* Chicago: University of Chicago Press.

Schenk, A., and K. M. Kovacs. 1995. Multiple mating between black bears revealed by DNA fingerprinting. *Animal Behaviour* 50: 1483–1490.

Schöfl, G., and M. Taborsky. 2002. Prolonged tandem formation in firebugs (*Pyrrhocoris apterus*) serves mate-guarding. *Behavioral Ecology and Sociobiology* 52:426–433.

Schwartz, J. J., B. W. Buchanan, and H. C. Gerhardt. 2001. Female mate choice in the gray treefrog (*Hyla versicolor*) in three experimental environments. *Behavioral Ecolology and Sociobiology* 49: 443–455.

Sillén-Tullberg, B. 1981. Prolonged copulation: A male "postcopulatory" strategy in a promiscuous species, *Lygaeus equestris* (Heteroptera: Lygaeidae). *Behavioral Ecology and Sociobiology* 9: 283–289.

Smedley, S. R., and T. Eisner. 1996. Sodium: A male moth's gift to its offspring. *Proceedings of the National Academy of Sciences USA* 93: 809–813.

Smith, R. L. 1984. Human sperm competition. In R. L. Smith, ed. *Sperm Competition and the Evolution of Animal Mating Systems.* Orlando: Academic Press, 601–659.

Stanton, M. L., A. A. Snow, and S. N. Handel. 1986. Floral evolution: Attractiveness to pollinators increases male fitness. *Science* 232: 1625–1627.

Tallamy, D. W., B. E. Powell, and J. A. McClafferty. 2002. Male traits under cryptic female choice in the spotted cucumber beetle (Coleoptera: Chrysomelidae). *Behavioral Ecology* 13: 511–518.

Tallamy, D. W., M. B. Darlington, J. D. Pesek, and B. E. Powell. 2003. Copulatory courtship signals male genetic quality in cucumber beetles. *Proceedings of the Royal Society of London B* 270: 77–82.

Tang-Martinez, Z., and T. B. Ryder. 2005. The problem with paradigms: Bateman's worldview as a case study. *Integerative and Comparative Biology* 45: 821-830.

Thornhill, R. 1976. Sexual selection and nuptial feeding behavior in *Bittacus apicalis* (Insecta: Mecoptera). *American Naturalist* 110: 529–548.

Thornhill, R., and J. Alcock. 1983. *The Evolution of Insect Mating Systems.* Cambridge, MA: Harvard University Press.

Trillmich, K. G. K. 1983. The mating system of the marine iguana (*Amblyrhynchus cristatus*). *Zeitschrift für Tierpsychologie* 63: 141–172.

Trivers, R. L. 1972. Parental investment and sexual selection. In B. Campbell, ed. *Sexual Selection and the Descent of Man 1871–1971.* Chicago: Aldine, 136–179.

Trivers, R. 1985. *Social Evolution.* Menlo Park, CA: Benjamin/Cummings.

Vaughton, G., and M. Ramsey. 1998. Floral display, pollinator visitation, and reproductive success in the dioecious perennial herb *Wurmbea dioica* (Liliaceae). *Oecologia* 115: 93–101.

Waage, J. K. 1984. Sperm competition and the evolution of Odonate mating systems. In R. L. Smith, ed. *Sperm Competition and the Evolution of Animal Mating Systems.* Orlando: Academic Press, 251–290.

Waage, J. K. 1986. Evidence for widespread sperm displacement ability among Zygoptera (Odonata) and the means for predicting its presence. *Biological Journal of the Linnean Society* 28: 285–300.

Walther, B. A., and D. H. Clayton. 2005. Elaborate ornaments are costly to maintain: Evidence for high maintenance handicaps. *Behavioural Ecology* 16:89-95.

Watson, P. J. 1991. Multiple paternity as genetic bet-hedging in female sierra dome spiders, *Linyphia litigiosa* (Linyphiidae). *Animal Behaviour* 41: 343–360.

Welch, Allison, R. D. Semlitsch, and H. Carl Gerhardt. 1998. Call duration as an indicator of genetic quality in male gray tree frogs. *Science* 280: 1928–1930.

Wikelski, M., and S. Bäurle. 1996. Precopulatory ejaculation solves time constraints during copulations in marine iguanas. *Proceedings of the Royal Society of London B* 263: 439–444.

Wikelski, M., C. Carbone, and F. Trillmich. 1996. Lekking in marine iguanas: Female grouping and male reproductive strategies. *Animal Behaviour* 52: 581–596.

Wikelski, M., V. Carrillo, and F. Trillmich. 1997. Energy limits to body size in a grazing reptile, the Galápagos marine iguana. *Ecology* 78: 2204–2217.

Wikelski, M., and C. Thom. 2000. Marine iguanas shrink to survive El Niño. *Nature* 403: 37.

Wikelski, M., and F. Trillmich. 1997. Body size and sexual size dimorphism in marine iguanas fluctuate as result of opposing natural and sexual selection: An island comparison. *Evolution* 51: 922–936.

Wilkinson, G. S., H. Kahler, and R. H. Baker. 1998. Evolution of female mating preferences in stalk-eyed flies. *Behavioral Ecology* 9: 525–533.

Wilkinson, G. S., D. C. Presgraves, and L. Crymes. 1998. Male eye span in stalk-eyed flies indicates genetic quality by meiotic drive suppression. *Nature* 391: 276–279.

Wilkinson, G. S., and P. R. Reillo. 1994. Female choice response to artificial selection on an exaggerated male trait in a stalk-eyed fly. *Proceedings of the Royal Society of London B* 255: 1–6.

Wilkinson, G. S., and M. Taper. 1999. Evolution of genetic variation for condition-dependent traits in stalk-eyed flies. *Proceedings of the Royal Society of London B* 266: 1685–1690.

Williams, G. C. 1966. *Adaptation and Natural Selection: A Critique of Some Current Evolutionary Thought.* Princeton University Press, Princeton, NJ.

Willson, M. F. 1979. Sexual selection in plants. *American Naturalist* 113: 777–790.

Woodroffe, R., and A. Vincent. 1994. Mother's little helpers: Patterns of male care in mammals. *Trends in Ecology and Evolution* 9: 294–297.

Zahavi, A. 1975. Mate selection—A selection for a handicap. *Journal of Theoretical Biology* 53: 205–214.

Genetically related female banded mongooses live and breed in groups, and care for each other's young (Gilchrist 2004; Gilchrist et al. 2004; Hodge 2005).

12

Kin Selection and Social Behavior

Social interactions create the possibility for conflict and cooperation. Consider two American crows (*Corvus brachyrhynchos*) patrolling the edge of their adjacent nesting territories. If one moves across the established boundary, its action may trigger aggressive calls, a flight chase, or even physical combat. But if a hawk flies by, the two antagonists will cooperate in chasing the predator away. Later in the day, these same individuals may spend considerable time and effort feeding the young birds in the nests in their respective territories, even though the nestlings are the crows' siblings or half-siblings and not their own offspring.

When and why do these individuals cooperate with each other, and why do they help their parents raise their siblings instead of leaving home to rear their own offspring? What conditions lead to conflicts with each other and with their parents, and how are these conflicts resolved? These are the types of questions addressed in this chapter.

In fitness terms, an interaction between individuals has four possible outcomes (Table 12.1). Cooperation (or **mutualism**) is the term for actions that result in fitness gains for both participants. **Altruism** represents cases in which the individual instigating the action pays a fitness cost and the individual on the receiving end benefits. In other words, altruism entails a sacrifice on behalf of another.

Table 12.1 Types of social interactions

The "actor" in any social interaction affects the recipient of the action as well as itself. The costs and benefits of interactions are measured in units of surviving offspring (fitness).

	Actor benefits	Actor is harmed
Recipient benefits	**Cooperative**	**Altruistic**
Recipient is harmed	**Selfish**	**Spiteful**

Selfishness is the opposite: The actor gains and the recipient loses. **Spite** is the term for behavior that results in fitness losses for both participants.

Understanding the evolution of these four interactions is simplified because spite, if it exists in nature, is rare (Keller et al. 1994; for discussion of possible examples see Hurst 1991, Foster et al. 2000, Foster et al. 2001, Gardner and West 2004, Johnstone and Bshary 2003). The rarity of spite is readily explained: An allele that results in fitness losses for both actor and recipient would quickly be eliminated by natural selection. Intuition might suggest that a spiteful actor could come out ahead if the cost he imposes on a competitor is greater than cost he suffers himself. Consider, however, the fitness of a third party who avoids involvement in spiteful acts and pays neither the cost of doling them out nor the cost of receiving them.

Altruism would seem equally difficult to explain because the actor suffers a fitness loss. Altruistic behavior appears to be common, however. Examples range from the crows that help at their parents' nests to a human who dives into a river and saves a drowning child. This is the first question we need to address: Why does altruism exist in nature?

12.1 Kin Selection and the Evolution of Altruism

Explaining altruistic behavior is a challenge for the Theory of Evolution by Natural Selection.

Altruism is a central paradox of Darwinism. It would seem impossible for natural selection to favor an allele that results in behavior benefiting other individuals at the expense of the individual bearing the allele. For Darwin (1859: 236), the apparent existence of altruism presented a "special difficulty, which at first appeared to me insuperable, and actually fatal to my whole theory." Fortunately he was able to hint at a resolution to the paradox: Selection could favor traits that result in decreased personal fitness if they increase the survival and reproductive success of close relatives. Over a hundred years passed, however, before this result was formalized and widely applied.

Inclusive Fitness

In 1964, William Hamilton developed a genetic model showing that an allele that favors altruistic behavior could spread under certain conditions (Hamilton 1964a). The key parameter in Hamilton's formulation is the **coefficient of relatedness, *r*.** This is the probability that the homologous alleles in two individuals are **identical by descent** (Box 12.1). The parameter is closely related to F, the coefficient of inbreeding, which we introduced in Chapter 6. F is the probability that homologous alleles in the same individual are identical by descent.

Box 12.1 Calculating coefficients of relatedness

Calculating r, the coefficient of relatedness, requires a pedigree that includes the actor (the individual dispensing the behavior) and the recipient (the individual receiving the behavior). Starting with the actor, all paths of descent are traced through the pedigree to the recipient. For example, half-siblings share one parent and have two genealogical connections, as indicated in Figure 12.1a. Parents contribute half their genes to each offspring, so the probability that genes are identical by descent (ibd) in each step in the path is 1/2. Put another way, the probability that a particular allele was transmitted from parent to actor is 1/2. The probability that the same allele was transmitted from parent to recipient is 1/2. The probability that this same allele was transmitted to both the actor and the recipient (meaning that the alleles in actor and recipient are ibd) is the product of these two independent probabilities, or 1/4.

Full siblings, on the other hand, share genes inherited from both parents. To calculate *r* when actor and recipient are full-siblings, we have to add the probabilities that genes are ibd through each path in the pedigree. In this case, we add the probability that genes are ibd through the mother to the probability that they are ibd through the father (see Figure 12.1b). This is $\frac{1}{4} + \frac{1}{4} = \frac{1}{2}$.

Using this protocol results in the following coefficients:

- First cousins, $\frac{1}{8}$ (Figure 12.1c)
- Parent to offspring, $\frac{1}{2}$
- Grandparent to grandchild, $\frac{1}{4}$
- Aunt or uncle to niece or nephew, $\frac{1}{4}$

The analyses we have just performed work for autosomal loci in sexual organisms and assume that no inbreeding has occurred. If the population is inbred, then coefficients will be higher. But when studying populations in the field, investigators usually have no data on inbreeding and have to assume that individuals are completely outbred. On this basis, coefficients of relationship that are reported in the literature should be considered minimum estimates. Another uncertainty in calculating values of *r* comes in assigning paternity in pedigrees. As we indicated in Chapter 10, extra-pair copulations are common in many species. If paternity is assigned on the basis of male-female pairing relationships and extra-pair copulations go undetected, estimates of *r* may be inflated.

When constructing genealogies is impractical, coefficients of relatedness can be estimated directly from genetic data (Queller and Goodnight 1989). Microsatellites and other marker loci are proving to be extremely useful for calculating *r* in a wide variety of social insects (e.g., Peters et al. 1999).

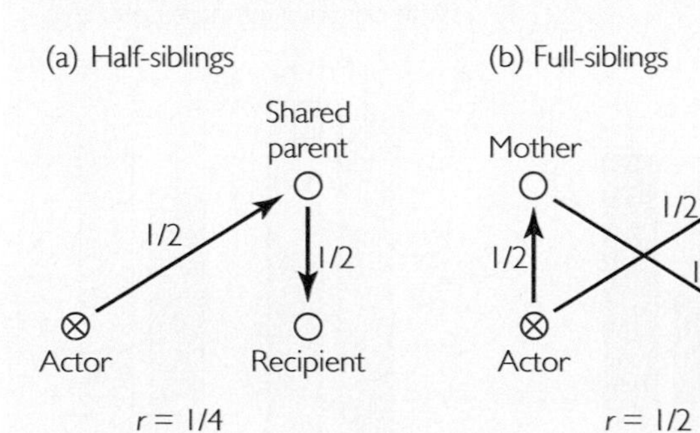

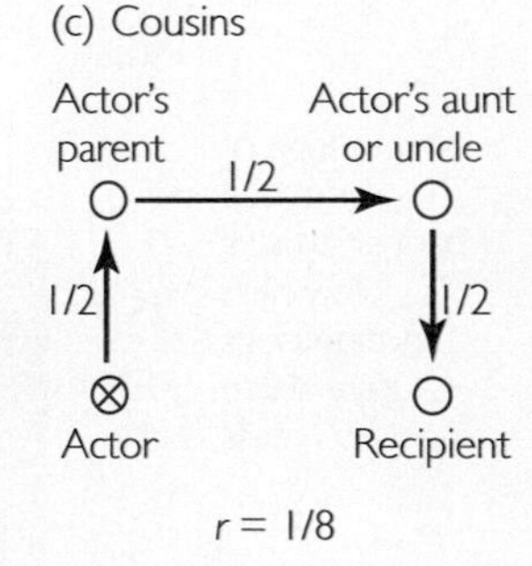

Figure 12.1 Computing relatedness with pedigrees The arrows describe paths by which genes can be identical by descent. The text explains how these paths are used to calculate *r*, the coefficient of relatedness.

Given *r*, the coefficient of relatedness between the actor and the recipient, **Hamilton's rule** states that an allele for altruistic behavior will spread if

$$Br - C > 0$$

where B is the benefit to the recipient and C is the cost to the actor. Both B and C are measured in units of surviving offspring. This simple law means that altruism is more likely to spread when the benefits to the recipient are great, the cost to the actor is low, and the participants are closely related.

Inclusive fitness consists of direct fitness due to personal reproduction and indirect fitness due to additional reproduction by relatives. Behavior that results in indirect fitness gains is favored by kin selection.

To generalize this result, Hamilton offered the concept of **inclusive fitness.** He pointed out that an individual's fitness can be partitioned into two components. **Direct fitness** results from personal reproduction. **Indirect fitness** results from additional reproduction by relatives that is made possible by an individual's actions. Indirect fitness accrues when relatives achieve reproductive success above and beyond what they would have achieved on their own—that is, without aid. Natural selection favoring the spread of alleles that increase the indirect component of fitness is called **kin selection.** As we will see, most instances of altruism in nature are the result of kin selection.

Robert Trivers (1985:47) called Hamilton's rule and the concept of inclusive fitness "the most important advance in evolutionary theory since the work of Charles Darwin and Gregor Mendel." To see why, we will apply the theory by venturing to the Sierra Nevada of California and observing a social mammal: Belding's ground squirrel (*Spermophilus beldingi*).

Alarm Calling in Belding's Ground Squirrels and Prairie Dogs

Explaining alarm calling in birds and mammals is a classical application of inclusive fitness theory. When flocks or herds are stalked by a predator, prey that notice the intruder sometimes give loud, high-pitched calls. These warnings alert nearby individuals and allow them to flee or dive for cover. They may also expose the calling individual to danger.

The first question to ask is whether alarm calls are genuinely altruistic. Research by Paul Sherman (1985) on Belding's ground squirrels shows why. Belding's ground squirrels give two kinds of alarms: they trill in response to predatory mammals and whistle in response to flying hawks. During 14 years of observation, Sherman and his assistants witnessed 30 natural predator attacks in which ground

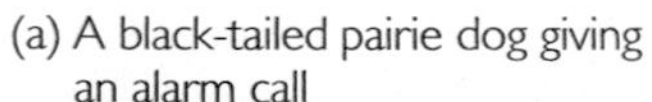
(a) A black-tailed pairie dog giving an alarm call

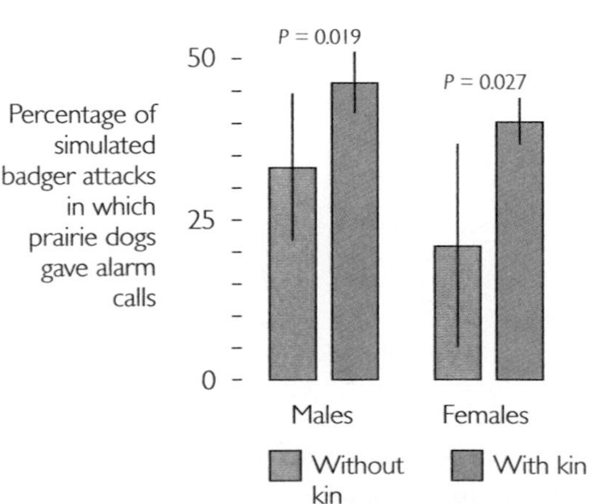

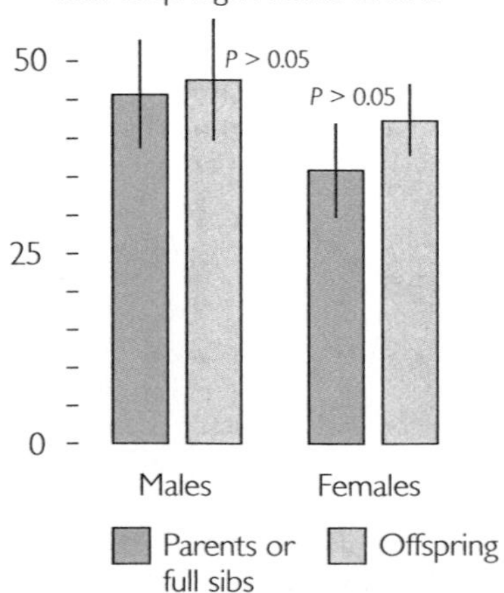

Figure 12.2 Black-tailed prairie dogs give more alarm calls when kin are nearby (a) This black-tailed prairie dog is giving an alarm call. (b) This bar chart reports the rates of alarm calling by prairie dogs who have vs. do not have kin living with them. Both sexes call more often when kin are near. (c) This chart reports the rates of alarm calling by males and females living with non-offspring kin vs. offspring. Both sexes call nearly as often when non-offspring kin are near as when offspring are near. Redrawn from Hoogland (1995).

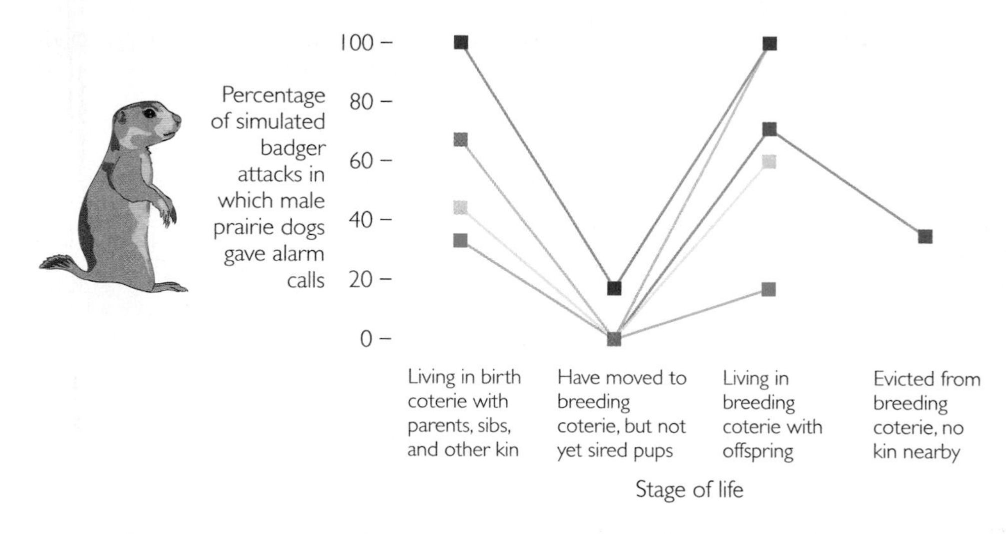

Figure 12.3 Male black-tailed prairie dogs change their alarm calling behavior when their living situation changes This graph plots the rate of alarm calling by five individual males at different stages of life. The males increased or decreased their rate of calling according to whether kin were nearby. Redrawn from Hoogland (1995).

squirrels were captured and killed. It turns out that when squirrels spot an attacking hawk and whistle, the whistling squirrel is captured only 2% of the time while nonwhistling squirrels are captured 28% of the time. The squirrel raising the alarm reduces its own chances of dying, perhaps by informing the hawk that the caller has seen it and at the same time sowing panic and confusion among the other squirrels. When squirrels spot an attacking mammal and trill, however, the trilling squirrel is killed 8% of the time while non-trilling squirrels are killed only 4% of the time. The squirrel raising the alarm increases its own peril to the benefit of other squirrels nearby. Whistles are selfish, but trills are altruistic.

John Hoogland (1983, 1994, 1995) studied alarm calls given by black-tailed prairie dogs (Figure 12.2a). Praire dogs live family groups called coteries; each coterie occupies a territory within a praire dog town. Females typically remain in their birth coterie for life, whereas males usually disperse when sexually mature. Despite logging over 50,000 person-hours watching individually marked prairie dogs in a town in South Dakota, Hoogland and his assistants were unable to document for certain whether prairie dog alarms are selfish or altruistic. Hoogland suspected that they are altruistic, however, and sought to determine whether the pairie dogs' calling behavior was consistent with the hypothesis that it evolved as a result of kin selection. Hoogland simulated predator attacks by having an assistant pull a stuffed badger through a praire dog town on a sled, and watched to see who gave alarm calls and who just dove for cover.

Individuals are more likely to give alarm calls when close relatives are nearby.

Both male and female prairie dogs are more likely to give alarm calls if their coterie includes genetic kin (Figure 12.2b). These calls are not simply a form of parental care, as individuals give calls nearly as often when the kin they live with are parents and siblings as when they are offspring (Figure 12.2c). Hoogland was even able to follow individual males across different stages of life, and saw them modify their rate of calling with changes in their proximity to kin (Figure 12.3).

The data show that apparently altruistic alarm-calling behavior is not dispensed randomly. It is nepotistic. Self-sacrifice is directed at close relatives and should result in indirect fitness gains.

White-Fronted Bee-Eaters

Another classical system for studying kin selection in vertebrates is helping behavior in birds (see Brown 1987; Stacey and Koenig 1990). In species from

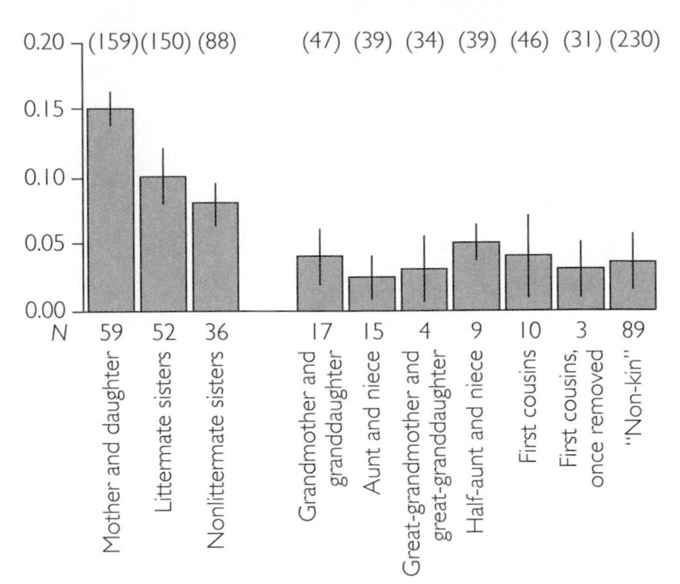

Figure 12.4 Closely related female ground squirrels are more likely to cooperate than distant kin This bar chart reports how frequently territory-owning females were joined by different categories of relatives and nonrelatives in chasing away trespassing ground squirrels. The number in parentheses indicates the number of chases that occurred when both types of individuals were present. N gives the number of different dyads of each kind that were observed. The vertical bars give the standard deviations of the frequencies. There is a significant difference in the frequency of cooperation between the three categories on the left and the seven categories on the right ($P < 0.01$), but no significant difference among the seven kinship categories on the right ($P > 0.09$). From Sherman (1981).

a wide variety of bird families, young that are old enough to breed on their own will instead remain and help their parents rear their brothers, sisters, or half-siblings. Helpers assist with nest building, nest defense, or food delivery to incubating parents and the nestlings.

Helping at the nest is usually found in species where breeding opportunities are extremely restricted, either because habitats are saturated with established breeding pairs or because suitable nest sites are difficult to obtain. In these cases, gaining direct fitness is almost impossible for young adults. Gaining indirect fitness by helping becomes a best-of-a-bad-job strategy.

Steve Emlen, Peter Wrege, and Natalie Demong have completed an intensive study of helping behavior in the white-fronted bee-eater (*Merops bullockoides*). This colonial species, native to eastern and central Africa, breeds in nesting chambers excavated in sandy riverbanks (Figure 12.5). The 40–450 individuals in a colony are subdivided into groups of 3 to 17, each of which defends a feeding territory up to 7 km away. These clans may include several sets of parents and offspring.

Many year-old bee-eaters stay to help at the nest during what would otherwise be their first breeding season. Clan members are related, so helpers usually have a choice of nestlings with different degrees of kinship as recipients of their helping behavior (Hegner et al. 1982; Emlen et al. 1995). This choice is a key

Figure 12.5 White-fronted bee-eaters The individual in the middle is performing a wing-waving display and may be soliciting a feeding.

point: Because kinship varies among the potential recipients of altruistic behavior, white-fronted bee-eaters are an excellent species for researchers to use in testing theories about kin selection.

After marking large numbers of individuals and working out genealogies over an eight-year study period, Emlen and Wrege (1988, 1991) found that bee-eaters conform to predictions made by Hamilton's rule. They determined, for example, that the coefficient of relatedness with recipients has a strong effect on whether a nonbreeding member of the clan helps (Figure 12.6a). Further, nonbreeders actively decide to help the most closely related individuals available (Figure 12.6b). That is, when young with different coefficients of relatedness are being reared within their clan, helpers almost always chose to help those with the highest r (Box 12.2). Their assistance is an enormous benefit to parents. More than half of bee-eater young die of starvation before leaving the nest. On average, the presence of each helper results in an additional 0.47 offspring being reared to fledging (Figure 12.7). For young birds, helping at the nest results in clear benefits for inclusive fitness.

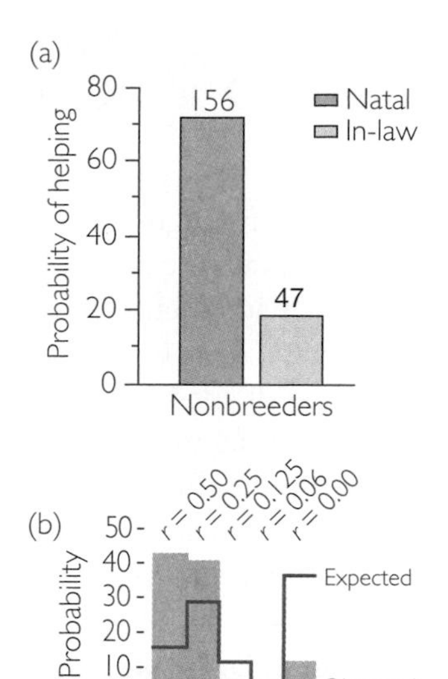

Figure 12.6 In bee-eaters, helpers assist close relatives (a) Bee-eater clans often contain nonbreeders that have paired with members of the clan. Their r with the offspring being raised that season is 0. This bar chart shows that they are much less likely to help than are clan members ($P < 0.01$). (b) In this bar chart, the expected probability of helping is calculated by assuming that helpers assist clan members randomly, in proportion to the r's of nestlings in clan nests. A G test rejects the null hypothesis that helping is directed randomly with respect to kinship ($P < 0.01$). From Emlen and Wrege (1988).

Kin Selection in Other Contexts

Recent research has revealed that kin selection is responsible for the evolution of altruistic behavior in a great variety of other situations beyond the classical examples we have reviewed. We briefly consider three such cases here.

Cannibalistic Tadpoles

The tadpoles of spadefoot toads (*Spea bombifrons*) can develop into either of two morphs. The typical morph is omnivorous and feeds mostly on decaying plant matter. The alternative morph has enlarged jaw muscles that allow it to eat big prey, including other spadefoot toad tadpoles (Figure 12.8a). David Pfennig (1999) tested whether cannibalistic tadpoles distinguish between kin and non-kin. He put 28 cannibalistic tadpoles into separate containers. Each tadpole shared its container with two omnivorous tadpoles it had never seen before: One was a full sibling; the other was non-kin. Pfennig then waited until each cannibal had eaten one of its companions. If cannibalistic tadpoles are indiscriminate, then they would be as likely to eat their siblings as they are to eat the non-kin tadpoles. In fact, only 6 of 28 cannibals ate their siblings, a significantly lower percentage than expected by chance (Figure 12.8b).

Pfennig and colleagues (1999) tested whether the discriminating behavior of cannibalistic tadpoles satisfies the requirements of Hamilton's inequality. The researchers used tiger salamanders (*Ambystoma tigrinum*), another species in which tadpoles can develop either into a typical morph or a cannibal morph. Pfennig and colleagues kept 18 cannibal tadpoles in separate cages placed in a natural lake. Each cannibal shared its cage with 6 of its typical-morph siblings, and 18 non-kin typical-morph tadpoles. As in spadefoot toads, some of the cannibalistic tadpoles discriminated between kin versus non-kin—eating kin less often than would be expected by chance—and others did not.

The degree of relatedness between the cannibals and their siblings was 1/2. Thus, by Hamilton's rule discrimination is favored if $B(1/2) - C > 0$, where B is the benefit to kin of avoiding eating them and C is the cost to the cannibal of missing a chance for a meal. Pfennig and colleagues estimated the benefit by counting the number of each cannibal's siblings that survived. They found that the siblings of discriminating cannibals were twice as likely to survive as the siblings of

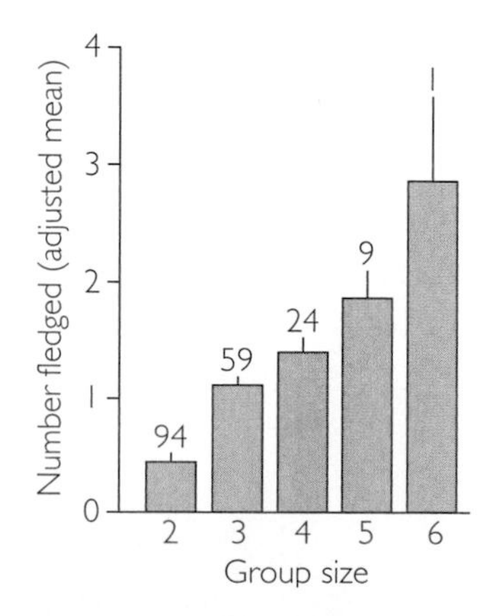

Figure 12.7 Fitness gains due to helping From Emlen and Wrege (1991).

Box 12.2 Kin recognition

The data in Figures 12.2–6 suggest that individuals have accurate mechanisms for assessing their degree of kinship with members of their own species, or conspecifics. This phenomenon, called kin recognition, has been divided into two broad categories: direct and indirect (Pfennig and Sherman 1995). Indirect kin recognition is based on cues like the timing or location of interactions. Many species of adult birds rely on indirect kin recognition when their chicks are young and will feed any young bird that appears in their nest. Direct kin recognition, in contrast, is based on specific chemical, vocal, or other cues.

There is currently a great deal of interest in determining whether loci in the major histocompatibility complex (MHC) function in direct kin recognition. In Chapter 6, we introduced the MHC and its role in self–nonself recognition by cells in the immune system. Although loci in the MHC clearly evolved to function in disease prevention, polymorphism is so extensive that non-kin share very few alleles. As a result, these genes can serve as reliable markers of kinship (see Brown and Eklund 1994). Could similarity in MHC provide a reliable cue of kinship and offer a criterion for dispensing altruistic behavior?

Jo Manning and colleagues (1992) addressed this question in a population of house mice (*Mus musculus domesticus*). A kin recognition system requires three components: production of the signal, recognition of the signal by conspecifics, and action based on that recognition. Previous work had shown that glycoproteins coded for by MHC loci are released in the urine of mice, and that mice can distinguish these molecules by smell. Mice are, for example, able to distinguish full siblings from half-siblings on the basis of their MHC genotypes. But do mice dispense altruistic behavior accordingly?

House mice form communal nests and nurse each others' pups. Because individuals could take advantage of this cooperative system by contributing less than their fair share of milk, Manning et al. predicted that mothers would prefer to place their young in nests containing close relatives. The logic here is that close kin should be less likely to cheat on one another because of the cost to their indirect fitness. Through a program of controlled breeding in which wild-caught mice were crossed with laboratory strains, Manning et al. created a population of mice with known MHC genotypes. This population was allowed to establish itself in a large barn. The researchers then recorded where mothers in this population placed their newborn pups after birth. The null hypothesis was that mothers would choose to rear their offspring randomly with respect to the MHC genotypes present in the communal nests available at the time. Contrary to the null expectation, mothers showed a strong preference for rearing their young in nests containing offspring with similar MHC genotypes. This result confirms a role for MHC as a signal used in direct kin recognition and shows that mice are capable of dispensing altruistic behavior on the basis of MHC genotypes.

nondiscriminating cannibals (Figure 12.8c). Thus the benefit in this system is roughly 2. The researchers assessed the cost to a discriminating cannibal of missing a chance for a meal by comparing the growth rate and age at metamorphosis of discriminating versus nondiscriminating cannibals. In neither case was there a significant difference. Thus the cost in this system is nearly zero. Putting these results together, we have $B(1/2) - C =$ approximately 1. Pfennig and colleagues concluded that not eating siblings is favored by kin selection in tiger salamander tadpoles.

Altruistic Sperm

Harry Moore and colleagues (2002) documented altruistic behavior in the sperm of the common European wood mouse (*Apodemus sylvaticus*). Female wood mice are highly promiscuous (Short 2002). When females mate with more than one male in short succession, the sperm from the two males are in a race to the egg (see Chapter 10). This appears to explain why male wood mice have such large testes

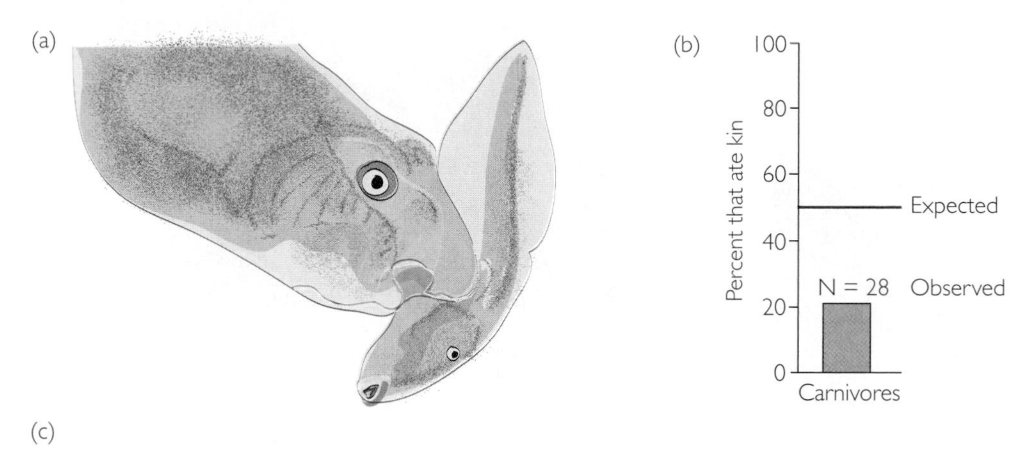

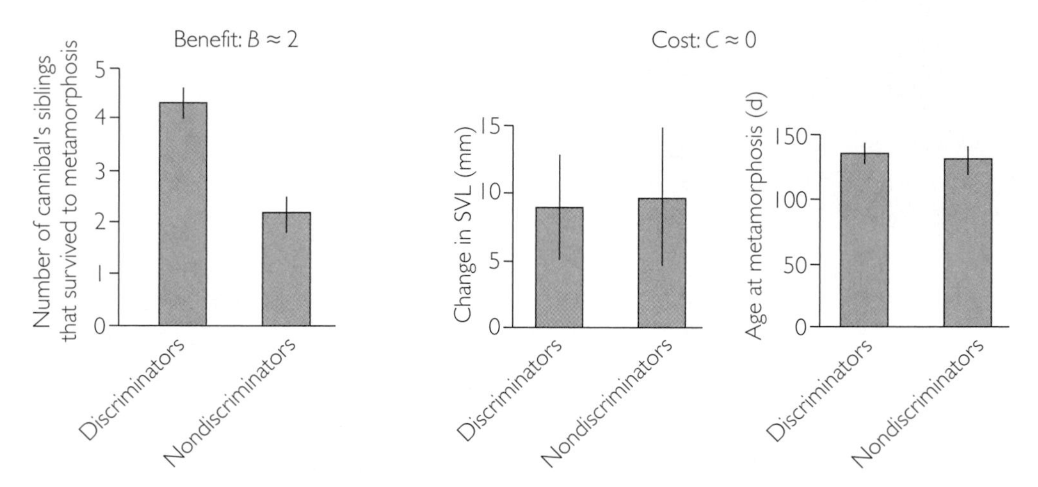

Figure 12.8 Kin-selected discrimination in cannibalistic tadpoles (a) In spadefoot toads some tadpoles develop into carnivores and become cannibals. (b) Given a choice between eating siblings versus non-kin, most cannibals are discriminating and preferentially eat non-kin. From Pfennig (1999). (c) The same phenomenon occurs in tiger salamanders. Under semi-natural conditions, discriminating cannibals see twice as many siblings survive as nondiscriminating cannibals (left) but suffer negligible costs (center and right). The costs and benefits of discriminating tastes thus satisfy Hamilton's equation for kin selection. From Pfennig et al. (1999).

(Figure 12.9a). Large testes are an adaptation for sperm competition, because the more sperm a male enters in the race the greater his chances of winning. Wood mice show other adaptations for sperm competition, too, as Moore and colleagues discovered when they examined wood mouse sperm under a microscope.

Wood mouse sperm have hooks on their heads (Figure 12.9b). Using these hooks to hold other sperm by their heads or their flagella, wood mouse sperm form trains, composed of tens or thousands of individual cells (Figure 12.9c). Swimming together, the sperm in a train can move about twice as fast as sperm swimming alone (Figure 12.9d). This cooperation would seem to improve their chances in competition with sperm from other males, but there is a catch. Before any of the sperm in a train can fertilize an egg, the train must break up. Breakup occurs when many of the sperm in the train undergo acrosome reaction, an event that releases enzymes that usually help a sperm fertilize an egg. By releasing these enzymes before contacting an egg, many of the sperm in the train sacrifice

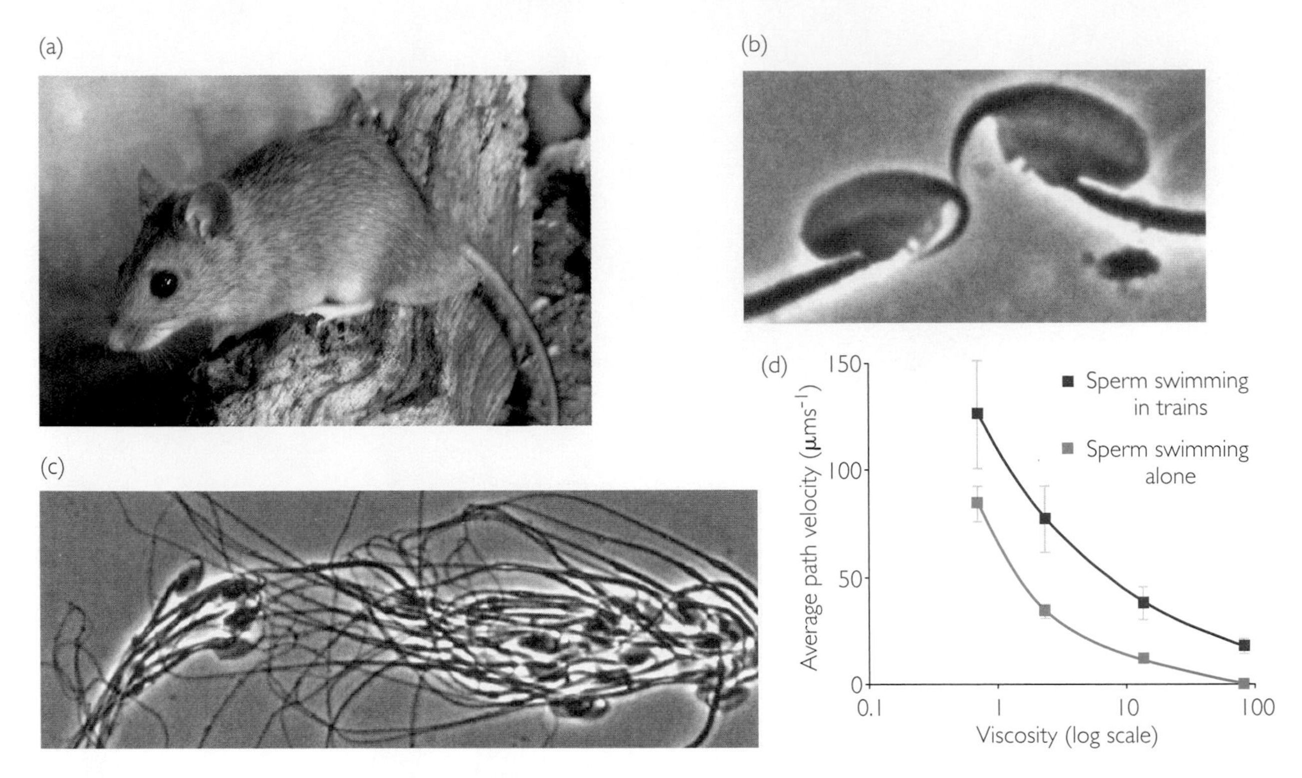

Figure 12.9 Altruistic sperm (a) Male wood mice have large testes, which suggests that sperm competition is common. From Short (2002). (b) Wood mouse sperm have hooks on their heads. (c) The sperm use these hooks to join together in large trains. (d) Trains of sperm swim about twice as fast as lone sperm. Before fertilization, however, the sperm in a train must disengage. They accomplish this via the self-sacrifice of many of the train's members. (b) (c) and (d) from Moore et al. (2002).

themselves on behalf of their siblings. Given that sibling sperm share half of their alleles, this altruistic sacrifice appears to be consistent with kin selection.

Clever Coots

The flip side of incurring costs on behalf of kin is avoiding paying costs on behalf of non-kin. Bruce Lyon (2003) documented a defense against parasitized altruism in an aquatic bird called the American coot (*Fulica americana*). Female coots often try to increase their reproduce success by laying eggs in other females' nests, a phenomenon known as conspecific nest parasitism. Accepting a parasitic egg and rearing the chick that hatches from it comes at a substantial cost. About half of the chicks in a typical nest die of starvation and the number of chicks that survive is the same for parasitized and non-parasitized nests. These facts suggest that a foster parent loses one of its biological offspring for every parasitic offspring it rears. Assuming that the coefficient of relatedness is zero between the foster parent and the parasitic chick, Hamilton's rule predicts that coots should have evolved defenses against parasitism.

Selection can favor making sacrifices for kin; it should also favor avoiding sacrifices for non-kin.

During four field seasons, Lyon monitored more than 400 nests in a wild coot population. Coot eggs show considerable variation in appearance (Figure 12.10a). If two new eggs appeared in a nest within a 24-hour period, Lyon could infer that the nest had been parasitized. And if one of the eggs did not match the others, Lyon could infer that it was the impostor.

The coots themselves apparently pay attention to similar clues. Among 133 hosts, 43% rejected one or more parasitic eggs. The rejected eggs differed from the host's eggs significantly more than did the accepted eggs (Figure 12.10b). This

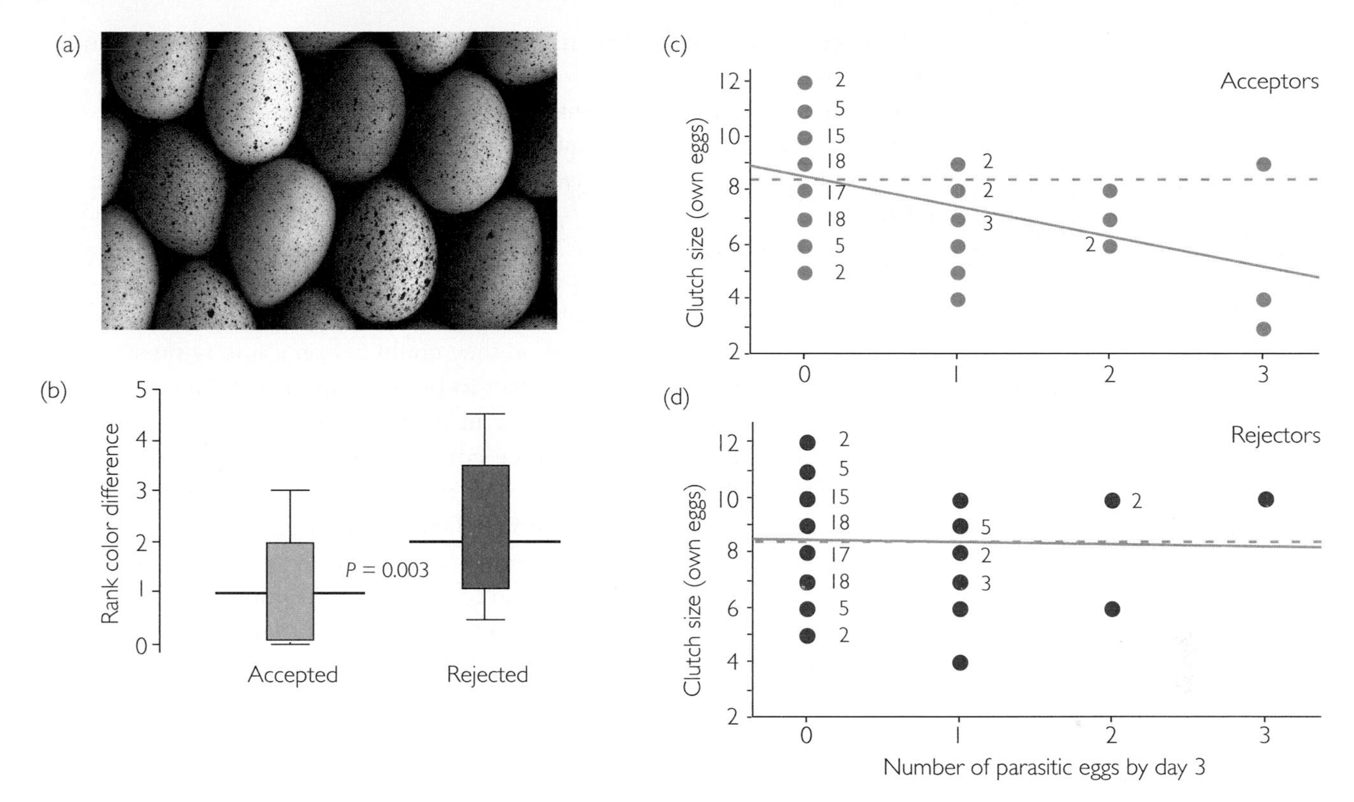

Figure 12.10 Clever coots protect themselves against parasitized altruism (a) Coot eggs vary in both background color and distribution of spots. As a result, when an extra egg appeared in a nest Lyon (2003) was able to tell which egg had been laid by a parasite. Lyon developed a scale, color rank, that allowed him to quantify the difference in appearance between host eggs and parasitic eggs. (b) Many hosts reject parasitic eggs. These box plots show that rejected eggs differ from the host's eggs more than do accepted eggs. The large horizontal lines indicate the median rank color difference, the tops and bottoms of the boxes indicate the 25th and 75th percentiles, and the thin ticks indicate the 10th and 90th percentiles. (c) Females that accepted parasitic eggs laid fewer of their own eggs. The dashed horizontal line indicates the mean clutch size for unparasitized females, the solid line is the best-fit line through the data, and the numbers next to data points indicate multiple females. (d) Females that rejected parasitic eggs laid the same number of their own eggs as unparasitized females, even though they decided how many eggs to lay before they disposed of the eggs they would reject. From Lyon (2003).

indicates that coots can distinguish their own eggs from those of other females based on appearance.

Females that accepted parasitic eggs laid one fewer egg of their own for every parasitic egg they accepted (Figure 12.10c). This meant that the acceptors reared an average of 8 eggs total, the same as the average clutch size for non-parasitized females. Females that rejected parasitic eggs laid an average of 8 of their own eggs, even though they decided how many of their own eggs to lay before they disposed of the eggs they would reject (Figure 12.10d). This gave them a final clutch size of 8 eggs. And it shows that coots can count.

By counting their eggs and rejecting the extras that do not look right, coots prevent themselves from being duped into unwitting maladaptive altruism.

Greenbeard Alleles

In the examples of kin selection we have discussed so far, the alleles responsible for altruistic behavior have risen to high frequency by playing the odds. Some of the kin that the alleles influence their carriers to help contain copies of the alleles; others do not. But so long as the alleles for altruism induce their carriers to obey Hamilton's rule—helping kin only when the cost/benefit ratio is sufficiently low and the coefficient of relatedness is sufficiently high—then the alleles win more often than they lose.

Most alleles favored by kin selection rise to high frequency by inducing altruism toward individuals likely to be carrying copies of the same allele. Greenbeard alleles would rise to high frequency by inducing altruism toward individuals certain to be carrying copies of the same allele.

Hamilton (1964b) recognized that there is another mechanism that could drive an allele for altruism to high frequency, at least in principle. Richard Dawkins (1976) called this mechanism the *greenbeard effect*. Dawkins imagined an allele that simultaneously causes its carriers to grow green beards, to recognize green beards on others, and to behave altruistically towards them. Carriers of such an allele would not have to distinguish full-siblings versus half-siblings versus cousins and adjust their behavior accordingly. Under the greenbeard effect, alleles for altruism would not have to play the odds. Instead they could bet on a sure winner.

We might expect the greenbeard effect to be nothing more than a theoretical curiosity, interesting in principle but rare in nature. The reason is that the effect requires a single allele to generate three complex and distinct phenotypes: the beard, the ability to recognize it, and the discriminating altruism. Rare the greenbeard effect may be, but it is not unknown. David Quellar and colleagues (2003) recently reported an example that comes closer than any yet discovered to matching the scenario that Hamilton and Dawkins had in mind.

Quellar and colleagues studied the slime mold *Dictyostelium discoideum*. Individuals in this soil-dwelling species germinate from spores and spend most of their lives as free-living single-celled amoebae (Figure 12.11a). When food runs short, however, these amoebae send each other chemical signals and under their influence stream together to form a mass containing thousands of cells. This mass differentiates into a slug that travels about for a time, then transforms into a tall thin stalk that supports a fruiting body. The cells in the fruiting body form spores, which disperse to new locations and begin the cycle anew. The cells in the stalk, some 20% of the individuals that came together to form the collective, altruistically sacrifice themselves to support the reproduction and dispersal of the rest.

The particular allele Quellar and colleagues studied is the wild-type allele of a gene called *csA*. Relative to loss-of-function mutations, the wild-type allele exhibits all the properties of a greenbeard allele. The protein encoded by *csA* sits on the surface of slime mold amoebae and sticks to other copies of itself on the surface of other cells. Thus the wild-type *csA* allele simultaneously specifies a trait (the protein) and the ability to recognize the trait in others (by adhesion). The remaining greenbeard trait is discriminating altruism.

Quellar and colleagues mixed wild-type amoebae with amoebae carrying a knock-out allele of the *csA* gene, grew them on agar plates in the lab, then starved them to encourage them to stream together and make fruiting bodies. The researchers found that wild-type cells were disproportionately represented in the *stalks* (Figure 12.11b). The wild-type cells apparently ended up on the bottom, delegated to a supporting and non-reproductive role, because they stuck to each other more strongly. So far, the wild-type allele appears to be not a greenbeard but a sucker. In lab cultures growing on agar plates it should quickly disappear.

The situation is reversed, however, when mixed cultures are grown in soil, their natural environment (Figure 12.11c). It is more difficult for amoeba to stream together in soil than it is on agar plates. Wild-type cells can stick to each other and pull each other along. Now wild-type cells are disproportionately represented among the spores in fruiting bodies as well as in the stalks. Knock-out cells are less adhesive and tend to get left out of aggregations altogether. Under natural conditions, then, the wild-type allele of *csA* renders its carriers preferen-

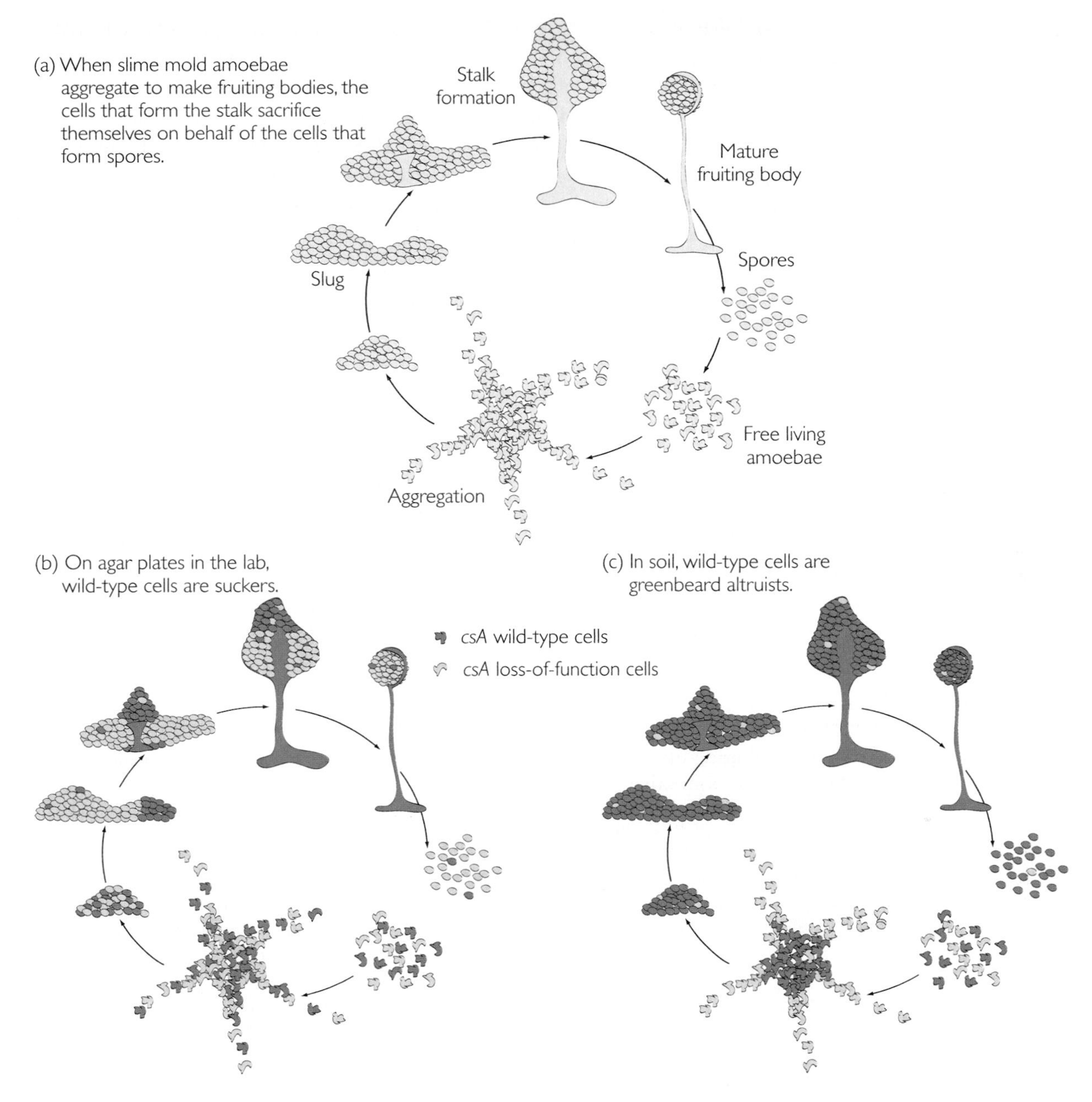
(a) When slime mold amoebae aggregate to make fruiting bodies, the cells that form the stalk sacrifice themselves on behalf of the cells that form spores.
Stalk formation
Mature fruiting body
Spores
Free living amoebae
Aggregation
Slug
(b) On agar plates in the lab, wild-type cells are suckers.
(c) In soil, wild-type cells are greenbeard altruists.
csA wild-type cells
csA loss-of-function cells

Figure 12.11 Life cycle of the slime mold *Dictyostelium discoideum* (a) When free-living amoebae aggregate to form a fruiting body, 20% of them sacrifice themselves on behalf of the others. The altruists form a stalk that supports the fruiting body proper. Only the amoebae in the fruiting body make spores. The spores disperse, germinate, and start a new life cycle. (b and c) The outcome in genetically mixed aggregates depends on the environment.

tially altruistic toward other wild-type cells. Kin selection works not just at the level of individual organisms, but also at the level of individual alleles.

12.2 Evolution of Eusociality

Darwin (1859) recognized that social insects represent the epitome of altruism, and thus a special challenge to the theory of evolution by natural selection. Many worker ants and bees, for example, do not reproduce at all. They are helpers at the nests of their parents, for life. This is an extreme form of reproductive altruism.

Eusociality (true sociality) is used to describe social systems with three characteristics (Michener 1969; Wilson 1971; Alexander et al. 1991): (1) overlap in generations between parents and their offspring, (2) cooperative brood care, and (3) specialized castes of nonreproductive individuals. Eusocial species are found in a variety of insect orders (Table 12.2), snapping shrimp (Duffy 1996; Duffy et al. 2000; Duffy et al. 2002), and one family of rodents (the mammal family Bathyergidae, or mole-rats).

As an entree to the extensive literature on eusociality, in this section we consider how reproductive altruism evolved in two very different groups: the Hymenoptera (ants, bees, and wasps) and mole-rats.

Haplodiploidy and Eusocial Hymenoptera

The Hymenoptera represent the pinnacle of social evolution. A single ant colony may number millions of individuals, each appearing to function more like a cell in a superorganism than an individual pursuing its own reproductive interests. Worker, soldier, and reproductive castes, which seem analogous to tissues in a body, can be identified on the basis of their morphology and the tasks they perform. But unlike cells and tissues, individuals in the colony are not genetically identical. What factors laid the groundwork for such extensive altruism? Why is eusociality so widespread in Hymenoptera?

In haplodiploid species, females are more closely related to their sisters than they are to their own offspring.

William Hamilton (1972) proposed that the unusual genetic system of ants, wasps, and bees predisposes them to eusociality. Hymenoptera have an

Table 12.2 Sociality in insects

This table summarizes the taxonomic distribution of eusociality in insects. Species are called "primitively eusocial" if queens are not morphologically differentiated from other individuals.

Order	Family	Subfamily	Eusocial species
Hymenoptera	Anthophoridae (carpenter bees)		In seven genera
	Apidae	Apinae (honeybees)	Six highly eusocial species
		Bombinae (bumble bees)	300 primitively eusocial species
		Euglossinae (orchid bees)	None
		Meliponinae (stingless bees)	200 eusocial species
	Halictidae (sweat bees)		In six genera
	Sphecidae (sphecoid wasps)		In one genus
	Vespidae (paper wasps, yellow jackets)	Polistinae	Over 500 species, all eusocial
		Stenogastrinae	Some primitively eusocial species
		Vespinae	About 80 species, all eusocial
	Formicidae (ants)	11 subfamilies	Over 8,800 described species, all eusocial or descended from eusocial species
	Many other families		None
Isoptera (termites)	Nine families		All species (over 2,288) are eusocial
Homoptera (plant bugs)	Pemphigidae		Sterile soldiers found in six genera
Coleoptera (beetles)	Curculionidae		Austroplatypus incompertus
Thysanoptera (thrips)	Phlaeothripidae		Subfertile soldiers are found in Oncothrips

Source: From Crozier and Pamilo (1996).

unusual form of sex determination: Males are haploid and females are diploid. Males develop from unfertilized eggs; females develop from fertilized eggs. As a result of this system, called **haplodiploidy**, female ants, bees, and wasps are more closely related to their sisters than they are to their own offspring. This follows because sisters share all of the genes they inherited from their father, which is half their genome, and half the genes they inherited from their mother (the colony's queen), which is the other half of their genome. Thus, the probability that homologous alleles in hymenopteran sisters are identical by descent is $(1 \times 1/2) + (1/2 \times 1/2) = 3/4$. To their own offspring, however, females are related by the usual r of 1/2. This unique system favors the production of reproductive sisters over daughters, sons, or brothers. (Females are related to their brothers by $r = 1/4$; see Figure 12.12). Thus, females will maximize their inclusive fitness by acting as workers rather than as reproductives (Hamilton 1972). Specifically, their alleles will increase in the population faster when they invest in the production of sisters rather than producing their own offspring. This is the haplodiploidy hypothesis for the evolution of eusociality in Hymenoptera.

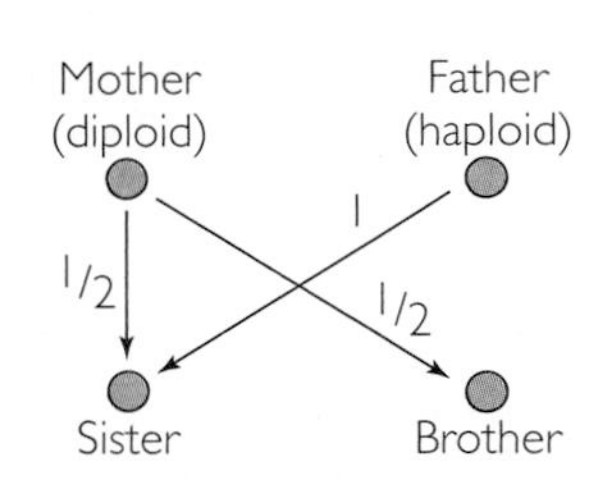

Figure 12.12 Haplodiploidy produces unusual coefficients of relationship The arrows describe the paths by which genes can be identical by descent in hymenopterans. Note that there is no path of shared descent between sisters and brothers through their father, because males have no father.

Testing the Haplodiploidy Hypothesis

In addition to offering an explanation for why workers prefer to invest in sisters rather than their own offspring, the haplodiploidy hypothesis predicts that workers prefer to invest in sisters over brothers. Because their r with sisters is 3/4 and only 1/4 with brothers, workers should favor a 3:1 female-biased sex ratio in reproductive offspring (meaning, offspring that are not destined to become sterile workers or soldiers; Trivers and Hare 1976; but see Alexander and Sherman 1977). Queens, in contrast, are equally related to their sons and daughters and should favor a 1 : 1 sex ratio in the reproductives produced (see Box 12.3). The fitness interests of workers and queens are not the same. The question is: Who wins the conflict? Do queens or workers control the sex ratio of reproductive offspring?

Liselotte Sundström and coworkers (1996) set out to answer this question by determining the sex ratio of reproductive offspring in wood ant (*Formica exsecta*) colonies. They found that queens laid a roughly equal number of male and female eggs, but that sex ratios were heavily female-biased at hatching. To make sense of this result, the researchers hypothesize that workers are able to determine the sex of eggs and that they selectively destroy male offspring.

Based on results from similar studies, most researchers acknowledge that female-biased sex allocation is widespread among eusocial hymenoptera. (For reviews, see: Bourke and Franks 1995; Crozier and Pamilo 1996; Chapuisat and Keller 1999; Sundström and Boomsma 2001.) In the tug-of-war over the fitness interests, workers appear to have the upper hand over queens.

Perhaps the more important general message of this work, however, is that colonies of ants, bees, and wasps are not harmonious "superorganisms." The asymmetry in relationship between queens and offspring versus workers and offspring produces a sharp conflict of interest.

Does the Haplodiploidy Hypothesis Explain Eusociality?

The prediction and affirmation of 3:1 sex ratios in reproductive offspring, at least in some hymenopterans, confirms that the haplodiploid system of sex determination has a strong effect on how workers behave. But is haplodiploidy the reason

Box 12.3 The evolution of the sex ratio

In many species, the sex ratio at hatching, germination, or birth is 1 : 1. Ronald Fisher (1930) explained why this should be so. Fisher pointed out that if one sex is in short supply in a population, then an allele that leads to the production of the rarer sex will be favored. This is because individuals of the rarer sex will have more than one mate on average when they mature, simply because in a sexual species every individual has one mother and one father. Members of the rarer sex will experience increased reproductive success relative to individuals of the more common sex. Indeed, whenever the sex ratio varies from 1 : 1, selection favoring the rarer sex will exist until the ratio returns to unity. Fisher's explanation is a classic example of frequency-dependent selection—a concept we introduced in Chapter 6 with the example of yellow and purple Elderflower orchids.

Fisher's argument is based on an important assumption: that parents invest equally in each sex. When one sex is more costly than the other, parents should adjust the sex ratio to even out the investment in each. For this reason evolutionary biologists distinguish the numerical sex ratio from the investment sex ratio and speak, in general terms, about the issue of sex allocation (Charnov 1982).

Robert Trivers and Dan Willard (1973) came up with an important extension to Fisher's model. They suggested that when females are in good physiological condition and are better able to care for their young, and when differences in the condition of young are sustained into adulthood, then they should preferentially invest in male offspring. This is because differences in condition affect male reproductive success RS more than female RS (see Chapter 11). This prediction, called condition-dependent sex allocation, has been confirmed in a wide variety of mammals, including humans. (For examples, see Clutton-Brock et al. 1984; Betzig and Turke 1986.)

A third prominent result in sex-ratio theory is due to William Hamilton (1967). In insects that lay their eggs in fruit or other insects, the young often hatch, develop, and mate inside the host. Frequently, hosts are parasitized by a single female. Given this situation, Hamilton realized that selection should favor females that produce only enough males to ensure fertilization of their sisters, resulting in a sex ratio with a strong female bias. This phenomenon, known as local mate competition, has been observed in a variety of parasitic insects. (For examples, see Hamilton 1967; Werren 1984.)

that so many hymenopteran species are eusocial? Most researchers are concluding that the answer is no. There are several reasons for this.

First, the prediction that workers favor the production of sisters over the production of their own offspring is based on an important assumption: that all of the female workers in the colony have the same father. In many species this is not true. Multiple mating is common in certain groups of eusocial Hymenoptera. Honeybee queens, for example, mate an average of 17.25 times before founding a colony (Page and Metcalf 1982). As a result, it is common to find that the average coefficient of relatedness among honeybee workers is under 1/3 (Oldroyd et al. 1997, 1998). In these colonies, workers are *not* more closely related to their sisters than they are to their own offspring.

Haplodiploidy affects the behavior of eusocial species, but it is not the most important factor leading to the evolution of eusociality.

Second, in many species more than one queen is active in founding the nest. If they have neither parent in common, then workers in these colonies have a coefficient of relatedness of 0.

Third, many eusocial species are not haplodiploid, and many haplodiploid species are not eusocial. Nonreproductive castes are found in all termite species, for example, even though termites are diploid and have a normal chromosomal system of sex determination (Thorne 1997). And although eusociality is common among hymenopterans, it is by no means universal. Reviewing recent work on the phylogeny of the hymenoptera will help drive this last point home.

Using Phylogenies to Analyze Social Evolution

To understand which traits are most closely associated with the evolution of eusociality in hymenoptera, James Hunt (1999) analyzed the evolutionary tree shown in Figure 12.13. Because all hymenopterans are haplodiploid, Hunt could infer that this system of sex determination evolved early in the evolution of the group, at the point marked A on the tree. Eusociality is found in just a few families of hymenopterans, however. Because these families are scattered around the tree, it is likely that eusociality evolved not just once, but several times independently. Most

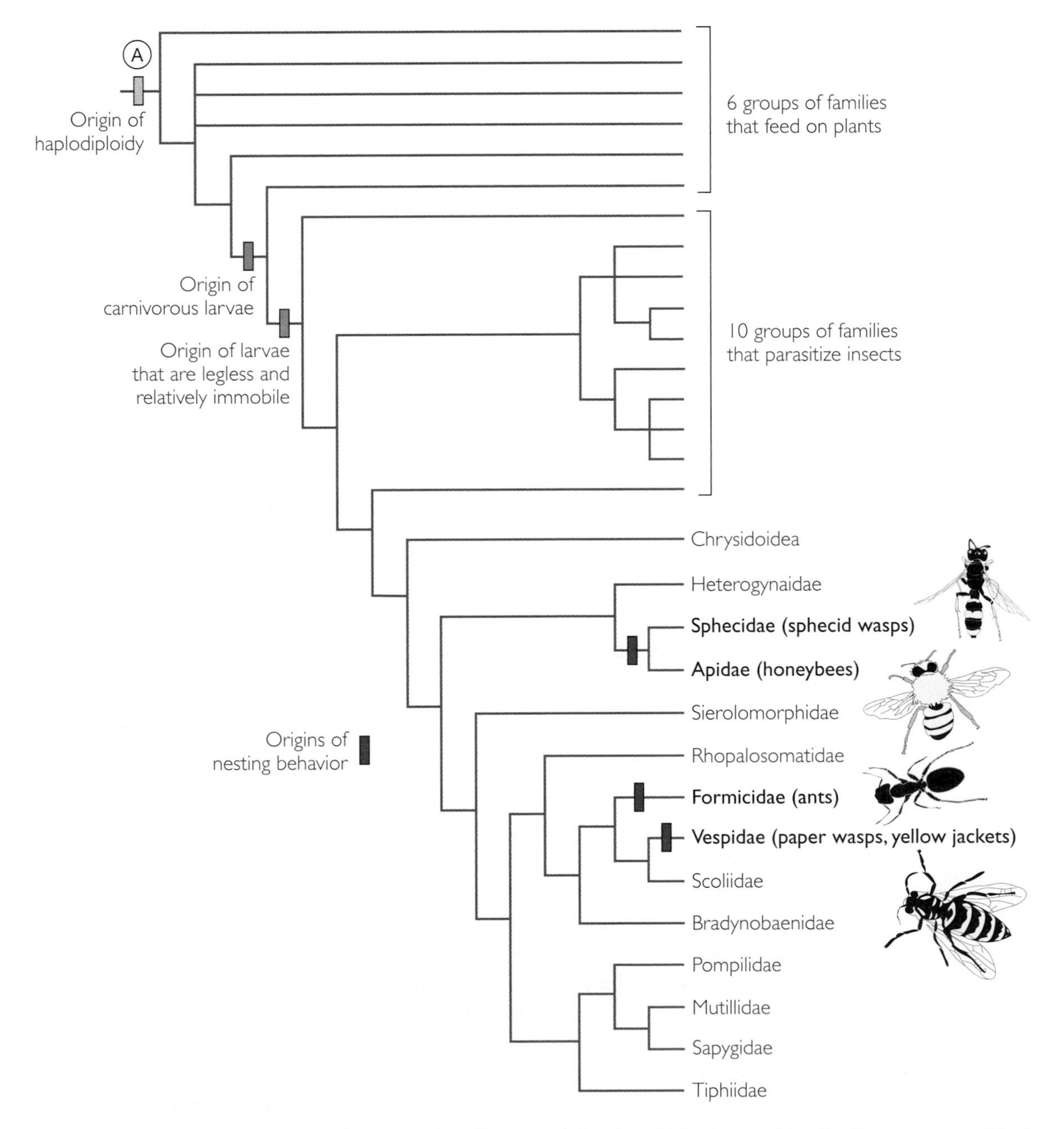

Figure 12.13 **A phylogeny of the hymenoptera** The taxa at the tips of this tree are either families or groups of families. Families that include eusocial species are indicated in bold type. (Not all of the species in these families are eusocial, however.) The colored boxes indicate points where certain key traits evolved. Modified from Hunt (1999).

importantly, Hunt noted that eusociality only evolved in groups that build complex nests and that care for their larvae for extended periods.

The association between nest building, care of larvae, and eusociality is important because it suggests that the primary agent favoring reproductive altruism in insects is ecological in nature—not genetic as proposed by the haplodiploidy hypothesis. The logic here is similar to the "best-of-a-bad-job" explanation for helping behavior in birds reviewed in Section 12.1. Nest building and the need to supply larvae with a continuous supply of food make it difficult or impossible for a female to breed on her own (see Alexander et al. 1991). Also, when predation rates are high but young are dependent on parental care for a long period, then individuals who breed alone are unlikely to survive long enough to bring their young to adulthood (Queller 1989; Queller and Strassmann 1998). In short, to explain the evolution of eusociality we clearly need to consider ecological factors that affect B and C in Hamilton's inequality, as well as genetic factors that dictate r.

Facultative Strategies in Paper Wasps

Paper wasps in the genus *Polistes* have been an especially productive group for research into the costs and benefits of reproductive altruism. Unlike workers and soldiers in ants and termites, paper wasp workers are not sterile. Instead of being obligate helpers, *Polistes* females are capable of reproducing on their own. This contrast is important. To achieve reproductive success, worker and soldier ants and termites have no choice but to assist relatives—the nutrition they received as larvae guarantees that they are sterile. But in paper wasps, females have the option of helping relatives or breeding on their own.

In paper wasps, reproductive altruism is facultative. Females can choose between helping at a nest or breeding on their own.

In *Polistes dominulus,* Peter Nonacs and Hudson Reeve (1995) found that females pursue one of three distinct strategies: They either initiate their own nest, join a nest as a helper, or wait for a breeding opportunity. Each option is associated with costs and benefits.

What are the costs and benefits of founding a nest? In the population that Nonacs and Reeve studied, nests were founded by single females or by multifemale groups. Earlier studies had shown that single foundresses are at a distinct disadvantage compared to multifemale coalitions. Adult mortality is high, and nests with multiple foundresses are less likely to fail because surviving females keep the nest going. Nonacs and Reeve also found that multifoundress coalitions are more likely to renest after a nest is destroyed. When they analyzed 106 instances of nest failure due to predation or experimental removal, they determined that only 5 of 54 single foundresses rebuilt while 21 of 51 multifoundress groups did.

Although the success rate of multifemale nests is high when compared to single-foundress nests, multifemale coalitions are not free of conflicts. Fights between wasps are decided by body size. As the graph in Figure 12.14 shows, multifoundress nests grew fastest when there was a large difference in the body size of the dominant female and her subordinate helpers. To interpret this result, Nonacs and Reeve suggest that productivity is low in coalitions where body size is similar because subordinate females frequently challenge for control of the nest and the right to lay most of the eggs.

Why would females join a coalition and help rear offspring that are not their own? Subordinates gain indirect fitness benefits because they are usually closely related to the dominant female. They may also gain direct fitness benefits if the

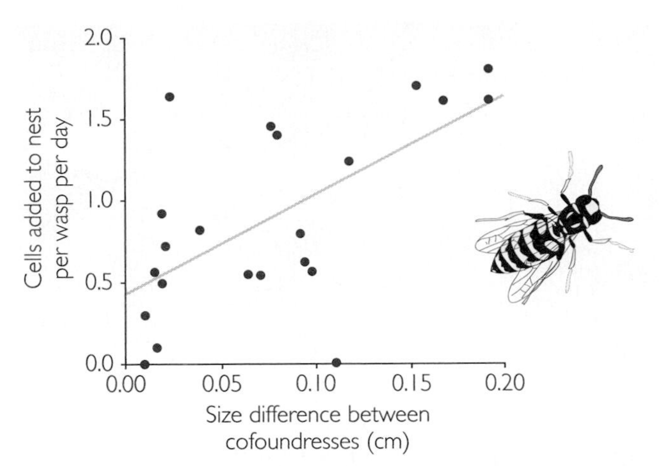

Figure 12.14 In paper wasps, the success of female coalitions varies This graph plots the growth rate of *Polistes dominulus* nests as a function of the size difference between cofoundresses. Nests grow fastest when the founding females are markedly different in size. Modified from Nonacs and Reeve (1995).

dominant individual dies and they are able to take over the nest. Thus, the costs and benefits of helping depend on a female's body size and her coefficient of relatedness with other members of the coalition.

Nonacs and Reeve also found that a "sit-and-wait" strategy could make sense in fitness terms because females that do not participate in nest initiation are able to adopt orphaned nests (after all of the adult females have died) or usurp small nests later in the season by defeating the attending female(s) in combat. In *Polistes fuscatus,* Reeve and colleagues (1998) found that some females pursue an interesting twist on these sit-and-wait tactics: They leave the nest early in the spring, enter a dormant state in a sheltered location, and wait until the following breeding season before attempting to nest.

For female paper wasps, the decision to join an existing nest or to breed independently hinges on a series of costs of benefits. These costs and benefits are dictated by environmental and social conditions, and may change through time.

The fundamental message of these studies is that reproductive altruism is facultative. It is an adaptive response to environmental conditions. In the case of the population studied by Nonacs and Reeve, the conditions that are relevant to a female are her body size relative to her competitors, her coefficient of relatedness to members of a nesting coalition, and the availability of other nests or nest sites. Genetic, social, and ecological factors have also been invoked to explain the evolution of eusociality in naked mole-rats.

Naked Mole-Rats

Naked mole-rats (*Heterocephalus glaber*) are one of the great oddities of the class Mammalia (Figure 12.15). They are neither moles nor rats, but are members of the family Bathyergidae, native to desert regions in the Horn of Africa. They eat tubers, live underground in colonies of 70–80 members, and construct tunnel systems up to two miles long by digging cooperatively in the fashion of a bucket brigade. Mole-rats are nearly hairless and ectothermic ("cold-blooded") and, like termites, can digest cellulose with the aid of specialized microorganisms in their intestines.

Naked mole-rats are also eusocial. All young are produced by a single queen and all fertilizations are performed by a group of two to three reproductive males. As other members of the colony grow older and increase in size, their tasks change from tending young and working in the tunnels to specializing in colony defense. For unknown reasons, there is a slight male bias in the colony sex ratio, of 1.4 : 1. Naked mole-rats are diploid and have an XY-chromosome system of sex determination.

Figure 12.15 **Naked mole-rats** This photo shows a naked mole-rat queen threatening a worker. For superb introductions to the biology of naked mole-rats, see Honeycutt (1992) and Sherman et al. (1992).

In naked mole-rats, helpers gain indirect fitness benefits because they are very closely related to the queen's offspring.

The leading hypothesis to explain why naked mole-rats are eusocial centers on inbreeding. Analyses of microsatellite loci confirm that colonies are highly inbred (Reeve et al. 1990). Researchers studying colonies established in the lab have determined that approximately 85% of all matings are between parents and their offspring or between full siblings and that the average coefficient of relationship among colony members is 0.81 (Sherman et al. 1992). These are among the highest coefficients ever recorded in animals.

Even extensive inbreeding does not mean that the reproductive interests of workers and reproductives are identical, however. Conflicts exist because workers are still more closely related to their own offspring than they are to their siblings and half-siblings. Queens are able to maintain control, however, through physical dominance (Reeve and Sherman 1991; Reeve 1992; see also Clarke and Faulkes 2001). If nonreproductives slow their pace of work, mole-rat queens push them. These head-to-head shoves are aggressive and can move a worker more than a meter backward through a tunnel. Shoves are directed preferentially toward nonrelatives and toward relatives more distant than offspring and siblings of the queen (Figure 12.16). Workers respond to shoves by nearly doubling their work

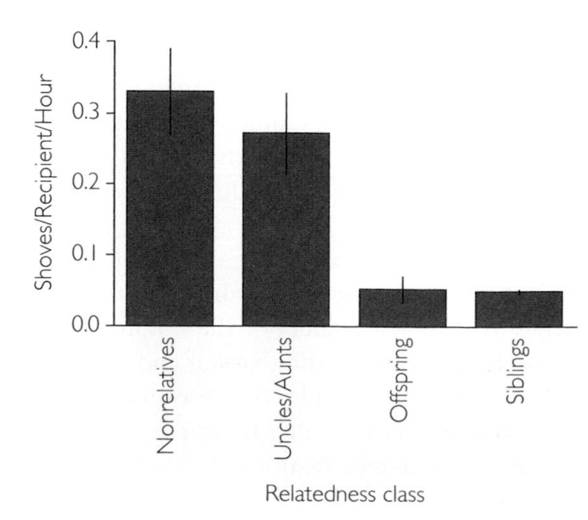

Figure 12.16 **Naked mole-rat queens preferentially shove nonrelatives** These data were collected from a captive colony of naked mole-rats. The bars indicate the average and standard errors of shoves given to different kin classes by a queen called three-bars. There is a statistically significant difference between shoving rates for nonrelatives and uncles/aunts versus the two closer kin classes. From Reeve and Sherman (1991).

Table 12.3 Shoves from naked mole-rat queens motivate workers

Working in captive colonies, Hudson Reeve (1992) performed regular scans and recorded the activity of all individuals in the colony. The work level reported in this table represents the proportion of scans during which individuals were performing work, with standard errors given in parentheses. Shoving events by queens that took place in the nestbox or tunnels were also recorded. There is a statistically significant difference between work rates before and after shoves, for both types of shoves ($P < 0.01$).

Shove recipient's work level	Before shove	After shove
All shoves	0.14 (0.03)	0.25 (0.06)
Tunnel shoves only	0.34 (0.05)	0.58 (0.07)

rate (Table 12.3). These data suggest that queens impose their reproductive interests on subordinates through intimidation.

Inbreeding has also been hypothesized as a key factor predisposing termites to eusociality (Bartz 1979; but see Pamilo 1984; Roisin 1994). Not all inbred species are eusocial, however. This means that inbreeding is just one of several factors that contribute to eusociality in naked mole-rats and termites. Ecological factors such as extended parental care, group defense against predation, and severely constrained breeding opportunities are also important in explaining the evolution of reproductive altruism.

12.3 Parent–Offspring Conflict

The theory of kin selection has been remarkably successful in explaining the structure and dynamics of social groups such as bee-eater clans and wasp colonies. Now we consider how the theory might inform questions about a more fundamental social unit: parents and offspring.

Parental care is a special case of providing fitness benefits for close relatives. Although kin selection can lead to close cooperation between related individuals such as parents and offspring, even close kin can be involved in conflicts when the costs and benefits of altruism change or when degrees of relatedness are not symmetrical. Robert Trivers (1974) was the first to point out that parents and offspring are *expected* to disagree about each other's fitness interests. Because parental care is so extensive in birds and mammals, conflicts over the amount of parental investment should be especially sharp in these taxa.

Weaning Conflict

Weaning conflict is a well-documented example of parent–offspring strife. Aggressive and avoidance behaviors are common toward the end of nursing in a wide variety of mammals. Mothers will ignore or actively push young away when they attempt to nurse, and offspring will retaliate by screaming or by attacking their mothers (Figure 12.17).

Parents maximize their fitness by investing in all of their offspring equally. Offspring, in contrast, maximize their fitness by receiving more parental investment than their siblings.

The key to explaining weaning conflict is to recognize that the fitness interests of parents and offspring are not symmetrical. Offspring are related to themselves with $r = 1$, but parents are related to their offspring with $r = 1/2$. Further, parents are equally related to all their offspring and are expected to equalize their

Figure 12.17 Weaning conflict (a) The infant langur monkey on the left has just attempted to nurse from its mother, at the right. The mother refused to nurse. In response, the infant is screaming at her. (b) The infant then dashes across the branch and slaps its mother. From Trivers, (1985).

investment in each. Siblings, in contrast, are related by 1 to themselves but 1/2 to each other. The theory of evolution by natural selection predicts that each offspring will demand an unequal amount of parental investment for itself.

When these asymmetries are applied to nursing, conflicts arise. At the start of nursing, the benefit to the offspring is high relative to the cost to the parent (Figure 12.18a). As nursing proceeds, however, this ratio declines. Young grow and demand more milk, which increases the cost of care. At the same time, they are increasingly able to find their own food, which decreases the benefit. Natural selection should favor mothers who stop providing milk when the bene-

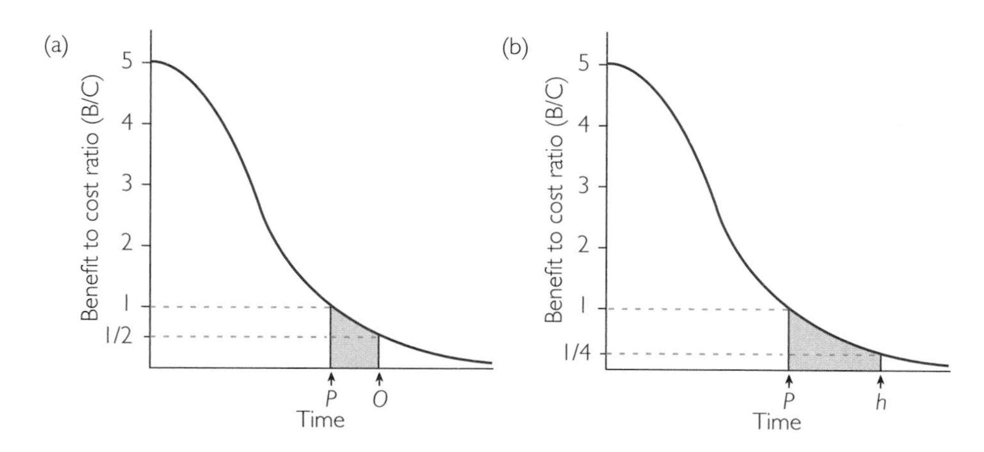

Figure 12.18 Parent–offspring conflict results from changes in the costs and benefits of parental care and asymmetries in relationship (a) This graph illustrates why parent–offspring conflict occurs. The *y*-axis plots the benefit-to-cost ratio (*B/C*) for an act of parental care such as providing milk. Benefit is measured in terms of increased survival by the offspring receiving the care, while cost is measured as decreased production of additional offspring by the parent. Time is plotted along the *x*-axis. The curve drawn here is hypothetical; its shape will vary from species to species. See the text for an explanation of how this curve is interpreted. (b) This is the same graph as in (a), modified to illustrate how the period of parent–offspring conflict is extended when parents produce half-siblings instead of full siblings. From Trivers (1985).

fit-to-cost ratio reaches 1 (this is time P in Figure 12.18a). From the mother's perspective, this is when weaning should occur. Offspring, on the other hand, devalue their mother's cost of providing care. They do this because the "savings" that a parent achieves through weaning will be invested in brothers or sisters with $r = 1/2$ instead of in themselves with $r = 1.0$. Natural selection should favor offspring who try to coerce continued parental investment until the benefit-cost ratio is 1/2 (this is time O on Figure 12.18a). The period between time P and time O defines the interval of weaning conflict. Avoidance and aggressive behavior should be observed throughout this period. If mating systems are such that mothers routinely remate and produce half-siblings, the period of weaning conflict will extend to time h, when the ratio of benefit to cost $(B/C) = r = 1/4$ (Figure 12.18b). For field studies that confirm weaning conflict, see Trivers (1985, Chapter 7); for a theoretical treatment of other types of parent–offspring strife, see Godfrey (1995).

Harassment in White-Fronted Bee-Eaters

Another dramatic example of parent–offspring conflict occurs in the white-fronted bee-eaters introduced earlier. Steve Emlen and Peter Wrege (1992) have collected data suggesting that fathers occasionally coerce sons into helping to raise their siblings. They do this by harassing sons who are trying to raise their own young.

A variety of harassment behaviors are observed at bee-eater colonies. Individuals chase resident birds off their territory, physically prevent the transfer of food during courtship feeding, or repeatedly visit nests that are not their own before egg laying or hatching. During the course of their study, Emlen and Wrege observed 47 cases of harassment. Over 90% of the instigators were male and over 70% were older than the targeted individual. In 58% of the episodes, the instigator and victim were close genetic kin. In fact, statistical tests show that harassment behavior is not targeted randomly, but is preferentially directed at close kin ($P < 0.01$; χ^2 test).

Emlen and Wrege interpret this behavior by proposing that instigators are actively trying to break up the nesting attempts of close kin. Furthermore, they suggest that instigators do this to recruit the targeted individuals as helpers at their own (the instigator's) nest.

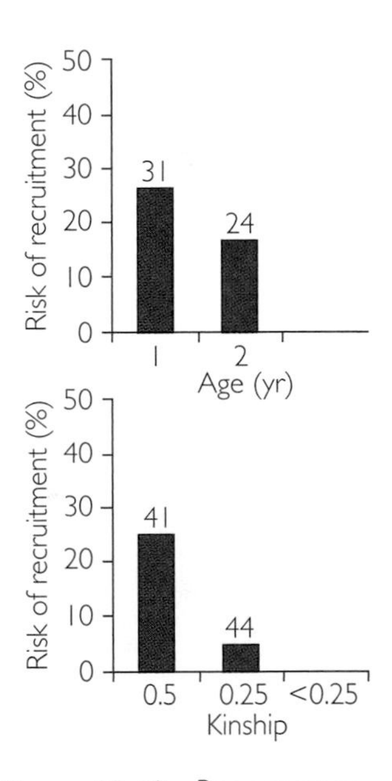

Figure 12.19 Bee-eaters recruit helpers who are younger and closely related Emlen and Wrege (1992) considered each paired male in the colony who had a same-age or older male present in its clan, and calculated the percentage of these individuals who were recruited as helpers after experiencing harassment. This probability is plotted on the *y*-axis of these two bar charts. Different age and kinship categories of targeted individuals are plotted on the *x*-axis. Here kinship represents the r value between the targeted individual and the offspring in the instigator's nest.

What evidence do Emlen and Wrege present to support this hypothesis? In 16 of the 47 harassment episodes observed, the behavior actually resulted in recruitment: The harassed individuals abandoned their own nesting attempts and helped at the nest of the instigator. Of these successful events, 69% involved a parent and offspring and 62% involved a father and son. The risk of being recruited is clearly highest for younger males and for males with close genetic relatives breeding within their clan (Figure 12.19).

These data raise the question of why sons do not resist harassment more effectively. Emlen and Wrege suggest that harassment can be successful because sons are equally related to their own offspring and to their siblings. Parents, in contrast, are motivated to harass because they are more closely related to their own offspring ($r = 1/2$) than they are to their grandchildren ($r = 1/4$). We have already mentioned that on average, each helper is responsible for an additional 0.47 offspring being raised. In comparison, each parent at a nest unaided by relatives is able to raise 0.51 offspring. This means that for a first-time breeder, the fitness payoff from breeding on its own is only slightly greater than the fitness payoff from helping. The payoffs are close enough to suggest that parents can change the bottom line of the fitness accounting. Perhaps harassing a son tips the balance by increasing his cost of rearing young. Then helping becomes a more favorable strategy for the son than raising his own young. Emlen and Wrege's data imply that bee-eater fathers recognize sons and coerce them into serving the father's reproductive interests.

Siblicide

In certain species of birds and mammals, it is common for young siblings to kill each other while parents look on passively. How can this behavior be adaptive, given that the parents and siblings are related by an r of 1/2?

Lynn Lougheed and David Anderson (1999) took an experimental approach to answering this question. Their research subjects were species of seabirds called the blue-footed booby and the masked booby (Figure 12.20). In both taxa, females normally lay a two-egg clutch. Because the eggs are laid 2–10 days apart, one chick hatches before the other. In the masked booby, the older offspring pushes its younger sibling from the nest within a day or two of hatching. There the smaller chick quickly dies of exposure or is taken by a predator.

Siblicide is more complex in the blue-footed booby, however. The older nestling does not always kill its younger sibling right after it hatches. During short-term food shortages, Anderson and Robert Ricklefs (1995) actually found that older chicks reduce their food intake. By doing so, they help their younger siblings survive. But if food shortages continue, the older chick attacks and kills its sibling. Presumably, this enhances the older chick's chance of survival by removing competition for food.

Lougheed and Anderson (1999) wanted to understand whether parents play a role in these events. In both masked boobies and blue-footed boobies, siblicide is sensible in light of the relatedness asymmetry between individuals (where $r = 1$) and their siblings (where $r = 1/2$). But parents are equally related to each chick and would be expected to intervene and prevent attacks.

Figure 12.20 Masked and blue-footed boobies These photos show masked boobies (left) and blue-footed boobies (right). The two species nest in adjacent colonies at the study site established by Lougheed and Anderson in the Galápagos Islands off the northwest coast of South America.

Table 12.4 In boobies, the probability of siblicide varies with parent species and nestmate species

These data show that siblicide is much more common in nests with masked booby parents or masked booby nestlings than in nests with blue-footed booby nestlings or blue-footed booby parents. To explain these patterns, Lougheed and Anderson hypothesize that masked booby nestlings are more siblicidal than blue-footed booby nestlings, and that blue-footed booby parents attempt to intervene and prevent siblicide.

Treatment	No siblicide	Siblicide
Masked booby nestlings with masked booby parents	0	25
Blue-footed booby nestlings with masked booby parents	12	8
Masked booby nestlings with blue-footed booby parents	4	16
Blue-footed booby nestlings with blue-footed booby parents	17	0

Source: Lougheed and Anderson (1999).

To explore whether parental behavior differs between masked boobies and blue-footed boobies, Lougheed and Anderson performed a reciprocal transplant experiment. They placed newly hatched masked booby chicks in blue-footed booby nests, and vice versa. As controls, they also monitored the fate of masked booby chicks transferred to other masked booby nests and blue-footed booby broods transferred to other blue-footed booby nests.

As the data in Table 12.4 show, the fate of the chicks varied dramatically in the four treatments. Chicks were much more likely to die if they had a masked booby nestmate. This is consistent with the observation that siblicide is virtually universal in this species. But nestlings were also much more likely to die if they had masked booby parents. To explain this result, Lougheed and Anderson proposed that masked booby parents tolerate siblicidal chicks, while blue-footed booby parents attempt to intervene and prevent the death of their younger offspring. Their data set is the first to indicate that in some siblicidal species, parents act to defend their reproductive interests.

Siblicide may increase the fitness of parents as well as the siblicidal offspring if the offspring that is killed is likely to die anyway.

Why do masked booby and blue-footed boobies differ so strongly in their response to parent–offspring conflict? The leading hypothesis is that food shortages are much more likely in masked boobies, and that second chicks almost always starve to death even without siblicide. This hypothesis is still untested, however. Research into the dynamics of siblicide and other forms of parent–offspring conflict continues (see Mock and Parker 1997).

12.4 Reciprocal Altruism

Inclusive fitness theory has been remarkably successful in explaining a wide range of phenomena in social evolution. In many cases, altruistic acts can be understood in light of Hamilton's rule, and conflicts can be understood by analyzing asymmetries in coefficients of relatedness and differences in fitness payoffs. But the theory and data we have reviewed thus far are only relevant to interactions among kin. What about the frequent occurrence of cooperation among unrelated individuals?

Reciprocal altruism provides one theoretical framework for studying cooperation among non-kin. Robert Trivers (1971) proposed that individuals can be selected to dispense altruistic acts if equally valuable favors are later returned by the beneficiaries. According to Trivers, natural selection can favor altruistic behavior if the recipients reciprocate.

Two important conditions must be met for reciprocal altruism to evolve. First, selection can favor altruistic acts only if the cost to the actor is smaller than or equal to the benefit to the recipient. Alleles that lead to high-cost, low-benefit behavior cannot increase in the population even if this type of act is reciprocated. (As with kin selection, the costs and benefits of altruistic acts are measured by the numbers of surviving offspring.) Second, individuals that fail to reciprocate must be punished in some way. If they are not, then altruistic individuals suffer fitness losses with no subsequent return. Alleles that lead to cheating behavior would increase in the population and altruists would quickly be eliminated. As a result, the theory predicts that altruists will be selected to detect and punish cheaters by physically assaulting them or by withholding future benefits (Box 12.4).

Trivers pointed out that reciprocal altruism is most likely to evolve when

Reciprocal altruism can evolve only under a restricted set of conditions.

- each individual repeatedly interacts with the same set of individuals (groups are stable);
- many opportunities for altruism occur in an individual's lifetime;
- individuals have good memories; and
- potential altruists interact in symmetrical situations.

This means interacting individuals sometimes need and sometimes can offer favors, and are able to dispense roughly equivalent benefits at roughly equivalent costs.

Accordingly, we expect reciprocal altruism to be characteristic of long-lived, intelligent, social species with small group size, low rates of dispersal from the group, and a high degree of mutual dependence in group defense, foraging, or other activities. Reciprocal altruism should be less likely to evolve in species where strong dominance hierarchies are the rule. In these social systems, subordinate individuals are rarely able to provide benefits in return for altruistic acts dispensed by dominant individuals.

Based on these characteristics, Trivers (1971, 1985; see also Packer 1977) has suggested that reciprocal altruism is responsible for much of the cooperative behavior observed in primates like baboons, chimpanzees, and humans. Indeed, Trivers has proposed that human emotions like moralistic aggression, gratitude, guilt, and trust are adaptations that have evolved in response to selection for reciprocal altruism. He suggests that these emotions function as "scorekeeping" mechanisms useful in moderating transactions among reciprocal altruists. Recent observational and experimental studies indicate that humans have a specialized mental apparatus that is particularly good at detecting cheaters in social exchanges (see Stone et al. 2002; Sugiyama et al. 2002).

Studying reciprocal altruism in natural animal populations is exceptionally difficult, however. For example, it is likely that kin selection and reciprocal altruism interact and are mutually reinforcing in many social groups. This makes it difficult for researchers to disentangle the effect that each type of selection has independently of the other. Also, the fitness effects of some altruistic actions can be difficult to quantify. When a young male baboon supports an older, unrelated male in his fight over access to a female, or when a young lioness participates in group defense of the territory, how do we quantify the fitness costs and benefits

Box 12.4 Prisoner's dilemma: Analyzing cooperation and conflict using game theory

Robert Trivers (1971) recognized that a classical problem from the branch of mathematics called game theory closely simulates the problems faced by nonrelatives in making decisions about their interactions. The central idea in game theory is that the consequences of any move in a game are contingent: The result or payoff from an action depends on the move made by the opponent. When players in a game pursue contrasting strategies, game theory provides a way to quantify the outcomes and decide which strategy works best.

Game theory was invented in the 1940s to analyze contrasting strategies in games like poker and chess. Later, the approach was applied by economists to a variety of problems in market economics and business competition. John Maynard Smith (1974, 1982) pioneered the use of game theory in evolutionary biology in analyses of animal contests. His work inspired a series of productive studies on the evolution of display behavior and combat (e.g., Sigurjónsdóttir and Parker 1981; Hammerstein and Reichert 1988).

The game that Trivers employed to analyze cooperation is called Prisoner's Dilemma. Prisoner's Dilemma models the following situation: Two prisoners who have been charged as accomplices in a crime are locked in separate cells. The punishment they suffer depends on whether they cooperate with one another in maintaining their innocence or implicate the other in the crime. Each prisoner has to choose his strategy without knowing the other prisoner's choice. The payoffs to Player A in this game are as follows:

		Player B's action	
		C Cooperation	**D Defection**
Player A's action	**C Cooperation**	R (reward for cooperation—both receive light sentences)	S (sucker gets longer sentence if partner defects)
	D Defection	T (temptation—reduced sentence for defector)	P (punishment for mutual defection—both receive intermediate sentences).

The values are such that $T > R > P > S$ and $R > (S + T)/2$. If A assumes B has cooperated, A's best strategy is to defect. If A assumes B has defected, A's best strategy is still to defect. When the players interact just once, the best plan for each is to defect. Yet if both cooperated they would get lighter sentences.

What happens when the two players interact repeatedly? Now they can reward—and punish—each other for past behavior. This allows them to negotiate a long-term cooperative relationship. Robert Axelrod and William Hamilton (1981) performed a widely cited analysis of an iterated Prisoner's Dilemma. They invited game theorists from all over the world to submit strategies for players competing in a computerized simulation of the game. Each round in this tournament had the following payoffs: $R = 3$, $T = 5$, $S = 0$, and $P = 1$. Axelrod and Hamilton let each strategy play against all of the other strategies submitted and computed the outcome of every one-on-one game over many interactions. Most of the theorists submitted complicated decision-making algorithms, but the winner was always the simplest strategy of all, called tit for tat, TFT. An individual playing TFT starts by cooperating, then simply does whatever the opponent did in the previous round. This strategy has three prominent features: A player using it is (1) never the first to defect; (2) provoked to immediate retaliation by defection; and (3) willing to cooperate again after just one act of retaliation for a defection.

In analyzing the outcome of games like this, researchers use the concept of an **evolutionarily stable strategy**, or ESS. A strategy is an ESS if a population of individuals using it cannot be invaded by a rare mutant adopting a different strategy. Axelrod and Hamilton's tournament showed that TFT is an evolutionarily stable strategy with respect to other strategies employed in the tournament. Their result offers an explanation for the evolution of cooperative behavior in unrelated individuals. Laboratory experiments with guppies and sticklebacks suggest that animals may actually play TFT when they interact (see Milinski 1996; Dugatkin 1998).

for each participant? What are our chances of observing the return behavior, and quantifying its costs and benefits as well? Finally, it can be difficult to distinguish reciprocal altruism from what biologists call by-product mutualism. This is cooperative behavior that benefits both individuals more or less equally. The critical difference between reciprocity and mutualism is that there is a time lag between the exchange of benefits in reciprocal altruism.

For these reasons, it has been difficult for evolutionary biologists to document reciprocal altruism—difficult enough, in fact, to make many biologists suspect that this type of natural selection is rare. Here we will examine one of the most robust studies done to date: food sharing in vampire bats. In this system, the altruistic act is regurgitating a blood meal. The cost of altruism can be measured as an increase in the risk of starvation and the benefit as a lowered risk of starvation.

Blood-Sharing in Vampire Bats

Gerald Wilkinson (1984) worked on a population of about 200 vampire bats (*Desmodus rotundus*) at a study site in Costa Rica (Figure 12.21). The basic social unit in this species consists of 8–12 adult females and their dependent offspring. Members of these groups frequently roost together in hollow trees during the day, although subgroups often move from tree to tree. (There was a total of 14 roosting trees at Wilkinson's study site.) As a result of this social structure, many individuals in the population associate with one another daily. The degree of association between individuals varies widely, however. Wilkinson quantified the degree of association between each pair of bats in the population by counting the number of times they were seen together at a roost and dividing by the total number of times they were observed roosting.

Wilkinson was able to capture and individually mark almost all of the individuals at his study site over a period of four and a half years, and estimate coefficients of relatedness through analysis of pedigrees. In the female group for which he had the most complete data, the average r between individuals was 0.11. (Recall that cousins have an r of 0.125.)

The combination of variability in association and relatedness raises interesting questions about the evolution of altruism. Vampire bats dispense altruistic behavior by regurgitating blood meals to one another. This food sharing is important because blood meals are difficult to obtain. The bats leave their roosts at night to

Figure 12.21 Vampire bats This photo shows a group of vampire bats roosting in a hollow tree.

search for large mammals—primarily horses and cattle—that can provide a meal. Prey are wary, however, and 33% of young bats and 7% of adults fail to feed on any given night. By studying weight loss in captive bats when food was withheld, Wilkinson was able to show that bats who go three consecutive nights without a meal are likely to starve to death.

Vampire bats reciprocate by sharing blood meals. They usually share with close relatives or nonrelatives who are roostmates and may later reciprocate.

Because the degree of relatedness and degree of association varied among individuals in the population, either kin selection or reciprocal altruism could operate in this system. Wilkinson was able to show that both occur. Over the course of the study, he witnessed 110 episodes of regurgitation. Seventy-seven of these were between mother and child and are simply examples of parental care. In 21 of the remaining 33 cases, Wilkinson knew both the r and the degree of association between the actor and beneficiary and could examine the effect of both variables. Wilkinson discovered that both degree of relatedness and degree of association have a statistically significant effect in predicting the probability of regurgitation (Figure 12.22). Bats do not regurgitate blood meals to one another randomly. They are much more likely to regurgitate to relatives and to nonrelatives who are frequent roostmates.

To confirm that bats actually do reciprocate, Wilkinson held nine individuals in captivity, withheld food from a different individual each night for several weeks, and recorded who regurgitated to whom over the course of the experiment. Statistical tests rejected the null hypothesis that hungry individuals received blood randomly from cagemates. Instead, hungry individuals were much more likely to receive blood from an individual they had fed before. This confirms that vampire bats are reciprocal altruists.

Territory Defense in Lions

Lions (*Panthera leo;* Figure 12.23) are the only species of cat that lives in social groups. These groups, called prides, consist of three to six related females, their offspring, and a coalition of males. Males are often related to one another but are

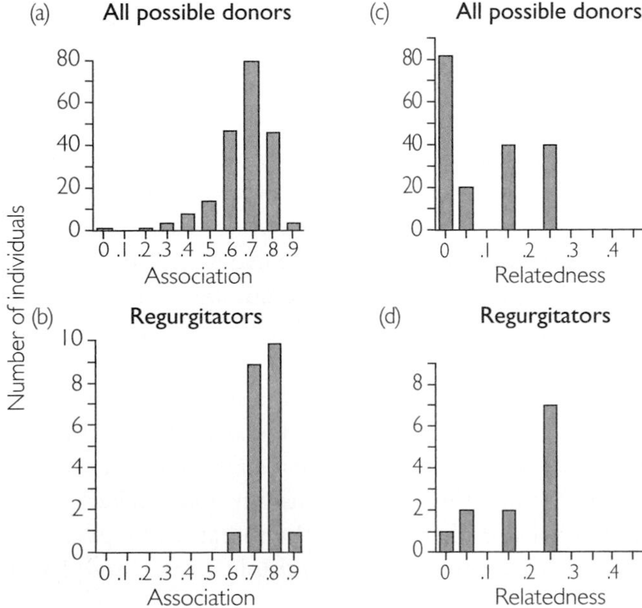

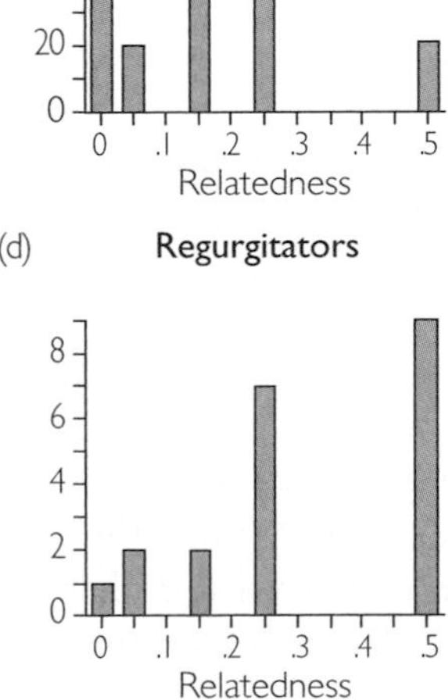

Figure 12.22 Association, relatedness, and altruism in vampire bats These histograms plot the total number of bat–bat pairs at Gerald Wilkinson's study site versus their (a) degree of association for all potential blood donors in the roost, (b) degree of association for blood-sharing pairs who are not related as mother–offspring, (c) coefficient of relatedness for all potential blood donors in the roost, and (d) coefficient of relatedness for blood-sharing pairs who were not related as mother–offspring. The visual impression in these figures is that regurgitators are more likely to be related and more likely to be roostmates than the general population. This is confirmed by a statistical procedure called a stepwise logistic regression. Both relatedness and association affect the probability of blood-sharing. From Wilkinson (1984).

Figure 12.23 **African lions** When female lions hear roars made by unfamiliar females, they move across their territory toward the source. Females that lead the movement frequently glance back at female pride members that are lagging behind.

unrelated to pride females (Packer and Pusey 1982; Packer et al. 1991). Males cooperate in defending themselves against other coalitions of males that attack and try to take over the pride. If a pride takeover occurs, the incoming males often kill the young cubs present (see Chapter 11).

Females cooperate in defending their young against infanticidal attacks, in nursing young, in hunting prey that are difficult to capture (females do the vast majority of hunting in lions), and in defending the pride's territory against incursions by females in neighboring prides (Packer and Pusey 1983, 1997). Battles with intruding females are dangerous, especially if pride females do not cooperate in defense. Solitary lions are often killed in same-sex encounters.

To study cooperation during incursions by strange females who threaten the territory, Robert Heinsohn and Craig Packer (1995) placed speakers near the edge of pride territories and played tape recordings of roars given by unfamiliar females. Pride females respond to these roars by approaching the speakers; they even attack if a stuffed female lion is placed near the speaker. They also do not habituate to this stimulus—they continue responding when the experiment is repeated. As a result, Heinsohn and Packer were able to quantify how females responded to threats over multiple trials. In addition to recording how long it took an individual to reach the midpoint between the pride's original position and the speaker, they also calculated the difference between each individual's time-to-midpoint and the leader's time, documented the order within the group that each female reached the midpoint, and counted the number of glances a female made back at lagging pride members. They collected data on female responses in eight different prides.

Statistical analyses showed significant differences in the strength of responses among females within prides. Some females in each pride always led; others always lagged behind. Other females adopted conditional strategies: They would assist the lead female(s) more frequently when Heinsohn and Packer played tapes of more than one female roaring (that is, when the threat to the pride was greater). Still others would lag behind more when multiple roars were played.

These differences were not correlated with the age or body size of females or with their coefficient of relationship with other pride members.

These results are paradoxical in the context of both inclusive fitness and reciprocity theory. Why do the leaders tolerate the laggards? According to the theory we developed earlier, they should be punished. But leaders were never observed to threaten laggards or to withhold benefits. For example, they did not stop leading until laggards caught up. They glanced back and appeared to recognize that laggards were lagging, but continued approaching the speaker.

The dynamics of group defense in lions have yet to be explained by kin selection or reciprocal altruism.

A lack of dominance and coercion among females is a general feature of lion society. Packer, Anne Pusey, and Lynn Eberly (2001) analyzed an extensive data set on female reproductive success. They found no evidence that the variation in reproductive success among females within prides was due to anything other than chance. This is in contrast to many other cooperatively breeding animals, in which dominant individuals do most of the breeding and subordinates provide assistance.

The simple answer is that we do not know how a mixture of altruist and nonreciprocator strategies can persist among lions. There are a number of possibilities. A leading hypothesis, still to be tested, is that females who contribute little to territory defense reciprocate in other ways—through exceptional hunting prowess or milk production, for example. Another idea is that even cowardly lions are so well armed in tooth and claw that it is just too dangerous to punish cheaters. Finally, having even a laggard for an ally may be better than having no ally at all. If so, tolerating a certain amount of social parasitism may be a superior strategy to chasing the offender away. Clearly, social interactions among lions are complex.

Summary

When individuals interact, four outcomes are possible with respect to fitness: cooperation, altruism, selfishness, and spite. The evolution of altruism was one of the great paradoxes of evolutionary biology until it was resolved by two important advances:

1. William Hamilton showed mathematically that a gene for altruism will spread when $Br - C > 0$ where B is the benefit to the recipient in units of surviving offspring, r is the coefficient of relatedness between actor and recipient, and C is the cost to the actor. When Hamilton's rule holds, kin selection results in altruistic behavior.
2. Robert Trivers developed the theory of reciprocal altruism. Altruism among unrelated individuals can evolve if the benefits of an altruistic act to the recipient are large and the cost to the actor is small, and the benefits are later returned to the actor by the recipient.

Kin selection explains a great variety of phenomena such as alarm calling in ground squirrels and helping behavior in birds. In Belding's ground squirrels, individuals are more willing to risk giving an alarm call if close relatives are nearby. In white-fronted bee-eaters, 1-year-old individuals preferentially help at the nests of closely related individuals (often their parents).

A variety of other behaviors are likewise illuminated by kin selection. It explains why cannibalistic tadpoles spit out their siblings, why wood mouse sperm sacrifice themselves to help their brothers win the big race, and why coots count their eggs.

Kin selection has been important in explaining the evolution of eusociality in Hymenoptera and in naked mole-rats. Phylogenetic analyses show that nest-building and extensive care of young was a precondition for the evolution of reproductive altruism in ants, wasps, and bees. In these groups, individuals are unlikely to reproduce on their own successfully, so reproductive altruism is favored by kin selection. In obligately eusocial groups like ants and bees, asymmetries in relatedness between egg-laying queens and nonreproductive workers result from the haplodiploid system of sex determination and produce sharp conflicts over the sex ratio of offspring. Helping behavior is facultative in *Polistes* wasps, however. In this group, females may or may not help at the nest depending on their body size and their coefficient of relatedness relative to nestmates. In naked mole-rats, eusociality is supported by the queen's physical dominance and by extensive inbreeding that leads to high coefficients of relatedness among individuals.

Parent–offspring conflict and sibling conflict occur when the fitness interests of individuals within families clash. Parents are equally related to each of their offspring, but offspring are more closely related to themselves than to their siblings. Weaning conflict occurs because young mammals demand more resources for themselves than for their siblings; siblicide occurs if perpetrators gain enough direct fitness benefits by removing competition for food to outweigh the indirect fitness cost of killing a sibling. Parents may acquiesce to siblicide if it increases the probability that at least one offspring will survive.

Reciprocal altruism has been successful as an explanation for food-sharing between vampire bats, and may be involved in interactions among females in lion prides. It is often difficult, however, to distinguish when cooperation is due to reciprocal altruism and when it results from kin selection or simple mutualism.

Questions

1. Suppose adult bee-eaters could raise only 0.3 more offspring with a helper than without a helper. Would you still expect male bee-eaters to give in to the harassment of their fathers, or would male bee-eaters tend to fight off their fathers? Explain your reasoning.
2. When a Thomson's gazelle detects a nearby stalking cheetah, the gazelle often begins bouncing up and down with a stiff-legged gait called *stotting* (see Figure 12.24). Stotting was originally assumed to be an altruistic behavior that distracts the cheetah from the gazelle's kin and also alerts the gazelle's kin to the presence of the predator, at considerable risk to the stotting gazelle. However, Caro reports that stotting does not seem to increase the gazelle's risk of being attacked. In fact, once a gazelle begins to stott, the cheetah often gives up the hunt.
   - **a.** If Caro is right, how does C (the cost of stotting) for a gazelle compare to C (the cost of trilling) for a ground squirrel?
   - **b.** Do you think stotting is altruistic, selfish, spiteful, or cooperative (mutualistic)? If you are not sure, what further studies could you do to answer this question?
   - **c.** With this in mind, make a prediction about whether a gazelle will stott when there are no other gazelles around, and then look up Caro's papers to see if you are right.

Caro, T. M. 1986. The function of stotting in Thomson's gazelles: Some tests of the hypotheses. *Animal Behaviour* 34: 663–684.

Caro, T. M. 1994. Ungulate antipredator behaviour: Preliminary and comparative data from African bovids. *Behaviour* 128: 189–228.

Figure 12.24 A Thomson's gazelle stotting

3. The cubs of spotted hyenas often begin fighting within moments of birth, and often one hyena cub dies. The mother hyena does not interfere. How could such a behavior have evolved? For instance:
   a. From the winning sibling's point of view, what must B (benefit of siblicide) be, relative to C (cost of siblicide), to favor the evolution of siblicide?
   b. From the parent's point of view, what must B be, relative to C, for the parent to watch calmly rather than to interfere?
   c. In general, under what conditions would you expect parents to evolve "tolerance of siblicide" (watching calmly while siblings kill each other, and not interfering).

   For more about the unusual social system of spotted hyenas and for new information from studies of wild hyenas, see:

   Frank, L G., S. E. Glickman, and P. Licht. 1991. Fatal sibling aggression, precocial development, and androgens in neonatal spotted hyenas. *Science* 252: 702–4.

   Frank, Laurence G. 1997. Evolution of genital masculinization: Why do female hyaenas have such a large "penis"? *Trends in Ecology and Evolution* 12: 58–62.

   Golla, W., H. Hofer, and M. L. East. 1999. Within-litter sibling aggression in spotted hyaenas: Effect of maternal nursing, sex and age. *Animal Behaviour* 58: 715–726.

4. Blue jays (*Cyanocitta cristata*) seem to be better than American robins (*Turdus migratorius*) at recognizing individuals. In one study, (Schimmel and Wasserman 1994), blue jays raised with American robins could distinguish strange from familiar robins better than the robins themselves. Do you think these species differ in occurrence of kin selection or reciprocal altruism (or both)? Why?

5. When an interviewer asked evolutionary biologist J. B. S. Haldane if he would risk his life to save a drowning man, he reportedly answered, "No, but I would for two brothers or eight cousins." Explain his reasoning.

6. a. Look at Figure 12.18 on parent–offspring conflict. Explain, in general terms, why the behavior of females should evolve so that mothers start weaning when B/C falls below 1. (*Hint:* Consider the reproductive success of mothers who wean very early and of mothers who wean very late.)
   b. If a mother could have only one litter of young in her lifetime, or if an old mother could expect to produce only one more litter of young, how would the period of weaning conflict change?

7. The text claims that eusociality has evolved several times independently within the hymenoptera. What is the evidence for this statement? If it is true, in what sense is eusociality in ants, bees, and wasps an example of convergent evolution (see Chapter 14)?

8. How would you go about testing the hypothesis that female lions who do not participate in territory defense reciprocate by providing milk to the offspring of territory defenders? List the predictions made by the hypothesis and the types of data you would have to collect.

9. House sparrows often produce two successive broods of young. Males feed their first brood only briefly, but feed their second brood for much longer. Why do males feed first broods less than second broods? (*Hint:* Consider how C, the cost of feeding the current brood, changes). How could you test your hypothesis? How is this situation analogous to weaning conflict in mammals?

10. Which is more common in human cultures—eusociality (look back at the three requirements of eusociality; can you think of any human cultures that fit?) or a helper-at-the-nest social system? Which do you think is generally more common in social animals? Why?

11. Human siblings often show intense sibling rivalry that typically declines during the teenage years. Suggest an evolutionary explanation for this pattern.

Exploring the Literature

12. Throughout this chapter, we concentrated on the fitness consequences of social interactions and paid little attention to the issue of why organisms live in groups in the first place. To learn how social living has been favored by factors such as a requirement for group defense against predators, benefits from group foraging, and a need for long-term care of dependent young, here are two classic papers from the 1970s and two recent updates:

    Alexander, R. D. 1974. The evolution of social behavior. *Annual Review of Ecology and Systematics* 5: 325–383.

    Hoogland, J. L., and P. W. Sherman. 1976. Advantages and disadvantages of bank swallow (*Riparia riparia*) coloniality. *Ecological Monographs* 46: 33–58.

    Sachs, J. L., U.G. Mueller, T. P. Wilcox, and J. Bull. 2004. The evolution of cooperation. *Quarterly Review of Biology* 79: 135–160.

    Whitehousel, M. E., and Y. Lubin. 2005. The function of societies and the evolution of group living: Spider societies as a test case. *Biological Revoews of the Cambridge Philosophical Society* 80: 347–361.

13. Because this chapter emphasizes theories that explain why cooperative behavior can evolve when a fitness cost is involved, we spent little time on the evolution

of mutualism. For examples of cooperative behavior that are not caused by kin selection or reciprocal altruism, see:

McDonald, D. B., and W. K. Potts. 1994. Cooperative display and relatedness among males in a lek-breeding bird. *Science* 266: 1030–1032.

Watts, D. P. 1998. Coalitionary mate guarding by male chimpanzees at Ngogo, Kibale National Park, Uganda. *Behavioral Ecology and Sociobiology* 44: 43–55.

Gibson, R. M., D Pires, K. S. Delaney, and R. K. Wayne. 2005. Microsatellite analysis shows that greater sage grouse leks are not kin groups. *Molecular Ecology* 14: 4453–4459.

14. The classic model of a "greenbeard allele" assumes that the same allele controls both a phenotypic trait (the green beard) and the ability to recognize that trait. For a recent model indicating that a greenbeard effect can evolve even with two different genes involved, see:

Jansen, V. A., and M. van Baalen. 2006. Altruism through beard chromodynamics. *Nature* 440: 663–666.

15. A variety of models and experiments (some using college students as experimental subjects) have shown that certain variations on the tit-for-tat strategy are extremely successful in interactions among individuals. To begin reviewing this literature, see:

Wedekind, C., and M. Milinski. 1996. Human cooperation in the simultaneous and the alternating Prisoner's Dilemma: Pavlov versus generous tit-for-tat. *Proceedings of the National Academy of Sciences USA* 93: 2686–2689.

Roberts, G., and T. N. Sherratt. 1998. Development of cooperative relationships through increasing investment. *Nature* 394: 175–179.

Imhof, L. A., D. Fudenberg, and M. A. Nowak. 2005. Evolutionary cycles of cooperation and defection. *Proceedings of the National Academy of Sciences USA* 102: 10797–10800.

16. Researchers seeking to explain cooperative breeding, such as we discussed in white-fronted bee-eaters, have tended to rely on kin selection and its indirect benefits (that is, inclusive fitness) to the exclusion of considering other ideas. For a review of some of these other ideas see:

Konig, B. 1997. Cooperative care of young in mammals. 1997. *Naturwissenschaften* 84: 95–104.

Clutton-Brock, T. 2002. Breeding together: Kin selection and mutualism in cooperative vertebrates. *Science* 296: 69–72.

For Clutton-Brock's own research on meerkats, see:

Clutton-Brock, T. H., M. J. O'Rlain, et al. 1999. Selfish sentinels in cooperative animals. *Science* 284: 1640–1644.

Clutton-Brock, T. H., A. F. Russell, et al. 2001. Effects of helpers on juvenile development and survival in meerkats. *Science* 293: 2446–2449.

17. The evolutionary origin of eusociality has turned out to be more complex than originally thought. See these papers for some recent analyses of the roles of group selection, male parentage, and other factors:

Wilson, E. O., and B. Hölldobl. 2005. Eusociality: Origin and consequences. *Proceedings of the Natlional Academy of Sciences USA* 102: 13367–13371.

R. L. Hammond and L. Keller. 2004. Conflict over male parentage in social insects. *PLoS Biology* 2: e248.

18. Kin selection should be especially effective in insects like aphids, which reproduce asexually. For evidence on how individuals in one aphid species adjust their behavior depending on whether they are living among clone-mates or strangers, see:

Abbot, P., J. H. Withgott, and N. A. Moran. 2001. Genetic conflict and conditional altruism in social aphid colonies. *Proceedings of the National Academy of Sciences USA* 98: 12068–12071.

19. Humans often exhibit high levels of nonreciprocal altruism—altruistic behavior directed toward nonrelated individuals who do not reciprocate. However, the altruist may gain help from other individuals. This has been described as "I help you, and somebody else will help me later." Humans also appear to have neural mechanisms that are extremely good at detecting "cheaters" and also have a tendency to punish cheaters strongly, even at a high cost for the punisher. Recent models show that alleles promoting nonreciprocal altruism can evolve in small societies and may also promote the evolution of neural mechanisms involved in cheater detection, punishment, and other features of complex human behavior. For recent developments in this interesting field, see:

Sanchez, A., and J. A. Cuesta. 2005. Altruism may arise from individual selection. *Journal of Theoretical Biology* 235: 233–240.

Pfeiffer, T., C. Rutte, T. Killingback, M. Taborsky, and S. Bonhoeffer. 2005. Evolution of cooperation by generalized reciprocity. *Procceedings of the Royal Society of London B* 272: 1115–1120.

Nowak, M. A., and K. Sigmund. 2005. Evolution of indirect reciprocity. *Nature* 437 (7063): 1291–1298.

Fehr, E., and B. Rockenbach. 2004. Human altruism: Economic, neural, and evolutionary perspectives. *Current Opinion in Neurobiology* 14: 784–790.

Citations

Alexander, R. D., K. M. Noonan, and B. J. Crespi. 1991. The evolution of eusociality. In P. W. Sherman, J. U. M. Jarvis, and R. D. Alexander, eds. *The Biology of the Naked Mole Rat*. Princeton, NJ: Princeton University Press, 3–44.

Alexander, R. D., and P. W. Sherman. 1977. Local mate competition and parental investment in social insects. *Science* 196: 494–500.

Anderson, D. J., and R. E. Ricklefs. 1995. Evidence of kin-selected tolerance by nestlings in a siblicidal bird. *Behavioral Ecology and Sociobiology* 37: 163–168.

Axelrod, R., and W. D. Hamilton. 1981. The evolution of cooperation. *Science* 211:1390–1396.

Bartz, S. H. 1979. Evolution of eusociality in termites. *Proceedings of the National Academy of Sciences USA* 76:5764–5768.

Betzig, L. L., and P. W. Turke. 1986. Parental investment by sex on Ifaluk. *Ethology and Sociobiology* 7:29–37.

Bourke, A. F. G., and N. R. Franks. 1995. *Social Evolution in Ants.* Princeton, NJ: Princeton University Press.

Brown, J. L. 1987. *Helping and Communal Breeding in Birds.* Princeton, NJ: Princeton University Press.

Brown, J. L., and A. Eklund. 1994. Kin recognition and the major histocompatibility complex: An integrative review. *American Naturalist* 143: 435–461.

Chapuisat, M., and L. Keller. 1999. Testing kin selection with sex allocation data in eusocial Hymenoptera. *Heredity* 82: 473–478.

Charnov, E. L. 1982. *The Theory of Sex Allocation.* Princeton, NJ: Princeton University Press.

Clarke, F. M., and C. G. Faulkes. 2001. Intracolony aggression in the eusocial naked mole-rat, *Hetercephalus glaber. Animal Behaviour* 61: 311–324.

Clutton-Brock, T. H., S. D. Albon, and F. E. Guinness. 1984. Maternal dominance, breeding success and birth sex ratios in red deer. *Nature* 308: 358–360.

Crozier, R. H., and P. Pamilo. 1996. *Evolution of Social Insect Colonies.* Oxford: Oxford University Press.

Darwin, C. 1859. *On the Origin of Species.* London: John Murray.

Dawkins, R. 1976. *The Selfish Gene.* Oxford: Oxford University Press.

Duffy, J. E. 1996. Eusociality in a coral-reef shrimp. *Nature* 381: 512–514.

Duffy, J. E., C. L. Morrison, and K. S. Macdonald. 2002. Colony defense and behavioral differentiation in the eusocial shrimp *Synalpheus regalis. Behavioral Ecology and Sociobiology* 51: 488–495.

Duffy, J. E., C. L. Morrison, and R. Ríos. 2000. Multiple origins of eusociality among sponge-dwelling shrimps (*Synalpheus*). *Evolution* 54: 503–516.

Dugatkin, A. L. 1998. *Cooperation among Animals.* Oxford: Oxford University Press.

Emlen, S. T., and P. H. Wrege. 1988. The role of kinship in helping decisions among white-fronted bee-eaters. *Behavioral Ecology and Sociobiology* 23: 305–315.

Emlen, S. T., and P. H. Wrege. 1991. Breeding biology of white-fronted bee-eaters at Nakuru: The influence of helpers on breeder fitness. *Journal of Animal Ecology* 60: 309–326.

Emlen, S. T., and P. H. Wrege. 1992. Parent–offspring conflict and the recruitment of helpers among bee-eaters. *Nature* 356:331–333.

Emlen, S. T., P. H. Wrege, and N. J. Demong. 1995. Making decisions in the family: An evolutionary perspective. *American Scientist* 83: 148–157.

Fisher, R. A. 1930. *The Genetical Theory of Natural Selection.* Oxford: Clarendon Press.

Foster, K. R., F. L. W. Ratnieks, and T. Wenseleers. 2000. Spite in social insects. *Trends in Ecology and Evolution* 15: 469–470.

Foster, K. R., T. Wenseleers, et al. 2001. Spite: Hamilton's unproven theory. *Annales Zoologici Fennici* 38: 229–238.

Gardner, A., and S. A. West. 2004. Spite among siblings. *Science* 305: 1413–1414.

Gilchrist, J. S. 2004. Pup escorting in the communal breeding banded mongoose: behavior, benefits, and maintenance. *Behavioral Ecology* 15: 952–960.

Gilcrhist, J. S., E. Otali, and F. Mwanguhya. 2004. Why breed communally? Factors affecting fecundity in a communal breeding mammal: the banded mongoose (*Mungos mungo*). *Behavioral Ecology and Sociobiology* 57: 119–131.

Godfrey, H. C. J. 1995. Evolutionary theory of parent–offspring conflict. *Nature* 376: 133–138.

Hamilton, W. D. 1964a. The genetical evolution of social behaviour. I. *Journal of Theoretical Biology* 7: 1–16.

Hamilton, W. D. 1964b. The genetical evolution of social behaviour. II. *Journal of Theoretical Biology* 7: 17–52.

Hamilton, W. D. 1967. Extraordinary sex ratios. *Science* 156: 477–488.

Hamilton, W. D. 1972. Altruism and related phenomena, mainly in the social insects. *Annual Review of Ecology and Systematics* 3: 193–232.

Hammerstein, P., and S. E. Riechert. 1988. Payoffs and strategies in territorial contests: ESS analyses of two ecotypes of the spider *Agelenopsis aperta. Evolutionary Ecology* 2: 115–138.

Hegner, R. E., S. T. Emlen, and N. J. Demong. 1982. Spatial organization of the white-fronted bee-eater. *Nature* 298: 264–266.

Heinsohn, R., and C. Packer. 1995. Complex cooperative strategies in group-territorial African lions. *Science* 269: 1260–1262.

Hodge, S. J. 2005. Helpers benefit offspring in both the short and long-term in the cooperatively breeding banded mongoose. *Proceedings of the Royal Society of London B* 272: 2479–2484.

Honeycutt, R. L. 1992. Naked mole-rats. *American Scientist* 80: 43–53.

Hoogland, J. L. 1983. Nepotism and alarm calling in the black-tailed prairie dog (*Cynomys ludovicianus*). *Animal Behaviour* 31: 472-479.

Hoogland, J. L. 1994. Nepotism and infanticide among prairie dogs. Pages 321-337 in S. Parmigiani and F. S. vom Saal, eds. *Infanticide and Parental Care.* Chur, Switzerland: Harwood Academic Publishers.

Hoogland, J. L. 1995. *The Black-Tailed Prairie Dog: Social Life of a Burrowing Mammal.* Chicago: University of Chicago Press.

Hunt, J. H. 1999. Trait mapping and salience in the evolution of eusocial vespid wasps. *Evolution* 53: 225–237.

Hurst, L. D. 1991. The evolution of cytoplasmic incompatibility or when spite can be successful. *Journal of Theoretical Biology* 148: 269–277.

Johnstone, R., A., and R. Bshary. 2004. Evolution of spite through indirect reciprocity. *Proceedings of the Royal Society of London B* 271: 1917–1922.

Keller, L., M. Milinski, M. Frischknecht, N. Perrin, H. Richner, and F. Tripet. 1994. Spiteful animals still to be discovered. *Trends in Ecology and Evolution* 9: 103.

Lougheed, L. W., and D. J. Anderson. 1999. Parent blue-footed boobies suppress siblicidal behavior of offspring. *Behavioral Ecology and Sociobiology* 45: 11–18.

Lyon, B. E. 2003. Egg recognition and counting reduce costs of avian conspecific brood parasitism. *Nature* 422: 495–499.

Manning, C. J., E. K. Wakeland, and W. K. Potts. 1992. Communal nesting patterns in mice implicate MHC genes in kin recognition. *Nature* 360: 581–583.

Maynard Smith, J. 1974. The theory of games and the evolution of animal conflicts. *Journal of Theoretical Biology* 47: 209–221.

Maynard Smith, J. 1982. *Evolution and the Theory of Games.* Cambridge: Cambridge University Press.

Michener, C. D. 1969. Comparative social behavior of bees. *Annual Review of Entomology* 14: 299–342.

Milinski, M. 1996. By-product mutualism, tit-for-tat reciprocity and cooperative predator inspection: A reply to Connor. *Animal Behaviour* 51: 458–461.

Mock, D. W., and G. A. Parker. 1997. *The Evolution of Sibling Rivalry.* Oxford: Oxford University Press.

Moore, H., K. Dvořáková, N. Jenkins, and W. Breed. 2002. Exceptional sperm cooperation in the wood mouse. *Nature* 418: 174–177.

Nonacs, P., and H. K. Reeve. 1995. The ecology of cooperation in wasps: Causes and consequences of alternative reproductive decisions. *Ecology* 76: 953–967.

Oldroyd, B. P., M. J. Clifton, K. Parker, S. Wongsiri, T. E. Rinderer, and R. H. Crozier. 1998. Evolution of mating behavior in the genus *Apis* and an estimate of mating frequency in *Apis cerana* (Hymenoptera: Apidae). *Annals of the Entomological Society of America* 91: 700–709.

Oldroyd, B. P., M. J. Clifton, S. Wongsiri, T. E. Rinderer, H. A. Sylvester, and R. H. Crozier. 1997. Polyandry in the genus *Apis*, particulary *Apis andreniformis. Behavioral Ecology and Sociobiology* 40: 17–26.

Packer, C. 1977. Reciprocal altruism in *Papio anubis. Nature* 265: 441–443.

Packer, C., D. A. Gilbert, A. E. Pusey, and S. J. O'Brien. 1991. A molecular genetic analysis of kinship and cooperation in African lions. *Nature* 351: 562–565.

Packer, C., and A. E. Pusey. 1982. Cooperation and competition within coalitions of male lions: Kin selection or game theory? *Nature* 296: 740–742.

Packer, C., and A. E. Pusey. 1983. Adaptations of female lions to infanticide by incoming males. *American Naturalist* 121: 716–728.

Packer, C., and A. E. Pusey. 1997. Divided we fall: Cooperation among lions. *Scientific American* 276 (May): 52–59.

Packer, C., A. E. Pusey, and L. E. Eberly. 2001. Egalitarianism in female African lions. *Science* 293: 690–693.

Page, R. E., Jr., and R. A. Metcalf. 1982. Multiple mating, sperm utilization, and social evolution. *American Naturalist* 119: 263–281.

Pamilo, P. 1984. Genetic relatedness and evolution of insect sociality. *Behavioral Ecology and Sociobiology* 15: 241–248.

Peters, J. M., D. C. Queller, V. L. Imperatriz-Fonseca, D. W. Roubik, and J. E. Strassmann. 1999. Mate number, kin selection and social conflicts in stingless bees and honeybees. *Proceedings of the Royal Society of London B* 266: 379–384.

Pfennig, D. W. 1999. Cannibalistic tadpoles that pose the greatest threat to kin are most likely to discriminate kin. *Proceedings of the Royal Society of London B* 266: 57–61.

Pfennig, D. W., J. P. Collins, and R. E. Ziemba. 1999. A test of alternative hypotheses for kin recognition in cannibalistic tiger salamanders. *Behavioral Ecology* 10: 436–443.

Pfennig, D. W., and P. W. Sherman. 1995. Kin recognition. *Scientific American* 272: 98–103.

Queller, D. C. 1989. The evolution of eusociality: Reproductive head starts of workers. *Proceedings of the National Academy of Sciences USA* 86: 3224–3226.

Queller, D. C., and K. F. Goodnight. 1989. Estimating relatedness using genetic markers. *Evolution* 43: 258–275.

Quellar, D. C., E. Ponte, S. Bozzaro, and J. E. Strassmann. 2003. Single-gene greenbeard effects in the social amoeba *Dictyostelium discoideum*. *Science* 299: 105–106.

Queller, D. C., and J. E. Strassmann. 1998. Kin selection and social insects. *BioScience* 48: 165–175.

Reeve, H. K. 1992. Queen activation of lazy workers in colonies of the eusocial naked mole-rat. *Nature* 358: 147–149.

Reeve, H. K., J. M. Peters, P. Nonacs, and P. T. Starks. 1998. Dispersal of first "workers" in social wasps: Causes and implications of an alternative reproductive strategy. *Proceedings of the National Academy of Sciences USA* 95: 13737–13742.

Reeve, H. K., and P. W. Sherman. 1991. Intracolonial aggression and nepotism by the breeding female naked mole-rat. In P. W. Sherman, J. U. M. Jarvis, and R. D. Alexander, eds. *The Biology of the Naked Mole Rat*. Princeton, NJ: Princeton University Press, 337–357.

Reeve, H. K., D. F. Westneat, W. A. Noon, P. W. Sherman, and C. F. Aquadro. 1990. DNA "fingerprinting" reveals high levels of inbreeding in colonies of the eusocial naked mole-rat. *Proceedings of the National Academy of Sciences USA* 87: 2496–2500.

Roisin, Y. 1994. Intragroup conflicts and the evolution of sterile castes in termites. *American Naturalist* 143: 751–765.

Schimmel, K. L, and F. E. Wasserman. 1994. Individual and species preference in two passerine birds: Auditory and visual cues. *Auk* 111: 634–642.

Sherman, P. W. 1985. Alarm calls of Belding's ground squirrels to aerial predators: Nepotism or self-preservation? *Behavioral Ecology and Sociobiology* 17: 313–323.

Sherman, P. W., J. U. M. Jarvis, and S. H. Braude. 1992. Naked mole rats. *Scientific American* 267: 72–78.

Short, R. V. 2002. Do the locomotion. *Nature* 418: 137.

Sigurjónsdóttir, H., and G. A. Parker. 1981. Dung fly struggles: Evidence for assessment strategy. *Behavioral Ecology and Sociobiology* 8: 219–230.

Stacey, P. B., and W. D. Koenig, eds. 1990. *Cooperative Breeding in Birds: Long-term Studies of Ecology and Behaviour*. Cambridge: Cambridge University Press.

Stone, V. E., L. Cosmides, J. Tooby, N. Kroll, and R. T. Knight. 2002. Selective impairment of reasoning about social exchange in a patient with bilateral limbic system damage. *Proceedings of the National Academy of Sciences USA* 99: 11531–11536.

Sugiyama, L. S., J. Tooby, and L. Cosmides. 2002. Cross-cultural evidence of cognitive adaptations for social exchange among the Shiwiar of Ecuadorian Amazonia. *Proceedings of the National Academy of Sciences USA* 99: 11537–11542.

Sundström, L., and J. J. Boomsma. 2001. Conflicts and alliances in insect families. *Heredity* 86: 515–521.

Sundström, L., M. Chapuisat, and L. Keller. 1996. Conditional manipulation of sex ratios by ant workers: A test of kin selection theory. *Science* 274: 993–995.

Thorne, B. L. 1997. Evolution of eusociality in termites. *Annual Review of Ecology and Systematics* 28: 27–54.

Trivers, R. L. 1971. The evolution of reciprocal altruism. *Quarterly Review of Biology* 46: 35–57.

Trivers, R. L. 1974. Parent–offspring conflict. *American Zoologist* 14: 249–264.

Trivers, R. L. 1985. *Social Evolution*. Menlo Park, CA: Benjamin Cummings.

Trivers, R. L., and H. Hare. 1976. Haplodiploidy and the evolution of the social insects. *Science* 191: 249–263.

Trivers, R. L., and D. E. Willard. 1973. Natural selection of parental ability to vary the sex ratio of offspring. *Science* 179: 90–92.

Werren, J. H. 1984. A model for sex ratio selection in parasitic wasps: Local mate competition and host quality effects. *Netherlands Journal of Zoology* 34: 81–96.

Wilkinson, G. S. 1984. Reciprocal food sharing in the vampire bat. *Nature* 308: 181–184.

Wilson, E. O. 1971. *The Insect Societies*. Cambridge, MA: Harvard University Press.

13

Aging and Other Life History Characters

After female sockeye salmon (*Onchorhynchus nerka*) lay their eggs, they defend their nests against disruption by later-spawning salmon for as long as they have sufficient energy to stay alive. Andrew Hendry and colleagues (2004) found that early spawners reserve more energy for nest defense, and live longer, than late spawners.

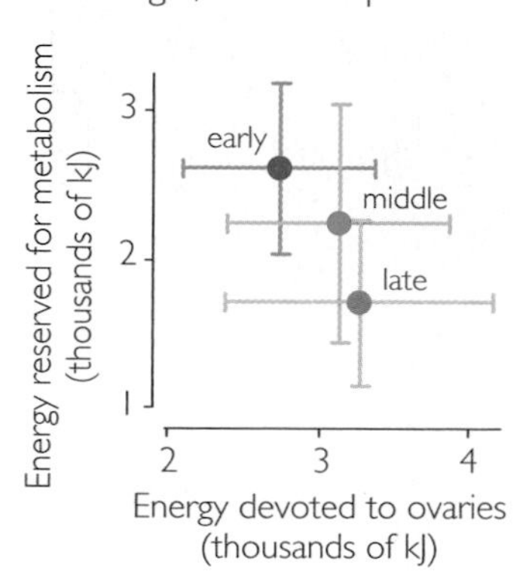

Evolution by natural selection has engineered all organisms to perform the same single ultimate task: to reproduce. How organisms go about the business of reproducing, however, is enormously variable. A few examples illustrate the diversity:

- Some mammals mature early and reproduce quickly, whereas others mature late and reproduce slowly. For example, female deer mice (*Peromyscus maniculatus*) mature at about seven weeks and have three or four litters of pups each year, whereas female black bears (*Ursus americanus*) mature at four or five years and produce cubs only once every two years (Nowak 1991).
- Plants have a wide range of reproductive life spans. Some, like the California poppy (*Eschscholtzia californica*), live and flower for just a single season. Others, like the black cherry (*Prunus serotina*), flower yearly for decades.
- Some bivalves produce enormous numbers of tiny eggs, whereas others produce small numbers of large eggs (Strathmann 1987). The oyster *Crassostrea gigas*, for example, releases 10 to 50 million eggs in a single spawn, each 50–55 micrometers in diameter. The clam *Lasaea subviridis*, in contrast, broods fewer than 100 eggs at a time, each some 300 micrometers in diameter.

The branch of evolutionary biology that attempts to make sense of the diversity in reproductive strategies is called life history analysis.

An organism truly perfected for reproduction would mature at birth, continuously produce high-quality offspring in large numbers, and live forever. Richard Law (1979) called such an organism a Darwinian demon: It would bedevil all other organisms and eventually monopolize life on the planet. No such organism exists, even after 3.5 billion years of evolution by natural selection. The reason is that such an organism is impossible. Some actual organisms come close to realizing one or another of the traits of an ideal reproducer, but all such organisms fall strikingly short by one or more of the remaining measures.

For example, the female thrips egg mite (*Adactylidium* sp.) is mature at birth. Furthermore, she is already inseminated, having hatched inside her mother's body and mated with her brother (Elbadry and Tawfik 1966; see also Gould 1980, essay 6). But she produces just one clutch of offspring and her life is brief: She dies at the age of just four days, when her own offspring eat her alive from the inside (Figure 13.1a).

Another example, the brown kiwi (*Apteryx australis mantelli*), produces high-quality offspring (Taborsky and Taborsky 1993). Female kiwis weigh about six pounds and lay eggs that weigh one pound (Figure 13.1b). The chicks that hatch from these huge eggs become largely self-reliant within a week. However, kiwi parents cannot produce these chicks continuously, and they cannot produce them in large numbers. It takes the female over a month to make each of the eggs in a typical two-egg clutch. The male has to incubate the eggs for about three months, during which time he loses some 20% of his body weight.

Organisms face fundamental trade-offs in their use of energy and time.

As the egg mite and the kiwi suggest, the laws of physics and biology impose fundamental trade-offs. The amount of energy an organism can harvest is finite, and biological processes take time. Energy and time devoted to one activity are energy and time that cannot be devoted to another. For example, an individual can allocate energy to growth for a long time, which may enable it to reach a larger size and ultimately enable it to produce more offspring. This benefit of

(a)

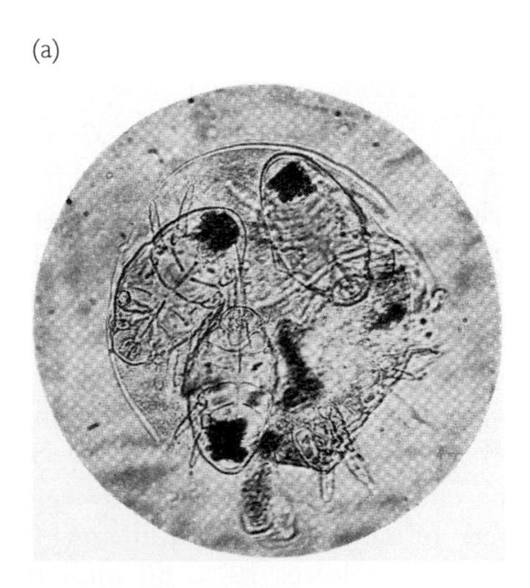

(b)

Figure 13.1 Extreme reproductive strategies (a) Having devoured their mother from the inside, three thrips egg mites (*Adactylidium* sp.) prepare to depart her empty cuticle. The mother's legs are visible at lower right (180×). Reproduced by permission from Elbadry and Tawfik (1966). (b) An X ray of a female brown kiwi (*Apteryx australis mantelli*) ready to lay an egg.

large size, however, is balanced by a cost. The time required to grow to a large size is time during which predators, diseases, or accidents may strike. An individual that takes the time to grow to a large size thus incurs a greater risk of dying without ever having reproduced at all. We introduced the concept of trade-offs in Chapter 10, where we discussed how trade-offs constrain the evolution of adaptations. Whenever there is a trade-off between different components of fitness, we expect natural selection to favor individuals that allocate energy and time with an optimal balance between benefits and costs, thereby maximizing lifetime reproductive success. Different balances are optimal in different environments. Environmental variation is undoubtedly the source of much of the life history variation seen among living organisms.

In exploring the evolution of life histories, we analyze costs and benefits, and fitness trade-offs, as they apply to the following questions:

- Why do organisms age and die?
- How many offspring should an individual produce in a given year?
- How big should each offspring be?

These questions focus on the balance among aspects of fitness for individual organisms. One emerging new area of life history studies is the analysis of conflicts of interest between organisms of the same species and the evolution of strategies for managing these conflicts. We briefly discuss two of these conflicts between the interests of male and female parents. In the final section, we place life history analysis in a broader evolutionary context by considering the maintenance of genetic variation and evolutionary transitions in life history.

13.1 Basic Issues in Life History Analysis

An example of a life history, one that we will return to near the end of Section 13.2, appears in Figure 13.2. The figure follows the career of a hypothetical female Virginia opossum (*Didelphis virginiana*). As a baby, this female nursed for a little more than 3 months, was then weaned, and became independent. She continued to grow for another several months, reaching sexual maturity at an age of about 10 months. Shortly thereafter the female had her first litter consisting of eight offspring. A few months later, she had a second litter, this time with seven offspring. At the age of 20 months, the female was killed by a predator.

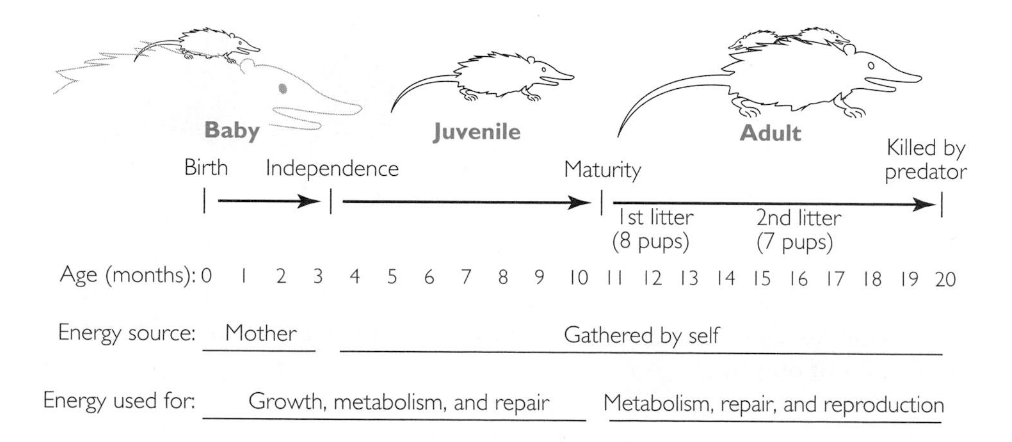

Figure 13.2 The life history of a hypothetical female Virginia opossum (*Didelphis virginiana*) This hypothetical female has a life history typical of female Virginia opossums living in the mainland United States (Austad 1988; 1993). Figure designed after Charnov and Berrigan (1993).

The figure also indicates where the female opossum got her energy at different stages of her life, and the functions to which she allocated that finite energy supply. Before she became sexually mature, the female used her energy for growth, metabolic functions like thermoregulation, and the repair of damaged tissues. After she became sexually mature, the female stopped growing, thereafter using her energy for metabolism, repair, and reproduction.

Changes in life history are caused by changes in the allocation of energy.

Fundamentally, differences among life histories concern differences in the allocation of energy. For example, a different female opossum than the one shown in Figure 13.2 might stop allocating energy to growth at an earlier age, thereby reaching sexual maturity more quickly. This strategy involves a trade-off: The female also matures at a smaller size, which means that she will produce smaller litters. Still another female might, after reaching sexual maturity, allocate less energy to reproduction and more to repair, thereby keeping her tissues in better condition. Again there is a trade-off: Allocating less energy to reproduction means having smaller litters.

Anthony Zera and Zhangwu Zhao have investigated a trade-off in the allocation of energy by female sand crickets (Zera and Zhao 2003; Zhao and Zera 2002). Natural populations of this insect contain two kinds of females: long-winged and short-winged (Figure 13.3). When long-winged females emerge as adults they have well-developed flight muscles and a considerable store of triglycerides, the energy source used by the flight muscles as fuel. Some of these long-winged females can fly. The ability to fly is presumably adaptive under some circumstances. If a female finds herself in a poor environment, she can seek a better one elsewhere. The ability to fly comes at a cost, however. The ovaries of long-winged females grow much more slowly than those of short-winged females. Short-winged females have poorly developed flight muscles and lower stores of triglycerides. They use the energy they save to make phospholipids, which they put into their eggs. Female sand crickets thus face a trade-off between dispersal ability versus early reproduction.

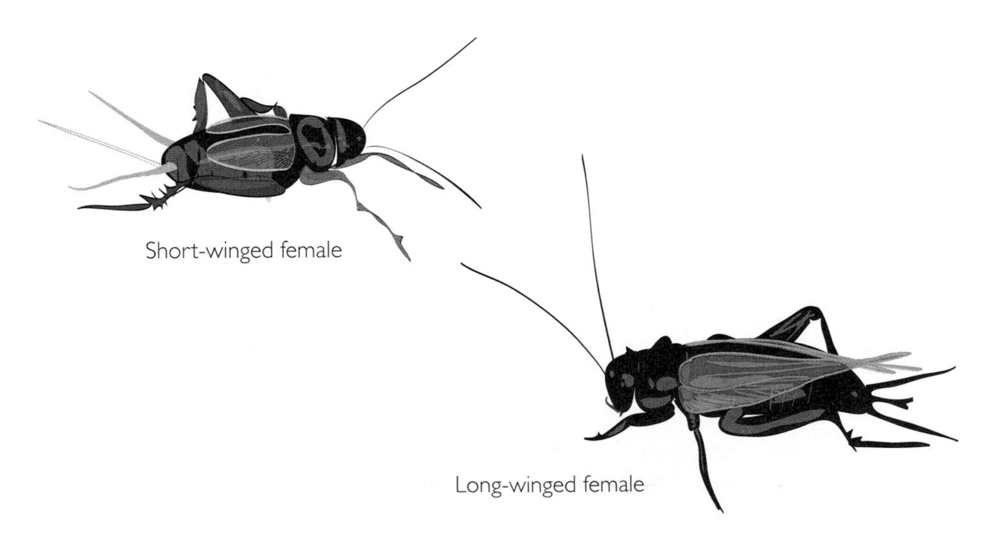

Figure 13.3 Short-winged and long-winged morphs of female sand crickets (*Gryllus firmus*)

Many trade-offs in energy allocation are inescapable. Natural selection acts on life histories to adjust energy allocation in a way that maximizes the total lifetime production of offspring.

13.2 Why Do Organisms Age and Die?

Aging, or **senescence**, is a late-life decline in an individual's fertility and probability of survival (Partridge and Barton 1993). Figure 13.4 documents aging in three animal species: a bird, a mammal, and an insect. All three show declines in both fertility and survival. All else being equal, aging reduces an individual's fitness. Aging should therefore be opposed by natural selection.

Aging should be opposed by natural selection.

We consider two theories on why aging persists. The first, called the rate-of-living theory, invokes an evolutionary constraint (see Chapter 10); it posits that populations lack the genetic variation to respond any further to selection against aging. The second, called the evolutionary theory, invokes, in part, a trade-off between the allocation of energy to reproduction versus repair.

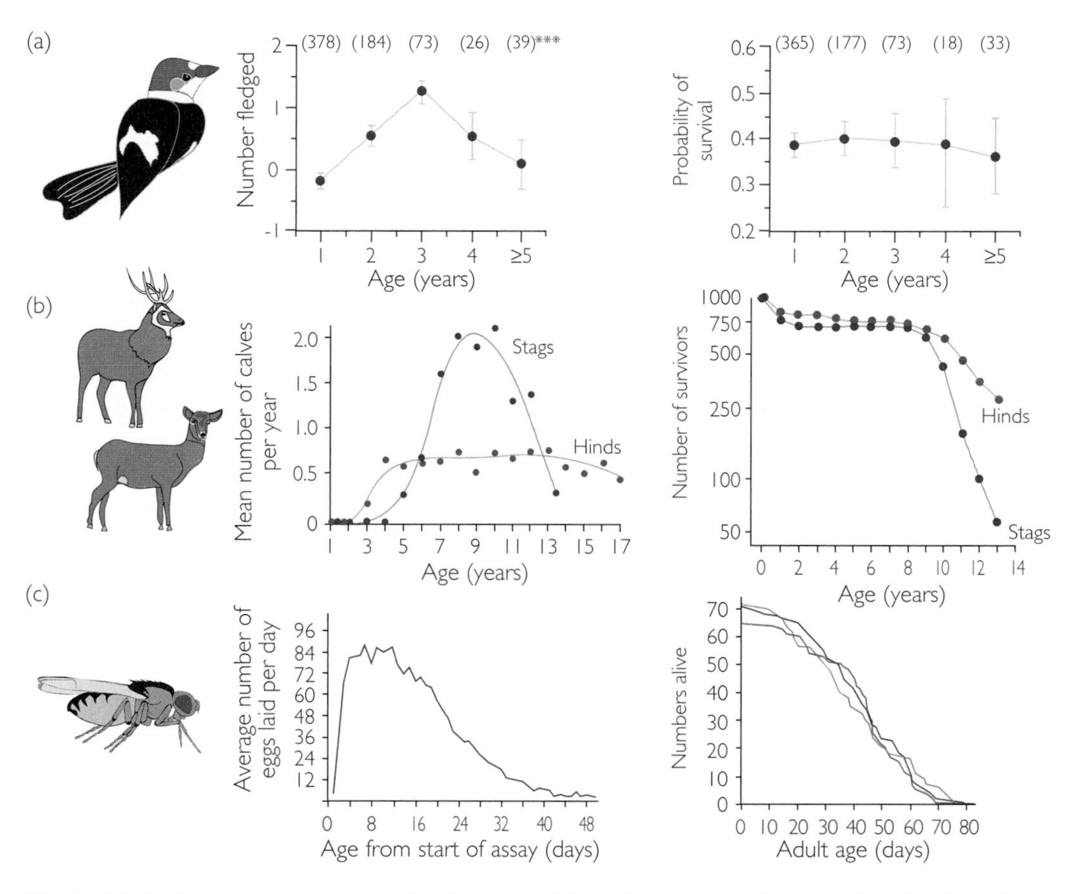

Figure 13.4 **Aging in three animals** (a) Aging in a bird. For females in a wild population of collared flycatchers (*Ficedula albicollis*), the number of young fledged each year declines after age 3 (age-related differences in number fledged are significant at $P < 0.001$). The probability of survival from one year to the next declines slightly, but not significantly, after age 2. Sample sizes in parentheses. From Gustafsson and Pärt (1990). (b) Aging in a mammal. For males in a wild population of red deer (*Cervus elaphus*), the number of calves fathered each year declines sharply after a peak at about age nine; for females, the number of calves produced each year declines gradually starting at about age 13. For both sexes, the probability of surviving from one year to the next is nearly 100% from age 2 to 9. After age 9, the probability of surviving plummets. From Clutton-Brock et al. (1988). (c) Aging in an insect. For females in a laboratory population of fruit flies (*Drosophila melanogaster*), the average number of eggs laid per day declines after the age of about 12 days. In three laboratory populations, the probability of surviving from one day to the next falls at the age of about 20 days. Modified from Rose (1984).

The Rate-of-Living Theory of Aging

The rate-of-living theory of senescence holds that aging is caused by the accumulation of irreparable damage to cells and tissues (reviewed in Austad and Fischer 1991). Damage to cells and tissues is caused by errors during replication, transcription, and translation, and by the accumulation of poisonous metabolic by-products. Under the rate-of-living theory, all organisms have been selected to resist and repair cell and tissue damage to the maximum extent physiologically possible. They have reached the limit of biologically possible repair. In other words, populations lack the genetic variation that would enable them to evolve more effective repair mechanisms than they already have.

One theory holds that aging is a function of metabolic rate . . .

The rate-of-living theory makes two predictions: (1) Because cell and tissue damage is caused in part by the by-products of metabolism, the aging rate should be correlated with the metabolic rate; and (2) because organisms have been selected to resist and repair damage to the maximum extent possible, species should not be able to evolve longer life spans, whether subjected to natural or artificial selection.

Steven Austad and Kathleen Fischer (1991) tested the first prediction, that aging rate will be correlated with metabolic rate, with a comparative study of mammals. Using data from the literature, Austad and Fischer calculated the amount of energy expended per gram of tissue per lifetime for 164 mammal species in 14 orders. According to the rate-of-living theory, all species should expend about the same amount of energy per gram per lifetime, whether they burn it slowly over a long life or rapidly over a short one. In fact, there is wide variation among mammals (Figure 13.5). Across all 164 species Austad and Fischer surveyed, energy expenditure ranges from 39 kcal/g per lifetime in an elephant shrew to 1,102 kcal/g per lifetime in a bat. Even within orders, energy expenditure varies greatly. The range for bat species runs from 325 to 1,102 kcal/g per lifetime. As a group, bats have metabolic rates that are similar to those of other mammals of the same size, but life spans that average nearly three times longer. Energy expenditure for marsupial species ranges from 43 to 406 kcal/g per lifetime. As a group, marsupials have metabolic rates that are significantly lower than those of other mammals of the same size but life spans that are significantly shorter. These patterns are not consistent with the rate-of-living theory of aging.

. . . but data on variation in metabolic rate and aging among mammals refute this theory.

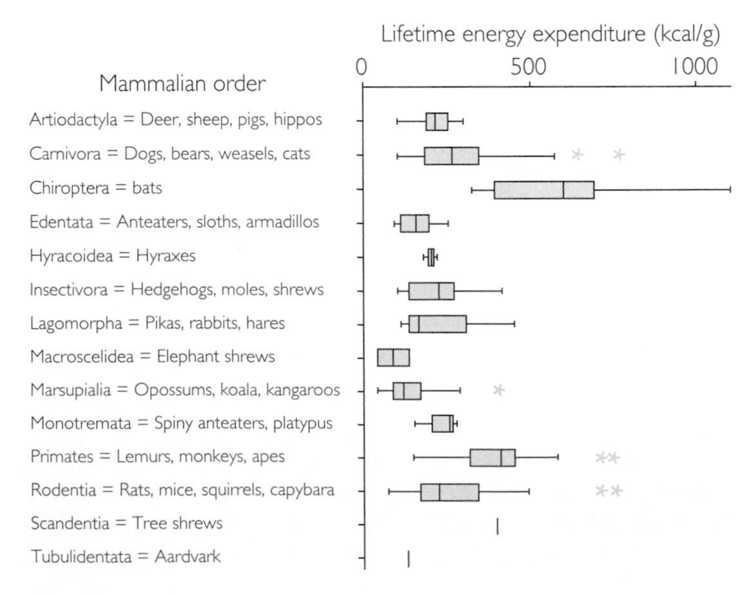

Figure 13.5 Variation among mammals in lifetime energy expenditure This box plot represents the range of lifetime energy expenditure within each of 14 orders of mammals. The vertical line dividing each box represents the median value for that order. The right and left ends of each box represent the 75th and 25th percentiles. The horizontal lines extending to the right and left of each box represent the range of values; the asterisks represent statistical outliers. From Austad and Fischer (1991).

Leo Luckinbill and colleagues (1984) tested the second prediction, that species cannot evolve longer life spans, by artificially selecting for longevity in laboratory populations of fruit flies (*Drosophila melanogaster*). Luckinbill et al. collected wild flies and used them to establish four laboratory populations. In two populations, the researchers selected for early reproduction by collecting eggs from young adults (two to six days after eclosion) and using the individuals that hatched from these eggs as the next generation's breeders. Longevity in these populations did not change significantly during 13 generations of selection (Figure 13.6). In the other two populations, Luckinbill and colleagues selected for late reproduction by collecting eggs from old adults. "Old" meant 22 days after eclosion at the beginning of the experiment and 58 days after eclosion by the end. Longevity in these populations increased dramatically during 13 generations of selection (Figure 13.6). At the beginning of the experiment, the average life span of the flies in these populations was about 35 days; by the end, the average life span was about 60 days. Other researchers conducting similar experiments have confirmed that average life span increases in *Drosophila* populations in response to selection for late-life reproduction (Rose 1984; Partridge 1987; Partridge and Fowler 1992; Roper et al. 1993). These results are consistent with the rate-of-living theory of aging only if the long-lived populations have evolved lower metabolic rates. Phillip Service (1987) found that fruit flies selected for long life span indeed had lower metabolic rates than controls, but only in the first 15 days of life. It is not clear that an evolved difference in metabolic rate can explain an evolved difference in life span as large as that obtained by Luckinbill and colleagues.

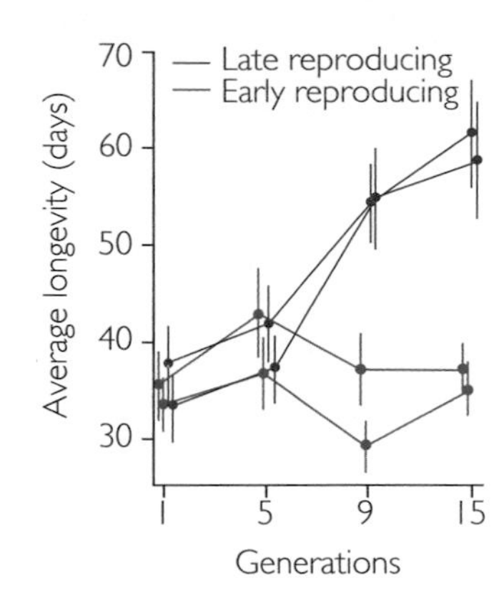

Figure 13.6 Artificial selection increases life span in fruit flies The graph shows average life span in each of four laboratory populations of *Drosophila melanogaster* over 13 generations of selection. The vertical lines show the 95% confidence intervals for the estimated population averages. From Luckinbill et al. (1984).

These experiments and observations seem to largely contradict the predictions of the rate-of-living theory. However, the general idea that organisms age and die as a result of intrinsic physiological limits on cells and tissues has persisted. Its tenacity is due, in part, to the discovery of cellular and genetic mechanisms that have appeared to link the senescence of cells to the senescence of organisms.

One such mechanism in animals is based on the cumulative effects, not of energy expenditure, but of cell division. Normal animal cells are capable of a limited number of divisions (and duplications of their chromosomes), after which the cells cease dividing and eventually die. This pattern occurs in all cells except germ line cells, cancer cells, and some embryonic and blood stem cells, and may be caused by damage to chromosomes (Campisi 1996; Reddel 1998). Each end—or telomere—of a eukaryotic chromosome consists of many copies of a repetitive DNA sequence (in humans the repeat is TTAGGG) that is tagged onto the end of the chromosome by a DNA polymerase enzyme called telomerase. Telomerase is strongly expressed in cancers and germ line cells but not in most other cells. A portion of the telomere is lost with each cycle of DNA replication and cell division (Harley et al. 1990). The progressive loss of part of the telomere with each cell division is associated with senescence and death of the cell (Harley et al. 1990; Reddel 1998). The mechanism causing this association is complex, and is still being elucidated (Karlsleder et al. 2002), but it involves the protein p53 (see Patil et al. 2005; Rodier et al. 2005). p53 is a transcription regulator and tumor suppressor that appears, among other functions, to mediate a trade-off between cancer risk and the rate of aging (Box 13.1).

The observation that telomere shortening is associated with the senescence of cells suggests a simple answer to the question, Why do organisms age and die? Perhaps they die, in part, because their telomeres are lost and their

Box 13.1 A trade-off between cancer risk and aging

The protein p53 is known as a tumor suppressor because of its role in preventing cancer (see Ferbeyre and Lowe 2002). p53 appears to monitor cells for damage to their DNA. Such damage can lead to alterations of gene expression and to uncontrolled cell division—that is, to the development of tumors. When p53 detects DNA damage, it signals cells to stop dividing or even to undergo programmed cell death. Individuals that carry germ-line mutations in the p53 gene have an elevated susceptibility to a variety of cancers. And more cancers carry mutations in the p53 gene, whether transmitted via the germ line or acquired somatically, than in any other gene.

Recent work by Stuart Tyner and colleagues (2002) has revealed that p53 mediates a trade-off between the risk of getting cancer and the rate of aging. Tyner and colleagues were using genetic engineering techniques to make mice deficient in p53 activity. Compared to normal mice, p53-deficient mice are highly susceptible to cancer. As a result, such mice tend to die before showing other obvious signs of old age. The difference in life span and cancer susceptibility of normal versus p53-deficient mice can be seen by comparing the green versus red curves in Figure 13.8a.

During their research, Tyner and colleagues experienced a lucky accident. Some of the mice whose p53 genes they had manipulated showed an increase, not a decrease, in p53 activity. These mice were heterozygotes, with one copy of the normal allele for p53 and one copy of a mutant allele that the researchers named *m*. Compared to normal mice, $p53^{+/m}$ mice have a dramatically reduced susceptibility to cancer (Figure 13.8a). Having elevated p53 activity would appear to be advantageous. Tyner and colleagues discovered, however, that it carries a steep cost. As the green versus blue curves in Figure 13.8a show, $p53^{+/m}$ mice have shorter life spans than normal mice. The reason is that they age more rapidly. The photos in Figure 13.8b and (c) show a normal male mouse and a $p53^{+/m}$ male mouse, both 2 years old. The normal mouse still looks healthy, whereas the $p53^{+/m}$ mouse has a hunched back, emaciated appearance, and other symptoms characteristic of old age.

Why does elevated p53 activity lead to rapid aging? Tyner and colleagues think that the answer lies in p53's effects on stem cells. Stem cells play a key role in the maintenance and repair of organs and tissues. Their job is to divide and produce daughter cells that serve as replacements for mature cells that have worn out and died or been destroyed by illness or injury. If p53 is too active—if it is overly sensitive to signs of even minor damage—it may cause stem cells to stop dividing and

chromosomes become too damaged to function. But this will only be true if there is an association between senescence of individual cells and the longevity of whole organisms.

Dan Röhme (1981) reported that across the eight mammal species he tested, the longevity of whole animals (measured in years) is correlated with the replicative life spans of their fibroblast cells growing in culture (measured in days). This suggests that organisms live longer if their cells are capable of more cell divisions. Does this correlation rest on a causal connection between the decay of telomeres, the senescence of cells, and the aging the whole organism? Kyu Sang Joeng and colleagues (2004) genetically engineered nematode worms (*Caenorhabditis elegans*) to overexpress a telomere-binding protein, thereby increasing their telomere length. As the researchers predicted, worms with longer telomeres outlived control worms (Figure 13.7).

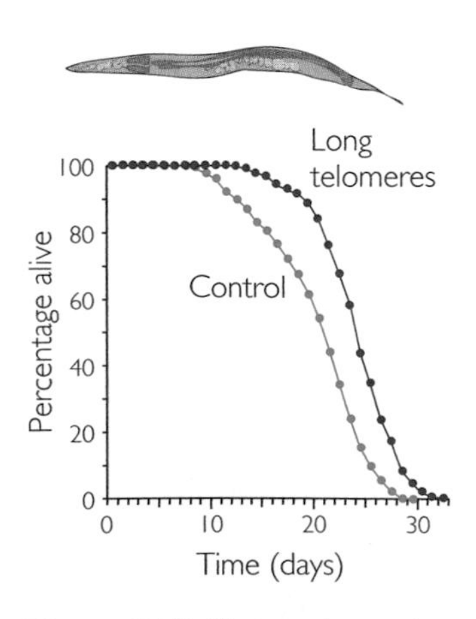

Figure 13.7 Nematodes engineered to have longer telomeres live longer From Joeng et al. (2004).

On the other hand, Marcela Raices and colleagues (2005) found that across *C. elegans* strains longevity is not closely associated with telomere length (Figure 13.9). This is perhaps not surprising, as cells in adult *C. elegans* do not di-

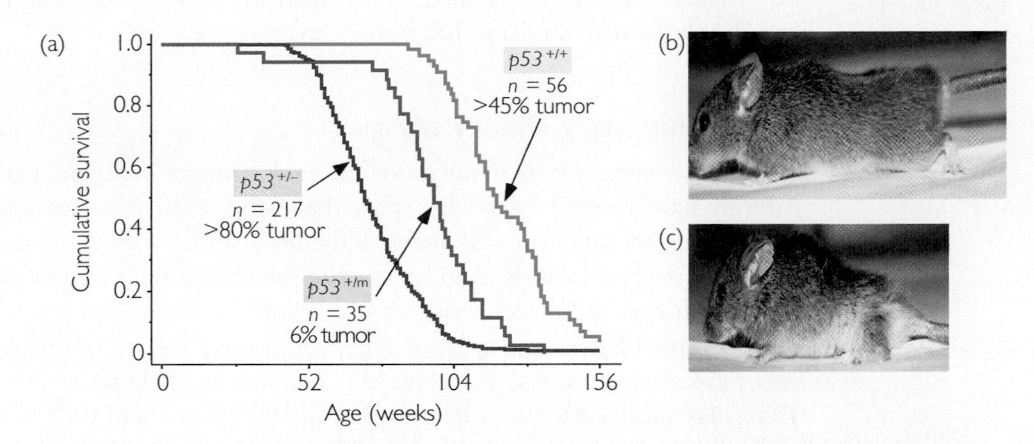

Figure 13.8 The tumor suppressor p53 trades-off cancer risk with rate of aging The graph in (a) tracks the cumulative survival of mice with three p53 genotypes: $p53^{+/-}$ mice (red) have one normal allele of the p53 gene and one loss-of-function allele, a genotype with reduced p53 activity; $p53^{+/m}$ mice (blue) have one normal allele and one *m* allele, a genotype with increased p53 activity; $p53^{+/+}$ mice (green) have two normal alleles, a genotype with normal p53 activity. $p53^{+/-}$ mice die young because they are highly susceptible to cancer. $p53^{+/m}$ mice have a low susceptibility to cancer, but die young anyway as a result of premature senescence. $p53^{+/+}$ mice have an intermediate susceptibility to cancer, a normal rate of senescence, and the longest average life spans. The photo in (b) shows a $p53^{+/+}$ mouse at the age of 104 weeks, when mice with his genotype have just begun to die of old age. He still looks healthy. The photo in (c) shows a $p53^{+/m}$ mouse at the same age, when half the mice with his genotype have already died. He has a hunched back, emaciated appearance, and other symptoms of aging. From Tyner et al. (2002).

die earlier than they should. As they lose their ability to maintain and repair their body parts, mice with overactive p53 age prematurely. As the curves in Figure 13.8a show, the level of p53 activity in normal mice appears to strike an optimal balance between cancer risk and aging rate.

vide. Working with mice, Michael Hemann and Carol Greider (2000) found that wild-derived strains have shorter telomeres than laboratory strains, but similar life spans. And Claus Bischoff and colleagues (2006), following 812 men and women that ranged in age from 73 to 101 at the start of their study, found no association between telomere length and subsequent survival. Finally, Antonello Lorenzini and colleagues (2005) repeated Röhme's study with a wider diversity of mammals and better data on longevity. They found that both longevity and fibroblast replicative capacity were correlated with body mass. After controlling statistically for the effect of mass, there was no association across species between longevity and replicative capacity. Collectively, these results demonstrate that there is no simple relationship between telomere length, the replicative senescence of cells and whole-organism aging and death.

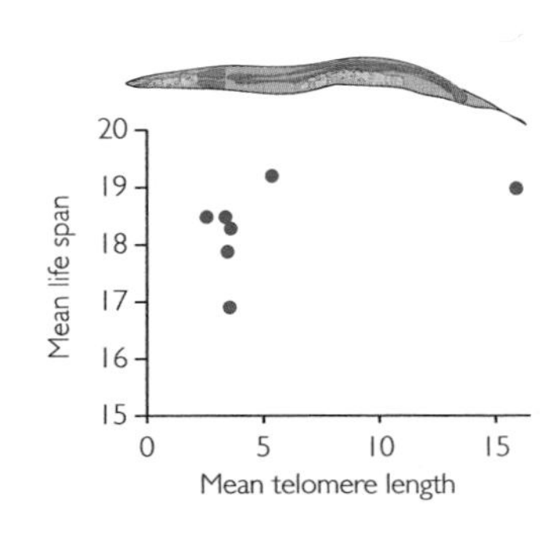

Figure 13.9 Longevity and telomere length across worm strains The association is weak at best, and not statistically significant.. Plotted from data in Raices et al. (2005)

To summarize, the rate-of-living theory and related hypotheses about aging have been scientific successes in the sense that they have stimulated considerable research and pointed the way to important discoveries. Among these discoveries, however, is a paradox. Many populations are not, in fact, up against

Many populations harbor genetic variation for longevity, yet longer life spans have not evolved.

instrinsic limits to longevity. They harbor genetic variation that would allow the evolution of longer life spans. And yet, longer life spans have not evolved. For a resolution to this paradox, we turn to the evolutionary theory of aging and the fundamental idea in life history analysis: trade-offs.

The Evolutionary Theory of Aging

As we have seen, fruit fly populations can evolve longer life spans, bat species apparently have evolved longer life spans than other eutherian mammals, and genetic engineers can artificially increase the life span laboratory strains of worms. If natural selection can lead to longer life spans, and the physiological mechanisms for longer life already exist, why has natural selection not produced this result in all populations and species? The evolutionary theory of senescence offers two related mechanisms to resolve this conundrum (Medawar 1952; Williams 1957; Hamilton 1966; Partridge and Barton 1993; Nesse and Williams 1995; Partridge 2001). Under the evolutionary theory, aging is caused not so much by cell and tissue damage itself as by the failure of organisms to completely repair such damage. This failure to fully repair leads to gradual decay and ultimate collapse. George C. Williams argues that complete repair ought to be physiologically possible (Williams 1957; Nesse and Williams 1995). Given that organisms are capable of constructing themselves from scratch, they should also be capable of maintaining their organs and tissues once formed. Upkeep is, in principle, easier than manufacture. Indeed, organisms do have remarkable abilities to replace or repair damaged parts. Yet in many organisms repair is incomplete. Under the evolutionary theory of senescence, the failure to completely repair damage is ultimately caused by either (1) deleterious mutations, or (2) trade-offs between repair and reproduction.

To understand why populations have not evolved longer life spans, researchers explore how natural selection varies as a function of an individual's age.

Figure 13.10 uses a simple genetic and demographic model of a hypothetical population to show how deleterious mutations or trade-offs can lead to the evolution of senescence. The figure follows the life histories of individuals in the population from birth until death. Individuals in the population are always at risk of death due to accidents, predators, and diseases. Except where noted, the probability that an individual will survive from one year to the next is 0.8. This leads to an exponential decline over time in the fraction of individuals that are still alive.

Figure 13.10a tracks the life histories of individuals with the wild-type genotype. These individuals mature at age 3 and die at age 16. The columns of the table are as follows:

- The first column lists ages.
- The second column indicates the fraction of all wild-type individuals born that are still alive at each age. From age 1 onward, each number in the second column is simply the number immediately above it multiplied by 0.8. Few individuals survive to the age of 15. To keep the size of the table reasonable, we assume that all individuals that survive until their 16th birthday die before reproducing that year (this assumption affects only about 3% of the population).
- The third column shows that wild-type individuals reach reproductive maturity at age 3. Once the organisms start to reproduce at age 3, they each have one offspring every year (for as long as they survive).

(a) Wild type matures at age 3 and dies at age 16; prior to age 16 annual rate of survival = 0.8

Age	Fraction of individuals surviving	RS of survivors	Expected RS for individuals
0	1.000	0	0.000
1	0.800	0	0.000
2	0.640	0	0.000
3	0.512	1	0.512
4	0.410	1	0.410
5	0.328	1	0.328
6	0.262	1	0.262
7	0.210	1	0.210
8	0.168	1	0.168
9	0.134	1	0.134
10	0.107	1	0.107
11	0.086	1	0.086
12	0.069	1	0.069
13	0.055	1	0.055
14	0.044	1	0.044
15	0.035	1	0.035
		Expected lifetime RS:	2.419

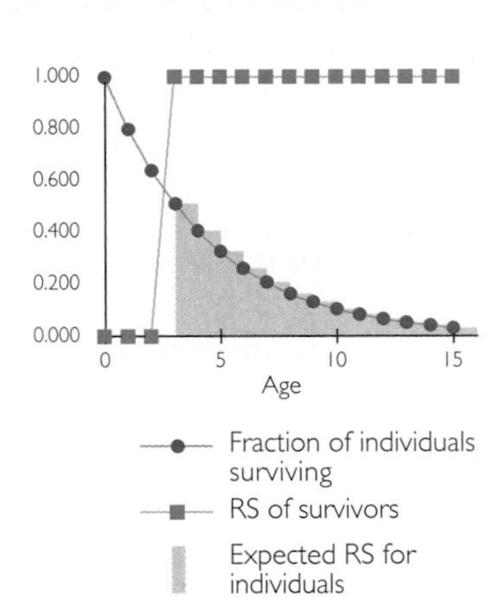

(b) Mutation from wild type that causes death at age 14; prior to age 14 annual rate of survival = 0.8

Age	Fraction of individuals surviving	RS of survivors	Expected RS for individuals
0	1.000	0	0.000
1	0.800	0	0.000
2	0.640	0	0.000
3	0.512	1	0.512
4	0.410	1	0.410
5	0.328	1	0.328
6	0.262	1	0.262
7	0.210	1	0.210
8	0.168	1	0.168
9	0.134	1	0.134
10	0.107	1	0.107
11	0.086	1	0.086
12	0.069	1	0.069
13	0.055	1	0.055
14	0.000	1	0.000
15	0.000	1	0.000
		Expected lifetime RS:	2.340

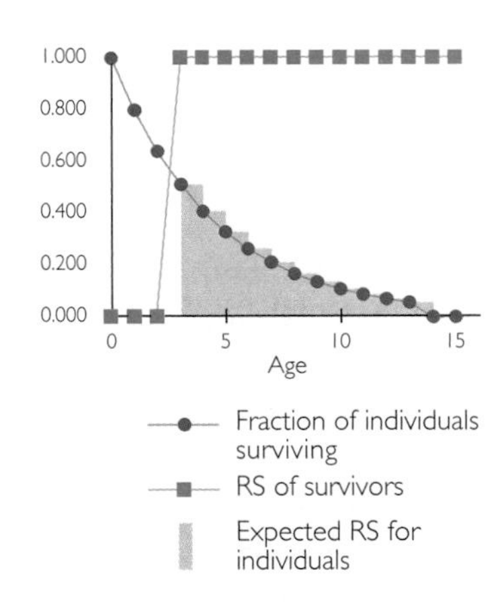

(c) Mutation from wild type that causes maturation at age 2 and death at age 10; prior to age 10 annual rate of survival = 0.8

Age	Fraction of individuals surviving	RS of survivors	Expected RS for individuals
0	1.000	0	0.000
1	0.800	0	0.000
2	0.640	1	0.640
3	0.512	1	0.512
4	0.410	1	0.410
5	0.328	1	0.328
6	0.262	1	0.262
7	0.210	1	0.210
8	0.168	1	0.168
9	0.134	1	0.134
10	0.000	1	0.000
11	0.000	1	0.000
12	0.000	1	0.000
13	0.000	1	0.000
14	0.000	1	0.000
15	0.000	1	0.000
		Expected lifetime RS:	2.663

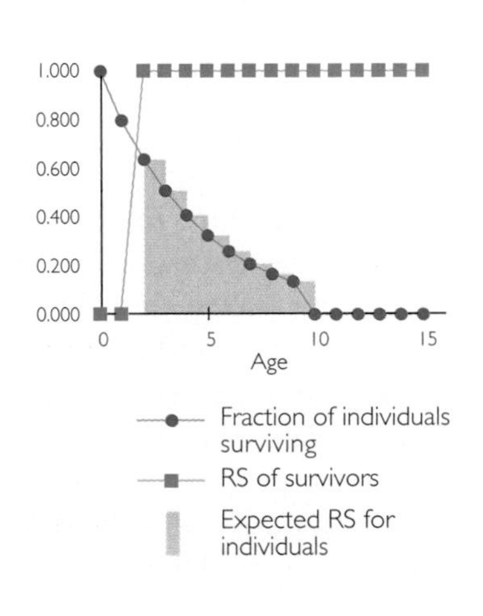

Figure 13.10 **A simple genetic model reveals two mechanisms through which aging can evolve** (a) The life history of individuals with the wild-type genotype. The sum of the values in column 4 gives the expected lifetime reproductive success of wild-type individuals: 2.419. (b) The life history of individuals with a mutation that causes death at age 14. The expected lifetime reproductive success of individuals with this mutation is 2.340. The mutation is deleterious, but not extremely so. (c) The life history of individuals with a mutation that causes maturation at age 2 and death at age 10. The expected lifetime reproductive success of individuals with this mutation is 2.663. The mutation is advantageous. See text for more details.

- The fourth column shows the expected reproductive success at each age for wild-type individuals. The expected reproductive success at age 5, for example, is simply the fraction of individuals that will survive to age 5 multiplied by the number of offspring each survivor will have at age 5. The sum of the numbers in this column gives the expected lifetime reproductive success of wild-type individuals.

The numbers in the table are plotted in the graph; the expected lifetime reproductive success of the wild-type individuals is equal to the area of the shaded region. The expected lifetime reproductive success of wild-type individuals is about 2.42.

We now consider two mutations that change the life histories of the individuals that carry them. If we imagine that these mutations are dominant in their effects, then our consideration covers both homozygotes and heterozygotes. If we imagine that the mutations are recessive, then our consideration covers only homozygotes.

Deleterious Mutations and Aging: The Mutation Accumulation Hypothesis

Figure 13.10b depicts a mutation that causes death at age 14. In other words, the mutation causes premature senescence. All other aspects of life history are unchanged. The mutation is obviously deleterious, but how strongly will it be selected against? As shown in the table and graph, the expected lifetime reproductive success of individuals with the mutation is about 2.34. This is over 96% of the fitness of wild-type individuals. Because few individuals survive to age 14 anyway, individuals carrying the mutation causing death at 14 do not, on average, suffer much of a penalty. The mutation is not selected against very strongly.

Natural selection is weak late in life, so alleles that cause aging are only mildly deleterious. They may persist in mutation-selection balance or rise to high frequency by drift.

At first glance, it is a bit surprising that a mutation causing death is only mildly deleterious. Many mutations that cause death are, in fact, highly deleterious. A mutation causing death at age 2, for example, would be selected against strongly. Individuals carrying such a mutation would have an expected lifetime reproductive success of zero. But mutations causing death after reproduction has begun are selected against less strongly. The later in life that such mutations exert their deleterious effects, the more weakly they are selected against. Mutations that are selected against only weakly can persist in mutation–selection balance (see Chapter 6). The accumulation in populations of deleterious mutations whose effects occur only late in life is one evolutionary explanation for aging (Medawar 1952).

What kind of mutation could cause death, but only at an advanced age? One possibility is a mutation that reduces an organism's ability to maintain itself in good repair. Humans provide an example. Among the kinds of cellular damage that humans (and other organisms) must repair are DNA mismatch errors. Mismatched nucleotide pairs can be created by mistakes during DNA replication, or they can be induced by chemical damage to DNA (Vani and Rao 1996). Repair of these errors is performed by a suite of special enzymes. Germ-line mutations in the genes that code for these enzymes can result in the accumulation of mismatch errors, which in turn can result in cancer.

Germ-line mutations in DNA mismatch repair genes cause a form of cancer in humans called hereditary nonpolyposis colon cancer (Eshleman and Markowitz 1996; Fishel and Wilson 1997). In one study, the age at which individuals were diagnosed with hereditary nonpolyposis colon cancer ranged from 17 to 92 years. The median age of diagnosis was 48 (Rodriguez Bigas et al. 1996). Thus, most people carrying mutations in the genes for DNA mismatch repair enzymes do not suffer the deleterious consequences of the mutations until well after the age at which reproduction begins. In an evolutionary sense, hereditary nonpolyposis colon cancer is a manifestation of senescence that is caused by deleterious mutations. These deleterious mutations persist in populations because they reduce survival only late in life.

Kimberly Hughes and colleagues (2002) used inbreeding depression to detect deleterious mutations associated with aging in fruit flies (*Drosophila melanogaster*). If inbreeding depression is caused by deleterious recessive alleles, and if late-acting deleterious alleles are maintained at higher frequency under mutation-selection balance than early-acting deleterious alleles, then the severity of inbreeding depression should increase with age. Hughes and colleagues tested this prediction by preparing 10 inbred stocks of fruit flies. Within each stock, all individuals were homozygous at most loci. The researchers then performed all 100 possible crosses among their inbred stocks. Ten of these crosses involved inbred lines crossed with themselves. The progeny of these crosses were inbred. The other 90 crosses involved crosses among inbred lines. The progeny of these crosses were outbred. That is, they were heterozygous at most loci. The researchers measured the reproductive success of the progeny at various ages. They calculated inbreeding depression as the difference in fitness between outbred versus inbred lines, divided by the fitness of outbred lines. The results, shown in Figure 13.11, reveal that, as predicted, inbreeding depression increases with age. The fruit fly stocks that Hughes and colleagues studied harbor deleterious mutations that contribute to senescence.

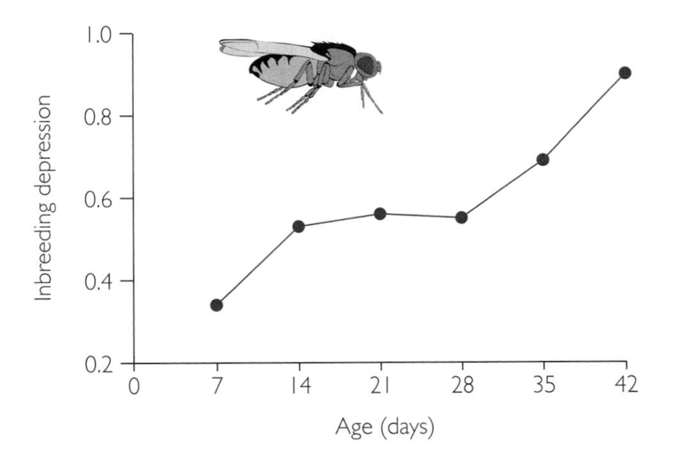

Figure 13.11 An increase in inbreeding depression with age consistent with the mutation accumulation hypothesis of senescence If inbreeding depression is caused by recessive deleterious alleles and if natural selection acts more weakly against late-acting deleterious alleles, then inbreeding depression should increase with age. This graph, tracking inbreeding depression as a function of age in laboratory fruit flies, shows that it does. Drawn from data in Hughes et al. (2002).

Deleterious mutations that contribute to aging can accumulate rapidly. David Reed and Edwin Bryant (2000) documented the rapid accumulation of late-acting deleterious mutations in populations of houseflies (*Musca domestica*). Reed and Bryant established their laboratory populations with wild houseflies. Each generation, the researchers allowed the adult flies to reproduce for only four or five days, then used the offspring they produced during this time to establish the next generation. This procedure in essence limited the adult life span of every fly to less than a week. Reed and Bryant reasoned that any late-acting deleterious mutations present in their populations would be rendered neutral. Some of these now-neutral alleles should drift to high frequency. Because neutral evolution proceeds as rapidly in large populations as it does in small ones (see Chapter 7), the researchers expected to see this effect proceed at the same pace regardless of population size.

Reed and Bryant monitored the accumulation of late-acting deleterious mutations in their housefly populations by periodically allowing the flies to live out their natural life spans. Over 24 generations, in both large populations and

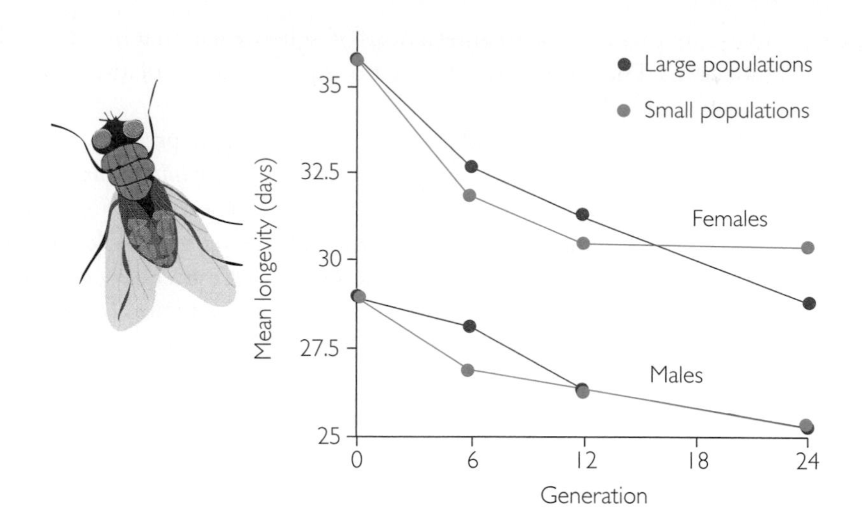

Figure 13.12 Evolution of shorter life spans consistent with the accumulation of deleterious mutations In populations of individuals allowed to live only a short time, late-acting mutations are effectively neutral. Some of these alleles will rise to high frequency by drift, causing a decline in natural longevity. Such a decline is evident in this graph, which tracks the natural longevity over 24 generations in populations of houseflies in which the adults were allowed to live only four days. Redrawn from Reed and Bryant (2000).

small populations, the natural longevity of the houseflies declined substantially (Figure 13.12). These results are consistent with the mutation accumulation hypothesis of senescence.

Trade-Offs and Aging: The Antagonistic Pleiotropy Hypothesis

Figure 13.10c, on page 493, depicts a mutation that affects two different life history characters. That is, the mutation is pleiotropic. The mutation causes reproductive maturation at age 2 instead of age 3, and the mutation causes death at age 10. In other words, the mutation involves a trade-off between reproduction early in life and survival late in life; its pleiotropic effects are antagonistic.

Because natural selection is weaker late in life, alleles that enhance early-life reproduction may be favored even if they also hasten death.

As shown in the table and graph, the expected lifetime reproductive success of individuals with the mutation is about 2.66. This is 1.1 times the expected lifetime reproductive success of wild-type individuals. Most of the individuals born with the mutation will live long enough to reap the benefit of earlier reproduction, but few will survive long enough to pay the cost of early aging. This mutation that causes both early maturation and early senescence is therefore favored by selection. Selection for alleles with pleiotropic effects that are advantageous early in life and deleterious late in life is a second evolutionary explanation for aging (Williams 1957; Rose 1991).

What kind of mutation could increase reproduction early in life at the same time it reduced reproduction or survival late in life? Perhaps a mutation that causes less energy to be allocated to repair early in life and more energy to be allocated to reproduction (see Figure 13.2 on page 485). Until recently, specific genes with this kind of pleiotropic action had not been identified. Now, however, several genes that appear to act in this way have been found (Leroi et al. 2005).

Research by David Walker and colleagues (2000) on the *age-1* gene in *Caenorhabditits elegans* provides an example. *C. elegans,* mentioned earlier, is a tiny nematode worm, about 1 mm in length, that lives in soil and eats bacteria. In *C. elegans* the protein encoded by *age-1* plays a role in an intracellular signaling pathway involved in the control of development and the determination of stress resistance. The *age-1* gene product also plays a role in senescence. Mutations in the gene increase life span by as much as 80%. Carriers of such mutations appear to be otherwise normal. They develop at the same rate as wild-type worms, have similar activity levels, and achieve comparable total fertility.

Walker and colleagues sought more subtle effects on fitness of a mutant *age-1* allele called *hx546*. The researchers established laboratory populations of worms in which the individuals were genetically identical except that some were homozygous for the normal *age-1* allele whereas others were homozygous for *hx546*. All the worms were hermaphrodites and reproduced by self-fertilization. The researchers tracked the frequency of the *hx546* allele over 10–12 generations. If the allele were beneficial its frequency would rise; if it were deleterious, its frequency would fall. Given that the only obvious difference between *hx546* worms and normal worms is that the *hx546* worms live considerably longer, one might expect that the allele would be advantageous.

Walker and colleagues first reared populations in which they gave the worms ample food. The researchers established two populations in which the starting frequency of *hx546* was 0.9, two in which it was 0.5, and two in which it was 0.1. Surprisingly, the frequency of the allele changed little over 10 generations, regardless of its starting frequency (Figure 13.13a). The *hx546* allele was not advantageous, but neither was it deleterious. This suggests that the benefit of longer life span was balanced by a roughly equivalent cost.

The true cost of carrying the *hx546* allele was revealed when Walker and colleagues reared populations under conditions that more closely resemble what *C. elegans* experiences in nature. Each time they established a new culture, they let the worms eat all of the bacteria in their petri dish. The researchers let the worms starve for four days, then gave them more bacteria to eat. Finally, the researchers used only eggs the worms produced in the first 24 hours after feeding resumed to establish the next culture. From start to finish each starvation cycle lasted about 2 generations. Walker and colleagues tracked the frequency of *hx546* for six starvation cycles in five populations, all of which had a starting frequency of 0.5. The result was dramatic. The frequency of *hx546* plummeted to an average of 0.06 in 12 generations (Figure 13.13b). For the frequency of *hx546* to fall that far that fast, the fitness of *hx546* worms must have been less than 80% that of normal worms.

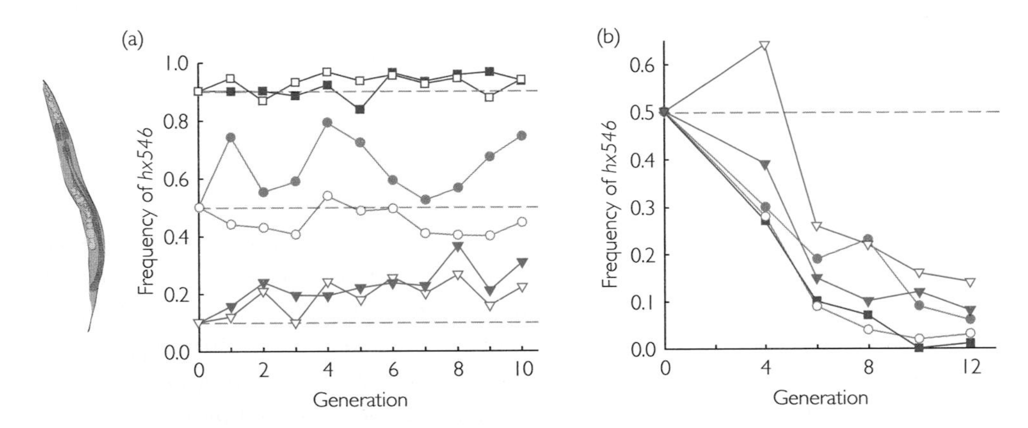

Figure 13.13 Under semi-natural conditions, the age-1 gene in the nematode worm *C. elegans* exhibits antagonistic pleiotropy (a) This graph tracks the frequency of the *hx546* allele of the *age-1* gene, which causes increased longevity, in six experimental populations reared in the lab with ample food. The frequency of *hx546* changed little over 10 generations, regardless of its starting frequency, indicating that the allele was effectively neutral. (b) This graph tracks the frequency of the *hx546* allele in five populations reared under semi-natural conditions characterized by periodic bouts of starvation. The frequency of *hx546* declined rapidly over 12 generations in all five populations, indicating that the allele was strongly selected against. From Walker et al. (2000).

Walker and colleagues isolated worms from cultures that had been starved for four days, then fed. They examined the worms at 12 and 24 hours after feeding resumed. They found that only young adults laid eggs during this window. The implication is that, compared to the *hx546* allele, the normal allele acts in just the way the antagonistic pleiotropy hypothesis of senescence predicts: It increases the reproductive success of its carriers in young adulthood, at the cost of a shorter life span.

A gene in the fruit fly, *Drosophila melanogaster*, has similar effects. When Yi-Jyun Lin and colleagues (1998) discovered this gene they named it *methuselah* because homozygotes for a mutant allele with reduced expression live 35% longer than normal flies. The *methuselah* mutation attracted notice because, in addition to increasing longevity, it enhances resistance to starvation, heat, and the herbicide paraquat (see Pennisi 1998). A cost-free mutation that extends life span and increases tolerance to a variety of stresses would challenge the evolutionary theory of senescence.

But is the *methuselah* mutation free of costs? To find out, Robin Mockett and Rajindar Sohal (2006) reared normal and homozygous mutant females and counted the offspring each fly produced over her life span. The researchers confirmed that mutant flies live longer (Figure 13.14a). However, the mutants also lay fewer eggs during early adulthood resulting in lower lifetime reproductive success (Figure 13.14b). Compared to the mutation, the normal allele of the *methuselah* gene thus appears to trade stress resistance and longevity for reprodutive fitness. Intriguingly, the effects of genotype depend on temperature. The flies tracked in Figure 13.14 lived at 29°C. At 18°C, normal females lived just as long as mutants but had *lower* reproductive success. This may explain why alleles of the *methuselah* gene vary in frequency among natural populations (Schmidt et al. 2000).

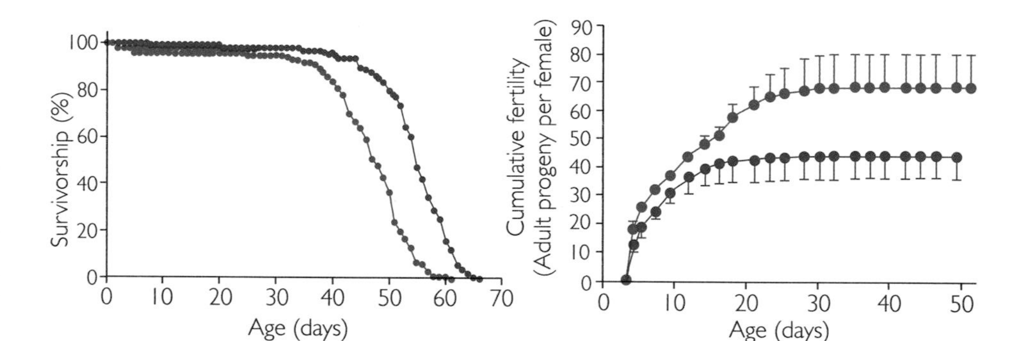

Figure 13.14 The *methuselah* gene controls a trade-off between reproductive success versus longevity and stress resistance Normal female fruit-flies die yougner than homozygous *methuselah* mutants (a), but they have considerably higher lifetime reproductive success (b). These patterns, detected at 29°C, change at other temperatures. Redrawn from Mockett and Sohal (2006).

In addition to identifying specific genes exhibiting antagonistic pleiotropy, evolutionary biologists have found widespread evidence that there is indeed a trade-off between reproduction early in life and reproduction or survival late in life. Most of this evidence comes from the analysis of quantitative genetic or phenotypic trade-offs between traits. In the paragraphs that follow, we review two examples.

Researchers have documented a trade-off between reproductive effort early in life and reproductive success late in life.

Lars Gustafsson and Tomas Pärt (1990) studied trade-offs in a bird, the collared flycatcher (*Ficedula albicollis*). Working for 10 years with a flycatcher population on the Swedish island of Gotland, Gustafsson and Pärt followed the life histories of individual birds from hatching to death. The researchers found that some female flycatchers begin breeding at age 1, whereas others wait until age 2. The females that breed at age 1 have smaller clutch sizes throughout life (Figure 13.15a), indicating that there is a cost later in life to breeding early. To further investigate this interpretation, Gustafsson and Pärt manipulated early reproductive effort by giving some first-year breeders extra eggs. The females given extra eggs had progressively small-

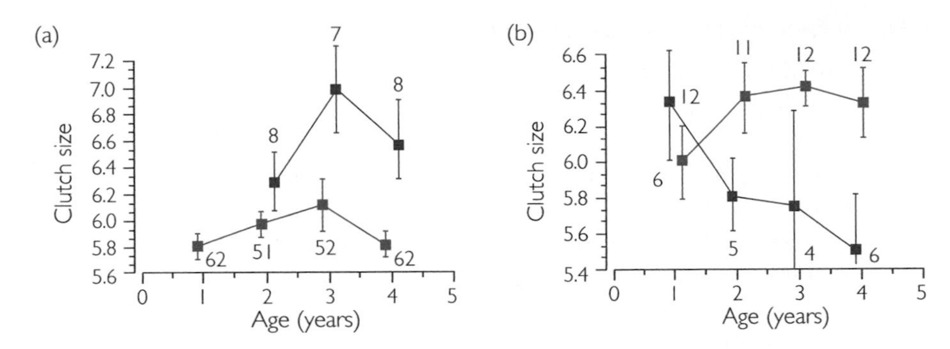

Figure 13.15 **In collared flycatchers (*Ficedula albicollis*), natural variation and experimental manipulations demonstrate a trade-off between reproduction early in life versus reproduction late in life** (a) Females that first breed at age 1 (blue boxes) have smaller clutches at ages 2, 3, and 4 than females that do not breed until age 2 (red boxes). The error bars indicate the standard errors of the estimated mean clutch sizes; the numbers above the symbols give the sample sizes. (b) Females given extra eggs at age 1 (red boxes) have progressively smaller clutches each year at ages 2, 3, and 4. In contrast, control females (blue boxes) do not begin to show a decline in clutch size until age 4. Note that the vertical scales are different. From Gustafsson and Pärt (1990).

er clutch sizes in subsequent years, whereas control females did not begin to show reproductive senescence until age 4 (Figure 13.15b). Gustafsson and Pärt conclude that there is a trade-off in collared flycatchers between early-life and late-life reproduction. The researchers note that despite this trade-off, first-year breeders had higher lifetime reproductive success than second-year breeders (1.24 ± 0.08 versus 0.90 ± 0.14 offspring surviving to adulthood $P < 0.05$).

Truman Young (1990) studied trade-offs in plants. Young reviewed data from the literature on the energy allocated to reproduction by closely related pairs of annuals versus perennials (Table 13.1). Annuals, which reproduce once and die, always allocate more energy to their sole bout of reproduction than perennials allocate to any given bout. This indicates that there is a trade-off in plants between reproduction and survival. Annual plants enjoy enhanced reproduction in their first reproductive season at the expense of drastically accelerated senescence.

Table 13.1 **Reproductive allocation in annual versus perennial plants**

Each line in the table gives the amount of energy allocated to reproduction by annual plants during their only bout of reproduction, divided by the amount of energy allocated to reproduction by closely related perennial plants during any single bout of reproduction. All the values are larger than one, indicating that annuals always allocate more per reproductive episode. The comparisons for *Oryza perennis* (two studies) and *Ipomopsis aggregata* are within species; all other comparisons are between species. From Young (1990); see this source for citations to individual studies.

Species	Allocation by annuals/allocation by perennials
Oryza perennis	2.9
Oryza perennis	5.3
Gentiana spp.	2.2–3.5
Lupinus spp.	2.2–3.2
Helianthus spp.	1.7–4.0
Temperate herbs	2.8–2.9
Old field herbs	1.7
Ipomopsis aggregata	1.5–2.3
Sesbania spp.	2.1–2.3
Hypochoeris spp.	2.4–3.7

A Natural Experiment on the Evolution of Aging

Steven Austad took advantage of a natural experiment in order to compare populations that had been historically exposed to different rates of mortality caused by extrinsic factors such as predators, diseases, and accidents. We will call this kind of mortality *ecological mortality*, in contrast to mortality caused by processes intrinsic to the organism, like the wearing out of body parts. (Mortality caused by intrinsic processes might be called physiological mortality.)

The evolutionary theory of senescence predicts that populations with lower rates of ecological mortality will evolve delayed senescence (Austad 1993). What is the logic behind this prediction? Both of the evolutionary mechanisms that lead to senescence have reduced effectiveness in populations with lower ecological mortality rates. In the case of late-acting deleterious mutations (Figure 13.10b, page 493), lower ecological mortality means that a higher fraction of zygotes will live long enough to experience the deleterious effects. Late-acting deleterious mutations are thus more strongly selected against and will be held at lower frequency in mutation–selection balance. In the case of mutations with pleiotropic effects (Figure 13.10c), lower ecological mortality means that a higher fraction of zygotes will live long enough to experience both the early-life benefits and the late-life costs. The change in the fraction of zygotes experiencing the benefits and costs is more pronounced, however, for the costs. Thus mutations with pleiotropic effects are less strongly favored by selection. All else being equal, then, individuals in populations with lower ecological mortality should show later senescence.

Austad (1993) studied the Virginia opossum (Figure 13.16). He compared a population living in the mainland southeastern United States to a population living on Sapelo Island, located off the coast of Georgia. In the mainland population, opossums have high ecological mortality rates. In one study reviewed by Austad, more than half of all naturally occurring opossum deaths were caused by

Figure 13.16 A Virginia opossum and her young Opossums, like other marsupials, have relatively short life spans for mammals. But opossums in some populations have shorter life spans than opossums in others. This suggest that rates of aging evolve in nature.

predators. When identifiable, two-thirds of the predators were mammals, including bobcats and feral dogs. Mammalian predators are absent on Sapelo Island, however. Sapelo Island supports an opossum population that has been isolated from the mainland population for 4,000–5,000 years. Other than the difference in mammalian predators, Sapelo Island differs little from Austad's mainland study site at Savanna River, South Carolina. The two habitats are similar in temperature, rainfall, opossum ectoparasite loads, and food available per opossum. The evolutionary theory of senescence predicts that the Sapelo Island opossums will show delayed senescence relative to the mainland opossums.

To test this prediction, Austad put radio collars on 34 island females and 37 mainland females and followed their life histories from birth until death. By three different measures, island females indeed show delayed senescence:

In populations where mortality rates are high, individuals tend to breed earlier in life.

- Island females show delayed senescence in month-to-month probability of survival. This is indicated by the shapes of the curves in Figure 13.17a (not the fact that the island curve is higher than the mainland curve). For island opossums, ln(proportion surviving) versus age traces a relatively straight line. This means that their risk of dying in their 40th month of life is not much higher than their risk of dying in their 20th month. In other words, the island opossums do not seem to accumulate much wear and tear. For mainland opossums, however, ln(proportion surviving) versus age traces a precipitous downward curve. Their risk of dying in their 28th month is much higher (approximately 100%) than their risk of dying in their 20th month. The immediate cause of death for most mainland opossums is predation. But older opossums are much more vulnerable to predation, because they are getting stiff and slow. As a result, the average life span of island females is significantly longer than the average life span of mainland females (24.6 versus 20.0; $P < 0.02$).
- Island females show delayed senescence in reproductive performance (Figure 13.17b). Austad measured reproductive performance by monitoring the

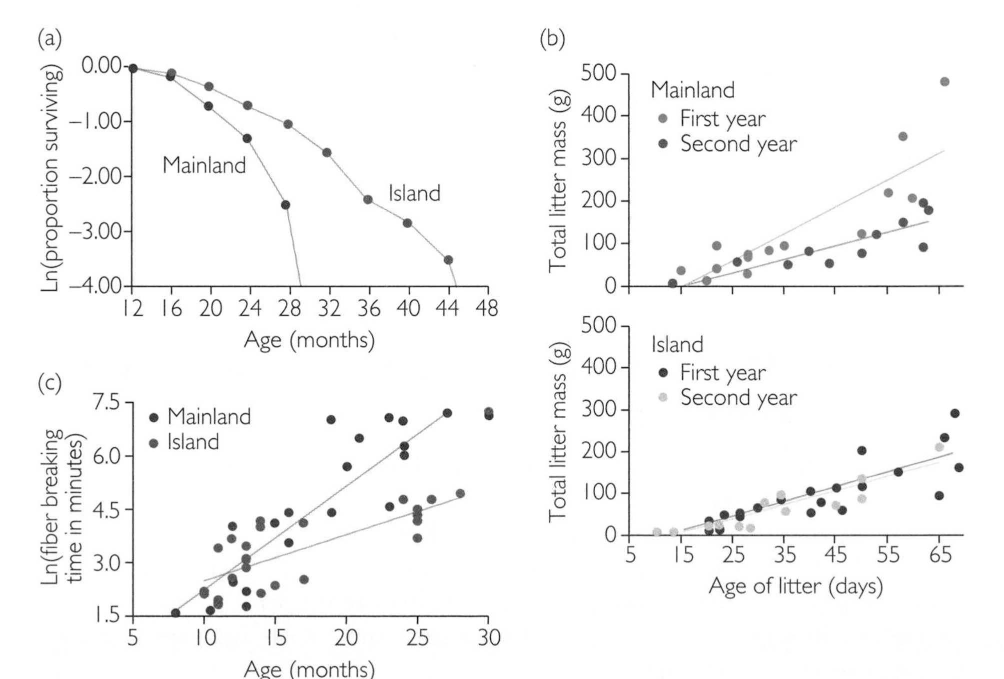

Figure 13.17 Female opossums on Sapelo Island age more slowly than female opossums on the mainland (a) The rate of survival from one month to the next falls more rapidly with age for mainland opossums than for island opossums. (b) Each graph shows total litter mass as a function of litter age for females in their first year of reproduction (green or red) and for females in their second year of reproduction (blue or yellow). In mainland females (top graph), offspring in second-year litters grow more slowly than offspring in first-year litters ($P < 0.001$). In island females, second-year litters grow just as fast as first-year litters. (c) Tail collagen fiber-breaking time increases more slowly with age in island females than in mainland females ($P < 0.001$). From Austad (1993).

growth rates of litters of young. For mainland females, litters produced in the mother's second year of reproduction grew more slowly than litters produced in the mother's first year of reproduction. This difference indicates that second-year mothers are less efficient at nourishing their young. Island females show no such decline in performance with age.

- Island females show delayed senescence in connective tissue physiology (Figure 13.17c). As mammals age, the collagen fibers in their tendons develop cross-links between protein molecules. These cross-links reduce the flexibility of the tendons. The amount of cross-linking in a tendon can be determined by measuring how long it takes for collagen fibers from the tendon to break. In both island and mainland opossums, breaking time in tail-tendon collagen increases with age, but breaking time increases less rapidly with age in island opossums. In other words, island opossums have slower rates of physiological aging.

These results are all consistent with the evolutionary theory of senescence.

While they support the conclusion that ecological mortality is an important factor in the evolution of senescence, Austad's results do not allow us to determine which of the evolutionary theory's two hypotheses is more important. Is the more rapid aging of mainland opossums due to late-acting deleterious mutations, or to trade-offs between early reproduction and late reproduction and survival? At least part of the difference in rates of senescence appears to be due to trade-offs. Island opossums have, on average, significantly smaller litters (5.66 versus 7.61; $P < 0.001$). This suggests that mainland opossums are, physiologically and evolutionarily, trading increased early reproduction for decreased later reproduction and survival.

Populations have not evolved longer life spans because the power of natural selection declines with age.

In summary, the evolutionary theory of senescence hinges on the observation that the power of natural selection declines late in life. This is because most individuals die—due to predators, diseases, or accidents—before reaching late life. Two mechanisms can lead to the evolution of senescence: (1) Deleterious mutations whose effects occur late in life can accumulate in populations, and (2) when there are trade-offs between reproduction and maintenance, selection may favor investing in early reproduction even at the expense of maintaining cells and tissues in good repair. The evolutionary theory of senescence has been successful in explaining variation in life history among populations and species. Among the questions that remain are these: What is the relative importance of deleterious mutations and trade-offs in the evolution of senescence (Partridge and Barton 1993)? Can evolutionary theory explain unusual reproductive life histories, such as menopause in human females (Box 13.2)?

13.3 How Many Offspring Should an Individual Produce in a Given Year?

Section 13.2 dealt, in part, with the optimal allocation of energy to reproduction versus repair over an organism's entire life history. In this section, we turn to the related issue of how much an organism should invest in any single episode of reproduction. Again we are concerned with trade-offs. Primary among them is this intuitively straightforward constraint: The more offspring a parent (or pair of parents) attempts to raise at once, the less time and energy the parent can devote to caring for each one.

Box 13.2 Is there an evolutionary explanation for menopause?

In humans, reproductive capacity declines earlier and more rapidly in women than in men (Figure 13.18a). The early decline in the reproductive capacity of women is puzzling, especially given that other measures of women's physiological capacity decline much more slowly (Figure 13.18b). Why should women's reproductive systems shut down by age 50, while the rest of their organs and tissues are still in good repair?

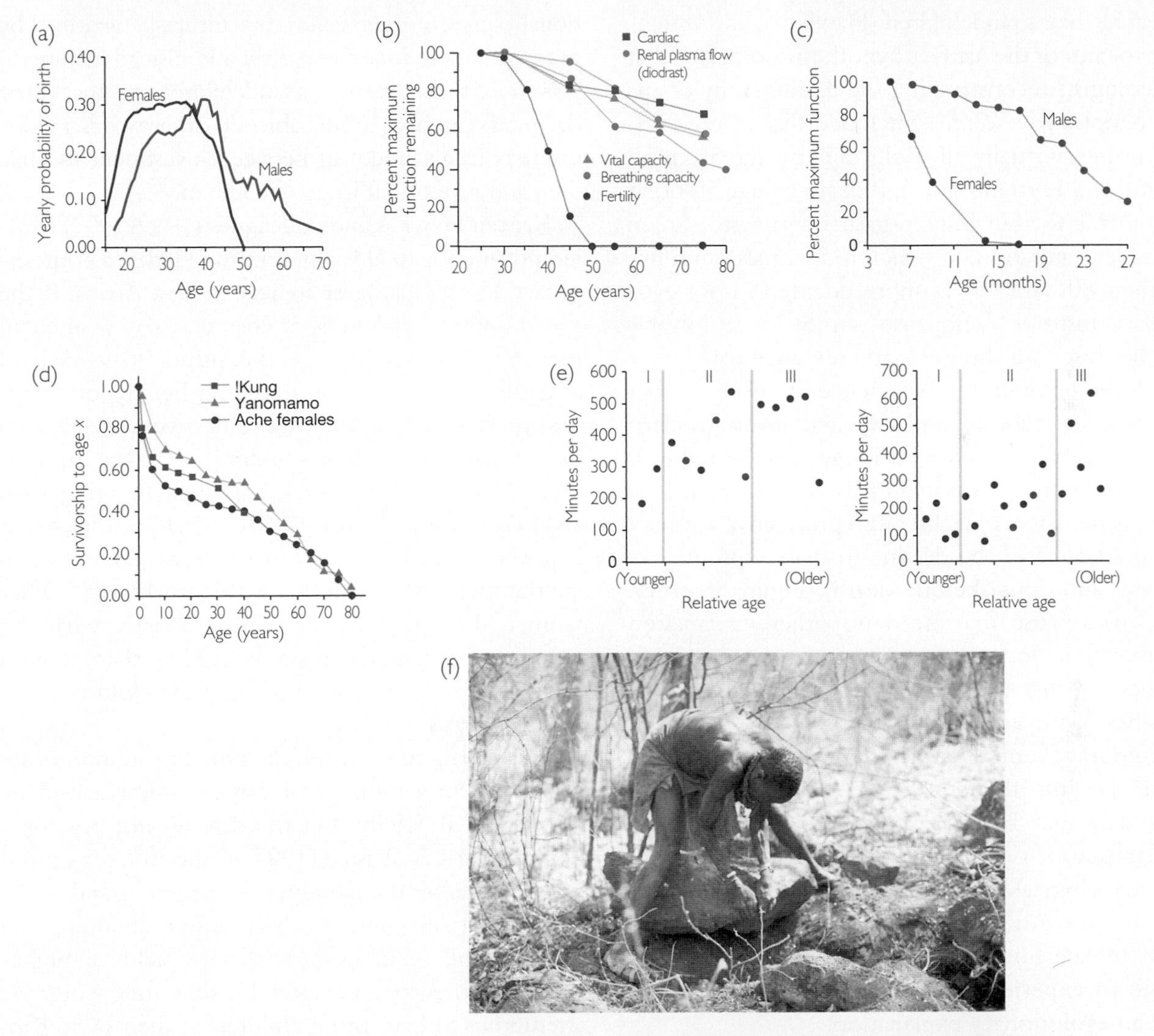

Figure 13.18 Data on menopause (a) Reproductive senescence in women and men. These data are from the Ache, hunter–gatherers of Paraguay. The graph shows the probability that women and men will have a child born during the year that they are any given age. From Hill and Hurtado (1996). (b) This graph shows the functional capacity of various physiological systems in women as a function of age. Capacity is calculated as the fraction of youthful capacity still remaining. From Hill and Hurtado (1991); see citations therein for sources of data. (c) Reproductive capacity as a function of age in captive rats. From Austad (1994); see citations therein for sources of data. (d) Fraction of individuals surviving as a function of age in three hunter–gatherer cultures. From Hill and Hurtado (1991); see citations therein for sources of data. (e) Time spent foraging each day during the wet season (left) and the dry season (right) by Hadza women of different ages: I = women who have reached puberty, but not yet married or begun to have children; II = women who are pregnant or have young children; III = women who are past childbearing age and have no children younger than 15. Blue circles represent women nursing young children. From Hawkes et al. (1989). (f) This Hadza woman is approximately 65 years old. She is using a digging stick and muscle power to dig tubers from underneath large rocks. Digging tubers requires knowledge, skill, patience, strength, and experience, making Hadza grandmothers the most productive foragers.

Box 13.2 (Continued)

We consider two hypotheses. One hypothesis suggests that menopause is a nonadaptive artifact of our modern lifestyle (see Austad 1994). The other hypothesis suggests that menopause is a life history adaptation associated with the contribution grandmothers make to feeding their grandchildren (Hawkes at al. 1989).

Advocates of the artifact hypothesis point out that archaeologists reconstructing the demography of ancient peoples have often concluded that in premodern cultures, virtually all adults died by age 50 or 55 (see Hill and Hurtado 1991). If death by age 50 or 55 was the rule for our hunter–gatherer ancestors, then the modern situation, in which individuals often live into their 80s and 90s, is unprecedented in our evolutionary history. Menopause cannot be an adaptation, because our hunter–gatherer ancestors never lived long enough to experience it. When other mammals are kept in captivity and given modern medical care, they too live far longer than individuals do in nature. In captive mammals, females in at least some species show a decline in reproductive capacity well in advance of the decline in male reproductive capacity, and long before death (Figure 13.18c). Thus, menopause may need no other explanation than that our modern lifestyle has extended our life span beyond that experienced by our ancestors.

Critics of the artifact hypothesis point out that in contemporary hunter–gatherer societies, many individuals live into their 60s and 70s (Figure 13.18d). These data may be more reliable indicators of the demography of our hunter–gatherer ancestors than are archaeological reconstructions (Hill and Hurtado 1991; see also Austad 1994). If a substantial fraction of our female hunter–gatherer ancestors lived long enough to experience menopause, then menopause needs an evolutionary explanation.

Advocates of the grandmother hypothesis note that human children are dependent on their mothers for food for several years after weaning. This is true in contemporary hunter–gatherer cultures, particularly when mothers harvest foods that yield a high return for adults, but are difficult for children to process (Hawkes et al. 1989). Thus, a woman's ability to have additional children may be substantially limited by her need to provision her older, still-dependent children. Furthermore, as a woman gets older, several relevant trends are likely to occur: (1) The probability that she will live long enough to be able to nurture another baby from birth to independence declines, (2) the risks associated with pregnancy and childbirth rise, and (3) her own daughters will themselves start to have children. The grandmother hypothesis suggests that older women may reach a point at which they can get more additional copies of their genes into future generations by ceasing to reproduce themselves and instead helping to provision their weaned grandchildren so that their daughters can have more babies. In other words, grandmothers face a trade-off between investment in children and investment in grandchildren.

Kristen Hawkes and colleagues (1989; 1997) studied postmenopausal women in the Hadza, a contemporary hunter–gatherer society in East Africa. If the grandmother hypothesis is correct, then women in their 50s, 60s, and 70s should continue to work hard at gathering food. If the grandmother hypothesis is wrong, then we might expect older women (who no longer have dependent children) to relax. In fact, older Hadza women work harder at foraging than any other group (Figure 13.18e). Furthermore, for at least some crops at some times of year, older women are the most effective foragers (Figure 13.18f). Older women do with their extra food exactly what the grandmother hypothesis predicts: They share it with young relatives, thereby improving the children's nutritional status.

These data are consistent with the grandmother hypothesis in a number of ways, as summarized by Hawkes et al. (1998), but the data do not provide a definitive test. As Austad (1994) points out, the crucial issue is whether the daughters of helpful grandmothers are actually able to have more children, and whether the grandmothers thereby achieve higher inclusive fitness (see Chapter 12) than they would by attempting to have more children of their own. Kim Hill and Magdalena Hurtado (1991; 1996) addressed this issue with data on the Ache hunter–gatherers of Paraguay. Hill and Hurtado's data show that the average 50-year-old woman has 1.7 surviving sons and 1.1 surviving daughters. The researchers calculate that by helping these children reproduce, the average Ache grandmother can gain the inclusive fitness equivalent of only 5% of an additional offspring of her own. This conclusion offers little support for the grandmother hypothesis. More complete data on more cultures are needed for definitive evaluation of both the artifact hypothesis and the grandmother hypothesis.

Questions about the optimal number of offspring have been addressed most thoroughly by biologists studying clutch size in birds. The reason is probably that it is easy to count the eggs in a nest, and it is easy to manipulate clutch size by adding or removing eggs. Assuming that the size of individual eggs is fixed, how many eggs should a bird lay in a single clutch?

Clutch Size in Birds

The simplest hypothesis for the evolution of clutch size, first articulated by David Lack (1947), is that selection will favor the clutch size that produces the most surviving offspring. Figure 13.19 shows a simple mathematical version of this hypothesis (for a more detailed mathematical treatment, see Stearns 1992). The model assumes a fundamental trade-off in which the probability that any individual offspring will survive decreases with increasing clutch size. Many researchers have tested this assumed trade-off by adding eggs to nests; in the majority of cases they have found that adding eggs indeed reduces the survival rate for individual chicks (see Stearns 1992). One explanation could be that the ability of the parents to feed any individual offspring declines as the number of offspring increases. In Figure 13.19a, we assume that the decline in offspring survival is a linear function of clutch size, but the model depends only on survival being a decreasing function. Given a function describing offspring survival, the number of surviving offspring from a clutch of a given size is just the product of the clutch size and the probability of survival (Figure 13.19b). The number of surviving offspring reaches a maximum at an intermediate clutch size. It is this most-productive clutch size that Lack's hypothesis predicts will be favored by selection.

Mark Boyce and C. M. Perrins (1987) tested Lack's hypothesis with data from a long-term study of great tits (*Parus major*) nesting in Wytham Wood, a research site near Oxford, England. Combining data for 4,489 clutches monitored over the years 1960 through 1982, Boyce and Perrins plotted a histogram showing the distribution of clutch sizes in the Wytham Wood tit population (Figure 13.20). The mean clutch size was 8.53. Boyce and Perrins also determined the average number of surviving offspring from clutches of each size (Figure 13.20). This number was highest for clutches of 12 eggs. When researchers added 3 eggs to each of a large number of clutches, the most productive clutch size was still 12 (but see below). In other words, birds that produced smaller clutches apparently could have increased their reproductive success for the year by laying 12 eggs. Taken at face value, these data indicate that natural selection in Wytham Wood

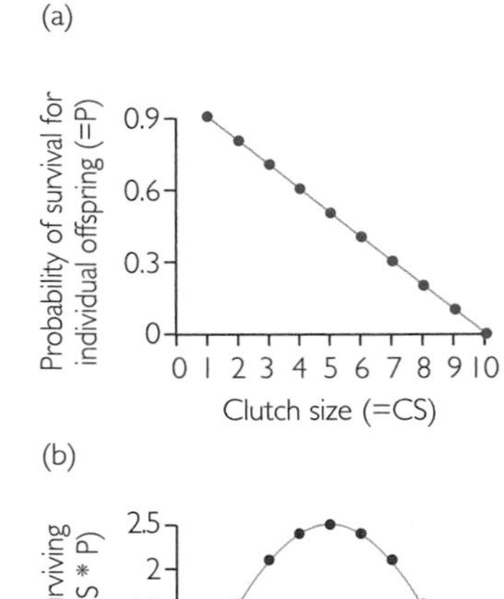

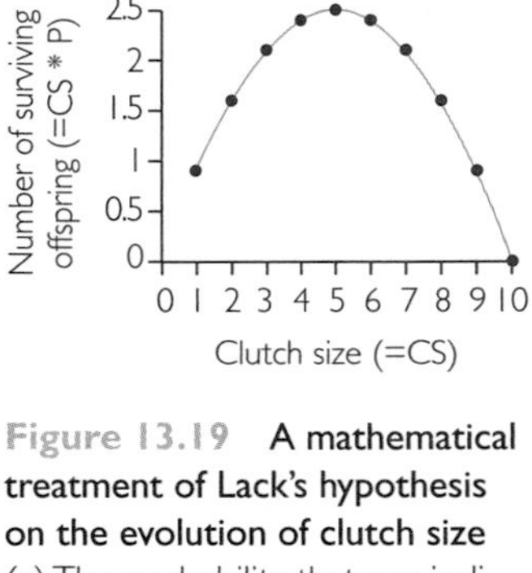

Figure 13.19 A mathematical treatment of Lack's hypothesis on the evolution of clutch size (a) The probability that any individual offspring will survive declines with increasing clutch size. Here, the probability of survival starts at 0.9 for the single offspring in a one-egg clutch, then declines by 0.1 for each one-egg increment in clutch size. (b) The number of surviving offspring per clutch is the number of eggs multiplied by the probability that any individual offspring will survive. The optimal clutch size is the one that produces the maximum number of survivors. Here, the optimal clutch size is five.

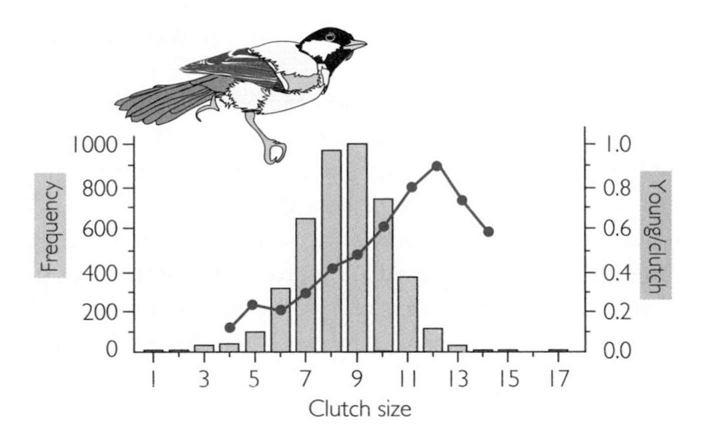

Figure 13.20 Lack's hypothesis tested with data on great tits (*Parus major*) The bars and left vertical axis form a histogram showing the variation in clutch size. The average clutch size for 4,489 clutches was 8.53. The blue dots connected by line segments and the right vertical axis show the number of surviving young per clutch as a function of clutch size. The number of surviving young per clutch was highest for clutches of 12 eggs. From Roff (1992); redrawn from Boyce and Perrins (1987).

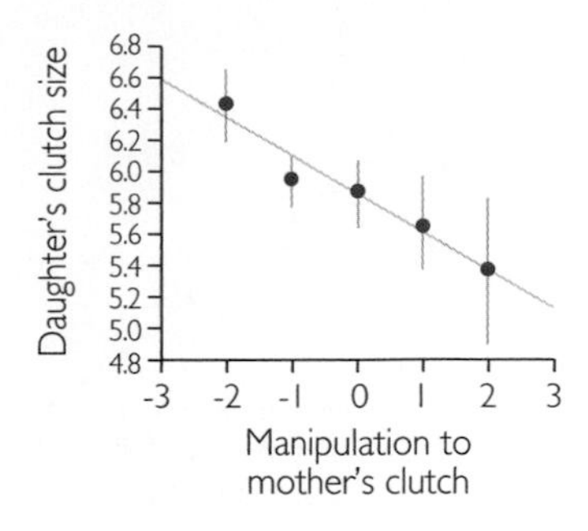

Figure 13.21 Does clutch size affect offspring reproductive performance? The graph shows the relationship between daughters' clutch sizes and the number of eggs added to or removed from the mothers' nests in which the daughters were reared. From Schluter and Gustafsson (1993).

favors larger clutches than the birds in the population actually produce. Because the average clutch size was less than the most productive clutch size, these results are not consistent with Lack's hypothesis.

The literature on Lack's hypothesis is extensive, and many researchers have done studies similar to that of Boyce and Perrins (see reviews in Roff 1992 and Stearns 1992). The results of Boyce and Perrins are typical: The majority of studies have shown that birds lay smaller clutches than predicted. How can we explain this discrepancy? The mathematical logic of Lack's hypothesis is correct. The hypothesis must therefore make one or more implicit assumptions that often turn out to be wrong. Evolutionary biologists have identified and tested several assumptions implicit in Lack's hypothesis. We discuss three of them now.

First, Lack's hypothesis assumes that there is no trade-off between a parent's reproductive effort in one year and its survival or reproductive performance in future years. As we discussed in Section 13.2, however, reproduction often entails exactly such costs. The data in Figure 13.15b, page 499, demonstrated that when female collared flycatchers are given an extra egg in their first year, their clutch size in future years is lower than that of control females. In a review of the literature on reproductive costs in birds, Mats Lindén and Anders Møller (1989) found that 26 of the 60 studies that looked for trade-offs between current reproductive effort and future reproductive performance found them. In addition, Lindén and Møller found that 4 of the 16 studies that looked for trade-offs between current reproductive effort and future survival found them. When reproduction is costly and selection favors withholding some reproductive effort for the future, the optimal clutch size may be less than the most productive clutch size.

Lack's hypothesis predicts that parents will attempt to rear that number of young that maximizes the number of surviving offspring. Data indicate that parents often rear fewer offspring. Efforts to identify which of Lack's assumptions are violated have led to the discovery of additional trade-offs and improved estimates of lifetime fitness.

Second, Lack's hypothesis assumes that the only effect of clutch size on offspring is in determining whether the offspring survive. Being part of a large clutch may, however, impose other costs on individual offspring than just reducing their probability of survival. Dolph Schluter and Lars Gustafsson (1993) added or removed eggs from the nests of collared flycatchers, put leg bands on the chicks that hatched from the nests, and then monitored the chicks' subsequent life histories. When the female chicks matured and built nests of their own, there was a strong relationship between the size of the clutches they produced and how much the clutch they were reared in had been manipulated (Figure 13.21). Females reared in nests from which eggs had been removed produced larger clutches, whereas females reared in nests to which eggs had been added produced smaller clutches. This indicates that clutch size affects not only offspring survival, but also offspring reproductive performance. These data suggest that there is a trade-off between the quality and quantity of offspring produced. When larger clutches entail lower offspring reproductive success, the optimal clutch size will be smaller than the most numerically productive clutch size.

Third, the discrepancy between Lack's hypothesis and the behavior of individual birds may sometimes be more apparent than real. When Richard Pettifor, Perrins, and R. H. McCleery (2001) reanalyzed the data on egg addition experiments used by Boyce and Perrins (1987), they concluded that Boyce and Perrins had compared their experimental birds to an inappropriate control group. Pettifor, Perrins, and McCleery found that when they used an appropriate control group, there was in fact no evidence that the birds that received extra eggs produced more surviving young than they would have had they been left alone. This suggests that, in the observational data represented in Figure 13.17, the birds that laid fewer than 12 eggs did so because they had lower reproductive capacities—and that each bird was producing a clutch size that would optimize its own reproductive success.

Note that we have been assuming that clutch size is fixed for any given genotype. In fact, clutch size is often phenotypically plastic (see Chapter 10). If clutch size is plastic, and if birds can predict whether they are going to have a good year or a bad year, then we would predict that individuals will adjust their clutch size to the optimum value for each kind of year (for example, see Sanz and Moreno 1995).

Lack's Hypothesis Applied to Parasitoid Wasps

Although Lack's hypothesis often proves to be too simple to accurately predict clutch size, the examples we have reviewed demonstrate that it offers a useful null model. By explicitly specifying what we should expect to observe under minimal assumptions, Lack's hypothesis alerts us to interesting patterns we might not otherwise have noticed. This application of Lack's hypothesis is not limited to birds.

Lack's hypothesis is a useful null model for other organisms in addition to birds.

Eric Charnov and Samuel Skinner (1985) used Lack's hypothesis to explore the evolution of clutch size in parasitoid wasps. Parasitoid wasps use a stingerlike ovipositor to inject their eggs into the eggs or body cavity of a host insect. When the larval parasitoids hatch, they eat the host alive from the inside. The larvae then pupate inside the empty cuticle of the host, finally emerging as adults to mate and repeat the life cycle.

For a parasitoid, a host is analogous to a nest. A female parasitoid can lay a clutch of one or more eggs in a single host. The larvae compete among themselves for food, so there is a trade-off between clutch size and the survival of individual larvae. An added twist with insects is that adult size is highly flexible. In addition to reducing offspring survival, competition for food may result in larvae simply becoming smaller adults. The maternal fitness associated with a given clutch size must therefore be calculated as the product of the clutch size, the probability of survival of individual larvae, and the expected lifetime egg production by offspring of the size that will emerge.

Charnov and Skinner used this modified version of Lack's hypothesis to analyze the oviposition behavior of female parasitoid wasps in the species *Trichogramma embryophagum*. This wasp deposits its eggs in the eggs of a variety of host insects. Using data from the literature, Charnov and Skinner calculated maternal fitness as a function of clutch size for three different host species (Figure 13.22). Table 13.2 gives

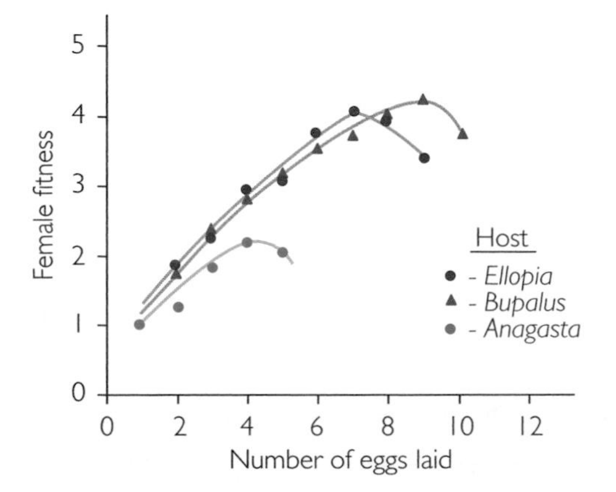

Figure 13.22 Lack's hypothesis applied to a parasitoid wasp This graph shows the fitness of female parasitoid wasps (*Trichogramma embryophagum*) as a function of clutch size. Each curve represents wasp clutches deposited in a different species of host egg. From Charnov and Skinner (1985).

Table 13.2 Parasitoid wasp (*Trichogramma embryophagum*) clutch sizes predicted by Lack's hypothesis versus actual clutch sizes for three host species

Host species	Lack's optimal clutch size	Actual clutch size
Anagasta	4	1–2
Ellopia	7	5–8
Bupalus	9	5–8

Source: From Charnov and Skinner (1985).

the most productive clutch sizes and the actual clutch sizes female wasps lay in each species of host egg. The data indicate that female wasps shift their behavior in a manner appropriate to different hosts. Females lay fewer eggs in the relatively poor hosts and more eggs in the relatively good hosts. As with many birds, however, female wasps lay smaller clutches than those predicted by Lack's hypothesis.

Why do female wasps typically lay clutches smaller than the predicted sizes? Charnov and Skinner consider three reasons. Two of the three are similar to factors we discussed for birds: Larger clutch sizes may reduce offspring fitness in ways that Charnov and Skinner did not include in their calculations; and there may be trade-offs between a female's investment in a particular clutch and her own future survival or reproductive performance. Charnov and Skinner's third hypothesis is novel to parasitoid wasps. Unlike birds, female parasitoid wasps may produce more than one clutch in rapid succession. Soon after she has laid one clutch, a female wasp may begin looking for another host to parasitize. The appropriate measure of a wasp's fitness with regard to clutch size may not be the discrete fitness she gains from a single clutch. Instead, it may be the rate at which her fitness rises as she searches for hosts and lays eggs in them. Readers familiar with behavioral ecology may recognize this as an optimal foraging problem.

Figure 13.23 presents a graphical analysis of a female's rate of increase in fitness over time. The figure follows the female from the time she sets out to find a host egg until she leaves that host egg to look for another. While she is searching, the female gains no fitness. Once she finds a host and begins to lay eggs in it, however, her fitness begins to rise. The fitness she gets from a clutch of any given size is determined by a parabolic function, as in our original depiction of Lack's hypothesis (Figure 13.19). In this example, if a female leaves to look for a new host

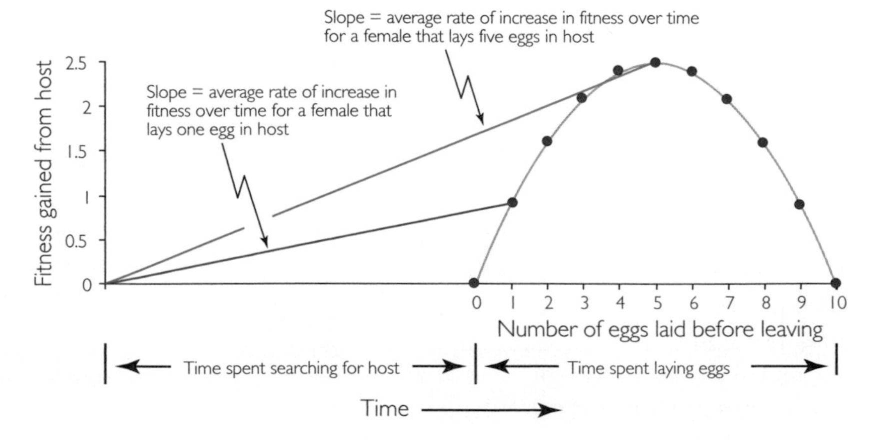

Figure 13.23 Rate of increase in parasitoid maternal fitness with time spent searching for hosts and laying eggs The horizontal axis represents the time spent by a female parasitoid in searching for a host egg and depositing a clutch. The vertical axis represents female fitness, in units of surviving offspring. The red dots show the relationship between number of surviving offspring and clutch size, as in Lack's hypothesis. After Charnov and Skinner (1985).

after laying just one egg, her total fitness gain from the first host is 0.9. Her average rate of fitness gain from the time she set out looking for the first host to the time she leaves to look for a second is given by 0.9 divided by the total elapsed time. This rate of fitness gain is equal to the slope of the diagonal line from the origin to the point representing a clutch size of one. Likewise, if the female stays to lay five eggs, her average rate of fitness gain for the whole trip is given by the slope of the upper diagonal line. In this example, the female would get the highest rate of fitness gain from this host if she left after laying four eggs. This is one egg less than the most productive clutch size. Thus, if female parasitoids are selected to maximize their rate of fitness increase, they may produce smaller clutches than those predicted by Lack's hypothesis.

To summarize, Lack's hypothesis is a useful starting point for the evolutionary analysis of clutch size. Assuming only that there is a trade-off between the number of offspring in a clutch and the survival of individual offspring, Lack's hypothesis predicts that parents will produce clutches of the size that maximizes the number of surviving offspring. This prediction is often violated, with actual clutches typically smaller than expected. These violations indicate the possible presence of other trade-offs. Current parental reproductive effort may be negatively correlated with future parental survival or reproductive performance, or clutch size may be negatively correlated with offspring reproductive performance. Alternatively, a violation of predicted clutch size may indicate that we have chosen the wrong measure of parental fitness.

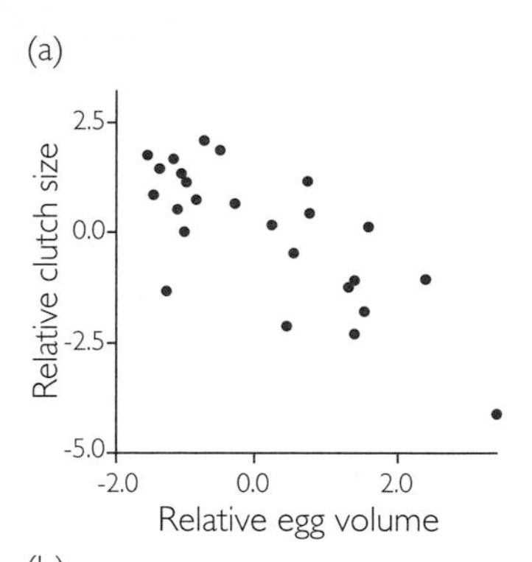

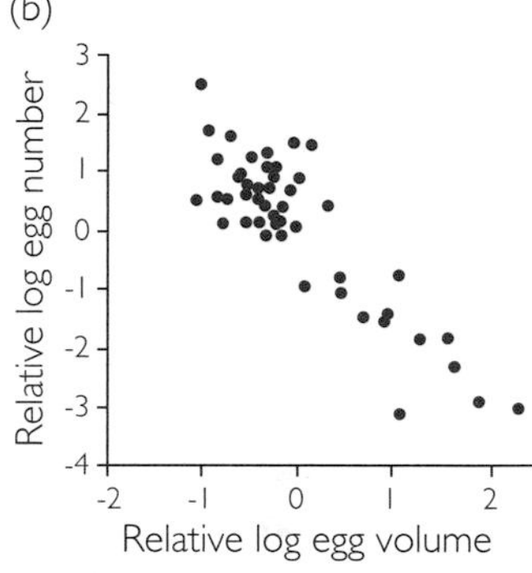

Figure 13.24 Trade-offs across taxa in size and number of offspring (a) A trade-off across 26 fish families. Larger fish produce bigger clutches, so Elgar had to use statistical techniques to remove the effect of variation among fish families in body size. The vertical axis shows relative clutch size,or the number of eggs per clutch adjusted for differences in body size among families. The horizontal axis shows relative egg volume, or egg size adjusted for differences in body size among families. The negative correlation between size and number of eggs is statistically significant ($P < 0.001$). From Elgar (1990). (b) A trade-off across fruit fly species. Larger fruit flies produce more and larger eggs, so Berrigan also had to use statistical techniques to remove the effect of variation in body size. The vertical axis shows relative egg number; the horizontal axis shows relative egg volume. The negative correlation is significant ($P < 0.001$). Berrigan found similar patterns in wasps and beetles. Provided by David Berrigan using data analyzed in Berrigan (1991).

13.4 How Big Should Each Offspring Be?

In Section 13.3, we assumed that the size of individual offspring was fixed. We now relax that assumption. Given that an organism will invest a particular amount of energy in an episode of reproduction, we can ask whether that energy should be invested in many small offspring or a few large offspring.

A trade-off between the size and number of offspring should be fundamental. A pie can be sliced into many small pieces or a few large pieces, but it cannot be sliced into many large pieces. Biologists have found empirical evidence for a size-versus-number trade-off in a variety of taxa. Mark Elgar (1990), for example, analyzed data from the literature on 26 families of fish. Elgar found a clear negative correlation between clutch size and egg size: Fish that produce larger eggs produce fewer eggs per clutch (Figure 13.24a). David Berrigan (1991) performed a similar analysis of variation in egg size and number among species in three orders of insects. In each case, Berrigan found a clear negative correlation between egg number and egg size. Data for one of his insect orders appear in Figure 13.24b.

Selection on Offspring Size

If selection on parents is forced by a fundamental constraint to strike a balance between the size and number of offspring, what is the optimal compromise? Christopher Smith and Stephen Fretwell (1974) offered a mathematical analysis of this question. Smith and Fretwell's analysis is based on two assumptions. The first assumption is the trade-off between size and number of offspring (Figure 13.25a). The second assumption is that individual offspring will have a better chance of surviving if they are larger (Figure 13.25b). There must be a minimum

(a)

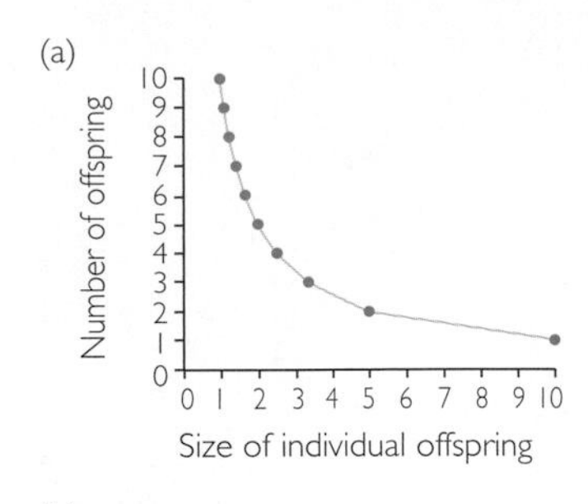

(b)

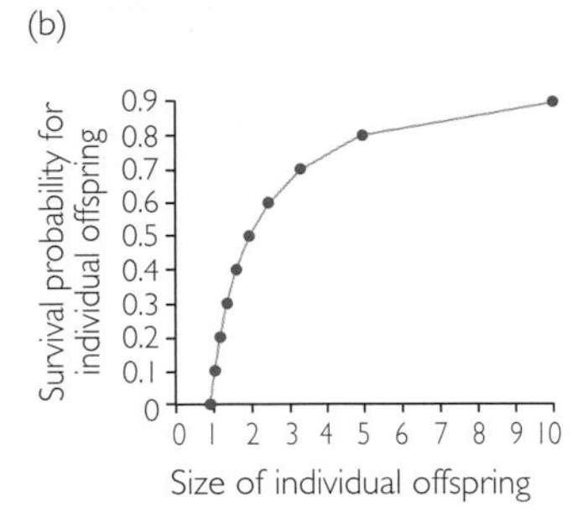

(c)

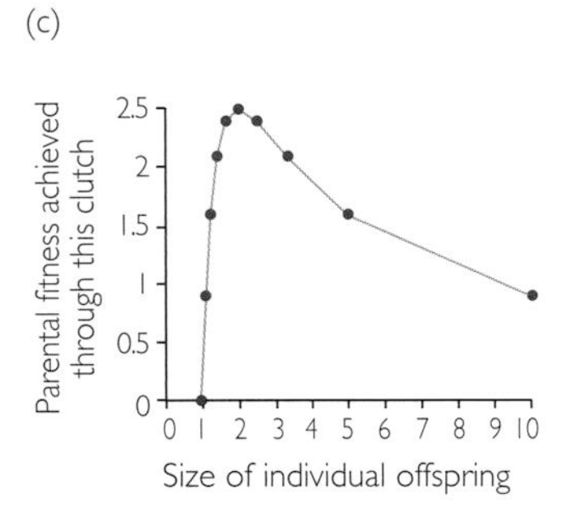

Figure 13.25 The optimal compromise between size and number of offspring (a) Assumption 1: There is a trade-off between size and number of offspring. The units we have used are arbitrary. The shape of the curve may vary from species to species. Here we have used the equation: Number = 10/Size. (b) Assumption 2: Above a minimum size, the probability that any individual offspring will survive is an increasing function of its size. Again, we have used arbitrary units. The shape of the curve may vary from species to species. Here we have used the equation: Survival = 1 − (1/size). (c) Analysis: The parental fitness gained from a single clutch of offspring of a given size is the number of offspring in the clutch multiplied by the probability that any individual offspring will survive. For example, given the equations and units used here, if a parent makes offspring of size five it can make two of them, and each has a probability of survival of 0.8. Thus, the expected fitness gained by the parent from this clutch is 2 × 0.8 = 1.6. For some (but not all) combinations of the trade-off function (a) and the survival function (b), parents achieve maximum fitness through offspring of intermediate size (as in c). After Smith and Fretwell (1974).

size below which offspring have no chance of survival. As offspring get larger, their probability of surviving rises. If survival probability approaches one, it must do so in a saturating fashion, because survival probability cannot exceed one. Given the two assumptions, the analysis is simple: The expected fitness of a parent producing offspring of a particular size is the number of such offspring the parent can make, multiplied by the probability that any individual offspring will survive. A plot of expected parental fitness versus offspring size (Figure 13.22c) reveals the size of offspring that gives the highest parental fitness.

The optimal offspring size depends on the shapes of the relationships for offspring number versus size and offspring survival versus size. In many cases, as in Figure 13.25, the optimal offspring size is intermediate. The important point here is that selection on parental fitness often favors offspring smaller than the size favored by selection on offspring fitness. This identification of a potential conflict of interest between parents and offspring is the primary contribution of Smith and Fretwell's model.

The shape of the offspring survival curve is particularly important (Figure 13.25b). In Smith and Fretwell's model, survival probability increases with offspring size, but the rate of increase declines: That is, increasingly large offspring gain a progressively smaller survival benefit. This leads directly to the prediction of an intermediate offspring size that gives the highest parental fitness (Figure 13.25c). If the offspring survival curve were a linear relationship instead of a concave curve (Vance 1973), the model would predict selection favoring extremes of offspring size: the smallest offspring capable of development, or the largest offspring that a female could manufacture, rather than some optimal intermediate offspring size (see Levitan 1993, 1996; Podolsky and Strathmann 1996).

Organisms face a trade-off between making many low-quality offspring or a few high-quality offspring. Selection on parents favors a compromise between the quality and quantity of offspring, but selection on individual offspring favors high quality.

It is possible to test Smith and Fretwell's analysis empirically only if there is substantial variation in offspring size among parents within a population. Variation in offspring size is relatively small in most species (Stearns 1992). We review two recent studies that have confirmed both the assumptions and the conclusion of Smith and Fretwell's analysis. In one study, researchers took advantage of the large variation in egg size in a population of fish. In the other study, researchers took advantage of phenotypic plasticity in egg size in a beetle.

Selection on Offspring Size in a Population of Fish

Daniel Heath and colleagues (2003) studied chinook salmon (*Oncorhynchus tshawytscha*) at a commercial salmon hatchery in British Columbia, Canada. When adult salmon return to the hatchery, workers harvest eggs from the females and fertilize them with sperm from the males. After the fry have hatched and grown for a time, the hatchery workers release them into natural rivers. When the fry grow up, they return to the hatchery to continue the cycle. Eggs produced by female salmon at the hatchery range in mass from less than 0.15 g to more than 0.30 g.

Heath and colleagues tested the first assumption of Smith and Fretwell at the level of individual females. The plot in Figure 13.26a shows the relationship between female relative fecundity—that is, the number of eggs a female lays per kg of her body mass—and the average size of a female's eggs. For the year shown, and for the three other years the researchers analyzed, there was a trade-off between size and number of eggs.

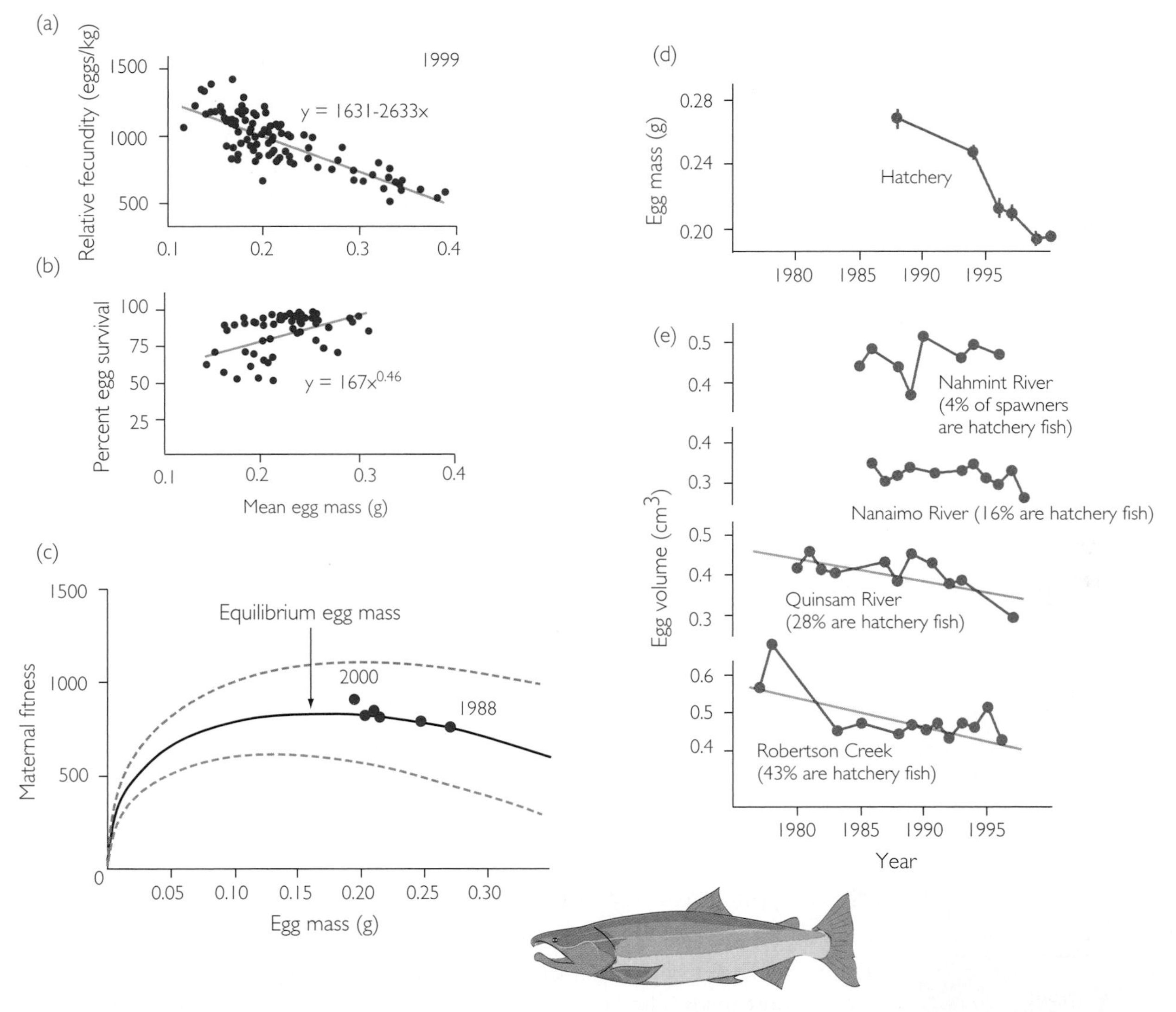

Figure 13.26 Evolution of reduced egg size in hatchery salmon (a) There is a trade-off, for female salmon of a given body size, between size and number of eggs. (b) The probability that a given offspring will survive increases with egg size. (c) Maternal fitness, calculated as the product of relative fecundity and the probability of offspring survival, is maximized at intermediate egg size. The optimal egg size for hatchery salmon, however, is smaller than that for wild salmon because offspring emerging from small eggs have a better chance of surviving in the hatchery than in natural streams. The data points show mean values for salmon in a recently established hatchery population for each of several years. (d) The average egg size in the hatchery population evolved toward smaller sizes over a period of just a few years. (e) Egg size also declined in wild populations receiving substantial numbers of immigrants from the hatchery population (Quinsam River and Robertson Creek) but remained stable in wild populations receiving relatively few immigrants from the hatchery (Nahmint River and Nanaimo River). From Heath et al. (2003).

Heath and colleagues tested the second assumption of Smith and Fretwell by tracking the fates of individual fry. The plot in Figure 13.26b shows the probability of survival as a function of the mass of the egg a fry hatches from. For the year shown and for the three other years the researchers analyzed, selection favored larger offspring. Furthermore, the relationship between survival and size followed a concave curve, just as Smith and Fretwell assumed.

How do the egg size versus number trade-off and selection on fry combine to select on the egg size produced by the mothers? Heath and colleagues estimated the relationship between maternal fitness and egg size by multiplying the fitted curve in Figure 13.26a and the fitted curve in Figure 13.26b. The result appears in Figure 13.26c. It shows that female fitness is maximal at intermediate egg sizes. The optimal egg size for females breeding at the hatchery was just over 0.15 g.

The optimal egg size for hatchery females turns out to be lower than the optimal egg size for females in the wild. This is because the hatchery provides a safe environment for young fry. Small fry, in particular, are more likely to survive in the hatchery than they are in natural rivers. When small fry are more likely to survive, females that make more and smaller fry have higher reproductive success. This enabled Heath and colleagues to predict that the hatchery population, which was founded in the late 1980s from wild stock, should be evolving toward smaller average egg sizes. Data from the hatchery confirm this prediction, as can be seen from the data points in Figure 13.26c and the time series in Figure 13.26d. In other words, the commercial hatchery has been running an unintentional experiment that confirms the predictions of Smith and Fretwell's analysis.

The evolution of hatchery populations toward smaller egg sizes has implications for the conservation of wild salmon stocks. A widespread conservation strategy for salmon is to supplement wild populations with fish from hatcheries. This amounts to migration from hatchery populations into wild populations. Heath and colleagues analyzed data from four rivers on Vancouver Island in which chinook stocks are being supplemented with hatchery fish. The amount of supplementation varies among the four rivers: 4% of the fish spawning in the Nahmint River are migrants from the hatchery, as are 16% of the fish in the Nanaimo River, 28% of the fish in Roberston Creek, and 43% of the fish in Quinsam River. As the time series in Figure 13.26e show, the chinook populations in Quinsam River and Robertson Creek have been evolving toward smaller egg sizes since at least 1980. Gene flow from the hatchery appears to be driving the evolution of suboptimal egg size in heavily supplemented wild chinook populations.

Phenotypic Plasticity in Egg Size in a Beetle

Charles Fox and colleagues (1997) studied the seed beetle *Stator limbatus* (see Figure 13.27. The females of this small beetle lay their eggs directly onto the surface of host seeds. The larvae hatch and burrow into the seed. Inside, the larvae feed, grow, and pupate. They emerge from the seed as adults. *S. limbatus* is a generalist seed predator; it has been reared on the seeds of over 50 different host species.

Fox and colleagues studied *S. limbatus* on two of its natural hosts: an acacia (*Acacia greggii*) and a palo verde (*Cercidium floridum*). The acacia is a good host; most larvae living in its seeds survive to adulthood. The palo verde is a poor host; fewer than half the larvae living in its seeds survive. We can easily add hosts of different quality to the Smith and Fretwell analysis (Figure 13.28). When we do so,

Figure 13.27 Athe seed beetle, *Stator limbatus* (a) This female is looking for a place to lay her eggs on the seeds of catclaw acacia (*Acacia greggii*) and blue palo verde (*Cercidium floridum*).

we get a clear prediction: Females should lay larger eggs on the poor host than on the good host. Recall from Chapter 10 that when selection favors different phenotypes at different times or in different places, organisms sometimes evolve phenotypic plasticity. The analysis in Figure 13.28 predicts that *S. limbatus* should exhibit phenotypic plasticity in egg size.

Fox and colleagues found that, as predicted, female *S. limbatus* adjust the size of the eggs that they lay to the host on which they deposit them. When the researchers took newly emerged females from the same population and gave them only one kind of seed, females given palo verde seeds (the poor host) laid significantly larger eggs than females given acacia seeds (Figure 13.29a). Confirming assumption 1 of Smith and Fretwell, these larger eggs came at the cost of fewer eggs produced over a lifetime (Figure 13.25b).

For females laying on palo verde seeds, the production of large eggs is adaptive. Fox et al. manipulated females into laying small eggs on palo verde seeds by keeping the females on acacia seeds until they laid their first egg, then moving them to palo verde seeds. Only 0.3% of the larvae hatching from small eggs on palo verde seeds survived to adulthood, whereas 24% of the larvae hatching from

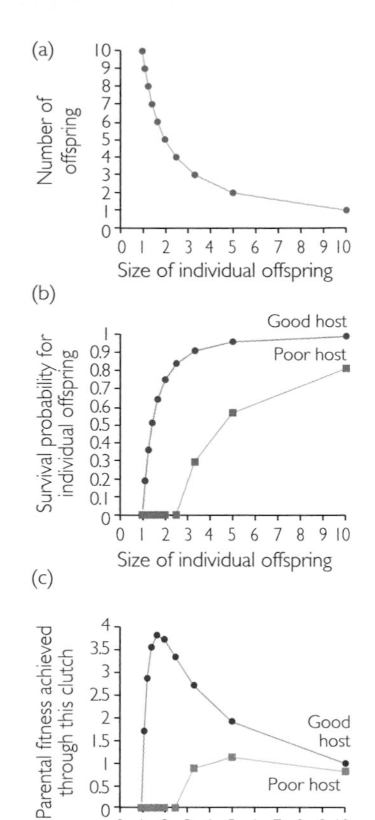

Figure 13.28 The optimal compromise between size and number of offspring on a good host and a poor host (a) As in Figure 13.22, we assume a trade-off between size and number of offspring. (b) As in Figure 13.22, we assume that there is a minimum size below which individual offspring do not survive; above this minimum size, the probability that any individual offspring will survive is an increasing function of offspring size. The minimum size for offspring survival is smaller on the good host. Furthermore, survival is higher on the good host at all sizes above the minimum. (c) Analysis: The parental fitness gained from a single clutch of offspring of a given size is the number of offspring in the clutch multiplied by the probability that any individual offspring will survive. The optimal offspring size (for the mother) is bigger on the poor host than on the good host. After Smith and Fretwell (1974).

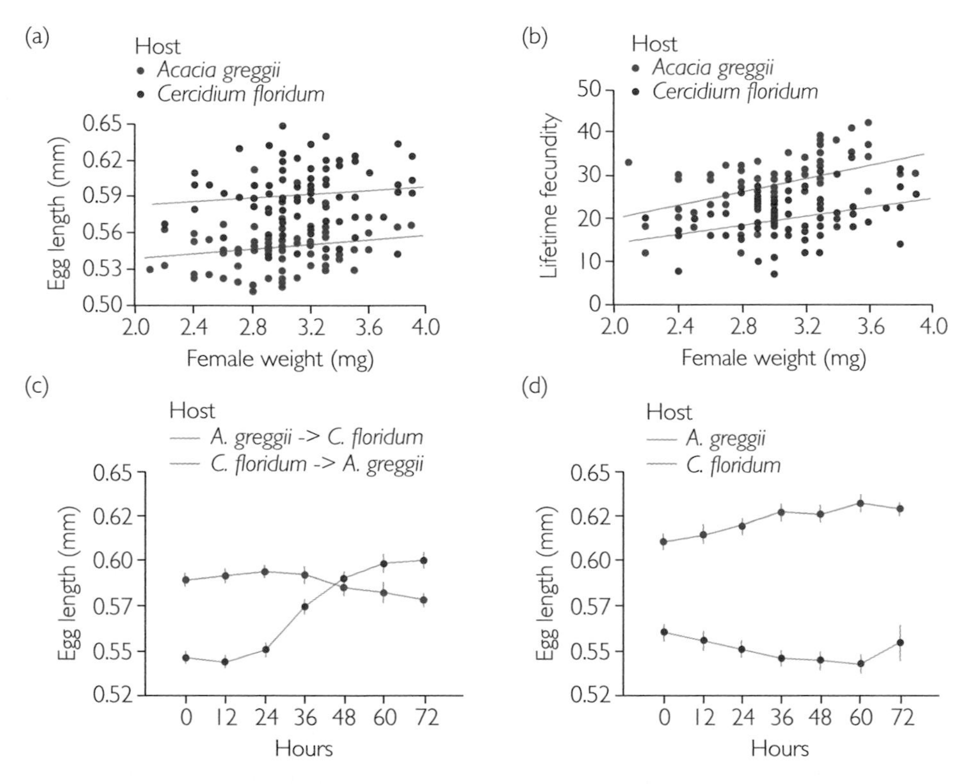

Figure 13.29 Phenotypic plasticity in egg size in the seed beetle, *Stator limbatus* (a) Larger females lay slightly larger eggs, but females laying on *Cercidium floridum* lay larger eggs for their size ($P < 0.001$). (b) Larger females have slightly higher fecundity (number of eggs), but females laying on *A. greggii* have higher fecundity for their size ($P < 0.001$) (c) In a host switch experiment, Fox and colleagues placed newly emerged females on each kind of seed, let them lay their first egg, and then switched them to the other kind of seed. The graph shows egg size as a function of time for the subsequent 72 hours. Females that laid their first egg on acacia (*A. greggi*, the good host) produced small eggs at first, but gradually switched to large eggs. Females that laid their first egg on palo verde (*C. floridum*, the poor host) produced large eggs at first, but gradually switched to smaller eggs. (d) In a control experiment for (c), Fox and colleagues left females on each kind of seed for life. Each group laid large or small eggs for life. From Fox et al. (1997).

large eggs on palo verde seeds survived ($P < 0.0001$). Confirming assumption 2 of Smith and Fretwell, even among the large eggs on palo verde seeds, the probability of survival from egg to adult was positively correlated with egg size. For females laying on acacia seeds, the production of small eggs is adaptive. Given that nearly all larvae hatching on acacia seeds survive, females producing more and smaller eggs have higher lifetime reproductive success.

Fox and colleagues even showed that individual females that had started to lay size-appropriate eggs on one host could readjust their egg size when switched to the other host (Figure 13.29d). Control females left on one kind of seed consistently produced large or small seeds for life (Figure 13.29e).

In summary, selection on offspring size often involves a conflict of interest between parents and offspring. Because making larger offspring also means making fewer offspring, selection on parents can favor smaller offspring sizes than are optimal for offspring survival. The exact balance between size versus number of offspring depends on the relationship between offspring size and survival. Poor environments pose a greater obstacle to offspring survival and thus favor larger offspring.

13.5 Conflicts of Interest between Life Histories

Analyzing trade-offs among life history traits has helped to explain many aspects of the extraordinary life history diversity known among living organisms. However, this view can sometimes obscure the fact that the life history of each organism unfolds in an ecological context that includes other individuals. For example, the opossum life history in Figure 13.2 shows reproduction by a hypothetical female but does not show the males with which she mated to produce offspring. This simplification might imply that the males are mere sperm providers and that their interests in the production of offspring are the same as the interests of the female shown in the figure. In fact, the reproductive interests of males and females will often be different. In this section, we discuss two such conflicts of interest and their evolutionary consequences.

Genetic Conflict between Mates: Genomic Imprinting

When different males father offspring within the same litter or clutch, the reproductive interests of the fathers and the mother conflict.

Opossums and other mammals that nourish their offspring through a placenta offer a surprising opportunity for conflict between the reproductive interests of females (which brood the offspring) and males (which do not). Consider the copies of a mammalian gene inherited from the father versus the mother. Why might these alleles be in conflict? In most mammals, females carry offspring from many different males in the course of a lifetime. Indeed, offspring with different fathers are frequently found in the same litter. Because the mother is related to each of these offspring equally, natural selection should act to equalize her physiological investment in each. Natural selection should, on the other hand, favor a father that can coerce the mother into investing more heavily in his offspring at the expense of offspring from other males.

Consistent with this prediction, at least some loci are biochemically marked (or imprinted) in mammals, to distinguish paternal and maternal alleles (Barlow 1995). This marking of alleles occurs in the testis and ovary during production of gametes. Imprinting affects the subsequent transcription of the marked genes within cells of the embryo after fertilization. The paternal allele of a hormone called insulin-like growth factor II (IGF-II), for example, is widely expressed in mice, while the maternal copy is hardly transcribed. This is a surprising pattern of gene expression for a diploid organism, because natural selection should favor the equal expression of both alleles. Equal expression protects the offspring against the effects of deleterious recessive mutations that interfere with the function of one allele (Hurst 1999). Why should a mother imprint her IGF-II alleles to reduce transcription of this gene in her offspring, especially when the paternal allele of the same gene is actively transcribed?

Genomic imprinting occurs when male and female alleles contain distinct chemical markers and are transcribed differently.

The answer hinges on the function of IGF-II and its interaction with other molecules. This hormone is a general stimulant to cell division and acts through a cell surface protein called the type-1 IGF-II receptor. However, it happens that another abundant cell surface protein in mice, called the cation-independent mannose-6-phosphate receptor (CI-MPR), also has a binding site for IGF-II (this alternative binding site is called the type-2 receptor). CI-MPR's function is completely unrelated to growth, and in mouse embryos it is transcribed only from the maternal allele.

David Haig and colleagues have proposed that this bizarre arrangement of hormones, receptors, and transcription patterns results from a tug-of-war between the interests of maternal and paternal alleles within the uterus. According to this interpretation, the paternally transcribed IGF-II is selected to maximize rates of cell division in the developing embryo. This increases growth rates and monopolizes the flow of maternal resources to the embryo through the placenta. The maternal IGF-II allele is turned off in order to conserve resources for future reproduction. In contrast, the maternally transcribed type-2 receptor is selected to bind excess paternal hormone, mitigate the effects of IGF-II overtranscription, and equalize the flow of resources to different embryos, while the paternal type-2 receptor allele is turned off in order to maximize the influence of the paternal IGF-II hormone on the mother (Haig and Graham 1991; Moore and Haig 1991).

Consistent with this interpretation, CI-MPR does not bind IGF-II in chickens and frogs; their embryos are provisioned before fertilization. Chicken and frog fathers have no opportunity to manipulate the distribution of maternal resources among offspring. This is a hint that the type-2 receptor of mammals evolved after the advent of the placenta, in response to selection that favored equalization of maternal resources among all offspring. Genomic imprinting has also been confirmed in flowering plants and may have been important in the evolution of the nutritive tissue called endosperm (see Haig and Westoby 1989, 1991).

The qualitative predictions of Haig's hypothesis for genomic imprinting have generally been confirmed, though there is some debate over whether quantitative variation in imprinting occurs and whether multiple paternity is necessary for imprinting to arise (Haig 1999; Hurst 1999; Spencer et al. 1999). For example, alleles could vary in the amount of transcription rather than being turned "on" or "off." Imprinting is known to be widespread in mammalian genomes (see references in Spencer et al. 1999), and the details of the imprinting mechanism and interactions among imprinted alleles could vary among genes and species.

Finally, it is important to note that mammals are not the only animals that have evolved placental development. For example, lizards (Guillette and Jones 1985), sharks (Wourms 1993), and numerous marine invertebrate groups (Strathmann 1987) have evolved structures like a placenta that transfer materials between the maternal body and the internally brooded offspring. Haig's hypothesis predicts that imprinted genes should be found in these groups, and that these will be genes that moderate the conflict among offspring within a brood as they compete for maternal resources. The hypothesis has not yet been tested in these groups, however (Spencer et al. 1999).

Physiological Conflict between Mates: Sexual Coevolution

In Chapter 8, we introduced the idea of adaptations arising in competing species, such as hosts and pathogens, that counteract each other's effects so that neither lineage shows a net gain in fitness. In these circumstances, fitness evolves around a kind of dynamic equilibrium in which the environment, and thus the nature of selection acting on a population of organisms, is largely determined by interactions with other organisms and their adaptations (Van Valen 1973).

This idea can be extended to life history adaptations arising within species as well. Experiments by William Rice and colleagues show that, where the reproductive interests of male and female fruit flies differ, sexual selection may favor adaptations that arise in one sex but are actually detrimental to the other sex. One of these adaptations involves the biochemistry of male seminal fluid, which has evolved to influence female behavior, such as egg laying rate or the tendency to remate with another male (Fowler and Partridge 1989). These effects are beneficial to a male if his mate is likely to have multiple partners because these adaptations will tend to increase the number of eggs that are fertilized by his sperm. Such seminal fluid is toxic and increases mortality of females, however (Fowler and Partridge 1989). Toxic effects favor the subsequent evolution of resistance among females, followed by more extreme adaptations among males to overcome female resistance. This iterated process has been called chase-away sexual selection (Rice 1987; Rice and Holland 1997; Holland and Rice 1998).

When mates are not monogamous, the life history strategy that is optimal for one sex may be suboptimal for the other.

Direct evidence for this kind of antagonistic sexual adaptation comes from experiments conducted by Rice (1996). In these experiments, male flies competed with each other for matings with females. Females, in turn, were able to mate with multiple partners. The competition among males resulted in selection for traits such as high rate of remating with the same female and highly toxic seminal fluid. However, only male offspring were retained from each generation of experimental mating. After each round of selection, the selected males were mated to females from a control group in which competition for mates was not occurring. In this way, Rice kept the female response to male sexual adaptations static while the selected males competed with one another to overcome female defenses.

The results of 31–41 generations of such selection are shown in Figure 13.30. Compared to males in the control group, selected males had higher fitness (more sons born per male, shown as the *net fitness* assays in Figure 13.30a). The data on the right-hand side of Figure 13.26a suggest that two traits contributed to the higher fitness of selected males. They were more likely to remate with the same female, and they fertilized a much higher proportion of eggs when the female was remated to another male (*defense* assays, Fig. 13.30a). These benefits to male reproductive success came at a cost to females, however: After 41 generations of selection, the mortality rate for females mated to selected males was about 50% higher than the mortality rate for females mated to unselected males (Figure 13.30b). The experiment suggests that males and females are engaged in a reproductive "arms race," which males win if female countermeasures are prevented.

It is important to recognize, however, that this result is based on a particular mating system. By enforcing monogamous mating upon flies from the same source population for many generations, Holland and Rice (1999) showed that

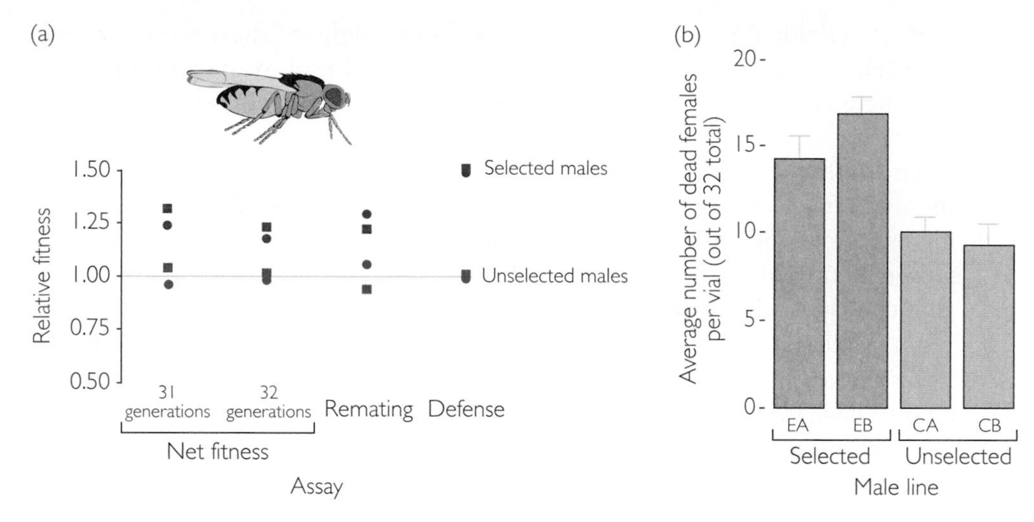

Figure 13.30 Antagonistic sexual selection between male and female fruit flies (a) Males selected for success in mating competition have higher relative fitness, measured as the number of sons per father, than unselected males. Net fitness was assayed after 31 and 32 generations in two populations of selected males (red circle or square) and compared to net fitness of unselected males (blue circle or square). Two components of this increased fitness are the rate of Remating with females to ensure paternity and Defense—measured as the relative number of offspring sired by a selected male after his mate was given a chance to mate with a second male. (b) When females are prevented from evolving compensatory life history traits, the cost of these male adaptations is higher female mortality. After 41 generations of selection on males, females mated to two different lines of selected males (EA and EB) had much higher mortality rates than females mated to unselected males (CA and CB) ($P = 0.0194$). Mortality is shown as the number of dead females out of initial groups of 32. After Rice (1996).

the effects of antagonistic sexual adaptation could actually be reversed: Monogamous lines of flies evolved lower sperm toxicity in males and lower resistance to sperm toxicity in females. These results make sense in light of the new relationship between male and female fitness: Males with only one lifetime mate depend on the fitness of that female alone to produce offspring, and they have no fear of being cuckolded by another male. These males should evolve less harmful life history traits in order to increase their own fitness. The defenses evolved by females against toxic male seminal fluid (or other male life history traits) become less beneficial as males evolve more benign traits of their own. If female resistance is expensive in time or energy, then resistance traits should be selected against (Holland and Rice 1999).

13.6 Life Histories in a Broader Evolutionary Context

In this final section of the chapter, we place life histories in a broader evolutionary context. We briefly consider examples of research addressing two general questions: What forces maintain genetic variation in populations? How do novel traits evolve?

The Maintenance of Genetic Variation

Natural selection on a trait should reduce the genetic variation for the trait (Fisher 1930). A simple example illustrates why (Roff 1992). Imagine a series of loci, each with two alleles, that affect a single trait correlated with fitness. At each locus, one allele ("0") contributes to the trait in such a way as to add zero units to the fitness of individuals, whereas the other allele ("1") contributes one unit to individual fitness. The genotype with the highest fitness is homozygous for allele "1" at all loci. Over time, selection should lead to the fixation of the "1" allele at each locus, and there will no longer be genetic variation in the trait.

Life history traits are closely correlated with fitness and have relatively low heritabilities.

Life history traits, because of their intimate connection with reproduction, should be more closely correlated with fitness than other kinds of traits, including behavioral, physiological, or morphological traits (Mousseau and Roff 1987). Consequently, life history traits should show less genetic variation—lower heritability—than other kinds of traits (for a discussion of heritability, see Chapter 9). From the literature, Mousseau and Roff (1987) assembled a sample of 1,120 estimates of the heritability of various traits. They found that life history traits indeed tend to have the lowest heritabilities (Figure 13.31). This result is consistent with the expectation from our simple theoretical treatment (for an alternative interpretation, see Price and Schluter 1991).

Nonetheless, Mousseau and Roff's review documents that life history traits typically have substantial genetic variation. What evolutionary forces maintain genetic variation in populations? The list of possibilities includes mutation, heterozygote advantage, frequency-dependent selection, and genotype-by-environment interaction in which different genotypes have higher fitness in different environments or at different times (see Chapters 6–10).

Richard Grosberg (1988) studied the maintenance of genetic variation for life history traits in a population of the sea squirt *Botryllus schlosseri*. *Botryllus schlosseri* is a colonial animal that lives attached to hard surfaces in shallow marine waters of the temperate zone. Colonies consist of a number of identical modules. The modules in a colony are physiologically connected, and their life histories are synchronous. The population Grosberg studied contains two distinct life history morphs. One morph is semelparous: Upon reaching sexual maturity, the modules in a colony reproduce once and die. The other morph is iteroparous: Colonies have at least three episodes of sexual reproduction before they die. In a series of experiments in which he grew sea squirts in a common environment and bred the morphs with each other, Grosberg demonstrated that the two morphs are genetically determined.

What maintains genetic variation for life history morphs in this sea squirt population? Grosberg tracked the seasonal frequency of the two morphs over two years (Figure 13.32). In both years, the semelparous morph dominated the population in the spring and early summer, whereas the iteroparous morph dom-

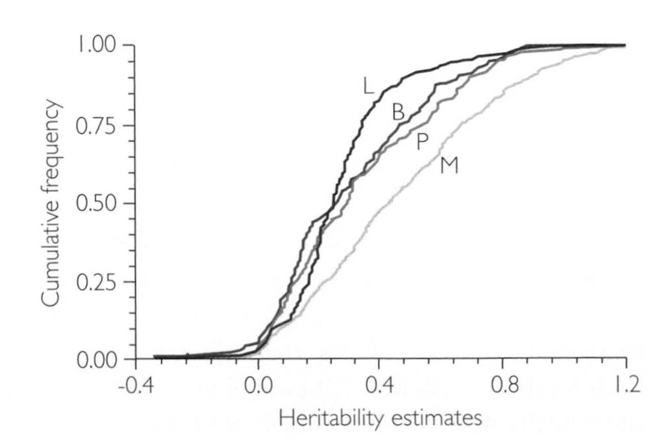

Figure 13.31 Life history traits have lower heritabilities than other kinds of traits This plot shows four cumulative frequency distributions. A cumulative frequency distribution is a running sum, moving across a histogram, of the heights of the bars. The more rapidly the curve in a cumulative frequency distribution rises to 1, the lower the mean of the histogram. The line marked L is the cumulative frequency distribution for estimates of the heritability of life history traits; *B* = behavioral traits; *P* = physiological traits; M = morphological traits. Life history traits tend to have the lowest heritabilities. From Mousseau and Roff (1987).

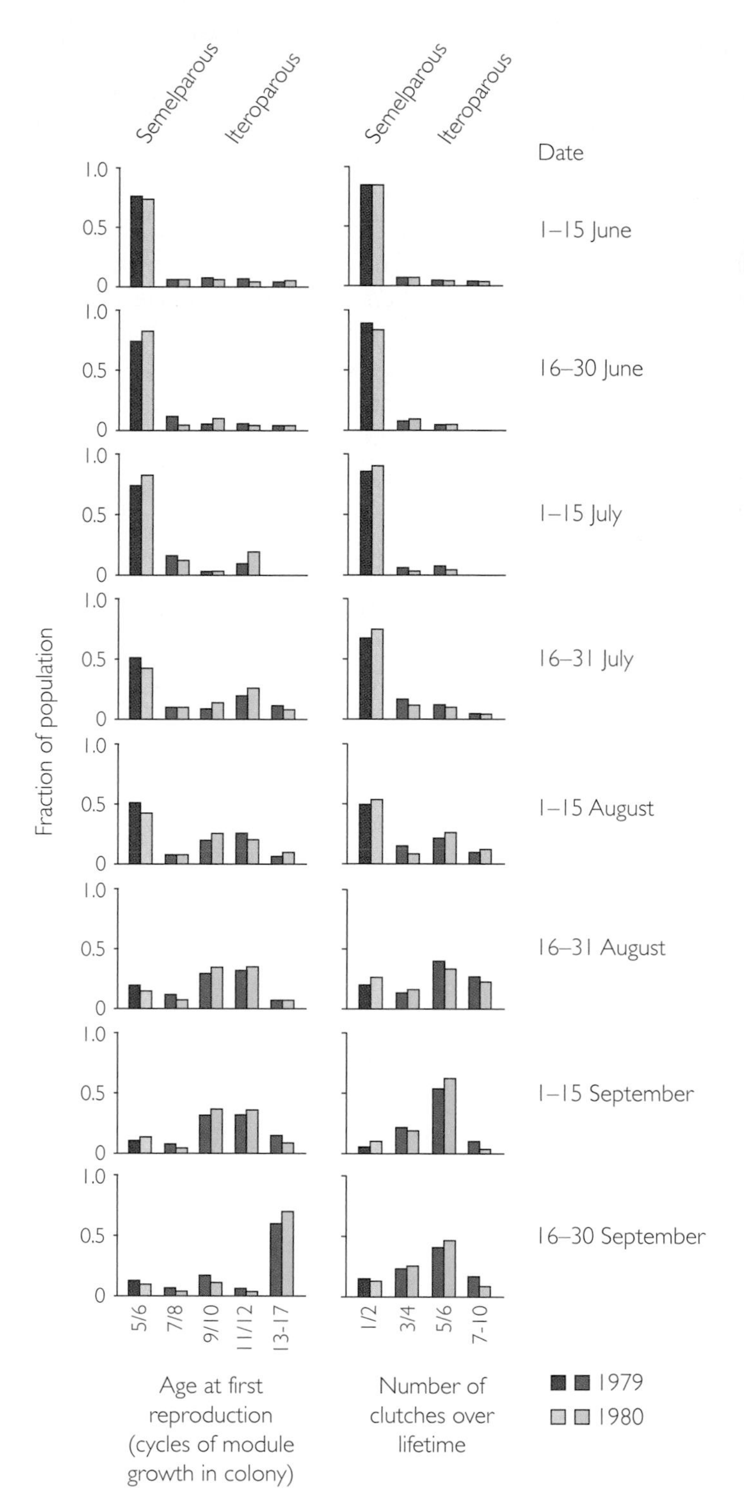

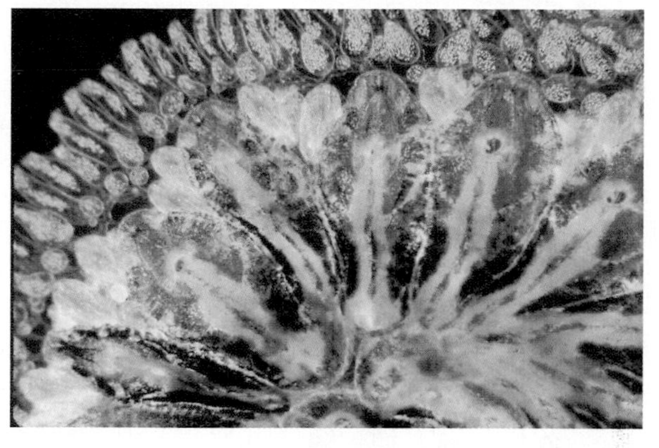

Figure 13.32 Annual cycles in the frequencies of two life history morphs in a population of sea squirts The photo shows a colony of the sea squirt, *Botryllus schlosseri*. Each of the bar graphs shows a frequency distribution for the population during a two-week period. Colonies of the semelparous morph (brown and tan bars) reproduce at an early age (age cycles of module growth in the colony) and produce only a single clutch of offspring. Colonies of the iteroparous morph (blue and green bars) reproduce at a late age and produce at least three clutches of offspring. From Grosberg (1988).

inated in the late summer. This indicates that the two morphs are maintained in the population by seasonal variation in selection. One important selective factor may be competitive interactions with another sea squirt (*Botryllus leachi*). This competitor, which becomes more abundant late in the summer, overgrows colonies of the semelparous *B. schlosseri* morph but not the iteroparous morph—a genotype-by-environment interaction.

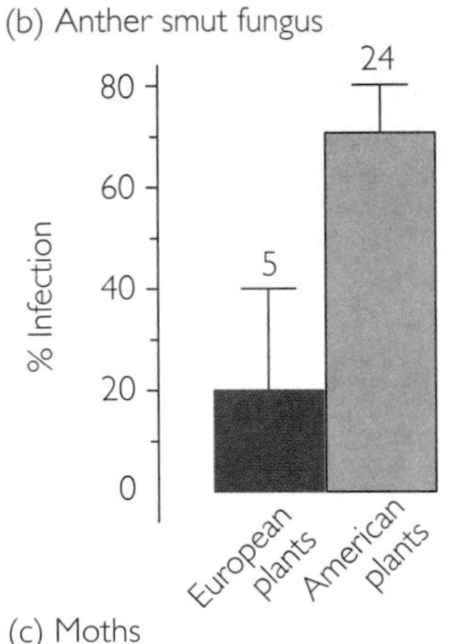

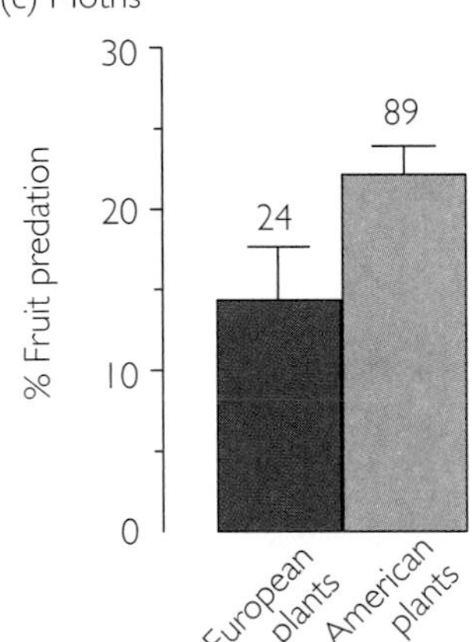

Figure 13.33 Life history evolution in a weed (a) Snowy campion is benign in its native Europe but weedy in North America. (b) and (c) Grown together in Europe, American campions were more vulnerable to diseases and predators than European campions. From Wolfe et al. (2004). Note that Wolfe divided plants by size for analysis; we have included only the largest. Overall, male American plants were significantly more vulnerable to infection (P = 0.008) and all American plants were significantly more vulnerable to predation (P = 0.042).

Biological Invasions

How does a species that is benign on one continent transmogrify into a pest when transported to another continent? Research by Lorne Wolfe and colleagues on the snowy campion, *Silene latifolia,* provides an example. Snowy campion is a small perennial herb native to Europe, where it is innocuous (Figure 13.33a). Accidentally introduced into North America some 200 years ago, it has there become an agricultural weed. The traditional explanation is that when the plant moved to the New World, it left all of its natural enemies behind. These include, among many others, a fungus that attacks the plant's anthers and a seed-eating caterpillar. Freed from the burdens imposed by these enemies, snowy campion thrived in America.

Amy Blair and Wolfe (2004) suspected that there was more to the story; that snowy campion had not just escaped its enemies but had also evolved. They tested their hypothesis by planting seeds from European and North American snowy campions together in a common garden in the United States. Since the plants would all experience the same environment, any differences in phenotype must be due to genotype. Consistent with the reseachers' prediction, the European and American plants were not the same. The American plants germinated earlier, grew faster, made more flowers, and survived at higher rates than the European plants.

How could a plant that had been evolving in Europe for millions of years suddenly become so superior in North America? Blair and Wolfe had evidence to suggest that the answer involved a life-history trade-off. With few enemies, a change in energy budget was possible, even adaptive. Mutants that skimped on defense to invest more heavily in reproduction should enjoy higher fitness.

Wolfe and colleagues (2004) planted a second common garden experiment in Europe. Consistent with the results from the American garden, the campions from the two continents were strikingly different. This time, however, the difference was that the American plants were easy pickings for predators and pathogens. Figure 13.33b shows the American plants' greater susceptibility to anther smut fungus. Figure 13.33c shows their greater vulnerability to seed-eating moths. Upon repatriation to Europe, the snowy campion's evolved defenselessness was a grave liability. Wolfe and colleagues concluded that what had turned snowy campion into a weed was life history evolution toward an optimal energy budget in a new habitat.

The Evolution of Novel Traits

The evolution of novel traits represents a challenge for the theory of evolution by natural selection (see Chapter 3). Providing an example that is the focus of current research, closely related species of sea urchins can have strikingly different larval forms. Figure 13.34a shows the larvae of two urchins in the same genus: one is a pluteus; the other is a schmoo. Pluteus larvae hatch from small eggs. Before metamorphosis to the adult form, pluteus larvae live, feed, and grow in the plankton. Schmoo larvae hatch from large eggs. Schmoo larvae undergo metamorphosis earlier than do pluteus larvae, and they do not feed.

The process of development in pluteus larvae versus schmoo larvae is as strikingly different as their morphology (Wray and Bely 1994; Wray 1995; 1996; Raff 1996). The two larval forms differ in the pattern of the earliest cell divisions (Figure 13.34b). Likewise, the larval forms differ in the expression of a variety of genes.

(a)

(b)

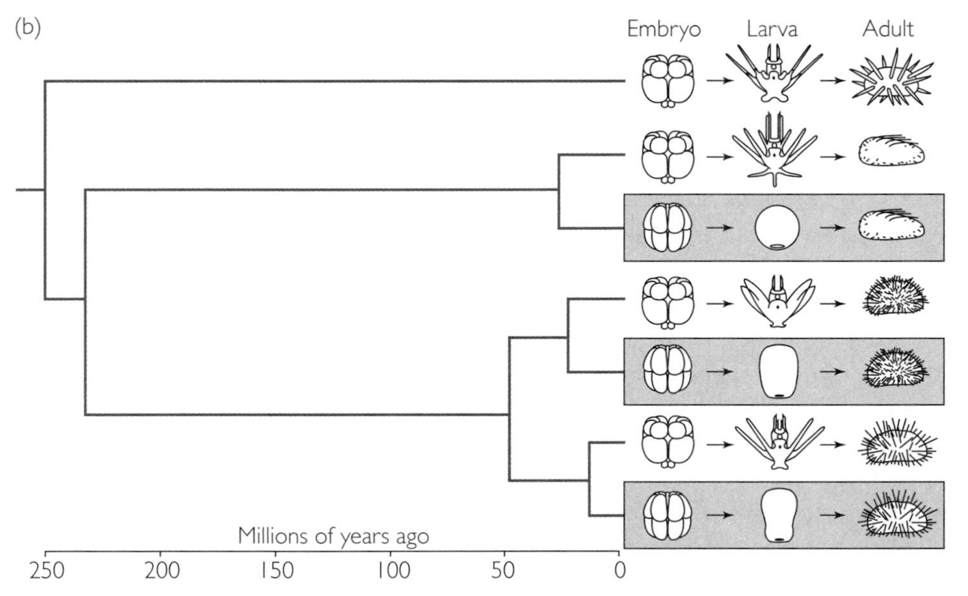

Figure 13.34 The evolution of nonfeeding larvae in sea urchins (a) The larva on the left is a form called a pluteus. The larva on the right is a form called a schmoo. These two larvae belong to closely related species of sea urchin: The pluteus is from *Heliocidaris tuberculata*; the schmoo is from *H. erythrogramma*. Both larvae in this photo are about three days old. (b) The pluteus larval form is ancestral and ancient in sea urchins. In the urchin phylogeny shown here, the pluteus form is at least 250 million years old. More extensive phylogenies have shown the pluteus form to be about 500 million years old. Extant species with schmoo larvae appear to be relatively recently derived from ancestors with pluteus larvae. As the diagrams of early embryos included in the figure show, the pattern of the earliest cell divisions is different in species with pluteus versus schmoo larvae. Reprinted with permission from Wray (1995).

Although the schmoo form is simpler than the pluteus, the pluteus form is probably ancestral (Strathmann 1978). Strathmann argued that the pluteus form is so complex, and is shared among such distantly related sea urchin species, that the same larval form is unlikely to have evolved twice. Strathmann argued instead that pluteus larvae evolve into schmoo larvae, but this change is not reversible. From their morphological features, schmoo larvae appear to have evolved by simplification and loss of the complex features used by a pluteus to feed and swim in the plankton.

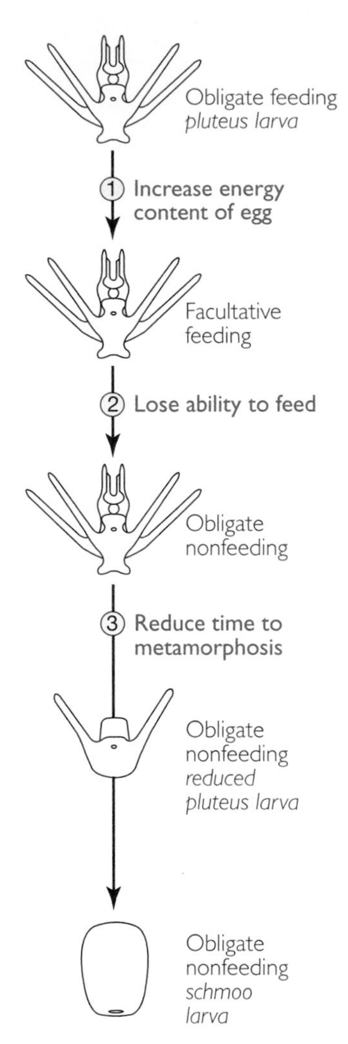

Figure 13.35 Hypothesized steps in the transition from pluteus larva to schmoo From Wray (1996).

How can this hypothesis be tested? The alternative hypothesis is that the larval forms of sea urchins can switch from pluteus to schmoo and back again. One approach to compare these hypotheses is to build a phylogeny for sea urchin groups and ask whether such reversibility seems likely from the distribution of larval forms on the phylogeny (Figure 13.34b). We introduced this idea of using phylogenetic methods to reconstruct evolutionary history in Chapter 1, and we analyzed numerous examples of this approach in Chapter 4. In the case of sea urchins, reconstructed phylogenies indicate that the pluteus form is hundreds of millions of years old. In contrast, extant species with schmoo larvae appear to have been derived within the last 50 million years from ancestors with pluteus larvae. In addition, Jurassic-aged fossils of pluteus skeletons also support an ancient origin of the pluteus form (see Wray 1996). If a pluteus is the ancestral form for the sea urchins in Figure 13.34b, then a schmoo evolved three times independently. On the other hand, if a schmoo can reversibly give rise to a pluteus, then four different origins of a pluteus could have occurred among these species. By the parsimony method that we introduced in Chapters 1 and 4, we would conclude that three changes are more likely to have happened than four changes, and by this method we would conclude that the transition from pluteus larvae to schmoo larvae appears to be one-way. Larger phylogenies for more species indicate that schmoo larvae have evolved from pluteus ancestors at least 20 times in the sea urchins (Wray and Bely 1994; Wray 1995).

Figure 13.35 shows a hypothesis for how the schmoo larval form evolves from a pluteus ancestor. In the first step in this scheme, selection favors the production of larger, more energy-rich eggs. This makes feeding optional for the derived larva. In the second step, the derived larva loses the ability to feed. Finally, selection for earlier metamorphosis causes the loss of all feeding structures, ultimately resulting in the most derived larval form, the schmoo. Michael Hart (1996) studied the larval form of one sea urchin (*Brisaster latifrons*) that appears to represent the first transitional step in this scenario. *Brisaster latifrons* larvae hatch from large eggs. Although they can feed, *B. latifrons* larvae do not have to feed in order to complete metamorphosis to the adult form. If the scenario outlined in Figure 13.30 is both correct and general, then the evolution of the novel developmental and morphological features of the schmoo form has been repeatedly initiated by selection for a simple life history trait: larger eggs.

Suboptimal Life Histories

A crucial assumption behind many of the models and experiments reviewed in this chapter is that populations have had both the time and the additive genetic variation in life history traits to enable evolution toward an optimum. We end this discussion with an important caveat: Life histories (like other kinds of adaptations) are not perfect, and they need not be optimal. Suboptimal life histories may be found because populations lack the time or variation to evolve toward an optimum, or there may be fundamental limits on the ability of populations to evolve an optimal life history (other than the limits imposed by time and energy trade-offs).

Douglas Gill (1989) described one kind of apparently suboptimal life history in a population of pink lady's slipper orchids, *Cypripedium acaule*. These orchids can live more than 20 years, have high adult survival rates, and produce large, showy flowers to attract pollinating insects (usually bumblebees in the genus *Bombus*). The flowers do not self-fertilize, so they require pollinators to deliver pollen from another *C. acaule*. However, the orchids deceive the pollinator because they offer no nectar reward in return for pollination. As a result, only naïve pollinators (newly emerged queen bees) visit the orchids, and even these pollinators do so only once.

Gill measured reproductive success of hundreds of these orchids in one study plot in the mountains of Virginia. In 10 years, this population produced 895 flowers, of which only 20 produced a natural fruit capsule. Pollinators are common at the study site, and this very low rate of fruiting is typical of other *C. acaule* populations. Gill concluded that such a life history is suboptimal for these orchids because this population could rapidly be invaded by several kinds of life history mutations that would increase fruiting success, as follows:

- The flowers are self-fertile. When Gill self-pollinated flowers by hand they produced abundant fruits. A self-pollinating mutant would be an improved life history in this population.
- Many other orchids do offer nectar rewards. A mutant that provides nectar to attract repeated visits by bees would also be an improved life history.
- New queen bees are more abundant later in the summer. A mutant that flowered later, even without a nectar reward, would be more likely to attract repeated visits (especially by bees carrying pollen from another *C. acaule*) and would be an improved life history.

Not all life history traits are optimized.

Given that such life histories can and do evolve in other orchids, why does this suboptimal life history persist in *C. acaule* and evolve repeatedly among other orchids? Gill could not explain this, and the life history of this plant remains an enigma.

Richard Strathmann and colleagues (1981) described another apparently suboptimal life history that gives some clues to potential constraints on the evolution of an optimal life history. They studied the reproduction and dispersal of the larvae of an intertidal barnacle, *Balanus glandula*. Adult barnacles release their larvae into the plankton, where they must develop for about two weeks before they are able to settle and become benthic juveniles. The adults tend to live in the low rocky intertidal zone, and settling larvae preferentially look for areas low in the intertidal zone, probably by evaluating the algae and other organisms growing on the rocks (Strathmann et al. 1981). Unfortunately for the barnacles, this preference sometimes misleads them to settle in low intertidal areas that are unsuitable for later adult life. In some parts of the geographic range of this species, the low intertidal is an ideal place to live, while in other parts of the range settlement in the low intertidal zone results in massive mortality of young barnacles (Strathmann and Branscomb 1979). In this case, the limitation on a more optimal life history for *B. glandula* is the length of the early larval life. Barnacles in the low intertidal zone of a good habitat might benefit by retaining their larvae in that habitat, but these barnacles have an obligate period of about two weeks as a planktonic larva. In those two weeks, larvae are so widely dispersed that it is impossible for a single larva to predict whether the low intertidal habitat it has selected for settlement will be a good or poor habitat as an adult. This obligate dispersal of the larva prevents natural selection from optimizing the life history to keep offspring living in the same neighborhood as their parents.

Summary

Organisms face fundamental trade-offs. The amount of energy available is finite, and energy devoted to one function—such as growth or tissue repair—cannot be devoted to other functions—such as reproduction. Furthermore, biological processes take time. An individual growing to a large size before maturing risks dying from disease or predation before ever reproducing. Fundamental trade-offs involving energy and time mean that every organism's life history is an evolutionary compromise.

Senescence evolves because natural selection is weaker late in life. Late-acting deleterious mutations can persist in populations under mutation–selection balance. Selection may favor increased investment in reproduction early in life at the expense of repair. Both mechanisms can result in a decline in reproductive performance and survival with age.

A trade-off between the number of offspring in a clutch and the survival of individual offspring constrains the evolution of clutch size. Additional constraints may involve trade-offs between present parental reproductive effort and future reproductive performance or survival, as well as trade-offs between clutch size and offspring reproductive performance.

A trade-off between the size and number of offspring constrains the evolution of offspring size. Selection on parents may favor smaller offspring than does selection on the offspring themselves.

Life history traits may reflect conflicts of interest between individuals. These conflicts have led to the evolution of differential gene expression (imprinting) and sexually antagonistic traits in males and females.

Theory predicts that life history traits should have low heritability because they are closely related to fitness. Life history traits do tend to have lower heritability than other kinds of traits, but nonetheless they typically show substantial genetic variation. One mechanism demonstrated to maintain genetic variation in life history traits is temporally varying selection.

Selection on life history traits can have dramatic consequences for other aspects of an organism's biology. Selection for larger eggs in echinoderms appears to have initiated, in numerous independent lineages, dramatic rearrangements in larval form and development. However, not all life history traits have evolved toward a stable optimum, and there are some examples of apparently suboptimal life histories.

Questions

1. Look again at Figure 13.4, which illustrates fertility and survival as a function of age in three different species.
   a. What similarities are there across all three species? What is the general trend in fertility and in annual probability of survival? Why are these trends considered to be an evolutionary puzzle?
   b. Which species has the best probability of survival even in old age? This is a characteristic of this taxon. Do you remember another animal (a mammal, discussed later in this chapter) that has a similarly high probability of survival in old age? What trait do these long-lived animals have in common?
   c. In the red deer, how do patterns of survival and reproduction vary with the two sexes? Why do you think these differences occur between males and females?
2. What are the two predictions of the rate-of-living theory of aging? What data exist to support or refute the two predictions?
3. What is a telomere? Describe the telomere-shortening theory of aging. Is telomere length associated with life span in:
   - *C. elegans* that have been genetically engineered to have longer-than-normal telomeres?
   - different strains of *C. elegans*?
   - different mammal species, when controlled for body mass?
   - wild mice as compared to laboratory mice?
   - elderly humans?
4. What is the evolutionary theory of aging? What are the two major mechanisms that are associated with it? Is natural selection the major evolutionary force for both mechanisms?
5. Listed below are four possible causes of aging that were discussed in the text and in the questions above. As a review, name the theory that is associated with each cause, and describe whether or not selection for a longer life span is possible under each theory. What

predictions does each theory make about the effect of ecological mortality (death due to external causes—predators, starvation, etc.) on aging rate?
- "Wearing out" due to metabolic activity
- Reduction in size of telomeres with each cell division
- Mutations that have negative effects late in life
- Mutations that have positive effects early and negative effects late in life.

6. Most domestic female rabbits will get uterine cancer if they are not spayed. The uterine cancer usually appears sometime after the age of 2 years. Describe a hypothesis for why rabbits have not evolved better defenses against uterine cancer. What do you think the average life span of a wild female rabbit might be? What do you think is a typical cause of death in wild rabbits? Why do you think that uterine cancer, and not (say) pancreatic cancer or throat cancer, is the most common cancer in female rabbits?

7. We have seen how aging can evolve due to two different phenomena: First, aging may evolve due to mutations that have deleterious effects only late in life. As a review, explain how such mutations could ever become common in a population. Second, aging may evolve due to mutations with pleiotropic effects that cause "trade-offs"—positive effects early and negative effects late. What would happen if a mutation arose with a reverse trade-off, that is, a mutation with negative effects early and positive effects late in life? Could such a mutation ever be selected for?

8. Look again at Figure 13.10, which shows life history trade-offs for a hypothetical species. Suppose you are studying these animals, and you notice the appearance of a new mutation from the wild type that causes its carriers to have two offspring per year instead of one. The new mutation does not alter the age of maturation, which still occurs at 3 years. Your initial observations indicate that the new mutation may cause an early death, but you are not certain exactly how early the deaths occur. You do notice, however, that the new mutation is increasing in frequency and the wild-type allele is decreasing. Make a prediction about the minimum possible age of death of organisms that carry this mutation and explain your reasoning.

9. Now suppose that during your research, you bring a large population of these animals into captivity. You notice that their annual survival rate immediately jumps from 0.80 to 0.95 once they are in captivity, primarily due to protection from predators. Make a prediction about whether the captive population will evolve changes in fertility or life span, simply due to this reduction in predation. Could this same process be occurring in zoo populations of captive animals today? Explain your reasoning.

10. Assuming that the grandmother hypothesis of menopause is correct, speculate on what aspects of a species' behavior and sociality may make menopause likely to evolve. For instance, is it important whether or not the species is highly social, or whether or not the species lives in kin groups? Might the age of independence of the young be important? Could menopause ever evolve in a species without parental care, such as aphids or willow trees? As fuel for thought, consider the likelihood of evolution of menopause in (1) orangutans, who live in small groups consisting simply of a female and her dependent young; (2) lions, in which females are very social and remain with their female kin for most of their lives; and (3) Arabian oryx, a species of antelope that lives in small family groups in extremely arid deserts and must sometimes find distant waterholes known only to the older oryx.

11. As a review, describe why hatchery salmon may be evolving smaller egg size, and the implications for wild populations. What could hatchery managers do to reverse the effects on wild populations?

12. The examples of the chinook salmon and seed beetles indicate that females, in general, cannot produce many large eggs. Instead, they must choose between producing many small eggs or producing a few large eggs (and sometimes, in unfortunate cases, just a few small eggs). Explain, then, how it is possible for a queen honeybee to produce a very large number of relatively large eggs. (*Hint*: Consider what the other bees are doing.) Does this suggest a general way in which a female can escape from the size–number trade-off?

13. The 1998 movie *Godzilla* and the 1986 movie *Aliens* each depict a fictional large female carnivore. The *Godzilla* female lives off a large prey population of humans and fishes but has no assistance from others of her kind. In a few days she produces hundreds of 7-ft-tall eggs, enough to fill Madison Square Garden. The *Aliens* female lives off a small prey population of a few dozen humans, is assisted by nonreproducing workers, and produces hundreds of large eggs in a few weeks. Comment on what is realistic and unrealistic about the life history traits and egg production abilities of each of these fictional animals. If they were real, would they have long or short life spans? Why?

14. Dairy farmers are sometimes frustrated in their attempts to breed a better milk cow because heritability values for milk production and reproductive traits are low—generally below 0.10. In addition, those cows that produce the most milk tend to have longer intervals between birth of successive calves, and require more breedings to a bull before the cow will conceive. Do these patterns make sense in light of evolutionary life history theory? Explain.

15. What is Lack's hypothesis? Is it supported by most experimental data? If not, why not?

16. Many human generations ago, most women worldwide began childbearing in their mid-teens. Today, a large proportion of women worldwide delays childbearing until their 20s. Among college-educated women in developed nations, the trend in delaying reproduction has been taken even further; with childbearing often delayed until past age 30 due to education and career pressures. Suppose that the majority of women worldwide were to delay child-bearing until age 30, and that women were to continue to make this choice for many human generations. Make a prediction about how human life span and fertility might evolve in response.

Exploring the Literature

17. A highly restricted, near-starvation diet will prolong life span markedly in almost all mammals. The mechanism is suspected to involve reduced free radical generation by mitochondria. See the following papers for a review and for an interesting new search for drugs that can mimic the longevity-promoting effects of caloric restriction without actually restricting calories:

Gredilla, Ricardo, and Gustavo Barja. 2005. Minireview: The role of oxidative stress in relation to caloric restriction and longevity. *Endocrinology* 146 (9): 3713–3717.

Roth, G. S., M. A. Lane, and D. K. Ingram. 2006. Caloric restriction mimetics—the next phase. *Annals of the New York Academy of Sciences* 1057: 365–371.

18. Genes that might produce antagonistic pleiotropic effects on life history traits are difficult to identify. One approach is to predict which reproductive functions are important for organisms and identify the genes that may regulate those functions. For an example of this approach, see this study of wild swans:

Charmantier, A., C. Perrins, R. H. McCleery, and B. C. Sheldon. 2006. Quantitative genetics of age at reproduction in wild swans: Support for antagonistic pleiotropy models of senescence. *Proceedings of the National Academy of Sciences USA* 103: 6587–6592.

An alternative approach is to perform massive mutation experiments and analyze the mutant strains for different life history traits. This approach resulted in the discovery of a new mutant, *methuselah*, that lives much longer than normal flies. See these papers for further information on *methuselah* and how it was discovered:

Lin, Y. J., L. Seroude, and S. Benzer. 1998. Extended life span and stress resistance in the *Drosophila* mutant *methuselah*. *Science* 282: 943–946.

Wang, H. D., P. Kazemi-Esfarjani, and S. Benzer. 2004. Multiple-stress analysis for isolation of *Drosophila* longevity genes. *Proceedings of the National Academy of Sciences USA* 101: 12610–12615.

How would you determine whether the trade-off between life span and early fecundity in *Drosophila* is influenced by the *methuselah* mutation? See:

Mockett, R. J., and R. S. Sohal. 2006. Temperature-dependent trade-offs between longevity and fertility in the *Drosophila* mutant, *methuselah*. *Experimental Gerontology* 41: 566–573

19. Graham Bell distinguished between the rate-of-living versus evolutionary theories of aging by comparing invertebrates that lay eggs (and thus have a distinct soma and germ line) with invertebrates that reproduce by fission (and thus have no soma/germ line division). According to the rate-of-living theory, both kinds of organisms will inevitably accumulate irreparable damage. According to the evolutionary theory, selection would allow genes responsible for senescence to accumulate only in organisms with a disposable soma. See:

Bell, G. 1984. Evolutionary and nonevolutionary theories of senescence. *American Naturalist* 124: 600–603.

20. For tests of the evolutionary theory of aging employing comparisons between eusocial versus non-eusocial insects and comparisons between castes of worker ants, see:

Keller, L., and M. Genoud. 1997. Extraordinary life spans in ants: A test of evolutionary theories of ageing. *Nature* 389: 958–960.

Chapuisat, M., and L. Keller. 2002. Division of labour influences the rate of ageing in weaver ant workers. *Proceedings of the Royal Society of London B* 269: 909–913.

21. For a study in which researchers tested the assumptions of Smith and Fretwell's analysis by surgically manipulating female lizards to increase the variation in egg size, see:

Sinervo, B., and P. Doughty. 1996. Interactive effects of offspring size and timing of reproduction on offspring reproduction: Experimental, maternal, and quantitative genetic aspects. *Evolution* 50: 1314–1327.

Sinervo, B., P. Doughty, R. B. Huey, and K. Zamudio. 1992. Allometric engineering: A causal analysis of natural selection on offspring size. *Science* 258: 1927–1930.

Sinervo, B., and P. Licht. 1991. Proximate constraints on the evolution of egg size, number, and total clutch mass in lizards. *Science* 252: 1300–1302.

22. It now appears that not all populations with high ecological mortality will evolve short life spans. In guppies, populations with the highest ecological mortality also have the longest life span, and some other studies have found similar results. Under what conditions does ecological mortality affect the evolution of senescence? See these papers for more information about the puzzle of the long-lived guppies:

Reznick, David N., M. J. Bryant, D. Roff, C. K. Ghalambor, and D. E. Ghalambor. 2004. Effect of extrinsic mortality on the evolution of senescence in guppies. *Nature* 431: 1095–1099.

Bronikowski, Anne M., and Daniel E. L. Promislow. 2005. Testing evolutionary theories of aging in wild populations. *Trends in Ecology and Evolution* 20: 271–273.

Williams, Paul D., T. Day, Q. Fletcher, and L. Rowe. 2006. The shaping of senescence in the wild. *Trends in Ecology and Evolution* 21: 458-463.

Citations

Austad, S. N. 1988. The adaptable opossum. *Scientific American* (February): 98–104.

Austad, S. N. 1993. Retarded senescence in an insular population of Virginia opossums (*Didelphis virginiana*). *Journal of Zoology, London* 229: 695–708.

Austad, S. N. 1994. Menopause: An evolutionary perspective. *Experimental Gerontology* 29: 255–263.

Austad, S. N., and K. E. Fischer. 1991. Mammalian aging, metabolism, and ecology: Evidence from the bats and marsupials. *Journal of Gerontology* 46: B47–53.

Barlow, D. P. 1995. Gametic imprinting in mammals. *Science* 270: 1610–1613.

Berrigan, D. 1991. The allometry of egg size and number in insects. *Oikos* 60: 313–321.

Bischoff, C., H. C. Petersen, et al. 2006. No association between telomere length and survival among the elderly and oldest old. *Epidemiology* 17: 190-194.

Blair, A. C., and L. M. Wolfe. 2004. The evolution of an invasive plant: an experimental study with Silene latifolia. *Ecology* 85: 3035-3042.

Boyce, M. S., and C. M. Perrins. 1987. Optimizing great tit clutch size in a fluctuating environment. *Ecology* 68: 142–153.

Campisi, J. 1996. Replicative senescence: an old lives' tale? *Cell* 84: 497–500.

Charnov, E. L., and D. Berrigan. 1993. Why do female primates have such long life spans and so few babies? or Life in the slow lane. *Evolutionary Anthropology* 1: 191–194.

Charnov, E. L., and S. W. Skinner. 1985. Complementary approaches to the understanding of parasitoid oviposition decisions. *Environmental Entomology* 14: 383–391.

Clutton-Brock, T. H., S. D. Albon, and F. E. Guinness. 1988. Reproductive success in male and female red deer. In T. H. Clutton-Brock, ed. *Reproductive Success*. Chicago: University of Chicago Press, 325–343.

Elbadry, E. A., and M. S. F. Tawfik. 1966. Life cycle of the mite *Adactylidium* sp. (Acarina: Pyemotidae), a predator of thrips eggs in the United Arab Republic. *Annals of the Entomological Society of America* 59: 458–461.

Elgar, M. A. 1990. Evolutionary compromise between a few large and many small eggs: Comparative evidence in teleost fish. *Oikos* 59: 283–287.

Eshleman, J. R., and S. D. Markowitz. 1996. Mismatch repair defects in human carcinogenesis. *Human Molecular Genetics* 5: 1489–1494.

Ferbeyre, G., and S. W. Lowe. 2002. The price of tumor suppression? *Nature* 415: 26–27.

Fishel, R., and T. Wilson. 1997. *MutS* homologs in mammalian cells. *Current Opinion in Genetics and Development* 7: 105–133.

Fisher, R. A. 1930. *The Genetical Theory of Natural Selection*. Oxford: Clarendon Press.

Fowler, K., and L. Partridge. 1989. A cost of mating in female fruit flies. *Nature* 338: 760–761.

Fox, C. W., M. S. Thakar, and T. A. Mousseau. 1997. Egg size plasticity in a seed beetle: An adaptive maternal effect. *American Naturalist* 149: 149–163.

Gill, D. E. 1989. Fruiting failure, pollinator inefficiency, and speciation in orchids. In D. Otte and J. A. Endler, eds. *Speciation and Its Consequences*. Sunderland, MA: Sinauer Associates.

Gould, S. J. 1980. *The Panda's Thumb: More Reflections in Natural History*. New York: W. W. Norton.

Grosberg, R. K. 1988. Life-history variation within a population of the colonial ascidian *Botryllus schlosseri*. I. The genetic and environmental control of seasonal variation. *Evolution* 42: 900–920.

Guillette, L. J., Jr., and R. E. Jones. 1985. Ovarian, oviductal and placental morphology of the reproductively bimodal lizard, *Sceloporus aeneus*. *Journal of Morphology* 184: 85–98.

Gustafsson, L., and T. Pärt. 1990. Acceleration of senescence in the collared flycatcher (*Ficedula albicollis*) by reproductive costs. *Nature* 347: 279–281.

Haig, D. 1999. Multiple paternity and genomic imprinting. *Genetics* 151: 1229–1231.

Haig, D., and C. Graham. 1991. Genomic imprinting and the strange case of the insulin-like growth factor II receptor. *Cell* 64: 1045–1046.

Haig, D., and M. Westoby. 1989. Parent-specific gene expression and the triploid endosperm. *American Naturalist* 134: 147–155.

Haig, D., and M. Westoby. 1991. Genomic imprinting in endosperm: Its effect on seed development in crosses between species, and between different ploidies of the same species, and its implications for the evolution of apomixis. *Philosophical Transactions of the Royal Society of London B* 333: 1–13.

Hamilton, W. D. 1966. The moulding of senescence by natural selection. *Journal of Theoretical Biology* 12: 12–45.

Harley, C. B., A. B. Futcher, and C. W. Greider. 1990. Telomeres shorten during ageing of human fibroblasts. *Nature* 345: 458–460.

Hart, M. W. 1996. Evolutionary loss of larval feeding: Development, form, and function in a facultatively feeding larva, *Brisaster latifrons*. *Evolution* 50: 174–187.

Hawkes, K., J. F. O'Connell, and N. G. Blurton Jones. 1989. Hardworking Hadza grandmothers. In V. Standen and R. A. Foley, eds. *Comparative Socioecology*. Oxford: Blackwell Scientific Publications, 341–366.

Hawkes, K., J. F. O'Connell, and N. G. Blurton Jones. 1997. Hadza women's time allocation, offspring provisioning, and the evolution of long postmenopausal life spans. *Current Anthropology* 38: 551–577.

Hawkes, K., J. F. O'Connell, N. G. Blurton Jones, H. Alvarez, and E. L. Charnov. 1998. Grandmothering, menopause, and the evolution of human life histories. *Proceedings of the National Academy of Sciences USA* 95: 1336–1339.

Heath, D. D., J. W. Heath, C. A. Bryden, et al. 2003. Rapid evolution of egg size in captive salmon. *Science* 299: 1738–1740.

Hemann, M. T., and C. W. Greider. 2000. Wild-derived inbred mouse strains have short telomeres. *Nucleic Acids Research* 28: 4474-4478.

Hendry, A. P., Y. E. Morbey, et al. 2004. Adaptive variation in senescence: reproductive lifespan in a wild salmon population. *Proceedings of the Royal Society of London B* 271: 259–266.

Hill, K., and A. M. Hurtado. 1991. The evolution of premature reproductive senescence and menopause in human females: An evaluation of the "grandmother hypothesis." *Human Nature* 2: 313–351.

Hill, K., and A. M. Hurtado. 1996. *Ache Life History: The Ecology and Demography of a Foraging People*. New York: Aldine de Gruyter.

Holland, B., and W. R. Rice. 1998. Perspective: Chase-away sexual selection: antagonistic seduction versus resistance. *Evolution* 52: 1–7.

Holland, B., and W. R. Rice. 1999. Experimental removal of sexual selection reverses intersexual antagonistic coevolution and removes a reproductive load. *Proceedings of the National Academy of Sciences USA* 96: 5083–5088.

Hughes, K. A., J. A. Alipaz, J. M. Drnevich, and R. M. Reynolds. 2002. A test of evolutionary theories of aging. *Proceedings of the National Academy of Sciences USA* 99: 14286–14291.

Hurst, L. D. 1999. Is multiple paternity necessary for the evolution of genomic imprinting? *Genetics* 153: 509–512.

Joeng, K. S., E. J. Song, et al. 2004. Long lifespan in worms with long telomeric DNA. *Nature Genetics* 6: 607-611.

Karlsleder, J., A. Smogorzewska, and T. de Lange. 2002. Senescence induced by altered telomere state, not telomere loss. *Science* 295: 2446–2449.

Lack, D. 1947. The significance of clutch size. *Ibis* 89: 302–352.

Law, R. 1979. Ecological determinants in the evolution of life histories. Pages 81-103 in R. M. Anderson, B. D. Turner, and L.R. Taylor, eds. *Population Dynamics*. Oxford: Blackwell Scientific.

Leroi, A. M., A. Bartke, et al. 2005. What evidence is there for the existence of individual genes with antagonistic pleiotropic effects? *Mechanisms of Ageing and Development* 126: 421–429.

Levitan, D. R. 1993. The importance of sperm limitation to the evolution of egg size in marine invertebrates. *American Naturalist* 141: 517–536.

Levitan, D. R. 1996. Predicting optimal and unique egg sizes in free-spawning marine invertebrates. *American Naturalist* 148: 174–188.

Lin, Y.-J., L. Seroude, and S. Benzer. 1998. Extended life-span and stress resistance in the *Drosophila* mutant *methuselah*. *Science* 282: 943–946.

Lindén, M., and A. P. Møller. 1989. Cost of reproduction and covariation of life history traits in birds. *Trends in Ecology and Evolution* 4: 367–371.

Lorenzini, A., M. Tresini, et al. 2005. Cellular replicative capacity correlates primarily with species body mass not longevity. *Mechanisms of Ageing and Development* 126: 1130–1133.

Luckinbill, L. S., R. Arking, M. J. Clare, W. C. Cirocco, and S. A. Buck. 1984. Selection for delayed senescence in *Drosophila melanogaster. Evolution* 38: 996–1003.

Medawar, P. B. 1952. *An Unsolved Problem in Biology.* London: H. K. Lewis.

Mockett, R. J., and R. S. Sohal. 2006. Temperature-dependent trade-offs between longevity and fertility in the *Drosophila* mutant, *methuselah. Experimental Gerontology* 41: 566–573.

Moore, T., and D. Haig. 1991. Genomic imprinting in mammalian development—a parental tug-of-war. *Trends in Genetics* 7: 45–49.

Mousseau, T. A., and D. A. Roff. 1987. Natural selection and the heritability of fitness components. *Heredity* 1987: 181–197.

Nesse, R. M., and G. C. Williams. 1995. *Why We Get Sick: The New Science of Darwinian Medicine.* New York: Vintage Books.

Nowak, R. M. 1991. *Walker's Mammals of the World*, 5th ed. Baltimore: The Johns Hopkins University Press.

Partridge, L. 1987. Is accelerated senescence a cost of reproduction? *Functional Ecology* 1: 317–320.

Partridge, L. 2001. Evolutionary theories of ageing applied to long-lived organisms. *Experimental Gerontology* 36: 641–650.

Partridge, L., and N. H. Barton. 1993. Optimality, mutation, and the evolution of ageing. *Nature* 362: 305–311.

Partridge, L., and K. Fowler. 1992. Direct and correlated responses to selection on age at reproduction in *Drosophila melanogaster. Evolution* 46: 76–91.

Patil, C. K., I. S. Mian, and J. Campisi. 2005. The thorny path linking cellular senescence to organismal aging. *Mechanisms of Ageing and Development* 126: 1040–1045.

Pennisi, E. 1998. Singe gene controls fruit fly life-span. *Science* 282: 856.

Pettifor, R. A., C. M. Perrins, and R. H. McCleery. 2001. The individual optimization of fitness: Variation in reproductive output, including clutch size, mean nestling mass and offspring recruitment, in manipulated broods of great tits *Parus major. Journal of Animal Ecology* 70: 62–79.

Podolsky, R. D., and R. R. Strathmann. 1996. Evolution of egg size in free-spawners: Consequences of the fertilization-fecundity trade-off. *American Naturalist* 148: 160–173.

Price, T., and D. Schluter. 1991. On the low heritability of life-history traits. *Evolution* 45: 853–861.

Raff, R. A. 1996. *The Shape of Life: Genes, Development, and the Evolution of Animal Form.* Chicago: University of Chicago Press.

Raices M, H., Maruyama et al. 2005. Uncoupling of longevity and telomere length in *C. elegans. PLoS Genetics* 1: e30.

Reddel, R. R. 1998. A reassessment of the telomere hypothesis of senescence. *BioEssays* 20: 977–984.

Reed, D. H., and E. H. Bryant 2000. The evolution of senescence under curtailed life span in laboratory populations of *Musca domestica* (the housefly). *Heredity* 85: 115–121.

Rice, W. R. 1987. The accumulation of sexually antagonistic genes as a selective agent promoting the evolution of reduced recombination between primitive sex chromosomes. *Evolution* 41: 911–914.

Rice, W. R. 1996. Sexually antagonistic male adaptation triggered by experimental arrest of female evolution. *Nature* 381: 232–234.

Rice, W. R., and B. Holland. 1997. The enemies within: Intergenomic conflict, interlocus contest evolution (ICE), and the intraspecific Red Queen. *Behavioral Ecology and Sociobiology* 41: 1–10.

Rodier, F., S.-H. Kim, et al. 2005. Cancer and aging: the importance of telomeres in genome maintenance. *International Journal of Biochemistry & Cell Biology* 37: 977-990.

Rodriguez Bigas, M. A., P. H. U. Lee, L. O'Malley, T. K. Weber, O. Suh, G. R. Anderson, and N. J. Petrelli. 1996. Establishment of a hereditary nonpolyposis colorectal cancer registry. *Diseases of the Colon & Rectum* 39: 649–653.

Roff, D. A. 1992. *The Evolution of Life Histories.* New York: Chapman & Hall.

Röhme, D. 1981. Evidence for a relationship between longevity of mamamlian species and life spans of normal fibroblasts *in vitro* and erythrocytes *in vivo. Proceedings of the National Academy of Sciences USA* 78: 5009–5013.

Roper, C., P. Pignatelli, and L. Partridge. 1993. Evolutionary effects of selection on age at reproduction in larval and adult *Drosophila melanogaster. Evolution* 47: 445–455.

Rose, M. R. 1984. Laboratory evolution of postponed senescence in *Drosophila melanogaster. Evolution* 38: 1004–1010.

Rose, M. R. 1991. *Evolutionary Biology of Aging.* New York: Oxford University Press.

Sanz, J. J., and J. Moreno. 1995. Experimentally induced clutch size enlargements affect reproductive success in the pied flycatcher. *Oecologia* 103: 358–364.

Schluter, D., and L. Gustafsson. 1993. Maternal inheritance of condition and clutch size in the collared flycatcher. *Evolution* 47: 658–667.

Schmidt, P. S., D. D. Duvernell, and W. F. Eanes. 2000. Adaptive evolution of a candidate gene for aging in *Drosophila. Proceedings of the National Academy of Sciences USA* 97: 10861–10865.

Service, P. 1987. Physiological mechanisms of increased stress resistance in *Drosophila melanogaster* selected for postponed senescence. *Physiological Zoology* 60: 321–326.

Smith, C. C., and S. D. Fretwell. 1974. The optimal balance between size and number of offspring. *American Naturalist* 108: 499–506.

Spencer, H. G., A. G. Clark., and M. W. Feldman. 1999. Genetic conflicts and the evolutionary origin of genomic imprinting. *Trends in Ecology and Evolution* 14: 197–201.

Stearns, S. C. 1992. *The Evolution of Life Histories.* Oxford: Oxford University Press.

Strathmann, M. F. 1987. *Reproduction and Development of Marine Invertebrates of the Northern Pacific Coast.* Seattle, WA: University of Washington Press.

Strathmann, R. R. 1978. The evolution and loss of feeding larval stages of marine invertebrates. *Evolution* 32: 894–906.

Strathmann, R. R., and E. S. Branscomb. 1979. Adequacy of cues to favorable sites used by settling larvae of two intertidal barnacles. In S. E. Stancyk, ed. *Reproductive Ecology of Marine Invertebrates.* Columbia, SC: University of South Carolina Press.

Strathmann, R. R., E. S. Branscomb, and K. Vedder. 1981. Fatal errors in set as a cost of dispersal and the influence of intertidal flora on set of barnacles. *Oecologia* 48: 13–18.

Taborsky, M., and B. Taborsky. 1993. The kiwi's parental burden. *Natural History* 1993: 50–56.

Tyner, S. D., S. Venkatachalam, J. Choi, et al. 2002. p53 mutant mice that display early ageing-associated phenotypes. *Nature* 415: 45–53.

Vance, R. R. 1973. On reproductive strategies in marine benthic invertebrates. *American Naturalist* 107: 339–352.

Vani, R. G., and M. R. S. Rao. 1996. Mismatch repair genes of eukaryotes. *Journal of Genetics* 75: 181–192.

Van Valen, L. 1973. A new evolutionary law. *Evolutionary Theory* 1: 1–30.

Walker, D. W., G. McColl, N. L. Jenkins, et al. 2000. Evolution of lifespan in *C. elegans. Nature* 405: 296–297.

Williams, G. C. 1957. Pleiotropy, natural selection, and the evolution of senescence. *Evolution* 11: 398–411.

Wolfe, L. M., J. A. Elzinga, and A. Biere. 2004. Increased susceptibility to enemies following introduction in the invasive plant Silene latifolia. *Ecology Letters* 7: 813-820.

Wourms, J. P. 1993. Maximization of evolutionary trends for placental viviparity in the spadenose shark, *Scoliodon laticaudus. Environmental Biology of Fish* 38: 269–294.

Wray, G. A. 1995. Punctuated evolution of embryos. *Science* 267: 1115–1116.

Wray, G. A. 1996. Parallel evolution of nonfeeding larvae in echinoids. *Systematic Biology* 45: 308–322.

Wray, G. A., and A. E. Bely. 1994. The evolution of echinoderm development is driven by several distinct factors. *Development 1994 Supplement:* 97–106.

Young, T. P. 1990. Evolution of semelparity in Mount Kenya lobelias. *Evolutionary Ecology* 4: 157–171.

Zera, A. J. and Z. Zhao. 2003. Life-history evolution and the microevolution of intermediary metabolism: Activities of lipid-metabolizing enzymes in life-history morphs of a wing-dimorphic cricket. *Evolution* 57: 586–596.

Zhao, Z., and A. J. Zera. 2002. Differential lipid biosynthesis underlies a trade-off between reproduction and flight capability in a wing-polymorphic cricket. *Proceedings of the National Academy of Sciences USA* 99: 16829–16834.

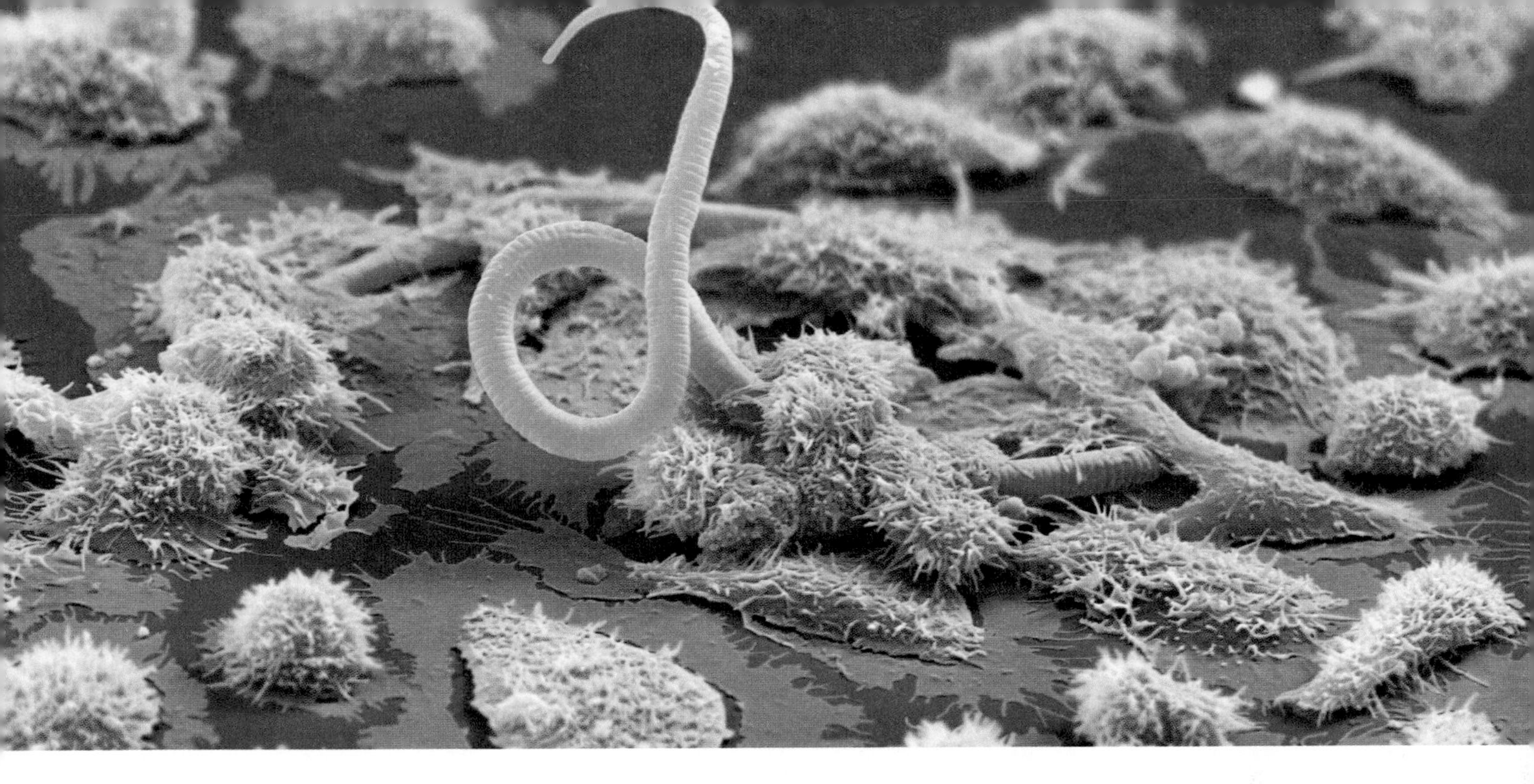

14

Evolution and Human Health

In 1854 a cholera epidemic struck central London. The disease, which causes severe diarrhea and dehydration, killed more than 500 people. In a famous act of medical detection, John Snow prepared a map of the affected neighborhood (see Tufte 1997). On it, he plotted the homes of the victims and the locations of the area's 11 water pumps (Figure 14.1, next page). The fatalities clustered around the Broad Street pump, at the center of Snow's map. Sealing the case were the deaths of two women in distant neighborhoods, who fell ill shortly after drinking water delivered by special arrangement from Broad Street. Although cholera's cause remained to be discovered, it was clearly associated with contaminated water.

In 1858, Louis Pasteur proposed that contagious diseases like cholera are caused by germs. Pasteur had been studying the fermentation of beer, wine, and milk, and had also been working to stop an epidemic of childbirth fever in a Paris maternity hospital. In a paper on lactic acid fermentation, Pasteur suggested that just as a particular microorganism is the cause of each kind of fermentation, so too might a particular microorganism be the cause of each infectious illness. Inspired by Pasteur, Robert Koch and others soon discovered the bacteria responsible for anthrax, wound infections, gonorrhea, typhoid fever, and tuberculosis. In 1883, Koch showed that cholera is caused by the bacterium *Vibrio cholerae*.

Macrophages attack a parasitic worm. Parasites that can evade their host's immune system are more likely to survive and reproduce than parasites that cannot.

The germ theory of disease was arguably the most important breakthrough in the development of modern medicine. It laid the foundation not only for the identification of numerous pathogens, but also for the development of antiseptic

Figure 14.1 John Snow's map of central London The cholera deaths during the 1854 epidemic were concentrated around the Broad Street pump.

surgery by Joseph Lister, the discovery of antibiotics by Alexander Fleming and others, and dramatic improvements in sanitation. The impact of sanitation and antibiotics on public health can be seen in Figure 14.2. The figure plots the death rate due to tuberculosis in the United States from 1900 through 1997. From 1900 to 1945, the tuberculosis death rate dropped from nearly 200 per 100,000 to about 40 per 100,000. This decline was largely due to improvements in sanitation, housing conditions, and nutrition. Then, in 1945, the death rate began falling more sharply still. The accelerated decline was due to the introduction of antibiotics, including streptomycin and isoniazid. By 1997, the tuberculosis death rate was fewer than 0.4 per 100,000, less than two-tenths of 1% what it was in 1900.

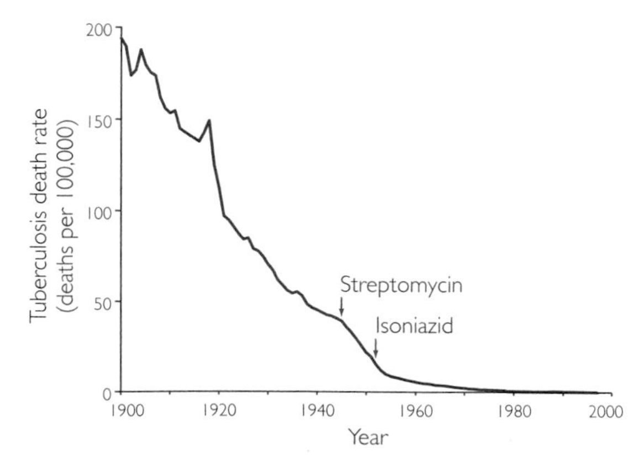

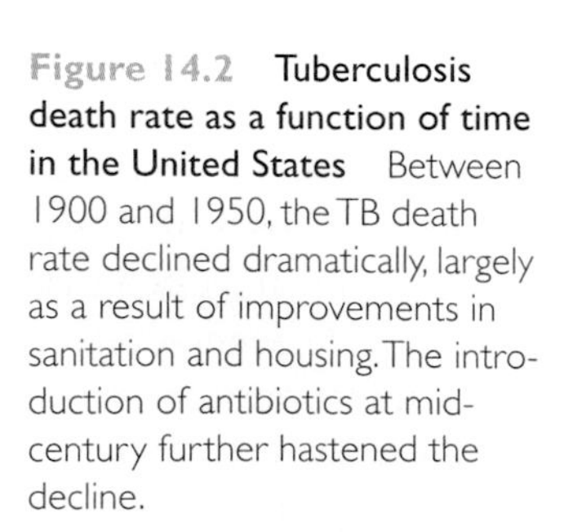
Figure 14.2 Tuberculosis death rate as a function of time in the United States Between 1900 and 1950, the TB death rate declined dramatically, largely as a result of improvements in sanitation and housing. The introduction of antibiotics at mid-century further hastened the decline.

Charles Darwin published *On the Origin of Species* in 1859, the year after Pasteur proposed the germ theory of disease. Evolutionary biology and modern medicine were born at the same time and have grown up in parallel. The relevance of evolutionary biology to medicine is deep and, in some ways, has only recently begun to be appreciated. George C. Williams and Randolph Nesse have been leaders in this field, which they call Darwinian medicine (Williams and Nesse 1991; Nesse and Williams 1994, 1998). Throughout this book we have highlighted medical applications of evolutionary analysis. In Chapter 1, for example, we discussed the evolution of HIV. In Chapters 5 through 8 we considered the impact of infectious diseases such as AIDS, sickle-cell anemia, and cystic fibrosis on the evolution of human populations. In Chapter 13 we explored the evolution of senescence and menopause. Here we devote an entire chapter to medical applications we have not had the chance to address elsewhere.

The chapter is divided into two parts. In Sections 14.1 through 14.4, we consider medical consequences of the fact that populations evolve. Examples include the evolution of pathogen populations and the evolution of cell populations within individual patients. In Sections 14.5 through 14.7, we turn our attention from pathogen and cell populations to the human animal as it has been shaped by natural selection. We consider applications of the adaptationist program, introduced in Chapter 10, in understanding puzzling aspects of human physiology and behavior. The message throughout is that evolutionary analysis is an invaluable tool for researchers and clinicians seeking to improve public health.

14.1 Evolving Pathogens: Evasion of the Host's Immune Response

The fundamental event in evolution is a change in the frequencies of genotypes within a population. It is with this phenomenon that we begin our discussion of evolution and human health. There are two kinds of populations whose ongoing evolution is important in medicine: populations of pathogens, and populations of human cells within individual patients. We consider evolving pathogen populations first, then evolving populations of cells.

A population of pathogens and their host are, by definition, in conflict. The pathogens attempt to consume the host's tissues and convert them into more pathogens; the host attempts to limit the damage by slowing or killing the pathogens. When we become hosts, our bodies employ an impressive array of weapons against the invaders. Our immune systems are capable of recognizing billions of foreign proteins, of mounting an aggressive and multifaceted response, and of remembering the structure of the pathogen's proteins so that the response will be mobilized more quickly in the event of future invasions. Pathogens are formidable enemies, however. Many pathogens have large population sizes, short generation times, and high mutation rates. These traits mean that pathogen populations evolve quickly. Any mutation that enables its possessors to evade or withstand the host's immune response should be strongly selected and should quickly increase in frequency. If we can understand how particular pathogens evolve in response to attack by the human immune system, we should be better able to intervene in the conflict and improve the odds of a favorable outcome.

Conflicts among organisms are inevitable. In the conflict between a parasite and its host, the host's immune system selects for parasites that can evade detection.

Walter Fitch and colleagues (1991) investigated whether selection imposed by the human immune system is responsible for detectable evolution in populations of influenza A viruses. If they could discern how and why flu populations have evolved in the past, perhaps they would be able predict how currently circulating populations will evolve in the future. Such predictions would be of great help to the developers of flu vaccines, which have to be redesigned every year.

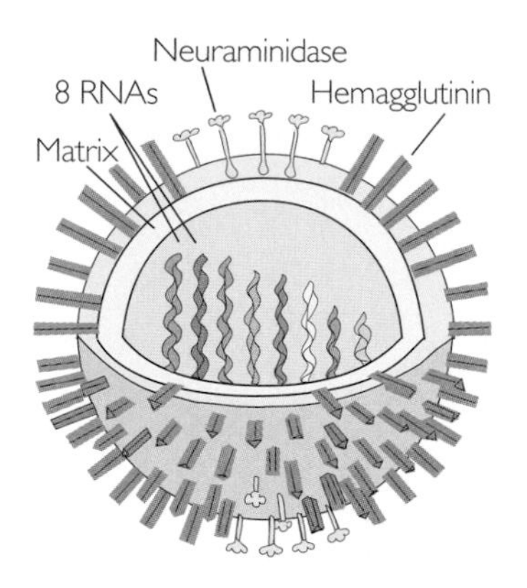

Figure 14.3 The influenza A virus The flu virus has two major surface proteins, hemagglutinin and neuraminidase. The viral genome, consisting of 10 genes, is carried on eight separate pieces of RNA. Redrawn from Webster et al. (1992).

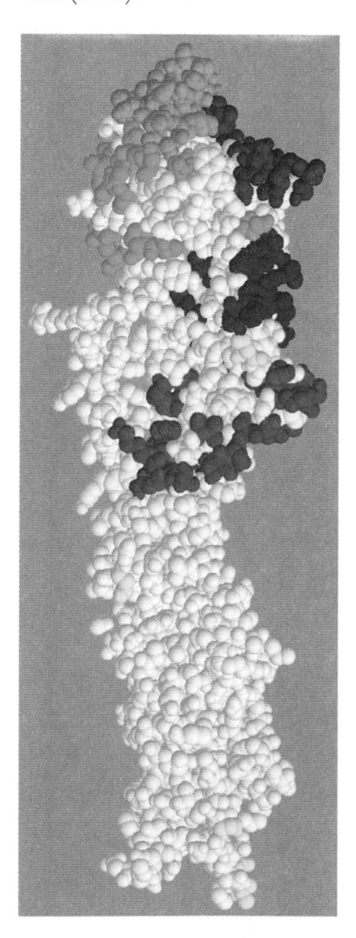

Figure 14.4 Hemagglutinin The five antigenic sites, regions recognized by the immune system, are highlighted with colors.

Flu Virus Evolution

Influenza A is responsible for annual flu epidemics and for occasional global pandemics, such as occurred in 1918, 1957, and 1968. Most of us think of flu as merely an annoyance—worse than a cold, certainly, but not as bad as chickenpox. In fact, flu can be deadly. In an ordinary flu season, the disease kills about 20,000 Americans. The 1918 pandemic flu was among the most devastating plagues in history. Within a period of months, it sickened some 20% of the world's population and killed between 50 and 100 million people (Kolata 1999, Johnson and Mueller 2002).

Influenza A has a genome composed of eight RNA strands that encode a total of 11 proteins (Figure 14.3). These proteins include polymerases, structural proteins, and coat proteins (Webster et al. 1992). The predominant coat protein is called hemagglutinin. Hemagglutinin initiates an infection by binding to sialic acid on the surface of a host cell (Laver et al. 1999). Hemagglutinin is also the primary protein recognized, attacked, and remembered by the host's immune system. To stay alive, any given strain of influenza A must either find a steady supply of naive hosts who have never been exposed to its version of hemagglutinin or alter its hemagglutinin so that previously exposed hosts no longer recognize it. Fitch and colleagues focused on mutations that alter the amino acids in hemagglutinin's antigenic sites (Figure14.4). **Antigenic sites** are the specific parts of a foreign protein that the immune system recognizes and remembers. Fitch and colleagues hypothesized that flu strains with novel antigenic sites would enjoy a selective advantage.

To test their hypothesis, the researchers examined the hemagglutinin genes of influenza A viruses that had been isolated from infected humans, and stored in freezers, between 1968 and 1987. Flu viruses evolve a million times faster than mammals, so the 20 years spanned by the frozen virus samples is equivalent to roughly four times the duration that separates humans from our common ancestor with the chimpanzees. In other words, the frozen flu samples constitute a fossil record—but one from which we can sequence genes.

From the sequences of the hemagglutinin genes, Fitch and colleagues estimated rate of evolution and the phylogeny of the frozen flu samples. The results appear in Figure 14.5. Two patterns are apparent. First, the flu strains accumulated nucleotide substitutions in their hemagglutinin genes at a steady rate, about 6.7×10^{-3} per nucleotide per year (Figure 14.5a). Second, most of the flu samples represent extinct side branches on the evolutionary tree (Figure 14.5b). The flu lineages that persisted into the 1980s were not a diverse assembly of strains descended from a variety of ancestors from the late 1960s and early 1970s. Instead the strains alive in the 1980s were close relatives, and all were descended from a single one of the late 1960s strains. The progeny of the other late-sixties and early-seventies strains had all died out.

What allowed the surviving lineage to endure while the other lineages perished? According to the researchers' hypothesis, it was nucleotide substitutions resulting in amino acid replacements in hemagglutinin's antigenic sites. From the nucleotide sequences, the researchers inferred all of the amino acid replacements that had occurred in the surviving lineage and in the extinct lineages. Then they noted whether each replacement had occurred in an antigenic site or a nonantigenic site. Fitch and colleagues predicted that, compared with the extinct lineages, the surviving lineage would have a higher fraction of its amino acid

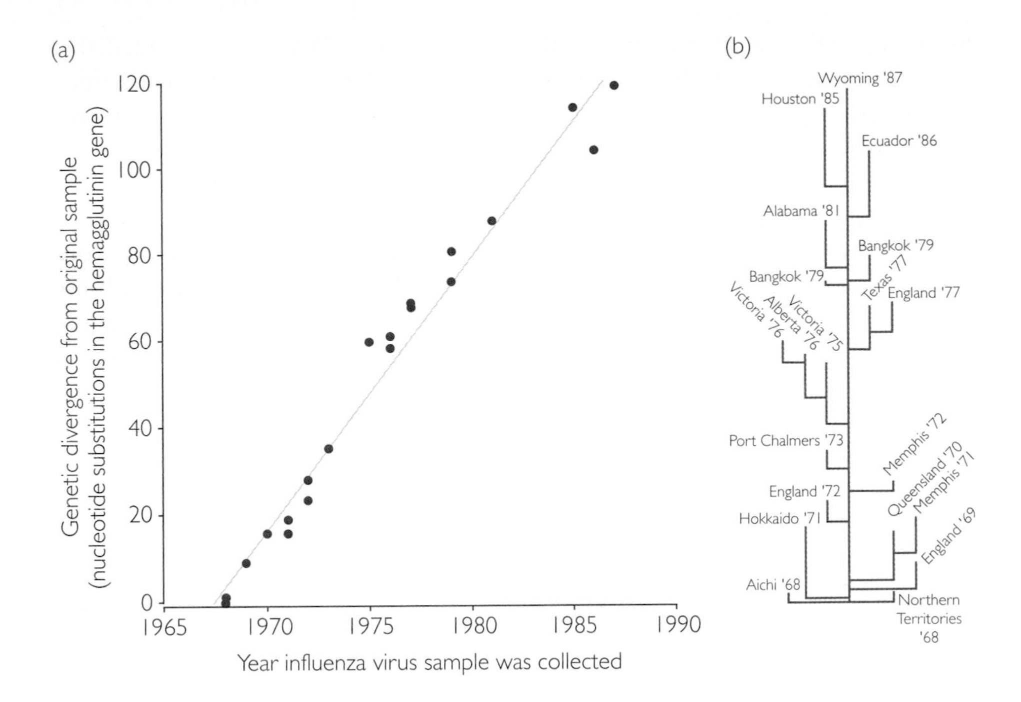

Figure 14.5 A phylogenetic analysis of frozen flu samples (a) The molecular evolution of the influenza A hemagglutinin gene as a function of time. The surviving lineage accumulated nucleotide substitutions at a constant rate. (b) A phylogeny of flu viruses isolated between 1968 and 1987. From Fitch et al. (1991).

replacements in its antigenic sites. The amino acid replacements, 109 in total, were distributed as follows:

	In antigenic sites	*In nonantigentic sites*
Surviving lineage	33	10
Extinct lineages	31	35

Consistent with the researchers' prediction, more than three-quarters of the surviving lineage's replacements had occurred in regions of hemagglutinin recognized by the immune system, compared with fewer than half of the extinct lineages' replacements. This association between a lineage's fate and the location of its replacements is statistically significant ($P = 0.002$).

Robin Bush, Walter Fitch, and colleagues (1999) followed up on this result by examining nucleotide substitutions in a phylogeny of hemagglutinin genes from 357 influenza A strains isolated between 1985 and 1996. The researchers took as their null hypothesis the neutral theory of molecular evolution (see Chapter 7). Recall that under the neutral theory two processes dominate molecular evolution: (1) Mutations resulting in amino acid replacements are typically deleterious and are eliminated by selection, and (2) mutations to synonymous codons are neutral and may become fixed in the population by genetic drift. According to the neutral theory, when we look at the nucleotide substitutions that have occurred on an evolutionary tree, silent substitutions should outnumber replacement substitutions. Of the 331 nucleotide substitutions that Bush, Fitch, and colleagues analyzed, 191 (58%) were silent and 140 (42%) were replacement substitutions. This result is consistent with the neutral theory (see Figure 7.21, page 253).

However, the researchers also identified 18 codons in the hemagglutinin gene in which there had been significantly more replacement substitutions than silent substitutions. The ratios in these 18 codons ranged from 4 replacement substitutions and 0 silent substitutions to 20 replacement substitutions and 1 silent substitution. An excess of replacement substitutions over silent substitutions is not

Phylogenetic analyses show that flu strains are more likely to survive if they have novel amino acid sequences in proteins recognized by the host's immune system.

consistent with the neutral theory. Bush, Fitch, and colleagues concluded that these 18 codons had been under positive selection for changes in the encoded amino acid. All 18 of the positively selected codons were for amino acids in antigenic sites of the hemagglutinin protein. It appears that the human immune system does, indeed, exert strong selection on flu virus hemagglutinin genes and that virus populations evolve in response.

This result is potentially useful to the makers of flu vaccines. Flu vaccines work by exposing the patient's immune system to killed flu viruses. Even though the viruses are dead, the immune system recognizes the viral proteins as foreign, mounts a response against them, and remembers their structure. In the event of a later infection by live viruses, the immune system can respond immediately. It can respond immediately, that is, as long as the hemagglutinin on the live invaders is similar enough to the hemagglutinin on the dead viruses that were in the vaccine. The problem is that flu populations evolve rapidly and vaccines take months to prepare in large quantities. Vaccine makers must begin production well in advance of the flu season. That means their scientific advisors must try to predict which among recently circulating flu strains are most likely to be responsible for next season's epidemic, so that they know which strains to include in the vaccine.

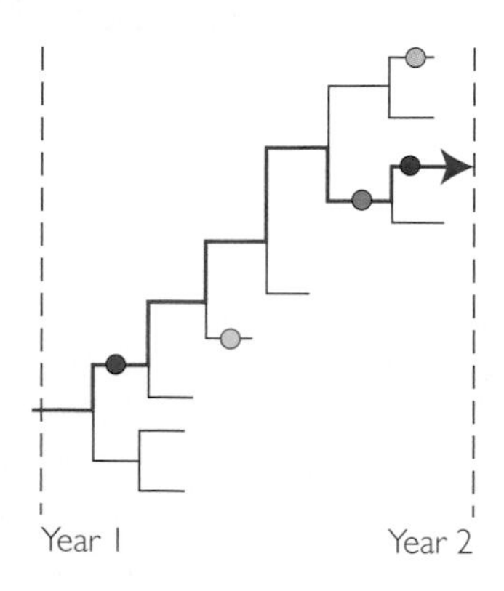

Figure 14.6 Predicting which lineages of flu will survive to cause future epidemics Mutations continually generate new lineages of flu, represented on an evolutionary tree as new branches. Among the lineages alive at any given time, one (red line) will ultimately survive; the rest (black lines) will eventually go extinct. Which lineage will survive? Usually, it is the one with the most amino-acid replacements in its hemagglutinin antigenic sites (indicated by colored dots). Redrawn from Bush (2001).

Robin Bush, Catherine Bender, and colleagues (1999; see also Bush 2001) devised a way to predict which of the currently circulating flu strains is most likely to have surviving descendents in the future. The survivor, they reasoned, is most likely to be the currently circulating strain with the most mutations in the 18 codons known to be under positive selection (Figure 14.6). On this basis, the researchers were able to accurately "predict," for 9 of 11 recent flu seasons, which of each season's strains would be the one to survive while the rest became extinct. Bush, Bender, and colleagues are careful to note that predicting which of this season's flu strains will be the ancestor of future lineages is not the same as predicting which, if any, of this season's strains will be responsible for next season's epidemic. Nonetheless, the predictive technique devised by Bush, Bender, and colleagues should be a valuable addition to the forecasting methods already in use. And it has spurred other researchers to look for ways to refine the technique (Plotkin et al. 2002; see also Fergusson and Anderson 2002).

The Origin of Pandemic Flu Strains

The fact that flu viruses with novel hemagglutinin genes appear to be at a selective advantage in evading their hosts' immune systems suggests a mechanism by which a strain could gain the ability to cause a pandemic. If a flu strain could somehow radically alter the structure of its hemagglutinin so that it was different from any hemagglutinin that had ever been seen by any human's immune system, then the strain could sweep the world and potentially infect everyone alive.

How could a flu strain radically alter the structure of its hemagglutinin? The organization of the influenza genome indicates a way (Figure 14.3). Recall that the flu genome has eight different RNA strands that encode a total of 11 different genes. If two flu strains simultaneously infect the same host cell, their genomes can recombine. That is, when new virions form, they may contain some RNA strands from Strain 1 and other RNA strands from Strain 2. Strain 1, for example, might produce offspring carrying Strain 2's hemagglutinin gene.

Phylogenetic analyses shed light on where, when, and how epidemics emerge.

The phylogeny in Figure 14.7 provides evidence that flu strains do, in fact, swap genes. This phylogeny, by Owen Gorman and colleagues (1991), is based on nucleotide sequences of influenza nucleoprotein genes. Nucleoprotein is thought to be the viral protein most responsible for host specificity. That is, the structure of a strain's nucleoprotein both enables the strain to infect particular species of hosts and tends to confine the strain to those species. Phylogenies based on the nucleoprotein gene should therefore be reliable indicators of strain history. Notice first that the nucleoprotein phylogeny has several distinct **clades**. A clade, in this context, is a set of strains that are all derived from a particular common ancestor. The distinct clades on the flu phylogeny include one that infects mainly horses, one that infects mainly humans, another that infects mainly pigs, and two that infect mainly birds.

Now look at the branch tips, their colors, and their labels. The colors indicate the species from which each strain was isolated. The labels give the year of isolation and the viral subtype. The subtype H3N2, for example, means *hemagglutinin-3, neuraminidase-2.* Neuraminidase, like hemagglutinin, is a coat protein. The numbers refer to groups of hemagglutinins or neuraminidases, defined by the ability of host antibodies to recognize them. The most important point for our purposes is that each hemagglutinin group constitutes a clade. That is, all H1s are more closely related to each other than to any H2 or H3 or H4. The same is true of the neuraminidases.

Find the human strains Human/Victoria/1968 (H2N2) and Human/Northern Territory/60–1968 (H3N2); they are set in boldface type. These two strains have nucleoproteins that are very closely related; that is why they are on nearby branches of the nucleoprotein phylogeny. They have neuraminidases that are closely related; both carry neuraminidase N2. But they have hemagglutinins that are distantly related; one carries H2, the other H3. How is it possible that two flu strains can have some genes that are closely related and others that are distantly related? The simplest explanation is that flu strains can trade genes. An examination of the phylogeny will reveal numerous additional examples.

Prior to the global pandemic of 1968, human flu viruses had never carried H3. This suggests that it was the acquisition of H3 from a nonhuman strain that allowed the 1968 flu to infect huge numbers of people

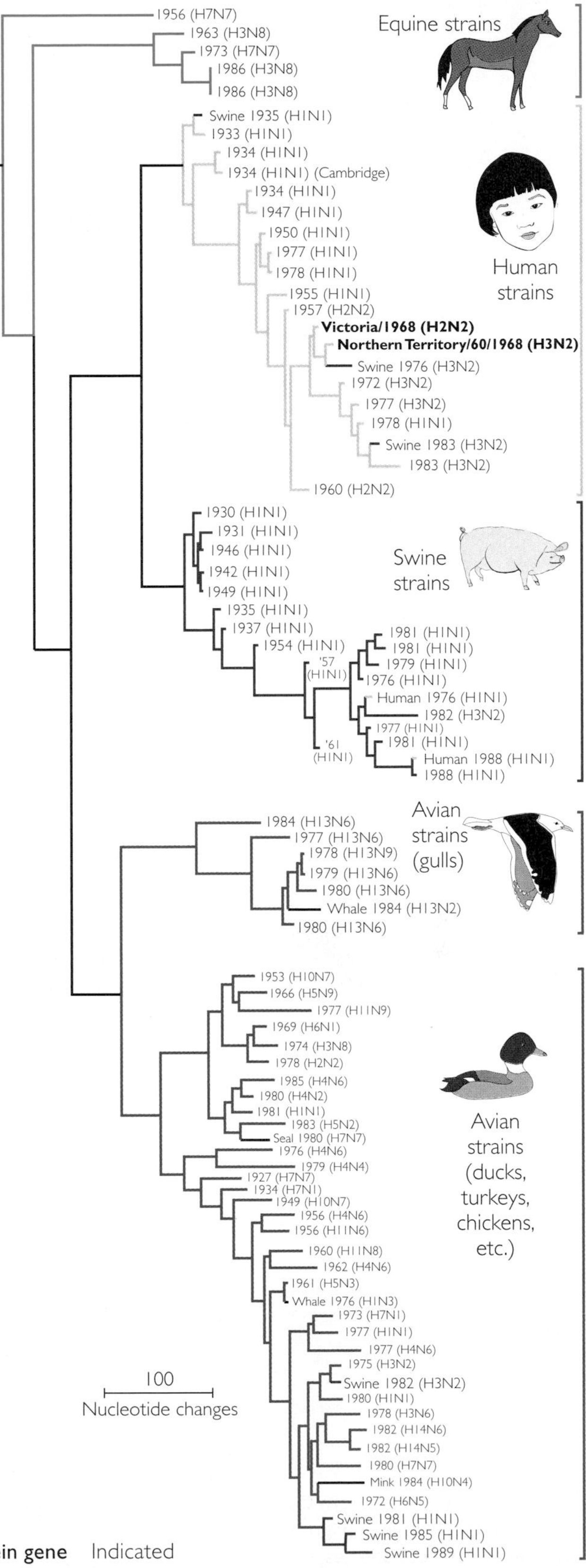

Figure 14.7 **A phylogeny of flu viruses, based on the nucleoprotein gene** Indicated for each viral strain is the host species from which it was isolated, the year, and the type of hemagglutinin and neuraminidase it carries. Redrawn from Gorman et al. (1991).

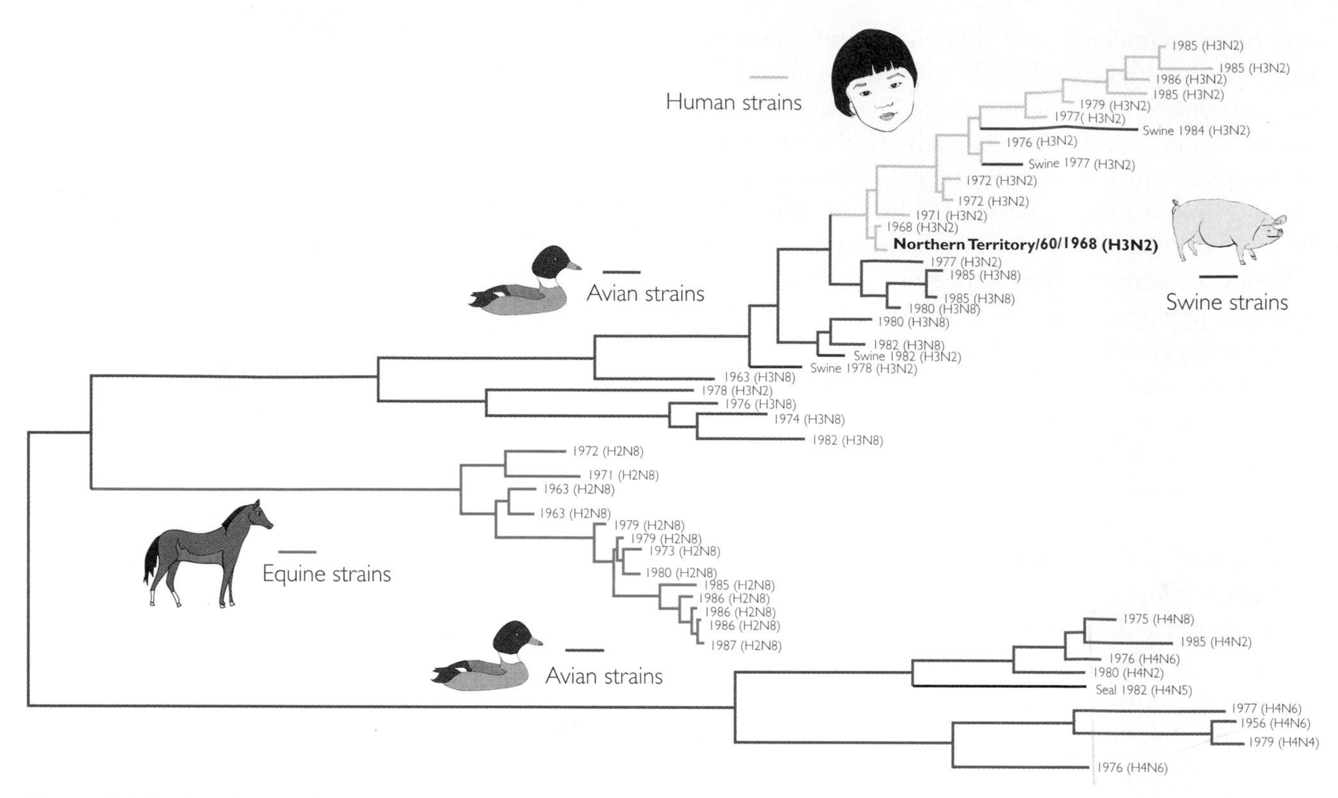

Figure 14.8 A phylogeny of flu virus hemagglutinin genes Indicated for each viral strain is the species from which it was isolated, the year, and the type of neuraminidase and hemagglutinin it carries. The 1968 human flu appears to have acquired its hemagglutinin gene from a bird flu strain. Redrawn from Bean et al. (1992).

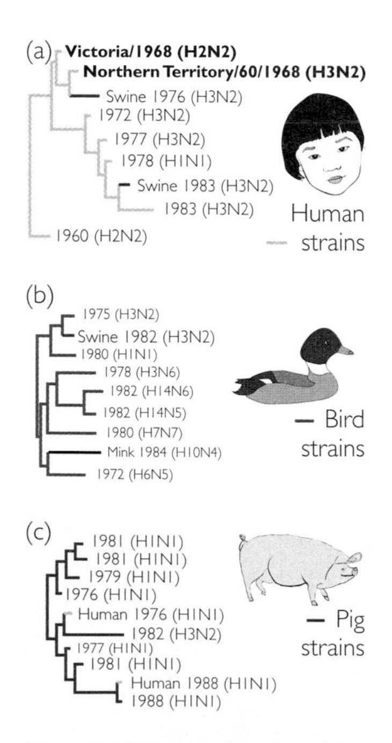

Figure 14.9 Branches from the nucleoprotein phylogeny (Figure 14.7) revealing cross-species transmissions (a) Human flu can infect pigs. (b) Bird flu can infect pigs. (c) Pig flu can infect people.

worldwide. What was the source of the H3 gene? Figure 14.8 shows a phylogeny from W. J. Bean and colleagues (1992) of human and nonhuman H3 genes, with equine H2 genes and bird H4 genes as the outgroups. The human H3 genes branch from within the bird H3s, with Northern Territory/60–1968's H3 near the base of the human clade. Apparently, the 1968 human pandemic flu strain acquired its H3 gene from a bird flu virus.

Similar evidence indcates that the 1968 pandemic strain also picked up a new version of the gene for a component of its polymerase enzyme called PB1. Again, the source was a bird virus (Parrish and Kawaoka 2005). Eleven years earlier, the 1957 pandemic was caused by a human strain that had replaced genes for three of its proteins—hemagglutinin, neuraminidase, and PB1, with copies it took up from a bird virus.

How do human flu strains acquire genes from bird strains? The nucleoprotein phylogeny in Figure 14.7, key parts of which are reproduced in Figure 14.9, contains clues. The nucleoprotein phylogeny reveals that human flu strains sometimes infect pigs (for example, Swine 1976, which was isolated in Hong Kong; Figure 14.9a). It reveals that bird strains sometimes infect pigs (Swine 1982, also from Hong Kong; Figure 14.9b). And that pig strains sometimes infect humans (Human 1976, from New Jersey; Figure 14.9c). A popular hypothesis among flu researchers is that pandemics begin when human strains and bird strains simultaneously infect a pig, swap genes, and later move from pigs to people (Webster et al. 1992).

Of course, the pandemic flu we most need to understand is the scourge of 1918. Toward this end Jeffery Taubenberger, Ann Reid, and colleagues (2005) have sequenced the genome of a flu strain recovered from the body of an Inuit

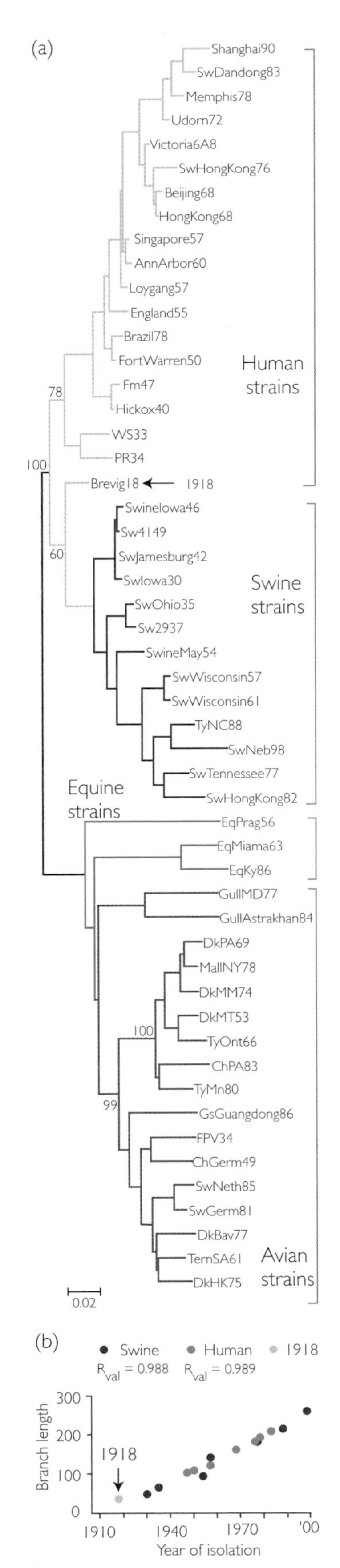

Figure 14.10 Evolutionary analysis of the nucleoprotein gene from the 1918 flu (a) An evolutionary tree of nucleoprotein genes from a variety of flu strains, including the one that killed an Inuit woman in 1918. The 1918 virus is near the base of the branch that includes all more recent human and swine strains. (b) A molecular clock plot for human and swine strains. Each point shows the genetic distance between the reconstructed human-swine common ancestor and one of its descendents, as a function of the year in which the descendent was collected. The point where the best-fit line crosses the horizontal axis estimates the year in which the common ancestor lived. From Reid et al. (2004).

woman who died in 1918 at Brevig Mission, Alaska, and was buried in permafrost. Terrence Tumpey and colleagues (2005) recreated live flu virus from the sequence, tested it in mice, and confirmed that it is extraordinarily deadly.

Reid, Taubenberger, and coworkers have analyzed the evolutionary history of each of the 1918 virus' genes (Reid et al. 1999, 2000, 2002, 2004a; Basler et al. 2001; Taubenberger et al. 2005). The results for the nucleoprotein gene are typical (see Reid et al. 2004b) and it is these we will focus on here.

An evolutionary tree of nucleoprotein genes, including the one from 1918, appears in Figure 14.10a. Like the nucleoprotein phylogeny in Figure 14.7, the genes on this tree form clades based on the animals they infect. The gene from the 1918 flu branches near the base of the clade that infects humans and pigs. This implies that all more recent human and pig strains are descended from the 1918 flu or one of its close relatives. The molecular clock plot in Figure 14.10b (like the one in Box 1.2, pages 28-29) allows us to estimate the age of the common ancestor of the human and pig strains in the clade. This common ancestor lived shortly before the great pandemic.

Where did the virus that wrought havoc in 1918 and begat all subsequent human influenzas come from? Alas, the data yield no definitve answer. The problem is that the human-swine clade does not arise from within any other clade on the tree. The founder of the human-swine clade could have come from an earlier human strain, a horse strain, a bird strain, or a strain that infects some other animal not represented in the tree at all.

Given currently available evidence, the smart money is on birds. Bird influenzas are, in general, much more diverse than those of other hosts, and birds are thought to be the ultimate reservoir from which all other influenzas derive. Reid and colleagues note that the nucleoprotein gene of the 1918 flu encodes relatively few amino acid changes versus the genes of typical bird flus. This suggests that the 1918 flu was recently derived from a strain adapted to living in birds. At the same time, however, the gene of the 1918 flu contains a great many synonymous nucleotide substitutions. This indicates that the 1918 flu was only distantly related to the bird strains presently known. Analyses of the remainder of the 1918 flu's genome provide no convincing evidence that it arose via recombination, like the 1957 and 1968 pandemic strains. Reid, Taubenberger, and colleagues hypothesize that the 1918 pandemic began when a virus jumped directly into humans from a group of birds whose viruses remain to be discovered and studied.

As a result of the discoveries discussed in this section—the work of evolutionary biologists applying the tools of their discipline to a practical problem—an international team of researchers maintains constant surveillance of flu strains circulating in pigs, birds, and humans. Their goal is to spot new pandemic strains early enough to allow the production and distribution of large quantities of vaccine. The flu surveillance researchers keep an especially keen watch for recombinant strains and for strains that are moving from species to species.

14.2 Evolving Pathogens: Antibiotic Resistance

Antibiotics are chemicals that kill bacteria by disrupting particular biochemical processes. For human patients, antibiotics are lifesaving drugs. For populations of bacteria, however, antibiotics are powerful agents of selection. When applied to a population of bacteria, an antibiotic quickly sorts the resistant individuals (those that can tolerate the drug) from the susceptible ones (those that cannot). An evolutionary perspective suggests that antibiotics should be used judiciously; otherwise, these miracle drugs may undermine their own effectiveness.

There are dozens of antibiotics and dozens of molecular mechanisms whereby bacteria can become resistant [see Baquero and Blázquez (1997) for a review]. Some of these mechanisms of resistance involve losses of function. Resistance to isoniazid in the tuberculosis pathogen *Mycobacterium tuberculosis* provides an example. Isoniazid poisons bacteria by interfering with the production of components of the cell wall (Rattan et al. 1998). Before it can do so, however, isoniazid must be converted by a bacterial enzyme into its biologically active form. The conversion is performed by the enzyme catalase/peroxidase, encoded by a gene called *KatG*. Mutations in *KatG* that reduce or eliminate catalase/peroxidase activity render bacteria tolerant or immune to isoniazid's effects.

Other mechanisms of resistance involve gains of function. Numerous extrachromosomal elements of bacteria, such as plasmids and transposons, carry genes conferring resistance to one or more antibiotics. The plasmid *Tn3*, for example, found in *Escherichia coli*, contains a gene called *bla*. This gene encodes an enzyme, β-lactamase, that breaks down the antibiotic ampicillin.

Evidence That Antibiotics Select for Resistant Bacteria

Several lines of evidence show that antibiotics select in favor of resistant bacteria, and that bacterial populations evolve rapidly in response.

Evidence that antibiotics select in favor of resistant bacteria comes from a variety of studies. On the smallest scale are studies of bacterial evolution within individual patients. William Bishai and colleagues (1996) monitored an AIDS patient with tuberculosis. When they initially determined that the patient had tuberculosis, the researchers cultured bacteria from the patient's lungs and found them sensitive to a variety of antibiotics, including rifampin. They and other doctors treated the patient with rifampin in combination with several other drugs. The patient responded well to treatment. At one point, the patient was so nearly recovered that the researchers were unable to culture tuberculosis bacteria from his lungs. Soon, however, the patient relapsed and died. After his death, the researchers found that the tuberculosis bacteria in the patient's lungs had resurged. They screened these bacteria for resistance to antibiotics. The bacteria were still susceptible to most drugs, but they were resistant to rifampin. The researchers sequenced the *rpoB* gene from some of the resistant bacteria. In the gene, they found a point mutation known to confer rifampin resistance.

Did the rifampin-resistant strain of tuberculosis evolve in the patient's lungs, or had he been infected with a new strain that was already resistant when he got it? The researchers prepared genetic fingerprints of rifampin-sensitive bacteria from the patient's initial infection and rifampin-resistant bacteria from the patient's autopsy. Other than the *rpoB* point mutation, the genetic fingerprints of the two groups of bacteria were identical. The researchers examined over 100 other strains of bacteria from patients living in the same city at the same time. Only two of the strains had genetic fingerprints matching the strain that killed the patient, and neither of these was rifampin-resistant. The simplest explanation for these results is that the *rpoB* point mutation occurred in bacteria living in the patient's lungs and ultimately rose to high frequency due to selection imposed by treatment with rifampin.

On a larger scale, researchers can compare the incidence of susceptible versus resistant bacterial strains among patients who are newly diagnosed, and thus have not been previously treated with antibiotics, versus patients who have relapsed after antibiotic treatment. If antibiotics select in favor of drug resistance, then a higher fraction of relapsed patients should harbor antibiotic-resistant bacteria. Alan Bloch and colleagues (1994) reported the results of a survey of tuberculosis patients conducted by the Centers for Disease Control. The results for bacterial susceptibility to isoniazid are as follows:

	New cases	*Relapsed cases*
Number with resistant bacteria	**243**	**41**
Number with susceptible bacteria	**2728**	**150**
Fraction resistant	**8.2%**	**21.5%**

These numbers are consistent with the notion that populations of bacteria within patients evolve in response to treatment.

Finally, on the largest scale, researchers can examine the relationship over time between the fraction of patients with resistant bacteria and the societywide level of antibiotic use. If antibiotics select in favor of resistance, then the level of resistance should track antibiotic consumption. D. J. Austin and colleagues (1999) plotted data on penicillin resistance among *Pneumococcus* bacteria in children in Iceland (Figure 14.11). In the late 1980s and the early 1990s, the fraction of children whose bacteria were resistant to penicillin rose dramatically. In response, Icelandic public health authorities waged a campaign to reduce the use of penicillin. Between 1992 and 1995, the per capita consumption of penicillin by children dropped by about 13%. The level of penicillin resistance peaked at just under 20% in 1993, then fell below 15% by 1996. Again, the data are consistent with the hypothesis that bacterial populations evolve in response to selection imposed by antibiotics.

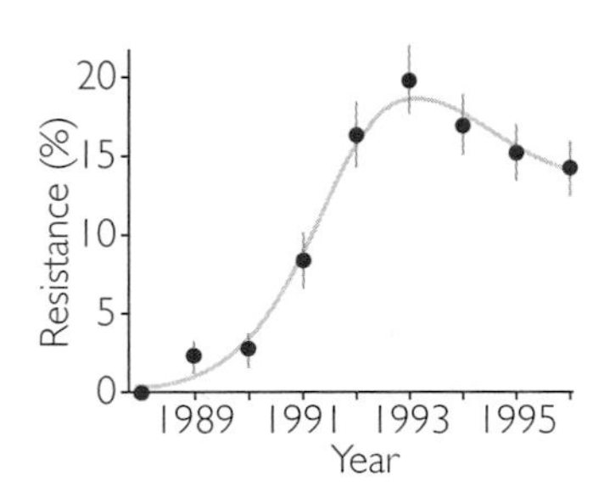

Figure 14.11 The frequency of penicillin resistance among *Pneumococcus* bacteria in Icelandic children as a function of time When public health officials waged a public campaign to reduce the use of penicillin, the incidence of resistant strains declined. Redrawn from Austin et al. (1999).

Evaluating the Costs of Resistance to Bacteria

Presumably, penicillin resistance in Iceland fell with declining penicillin use because penicillin resistance is costly to bacteria. If resistance comes at a cost, then when antibiotics are absent, sensitive bacteria will have higher fitness.

Costs of resistance are generally assumed to be common. When antibiotic resistance is conferred by loss-of-function mutations, costs might be levied by the loss-of-function itself. When antibiotic resistance is conferred by gains of function, costs might be levied by the expense of maintaining new genes and their associated proteins.

Costs to bacteria associated with antibiotic resistance would be good news for public health. If an antibiotic begins to lose its effectiveness because too many bacteria are resistant, doctors and patients could simply make a collective agreement to suspend the use of the antibiotic until bacterial populations have evolved back to the point that they are again dominated by susceptible strains.

Antibiotic resistance is generally assumed to be costly to bacteria, but over the long term costs of resistance can be eliminated by natural selection.

While resistance may, indeed, typically come at a cost, the cost may not always persist. Additional mutations elsewhere in the bacterial genome may compensate for the costs, making resistant bacteria equal in fitness to sensitive bacteria, even when antibiotics are absent. Stephanie Schrag and colleagues (1997) investigated the

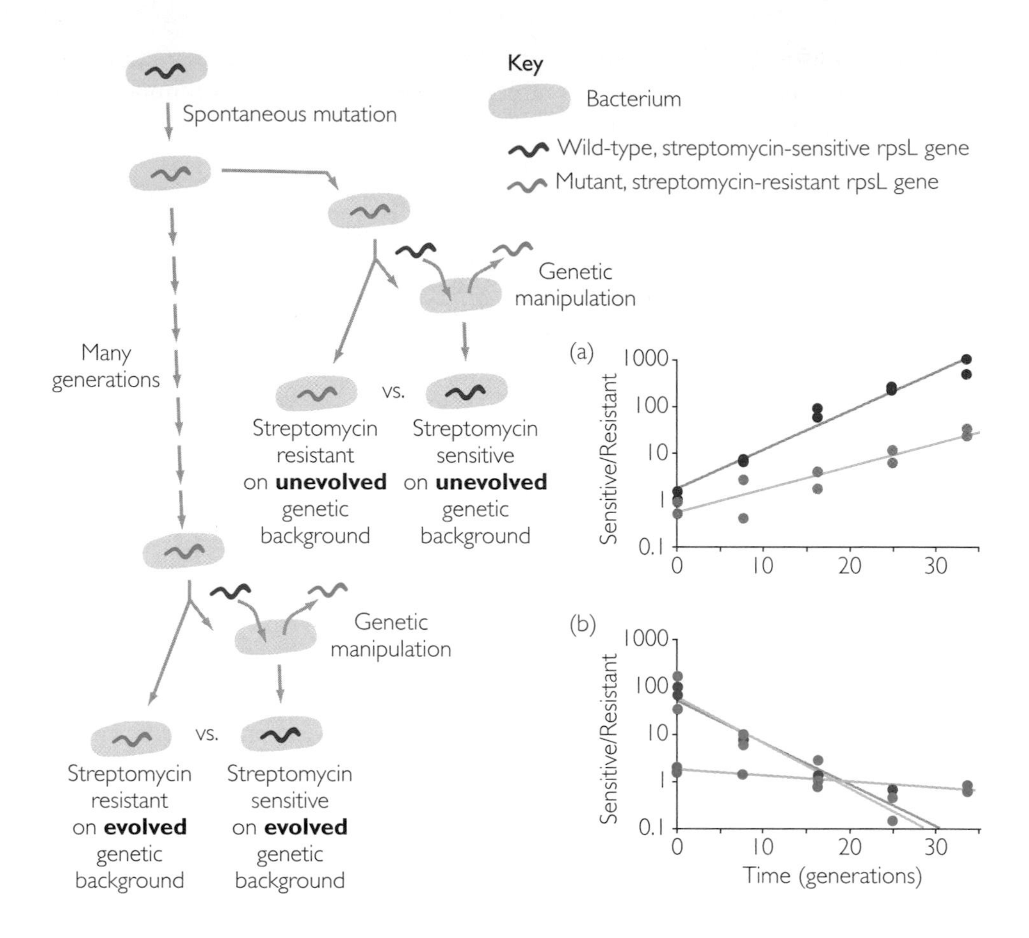

Figure 14.12 An assessment of the costs of antibiotic resistance to bacteria, over the short term and over the long term See text for explanation. Scatterplots from Schrag et al. (1997).

possibility that compensatory mutations could alleviate the cost of streptomycin resistance in *E. coli*. Their experimental design and results appear in Figure 14.12.

Schrag and colleagues started with a population of streptomycin-sensitive *E. coli* and screened for new streptomycin-resistant mutants. Streptomycin interferes with protein synthesis by binding to a ribosomal protein encoded by the *rpsL* gene. Point mutations in the *rpsL* gene render bacteria resistant to streptomycin. In one set of experiments, the researchers competed streptomycin-resistant strains against identical strains restored to sensitivity by replacement of the mutant version of *rpsL* with the wild-type version. If streptomycin resistance comes at a cost, at least in the short-term, then in mixed cultures the streptomycin-sensitive bacteria should increase in frequency over time. This is exactly what happened (Figure 14.12a).

In a second set of experiments, Schrag and colleagues let streptomycin-resistant strains evolve for many generations in the lab. Then the researchers again competed streptomycin-resistant strains against identical strains restored to sensitivity by genetic manipulation. If compensatory mutations had occurred and become fixed while the resistant strains were allowed to evolve, then in mixed cultures the streptomycin-sensitive strains should fail to increase in frequency over time. In fact, the result was even more dramatic: The streptomycin-sensitive strains decreased in frequency over time (Figure 14.12b). Schrag and colleagues concluded that compensatory evolution had not only alleviated the cost of streptomycin resistance, it had created a multilocus genotype, or genetic background, on which the resistant allele of *rpsL* enjoyed a fitness advantage over the sensitive allele.

Judicious Use of Antibiotics

If Schrag et al.'s results are general, there is no guarantee that an antibiotic can be restored to medical effectiveness simply by withdrawing it from use until bacterial populations have evolved back to sensitivity. If we want to maintain an arsenal of potent antibiotics that we can use when patients' lives are at stake, it seems that our best chance is to try to avoid letting bacterial populations evolve resistance in the first place.

Bacteriologist Stuart Levy (1998) recommends guidelines for the limitation of antibiotic resistance. The guidelines are intended to prevent people from contracting bacterial infections in the first place, restrict unnecessary uses of antibiotics that may select for resistance in potentially pathogenic bacterial bystanders, and ensure that, when antibiotics are used, they exterminate the targeted bacterial population before resistance evolves. Among Levy's guidelines are these:

The best defense against antibiotic-resistant bacteria is to avoid letting bacterial populations evolve resistance in the first place.

- To avoid contracting foodborne bacteria, consumers should wash fruits and vegetables and avoid raw eggs and undercooked meat.
- Consumers should use antibacterial soaps and cleaners only when they are needed to prevent infection in patients with compromised immune systems.
- Patients should not request antibiotics for viral infections, such as colds or flu.
- When they do take antibiotics, patients should complete the full course of treatment. Patients should never save antibiotics prescribed for one infection and use them to treat another.
- To avoid spreading infections from patient to patient, doctors should wash their hands thoroughly between patients.
- Doctors should never prescribe unneeded antibiotics, even when patients request them.
- When they do prescribe antibiotics, doctors should use drugs that target the narrowest possible range of bacterial species.
- Doctors should isolate patients infected with bacteria resistant to several drugs to reduce the risk that such bacteria will spread.

14.3 Evolving Pathogens: Virulence

The final issue we will consider relating to the evolution of pathogen populations is virulence. Virulence is the harm done by a pathogen to the host during the course of an infection. Virulence varies dramatically among human pathogens. Some pathogens, like cholera and smallpox, are often lethal; others, like herpes viruses and cold viruses, may produce no symptoms at all. Evolutionary biologists investigating virulence seek to explain this diversity.

Virulence, the harm done by a parasite to its host, is a trait that can evolve.

How Virulence Evolves

There are three general models to explain the evolution of virulence (Bull 1994, Ewald 1994, Levin 1996):

1. **The coincidental evolution hypothesis.** The virulence of many pathogens in humans may not be a target of selection itself, but rather an accidental by-product of selection on other traits. For example, tetanus is caused by a soil bacterium, *Claustridium tetanae*. When tetanus bacteria find themselves

inside a human wound, they can grow and divide. They also produce a potent neurotoxin, making tetanus infections highly lethal. However, tetanus bacteria ordinarily do not live in humans and are not transmitted by humans. The ability of these bacteria to produce tetanus toxin is probably the result of selection during their ordinary life in the soil, not selection inside human hosts.

2. **The short-sighted evolution hypothesis.** Pathogens may experience many generations of evolution by natural selection within an individual host before they have the opportunity to move to a new host. As a result, traits that enhance the within-host fitness of pathogen strains may rise to high frequency even if they are detrimental to transmission of the pathogen to new hosts. Poliovirus may provide an example. Ordinarily, polioviruses infect only cells that line the digestive tract, produce no symptoms, and are transmitted via feces. Occasionally, however, polio virions invade the cells of the nervous system. Acquiring the ability to invade the nervous system probably increases within-host fitness, because the virions that can do so have fewer intraspecific competitors. But the virions living in the nervous system are unlikely to ever be transmitted to a new host.
3. **The trade-off hypothesis.** Biologists traditionally believed that all pathogen populations would evolve toward ever-lower virulence. The reasoning was that damage to the host must ultimately be detrimental to the interests of the pathogens that live inside it. If the host dies, for example, the pathogens die with it. Thus, it was thought, more benign pathogens should enjoy higher lifetime reproductive success. This view was naive. Recall our discussion of the evolution of aging in Chapter 13. There we concluded that genes that hasten the death of their carriers can nonetheless rise in frequency if they confer a sufficient enhancement of early-life reproductive success. As with genes, so too with pathogens. A strain can be virulent but nontheless increase in frequency in the total pathogen population if, in the process of killing its hosts, it sufficiently increases its chances of being transmitted (Figure 14.14). Natural selection should favor pathogens that strike an optimal balance between the costs and benefits of harming their hosts.

Figure 14.13 A parasite that kills its host before reproducing The fungus shown here has invaded the body of a weevil and grown inside. Finally, the fungus killed the weevil and sprouted fruiting bodies that will release spores. Killing its host is not necessarily detrimental to the interests of a parasite.

We explored the short-sighted evolution hypothesis, as it applies to HIV, in Chapter 1. It is the trade-off hypothesis that we will focus on here. A key assumption of the trade-off hypothesis is that a pathogen cannot reproduce inside its host without doing the host some harm. Every offspring that the pathogen makes is constructed with energy and nutrients stolen from the host. In addition, the pathogen produces metabolic wastes that the host must detoxify and eliminate. These are the reasons the host mounts an immune response against the pathogen—an expensive endeavor that compounds the host's costs but that may be better for the host than the alternative.

All else being equal, pathogens with higher within-host reproductive rates should be transmitted to new hosts at higher rates as well. But all else is equal only up to a point. Because reproducing faster within the host necessarily means harming the host more severely, it is possible for the pathogen to reproduce too fast. Reproducing too fast may mean debilitating the host so severely, or killing it so quickly, that the rate of transmission to new hosts is reduced.

According to the trade-off hypothesis, selection favors parasites that reproduce more quickly within their hosts—until the parasites begin to harm the hosts so severely that the probability of transmission begins to fall.

Sharon Messenger, Ian Molineux, and James Bull (1999) tested the trade-off hypothesis by using *E. coli* as the host and a virus, bacteriophage f1, as the pathogen. Phage f1 produces lasting, nonlethal infections in *E. coli*. The phage invades a bacterium and lives inside it as a plasmid. It induces the machinery of the host cell to produce new phage copies that are secreted from the cell as phage chromosomes encased in protein filaments. Production of new phage copies slows the growth rate of the host bacterium to about one-third of normal. But when the host bacterium does divide, copies of phage f1 typically travel with both daughter cells. Thus phage f1 has two modes of transmission: It is transmitted vertically, from one host generation to the next, when the host cell divides, and it is transmitted horizontally, from one host to another, when secreted virions invade new hosts.

Messenger and colleagues maintained cultures of phage f1 in which they forced the viruses to alternate between the two modes of transmission. During the vertical-transmission phase, the researchers prevented secreted virions from infecting new bacterial cells. The only way phages could spread was via the reproduction of their hosts. During the horizontal-transmission phase, the researchers harvested secreted virions and introduced them to cultures of uninfected bacteria. Now the only way the phages could spread was via secretion.

The researchers maintained two sets of cultures for 24 days. For one set of cultures, they alternated 1-day-long vertical-transmission phases with brief horizontal-transmission phases. For another set of cultures they alternated 8-day-long vertical-transmission phases with brief horizontal-transmission phases. At the end of 24 days, the researchers measured phage virulence and phage reproductive rate. They measured virulence as the growth rate of infected hosts, with lower host growth rates indicating more virulent viruses. They measured phage reproductive rate as the rate of virion secretion from infected hosts, with more rapid secretion indicating faster phage reproduction.

Messenger and colleagues made two predictions. First, they predicted that, across their cultures, they would find a correlation between phage virulence and phage reproduction rate. In other words, phages that induced their hosts to produce and secrete more phage copies would slow the growth of their hosts more severely. Second, they predicted that the cultures subjected to eight-day vertical-transmission phases would evolve lower reproductive rates and lower virulence than the cultures subjected to one-day vertical-transmission phases. Their reasoning here was that during the vertical-transmission phase natural selection should favor viral strains that allow their host bacteria to divide more quickly, whereas during the horizontal transmission phase selection should favor viral stains that induce their host bacteria to secrete more viral copies. The eight-day cultures experienced more selection to allow their hosts to divide and less selection to induce secretion, so they should evolve lower virulence.

The results appear in Figure 14.14. The figure is a scatterplot showing the reproductive rate and virulence of each of 13 one-day cultures and 13 eight-day cultures. First, note the strong correlation across both experiments between viral reproductive rate and virulence. As Messenger and colleagues predicted, the phage strains that slow their hosts' growth rate most dramatically are the strains that reproduce more quickly within their hosts. Second, note that the eight-day cultures had lower reproductive rates and lower virulence than the one-day cultures. As the researchers predicted, different patterns of selection favor different levels of virulence. These results are consistent with the trade-off hypothesis.

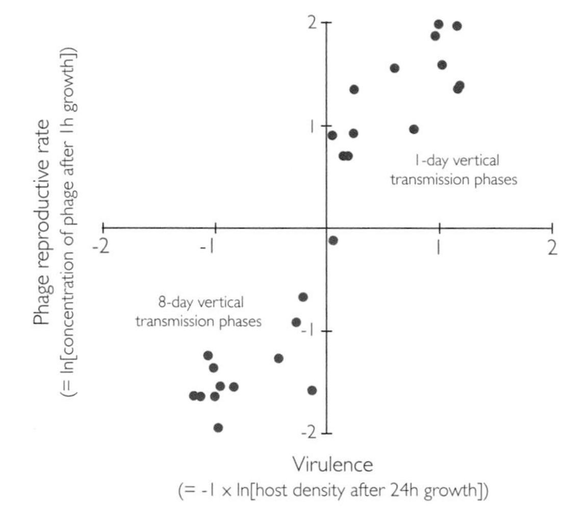

Figure 14.14 A trade-off between the virulence and within-host reproductive rate of a virus that infects *E. coli* When researchers gave the viruses more opportunities for horizontal transmission (red dots), the viruses evolved higher virulence and higher reproductive rates than viruses given fewer opportunities for horizontal transmission (blue dots). Redrawn from Messenger et al. (1999).

Other researchers have also found experimental support for the trade-off hypothesis. Vaughn Cooper and colleagues (2002), for example, selected for rapid replication in populations of nuclear polyhedrosis virus by allow transmission from host to host only during the early stages of infection. Compared to viral strains transmitted late, the early transmission strains not only had higher replication rates, they were also significantly more likely to kill their gypsy moth hosts.

Virulence in Human Pathogens

Paul Ewald (1993, 1994) considered how the trade-off hypothesis might apply to human pathogens. He used the hypothesis to guide his thinking about how the details of disease transmission should select for different levels of virulence. His key insight was that some pathogens thrive only so long as their hosts are reasonably healthy, whereas other pathogens thrive when even when their hosts are severely ill. Here we will discuss two of Ewald's specific predictions, along with the tests he devised using comparative data compiled from the literature.

Parasites transmitted by insect vectors or water can thrive even when their host is severely debilitated. As a result they tend to be more virulent than parasites that are transmitted by direct contact.

Ewald's first prediction concerns the virulence of diseases like colds and flu that are transmitted by direct contact between an infected person and an uninfected person, versus diseases like malaria, that are transmitted by insect vectors. Ewald reasoned that diseases transmitted by direct contact cannot afford to be virulent. If a host is so incapacitated by illness that he or she stays home in bed and avoids contact with uninfected individuals, the pathogen has no chance to be transmitted. Vectorborne diseases, on the other hand, can afford to be highly virulent. An insect vector can carry pathogens away from even a severely debilitated host, and might, in fact, be at less risk of being killed in the process. Ewald compiled data on the mortality rates of a variety of vectorborne and directly transmitted diseases. These data are consistent with Ewald's prediction (Figure 14.15). The vast majority of directly transmitted diseases have mortality rates under 0.1%, whereas more than 60% of vectorborne diseases have mortality rates of 0.1% or higher.

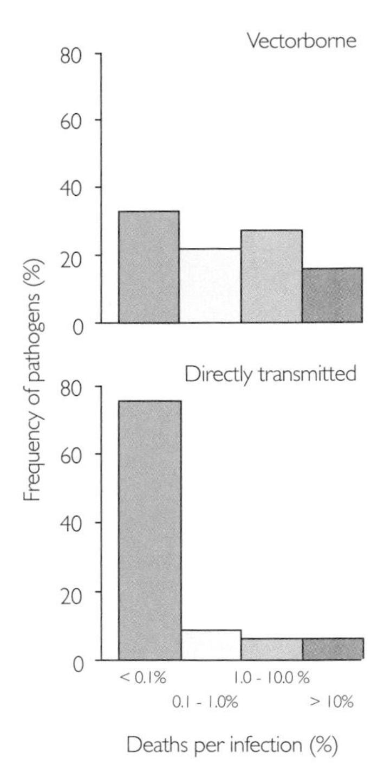

Figure 14.15 The virulence of vectorborne versus directly transmitted diseases Diseases carried from host to host by insects (top) are, on average, more virulent. From Ewald (1994).

Ewald's second prediction concerns bacteria that infect the digestive tract and cause diarrhea. These bacteria can typically be transmitted both directly from person to person and via contaminated water. Ewald reasoned that contaminated water can play the same role that insect vectors did in his first prediction. That is, when sewage enters the drinking water supply, even a severely incapacitated host can transmit bacteria to remote individuals over long distances. Ewald assembled data on roughly 1,000 outbreaks of illness caused by nine different kinds of bacteria. Some of these bacteria have a stronger tendency to be spread by direct contact, others have a stronger tendency to be spread by contaminated water. For each kind of bacteria, Ewald calculated both the fraction of outbreaks attributed to contaminated water and the victim mortality rate. He predicted that diseases with a higher frequency of waterborne transmission would be more virulent. The data, plotted in Figure 14.16, are consistent with Ewald's prediction. The most virulent of the nine bacteria in the study is classical *Vibrio cholerae*, the pathogen responsible for the waterborne, and lethal, cholera outbreak in London during 1854.

The trade-off hypothesis for the evolution of virulence implies that human behavior can affect the severity of human diseases. For example, when people dump untreated sewage directly into rivers, or when health-care workers fail to wash their hands thoroughly between patients, they create conditions that may select for increased virulence in human pathogens. Conversely, when people keep their drinking water supplies pure, and when health-care workers avoid becoming inadvertent vectors, they create conditions that may select for reduced virulence in human pathogens.

We now turn our attention from evolving populations of pathogens to another group of evolving populations important to human health: populations of humans cells inside individual humans. It may seem surprising to even propose that human cell populations can evolve because we are used to thinking of all of the cells in a human body as being genetically identical. In fact, however, there are mechanisms that produce genetic diversity among somatic cells and conditions under which somatic cell populations can evolve.

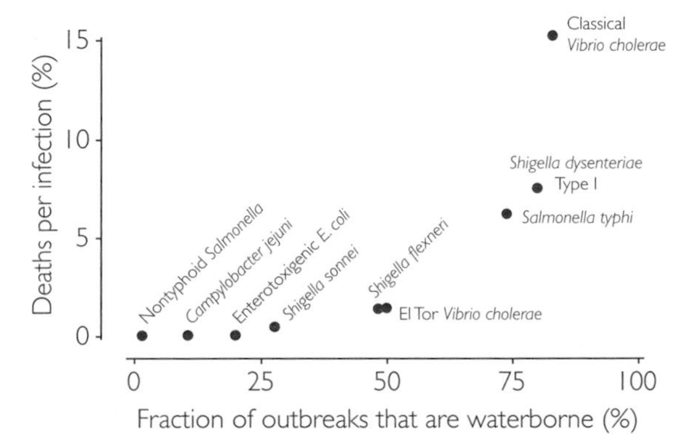

Figure 14.16 The virulence of intestinal bacteria, as a function of tendency toward waterborne transmission The higher the fraction of a disease's outbreaks that are waterborne, the more virulent the disease. From Ewald (1991).

14.4 Tissues as Evolving Populations of Cells

All of the cells in an individual's body are descended from a common ancestor, the zygote. If, during the development of a tissue, a mutation occurs in a cell still capable of continued division, then we can think of the tissue as a population of reproducing cells with heritable genetic variation. If one of the genetic variants leads to increased cell survival or faster reproduction, then the tissue will evolve by natural selection, just like a population of free-living organisms.

A Patient's Spontaneous Recovery

Rochelle Hirschhorn and colleagues (1996) documented a case in which tissue evolution saved the life of a boy with a serious genetic disease. The disease is called adenosine deaminase deficiency. Adenosine deaminase (ADA), encoded by a locus on chromosome 20, is a housekeeping enzyme normally made in all cells of the body. ADA's job is to recycle purines. Cells lacking ADA accumulate two poisonous metabolites—deoxyadenosine and deoxyadenosine triphosphate. The cells in the body most susceptible to these poisons are the lymphocytes, including the T cells and B cells vital to the immune system (Youssoufian 1996). Individuals who inherit loss-of-function mutations in both copies of the ADA gene have no T cells and have B cells that are nonfunctional or absent (Klug and Cummings 1997). In consequence, these individuals suffer from severe combined immunodeficiency. Without treatment, they usually die of opportunistic infections at an early age.

Both parents of the boy that Hirschhorn and colleagues studied carry, in heterozygous state, a recessive loss-of-function allele for ADA. One of the boy's older siblings inherited two loss-of-function alleles, made no ADA, and died of severe combined immunodeficiency at age two. The boy himself also inherited both his parents' mutant alleles, and during his first 5 years he suffered the recurrent bacterial and fungal infections characteristic of severe combined immunodeficiency. Between the ages of 5 and 8, however, the boy spontaneously and mysteriously recovered. He was 12 years old when Hirschhorn and colleagues published their paper, and he had been clinically healthy for 4 years.

With a careful genetic analysis of the mother, father, and son, Hirschhorn and colleagues were able to reconstruct a plausible explanation for the boy's recovery. Although the boys' parents are both carriers for ADA deficiency, they are carriers for different loss-of-function alleles. Hirschhorn and colleagues showed that the son's blood cells are a genetic mosaic (Box 14.1). The father's mutation is present in all of the boy's peripheral leukocytes (white blood cells) and lymphoid B cells. The mother's mutation is present in all of the boy's peripheral leukocytes, but absent in most of his B cells.

How could this happen? Hirschhorn and colleagues found evidence that the cell ancestral to most of the boy's existing lymphoid B cells had sustained a lucky back mutation in the allele the boy inherited from his mother, thus spontaneously reverting to wild type, or normal (Box 14.1). Over time, the descendants of this reverted cell apparently became more and more abundant in the boy's B-cell population. Eventually, reverted B cells became abundant enough, and made and released enough ADA, that the boy's clinical symptoms of ADA deficiency vanished.

Populations of cells inside an individual's body may exhibit genetic variation and differential fitness and thus may evolve.

Hirschhorn and colleagues believe that the increase in frequency of reverted B cells in the boy's B-cell population happened by natural selection. It is also possible that the increase happened by drift. It is likely, however, that reverted cells

Box 14.1 | Genetic sleuthing solves a medical mystery

Figure 14.17 shows the nucleotide sequence of a short piece of the ADA gene. Both parents have point mutations substituting an A for a G. Dad's mutation is in an intron/exon splice site, resulting in the deletion of an entire exon from the processed ADA transcript. Mom's mutation results in the substitution of one amino acid for another in the primary structure of the ADA enzyme. Both mutations are known, from other individuals who have them, to virtually destroy the enzymatic activity of ADA.

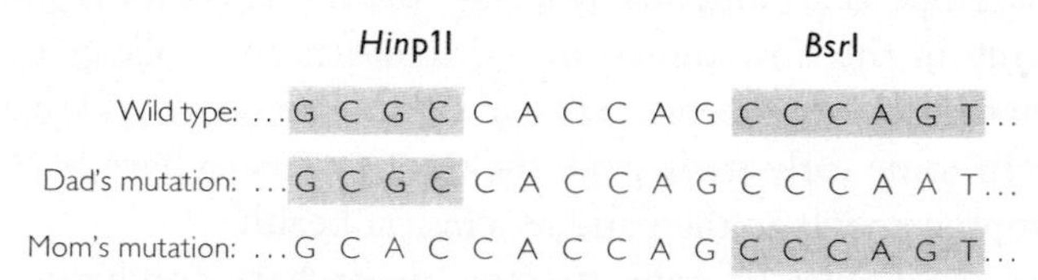

Figure 14.17 A short piece of the gene for adenosine deaminase Point mutations are shown in red; the gray boxes indicate sites recognized by the restriction enzymes *Hinp*1I and *Bsr*I. From Hirschhorn et al. (1996).

Notice that the wild-type, or normal, sequence contains a cutting site for the restriction enzyme *Hinp*1I and a cutting site for the restriction enzyme *Bsr*I. The father's mutation eliminates the *Bsr*I cutting site, whereas the mother's mutation eliminates the *Hinp*1I cutting site. Hirschhorn and colleagues amplified a 254-bp-long sequence from the ADA genes in both parents and in different populations of the boy's blood cells. When digested with *Bsr*I, the wild-type 254-bp fragment yields a 182-bp fragment and a 72-bp fragment. Figure 14.18a illustrates a *Bsr*I restriction fragment analysis of the family. Look first at the mother's lane on the gel. Both her wild-type allele and her mutant allele have the *Bsr*I cutting site, so her gel shows only two bands: the 182-bp fragment and the 72-bp fragment. Now look at the father's lane. His wild-type allele has the *Bsr*I cutting site, so his lane shows the 182-bp and 72-bp bands. His mutant allele, however, lacks the *Bsr*I cutting site, so his lane also shows a 254-bp fragment. Finally, look at the boy's lanes. They show that the father's mutant allele is present in both the boy's peripheral leukocytes (white blood cells) and his lymphoid B cells.

Figure 14.18b illustrates a *Hinp*1I restriction fragment analysis. This time, the mother's mutant allele shows up as an uncut 254-bp fragment. The son's lanes show something unexpected: His mother's mutant allele is present in his peripheral leukocytes, but absent in most of his lymphoid B cells. We say most because there is a faint band from a 254-bp fragment in the boy's B-cell lane. Closer examination of 15 different B-cell lines revealed that all carry the father's mutant allele, but only 2 carry the mother's mutant allele.

Hirschhorn and colleagues discovered that the mother's mutant allele also contains a unique neutral marker. When they checked the boy's lymphoid B cell lines that appear to lack the mother's mutant allele, the researchers found that these cells in fact carry the mother's neutral marker. Hirschhorn and colleagues concluded that the ancestral cell to most of the boy's existing lymphoid B cells had sustained a lucky back mutation in the allele from the mother, thus spontaneously reverting to wild type.

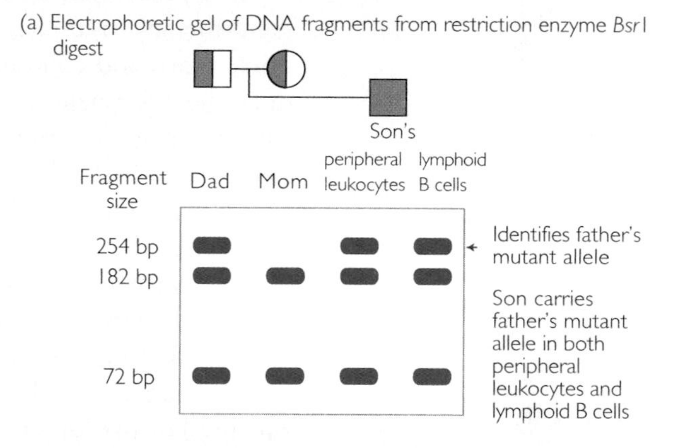

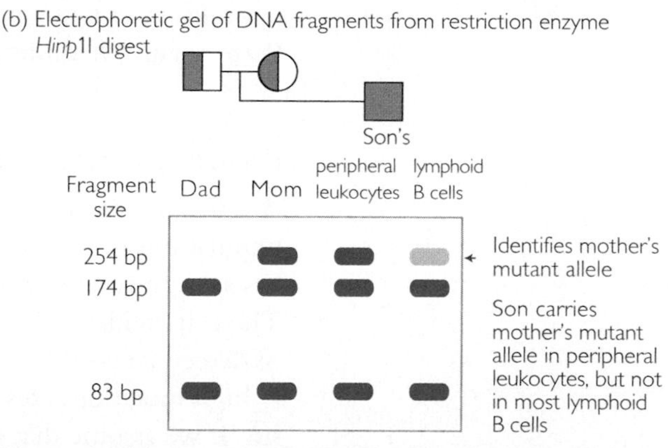

Figure 14.18 Restriction fragment analysis of the genetics of ADA deficiency in a family (a) A *Bsr*I restriction fragment assay for the father's mutant allele. The 254-bp band at the top of the gel is a marker for the presence of the father's mutation, which eliminates a cutting site recognized by *Bsr*I. (b) A *Hinp*1I restriction fragment assay for the mother's mutant allele. The 254-bp band at the top of the gel is a marker for the presence of the mother's mutation, which eliminates a cutting site recognized by *Hinp*1I. After Hirschhorn et al. (1996).

are at a distinct selective advantage. Because they make their own supply of a crucial housekeeping enzyme, they should live longer than cells that have to pick up the enzyme after it has been released by the cells that make it.

The boy's story may have implications for the treatment of other individuals with ADA deficiency. The outlook for patients with ADA deficiency has improved in recent years. In 1987, researchers developed a form of injectable ADA that is an effective enzyme replacement treatment for many ADA patients (Hershfield et al. 1987). In the early 1990s, researchers began the first clinical experiments with somatic-cell gene therapy (Blaese et al. 1995, Bordignon et al. 1995). Somatic-cell gene therapy involves removing lymphocytes and/or bone marrow cells from the patient, inserting a functioning version of the ADA gene with its own promoter into their chromosomes, and returning the cells to the patient's body. In other words, gene therapy is an attempt to accomplish by design the reversion mutation that happened spontaneously in the boy studied by Hirschhorn and colleagues. The early trials have shown that the engineered cells can survive for years, and that they can grow and divide. In some early trials, gene therapy appears to have been responsible for dramatic improvements in the patients' clinical health.

As a precaution against the failure of gene therapy, researchers conducting gene-therapy trials have kept their patients on enzyme replacement therapy. Hirschhorn and colleagues suggest, however, that enzyme replacement may reduce the effectiveness of gene therapy. If enzyme replacement reduces the selective advantage the engineered cells have over ADA-deficient cells, it will slow the rate at which the patients' blood cell population evolves by natural selection. Consistent with this idea, Alessandro Aiuti and colleagues (2002a) report on an ADA gene therapy patient in which the genetically modified cells appeared to enjoy a stronger selective advantage when enzyme replacement therapy was stopped. For the short term, gene therapists will have to balance the benefits of encouraging rapid selective fixation of engineered cells against the risks of depriving patients of the insurance provided by continued enzyme replacement therapy. For the long term, researchers are working on new methods to give genetically modified cells a selective advantage inside patients (Aiuti et al. 2002b, Persons and Nienhuis 2002).

Reconstructing the History of a Cancer

Another medical context in which it is productive to view tissues as evolving populations of cells is cancer (Shibata et al. 1996). A cancer starts with a cell that has accumulated mutations that free it from the normal controls on cell division. The cell divides, and its offspring divide, and so on, to produce a large population of descendants—that is, a tumor. The cells in some kinds of cancer have extremely high mutation rates, allowing tumors to accumulate measurable genetic diversity. If we assume that the mutation rate per cell division is constant, and that the mutations are neutral, then the genetic diversity within a particular tumor is a measure of the tumor's age.

Occasionally, a cell may leave the tumor it was born in and migrate elsewhere to initiate a new tumor. (This process is called metastasis.) This new tumor represents a new population of cells. Because it was founded by a single individual, this new population will have low genetic diversity. As it grows, however, the population will evolve. Like the population from which its founder came, the new tumor will accumulate genetic diversity as a result of mutation and genetic drift.

Darryl Shibata and colleagues (1996) use the amount of genetic variation in tumors to reconstruct the history of cancers within patients. Their approach is similar to the one used by Sarah Tishkoff and colleagues to reconstruct the history of human populations (see Chapter 20). Non-African populations have lower genetic diversity than African populations. With other evidence, this pattern indicates that non-African populations were founded by emigrants from Africa.

The analytical tools evolutionists use to reconstruct the history of populations can also be used to reconstruct the history of tissues and tumors within a patient's body.

Figure 14.19a shows the allelic diversity at a tandem repeat locus in subpopulations of two adjacent tumors in a 43-year-old patient. The patient has a type of cancer called a colorectal adenocarcinoma. One tumor, called an adenoma, is in glandular tissue; the other tumor, called a carcinoma, is in colorectal tissue. Three of the four adenoma subpopulations show substantially greater genetic diversity than the carcinoma subpopulations. Figure 14.19b shows estimates of the ages of the eight subpopulations, based on the estimated mutation rate for this kind of cancer and the assumption that each cancer cell divides once per day. It appears that the adenoma grew benignly for up to 10 years before it spawned a cell that migrated into the patient's colon to initiate the malignant carcinoma.

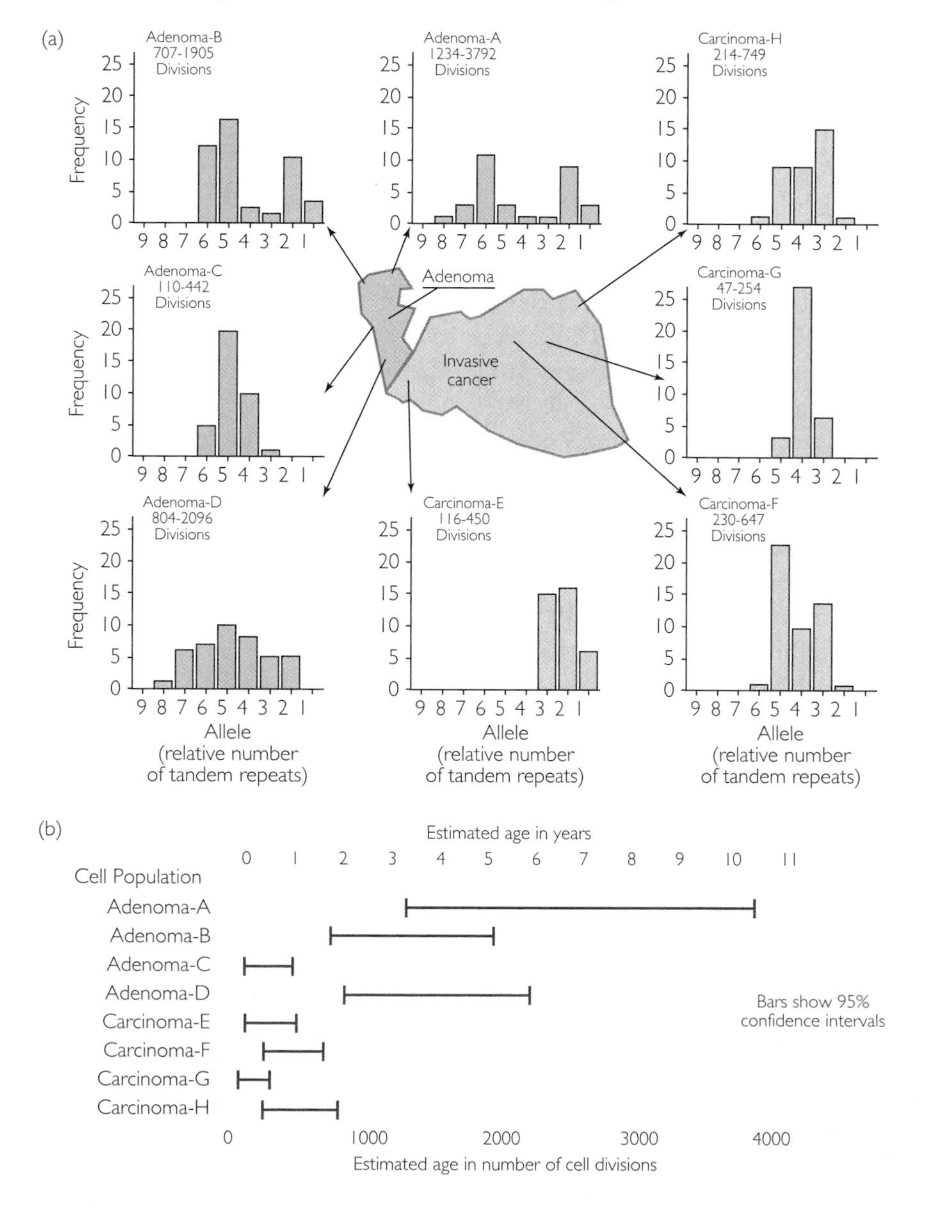

Figure 14.19 An evolutionary reconstruction of the history of a cancer within an individual patient (a) The diagram at the center is a map of a pair of neighboring tumors in a 43-year-old patient. At left is an adenoma, a tumor of glandular tissue that is usually benign. At right is a colorectal carcinoma, an invasive malignant tumor. The histograms that surround the map show the genetic diversity within samples of cells taken from the locations indicated by the arrows. Each histogram shows the frequencies of various alleles at a tandem repeat locus. Different alleles are identified by the number of tandem repeats they carry. Some cell samples, such as Adenoma-D, show high genetic diversity (many different alleles present). Other cell samples, such as Carcinoma-E, show low genetic diversity. From Shibata et al. (1996). (b) Estimates of the age of the different tumor subpopulations shown in (a). The higher genetic diversity of the samples from the adenoma indicates that the adenoma is older than the carcinoma. Redrawn from Shibata et al. (1996).

14.5 The Adaptationist Program Applied to Humans

We now shift our focus from pathogen and cell populations that evolve within humans to the human animal itself. Our goal is to illustrate how researchers use the analytical tools of the adaptationist program to understand aspects of human physiology and behavior relevant to medicine and public health. Researchers following the adaptationist program identify traits that appear to be adaptive (see Chapter 10). On the assumption that these traits were produced by natural selection, the researchers formulate and test hypotheses about how the traits enhance fitness.

The analytical tools evolutionists use to study form and function in other organisms can, with appropriate caution, also be used to study form and function in humans.

Section 14.5 considers complications that arise when applying the adaptationist program to a species that alters its own environment at a rate that outpaces evolution by natural selection. Section 14.6 uses fever to show how the adaptationist program can be applied to physiological puzzles. Finally, Section 14.7 explores parenting as an example of how the adaptationist program can be used to analyze human behavior. We should warn the reader now that few of the hypotheses we discuss have been adequately tested. Our coverage of the adaptationist approach to human health is partly a plea for more research.

Adaptation to What Environment?

Figure 14.20 A Yagua hunter–gatherer from the Amazon rain forest of Peru Dressed for hunting, he is leaning on his blow gun. He uses the blow gun to shoot poisoned darts at game animals.

Before any attempt to apply the adaptationist program to humans, it is crucial to ask: To what environment are humans adapted? Until the advent of agriculture some 10,000 years ago, all humans lived as hunter–gatherers (Figure 14.20). Hunter–gatherers occupied, and still do occupy, a wide variety of habitats, ranging from deserts to Arctic tundra. But none of these environments resembles that of a modern urbanite. In other words, our lives are not like the lives of our ancestors.

S. Boyd Eaton and various colleagues have attempted to reconstruct some of the basic features of our ancestral Stone Age lifestyle (Eaton et al. 1997; Eaton and Cordain 1997; Cordain et al. 1997; Cordain, Eaton, et al. 2002). Their evidence comes from observations of present-day hunter–gatherers, archaeological remains, and analyses of uncultivated edible plants and wild game animals. Figure 14.21 shows their estimate of the energy sources in a typical hunter–gatherer diet, compared with the energy sources in a typical, modern American diet. While hunter–gatherers get more of their energy from meat, they eat far leaner meat, more fruits and vegetables, fewer cereal grains, and fewer milk products. Hunter–gatherers also get considerably more exercise. Among the !Kung hunter–gatherers of the Kalahari Desert, for example, a typical individual walks 10 to 15 kilometers each day (6 to 9 miles), and uses energy at more than 1.5 times his or her resting metabolic rate. A sedentary modern American, in contrast, may walk effectively no distance each day and uses energy at less than 1.2 times his or her resting metabolic rate.

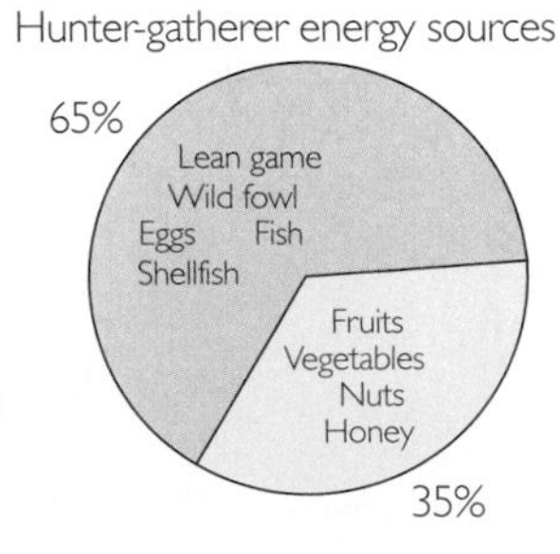

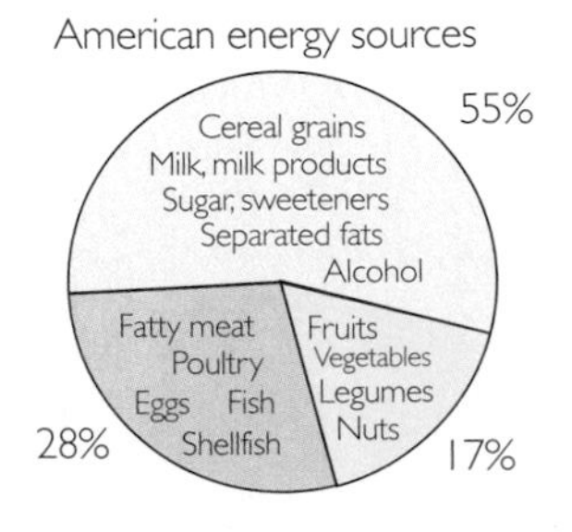

Figure 14.21 A typical hunter–gatherer diet versus a typical modern American diet Modified from Eaton and Cordain (1997), after Cordain et al. (2002).

Eaton points out that the impact of our novel lifestyle on health has not been assessed through rigorous controlled experiments. However, in observational studies, features of the modern lifestyle, such as a high-fat diet, are associated with a variety of diseases, including heart disease, strokes, and cancer. These conditions, rare in hunter–gatherers, are diseases of civilization (Nesse and Williams 1994).

There are a few cases in which researchers have shown that human populations have evolved in response to selection imposed by a modern lifestyle. Production of lactase provides an example. Lactase is the enzyme that enables us to digest milk sugar, or lactose. The only source of milk sugar in the diet of most mammals is mother's milk, so there would be no advantage, and probably some cost, in continuing to produce lactase after weaning. Indeed, individuals stop producing lactase around the age of weaning in most mammal species and in many populations of humans. However, for modern humans with lifelong access to cows' milk, continuing to produce lactase after weaning can be advantageous. In human populations with a long history of drinking fresh milk, and only in these populations, many individuals have a heritable ability to continue producing lactase as adults (Durham 1991).

In a milk-drinking culture, the ability to produce lactase is probably beneficial throughout life. Many diseases of civilization, however, strike only late in life. As a result, selection on genetic susceptibility to these diseases is probably weak (see Chapter 13). This consideration, combined with the fact that a modern lifestyle dates back at most a few hundred generations, implies that we should not expect that evolution by natural selection in human populations will have kept pace with our changing lifestyle. In other words, even the most sophisticated modern urbanites have bodies and brains that are largely designed for life in the Stone Age.

The pace of cultural change has been so much faster than the pace of evolution by natural selection that modern humans are largely still adapted to life in the Stone Age.

The recognition that the environment we live in is different from the environment we are adapted to has at least two implications. It helps make sense of some otherwise puzzling features of our physiology, and it suggests ways to reduce some of the risks associated with modern life. Myopia provides an example of the former, and breast cancer an example of the latter.

Myopia

In many populations the incidence of myopia, or nearsightedness, is 25% or more. Researchers have used twin studies to determine whether variation in vision has a genetic basis. For example, J. M. Teikari and colleagues (1991) assessed the similarities in vision between monozygotic versus dizygotic twins. As discussed in Chapter 9, if a trait is heritable, then monozygotic twins should resemble each other more strongly than dizygotic twins (see Figure 9.16, page 337). Teikari et al.'s data are as follows:

	Monozygotic twins	*Dizygotic twins*
Concordant pairs	**36**	**19**
Discordant pairs	**18**	**36**

Two-thirds of the monozygotic twin pairs were concordant—that is, both nearsighted or both normal—versus only one-third of the dizygotic twin pairs. These data suggest that nearsightedness is partially heritable.

But how could nearsightedness be even partially heritable? Many Americans are legally blind without their corrective lenses. These people would be at a serious disadvantage if forced to live as hunter–gatherers. Surely natural selection among hunter–gatherers would quickly eliminate alleles associated with myopia. And if natural selection eliminated alleles that caused myopia among our hunter–gatherer ancestors, they cannot have passed such alleles to us.

The fact that human populations have not had time to adapt to the environment we live in can help explain some of our apparently maladaptive traits.

The solution to the puzzle of myopia is to recognize that modern humans live a lifestyle that is as novel in its visual demands as in its diet and activity levels. Hunter–gatherers do not spend their childhoods indoors reading under artificial light. Perhaps the alleles that predispose some of us to myopia actually cause myopia only in a modern environment.

Evidence to evaluate this hypothesis comes from populations of people who have only recently adopted a modern lifestyle. Francis Young and colleagues (Young et al. 1969, Sorsby and Young 1970) went to Barrow, Alaska, to measure the incidence of myopia among Inuit. The researchers chose this population because most of the families in it had moved to Barrow from isolated communities during and after World War II, drawn by the economic activity associated with a naval research laboratory, a radar station, and oil exploration. As the town of Barrow grew, a school system grew with it. Most of the children Young and colleagues examined attended formal, American-style schools and did a great deal of reading. Most of the adults over 35 had attended less formal, ungraded schools for a maximum of six years. Young et al.'s data on the incidence of myopia in younger versus older individuals are as follows:

Age	*Number myopic*	*Number not myopic*	*Fraction myopic*
6–35	146	202	42%
36–88	8	152	5%

The children, who had been much more strongly exposed to a modern visual lifestyle, had a substantially higher incidence of myopia.

Young et al.'s study was observational, not experimental, and the amount of schooling was just one of many differences between the environments of the children versus the adults. Nonetheless, the data are consistent with a variety of studies on humans and animals indicating that the shape of the growing eye is molded by visual experience (see Norton and Wildsoet 1999 for a review). This body of human and animal research suggests that myopia is caused by a combination of genetic susceptibility and close-in visual work. The older Inuit in Barrow had grown up in a visual environment more like that of our hunter–gatherer ancestors. The older adults had the same alleles they passed to their children and grandchildren, but they themselves were not myopic. In other words, myopia can be partially heritable because the alleles that predispose some modern humans to myopia do not cause myopia in a hunter–gatherer environment.

Breast Cancer

About one in eight North American women gets breast cancer during her lifetime. Some of these women die while still in their child-bearing years. Breast cancer is commonly thought to result from a combination of genetic susceptibility and environmental factors. But if we take a Darwinian view, breast cancer presents a puzzle (Cochran et al. 2000). If genes are responsible for a substantial fraction of breast cancers, natural selection should have eliminated breast cancer alleles from our ancestors' populations. And if breast cancer is caused by environmental factors to which our ancestors were long exposed, evolution by natural selection should have favored individuals immune to their effects. So why is the rate of breast cancer so high?

Among the possible solutions to this puzzle are these:

- Breast cancer may be caused by a pathogen, such as a virus or bacterium. Viruses and bacteria are living organisms. Their populations evolve in response to selection imposed by the host's immune system. Thus we do not expect host populations to be able to evolve complete immunity to all diseases.
- Breast cancer may be a disease of civilization, like myopia. That is, it may be caused by the interaction between genes and novel environments to which our ancestors were never exposed.

Breast Cancer as a Viral Disease

Mice carry a virus, called Mouse Mammary Tumor Virus (MMTV), that causes the mouse equivalent of breast cancer. A small group of researchers has long suspected that MMTV, or something like it, can cause breast cancer in humans. We will briefly consider two suggestive pieces of evidence.

The first piece of evidence comes from Yue Wang and colleagues (1995), working in the lab of Beatriz G.-T. Pogo. Wang and colleagues tried to extract from human tissue samples DNA sequences similar to a piece of one of MMTV's genes. The researchers analyzed 314 tissue samples from breast cancers and 107 tissue samples from normal breasts. Their results were as follows:

	Normal tissue	*Cancer tissue*
Number of samples with a sequence similar to MMTV's:	**2**	**121**
Total number of samples examined:	**107**	**314**
Fraction of samples with MMTV-like sequences:	**1.9%**	**38.5%**

Sequences similar to MMTV's lurked in over a third of the samples from breast cancers but almost none of the samples from normal tissues.

The second piece of evidence comes from T. H. M. Stewart and colleagues (2000). These researchers looked at rates of breast cancer in various countries in Europe. The researchers knew that the species of house mouse found in western Europe is *Mus domesticus*, whereas the species of house mouse found in eastern Europe is *Mus musculus*. And they knew that *Mus domesticus* tends to be more heavily infected with MMTV. The researchers reasoned that if MMTV causes breast cancer in humans, than the rate of breast cancer should be higher in countries with *Mus domesticus* than in countries with *Mus musculus*. The data, shown in Figure 14.22, are consistent with this prediction.

Most cancer researchers view the evidence of MMTV's involvement in breast cancer as only suggestive. They note that no one has found MMTV virus particles in a breast tumor, and that no one has been able to explain how this mouse virus gets into people. Furthermore, on the evidence we have seen, MMTV can explain fewer than 40% of breast cancers. These considerations indicate that we should also consider the hypothesis that breast cancer is, in part, a disease of civilization.

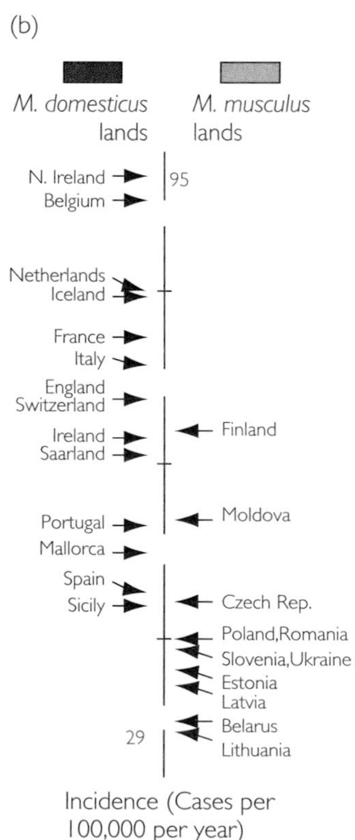

Figure 14.22 **The incidence of breast cancer is higher where *Mus domesticus* lives than where *Mus musculus* lives** (a) This map of Europe shows the geographic range of *Mus domesticus* (blue) and *Mus musculus* (green). (b) On average, rates of breast cancer are higher in countries occupied by *Mus domesticus*. The rate of breast cancer is 95 cases per 100,000 women per year in Northern Ireland; it is 29 cases per 100,000 women per year in Lithuania. From Stewart et al. (2000).

Menstrual Cycling and Breast Cancer

The monthly menstrual cycling experienced by most modern women is usually considered normal. However, a growing body of epidemiological evidence suggests that continuous menstrual cycling increases a woman's risk of breast cancer. A woman's risk of cancer is higher the earlier she begins to menstruate, the later she has her first child, and the less time she spends nursing (see, for example, Layde et al. 1989, Berkey et al. 1999). Menstrual cycling appears to elevate the risk of breast cancer because the combination of estrogen and progesterone present during the postovulatory phase of the cycle stimulates cell division in the lining of the milk ducts (Henderson et al. 1993). With more cell divisions come more opportunities for mutations that may create cancers. Given the high incidence of breast cancer among modern women—about one in eight in North America—it is worth knowing whether continuous menstrual cycling really is normal.

Beverly Strassmann (1999) spent two years observing menstrual cycling among the Dogon of Mali (Figure 14.23). The Dogon are a premodern people who use no contraceptives. Their culture is an easy one within which to study menstrual cycling because custom dictates that the women sleep in special menstrual huts while they are menstruating. Strassmann confirmed that menstruating women do, in fact, sleep in the huts, and that women sleeping in the huts are menstruating, by regularly collecting urine samples from 93 women for two and one-half months and checking for metabolites of estrogen and progesterone. Strassmann then tracked visits to the huts by the Dogon women over a period of two years. She found that women between the ages of 20 and 35 spend little time cycling (Figure 14.24a). Instead, they are usually either pregnant or experiencing lactational amenorrhea—a suppression of cycling due to nursing. On any given day, about 25% of adult Dogon women are cycling, about 15% are pregnant, about 30% are in lactational amenorrhea, and about 30% are past menopause

Figure 14.23 Dogon women grinding millet with a mortar and pestle The Dogon are a traditional society located in Mali, northwestern Africa.

(Figure 14.24b). Strassmann estimates that the average Dogon woman has a total of about 100 menstrual cycles during her lifetime. This is less than one-third the number for a typical modern woman.

Strassmann's data on the Dogon suggest that women's bodies may not have been designed by natural selection to tolerate long periods of continuous menstrual cycling. If continuous menstrual cycling is not normal for women, then we can think of the high rates of breast cancer among modern women as another maladaptive consequence of life in a novel environment. Strassmann does not have data on the incidence of breast cancer among Dogon women, but she notes that among urban West African women, whose menstrual patterns parallel those of the Dogon, the breast cancer rate is about one-twelfth that among North American women.

The fact that human populations have not had time to adapt to the lifestyle we live may explain why we lack physiological defenses against many diseases of civilization.

Perhaps modern women should consider using hormonal treatments that maintain their bodies in a hormonal state more consistent with the state experienced by our ancestors. Oral contraceptives reduce the risk of endometrial and ovarian cancer among modern women who use them, but they do not reduce the risk of breast cancer (Henderson et al. 1993). D.V. Spicer and colleagues are developing an oral contraceptive that suppresses ovarian function, does not stimulate cell division in the breast, and contains sufficient concentrations of sex steroids to avoid adverse side effects such as accelerated osteoporosis (Spicer et al. 1991, Henderson et al. 1993). They predict that their new strategy will be equally as effective as present oral contraceptives at reducing the risk of ovarian cancer, about half as effective at reducing the risk of endometrial cancer, and much better at reducing the risk of breast cancer.

Keeping in mind that the environment modern humans live in may not be the environment to which we are adapted, we devote the remaining two sections of this chapter to examples showing how researchers use an adaptationist framework to develop and test hypotheses about medical physiology and hypotheses about fundamental aspects of human behavior.

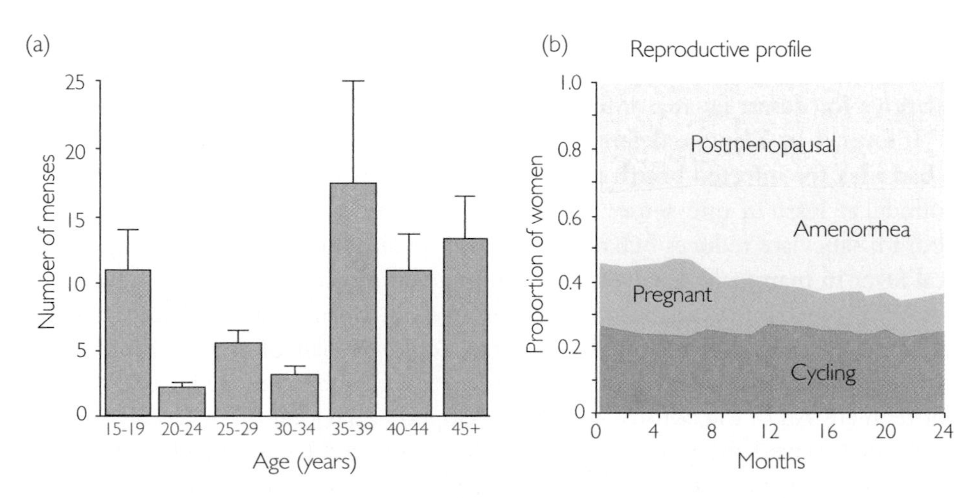

Figure 14.24 Menstrual cycling in Dogon women (a) Most Dogon women had relatively few menstrual cycles during a two-year period. (b) This is because, at any given time, a relatively high fraction of the Dogon women were pregnant or experiencing lactational amenorrhea. From Strassmann (1999).

14.6 Adaptation and Medical Physiology: Fever

Many people consider the symptoms that accompany illness to be a nuisance. A common response to the fever associated with a cold or flu, for example, is to take aspirin, acetaminophen, or ibuprofen. These drugs reduce the fever, but they do not combat the virus that is causing the cold or flu. Here we ask whether taking drugs to reduce fever is a good idea. To answer the question, we need to know why people run a fever when they are sick.

An evolutionary perspective can help medical researchers develop hypotheses about physiological function.

An evolutionary perspective suggests two interpretations of fever. One interpretation is that fever may represent manipulation of the host by the pathogen. Viruses or bacteria may release chemicals that cause the host to elevate its body temperature so as to increase the pathogen's growth rate or reproductive rate. If this hypothesis is correct, then reducing the fever would probably help the host combat the infection. The second interpretation is that fever may be an adaptive defense against the pathogen. The pathogen may grow and reproduce more slowly at higher temperatures, or the host's immune response may be more effective at higher temperatures. If this hypothesis is correct, then taking drugs that alleviate fever might be counterproductive to recovery.

Matthew Kluger has, for over 20 years, advocated the second hypothesis—that fever is an adaptive defense against disease. In 1974, Linda Vaughn, Harry Bernheim, and Kluger discovered that the desert iguana (*Dipsosaurus dorsalis*, Figure 14.25a) develops a behavioral fever in response to infection with a bacterium called *Aeromonas hydrophila*. Recall from Chapter 10 that iguanas, being ectotherms, use behavior instead of physiology to regulate their body temperatures. They move to hot spots to warm themselves and to cold spots to cool themselves. Vaughn et al. found that when they injected desert iguanas with dead bacteria, the lizards chose body temperatures about 2°C higher than they normally choose (Figure 14.25b).

Is behavioral fever an adaptive response to infection, or are the bacteria manipulating the iguanas? To distinguish between these hypotheses, Kluger, Daniel Ringler and Miriam Anver (1975) infected desert iguanas with live bacteria, then prevented the lizards from thermoregulating by keeping them in fixed-temperature incubators. Most of the iguanas kept at temperatures mimicking behavioral fever survived, whereas most of the iguanas kept at lower temperatures died (Figure 14.25c). This result suggests that behavioral fever is, in fact, adaptive for desert iguanas infected with *A. hydrophila*.

If fever is an adaptive defense against *A. hydrophila*, then it would probably be a bad idea for infected lizards to take aspirin. This is not as silly a statement as it sounds, at least in one sense: The researchers found that the aspirin-like drug sodium salicylate reduces behavioral fever in iguanas just as it reduces physiological fever in mammals. Apparently thermoregulation is controlled by similar neurological mechanisms in both groups of animals. Bernheim and Kluger (1976) infected 24 desert iguanas with bacteria, then gave half of the infected lizards sodium salicylate. The researchers allowed all the iguanas to behaviorally thermoregulate. All of the control iguanas developed behavioral fever, and all but one of them survived the infection. Five of the medicated lizards developed behavioral fever in spite of the medication, and all of them survived. The other seven medicated iguanas failed to develop behavioral fever, and all of them died.

Since the mid-1970s, researchers have documented behavioral fever in a wide variety of reptiles, amphibians, fishes, and invertebrates. In several animal studies researchers have shown that fever increases survival (see Kluger 1992 for a re-

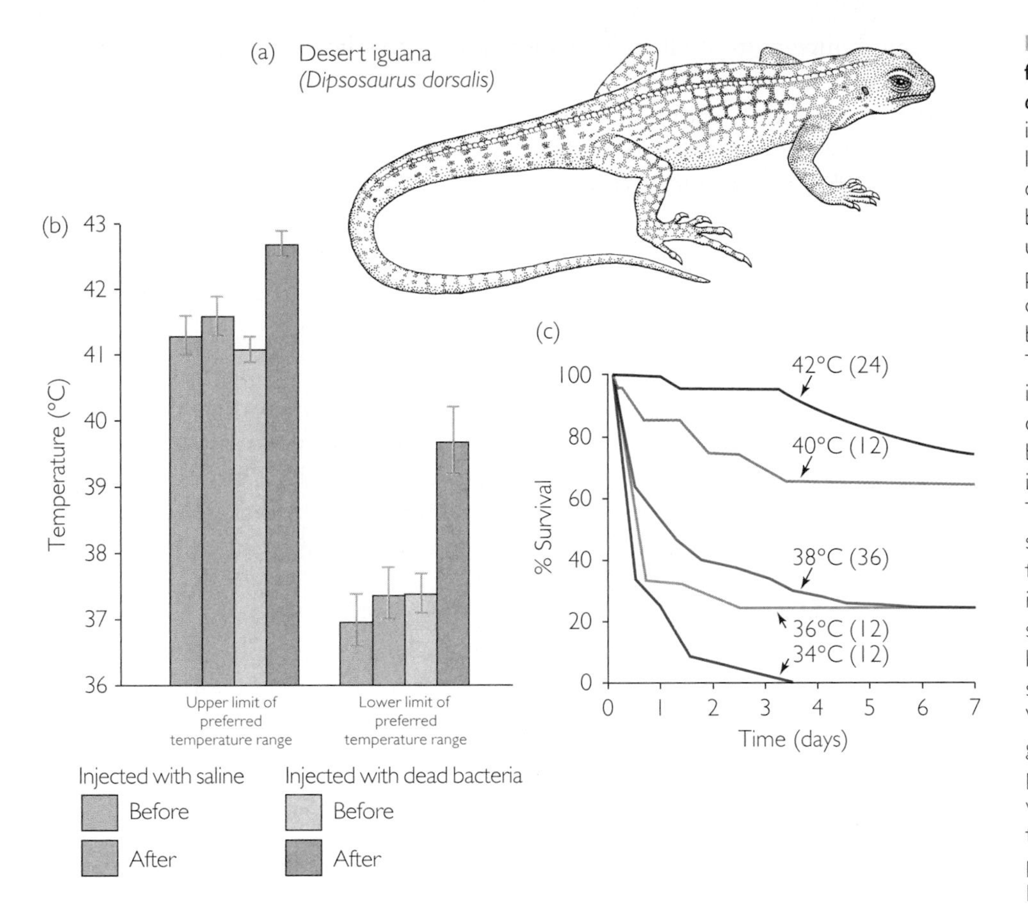

Figure 14.25 Behavioral fever in the desert iguana (*Dipsosaurus dorsalis*) (a) A desert iguana. (b) Linda Vaughn and colleagues injected saline into nine control iguanas. The gray and blue bars show the average upper and lower preferred temperatures, or set-points, (± standard error) for these lizards before and after the injection. There was no significant change in either set-point. Vaughn and colleagues then injected killed bacteria (*Aeromonas hydrophila*) into 10 experimental iguanas. The light and dark orange bars show the average set-points for these lizards before and after the injection. The increase in both set-points after injection with killed bacteria was statistically significant ($P < 0.001$). From Vaughn et al. (1974). (c) This graph shows the fraction of experimentally infected lizards surviving at each temperature over time (numbers of lizards in parentheses). From Kluger, Ringler, and Anver (1975).

view). These results broadly support the hypothesis that fever is an adaptive response to infection.

Fever is much harder to study in endotherms than it is in ectotherms. Researchers cannot force endotherms to be at an arbitrary body temperature simply by putting them in an incubator. And as we will see shortly, drugs that reduce fever have effects on the immune system that are independent of fever.

In an attempt to disentangle the effects of the increased metabolic rate that accompanies a fever from the increase in body temperature per se, Manuel Banet used rats implanted with cooling devices and infected with *Salmonella enteritidis*. First, Banet (1979) implanted cooling devices in the brains of rats and used them to chill the hypothalamus. This induced the infected rats to develop extra high fevers, without dramatically elevating their metabolic rates. The extra-high-fever rats survived the bacterial infection at much lower rates than control rats. Second, Banet (1981a) cooled the spinal cords of infected rats. This induced the rats to greatly increase their metabolic rates, while preventing them from elevating their body temperatures. The high-metabolic-rate rats survived the infections at somewhat higher rates than normal-metabolic-rate rats. Finally, Banet (1981b) carefully monitored the body temperatures and metabolic rates of a group of infected rats, some of which had implants, but none of which were heated or cooled. Banet found that the rats that ran the highest fevers had the lowest rates of survival, but that the rats that showed the highest metabolic rates had the highest rates of survival. Together, Banet's results suggest that moderate fever is beneficial to infected rats, that the benefits of fever may be associated not so much with elevated temperature itself as with increased metabolic rate or other effects on the immune system, and that high fever in rats is deleterious to survival.

It is unclear how the results with iguanas and rats might apply to humans. Fewer clinical studies have been done on this question than one might expect (Kluger 1992; Green and Vermeulen 1994). We review three studies, none of which is conclusive.

Fever and Chickenpox

Timothy Doran and colleagues (1989) studied 68 children with chickenpox. After getting the informed consent of the children's parents, the researchers divided the children at random into two groups. The experimental group took acetaminophen, a common over-the-counter antifever medicine. The control group took a placebo (pills that looked like acetaminophen, but contained no medicine). The assignment of children to study groups was double blind: Neither the researchers nor the parents (nor the children) knew which children were in which group until after the study was over.

By most measures of the duration and severity of illness, there was no difference between the acetaminophen and placebo groups. Where the results hint at a difference, the children taking the placebo recovered faster. It took less time, on average, for all of the vesicles to scab over in the placebo children (5.6 ± 2.5 days) than in the acetaminophen children (6.7 ± 2.3 days). By itself, this result is statistically significant at $P = 0.048$; but given that several other measures failed to turn up any difference, it is not persuasive. (Remember that if we do 20 statistical tests, one of them is likely to be significant at $P < 0.05$ simply due to chance.) The children's itching seemed to subside faster in the placebo group (Figure 14.26), but this pattern is not statistically significant.

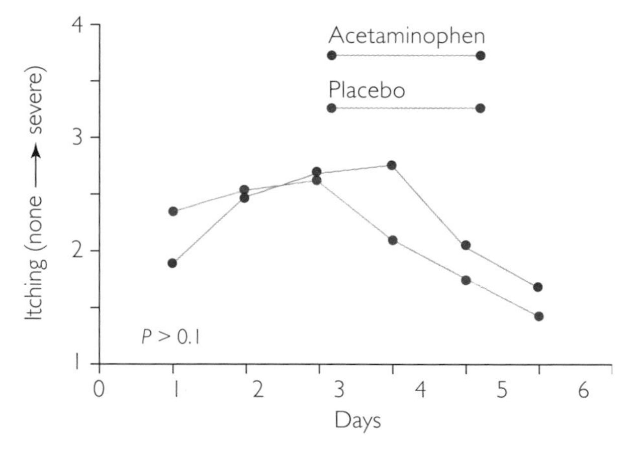

Figure 14.26 Does the antifever medicine acetaminophen have any effect on the course of chickenpox? The graph shows the severity of itching as a function of time (as judged by the children's parents) in 37 children taking acetaminophen versus 31 children taking a placebo. The placebo group appears to have recovered faster, but the difference is not significant. From Doran et al. (1989).

The simple interpretation is that acetaminophen has little or no effect on the course of chickenpox, and therefore that fever is neither adaptive nor maladaptive as a defense against the disease. Kluger (1992) points out, however, that only slightly over half of the children in the study ran a fever (defined by Doran et al. as a temperature of 100.4°F or higher) and that the fraction of children who did run a fever was the same in the acetaminophen group (57%) and the placebo group (55%). Kluger's interpretation is that the acetaminophen children were not given enough medicine to reduce fever, and thus that the study did not test the hypothesis that fever is adaptive. Kluger's analysis illustrates that it is important to critically examine a study's methods and results before accepting the conclusions.

Fever and the Common Cold

Neil Graham and colleagues (1990) intentionally infected 56 consenting adult volunteers with rhinovirus type 2, one of the viruses that can cause the common cold. Assigned to groups double blind and at random, 14 subjects took a placebo. The rest took common over-the-counter antifever medications: 13 took ibuprofen, 15 took aspirin, and 14 took acetaminophen. The volunteers taking the placebo suffered less nose stuffiness (Figure 14.27a) and made more antibodies against the rhinovirus (Figure 14.27b) than did the volunteers taking antifever medicines. The reason for the reduced antibody response in volunteers taking medicine may be that the medicines prevented monocytes, a class of white blood cells, from moving from the blood to the infected tissues (Figure 14.27c). Once in the infected tissues, monocytes differentiate into macrophages, which help mount an immune response against the virus (Graham et al. 1990).

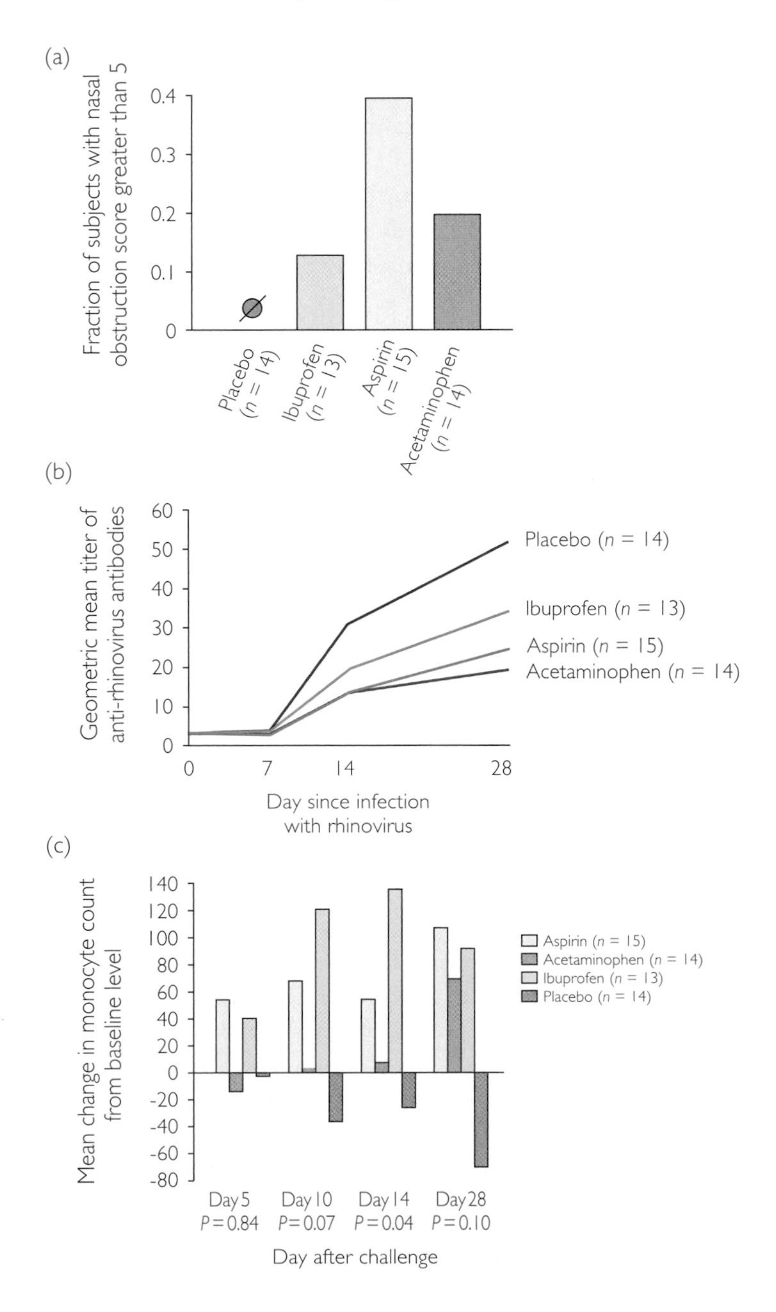

Figure 14.27 Do antifever medicines have any effect on the course of the common cold? (a) Volunteers taking a placebo had less-stuffy noses (fewer had "nasal obstruction scores" over 5) than volunteers taking one of three antifever medicines ($P = 0.02$). (b) Volunteers taking a placebo made more antibodies against the rhinovirus that caused their cold than did volunteers taking one of three antifever medicines. On day 28, the difference between the placebo group and the other three groups combined was significant at $P = 0.03$. (c) Monocytes are white blood cells that circulate to infected tissues, then leave the blood and differentiate into macrophages. Volunteers taking a placebo showed a drop in the concentration of monocytes in their blood over time (indicating that the cells had moved into the tissues), whereas volunteers taking one of three antifever medicines showed an increase in the concentration of monocytes. From Graham et al. (1990).

This time, the simple interpretation is that the antifever medications interfered with the immune response to the common cold and therefore that fever is an adaptive defense against the disease. Kluger (1992) points out, however, that few of the subjects in the study ran a fever. Furthermore, the fraction of subjects taking the placebo who did run a fever (14%) was not significantly higher than the fraction of subjects taking medicine who ran a fever (7%). Kluger's interpretation is that few people infected with rhinovirus type 2 run a fever and thus that the study did not test the hypothesis that fever is adaptive.

The study by Graham and colleagues did show, however, that antifever medicines interfered with the immune response to the virus. This demonstrates that antifever medicines have multiple physiological effects. As we noted above, this fact makes it extremely difficult to design studies on the adaptive significance of fever in mammals. Studies using traditional antifever medicines cannot separate fever from other aspects of the immune response.

Fever and Sepsis

Gordon Bernard and colleagues (1997) studied human patients with severe systemic bacterial infections, or sepsis. They randomly assigned 455 patients to receive either ibuprofen or a placebo, in addition to standard treatments for sepsis. The ibuprofen group had significantly lower body temperatures and metabolic rates. However, both groups had about the same mortality rate, roughly 40%. Because all of the patients in the study were already gravely ill when they were enrolled, and because virtually all of them received aggressive treatment with antibiotics and other medications, it is difficult to interpret the results with regard to our two evolutionary hypotheses about fever.

Fever and Medical Practice

Even if researchers find clear evidence that fever in humans is, indeed, an adaptive response to some infections, no responsible doctor (or evolutionary biologist) would suggest that it is always a bad idea to suppress fever. There are several reasons:

- Fever may be an adaptive response against some pathogens, but not against others. Some bacteria or viruses may grow and reproduce faster at fever temperatures than at normal temperatures. In other words, the adaptive-response and pathogen-manipulation-of-host hypotheses may be mutually exclusive for any particular pathogen, but they are not mutually exclusive across all pathogens.
- Even when fever is beneficial, it carries costs as well (Nesse and Williams 1994). In the case of mild illness and low fever, sometimes the benefits of antifever medicines in alleviating symptoms and allowing people to continue their normal activities outweigh the costs of a somewhat diminished immune response. In the case of serious illness and high fever, fever itself can deplete nutrient reserves and can even cause temporary or permanent tissue damage.
- There are circumstances in which fever may cause damage directly, unconnected with its role in infections. For example, experiments with animals and observational studies of humans suggest that fever following a stroke causes neurological damage and reduces the likelihood of survival (Azzimondi et al. 1995).

More research is needed on the adaptive significance of fever in humans and on the costs and benefits of using various antifever medications.

14.7 Adaptation and Human Behavior: Parenting

Using an adaptationist approach to understand human behavior, evolutionary psychologists assume that the brain is an organ whose properties as a regulator of behavior have been shaped by natural selection. As a regulator of behavior, the brain is a flexible machine, not a computer slavishly converting input to output according to some fixed program. The human brain runs on a complex mix of conscious and unconscious perception, emotion, experience, and calculation, in pursuit of a variety of goals. But in the view of evolutionary psychologists,

> The ultimate objective of our conspicuously purposive physiology and psychology is not longevity or pleasure or self-actualization or health or wealth or peace of mind. It is fitness. Our appetites and ambitions and intellects and revulsions exist because of their historical contributions to this end. Our perceptions of self-interest have evolved as proximal tokens of expected gains and losses of fitness, "expected" being used here in its statistical sense of what would be anticipated on average from the cumulative evidence of the past (Daly and Wilson 1988a, page 10).

This adaptationist approach to human behavior requires caution. In its capacity as a regulator of behavior, the human brain is influenced by culture as well as by evolutionary history. Culture evolves by its own set of rules (see Box 14.2). Furthermore, culture can manifestly induce individuals to behave in ways contrary to the interests of their genetic fitness. The mass suicide of 39 members of the Heaven's Gate cult in March of 1997, for example, defies adaptationist explanation.

The influence of culture on human behavior means that studies of behavior within a single society cannot disentangle the effects of culture from those of evolutionary history. To make a plausible claim that a psychological trait or pattern of behavior is a product of natural selection, evolutionary psychologists must show that the trait or pattern is broadly cross-cultural. Cross-cultural diversity has fallen dramatically during the last century. All but the most remote and isolated traditional societies have been contacted, and Western ideas and artifacts have spread virtually everywhere (see Diamond 1992). Some biologists feel it is no longer possible to conduct a genuine cross-cultural study. Others feel that such studies are still worth pursuing, particularly when new findings are combined with information extracted from databases of earlier anthropological research.

Another caveat for the study of human behavior is one we have already discussed. The environments most humans live in today are strikingly different from the environments all humans lived in for most of our evolutionary history. From the time of the earliest members of the genus *Homo*, over 2 million years ago (see Chapter 20), until the advent of agriculture, at about 8000 B.C., all humans lived in small groups and made their living by hunting and gathering. The rapid pace of cultural change in the last 10,000 years has been largely too fast for genetic evolution to keep up. As a result, it is of little use to ask why natural selection would have produced human behaviors, such as a willingness to ski down a mountainside at 75 miles per hour, that can only occur in a modern context. So long as we are careful to allow for our incomplete understanding of the hunter–gatherer lifestyle, however, it may make sense to ask why natural selection would have built into us a desire for the social rewards that we can attain, under the right circumstances, through dramatic demonstrations of superior

Box 14.2 Is cultural evolution Darwinian?

A treatment of the mechanisms of cultural evolution is beyond the scope of this chapter. In fact, the mechanisms of cultural evolution are probably beyond the scope of evolutionary biology altogether.

Richard Dawkins, in his 1976 book *The Selfish Gene*, suggested that we might develop a theory of cultural evolution by natural selection that works exactly like our theory of biological evolution. Central to this suggestion is the idea that natural selection is a generalizable process. Natural selection works on organisms because organisms have four key features: mutation, reproduction, inheritance, and differential reproductive success. In principle, natural selection should operate on any class of entities that have the same four properties.

Dawkins noted that elements of culture have these four properties, and thus should evolve by natural selection. A new word, song, idea, or style is analogous to a new allele created by mutation. The Shakers' austere and beautiful style of furniture design, for example, is an element of culture. A new piece of culture reproduces when other people adopt it and pass it on, as when a woodworker admires a Shaker table, then goes home to her shop and imitates the design. Some pieces of culture are more successful than others at getting themselves transmitted from person to person. For example, Shaker-style furniture has achieved much wider adoption than Shaker-style lifelong celibacy. Culture evolves as the relative frequencies of styles and ideas change.

Dawkins coined the term *meme* for the fundamental unit of cultural evolution. He saw the meme as analogous to the gene, the fundamental unit of biological evolution. Dawkins envisioned a detailed theory of population memetics that would be similar to the theory of population genetics we covered in Chapters 5 through 8. [For a more recent exposition on the potential explanatory power of this idea, see Dennett (1995).]

The trouble with Dawkins' suggestion, noted by Dawkins himself (see also the 1989 edition of his book), is that the effectiveness of natural selection as a mechanism of evolution depends not just on the property of inheritance, but also on the details of how inheritance works. This fact was first recognized by Fleeming Jenkin, one of Darwin's critics, in 1867. In Darwin's time, the prevailing model of inheritance involved the blending in the offspring of infinitely divisible particles contributed by the parents. Jenkin pointed out that blending inheritance undermines evolution by natural selection because of the fate it implies for new variations. In a sexual population with blending inheritance, any new variation would quickly vanish, like a single drop of black paint dissolving into a bucket of white. Mendelian genetics rescues Darwin's theory, because Mendelian inheritance is particulate. Genes do not blend. A new recessive mutation can remain hidden in a population for generations. Eventually, the mutant allele may reach a high enough frequency that heterozygotes start to mate with each other, producing among their offspring a few homozygous recessives.

In its correct form, then, the generalizable theory of evolution by natural selection applies to any class of entities that has the properties of mutation, reproduction, *particulate* inheritance, and differential reproductive success. The crucial question for the theory of cultural evolution by natural selection is whether memes are transmitted by particulate or blending inheritance. As Allen Orr (1996) puts it, "Do street fashion and high fashion segregate like good genes, or do they first mix before replicating in magazines or storefronts?" Nobody knows. If memes are transmitted by blending inheritance, then natural selection is, at best, a weak mechanism of cultural evolution. We need other mechanisms to explain cultural evolution—perhaps mechanisms entirely different from those responsible for biological evolution.

Although biological evolution and cultural evolution may proceed by different mechanisms, this does not mean that either is irrelevant to the other. Cultural evolution can set the stage for biological evolution. Most humans, for example, stop producing the enzyme lactase in childhood, but the cultural practice of dairy farming led to the evolution of lifelong lactase production in many human populations (Durham 1991). Likewise, biological evolution can influence cultural evolution. For example, the division of the visible light spectrum into verbally distinguished colors follows cross-culturally universal patterns (Durham 1991). These patterns are determined by the way our eyes and brains encode visual information, indicating that the structure of our nervous systems has constrained cultural variation in color terminology. Cultural and biological evolution are distinct but interdependent.

athleticism and bravery. Scientists pursuing such an inquiry would formulate and test hypotheses about how a recipient of these social rewards, living in a hunter–gatherer society, might convert them into reproductive success.

Evolution and Parenthood

An evolutionary perspective can help researchers develop hypotheses about patterns of human behavior.

We now explore evolutionary psychology by considering aspects of parenthood. We begin with a prediction. On the assumption that the psychology of parenting has been shaped by natural selection, we can predict that human adults should direct more of their parental caregiving to their own genetic offspring than to the genetic offspring of others. We would make the same prediction about any organism that provides parental care. Care is expensive to the caregiver, and caregivers who reserve their efforts for their own genetic young should enjoy higher lifetime reproductive success than caregivers who are indiscriminate. The generality of this prediction gives us confidence that it is legitimate for human hunter–gatherers. And it has hidden subtleties, as an animal example will show.

Reed buntings (*Emberiza shoeniclus*) are small ground-nesting birds in which both males and females provide parental care. Most nesting pairs are socially monogamous: Each partner tends no other nest than the one they tend together. Genetic testing by Andrew Dixon and colleagues (1994) revealed that there is more to the reed bunting mating system than meets the eye. They found that 55% of chicks were sired by males other than their mother's social mate and that 86% of all nests included at least one such chick. Dixon predicted that males, if they could tell what fraction of the chicks they had sired in any given nest, would adjust their parental effort accordingly. Dixon looked at the chick-feeding behavior of 13 pairs of buntings that raised two clutches of chicks in a single season. The males fed the chicks more often in the nest in which they had sired a higher proportion of the chicks (Figure 14.28a). The females, who were the genetic mothers of all the chicks in both nests, showed no such pattern (Figure 14.28b).

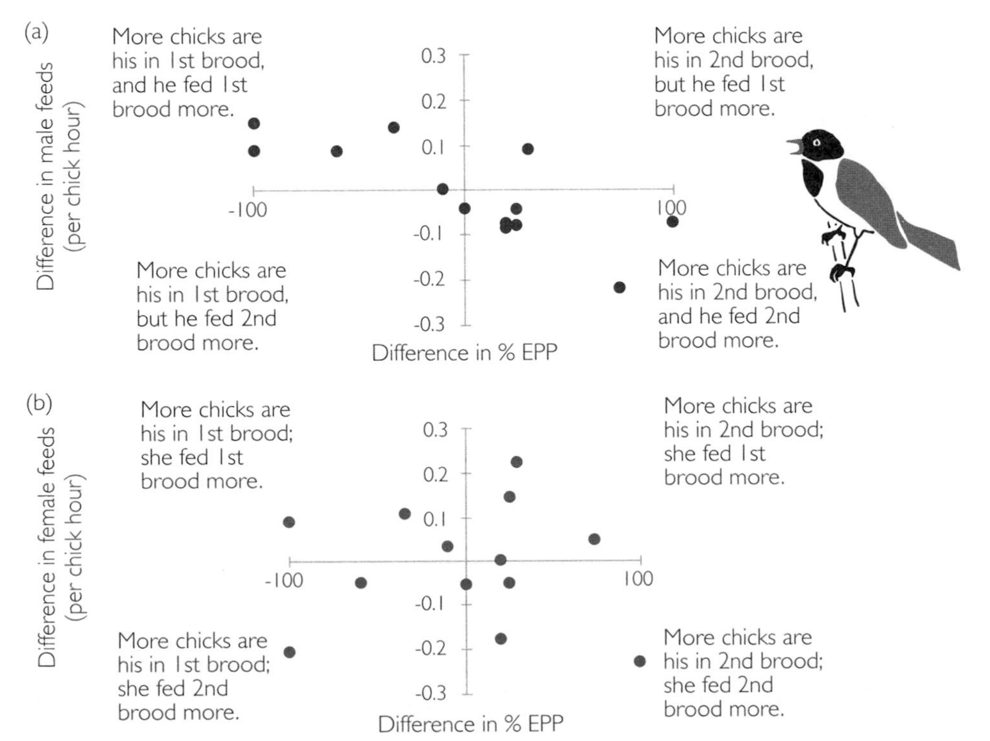

Figure 14.28 Male reed buntings adjust their parental effort depending on whom they are feeding (a) Each dot represents a single male who raised two broods of chicks in a season. The horizontal axis represents the difference between the two broods in the percentage of extra-pair paternity (% EPP), or fraction of chicks sired by some other male. The vertical axis represents the difference between how frequently the male fed chicks in the first brood and how frequently he fed chicks in the second brood. Most males fed chicks more often in the nest in which they had sired a higher fraction of chicks. This association was significant at $P = 0.0064$. (b) Each dot represents a single female who raised two broods of chicks in a season. The horizontal axis represents the difference between the percentage of chicks in the first brood sired by extra-pair males and the percentage of chicks in the second brood sired by extra-pair males (% EPP). The vertical axis represents the difference between how frequently the female fed chicks in the first brood and how frequently she fed chicks in the second brood. Females showed no relationship between the number of feedings they provided and the relative number of chicks sired by extra-pair males. Redrawn from Dixon et al. (1994).

We have presented the reed bunting example because evolutionary biologists using Darwinism to understand human behavior are often accused of genetic determinism (see, for example, Lewontin 1980). Genetic determinism is the notion that fundamental characteristics of human societies are unchangeably programmed into our genes. Note, however, the sense in which genes do and do not determine the parental behavior of male reed buntings. A male's genotype does not specify a particular level of parental care that the male will provide at all times, no matter what. Instead, each male's genotype specifies a range of phenotypic plasticity in parental care (see Chapter 10). That is, the bird's brain has a mechanism that adjusts the effort the male expends in caring for a brood, based on cues that indicate his probable level of paternity in that brood. If a male's social or biological environment changes, he alters his level of parental care accordingly, as Figure 14.28a shows. The pattern of phenotypic plasticity in a trait is called the trait's reaction norm. Reaction norms for reed bunting parental care presumably vary from male to male—or at least did vary in ancestral populations. It is this genetic variation in reaction norms that provides the raw material for the evolution of parental behavior. The average reaction norm of today's reed buntings appears to be adaptive. The average reaction norm might be described as "reed bunting nature."

Evolutionary psychologists studying human behavior are likewise interested in phenotypic plasticity—that is, reaction norms. They recognize that human reaction norms allow wide latitude for social and environmental circumstances to modify human behavior, and they recognize that reaction norms vary from person to person. What evolutionary psychologists do is formulate and test hypotheses about average human reaction norms.

Are humans as discriminating as male reed buntings in adjusting their provision of parental care? The question is difficult to study directly, at least in modern Western cultures, in which most of the interactions between parents and children take place in private. Other cultures, however, are less private and therefore more amenable to study.

Mark Flinn (1988) conducted an extensive and detailed observational study of the interactions between parents and offspring in a small village in rural Trinidad. Flinn interviewed all the residents of the village to determine which people were genetically related and which people were living together. Then, once or twice a day for six months, Flinn walked a standard route through the village that took him within 20 meters of every house and public building. He started at a different, randomly determined point each day, so as not to regularly pass particular places at particular times. Each time he saw any one of the village's 342 residents, Flinn recorded what the person was doing, who he or she was with, and the nature of the interaction they were having. The houses and buildings in the village are all rather open, so Flinn was able to see much that went on inside the houses as well as outside

Fourteen of the village's 112 households included mothers that were the genetic mothers of all the resident children and fathers that were the genetic fathers of some of the resident children and the stepfathers of others. These 14 families included 28 genetic offspring and 26 stepoffspring of the fathers. There can be no hidden differences between the genetic fathers and the stepfathers because the genetic fathers and the stepfathers are the same men.

Flinn calculated the amount of time the fathers spent with their children and the fraction of their interactions with their children that were agonistic. An agonistic interaction was one that "involved physical or verbal combat (e.g., spanking or arguing) or expressions of injury inflicted by another individual (e.g., screaming in pain or anguish or crying)" (Flinn 1988). Note that, overall, only 6% of the parent–offspring interactions Flinn saw were agonistic, and 94% of these involved only verbal exchanges. During his study, Flinn was not aware of any interactions between parents and children that would be considered physical child abuse. ("Screaming in pain or anguish" may sound like evidence of physical child abuse, but anyone who has spent time with a two-year-old knows that this is not necessarily so.) In other words, Flinn's research concerns parent–offspring interactions that most anyone would consider normal.

Flinn found that the 14 fathers with both genetic offspring and stepoffspring spent more of their time with their genetic offspring (Figure 14.29a). Furthermore, a smaller fraction of the father–genetic offspring interactions were agonistic (Figure 14.29b). These results are consistent with the prediction that parents discriminate among children on the basis of their genetic relationship with those children.

This is an observational study, however, and there is a potentially confounding variable. The pattern in Flinn's data could be explained by the late arrival of the stepfathers in the lives of their stepchildren. Men might feel less affection and concern for their stepchildren simply because they joined the family when the stepchildren were older, whereas the men were already in the family when all of their own genetic children were born.

Flinn's data set for this village, remarkably, includes 23 stepchildren who were born when their mothers and stepfathers were already living together, as well as 11 stepchildren born before their mothers and stepfathers moved in together. (This sample includes all the stepfathers in the village, not just those who also have genetic children living in the same house.) If a man's parental affection is simply a function of the fraction of the child's life during which the man has lived with the child, then the stepfathers who have lived with their stepchildren from the children's births should be more affectionate. In fact, the opposite appears to be true. The stepfathers in Flinn's study spent more of their time, and had a lower fraction

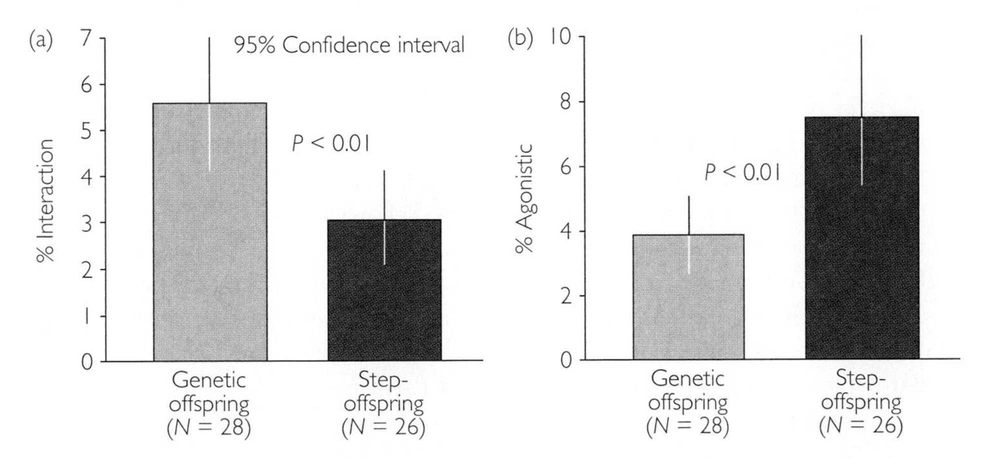

Figure 14.29 Fathers with both step- and biological children spend more time, and get along better, with their biological children (a) The bars show the fraction of their time (% of all observed interactions) that 14 fathers spent with their biological children versus their stepchildren. The 95% confidence interval is an estimate of the certainty that Flinn's estimated percentage is close to the truth. Roughly speaking, we can be 95% certain that the true number is within the 95% confidence interval. (b) The bars show the fraction of interactions between 14 fathers and their children that were agonistic (see text for definition). From Flinn (1988).

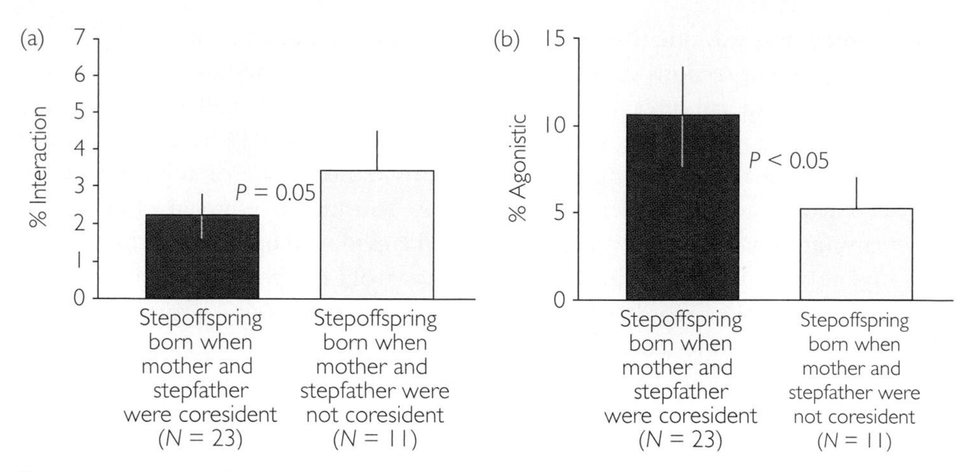

Figure 14.30 Stepfathers spend more time with their stepchildren, and get along better with them, when the stepchildren are born before the stepfather joins the family (a) Fraction of time (% of all observed interactions) that stepfathers spent with their stepchildren, with the data separated by whether the stepfather was living in the house when the stepchild was born. (b) Fraction of interactions between stepfathers and their stepchildren that were agonistic (see text for definition). From Flinn (1988).

of agonistic interactions, with stepchildren born before the stepfathers joined the family (Figure 14.30).

We noted earlier that studies within a single society offer no means of disentangling the influences of culture and evolutionary history. We could argue that the pattern of discrimination revealed in Flinn's study is simply a product of culture and has nothing to do with our species' adaptive history. Evidence is accumulating, however, that parental discrimination between own and others' genetic offspring is a cross-cultural phenomenon. For example, Kim Hill and Hillard Kaplan (1988) studied the survival of biological children versus stepchildren in the Ache Indians, a traditional foraging culture in Paraguay. Hill and Kaplan found that 81% of children raised by both biological parents survived to their 15th birthday, whereas only 57% of children raised by one biological parent and one stepparent survived. Napoleon Chagnon (1992; see also 1988) studied the Yąnomamö Indians, a traditional hunting, gathering, and gardening culture in Venezuela and Brazil. The Yąnomamö are polygynous, which means that women have little trouble finding husbands, but men often have difficulty finding wives. Chagnon reports that men work harder to find wives for their biological sons than for their stepsons. Frank Marlowe (1999) studied Hadza hunter–gatherers in Tanzania. He found that compared with stepfathers, genetic fathers spend more time near their children, and play with them, talk with them, and nurture them more. Kermyt Anderson and colleagues (1999) studied modern American men living in Albuquerque, New Mexico. They found that men invest more toward the college education of their genetic children than that of their stepchildren.

Parental Discrimination and Children's Health

Discrimination by parents against stepchildren becomes a public-health issue when we consider its impact on the childrens' physiological state. Mark Flinn and Barry England (1995, 1997), working in another rural Caribbean village, this time in Dominica, gave chewing gum to children, and then asked the children to provide saliva samples. In the saliva samples, the researchers measured the concentration of cortisol. Cortisol is a hormone that animals produce when they are under stress. Over the short term, high levels of cortisol cause an animal to divert resources to immediate demands, for example by increasing metabolic rate and alertness and inhibiting growth and reproduction. Over the long term, chronically high levels of cortisol can inhibit the immune system, deplete energy stores, and induce social withdrawal.

Flinn and England found that among the children in the village they studied, individuals with relatively high concentrations of cortisol in their saliva were sick more often (Figure 14.31a). Not surprisingly, it was the stepchildren who had the highest concentrations of cortisol (Figure 14.31b) and higher frequencies of illness (Figure 14.31c). Ultimately, stepchildren had lower reproductive success during early adulthood (Figure 14.31d) and were more likely to leave town.

Researchers investigating Darwinian predictions have discovered patterns of human behavior with profound consequences for public health.

Martin Daly and Margo Wilson approached the public-health consequences of parental discrimination by analyzing case files of homicides in which parents killed their children (Daly and Wilson 1988a; see also Daly and Wilson 1988b; 1994a; 1994b). Daly and Wilson predicted that children would be killed at a higher rate by stepparents than by biological parents.

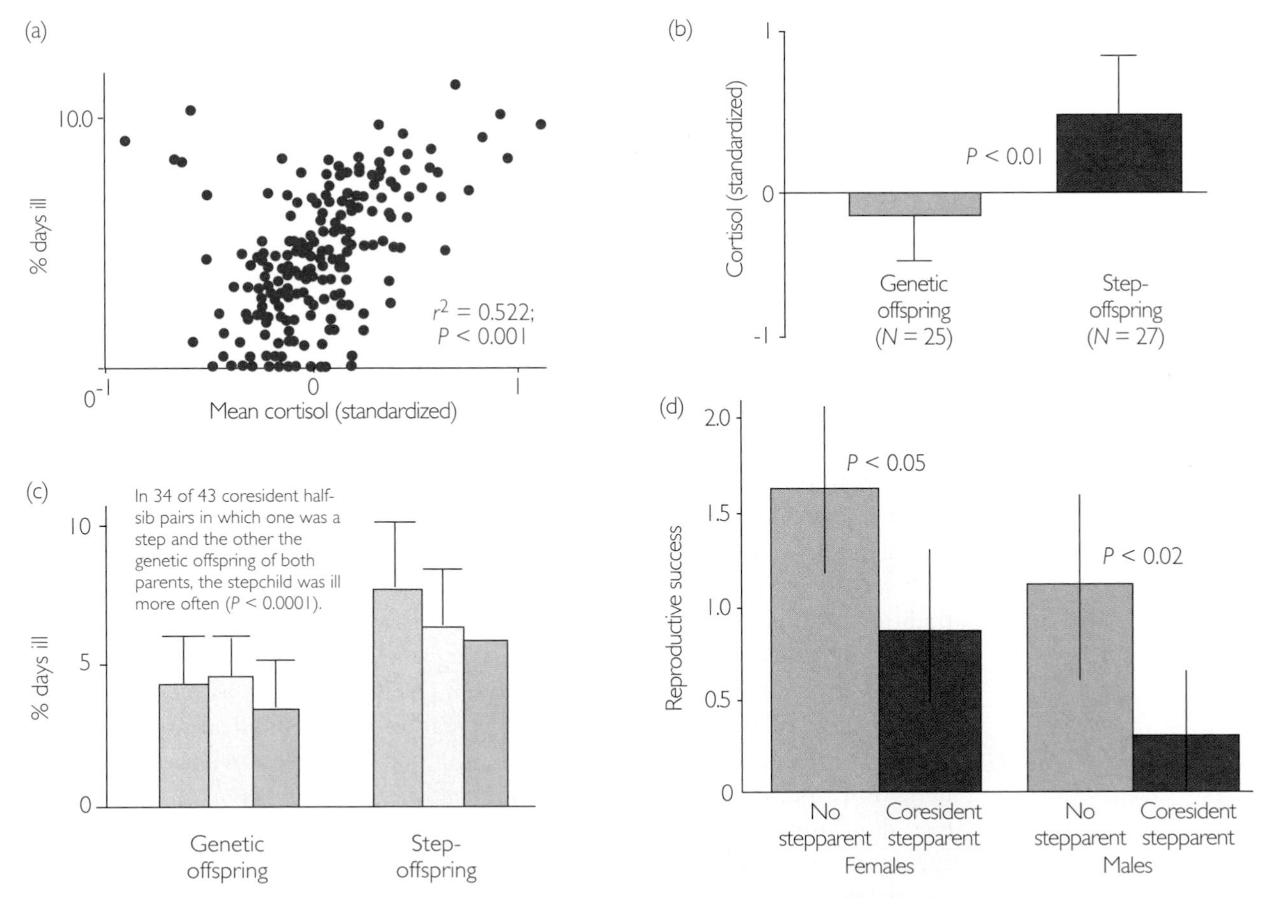

Figure 14.31 Stress, cortisol levels, illness, and reproductive success for stepchildren versus genetic children (a) Children with higher levels of the stress hormone cortisol in their blood get sick more often. Negative numbers indicate lower than average cortisol concentrations; positive numbers indicate higher than average concentrations. (b) Stepoffspring have higher concentrations of cortisol in their blood than do biological children. (a, b) From Flinn and England (1995). (c) The difference in health between biological children and stepchildren is larger than the differences attributable to socioeconomic status (different colors indicate different levels of status). From Flinn and England (1997). (d) Biological children have higher reproductive success during early adulthood (ages 18–28 for women, 20–30 for men) than do stepchildren. From Flinn (1988).

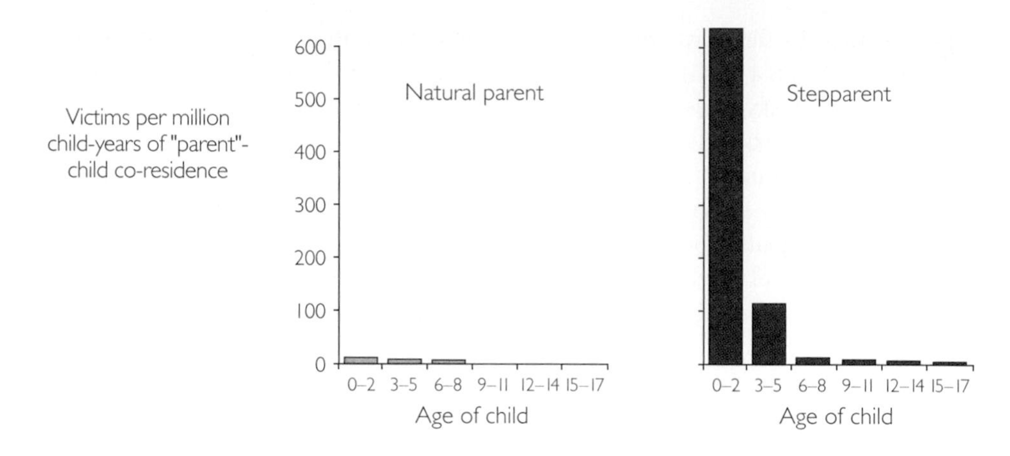

Figure 14.32 The risk to children of being killed by a biological parent versus a stepparent The graphs show, for biological parents (left) and stepparents (right), the rate at which parents killed children (number of homicides per million child-years that parents and children spent living in the same house). Children aged two or younger are killed by stepparents at a rate about 70 times higher than such children are killed by biological parents. The data are for homicides in Canada, 1974 through 1983. From Daly and Wilson (1988a; 1988b).

Data on murders of children in Canada dramatically confirm Daly and Wilson's prediction: Stepparents kill stepchildren at a much higher rate than biological parents kill biological children (Figure 13.26).

It is worth discussing this result a bit. In absolute numbers (that is, simply counting up homicides) more children are killed by biological parents than by stepparents (341 versus 67 in Daly and Wilson's study). But this is because only a small minority of children have stepparents. This is especially true for young children, the most common victims of parental homicide. In 1984, only 0.4% of Canadian children one to four years old lived with a stepparent. To adjust for the fact that few young children live with stepparents, Daly and Wilson reported the data in Figure 13.26 as rates: the number of homicides per million child–years that parents or stepparents and children spend living together. Epidemiologists often summarize the results of such a study by reporting a relative risk. Here the relative risk of homicide in stepchildren versus biological children is the rate at which stepparents kill stepchildren divided by the rate at which biological parents kill biological children. For children zero to two years old, the relative risk of parental homicide for stepchildren versus biological children is about 70. This is an extraordinarily high relative risk. For comparison, the relative risk of lung cancer in smokers versus nonsmokers is about 11.

Daly and Wilson do not suggest that killing stepchildren, in and of itself, is or ever was adaptive for humans. Anyone who kills someone else's child, even in a traditional hunter–gatherer society, is likely to suffer social penalties that outweigh any potential benefits of eliminating an unwelcome demand for stepparental investment. Instead, what Daly and Wilson suggest is adaptive is the combination of two traits: (1) an intellectual and psychological apparatus that perceives a personal interest in the distinction between one's own and others' genetic offspring, and (2) the emotional motivation to turn this perception into active discrimination between the two kinds of children. Whenever such an apparatus exists, individuals will, rarely, commit errors of excess. These errors of excess become Daly and Wilson's data.

Daly and Wilson's data come from an observational study in which it was impossible to control, as Flinn (1988) was able to, for differences between biological parents and stepparents. Nonetheless, they provide an argument that research conducted within a Darwinian framework can yield insights useful to public-health workers and providers of social services.

Summary

Evolutionary biology has numerous applications in medicine. This chapter has considered two general ways in which evolutionary analysis improves our understanding of issues relating to human health. First, we used our knowledge of the mechanisms of evolution to study pathogens and tumors. Second, we used the methods of the adaptationist program to address questions about human physiology and behavior.

Pathogens and their hosts are locked in a perpetual evolutionary arms race. Our immune systems and the drugs we take impose strong selection on the viruses and bacteria that infect our tissues. Because pathogens have short generation times, large population sizes, and often high mutation rates, populations of viruses and bacteria evolve quickly. Phylogenetic analysis helps us reconstruct the history of pathogen evolution, and in the case of flu, understand some of the mechanisms that create pathogen strains capable of causing epidemics. Selection thinking also helps us predict when pathogen populations will become resistant to drugs, whether drug resistance will persist in pathogen populations if drug use is suspended, and what makes some diseases virulent and others benign.

Humans, like other organisms, are a product of evolution by natural selection. As a result, the tools of the adaptationist program can help us understand aspects of our own form and function. Selection thinking suggests, for example, that symptoms of disease, such as fever, may be adaptive facets of our immune response. And aspects of our behavior with significant public-health consequences, such as cross-culturally consistent patterns in the way we treat children, may be interpretable as psychological adaptations. It is also important to keep in mind that change in our environment in recent centuries has far outpaced the rate of adaptive evolution. Modern epidemics of myopia and breast cancer may be the result of our exposure to novel environments.

Recognition that evolutionary biology is a medical science has, in some respects, been slow in arriving. We expect that interactions between evolutionary biologists and medical researchers will become more frequent and more productive in the years to come.

Questions

1. a. As a review, summarize the evidence discussed in this chapter that antibiotic resistance is due to evolution (i.e., due to new mutations that increase in frequency due to antibiotic exposure).

b. What would health-care workers, patients, and healthy people do if they wanted antibiotic resistance to evolve as *quickly* as possible? Do you know of any cases where humans are (unintentionally) doing this?

2. Some biologists regard our bodies as small ecosystems that exert selective pressure for the evolution of invasive metastatic cancer. If this is true, why don't we all get cancer? (*Hint*: Consider the speed of evolution.) However, these same biologists believe that humans have certain genes that have evolved specifically to prevent cancer. How is it possible to have both strong selection for cancer and strong selection for anticancer genes? (*Hint*: Consider which population is under selection in each case.)

3. We have seen how the genetic diversity within a tumor can be used to estimate the tumor's age (see Figure 13.14). This analysis depends on mutation rate per cell division being constant.

a. If a cancerous tumor has evolved a high mutation rate, how will this bias the results?

b. Do the genetic markers we use to estimate a tumor's age need to be selectively neutral? Why or why not?

4. Pathogens require a minimum population size of potential hosts. If the host population is too small, in a short time the entire population has either been killed by the pathogen, or has survived the initial infection and become immune. If this occurs, the pathogen dies out. What evolutionary changes in a pathogen might increase its ability to survive in a smaller population? For example, measles requires a host population of about 500,000 humans, while diphtheria can get by with only about 50,000 humans. Develop some hypotheses for why diphtheria can survive with just one-tenth the number of hosts. For example, how might these two diseases differ from each other in transmission rate, virulence, latency to infection, or mutation rate?

5.a. In the study of streptomycin resistance, why did Shrag and colleagues use genetically manipulated bacteria, instead of the original wild-type bacteria, to compare streptomycin-sensitive to streptomycin-resistant strains?

b. Summarize the key finding of Shrag et al.'s study. Why are these results worrying to the medical and veterinary professions?

6. Review the studies on fever that were presented in this chapter, and summarize each in a sentence or two. Do you agree with Kluger that several of the studies did not really test the adaptive fever hypothesis? If so, can you design an experiment that will truly test the hypothesis? Is your experiment ethical?

7. The male reed buntings in Dixon et al.'s study (Figure 13.22) seem to be consciously aware of genetic relationships and "trying" to increase their reproductive success. Can evolution cause reed buntings (and other animals) to behave as if they are aware of the evolutionary consequences of their actions, without actually being aware of those consequences? Does your answer also apply to humans?

8. Daly and Wilson's data on infanticide risks might be explained by stepfathers having, on average, more violent personalities than biological fathers. Could this "violent personality" explanation also apply to Flinn's data from the Trinidad village? Why or why not? Daly and Wilson's study involved general data about a large number of families, whereas Flinn's study involved detailed data on a small number of families. What are the advantages and disadvantages of each kind of study?

9. An evolutionary biologist once hypothesized that if evolution has affected human social behavior, then a mother's brothers should take a particular interest in her children—more so than the father's brothers, and perhaps even more so than the father himself. Why did he hypothesize this? (As it turns out, there are many cultures in which men do, in fact, direct parental care primarily toward their sisters' children.)

10. In 1999, a mysterious outbreak of human encephalitis occurred in the northeastern United States. The cause was tentatively identified as St. Louis encephalitis virus. At the same time, an unusual number of dead birds were noticed along the northeastern Atlantic coast. Figure 14.33 shows genetic relationships of three known encephalitis viruses (St. Louis, Japanese, and West Nile), with viruses isolated from the birds, from two human patients that died, from one dead horse, and from mosquitoes. (Data compiled from Anderson et al. 1999 and Lanciotti et al. 1999.)

    Were the birds, the horse, and the humans all suffering from the same disease?

    Do you think the outbreak was caused by St. Louis encephalitis virus?

    Does this cladogram suggest how the disease might be spread?

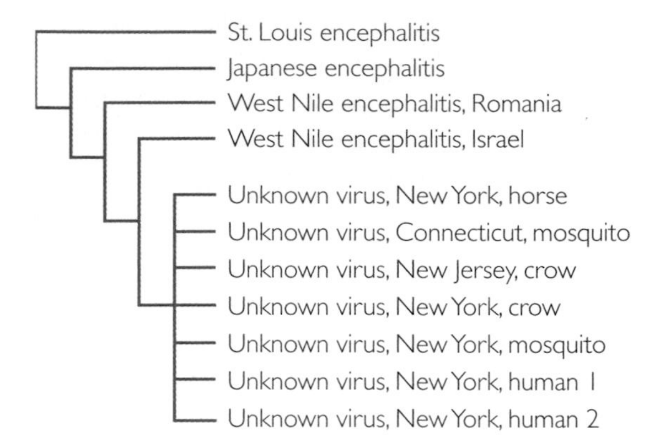

Figure 14.33 **A phylogeny of encephalitis viruses isolated from various hosts in the northeastern United States during 1999** Based on data and analyses in Anderson et al. (1999) and Lanciotti et al. (1999).

11. It has now become clear that birds are the primary host of West Nile virus. If the virus reaches a human (or horse) it is not spread from human to human (nor from horse to horse) and is unlikely to be transferred back to birds. Is the virulence of the virus in humans and horses an example of coincidental evolution or short-sighted evolution? Explain your reasoning.

12. In a review on oral contraceptives' (OCs) effect on various cancers, Spicer et al. (2000) stated: "Direct observational studies of breast-cell proliferation in women taking OCs suggest that the total breast-cell proliferation is very similar over an OC cycle and a normal menstrual cycle. These results predict that breast cancer risk should not be substantially affected by OC use, as is observed."

    If it is correct that oral contraceptives have no more effect on breast cancer than does a normal menstrual cycle, does it follow that OCs do not affect the risk of breast cancer? [*Hint*: the risk of breast cancer *as compared to what*?]

13. An avian influenza virus of type H5N1 has recently evolved a "high pathogenicity" (hp) strain that causes severe illness in most wild birds (except ducks) as well as in domestic poultry. A few humans have been infected. The World Health Organization (WHO) currently inspects every human case with particular attention to how the patients contracted the virus. Why is this virus a cause for concern, and why are WHO officials so interested in each patient's source of infection? (See Question 21 for further reading.)

Exploring the Literature

14. For another study on the effects of antifever medication on human illness, see:

Sugimura, T., T. Fujimoto, H. Motoyama, T. Maruoka, and S. Korematu, et al. 1994. Risks of antipyretics in young children with fever due to infectious disease. *Acta Paediatrica Japonica* 36: 375–378.

How strong is the evidence in this paper that fever is an adaptive response to bacterial respiratory infection? Consider that acetaminophen affects aspects of the immune response other than fever ("Fever and the Common Cold" in Section 13.6). Also consider that Sugimura and colleagues conducted an observational study, not an experimental one. That is, the researchers did not randomly assign their subjects to acetaminophen versus placebo groups. Instead, the researchers asked parents to keep a diary of the number of doses of acetaminophen they gave their children.

15. See these papers for recent studies on fever in critically ill adults:

J. R. Ostberg and E. A. Repasky. 2006. Emerging evidence indicates that physiologically relevant thermal stress regulates dendritic cell function. *Cancer Immunol. Immunother.* 55 (3): 292–298.

Su, F., N. D. Nguyen, Z. Wang, Y. Cai, P. Rogiers, and J. L. Vincent. 2005. Fever control in septic shock: beneficial or harmful? *Shock* 23 (6): 516–20.

M. Ryan and M. M. Levy. 2003. Clinical review: Fever in intensive care unit patients. *Critical Care* 7 (3): 221–225.

16. In many countries, meat producers routinely feed antibiotics to livestock to promote growth. The reasons why antibiotics promote growth are unclear, but this use accounts for a large fraction of global antibiotic production. For strong circumstantial evidence that antibiotics fed to livestock select for resistant bacteria that later infect people, see:

Wegener, H. C., et al. 1999. Use of antimicrobial growth promoters in food animals and *Enterococcus faecium* resistance to therapeutic antimicrobial drugs in Europe. *Emerging Infectious Diseases* 5: 329–325. Available from *http://www.cdc.gov/ncidod/EID/eid.htm.*

17. A crucial question in deciding whether modern women should use hormonal treatments that suppress menstruation is whether menstruation itself is adaptive. The issue is controversial. See:

Profet, M. 1993. Menstruation as a defense against pathogens transported by sperm. *Quarterly Review of Biology* 68: 335–381.

Strassmann, B. I. 1996. The evolution of endometrial cycles and menstruation. *Quarterly Review of Biology* 71: 181–220.

18. We presented evidence in Section 13.5 that myopia is a disease of civilization, and that the crucial change in lifestyle responsible for myopia is close visual work in childhood. For evidence that modern diets may also be involved in myopia, see:

Cordain, L., S. B. Eaton, et al. 2002. An evolutionary analysis of the aetiology and pathogenesis of juvenile-onset myopia. *Acta Ophthalmologica Scandinavica* 80: 125–135.

19. For evidence that acne is a disease of civilization, see:

Cordain, L., S. Lindeberg, et al. 2002. Acne vulgaris: A disease of Western civilization. *Archives of Dermatology* 138: 1584–1590.

Maziak, W. 2005. The asthma epidemic and our artifical habitats. *BMC Pulmonary Medicine* 5: 5.

20. The quest for the perfect oral contraceptive continues, particularly a contraceptive that will reduce the risks of several types of reproductive cancers. See:

Pike, M. C., and D. V. Spicer. 2000. Hormonal contraception and chemoprevention of female cancers. *Endocrine-Related Cancer* 7: 73–83.

Gardner, J., L. Miller. 2005. Promoting the safety and use of hormonal contraceptives. *Journal of Women's Health* 14 (1): 53–60.

21. Birds are the major hosts for several avian pathogens that can also, on occasion, infect humans. The combination of the density of the commercial poultry industry with the huge distances that wild birds can travel has made birds a major factor in the spread of several emerging human diseases. See these papers for recent information on the role of birds in human health:

Reed, K.D., J.K. Meece, et al. 2003. Birds, migration, and emerging zoonoses: West Nile virus, Lyme disease, influenza A, and enteropathogens. *Clinical Medicine & Research* 1: 5–12.

Riedel, S. 2006. Crossing the species barrier: The threat of an avian influenza pandemic. *Baylor University Medical Center Proceedings* 19: 16–20.

Citations

Aiuti, A., S. Vai, A. Mortellaro, et al. 2002a. Immune reconstitution in ADA-SCID after PBL gene therapy and discontinuation of enzyme replacement. *Nature Medicine* 8: 423–425.

Aiuti, A., S. Slavin, M. Aker, et al. 2002b. Correction of ADA-SCID by stem cell gene therapy combined with nonmyeloablative conditioning. *Science* 296: 2410–2413.

Anderson, J. F., T. G. Andreadis, et al. 1999. Isolation of West Nile virus from mosquitoes, crows, and a Cooper's hawk in Connecticut. *Science* 286: 2331–2333.

Anderson, K., H. Kaplan, and J. Lancaster. 1999. Paternal care by genetic fathers and stepfathers I: Reports from Albuquerque men. *Evolution and Human Behavior* 20: 405–431.

Austin, D. J., K. G. Kristinsson, and R. M. Anderson. 1999. The relationship between the volume of antimicrobial consumption in human communities and the frequency of resistance. *Proceedings of the National Academy of Sciences USA*: 96: 1152–1156.

Azzimondi, G., L. Bassein, F. Nonino, L. Fiorani, L. Vignatelli, et al. 1995. Fever in acute stroke worsens prognosis: A prospective study. *Stroke* 26: 2040–2043.

Banet, M. 1979. Fever and survival in the rat. *Pflügers Archive* 381: 35–38.

Banet, M. 1981a. Fever and survival in the rat. The effect of enhancing the cold defense response. *Experientia* 37: 985–986.

Banet, M. 1981b. Fever and survival in the rat. Metabolic rate versus temperature response. *Experientia* 37: 1302–1304.

Baquero, F., and J. Blázquez. 1997. Evolution of antibiotic resistance. *Trends in Ecology and Evolution* 12: 482–487.

Basler, C. F., A. H. Reid, et al. 2001. Sequence of the 1918 pandemic influenza virus nonstructural gene (NS) segment and characterization of recombinant viruses bearing the 1918 NS genes. *Proceedings of the National Academy of Sciences USA* 98: 2746-2751.

Bean, W. J., Schell, M., et al. 1992. Evolution of the H3 influenza virus hemagglutinin from human and nonhuman hosts. *Journal of Virology* 66: 1129–1138.

Berkey, C. S., A. L. Frazier et al. 1999. Adolescence and breast carcinoma risk. *Cancer* 85: 2400–2409.

Bernard, G. R., A. P. Wheeler, et al. 1997. The effects of ibuprofen on the physiology and survival of patients with sepsis. *New England Journal of Medicine* 336: 912–918.

Bernheim, H. A., and M. J. Kluger. 1976. Fever: Effect of drug-induced antipyresis on survival. *Science* 193: 237–239.

Bishai, W. R., N. M. H. Graham, et al. 1996. Rifampin-resistant tuberculosis in a patient receiving rifabutin prophylaxis. *New England Journal of Medicine* 334: 1573–1576.

Blaese, R. M., K. W. Culver, A. D. Miller, C. S. Carter, T. Fleisher, et al. 1995. T. lymphocyte-directed gene therapy for ADA-SCID: Initial trial results after 4 years. *Science* 270: 475–480.

Bloch, A. B., G. M. Cauthen, et al. 1994. Nationwide survey of drug-resistant tuberculosis in the United States. *Journal of the American Medical Association* 271: 665–671.

Bordignon, C., L. D. Notarangelo, N. Nobili, G. Ferrari, and G. Casorati, et al. 1995. Gene therapy in peripheral blood lymphocytes and bone marrow for ADA-immunodeficient patients. *Science* 270: 470–575.

Bull, J. J. 1994. Virulence. *Evolution* 48: 1423–1437.

Bush, R. M. 2001. Predicting adaptive evolution. *Nature Reviews Genetics* 2: 387-392

Bush, R. M., C. A. Bender, et al. 1999. Predicting the evolution of human influenza A. *Science* 286: 1921–1925.

Bush, R. M., W. M. Fitch, et al. 1999. Positive selection on the H3 hemagglutinin gene of human influenza virus A. *Molecular Biology and Evolution* 16: 1457–1465.

Chagnon, N.A. 1988. Male manipulations of kinship classifications of female kin for reproductive advantage. In L. Betzig, M. Borgerhoff Mulder, and P. Turke, eds. *Human Reproductive Behavior: A Darwinian Perspective*. Cambridge: Cambridge University Press, 23–48.

Chagnon, N.A. 1992. *Yąnomamö*. Fort Worth: Harcourt Brace College Publishers.

Cochran, G. M., P. W. Ewald, and K. D. Cochran. 2000. Infectious causation of disease: An evolutionary perspective. *Perspectives in Biology and Medicine* 43: 406–448.

Cooper, V. S., M. H. Reiskind, J. A. Miller, et al. 2002. Timing of transmission and the evolution of virulence of an insect virus. *Proceedings of the Royal Society of London B* 269: 1161–1165.

Cordain, L., S. B. Eaton, et al. 2002. The paradoxical nature of hunter-gatherer diets: meat-based, yet non-atherogenic. *European Journal of Clinical Nutrition* 56: S42–S52.

Cordain, L., R. W. Gotshall, and S. B. Eaton. 1997. Evolutionary aspects of exercise. *World Review of Nutrition and Dietetics* 81: 49–60.

Daly, M., and M. Wilson. 1988a. *Homicide*. New York: Aldine de Gruyter.

Daly, M., and M. Wilson. 1988b. Evolutionary social psychology and family homicide. *Science* 242: 519–524.

Daly, M., and M. I. Wilson. 1994a. Some differential attributes of lethal assaults on small children by stepfathers versus genetic fathers. *Ethology and Sociobiology* 15: 207–217.

Daly, M., and M. Wilson. 1994b. Stepparenthood and the evolved psychology of discriminative parental solicitude. In S. Parmigiani and F. S. vom Saal, eds. *Infanticide and Parental Care*. London: Harwood Academic Publishers, 121–133.

Dawkins, R. 1976. *The Selfish Gene*, 1st ed. Oxford: Oxford University Press.

Dawkins, R. 1989. *The Selfish Gene*, 2nd ed. Oxford: Oxford University Press.

Dennett, D. C. 1995. *Darwin's Dangerous Idea*. New York: Simon and Schuster.

Diamond, J. 1992. *The Third Chimpanzee*. New York: Harper Collins.

Dixon, A., D. Ross, S. L. C. O'Malley, and T. Burke. 1994. Paternal investment inversely related to degree of extra-pair paternity in the reed bunting. *Nature* 371: 698–700.

Doran, T. F., C. De Angelis, R. Baumgardner, and E. D. Mellits. 1989. Acetaminophen: More harm than good for chickenpox? *Journal of Pediatrics* 114: 1045–1048.

Durham, W. H. 1991. *Coevolution: Genes, culture, and human diversity*. Stanford University Press, Stanford.

Eaton, S. B., and L. Cordain. 1997. Evolutionary aspects of diet: Old genes, new fuels. *World Review of Nutrition and Dietetics* 81: 26–37.

Eaton, S. B., S. B. Eaton III, and M. J. Konner. 1997. Paleolithic nutrition revisited: A twelve-year retrospective on its nature and implications. *European Journal of Clinical Nutrition* 51: 207–216.

Ewald, P. W. 1991. Waterborne transmission and the evolution of virulence among gastrointestinal bacteria. *Epidemiology and Infection* 106: 83–119.

Ewald, P. W. 1993. The evolution of virulence. *Scientific American* (April): 86–93.

Ewald, P. W. 1994. *Evolution of Infectious Disease*. Oxford: Oxford University Press.

Ferguson, N. M., and R. M. Anderson. 2002. Predicting evolutionary change in the influenza A virus. *Nature Medicine* 8: 562–563.

Fitch, W. M., J. M. Leiter, et al. 1991. Positive Darwinian evolution in human influenza A viruses. *Proceedings of the National Academy of Sciences USA* 88: 4270–4274.

Flinn, M. V. 1988. Step- and genetic-parent-offspring relationships in a Caribbean village. *Ethology and Sociobiology* 9: 335–369.

Flinn, M.V., and B. G. England. 1995. Childhood stress and family environment. *Current Anthropology* 36: 854–866

Flinn, M.V., and B. G. England. 1997. Social economics of childhood glucocorticoid stress response. *American Journal of Physical Anthropology* 102: 33–53.

Gorman, O.T., W. J. Bean, et al. 1991. Evolution of influenza A virus nucleoprotein genes: Implications for the origins of H1N1 human and classical swine viruses. *Journal of Virology* 65: 3704–3714.

Graham, N. M. H., C. J. Burrell, R. M. Douglas, P. Debelle, and L. Davies. 1990. Adverse effects of aspirin, acetaminophen, and ibuprofen on immune function, viral shedding, and clinical status in rhinovirus-infected volunteers. *Journal of Infectious Diseases* 162: 1277–1282.

Green, M. H., and C. W. Vermeulen. 1994. Fever and the control of gram-negative bacteria. *Research in Microbiology* 145: 269–272.

Henderson, B. E., P. K. Ross, and M. C. Pike. 1993. Hormonal chemoprevention of cancer in women. *Science* 259: 633–638.

Hershfield, M. S., R. H. Buckley, M. L. Greenberg, A. L. Melton, L. Schiff, et al. 1987. Treatment of adenosine deaminase deficiency with polyethylene glycol-modified adenosine deaminas. *New England Journal of Medicine* 316: 589–596.

Hill, K., and H. Kaplan. 1988. Trade-offs in male and female reproductive strategies among the Ache, part 2. In L. Betzig, M. Borgerhoff Mulder, and P. Turke, eds. *Human Reproductive Behavior: A Darwinian Perspective*. Cambridge: Cambridge University Press, 291–305.

Hirschhorn, R., D. R. Yang, et al. 1996. Spontaneous in vivo reversion to normal of an inherited mutation in a patient with adenosine deaminase deficiency. *Nature Genetics* 13: 290–295.

Johnson, N. P. A. S., and J. Mueller. 2002. Updating the accounts: Global mortality of the 1918–1920 "Spanish" influenza pandemic. *Bulletin of the History of Medicine* 76: 105–115.

Klug, W. S., and M. R. Cummings. 1997. *Concepts of Genetics*, 5th ed. Upper Saddle River, NJ: Prentice Hall.

Kluger, M. J. 1992. Fever revisited. *Pediatrics* 90: 846–850.

Kluger, M. J., D. H. Ringler, and M. R. Anver. 1975. Fever and survival. *Science* 188: 166–168.

Kolata, Gina. 1999. *Flu*. New York. Farrar, Straus, and Giroux.

Lanciotti, R. S., J. T. Roehrig, et al. 1999. Origin of West Nile virus responsible for an outbreak of encephalitis in the northeastern United States. *Science* 286: 2333–2337.

Laver, W. G., N. Bischofberger, and R. G. Webster. 1999. Disarming flu viruses. *Scientific American* 280 (January): 78–87.

Layde P. M., L. A. Webster, et al. 1989. The independent associations of parity, age at first full term pregnancy, and duration of breastfeeding with the risk of breast cancer. *Journal of Clinical Epidemiology* 42: 963–73.

Levin, B. R. 1996. The evolution and maintenance of virulence in microparasites. *Emerging Infectious Diseases* 2: 93–102.

Levy, S. B. 1998. The challenge of antibiotic resistance. *Scientific American* 278 (March): 46–53.

Lewontin, R. C. 1980. Sociobiology: Another biological determinism. *International Journal of Health Services* 10: 347–363.

Marlowe, F. 1999. Showoffs or providers? The parenting effort of Hadza men. *Evolution and Human Behavior* 20: 391–404.

Messenger, S. L., I. J. Molineux, and J. J. Bull. 1999. Virulence evolution in a virus obeys a trade off. *Proceedings of the Royal Society of London B* 266: 397–404.

Nesse, R. M., and G. C. Williams. 1994. *Why We Get Sick*. New York: Vintage Books.

Nesse, R. M., and G. C. Williams. 1998. Evolution and the origins of disease. *Scientific American* (November): 86–93.

Norton, T. T., and C. Wildsoet. 1999. Toward controlling myopia progression? *Optometry and Vision Science* 76: 341–342.

Orr, H. A. 1996. Dennett's dangerous idea. *Evolution* 50: 467–472.

Parrish, C. R., and Y. Kawaoka. 2005. The origins of new pandemic viruses: The acquisition of new host ranges by canine parvovirus and influenza A viruses. *Annual Review of Microbiology* 59: 553–586.

Persons, D. A., and A. W. Nienhuis. 2002. In vivo selection to improve gene therapy of hematopoietic disorders. *Current Opinion in Molecular Therapeutics* 4: 491–498.

Plotkin, J. B., J. Dushoff, and S. A. Levin. 2002. Hemagglutinin sequence clusters and the antigenic evolution of influenza A virus. *Proceedings of the National Academy of Sciences USA* 99: 6263–6268.

Rattan, A., A. Kalia, and N. Ahmad. 1998. Multidrug-resistant Mycobacterium tuberculosis: Molecular perspectives. *Emerging Infectious Diseases* 4: 195–209.

Reid, A. H., T. G. Fanning, et al. 1999. Origin and evolution of the 1918 "Spanish" influenza virus hemagglutinin gene. *Proceedings of the National Academy of Sciences USA* 96: 1651–1656.

Reid, A. H., T. G. Fanning, et al. 2000. Characterization of the 1918 "Spanish" influenza neuraminidase gene. *Proceedings of the National Academy of Sciences USA* 97: 6785–6790.

Reid, A. H., T. G. Fanning, et al. 2002. Characterization of the 1918 "Spanish" influenza virus matrix gene segment. *Journal of Virology* 76: 10717–10723.

Reid, A. H., T. G. Fanning, et al. 2004. Novel origin of the 1918 pandemic influenza virus nucleoprotein gene. *Journal of Virology* 78: 12462–12470.

Reid, A. H., J. K. Taubenberger, and T. G. Fanning. 2004. Evidence of an absence: The genetic origins of the 1918 pandemic influenza virus. *Nature Reviews Microbiology* 2: 909–914.

Schrag, S. J., V. Perrot, and B. R. Levin. 1997. Adaptation to the fitness costs of antibiotic resistance in *Escherichia coli*. *Proceedings of the Royal Society of London B* S264: 1287–1291.

Shibata, D., W. Navidi, R. Salovaara, Z.-H. Li, and L. A. Aaltonen. 1996. Somatic microsattelite mutations as molecular tumor clocks. *Nature Medicine* 2: 676–681.

Sorsby, A., and F. A. Young. 1970. Transmission of refractive errors within Eskimo families. *American Journal of Optometry* 47: 244–249.

Spicer, D. V., D. Shoupe, and M. C. Pike. 1991. GnRH agonists as contraceptive agents: Predicted significantly reduced risk of breast cancer. *Contraception* 44: 289–310.

Stewart, T. H. M., R. D. Sage, A. F. R. Stewart, and D. W. Cameron. 2000. Breast cancer incidence highest in the range of one species of house mouse, *Mus domesticus*. *British Journal of Cancer* 82: 446–451.

Strassmann, B. I. 1999. Menstrual cycling and breast cancer: An evolutionary perspective. *Journal of Women's Health* 8: 193–202.

Taubenberger, J. K., A. H. Reid, et al. 2005. Characterization of the 1918 influenza virus polymerase genes. *Nature* 437: 889–893

Teikari, J. M., J. O'Donnell, J. Kaprio, and M. Koskenvuo. 1991. Impact of heredity in myopia. *Human Heredity* 41: 151–156.

Tufte, E. R. 1997. *Visual Explanations*. Cheshire, CT: Graphics Press.

Tumpey, T. M., C. F. Basler, et al. 2005. Characterization of the reconstructed 1918 Spanish influenza pandemic virus. *Science* 310: 77–80.

Vaughn, L. K., H. A. Bernheim, and M. Kluger. 1974. Fever in the lizard *Dipsosaurus dorsalis*. *Nature* 252: 473–474.

Wang, Y., J. F., Holland, I. J. Bleiweiss, et al. 1995. Detection of mammary tumor virus ENV gene-like sequences in human breast cancer. *Cancer Research* 55: 5173–5179.

Webster, R. G., W. J. Bean, et al. 1992. Evolution and ecology of influenza A viruses. *Microbiological Reviews* 56: 152–179.

Williams, G.C., and R. M. Nesse. 1991. The dawn of Darwinian medicine. *The Quarterly Review of Biology* 66: 1–22.

Young, F. A., W. R. Baldwin, et al. 1969. The transmission of refractive errors with Eskimo families. *American Journal of Optometry* 46: 676–685.

Youssoufian, H. 1996. Natural gene therapy and the Darwinian legacy. *Nature Genetics* 13: 255–256.

15

Phylogenomics and the Molecular Basis of Adaptation

For generations, biologists were restricted to studying adaptation at the level of the phenotype. The goal was to understand which morphological or behavioral traits allowed individuals to achieve higher fitness in particular environments, and why. The situation was similar in cell biology, developmental biology, genetics, and other fields, where researchers focused on studying the phenotypes of cells, embryos, and offspring.

All that has changed. A revolution began with the discovery of DNA as the hereditary material and data indicating that genes code for proteins and for RNA molecules that perform specific functions in the cell. These insights inspired the development of techniques for studying the makeup of nucleic acids and proteins. The ability to sequence and manufacture specific polypeptides and nucleic acids allowed biologists to document the genotypes of individuals and even manipulate them. Molecular biologists began tracking the products of specific alleles in organisms and recording the effect on cells, development, or inheritance.

Much of the data discussed in this chapter was gathered and analyzed at genome sequencing centers like this one.

In a similar way, evolutionary biologists seized the opportunity to study adaptation at the level of the gene. Molecular evolutionists, in particular, delved into questions about the rates and patterns of change in nucleotide and amino acid sequences. That work contributed new techniques for inferring phylogenies (Chapter 4), the use of molecular clocks to infer divergence times that are not documented in the fossil record (Chapter 4), insights into the molecular basis of mutation (Chapter 5), and tests to distinguish the effects of natural selection and genetic drift at the level of the nucleotide (Chapter 7).

Now laboratories around the world are sequencing the entire genomes of an increasing number and diversity of species. These data sets estimate total gene number, document gene order, characterize the function of at least some genes, and make it possible to fully describe the size and composition of gene families (see Chapter 4). Follow-up studies in the new field of proteomics are documenting which genes are expressed at different times during development or in response to different environmental challenges.

Genome sequencing has also made it possible to ask entirely new types of questions about adaptation at the level of the gene. This chapter explores some of these questions, which lie at the intersection of genomics and evolutionary analysis—an emerging field called **phylogenomics** (Eisen and Fraser 2003).

Phylogenomics focuses on the evolutionary analysis of genome sequence data.

Let's begin by examining the genomic parasites called **transposable elements**, which transmit copies of themselves at the expense of their host genome. Then we'll consider evidence that genes have been transferred between lineages on the tree of life, investigate patterns in the types of genes that various organisms have, and consider the sorts of problems that future research in phylogenomics might address.

15.1 Transposable Elements and Levels of Selection

Genomes are not simple collections of sequences that code for proteins, rRNAs, and tRNAs. Instead they contain a bestiary of sequence types. In humans, for example, only about 1.2% of the genome codes for proteins. Our genome, like the genomes of many other eukaryotes, is dominated by parasitic sequences that do not code for products that help cells function efficiently.

Two pioneering observations hinted at this conclusion and launched research on how genomes evolve. One observation was called the C-value paradox. In eukaryotes the total amount of DNA found in a cell, also known as its C-value, does not correlate with the organism's size or perceived morphological complexity or phylogenetic position. The single-celled eukaryote *Amoeba dubia* has 670 billion kilobases (kb) of DNA per cell, whereas humans have only 3.4 billion kb and the fruit fly *Drosophila melanogaster* has only 180,000 kb. This finding suggests that much or most of the DNA in eukaryotes is functionless from the cell's viewpoint. A second important observation was Barbara McClintock's discovery of transposable genetic elements, or "jumping genes." While studying the inheritance of kernel color in corn, McClintock found genes that produced novel color patterns by moving to new locations in the genome.

Sequencing studies later revealed that much of the "extra" DNA responsible for the C-value paradox consists of transposable genetic elements, also called mobile genetic elements. In the human genome, for example, over 44% of the DNA present is derived from transposable elements (International Human Genome Sequencing Consortium 2001). It's important to consider three basic questions about these sequences: What are they? Where did they come from? And what effect do they have on the genomes that host them?

Transposable Elements Are Genomic Parasites

As the ensuing discussion will show, transposable element is really an umbrella term for genes with a diverse set of characteristics. The majority of transposable elements contain only the sequences required for transposition, and all of them share the ability to transpose, or move, from one location to another in the genome. Most leave a copy of themselves behind when they move. In this way, transposition events lead to an increase in the number of transposable elements in the host genome. This is a key observation. A transposable element is an allele that can increase in frequency in a population by moving to a new location in an individual's genome, and then being passed on to offspring.

Transposition can increase the fitness of a transposable element, but what about the rest of the genome? If a transposable element disrupts an important coding sequence when it inserts into a new location in the genome, deleterious "knock-out" mutations result. In humans, transposition events have resulted in tumor formation and cases of hemophilia (see Hutchison et al. 1989). The sheer bulk of transposable element DNA should also have deleterious consequences for the host genome. The time, energy, and resources required to replicate a genome burdened by parasitic DNA could place a limit on growth rates, particularly in small, rapidly dividing organisms. Because they may disrupt coding sequences and place an energetic burden on the cell, transposable elements are most accurately characterized as genome parasites. Carrying them in the genome—often in appallingly large numbers—appears to be maladaptive. Yet surveys done to date suggest that transposable elements exist in virtually every organism.

Transposable elements are considered genomic parasites because they may cause deleterious mutations when they transpose and because they add DNA that requires time and resources to be copied.

What is the key to their success? The answer is that while selection at the level of host organisms may select against transposable elements, selection at the level of the elements themselves favors their spread. Even if a transposition event reduces the survival and reproductive capacity of the host slightly—meaning that the transposable element is also a little less likely to be transmitted from that host to the next generation—the extra copies of the transposable element now present in the gene pool can make up the deficit and result in the parasite's spread throughout the population. According to models developed by Brian Charlesworth and Charles Langley (1989), transposable elements that replicate themselves most efficiently and with the least fitness cost to the host genome are, on balance, favored by natural selection and tend to spread.

Transposable elements should also spread most efficiently in organisms that undergo sexual reproduction (Hickey 1992). In eukaryotes, sex results in haploid genomes being mixed. This presents transposable elements with new targets for transposition and allows them to spread through a population faster. In bacteria and archaea, however, gene transfer is one-way and most reproduction is by fission. In these species, transposable elements in the main chromosome tend to be eliminated by selection or drift. Based on these observations it is logical to find that the transposable elements found in bacterial and archaea tend to reside on the circular extrachromosomal elements called **plasmids**, which can be transmitted from one cell to another.

Box 15.1 provides an overview of the variety of transposable elements that have been discovered to date.

Box 15.1 Categories of transposable elements

Transposable genetic elements are grouped into two broad classes, based on whether they move via an RNA or DNA intermediate sequence. These Class I and Class II elements come in an enormous array of sizes, copy numbers, and structurally related families. The literature on mobile genetic elements is vast and growing rapidly. Although we can only touch on this body of knowledge here, it is helpful to understand a few of the basic types of transposable elements.

Class I elements

Class I elements, also called **retrotransposons**, are the products of reverse transcription events. Work on the molecular mechanism of transposition has confirmed that movement of Class I elements occurs through a ribonucleic acid (RNA) intermediate (Cosineau et al. 2000). Transposition is also replicative, meaning that the original copy of the sequence is intact after the event.

The long interspersed elements (LINEs) are retrotransposons that contain the coding sequence for reverse transcriptase and are thought to catalyze their own transposition. In mammals, LINEs are typically 6–7 kb in length (Hutchison et al. 1989; Wichman et al. 1992).

Another important category of retrotransposons is distinguished by the presence of long terminal repeats (LTRs). LTRs are one of the hallmarks of retroviral genomes. When retroviruses insert themselves into a host DNA to initiate an infection, LTRs mark the insertion point. In corn, 10 different families of retrotransposons with LTRs have been identified, each of which is found in 10 to 30,000 copies per haploid genome (SanMiguel et al. 1996). The complete genome of baker's yeast, *Saccharomyces cerevisiae*, has been sequenced and found to contain 51 complete LTR-containing sequences called Ty elements, and 264 "naked" LTRs that lack the coding regions of normal retrotransposons. These empty LTRs are interpreted as transposition footprints, meaning that they are sequences left behind when Ty elements were somehow excised from the genome (see Boeke 1989; Goffeau et al. 1996).

Where did retrotransposons come from? One hypothesis is that LINEs and the LTR-containing retrotransposons evolved from retroviruses. Retrotransposons resemble retroviruses that have lost the coding sequences required to make capsule proteins. This hypothesis proposes that retrotransposons have adopted a novel evolutionary strategy. In contrast to retroviruses, which replicate in their host cell, move on to infect new cells, and eventually infect new host individuals of the same generation, retrotransposons replicate by infecting the germline and being passed along to subsequent generations. Instead of being transmitted horizontally—meaning, from host to host in the same generation—they replicate by being transmitted vertically to the next generation of hosts. Their transmission is much slower than that of conventional retroviruses, but retrotransposons also escape detection and destruction by the immune system.

The other important type of Class I element is called a retrosequence. Retrosequences do not contain the coding sequence for reverse transcriptase, but amplify via RNA intermediates that are reverse transcribed and inserted in the genome. The short interspersed elements (SINEs) of mammals are among the best-studied examples. SINEs are grouped into several different families, each of which is distinguished by its sequence homology with a different functional gene. The Alu family of sequences in primates, for example, is about 90% identical to the 7SL RNA gene, which is involved in transmembrane transport; other families of SINEs are homologous with various tRNA genes. SINEs are typically under 500 base pairs (bp) in length and lack the sequences necessary for translation of a transcribed RNA message. They are particularly abundant in mammals, and especially in the primates. The human genome, for example, contains over one million Alu elements (International Human Genome Sequencing Consortium 2001).

Where did SINEs and other retrosequences come from, and how are they replicated? In most SINE families it appears that only one or a very few master copies of the locus are actively transposing, and that the remainder represent inactive copies analogous to pseudogenes (see Shen et al. 1997). The mechanism of transposition in retrosequences is not known, however. We do not know how transcription of the master gene locus is regulated, where the reverse transcriptase comes from, or how insertion of the resulting DNA copy proceeds.

Class II elements

Class II transposable elements replicate via a DNA intermediate and are the dominant type of transposable genetic element in bacteria. Their transposition can be replicative, as in Class I elements, or conservative. In conservative transposition, the element is excised during the move so that copy number does not increase. The first Class II elements ever described were the insertion sequences, or IS elements, discovered in the bacterium *Escherichia coli*. When insertion sequences contain one or more coding sequences, they are called **transposons**. In addition to being inserted in to the main bacterial chromosome, however, transposons are commonly inserted into plasmids. Plasmids are loops of double-stranded DNA that replicate independently of the main chromosome. Copies of plasmids are readily transferred from one bacterial cell to another during a process called conjugation.

Transposons encode a protein, called a transposase, that catalyzes transposition. In some bacterial transposons, the coding region also codes for a protein that confers resistance to an antibiotic. As a result, plasmid-borne transposons have been responsible for the rapid evolution of drug resistance in disease-causing bacteria. Transposons that confer antibiotic resistance create a fitness advantage for its host. This type of transposable element is not parasitic.

Class II transposable elements are also found in eukaryotes. The *Ac* and *Ds* elements of corn, discovered by Barbara McClintock in the 1950s, belong to this group. These Class II sequences code for a transposase as well as other proteins. The P elements found in *Drosophila melanogaster* are another example. A typical fly genome contains 30-50 copies of these elements, which have insertion-sequence-like repeats at their ends and as many as three coding regions for protein products (Ajioka and Hartl 1989).

Table 15.1 summarizes data on the types and frequency of various classes of transposable elements in some eukaryotic species, based on genome sequencing. The variation in types of sequences present and in the overall presence of transposable elements is remarkable. Explaining this variation is one of the great challenges facing biologists interested in the evolution of genetic parasites.

Table 15.1 Presence of transposable elements in some eukaryotic genomes

The percentage of the genome made up of different types of transposable elements in selected eukaryotes. The entries in the "Total" row include types of transposable elements other than LINEs, SINEs, LTRs, and Class II sequences. Data from IHGSC 2001, ICGSC 2004, RGSCP 2004.

Type of Transposable Element	Human	Rat	Rice	*Arabidopsis*	Chicken	*Caenorhabditis*	*Drosophila*
LINE/SINE	33.4	30.2	1.18	0.5	6.5	0.4	0.7
LTR	8.1	9.0	14.8	4.8	1.3	0.0	1.5
Class II	2.8	0.8	13.0	5.1	0.8	5.3	0.7
Total	44.4	40.3	35.0	10.5	8.6	6.5	3.1

Selfish Genes and Levels of Selection

Research on transposable elements has delivered an important message: Genomes are not cohesive communities of sequences that contribute to the fitness of the individual. Instead, they are riddled with "selfish genes"—parasites that transmit themselves at their host's expense.

Selfish sequences illustrate a more general point: Natural selection is not limited to acting on individual organisms within populations. It can also act at the level of genes within organisms. In fact, evolution by natural selection can act at any level of organization where heritable variation and differential success occur.

Natural selection does not just act on individuals. It can act at any level where heritable variation and differential success occur.

Let's consider this point in more detail. When selection acts at the level of an individual in a population of individuals, the fitness of those individuals determines which alleles increase in frequency and which decline or disappear. If an allele influences the phenotype in a way that leads an average individual carrying that allele to have higher than average reproductive success, then that allele will increase in frequency; otherwise, it will decrease.

Selection can act at the level of the gene if the fitness of a particular sequence is at least partially independent of the fate of other sequences in the same genome. If transposable alleles vary in their ability to spread within genomes, and if some transposable alleles are more successful at spreading than others, then the more successful transposable alleles will increase in frequency in genomes over time. In this way, natural selection can act below the level of the individual.

There is no question that selection at the level of the gene is an important phenomenon. In terms of sheer numbers, transposable elements represent some of the most successful genes in the history of life.

Evolutionary Impact of Transposable Elements

Once biologists recognized the nature of transposable elements and had characterized their diversity and distribution, they turned to understanding their dynamics. Let's consider three questions, in turn. If transposable elements are parasitizing hosts, does natural selection limit their spread? Do host genomes have mechanisms to counter them? And are transposable elements ever beneficial to their hosts?

Defending against the Spread of Transposable Elements

When a transposable element inserts into the coding region of a gene, the mutation that results should quickly be eliminated by natural selection. Insertion events in coding regions are highly deleterious mutations. Based on this observation, biologists predicted that purifying selection (see Chapter 7) should limit transposition events to portions of the genome that do not contain large numbers of important coding sequences. Genome sequence data confirm that transposable elements are found in highest density in heterochromatic regions near the centromere. Heterochromatin contains few expressed genes and remains in a highly condensed state at all times. These data support the hypothesis that transposable elements are frequently subject to purifying selection after insertion.

Do host genomes have mechanisms to keep transposition from occurring in the first place? Rachel Waugh O'Neill and coworkers (1998) suggest that the answer may be yes. These biologists conducted a study inspired by observation that eukaryotes contain enzymes that selectively add methyl ($—CH_3$) groups to certain genes in the genome. This process is called methylation and is associated with lack of expression. Stated another way, methylation is a mechanism to shut down certain genes. Waugh O'Neill and colleagues hypothesized that organisms add methyl groups to DNA as a way to thwart mobile genetic elements (Bester and Tycko 1996; Yoder et al. 1997). To test the prediction that there should be an association between DNA methylation and the spread of transposable elements, the researchers analyzed the chromosomes found in the hybrid offspring of a mating between a tammar wallaby and a swamp wallaby. For unknown reasons, the DNA of the hybrid individual was virtually unmethylated. The biologists found that in many of this individuals' chromosomes, a retrotransposon called KERV-1 virtually exploded in copy number (see Figure 15.1). According to Waugh O'Neill and colleagues, this correlation is strong support for the hypothesis that methylation protects host DNA from insertion by parasites. Work on this topic continues, however. It has yet to be firmly established that methylation functions as adaptation for controlling transposable elements.

Researchers are currently exploring whether genomes contain defense mechanisms that limit the spread of transposable elements.

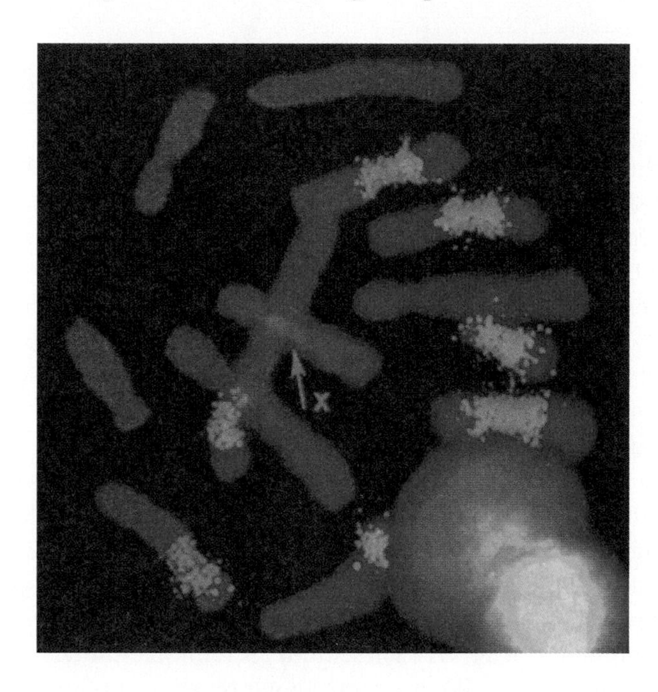

Figure 15.1 When DNA is unmethylated, transposable elements explode in number To produce this photograph, Waugh O'Neill and co-workers labeled single-stranded DNA from the KERV-1 transposable element with a fluorescent molecule, then allowed the labeled DNA to hybridize to KERV-1 sequences in the chromosomes of a wallaby hybrid individual. The pink dots indicate the location of KERV-1 elements. (Note that they cluster near the centromeres of certain chromosomes.) No hybridization was observed in the parental wallaby species. From Waugh O'Neill et al. (1998).

Positive Impacts of Transposable Elements?

For decades, the only example of a positive fitness benefit provided by transposable elements was the trait of antibiotic resistance, conferred to certain species of bacteria by plasmid-borne transposons (see Box 15.1). But then work by John Moran and colleagues (1999) suggested that transposition events in eukaryotes may occasionally result in mutations that confer a fitness benefit. This conclusion is based on experiments with the LINE elements found in humans. To analyze how these genes move from one location to another, Moran and coworkers used recombinant DNA techniques to attach a marker gene to LINE-1 sequences. They introduced the engineered sequence into human cells growing in culture, allowed the parasitic sequences to insert themselves into the genome, and sequenced the LINE-1 elements that succeeded in transposing to new locations in the genome. In several cases, the sequence data showed that the mobile elements had carried a chunk of host DNA along with it and the marker gene during transposition. In essence, the LINEs had duplicated segments of the host DNA

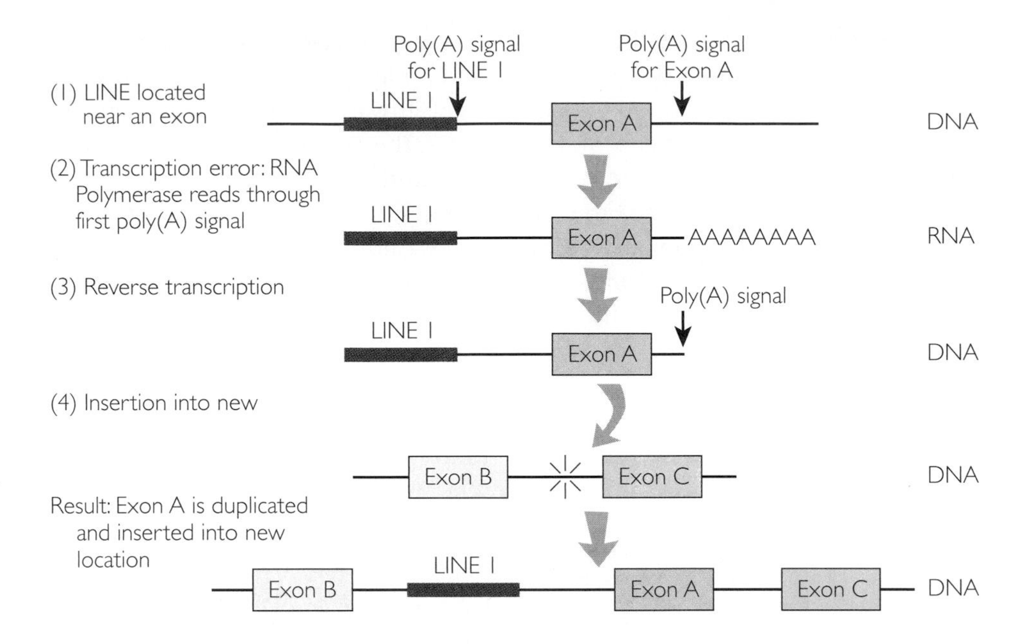

Figure 15.2 Exon shuffling via transposition events Based on the series of events diagrammed here, transposition by LINE elements can result in exons or regulatory sequences being moved to new locations in the genome. This phenomenon is known as exon shuffling. The experiment by Moran et al. (1999), described in the text, shows that each of the steps illustrated here can actually occur.

and moved them to new locations. Figure 15.2 illustrates how this happened. The diagram also shows that, if the transposed host DNA segment happens to contain an exon or regulatory sequences, the transposition results in a novel gene.

Moran and coworkers contend that these types of transposition events are important in the evolution of genomes. More specifically, they furnish a mechanism for a hypothesis called **exon shuffling**. An exon is a portion of a protein-coding gene; exon shuffling is the idea that functional portions of proteins can be recombined in novel ways.

Transposable elements may have a positive impact on fitness if they move exons with them, and put them in locations where the exons become parts of valuable new genes.

A causal connection between transposition and exon shuffling was supported by a recent analysis of the rice genome by Ning Jiang and coworkers (2004). After searching the rice database for Class II transposable elements called MULEs, Jiang et al. found 3,000 copies that contained fragments from cellular genes, including some that contained fragments of several cellular genes that had been spliced together. Some of these fragments appear to be expressed and produce working proteins or RNAs. These findings reinforce the view that transposition events can create advantageous mutations by mixing and matching fragments of genes and producing novel combinations that benefit the organism.

In a similar vein, work by Alka Agrawal and colleagues (1998) suggest that a key gene in the vertebrae immune system was originally part of a transposable element. As Figure 15.3a shows, the proteins that serve as antigen-recognition sites on the surface of immune systems cells are encoded by three gene segments. As immune systems cells develop in an embryo, a series of reactions take place that result in portions of the V (variable), D (diversity), and J (joining) gene segments being excised and recombined (see Figure 15.3b). These reactions are catalyzed by proteins called RAG1 and RAG2. Using the experimental design diagrammed in Figure 15.3c, Agrawal et al. showed that RAG1 and RAG2 can also catalyze the transposition of gene constructs that are unrelated to the V, D, and J regions. The reaction mechanism that is involved turns out to be identical to the chemical events that take place during movement by transposable elements (see Zhao et al. 2004). To make sense of this result, the researchers propose that the RAG proteins are transposases homologous to those found in present-day transposable elements.

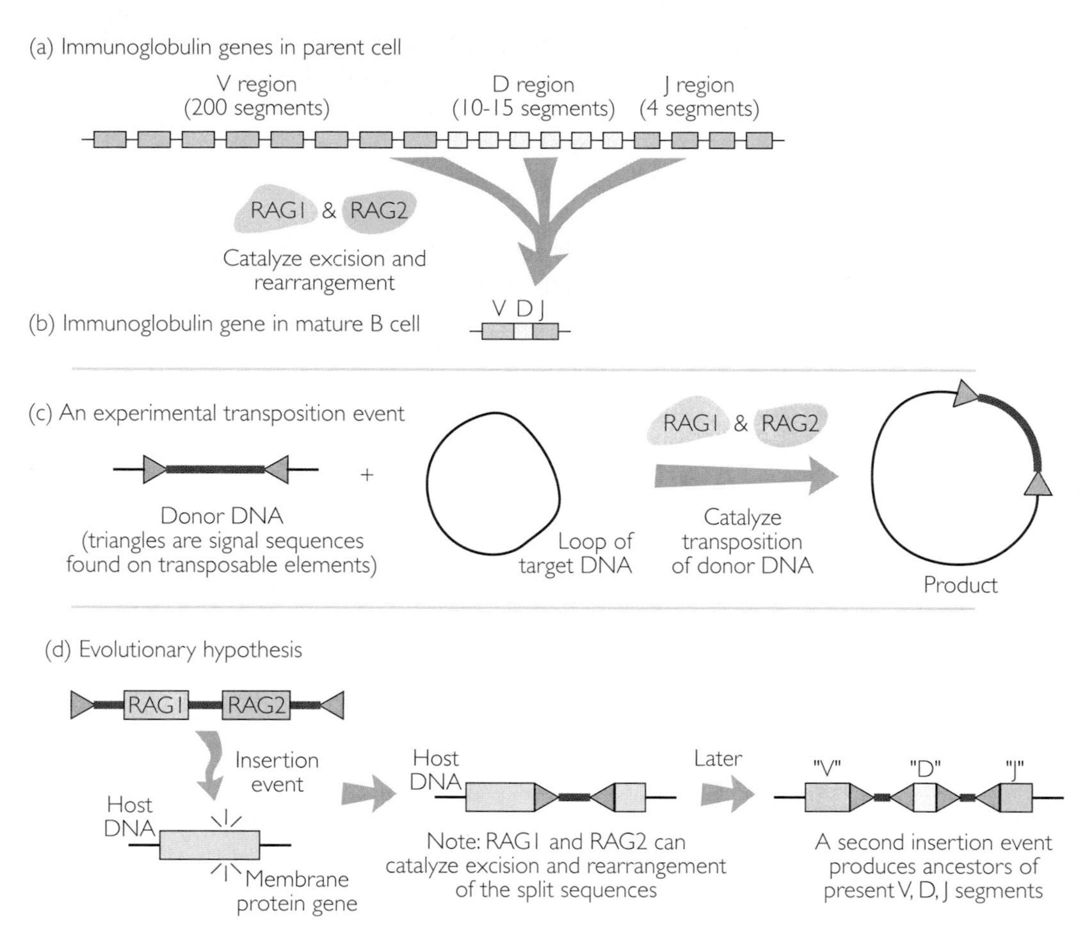

Figure 15.3 Did the vertebrate immunoglobulin genes originate in a transposition event? (a) As a human embryo develops, the cells that serve as precursors to immune system cells have immunoglobulin genes with many different V, D, and J regions. (b) As individual immune system cells mature, RAG1 and RAG2 catalyze reactions that results in a single V, D, and J segment being combined and the others being excised. (c) The experiment by Agrawal et al. (1998) showed that RAG1 and RAG2 can catalyze transposition events. (d) According to the evolutionary hypothesis outlined here, insertion events by transposable elements containing the RAG1 and RAG2 genes created the basic structure of the immunoglobulin genes. Gene duplication events could later produce the variety of V, D, and J segments observed today.

The implications of this work are important. Agrawal and associates hypothesize that the V(D)J excision and rearrangement reactions observed in today's vertebrates are possible because of an insertion event by a transposable element several hundred million years ago. According to the hypothesis outlined in Figure 15.3d, a transposable element bearing RAG1 and RAG2 inserted into a gene for a membrane protein early in the evolution of vertebrates. The transposase could catalyze the recombination of the resulting receptor gene segments into a functional gene. Gene duplication events later in evolution could have expanded the membrane receptor locus and resulted in the extensive V, D, and J regions observed today, which are still recombined by RAG1 and RAG2-enzymes that trace their ancestry back to the original transposase. The key take-home message here is that an important adaptation at the level of individuals may derive from the activity of selfish genomic parasites.

Key proteins in the vertebrate immune system may be the descendants of proteins in transposable elements.

The other well-documented example of a "domesticated" transposable element—meaning, one that serves a positive function in the cell—is the reverse transcriptase called telomerase that copies the ends of linear chromosomes in eukaryotes (see Pardue and DeBaryshe 2003; Abad et al. 2004). Several other candidates for positive function are under investigation, and it is likely that further research will show that additional sequences associated with transposition increase the fitness of the host genome and qualify as adaptations (e.g., Cowan et al. 2005; Gao and Voytas 2005). Even though most transposable elements function as genomic parasites and most transposition events result in deleterious mutations, it is increasingly clear that at least some transposition events result in important new genes or other changes that have a positive impact on the fitness of organisms (Britten 2006).

15.2 Lateral Gene Transfer

If studies of transposable elements have one overriding message, it is this: Genomes are dynamic. Studies of chromosome inversions and gene and genome duplication, highlighted in Chapter 5, reinforce this theme. Transposable elements move themselves and other genes around within genomes, chromosome segments can flip or move to new positions, segments of genes can be added or subtracted from chromosomes during crossing-over, and entire chromosome sets can double.

But the most dramatic example of dynamism in genomes may be the phenomenon known as **lateral gene transfer** (**LGT**), also known as **horizontal gene transfer**. When LGT occurs, one or several genes move from one species to another. The lateral or horizontal in the name alludes to the transmission of alleles across species and within the same generation, as opposed to vertical transmission, in which alleles are transmitted within species and from one generation to the next. In some or even most cases, the species involved in lateral gene transfer are not at all closely related. LGT is the acquisition of foreign DNA.

When LGT occurs, genes move between species.

As an example, consider the phylogenetic tree in Figure 15.4. The relationships shown are based on an analysis of the coding sequences for an enzyme called HMGCoA reductase (Doolittle and Logsdon 1998). Now look at the location on the tree for the HMGCoA reductase gene from *Archaeoglobus fulgidus*, highlighted in tan. *A. fulgidus* is unquestionably a member of the domain Archaea: it has a small-subunit rRNA gene that it is clearly archaean, machinery for transcription and translation that are typical of archaeans, and lipids in its plasma membrane that are diagnostic of archaeans. But its gene for HMGCoA reductase is closely related to homologs in bacteria—much more so, in fact, than to homologs of HMGCoA in other archaeans. How can this be?

The most likely answer is that the HMGCoA reductase gene in *Archaeoglobus fulgidus* is bacterial. In other words, the data are consistent with the hypothesis that an ancestor of *A. fulgidus* lost its native archaean HMGCoA reductase gene and replaced it with a gene picked up from a bacterium. This is an example of LGT.

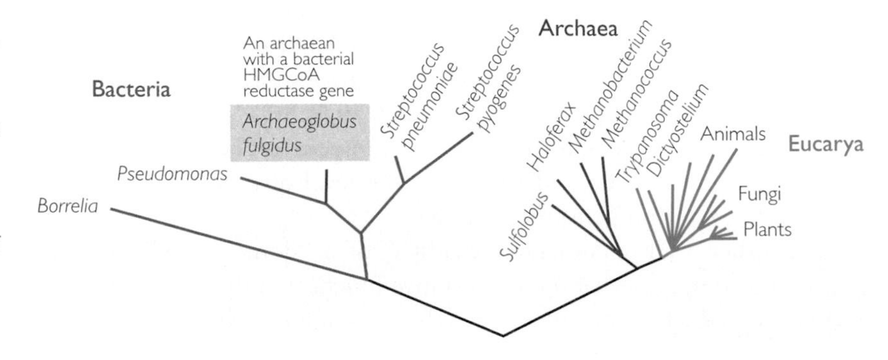

Figure 15.4 **Phylogenetic evidence for lateral transfer of the HMGCoA reductase gene** *Archaeoglobus fulgidus* is an archaean, yet its gene for HMGCoA reductase branches from within the Bacteria. The most reasonable explanation is that *Archaeoglobus* has lost its native archaean version of the HMGCoA reductase gene and replaced it with a version of the gene acquired from a bacterium. Modified from Doolittle and Logsdon (1998), after Doolittle (2000).

Mechanisms of Gene Transfer

For LGT to occur, a gene or gene fragment from one organism has to enter a cell from a different species and become integrated into its chromosome. If the gene is going to increase in the population due to natural selection, it must also be expressed and contribute a product that increases the individual's fitness. How does all this get started?

Genes can move from one species to another through several different mechanisms:

- Viruses are responsible for the first gene-transfer process ever described. In the course of infecting a host cell, viruses that parasitize bacteria insert their DNA directly into the host cell chromosome via specialized sections of DNA called insertion sequences (ISs). When copies of the viral DNA leave the host chromosome to make a new generation of virus particles, they can pick up DNA segments from the infected host. When the next generation of viruses leaves that host, some of the particles carry the bacterial sequences to the next host they parasitize. In this way, viruses can transfer DNA from one species into the chromosome of another species. This process creates LGT and is called **transduction**. Even though it was discovered in the early 1950s, it is still unclear exactly how the bacterial host cell genes initially become attached to the viral DNA.
- Plasmids are small loops of DNA found in bacteria and archaea that replicate independently of the main chromosome. Copies of plasmids can move from one bacterial cell to another via a process called **conjugation**; it is possible that conjugation occasionally occurs between members of different bacterial or archaeal species and results in LGT.
- At least some bacteria and archaea can take nucleic acids up directly from the environment (Meibom et al. 2005). If these foreign genes are not digested and used as nutrients, they may become incorporated into the cell's chromosome. This process is known as **transformation**.
- In eukaryotes, the major cause of LGT is **endosymbiosis**. When endosymbiosis occurs, a cell from a different species begins living inside a host cell (Margulis 1970; 1993). Many contemporary examples of endosymbiosis exist (e.g., Nakabachi et al. 2005; Okamoto and Inouye 2005; Partida-Martinez and Hartweck 2005) and there is strong evidence, reviewed later in the chapter, that the eukaryotic organelles called mitochondria and chloroplasts originated in endosymbioses involving bacteria. Stated another way, mitochondria and chloroplasts started out as bacteria living symbiotically inside cells that are the ancestors of today's eukaryotes. Because bacterial cells typically have about 5,000 genes, the origin of mitochondria and chloroplasts represent the largest LGT events in the history of life and possibly the most important, given their role in the origin and diversification of an entirely new domain of life: the Eukarya.

Genes can move between species via viruses, plasmids, or cells that take up residence inside a host. Genes can also be taken up directly, like food.

Diagnosing LGT

Given the array of processes that can cause lateral gene transfer, it should not be surprising to find that it has occurred numerous times during the history of life. The question is: How do researchers know it when they see it?

Confirming that LGT is responsible for the presence of a particular gene can be difficult (Stanhope et al. 2001; Eisen and Fraser 2003), but researchers have an array of techniques at their disposal. One approach relies on documenting the relative abundance of guanine-cytosine (GC) versus adenine-thymine (AT) pairs in a genome. Because overall GC content varies widely among organisms, a gene with an anomalously high or low proportion of GC pairs can indicate that it originated from a different species. Similarly, researchers routinely document which codons are used to specify the same amino acid in various parts of the genome. If a particular gene contains codons that are not commonly used elsewhere in the genome, it suggests that the gene may have originated in LGT.

Scanning a genome for segments identical to the transposases found in transposable elements or the insertion sequences of viruses is another way to infer that a gene arrived via LGT. If a foreign gene was delivered via a transposable element that was transferred between species or when an invading virus integrated its genome into a host cell chromosome, then it is likely that at least part of the original transposase or viral insertion sequence is still associated with the inserted gene.

Researchers suspect that a gene originated in LGT if it has an unusual GC content or codon use, or if the gene sequence is extremely similar to homologous genes in other species.

Doolittle and Logsdon's technique for inferring the lateral transfer of the gene for HMGCoA reductase—constructing a phylogenetic tree for the gene in question and finding that it comes out in an anomalous location—is another classic way to identify LGT. For example, phylogenetic analyses were the key evidence linking today's mitochondria and chloroplasts with endosymbiosis. When researchers sequenced genes found in the DNA inside mitochondria and chloroplasts and compared them to homologs in a wide array of organisms, they found that mitochondrial genes are most closely related to homologs in a bacterial clade called the α-proteobacteria (Yang et al. 1985). Chloroplast genes are most closely related to the bacterial group called cyanobacteria (Giovannoni et al. 1988). In a phylogenetic sense, mitochondria and chloroplasts are bacteria (see Figure 15.5). Anomalous placement of particular genes on a phylogenetic tree can furnish strong support for LGT. The most convincing evidence of all is when analyses of GC content, codon usage, transpose-sequences or ISs, and phylogenetic analyses are all consistent with the hypothesis of lateral gene transfer.

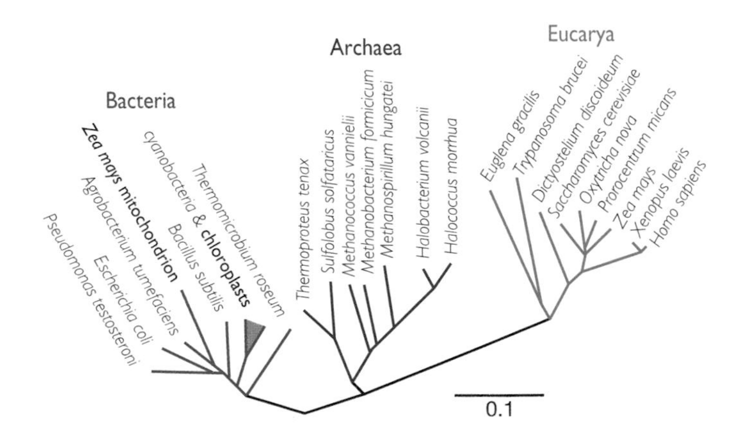

Figure 15.5 Phylogenetic evidence for lateral transfer of genes in mitochondrial DNA and chloroplast DNA This tree is based on sequences of small-subunit rRNA genes. The mitochondrial DNA sequences are from corn (*Zea mays*) and the chloroplast DNA sequences are represented by a variety of species, all of which branch within the region indicated. From Giovannoni et al. (1988).

Lateral Gene Transfer in Eukaryotes: Evolution of Mitochondria and Chloroplasts

Although the phylogeny of mitochondrial DNA and of chloroplast DNA furnished the strongest data in favor of the endosymbiosis hypothesis, several other lines of evidence support the hypothesis that mitochondria and chloroplasts originated in endosymbiosis. For example, the organelles are also about the size of a bacterium and have a bacteria-like circular chromosome with supercoiled DNA but without chromosomal proteins. In addition, they have a double membrane—presumably, one derived from the original bacterium and one de-

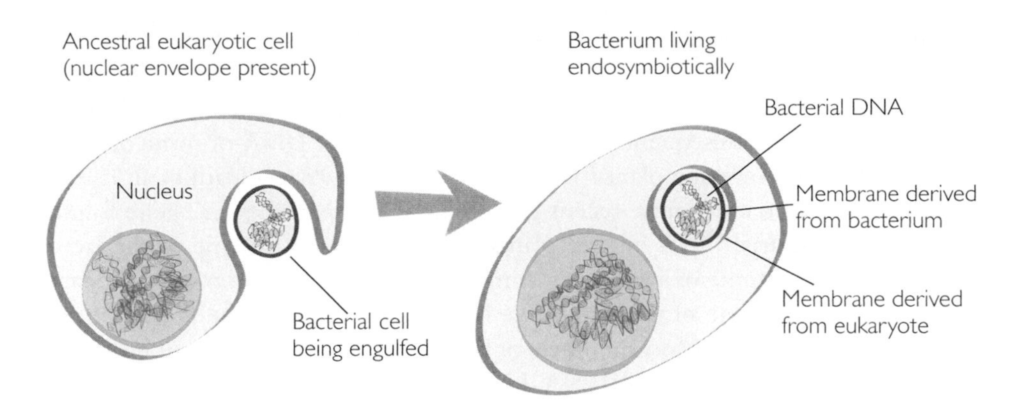

Figure 15.6 Endosymbiosis and the origin of mitochondria and chloroplasts If an ancestral eukaryote engulfed a bacterial cell which then began living endosymbiotically, the bacterial cell would have a double membrane and its own DNA.

rived from the plasma membrane from the ancient eukaryote that engulfed the bacterial cell (Figure 15.6).

Present-day mitochondria and chloroplasts have just a tiny fraction of the genetic information present in a bacterium, however (Gray 1992). Chloroplast DNA, for example, typically contains slightly over 100 genes, most of which code for proteins required to express the chloroplast genome or perform photosynthesis (Wolfe et al. 1991; Clegg et al. 1994). The mitochondrial DNA of most eukaryotes codes for rRNAs, tRNAs, and ribosomal proteins, along with proteins involved in cellular respiration. The question is: What happened to the rest of the bacterial genes that arrived via LGT?

Evolution of Organelle Genomes

If an ancient eukaryote gained about 4,300 new genes via LGT, and if today's mitochondria only contain a few hundred genes, then more than 4,000 bacterial genes ended up in one of two places: They were either lost completely or transferred to the nucleus. A search of the complete genome sequences of two eukaryotes—humans and yeast—suggests that about 630 α-proteobacterial genes exist in these species (Gabaldon and Huynen 2003). Based on these results, it is likely that a few hundred of the laterally transferred genes still reside inside the mitochondrion, another 400–500 have been transferred to the nucleus, and more than 3,500 were deleted entirely.

Mitochondria and chloroplasts originated as bacterial cells that were living inside eukaryotic cells.

To understand the evidence that organelle-to-nucleus transfers have taken place, consider the genes that encode what may be the most abundant protein in nature, ribulose bisphosphate carboxylase (RuBPCase). This enzyme catalyzes the fixation of CO_2 during the Calvin-Benson cycle, which is the central pathway in the light-independent reactions of photosynthesis. RuBPCase is made up of two subunits. The gene for the protein's small subunit is found in the nuclear genome, while the gene for the large subunit is part of the chloroplast genome (Gillham et al. 1985). Analogous evidence for gene transfers exists in mitochondria. The organelle's ribosomes, for example, are composed of rRNAs encoded by mitochondrial DNA and proteins encoded by nuclear DNA.

Because the gene content of mitochondria and chloroplasts is similar among most of today's eukaryotes, it is logical to conclude that many gene transfers and losses occurred early in the history of endosymbiosis (Gillham et al. 1985; Clegg et al. 1994). Recent gene transfers have also been documented, however. For example, the gene *tufA*, which codes for a translation factor active only in the chloroplast, is found in the chloroplast DNA of green algae but in the nuclear genome of the flowering plant *Arabidopsis*. This observation suggests that *tufA*

Once the ancestors of today's mitochondria and chloroplasts took up residence inside a eukaryote, most of their genes were lost or transferred to the nucleus.

was transferred to the nucleus after the green algae and land plants diverged. But because copies of the gene exist in both genomes of some green algae, it is more likely that the gene was duplicated and transferred to the nucleus early in plant evolution and then subsequently lost from the chloroplast DNA of some derived lineages, like the flowering plants (Baldauf and Palmer 1990; Baldauf et al. 1990).

Evidence for an even more recent gene transfer involves the *cox2* gene found in the mitochondrial DNA of land plants. This gene codes for one of the large subunits of cytochrome oxidase, a key component of the electron transport chain in the inner membrane of mitochondria. In most plants, this gene is part of the mitochondrial genome. Most members of the pea family have a copy in both nuclear DNA and mitochondrial DNA, however, and in mung bean and cowpea the only copy is located in the nucleus. In these species, the structure of the gene closely resembles the structure of an edited mRNA transcript. Because RNAs can be reverse transcribed to DNA, these facts suggest that gene transfer from the mitochondrion to the nucleus took place recently via the reverse transcription of an edited mRNA intermediate (Nugent and Palmer 1991; Covello and Gray 1992). This mechanism is reminiscent of the way that the transposable elements called SINEs are inserted into genomes (Figure 15.7). In effect, the massive LGT event that began with endosymbiosis has been followed by slower, and still continuing, lateral transfers of genes from mitochondria and chloroplasts to the chromosomes in the nucleus.

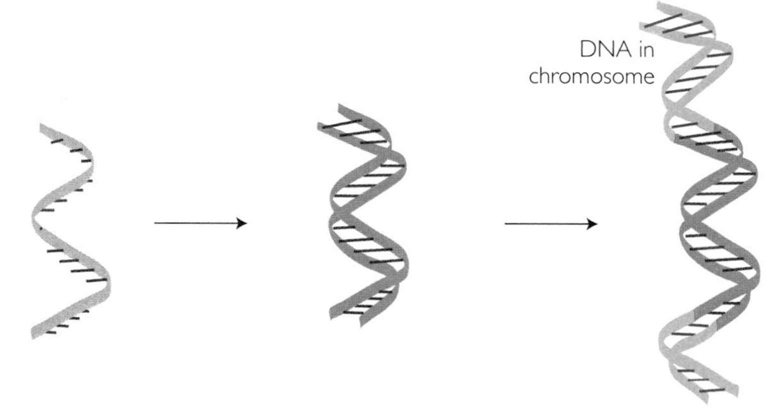
DNA in chromosome
(1) A mature mRNA (no introns)
(2) Reverse transcriptase catalyses synthesis of a double-stranded DNA complementary to the mRNA
(3) Via unknown mechanisms, the complementary DNA becomes inserted into the main chromosome.

Figure 15.7 SINEs originate from reverse-transcribed mRNAs SINEs are transposable elements that arise from mature mRNAs. They are reverse-transcribed into a complementary DNA (cDNA) and then inserted into the main chromosome.

Many other questions remain about the evolution of these cohabiting genomes. Why have some genes, but not all, been transferred from organelles to the nucleus? Is there a selective advantage to having certain genes located in each genome, or were the movements merely chance events? Is reverse transcription of mRNAs the usual mechanism of transfer, or are other processes involved? Work on the relationship between nuclear and organelle genomes continues.

Secondary Endosymbiosis

The acquisitions of mitochondria and chloroplasts were seminal events in the evolution of eukaryotes, but LGT via endosymbiosis has continued to occur throughout the history of the lineage. In an array of eukaryotes, genes have been acquired through LGT events known as **secondary endosymbioses**.

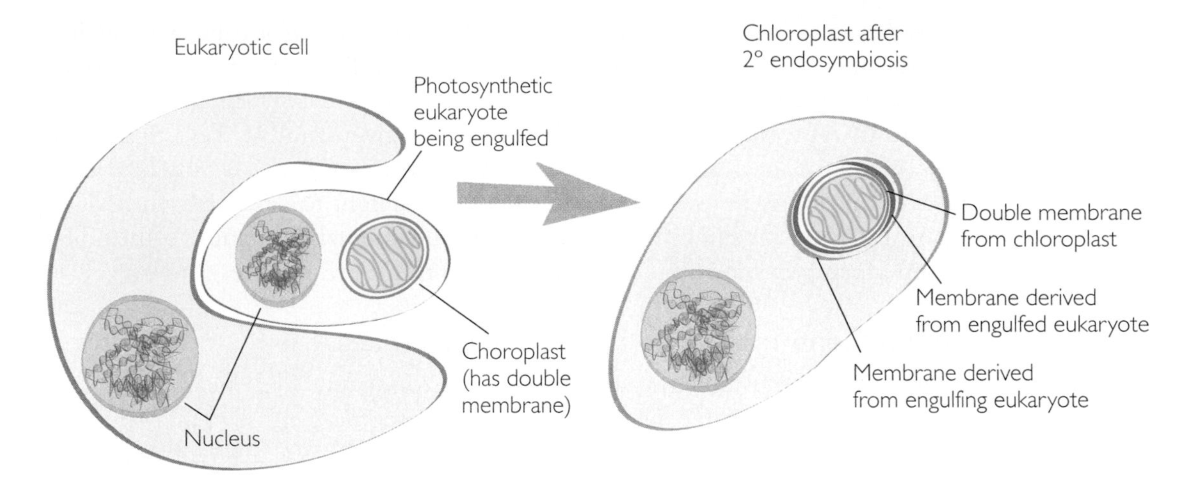

Figure 15.8 Secondary endosymbiosis If a eukaryote engulfs a photosynthetic eukaryote containing a chloroplast, and if the chloroplast is retained as an endosymbiont while the rest of the engulfed cell is ingested, then the resulting chloroplast would have four membranes.

Figure 15.8 shows how secondary endosymbiosis occurs. If a eukaryote containing a chloroplast is engulfed by a eukaryote that lacks a chloroplast, and if the chloroplast is subsequently retained as an active, functional organelle, it has been acquired via secondary endosymbiosis. Note that organelles acquired in this way should have four membranes. *Cryptomonas F*, for example, is an alga whose chloroplasts have two pairs of envelope membranes, for a total of four. Inside the inner membrane pair, the *Cryptomonas* chloroplast has a typical circular chloroplast chromosome; between the inner and outer membrane pairs, it also has a small nucleus-like organelle, called the nucleomorph. The nucleomorph makes a functional ribosome, which remains between the membrane pairs. Susan Douglas and colleagues (1991) sequenced the small-subunit ribosomal RNA from both the nucleomorph and the nucleus and then placed these sequences on a phylogeny of the eukaryotes. Both rRNAs are clearly of eukaryotic origin, but they are not closely related to each other. The implication is that the outer chloroplast membrane pair and the nucleomorph are vestiges of a eukaryotic ancestor. This ancestor, itself host to a chloroplast, became the secondary endosymbiont of another eukaryotic host.

When secondary endosymbiosis occurs, a cell gains a chloroplast by engulfing a cell that already contains a chloroplast.

The Adaptive Significance of LGT in Eukaryotes

The take-home message of research on organelle genomes is that eukaryotes are chimeras: a blend of organisms akin to the mythical beast with a lion's head, a goat's body, and a snake's tail. All eukaryotes have either a fully functional mitochondrion or a vestigial one (e.g., Williams et al. 2002; Tovar et al. 2003), suggesting that endosymbiosis occurred early in the history of the lineage. The phylogeny of chloroplast genes suggests that the original endosymbiosis event took place in an ancestor of today's red algae, green algae, and land plants. All other photosynthetic eukaryotes—including the brown algae, euglenids, diatoms, dinoflagellates, and golden-brown algae—acquired their photosynthetic machinery via secondary endosymbiosis or even tertiary endosymbiosis (Falkowski et al. 2004; Keeling 2004; Yoon et al. 2005).

Lateral gene transfer has played a key role in the history of the Eukarya for a simple reason: The information acquired through LGT gave eukaryotes new metabolic capabilities, which allowed them to use existing resources more efficiently or

Eukaryotes can be interpreted as "cut-and-paste jobs." They are a lineage that diversified after cells combined via endosymbiosis.

to obtain entirely new types of resources. The acquisition of a mitochondrion gave the ancestral eukaryote the ability to perform cellular respiration with oxygen as the final electron acceptor—an innovation that vastly increased the amount of ATP that could be produced from a given molecule of sugar, fatty acids, or amino acids. The acquisition of a chloroplast gave an array of eukaryotic lineages the ability to use the kinetic energy in sunlight to drive the synthesis of sugars. Eukaryotes gained two of their most fundamental adaptations—the ability to perform aerobic respiration and photosynthesis—via LGT.

Lateral Gene Transfer in Bacteria and Archaea

When evolutionary biologists compared the history of LGT in eukaryotes versus bacteria and archaea, they discovered a major contrast as well as a major similarity. The contrast is that instead of occurring in just a few events that transferred thousands of genes at a time, LGT in bacteria and archaea has occurred many times but transferred a small number of genes each time. The similarity lies in the adaptive significance of lateral gene transfer: As in eukaryotes, LGT has given a wide array of bacterial and archaeal lineages new metabolic capabilities.

When researchers analyze the make-up of bacterial and archaeal genomes, they almost always find evidence of lateral gene transfer. The first bacterial genome to be completely sequenced was the laboratory strain of *Escherichia coli* (Blattner et al. 1997). This organism has an estimated 4,288 genes, of which 755 or 18% are thought to have been acquired via LGT (Lawrence and Ochman 1998). When Nicole Perna and colleagues (2001) sequenced the genome of the closely related strain O157:H7 of *E. coli,* they found that it has an estimated 5,361 genes. Of these, 1,387 are not present in K-12. In turn, K-12 has 528 genes not present in O157:H7. Many or most of these differences in gene content are thought to be due to lateral transfer.

What types of genes have been transferred between more distantly related species of bacteria and archaea? A few examples will illustrate a general conclusion: Most laterally transferred genes that have been retained in bacteria and archaea provide information for novel types of metabolism or other adaptations to specific environments.

- The archaean *Thermoplasma acidophilum* lives in environments where temperatures routinely approach 60°C and pH is as low as 2. About 17% of its genome is extremely similar to genes found in the distantly related archaean *Sulfolobus solfataricus*, which thrives in the same types of habitats. Most of the shared genes code for proteins involved in the transport or processing of nutrients. To explain these observations, Ruepp et al. (2000) hypothesize that genes adapted to the extreme environment have been swapped between species. In this way, LGT has been responsible for convergent evolution in habitat use.
- In the cyanobacterium *Synechococcus,* which lives in the extremely nutrient-poor habitats of the open ocean, genes that arrived via LGT are involved in transport of nutrients across the plasma membrane and in obtaining nitrogen atoms from an array of source ions and molecules (Palenik et al. 2003). In addition, the cells have an unusual and poorly understood swimming mechanism that does not involve flagella. At least some of the genes involved in motility arrived via LGT. Thus, two key attributes of these *Synechococcus*—their ability to take up rare nutrients and swim—appear to have been gained via LGT.
- The bacterium *Ralstonia solanacearum* lives in soil and infects the xylem tissue of plants, causing devastating diseases in potatoes, bananas, plaintains, and a

wide range of other species. The team that sequenced its genome found so many genes traced to LGT that they refer to the genome as a mosaic (Salanoutbat et al. 2002). Several of the foreign genes code for proteins linked to pathogenicity, or the ability to cause disease. These molecules include membrane transport proteins responsible for secreting toxins into the host. This result suggests that this bacterium's ability to feed off an array of plant species depends in part on genes acquired through lateral gene transfer.

The general conclusion from a wide array of similar studies is that bacteria and archaeal species have a "core genome" that is rarely affected by lateral transfer, consisting of genes involved in information-processing events like DNA replication, gene transcription, and protein synthesis. In contrast, the enzymes and membrane proteins involved in metabolizing certain nutrients are frequently transferred between species (Coleman et al. 2006). The take-home message is clear: Genes that allow bacteria and archaea to adapt to certain habitats are subject to frequent LGT.

LGT is common in bacteria and archaea. The laterally transferred genes usually code for proteins needed to exploit new food resources or occupy new habitats.

Let's take a closer look at the general phenomenon of how specific genes or sets of genes increase fitness in specific environments. In particular, what are genome-sequencing data revealing about adaptation at the level of the gene?

15.3 The Molecular Basis of Adaptation

Genome sequences from bacteria, archaea, and eukaryotes have been available for less than a decade, so it is not surprising that research on how gene content, gene organization, and gene expression affect adaptation is in its infancy. Most of the data that have accumulated to date are descriptive and correlative in nature. Still, they are fascinating. For the first time in history, biologists can examine the complete catalog of genes in an organism and ask questions about how those genes affect fitness.

Most genome sequencing projects to date have focused on species that have a direct impact on humans. Genome sequencers have been particularly interested in species that make us or our crop plants sick. Let's begin with some general observations about the nature of parasite genomes before going on to study adaptation at the level of the gene in free-living organisms.

Patterns in Parasite Genomes

Complete genome sequences are now available from dozens of bacteria and eukaryotes that parasitize hosts and cause disease (see Raskin et al. 2006). The first observation that jumps out from these data sets is that parasites have extremely small genomes compared to free-living organisms (Figure 15.9). As an example,

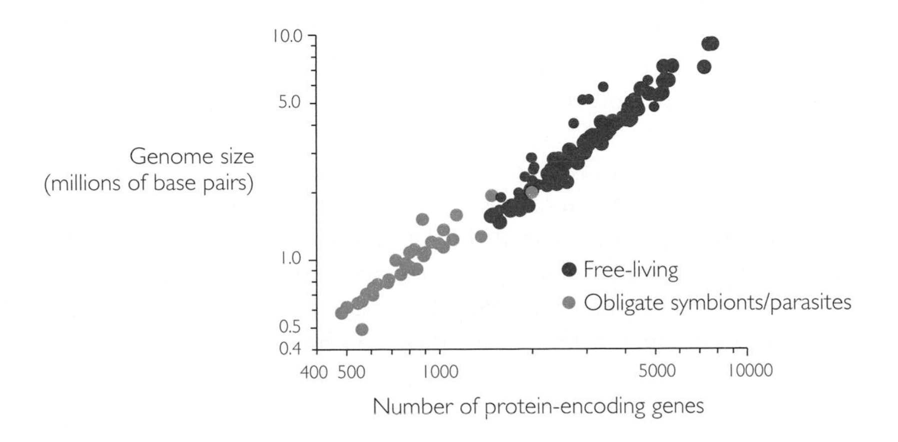

Figure 15.9 Parasites have small genomes This graph plots the number of protein-encoding genes versus total genome size. Each datapoint represents a species or distinctive population. Note the log-log scale.

consider the eukaryote *Cryptosporidium parvum,* which lives inside intestinal cells in humans and other mammals. Healthy humans that become infected with *C. parvum* get diarrhea that resolves in a few days, but AIDS patients and other immunocompromised individuals may die (Abrahamson et al. 2004). *Cryptosporidium parvum* lacks the chloroplast-like organelle found in close relatives that are not parasitic, and it has a vestigial mitochondrion that lacks DNA. As a result, it cannot perform cellular respiration and relies exclusively on glycolysis to make ATP and stay alive. In addition, its nuclear genome lacks the genes required to synthesize nucleotides or amino acids. In sum, this organism has lost many or most of the basic genes required for normal metabolism. A relative that parasitizes only humans has similar characteristics (see Xu et al. 2004).

To interpret extreme cases of gene loss like this, biologists sometimes employ the quip, "Use it or lose it." The logic here is that parasites gain the nutrients they need, including nucleotides and amino acids, directly from their hosts. Instead of synthesizing their own compounds, parasites simply absorb them from their surroundings. If a parasite has a gene for an enzyme that is active in pathways for synthesizing nucleotides or amino acids, the gene product is likely to be unnecessary. When a mutation removes an unneeded gene, natural selection will favor the mutant individuals because they no longer have to commit resources to copying and expressing the superfluous sequence. In this way, selection quickly reduces genome size in organisms that adopt a parasitic lifestyle. In many parasites, gene loss has been massive.

Although the genomes of parasites tend to be small, they code for proteins that allow the parasite to enter or stick to a host and acquire nutrients from it.

The second major observation about parasite genomes is that they contain "virulence genes." These are sequences that code for proteins required by the parasitic mode of life. In the bacterium *Erwinia carotovora*, genome sequencing identified an array of genes for membrane proteins involved in secreting substances into a host, along with several compounds that are toxic to its plant hosts. To test the hypothesis that these genes were required for successful parasitism, K.S. Bell and associates (2004) created individuals with mutations that incapacitated one of the putative virulence genes. The logic here is simple: If a gene is hypothesized to have a particular function, then you should not see that function when the gene is knocked out. As the data in Figure 15.10 show, bacteria that lacked these genes were much less virulent than normal bacteria. This is strong evidence that the bacteria require the genes to successfully parasitize a host.

Another research strategy for locating virulence genes is to compare the genome sequence of a parasitic bacterium to a closely related non-parasitic bacterium. Studies like these have implicated a variety of genes in virulence, including sequences involved in adhesion (so the parasite can stick to host cells), toxin production, enzymes that break down host cell walls or proteins, and membrane transporters that secrete toxins or allow the parasitic cell to rid itself of antibiotics (Perna et al. 2001; Salanoutbat et al. 2002; Dean et al. 2005; Feil et al. 2005; Kuroda et al. 2005).

In a similar vein, comparisons of drug-resistance and drug-susceptible bacterial strains have shown that in most cases, genes that confer drug resistance are carried on plasmids (e.g., Holden et al. 2004). This observation explains why drug resistance has been passed among disease-causing bacterial lineages so efficiently over the past several decades. The antibiotic resistance genes are moving rapidly among lineages via lateral gene transfer, mediated by plasmids.

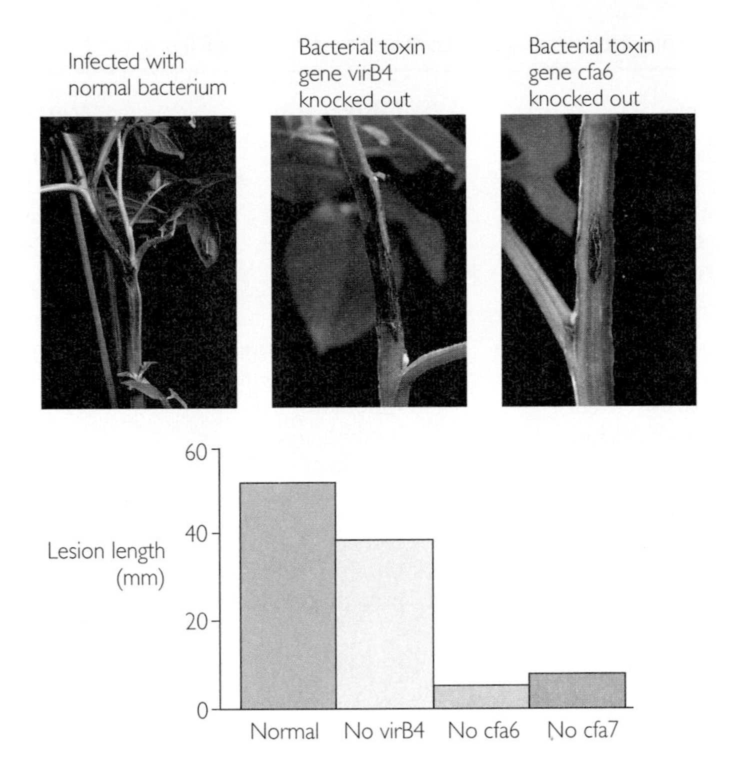

Figure 15.10 Knocking out virulence genes makes a parasite less virulent The photo on the top left shows a host plant infected with normal cells of the bacterial parasite *Erwinia carotovora*. Note the large lesion consisting of dead host cells. The other photos show the smaller lesions caused by infections with strains of *Erwinia carotovora* that have a specific "virulence gene" knocked out. In this case, the virulence genes code for compounds that are toxic to plants. The graph at the bottom shows the average lengths of lesions caused by normal bacteria versus bacteria with various toxin-coding genes knocked out.

A final generalization about gene-level adaptations in pathogens concerns organisms that parasitize vertebrates. When researchers sequence the genomes of vertebrate parasites, the parasite genomes routinely contain large numbers of genes that code for variants of membrane proteins (Gardner et al. 2002; Paulson et al. 2003; Berriman et al. 2005). The eukaryote that causes malaria in mice, for example, has 806 genes that code for glycoproteins found on the surface of the cell (Carlton et al. 2002). Why would almost 9% of this species' entire genome code for membrane proteins that have carbohydrate groups attached? To explain this observation, biologists point out that the vertebrate immune system is activated when host cells recognize particular proteins as foreign. Once activated, immune system cells then destroy the cell with the foreign protein. But if a population of parasitic cells displays an array of proteins on their surfaces, it is unlikely that the immune system will be able to recognize all of the cells and eliminate them. Thus, large numbers of membrane-protein genes are interpreted as an adaptation for parasitic cells to avoid immune detection.

The genomes of vertebrate parasites usually contain sequences that help the parasite evade detection by the immune system.

Patterns in the Genomes of Free-Living Organisms

If the nature of genomes in parasitic organisms correlates closely with their way of life, it should come as no surprise that analogous types of correlations exist among free-living organisms. When researchers compared the complete genome sequences of the mosquito *Anopheles gambiae* with the genome of the fruit fly *Drosophila melanogaster*, they found that the mosquito has 58 fibrinogen-like proteins that are not found in fruit flies. Adult female mosquitoes feed on blood, while adult fruit flies feed on rotting fruit. Fibrinogens function as blood-clotting proteins in humans, and the fibrinogen-like proteins in mosquitoes probably function as an anti-coagulant, to keep blood flowing smoothly while a female mosquito is taking a meal (De Gregorio and Lemaitre 2002; Holt et al. 2002).

Table 15.2 Examples of correlations between gene content and mode of living

Species studied	Description	Observation	Reference
Galderia sulfuria & *Cyanidioschyzon meroloae*	Closely related unicellular red algae. Both are photosynthetic, but *G. sulfuria* can also live on over 50 external carbon sources.	*G. sulfuria* has many genes for uptake of carbohydrates.	Barbier et al. 2005. *Plant Physiology* 137: 460–474.
Prochlorococcus sp.	Abundant in marine plankton; compare populations found at different depths (high light, medium light, low light).	Each population has unique genes for chlorophyll-binding antenna proteins.	Bibby et al. 2003. *Nature* 424:1051–1054; Rocap et al. 2003. *Nature* 424: 1042–1047.
Pseudomonas syringae	Compare populations of this species that live in specialized habitats versus many different habitats.	Generalist population has 976 additional protein-coding genes.	Fiel et al. 2005. *PNAS* 102: 11064–11069.
Dehalococcoides ethenogenes	This bacterial species can dechlorinate the pollutants PCE and trichloroethene.	Has 17 genes coding for dehalogenase enzymes; each is next to a transcription regulatory sequence.	Seshadri et al. 2005. *Science* 307: 105–108.

Strong correlations usually exist between the genes present in a genome and how the organism makes its living. But researchers must rigorously test the hypothesis that a certain gene functions as an adaptation to a certain environment.

Table 15.2 summarizes other examples of correlations between the way free-living organisms make a living and the contents of their genomes. In each case, the logical hypothesis is that the observed genetic differences correspond to adaptations important to the organism's fitness. The carbohydrate transporters found in *Galderia sulfuria* should allow it to live on a diversity of carbon sources; the distinctive chlorophyll-binding proteins in populations of *Prochlorococcus* should allow them to harvest light efficiently at different depths; the extra genes in some populations of *Pseudomonas* should help them exploit a wide array of habitats; the dehalogenase enzymes in *Dehalococcoides* are probably involved in using chlorine-containing molecules as food. These hypotheses still need to be tested, however, using targeted mutations—like those used to evaluate the function of suspected virulence genes in *Erwinia carotovora*—or other experimental approaches.

The other important thing to note about the observations in Table 15.2 is that most depend on comparing the genomes of closely related species. Once researchers had worked out the technical issues involved in sequencing complete genomes, they quickly realized that the key to interpreting the data sets was the ability to make comparisons across species. These types of cross-species comparisons are already moving beyond comparisons of protein-coding sequences, however, into an exciting new frontier: looking at differences in gene regulation. For example, comparing the recently completed human and chimp genome is inspiring research on a key aspect of understanding adaptation at the molecular level: the importance of changes in the DNA sequences and proteins that regulate gene expression.

Evolution of Regulatory Sequences

One of the most striking observations to come out of genome sequencing studies emerged when researchers compared the complete gene set in humans and common chimpanzees (International Human Gene Sequencing Consortium 2001; Venter et al. 2001; International Human Gene Sequencing Consortium 2004; Chimpanzee Sequencing and Analysis Consortium 2005). When researchers identified sequences that code for proteins and compared the DNA sequences of homologous genes in chimps and humans, they found that homologs differ by only about 1%. When the DNA sequences are used to predict the amino acid sequences of the resulting proteins, 29% of the homologous proteins in chimps and humans are identical. When all homologous proteins in humans and chimps are compared, the average difference is a grand total of two amino acids.

This result confirms a tentative conclusion initially made on the basis of similarities in individual human and chimp genes and chromosomes. At the level of DNA sequences and amino acid sequences, humans and chimps are extremely similar. The obvious question is: If they are so similar genetically, why are they so different phenotypically? Mary-Claire King and Allan Wilson (1975) proposed an answer more than three decades ago. King and Wilson suggested that if chimps and humans are producing nearly identical proteins, then their phenotypic differences must result from differences in the amount or timing of gene expression.

King and Wilson were focusing on the distinction between structural genes, which code for RNAs and proteins that perform functions in the cell, and regulatory genes and sequences. Regulatory genes code for transcription factors and other proteins that bind to DNA or to other proteins. These binding events help turn gene expression on and off or up and down. Regulatory sequences are segments of DNA where regulatory proteins bind to influence gene expression. According to King and Wilson, at least some of the key phenotypic differences between chimps and humans should result from divergence in regulatory genes and proteins.

Research on genetic changes responsible for human traits like hair loss, increased brain size, language, and remodeling of the face and jaw is now occurring at a furious pace (e.g., Enard et al. 2002; Carroll 2003; Stedman et al. 2004; Khaitovich et al. 2005). But one of the best current examples of a regulatory change that produced a dramatic change in phenotype comes from a plant: the ground cherry or husk tomato *Physalis pubescens*. This species is distinguished by sepals that grow up and around the maturing fruit to form a structure called a "Chinese lantern." In close relatives including the potato, *Solanum tuberosoma,* the sepals do not continue growing after pollination and the fruit remains exposed as it matures (Figure 15.11).

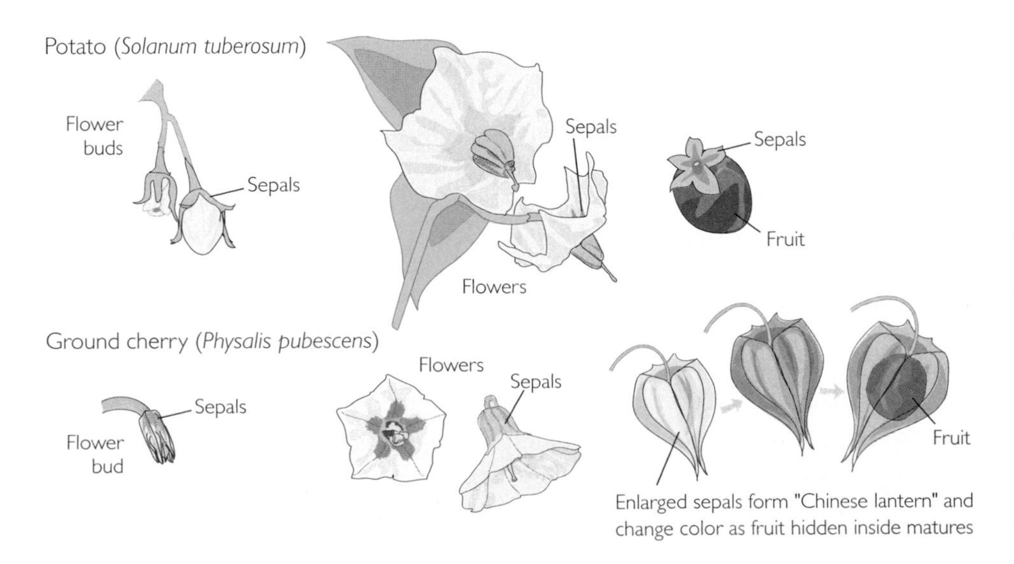

Figure 15.11 The "Chinese lantern" is an innovation found in certain members of the plant family *Solanaceae* These drawings show the development of a flower and mature fruit of the closely related potato (top) and ground cherry (bottom). The "Chinese lantern" structure forms around the fruits of ground cherries because the sepals keep growing instead of staying small, as they do in most members of this plant family.

Chaoying He and Heinz Saedler (2005) set out to characterize the genetic changes responsible for the evolution of the lantern structure. They focused on a gene in corn that produces structures reminiscent of the *Physalis* lantern when it is expressed in floral tissue instead of vegetative tissue. He and Saedler found a homolog of the corn gene in *Physalis* and *Solanum* and named it MPF2. They showed that MPF2 codes for a transcription factor that stimulates cell division, that the gene is expressed only in vegetative tissue in *Solanum* but in floral *and* vegetative cells of *Physalis*, and that the lantern structure in *Physalis* diminishes or disappears if MPF2 is knocked out experimentally. These results provided strong support for the hypothesis that MPF2 is responsible for the formation of the lantern, and that lanterns form because a mutation made MPF2 active in floral organs as well as vegetative cells.

To understand the nature of this evolutionary change, He and Saedler sequenced the genes and flanking regions of MPF2 in *Physalis* and *Solanum*. The data showed that the coding regions of the genes were 86% identical in DNA sequence, but that the promoter regions—where transcription factors and other proteins bind to initiate gene expression—are only 42% identical. Based on this observation, they claim that evolutionary change in a regulatory sequence is responsible for the change in where the gene is expressed.

Researchers are moving beyond comparisons of protein-coding "structural genes" to analyze evolutionary change in the proteins and DNA sequences responsible for regulating gene epxression.

Studies like these bring us to the frontier of research in phylogenomics. After comparing gene content and gene expression in closely related species, researchers create hypotheses about which genes are responsible for the phenotypic differences in those species. These hypotheses, in turn, inspire rigorous experimental tests. The goal is to identify which alleles are responsible for evolutionary change.

15.4 Frontiers in Phylogenomics

Trying to predict the course of scientific research is notoriously difficult. This is especially so in a field like phylogenomics, which is still in an emerging, rapidly growing phase. Even so, it can be helpful to take a look at the broad picture of research in new areas of evolutionary analysis like phylogenomics. The idea is try to pick out patterns, as a sign of where the most exciting research frontiers might lie.

Let's consider two broad themes that are emerging from studies at the interface of phylogenetic studies and genomics. Each touches on a fundamental characteristic of good science: Research should inspire more research, and new fields should "cross-pollinate," or contribute new insights to other areas of investigation.

New Data, New Questions

In the late 1980s and early 1990s, breakthroughs in gene sequencing technology and in methods for inferring phylogenetic trees triggered an explosion of work on the history of lineages. For the first time, evolutionary biologists could answer questions about the origins of pathogenic viruses, cospeciation between parasites and hosts, the timing of key events in the history of life that are not documented in the fossil record, and correlations between phylogeny and geography (see Chapter 4). In a similar way, genome-sequencing projects are contributing a fundamentally new type of data in biology and triggering research on an array of new questions.

When a genome sequence is complete, biologists have the comprehensive catalog of genes present in a species. This makes it possible to study gene loss and trait loss—a topic highlighted earlier in the chapter—in a rigorous way. Until a complete genome sequence is available, a critic can take issue with claims that a particular gene has been lost in a certain lineage by championing an alternative hypothesis: The gene hasn't been lost; it simply hasn't been found yet.

Having a complete gene catalog also promises surprises, including the possibility of finding genes that researchers didn't expect to find. A recent analysis of a diatom genome, for example, led to the discovery of genes for the complete urea cycle (Armbrust et al. 2004). Diatoms are single-celled eukaryotes and are extremely abundant in marine plankton. They secrete a glass-like case around their cells and make their living performing photosynthesis (Figure 15.12). Prior to this study, the genes required for the urea cycle had never been found in a photosynthetic eukaryote. In mammals, the urea cycle is responsible for processing waste molecules that contain nitrogen. The urea that is produced by the cycle is excreted in the urine. Instead of processing waste, however, it is likely that diatoms use the urea cycle to supply precursors to other biosynthetic pathways. Why diatoms have these genes, while other photosynthetic eukaryotes do not, is still a mystery. For our purposes, the important point is that new data from genome sequences are inspiring new questions about what certain genes are doing in certain species. New sources of data are triggering new questions in phylogenomics.

Throughout the history of biology, the availability of new types of data has inspired new types of questions and new fields of research.

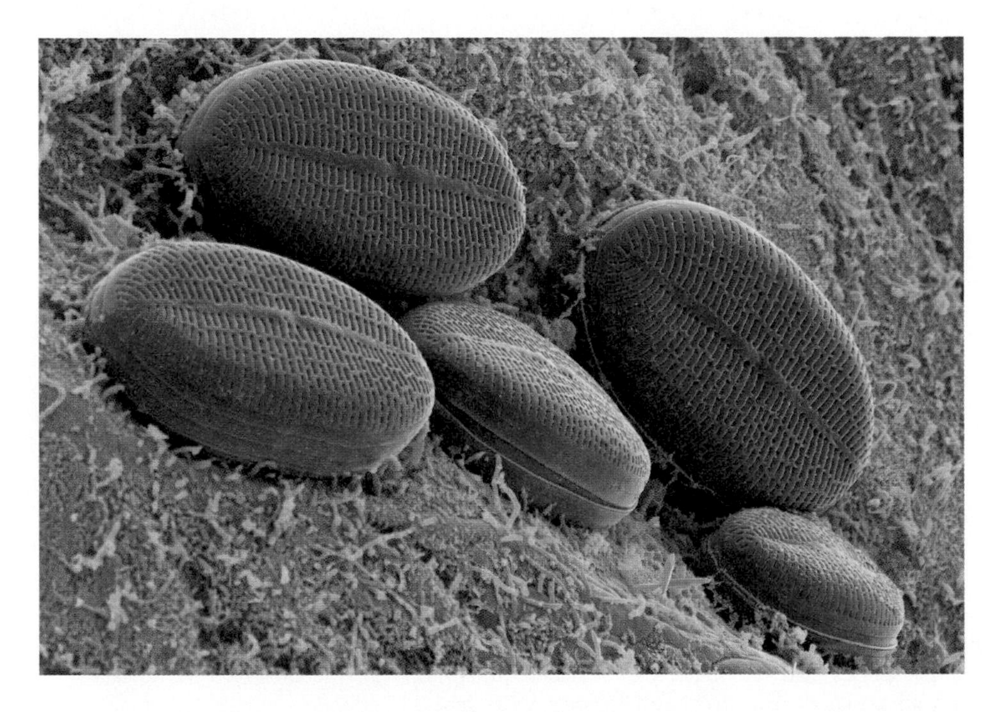

Figure 15.12 Diatoms are photosynthetic organisms that are common in marine plankton Diatoms have a glassy, silica-rich shell that comes in a "box and lid" arrangement, like a petri plate. These shells have been colored purple to make them easier to see.

Another novel source of data is springing from a novel research approach. Instead of sequencing the genomes of individual organisms, researchers are beginning to sequence the genomes present in a particular environment. This tactic was inspired by a research strategy called **environmental sequencing** or **direct sequencing** (Giovannoni et al. 1990; Ward et al. 1990). Environmental sequencing, in turn, was inspired by the realization that only a few percent of the total number of bacteria and archaea on Earth have been cultivated and studied to date. To get a better idea of which bacteria and archaea are actually present in a particular habitat, research teams began collecting DNA samples, sequencing small subunit rRNA genes or other specific DNA segments, and

(a) Environmental sequencing

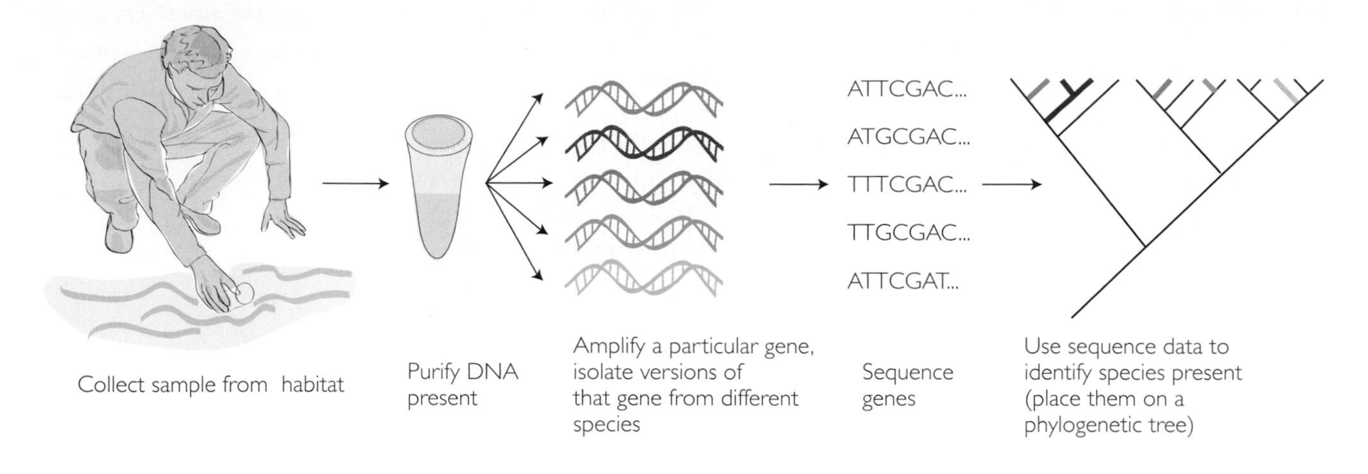

Collect sample from habitat

Purify DNA present

Amplify a particular gene, isolate versions of that gene from different species

Sequence genes

Use sequence data to identify species present (place them on a phylogenetic tree)

(b) Environmental genomics

Lots of sequences

Are there genes for:

Photosynthetic pigments, enzymes?

Nitrogen fixation?

Using certain electron donors in cellular respiration?

Fermenting certain substrates?

Absorbing or secreting certain organic or inorganic compounds?

Collect samples from different habitats

Purify DNA present

Sequence all genes present

Analyze genes present in each habitat—identify function and compare

Figure 15.13 Environmental sequencing and environmental genomics (a) Environmental sequencing is a way to identify which organisms are present in a habitat, even though the species have never been seen or studied. (b) Environmental genomics is a way to catalog the genes present in a habitat; even investigators do not know which species contain those genes.

Environmental sequencing is based on extracting DNA from a particular gene directly from an environment, then using the sequence data to identify the organisms present.

using the sequence data to place the organisms on the tree of life (Figure 15.13a). In this way, researchers can describe new species without ever actually seeing them. Environmental sequencing has deepened our understanding of bacterial and archaeal diversity and led to the discovery of a previously unknown, exceptionally abundant lineage of marine bacteria called the SAR11 clade.

Environmental sequencing has recently been extended from its initial focus on using a particular gene to identify unknown species to sequencing the entire genomes present in a sample. This approach is called **environmental genomics**. The goal here is to understand the genes present in a habitat without necessarily knowing which species those genes belong to (Figure 15.13b). In this way, researchers hope to infer how the organisms present are making a living—how they are adapted to the habitat being studied.

As an example of environmental genomics, consider work by Craig Venter and coworkers (2004) on the complete genomes present in the Sargasso Sea off the coast of Bermuda. The sequences represented an estimated 1,800 species, including almost 150 that had never been described previously. The biggest surprise in the data, however, was finding over 780 distinct genes for a protein called proteorhodopsin. Proteorhodopsin functions as a light-driven protein pump in the membranes of certain bacteria (Beja et al. 1980). Using the energy in light, proteorhodopsin sets up a proton gradient that allows cells to synthesize ATP. Early work on proteorhodopsin indicated that different species have different versions of the protein, and that these different versions might be adapted to specific light regimes (Beja et al. 1981). The 780 proteorhodopsin sequences in the recent Sargasso Sea sample suggest that hundreds of slightly different proton pumps exist in species that live virtually side by side. Further work will be needed to test the hypothesis that the observed diversity in proteorhodopsins is a result of natural selection to increase the efficiency of ATP production in different species or reduce competition for light. But without environmental genomics data, evolutionary biologists never would have thought to ask the question.

Environmental genomics is based on sequencing entire genomes from an environment and analyzing which genes are present.

In the meantime, environmental genomics is now being done in a comparative framework, with researchers comparing the genes present in distinctive habitats. Susannah Tringe and colleagues (2005), for example, recently compared the protein-coding sequences present in a sample of agricultural soil versus rotting whale carcasses in deep ocean water. Many of the observations they made were logical. The soil sample contained an array of genes responsible for degrading the cellulose found in plant cell walls, but the whale-carcass sample lacked these genes entirely. Neither of the light-poor environments contained proteorhodopsin genes. Perhaps their most intriguing finding, though, was a large number of genes in each habitat type whose function is unknown. Time and further research will tell whether these uncharacterized genes represent adaptations specific to each habitat.

Functional Genomics and Evo-Devo

The "mystery genes" that are being found in environmental genome sequences highlight an important point: Evolutionary analyses of genome sequences are much more tractable when the function of the genes involved is known (Raskin et al. 2006). Within the field of genetics, the research program called **functional genomics** seeks not only to characterize the function of genes, but to document when they are expressed and in what quantities. In this way, basic research in genetics is fueling advances in phylogenomics. For example, if changes in regulatory sequences or regulatory proteins are responsible for phenotypic differences between species, and if functional genomics can specify how gene expression differs between species, then an evolutionary analysis should identify at least some of the regulatory changes involved in phenotypic differentiation. In chimpanzees and humans, researchers are now comparing differences in protein coding sequences and in expression patterns due to regulatory changes (Khaitovich et al. 2005). Comparative data from functional genomics should be a treasure trove for studying adaptation at the level of the gene.

Phylogenomics draws on data from the research area called functional genomics, and in turn provides insights useful to the field called evo-devo.

To summarize, genome sequencing and functional genomics allow researchers to compare the entire gene sets of different species and how they are expressed.

If progress continues, these fields of study should begin to complement research in the field called evolution and development, or **evo-devo**. Evo-devo seeks to understand the genetic changes responsible for evolutionary innovations like the insect wing, the tetrapod limb, and the flower. The focus is on individual traits and single genes or small sets of genes. As Chapter 19 will show, this research program has been tremendously productive. With the addition of genomic perspectives on gene content and gene expression, our understanding of the genetic changes responsible for phenotypic evolution should continue to accelerate.

Summary

Phylogenomics is based on evolutionary analyses of genome sequences, amino acid sequences, gene expression profiles, and other types of molecular data sets. The development of this field is extending our ability to study adaptation at the level of the genotype.

Early studies of gene content documented the diversity and abundance of transposable elements in bacteria, archaea, and eukaryotes. Transposable elements are DNA sequences that encode the information required for that sequence to be copied and moved to a new location in the genome. Because they may cause deleterious mutations when they insert in a new location and because they must be copied by the host cell, they are more accurately characterized as genomic parasites. They illustrate the principle that natural selection can act at the level of single genes as well as at the level of an individual's phenotype. At least on occasion, however, transposition events result in the formation of new genes that serve a positive function in the organism.

Lateral gene transfer occurs when transduction, transformation, plasmid transfer, or other processes result in a movement of genes between species. LGT has been a prominent feature of eukaryote evolution-especially during the acquisition of mitochondria and chloroplasts—and has been rampant during the evolution of bacteria and archaea. In most cases, LGT has allowed organisms to acquire the genes required for novel metabolic adaptations, such as the ability to use certain nutrients.

Researchers are now comparing the gene catalogs available in whole-genome sequence data to study topics like gene and trait loss and to look for correlations between the genes present in organisms and how they exploit particular habitats. The new types of data emerging from genome sequencing projects, environmental sequencing and environmental genomics, and from functional genomics studies of gene expression are also inspiring new questions. These questions include why certain genes exist in certain organisms and how changes in regulatory sequences and regulatory proteins have influenced phenotypic divergence in closely related species.

Questions

1. In interactions between species, parasitism occurs when one species gains a fitness benefit and another suffers a fitness loss. In contrast to predation, parasites are small relative to their hosts and kill the host slowly if at all. Based on this definition, should transposable elements be considered parasites? Should biologists still use the original term used to describe them, which is "junk DNA?"
2. Explain how the movement of transposable elements can create mutations that are beneficial to hosts as well mutations that are deleterious.
3. Transposable elements are considered the best example of the prediction that natural selection can act on other levels besides individual organisms. Explain how heritable variation and differential success among transposable sequences can lead to evolution by natural selection.
4. Why are lateral gene transfer and horizontal gene transfer appropriate names for the phenomenon they describe?
5. Genes that confer antibiotic resistance in disease-causing bacteria are usually located on plasmids and can pass from one species to another via lateral gene transfer. Now suppose that LGT never occurred. Would drug resistance in bacteria evolve more rapidly or more slowly in the absence of LGT?
6. Why is it reasonable to characterize eukaryotes as "chimeras?"
7. Describe three ways that researchers can diagnose LGT.
8. Recall that most LGT events involve genes that confer new metabolic abilities. Which genes should researchers sequence to infer the actual phylogeny of organisms where LGT has been common?

9. Why is it logical to observe that parasites tend to have much smaller genomes than free-living organisms?
10. Give an example of a correlation between the gene content of an organism and its way of making a living.
11. How is it possible for humans and chimpanzees to be so dissimilar at the phenotypic level if they are so similar at the genotypic level?
12. Suppose a genome sequencing project set out to compare the gene content of two closely related bacteria. One is a highly virulent pathogen that lives in the lung tissue of mammals while the other is a harmless commensal that lives in the nasal passages of mammals. How would you expect the genomes to differ?
13. Suppose that an environmental genomics project set out to compare the gene content in a rich agricultural soil versus soil at an abandoned oil refinery. How would you expect the two habitats to compare?

Exploring the Literature

14. Extensive gene loss or highly streamlined genomes have been observed in some free-living organisms as well as in many parasites. In many cases, the missing or non-functional genes or small genomes are thought to be adaptive. To investigate possible examples of adaptive gene loss in free-living species, see:

    Giovannoni, S.J., H.J. Tripp, et al. 2005. Genome streamlining in a cosmopolitan oceanic bacterium. *Science* 309: 1242–1245.

    Go, Y., Y. Satta, O. Takenaka, and N. Takahata. 2005. Lineage-specific loss of function of bitter taste receptor genes in humans and non-human primates. *Genetics* 170: 313–326.

    Hughes, A.L., and Friedman, R. 2005. Loss of ancestral genes in the genomic evolution of *Ciona intestinalis*. *Evolution and Development* 7: 196–200.

15. Compared to bacteria and archaea, eukaryotic genomes carry a large number of transposable elements. Why don't eukaryotes do a better job of defending themselves against these parasites?

    Galagan, J.E., S.E. Calvo, et al. 2003. The genome sequence of the filamentous fungus *Neurospora crassa*. *Nature* 422: 859–868.

    Lynch, M. and J.S. Conery. 2003. The origins of genome complexity. *Science* 302: 1401–1404.

Citations

Abad, J.P., B. De Pablos, K. Osoegawa, P.J. De Jong, A. Martin-Gallardo, and A. Villasante. 2004. TAHRE, a novel telomeric retrotransposon from *Drosophila melanogaster*, reveals the origin of *Drosophila telomeres*. *Molecular Biology and Evolution*. 21: 1620–4.

Abrahamson, M.S., T.J. Templeton, et al. 2004. Complete genome sequence of the apicomplexan, *Cryptosporidium parvum*. *Science* 304: 441–445.

Agrawal, A., Q.M. Eastman, and D.G. Schatz. 1998. Transposition mediated by RAG1 and RAG2 and its implications for the evolution of the immune system. *Nature* 394: 744–751.

Armbrust, E.V., J.A. Berges, et al. 2004. The genome of the diatom *Thalassiosira pseudonana*: ecology, evolution, and metabolism. *Science* 306: 79–86.

Baldauf, S.L. and J.D. Palmer. 1990. Evolutionary transfer of the chloroplast *tufA* gene to the nucleus. *Nature* 344: 262–265.

Baldauf, S.L. J.R. Manhart, and J.D. Palmer. 1990. Different fates of the chloroplast *tufA* gene following its transfer to the nucleus in green algae. *Proceedings of the National Academy of Sciences USA* 87: 5317–5321.

Bell, K.S., M. Sebaihia, et al. 2004. Genome sequence of the enterobacterial phytopathogen *Erwinia carotovova* subsp. *atroseptica* and the characterization of virulence factors. *Proceedings of the National Academy of Sciences USA* 101: 11105–11110.

Beja, O., L. Aravind, et al. 2000. Bacterial rhodopsin: evidence for a new type of phototrophy in the sea. *Science* 289: 1902–1906.

Beja, O., E.N. Spudich, J.L. Spudich, M. Leclerc, and E.F. DeLong. 2001. Proteorhodopsin phototrophy in the ocean. *Nature* 411: 786–789.

Berriman, M., E. Ghedin, et al. 2005. The genome of the African trypanosome *Trypanosoma brucei*. *Science* 309: 416–422.

Bester, T.H. and B. Tycko. 1996. Creation of genomic methylation patterns. *Nature Genetics* 12: 363–367.

Blattner, F.R. et al. 1997. The complete genome sequence of *Escherichia coli* K-12. *Science* 277: 1453–1474.

Britten, R. 2006. Transposable elements have contributed to thousands of human proteins. *Proceedings of the National Academy of Sciences USA* 103: 1798–1803.

Carlton, J.M., S.V. Angiuoli, et al. 2002. Genome sequence and comparative analysis of the model rodent malaria parasite *Plasmodium yoelii yoelii*. *Nature* 419: 512–519.

Carroll, S.B. 2003. Genetics and the making of *Homo sapiens*. *Nature* 422: 849–857.

Charlesworth, B., and C.H. Langley. 1989. The population genetics of *Drosophila* transposable elements. *Annual Review of Genetics* 23: 251–287.

Chimpanzee Sequencing and Analysis Consortium. 2005. Initial sequence of the chimpanzee genome and comparison with the human genome. *Nature* 437: 69–87.

Clegg, M.T., B.X. Gaut, G.H. Learn, Jr., and B.R. Morton. 1994. Rates and patterns of chloroplast DNA evolution. *Proceedings of the National Academy of Sciences USA* 91: 6795–6801.

Coleman, M.L., M.B. Sullivan et. al. 2006. Genomic islands and the ecology and evolution of *Prochlorococcus*. *Science* 311: 1768–1770.

Covello, P.S. and M.W. Gray 1992. Silent mitochondrial and active nuclear genes for subunit 2 of *cytochrome c oxidase* (*cox2*) in soybean: evidence for RNA-mediated gene transfer. *EMBO Journal* 11: 3815–3820.

Cowan, R.K., D.R. Hoen, D.J. Schoen, and Thomas E. Bureau. 2005. MUSTANG is a novel family of domesticated transposase genes found in diverse angiosperms. *Molecular Biology and Evolution* 22: 2084–2089.

De Gregorio, E. and B. Lemaitre. 2002. The post-genomic era opens. *Nature* 419: 496–497.

Dean, R.A., N.J. Talbot, et al. 2005. The genome sequence of the rice blast fungus *Magnaporthe grisea*. *Nature* 434: 980–986.

DeLong, E.F., C.M. Preston, et al. 2006. Community genomics among stratified microbial assemblages in the ocean's interior. *Science* 311: 496–503.

Doolittle, W.F. and J.M. Logsdon, Jr. 1998. Archaeal genomics: Do archaea have a mixed heritage? *Current Biology* 8: R209–R211.

Douglas, S.E., C.A. Murphy, D.F. Spencer, and M.W. Gray 1991. Cyrptomonad algae are evolutionary chimaeras of two phylogenetically distinct unicellular eukaryotes. *Nature* 350: 148–151.

Eisen, J.A. and C. Fraser. 2003. Phylogenomics: intersection of evolution and genomics. *Science* 300: 1706–1707.

Enard, W., M. Przeworski, et al. 2002. Molecular evolution of FOXP2, a gene involved in speech and language. *Nature* 418: 869–872.

Falkowski, P.G., M.E. Katz, et al. 2004. The evolution of modern eukaryotic phytoplankton. *Science* 305: 354–360.

Feil, H., W.S. Feil, et al. 2005. Comparison of the complete genome sequences of *Pseudomonas syringae pv. syringae* B728a and *pv. tomato* DC 3000. *Proceedings of the National Academy of Sciences USA* 102: 11064–11069.

Gabaldon, T. and M.J. Huynen. 2003. Reconstruction of the proto-mitochondrial metabolism. *Science* 301: 609.

Gao, X. and D.F. Voytas. 2005. A eukaryotic gene family related to retroelement integrases. *Trends in Genetics* 21: 133–137.

Gardner, M.J., N. Hall, et al. 2002. Genome sequence of the human malarial parasite *Plasmodium falciparum*. *Nature* 419: 498–511.

Gillham, N.W., J.E. Boynton, and E.H. Harris. 1985. Evolution of plastid DNA. In T. Cavalier-Smith, ed. *The Evolution of Genome Size*. New York: John Wiley & Sons, 299–351.

Giovannoni, S.J. S. Turner, G.J. Olsen, S. Barns, D.J. Lane, and N.R. Pace. 1988. Evolutionary relationships among cyanobacteria and green chloroplasts. *Journal of Bacteriology* 170: 3584–3592.

Giovannoni, S.J., T.B. Britschgi, C.L. Moyer, and K.G. Field. 1990. Genetic diversity in Sargasso Sea bacterioplankton. *Nature* 345: 60–63.

Giovannoni, S.J., H.J. Tripp, et al. 2005. Genome streamlining in a cosmopolitan oceanic bacterium. *Science* 309: 1242–1245.

Gray, M.W. 1992. The endosymbiont hypothesis revisted. In D.R. Wolstenholme and K.W. Jeon, eds. *Mitochondrial Genomes*. San Diego, CA: Academic Press, 233–257.

He, C. and H. Saedler. 2005. Heterotopic expression of MPF2 is the key to the evolution of the Chinese lantern of *Physalis*, a morphological novelty in the Solanaceae. *Proceedings of the National Academy of Sciences USA* 102: 5779–5784.

Hickey, D.A. 1992. Evolutionary dynamics of transposable elements in prokaryotes and eukaryotes. *Genetica* 86: 269–274.

Holden, M.T.G., E.J. Feil, et al. 2004. Complete genomes of two clinical *Staphylococcus aureus* strains: evidence for the rapid evolution of virulence and drug resistance. *Proceedings of the National Academy of Sciences USA* 101: 9786–9791.

Holt, R.A., G.M. Subramanian, et al. 2002. The genome sequence of the malaria mosquito *Anopheles gambiae*. *Science* 298: 129–149.

Hutchison, C.A. III, S.C. Hardies, D.D. Loeb, W.R. Shehee, and M.H. Edgell. 1989. LINEs and related retroposons: Long interspersed repeated sequences in the eukaryotic genome. In D.E. Berg and M.M. Howe, eds. *Mobile DNA*. Washington, D.C.: American Society of Microbiology, 593–617.

International Chicken Genome Sequencing Consortium. 2004. Sequence and comparative analysis of the chicken genome provide unique perspectives and vertebrate evolution. *Nature* 432: 695–716.

International Human Gene Sequencing Consortium. 2001. Initial sequencing and analysis of the human genome. *Nature* 409: 860–921.

International Human Gene Sequencing Consortium. 2004. Finishing the euchromatic sequence of the human genome. *Nature* 431: 931–945.

Jiang, N., Z. Bao, X. Zhang, S.R. Eddy, and S.R. Wessler. 2004. Pack-MULE transposable elements mediate gene evolution in plants. *Nature* 431: 569–573.

Keeling, P.J. 2004. Diversity and evolutionary history of plastids and their hosts. *American Journal of Botany* 91: 1481-1493.

Khaitovich, P., I. Hellmann, et al. 2005. Parallel patterns of evolution in the genomes and transcriptomes of humans and chimpanzees. *Science* 309: 1850–1854.

King, M.-C. and A.C. Wilson. 1975. Evolution at two levels in humans and chimpanzees. *Science* 188: 107–116.

Kuroda, M. and A. Yamashita, et al. 2005. Whole genome sequence of *Staphylococcus saprophyticus* reveals the pathogenesis of uncomplicated urinary tract infection. *Proceedings of the National Academy of Sciences USA* 102: 13272–13277.

Lawrence, J.G. and H. Ochman. 1998. Molecular archaeology of the *Escherichia coli* genome. *Proceedings of the National Academy of Sciences USA* 95: 9413–9417.

Meibom, K.L., M. Blokesch, N.A. Dolganov, C.-Y. Wu, and G.K. Schoolnik. 2005. Chitin induces natural competence in *Vibrio cholerae*. *Science* 310: 1824–1827.

Moran, J.V., R.J. Debarardinis, H.H. Kazazian, Jr. 1999. Exon shuffling by L1 retrotransposition. *Science* 283: 1530–1534.

Nugent, J.M. and J.D. Palmer. 1991. RNA-mediated transfer of the gene *coxII* from the mitochondrion to the nucleus during flowering plant evolution. *Cell* 66: 473–481.

Palenik, B., B. Brahamsha, et al. 2003. The genome of a motile marine *Synechococcus*. *Nature* 424: 1037–1042.

Pardue, M.L. and P.G. DeBaryshe. 2003. Retrotransposons provide an evolutionarily robust non-telomerase mechanism to maintain telomeres. *Annual Review of Genetics* 37: 485–511.

Paulsen, I.T., L. Banerji, et al. 2003. Role of mobile DNA in the evolution of vancomycin-resistant *Enterococcus faecalis*. *Science* 299: 2071–2074.

Perna, N.T., G. Plunkett III, et al. 2001. Genome sequence of enterohaemorrhagic *Escherichia coli* O157:H7. *Nature* 409: 529–533.

Raskin, D.M., R. Seshadri, S.U. Pukatzki, and J.J. Mekalanos. 2006. Bacterial genomics and pathogen evolution. *Cell* 124: 703–714.

Rat Genome Sequencing Project Consortium 2004. Genome sequence of the Brown Norway rat yields insights into mammalian evolution. *Nature* 428: 493–521.

Ruepp, A., W. Grami, et al. 2000. The genome sequence of the thermoacidophilic scavenger *Thermoplasma acidophilum*. *Nature* 407: 508–513.

Salanoubat, M., S. Genin, et al. 2002. Genome sequence of the plant pathogen *Ralstonia solanacearum*. *Nature* 415: 497–502.

Stanhope, M.J., A. Lupas, M.J. Italia, K.K. Koretke, C. Volker, and J.R. Brown. 2001. Phylogenetic analyses do not support horizontal transfers from bacteria to vertebrates. *Nature* 411: 940–944.

Stedman, H.H., B.W. Kozyak, et al. 2004. Myosin gene mutation correlates with anatomical changes in the human lineage. *Nature* 428: 415–418.

Tovar, J., G. Leo'n-Avila, et al. 2003. Mitochondrial remnant organelles of *Giardia* function in iron-sulphur protein maturation. *Nature* 426: 172–176.

Tringe, S.G., C. von Mering, et al. 2005. Comparative metagenomics of microbial communities. *Science* 308: 554–557.

Venter, J.C., M.D. Adams, et al. 2001. The sequence of the human genome. *Science* 291: 1304–1351.

Venter, J.C., K. Remington, et al. 2004. Environmental genome shotgun sequencing of the Sargasso Sea. *Science* 304: 66–74.

Ward, D.M., R. Weller, and M.M. Bateson. 1990. 16S rRNA sequences reveal numerous uncultured microorganisms in a natural community. *Nature* 345: 63–65.

Waugh O'Neill, R.J., M.J. O'Neill, and J.A. Marshall Graves. 1998. Undermethylation associated with retroelement activation and chromosome remodeling in an interspecific mammalisn hybrid. *Nature* 393: 68–72.

Williams, B.A.P., R.P. Hirt, J.M. Lucocq, and T.M. Embley. 2002. A mitochondrial remnant in the microsporidian *Trachipleistophora hominis*. *Nature* 418: 865–869.

Wolfe, K.H., C.W. Morden, and J.D. Palmer. 1991. Ins and outs of plastid genome evolution. *Current Opinions in Genetics and Development* 1: 523–529.

Xu, P., G. Widmer, et al. 2004. The genome of *Crytosporidium hominis*. *Nature* 431: 1107–1112.

Yang, D., Y. Oyaizu, H. Oyaizu, G.J. Olsen, and C.R. Woese. 1985. Mitochondrial origins. *Proceedings of the National Academy of Sciences USA* 82: 4443–4447.

Yoder, J.A., C.P. Walsh, and T.H. Bestor. 1997. Cytosine methylation and the ecology of introgenomic parasites. *Trends in Genetics* 13: 335–340.

Yoon, H.S., J.D. Hackett, F.M. Van Dolah, T. Nosenko, K.L. Lidie, and D. Bhattacharya. 2005. Tertiary endosymbiosis driven genome evolution in dinoflagellate algae. *Molecular Biology and Evolution* 22: 1299–1308.

Zhao, L., R. Mitra, P.W. Atkinson, A.B. Hickman, F. Dyda, and N.L. Craig. 2004. Transposition of hAT elements links transposable elements and V(D)J recombination. *Nature* 432: 995–1001.

PART IV
THE HISTORY OF LIFE

The fighting dinosaurs. Eighty million years ago, in what is now Mongolia's Gobi desert, a young *Velociraptor* (right) attacked a *Protoceratops* (left). As the predator sunk its foot claw into its prey's neck, the *Protoceratops* bit down hard on the *Velociratpor's* right arm. At that moment the pair were buried alive in a sand slide. Their fossils preserve their final instant of mortal combat.

Part II and Part III of this text detailed how allele frequencies change within populations. Now our focus shifts to questions about evolutionary change between populations and species. Instead of considering changes that occur in a population from one generation to the next and thinking in time scales of a single breeding season or a few years, we need to consider the entire sweep of Earth's 4.6-billion-year history and ask what happened and why. How did life begin? How are new species produced? What changes in allele frequencies were responsible for major evolutionary innovations like the limbs of animals and the flowers of angiosperms?

To launch this investigation, Chapter 16 analyzes how populations diverge to become distinct species. The rest of Part IV focuses on major events in the history of life. Chapter 17 introduces the earliest traces of life in the fossil record and recent experimental work on the origin of life. Chapter 18 investigates the evolution of multicellular life with an emphasis on the initial diversification of animals and on catastrophic mass extinctions. Chapter 19 focuses on how major innovations occurred during the history of life and introduces research that integrates experimental results from molecular genetics with data in the fossil record. The unit closes with a look at the evolution of our own species. ■

In North America, *Rhagoletis pomonella* populations are diverging into species that are specialized for parasitizing fruits of apple (left) versus hawthorn (right). They are just one example where biologists are documenting speciation in nature.

16 Mechanisms of Speciation

No one knows how many different species are alive today. Slightly over 1.5 million species have been described thus far; conservative estimates propose that the actual total is 3–5 million and more aggressive analyses suggest that the total might be as high as 100 million. This chapter focuses on how these species came to be. More specifically, we'll explore how mutation, natural selection, migration, and drift can cause populations to diverge and form new, independent species.

Speciation is among the most fundamental events in the history of life. It has occurred millions, if not billions, of times since life originated more than 3 billion years ago. In addition to its intrinsic importance, though, studying speciation has interesting practical applications. Understanding what species are and how they form is central to efforts to preserve biodiversity. To begin, let's start with the field's most fundamental question: What is a species?

16.1 Species Concepts

All human cultures recognize different types of organisms in nature and name them. These taxonomic or naming systems are based on judgments about the degrees of similarity among organisms. People intuitively group like with like. The challenge to biologists has been to move beyond these informal judgments to a definition of species that is mechanistic and testable and to a system for naming and classifying the diversity of life that accurately reflects the evolutionary history of organisms.

These goals have been difficult to achieve, even though most biologists agree on what a species is: It is the smallest evolutionarily independent unit. Evolutionary independence occurs when mutation, selection, gene flow, and drift operate on populations separately. Evolution consists of changes in allele frequencies, and species form a boundary for the spread of alleles. As a result, different species follow different evolutionary trajectories. Based on this definition, it is clear that the essence of speciation is lack of gene flow.

Species consist of interbreeding populations that evolve independently of other populations.

Although this definition sounds straightforward, it is often difficult to put it into practice. The challenge is to establish practical criteria for identifying when populations are actually evolving independently. To drive this point home, consider the three most important "species concepts" currently in use. Each of the three agrees that species are evolutionarily independent units that are isolated by lack of gene flow, but each employs a different criterion for determining that independence is actually in effect.

The Morphospecies Concept

In traditional cultures, people name species based on morphological similarities and differences. In biology, careful analyses of phenotypic differences are the basis of identifying morphospecies.

The great advantage of the morphospecies concept is that it is widely applicable. Morphospecies can be identified in species that are extinct or living, and in species that reproduce sexually or asexually. The most troublesome disadvantage of the morphospecies concept is that when it is not applied carefully, species definitions can become arbitrary and idiosyncratic. In the worst-case scenario, species designations made by different investigators are not comparable. In addition, the concept can be difficult to apply in groups like bacteria, archaea, and many fungi that are small and have few measurable morphological characters to assess.

Paleontologists have to work around other restrictions when identifying species. Fossil species that differed in color or the anatomy of soft tissues cannot be distinguished. Neither can populations that are similar in morphology but were strongly divergent in traits like songs, temperature or drought tolerance, habitat use, or courtship displays. Whether living or fossil, populations like these are called **cryptic species**. The adjective cryptic is appropriate because groups that were or are actually independent of one another appear to be members of the same species based on morphological similarity.

The Biological Species Concept

Under the biological species concept (BSC), the criterion for identifying evolutionary independence is reproductive isolation. Specifically, if populations of organisms do not hybridize regularly in nature, or if they fail to produce fertile offspring when they do, then they are reproductively isolated and considered good species. The biological species concept has been widely accepted since Ernst Mayr proposed it in 1942. It is used in practice by many zoologists and is the legal definition employed in the Endangered Species Act, which is the flagship biodiversity legislation in the United States.

The great strength of the BSC is that reproductive isolation is a meaningul criterion for identifying species because it confirms lack of gene flow. Lack of gene flow is the litmus test of evolutionary independence in organisms that reproduce sexually. Although this criterion is compelling in concept and useful in some situations, it is often difficult to apply. For example, if nearby populations do not actually overlap, researchers have no way of knowing whether they are reproductively isolated. Instead, biologists have to make subjective judgments to the effect that, "If these populations were to meet in the future, we believe that they are divergent enough already that they would not interbreed, so we will name them different species." In these cases, species designations cannot be tested with data. Furthermore, the biological species concept can never be tested in fossil forms, is irrelevant to asexual populations (see Box 16.1), and is difficult to apply in the many plant groups where hybridization between strongly divergent populations is routine (but see Rieseberg et al. 2006).

Box 16.1 What about bacteria and archaea?

Much of the research reviewed in this chapter focuses on events that lead to reproductive isolation and result in lack of gene flow. Indeed, the BSC treats reproductive isolation as the criterion of speciation. But in many eukaryotes and in all bacteria and archaea, reproduction takes place asexually. Thus, there is no exchange of genetic material when bacteria and archaea reproduce. When gene flow does occur between bacterial cells, it is limited to small segments of the genome. Gene flow in bacteria and archaea is also unidirectional, occurs in the absence of reproduction, and may result in genetic recombination—meaning, the creation of a chromosome with a new combination of alleles.

Although biologists have confirmed that several processes can result in gene transfer between bacterial or archaeal species, researchers are just beginning to quantify the extent of gene flow in nature (Nesbo et al. 2006). One thing is certain, though: alleles are routinely transferred between members of widely diverged bacterial and archaeal lineages. In some cases the species involved have genomes whose base sequences have diverged up to 16% (Cohan 1994, 1995). In contrast, genetic exchange between eukaryotes is generally limited to organisms whose genomes have diverged a total of 2% or less.

A key point here is that what most of us consider normal sex—meaning meiosis followed by the reciprocal exchange of homologous halves of genomes, among members of the same species—is unheard of in most organisms. As a result, gene flow plays a relatively minor role in homogenizing allele frequencies among bacteria populations. In these organisms the primary consequence of gene flow is that certain cells acquire alleles—via one-way flow from other cells—with high fitness advantages, such as sequences that confer antibiotic resistance or give the recipient cell the ability to use a new type of sugar or other energy source.

Based on these observations, Lawrence and Ochman (1998) have proposed that acquiring novel alleles through lateral gene transfer is the primary mechanism for speciation in bacteria. Their hypothesis is that gene flow triggers divergence among bacteria populations, even though it prevents divergence among eukaryotes. If their hypothesis is correct, it means that bacterial species may consist of cells that recently descended from a common ancestor and that have not experienced gene flow via lateral transfer.

More recent work, based on comparing the complete genome sequences of strains of the same bacterial species, has come to a similar conclusion—specifically, that bacterial and archaeal species should be identified on the basis of gene content, or which genes are present (Constantinidis and Tiedje 2005). The logic here is that bacterial and archaeal species are best defined in an ecological context, based on their ability to thrive in a particular environment. This ability, in turn, is dependent on which genes are present in the genomes of these organisms. Because the genes required for using particular sources of food are often acquired through gene flow, the ecological view coincides nicely with the idea that gene flow triggers speciation in bacteria and archaea, instead of impeding speciation as it does in eukaryotes.

The Phylogenetic Species Concept

Systematists are biologists who are responsible for classifying the diversity of life. A growing number of systematists and other evolutionary biologists are promoting an alternative to the biological species concept called the phylogenetic species concept (PSC)—also known as the genealogical species concept. This approach focuses on a criterion for identifying species called monophyly. You might recall from Chapter 4 that monophyletic groups are defined as lineages that contain all of the known descendants of a single common ancestor. Under the phylogenetic species concept, species are identified by estimating the phylogeny of closely related populations and finding the smallest monophyletic groups. On a tree like this, species form the tips. For example, if the taxa labeled A–G in Figure 16.1 represent populations—as opposed to genera, families, orders, or other types of **taxa**, or named groups—then they are the smallest monophyletic groups on the tree and represent distinct species. In contrast, if populations cannot be clearly distinguished in a phylogeny by unique, derived characters, then they will form clusters like the populations designated B, E, and G in Figure 16.1. The populations that make up these clusters would be considered part of the same species.

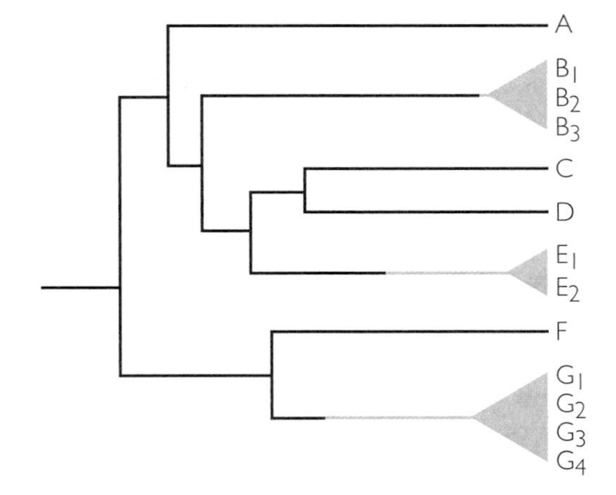

Figure 16.1 Phylogenetic species The taxa labeled A–G on the tips of this phylogeny represent distinct species. Groups labeled G_1, G_2, etc. represent populations of the same species.

The rationale behind the phylogenetic species concept is that traits can only distinguish populations on a phylogeny if the populations have been isolated in terms of gene flow and have diverged genetically, and possibly morphologically as well. Put another way, to be called separate phylogenetic species, populations must have been evolutionarily independent long enough for the diagnostic traits to have evolved. Populations within species have shared, derived traits that distinguish them from populations of other species (see Chapter 4).

The appeals of this approach are that it can be applied to any type of organism—sexually reproducing, asexually reproducing, or fossil—and that it is testable: Species are named on the basis of statistically significant differences in the traits used to estimate the phylogeny.

The challenge comes with putting this criterion into practice. As Chapter 4 indicated, it takes a significant amount of time, money, and careful analysis to estimate evolutionary relationships. As a result, well-supported, carefully constructed phylogenies are available for a relative handful of groups thus far. In addition, it is widely recognized that instituting the phylogenetic species concept could easily double the number of named species and might create a great deal of confusion if traditional names and species identities are changed.

Proponents of the concept are not bothered by the prospect of recognizing many additional species, however. They claim that if a dramatic increase in the number of named species did occur, it is necessary to reflect biological reality. As predicted, recent analyses have found that the PSC often distinguishes a series of cryptic species in populations that were formerly considered a single species (e.g., Dettman et al. 2003; Gaines et al. 2005; Hebert et al. 2003; Pringle et al. 2005). For an exception to this generalization, see Johnson et al. 2005.

Applying Species Concepts: Two Case Histories

Species can be identified by distinctive morphological traits, reproductive isolation, and/or phylogenetic independence. Each species concept has advantages and disadvantages.

Although it is probably unrealistic to insist on a single, all-purpose criterion for identifying species (Endler 1989), the major species concepts that have been proposed are productive when applied in appropriate situations. Consider, for example, how recent efforts to apply more than one species concept have improved our understanding of diversity in the most abundant group of ocean-dwelling animals and informed efforts to preserve African elephants.

Diversification in Marine Copepods

Copepods are small crustaceans that are extremely abundant in the world's oceans. The species *Eurytemora affinis,* for example, is only 1–2 mm long but is the most important grazing animal in many of the world's largest estuaries (Figure 16.2). Estuaries, in turn, are ecosystems that form where rivers flow into the sea. Because they tend to be nutrient rich, estuaries are among the most productive habitats in marine environments. In many cases, fish that spend their juvenile stage feeding on *Eurytemora* and other small inhabitants of estuarine environments spend their adult lives in the open ocean.

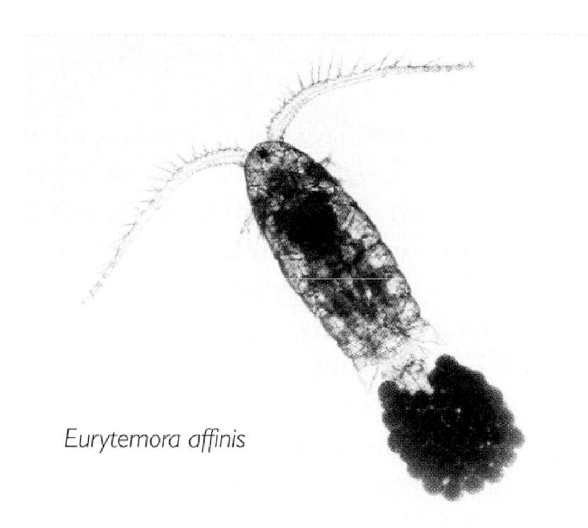

Figure 16.2 A cosmopolitan copepod *Eurytemora affinis* is a common inhabitant of coastal environments throughout the world.

Although *E. affinis* is found along the coasts of Asia, Europe, and North America, traditional analyses based on the morphospecies concept had grouped all of the populations into the same species. To test this hypothesis, Carol Eunmi Lee (2000) collected *E. affinis* from a wide array of locations throughout the Northern Hemisphere. To assess whether some of the 38 populations in her sample represent separate species under the biological species concept, she tested individuals from several populations for the ability to mate and produce fertile offspring. To address the same question with the phylogenetic species concept, she sequenced two genes and used similarities and differences in the base sequences observed in the 38 populations to estimate their evolutionary relationships.

Her results? The phylogeny clearly showed that at least eight independent species exist, with each occupying a distinctive geographic area (Figure 16.3). These results were supported by the results of the mating tests, which showed that populations from different phylogenetic species are unable to produce fertile offspring and are thus reproductively isolated.

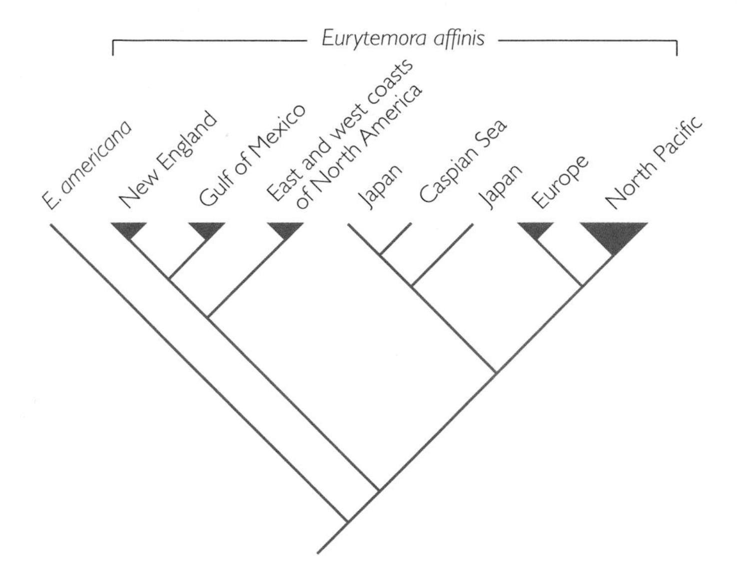

Figure 16.3 Copepod diversity A phylogeny of *E. affinis* populations shows that at least eight phylogenetic species exist, each in a different geographic area.

The take-home message of this study is that species diversity in copepods is likely to be far greater than previously thought. Employing more than one species concept can help biologists recognize diversity and organize research on its consequences.

How Many Species of Elephant Live in Africa?

Traditionally, the world's elephants have been classified as two species: African (*Loxodonta africana*) and Asian (*Elephas maximus*). Recent morphological analyses began to challenge this view, however, by pointing out that African elephants that live in forest habitats versus savanna or grassland habitats have distinctive morphological features (Figure 16.4a). But because the populations that inhabit the two habitat types don't interact regularly, individuals rarely have a chance to interbreed. As a result, it has been difficult to assess whether forest and savanna elephants qualify as separate species under the BSC.

To clarify the situation, Alfred Roca and coworkers (2001) applied the PSC. They began by collecting tissue samples from 195 elephants in 21 populations throughout central and south Africa. They then isolated DNA from the tissues, sequenced four genes from each individual, and used the resulting data to estimate which populations were closely or more distantly related. The phylogeny that resulted from this analysis clearly showed that forest and savanna elephants qualify as distinct phylogenetic species (Figure 16.4b). The group has proposed naming the forest elephants *Loxodonta cyclotis* and retaining the name *L. africana* for the savanna/grassland populations.

Reliable criteria for identifying species are essential for preserving biodiversity.

The result has a sense of urgency because many elephant populations in Africa are declining due to habitat loss and illegal hunting. Based on this work, it is clear that conservation programs should be focused on preserving both forest and savanna populations as distinct entities. In this case, employing several criteria for

Figure 16.4 **Elephant diversity** (a) In West Africa, elephants that live in forest habitats (left) have morphological characteristics that distinguish them from savanna-dwelling elephants (right) from west, central, and east Africa. (b) This evolutionary tree indicates that forest-dwelling elephants are a distinct phylogenetic species.

identifying species proved to be a productive approach in clarifying both conservation and evolutionary issues.

16.2 Mechanisms of Genetic Isolation

Given that several tools are available for identifying species, let's turn our attention now to the question of how species form. Classically, speciation has been hypothesized to be a three-stage process: an initial step that isolates populations, a second step that results in divergence in traits such as mating system or habitat use, and a final step that produces reproductive isolation. According to this model, the isolation and divergence steps were thought to take place over time and to occur while populations were located in different geographic areas. The final phase was hypothesized to occur when these diverged populations came back into physical contact—an event known as **secondary contact**.

Recent research has shown that this view was oversimplified. For example, it is now clear that the isolation and divergence steps that initiate speciation frequently take place at the same time and in the same place. In addition, it appears likely that in a significant number of speciation events or even a majority, the third phase never occurs. Even so, the isolation/divergence/secondary contact hypothesis provides a useful framework for analyzing how speciation takes place.

The focus of this section is on the first step in speciation—genetic isolation. We'll analyze how physical separation or changes in chromosome complements can reduce gene flow between populations. Once gene flow is dramatically reduced or ceases, evolutionary independence begins and speciation is underway. Section 16.3 addresses how genetic drift, natural selection, and sexual selection act on mutations and cause genetically isolated populations to diverge. In Section 16.4, we consider what happens if secondary contact occurs.

The speciation process begins when gene flow is disrupted and populations become genetically isolated.

Physical Isolation as a Barrier to Gene Flow

Chapter 7 introduced models showing that gene flow tends to homogenize gene frequencies and reduce the differentiation of populations. You might recall the example of water snakes from mainland and island habitats in Lake Erie, and experiments that showed a selective advantage for unbanded snakes on island habitats. But because migration of banded forms from the mainland to the islands occurs regularly, and because banded and unbanded forms subsequently interbreed, the island populations did not completely diverge from mainland forms. Migration continually introduced alleles for bandedness, even though selection tended to eliminate them from the island populations.

Now consider a thought experiment: What would happen if lake currents changed in a way that effectively stopped the migration of banded forms from the mainland to the islands? Gene flow between the two populations would end and the balance between migration and natural selection would tip. The island population would be free to differentiate as a consequence of mutation, natural selection, and drift. These forces would act on them independently of the forces acting on mainland forms.

Geographic isolation produces reproductive isolation, and thus genetic isolation.

This scenario illustrates a classical theory for how speciation begins, called the **allopatric model** (Mayr 1942, 1963). Translated literally, "allopatric" means different country or homeland. The essence of allopatric speciation is that physical isolation creates an effective barrier to gene flow. Research has shown that in many cases, geographic isolation has been an important trigger for the second stage in the speciation process: genetic and ecological divergence. Geographic isolation can come about through dispersal and colonization of new habitats or through vicariance events, where an existing range is split by a new physical barrier (Figure 16.5).

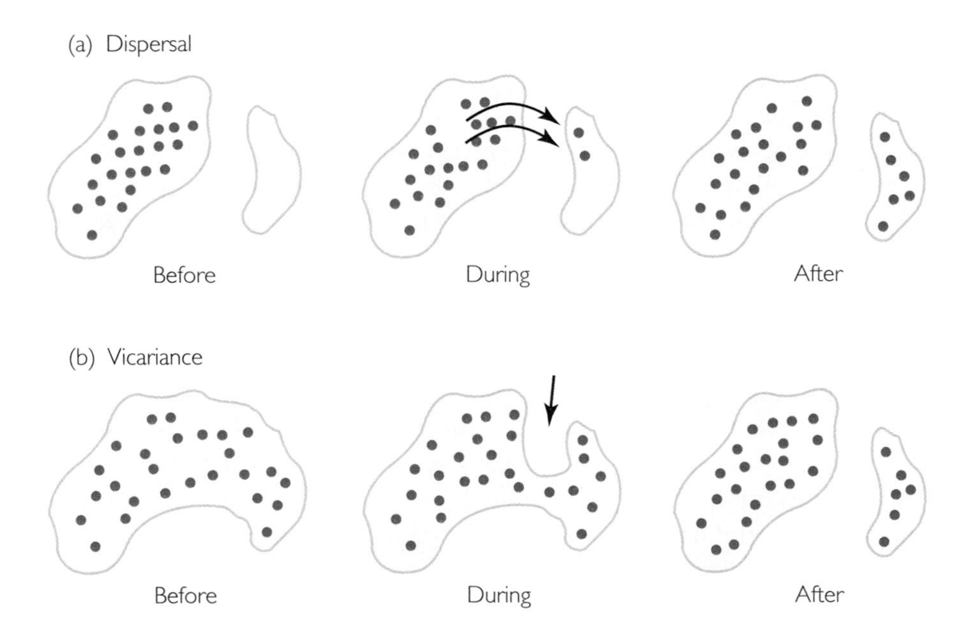

Figure 16.5 Isolation by dispersal and vicariance In the diagram of dispersal (a), the arrows indicate movement of individuals. In the diagram of vicariance (b), the arrows indicate an encroaching physical feature such as a river, glacier, lava flow, or new habitat.

Geographic Isolation through Dispersal and Colonization

One of the most spectacular adaptive radiations among insects is also a superb example of geographic isolation through dispersal. The Hawaiian drosophilids, close relatives of the fruit flies we have encountered before, include an estimated 1,000 species and are renown for their exceptional ecological diversity. Hawaiian flies can be found from sea level to montane habitats and from dry scrub to rain forests. Food sources, especially the plant material used as the medium for egg laying and larval development, vary widely among species. One of the Hawaiian flies even lays its eggs in spiders, while another has aquatic larvae. In addition, many species have elaborate traits, such as patterning on their wings or modified head shapes, that are used in combat or in courtship displays (Figure 16.6).

The leading explanation to explain this diversity begins with dispersal and colonization. Many of the Hawaiian flies are island endemics, meaning that their range is restricted to a single island in the archipelago. If small populations of flies, or perhaps even single gravid females, disperse to new habitats or islands, then the colonists found new populations that are physically cut off from the ancestral species. Divergence begins after the founding event, resulting from genetic drift and natural selection acting on the genes responsible for courtship displays and habitat use.

The logic of the dispersal-and-colonization hypothesis is compelling, but do we have evidence, other than endemism, that these events actually occurred? Because the geology of the Hawaiian islands is well known, the hypothesis makes a strong prediction about speciation patterns in flies. The Hawaiian islands are produced by a volcanic hotspot under the Pacific Ocean. The hotspot is stationary, but periodically spews magma up and out onto the Pacific plate, forming islands. After islands form, continental drift carries them with the Pacific plate to the north and west (see Figure 16.7a). As time passes, the volcanic cones gradually erode down to atolls and submarine mountains.

The dispersal-and-colonization hypothesis makes two predictions based on these facts: (1) Closely related species should almost always be found on adjacent islands, and (2) at least some sequences of branching events should correspond to the sequence in which islands were formed. James Bonacum and coworkers (2005) used sequence differences in a series of mitochondrial and nuclear genes to estimate the phylogeny of closely related Hawaiian flies and found exactly these patterns (Figure 16.7b). This is strong evidence that dispersal to new habitats triggered speciation. Similar patterns have been observed in the phylogenies

Figure 16.6 Hawaiian *Drosophila* As these photos of *Drosophila suzukii, D. macrothrix, and D. nigribasis* (top to bottom) show, the *Drosophila* found in Hawaii are remarkably diverse in body size, wing coloration, and other traits.

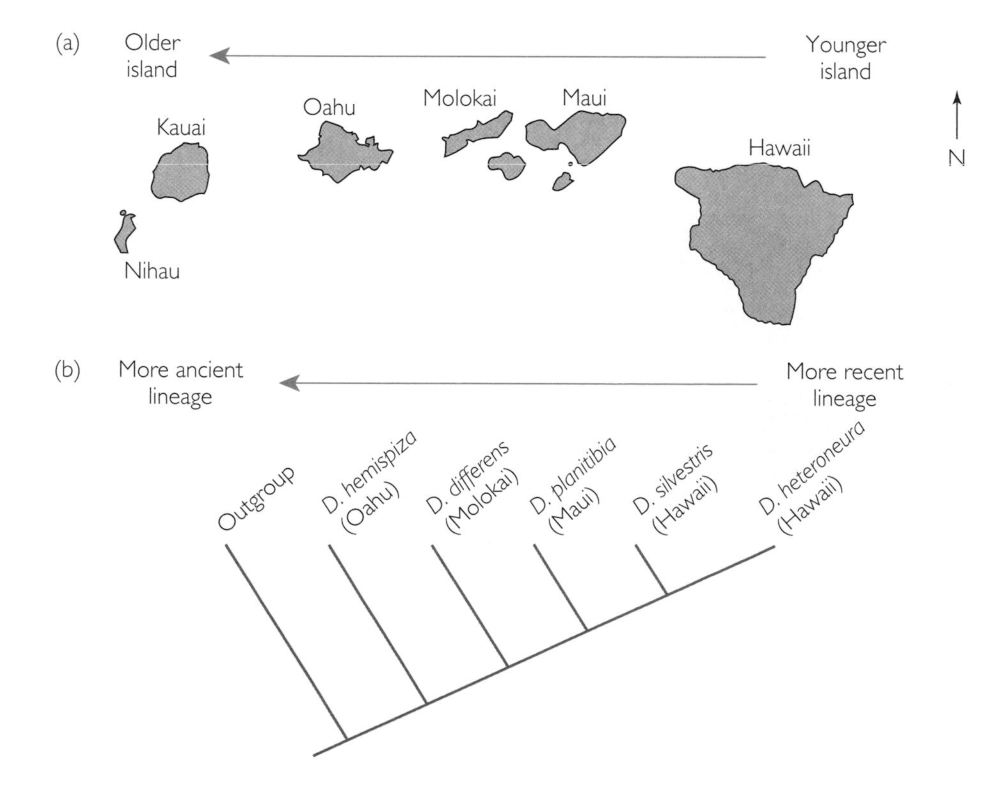

Figure 16.7 Evidence for speciation by dispersal and colonization events (a) The Hawaiian islands are part of an archipelago that stretches from the island of Hawaii to the Emperor Seamounts near Siberia. The youngest landform in the chain is the island of Hawaii, which still has active volcanoes. (b) The five *Drosophila* species on the tree are a closely related group. Note that the older-to-younger sequence of branches on the phylogeny corresponds to the older-younger sequence of island formation shown in part (a). This pattern is consistent with the hypothesis that at least some of the speciation events in this group were the result of island hopping. See Bonacum et al. (2005).

of Hawaiian crickets (Mendelson and Shaw 2005) and Galápagos turtles (Beheregaray et al. 2004).

Populations can become geographically isolated when individuals colonize a new habitat.

As a mechanism for producing physical isolation and triggering speciation, the dispersal-and-colonization hypothesis is relevant to a wide variety of habitats in addition to oceanic islands. Hot springs, deep sea vents, fens, bogs, caves, mountaintops, and lakes or ponds with restricted drainage also represent habitat islands (e.g., Dawson and Hammer 2005). Dispersal to novel environments has proven to be a general mechanism for initiating speciation.

Geographic Isolation through Vicariance

Populations can also become geographically isolated when a species' former range is split into two or more distinct areas.

Vicariance events split a species' distribution into two or more isolated ranges and discourage or prevent gene flow between them. There are many possible mechanisms of vicariance, ranging from slow processes such as the rise of a mountain range or a long-term drying trend that fragments forests, to rapid events such as a mile-wide lava flow that bisects a snail population.

Nancy Knowlton and colleagues studied a classical vicariance event: the recent separation of marine organisms on either side of Central America. Geological evidence has established that the Isthmus of Panama closed about 3 million years ago. As the isthmus rose and created a land bridge between North and South America, populations of marine organisms became separated on the Atlantic and Pacific sides. When the oceans were separated in this way, did the populations that ended up on either side speciate?

To address this question, Knowlton and coworkers (1993) analyzed a series of snapping shrimp (*Altheus*) populations from either side of the isthmus (Figure 16.8a). Based on the morphospecies concept, the populations that they sampled appeared to represent seven pairs of closely related species pairs, or **sister species**, with one member of each pair found on each side of the land bridge. The phylogeny of these shrimp, estimated from data on DNA sequences, confirms this hypothesis (Figure 16.8b). The species pairs from either side of the isthmus, reputed to be sisters on the basis of morphology, are indeed each others' closest relatives. This result is consistent with the prediction from the vicariance hypothesis.

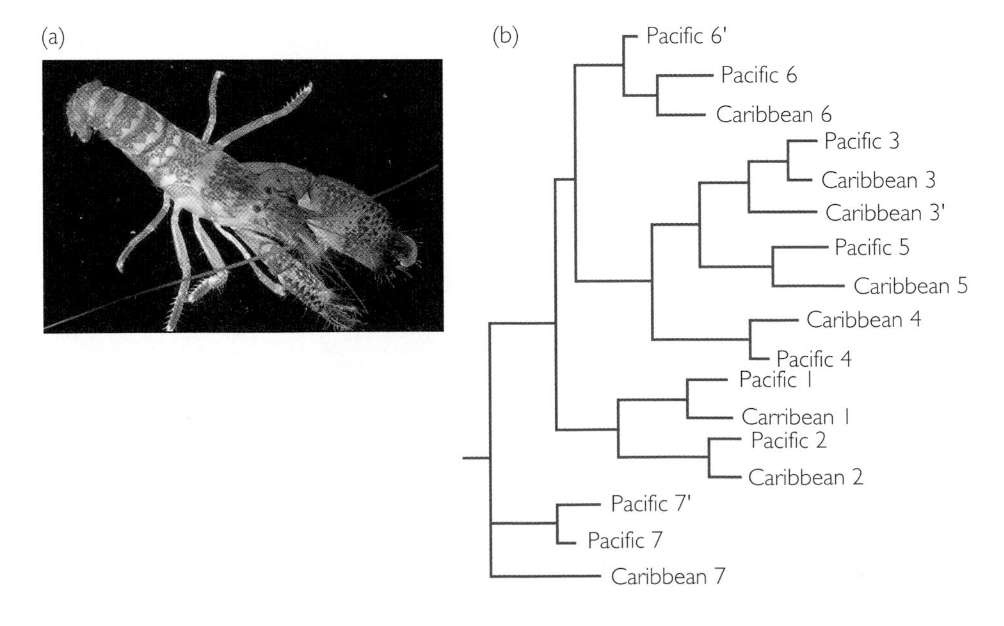

Figure 16.8 Snapping shrimp speciated due to vicariance (a) This is *Alpheus malleator*, found on the Pacific side of the Panamanian isthmus. (b) This tree of unnamed species was estimated from sequence divergence in mitochondrial DNA. Reprinted with permission from Knowlton et al. (1993). Morphological sister species are numbered and identified by location. The prime marks after some letters indicate cryptic species distinguished by sequence differences. In every case, the putative sister species are indeed each others' closest relative.

Furthermore, when Knowlton and coworkers put males and females of various species pairs together in aquaria and watched for aggressive or pairing interactions, the researchers found a strong correlation between the degree of genetic distance between species pairs and how interested the shrimp were in courting. Males and females from species with greater genetic divergence, indicative of longer isolation times, were less interested in one another. Finally, almost none of the pairs that formed during the courtship experiments produced fertile clutches. This last observation confirms that the Pacific and Caribbean populations are indeed separate species under all three of the species concepts we have reviewed.

One of the most interesting aspects of the study, though, was that the data contradicted an original prediction made by the vicariance hypothesis. If the land bridge had formed rapidly, we would expect that genetic distances and degrees of reproductive isolation would be identical in all seven species pairs. This is not the case. For example, DNA sequence divergence between species pairs varied from about 6.5% to more than 19%.

What is going on? Upon reflection, it became clear that a prediction of identical divergence was naïve, because it is unlikely that the land bridge popped up all at once. Instead, as the land rose and the ocean gradually split and retreated on either side, different shrimp populations would become isolated in a staggered fashion, depending on the depth of water each species occupied and how efficiently their larvae dispersed. The ranges of deeper-water species, or those with less-motile larvae, would be cut in two first. Consistent with this logic, the species numbered 6 and 7 in Figure 16.8b inhabit deep water, while the species numbered 1–4 live in shallower water along the coast (see Knowlton and Weigt 1998). Note also that the degree of genetic divergence in the species pairs numbered 1–4 is very similar. These are the lowest values observed, perhaps indicating "the final break" between the two oceans 3 million years ago.

An array of similar studies has convinced biologists that vicariance has been an important isolating mechanism and trigger for speciation in a wide variety of groups. Other well-studied examples include a seaway that separated the northern and southern portions of the Baja peninsula about 1 million years ago (see Riginos 2005) and the fragmentation of habitats by glacial advances during the Pleistocene (e.g., Weir and Schluter 2004; Hoskin et al. 2005).

A Role for Mutation: Polyploidy and Other Chromosome Changes as a Barrier to Gene Flow

Theory predicts that populations may speciate after becoming physically isolated due to dispersal or vicariance, and data have confirmed that these events are common triggers for speciation. But as examples here and in the next section will show, it is entirely possible for speciation to occur in the absence of physical isolation between populations. For example, Chapter 5 pointed out that mutations that result in polyploidization can produce instant reproductive isolation between parental and daughter populations. To recall why this is so, consider a tetraploid population newly created by mutation. These individuals produce diploid gametes. If they mate with members of the diploid population, which make haploid gametes, the offspring will be triploid. Triploid individuals are rarely able to make viable gametes, because homologs do not pair properly at meiosis, leading to cells with dysfunctional chromosome complements.

Changes in chromosome number isolate populations genetically.

In addition to problems with meiosis and gamete formation, researchers have documented that flower shape and the timing of flowering is often different in tetraploid populations that are derived from diploid populations. Because different pollinators tend to visit individuals with different ploidy levels, the potential for gene flow between diploid and tetraploid populations is low, even when individuals are growing side by side (e.g., Segraves and Thompson 1999; Husband and Schemske 2000).

Speciation triggered by changes in chromosome number has been especially important in plants.

How important is polyploidization as a mechanism of speciation's first stage? Biologists estimate that among the estimated 300,000 species of land plants, at least 2–4% are derived directly from polyploidization events. In addition, biologists have documented the origin of new plant species via polyploidization in Tragopogon, Senecio, and Spartina within the past 200 years (Novak et al. 1991; Kadereit et al. 2006; Salmon et al. 2005). Although speciation by polyploidy is much less common in animals than in plants, a phylogeny of 25 species in the fish genus *Barbus*, estimated by Annie Marchordum and Ignacio Doadrio (2001), showed that polyploidization initiated a speciation event at least three times.

Changes in chromosome number less drastic than polyploidization may also be important in speciation. For example, Oliver Ryder and colleagues (1989) studied chromosome complements in a series of small African antelopes—called dik-diks—that were being displayed in North American zoos. Although zookeepers traditionally recognized just two species, Ryder's team distinguished three on the basis of chromosome number and form. Furthermore, they were able to show that hybrid offspring between unlike karyotypes are infertile. Their research revealed a cryptic species. Two of the dik-dik species have apparently differentiated on the basis of karyotype.

It is extremely common to find small-scale chromosomal changes like these when the karyotypes of closely related species are compared. Although these mutations could be important in causing genetic divergence between populations (White 1978), much of the extensive work on chromosomal differentiation done to date is only correlative. That is, many studies have measured chromosome differences in related species and claimed that chromosomal incompatibilities are responsible for genetic isolation. But in many cases, it is likely that the chromosome differences arose after speciation had occurred due to other causes (Patton and Sherwood 1983). Work continues on establishing causative links between small-scale karyotype differences and speciation (e.g., Noor et al. 2001; Navarro and Barton 2003).

16.3 Mechanisms of Divergence

Dispersal, vicariance, and polyploidization only create the conditions for speciation. For the event to continue, genetic drift and natural selection have to act on mutations in a way that creates divergence in the isolated populations. This section focuses on how drift and selection act on closely related populations once gene flow between them has been reduced or eliminated.

Genetic Drift

Population genetic models, like those analyzed in Chapter 7, have quantified the major effects of genetic drift. These effects are random fixation of alleles and random loss of alleles. We also reviewed data from a dispersal-and-colonization event in small birds called silvereyes. You might recall that measurements of genotypes confirmed that the founding population was a nonrandom sample of the source population. Drift had produced a colonizing population that was genetically distinct from the original population.

Because genetic drift is a sampling process, its effects are most pronounced in small populations. This is an important observation because traditionally, most species are thought to have originated with low population sizes. Normally, only tiny numbers of individuals are involved in colonization events; vicariance events fragment large populations into two or more smaller ones; and polyploidization initially produces only a handful of individuals. As a consequence, genetic drift has long been hypothesized as the key to speciation's second stage.

Drift can produce rapid genetic divergence in small, isolated populations.

A variety of genetic models have examined how drift might lead to rapid genetic differentiation in small populations (see Templeton 1996). The general message of these models is that small populations that become isolated start out as a nonrandom sample of the ancestral population. As drift continues to occur in the small, derived population, it leads to a random loss of alleles and the random fixation of existing and new alleles. As a result, the isolated population should undergo rapid genetic divergence from the ancestral population.

The role of drift in speciation events is controversial, however. Russell Lande (1980, 1981) has shown that when a population is reduced to a small size for a short period of time—the phenomenon known as **bottlenecking**—only very rare alleles tend to be lost due to drift. For drift to change allele frequencies dramatically, the founding population has to be extremely small and remain small for a significant period of time. Peter Grant and Rosemary Grant (1996) have also pointed out that hundreds of small populations have been introduced to new habitats around the world in the last 150 years due to the action of humans, but that few, if any, dramatic changes in genotypes and no speciation events have resulted. Although genetic drift once dominated discussions of speciation mechanisms, most evolutionary biologists now take a much more balanced view. Natural selection is becoming recognized as the most important force promoting the divergence of populations (Schulter 2001; Via 2002).

Natural Selection

Marked genetic differences have to emerge between closely related populations for speciation to occur. Drift almost always plays a role in creating these genetic differences when at least one of the populations is small. But natural selection can also lead to divergence if one of the populations occupies a novel environment or uses a novel resource.

Selection's role in speciation is clearly illustrated by recent research on apple and hawthorn flies. These closely related insect populations are diverging because of natural selection on preferences for a crucial resource: food. The work is important because it focuses on insects that specialize on parasitizing a specific host plant, and because "there are more host-specific phytophagous insects than any other life form on Earth" (Dambrowski et al. 2005, p. 1963).

The apple maggot fly, *Rhagoletis pomonella,* is found throughout the northeastern and north central United States (see photos at the start of the chapter). The species is a major agricultural pest, causing millions of dollars of damage to apple crops each year. The flies also parasitize the fruits of trees in the hawthorn group (species of *Crataegus*), which are closely related to apples.

Male and female *Rhagoletis* identify their host trees by sight, touch, and smell. Courtship and mating occur on or near the fruits. Females lay eggs in the fruit while it is still on the tree, and the eggs hatch within two days. The larvae begin eating their way through the fruit and then continue developing through three larval stages in the same fruit. This takes about a month. After the fruit falls to the ground, the fully developed larvae leave and burrow a few inches into the soil. After 3–4 days there they pupate—meaning that they secrete a tough case, inside of which they will undergo metamorphosis. They spend the winter underground as pupae, in a resting state called diapause. Most leave this resting stage and emerge as adults the next summer, starting the cycle anew.

Apple trees clearly represent a novel food source for *Rhagoletis*. Hawthorn trees and *Rhagoletis* are native to North America, but apple trees were introduced from Europe less than 300 years ago. *Rhagoletis* were first observed parasitizing apples in the mid-1800s. Thus, *Rhagoletis* has recently made a host switch. The question is: Are the flies that parasitize apple fruits and hawthorn fruits distinct populations?

The question could be answered by two contrasting hypotheses. The first is that natural selection, based on a preference for different food sources, has created two genetically distinct groups of flies. The alternative is that flies parasitizing hawthorns and apples are members of the same population. If so, then flies on hawthorns and apples interbreed freely and are genetically indistinguishable, and selection for exploiting different hosts has not occurred.

The hypothesis of no differentiation actually appears much more likely, because the two host trees occur together throughout their ranges. Far from being physically isolated, at some sites hawthorn and apple trees are almost in physical contact. Marked flies have been captured over a mile from the site where they were originally captured, suggesting that individuals may search widely for appropriate fruit to parasitize. Thus, flies from the same population might simply switch from apple to hawthorn trees and back, based on fruit availability.

To test these contrasting hypotheses, Jeff Feder and colleagues analyzed the genetic makeup of flies collected from hawthorns versus apples. Remarkably, they found a clear distinction in the two samples: Flies collected from hawthorns versus apples have statistically significant differences in the frequencies of alleles for six different enzymes (Feder et al. 1988, 1990). This is strong support for the hypothesis that hawthorn and apple flies have diverged and now form distinct populations. Even though the two "races" look indistinguishable, they are easily differentiated on the basis of their genotypes.

Natural selection can cause populations to diverge based on food preferences, habitats used, or other ecological differences

How could this have occurred? The key is that instead of being isolated by geography or by chromosomal incompatibilities, apple and hawthorn flies are isolated on different host species. In experiments where individuals are given a choice of host plants, apple and hawthorn flies show a strong preference for their own fruit type (Prokopy et al. 1988) and avoid the other fruit type (Forbes et al. 2005). These preferences are heritable (Dambrowski et al. 2005). Because mating takes place on the fruit, this habitat preference should result in strong nonrandom mating. Feder and colleagues (1994) confirmed this prediction by following marked individuals in the field. They found that matings between hawthorn and apple flies accounted for just 6% of the total observed.

Although host plant fidelity serves as an important barrier to mating, the two populations continue to exchange alleles. To cause the genetic divergence that has been observed, natural selection must overwhelm this gene flow in some way.

Feder and coworkers (1997; Filchak et al. 2000) hypothesized that natural selection for divergence is triggered by a marked difference in when apple and hawthorn fruits ripen. Hawthorn fruits ripen 3–4 weeks after apples do. As a result, hawthorn fly larvae experience cool temperatures while feeding prior to pupating, while apple fly larvae experience warm temperatures before pupating. Feder and colleagues suggested that certain alleles are favored in cool prewinter temperatures versus warm prewinter temperatures.

To test this hypothesis, the researchers collected a large number of pupae from hawthorn fruits, split them into groups, and exposed each sample to 1, 2, 3, 4, or 5 weeks of warm weather. In each group of pupae, the warm period was followed by a period of freezing to simulate winter and then a period of warm temperatures to simulate spring. The last step in the experiment was to collect the individuals that emerged as adults and assay the frequencies of the six enzyme alleles that differ between the hawthorn and apple races. The results for the allele called Acon-2 95, shown in Figure 16.9, are typical. This graph shows that the hawthorn race individuals that were exposed to a month of warm days as pupae, and that survived to develop into adults during the spring, had enzyme allele frequencies similar to those found in apple flies. The group got similar results when fly larvae were exposed to cool versus warm temperatures prior to pupation (Filchak et al. 2000).

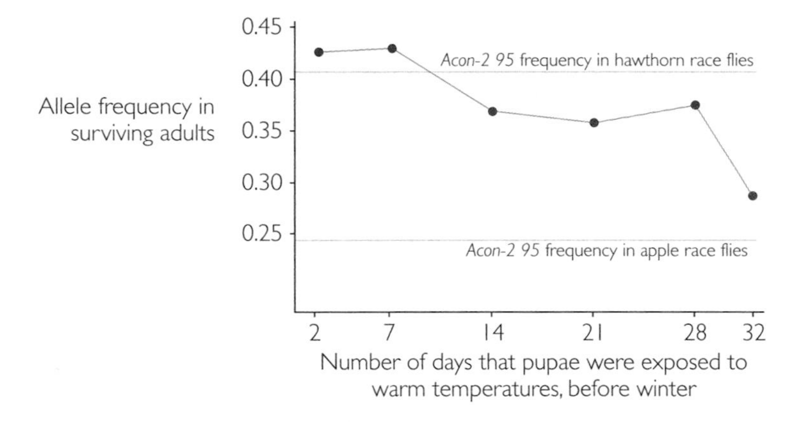

Figure 16.9 Allele frequency changes caused by differences in temperatures experienced by apple maggot flies This graph plots the frequencies of an allele called *Acon-2 95* in populations of hawthorn maggot fly pupae that survived to emerge as adults, as a function of the number of days of warm temperatures the pupae experienced. In the population that experienced an extended period of warm days—similar to the regime experienced by apple flies in nature—allele frequencies approximated those observed in natural populations of apple flies. From Feder et al. (1997).

These data are remarkable. They support the hypothesis that hawthorn flies making the switch to apples had alleles that enabled them to thrive in warm temperatures and then stay in diapause long enough to emerge at the correct time in spring. The experiment also suggests that the six alleles surveyed, or genes that are closely linked to these loci, are responsible for the adaptation to apples. In a single generation, then, the researchers succeeded in replicating the selection events that have produced divergence between the apple race and the hawthorn race in nature over the past 200 years.

Natural selection can cause populations to diverge even when a small amount of gene flow occurs.

The diapause experiment demonstrates that natural selection is responsible for strong divergence between *R. pomonella* populations even in the face of gene flow. Follow-up work has shown that the warm-adapted and cold-adapted alleles existed in *R. pomonella* populations from southern and northern regions of North America prior to the introduction of apples, and that the alleles are located inside chromosomal inversions (reviewed in Jiggins and Bridle 2004).

Many biologists now consider hawthorn and apple flies to be incipient species. This means that the populations have clearly diverged and are largely, but not completely, isolated in terms of gene flow.

Table 16.1 Speciation by natural selection

Recent studies on populations that are diverging due to differences in habitat or resource use.

Species	Type of divergence that is underway	Citation
Flowering dogwood flies	Different host plants	Dambrowski et al. 2005. *Evolution* 59: 1953–1964.
European corn borer	Different host plants	Malausa et al. 2005. *Science* 308: 258–260.
Treehoppers	Different host plants	Rodríguez et al. 2004. *Evolution* 58: 571–578.
Seed beetles	Different host plants	Messina. 2004. *Evolution* 58: 2788–2797.
Leaf beetles	Different host plants	Funk, D. J. 1998. *Evolution* 52: 1744–1759.
Larch budmoth	Different host plants	Emelianov et al. 2004. *Proceedings of the Royal Society of London B* 271: 97–105.
Pea aphids	Different host plants	Via, S. 1999. *Evolution* 53: 1446–1457.
Army worms	Different host plants	Pashley, D. P. 1988. *Evolution* 42: 93–102.
Soapberry bugs	Different host plants	Carroll, S., et al. 1997. *Evolution* 51: 1182–1188.
Goldenrod ball gallmakers	Different host plants	Abrahamson, W. G. et al. 2001. *American Zoologist* 41: 928–938.
Blueberry and apple maggot flies	Different host plants	Feder, J. L. et al. 1989. *Entomological Experiments and Applications* 51: 113–123.
Walking stick insects	Different host plants	Nosil, P. et al. 2002. *Nature* 417: 440–443.
Skipper butterflies	Different host plants	Hebert et al. 2004. *Proceedings of the National Academy of Sciences USA* 101: 14812–14817.
Lake whitefish	Populations of dwarf versus normal-sized individuals	Lu, G. and L. Bernatchez. 1999. *Evolution* 53: 1491–1505.
Three-spine sticklebacks (freshwater fish)	Benthic (bottom-dwelling) versus limnetic (open-water-dwelling) forms	Rundle, H. D. et al. 2000. *Science* 287: 306–308.
Lake Nicaragua cichlids	Benthic versus limnetic forms	Barluenga et al. 2006. *Nature* 439: 719–723.
Sockeye salmon	Sea-run versus lake-dwelling populations	Wood, C. C., and C. J. Foote. 1996. *Evolution* 50: 1265–1279.
Sockeye salmon	Populations that breed in beach versus river habitats	Hendry, A. P. et al. 2000. *Science* 290: 516–518.
Blue tits (birds)	Populations that breed in evergreen versus deciduous oak forests	Blondel, J. et al. 1999. *Science* 285: 1399–1402.
Warbler finches	Habitat selection (high vs. low elevation)	Tonnis et al. 2003. *Proceedings of the Royal Society of London B* 272: 819–826.
Crossbills (birds)	Food source (type of conifer tree seed)	Benkman, 2003. *Evolution* 57: 1176–1181.
Indigobirds	Different hosts of nest parasites	Sorenson et al. 2003. *Nature* 928–931
Mycorrhizal fungi	Different host species (orchids)	Taylor et al. 2004. *Proceedings of the Royal Society of London B* 271: 35–43.
Heliconius butterflies	Different habitats and types of warning coloration	Jiggins, C. D. et al. 2001. *Nature* 411: 302–305.

Finally, it is important to reemphasize that apple maggot flies are by no means an isolated case. As Table 16.1 shows, many other examples of divergence due to host switching in insects are being investigated. Researchers inspired by these results are even exploring the hypothesis that wasps and other species that parasitize host-specific herbivores also speciate when their hosts diverge to feed on new plants (Stireman et al. 2006). In addition, Table 16.1 highlights examples in species other than insects where populations are diverging due to natural selection on food or habitat choice (for a review, see Drès and Mallet 2002). Natural selection is a potent force for divergence.

Sexual Selection

Sexual selection acts on characters involved in mate choice. Changes in sexual selection can cause reproductive isolation and trigger rapid divergence.

Sexual selection is a form of natural selection that results from differences among individuals in their ability to obtain mates (see Chapter 11). Population genetic models have shown that changes in the way that a population of sexual organisms chooses or acquires mates can lead to rapid differentiation from ancestral populations (e.g., Fisher 1958; Lande 1981, 1982; Higashi et al. 1999). For example, if a new mutation led females in a certain population of barn swallows to prefer males with iridescent feathers instead of preferring males with long tails, then sexual selection would trigger rapid divergence. The key point is that sexual selection promotes divergence efficiently because it affects gene flow directly.

In the Hawaiian *Drosophila*, for example, sexual selection is thought to have been a key factor in promoting divergence among isolated populations. Many of these flies court and copulate in aggregations called leks. Males fight for small display territories and dance or sing for females, who visit the lek to select mates. Lek breeding is often associated with elaborate male characters, which vary widely among Hawaiian flies. Does this imply that sexual selection has been important in speciation?

The evidence in favor of the hypothesis is tantalizing, though not yet conclusive. For example, males of *Drosophila heteroneura* have wide, hammer-shaped heads (Figure 16.10a). Because males butt heads when fighting to stake out a courting arena on the lek, the unusual head shape appears to be a product of sexual selection for success in fighting while minimizing damage to the eyes (Kaneshiro and Boake, 1987). In contrast, males and females of *D. heteroneura's* closest relative, *Drosophila silvestris,* don't have hammer-shaped heads. Their heads are similar in size and shape to female *D. heteroneura* (Figure 16.10b). Instead of head-butting, *D. silvestris* males fight on the lek by rearing up and grappling with one another. Both species are endemic to the island of Hawaii.

These facts are consistent with the following scenario:

1. In the ancestor to *D. silvestris* and *D. heteroneura,* males had normal heads, courted on leks, and fought for display territories by rearing up and grappling. Females chose the males who were most successful in combat.
2. A mutation occurred in an isolated subpopulation, which led to males with a new fighting behavior: head-butting.
3. The mutant males were more efficient in combat on leks and experienced increased reproductive success because females still preferred to mate with males who won the most contests.
4. The mutation increased to fixation and additional mutations led to the elaboration of the trait over time. For example, it is possible that mutations leading to widely spaced eyes were favored because they made the eyes less prone to damage during head-butting fights.

(a) *Drosophila heteroneura*

(b) *Drosophila silvestris*

Figure 16.10 Contrasting head shapes and fighting strategies in Hawaiian *Drosophila* (a) Male *Drosophila heteroneura* have wide heads; male *Drosophila silvestris* have narrow heads. (b) *D. heteroneura* establish display territories on a lek by butting heads;. *D. silvestris* fight over display territories by rearing up and grappling with one another.

As a result of this sequence of events, the two populations would diverge due to sexual selection. The differentiation between populations would be based on the strategies and weaponry employed in male-male combat and female choice.

Formulating this type of plausible sequence is a productive way to generate testable hypotheses, but it does not substitute for genetic models, experiments, or other types of evidence. Fortunately, Christine Boake and associates (1997) have directly tested two assumptions of the sexual selection scenario outlined above. These researchers staged a series of tests in the laboratory to assess whether female *D. heteroneura* prefer to mate with especially wide-headed males. They also staged male-male contests, to test the prediction that males with wider heads are more likely to win fights. As Figure 16.11 shows, both patterns were strongly supported by their data. The results increase our confidence that sexual selection has been a prominent cause of divergence in these populations (see Boake 2005).

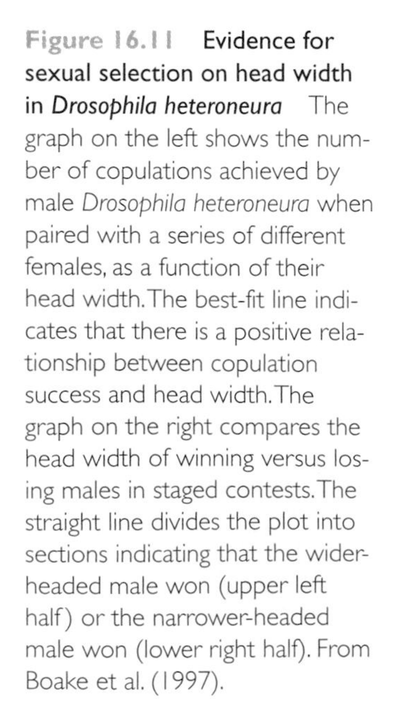

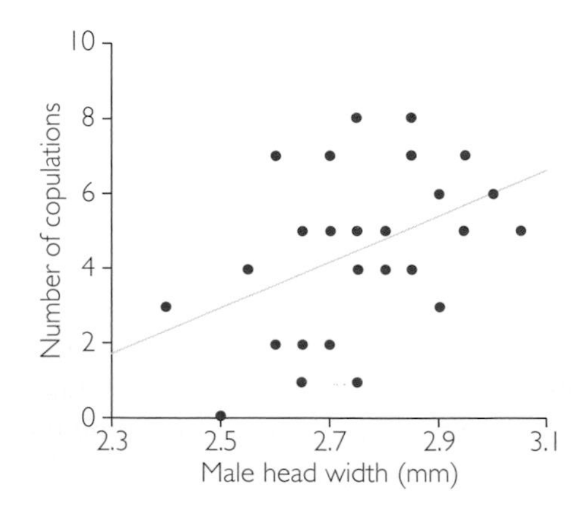

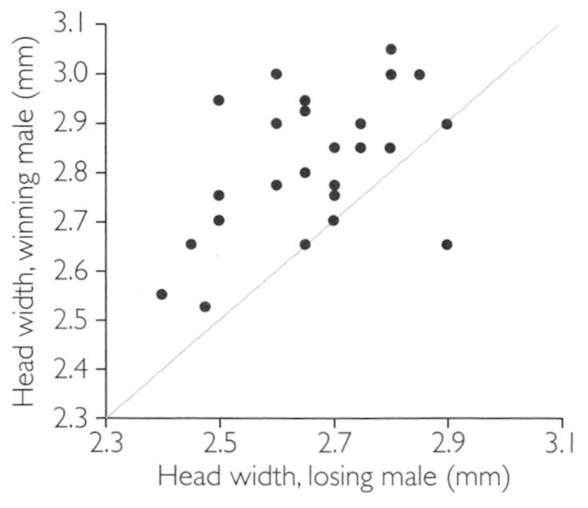

Figure 16.11 Evidence for sexual selection on head width in *Drosophila heteroneura* The graph on the left shows the number of copulations achieved by male *Drosophila heteroneura* when paired with a series of different females, as a function of their head width. The best-fit line indicates that there is a positive relationship between copulation success and head width. The graph on the right compares the head width of winning versus losing males in staged contests. The straight line divides the plot into sections indicating that the wider-headed male won (upper left half) or the narrower-headed male won (lower right half). From Boake et al. (1997).

Table 16.2 Speciation by sexual selection

Some recent studies on populations that are diverging due to differences in how females choose mates, how males attract females, or how males compete for females.

Species	Character under selection	Citation
Hawaiian crickets	Male courtship song	Shaw, K. L., and E. Lugo. 2001. *Molecular Ecology* 10: 751–759.
Field crickets	Male courtship song	Gray, D. A., and W. H. Cade. 2000. *Proceedings of the National Academy of Sciences USA* 97: 14449–14454.
Ground crickets	Male courtship song	Mousseau, T. A., and D. J. Howard. 1998. *Evolution* 52: 1104–1110.
Heliconius butterflies	Wing pattern and coloration	Jiggins, C. D. et al. 2001. *Nature* 411: 302–305.
Sepsid flies	Male genitalia	Eberhard, W. G. 2001. *Evolution* 55: 93–102.
Treehoppers	Male courtship vibrations	Rodríguez, R. L. et al. 2004. *Evolution* 58: 571–578.
Threespine sticklebacks	Body size	McKinnon, J. S. et al. 2004. *Nature* 429: 294–298.
Lake Victoria cichlids	Male coloration	Haesler, M. P. and O. Seehausen. 2005. *Proceedings of the Royal Society of London B* 272: 237–245.
Lake Malawi cichlids	Male coloration	Knight, M. E. and G. F Turner. 2004. *Proceedings of the Royal Society of London B* 271: 675–680.
Bowerbirds	Male courtship structures	Uy, J. A. C., and G. Borgia. 2000. *Evolution* 54: 273–278.

Research on speciation by sexual selection is accelerating. Tami Panhuis and coworkers (2001) have provided a recent review of the field, and Table 16.2 lists a few of the model systems that are currently being analyzed. Most of the papers cited in the table provide experimental evidence, analogous to the data on Hawaiian *Drosophila* in Figure 16.11, that females of extremely closely related species or populations use different characteristics to choose mates. This is important support for the hypothesis that the populations involved are diverging due to sexual selection. The citations also focus exclusively on animals—ignoring extensive work on how changes in pollinator attraction lead to speciation in plants.

16.4 Secondary Contact

Suppose a particular speciation event begins with the geographic isolation of two populations and a corresponding reduction in gene flow, and then continues as selection, mutation, and drift cause genetic divergence between the two groups. What happens if the recently diverged populations come back into contact and have the opportunity to interbreed? Hybridization events between these types of recently diverged species are especially common in plants. For example, over 700 of the plant species that have been introduced to the British Isles in the recent past have hybridized with native species at least occasionally, and about half of these native/nonnative matings produce fertile offspring (Abbott 1992). Ten percent of all bird species also hybridize regularly and produce fertile offspring (Grant and Grant 1992).

Hybridization occurs when recently diverged populations interbreed.

In at least some cases, the fate of these hybrid offspring determines the course of speciation. Will the hybrids thrive, interbreed with each of the parental populations, and eventually erase the divergence between them? Or will hybrids have new characteristics and create a distinct population of their own? What happens if hybrid offspring have reduced fitness relative to the parental populations?

Reinforcement

The geneticist Theodosius Dobzhansky (1937) reasoned that if populations have diverged during a period when the groups lived in different geographic areas, then any hybrid offspring that are produced should have markedly reduced fitness relative to individuals in each of the parental populations. The logic here is that if natural selection produced adaptations to distinct habitats, if sexual selection produced changes in the mating system, or if genetic drift led to the fixation of alleles that do not work well together when heterozygous, then hybrid offspring would have low fitness. As a result, there should be strong natural selection in favor of assortative mating—meaning that selection should favor individuals that choose mates only from the same population. Selection that reduces the frequency of hybrids in this way is called **reinforcement**. Reinforcement should finalize the speciation process by producing complete reproductive isolation.

Reinforcement occurs when hybrid offspring have low fitness, and natural selection leads to assortative mating and the prezygotic isolation of populations.

The reinforcement hypothesis predicts that when closely related species come into contact and hybridize, a mechanism that prevents hybridization will evolve. For example, selection might favor mutations that alter aspects of mate choice or life history (such as the timing of breeding). Divergence in these traits prevents fertilization from occurring and results in **prezygotic isolation** of the two species. But populations can also remain genetically isolated in the absence of reinforcement if hybrid offspring are sterile or infertile. This possibility is known as **postzygotic isolation**.

Some of the best data on prezygotic isolation and the reinforcement hypothesis have been assembled and analyzed by Jerry Coyne and Allen Orr (1997). Coyne and Orr examined data from a large series of sister-species pairs in the genus *Drosophila*. Some of these species pairs live in allopatry and others in sympatry. Coyne and Orr's data set included estimates of genetic distance, calculated from differences in enzyme allele frequencies, along with measurements of the degree of pre- and postzygotic isolation. When they plotted the degree of prezygotic isolation against genetic distance, which they assumed correlates at least roughly with time of divergence, they found a striking result: Prezygotic isolation evolves much faster in sympatric species pairs than it does in allopatric species pairs (Figure 16.12). This is exactly the prediction made by the reinforcement hypothesis.

These laboratory experiments with *Drosophila* are some of the best evidence available in support of the reinforcement hypothesis. In contrast, most field studies that have looked for evidence of reinforcement in hybridizing populations have produced mixed results.

Work by Megan Higgie and colleagues (2000) is an important exception, however, because it provides direct experimental evidence of reinforcement. These researchers investigated the chemical composition of mating signals called pheromones in two species of *Drosophila* that are native to northeast Australia. In regions where populations of *D. birchii* and *D. serrata* overlap, *D. serrata* individuals produce pheromones that are chemically very different from those produced in regions where the two species do not overlap. This is the pattern that should occur if natural selection has favored *D. serrata* individuals with pheromones that preclude matings with *D. birchii* individuals.

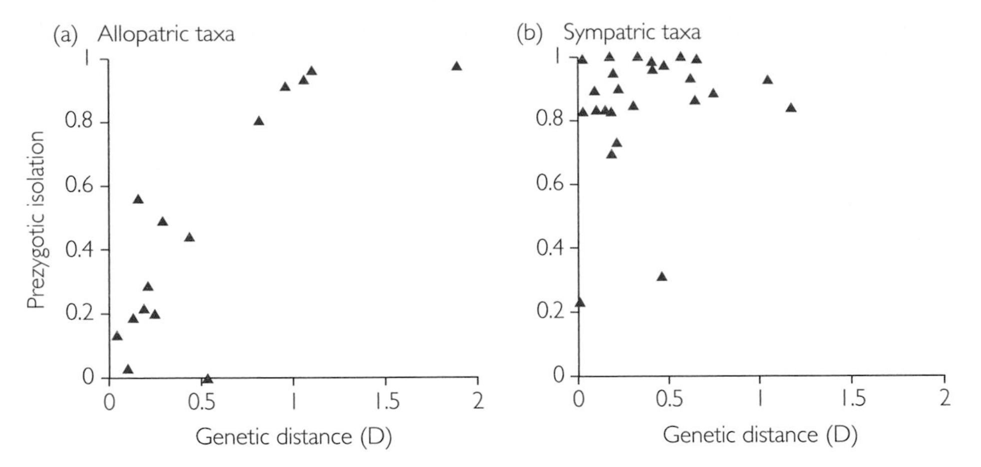

Figure 16.12 Prezygotic isolation in allopatric versus sympatric species pairs of *Drosophila* These graphs (Coyne and Orr 1997) plot degree of prezygotic isolation versus genetic distance in a variety of sister-species pairs from the genus *Drosophila*. Prezygotic isolation is estimated from mate-choice tests performed in the laboratory. A value of 0 indicates that different populations freely interbreed (0% prezygotic isolation) and 1 indicates no interbreeding (100% prezygotic isolation). Genetic distance is estimated from differences in allele frequencies found in allozyme surveys. Sibling species with the same degree of overall genetic divergence show much more prezygotic isolation if they live in sympatry.

To test this hypothesis more rigorously, Higgie and colleagues collected individuals of each species from areas where they do not overlap, mixed the populations in the lab, and reared them for nine generations. By the end of the experiment, the pheromones produced by *D. serrata* individuals had changed from those present originally. More specifically, the experimental *D. serrata* were now producing pheromones like those observed in natural *D. serrata* populations that live in sympatry with *D. birchii*. In effect, Higgie and coworkers had observed the evolution of reinforcement in just nine generations.

Reinforcement can play an important part in the speciation process, but it is not essential.

Even with this strong experimental evidence of reinforcement in nature, the classical view of reinforcement as a universal and required stage in speciation is overstated. Reinforcement can and does occur, but it is not essential for speciation (Butlin 1995; Noor 1995).

Hybridization

Reinforcement should occur when hybrid offspring have reduced fitness. But what happens to hybrid offspring that survive and reproduce well? Their fate has important consequences for speciation.

Creation of New Species through Hybridization

If hybrid offspring occupy habitats that are different from those occupied by either parental population, it is entirely possible that the hybrid offspring will have higher fitness in the novel habitat than either of the parental species. If so, will these hybrid populations occupy the new environment and become distinct species?

An experimental study of plant hybridization, conducted in Loren Rieseberg's lab (1996), suggests that the answer to this question is yes. These researchers worked with three annual sunflower species native to the American southwest-*Helianthus annuus, H. petiolaris, and H. anomalous*—and experimentally duplicated a natural hybridization event that led to speciation.

Based on morphological and chromosome studies, it had long been thought that *H. anomalous* originated in a hybridization event between *H. annuus* and *H. petiolaris. H. anomalous* lives on sand dunes, where *H. annuus* and *H. petiolaris* do not grow well.

To test the hybrid-origin hypothesis rigorously, Rieseberg and coworkers crossed individuals of *H. annuus* and *H. petiolaris* to produce three lines of hybrids. They then either mated individuals from each of these lines back to *H. annuus* (this is called a backcross) or sib-mated the individuals within each line for four additional generations. As a result, each experimental line underwent a different sequence and combination of backcrossing and sib-mating. This protocol simulated different types of matings that may have occurred when populations of *H. annuus* and *H. petiolaris* hybridized naturally.

At the end of the experiment, Rieseberg and colleagues surveyed the three hybrid populations genetically. Their goal was to determine how similar the hybrids were to each other and to naturally occurring *H. anomalous* individuals. To make this comparison possible, the researchers mapped a large series of species-specific DNA sequences, called randomly amplified polymorphic DNA (RAPD) markers, in the two parental species. Some of these markers were unique to *H. petiolaris,* while others were unique to *H. annuus.* These markers allowed the researchers to determine which alleles from the two parental species were present in the three experimental hybrid populations and compare them to alleles present in naturally occurring *H. anomalous.* The results were striking: The three independently derived experimental hybrids and the natural hybrid shared an overwhelming majority of markers. The experimental and natural hybrids were genetically almost identical.

Researchers have experimentally re-created a speciation event that occurred naturally via hybridization.

To interpret this result, Rieseberg and coworkers contend that only certain alleles from *H. petiolis* and *H. anuus* work in combination and that other hybrid types are inviable or have reduced fitness. The idea here is that the genetic composition of the experimental hybrids quickly sorted out into a favorable combination. Even more remarkable, this combination of alleles was nearly the same as that produced by a natural hybridization event that occurred thousands of years ago. This puts an interesting twist on speciation's third stage: Secondary contact and gene flow between recently diverged species can result in the formation of a new, third, species. For other examples of this phenomenon, see Lexer et al. (2003) and Schwarz et al. (2005).

Hybrid Zones

A **hybrid zone** is a region where interbreeding between diverged populations occurs and hybrid offspring are frequent. Hybrid zones are usually produced when secondary contact occurs between species that have diverged in allopatry.

Data from *Drosophila* and *Helianthus* show that it is possible for hybrid offspring to have lower or higher fitness than purebred offspring, with very different consequences: reinforcement of parental forms or the formation of a new species. But frequently no measurable differences can be found between the fitness of hybrid and pure offspring. In cases like this, a new outcome is possible: the formation of a stable hybrid zone, or region where hybridization is ongoing and hybrid offspring are common.

The following three possibilities dictate the size, shape, and longevity of hybrid zones (Endler 1977; Barton and Hewitt 1985; see Table 16.3):

- When hybrid and parental forms are equally fit, the hybrid zone is wide. Individuals with hybrid traits are found at high frequency at the center of the zone

Table 16.3 Outcomes of secondary contact and hybridization

When populations hybridize after diverging in allopatry, several different outcomes are possible. The type of hybrid zone formed and the eventual outcome depend on the relative fitness of hybrid individuals.

Fitness of hybrids	Hybrid zone	Eventual outcome
Lower than parental forms	Relatively narrow and short lived	Reinforcement (differentiation between parental populations increases)
Equal to parental forms	Relatively wide and long lived	Parental populations coalesce (differentiation between parental populations decreases)
Higher than parental forms	Depends on whether fitness advantage occurs in ecotone or new habitat	Stable hybrid zone or formation of new species

and progressively lower frequencies with increasing distance. In this type of hybrid zone, the dynamics of allele-frequency change are dominated by drift. The width of the zone is a function of two factors: how far individuals from each population disperse each generation, and how long the zone has existed. The farther the individuals move each generation and the longer the populations are in contact, the wider the zone.

- When hybrids are less fit than purebred individuals, the fate of the hybrid zone depends on the strength of selection against them. If selection is very strong and reinforcement occurs, then the hybrid zone is narrow and short-lived. If selection is weak, then the region of hybridization is wider and longer-lived. These types of hybrid zones are an example of a selection-migration balance, analogous to the situation with water snakes in Lake Erie (see Chapter 7).
- When hybrids are more fit than purebreds, the fate of the hybrid zone depends on the extent of environments in which hybrids have an advantage. If hybrids achieve higher fitness in environments outside the ranges of the parental species, then a new species may form in the new habitat. This was the case in the evolution of *Helianthus anomalus*. If hybrids have an advantage at the boundary of each parental population's range, then a stable hybrid zone may form. For example, many hybrid zones are found in regions called ecotones, where markedly different plant and animal communities meet. In this case, two closely related species or populations are found on either side of the ecotone, with a hybrid zone between them. To explain this pattern, researchers hypothesize that hybrid individuals with intermediate characteristics have a fitness advantage in these transitional habitats.

Hybridization can have a variety of outcomes, depending on the fitness of the hybrids relative to the parental forms. The outcomes include creation of a new species made up of hybrid individuals, formation of a stable hybrid zone, and reinforcement.

To illustrate how biologists go about distinguishing between these possibilities, consider recent work on what may be the most widespread and economically important plant in the American West: big sagebrush (*Artemesia tridentata*). A total of four distinctive populations or subspecies of big sagebrush have been described, including two that hybridize in the Wasatch Mountains of Utah (Freeman et al. 1995). Basin big sagebrush (*A.t. tridentata*) is found at low elevations in river flats, while mountain sagebrush (*A.t. vaseyana*) grows at higher elevations in upland habitats. The two subspecies hybridize where they make contact at intermediate elevations.

The first task in analyzing a hybrid zone is to describe the distribution and morphology of hybrids relative to parental populations. Previous work had shown that hybrid zones between sagebrush populations are narrow—often less

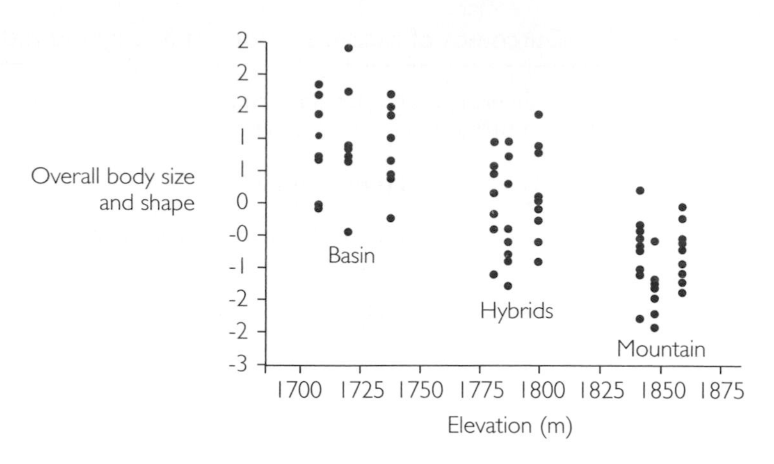

Figure 16.13 Hybrid sagebrush are intermediate in form between parental subspecies On this graph, a quantity called the principal component score is plotted against the elevation where sagebrush plants were sampled. Principal component analysis (PCA) is a statistical procedure for distilling information from many correlated variables into one or two quantities that summarize the variation measured among individuals in the study. In this case, Carl Freeman and colleagues (1991) measured a large series of morphological traits in sagebrush such as height, circumference, crown diameter, and branch length. The PCA was performed to combine these many variables into a single quantity, the PCA score, that summarizes overall size and shape. Each data point represents an individual.

than the length of a football field—and Carl Freeman and colleagues (1991) found that hybrids are intermediate in form between basin and mountain subspecies (Figure 16.13). Historical records indicate that the size and distribution of hybrid zones have been stable in extent for at least 2–3 sagebrush generations.

To assess the relative fitness of hybrid offspring versus pure forms, John Graham et al. (1995) compared a variety of fitness components in individuals sampled along an elevation gradient. These fitness components included seed and flower production, seed germination, and extent of browsing by mule deer and grasshoppers. Table 16.4 shows some data representative of their results. In general, hybrids show equal or even superior production of flowers and seeds and resistance to herbivores that is equal to the mountain forms. Thus, hybrid offspring do not appear to be less fit than offspring of the parental populations when each is growing in its natural habitat.

This leaves two possibilities: The hybrid zone could be maintained by selection-migration balance or by positive selection on hybrids in the ecotone. To test these alternative hypotheses, the research group studied growth rates and other components of fitness in basin, hybrid, and mountain seedlings that were transplanted to habitats at low, intermediate, and high elevation (Wang et al. 1997; Freeman et al. 1999). These reciprocal transplant experiments showed that each form performed best in its native habitat (Figure 16.14). These data suggest that the hybrid zone is maintained because hybrid offspring have superior fitness in a transitional habitat.

Table 16.4 Fitness of sagebrush hybrids

The rows in this table, listing basin through mountain populations, represent big sagebrush plants sampled along a lower-to-higher elevational gradient. *N* is the sample size, and the numbers in parentheses are standard deviations—a measure of variation around the average value. Differences among these populations in the number of inflorescences (flowering stalks) are not statistically significant. But hybrids have significantly more flowering heads per unit inflorescence length—a measure of flower density.

Population	*N*	Number of inflorescences		Number of flowering heads	
Basin	25	19.92	(6.16)	175.1	(124.9)
Near basin	25	17.72	(6.59)	174.4	(92.5)
Hybrid	27	20.11	(6.75)	372.7	(375.9)
Near mountain	25	17.04	(6.50)	153.7	(75.2)
Mountain	25	16.80	(6.34)	102.0	(59.4)

Source: Adapted from Table 2 in Graham et al. (1995).

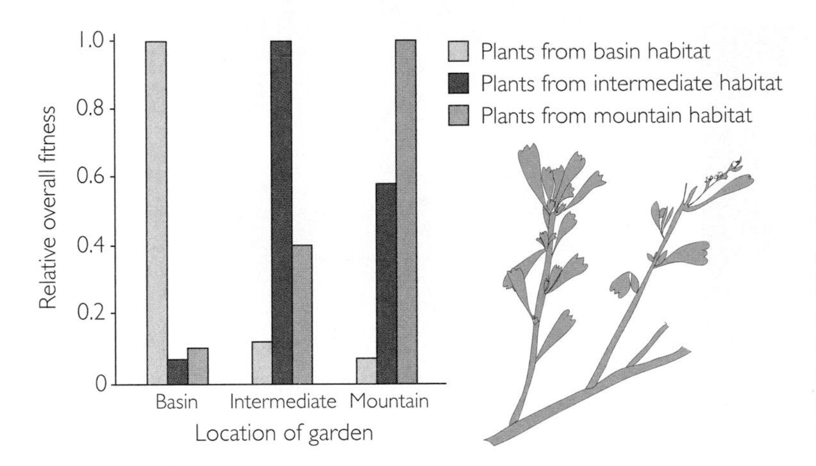

Figure 16.14 Relative fitness of big sagebrush taxa The vertical axis on this graph plots an overall measure of fitness that combines data on survivorship, flowering, seed production, and seed germination rate. The data are expressed as relative fitness by assigning a value of 1.0 to the group that had the highest fitness in each of three experimental gardens and then expressing the fitness of the other groups as a percentage of that group's fitness. The horizontal axis indicates whether the data come from gardens at the basin, intermediate, or mountain elevations. From Wang et al. (1997).

This result suggests that the hybrid zone should continue to be stable, and should correspond geographically to the intermediate, mid-elevation environments in these mountains.

In intermediate or transitional habitats, hybrid populations may be more fit than either parental population.

Having reviewed isolation, divergence, and secondary contact, we can move on to consider the genetic mechanisms responsible for these events. Understanding the genetic basis of speciation is the focus of Section 16.5.

16.5 The Genetics of Speciation

What degree of genetic differentiation is required to isolate populations and produce new species? The traditional view was that some sort of radical reorganization of the genome, called a genetic revolution, was necessary (Mayr 1963). This hypothesis was inspired by a strict interpretation of the BSC. The logic went as follows: Under the BSC, species are reproductively isolated if and only if hybrids are inviable or experience dramatic reductions in fitness. For this to happen, sister species would have to be genetically incompatible. Combining their alleles would produce dysfunctional development, morphology, or behavior.

Genetic models have shown that these types of large-scale changes in the genome are not only unlikely but unnecessary for divergence and speciation to occur (Lande 1980; Barton and Charlesworth 1984). These theoretical results have been verified by the work reviewed in Section 16.4, which demonstrates that marked differentiation can occur between populations of sunflowers and sagebrush that still produce fertile hybrid offspring. As a result, the questions that motivate current research in the genetics of speciation are focused on the number and nature of alleles that distinguish closely related species. Let's consider two cases.

Pea Aphids

Pea aphids (*Acyrthosiphon pisum*) are small insects that make their living by sucking sap from plants in the pea family (Figure 16.15a). Sara Via and colleagues have been studying two populations of this species. One of these populations lives on red clover plants and the other is found on alfalfa. A series of experiments have documented that individuals derived from red clover populations actively prefer to settle and feed on red clover when given a choice, and that individuals from alfalfa populations prefer alfalfa (Via 1999; Caillaud and Via 2000). Alfalfa individuals transferred to red clover have low fitness, while red clover individuals

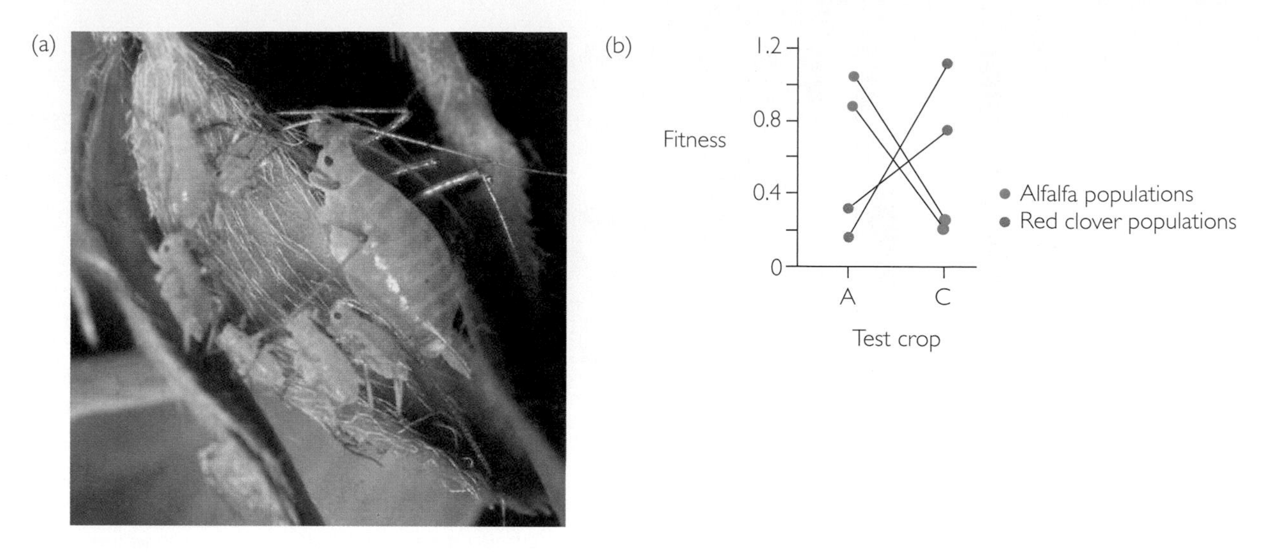

Figure 16.15 Speciation in pea aphids (a) Pea aphids settle on host plants, insert a mouthpart called a stylet into the stem or leaves, and suck sap from phloem tissues. In North America, genetically distinct populations specialize on red clover or on alfalfa. (b) These data show the relative fitness of pea aphid individuals from alfalfa (A) and red clover (C) populations when grown on alfalfa plants or red clover plants.

raised on alfalfa have low fitness (Figure 16.15b). In addition, when the group crossed individuals from the two populations and raised the offspring on both types of host plants, they found that the hybrids had lower fecundity and other measures of fitness than individuals from the natural populations on both hosts (Via et al. 2000).

Based on these observations, it appears that the red clover and alfalfa populations are beginning to diverge and become separate species. What alleles are changing in frequency as the populations diverge? To answer this question, Via and coworkers created a genetic map of the pea aphid genome using the techniques introduced in Chapter 9 (see Hawthorne and Via 2001; Via and Hawthorne 2002). Briefly, they were able to identify a series of detectable gene sequences ("markers") that are specific to the genomes of aphids found on red clover or on alfalfa. They then mated clover and alfalfa individuals to produce individuals that were heterozygous at each gene in the genome. These F_1 offspring contained a clover-derived and an alfalfa-derived allele at each gene. Finally, they mated these F_1 individuals with each other to produce a large number of aphids with a complex array of genotypes. They concluded their analysis by testing individuals in the F_2 generation to determine their preference for alfalfa or clover plants and their fitness on each type of host.

To find genes that have been under selection as populations diverged to form new species, researchers look for alleles associated with species-specific traits like food or habitat preferences.

The results of these experiments were remarkable. At several locations in the genome, there are alleles that increase fecundity on clover while decreasing fitness on alfalfa, or vice versa. These data suggest a genetic trade-off—that alleles responsible for high fitness on one host lead to low fitness on the other host. Similarly, there appear to be several genes with alleles that lead to a preference for settling on clover and a disinclination to feed on alfalfa, or vice versa. What really strikes home about the data, though, is that at least some of these host-preference genes appear to be the same or very closely linked to the genes that affect fitness on the two hosts. Although finer-resolution studies still need to be done, the data suggest that either genes with pleiotropic effects or genes that are tightly linked to each other simultaneously cause increased fitness on one host and a preference for that host. Recall that pleiotropy occurs when a single gene affects more than one trait, and that linked alleles are in close physical proximity and thus rarely separated by recombination. If selection favors individuals with certain combinations of linked genes, those alleles are in linkage disequilibrium or LD (see Chapter 9.)

If further studies confirm these results, it would suggest a genetic mechanism responsible for at least some of the speciation-in-action examples listed in Table 16.1. If the same or closely linked genes are responsible for both host choice and fitness in insects that feed on plants, and if mutations in these genes lead to a higher fitness and a preference for a different host plant, then the new alleles should spread rapidly and lead to speciation based on host-plant use. Because there are several million species of plant-feeding insects alive today, this may be the genetic basis for an exceptionally important mechanism of speciation.

Research on insects suggests that speciation can occur rapidly when a population switches to a new food plant and when alleles for choosing that plant and eating it successfully are identical or closely linked.

Threespine Sticklebacks

Threespine sticklebacks (*Gasterosteus aculeatus*) are small fish found in marine environments throughout the northern hemisphere (Figure 16.16a). They are also common in some freshwater environments, including lakes along the western coast of British Columbia, Canada. In at least six of these lakes, two closely related species of stickleback exist. A relatively small, slim, limnetic species lives in open water (Figure 16.16b) while a benthic species consists of larger, deeper-bodied individuals that live in shallow water near the shore (Figure 16.16c). Limnetic refers to the deeper, open water of a lake or pond; benthic refers to the bottom of a lake or sea. The limnetic species feeds by sieving microscopic prey from the plankton using barb-like "rakers" on its gills. The benthic species feeds by biting invertebrates.

Marine sticklebacks are larger in size than the freshwater species. They have longer spines along their dorsal and ventral surfaces and more numerous and

(a) Marine species

(b) Limnetic species in freshwater

(c) Benthic species in freshwater

Figure 16.16 Speciation in threespine sticklebacks (a) Stickleback populations that occupy marine environments are large and have prominent spines along their dorsal side. (b) Stickleback species that live in limnetic or open-water lake habitats consist of individuals that are small and slim with relatively large eyes and spines. (c) Lake populations that occupy benthic habitats near the shores of a lake consist of larger, deeper-bodied individuals.

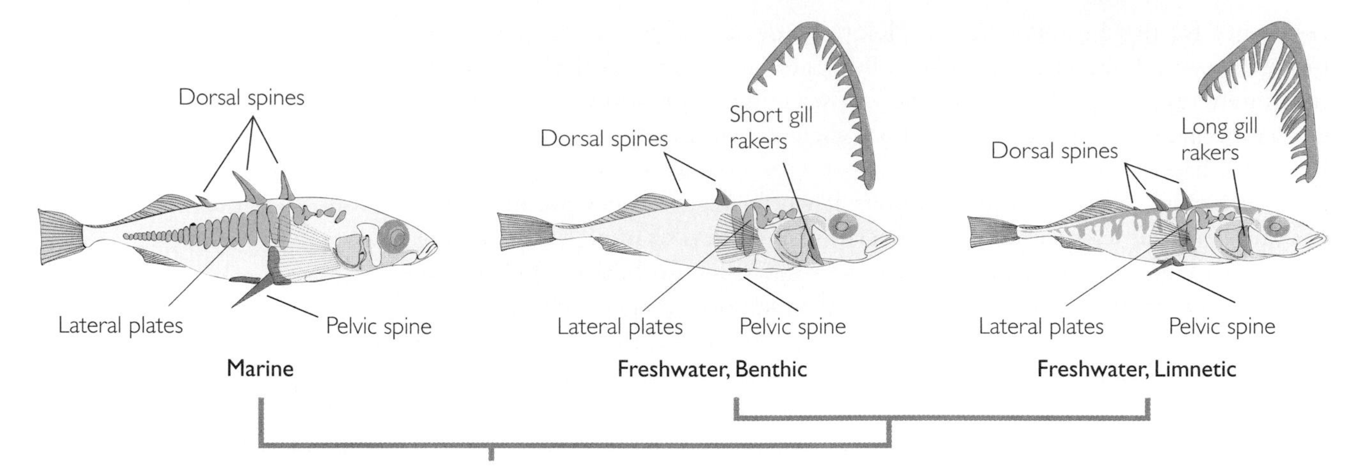

Figure 16.17 Morphological characters that vary among stickleback species These drawings detail the traits that vary between marine and freshwater sticklebacks, and between the limnetic and benthic sticklebacks found in freshwater. The key traits are the size of gill rakers, and the size and extent of the dorsal spines, lateral plates, and pelvic girdle and pelvic spines. After Schluter (1993); Peichel et al. (2001); Cole et al. (2003); Peichel (2005); Craig Miller (personal communication).

relatively larger bony structures along their sides (Figure 16.17). In addition to inspiring the name "stickleback," the spines and the bony lateral plates are thought to offer protection from large fish that prey on sticklebacks by biting them.

The limnetic and benthic species, in turn, are distinguished by several morphological traits other than overall body size and shape:

1. Compared with benthic individuals, the limnetic species has longer spines along its dorsal side. Experiments have shown that the major fish predator in the lakes, cutthroat trout, prefer to feed in open-water habitats where limnetics live (Vamosi and Shluter 2002).
2. The limnetic species has numerous and relatively large bony structures along its sides. These lateral plates are thought to support the long spines and make the limnetic individuals more difficult for predators to bite and swallow.
3. The number and size of gill rakers is dramatically reduced in the benthic species compared to the limnetic species. The reduction correlates with their contrasting feeding methods.

Sticklebacks have speciated repeatedly as populations moved from marine environments to lakes. Two stickleback species are usually found in each lake, with the two species having traits that are distinct from each other and from the marine forms.

Experimental tests conducted in Dolph Schluter's laboratory have shown that benthic and limnetic species mate assortatively and that hybrids between the two species have lower fitness than either parental form (Hatfield and Schulter 1999; Rundle et al. 2000). Phylogenetic studies have shown that in each of the six lakes, the benthic and limnetic forms have speciated independently, or in parallel, since glaciers retreated from the region about 15,000 years ago. According to current data, it is likely that each benthic and limnetic species is derived from an independent influx of marine-dwelling individuals (Schluter et al. 2001). Each lake, then, represents a replicated experiment in speciation.

What alleles changed in frequency as ancestral, marine-dwelling stickleback populations speciated into the benthic and limnetic freshwater forms? To address this question, Catherine Peichel and colleagues (2001) created a genetic map for sticklebacks, based on a large number of microsatellite markers. (For details on how genetic maps are constructed, see Chapter 9.) By analyzing correlations between the morphology of hybrid offspring and the limnetic- or benthic-specific genetic markers they carried, the group was able to locate a series of genes, scattered among the 21 chromosomes found in these species, that affect spine length, gill raker number, and the distribution of lateral plates. The group is still analyzing these "candidate genes." They are trying to identify the specific alleles responsible for morphological divergence in the limnetic and benthic forms.

In the meantime, several teams of researchers have used the stickleback genetic map to identify alleles responsible for the contrasting morphology in marine versus freshwater forms. By crossing individuals from marine and freshwater populations and then correlating the genetic markers found in hybrid offspring with their morphology, biologists have been able to pinpoint two genes responsible for the dramatic morphological divergence that has occurred:

Some of the alleles responsible for morphological divergence in marine and freshwater stickleback species are now known.

- Michael Shapiro and colleagues (2004; 2006) have determined that alleles of the *Pitx1* gene are responsible for the loss of pelvic spines and supporting bones, called the pelvic girdle (Figure 16.17) in freshwater forms. Although the sequence of the *Pitx1* gene is identical in marine and freshwater species, a regulatory change prevents the gene from being expressed as pelvic structures are developing in freshwater species. When normal *Pitx1* expression is missing in this region, the pelvic bones and spines fail to form.
- Pamela Colosimo and coworkers (2005) have shown that alleles of the *Eda* gene are responsible for the loss of extensive bony plates on the sides of the freshwater species (Figure 16.18). Further, they found that the alleles for reduced lateral plates exist at low frequencies in marine populations—meaning that they were probably introduced to the freshwater populations when the initial colonization of the lakes occurred. If so, then rapid morphological divergence occurred through natural selection favoring alleles that led to loss of lateral plates. The logic here is that large predators are rarer in freshwater than in marine habitats, and so selection favors reduced investment in extensive defenses.

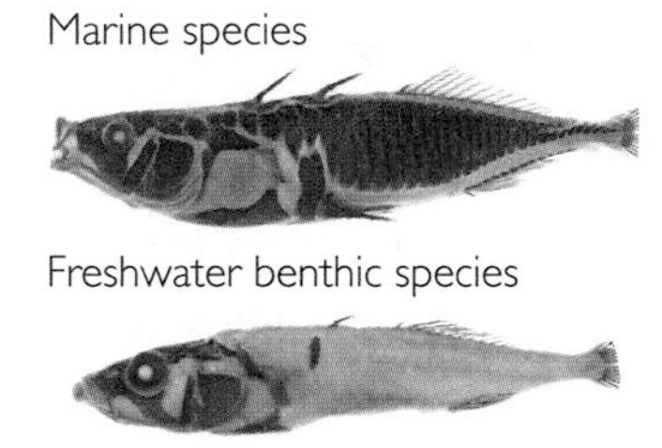

Figure 16.18 Marine and freshwater sticklebacks: Contrasts in extent of lateral plates These fish have been stained with the dye alizarin-red, which highlights bony tissues.

Taken together, the work on sticklebacks furnishes one of the best-understood examples of genes that are responsible for morphological divergence during speciation. What genes influence the isolation step in stickleback speciation? Recent work has shown that benthic and limnetic forms rarely hybridize in nature because females choose mates on the basis of body size and courtship coloration (Boughman et al. 2005). If the genetic basis of these mating preferences were known, sticklebacks would furnish a comprehensive story in the genetics of speciation.

Summary

Although a wide variety of species concepts have been proposed, all agree that the distinguishing characteristic of a species is evolutionary independence. The various species concepts differ in the criteria that are employed for recognizing evolutionary independence.

Speciation can be analyzed as a three-step process: (1) Isolation of populations caused by dispersal, vicariance, or large-scale chromosome changes such as polyploidization; (2) divergence based on genetic drift or natural selection; and (3) completion or elimination of divergence upon secondary contact. There are numerous exceptions to this sequence, however. In some cases, selection for divergence is strong enough that populations can differentiate without physical isolation, as research on apple and hawthorn maggot flies illustrates. It may also be possible for sexual selection to cause genetic isolation and divergence in traits related to courtship or mating. Furthermore, a variety of outcomes are possible after secondary contact. These include formation of stable hybrid zones and creation of a new species containing genes from each of the parental forms.

The primary strategy employed in genetic analyses of speciation is to look for correlations between mapped phenotypic or molecular markers and the distribution of traits in offspring of recently diverged species. These strategies have confirmed that in pea aphids, the same or closely linked alleles influence host plant choice and fitness on clover versus alfalfa, and have identified several genes responsible for changes in the morphology of marine-dwelling sticklebacks and the benthic and limnetic species found in freshwater lakes.

Questions

1. What does it mean to say that species are "evolutionarily independent" or that "species form a boundary for gene flow?"
2. Compare and contrast the morphospecies concept, the biological species concept, and the phylogenetic species concept. What criterion does each use to identify species? What are the pros and cons of each?
3. The text introduces work on speciation in marine copepods and African elephants. In each case, did application of the phylogenetic species concept lead to the recognition of fewer, the same, or more species? Explain.
4. What does "sex" consist of in bacteria? How do biologists define species in bacteria and archaea?
5. Explain the difference between dispersal and vicariance. Why might dispersal or vicariance events initiate speciation?
6. When the Panama land bridge between North and South America was uncovered, some North American mammal lineages crossed to South America and underwent dramatic radiations. For terrestrial species, did the completion of the land bridge represent a vicariance or dispersal event? Does the recent building of the Panama Canal represent a vicariance or dispersal event for terrestrial organisms? For marine organisms?
7. Morphologically similar species of snapping shrimp that are each others' closest relatives live on either side of the Isthmus of Panama. Why does this observation support the hypothesis that speciation occurred via vicariance?
8. Phylogenetic analyses show that in many cases, closely related Hawaiian species occur in the following pattern: The more ancestral groups occupy older islands in the archipelago, while the more derived populations occupy younger islands. Why does this observation support the hypothesis that speciation has occurred via dispersal?
9. Would glaciation in northern Europe and North America have created vicariance events over the past 150,000 years? If so, how? Which organisms might have been affected? For example, consider the different effects glaciation might have on small mammals, migratory birds, and trees.
10. Why do tetraploid populations tend to be reproductively isolated from closely related diploid populations?
11. What are the possible outcomes when species that have long been separated geographically come back into contact and begin hybridizing, and under what conditions does each outcome occur?
12. What is reinforcement? Is it an example of genetic drift, natural selection, or sexual selection?
13. Why is sexual selection considered a particularly effective process in producing speciation?
14. Researchers are claiming that benthic and limnetic three-spine sticklebacks and red clover and alfalfa populations of pea aphids are speciating due to natural selection. Do you agree with this hypothesis? Why or why not?
15. Within the past 50 years, soapberry bug populations in the United States have diversified into populations distinguished by markedly different beak lengths. These bugs eat the seeds at the center of soapberry fruits. Native and recently introduced varieties of soapberries differ greatly in fruit size. Describe the experiments or observations you would make to launch an in-depth study of speciation in these bugs. What data would tell you whether they are separate populations evolving independently, or a single interbreeding population? Many museums contain insect specimens from decades ago. What would you examine in these old specimens? What information about the host plants would be useful?
16. Red crossbills are small finches specialized for eating seeds out of the cones of conifer trees. They fly thousands of kilometers each year in search of productive cone crops. Despite their mobility, crossbills have diverged into several "types" that differ in bill shape, body size, and vocalizations. Each type prefers to feed on a different species of conifer, and each species of conifer is found only in certain forests. Bill size and shape affects how efficiently a bird can open cones of a certain conifer species. Explain how a highly mobile animal such as the red crossbill could have diverged into different types in the absence of any geographic barrier. If crossbills could not fly, do you think speciation would occur more quickly or more slowly? If conifer species were not patchily distributed (i.e., in different forests), do you think crossbill speciation would occur more quickly or more slowly? Compare your answers to the

analyses and data in Benkman (2003) *Evolution* 57: 1176–1181.

17. Ellen Censky and coworkers (1998) documented the arrival of a small group of iguanas to the Caribbean island of Anguilla, that previously had no iguanas. The animals were carried there on a raft of fallen trees and other debris during a hurricane. Outline a long-term study that would document whether this newly isolated population diverges from iguanas on nearby islands to form a new species.

18. A total of 30 different species of yuccas are found in the deserts of the American Southwest. Yuccas are pollinated when female moths lay their eggs in the flowers of yucca plants and male moths visit to fertilize the moth eggs. Traditionally, just one species of moth was thought to pollinate all 30 species of yucca. Jim Leebens-Mack and coworkers (1998) studied the moths that pollinate two different yucca species, however, and found that the moths are morphologically distinct and form independent tips on a phylogeny. Are these moths different species? How would you go about testing the hypothesis that each species of yucca is pollinated by a different species of moth?

19. At least eight geographically isolated populations of the spiny lizard *Sceloporus jarrovii* exist in the mountains and deserts of north and central Mexico. Although females look similar, males in each population are strikingly different—ranging in color from jet black to gold to green, orange, and blue. An analysis by John Wiens and colleagues (1999) has shown that at least five of the populations are phylogenetic species. Would they be considered species under the morphospecies and biological species concept as well? Wiens and coworkers suggest that sexual selection has been responsible for speciation in this group. Outline a study that would test this hypothesis.

Exploring the Literature

20. Biologists frequently use the word spectacular to describe the species numbers and ecological and morphological diversity of cichlid fish found in east Africa's lakes Malawi, Tanganyika, and Victoria. Males are brightly colored in many of the 1,000 species, and recent work has shown that sexual selection may be intense. Phylogenies and geologic data indicate that the 300 species found in Lake Victoria are derived from a single founding population that arrived about 100,000 years ago. Furthermore, each lake contains species that eat fish, mollusks, insect larvae, algae, zooplankton, or phytoplankton. To learn more about this dramatic example of rapid speciation, see:

Galis, R., and J. A. J. Metz. 1998. Why are there so many cichlid species? *Trends in Ecology and Evolution* 13: 1–2.

Haesler, M.P. and O. Seehausen. 2005. Inheritance of female mating preference in a sympatric sibling species pair of Lake Victoria cichlids: implications for speciation. *Proceedings of the Royal Society of London B* 272: 237–245.

Verheyen, E., W. Salzburger, J. Snoeks, and A. Meyer. 2003. Origin of the superflock of cichlid fishes from Lake Victoria, East Africa. *Science* 300: 325–329.

Citations

Abbott, R. J. 1992. Plant invasions, interspecific hybridization and the evolution of new plant taxa. *Trends in Ecology and Evolution* 7: 401–405.

Barton, N. H., and B. Charlesworth. 1984. Genetic revolutions, founder effects, and speciation. *Annual Review of Ecology and Systematics* 15: 133–164.

Barton, N. H., and G. M. Hewitt. 1985. Analysis of hybrid zones. *Annual Review of Ecology and Systematics* 15: 133–164.

Beheregaray, L. B., J. P. Gibbs, N. Havill, J. R. Powell, and A. Caccone. 2004. Giant tortoises are not so slow: rapid diversfication and biogeographic consensus on the Galapagos. *Proceedings of the National Academy of Sciences USA* 101: 6514–6519.

Boake, C. R. B. 2005. Sexual selection and speciation in Hawaiian *Drosophila*. *Behavioral Genetics* 35: 297–303.

Boake, C. R. B., M. P. DeAngelis, and D. K. Andreadis. 1997. Is sexual selection and species recognition a continuum? Mating behavior of the stalk-eyed fly *Drosophila heteroneura*. *Proceedings of the National Academy of Sciences USA* 94: 12442–12445.

Bonacum, J., P .M. O'Grady, M. Kambysellis, and R. DeSalle. 2005. Phylogeny and age of diversification of the planitibia species group of the Hawaiian *Drosophila*. *Molecular Phylogenetics and Evolution* 37: 73–82.

Boughman, J. W., H. D. Rundle, and D. Schluter. 2005. Parallel evolution of sexual isolation in sticklebacks. *Evolution* 59: 361–373.

Butlin, R. 1995. Reinforcement: An idea evolving. *Trends in Ecology and Evolution* 10: 432–434.

Caillaud, M. C., and S. Via. 2000. Specialized feeding behavior influences both ecological specialization and assortative mating in sympatric host races of pea aphids. *American Naturalist* 156: 606–621.

Censky, E. J., K. Hodge, and J. Dudley. 1998. Over-water dispersal of lizards due to hurricanes. *Nature* 395: 556.

Cohan, F. M. 1994. Genetic exchange and evolutionary divergence in prokaryotes. *Trends in Ecology and Evolution* 9: 175–180.

Cohan, F. M. 1995. Does recombination constrain neutral divergence among bacterial taxa? *Evolution* 49:164–175.

Cole, N. J., M. Tanaka, et al. 2003. Expression of limb initiation genes and clues to the morphological diversification of threespine stickleback. *Current Biology* 13: R951-R952.

Colosimo, P. F., K. E. Hosemann, et al. 2005. Widespread parallel evolution in sticklebacks by repeated fixation of ectodysplasin alleles. *Science* 307: 1928–1933.

Coyne, J. A., and H. A. Orr. 1997. "Patterns of speciation in *Drosophila*" revisited. *Evolution* 51: 295–303.

Dambrowski, H. R., C. Linn, Jr., S. H. Berlocher, A. A. Forbes, W. Roelofs, and J. L Feder. 2005. The genetic basis for fruit odor discrimination in *Rhagoletis* flies and its significance for sympatric host shifts. *Evolution* 59: 1953–1964.

Dawson, M. N. and W. M. Hammer. 2005. Rapid evolutionary radiation of marine zooplankton in peripheral environments. *Proceedings of the National Academy of Sciences USA* 102: 9235–9240.

Dettman, J. R., D. J. Jacobson, and M. W. Taylor. 2003. A multilocus genealogical approach to phylogenetic species recognition in the model eukaryote *Neurospora*. *Evolution* 57: 2703–2720.

Dobzhansky, T. 1937. *Genetics and the Origin of Species*. New York: Columbia University Press.

Drès, M. and J. Mallet. 2002. Host races in plant-feeding insects and their importance in sympatric speciation. *Philosophical Transactions of the Royal Society of London B* 357: 471–492.

Endler, J. A. 1977. *Geographic Variation, Speciation, and Clines*. Princeton, NJ: Princeton University Press.

Endler, J. A. 1989. Conceptual and other problems in speciation. In D. Otte and J. A. Endler, eds. *Speciation and Its Consequences*. Sunderland, MA: Sinauer, 625–648.

Feder, J. L., C. A. Chilcote, and G. L. Bush. 1988. Genetic differentiation between sympatric host races of the apple maggot fly *Rhagoletis pomonella*. *Nature* 336: 61–64.

Feder, J. L., C. A. Chilcote, and G. L. Bush. 1990. The geographic pattern of genetic differentiation beween host associated populations of *Rhagoletis pomonella* (Diptera: Tephritidae) in the eastern United States and Canada. *Evolution* 44: 570–594.

Feder, J. L., S. B. Opp, B. Wlazlo, K. Reynolds, W. Go, and S. Spisak. 1994. Host fidelity is an effective premating barrier between sympatric races of the apple maggot fly, *Rhagoletis pomonella*. *Proceedings of the National Academy of Sciences USA* 91: 7990–7994.

Feder, J. L., J. B. Roethele, B. Wlazlo, and S. H. Berlocher. 1997. Selective maintenance of allozyme differences among sympatric host races of the apple maggot fly. *Proceedings of the National Academy of Sciences USA* 94: 11417–11421.

Filchak, K. E., J. B Roethele, and J. L. Feder. 2000. Natural selection and sympatric divergence in the apple maggot *Rhagoletis pomonella*. *Nature* 407: 739–742.

Fisher, R. A. 1958. *The Genetical Theory of Natural Selection*. New York: Dover.

Forbes, A. A., J. Fisher, and J. L. Feder. 2005. Habitat avoidance: overlooking an important aspect of host-specific mating and sympatric speciation? *Evolution* 59: 1552–1559.

Freeman, D. C., W. A. Turner, E. D. McArthur, and J. H. Graham. 1991. Characterization of a narrow hybrid zone between two subspecies of big sagebrush (*Artemisia tridentata*: Asteraceae). *American Journal of Botany* 78: 805–815.

Freeman, D. C., J. H. Graham, D. W. Byrd, E. D. McArthur, and W. A. Turner. 1995. Narrow hybrid zone between two subspecies of big sagebrush *Artemisia tridentata* (Asteraceae). III. Developmental instability. *American Journal of Botany* 82: 1144–1152.

Freeman, D. C., K. J. Miglia, E. D. McArthur, J. H. Graham, and H. Wang. 1999. Narrow hybrid zone between two subspecies of big sagebrush (*Artemesia tridentata*: Asteraceae): X. Performance in reciprocal transplant gardens. *USDA Forest Service Proceedings RMRS-P* 11: 15–23.

Gaines, C. A., M. P. Hare, S. E. Beck, and H. C. Rosenbaum. 2005. Nuclear markers confirm taxonomic status and relationships among highly endangered and closely related right whale species. *Proceedings of the Royal Academy of London B* 272: 533–542.

García-Ramos, G., and M. Kirkpatrick. 1997. Genetic models of adaptation and gene flow in peripheral populations. *Evolution* 51: 21–28.

Graham, J. H., D. C. Freeman, and E. D. McArthur. 1995. Narrow hybrid zone between two subspecies of big sagebrush. II. Selection gradients and hybrid fitness. *American Journal of Botany* 82: 709–716.

Grant, P. R., and B. R. Grant. 1992. Hybridization of bird species. *Science* 256: 193–197.

Grant, P. R., and B. R. Grant. 1996. Speciation and hybridization in island birds. *Philosophical Transactions of the Royal Society of London B* 351: 765–772.

Hatfield, T., and D. Schluter. 1999. Ecological speciation in sticklebacks: environment-dependent hybrid fitness. *Evolution* 53: 866–873.

Hawthorne, D. J., and S. Via. 2001. Genetic linkage of ecological specialization and reproductive isolation in pea aphids. *Nature* 412: 904–907.

Hebert, P. D. N., E. H. Penton, J. M. Burns, D. H. Janzen, and W. Hallwachs. 2003. Tens species in one: DNA barcoding reveals cryptic species in the neotropical skipper butterfly *Astraptes fulgerator*. *Proceedings of the National Academy of Sciences USA* 101:14812–14817.

Higashi, M., G. Takimoto, and N. Yamamura. 1999. Sympatric speciation by sexual selection. *Nature* 402: 523–526.

Higgie, M., S. Chenoweth, and M. W. Blows. 2000. Natural selection and the reinforcement of mate recognition. *Science* 290: 519–521.

Hoskin, C. J., M. Higgie, K. R. McDonald, and C. Moritz. 2005. Reinforcement drives rapid allopatric speciation. *Nature* 437: 1353–1356.

Husband, B. C., and D. W. Schemske. 2000. Ecological mechanisms of reproductive isolation between diploid and tetraploid *Chamerion angustifolium*. *Journal of Ecology* 88: 689–701.

Jiggins, C. D. and J. R. Bridle. 2004. Speciation in the apple maggot fly: a blend of vintages? *Trends in Ecology and Evolution* 19: 111–114.

Johnson, J. A., R. T. Watson, and D.P. Mindell. 2005. Prioritizing species conservation: does the Cape Verde kite exist? *Proceedings of the Royal Society of London B* 272: 1365–1371.

Kadereit, J. W., S. Uribe-Convers, E. Westberg, and H.P. Comes. 2006. Reciprocal hybridization at different times between *Senecio flavus* and *Senecio glaucus* gave rise to two polyploid species in north Africa and south-west Asia. *New Phytologist* 169:431–41.

Kaneshiro, K.Y., and C. R. B. Boake. 1987. Sexual selection and speciation: Issues raised by Hawaiian *Drosophila*. *Trends in Ecology and Evolution* 2: 207–213.

Knowlton, N., L. A. Weigt, L. A. SolÛrzano, D. K. Mills, and E. Bermingham. 1993. Divergence in proteins, mitochondrial DNA, and reproductive incompatibility across the isthmus of Panama. *Science* 260: 1629–1632.

Knowlton, N., and L. A. Weigt. 1998. New dates and new rates for divergence across the Isthmus of Panama. *Proceedings of the Royal Society of London B* 265: 2257–2263.

Konstantinidis, K. T. and J. M. Tiedje. 2005. Genomic insights that advance the species definition for prokaryotes. *Proceedings of the National Academy of Scienes USA* 102: 2567–2572.

Lande, R. 1980. Genetic variation and phenotypic evolution during allopatric speciation. *American Naturalist* 116: 463–479.

Lande, R. 1981. Models of speciation by sexual selection on polygenic traits. *Proceedings of the National Academy of Sciences USA* 78: 3721–3725.

Lande, R. 1982. Rapid origin of sexual isolation and character divergence in a cline. *Evolution* 36: 213–223.

Lawrence, J. G., and H. Ochman. 1998. Molecular archaeology of the Escherichia coli genome. *Proceedings of the National Academy of Sciences USA* 95: 9413–9417.

Lee, C. E. 2000. Global phylogeography of a cryptic copepod species complex and reproductive isolation between genetically proximate "populations." *Evolution* 54: 2014–2027.

Leebens-Mack, J., O. Pellmyr, and M. Brock. 1998. Host specificity and the genetic structure of two yucca moth species in a yucca hybrid zone. *Evolution* 52: 1376–1382.

Lexer, C., M. E. Welch, O. Raymond, and L. H. Rieseberg. 2003. The origin of ecological divergence in *Helianthus paradoxus* (Asteraceae): selection on transgressive characters in a novel hybrid habitat. *Evolution* 57: 1989–2000.

Machordom, A., and I. Doadrio. 2001. Evolutionary history and speciation modes in the cyprinid genus *Barbus*. *Proceedings of the Royal Society of London B* 268: 1297–1306.

Mayr, E. 1942. *Systematics and the Origin of Species.* New York: Columbia University Press.

Mayr, E. 1963. *Animal Species and Evolution*. Cambridge, MA: Harvard University Press.

Mendelson, T. C. and K. L. Shaw. 2005. Rapid speciation in an arthropod. *Nature* 433: 375.

Navarro, A. and N. H. Barton 2003. Chromosomal speciation and molecular divergence-accelerated evolution in rearranged chromosomes. *Science* 300: 321–324.

Nesbo, C. L., M. Dlutek, and W. F. Doolittle. 2006. Recombination in *Thermotoga*: implications for species concepts and biogeography. *Genetics* 172: 759–769.

Noor, A. M. 1995. Speciation driven by natural selection in *Drosophila. Nature* 375: 674–675.

Noor, A. M., K. L. Grams, L. A. Bertucci, and J. Reiland. 2001. Chromosomal inversions and the reproductive isolation of species. *Proceedings of the National Academy of Sciences USA* 98: 12084–12088.

Novak, S. J., D. E. Soltis, and P.E. Soltis. 1991. Ownbey's Tragopogons: 40 years later. *American Journal of Botany* 78: 1586–1600.

Panhuis, T. M., R. Butlin, M. Zuk, and T. Tregenza. 2001. Sexual selection and speciation. *Trends in Ecology and Evolution* 16: 364–371.

Patton, J. L., and S. W. Sherwood. 1983. Chromosome evolution and speciation in rodents. *Annual Review of Ecology and Systematics* 14: 139–158.

Peichel, C. L. 2005. Fishing for the secrets of vertebrate evolution in threespine sticklebacks. *Developmental Dynamics* 234: 815–823.

Peichel, C. L., K. S. Nereng, K. A. Ohgi, B. L. E. Cole, P. F. Colosimo, C. A. Buerkle, D. Schluter, and D. M. Kingsley. 2001. The genetic architecture of divergence between threespine stickleback species. *Nature* 414: 901–905.

Pringle, A., D. M. Baker, J. L. Platt, J. P. Wares, J. P. Latgé, and J. W. Taylor. 2005. Cryptic speciation in the cosmopolitan and clonal human pathogenic fungus *Aspergillus fumigatus*. *Evolution* 59: 1886–1899.

Prokopy, R. J., S. R. Diehl, and S. S. Cooley. 1988. Behavioral evidence for host races in *Rhagoletis pomonella* flies. *Oecologia* 76: 138–147.

Rieseberg, L. H., B. Sinervo, C. R. Linder, M. C. Ungerer, and D. M. Arias. 1996. Role of gene interactions in hybrid speciation: Evidence from ancient and experimental hybrids. *Science* 272: 741–745.

Rieseberg, L. H., T. E. Wood, and E. J. Baack. 2006. The nature of plant species. *Nature* 440: 524–527.

Riginos, C. 2005. Cryptic vicariance in Gulf of California fishes parallels vicariant patterns found in Baja California mammals and reptiles. *Evolution* 59: 2678–2690.

Roca, A. L. 2001. Genetic evidence for two species of elephant in Africa. *Science* 293: 1473–1477.

Rundle, H. D., L. Nagel, J. W. Boughman, and D. Schluter. 2000. Natural selection and parallel speciation in sympatric sticklebacks. *Science* 287: 306–308.

Ryder, O. A., A. T. Kumamoto, B. S. Durrant, and K. Benirschke. 1989. Chromosomal divergence and reproductive isolation in dik-diks. In D. Otte and J. A. Endler, eds. *Speciation and Its Consequences*. Sunderland, MA: Sinauer, 180–207.

Salmon, A., M.L. Ainouche, and J.F. Wendel. 2005. Genetic and epigenetic consequences of recent hybridization and polyploidy in *Spartina* (Poaceae). *Molecular Ecology* 14: 1163–1175.

Schluter, D. 1993. Adaptive radiation in sticklebacks: Size, shape, and habitat use efficiency. *Ecology* 73: 699–709.

Schluter, D. 2001. Ecology and the origin of species. *Trends in Ecology and Evolution* 16: 372–380.

Schluter, D., J. W. Boughman, and H. D. Rundle. 2001. Parallel speciation with allopatry. *Trends in Ecology and Evolution* 16: 283–284.

Schwarz, D., B. M. Matta, N. L. Shakir-Botteri, and B. A. McPheron. 2005. Host shift to an invasive plant triggers rapid animal hybrid speciation. *Nature* 436: 546–549.

Segraves, K. A., and J. N. Thompson. 1999. Plant polyploidy and pollination: Floral traits and insect visits to diploid and tetraploid *Heuchera grossulariifolia*. *Evolution* 53: 1114–1127.

Shapiro, M. D., M. E. Marks, et al. 2004. Genetic and developmental basis of evolutionary pelvic reduction in threespine sticklebacks. *Nature* 428: 717–723.

Shapiro, M. D., M. E. Marks, et al. 2006. Genetic and developmental basis of evolutionary pelvic reduction in threespine sticklebacks; Corrigendum. *Nature* 439: 1014.

Smith, M. A., N. E. Woodley, D. H. Janzen, W. Hallwachs, and P. D. N. Hebert. 2006. DNA barcodes reveal cryptic host-specificity within the presumed polyphagous members of a genus of parasitoid flies (Diptera: Tachinidae). *Proceedings of the National Academy of Sciences USA* 103: 3657–3662.

Stireman, J. O. III, J. D. Nason, S. B Heard, and J. M. Seehawer. 2006. Cascading host-associated genetic differentiation in parasitoids of phytophagous insects. *Proceedings of the Royal Society of London B* 273: 523–530.

Templeton, A. R. 1996. Experimental evidence for the genetic-transilience model of speciation. *Evolution* 50: 909–915.

Vamosi, S. M. and D. Schluter. 2002. Impacts of trout predation on fitness of sympatric sticklebacks and their hybrids. *Proceedings of the Royal Society of London B* 269: 923–930.

Via, S. 1999. Reproductive isolation between sympatric races of pea aphids. I. Gene flow restriction and habitat choice. *Evolution* 53: 1446–1457.

Via, S. 2002. The ecological genetics of speciation. *American Naturalist* 159: S1–S7.

Via, S., A. C. Bouck, and S. Skillman. 2000. Reproductive isolation between divergent races of pea aphids on two hosts. II. Selection against migrants and hybrids in the parental environments. *Evolution* 54: 1626–1637.

Via, S., and D. J. Hawthorne. 2002. The genetic architecture of ecological specialization: Correlated gene effects of host use and habitat choice in pea aphids. *American Naturalist* 159: S76–S88.

Wang, H., E. D. McArthur, S. C. Sanderson, J. H. Graham, and D. C. Freeman. 1997. Narrow hybrid zone between two subspecies of big sagebrush (*Artemisia tridentata*: Asteraceae). IV. Reciprocal transplant experiments. *Evolution* 51: 95–102.

Weir, J. T. and D. Schluter. 2004. Ice sheets promote speciation in boreal birds. *Proceedings of the Royal Society of London B* 271: 1881–1887.

White, M. J. D. 1978. *Modes of Speciation*. San Francisco: W. H. Freeman.

Wiens, J. J., T. W. Reeder, and A. N. Montes de Oca. 1999. Molecular phylogenetics and evolution of sexual dichromatism among populations of the Yarrow's spiny lizard (*Sceloporus jarrovii*). *Evolution* 53: 1884–1897.

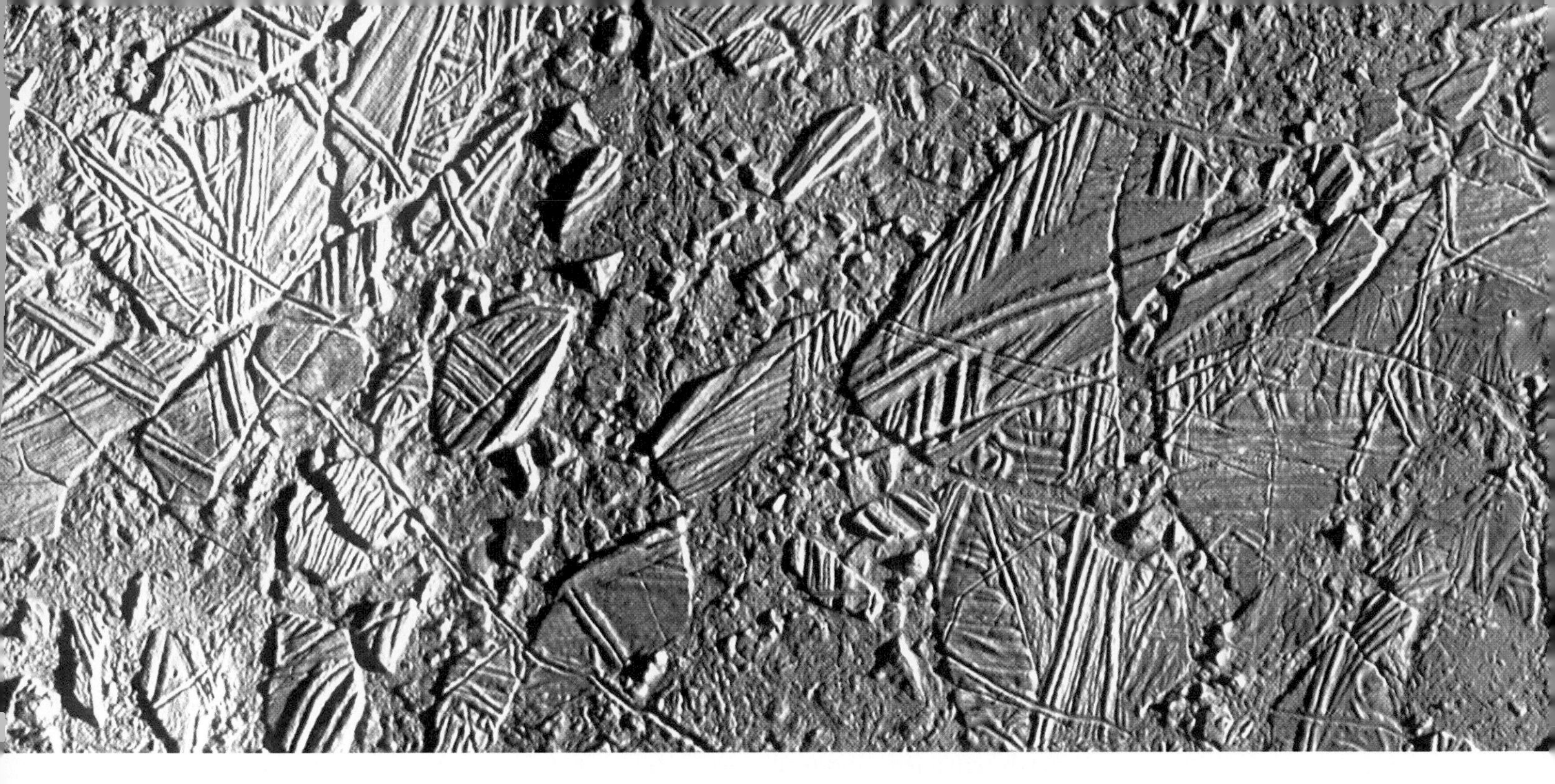

17

The Origins of Life and Precambrian Evolution

"There is grandeur in this view of life, with its several powers, having been originally breathed into a few forms or into one; and that, whilst this planet has gone cycling on according to the fixed law of gravity, from so simple a beginning endless forms most beautiful and most wonderful have been, and are being, evolved."

Charles Darwin, 1859,
On the Origin of Species,
page 490

Jupiter's moon Europa, visible below as the white speck on the right casting a small dark shadow on the lower portion of its mother planet, has a scarred icy surface (above) suggesting the presence a liquid water ocean underneath. This raises at least the possibility of life.

In this chapter we turn, as Darwin did at the end of *The Origin*, to the big picture of life on Earth. We review work by scientists trying to answer some of the most intriguing, profound, and difficult questions in biology:

- What was the first living thing?
- Where did it come from?
- What was the last common ancestor of today's organisms and when did it live?
- What is the shape of the tree of life?
- How did the last common ancestor's descendants evolve into modern life forms?

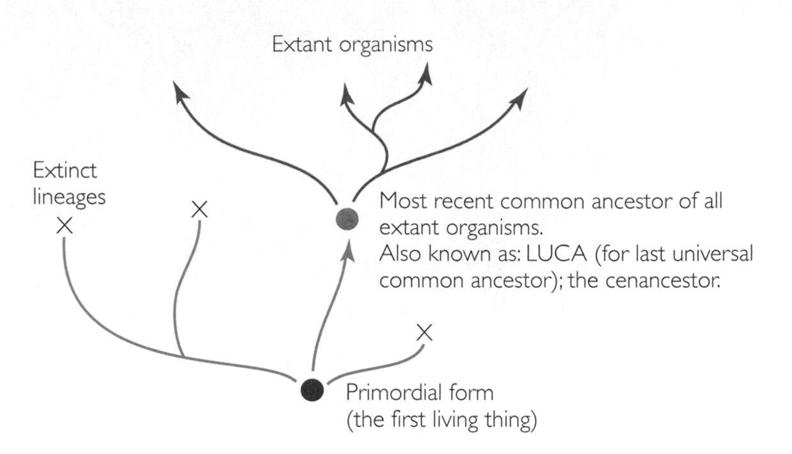

Figure 17.1 Cartoon of the tree of life Presumably, the first living organism had several descendant lineages, all but one of which eventually died out. The most recent common ancestor of all living things is the organism whose immediate descendants diverged to found the lineages that ultimately became all extant organisms. The whole-life phylogeny drawn here does not include the viruses, whose position on the tree is unclear.

These questions concern events of the far distant past. How far distant? Rocks dating from the time of Earth's formation do not exist on the planet's surface, but radiometric dating of meteorites yields an estimated age for the solar system, and hence Earth, of 4.5 to 4.6 billion years (see Badash 1989). The newborn Earth remained inhospitable for at least a few hundred million years. At first it was simply much too hot. This is because the collisions of the planetesimals that coalesced to form Earth released enough heat to melt the entire planet all the way through (Wetherill 1990). Eventually, Earth's outer surface cooled and solidified to form a crust, and water vapor released from the planet's interior cooled and condensed to form the oceans. By the best estimates, life arose on the Earth a bit less than 4 billion years ago.

No physical record of the first biological events has survived. In contrast to the evolutionary processes we have investigated so far in this book, the origins of life must be reconstructed using indirect evidence alone. Consequently, biologists have turned to gathering disparate bits of information and fitting them together like pieces of a jigsaw puzzle. When more complete, this puzzle should present a clearer picture of life's origins.

Figure 17.1 shows a hypothetical history of life on Earth that will serve to organize our discussion. Given that we are here to wonder what it was, there must have been a **primordial form**, or first living thing, represented by the red dot at the bottom of the figure. Presumably the primordial form beget a diversity of descendant lineages (green branches), most of which are long since extinct. Among the descendants of the primordial form was the last common ancestor of all extant organisms (orange dot), sometimes referred to as the **cenancestor** (Fitch and Upper 1987) or as **LUCA**, for last universal common ancestor (Forterre and Philippe 1999). The evolutionary history of LUCA's descendants (blue branches) constitutes the tree of life.

It is important to keep in mind that the history in Figure 17.1 is hypothetical, and subject to revision. For example, as we will see in Section 17.3, recent discoveries suggest that LUCA might not have been a single species, but instead a community of interbreeding forms.

17.1 What Was the First Living Thing?

In the early 1980s, two teams of scientists independently discovered small enzymes that could break and reform the chemical bonds that link nucleic acids together in chains. The enzymes did their job poorly. Compared to the hundreds of other such enzymes already known, they were slow at their catalytic task and showed little versatility. Yet the discovery has been recognized as among the most significant biological breakthroughs of the era. In 1989, the teams' leaders, Sidney Altman and Thomas Cech, shared the Nobel prize.

Why were biologists so excited by these new enzymes? The answer is that the enzymes were made not of protein, but of nucleic acid—specifically RNA. Until 1982, all known enzymes were proteins. RNA was often considered to be DNA's poor cousin, relegated to the task of shuttling genetic information from DNA, where the information is stored, to proteins, which carry out all the actual work of the cell. But Altman and Cech's discovery of RNA enzymes, or **ribozymes**, has changed how biologists view the operations of the cell. Perhaps more importantly, the existence of ribozymes has forever changed how biologists view the origin of the primordial form—how they believe life originated and evolved on the early Earth.

The origin of life has been under investigation, via observation and experimentation, for over 80 years (see Fry 2006). Biologists have made artificial cells and artificial cell membranes, and have zeroed in on chemical reactions that could have built cellular materials from nonliving sources. Early in the course of these studies, however, a quandary became apparent. Which of its two most vital substances did life acquire first, proteins or DNA? Proteins can perform all sorts of complicated biological tasks, but there is no evidence that proteins can propagate themselves. They cannot store and transmit the information needed to replicate. DNA, on the other hand, is perfectly suited to store and transmit genetic information by complementary base pairing, but it was not known to be able to perform any biological work. Neither DNA nor proteins seems to be of any use without the other, but it is implausible that they appeared simultaneously.

This chicken-and-egg problem was essentially resolved with the discovery of catalytic RNA. Because RNA has both a capacity for information storage and transmission and the ability to perform biological work, researchers now think that it preceded both proteins and DNA in the origin of life. Was there once a time when life was based entirely on RNA—an RNA World (Gilbert 1986)? This question is the topic of Section 17.1.

The RNA World hypothesis proposes that catalytic RNA molecules were a transitional form between nonliving matter and the earliest cells.

An RNA World appeals to scientists because it would possess many of the characteristics of modern life without the need for much more than a few organic molecules in solution. The hypothesis of an RNA World is based on the realization, since the discovery of ribozymes, that RNA can simultaneously possess both a genotype and a phenotype (Joyce 1989). The genotype is the primary sequence of nucleotides along the RNA (Figure 17.2a), much like the genotype of a modern organism is the sequence of nucleotides along the DNA in the chromosome. Catalytic RNA, for example, contains between 30 and 1,000 ribonucleotides that form its primary sequence, and hence its genotype. The *Tetrahymena* ribozyme discovered by Cech and colleagues (Kruger et al. 1982; Zaug and Cech 1986) stretches some 400 nucleotides from head (the 5′ end) to tail (the 3′ end). However, unlike genomic DNA, which is usually double-stranded (see Figure 5.1, page 144), RNA usually exists as a single-stranded molecule that folds back on itself many times to form a three-dimensional structure. In the case of ribozymes, this folded state can have an active site that enables the RNA molecule to catalyze a chemical reaction on a substrate, like a protein enzyme. This reactivity gives RNA its phenotype (Figure 17.2b).

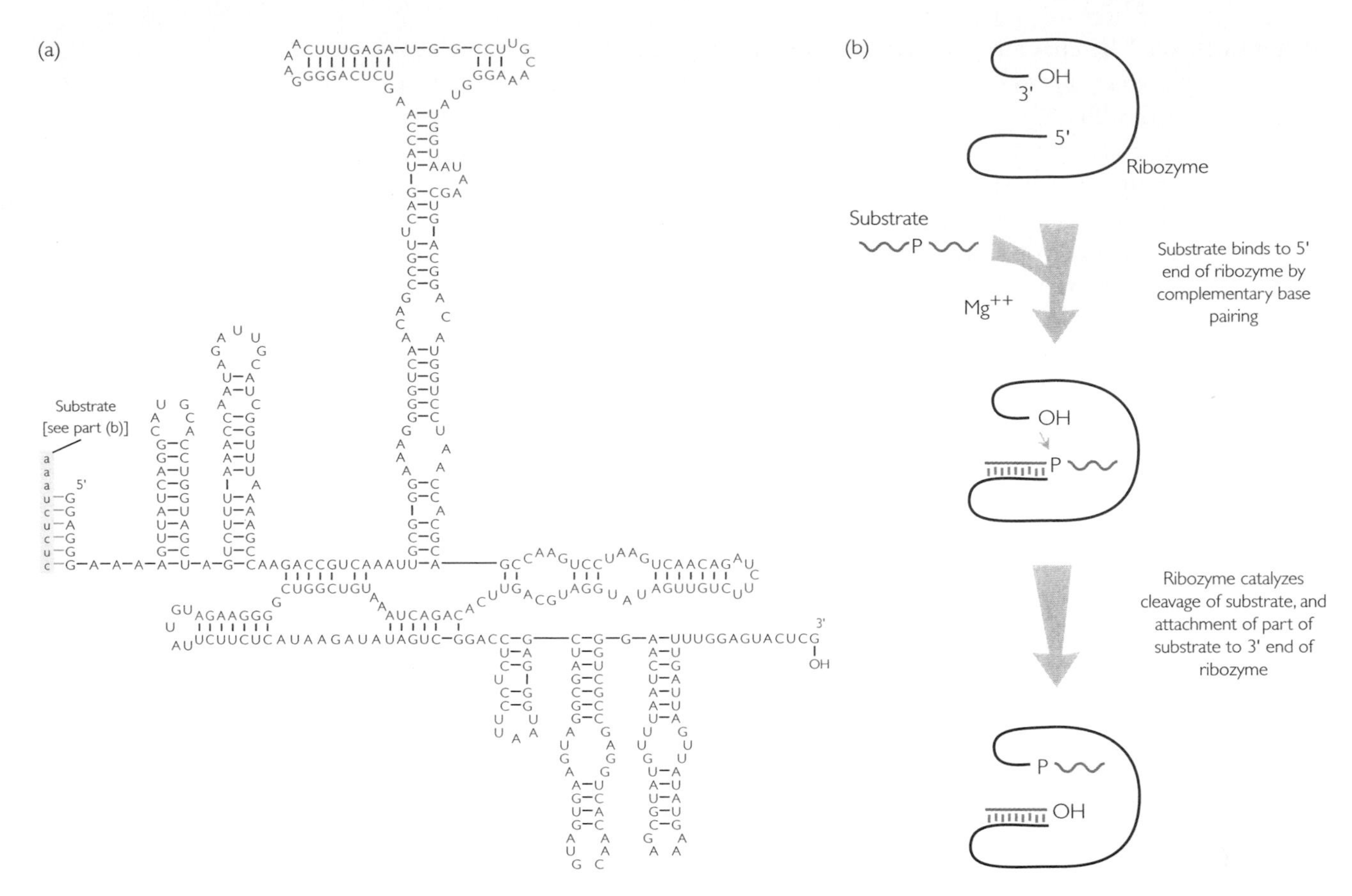

Figure 17.2 The ribozyme from *Tetrahymena thermophila* (a) The primary nucleotide sequence, which is the genotype of this catalytic RNA. This RNA is an intron (an intervening sequence between two genes) that separates two regions of the *Tetrahymena* genome that code for ribosomal RNA (rRNA) genes. Tom Cech and colleagues found that this sequence has the catalytic capability to splice itself out from between the two adjacent rRNAs after they have been transcribed (Kruger et al. 1982). The sequence shown here is a 413-nucleotide version, shortened from the naturally occurring form by the use of a restriction enzyme. A secondary structure of this molecule is formed when nucleotides base pair with each other as the molecule folds back upon itself during transcription from DNA. Note that in RNA, uridine (U) replaces thymidine (T), and that three types of base pairs are common: A–U, G–U, and G–C. The secondary structure shown here is drawn such that no RNA strands cross each other and, as such, does not accurately reflect how the molecule folds further into a tertiary (three-dimensional) structure. (b) A cartoon of the catalysis performed by the *Tetrahymena* ribozyme *in vitro,* which is its phenotype. A short oligonucleotide substrate (orange) binds to the 5′ end of the ribozyme (red) through complementary base pairing (green ticks). In the presence of a divalent cation, such as Mg^{2+}, the ribozyme catalyzes the breakage of a phosphoester bond in the substrate and the ligation of the 3′ fragment to its own 3′ end. This "pick-up-the-tail" event can be used to discriminate catalytically active mutant sequences from less active or inactive mutants during *in vitro* evolution experiments (see Figure 17.4).

Defining Life

Here is one way we might define life: If it forms populations capable of evolving by natural selection, then it is alive.

All living organisms possess both a genotype and a phenotype. In fact, when we consider what life really is, and how living systems can be distinguished from nonliving ones, the ability to store and transmit information (a genotype) and the ability to express that information (a phenotype) are perhaps the most important criteria that set life apart from nonlife. There is no neat list of characteristics that define life. Most biologists would include traits like growth and reproduction on such a list, but they cannot agree on what else should be used to exclude such lifelike systems as a growing salt crystal or a computer virus (if, indeed, these should be excluded). However, many now agree that the ability to evolve is a crucial component of any definition of life. Evolution—descent with modification—requires both the ability to record and make alterations in heritable information and a sorting process that distinguishes valuable changes from detrimental ones. The former is a property of genotype, while the latter occurs as a result of variation among individuals in phenotype.

Dozens of naturally occurring ribozymes have been discovered (Gesteland et al. 1999), and the phenotypes of most of them involve the formation and breaking of phosphoester bonds in RNA or DNA (Figure 17.2b). The chemistry of these reactions is precisely what is needed to replicate nucleic acids. This observation gives support to the idea of a primordial RNA World, where RNA would be responsible for replicating itself in order to persist. If a molecule of RNA could make a copy of itself while accommodating the possibility of mistakes—mutations—then that molecule would exhibit many of the characteristics of modern life and could therefore be considered alive.

The Case for RNA as an Early Life Form

The RNA World hypothesis posits that the primordial form was an RNA-based living system that later evolved into life forms like those we see today, in which DNA stores biological information and proteins manifest this information. DNA is better suited as an information repository because it is chemically more stable than RNA. Especially when double-stranded, DNA can better withstand high temperatures and spontaneous degradation by acids or bases.

What is the evidence that RNA is ancient? The existence of catalytic RNA is critical, but there are other indicators as well. One clue that RNA was involved in early life forms is its role in the machinery cells use for replication and metabolism (Crick 1966; White 1976). The most conserved and universal component of the information-processing machinery, for example, is the apparatus for translation of genetic information into protein: the ribosome (Harris et al. 2003; Koonin 2003). This apparatus, while it incorporates proteins, is built on a frame of RNA (rRNA). Ribosomes not only contain RNA themselves, they require RNA adaptors (tRNAs) to do their job. Furthermore, it is the RNA portion of ribosomes that actually carries out the catalytic steps in protein synthesis (Nissen et al. 2000; Steitz and Moore 2003). Another argument for the antiquity of RNA is that the basic currency for biological energy is ribonucleoside triphosphates, such as ATP and GTP (Joyce 1989). These molecules are involved in almost every energy-transfer operation of all cells and are even components of electron-transfer cofactors such as NAD (nicotinamide adenine dinucleotide), FAD (flavin adenine dinucleotide), and SAM (S-adenosyl methionine). With these ghosts of an RNA World in mind, we turn next to the question: Can RNA evolve?

The Experimental Evolution of RNA

RNA sequences can provide a blueprint for their own replication. For any RNA sequence, we can build a complementary sequence by base pairing. Thus RNA, like DNA, has the capacity to store heritable information that can be propagated. A good example is the life cycle of the HIV virus we tracked in Chapter 1. HIV uses the protein enzyme reverse transcriptase to copy its RNA strand into a DNA complement, which can then be converted into double-stranded DNA (see Figure 1.5, page 7). Given that RNA can store genetic information, populations of RNA molecules should be able to evolve.

One way researchers have tested the RNA World hypothesis is to check whether populations of RNA molecules can evolve by natural selection.

Donald Mills, Roger Peterson, and Sol Spiegelman (1967) used RNA from a virus called bacteriophage Q_β to demonstrate that populations of RNA can evolve in test tubes. Their experiment also employed Q_β's replicase protein.

Replicase is a general term to describe any enzyme that can make a copy of another molecule. When a small amount of Q_β RNA is incubated with Q_β replicase for a few minutes, the replicase makes copies of the RNA, and copies of the copies, and so on. With each copying, there exists a small, but finite, probability that the replicase will make a mistake and miscopy a nucleotide, by putting an A instead of a G across from a C, for example. This mutability provides the raw material for evolution. After only four serial transfers of the Q_β RNA solution (removing a small amount of RNA after a prespecified time and using it to seed a fresh test tube containing fresh replicase), the phenotypic composition of the RNA population had already changed: The Q_β RNAs had, on average, a dramatically reduced ability to infect bacteria (Figure 17.3).

An important aspect of test-tube evolution experiments is to demonstrate that any observed phenotypic change has an underlying genotypic basis. Although the Q_β RNA experiment was performed before technology existed to determine the nucleotide sequences of large numbers of nucleic acids, Mills and colleagues were able to infer a genotypic shift in the RNA population after 74 transfers by digesting the RNA into its composite nucleotides. The researchers determined the percentages of all four nucleotides by chromatographic analysis and found that values from the descendant RNA population differed from those of the original Q_β RNA by as much as 5%. The population of Q_β RNAs had, indeed, evolved.

The experiments with Q_β RNA showed that mutations could arise and spread in a test-tube RNA population, but they were not designed to correlate a specific phenotypic trait of RNA with a particular genotype during evolution. The novel genotypes that rose to high frequency in the Mills experiment were simply favored by selection for an RNA sequence that could be more quickly replicated by the Q_β replicase protein. In fact, a typical RNA from the 75th transfer was 83% shorter than one from the original Q_β RNA population, and it could be replicated about 15 times faster.

Although Mills' experiment demonstrated that populations of RNA molecules can evolve, the traits that changed across generations—the length of the RNA sequences and the speed with which they could be copied by Q_β replicase—are not

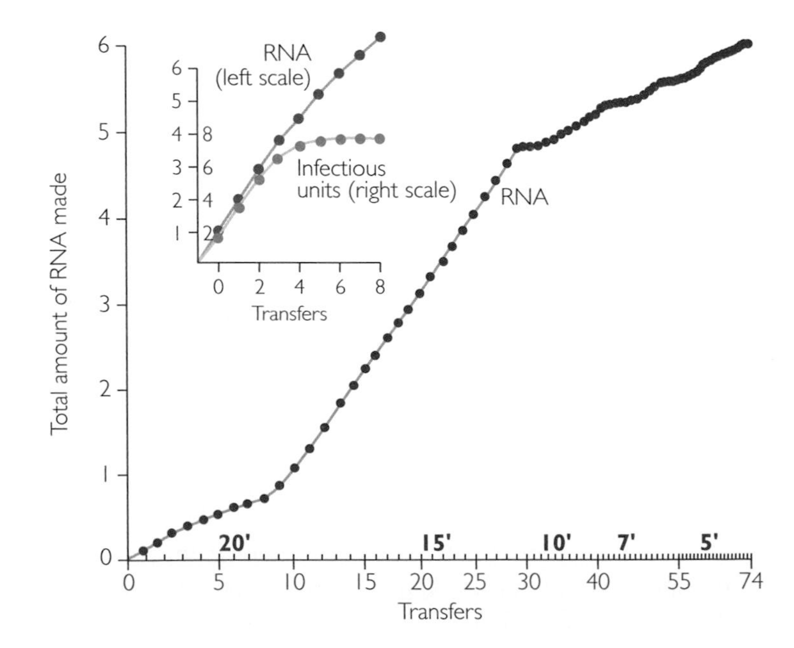

Figure 17.3 Mills, Peterson, and Spiegelman's experiment with Q_β RNA The graphs show the accumulation of Q_β RNA over many serial transfers from one test tube to another. In each test tube, a small amount of Q_β RNA from the previous tube was mixed with some purified Q_β replicase protein and the four ribonucleotide triphosphates, incubated, then transferred to a new tube. The incubation times, shown in red on the x-axis, were reduced during the experiment to maintain selection for rapid replication. The researchers assessed the amount of RNA made as the cumulative amount of radioactive UTP incorporated into RNA molecules. (For the main graph, the units are counts per minute $\times 10^{-5}$ of radioactive phosphorus per 0.25 ml). The inset graph (in which the units for amount of RNA made are a tenth as large) shows that after the 4th transfer, the RNA had evolved a new phenotype: It could no longer infect *E. coli* bacteria. The RNA from the 74th transfer had evolved a base composition that was about 5% different from the original Q_β RNA and was only 17% the length. From Mills et al. (1967).

associated with the ability to do the kinds of useful biological work that would have been required of the RNAs that would have lived in RNA World. With the discovery of ribozymes came a more powerful way to perform test-tube evolution of RNA, and with that, a more convincing argument that life may have passed through an RNA-based period. If an RNA molecule has a phenotype that involves catalyzing a specific chemical reaction, can we apply selection to improve or modify this phenotype and observe a heritable change?

Beaudry and Joyce (1992) exploited the catalytic capacity of the *Tetrahymena* ribozyme to address this question (Figure 17.4). Researchers had previously determined that a shortened form of the *Tetrahymena* ribozyme could catalyze a phosphoester transfer reaction on a short RNA substrate, called an oligonucleotide (a piece of single-stranded nucleic acid between about 5 and 30 nucleotides in length). In the reaction, the 3′ half of the substrate is broken off by the ribozyme and attached to the ribozyme's own 3′ end (Kruger et al. 1982). If this 3′ "tail" could be used as a tag, then ribozymes that performed the catalysis could be distinguished from those that did not. Beaudry and Joyce (1992) first made a large population of RNA molecules by sprinkling random mutations throughout the

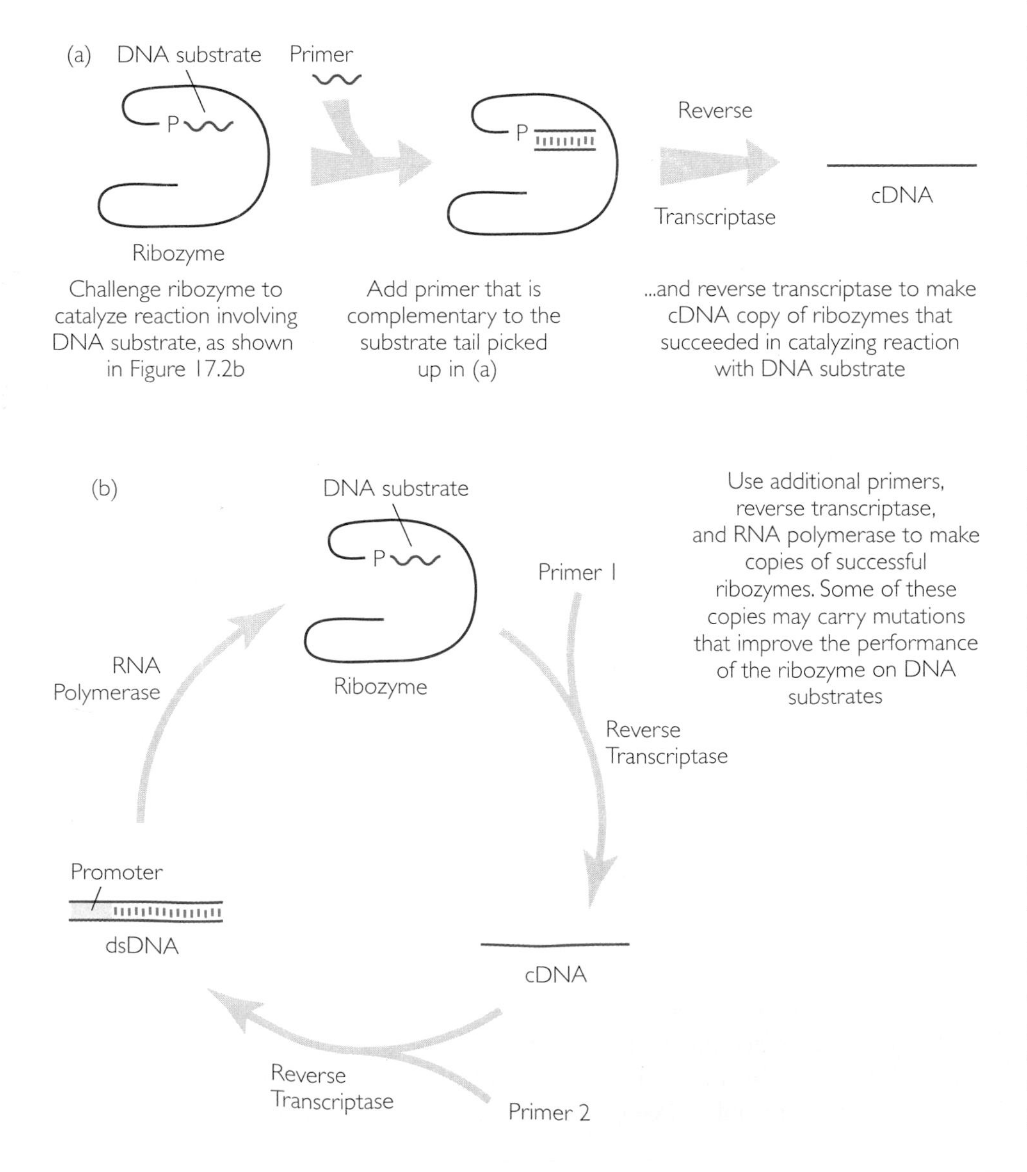
(a)
DNA substrate
Primer
P
Ribozyme
P
Reverse
Transcriptase
cDNA
Challenge ribozyme to catalyze reaction involving DNA substrate, as shown in Figure 17.2b
Add primer that is complementary to the substrate tail picked up in (a)
...and reverse transcriptase to make cDNA copy of ribozymes that succeeded in catalyzing reaction with DNA substrate
(b)
DNA substrate
P
Ribozyme
Primer 1
Use additional primers, reverse transcriptase, and RNA polymerase to make copies of successful ribozymes. Some of these copies may carry mutations that improve the performance of the ribozyme on DNA substrates
RNA Polymerase
Reverse Transcriptase
Promoter
dsDNA
cDNA
Reverse Transcriptase
Primer 2

Figure 17.4 Test-tube selection and reproduction of RNA (a) ***Selection.*** A pool of RNA sequences (red), made by random mutagenesis of a ribozyme such as the *Tetrahymena* ribozyme (Figure 17.2a), is challenged to perform a desired chemical reaction. Only those that can perform the reaction acquire a short "tail" of DNA nucleotides attached to their 3′ end (blue). This tail of nucleotides is complementary to a primer that is required for copying the ribozyme's RNA into cDNA by reverse transcriptase.
(b) ***Reproduction.*** RNA sequences that have successfully acquired a 3′ tail (top) can bind primer 1 by complementary base pairing and be copied by reverse transcriptase into complementary DNA (cDNA). A second primer (primer 2) then binds to the cDNA so that the reverse transcriptase can make the DNA double-stranded. Primer 2 contains the promoter region for RNA polymerase, so that in the last step, RNA polymerase can bind to the double-stranded DNA and copy it back into RNA many times. The overall effect of this cycle is that successful RNA sequences, those that can perform the initial chemical reaction, are able to reproduce themselves into thousands of copies. Additional variation is introduced into the RNA population during this cycle because the two protein enzymes, reverse transcriptase and RNA polymerase, can make mistakes during copying and create mutations, which themselves are subject to selection. Reprinted with permission from Beaudry and Joyce (1992).

Tetrahymena ribozyme at a rate of 5% per position. Then, this mutant population was challenged with a novel task to select certain genotypes. The task in this case was that the substrate oligonucleotide was provided in the form of DNA, not RNA. The naturally occurring sequence of the *Tetrahymena* ribozyme (the "wild type") used to start these experiments could cleave a DNA substrate only at a miserably slow rate. Beaudry and Joyce hoped that in the mutant pool there were variant sequences that, by chance, had an increased capacity for DNA cleavage.

The researchers incubated the mutant RNA population with a DNA substrate for an hour, and then amplified the ribozyme RNA into many more copies by adding two protein enzymes—reverse transcriptase and RNA polymerase. Because ribozymes pick up a 3′ tail as a consequence of cleaving the substrate, a DNA primer for reverse transcriptase that is complementary to the 3′ tail can be used to discriminate sequences that catalyzed a reaction with the DNA substrate from sequences that did not. A 3′ tail is necessary to bind the primer, which in turn is necessary to initiate reverse transcription, which is in turn necessary to make more RNA. The RNA that results from this cycle of events can be used to seed a completely new cycle (a new generation) to continue and refine the selection process. After 10 such generations, the activity of the average RNA in the population in cleaving DNA substrates and attaching one of the resulting fragments to its own 3′ end had improved by a factor of 30. Importantly, this phenotypic enhancement could be traced to specific changes in nucleotide sequence (Figure 17.5). Specific mutations at four nucleotide positions in the ribozyme's sequence were responsible for the majority of the catalytic improvement. Individual ribozymes carrying mutations at positions 94, 215, 313, and 314 proved to have a catalytic efficiency over 100 times greater than the ancestral sequence.

Populations of catalytic RNA molecules exhibit variation in nucleotide sequence. This variation is heritable when RNA is replicated. And researchers have devised experimental conditions under which sequence variation results in differences in survival.

This experiment demonstrated that RNA molecules in solution can possess features of living organisms that allow them to evolve. Each RNA can be ascribed a particular fitness, which is a function of both survival (substrate catalysis) and reproduction (ability to be reverse and forward transcribed). The fitness of the molecule is a reflection of its phenotype, which, in the case of ribozymes, is immediately specified by their primary sequence. Variation in an RNA population can be introduced at the outset by the randomization of a wild-type sequence, as was the case with the Beaudry and Joyce (1992) experiment. Alternatively, an investigator can rely on the intrinsic error rates of the protein enzymes used in RNA amplification and can even alter the chemical environment to make the error rates higher. With such online mutagenesis the system becomes truly evolutionary, and selection can operate on variants of variants over many generations. Thus, it is easy to see a parallel between an evolving population of RNA in a test tube and an evolving population of modern organisms in the natural environment (Lehman and Joyce 1993).

In test-tube experiments like these, researchers have evolved many ribozymes with either improved function or an entirely new function. The catalytic repertoire of RNA has greatly expanded (Joyce 1998), and we now know that RNA can catalyze such reactions as phosphorylation (Lorsch and Szostak 1994), aminoacyl transfer (Illangasekare et al. 1995), peptide-bond formation (Zhang and Cech 1997), and carbon–carbon bond formation (Tarasow et al. 1997; Fusz et al. 2005). Ribozymes have been designed that are allosteric, requiring a small-molecule cofactor to carry out catalysis (Tang and Breaker 1997). Ribozymes can be selected that can play a role in ribonucleotide synthesis (Unrau and Bartel 1998), retain activity with only three of four nucleotides (Rogers and Joyce 1999), and

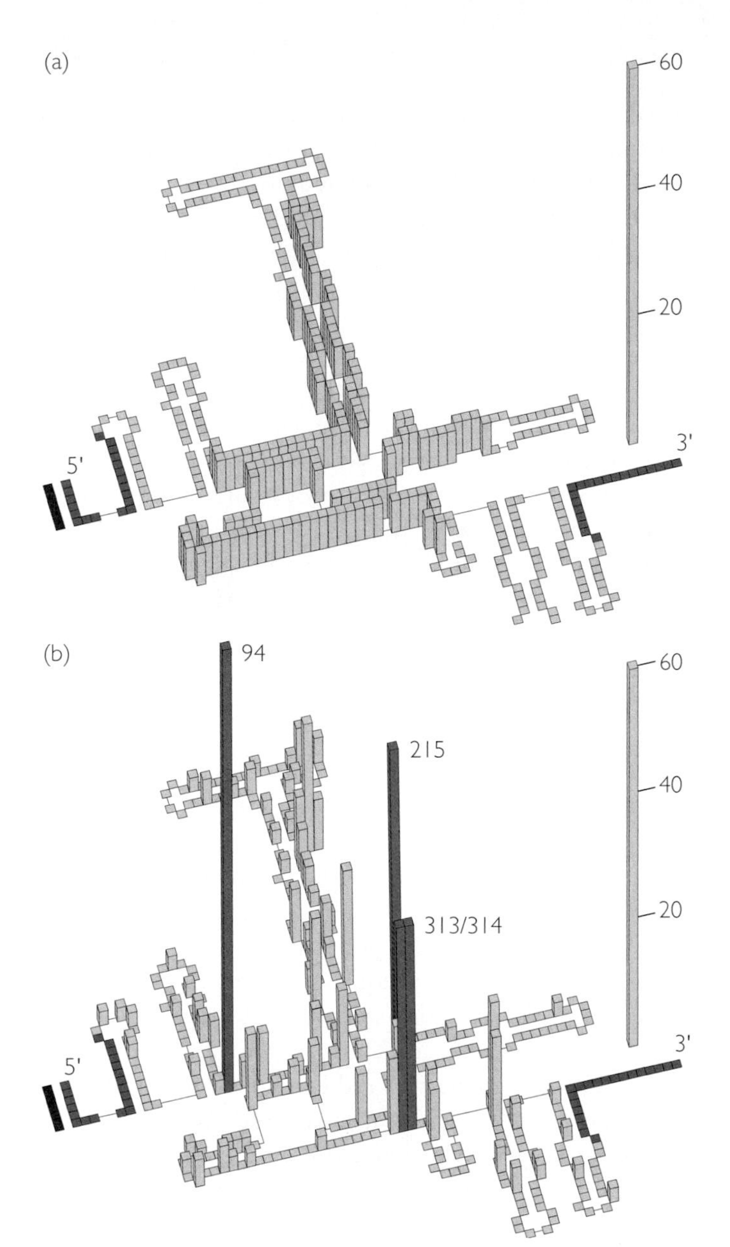

Figure 17.5 Genotypic changes in an evolving RNA population These histograms depict the genotypic changes that occurred during the test-tube evolution experiment of Beaudry and Joyce (1992) with variants of the *Tetrahymena* ribozyme. (a) Mutations were introduced randomly throughout the middle portion of the ribozyme. A crude representation of the folded secondary structure of the 413-nucleotide ribozyme is shown as a base from which the heights of each bar depict the frequencies of mutations in the test-tube population relative to the wild-type *Tetrahymena* sequence. In the starting generation, 140 of the nucleotides were randomly mutated such that each nucleotide had a 5% chance of not being the wild-type nucleotide. This pool of variants (about 10^{13} molecules) was challenged such that only those sequences that could catalyze the cleavage of a DNA oligonucleotide substrate (black boxes) would be allowed to reproduce (Figure 17.4b). (b) The genotypic composition of the population after nine rounds of selection and reproduction. Four mutations, at nucleotides 94, 215, 313, and 314, have increased in frequency over 50% and are mainly responsible for the new phenotypic characteristics of the population. Reprinted with permission from Beaudry and Joyce (1992).

operate without divalent metal–ion cofactors (Geyer and Sen 1997). RNA sequences, called aptamers, can be selected to bind tightly to almost any other molecule desired (Tuerk and Gold 1990; Ellington and Szostak 1990), much like the immunoglobulin proteins of the mammalian immune system. Together, these developments implicate RNA as a possible living system that preceeded cells.

As discussed in Chapter 3, Darwin deduced that when the individuals in a population exhibit (1) variation, (2) inheritance, (3) excess reproduction, and (4) variation in survival or reproductive success, populations will evolve. When stripped of the particular characteristics of an intact, complex organism, we realize that traits (1) and (2) are about having a genotype, trait (3) is about being self-replicating, and trait (4) is about having a phenotype that makes a difference. Consequently, a self-replicating population of RNA would have the essence of life, even without the cells or organelles or tissues or leaves or fur or behavioral characteristics, and so on, that we are accustomed to seeing in living creatures.

Self-Replication

Although populations of catalytic RNA molecules have most of the properties required for evolution by natural selection, they still cannot evolve on their own without considerable help from human researchers. This is because catalytic RNA molecules cannot yet copy themselves.

From what we have discussed so far, there is a crucial piece conspicuously missing from the evidence that today's organisms could be descended from inhabitants of an RNA world. We know that RNA is an incredibly diverse molecule, and that it can evolve under the right circumstances. In all the experiments previously described, however, RNA was copied by protein enzymes. These proteins, of course, would not have existed in the RNA World. A main premise of the RNA World hypothesis is that RNA predates the time when life used proteins to do most of the biological work. The piece of evidence that we lack for the RNA World is the demonstration that RNA can copy itself. The as-yet-undiscovered "RNA-dependent RNA autoreplicase" remains a Holy Grail for origins-of-life research (Bartel and Unrau 1999; Müller 2006). Whether the RNA World used only one type of self-replicating RNA or a suite of interacting RNAs, an RNA with a replicase phenotype would be necessary (Bartel 1999). The acquisition of the ability to self-replicate by a collection of organic molecules, such as RNA, is arguably the point at which nonliving matter came to life.

The hypothesis that an RNA molecule could replicate itself, serving as a simple proto-organism, is testable. If the hypothesis is correct, then we should be able to make a self-replicating RNA molecule in the lab. Although this has not been achieved to date, researchers have made significant advances. David Bartel and coworkers, for example, are using test-tube evolution to search for ribozymes capable of synthesizing RNA (Bartel and Szostak 1993; Ekland et al. 1995).

Figure 17.6 shows the selection scheme Bartel and Szostak (1993) used to make ribozymes that catalyze the formation of a phosphoester bond to link a pair of adjacent RNA nucleotides. The researchers started with a large pool of RNA polynucleotides. These pool RNAs constituted the population to be subjected to selection. Every RNA in the pool had the same sequence on its 5′ and 3′ ends (represented by lines), plus a unique 220-nucleotide stretch of random sequence in the middle (represented by the box labeled *Random 220*). Figure 17.6 follows two RNAs from the pool: Random 220 A (left column) and Random 220 B (right).

Bartel and Szostak bound the pool RNAs to agarose beads by means of a base-pairing interaction on their 3′ ends. The scientists then bathed the pool RNAs in a solution containing many copies of a specific substrate polynucleotide (row 1). This short RNA molecule had, on its 5′ end, a sequence of nucleotides forming a tag, whose function will soon become clear. On its 3′ end, the substrate RNA had a sequence of nucleotides complementary to the free end of the pool RNA molecules. The substrate molecules quickly became bound, by base-pairing hydrogen bonds, to the pool RNAs (row 2).

Researchers are performing selective breeding experiments with populations of catalytic RNAs in an effort to develop catalytic RNAs that can replicate themselves. If the researchers succeed, then by the definition we proposed earlier, they will have created life.

This annealing brought into adjacent position the triphosphate group (PPP) on the 5′ end of the pool RNA and the hydroxyl group (OH) on the 3′ end of the substrate RNA. If, by chance, the 220-nucleotide stretch of random sequence in a pool RNA molecule had some ability to catalyze the formation of phosphoester bonds, then it catalyzed the formation of such a bond between the substrate and pool RNA molecules. In row 3, Random 220 A has catalyzed such a reaction, liberating a diphosphate molecule, whereas Random 220 B has not.

Bartel and Szostak then rinsed the pool RNAs under conditions that washed away any substrate RNAs not covalently bound (by phosphoester bonds) to pool RNAs, and liberated the pool RNAs from their agarose beads (row 4). Random 220 A still has its substrate (with tag); Random 220 B does not.

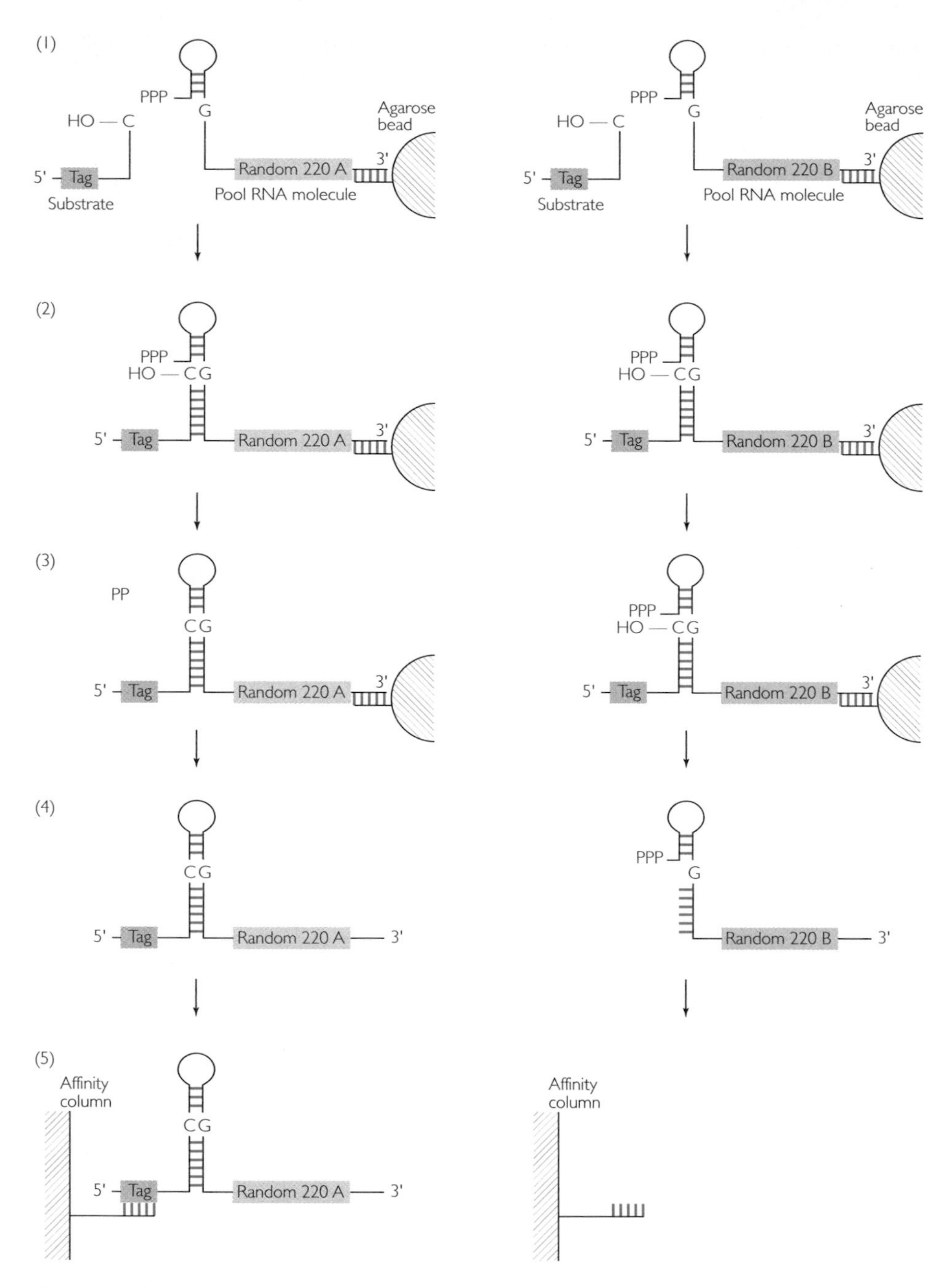

Figure 17.6 Test-tube selection scheme for identifying ribozymes that can synthesize RNA See text for explanation. After Bartel and Szostak (1993).

Finally, the scientists ran the pool RNAs through an affinity column (row 5). The affinity column caught hold of the tag sequence on the substrate RNA by base pairing. The column thus captured any pool RNA whose 220-nucleotide stretch of random sequence had catalytic activity (like Random 220 A) and let pass any pool RNA whose 220-nucleotide sequence did not. This selection step is analogous to the discrimination between tailed and untailed *Tetrahymena* ribozymes in the experiment of Beaudry and Joyce (1992).

Now Bartel and Szostak released the captured pool RNAs from the affinity column, made many copies of each by using replication enzymes that allowed

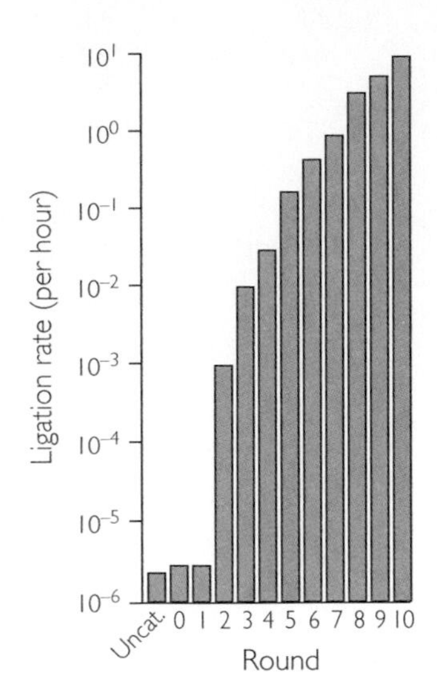

Figure 17.7 Evolution of catalytic ability in a laboratory population of ribozymes The graph shows the average rate at which the members of Bartel and Szostak's (1993) RNA pool catalyzed the formation of phosphoester bonds (ligation rate) as a function of round of selection. Note the logarithmic scale on the vertical axis. Over the course of the experiment, the catalytic activity of the molecules in the RNA pool increased by several orders of magnitude. From Bartel and Szostak (1993).

some mutations, and repeated the whole process again. Notice that Bartel and Szostak's protocol also gives, to the pool RNAs, all the properties necessary and sufficient for evolution by natural selection. The RNAs have reproduction with heritability (via the copying process), variation (due to mutation), and differential survival (in the affinity column). The RNAs most likely to survive from one generation to the next are the ones that are most efficient at catalyzing phosphoester bonds. After 10 rounds of selection, the RNA pool had evolved ribozymes that could catalyze the formation of phosphoester bonds at a rate several orders of magnitude faster than such bonds form without a catalyst (Figure 17.7).

More recently, Wendy Johnston and colleagues (2001; see also Lawrence and Bartel 2005), working in Bartel's laboratory, reported that they had used a similar scheme to evolve a catalytic RNA that can add up to 14 nucleotides to a growing RNA chain (Figure 17.8). This ribozyme uses template RNA and nucleoside triphosphates, and catalyzes RNA polymerization by the same chemical reaction promoted by the protein-based RNA polymerase enzymes used by living organisms. This laboratory-evolved ribozyme is not yet capable of self-replication. It cannot copy template strands that are even close to its own length; it too readily falls off the template strand it is copying, and it synthesizes new RNAs more slowly than it decomposes (see Strobel 2001). But it appears that biochemists are homing in on an RNA sequence, or set of sequences, capable of self-replication.

As Gerald Joyce (1996) put it, "Once an RNA enzyme with RNA replicase activity is in hand, the dreaming stops and the fun begins." If given the right organic molecules to feed on, a population of self-replicating RNAs should be able to evolve on its own by mutation and natural selection. The population would not require the generation-by-generation management practiced by Beaudry and Joyce or Bartel and Szostak. Would a species of self-replicating RNA evolve a DNA genome with DNA replication and transcription? Would it invent proteins and translation? Would its machinery be anything like the machinery in naturally evolved organisms? Would one of its descendants resemble the cellular life we see today? Perhaps one day soon we will have more pieces to fit into the puzzle.

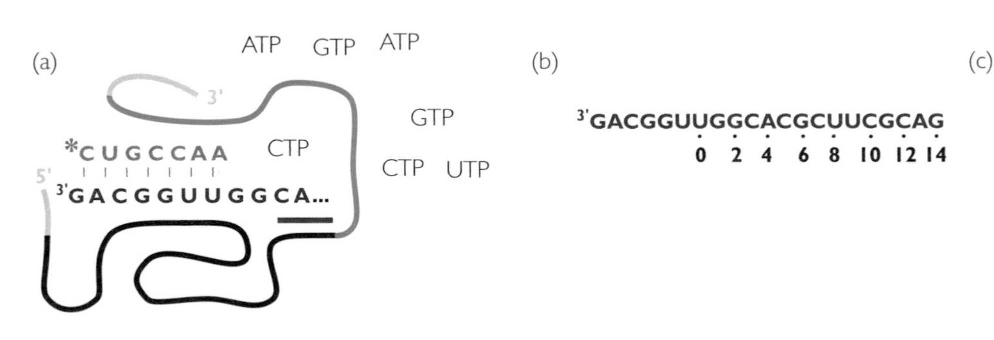

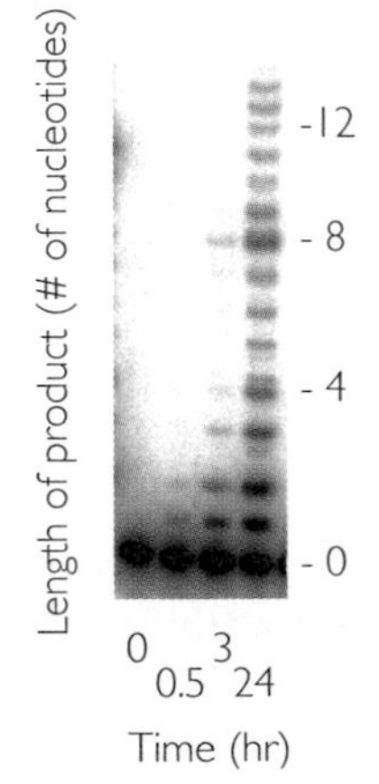

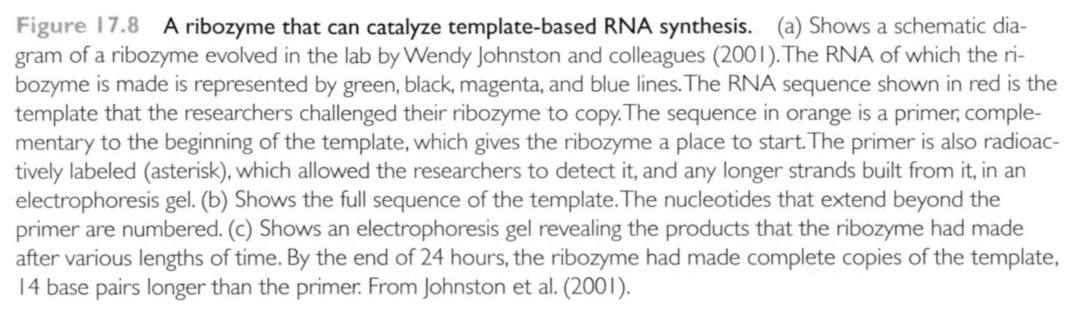

Figure 17.8 A ribozyme that can catalyze template-based RNA synthesis. (a) Shows a schematic diagram of a ribozyme evolved in the lab by Wendy Johnston and colleagues (2001). The RNA of which the ribozyme is made is represented by green, black, magenta, and blue lines. The RNA sequence shown in red is the template that the researchers challenged their ribozyme to copy. The sequence in orange is a primer, complementary to the beginning of the template, which gives the ribozyme a place to start. The primer is also radioactively labeled (asterisk), which allowed the researchers to detect it, and any longer strands built from it, in an electrophoresis gel. (b) Shows the full sequence of the template. The nucleotides that extend beyond the primer are numbered. (c) Shows an electrophoresis gel revealing the products that the ribozyme had made after various lengths of time. By the end of 24 hours, the ribozyme had made complete copies of the template, 14 base pairs longer than the primer. From Johnston et al. (2001).

17.2 Where Did the First Living Thing Come From?

The RNA World has many attractive features, and it solves the problem of having to propose the advent of proteins before DNA existed to encode them. But an RNA World comes with troubles of its own. The fundamtental problem is a simple one: How could RNA sequences of any kind arise in an abiotic environment?

The proposition that catalytic RNAs were a transitional form between nonliving matter and cellular life leaves many gaps. We must still explain where the first RNA molecules came from, and how a population of self-replicating RNA molecules evolved into DNA- and protein-based cells.

Chemists have studied the ways in which nucleic acids could be made without the aid of living systems. Certain aspects of the abiotic synthesis of nucleic acids turn out to be surprisingly easy, while others turn out to be dauntingly difficult. The general consensus is that the RNA World was probably not the first self-replicating system. This is because the likelihood of making RNA abiotically is too minute. Later, we will talk about the challenges of RNA synthesis, but for now we will note that RNA was probably derived from a more primitive chemical system. In other words, the primordial form was not made of RNA, but of something else that preceded RNA.

Regardless of what the primordial form was made of, in order to reconstruct the advent of any information-containing organic molecule with the properties of self-replication, the following four issues need to be addressed:

1. Information-containing biomolecules need to be made from simple inorganic compounds. Where did these compounds come from?
2. The chemical reactions that construct larger molecules from simple inorganics must be favorable and have a source of energy. What were these reactions?
3. The building blocks must be able to self-assemble into polymers such as RNA and polypeptides. How did this happen?
4. Larger biomolecules must be protected from harsh environmental conditions. How was this accomplished?

We will assume, following most researchers in the field, that the primordial form arose on Earth. As discussed in Box 17.1, however, this is not necessarily correct.

Where Did the Stuff of Life Come From?

On September 28, 1969 at about 11 o'clock in the morning, a meteor entered Earth's atmosphere over the town of Murchison, Australia, and broke up into several meteorites that scattered over a 5-square-mile area of the ground (Figure 17.9). Soon after, scientists collected some of the meteorites and carefully brought them back to the laboratory for chemical analysis (Kvenvolden et al. 1970). To their astonishment, the analyses showed that organic compounds were present in the interior of the rocks. In particular, the amino acids glycine, alanine, glutamic acid, valine, and proline were found in significant concentrations (1–6 micrograms of amino acid per gram of meteorite). These amino acids are among the ones used by modern organisms to make proteins. Amino acids had been found in meteorites before, but their presence was more than likely the result of contamination from human handling. The scientists who studied the Murchison meteorites fractured them in the lab and analyzed only the interior portions. In addition, the amino acids they found in the Murchison stones were racemic; that is, they included roughly equal proportions of the D- and L-stereoisomers (mirror-image forms). By contrast, biological amino acids are almost purely of the L-form, and thus terrestrial life could not be the source of the compounds the researchers found in the meteorites.

Figure 17.9 The Murchison Meteorite This photo shows one fragment of the 100 kg of meteorites that fell near Murchison, Australia, in 1969. These meteorites contain dozens of amino acids.

Box 17.1 The Panspermia Hypothesis

Most contemporary specialists assume that life originated on Earth, but there is an alternative: Life could have originated elsewhere and traveled here. This suggestion, the Panspermia Hypothesis, invites the complaint that it merely shifts the problem of life's origin to a remote location where it is even harder to study. Francis Crick and Leslie Orgel (1973) pointed out, however, that such criticism is not only unfair, it could even prevent us from discovering the truth:

> For all we know there may be other types of planet on which the origin of life ... is greatly more probable than on our own. For example, such a planet may possess a mineral, or compound, of crucial catalytic importance, which is rare on Earth.

One version of the Panspermia Hypothesis is that life originated on another planet within our own solar system. Microbes could then have been dislodged from their home world by a meteor impact, carried through space on a chunk of debris, and dropped to Earth in another meteor impact. The field of exobiology, or astrobiology, is centered around the study of life elsewhere in the solar system. Two crucial questions are: (1) Do (or did) microbes exist on other planets in our solar system? and (2) Could they survive such a trip?

To address the first question, the United States sent three experiments aboard two *Viking* spacecraft that landed on Mars in 1976. Designed to determine whether microbes were living in the Martian soil, the experiments attempted to detect gases released as byproducts of metabolism. None yielded positive results. They did not, however, rule out the possibility that life existed on Mars at other places or times.

David McKay and colleagues (1996) reported evidence suggesting that life did indeed exist on Mars some 4 billion years ago. McKay's team studied a rock from Mars that had fallen as a meteorite onto Antarctica. On freshly chipped pieces of the rock, the team found globules of carbonate ($—CO_3$) in close association with magnetite (Fe_3O_4), iron sulfide (FeS_2), and organic molecules called polycyclic aromatic hydrocarbons. All of these chemicals can be produced by either biological or nonbiological processes. Carbonate crystals, however, form in the presence of water, a requisite for life as we know it.

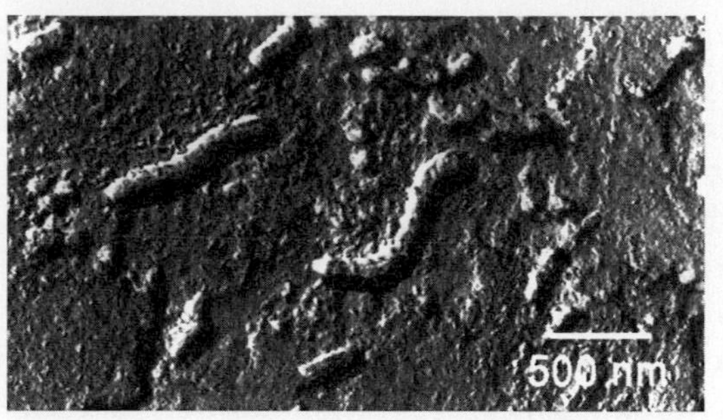

Figure 17.10 Fossils of Martian bacteria? This scanning electron microscope image of a Martian meteorite shows objects that resemble tiny bacteria.

In addition, McKay and colleagues found objects on the Martian rock that resemble tiny bacteria (Figure 17.10). McKay et al. concluded that the most plausible explanation for their findings is that the objects are fossils of biological processes that produced the minerals and chemicals. Other scientists are not convinced (see Anders et al. 1996; Kerr 1997a; Weiss et al. 2004). But insufficient evidence is not evidence of absence. As Bill Schopf noted, "This attempt failed to find life on Mars. That does not mean Mars contained no life—just that these scientists didn't find any." A new series of expeditions to Mars is under way and promises eventually to yield more decisive

Why were the Murchison meteorites significant? The biomolecules of life, as well as their likely precursors, all require the elements carbon, hydrogen, oxygen, nitrogen, sulfur, and phosphorus in large amounts, plus trace quantities of other elements such as magnesium, calcium, and potassium. Moreover, these elements must be in a chemical form that allows them to be used in the construction of biological building blocks like amino acids, sugars, and carbohydrates. If these building blocks could have been synthesized on the primitive Earth, then presumably they would have been available for condensation into larger biomolecules. But if they could not have been made on Earth, we would have to look to extraterrestrial sources, such as meteors, to account for their presence.

information about past or present life on the red planet.

There are other candidate locations for extraterrestrial life in the solar system. One is Jupiter's moon, Europa. Recent photographs taken by the spacecraft *Galileo* suggest that Europa has abundant liquid water (see photo on page 639) and active volcanoes. Chyba (2000) argues that energy capable of sustaining life on this moon could be generated by charged particles provided in Jupiter's magnetosphere. Together, these data offer the possibility that life could have evolved in a liquid ocean hidden under Europa's icy surface (Belton et al. 1996; Kerr 1996, 1997b).

What about a microbe's chances of surviving a trip to Earth? For microbes on meteorites knocked loose from Mars, the trip to Earth would typically take several million years, but a lucky few would reach Earth in less than a year (Gladman et al. 1996; Gladman and Burns 1996). Spacefaring microbes would be exposed to cold, vacuum, and both UV and ionizing radiation. Two teams, Peter Weber and J. Mayo Greenberg (1985) and Klaus Dose and Anke Klein (1996), have measured the survival rates of spores of the bacterium *Bacillus subtilis* exposed to various combinations of these conditions; Jeff Secker and colleagues (1994) have made theoretical calculations. The consensus is that bare spores could not survive an interplanetary trip. With some sort of shield against radiation, however, spores would have some chance of making it. A shield could be provided by ice, rock, or carbon.

To our knowledge, only once has the long-term survival of microbes in space been directly tested. In November of 1969, astronauts from *Apollo 12* recovered a camera from the unmanned lunar lander *Surveyor 3*, which had touched down on the Moon two and a half years earlier. Back on Earth, NASA scientists opened the camera in a sterile environment and found that the foam insulation inside harbored viable bacteria (*Streptococcus mitis*). The microbes had apparently stowed away in the camera before it left Earth, and, shielded from radiation, survived their stay in the vacuum and cold of space (Mitchell and Ellis 1972).

To find out whether spacefaring microbes could survive a fall to Earth, Robert McLean and colleagues (2006) opened a sealed microbiology experiment that was recovered intact from the crash of the space shuttle Columbia. Inside, they found viable bacteria in the genus *Microbispora*. Although these microbes were not supposed to be there, the researchers think they got into the experiment before it was sent on the ill-fated flight. If so, they survived Columbia's reentry and breakup.

A second possibility under panspermia is that life originated in another solar system and traveled to Earth through interstellar space. Spores embarking on such a voyage would need a force to accelerate them to sufficient velocity to escape the gravitational field of their home star. This force can be supplied by radiation pressure (Arrhenius 1908; Secker et al. 1994). Sailing on radiation pressure severely limits the mass of the shielding a spore can carry. Secker and colleagues suggest that spores encased in a film of carbonaceous material could both achieve escape velocity and survive the radiation they would encounter on a trip between stars.

Finally, Crick and Orgel (1973) suggest a third possibility, which they call directed panspermia: Earth's founding microbes were sent here intentionally, aboard a spacecraft, by intelligent extraterrestrials bent on seeding the galaxy with life. Crick and Orgel argue that, within the foreseeable future, it will probably be possible for us to launch such a mission. Therefore, it is at least conceivable that some other civilization actually did so 4 billion years ago.

The problem with terrestrial sources is that, 4 billion years ago, Earth's environment might or might not have been permissive for the synthesis of life's building blocks. In addition to temperature and pressure, a key feature of the environment is whether it was primarily oxidizing, with high abundances of molecular oxygen (O_2) and carbon dioxide (CO_2), or primarily reducing, with high concentrations of hydrogen (H_2), methane (CH_4), and ammonia (NH_3). Or it could have been intermediate in oxidizing activity. Which state it was in would have determined which chemical reactions were possible.

The composition of the early atmosphere remains a subject of debate (Lazcano and Miller 1996; Chyba 2005), and atmospheric chemists are looking for mechanisms by which organic molecules could have been synthesized, even in relatively impermissive mixtures of gases (Kasting 1993). Some feel that geochemical evidence points to an atmosphere that would not be favorable for the generation of biologically important molecules, or at least would not produce them in the concentrations needed for the origins of life. Thus, many have explored an alternative hypothesis that certain critical biochemicals were made elsewhere in the solar system and delivered to Earth in vehicles such as the Murchison meteorite.

The simple organic molecules from which life was built may have formed in space and then fallen to Earth. Researchers have tested this idea by looking for amino acids and other organic molecules inside meteorites.

The young Earth experienced heavy bombardment by meteors and comets. Figure 17.11 shows the history of very large impacts on both Earth and the Moon. Like the Murchison meteorite, several carbonaceous chondrite meteorites, believed to be fragments of asteroids, have proven to contain an abundance of organic molecules (see Chyba et al. 1990; Lazcano and Miller 1996). Many comets also contain a variety of organic molecules (Chyba et al. 1990; Cruikshank 1997). It is thus possible, at least in principle, that the stuff of life was delivered from space.

There is at least one crucial difficulty with the hypothesis that life's building blocks came from space: When meteors and comets crash to Earth, friction with the atmosphere and collision with the ground generate tremendous heat (Anders 1989). This heat may destroy most or all of the organic molecules the meteors and comets carry (Chang 1999). Edward Anders (1989) notes that very small incoming particles are slowed gently enough by the atmosphere to avoid incinerating all of their organics; he suggests that cometary dust may have been the primary source of the young Earth's organic molecules. Christopher Chyba (1990) and colleagues look instead to the possibility that the early atmosphere was dense with carbon dioxide. A dense CO_2 atmosphere may have provided a soft enough landing, even for large meteors and comets, for some of their organics to survive. The Murchison meteorites certainly provide direct evidence that at least some organic molecules can survive a descent to Earth.

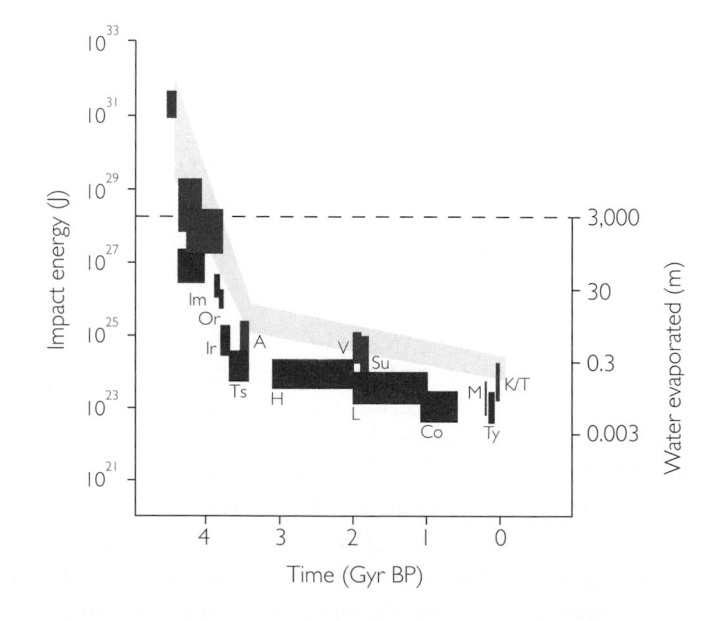

Figure 17.11 The history of large impacts on Earth and the Moon The horizontal axis represents time (billions of years before present). The left vertical axis represents the energy of impact (joules). The right vertical axis shows the depth (m) to which the oceans would be vaporized by an impact with a given energy. The dashed line represents vaporization of the entire global ocean. Each box encloses the range of times during which a particular impact is estimated to have occurred, and the range of energies the impact is estimated to have fallen within. The red boxes are for the Moon; the blue boxes are for Earth. Boxes with labels represent impacts documented by craters or other geological evidence. The unlabeled box at the upper left represents a large impact thought to be responsible for the formation of the Moon. The other unlabeled boxes are for hypothetical impacts. The gray band shows the largest impacts likely to have hit Earth at any given time. Lunar craters—Im = Imbrium, Or = Orientale, Ir = Iridium, Ts = Tsiolkovski, H = Hausen, L = Langrenus, Co = Copernicus, Ty = Tycho. Terrestrial craters—A = Archaean spherule beds, V = Vredevort, Su = Sudbury, M = Manicougan, KT = Cretaceous–Tertiary impact (crater located off the Yucatan peninsula). From Sleep et al. (1989).

The Oparin–Haldane Model

Originally, there was great hope that Earth itself could provide the "right stuff" for prebiotic synthesis. In 1953, Stanley Miller, then a graduate student in Harold Urey's laboratory at the University of Chicago, reported a simple and elegant experiment. He built an apparatus that boiled water and circulated the hot vapor through an atmosphere of methane, ammonia, and hydrogen, past an electric spark, and finally through a cooling jacket that condensed the vapor and directed it back into the boiling flask. Miller let the apparatus run for a week; the water inside turned deep red and cloudy. Using paper chromatography, Miller identified the cause of the red color as a mixture of organic molecules, most notably the amino acids glycine, α-alanine, and β-alanine. Since 1953, chemists working on the prebiotic synthesis of organic molecules have, in similar experiments, documented the formation of a tremendous diversity of organic molecules, including amino acids, nucleotides, and sugars (see Fox and Dose 1972; Miller 1992).

Miller used methane, ammonia, and hydrogen as his atmosphere; in the 1950s, this highly reducing mixture was thought to model the atmosphere of the young Earth. The implication of Miller's result was that if lightning or UV radiation could have played the role that the spark did in his experiment, then the young Earth's oceans would have quickly become rich in biological building blocks.

The simple organic molecules from which life was built may also have formed on Earth. Researchers have tested this idea by trying to re-create the chemical conditions on the early Earth and replicate the chemical reactions that might have created amino acids and nucleotides.

Many atmospheric chemists now believe that Earth's early atmosphere was not so reducing, being dominated by carbon dioxide rather than methane, and molecular nitrogen (N_2) rather than by ammonia (Kasting 1993). This conclusion is based on the mixture of gases released by contemporary volcanoes, and on improved knowledge of the chemical reactions that occur in the upper atmosphere. Reaching a consensus on the prebiotic environment is important, because an atmosphere dominated by carbon dioxide and molecular nitrogen appears to be much less conducive to the formation of certain organic molecules. However, the formation of aldehydes, especially formaldehyde (H_2CO), from carbon dioxide has been deemed plausible by several researchers—particularly in light of work by Feng Tian and colleagues (2005) suggesting that the early atmosphere might have contained as much as 30% molecular hydrogen (H_2). Aldehydes are necessary in the construction of the ribose sugars needed to make the nucleotides of an RNA World (Mojzsis et al. 1999).

The view that Earth possessed all the necessary ingredients for the origins of life is perhaps the most thoroughly investigated hypothesis and holds great appeal for many scientists even today. This opinion dates back to the efforts of A. Oparin and J. B.S. Haldane, working in the first half of the 20th century, to reconstruct how life may have begun. These scientists and others (including Charles Darwin) created a lasting image of life arising in an aqueous environment brimming in high concentrations of biological building blocks. This was Darwin's "warm little pond" (Darwin 1887), the famous "prebiotic soup." There are many severe criticisms of this vision, not the least of which is whether liquid water existed on Earth at the time of life's origin. Nonetheless, this view remains as sort of a null model against which deviations can be tested, much like the Hardy–Weinberg equilibrium principle in population genetics. This scenario is often referred to as the Oparin–Haldane model.

We can break the Oparin–Haldane model into a series of steps that occurred sequentially in the waters or moist soil of the young Earth (Figure 17.12). First, nonbiological processes synthesized organic molecules, such as amino acids and

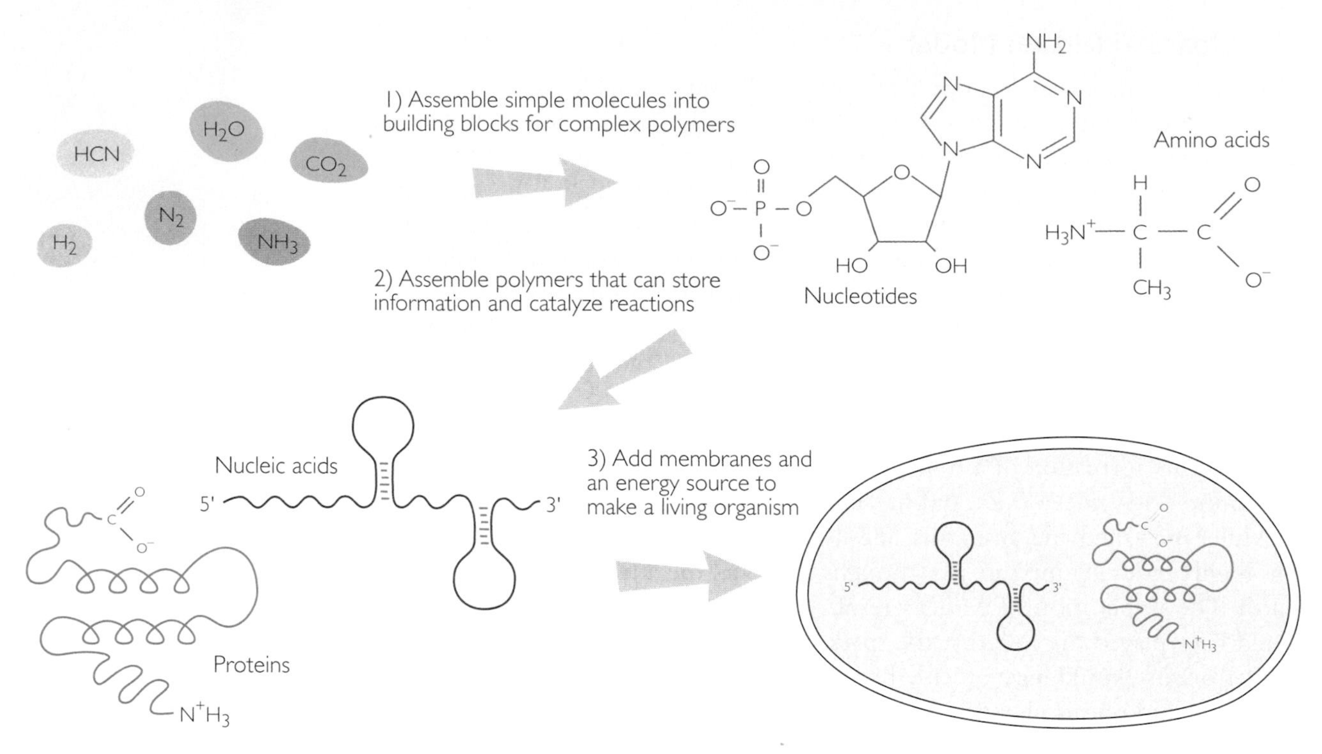

Figure 17.12 Synopsis of the Oparin–Haldane model The first stage would have been the formation of biological building blocks (nucleotides and amino acids) from existing inorganic material on the early Earth. The second stage would have been the polymerization of these building blocks to form biological macromolecules (nucleic acids and proteins). Finally, these macromolecules would have directed the formation of other biological structures, such as cell membranes.

nucleotides, that would later serve as the building blocks of life. Then, the organic building blocks in the prebiotic soup were assembled into biological polymers, such as proteins and nucleic acids. Last, some combination of biological polymers was assembled into a self-replicating organism that fed off of the existing organic molecules, much as we discussed earlier for the RNA World.

From Simple Inorganics to the Building Blocks of Life

Previously, we saw the ease with which amino acids can be made from simple inorganics like methane, ammonia, and hydrogen. What about nucleotides? A second monumental achievement in origins-of-life research was the demonstration by Juan Oró (1961) that the nitrogenous base adenine (a purine) could be readily made from a thermodynamically favorable reaction involving only ammonia and hydrogen cyanide (HCN). When these two compounds are heated in water, adenine is produced in yields as high as 0.5%, which is significant if the early atmosphere was reducing and contained large amounts of ammonia and hydrogen cyanide. Miller refers to this reaction as the "rock of faith" for terrestrial prebiotic synthesis. Other chemists have had similar results for other purine bases. Pyrimidines (C, U, and T) are slightly more difficult to construct abiotically, but chemists have had some successes (Voet and Schwartz 1982). Finally, the ribose sugars that form nucleotides can, at least under the right environmental conditions, be derived from a cascade of condensation reactions that begin only with formaldehyde.

Unfortunately, the description of several mostly independent, plausible chemical pathways that could have produced amino acids, nucleotides, and sugars leaves us a long way from fully formed building blocks that are on the verge of becoming a self-replicating system. A major obstacle that has plagued biochemists for decades is the origin of chirality, or handedness. As noted above, living systems today use only one stereoisomer, or mirror-image form, of the amino acids in their proteins, and the same is true of nucleotides. In many of the chemical syntheses described by adherents to the Oparin–Haldane model, both mirror images of the building blocks are made in roughly equal quantities, and it is difficult to devise mechanisms that produce only one or the other. Exacerbating this problem is the fact that one mirror-image form would inhibit the polymerization of the other during any type of polymer self-replication (see Joyce et al. 1987).

Another problem is exemplified by the case of sugar formation. Not only does the sugar that we see today in nucleic acids (ribose) constitute a very small percentage of all the sugars produced by formaldehyde condensation, but there also exist multiple equally probable ways that the nitrogenous bases could be attached to the sugar. Each of these combinations produces a subtly, but importantly, different nucleotide isomer than that used by contemporary RNA. To make matters worse, each building block needs to be activated, or chemically charged, before it can be incorporated into a polymer. Activation requires a preexisting source of chemical energy. Without cell membranes to concentrate this energy, it is challenging to understand how building blocks became activated in the RNA World (Orgel 1986). These problems have been described by Joyce and Orgel (1999) as turning the "molecular biologist's dream" (. . . once upon a time there was a prebiotic pool full of β-D-nucleotides . . .) into the "prebiotic chemist's nightmare."

Where does this leave the RNA World? Staggering, perhaps, but still standing. Many researchers now think that the RNA World did not arise *de novo* from a warm little pond. Instead, RNA is likely to have been a later stage in an evolutionary lineage that derived from a simpler genetic system. Several non-RNA self-replicating systems have been proposed (see Orgel 2000). Among them are: polymers made up not of ribonucleotides as we know them today, but of ribonucleotide analogs that have only one stereoisomer (Joyce et al. 1987); polymers made up of a hybrid between peptides and nucleic acids (Egholm et al. 1992); polymers made up of nucleotides composed of pyranose or tetrose sugar (Eschenmoser 1999; Schöning et al. 2000) or propylene glycol (Zhang et al. 2005) instead of ribose sugar; and even polymers made up of inorganic substances such as clay (Cairns-Smith et al. 1992). Christian deDuve (1991) has outlined a "Thioester World" in which information transfer is intricately linked to the metabolic turnover of thioester linkages in a complex chemical milieu. All of these scenarios are based on the presumption that another self-replicating system could arise abiotically with a higher probability than RNA. Some of these are envisioned such that RNA could develop from them; presumably, the preexistence of a self-replicator could allow for a bias in the way RNA is synthesized, such that the problem of chirality could be overcome, for example. Others are envisioned as alternatives to an RNA World, many formulated in a way that would favor the construction and use of catalysts other than RNA.

The Assembly of Biological Polymers

The second step in the Oparin–Haldane theory, the formation of biological polymers from the building blocks in the prebiotic soup, has presented other theoretical and practical challenges. The prebiotic soup would contain organic building blocks dissolved in water, and although biological polymers can readily be synthesized in water, they also break down by hydrolysis even as they are

The building blocks of life may have been assembled into polymers while stuck to the surface of clay crystals. Adhering to clay helps a growing polymer avoid being broken apart by hydrolysis.

being built. This problem raises doubts that polymers sufficiently long to serve as the basis of a self-replicating primordial organism would ever have formed in a simple organic soup (Ferris et al. 1996).

James Ferris and colleagues (1996), extending a tradition that dates from the 1940s and 1950s (see Ferris 1993), demonstrated a plausible mechanism to overcome the hydrolysis problem. Ferris et al. prepared a simple prebiotic soup in the lab and added the common clay mineral montmorillonite. Montmorillonite is a naturally occurring aluminum-silicate clay to which organic molecules readily adhere. When activated nucleotides (that is, nucleoside triphosphates) stick to montmorillonite, the clay acts as a catalyst and will join them together in a polynucleotide chain. While bound to the clay, the polynucleotides form more rapidly than they are hydrolyzed, and the researchers succeeded in encouraging the formation of polynucleotide chains containing 8–10 nucleotides in a row.

Ferris and colleagues then demonstrated that it was possible to prepare much longer polynucleotides by the daily addition of activated nucleotides to a premade oligonucleotide primer. They started with a polyadenylate primer 10 nucleotides long and let it bind to the montmorillonite. The scientists then added to the polyadenylate/clay solution a bath of activated adenosine nucleotides. The activated nucleotides reacted with the polyadenylate primers, adding themselves to the nucleotide chains. Ferris and colleagues then used a centrifuge to spin down the clay (and its attached nucleotide chains), poured off the spent solution, and added a fresh bath of activated nucleotides. By repeating this process, adding a fresh bath of activated nucleotides once each day, Ferris et al. synthesized polyadenylates over 40 nucleotides long (Figure 17.13). Ferris and his colleagues have since refined their recipe to the point that in one step, run over a single day, it can produce polynucleotides up to 50 nucleotides long (Huang and Ferris 2006).

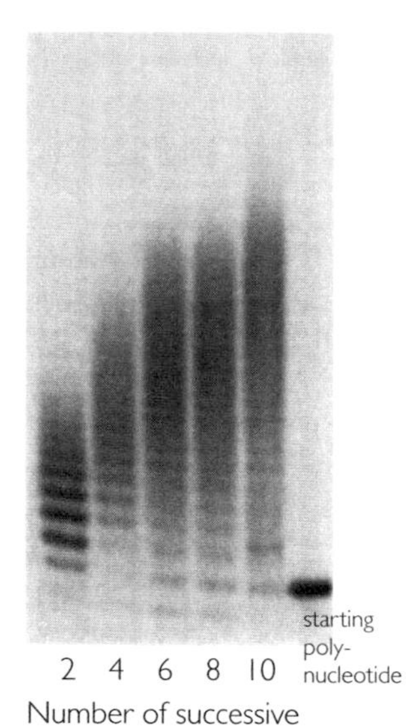

Figure 17.13 Synthesis of long nucleotide chains on clay This electrophoresis gel has separated mixtures of polyadenylates by size. The right lane contains a single band that corresponds to nucleotide chains 10 bases long; this was the starting point for Ferris et al.'s (1996) experiment. The left lane contains the mixture of polynucleotides produced when 10-nucleotide polyadenylates were allowed to bind to montmorillonite, then given two successive baths with activated adenosine nucleotides. Each successive band represents a one-nucleotide difference in length. The leftmost lane thus contains polyadenylates ranging from 11 to 20 nucleotides in length. The second lane from the left shows the results of four successive baths with activated nucleotides, and so on. Reactions run without montmorillonite failed to produce elongated nucleotide chains.

Ferris and Orgel have used repeated-bathing procedures to grow polypeptides up to 55 amino acids long on the minerals illite and hydroxylapatite (Ferris et al. 1996; Hill et al. 1998). The teams assert that their method models a mechanism by which biological polymers could have grown on the early Earth. Minerals in sediments that were repeatedly splashed with the pre-biotic soup, or continuously bathed by it, could have nursed the formation of polymers that were long enough to form a self-replicating primordial form. This view has its critics (see Shapiro 2006), but the clay-catalysis research has given the second step of the Oparin–Haldane model at least some experimental support. We will briefly discuss the third step in Section 17.3.

Protecting Life from the Environment

At this point, one can at least grapple with the possibility that we can discover a logical chain of events that led from simple inorganics, such as carbon dioxide, ammonia, and hydrogen cyanide, to fully formed nucleic acids. These events could have all taken place on Earth, or some of them could have taken place on extraterrestrial bodies. Some researchers have even suggested that certain chemical reactions might have occurred in the atmosphere itself, perhaps suspended in water droplets that rose and fell with the temperature. Regardless of the chemistry of the early atmosphere, early Earth probably offered many local opportunities for organic synthesis—hydrothermal environments, ocean water rich in ferrous iron, or the caldera of volcanoes, just to name a few. However the final challenge for any model of the origins of

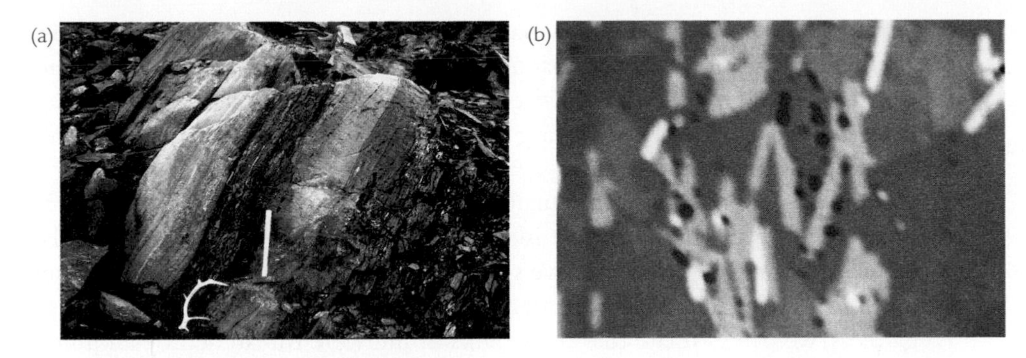

Figure 17.14 **3.7-billion-year-old sedimentary rocks contain chemical evidence suggesting life** (a) This sedimentary rock, from Isua, Greenland, is 3.7 billion years old. Note the geologist's hammer for scale. (b) Examination under a microscope reveals that the rock shown in (a) contains microscopic graphite particles, which appear as black dots in this photo. The graphite particles contain ratios of carbon isotopes that suggest they are derived from living cells. From Rosing (1999).

life is not whether the early Earth would have provided for the needs of life, but whether it would have been hospitable enough to allow life to evolve.

Sedimentary rocks from Isua, Greenland, contain evidence suggesting that life was already established on Earth by 3.7 billion years ago (Figure 17.14a). The rocks, once part of the seafloor, have been exposed to high temperatures and pressures, which have compacted the rocks and crystallized many of the minerals they contain. This transformation would have destroyed any microfossils the rocks might have originally harbored. The rocks do contain microscopic graphite globules, however (Figure 17.14b). Graphite is a mineral form of carbon.

Minick Rosing (1999) hypothesized that the graphite globules in the Greenland rocks are chemical fossils of ancient organisms. Rosing tested this hypothesis by measuring the isotopic composition of the graphite. Carbon has two stable isotopes: ^{12}C and ^{13}C. When organisms capture and fix environmental carbon, during photosynthesis for example, they harvest ^{12}C at a slightly higher rate than ^{13}C. As a result, carbonaceous material produced by biological processes has a slightly higher ratio of ^{12}C to ^{13}C than does carbonaceous material produced by nonbiological processes. When Rosing analyzed the graphite globules, he found carbon isotope ratios characteristic of life. The results of subsequent analyses are consistent with this interpretation (Rosing and Frei 2004; Fedo et al. 2006).

Other researchers have examined rocks from Greenland that may be as much as 3.85 billion years old. Some have concluded that these rocks, too, contain chemical fossils of ancient life (Schidlowski 1988; Mojzsis et al. 1996). These conclusions have proven controversial (see: Fedo and Whitehouse 2002a, b; Friend et al. 2002; Lepland et al. 2002; Mojzsis and Harrison 2002; van Zuilen et al. 2002; Moorbath 2005).

Are we likely to find evidence of life much earlier than the 3.7 (or 3.85) billion years ago demonstrated by the Greenland rocks? Probably not, for at least two reasons. First, erosion, plate tectonics, and volcanic eruptions have obliterated all rocks from any crust that might have existed earlier and thus have eliminated all direct evidence. Second, even if crust and oceans did exist earlier, continued bombardment of the planet by large meteors could have prevented life from being established much earlier than 3.7 to 3.85 billion years ago (see Figure 17.11). Large meteor impacts generate heat, create sun-blocking dust, and produce a blanket of debris. As time passed, and the largest planetesimals

Although we still lack a complete scenario for how the first living things arose from nonliving matter, it appears that they did so quickly—almost as soon as the early Earth was habitable.

got swept up by Earth and other planets, the sizes of the largest impacts decreased. Norman Sleep and colleagues (1989) estimated that the last impact with sufficient energy to vaporize the entire global ocean, and thereby frustrate the emergence of any self-replicating system, probably happened between 4.44 and 3.8 billion years ago.

Life's development may have actually been initiated many times over, if sterilizing events allowed enough time in between for self-replication to re-evolve each time. Alternatively, life may have survived some of these impacts, sequestered in protective niches in the environment, such as deep-sea hydrothermal settings. Regardless, we can estimate that the origins of life were threatened by an inhospitable environment until about 4 billion years ago. Whether the Oparin–Haldane model leading to an RNA World is correct, or whether some other scenario turns out to be more plausible, we should note one last point before we consider cellular life: The origins of life occurred in a tumultuous abiotic environment. Paradoxically, Earth today is even more inhospitable to the origins of life. Life has become so successful at exploiting extreme niches that there are no places left for inorganic molecules to reinvent self-replication before the first stages of these attempts would be gobbled up by extant creatures.

17.3 What Was the Last Common Ancestor of All Extant Organisms and What Is the Shape of the Tree of Life?

Once self-replicating systems evolved on Earth, at least one of them adapted to the use of DNA to store heritable information and to the use of proteins to express that information. This system eventually gave rise to all lineages of life on the planet today. We draw this conclusion because all life forms (except some viruses) use DNA and proteins. In fact, all modern organisms use them in the same way; the same 20 amino acids and the same basic structure of the genetic code have been found in all creatures studied to date. Thus, we apply the principle of parsimony to infer that all organisms share a common ancestor.

What Was the Most Recent Common Ancestor of All Living Things?

Because another shared feature of all extant life is the existence of cells, we also infer that the common ancestor was a cellular form. Technically speaking, we need to say that all life has descended from a population of interbreeding cells, because if portions of the primitive genome could be readily swapped, then life today cannot trace its ancestry to a single organism. The picture that emerges of the origins and early evolution of life on Earth can be diagrammed as in Figure 17.15. The first cellular life whose descendants ultimately survived, the cenancestor (or cenancestors), appeared at least 2 billion years ago and probably much earlier. The advantages of cellular membranes, as well as internal organellar membranes would have been enormous. Cells allow for compartmentalization. Certain chemicals can be concentrated inside the cell, while others can be pumped outside the cell. This allowed life to accumulate its necessary constituents in much higher concentrations than they are found free in solution—activated nucleotides, for example. Cells also allowed genotypes and phenotypes to be linked, even after the latter had become the domain of proteins and not of the genetic material itself. It does a genotype little (evolutionary) good if the phenotype it encodes is free to diffuse to other genotypes.

It is a long way from a self-replicating RNA molecule to the cenancestors, and many questions remain. For example, how did the earliest organisms acquire cellular form? One potential answer has come from the work of Sidney Fox and colleagues, who found that mixtures of polyamino acids in water or salt solution spontaneously organize themselves into microspheres with properties reminiscent of living cells (see Fox and Dose 1972; Fox 1988; Fox 1991). Another potential answer holds that the precursors to cells were tiny compartments formed by inorganic mineral deposits at the mouths of deep-sea hydrothermal vents (Martin and Russell 2003).

Little is known about how the first self-replicating molecules evolved into cellular life forms, although researchers have shown that structures reminiscent of cell membranes form spontaneously.

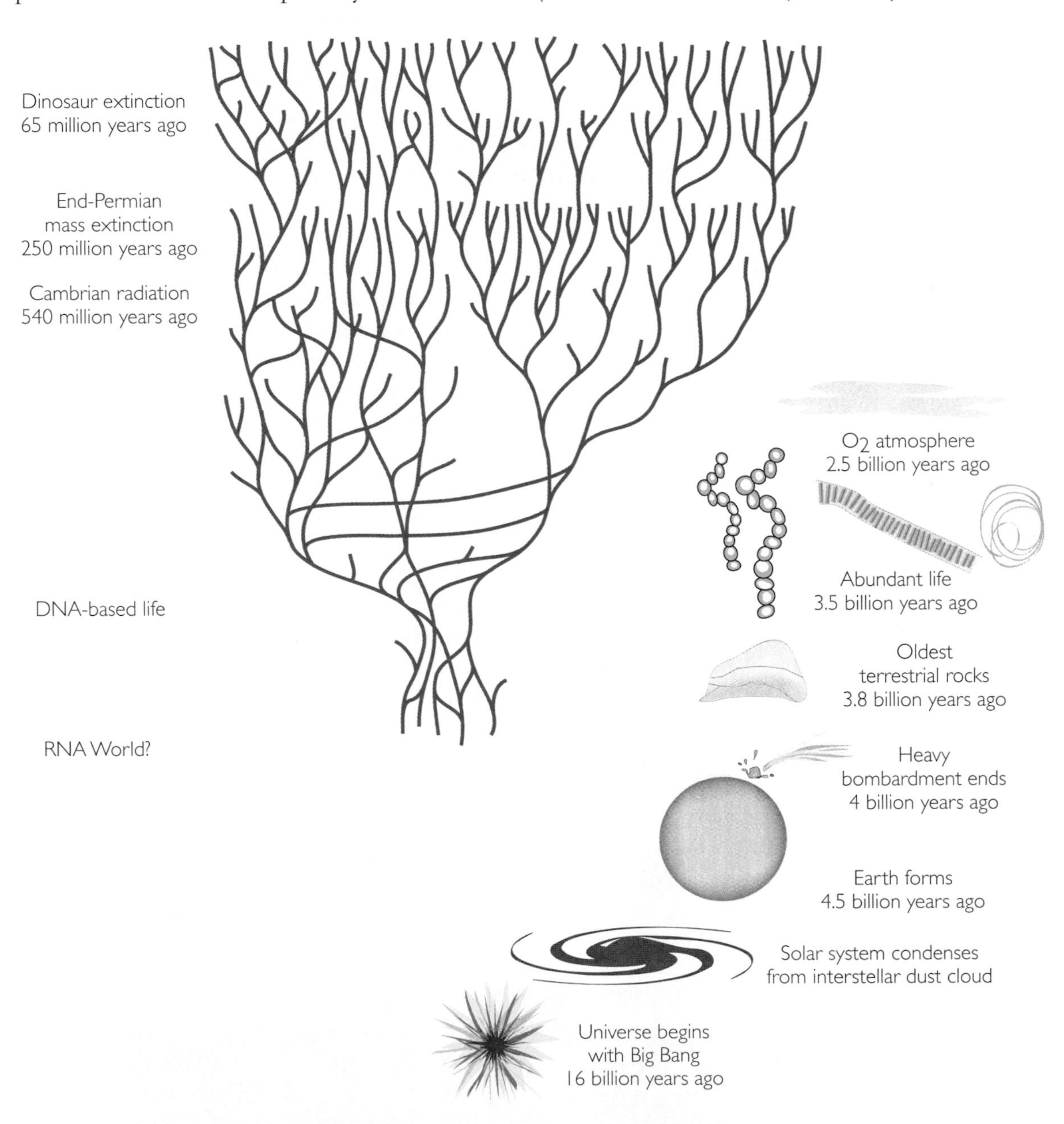

Figure 17.15 Overview of the evolution of life In the tree of life shown here (blue), the fusion of branches represents the acquisition of symbionts and other forms of lateral gene transfer. These phenomena will be discussed later in the chapter. After Atkins and Gesteland (© 1998) in Gesteland et al. (1999); W. F. Doolittle (2000).

About the ancestral cellular lineage, like the first self-replicating system, we can ask what its general characteristics were, when it lived, and by what route its descendants evolved into today's orchids, ants, mushrooms, amoebae, and bacteria. Again, these events occurred early in Earth's history and much direct information has been lost. But if we know what questions to ask, the available data in the geological record can begin to remove the mystery of the first cellular life (Schopf 1994b).

One way to learn about the characteristics of the earliest cells is to look for their fossils.

The first place we might look in trying to identify the ancestral cells is the fossil record. In principle, a complete fossil record would allow us to trace lines of descent from present-day organisms all the way back to the cenancestors. However, it does not appear that the fossil record so far assembled can take us that deep into the past.

Several researchers have reported fossil cells preserved in rocks that are 3.2 to 3.5 billion years old (see Schopf 2006 for a review). For example, the fossils in Figure 17.16a–d, discovered by Andrew Knoll and Elso Barghoorn (1977; see also Westall et al. 2001), are from a geological formation called the Swartkoppie chert in South Africa. Originally thought to be 3.4 billion years old, they are now estimated to be slightly younger. Among other reasons, Knoll and Barghoorn identified them as cells based on their carbon content, size distribution, location in sedementary rocks, and resemblance to dividing bacteria (Figure 17.16e–h). William Schopf (1993) has reported fossils of what he believes are cyanobacteria from the slightly older Apex chert of Western Australia. Shopf's evidence has been the subject of recent controversy (see Brasier et al. 2002; Dalton 2002; Kázmierczak and Kremer 2002; Kempe et al. 2002; Schopf et al. 2002a, 2002b; Pasteris and Wopenka 2002). Indeed, skeptics have questioned the biological origin of virtually all purported fossils more than 3 billion years old (Brasier et al. 2006).

Unfortunately, even if some or all of these purported fossils prove genuine, they will not answer our present question. The fossil record for times earlier than 2.5 billion years ago is too spotty to allow paleontologists to trace lines of evolutionary descent from present-day organisms back to the fossils in the Swartkoppie or Apex cherts (Altermann and Schopf 1995). As a result, we have no direct way of knowing whether the organisms recorded in these rocks represent extinct or living branches of the tree of life, or whether they lived before or after the last common ancestor. If we want to discover the characteristics of ancestral cell lineages, we must use methods other than examination of the fossil record.

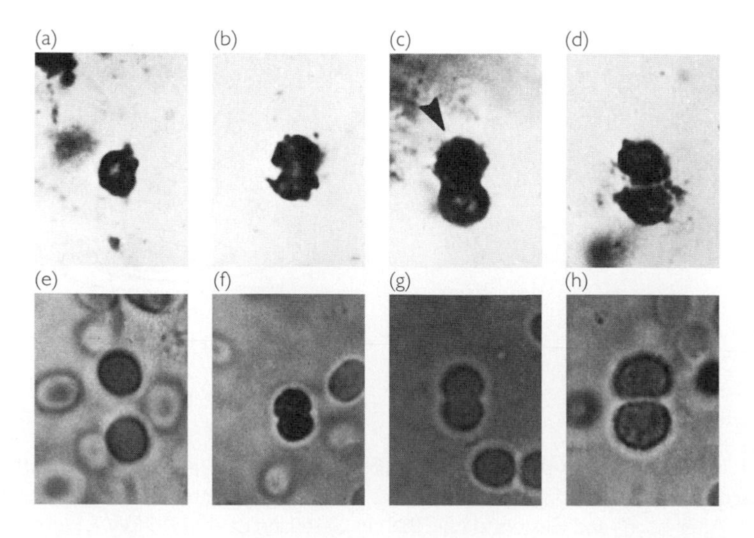

Figure 17.16 3.26-billion-year-old fossils of dividing cells The top row (a–d) shows microscopic fossils in 3.26 billion-year-old rocks from South Africa. The bottom row (e–h) shows living bacterial cells in various stages of division. Note the striking resemblance between the fossils and the living cells. From Knoll and Barghoorn (1977).

The Phylogeny of All Living Things

Another way to study the ancestral lineage is to reconstruct the phylogeny of all living things. A universal phylogeny should allow us to infer additional characteristics of the earliest life forms (see Chapter 4), beyond just their cellular nature. The first attempts to reconstruct the phylogeny of everything were based on the morphologies of organisms (see reviews in Woese 1991; Doolittle and Brown 1994). The morphological approach was productive for biologists interested in the branches of the tree of life that contain eukaryotes. Morphology was, historically, the basis of the phylogeny of many taxonomic groups. The morphological approach led only to frustration, however, for biologists interested in the branches of the universal phylogeny containing prokaryotes. Prokaryotes lack sufficient structural diversity to allow the reconstruction of morphology-based evolutionary trees.

When biologists developed methods for reading the sequences of amino acids in proteins, and the sequences of nucleotides in DNA and RNA, a new technique for estimating phylogenies quickly became established (Zuckerkandl and Pauling 1965). Some of the details of this technique are devilish (see Chapter 4), but the basic idea is straightforward. Imagine that we have a group of species, all carrying in their genomes a particular gene. We can read the sequence of nucleotides in this gene in each of the species, then compare the sequences. If species are closely related, their sequences ought to be fairly similar. If species occupy distant branches on the evolutionary tree, then their sequences ought to be less similar. As a result, we can use the relative similarity of the sequences of species to infer their evolutionary relationships. We place species with more similar sequences on neighboring branches of the evolutionary tree and species with less similar sequences on more distant branches.

Another way to learn about the earliest cells is to estimate the phylogeny of all living things, then infer the characteristics of the common ancestors.

The challenge in using sequence data to estimate the evolutionary tree for all living things is to find a gene that shows recognizable sequence similarities even between species that are as distantly related as *Escherichia coli* and *Homo sapiens* (Woese 1991). We need a gene that is present in all organisms and that encodes a product whose function is essential and thus subject to strong stabilizing selection. Without strong stabilizing selection, billions of years of genetic drift will have obliterated any recognizable similarities in the sequences of distantly related organisms. Additionally, the function of the gene must have remained the same in all organisms. This is because when a gene product's function shifts in some species but not in others, selection on the new function can cause a rapid divergence in nucleotide sequence that makes species look more distantly related than they actually are.

One gene that meets all the criteria for use in reconstructing the universal phylogeny is the gene that codes for the small-subunit ribosomal RNA (Woese and Fox 1977; Woese 1991). All organisms have ribosomes, and in all organisms the ribosomes have a similar composition, including both rRNA and protein. All ribosomes have a similar tertiary structure, including small and large subunits. In all organisms the function of the ribosomes is the same: They are the machines responsible for translation. Translation is so vital, and organisms are under such strong natural selection to maintain it, that the ribosomal RNAs of humans and their intestinal bacteria show recognizable similarities in nucleotide sequence, even though humans and bacteria last shared a common ancestor billions of years ago. The small-subunit rRNA was the molecule chosen by Carl R. Woese, the chief pioneer of the use of molecular sequences in estimating the universal phylogeny (Fox et al. 1977; Woese and Fox 1977; see also Doolittle and Brown 1994). Though it is not a perfect solution, the small-subunit rRNA remains an informative resource for whole-life phylogenies.

Before presenting the tree of life as revealed by small-subunit rRNA sequences, it is worth recalling what biologists thought it looked like when Woese embarked on his project (Figure 17.17). According to the five-kingdom model (Whittaker 1969), the first split in the tree separates what will become the prokaryotes—the bacteria—on the left from what will become the eukaryotes on the right. The eukaryotes comprise three kingdoms containing the large multicellular organisms we are familiar with in daily life, plus a fourth kingdom of microorganisms.

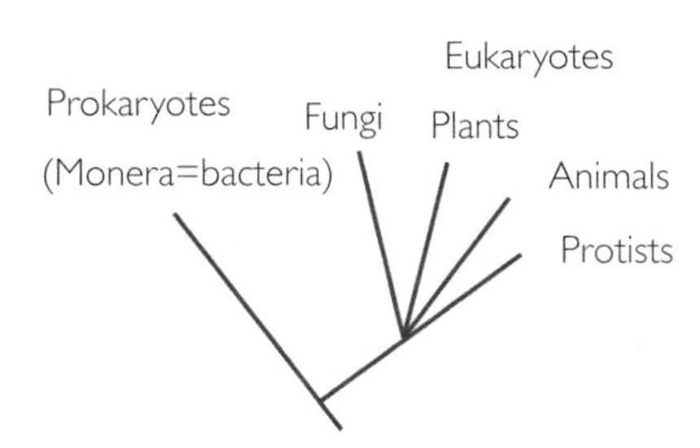

Figure 17.17 The tree of life, according to the five-kingdom scheme According to this scheme, the deepest node on the universal phylogeny is the split between the lineages that evolved into today's prokaryotes (bacteria) versus eukaryotes (fungi, plants, animals, and protists).

An estimate of the universal phylogeny based on sequences of the small-subunit rRNA appears in Figure 17.18. This whole-life rRNA phylogeny has prompted a dramatic revision of our traditional view of the organization of life, because it reveals that the five-kingdom system of classification bears only a limited resemblance to actual evolutionary relationships (Woese et al. 1990; for a contrary view, see Margulis 1996).

The first whole-life phylogenies based on sequence data were estimated on the basis of small-subunit rRNA genes. These rRNA phylogenies revealed that the traditional five-kingdom system of classification offers a misleading view of evolutionary relationships.

The prokaryotes, for example, which are all grouped in the kingdom Monera in the traditional classification, occupy two of the three main branches of the rRNA phylogeny. One of these two branches, the Bacteria, includes virtually all of the well-known prokaryotes. The Gram-positive bacteria, for instance, include *Mycobacterium tuberculosis,* the pathogen that causes tuberculosis. The purple bacteria include *E. coli.* (The purple bacteria are so named because some of them are purple and photosynthetic, although *E. coli* is neither.) The cyanobacteria, all of which are photosynthetic, include *Nostoc,* an organism often seen in introductory biology labs.

The other prokaryote branch, the Archaea, is not as well known. Many of the Archaea live in physiologically harsh environments, are difficult to grow in culture, and were discovered only recently (see Madigan and Marrs 1997). Most of the Crenarchaeota, for example, are hyperthermophiles, living in hot springs at temperatures as high as 110°C. Many of the Euryarchaeota are anaerobic methane producers. Another group in the Euryarchaeota, the Haloarchaea, are highly salt dependent and are thus referred to as extreme halophiles.

Because of their prokaryotic cell structure, the Archaea were originally considered bacteria. When Woese and colleagues discovered that these organisms were only distantly related to the rest of the bacteria, they renamed them the archaebacteria (Fox et al. 1977; Woese and Fox 1977). Eventually biologists realized that, as the phylogeny in Figure 17.18 shows, the archaebacteria are in fact more closely related to the eukaryotes than they are to the true bacteria (see Bult et al. 1996; Olsen and Woese 1996). In recognition of this, Woese and colleagues (1990) proposed the new classification used in Figure 17.18. Woese and colleagues dropped the *bacteria* from "archaebacteria," renaming this group the Archaea. Given that the Bacteria and the Archaea do not form a monophyletic group, some biologists feel the term "prokaryote" should be dropped as well

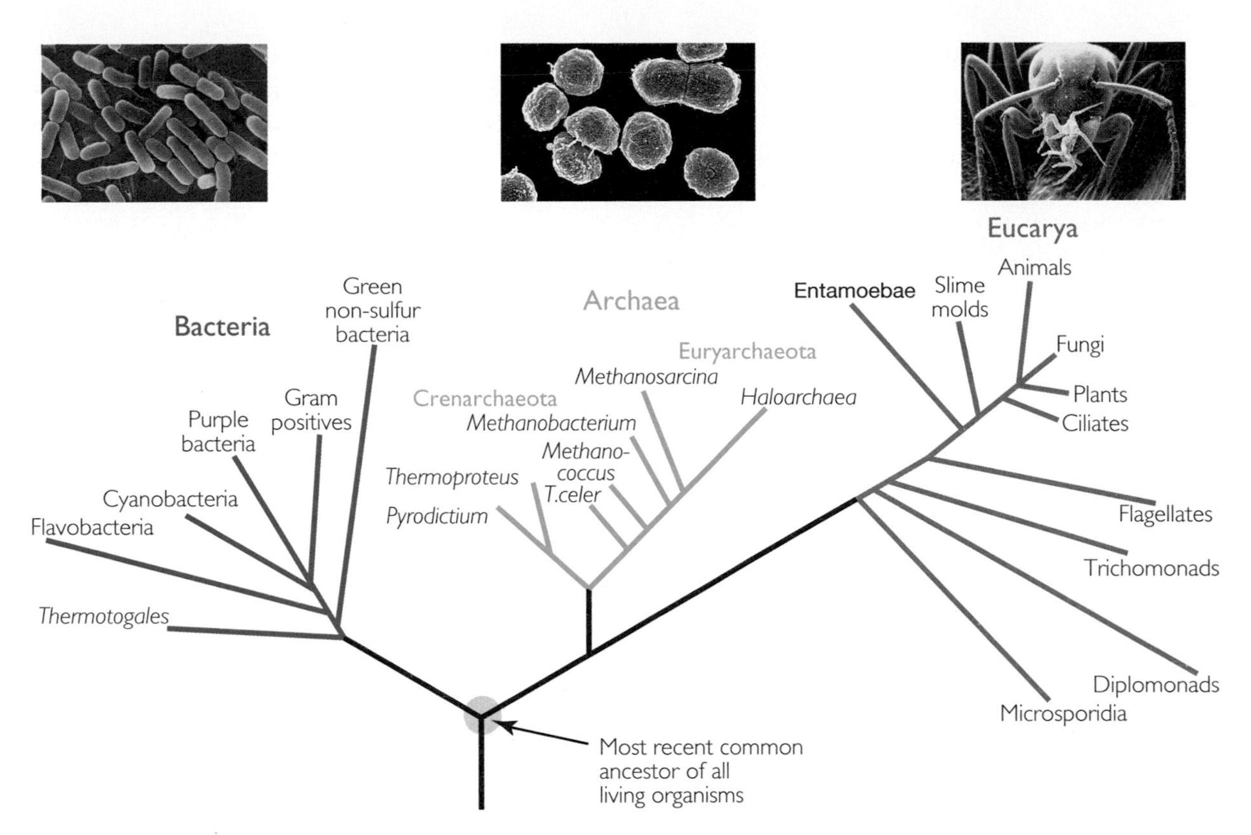

Figure 17.18 An estimate of the phylogeny of all living organisms This tree is based on the analysis of nucleotide sequences of small-subunit rRNAs. The photos show *E. coli* representing the Bacteria, *Methanococcus jannaschii* representing the Archaea, and an ant and an aphid representing the Eucarya. From Woese (1996).

(Pace 2006). The most inclusive taxonomic units in the new classification are three domains corresponding to the three main branches on the tree of life: the Bacteria, the Archaea, and the Eucarya. Woese and colleagues proposed that the two fundamental branches of the Archaea, the Crenarchaeota and the Euryarchaeota, be designated as kingdoms.

Woese et al. (1990) declined to offer a detailed proposal on how to divide the Eucarya into kingdoms. The Protista, a single kingdom in the traditional classification, are scattered across several fundamental limbs on the eukaryotic branch of the tree of life. The diplomonads, for example, which include the intestinal parasite *Giardia lamblia,* represent one of the deepest branches of the Eucarya. They are well separated from such other protists as the flagellates, which include *Euglena,* and the ciliates, which include *Paramecium*. If we want our kingdoms to be natural evolutionary groups, they should be monophyletic. That is, each kingdom should include all the descendants of a single common ancestor. Unless we want the kingdom Protista to include the animals, plants, and fungi, it will have to be disbanded and replaced by several new kingdoms.

The remaining three kingdoms in the traditional classification, the Animals, Plants, and Fungi, require only minor revision. To make the Fungi a natural group, for example, the cellular slime molds (such as *Dictyostelium,* a favorite of developmental biologists) will have to be removed.

The universal rRNA phylogeny demonstrates, however, that the Animals, Plants, and Fungi, the kingdoms that have absorbed most of the attention of evolutionary biologists (and represent most of the examples in this book), are mere twigs on the tip of one branch of the tree of life. The multicellular, macroscopic organisms in these three kingdoms are newcomers on the evolutionary scene, with a relatively recent last common ancestor. For genes shared among all organisms, such as the gene for the small-subunit rRNA, Animals, Plants, and Fungi appear to possess less than 10% of the nucleotide-level diversity observed on Earth (Olsen and Woese 1996).

An Examination of Early Cellular Life

Now that we have a universal phylogeny, what does it tell us about the earliest cellular life forms? The arrow in Figure 17.18 points to the last common ancestor of all extant organisms. According to this tree, the common ancestor's descendants diverged to become the Bacteria on one side and the Archaea–Eucarya on the other. Rooting the tree of life in this way was, and remains, a challenge because there is no outgroup to work with. The position of the root shown in Figure 17.18 is based on the work of several groups of researchers who used different analytical tricks (see Figure 17.19), but who all came up with approximately the same answer: The Archaea and Eucarya are more closely related to each other than either is to the Bacteria (Gogarten et al. 1989; Iwabe et al. 1989; Brown and Doolittle 1995; Baldauf et al. 1996). More recent data have yielded surprises, as we will discuss shortly. Estimating the location of the root remains an active area of research (see Zhaxybayeva et al. 2005).

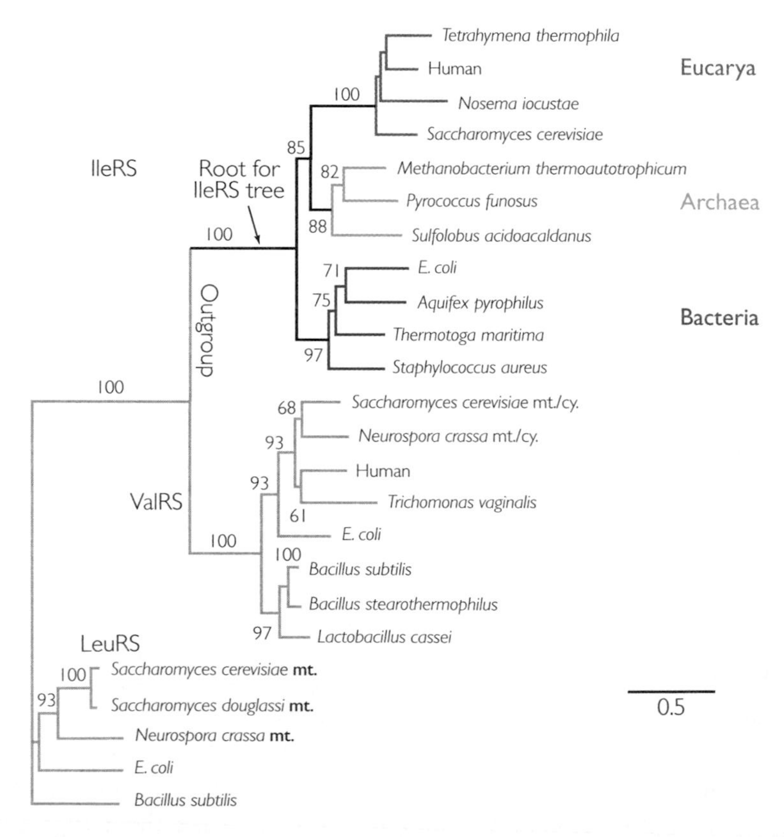

Figure 17.19 Rooting the universal phylogeny To root the tree of life, which has no organism that can serve as an outgroup, an analytical trick must be used. Here, the observation that gene families exist as a result of ancient duplications can provide a molecular outgroup. The aminoacyl-tRNA synthetase gene family serves as an example. The genes in this family arose in a series of ancient duplications. Keeping in mind that the phylogeny shown here is a gene tree, not an organism tree, look at the top portion, shown in black and colored lines. This portion is a phylogeny of the isoleucine aminoacyl-tRNA synthetase (IleRS) genes of organisms representing all three domains (the Bacteria, the Archaea, and the Eucarya). Genes for valine (ValRS) and leucine (LeuRS) aminoacyl-tRNA synthetase from a variety of bacteria and eukaryotes are the outgroups that root the IleRS tree. The outgroup phylogeny is shown in gray lines. From Brown and Doolittle (1995).

Assuming that we can accurately estimate a phylogeny that extends so far back in time, and that our placement of the root is reasonable, we can make inferences about when certain fundamental cellular traits evolved, and in which lineages. Recall from Chapter 4 that we can map character-state changes onto phylogenies using the principle of parsimony. For examples of how we might do this on the universal phylogeny look at Figure 17.20. If a trait occurs in all three domains

(Figure 17.20a), or if it occurs in the Bacteria and the Archaea but not in the Eucarya (Figure 17.20b), or if it occurs in the Bacteria and the Eucarya but not in the Archaea (Figure 17.20c), we can infer that the trait was present in the common ancestor and was lost on the lineage (if there is one) that lacks it. Alternative scenarios would require that the trait arose independently two or three times and would therefore require more evolutionary transitions.

We have already noted that the most conserved pieces of machinery inside cells function in the translation of genetic information from nucleic acids into proteins. Of the roughly 60 genes that occur in the genomes of all cellular organisms in all domains, 30 are ribosomal proteins and 15 are aminoacyl-tRNA synthetases—enzymes that attach amino acids to their tRNAs (Koonin 2003). We can infer that the last common universal ancestor had enzymes made of protein and a well-elaborated capacity for manufacturing them.

Can parsimony tell us whether the common ancestor of all extant organisms was already storing its genetic information in DNA? The fact that all extant organisms use DNA suggests that the common ancestor did the same. An alternative possibility is that the common ancestor stored its genetic information in some other molecule, such as RNA, but that storage in DNA was favored so strongly by natural selection that a conversion from RNA storage to DNA storage occurred independently in more than one domain. Use of DNA by the common ancestor appears more likely than this scenario of convergent evolution. One clue is that the DNA-dependent RNA polymerases used in transcription show strong similarities across all three domains. This suggests that a DNA-dependent RNA polymerase was present in the last common ancestor. The possession of a DNA-dependent RNA polymerase implies the possession of DNA (Benner et al. 1989). Likewise, DNA polymerases found in all three domains show enough similarities to suggest that the common ancestor also had a DNA polymerase. And where there was a DNA polymerase, again, there was probably DNA. On the other hand, some components of the machinery for DNA replication are so different in Bacteria versus Archaea and Eucarya that we can infer that they evolved independently (Leipe et al. 1999). Perhaps the last universal common ancestor stored its genetic information in DNA, but copied it differently than modern organisms.

Based on similar kinds of evidence and reasoning, many researchers have tentatively concluded that the most recent common ancestor was highly evolved and biologically sophisticated. Overall, the common ancestor appears in many ways to have been rather like a modern bacterium (Ouzounis et al. 2006).

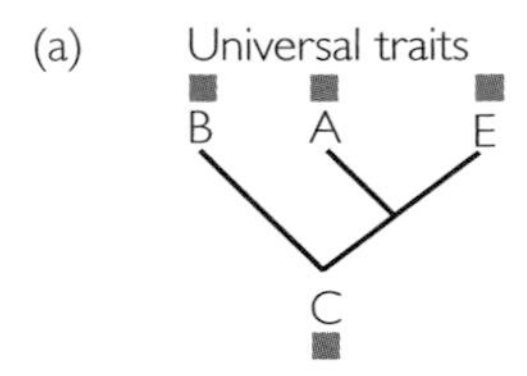

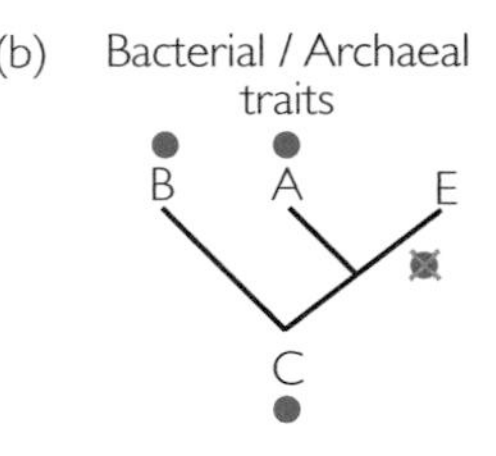

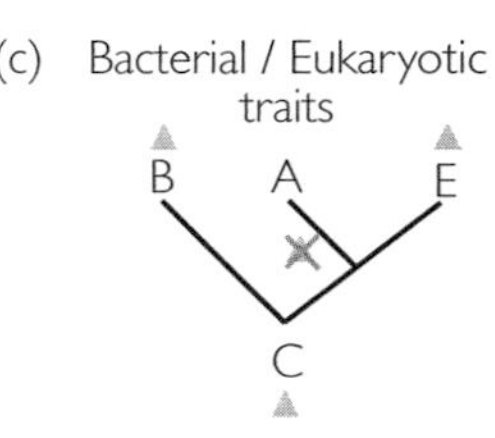

Figure 17.20 Three possible distributions of complex traits among the three domains of life The first appearance of a symbol on a tree represents the origin of a trait. A crossed-out symbol represents the loss of a trait.

Whole-life phylogenies based on molecular data suggest that the most recent common ancestor of all extant life was a sophisticated organism, with a DNA genome and much of the machinery of modern cells. . . .

Our Picture of the Tree of Life, and of the Earliest Cells, Continues to Evolve

Our understanding of the tree of life, and of the earliest cells whose descendants survive today, depends crucially on genetic sequence data. Such data allow us to estimate the universal phylogeny. Furthermore, sequence data provide much of the information about the traits of organisms that, when placed on the universal phylogeny, allow us to make inferences about the common ancestors. The amount of sequence data we have is still limited, but is growing explosively. Two trends in particular promise to yield many new insights.

First, our knowledge of the Archaea is increasing dramatically. As we mentioned earlier, many archaeans live in harsh and unusual environments. *Methanococcus jannaschii,* for example, lives anaerobically in deep-sea hydrothermal vents, at temperatures near 85°C and depths of at least 2,600 m (Jones et al. 1983). Not surprisingly, most known archaeans are difficult or impossible to grow in culture and thus are hard to study.

In 1984, a team of biologists working in the laboratory of Norman Pace pioneered a new approach to studying the environmental distribution of the Archaea. The researchers extracted DNA directly from mud and water samples collected in nature, then amplified and sequenced the DNA in the lab (Stahl et al. 1984). Following this approach, Edward DeLong and colleagues examined ribosomal RNA genes extracted from seawater collected in the Antarctic and off the coast of North America. DeLong and colleagues found many genes that were recognizable, based on their sequences, as belonging to previously unknown archaeans (DeLong 1992; DeLong et al. 1994). Susan Barns and colleagues (1994) likewise looked at rRNA genes extracted directly from mud in a hot spring in Yellowstone National Park. They also detected several rRNAs from previously unknown archaeans. Researchers in several laboratories are now pursuing similar studies (Service 1997; Schleper et al. 2005).

These environmental sequencing surveys have established that Archaea live not only in extreme envirnoments, but in moderate ones as well—including saltwater, freshwater, and soil. They are sufficiently abundant that they may turn out to play a substantial role in global energy and chemical cycles. And they include the only organisms capable of converting hydrogen and carbon dioxide into methane.

Barnes and colleagues (1996) used several new archaean rRNA sequences in the estimate of the whole life phylogeny shown in Figure 17.21. This tree suggests the existence of a previously unknown kingdom of archaeans, the Korarchaeota. Given that the Achaea are one of the three fundamental groups of organisms, everything we learn about them improves our our understanding of the universal phylogeny (Gribaldo and Brochier-Armanet 2006).

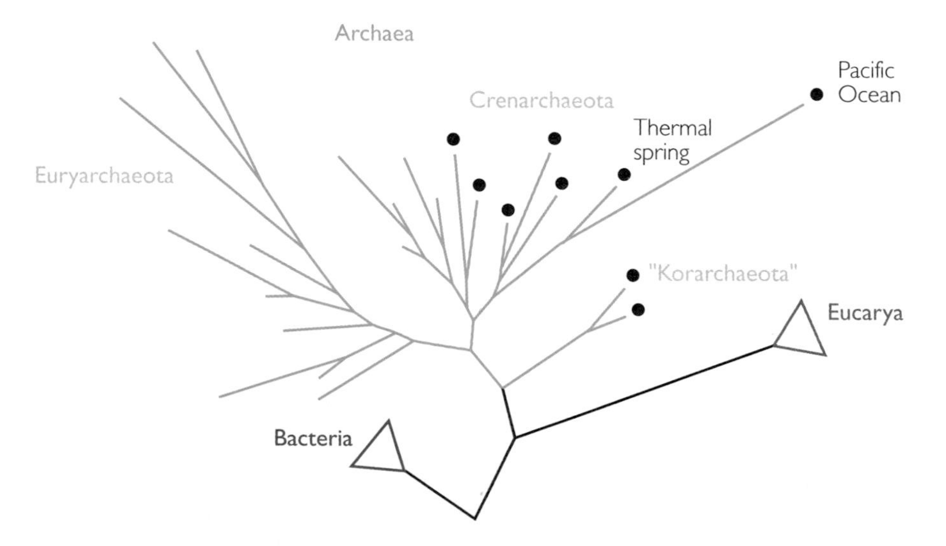

Figure 17.21 An estimate of the phylogeny of all extant organisms based on rRNA sequences Black dots indicate lineages known only from their rRNA genes. Branch lengths within the Archaea and among the domains are proportional to genetic distances; the genetic diversities among Bacteria and Eucarya are not to scale. Redrawn from Barnes et al. (1996).

The second trend that will improve our understanding of both the universal phylogeny and the biology of the most recent common ancestor is the advent of whole-genome sequencing. When we wrote the first edition of this book, complete genomes had been sequenced for five organisms. When we wrote the second edition, 27 genomes were available; when we wrote the third edition, 114. As of this writing, complete genomes have been sequenced for 389 organisms: 339 bacteria, 28 archaeans, and 22 eukaryotes—including a human, a mouse, a fruitfly, a roundworm, three plants, several fungi, and several protists. Rough drafts are complete for an additional 345 species, and genome projects are under way for 483 more (NCBI 2006). The availability of whole genome sequences gives researchers the opportunity to estimate the universal phylogeny based on information from a great variety of different genes, and on whole genomes themselves. And this has produced some surprises.

We would expect estimates of the universal phylogeny based on different genes to be broadly congruent. In fact, however, they are not. James R. Brown and W. Ford Doolittle (1997) estimated whole-life phylogenies based on some four dozen genes (Figure 17.22). Genes for proteins involved in the storage and processing of genetic information often give a tree consistent with the small-subunit rRNA tree (Figure 17.22a). Genes for proteins involved in metabolism, however, often give a tree in which the Bacteria and the Archaea are closest relatives (Figure 17.22b). Still other genes give a tree in which the Bacteria and the Eucarya are closest relatives (17.22c), or in which there is an unresolved trichotomy of the three domains (Figure 17.22d).

... However, whole-life phylogenies based on molecular data have also yielded rude surprises....

How can we explain the discordance among the whole-life phylogenies estimated from different genes? Many researchers, Carl Woese included (1998, 2000, 2002), argue that the conflicts among data sets are too numerous and persistent to ignore. They believe the explanation is the lateral movement of genes among taxa, a process known as **lateral** (or **horizontal) gene transfer**.

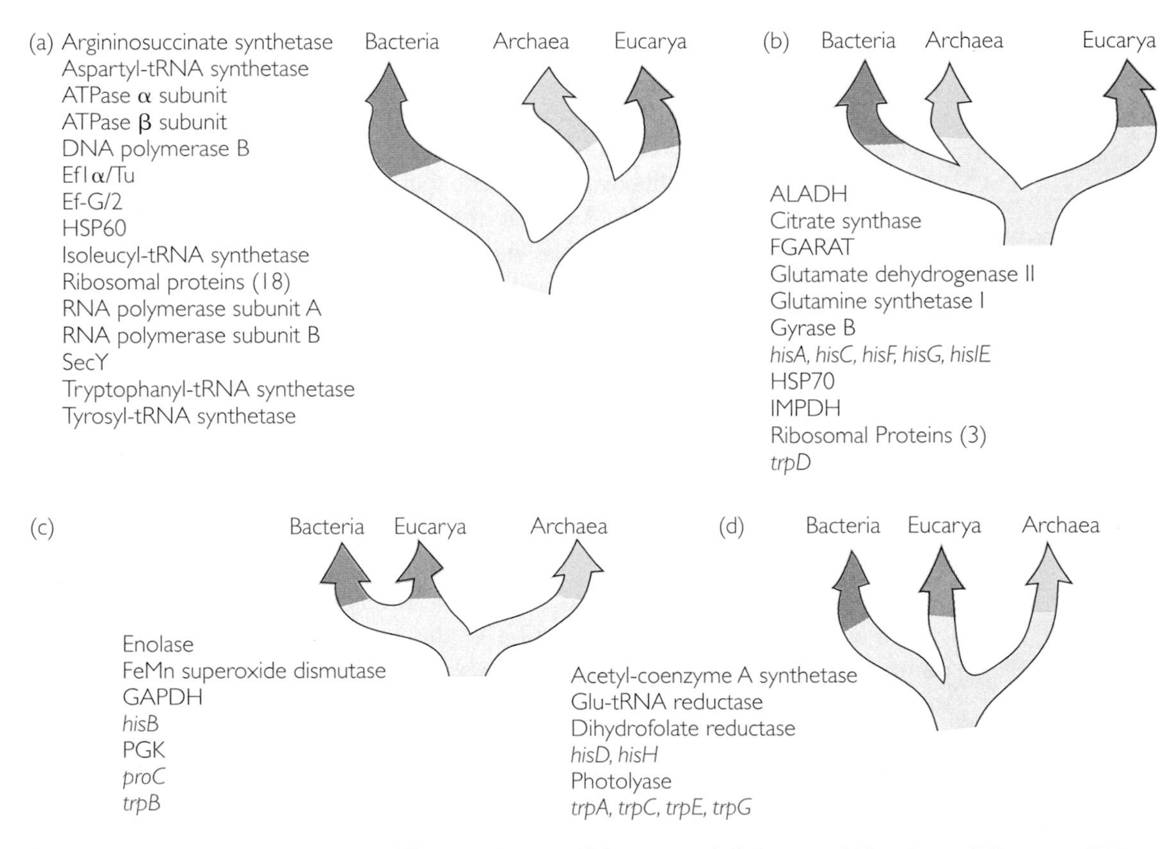

Figure 17.22 Different genes give different estimates of the universal phylogeny When James R. Brown and W. Ford Doolittle (1997) reconstructed the tree of life using a variety of genes, they found that different genes give fundamentally different phylogenies. Some genes give trees in which Archaea and Eucarya are closest relatives (a), others give trees in which Bacteria and Archaea are closest relatives (b). Still other genes give trees in which Bacteria and Eucarya are closest relatives (c), or in which the relationships among the three domains are unresolved (d). From Brown and Doolittle (1997).

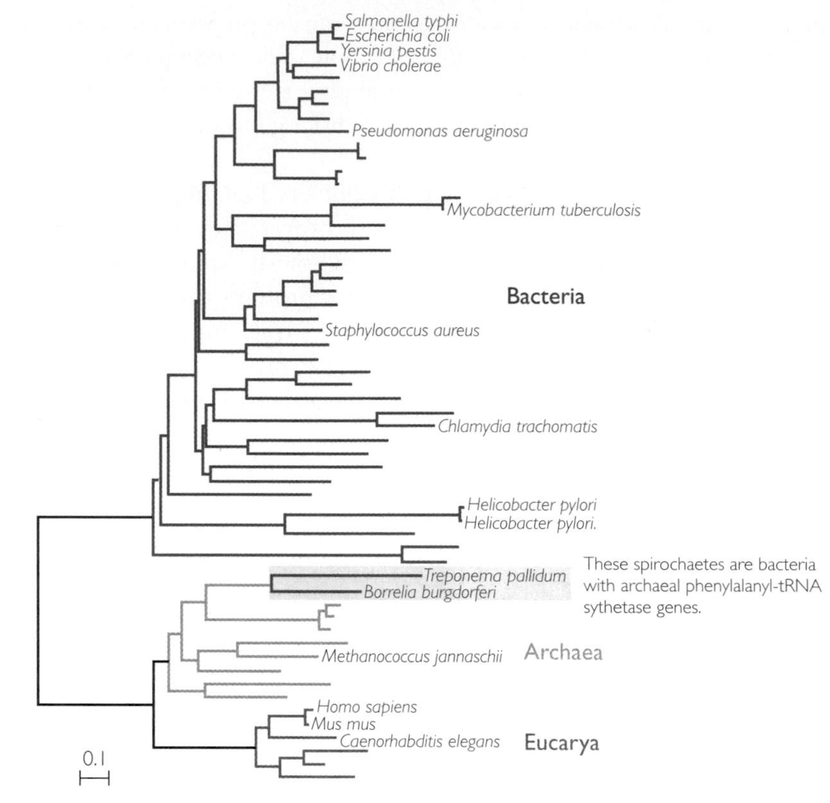

Figure 17.23 A universal phylogeny based on the phenylalanyl-tRNA synthetase β-subunit gene reveals lateral gene transfer The spirochaetes *Treponema pallidum* and *Borrelia burgdorferi* are bacteria, yet their genes for the β-subunit of the phenylalanyl-tRNA synthetase enzyme branch from within the archaea. The likely explanation is that the spirochaetes have lost their native bacterial version of the ß-subunit gene and replaced it with a version of the gene picked up from an archaean. From Brown (2001).

... Chief among the rude surprises from whole-life phylogenies is that organisms appear to have swapped their genes much more readily than anyone suspected. This means that the phylogenies of genes may be different from the phylogenies of the organisms that harbor them.

An example of lateral gene transfer appears in Figure 17.23. This whole life phylogeny, by James R. Brown (2001), is based on the gene for the β-subunit of phenylalanyl-tRNA synthetase, the enzyme that attaches the amino acid phenylalanine to its transfer RNA. Look at the location on the phylogeny of the β-subunits from *Treponema pallidum* and *Borrelia burgdorferi*. These organisms are pathogenic spirochaetes. *Treponema pallidum* causes syphilis; *Borrelia burgdorferi* causes Lyme disease. They are unambiguously bacteria, and phylogenies based on most other components of the translation machinery put them where they belong. In Figure 17.24, for example, they appear solidly within the bacterial domain (at about 5 o'clock in the diagram). And yet on the tree in Figure 17.23 their genes for the β-subunit of phenylalanyl-tRNA synthetase appear to be archaeal. How could this be? The likely answer is that their β-subunit genes *are* archaeal. A common ancestor of the two spirochaetes lost its native bacterial β-subunit gene and replaced it with a gene picked up from an archaean. We discussed additional examples of lateral gene transfer, and the mechanisms responsible, in Chapter 15.

For any given organism, the fraction of the genome acquired by lateral gene transfer may be startlingly high. Jeffrey Lawrence and Howard Ochman (1998) estimated, for example, that 18% of the genes carried by *E. coli* strain MG1655 were picked up laterally within the last 100 million years. And Karen E. Nelson and colleagues (1999) estimated that 24% of the genes in the bacterium *Thermotoga maritima* were picked up by lateral transfer from archaeans. Numbers this high, if they are accurate, are probably atypical. Fen Ge and colleagues (2005) developed stringent methodological and statistical criteria for identifying laterally transferred genes, then applied them to the genomes of 40 microbes. They estimated that among these genomes, between 0 and 6.74% of the genes were aquired by lateral transfer, with a mean of 2%. Nonetheless, lateral gene transfer is rampant enough to raise questions about the very enterprise of reconstructing evolutionary trees from genetic data.

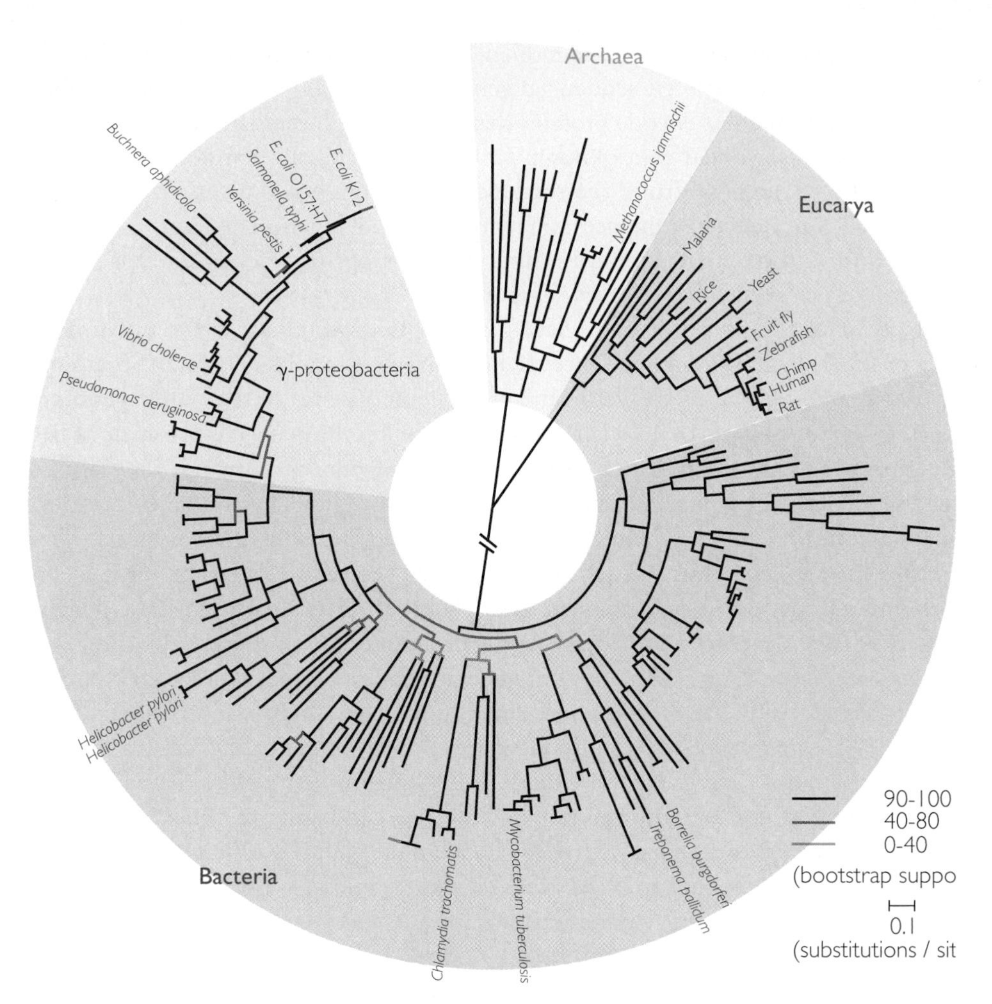

Figure 17.24 A universal phylogeny based on concatenated sequences of 31 universal genes, most for ribosomal proteins The tree resolves the Archaea, Bacteria, and Eucarya into monophyletic domains, and is in good overall agreement with the rRNA tree in Figure 17.18. From Ciccarelli et al. (2006).

Some researchers doubt that deep evolutionary history can be accurately characterized by trees at all (Doolittle 1999; Gogarten et al. 2002; Bapteste et al. 2005). Christopher Creevey and colleagues (2004) investigated this issue by using hundreds of different genes to reconstruct bacterial phylogenies. Once they had phylogenies based on individual genes, they constructed a single "supertree" that minimized the summed differences between the supertree and the individual gene trees. They then assessed the statistical significance of this supertree by comparing the summed differences between the supertree and the real gene trees versus the supertree and randomly generated gene trees. When Creevey and colleagues completed this exercise using genes from the genomes of ten bacterial species belonging to a single clade (the γ-proteobacteria, 10 o'clock in Figure 17.24), they found that the supertree fit the real gene trees significantly better than it fit randomly generated trees. That is, there is detectable consensus among the phylogenies reconstructed from individual genes. When Creevey and colleagues completed the exercise using genes from the genomes of 11 bacterial species spanning the entire bacterial domain, however, the supertree fit the real gene trees no better than it fit random ones. Apparently, most individual genes have either evolved so rapidly that they no longer retain any record of their deep history, or most individual genes have such different histories that they cannot be described by a single phylogeny. Deep evolutionary history, under the latter interpretation, is not a tree but a web.

Is the universal tree of life an idea worth keeping? It is, as shown by reconstructions based not on individual genes but on whole genomes (Brown 2003; Doolittle 2005). The tree in Figure 17.24 comes from one such reconstruction. Francesca

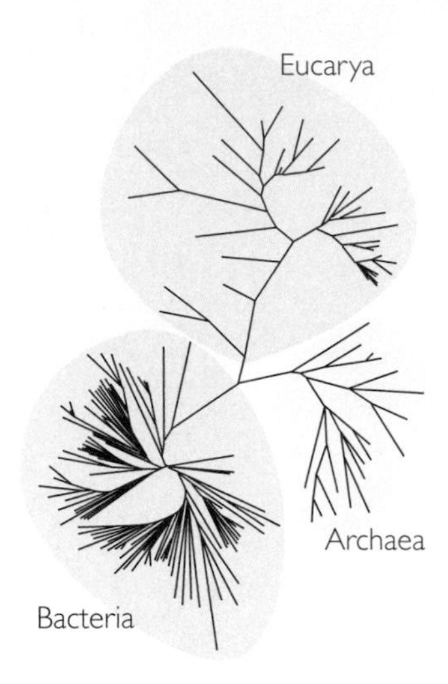

Figure 17.25 A universal phylogeny based on the presence/absence of protein superfamilies This tree, too, is in good overall agreement with the rRNA tree in Figure 17.18. From Yang et al. (2005).

Ciccarelli and colleagues (2006) identified 31 genes present in every one of nearly 200 species with completely sequenced genomes. Most of these genes encode ribosomal proteins; the rest encode proteins that participate in translation in other ways. The researchers avoided genes known to have been laterally transferred. For each species, the researchers strung the sequences of the 31 genes together. They then used these concatenated sequences to reconstruct a phylogeny. The resulting tree resolves the Bacteria, Archaea, and Eucarya into monophyletic clades, and is largely congruent with Woese's rRNA tree (Figure 17.18, page 665).

Researchers using a different whole-genome method produced the tree in Figure 17.25. Instead of using sequence data on a limited number of genes, Song Yang and colleagues (2005) used data on the presence or absence in the whole genomes of 174 species of each of 1,294 protein superfamilies. Proteins are classified as belonging to the same superfamily if they show sufficient similarity in their amino acid sequence, higher-level structure, and/or function to indicate that they probably share a common ancestry (Murzin et al. 1995). Yang and colleagues assigned every possible pair of organisms a genetic distance based on the overlap in the protein superfamilies represented or lacking in their genomes. They then used these genetic distances to reconstruct a phylogeny. Again, the resulting tree features the three domains as monophyletic clades, and is in general agreement with Woese's rRNA tree.

Ciccarelli's study, Yang's study, and other studies like them confirm that whole genomes can give us at least a misty view into the far distant past. What we can see, however, it not exactly what we expected. Life's history is treelike (Kurland et al. 2003; Delsuc et al. 2005). But it is not exclusively so, and any two genes in the same genome at the tip of a twig may have ascended there there by different routes.

An important implication is that when trying to infer the nature of the last universal common ancestor, the best metaphor for life's history may be neither a simple tree nor a web, but instead a set of interconnected roots like those shown in Figure 17.26. It may be incorrect, therefore, to think of the last common ancestor of all extant organisms as having been a single species. A more accurate depiction of the common ancestor may be that it was a community of interacting species that readily traded their genes (Woese 1998; Doolittle 2000; Whitfield 2004). How the Bacteria, Eucarya, and Archaea might have emerged from this tangled base is a topic we will consider in the final section of the chapter.

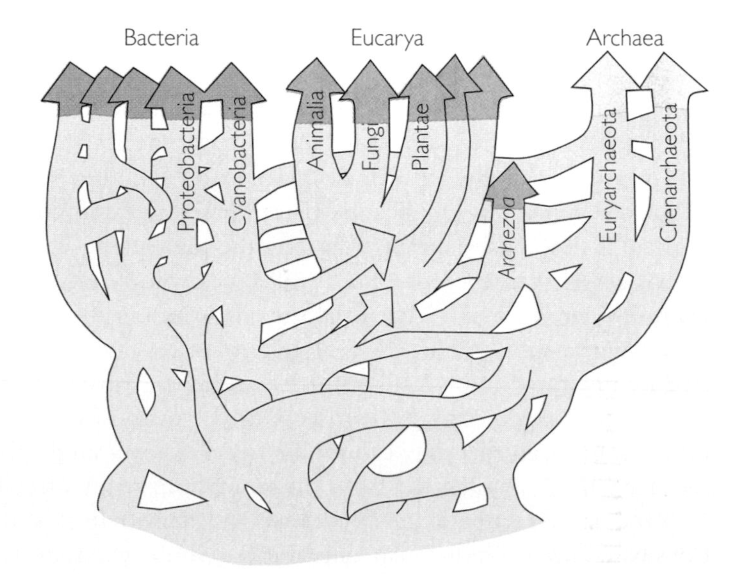

Figure 17.26 The cenancestor was not a single species, but a community Given the evidence for extensive horizontal gene transfer (documented in Figures 17.22 and17.23), many researchers have concluded that we can no longer think of the last common ancestor of all extant organisms as a single species. Instead, the cenancestor was a community of species that readily traded genes. Eventually, the three main branches on the tree of life emerged from this tangled base. After Doolittle (1999).

The Latest Possible Date for the Root of the Tree of Life

The organisms at the base of the universal phylogeny could not have lived more recently than any of the branch points above them. Attempts have been made to date the branch points using sequence data and molecular clocks (for example, see Doolittle et al. 1996, but also Hasegawa et al. 1996). However, the most definitive information about branching times comes from fossils. The fossils useful in this regard are those that can be confidently identified as belonging to a particular group of organisms. If we can place a fossil in one of the three domains in the whole-life phylogeny, then we know that the deepest branch point is older than the fossil.

Fossils of single-celled organisms can be identified as eukaryotes if they show sufficient structural complexity. The 590-million-year-old fossil in Figure 17.27a, for example, is clearly eukaryotic: It is about 250 μm in diameter and, unlike any known archaean or bacterium, it is covered with spikes. The 850-to-950-million-year-old fossil in Figure 17.27b is also eukaryotic: It is about 40 μm in diameter and covered with knobs. The 1.4- to 1.5-billion-year-old fossil in Figure 17.27c is about 60 μm in diameter, but simple in structure. It is probably, but not certainly, eukaryotic.

(a)

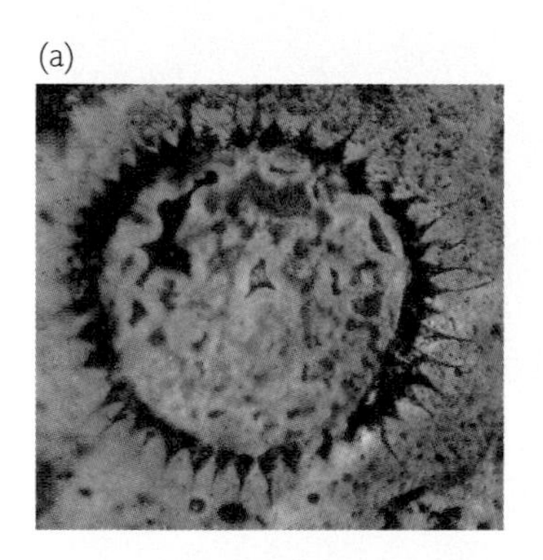

(b)

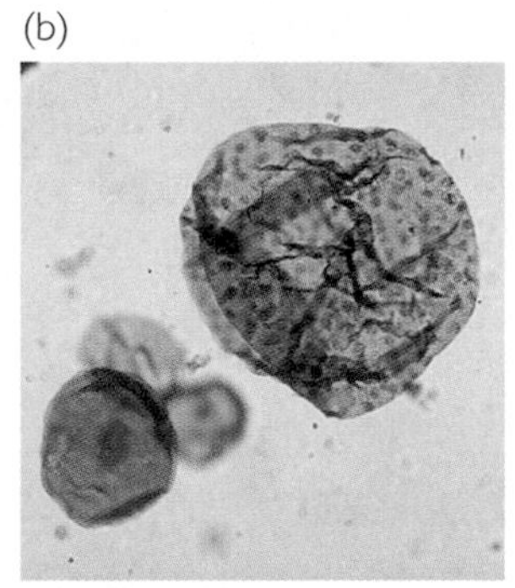

(c)

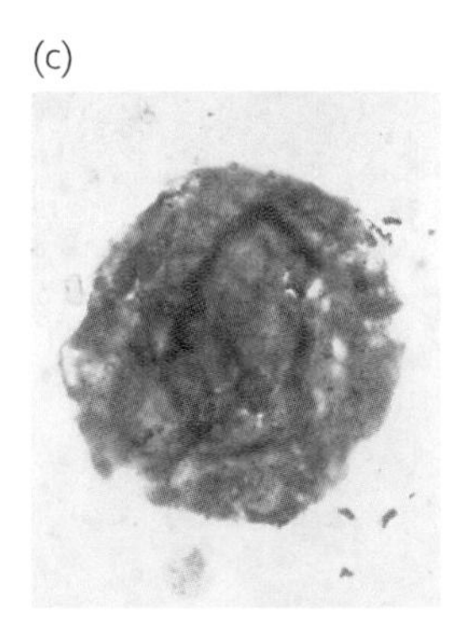

Figure 17.27 Fossils of single-celled Eucarya (a) A spiny fossil from the Doushantuo Formation, China. This 590-million-year-old fossil represents either the preserved cell wall of a single-celled eukaryote, the reproductive cyst of a multicellular alga, or the egg case of an early animal. The fossil has a diameter of about 250 μm. From Knoll (1994); see also Knoll (1992). (b) A structurally complex fossil from the Miroyedicha Formation, Siberia. This 850- to 950-million-year-old fossil is clearly eukaryotic, but like that shown in (a), its exact identification is unclear. It probably represents a single-celled organism. It has a diameter of about 40 μm. From Knoll (1994); see also Knoll (1992). (c) Fossil cell from the Roper Group, Australia. This cell is 1.4 to 1.5 billion years old; it probably represents a eukaryote, but it lacks sufficient complexity for a definitive identification. The fossil is about 60 μm in diameter.

The oldest known fossils that are probably those of eukaryotes are 1.85–2.1 billion years old (Figure 17.28). Found by Tsu-Ming Han and Bruce Runnegar (1992) at the Empire Mine in Michigan, these fossils show a spiral-shaped organism similar to a more recent fossil named *Grypania spiralis*. *G. spiralis* is known from fossils in Montana, China, and India that range in age from 1.1 to 1.4 billion years old (see Han and Runnegar 1992). Because of its large size and structural complexity, paleontologists believe that *Grypania* was a eukaryote—probably an alga.

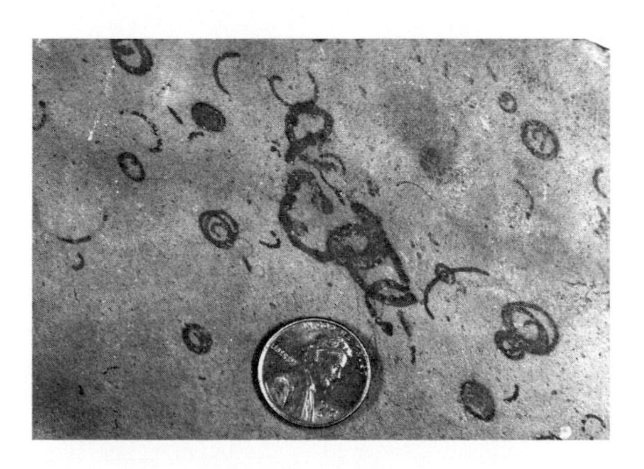

Figure 17.28 2-billion-year-old fossils from Michigan Paleontologists believe these fossils represent eukaryotic algae. The penny is 18.5 mm in diameter.

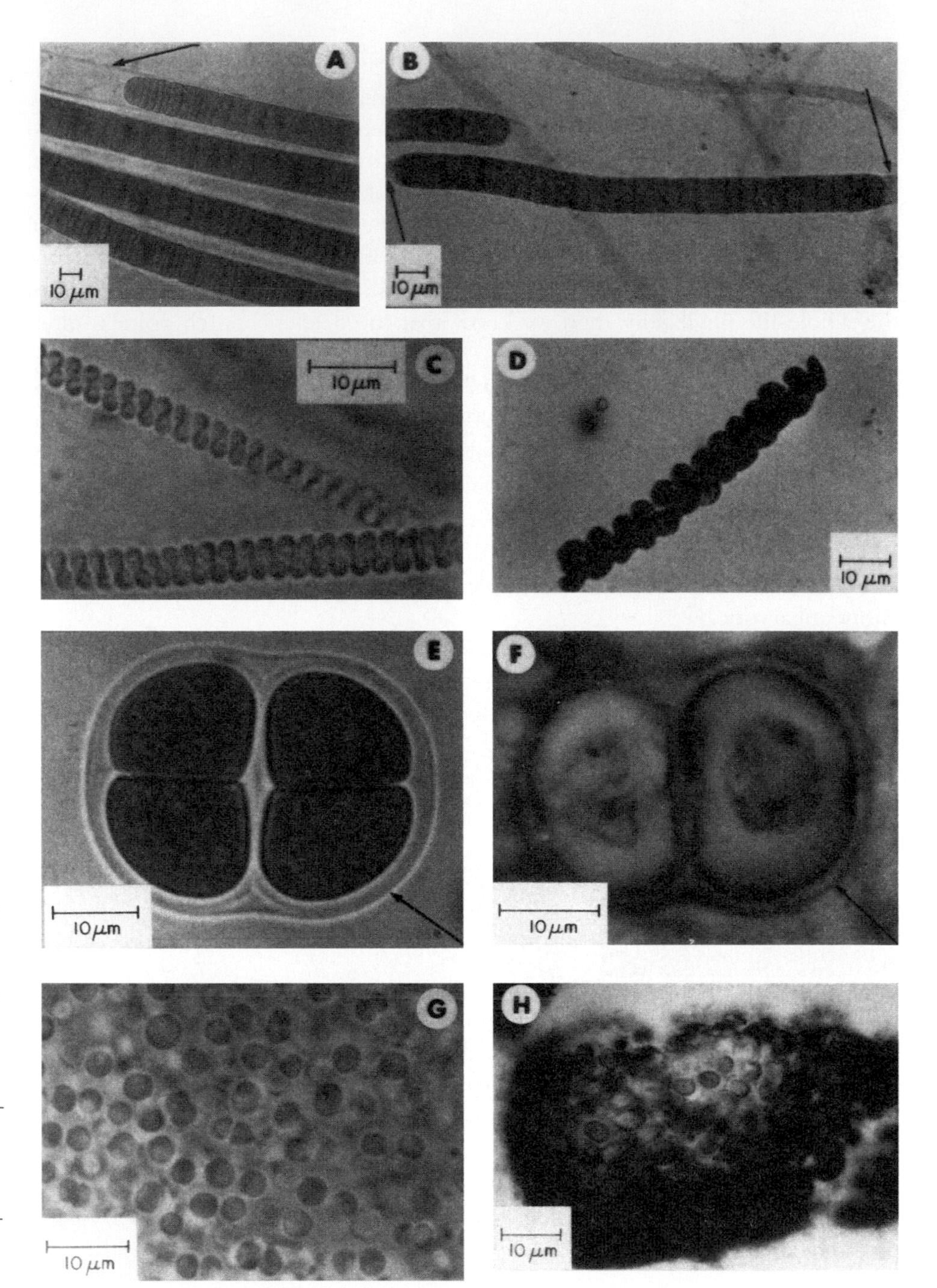

Figure 17.29 Fossil cyanobacteria and their extant relatives
(a) *Lyngbya* (extant); (b) *Paleolyngbya* fossil from the 950-million-year-old Lakhanda Formation, Siberia; (c) *Spirulina* (extant); (d) *Heliconema* fossil from the 850-million-year-old Miroedikha Formation, Siberia; (e) *Gloeocapsa* (extant); (f) *Gloeodiniopsis* fossil from the 1.55-billion-year-old Satka Formation, Bashkiria; (g) *Entophysalis* (extant); (h) *Eoentophysalis* fossil from the 2-billion-year-old Belcher Group, Canada. For more details see Schopf (1994a).

Fossil cyanobacteria also suggest that the root of the universal phylogeny predates 2 billion years (Schopf 1994a). In a testament to the length of time that successful organisms can remain at least superficially unchanged, many fossil cyanobacteria are identifiable based on their structural similarity to extant forms. Each row of Figure 17.29 shows an extant species of cyanobacteria on the left, and a similar fossil form on the right. The fossils range in age from 850 million to 2 billion years old. The extant cyanobacteria occupy a limb of the universal phylogeny that is, like those occupied by the extant algae, several branch points above the last common ancestors (Figure 17.18). Again, we can conclude that the last common ancestors lived at least 2 billion years ago, and probably earlier.

Fossils that belong to identifiable taxa can give us a minimum age for the last common ancestor of all living things.

In summary, we can use the estimated universal phylogeny, along with geological and paleontological data, to bracket the time when the first branching in the universal phylogeny took place. The earliest possible date is when life on Earth began between 4.4 and 3.7 billion years ago; the most recent possible date, set by the oldest identifiable fossils, is at least 2 billion years ago.

17.4 How Did the Last Common Ancestor's Descendants Evolve into Today's Organisms?

In the preceeding sections of this chapter we have explored a hypothesized early stage of life, the RNA World. We have surveyed ideas on how the inhabitants of RNA World might have arisen from nonliving matter. And we have looked at reconstructions of the tree of life to see what they show us about the nature of its root. The researchers engaged in this last pursuit had set out to find LUCA, the last universal common ancestor, whch they thought would be a single species of microbe. What they discovered instead was evidence of lateral gene transfer extensive enough to suggest that last universal common ancestor was, in fact, a community.

In this last section of the chapter we review a variety of ideas about how this ancestral community gave rise to the bacteria, archeans, and eukaryotes that populate the Earth today. Each of the hypotheses we discuss offers a different scenario of lateral gene transfer to resolve apparent conflicts among the evolutionary histories of different genes (as illustrated in Figure 17.22). All of the hypotheses are speculative and all are controversial.

The Universal Gene-Exchange Pool Hypothesis

The first biologist to reconstruct a universal phylogeny based on rRNA genes was, as we have discussed, Carl Woese. In building the first whole life phylogeny, Woese discovered the Archaea. He was also among the first to recognize that the conflicts among universal phylogenies based on different genes were revealing something unexpected about the importance of lateral gene transfer in early evolution (Woese 1998).

In recent years Woese (2002, 2004) has been outlining a scenerio of early evolution in which lateral gene transfer was so rampant that it overshadowed vertical inheritance. Genomes, such as they existed, were modular in nature. That is, most ribozymes and proteins functioned independently of other ribozymes proteins, and the genes encoding them could readily move from genome to genome. Organisms were assembled more by drawing genes from a universal-gene exchange pool (Figure 17.30), than by self-replication. Geneological lineages, as we think of them today, did not exist. Nor did evolutionary trees.

As Woese himself asserts, the situation he describes is not conducive to evolution by natural selection. When genotypes and phenotypes are acquired rather than inherited, differential reproductive success is of limited consequence. Instead, Woese postulates a non-Darwinian mechanism of communal evolution. Gradually, as proteins became more interdependent, the modularity of genomes gave way to a more integrated and stable format. Individual genes could no longer move so easily from one genome to another. Self-replication now had the more prominent role in the generation of new organisms. At this point, which Woese calls the Darwinian threshold, populations began to evolve by natural selection.

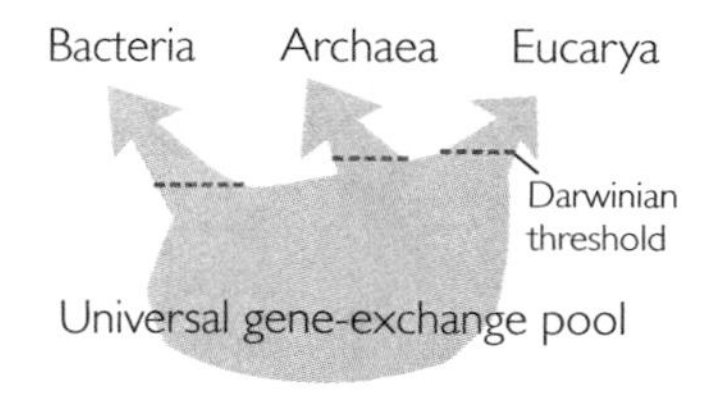

Figure 17.30 Carl Woese's conjecture on the origin of the three domains of life The last universal common ancestor was not a single species but rather a pool of readily exchanged and largely independent genes. Eventually, three cellular forms emerged with genomes stable enough to establish persistent lineages. At this point, which Woese calls the Darwinian threshold, populations began to evolve by natural selection. The three lineages became the three domains of life.

Woese believes that at least three stable lineages emerged independently from the universal gene-exchange pool. These were the ancestors of today's Bacteria, Archaea, and Eucarya. It is the order in which the three domains crossed their Darwinian thresholds that makes Archaea and Eucarya appear to be each other's closest kin in universal phylogenies based on rRNA (Figure 17.18), ribosomal proteins (Figure 17.24), and protein families (Figure 17.25). Bacteria crossed first, followed by Archaea, then Eucarya. Because they continued to draw from the universal gene exchange after the Bacteria had separated from it, the Archaea and Eucarya have more similar genes across most, but not all, of their genomes. But a deep phylogeny with a single unique root is a pattern we impose on the data, rather than a pattern that emerges from the data.

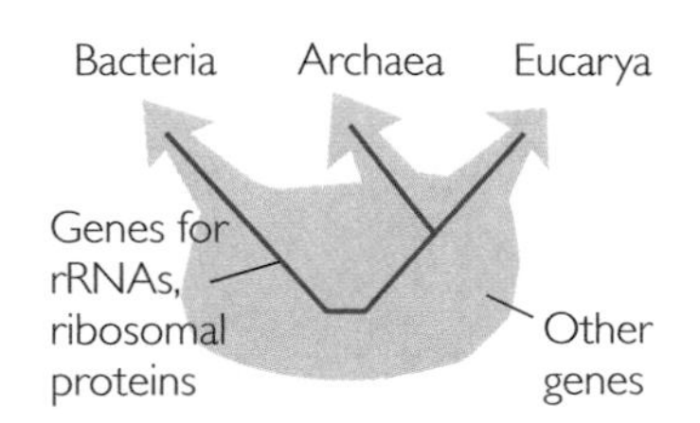

Figure 17.31 **The deepest node on the universal phylogeny as a community** We can reconstruct evolutionary trees from various kinds of data, as shown here for rRNA and ribosomal protein genes, but the most biologically accurate interpretation of the data may be to view the last common ancestor of the three domains as a community.

It is the non-Darwinian communal evolution at the heart of Woese's hypothesis that elicits skepticism from other biologists (see Whitfield 2004). What is the mechanism responsible? Peter Antonelli and colleagues (2003) developed a mathematical model of the universal gene-exchange pool and demonstrated that it was unstable. Woese asserts that other mathematical formulations of his verbal argument are possible. He and colleagues have developed a model showing that lateral gene transfer leads to convergence on a universal genetic code (Vestigian et al. 2006). But until a full quantitative model of the universal gene-exchange is developed and shown to be workable, Woese's ancestral gene exchange will remain a rather abstract conjecture.

While they may not agree with the notion of non-Darwinian communal evolution, many other researchers do concur with Woese that lateral gene transfer was rampant enough during life's early history that we have to think of the last common ancestor of the three domains as a community rather than a single species (Figure 17.31; see Kurland et al. 2006).

The Ring of Life Hypothesis

Like researchers reconstructing the whole life phylogeny from different genes, researchers comparing the genes of eukaryotes to those of bacteria and archaea have discovered a curious pattern. Christian Esser and colleagues (2004), for example, compared the amino acid sequences specified by more than 6,000 yeast genes to those encoded by more than 175,000 bacterial and archaean genes. For some 75% of the yeast genes the most similar non-eukaryotic gene came from a bacterium; for the rest it came from an archaean. In general, eukaryotic genes involved in the storage and use of genetic information, in processes such as transcription and translation, tend to be more simliar to archaean genes. Eukaryotic genes involved in metabolic processes, such as the synthesis of amino acids, tend to be more similar to bacterial genes (Simonson et al. 2005).

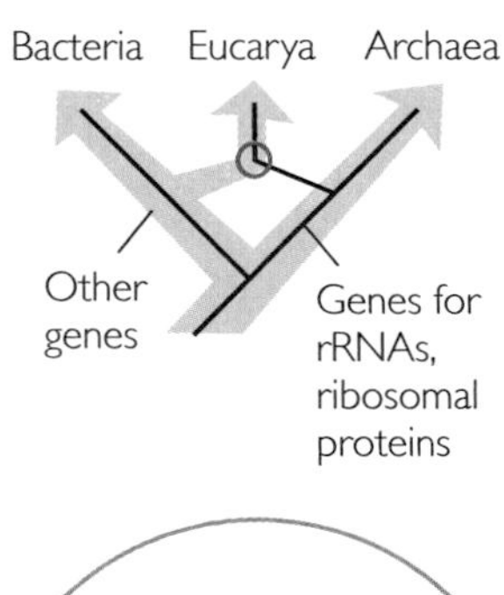

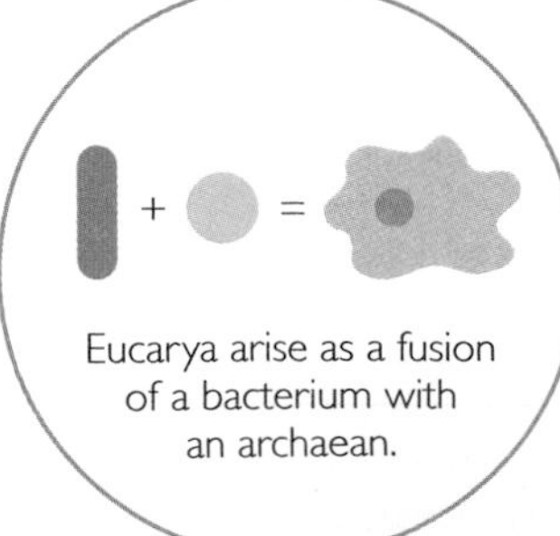

Figure 17.32 **The ring of life** According to this hypothesis, the Eucarya arose from the fusion of a bacterium with an archaean.

This pattern suggested to Maria Rivera and James Lake (2004) that the first eukaryote arose when a bacterium fused with an archean (see also Horiike et al. 2002, 2004). The lineage that arose from this union retained the informational genes from the archaean and the metabolic genes from the bacterium. The whole-life phylogeny, as shown in Figure 17.32, has a ring at its center. In some versions of this hypothesis the bacterial partner in the fusion that created the first eukaryote eventually became the mitochondrion (see Chapter 15); in other versions the eukaryote lineage acquired the mitochondrion later.

If the bacterial partner in the fusion that created the first eukaryote was the ancestor of the mitochondrion, then the metabolic genes of eukaryotes should arise from within the α-proteobacteria, the bacterial clade known from rRNA phylogenies to be the source of the mitochondrion. Bjorn Canback and colleagues (2002) tested this prediction by reconstructing deep phylogenies for eight enzymes involved in glycolysis. In fact, none of the glycolytic enzymes of Eucarya are closely related to their α-proteobacterial homologs. If the Eucarya were born from the union of a bacterium and an archaean, that union long predated the acquisition of the mitochondrion.

An early fusion of a bacterium and an archaean is still a possibility. Critics argue, however, that the bacterial and archaeal genes present in the eukaryotic genome arose at different times from different lineages (Lester et al. 2005), that the ring of life hypothesis provides no explanation of where the hundreds of proteins found only in Eucarya came from (Kurland et al. 2006), and that because Archaea and Bacteria lack a cytoskeleton that enables phagocytosis, it is difficult to see how they could have fused in the first place (Kurland et al. 2006, including supporting online material).

The Chronocyte Hypothesis

Russell Doolittle (2000) is among the advocates of a scenario that offers a solution to the phagocytosis problem. In this scenario, outlined in Figure 17.33, the deepest fork in the tree of life separates a lineage that will become the Bacteria and the Archaea from the lineage that will become the Eucarya. Hyman Hartman calls this lineage the chronocytes (Hartman and Fedorov 2002). The chronocyte lineage evolved a cytoskeleton and the ability to eat other microbes by phagocytosis. A chronocyte then ate an archaean that resisted digestion and became an endosymbiont. This endosymbont eventually evolved into an organelle: the nucleus. The nucleus preserved the information-processing genes from its archaeal ancestor but incorporated cytoskeletal genes from its host. The chronocytes had spawned the Eucarya. The Eucarya later acquired the mitochondrion and the chloroplast in the same way.

One way to test the chronocyte hypothesis is to look for a living chronocyte. Such a creature would have a cytoskelton and feed on other cells, but it would lack a nucleus and mitochondria. To date, no such beast has been found.

Hyman Hartman and Alexei Fedorov (2002) assert, however, that they have found the next best thing. In an exhaustive search of whole genomes representing all three domains, Hartman and Fedorov identified 347 genes found in all eukaryotes but completely absent in the genomes of bacteria and archaeans. Among the 347, those with known function encode proteins that build and operate the cytoskeleton and inner membranes, modify RNA, and control various aspects of cellular physiology. Hartman and Fedorov believe that today's eukaryotes inherited these genes from their chronocyte ancestors.

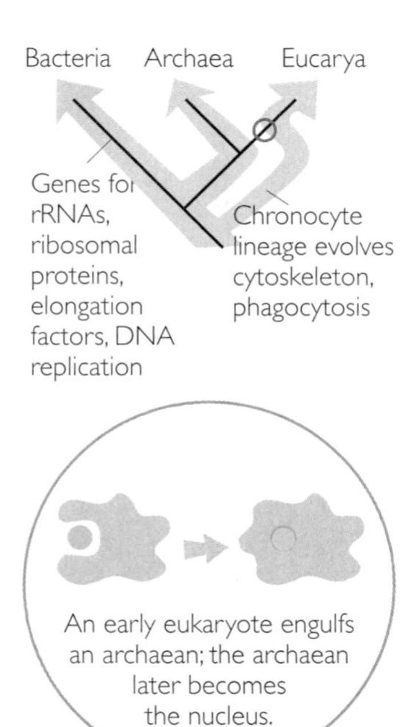

Figure 17.33 The chronocyte hypothesis According to this scheme, the deepest split in the tree of life separates the Bacteria and Archaea from the Chronocytes, which will eventually become the Eucarya. After evolving a cytoskeleton and the ability to ingest other cells, a member of this lineage engulfed an archaean that resisted digestion and eventually became the eukaryotic nucleus. Later still, eukaryotes acquired the mitochondrion and chloroplast in a similar fashion.

The Three Viruses, Three Domains Hypothesis

Among the proteins Hartman and Fedorov found to be nearly universal among the Eucarya, but missing from Bacteria and Archaea, is an RNA-dependent RNA polymerase. Eukaryotic cells use this polymerase to replicate RNAi, a form of RNA involved in posttranscriptional gene regulation. On this and other evidence, Hartman and Fedorov suggest that chronocytes had RNA-based genomes. This contradicts the tentative inference, discussed in Section 17.3, that the last universal common ancestor stored its genetic information in DNA. And it raises the question of how the three domains made the transition to DNA from RNA.

Patrick Forterre has suggested an answer. This answer hinges on Forterre's view of where viruses came from. Viruses, the reader may have noticed, have been conspicuously absent from our discussion of the origin and history of life. This is a glaring omission, given that viruses vastly outnumber all other forms of life (Hamilton 2006). Virologists estimate, for example, that 1,200 different kinds of viruses inhabit the human gut, that a kilogram of marine sediment harbors a million distinguishable viral genotypes, that the Earth is home to 10^{31} individual virions, and that most of the genetic diversity among viruses remains undiscovered. To connect viruses to the tree of life, researchers have offered a full range of hypotheses on their origin (Forterre 2006a): Viruses are genes that have escaped from the genomes of cellular organisms; viruses are descended from cellular organisms that have evolved reduced genomes in association with a parasitic lifestyle; viruses are remnants of the earliest eras of life on Earth, including the RNA World and the early DNA world. Forterre believes the balance of recent evidence favors the last of these hypotheses. Among other reasons, viral genes are often not closely related to the homologous genes of their hosts—indeed, many viral genes have no known homologs—and structural similarities among viruses infecting all three domains suggest that they derive from a common ancestor that lived before the last universal common ancestor of cellular life.

If viruses evolved early, Forterre (2005, 2006a, 2006b) maintains, they provide a plausible explanation for how and why DNA-based life evolved from RNA-based life. The traditional view is that a switch to DNA was adaptive because DNA is more chemically stable and because mutations converting cytosine to uracil can be recognized and repaired in DNA but not in RNA. The trouble with this explanation is that the advantages it cites acrue over the long term, whereas natural selection happens in the short term. A more plausible scenario, according to Forterre, is that the switch from RNA to DNA first happened in viruses that made their living parasitizing cells with RNA-based genomes.

Cells that are parasitized by viruses evolve defenses. Among these defenses are enzymes that recognize viral genomes and chop them up. In turn, viruses that parasitze cells evolve counterdefenses. These include chemical modifications of the parasite's nucleic acids that prevent the host's defensive enzymes from recognizing and destroying the parasite's genome. Given that DNA is a chemically modified form of RNA, it seems plausible that DNA first appeared as an adaptation in a previously RNA-based virus engaged in an evolutionary arms race with an RNA-based host. Consistent with this scenario, extant viruses illustrate many of the required transitional forms. There are viruses with purely RNA-based genomes (and other means of defending themselves against their hosts), RNA viruses that replicate their genomes via DNA intermediates, DNA viruses that replicate their genomes through RNA-based intermediates, and viruses with purely DNA-based genomes. There are even viruses with DNA-based genomes that use uracil instead of thymidine.

Finally, Forterre uses this scenario to explain how the cells that were the ancestors of the Bacteria, Archaea, and Eucarya themselves made the transition from RNA to DNA. Imagine that a DNA-based virus invades an RNA-based cell, loses the genes that encode its coat proteins, and thereby becomes an obliglately intracellular extrachromosomal element. If the DNA virus carries a gene for reverse transcriptase, it may occasionally copy one of its host's genes into DNA and incorporate the gene into its own genome. Eventually the DNA genome will absorb all of the genes from the RNA genome, along the way ceasing to be a parasite and instead transmogrifying into a component of the host cell. Because the DNA genome replicates more efficiently, it will outcompete the RNA genome and ultimately cause its extinction. The RNA-based host cell has been converted to a DNA-based cell with an expanded genetic repertoire.

To explain the phylogenetic distributions of various cellular genes, Forterre postulates that the ancestors of the three domains of cellular life first diverged while still carrying their genetic information in RNA, and that each was converted to DNA by a separate virus (Figure 17.34). In particular, his hypothesis of three viruses for three domains accounts for the fact, mentioned in Section 17.3, that much of the machinery bacteria use to replicate their DNA appears unrelated to the machinery used by archaeans and eukaryotes. The viruses that carried DNA into the Archaea and Eucarya happened to be related to each other, but distantly related or unrelated to the virus that carried DNA into the Bacteria.

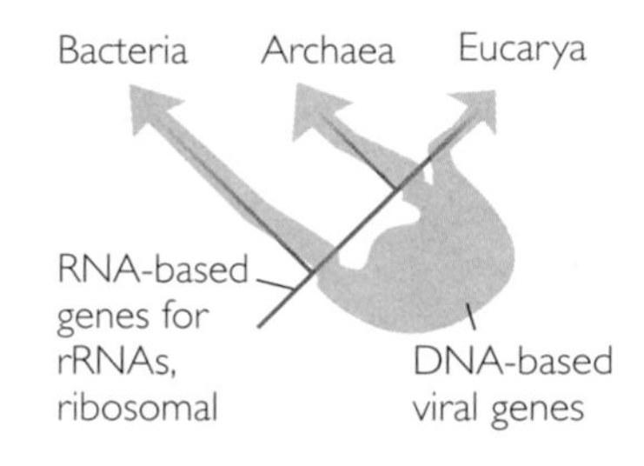

Figure 17.34 Three viruses, three domains According to this hypothesis, viruses infecting RNA-based cells first evolved DNA to counter their hosts' defenses. DNA was then transfered to cellular life when DNA-based viruses took up permanent residence inside their hosts.

Forterre asserts that his hypothesis also explains why there are only three domains of life. Once three lineages of DNA cells had evolved, they outcompeted and eliminated all other lineages of RNA-based cells. If, on the other hand, new domains of life can be generated by one of the endosymbiosis or fusion mechanisms discussed above, they should be appearing all the time.

Forterre notes that the best way to test his hypothesis would be to infect an RNA-based cell with a DNA-based virus and see if the host's descendants are ultimately transformed into DNA-based cells. Unfortunately, there are no known cellular organisms with RNA genomes. Forterre suggests, however, that it might be possible to use genetic engineering to make an RNA plasmid from the genome of an RNA virus, then insert it into a host cell whose genome encodes reverse transcriptase. If his hypothesis is correct, then RNA-based genes from the plasmid ought to turn up as DNA-based genes in the host cell's genome.

Researchers have proposed many hypotheses on how the three domains of life emerged. The most productive hypotheses make specific testable predictions.

Another way to test the three viruses, three domains hypothesis is by reconstructing phylogenies of genes involved in managing DNA-based genomes. Forterre's hypothesis predicts that in such phylogenies, genes from the cellular domains of life will be derived from, and thus nested within, genes from viruses.

Figure 17.35 shows a phylogeny of DNA-dependent RNA polymerases from a sample of bacteria, archaeans, and eukaryotes, plus a variety of viruses. The tree

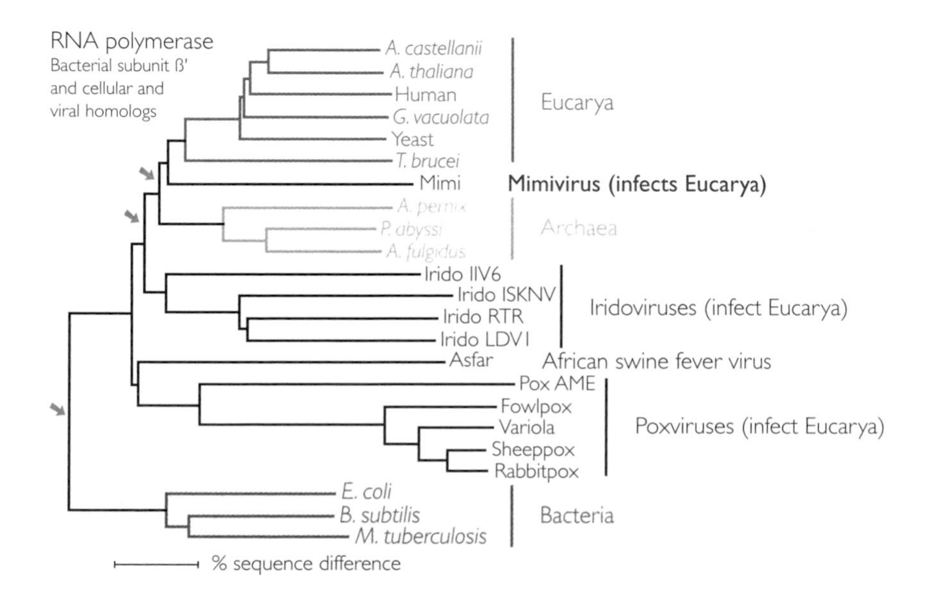

Figure 17.35 A phylogeny of cellular and viral DNA-dependent RNA polymerases Note that the cellular RNA polymerases are interspersed with the viral RNA polymerases. The tree is consistent with the hypothesis that the cellular genes are derived from viral genes. It is also consistent with the hypothesis that the viral genes are derived from cellular genes. The gray arrows mark some of the plausible roots. Regardless of whether we assume the root is cellular or viral, the tree requires a minimum of three transitions. Redrawn from supplement to Raoult et al. (2004).

was prepared by Didier Raoult and colleagues (2004) as a supplement to their report on the complete genome of Mimivirus. Mimivirus, which infects amoebae, is the largest virus yet discovered. Its genome is nearly 1.2 million base-pairs long and appears to encode well over 1,000 genes. Mimivirus is, in fact, a virus. It lacks a ribozome and thus cannot reproduce without infecting a host cell. But it is more genetically complex than many bacteria.

The phylogeny is consistent with Forterre's hypothesis. The clades containing the three domains are interspersed with the viral clades. Mimivirus branches between the Eucarya and the Achaea. The iridoviruses, African swine fever virus, and the poxviruses branch in between the bacteria and the other two cellular domains. One explanation for this pattern is that cellular RNA polymerases emerged, three times independently, from viral RNA polymerases.

Unfortunately, as Forterre (2005) points out, this is not the only possible explanation. The tree is also consistent with viral polymerases emerging three times independently from cellular polymerases. Note, for example, that among the places we could root the tree are the three gray arrows in the figure. Whether we assume the ancester at root was viral or cellular, the tree requires a minimum of three transitions.

Broader surveys of viral genomes might turn up genes that would allow us to reconstruct more extensive trees. These, in turn, might allow us to determine whether cellular genes evolved from viral ones or vice versa. In the meantime, the three viruses, three domains hypothesis will remain controversial (Whitfield 2006; Zimmer 2006)—just like the other hypotheses we have reviewed in this section.

Summary

Life arose from an abiotic environment a bit less than 4 billion years ago. As a consequence of its extreme antiquity, the reconstruction of this event poses many challenges. Life may have begun only once and spread quickly over Earth. It may have arisen several times, each time only to be extinguished by the vaporization of Earth's water by the impact of meteorites. It may have evolved entirely on Earth or had its origins elsewhere in the solar system.

Scientists have broken down the origins of life, regardless of its particulars, into three phases. The first phase would have been the synthesis of the building blocks of life, such as amino acids, nucleotides, and simple carbohydrates, from small inorganic molecules. Many plausible scenarios for these reactions exist, but significant uncertainties remain. The second phase would be the assembly of building blocks into a polymer, such as RNA, that contains and transmits information. Again, researchers have demonstrated that many of the details of such polymerization may be possible. And the third phase would be the advent of cellular compartmentalization, which would allow significant advances in phenotypic evolution and lead to the community of cells from which all current life is descended—the last universal common ancestors.

The study of life's origins is a highly collaborative venture, drawing on expertise from such diverse fields as astronomy, geology, chemistry, molecular biology, and evolutionary biology. It has forced us to consider exactly what "life" means. It is an excellent example of how science works, by formulating and testing hypotheses. It also reveals how great progress can be made in the absence, at this time, of a general-consensus viewpoint. Notably, the editors of *Chemical and Engineering News* (December 6, 1999) asked prominent chemists what will be the major scientific questions for the next hundred years. Three responded that the origins of life would be one of the major topics of study. Rita R. Colwell, director of the National Science Foundation, remarked, "Chemists also will develop self-replicating molecular systems to provide insights into the molecular origins of life." This would be a milestone achievement, and yet it would merely be another piece in an elusive puzzle.

Evolutionary biologists attempt to assemble the big picture of evolution since the last common ancestors by reconstructing the universal phylogeny based on genomic sequence data. Sequence-based universal phylogenies have forced a dramatic revision of the fundamental organization of living things. Instead of five kingdoms of life, there are three domains: Bacteria, Archaea, and Eucarya. The first universal phylogeny indicated that among the three domains Archaea and Eucarya are closest relatives. Comparison of phylogenies based on a variety of genes reveals, however, that there has been considerable horizontal gene transfer. Horizontal gene transfer may have been so rampant during life's history that it will force us to give up the idea of a single tree of life, and with it the idea of a single last common ancestor of all extant organisms.

How three domains of life emerged from the community of gene-exchanging organisms that now appears to have been the last universal common ancestor is the subject of much speculation and little consensus. As Russell Doolittle (2000) put it, "Vast amounts of sequence data notwithstanding, there are many things about early life on Earth that are not yet known."

Questions

1. The genesis of life is sometimes said to have required four things: energy, concentration, protection, and catalysis (for example, Cowen 1995). Explain why each of these four things was necessary for the generation of the primordial form.
2. What clues in the way RNA is used in modern cells hint that RNA may have an ancient role in cellular metabolism?
3. Briefly summarize two studies on evolution of RNA populations in the lab. In each experiment, what ability(s) did the RNA population develop during evolution (i.e., what was the change in phenotype)? Do you think these RNA populations qualify as "life"? Do you think that a self-replicating RNA population will be developed in the lab in your lifetime?
4. Why was the gene for small-subunit RNA particularly well suited for studies of the phylogeny of all living things? Do you think this gene is also useful for studying relationships among living mammals, such as for elucidating the family tree of humans, chimpanzees, and gorillas? Why or why not?
5. Consider the classic five-kingdom model of life:

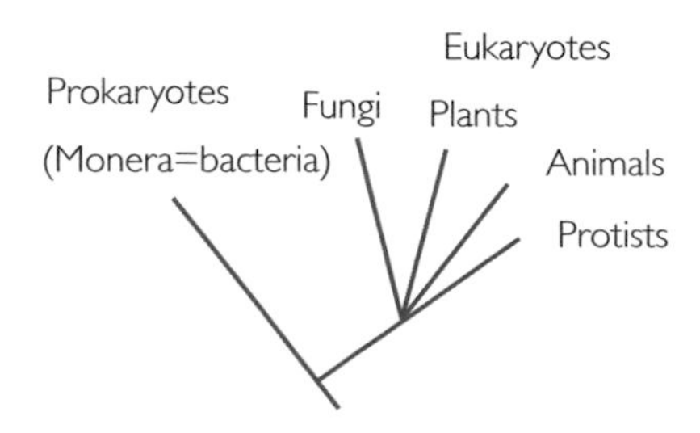

According to the "tree of life" first described by small-subunit rRNA analysis, which of the original "five kingdoms" need to be revised and which are still valid? Has this new "tree of life" stood the test of time, as other genes have been studied?

6. Briefly outline four possible hypotheses for the emergence of the three domains of life. Which is best supported (at present) from the evidence? Which is your favorite hypothesis (this need not be the one you think is most likely to be true!), and why?
7. It has been said that life could develop on Earth only because Earth is just the right distance from the sun. Any closer, and Earth would have been too hot (like Mercury or Venus); any farther away, there would not have been sufficient solar energy for the evolution of living things. Recently, communities of organisms have been found in deep-sea vents on Earth. These communities seem to get all of their energy from the vents rather than the sun. That is, the vent communities derive energy from the inner heat of Earth (which is provided ultimately by radioactivity). Even more recently, communities of bacteria have been found deep in subsurface rock. How does this discovery inform consideration of whether life might exist on other planets or moons that are not at "the right distance" from the sun?
8. The SETI project (Search for Extraterrestrial Intelligence) is a research program that searches for intelligent life on other planets, using the assumptions that (a) intelligent life has probably evolved elsewhere in the universe, and (b) it should be detectable by scanning regions of the sky for anomalous radio signals. One of

the fundamental uncertainties in this endeavor is the probability that any life at all will evolve on a planet, and if so, whether it will develop a civilization that uses radio waves. On Earth, how soon after Earth became habitable did life appear? How long did it take until eukaryotes appeared? How long until intelligent life appeared? In your opinion, do the answers indicate that the evolution of life (of any kind) on other Earth-like planets is probable or improbable? How about the evolution of intelligent life? How about advanced civilization and radio communications?

9. Leslie Orgel admitted to John Horgan (1991) that he and Crick intended their directed panspermia hypothesis as "sort of a joke." In their 1973 paper, however, Crick and Orgel treat the idea seriously enough to consider biological patterns that might serve as evidence. They point out, for example, that it is a little surprising that organisms with somewhat different [genetic] codes do not exist. The universality of the code follows naturally from an "infective" theory of the origins of life. Life on Earth would represent a clone derived from a single extraterrestrial organism.

Since 1973, biologists have discovered that the genetic code is not universal and that organisms with "somewhat different codes" do, in fact, exist. Our mitochondria, for example, use a code slightly different from that used by our nuclei (see Knight et al. 1999). Many ciliates and other organisms also have slightly deviant codes (see Osawa et al. 1992). How strongly does the discovery that the genetic code is not universal refute the directed panspermia hypothesis? How strongly does it refute other versions of panspermia? Explain your reasoning. Can you think of other kinds of evidence that could (or do) either support or refute some version of panspermia?

10. Examine closely Figure 17.5. Recall that mutations were introduced at 140 randomized nucleotide positions. By the ninth round of selection, 4 nucleotides were responsible for most of the evolutionary change. Look at the other 136 nucleotides. Many of these have reverted to their original state. Why?

11. In the experiment diagrammed in Figure 17.6, why was it important for the researchers to include a tag on the end of the substrate RNAs?

12. In the chain of events leading from the abiotic synthesis of biological building blocks to the evolution of eukaryotes (Figure 17.12), which transition appears to be the least characterized? Why do you think this is the case?

13. Imagine an extremely primitive organism that has very primitive ribosomes with no proteins. Would it be possible to place this organism on the "tree of life" shown in Figure 17.18? Why or why not? How about an organism with no ribosomes? (Can you think of such an organism?) Is it conceivable that there are some as-yet-undiscovered primitive organisms that cannot be placed on these phylogenies? How would the discovery of such organisms affect our reconstruction of the cenancestor?

14. When biologists worked out the details of DNA replication in bacteria and eukaryotes, many researchers were surprised to discover that there are several different DNA polymerases, each with a different role. The machinery for replication seemed enormously complex, and every piece seemed essential if the whole system was to function at all. Many people found it hard to imagine how such a complex system of interdependent parts could have evolved by natural selection. Does the discovery of organisms with only one DNA polymerase (such as *Methanococcus jannaschii*) offer new insight into the evolution of replication? Why or why not?

15. Suppose you are trekking through remote Greenland on a day off from your summer job at a scientific camp and you find an unusual layer of sedimentary rock that is not mapped on your geological charts. You suspect this rock might be even older than the 3.7-billion-year-old rocks from Isua (Figure 17.14). What would you do to determine whether these rocks have any evidence of ancient life? What results would show that life was indeed present before 3.7 billion years ago?

16. A recurring theme in literature of the past two centuries is that scientists should not try to "play God by trying to create life in the lab." Until recently, this phrase was just an unrealistic exaggeration used to make a point. Now, however, it appears that some scientists may be getting close to doing exactly that, by evolving self-replicating entities from abiotic molecules. Generally speaking, do you think these projects are worthwhile? What have they taught us about life and how it appeared on Earth?

17. **a.** A common objection to genetically modified food, and genetic engineering in general, is that it is "not natural for genes to cross the species barrier." Comment on whether this argument is logically sound.

b. Whether or not it is "natural" for genes to cross species barriers, many people have additional worries about genetically modified food. One such concern is the possibility that the genetically modified organisms might escape into the natural environment, where they could, conceivably, compete with other organisms and cause problems. Is this also a concern for research projects aimed at evolving self-replicating RNA populations? Which are more likely to survive if they escape into the natural environment: genetically engineered modern organisms or self-replicating RNA populations? Why?

c. Do you think either of these research endeavors is unwise in any way? Why?

Exploring the Literature

18. For a thorough review of recent discoveries on the role of RNA in the origin of life, see this recent compilation of papers on the subject:

Gesteland, R. F., T. R. Cech, and J. F. Atkins (eds.). 2006. *The Nature of Modern RNA Suggests a Prebiotic RNA World.* Cold Spring Harbor Monograph Series: 43.

And see this paper for an interesting computer model of what might happen once an RNA population manages to achieve self-replication:

Kuhn, C. 2005. A computer-glimpse of the origin of life. *Journal of Biological Physics* 31: 571–585

19. During their first 2 billion years on Earth, organisms wrought dramatic changes in the chemical composition of the atmosphere and oceans. For an introduction, see:

Schopf, J. W. 1992. The oldest fossils and what they mean. In J. W. Schopf, ed. *Major Events in the History of Life.* Boston: Jones and Bartlett Publishers, 29–63.

20. Periodic meteor impacts were not the only hazard faced by early life on Earth. The Sun was much less luminous, which might have resulted in the entire Earth being frozen solid during the period that life apparently first appeared (this is known as the Faint Young Sun Paradox). Yet at the same time, the Sun may also have produced more intense UV radiation, with UV doses at sea level on Earth more than 400 times as intense as today. See these papers for some ideas about these solar effects on the origin of life:

Karam, P. A. 2003. Inconstant sun: How solar evolution has affected cosmic and ultraviolet radiation exposure over the history of life on Earth. *Health Physics* 84: 322–333.

Pavlov, A. A., O. B. Toon, and T. Feng. 2006. Methane runaway in the early atmosphere—Two stable climate states of the Archean? *Astrobiology* 6:161.

Bada, J. L., C. Bigham, and S. L. Miller. 1994. Impact melting of frozen oceans on the early earth: Implications for the origin of life. *Proceedings of the National Academy of Sciences USA* 91: 1248–2350.

21. The conditions present in our solar system that allowed the evolution of life might be rare in the universe at large. For one thing, our solar system lies in a region of the galaxy that experiences relatively little bombardment from solid debris and a minimum of ionizing radiation. For another, our solar system's arrangement of inner small, rocky planets surrounded by outer gas-giant planets appears to be crucial. The inner rocky planets can receive enough solar radiation to support life, while the outer gas giants, though inhospitable to life themselves, tend to protect the inner planets from bombardment from loose material. (The reason for the cessation of cosmic bombardment after 3.5 billion years on Earth is thought to be that Jupiter eventually attracted most of the loose material in the solar system; it continues to protect us today.) Does this mean that life is unlikely to have evolved elsewhere? See the following book by a paleontologist and an astronomer for an interesting discussion of recent research on this issue.

Ward, P., and D. Brownlee. 2003. *Rare Earth: Why Complex Life Is Uncommon in the Universe.* New York: Springer Publishing Co.

22. The possibility that life exists on Mars has engendered a great debate, and all evidence for or against is heavily scrutinized. The argument in favor is clearly presented in

Gibson, E. K., Jr., et al. 1997. The case for relic life on Mars. *Scientific American* 277: 58–65.

Recent discoveries of bacteria and archaea on Earth that live deep in subsurface rock and in deep ice sheets have broadened our conceptions of what form life on Mars might take. See the February 2006 issue of the journal *Astrobiology* (vol. 6, issue 1) for a sampling of research on this topic (abstracts from the 2006 Astrobiology Science Conference).

23. Currently, scientists are in the process of seeking and judging the utility of several "biomarkers" that are chemical traces that life exists or existed in a particular environment or in a fossil. Some biomarkers may even indicate the presence of particular taxa. For a discussion of the use of biomarkers to date eukaryotic fossils to at least 2.7 billion years ago, see:

Brocks, J. J. et al. 1999. Archean molecular fossils and the early rise of eukaryotes. *Science* 285: 1033–1036.

J. J. Brocks, G. D. Love, R. E. Summons, A. H. Knoll, G. A. Logan, and S. A. Bowden. 2005. Biomarker evidence for green and purple sulphur bacteria in a stratified Palaeoproterozoic sea. *Nature* 437: 866–870.

In contrast, other teams continue to look for physical fossils that represent the shapes of extinct organisms. Zhu Shixing and Chen Huineng reported 1.7-billion-year-old fossils that they interpreted as representing multicellular algae with differentiated tissues. See:

Zhu S. and Chen H. 1995. Megascopic multicellular organisms from the 1700-million-year-old Tuanshanzi Formation in the Jixian area, North China. *Science* 270: 620–622.

Examine Zhu and Chen's photographs. Which do you find more convincing, the evidence from the biomarkers, or that of the fossils?

24. The organisms on the deepest eukaryotic branches in Figure 17.21 were long thought to lack mitochondria. For a review of evidence suggesting that this belief is mistaken, see:

Palmer, J. D. 1997. Organelle genomes: Going, going, gone! *Science* 275: 790–791.

25. For a startling example in which a human parasite appears to have evolved from an ancestor that could photosynthesize, see:

Hannaert, V., E. Saavedra, et al. 2003. Plantlike traits associated with metabolism of *Trypanosoma* parasites. *Proceedings of the National Academy of Sciences USA* 100: 1067–1071.

Citations

Altermann, W., and J. W. Schopf. 1995. Microfossils from the Neoarchean Campbell Group, Griqualand West Sequence of the Transvaal Supergroup, and their paleoenvironmental and evolutionary implications. *Precambrian Research* 75: 65–90.

Anders, E. 1989. Prebiotic organic matter from comets and asteroids. *Nature* 342: 255–257.

Anders, E., C. K. Shearer, J. J. Papike, J. F. Bell, S. J. Clemett, R. N. Zare, D. S. McKay, K. L. Thomas-Keprta, C. S. Romanek, E. K. Gibson, Jr., and H. Vali. 1996. Evaluating the evidence for past life on Mars. *Science* 274: 2119–2125.

Antonelli, P. L., L. Bevilacqua, and S. F. Rutz. 2003. Theories and models in symbiogenesis. *Nonlinear Analysis: Real World Applications* 4: 743–753.

Arrhenius, S. 1908. *Worlds in the Making*. New York: Harper and Row.

Badash, L. 1989. The age-of-the-Earth debate. *Scientific American* 261: 90–96.

Baldauf, S. L., J. D. Palmer, and W. F. Doolittle. 1996. The root of the universal tree and the origin of eukaryotes based on elongation factor phylogeny. *Proceedings of the National Academy of Sciences USA* 93: 7749–7754.

Bapteste, E., E. Susko, et al. 2005. Do orthologous gene phylogenies really support tree-thinking? *BMC Evolutionary Biology* 5: 33.

Barns, S. M., C. F. Delwiche, J. D. Palmer, and N. R. Pace. 1996. Perspectives on archaeal diversity, thermophily and monophyly from environmental rRNA sequences. *Proceedings of the National Academy of Sciences USA* 93: 9188–9193.

Barns, S. M., R. E. Fundyga, M. W. Jeffries, and N. R. Pace. 1994. Remarkable archaeal diversity detected in a Yellowstone National Park hot spring environment. *Proceedings of the National Academy of Sciences USA* 91: 1609–1613.

Bartel, D. P. 1999. Recreating an RNA replicase. In R. F. Gesteland, T. R. Cech, and J. F. Atkins, eds. *The RNA World*, 2nd ed. Cold Spring Harbor, NY: Cold Spring Harbor Laboratory Press, 143–162.

Bartel, D. P., and J. W. Szostak. 1993. Isolation of new ribozymes from a large pool of random sequences. *Science* 261: 1411–1418.

Bartel, D. P., and P. J. Unrau. 1999. Constructing an RNA World. *Trends in Cell Biology* 9: M9–M13.

Beaudry, A. A., and G. F. Joyce. 1992. Directed evolution of an RNA enzyme. *Science* 257: 635–641.

Belton, M. J. S., et al. 1996. Galileo's first images of Jupiter and the Galilean satellites. *Science* 274: 377–385.

Benner, S. A., A. D. Ellington, and A. Tauer. 1989. Modern metabolism as a palimpsest of the RNA world. *Proceedings of the National Academy of Sciences USA* 86: 7054–7058.

Brasier, M. D., O. R. Green, et al. 2002. Questioning the evidence for Earth's oldest fossils. *Nature* 416: 76–81.

Brasier, M., N. McLoughlin, et al. 2006. A fresh look at the fossil evidence for early Archaean cellular life. *Philosophical Transactions of the Royal Society B* 361: 887–902.

Brown, J. R. 2001. Genomic and phylogenetic perspectives on the evolution of prokaryotes. *Systematic Biology* 50: 497–512.

Brown, J. R. 2003. Ancient horizontal gene transfer. *Nature Reviews Genetics* 4: 121–132.

Brown, J. R., and W. F. Doolittle. 1995. Root of the universal tree of life based on ancient aminoacyl-tRNA synthetase gene duplications. *Proceedings of the National Academy of Sciences USA* 92: 2441–2445.

Brown, J. R., and W. F. Doolittle. 1997. *Archaea* and the prokaryote-to-eukaryote transition. *Microbiology and Molecular Biology Reviews* 61: 456–502.

Bult, C. J., et al. 1996. Complete genome sequence of the methanogenic archaeon, *Methanococcus jannaschii*. *Science* 273: 1058–1073.

Cairns-Smith, A. G., A. J. Hall, and M. J. Russell. 1992. Mineral theories of the origin of life and an iron sulfide example. *Origins of Life and Evolution of the Biosphere* 22: 161–180.

Canback, B, S. G. E. Andersson, and C. G. Kurland. 2002. The global phylogeny of glycolytic enzymes. *Preceedings of the National Academy of Sciences USA* 99: 6097–6102.

Chang, S. 1999. Planetary environments and the origin of life. *Biological Bulletin* 196: 308–310.

Chyba, C. F. 2000. Energy for microbial life on Europa. *Nature* 403: 381–382.

Chyba, C. F. 2005. Rethinking Earth's early atmosphere. *Science* 308: 962–963.

Chyba, C. F., P. J. Thomas, L. Brookshaw, and C. Sagan. 1990. Cometary delivery of organic molecules to the early Earth. *Science* 249: 366–373.

Ciccarelli, F. D., T. Doerks, et al. 2006. Toward automatic reconstruction of a highly resolved tree of life. *Science* 311: 1283–1287.

Cowen, R. 1995. *History of Life*, 2nd ed. Cambridge: Blackwell Scientific Publications.

Creevey, C. J., D. A. Fitzpatrick, et al. 2004. Does a tree-like phylogeny only exist at the tips in the prokaryotes? *Proceedings of the Royal Society of London B* 271: 2551–2558.

Crick, F. H. C. 1966. The genetic code—yesterday, today, and tomorrow. *Cold Spring Harbor Symposia on Quantitative Biology* 31: 1–9.

Crick, F. H. C., and L. E. Orgel. 1973. Directed panspermia. *Icarus* 19: 341–346.

Cruikshank, D. P. 1997. Stardust memories. *Science* 275: 1895–1896.

Dalton, R. 2002. Squaring up over ancient life. *Nature* 417: 782–784.

Darwin, C. 1859. *On the Origin of Species by Means of Natural Selection*. London: John Murray.

Darwin, F. 1887. *The Life and Letters of Charles Darwin*, vol. 2. New York: Appleton.

DeDuve, C. 1991. *Blueprint for a Cell: The Nature and Origin of Life*. Burlington, NC: Patterson.

DeLong, E. F. 1992. Archaea in coastal marine environments. *Proceedings of the National Academy of Sciences USA* 89: 5685–5689.

DeLong, E. F., K. Y. Wu, B. B. Prézelin, and R. V. M. Jovine. 1994. High abundance of Archaea in Antarctic marine picoplankton. *Nature* 371: 695–697.

Delsuc, F., H. Brinkman, and H. Philippe. 2005. Phylogenomics and the reconstruction of the tree of life. *Nature Reviews Genetics* 6: 361–375.

Doolittle, W. F. 1999. Phylogenetic classification and the universal tree. *Science* 284: 2124–2128.

Doolittle, W. F. 2000. Uprooting the tree of life. *Scientific American* (February): 90–95.

Doolittle, R. F. 2000. Searching for the common ancestor. *Research in Microbiology* 151: 85–89.

Doolittle, R. F. 2005. Evolutionary aspects of whole-genome biology. *Current Opinion in Structural Biology* 15: 248–253.

Doolittle, W. F., and J. R. Brown. 1994. Tempo, mode, the progenote, and the universal root. *Proceedings of the National Academy of Sciences USA* 91: 6721–6728.

Doolittle, R. F., D.-F. Feng, S. Tsang, G. Cho, and E. Little. 1996. Determining divergence times of the major kingdoms of living organisms with a protein clock. *Science* 271: 470–477.

Dose, K., and A. Klein. 1996. Response of Bacillus subtilis spores to dehydration and UV irradiation at extremely low temperatures. *Origins of Life and Evolution of the Biosphere* 26: 47–59.

Egholm, M., O. Buchardt, L. Christensen, P. E. Nielson, and R. H. Berg. 1992. Peptide nucleic acids (PNA). Oligonucleotide analogues with an achiral peptide backbone. *Journal of the American Chemical Society* 114: 1895–1897.

Ekland, E. H., J. W. Szostak, and D. P. Bartel. 1995. Structurally complex and highly active RNA ligases derived from random RNA sequences. *Science* 269: 364–370.

Ellington, A. D., and J. W. Szoztak. 1990. *In vitro* selection of RNA molecules that bind specific ligands. *Nature* 346: 818–822.

Eschenmoser A. 1999. Chemical etiology of nucleic acid structure. *Science* 284: 2118–2124.

Esser, C., N. Ahmadinejad, et al. 2004. A genome phylogeny for mitochondria among α-proteobacteria and a predominantly eubacterial ancestry of yeast nuclear genes. *Molecular Biology and Evolution* 21: 1643–1660.

Fedo, C. M., and M. J. Whitehouse. 2002a. Quartz-pyroxene rock, Akilia, Greenland, and implications for Earth's earliest life. *Science* 296: 1448–1452.

Fedo, C. M., and M. J. Whitehouse. 2002b. Origin and significance of archean auartzose rocks at Akilia, Greenland. *Science* 298: 917a.

Fedo, C. M., M. J. Whitehouse, and B. S. Kamber. 2006. Geological constraints on detecting the earliest life on Earth: a perspective from the Early Archaean (older than 3.7 Gyr) of southwest Greenland. *Philosophical Transactions of the Royal Society B* 361: 851–867.

Ferris, J. P. 1993. Catalysis and prebiotic RNA synthesis. *Origins of Life and Evolution of the Biosphere* 23: 307–315.

Ferris, J. P., A. R. Hill, Jr., R. Liu, and L. E. Orgel. 1996. Synthesis of long prebiotic oligomers on mineral surfaces. *Nature* 381: 59–61.

Fitch, W. M., and K. Upper. 1987. The phylogeny of tRNA sequences provides evidence for ambiguity reduction in the origin of the genetic code. *Cold Spring Harbor Symposia on Quantitative Biology* 52: 759–767.

Forterre, P. 2005. The two ages of the RNA world, and the transition to the DNA world: a story of viruses and cells. *Biochimie* 87: 793–803.

Forterre, P. 2006a. The origin of viruses and their possible roles in major evolutionary transitions. *Virus Research* 117: 5–16.

Forterre, P. 2006b. Three RNA cells for ribosomal lineages and three DNA viruses to replicate their genomes: A hypothesis for the origin of cellular domain. *Proceedings of the National Academy of Sciences USA* 103: 3669–3674.

Forterre, P., and H. Philippe. 1999. The last universal common ancestor (LUCA), simple or complex? *Biological Bulletin* 196: 373–377.

Fox, S. W. 1988. *The Emergence of Life: Darwinian Evolution from the Inside.* New York: Basic Books.

Fox, S. W. 1991. Synthesis of life in the lab? Defining a protoliving system. *Quarterly Review of Biology* 66: 181–185.

Fox, S. W., and K. Dose. 1972. *Molecular Evolution and the Origin of Life.* San Francisco: W. H. Freeman.

Fox, G. E., L. J. Magrum, W. E. Balch, R. S. Wolfe, and C. R. Woese. 1977. Classification of methanogenic bacteria by 16S ribosomal RNA characterization. *Proceedings of the National Academy of Sciences USA* 74: 4537–4541.

Friend, C. R. L., A. P. Nutman, and V. C. Bennett. 2002. Origin and significance of archean auartzose rocks at Akilia, Greenland. *Science* 298: 917a.

Fry, I. 2006. The origins of research into the origins of life. *Endeavour* 30: 24–28.

Fusz, S., A. Eisenführ, et al. 2005. A ribozyme for the aldol reaction. *Chemistry & Biology* 12: 941–950.

Ge, F., L.-S. Wang, and J. Kim. 2005. The cobweb of life revealed by genome-scale estimates of horizontal gene transfer. *PLoS Biology* 3: e316.

Gesteland, R. F., T. R. Cech, and J. F. Atkins, eds. 1999. *The RNA World*, 2nd ed. Cold Spring Harbor, NY: Cold Spring Harbor Laboratory Press.

Geyer, C. R., and D. Sen. 1997. Evidence for the metal-cofactor independence of an RNA phosphodiester-cleaving DNA enzyme. *Chemistry & Biology* 4: 579–593.

Gilbert, W. 1986. The RNA world. *Nature* 319: 618.

Gladman, B. J., and J. A. Burns. 1996. Mars meteorite transfer: Simulation. *Science* 274: Letters.

Gladman, B. J., J. A. Burns, M. Duncan, P. Lee, and H. F. Levison. 1996. The exchange of impact ejecta between terrestrial planets. *Science* 271: 1387–1392.

Gogarten, J. P., et al. 1989. Evolution of the vacuolar H1-ATPase: Implications for the origin of eukaryotes. *Proceedings of the National Academy of Sciences USA* 86: 6661–6665.

Gogarten, J. P., W. F. Doolittle, and J. G. Lawrence. 2002. Prokaryotic evolution in light of gene transfer. *Molecular Biology and Evolution* 19: 2226–2238.

Gribaldo, S., and C. Brochier-Armanet. 2006. The origin and evolution of the Archaea: a state of the art. *Philosophical Transactions of the Royal Society B* 361: 1007–1022.

Hamilton, G. 2006. The gene weavers. *Nature* 441: 683–685.

Han, T.-M., and B. Runnegar. 1992. Megascopic eukaryotic algae from the 2.1-billion-year-old Negaunee Iron-Formation, Michigan. *Science* 257: 232–235.

Harris, J. K., S. T. Kelley, et al. 2003. The genetic core of the universal ancestor. *Genome Research* 13: 407–412.

Hartman, H., and A. Fedorov. 2002. The origin of the eukaryotic cell: A genomic investigation. *Proceedings of the National Academy of Sciences USA* 99: 1420–1425.

Hasegawa, M., W. M. Fitch, J. P. Gogarten, L. Olendzenski, E. Hilario, C. Simon, K. E. Holsinger, R. F. Doolittle, D.-F. Feng, S. Tsang, G. Cho, and E. Little. 1996. Dating the cenancestor of organisms. *Science* 274: 1750–1753.

Hill, A. R., Jr., C. Böhler, and L.E. Orgel. 1998. Polymerization on the rocks: Negatively-charged α-amino acids. *Origins of Life and Evolution of the Biosphere* 28: 235–243.

Horgan, J. 1991. In the beginning ... *Scientific American* 264(2): 116–125.

Horiike, T., K. Hamada, and T. Shinozawa. 2002. Origin of eukaryotic cell by symbiosis of archaea in bacteria supported by the newly clarified origin of functional genes. *Genes and Genetic Systems* 77: 369–376.

Horiike, T., K. Hamada, et al. 2004. The origin of eukaryotes is suggested as the symbiosis of *Pyrococcus* into γ-proteobacteria by phylogenetic tree based on gene content. *Journal of Molecular Evolution* 59: 606–619.

Huang, W., and J. P. Ferris. 2006. One-step, regioselective synthesis of up to 50-mers of RNA oligomers by montmorillonite catalysis. *Journal of the American Chemical Society* 128: 8914–8919.

Illangasekare, M., G. Sanchez, T. Nickles, and M. Yarus. 1995. Aminoacyl-RNA synthesis catalyzed by an RNA. *Science* 267: 643–647.

Iwabe, N., K.-i. Kuma, M. Hasegawa, S. Osawa, and T. Miyata. 1989. Evolutionary relationship of archaebacteria, eubacteria, and eukaryotes inferred from phylogenetic trees of duplicated genes. *Proceedings of the National Academy of Sciences USA* 86: 9355–9359.

Johnston, W. K., P. J. Unrau, et al. 2001. RNA-catalyzed RNA polymerization: Accurate and general RNA-templated primer extension. *Science* 292: 1319–1325.

Jones, W. J., J. A. Leigh, F. Mayer, C. R. Woese, and R. S. Wolfe. 1983. *Methanococcus jannaschii sp. nov.*, an extremely thermophilic methanogen from a submarine hydrothermal vent. *Archives of Microbiology* 136: 254–261.

Joyce, G. F. 1989. RNA evolution and the origins of life. *Nature* 338: 217–224.

Joyce, G. F. 1996. Ribozymes: Building the RNA world. *Current Biology* 6: 965–967.

Joyce, G. F. 1998. Nucleic acid enzymes: Playing with a fuller deck. *Proceedings of the National Academy of Sciences USA* 95: 5845–5847.

Joyce, G. F., and L. E. Orgel. 1999. Prospects for understanding the RNA world. In R. F. Gesteland, T. R. Cech, and J. F. Atkins, eds. *The RNA World,* 2nd ed. Cold Spring Harbor, NY: Cold Spring Harbor Laboratory Press, 49–77.

Joyce, G. F., A. Schwartz, S. L. Miller, and L. Orgel. 1987. The case for an ancestral genetic system involving simple analogues of the nucleotides. *Proceedings of the National Academy of Sciences USA* 84: 4398–4402.

Kasting, J. F. 1993. Earth's early atmosphere. *Science* 259: 920–926.

Kázmierczak, J., and B. Kremer. 2002. Thermal alteration of the Earth's oldest fossils. *Nature* 420: 477–478.

Kempe, A., Schopf, J. W., et al. 2002. Atomic force microscopy of Precambrian microscopic fossils. *Proceedings of the National Academy of Sciences USA* 99: 9117–9120.

Kerr, R. A. 1996. Galileo turns geology upside down on Jupiter's icy moons. *Science* 274: 341.

Kerr, R. A. 1997a. Martian "microbes" cover their tracks. *Science* 276: 30–31.

Kerr, R. A. 1997b. An ocean emerges on Europa. *Science* 276: 355.

Knight, R. D., S. J. Freeland, and L. F. Landweber. 1999. Selection, history, and chemistry: The three faces of the genetic code. *Trends in Biochemical Sciences* 24: 241–247.

Knoll, A. H. 1992. The early evolution of eukaryotes: A geological perspective. *Science* 256: 622–627.

Knoll, A. H. 1994. Proterozoic and Early Cambrian protists: Evidence for accelerating evolutionary tempo. *Proceedings of the National Academy of Sciences USA* 91: 6743–6750.

Knoll, A. H., and E. S. Barghoorn. 1977. Archean microfossils showing cell division from the Swaziland system of South Africa. *Science* 198: 396–398.

Koonin, E. V. 2003. Comparative genomics, minimal gene-sets, and the last universal common ancestor. *Nature Reviews Microbiology* 1: 127–136.

Kruger, K., Grabowski, P. J., Zaug, A. J., Sands, J., Gottschling, D. E., and T. R. Cech. 1982. Self-splicing RNA: Autoexcision and autocatalyization of the ribosomal RNA intervening sequence of *Tetrahymena*. *Cell* 31: 147–157.

Kurland, C. G., C. Canback, and O. G. Berg. 2003. Horizontal gene transfer: A critical review. *Proceedings of the National Academy of Sciences USA* 100: 9658–9662.

Kurland, C. G., L. J. Collins, and D. Penny. 2006. Genomes and the irreducible nature of Eukaryote cells. *Science* 312: 1011– 1014.

Kvenvolden, K., et al. 1970. Evidence for extraterrestrial amino acids and hydrocarbons in the Murchison meteorite. *Nature* 228: 923–926.

Lawrence, J. G., and H. Ochman. 1998. Molecular archaeology of the *Escherichia coli* genome. *Proceedings of the National Academy of Sciences USA* 95: 9413–9417.

Lawrence, M. S., and D. P. Bartel. 2005. New ligase-derived RNA polymerase ribozymes. *RNA* 11: 1173–1180.

Lazcano, A., and S. L. Miller. 1996. The origin and early evolution of life: Prebiotic chemistry, the pre-RNA world, and time. *Cell* 85: 793–798.

Lehman, N., and G. F. Joyce. 1993. Evolution *in vitro* of an RNA enzyme with altered metal dependence. *Nature* 361: 182–185.

Leipe, D. D., L. Aravind, and E. V. Koonin. 1999. Did DNA replication evolve twice independently? *Nucleic Acids Research* 27: 3389–3401.

Lepland, A., G. Arrhenius, and D. Cornell. 2002. Apatite in early Archean Isua supracrustal rocks, southern West Greenland: Its origin, association with graphite and potential as a biomarker. *Precambrian Research* 118: 221–241.

Lester, L., A. Meade, and M. Pagel. 2005. The slow road to the eukaryotic genome. *BioEssays* 28: 57–64.

Lorsch, J. R., and J. W. Szostak. 1994. *In vitro* evolution of new ribozymes with polynucleotide kinase activity. *Nature* 371: 31–36.

Madigan, M. T., and B. L. Marrs. 1997. Extremophiles. *Scientific American* 276(4): 82–87.

Margulis, L. 1996. Archaeal-eubacterial mergers in the origin of Eukarya: Phylogenetic classification of life. *Proceedings of the National Academy of Sciences USA* 93: 1071–1076.

Martin, W., and M. J. Russell. 2003. On the origins of cells: A hypothesis for the evolutionary transitions from abiotic geochemistry to chemoautotrophic prokaryotes, and from prokaryotes to nucleated cells. *Philosophical Transactions of the Royal Society of London B* 358: 59–85.

McKay, D. S., et al. 1996. Search for past life on Mars: Possible relic biogenic activity in Martian meteorite ALH84001. *Science* 273: 924–930.

McLean, R. J. C., A. K. Welch, and V. A. Casasanto. 2006. Microbial survival in space shuttle crash. *Icarus* 181: 323–325.

Miller, S. L. 1953. A production of amino acids under possible primitive Earth conditions. *Science* 117: 528–529.

Miller, S. L. 1992. The prebiotic synthesis of organic compounds as a step toward the origin of life. In J. W. Schopf, ed. *Major Events in the History of Life*. Boston: Jones and Bartlett, 1–28.

Mills, D. R, R. L. Peterson, and S. Spiegelman. 1967. An extracellular Darwinian experiment with a self-duplicating nucleic acid molecule. *Proceedings of the National Academy of Sciences USA* 58: 217–220.

Mitchell, F. J., and W. L. Ellis. 1972. *Surveyor 3*: Bacterium isolated from lunar-retrieved television camera. In *Analysis of Surveyor 3 Material and Photographs Returned by Apollo 12*. Washington, D.C.: National Aeronautics and Space Administration, 239–248.

Mojzsis, S. J., G. Arrhenius, K. D. McKeegan, T. M. Harrison, A. P. Nutman, and C. R. L. Friend. 1996. Evidence for life on Earth before 3,800 million years ago. *Nature* 384: 55–59.

Mojzsis, S. J., and T. M. Harrison. 2002. Origin and significance of archean auartzose rocks at Akilia, Greenland. *Science* 298: 917a.

Mojzsis, S. J., R. Krishnamurthy, and G. Arrhenius. 1999. Before RNA and after: Geophysical and geochemical constraints on molecular evolution. In R. F. Gesteland, T. R. Cech, and J. F. Atkins, eds. *The RNA World*, 2nd ed. Cold Spring Harbor, NY: Cold Spring Harbor Laboratory Press. 1–47.

Moorbath, S. 2005. Dating earliest life. *Nature* 434: 155.

Müller, U. F. 2006. Re-creating an RNA world. *Cellular and Molecular Life Sciences* 63: 1278–1293.

Murzin, A. G., S. E. Brenner, et al. 1995. SCOP: A structural classification of proteins database for the investigation of sequences and structures. *Journal of Molecular Biology* 247: 536–540.

National Center for Biotechnology Information. 2006. Genome sequencing projects statistics. *http://www.ncbi.nlm.nih.gov/genomes/static/gpstat.html*. Accessed 29 August 2006.

Nelson, K. E., R. A. Clayton, et al. 1999. Evidence for lateral gene transfer between Archaea and Bacteria from genome sequence of *Thermotoga maritima*. *Nature* 399: 323–329.

Nissen, P., J. Hansen, et al. 2000. The structural basis of ribosome activity in peptide bond synthesis. *Science* 289: 920–930.

Olsen, G. J., and C. R. Woese. 1996. Lessons from an Archaeal genome: What are we learning from *Methanococcus jannaschii*? *Trends in Genetics* 12: 377–379.

Orgel, L. E. 1986. RNA catalysis and the origins of life. *Journal of Theoretical Biology* 123: 127–149.

Orgel, L. 2000. A simpler nucleic acid. *Science* 290: 1306–1307.

Oró, J. 1961. Mechanism of synthesis of adenine from hydrogen cyanide under plausible primitive earth conditions. *Nature* 191: 1193–1194.

Osawa, S., T. H. Jukes, K. Watanabe, and A. Muto. 1992. Recent evidence for evolution of the genetic code. *Microbiological Reviews* 56: 229–264.

Ouzounis, C. A., V. Kunin, et al. 2006. A minimal estimate for the gene content of the last universal common ancestor—exobiology from a terrestrial perspective. *Research in Microbiology* 157: 57–68.

Pace, N. R. 2006. Time for a change. *Nature* 441: 289.

Pasteris, J. D., and B. Wopenka. 2002. Images of the Earth's oldest fossils? *Nature* 420: 476–477.

Raoult, D., S. Audic, et al. 2004. The 1.2-megabase genome sequence of Mimivirus. *Science* 306: 1344–1350.

Rivera, M. C., and J. A. Lake. 2004. The ring of life provides evidence for a genome fusion origin of eukaryotes. *Nature* 431: 152–155.

Rogers, J., and G. F. Joyce 1999. A ribozyme that lacks cytidine. *Nature* 402: 323–325.

Rosing, M. T. 1999. ^{13}C-depleted carbon microparticles in >3700-Ma sea-floor sedimentary rocks from West Greenland. *Science* 283: 674–676.

Rosing, M. T., and R. Frei. 2004. U-rich Archaean sea-floor sediments form Greenland—indications of >3700 Ma oxygenic photosynthesis. *Earth and Planetary Science Letters* 217: 237–244.

Schidlowski, M. 1988. A 3,800-million-year isotopic record of life from carbon in sedimentary rocks. *Nature* 333: 313–318.

Schleper, C., G. Jurgens, and M. Jonusscheit. 2005. Genomic studies of uncultivated Archaea. *Nature Reviews Microbiology* 3: 479–488.

Schöning, K.-U., P. Scholz, et al. 2000. Chemical etiology of nucleic acid structure: The α-threofuranosyl-(3' $\rightarrow$ 2') oligonucleotide system. *Science* 290: 1347–1351.

Schopf, J. W. 1993. Microfossils of the early Archean apex chert: New evidence of the antiquity of life. *Science* 260: 640–646.

Schopf, J. W. 1994a. Disparate rates, differing fates: Tempo and mode of evolution changed from the precambrian to the phanerozoic. *Proceedings of the National Academy of Sciences USA* 91: 6735–6742.

Schopf, J. W. 1994b. The early evolution of life: Solution to Darwin's dilemma. *Trends in Ecology and Evolution* 9: 375–378.

Schopf, J. W. 2006. Fossil evidence of Archaean life. *Philosophical Transactions of the Royal Society B* 361: 869–885.

Schopf, J. W., A. B. Kudryavtsev, et al. 2002a. Laser-Raman imagery of Earth's earliest fossils. *Nature* 416: 73–76.

Schopf, J. W., A. B. Kudryavtsev, et al. 2002b. Images of the Earth's oldest fossils? *Nature* 420: 477.

Secker, J., J. Lepock, and P. Wesson. 1994. Damage due to ultraviolet and ionizing radiation during the ejection of shielded micro-organisms from the vicinity of 1 M main sequence and red giant stars. *Astrophysics and Space Science* 219: 1–28.

Service, R. F. 1997. Microbiologists explore life's rich, hidden kingdoms. *Science* 275: 1740–1742.

Shapiro, R. 2006. Small molecule interactions were central to the origin of life. *The Quarterly Review of Biology* 81: 105–125.

Simonson, A. B., J. A. Servin, et al. 2005. Decoding the genomic tree of life. *Proceedings of the National Academy of Sciences USA* 102: 6608–6613.

Sleep, N. H., K. J. Zahnle, J. F. Kasting, and H. J. Morowitz. 1989. Annihilation of ecosystems by large asteroid impacts on the early Earth. *Nature* 342: 139–142.

Stahl, D. A., D. J. Lane, G. J. Olsen, and N. R. Pace. 1984. Analysis of hydrothermal vent-associated symbionts by ribosomal RNA sequences. *Science* 224: 409–411.

Steitz, T. A., and P. B. Moore. 2003. RNA, the first macromolecular catalyst: the ribosome is a ribozyme. *Trends in Biochemical Sciences* 28: 411–418.

Strobel, S. A. 2001. Repopulating the RNA world. *Nature* 411: 1003–1006.

Tang, J., and R. R. Breaker. 1997. Rational design of allosteric ribozymes. *Chemistry and Biology* 4: 453–459.

Tarasow, T. M., S. L. Tarasow, and B. E. Eaton. 1997. RNA-catalyzed carbon–carbon bond formation. *Nature* 389: 54–57.

Tian, F., O. B. Toon, et al. 2005. A hydrogen-rich early atmosphere. *Science* 308: 1014–1017.

Tuerk, C., and L. Gold. 1990. Systematic evolution of ligands by exponential enrichment: RNA ligands to bacteriophage T4 DNA polymerase. *Science* 249: 505–510.

Unrau, P. J., and D. P. Bartel. 1998. RNA-catalyzed nucleotide synthesis. *Nature* 395: 260–263.

van Zuilen, M. A., A. Lepland, and G. Arrhenius. 2002. Reassessing the evidence for the earliest traces of life. *Nature* 418: 627–630.

Vestigian, K., C. Woese, and N. Goldenfeld. 2006. Collective evolution of the genetic code. *Proceedings of the National Academy of Sciences USA* 103: 10696–10701.

Voet, A. B., and A. W. Schwartz. 1982. Uracil synthesis via HCN oligomerization. *Origins of Life* 12: 45–49.

Ward, P. D., and D. Brownlee. 2000. *Rare Earth: Why Complex Life Is Uncommon in the Universe.* New York: Copernicus Books.

Weber, P., and J. M. Greenberg. 1985. Can spores survive in interstellar space? *Nature* 316: 403–407.

Weiss, B. P., S. S. Kim, et al. 2004. Magnetic tests for magnetosome chains in Martian meteorite ALH84001. *Proceedings of the National Academy of Sciences USA* 101: 8281–8284.

Westall, F., M. J. de Wit, et al. 2001. Early Archean fossil bacteria and biofilms in hydrothermally-influenced sediments from the Barberton greenstone belt, South Africa. *Precambrian Research* 106: 93–116.

Wetherill, G. W. 1990. Formation of the Earth. *Annual Review of Earth and Planetary Science* 18: 205–256.

White, H. B. 1976. Coenzymes as fossils of an earlier metabolic state. *Journal of Molecular Evolution* 7: 101–104.

Whitfield, J. 2004. Born in a watery commune. *Nature* 427: 674–676.

Whitfield, J. 2006. Base invaders. *Nature* 439: 130–131.

Whittaker, R. H. 1969. New concepts of kingdoms of organisms. *Science* 163: 150–160.

Woese, C. R. 1991. The use of ribosomal RNA in reconstructing evolutionary relationships among bacteria. In R. K. Selander, A. G. Clark, and T. S. Whittam, eds. *Evolution at the Molecular Level.* Sunderland, MA: Sinauer, 1–24.

Woese, C.R. 1996. Phylogenetic trees: Whither microbiology? *Current Biology* 6: 1060–1063.

Woese, C. R. 1998. The universal ancestor. *Proceedings of the National Academy of Sciences USA* 95: 6854–6859.

Woese, C. R. 2000. Interpreting the universal phylogenetic tree. *Proceedings of the National Academy of Sciences USA* 97: 8392–8396.

Woese, C. R. 2002. On the evolution of cells. *Proceedings of the National Academy of Sciences USA* 99: 8742–8747.

Woese, C. R. 2004. A new biology for a new century. *Microbiology and Molecular Biology Reviews* 68: 173–186.

Woese, C. R., and G. E. Fox. 1977. Phylogenetic structure of the prokaryotic domain: The primary kingdoms. *Proceedings of the National Academy of Sciences USA* 74: 5088–5090.

Woese, C. R., O. Kandler, and M. L. Wheelis. 1990. Towards a natural system of organisms: Proposal for the domains Archaea, Bacteria, and Eucarya. *Proceedings of the National Academy of Sciences USA* 87: 4576–4579.

Yang, S., R. F. Doolittle, and P. E. Bourne. 2005. Phylogeny determined by protein domain content. *Proceedings of the National Academy of Sciences USA* 102: 373–378.

Zaug, A. J., and T. R. Cech. 1986. The intervening sequence RNA of *Tetrahymena* is an enzyme. *Science* 231: 470–475.

Zhang, B., and T. R. Cech. 1997. Peptide bond formation by *in vitro* selected ribozymes. *Nature* 390: 96–100.

Zhang, L., A. Peritz, and E. Meggers. 2005. A simple glycol nucleic acid. *Journal of the American Chemical Society* 127: 4174–4175.

Zhaxybayeva, O., P. Lapierre, and J. P. Gogarten. 2005. Ancient gene duplications and the root(s) of the tree of life. *Protoplasma* 227: 53–64.

Zimmer, C. 2006. Did DNA come from viruses? *Science* 312: 870–872.

Zuckerkandl, E., and L. Pauling. 1965. Molecules as documents of evolutionary history. *Journal of Theoretical Biology* 8: 357–366.

18

The Cambrian Explosion and Beyond

Once the fundamental life processes of DNA replication, protein synthesis, respiration, and cell division had evolved, a spectacular diversification of life ensued. Innovations like photosynthesis and the nuclear envelope evolved. These events, along with others reviewed in Chapter 17, spanned some 3.2 billion years and created the deep branches on the tree of life. During this interval all organisms, with the exception of some small red, brown, and green algae, were unicellular.

The first animals appear in the fossil record 565 million years ago (mya). The jellyfish and sponges in these rocks are multicellular, but small in size and morphologically simple. Similar organisms are found in deposits created over the subsequent 20 million years. But then, in sediments dated from 543 to about 506 mya, most of the animal phyla living today appear: crustaceans and other arthropods, onychophorans, sipunculid worms, segmented worms, molluscs, and chordates. This time interval is called the Cambrian period and the events are known as the Cambrian explosion. The Cambrian represents less than 1% of Earth's history, but the relatively rapid appearance of so many large, complex animals ranks as one of the great events in the history of life.

The first large and morphologically complex animals appear in the fossil record during the Cambrian period, 543 to 495 million years ago.

The fossil record and phylogenetic analyses have confirmed many other periods of rapid species diversification over the past 543 million years, as well as five episodes of cataclysmic extinction. The time interval between the start of the

(a)

(b)

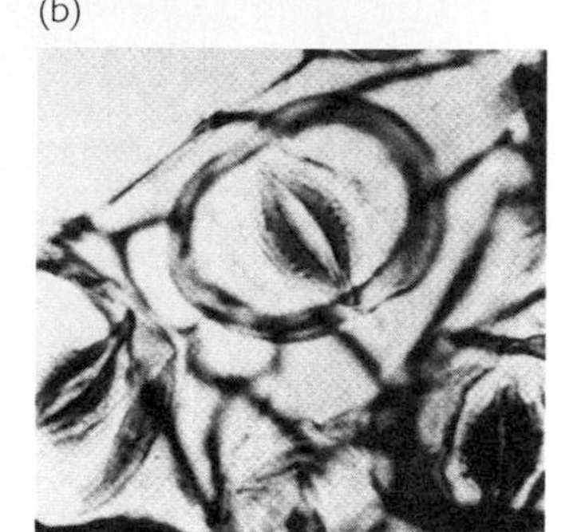

Figure 18.1 Compression fossils These are two-dimensional fossils, usually found by splitting sedimentary rocks along the bedding plane.
(a) A compression fossil of a Paleocene leaf, about 60 million years old, found near Alberta, Canada. (b) A close-up from the leaf pictured in (a), showing a stoma—the guard cells and pore where gas exchange occurs.

Figure 18.2 Casts and molds This is a cast of a horsetail stem from the Carboniferous, about 310 million years old.

Cambrian and the present is called the Phanerozoic ("visible life") eon. When did the major groups of plants and animals and fungi appear, and why? Why did some groups diversify quickly at certain places or times during the Phanerozoic, while others show little change over time? Why did so many mass extinctions occur? This chapter is focused on questions like these. Before addressing them, however, we need to have a look at the basics: how paleontologists read the fossil record and document the history of life.

18.1 The Nature of the Fossil Record

Chapter 2 introduced the geological time scale that was established by paleontologists in the early 19th century. You might also recall from that chapter that 20th-century geochronologists are using radioactive isotopes to estimate the absolute age of each eon, era, period, and epoch. To get a deeper understanding of how life has changed through time, let's review the process of fossilization, examine the strengths and weaknesses of the fossil record, and present a time line of major events in evolution.

How Organic Remains Fossilize

A fossil is any trace left by an organism that lived in the past. Fossils are enormously diverse, but four general categories can be defined by method of formation. In the list that follows, there are two important issues to focus on: Which part of the organism is preserved and available for study? What kinds of habitats produce fossils?

- Compression fossils (Figure 18.1) can result when organic material is buried in water- or wind-borne sediment before it decomposes. Under the weight of sand, mud, ash, or other particles deposited above, a structure can leave an impression in the material below. The resulting fossil is analogous to the record left by footprints in mud or leaves in wet concrete.
- Casts and molds (Figure 18.2) originate when remains decay after being buried in sediment. Molds consist of unfilled spaces, while casts form when new material infiltrates the space, fills it, and hardens into rock. This process is analogous to the lost-wax casting technique used by sculptors. Molds and casts preserve information about external and internal surfaces.
- Permineralized fossils (Figure 18.3) can form when structures are buried in sediments and dissolved minerals precipitate in the cells. This process, which is analogous to the way a microscopist embeds tissues in resin before sectioning, can preserve details of internal structure.
- Unaltered remains (Figure 18.4) are sometimes preserved in environments that discourage loss from weathering, consumption by scavenging animals, and decomposition by bacteria and fungi. How long can organic material remain unaltered? Two-thousand-year-old human cadavers from the Iron Age, buried in the highly acidic environment of peat bogs, have been recovered with flesh still intact. Woolly mammoths dug out of permafrost have fur and many tissues preserved. Dried but otherwise unaltered 20,000-year-old dung from giant ground sloths can be found in protected, desiccating environments such as desert caves. Viscous plant resins can harden into amber, preserving the insects trapped inside so well that wing veins are visible. Paleobotanists have driven nails into 100-million-year-old logs recovered from oil-saturated tar sands.

Though spectacular, intact remains are so rare that they represent only a small fraction of the total fossil record. Compression, impression, casting, molding, and permineralization are far more common. All of these processes depend on three key features of the specimen: durability, burial (usually in a water-saturated sediment), and lack of oxygen. Each of these factors slows decomposition and makes fossilization more likely. As a result, the fossil record consists primarily of hard structures left in depositional environments such as river deltas, beaches, floodplains, marshes, lakeshores, and seafloors.

It is not hard to understand why these structures predominate, while soft tissues and organisms from upland habitats are rarely preserved. Marine bivalves that burrow are automatically buried in saturated sediments after death. Because enamel is one of the highest-density substances known in nature, teeth decay slowly and have more time to fossilize. Trees that grow in floodplain forests drop their leaves and seeds onto substrates that are frequently washed over with sediment, and the saturated soils of swamps are routinely anoxic and permit only slow decomposition. Shelled organisms in the marine plankton drift to the ocean floor after death and are likewise buried.

(a)

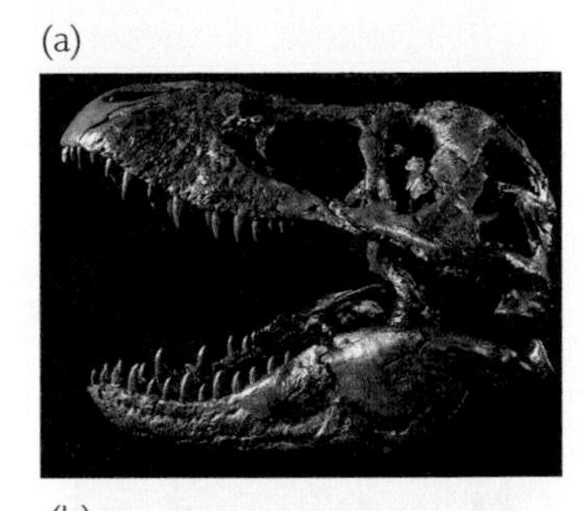

(b)

Figure 18.3 Permineralized fossils Permineralized fossils are usually found in rock outcrops after they have been partially exposed by natural weathering. (a) The skull of *Tyrannosaurus rex*, a predatory dinosaur; (b) Petrified wood.

Strengths and Weaknesses of the Fossil Record

Understanding the nature of fossilization processes is important because it results in three types of sampling bias in the fossil record: geographic, taxonomic, and temporal (see Donovan and Paul 1998). The geographic bias is produced by the propensity for fossils to come from lowland and marine habitats. To appreciate the taxonomic bias more fully, consider this: Marine organisms dominate the fossil record, but make up only 10% of extant species. A full two-thirds of animal phyla living today are underrepresented in the fossil record because they lack any sort of mineralized hard parts, such as bone or shell, that are amenable to fossilization. Likewise, critical parts of plants, including reproductive structures such as flowers, are seldom preserved. The temporal bias results because Earth's crust is constantly being recycled. When tectonic plates subduct or mountains erode, their fossils go with them. As a result, old rocks are rarer than new rocks, and our ability to sample life forms should decline with time. An analysis by Benton and coworkers (2000), however, suggests that older rocks still contain enough fossils to accurately record the order of branching events implied by molecular phylogenies of living groups—meaning that the temporal bias does not doom our ability to understand life's diversity.

Figure 18.4 Unaltered remains This is a winged male termite, preserved in amber, from the Upper Cretaceous of Canada. It is about 125 million years old.

It is also important to realize that taxonomic and other types of sampling bias are by no means unique to paleontology. Advances in developmental genetics depend on the generality of a few model systems, such as *Drosophila melanogaster,* the roundworm *Caenorhabditis elegans*, the annual flower Arabidopsis thaliana, corn (*Zea mays*), and zebrafish. Most work in molecular genetics is done on a few bacteriophages, the yeast *Saccharomyces cerevisieae*, and *Escherichia coli*. The vast majority of research in behavioral ecology is done on birds and mammals; community ecology has historically focused on upland habitats in North America and Europe.

The important point is that the fossil record, like any source of data, has characteristics that limit the types of information that can be retrieved and how broadly the data can be interpreted. The goal for paleontologists is to recognize the constraints and work creatively within them.

Like all sources of data, the fossil record has inherent strengths and limitations.

With these caveats in place, let's begin our intensive use of the fossil record with a broad look at the sequence of events during the Phanerozoic.

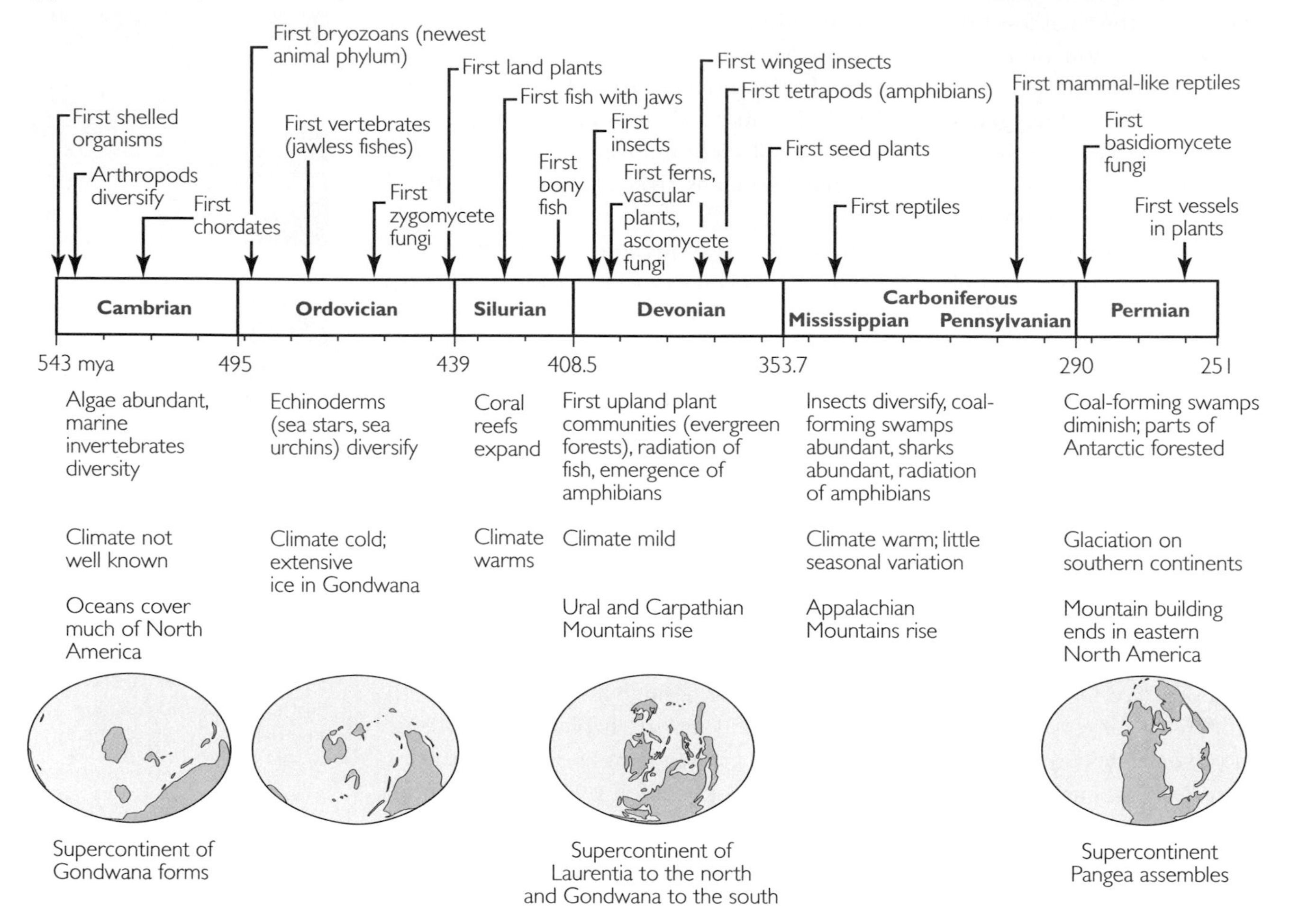

Figure 18.5 The Phanerozoic eon The diagrams show a selection of events from the three eras that make up the Phanerozoic. The labels indicate the first appearances of life forms discussed in this and other chapters, the names of periods or epochs within the era, absolute ages assigned by radioactive dating, notes on prominent plant communities, and the climate and important geological events. The maps show the estimated positions of major landmasses.

(a) The Paleozoic begins with the radiation of animals and ends with a mass extinction at the end of the Permian. Each hash mark on the time bar represents about 12 million years.

(b) The Mesozoic, sometimes nicknamed the age of reptiles, begins after the end-Permian extinction event and ends with the extinction of the dinosaurs and other groups at the Cretaceous-Tertiary boundary. Each hash mark on the time bar represents about 7.5 million years.

(c) The Cenozoic is divided into the Tertiary and Quaternary periods. The Tertiary includes the Paleocene, Eocene, Oligocene, Miocene, and Pliocene epochs. The Quaternary includes the Pleistocene and Holocene epochs. The Cenozoic is sometimes nicknamed the age of mammals. Each hash mark on the time bar represents about 2.8 million years.

Life through Time: An Overview

The geologic time scale is a hierarchy divided into eons, eras, periods, epochs, and stages. Each named interval is defined by a suite of diagnostic fossils. When the scale was first being formulated, in the early 1800s, the intervals were arranged by relative age only, with rocks placed in younger-to-older sequence relative to each other. It was only much later, after the discovery of radioisotopes and the development of accurate dating techniques, that absolute times were assigned to each interval. Consequently, levels in the hierarchy are not equivalent in terms of time. The Paleozoic era lasted 292 million years and the Mesozoic era 186 million, for example. Also, the geologic time scale is a work in progress. Esti-

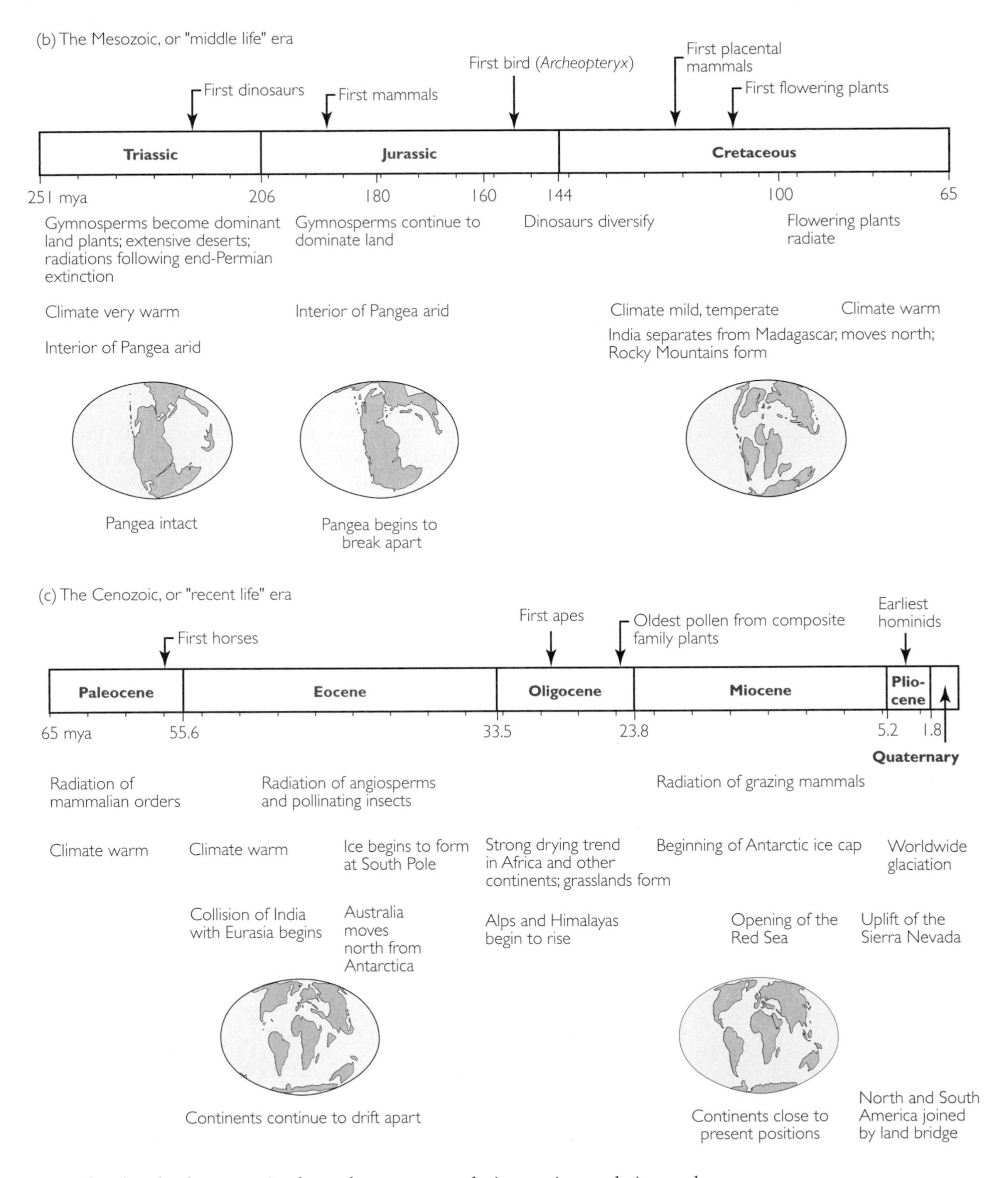

mates for the absolute ages in the scale are constantly improving as dating techniques become more sophisticated and more rocks are sampled.

Figure 18.5 presents time lines for the three eras that make up the Phanerozoic. The eon begins with the Cambrian explosion and ends in the present. Its three component eras are the Paleozoic (ancient life), Mesozoic (middle life), and Cenozoic (recent life). In addition to offering a compact overview of the history of multicellular life, the diagrams in Figure 18.5 should inspire questions. For example, each of the fossil "firsts" mapped along the top of the time lines represents a suite of new traits. What are these? Why did selection favor them? How long did they persist? Note, too, that many of these events led to the recognition of some organisms as new orders, classes, and phyla. Why did some of them become extinct?

Given enough time and space, we could explore these questions for any of the many events diagrammed on the time lines. However, the history of Phanerozoic evolution includes a smaller number of broad patterns. Because our goal is to introduce how research in contemporary paleontology is done and to illustrate the most important concepts in the field, let's focus on a few of these patterns to review in detail.

18.2 The Cambrian Explosion

Almost all of the animal phyla currently recognized by biologists make their first appearance in the fossil record during the Cambrian period—a span of just 40 million years. This amount of time is little more than the blink of an eye, geologically speaking. To appreciate just how much evolutionary change occurred during the interval, consider the fossil record of animals that existed before and after the explosion.

The Ediacaran Fauna

The first unequivocal evidence for animals in the fossil record comes from the Ediacaran faunas. The earliest members of this fauna are dated at or around 565 mya and the youngest at 544 mya, placing them at the end of the Proterozoic (early life) era. The initial specimens were found in the 1940s in the Ediacara Hills of south Australia, but similar fossils have now been found at some 20 sites around the world. Most are preserved as compression and impression fossils, and virtually none have shells or any other type of hard part. The fossils are often difficult to identify, but most experts now agree that sponges, jellyfish, and comb jellies were present (Figure 18.6; Conway Morris 1989; Zhang et al. 2001; Xiao et

Figure 18.6 The Ediacaran fauna (a) This is *Dickinsonia*, a radially symmetric animal of uncertain identity that is common in Ediacaran deposits. Individuals were about 3.5 cm across. (b) These "frondlets" belong to animals of uncertain identity that are generally referred to as Rangea. The scale bar is 0.25 cm.

al. 2002). Ediacaran animals are small in size, typically reaching only a few centimeters across. The groups that are represented are also relatively simple in terms of their morphology. More specifically, the bodies of adult sponges, jellyfish, and comb jellies are either asymmetrical or have radial symmetry—meaning that the body has many planes of symmetry.

Recent analyses of the Ediacaran fauna have focused on a single question: Were complex, bilaterally symmetric animals, like the molluscs and worms and crustaceans that dominant today's oceans and continents, present this early in animal evolution? This question has been difficult to answer. Recently discovered fossilized embryos support the hypothesis that bilaterians evolved prior to the Cambrian. These specimens come from the Doushantuo formation of southern China, which is dated from 635–551 mya (Condon et al. 2005). In the Doushantuo rocks, phosphate minerals replaced soft tissue when organisms died and created fossils that show the finest anatomical details. Li et al. (1998) have described beautifully preserved sponge specimens with larvae similar to the forms found in living sponges. Xiao and coworkers (1998) have found fossils that are about half a millimeter in diameter and consist of two, four, eight, or more round structures that appear to be blastomeres in a cleaving embryo (Figure 18.7). Because these structures resemble the first cells observed in a developing arthropod, the fossils have been interpreted as the embryos of a bilaterally symmetric species (Xiao et al. 1998). Other researchers disagree with this interpretation, however (see Conway Morris 1998a). Controversy also exists over Ediacaran specimens that qualify as trace fossils. Trace fossils are the remnants of burrows, fecal pellets, tracks, or

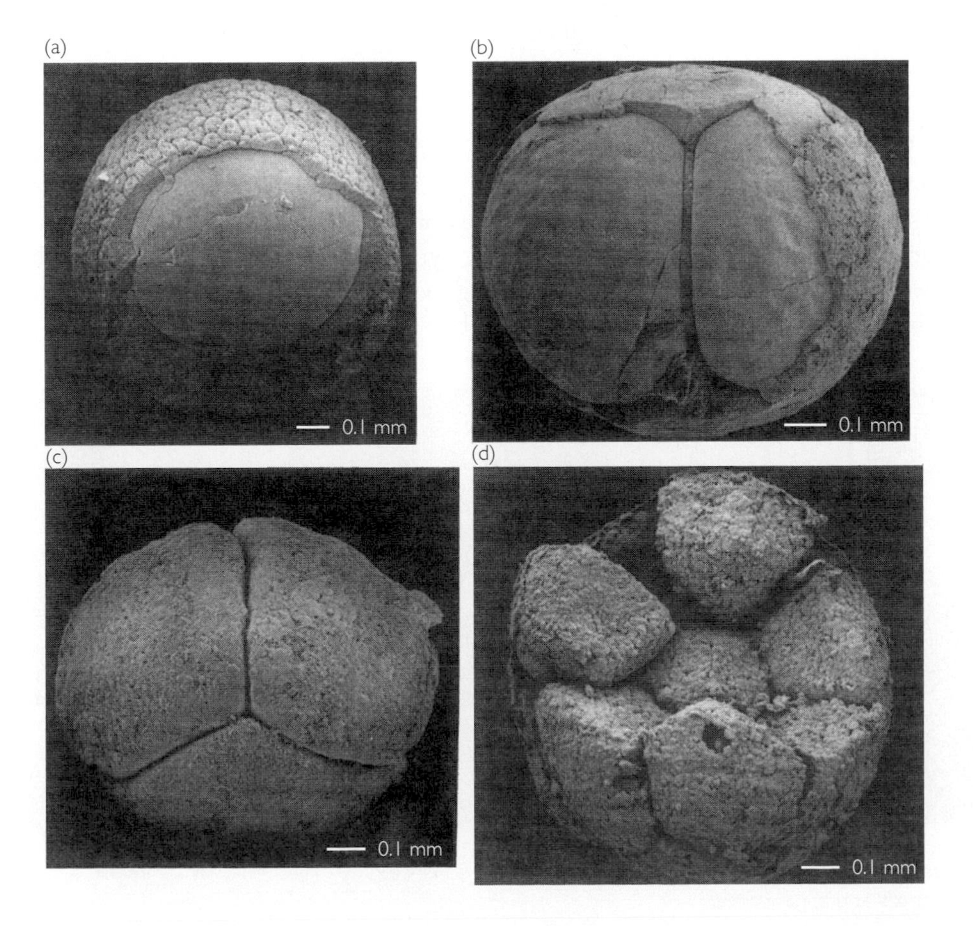

Figure 18.7 Precambrian embryos? These microfossils from the Doushantuo formation may be (a) zygotes and (b–d) cleavage-stage embryos of a Proterozoic bilateral animal, possibly an arthropod or a flatworm. Notice that a structure resembling an egg envelope surrounds each embryo, that the cells occur in multiples of two, and that they are arranged geometrically in a pattern similar to the embryos of some living arthropods and flatworms.

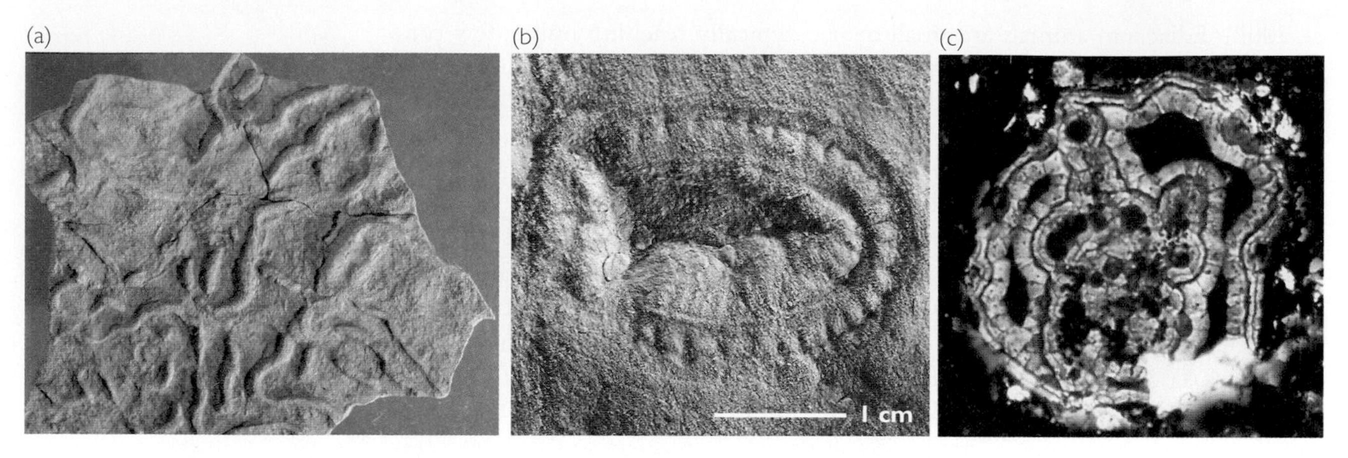

Figure 18.8 Evidence for bilaterally symmetric animals in the Precambrian (a) Trace fossils do not contain body parts but instead consist of tracks or burrows or feces. These tracks may have been made by a worm-like burrowing animal that was bilaterally symmetric. Each track is about 0.5 cm wide. Only two whole-body fossils of bilaterally symmetric animals are known from the Precambrian: (b) the mollusc-like *Kimberella*, and (c) the tiny *Vernanimacula guizhouena*, shown here in cross section.

There is increasingly strong evidence that small-bodied, bilaterally symmetric animals were present prior to the Cambrian explosion.

other traces left by organisms (Figure 18.8a). Trace fossils are notoriously difficult to interpret, but at least some researchers have argued that the linear burrows and tracks that are present must have been made by bilaterally symmetric organisms that had a head and tail region and moved in a line (Waggoner 1998).

Recent fossil finds provide stronger evidence for bilaterians in the late Precambrian: the mollusc-like *Kimberella* (Figure 18.8b; Fedonkin and Waggoner 1997) and the tiny *Vernanimacula guizhouena* (Figure 18.8c; Chen et al. 2004). Taken together, the data argue that bilaterally symmetric animals were small in size but definitely present prior to the Cambrian period.

The Burgess Shale Fauna

In sharp contrast to the paucity of bilaterians among Ediacaran fossils in the Precambrian, the Burgess Shale fauna of the Cambrian period records an astonishing variety of large, complex, and bilaterally symmetric forms (Figure 18.9). The most species-rich lineages of animals living today—the arthropods, mollusks, vertebrates, and echinoderms—are all present.

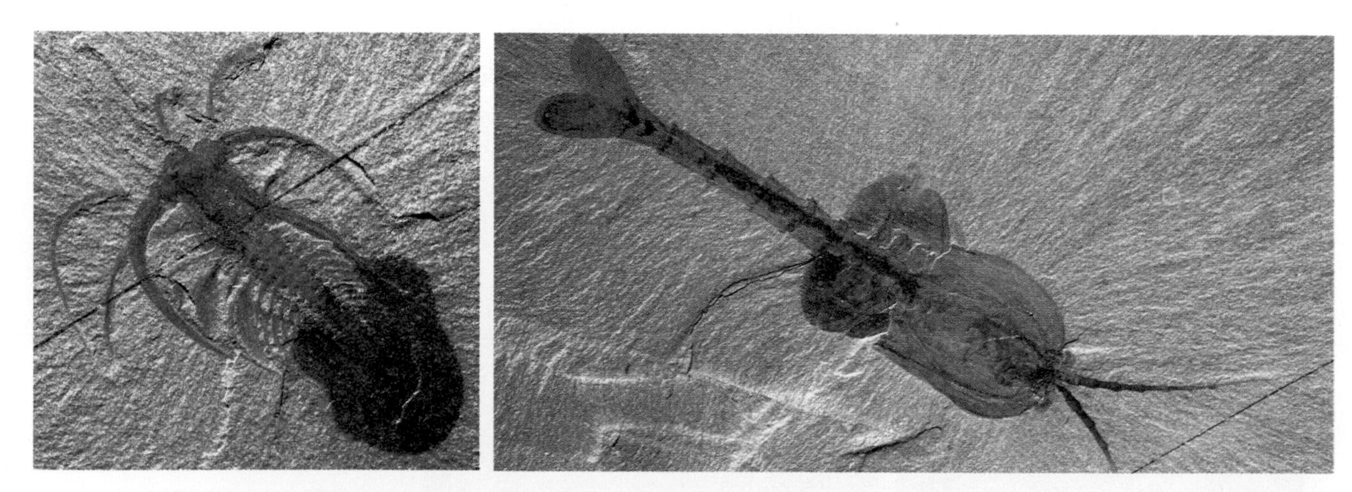

Figure 18.9 The Burgess Shale faunas The Burgess Shale and Chengjiang faunas are dominated by large, bilaterally symmetric animals with well-developed segmentation, heads, and appendages.

(a)

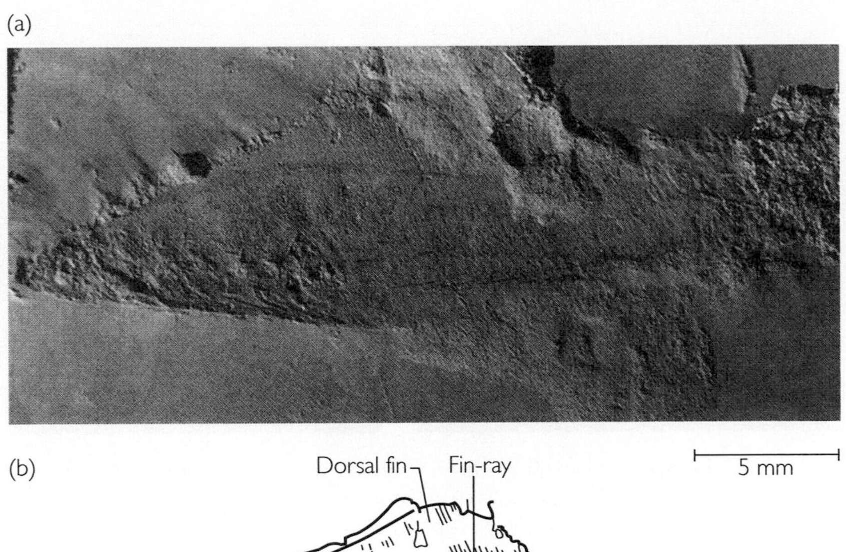

(b)

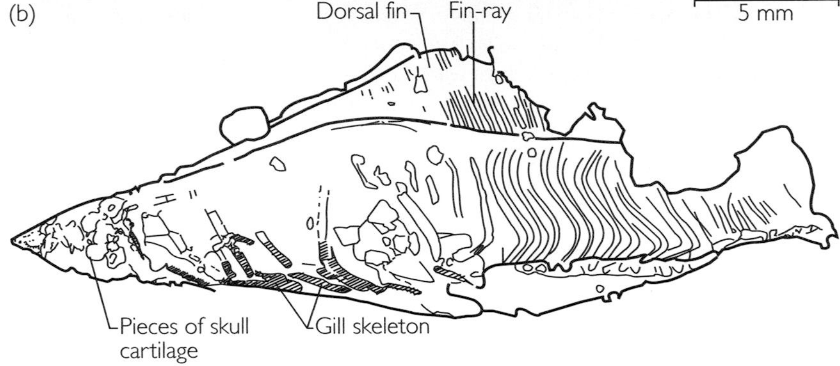

Figure 18.10 A vertebrate from the Cambrian: *Haikouichthys ercaicunensis* The photograph in part (a) shows the whole specimen. The interpretive drawing below (b) shows some distinctive vertebrate characters, including a cartilaginous skull, gill skeleton, and fin-rays. From Shu et al. (1999).

The fossils were initially discovered early in the 1900s near the town of Field, British Columbia. These Burgess Shales, dated to 520–515 mya, along with the Chengjiang biota from Yunnan Province in China (525–520 mya) are arguably the most spectacular fossil deposits ever found (Conway Morris 1998b; Zhang et al. 2001). They are primarily impression and compression fossils and are extraordinary for the detail they preserve and the story they tell.

There is little overlap between species found in the Ediacaran and Burgess Shale deposits, but at least a few organisms, such as large, colonial cnidarians that are similar to modern sea pens, are present in both (Conway Morris 1998b). The Cambrian specimens include a wide array of complex and unusual arthropods, including trilobites, as well as segmented worms, wormlike priapulids and sipunculids, and a diversity of molluscs. Remarkably, the deposits also harbor several chordates, including species of jawless vertebrates (Figure 18.10). These early chordates had segmented trunk muscles and a skeletal rod called a notochord. In overall morphology they resembled the jawless vertebrates living today—the hagfishes and lampreys (Shu et al. 1999; Chen et al. 1999; Shu et al. 2003).

Compared to animals present earlier in the fossil record, the hallmarks of the Cambrian fauna are a dramatic increase in body size, the origin of hard exoskeletons and complex body parts like limbs, and a diversification in basic body shapes and organization.

Initially, though, many members of the Burgess Shale and Chengjiang faunas looked so unusual that biologists were perplexed. Seemingly bizarre fossils like those in Figure 18.11 were assigned to a jumbled group referred to as Problematica. Some observers suggested that these species represented unique phyla, unlike any organisms living today. Further study revealed that most or all of the problematica are in fact members or close relatives of living phyla, however. *Opabinia regalis* (Figure 18.11a), for example, was first described as an elongate bilaterian with serially repeated lateral plates, five dorsal eyes, and an elongate "nozzle" on the anterior end. This description left *Opabinia* out of the classification of living phyla. More recent specimens, though, appear to have legs with a terminal

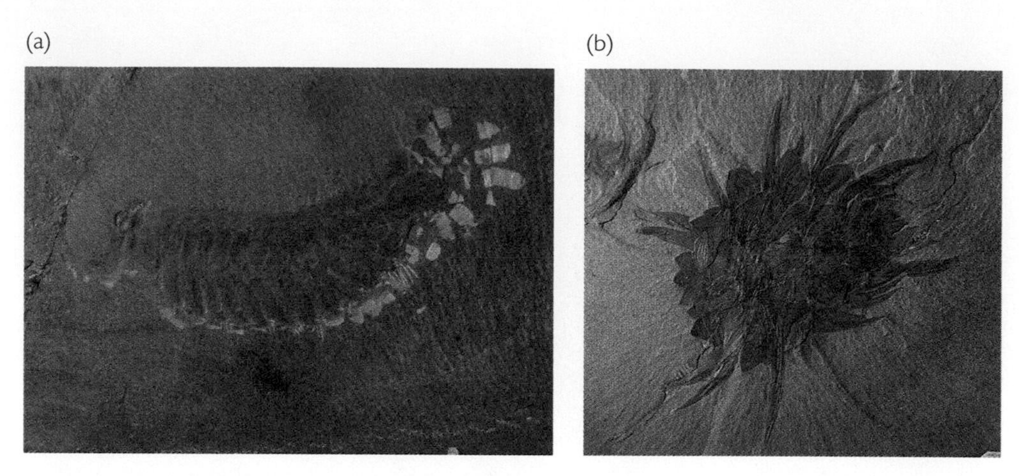

Figure 18.11 Problematica? About 60 species in the Burgess Shale fauna are so unusual that they were initially left out of the classification of known living phyla. Many of these problematica are now more reliably grouped with living phyla, or at least with early Phanerozoic fossils of known affinity. (a) *Opabinia* is related to arthropods (or may belong within the arthropods); (b) *Wiwaxia* is most likely a polychaete worm.

claw (Budd 1996). This redescription suggests that *Opabinia* is actually a close relative or a member of the arthropods (Conway Morris 1998b). Similarly, *Wiwaxia corrugata* (Figure 18.11b) was formerly given phylum status as a problematic fossil consisting of spines and plates, but further study revealed that it is almost certainly a polychaete worm or a member of the stem group that gave rise to the annelids (Butterfield 1990; Conway Morris 1998b). As a result of these analyses, the number of phyla that existed during the Cambrian is now recognized as roughly equivalent to the diversity observed today (Briggs et al. 1992).

Most of the animal phyla living today make their first appearance in the fossil record during the Cambrian.

This is a remarkable conclusion. The earliest members of virtually all major animal lineages appeared relatively suddenly in the fossil record, at the same time, in geographically distant parts of the globe (Conway Morris 2000; Valentine 2002). The Burgess Shale fauna records an amazing variety and number of major morphological innovations, including large body size and the first segmented body plans, limbs, antennae, shells, external skeletons, and notochords. These animals sat, swam, burrowed, crawled, floated, and walked. They found food in almost every conceivable way, from filtering it out of the water to hunting it down. To get a more precise view of these remarkable changes, we need to look at the relationship between morphological diversity and the phylogeny of animals.

Phylogeny and Morphology

Our understanding of animal phylogeny has deepened considerably over the past few years. Beginning with a breakthrough analysis of the gene that encodes the small subunit of the ribosome, a series of follow-up studies have analyzed DNA sequences from a variety of genes and reorganized our picture of animal evolution. Although several important issues remain controversial, researchers are increasingly confident that the pattern of branching events in Figure 18.12 accurately reflects actual evolutionary history (e.g., Philippe et al. 2005).

In examining Figure 18.12, note that the blue bars to the right of the tree indicate when lineages appeared in the fossil record. The key observation is that the lineages at the base of the tree—the Porifera (sponges), Cnidaria (jellyfish), and Ctenophora (comb-jellies)—predominate in the Ediacaran fauna while more de-

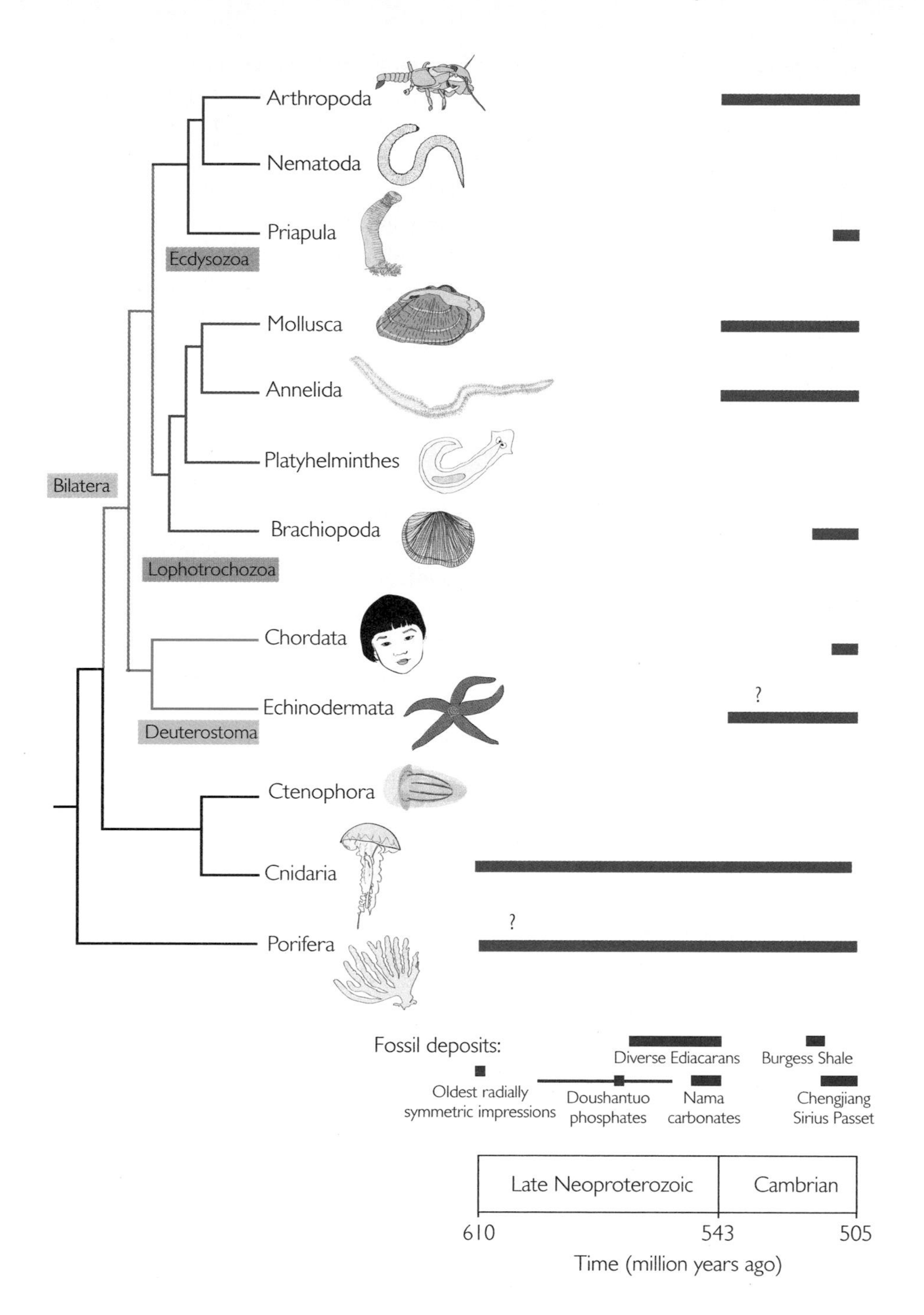

Figure 18.12 The first animals: phylogeny and fossils The phylogeny on the left shows the relationships among some of the major taxa represented in the earliest fossil faunas. The phylogeny is based on molecular data from living members of these phyla (Philippe et al. 2005). The blue bars to the right of the tree indicate which phyla are represented in the fossil faunas identified by the red bars at the bottom of the figure; question marks indicate that the presence of a group at a certain time is controversial.

rived groups such as the Arthropoda (crustaceans, millipedes, and trilobites) and annelids (segmented worms) appear in the Burgess Shale fauna. To extend this observation, consider several key points about how animals develop as embryos and how their adult body plan is organized.

- **Diploblasts and triploblasts.** The Cnidaria and Ctenophora are termed diploblastic because they have two embryonic tissue types. The remaining animals are called triploblasts because they have three embryonic tissues. The tissue types present in the embryos of both groups are the ectoderm and endoderm. Ectodermal cells produce the adult skin and nervous system, and endodermal cells produce the gut and associated organs. The embryonic tissue that is unique

The bilaterally symmetric animals, or Bilateria, have three embryonic tissues. Many of the bilaterians can move efficiently using a hydrostatic skeleton..

Within the Bilateria, the protostomes and deuterostomes form distinct groups.

to triploblasts is called mesoderm and develops into the gonads, heart, muscle, connective tissues, and blood. The vast majority of triploblasts have one plane of symmetry, meaning that they are bilaterally symmetric. In contrast, some diploblasts are radially symmetrical in whole or in part while others are asymmetrical. The fossil record and molecular phylogeny agree that diploblasty and radial symmetry evolved before triploblasty and bilateral symmetry. The origin of mesoderm and triploblasty was important because it made it possible for muscle-lined, fluid-filled cavities to evolve. These structures furnished hydrostatic skeletons that made directed movement much more efficient.

- **Protostomes and deuterostomes.** The bilaterians consist of protostomes and deuterostomes. These two major animal groups differ in the way that they undergo important developmental processes, particularly gastrulation. Gastrulation is the mass movement of cells that rearranges the embryonic cells after cleavage and defines the ectoderm, endoderm, and mesoderm. In protostomes, gastrulation forms the mouth region first. But in deuterostomes, gastrulation forms the anal region first and the mouth later. This contrast inspired the use of the Greek roots *proto* (first), *deutero* (second), and *stoma* (mouth). In essence, animals evolved two basic methods for producing a multicellular body with three types of embryonic tissues. Because both protostomes and deuterostomes occur in the Burgess Shale fauna, it appears that natural selection resulted in the rapid elaboration of a wide array of body types and structures in each lineage.
- **Lophotrochozoans and ecdysozoans.** The protostomes consist of two major groups, just as bilaterians do. The primary lineages of protostomes are called Ecdysozoa and Lophotrochozoa. As suggested by the Greek root *ecdysis*, meaning to slip out, the Ecdysozoa are molting animals. The Lophotrochozoa includes animals that have a feeding structure called a lophophore. Both ecdysozoans and lophotrochozoans are represented in the Burgess Shale fauna. Based on these observations, it appears that molting and the lophophore were evolutionary innovations that appeared in the Cambrian explosion.

To summarize, the fossil record and molecular phylogeny agree that the most basal groups of animals populate the Ediacaran fauna, while the vast majority of more derived groups first appear in the Cambrian. The radiation of triploblasts into major lineages such as deuterostomes, lophotrochozoans, and ecdysozoans emphasizes just how many types of fundamental changes occurred. Now the question is: Just how rapidly did all this change occur? Did the major groups of bilaterians all evolve between 543–505 million years ago?

Was the Cambrian Explosion Really Explosive?

Many phyla and morphological innovations first appear in the Cambrian. But it is important to recognize that these species and traits had to have existed for some time before being immortalized in the Burgess Shale and Chengjiang deposits. Just how long? To answer this question, evolutionary biologists have used molecular clocks to estimate when the earliest branches on the animal phylogeny occurred. As Chapter 4 and Chapter 7 explained, changes in DNA or protein sequences that are neutral with respect to selection should arise by mutation and then drift to fixation at a steady, clocklike rate. By observing the amount of selectively neutral genetic change that has occurred between taxa whose divergence is dated in the fossil or geological record, a molecular clock can be calibrated in terms of the amount of change expected per million years. This calibration, in turn, can then be used to date events that are not dated in the fossil record.

To date the origin of the bilaterians, Bruce Runnegar (1982) analyzed differences among hemoglobin amino acid sequences in vertebrates and various invertebrate phyla. To translate these genetic distances into times of divergence, Runnegar used estimates of the rate of hemoglobin evolution among vertebrate groups with known fossil ages. He concluded that the earliest branches in Figure 18.12 occurred about 900 mya—long before the Cambrian explosion. Greg Wray, Jeff Levinton, and Leo Shapiro (1996) came to a similar conclusion using a different data set. Wray's group estimated that chordates and echinoderms diverged about 1,000 mya, while protostomes and deuterostomes diverged about 1,200 mya. This study included more genes and more taxa, but used the same vertebrate fossil record for calibration as the Runnegar study. Both of these analyses admitted some uncertainty about the exact ages of the divergences among the bilaterally symmetric organisms, but agreed that these divergences occurred hundreds of millions of years before their first appearance in the fossil record.

These papers generated a storm of controversy, however, because such early divergence dates imply a long history of animal evolution prior to the Cambrian explosion. If the dates are correct, then Proterozoic rocks should eventually yield fossils of deuterostomes, ecdysozoans, and lophotrochozoans. But with the exception of the Doushantuo embryos, *Kimberella*, and *Ventanimaculan*, they have not.

To resolve the discrepancy between the fossil record and the predictions of the molecular clock, Andrew Smith (1999) suggests that the lineages leading to the living Bilateria diverged from each other over a prolonged period in the Proterozoic, but that the vast majority of resulting species existed as small, larvalike organisms that left no trace in the fossil record (see also Erwin and Davidson 2002).

According to this point of view, the Cambrian explosion is an explosion of morphological diversity, but not necessarily an explosion of lineages, which occurred much earlier. The idea that the major animal lineages existed long before they diversified and produced large-bodied forms is captured in the quip that the Cambrian explosion had a "long fuse." Even if this hypothesis is correct, however, we are left with the question of why the explosion occurred. Why did so many lineages evolve dramatic changes in body size and body plan during the same brief period? As a last question about the Cambrian explosion, let's consider some mechanisms behind the rapid morphological diversification that occurred.

What really exploded during the Cambrian was not lineages, but morphologies and ways of making a living.

What Caused the Cambrian Explosion?

An astonishing variety of body plans, cell types, and developmental patterns evolved during the Cambrian explosion. But it is critical to recognize that the radiation of bilaterally symmetric animals was really driven by how the organisms made a living. Most of the Ediacaran animals are either sessile filter feeders or predators that floated high in the water column and fed on planktonic organisms. But the Burgess Shale fauna introduces a huge variety of benthic and pelagic predators, filter feeders, grazers, scavengers, and detritivores, most of which actively chased their prey. The Cambrian explosion filled many of the ecological niches present in shallow marine habitats. Based on this observation, the question of why the Cambrian explosion occurred turns on what environmental changes made all of these novel ways of life possible.

Rising oxygen concentrations in seawater, due to an increase in photosynthetic algae during the Proterozoic, was clearly a key to the origin of multicellularity and large size (Valentine 1994). Increased availability of oxygen makes higher

The leading hypothesis to explain the Cambrian explosion is based on a dramatic rise in atmospheric oxygen concentration, which would have made higher activity levels and more rapid growth possible.

metabolic rates and bigger bodies possible. Larger size is a prerequisite for the evolution of tissues, and higher metabolic rates are required for powered movement. The beginnings of both of these traits are recorded in the Ediacaran faunas.

To explain the Cambrian explosion, Andrew Knoll and Sean Carroll (1999) have suggested that a sudden and dramatic increase in atmospheric oxygen occurred about 543 mya and that this environmental change made large size and rapid movement possible. Knoll and Carroll also posit that a mass extinction event eliminated much of the Ediacaran fauna at the end of the Proterozoic, creating an opportunity for the tiny deuterostomes and protostomes present at the time to evolve in response to the changed conditions.

These hypotheses are provocative and are currently being tested. If they are valid, then a series of predictions should prove to be correct. Further analyses of molecular clocks should be consistent with the claim that bilaterians arose hundreds of millions of years before the Cambrian explosion. Additional fossil evidence of small-bodied protostomes and deuterostomes should eventually be found in the Proterozoic, and fossil or geological evidence of a mass extinction event and a rise in available oxygen should come to light. Only time and further research will tell if the evolution of animals really was explosive and triggered by dramatic environmental change in the form of increased oxygen availability.

18.3 Macroevolutionary Patterns

Documenting the Cambrian explosion and evolution's other "greatest hits" is only part of historical biology's portfolio. Searching for broad patterns in the fossil record is an equally important research program. The fossils document that the types of species on Earth have changed radically over vast reaches of time. As biologists analyze these changes, what patterns come to light?

The literature on patterns of change through time is enormous, so we can only touch on a few of the major results here. Let's start by summarizing a classical pattern: The rapid diversification of species in response to a morphological innovation or ecological opportunity.

Adaptive Radiations

An **adaptive radiation** occurs when a single or small group of ancestral species rapidly diversifies into a large number of descendant species that occupy a wide variety of ecological niches (Figure 18.13). The Galápagos finches and Hawaiian *Drosophila*, which figure prominently in earlier chapters, are well-studied examples. But adaptive radiations have occurred in a wide array of groups at intervals throughout the history of life. They represent a prominent pattern. It's as if the tree of life suddenly sprouts a large number of highly diverse branches.

What factors trigger adaptive radiations? Why do only certain lineages diversify broadly and rapidly? The answers vary from time to time and clade to clade.

Ecological Opportunity as a Trigger

An ecological opportunity occurs when a small number of individuals or species is suddenly presented with a wide and abundant array of resources to exploit. The ancestors of the Hawaiian *Drosophila* and Galápagos finches, for example, colonized islands that had few competitors and a wide variety of resources and habitats to use. Such conditions favor rapid diversification and speciation.

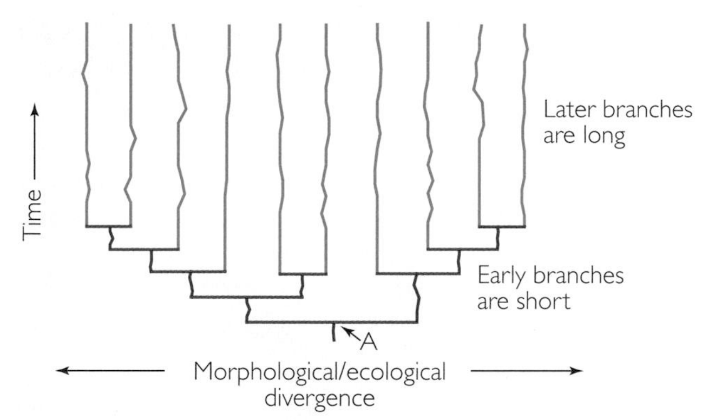

Figure 18.13 Adaptive radiation This diagram shows the branching pattern produced by a hypothetical adaptive radiation. Note that the initial radiation is rapid, producing lineages with widely divergent forms. When phylogenetic trees are estimated for lineages that have undergone adaptive radiation, the early branches on the tree are often extremely short and later branches much longer. The pattern is "bushy" or "stemmy." The event that triggered the radiation occured at node A.

Why certain populations colonize an area and undergo adaptive radiation is largely a matter of blind luck, however. For example, recent phylogenies of the Galápagos finches have shown that most of the closest living ancestors of the group live in the Caribbean (Sato et al. 2001; Burns et al. 2002). Thus, the leading hypothesis to explain the radiation is that a small population of birds happened to move from the Caribbean to the Galápagos and take up residence. Similarly, the ancestors of the Hawaiian *Drosophila* was a fruit-fly species that happened to be blown onto the islands millions of years ago; the ancestor of today's diverse array of Hawaiian silverswords was a tarweed native to California, that probably arrived in Hawaii as a seed hitchhiking on a bird's foot or in its digestive tract.

A lineage can diversify into many different species with divergent ways of life in response to the availability of new habitats and resources . . .

Ecological opportunity is not created solely through colonization events, however. In the aftermath of a mass extinction at the end of the Cretaceous period, mammals diversified rapidly. The leading hypothesis for why they did so was that they lacked competition, not that they had superior adaptations. The extinction of the dinosaurs created an ecological opportunity for mammals. Ecological opportunities can be created by dispersal and colonization or extinction of competitors.

Morphological Innovation as a Trigger

Not all adaptive radiations are associated with ecological opportunity; many are correlated with morphological innovations that represent important new adaptations. The diversification of arthropods is a prime example. The variety of ecological niches occupied by insects, crustaceans, and spiders and the number of species in these lineages are remarkable. Their success is closely associated with modifications and elaborations of their jointed limbs, which allowed species in these groups to move efficiently and find food. Chapter 19 explores the genetic mechanisms responsible for the elaboration of arthropod limbs; here the central point is that jointed limbs were a morphological innovation that is correlated with an adaptive radiation.

. . . or in response to a newly evolved morphological trait that allows individuals to exploit resources in a new way.

Other Examples: Adaptive Radiations in Land Plants

Adaptive radiations have occurred at several different taxonomic levels during the evolution of land plants. Two of the most notable ones were unique events, similar to the Cambrian explosion of animal diversity. The first was the radiation of terrestrial plants from aquatic ancestors in the early Devonian, about 400 mya. During this period, early terrestrial plants evolved key morphological features such as a waxy cuticle and the surface openings called stomata. They also evolved the life history, characterized by alternating gametophyte and sporophyte generations, observed in their living descendants (Bateman et al.

Figure 18.14 Like the ancestral flower? The tropical shrub *Amborella* is the sister group to all of the other living flowering plants. If it has undergone less evolutionary change from the ancestral condition compared to other flowering plants, then *Amborella* may provide clues to the nature of the ancestral angiosperm.

1998). These innovations are associated with the transition to terrestrial life and the adaptive radiation of land plants.

The second radiation in plant evolution was the Cretaceous explosion of flowering plants or angiosperms about 110 mya. Work on the phylogeny of angiosperms and their close relatives produced a surprising result: A little-known shrub from the island of New Caledonia called *Amborella* (Figure 18.14) is the sister group to all other flowering plants (Qiu et al. 1999; Soltis et al. 1999). Identifying *Amborella* as the direct descendant of the angiosperm ancestor is an important clue to what that ancestor might have looked like and, by extension, what traits of that ancestor might have contributed to the early evolutionary success of flowering plants (Brown 1999). More than 250,000 species of angiosperms are alive today. They occupy a spectacular diversity of habitats and range from sprawling inhabitants of arctic tundra to the trees that dominate tropical rain forests. Because it made pollination so efficient, the flower is thought to be a morphological innovation that made this radiation possible.

Adaptive radiation has been a popular area of evolutionary analysis because it is spectacular: A great deal of evolution takes place in a relatively short amount of time. Another prominent pattern in the history of life is the opposite: lack of appreciable morphological change or speciation over long periods of time.

Stasis

In contrast to adaptive radiations, the fossil record contains many cases of new species that appear and then persist for millions of years without apparent change. Stated another way, evolution in some groups seems to consist of long periods of stasis that are occasionally punctuated by speciation events that appear instantaneously in geological time. There is no burst of speciation and morphological change as in an adaptive radiation, and no gradual change over time in response to environmental changes.

Darwin (1859) was well aware of these observations and considered them a problem for his theory. Because his ideas were presented in opposition to the theory of special creation, which predicts the instantaneous creation of new forms, Darwin repeatedly emphasized the gradual nature of evolution by natural selection. He attributed the sudden appearance of new taxa to the incompleteness of the fossil record and predicted that as specimen collections grew, the apparent gaps between fossil forms caused by stasis and punctuated by sudden jumps would be filled in by forms showing gradual transitions between species. For a century thereafter, most paleontologists followed his lead.

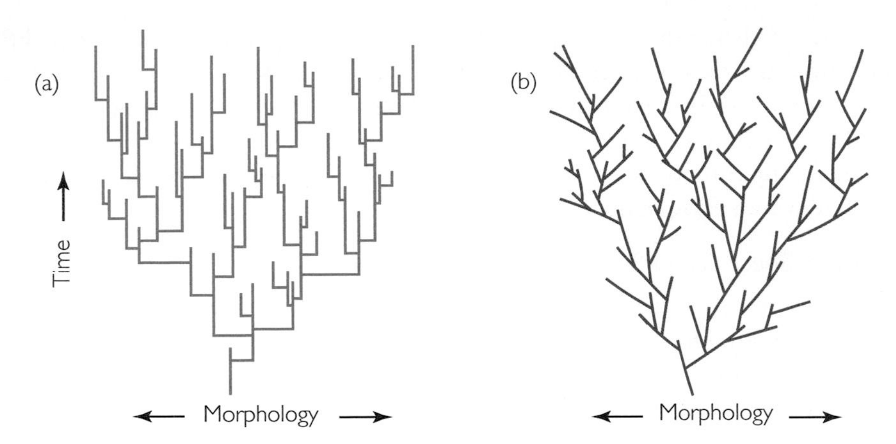

Figure 18.15 Patterns of morphological change: stasis and gradualism When time is plotted against morphology, two extreme patterns are possible, along with many intermediate or mixed patterns. (a) In punctuated equilibrium, all morphological variation occurs at the time of a speciation (branching) event; otherwise there is stasis. (b) In phyletic gradualism, morphological change occurs gradually and is unrelated to speciation events. Eldredge and Gould (1972) claimed that stasis dominates the history of species; Darwin claimed that the gradual pattern does.

In 1972, however, Niles Eldredge and Stephen Jay Gould broke with this tradition by claiming that stasis is a real pattern in the fossil record and that most morphological change occurs during speciation. They called their proposal the theory of punctuated equilibrium (Figure 18.15). The theory and its implications were hotly debated for 20 years.

Demonstrating Stasis

One benefit of the debate over punctuated equilibrium was that it spurred paleontologists to ask whether stasis is in fact real. Do the data support the claim that stasis, punctuated by morphological change at speciation events, is the predominant feature of species histories through time?

Before looking at some tests of this hypothesis, it is important to clarify the requirements for testing the pattern. The goal is to follow changes in morphology in speciating clades through time and determine whether change occurs in conjunction with speciation events or independently, and whether rapid change is followed by stasis or continued change. As critics of the theory have emphasized, a rigorous test for stasis vs. gradualism is exceptionally difficult. This is because the theory of punctuated equilibrium can become tautological. Fossil species are defined on the basis of morphology, so it might be trivial to observe a strong correlation between speciation and morphological change. To avoid circularity, an acceptable test requires that

1. the phylogeny of the clade is known, so researchers can identify which species are ancestral and which descendant, and
2. ancestral species survive long enough to co-occur with the new species in the fossil record.

The second criterion is critical. If it is not fulfilled, it is impossible to know whether the new morphospecies is indeed a product of a splitting event or whether it is the result of rapid evolution in the ancestral form without speciation taking place. This second possibility is called **phyletic transformation**, or **anagenesis**.

These are demanding criteria, especially when compounded with other difficult practical issues: the problem of misidentifying cryptic species in the fossil record, the need for analyzing change at the level of species, the requirement of sampling at frequent time intervals, and the necessity of sampling multiple localities to distinguish normal, within-species geographic variation from authentically different morphospecies.

Stasis and Speciation in Bryozoans

There are relatively few fossil series that meet these stringent requirements (Jablonski 2000). One of them is a series of late Cenozoic fossils of the marine invertebrate phylum Bryozoa. Experimental studies on cheilostome bryozoans that are alive today established that bryozoans that are identified as morphospecies also qualify as phylogenetic species (Jackson and Cheetham 1990, 1994). In addition to being abundant in the fossil record of the past 100 million years, then, we can also be confident that species designations in this group actually reflect phylogeny.

Cheetham (1986) and Jackson and Cheetham (1994) performed a high-resolution analysis of speciation and morphologic change in cheilostomes from the Caribbean, starting about 15 mya in the Miocene and ending with living taxa. They began the study by defining 19 morphospecies in the genus *Stylopoma*, based on an analysis of 15 skeletal characters. Then they estimated the phylogeny of the 19 morphospecies from differences in skeletal characters and scaled the tree so that the branch points and branch tips lined up with the dates of first and last appearance for fossil forms. They did a similar analysis for 19 living or extinct morphospecies in the genus *Metrarabdotos*.

Although some lineages show rapid morphological change at speciation followed by stasis, this is not the predominant pattern in the history of life.

The trees generated by the study are pictured in Figure 18.16. The phylogenies show an unequivocal pattern of stasis punctuated by rapid morphological change. The fact that ancestral and descendant species co-occur defends the idea that morphologic change was strongly associated with speciation events. This is an almost flawless example of stasis punctuated by evolutionary change at speciation.

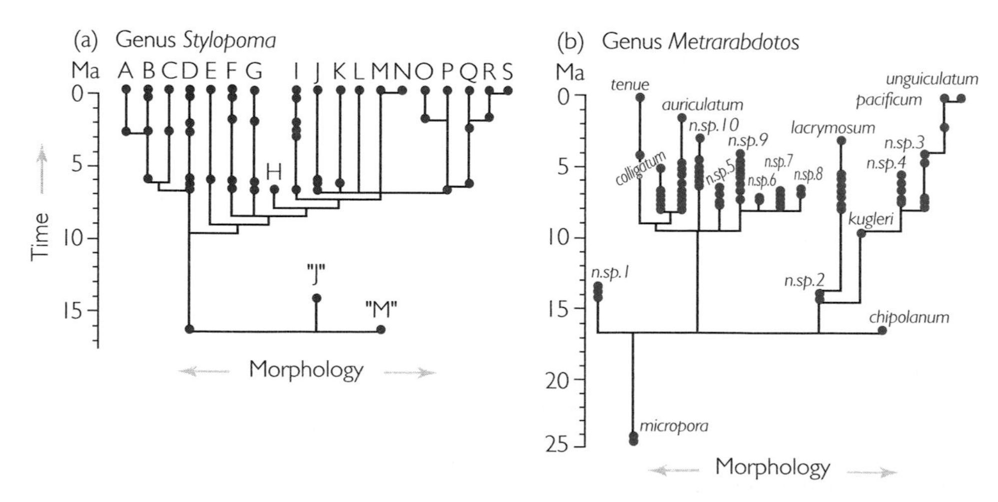

Figure 18.16 Punctuated change in cheilostome Bryozoa These phylogenies, for 19 living and fossil morphospecies in the genus *Stylopoma* (a) and 19 living and fossil morphospecies in the genus *Metrarabdotos* (b), were estimated from differences in skeletal characters (Jackson and Cheetham 1994). Each dot indicates a population that was sampled. None of the populations in the study showed skeletal traits that were intermediate between species, and the characteristics of species were stable through time. As a result, the pattern of change is strong stasis punctuated by speciation events.

What Is the Relative Frequency of Stasis and Gradualism?

How common is the pattern observed in cheilostome bryozoans? Doug Erwin and Robert Anstey (1995a, b) reviewed a total of 58 studies conducted to test the theory of punctuated equilibrium. The analyses represent a wide variety of taxa and periods. Although the studies varied in their ability to meet the strict criteria required for a rigorous test of the theory, their sheer number may compensate somewhat. Erwin and Anstey's conclusion was that "Paleontological evidence overwhelmingly supports a view that speciation is sometimes gradual and sometimes punctuated, and that no one mode characterizes this very complicated process in the history of life." Furthermore, Erwin and Anstey noted that a quarter of the studies reported a third pattern: gradualism and stasis.

Once the controversy over punctuated equilibrium was resolved, biologists turned to other questions. Is it possible that different types of organisms exhibit distinct patterns of change through time? Researchers who have worked on the problem are beginning to argue that gradualist patterns tend to predominate in foraminifera, radiolarians, and other microscopic marine forms, while stasis occurs more often in macroscopic fossils such as marine arthropods, bivalves, corals, and bryozoans (Hunter 1988; Benton and Pearson 2001). If so, why? Research continues.

Why Does Stasis Occur?

One of Eldredge and Gould's prominent claims about the fossil record is that "stasis is data." That is, lack of change is a pattern that needs to be explained. Jackson and Cheetham's study of bryozoans showed that virtually no change occurred in these sessile invertebrates over millions of years. Why would morphology remain unchanged for so long in these lineages? One approach to answering this question has focused on the so-called living fossils. These are species or clades that show little or no measurable morphological change over extended periods. The leaves of the living ginko tree in Figure 18.17a, for instance, are similar to the impression fossils of ginko leaves in Figure 18.17b, which are 40 million years old. The living stromatolites pictured in Figure 18.18a were made by intertidal bacteria. They resemble the fossil stromatolites in shown in Figure 18.18b, which are 1.8 billion years old.

(a)

(b)

Figure 18.17 "Living fossils" Leaves from a living gingko tree (a) are similar to 40-million-year-old impression fossils made by gingko leaves (b). We put "living fossils" in quotes because it is an oxymoron.

(a)

(b)

Figure 18.18 More "living fossils" Contemporary stromatolite-forming bacteria from Australia (a) are similar to 1,800-million-year-old fossil forms from the Great Slave Lake area of Canada (b).

Horseshoe crabs are a spectacular example of an entire clade of "living fossils." The extant species, in the genus *Limulus*, are virtually identical in morphology to fossil species in a different family that existed 150 mya. While some horseshoe crab lineages stayed virtually unchanged, the entire radiation of birds, mammals, and flowering plants took place.

Have these species failed to change simply because they lack genetic variation? John Avise and colleagues (1994) answered this question by sequencing several genes in the mitochondrial DNA of horseshoe crabs and comparing the amount of genetic divergence they found to a previously published study of genetic distances in another arthropod clade: the king and hermit crabs (Cunningham et al. 1992). The result is striking: The horseshoe crabs show just as much genetic divergence as the king-hermit crab clade, even though far less morphological change has occurred (Figure 18.19). This is strong evidence that stasis is not from a lack of genetic variability.

Although it might appear static, morphology in a lineage may actually fluctuate over time around a long-term average.

What about some of the other possibilities? When Steve Stanley and Xiangning Yang (1987) looked at bivalve species that have shown remarkably little change over the past 15 million years, they discovered an interesting pattern. When they mapped change in 24 different shell characters over this interval, they found that most of the traits showed little net change within species, but that many had undergone large fluctuations, or what Stanley and Yang called "zigzag evolution" (Figure 18.20). These clam populations probably changed in response to environmental changes over time. But because these changes tended to fluctuate about a mean value, we perceive stasis as a result. This phenomenon has also been called **habitat tracking** or **dynamic stasis.**

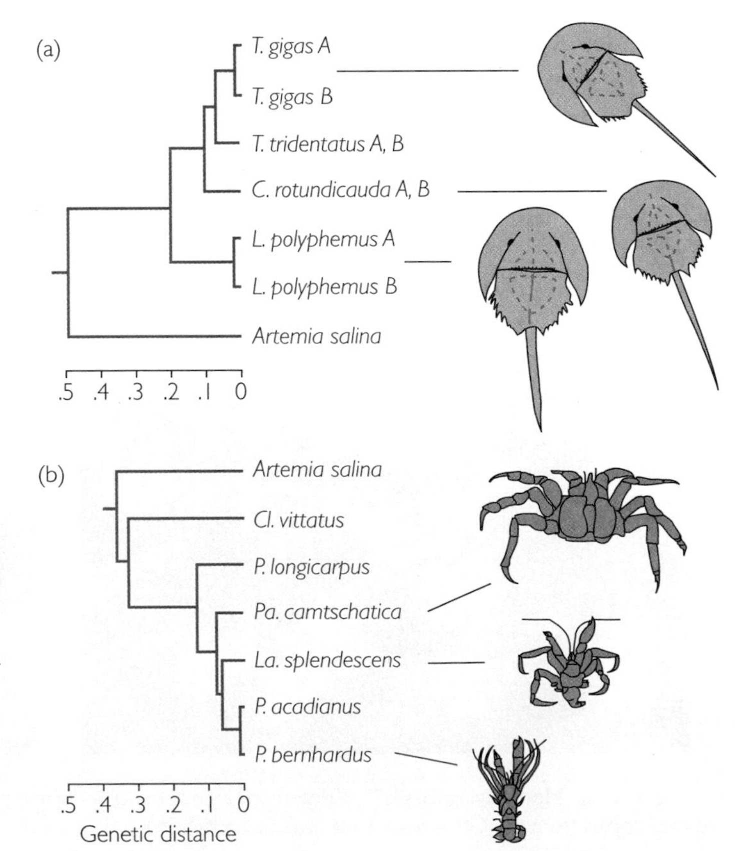

Figure 18.19 Genetic and morphological change in two arthropod clades The length of each branch on these phylogenies represents a genetic distance, measured as the percentage difference in 16S rRNA sequences in mtDNA. The scale is the same for both trees. The fairy shrimp *Artemia salina* was used as the outgroup to root each of these trees. Slightly more genetic divergence in 16S rRNA sequences has occurred in the horseshoe crab clade (a), even though much more morphological divergence has occurred in the clade that includes hermit crabs and allies (b). From Avise et al. (1994).

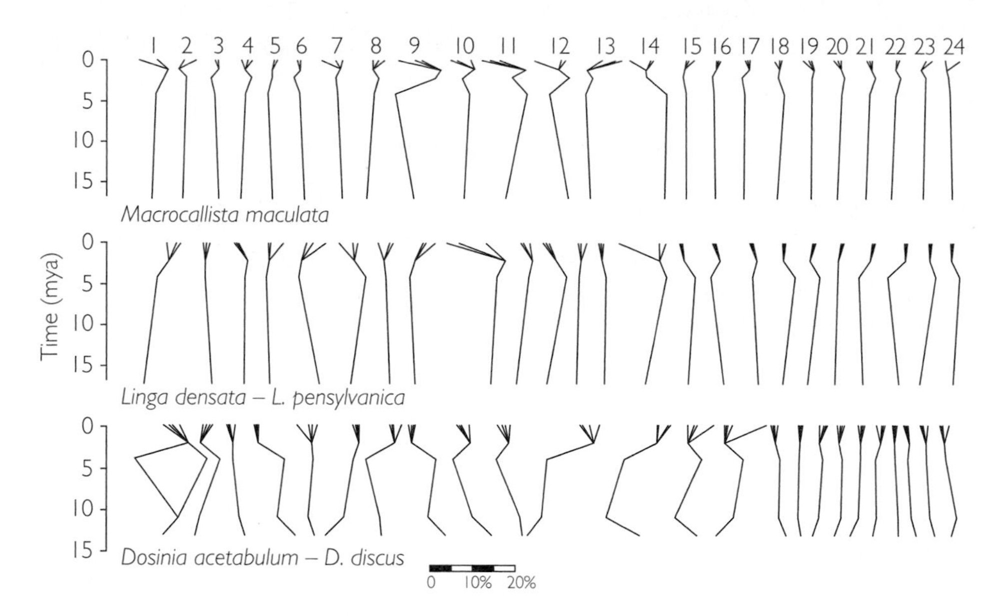

Figure 18.20 Zigzag evolution in Pliocene bivalves results in stasis This diagram shows how 24 different morphological characters change through time in three lineages. The horizontal axis plots percent change in shell morphology between each time interval sampled in the study. From Stanley and Yang (1987).

As a consequence of studies like these, the current view is that there is no single and general explanation for the low rates of morphological change that occur in particular lineages. Stasis is best tested and explained case by case.

18.4 Mass Extinctions

Extinction is the ultimate fate of all species. Over the course of Earth's history, what patterns occur in the rate of extinction? To answer this question, consider a plot that David Raup (1991, 1994) constructed by calculating, for each one-million-year interval over the last 543 million years, the percentage of taxa that went extinct in that interval (Figure 18.21). The histogram has a pronounced right skew, created by a few particularly large events. The most extreme of these intense periods are commonly referred to as **mass extinctions**. They represent intervals in which over 60% of the species that were alive went extinct in the span of a million years. Because of their speed and magnitude, they qualify as biological catastrophes.

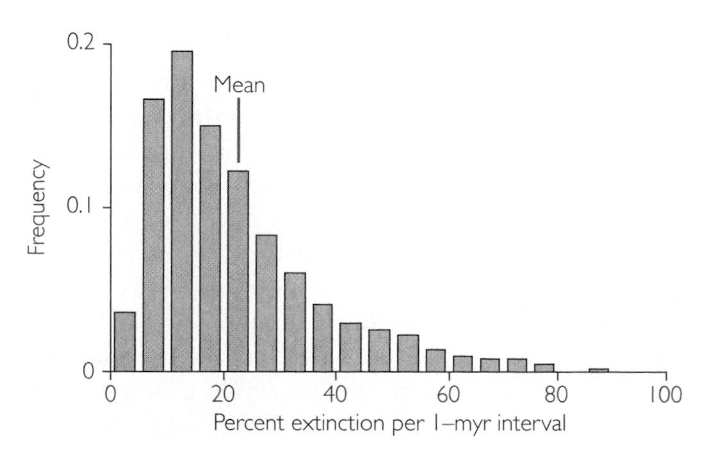

Figure 18.21 Distribution of extinction intensities David Raup (1994) broke the 543-million-year fossil record for the Phanerozoic into 1-million-year intervals and calculated the percentage of species that went extinct during each such interval. Many of the larger events plotted in this figure were recognized early in the 19th century and were used to define boundaries for the eras, periods, and epochs that make up the geologic time scale.

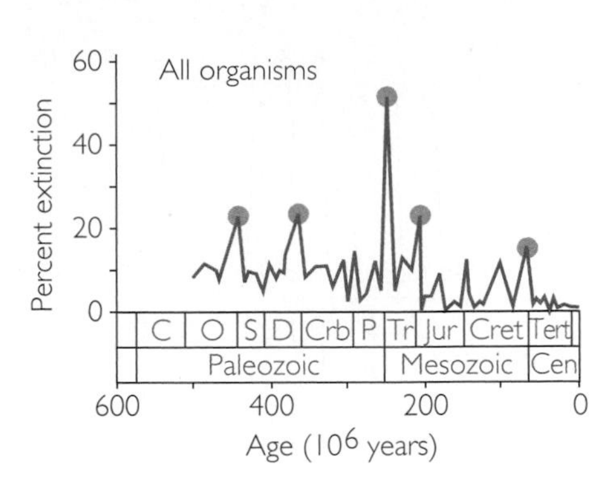

Figure 18.22 Patterns of extinctions of families through time This diagram shows the distribution of mass extinctions events throughout the Phanerozoic. Families are groups of closely related genera. The Big Five extinctions are indicated by red dots. Redrawn with permission from Benton (1995).

How many mass extinctions have occurred during the Phanerozoic? Figure 18.22 plots the percentage of families that died out during each stage in the fossil record over the past 510 million years (Benton 1995). The five prominent spikes that are circled on the graph are traditionally recognized as mass extinctions and are referred to as the Big Five. On the geologic time scale, these events occurred at the terminal-Ordovician (ca. 440 mya), late-Devonian (ca. 365 mya), end-Permian (250 mya), end-Triassic (ca. 215 mya), and Cretaceous-Tertiary, or K-T (65 mya). (Note that the Cretaceous is routinely symbolized with a K to distinguish it from other eras and periods that start with C.)

The five largest extinction events in the Phanerozoic are known as the Big Five.

It is important to recognize, however, that the Big Five are responsible for perhaps 4% of all extinctions during the Phanerozoic. The other 96% of extinctions recorded in Figure 18.21 and 18.22 are referred to as **background extinctions**-meaning that they occurred at normal rates. To distinguish mass extinctions from background extinctions, biologists point out that a mass extinction is global in extent, involves a broad range of organisms, and is rapid relative to the expected life span of taxa that are wiped out (Jablonski 1995). It is difficult to differentiate the two categories of extinction more precisely than this, however. As Raup's analysis makes clear, mass extinctions simply represent the tail of a continuous distribution of extinction events over time.

In this section we look at patterns that occur during times of background extinctions, delve into the causes of a particularly spectacular mass extinction, and ask whether a mass extinction event, caused by human beings, is now underway.

Background Extinction

Several interesting patterns have been resolved from data on background extinctions. First, within any particular group of organisms, the likelihood of particular lineages becoming extinct is constant and independent of how long the taxa have been in existence. Leigh Van Valen (1973) discovered this when he plotted simple survivorship curves for a wide variety of fossil groups. Survivorship curves show the proportion of an original sample that survives for a particular amount of time. For fossil taxa, Van Valen plotted the number of species, genera, or families from an order or phylum of fossil animals that survived for different intervals. He put the number surviving on a logarithmic scale, so the slope of the curve at any point equaled the probability of becoming extinct at that time. Virtually every plot he constructed, from many different fossil groups and eras, produced a straight line. This means that the probability of subgroups becoming extinct was constant over the life span of the larger clade. The data in Figure 18.23 are typical. Note that the slopes of the lines vary from taxon to taxon, meaning that rates of extinction vary

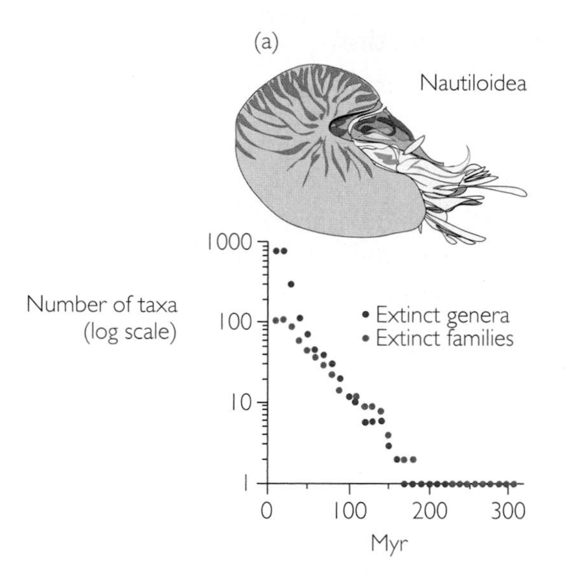

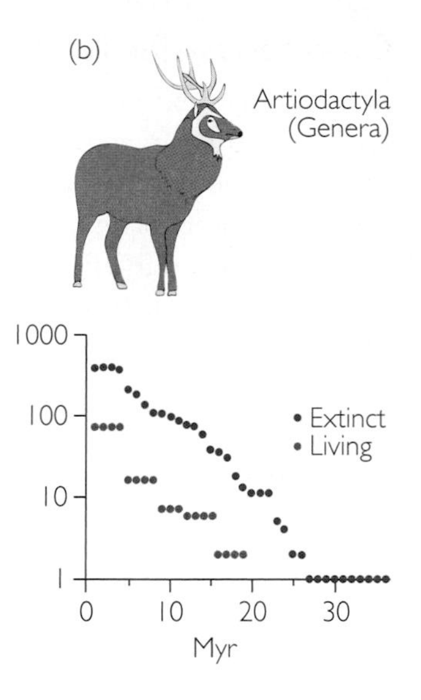

Figure 18.23 Lineage survivorship curves The first step in constructing these curves is to select a random sample of taxa from the fossil record of a particular clade—say, a family or class. The taxa included in the sample can come from any period. The logarithm of the number of genera or families in the clade that survive for different intervals is then plotted. The curves reproduced here are typical (Van Valen 1973): (a) is for genera and families of fossil marine invertebrates called nautiloids, and (b) is for genera in the deer family.

dramatically between lineages. These data indicate that during background times, extinction rates are constant within clades, but highly variable across clades. Why background extinction rates vary among lineages is still a mystery.

Second, in marine organisms, extinction rates vary with how far larvae disperse after eggs are fertilized and begin development. David Jablonski (1986) came to this conclusion by studying extinction patterns in bivalve (clams and mussels) and gastropod (slugs and snails) species from the Gulf of Mexico and the Atlantic coastal plain region over the last 16 million years of the Cretaceous period. Jablonski found that marine invertebrate species with a planktonic larval stage survived longer, on average, than species whose young develop directly from the egg (Figure 18.24). In living species, planktonic larvae are carried on currents and often disperse long distances. This gives them greater colonizing ability, which might reduce the frequency of extinction. Populations with this life history also tend to have larger ranges. Indeed, Jablonski confirmed that geographic range also influences extinction rates: Species with large ranges survived longer than those with more limited ranges (Figure 18.25). Taxa found in small areas are less likely to survive sea-level changes, new predation, new diseases, and other stresses that can lead to extinction.

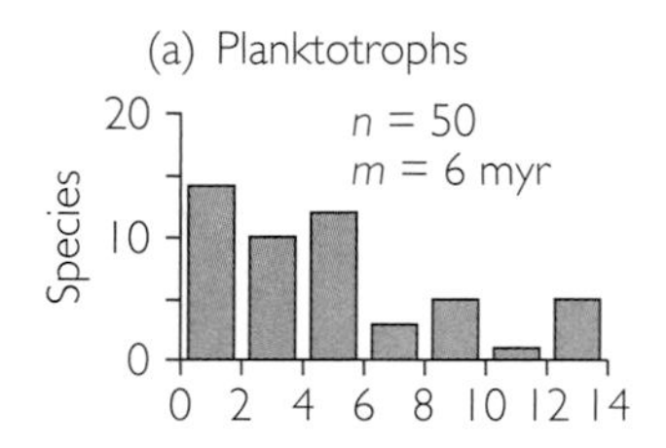

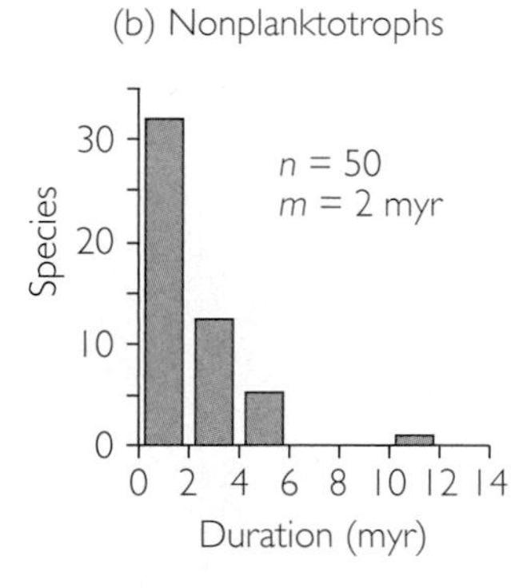

Figure 18.24 How long does a species of marine bivalve exist? Planktotrophs are species with larvae that spend at least some time floating and feeding in the plankton. Nonplanktotrophs are species with larvae that develop directly from the egg. The number of species plotted in each histogram is given by n; m stands for mean and myr stands for million years. Reprinted by permission from Jablonski (1986).

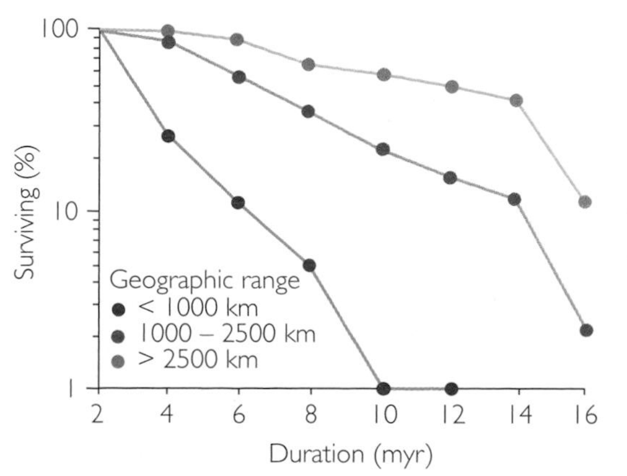

Figure 18.25 Geographic range affects the survivorship of species Jablonski (1986) broke the fossil bivalve and gastropod species in his study into three groups—those with broad, intermediate, and narrow geographic ranges along the Atlantic coast of North America—and created separate survivorship curves. The slopes of these curves give the extinction rate, as in Figure 18.23.

Cretaceous–Tertiary: High-Impact Extinction

Why do mass extinctions occur? The short answer is that they result from short-term, catastrophic episodes of environmental change. But current research suggests that the type of environmental change that occurred, and the underlying cause, are different for each of the Big Five.

Here we examine the impact hypothesis for the extinction at the K-T boundary (Alvarez et al. 1980). The K-T extinction is the best understood of the Big Five, and the impact hypothesis was successful in firing the imaginations of scientists and the lay public. The idea that an enormous asteroid hit Earth and caused widespread extinctions, including the demise of the dinosaurs, provoked intense debate and research.

Evidence for the Impact Event

The discovery of anomalous concentrations of the element iridium in sediments that were laid down at the K-T boundary (Figure 18.26a) was the first clue that an asteroid hit Earth 65 mya. Iridium is rare in Earth's crust but abundant in meteorites and other extraterrestrial objects. Figure 18.26b shows a typical iridium spike found in strata that were laid down over the Cretaceous-Tertiary boundary. Glen (1990) counted 95 different sampling localities from all over the globe that had been found with iridium anomalies dating to the K-T boundary. On the basis of estimates for the amount of iridium needed to produce the anomalies and the density of iridium in typical meteorites, Alvarez et al. (1980) suggested that the asteroid was on the order of 10 km wide; more recent estimates suggest that it may have been as large as 15 km wide. It was, quite literally, the size of a mountain. It would also have been intensely hot, from friction with the atmosphere.

(a) Extinction of shelled organisms caused a clay layer to form

(b) The clay layer contains high concentrations of iridium

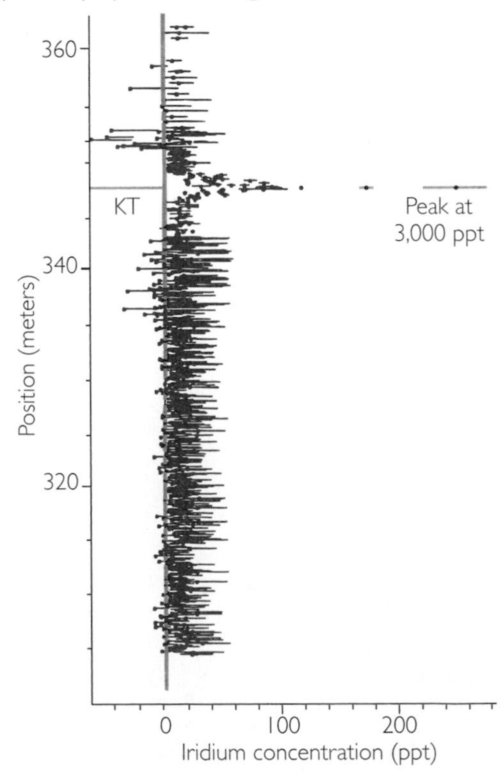

Figure 18.26 Evidence for an impact event at the K–T boundary (a) The dark band is a clay layer that was laid down at the K-T boundary. The limestones on either side are made up of shells from marine invertebrates. A mass extinction event stopped limestone formation and allowed clay to build up. (b) This graph shows the concentration of the element iridium, in parts per trillion (ppt), found in the strata pictured in part (a). The large spike is in the clay layer.

The discovery of two unusual minerals in K-T boundary layers provides additional support for the hypothesis. Shocked quartz particles (Figure 18.27a), produced by intense, short-term pressure, had been found only on the margins of well-documented meteorite impact craters until they were discovered at K-T boundary sites. The other unusual structures are tiny glass particles called **microtektites** (Figure 18.27b). Microtektites can have a variety of mineral compositions, depending on the source rock, but all originate as grains melted by the heat of an impact. If the melted particles are ejected from the crash site instead of being cooled in place, they are often teardrop or dumbbell shaped—a result of solidifying in flight.

(a) Quartz grains shocked by intense pressure

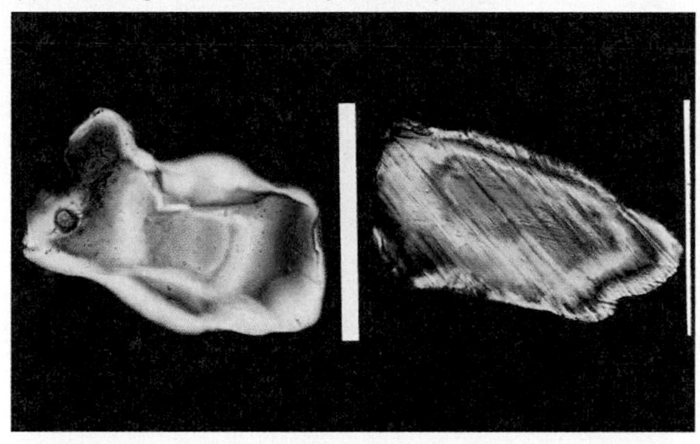

(b) Quartz grains melted into glassy microtektites

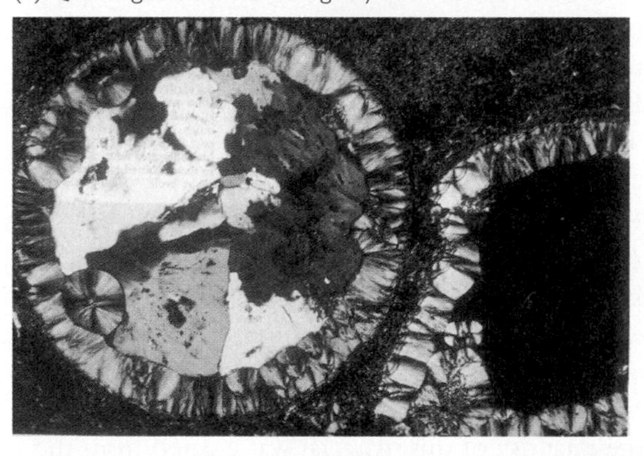

Figure 18.27 More evidence for an impact event at the K-T boundary (a) Small quartz grains (1-2 mm across) with parallel planes called lamellae are routinely found near meteorite strikes. The deformation is thought to be caused by the shock of impact. A shocked quartz grain is shown on the right, a normal grain on the left. (b) Microtektites are spherical or teardrop-shaped particles of glass associated with impact sites. The tektites pictured here have been sectioned to show the interior.

The discovery of abundant shocked quartz and microtektites in K-T boundary layers from Haiti and other localities in the Caribbean helped investigators narrow the search for the crater. Then, in the early 1990s, a series of papers on magnetic and gravitational anomalies confirmed the existence of a crater 180 kilometers in diameter, centered near a town called Chicxulub (cheek-soo-LOOB) in the northwest part of Mexico's Yucatán peninsula (Figure 18.28). The shape of the crater suggested that it was created by an oblique impact-meaning that the asteroid hit at an angle and splashed material to the north and west.

Subsequent dating work confirmed that microtektites from the wall of the crater, recovered from cores drilled in the ocean floor, were 65 million years old (Swisher et al. 1992; but see Keller et al. 2004). This is an almost exact match to dates for glasses ejected from the site and recovered in the Haitian K-T boundary layer.

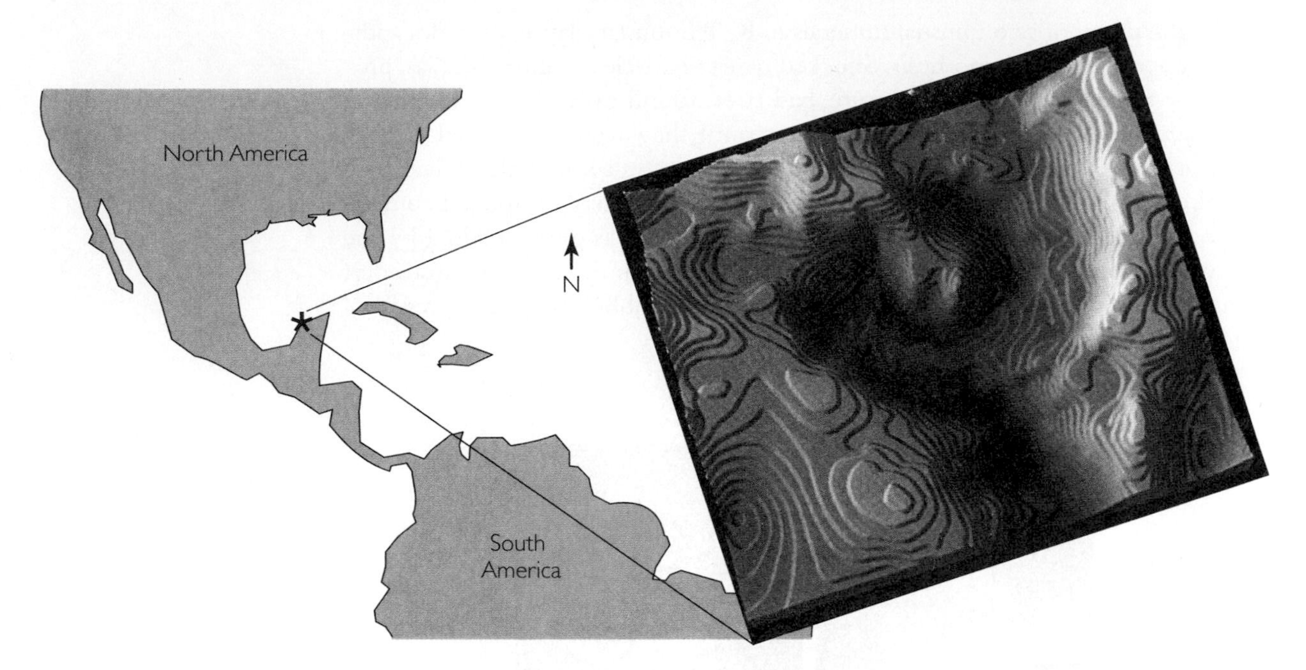

Figure 18.28 Location and shape of the Chicxulub crater Gravitational and magnetic anomalies outline the rim and walls of a 180-km diameter crater, buried beneath sediments near Chicxulub on the Yucatán peninsula. The inset is a relief map of the impact site, showing the gravity field density (Schultz and D'Hondt 1996).

Evidence for an asteroid impact at the end-Cretaceous is now overwhelming.

The discovery of the crater was the long-sought smoking gun. It solidified a consensus among paleontologists, physicists, geologists, and astronomers that a large meteorite struck Earth 65 mya (Shukolyukov and Lugmair 1998; Kyte 1998). The existence of the impact is no longer controversial; the consequences of the impact are.

Killing Mechanisms

The mountain-sized asteroid that struck the ocean would have produced a series of events capable of affecting climate and atmospheric and oceanic chemistry all over the globe. The ocean floor near the impact site at Chicxulub consisted of carbonates, including large beds of anhydrite ($CaSO_4$). The distribution of shocked quartz and microtektites, far to the north and west of Chicxulub, confirms that a large quantity of this material was ejected from the site and that significant amounts were melted or vaporized by the heat generated at impact.

What consequences did the ejected material have? Vaporization of anhydrite and seawater would have contributed an enormous influx of sulfur dioxide (SO_2) and water vapor to the atmosphere. These two molecules would react to form sulfuric acid (H_2SO_4) and produce intense acid rain. Sulfur dioxide is also a strong scatterer of solar radiation in the visible spectrum, which would lead to global cooling (McKinnon 1992). The cooling effect would have been enhanced by dust-sized carbonate, granitic, and other particles. These were ejected in quantities large enough to block incoming solar radiation. Furthermore, soot deposits at numerous K-T boundary localities (Wolbach et al. 1988; Heymann et al. 1994; Vajda et al. 2001) suggest that widespread wildfires occurred, perhaps triggered by the fireball of hot gas and particulates expelled from the impact site. The ash and soot produced by the fires would have added to a widespread smog layer and the accentuation of global cooling. All these data indicate that once the fires triggered by the impact had burned themselves out, Earth became cold and dark.

A variety of models suggest that the force of the impact was also sufficient to trigger massive earthquakes, perhaps as large as magnitude 13 on the Richter scale, and to set off volcanoes. The second largest magma deposits of the Phanerozoic, the Deccan Traps of India, were closely contemporaneous with the K-T extinctions. It is not clear that the Deccan Traps began forming at precisely the same time as the impact, however, and no causal connection has yet been established to directly tie their formation to the impact (Mukhopadhyay et al. 2001; Ravizza and Peucker-Ehrenbrink 2004). If widespread volcanism did occur, it would have added sulfur dioxide, carbon dioxide, and ash to the atmosphere. The ash and sulfur dioxide would accentuate global cooling, while carbon dioxide would contribute to a longer-term greenhouse effect and warming.

Finally, the impact would have created an enormous tidal wave, or tsunami, in the Atlantic Ocean. If the asteroid was indeed 10 km wide, models suggest that the wave produced by the strike would have been as large as 4 km high. The mountain of rock made a mountain-sized splash. Joanne Bourgeois and colleagues (1988) provided evidence of the tsunami when they discovered a huge sandstone deposit along the Brazos River in Texas, which has been mapped throughout northeast Mexico. It is 300 km long and several meters thick and is now interpreted by most geologists as a product of the rapid and massive deposition typical of tsunamis. Additional evidence for a tsunamai is found in the 65-mya sediments of Haiti, where there is a thick jumble of coarse- and fine-grained particles sandwiched between the iridium-enriched clay layer above and extensive tektite deposits below. Florentin and Sen (1990) interpret this middle stratum as the product of tsunami-induced mixing and deposition. This occurred after the initial splash of microtektites, but before the fallout of iridium-enriched particulates from the atmosphere.

The asteroid impact had profound effects on both marine and terrestrial ecosystems.

The asteroid strike would have affected the world's oceans in two ways. Globally, the primary productivity of phytoplankton would have been dramatically reduced by the cooling and darkening of the atmosphere. Locally, temperature regimes and chemical gradients in the Atlantic would have been disrupted by the largest tsunami ever recorded.

These physical consequences of the asteroid impact are dramatic, and they undoubtedly led directly to the rapid elimination of marine and terrestrial biotas in the days or months immediately following the blow. However, a large proportion of the end-Cretaceous extinctions must have been caused by ecological interactions between organisms and their traumatized environment. The decline of many groups was not instantaneous, but rather was drawn out over the 500,000 years following the impact. These extinctions were probably due to the disruption of ecological processes, biogeochemical cycles of nutrients, and interactions among species.

Among calamities such as acid rain, widespread wildfires, intense cooling, extensive darkness, an enormous tsunami, and subsequent ecological disruption, there is no shortage of plausible killing mechanisms for both terrestrial and marine environments. The question now becomes, Are there any patterns indicating which groups died out? If so, do these patterns tell us anything about which of the possible killing mechanisms were most important?

Extent and Selectivity of Extinctions

To date, our best estimate is that 60% to 80% of all species became extinct at the end-Cretaceous (Jablonski 1991). As in other mass extinctions, however, the losses were not distributed evenly across taxa. Among vertebrates, for example, amphibians, crocodilians, mammals, and turtles were little affected. Prominent terrestrial groups like the dinosaurs and pterosaurs (flying reptiles) were wiped out, and large-bodied marine reptiles like the ichthyosaurs, plesiosaurs, and mosasaurs disappeared. Only a few groups of birds survived, while insects escaped virtually unscathed. Among marine invertebrates, the ammonites and the rudists (a group of clams) were obliterated. Marine plankton became so scarce that micropaleontologists describe a "plankton line" at the K-T boundary. At some localities in North America, more than 35% of land plant species became extinct (Schultz and D'Hondt 1996). Pollen and spore deposits dated to just after the boundary show a prominent "fungal spike" followed by a "fern spike." These data suggest that forest communities were dramatically reduced and that forest communities were replaced by widespread stands of ferns once the dead trees had been rotted away by fungi (Nichols et al. 1992; Sweet and Braman 1992; Vajda et al. 2001; Vajda and McGloughlin 2004).

Initially, data on the nature of extinct groups supported the hypothesis that the K-T extinction was size selective, with large-bodied organisms suffering most. The logic was that an extended period of cold and dark after the impact would have affected large animals and plants the most, because of their higher nutritional requirements. But the analyses done to date have shown no difference in extinction rates between small- and large-bodied forms of marine bivalves and gastropods (Figure 18.29). Jablonski (1996) and others have pointed out that large crocodilians survived, while small-bodied and juvenile dinosaurs did not. The selectivity of the vertebrate extinctions is unresolved and still the focus of vigorous debate and research. One hypothesis attracting interest is that large species capable of extended movement to new habitats or extended inactivity, through hibernation or dormancy, were more likely to survive the consequences of the impact.

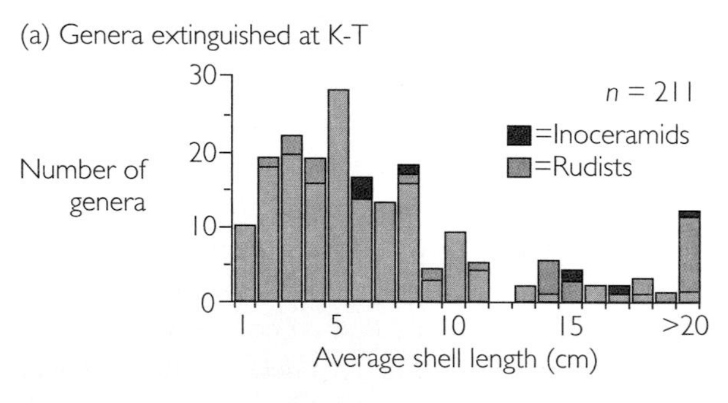

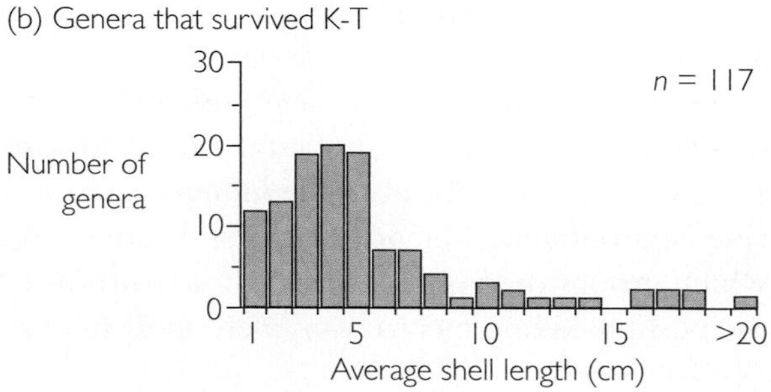

Figure 18.29 Bivalve extinctions at K–T were not size dependent Groups that went extinct at the K-T boundary are shown in (a) and survivors in (b); n is the total number of genera. When the group of unusually large bivalves called rudists are excluded from the analysis, there is no significant difference between victims and survivors (Jablonski and Raup 1995; see also Jablonski 1986, 1996). Reprinted by permission from Jablonski and Raup (1995).

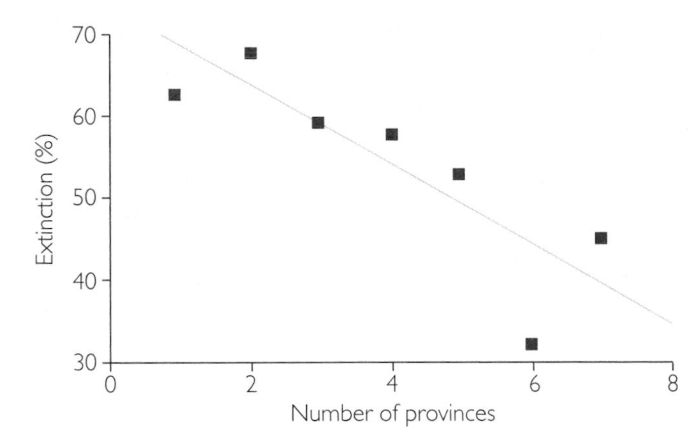

Figure 18.30 Which marine bivalves are most likely to become extinct at the K–T boundary? On this graph, the percentage of bivalve genera that went extinct at the K–T boundary is plotted as a function of the number of biogeographic provinces occupied by those genera. (Biogeographic provinces are regions of the world that share similar floras and faunas.) Genera that occupied a smaller geographic area were more likely to be wiped out during the mass extinction. Reprinted from Jablonski and Raup (1995).

There has been more substantial progress in understanding the selectivity of the marine invertebrate extinctions, largely because the fossil record is so much more extensive (Jablonski 1986, 1991). For example, the probability that a bivalve genus survived the K-T boundary had nothing to do with whether it lived by burrowing or in exposed positions, whether it lived close to shore or far offshore, or whether it occupied tropical or polar latitudes (Raup and Jablonski 1993; Jablonski and Raup 1995). The outstanding pattern that has emerged in studies of bivalves and gastropods is that genera with wide geographic ranges were less susceptible to being eliminated than genera with narrower ranges (Figure 18.30). This is exactly the same result in earlier mass extinction events and for bivalves and gastropod species undergoing background extinction in the interval just before the asteroid impact at the end-Cretaceous (Figure 18.25). Wide geographic ranges clearly buffer marine invertebrate clades against extinction. To date, this pattern represents the most robust result to emerge from the study of both background and mass extinctions.

Extinctions of marine invertebrates at K–T were selective: Genera with broad geographic ranges survived better.

Recent Extinctions: The Human Meteorite

Concern about widespread extinction is on the minds of people from all walks of life, from grade-school children to heads of state. But despite celebrity examples like the dodo, passenger pigeon, and Carolina parakeet, is anything of special evolutionary significance going on now? That is, are we currently experiencing or contemplating an event anything like the Big Five in scale and speed? To answer this question, let's examine data on extinctions that have occurred over the past 2,000 years.

Polynesian Avifauna

David Steadman (1995) has amassed compelling evidence that an important extinction has just occurred among birds. Steadman estimates that 2,000 avian species have been extinguished over the past two millennia in the Pacific region alone, as a result of human colonization of islands. Because slightly over 9,600 species of birds exist today, Steadman's work means that the clade called Aves has recently lost over 15% of its species.

The evidence for this claim comes from archaeological digs throughout the Pacific islands, conducted over the past 20 years by Steadman, Storrs Olson, Helen James, and colleagues. These researchers sampled sites that record bird species present before and after human colonization.

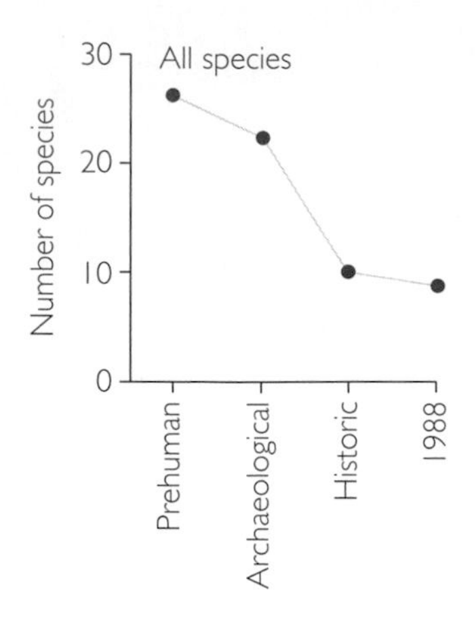

Figure 18.31 Extinction of forest birds on the island of 'Eua This graph plots the number of species present on an island in the South Pacific at four different time intervals. "Archaeological" indicates data from cultures present on the island before contact with Europeans ("Historic"). Reprinted with permission from Steadman (1995).

Archaeological evidence suggests that humans were responsible for the recent extinction of at least 2,000 bird species on islands in the Pacific.

- Sixty bird species endemic to the Hawaiian archipelago became extinct after the arrival of settlers there about 1,500 years ago. These species begin dropping out of the fossil record soon after the first fire pits, middens (trash piles), and tools appear in the digs.
- In New Zealand, 44 bird species became extinct after human colonization, but before historical times. The losses included 8 species of gigantic flightless birds called moas.
- On the island of 'Eua in the Kingdom of Tonga, only 6 of the 27 land birds represented in the prehuman fossil record are still extant (Figure 18.31).
- On each of the seven best-studied islands in central Polynesia, where research has been intense enough to recover and identify at least 300 bones and thus provide a reasonable sample of the fossil avifauna, at least 20 endemic species or populations were wiped out after human settlement.

The total losses are staggering. Extrapolating on the basis of data from the best-studied sites, Steadman suggests that a minimum of 10 species or populations has been lost on each of Oceania's 800 major islands. This estimate is probably conservative. Consider, for example, the fossil record for the small, forest-dwelling, flightless birds called rails. One to four endemic species have been recovered as fossils from each of the 19 islands on which enough research has been done to uncover at least 50 land-bird bones. Extrapolating from this well-studied subset implies that there may have been 2,000 species of rail in the Pacific alone. But just 4 species are left in Oceania. As Steadman (1995: 1127) has written: "Only the bones remain as evidence of one of the most spectacular examples of avian speciation."

Several possible agents of human-caused extinction have been documented in the digs (Olson and James 1982; James et al. 1987; Steadman 1995). In Hawaii, for example, bird bones have been found that are split or charred by fire. These observations suggest direct predation by humans. The presence of pig, dog, and rat remains in the human-associated deposits indicates that these animals were imported by the colonists. Because most islands in the world lacked mammalian predators before the arrival of people, the introduction of rats and dogs was devastating for ground-nesting birds and for the many species, like rails, that had evolved flightlessness. Habitat destruction was also a factor. Records from early European visitors to Hawaii indicate that slash-and-burn agriculture, or permanent irrigated fields for the cultivation of taro, had virtually wiped out the lowland forests of Hawaii before the 1700s.

As a control for the hypothesis that humans were the causative agent, Steadman offers the Galápagos archipelago. The Galápagos lacked any permanent human settlement until 1535 and had a tiny human population until 1800. Archaeological and paleontological work has been extensive on the five largest islands in the group: More than 500,000 vertebrate bones have been excavated. The extensive fossil record documents the loss of only three populations in the 4,000 to 8,000 years preceding the arrival of humans. More than 20 taxa, however, have been eliminated since humans arrived.

Is a Mass Extinction Event Currently Underway?

As impressive (and discouraging) as the data are for recent Polynesian birds, they do not approach the intensity or geographic scope of the Big Five, when over 60% of species were lost. What about more recent rates? About 1.5 million species have been studied and named thus far, but only about 1,100 species have become extinct since 1600. Is concern about an impending mass extinction overblown?

To answer this question, it is important to note that the majority of recently extinct species inhabited islands (Smith et al. 1993b). These island extinctions, in turn, usually resulted from human hunting or the introduction of nonnative predators or competitors. This introduction process has probably peaked in intensity and should be less important in the future.

Instead, current concern is focused on a different agent of extinction: habitat loss due to expanding human populations. The current human population is about 6.5 billion and is growing at the rate of 1.2% per year—an addition of 77 million people annually (United Nations Population Division 2004). If this rate continues, world population will exceed 12 billion by the year 2050. Unless human population growth declines rapidly, threats to natural habitats will grow in intensity over the next several decades.

Biologists have employed three types of approaches to predict how continued habitat destruction will impact extinction rates (May et al. 1995):

1. Multiply the number of species found per hectare in different environments by rates of habitat loss measured from satellite photos.
2. Quantify the rate that well-known species are moving from threatened to endangered to extinct status in the lists maintained by conservation groups.
3. Estimate the probability that all species currently listed as threatened or endangered will actually go extinct over the next 100 or 200 years.

All of these approaches suggest that extinctions are now occurring at 100 to 1,000 times the normal, or background, rate of extinction (May et al. 1995; Pimm et al. 1995). For example, the International Union for the Conservation of Nature estimates that the number of threatened and critically endangered species grew from slightly over 10,000 in 1996 to over 15,000 in 2004. If this rate continued, and if all of these rare species were to go extinct, then it would take less than 100 years for 60% of all species living today to be wiped out. These data suffer from an ascertainment bias, however: the 2004 number is much higher than the 1996 number simply because much more effort has been expended recently in studying endangered populations. Taken together, though, the analyses that have been done to date suggest that if current rates of habitat destruction continue, the coming centuries or millennia will see a mass extinction on the same scale as the Big Five documented in the fossil record. A human meteor is hitting the Earth.

Most observers agree that a mass extinction event is currently underway, caused by expanding human populations destroying or altering habitats.

Where Is the Problem Most Acute?

If you asked a biologist where habitat destruction is causing the most severe threat to biodiversity, the answer is likely to be tropical rain forests. There are two reasons:

- Tropical rain forests are extraordinarily rich in species. E. O. Wilson (1988) recounts that he once collected 43 species of ants belonging to 26 genera from a single tree in a Peruvian rain forest. These numbers are roughly equivalent to the entire ant fauna of the British Isles. Similarly, Peter Ashton identified 700 different species of trees—the same number found in all of North America—in just ten 1-hectare sample plots from a rain forest in Borneo. With the exception of conifers, salamanders, and aphids, nearly every well-studied lineage on the tree of life shows a latitudinal gradient in diversity, with the largest number of species residing in the tropics. Why this pattern occurs is not clear, but the results are striking. Tropical forests occupy less than 7% of Earth's land area, but contain at least 50% of all plant and animal species.

- Tropical rain forests are presently under acute threat. Many nontropical habitats in the Northern Hemisphere, as well as most oceanic islands, have been under continuous occupation by high densities of humans for several hundred years. As a result, the flora and fauna of these nontropical habitats have already sustained numerous extinctions. Andrew Balmford (1996) suggests that the long history of dense human occupation, combined with extinctions caused by radical climate change during the ice ages of the Pliocene and Pleistocene, have put nontropical biomes through an "extinction filter." The plant and animal communities now living in these regions are expected to be relatively resilient in the face of continued human impact. In contrast, many areas of the tropics have been relatively unaffected by humans in recent history and were less affected by glaciation and sea-level changes in the Pleistocene. The tropics are now experiencing the highest rates of growth in human populations and the highest rates of habitat loss.

The threat to these forests is grave. According to the United Nations Food and Agriculture Organization (FAO), total forest loss currently averages about 7 million hectares per year—an area about the size of Scotland. Most of these losses are occurring in the tropics. South and southeast Asia is an area of particular concern because the total forest area is relatively small. In this region, forest loss is averaging about 1.1% per year compared to about 0.7% per year in Africa and the Americas (Laurance 1999). The Brazilian Amazon is also a region of interest, because it is the largest continuous tropical forest in the world. Using satellite photographs, Brazil's National Institute of Space Research estimates that an average of about 25,000 square kilometers of forest was lost each year from 2002–2005—an annual loss equivalent to the state of Massachusetts. The recent pace of forest destruction is up from a rate of about 15,000 square kilometers lost annually in the Brazilian Amazon from 1978–1988 (Skole and Tucker (1993). Biologists also maintain that more than double the amount of cleared forest is adversely affected each year because of edge effects. Forested areas adjacent to clearings undergo dramatic changes: Light levels increase, soils dry, daily temperature fluctuations increase markedly, domestic livestock encroach, and hunting pressure by humans heightens. As Skole and Tucker (1993: 1909) note: the "Implications for biological diversity are not encouraging." Stopping the human meteor will take a combination of lower human population growth rates and sustainable development that preserves tropical forests.

Two factors make the fate of the world's tropical rainforests of particular concern; the high rate of deforestation and the high species diversity.

Summary

The most efficient fossilization processes are compression, impression, casting, and permineralization. Because these events depend on the rapid burial of organic remains in water-saturated sediments, the fossil record is dominated by organisms with hard parts that lived in lowland or shallow-water marine environments. Thanks to new fossil finds and increasingly high-resolution dating techniques, the geological record of life on Earth is steadily improving.

Although some bilaterally symmetric animals were present in the Precambrian, most of the major animal lineages present today first appear in the fossil record during the Cambrian. The Cambrian explosion was characterized by the relatively sudden appearance of large and morphologically diverse animals that swam, crawled, or burrowed and that filled an array of ecological niches in shallow-water marine communities.

The Cambrian explosion is just the most spectacular of a series of adaptive radiations that characterize the rise of morphological complexity and diversity through the Phanerozoic. Adaptive radiations can be triggered by key morphological adaptations or chance events that create an ecological opportunity, such as colonizing a new habitat or surviving a mass extinction. Prolonged stasis is another pattern in evolutionary history. In some lineages, morphological stasis is punctuated by rapid change that occurs during speciation events.

The eventual fate of both new taxa and new morphological traits is extinction. The five most intense extinctions are designated as mass extinctions and are commonly distinguished from background extinctions. The K–T extinction is the best-understood of the Big Five mass extinctions and was caused by an asteroid that slammed into Earth near Mexico's Yucatán peninsula. During both background and mass extinctions, geographically widespread species are less likely to go extinct. During recent times, a prominent extinction has been documented in the loss of bird species on Polynesian islands. Although dramatic, this event was too local to qualify as a mass extinction. But current projections of species loss due to rapid habitat destruction indicate that a mass extinction, caused by humans, may now be underway.

Questions

1. Explain why the following are relatively common in the fossil record:
   - marine-dwelling forms
   - burrowing species
   - recent organisms
   - pollen grains
2. Explain why the following are relatively rare in the fossil record:
   - desert-dwelling forms
   - species that were capable of flight
   - skeletal elements from sharks and rays
   - flowers
3. Define the Phanerozoic and Proterozoic eras.
4. What important events occurred during the following time intervals?
   - the boundary between the Silurian and Devonian periods
   - the Cambrian period
   - the boundary between the Permian and Triassic periods
   - the Cretaceous period
5. In what sense is the Cambrian period "explosive?" In what sense is the term "explosion" misleading?
6. Compare and contrast the Ediacaran and Burgess Shale faunas. What phyla are found in each? How did the species that were present differ in terms of their morphology and ecology?
7. What is an adaptive radiation? State two hypotheses to explain why adaptive radiations occur.
8. Compare and contrast mass extinctions and background extinctions. How do their size and geographic and taxonomic extent differ?
9. Suppose that a species first appears in the fossil record 350 mya. Why is it logical to argue that this species actually existed before this date?
10. What data support the hypothesis that the origin of bilaterians, the deuterostome-protostome split, and the lophotrochozoan-ecdysozoan split all occurred long before the Cambrian explosion?
11. If data confirm that a mass extinction event occurred at the end of the Proterozoic era, what would be the consequences for our understanding of why the Cambrian explosion occurred?
12. Why would a rise in the availability of oxygen help explain why the Cambrian explosion occurred?
13. Give an example of an adaptive radiation. Provide evidence to support the claim that the radiation originated with one or a few species, that it was rapid, and the descendant groups occupy a wide array ecological niches. Suggest a hypothesis to explain why the radiation occurred.
14. List the evidence in favor of the impact hypothesis for the K–T extinction. Which piece of evidence do you find most persuasive, and why?
15. Why would a meteor strike lead to global cooling, or what researchers call an "impact winter?"
16. Do you accept the hypothesis that a mass extinction event is currently underway? Why or why not?
17. Terrestrial fossils from a particular time (say, 230 mya) are patchily distributed around the world. Instead of being evenly distributed over the continents in a continuous thin layer, they often occur in narrow strips or pockets a few miles wide. Why is this?

18. Most fossils of Mesozoic birds are from marine diving birds. Relatively few terrestrial species are known. Does this mean that most Mesozoic birds were, in fact, marine diving birds? Explain your reasoning.

19. One of the (many) mysteries of the K–T extinction is the different fate of ammonites and nautiloids. These were molluscs with buoyant, chambered shells that lived in open-water habitats. Ammonites went extinct during the K–T extinction, but some nautiloids survived. The two groups had different reproductive strategies. Ammonites are thought to have produced many free-swimming young each year that fed near the ocean surface and grew rapidly. In contrast, a female nautilus produces just a few large eggs each year, each of which rests quietly in the depths for up to a year before hatching into a small, slow-growing nautilus. Based on these different reproductive strategies, suggest a possible hypothesis for why the nautiloids, but not the ammonites, might have been able to survive an asteroid impact.

20. In 1996, Gregory Retallack announced that he had found shocked quartz crystals dating from the same time as the end-Permian extinction. What is the implication of this finding? Other geologists point out that no one has found any evidence of elevated iridium from these strata (despite much searching). What is the significance of the lack of iridium?

21. Suppose you are talking to a friend about extinctions, and you mention that humans are known to have caused thousands of extinctions in the last few millennia. Your friend responds "So? Extinction is natural. Species have always gone extinct. So it's really not something we need to worry about." Is your friend correct that extinction is natural? Is the current rate of extinction typical? Is your friend correct that if extinctions are natural, then they are not a problem for the dominant life forms on Earth?

Exploring the Literature

22. Recent research has implicated human activities in the exinction of large animals during the Pleistocene era, including mammoths, giant ground sloths, the birds called moas, and saber-toothed cats. Read the following papers and decide if you agree with this hypothesis.

Barnosky, A. D., P. L. Koch, R.S. Feranec, S. L. Wing, and A. B. Shabel. 2004. Assessing the causes of late Pleistocene extinctions on the continents. *Science* 306: 70–75.

Burney, D. A. and T .F. Flannery. 2005. Fifty millennia of catastrophic extinctions after human contact. *Trends in Ecology and Evolution* 20: 395–401.

Steadman, D.W., P .S. Martin, et al. 2005. Asynchronous extinction of late Quaternary sloths on continents and islands. *Proceedings of the National Academy of Sciences USA* 102: 11763–11768.

Trueman, C. N. G., J. H. Field, J. Dortch, B. Charles, and S. Wroe. 2005. Prolonged coexistence of humans and megafauna in Pleistocene Australia. *Proceedings of the National Academy of Sciences USA* 102: 8381–8385.

Citations

Alvarez, L. W., W. Alvarez, F. Asaro, and H.V. Michel. 1980. Extraterrestrial cause for the Cretaceous–Tertiary extinction. *Science* 208: 1095–1108.

Alvarez, W., F. Asaro, and A. Montanari. 1990. Iridium profile for 10 million years across the Cretaceous–Tertiary boundary at Gubbio (Italy). *Science* 250: 1700–1702.

Avise, J. C., W. S. Nelson, and H. Sugita. 1994. A speciational history of "living fossils:" Molecular evolutionary patterns in horseshoe crabs. *Evolution* 48: 1986–2001.

Balmford, A. 1996. Extinction filters and current resilience: The significance of past selection pressures for conservation biology. *Trends in Ecology and Evolution* 11: 193–196.

Bateman, R. M., P. R. Crane, W. A. DiMichele, P. R. Kenrick, N. P. Rowe, T. Speck, and W. E. Stein. 1998. Early evolution of land plants: Phylogeny, physiology, and ecology of the primary terrestrial radiation. *Annual Review of Ecology and Systematics* 29: 263–292.

Benton, M. J. 1995. Diversification and extinction in the history of life. *Science* 268: 52–58.

Benton M. J., and P .N. Pearson. 2001. Speciation in the fossil record. *Trends in Ecology and Evolution* 16: 405–411.

Benton, M. J., M. A. Wills, and R. Hitchin. 2000. Quality of the fossil record through time. *Nature* 403: 534–537.

Bourgeois, J., T. A. Hansen, P. L. Wiberg, and E. G. Kauffman. 1988. A tsunami deposit at the Cretaceous–Tertiary boundary in Texas. *Science* 241: 567–570.

Briggs, D. E. G., R. A. Fortey, and M. A. Wills. 1992. Morphological disparity in the Cambrian. *Science* 256: 1670–1673.

Brown, K. S. 1999. Deep Green rewrites evolutionary history of plants. *Science* 285: 990–991.

Budd, G. E. 1996. The morphology of *Opabinia regalis* and the reconstruction of the arthropod stem-group. *Lethaia* 29: 1–14.

Burns, K. J., S. J. Hackett, and N. K. Klein. 2002. Phylogenetic relationships and morphological diversity in Darwin's finches and their relatives. *Evolution* 56: 1240–1252.

Butterfield, N. J. 1990. A reassessment of the enigmatic Burgess Shale [British Columbia, Canada] fossil *Wiwaxia corrugata* (Matthew) and its relationship to the polychaete *Canadia spinosa* Walcott. *Paleobiology* 16: 287–303.

Cheetham, A. H. 1986. Tempo of evolution in a Neogene bryozoan: Rates of morphologic change within and across species boundaries. *Paleobiology* 12: 190–202.

Chen, J.-Y., D.-Y. Huang, and C-W. Li. 1999. An early Cambrian craniate-like chordate. *Nature* 402: 518–522.

Chen, J.-Y., D. J. Bottjer, et al. 2004. Small bilaterian fossils from 40 to 55 million years before the Cambrian. *Science* 305: 218–222.

Condon, D., M. Zhu, S. Bowring, W. Wang, A. Yang, and Y. Jin. 2005. U-Pb ages from the Neoproterozoic Doushantuo formation, China. *Science* 308: 95–98.

Conway Morris, S. 1989. Burgess Shale faunas and the Cambrian explosion. *Science* 246: 339–346.

Conway Morris, S. 1998a. Early metazoan evolution: Reconciling paelontology and molecular biology. *American Zoologist* 38: 867–877.

Conway Morris, S. 1998b. *The Crucible of Creation*. Oxford: Oxford University Press.

Conway Morris, S. 2000. The Cambrian "explosion": Slow-fuse or megatonnage? *Proceedings of the National Academy of Sciences USA* 97: 4426–4429.

Cunningham, C. W., N. W. Blackstone, and L. W. Buss. 1992. Evolution of king crabs from hermit crab ancestors. *Nature* 355: 539–542.

Darwin, C. 1859. *On the Origin of Species*. London: John Murray.

De Rosa, R., J. K. Grenier, T. Andreeva, C. E. Cook, A. Adoutte, M. Akam, S. B. Carroll, and G. Balavoine. 1999. *Hox* genes in brachiopods and priapulids and protostome evolution. *Nature* 399: 772–776.

Donovan, S. K., and C. R. C. Paul, eds. 1998. *The Adequacy of the Fossil Record*. Chichester, UK. John Wiley.

Eldredge, N., and S. J. Gould. 1972. Punctuated equilibria: An alternative to phyletic gradualism. In T. J. M. Schopf, ed. *Models in Paleobiology*. San Francisco: Freeman, Cooper & Company, 82–115.

Erwin, D. H., and R. L. Anstey. 1995a. Introduction. In D. H. Erwin and R. L. Anstey, eds. *New Approaches to Speciation in the Fossil Record*. New York: Columbia University Press, 1–8.

Erwin, D. H., and R. L. Anstey. 1995b. Speciation in the fossil record. In D. H. Erwin and R. L. Anstey, eds. *New Approaches to Speciation in the Fossil Record*. New York: Columbia University Press, 11–38.

Erwin, D. H., and E. H. Davidson. 2002. The last common bilaterian ancestor. *Development* 129: 3021–3032.

Fedonkin, M. A., and B. M. Waggoner. 1997. The late Precambrian fossil *Kimberella* is a mollusc-like bilaterian organism. *Nature* 388: 868–871.

Florentin, J-M. R. M., and G. Sen. 1990. Impacts, tsunamis, and the Haitian Cretaceous–Tertiary boundary layer. *Science* 252: 1690–1693.

Glen, W. 1990. What killed the dinosaurs? *American Scientist* 78: 354–370.

Heymann, D., L. P. F. Chibante, R. R. Brooks, W. S. Wolbach, and R. E. Smalley. 1994. Fullerenes in the Cretaceous–Tertiary boundary layer. *Science* 265: 645–647.

Hunter, R. S. T., A. J. Arnold, and W. C. Parker. 1988. Evolution and homeomorphy in the development of the Paleocene *Planorotalites pseudomenardii* and the Miocene *Globorotalia (Globorotalia) maragritae* lineages. *Micropaleontology* 34: 181–192.

Jablonski, D. 1986. Background and mass extinctions: The alternation of evolutionary regimes. *Science* 231: 129–329.

Jablonski, D. 1991. Extinctions: A paleontological perspective. *Science* 253: 754–757.

Jablonski, D. 1995. Extinctions in the fossil record. In J. H. Lawton and R. M. May, eds. *Extinction Rates*. Oxford: Oxford University Press, 25–44.

Jablonski, D. 1996. Body size and macroevolution. In D. Jablonski, D. H. Erwin, and J. H. Lipps, eds. *Evolutionary Paleobiology*. Chicago: University of Chicago Press, 256–289.

Jablonski, D. 2000. Micro- and macroevolution: Scale and hierarchy in evolutionary biology and paleobiology. *Paleobiology* 26: S15–52.

Jablonski, D., and D. M. Raup. 1995. Selectivity of end-Cretaceous marine bivalve extinctions. *Science* 268: 389–391.

Jackson, J. B. C., and A. H. Cheetham. 1990. Evolutionary significance of morphospecies: A test with cheilostome bryozoa. *Science* 248: 579–583.

Jackson, J. B. C., and A. H. Cheetham. 1994. Phylogeny reconstruction and the tempo of speciation in the cheilostome Bryozoa. *Paleobiology* 20: 407–423.

James, H. F., T. W. Stafford, Jr., D. W. Steadman, S. L. Olson, P. S. Martin, A. J. T. Jull, and P. C. McCoy. 1987. Radiocarbon dates on bones of extinct birds from Hawaii. *Proceedings of the National Academy of Sciences USA* 84: 2350–2354.

Keller, G., T. Adatte, et al. 2004. Chicxulub impact predates the K-T boundary mass extinction. *Proceedings of the National Academy of Sciences USA* 101: 3753–3758.

Knoll, A. H. and S. B. Carroll. 1999. Early animal evolution: Emerging views from comparative biology and geology *Science* 284: 2129–2137.

Kyte, F.T. 1998. A meterorite from the Cretaceous/Tertiary boundary. *Nature* 396: 237–239.

Laurance, W. F. 1999. Reflections on the tropical deforestation crisis. *Biological Conservation* 91: 109–118.

Li, C.-W., J.-Y. Chen, and T-E. Hua. 1998. Precambrian sponges with cellular structures. *Science* 279: 879–882.

May, R. M., J. H. Lawton, and N. E. Stork. 1995. Assessing extinction rates. In J. H. Lawton and R. M. May, eds. *Extinction Rates*. Oxford: Oxford University Press, 1–24.

McKinnon, W. B. 1992. Killer acid at the K–T boundary. *Nature* 357: 15–16.

Mukhopadhyay, S. K., A. Farley, and A. Montanari. 2001. A short duration of the Cretaceous-Tertiary boundary event: Evidence from extra-terrestrial helium-3. *Science* 291: 1952–1955.

Nichols, D. J., J. L. Brown, M. A. Attrep, Jr., and C. J. Orth. 1992. A new Cretaceous–Tertiary boundary locality in the western Powder River basin, Wyoming: Biological and geological implications. *Cretaceous Research* 13: 3–30.

Olson, S. L., and H. F. James. 1982. Prodromus of the fossil avifauna of the Hawaiian islands. *Smithsonian Contributions in Zoology* 365: 1–59.

Philippe, H., N. Lartillot, and H. Brinkmann. 2005. Multigene analyses of bilaterian animals corroborate the monophyly of Ecdysozoa, Lophotrochozoa, and Protostomia. *Molecular Biology and Evolution* 22: 1246-1253.

Pimm, S. L., G. J. Russell, J. L. Gittleman, and T. M. Brooks. 1995. The future of biodiversity. *Science* 269: 347–350.

Qiu, Y.-L., J. Lee, F. Bernasconi-Quadroni, D. E. Soltis, P. S. Soltis, M. Zanis, E. A. Zimmer, Z. Chen, V. Savolainen, and M. W. Chase. 1999. The earliest angiosperms: Evidence from mitochondrial, plastid and nuclear genes. *Nature* 402: 404–407.

Raup, D. M. 1991. A kill curve for Phanerozoic marine species. *Paleobiology* 17: 37–48.

Raup, D. M. 1994. The role of extinction in evolution. *Proceedings of the National Academy of Sciences USA* 91: 6758–6763.

Raup, D. M., and D. Jablonski. 1993. Geography of end-Cretaceous marine bivalve extinctions. *Science* 260: 971–973.

Runnegar, B. 1982. A molecular-clock date for the origin of the animal phyla. *Lethaia* 15: 199–205.

Sato, A., H. Tichy, C. O'hUigin, P. R. Grant, B. R. Grant, and J. Klein. 2001. On the origin of Darwin's finches. *Molecular Biology and Evolution* 18: 299–311.

Schultz, P. H., and S. D'Hondt. 1996. Cretaceous–Tertiary (Chicxulub) impact angle and its consequences. *Geology* 24: 963–967.

Shu, D.-G., H.-L. Luo, S. Conway Morris, X.-L. Zhang, S.-X. Hu, L. Chen, J. Han, M. Zhu, Y. Li, and L.-Z. Chen. 1999. Lower Cambrian vertebrates from south China. *Nature* 402: 42–46.

Shu, D.-G., S. Conway-Morris, et al. 2003. Head and backbone of the early Cambrian vertebrate Haikouichthys. *Nature* 421: 526-529.

Shukolyukov, A., and G. W. Lugmair. 1998. Isotopic evidence for the Cretaceous-Tertiary impactor and its type. *Science* 282: 927–929.

Skole, D., and C. Tucker. 1993. Tropical deforestation and habitat fragmentation in the Amazon: Satellite data from 1978 to 1988. *Science* 260: 1905–1920.

Smith, F. D. M., R. M. May, R. Pellew, T. H. Johnson, and K. S. Walter. 1993a. Estimating extinction rates. *Nature* 364: 494–496.

Smith, F. D. M., R. M. May, R. Pellew, T. H. Johnson, and K. R. Walter. 1993b. How much do we know about the current extinction rate? *Trends in Ecology and Evolution* 8: 375–378.

Smith, A. B. 1999. Dating the origin of metazoan body plans. *Evolution and Development* 1: 138–142.

Soltis, P. S., D. E. Soltis, and M. W. Chase. 1999. Angiosperm phylogeny inferred from multiple genes as a tool for comparative biology. *Nature* 402: 402–404.

Stanley, S.M., and X. Yang. 1987. Approximate evolutionary stasis for bivalve morphology over millions of years: A multivariate, multilineage study. *Paleobiology* 13: 113–139.

Stanley, S. M., and X. Yang. 1994. A double mass extinction at the end of the Paleozoic era. *Science* 266: 1340–1344.

Steadman, D. W. 1995. Prehistoric extinctions of Pacific island birds: Biodiversity meets zooarchaeology. *Science* 267: 1123–1131.

Sweet, A. R., and D. R. Braman. 1992. The K–T boundary and contiguous strata in western Canada: Interactions between paleoenvironments and palynological assemblages. *Cretaceous Research* 13: 31–79.

Swisher, C. C. III, J. M. Grajales-Nishimura, A. Montanari, S. V. Margolis, P. Claeys, W. Alvarez, P. Renne, E. Cedillo-Pardo, F. J-M. R. Maurrasse, G. H. Curtis, J. Smit, and M. O. McWilliams. 1992. Coeval $^{40}Ar/^{39}Ar$ ages of 65.0 million years ago from Chicxulub crater melt rock and Cretaceous–Tertiary boundary tektites. *Science* 257: 954–958.

United Nations Population Division. 2004. World Population Prospects, the 2000 Revision: Highlights. *http://www.un.org/esa/population/publications/wpp2004/2004highlights-finalrevised.pdf.*

Vajda, V., J. I. Raine, and C. J. Hollis. 2001. Indication of global deforestation at the Cretaceous–Tertiary boundary by New Zealand fern spike. *Science* 294: 1700–1702.

Vajda, V. and S. McGloughlin. 2004. Fungal proliferation at the Cretaceous-Tertiary boundary. *Science* 303: 1489.

Valentine, J. W. 1994. The Cambrian explosion. In S. Bengtson, ed. *Early Life on Earth*. Nobel Symposium No. 84. New York: Columbia University Press, 401–411.

Valentine, J. W. 2002. Prelude to the Cambrian explosion. *Annual Review of Earth and Planetary Sciences* 30: 285–306.

Van Valen, L. 1973. A new evolutionary law. *Evolutionary Theory* 1: 1–30.

Waggoner, B. 1998. Interpreting the earliest metazoan fossils: What can we learn? *American Zoologist* 38: 975–982.

Wilson, E. O. 1988. The current state of biodiversity. In E. O. Wilson, ed. *Biodiversity*. Washington DC: National Academy Press, 3–18.

Wolbach, W. S., I. Gilmour, E. Anders, C. J. Orth, and R. R. Brooks. 1988. Global fire at the Cretaceous–Tertiary boundary. *Nature* 334: 665–669.

Wray, G. A., J. S. Levinton, and L. H. Shapiro. 1996. Molecular evidence for deep Precambrian divergences among metazoan phyla. *Science* 274: 568–573.

Xiao, S., Y. Zhang, and A. H. Knoll. 1998. Three-dimensional preservation of algae and animal embryos in a Neoproterozoic phosphorite. *Nature* 391: 553–558.

Xiao, S., X. Yuan, M. Steiner, and A. Knoll. 2002. Macroscopic carbonaceous compressions in a terminal Proterozoic shale: A systematic reassessment of the Miaohe biota, South China. *Journal of Paleontology* 76: 347–376.

Zhang, X., D. Shu, Y. Li, and J. Han. 2001. New sites of Chengjiang fossils: Crucial windows on the Cambrian explosion. *Journal of the Geological Society, London* 158: 211–218.

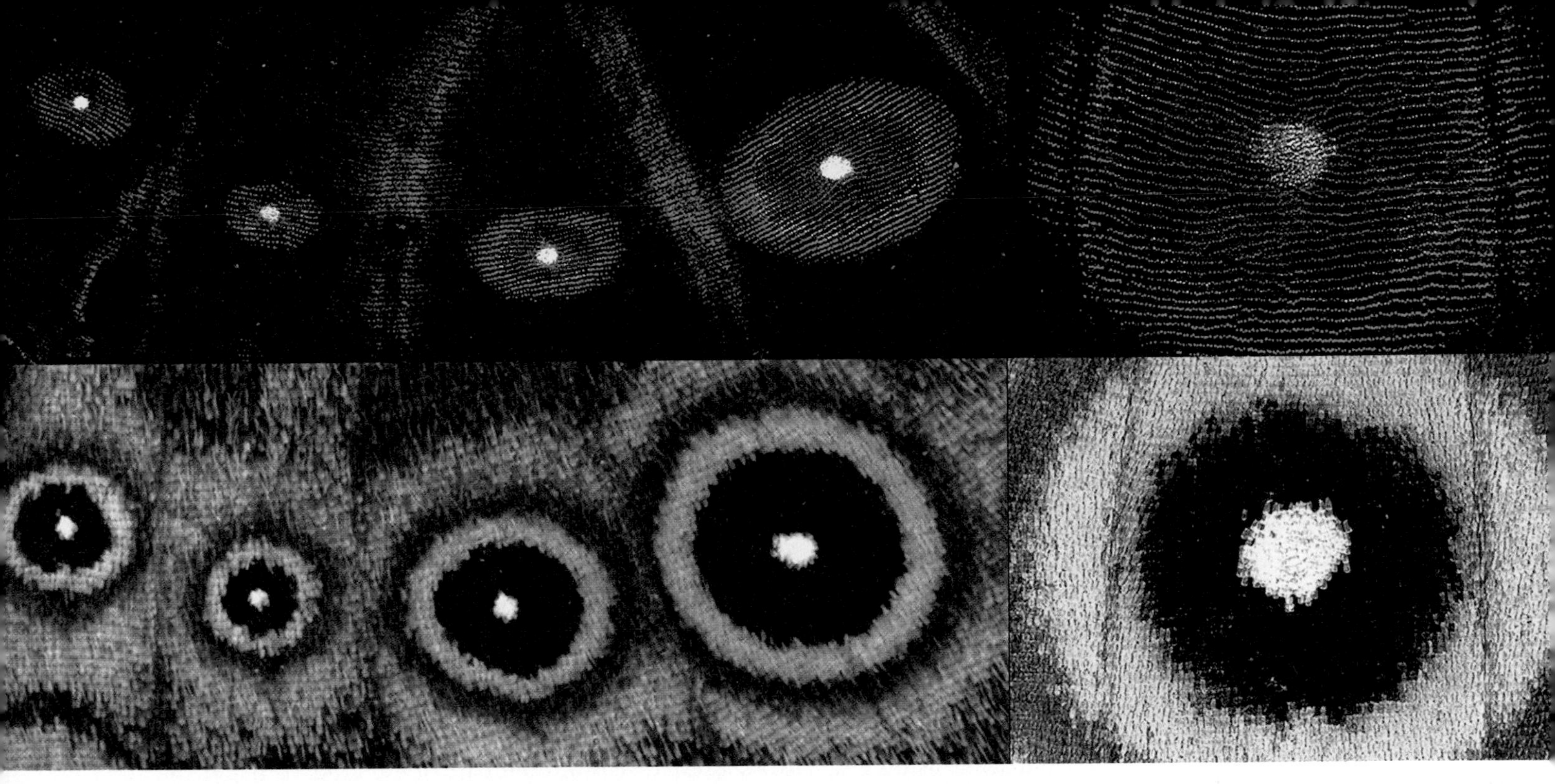

The lower photos show the eyespots on a portion of a *Bicyclus* butterfly wing. The upper photos show where the *En/Inv* (purple) and *Dll* (green) genes are expressed as the wing is developing. The presence, size, and shape of eyespots depend on the expression of *En/Inv* and *Dll*.

19

Development and Evolution

The history of life is marked by great events: the evolution of the first cell, a massive diversification of unicellular forms, the rise of multicellular organisms in an array of eukaryotic lineages, the movement of plants and animals and fungi onto the land. Studying the fossil record and analyzing phylogenies can tell us what has happened over the past 4 billion years, and evolution by natural selection can explain why changes occurred. But what about the mechanics? How did the great events happen? To put the question more specifically: What genetic changes were responsible for key morphological innovations during the history of life—for the evolution of traits like limbs and segmented bodies in animals and flowers in land plants?

Until recently, the answer to these questions was, "Nobody knows." What's more, evolutionary biologists had few tools to use in addressing the problem. Then starting in the late 1980s, a remarkable collaboration involving developmental biologists, systematists who work on phylogenies, and paleontologists created a new research program aimed at understanding the genetic changes that led to great events in evolution. The field became known as molecular paleontology or evo-devo-short for "evolution and development."

This chapter explores what evo-devo is all about, starting with an overview of how molecules control the development of multicellular organisms, a brief look at key advances that launched evo-devo research, and an introduction to some themes that have emerged from the field's early years. The chapter continues with an in-depth look at three case studies in evo-devo and ends with some thoughts about where the field is headed. The overall message is simple: Thanks to two decades of successful research on development and evolution, biologists understand the molecular basis of some of the great events in the history of life.

19.1 The Foundations of Evo-Devo Research

If you ask a developmental biologist to outline how a zygote is transformed into an adult organism, the answer might run something like the following. An embryo starts out as a mass of cells, each of which contains the same genes. Depending on its position in the embryo, a cell receives a specific combination of signaling molecules. These signals may have been sequestered at particular locations in the egg or received from other cells after fertilization occurred. The signals activate proteins called transcription factors that turn certain genes in each cell on or off. Some of the genes that are activated code for other transcription factors that turn still other genes on or off. As development continues, this cascading network of gene activity is constantly modified. New genes are expressed or shut down as the cell divides, receives signals from other parts of the embryo, and sends its own signals. As a result, cells in each part of the embryo begin expressing distinct subsets of the genome and differentiate into recognizable and functioning muscle, vascular tissue, bone, epidermis, cartilage, or parenchyma.

If you asked an evolutionary biologist how phenotypes change over time, the answer would probably center around changes in development. If the form of an animal or plant or multicellular fungus or alga is going to change over time, something about the signaling networks and transcription-factor cascades that run development has to change. Adult phenotypes vary because of variations in development; development varies if a particular signal or transcription factor is produced in a new location, in a different amount, or at a different time.

Morphology changes over time because mutation, selection, and drift act on genes that control embryonic development.

Based on these insights, it follows that the great changes in evolutionary history were caused by mutations that changed developmental sequences and pathways. This is the fundamental premise behind evo-devo research.

Launching the Field

A key paper in the founding of evo-devo earned a Nobel prize for Christiane Nüsslein-Volhard and Eric Weischaus. Their work was based on a mutant screen in the fruit fly *Drosophila melanogaster*: a search among tens of thousands of larvae for mutants with changes in body polarity or segmentation (Nüsslein-Volhard and Weischaus 1980). The mutant hunt allowed them to name and characterize 15 genes involved in the specification of the anterior-posterior axis and the organization of the segmented body plan. Follow-up work in a series of labs identified a large catalog of genes responsible for the formation of structures found in the head or tail or inspecific segments. The work by Nüsslein-Volhard and Weischaus inspired gene hunts in the mustard plant *Arabidopsis thaliana*, which allowed Gerd Jürgens and colleagues (1991) to identify a series of genes involved in the formation of the root-to-shoot axis and Eliot Meyerowitz's and other research groups to characterize genes responsible for specifying flower parts (e.g., Bowman et al. 1989; Bowman et al. 1991).

These early findings in developmental genetics led biologists to wonder: If the gene I'm studying is involved in specifying the nature of body segments in flies or the structure of an *Arabidopsis* embryo or flower, was the origin of this gene partly responsible for the origin of segmented bodies in animals, or the root-shoot distinction in plants, or the flower in angiosperms? Might different alleles of this gene be responsible for variation in body segmentation among arthropods or variation in body shape and flower form among plants? A research program inspired by these types of questions is called a candidate gene approach. Early work in developmental genetics gave evo-devo researchers an array of candidate genes to analyze.

A technical breakthrough—the development of the polymerase chain reaction (PCR)—made it possible to start studying candidate genes in an array of different species. You might recall from Chapter 5 that PCR generates many copies of a specific DNA sequence using an in vitro DNA synthesis protocol. The technique requires that sequences on either side of the target gene are known. Once researchers had sequenced the genes responsible for segmentation in fruit flies or petal formation in *Arabidopsis*, they could use PCR to look for homologous genes in other species. If the gene sequence was different in a species, or if a candidate gene was expressed at a different time or a different location, researchers would begin to suspect that some of the phenotypic differences observed between the species being studied might be due to differences in the candidate gene.

Thanks to these early findings, developmental biologists began to study phylogenies and the fossil record. Systematists used phylogenies to map which taxa contained homologous genes that are important in development and to infer where these genes might have originated. Paleontologists looked to the fossil record to estimate when these genes arose, and began analyzing how they could have led to the innovations documented in the rocks. Research on evo-devo exploded.

A Theme and Variations

Although evo-devo is still considered a young field in evolutionary biology, a major theme is already unifying results on a wide array of genes and traits. As evolution proceeds, the cascade of regulatory genes that direct the development of a trait functions like a groove in jazz, a tala in Indian music, or a beat in hip-hop. It is the basic framework that makes a wide range of variations possible. Once a developmental pathway exists for a particular structure, it is elaborated and modified during evolution to produce an array of phenotypes.

The emergence of this theme should come as no surprise. As you might recall from earlier chapters, natural selection can only act on variation in existing traits. As a result, it is logical to observe that the limbs of bats, seals, humans, and horses all contain variations on an arrangement of bones that existed in the fins of lungfish-like organisms 375 million years ago. That arrangement of bones was encoded by a cascade of developmental signals and transcription factors that were modified over time, by mutation and natural selection, to produce the diversity of limbs observed today. Stated another way, the underlying similarity we see in homolgous structures is due to homologies in the genes that are active as those structures develop. Evo-devo researchers like to call this phenomenon 'deep homology." The *Pax6* genes found in mice and fruit flies are a particularly good example. These genes are similar in structure, and function in the early development of the camera eye of vertebrates and the compound eye of insects. The experimental evidence for homology is strong: When a mouse gene is inserted into a fruit fly embryo, a fly eye forms wherever the mouse gene happens to be expressed during early development (Halder et al. 1995). *Pax6* is part of a regulatory cascade involved in forming eyes. The expression of *Pax6* and other members of this cascade has been modified by evolution to produce the wide array of eyes observed in animals today.

Development is controlled by networks of interacting genes. These networks are often shared by many species and are modified to produce variations of the same trait, such as an eye.

In addition, traits that serve one function can be selected to serve an entirely new function. Like the wrist bone that became elaborated by mutation and natural selection into the panda's thumb (see Chapter 3), genes that code for one regulatory pathway can be expressed in new regions of the body and influence the development of completely new traits. In evo-devo research, biologists refer to this phenomenon as "re-purposing" or "co-opting" of genes to function in the development of new characteristics. The *Distal-less* (*Dll*) gene, which figures prominently later in the chapter, is a good example. *Dll* is expressed early in the development of a wide array of animal species and in an impressive variety of structures. Its protein product is a signal that directs other cells to grow out and away from the main body axis, or distally. *Dll* helps trigger the development of diverse appendages, ranging from flippers to legs to antennae and other sensory structures. It is also involved in the development of the vertebrate forebrain. From its original function, *Dll* has been co-opted to function in a variety of structures that require cells to grow distally.

Gene networks that control the development of a particular trait can also be co-opted or re-purposed to influence the development and diversification of a new trait.

Again and again, evo-devo researchers see homologous genes involved in the formation of homologous structures or co-opted to direct the development of new structures. Evolutionary change occurs when genes involved in regulatory cascades are expressed at new times or location or in new amounts. With the theme of deep homology and re-purposing providing a steady riff, let's delve into some case histories of evo-devo research in action.

19.2 Homeotic Genes and Diversification in Animal Body Plans

The cells that make up a multicellular organism develop in four dimensions: the three spatial dimensions and time. For tissues and organs to form properly, every cell in the growing body has to know where it is relative to other cells and relative to time—meaning, which phase is occuring in the developmental sequence. Specific molecular signals are responsible for telling cells where they are in the embryo at each stage. If these spatial and temporal signals change, then the activity and fate of cells in the embryo change. If cells in the embryo change, then the adult phenotype changes.

Many of the sequences involved in providing locational information are classified as **homeotic genes**. When homeotic genes are knocked out by mutation, structures that usually appear at the location signaled by that gene are replaced by structures normally found elsewhere in the body. When these map molecules are missing, cells think they are in the wrong place and form structures appropriate to the wrong location.

Homeotic genes are active in gene networks that control the overall organization of an animal or plant body, including its size and shape.

In plants, the most important homeotic loci belong to a family called the *MADS*-box genes. In animals, the key homeotic sequences are called the *HOM* or *Hox* genes, depending on whether they are found in invertebrates or vertebrates. Because the *HOM* and *Hox* genes are similar in terms of their DNA sequence, organization on the chromosome, and function, they are considered homologous and are collectively referred to as *Hox* genes.

Soon after their discovery, evo-devo researchers began exploring how changes in *MADS*-box and *Hox* genes could lead to changes in the organization of plant and animal bodies and thus important new phenotypes. In this section we focus on the *Hox* genes of animals; Section 19.3 will explore the function of *MADS*-box genes in flowering plants.

Hox Gene Structure and Function: An Overview

Hox genes have been found in all major animal phyla and share three key traits (Carroll 1995):

- They occur in groups. To explain why similar genes are found in close proximity on the chromosome, researchers hypothesize that the original *Hox* sequence was elaborated by gene duplication events (see Chapter 5) as animals diversified.
- There is a perfect correlation between the 3′–5′ order of genes along the chromosome and the anterior-to-posterior location of gene products in the embryo (Lewis 1978). Genes that are located at the 3′ end of the complex are expressed in the head region of the embryo, while genes located at the 5′ end are expressed toward the posterior of the embryo. In addition, genes in the 3′ end of *Hox* complexes are expressed earlier in development than genes located upstream, and genes at the 3′ end are expressed in higher quantity than genes toward the 5′ end. This phenomenon is called temporal, spatial, and quantitative colinearity and appears to be unique to *Hox* sequences. According to Kmita and coworkers (2002), colinearity occurs because a gene product from one end of the sequence is required to express all of the other genes downstream, and because the regulatory molecule's effectiveness in starting gene expression declines with distance.
- Each gene within the complex contains a highly conserved 180-bp sequence called the homeobox (McGinnis et al. 1984a, b). These bases code for a DNA-binding segment in the proteins encoded by *Hox* genes. The discovery of the homeobox confirmed that *Hox* gene products are transcription factors that bind to DNA and control the transcription of other genes.

Homeotic genes code for regulatory proteins that control the transcription of other genes.

The conclusion that *Hox* genes tell cells where they are in the embryo emerged from experiments with fruit fly mutants lacking specific *Hox* genes (e.g., Lewis 1978; Nüsslein-Volhard et al. 1987; McGinnis and Kuziora 1994). In *Drosophila*, homeotic genes are found in two main clusters called the *bithorax* and *Antennapedia* complexes (Figure 19.1). Mutations in *bithorax* complex genes

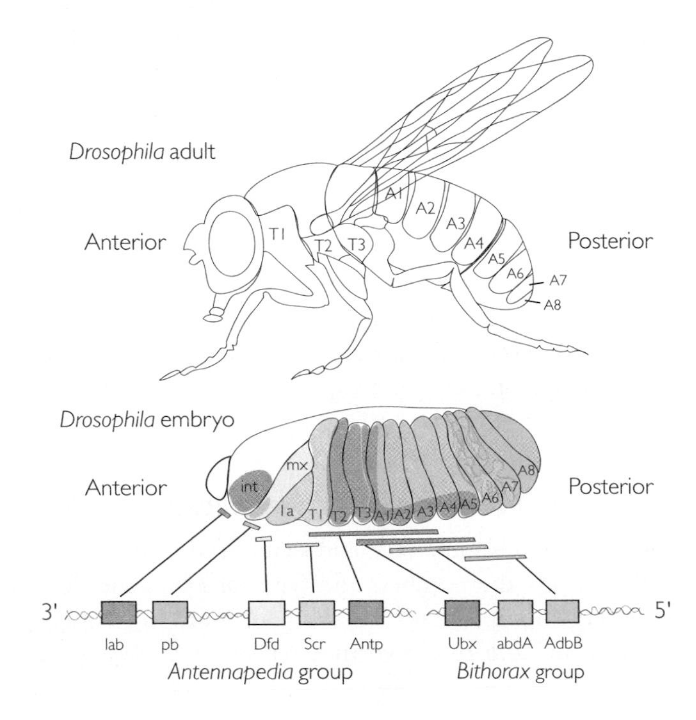

Figure 19.1 ***Hox* genes in *Drosophila*** In these diagrams, T1–T3 and A1–A8 indicate the three thoracic and eight abdominal segments, respectively. The int, mx, and 1a regions of the embryo form head structures. The bottom part of the figure shows the relative locations of genes in the *HOM* cluster. Each gene is color coded to indicate where it is expressed. Expression of *pb*, for example, influences the identity of cells that make the proboscis or mouthparts of the fly; both are shaded green. From Gerhart and Kirschner (1997).

(a)

(b)

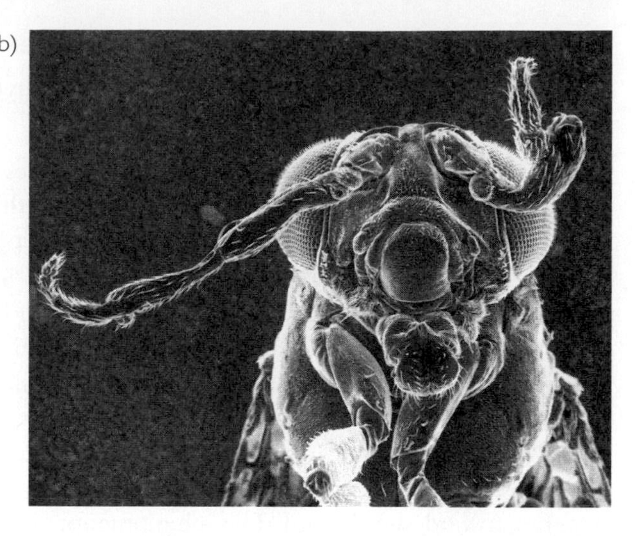

Figure 19.2 **Homeotic mutants in *Drosophila*** (a) Mutations in the *Hox* genes *bx*, *pbx*, and *abx* produce four-winged flies. The phenotype occurs because appendages that should appear on thoracic segment 2 (T2) also appear on T3. In effect, the mutations have changed the identity of T3 to equal that of T2. (b) Mutations in the *antp* gene can produce adult flies with legs growing from the head. In this case, the identity of one head segment has been changed to that of a thoracic segment.

tended to cause defects in the posterior half of embryos, while mutations in *Antennapedia* genes affected the anterior. The key observation was that flies missing one or more *Hox* gene products tended to produce segment-specific appendages such as legs or antennae in the wrong place. The cells that produced the misplaced structures acted as if they misunderstood their location. The general message was that rather than specifying a particular structure, gene products from *Hox* genes demarcate relative positions in the embryo. That is, instead of signaling "make wing," a *Hox* protein indicates "this is thoracic segment 2."

The function of homeotic genes is to define where cells are within the embryo. As a result, changes in the expression of homeotic genes can lead to changes in the organization of the body.

These results helped researchers figure out where *Hox* genes fit in the regulatory cascade responsible for organizing a fly embyro's body. The polarity and segmentation genes that Nüsslein-Volhard and Weischaus discovered turn on certain *Hox* genes in certain locations. Once activated, a *Hox* gene product activates genes responsible for making the structures appropriate for each location in the embryo.

By mixing and matching mutations in the two *Hox*-gene complexes of fruit flies, researchers have been able to change the signals that particular groups of cells in the embryo receive. Because these mutations result in cells that act as if they are in a different location than they actually are, it has been possible to produce adults with four wings instead of two or with legs growing from the head region (Figure 19.2). The four-winged fly is particularly interesting, as data from the fossil record and phylogenetic analyses indicate that flies evolved from ancestors with four wings. The experimental production of a four-winged fly supports the hypothesis that changes in *Hox* gene expression were responsible for a major change in the morphology of insects.

Changes in *Hox* Gene Number: The Diversification of Animals

Once *Hox* genes had been characterized and sequenced in fruit flies, researchers began using PCR to study the numbers and types of *Hox* genes found in other types of organisms. *Hox* homologs have now been found in nonsegmented animals such as jellyfish and the roundworm *Caenorhabditis elegans*, in sponges—

which lack an anterior-to-posterior body axis altogether—and in green plants and fungi (see Kenyon and Wang 1991; Kruse et al. 1994; Seimiya et al. 1994; Finnerty and Martindale 1997). These observations push the origin of the *Hox* complex prior to the evolution of animals.

Within animals, the number of genes within each *Hox* cluster and the total number of *Hox* clusters varies widely (see Kenyon and Wang 1991; Balavoine and Telford 1995; Valentine et al. 1996; Brook et al. 1998; de Rosa et al. 1999). This observation raised an important question: If *Hox* genes help organize the body, and if the number of *Hox* genes has increased over time due to gene duplication events, were changes in the number and function of these genes responsible for at least some of the major evolutionary changes that occurred as animals diversified?

Because Hox *genes are so important for organizing the body, and because the number of* Hox *genes varies among animals, it was logical to hypothesize that changes in* Hox *gene numbers triggered evolutionary changes in morphology.*

Several research groups have tried to answer this question by looking for correlations between the morphology of different groups and the number and types of *Hox* genes they contain. The approach here is to map the presence and absence of *Hox* sequences onto the phylogeny of animals and then use parsimony to infer which genes in the *Hox* complex were gained or lost at key branching points (e.g., Valentine et al. 1996; Erwin et al. 1997; de Rosa et al. 1999; Hoegg and Meyer 2005). An analysis of the data in Figure 19.3 makes several important points:

- The genes called *labial* (*lab*), *proboscipedea* (*pb*), *Deformed* (*Dfd*), *Antennapedia* (*Antp*), and *Abdominal-B* (*Abd-B*) appear to be ancestral to many other genes in the complex. Other sequences within the cluster are probably descended by gene duplication from these original genes.
- A parsimony analysis suggests that 8–10 *Hox* genes existed in the common ancestor of all bilaterally symmetric animals. Because sponges and cnidarians have 5 or fewer *Hox* genes that are not clustered (Kamm et al. 2006), it appears that a dramatic change in the number and organization of *Hox* sequences occurred sometime before the Cambrian explosion. In this case, the number and the organization of genes correlates roughly with the overall complexity of metazoan body plans (Knoll and Carroll 1999). In particular, the addition of the *Abdominal-B* locus appears to be associated with the evolution of bilaterally symmetric animals.
- The entire *Hox* complex was duplicated several times during vertebrate evolution. The closest living relatives of vertebrates and other chordates are the echinoderms, which have just one *Hox* cluster. Most vertebrates have four clusters, however, and ray-finned fish have eight (see Hoegg and Meyer 2005). To explain these observations, researchers suggest that the *Hox* cluster was duplicated twice early in vertebrate evolution, and then again in an ancestor of the ray-finned fish.

Based on the data in Figure 19.3, it is clear that the number of *Hox* genes has increased dramatically over the course of animal evolution. This is an important point, because duplications of individual Hox genes and the entire gene complex may have provided the raw material required for natural selection to produce the diversity of body plans observed today. If mutation creates larger suites of genes involved in specifying locations in the body, then it is possible for larger and more complex bodies to evolve.

In some cases, then, it appears that the number of *Hox* genes present correlates closely with morphological diversification and speciation. Molluscs are an important exception, however. This lineage includes over 100,000 species, ranging from slugs and snails to clams, mussels, squid, and octopuses. Molluscs have just

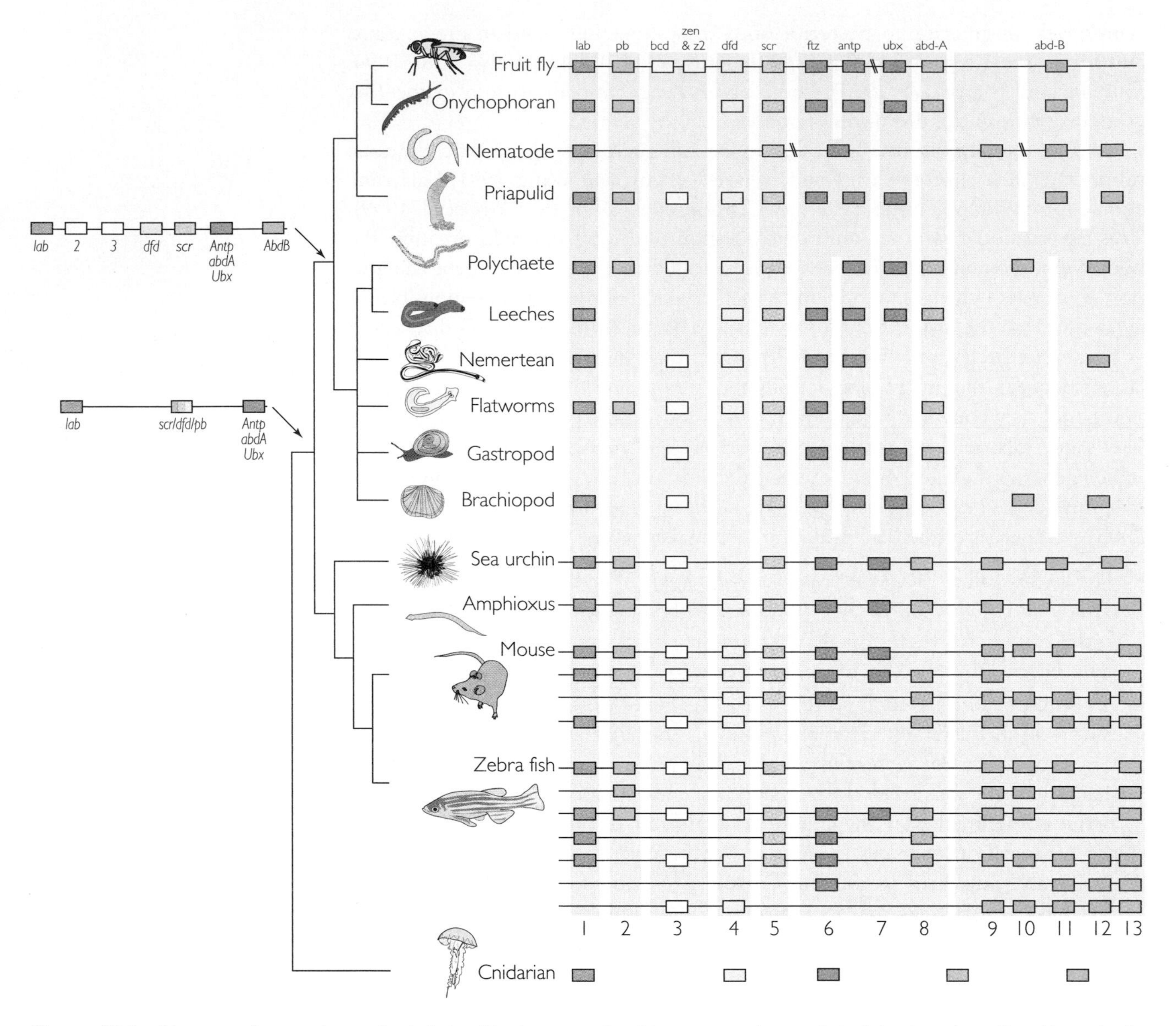

Figure 19.3 ***Hox* genes from various animal phyla** The boxes on the right represent homeotic loci that have been found in each of the taxa on the tree. Genes that are thought to be homologous are lined up vertically; the vertical white lines group genes that are clearly homologous. Modified from Valentine et al. (1996), de Rosa et al. (1999), and Hoegg and Meyer (2005).

3–6 *Hox* genes, however, while tubeworms have 10, mice have 39, and zebrafish have 52.

The current consensus is that changes in the number of *Hox* genes were important during the diversification of animals, but that in some or many lineages, changes in how the available genes were regulated were even more significant (Erwin and Davidson 2002). To get a better idea of how changes in the timing or location of *Hox* gene expression could affect the course of evolution, let's consider recent work on the segmentation of arthropods.

Changes in *Hox* Expression: Arthropod Segmentation

Arthropods are far and away the most diverse animal phylum. Over a million species have been described, and perhaps 50 times that many are still to be named. Members of the group are distinguished by having an exoskeleton, a segmented body organized into head, thoracic, and abdominal regions, pairs of appendages on each body segment, and an open circulatory system. The initial diversification of arthropods occurred during the Cambrian explosion, but many of the lineages living today originated and diversified later. Insects, for example, do not appear in the fossil record until early in the Devonian, about 400 million years ago.

Changes in Hox gene number have been important during animal evolution, but changes in the regulation of Hox gene expression have also been significant.

Although estimating the phylogeny of arthropods is notoriously difficult and controversial (see Nardi et al. 2003; Cook et al. 2005; Regier et al. 2005), current data recognize four major groups: The hexapods include the insects (Figure 19.4a); crustaceans include the shrimp, copepods, barnacles, crabs, lobsters, crayfish, and pill bugs (Figure 19.4b); myriapods encompass the centipedes and millipedes (Figure 19.4c); and chelicerates include scorpions, spiders, mites, horseshoe crabs, and a variety of extinct groups (Figure 19.4d). Prominent among the extinct chelicerates are the eurypterids, or sea spiders, which occasionally reached 3 meters in length. The onychophorans, or velvet worms, are probably the closest living relative of the arthropods. These animals are burrowers that are restricted to the humid tropics (Figure 19.4e).

The key to understanding morphological diversity in arthropods is to recognize how their body segments vary. For example, centipedes have a pair of legs on each of many similar body segments. Insects, in contrast, have three pairs of legs on their thoracic segments, while spiders have four pairs of legs on the thoracic segments. As a centipede or insect or spider embryo develops, the differentiation of these segments is controlled by *Hox* gene expression (Nagy 1998). But despite the diversity in segment size, shape, and appendages observed in arthropods and

Figure 19.4 The arthropod radiation Note the diversity of form and function in the segments and limbs of these representative taxa. Onychophorans are the closest living relative of arthropods.

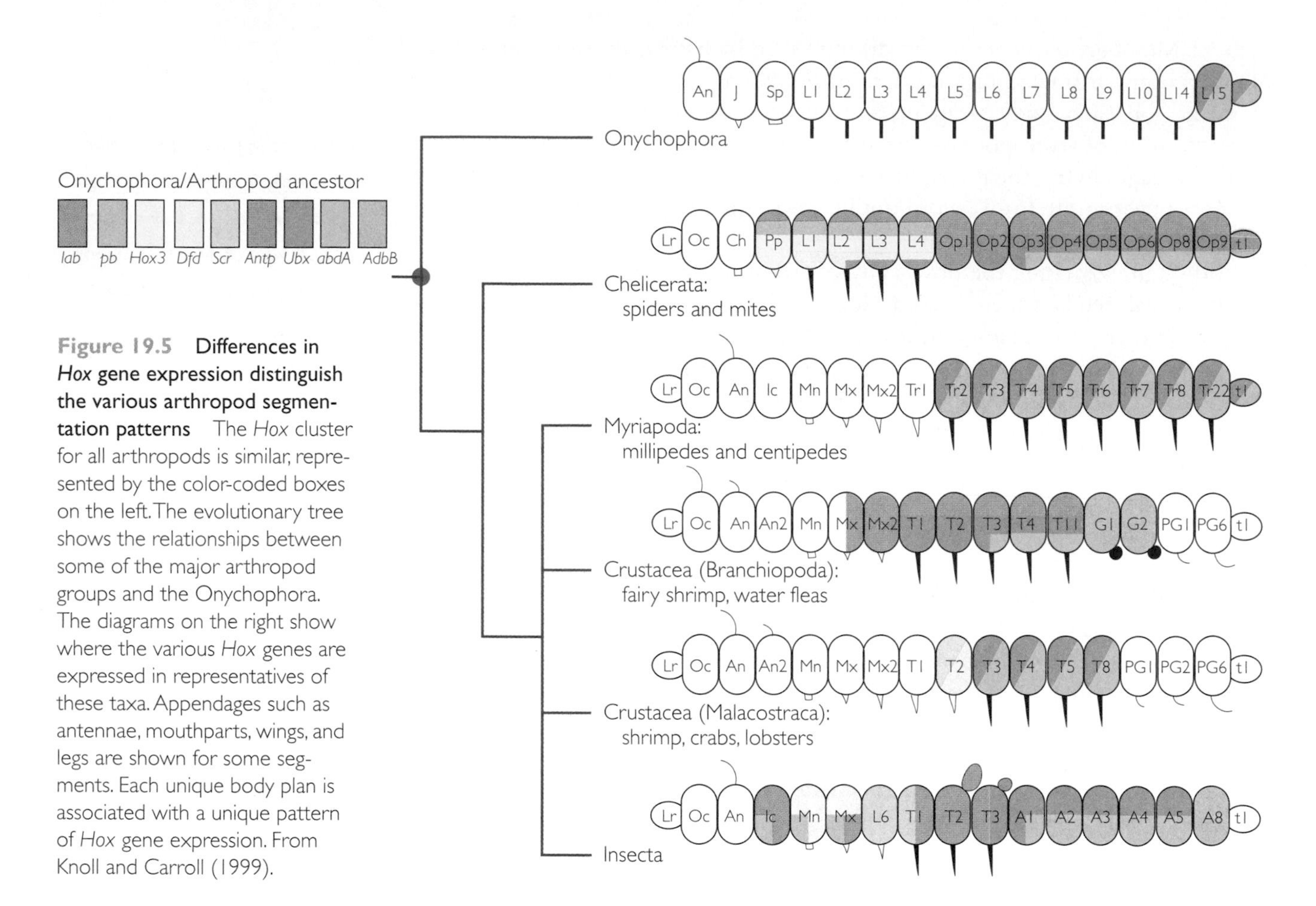

Figure 19.5 Differences in *Hox* gene expression distinguish the various arthropod segmentation patterns The *Hox* cluster for all arthropods is similar, represented by the color-coded boxes on the left. The evolutionary tree shows the relationships between some of the major arthropod groups and the Onychophora. The diagrams on the right show where the various *Hox* genes are expressed in representatives of these taxa. Appendages such as antennae, mouthparts, wings, and legs are shown for some segments. Each unique body plan is associated with a unique pattern of *Hox* gene expression. From Knoll and Carroll (1999).

onychophorans, they all have the same complement of nine *Hox* loci (Grenier et al. 1997). Thus, the morphological variation that we observe must result from changes in the expression of these sequences, rather than from the presence or absence of particular genes.

These changes in gene expression have been summarized by Andrew Knoll and Sean Carroll (1999). Let's begin with onychophorans, which have a pair of unjointed legs on each of many similar body segments. *Hox* expression in their trunk segments is limited to just two genes, called *Ubx* and *abd-A*, which are transcribed only in the most posterior segments (Figure 19.5).

All arthropods have the same suite of homeotic genes. Among species, however, the genes are expressed at different times and in different locations. The variation in homeotic gene expression correlates with variation in the arthropod body plan.

Arthropods share two important aspects of *Hox* gene expression with onycophorans. *Ubx* and *abd-A* are expressed in at least some posterior trunk segments in all of the arthropod groups, and *abd-A* transcripts always appear on the ventral side of a segment relative to *Ubx*.

Other patterns of *Hox* gene expression distinguish the four arthropod groups, however. For example, *Ubx/abd-A* expression does not occur in the most posterior trunk segments of insects and crustaceans. In insects, the gene called *pb* is expressed in the ventral and anterior part of three head segments, where it helps to define the mouthparts. Observations like these support the hypothesis that evolutionary diversification in arthropods is at least partially the result of changes in where shared *Hox* genes are expressed (Nagy 1998; Knoll and Carroll 1999).

Recent experimental work also suggests that at least some key events in the diversification of the arthropod body resulted from changes in the DNA sequence of certain *Hox* genes. In insects, for example, the *Ubx* gene contains a region that codes for a string of alanine residues in the protein product. *Ubx* genes from onycophorans and non-insect arthropods lack this alanine-rich region. To investigate whether this sequence affects how *Ubx* affects body segments in the different groups, Ron Galant and Sean Carroll (2002) created genes that mixed and matched different segments from the *Ubx* sequences of an insect and an onycophoran. One of these mix-and-matched *Ubx* genes was an onycophoran *Ubx* that contained the alanine-rich regions from the fly *Ubx*. When this gene was expressed in fly embryos, they found that the embryo's thoracic segments, which normally bear legs, were transformed to abdominal segments which do not bear legs. In a similar set of results, Matthew Ronshaugen and colleagues (2002) expressed the *Ubx* gene from flies and a crustacean in fly embryos. They found that fly *Ubx*, with its alanine-rich region, was able to repress limb development in the abdomen, while crustacean *Ubx* was not. Follow-up work has shown that efficient blockage of limb development requires reduced expression of both the alanine-rich *Ubx* and *abd-A* (Hittinger et al. 2005).

We now have a specific example of a change in Hox *gene expression that is associated with a dramatic change in animal body plans.*

Taken together, these results support the hypothesis that as insects evolved, mutations created an alanine-rich region in the *Ubx* gene, and that the resulting protein acted as an efficient repressor of leg development in combination with reduced expression of *abd-A*. When the mutant protein was expressed in posterior segments along with reduced *abd-A*, it contributed to the evolution of the legless abdomen observed in insects.

To summarize how mutations in the *Hox* gene complex have affected the diversification of animals, current data suggest that dramatic changes in *Hox* gene number were correlated with changes in morphological complexity on three occasions: early in the evolution of animals, early in the evolution of vertebrates, and early in the evolution of ray-finned fishes. Within the arthropods, changes in where and when the same *Hox* genes are expressed appear to be correlated with changes in the structure and function of particular body segments or regions. Finally, changes in the *Ubx* gene sequence may have contributed to the evolution of the legless abdomen observed in insects. Research continues on how variations in *Hox* gene structure and expression led to the morphological diversification of animals.

19.3 Deep Homology and Diversification in Animal Limbs

The evolution of limbs in animals was one of the most striking innovations in the Cambrian explosion, and diversity in limb structure and function is widely viewed as a key to the ecological and evolutionary success of arthropods and the land-dwelling vertebrates. Among the vertebrates, limbs range from the supple wings of bats to the powerful legs of horses. Among the arthropods, limbs range from filmy fly wings to sturdy crab claws. What genes were involved in the origin and elaboration of these structures? Let's first consider data on this question from the land-dwelling vertebrates, then move on to arthropods.

The Tetrapod Limb

The terrestrial vertebrates include the amphibians, crocodiles, birds, and mammals. The signature adaptation of this lineage, called the Tetrapoda, is the limb. This structure distinguished the early tetrapods from their closest relatives, the ancestors of today's lungfish, and eventually allowed tetrapods to crawl about on

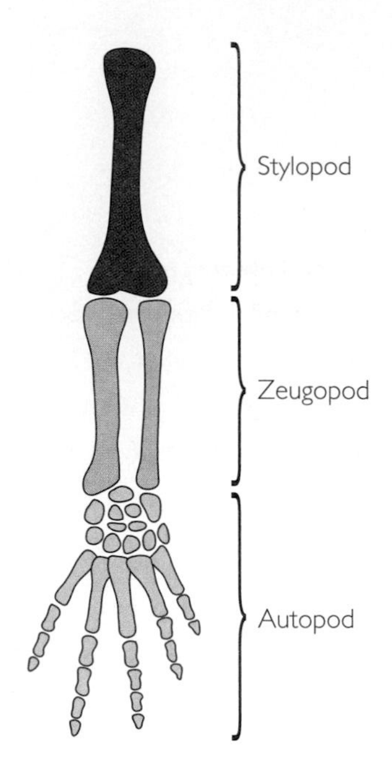

Figure 19.6 The tetrapod limb has 3 basic elements Tetrapod limbs have a single proximal element, or stylopod (called the humerus in the forelimb and femur in the hindlimb), followed by an element with two bones (radius and ulna or tibia and fibula), and an autopod consisting of small elements (carpus or wrist, tarsus or ankle) and the digits (fingers or toes).

land. Later, this innovation was modified into an enormous variety of shapes and sizes, with functions ranging from digging to flying. But in general, tetrapod limbs are variations on the same theme. From frogs to foxes, the number and arrangement of limb bones is similar. A single bony element called the stylopod is closest to the body, followed by paired bones that make up an element called the zeugopod further away from the body, with an autopod consisting of carpels and digits (Figure 19.6).

What Structure Gave Rise to the Limb?

Establishing homologies between the tetrapod limb and ancestral forms is the key to understanding its origin. Phylogenetic analyses show that the sister group of the tetrapods is an extinct lineage of lobe-finned fish from the late Devonian represented by the species *Tiktaalik roseae*, and that the closest living relatives to tetrapods are the lungfishes found today in Africa, South America, and Australia (Daeschler et al. 2006; Shubin et al. 2006). Lobe-finned fish have fleshy fins, in contrast to the flattened structures observed in the more familiar ray-finned fish.

Based on recent analyses of the earliest tetrapod fossils, it is likely that the early tetrapods and their lobe-finned-fish ancestors had internal gills to breathe underwater as well as lungs to breathe air. The skeletons of both groups suggest that they were large predators in shallow, freshwater habitats. Recent analyses suggest that *Tiktaalik roseae* could prop itself up on its limbs and walk or crawl around its shallow-water, floodplain habitat (Daeschler et al. 2006; Shubin et al. 2006).

As Figure 19.7 illustrates, there are important structural homologies between the pectoral fin of *Tiktaalik* and the forelimbs of the earliest tetrapods. Specifically, both appendages are supported by a single element near the body followed by a pair of bones further down the structures. The humerus, radius, and ulna in the tetrapod forelimb and the femur, tibia, and fibula of the hindlimb are homologous with bones observed in ancient lobe-finned fish. In addition, *Tiktaalik* has bones at the end of its fin that resemble the carpels and digits (wrist bones and hand bones) found in the earliest tetrapods.

To summarize, phylogenetic and morphological analyses support the hypothesis that the tetrapod limb is derived from the fins of predatory lobe-finned fish that had lungs as well as gills and that lived in shallow water habitats. The early lobe-finned fishes had a stylopod- and zygopod-like structure as well as some of the elements of an autopod. These results prompt two questions: How was the prototypical tetrapod limb elaborated into the array of appendages observed today, and which genes were responsible for the origin and elaboration of the hand and foot?

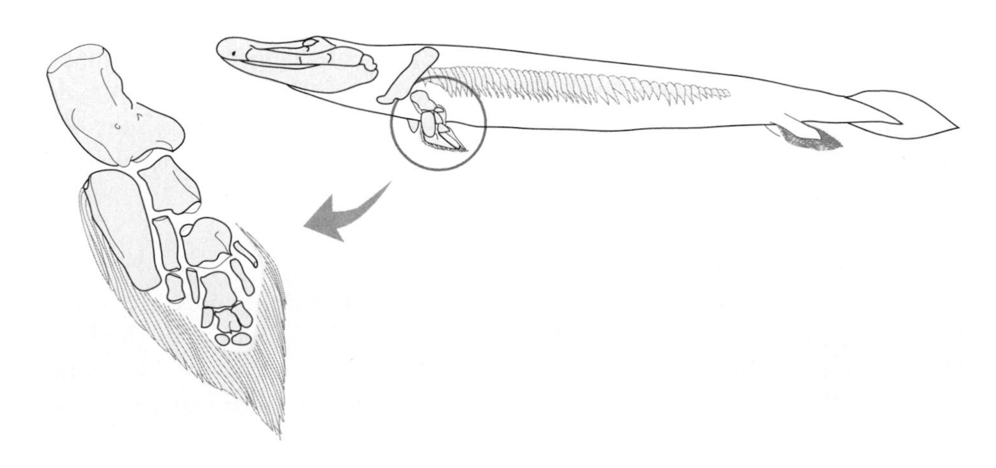

Figure 19.7 Fin bones in lobe-finned fish from the Devonian *Tiktaalik*, the genus pictured here, had fins containing bones similar to the tetrapod limb arrangement. After Daeschler et al. (2006); Shubin et al. (2006).

What Developmental and Genetic Changes Were Responsible for the Origin of the Tetrapod Limb and Its Subsequent Modification?

To understand the genetic changes responsible for the evolution of the tetrapod limb, it's important to understand how limbs develop in contemporary tetrapods. Let's start by investigating the mechanisms that create the shared structure of the tetrapod limb, then consider how these mechanisms first appeared in an ancestor.

Every tetrapod investigated to date shares the same basic features of limb development. Specifically, all tetrapod limbs originate with a bud formed from mesodermal cells (Figure 19.8a). At the tip of this bud is a structure called the apical ectodermal ridge, or AER (Figure 19.8b and c). Cells in the AER secrete a signal that keeps the underlying cells in a growing and undifferentiated state. This population of cells is called the progress zone. Its descendants will eventually be instructed by other signals to form specific elements of the limb such as skeleton, muscle, and connective tissue. The progress zone grows outward and defines the long axis of the developing limb (Figure 19.8c).

Tetrapod limbs have a common ground plan. This results from a shared developmental program.

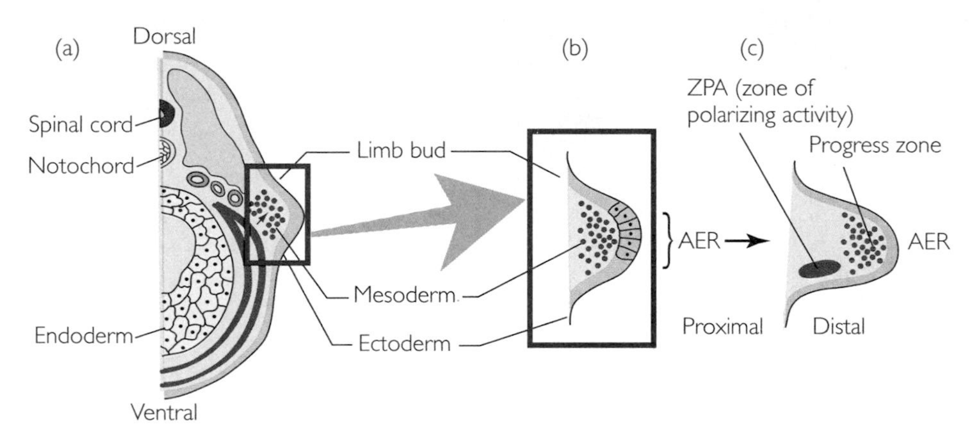

Figure 19.8 The developing limb bud (a) In tetrapods, the limb bud originates from mesodermal cells. (b) The mesodermal cells induce the formation of a structure called the apical ectodermal ridge (AER). (c) Cells in the AER secrete molecules that induce growth.

At the base of the bud in Figure 19.8c is a group of cells called the zone of polarizing activity (ZPA). A molecule secreted from the ZPA diffuses into the surrounding tissue and establishes a gradient that supplies positional information to cells in the structure: The molecule is found at highest concentration near the ZPA and at lower concentration in the parts of the limb bud farthest from the ZPA.

The signals emanating from the AER and ZPA form concentration gradients that provide cells with information about where they are in three-dimensional space. As Figure 19.9 shows, these three axes are described as anterior to posterior, dorsal to ventral, and proximal to distal (Figure 19.9). On your arm, the axes run from the thumb to the last finger (anterior to posterior), from the back of the hand and arm to the palm and underarm (dorsal to ventral), and from the shoulder to the fingertips (proximal to distal).

Thanks to a series of ingenious experiments carried out in a wide array of labs, the molecules responsible for setting up the three-dimensional axis of the developing limb are now known (see Cameron et al. 1998; Tabin et al. 1999):

1. A gradient of the *Sonic hedghog* (*shh*) gene product from the ZPA establishes the anterior-posterior axis of the limb;
2. Proteins called *fibroblast growth factors 4* and *8* (*FGF-4* and *FGF-8*) are secreted by the AER and establish the proximal-distal axis; and
3. The protein product of the *Wnt7a* gene, expressed in cells along the limb bud's dorsal surface, establishes the dorsal-ventral axis.

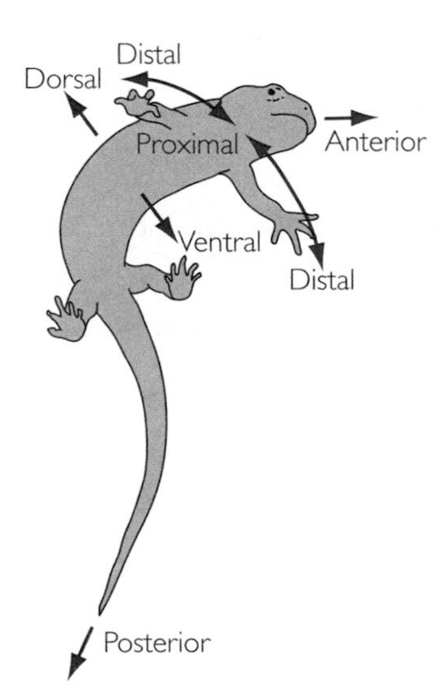

Figure 19.9 Developmental and morphological axes Dorsal = toward the back; ventral = toward the belly; anterior = toward the front; posterior = toward the rear; proximal = toward the main body axis; distal = away from the main body axis.

In addition, distinctive sets of genes from the *Hox* complex respond to these signals as the limb grows distally. Figure 19.10 summarizes how *Hox* gene expression changes as the stylopod, zygopod, and autopod develop. It is clear from these data that *Hox* genes play a large role in identifying where structures inside the limb will form, as well as in patterning the anterior-to-posterior axis of the entire body and marking where limbs will develop. This observation supports the hypothesis that *Hox* genes were co-opted, early in vertebrate development, to organize the structure of limbs.

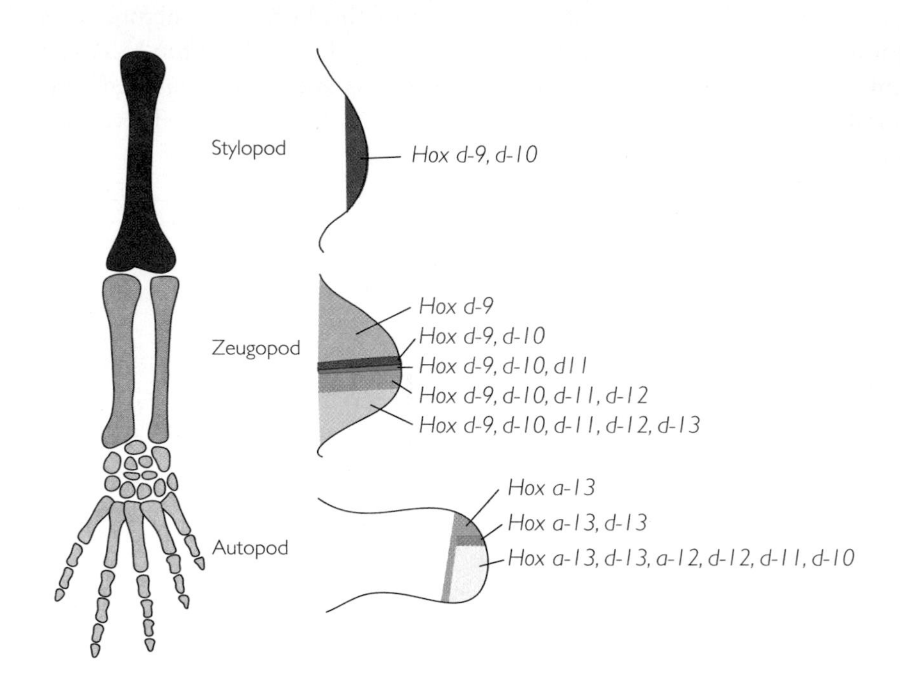

Figure 19.10 **Distinct sets of *Hox* genes are expressed as the stylopod, zeugopod, and autopod form** After Shubin et al. 1997.

The shared developmental program observed in tetrapod limbs is produced by homologous genes.

This brief introduction to the developmental genetics of the vertebrate limb carries two messages. First, homologous genes and developmental pathways underlie the structural homology of amphibian, reptile, bird, and mammal limbs. Second, changes in the timing, location, or level of expression in the pattern-forming genes we have reviewed—the *Fgf*, *shh*, *Wnt*, or *Hox* sequences—could be responsible for adaptive changes in the limb. The evolution of longer limbs, for example, might be based on variation in *Fgf-8* gene expression and how long the progress zone is maintained.

Just this type of genetic connection has been made concerning the origin of the unique portion of the tetrapod limb: the autopod, or hand and foot. To investigate which genes were involved in the evolution of the autopod, researchers have relied on a technique called in situ hybridization. The protocol is based on labeling single-stranded copies of a gene that is involved in limb formation with a dyed or fluorescing compound and then exposing the labeled DNA to fish fins in various stages of development. Cells in the fin that are actively expressing the gene in question contain copies of the gene's messenger RNA. The labeled DNA binds to this mRNA, making it visible. As a result, researchers can determine when and where various genes are expressed as fins develop and compare them with similar experiments in vertebrate limbs.

The pattern of expression of *shh* and various *Hox* genes in fish fins, when contrasted with results for the same genes in a mouse or chicken, is striking. Early in development, gene expression in tetrapods and fish is similar. *Hoxd-11* transcripts, for example, localize to the posterior margin of both elongating zebrafish fins and limb buds (Figure 19.11a; see Sordino et al. 1995; Sordino and Duboule 1996). But in tetrapods a switch occurs. Later in development, *Hoxd-11* gene transcripts are found in the anterior and distal parts of the mouse limb bud. There is also a late expression of *shh*. Why are these differences interesting? Because they do not happen in zebrafish. There is no late expression of *Hoxd-11* or *shh* in the limb bud of zebrafish. Researchers also found striking differences in the way that *Hoxa-11* and *Hoxa-13* are expressed in zebrafish versus tetrapods versus paddlefish, which are descended from a lineage that split off prior to the group containing zebrafish. In zebrafish and paddlefish, *Hoxa-11* and *Hoxa-13* are usually expressed in the same parts of the limb bud. But in tetrapods, the expression patterns of the two genes do not overlap at all. Only *Hoxa-13* is expressed on the portion of the limb that becomes the autopod (Figure 19.11b; see Metscher et al. 2005).

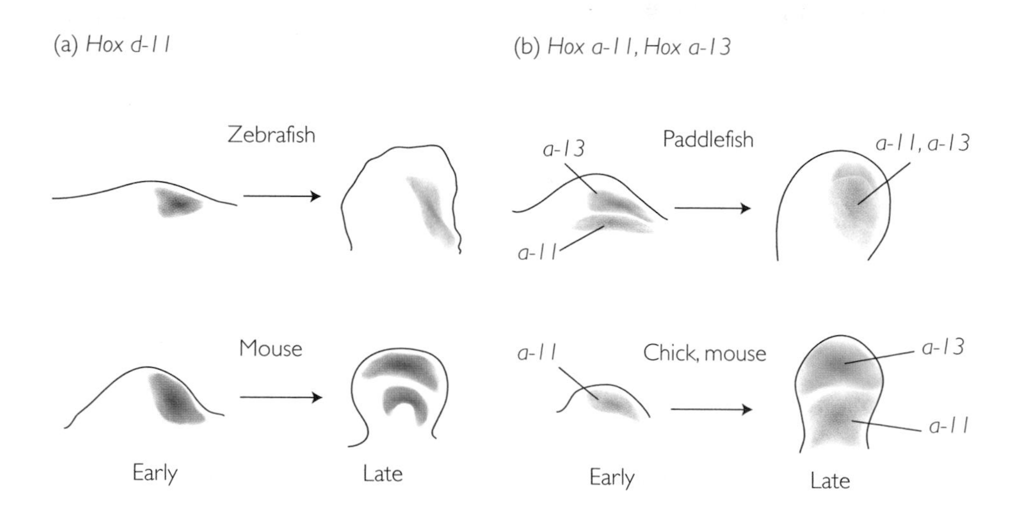

Figure 19.11 Unique patterns of *Hox*-gene expression are associated with the evolution of the autopod These diagrams show patterns of *Hox*-gene expression in several species of ray-finned fish versus tetrapods. (a) *Hoxd-11* is expressed late in limb bud development in tetrapods, but not in zebrafish. After Sordino et al. 1995. (b) *Hoxa-13* is expressed by itself in the distal part of the late limb bud, in contrast to the pattern in paddlefish. After Metscher et al. 2005.

These data support the hypothesis that changes in the timing and location of *shh* and *Hox* gene expression were responsible for producing tissues that became the first hands and feet in the history of life. Mutation caused genes that are active in limb development to be expressed at a different time or location. In this way, genes in established regulatory cascades were co-opted to direct the formation of a novel structure: the autopod. The regulatory changes responsible for the difference in timing and location of expression are still a mystery, however. Research continues.

Tetrapods may have gained hands and feet because of a change in the timing and location of homeotic gene expression. Mutation and selection re-purposed existing genes to organize the tissues in a novel structure.

The Arthropod Limb

Limbs are a hallmark adaptation of arthropods, just as they are in tetrapods. Arthropod appendages come in two basic forms: uniramous (one branch) and biramous (two branches). Chelicerates and hexapods have unbranched appendages, while trilobite and most crustacean limbs have two elements (Figure 19.12). In addition, several crustacean groups have multiply branched or phyllopodous (leafy feet) appendages that are used for swimming. The position, size, and function of biramous, uniramous, and phyllopodous limbs vary widely. They may be located on the head, thoracic, or abdominal segments and can be used for swimming,

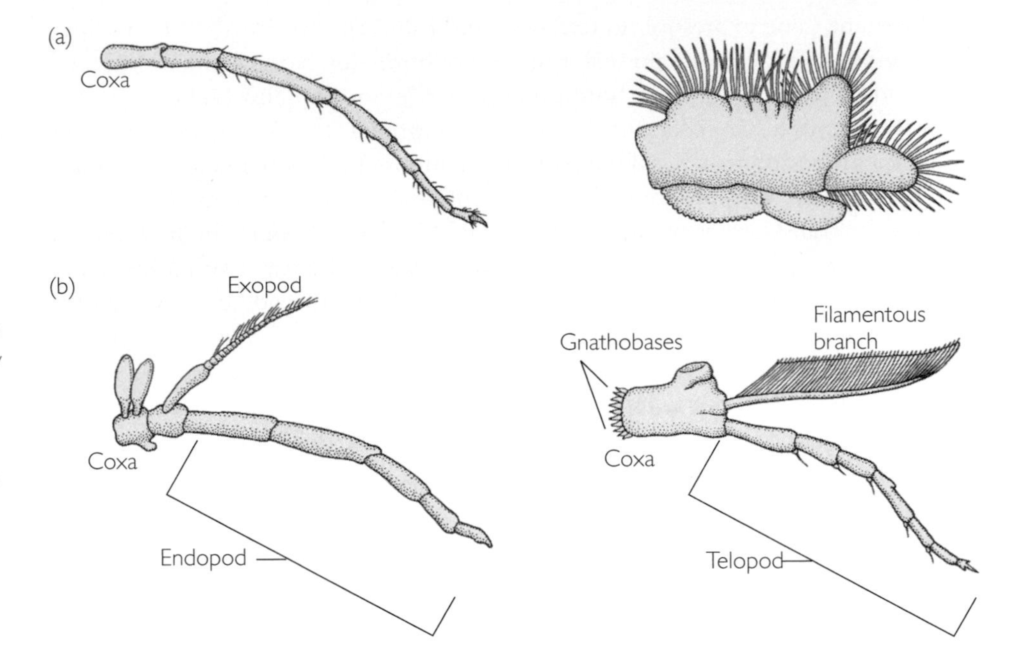

Figure 19.12 Unbranched and branched limbs of arthropods This figure shows just a few of the many variations on the jointed limbs of arthropods. A key distinction among arthropod limbs is whether they are uniramous (a) or biramous (b). Major groups within the phylum tend to have one type of limb or the other, except that some crustaceans have secondarily derived uniramous limbs. Modified from Brusca and Brusca (2002).

Because their limbs are so diverse, arthropods are able to move in different ways and capture many different types of food. The development of these diverse arthropod limbs is based on a common underlying genetic program.

walking, breathing, fighting, or foraging. The ecological diversification of arthropods goes hand in hand with the elaboration of this morphological innovation.

What do we know about the genes involved in the evolution of arthropod appendages? Research on *Drosophila melanogaster* has established three major types of genetic control over limb formation in arthropods (for a recent review, see Angelini and Kaufmann 2005):

- The decision whether to make a limb depends on a gene called *wingless* (*wg*). Mutant flies that lack *wg* fail to make any limb primordia at all. *Wg* is expressed in the anterior section of limb primordia; the protein products of a gene called *engrailed* (*en*) show up in the posterior. These observations suggest that the two genes may be responsible for defining the anterior-posterior axis of the early limb.
- The decision to extend the limb primordium distally hinges on the expression of a gene called *Distal-less* (*Dll*). This is the first gene activated specifically in limb primordia.
- The decision on which type of limb will develop is controlled by homeotic genes. This is not surprising because the fate of a limb primordium should depend on its position in the embryo, and homeotic genes specify location.

What do these observations have to do with the radiation of arthropods? There are hints, from studies on *Drosophila melanogaster*, that changes in these genes correlate with significant evolutionary events. Specifically, abnormal expression of *Dll* can lead to branched limbs in fruit flies, even though their appendages normally have just one element. This result suggests that changes in the regulation of the gene could have played a role in the evolution of uniramous, biramous, and phyllopodous appendages. Variation in when or where *Dll* is turned on or off might lead to new or modified limb outgrowths.

Grace Panganiban and colleagues (1995, 1997) were able to test this hypothesis by making an antibody to the *Dll* protein product and using it to find out where *Dll* is expressed in arthropod embryos. (Antibodies are proteins that bind

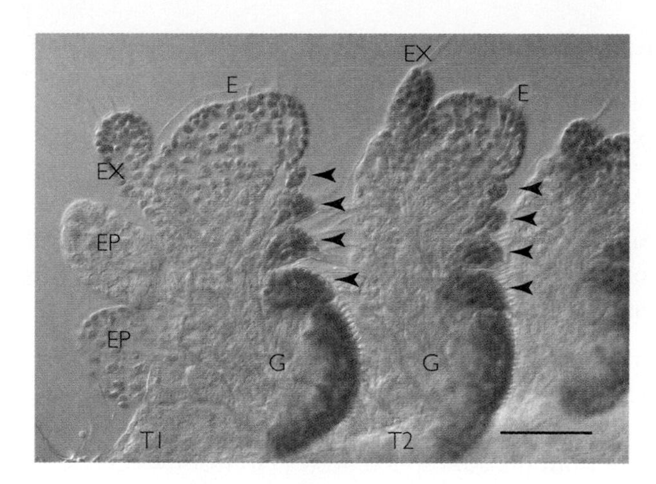

Figure 19.13 ***Distal-less* is expressed in each branch of crustacean phyllopodous limbs** The dark staining grains indicate the protein product of *Dll*. In both the first and second thoracic segments of the common brine shrimp (*Artemia franscicana*) pictured here, marked T1 and T2, all of the limb branches (indicated by the letters and arrowheads) contain *Dll* product. The scale bar represents 0.1 mm. From Panganiban et al. (1995).

to a specific site on a target molecule.) One set of experiments involved the common brine shrimp *Artemia franscicana* and the opossum shrimp *Mysidopsis bahia*. *Artemia* has phyllopodous appendages on its thorax that are used for swimming. *Mysidopsis* has several types of biramous appendages: antennae on its head, walking legs on its thorax, and pleopods used for swimming on its abdomen. In *Artemia*, *Dll* is expressed in each of the outgrowths present in phyllopods (Figure 19.13). In *Mysidopsis*, the anti-*Dll* antibody localizes to each branch of the biramous appendages (Figure 19.14). *Dll* seems to be regulated in different ways in different body regions of *Mysidopsis*, however. In the thorax, the two limb branches grow out from two independent clusters of cells, each of which expresses *Dll*. In the thorax, then, the expression of the gene varies spatially. But in the abdomen, the branches of pleopods form sequentially. In this case, the gene is turned on at two different times in the same location. This means that *Dll* is regulated temporally. In both cases, there is a close correlation between when and where *Dll* is turned on or off and where limbs are produced. The message here is that changes in when and where the gene is expressed appear to affect the branching pattern of arthropod limbs.

Early in development

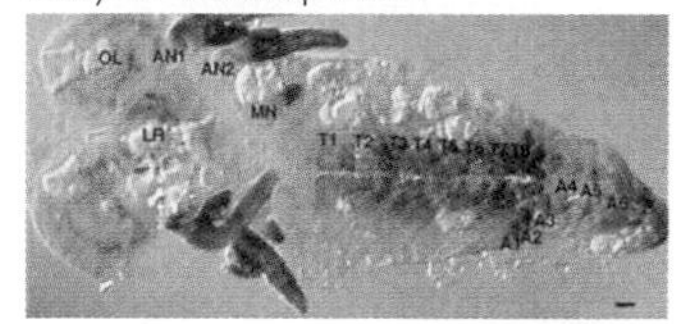

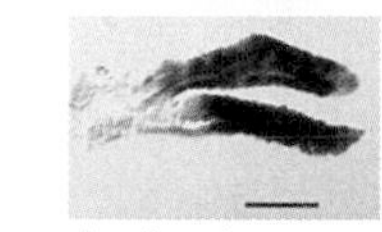

Later in development

Figure 19.14 ***Distal-less* is expressed in both branches of crustacean biramous limbs** The dark-staining grains in these photos indicate the protein product of *Dll*. The photos show how biramous limbs in the first and second thoracic segments (T1 and T2) of the opossum shrimp *Mysidopsis bahia* develop through time. The scale bars represent 0.1 mm. From Panganiban et al. (1995).

Panganiban and coworkers (1997) also used their anti-*Dll* antibody to stain embryos from bilaterally symmetric species that have all sorts of different limb-like appendages, ranging from the saclike outgrowths of onycophorans to the hairlike parapodia of segmented worms. Their results were striking: *Dll* is expressed in the body wall and in the distal parts of the lobopods of the onychophoran *Peripatopsis capensis* as well as in the simple parapodia of the annelid worm *Chaetopterus variopedatus*. In addition, their data show that *Dll* is expressed in the tube feet of a sea urchin and in the ampullae (attachment structures) of a tunicate. In each case, they detected *Dll* expression before these body-wall outgrowths are formed and then again later, in the cells of the more distal parts of the outgrowth.

In all of the bilaterians that have been examined to date, then, *Distal-less* seems to instruct cells to form an outgrowth with proximal-distal polarity. Stated another way, every type of appendage-like "sticky-outey" (Nagy 1998) observed in animals appears to be homologous at a genetic level. The role of *Dll* in the formation of arthropod appendages has become a classic example of deep homology. In this case, the same regulatory gene directs the development of an array of appendages with diverse structures and functions, in a wide variety of animal lineages.

Variation in the timing or location of Dll *expression appears to correlate with variation in the number, location, and shape of arthropod limbs.*

19.4 Homeotic Genes and the Evolution of the Flower

Our third and final example of "molecular paleontology" concerns the most species-rich group of land plants. Multicellular organisms are found in terrestrial environments starting about 440 million years ago, when a species derived from green algae first made the transition to land. There have been four major radiations of land plants since then: early groups called the Prilophyta and Rhyniophyta, ferns, early seed plants without flowers, and the angiosperms or flowering plants. Each of these radiations was associated with one or more evolutionary innovations:

- Cuticle, a waxy material that covers epidermal cells, cuts down on water loss through transpiration and allowed early land plants to tolerate drying. The evolution of guard cells and pores provided breaks in the cuticle and a mechanism for acquiring carbon dioxide from the atmosphere instead of from water.
- Vascular tissues provided structural support and efficient water transport, allowing land plants to grow beyond small size and setting the stage for a radiation of ferns and other large plants about 390 million years ago. These cells provide a mechanism for conducting water along the potential gradient created by water loss at the leaves and uptake at the roots.
- Pollen and seeds are multicellular structures with a protective coat that increased the efficiency of reproduction on land. Pollen can be transported by wind while sperm are restricted to swimming; seeds can be transported far from the parent plant yet contain a nutrient supply provided by the parent. Gymnosperms were among the first plants with pollen and seeds and radiated from 250–150 million years ago, during the Triassic and Jurassic periods. The evolution of wind pollination made it possible for plants to grow in drier habitats than ever before.
- Flowers protect egg-bearing ovules in a structure called a carpel. The fertilized ovule develops into a seed while the carpel develops into the wall of a fruit. In many species, other parts of the flower function in attracting animal pollinators.

The adaptive radiation of land plants was based on a series of morphological innovations, the most recent of which is the flower.

Angiosperms have all four of the major plant innovations: cuticle and stomata, vascular tissue, pollen and seeds, and flowers. The success of this suite of adaptations is evident in the number of angiosperm species living today: an estimated

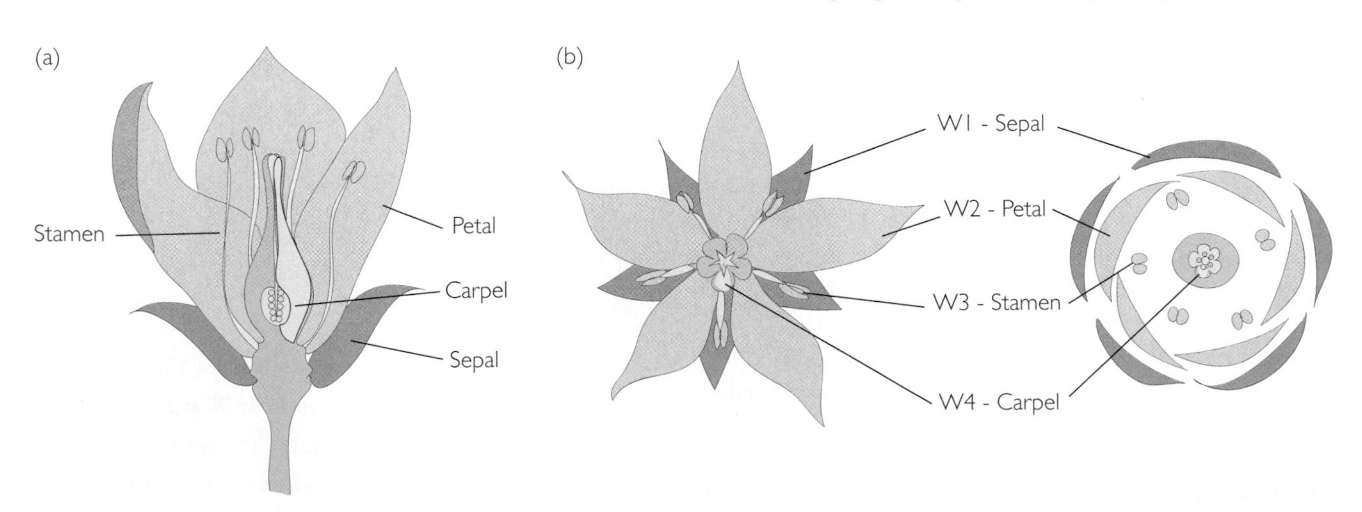

Figure 19.15 **Parts of a flower** (a) A long section through an idealized flower; (b) A top-down view showing how the four major organs are arranged in whorls.

250,000 to 300,000, making them far and away the most species-rich plant group in the history of life.

The hallmark trait of angiosperms, the flower, consists of four concentric whorls of repeated organs. From outermost to innermost, these are the sepals, petals, stamens, and carpels (Figure 19.15). The sepals are sometimes fused together to form a calyx that encloses the developing flower bud. The petals are sometimes fused together to form a corolla. Stamens are organs that produce pollen, and the carpels are sometimes fused together to form a single, central pistil. All of these organs are thought to have evolved from leaflike branching structures—a hypothesis that can be traced back to the 18th-century poet and naturalist Johann Goethe (Coen 1999).

Early evo-devo research in plants focused on documenting the genetic changes responsible for changing Goethe's leaflike branches into sepals, petals, stamens, and carpels. Researchers wanted to understand the genes that direct the development of the flower, and then investigate how those genes were modified to produce the fabulous array of flower structures seen today.

Homeosis in the Flower

Genetic research on flower development began with a mutant screen in *Arabidopsis thaliana*, just as studies on the genetic control of animal development began with an analysis of *Drosophila* mutants. By exposing plants to mutagens and examining thousands of offspring for unusual floral phenotypes, researchers found mutants in which the normal arrangement of sepals, petals, stamens, and carpels is altered. The most interesting of these mutations were homeotic: they produced phenotypes where structures developed in the wrong place. Three main classes of homeotic floral mutations emerged (Figure 19.16a):

- Class A mutants, in which the outer whorls are replaced by sex organs, leaving carpels, stamens, stamens, and carpels, in that order;
- Class B mutants, in which the middle two whorls are altered, leaving sepals, sepals, carpels, and carpels; and
- Class C mutants, in which the inner two whorls are altered, leaving sepals, petals, petals, and sepals. Note that this is the opposite pattern to class A mutants, in which sex organs replace the outer vegetative organs.

The expression of three genes specify the four structures found within a flower.

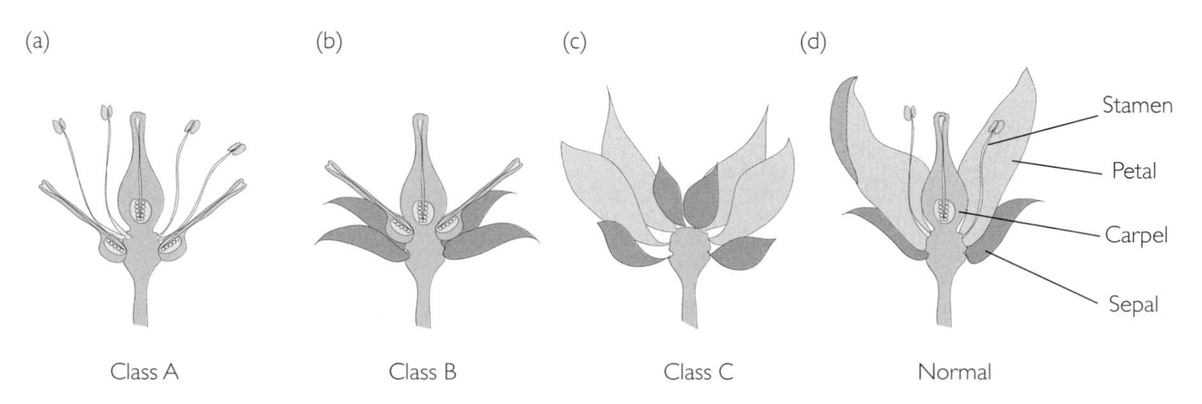

Figure 19.16 The ABCs of flower development Class A, B, and C floral homeotic mutants have two of the four flower whorls replaced by other parts: (a) Sepals and petals replaced by carpels and stamens; (b) petals and stamens replaced by sepals and carpels; (c) stamens and carpels replaced by petals and sepals. (d) A normal flower. See Coen (1999).

Combinations of these mutations in the same individual lead to the replacement of floral organs by leaflike structures (Figure 19.17). Like the four-winged fly of Figure 19.2, this mutant appears to "phenocopy" the ancestral state of the flower predicted by Goethe. Honma and Goto (2001) have also succeeded in converting leaf buds into floral organs by manipulating the expression of the ABC genes. Taken together, these data strongly support the hypothesis that flowers are derived from whorls of leaves (Weigel and Meyerowitz 1994).

Figure 19.17 Null mutants in floral homeotic genes (a) A wild-type flower of *Arabidopsis thaliana*. (b) In a triple mutant lacking the protein product of the AP2, API, and AG loci, sepals, petals, stamens, and ovules are replaced by leaflike structures.

The discovery of homeotic floral mutants in Arabidopsis and in the snapdragon *Antirrhinum majus* led to Enrico Coen and Eliot Meyerowitz (1991) to propose the ABC model of flower development, summarized in Figure 19.18. The key idea is that the protein products of A, B, and C-class genes interact, and that specific combinations of the three types of proteins direct the development of each of the four types of floral organs. To test this hypothesis, researchers had to hunt down the genes responsible for the A, B, and C-class mutations.

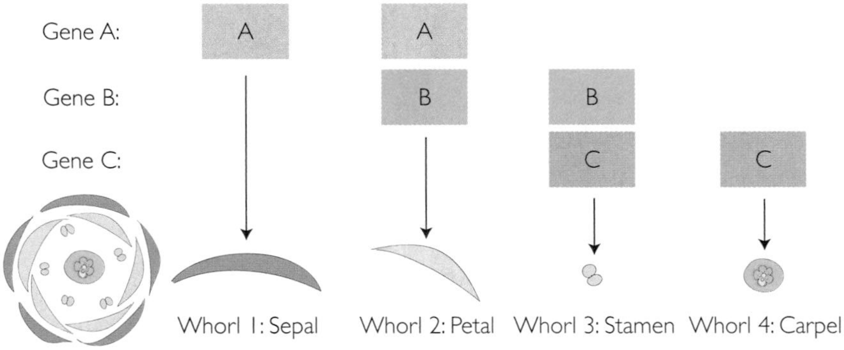

Gene A: A A
Gene B: B B
Gene C: C C
Whorl 1: Sepal Whorl 2: Petal Whorl 3: Stamen Whorl 4: Carpel

Figure 19.18 The ABC model of flower development According to the ABC model, each whorl of floral organs is specified by a different combination of the A,B, and C *MADS*-box genes.

The Discovery of *MADS*-box Genes

When the genes responsible for the homeotic transformations in flowers were sequenced, they turned out to have a common structure. The floral homeotic genes are transcription factors with a DNA-binding region called the *MADS* box. The *MADS* domain is analogous to the DNA-binding homeobox found in *Hox* genes.

Specific genes within the *MADS*-box family have now been associated with each of the class A, B, and C mutants (Theissen and Saedler 1999). *APETALA1* and *APETALA2*, or *AP1* and *AP2*, are class A genes. *APETALA3* and *PISTIL-*

LATA, or *AP3* and *PI*, are class B genes, while *AGAMOUS* or *AG* is a class C gene. Homologs of these genes have now been found in maize and other species, suggesting that they are widespread or even nearly universal in flowering plants (Ambrose et al. 2000).

As the ABC model predicts, failure to express *AP1* results in class A mutant phenotypes that lack sepals and petals. *AP3* and *PI* are expressed later and mainly in the middle whorls of the flower bud. Inactivation of *AP3* prevents normal development of petals plus stamens. *AG* also is expressed late, in the center of the flower bud. Inactivation of *AG* produces class C mutant phenotypes lacking sex organs.

Different combinations of *AP1*, *AG*, and *AP3* proteins in cells of the flower bud provide enough coordinate information to instruct each cell to participate in the construction of a particular floral organ. To summarize:

- *AP1* without *AG* or *AP3* induces sepals,
- *AP1* with *AP3* induces petals,
- *AP3* with *AG* induces stamens, and
- *AG* without *AP1* or *AP3* induces carpels.

Follow-up work showed that the ABC genes are expressed in response to a master control gene called *LEAFY* (Weigel and Meyerowitz 1993). The LFY protein activates *AP1*, *AP3*, and *AG* expression in the appropriate cells of the four whorls in the flower bud, beginning the specification of sepals, petals, stamens, and carpels. In many cases, though, the LFY protein acts on the ABC genes indirectly, by activating transcription factors that stimulate or repress expression. As Figure 19.19 shows, the floral homeotic genes, like the *Hox* genes of animals, are part of a complex regulatory cascade.

A common gene regulatory network controls the development of the diverse array of flowers.

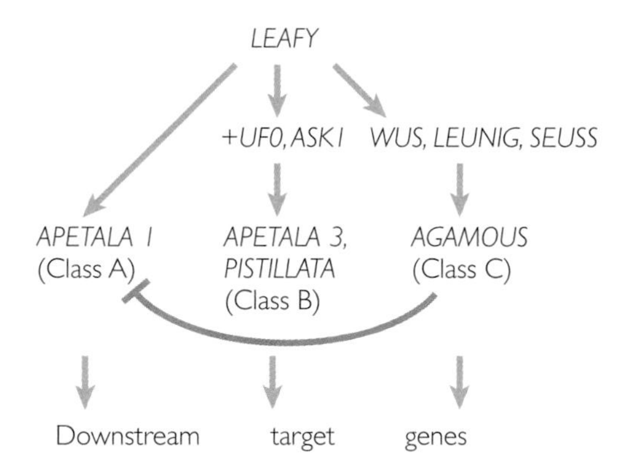

Figure 19.19 A regulatory gene network directs flower development The regulatory proteins and signals produced by these genes interact to control floral morphology. From Parcy et al. 1998; Buzgo et al. 2004.

Researchers continue to identify genes involved in the flower developmental network and explore their relationships. For example, Takashi Honma and Koji Goto (2001) have found that multiple genes are required to produce most of the floral organs because the protein products of genes in the network physically bind to one another before contacting DNA and directing gene expression. In addition, the downstream targets of *AP1*, *AP3*, *AG*, and other *MADS*-box genes are still in the early stages of investigation, much as the downstream targets of animal *Hox* genes are (e.g., Ito et al. 2004; Gomez-Mena et al. 2005; Cobb and Duboule 2005). Enough is known about how flowers develop, though, that evo-devo researchers have been able to wade in and explore the genetic basis for the origin and diversification of this key land plant innovation.

Where Did the Floral Genes Come From?

The genes that direct the development of the flower evolved long before flowers did. Homologs of *LFY*, *AG*, and other genes with *MADS* boxes have been discovered in pines, ferns, and other plant groups that don't form flowers, as well as in green algae (Tandre et al. 1995; Munster et al. 1997; Mouradov et al. 1998; Tanabe et al. 2005). In land plants that don't have flowers, the genes are involved in the formation of reproductive structures. In green algae, they direct the differentiation of reproductive cells.

These results have two important implications. First, the *MADS*-box genes do not produce proteins that form flower parts. Instead, they signal where cells are in the flower bud. Like *Hox* genes, they indicate location in an embryonic structure. Their proteins carry messages like "this is whorl 2," not "make the color needed for a petal." Second, deep homology exists in the reproductive structures of green plants. Homologous genes are involved in the formation of reproductive structures in all of the green plants. Like the *Hox* genes of animals, the *MADS*-box genes of angiosperms were re-purposed into directing flower formation.

The parallels between the *Hox* genes of animals and the *MADS*-box genes of plants are striking, and carry an important message. The genetic mechanisms responsible for organizing the plant and animal body are a spectacular example of evolutionary convergence. Multicellularity evolved independently in the land plants and in the animals, but the development of a plant or animal structure is directed in similar ways at the molecular level (Meyerowitz 2002). Complex series of regulatory genes interact in a cascade or hierarchy; signaling molecules originate at a point and are active along a concentration gradient; different signaling molecules are responsible for demarcating each of the three-dimensional axes. Although the logic of how development is controlled in plants and animals is similar, the molecules involved are different. Plants have homologs to some of the homeobox-containing genes found in animals, but it is the *MADS*-box genes that actually do the work of signaling where cells are in space, so that they develop into the correct cell and tissue types.

Diversification in Floral Genes and Structures

Thanks to the success of the ABC model, the basic framework of how sepals, petals, stamens, and carpels develop is well understood. What is much less well-understood is how mutation and selection produced the diversity of flower shapes and sizes observed today. Researchers are pursuing several different approaches to make progress on this question. Using a candidate gene approach in *Arabidopsis thaliana*, Michael Purugannan and coworkers have confirmed that (1) alleles of homeotic genes vary in sequence and in their effect on the phenotype of *Arabidopsis* flowers, and (2) certain homeotic alleles have recently been under selection (Purugganan and Suddith 1998, 1999; Purugganan et al. 2000). These studies were an encouraging early step in the effort to link particular alleles with changes in floral morphology.

Diversification in floral morphology over time is based on mutation and selection acting on elements in the genetic network that controls flower development.

The candidate gene approach has also been used in between-species comparisons (e.g., Lawton-Rauh et al. 1999; Hileman et al. 2003; Costa et al. 2005). The idea here is to choose one or two components of the floral regulatory network and compare how homologous genes are expressed in closely related species with different flower shapes. These studies suggest that changes in the level of expression of the *CYCLOIDEA* and *DICHOTOMA* genes—possibly in combination with changes in their downstream targets—have been responsible for evolutionary changes in flower symmetry.

A third research strategy has been particularly productive. Instead of starting with candidate genes identified through studies of floral development, some research groups are looking for candidate genes by arranging matings between closely related species with distinctive floral characteristics and then documenting associations between specific genetic markers and the phenotypes of the offspring. Chapter 9 detailed how this work is done, using research on monkeyflowers as an example. Similar research programs are beginning to identify the alleles responsible for divergence in the size and shape of grass flowers (Doust et al. 2005) and petunia flowers (Stuurman et al. 2004).

Although research on the genetic mechanisms of floral diversification is still in its infancy, growth has been rapid. With an array of research strategies and candidate genes available, biologists are optimistic about achieving a much deeper understanding of the genetic changes responsible for remarkable array of flowers observed today.

19.5 Frontiers in Evo-Devo Research

This chapter has explored the genetic basis for three great events in the history of life: the diversification of body plans in animals, the evolution of limbs in tetrapods and insects, and the origin and diversification of the flower. The genes involved in the origin of these traits are no longer a mystery. Current research is beginning to reveal how changes in the structure and regulation of these genes helped produce the diversity of forms we see today.

Before leaving the topic of evo-devo, let's briefly examine some broad lessons from the past 25 years of research in the field and consider where research might be headed now.

Extending Evo-Devo Research to Non-Model Systems and Structures

The first generation of evo-devo research focused on model organisms such as *Drosophila melanogaster*, chicken and mouse, and *Arabidopsis thaliana*, and model structures such as body segments, limbs, and flowers. Developmental biologists knew the most about these organisms and structures, so evolutionary biologists had a knowledge base from which to start.

Current research is branching out to investigate the evolution of a range of structures in an array of lineages. A few examples will highlight this development:

- Research is beginning to clarify how changes in gene expression have led to evolutionary changes in leaf shape (e.g., Byrne et al. 2000; Bharathan et al. 2002; Friedman et al. 2004);
- Allelic variation in the *MADS*-box gene *AP2* affects seed size (Jofuku et al. 2005; Ohta et al. 2005);
- Mutations in the coding regions of two genes affect the size and shape of the dog skull and jaws (Fondon and Garner 2004);
- Changes in the timing and level of expression of a gene called *Bmp4* are associated with evolutionary changes in the size and shape of the beak in Galapagos finches and other bird species (Abzhanov et al. 2004; Wu et al. 2004);

- Changes in the regulation of pigment genes led to the evolution of dark spots in the wings of some fruit flies (Gompel et al. 2005);
- Defects in a gene regulatory network led to the loss of body stripes in a zebrafish (Quigley et al. 2005).

The take-home message of these examples is simple. Evo-devo researchers have an increasingly sophisticated array of research techniques at their disposal, along with a broadening catalog of candidate genes. These advances have made it possible to explore the genetic basis of evolutionary change in almost any trait, in almost any plant or animal lineage.

Using Deep Homology to Reconstruct Ancestors

The discovery of deep genetic homology is one of the leading results from evo-devo research. Among the bilaterally symmetric animals, for example, there is now good evidence that homologous genes are involved in the formation of eyes, hearts, nerve chords, and segments. These findings suggest that the common ancestor of these species also had these genes. In this way, understanding deep homology can help us reconstruct what that ancestor looked like and perhaps how it lived—even though it is not yet present in the fossil record.

In the case of bilaterally symmetric animals, deep homologies suggest that their common ancestor had the following traits (Arthur et al. 1999; Holland 1999; Knoll and Carroll 1999; Erwin and Davidson 2002; see also Finnerty et al. 2004):

- Serial repetition of some body parts without complex differentiation of segments;
- A simple type of appendage, but not a complex limb;
- Clusters of simple photoreceptors, but not an elaborate image-forming eye;
- A contractile blood vessel rather than a heart;
- Condensations of nerve cell bodies into nerve cords, but not a brain.

Paleontologists can test these predictions by looking for fossils of ancient bilaterally symmetric animals in Precambrian rocks. The deep homology underlying many organ systems in its descendants provides a reasonable sketch of what the ancestor of today's bilaterians looked like. If this sketch is correct, fossils resembling that ancestor eventually should be found.

Microevolution and Macroevolution

One of the long-standing debates in evolutionary biology has been whether changes that take place within populations due to mutation, selection, drift, and gene flow are responsible for the types of changes that distinguish groups like chordates versus crustaceans or mosses versus angiosperms. Traditionally, changes that take place on short time scales within populations have been referred to as **microevolution**, while changes that take place on long time scales between broader taxonomic groupings have been termed **macroevolution**. Some workers have argued that microevolutionary processes are not sufficient to explain macroevolutionary change, and that other—often unnamed—evolutionary processes must be at work.

Sean Carroll has proposed that the debate over micro- and macroevolution should now be put to rest (Carroll 2001). Based on the work done to date on evolution and development, Carroll claims that mutation, selection, drift, and gene flow can explain all of the great innovations and changes that have occurred over the history of life. Thanks to progress in evo-devo research, we can now infer the types of mutations and selection pressures that led to the evolution of the animal phyla, as well as the origin and diversification of body segmentation and limbs in animals and the flower in angiosperms.

Evo-devo research is allowing investigators to understand the genetic changes responsible for key innovations during evolution—to identify the alleles that arose by mutation and that responded to selection as important new lineages diversified.

Although this chapter focused on genetic changes responsible for macroevolutionary events, earlier chapters introduced several areas where researchers are investigating the genetic changes responsible for microevolutionary change, including divergence in *Mimulus* flowers (Chapter 9), the preferred host plants of apple maggot flies and pea aphids (Chapter 16), and threespine stickleback morphology (Chapter 16). Investigators are also exploring within-population variation in genes responsible for traits such as the shape of leaves in *Arabidopsis* and broccoli or the eyespots observed in some butterflies, in addition to the studies of flower-shape variation mentioned earlier. The idea is to understand which alleles are responding to natural selection and creating morphological change within populations. For more on this work, see the Exploring the Literature section at the end of this chapter.

If future research continues to support Carroll's viewpoint, it means that recent collaborations between paleontologists, developmental biologists, and systematists have led to an important conceptual unification in these disparate fields. A grand synthesis of micro- and macroevolution is underway.

Summary

The diversity of animal and plant form is largely due to evolutionary changes in the genes that control embryonic development. Development is coordinated by intercellular signals and transcription factors that interact to form regulatory networks. These networks are responsible for telling cells where they are in the multicellular body, and directing the expression of genes that make the cell differentiate into the type appropriate for that location: muscle, tracheid, or bone, for example. If mutation causes a gene in one of these regulatory networks to be expressed in a different location, amount, or time, the result is a change in morphology.

The existence of a regulatory network creates a theme; mutation, drift, and natural selection create variations. When components of a regulatory network are conserved over time, the phenomenon of deep homology results. Mutation can also result in regulatory networks being co-opted to direct the formation of entirely new structures.

Homeotic genes—the *Hox* genes in animals and the *MADS*-box genes in plants—are particularly prominent components of important regulatory networks. In animals, changes in the number of genes in the homeotic complexes may have been associated with the origin of bilateral symmetry and the vertebrates, and the diversification of ray-finned fish. Changes in when and where these genes are expressed were important in the diversification of segmented body plans and in the elaboration of the tetrapod and arthropod limbs. Changes in the sequence of certain Hox genes may have been important in changing the number and arrangement of limbs in arthropods. In plants, changes in the location and timing of expression *MADS*-box genes are associated with evolutionary changes in flower size and shape.

Questions

1. What do homeotic genes do?
2. Why is it significant that *Hox* and *MADS*-box genes contain a DNA-binding domain?
3. Is it legitimate to claim that changes in Hox-gene number were correlated with increased morphological complexity during animal evolution? Explain why or why not.
4. Why is it logical for researchers to infer that the ancestor of all bilaterally symmetric animals had about 10 *Hox* genes?

5. What does the animal gene called *Distal-less* do?
6. Are the same *MADS*-box genes involved in the formation of sepals, petals, stamens, and carpels, or is each floral organ coded for by unique genes?
7. Define microevolution and macroevolution. How does evo-devo research unite these fields?
8. What is "deep homology?"
9. During the evolution of animals, when did changes in the timing and spatial location of *Hox* gene expression appear to correlate with important morphological innovations?
10. During the evolution of animals, when did changes in the sequence of a *Hox* gene appear to correlate with an important morphological innovation?
11. What did the ancestor of the bilateral animals look like? Was it segmented? Did it have limbs, eyes, and a heart?
12. In fruit flies, the protein product of the *Ubx* gene turns off the gene called *Distal-less*. Go back to Section 19.3 and review the evolutionary changes that have occurred in *Ubx* sequences and the function of *Dll* in limb development. Why is the observation that *Ubx* represses *Dll* important?
13. How does the finding that different MADS-box proteins physically bind to one another relate to the ABC model of flower development?
14. *Hox* and *MADS*-box genes are often called "selector genes." Why?

Exploring the Literature

15. Researchers are finding many new examples of how genes in established regulatory networks have been co-opted to direct the development of novel structures, including traits that are unique to cephalopods (octopus and squid), the eyespots found on some butterfly wings, and beetle horns.

    Lee, P.N., P. Callaerts, H.G. de Couet, and M.Q. Martindale. 2003. Cephalopod *Hox* genes and the origin of morphological novelties. *Nature* 424: 1061–1065.

    Keys, D. N., D. L. Lewis, et al.. 1999. Recruitment of a *hedgehog* regulatory circuit in butterfly eyespot evolution. *Science* 283: 532–534.

    Moczek, A.P. and L.M. Nagy. 2005. Diverse developmental mechanisms contribute to different levels of diversity in horned beetles. *Evolution and Development* 7: 175–185.

16. A "deep homology" hypothesis claims that the ciliary bands observed in echinoderm embryos and the chordate central nervous system are homologous. For experimental support of this idea see:

    Poutstka, A.J., A Ku:hn, V. Radosavljevic, R. Wellenruether, H. Lehrach, and G. Panopoulou. 2004. On the origin of the chordate nervous system: expression of onecut in the sea urchin embryo. *Evolution and Development* 6: 227–236.

17. Understanding how intraspecific variation in regulatory genes affects phenotypic variation within populations is an active area of research. For interesting work on eyespot variation in butterflies and bird coloration, see:

    Beldade, P., P. M. Brakefield, and A. D. Long. 2002. Contribution of *Distal-less* to quantitative variation in butterfly eyespots. *Nature* 415: 315–318.

    Mundy, N.I., N.S. Badcock, T. Hart, K. Scribner, K. Janssen, and N.J. Nadeau. 2004. Conserved genetic basis of a quantitative plumage trait involved in mate choice. *Science* 303: 1870–1873.

Citations

Abzhanov, A., M. Protas, B. R. Grant, P. R. Grant, C. J. Tabin. 2004. *Bmp4* and morphological variation of beaks in Darwin's finches. *Science* 305: 1462–1465.

Ambrose, B. A., D. R. Lerner, P. Ciceri, C. M. Padilla, M. F. Yanofsky, and R. J. Schmidt. 2000. Molecular and genetic analysis of the *Silky1* gene reveal conservation in floral organ specification between Eudicots and Monocots. *Molecular Cell* 5: 569–579.

Angelini, D.R. and T.C. Kaufman. 2005. Insect appendages and comparative genomics. *Developmental Biology* 286: 57–77.

Arthur, W., T. Jowett, and A. Panchen. 1999. Segments, limbs, homology, and co-option. *Evolution and Development* 1: 74–76.

Balavoine, G., and M. J. Telford. 1995. Identification of planarian homeobox sequences indicates the antiquity of most *Hox*/homeotic gene subclasses. *Proceedings of the National Academy of Sciences USA* 92: 7227–7231.

Bharathan, G., T. E. Goliber, C. Moore, S. Kessler, T. Pham, and N. R. Sinha. 2002. Homologies in leaf form inferred from *KNOXI* gene expression during development. *Science* 296: 1858–1860.

Bowman, J. L., Smyth, D. R., and Meyerowitz, E. M. 1989. Genes directing flower development in *Arabidopsis*. *The Plant Cell* 1:, 37–52.

Bowman, J. L., Smyth, D. R., and Meyerowitz, E.M. 1991. Genetic interactions among floral homeotic genes of *Arabidopsis*. *Development* 1: 12, 1–20.

Brooke, N. M., J. Garcia-Fernandez, and P. W. H. Holland. 1998. The *Para-Hox* gene cluster is an evolutionary sister of the *Hox* gene cluster. *Nature* 392: 920–922.

Brusca, R. C., and G. J. Brusca. 2002. *Invertebrates*, 2nd ed. Sunderland, MA: Sinauer.

Buzgo, M., D. E. Soltis, P. S. Soltis, and H. Ma. 2004. Towards a comprehensive integration of morphological and genetic studies of floral development. *Trends in Plant Science* 9: 164–173.

Byrne, M. E., R. Barley, M. Curtis, J. M. Arroyo, M. Dunham, A. Hudson, and R. A. Martienssen. 2000. Asymmetric leaves1 mediates leaf patterning and stem cell function in *Arabidopsis*. *Nature* 408: 967-971.

Cameron, R. A., K. J. Peterson, and E. H. Davidson. 1998. Developmental gene regulation and the evolution of large animal body plans. *American Zoologist* 38: 609–620.

Carroll, S. B. 1995. Homeotic genes and the evolution of the chordates. *Nature* 376: 479–485.

Carroll, S. B. 2001. The big picture. *Nature* 409: 669.

Cobb, J. and D. Duboule. 2005. Comparative analysis of genes downstream of the *Hoxd* cluster in developing digits and external genetalia. *Development* 132: 3055–3067.

Coen, E. 1999. *The Art of the Gene*. Oxford: Oxford University Press.

Coen, E. S. and E. M. Meyerowitz. 1991. The war of the whorls: Genetic interactions controlling floral development. *Nature* 353: 31–37.

Cook, C. E., Q. Ye, and M. Akam. 2005. Mitochondrial genomes suggest that hexapods and crustaceans are mutually paraphyletic. *Proceedings of the Royal Society of London B* 272: 1295–1304.

Costa, M. M. R., S. Fox, A. I. Hanna, C. Baxter, and E. Coen. 2005. Evolution of regulatory interactions controlling floral asymmetry. *Development* 132: 5093–5101.

Daeschler, N. H. Shubin, and F. A. Jenkins Jr. 2006. A Devonian tetrapod-like fish and the evolution of the tetrapod body plan. *Nature* 440: 757–763.

de Rosa, R., J. K. Grenier, T. Andreeva, C. E. Cook, A. Adoutte, M. Akam, S. B. Carroll, and G. Balavoine. 1999. *Hox* genes in brachiopods and priapulids and protostome evolution. *Nature* 399: 772–776.

Doust, A. N., K. M. Devos, M. D. Gadberry, M. D. Gale, and E. A. Kellog. 2005. The genetic basis for inflorescence variation between foxtail and green millet (Poaceae). *Genetics* 169: 1659–1672.

Erwin, D. H., and E. H. Davidson. 2002. The last common bilaterian ancestor. *Development* 129: 3021–3032.

Erwin, D., J. Valentine, and D. Jablonski. 1997. The origin of animal body plans. *American Scientist* 85: 126–137.

Finnerty, J. R., and M. Q. Martindale. 1997. Homeoboxes in sea anemones (Cnidaria; Anthozoa): A PCR-based survey of *Nematostella vectensis* and *Metridium senile*. *Biological Bulletin* 193: 62–76.

Finnerty, J. R., K. Pang, P. Burton, D. Paulson, and M. Q. Martindale. 2004. Origins of bilateral symmetry: *Hox* and *Dpp* expression in a sea anemone. *Science* 304: 1335–1337.

Fondon, J. W. III and H. R. Garner. 2004. Molecular origins of rapid and continuous morphological evolution. *Proceedings of the National Academy of Sciences USA* 101: 18058–18063.

Friedman, W. E., R. C. Moore, and M. D. Purugganan. 2004. The evolution of plant development. *American Journal of Botany* 91: 7126–7141.

Galant, R., and S. B. Carroll. 2002. Evolution of a transcriptional repression domain in an insect *Hox* protein. *Nature* 415: 910–913.

Gerhart, J. and M. Kirschner. 1997. *Cells, Embryos, and Evolution*. Malden, MA: Blackwell Science.

Gómez-Mena, C., S. de Folter, M. M. R. Costa, G. C. Angenent, and R. Sablowski. 2005. Transcriptional program controlled by the floral homeotic gene *AGAMOUS* during early organogenesis. *Development* 132: 429–438.

Gompel, N., B. Prud'homme, P. J. Wittkopp, V. A. Kassner, and S. B. Carroll. 2005. Chance caught on the wing: cis-regulatory evolution and the origin of pigment patterns in *Drosophila*. *Science* 433: 481–487.

Grenier, J. K., T. L. Garber, R. Warren, P. M. Whitington, and S. B. Carroll. 1997. Evolution of the entire arthropod *Hox* gene set predated the origin and radiation of the onychophoran/arthropod clade. *Current Biology* 7: 547–553.

Halder, G., P. Callaerts, and W. J. Gehring. 1995. Induction of ectopic eyes by targeted expression of th eyeless gene in *Drosophila*. *Science* 267: 1788–1792.

Hileman, L. C., E. M. Kramer, and D. A. Baum. 2003. Differential regulation of symmetry genes and the evolution of floral morphologies. *Proceedings of the National Academy of Sciences USA* 100: 12814–12819.

Hittinger, C. T., D. L. Stern, and S. B. Carroll. 2005. Pleiotropic functions of a conserved insect-specific *Hox* peptide motif. *Development* 132: 5661–5670.

Hoegg, S. and A. Meyer. 2005. *Hox* clusters as models for vertebrate genome evolution. *Trends in Genetics* 21: 421–424.

Holland, P. W. H. 1999. The future of evolutionary developmental biology. *Nature* 402 (Supplement): C41–C44.

Honma, T., and K. Goto. 2001. Complexes of MADS-box proteins are sufficient to convert leaves into floral organs. *Nature* 409: 525–528.

Ito, T., F. Wellmer, et al. 2004. The homeotic gene *AGAMOUS* controls microsporogenesis by regulation of *SPOROCYTELESS*. *Nature* 430: 356–360.

Jarvik, E. 1980. *Basic Structure and Evolution of Vertebrates*, vol. 2. London: Academic Press.

Jofuku, K. D., P. K. Omidyar, Z. Gee, and J. K. Okamuro. 2005. Control of seed mass and seed yield by the floral homeotic gene *APETALA2*. *Proceedings of the National Academy of Sciences USA* 102: 3117–3122.

Jürgens, G., U. Mayer, R. A. Torres Ruiz, T. Berleth, and S. Misera. 1991. Genetic analysis of pattern formation in the *Arabidopsis* embryo. *Development* 1 (suppl 1): 27–38.

Kamm, K., B. Schierwater, and W. Jakob. 2006. Axial patterning and diversification in the Cnidaria predate the *Hox* system. *Current Biology* 16: 1–7.

Kenyon, C., and B. Wang. 1991. A cluster of *Antennapedia*-class homeobox genes in a nonsegmented animal. *Science* 253: 516–517.

Kmita, M., N. Fraudeau, Y. Herault, and D. Duboule. 2002. Serial deletions and duplications suggest a mechanism for the collinearity of *Hoxd* genes in limbs. *Nature* 420: 145–150.

Knoll, A. H., and S. B. Carroll. 1999. Early animal evolution: Emerging views from comparative biology and geology. *Science* 284: 2129–2137.

Kruse, M., A. Mikoc, H. Cetkovic, V. Gamulin, B. Rinkevich, I. M. Müller, and W. E. G. Müller. 1994. Molecular evidence for the presence of a developmental gene in the lowest animals: Identification of a homeobox-like gene in the marine sponge *Geodia cydonium*. *Mechanisms of Aging and Development* 77: 43–54.

Lawton-Rauh, A. L., E. S. Buckler, IV, and M. D. Purugganan. 1999. Patterns of molecular evolution among paralogous floral homeotic genes. *Molecular Biology and Evolution* 16: 1037–1045.

Lewis, E. B. 1978. A gene complex controlling segmentation in *Drosophila*. *Nature* 276: 565–570.

McGinnis, W., R. L. Garber, J. Wirz, A. Kuroiwa, and W. J. Gehring. 1984. A homologous protein-coding sequence in *Drosophila* homeotic genes and its conservation in other metazoans. *Cell* 37: 403–408.

McGinnis, W., and M. Kuziora. 1994. The molecular architects of body design. *Scientific American* 270: 58-66.

McGinnis, W., M. S. Levine, E. Hafen, A. Kuroiwa, and W. J. Gehring. 1984a. A conserved DNA sequence in homoeotic genes of the *Drosophila Antennapedia* and *bithorax* complexes. *Nature* 308: 428–433.

Metscher, B. D., K. Takahashi, et al. 2005. Expression of *Hoxa-11* and *Hoxa-13* in the pectoral fin of a basal ray-finned fish, *Polyodon spathula*: implications for the origin of tetrapod limbs. *Evolution and Develoment* 7: 186–195.

Meyerowitz, E. M. 2002. Plants compared to animals: The broadest comparative study of development. *Science* 295: 1482–1485.

Mouradov, A., T. Glassick, B. Hamdorf, L. Murphy, B. Fowler, S. Maria, and R. D. Teasdale. 1998. *NEEDLY*, a *Pinus radiata* ortholog of *FLORICAULA/LEAFY* genes, expressed in both reproductive and vegetative meristems. *Proceedings of the National Academy of Sciences USA* 95: 6537–6542.

Munster, T., J. Pahnke, A. DiRosa, J. T. Kim, W. Martin, H. Saedler, and G. Theissen. 1997. Floral homeotic genes were recruited from homologous *MADS*-box genes preexisting in the common ancestor of ferns and seed plants. *Proceedings of the National Academy of Sciences USA* 94: 2415–2420.

Nagy, L. 1998. Changing patterns of gene regulation in the evolution of arthropod morphology. *American Naturalist* 38: 818–828.

Nardi, F., G. Spinsanti, et al. 2003. Hexapod origins: monophyletic or paraphyletic? *Science* 299: 1887–1889.

Nüsslein-Volhard, C., H. G. Frohnhofer, and R. Lehmann. 1987. Determination of anterioposterior polarity in *Drosophila*. *Science* 238: 1675–1681.

Nüsslein-Volhard, C., and E. Wieschaus. 1980. Mutations affecting segment number and polarity in *Drosophila*. *Nature* 287: 795–801.

Ohta, M., R.L. Fischer, R.B. Goldberg, K. Nakamura, and J.J. Harada. 2005. Control of seed mass by *APETALA2*. *Proceedings of the National Academy of Sciences USA* 102: 3123–3128.

Panganiban, G., et al. 1997. The origin and evolution of animal appendages. *Proceedings of the National Academy of Sciences USA* 94: 5162–5166.

Panganiban, G., A. Sebring, L. Nagy, and S. Carroll. 1995. The development of crustacean limbs and the evolution of arthropods. *Science* 270: 1363–1366.

Parcy, F., O. Nilsson, M. A. Busch, I. Lee, and D. Weigel. 1998. A genetic framework for floral patterning. *Nature* 395: 561–566.

Purugganan, M. D., and J. I. Suddith. 1998. Molecular population genetics of the *Arabidopsis CAULIFLOWER* regulatory gene: Nonneutral evolution and naturally occurring variation in floral homeotic function. *Proceedings of the National Academy of Sciences USA* 95: 8130–8134.

Purugganan, M. D., and J. I. Suddith. 1999. Molecular population genetics of floral homeotic loci: Departures from the equilibrium-neutral model at the *APETALA3* and *PISTILLATA* genes of *Arabidopsis thaliana*. *Genetics* 151: 839–848.

Purugganan, M.D., A.L. Boyles, and J.I. Suddith. 2000. Variation and selection at the *CAULIFLOWER* floral homeotic gene accompanying the evolution of domesticated *Brassica oleracea*. *Genetics* 155: 855–62.

Quigley, I.K.J.L. Manuel, et al. 2005. Evolutionary diversification of pigment pattern in *Danio* fishes: differential fms dependence and stripe loss in *D. albolineatus*. *Development* 132: 89–104.

Regier, J. C., J. W. Schultz, and R. E. Kambic. 2005. Pancrustacean phylogeny: hexapods are terrestrial crustaceans and maxillopods are not monophyletic. *Proceedings of the Royal Society of London B* 272: 395–401.

Ronshaugen, M., N. McGinnis, and W. McGinnis. 2002. Hox protein mutation and macroevolution of the insect body plan. *Nature* 415: 914–917.

Seimiya, M., H. Ishiguro, K. Miura, Y. Watanabe, and Y. Kurosawa. 1994. Homeobox-containing genes in the most primitive metazoa, the sponges. *European Journal of Biochemistry* 221: 219–225.

Shubin, N., C. Tabin, and S. Carroll. 1997. Fossils, genes, and the evolution of animal limbs. *Nature* 388: 639–648.

Shubin, N.H., E.B. Daeschler, and M.I. Coates. 2004. The early evolution of the tetrapod humerus. *Science* 304: 90–93.

Shubin, N. H., E. B. Daeschler, and F. A. Jenkins Jr.. 2006. The pectoral fin of *Tiktaalik roseae* and the origin of the tetrapod limb. *Nature* 440: 764–771.

Sordino, P., and D. Duboule. 1996. A molecular approach to the evolution of vertebrate paired appendages. *Trends in Ecology and Evolution* 11: 114–119.

Sordino, P., F. van der Hoeven, and D. Duboule. 1995. *Hox* gene expression in teleost fins and the origin of vertebrate digits. *Nature* 375: 678–681.

Stuurman, J., M.E. Hoballah, L. Broger, J. Moore, C. Basten, and C. Kuhlemeier. 2004. Dissection of floral pollination syndromes in *Petunia*. *Genetics* 168: 1585–1599.

Tabin, C. J., S. B. Carroll, and G. Panganiban. 1999. Out on a limb: Parallels in vertebrate and invertebrate limb patterning and the origin of appendages. *American Zoologist* 39: 650–663.

Tanabe, Y, M. Hasebe, et al. 2005. Characterization of *MADS*-box genes in charophycean green algae and its implication for the evolution of *MADS*-box genes. *Proceedings of the National Academy of Sciences USA* 102: 2436–2441.

Tandre, K., V. A. Albert, A. Sundas, and P. Engstrom. 1995. Conifer homologs to genes that control floral development in angiosperms. *Plant Molecular Biology* 27: 69–78.

Theissen, G., and H. Saedler. 1999. The golden decade of molecular floral development (1990–1999): A cheerful obituary. *Developmental Genetics* 25: 181–193.

Valentine, J. W., D. H. Erwin, and D. Jablonski. 1996. Developmental evolution of metazoan body plans: The fossil evidence. *Developmental Biology* 173: 373–381.

Weigel, D., and E. M. Meyerowitz. 1993. Activation of floral homeotic genes in *Arabidopsis*. *Science* 261: 1723–1726.

Weigel, D., and E. M. Meyerowitz. 1994. The ABC's of floral homeotic genes. *Cell* 78: 203–209.

Wu, P., T.-X. Jiang, S. Suksaweang, R. B. Widelitz, and C.-M. Chuong. 2004. Molecular shaping of the beak. *Science* 305: 1465–1466.

20

Human Evolution

Lucy's daughter: This 3-year-old *Australopithecus afarensis,* shown by expedition leader Zeresenay Alemseged, lived 3.3 million years ago in what is now Dikika, Ethiopia. Her legs, including the knee below, are those of a biped. Her shoulder blades are like a gorilla's, however, hinting that she may also have been a climber (Alemseged et al. 2006).

The first printing of *On the Origin of Species* sold out on November 22, 1859, the first day Darwin's publisher, John Murray, offered it to booksellers. Among the profound implications that attracted such attention was what the book told its readers about themselves. Although Darwin saw this as clearly as anyone, his only explicit treatment of human evolution was a single paragraph in the last chapter, in which his strongest claim was that "Light will be thrown on the origin of man and his history" (Darwin 1859, page 488).

Not until 12 years later did Darwin reveal the depth and breadth of his thinking about humans. In 1871 he published a two-volume work, *The Descent of Man, and Selection in Relation to Sex*. In the introduction, Darwin explained his initial reticence on the subject of human evolution: "During many years I collected notes on the origin or descent of man, without any intention of publishing on the subject, but rather with the determination not to publish, as I thought that I should thus only add to the prejudices against my views" (Darwin 1871, page 1).

Darwin's apprehensions were well founded. The human implications of evolutionary biology have been, and remain, a cause of heated controversy. In 1925, Tennessee schoolteacher John T. Scopes was convicted of violating a new state law prohibiting the teaching of evolution (see Chapter 3). The Scopes case was popularly known as the Monkey Trial, indicating that for many observers the central issue at stake was the origin of the human species. In 2004, the school board in Dover, Pennsylvannia, adopted a policy requiring that a disclaimer be read to students in ninth grade biology classes. The disclaimer admonished students to consider evolution as theory, not fact, and notified them that a book espousing Intelligent Design creationism was available as a reference. The Dover disclaimer became the subject of a widely publicized trial, mentioned in Chapter 3, in which teaching Intelligent Design in public schools was ruled unconstitutional. Again, the origin of our species was a key issue for many involved. In a meeting leading up to the adoption of the disclaimer policy, board member William Buckingham declared, "It's inexcusable to have a book that says man descended from apes with nothing to counterbalance it" (Maldonado 2004; Jones 2005).

In this chapter, we explore research on the evolutionary history of our species. We start by reviewing attempts to determine the evolutionary relationships among humans and the extant apes. Then we consider the fossil evidence bearing on the course of human evolution following the split between our lineage and the lineage of our closest living relatives. Next, we look at fossil and molecular evidence on the emergence of *Homo sapiens*. Finally, we consider the evolutionary origins of some of our species' defining characteristics, including tool use and language. Our exploration illustrates that the subject of human evolution generates controversies within the scientific community that, while different in focus, are as heated as those it generates among the lay public.

20.1 Relationships among Humans and Extant Apes

Humans (*Homo sapiens*) belong to the primate taxon Catarrhini (Goodman et al. 1998), which includes the Old World monkeys, such as the baboons and macaques, and the apes (Figure 20.1). The apes include the gibbons (*Hylobates*) of southeast Asia and the great apes. The great apes include the orangutan (*Pongo pygmaeus*), also of southeast Asia, and three African species: the gorilla (*Gorilla gorilla*), the common chimpanzee (*Pan troglodytes*), and the bonobo, or pygmy chimpanzee (*Pan paniscus*).

Humans Belong to the Same Clade as the Apes

Scientists universally agree that humans evolved from within the apes. Humans share with the apes numerous derived characteristics (synapomorphies). These evolutionary innovations distinguish the apes from the rest of the Catarrhini and indicate that the apes are descended from a common ancestor (see Chapter 4). The shared derived traits of the apes include relatively large brains, the absence of a tail, a more erect posture, greater flexibility of the hips and ankles, increased flexibility of the wrist and thumb, and changes in the structure and use of the arm and shoulder (Andrews 1992; see also Groves 1986; Andrews and Martin 1987; Begun et al. 1997). In addition to this morphological evidence, the molecular analyses described later in this chapter also unequivocally demonstrate that humans are apes.

Humans Belong to the Same Clade as the African Great Apes

Figure 20.1 includes a reconstruction of the phylogenetic relationships among the apes. This reconstruction places humans with the great apes, and more specifically with the African great apes. The reconstruction was first proposed by

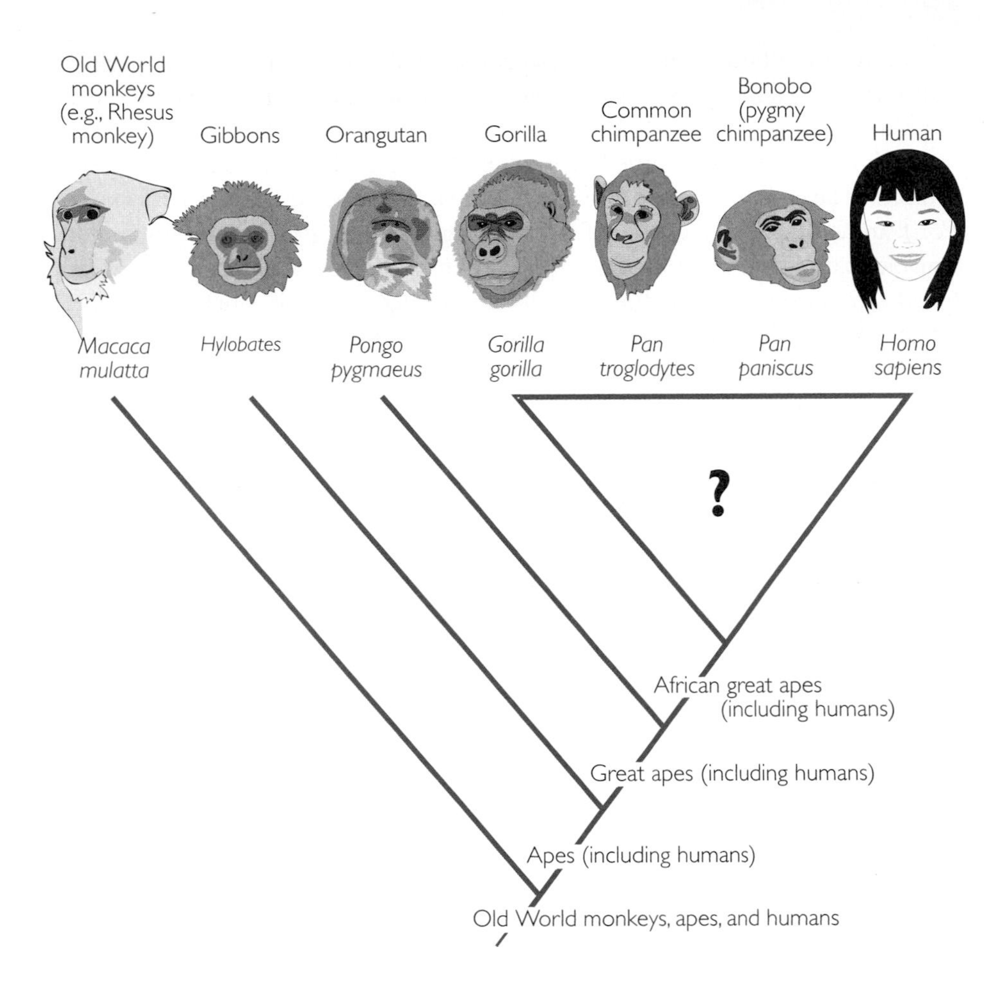

Figure 20.1 Phylogeny of the apes This evolutionary tree shows the relationships among the Old World monkeys, represented by a rhesus monkey, and the apes and humans. Among the apes, the gibbons branch off first, followed by the orangutan. The evolutionary relationships among the gorilla, the two chimpanzees, and humans (triangle with question mark) were long the subject of considerable dispute.

Thomas Henry Huxley (Huxley 1863). Huxley's proposal raised dispute, but in recent years, as more data have been collected and analyzed, scientists in all fields have accepted the tree in Figure 20.1.

Cladistic analyses of morphology support the tree. Humans and the African great apes share a number of derived traits that distinguish them from the rest of the apes. These synapomorphies include elongated skulls, enlarged brow ridges, shortened but stout canine teeth, changes in the front of the upper jaw (premaxilla), fusion of certain bones in the wrist, enlarged ovaries and mammary glands, changes in muscular anatomy, and reduced hairiness (Ward and Kimbel 1983; Groves 1986; Andrews and Martin 1987; Andrews 1992; Begun et al. 1997).

Morphological and molecular analyses demonstrate that humans are closely related to gorillas and chimpanzees.

Molecular analyses concur. They have, in fact, indicated a close relationship between humans and the African great apes since the beginnings of modern molecular systematics. Using a technique pioneered by George H. F. Nuttall (1904) and Morris Goodman (1962), Vincent Sarich and Allan Wilson (1967) took purified human serum albumin, a blood protein, and injected it into rabbits. After giving the rabbits time to make antibodies against the human albumin protein, Sarich and Wilson took blood serum from the rabbits. This serum contained rabbit antihuman antibodies. The researchers mixed the rabbit serum with purified serum albumin from a variety of apes and Old World monkeys. Sarich and Wilson used the strength of the immune reaction between the rabbit antihuman antibodies and the primate albumins as a measure of similarity among the albumins they tested, and they assumed that the similarity of two species' serum albumin proteins reflects the species' evolutionary kinship. The resulting phylogeny shows that humans are closely related to gorillas and the two chimpanzees (Figure 20.2).

Sarich and Wilson put a time line on their phylogeny by assuming that (1) serum albumin evolves at a constant rate and (2) the split between the apes and the Old World monkeys occurred 30 million years ago, as indicated by the fossil record available at the time. The time line suggests that humans and the African great apes shared a common ancestor about 5 million years ago, which is much more recently than scientists had previously suspected (see Lowenstein and Zihlman 1988). We will review other, more recent molecular phylogenies shortly; all are consistent with Sarich and Wilson's conclusions in showing a close kinship between humans and the African great apes.

The phylogenies in Figure 20.1 and Figure 20.2 show that humans, gorillas, and the two chimpanzees are close relatives, but they do not resolve the evolutionary relationships among these four species. The true phylogeny for humans, gorillas, and the two chimpanzees could be any one of the four trees shown in Figure 20.3. It is probably safe to say that more scientists have invested more effort in attempting to determine which of these trees is correct than has been invested in any other species-level problem in the history of systematics.

Humans, Gorillas, Chimpanzees, and Bonobos

After decades of debate, researchers have come to a consensus that the evolutionary relationships among humans and the African great apes are best characterized by the tree in Figure 20.3a. Humans and the chimpanzees are more closely related to each other than either is to gorillas. This consensus was slow in forming for

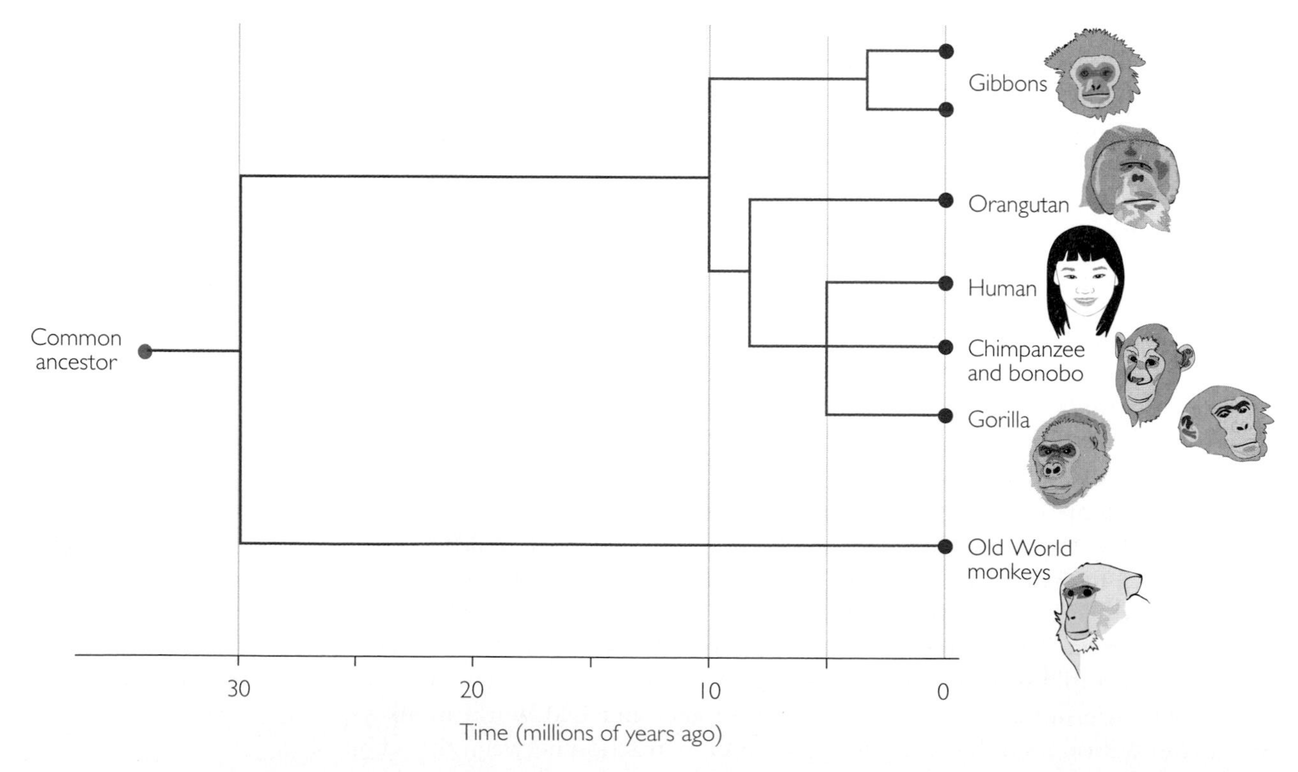

Figure 20.2 Sarich and Wilson's phylogeny of the apes The time line on the bottom is in millions of years before the present. Reprinted with permission from Sarich and Wilson (1967).

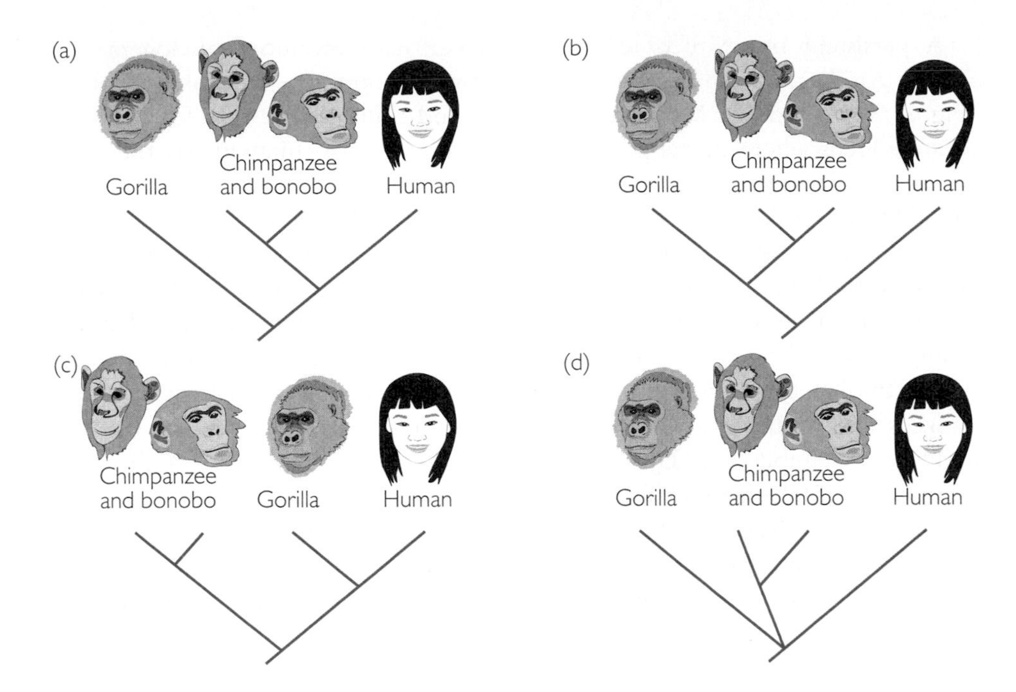

Figure 20.3 Possible phylogenies of humans and the African great apes The figure shows four possible resolutions of the evolutionary relationships among humans and the African great apes. All assume that the two species of chimpanzee are closest relatives. The true tree could have (a) chimpanzees and humans as closest relatives, (b) gorillas and chimpanzees as closest relatives, (c) gorillas and humans as closest relatives, or (d) a genuine simultaneous three-way split (trichotomy).

at least two reasons: There were conflicts among molecular data sets, and there were conflicts between molecular evidence and morphological evidence.

Molecular Evidence

Molecular biologists sought to resolve the human/African great ape phylogeny by analyzing DNA sequences. Maryellen Ruvolo and colleagues (1994), for example, reconstructed the evolutionary tree of the apes based on sequence data for a mitochondrial gene (Figure 20.4). On the evidence of these data, humans and the chimpanzees diverged from each other only after their common ancestor diverged from the gorillas. Researchers have reconstructed the evolutionary history of the apes with data from a great variety of loci. Most analyses have produced trees like the one in Figure 20.4, in which humans and the chimpanzees are closest relatives (for examples, see: Horai et al. 1992; Goodman et al. 1994; Kim and Takenaka 1996).

Molecular analyses indicate that humans and chimpanzees are closest relatives . . .

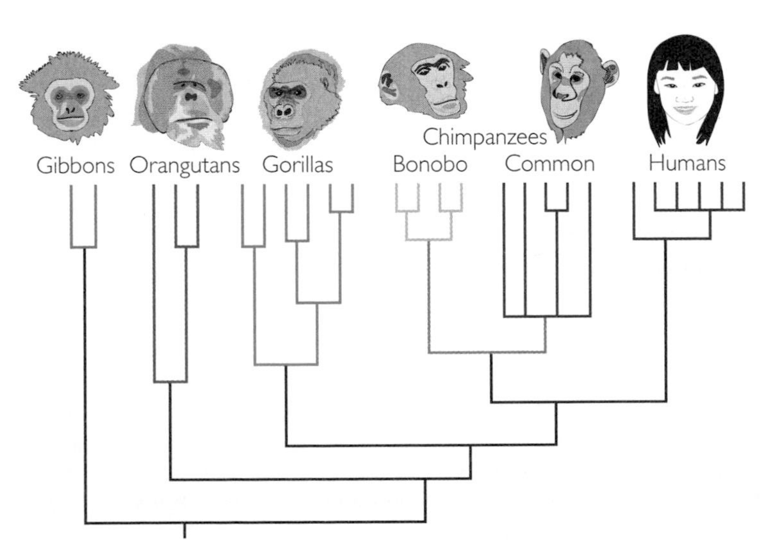

Figure 20.4 Phylogeny of mitochondrial cytochrome oxidase II alleles in humans and the African great apes Ruvolo and colleagues estimated this tree using the maximum parsimony method. From Ruvolo et al. (1994).

A persistent minority of analyses, however, have produced phylogenies in which gorillas and chimps, or even gorillas and humans, are closest relatives [Figure 20.3b and (c)]. Madalina Barbulescu and colleagues (2001), for example, found a locus where, in the genome of gorillas and both chimpanzees, there exists an inserted nucleotide sequence. This insert is the genome of a retrovirus called human endogenous retrovirus K, or HERV-K. The implication is that this particular retroviral invasion happened long ago, in the common ancestor of gorillas and the chimpanzees. The same locus in humans lacks the HERV-K insert and appears never to have had it. This locus, taken on its own, suggests that humans diverged from the lineage that would give rise to gorillas and the chimps before the latter acquired their HERV-K insertion, and that gorillas and chimpanzees are thus more closely related to each other than either is to humans (for additional examples see: Djian and Green 1989; Marks 1993, 1994 [but also Borowik 1995]; Deinard et al. 1998).

. . . but the phylogenies of genes and the phylogenies of species are not necessarily the same.

How can we reconcile the conflicting implications of these molecular analyses? The answer is that we should not necessarily have expected all molecular analyses to agree with each other in the first place. Phylogenies like the one in Figure 20.4 are gene trees, not species trees. If the ancestral species was genetically variable for the gene under study, then the gene tree estimated from sequence data may differ from the true species tree. Figure 20.5 illustrates the reasoning. If different descendant species lose different ancestral alleles (Figure 20.5a), then we can end up reconstructing only a portion of the original gene tree. This portion may imply a different branching pattern than that of the true species tree (Figure 20.5b).

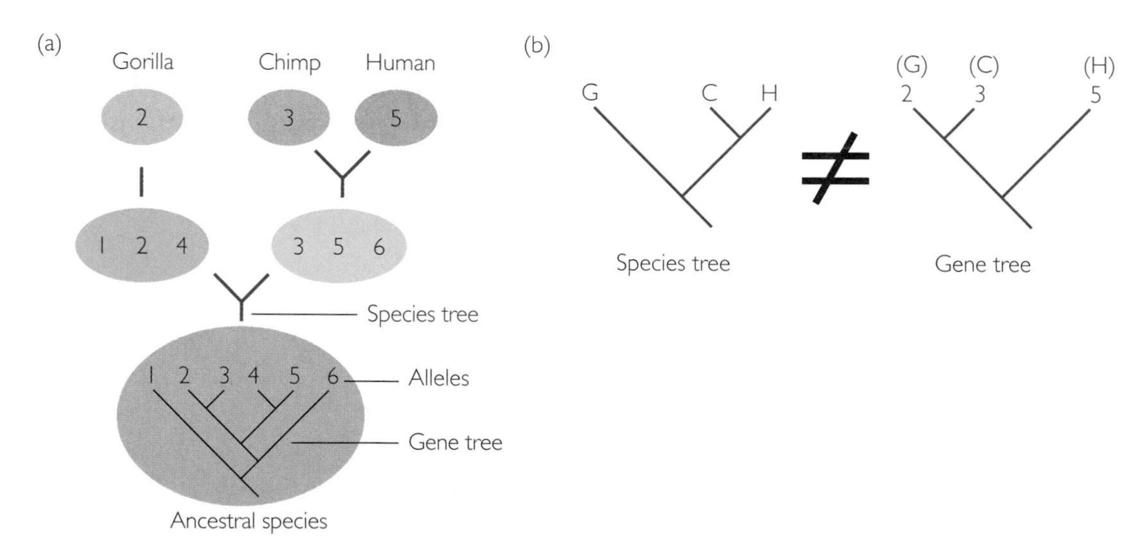

Figure 20.5 Gene trees versus species trees (a) Read the figure from the bottom to the top. The oval at the bottom represents an ancestral species containing six alleles for a gene. The alleles are related to each other as shown by the gene phylogeny inside the oval: Each allele was derived from its ancestors via a series of mutations. Moving up the figure, we see that a speciation event produces two sister species from the ancestral species. By selection or drift, one species loses alleles 3, 5, and 6, while the other species loses alleles 1, 2, and 4. Moving further up the figure, we find that another speciation event occurs, followed by the loss of alleles in its descendant species. We now have gorillas, containing allele 2; chimps, containing allele 3; and humans, containing allele 5. Now imagine that we sequence the allele from each species and reconstruct a phylogeny (b). In the true species tree, chimps and humans are closest relatives, but in the gene tree, alleles 2 and 3 are closest relatives. The gene tree and the species tree show different branching patterns. After Ruvolo (1994).

To determine the species tree from gene trees, we need to reconstruct the phylogeny using many independent genes. The cases in which different alleles have persisted at random from a genetically diverse ancestral species will be equally likely to support a human–chimpanzee pairing, a human–gorilla pairing, and a chimpanzee–gorilla pairing. Unless there has been a true three-way split (Figure 20.3d), however, the cases in which the gene tree and the species tree are the same should agree with each other and produce a clear signal against this random background. Ruvolo (1995, 1997) reviewed and tallied independent data sets of DNA sequences that were informative about the human/African great ape phylogeny. She counted all mitochondrial DNA studies as a single data set, because all mitochondrial genes are linked and are thus not independent of each other. Likewise, any groups of nuclear genes that are near each other on the same chromosome counted as a single data set, because the genes are linked. In all, Ruvolo found 14 independent data sets. Eleven of these show humans and the chimpanzees as closest relatives, two show gorillas and the chimpanzees as closest relatives, and one shows humans and gorillas as closest relatives. Ruvolo calculated that under the null hypothesis of a true trichotomy (Figure 20.3d), this distribution of results has a probability of only 0.002. Ruvolo concluded that the molecular phylogeny data reject the trichotomous tree and favor the tree in which humans and the chimpanzees are closest relatives (Figure 20.3a).

Combined analyses of several molecular data sets strongly support the hypothesis that humans and chimps are closest relatives.

This resolution of conflicting molecular data sets is consistent with more recent analyses of even larger collections of loci (Satta et al. 2000; O'hUigin et al. 2002; Wildman et al. 2003; Rauum et al. 2005; Elango et al. 2006). In a notable study, Abdel-Halim Salem and colleagues (2003) reconstructed the phylogeny of the apes using Alu elements. Alu elements are a type of short interspersed element, or SINE. As we discussed in Chapter 4, SINEs are selfish DNA sequences that occassionally insert themselves into the chromosomes of their hosts. Because insertions are infrequent, and because deletion of a SINE is usually detectable from the sequence left behind, SINEs are nearly ideal derived traits for reconstructing evolutionary history. Salem and colleagues found one Alu element insertion that was shared by humans and gorillas, but absent in the chimpanzees, versus seven insertions shared by humans and the chimpazees but absent in gorillas. Judging from the mutations it has accumulated, the insertion shared by humans and gorillas is older that the seven shared by humans and the chimps. Salem and colleagues infer that it reflects an ancestral polymorphism of the kind diagrammed in Figure 20.5. Overall, Salem's data strongly support the concensus that humans and the chimps form a monophyletic clade.

Perhaps the most elegant summary of the evidence provided by molecular analyses is Svante Pääbo's (2003). Pääbo describes the human genome, and by implication the genomes of the African great apes, as mosaics. For each genomic segment the homologous components from humans and the great apes have their own phylogeny. For some genomic segments gorillas and the chimpanzees, or gorillas and humans, are closest relatives. For most genomic segments, however, the closest kin are humans and the chimpanzees. Indeed, Wildman and colleagues (2003) argue that humans and the chimpanzees are so closely related genetically that they all belong together in genus *Homo*.

Morphological Evidence

Paleontologists sought to solve the human/chimp/gorilla puzzle with cladistic analyses of morphology. These researchers pointed out several features that are shared by gorillas and the two chimpanzees but absent in humans. These features mainly include skeletal traits associated with knucklewalking (Andrews and Martin 1987). Knucklewalking is moving on four legs with the upper body supported on the backs of the middle phalanges—the middle of the three bones in each finger.

Knucklewalking is derived in the African great apes. The Asian great apes, the orangutans, do walk on their knuckles occasionally, but typically they walk on their fists. That is, they support their weight on the backs of the proximal phalanges—the bone in each finger that is directly connected to the hand. The African great apes, the gorillas and both species of chimpanzee, are dedicated knucklewalkers. They have specialized hand and wrist anatomy to go with the habit. Humans, of course, are not knucklewalkers.

Considering knucklewalking in isolation, the simplest explanation for its distribution is that humans diverged first from the lineage that would later produce the gorilla and the two chimpanzees (Figure 20.3b). This scenario requires only one appearance of knucklewalking, in a common ancestor of the gorilla and the chimps, and no losses. There is a catch, however. While the tree in Figure 20.3b gives a parsimonious explanation for knucklewalking, it requires that several other traits shared only by humans and the two chimps be interpreted either as ancestral traits that were lost in gorillas or as convergent derived traits that evolved independently in humans and the chimps. These traits include features of the teeth, skull, and limbs, delayed sexual maturity, and prominent labia minora in females and a pendulous scrotum in males (Groves 1986; Begun 1992). The tree in Figure 20.3b also conflicts with the consensus result from molecular evidence, which shows that humans and the chimpanzees are closest relatives.

Resolution of the human/chimp/gorilla evolutionary tree on morphological grounds thus depends on the identification of which traits are ancestral and which are derived. David R. Begun classified characteristics of the skulls of the great apes by including in his analysis an extinct European ape called *Dryopithecus,* known only from fossils about 10 million years old (Begun 1992; see also Begun 1995). *Dryopithecus* shares several cranial traits with gorillas that are absent in the two chimpanzees and humans. These traits might previously have been classified as uniquely derived in gorillas, but given their presence in *Dryopithecus,* the traits now appear to be ancestral. This, in turn, means that some of the traits thought to be ancestral or convergent in humans and chimpanzees now appear to be derived. When Begun reconstructed the ape evolutionary tree with the new classification of traits, he concluded that humans and chimpanzees are closest relatives (Figure 20.3a). This implies either (1) that the most recent common ancestor of humans, gorillas, and the chimpanzees was a knucklewalker, and that knucklewalking was subsequently lost in the human lineage, or (2) that knucklewalking evolved independently in gorillas and the chimpanzees. It also implies that a number of characters of the teeth, skull, and limbs, as well as the delayed sexual maturity and shared genital anatomy of humans and chimpanzees, need have evolved only once.

Some researchers were not convinced by Begun's reasoning, arguing that some of the skull features that Begun believes are shared derived traits in humans and chimpanzees may be ancestral or convergent and that knucklewalking may not be so readily evolved or lost as Begun's phylogeny requires (Dean and Delson 1992; Andrews 1992). However, more recent analyses of much expanded data sets seem to confirm the close relationship among chimpanzees and humans (Shoshani et al. 1996; Begun et al. 1997; Gibbs et al. 2000; Gibbs et al. 2002; Lockwood et al. 2004; Strait and Grine 2004).

Recent morphological analyses also suggest that humans and chimpanzees are closest relatives.

Furthermore, Brian Richmond and David Strait (2000) recently compared the wrist bones of African fossils with those of living primates. They found evidence that at least two extinct species thought to be more closely related to humans than to chimpanzees or gorillas had anatomical features associated with knucklewalking (see also Collard and Aiello 2000; Corruccini and McHenry 2001; Dainton 2001; Lovejoy et al. 2001; Richmond and Strait 2001a, 2001b). This interpretation is consistent with the hypothesis that humans evolved from a knucklewalking ancestor, and that humans and the chimpanzees are each other's closest living relatives (Figure 20.3a). Thus it appears that the morphological evidence is now converging on the same conclusion as the molecular analyses.

Estimating the Divergence Times for Humans and the Apes

Working in the laboratory of S. Blair Hedges, a team led by undergraduate Rebecca L. Stauffer used molecular clocks to estimate the divergence times for humans and the apes (Stauffer et al. 2001). Based on the fossil record, the Old World monkeys diverged from the apes 23.3 million years ago. From comparisons of sequence differences in a variety of protein-coding genes from apes versus Old World monkeys, the researchers estimated the rate at which the genes have evolved since the two lineages diverged. Then, by counting the sequence differences in the same genes from, for example, humans versus chimpanzees, the team estimated the divergence times among the humans and the apes. To improve the accuracy of their estimates Stauffer and colleagues combined data from dozens of genes. Their estimates are summarized in Figure 20.6. The lineage that would become today's gorillas diverged from the lineage that would become humans and the chimpanzees 6.4 ± 1.5 million years ago. The human and chimpanzee lineages split 5.4 ± 1.1 million years ago. Using similar methods, an even larger dataset, and an ape–Old World monkey divergence time of 23.8 million years ago, Sudhir Kumar and colleagues (2005) estimated that the human and chimpanzee lineages diverged 4.98 million years ago, with a 95% confidence interval of 4.38 to 5.94 million years ago.

How large are the genetic differences that have evolved in the 4–8 million years since humans and the African great apes diverged? Box 20.1 summarizes evidence from some recent assessments.

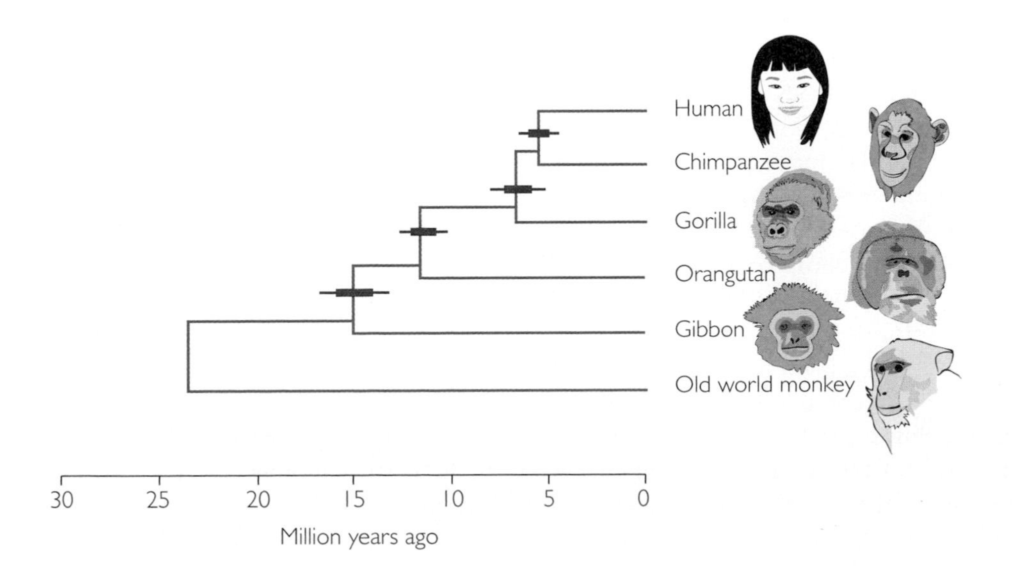

Figure 20.6 Divergence times for the apes Stauffer and colleagues estimated the dates of the common ancestors on this phylogeny by combining data from dozens of proteins used as molecular clocks. The heavy bars show ±1 standard error around the time estimates; the lighter bars show 95% confidence intervals. From Stauffer et al. (2001).

Box 20.1 Genetic differences between humans, chimpanzees, and gorillas

The most obvious genetic difference between humans and the African great apes is in their karyotypes. Gorillas and chimpanzees have 24 pairs of chromosomes, whereas humans have only 23. The reason is that in the ancestors of humans, sometime after our lineage split from that of the chimpanzees, two chromosomes fused to become what we now know as chromosome 2 (Figure 20.7).

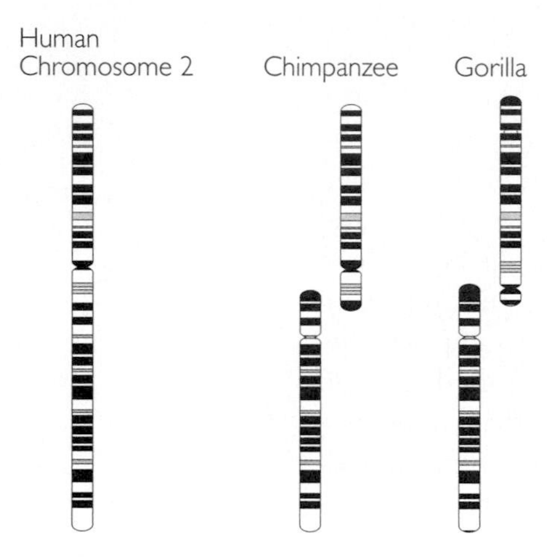

Figure 20.7 Human chromosome 2 and its homologues in chimpanzees and gorillas The banding patterns on stained chromosomes reveal that human chromosome 2 is derived from the fusion of two chromosomes that remain separate in the other great apes. From Hacia (2001).

The human genome project, and the technologies and research it has fostered, have produced data that are allowing researchers to assess the genetic differences between ourselves and our closest kin on a finer scale. Some of the differences are striking. In humans but not chimpanzees and gorillas, for example, a frameshift mutation in the gene for an enzyme called CMP-sialic acid hydroxylase has resulted in the complete loss of a cell-surface sugar common in mammals (see Gagneux and Varki 2001). Most of the differences, however, are more subtle.

Summarizing data from Feng-Chi Chen and Wen-Hsiung Li (2001), Joseph Hacia (2001) compiled the numbers in Table 20.1. Humans, chimps, and gorillas have diverged more in some kinds of DNA than in others. In pseudogenes—nontranscribed copies of functional genes—human and chimp sequences differ

Table 20.1 Genetic distances between humans, chimpanzees, and gorillas

This table gives estimates of the percentage sequence divergence between species pairs, based on examinations of 4.8–29.3 kb of sequence.

Sequence type	Chimp–Human	Gorilla–Human	Chimp–Gorilla
Noncoding intergenic	1.24	1.62	1.63
Intronic	0.93	1.23	1.21
Pseudogenes	1.64	1.87	2.14
X chromosome noncoding	1.16	1.47	1.50
Y chromosome	1.68	2.33	2.78
Coding sequences			
Synonymous	1.11	1.48	1.64
Nonsynonymous	0.80	0.93	0.90
Amino acid divergence	1.34	1.58	1.65

at an average of 1.64% of their nucleotides. In coding sequences, on the other hand, the two species differ at only 1.11% of synonymous nucleotides and just 0.80% of nonsynonymous nucleotides.

Roy Britten (2002) argues that comparisons like those in Table 20.1, which only count nucleotide substitutions, underestimate the true genetic divergence among species. The reason is that they ignore larger insertions and deletions. Comparing 779 kb of chimpanzee sequence with the most closely matching regions of the human genome, Britten found a 1.4% difference due to nucleotide substitution and an additional 3.4% difference due to insertions and deletions. Britten estimates that overall about 95% of the base pairs in the human and chimpanzee genomes are identical.

To put all of these numbers in perspective, we can measure the human–chimpanzee genetic difference against the genetic differences that separate other species pairs. Mitochondrial DNA sequences are a useful point of comparison, because they are known for a great variety of species. John Klicka and Robert Zink (1997), for example, sequenced mitochondrial genes from a number of bird species. The mitochondrial genomes in the great apes have evolved more rapidly than the nuclear genomes. In coding sequences of their mitochondrial genomes, humans and chimps differ at just under 10% of their nucleotides (see Hacia 2001). By Klicka and Zink's data, this means that humans are not as closely related to chimps as eastern bluebirds are to western bluebirds. We are, however, more closely related to chimps than blue jays are to Steller's jays. By just about any standard, blue jays and Steller's jays are similar birds. The clear message is that the genetic divergence between humans and chimpanzees is rather small.

Our genomes are large, however. Upon completing a draft sequence of a chimpanzee genome, Tarjei Mikkelsen and colleagues (2005) reported that humans and chimpanzees are distinguished by about 35 million single-nucleotide substitutions, 5 million insertions and deletions, and an assortment of chromosomal rearrangements. About 29% of the proteins encoded in our genome are identical to the homologous protein in chimps. For the remaining proteins, the typical difference is 2 amino acid substitutions.

The genetic difference between humans and chimpanzees varies across the genome. Nick Patterson and colleagues (2006) interpret this as evidence of that the human and chimp lineages continued to exchange genes for an extended period before diverging for good. Nick Barton (2006) argues, however, that an abrupt split followed by the random loss of alleles from polymorphic loci provides a simpler explanation.

Which of the genetic differences between us and our closest relatives are the ones that make us human and them chimps? We do not yet know (Sikela 2006).

They could be among the relatively small number of genes that have been gained or lost in one lineage or the other. Wang and colleagues (2006) identified 80 genes that are active in chimps but disabled in humans by loss-of-function mutations. These genes encode olfactory, taste, and other chemoreceptors, and proteins involved in the immune response. Magdalena Popesco and colleagues (2006) found a family of duplicated genes with far more copies in the human genome than in the chimp genome. The function of the encoded proteins is unknown, but they are expressed in the brain.

The crucial differences could also be among the amino acid substitutions in proteins that are encoded by both gemones. Rasmus Nielsen and colleagues (2005) combed the human and chimp genomes for genes with high ratios of nonsynonymous to synonymous substitutions. Many of the genes they turned up function in sensory perception, immune defense, tumor suppression, and spermatogenesis.

Finally, the genetic differences that make us human could, as predicted decades ago by Marie-Claire King and Allan Wilson (1975), lie in the regulatory regions that control when, where, and in what amount each protein is made. Yoav Gilad and colleagues (2006) measured the transcriptional expression of genes in the livers of humans, chimpanzees, orangutans, and rhesus macaques. Among the genes with elevated expression in humans, a higher than expected number were themselves transcription factors. Wolfgang Enard and colleagues (2002) compared overall patterns of gene expression in the livers, blood, and brains of humans, chimpanzees, and rhesus macaques. They found that in their patterns of gene expression in blood and liver, humans and chimps have diverged about equally from their common ancestor. In their brain gene expression, however, humans have diverged substantially more (Figure 20.8). The proper interpretation of this pattern is a topic of ongoing research (see, for example, Gu and Gu 2003; Khaitovich et al. 2006).

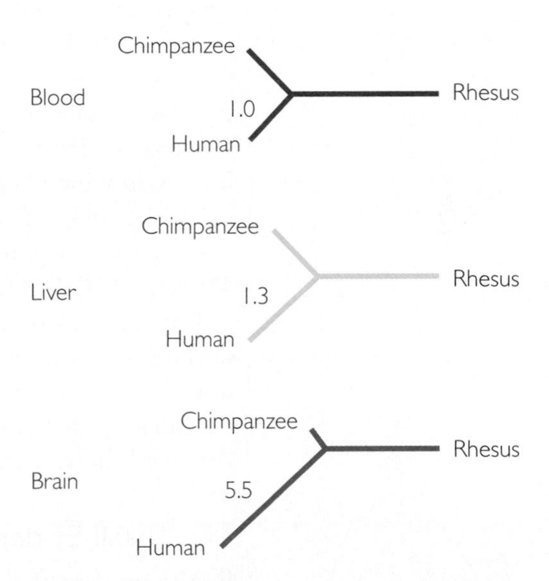

Figure 20.8 Differences in gene expression patterns in different tissues of humans, chimps, and rhesus macaques These unrooted trees represent the divergence in overall patterns of gene expression in humans versus chimpanzees versus rhesus macaques. The numbers on the human branches represent the ratio of the human divergence versus the chimp divergence. In blood and liver, humans have diverged from the common pattern about as much as chimps have. In brain, however, humans have diverged considerably more. From Enard et al. (2002).

20.2 The Recent Ancestry of Humans

According to the evidence presented in Figure 20.6, humans and the two chimpanzees last shared an ancestor about 5.4 million years ago. With appropriate caution, we can use what we know about humans, chimpanzees, and bonobos to infer something of the nature of that last common ancestor. It is probable that we inherited from it at least some of the behaviors that are shared by its three descendants today. If this is the case, then the last common ancestor, in addition to being a knucklewalker, would have had a broad, fruit-based diet and lived in a range of different habitats. It may have used tools to obtain and process food, and it may have hunted, as do living bonobos, chimpanzees, and humans.

The last common ancestor also may have had culture—behavior that is taught and learned and varies among populations. Chimpanzees, like humans, exhibit cultural variation today (de Waal 1999; Whiten et al. 1999, Whiten 2005). Indeed, culture may have appeared in our lineage long before our last common ancestor with chimps and bonobos, as it is also present in orangutans (van Schaik et al. 2003).

Other aspects of the last common ancestor's behavior are more elusive. Bonobos and common chimpanzees are equally close kin to humans, yet show striking differences in behavior (Parish and de Waal 2000). Elements of each's behavior resonate, for some observers, with the behavior shown by humans in at least some cultures. Chimpanzee societies, for example, are dominated by males that form strategic alliances, fight viciously, and sometimes stalk and kill their rivals. Bonobo societies, in contrast, are dominated by females that form strong bonds with each other, even when they are not kin, and that, while aggressive toward males, are less violent than chimpanzees. While chimpanzees sometimes engage in homosexual behaviors, they are primarily heterosexual. Bonobos, in contrast, have sex in all possible combinations, and for a variety of reasons not directly connected with procreation. Perhaps the safest inference to draw is that we humans belong to a lineage in which behavior is flexible, both culturally and evolutionarily. For more on the behavior of chimpanzees and bonobos, and its possible relevance to human behavior, see Begun (1994), Boesch and Tomasello (1998), de Waal (1997, 2005), Manson et al. (1997), Parish (1994, 1996), and Wrangham (1999).

Our goal in this section is to review evidence on the pattern of evolution leading from our last common ancestor with the chimpanzees to ourselves. Has our history involved only the steady transformation of a single lineage, finally culminating in *Homo sapiens,* or have there been splits and extinctions in our recent evolutionary tree?

The Fossil Evidence

Fossils provide the only data available for answering this question. The fossil record for early humans and their kin is frustratingly sparse, but steadily improving (see Tattersall 1995; Johanson et al. 1996; Tattersall 1997). Illustrations and photos of some of the key specimens appear in Figures 20.9 through 20.14.

The fossil record includes a diversity of hominins—species that lived after the human and chimpanzee lineages separated and are more closely related to humans than to chimpanzees.

Paleoanthropologists disagree about the most appropriate names for many of these specimens. With some exceptions, we use the names used by Johanson et al. (1996) in the belief that they will be the names most familiar to readers. In several cases we note alternative names. We use the term *hominin* to describe any species more closely related to humans than to chimpanzees, but note that some paleoanthropologists still prefer the more traditional term *hominid* (see Box 1 in Gee 2001).

Likewise, paleoanthropologists disagree about the number of species represented by the specimens in the figures (see Tattersall 1986, 1992). For example, the specimens of *Homo habilis* and *Homo rudolfensis* in Figures 20.13b and (c) are both from Koobi Fora, Kenya, and are both about 1.9 million years old. Some researchers consider them to be variants of the same species, whereas others consider them different species. As with the names, we have largely followed the classification used by Johanson et al. (1996).

Most of the time ranges noted in the figures are also those given in Johanson et al. (1996); they differ somewhat from the estimates of other researchers, including those given by Strait et al. (1997) and used in Figure 20.16, and those used in Figure 20.15.

We start with *Sahelanthropus tchadensis* (Figure 20.9). Found in the Djurab Desert of Chad in July 2001 by Djimdoumalbaye Ahounta, a member of a team led by Michel Brunet, this nearly complete cranium stunned paleoanthropologists (Brunet et al. 2002; Gibbons 2002a; see also Brunet et al. 2005). For one thing, it is 6–7 million years old. This places it at the older end of the window during which molecular biologists estimate humans diverged from chimpanzees. For another, it shows a curious mixture of traits. As Bernard Wood (2002) describes it, its small braincase (320–380 cm^3) makes it look, from the back, like a chimpanzee. From the front, however, its relatively flat face makes it look like an *Australopithecus, Kenyanthropus*, or *Homo* from as recently as 1.75 million years ago. In other words, it looks like a closer relative of humans than anyone expected in a fossil so old. *Sahelanthropus tchadensis* could be a close relative of the last common ancestor—or even, in principle, the last common ancestor itself.

Name: ***Sahelanthropus tchadensis***
Also known as: "*Toumaï*"
Specimen: TM 266-01-060-1
Age: 6–7 million years
Found by: Djimdoumalbaye Ahounta
Location: Djurab Desert, Chad

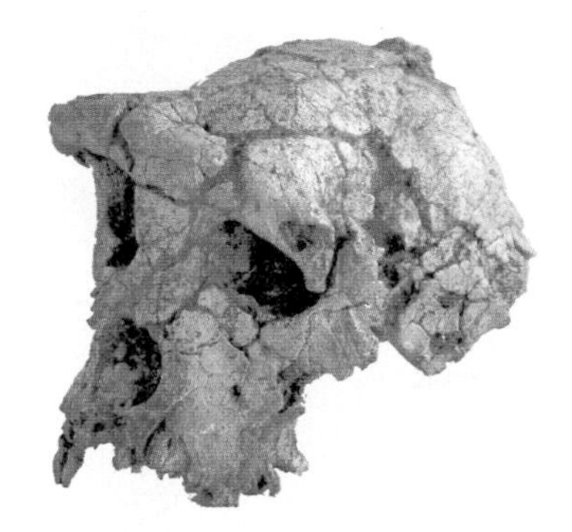

Figure 20.9 Sahelanthropus tchadensis This 6–7 million-year-old skull, found by a member of a team led by Michel Brunet, may represent a close relative of our common ancestor with the chimpanzees. From Wood (2002).

Brunet and colleagues believe that *Sahelanthropus* is, indeed, an early hominin, a descendant of the last common ancestor on the human side of the evolutionary tree. Many other paleoanthropologists are inclined to agree (see Cela-Conde and Ayala 2003; Gibbons 2005; Wilford 2005). The view is not unanimous, however.

Among the dissenters are Brigitte Senut and Martin Pickford, discoverers of a rival candidate for the title of oldest known hominin. Their discovery, *Orrorin tugenensis*, lived about 6 million years ago in what is now Kenya. It is known primarily from three thighbones (Aiello and Collard 2001; Senut et al. 2001; Gibbons 2002b). Shortly after Brunet and colleagues reported the discovery of *Sahelanthropus*, Senut and Pickford, along with Milford Wolpoff and John Hawks, suggested that *Sahelanthropus* belongs to the lineage that produced chimpanzees, or even to the lineage that led to gorillas (Wolpoff et al. 2002). Brunet (2002) rejected this suggestion.

Part of the difficulty in interpreting *Sahelanthropus* is due to the fact that its skull is crushed and distorted. To overcome this problem, Brunet enlisted the

(a) Name: **Australopithecus garhi**
Specimen: BOU-VP-12/130
Age: 2.5 million years
Found by: Yohannes Haile-Selassie
Location: Bouri Formation, Ethiopia

Species Time Range: ~2.5 mya

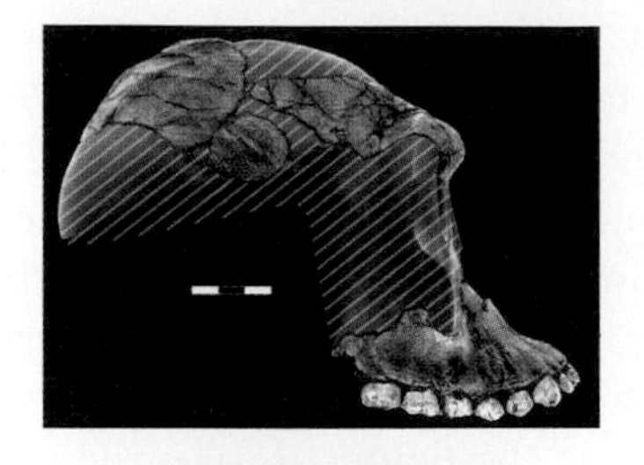

(b) Name: **Australopithecus africanus**
Specimen: Sts 5
Age: 2.5 million years
Found by: Robert Broom and John T. Robinson
Location: Sterkfontein, South Africa
Color photo: Johanson et al. (1996) pages 3; 135

Species Time Range: ~2.4–2.8 mya

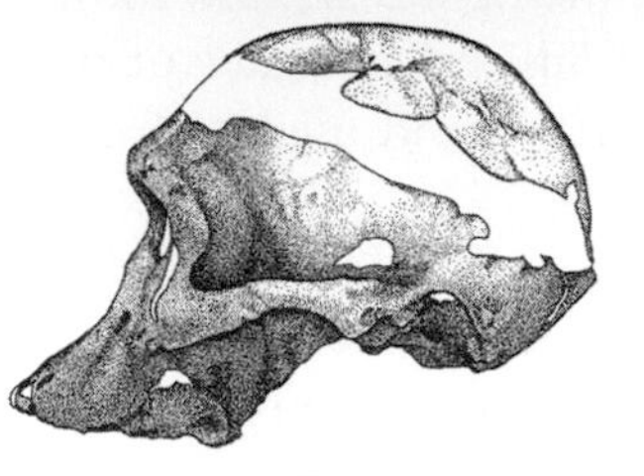

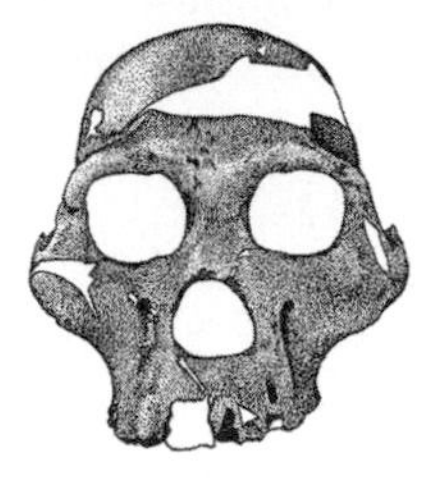

(c) Name: **Australopithecus afarensis**
Also known as: *Praeanthropus africanus*
Specimen: Reconstruction from fragments
Color photo of same species: Johanson et al. (1996) page 129

Species Time Range: ~3.0–3.9 mya

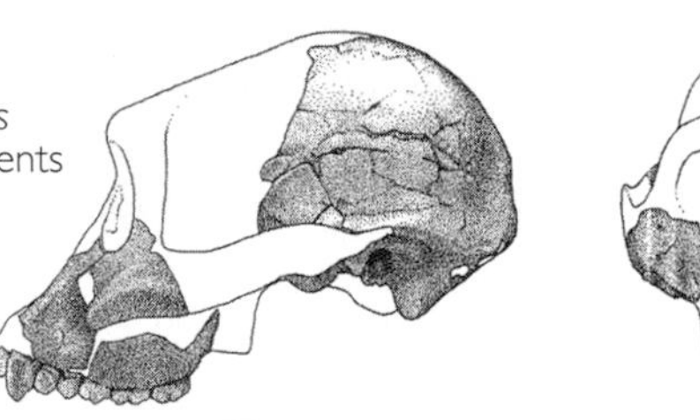

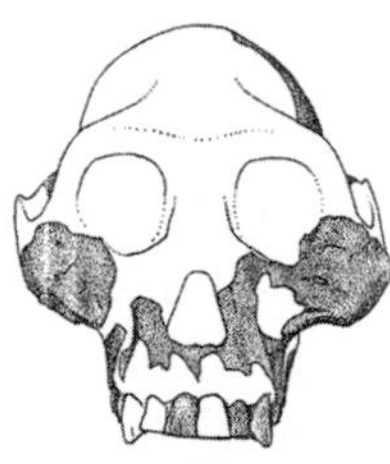

(d) Name: **Kenyanthropus platyops**
Specimen: KNM-WT 40000
Age: 3.5 million years
Found by: J. Erus
Location: Lake Turkana, Kenya

Species Time Range: ~3.5 mya

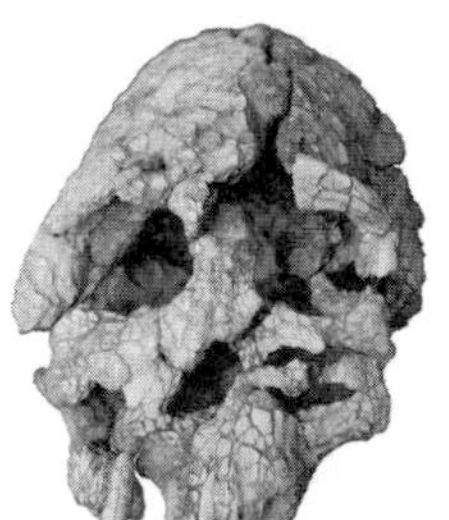

(e) Name: **Australopithecus anamensis**
Specimen: KNM-KP 29281
Age: 4.1 million years
Found by: Peter Nzube
Location: Kanapoi, Kenya
Color photo: Johanson et al. (1996) page 123

Species Time Range: ~3.9–4.2 mya

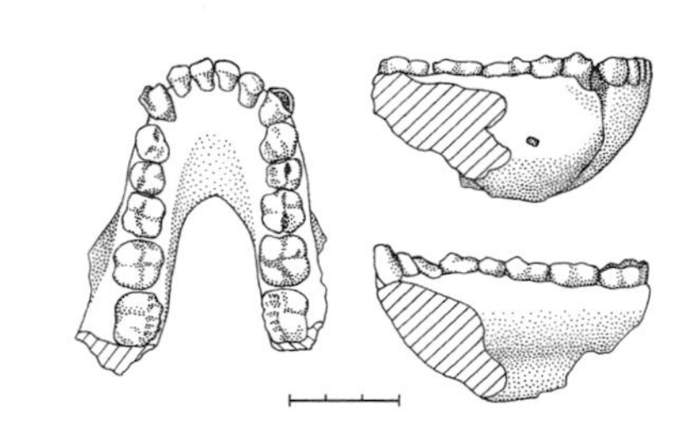

(f) Name: **Ardipithecus ramidus**
Originally named as: *Australopithecus ramidus*
Specimen: ARA-VP-1/128
Age: 4.4 million years
Found by: T. Assebework
Location: Aramis, Ethiopia
Color photo of same species: Johanson et al. (1996) page 116

Species Time Range: ~4.4 mya

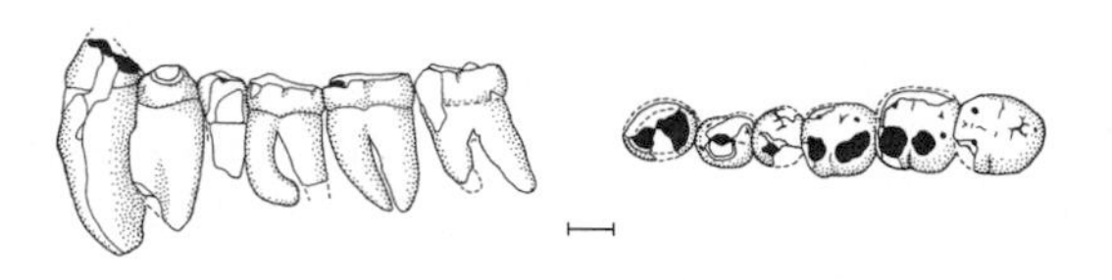

Figure 20.10 **Gracile australopithecines, *Kenyan-thropus*, and *Ardipithecus*** (a) Photo by David Brill. Scale unit = 1 cm. (b) By Don McGranaghan, page 70 in Tattersall (1995). Scale bar = 1 cm. (c) By Don McGranaghan, page 146 in Tattersall (1995). Scale bar = 1 cm. (d) Figure 1 in Leakey et al. (2001). Scale bar = 3 cm. (e) By Laszlo Meszoly, after Figure 1a in Leakey et al. (1995). Scale unit = 1 cm. (f) By Laszlo Meszoly, after Figure 3b in White et al. (1994); see also White et al. (1995). Scale bar = 1 cm.

collaboration of a team led by Chistoph Zollikofer and Marcia Ponce de León. The researchers X-rayed the skull with a CT scanner and used the resulting images to prepare a corrected three-dimensional reconstruction. The reconstructed *Sahelanthropus* resembles known hominins more strongly than it does chimpanzees or gorillas, bolstering Brunet's view that his fossil represents the human side of the family tree (Guy et al. 2005; Zollikofer et al. 2005). The dissenters have declined to concede (Wolpoff et al. 2006). The issue is likely to be resolved only by the discovery of additional fossils, including post-cranial remains.

Figure 20.10 shows examples of undisputed early hominins: the gracile australopithecines, *Kenyanthropus* and *Ardipithecus.* The species depicted in Figure 20.10a, (b), and (c), *Australopithecus gahri, Australopithecus africanus*, and *Australopithecus afarensis,* had skulls with small braincases (400 to just over 500 cm^3) and relatively large, projecting faces (Johanson et al. 1996; Asfaw et al. 1999). Female *Australopithecus africanus* and *Australopithecus afarensis* grew to heights of about 1.1 meters (3′7″), whereas the males were some 1.4–1.5 meters (4′7″–4′11″) tall (but see Reno et al. 2003). Both species walked on two legs. Evidence for their erect posture comes from many bones of the skeleton, including the hips, knees, feet, limb proportions, and vertebral column, all of which are anatomically modified to permit upright posture and the support of the body mass on two rather than four feet. Other evidence for bipedal locomotion appears in the photo in Figure 20.11: fossilized footprints at Laetoli, Tanzania, of a pair of *A. afarensis* that walked side by side through fresh ash from the Sadiman volcano 3.6 million years ago (Stern and Susman 1983; White and Suwa 1987).

Kenyanthropus platyops (Figure 20.10d), 3.5 million years old, was discovered in August 1999 by J. Erus, an assistant to Meave Leakey and colleagues (2001). *Kenyanthropus platyops* has a brain the same size as that of *Australopithecus afarensis* (Figure 20.10c), which lived at the same time, and a variety of other ancestral skull characters. At the same time, *K. platyops* has smaller teeth and a flatter and more human-looking face than *A. afarensis*, or any other species traditionally

Figure 20.11 Footprints of a pair of *Australopithecus afarensis* These 3.6-million-year-old footprints from Laetoli, Tanzania were made by a pair of individuals who walked side-by-side through fresh ash from a volcanic eruption.

classified as *Australopithecus*. Feeling that *K. platyops* does not fit into either *Australopithecus* or *Homo*, Leakey and colleagues assigned it to a new genus. Tim White (2003), in contrast, argues that the more human appearance of *K. platyops* is an illusion resulting from the fact that the skull has been fragmented and distorted by the rock it is preserved in. Were the skull not so distorted, White believes it would fall within the range of variation already known for other fossils of similar age that are assigned to genus *Australophithecus*. White maintains that all these fossils, *K. platyops* included, belong to a single lineage connecting *A. anamensis* (Figure 20.10e) to *A. afarensis* (Figure 20.10c). The fossils show considerable structural variation, but given the variation we can observe among today's humans, among bonobos, and among chimpanzees, White asserts that we should expect the recent ancestors of these species to have been variable, too.

The species depicted in Figures 20.10e and (f), *Australopithecus anamensis* and *Ardipithecus ramidus,* are less well known than the more recent species in the figure. The structure and size of a tibia from *A. anamensis* indicates that its owner was a biped somewhat larger than *A. afarensis* (Leakey et al. 1995). *Ardipithecus ramidus* is the least derived of these species, with teeth in many ways intermediate between chimpanzees and humans. It has skeletal features that suggest it was a biped, but its discoverers are withholding final judgment until they can complete a more thorough analysis (White et al. 1994; Johanson et al. 1996).

Figure 20.12 shows examples of three species formerly known as the robust australopithecines, now called *Paranthropus*. Like the gracile australopithecines, these species had relatively small braincases (in most instances between those of the gracile australopithecines and early *Homo* in relative size) and very large faces.

(a) Name: ***Paranthropus robustus***
Also known as: *Australopithecus robustus*
Specimen: SK 48
Age: 1.5–2.0 million years
Found by: Fourie
Location: Swartkrans, South Africa
Color photo: Johanson et al. (1996) pages 108; 150

Species Time Range: ~1.0–2.0 mya

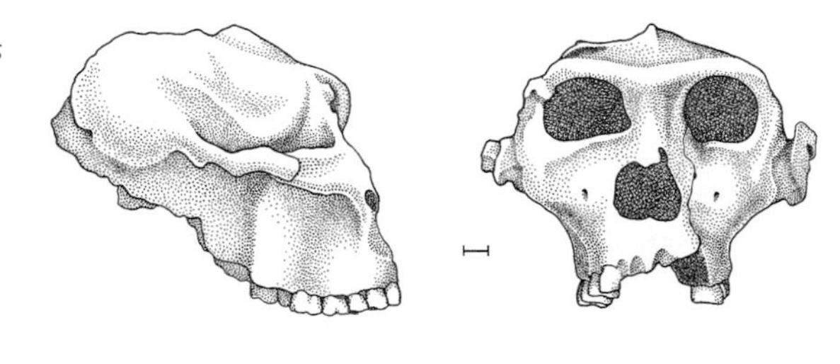

(b) Name: ***Paranthropus boisei***
Also known as: *Australopithecus boisei*
Specimen: KNM-ER 406
Age: 1.7 million years
Found by: Richard Leakey and H. Mutua
Location: Koobi Fora, Kenya
Color photo: Johanson et al. (1996) pages 54; 159; 160

Species Time Range: ~1.4–2.3 mya

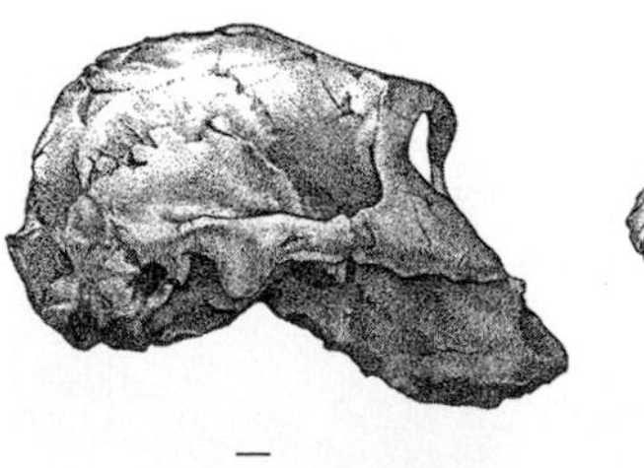

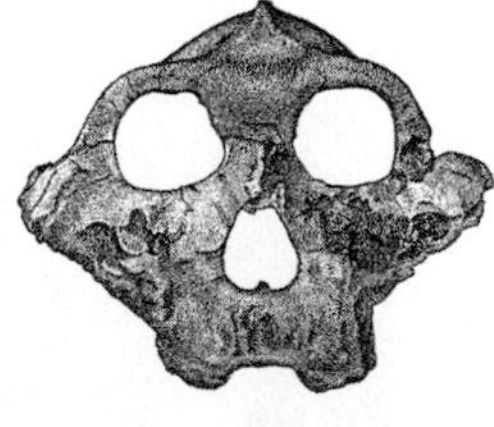

(c) Name: ***Paranthropus aethiopicus***
Also known as: *Australopithecus aethiopicus*
Specimen: KNM-WT 17000 (Black Skull)
Age: 2.5 million years
Found by: Alan C. Walker
Location: Lake Turkana, Kenya
Color photo: Johanson et al. (1996) pages 153; 154

Species Time Range: ~1.9–2.7 mya

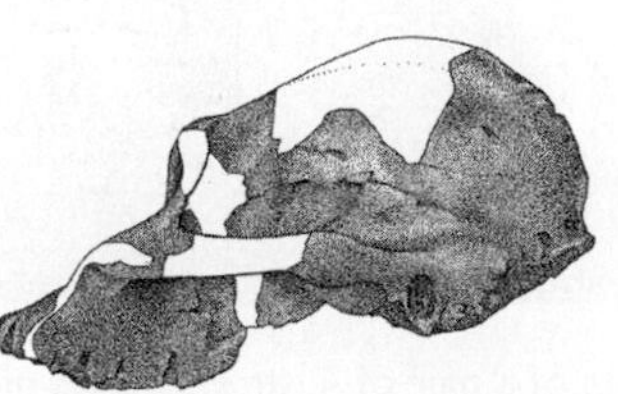

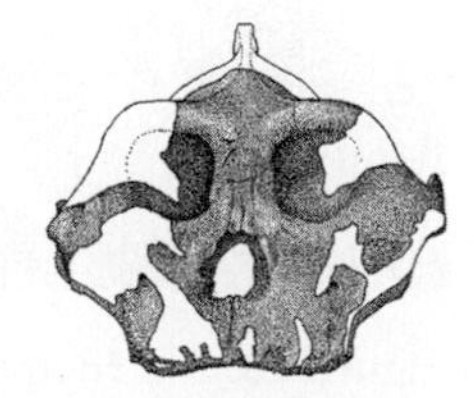

Figure 20.12 ***Paranthropus* (robust australopithecines)** (a) By Laszlo Meszoly, after pages 108 and 150 in Johanson et al. (1996). Scale bar = 1 cm. (b) By Don McGranaghan, page 131 in Tattersall (1995). Scale bar = 1 cm. (c) By Don McGranaghan, page 195 in Tattersall (1995). Scale bar = 1 cm.

Unlike the gracile australopithecines, they had enormous cheek teeth, robust jaws, and massive jaw muscles, sometimes anchored to a bony crest running along the centerline on the top of the skull (Johanson et al. 1996). These adaptations for powerful chewing have given one of the species, *P. boisei,* the nickname "nutcracker man." The robust australopithecines were about the same size as the gracile forms, and all were bipeds.

Figure 20.13 shows examples of an undisputed early member of the genus *Homo*, and two forms whose taxonomic status is controversial, all from Africa. That *Homo ergaster* (Figure 20.13a) is human is undisputed. Its braincase volume is 850 cm^3 (Johanson et al. 1996). In present-day humans the average is about 1,200 cm^3; some examples of *H. sapiens* have braincase volumes as large as 2,000 cm^3. Even at two-thirds today's average, however, *H. ergaster* has a much larger brain than any of the other fossils we have discussed so far. Furthermore, compared to the fossils in Figures 20.9, 20.10, and 20.12, it has a number of other features characteristic of humans. These include a relatively smaller, flatter face, smaller teeth and jaws, greater height, longer legs, and reduced sexual size dimorphism.

Homo habilis (Figure 20.13b) and *Homo rudolfensis* (Figure 20.13c) lived in the same place at about the same time, a factor that contributes to a dispute, mentioned above, over whether they are different species or just large and small individuals of a single species. Robert Blumenschine and colleagues (2003) report their discovery of a set of jaws and teeth from Olduvai Gorge, Tanzania. This specimen is 1.8 million years old and was found in association with stone tools and bones bearing the marks of butchery. The researchers assign their fossil to the species *H. habilis*, but note that it bears a decided resemblance to the *H. rudolfensis* shown in Figure 20.13c. Blumenschine and colleagues believe that their new

(a) Name: **Homo ergaster**
Also known as: (African) *Homo erectus*
Specimen: KNM-ER 3733
Age: 1.75 million years
Found by: Bernard Ngeneo
Location: Koobi Fora, Kenya
Color photo: Johanson et al. (1996) pages 180; 181

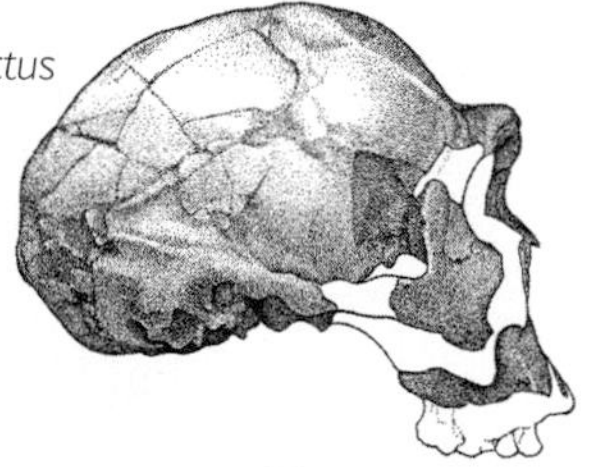

Species Time Range: ~1.5–1.8 mya

(b) Name: **Homo habilis**
Also known as: *Australopithecus habilis*
Specimen: KNM-ER 1813
Age: 1.9 million years
Found by: Kamoya Kimeu
Location: Koobi Fora, Kenya
Color photo: Johanson et al. (1996) pages 6; 175

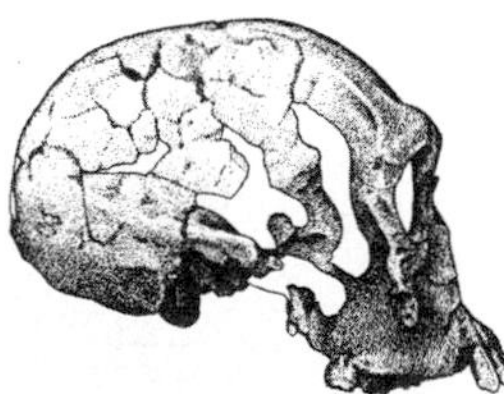

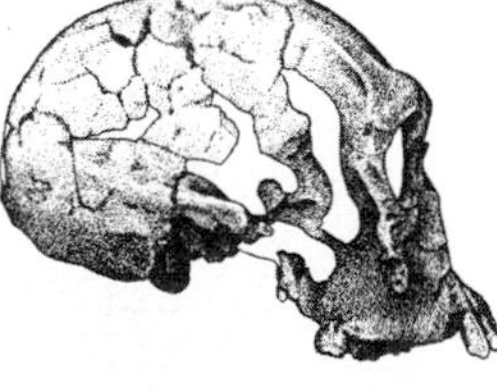

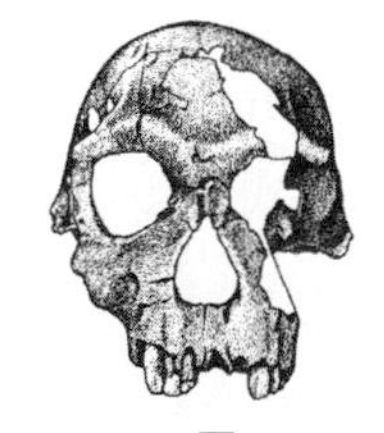

Species Time Range: ~1.6–1.9 mya

(c) Name: **Homo rudolfensis**
Also known as: *H. habilis; A. rudolfensis; K. rudolfensis*
Specimen: KNM-ER 1470
Age: 1.8–1.9 million years
Found by: Bernard Ngeneo
Location: Koobi Fora, Kenya
Color photo: Johanson et al. (1996) pages 178; 179

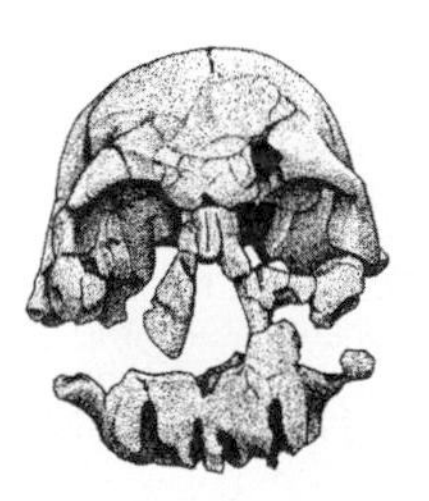

Species Time Range: ~1.8–2.4 mya

Figure 20.13 Early humans (a) By Don McGranaghan, page 138 in Tattersall (1995). Scale bar = 1 cm. (b) By Don McGranaghan, page 134 in Tattersall (1995). Scale bar = 1 cm. (c) By Don McGranaghan, page 133 in Tattersall (1995). Scale bar = 1 cm .

discovery argues against the designation of *H. rudolfensis* as a separate species, but it is unlikely to settle the issue (see Tobias 2003).

Are *Homo habilis* and *Homo rudolfensis* human, as their traditional genus name *Homo* implies? The *H. habilis* specimen shown here has a braincase volume of just 510 cm^3; the *H. rudolfensis* has a cranial capacity of 775 cm^3 (Johanson et al. 1996). This means both forms have larger brains than the australopithecines, though just barely larger in *H. habilis*'s case. Both have somewhat flatter faces than the australopithecines, but they overlap the australopithecines in tooth and body size. Bernard Wood and Mark Collard (1999) argue that *H. habilis* and *H. rudolfensis* should not be considered human and assign them instead to genus *Australopithecus*. Many other paleoanthropologists have adopted this view.

Given that the *Homo rudolfensis* specimen shown in Figure 20.13c, KNM-ER 1470, resembles *Kenyanthropus platyops* (Figure 20.10d), Maeve Leakey and colleagues (2001) suggest that the specimen should be rechristened *Kenyanthropus rudolfensis*. This proposal, too, has attracted adherents (see Aiello and Collard 2001; Lieberman 2001). KNM-ER 1470, whatever it is, seems to bear the brunt of considerable pushing and shoving among paleoanthropologists.

(a) Name: ***Homo sapiens***
Specimen: Cro-Magnon I
Age: 30,000 to 32,000 years
Found by: Louis Lartet and Henry Christy
Location: Abri Cro-Magnon, Les Eyzies, France
Color photo: Johanson et al. (1996) pages 245; 246

Species Time Range: ~0.1 mya–Present

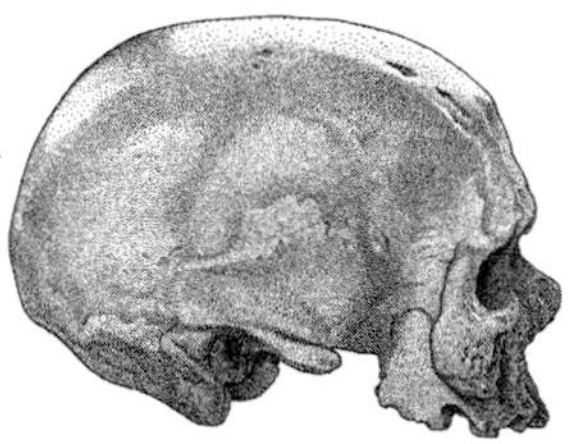

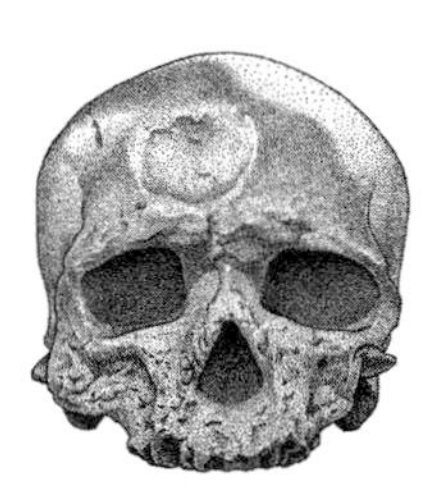

(b) Name: ***Homo neanderthalensis***
Specimen: Saccopastore I
Age: ~120,000 years
Found by: Mario Grazioli
Location: Saccopastore quarry, Rome, Italy
Color photo: Johanson et al. (1996) pages 213; 214

Species Time Range: ~0.03–0.3 mya

(c) Name: ***Homo heidelbergensis***
Specimen: Broken Hill I
Age: ~300,000 years
Found by: Tom Zwigelaar
Location: Kabwe, Zambia
Color photo: Johanson et al. (1996) pages 209; 210

Species Time Range: ~0.2–0.6 mya

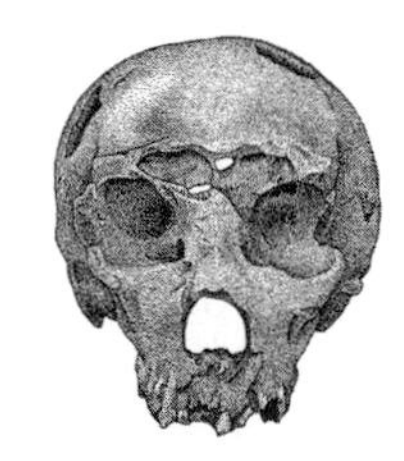

(d) Name: ***Homo erectus***
Specimen: Sangiran 17
Age: ~800,000 years
Found by: Mr. Towikromo
Location: Sangiran, Java, Indonesia
Color photo: Johanson et al. (1996) pages 192; 193

Species Time Range: ~0.4–1.2 mya

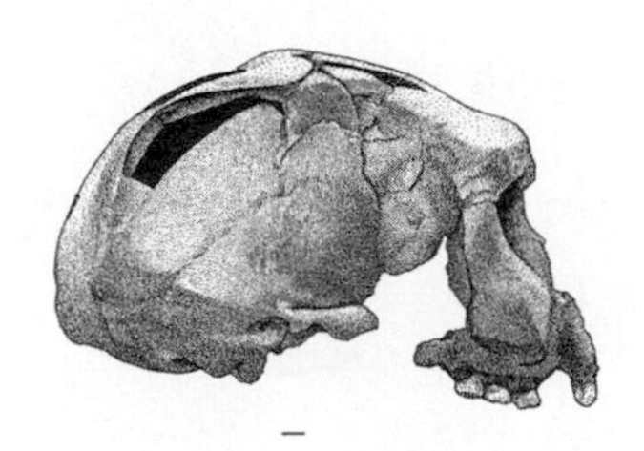

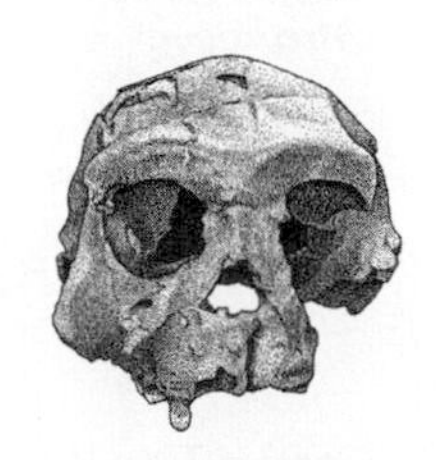

Figure 20.14 Recent humans (a) By Don McGranaghan, page 25 in Tattersall (1995). Scale bar = 1 cm. (b) By Don McGranaghan, page 83 in Tattersall (1995). Scale bar = 1 cm. (c) By Don McGranaghan, page 54 in Tattersall (1995). Scale bar = 1 cm. (d) By Don McGranaghan, page 172 in Tattersall (1995). Scale bar = 1 cm.

Figure 20.14 shows examples of more recent members of the genus *Homo* from Africa, Europe, and Java. Modern *H. sapiens,* including Cro-Magnon I, whose skull appears in Figure 20.14a, differ from earlier forms in a variety of traits (Johanson et al. 1996). Modern humans have very large braincases. Cro-Magnon I's is over 1,600 cm^3, substantially larger than the present-day average. Associated with their large braincases, modern humans have high, steep foreheads. They also have relatively short, flat, vertical faces and prominent noses. Cro-Magnon I was a man who died in middle age about 30,000 years ago. His skeleton was found in a single prepared grave along with those of two other adult men, an adult woman, and an infant. The group had been buried with an assortment of animal bones, jewelry, and stone tools.

Interpreting the Fossil Evidence

Figure 20.15 summarizes the fossil evidence we have discussed. On the lower right of the figure are the intriguing and nearly too-ancient *Sahelanthropus tchadensis* and *Orrorin tugenensis*, apparently close relatives to our last common ancestor with the chimpanzee. In the center is a rather confusing suite of fossils assigned, in different combinations by different researchers, to the genera *Australopithecus, Kenyanthropus*, and *Homo*. Out of this confusion emerge two distinct hominin groups: later members of genus *Homo* (that is, humans) and *Paranthropus* (also known as the robust australopithecines). Can we organize these fossils any more coherently, by arranging them on an evolutionary tree?

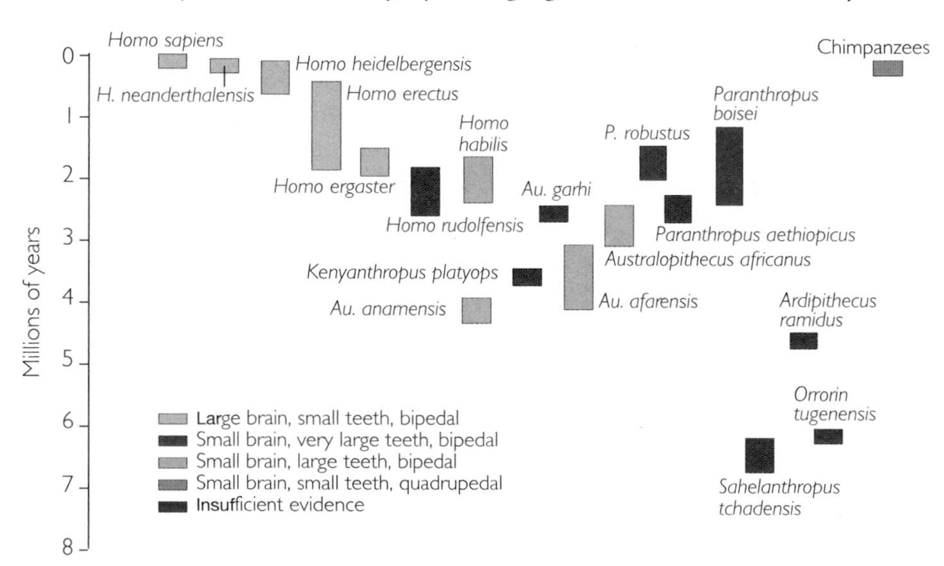

Figure 20.15 Summary of fossil evidence on the recent ancestry of humans The vertical axis gives approximate time ranges for the species we have mentioned. Horizontally, the hominin species are grouped roughly by morphological similarity. Chimpanzees are the outgroup. From Wood (2002).

Paleoanthropologists have tried to reconstruct the evolutionary history of the hominins with a two-step process (Strait et al. 1997). First, the researchers use a cladistic analysis to estimate the evolutionary relationships among the various fossil species. Then they make educated guesses about which fossil species represent ancestors that lived at the branch points of the cladogram and which fossil species represent extinct side branches.

Results of one such study, by David S. Strait and colleagues (1997), appear in Figure 20.16. Included are Strait et al.'s cladogram (Figure 20.16a) and a hypothesis about what the cladogram tells us concerning the phylogenetic relationships among the various species (Figure 20.16b). The cladogram is based on a variety of skull and tooth characters. Note that in the cladogram, the lengths of the

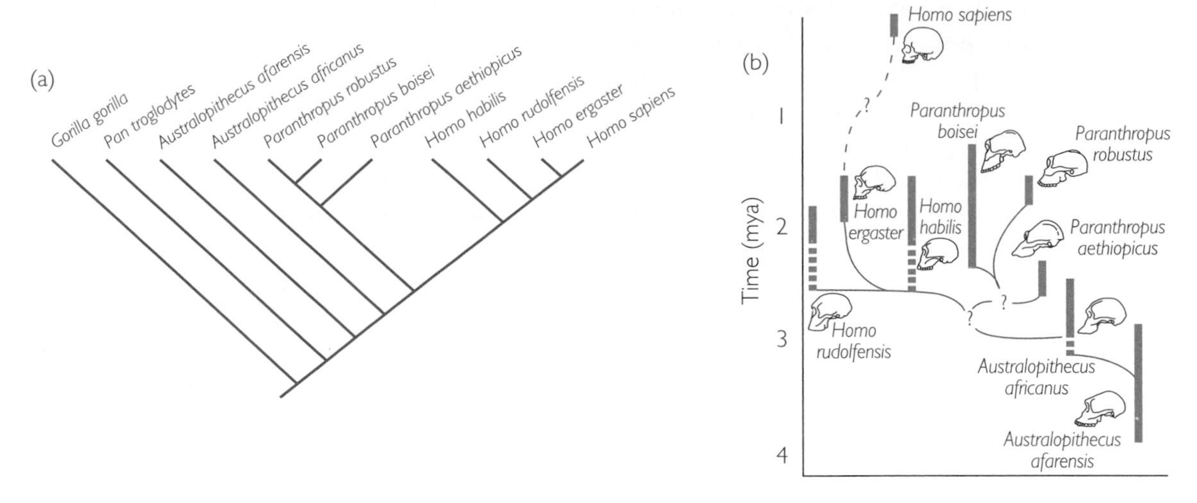

Figure 20.16 **Cladogram and phylogeny of *Homo sapiens* and its recent ancestors and extinct relatives** (a) A cladogram of three extant hominins (the gorilla, the common chimpanzee, and the modern human), and several extinct hominins known only from fossils. (b) A hypothesis about the ancestor–descendant relationships implied by the cladogram in (a). The heavy green vertical bars indicate the known range of times over which each species existed, whereas the heavy green dashes represent the suspected range of times over which each species existed.

The phylogenetic relationships among the species of fossil hominins have not been definitively established.

branches are meaningless; the only information encoded in the cladogram is in the order of branching. The hypothesized phylogeny (Figure 20.16b) makes educated guesses about the actual lengths ascribed to the branches in the cladogram. For example, the branch leading to *Paranthropus aethiopicus* is relatively long, so that *P. aethiopicus* is a sister species to *P. boisei* and *P. robustus*. Another possibility, not shown here, is that the length of the branch leading to *P. aethiopicus* is zero, so that *P. aethiopicus* is the common ancestor of *P. boisei* and *P. robustus*.

Nearly all such analyses performed to date rely heavily on characters of skulls and teeth. The reason is simply that the fossil record is more complete for skulls and teeth than for other parts of the skeleton. To evaluate the value of skull and tooth characters for reconstructing the evolutionary history of hominins, Mark Collard and Bernard Wood (2000) attempted to reconstruct the phylogeny of the living apes using a cladistic analysis of skull and tooth characters that are equivalent to the ones typically used for hominin fossils. Collard and Wood took the well-established molecular phylogeny of the apes (discussed in Section 20.1 and illustrated in Figure 20.4) as the "truth." If their analysis could reconstruct the "truth" as known from the molecules, then we can have some confidence in phylogenies of hominin fossils reconstructed by the same method.

The startling and depressing result was that Collard and Wood's cladistic analysis of skull and tooth characters in living apes produced a phylogeny in which gorillas and orangutans are closest relatives, chimpanzees are their next of kin, and humans branch first from the lineage that will become the other great apes. In other words, the analysis failed completely to recover the known phylogeny. This does not mean that cladistic analyses of morphology are a poor method in general for reconstructing the phylogeny of the apes or the hominins. Indeed, cladistic analyses of soft tissue characters yield ape phylogenies that match the molecular phylogeny exactly (Gibbs et al. 2000, 2002). Nor does it mean that cladistic analyses of skull and tooth characters are unreliable for vertebrates in general. It simply suggests that cladistic analyses of skull and tooth characters may not produce a reliable answer to the question we are interested in here.

David Strait and Frederick Grine (2004) felt that this conclusion was too pessimistic. They suspected that a cladistic reconstruction using skulls and teeth could recover the true phylogeny of the living apes if extinct species were added to the analysis. They used methods similar to Collard and Wood's to infer the evolutionary relationships among gibbons, orangutans, the chimpanzees, humans, and most of the fossil taxa we have discussed in this section. The results were heartening. The relationships among the living species matched the molecular phylogeny exactly. This suggests we can have some confidence in what the tree suggests about the relationships among the fossil taxa, which was largely consistent with Figure 20.16. And it indicates that gathering more evidence will yield better answers. As David Begun (2004) observed, the mantra of every paleontologist is: "We need more fossils!"

Some Answers

Although the phylogeny of the fossil hominins is not yet known with certainty, the evidence we have reviewed gives a general answer to the question we posed at the beginning of this section. The pattern of evolution leading from our common ancestor with the chimpanzees to ourselves has not been simple. Instead, speciation has produced a diversity of lineages. Throughout most of the last 4 million years, multiple species, perhaps even as many as five at a time, have coexisted in Africa (see Tattersall 2000). For example, specimen KNM-ER 406 (Figure 20.12b) and specimen KNM-ER 3733 (Figure 20.13a) clearly represent different species (Figure 20.17). Both were found at Koobi Fora, Kenya, in sediments of nearly the same age. *Paranthropus boisei* and *Homo ergaster* knew each other, but only one belonged to a lineage that persists today. We *Homo sapiens* are the lone survivors of an otherwise extinct radiation of bipedal African hominins.

The hominin fossil record is sufficiently detailed to allow us to conclude that Homo sapiens *is the sole survivor among a diversity of species.*

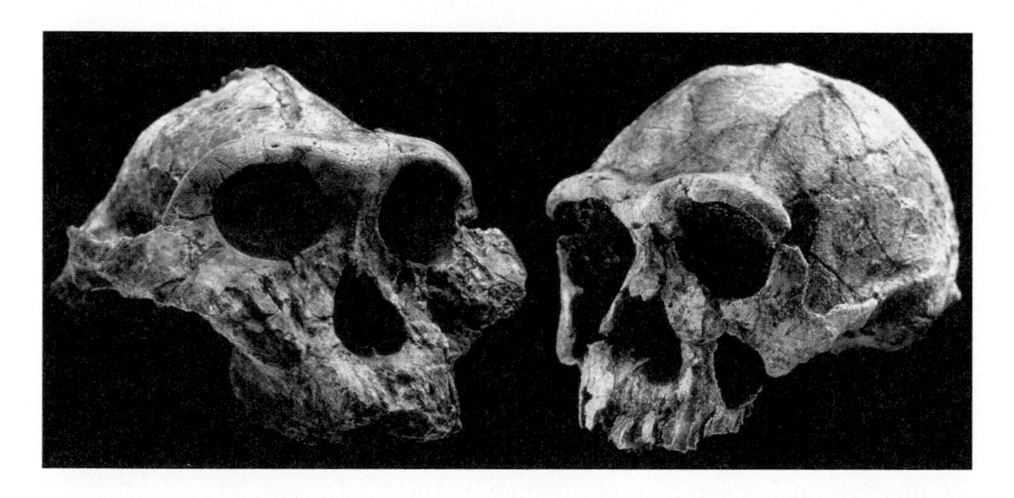

Figure 20.17 Evidence of a hominin radiation *Paranthropus boisei* (specimen KNM-ER 406, left) and *Homo ergaster* (specimen KNM-ER 3733) both lived in what is now Koobi Fora, Kenya, about 1.7 million years ago. From Johanson, Edgar, and Brill (1996).

20.3 The Origin of the Species *Homo sapiens*

Figure 20.13a and Figure 20.14a–d show five specimens that are uncontroversially human. We have used for them the names *Homo ergaster, H. erectus, H. heidelbergensis, H. neanderthalensis,* and *H. sapiens*. There is considerable uncertainty and debate, however, over how many species they actually represent and how modern humans, *Homo sapiens*, emerged from among the others.

Controversies over the Origin of Modern Humans

Paleoanthropologists are split on the taxonomic status of *H. ergaster* (Figure 20.13a) and *H. erectus* (Figure 20.14d). Some researchers consider these two forms to be regional variants of a single species (*H. erectus*), whereas others consider *H. erectus* to be a distinct Asian species descended from the African species *H. ergaster.* Likewise, some researchers consider *H. neanderthalensis* (Figure 20.14b) and *H. heidelbergensis* (Figure 20.14c) to be regional variants of transitional forms between *H. erectus* and modern *H. sapiens.* Others consider them to be distinct species, with *H. heidelbergensis* descended from *H. ergaster* and *H. neanderthalensis* descended from *H. heidelbergensis* (see Tattersall 1997). Recently, a new species, *Homo antecessor,* has been suggested to be the common ancestor of both Neandertals and modern humans (Bermúdez de Castro et al. 1997; Arsuaga et al. 1999). Paleoanthropologists generally agree that modern humans are the descendants of some or all of the populations in the *H. ergaster/erectus* group. However, how and where the transition from *H. ergaster/erectus* to *H. sapiens* took place is a matter of debate.

All hominins prior to *H. ergaster/erectus* were confined to Africa. The oldest examples of *H. ergaster/erectus*, however, appear in the fossil record nearly simultaneously at Koobi Fora in Africa, at Dmanisi in the Caucasus region of eastern Europe, at Longgupo Cave in China, and at Sangiran and Mojokerto in Java—all 1.6 to 1.9 million years ago (Gibbons 1994; Swisher et al. 1994; Gabunia and Vekua 1995; Huang Wanpo et al. 1995; Wood and Turner 1995; Gabunia et al. 2000). Because its immediate ancestors and closest relatives appeared to be restricted to Africa, most paleontologists had assumed that *H. erectus* evolved in Africa and then moved to Asia. The fossils at Longgupo Cave, China, however, are similar enough to African *H. habilis* and *H. ergaster* to suggest that *H. erectus* may have evolved in Asia from earlier migrants (Huang et al. 1995). Either way, prior to 2 million years ago the ancestors of our species within the genus *Homo* almost certainly lived in Africa.

The origin of modern Homo *sapiens is controversial.*

Anatomically modern *H. sapiens* first appear in the fossil record about 100,000 years ago in Africa and Israel and somewhat later throughout Europe and Asia (Stringer 1988; Valladas et al. 1988; Aiello 1993; White et al. 2003; but see McDougall et al. 2005). The range of hypotheses concerning the evolutionary transition from *H. ergaster/erectus* to *H. sapiens* is illustrated in Figure 20.18. At one extreme, the African replacement (or out-of-Africa) model (Figure 20.18a) posits that *H. sapiens* evolved in Africa, then migrated to Europe and Asia, replacing *H. erectus* and *H. neanderthalensis* without interbreeding. At the other extreme, the candelabra model (Figure 20.18d) holds that *H. sapiens* evolved independently in Europe, Africa, and Asia, without gene flow between regions. Between the extremes are hypotheses that postulate different combinations of migration, gene flow, and local evolutionary transition from *H. ergaster/erectus* to *H. sapiens.* These intermediates are the hybridization and assimilation model (Figure 20.18b), and the multiregional evolution model (Figure 20.18c).

At stake in the debate is the nature and antiquity of the present-day geographic races of humans. If the African replacement model is correct, then present-day racial variation is the result of recent geographic differentiation that occurred within the last 100,000 to 200,000 years, after anatomically modern *H. sapiens* emerged from Africa. If one of the intermediate models is correct, then present-day racial variation represents some mixture of recent and ancient geographic differentiation. If the candelabra model is correct, then present-day racial variation derives from geographic differentiation among *H. ergaster/erectus* populations and may be as much as 1.5 to 2 million years old.

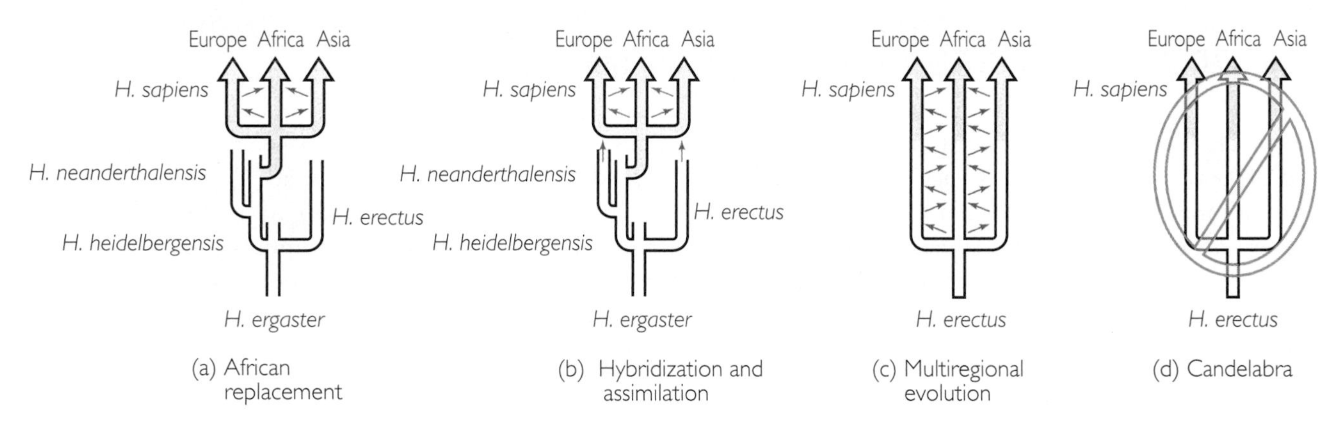

Figure 20.18 Hypotheses concerning the transition from *Homo ergaster/erectus* to *Homo sapiens* The white portions of the phylogenies represent various archaic forms of *Homo,* including *H. ergaster, H. erectus, H. heidelbergensis,* and *H. neanderthalensis.* The colored portions represent modern *H. sapiens.* The small blue arrows represent gene flow. Note that specimens identified as *H. heidelbergensis* have been found in Europe and Africa, and specimens identified as *H. neanderthalensis* have been found in Europe and the Middle East. (a) The African replacement model. According to this model, modern *H. sapiens* evolved in Africa and then migrated to Europe and Asia. *H. sapiens* replaced the local forms without hybridization. No genes from these earlier forms persist in modern human populations. (b) The hybridization and assimilation model. According to this model, modern *H. sapiens* evolved in Africa and then migrated to Europe and Asia. *H. sapiens* largely replaced the local populations, but there was hybridization between the newcomers and the established residents. As a result, some genes from the archaic local populations were assimilated and persist in modern human populations. (c) The multiregional evolution model. According to this model, *H. sapiens* evolved concurrently in Europe, Africa, and Asia, with sufficient gene flow among populations to maintain their continuity as a single species. Gene pools of all present-day human populations are derived from a mixture of distant and local archaic populations. (d) The candelabra model: *H. sapiens* evolved independently in Europe, Africa, and Asia, without gene flow among populations. All genes in present-day European and Asian populations are derived from local archaic populations. The model names and characterizations we use are based on those used by Aiello (1993), Ayala et al. (1994), and Tattersall (1997). Not all authors would agree with our characterizations and names. Frayer et al. (1993), for example, apparently consider both models (b) and (c) to be variations of the multiregional evolution model they favor.

The candelabra model has been widely and thoroughly rejected by scientists in all fields (see Frayer et al. 1993; Ayala et al. 1994). It is flatly implausible that the same single descendant species, *H. sapiens,* could emerge in parallel in three different regions with no gene flow to maintain its continuity (see Chapters 7 and 16). Arguments over the remaining three models are based on archaeological and paleontological evidence and on genetic analyses. In much of the discussion that follows, we focus on distinguishing between the remaining extremes (the African replacement model of Figure 20.18a versus the multiregional evolution model of Figure 20.18c), but it is useful to keep in mind that these two models fall at the ends of a continuum of possibilities. It is also useful to keep in mind that whatever their origins, all living humans are extremely closely related to each other (Box 20.2).

African Replacement versus Multiregional Evolution: Archaeological and Paleontological Evidence

David Frayer and colleagues (1993) use archaeological and paleontological data to argue against the African replacement model (Figure 20.17a) and in favor of some alternative [Figure 20.17b or (c)]. The researchers note that the African replacement model holds that long-established populations of one or more tool-using, hunter–gatherer species (*H. erectus* and other archaic forms of *Homo*) were supplanted wholesale throughout Europe and Asia by populations of another tool-using, hunter–gatherer species (modern *H. sapiens* emerging from Africa). It is hard to imagine how this could have happened, except by direct competition

Box 20.2 | Genetic diversity among living humans

How closely related are living humans? One way to address this question is to compare the genetic diversity among humans to the genetic diversity among other great apes. Pascal Gagneux and colleagues (1999) examined mitochondrial DNA nucleotide sequences of 811 humans, 292 chimpanzees, 24 bonobos, and 26 gorillas. For each species, the researchers sorted all possible pairs of individuals by the percentage of nucleotides that were different between their sequences. The graphs in Figure 20.19 show the distributions of pairwise differences. A randomly chosen pair of humans is likely to be about as different as a randomly chosen pair of eastern chimps, and substantially less different than a randomly chosen pair of central chimps, western chimps, bonobos, or gorillas. Indeed, the researchers found several cases in which bonobos or western chimpanzees living in the same social group were more genetically different than any two humans from anywhere in the world. By the standards of the other African great apes, all living humans are closely related. Furthermore, most of the genetic diversity that exists among living humans occurs as differences among individuals within populations, rather than as differences between populations (Jorde et al. 2000).

Are there any genetic differences at all distinguishing present-day human populations? To find out, Noah Rosenberg and colleagues (2002) analyzed data on the genotypes, at 377 variable short tandem repeat loci, of 1,056 individuals from 52 populations. They used a computer program to (1) define, using only the genotype data on the 1,056 individuals, five groups with distinct allele frequencies, and (2) assign each individual a partial membership in each group determined by how closely his or her genotype matches each group's membership criteria. If there are genetic differences that distinguish human populations, then individuals from the same geographic region should have similar group-membership profiles.

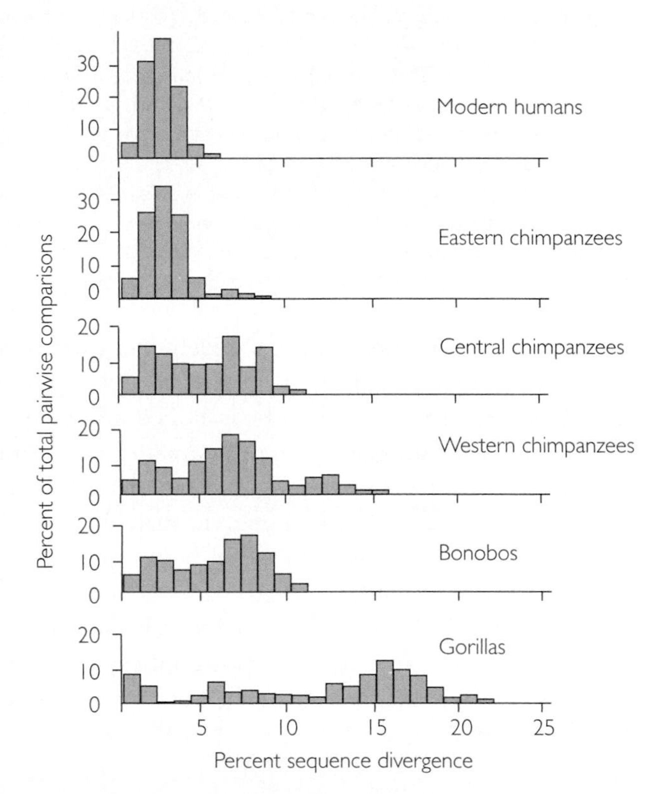

Figure 20.19 Genetic diversity in humans versus African great apes These histograms show the distribution of mitochondrial DNA sequence differences between all possible pairs of individuals in large samples of humans and several African great apes. A randomly chosen pair of humans are, on average, much more genetically similar than a randomly chosen pair of central or western chimps, bonobos, or gorillas. From Gagneux et al. (1999).

between the invaders and the established residents. It is implausible that modern *H. sapiens* could have been such a relentlessly superior competitor without a substantial technological advantage in the form of better tools or weapons. Thus, Frayer and colleagues conclude, the African replacement model predicts that the archaeological record will show evidence of abrupt changes in the level of technology in Europe and Asia as modern *H. sapiens* replaced archaic *Homo*. In fact, the researchers say, there is no evidence of any such abrupt technological changes.

Frayer and colleagues (1993) also argue that the African replacement model predicts that fossils of *Homo* populations in any given non-African region should

The results appear in Figure 20.20. Each individual is represented by a thin horizontal line. Each line is composed of five segments with different colors, representing the individual's partial membership in each of the five groups. It turns out that individuals from different geographic regions do, indeed, tend to have similar group-membership profiles.

Other teams of researchers conducting similar analyses have found similar patterns. Richard Redon and colleagues (2006) analyzed the genotypes of 210 individuals at 67 loci polymorphic for insertions, deletions, or duplications. Based on these genotypes, the researchers asked a computer to sort the individuals into groups derived from three ancestral populations. Nearly all the individuals in the sample ended up in a group with others from the same continent of origin. Michael Bamshad and colleagues (2003) analyzed the genotypes of 206 individuals at 60 polymorphic short tandem repeat loci and 100 loci polymorphic for Alu transposable elements (see Box 15.1, page 578). Based solely on genotype, their computer assigned individuals to the correct continent of origin with 99% accuracy.

The take-home message is that genetic differences among modern human populations do exist, but they are so subtle that it takes an extraordinarily large data set and considerable computational effort to find them.

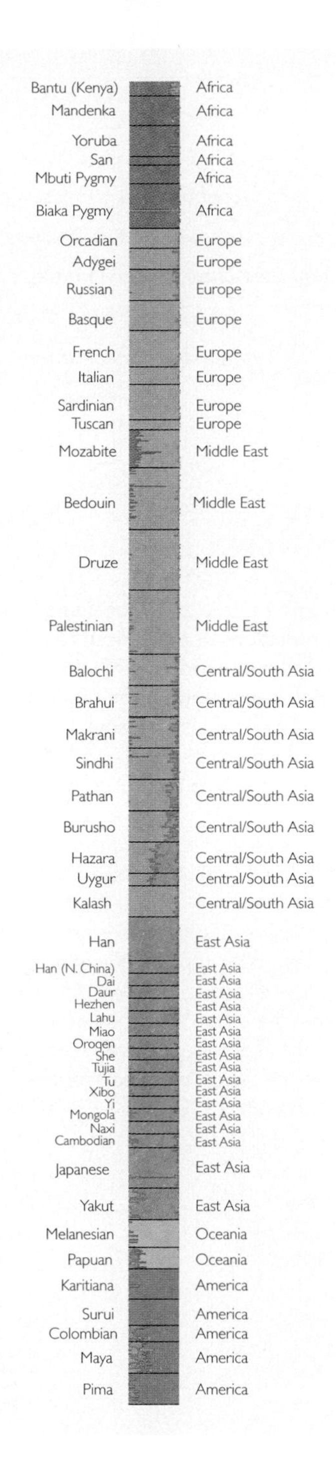

Figure 20.20 Evidence of geographic structure in living human populations Noah Rosenberg and colleagues (2002) programmed a computer to define five maximally distinct human groups based on allele frequencies at 377 marker loci, then assign each of over a thousand individuals a partial membership in each group. As depicted in this illustration, the computer defined groups in which people from the same geographic region tended to have similar partial membership profiles. See box text for details. From Rosenberg et al. (2002).

show distinct changes in morphology when modern *H. sapiens* migrating from Africa replaced local archaic *Homo.* To refute this prediction, Frayer and colleagues point to distinctive traits of regional populations that have persisted from the distant past to the present. One-million-year-old fossils of *H. erectus* from Java, for example, have a straighter, more prominent browridge than their contemporaries elsewhere in the world. This strong browridge remains a distinctive feature of present-day Australian aborigines, whose ancestors may have arrived by boat from Java up to 60,000 years ago. Likewise, many present-day Asians have shovel-shaped upper front teeth, a trait that characterizes virtually all fossil specimens of

Asian *H. erectus* and *H. sapiens.* (For other examples of the continuity of distinctive regional traits, see Thorne and Wolpoff 1981; Li Tianyuan and Etler 1992; and Frayer et al. 1993.) On these and other grounds, Frayer and colleagues reject the African replacement model.

Morphological analyses, though not definitive, suggest that modern humans evolved in Africa, then replaced archaic humans elsewhere.

Diane Waddle (1994) and Daniel Lieberman (1995) use statistical and cladistic approaches, respectively, to evaluate predictions from the African replacement model (Figure 20.18a) and the multiregional evolution model (Figure 20.18c). While Lieberman stresses that his data set of only 12 characters is too small to produce reliable inferences, both he and Waddle tentatively conclude that all modern humans are more closely related to archaic forms from Africa (Figure 20.21a) than regional groups are to local archaic forms (Figure 20.21b). If Lieberman and Waddle are correct, then the examples described by Frayer and colleagues of apparent long-term continuity of regionally distinctive traits must be the result of convergent evolution in *H. erectus* and *H. sapiens.*

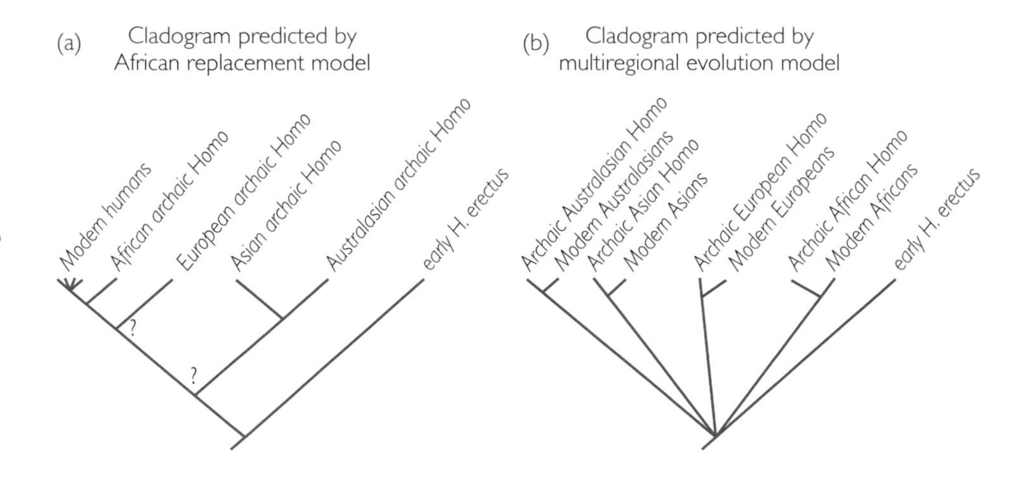

Figure 20.21 **Phylogenetic predictions of the African replacement model versus the multiregional evolution model** (a) The African replacement model predicts that all modern humans will be more closely related to each other than any is to any archaic species and that, among the archaic species, those from Africa will be the most closely related to modern humans. (b) In contrast, the multiregional evolution model predicts that the archaic and modern humans in each region will be each other's closest relatives. From Lieberman (1995).

The most recent analyses of Neandertal and pre-Neandertal fossils indicate that the story of modern human origins may be very complex indeed. Bermúdez de Castro and colleagues recently described a newly discovered set of human fossils and associated artifacts and animal remains from the Gran Dolina section of the Atapuerca locality in Spain (Carbonell et al. 1999 and papers cited therein). The researchers have attributed the human fossils from this locality, which is dated to somewhere between 780,000 and 980,000 years ago, to a new species, *Homo antecessor.* Their analysis concludes that the Gran Dolina specimens share features of both modern humans and Neandertals and may be the common ancestor of both. At the opposite end of the time range for Neandertals, Trinkaus and colleagues recently found evidence of mixed Neandertal and modern human charactersistics in two different specimens, one from France and the other from Portugal, both roughly 30,000 years old. Trinkaus and colleagues have interpreted these fossils as possible evidence of hybridization between Neandertals and modern humans (Trinkaus et al. 1998; Duarte et al. 1999; see, though, Tattersall and Schwartz 1999 for a dissenting opinion). *Homo antecessor* and possible Neandertal–modern human hybrids, all from western Europe, may yet breathe new life into the multiregional model, at least insofar as it relates to Europe.

African Replacement versus Multiregional Evolution: Molecular Evidence

In principle, we could take Lieberman's cladistic approach and use it with DNA sequence data. If we had sequences of genes from both modern and archaic humans in all regions, we could estimate their phylogeny and see whether it most closely matches the tree predicted by the African replacement model or the tree predicted by the multiregional evolution model.

A team led by Svante Pääbo recovered sequences of mitochondrial DNA from the skeletons of two *Homo neanderthalensis*: one that lived in Germany some 30,000 to 100,000 years ago, and another that lived in Croatia 42,000 years ago (Krings et al. 1997, 1999, 2000). Igor Ovchinnikov and colleagues (2000) recovered a sequence from a third Neandertal that lived in the northern Caucasus 29,000 years ago. The researchers compared the Neandertal sequences with several hundred modern human mtDNA sequences. The cartoon in Figure 20.22 summarizes their results. The researchers found that modern humans from Europe, Africa, Asia, America, Australia, and Oceania are all more closely related to each other than any of them is to the archaic Europeans (that is, the Neandertals). Likewise, the Neandertal sequences are all more closely related to each other than any of them is to modern humans. The Neandertals have, on average, more than three times as many sequence differences from a typical modern human as are found among modern humans themselves. The last common ancestor of all modern humans lived roughly 170,000 years ago (Ingman et al. 2000). The last common ancestor of the three Neandertals lived about 250,000 years ago. The last common ancestor of the Neandertals and modern humans, in contrast, lived about 500,000 years ago. These results are consistent with the African replacement model (Figure 20.18a), in which modern humans were a distinct lineage from the Neandertals and supplanted them without interbreeding.

Analyses of a limited sample of ancient DNAs suggest that modern humans and Neandertals were distinct lineages, and that modern humans emerging from Africa replaced Neandertals in Europe.

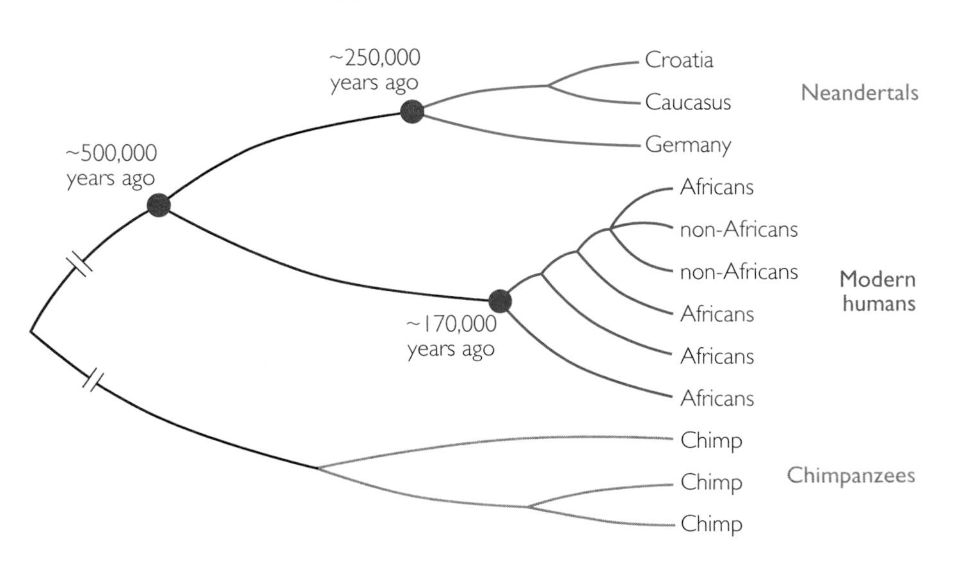

Figure 20.22 Phylogeny of Neandertals and modern humans This cartoon summarizes evidence from analyses of mitochondrial DNA sequences from several hundred modern humans and three Neandertals. The split between Neandertals versus humans predates the diversification within each lineage by a substantial margin. This suggests that modern humans are a distinct lineage from Neandertals and replaced them without hybridization. After Ingman et al. (2000), Krings et al. (2000), Ovchinnikov et al. (2000), and Hoffreiter et al. (2001).

This conclusion is not definitive, however. It is based on just three Neandertal sequences, and leaves open the possibility that modern Europeans inherited nuclear genes, if not mitochondrial genes, from archaic Europeans (see Nordborg 1998). Furthermore, Gabriel Gutiérrez and colleagues (2002) argue that the statistical support for distinct Neandertal versus modern human lineages is not as strong as the initial analyses indicated and suggest that the apparent divergence between them may be due, in part, to chemical degradation of the Neandertal DNA. What we need is more ancient DNA sequences from a variety of time depths. Unfortunately, retrieving DNA from fossils is difficult, and may prove impossible for bones older than 100,000 years.

Working with DNA sequences of present-day humans only, researchers find it more difficult to design tests that distinguish the African replacement model from the multiregional evolution model [see, for example, the exchange between Sarah Tishkoff and colleagues (1996a) and Milford Wolpoff (1996)]. The trouble is that from a genetic perspective, the two models are identical in most respects. Both describe a species originating in Africa, spreading throughout Europe and Asia, and then differentiating into regionally distinct populations that nonetheless remain connected by gene flow [Figures 20.18a and (c)]. The only difference is that under the African origin model this process began 200,000 years ago or less, whereas under the multiregional evolution model it began some 1.8 million years ago. This means that any genetic patterns that might allow us to distinguish between the two models will involve quantitative differences rather than qualitative differences. Table 20.2 lists four criteria that molecular geneticists have used in efforts to distinguish between the African replacement versus multiregional evolution models. We will refer to the table in the paragraphs that follow.

S. Blair Hedges and colleagues (1992) attempt to distinguish between the African replacement model and the multiregional evolution model by analyzing sequence data for mitochondrial DNA. The data they use, collected by Linda Vigilant and colleagues (1991), are for sequences of noncoding mitochondrial DNA from 189 people from various geographic regions. Hedges and colleagues use a

Table 20.2 Genetic predictions distinguishing African replacement versus multiregional evolution

Each of the criteria in the first column is a category of data we might use to distinguish between the African replacement model and the multiregional evolution model. The next two columns predict the patterns in each type of data under each model. The last column gives reasons why the distinctions implied by the predictions are not definitive. See text for more details.

	Predictions		
Criteria	**African replacement**	**Multiregional evolution**	**Caveat**
1. Location of ancestor of neutral alleles	Mostly Africa	Random	African origin of *H. ergaster/erectus* may bias the location of alleles toward African even under multiregional evolution.
2. Divergence time of African vs. non-African populations	200,000 years or less	1 million years or more	Gene flow among regional populations can reduce the apparent age of population divergence under multiregional evolution.
3. Genetic diversity	Genetic diversity is greater in Africa.	Genetic diversity is roughly equal in all regions.	African origin of *H. ergaster/erectus* and gene flow or selection may lead to greater diversity in Africa, even under multiregional evolution.
4. Sets of neutral alleles	Alleles present in Europe and Asia are subsets of those in Africa.	Each region has some unique alleles; no region's alleles are a subset of another region's.	African origin of *H. ergaster/erectus* may mean that alleles present in Europe and Asia are subsets of those in Africa, even under multiregional evolution.

neighbor-joining technique to estimate the evolutionary tree linking the 189 mitochondrial sequences. Recall that mtDNA carries a record of direct maternal ancestry. By direct maternal ancestry we mean a person's mother, mother's mother, and so on. (A person's mother's father is part of the person's indirect maternal ancestry). Hedges et al.'s mitochondrial tree traces the direct maternal ancestries of the 189 people back to a point at which they all converge to a single woman. (Note that the woman, often called by the misleading name Mitochondrial Eve, is not, in her generation, the sole female ancestor of the 189 present-day people. The 189 present-day people undoubtedly have a large number of indirect female ancestors who were contemporaries of their common direct maternal ancestor and from whom they inherited many of their nuclear genes. See Ayala et al. 1994; Ayala 1995.) Because the deepest branches in the phylogeny involve splits within African lineages, the tree suggests that this woman lived in Africa. Hedges and colleagues note, however, that their mitochondrial tree is poorly supported statistically.

Max Ingman and colleagues (2000) perform a similar analysis, this time using sequences of the entire mitochondrial genomes of 53 individuals. Their tree has strong statistical support and, like that of Hedges et al., shows all non-African sequences branching from within the African sequences (Figure 20.23). It appears that the common ancestor of all present-day human mtDNAs did, in fact, live in

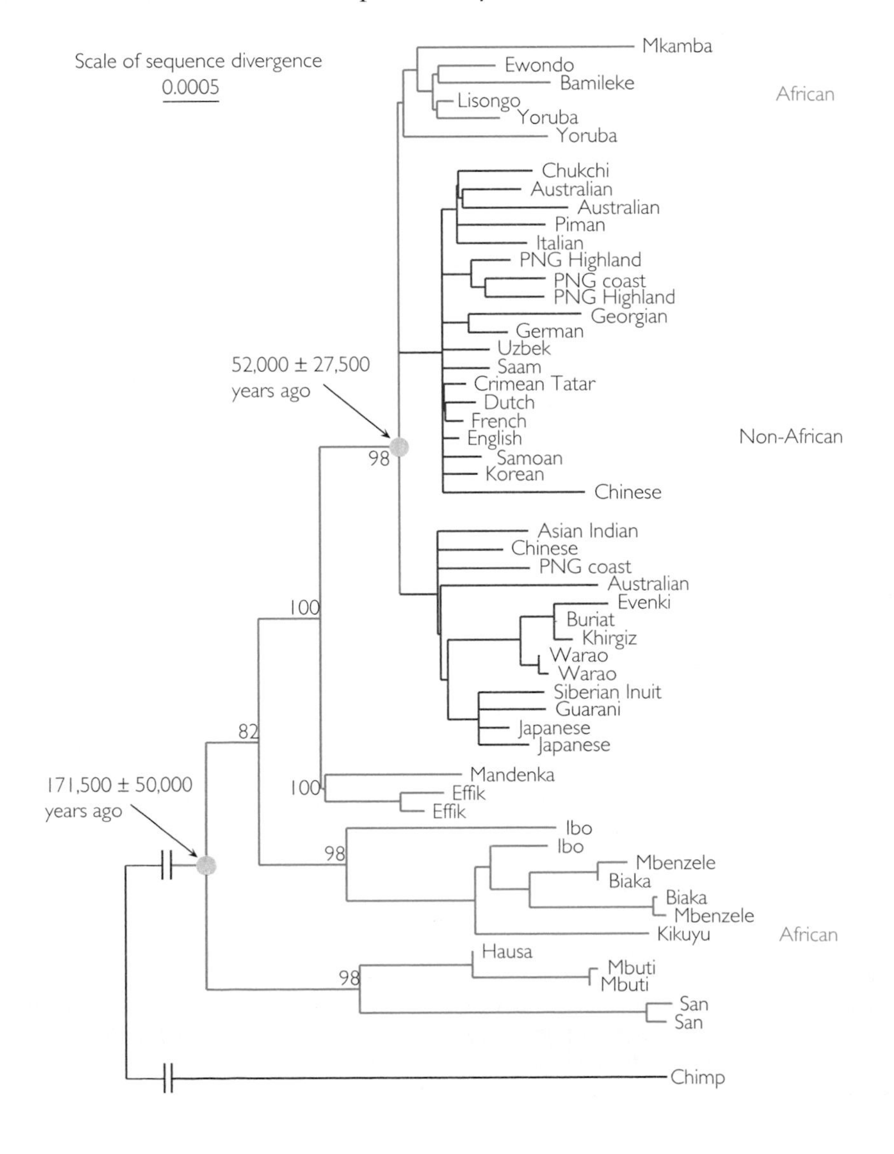

Figure 20.23 An evolutionary tree of complete mitochondrial DNAs of 53 humans Each branch tip represents a single individual, identified by his or her population. The tree was rooted by using a chimpanzee sequence as the outgroup. All non-African individuals branch from within the Africans. Numbers at nodes represent the strength of statistical support (percentage of 1,000 bootstrap replicates in which the node was present). From Ingman et al. (2000).

Africa. Does that help us decide, by criterion 1 of Table 20.2, between African replacement and multiregional evolution? Not by itself, because both models include a common ancestry for all present-day humans that traces back to Africa. We need to know when the common ancestor of all human mtDNAs lived.

Ingman and colleagues estimate that the most recent common ancestor of all present-day mitochondrial DNAs, at the node highlighted in yellow in Figure 20.23, lived between 120,000 and 220,000 years ago. The researchers arrive at this estimate by using a molecular clock. They can verify for their data set that mutations have accumulated in their mtDNAs at a constant, clocklike rate. They assume a human-chimpanzee divergence time of 5 million years ago and estimate the divergence time of the most distant human sequences by comparing them to the sequence from the chimp. By the same method, Ingman and colleagues estimate that the most recent common ancestor of Africans and non-Africans, at the node highlighted in orange in Figure 20.23, lived just 25,000 to 80,000 years ago (see Hedges 2000 for a commentary).

At first glance, these dates, which are consistent with the African replacement hypothesis, appear to refute the multiregional origin hypothesis by criterion 2 of Table 20.2. The dates suggest that non-African populations of humans diverged from African populations no more than a hundred thousand years ago and certainly nothing like a million years ago. Examination of Figure 20.24, however, shows that this is not necessarily the case. It is true that species cannot diverge any earlier than the divergence of any of their alleles, but populations connected by gene flow can. It is possible that the mitochondrial clock, which is effectively based on a single gene, makes the split between African and non-African populations look more recent than it actually was.

What we need to do is look at many loci at once and see whether, taken together, they tell the same story of a recent divergence between African and non-African populations. A. M. Bowcock and colleagues (1994) look at 30 nuclear microsatellite loci from people in each of 14 populations. Microsatellite loci are places in the genome where a short string of nucleotides, usually two to five bases long, is repeated in tandem. The number of repeats at any given locus is highly variable among individuals, meaning that each microsatellite locus has many alleles.

(a) Species or populations can diverge simultaneously with a pair of alleles.

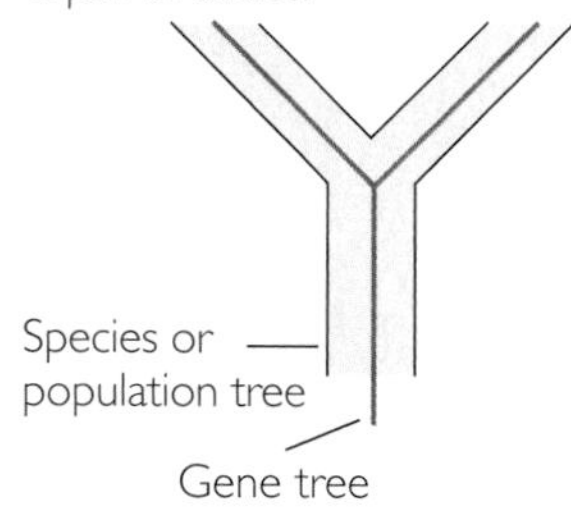

(b) Species or populations can diverge after a pair of alleles diverge.

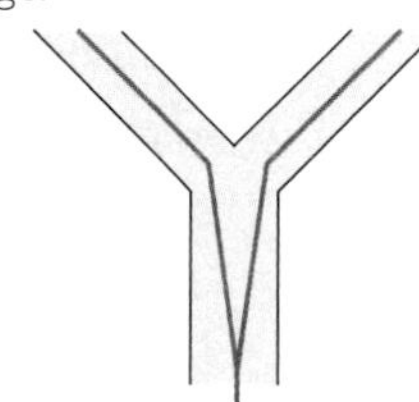

(c) Species cannot diverge before a pair of alleles diverge...

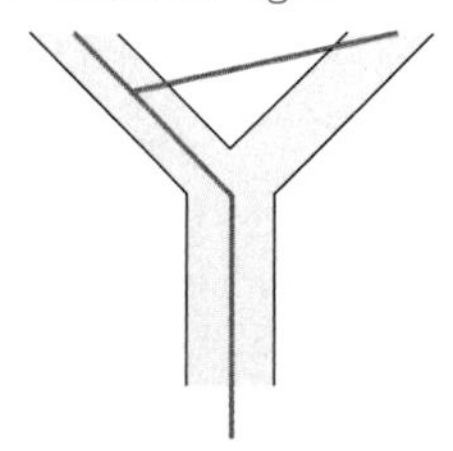

(d) ...but populations connected by gene flow can.

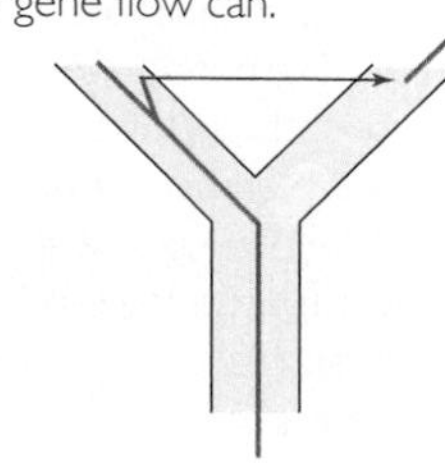

Figure 20.24 Divergence times of species trees, population trees, and gene trees These figures illustrate a hypothetical gene phylogeny embedded within a species or population phylogeny. (a) In this scenario, a mutation creates a new allele to produce a split in the gene tree at the same time the species or population splits into two. One allele is then lost by drift or selection in each descendant species or population. The result is a gene tree that is exactly congruent with the species or population tree. If we use a molecular clock to estimate the divergence time for the species or populations, we will get the right answer. (b) Here, a mutation creates a new allele, producing a split in the gene tree. Some time later, the species or population splits into two. One allele is then lost by drift or selection in each descendant species or population. If we use a molecular clock for the gene tree to estimate the divergence time for the species or populations, the species or populations will appear to have diverged earlier than they actually did. (c) First, a species splits into two. Some time later, a mutation creates a new allele, producing a split in the gene tree. Finally, one of the alleles moves from one species to the other. This last step is impossible; thus a molecular clock will not make a split between species appear more recent than it was. (d) A scenario like that in (c) is, however, possible for populations within a species. First, a population splits into two. Some time later, a mutation creates a new allele, producing a split in the gene tree. Then a migrant carries the new allele to the other population (red arrow). Finally, the new allele is lost in one population, and the ancestral allele is lost in the other population. If we use a molecular clock for the gene tree to estimate the divergence time for the population tree, the populations will appear to have diverged more recently than they actually did.

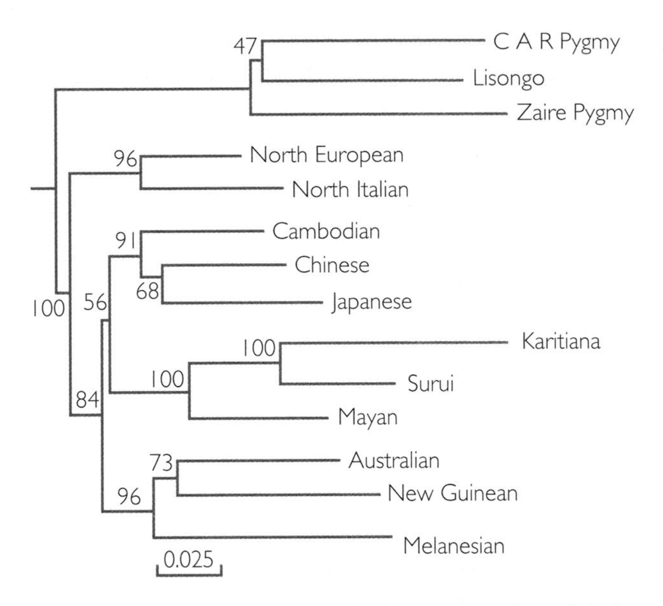

Figure 20.25 Phylogenetic tree for 14 human populations based on allele frequencies at 30 microsatellite loci The number at a node indicates the percentage of times that the node was present in 100 bootstrap replicates (see Chapter 14). The deepest split in the tree was present in 100% of the bootstrap replicates, indicating strong statistical support for the conclusion that African (Central African Republic Pygmy, Lisongo, Zaire Pygmy) versus non-African is the most fundamental division in the population phylogeny. From Bowcock et al. (1994).

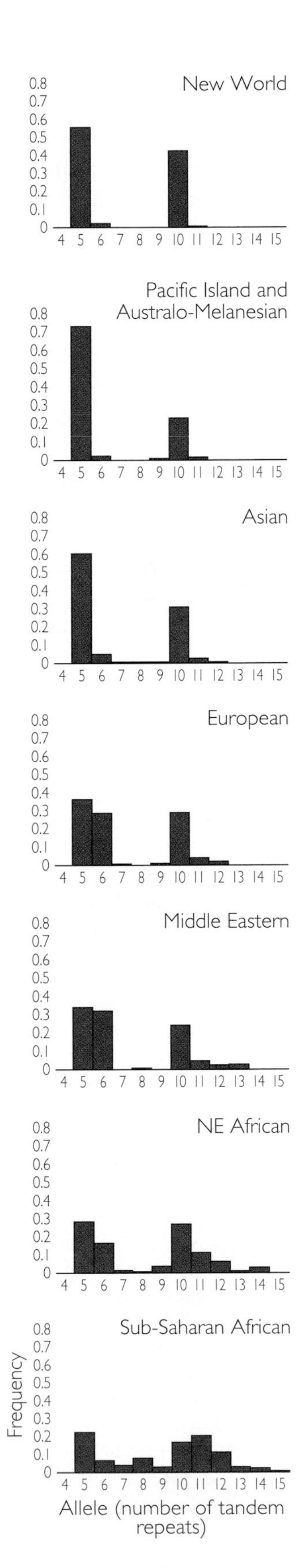

Bowcock and colleagues calculate multilocus genetic distances among the 14 populations on the basis of the allele frequencies at each of the 30 microsatellite loci. They then use the genetic distances among populations to estimate the population phylogeny (Figure 20.25).

On Bowcock et al.'s phylogenetic tree, geographically neighboring populations cluster together. Furthermore, the deepest branch point separates African from non-African populations. Analyzing the same data, D. B. Goldstein and colleagues (1995) estimate that the split between African and non-African populations occurred 75,000 to 287,000 years ago. This time range, which is consistent with the African replacement model, makes a more persuasive case than the mitochondrial clock date that the multiregional evolution model can be rejected by criterion 2 of Table 20.2. It is still possible under the multiregional evolution model to argue that there was enough gene flow to make the population split look more shallow than it was. But if there was that much gene flow, then it becomes hard to explain how any regional differentiation of characters could be maintained for a million years or more (Nei 1995).

Finally, we consider a study by Sarah Tishkoff and colleagues (1996b). These researchers examined allelic variation at a locus on chromosome 12 that is the site of a short tandem-repeat polymorphism. This is a region of noncoding DNA in which the sequence TTTTC is repeated between 4 and 15 times, producing a total of 12 alleles. Tishkoff and colleagues determined the genotypes of more than 1,600 people from seven different geographic regions. Figure 20.26 shows plots of the allele frequencies in each of the seven regions.

Figure 20.26 Genetic diversity at a single locus among the people of seven geographic regions Each plot shows, for the people of a particular geographic region, the frequencies of the various alleles (numbered 4–15) at a short tandem-repeat locus on chromosome 12. Plotted from tables in Tishkoff et al. (1996b).

Box 20.3 Using linkage disequilibrium to date the divergence between African and non-African populations

Near the short-tandem-repeat locus analyzed in Figure 20.26 is another locus, also noncoding, with a nucleotide sequence known as an Alu element. This Alu locus has two alleles: the ancestral, or *Alu*(+) allele; and a derived *Alu*(−) allele with a 256-base-pair deletion. Gorillas and chimpanzees lack the *Alu*(−) allele, so it probably arose in the human lineage after the split between the human lineage and the chimpanzee lineage.

The deletion mutation that created the *Alu*(−) allele probably occurred only once, in Africa, most likely in a chromosome that carried the six-repeat allele at the short-tandem-repeat locus (Tishkoff et al. 1996b). Upon its appearance in the population, the *Alu*(−) allele was in linkage disequilibrium with the short-tandem-repeat allele (see Chapter 8). The only kind of *Alu*(−) chromosome 12 in the population had the haplotype short-tandem-repeat-6-*Alu*(−).

Recall from Chapter 8 that sexual reproduction in a population reduces linkage disequilib-

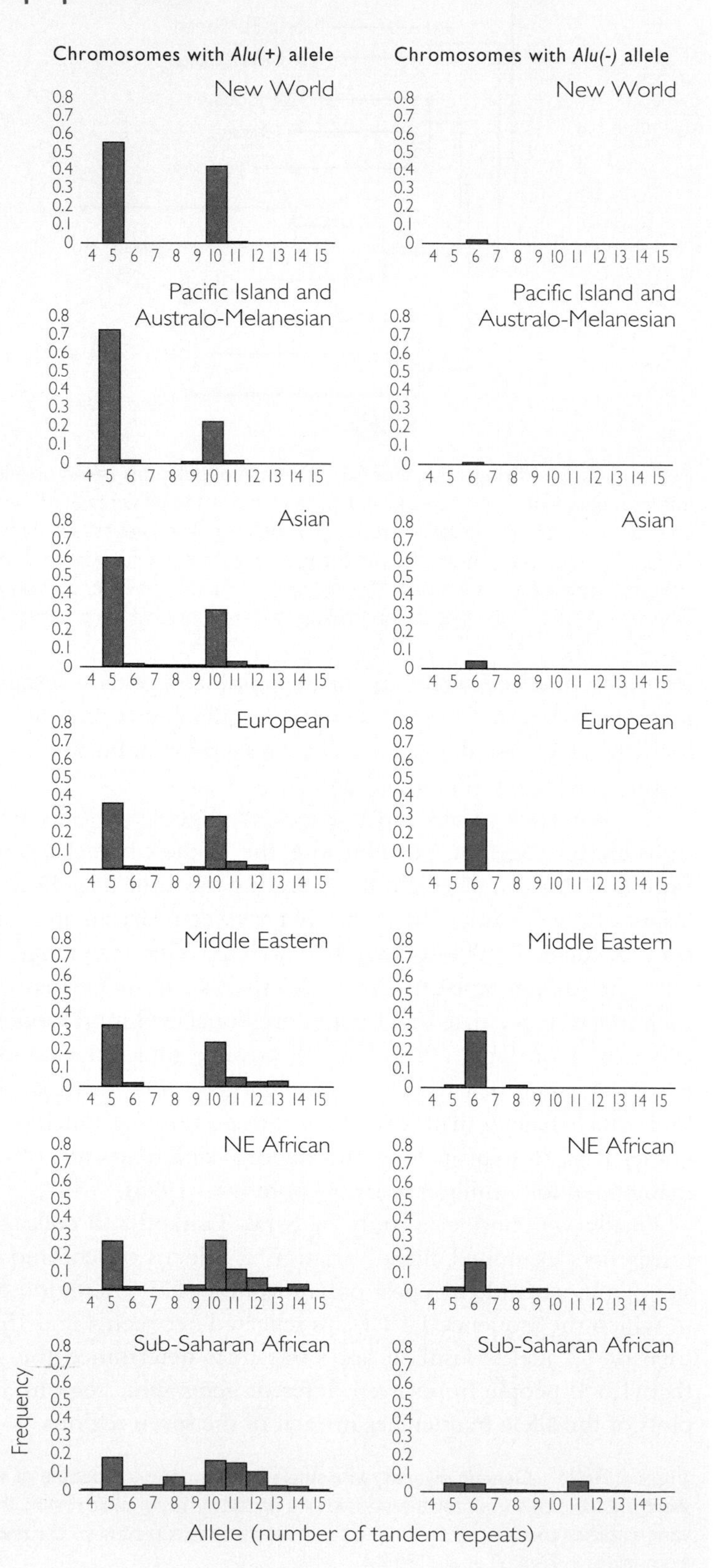

Figure 20.27 Genetic diversity at two linked loci in the people of seven geographic regions These graphs represent the same populations and the same alleles represented in Figure 20.25. Near the short-tandem-repeat locus is an Alu element polymorphic for a deletion. There are two alleles at this locus: *Alu*(+) and *Alu*(−). The left column shows allele frequencies at the short-tandem-repeat locus among chromosomes that carry the *Alu*(+) allele. The right column shows allele frequencies at the short-tandem-repeat locus among chromosomes that carry the *Alu*(−) allele. If a population is at linkage equilibrium for the short-tandem-repeat locus and the Alu locus, then the shape of the distributions of alleles in the two columns (but not necessarily the height of the distributions) will be the same. In sub-Saharan Africa, the distribution of alleles is roughly the same for *Alu*(+) and *Alu*(−) chromosomes. This pattern indicates that the short- tandem-repeat locus and the Alu locus are near linkage equilibrium in this population. In the people of other regions, the distribution of alleles is dramatically different in *Alu*(+) versus *Alu*(−) chromosomes, indicating that the short-tandem-repeat locus and the Alu locus are in linkage disequilibrium. Plotted from tables in Tishkoff et al. (1996b).

Box 20.3 (Continued)

rium. After the *Alu*(−) allele appeared in Africa, mutations at the short-tandem-repeat locus and genetic recombination between it and the Alu locus created a great variety of other genotypic combinations on chromosome 12 (Figure 20.27, African populations). The sub-Saharan African population is near linkage equilibrium at the two loci.

When the first migrants left Africa, it appears that they carried with them at appreciable frequency only three 2-locus genotypes: 5-*Alu*(+); 10-*Alu*(+); and 6-*Alu*(−). In other words, genetic drift in the form of the founder effect put the migrant populations in linkage disequilibrium. The time since the migrants' departure from Africa has not been sufficient for mutation and recombination to create new two-locus genotypes and replenish the haplotype diversity to the levels seen in Africa. In other words, the non-African populations are still in linkage disequilibrium, a population-genetic legacy from their ancient ancestors who left Africa. Using estimates of the rates of mutation and recombination, both of which affect the rate at which populations approach linkage equilibrium (see Chapter 8), Tishkoff and colleagues estimate that the founders of the non-African populations left Africa not more than 102,000–450,000 years ago (see also Pritchard and Feldman 1996; Risch et al. 1996).

The African populations show much greater allelic diversity than non-African populations. This pattern is consistent with the African replacement model. If non-African populations were founded by small bands of people migrating out of Africa, then non-African populations should have reduced genetic diversity because of the founder effect (see Chapter 7).

Notice also that the graphs in Figure 20.26 are arranged by travel distance from sub-Saharan Africa, with the closest regions near the bottom and the most distant regions near the top. Moving up the figure from sub-Saharan Africa to northeast Africa, then to the Middle East and beyond, we see that each region shows a set of alleles that is a subset of those present in the region below. Again, this is consistent with African replacement. It is what we would expect if each more-distant region were founded by a small band of people picking up from where their ancestors had settled and moving on. Not only is the pattern of allelic diversity consistent with African replacement, but it also tends to refute multiregional evolution by criteria 3 and 4 of Table 20.2. This refutation is not definitive, however, because multiregional evolution postulates the same pattern of migration and settlement, just earlier.

Tishkoff and colleagues can estimate when the founders of the non-African populations left Africa. They use a method based on linkage disequilibrium between the short tandem-repeat locus and a second locus nearby (see Box 20.3). Tishkoff and colleagues estimate that the founders of the non-African populations left Africa not more than 102,000–450,000 years ago. These dates are consistent with the African replacement model and tend to refute the multiregional evolution model by criterion 2 of Table 20.2.

Analysis of modern DNAs, though not definitive, suggest that modern humans evolved in Africa, then replaced archaic humans elsewhere.

The balance of evidence we have reviewed appears to favor the African replacement model for the origin of *H. sapiens*. None of the tests is definitive, so some form of intermediate model (Figure 20.18b) cannot be ruled out. But taken together, the genetic data and at least some of the morphological data suggest that (1) all present-day people are descended from African ancestors and (2) all present-day non-African people are descended from *H. sapiens* ancestors who left Africa within the last few hundred thousand years. Present-day differences among races must have arisen since then.

20.4 The Evolution of Uniquely Human Traits

Humans have a number of traits that are unique among extant primates: We walk bipedally, we have very large brains, we manufacture and use complex tools, and we use language. We discussed bipedal locomotion and brain size briefly in Section 20.2. Here we consider evidence on the origin of tools and language.

Which of Our Ancestors Made and Used Stone Tools?

Chimpanzees make and use simple tools. They strip stems and twigs of leaves and use the resulting tools to fish termites out of termite mounds; they use leaves as umbrellas; they use rocks and sticks to hammer open nuts (see Mercader et al. 2002, Vogel 2002). Other animals use tools as well. One species of Darwin's finch, the woodpecker finch (*Camarhynchus pallidus*), uses cactus spines to extract insects from bark. So making and using tools is not, in itself, unique to humans. What is unique to humans is making and using complex tools.

The earliest uniquely complex tools that appear in the archaeological record are sharp-edged stone flakes and handheld chopping tools (Figure 20.28). A stone knapper making such tools began by selecting an appropriate cobble from a riverbed, preferably one of fine-grained volcanic rock (Schick and Toth 1993). He or she then struck the cobble with a second rock to chip off flakes. The flakes themselves were usable as cutting tools. Chipping numerous flakes off a cobble in an appropriate pattern produced a chopper. Tools of this style are said to belong to the Oldowan industrial complex because they were first discovered at Olduvai Gorge, Tanzania. Archaeologists have learned firsthand that making Oldowan-style stone tools requires skill and experience.

The oldest known Oldowan tools are from Gona, Ethiopia. Based on the ages of the strata just above and below the sediment layer that contains the tools, Sileshi Semaw and colleagues (1997) established that the tools are 2.5 to 2.6 million years old. Who were the stone knappers who made them?

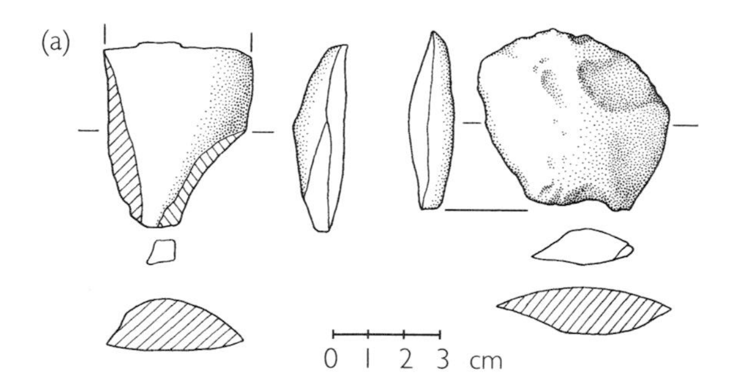

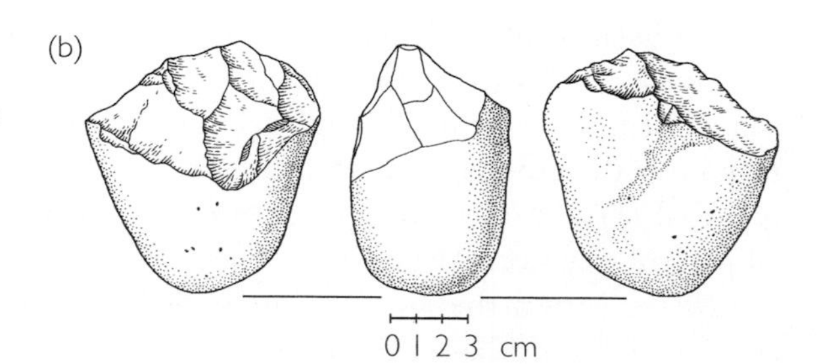

Figure 20.28 Oldowan stone tools from Hadar, Ethiopia These 2.3-million-year-old stone tools are among the oldest known. (a) Two sharp-edged flakes. Each is shown from various sides. (b) A handheld chopper, shown from three sides. After Kimbel et al. (1996).

An obvious answer is, some early member of the genus *Homo*. The trouble with this answer is that we have no definitive evidence that any species of *Homo* had appeared by 2.5 million years ago. The oldest reliably dated *Homo* fossil is a 2.3-million-year-old upper jaw (maxilla) from Hadar, Ethiopia (Gibbons 1996; Kimbel et al. 1996). Which species this fossil represents is unclear; it could be *H. habilis* (Figure 20.13b), *H. rudolfensis* (Figure 20.13c), or some heretofore unknown species.

The earliest known stone tools predate the earliest known Homo *specimens.*

Circumstantial evidence certainly suggests that the Hadar fossil represents the same species that made the 2.5-million-year-old Gona tools. Hadar is geographically close to Gona, 2.3 million years ago is geologically close to 2.5 million years ago, and the Hadar fossil was found near 34 Oldowan tools. It is possible that 2.5-million-year-old *Homo* fossils will eventually be found at Gona and that these early humans were the Gona stone knappers. However, as Bernard Wood (1997) points out, circumstantial evidence is not proof. If other hominins were present at the same time and place, then they are suspects, too. This is true even at Hadar, where the 2.3-million-year-old *Homo* jaw was found near Oldowan tools.

Wood notes that *Paranthropus* (the robust australopithecines) existed with early *Homo* in the same part of Africa over approximately the same time span as the Oldowan industrial complex. While there is no good circumstantial evidence to indicate that *Paranthropus* may have been responsible for Oldowan tools, some indirect evidence of their tool-using capabilities comes from their anatomy. Randall L. Susman (1994) makes an argument based on the anatomy of opposable thumbs. He starts by comparing the bones and muscles of the thumb in humans versus chimpanzees (Figure 20.29). Humans have three muscles that chimpanzees lack. Associated with these extra muscles, humans have thicker metacarpals with broader heads (Figure 20.30a). These differences in thumb anatomy make the human hand more adept at precision grasping than the chimpanzee hand. Susman argues that the modified anatomy of the human thumb evolved in response to selection pressures associated with the manufacture and use of complex tools.

Susman then compares the relative thickness of the thumb metacarpals in humans and chimpanzees with that in a variety of hominin fossils (Figure 20.30b). *H. neanderthalensis, H. erectus*, and *P. robustus* resemble *H. sapiens* in having thumb

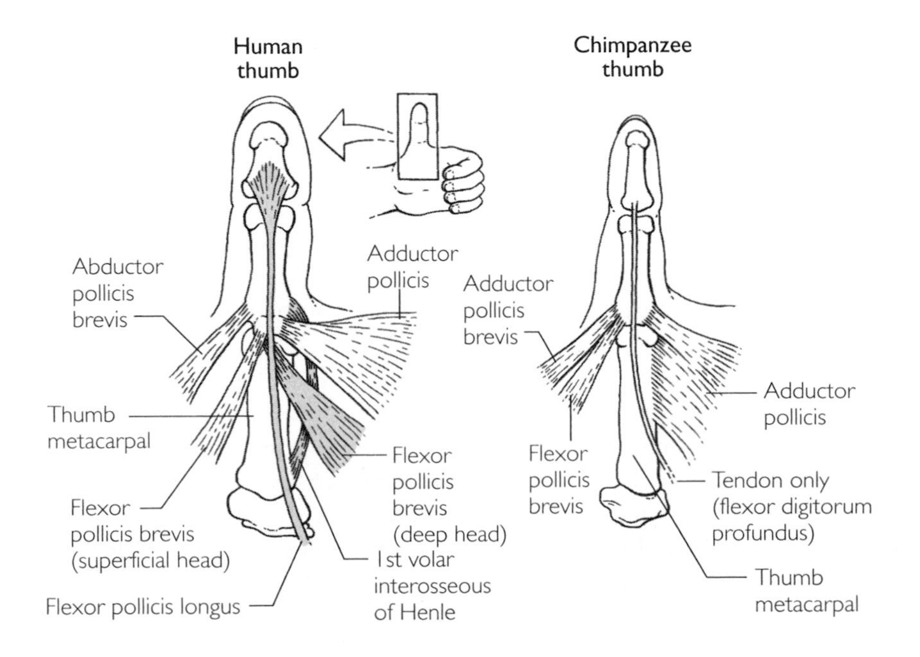

Figure 20.29 Human versus chimpanzee thumbs The three muscles in a human thumb that chimpanzees lack are highlighted with color. Reprinted with permission from Susman (1994).

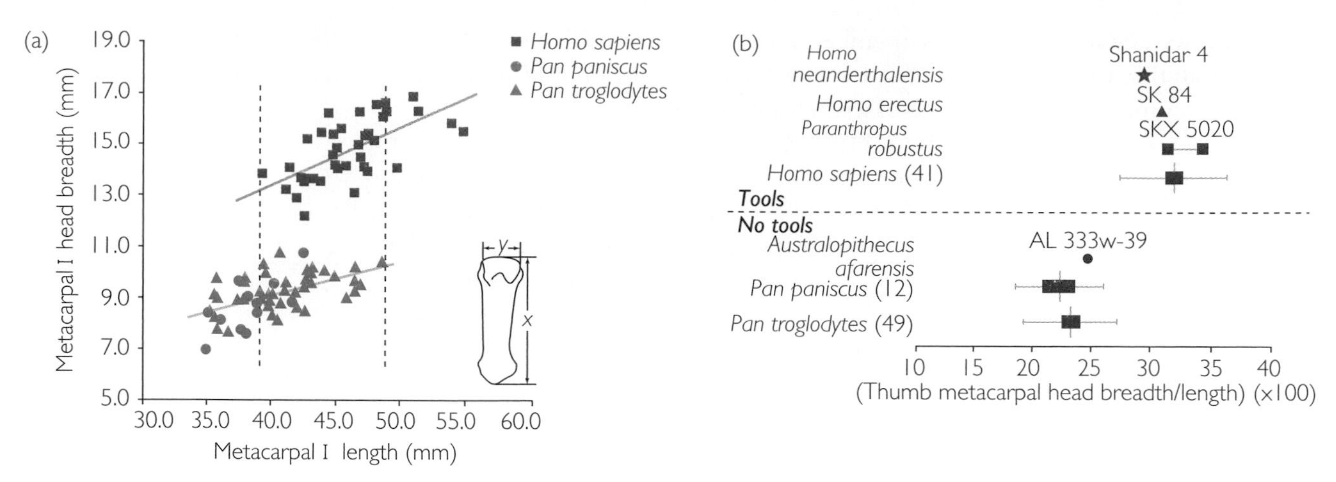

Figure 20.30 Thumb metacarpal bones in a variety of hominins (a) This graph plots the breadth of the thumb metacarpal head (the end of the bone pointing out along the thumb) against the length of the thumb metacarpal for a sample of humans, bonobos, and common chimpanzees. Longer metacarpals have broader heads, but human metacarpals have broader heads for their size than chimpanzee metacarpals. In other words, the ratio of metacarpal head breadth to length is greater in humans. (b) This plot shows the range of ratios (metacarpal head breadth to length) for several hominin fossils (the labels on the data points are the specimen numbers) and samples of present-day humans, bonobos, and common chimpanzees. For *Paranthropus robustus*, the ratio is reported as an estimated range because the bone in question is not quite complete. Species above the dashed line are associated (at least temporally) with manufactured stone tools in the archaeological or fossil record or the present. Species below the dashed line are not. Reprinted with permission from Susman (1994) with some changes in the species names.

metacarpals with broad heads for their length. *A. afarensis*, a gracile australopithecine that disappears from the fossil record before Oldowan tools appear, is like the chimpanzees in having thumb metacarpals with narrow heads for their length. Susman asserts that we can use the thumb metacarpals to diagnose whether extinct hominins were makers and users of stone tools. Susman concludes that both *H. erectus* and *P. robustus* were toolmakers. Susman's argument has been controversial (McGrew et al. 1995). More recently, Hamrick et al. (1998) have argued that several australopithecine species possessed powerful grasping thumbs, including *A. africanus* (the appropriate bones are not known for *A. afarensis*), and that they may have used unmodified stones as tools.

Some morphological analyses suggest that both early Homo species and the robust australopithecines could have been toolmakers.

None of the evidence we have discussed establishes for certain whether the Oldowan tools at Gona were made by a species of *Homo* or a *Paranthropus*. Instead, the evidence argues that they could have been made by either or both. If we accept Susman's conclusion that the robust australopithecines were toolmakers, but the gracile australopithecines were not, and if we accept that the robust australopithecines were a sister lineage to *Homo* (Figures 20.15 and 20.16), then we must make one of two inferences: (1) The manufacture and use of complex stone tools originated in an undiscovered common ancestor of *Homo* and *Paranthropus*, or (2) the manufacture and use of complex stone tools originated independently in at least two hominin lineages.

Most paleoanthropologists, however, still believe that most, if not all, Oldowan stone tools are the handiwork of Homo.

In the absence of definitive proof in the form of a fossil hand grasping a stone tool, we may never know the answer to these questions. Currently, however, most paleoanthropologists favor the view that early *Homo* is responsible for most if not all Oldowan tools. Whenever stone tools are found in association with fossil humans, *Homo* is always there, but *Paranthropus* are often absent. (This includes Susman's fossils from Swartkrans, South Africa, which also contains *Homo erectus*.) And *Homo habilis*, if it is present at Hadar at 2.3 million years ago, comes very close to matching the time span of the Oldowan industrial complex. Three different species of robust australopithecine span the time from 2.5 to 1.0 million years ago, and there are no Oldowan tools after about 1.5 million years ago.

Which of Our Ancestors Had Language?

If the history of hominin tool use is murky, the history of hominin language is even murkier. Like tool use, language is a behavior. Because behaviors do not fossilize, we have no direct evidence of their history. We are left to examine circumstantial evidence in the archaeological and fossil record. Before the invention of writing, language left even less circumstantial evidence than tool use.

Language is a complex adaptation located in the neural circuitry of the brain. The vocabulary and particular grammatical rules of any given language are transmitted culturally, but the capacity for language and a fundamental grammar are, in present-day humans, both innate and universal (see Pinker 1994). Among the evidence for this assertion is the observation that communities of deaf children, if isolated from native signers, invent their own signed languages from scratch. By the end of two generations of transmission to young children within the new deaf culture, these new sign languages develop all the hallmarks of genuine language. They have a standardized vocabulary and grammar, and their fluent users can efficiently communicate the full range of human ideas and emotions. Each of these new sign languages is unique, but all reflect the same universal grammar that linguists have identified in spoken languages.

Many of the brain's language circuits are concentrated in an area called the perisylvian cortex, usually in the brain's left hemisphere (see Pinker 1994). These language circuits include Broca's area and Wernicke's area. Homologous structures exist in the brains of monkeys (Galaburda and Pandya 1982). The monkey homologues of Wernicke's area function in the recognition of sounds, including monkey calls. The monkey homologues of Broca's area function in controlling the muscles of the face, tongue, mouth, and larynx. However, neither of these structures plays a role in the production of monkey vocal calls. Instead, vocal calls are generated by circuits in the brain stem and limbic system. These same structures control nonlinguistic vocalizations in humans, such as laughing, sobbing, and shouting in pain. Thus, the human language organ appears to be a derived modification of neural circuits common to all primates. But among extant species the nature of this modification—its specialization for linguistic communication—appears to be unique to humans. The implication is that the language organ, as such, evolved after our lineage split from the lineage of chimpanzees and bonobos.

Spoken language also relies on derived modifications of the larynx that are unique to humans. In modern human newborns and in other mammals, the larynx is high enough in the throat that it can rise to form a seal with the back opening of the nasal cavity (see Pinker 1994). This allows air to bypass the mouth and throat on its way from the nose to the lungs and prevents the infant from choking on accidentally inhaled food or water. When human babies are about three months old, the larynx descends to a lower position in the throat. This clears more space in which the tongue can move, changes the shape of a pair of resonating chambers, and makes it possible for humans to articulate a much greater diversity of vowel sounds than, for example, chimpanzees.

How far back in our evolutionary lineage can we trace the existence of language, and on what evidence? William Noble and Iain Davidson (1991) assert that the only reliable evidence is the archaeological record. In their view, the hallmark of language is the use of arbitrary symbols, standardized within a culture, to represent objects and ideas. To find language, then, we must look for such symbols in the archaeological record. The first unequivocally arbitrary symbols occur in cave paintings in Germany and France that are about 32,000 years old. Even Noble and Davidson cannot quite accept that language is as recent an innovation as that. They note that *Homo sapiens* had colonized Australia by 40,000 years ago (as early as 60,000, according to some) and confess that they cannot imagine how people could build boats and cross the open ocean without language to help them conceive and coordinate the expedition. However, Noble and Davidson hold the line at about 40,000 years. This would imply that *H. sapiens* is the only species ever to use language.

Early studies suggested that Neandertals could not talk . . .

Other support for the view that language is a recent innovation of *H. sapiens* came from anatomical studies of the skulls of *H. neanderthalensis.* These studies were attempts to reconstruct the Neandertal larynx. Analyses of the shape of the base of the skull and the position of its muscle attachment sites were used to argue that Neandertals had an undescended larynx that would have limited their ability to articulate vowels. This, it was argued, would have prevented them from developing language.

. . . but this conclusion was contradicted by the discovery of a key Neandertal bone.

B. Arensburg and colleagues (1989; 1990) refuted these arguments with a rare paleontological find. A 60,000-year-old Neandertal skeleton from Israel included an intact hyoid bone (Figure 20.31). The hyoid is located in the larynx and serves as the anchor for throat muscles that, in humans, are important in speaking. Arensburg et al. analyzed the hyoid bone and compared it to those of chimpanzees and present-day humans. They found the Neandertal hyoid to be dramatically different from that of chimps and virtually identical to that of present-day humans. Based on this hyoid bone, Arensburg and colleagues suggest that Neandertals had a descended larynx. Given that a descended larynx entails an increased risk of choking, it is hard to imagine why it would evolve unless it carried a substantial benefit. An obvious candidate is an improved ability to speak.

If we accept the proposition that Neandertals could talk, can we trace language back any further in our lineage? Given that the language organ is in the brain, what can we glean from the fossil record about the brains of our ancestors? David Pilbeam and Stephen Jay Gould (1974) showed many years ago that even after taking body size into account, there has been something dramatically different about our brains since the first emergence of our genus. Figure 20.32 plots brain size as a function of body size in extant great apes, australopithecines (including *Paranthropus*), and three species of *Homo.* Inspection of the figure indi-

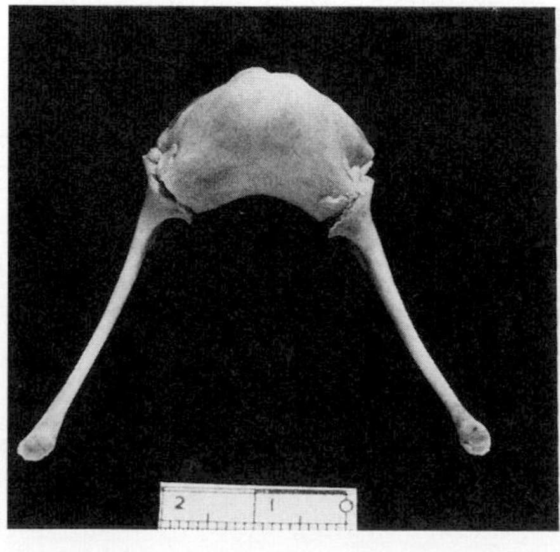

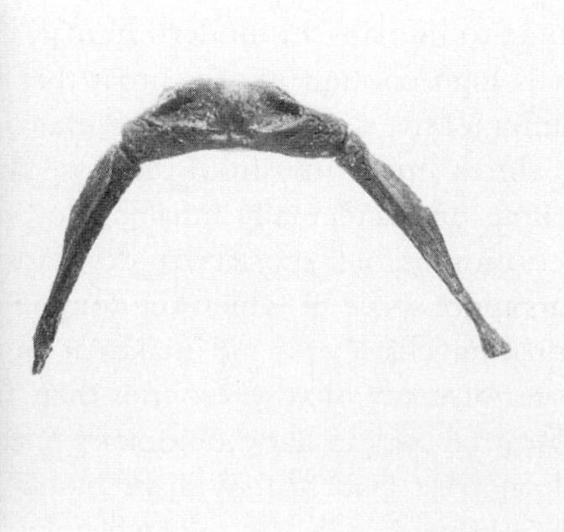

Figure 20.31 Hyoid bones from *Homo neanderthalensis* and a common chimpanzee The Neandertal hyoid is on the right. It is virtually identical to the hyoid of a modern human. The scale bar, which applies to both hyoids, is in centimeters. Reproduced by permission from Arensburg, Schepartz, Tillier, Vandermeersch, and Rak (1990).

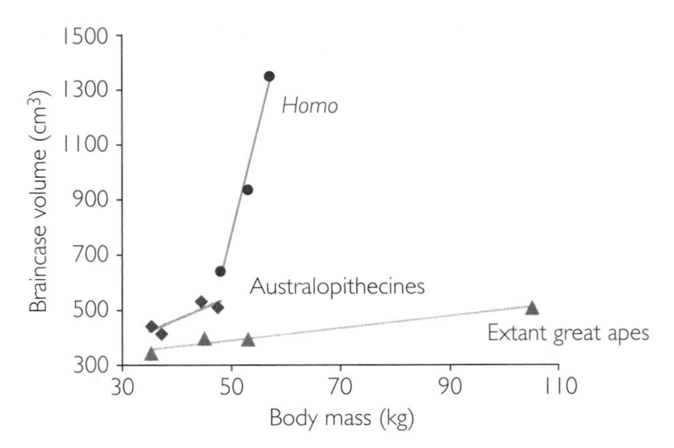

Figure 20.32 Brain size versus body size in a variety of hominins and great apes Data points represent species averages, with best-fit lines. In all three groups, species with larger brains have larger bodies. Australopithecines have larger brains for their size than extant great apes. *Homo* species have larger brains for their size, as well as a dramatically different relationship between brain size and body size. The extant great apes are the bonobo, common chimpanzee, orangutan, and gorilla; the australopithecines are *A. afarensis, A. africanus, P. boisei,* and *P. robustus*; the *Homo* species are *H. habilis, H. ergaster/erectus,* and *H. sapiens*. The data are from Tobias (1987) and Pilbeam and Gould (1974). See also McHenry (1994). After Pilbeam and Gould (1974).

cates that not only do *Homo* species have larger brains for their size, but brain size increases much more steeply with body size in *Homo* than in its hominin relatives. Nearly all the more recent analyses (for example, McHenry 1992; Kappelman 1996) benefiting from a much more complete fossil record of human brain and body size evolution, have essentially confirmed this result. Like a descended larynx, large brains come at a cost. They require a great deal of energy to maintain, and they generate considerable heat. What benefits could have compensated for these costs? Two possibilities are an increased capacity to make and use tools and an increased capacity for language.

Phillip Tobias (1987) examined casts of the insides of the braincases of specimens of *H. habilis.* In addition to their sheer size, these endocasts revealed the existence of derived structural traits unique to our genus. Among them are clear enlargments of Broca's and Wernicke's areas. By his own account, this discovery converted Tobias from a skeptic to an advocate of the hypothesis that language first emerged, at least in rudimentary form, in *H. habilis.* We do not know for certain, and perhaps never will, but language may be as much as 2 million years old.

An analysis of the skulls of Homo habilis suggest that they may have had at least rudimentary language.

Summary

As Darwin predicted, an evolutionary perspective throws light on the origin and nature of humans. Humans are relatives of the great apes. Although consensus was slow to arrive, morphological and molecular studies indicate that our closest living relatives are the chimpanzees. Our most recent common ancestor with the chimpanzees lived about 5 or 6 million years ago.

Following its split from the chimpanzee lineage, our own lineage gave rise to multiple species of bipedal African hominins. Fossils provide strong evidence for the coexistence of at least two of these species and perhaps as many as five. We are the sole survivors of this evolutionary radiation.

The first members of genus *Homo* left Africa nearly 2 million years ago. Whether these populations ultimately contributed genes to present-day populations of humans is the subject of debate. No definitive tests have been performed, but the balance of the evidence suggests that all present-day non-African populations are descended from a more recent wave of emigrants that left Africa within the last 200,000 years. This implies that the present-day geographic variation among populations of humans is of relatively recent origin.

Among the derived traits unique to our species are the manufacture and use of complex tools and the capacity for language. Because behavior does not fossilize, researchers have to rely on circumstantial evidence to reconstruct the history of these traits. Tool use appeared at least 2.5 million years ago. It is most likely to have arisen in an early species of *Homo,* although robust australopithecines may also have used stone tools. The evidence on language is more tenuous, but it suggests that language may have emerged nearly as early as tool use.

Questions

1. What is the difference between a gene tree and a species tree? Explain in your own words how it is possible for gorillas and humans to share a genetic trait (such as a retroviral insertion) that chimpanzees do not share, if chimpanzees and humans are really each other's closest relatives. Given the lack of agreement between gene trees and species trees, how is it possible to reconstruct the true species tree?
2. Just a generation ago, almost none of the fossils described in this chapter had been found. At that time, it was expected that when early hominin fossils were finally found, the first distinctly human feature—that is, the first derived trait of hominins that would distinguish them morphologically from the chimpanzee lineage—would prove to be an enlarged brain. It was thought that a large, human-sized brain must have evolved either before or simultaneously with bipedality. Now that we have the fossils to test this question, what do the fossils show? Which came first, large brains or bipedality?
3. How were Sarich and Wilson able to test the genetic relationships of humans to great apes when at the time (1967) it was not possible to sequence DNA?
4. Explain why Ruvolo (1995, 1997) thought it was important to look at several nuclear genes, and not just the mtDNA genes, to study the relationships of humans and the great apes.
5. The data in this chapter show that humans and the chimpanzees are each other's closest relatives. Is it accurate to say that humans evolved from chimpanzees? How about that chimpanzees evolved from humans? Is it accurate to say that humans evolved from apes?
6. In a study of the phylogeny of Old World monkeys (Hayasaka et al. 1996), the three individual rhesus macaques that were studied did not form a monophyletic group. Instead, the mtDNA of one of the rhesus macaques was more similar to the mtDNA of Japanese and Taiwanese macaques (which are different species) than it was to the other rhesus macaques. How might this have happened? (There are at least two possibilities.) With this as background, explain why it is useful that the phylogeny in Figure 20.4 includes several individuals from each species.
7. Look again at Figure 20.4, and this time focus on the diversity of sequences within each species. The length of the colored lines indicates the degree of genetic diversity within each species.
   - **a.** Do humans have a large or small amount of genetic diversity, compared to the other primates studied? Which other species show similar patterns to humans? Why do you think some species have greater diversity than others?
   - **b.** One of the human sequences is distinct from the other five. Can you guess the geographic location of this person?
8. Knuckle-walking and fist-walking are exceedingly rare methods of locomotion among animals. A similar type of locomotion occurs in a few other unusual cases such as the giant anteater and a few sloths. Why do you think it evolved in only these animals—what advantage does it bring? Why do you think humans lost this trait? What does this trait indicate about the different uses to which forelimbs may be put?
9. Briefly describe (one sentence each) the four main models of the origin of *Homo sapiens* from *Homo erectus*. Which one has been rejected, and why? Why has it been difficult to determine which of the other three models is correct? Which model does the most recent evidence support?
10. What's in a name? Jared Diamond (1992) suggests that if we follow the naming traditions of cladistic taxonomy, then humans, chimpanzees, and bonobos should all be considered members of a single genus. Diamond proposes calling these species, respectively, *Homo sapiens, Homo troglodytes*, and *Homo paniscus*. Jonathan Marks (1994) objects to Diamond's taxonomic reasoning. Concerning the nature of humans and apes, Marks asserts that "Popular works tell us we are not merely genetically apes but that we are literally apes (e.g., Diamond 1992). Sometimes there is profundity in absurdity, but I don't think this is one of those times. It merely reflects the paraphyletic nature of the category 'apes'—humans are apes, but only in the same sense that pigeons are reptiles and horses are fish.... Focusing on the genetic relations obscures biologically significant patterns of phenotypic divergence." Do you think humans, chimpanzees, and bonobos should all be classified as members of the same genus? Is there more at stake in the disagreement between Diamond and Marks than just Latin names? If so, what?
11. Jared Diamond finds ethical dilemmas in the close kinship between humans and chimpanzees: "It's considered acceptable to exhibit caged apes in zoos, but it's not acceptable to do the same with humans. I wonder how the public will feel when the identifying label on the chimp cage in the zoo reads *Homo troglodytes*" (Diamond 1992, p. 29). Diamond finds the use of chimpanzees in medical research even more problematic. The scientific justification for the use of chimpanzees is that chimpanzee physiology is extremely similar to human physiology, so chimpanzees are the best substitute for human subjects. Diamond notes that jails are a very rough analogue to zoos, in the sense that they rep-

resent conditions under which we do consider it acceptable to keep people in cages without their consent (if not to display them). But there is no human analogue to research on chimpanzees: There are no conditions under which we consider it acceptable to do medical experiments on humans without their consent. Is it ethically justified to keep animals in zoos? To use animals in medical research? Does the phylogenetic relationship between ourselves and the animals in question matter? If so, how and why?

12. Section 20.2 mentions that there is debate over the evolutionary affinities of *Sahelanthropus tchadensis*. Given the age and appearance of the skull in Figure 20.9, there are many possibilities: It could be our common ancestor with the chimpanzees; it could be a species more closely related to us than to the chimpanzees; it could be a species more closely related to the chimpanzees than to us; it could even be a species more closely related to gorillas than to either the chimpanzees or us. Suppose you are a paleoanthropologist who wants to figure out which of these possibilities is correct. What strata would you choose to search for more fossils? If you are lucky enough to find a skull, what features would tell you which hypothesis is closest to the truth?

13. One of the most heated aspects of human racial politics is the contention that human races are genetically distinct. How is this issue addressed by the African replacement model versus the multiregional evolution model? That is, which model predicts that human races are more genetically different from each other? How different are people from different geographic regions?

14. We reviewed genetic studies showing that non-African human populations are descended from African populations. Some people might conclude from these data that modern African people are in some sense "primitive." What is the logical flaw in this thinking?

15. Different ethnic groups within Africa are more diverse than are the ethnic groups on all other continents taken together. What does this imply about the common U.S. practice of categorizing people into "African," "Caucasian," "Asian," "Hispanic," and "Native American"?

16. Work by C. Swisher and colleagues (1996) indicates that *Homo erectus* may have persisted in Java until 53,000 years ago at Sambungmachan and until 27,000 years ago at Ngandong. If correct, these dates imply that *H. erectus* and *H.sapiens* coexisted in Java. Will this finding help settle the debate over the out-of-Africa model versus the multiregional origin model? Why or why not?

17. Describe the two lines of morphological evidence that have been used to infer whether extinct hominins like Neandertals or *Homo erectus* might have used spoken language. What did the data show? Do you find this to be convincing evidence?

18. For the sake of argument, adopt the proposition implied by Wood (1997) that early species of *Homo* did not participate in the production of Oldowan stone tools. What other puzzles must we now solve? Note that after the invention of Oldowan tools, the next advance in toolmaking is marked in the archaeological record by the appearance of Acheulean tools, which are substantially more sophisticated than Oldowan tools (Johanson et al. 1996). Acheulean tools appeared about 1.4 million years ago and persisted until less than 200,000 years ago.

19. Derek Bickerton (1995) and Charles Catania (1995) object to the suggestion that *Homo habilis* had language. Bickerton writes, "If *H. habilis* already had all the necessary ingredients for language, what happened during the next million years?" And Catania writes, "I [deduce] that our hominin ancestors should have taken over the world 950,000 years ago if a million years ago they were really like us in their language competence. But they did not, so they were not; if they had been, those years would have been historic instead of prehistoric." How do you think Phillip Tobias would respond to Bickerton and Catania? Who do you think is right?

20. The ancestors of horses are each known from dozens, hundreds, or in some cases even thousands, of virtually complete specimens. Yet hominin species of the same age are often known only from one or a very few partial specimens, such as the crushed partial skull of *Sahelanthropus*. As a result, we understand equine evolution much better than we do hominid evolution. Speculate as to why this is so: Why are hominin fossils rare? Does the scarcity of hominin fossils invalidate the conclusions of paleontologists? Put another way, is it really possible to learn anything useful from a single bone fragment?

Exploring the Literature

21. In Section 20.3 we discussed sequence data from Neandertal mitochondrial genomes and what they tell us about the relathionship between Neandertals and modern humans. For preliminary reports on efforts to sequence the nuclear genome of a Neandertal see:

Green, R. E., J. Krause, et al. 2006. Analysis of one million base pairs of Neanderthal DNA. *Nature* 444: 330–336.

Noonan, J. P., G. Coop, et al. 2006. Sequencing and analysis of Neanderthal genomic DNA. *Science* 314: 1113–1118.

22. In 2004, Peter Brown, Mike Morwood, and colleagues reported their discovery, in a cave on Flores Island, Indonesia, of a tiny adult hominin that lived just 18,000 years ago. The researchers assigned the fossil to a new human species, *Homo floresiensis,* which they initially

conjectured was a descendent of *Homo erectus* that persisted long enough to overlap with modern humans. For a start on the contentious literature surrounding the Flores hominin, see:

Brown, P., T. Sutikna, et al. 2004. A new small-bodied hominin from the Late Pleistocene of Flores, Indonesia. *Nature* 431: 1055–1061.

Brumm, A., F. Aziz, et al. 2006. Early stone technology on Flores and its implications for *Homo floresiensis*. *Nature* 441: 624–628.

Falk, D., C. Hildebolt, et al. 2005. The brain of LB1, *Homo floresiensis*. *Science* 308: 242–245.

Falk, D., C. Hildebolt, et al. 2005. Response to Comment on "The Brain of LB1, *Homo floresiensis*." *Science* 310: 236c.

Falk, D., C. Hildebolt, et al. 2006. Response to Comment on "The Brain of LB1, *Homo floresiensis*." *Science* 312: 999c.

Jacob, T., E. Indiriati, et al. 2006. Pygmoid Australomelanesian *Homo sapiens* skeletal remains from Liang Bua, Flores: Population affinities and pathological abnormalities. *Proceedings of the National Academy of Sciences USA* 103: 13421–13426.

Lahr, M. M., and R. Foley. 2004. Human evolution writ small. *Nature* 431: 1043–1044.

Martin, R. D., A. M. MacLarnon, et al. 2006. Comment on "The Brain of LB1, *Homo floresiensis*." *Science* 312: 999b.

Morwood, M. J., P. Brown, et al. 2005. Further evidence for small-bodied hominins from the Late Pleistocene of Flores, Indonesia. *Nature* 437: 1012–1017.

Morwood, M. J., R. P. Soejono, et al. 2004. Archaeology and age of a new hominin from Flores in eastern Indonesia. *Nature* 431: 1087–1091.

Weber, J., A. E., Czarnetzki, and C. M. Pusch. 2005. Comment on "The brain of LB1, *Homo floresiensis*." *Science* 310: 236.

23. Read Susman's (1994) paper on opposable thumbs and tool use. What weaknesses can you identify in Susman's argument? What additional data would you like to see? If you were Susman, how would you respond to these critiques? Read McGrew et al. (1995) to see if the critiques and responses therein are similar to your own.

McGrew, W. C., M. W. Hamrick, S. E. Inouye, J. C. Ohman, M. Slanina, G. Baker, R. P. Mensforth, and R. L. Susman. 1995. Thumbs, tools, and early humans. *Science* 268: 586–589.

Susman, R. L. 1994. Fossil evidence for early hominid tool use. *Science* 265: 1570–1573.

For an additional approach to the problem of tool use, see:

Susman, R. L. 1998. Hand function and tool behavior in early hominids. *Journal of Human Evolution* 35: 23–46.

24. Is tool use in animals hard-wired or learned? See:

Matsusaka, T., H. Nishie, M. Shimada, N. Kutsukake, K. Zamma, M. Nakamura, and T. Nishida. 2006. Tool-use for drinking water by immature chimpanzees of Mahale: prevalence of an unessential behavior. *Primates* 42 (2): 113–122.

Kruetzen, M., J. Mann, M. R. Heithaus, R. C. Connor, L. Bejder, and W. B. Sherwin. 2005. Cultural transmission of tool use in bottlenose dolphins. *Proceedings of the National Academy of Sciences USA* 102 (25): 8939–8943

Recent evidence suggests that certain birds not only make and use tools, but appear to understand what it is they are doing. See:

Weird, A. A. S., J. Chappell, and A. Kacelnik. 2002. Shaping of hooks in New Caledonian crows. *Science* 297:981.

Videos of the New Caledonian crows making and using hook tools can be seen at:

http://users.ox.ac.uk/~kgroup/tools/movies.shtml.

25. There were hints in this chapter that present-day humans have smaller bodies and brains than some of our recent ancestors. For documentation that this is the case, see:

Gibbons, A. 1997. Bone sizes trace the decline of man (and woman). *Science* 276: 896–897.

Ruff, C. B., E. Trinkaus, and T. W. Holliday. 1997. Body mass and encephalization in Pleistocene *Homo*. *Nature* 387: 173–176.

26. See the following sequence of studies for a valiant effort to find another morphological marker of human speech. What was the final conclusion?

Kay, R. F., M. Cartmill, and M. Balow. 1998. The hypoglossal canal and the origin of human vocal behavior. *Proceedings of the National Academy of Sciences USA* 95: 5417–5419

DeGusta, D., W. H. Gilbert, and S. P. Turner. 1999. Hypoglossal canal size and hominid speech. *Proceedings of the National Academy of Sciences USA* 96: 1800–1804.

Jungers, W. L., A. A. Pokempner, R. F. Kay, and M. Cartmill. 2003. Hypoglossal canal size in living hominoids and the evolution of human speech. *Human Biology* 75: 473–484.

27. Might any non-human animals use a simple language? Most animals can produce vocalizations, but few show two fundamental aspects of human language: learned vocalizations, and the use of certain sounds to symbolize a concept. The ability to learn vocalizations is common in many birds, but rare in mammals. The use of sounds to symbolize concepts is rarer still. See these papers for two examples that might qualify:

Janik, V. M., L. S. Sayigh, and R. S. Wells. 2006. Signature whistle shape conveys identity information to bottlenose dolphins. *Proceedings of the National Academy of Sciences USA* 103 (21): 8293–8297.

Seyfarth, R. M., D. L. Cheney, and P. Marler. 1980. Vervet monkey alarm calls—semantic communication in a free-ranging primate. *Animal Behaviour* 28: 1070–1094.

This paper reports evidence that dogs, though they obviously cannot produce language themselves, may have the capability to understand the symbolic meaning of up to 200 words, and can rapidly learn more—at least in the case of this border collie:

Kaminski, J., J. Call, and J. Fischer. 2004. Word learning in a domestic dog: Evidence for "fast mapping." *Science* 304: 1682–1683.

28. Why did hominins evolve bipedality? See this paper for evidence that endurance running may have played a major role in human evolution:

Bramble, D. M., and D. E. Lieberman. 2004. Endurance running and the evolution of *Homo*. *Nature* 432: 345–352.

29. We noted in Box 20.1 that, genetically, humans and chimpanzees are startlingly similar. Are they similar enough that they could produce viable hybrids? How could we identify a hybrid if it existed? Claims that a sideshow performer named Oliver was a human-chimpanzee hybrid were credible enough to warrant scientific investigation. See:

Ely, J. J., M. Leland, et al. 1998. Technical note: Chromosomal and mtDNA analysis of Oliver. *American Journal of Physical Anthropology* 105: 395–403.

Citations

Aiello, L. C. 1993. The fossil evidence for modern human origins in Africa: A revised review. *American Anthropologist* 95: 73–96.

Aiello, L. C., and M. Collard. 2001. Our newest oldest ancestor? *Nature* 410: 526–527.

Alemseged, Z., F. Spoor, et al. 2006. A juvenile early hominin skeleton from Dikika, Ethiopia. *Nature* 443: 296–301.

Andrews, P. 1992. Evolution and environment in the Hominoidea. *Nature* 360: 641–646.

Andrews, P., and L. Martin. 1987. Cladistic relationships of extant and fossil hominoids. *Journal of Human Evolution* 16: 101–118.

Arensburg, B., A. M. Tillier, et al. 1989. A middle paleolithic human hyoid bone. *Nature* 338: 758–760.

Arensburg, B., L. A. Schepartz, et al. 1990. A reappraisal of the anatomical basis for speech in Middle Paleolithic hominids. *American Journal of Physical Anthropology* 83: 137–146.

Arsuaga, J. L., I. Martínez, et al.. 1999. The human cranial remains from Gran Dolina lower-Pleistocene site (Sierra de Atapuerca, Spain). *Journal of Human Evolution* 37: 431–457.

Asfaw, B., T. White, et al. 1999. *Australopithecus gahri*: A new species of early hominid from Ethiopia. *Science* 284: 629–635.

Ayala, F. J. 1995. The myth of Eve: Molecular biology and human origins. *Science* 270: 1930–1936.

Ayala, F. J., A. Escalante, et al. 1994. Molecular genetics of speciation and human origins. *Proceedings of the National Academy of Sciences USA* 91: 6787–6794.

Bamshad, M. J., S. Wooding, et al. 2003. Human population genetic structure and inference of group membership. *American Journal of Human Genetics* 72: 578–589.

Barbulescu, M., G. Turner, et al. 2001. A HERV-K provirus in chimpanzees, bonobos, and gorillas, but not humans. *Current Biology* 11: 779–783.

Barton, N. H. 2006. Evolutionary biology: How did the human species form? *Current Biology* 16: R647–R650.

Begun, D. R. 1992. Miocene fossil hominids and the chimp–human clade. *Science* 257: 1929–1933.

Begun, D. R. 1994 Relations among the great apes and humans: New interpretations based on the fossil great ape *Dryopithecus*. *Yearbook of Physical Anthropology* 37: 11–63.

Begun, D. R. 1995. Late-Miocene European orangutans, gorillas, humans, or none of the above? *Journal of Human Evolution* 29: 169–180.

Begun, D. R. 2004. The earliest hominins—Is less more? *Science* 303: 1478–1480.

Begun, D. R., C. V. Ward, and M. D. Rose. 1997. Events in hominoid evolution. In D. R. Begun, C. V. Ward, and M. D. Rose, eds. *Function, Phylogeny and Fossils: Miocene Hominoid Evolution and Adaptations.* New York: Plenum Publishing Company, 389–415.

Bermúdez de Castro, J. M., J. L. Arsuaga, et al. 1997. A hominid from the lower Pleistocene of Atapuerca, Spain: Possible ancestor to Neandertals and modern humans. *Science* 276: 1392–1395.

Bickerton, Derek. 1995. Finding the true place of *Homo habilis* in language evolution. Open peer commentary. In: Wilkins, W. K., and J. Wakefield, Brain evolution and preconditions. *Behavioral and Brain Sciences* 18: 161–226.

Blumenschine, R. J., C. R. Peters, et al. 2003. Late Pliocene *Homo* and hominid land use from western Olduvai Gorge, Tanzania. *Science* 299: 1217–1221.

Boesch, C., and M. Tomasello. 1998 Chimpanzee and human cultures. *Current Anthropology* 39: 591–614.

Borowik, O. A. 1995. Coding chromosomal data for phylogenetic analysis: Phylogenetic resolution of the *Pan-Homo-Gorilla* trichotomy. *Systematic Biology* 44: 563–570.

Bowcock, A. M., A. Ruiz-Linares, et al. 1994. High resolution of human evolutionary trees with polymorphic microsatellites. *Nature* 368: 455–457.

Britten, R. J. 2002. Divergence between samples of chimpanzee and human DNA sequences is 5%, counting indels. *Proceedings of the National Academy of Sciences USA* 99: 13633–13635.

Brunet, M. 2002. *Sahelanthropus* or '*Salhelpithecus*'? *Nature* 419: 582.

Brunet, M., F. Guy, et al. 2002. A new hominid from the upper Miocene of Chad, central Africa. *Nature* 418: 145–151.

Brunet., M., F. Guy, et al. 2005. New material from the earliest hominid from the Upper Miocene of Chad. *Nature* 434: 752–755.

Carbonell, E., J. M., Bermúdez de Castro, and J. L. Arsuaga. 1999. Preface: Special issue on Gran Dolina site: TD6 Aurora Stratum (Burgos, Spain). *Journal of Human Evolution* 37: 309–311.

Catania, A. C. 1995. Single words, multiple words, and the functions of language. Open peer commentary. In W. K. Wilkins and J. Wakefield. Brain evolution and neurolinguistic preconditions. *Behavioral and Brain Sciences* 18: 161–226.

Cela-Conde, C. J., and F. J. Ayala. 2003. Genera of the human lineage. *Proceedings of the National Academy of Sciences USA* 100: 7684–7689.

Chen, F.-C., and W.-H. Li. 2001. Genomic differences between humans and other hominoids and the effective population size of the common ancestor of humans and chimpanzees. *American Journal of Human Genetics* 68: 444–456.

Collard, M., and L. C. Aiello. 2000. From forelimbs to two legs. *Nature* 404: 339–340.

Collard, M., and B. Wood. 2000. How reliable are human phylogenetic hypotheses? *Proceedings of the National Academy of Sciences USA* 97: 5003–5006.

Corruccini, R. S., and H. M. McHenry. 2001. Knuckle-walking hominid ancestors. *Journal of Human Evolution* 40: 507–511.

Dainton, M. 2001. Did our ancestors knuckle-walk? *Nature* 410: 324–325.

Darwin, C. 1859. *On the Origin of Species by Means of Natural Selection, Or the Preservation of the Favoured Races in the Struggle for Life.* London: John Murray.

Darwin, C. 1871. *The Descent of Man, and Selection in Relation to Sex.* London: John Murray.

Dean, D., and E. Delson. 1992. Second gorilla or third chimp? *Nature* 359: 676–677.

Deinard, A. S., G. S. Sirugo, and K. K. Kidd. 1998. Hominoid phylogeny: Inferences from a sub-terminal minisatellite analyzed by repeat expansion detection (RED). *Journal of Human Evolution* 35: 313–317.

de Waal, F. B. M. 1997. Bonobo: *The Forgotten Ape.* Berkeley: University of California Press.

de Waal, F. B. M. 1999. Cultural primatology comes of age. *Nature* 399: 635–636.

de Waal, F. B. M. 2005. A century of getting to know the chimpanzee. *Nature* 437: 56–59.

Diamond, J. 1992. *The Third Chimpanzee.* New York: Harper Collins.

Djian, P., and H. Green. 1989. Vectorial expansion of the involucrin gene and the relatedness of the hominoids. *Proceedings of the National Academy of Sciences USA* 86: 8447–8451.

Duarte, C., J. Maurício, et al. 1999. The early upper Paleolithic human skeleton from the Abrigo do Laga Velho (Portugal) and modern human emergence in Iberia. *Proceedings of the National Academy of Sciences USA* 96: 7604–7609.

Elango, N, J. W. Thomas, et al. 2006. Variable molecular clocks in hominoids. *Proceedings of the National Academy of Sciences USA* 103: 1370–1375.

Enard, W., P. Khaitovich, et al. 2002. Intra- and interspecific variation in primate gene expression patterns. *Science* 296: 340–343.

Frayer, D. W., M. H. Wolpoff, et al. 1993. Theories of modern human origins: The paleontological test. *American Anthropologist* 95: 14–50.

Gabunia, L., and A. Vekua. 1995. A Plio-Pleistocene hominid from Dmanisi, East Georgia, Caucasus. *Nature* 373: 509–512.

Gabunia, L., A. Vekua, et al. 2000. Earliest Pleistocene hominid cranial remains from Dmanisi, Republic of Georgia: Taxonomy, geological setting, and age. *Science* 288: 1019–1025.

Gagneux, P., and A. Varki. 2001. Genetic differences between humans and great apes. *Molecular Phylogenetics and Evolution* 18: 2–13.

Gagneux, P., C. Wills, et al. 1999. Mitochondrial sequences show diverse evolutionary histories of African hominoids. *Proceedings of the National Academy of Sciences USA* 96: 5077–5082.

Galaburda, A. M., and D. N. Pandya. 1982. Role of architectonics and connections in the study of primate brain evolution. In E. Armstrong and D. Falk, eds. *Primate Brain Evolution.* New York: Plenum.

Gee, H. 2001. Return to the planet of the apes. *Nature* 412: 131–132.

Gibbons, A. 1994. Rewriting—and redating—prehistory. *Science* 263: 1087–1088.

Gibbons, A. 1996. A rare glimpse of an early human face. *Science* 274: 1298.

Gibbons, A. 2002a. One scientist's quest for the origin of our species. *Science* 298: 1708–1711.

Gibbons, A. 2002b. In search of the first hominids. *Science* 295: 1214–1219.

Gibbons, A. 2005. Facelift supports skull's status as oldest member of the human family. *Science* 308: 179–180.

Gibbs, S., M. Collard, and B. Wood. 2000. Soft-tissue characters in higher primate phylogenetics. *Proceedings of the National Academy of Sciences USA* 97: 11130–11132.

Gibbs, S., M. Collard, and B. Wood. 2002. Soft-tissue anatomy of the extant hominoids: A review and phylogenetic analysis. *Journal of Anatomy* 200: 2–40.

Gilad, Y., A. Oshlack, et al. 2006. Expression profiling in primates reveals a rapid evolution of human transcription factors. *Nature* 440: 242–245.

Goldstein, D. B., A. Ruiz Linares, et al. 1995. Genetic absolute dating based on microsatellites and the origin of modern humans. *Proceedings of the National Academy of Sciences USA* 92: 6723–6727.

Goodman, M. 1962. Evolution of the immunologic species specificity of human serum proteins. *Human Biology* 34: 104–150.

Goodman, M., W. J. Bailey, et al. 1994. Molecular evidence on primate phylogeny from DNA sequences. *American Journal of Physical Anthropology* 94: 3–24.

Goodman, M., C. A. Porter, J. Czelusniak, et al. 1998. Toward a phylogenetic classification of primates based on DNA evidence complemented by fossil evidence. *Molecular Phylogenetics and Evolution* 9: 585–598.

Green, H., and P. Djian. 1995. The involucrin gene and Hominoid relationships. *American Journal of Physical Anthropology* 98: 213–216.

Groves, C. P. 1986. Systematics of the great apes. In D. R. Swindler and J. Erwin, eds. *Comparative Primate Biology*, Volume 1: Systematics, Evolution, and Anatomy. New York: Alan R. Liss, Inc., 187–217.

Gu, J., and X. Gu. 2003. Induced gene expression in human brain after the split from chimpanzee. *Trends in Genetics* 19: 63–65.

Gutiérrez, G., D. Sánchez, and A. Marín. 2002. A reanalysis of the ancient mitochondrial DNA sequences recovered from Neandertal bones. *Molecular Biology and Evolution* 19: 1359–1366.

Guy, F., D. E. Lieberman, et al. 2005. Morphological affinities of the *Sahelanthropus tchadensis* (Late Miocene hominid from Chad) cranium. *Proceedings of the National Academy of Sciences USA* 102: 18836–18841.

Hacia, J. G. 2001. Genome of the apes. *Trends in Genetics* 17: 637–645.

Hamrick M. W., S. E. Churchill, et al. 1998. EMG of the human flexor pollicis longus muscle: Implications for the evolution of Hominid tool use. *Journal of Human Evolution* 34: 123–136.

Hayasaka, K., K. Fujii, and S. Horai. 1996. Molecular phylogeny of macaques: Implications of nucleotide sequences from an 896-base-pair region of mitochondrial DNA. *Molecular Biology and Evolution* 13: 1044–1053.

Hedges, S. B. 2000. A start for population genomics. *Nature* 408: 652–653.

Hedges, S. B., S. Kumar, et al. 1992. Human origins and analysis of mitochondrial DNA sequences. *Science* 255: 737–739.

Hofreiter, M., D. Serre, et al. 2001. Ancient DNA. *Nature Reviews Genetics* 2: 353–359.

Horai, S., Y. Satta, et al. 1992. Man's place in the Hominoidea revealed by mitochondrial DNA genealogy. *Journal of Molecular Evolution* 35: 32–43.

Huang W. P., R. G. Ciochon, et al. 1995. Early *Homo* and associated artifacts from Asia. *Nature* 378: 275–278.

Huxley, T. H. 1863. *Evidence as to Man's Place in Nature.* New York: D. Appelton and Company.

Ingman, M., H. Kaessmann, et al. 2000. Mitochondrial genome variations and the origin of modern humans. *Nature* 408: 708–713.

Johanson, D. C., B. Edgar, and D. Brill. 1996. *From Lucy to Language.* New York: Simon & Schuster Editions.

Jones, J. E., III. 2005. *Tammy Kitzmiller v. Dover Area School District, Memorandum Opinion.* US District Court for the Middle District of Pennsylvannia, Case No. 04cv2688.

Jorde, L. B., W. S. Watkins, et al. 2000. The distribution of human genetic diversity: A comparison of mitochondrial, autosomal, and Y-chromosome data. *American Journal of Human Genetics* 66: 979–988.

Kappelman, J. 1996. The evolution of body mass and relative brain size in fossil hominids. *Journal of Human Evolution* 30: 243–276.

Khaitovich, P., W. Enard, et al. 2006. Evolution of primate gene expression. *Nature Reviews Genetics* 7: 693–702.

Kim, H.-S., and O. Takenaka. 1996. A comparison of TSPY genes from Y-chromosomal DNA of the great apes and humans: Sequence, evolution, and phylogeny. *American Journal of Physical Anthropology* 100: 301–309.

Kimbel, W. H., R. C. Walter, et al. 1996. Late Pliocene *Homo* and Oldowan tools from the Hadar Formation (Kada Hadar member), Ethiopia. *Journal of Human Evolution* 31: 549–561.

King, M.-C., and A. C. Wilson. 1975. Evolution at two levels in humans and chimpanzees. *Science* 188: 107–116.

Klicka, J., and R. M. Zink. 1997. The importance of recent ice ages in speciation: A failed paradigm. *Science* 277: 1666–1669.

Krings, M. H., C. Capelli, et al. 2000. A view of Neandertal genetic diversity. *Nature Genetics* 26: 144–146.

Krings M., H. Geisert, et al. 1999. DNA sequence of the mitochondrial hypervariable region II from the Neandertal type specimen. *Proceedings of the National Academy of Sciences USA* 96: 5581–5585.

Krings, M., A. Stone, et al. 1997. Neandertal DNA sequences and the origin of modern humans. *Cell* 90: 19–30.

Kumar, S., A. Filipski, et al. 2005. Placing confidence limits on the molecular age of the human-chimpanzee divergence. *Proceedings of the National Academy of Sciences USA* 102: 18842–18847.

Leakey, M. G., C. S. Feibel, et al. 1995. New 4-million-year-old hominid species from Kanapoi and Allia Bay, Kenya. *Nature* 376: 565–571.

Leakey, M. G., F. Spoor, et al. 2001. New hominin genus from eastern Africa shows diverse middle Pliocene lineages. *Nature* 410: 433–440.

Li, T. Y., and D. A. Etler. 1992. New middle pleistocene hominid crania from Yunxian in China. *Nature* 357: 404–407.

Lieberman, D. E. 1995. Testing hypotheses about recent human evolution from skulls: Integrating morphology, function, development, and phylogeny. *Current Anthropology* 36: 159–197.

Lieberman, D. E. 2001. Another face in our family tree. *Nature* 410: 419–420.

Lockwood, C. A., W. H. Kimbel, and J. M. Lynch. 2004. Morphometrics and hominoid phylogeny: Support for a chimpanzee-human clade and differentiation among great ape subspecies. *Proceedings of the National Academy of Sciences USA* 101: 4356–4360.

Lovejoy, C. O., K. G. Heiple, and R. S. Meindl. 2001. Did our ancestors knuckle-walk? *Nature* 410: 325–326.

Lowenstein, J., and A. Zihlman. 1988. The invisible ape. *New Scientist* 3 (December): 56–59.

Maldonado, J. 2004. Dover schools still debating biology text. *York Daily Record*: 9 June.

Manson, J. H., S. Perry, and A. R. Parish. 1997. Nonconceptive sexual behavior in bonobos and capuchins. *International Journal of Primatology* 8: 767–786.

Marks, J. 1993. Hominoid heterochromatin: Terminal C-bands as a complex genetic trait linking chimpanzee and gorilla. *American Journal of Physical Anthropology* 90: 237–246.

Marks, J. 1994. Blood will tell (won't it?): A century of molecular discourse in anthropological systematics. *American Journal of Physical Anthropology* 94: 59–79.

McDougall, I., F. H. Brown, and J. G. Fleagle. 2005. Stratigraphic placement and age of modern humans from Kibish, Ethiopia. *Nature* 433: 733–736.

McGrew, W. C., M. W. Hamrick, et al. 1995. Thumbs, tools, and early humans. *Science* 268: 586–589.

McHenry, H. M. 1992. Body size and proportions in early hominids. *American Journal of Physical Antropology* 87: 407–430.

McHenry, H. 1994. Tempo and mode in human evolution. *Proceedings of the National Academy of Sciences USA* 91: 6780–6784.

Mercader, J., M. Panger, and C. Boesch. 2002. Excavation of a chimpanzee stone tool site in the African rainforest. *Science* 296: 1452–1455.

Mikkelsen, T. S., L. W. Hillier, et al. 2005. Initial sequence of the chimpanzee genome and comparison with the human genome. *Nature* 437: 69–87.

Nei, M. 1995. Genetic support for the out-of-Africa theory of human evolution. *Proceedings of the National Academy of Sciences USA* 92: 6720–6722.

Nielsen, R., C. Bustamante, et al. 2005. A scan for positively selected genes in the genomes of humans and chimpanzees. *PLoS Biology* 3: e170.

Noble, W., and I. Davidson. 1991. The evolutionary emergence of modern human behavior: Language and its archaeology. *Man* 26: 223–253.

Nordborg, M. 1998. On the probability of Neanderthal ancestry. *American Journal of Human Genetics* 63: 1237–1240.

Nuttall, G. H. F. 1904. *Blood, Immunity, and Blood Relationship: A Demonstration of Certain Blood-Relationships amongst Animals by Means of the Precipitin Test for Blood.* London: Cambridge University Press.

O'hUigin, C., Y. Satta, et al. 2002. Contribution of homoplasy and of ancestral polymorphism to the evolution of genes in anthropoid primates. *Molecular Biology and Evolution* 19: 1501–1513.

Ovchinnikov, I. V., A. Götherström, et al. 2000. Molecular analysis of Neanderthal DNA from the northern Caucasus. *Nature* 404: 490–493.

Pääbo, S. 2003. The mosaic that is our genome. *Nature* 421: 409–412.

Parish, A. R. 1994. Sex and food control in the "uncommon chimpanzee": How bonobo females overcame a phylogenetic legacy of male dominance. *Ethology and Sociobiology* 15: 157–179.

Parish, A. R. 1996. Female relationships in bonobos (*Pan paniscus*): Evidence for bonding, cooperation, and female dominance in a male-philopatric species. *Human Nature* 7: 61–96.

Parish, A. R., and F. B. M. de Waal. 2000. The other "closest living relative": How bonobos (*Pan paniscus*) challenge traditional assumptions about females, dominance, intra- and intersexual interactions, and Hominid evolution. *Annals of the New York Academy of Sciences* 907: 97–113.

Patterson, N., D. J. Richter, et al. 2006. Genetic evidence for complex speciation of humans and chimpanzees. *Nature* 441: 1103–1108.

Pilbeam, D., and S. J. Gould. 1974. Size and scaling in human evolution. *Science* 186: 892–901.

Pinker, S. 1994. *The Language Instinct.* New York: Harper Collins.

Popesco, M. C., E. J. MacLaren, et al. 2006. Human lineage-specific amplification, selection, and neuronal expression of DUF1220 domains. *Science* 313: 1304–1307.

Pritchard, J. K., and M. W. Feldman. 1996. Genetic data and the African origin of humans. *Science* 274: 1548.

Rauum, R. L., K. N. Sterner, et al. 2005. Catarrhine primate divergence dates estimated from complete mitochondrial genomes: concordance with fossil and nuclear DNA evidence. *Journal of Human Evolution* 48: 237–257.

Redon, R., S. Ishikawa, et al. 2006. Global variation in copy number in the human genome. *Nature* 444: 444–454.

Reno, P. L., R. S. Meindl, et al. 2003. Sexual dimorphism in *Australopithecus afarensis* was similar to that of modern humans. *Proceedings of the National Academy of Sciences USA* 100: 9404–9409.

Richmond, B. G., and D. S. Strait. 2000. Evidence that humans evolved from a knuckle-walking ancestor. *Nature* 404: 382–385.

Richmond, B. G., and D. S. Strait. 2001a. Did our ancestors knuckle-walk? *Nature* 410: 326.

Richmond, B. G., and D. S. Strait. 2001b. Knuckle-walking hominid ancestor: A reply to Corruccini and McHenry. *Journal of Human Evolution* 40: 513–520.

Risch, N., K. K. Kidd, and S. A. Tishkoff. 1996. Genetic data and the African origin of humans—reply. *Science* 274: 1548–1549.

Rogers, J., and A. G. Comuzzie. 1995. When is ancient polymorphism a potential problem for molecular phylogenetics? *American Journal of Physical Anthropology* 98: 216–218.

Rosenberg, N. A., J. K. Pritchard, et al. 2002. Genetic structure of human populations. *Science* 298: 2381–2385.

Ruvolo, M. 1994. Molecular evolutionary processes and conflicting gene trees: The Hominoid case. *American Journal of Physical Anthropology* 94: 89–113.

Ruvolo, M. 1995. Seeing the forest and the trees: Replies to Marks; Rogers and Commuzzie; Green and Djian. *American Journal of Physical Anthropology* 98: 218–232.

Ruvolo, M. 1997. Molecular phylogeny of the hominoids: Inferences from multiple independent DNA sequence data sets. *Molecular Biology and Evolution* 14: 248–265.

Ruvolo, M., D. Pan, et al. 1994. Gene trees and hominoid phylogeny. *Proceedings of the National Academy of Sciences USA* 91: 8900–8904.

Salem, A.-H., D. A. Ray, et al. 2003. Alu elements and hominid phylogenetics. *Proceedings of the National Academy of Sciences USA* 100: 12787–12791.

Sarich, V. M., and A. C. Wilson. 1967. Immunological time scale for Hominid evolution. *Science* 158: 1200–1203.

Satta, Y., J. Klein, and N. Takahata. 2000. DNA archives and our nearest relative: The trichotomy revisited. *Molecular Phylogenetics and Evolution* 14: 259–275.

Schick, K. D., and N. P. Toth NP. 1993. *Making Silent Stones Speak: Human Evolution and the Dawn of Technology.* New York: Simon & Schuster.

Semaw, S., P. Renne, et al. 1997. 2.5-million-year-old stone tools from Gona, Ethiopia. *Nature* 385: 333–336.

Senut, B. M. Pickford, et al. 2001. First hominid from the Miocene (Lukeino Formation, Kenya). *Comptes Rendus de l'Academie des Sciences Serie II Fascicule A-Sciences de la Terre et des Planetes* 332: 137–144.

Shoshani, J., C. P. Groves, et al. 1996. Primate phylogeny: Morphological vs. molecular results. *Molecular Phylogenetics and Evolution* 5: 101–153.

Sikela, J. M. 2006. The jewels of our genome: The search for the genomic changes underlying the evolutionarily unique capacities of the human brain. *PLoS Genetics* 2: e80.

Stauffer, R. L., A. Walker, et al. 2001. Human and ape molecular clocks and constraints on paleontological hypotheses. *Journal of Heredity* 92: 469–474.

Stern, J. T., and R. L. Susman. 1983. The locomotor anatomy of *Australopithecus afarensis. American Journal of Physical Anthropology* 60: 279–317.

Strait, D. S., F. E. Grine, and M. A. Moniz. 1997. A reappraisal of early hominid phylogeny. *Journal of Human Evolution* 32: 17–82.

Strait, D. S., and F. E. Grine. 2004. Inferring hominoid and early hominid phylogeny using craniodental characters: the role of fossil taxa. *Journal of Human Evolution* 47: 399–452.

Stringer, C. 1988. The dates of Eden. *Nature* 331: 565–566.

Susman, R. L. 1994. Fossil evidence for early hominid tool use. *Science* 265: 1570–1573.

Swisher, C. C., III, G. H. Curtis, et al. 1994. Age of the earliest known hominids in Java, Indonesia. *Science* 263: 1118–1121.

Swisher, C. C., III, W. J. Rink, et al. 1996. Latest *Homo erectus* of Java: Potential contemporaneity with *Homo sapiens* in Southeast Asia. *Science* 274: 1870–1874.

Tattersall, I. 1986. Species recognition in human paleontology. *Journal of Human Evolution* 15: 165–175.

Tattersall, I. 1992. Species concepts and species identification in human evolution. *Journal of Human Evolution* 22: 341–349.

Tattersall, I. 1995. *The Fossil Trail.* Oxford: Oxford University Press.

Tattersall, I. 1997. Out of Africa again . . . and again? *Scientific American* 276 (April): 60–67.

Tattersall, I. 2000. Once we were not alone. *Scientific American* 282 (January): 56–62.

Tattersall, I., and J. H. Schwartz. 1999. Hominids and hybrids: The place of Neanderthals in human evolution. *Proceedings of the National Academy of Sciences USA* 96: 7117–7119.

Thorne, A. G., and M. H. Wolpoff. 1981. Regional continuity in Australasian pleistocene hominid evolution. *American Journal of Physical Anthropology* 55: 337–349.

Tishkoff, S. A., E. Dietzsch, et al. 1996b. Global patterns of linkage disequilibrium at the CD4 locus and modern human origins. *Science* 271: 1380–1387.

Tishkoff, S. A., K. K. Kidd, and N. Risch. 1996a. Interpretations of multiregional evolution—reply. *Science* 274: 706–707.

Tobias, P. V. 1987. The brain of *Homo habilis*: A new level of organization in cerebral evolution. *Journal of Human Evolution* 16: 741–761.

Tobias, P. V. 2003. Encore Olduvai. *Science* 299: 1193–1194.

Trinkaus, E., C. B. Ruff, et al. 1998. Locomotion and body proportions of the Saint-Césaire 1 Châtelperronian Neandertal. *Proceedings of the National Academy of Sciences USA*. 95: 5836–5840.

Valladas, H., J. L. Reyss, et al. 1988. Thermoluminescence dating of Mousterian "Proto-Cro-Magnon" remains from Israel and the origin of modern man. *Nature* 331: 614–616.

van Schaik, C. P., M. Ancrenaz, et al. 2003. Orangutan cultures and the evolution of material culture. *Science* 299: 102–105.

Vigilant, L., M. Stoneking, et al. 1991. African populations and the evolution of human mitochondrial DNA. *Science* 253: 1503–1507.

Vogel, G. 2002. Can chimps ape ancient hominid toolmakers? *Science* 296: 1380.

Waddle, D. M. 1994. Matrix correlation tests support a single origin for modern humans. *Nature* 368: 452–454.

Wang, X., W. E. Grus, and J. Zhang. 2006. Gene losses during human origins. *PLoS Biology* 4: e52.

Ward, S. C., and W. H. Kimbel. 1983. Subnasal alveolar morphology and the systematic position of *Sivapithecus*. *American Journal of Physical Anthropology* 61: 157–171.

White, T. 2003. Early hominids—diversity or distortion? *Science* 299: 1994–1997.

White, T. D., B. Asfaw, et al. 2003. Pleistocene *Homo sapiens* from Middle Awash, Ethiopia. *Nature* 423: 742–747.

White, T. D., and G. Suwa. 1987. Hominid footprints at Laetoli: Facts and interpretations. *American Journal of Physical Anthropology* 72: 485–514.

White, T. D., G. Suwa, and B. Asfaw. 1994. *Australopithecus ramidus*, a new species of early hominid from Aramis, Ethiopia. *Nature* 371: 306–312.

White, T. D., G. Suwa, and B. Asfaw. 1995. Corrigendum: *Australopithecus ramidus*, a new species of early hominid from Aramis, Ethiopia. *Nature* 375: 88.

Whiten, A., J. Goodall, et al. 1999. Cultures in chimpanzees. *Nature* 399: 682–685.

Whiten, A. 2005. The second inheritance system of chimpanzees and humans. Nature 437: 52–55.

Wildman, D. E., M. Uddin, et al. 2003. Implications of natural selection in shaping 99.4% nonsynonymous DNA identity between humans and chimpanzees: Enlarging genus *Homo*. *Proceedings of the National Academy of Sciences USA* 100: 7181–7188.

Wilford, J. N. 2005. Fossils of apelike creature still stir lineage debate. *The New York Times*: 12 April.

Wilkins, W. K., and J. Wakefield. 1995. Brain evolution and neurolinguistic preconditions. *Behavioral and Brain Sciences* 18: 161–226.

Wolpoff, M. H. 1996. Interpretations of multiregional evolution. *Science* 274: 704–706.

Wolpoff, M., B. Senut, et al. 2002. *Sahelanthropus* or '*Sahelpithecus*'? *Nature* 419: 581–582.

Wolpoff, M. H., J. Hawks, et al. 2006. An ape or the ape: Is the Toumaï cranium TM 266 a hominid? *PaleoAnthropology* 2006: 36–50.

Wood, B. 1992. Origin and evolution of the genus *Homo*. *Nature* 355: 783–790.

Wood, B. 1997. The oldest whodunnit in the world. *Nature* 385: 292–293.

Wood, B. 2002. Hominid revelations from Chad. *Nature* 418: 133–135.

Wood, B., and M. Collard. 1999. The human genus. *Science* 284: 65–71.

Wood, B., and A. Turner. 1995. Out of Africa and into Asia. *Nature* 378: 239–240.

Wrangham, R. W. 1999. Evolution of coalitionary killing. *Yearbook of Physical Anthropology* 42: 1–30.

Zollikofer, C. P. E., M. S. Ponce de León, et al. 2005. Virtual cranial reconstruction of *Sahelanthropus tchadensis*. *Nature* 434: 755–759.

Glossary

adaptation A trait that increases the ability of an individual to survive or reproduce compared with individuals without the trait.
adaptive radiation The divergence of a clade into populations adapted to many different ecological niches.
adaptive trait A trait that increases the fitness of its bearer.
additive effect The contribution an allele makes to the phenotype that is independent of the identity of the other alleles at the same or different loci.
additive genetic variation Differences among individuals in a population that are due to the additive effects of genes.
agent of selection Any factor that causes individuals with certain phenotypes to have, on average, higher fitness than individuals with other phenotypes.
alleles Variant forms of a gene, or variant nucleotide sequences at a particular locus.
allopatric model The hypothesis that speciation occurs when populations become geographically isolated and diverge because selection and drift act on them independently.
allopatry Living in different geographic areas.
allozymes Distinct forms of an enzyme, encoded by different alleles at the same locus.
altruism Behavior that decreases the fitness of the actor and increases the fitness of the recipient.
anagenesis Descent with modification but no speciation.
ancestral Describes a trait that was possessed by the common ancestor of the species on a branch of an evolutionary tree; used in contrast with *derived*.
antibiotic A chemical, typically extracted from a microorganism, that kills bacteria by disrupting a particular biochemical process.
antigenic site A portion of a protein that is recognized by the immune system and initiates a response.
assortative mating Occurs when individuals tend to mate with other individuals with the same genotype or phenotype.
average excess The average excess for allele *a* is the difference between the mean fitness of individuals carrying allele *a* and the mean fitness of the entire population. If the average excess for an allele is positive, then the allele will rise in frequency.

back mutation A mutation that reverses the effect of a previous mutation; typically a mutation that restores function after a loss-of-function mutation.
background extinction Extinctions that are not part of mass extinction events; thought to be due to typical types and rates of environmental change or species interactions as opposed to the extraordinary environmental changes that occur during mass extinctions.
Bateman gradient The slope of the best-fit line relating reproductive success to mating success. Measures the strength of sexual selection.
Bayesian Markov Chain Monte Carlo An approach to phylogeny inference based on computing the probability that a particular tree is correct, given a specific model of evolution for the characters being analyzed and the data observed.
best-fit line The line that most accurately represents the trend of the data in a scatterplot; typically best-fit lines are calculated by least-squares linear regression.
blending inheritance The hypothesis that heritable factors blend to produce a phenotype and are passed on to offspring in this blended form.
bootstrapping In phylogeny reconstruction, a technique for estimating the strength of the evidence that a particular node in a tree exists. Bootstrap values range from 0% to 100%, with higher values indicating stronger support.
bottleneck A large-scale but short-term reduction in population size followed by an increase in population size.
branch (of a phylogenetic tree) Lines that indicate a specific population or taxonomic group through time.
broad-sense heritability That fraction of the total phenotypic variation in a population that is caused by genetic differences among individuals.

catastrophism In geology, the view that most or all landforms are the product of catastrophic events, such as the flood at the time of Noah described in the Bible. See *uniformitarianism*.
cenancestor The last common ancestor of all extant organisms.
chromosome inversion A region of DNA that has been flipped, so that the genes are in reverse order; results in lower rates of crossing-over and thus tighter linkage among loci within the inversion.
clade The set of species descended from a particular common ancestor; synonymous with *monophyletic group*.
cladistics A classification scheme based on the historical sequence of divergence events (phylogeny); also used to identify a method of inferring phylogenies based on the presence of shared derived characters (synapomorphies).
cladogram An evolutionary tree reflecting the results of a cladistic analysis.
cline A systematic change along a geographic transect in the frequency of a genotype or phenotype.
clone An individual that is genetically identical to its parent, or a group of individuals that are genetically identical to each other.
codon A set of three bases in DNA that specifies a particular amino acid–carrying tRNA.
codon bias A nonrandom distribution of codons in a DNA sequence.
coefficient of inbreeding (F) The probability that the alleles at any particular locus in the same individual are identical by descent from a common ancestor.
coefficient of linkage disequilibrium (D) A calculated value that quantifies the degree to which genotypes at one locus are nonrandomly associated with genotypes at another locus.

coefficient of relatedness (*r*) The probability that the alleles at any particular locus in two different individuals are identical by descent from a common ancestor.
coevolution That which occurs when interactions between species over time lead to reciprocal adaptation.
common garden experiment An experiment in which individuals from different populations or treatments are reared together under identical conditions.
comparative method A research program that compares traits and environments across taxa and looks for correlations that test hypotheses about adaptation.
complementary base pairs Nucleotides that match up and form hydrogen bonds on opposite strands of a DNA molecule or DNA-RNA duplex. C complements G; A complements T or U.
confidence interval An indication of the statistical certainty of an estimate; if a study yielding an estimate is done repeatedly, and a 95% confidence interval is calculated for each estimate, the confidence interval will include the true value 95% of the time.
conjugation In bacterial genetics, the transfer of one or more genes from one cell to another via a plasmid that travels through a conjugation tube.
constraint Any factor that tends to slow the rate of adaptive evolution or prevent a population from evolving the optimal value of a trait.
control group A reference group that provides a basis for comparison; in an experiment, the control group is exposed to all conditions affecting the experimental group except one—the potential causative agent of interest.
convergent evolution Similarity between species that is caused by a similar, but evolutionarily independent, response to a common environmental problem.
cryptic species Species that are indistinguishable morphologically, but divergent in songs, calls, odor, or other traits.

Darwinian fitness The extent to which an individual contributes genes to future generations, or an individual's score on a measure of performance expected to correlate with genetic contribution to future generations (such as lifetime reproductive success).
derived Describes a trait that was not possessed by the common ancestor of the species on a branch of an evolutionary tree; an evolutionary novelty; used in contrast with *ancestral.*
deuterostome A lineage of animals that share a pattern of development, including radial cleavage and formation of the anus before the mouth. Includes echinoderms and chordates.
differential success A difference between the average survival, fecundity, or number of matings achieved by individuals with certain phenotypes versus individuals with other phenotypes.
dioecious Describes a species in which male and female reproductive function occurs in separate individuals; usually used with plants.
diploblast An animal that develops from two basic embryonic cell layers (endoderm and ectoderm).
direct fitness Fitness that is due to the production of offspring. See *indirect fitness.*
direct sequencing A research program aimed at understanding which species are present in a particular environment, based on sequencing one or more genes directly from an environmental sample and using the data to place the organisms present on a phylogenetic tree. In most cases, the organisms that are identified have never been seen. Synonymous with environmental sequencing.
directional selection That which occurs when individual fitness tends to increase or decrease with the value of phenotypic trait; can result in steady evolutionary change in the mean value of the trait in the population.
disruptive selection Occurs when individuals with more extreme values of a trait have higher fitness; can result in increased phenotypic variation in a population.
dominance genetic variation Differences among individuals in a population that are due to the nonadditive effects of genes, such as dominance; typically means the genetic variation left over after the additive genetic variation has been taken into account.
drift Synonym for *genetic drift.*
dynamic stasis Lack of morphological change over a long interval of evolution, despite many short-term changes during the same interval. No or little net evolution.

ecdysozoan An lineage of protostome animals distinguished by the presence of molting.
effective population size (N_e) The size of an ideal random mating population (with no selection, mutation, or migration) that would lose genetic variation via drift at the same rate as is observed in an actual population.
endosymbiosis A relationship where one organism lives inside the body or within the cells of another organism.
environmental genomics A research program aimed at understanding which genes are present in a particular environment, based on sequencing the entire genomes present. In most cases, the genes studied come from organisms that have never been identified or seen.
environmental sequencing A research program aimed at understanding which species are present in a particular environment, based on sequencing one or more genes directly from an environmental sample and using the data to place the organisms present on a phylogenetic tree. In most cases, the organisms that are identified have never been seen. Synonymous with direct sequencing.
environmental variation Differences among individuals in a population that are due to differences in the environments they have experienced.
epitope The specific part of a protein that is recognized by the immune system and initiates a response. Synonymous with *antigenic site* and *antigenic determinant.*
eugenics The study and practice of social control over the evolution of human populations; positive eugenics seeks to increase the frequency of desirable traits, whereas negative eugenics seeks to decrease the frequency of undesirable traits.
eusocial Describes a social system characterized by overlapping generations, cooperative brood care, and specialized reproductive and nonreproductive castes.
eusociality A social system characterized by overlapping generations, cooperative brood care, and specialized reproductive and nonreproductive castes.
evo-devo The study of how changes in genes that affect embryonic development could lead to important evolutionary changes; short for "evolution and development."
evolution Originally defined as descent with modification, or change in the characteristics of populations over time. Currently defined as changes in allele frequencies over time.
evolutionarily stable strategy (ESS) In game theory, a strategy or set of strategies that cannot be invaded by a new, alternative strategy.
evolutionary arms race That which occurs when an adaptation in one species (a parasite, for example) reduces the fitness of

individuals in a second species (such as a host), thereby selecting in favor of counter-adaptations in the second species. These counter-adaptations, in turn, select in favor of new adaptations in the first species, and so on.

evolutionary tree A diagram (typically an estimate) of the relationships of ancestry and descent among a group of species or populations; in paleontological studies the ancestors may be known from fossils, whereas in studies of extant species the ancestors may be hypothetical contructs. Also called a *phylogenetic tree*.

exon A nucleotide sequence that occurs between introns and that remains in the messenger RNA after the introns have been spliced out.

extant Living today.

extended haplotype homozygosity (EHH) A measure of the linkage disequilibrium between an allele at a locus of interest and alleles at other loci on the same chromosome. Allele a's EHH to a particular distance x is the probability that two randomly chosen chromosomes carrying a will also have the same genotype at all marker loci between a and x.

fecundity The number of gametes produced by an individual; usually used in reference to the number of eggs produced by a female.

fitness The extent to which an individual contributes genes to future generations, or an individual's score on a measure of performance expected to correlate with genetic contribution to future generations (such as lifetime reproductive success).

fixation The elimination from a population of all the alleles at a locus but one; the one remaining allele, now at a frequency of 1.0, is said to have achieved fixation, or to be fixed.

fossil Any trace of an organism that lived in the past.

fossil record The complete collection of fossils, located in many institutions around the world.

founder effect A change in allele frequencies that occurs after a founder event, due to genetic drift in the form of sampling error in drawing founders from the source population.

founder event The establishment of a new population, usually by a small number of individuals.

founder hypothesis The hypothesis that many speciation events begin when small populations colonize new geographic areas.

frameshift mutation An insertion or deletion in a coding region of a gene in which the length of the inserted or deleted sequence is not a multiple of three; causes the codons downstream of the mutation to be translated in the wrong reading frame.

frequency The proportional representation of a phenotype, genotype, gamete, or allele in a population; if 6 out of 10 individuals have brown eyes, the frequency of brown eyes is 60%, or 0.6.

frequency-dependent selection Occurs when an individual's fitness depends on the frequency of its phenotype in the population; typically occurs when a phenotype has higher fitness when it is rare and lower fitness when it is common.

functional genomics The study of gene function—understanding the role of gene products in the organism and how the timing and quantity of gene expression is controlled.

gamete pool The set of all copies of all gamete genotypes in a population that could potentially be contributed by the members of one generation to the members of the next generation.

gene duplication Generation of an extra copy of a locus, usually via unequal crossing-over.

gene family A group of loci related by common descent and sharing identical or similar function.

gene flow The movement of alleles from one population to another population, typically via the movement of individuals or via the transport of gametes by wind, water, or pollinators.

gene pool The set of all copies of all alleles in a population that could potentially be contributed by the members of one generation to the members of the next generation.

genetic distance A statistic that summarizes the number of genetic differences observed between populations or species.

genetic drift Change in the frequencies of alleles in a population resulting from sampling error in drawing gametes from the gene pool to make zygotes and from chance variation in the survival and/or reproductive success of individuals; results in nonadaptive evolution.

genetic load Reduction in the mean fitness of a population due to the presence of deleterious alleles.

genetic recombination The placement of allele copies into multilocus genotypes (on chromosomes or within gametes) that are different from the multilocus genotypes they belonged to in the previous generation; results from meiosis with crossing-over and sexual reproduction with outcrossing.

genetic variation Differences among individuals in a population that are due to differences in genotype.

genotype-by-environment interaction Differences in the effect of the environment on the phenotype displayed by different genotypes; for example, among people living in the same location some change their skin color with the seasons and others do not.

geologic column A composite, older-to-younger sequence of rock formations that describes geological events at a particular locality.

geologic time scale A sequence of eons, eras, periods, epochs, and stages that furnishes a chronology of Earth history.

h^2 Symbol for the narrow-sense heritability (see *heritability*).

habitat tracking Morphological evolution in response to short-term environmental change that in the long term results in variation around a mean value. Also called *dynamic stasis*.

half-life The time required for half of the atoms of a radioactive material, present at any time, to decay into a daughter isotope.

Hamilton's rule An inequality that predicts when alleles for altruism should increase in frequency.

haplodiploidy A reproductive system in which males are haploid and develop from unfertilized eggs, while females are diploid and develop from fertilized eggs.

haplotype Genotype for a suite of linked loci on a chromosome; typically used for mitochondrial genotypes, because mitochondria are haploid and all loci are linked.

Hardy–Weinberg equilibrium A situation in which allele and genotype frequencies in an ideal population do not change from one generation to the next, because the population experiences no selection, no mutation, no migration, no genetic drift, and random mating.

heritability In the broad sense, that fraction of the total phenotypic variation in a population that is caused by genetic differences among individuals; in the narrow sense, that fraction of the total variation that is due to the additive effects of genes.

hermaphroditic In general, describes a species in which male and female reproductive function occur in the same individual; with plants, describes a species with perfect flowers (that is, flowers with both male and female reproductive function).

heterozygosity That fraction of the individuals in a population that are heterozygotes.
heterozygote inferiority (underdominance) Describes a situation in which heterozygotes at a particular locus tend to have lower fitness than homozygotes.
heterozygote superiority (overdominance) Describes a situation in which heterozygotes at a particular locus tend to have higher fitness than homozygotes.
histogram A bar chart that represents the variation among individuals in a sample; each bar represents the number of individuals, or the frequency of individuals, with a particular value (or within a particular range of values) for the measurement in question.
hitchhiking Change in the frequency of an allele due to selection on a closely linked locus. Also called a *selective sweep*.
homeotic loci Genes whose products provide positional information in a multicellular embryo.
homology Classically defined as curious structural similarity between species despite differences in function. Today defined as similarity between species that results from inheritance of traits from a common ancestor.
homoplasy Similarity in the characters found in different species that is due to convergent evolution, parallelism, or reversal—not common descent.
horizontal gene transfer The movement of genetic material across species barriers.
hybrid zone A geographic region where differentiated populations interbreed.

identical by descent Describes alleles, within a single individual or different individuals, that have been inherited from the same ancestral copy of the allele.
inbreeding Mating among kin.
inbreeding depression Reduced fitness in individuals or populations resulting from kin matings; often due to the decrease in heterozygosity associated with kin matings, either because heterozygotes are superior or because homozygotes for deleterious alleles become more common.
inclusive fitness An individual's total fitness; the sum of its indirect fitness, due to reproduction by relatives made possible by its actions, and direct fitness, due to its own reproduction.
indel A type of mutation based on the insertion or deletion of one or more deoxyribonucleotides (bases).
independence (statistical) Lack of association among data points, such that the value of a data point does not affect the value of any other data point.
indirect fitness Fitness that is due to increased reproduction by relatives made possible by the focal individual's actions. See *direct fitness*.
inheritance of acquired characters The hypothesis that phenotypic changes in the parental generation can be passed on, intact, to the next generation.
interaction In genetics, occurs when the effect of an allele on the phenotype depends on the other alleles present at the same or different loci; in statistics, occurs when the effect of a treatment depends on the value of other treatments.
intersexual selection Differential mating success among individuals of one sex due to interactions with members of the other sex; for example, variation in mating success among males due to female choosiness.
intrasexual selection Differential mating success among individuals of one sex due to interactions with members of the same sex; for example, differences in mating success among males due to male–male competition over access to females.
intron (intervening sequence) A noncoding stretch of DNA nucleotides that occurs between the coding regions of a gene and that must be spliced out after transcription to produce a functional messenger RNA.
iteroparous Describes a species or population in which individuals experience more than one bout of reproduction over the course of typical lifetime; humans provide an example.

kin recognition The ability to discern the degree of genetic relatedness of other individuals.
kin selection Natural selection based on indirect fitness gains.

lateral gene transfer Transfer of genetic material across species barriers.
law of succession The observation that fossil types are succeeded, in the same geographic area, by similar fossil or living species.
life history An individual's pattern of allocation, throughout life, of time and energy to various fundamental activities, such as growth, repair of cell and tissue damage, and reproduction.
likelihood The probability of a particular outcome given a model of the process that produced it. For example we might calculate the probability that a pair of parents will have an offspring with a particular multilocus genotype given a model specifying how closely the loci in question are linked. Or we might calculate the probability of getting a particular set of sequences given a phylogeny of the species from which we sampled them.
likelihood ratio Literally, the ratio of two likelihoods. Typically, the probability of a particular outcome given a model we are evaluating divided by the probability of the same outcome under a null model.
lineage A group of ancestral and descendant populations or species that are descended from a common ancestor. Synonymous with *clade*.
linkage The tendency for alleles at different loci on a chromosome to be inherited together. Also called *genetic linkage*.
linkage (dis)equilibrium If, within a population, genotypes at one locus are randomly distributed with respect to genotypes at another locus, then the population is in linkage equilibrium for the two loci; otherwise, the population is in linkage disequilibrium.
LOD Literally, logarithm of the odds. The logarithm of a likelihood ratio.
lophotrochozoan A lineage of protostome animals, many of which have a feeding structure called a lophophore.
loss-of-function mutation A mutation that incapacitates a gene so that no functional product is produced; also called a *forward, knock-out*, or *null mutation*.

macroevolution Large evolutionary change, usually in morphology; typically refers to the evolution of differences among populations that would warrant their placement in different genera or higher-level taxa.
mass extinction A large-scale, sudden extinction event that is geographically and taxonomically widespread.
maternal effect Variation among individuals due to variation in nongenetic influences exerted by their mothers; for example, chicks whose mothers feed them more may grow to larger sizes,

and thus be able to feed their own chicks more, even when size is not heritable.

maximum likelihood In phylogeny inference, a method for choosing a preferred tree among many possible trees. In using maximum likelihood, a researcher asks how likely a particular tree is given a particular data set and a specific model of character change.

Mendelian gene A locus whose alleles obey Mendel's laws of segregation and independent assortment.

methodological naturalism The convention, adopted by scientists, that natural phenomena are to be explained by natural causes. Compare to *ontological naturalism*

microevolution Changes in gene frequencies and trait distributions that occur within populations and species.

microtektites Tiny glass particles created when minerals are melted by the heat generated in a meteorite or asteroid impact.

midoffspring value The mean phenotype of the offspring within a family.

midparent value The mean phenotype of an individual's two parents.

migration In evolution, the movement of alleles from one population to another, typically via the movement of individuals or via the transport of gametes by wind, water, or pollinators.

Modern Synthesis The broad-based effort, accomplished during the 1930s and 1940s, to unite Mendelian genetics with the theory of evolution by natural selection; also called the *Evolutionary Synthesis*.

molecular clock The hypothesis that base substitutions accumulate in populations in a clock-like fashion; that is, as a linear function of time.

monoecious Typically used for plants, to describe either: (1) a species in which male and female reproductive function occur in the same individual; or (2) a species in which separate male and female flowers are present on the same individual (see also *hermaphroditic*).

monophyletic group The set of species (or populations) descended from a common ancestor.

morphology Structural form, or physical phenotype; also the study of structural form.

morphospecies Populations that are designated as separate species based on morphological differences.

mutation-selection balance Describes an equilibrium in the frequency of an allele that occurs because new copies of the allele are created by mutation at exactly the same rate that old copies of the allele are eliminated by natural selection.

mutualism An interaction between two individuals, typically of different species, in which both individuals benefit.

narrow-sense heritability That fraction of the total phenotypic variation in a population that is due to the additive effects of genes.

natural selection A difference, on average, between the survival or fecundity of individuals with certain phenoypes compared with individuals with other phenotypes.

negative selection Selection against deleterious mutations. Also called *purifying selection*.

neutral (mutation) A mutation that has no effect on the fitness of the bearer.

neutral evolution (neutral theory) A theory that models the rate of fixation of alleles with no effect on fitness; also associated with the claim that the vast majority of observed base substitutions are neutral with respect to fitness.

node A point on an evolutionary tree at which a branch splits into two or more subbranches.

nonsynonymous substitution A DNA substitution that changes the amino acid sequence specified by a gene.

null hypothesis The predicted outcome, under the simplest possible assumptions, of an experiment or observation; in a test of whether populations are different, the null hypothesis is typically that they are not different and that apparent differences are due to chance.

null model The set of simple and explicit assumptions that allows a researcher to state a null hypothesis.

ontological naturalism The philosophical position that the natural world is all that exists. Compare to *methodological naturalism*.

outbreeding Mating among unrelated individuals.

orthologous Genes that diverged after a speciation event; describes the relationship among homologous genes found in different species.

outgroup A taxonomic group that diverged prior to the rest of the taxa in a phylogenetic analysis.

overdominance Describes a situation in which heterozygotes at a particular locus tend to have higher fitness than homozygotes.

***P* value** An estimate of the statistical support for a claim about a pattern in data, with smaller values indicating stronger support; an estimate of the probability that apparent violations of the null hypothesis are due to chance (see *statistically significant*).

paleontology The study of fossil organisms.

paralogous Duplicated genes found in the same genome; describes the relationship among members of the same gene family. A type of genetic homology.

paraphyletic group A set of species that includes a common ancestor and some, but not all, of its descendants.

parental investment Expenditure of time and energy on the provision, protection, and care of an offspring; more specifically, investment by a parent that increases the fitness of a particular offspring and reduces the fitness the parent can gain by investing in other offspring.

parsimony A criterion for selecting among alternative patterns or explanations based on minimizing the total amount of change or complexity.

parthenogenesis A reproductive mode in which offspring develop from unfertilized eggs.

phenetics, phenetic approach A classification scheme based on grouping populations according to their similarities.

phenotypic plasticity Variation, under environmental influence, in the phenotype associated with a genotype.

phenotypic variation The total variation among the individuals in a population.

phyletic transformation The evolution of a new morphospecies by the gradual transformation of an ancestral species, without a speciation or splitting event taking place. Also called *anagenesis*.

phylogeny The evolutionary history of a group.

phylogenetic tree A diagram (typically an estimate) of the relationships of ancestry and descent among a group of species or populations; in paleontological studies the ancestors may be known from fossils, whereas in studies of extant species the ancestors may be hypothetical constructs. Also called an *evolutionary tree*.

phylogenomics The use of data from genome sequencing to answer questions about evolution.

phylogeography The use of evolutionary trees in answering questions about the geographic distribution of organisms.
plasmids Small loops of DNA that can replicate themselves; common in bacteria and observed in a small number of eukaryotes.
point mutation Alteration of a single base in a DNA sequence.
polyandry A mating system in which at least some females mate with more than one male.
polygyny A mating system in which at least some males mate with more than one female.
polymorphic Describes a population, locus, or trait for which there is more than one phenotype or allele; variable.
polymorphism The existence within a population of more than one variant for a phenotypic trait, or of more than one allele.
polyphyletic group A set of species that are grouped by similarity, but not descended from a common ancestor.
polyploid Having more than two haploid sets of chromosomes.
polytomy A node, or branch point, on a phylogeny with more than two descendent lineages emerging.
population For sexual species, a group of interbreeding individuals and their offspring; for asexual species, a group of individuals living in the same area.
population genetics The branch of evolutionary biology responsible for investigating processes that cause changes in allele and genotype frequencies in populations.
positive selection Selection in favor of advantageous mutations.
postzygotic isolation Reproductive isolation between populations caused by dysfunctional development or sterility in hybrid forms.
preadaptation A trait that changes due to natural selection and acquires a new function.
prezygotic isolation Reproductive isolation between populations caused by differences in mate choice or timing of breeding, so that no hybrid zygotes are formed.
primordial form The first organism; the first entity capable of (1) replicating itself through the directed chemical transformation of its environment, and (2) evolving by natural selection.
processed pseudogene A pseudogene that originated when a messenger RNA from which the introns had already been removed was reverse-transcribed and inserted into the genome.
protostome A lineage of animals that share a pattern of development, including spiral cleavage and formation of the mouth before the anus. Includes arthropods, mollusks, and annelids.
proximate causation Explanations for how, in terms of physiological or molecular mechanisms, traits function.
pseudogene DNA sequences that are homologous to functioning genes, but are not transcribed.
purifying selection Selection against deleterious mutations. Also called *negative selection.*

quantitative genetics The branch of evolutionary biology responsible for investigating the evolution of continuously variable traits that are influenced by the combined effects of genotype at many loci and the environment. That is, for investigating the evolution of traits not controlled by genotype at a single locus.
qualitative trait A trait for which phenotypes fall into discrete categories (such as affected versus unaffected with cystic fibrosis).
quantitative trait A trait for which phenotypes do not fall into discrete categories, but instead show continuous variation among individuals; a trait determined by the combined influence of the environment and many loci of small effect. See *qualitative trait.*
quantitative trait locus A locus at which there is genetic variation that contributes to the phenotypic variation in a quantitative trait.
QTL Quantitative trait locus.
QTL mapping A collection of techniques that allow researchers to identify chromosomal regions containing loci that contribute to quantitative traits.

radiometric dating Techniques for assigning absolute ages to rock samples, based on the ratio of parent-to-daughter radioactive isotopes present.
reaction norm The pattern of phenotypic plasticity exhibited by a genotype.
reciprocal altruism An exchange of fitness benefits, separated in time, between two individuals.
recombination rate (*r*) The frequency, during meiosis, of crossing-over between two linked loci; ranges from 0 to 0.5.
reinforcement Natural selection that results in assortative mating in recently diverged populations in secondary contact; also known as reproductive character displacement.
relative dating Techniques for assigning relative ages to rock strata, based on assumptions about the relationships between newer and older rocks.
relative fitness The fitness of an individual, phenotype, or genotype compared with others in the population; can be calculated by dividing the individual's fitness by either (1) the mean fitness of the individuals in the population, or (2) the highest individual fitness found in the population; method 1 must be used when calculating the selection gradient.
replacement substitution A DNA substitution that changes the amino acid or RNA sequence specified by a gene. Also called a *nonsynonymous substitution.*
reproductive success (RS) The number of viable, fertile offspring produced by an individual.
response to selection (*R*) In quantitative genetics, the difference between the mean phenotype of the offspring of the selected individuals in a population and the mean phenotype of the offspring of all the individuals.
retrotransposons Transposable elements that move via an RNA intermediate and contain the coding sequence for reverse transcriptase; closely related to retroviruses.
retrovirus An RNA virus whose genome is reverse transcribed to DNA by reverse transcriptase.
reversal An event that results in the reversion of a derived trait to the ancestral form.
ribozyme An RNA molecule that has the ability to catalyze a chemical reaction.
root The location on a phylogeny of the common ancestor of a clade.

sampling error A chance difference between the frequency of a trait in a subset of individuals from a population versus the frequency of the trait in the entire population. Sampling error is larger for small samples than for large ones.
secondary contact When two populations that have diverged in isolation from a common ancestor are reunited geographically.
secondary endosymbiosis A relationship where an endosymbiotic cell or organelle is taken up by another organism and lives inside it. The leading explanation for how several lineages of eukaryotes acquired chloroplasts.

selection Synonym for *natural selection*.
selection coefficient A variable used in population genetics to represent the difference in fitness between one genotype and another.
selection differential (*S*) A measure of the strength of selection used in quantitative genetics; equal to the difference between the mean phenotype of the selected individuals (for example, those that survive to reproduce) and the mean phenotype of the entire population.
selection gradient A measure of the strength of selection used in quantitative genetics; for selection on a single trait, it is equal to the slope of the best-fit line in a scatterplot showing relative fitness as a function of phenotype.
selectionist theory The viewpoint that natural selection is responsible for a significant percentage of substitution events observed at the molecular level.
selective sweep Change in the frequency of an allele due to selection on a closely linked locus. Also called *hitchhiking*.
selfishness An interaction between individuals that results in a fitness gain for one individual and a fitness loss for the other.
semelparous Describes a species or population in which individuals experience only one bout of reproduction over the course of typical lifetime; salmon provide an example.
senescence A decline with age in reproductive performance, physiological function, or probability of survival.
sexual dimorphism A difference between the phenotypes of females versus males within a species.
sexual selection A difference, among members of the same sex, between the average mating success of individuals with a particular phenotype versus individuals with other phenotypes.
significant In scientific discussions, typically a synonym for *statistically significant*.
silent substitution (or **silent-site substitution)** A DNA substitution that does not change the amino acid or RNA sequence specified by the gene. Also called a *synonymous substitution*.
sister species The species that diverged from the same ancestral node on a phylogenetic tree.
sister taxa Lineages that diverged from the same ancestral node on a phylogenetic tree. See *sister species*.
species Groups of interbreeding populations that are evolutionarily independent of other populations.
spite Behavior that decreases the fitness of both the actor and the recipient.
stabilizing selection That which occurs when individuals with intermediate values of a trait have higher fitness; can result in reduced phenotypic variation in a population and can prevent evolution in the mean value of the trait.
standard deviation A measure of the variation among the numbers in a list; equal to the square root of the variance (see *variance*).
standard error The likely size of the error due to chance effects in an estimated value, such as the average phenotype for a population.
stasis Lack of change.
statistically significant Describes a claim for which there is a degree of evidence in the data; by convention, a result is considered statistically significant if the probability is less than or equal to 0.05 that the observed violation of the null hypothesis is due to chance effects.
substitution Fixation of a new mutation in a population.
sympatric Living in the same geographic area.
synapomorphy A shared, derived character; in a phylogenetic analysis, synapomorphies are used to define clades and distinguish them from outgroups.
synonymous substitution A DNA substitution that does not change the amino acid or RNA sequence specified by the gene. Also called a *silent* (or *silent-site*) *substitution*.
systematics A scientific field devoted to the classification of organisms.

taxon Any named group of organisms (the plural form is *taxa*).
Theory of Evolution by Natural Selection The hypothesis that descent with modification is caused in large part by the action of natural selection.
tip (of a phylogenetic tree) The ends of the branches on a phylogenetic tree, which represent extinct or living taxa.
trade-off An inescapable compromise between one trait and another.
transformation In genetics, the acquisition of DNA from the environment or another organism that becomes incorporated into an organism's genome.
transition In DNA, a mutation that substitutes a purine for a purine or a pyrimidine for a pyrimidine.
transitional form A species that exhibits traits common to ancestral and derived groups, especially when the groups are sharply differentiated.
transposable elements Any DNA sequence capable of transmitting itself or a copy of itself to a new location in the genome.
transposons Transposable elements that move via a DNA intermediate, and contain insertion sequences along with a transposase enzyme and possibly other coding sequences.
transversion In DNA, a mutation that substitutes a purine for a pyrimidine, or a pyrimidine for a purine.
triploblast An animal that develops from three basic embryonic cell layers (endoderm, mesoderm, and ectoderm).

ultimate causation Explanations for why, in terms of fitness benefits, traits evolved.
underdominance Describes a situation in which heterozygotes at a particular locus tend to have lower fitness than homozygotes.
unequal cross-over A crossing-over event between mispaired DNA strands that results in the duplication of sequences in some daughter strands and deletions in others.
uniformitarianism The assumption (sometimes called a "law") that processes identical to those at work today are responsible for events that occurred in the past; first articulated by James Hutton, the founder of modern geology.

variance A measure of the variation among the numbers in a list; to calculate the variance of a list of numbers, first square the difference between each number and the mean of the list, then take the sum of the squared differences and divide it by the number of items in the list. (For technical reasons, when researchers calculate the variance for a sample of individuals, they usually divide the sum of the squared differences by the sample size minus one).
vestigial traits (or **structures)** Rudimentary traits that are homologous to fully functional traits in closely related species.
vicariance Splitting of a population's former range into two or more isolated patches.
virulence The damage inflicted by a pathogen on its host; occurs because the pathogen extracts energy and nutrients from the host and because the pathogen produces toxic metabolic wastes.
wild type A phenotype or allele common in nature.

Illustration Credits

PART I © Frans Lanting/CORBIS All Rights Reserved.

CHAPTER 1 C0.1 © Jayanta Shaw/CORBIS All Rights Reserved. **1.1b** From World Health Organization (WHO), 2004, The world health report 2004: Changing history. Reprinted with permission. **1.4** From N. Macdonald, S. Dougan, et al., 2004, Recent trends in diagnoses of HIV and other sexually transmitted infections in England and Wales among men who have sex with men. *Sexually Transmitted Infections* 80: 492–497, Figure 1b. Reprinted with permission. **1.7b** Reprinted by permission from Macmillan Publishers Ltd.: *Nature Medicine*, Z. Grossman, M. Meier-Schellersheim, et al., 2002, CD4 T-cell depletion in HIV infection: Are we closer to understanding the cause? *Nature Medicine* 8: 319–323, Figure 1a, copyright © 2002. **1.8(2)** Reprinted by permission from Macmillan Publishers Ltd.: *Nature Medicine*, Z. Grossman, M. Meier-Schellersheim, et al., 2006, Pathogenesis of HIV infection: what the virus spares is as important as what it destroys. *Nature Medicine* 12: 289–295, Figure 3, copyright © 2006. **1.8(3)** Reprinted by permission from Macmillan Publishers Ltd.: *Nature Medicine*, J. M. Brenchley, D. A. Price, and D. C. Douek, 2006, HIV disease: Fallout from a mucosal catastrophe? *Nature Immunology* 7: 235–239, Figure 3, copyright © 2006. **1.11 & 1.12** From B. A. Larder, G. Darby, and D. D. Richman, 1989, HIV with reduced sensitivity to Ziodvudine (AZT) isolated during prolonged therapy, *Science*, 243: 1731–1734, Figures 1c & d, 2b. Reprinted with permission from American Association for the Advancement of Science.* **1.13b** Thomas A. Steitz, Yale University. **1.13c** Courtesy of Lori A. Kohlstaedt, University of California at Santa Barbara. Reproduced from L. Kohlstaedt and J. Cohen, AIDS Research: The Mood Is Uncertain, *Science* 260: 1254–1258, May 28, 1993. **1.15** From F. J. Palella Jr., J. S. Chmiel, A. C. Moorman, et al., 2002, Durability and active antiretroviral predictors of success of highly therapy for ambulatory HIV-infected patients, *AIDS*, 16: 1617–1626, Figure 1. Reprinted by permission. **1.16a** Reprinted by permission from Macmillan Publishers Ltd.: *Nature Medicine*, A. J. Leslie, K. J. Pfafferott, et al., 2004, HIV evolution: CTL escape mutation and reversion after transmission, *Nature Medicine* 10: 282–289, Figure 1d, copyright © 2004. **1.17** From R. Shankarappa, J. B. Margolick, S. J. Gange, et al., 1999, Consistent viral evolutionary changes associated with the progression of human immunodeficiency virus type 1 infection," *Journal of Virology* 73: 10489–10502, Figures 1b, 2d. Reprinted with permission. **1.18** Redrawn from Figure 4a, page 9011, in R. M. Troyer, K. R. Collins, et al., 2005, Changes in human immunodeficiency virus type 1 fitness and genetic diversity during disease progression, *Journal of Virology* 79: 9006–9018. Reprinted with permission from the American Society for Microbiology and the *Journal of Virology*.

*Hereafter abbreviated as AAAS.

1.19 Redrawn from H. Blaak, A. B. van't Wout, M. Brouwer, et al., 2000, In vivo HIV-1 infection of T cells is established primarily by syncytium-inducing variants and correlates with the rate of T cell decline, *Proceedings of the National Academy of Sciences USA* 97(3): 1269–1274, Figure 4a. Copyright © 2000 National Academy of Sciences, U.S.A. **1.20** From S. A. Limborskaa, O. P. Balanovskya, E. V. Balanovskya, et al., 2002, Analysis of *CCR5-Δ32* geographic distribution and its correlation with some climatic and geographic factors," *Human Heredity* 53: 49–54. Reprinted with permission from S. Karger AG, Basel. **1.21** Reprinted from B. H. Hahn, G. M. Shaw, K. M. De Cock, and P. M. Sharp, 2000, AIDS as a zoonosis: Scientific and public health, *Science* 287: 607–614, Figure 1. Reprinted with permission from AAAS. **1.22** From B. Korber, M. Muldoon, J. Tehiler, et al., 2000, Timing the ancestor of the HIV-1 pandemic strains, *Science* 288: 1789–1796, Figure 1. Reprinted with permission from AAAS.

CHAPTER 2 CO.2a The Field Museum, Neg#GEO86481c, Chicago. **C0.2b** The Field Museum, Neg#GEO86483c, Chicago. **CO.2 (bottom)** Reprinted with permission from E. Tchernov, O. Rieppel, H. Zaher, et al., 2000, A fossil snake with limbs, *Science* 287: 2010–2012. Copyright © 2000 AAAS. **2.2** Drawings from Figure 1 and data from Figure 6, S. P. Carroll and C. Boyd, 1992, Host race radiation in the soapberry bug: Natural history with the history, *Evolution* 46: 1052–1069. Copyright © 1992 *Evolution*. Reprinted by permission. **2.3a** Drawings from Figures 3, 4, and 6 in S. P. Carroll and C. Boyd, 1992, Host race radiation in the soapberry bug: Natural history with the history, *Evolution* 46: 1052–1069. Copyright © 1992 *Evolution*. Reprinted by permission. **2.3b** Drawings from Figure 3 in S. P. Carroll. H. Dingle, and S. P. Klassen, 1997, Genetic differentiation of fitness-associated traits among rapidly evolving populations of the soapberry bug, *Evolution* 51: 1182–1188. Copyright © 1997 *Evolution*. Reprinted by permission. **2.4 (1 & 2)** Rogan Colbourne/Illustration & Graphics. **2.4 (3 & 4)** Ryan Hoyer/Ryan Hoyer. **2.5a** Vincent Zuber/Custom Medical Stock Photo, Inc. **2.5c** Mary Beth Angelo/Photo Researchers, Inc. **2.6 (top & bottom)** A.C. Burke/Alan Feduccia, Developmental patterns and the identification of homologies in the avian hand, *Science*, v. 278, 666–668m, Oct. 24, 1997. **2.7a–c** Armor loss in freshwater Alaskan threespine sticklebacks. This shot depicts a marine fish heavily armored with bony plates. Originally published as figure 1D in: William A. Cresko, et al, Parallel genetic basis for repeated evolution of armor loss in Alaskan threespine stickleback populations *Proceedings of the National Academy of Sciences USA* 101: 6050–6055. Copyright © 2007 National Academy of Sciences, U.S.A. **2.8** A. J. Copley/Visuals Unlimited. **2.9 (left top)** Tom McHugh/Photo Researchers, Inc. **2.9 (left bottom)** © The Natural History Museum, London. **2.9 (right top)** Tom Brakefield/Corbis/Bettmann. **2.9 (right bottom)**

Neg./Transparency no. 310314. Photo by: E. M. Fulda. Courtesy Dept. of Library Services, American Museum of Natural History. **2.10a** Museum Fur Naturkunde; **2.10b** AP Wide World Photos. **2.11a** Reprinted by permission from: Qiang Ji, Philip J. Currie, Mark A. Norell & Shu-An Ji, *Nature:* 393: 753–761. Copyright © 1998 Macmillan Magazines Limited. **2.11b** © Reuters NewMedia Inc./Corbis. **2.12a** Mick Ellison. **2.12b** Sovfoto/Eastfoto. **2.13b** Reprinted with permission from P. D. Gingerich, B. H. Smith, E. L. Simons, 1990, Hind limbs of Eocene *Basilosaurus*: evidence of feet in whales, *Science* 249: 154–156. Copyright © 1990 AAAS. **2.13c** Reprinted with permission from J. G. M. Thewissen, S. T. Hussain, & M. Arif, 1994, Fossil evidence for the origin of aquatic locomotion in archaeocete whales, *Science* 263: 210–212. Copyright © 1994 AAAS. **2.15** Courtesy of the Library of Congress. **2.17a** Darren E. Irwin, Asst Professor. **2.17 (map)** From D. E. Irwin, S. Bensch, et al., 2005, Speciation by distance in a ring species, *Science* 307: 414–416, Figure 1a. Redrawn with permission from AAAS. **2.20** From: Richardson, M. K., et al. *Science*, Vol 280: Pg 983c, Issue # 5366, May 15, 1998. Embryo from Professor R. O'Rahilly. National Museum of Health and Medicine/Armed Forces Institute of Pathology. **2.24c** Reprinted from *Molecular Phylogenetics and Evolution*, 16, F. Friedberg & A. R. Rhoads, 2000, Calculation and verification of the ages of retroprocessed pseudogenes, 127–130, Copyright © 2000, with permission from Elsevier.

CHAPTER 3 CO.3 © *www.stevebloom.com*. **CO.3 (bottom)** Reprinted with permission from A. Pole, I. J. Gordon, & M. L. Gorman, 2003, African wild dogs test the "survival of the fittest" paradigm, *Proceedings of the Royal Society of London B* (Supplement) 270: S57, Figure 1. Published by the Royal Society of London. **3.1 & 3.2** Reproduced with permission from: Figures 1a and 1b on pg 86 in: Frary, A., T.C. Nesbitt, A. Frary, S. Grandillo, E. van der Knaap, B. Cong, J. Liu, J. Meller, R. Elber, K. Alpert, and S. D. Tanksley. Cloning and Transgenic Expression of fw2.2: A Quantative Trait Locus Key to the Evolution of Tomato Fruit. *Science* 289: 85–87. Copyright 2000 AAAS. **3.6** Photos from Peter R. Grant, Princeton University. **3.6 (mangrove finch)** Greg Lasley/Greg Lasley Nature Photography. **3.7 (top & bottom)** Peter R. Grant, Princeton University. **3.8b** Courtesy of Robert Podolsky. **3.11** Photos from Peter R. Grant, Princeton University. **3.11 (right)** Reproduced with permission: Abzhanov, A., Protas, M., Grant, R., Grant, P. and Tabin, C. J. Bmp4 and morphological variation of the beaks of Darwin's Finches. *Science* 2004, September; 305 (5689): 1462–1465. fig. 1 b & c; pg 1463. Copyright 2004 AAAS. **3.12** Reprinted with permission from P.T. Boag and P. T. Grant, 1981, Intense natural selection in a population of Darwin's finches (Geospizinae) in the Galapagos, *Science* 214: 82–85. Copyright © 1981 AAAS. **3.13** Reprinted with permission from P. T. Boag & P. R. Grant, 1984, Classical case of character release: Darwin's finches (Geospiza) on Isla Daphne Major, Galápagos, *Biological Journal of the Linnean Society* 22: 243–287, published by Blackwell Publishing. **3.14** Republished with permission of BioScience, from B. R. Grant and P. R. Grant, 2003, What Darwin's finches can teach us about the evolutionary origin and regulation of biodiversity, *BioScience* 53(10): 965–975, Figure 5; permission conveyed through Copyright Clearance Center, Inc. **3.15a** From P.R. Grant & B.R. Grant, 2006, Evolution of character displacement in Darwin's finches, *Science* 313: 224–226, Figure 2. Adapted with permission from AAAS. **3.15b & c** From P. R. Grant & B. R. Grant, 2002, Unpredictable evolution in a 30-year study of Darwin's finches, *Science* 296: 707–711, Figure 1. Adapted with permission from AAAS. **3.18** Reprinted from *Trends in Plant Science*, 9, S. P. Moose, J. W. Dudley, & T. R. Rocheford, Maize selection passes the century mark: A unique resource for the 21st century genomics, 358–364, Copyright © 2004, with permission from Elsevier. **3.19a** Bill Kamin/Visuals Unlimited. **3.19b** Reprinted with permission from *Nature*, H. Endo et al., 1999, Role of the giant panda's "pseudo-thumb," *Nature* 397: 309–310. Copyright © 1999 Macmillan Magazines Limited. **3.20** From: Fig. 3, pg 7622: Langerhans, R.B., Layman, C.A. and DeWitt, T.J. Male genital size reflects a tradeoff between attracting mates and avoiding predators in two livebearing fish species. *Proceedings of the National Academy of Sciences USA* 102: 7618–7623. Copyright © 2005 National Academy of Sciences, U.S.A. **3.21** From J.V. Scaffer and J. L. Bolognia, 2001, The melanocortin-1 receptor: Red hair and beyond, *Archives of Dermatology* 137: 1477–1485, Figure 3. © 2001, American Medical Association. All rights reserved. **3.22** © Bettmann/CORBIS All Rights Reserved. **3.24a–b** Reprinted from D. R. Mitchell, 2000, Chlamydomonas flagella, *Journal of Phycology* 36: 261–274, Figure 2. Reprinted with permission from the *Journal of Phycology*. **3.24c** From: Journal of Cell. *Science* 110: 85–91, Fig 3, pg 87. Woolley, D .M., 1997. Studies on the eel sperm flagellum. I. The structure of the inner dynein arm complex. **3.25** Adapted and reprinted, with permission, from the *Annual Review of Cell and Developmental Biology*, Volume 18 © 2002 by Annual Reviews, *www.annualreviews.org*.

CHAPTER 4 CO.4 Corbis Digital Stock. **4.4a** Lawson Wood/Corbis/Bettmann. **4.4b** Fred Bavendam/Minden Pictures. **4.4c** Joe McDonald/Corbis/Bettmann. **4.4d** Peter Johnson/Corbis/Bettmann. **4.7** From B. Schaeffer, 1948, The origin of a mammalian ordinal character, *Evolution* 2: 164–175. Copyright © 1948 *Evolution*. Reprinted by permission of *Evolution*. **4.12a** Reprinted with permission from *Nature*, J. G. M. Thewissen et al., 2001, Skeletons of terrestrial cetaceans and the relationship of whales to artiodactyles, *Nature* 413: 277–281, Figure 2. Copyright © 2001 Macmillan Magazines Limited. **4.12b** Reprinted from J. G. M. Thewissen, S. T. Hussain, & M. Arif, 1994, Fossil evidence for the origin of aquatic locomotion in archaeocete whales, *Science* 263: 210–212. Copyright © 1994 AAAS. **4.12c** Reprinted with permission from P. D. Gingerich et al., 2001, Origin of whales from early artiodactyls: Hands and feet of Eocene Protocetidae from Pakistan, *Science* 293: 2239–2242, Figure 3, illustration Douglas Boyer. Copyright © 2001 AAAS. **4.12d** Reprinted with permission from P. D. Gingerich, B. H. Smith, E. L. Simons, 1990, Hind limbs of Eocene Basilosaurus: evidence of feet in whales, *Science* 249: 154–156. Copyright © 1990 AAAS. **4.14** C. K. Lorenz/Photo Researchers, Inc. **4.15b–c** Reprinted with permission from *Nature*, C. J. Raxworthy et al., 2002, Chameleon radiation by oceanic dispersal, *Nature* 415: 784–787, Figure 2. Copyright © 2002 Macmillan Magazines Limited. **4.16a** Courtesy of Professor Nancy Moran/Dept of Ecology and Evolutionary/University of Arizona. Photo by Jonas Sandstrom. **4.16b** Courtesy of Dr. Angela Douglas, Department of Biology, University of York. **4.17** From J. A. Clark et al., 2000, Cospeciation between bacterial endosymbionts (*Buchnera*) and a recent readiation of aphids (*Uroleucon*) and pitfalls of testing for phylogenetic congruence," *Evolution* 54: 517–525, Figure 1. Copyright © 2000 *Evolution*.

Reprinted with permission. **4.18** Adapted with permission from W. J. Murphy et al., 2001, Resolution of the early placental mammal radiation using Bayesian phylogenetics, *Science* 294: 2348–2351. Copyright © 2001 AAAS.

PART II Top left David Scharf/Photo Researchers, Inc. **Top middle** Photo Researchers, Inc. **Bottom left** Carolina Biological Supply Company/Phototake NYC. **Right** Dr. David M. Phillips/Visuals Unlimited.

CHAPTER 5 CO.5 Julie Bruton/Ardea London Limited. **5.5a** From L. L.Vassilieva et al., 2000,The fitness effects of spontaneous mutations in *Caenorhabditis elegans, Evolution* 54: 1234–1246, Figure 2. Reprinted with permission. **5.5b** From M. Lynch et al., 1999, "Perspective: Spontaneous deleterious mutation, *Evolution* 53: 645–663, Figure 1. Reprinted with permission. **5.12** Reprinted by permission from Macmillan Publishers Ltd: *Nature*, M. Samson, M. F. Libert, et al., 1996, Resistance to HIV-1 infection in caucasian individuals bearing mutant alleles of the CCR-5 chemokine receptor gene, *Nature* 382: 722–725, Figure 3, copyright © 1996. **Table 5.3** Reprinted by permission from Macmillan Publishers Ltd.: *Nature Genetics*, J. J. Martinson, et al., Global distribution of the CCR5 gene 32-base-pair deletion, 1997, *Nature Genetics* 16: 100–103, copyright © 1997. **5.13** From J. C. Avise, *Molecular Markers, Natural History, and Evolution*, Figure 2.2, page 19. Copyright © 1994, Chapman and Hall. Reprinted with kind permission of Springer Science and Business Media. **5.14** Reprinted from *Trends in Genetics*, 8, L.-C Tsui, The spectrum of cystic fibrosis mutations, 392–398. Copyright © 1992, with permission from Elsevier.

CHAPTER 6 CO.6 Courtesy of Susan J. Brown, Assoc. Professor/Kansas State University, Kansas. **6.13** From D. R. Cavener and M.T. Clegg, 1981, Mulitgenic response to ethanol in *Drosophila melanogaster*, *Evolution* 39: 1–10. Copyright © 1991 *Evolution*. Reprinted by permission. **6.19** Redrawn from G. G. Foster, M. J. Whitten, T. Prout, and R. Gill, 1972, Chromosome rearrangements for the control of insect pests, *Science* 176: 875–880. Copyright © 1972 AAAS. **6.21a** Dr. Luc Gigord/Evolutionary Ecology & Genetics; **6.21b & c** From L. D. B. Gigord, M. R. McNair & A. Smithson, 2001, Negative frequency-dependent selection maintains a dramatic flower color polymorphism in the rewardless orchid *Dactylorhiza sambucina*, *Proceedings of the National Academy of Sciences USA* 98: 6253–6255, Figure 3. Copyright © 2001 National Academy of Sciences, U.S.A. **6.25** Reprinted from S. F. Elena, V. S. Cooper, and R. E. Lenski, 1996, Punctuated evolution caused by selection of rare beneficial mutations, *Science* 272: 1802–1804. Copyright © 1996 AAAS. **6.26** James R. MacFall, Duke University Medical Center. **6.27a** Reprinted by permission from Macmillan Publishers Ltd: *Nature*, G. Pier, B. M. Grout, et al., 1998, *Salmonella typhi* uses CFTR to enter intestinal epithelial cells, *Nature* 393: 79–82, copyright © 1998. **6.27b** From J. B. Lyczak, C. L. Cannon, & G. B. Pier, 2002, Lung infections associated with cystic fibrosis, *Clinical Microbiology Reviews* 15: 194–222, Figure 5. Reprinted with permission. **6.28** From K. Johannesson, B. Johannesson, and U. Lundgren, 1995, Strong natural selection causes microscale allozyme variation in a marine snail, *Proceedings of the National Academy of Sciences USA*, 92: 2602–2606, Figure 1. Copyright © 1995 National Academy of Sciences, U.S.A.

CHAPTER 7 CO.7 J. Paul Moore. **7.1** Richard Day/Daybreak Imagery. **7.2** Reprinted with permission from R. L. Westemeier, et al., 1998, Tracking the long-term decline and recovery of an isolated population, *Science* 282: 1695–1698, Figure 1. Copyright © 1998 AAAS. Derived from R. C. Anderson, *Transactions of the Illinois State Academy of Science* 63, 214 (1970). Reprinted by permission of the Illinois State Academy of Science. **7.3** Reprinted with permission from R. L. Westemeier, et al., 1998, Tracking the long-term decline and recovery of an isolated population, *Science* 282: 1695–1698, Figure 1. Copyright © 1998 AAAS. **7.6a** From R. B. King, et al., 1995, Color-pattern variation in Lake Erie water snakes: The role of gene flow, *Evolution* 49: 885–896. Copyright © 1995 *Evolution*. Reprinted by permission. **7.6b** Richard B. King, Northern Illinois University. **7.7** From J. H. Camin and P. R. Ehrlich, 1958, Natural selection in water snakes (*natrix sipedon L.*) on islands in Lake Erie, *Evolution* 12: 504–511. Copyright © 1958 *Evolution*. Reprinted by permission. **7.13a** Jason Elsworth Photography. **7.13b & c** From S. M. Clegg, S. M. Degnan, J. Kikkawa, et al., 2002, Genetic consequences of sequential founder events by an island-colonizing bird, *Proceedings of National Academy of Sciences USA* 99: 8127–8132, Figures 1a and 2a. Copyright © 2002 National Academy of Sciences USA. **7.18a** Alan R. Templeton. **7.18b–d** Copyright © 1990 Alan R. Templeton. Reprinted by permission of the author. **7.19** Reprinted from *Trends in Ecology and Evolution*, 11, A. Young, T. Boyle, and T. Brown, 1996, The population genetic consequences of habitat fragmentation for plants, 413–418. Copyright © 1996, with permission from Elsevier. **7.21c** From T. Gojoburi, E. N. Etsuko, & M. Kimura, 1990, Molecular clock of viral evolution, and the neutral theory, *Proceedings of the National Academy of Sciences USA* 87: 10015–10018, Figure 1. Copyright © 1990 National Academy of Sciences, U.S.A. **7.22** From L. Chao and D. E. Carr, 1993, The molecular clock and the relationship between population size and generation time, *Evolution* 47: 688–690, Figure 1. Copyright © 1993 *Evolution*. Reprinted by permission of *Evolution*. **7.23** Reprinted by permission from Macmillan Publishers Ltd.: *Nature Genetics*, G. A. Huttley, E. Easteat, M. C. Southby, et al., 2000, Adaptive evolution of the tumor supressor BRCA1 in humans and chimpanzees, *Nature Genetics* 25: 410–412, Figure 2, copyright © 2000. **7.24** Reprinted from W.-H Li and D. Graur, 1991, *Fundamentals of Molecular Evolution* (Sunderland, MA: Sinauer). Copyright © 1991 Sinauer Associates, Inc. Reprinted with permission. **7.26** Richard R. Hansen/Photo Researchers, Inc. **7.29** From L. M. Wolfe, 1993, Inbreeding depression in *Hydrophyllum appendicualum*: Role of maternal effects, crowding, and parental mating history, *Evolution* 47: 374–386, Figure 1. Copyright © 1993 *Evolution*. Reprinted by permission of *Evolution*. **7.30** From A. J. Van Noordwijk and W. Scharloo, 1981, Inbreeding in an island population of the Great Tit, *Evolution* 35: 674–688. Copyright © 1981 *Evolution*. Reprinted by permission. **7.31** Reprinted with permission from R. L. Westemeier, et al., 1998, Tracking the long-term decline and recovery of an isolated population, *Science* 282: 1695–1698. Copyright © 1998 AAAS. **7.32** Reprinted by permission from Macmillan Publishers Ltd: *Nature Genetics*, M. T. Seielstad, E. Minch, and L. Cavalli-Sforza,

1998, Genetic evidence for a higher female migration rate, *Nature Genetics* 20: 278–280, copyright © 1998.

CHAPTER 8 CO.8 Phil Savoie/Nature Picture Library. **8.7** From M. T. Clegg, J. F. Kidwell and C. R. Horch, 1980, Dynamics of correlated genetic systems. V. Rates of decay of linkage disequilibria in experimental populations of *Drosophila melanogaster*, *Genetics* 94: 217–234, Figure 3. Reprinted with permission from Genetics Society of America. **8.9** Plotted by the author from data in Table 1, based on the presentation in Figure 2, page 1628, in Schlenke, T.A., and D. J. Begun, 2004, Strong selective sweep associated with a transposon insertion in *Drosophila simulans, Proceedings of the National Academy of Sciences USA* 101: 1626–1631. Copyright © 2004 National Academy of Sciences, U.S.A. **8.10** Reprinted by permission from Macmillan Publishers Ltd: *Nature*, E. Dawson, et al., 2002, A first-generation linkage disequilibrium map of human chromosome 22, *Nature* 418: 544–548, copyright © 2002. **8.13** Reprinted with permission from L. Luzzatto & R. Notaro, 2001, Malaria: Protecting against bad air, *Science* 293: 442–443. Copyright © 2001 AAAS. **8.14** Reprinted with permission from *Nature*, P. C. Sabeti, D. E. Reich, J. M. Higgins, et al., 2002, Detecting recent positive selection in the human genome from haplotype structure, *Nature* 419: 832–837, Figures 2c, 2e. Copyright © 2002 Macmillan Magazines Limited. **8.15** From T. Bersaglieri, P. C. Sabeti, et al., 2004, Genetic signatures of recent positive selection at the lactase gene, *American Journal of Human Genetics* 74: 1111–1120, Figure 3b. Copyright © 2004 The University of Chicago Press. Reprinted by permission of the The University of Chicago Press. **8.16a** Peter J. Bryant/Biological Photo Service. **8.16b** Jon C. Herron. **8.16c** Photo by P. S. Tice from Buchsbaum and Pearse, *Animals Without Backbones*, 3rd ed., University of Chicago Press, 1987. **8.18** From R. L. Dunbrack, C. Coffin, and R. Howe, 1995, The cost of males and the paradox of sex: An experimental investigation of the short-term competitive advantages of evolution in sexual populations, *Proceedings of the Royal Society of London B* 262: 45–49. Copyright © 1995, The Royal Society. **8.19** Reprinted by permission from Macmillan Publishers Ltd: *Nature Reviews Genetics,* S. P. Otto & T. Lenormand, 2002, Resolving the paradox of sex and recombination, *Nature Reviews Genetics* 3: 252–261, copyright © 2002. **8.24** From M. H. Kohn, H-J. Peiz, R. K. Wayne, 2000, Natural selection mapping of the warfarin-resistance gene, *Proceedings of the National Academy of Sciences USA*, 97: 7911–7915, Figure 2. Copyright © 2000, National Academy of Sciences, U.S.A.

CHAPTER 9 CO.9 From: Fig 1a, pg 99 of Shikano, T. 2005. Marker-based estimation of heritability for body color variation in Japanese flounder *Paralichthys olivaceus. Aquaculture* 249: 95–105. **CO.9 (bottom)** Reprinted from *Aquaculture*, 249, T. Shikano, 2005. Marker-based estimation of heritability for body color variation in Japanese flounder *Paralichthys olivaceus. Aquaculture* 249: 95–105, Figure 1a. Copyright © 2005, with permission from Elsevier. **9.1a** Peter Morenus, University of Connecticut. **9.1b–c** Reprinted with permission from G. E. McClearn, B. Johansson, S. Berg, et al., 1997, Substantial genetic influence on cognitive abilities in twins 80 or more years old, *Science* 276: 1560–1563, Figure 1A. Copyright © 1997 AAAS. **9.4** From J. Clausen, D. D. Keck, & W. M. Hiesey, 1948, Experimental studies on the nature of the species, III, environmental responses of climatic races of achillea, Washington, DC: Carnegie Institution of Washington, Publication No. 581, 45–86, Figure 13. Reprinted with permission. **9.5** From P. M. Beardsley, A. Yen, and R. G. Olmstead, 2003, AFLP phylogeny of Mimulus section Erythranthe and the evolution of hummingbird pollination, Figure 5, *Evolution* 57: 1397–1410. Reprinted with permission. **9.6** Courtesy Douglas W. Schemske, Michigan State University, Michigan/W. K. Kellogg Biological Station. **9.8b** Reprinted with permission from *Nature*, A. H. Paterson, E. S. Lander, et al., 1988, Resolution of quantitative traits into Mendelian factors by using a complete linkage map of restriction fragment length polymorphisms, *Nature* 335: 721–726, Figure 3. Copyright © 1988 Macmillan Magazines Limited. **9.9** From H. D. Bradshaw, Jr., and K. G. Otto et al., 1998, Quantitative trait loci affecting differences in floral morphology between two species of monkeyflower (*mimulus*), *Genetics* 149: 367–382, Figure 5. Reprinted with permission of *Genetics*. **9.10a–b** Reprinted by permission from *Nature*: Monkeyflower. *M. lewisii with M. cardinalis* genotype at the UYP locus. About 70 times more attractive to hummingbirds than bees. Originally published as figure 1b on page 177 from: Allele substitution at a flower colour locus produces a pollinator shift in monkeyflower, by HD Bradshaw Jr. & D W Schemske. *Nature* 426: 176–178. Copyright © 2007 Macmillan Magazines Limited. **9.12** Reprinted by permission from *Nature*: Ogura et al., *Nature*. 411: 603–606, Fig 2; A frameshift mutation in NOD2 associated with susceptibility to Crohn's disease. Copyright © Macmillan Magazines Limited. **9.14** From J. M. N. Smith and A. A. Dhondt, 1980, Experimental confirmation of heritable morphological variation in a natural population of song sparrows, *Evolution* 34: 1155–1160. Copyright © 1980 *Evolution*. Reprinted by permission of *Evolution*. **9.19a** Richard Parker/Photo Researchers, Inc. **9.19b** Stephen Dalton/Photo Researchers, Inc. **9.24** From E. D. Brodie, III, 1992, Correlational selection for color pattern and antipredator behavior in the garter snake *thamnophis ordinoides*, *Evolution* 46: 1284–1298, Figure 1. Copyright © 1992 *Evolution*. Reprinted by permission of *Evolution*. **9.26** From A. E. Weis and W. G. Abrahamson, 1986, Evolution of host plant manipulation by gall makers: Ecological and genetic factors in the *solidago-eurosta* system, *American Naturalist* 127: 681–695, Figures 2 and 3. Copyright © 1986 The University of Chicago Press. Reprinted by permission of the The University of Chicago Press. **9.28** From J. Clausen, D. D. Keck, & W. M. Hiesey, 1948, Experimental studies on the nature of the species, III, environmental responses of climatic races of achillea, Washington, DC: Carnegie Institution of Washington, Publication No. 581, 45–86, Figure 13. Reprinted with permission. **9.32** Reprinted with permission from A. Caspi, J. McClay, et al. 2002. Role of genotype in the cycle of violence in maltreated children. *Science* 297: 851–854, Figure 1. Copyright 1996 © AAAS. **9.33** Reprinted with permission from K.-P. Lesch, D. Bengel, A. Heils, et al., 1996, Association of anxiety-related traits with a polymorphism in the serotonin transporter gene regulatory regions, *Science* 274: 1527–1531, Figure 3. Copyright 1996 © AAAS.

PART III Steven D. Johnson, University of KwaZulu-Natal, Pietermaritzburg, South Africa

CHAPTER 10 CO.10 © Johann Schumacher/VIREO. **CO.10 (bottom)** From D. H. Clayton, B. R. Moyer, et al. 2005. Adaptive significance of avian beak morphology for ectoparasite control, *Proceedings of the Royal Society of London B* 272: 811–817, Figure 1d. Copyright © 2005, The Royal Society. Reprinted by permission. **10.1** © Martin Harvey / CORBIS All Rights Reserved; **10.2, 10.3, & 10.4** Reprinted from P. Weeks, 2000, Red-billed oxpeckers: vampires or tickbirds? *Behavioral Ecology* 11(2): 154–160, Figure 1, by permission of the International Society for Behavioral Ecology. Copyright © 2000 International Society for Behavioral Ecology. **10.5** Erick Greene, University of Montana. **10.6 & 10.7** Adapted with permission from E. Greene, L. J. Orsak, and D. W. Whitman, 1987, A tephritid fly mimics the territorial displays of its jumping spider predators, *Science* 236: 310–312. Copyright © 1987 AAAS. **10.9** Reprinted from *Trends in Ecology and Evolution*, 4, R. B. Huey and J. G. Kingsolver, Evolution of thermal sensitivity of ectotherm performance, 131–135, Copyright © 1989, with permission from Elsevier. **10.10 & 10.11** From R. B. Huey, C. R. Peterson, S. J. Arnold, and W. P. Porter, 1989, Hot rocks and not-so-hot rocks: Retreat-site selection by garter snakes and its thermal consequences, *Ecology* 70: 931–944. Copyright © 1989 Ecological Society of America. Reprinted by permission. **Table 10.1** From R. B. Huey, C. R. Peterson, S. J. Arnold and W. P. Porter, 1989, Hot rocks and not-so-hot rocks: Retreat-site selection by garter snakes and its thermal consequences, *Ecology* 70: 931–944, Table 1. Copyright © 1989 Ecological Society of America. Reprinted by permission. **10.12a** Fritz Prenzel/Animals Animals/Earth Scenes. **10.12b** From D. J. Hosken, 1998, Testes mass in megachiropteran bats varies in accordance with sperm conpetition theory, *Behavioral Ecology and Sociobiology* 44: 169–177. Copyright © 1998 Springer-Verlag GmbH & Co. KG. Reprinted by permission. **10.13** After G. V. Lauder, R. B. Huey, et al., 1995, Systematics and the study of organismal form and function, *Bioscience* Vol. 45, No. 10, pp. 696–704, Figure A. Reprinted by permission of American Institute of Biological Sciences and the author. **10.15** From D. J. Hosken, 1998, Testes mass in megachiropteran bats varies in accordance with sperm conpetition theory, *Behavioral Ecology and Sociobiology* 44: 169–177. Copyright © 1998 Springer-Verlag GmbH & Co. KG. Reprinted by permission. **10.16** From T. Garland, Jr. and S. C. Adolph, 1994, Why not do two-species comparative studies: Limitations on inferring adaptation, *Physiological Zoology* 67: 797–828. Copyright © 1994 by The University of Chicago Press. Reprinted by permission of The University of Chicago Press. **10.17** Edward Kinsman/Photo Researchers, Inc. **10.18** From L. DeMeester, 1996, Evolutionary potential and local genetic differentiation in a phenotypically plastic trait of a cyclical parthenogen, *daphnia magna, Evolution* 50: 1293–1298. Copyright © 1996 *Evolution*. Reprinted by permission of *Evolution*. **10.19** From C. Cousyn, L. De Meester, et al., 2001, Rapid, local adaption of zooplankton behavior to changes in predation pressure in the absence of neutral genetic changes, *Proceedings of the National Academy of Sciences USA* 98(11): 6256–6260, Figure 2. Copyright © 2001 National Academy of Sciences USA. **10.20a & b** Courtesy of Douglas W. Schemske, Michigan State University and Jon Agren, Uppsala University. **10.21** From D. W. Schemske and J. Agren, 1995, Deceit pollination and selection on female flower size in *begonia involucrata*: An experimental approach, *Evolution* 49: 209–214. Copyright © 1995 *Evolution*. Reprinted by permission of *Evolution*. **10.22** Lynda Delph, Indiana University; **10.23** From L. F. Delph and C. M. Lively, 1989, The evolution of floral color change: Pollinator attraction versus physiological constraints in *fuchsia excorticata, Evolution* 43: 1252–1262. Copyright © 1989 *Evolution*. Reprinted by permission of *Evolution*. **10.24** From D. J. Futuyma, M. C. Keese, and D. J. Funk, 1995, Genetic constraints on macroevolution: The evolution of host affiliation in the leaf beetle genus *ophraella, Evolution* 49: 797–809. Copyright © 1995 *Evolution*. Reprinted by permission of *Evolution*. **Table 10.3** From D. J. Futuyma, M. C. Keese, and D. J. Funk, 1995, Genetic constraints on macroevolution: The evolution of host affiliation in the leaf beetle genus *ophraella, Evolution* 49: 797–809, Table 7. Copyright © 1995 *Evolution*. Reprinted by permission of *Evolution*. **10.25** From D. H. Clayton and K. P. Johnson, 2003, Linking coevolutionary history to ecological process: doves and lice, *Evolution* 57: 2335–2341, Figure 4. Copyright © 2003. Reprinted by permission of *Evolution*. **10.26** From D. H. S. Clayton, E. Bush, and K. P. Johnson, 2004, Ecology of congruence: past meets present, *Systematic Biology* 53: 165–173, Figure 1. Illustration by Sarah E. Bush. Reprinted with permission. **10.27** From D. R. Taylor, C. Zeyl, and E. Cooke, 2002, Conflicting levels of selection in the accumulation of mitochondrial defects in *Saccharomyces cerevisiae, Proceedings of the National Academy of Sciences USA* 99: 3690–3694, Figure 2. Copyright © 2002 National Academy of Sciences, U.S.A. **10.28** From A. Goriely, G. A. T. McVean, et al., 2003, Evidence for selective advantage of pathogenic FGFR2 mutations in the male germ line, *Science* 301: 643–646, Figure 3. Reprinted with permission from AAAS.

CHAPTER 11 CO.11 (both) Kim Hart Wildlife Photography. **11.1a** M. Hamblin/Animals Animals/Earth Scenes; **11.1b** Michel Gunther/Bios/Peter Arnold, Inc. **11.1c** Michael Fogden/Animals Animals/Earth Scenes; **11.5a** William Leonard/DRK Photo; **11.5b–c & 11.6b–c** Drawn from data from Adam Jones (Georgia Institute of Technology). **11.6a** Berglund Anders. **11.7** Anders Axelsson/S. J. O'Haren Photography. **11.8** Tui DeRoy/Minden Pictures; **11.10** Martin C. Wikelski, Princeton University; **11.12** From M. Wikelski and S. Bäurle, 1996, Pre-copulatory ejaculation solves time constraints during copulations in marine iguanas, *Proceedings of the Royal Society of London B* 263: 439–444. Copyright © 1996, The Royal Society. Reprinted by permission. **11.13** After M. R. Gross, 1991, Salmon breeding behavior and life history evolution in changing environments, *Ecology* 72: 1180–1186. Copyright © 1991 Ecological Society of America. Reprinted by permission. **11.16** George B. Schaller. **11.17** Charlie Moores. **11.18** From S. R. Pryke and S. Andersson, 2005, Experimental evidence for female choice and energetic costs of male tail elongation in red-collared widowbirds, *Biological Journal of the Linnean Society* 86: 35–43, Figure 3. Reprinted with permission from Blackwell Publishing. **11.19** From S. R. Pryke and S. Andersson, 2005, Experimental evidence for female choice and energetic costs of male tail elongation in red-collared widowbirds, *Biological Journal of the Linnean Society* 86: 35–43, Figure 1. Reprinted with permission from Blackwell Publishing. **11.20** David T. Roberts/Nature's Images Inc./Photo Researchers, Inc. **11.21c** Reprinted from *Animal Behaviour*, 63, S. L. Bush, H. C. Gerhardt, & J. Schul, Pattern recognition and call preferences in tree-frogs (*Anura: Hylidae*), 7–14, Copyright © 2002, with permission from Elsevier. **11.23** Randy Thornhill, University of New Mexico. **11.24** From R. Thornhill, 1976, Sexual selection and nuptial feeding behavior in *Bittacus Apicalis* (Insecta: *Mecoptera*),

American Naturalist 110: 529–548. Copyright © 1976 The University of Chicago Press. Reprinted by permission of The University of Chicago Press. **11.25** Rerendered from *Animal Behaviour*, 42, H. C. Proctor, Courtship in the water mite, *neumania papillator*: Males capitalize on female adaptations for predation, 589–598, Copyright © 1991, with permission from Elsevier. **11.26** Rerendered from H. C. Proctor, 1992, Sensory exploitation and the evolution of male mating behavior: A cladistic test using water mites (Acari: Parasitengona), *Animal Behaviour* 44: 745–752. Reprinted by permission of Academic Press, Ltd. **11.27** Courtesy of John H. Christy, Smithsonian Tropical Research Institute. **11.29** Gerald S. Wilkinson, University of Maryland. **11.31** From Gerald Wilkinson and P. R. Reillo, 1994, Female choice response to artificial selection on an exaggerated male trait in a stalk-eyed fly. *Proceedings of the Royal Society of London B* 255: 1–6. Copyright © 1994, The Royal Society. **11.32** Reproduced by permission from H. L. Gibbs,1990, Realized reproductive success of polygynous red-winged blackbirds revealed by DNA markers, *Science* 250:1394–1396, December 7, 1990, p. 1395, Fig. 1. © AAAS. **11.33** Reprinted from *Animal Behaviour*, 55, J. L. Hoogland, Why do female Gunnison's prairie dogs copulate with more than one male? 351–359, Copyright © 1998, with permission from Elsevier. **11.35a** Dowery Orchard Nursery. **11.35b** Sharon Dahl. **11.35c** From G. Vaughton and M. Ramsey, 1998, Floral display, pollinator visitation, and reproductive success in the dioecious perennial herb *Wurmbea dioica* (*Liliaceae*), *Oecologia* 115: 93–101. Copyright © 1998 Springer Verlag. Reprinted by permission. **11.36** From M. L. Stanton, A. A. Snow, and S. N. Handle, 1986, Floral evolution: Attractiveness to pollinators increases male fitness, *Science* 232: 1625–1727, Figure 1. Reprinted with permission from AAAS. **11.37** Drawn from data in L. F. Delph, F. L. Galloway, and M. L. Stanton, 1996, Sexual dimorphism in flower size, *American Naturalist* 148: 299–320. Copyright © 1996 *American Naturalist*. Reprinted by permission of The University of Chicago Press. **11.38** From Mulder M. Borgerhoff, 1988, Reproductive success in three Kipsigis cohorts. In T. H. Clutton-Brock, ed. *Reproductive Success*, pp. 419–435. Copyright © 1988 The University of Chicago Press. Reprinted by permission of The University of Chicago Press. **11.40** Reprinted with permission from *Nature*, B. Pawlowski, R. I. M. Dunbar, and A. Lipowicz, 2000, Tall men have more reproductive success, *Nature* 403: 156, Figure 1b. Copyright © 2000 Macmillan Magazines Limited. **11.41** From B. J. LeBoeuf and J. Reiter, 1988, Lifetime reproductive success in northern elephant seals. In T. H. Cutton-Brock, ed., *Reproductive Success* (Chicago: University of Chicago Press), pp. 344–362. Copyright © 1988 by The University of Chicago Press. Reprinted by permission of The University of Chicago Press.

CHAPTER 12 CO.12 © Peter Johnson/CORBIS All Rights Reserved. **12.2a** D. Robert & Lorri Franz/Corbis RF. **12.2b–c** From J. L. Hoogland, *The Black-Tailed Prairie Dog: Social Life of a Burrowing Mammal* (Chicago: University of Chicago Press), pp. 173, 174. Copyright © 1988 by The University of Chicago Press. Reprinted by permission of The University of Chicago Press. **12.3** From J. L. Hoogland, *The Black-Tailed Prairie Dog: Social Life of a Burrowing Mammal* (Chicago: University of Chicago Press), p. 177. Copyright © 1988 by The University of Chicago Press. Reprinted by permission of The University of Chicago Press. **12.5** Gerard Lacz/Animals Animals/Earth Scenes. **12.8b** Redrawn from Figure 1, D. W. Pfennig, 1999, Cannibalistic tadpoles that pose the greatest threat to kin are most likely to discriminate kin, *Proceedings of the Royal Society of London B* 266: 57–61. Reprinted with permission. **12.8c** From D. W. Pfennig, J. P. Collins, and R. E. Ziemba, 1999, A test of alternative hypotheses for kin recognition in cannibalistic tiger salamanders, *Behavioral Ecology* 10: 436–443, Figure 2, by permission of International Society for Behavioral Ecology and the Oxford University Press. **12.9a** From: Short, Roger V,. Reproductive biology: Do the locomotion. *Nature* 418 (July 11, 2002): 137. **12.9b–c** From: H. Moore, K. Dvorakova, N. Jenkins, W. Breed (2002). Exceptional sperm cooperation in the wood mouse, *Nature* 418: 174–177. **12.9d** Reprinted with permission from *Nature*, H. Moore, N. Dvoradova, Jenkins, and W. Breed, 2002, Exceptional sperm cooperation in the wood mouse, *Nature* 418: 174–177, Figure 3. Copyright © 2002 Macmillan Magazines Limited. **12.10a** Courtesy of Bruce E. Lyon, University of California. **12.10b–d** Reprinted with permission from *Nature*, B. E. Lyon, 2003, Egg recognition and counting reduce costs of avian conspecific brood parasitism, *Nature* 422: 495–499, Figures 1b, 3a, and 3b. Copyright © 2003 Macmillan Magazines Limited. **Table 12.2** From R. H. Crozier and P. Parrillo, 1996, *Evolution of Social Insect Colonies*. Copyright © 1996 Oxford University Press. Reprinted by permission of Oxford University Press. **12.13** Modified from J. H. Hunt, 1999, Trait mapping and salience in the evolution of eusocial vespid wasps, *Evolution* 53: 225–237. Copyright © 1999 *Evolution*. Reprinted by permission of *Evolution*. **12.14** Modified from P. Nonacs and H. K. Reeve, 1995, The ecology of cooperation in wasps: Causes and consequences of alternative reproductive decisions, *Ecology* 76: 953–967. Copyright © 1995 Ecological Society of America. Reprinted by permission. **12.15** Raymond A. Mendez/Animals Animals/Earth Scenes. **12.16** Sherman, Paul W.; *The Biology of the Naked Mole Rat*. © 1991 Princeton University Press. Reprinted by permission of Princeton University Press. **12.17a & b** Sarah Blaffer Hardy/Anthro-Photo File. **12.20a** D. Cavagnaro/Visuals Unlimited. **12.20b** Tui De Roy/Minden Pictures. **12.21** Gerald S. Wilkinson, University of Maryland. **12.23** Mitsuaki Iwago/Minden Pictures.

CHAPTER 13 CO.13 © Natalie Fobes/Corbis. **CO.13 (bottom)** Redrawn from A. P. Hendry, Y. E. Morbey, et al. 2004. Adaptive variation in senescence: reproductive lifespan in a wild salmon population, *Proceedings of the Royal Society of London B* 271: 259–266, Figure 2. Copyright © 2004, The Royal Society. Reprinted with permission. **13.1a** Reproduced by permission from E. A. Elbadry and M. S. F. Tawfik, 1966. Life cycle of the mite Adactylidium sp. (Acarina: Pyemotidae), a predator of thrips eggs in the United Arab Republic, *Annals of the Entomological Society of America* 59(3): 458–461, May 1966, p. 460, Fig. 6.; **13.1b** Otorohanga Kiwi House, New Zealand. **13.5** From S. N. Austad and K. E. Fischer, 1991, Mammalian aging, metabolism and ecology: Evidence from the bats and marsupials, *Journal of Gerontology* 46: B47–53, Figure 4, p. B51. Copyright © 1991 Gerontological Society of America. Reprinted by permission. **13.7** Reprinted by permission from Macmillan Publishers, Ltd: *Nature Genetics*, K. S. Joeng, K. S., E. J. Song, et al., 2004, Long lifespan in worms with long telomeric DNA, *Nature Genetics* 36(6): 607–611, Figure 3a, copyright © 2004. **13.8a** Reprinted with permission from *Nature*, Tyner, et al., 2002, p53 mutant mice that display early ageing-assocated phenotypes, *Nature* 415: 45–53, Figure 2a. Copyright © 2002 Macmillan Magazines Limited. **13.8b–c** From: Tyner, Donehower, et al (2002). Mutant mice that display early aging-associated phenotypes, *Nature* 415: 45–53. **13.12** Reprinted with permission from *Nature*, D. Reed and E. Bryant, 2000, The evalution of senescence under curtailed life span in laboratory populations of *Musca domestica* (the housefly), *Heredity* 85: 1115– 1121, Figure 3. Copyright © 2000 Macmillan Magazines Limited. **13.13** Reprinted with permission from *Nature*, D. W. Walker, et al., 2000, Evolution of life span in *C. elegans*, *Nature* 405: 296–297, Figure 1. Copyright © 2000 Macmillan Magazines Limited. **13.14** Reprinted from *Experimental Gerontology*, 41(6), R. J. Mockett and R. S. Sohal, Temperature-dependent trade-offs between longevity and fertility in the *Drosophila* mutant, *methuselah*, 566-673, Figures 1B, 2A. Copyright © 2006, with permission from Elsevier. **13.18a** Reprinted with permission from Kim Hill and A. Magdalena Hurtado. *Ache Life History: The Ecology and Demography of a Foraging People* (New York: Aldine de Gruyter). Copyright © 1996 Walter de Gruyter, Inc., New York. **13.18b & d** Reprinted from K. Hill and A. M. Hurtado, 1991, The evolution of premature reproductive senescence and menopause in human females: An evaluation of the "grandmother hypothesis," *Human Nature* 2(4): 313–351. Copyright © 1991 by Aldine Publishers. Reprinted by permission of AldineTransaction, a division of Transaction Publishers. **13.18f** Copyright James F. O'Connell, University of Utah. **13.26** From D. D. Heath, et al., 2003, Rapid evaluation of egg size in captive salmon, *Science* 299: 1738–1740, Figures 1, 2, and 3. Redrawn with permission from AAAS. **13.27** Timothy A. Mousseau, University of South Carolina. **13.29** From C. W. Fox, M. S. Thakar, and T. A. Mousseau, 1997, Egg size plasticity in a seed beetle: An adaptive maternal effect, *American Naturalist* 149: 149–163. Copyright © 1997 by The University of Chicago Press. Reprinted by permission of The University of Chicago Press. **13.30** Reprinted by permission from *Nature*, W. R. Rice, 1996, Sexually antagonistic male adaptation triggered by experimental arrest of female evolution, *Nature* 381: 232–234. Copyright © 1996 Macmillan Magazines Limited. **13.32** Richard K. Grosberg. **13.33b–c** From L. M. Wolfe, J. A. Elzinga, and A. Biere, 2004, Increased susceptibility to enemies following introduction in the invasive plant *Silene latifolia, Ecology Letters* 7(9): 813–820, Figure 2. Reprinted with permission by Blackwell Publishing. **13.34a & b** Reproduced by permission from G. A. Wray, 1995. Punctuated evolution of embryos, *Science* 267: 1115–1116, Figure 1. Reprinted with permission from AAAS.

CHAPTER 14 CO.14 Eye of Science/Photo Researchers, Inc. **14.1** From E. W. Gilbert, 1958, Pioneer maps of health and disease in England, *Geographical Journal* 124: 172–183. Reprinted by permission of the Royal Geographical Society. **14.3** From R. G. Webster and W. J. Bean, et al., 1992, Evolution and ecology of influenza A viruses, *Microbiological Reviews* 56: 152–179. Reprinted by permission of the American Society for Microbiology. **14.4** From: fig 3, p. 7156: Plotkin & Doshof. Codon bias and frequency-dependent selection on the hemagglutinin epitopes of influenza A virus. *Proceedings of the National Academy of Science USA* 100: 7152–7157. Copyright © 2003 National Academy of Sciences, U.S.A. **14.5** From W. M. Fitch and J. M. Leiter, et al., 1991, Positive Darwinian evolution in human influenza A viruses, *Proceedings of the National Academy of Sciences USA* 88: 4270–4274. Reprinted by permission of the author. **14.6** Adapted by permission from Macmillan Publishers Ltd.: *Nature Reviews Genetics*, R. M. Bush, 2001, Predicting adaptive evolution, *Nature Reviews Genetics* 2(5): 387–392. **14.7** From O. T. Gorman, W. J. Bean, et al., 1991, Evolution of influenza A virus nucleoprotein genes: Implications for the origins of H1N1 human and classical swine viruses, *Journal of Virology* 65: 3704–3714. Reprinted by permission of the American Society for Microbiology. **14.8** From W. J. Bean and M. Schell, et al., 1992, Evolution of the H3 influenza virus hemagglutinin from human and nonhuman hosts, *Journal of Virology* 66: 1129–1138. Reprinted by permission of the American Society for Microbiology. **14.11** From D. J. Austin, K. G. Kristinsson, and R. M. Anderson, 1999, The relationship between the volume of antimicrobial consumption in human communities and the frequency of resistance, *Proceedings of the National Academy of Sciences USA* 96: 1152–1156. Copyright © 1999 National Academy of Sciences, U.S.A. **14.12** From scatterplots in Figure 1 in S. J. Schrag, V. Perrot, and B. R. Levin, 1997, Adaptation to the fitness costs of antibiotic resistance in *Escherichia coli, Proceedings of the Royal Society of London B* 264: 1287–1291. Copyright © 1997, The Royal Society. **14.13** © Wolfgang Kaehler / CORBIS All Rights Reserved. **14.14** From Figure 3, S. L. Messenger, et al., 1999, Virulence evolution in a virus obeys a trade-off, *Proceedings of the Royal Society of London B* 366: 397–404. Copyright © 1999, The Royal Society. **14.15** From *Evolution of Infectious Disease* by Paul W. Ewald, copyright © 1994 by Paul W. Ewald. Used by permission of Oxford University Press, Inc. **14.16** From P. W. Ewald, 1991, Waterborne transmission and the evolution of virulence among gastrointestinal bacteria. *Epidemiology and Infection* 106: 83–119. © 1991 Cambridge University Press. Reprinted with the permission of Cambridge University Press. **14.20** James Holland/Stock Boston. **14.21** From S. B. Eaton and L. Cordain, 1997, Evolutionary aspects of diet: Oldgenes, new fuels, *World Review of Nutrition and Dietetics* 81: 26–37. Copyright © 1997 S. Karger AG. Reprinted by permission of S. Karger AG. **14.22** From T. H. M. Stewart, R. D. Sage, A. F. R. Stewart, & D. W. Cameron, 2000, "Breast cancer incidence highest in the range of one species of house mouse, *Mus domesticus, British Journal of Cancer* 82: 446–451, Figure 1. Reprinted with permission. **14.23** © Charles & Josette Lenars/CORBIS All Rights Reserved. **14.24** From B. I. Strassmann, 1999, Menstrual cycling and breast cancer: An evolutionary perspective, *Journal of Women's Health* 8: 193–202. Reprinted by permission of Mary Ann Liebert, Inc. **14.25b** Reprinted by permission from *Nature*. L. K. Vaughnn, H. A. Bernheim, and M. Kluger, 1974, Fever in the lizard *dipsosaurus dorsalis*, *Nature* 252: 473–474. Copyright © 1974 Macmillan Magazines Limited. **14.25c** Reprinted with permission from M. J. Kluger, D. H. Ringler, & M. R. Anver, 1975, Fever and survival. *Science*: 188: 166–168. Copyright © 1975 AAAS. **14.27** From N. M. H. Graham, C. J. Burrell, R. M. Douglas, P. Debelle, and L. Davies, 1990, Adverse effects of aspirin, acetaminophen, and ibuprofen on immune function, viral shedding, and clinical status in rhinovirus-infected volunteers, *Journal of Infectious Diseases* 162: 1277–1282. Copyright © 1990 *Journal of Infectious Diseases*. Reprinted by permission of The University of Chicago Press. **14.31a–b** From M. V. Flinn and B. G. England, 1995, Childhood stress and family environment, *Current Anthropology* 36: 854–866,

Figures 3, 7. Copyright © 1995 by The University of Chicago Press. Reprinted with permission. **14.31c** From Social economics of childhood gluocorticoid stress response, M. V. Flinn and B. G. England, *American Journal of Physical Anthropology* 102 (1997): 33–53, Figure 5. Copyright © 1997 John Wiley & Sons, Inc. Reprinted with permission of Wiley-Liss, Inc., a subsidiary of John Wiley & Sons, Inc. **14.31d** Reprinted from *Evolution & Human Biology*, 9, M. V. Flinn, Step and genetic parent/offspring relationships in a Caribbean village, 335–369, Copyright © 1988, with permission from Elsevier. **14.32** Reprinted with permission from Martin Daly and Margo Wilson. Homicide. (New York: Aldine de Gruyter). Copyright © 1988 by Aldine de Gruyter, Inc., New York.

CHAPTER 15 CO.15 © Karen Kasmauski/CORBIS All Rights Reserved. **15.1** From R. J. W. O'Neill, et al., Undermethylation associated with retroelement activation and chromosome remodelling in an interspecific mammalian hybrid, *Nature*, vol. 395, May 7, 1998, 68–72. **15.4** Reprinted from *Current Biology*, 8, W. F. Doolittle and J. M. Logsdon, Archaeal genomics: Do archaea have a mixed heritage? R209–R21, Copyright © 1998, with permission from Elsevier. From W. F. Doolittle, Uprooting the tree of life, *Scientific American* 90–95, Feb. 2000. Illustrated by C. Blumrich. Reprinted by permission. **15.5** From S. J. Giovannoni, S. Turner, et al., 1988, Evolutionary relationships among cyanobacteria and green chloroplasts, *Journal of Bacteriology* 170: 3584–3592, Figure 3. Reprinted by permission of the American Society of Microbiology. **15.9** Reprinted with permission from S. J. Giovannoni et al., 2005, Genome streamlining in a cosmopolitan oceanic bacterium, *Science* 309: 1242–1245. Copyright © 2005 AAAS. **15.10** Three photos depicting left to right: Plant infected with normal cells of the bacterial parasite *Erwinia carotovora*; Middle and right show smaller lesions caused by infections with strains of Erwinia carotovora that have a specific virulence gene knocked out. Originally published as figure 4, p. 11109 from: Bell et al, 2004, "Genome sequence of the enterobacterial phytopathogen *Erwinia carotovora* subsp. *atroseptica* and the characterization of virulence factors, *Proceedings of the National Academy of Sciences USA* 101: 11105–11110. Copyright © 2004 National Academy of Sciences, U.S.A. **15.12** Dennis Kunkel/ Phototake NYC.

PART IV © Louie Psihoyos/CORBIS All Rights Reserved.

CHAPTER 16 CO.16 (LEFT) Andrew A. Forbes. **CO.16 (right)** Tom Murray. **16.2** Courtesy of Dr. Carol Eunmi Lee, University of Wisconsin–Madison. **16.4a** Courtesy of Dr. Lori S. Eggert/Division of Biological Sciences/University of CA, San Diego. **16.4b** Mark Boulton/Photo Researchers, Inc. **16.6** Kenneth Y. Kaneshiro, University of Hawaii. **16.8a** Carl C. Hansen/Nancy Knowlton/ Smithsonian Institution Photo Services. **16.8b** Reprinted with permission from N. Knowlton, L. A. Weigt, L. A. Solorzano, D. K. Mills, and E. Bermingham, 1993, Divergence in proteins, mitochondrial DNA, and reproductive incompatibility across the isthmus of Panama. *Science* 260: 1629–1632. Copyright © 1993 AAAS. **16.9** From J. L. Feder, J. B. Roethele, B. Wlazlo, and S. H. Berlocher, 1997, Selective maintenance of allozyme differences among sympatric host races of the apple maggot fly, *Proceedings of the National Academy of Sciences USA* 94: 11417–11421. Copyright © 1997 National Academy of Sciences, U.S.A. **16.10ab** Kenneth Kaneshiro/University of Hawaii/CCRT. **16–10abR** Kenneth Y. Kaneshiro, University of Hawaii. **16.11** From D. R. B. Boake, M. P. DeAngelis and D. K. Andreadis, 1997, Is sexual selection and species recognition a continuum? Mating behavior of the stalk-eyed fly *drosophila heteroneura*, *Proceedings of the National Academy of Sciences USA* 94: 12442–12445. Copyright © 1997 National Academy of Sciences, U.S.A. **16.14** From H. Wang, E. D. McArthur, S. C. Sanderson, J. H. Graham, and D. C. Freeman, 1997, Narrow hybrid zone between two subspecies of big sagebrush (*Artemisia tridentata: asteraceae*). IV. Reciprocal Transplant Experiments, *Evolution* 51: 95–102. Copyright © 1997 *Evolution*. Reprinted by permission of *Evolution*. **16.15a** Prof. Sara Via/University of Maryland. **16.15b** From S. Via & D. J. Hawthorne, 2002, The genetic architecture of ecological specialization: Correlated gene effects of host use and habitat choice in pea aphids, *American Naturalist* 159: S76–S88, Figure 1. Copyright © 2002 by The University of Chicago Press. Reprinted with permission ofThe University of Chicago Press. **16.16a** Jerry Young © Dorling Kindersley. **16.16b–c** Courtesy of Ernie Cooper/WWF Canada & Dolph Schluter/University of British Columbia. **16.17** Adapted by permission from Macmillan Publishers, Ltd.: *Nature*, C. L.Peichel, et al., 2001, The genetic architecture of divergence between threespine stickleback species, 414 *Nature*: 901–905, Figure 3a, b, d, copyright © 2001. **16.18** Reproduced with permission from Japanese marine (top) and freshwater limnetic benthic Paxton Lake (bottom) species of sticklebacks. These fish have been stained with alizarin-red, which heightens bony tissue. Originally published as figure 1A from: Kingsley, D., et al. Widespread parallel evolution in sticklebackss by repeated fixation of ectodysplasin alleles. *Science*, 25 March 2005:Vol. 307. No. 5717, pp. 1928–1933. Copyright 2005 AAAS.

CHAPTER 17 CO.17 NASA/Jet Propulsion Laboratory. **CO-17 (bottom)** NASA/JPL/University of Arizona. **17.3** From D. R. Mills, R. I. Peterson, and S. Spiegelman, 1967, An extracellular Darwinian experiment with a self-duplicating nucleic acid molecule, *Proceedings of the National Academy of Sciences USA*, 58: 217–220, Figure 1. Reprinted with permission. **17.4 & 17.5** Reprinted with permission from A. A. Beaudry and G. F. Joyce, 1992, Directed evolution of an RNA enzyme, *Science* 257: 635–641, Figures 1 and 4. Copyright © 1992 AAAS. **17.8a–b** Redrawn with permission from W. K. Johnston, P. J. Unrau, et al., 2001, RNA-catalyzed RNA polymerization: Accurate and general RNA-templated primer extension, *Science* 292: 1319–1325, Figures 3 and 4. Copyright © 2001 AAAS. **17.8c** Reproduced with permission from: *Science* 292: Fig #4ABC, pp. 1319–1325. Johnston, W. K., et al. Copyright 2001 AAAS. **17.9** Lab Photo: Russell Kempton, New England Meteoritical Services. **17.10** NASA/Johnson Space Center. **17.13** Photo from James P. Ferris, NY Center for Studies on the Origins of Life, Rensselaer Polytechnic Institute, Troy, NY 12180. **17.14a–b** Geologisk Museum; **17.15** From R. F. Gesteland, T. R. Cech, and J. F. Atkins, eds., *The RNA World* 2nd ed. (Cold Spring Harbor, NY: Cold Spring Harbor Laboratory Press). Copyright © 1998 John F. Atkins and Raymond F. Gesteland. **17.16** Andrew H. Knoll, Harvard University; **17.18 (left)** Dennis Kunkel/Phototake NYC. **17.18 (middle)** B. Boonyartana/Visuals Unlimited. **17.18 (right)** A. Syred/Photo Researchers, Inc. **17.22** From J. R. Brown and W. F. Doolittle, 1997, Archaea and the prokaryote-to-eukaryote transition, *Microbiology and Molecular Biology Reviews* 61: 456–502, Figure 4. Reprinted with permission. **17.23** J. R. Brown,

2001, Genomic and phylogenetic perspectives on the evolution of prokaryotes, *Systematic Biology* 50: 497–512, Figure 2. Redrawn and reprinted with permission. **17.24** From F. D. Ciccarelli, T. Doerks, et al., 2006, Toward automatic reconstruction of a highly resolved tree of life, *Science* 311: 1283–1287, Figure 2. Reprinted with permission from AAAS. **17.25** From S. Yang, R. F. Doolittle, and P. E. Bourne, 2005, Phylogeny determined by protein domain content, *Proceedings of the National Academy of Sciences USA* 102: 373–378, Figure 4. Copyright © 2005 National Academy of Sciences, U.S.A. **17.26** Reprinted with permission from W. F. Doolittle, 1999, Phylogenetic classification and the universal tree, *Science* 284: 2124–2128, Figure 3. Copyright © 1999 AAAS. **17.27a–c** Andrew H. Knoll, Harvard University; **17.28** Tsu-Ming Han; **17.29a–h** J. William Schopf, University of California at Los Angeles; **17.35** Reprinted with permission from D. Raoult, S. Audic, et al., 2004, The 1.2-megabase genome sequence of Mimivirus, *Science* 306: 1344–1350, Figure S6. Copyright © 2004 AAAS.

CHAPTER 18 CO.18 Chip Clark. **18.1a** From: A. Chandrasekharam, 1974, Megafossil flora from the Genesee locality, Alberta, Canada. *Paleontographica B* 147: 1–41. Reproduced by permission of E. Schweizerbart'sche Verlagsbuchhandlung, Stuttgart, Germany. **18.1b** 0-521-38294-7. Stewart, W.N. and Rothwell, G.W. *Paleobotany and the Evolution of Plants*, 2nd ed.p. 15, Fig. 2.7D (Fig. 2.5D in 1st ed.) from Roth & Dilcher. Reproduced by permission of E. Schweizerbart'sche Verlagsbuchandlung, Stuttgart, Germany. **18.2** Ted Clutter/Photo Researchers, Inc. **18.3a** Francois Gohier/Photo Researchers, Inc. **18.3b** John D. Cunningham/Visuals Unlimited; **18.4** John Koivula/Science Source/Photo Researchers, Inc. **18.5** Paleogeographic Maps by Christopher R. Scotese, ©2003 PALEOMAP Project (*www.scotese.com*). **18.6a** Simon Conway Morris, University of Cambridge, Cambridge, United Kingdom. **18.6b** Reproduced with permission from 560–575 million-year-old fossil specimens of Rangeomorph frondlets from Ediacaran fauna, Spaniards Bay, Newfoundland. Originally published as the cover and figure 2A from Narbonne, G., Modular Construction of Early Ediacaran Complex Life Forms. *Science* 20 August 2004: Vol. 305. no. 5687, pp. 1141–1144. Copyright 2005 AAAS. **18.7** Reproduced by permission from S. Xiao, Y. Zhang, and A. H. Knoll, Three-dimensional preservation of algae and animal embryos in a Neoproterozoic phosphorite. *Nature* 391:553–558 (February 5, 1998), fig. 5. Copyright © 1998 Macmillan Magazines Limited. Image courtesy of Shuhai Xiao, Tulane University. **18.8a** Courtesy of Dr. Guy M. Narbonne/Queen's University, Department of Geological Sciences and Geological Engineering. **18.8b** Courtesy of Dr. Benjamin Martin Waggoner, Dept of Biology, University of Central Arkansas. **18.8c** Reproduced with permission from Fig B1: Jun-Yuan Chen, David J. Botjer, et al; Small bilaterian fossils from 40 to 55 million years before the Cambrian. *Science* 9: Vol. 305. no. 5681, pp. 218–222. Copyright 2000 AAAS. **18.9 (left & right)** Chip Clark. **18.10a** Photo by permission of Degan Shu, Northwest University, Xi'an, China. Reproduced from D.G. Shu et al., Lower Cambrian vertebrates from south China, *Nature* 402: 42, November 4, 1999. **18.10b** Reprinted with permission from *Nature*, D. G. Shu, et al., 1999, Lower Cambrian vertebrates from south China, *Nature* 402: 42–46, Figure 4. Copyright © 1999 Macmillan Magazines Limited. **18.11a–b** A. J. Copley/Visuals Unlimited. **18.12** Reprinted with permission from A. Knoll and S. Carroll, 1999, Early animal evolution: Emerging views from comparative biology and geology, *Science* 284: 2129–2137. Copyright © 1999 AAAS. **18.14** Sandra Floyd/USDA/ARS/Agricultural Research Service. **18.17a** Hugh Spencer/Photo Researchers, Inc. **18.17b** Sinclair Stammers/Science Photo Library/Photo Researchers, Inc. **18.18a** Biological Photo Service. **18.18b** Francois Gohier/Photo Researchers, Inc. **18.22** Reprinted with permission from M. J. Benton, 1995, Diviersification and extinction in the history of life, *Science* 268: 52–58. Copyright © 1995 AAAS. **18.24 & 18.25** Reprinted with permission from D. Jablonski, 1986, Background and mass extinctions: The alternation of evolutionary regimes, *Science* 231: 129–329. Copyright © 1986 AAAS. **18.26a** Alessandro Montanari/Geological Observatory of Coldigioco, Frontale di Apiro, Italy. **18.26b** Reprinted from permission from W. Alvarez, F. Asaro, and A. Montanari, 1990, Iridium profile for 10 million years across the Cretaceous-Tertiary boundary at Gubbio (Italy), *Science* 250: 1700–1702. Copyright © 1990 AAAS. **18.27a–b** Glen A. Izett/U.S. Geological Survey, Denver. **18.28** Peter H. Schultz, Brown University; Steven D'Hondt, University of Rhode Island Graduate School of Oceanography. **18.29& 18.30** Reprinted with permission from D. Jablonski and D. M. Raup, 1995, Selectivity of end-Cretaceous marine bivalve extinctions, *Science* 268: 389–391, Figures 1 and 3. Copyright © 1995 AAAS. **18.31** Reprinted with permission from D. W. Steadman, 1995, Prehistoric historic extinctions of Pacific island birds: Biodiversity meets zooarchaeology, *Science* 267: 1123–1131. Copyright © 1995 AAAS.

CHAPTER 19 CO.19 Reproduced by permission from From Brunetti, C. R. , Selegue, J. E., Monteiro, A., French, V., Brakefield, P. M., and Carroll, S. B. The generation and diversification of butterfly eyespot color patterns. *Current Biology* 11: 1578–1585. Copyright © 2001 by Elsevier Science Ltd. Image courtesy of Current Biology. **19.1** From J. Gerhart and M. Kirschner, 1997, *Cells, Embryos, and Evolution* (Malden, MA: Blackwell Science). Copyright © 1997 Blackwell Science, Inc. Reprinted by permission. **19.2a** Edward B. Lewis, California Institute of Technology. **19.2b** F. Rudolf Turner, Indiana University. **19.3** Reprinted with permission from *Nature*, R. de Rosa et al., 1999, Hox genes in brachiopods and priapulids and protostome evolution, *Nature* 399: 772–776. Copyright © 1999 Macmillan Magazines Limited. **19.4a** Rod Planck/Photo Researchers, Inc. **19.4b** Marty Snyderman/Visuals Unlimited. **19.4c** Tom McHugh/Photo Researchers, Inc. **19.4d** Jeffrey Howe/Visuals Unlimited. **19.4e** Dr. Morley Read/SPL/Photo Researchers, Inc. **19.5** Adapted with permission from A. H. Knoll and S. B. Carroll, 1999, Early animal evolution: Emerging views from comparative biology and geology, *Science* 284: 2129–2137. Copyright © 1999 AAAS. **19.13 & 19.14** Reproduced and reprinted by permission from Grace Panganiban et al., The development of crustacean limbs and the evolution of anthropods. *Science* 270:1363–1366 (November 24, 1995). © 1995 AAAS. **19.16** After E. Coen, 1999, *The Art of the Gene*, Oxford University Press, p. 62, Figure 4.4. Copyright © 1999 Oxford University Press. Reprinted by permission of Oxford University Press. **19.17a** Elliot M. Meyerowitz, California Institute of Technology. **19.17b** M.P. Running/Elliot M. Meyerowitz, California Institute of Technology. **19.19** Reprinted by permission from *Nature*, F. Parcy, et al., 1998, A genetic framework for floral patterning, *Nature* 395: 561–566, Figure 7. Copyright © 1998 Macmillan Magazines, Ltd.; reprinted from *Trends in Plant Science*, 9, M. Buzgo, D. E. Soltis, P. S. Soltis, and H. Ma, Towards a comprehensive integration of morphological and genetic studies of floral development, 164–173, Copyright © 2004, with permission from Elsevier.

CHAPTER 20 CO.20 Euan Denholm/Reuters/Landov LLC; **CO.20 (bottom)** Reprinted by permission from *Nature*. From: Figure 2b in Alemseged, Z., F. Spoor, et al. 2006. "A juvenile early hominin skeleton from Dikka, Ethipia" Nature 443: 296–301. Copyright © 2006 Macmillan Magazines Limited. Courtesy of Zeresenay Alemseged; Max Planck Institute for Evolutionary Anthropology, Germany. **20.2** Reprinted with permission from V. M. Sarich and A. C. Wilson, 1967, Immunological time scale for Hominid evolution, *Science* 158: 1200–1203. Copyright © 1967 AAAS. **20.6** R. L. Stauffer, A. Walker, O. A. Ryder et al., Human and ape molecular clocks and constraints on paleontological hypotheses, *Journal of Heredity*, 2001, 92: 469–474, by permission of Oxford University Press. **20.7** Reprinted from *Trends in Genetics*, 17, J. G. Hacia, Genome of the apes, 637–645, Copyright © 2001, with permission from Elsevier. **20.8** Reprinted with permssion from W. Enard, P. Khaitovich, et al., 2002, Intra- and interspecific variation in primate gene expression patterns, *Science* 296: 340–343, Figure 2. Copyright © 2002 AAAS. **20.9** Reprinted with permission from *Nature*, B. Wood, 2002, Hominid revelations from Chad," *Nature* 418: 133–135, Figure 1. Copyright © 2002 Macmillan Magazines Limited. **20.10a** David L. Brill/Brill Atlanta. **20.10b–f** From Don McGranaghan in I. Tattersall, 1995, *The Fossil Trail* (Oxford: Oxford University Press). Reprinted with permission of the Department of Anthropology, American Museum of Natural History. **20.11 (left & right)** John Reader/Science Photo Library/Photo Researchers, Inc. **20.12, 20.13, & 20.14** By Don McGranaghan in I. Tattersall, 1995, *The Fossil Trail* (Oxford: Oxford University Press), 131, 195. Reprinted with permission of Department of Anthropology, American Museum of Natural History. **20.15** Reprinted with permission from *Nature*, B. Wood, 2002, Hominid revelations from Chad, *Nature* 418: 133–135, modified from Figure 2. Copyright © 2002 Macmillan Magazines Limited. **20.17a** National Museums of Kenya, Nairobi. ©1985 David L. Brill. **20.17b** National Museums of Kenya, Nairobi. ©1994 David L. Brill. **20.19** From P. Gagneux, C. Wills, et al., 1999, Mitochondrial sequences show diverse evolutionary histories of African homindoids, *Proceedings of the National Academy of Sciences USA* 96: 5077–5082, Figure 2. Copyright © 1999 National Academy of Sciences, U.S.A. **20.20** Reprinted with permission from N. A. Rosenberg, J. K. Pritchard, et al. 2002, Genetic structure of human populations, *Science* 298: 2381–2385, Figure 1. Copyright © 2002 AAAS. **20.23** Reprinted with permission from *Nature*, M. Ingman, et al., 2000, Mitrochondrial genome variation and the origins of modern humans, *Nature* 408: 708–713. Copyright © 2000 Macmillan Magazines Limited. **20.29 & 20.30** Reprinted with permission from R. L. Susman, 1994, Fossil evidence forearly hominid tool use, *Science* 265:1570–1573. Copyright © 1994 AAAS. **20.31** Arensburg, Schepartz, Tillier, Vandermeersch, and Rak, A reappraisal of the anatomical basis for speech in middle paleolithic hominids. From *American Journal of Physical Anthropology* 83: 137–146. © 1990 American Association of Physical Anthropologists. Reprinted by permission of Wiley-Liss, Inc., a subsidiary of John Wiley & Sons, Inc.

Index

ABC model of flower development, 744, **744**, 745, 746
Abdominal-B (Abd-B) genes, 731, 734
Abrahamson, Warren, 348, **348**
Abzhanov, Arhat, 85
Acacia (*Acacia greggii*), 512–14, **512**, **513**
Acetaminophen, 558, **558**, 559
Ache Indians, **503**, 504, 566
Achillea, **354**
 use in common garden experiment, 351–52, **351**, 353–54
Achondroplasia, 148
Achromatopsia, 238
Acid rain, 714
Acon-2 95, 619, **619**
Acquired immune deficiency syndrome (AIDS), 3. *See also* Human immunodeficiency virus (HIV)
 in Asia, **5**
 CCR5-Δ32 allele frequency and, 161–62, 191–94, **191**
 combination therapies for, 16–17
 infection mechanism, 8–11
 natural history of epidemic of, 4–11, **5**
 in North America and Europe, 5, **5**
 prevention successes and failures, 6, **6**
 in sub-Saharan Africa, 5, **5**
Acute hemolytic anemia, 300
Acyrthosiphon pisum (pea aphids), 629–31, **630**
Adactylidium, 484, **484**
Adaptation(s), 93, 361, 363–400. *See also* Aging; Kin selection, evolution of altruism and; Sexual selection
 caveats in studying, 367
 comparative analysis of, 376–80, 396
 constraints on, 385–91
 defined, 77, 364
 experiments on, 367–72
  design issues, 367–71, **370**
  replication of, 370
 hypothesis-formation strategies, 395–96
 lateral gene transfer, significance of, 589–90
 medical/health, 550–68
  breast cancer, 552–55
  environment and, 550–51
  fever, 556–60
  menstrual cycle, 554–55, **555**
  myopia, 551–52
  parenting, 561–68
 molecular basis of, 591–96
  patterns in genomes of free-living organisms, 593–94
  patterns in parasite genomes, 591–93, **591**
  regulatory sequences, evolution of, 594–96
 mutualism, 447
 observational studies of, 370, 372–75
  behavioral thermoregulation, 372–75
  garter snake nighttime behavior, 373–75, **374**, **375**
 oxpeckers, 364–67, **364**
  mutualist relationship with cattle and, hypothesis of, 365–66, **365**, **366**
  parasitic relationship with cattle, hypothesis of, 366–67
 perceived perfection and complexity of, 98
 phenotypic plasticity, 380–83, 564
  in clutch size, 507
  in egg size in beetle, 510, 512–14, **513**
  in water flea behavior, 381–83, **381**, **383**
 selection and, 364

Note: Pages in **bold** locate figures; pages in *italic* locate tables.

 selection operating on different levels and, 392–95
  in Apert syndrome, 394–95, **395**
  demonstration of, 392–94
  transposable elements, 576–84
 trade-offs and constraints in, 383–91
  begonia flower size, 383–85, **384**
  fuchsia color change, 385–88, **386**
  host shifts in *Ophraella*, 388–90
Adaptationist program, 364, 550–55
Adaptive landscapes, 199
Adaptive radiations, 702–4, **703**
Adaptive significance of sex, 303–12
 environmentally-imposed selection and, 310–12
 genetic drift and, 308–10
 genetic recombination, 306–7
 reproductive modes and, 303–6
Additive genetic variation (V_A), 335, 336, **336**
Adenine, 144, **144**, 145, 656
Adenoma, 549
Adenosine deaminase deficiency, 546–48, **547**
Adenosine triphosphate (ATP), 590, 599
Adolph, S.C., 380
AER (apical ectodermal ridge), 737, **737**
Aeromonas hydrophila, 556–57
Africa, HIV/AIDS in sub-Saharan, 5
African elephants (*Loxodonta africana*), 610, **611**
African great apes, 754–56
African replacement model, 774–85, **775**, **784**
 archaeological and paleontological evidence of, 775–78, **778**
 cladogram predicted by, **778**
 genetic predictions distinguishing, 780–81, *780*
 molecular evidence of, 779–85
AGAMOUS (*AG*) gene, 745
Agassiz, Louis, 55
age-1 gene, 496–97, **497**
Aging, 487–502
 in birds, **487**
 cancer risk and, 490–91
 evolutionary theory of, 487, 492–502
  deleterious mutations and, 492, **493**, 494–96, 502
  natural experiments of, 500–502
  trade-offs and, 490, 496–99, 502
 in insects, **487**
 in mammals, **487**
 rate-of-living theory of, 487, 488–92
  trade-offs and, 490–91, **491**
Agonistic parent-child interaction, 564–65
Agrawal, Alka, 582–83
Ågren, Jon, 384–85
Ahounta, Djimdoumalbaye, 765
AIDS. *See* Acquired immune deficiency syndrome (AIDS)
AIDSVAX, 29–30
Aiello, L.C., 775
Aiuti, Alessandro, 548
Alanine, 655
Alarm calling in Belding's ground squirrels and prairie dogs, 450–51, **450**, **451**, **452**
Albumin, similarities between human and ape serum, 755–56
Alcohol dehydrogenase (ADH), 185, **185**, 187
Alcohol dehydrogenase (*Adh*) gene, 259
Alemseged, Zeresenay, **753**
Allele frequency(ies)
 calculating, 163
 CCR5-Δ32, 161–62, 182, 191–94, **191**
 chance events and, 232–35, **233**
 changes in, 90. *See also* Genetic drift; Migration; Mutation(s); Natural selection
 in general case, 177–80, **178**
 homogenization of, migration and, 230–32
 migration and, 225–32
 under mutation-selection balance, 215
 numerical calculation, 174–77, **174**, **176**, **177**
 predicting, in *CCR5-Δ32* allele, 194
 selection and changes in, 183–87, **183**, **185**
  empirical research on, 185–87
 selfing and, 264–66, **265**, *265*
 in simulation of life cycle, 170–73, **173**
 temperature and, **619**
Alleles
 defined, 146
 dominant and recessive
  mutation rates for, 216
  selection on, 195–97, 198–99, **199**
 greenbeard, 457–59
 neutral, 152
 new, 144–52
  from replacement substitutions, 147–48
 random fixation of, 238–49
  in natural populations, 245–49
  probability of allele to drift toward, 241
 resistance, 23–25
Allelic diversity, 163–66
Allelic richness, 248
Allelic state, correlation of, 290
Allopatric speciation, 612
Allozyme electrophoresis, 164
α-Tubulin, 100
Alpine skypilot (*Polemonium viscosum*), flower size in, 342–43, **343**
Altheus malleator, **614**
 geographic isolation through vicariance, **614**
Altman, Sidney, 641
Altruism, 94, 447, 448
 in birds, 451–53, 464
 as central paradox of Darwinism, 448
 kin selection and evolution of, 448–59
  inclusive fitness, 448–50, 471
  kin recognition, 454
 parasitized, **457**
 parent-offspring conflict and, 467–71, **468**
 reciprocal, 471–77
  blood-sharing in vampire bats, 474–75, **474**, **475**
  conditions for, 472
  territory defense in lions, 475–77, **476**
 reproductive, 459–67
Altruistic sperm, 454–56, **456**
Alu elements, 759, 784
Alu sequences, 578, 784–85, **784**
Alvarez, W., 712
Amazon region, deforestation in, 720, **720**
Amblyrhynchus cristatus. *See* Marine iguana (*Amblyrhynchus cristatus*)
Amborella, 704, **704**
Ambulocetus natans, 48–49, **49**
Ambystoma tigrinum, 453
Amenorrhea, lactational, 554, **555**
Amino acids in meteorites, 651
Ammonites, 716
Amniota, 114, **114**
Amoeba dubia, 576
Amphibia, 114, **114**
Amyloid plaques, 189
Anagenesis (phyletic transformation), 705
Anders, Edward, 654
Anderson, David, 470–71
Anderson, Kermyt, 566
Anderson, Virginia, 366
Andersson, Dan, 309
Andersson, Malte, 416

Andersson, Steffan, 416–17
Anemia, acute hemolytic, 300
Angiosperms
  Cretaceous explosion of, 704, **704**
  evolution of flower, 742–47
  inbreeding depression in, 271, **272**
Anhydrite, 714
Animals, **664**, 665–66
  diversification of. *See* Diversification of animals
  domestic, 74–75
  phylogeny and morphology of Cambrian, 698–700, **699**
Annelida, **699**
Anopheles gambiae (mosquito), genome sequences of, 593
Anstey, Robert, 707
Antagonistic pleiotropy hypothesis, 496–99
Ant and fungi symbionts, phylogenies of, **138**
Antennapedia (Antp) genes, 731
Antennapedia complex, 729, 730
Antibiotic resistance, 538–41
  costs to bacteria of, 539–40, **540**
  genes conferring, carried by plasmids, 592
  judicious use of antibiotics to inhibit, 541
  selection and, 538–39
Antibiotics, 530, **530**
  judicious use of, 541
Antibodies, 740–41
  anti-*Dll*, 740–41
Antifreeze proteins, 102
Antigenic sites, 532–33
Antigen recognition site (ARS), 257
Antonelli, Peter, 676
Anver, Miriam, 556
Apert syndrome, multilevel selection in, 394–95, **395**
Apes
  great, 754–56
    brain size in, 790–91, **791**
    genetic diversity among, **776**
  phylogeny of, **755**, **756**
  relationship between humans and extant, 754–63
    divergence times, 761, **761**
    molecular evidence of, 755–59
    morphological evidence of, 759–61
APETALA1 and *APETALA2* (*AP1* and *AP2*) genes, 744–45, 747
APETALA3 and *PISTILLATA* (*AP3* and *PI*) genes, 744–45
Apex chert fossils, 662
Aphids
  cospeciation in, 135–36, **135**, **136**
  pea, 629–31, **630**
  sexual and asexual reproduction in, **303**, 304
Apical ectodermal ridge (AER), 737, **737**
Apodemus sylvaticus, 454–56
Apple maggot fly (*Rhagoletis pomonella*), **605**, **619**
  divergence through natural selection, 617–19
Aptamers, 647
Apteryx australis mantelli (brown kiwi), 42, **42**, 484, **484**
Arabidopsis, 587–88
Arabidopsis thaliana (weedy mustard), 295, **295**, 691, 726, 727, 746
  mutant screen in, 743–44, **744**
Archaea, 664–65, 667–68, 669, 673
  hypotheses of origins of
    chronocyte hypothesis, 677, **677**
    ring of life hypothesis, 676–77, **676**
    three viruses, three domains hypothesis, 677–80, **679**
    universal gene-exchange pool hypothesis, 675–76, **675**
  lateral gene transfer in, 590–91
Archaeoglobus fulgidus, 584, **584**
Archaeological record. *See also* Fossil record
  African replacement model vs. multiregional origin model and, 775–78, **778**
  evidence of language in, 789–90
Archaeopteryx, 46–47, **46**
Ardipithecus, 767
  A. ramidus, **766**, 768
Arensburg, B., 790
Arg151 Cys, **95**
Argument from Design, 98
Arnold, S.J., 344, **427**
Artemesia tridentata tridentata (basin big sagebrush), 627, **629**
Artemesia tridentata vaseyana (mountain sagebrush), 627, *628*
  hybrid zone between subspecies of, 627–28
Artemia franscicana (brine shrimp), 741
Artemia salina (fairy shrimp), **708**
Arthropods, **699**
  diversification of, 703
  limb evolution in, 739–41, **740**, **741**
    forms of appendages, 739, **740**
  segmentation of, 732–35, **733**, **734**
Artifact hypothesis of menopause, 504
Artificial selection, 74–75
  genetic recombination and, **307**
Artiocetus clavis, 129
Artiodactyla, 120–22, **120**, 131
Artiodactyla hypothesis, 121, **121**, 123
Ascertainment bias in current extinction rate data, 719
Asexual reproduction
  changing vs. fixed environment and, 310
  comparison of sexual reproduction and, 303–6, **304**
  mutation and, 308–10, **308**, **309**
  in water flea, 381
Ashkenazi population, 295–98
  Δ32 allele in, 163
  GBA-84GG mutation in, 296–97
Ashton, Peter, 719
Asia
  AIDS epidemic in, **5**
  tropical forest destruction in, 720
Asian elephants (*Elephas maximus*), 610, **611**
Assortative mating, 426, **427**, 624
Astragalus, 120–22, **120**
Astrobiology, 652
Asymmetries in sexual reproduction, 403–4
  in fruit flies, 404
Asymmetry of sex, behavioral consequences of, 407–8
Atkins, John F., **661**
Atmosphere
  after asteroid impact at K-T boundary, 714, 715
  composition of the early, 653–54
Austad, Steven, 488, 500, 502, **503**, 504
Austin, A.J., 539
Australian aborigines, 777
Australopithecines
  brain size in, 790–91, **791**
  gracile, **766**, 767, 768, 788
Australopithecus, 767
  A. aethiopicus, **768**
  A. afarensis, **753**, **766**, 767, 768, 788
  A. africanus, **766**, 767
  A. anamensis, **766**, 768
  A. boisei, **768**
  A. gahri, **766**, 767
  A. robustus, **768**
Autopod, 736, **736**, 738, **738**, 739, **739**
Average excess, 186
Average fitness, 186
Aves, 717
Avifauna extinctions, Polynesian, 717–18
Avise, John, 708
Axelrod, Robert, 473
Axoneme, 100, 101
Ayala, F.J., **244**, 775
AZT (azidothymidine), 11–16, 77
  blockage of reverse transcriptase, **11**
  resistance to, evolution of, **12**, 13–15, **13**, **14**
  reverse transcriptase and, 11–15

Bacillus subtilis, 653
Backcross, 626
Background extinctions, 710–11
Background selection, 263
Back mutation, 15, 210, 308, 546–48
Bacteria, 664, **664**, 665, 667, 669, 673. *See also* Antibiotic resistance; *specific bacteria*
  disease caused by, 529
  endosymbiosis and, 585, 586–89
  genetic recombination in, 607
  hypotheses of origins of
    chronocyte hypothesis, 677, **677**
    ring of life hypothesis, 676–77, **676**
    three viruses, three domains hypothesis, 677–80, **679**
    universal gene-exchange pool hypothesis, 675–76, **675**
  intestinal, virulence of, 545, **545**
  lateral gene transfer in, 585, 590–91, 670, **670**
  Martian, 652, **652**
  purple, 664
  species concept in, 607
  stromatolite-forming, **707**
Bacteriocytes, 135
Bacteriophage f1, 543–44
Bacteriophage T4, 149
Baker, Richard, 428
Baker's yeast (*Saccharomyces cerevisiae*), 59, 151, 578, 691
  codon bias in, *261*, **262**
Balance (selectionist) theory, 165
Balanus glandula (barnacle), 523
Balmford, Andrew, 720
Bamshad, Michael, 777
Banded mongooses, **447**
Banet, M., 557
Barbulescu, Madalina, 758
Barbus, 616
Barghoorn, Elso, 662
Barnacle (*Balanus glandula*), 523
Barns, Susan, 668
Barn swallows, female choice in, 419
Bartel, D.P., 648–50
Barton, Nick, 763
Bases in DNA, 144–45, **144**
Base substitution, 147–48
Basilosaurus isis, 48, **49**
Basin big sagebrush (*Artemesia tridentata tridentata*), 627, *628*, **629**
Bateman, A.J., 403, 404, 407
Bates Smith, Thomas, 348, **349**
Bathyergidae, 465
Bats
  fruit, 376–79, **376**, **379**
  metabolic rates of, 488
  testes size in, 376–79, **376**, **379**
  vampire, 474–75, **474**, **475**
Bäurle, Silke, 410, 412
Bayesian Markov Chain Monte Carlo (BMCMC) methods, 124, 125
Beak shape in Galápagos finches, evolution of, 80–90, **83**, **85**, **87**, **88**, **89**
Bean, W.J., 536
Bears, phylogenetically independent contrasts among, 380–81, **381**
Beaudry, A.A., 645–47, 649
Beetles
  eusociality in, *460*
  flour beetle, 304–6, **306**
  host shifts in herbivorous, 388–90
  phenotypic plasticity in egg size in, 510, **513**
Begonia involucrata, 383–85, **384**
Begun, David R., 292–93, 760, 764, 773
Behavior
  asymmetry of sex and, 407–8
  social. *See* Social behavior
Behavioral fever, 556–57, **557**
Behavioral mechanisms of inbreeding avoidance, 272
Behavioral thermoregulation, 372–75
  behavior fever and, 556
  by garter snakes, 373–75, **374**, **375**, *375*
Behe, Michael, 100–101, 102
Belding's ground squirrels, alarm calling in, 450–51, **452**
Bell, K.S., 592
Bell, Michael, 43
Bell Curve, The (Murray and Herrnstein), critique of, 352–55, **352**, **353**
Bender, Catherine, 534
Benjamin, Jonathan, 331–32
Benton, M.J., 691, **710**
Berger, Edward, 23
Berglund, Anders, 432
Bermúdez de Castro, J.M., 778
Bernard, Gordon, 560
Bernheim, H.A., 556
Berrigan, David, 509
Berry, Andrew, 262–63

Bersaglieri, Todd, 302
Bertram, B.C.R., 414
Best-fit line, **83**, 336
  slope of, 339, **339**, 342, **342**
Beta-casein gene, 122
β-Tubulin, 100
Bias(es)
  experimental, **370**
  in fossil record, 691
  preexisting sensory, 422–24
Big Five extinctions, 710, **710**. *See also* Cretaceous-Tertiary (K-T) mass extinction
Big sagebrush (*Artemesia tridentata*), 627, *628*, **629**
Bilateria, **699**, 700, 701
Biodiversity. *See* Speciation
Biogeography, 133
Biological invasions, 520
Biological polymers, assembly of, 657–58, **658**
Biological species concept, 606–7, 629
Biotic potential (reproductive capacity), 86–87
Bipedal locomotion, 767
Biramous appendages, 739, **740**, 741
Birds. *See also specific species*
  aging in, **487**
  clutch size in, 498–99, **499**, 505–7, **506**
  Cretaceous-Tertiary extinction and, 716
  helping behavior in, 451–53, 464
  Polynesian avifauna extinction, 717–18
Bischoff, Claus, 491
Bishai, William, 538
Bithorax complex, 729–30
Bittacus apicalis (hangingfly), 421–22, **421**, **422**
Bittles, A.H., **271**
Bivalves
  Cretaceous-Tertiary extinction and, 716–17, **716**, **717**
  geographic range and survivorship of fossil, **711**
  zigzag evolution in Pliocene, 708, **709**
Blaak, Hetty, 20
Black bears (*Ursus americanus*), 483
Black-bellied seedcracker (*Pyrenestes o. ostrinus*), 348, **349**
Black cherry (*Prunus serotina*), 483
Black Death, 4, 299
Bla gene, 538
Blair, Amy, 520
Blending inheritance, 95, **95**
Bloch, Alan, 539
Blood-sharing in vampire bats, 474–75, **474**, **475**
Blue-footed booby, 470–71, **470**, *471*
Blumenschine, Robert, 769–70
BMP4 (bone morphogenic protein 4), beak development and, 85, **85**
Bmp4 gene, 747
Boag, P.T., 83, 86
Boake, Christine, 622
Bodmer, W.F., **347**
Body lice (*Pedicularis corporus*), 132–33
Body size
  in humans
    brain size vs., 790–91, **791**
    sexual dimorphism in, **402**, 438–40
  in marine iguanas, 408–12
  in paper wasp females, 464–65
Boesch, C., 764
Bombus sp. (bumblebee), 342–43, **342**, **343**
Bonacum, James, 613
Bone morphogenic protein 4 (BMP4), beak development and, 85, **85**
Bonobos (*Pan paniscus*), **1**, 754, 764
  relationship between humans and, 756–61
    molecular evidence, 757–59
Bootstrapping, 125–27
Borgerhoff Mulder, Monique, 438
Borrelia burgdorferi, 670, **670**
Botryllus leachi, 519
Botryllus scholosseri (sea squirt), 518–19, **519**
Botswana, AIDS and life expectancy in, **5**
Bottlenecking, 617
Bourgeois, Joanne, 715
Bouzat, Juan, 274–75
Bovine spongiform encephalopathy (mad cow disease), 189
Bowcock, A.M., 782–83
Boyce, Mark, 505–6
Brachiopoda, **699**
Bradshaw, H. D., Jr., 324, 325–26, 330–31
Brain
  language circuits in, 789
  spongy brain diseases, 189–91
Braincase volume in *Homo* species, 765
Brain size in hominoids, body size and, 790–91
Branches of phylogenetic tree, 50, 52, **52**
Branching events. *See* Speciation
Brassica oleracea (wild cabbage), 75, **75**
Brazilian Amazon, deforestation in, 720, **720**
BRCA1 gene, 258, **258**
Breast cancer, 552–55
  incidence of, 553, **553**, 554
  menstrual cycle and, 554–55
  as viral disease, 553
Brine shrimp (*Artemia franscicana*), 741
Brisaster latifrons, 522
Britten, Roy, 762
Broad-sense heritability, 334, 335
Broca's area, 789, 791
Broccoli (*Brassica oleracea italica*), **75**
Brodie, E.D., III, 344, 345–46
Bronze Age frequency of *CCR5-Δ32* allele, 299
Brown, D., **666**
Brown, James R., 669, 670
Brown kiwi (*Apteryx australis mantelli*), 42, **42**, 484, **484**
Brunet, Michel, 765–67
Brussels sprouts (*Brassica oleracea gemmifera*), **75**
Bryan, William Jennings, 97
Bryant, Edwin, 495–96
Bryozoans
  cheilostome, 706, 707
  punctuated change in, **706**
  stasis and speciation in, 706
BsrI enzyme, 547, **547**
Bubonic plague, 299
Buchanan, Bryant, 418
Buchnera, **136**
Buckingham, William, 754
Buck v. Bell, 194
Bufo periglenes (golden toads), **402**
Bull, James, 543
Bumblebee (*Bombus sp.*), 342–43, **342**, **343**
Burdick, Allan, 197–200
Burgess Shale faunas, 696–98, **696**, **698**, 700, 701
Buri, Peter, 242–45
Bush, Robin, 533–34
Bush, Sarah, 418
Bustamante, Carlos, 259
Butler Act (Tennessee), 97–98
Byers, David, 105

Caenorhabditis, 668
  C. elegans (nematode worms), 148–49, 150, 490–91, 496–97, **497**, 691, 730–31
California poppy (*Eschscholtzia californica*), 483
Calvin-Benson cycle, 587
Camarhynchus pallidus (woodpecker finch), 786
Cambrian explosion, 689, 694–702
  cause of, 701–2
  change in number and organization of *Hox* sequences before, 731
  Ediacaran and Burgess Shale faunas, 694–98, **696**, **698**, 700, 701
  evolution of limbs in animals in, 735–41
  morphology of, 698–700, **699**
  phylogeny of, 698–700, **699**
Camera eye, convergent evolution of, 115, **115**, 117–18, **117**
Camin, J.H., 227, **228**
Canback, Bjorn, 677
Cancer
  aging and risk of, 490–91
  breast, 552–55
  colon, 494–95
  endometrial, 555
  germ-line mutations in DNA mismatch repair genes causing, 494–95
  history of, 548–49, **549**
  ovarian, 555
Candelabra model, 774–75, **775**
Candidate gene approach, 727
  studying diversification of floral genes and structures using, 746–47
Cannibalism, kuru and ritual, 190
Cannibalistic tadpoles, 453–54, **455**
Cannon, Carolyn, 218
Carbonate ($—CO_3$), 652
Carcinoma, 549
Carpels, 742, **742**, 743, **743**
Carr, David, 256
Carroll, Sean, 702, 734, 735, 748–49
Caspi, Avshalon, 355, **356**
Castle, William, 179–80
Casts, 690, **690**, 691
Catalase/peroxidase, 538
Catalytic RNA. *See* Ribozymes (RNA enzymes)
Catarrhini, 754
Catasetum, 434, **435**, 437
Catasetum ochraceum, 437
Catastrophism, 60
Cation-independent mannose-6-phosphate receptor (CI-MPR), 515
Cats, evolutionary tree for, 52–53, **52**
Cattle-oxpecker relationship, hypotheses of, 364–67, **364**, **365**, **366**
Caudipteryx zoui, 47, **47**
Cauliflower (*Brassica oleracea botrytis*), **75**
Cavalli-Sforza, L.I., **347**
Cavener, Douglas, 185, 187
Cave paintings, symbols in, 790
CCR5, 10, 20, 162, **162**, *162*, 170
CCR5+, 161–62, **162**, *162*
CCR5-Δ32 allele, 23–25, **24**, **162**, *162*, 163, 170
  age of, 298–99
  changes in frequency of, 161–62, 182, 191–94, **191**
  evolution of, 282
  frequency in future generations, 194
  origin of, 299
CD4, 7, 10–11, 19
Cech, Thomas, 641
Cell division, senescence of cells and telomere shortening with each, 489–91
Cellular life, 641, 660–62
  advantages of, 660–61
  DNA and, 660
  examination of early, 666–67
  fossil record of, 662, **662**
  universal phylogeny, 663–66
    based on concatenated sequences of 31 universal genes, **671**, 672
    discordance of different estimates of, 669, **669**
    genetic sequencing approach to, 663–64, **665**, 667–69, **668**
    morphological approach to, 663
    rooting, 666, **666**, 673–75
    whole-genome sequencing in, 668–69, 671–72
Cenancestor, 640, 660–62, **672**
  as community, 672, **672**, 675, 676, **676**
Cenozoic era, **692**, **693**
Ceratitis capitata (Mediterranean fruit fly), 414
Cercidium floridum (Palo verde), 512–14, **513**
Cerrado, **720**
Certhidea fusca (warbler finch), 81
Certhidea heliobates (mangrove finch), 81
Certhidea olivacea (warbler finch), 81
Certhidea pallida (woodpecker finch), 81, 786
Cervus elaphus (red deer), **402**, **487**
Cetacea, 120
CFTR, 165, 214–18
Chaetopterus variopedatus, 741
Chagnon, Napoleon, 439, 566
Chamaeleonidae, 133, **133**
Chambers, Rober, 39
Chameleons, 133–34, **133**
Chance events
  allele frequencies and, 232–35, **233**
  genotype frequencies and, 232–35, **233**
Changing-environment theories for sex, 310–12
Chao, Lin, 256
Charcot-Marie-Tooth disease type 1A, 57
Charina bottae, 42, **42**
Charlesworth, Brian, 577
Charlesworth-Langley hypothesis, 580
Charnov, Eric, 507–8

Chase-away sexual selection, 516
Cheetham, Alan, 706, 707
Cheilostome Bryozoa, 706, 707
Chelicerates, 733, **733**, **734**
Chemical cues, kin recognition based on, 454
Chen, Feng-Chi, 762
Cheng, Ze, 153
Chengjiang biota from Yunnan Province China, **696**, 697
Chickenpox, fever and, 558, **558**
Chickens, vestigial traits in, 42, **43**
Chicxulub (Mexico), crater near, 713, **714**
Children, murder of, 567–68, **568**
Chimpanzees (*Pan troglodytes*), 754, 764–65
  genetic differences between humans, gorillas and, 762–63, *762*, **763**
  hyoid bone of, **790**
  relationship between humans and, 756–61
    molecular evidence, 757–59
    morphological evidence, 759–61
  subspecies (*P. t. troglodytes*), 27
  thumb in humans vs., **787**, 788
  tool making by, 786
Chinook salmon (*Oncorhynchus tshawytscha*), 510–12, **511**
Chirality (handedness), origin of, 657
Chi-square (χ^2) test, 192–93
Chlamydomonas, **100**, 101
Chloroplast DNA (cpDNA), 586–88
Chloroplasts
  evolution of, 585, 586–90, **586**
  secondary endosymbioses and, 588–89, **589**
Cholera, 529, **530**
Chordates, **699**, 701
Chromosome 17, 56–57
Chromosome frequencies, 282–83, 284, 286–87
Chromosome inversion, 156–58, **156**, **157**
Chromosome mutations, 156–60
  genome duplication, 159–60, *161*
  inversions, 156–58, **156**, **157**, 160, *161*
  polyploidy, 159–60
Chromosomes
  changes in, as barrier to gene flow, 615–16
  compound, 200–205, **201**
  polymorphic, 157
Chronic wasting disease, 189
Chronocyte hypothesis, 677, **677**
Chyba, Christopher F., 653, 654
Ciccarelli, Francesca, 671
Cilia, eukaryotic, 100–101, **100**
Ciliates, 665
Clades, **131**, 535
  on flu phylogeny, 535, 537
Cladistic analysis, 113, 131
  of African replacement vs. multiregional evolution model, 778, **778**
  of hominoid fossil evidence, 772–73
  of human/African ape relationship, 754–56
Cladogram, 25, 113. *See also* Phylogenetic tree(s)
  of *Homo sapiens* and recent ancestors, **772**
Clam (*Lasaea subviridis*), 483
Clark, Marta, 136
Clausen, J., 324, 351, 352, 353–54, **354**
Claustridium tetanae, 541
Clayton, Dale, 363–64, 390–91
Cleavage, 700
Clegg, Michael, 185, 187, 289–91
Clegg, Sonya, 236
Clift, William, 45
Climate after asteroid impact at K-T boundary, 714–15
Cline, 157–58
Clocks, molecular, 132–33, 700, 701
Clothing, divergence of body lice from head lice and wearing of, 132–33
Cluster analysis, genetic distances for, **126**
Clutch size
  in birds, 498–99, **499**, 505–7, **505**, **506**
  egg size and, 509–14, **513**
  in parasitoid wasps, 507–9
Clutton-Brock, T.H., **487**
CMAH (CMP-N-acetylneuraminic acid hydroxylase), 42–43
CMP-sialic acid hydroxylase, 762
CMT1A repeats, 56–57, **57**
Cnemidophorus uniparens (desert grasslands whiptails), **281**
CNGB3 gene, 238
Cnidarians, **699**
Coccyx, 42, **42**
Cocktails, multidrug, 16–17
Coding sequences. *See* Molecular evolution
Codon bias, 260–62, *261*, **262**
Codons, 56, 146
Coefficient of inbreeding (F), 268, 269–70, **270**
Coefficient of linkage disequilibrium (*D*), 284, 285
Coefficient of relatedness (r), 448–49
  altruistic behavior and, 449, 453
  blood-sharing among vampire bats and, 474, **475**
  in boobies, 470
  calculating, 449
  cannibalistic tadpoles, 453
  haploidiploidy and, **461**
  among Hymenoptera, 462
  among naked mole-rats, 466
  in paper wasps, 465
Coen, Enrico, 744
Coevolution, 135
  sexual, 516–17, **517**
Coho salmon (*Oncorhynchus kisutch*), 413, **413**
Coincidental evolution hypothesis of virulence, 541
Cold, fever and, 559–60, **559**
Coleoptera (beetles). *See* Beetles
Colinearity, temporal, spatial, and quantitative, 729
Collard, Mark, 770, 772–73
Collared flycatcher (*Ficedula albicollis*)
  aging in, **487**
  antagonistic pleiotropy in, 498–99, **499**
  clutch size in, impact of, 506
Collared lizard (*Crotaphytus collaris*), 246–47, **246**, **401**
Collinge, John, 190
Colon cancer, hereditary nonpolyposis, 494–95
Colonization, dispersal and, 612–14, **612**, **613**
Color terminology, 562
Colosimo, Pamela, 633
Columbia space shuttle, 653
Columbicola, 391
Combat, male-male
  intrasexual selection through, 408–12
  among stalk-eyed flies, 426
Combination therapies for AIDS, 16–17
Combined probabilities, 175
Comets, 654
Common cold, fever and, 559–60, **559**
Common garden experiments, 84, 351–52, **351**, 353–54
Communal evolution, non-Darwinian, 675–76
Comparative analysis of adaptation, 376–79, 396
Competition
  asexual vs. sexual reproduction and, 305
  local mate, 462
  male-male, 408–15
    alternative male mating strategies, 412–13
    combat, 408–12, 426
    homicides, 438–39
    infanticide, 94, 414–15
    sperm competition, 376–79, **376**, 413–14, **414**
    stalk-eyed flies, 426
Complementary base pairs, **144**, 145
Complexity in nature, 98–103
Compound chromosomes, 200–205, **201**
Compression fossils, 690, **690**, 691
Compsognathus, 46, 47
Compulsory sterilization, 194, 208–10
Comte de Buffon, 39
Condition-dependent sex allocation, 462
Condom use, HIV and, 3
Conery, John, 153
Conflict
  game theory to analyze, 473
  parent-offspring, 467–71, **468**
Conjugation, 585
Conservation genetics, 273–75
Constraints on adaptation, 385–91
  fuchsia color change, 385–88, **386**
  host shifts in feather lice, 390–91, **391**
  host shifts in *Ophraella*, 388–90, **389**
Continuous variation, **319**, 320
Contraceptives, oral, 555
Control groups, experimental, 370
Convergence in SINE or LINE characters, 128
Convergent evolution, 115–16, **115**, 117–18, **117**
  at molecular level, 116
Cooper, Vaughn, 544
Cooperation, 447. *See also* Altruism
  game theory to analyze, 473
Co-opting of genes, 728
Coots (*Fulica americana*), 456–57, **457**
Copepods, diversification in, 609–10, **609**, **610**
Copulations, extra-pair, 417, **417**, 430–31, **431**
Cordain, 550
Coreceptor, 7, 10, 19, 20, **21**
Corn (*Zea mays*), 691
Correlation of allelic state, 290
Cortisol, 567, **567**
Corvus brachyrhynchos (crow), 447
Cospeciation, 135–36, **136**
Courtship
  displays by males, evolution of, 415–16. *See also* Female choice
  in hangingflies, **421**, **422**
  in water mite, 423–24, **423**
COX10 protein, 56–57
cox2 locus, 588
Coyne, Jerry, 624
Crassostrea gigas (oyster), 483
Crater, from asteroid impact at end-Cretaceous, 713–14
Creationism, 37–38, 45, 97–105
  history of controversy, 97–98
  motivation for controversy, 105
  other objections to, 103–5
  perfection and complexity in nature, 98–103
Creevey, Christopher, 671
Crenarchaeota, 664
Cresko, William, 43
Cretaceous-Tertiary (K-T) mass extinction, 712–17
  extent and selectivity of, 716–17
  impact hypothesis of, 712–14
  killing mechanisms, 714–15
Creutzfeldt-Jakob disease, 189
  variant, 189, 190
Crick, Francis, 145, 652, 653
Crickets (*Gryllodes sigillatus*), 432
Crocodiles, eyes of, 115, **115**
Crohn's disease, 332, **332**
Cro-Magnon I, **770**, 771
Cross-cultural diversity, 561
Cross-cultural studies of human behavior, 438
Cross-cutting relationships, principle of, 61
Cross-fostering experiments, 84
Crossing over, 282, 286, 289, 296, 297
  inversions and, 157
Cross-over
  meiosis with, 306
  unequal, 152–53, **153**
Crotaphytus collaris (collared lizard), 246–47, **246**, **401**
Crouzon syndrome, 394, 395, **395**
Crow, J.F., *265*
Crows (*Corvus brachyrhynchos*), 447
Crustaceans, 733, **733**, **734**
Crymes, Lili, 428–29
Cryptic species, 606, **614**, 616
Cryptomonas, 589
Cryptosporidium parvum, 592
Crystallins, 102, **103**
Ctenophora, **699**
Cubitus interruptus dominant (*ciD*) gene, 263
Cultural evolution, 562
Curie, Marie, 63
Cuticle, 742
Cuvier, Georges, 44, 54
C-value paradox, 576
CXCR4, 20, 21, **21**
Cyanidioschyzon meroloae, *594*
Cyanobacteria, 586, 590, 662, 664
  fossil, 674, **674**
Cynomys gunnisoni (Gunnison's prairie dogs), 429–32, **431**
Cypripedium acaule (pink lady's slipper orchids), 522–23
Cyrtodiopsis dalmanni (stalk-eyed flies), 426–29, **427**, **428**
Cyrtodiopsis whitei (stalk-eyed flies), **426**

Cystic fibrosis, 165, **165**, 210
mutation rate for, 216, 217
mutation-selection balance in, 214–18
Cystic fibrosis transmembrane conductance regulator (CFTR), 165, 214–18
Cytochrome oxidase II, **757**
Cytosine, 144, **144**, 145

D1S305, linkage disequilibrium between GBA locus and, 296, 297
D4DR (D4 dopamine receptor), 331–32
Dactylorhiza sambucina (Elderflower orchid), 205–7, **206**
Daly, Martin, 438–39, 567–68
Damselflies, sperm competition in, 414, **414**
Daphne Major, 81–82, **81**
Geospiza fortis on, 81–90, **81**, **83**, **85**, **87**, **88**, **89**
Geospiza magnirostris on, 344–46, **345**
Daphnia magna (water flea), 381–83, **381**, **383**
Darrow, Clarence, 97
Darwin, Charles, 3, 38–40, 44, 45, 46, 49–50, 60, 61, 65, 73, 74, 86, 93, 97, 169, 402, 403, 448, 459, 530, 562, 639, 647, 655, 704. *See also* Darwinism
on altruism, 448
Descent of Man, and Selection in Relation to Sex, The, 416, 753
on organs of extreme perfection, 98, 99
On the Origin of Species by Means of Natural Selection, 3, 38, 39, 45, 46, 52, 73, 76, 77, 97, 530, 639, 753
on sexual selection, 403
Darwin, Erasmus, 39
Darwinian fitness, 77
Darwinian threshold, 675, 676
Darwinism. *See also* Evidence for evolution; Natural selection
altruism as central paradox of, 448
evolution of, 94–97
age of Earth and, 96
Modern Synthesis, 39, 96
mutation and, 94
view of life in, 97
Dating
radiometric, 61, 63–65, **64**, 104
relative, 61
Davidson, Iain, 789–90
Dawkins, Richard, 66, 98, 458, 562
Dawson, Elisabeth, 294
Dawson, Peter, 195–96, 197
Deacon, N.J., 21
Decay, radioactive, 63, **63**
Deccan Traps, 715
deDuve, Christian, 657
Deep homology, 727
diversification in animal limbs and, 735–41
arthropod limb, 739–41, **740**, **741**
tetrapod limb, 735–39, **736**, **737**, **738**, **739**
reconstructing ancestors using, 748
Deer mice (*Peromyscus maniculatus*), 483
Defense assays, 516, **517**
Deforestation, 720, **720**. *See also* Habitat destruction
Deformed (Dfd) genes, 731
Degree of genetic determination, 334
Degrees of freedom for chi-square, 193
Dehalococcoides ethenogenes, *594*
Deleterious mutations, **496**
evolutionary theory of aging and, 492, **493**, 494–96, 502
late-acting, 500
DeLong, Edward, 668
Delph, Lynda, 387, 436, **437**
Dembski, William, 104
De Meester, Luc, 381–82
Demong, Natalie, 452
Denver, Dee, 148, 149
Deoxyribonucleotides, 144–45, **144**
Deoxyribose, 144, **144**
Descent of Man, and Selection in Relation to Sex, The (Darwin), 416, 753
Descent with modification, 38, 39, 49, **51**
Desert grasslands whiptails (*Cnemidophorus uniparens*), **281**
Desert iguana (*Dipsosaurus dorsalis*), 372–73, **373**, 556–57, **557**
Desmodus rotundus, 474–75, **474**, **475**
Determinism, genetic, 564
Deuterostomes, **699**, 700, 701
Developmental genetics. *See* Evo-devo
Developmental homologies, 54–55, **55**
Dhondt, André, 335–37
Diabrotica undecimpunctata howardi (spotted cucumber beetle), 425, **425**
Diatoms, genes for urea cycle in, 597, **597**
Diaz, George, 296, 297, 298
Dickinsonia, *694*
Dicotyledenous plants, **131**
Dictyostelium, 665
Dictyostelium discoideum, 458–59, **459**
Didelphis virginiana (Virginia opossum), 485–86, 500–502, **500**, **501**
Dideoxy-3′-thiacytidine (3TC), 16
Diet, hunter-gatherer vs. modern American, 550–51, **550**
Differential success, measuring, 338–40
Differentiation, 629
Digital organisms, evolution of complexity in, 101–2
Dik-diks, 616
Di Masso, R.J., 338
Dimorphism, sexual. *See* Sexual dimorphism
Dinoflagellates, 99
Dioecious plants, 436
Diploblasts, 699–700
Dipsosaurus dorsalis (desert iguana), 372–73, **373**, 556–57, **557**
Directed panspermia, 653
Direct fitness, 450
Directional selection, 347, **347**, 348, 384, **385**
Direct kin recognition, 454
Directly transmitted vs. vector borne disease, virulence of, 544–45, **545**
Direct maternal ancestry, 781
Direct sequencing, 597–98
Disease
bacteria-caused, 529
germ theory of, 529–30
Hardy-Weinberg equilibrium principle and genetic variation for resistance to, 188–91
Dispersal ability, as constraint, 390–91, **391**
Dispersal and colonization, 612–14, **612**, **613**
Dispersal hypothesis, 134
Disruptive selection, 347–48, **347**, **349**
Distal-less (Dll) gene, 740–41, **741**
co-opting of, 728
Distance methods, 125, 126, **126**
Distribution, reference, 372
Divergence, 611
between African and non-African populations, 782–83, 784–85
dating events from estimates of genetic, 132–33
mechanisms of, 616–23, *620*
genetic drift, 616–17
natural selection, 617–21
sexual selection, 621–23, **622**
Diversification in floral genes and structures, 746–47
Diversification of animals, 730–32, **732**
arthropods, 703
body plans, 728–35
changes in *Hox* expression, 732–35
changes in *Hox* gene number and, 730–32, **732**
Hox gene structure and function, 729–30
Cambrian animal morphology, 698–700, **699**
limbs, 735–41
arthropod limb, 739–41, **740**, **741**
tetrapod limb, 735–39, **736**, **737**, **738**, **739**
marine copepods, 609–10, **610**
Diversity
cross-cultural, 561
genetic, 247–49, **248**
heterozygote inferiority and, 205
heterozygote superiority and, 200
among living humans, 776–77, **776**
population size and, 247–48, **248**
in reproductive strategies, 484
Dixon, Andrew, 563
Djurab Desert of Chad, 765
DNA, 144–50, 641, 643
bases in, 144–45, **144**
cellular life and, 660
chloroplast (cpDNA), 586–88
in common ancestor, 667
methylation of, 581, **581**
mismatch errors, 494–95
mitochondrial (mtDNA), 586–88
evolutionary tree of human, 779, 780–81, **781**
phylogenies of humans and African great apes based on, 757
repair systems, 149–50
single-base substitutions in, 147
structure of, 144–45, **144**
synthesis, **145**
three viruses, three domains hypothesis and transition to, 677–80, **679**
transcription of, 146, **146**
DNA fingerprints, assessing frequency of extra-pair copulation using, 430, **431**
DNA polymerases, 145, **145**, 147
variation in, 149–50
DNA sequences
divergence patterns, 252–54
phylogenetic analysis using, 119
Doadrio, Ignacio, 616
Dobzhansky, Theodosius, 3, 39, 94, 624
Dogon of Mali, 554–55, **555**
Dolphin (*Lagenorhynchus obscurus*), **122**
Domestic animals and plants, 74–75
Dominance genetic variation (VD), 335, 336, **336**
Dominant alleles, selection on, 195–97, 198–99
algebraic treatment of, 198–99, **199**
Doolittle, Russell, 677
Doolittle, W.F., **584**, 586, **661**, **666**, 669, **672**
Doran, Timothy, 558
Dose, Klaus, 653
Douglas, Susan, 589
Doushantuo Formation, China, **673**
Dover, Pennsylvania school board, 753–54
Doves, host shifts in feather lice infesting, 391, **391**
Drift. *See* Genetic drift
Dromaeosaur, 47–48, **47**, **48**
Drosophila, **141**, 157
chromosome inversions, 157
deleterious mutations and aging in, 495, **495**
fourth chromosome in, 263
Hawaiian
ancestors of, 702–3
contrasting fighting styles in, 621–22, **622**
geographic isolation through dispersal, 612–13, **613**
phylogeny of, **613**
sexual selection in, 621–23
Hox genes in, 729–30, **729**
mutants, **141**
pheromones in, 624–25
postzygotic isolation in, 624
prezygotic isolation in, 624
reinforcement and, 624, **625**
D. birchii, 624–25
D. heteroneura, **613**, 621–22, **622**
D. macrothrix, **613**
D. melanogaster, 151, 263, 264, 293, 593, 691
aging in, **487**
alcohol dehydrogenase (*Adh*) gene of, 259
allele frequency change by selection in, 185–87
antagonistic pleiotropy in, 498
artificial selection for longevity in, 489, **489**
asymmetry of sex in, 404
codon bias in, *261*
DNA bases in, 576
genetic control over limb formation, 740
heterozygote superiority in, 197–200, **200**
Indy mutation in, 498
mutant screen in, 726
P elements in, 579
random fixation of alleles in, 242–45, **244**, **245**
D. nigribasis, **613**
D. planitibia, **613**
D. serrata, 624–25
D. silvestris, **613**, 621, **622**
D. simulans, 259, 263, 264, 292–93
D. subobscura, 157, **157**
D. suzukii, **613**
D. yakuba, 259
Dryopithecus, 760

Dudash, Michele, 271–72
Dunbrack, R.L., 304, 305, 310
Duplication, gene, 152, *161*
  fate of duplicated genes, 153–56
  new genes from, 152–56, **153**
  origin of, **153**
  rates of, 153
Dynamic stasis, 708
Dyson, Miranda, 418

Earth
  age of, 60–65, 640
    creation scientists on, 104
    evolution of Darwinism and, 96
    geologic time scale, 61–63, **62**
    Kelvin's calculation of, 96
    radiometric dating, 61, 63–65, **64**
  history of large impacts on, 654, **654**
Earthquakes, asteroid impact at K-T boundary and, 715
East, Edward M., 208, 320, 321, 322–23, **323**, 325
Eaton, S. Boyd, 550–51
Eberly, Lynn, 477
Ecdysozoa, **699**, 700
Echinodermata (sea star), **699**
Echinoderms, 701, 731
Ecological factors in reproductive altruism, 464
Ecological mortality, 500, 502
Ecological opportunity, adaptive radiation and, 702–3
Ecotones, 627
Ectoderm, 699–700
Ectopic recombination, 580, **580**
Ectotherms, 372–73
Ediacaran faunas, 694–96
Edwards v. Aquillard, 98
Effective population size, 244, 245
Effector T cells, 9, 11
Eggs, female investment in, 404
Egg size, clutch size and, 509–14, **513**
Ehrlich, P.R., 227, **228**
Eiseley, Loren, 67
Ejaculation, prior, 412–13
Elderflower orchid (*Dactylorhiza sambucina*), 205–7, **206**
Eldredge, N., 705
Electrophoresis, allozyme, 164
Electrophoresis gel, 162
Elena, S.F., 349–50
Elephants, 610–11, **611**
Elephas maximus (Asian elephants), 610, **611**
Elgar, Mark, 509
El Niño event, 88
Emberiza shoeniclus (reed bunting), 563, **563**
Embryos
  Cambrian and Precambrian, 695, **695**, 699–700
  fossil, 695, **695**
  homologies among, 55
Emlen, Steve, 452, 453, 469, 470
Enard, Wolfgang, 763
Endangered Species Act, U.S., 606
Endemism, 613
Endoderm, 699–700
Endometrial cancer, 555
Endosperm, 515
Endosymbionts, obligate, 309
Endosymbiosis, 135, **136**, 585, 586–89, **586**, 677
  secondary, 588–89, **589**
Endosymbiosis theory, **587**
Energy
  body size in marine iguanas and, 409, **409**
  mammal expenditure of, 488, **488**
  trade-offs between time and, 484–85
  trade-offs in energy allocation, 486
England, Barry, 566–67
Engrailed (en) gene, 740
Enhydra lutris (sea otters), 266–68, **266**, *268*
Entophysalis, **674**
Entropy, 104
Environment
  changing-environment theories for sex, 310–12
  as complication in heritability estimates, 84
  genotype-by-environment interaction, 382
  medical/health adaptation and, 550–51
  origins of life and, 658–60
  phenotype and (phenotypic plasticity), 380–83
  prebiotic, 655
Environmental genomics, 598–99, **598**
Environmental sequencing, 597–98, **598**
Environmental variation (V^E), 334
Enzyme replacement therapy, 548
Eoentophysalis, **674**
Eons, **62**, 63
Epitope, 18–20, 28
Epochs, **62**, 63
Epperson v. *Arkansas*, 97–98
Equilibrium/equilibria. *See also* Hardy-Weinberg equilibrium principle; Linkage disequilibrium
  heterozygote fitness and, 202–3
  linkage, **283**, 284, 286, 287, **294**
  punctuated, 704, **705**, **706**, 707
Eras, **62**, 63
Erus, J., 767
Erwin, Douglas, 707
Erwinia carotovora, 592, **593**, 594
Escherichia coli, 59, 149, 150, 349–50, 691
  codon bias in, *261*, **262**
  Fisher's Fundamental Theorem hypothesis tested on, 350
  frequency-dependent selection in, 350
  insertion sequences in, 579
  lateral gene transfer in, 590, 670
  MG1655 strain, 670
  mutation in combination with selection in, 212–13, **213**
  mutation-selection balance hypothesis tested on, 350
  O157:H7 strain, 590
  phage f1 pathogen in, 543–44, **544**
  streptomycin resistance in, 540
Eschscholtzia californica (California poppy), 483
Esser, Christian, 676
Estuaries, 609
Ethnicity, IQ and, 350, 352–55, **352**, **353**
'Eua, avifauna extinctions in, 718, **718**
Eucarya, 665, 669
  fossils of single-celled, **673**
  hypotheses of origins of
    chronocyte hypothesis, 677, **677**
    ring of life hypothesis, 676–77, **676**
    three viruses, three domains hypothesis, 677–80, **679**
    universal gene-exchange pool hypothesis, 675–76, **675**
Eugenic sterilization, 194–95, 208–10
Euglena, 665
Euglenoids, 99
Eukaryotes
  genetic exchange between, 607
  lateral gene transfer in, 586–90
  transposable elements found in, 579, *579*
Eukaryotic cilium (flagellum), 100–101, **100**
Eumelanin, **95**
Euplectes ardens (red-collared widowbird), 416–17, **416**, **417**
Euplectes progne (long-tailed widow birds), 403, **403**
Europa (moon), **639**, 653
Europe, AIDS epidemic in, 5, **5**
European wood mouse (*Apodemus sylvaticus*), 454–56
Eurosta solidaginis, 348, **348**
Euryarchaeota, 664
Eurypterids (sea spiders), 733
Eurytemora affinis, 609, **609**, **610**
Eusociality, 459–67
  characteristics of, 460
  haplodiploidy and, 460–62, *460*
  in naked mole-rats, 465–67, **466**, *467*
  in paper wasps, 464–65
  phylogenies to analyze, 463–64, **463**
Evidence for evolution
  age of Earth
    geologic time scale, 61–63, **62**
    radiometric dating, 61, 63–65, **64**
  change through time, 40–49
    evidence from fossil record, 44–49
    evidence from living species, 40–43
  common ancestry, 50–60
    evolutionary trees, 50–53
    homology, 54–59
    interspecies relationship, 60
    ring species, 53–54
  descent with modification, 49, **51**
  relatedness of life forms
    interspecies relationships, 60
    phylogenetic trees and, 50–53, **52**
  religion-evolutionary biology conflict, 65–67
    Methodological Naturalism and Ontological Naturalism, 65–66
    religious evolutionary biologists and, 66–67
Evo-devo, 600, 725–52
  deep homology and diversification in animal limbs, 735–41
    arthropod limb, 739–41, **740**, **741**
    tetrapod limb, 735–39, **736**, **737**, **738**, **739**
  homeotic genes and diversification in animal body plans, 728–35
    changes in *Hox* expression, 732–35
    changes in *Hox* gene number, 730–32
    Hox gene structure and function, 729–30
  homeotic genes and the evolution of flower, 742–47
    discovery of *MADS*-box genes, 728, 744–47
    homeosis in flower, 743–44
  research
    deep homology to reconstruct ancestors, 748
    foundations of, 726–28
    frontiers in, 747–49
    microevolution and macroevolution, 748–49
    non-model systems and structures, 747–48
Evolution
  brief history of ideas on, 39
  convergent, 115–16, **115**, 117–18, **117**
    at molecular level, 116
  of Darwinism. *See under* Darwinism
  evidence for. *See* Evidence for evolution
  forces of. *See* Genetic drift; Migration; Mutation(s); Selection
  human. *See* Human evolution
  mechanisms of. *See* Mendelian genetics
  by natural selection, 15, 16–17, 40
  rate of
    dominance and allele frequency and, 196–97
    by genetic drift, 249–51
  reasons for studying, 3
Evolutionarily stable strategy (ESS), 473
Evolutionary history, reconstruction of, 25
Evolutionary independence, 606–7
Evolutionary species concept. *See* Phylogenetic species concept
Evolutionary Synthesis (Modern Synthesis), 96
Evolutionary theory of aging, 487, 492–502
  deleterious mutations and, 492, **493**, 494–96, 502
  natural experiments of evolution of, 500–502
  trade-offs and, 490, 496–99, 502
Evolutionary trees (phylogenies), 50–53, **52**. *See also* Phylogenetic tree(s)
Ewald, Paul, 544–45
Exhaustive search, 124
Exobiology, 652
Exons, 57
Exon shuffling, 582, **582**
Extended haplotype homozygosity (EHH), 301
Extinction(s), 709–20
  background, 710–11
  Cretaceous-Tertiary (K-T), 712–17
    extent and selectivity of, 716–17
    impact hypothesis of, 712–14
    killing mechanisms, 714–15
  as evidence for evolution, 44–45
  island, 717–18, 719
  mass, 702, 709–20, **710**
  during Phanerozoic, **710**
  rates of, 709, 710–11
  recent, 717–20
    mass extinction underway, 718–19
    Polynesian avifauna, 717–18
  underway
    most acutely affected regions, 719–20, **720**
    rate of species disappearance, 718–19
"Extinction filter," 720
Extinction vortex hypothesis, 274, 275
Extra-pair copulations, 417, **417**, 430–31, **431**

Eye(s), 364
camera, convergent evolution of, 115, **115**, 117–18, **117**
crocodile, 115, **115**
crystallins of lenses of, 102, **103**
mollusc, 99, **99**
octopus, 115, **115**, 117–18, **117**
Pax6 genes as part of regulatory cascade involved in forming, 727
variations in complexity of, 99, 102
Eyre-Walker, Adam, 259

*ΔF*508, 217–18, **218**
Factor XI deficiency, 298
Facultative strategies in paper wasps, 464–65
Fairy shrimp (*Artemia salina*), **708**
Falconer, D.S., 338, **427**
Family lineages, inbreeding depression varying among, 271–72
Fanconi anemia type C, 298
Fatal familial insomnia, 189
Faunal succession, principle of, 61
Feather lice, host shifts in, 390–91, **391**
Feathers, dinosaurs with, 46–48, **46**, **47**, **48**
Feder, J.L., 618–19
Fedorov, Alexei, 677
Feeblemindedness, compulsory sterilization to eliminate, 208–10
Felsenstein, Joseph, 215, 307, 377–78
Female choice, 407, 415–29
acquisition of resources and, 421–22
in barn swallows, 419
extra-pair copulations, 417, **417**, 430–31, **431**
in gray tree frogs, 417–21, **418**, **419**, **420**, *421*
offspring gene quality and, 419–21
preexisting sensory biases and, 422–24
in red-collared widowbirds, 416–17, **416**, **417**
sexy-son hypothesis of, 425
in stalk-eyed flies, 426–29, **426**, **427**, **428**
Fermentation, in yeast cells, 392, **393**
Ferns, 742
polyploidy in, 616
Fern spike, 716
Ferris, James, 658
Fever
behavioral, 556–57, **557**
chickenpox and, 558, **558**
common cold and, 559–60, **559**
medical practice and, 560
sepsis and, 560
Fibrinogens, 593
Fibroblast growth factor receptor 2 (FGFR2), 394–95, **395**
Fibroblast growth factors 4 and 8 (FGF-4 and FGF-8), 737, 738
Ficedula albicollis. *See* Collared flycatcher (*Ficedula albicollis*)
Finches, Galápagos (Darwin's finches), 80–90, **80**, **83**, **85**, **87**, **88**, **89**
Fischer, Kathleen, 488
Fish(es), **131**
correlation between clutch size and egg size in, 510–12
pipefish, 432–34, **433**
selection on offspring size in, 510–12, **511**
Fisher, Ronald, 208, 209, 325, 349, 426, 462
Fisher's Fundamental Theorem Hypothesis, 349, 350
Fitch, Walter, 531, 532–34
Fitness. *See also* Reproductive success
antagonistic sexual selection and, 516–17, **517**
asexual vs. sexual reproduction and, 308
asymmetric limits on, behavioral consequences of, 407–8
average, 186
Darwinian, 77
direct, 450
of genotype, 183, 186, 198
due to helping, **453**
of heterozygotes, 197–200
of homozygotes, 200–205, 249
of hybrid populations, 626
hybrid zone and, 626–29
inclusive, 448–50, 471, 504
in white-fronted bee eaters, 453, **453**
indirect, 450, 452
maternal
clutch size of parasitoid wasps and, 507–9, **508**, *508*
egg size and, **511**, 512
mean, of population, 199
as measurable in nature, 93
multiple traits and, 344, 345, **345**
mutation and, 150–52
natural selection and, 93, 98, 338–50
net, 516–17, **517**
as nontautological, 93
parent-offspring conflicts over interests of, 467
phenotype and, 346–47
relative, 339, 340
of RNA molecules, 646
of sagebrush hybrids, 628–29, *628*, **629**
selection modes and, 347–48
social behavior in terms of, 447–48, *448*
Fixation, random, 238–49
experimental study on, 242–45, **244**, **245**
in natural populations, 245–49
probability of allele to drift toward, 241
Flagellates, 665
Flagellum (eukaryotic cilium), 100–101, **100**
Flatworm, **699**
Fleming, Alexander, 530
Flinn, Mark, 564–67, 568
Florentin, J.-M.R., 715
Flour beetle (*Tribolium castaneum*), 195–97
sexual vs. asexual reproduction in, 304–6, **306**
Flower(s). *See also* Angiosperms
color evolution in, 78–80, **79**
homeotic genes and evolution of, 742–47
homeosis in flower, 743–44
MADS-box genes, discovery of, 728, 744–47
parts of, **742**, 743
Flu virus. *See* Influenza A virus
Flying fox (*Pteropus poliocephalus*), 376–79, **376**, **379**
Forced sterilization, 194, 208–10
Forelimbs, vertebrate, 54–55, **54**
Foré people of Papua New Guinea, kuru among, 188–91
Forterre, Patrick, 678–79, 680
Fossil record, 44–49, 132, 689, 690–94
of cellular life's origins, 662, **662**
methods of fossil formation, 690–91
overview of events during Phanerozoic eon, 692–94
strengths and weaknesses of, 691
Fossils, 606
date for root of tree of life and, 673–75, **673**, **674**
defined, 44, 690
hominid, 764–73
interpreting, 771–73
living, 707–9, **707**, **708**
methods of formation, 690–91
of single-celled Eucarya, **673**
trace, 695–96
transitional, 46–49, **49**
Foster, G.G., 200, 202–4, 205
Founder effect, 236–38, 298, 785
in silvereye, 236–37, **237**
Fox, Charles, 512–14
Fox, Sidney, 661
Frameshift mutations, 332
Frary, Anne, 74–75
Frayer, David, 775–78
Free-living organisms, patterns in genomes of, 593–94, *594*
Freeman, Carl, 628
Free-recombination model, 328–29
Frequency-dependent selection, 205–7, **206**, 350, 462
Fretwell, Stephen, 509–10, 511–12
Friedberg, Felix, 59
Fruit bats, 376–79, **376**, **379**
Fruit flies. *See Drosophila*
Fuchsia excorticata, 385–88, **386**
Fulica americana, 456–57, **457**
Functional constraints, 254
Functional genomics, 599–600
Fundamental Theorem of Natural Selection, 349, 350
Fungal spike, 716
Fungi, **664**, 665–66
coevolution of ants and, **138**
Fusion inhibitors, 16
Futuyma, D.J., 388, 389–90
fw2.2 gene, 74–75

G6PD-202A allele, 300–302, **301**
G6PD locus, 299–302
Gabriel, Wilfried, 273
Gage, Matthew, 414
Gagneux, Pascal, 776
Gain-of-function mutations, 395
Galant, Ron, 735
Galápagos archipelago, avifauna extinctions on, 718
Galápagos finches
ancestors of, 702–3
evolution of beak shape in, 80–90, **80**, **83**, **85**, **87**, **88**, **89**
Galápagos Islands, 60, **81**
Galderia sulfuria, 594, *594*
Galen, Candace, 342–43
Galileo spacecraft, 653
Galvani, Alison, 299
Gambusia affinis (mosquito fish), 92
Gametes, polyploid, 159
Game theory, analyzing cooperation and conflict using, 473
Garland, T., Jr., 380
Garter snake (*Thamnophis elegans*), 373–75, **374**, **375**, *375*
antipredator defenses, 346, **346**
Gasterosteus aculeatus (three-spined sticklebacks), 43, 631–33, **631**, **632**, **633**
Gastropods
Cretaceous-Tertiary extinction and, 716–17
geographic range and survivorship of fossil, **711**
Gastrulation in protostomes and deuterostomes, 700
Gatesy, John, 122, 123, 124
Gaucher disease, 295–96
GBA, 295–97
GBA-84GG mutation, 296–97, **296**
estimating age of, 296, 297
Ge, Fen, 670
Gene(s). *See also specific genes*
chromosome inversions and, 156–57
co-option, 102, **103**
homeotic. *See* Homeotic genes
homologous, 155
inactivation of, 148
jumping. *See* Transposable elements
molecular view of, 146
new, 152–56
from duplications, 152–56, **153**, *156*
orthologous, 155
paralogous, 155, 156
pseudogenes, 155
replicase, 643–44
repurposing of, 728
selfish, 580
variation among, 149
virulence, 592, **593**
Gene behavior, null model for, 170
Gene duplication, *161*
fate of duplicated genes, 153–56
new genes from, 152–56, **153**
origin of, **153**
rates of, 153
Gene expression, regulatory sequences and, 595–96
Gene families, size of, 155–56, **156**
Gene flow, barriers to
chromosome changes, 615–16
physical isolation, 611–15
Gene history, linkage disequilibrium and reconstruction of, 295–99
Gene loss in parasites, 592
Gene pool, 171, **172**, **174**, 175–76, 178–79, **178**
General Assembly of the United Presbyterian Church in the United States, 67
Gene therapy, enzyme replacement and, 548
Genetic code, homologies in, 56–58
Genetic determination, degree of, 334
Genetic determinism, 564
Genetic distance, 126, **126**, 624, **625**
Genetic diversity, 247–49
heterozygote inferiority and, 205
heterozygote superiority and, 200
among living humans, 776–77, **776**

population size and, 247–48, **248**
Genetic drift, 24, 132, 166, 173, 224, 232–64, 495, 785
adaptive significance of sex and, 308–10
cumulative effects of, **239**
defined, 234
in greater prairie chicken population, 274–75, *275*
Hardy-Weinberg equilibrium assumption of no, 181, **181**
linkage disequilibrium and, 288, **288**, 310, 312
as mechanism of divergence, 616–17
in Mendelian genetics of populations, 232–64
random fixation of alleles, 238–49
sampling error and founder effect, 236–38
model of, 232–35, **235**
in molecular evolution, 250, 251–64
DNA sequence divergence, 252–54
nearly neutral model of, 254–56
neutral theory of, 252, 256–64
in natural populations, 245–49
natural selection vs., 234
population size and, 235–36, 239, **240**, 617
probability of allele to drift, 241
random fixation of alleles, 238–49
rate of evolution by, 249–51
substitution rate under, 250
Genetic homologies, 56, **56**
Genetic innovation, genome duplication and, 159–60. *See also* Genetic variation
Genetic isolation, 611–16
Genetic linkage, 156–57. *See also* Linkage disequilibrium; Linkage equilibrium
Genetic load, 273, 308
Genetic polymorphism, 248
Genetic recombination
adaptive significance of sex and, 306–7
artificial selection and, **307**
in bacteria, 607
ectopic, 580, **580**
linkage disequilibrium reduced by, 289, 307
sex in population and, 306
Genetic revolution, 629
Genetics, population, 39
Genetics and the Origin of Species (Dobzyansky), 39
Genetic variation, 16, 18, 149, 334
additive vs. dominance, 335, 336, **336**
gene duplications and, 153, 156
genome duplication events and, 159–60
lack of, as constraint in adaptation, 388–90, **389**, *390*
for life history traits, 517–19
in natural populations
extent of, 163–66
measuring, 160–66
natural selection and, 517–18
from polyploidy, 159
reasons for, 165–66
for resistance to HIV infection, 23
selection modes and, 349–50
in tumors, 549
Gene transfers, lateral. *See* Lateral gene transfer (LGT)
Gene trees
divergence times of, **782**
species trees vs., 758–59, **758**
Genome duplication, *161*
genetic innovation and, 159–60
Genomes, 576
human, 294–95, 759
organelle, 586–89
selfish genes in, 580
Genome sequences. *See also* Phylogenomics
of free-living organisms, 593–94
new research strategies, 597–600
in parasite genomes, 591–93, **591**
regulatory sequences, evolution of, 594–96
Genomic imprinting, 514–16
Genotype(s)
determining, 161–62
fitness of, 183, 186, 198
RNA population evolution and, **647**
Genotype-by-environment interaction, 382
Genotype frequencies
chance events and, 232–35, **233**
in general case, 177–80, **178**
inbreeding and, **265**
linkage disequilibrium and, 282–84, **283**
migration and, 225–26
numerical calculation, 174–77, **174**, **176**, **177**
random mating and, 171, **172**, **174**, 175–76, 178–79, **178**
selection and, 187–91, **187**
empirical research on, 188–91
selfing and changes in, 264–66, **265**, *265*
in simulated life cycle, 170–73, **173**
Geographic bias in fossil record, 691
Geographic isolation
through dispersal and colonization, 612–14, **612**, **613**
through vicariance, **612**, 614–15
Geographic range, Cretaceous-Tertiary extinction and, 717
Geologic column, 61
Geologic time scale, 61–63, **62**, 690, 692–93, 710
Geospiza fortis (medium ground finch), 81–90, **81**, **83**, **85**, **87**, **88**, **89**
Geospiza magnirostris (large ground finch), 85, **85**, 89
Gerhardt, H. Carl, 417–18, **419**
Germ theory of disease, 529–30
Gerstmann-Sträussler-Scheinker disease, 189
Gesteland, Raymond F., **661**
Giant Panda's thumb, anatomy of, 92, **92**
Giardia lamblia, 665
Gibbons (*Hylobates*), 754
Gigord, Luc, 205, 207
Gilad, Yoav, 763
Giles, Barbara, 230, 231
Gill, Douglas, 522–23
Gillespie, John, 251
Gillin, Frances, 149
Gill rakers, 631, 632
Gingerich, Philip, 48
Gingko tree, 707, **707**
Giovannoni, S.J., **586**
Giraud, Antoine, 150
Gish, Duane, 104
Glades, 246–47
Glen, W., 712
Global cooling, asteroid impact at K-T boundary and, 714–15
Globin gene family
gene duplication events in, 154–56
timing of expression in, **154**
transcription units in, **155**
Gloeocapsa, **674**
Gloeodiniopsis, **674**
Glucocerebrosidase (acid β-glucosidase or GBA), 295–97
Glucose-6-phosphate dehydrogenase deficiency, 299–300, **300**
Glycine, 655
Glyptodont, 45
Goddard, Henry H., 208, 209
Goethe, Johann, 743
Golden toads (*Bufo periglenes*), **402**
Goldstein, D.B., 783
Gona, Ethiopia, Oldowan tools from, 786–87
Gondwanaland, 133–34, **134**
Goodman, Morris, 755
Goosebumps, 42
Goriely, Anne, 394–95
Gorilla (*Gorilla gorilla*), 754
genetic differences between humans, chimpanzees and, 762–63, *762*, **763**
relationship between humans and, 756–61
molecular evidence, 757–59
Gorman, Owen, 534
Goto, Koji, 744, 745
Goudet, Jérôme, 230, 231
Gould, Stephen Jay, 705, 790–91
gp120 protein, 28
Gracile australopithecines, **766**, 767, 768, 788
Gradualism, **705**
relative frequency of, 707
Graham, John, 628
Graham, Neil, 559, 560
Grammar, universal, 789
Gram positive bacteria, 664
Grandmother hypothesis of menopause, 504
Gran Dolina specimens, 778
Grant, Peter (P.R.), 80–82, 86, 88, 89–90, 104, 617, 344–46
Grant, Rosemary (B.R.), 80–82, 86, 88, 89–90, 104, 617, 344–46
Graphite, 659, **659**
Gray, Elizabeth, 430
Gray tree frogs (*Hyla versicolor*), 417–21, **418**, **419**, **420**, *421*
Great apes, 754–56
brain size in, 790–91, **791**
genetic diversity among, **776**
Greater prairie chicken. *See* *Tympanuchus cupido pinnatus* (greater prairie chicken)
Great tit (*Parus major*)
clutch sizes of, 505–6, **505**
inbreeding depression in, 272, **272**
Greenbeard effect, 457–59
Greenberg, J. Mayo, 653
Greene, Erick, 367–72
Greenish warblers
Siberian (*Phylloscopus trochiloides*), 53–54, **53**
speciation in, 104–5
Greenwood, Paul, 272
Greider, Carol, 491
Grine, Frederick, 773
Grosberg, Richard, 518–19, **519**
Gross, M.R., 413
Ground finches of Daphne Major, 344–46, **345**
Gryllodes sigillatus (crickets), 432
Gryllus firmus (sand crickets), 486, **486**
Grypania spiralis, 673
Guanine, 144, **144**, 145
Guard cells, 742
Gunnison's prairie dogs (*Cynomys gunnisoni*), 429–32, **431**
Guppies (*Poecilia reticulata*), **402**
Gustafsson, Lars, **487**, 498–99, 506
Gutiérrez, Gabriel, 779–80
Gymnosperms, 742

HAART (Highly Active Anti-Retroviral Therapy), 16–17
Haas's Holy Land snake (*Haasiophis terrasanctus*), **37**
Habitat destruction, **224**, 273, 718
expanding human population and, 719
Habitat tracking, 708
Hacia, Joseph, 762
Hadar, Ethiopia, oldest reliably dated *Homo* fossil from, **786**, 787
Hadza hunter-gatherers, 566
postmenopausal women, **503**, 504
Hahn, Beatrice, 25, 27
Haig, David, 515
Haigh, J., 309
Haikouichthys ercaicunensis, **697**
Haldane, J. B.S., 655
Haloarchaea, 664
Hamilton, William, 448–50, 458, 460–61, 462, 473
Hamilton's rule, 449–50, 453, 456, 457–58
Hamrick, M.W., 788
Han, Tsu-Ming, 673
Handedness (chirality), origin of, 657
Hangingfly (*Bittacus apicalis*), 421–22, **421**, **422**
Haplodiploidy, eusocial Hymenoptera and, 460–62, **461**
Haplotype, 282
Harassment in white-fronted bee-eaters, 469–70
Hardy, G.H., 177, 179, 180
Hardy-Weinberg equilibrium principle, 170–82
analysis using
adding inbreeding to, 269
adding migration to, 225–27
adding mutation to, 210–12
adding selection to, 183–85, 187–88
two-locus version of. *See* Two-locus version of Hardy-Weinberg analysis
crucial assumptions of, 180–82, **181**
fundamental conclusions of, 180
general case, 177–80, **178**
generating expected values using, 192–93
genetic variation for resistance to disease and, 188–91
with more than two alleles, 180
as null model, 181, 182
numerical calculation, 174–77, **174**, **176**, **177**
rate of evolution and, 196–97

Hardy-Weinberg equilibrium principle, (*continued*)
  simulation, 170–73, **173**
  use of, 180–82
Harris, H., 161
Hart, Michael, 522
Hartl, D.L., **245**
Hartman, Hyman, 677
Hawaii, avifauna extinctions in, 718
Hawaiian *Drosophila*. *See under Drosophila*
Hawkes, Kristen, **503**, 504
Hawks, John, 765
He, Chaoying, 596
Head lice (*Pedicularis capitus*), 132, 133
Health. *See* Medical applications
Heath, Daniel, 511–12
Hedges, S. Blair, 761, 780–81
Hedrick, P.W., 191, **289**
Height
  of offspring, parent height and, 333–34, **334**
  sexual dimorphism in, **402**
Heinsohn, Robert, 476–77
Helianthus annuus, 625–26
Helianthus anomalous, 625–26
Helianthus petiolaris, 625–26
Heliconema, **674**
Heliocidaris erythrogramma, **521**
Heliocidaris tuberculata, **521**
Helper T cells, response to viral infection, 8, **8**
  HIV infection, 10–11
Helping behavior. *See* Altruism
Hemagglutinin, 532–36, **532**, **536**
Hemagglutinin-3, neuraminidase-2 (H3N2), 535
Hemann, Michael, 491
Hemoglobin, 146–47
  gene duplication events in, 154
  sickle-cell, 146–47
Hemophilia A, 148
Hendry, Andrew, **483**
Hennig, Willi, 113
Herbivorous beetle, host shifts in, 388–90
Hereditary non-polyposis colon cancer, 494–95
Heritability, 78, 83, 333–38
  broad-sense, 334, 335
  defined, 334
  estimates of, 84
    from parents and offspring, 334–37
    from twins, 337–38, **337**
  of flower size in alpine skypilots, 342, **342**, **343**
  of IQ, 350, 352–55, **352**, **353**
  of life history traits, 518, **518**
  misinterpretations of, 350–55
  narrow-sense, 335
  natural selection acting on populations and, 90, **90**
  predicting evolutionary response to selection and, 341
  of variation, 83–85
Heritable variation, measuring, 333–38
Hermit crabs, 708, **708**
Herrnstein, Richard J., 352–55, **352**, **353**
HERV-K, 758
Heterocephalus glaber, 465–67, **466**, *467*
Heterozygosity, **164**, 268–69
  defined, 242
  inbreeding and, 269
  loss of, 241–49
    experimental study on, 242–45
    in natural populations, 245–49
    population size and, 245–49
  mean, 163–64
  in populations, 269
    random fixation and loss of, 245–49
Heterozygote inferiority (underdominance), 204
  genetic diversity and, 205
  unstable equilibria with, 202–3
Heterozygote superiority, 197–200
  cystic fibrosis and, 217–18, **218**
  in *Drosophila melanogaster*, 197–200, **200**
  genetic diversity and, 200
  stable equilibria with, 202–3
Hexapods, 733, **733**
Hiesey, W.M., 324, 351, 352, 353–54, **354**
Higgie, Megan, 624–25
Highly Active Anti-Retroviral Therapy (HAART), 16–17
Hill, Kim, **503**, 504, 566
Hillis, David, 128
*Hin*pII enzyme, 547, **547**
Hippopotamuses, 120
  eyes of, 115, **115**
Hirschhorn, Rochelle, 546–48
Histone proteins, 254
Hitchhiking (selective sweep), 262–63
HIV. *See* Human immunodeficiency virus (HIV)
HMGCoA reductase gene, 584, **584**, 586
Holland, B., 516–17
Homeobox, 729
Homeotic genes, 728–35
  diversification in animal body plans and, 728–35
    changes in *Hox* expression, 732–35
    changes in *Hox* gene number, 730–32, **732**
    Hox gene structure and function, 729–30
  evolution of flower and, 742–47
    homeosis in flower, 743–44
    MADS-box genes, discovery of, 728, 744–47
HOM genes. *See Hox* genes
Homicides
  parent-child, 567–68, **568**
  same-sex, 438–39
Hominids, 764
  thumb metacarpal bones in, **788**
Hominin, 764
Homogenization of allele frequencies, migration and, 230–32
Homologous genes, 155
Homology(ies)
  deep, 727
    diversification in animal limbs and, 735–41
    reconstructing ancestors using, 748
  developmental, 54–55, **55**
  distinguishing homoplasy from, 116–18
  as evidence for evolution, 54–59
    model organisms and, 59
    modern concepts, 59
    molecular homology, 56–59
    structural and developmental homology, 54–55, **54**, **55**, **56**
  genetic, 56, **56**
  model organisms and, 59
  structural, 54–55
  of traits, 112
Homoplasy, 116–18, 127, 128
  distinguishing homology from, 116–18
  in molecular characters, 119, 123
  in morphological characters, 119
Homoptera (plant bugs), *460*
Homo species, 765, 769–71, **770**. *See also* Human evolution
  brain case volume in, 765, 769
  brain size in, 790–91
Homo antecessor, 774, 778
Homo erectus, **770**, 773, 774, 778, 787–88
  hypotheses of transition to *H. sapiens* from, 774–75, **775**
  as toolmakers, 788
Homo ergaster, 769, **769**, 773, **773**, 774, **775**
Homo habilis, 769–70, 787, 788
  fossils, 765, **769**
  language in, 791
Homo heidelbergensis, **770**, 773, 774, **775**
Homo neanderthalensis, **770**, 773, 774, **775**, 779, 787–88
  hyoid bone in, 790, **790**
  larynx in, 790
Homo rudolfensis, 769–70, 787
  fossils, 765, **769**
Homo sapiens, 769, **770**, 771, 775
  cladogram of, **772**
  language in, 790
  origin of, 773–85
    African replacement model vs. multiregional evolution, 775–85, **778**, **784**
    controversies over, 774–75
    present-day geographic variation and, 785
  phylogeny of, **772**
Homozygotes
  fitness of, 200–205, 249
  inbreeding and, 264, 266–68, 269
  selection on, 200–205
  selfing and frequency of, 264
Honeybee queens, 462
Honma, Takashi, 744, 745
Hoogland, John, 429–32, **450**, 451
Hooper, R. E., 414
Horizontal gene transfer. *See* Lateral gene transfer (LGT)
Horizontality, principle of original, 61
Horseshoe crabs, 708, **708**
Hosken, David, 376, 377, 378–79
Host-parasite arms race, 311, **311**
Host plant fidelity, 618
Host shifts
  in feather lice, 390–91, **391**
  in herbivorous beetle, 388–90, **389**
Houseflies (*Musca domestica*), 495–96, **496**
House mice (*Mus musculus domesticus*), 454
Hoxa-13, 739
Hoxd-11 gene transcripts, 739
Hox genes, 728
  changes in expression, arthropod segmentation and, 732–35, **733**, **734**
  changes in number, diversification of animals and, 730–32, **732**
  parallels between *MADS*-box genes of plants and, 746
  structure and function, 729–30
  tetrapod limb development and, 738
Hubby, J.L., 161
Huey, Ray, 373–75
Hughes, Austin, 254, 257–58
Hughes, Diarmid, 309
Hughes, Kimberly, 495
Human endogenous retrovirus K (HERV-K), 758
Human evolution, 753–98
  human-ape relationship, 754–63
    divergence times, 761, **761**
    molecular evidence for, 755–59
    morphological evidence for, 759–61
  origin of *Homo sapiens*, 773–85
    African replacement vs. multiregional evolution models of, 775–85, **778**, **784**
    controversies over, 774–75
    present-day geographic variation and, 785
  recent ancestry, 764–73
    fossil evidence, 764–73
    interpreting fossil evidence, 771–73
  of uniquely human traits, 786–91
    language, 789–91
    stone tools, 786–89
Human genome
  linkage disequilibrium on, 294–95
  as mosaic, 759
Human immunodeficiency virus (HIV), 3–36, 170
  AIDS and, 3, 4–11
  AZT (azidothymidine) and temporary inhibition of, 11–16, 77
  AZT resistance, **12**, 15
  characteristics of, 7–8
  distribution of infection, **5**
  diversity of, 27–30
  escape mutation, **18**
  evolutionary history of, 25
  evolution within individual, 18–20, **18**, **19**, **20**
  HIV-1, 22, **26**, 27, 28–29
    CCR5 genotype and, 161–62, 170
  HIV-2, 22, 26–27, **26**
  human-chimpanzee molecular homologies and, 56
  infection mechanism, 8–11, **10**
  lethality of, 16–22
    short-sighted evolution and, 18–21
    transmission and, 21–22
  life cycle of, 7–8, **7**, 643
  movement from chimpanzee to humans, 27, 28–29
  mutation rate of, 14–15
  origin of, 25–27
  phases of infection, 10–11
  phylogeny of, 25, **26**
  prevention successes and failures, 6, **6**
  rates of infection by, 4–5, **5**
  replication of, 7–8
  resistance to, 12, 22–25
  transmission of, 5–6
  vaccine for, 27–30
  virions, 7–8, 10, 12–13

Human leucocyte antigen (HLA), 18
linkage disequilibrium in HLA loci, 294
Humans
extinctions caused by, 717–20
genetic differences between chimpanzees, gorillas and, 762–63, *762*, **763**
genetic diversity among living, 776–77, **776**
geographic structure in populations of, **777**
inbreeding depression in, 270–71, **271**
phylogeny of, **779**
relationship between extant apes and, 754–63
divergence times, 761, **761**
molecular evidence of, 755–59
morphological evidence of, 759–61
sexual dimorphism in, 401, **402**
body size, **402**, 438–40
Hummel, Susanne, 299
Hunt, James, 463
Hunter-gatherers, **503**, 504, 550–51, **550**
Hurtado, Magdalena, **503**, 504
Huttley, Gavin, 258, 294
Hutton, James, 60–61
Huxley, Thomas Henry, 46, 48, 755
hx546 allele, 497, **497**
Hybridization, 624, 625–29
creation of new species through, 625–26
secondary contact and, 623, 625–29, *627*
in situ, 738
Hybridization and assimilation model, 774, **775**
Hybrid zones, 626–29
Hydra, sexual and asexual reproduction in, 303, **303**, 304
Hydrolysis problem, 657–58
Hydrophyllum appendiculatum (waterleaf), 271
Hyla versicolor (gray tree frogs), 417–21, **418**, **419**, **420**, *421*
Hylobates (gibbons), 754
Hymenoptera
coefficient of relatedness among, 462
eusocial, 460–62, *460*
phylogeny of, 463–64, **463**
Hyoid bone, 790, **790**
Hypothesis/hypotheses
formation strategies, 395–96
null, 371–72
testing required by, 364–67

Ichthyolestes pinfoldi, 129
Iguanas, sexual dimorphism in, 401. *See also* Desert iguana (*Dipsosaurus dorsalis*); Marine iguana (*Amblyrhynchus cristatus*)
Immune response
pathogenic evasion of, 531–37
virulence, 541–45
to viral infection, 8–9, **8**
HIV infection, 10–11
Immunoglobulin, origin in transposition event, 582–83, **583**
Impact hypothesis of K-T mass extinction, 712–14
Impatiens capensis (jewelweed), 271
Imprinting, genomic, 514–16
Inbreeding, 264–73
coefficient of (F), 268, 269–70, **270**
empirical research on, 266–68
eusocial naked mole-rats and, 466–67
general analysis of, 268–69
genotype frequencies and, **265**
homozygotes and, 264, 266–68, 269
mechanisms to avoid, 272
selfing, 264–66, **265**, *265*, 268, 295
Inbreeding depression, 270–73, **272**, 495, **495**
genetic drift as, 274–75
Inclusions, principle of, 61
Inclusive fitness, 448–50, 471, 504
in white-fronted bee eaters, 453, **453**
Indels, 149
Independence, evolutionary, 606–7
Indinavir, 16
Indirect fitness, 450, 452
Indirect kin recognition, 454
Individuals
natural selection on, 90, **90**, 94
variation among, 83–87, 148–49
Individual variation, 78
Indy mutation, 498, **498**
Infanticide, 94, 414–15
Inference, phylogeny. *See* Phylogeny(ies)
Influenza A virus, 4, 531, 532–34, **532**
molecular evolution in, 253, **253**
mortality from, 532
pandemic strains of, 532, 534–37
phylogeny of, **533**, 535–36, **535**, **536**
vaccines against, 534
Ingman, Max, 781–82
Ingram, Vernon, 146
Ingroup, 113
Inheritance, blending, 95, **95**. *See also* Heritability
Insects, 734–35
aging in, **487**
Cretaceous-Tertiary extinction and, 716
Hox gene expression in, 734
sociality in, 460–62, *460*
Insertion sequences (IS elements), 579
In situ hybridization, 738
Insulin-like growth factor II (IGF-II), 515
Integrase inhibitors, 16
Intelligent Design Theory, 97–105, 754
International HapMap Consortium, 295
International Union for the Conservation of Nature, 719
Intersexual selection, 408, 415–29
Interspecies relationships, 60
Intestinal bacteria, virulence of, 545, **545**
Intrasexual selection, 407, 408–15
by combat, 408–12
by infanticide, 414–15
by sperm competition, 413–14, **414**
Intron, 57, **642**
Invasions, biological, 520
Inversions, chromosome, 156–58, **156**, **157**, 160, *161*
Invertebrates, marine, 716, 717
IQ, heritability of, 350, 352–55, **352**, **353**
Iridium anomaly at K-T boundary, 712, **712**
Irish elk, 44, **44**
Iron sulfide (FeS_2), 652
Irwin, Darren, 54
Island extinctions, 717–18, 719
Isolation
genetic, mechanisms of, 611–16
chromosomal changes, 615–16
dispersal and colonization, 612–14, **612**, **613**
migration, 612
physical isolation, 611–15
vicariance, **612**, 614–15
postzygotic, 624
prezygotic, 624, **625**
reproductive, 606, 607
Isoniazid, 538, 539
Isoptera (termites), eusociality in, *460*
Isotopes used in radiometric dating, 63, **64**
Isthmus of Panama, geographic isolation caused by, 614
Isua, Greenland, 659, **659**
Iteroparous morph, 518–19
Ivy, Tracie, 432

Jablonski, David, 711, 716, 717
Jackson, Jeremy, 706, 707
Jadera haematoloma, 40
James, Helen, 717
Janzen, F.J., 344
Jenkin, Fleeming, 95, 562
Jennings, H.S., 287
Jewelweed (*Impatiens capensis*), 271
Joeng, Kyu Sang, 490
Johanson, D.C., 764–65
Johansson, K., **433**
John Paul II, Pope, 105
Johnson, Kevin, 390–91
Johnson, Warren, 53
Johnston, Wendy, 650, **650**
Jones, Adam G., 405, 406, 407
Jones, Kristina Niovi, 78–80
Joyce, Gerald, 645–47, 649, 657
Jukes, 256
Jumping genes. *See* Transposable elements
Jumping spiders (*Phidippus apacheanus*), 367–72, **368**, **369**
Jupiter, **639**
Jürgens, Gerd, 726

Kahler, Heidi, 428
Kale (*Brassica oleracea acephala*), **75**
Kaplan, Hillard, 566
KatG gene, 538
Keck, D.D., 324, 351, 352, 353–54, **354**
Keightley, Peter, 310
Keller, Lukas, 83, 84, 272
Keller, Marcel, 57
Kelvin, Lord, 96
Kenyanthropus, 767
K. platyops, **766**, 767–68, 770
K. rudolfensis, 770
Kerr, W.E., 242
KERV-1 retrotransposon, 581, **581**
Kiger, J.A., Jr., **244**
Kimberella, *696*
Kimura, Motoo, 251, 252, 255, 256
King, Marie-Claire, 595, 763
King, Richard B., 227, **228**, 229–30, 256
King crabs, 708, **708**
Kingsolver, J.G., **373**
Kin recognition, 454
Kin selection, evolution of altruism and, 448–59
alarm calling and, 450–51, **450**, **451**
altruistic sperm, 454–56, **456**
in birds, 451–53, **452**
cannibalistic tadpoles, 453–54, **455**
coots, 456–57, **457**
defined, 450
greenbeard alleles, 457–59
inclusive fitness, 448–50
kin recognition, 454
parent-offspring conflict and, 467–71, **468**
siblicide and, 470–71
Kipsigis people, sexual selection among, 438, **438**
Kittler, Ralf, 132–33
Kitzmiller et al. v. *Dover Area School District*, 98
Kiwi, brown, 484, **484**
Klein, Anke, 653
Klicka, John, 762–63
Kluger, M., 556, 558, 560
Kmita, M., 729
Knock-out cells, 458–59
"Knock-out" mutations, 148, 577
Knoll, Andrew, 662, **673**, 702, 734
Knowlton, Nancy, 614
Knucklewalking, 760–61
Koch, Robert, 529
Kohlrabi (*Brassica oleracea gongylodes*), **75**
Korarchaeota, 668
Korber, Bette, 28–29
Kotukutuku (*Fuchsia excorticata*), 385–88, **386**
Kreitman, Martin, 259
K-T extinction. *See* Cretaceous-Tertiary (K-T) mass extinction
Kumar, Sudhir, 761
!Kung, 550
Kuru, 188–91, 192, 193

Labial (lab) genes, 731
Lack, David, 505
Lack's hypothesis, 505, **505**, 506, 507–9, **507**, *508*
Lactase, 551, 562
production, persistence beyond age of weaning, 302, **302**
Lactate dehydrogenase-B (Ldh-B), 102
Lactational amenorrhea, 554, **555**
Lagenorhynchus obscurus (dolphin), **122**
Lake, James, 676
Lake Erie water snakes, migration and selection in, 227–30
Lamarck, Jean-Baptiste, 39, 95
Lambert, J. David, 309–10
Lande, R., 344, 617
Langerhans, Brian, 92
Langley, Charles, 577
Language
origin of, 789–91
sign, 789
Langur monkey, weaning conflict in, **468**
Large ground finch (*Geospiza magnirostris*), 85, **85**, 89

Larson, Edward, 67
Larvae, fossil, 695
Larval growth rate and survival, 420, *421*
Larynx, spoken language and, 789, 790
Lasaea subviridis (clam), 483
Lateral gene transfer (LGT), 584–91, **584**, **661**, 669–70, **670**
  antibiotic resistance genes moving via, 592
  in bacteria and archaea, 585, 590–91, 670, **670**
  chronocyte hypothesis, 677, **677**
  diagnosing, 585–86
  in eukaryotes, 586–90
    adaptive significance of, 589–90
  mechanisms of, 585
  phylogenetic evidence for, 586, **586**
  ring of life hypothesis, 676–77, **676**
  three viruses, three domains hypothesis, 677–80, **679**
  universal gene-exchange pool hypothesis, 675–76, **675**
Lauder, G.V., **377**
Law, Richard, 484
Lawrence, J.G., 607, 670
Lawson, R., **228**, 229–30
Ldh-B (lactate dehydrogenase-B), 102
Leaf beetles (*Ophraella*), 388–90, **389**
LEAFY gene, 745
Leakey, Meave G., 767–68, 770
Least-squares linear regression, 334–35
Lee, Carol Eunmi, 609
Leg trembling in male water mites, 422–24
Leks, 223, 621, **622**
Lenormand, Thomas, 307
Lenski, R.E., 349–50
Lenski, Richard, 101–2, 212–13
Lesch, Klaus-Peter, 356, **356**
Leslie, A.J., 18
Leuba, James, 67
Levin, B.R., 306
Levinton, Jeff, 701
Levy, Stuart, 541
Lewontin, R.C., 161
Li, Wen-Hsiung, 762
Lice, 132–33
  body, 132–33
  head, 132, 133
  host shifts in feather, 390–91, **391**
  infestations in rock pigeons, beak hooks and, 363–64, **363**
  pubic, 132
Lidicker, William, 267, *268*
Lieberman, Daniel, 778, 779
Life, defining, 642–43
Life, origins of, 639–87
  biological polymers, 657–58, **658**
  building blocks, 652–57
    sources of, 652
  cellular life, 641, 660–62
    advantages of, 660–61
    DNA and, 660
    examination of, 666–67
    fossil record of, 662, **662**
    universal phylogeny, 663–66
  environment and, 658–60
  lateral gene transfer
    chronocyte hypothesis, 677, **677**
    ring of life hypothesis, 676–77, **676**
    three viruses, three domains hypothesis, 677–80, **679**
    universal gene-exchange pool hypothesis, 675–76, **675**
  nucleotides and, 656–57, **658**
  Oparin-Haldane model of, 655–56, **656**, 657
  overview of, **661**
  panspermia hypothesis of, 652–53
  RNA enzymes (ribozymes) and, 641
  RNA World hypothesis, 641–50
    case for, 643
  tree of life, **640**, 673–75
Life cycle, 170–73, **171**, **173**
  of HIV, 7–8, **7**, 643
Life histories, 483–528
  aging, 487–502, **487**
    deleterious mutations and, 492, **493**, 494–96, 502
    evolutionary theory of, 487, 492–502
    natural experiments of evolution of, 500–502
    rate-of-living theory of, 487, 488–92
    trade-offs and, 490–91, **491**, 496–99, 502
  basic issues in, 485–86
  in broader evolutionary context, 517–23
    biological invasions, 520
    maintenance of genetic variation, 517–19
    novel traits, 520–22
  conflicts of interest between, 514–17
    genomic imprinting, 514–16
    sexual coevolution, 516–17, **517**
  menopause, 503–4, **503**
  number of offspring produced in year, 502–9
    clutch size in birds, 498–99, **499**, 505–7, **506**
    clutch size in parasitoid wasps, 507–9
  size of offspring, 509–14
    phenotypic plasticity in egg size in beetle, 512–14, **513**
    selection on, 510–12
  suboptimal, 522–23
Likelihood
  maximum (ML), 124–25
  QTL mapping and calculating, 328–29
Likelihood ratio, 329
Limb evolution in animals, 735–41
  arthropod limb, 739–41, **740**, **741**
  tetrapod limb, 735–39, **736**, **737**, **738**, **739**
Limulus, 708
Lindén, Mats, 506
Lineages, 26
Linear regression
  best-fit line, **83**, 336
    slope of, 339, **339**, 342, **342**
  least-squares, 334–35
LINEs (Long INterspersed Elements), 128, **128**, 578
Linkage, genetic, 156–57
Linkage disequilibrium, 282, **283**, 284–302, 426
  coefficient of (*D*), 284, 285
  creation of, 285–89
    from genetic drift, 288, **288**, 309–10, 312
    from population admixture, 288–89, **289**
    from selection on multilocus genotypes, 285–88, **287**
  to date divergence between African and non-African populations, 784–85
  defined, 284–85
  elimination of, 289–91, **289**, **291**, 294
  genetic drift and, 288, **288**, 309–10
  genetic recombination and, 289, 307
  genotype frequencies and, 282–84, **283**
  in human leukocyte antigen (HLA) loci, 294
  practical reasons to study, 295–302
    gene and population history reconstruction, 295–99
    positive selection detection, 299–302
  selection and, 310
  significance of, 291–95
  single-locus studies and, 292, 294
Linkage equilibrium, **283**, 284, 286, 287
  on human chromosome 22, **294**
Linkage model with r = 0.1, 328
Lions (*Panthera leo*), 475–77, **476**
  infanticide among, 94, 414–15
  pride of, 414, 475–76, **476**
  territory defense in, 475–77, **476**
Lister, Joseph, 530
Liu, Rong, 23
Lively, Curtis, 311–12, 387
Living fossils, 707–9, **707**, **708**
Lizards, genetic imprinting in, 515
Lobe-finned fish, tetrapod limb derived from fins of, 736
Local mate competition, 462
Loci. *See* Gene(s)
LOD (logarithm of the odds) score, 329–30, **329**
Logsdon, John, **584**, 586
Longevity. *See* Aging
Longflower tobacco (*Nicotiana longiflora*), 320–23
Long INterspersed Elements (LINEs), 128, **128**, 578
Long-tailed widow birds (*Euplectes progne*), 403, **403**
Long terminal repeats (LTRs), 578
Lophotrochozoa, **699**, 700
Lorenzini, Antonello, 491
Loss-of-function mutations, 148, 165, **165**, 210, 212
  inbreeding depression and, 270
Lougheed, Lynn, 470–71
Loxodonta africana (African elephants), 610, **611**
Loxodonta cyclotis, 610
LTRs (long terminal repeats), 578
Luckinbill, Leo, 489
Lungfishes, 736
Lyczak, Jeffrey, 218
Lyell, Charles, 60–61
Lynch, Michael, 153, 273
Lyngbya, **674**
Lyon, Bruce, 456
Lysosomal storage disorder, Gaucher disease as, 295

McCall, C., 271
McClearn, Gerald, 337–38
McCleery, R. H., 506–7
McClintock, Barbara, 576, 579
McCollum, F.C., 267, *268*
McDonald, John, 259
McDonald-Kreitman (MK) test, 259
McElligott, Alan, 366–67
McGranaghan, Don, 768, 769, 770
McKay, David, 652
McLean, Robert, 653
McLennan, Deborah, 53
Macnair, Mark, 205, 206
Macroevolution, evo-devo research and, 748–49
Macroevolutionary patterns, 39, 40, 49, 702–9
  adaptive radiations, 702–4, **703**
  dynamic, 707–9
  relative frequency of stasis and gradualism, 707
  stasis, 704–9
    in Bryozoans, 706, **706**
    demonstrating, 705–6
Macrophages, 10, **529**
Madar, S., 122
Mad cow disease, 189
MADS-box genes, 728
  discovery of, 744–47
Magnetite (Fe_3O_4), 652
Major histocompatibility complex (MHC), 257, 454
Malaria, 135, 300, **300**
Malaria parasite, 135
Male-male competition, 407, 408–15
  alternative male mating strategies, 412–13
  combat, 408–12, 426
  homicides, 438–39
  infanticide, 94, 414–15
  sperm competition, 413–14, **414**
  stalk-eyed flies, 426
Mammals
  aging in, **487**
  phylogeny of, **137**
  variation in lifetime energy expenditure among, 488, **488**
Mangrove finch (*Certhidea heliobates*), 81
Manning, Jo, 454
Marchordum, Annie, 616
Marine iguana (*Amblyrhynchus cristatus*), 408–12, **408**
  body size of, 408–12
    natural selection on, 409
    sexual selection differentials for, 411, *411*
  male-male combat among, 408–12, **410**
Marine invertebrates, Cretaceous-Tertiary extinction and, 717
Marine organisms, extinction rates for, 711
Marlowe, Frank, 566
Marsupials, energy expenditure for, 488
Martian bacteria, 652, **652**
Martinson, Jeremy, 162, 163, 191
Masked booby, 470–71, **470**, *471*
Mass extinctions, 702, 709–20, **710**
  Cretaceous-Tertiary (K-T), 712–17
    extent and selectivity of, 716–17
    impact hypothesis of, 712–14
    killing mechanisms, 714–15
  recent, 717–20
    Polynesian avifauna extinctions, 717–18
Mass spectrometer, 64
Mate choice. *See* Sexual selection
Maternal ancestry, direct, 781
Maternal effects, 84

deleterious inbreeding effects masked by, 271
Maternal fitness
clutch size of parasitoid wasps and, 507–9, **508**, *508*
egg size and, **511**, 512
Mather, K., 310
Mating, 224. *See also* Sexual selection
alternative male strategies for, 412–13
assortative, 426, **427**, 624
asymmetry of sex and, 407–8
in hangingfly, 421–22, **421**, **422**
in marine iguanas, 410–12
multiple mating by females, 429–32
nonrandom. *See* Inbreeding
random, 171, **174**, 175–76, 178–79, **178**
in water mites, 422–24, **423**
Mating call, gray tree frog, 418–21, **418**, **419**, **420**, *421*
Mattern, Michelle, 53
Matthew, Patrick, 39
Maximum likelihood (ML), 124–25
Maynard Smith, John, 303–4, 305, **308**, 310, 473
Mazur, Allan, 440
MCR-1, **95**
Mead, Simon, 188, 190, 191, 192
Medical applications, 529–73
adaptations
breast cancer, 552–55
environment and, 550–51
fever, 556–60
menstrual cycle, 554–55, **555**
myopia, 551–52
parenting, 561–68
pathogens, 531–45
antibiotic resistance in, 538–41
evasion of immune response by, 531–37
virulence, 541–45
tissues as evolving populations of cells, 546–49
history of cancer, 548–49, **549**
patient's spontaneous recovery, 546–48
Medical practice, fever and, 560
Mediterranean fruit fly (*Ceratitis capitata*), 414
Medium ground finch (*Geospiza fortis*), 81–90, **81**, **83**, **85**, **87**, **88**, **89**
Megachiroptera, 376–79
Meiosis, with crossing over, 306
Melanin, 95
Melanocytes, **95**
Melospiza melodia (song sparrows), 335–37, **335**
Meme, 562
Memory T cells, 9, 10, 11
Mendel, Gregor, 39, 94–95, 196
Mendelian genetics, 169–280
genetic drift, 232–64
random fixation of alleles, 238–49
sampling error and founder effect, 236–38
Hardy-Weinberg equilibrium principle, 170–82
analysis using, 183–85, 187–88, 210–12, 225–27, 269
crucial assumptions of, 180–82, **181**
fundamental conclusions of, 180
general case, 177–80, **178**
generating expected values using, 192–93
genetic variation for resistance to disease and, 188–91
with more than two alleles, 180
as null model, 181, 182
numerical calculation, 174–77, **174**, **176**, **177**
rate of evolution and, 196–97
simulation, 170–73, **173**
use of, 180–82
migration, 225–32
defined, 225
as evolutionary force, 225–27, **226**, **228**, 230–32
homogenization across populations and, 230–32
migration-selection balance, 229–30, 627
mutation, 210–18
as evolutionary force, 210–12, **211**
mutation-selection balance, 213–18
selection and, 212–13, **213**
nonrandom mating (inbreeding), 264–73
coefficient of (F), 268, 269–70, **270**
empirical research on, 266–68
general analysis of, 268–69
homozygotes and, 264, 266–68, 269
inbreeding depression, 270–73, **271**
mechanisms to avoid, 272
quantitative traits and, 320–21, **321**, **323**
selection, 182–210
added to Hardy-Weinberg analysis, 183–85, 187–88
changes in allele frequencies and, 183–86, **183**, **185**
examples of, 182–83
frequency-dependent, 205–7, **206**
general treatment of, 186
genotype frequencies and, 187–91, **187**
Menopause, 503–4, **503**
Menstrual cycling, breast cancer and, 554–55
Merops bullockoides (white-fronted bee-eaters), 451–53, **452**, **453**, 469–70
Mesoderm, 700
Mesonychians, 119
Mesozoic era, 692, **692**, 693, **693**
Messenger, Sharon, 543
Messenger RNA (mRNA), 146, **146**
reverse-transcribed, SINEs originating from, 588, **588**
Metabolism
aging rate and rate of, 488
body size in marine iguanas and, 409, **409**
Metacarpals, human vs. chimpanzee thumb, 787–88, **787**
Metamorphosis, 420, *421*
Meteorites
amino acids in, 651
carbonaceous chondrite, 654
Murchison, 651–52, **651**
Methanococcus jannaschii, **665**, 667
Methodological Naturalism, 65–66
Methuselah gene, 498, **498**
Methylation, 580–81
Metrarabdotos, 706, **706**
Meyerowitz, Eliot, 726, 744
Mice, 553, **553**
deer, 483
kin recognition in, 454
Michod, R.E., 306
Microbispora, 653
Microevolution, 40, 49
evo-devo research and, 748–49
Microraptor gui, 48, **48**
Microsatellite, 782–83
as markers for calculating coefficients of relatedness, 449
Microtektites in K-T boundary layers, 713, **713**, 714
Migration, 224, 225–32
added to Hardy-Weinberg equilibrium analysis, 225–27
allele frequencies and, 225–32
defined, 225
genotype frequencies and, 225–26
Hardy-Weinberg equilibrium assumption of no, 181, **181**
as mechanism of isolation, 612
in Mendelian genetics of populations, 225–32
defined, 225
as evolutionary force, 225–27, **226**, **228**, 230–32
homogenization across populations and, 230–32
migration-selection balance, 229–30, 627
one-island model of, 225–27, **225**
Mikkelsen, Tarjei, 763
Milk sugar, ability to digest in adulthood, 302, **302**
Miller, Kenneth, 66–67
Miller, Stanley, 655
Mills, D.R., 643–45
Mimicry, 368–69, **369**
Mimivirus, 680
Mimulus cardinalis, 324–31, **326**
flowers of, *327*
phylogeny of, 325, **325**
QTLs for floral traits in, **330**
Mimulus guttatus, 272
Mimulus lewisii, 324–31, **326**
flowers of, *327*
phylogeny of, 325, **325**
QTLs for floral traits in, **330**
Miroyedicha Formation, Siberia, **673**
Mismatch errors, DNA, 494–95
Mitochondria
evolution of, 585, 586–90, **586**
selection at level of, 392–94, **393**
Mitochondrial DNA (mtDNA), 586–88
evolutionary tree of human, 779, 780–81, **781**
phylogenies of humans and African great apes based on, 757
Mitochondrial Eve, 781
Mitochondrial genomes, parasitic, 392–94, **393**
Mockett, Robin, 498
Mockingbird, homologies among, 60
Model organisms, homology and, 59
Modern Synthesis, 39, 96
Molds (fossils), 690, **690**, 691
Molecular basis of adaptation, 591–96
Molecular characters, parsimony with multiple, 122–23, **122**
Molecular clocks, 132–33, 700, 701
Molecular evidence of multiregional origin, 779–85
Molecular evolution
genetic drift and, 251–64
DNA sequence divergence, 252–54
nearly neutral model of, 254–56
neutral theory of, 252, 256–64
neutral theory of, 132, 250, 251, 252, 533
Molecular homology, 56–59
Molecular sequencing in phylogeny, 663–64, **668**
Molecular traits, 119–20
Molineux, Ian, 543
Møller, A.P., 506
Mollusca, **699**
Molluscs, 731–32
variation in eyes of, 99, **99**
Monera, **664**
Mongooses, banded, **447**
Monkeyflowers, 324–31, **326**
Monkeys, homologous language circuit structures in, 789
Monoecious plants, 383, 436
Monophyletic groups, 131, **131**
synapomorphies identifying, 112–14, **113**, **114**
Monophyly, 608, **608**
Montmorillonite, 658
Moon, history of large impacts on, 654, **654**
Moore, A.J., 344
Moore, Harry, 454–55, **456**
Morality, debate over scientific creationism and, 105
Moran, John, 581–82
Moran, Nancy, 309–10
Morgan, Thomas Hunt, 94, 426
Morphological approach to phylogeny, 663
Morphological change
adaptive radiations and, 703
patterns of, **705**
speciation and, 706
Morphological traits, 119–20
Morphology
of Cambrian animals, 698–700, **699**
human-ape relationship in, 759–61
parsimony with single character, 120–22
Morphospecies concept, 606
Mortality, 86
ecological, 500, 502
physiological, 500
Mosquito fish (*Gambusia affinis*), 92
Mosses, polyploidy in, 616
Mountain sagebrush (*Artemesia tridentata vaseyana*), 627
Mouse Mammary Tumor Virus (MMTV), 553
Mousseau, T.A., 518
MPF2, 596
mRNA, 146, **146**
reverse-transcribed, SINEs originating from, 588, **588**
mtDNA. *See* Mitochondrial DNA (mtDNA)
Mueller, Ulrich, 440
Mukai, Terumi, 197–200
Mukherjee, A., **402**
Muller, H.J., 308
Muller's ratchet, 308–10, **308**, **309**
Multiple loci, evolution at. *See* Linkage disequilibrium; Quantitative genetics; Sexual reproduction
Multiregional origin model, 774, 775–85, **775**
archaeological and paleontological evidence of, 775–78

Multiregional origin model, (*continued*)
cladogram predicted by, **778**
genetic predictions distinguishing, *780*
molecular evidence, 779–85
Murchison Meteorite, 651–52, **651**
Murder of children, 567–68, **568**
Murray, Charles, 352–55, **352**, **353**
Murray, John, 753
Musca domestica (houseflies), 495–96, **496**
Mus domesticus, 553, **553**
Mus musculus, 553, **553**
M. m. domesticus (house mice), 454
Mutation(s), 143–68, 210–18
added to Hardy-Weinberg equilibrium analysis, 210–12
asexual reproduction and, 308–9, **308**, **309**
back mutation, 15, 210, 308, 546–48
chromosome, 156–60
genome duplication, 159–60, **161**
inversions, 156–58, **156**, **157**, 160, *161*
polyploidy, 159–60
deleterious, **496**
evolutionary theory of aging and, 492, **493**, 494–96, 502
late-acting, 500
evolution of Darwinism and, 94
fitness effects of, 150–52
frameshift, 332
gain-of-function, 395
gene duplication, *161*
new genes from, 152–56, **153**
genome duplication, 159–60, *161*
Hardy-Weinberg equilibrium assumption of no, 180, **181**
in HIV genome, evolution of resistance and, 16–17
homeotic floral, 743–44, **743**, **744**
in *Hox* genes
changes in *Hox* expression, arthropod segmentation and, 732–35, **733**, **734**
changes in *Hox* gene number, animal diversification and, 730–32, **732**
homeotic mutants in *Drosophila*, 729–30, **730**
isolation from, 615–16
"knock-out," 148, 577
loss-of-function, 148, 165, **165**, 210, 212
inbreeding depression and, 270
in Mendelian genetics of populations, 210–18
as evolutionary force, 210–12, **211**
mutation-selection balance, 213–18
selection and, 212–13, **213**
in molecular evolution, 250
molecular view of, 146
nature of, 145–48
nearly neutral, **256**
neutral theory of, 132, 251, 252, 533
pleiotropic, 496–99
point, 147–48, 150, *161*, 253
as random, 232
rates of, 148–50
for cystic fibrosis, 216, 217
direct estimates of, 148–49
highest, 210, 212
of HIV, 14–15
molecular clock and, 132
natural selection and, 149–50
for recessive alleles, 216
for spinal muscular atrophy, 216
selectionist theory of, 250, 251
sex reproduction benefits from, 308–10
substitution vs., 249, **249**
Mutation accumulation hypothesis, 494–96, **495**
Mutational meltdown, 273
Mutation-selection balance, 213–18, 350
in cystic fibrosis, 214–18
Mutualism, 447
aphid-bacteria, 135, **135**
cattle-oxpecker, 365–66, **365**, **366**
Mycobacterium tuberculosis, 538, 664
Myopia, 551–52
Myriapods, 733, **733**, **734**
Mysidopsis bahia (opossum shrimp), 741

Naive T cells, 9, 11, 20
Naked mole-rats, eusociality in, 465–67, **466**, *467*
Narrow-sense heritability, 335
Natural history, 395–96
Natural selection, 15, 16–17, 39, 40, 73–109. *See also* Adaptation(s)
action on individuals, 90, **90**, 94
action on phenotypes, 90
action on populations, 90, **90**
backward-looking nature of, 91
beak shape in Galápagos finches, 80–90, **80**, **83**, **85**, **87**, **88**
on body size in marine iguanas, 409, **409**
cultural evolution by, 562
on DNA sequences, neutral theory as null hypotheses detecting, 256–64
falsifiability of, 103
fitness and, 93, 98, 338–50
genetic drift vs., 234
genetic variation and, 517–18
interspecies interactions and, 135
as mechanism of divergence, 617–21
mutation rates and, 149–50
new traits from, 91–92
non-perfect traits from, 92–93
as nonrandom and nonprogressive, 93
postulates of, 76–77, **76**
heritability of variation, 83–85
nonrandom survival and reproduction, 87–88
on "silent" substitutions, 260–63
in snapdragons, 78–80, **79**
speciation by, *620*
for translational efficiency, 261
Nature-nurture controversy, 333
Neandertals, phylogeny of, **779**
Nearly neutral model, 254–56, **256**
Nearsightedness (myopia), 551–52
Neel, J.V., **271**
Nef protein, 21
Negative (purifying) selection, 254, 261
Nei, Masatoshi, 257–58
Nelson, Craig, 437
Nematoda, **699**
Nematodes, pleitropic mutations in, 496–97, **497**
Nematode worms (*Caenorhabditis elegans*), 148–49, 150, 490–91, 496–97, **497**, 691, 730–31
Nerodia sipedon, 227–30, **228**
Nerophis ophidion, 432–34
Nesse, Randolph, 531
Nest parasitism, conspecific, 84
Nests, wasp, 464–65, **465**
Net fitness, 516–17, **517**
Net-stance, 422–24
Nettle, Daniel, 440
Neumania papillator (water mite), 422–24, **423**, **424**
Neuraminidase, 535
Neutral alleles, 152
Neutral evolution, canonical rate of, 252–53
Neutral theory, 165–66, 251, 252, 256–64
current status of, 263–64
of molecular evolution, 132, 250, 251, 252, 533
nearly neutral model, 254–56
as null hypotheses detecting natural selection on DNA sequences, 256–64
Newts (*Taricha granulosa*), 405–6, **405**
New Zealand, avifauna extinctions in, 718
Nicotiana longiflora, 320–23
Nielsen, Rasmus, 763
Nikaido, Masato, 128, 129
Noble, William, 789–90
NOD2 gene, 332
Node of phylogenetic tree, 25, **52**
No Free Lunch Theorems, 104
Nonacs, Peter, 464–65, **465**
Nonrandom mating, 264–73. *See also* Inbreeding
Nonrandom selection, 87–88, 93, 232
Nonsynonymous substitutions. *See* Replacement (nonsynonymous) substitutions
Nordborg, Magnus, 295
Normal distribution of trait, 333
Norms, reaction, 564
North America, AIDS epidemic in, 5, **5**
Nossal, Nancy, 149
Nostoc, 664
Novel traits, evolution of, 520–22
Novelty seeking, 331–32
Novembre, John, 299
Nucleic acids, abiotic synthesis of, 651
Nucleomorph, 589
Nucleoprotein, 535, **535**, 536
evolutionary analysis of, 537, **537**
Nucleosomes, 254
Nucleotides
origins of life and, 656–57, **658**
substitutions in antigenic sites, 532–33
Null hypothesis, 371–72
for gene behavior, 170
Hardy-Weinberg equilibrium principle as, 181, 182
neutral theory as, detecting natural selection on DNA sequences, 256–64
Nurture, nature vs., 333
Nüsslein-Volhard, Christiane, 726, 730
Nuttall, George H.F., 755

Obligate endosymbionts, 309
O'Brien, Stephen, 53
Observational studies
of adaptation, 370, 372–75
behavioral thermoregulation, 372–75
garter snake nighttime behavior, 373–75, **374**, **375**, *375*
of human behavior, 438
Oceanic chemistry, after asteroid impact at K-T boundary, 714, 715
Ochman, Howard, 607, 670
Octopuses, camera eye of, 115, **115**, 117–18, **117**
Offspring. *See also* Parenting
excess of, 86
female choice and quality of, 419–21
number produced in year, 502–9
clutch size in birds, 505–7
clutch size in parasitoid wasps, 507–9
parent-offspring conflict, 467–71, **468**
size of, 509–14
phenotypic plasticity in egg size in beetle, 512–14, **513**
selection on, 510–12
variations passed to, 94–95
Ogura, Yasunori, 332
Ohta, Tomoko, 255–56
Oldowan industrial complex, 786–89, **786**
Old World primates, 754, **755**
Oligonucleotide, 645
Olson, Storrs, 717
Olsson, Lennart, 52–53
Oncorhynchus kisutch (coho salmon), 413, **413**
Oncorhynchus nerka (sockeye salmon), **483**
Oncorhynchus tshawytscha (chinook salmon), 510–12, **511**
One-island model of migration, 225–27, **225**
On the Origin of Species by Means of Natural Selection (Darwin), 3, 38, 39, 45, 46, 52, 73, 76, 77, 97, 530, 639, 753
Ontological Naturalism, 66
Onychophorans (velvet worms), 733, **733**, 734–35, **734**
Opabinia regalis, 697, **698**
Oparin, A., 655
Oparin-Haldane model, 655–56, **656**, 657
Ophraella (leaf beetles), 388–90, **389**
Opossum. *See* Virginia opossum (*Didelphis virginiana*)
Opossum shrimp (*Mysidopsis bahia*), 741
Opposable thumb, anatomy of, 787
Optic cups, 99, **99**
Oral contraceptives, 555
Orangutan (*Pongo pygmaeus*), 404, 754
Orchids, 434, **435**, 437
homologies among flowers, 55, **56**
pink lady's slipper (*Cypripedium acaule*), 522–23
Order, 131
Organelle genomes, 586–89
chloroplast DNA (cpDNA), 586–88
evolution of, 587–88
mitochondrial DNA, 586–88
Organelles, 586–90
Organic molecules, prebiotic synthesis of, 655–56
Orgel, Leslie, 652, 653, 657, 658
Original horizontality, principle of, 61
Origin of Species. *See On the Origin of Species by Means of Natural Selection* (Darwin)
Oró, Juan, 656
Orr, H. Allen, 104, 325, 562, 624

Orrorin tugenensis, 765
Orthologous genes, 155
Otto, Sarah, 307, 310
Outbreeding, 289
Outgroup analysis, 113
Out of Africa model. *See* African replacement model
Ovarian cancer, 555
Ovchinnikov, Igor, 779
Overdominance. *See* Heterozygote superiority
Ovipositor, 135
Owen, Richard, **45**, 54
Oxpeckers, 364–67, **364**
  mutualist relationship with cattle, hypothesis of, 365–66, **365**, **366**
  parasitic relationship with cattle, hypothesis of, 366–67
Oxygen, Cambrian explosion and atmospheric, 701–2
Oyster (*Crassostrea gigas*), 483
Ozark Mountains, 245–46

p24 epitope, 18
p53 protein, 489, 490–91, **491**
$p53^{+/m}$ allele, 490, **491**
Pääbo, Svante, 759, 779
Pace, Norman, 668
Packer, Craig, 414, 415, 476–77
Paddlefish, 739
Pakicetus attocki, 129
Palella, Frank, 16
Paleolyngbya, **674**
Paleozoic era, 692, **692**, 693
Paley, William, 98
Palo verde (*Cercidium floridum*), 512–14, **513**
Panganiban, Grace, 740–41
Panhuis, Tami, 623
Pan paniscus. *See* Bonobos (*Pan paniscus*)
Panspermia hypothesis, 652–53
Panthera leo. *See* Lions (*Panthera leo*)
Pan troglodytes. *See* Chimpanzees (*Pan troglodytes*)
Paper wasps, 464–65, **465**
Papua New Guinea, kuru among Foré people of, 188–91
Parallel evolution, 293
Paralogous genes, 155, 156
Paramecium, 665
Paranthropus, 768, **768**, 771–72, 787, 790–91
  P. aethiopicus, 772
  P. boisei, **768**, 769, 772, 773, **773**
  P. robustus, 772, 787–88, **788**
    as toolmakers, 788
Paraphyletic group, 131, **131**
Parasite genomes, patterns in, 591–93, **591**
Parasites, **529**
  gene loss in, 592
  host-parasite arms race, 311, **311**
  malaria, 135
  parasitic oxpecker-cattle relationship, hypothesis of, 366–67
Parasitism, conspecific, 84
Parasitized altruism, **457**
Parasitoid wasps, 348, **348**
Parasitoid wasps, clutch size in, 507–9
Parental investment, 403–4, 409–10, 432–34
Parenting, 561–68
  of biological vs. step-children, 565–66, **565**, **566**
  children's health and, 566–68
  evolution and, 563–66
  male vs. female investment in, 403–4, 409–10
    sexual selection stronger for females than males and, 432–34
Parent-offspring conflict, 467–71
  siblicide and, 470–71
Parsimony, 117, 522
  with multiple molecular characters, 122–23, **122**
  phylogeny and, 111
  role in phylogeny inference, 118–19
  with single morphological character, 120–22
Pärt, Tomas, **487**, 498–99
Parthenogenesis, 159, 303, 304, 312
Parus major (great tit)
  clutch sizes of, 505–6
  inbreeding depression in, 272, **272**
Pasteur, Louis, 529
Paternity, misidentified, 84
Paterson, Andrew, 329–30
Path analysis with pedigrees, 449
Pathogens, 531–45
  antibiotic resistance in, 538–41
    costs to bacteria of, 539–40, **540**
    judicious use of antibiotics to inhibit, 541
    selection and, 538–39
  evasion of immune response by, 531–37
    virulence, 541–45
  evolution of, 541–44
    in human pathogens, 544–45, **545**
Patterson, Nick, 763
Pauling, Linus, 132, 146, 251
Pawlowski, B., 439–40
Pax6 genes, 727
PCR (polymerase chain reaction), 162, 727
Pea aphids (*Acyrthosiphon pisum*), 629–31, **630**
Pearson, Karl, 192
Pedicularis capitus (head lice), 132, 133
Pedicularis corporus (body lice), 132–33
Pedigree
  calculating coefficient of inbreeding (F) from, 269–70, **270**
  calculating coefficient of relatedness and, 449
  path analysis with, 449
Peichel, Catherine, 632
Penicillin-resistance in *Pneumococcus* bacteria, 539, **539**
Perfection in nature, 98–103
Perianth, 436, **437**
Periods, **62**, 63
Peripatopsis capensis, 741
Peripheral myelin protein-22 (PMP-22), 56–57
Perissodactyla, **121**, 122
Perisylvian cortex, 789
Permineralized fossils, 690, 691, **691**
Perna, Nicole, 590
Peromyscus maniculatus (deer mice), 483
Perrins, C.M., 505–6
Personality traits, quantitative traits influencing, **332**
Petals, **742**, 743, **743**
Pettifor, Richard, 506–7
Pf14, 101
Pfennig, David, 453–54
Phagocytosis problem, 677
Phanerozoic eon, 690
  extinctions during, **710**
  overview of, 692–94, **692–93**
Phenetic approach, 126, 131
Phenotypes
  blending inheritance and, 95
  fitness and, 346–47
  natural selection on, 90
  regulatory sequences and, 595–96
  of ribozymes, 643
  selection and, 182–83
Phenotypic plasticity, 564
  in clutch size, 507
  in egg size in beetle, 512–14, **513**
  environment and, 380–83
  in water flea behavior, 381–83, **381**, **383**
Phenotypic variation (*Vp*), 334
Phenylalanyl-tRNA synthetase, universal phylogeny based on gene for β-subunit of, 670, **670**
Pheomelanin, **95**
Pheromones, 624–25
Phidippus apacheanus, 367–72, **368**, **369**
Phillips, P.C., 344
Phosphate group, 144–45, **144**
Phosphodiester bonds, **144**, 145
Photoreceptor, 99
Photosynthesis, lateral gene transfer and, 589, 590
Phototactic behavior in *Daphnia magna*, 381–83, **383**
Phyletic transformation (anagenesis), 705
Phyllopodous appendages, 739, 741
Phylloscopus trochiloides (Siberian greenish warbler), 53–54, **53**
Phylogenetically independent contrasts, 378, **378**, 380–81, **381**
Phylogenetic approaches. *See* Cladistic analysis
Phylogenetic species concept, 608–9, 610
Phylogenetic tree(s), 25, 50–53, **52**, 111
  cladistic methods of building, 113
  cladogram, 25, 113
  evaluating, 124–27
  population, **782**, 783, **783**
  reading, 52
  searching, 123–24
Phylogenomics, 575–602
  frontiers in, 596–600
  lateral gene transfer (LGT), 584–91
    in bacteria and archaea, 590–91
    diagnosing, 585–86
    in eukaryotes, 586–90
    mechanisms of, 585
  molecular basis of adaptation, 591–96
    evolution of regulatory sequences, 594–96
    patterns in genomes of free-living organisms, 593–94
    patterns in parasite genomes, 591–93
  transposable elements and, 576–84, 586
    categories of, 578–79
    evolutionary impact of, 580–84
    as genomic parasites, 577
    selfish genes and levels of selection, 580
Phylogeny(ies), 25, 50, **52**, 111–40
  answering questions using, 130–36
    in biogeography, 133–34
    about classification and nomenclature, 130–31
    about coevolution, 135–36
    about rates of change, 132–33
  of apes, **755**, **756**
  of Cambrian explosion, 698–700, **699**
  of Catarrhini, 754
  of cellular and viral DNA-dependent RNA polymerases, 679–80, **679**
  of chimpanzees, 759
  of cytochrome oxidase II, **757**
  defined, 111
  of *Eurytemora affinis*, **610**
  of flu viruses, **533**, 535–36, **535**
  of fungi and ant symbionts, **138**
  of gorillas, **757**, 759
  of Hawaiian *Drosophila*, **613**
  of HIV, 25, **26**
  of *Homo sapiens*, **772**
  of humans, **757**, 759, **779**
  of hymenoptera, 463–64, **463**
  of mammals, **137**
  molecular sequencing in, 663–64, **668**
  of monkeyflowers, 325, **325**
  morphological approach to, 663
  of Neandertals, **779**
  parsimony and, 111
  problems in reconstructing, 114–19
  for sea urchin groups, **521**, 522
  of simian immunodeficiency viruses (SIVs), **26**, 27
  for species under study, 378
  universal, 663–66
    based on concatenated sequences of 31 universal genes, **671**, 672
    based on presence/absence of protein superfamilies, 672, **672**
    discordance of different estimates of, 669, **669**
    genetic sequencing approach to, 663–64, **665**, 668–69, **668**
    morphological approach to, 663
    rooting, 666, **666**, 673–75
    whole-genome sequencing in, 668–69, 671–72
  of water mite (*Neumania papillator*), 424, **424**
  of whales, 119–30, **121**
    choosing characters for data, 119–20
    competing ideas about, **121**
    distance methods, 125, 126, **126**
    nearly perfect phylogenetic character, **128**
    parsimony with multiple molecular characters, 122–23, **122**
    parsimony with single morphological character, 120–22
    resolving character conflict, 127–30
Phylogeography, **133**
Physalis pubescens, 595–96, **595**
Physconelloides, 391
Physical isolation as barrier to gene flow, 611–15
Physiological mortality, 500
Physiological performance, temperature and, 372–73
Phytoplankton, Cretaceous-Tertiary extinction and, 715
Pickford, Martin, 765

Pier, Gerald, 165, 217–18
Pigeons, 74
  lice infestations in rock pigeons, beak hooks and, 363–64, **363**
Pigment cup, 99, **99**
Pilbeam, David, 790–91
Pingelapese people, 238
Pink lady's slipper orchids (*Cypripedium acaule*), 522–23
Pipefish (*Syngnathus typhle*), 406–7, **406**, **407**
  sexual selection by male choice in, 432–34, **433**
PISTILLATA (*PI*) genes, 744–45
Placental development, genomic imprinting and, 515–16
Plague, 4, 299
Planetesimals, 640
Plankton, Cretaceous-Tertiary extinction and, 715, 716
Planktotrophs, **711**
Plant bugs (homoptera), *460*
Plants, **664**, 665–66. *See also specific species*
  Cretaceous-Tertiary extinction and, 716
  domestic, 74–75
  evolution of flower, 742–47
  host fidelity, 618
  hybridization, 625–26
  major plant innovations, 742
  monoecious, 383, 436
  polyploidy in, **158**, 159–60
  radiations among, adaptive, 703–4
  sexual selection in, 434–38, **436**
  trade-off between reproduction and survival in annual vs. perennial, 499, *499*
Plasmids, 577, 579
  conjugation, 585
  genes conferring drug resistance on, 592
Plasticity, phenotypic. *See* Phenotypic plasticity
Platyhelminthes (flatworms), **699**
Platyspiza crassirostris (vegetarian finch), 81
Pleiotropic mutation, 496–99
Pluteus larvae of sea urchin, 520–22, **521**, **522**
PMP-22 (peripheral myelin protein-22), 56–57
Pneumococcus, penicillin-resistant, 539, **539**
Poecilia reticulata (guppies), **402**
Pogo, Beatriz G.-T., 553
Point mutations, 147–48, 150, *161*, 253
Polemonium viscosum, 342–43, **342**, **343**
Polioviruses, 542
Polistes sp., 464–65, **465**
Pollen, 742
  reproductive success through donation of, **436**
Pollen tubes, growth of, 387, *387*
Polyadenylate primers, 658
Polyandry, 429–32
Polycyclic aromatic hydrocarbons, 652
Polymerase chain reaction (PCR), 162
  development of, 727
Polymers, biological, 657, **658**
Polymorphic chromosomes, 157
Polymorphism, 164, 254, 259
  genetic, 248
  short-tandem-repeat, 783, **784**
Polynesian avifauna extinctions, 717–18
Polyploidy, 159–60
  speciation and, 615–16
Polytomy, 127
Ponce de León, Marcia, 767
Pongo pygmaeus (orangutan), 404, 754
Popesco, Magdalena, 763
Population(s)
  defined, 170
  effective size, 244, 245
  heterozygosity in, 269
    random fixation and loss of, 245–49
  linkage disequilibrium from admixture of, 288–89, **289**
  natural selection consequences in, 90, **90**
  reconstructing history of, 295–99
  variability in, 82, 94
Population genetics, 39, 169. *See also* Mendelian genetics
  effect of sex in, 306–7
  model of varying selection, 310, 312
Population size
  genetic diversity and, 247–48, **248**
  genetic drift and, 235–36, 239, **240**, 617
  heterozygosity and, 245–49
  Muller's ratchet and, 308–10, **308**, **309**
  neutral theory and, 252
Population trees, 783, **783**
  divergence times of, **782**
Porifera, **699**
Positive feedback loop, 428
Positive selection, 254
  detecting, 299–302
  loci under, 259–60
  on replacement substitutions, *260*
  signature of, **301**, 302
Postzygotic isolation, 624
Potamopyrgus antipodarum, 311, **312**
Potassium-argon dating, 63, 64
Prairie dogs
  alarm calling in, 450–51, **450**, **451**
  Gunnison's (*Cynomys gunnisoni*), 429–32, **431**
Pratt, David, 366
Preadaptation, 92
Prebiotic synthesis of organic molecules, 655–56
Precision, experimental, **370**
Predation, trade-off between reproduction and survival with, 501
Preexisting sensory biases, 422–24
Presgraves, Daven, 428–29
Prezygotic isolation, 624, **625**
  in allopatric vs. sympatric species pairs, 624, **625**
Priapula, **699**
Price, Trevor, 86
Prilophyta, 742
Primates, Old World, 754, **755**
  phylogenies showing relationships of some, **138**
Primordial form, 640, **640**
Principal component analysis, **628**
Prion protein (PrP), 189, 190
Prior ejaculation in marine iguanas, 412–13
Prisoner's dilemma, 473
Pritchard, Jonathan, 302
Probabilities, combined, 175
Problematica, 697, **698**
Proboscipedea (pb) genes, 731, 734
Processed pseudogenes, 57–59, **58**
Prochlorococcus, 594, *594*
Proctor, Heather, 422–24
Progressivist view of evolution, 93
Progress zone, 737, **737**, 738
Prokaryotes, **131**, 663, 664, **664**
Promoter regions, 596
Protease inhibitors, 16
Proteins, 641
Protein superfamilies, 672, **672**
α-Proteobacteria, 586, 587
γ-Proteobacteria, 671
Proteorhodopsin, 599
Proterozoic Doushantuo formation, China, 695, **695**
Protista, **664**, 665
Protoceratops, **603**
Protostomes, 700, 701
Prunus serotina (black cherry), 483
Pryke, Sarah, 416–17
Pseudogenes, 155, 762
  canonical rate of neutral evolution established by, 252–53
  processed, 57–59, **58**
Pseudomonas, 594
Pseudomonas aeruginosa, 165, 210, 217
Pseudomonas syringae, *594*
Psychology, evolutionary, 563
Pteropus poliocephalus (flying fox), 376–79, **376**, **379**
Pubic lice (*Pthirus pubis*), 132
Punctuated equilibrium, 704–5, **705**, **706**, 707
Punnett, R.C., 174, 177, 208
Punnett square, 174–75, **174**, **178**
Purifying (negative) selection, 254, 261
Purines, 144–45, **144**
  prebiotic synthesis of, 656
Purple bacteria, 664
Purugganan, Michael, 746
Pusey, Anne, 414, 415, 477
P value, 372
Pygmy armadillo (*Zaedyus pichiy*), **45**
Pyrenestes o. ostrinus (black-bellied seedcracker), 348, **349**
Pyrimidines, 144–45, **144**
  prebiotic synthesis of, 656

Q_β RNA, 643–44, **644**
QTL mapping, 324–31
  logic of, 327–28, **328**
Qualitative traits, 320
  defined, 320
  loci contributing to, 324–33
    candidate loci, 331–33
    QTL mapping, 324–31, **328**, **329**
Quantitative genetics, 319–59, 389–90
  measuring differential success, 338–40
  measuring heritable variation, 333–38
  modes of selection, 346–50, **347**
  predicting evolutionary response to selection, 341–43, **343**
  selection on multiple traits, 344–46, **345**
Quantitative trait loci (QTL), 324
  QTL mapping, 324–31
Quantitative traits, 319–24
  environmental influence on, **324**
  in humans, **320**
  influencing personality trait, **332**
  Mendelian genetics and, 320–21, **321**, **323**
Quartz, shocked, 713, 714
Quaternary period, **692**, **693**
Queens
  honeybee, 462
  naked mole-rat, 466–67, **466**, *467*
Queller, David, 458
Questions about evolution, strategies for asking interesting, 395–96

Radiation
  adaptive, 702–4, **703**
  plant, 703–4
Radiation pressure, 653
Radioactive decay, 63, **63**
Radiometric dating, 61, 63–65, **64**, 104
RAG1 protein, 582–83
RAG2 protein, 582–83
Raices, Marcela, 490
Rails, extinction of, 718
Ralstonia solanacearum, 590–91
Ramsey, Justin, 159
Ramsey, Mike, **435**, 436
Random fixation of alleles, 238–49
  in natural populations, 245–49
  probability of allele to drift toward, 241
Randomization, experimental, 370
Randomly amplified polymorphic DNA (RAPD) markers, 626
Random mating, 171, **174**, 175–76, 178–79, **178**
  genotype frequencies produced by, 171, **174**, 175–76, 178–79, **178**
  Hardy-Weinberg equilibrium assumption of, 181
  linkage disequilibrium reduced by, **289**, 290, 294
Raoult, Didier, 680
Raphanus raphanistrum (wild radish), 434–36
Ratcliffe, Laurene, 86
Rate-of-living theory of aging, 487, 488–92
  trade-offs and, 490–91
Raup, David, 709, 710
Raxworthy, C. J., 134
Ray, John, 38
Ray-finned fish
  camera eye of, 115, **115**
  genome duplication event in early evolution of, 160
Reaction norm, 564
Recessive alleles
  equilibrium frequency, 214
  mutation rates for, 216
  selection on, 195–97, 198–99
    algebraic treatment of, 198–99, **199**
Reciprocal altruism, 471–77
  blood-sharing in vampire bats, 474–75, **474**, **475**
  conditions for, 472
  territory defense in lions, 475–77, **476**
Reciprocal-transplant experiments, 84
Recombination. *See* Genetic recombination
Red bladder campion (*Silene dioica*), 230–31, **231**

Red-collared widowbirds, **416**
  female choice in, 416–17, **416**, **417**
Red deer (*Cervus elaphus*), **402**, **487**
Redon, Richard, 777
Red Queen hypothesis, 311
Red-winged blackbirds, extra-pair copulations in, 430–31, **431**
Reed, David, 495–96
Reed bunting (*Emberiza shoeniclus*), 563, **563**
Reeve, Hudson, 464–65, **465**, *467*
Reference distribution, 372
Regression analysis
  best-fit line, **83**
    slope of, 339, **339**, 342, **342**
  least squares linear regression, 334–35
Regulatory cascade, 727, 728, 739, 746
  floral homeotic genes as part of, 745, **745**
Regulatory genome sequences, evolution of, 594–96
Reich, David, 299–300
Reid, Ann, 536–37
Reillo, Paul, 426–27, 428
Reinforcement, 624–25
Reithel, Jannifer, 78–80
Relatedness of life forms. *See also* Coefficient of relatedness (r)
  homology, 59
  interspecies relationships, 60
  phylogenetic trees and, 50–53, **52**
    reading, 52
Relative dating, 61
Relative fitness, 339, 340
Religion, conflict between evolutionary biology and, 65–67
Replacement (nonsynonymous) substitutions, 147–48, 253–54
  selection on, 257–60, *260*
Replicase gene, 644
Reproduction. *See also* Sexual reproduction; Sexual selection
  asexual, 308–9, **308**, **309**, 310
  comparison of modes of, 303–6, **304**
  nonrandom, 78–79, 87–88
  organisms with two modes of, 303–6, **303**
  trade-off between early- and late-life, 496–99
Reproductive altruism, 459–67
Reproductive capacity, 86–87
Reproductive isolation, 607
Reproductive strategies, diversity in, 484
Reproductive success, 78, 79
  asymmetric limits on, 405–7
  clutch size and offspring, 506
  measuring differences in, 338–40
  in plants, 434–38
  sexual dimorphism in body size in humans and, 438–40
  variation in, 86–87, **86**
Re-purposing of genes, 728
Resistance
  antibiotic. *See* Antibiotic resistance
  to AZT, **12**, 13–15, **13**, **14**
  to HIV, 12, 22–25
  molecular basis of, 23
  mutation in HIV genome and evolution of, 16–17
Resistance alleles, 23–25
Resources, female choice and acquisition of, 421–22
Respiration, in yeast cells, 392, **393**
Retrosequences, 578–79
Retrotransposition, 152, 153
Retrotransposons, 57–58, 578
Retroviruses, 8. *See also* Human immunodeficiency virus (HIV)
  retrotransposons and, 578
Reversal, 116, **116**
  in SINE or LINE characters, 128
Reverse transcriptase, 7–8, 58, **645**, 646
  AZT and, 11–15, 77
  retrotransposition and, 152, 153
  telomerase, 583–84
  transcription errors made by, 13–14, 15
Reverse transcriptase inhibitors, 16
Reverse transcription, 588
  retrotransposons and, 578
Reversion (back) mutation, 15, 210, 308, 546–48
Rhagoletis pomonella (apple maggot fly), **605**, **619**
  divergence through natural selection, 617–19
Rhesus macaques, **763**
Rhoads, Allen, 59
Rhyniophyta, 742
Ribonucleic acid (RNA) intermediate, 578
Ribonucleoside triphosphates, 643
Ribonucleotides, 641
Ribose sugars, prebiotic synthesis of, 656
Ribosomal RNA (rRNA), 663–64
  universal phylogeny of, 663, 668, **668**
Ribosomal RNA (rRNA) genes, 309–10
Ribosome, 643
Ribozymes (RNA enzymes), 641
  evolution of catalytic ability in laboratory population of, **650**
  experimental evolution of RNA and, **645**, 646–50, **647**
  phenotypes of, 643
  with RNA synthesizing ability, 648, **649**
  Tetrahymena, 641, **642**, 645–46, **647**
Ribulose bisphosphate carboxylase (RuBPCase), 587
Rice, William, 516–17
Richmond, Brian, 761
Ricklefs, Robert, 470
Rieseberg, Loren, 625–26
Rifampin, 538
Ringler, Daniel, 556
Ring of life hypothesis, 676–77, **676**
Ring species, 53–54
Risch, Neil, 298
Rivera, Maria, 676
RNA, **146**
  antiquity of, 643
  experimental evolution of, 643–47
    Q_{β} RNA, 643–44, **644**
    ribozymes and, **645**, 646–50, **647**
  genotype and phenotype possessed by, 641
  genotypic changes in population of, **647**
  messenger (mRNA), 146, **146**
    reverse-transcribed, SINEs originating from, 588, **588**
  precursor of, 651–60
  ribosomal (rRNA), 663–64, 668, **668**
  self-replication of, 648–50, **649**
  three viruses, three domains hypothesis and transition to DNA from, 677–80, **679**
  in vitro selection and reproduction of, **645**
RNA adaptors (tRNAs), 643
RNA enzymes. *See* Ribozymes (RNA enzymes)
RNA polymerases, 147, **645**, 646
  DNA-dependent, 667
    phylogeny of, 679–80, **679**
  RNA-dependent, 677
RNA synthesis, template-based, **650**
RNA World, 641–50
  activation of building blocks in, 657
  case for, 643
Robust australopithecines, **768**, 769, 771–72, 787
Roca, Alfred, 610
Rock pigeons, beak hooks and lice infestation in, 363–64, **363**
Rodhocetus kasrani, 129
Roff, D.A., 518
Rogers, A.R., **402**
Rogina, Blanka, 498
Röhme, Dan, 490
Romero, Gustavo, 437
Ronshaugen, Matthew, 735
Rosenberg, Noah, 776, 777
Rosenqvist, Gunilla, 432, **433**
Rose pinks (*Sabatia angularis*), 271
Rosing, Minick, 659
rRNA. *See* Ribosomal RNA (rRNA)
Rubber boa (*Charina bottae*), 42, **42**
Rudists, 716, **716**
Ruepp, A., 590
Runaway selection, 425, 426–29
Runnegar, Bruce, 673, 701
Ruvolo, Maryellen, 757–59
Ryder, Oliver, 616

Sabatia angularis (rose pinks), 271
Sabeti, Pardis, 299–300, 301, 302
Saccharomyces cerevisiae (baker's yeast), 59, 151, 578, 691
  codon bias in, *261*, **262**
Saedler, Heinz, 596
Sagebrush hybrids, fitness of, 628–29, *628*, **629**
Sahelanthropus tchadensis, 765–67, **765**
Salem, Abdel-Halim, 759
Salmonella enteritidis, 150, 557
Salmonella typhi, 217, 218
Salmonella typhimurium, 309
Salticidae, 367
Sampling error, 234
  founder effect and, 236–38. *See also* Genetic drift
Samson, Michel, 23, 24, 162
Sand crickets (*Gryllus firmus*), 486, **486**
Sanitation, 530
SAR11 clade, 598
Sarich, Vincent, 755, 756
Scharloo, W., 272
Schemske, Douglas, 159, 330–31, 384–85
Schlenke, Todd, 292–93
Schluter, Dolph, 506, 632
Schmoo larvae of sea urchin, 520–22, **521**, **522**
Schopf, J. William, 652, 662
Schrag, Stephanie, 539–40
Schul, Johannes, 418
Schwartz, Joshua, 418
Scientific creationism. *See* Creationism
Scientific theories, components of, 38
Scopes, John T., 97, **97**, 753
Scopes Monkey Trial of 1925, 97, 753–54
Scrapie, 189
Sea otters (*Enhydra lutris*), 266–68, **266**, *268*
Sea spiders (eurypterids), 733
Sea squirt (*Botryllus schlosseri*), 518–19, **519**
Sea star (Echinodermata), **699**
Sea urchins
  larval forms of, 520–22, **521**, **522**
  phylogeny for, **521**, 522
Secker, Jeff, 653
Secondary contact, 611, 623–29
  hybridization and, 623, 625–29, *627*
  reinforcement, 624–25
Secondary endosymbiosis, 588–89, **589**
Second Law of Thermodynamics, 104
Seed beetle (*Stator limbatus*), 512–14, **512**, **513**
Seeds, 742
Segmentation of arthropods, 732–35, **733**, **734**
Segregation, Mendel's law of, 171–72
Selection
  adaptation and, 364
  antibiotics as agents of, 538–39. *See also* Antibiotic resistance
  for antigenic sites, 532–34
  artificial, 74–75
  background, 263
  directional, 347, **347**, 349, 384, **385**
  disruptive, 347–48, **347**, 349, **349**
  empirical research on, 188–91
  environmentally-imposed, 310–12
  evolutionary response to, 87–88, 90
    predicting, 341–43, **343**
  frequency-dependent, 205–7, **206**, 350, 462
  genotype frequencies and, **187**
  Hardy-Weinberg equilibrium assumption of no, 180, **181**
  on heterozygotes and homozygotes, 197–205
  imposed by changing environment, sexual reproduction and, 310–12
  intersexual, 408, 415–29
  intrasexual, 407, 408–15
    by combat, 408–12
    by infanticide, 414–15
    by sperm competition, 413–14, **414**
  linkage disequilibrium and, 285–88, **287**, 310
  in Mendelian genetics, 182–210
    added to Hardy-Weinberg analysis, 183–85, 187–88
    changes in allele frequencies and, 183–85, **183**, **185**
    examples of, 182–83
    frequency-dependent, 205–7, **206**
    general treatment of, 186
    genotype frequencies and, 187–91
  migration-selection balance, 229–30, 627
  modes of, 346–50, **347**

Selection, (*continued*)
in molecular evolution, 250
on multiple traits, 344–46, **345**
mutation and, 212–13, **213**
mutation-selection balance, 213–18
natural. *See* Natural selection
negative (purifying), 254, 261
nonrandom, 87–88, 93, 232
on offspring size, 510–12
operation on different levels, 392–95
in Apert syndrome, 394–95, **395**
demonstration of, 392–94
selfish genes and, 580
transposable elements, 576–84
phenotypes and, 182–83
positive, 254, 259–60, 299–302, **301**
predicting evolutionary response to, 341–43
on recessive and dominant alleles, 195–97, 198–99
algebraic treatment of, 198–99, **199**
on replacement substitutions, 257–60
runaway, 426–29
sexual. *See* Sexual selection
stabilizing, 347, **347**, **348**, 384, **385**
strength of
hybrid zone and, 627
measuring, 338–40, **339**
Selection coefficient, 151, 198, 214, 215, 217
Selection differential (s), 338, 339, **339**, 340
in alpine skypilots pollinated by bumblebees, 342
predicting evolutionary response to selection and, 341, **341**, **343**
Selection gradient, 338–40, **339**, 342, **343**
Selectionist (balance) theory, 165
Selectionist theory of mutation, 250, 251
Selection modes, genetic variation and, 349–50
Selection-mutation balance hypothesis, 350
"Selection thinking," 16, 18, 22
Selective sweep (hitchhiking), 262–63
Self-fertilization (selfing), 264–66, **265**, *265*, 268, 295
Selfish Gene, The (Dawkins), 562
Selfish genes, 580
Selfishness, 448
Self-replication, 675
Semaw, Sileshi, 786
Semelparous morph, 518–19
Seminal fluid, toxic, 516–17
Sen, G., 715
Senescence. *See* Aging
Sensory biases, preexisting, 422–24
Sensory exploitation hypothesis, 422–24
Senut, Brigitte, 765
Sepals, **742**, 743, **743**
Sepsis, fever and, 560
Sequence(s). *See* DNA sequences; Molecular evolution
Serum albumin, human and ape, similarities between, 755–56
Service, Phillip, 489
Sex
adaptive significance of, 303–12
environmentally-imposed selection and, 310–12
genetic drift and, 308–10
genetic recombination, 306–7
reproductive modes and, 303–6, **304**
asymmetry of, 403–4
behavioral consequences of, 407–8
in fruit flies, 404
Sex ratio
evolution of, 462
among naked mole-rats, 465
Sex-role reversed species, 432–34
Sexual coevolution, 516–17, **517**
Sexual dimorphism, 401–7
in humans, 401, **402**
body size, **402**, 438–40
in plants, 434–38, **435**, **437**
in red-collared widowbirds, 416, **416**, 417
Sexual reproduction
adaptive significance of, 303–12
comparison of asexual and sexual reproduction, 303–6, **304**
linkage disequilibrium reduced by, **289**, 290, **291**, 294
as paradox in evolution, 304
Sexual selection, 401–46
antagonistic, 516–17, **517**
asymmetries in sexual reproduction, 403–4
asymmetry of sex
behavioral consequences of, 407–8
in fruit flies, 404
chase-away, 516
Darwin on, 403
female choice and, 407, 415–29
acquisition of resources and, 421–22
in barn swallows, 419
extra-pair copulations, 417, **417**, 430–31, **431**
in gray tree frogs, 417–21, **418**
offspring gene quality and, 419–21
preexisting sensory biases and, 422–24
in red-collared widowbirds, 416–17, **416**, **417**
sexy-son hypothesis of, 425
in stalk-eyed flies, 426–29, **426**, **427**, **428**
in Hawaiian *Drosophila*, 621–23, **622**
intersexual selection, 408
intrasexual selection, 407
by male-male competition, 408–15
alternative male mating strategies, 412–13
combat, 408–12, 426
homicides, 438–39
infanticide, 94, 414–15
sperm competition, 413–14, **414**
among stalk-eyed flies, 426
as mechanism of divergence, 621–23, **622**, *623*
in plants, 434–38
polyandry, 429–32
stronger for females than males, 432–34
Sexy-son hypothesis, 425
Shakers, 562
Shankarappa, Raj, 19
Shapiro, Leo, 701
Shapiro, Michael, 633
Sharks, 515
Sharp, Paul, 258
Sherman, Paul, 450–51, **452**
Shibata, Darryl, 549
Shocked quartz, 713, 714
Short INterspersed Elements (SINEs), 128, **128**, 578–79, 588, **588**, 759
Short-sighted evolution, 18–21
Short-sighted evolution hypothesis of virulence, 542
Short-tandem-repeat polymorphism, 783, **784**
Shu, D.-G., **697**
Sialic acid, 532
Siberian greenish warbler (*Phylloscopus trochiloides*), 53–54, **53**
Siblicide, 470–71
Sib-mating, 626
Sickle-cell anemia, 146, 147
Significance, statistical, 372
Sign language, 789
Silene dioica (red bladder campion), 230–31, **231**
Silene latifolia (snowy campion), 520, **520**
Silent-site (synonymous) substitutions, 148, 152, 253–54
selection on, 260–63
"Silent" substitutions, 260–63
Silvereyes (*Zosterops lateralis*), 236–37, **237**
Simian immunodeficiency viruses (SIVs), 9, 22–23, 27
SINEs (Short INterspersed Elements), 128, **128**, 578–79, 588, **588**, 759
Sinornithosaurus millenii, **47**
Sinosauropteryx prima, **46**, 47
Sister species, 614, **614**, 624
Sister taxa, **52**
Siva-Jothy, M.T., 414
Skeppsvik Archipelago, Sweden, 230
Skinner, Samuel, 507–8
Skole, David, 720
Slatkin, Montgomery, 299
Sleep, N.H., **654**, 660
Slime mold (*Dictyostelium*), 665
Slime mold (*Dictyostelium discoideum*), 458–59, **459**
Slope of line, calculation of, 342
Smallpox epidemic (c. 1520), 4
Smell, sense of, 454
Smith, Andrew, 701
Smith, Christopher, 509–10, 511
Smith, James, 335–37
Smith, Nick, 259
Smith, R.L., 414
Smithson, Ann, 205, 206
Snails, evolutionary tree for, 50, **51**
Snakes, migration and selection in, 227–30, **228**
Snapdragons, 78–80, **79**
Snapping shrimp (*Altheus*), 614–15, **614**
Sniegowski, Paul, 580
Snow, John, 529, **530**
Snowy campion (*Silene latifolia*), 520, **520**
Soapberry bugs, 40–41, **40**, 49
Social behavior, 447–82
eusociality, 459–67
characteristics of, 460
haplodiploidy and, 460–62
in naked mole-rats, 465–67, **466**, *467*
in paper wasps, 464–65
phylogenies to analyze, **463**
kin selection and evolution of altruism, 448–59
inclusive fitness, 448–50
kin recognition, 454
parent-offspring conflict, 467–71, **468**
siblicide and, 470–71
reciprocal altruism, 471–77
blood-sharing in vampire bats, 474–75, **474**, **475**
conditions for, 472
territory defense in lions, 475–77, **476**
in terms of fitness, 447–48, *448*
Sockeye salmon (*Oncorhynchus nerka*), **483**
Sohal, Rajindar, 498
Solanum lycopersicum (domestic tomato), 74, **74**
Solanum pimpinellifolium (wild tomato), 74, **74**
Solanum tuberosoma, 595–96, **595**
Solar system, age of, 640
Solidago altissima, 348
Somatic-cell gene therapy, 548
Song sparrows (*Melospiza melodia*), 335–37, **335**
Sonic hedghog (shh) gene, 737, 738, 739
Sooty mangabey, 26–27
Spea bombifrons, 453–54, **455**
Special Creation, Theory of, 38, 40, 42, 50, 60, 65, 98, 704
Speciation, 39, 104–5, 135, 605–37
allopatric, 612
in Bryozoans, 706, **706**
classical view of, 611
divergence, 611, 616–23, *620*
genetic drift, 616–17
natural selection, 617–21
sexual selection, 621–23, **622**, *623*
genetics of, 629–33
in pea aphids, 629–31, **630**
in three-spined sticklebacks, 631–33
isolation mechanisms, 611–16
chromosomal changes, 615–16
dispersal and colonization, 612–14, **612**, **613**
migration, 612
physical, 611–15
vicariance, **612**, 614–15
morphological change and, 706
by natural selection, *620*
polyploidy and, 159
secondary contact, 611, 623–29
hybridization, 623, 625–29, *627*
reinforcement, 624–25
by sexual selection, *623*
species concepts and, 605–11
synapomorphies and, 113, **113**
Species
cryptic, 606, **614**, 616
mutation rate variations in, 149
polyploidy and, 159
relationships among, 60
ring, 53–54
sister, 614, **614**, 624
Species concepts, 605–11
applying, 609–11
in bacteria, 607
biological, 606–7, 629
morphospecies concept, 606
phylogenetic, 608–9
Species trees
divergence times of, **782**
gene trees vs., 758–59, **758**
Sperm, 404

altruistic, 454–56, **456**
Spermatophores, 422–23
Sperm competition, 413–14, **414**
bat testes size and, 376–79, **376**
Spermophilus beldingi, 450–51, **452**
Spicer, D.V., 555
Spiders, jumping (*Phidippus apacheanus*), 367–72, **368**, **369**
Spiegelman, Sol, 643–45
Spinal muscular atrophy, 214, 216
Spirulina, **674**
Spite, 448
Splash zone of bolide impact at K-T boundary, 717
Spoken language, 789
Spongiform encephalopathies, or spongy brain diseases, 189–91
Spores, extraterrestrial, 653
Spotted cucumber beetle (*Diabrotica undecimpunctata howardi*), 425, **425**
Stabilizing selection, 347, **347**, **348**, 384, **385**
Stalk-eyed flies (*Cyrtodiopsis dalmanni*), 426–29, **427**, **428**
Malaysian (*C. whitei*), **426**
Stamens, **742**, 743, **743**
Standard error, **86**
Stanley, Steve, 708, **709**
Stanton, Maureen, 434–36
Stasis, 704–9
in Bryozoans, 706, **706**
demonstrating, 705–6
dynamic, 708
reasons for, 707–9
relative frequency of, 707
Statistical testing, 371–72
Stator limbatus (seed beetle), 512–14, **512**, **513**
Stauffer, Rebecca L., 761
Steadman, David, 717–18
Stem cells, 490–91
Stepchildren, discriminatory parenting of, **565**, 566–68, **566**
Stephens, J. Claiborne, 298, 299
Sterilization, compulsory, 194
Sterilization, eugenic, 194–95, 208–10
Stewart, T. H. M., 553
Sticklebacks, freshwater, with vestigial armor, 43, **43**
Stomata, 742
Stone tools, origin of, 786–89
Strait, David, 761, 765, 771, 773
Strassmann, Beverly, 554–55
Strathmann, Richard, 521, 523
Streptococcus mitis, 653
Streptomycin, 540
Stromatolite-forming bacteria, **707**
Structural homologies, 54–55, **54**, **56**
Sturtevant, A.H., 310
Stylet, **135**
Stylopod, 736, **736**, 738, **738**
Stylopoma, 706, **706**
Suboptimal life histories, 522–23
Sub-Saharan Africa, HIV/AIDS in, 5, **5**
Subspecies, 627–28
Substitution(s)
in antigenic sites, 532
base, 147–48
mutation vs., 249, **249**
rates of, *255*
under genetic drift, 250, 252
replacement (nonsynonymous), 147–48, 253–54
selection on, 257–60, *260*
"silent," 260–63
silent-site (synonymous), 148, 152, 253–54
selection on, 260–63
single-base DNA, 147
Succession, law of, 45, **45**
Sulfolobus solfataricus, 590
Sulfur dioxide, 714, 715
Sundström, Liselotte, 461
Superposition, principle of, 61
Survival, nonrandom, 87–88
Survival success, measuring differences in, 338–40
Survivorship curves, 710, **711**
Susman, Randall L., 787–88
Swartkoppie chert, 662
Sydney Bloodbank Cohort, 21
Synapomorphies, 112–14, **113**, **114**, 119, 123
of humans and African great apes, 755
Synechococcus, 590
Syngnathus typhle, 406–7, **406**, **407**, 432–34
Synonymous (silent-site) substitutions, 148, 152, 253–54
selection on, 260–63
Systematics, 130–31
Szostak, J.W., 648–50

T4 bacteriophage, 149
Tadpoles, cannibalistic, 453–54, **455**
Tailbone, 42, **42**
Tallamy, Douglas, 425
Tangle-veined fly, **361**
Tanksley, Steven, 74–75
Tanner, Steven, 418
Taper, Mark, 428
Tapeworms, 93
Taricha granulosa (newts), 405–6, **405**
Tattersall, I., 775
Taubenberger, Jeffery, 536–37
Taxonomic bias in fossil record, 691
Taxon/taxa, 52, 608
Taylor, Douglas, 392
Tay-Sachs disease, 298
T cells, 28
effector, 9, 11
life cycle, 9, **9**
memory, 9, 10, 11
naive, 9, 11, 20
Teikari, J.M., 551
Telomerase, 489, 583–84
Telomere, 489–91, **491**
Telomeric survival motor neuron gene (*telSMN*), 214
Temperature
allele frequency and, **619**
physiological performance and, 372–73
Templeton, Alan, 245, 246–47, 248
Temporal, spatial, and quantitative colinearity, 729
Temporal bias in fossil record, 691
Tendon, cross-linking in, 502
Tephritid fly (*Zonosemata vittigera*), 367–72, **368**, **369**
Termites, eusociality in, *460*
Territorial behavior in lions, 475–77, **476**
Tertiary period, **692**, **693**
Testes size in bats, 376–79, **376**, **379**
Testing, statistical, 371–72
Test statistic, 371
chi-square, 192–93
Tetanus, 541–42
Tetrahymena genome, **642**
Tetrahymena ribozyme, 641, **642**, 645–46
Tetrapoda, 735
Tetrapods, 42
limb evolution in, 735–39
basic elements in, **736**
developmental and genetic changes responsible for origin of, 737–39, **737**, **738**, **739**
structure giving rise to limb, 736, **736**
synapomorphies revealing relationships among, 114, **114**
Thamnophis elegans (garter snake), 373–75, **374**, **375**, *375*
antipredator defenses, 346, **346**
Theory of Special Creation, 38, 40, 42, 50, 60, 65, 98, 704
Thermal performance curve, 373
Thermoplasma acidophilum, 590
Thermoregulation, behavioral, 372–75
behavioral fever and, 556
by garter snakes, 373–75, **374**, **375**, *375*
Thermotoga maritima, 670
Thewissen, Johannes, 49, 122
"Thioester World," 657
Thornhill, Randy, 421
3020insC allele, 332
Three-spined sticklebacks (*Gasterosteus aculeatus*), 43, 631–33, **631**, **632**, **633**
3TC (dideoxy-3′-thiacytidine), 16
Three viruses, three domains hypothesis, 677–80, **679**
Thrips, eusociality in, *460*
Thrips egg mite (*Adactylidium*), 484, **484**
Thumb, opposable, 787
Thymidine, 12
Thymine, 144, **144**, 145
Thysanoptera, eusocial, *460*
Tiger salamanders (*Ambystoma tigrinum*), 453
Tiktaalik roseae, 736, **736**
Time scales, 710
Tips of phylogenetic tree, 52, **52**
Tishkoff, Sarah, 549, 780, 783, 784, 785
Tissues, evolution to resist disease, 546–49
cancer history, 548–49, **549**
spontaneous recovery, 546–48
Tit-for-tat (TFT) strategy, 473
Tn3 plasmid, 538
Tobias, Phillip, 791
Tomasello, M., 764
Tomato, domestication of, 74–75
Tools, origin of stone, 786–89
Trace fossils, 695–96
Trade-off hypothesis of virulence, 542–45, **544**
Trade-offs
in adaptation, 383–91
begonia flower size, 383–85, **384**, **385**
in allocation of energy, 486
between current reproductive effort and future reproductive performance, 506
between early- and late-life reproduction, 496–99
in energy and time, 484–85
in evolutionary theory of aging, 496–99, 502
optimal number of offspring and, 502–9
in pea aphids, 630–31
in rate-of-living theory of aging, 490–91, **491**
between reproduction and survival in plants, 499, *499*
between size and number of offspring, 509–10, **509**, **510**
Trait(s)
adaptive. *See* Adaptation(s)
convergent evolution of, 115–16, **115**
cross-species correlations among, 377–78, **377**, **379**
distinguishing homology from homoplasy, 116–18
evolution of uniquely human, 786–91
language, 789–91
stone tools, 786–89
heritability of. *See* Heritability
homologous, 112
molecular, 119–20
morphological, 119–20
natural selection acting on existing, 91–92
normally distributed variation in, 333
novel, evolution of, 520–22
qualitative, 320
loci contributing to, 324–33
quantitative, 319–24, **332**
selection on multiple, 344–46, **345**
Transcription factors, 726
Hox gene products as, 729
Transduction, 585
Transformation, 585
Transition, 147, **147**
Transitional forms, 46–49, **49**
Translational efficiency, natural selection for, 261
Transmission disequilibrium test, 332
Transposable elements, 576–84
categories of, 578–79
evolutionary impact of, 580–84
defending against spread of transposable elements, 580–81
positive, 581–84
as genomic parasites, 577
lateral gene transfer via, 586
Transposase, 579, 586
Transposons, 579
Transversion, 147, **147**
Tree of life, **640**, **661**, 673–75
five-kingdom scheme, **664**
Trematodes, 311–12
Treponema pallidum, 670, **670**
Tribolium castaneum (flour beetle), 195–97
sexual vs. asexual reproduction in, 304–6, **306**
Tribulus cistoides, 87, 89
Trichogramma embryophagum, clutch size of, 507–9, **507**, *508*
Trichotomy, human/African great ape phylogeny as, **757**, 759

Trillmich, Fritz, 409
Trillmich, Krisztina, 410–11
Tringe, Susannah, 599
Trinkaus, E., 778
Triploblasts, 699–700
Trivers, Robert, 403, 450, 462, 467, 471–72, 473
Tropical rain forests, species extinctions in, 719–20
Troyer, Ryan, 20
Tryannosaurus rex, 47
Tsunami, asteroid impact at K-T boundary and, 715
Tuberculosis, 530, **530**
Tucker, Compton, 720
tufA gene, 587–88
Tumors, genetic variation in, 549
Tumor suppressors, 490
Tumpey, Terrence, 537
Twin studies of heritability, 337–38, **337**
Two-locus version of Hardy-Weinberg analysis, 282–313
  adaptive significance of sex, 303–12
    environmentally-imposed selection and, 310–12
    genetic drift and, 308–10
    genetic recombination, 306–7
    reproductive modes and, 303–6
  chromosome frequencies and, 286–87
  linkage disequilibrium. *See* Linkage disequilibrium
  numerical example of, 282–84
Ty elements, 578
Tympanuchus cupido pinnatus (greater prairie chicken), 223–25, **224**, **225**
  conservation genetics of, 273–75
  declining hatching success in, 274, **274**
  genetic drift in, 274–75, *275*
  habitat destruction, **224**
Tyner, Stuart, 490–91
Type 1 Gaucher disease, 295–96
Typhoid fever, 217–18, **218**

Ubx gene, 734–35
Unaltered remains, 690, **691**
Underdominance (heterozygote inferiority), 204
  genetic diversity and, 205
  unstable equilibria with, 202–3
Unequal cross-over, 152–53, **153**
Ungulates, 120, **120**
Uniformitarianism, 60–61, 104
Uniramous appendages, 739, **740**
United Nations Food and Agriculture Organization (FAO), 720
United States Supreme Court, 97–98
Universal gene-exchange pool hypothesis, 675–76, **675**
Universal grammar, 789
Universal phylogeny, 663–66
  based on concatenated sequences of 31 universal genes, **671**, 672
  based on presence/absence of protein superfamilies, 672, **672**
  discordance of different estimates of, 669, **669**
  genetic sequencing approach to, 663–64, **665**, 668–69, **668**
  morphological approach to, 663
  rooting, 666, **666**, 673–75
  whole-genome sequencing in, 668–69, 671–72
Unokais, 439, **439**
Uranium-lead dating, 63
Urea cycle, 597
Ursus americanus (black bears), 483
Ussher, James, 38

Vaccine
  for flu, 534
  for HIV, 27–30
Vampire bats, blood-sharing in, 474–75, **474**, **475**
Van Noordwijk, A.J., 272
Van Valen, Leigh, 710
Variability in populations, 82, 94
Variant Creutzfeldt-Jakob disease, 189, 190
Variation. *See also* Genetic variation
  heritability of, 83–85
  individual, 78
  among individuals, 83–87, 148–49
  in reproductive success, 86–87
Varieties, 628
Varying-selection theory of sex, 310, 312
Vascular tissues in plants, 742
Vassilieva, Larissa, 150
Vaughton, Glenda, **435**, 436
Vector borne vs. directly transmitted disease, virulence of, 544–45, **545**
Vegetarian finch (*Platyspiza crassirostris*), 81
Velociraptor, **603**
Velvet worms (onychophorans), 733, **733**, 734–35, **734**
Venter, Craig, 599
Vertebrate parasites, genomes of, 593
Vertebrates
  Hox complex duplication during evolution of, 731
  limb, development genetics of, 735–39
Vestiges of the Natural History of Creation, The (Chambers), 39
Vestigial structures, 42–43, **42**
Via, Sara, 629–30
Vibrio cholerae, 529, 545
Vicariance, **612**, 614–15
Vigilant, Linda, 780
Viking spacecraft, 652
Viral disease, breast cancer as, 553
Viral infection, immune response to, 8–9, **8**
  HIV infection, 10–11
Virginia opossum (*Didelphis virginiana*), 485–86, 500–502, **500**, **501**
Virions
  HIV, 7–8, 10, 12–13
  X4, 20–21
Virulence, 541–45
  evolution of, 541–44
  in human pathogens, 544–45, **545**
Virulence genes, 592, **593**
Viruses, 7–8. *See also* Human immunodeficiency virus (HIV); Influenza A virus
  hypotheses on origin of, 678
  three viruses, three domains hypothesis, 677–80, **679**
  transduction, 585
Visser, Arjan de, 150
Voight, Benjamin, 302
Volcanism, asteroid impact at K-T boundary and, 715
Volvox, 303, **303**, 304

Waage, J.K., **414**
Waddle, Diane, 778
Walker, David, 496–98
Wallace, Alfred Russel, 39, 77
Wang, Yue, 553, 763
Warbler finches (*Certhidea olivacea* and *Certhidea fusca*), 81
Wasps, parasitoid, 507–9, **507**, *508*
Water flea (*Daphnia magna*), 381–83, **381**, **383**
Waterleaf (*Hydrophyllum appendiculatum*), 271
Water mite (*Neumania papillator*), 422–24, **423**, **424**
Water snakes, migration and selection in, 227–30, **228**
Watson, James, 145
Waugh O'Neill, Rachel, 580–81
Weaning conflict, 467–69, **468**
Weber, Peter, 653
Weedy mustard (*Arabidopsis thaliana*), 295, **295**, 691
Weeks, Paul, 365–66
Weinberg, Wilhelm, 179, 180
Weis, Arthur, 348, **348**
Weischaus, Eric, 726, 730
Welch, Allison, 419–21
Wells, W.C., 39
Werdelin, Lars, 52–53
Wernicke's area, 789, 791
Westemeier, Ronald, 273, 274, 275
Whale phylogenetic hypothesis, 120, 121, 123, 124, 128–30, **128**
Whales
  ancestors of, 48–49, **49**
  phylogeny of, 119–30, **121**
    choosing characters for data, 119–20
    competing ideas about, **121**
    distance methods, 125, 126, **126**
    nearly perfect phylogenetic character, **128**
    parsimony with multiple molecular characters, 122–23, **122**
    parsimony with single morphological character, 120–22
    resolving character conflict, 127–30
WHIPPO-1 data set, **122**, 123, 124
White, Tim D., 768
White-fronted bee-eaters (*Merops bullockoides*), 451–53, **452**, **453**, 469–70
Whiteman, Noah, 391
Whole-genome sequencing, 668–69, 671–72
Wikelski, Martin, 409, 410, 411, 412
Wild cabbage (*Brassica oleracea*), 75, **75**
Wildman, D.E., 759
Wild radish (*Raphanus raphanistrum*), 434–36
Wild tomato (*solanum pimpinellifolium*), 74, **74**
Wild type, 160–61, 458–59
Wilkinson, Gerald, 426–27, 428–29, 474–75
Willard, Dan, 462
Williams, George C., 492, 531
Wilson, Allan, 595, 755, 763
Wilson, D.E., 756
Wilson, E.O., 66, 719
Wilson, Margo, 438–39, 567–68
Wingless (wg) gene, 740
Wirth, Brunhilde, 214, 216
Witham, Larry, 67
Wiwaxia corrugata, 698, **698**
Wnt7a gene, 737
Woese, Carl, 663–65, 669, 675–76
Wolfe, Lorne, 271, 520
Wolpoff, Milford, 765, 780
Women, menopause in, 503–4, **503**
Wood, Bernard, 765, 770, 772–73, 787
Woodpecker finch (*Camarhynchus pallidus/Certhidea pallida*), 81, 786
Wrangham, R.W., 764
Wray, G.A., **521**, 701
Wrege, Peter, 452, 453, 469, 470
Wright, Sewall, 241, 242
Wurmbea dioica, 434, **435**, 436

X4 virions, 20–21
Xiao, S.Y., 695

Yagua hunter-gatherers, **550**
Yang, Song, 672
Yang, Xiangning, 708, **709**
Yąnomamö Indians, 439, **439**, 566
Yeast (*Saccaromyces cerevisiae*), 392–94
Young, Andrew, 247–48
Young, Francis, 552
Young, Truman, 499
Yule, G. Udny, 176–77

Zaedyus pichiy (pygmy armadillo), **45**
Zar, J.H., 193
Zea mays (corn), 691
Zebrafish, 691, 739
Zera, Anthony, 486
Zeugopod, 736, **736**
Zhao, Zhangwu, 486
Zhu, Tuofo, 29
Zigzag evolution, 708, **709**
Zink, Robert, 762–63
Zollikofer, C.P.E., 767
Zone of polarizing activity (ZPA), 737, **737**
Zonosemata vittigera (tephritid fly), 367–72, **368**, **369**
Zosterops lateralis (silvereyes), 236–37, **237**
ZPA (zone of polarizing activity), 737, **737**
Zuckerkandl, Emil, 132, 251
Zygopod, 738, **738**